The International Handbook on Innovation

The International Handbook on Innovation

Editor:

Larisa V. Shavinina

Département des Sciences Administratives,
Université du Québec en Outaouais, Canada

An imprint of Elsevier

AMSTERDAM – BOSTON – HEIDELBERG – LONDON – NEW YORK – OXFORD – PARIS
SAN DIEGO – SAN FRANCISCO – SINGAPORE – SYDNEY – TOKYO

Elsevier
The Boulevard, Langford Lane, Kidlington, Oxford OX5 1GB, UK
Radarweg 29, PO Box 211, 1000 AE Amsterdam, The Netherlands

First edition 2003
Reprinted 2005, 2006

British Library Cataloguing in Publication Data
The international handbook of innovation
1. Creative ability – Social aspects   2. Creative ability
3. Creative thinking   4. Creative ability in business
I. Shavinaina, Larissa V.
306 . 4'2

Library of Congress Cataloguing in Publication Data
A catalogue record for this book is available from the British Library

ISBN 008044198X

For information on all Elsevier publications
visit our website at books.elsevier.com

Working together to grow
libraries in developing countries

www.elsevier.com | www.bookaid.org | www.sabre.org

ELSEVIER   BOOK AID International   Sabre Foundation

Printed and bound in the Netherlands

Contents

This handbook is dedicated
to my parents
Anna Shavinina and Vladimir Shavinin,
to Professor Marina Kholodnaya,
and to my wonderful husband and son
Evgueni and Alexander

About the Authors

Mohi Ahmed is a visiting researcher at the Graduate School of International Corporate Strategy of Hitotsubashi University, Tokyo, working in conjunction with Professor Nonaka.

Susan G. Assouline is the Belin-Blank Center's Associate Director and Clinical Supervisor. She received her B.S. in general science with a teaching endorsement, her Ed.S. in School Psychology, and her Ph.D. in Psychological and Quantitative Foundations, all from The University of Iowa. She has an appointment as a Clinical Associate Professor in School Psychology. She is especially interested in identification of academic talent in elementary students and is co-author of the book *Developing Mathematical Talent: A Guide for Challenging and Educating Gifted Students* (Prufrock Press). Also, she is co-editor with Nicholas Colangelo of the series *Talent Development: Proceedings from the Wallace Research Symposia on Giftedness and Talent Development*, and co-developer of *The Iowa Acceleration Scale*.

James Bailey is an Associate Professor of social and organizational theory at George Washington University. Professor Bailey's scholarship examines how the nature and experience of change are manifested in crucial tensions between opposing values and how practitioners understand and adapt their competencies to environmental demands. His research has appeared in such outlets as *Organizational Science, Journal of International Business Studies,* and *Journal of Applied Behavioral Science.* Has worked extensively as a consultant on issues relating to conflict management, skill development and organizational change. In 2000 he was named the Associate Editor of the *Academy of Management Learning and Education.*

Clar M. Baldus is the administrator for Rural Schools Programs and Inventiveness Programs at the Belin-Blank Center and the Sate Coordinator of Invent Iowa. Her research has been guided by a passion for art, interest in creative processes, and commitment to talent development. Before joining the Belin-Blank Center, Dr Baldus was a resource specialist and assessment facilitator for elementary gifted and talented students. Her extensive, 25 years of teaching experience at the secondary, postsecondary, and professional development levels has also included working with students in the areas of art, creativity, and gifted education.

John Bessant is Professor of Innovation Management at Cranfield University and before that was Director of the Centre for Research in Innovation Management at the University of Brighton in the U.K. Originally a chemical engineer, he has been active in the field of research and consultancy in technology and innovation management for over 25 years. His areas of special interest include organizational development for innovation, inter-firm learning and enabling the emergence of agile enterprises. He is the author of 12 books (the latest—2001—J. Tidd, J. Bessant & K. Pavitt, *Managing Innovation* published by John Wiley) and many articles on the topic and has lectured and consulted widely on the topic around the world.

Vincent Boly is Professor of Innovation at Ecole Supérieure en Génie des Systèmes Industriels of Nancy, France. His research and teaching interests span the field of new product design, metrology and integration concepts. Before returning to University, Dr Boly worked as a consultant with major international companies as well as SMEs. He was a member of the board of a High Tech SME after being involved in its launch. Dr Boly is invited Professor at Quebec University in Trois Rivières.

John Callahan is cross-appointed in both the Department of Systems and Computer Engineering and the School of Business at Carleton University, Ottawa, Canada. He teaches graduate courses and supervises graduate research in telecommunications and product development management in Engineering, and in R&D management in the Sprott School of Business. He has published in journals such as *Management Science, Research Policy, IEEE Transactions on Engineering Management, Technovation* and *R&D Management.* Dr Callahan graduated from Carleton University with a B.Sc. (Hons.) in Mathematics and Physics. He then did an M.A. in Applied Mathematics and a Ph.D. in Industrial Engineering at the University of Toronto.

Elias G. Carayannis is an Associate Professor at the Management Science Department and Director of Research on Science, Technology, Innovation and Entrepreneurship of the European Union Center, School of Business and Public Management, George Washington University. He also co-founded the Environmental Studies Center and the Social and Organizational Learning Studies Center at the School. Dr Carayannis has published in *Research Policy, Journal of Engineering and Technology Management, International Journal of Technology Management, Technovation, Journal of Technology Transfer, R&D Management*, and others. He has consulted for the World Bank, USAID, NSF SBIR, NIST ATP, Sandia National Laboratories, the General Electric Corporate Training & Development Center, Cowen & Co, and others.

Hsing-Hsiung Chen received a B.Sc. in Electrical Engineering from the National Taiwan Institute of Technology in 1987, and an M.S. degree from the National Tsing Hua University in 1989. He then joined the Strategic Planning Department (SPD) of Electronic Research & Service Organization (ERSO), Industrial Technology Research Institute (ITRI), where he worked his way up from a Planning Engineer, Deputy Manager, to the Department Manager. In 2000, he transferred to the Industrial Economics & Knowledge Center (IEK), ITRI, where he currently serves as Deputy Director of Integrated Research Division. Mr Chen received his second M.S. degree in Information Management from the National Taiwan University in 2001 and is pursuing his Ph.D. in Management Science at the National Chiao Tung University.

Michelene Chi is a Senior Scientist at the Learning Research and Development Center and a Professor at the Department of Psychology, University of Pittsburgh. Her research interests include learning from tutoring, learning from collaboration, understanding emergent processes, and conceptual change.

Sunyang Chung is a Director of the Institute for Technological Innovation (ITI) and Professor of Technology Management at the School of Business Administration, Sejong University, Seoul, Korea. He received a Ph.D degree from the University of Stuttgart, Germany. He carried out several research projects at the Fraunhofer-Institute for Systems and Innovation Research (FhG-ISI) in Karlsruhe, Germany. He had worked about 13 years at the Science and Technology Policy Institute (STEPI), Korea. His research areas are technology management and policy, regional innovation strategies, environmental management and policy, and the integration of the South and North Korean innovation systems. He has published several books in Korea and Germany and has written over 50 articles in Korean and International journals.

Maria Clapham is an Associate Professor of Psychology at Drake University. She received her doctorate in Industrial and Organizational Psychology from Iowa State University. Her research interests include employee assessment and development, bias in assessment procedures, and factors that predict and affect creative performance. She has published numerous articles in professional journals such as the *Journal of Occupational and Organizational Psychology, Creativity Research Journal, Educational and Psychological Measurement,* and *Journal of Managerial Issues*.

Joseph F. Coates is Head of Joseph F. Coates, Consulting Futurist, Inc., also President Emeritus of Coates & Jarratt, Inc., a future research organization that he founded in 1979 and managed until May 2001. He consulted with 45 of the Fortune 100 companies, numerous smaller firms, trade, professional, and public interest groups, and all levels of government. Joe lectures to many groups each year about trends and future developments. He is the co-author of *2025: Scenarios of U.S. and Global Society Reshaped by Science and Technology* (Oakhill Press, 1997), *Futurework* (Jossey-Bass, 1990), *What Futurists Believe* (World Future Society, 1989), and *Issues Management: How Can You Plan, Organize, and Manage for the Future* (Lomond, 1986), and is the author of over 300 articles, chapters, papers, and other publications.

Nicholas Colangelo is the Myron and Jacqueline Blank Professor of Gifted Education and the Director of The Connie Belin & Jacqueline N. Blank International Center for Gifted Education and Talent Development. He has published extensively on the counseling and social-emotional needs of gifted students. He is co-author of the *Iowa Inventiveness Inventory* and the *Iowa Acceleration Scale*. He was presented with the Distinguished Scholar Award in 1991 by the National Association of Gifted Children. He is the senior editor of the *Handbook of Gifted Education* (3rd ed.).

Pedro Conceição is Assistant Professor at the Instituto Superior Tecnico in Lisbon, Portugal, at the Center for Innovation, Technology and Policy Research, http://in3.dem.ist.utl.pt/, and a Senior Research Fellow at the IC2 Institute, The University of Texas at Austin, USA. He is currently with the United Nations Development Program.

Robert G. Cooper is President of the Product Development Institute Inc., Professor of Marketing at the School of Business, McMaster University in Hamilton, Ontario, Canada, and ISBM Distinguished Research Fellow at Penn State University's Smeal College of Business Administration. Dr Cooper is the father of the *Stage-Gate*™

process, now widely used by leading firms around the world to drive new products to market. He is a thought-leader in the field of product innovation management, having published more than 90 articles and also written six books on the topic, including the best-seller, *Winning at New Products: Accelerating the Process from Idea to Launch.*

Martyn Cordey-Hayes is an Emeritus Professor at Cranfield University in the U.K. Previously he was Director of the Innovation and Technology Assessment Unit and Head of the International Ecotechnology Research Centre, Cranfield University.

Kevin G. Corley is an Assistant Professor of organizational behavior in the College of Business at the University of Illinois, Urbana-Champaign. His research interests focus on organizational change, especially as it pertains to issues of organizational identity, image, reputation, and learning. His research has recently appeared in the *Academy of Management Journal*, the *Academy of Management Review*, *Journal of Organizational Change Management*, and *Corporate Reputation Review.*

Laurie J. Croft is the administrator for Professional Development at the Belin-Blank Center, teaches and coordinates workshops for pre-service and in-service educators about various aspects of gifted education, as well as about the Advanced Placement Program. Formerly a teacher at the University of Tulsa School for Gifted Children, she supervises practicum experiences for educators of the gifted.

Subrata Dasgupta studies creativity, especially as it is manifested in science, technology, design, intellectual history, and art. He is presently Director, Institute of Cognitive Science, and holder of the Eminent Scholar Chair in Computer Science at the University of Louisiana at Lafayette, where he also has professorial appointments in the Department of History and the Center for Advanced Computer Studies. His most recent books include *Creativity in Invention and Design* (Cambridge, 1994), *Technology and Creativity* (Oxford, 1996), and *Jagadis Chandra Bose and the Indian Response to Western Science* (Oxford, 1999).

Jean Philippe Deschamps teaches technology and innovation management to executives at IMD in Lausanne (Switzerland). Prior to joining IMD, in 1996, he was a corporate vice-president at Arthur D. Little and the founder and Chairman of the firm's technology and innovation management practice. He has 30 years of international consulting and lecturing experience with a great variety of industries, and co-authored *Product Juggernauts—How Companies Mobilize to Generate Streams of Market Winners*, published in 1995 by the Harvard Business School Press. He graduated from HEC (Paris), INSEAD in Fontainebleau, and the Harvard Business School.

Icíar Domingue-Lacasa has a M.A. in economics from the University Complutense in Madrid, Spain. In 1996–1998 she was a postdoctoral fellow in international economics at the University of Mannheim, Germany. In 1998–1999 Lacasa worked at Boston Consulting Group in Munich. Since 2000 she has been a research fellow at the Fraunhofer Institute for Systems and Innovation Research (ISI) in Karlsruhe, Germany.

James L. Farr is Professor of Psychology at The Pennsylvania State University. He is the Editor of *Human Performance* and a former President of the Society for Industrial and Organizational Psychology. His research interests are in work group innovation, performance measurement and management, older workers, and work motivation.

Cameron Ford's interests are in the areas of organizational creativity, innovation, and new venture creation. His research has appeared in journals such as the *Academy of Management Review*, *Journal of Management*, *Journal of Applied Behavioral Science*, *IEEE Transactions on Engineering Management*, and *Journal of Organizational Change Management*. He has also published numerous book chapters related to creativity, edited the book Creative Action in Organizations, and has an edited book tentatively titled *The Handbook of Organizational Creativity* scheduled for publication in 2003. He is on the editorial boards of *Academy of Management Learning and Education, Journal of Creative Behavior* and *Creativity and Innovation Management.*

Gordon Foxall is a Distinguished Research Professor at Cardiff University. His chief research interests lie in psychological theories of consumer choice and consumer innovativeness and their relationships to marketing management and strategy. He has published some 16 books and over 250 articles and papers on these and related themes. His undergraduate work was at the University of Salford, and he is also a graduate of the Universities of

Birmingham (Ph.D. industrial economics and business studies) and Strathclyde (Ph.D. psychology). In addition, he holds a higher doctorate of the University of Birmingham (DSocSc).

Monika Friedrich-Nishio studied industrial engineering and management at Karlsruhe University, Germany, and graduated in informatics in 1999. Since 2000 she has been with the section System Dynamics and Innovation of the Institute of Economic Policy Research, Faculty of Economics, Karlsruhe University. With a Humboldt scholarship she was with the Institute of Innovation Research, Hitotsubashi University, Tokyo, from July to November, 2002.

Oliver Gassmann is Professor at the University of St Gallen, Switzerland, and Director of the Institute for Technology Management. In 1996–2002 he was Vice President responsible for Corporate Research and Strategic Technology Management at Schindler Corporation. He holds a Masters degree in economics and business administration from the University of Hohenheim-Stuttgart and a Ph.D. degree in business administration from the University of St Gallen. He is a member of several boards, such as the Board for Science and Research of Economiesuisse, the Comité Stratégique Euresearch, and the Editorial Board of the *R&D Management Journal*. His research focuses on international R&D and strategic innovation management. He has published five books and over 40 papers in the area of technology and innovation management.

Asta S. Georgsdottir is currently completing her doctoral studies at the University of Paris 5, France. She received her B.A. degree in Psychology from the University of Iceland, Reykjavik, and her D.E.A. (Master) degree from the University of Paris 5, France. Her research focuses on cognitive flexibility, with an emphasis on individual differences, and the development of cognitive flexibility in children and adolescents.

Isaac Getz is a Professor of Creativity and Innovation Management at the ESCP-EAP European School of Management, Paris, France. He was a Visiting Professor at Cornell and Stanford Universities and at the University of Massachusetts. Dr Getz's current research focuses on emotion, metaphor, creativity, and idea management. He is editor of a book *Organizational Creativity* (Vuibert, 2002) and co-author of another book *Your Ideas Change Everything* (Editions d'Organisation, 2003).

David V. Gibson is Director, Research and Global Programs, and The Nadya Kozmetsky Scott Centennial Fellow, IC^2 (Innovation, Creativity, Capital) Institute (www.ic2.org), The University of Texas at Austin. In 1983, he earned his Ph.D. from Stanford in organizational behavior and communication theory. He is a consultant to businesses and governments worldwide. His research and papers have been translated into Mandarin, Japanese, Korean, Russian, Spanish, Italian, French, German, Finnish, and Portuguese.

Dennis A. Gioia is Professor of Organizational Behavior in the Department of Management and Organization at the Smeal College of Business Administration, Penn State University. His current theory and research interests focus on organizational identity, image, reputation, learning, and sensemaking. His work has appeared in most of the major journals in the field, and he has co-edited two books: *The Thinking Organization* and *Creative Action in Organizations*. Prior to his academic career he worked for Boeing Aerospace at Kennedy Space Center during the Apollo lunar program and served as vehicle recall coordinator for Ford Motor Company.

Ronald E. Goldsmith is the Richard M. Baker Professor of Marketing at the Florida State University. He has published approximately 150 journal articles and conference papers as well as a book, *Consumer Psychology for Marketing*. His articles have appeared in the *Journal of the Academy of Marketing Science*, *Journal of Business Research*, *Service Industries Journal*, and others. He is currently the North American Editor of the *Service Industries Journal*.

Edgar Gonzalez holds an M.S. and M.D. degree from the London School of Economics, U.K. In Colombia he served as General Director of the Health Services Development Department in the Ministry of Health, Health Advisor to the President of the Republic, and founding Director of the Health Services Management Program at the National School of Public Management. He also worked as a consultant for the Pan-American Health Organization, World Health Organization, United Nations, and Multilateral Development Banks in the fields of public health and social and economic development.

Hariolf Grupp is a Professor of Systems Dynamics and Innovation and Director of the Institute for Economic Policy Research (IWW) in the Economics Faculty at the Karlsruhe Technical University. He graduated from Heidelberg University with Ph.D. in physics and mathematics. Since 1985 he has been at the Fraunhofer Institute for Systems and Innovation Research (ISI) and has been a Deputy Director of this institute since 1996.

Ronald J. Gutmann is a Professor of Electrical Engineering at Rensselaer Polytechnic Institute, where he is active in teaching and research activities in semiconductor devices and integrated circuit interconnect technology. He has authored approximately 300 technical papers in these and related fields and is a Fellow of IEEE for contributions in microwave semiconductor technology. He has co-authored three books in IC interconnect technology: *Chemical-Mechanical Planarization of Microelectronics Materials* (Wiley, 1997), *Copper — Fundamentals for Microelectronic Applications* (Wiley, 2000) and *Chemical-Mechanical Polishing of Low Dielectric Constant Polymers and Organosilicate Glasses* (Kluwer, 2002).

Athanasios Hadjimanolis obtained his first degree in Chemistry from the University of Athens (Greece). He then obtained the certificate and diploma of the Chartered Institute of Marketing (U.K.) and received his MBA and Ph.D. from Brunel University in London (U.K.). He has published several papers in international journals like *Research Policy* and *Technovation*. He has worked for many years in an industrial firm as a division head. He also teaches at the Business School at Intercollege in Nicosia, Cyprus. He is a member of the Chartered Institute of Marketing (U.K.) and other professional associations.

Rebecca Harding is a Senior Fellow of London Business School and the Chief Economist of The Work Foundation. Up until March 2002 she was a Senior Lecturer at the Science Policy Research Unit at the University of Sussex and a Research Associate of the Institute for Public Policy Research in London. Her work has focused on all areas of entrepreneurship and innovation management, including science-based entrepreneurship and international comparisons of venture capital structures and policy. She has published extensively in this area and has acted in an advisory capacity to the U.K., Danish and German regional governments on innovation, venture capital and entrepreneurship. Recently she was appointed as Research Director for the Entrepreneurial Working Party and she is on the Advisory Board of Inner City 100.

Jürgen Hauschildt is a Professor of Business Administration at the University of Kiel and Director of the Institute for Research in Innovation Management. Since 1999 he has been Managing Director of the 'Studienkolleg BWL' at Kiel University, responsible for education of technical and chemical engineers in business administration. Dr Hauschildt also coordinated the research program 'Theory of Innovation Management' sponsored by the German Science Foundation. He has published in the areas of organisation theory, financial management, controlling, financial statement analysis, decision theory, innovation management, and crisis management.

Robert Hausmann is a doctoral candidate in the cognitive program at the University of Pittsburgh. His current research interests include collaborative learning and collaborative scientific discovery.

Thomas E. Heinzen studies creativity among a variety of unlikely populations, provides clinical services to elderly individuals and their families, and writes extensively on creativity and aging. He is the author of six books and numerous scientific articles. He earned his Ph.D. at the State University of New York at Albany, studied the New York State workforce while affiliated with the Rockefeller College of Public Affairs and Policy, and now serves as Associate Professor of Psychology at William Paterson State University of New Jersey.

Manuel Heitor is Professor at the Instituto Superior Tecnico in Lisbon, Portugal, Director of the Center for Innovation, Technology and Policy Research, http://in3.dem.ist.utl.pt/, and a Senior Research Fellow at the IC2 Institute, The University of Texas at Austin, USA.

Gerald Holton is Mallinckrodt Professor of Physics and Professor of History of Science, Emeritus, at Harvard University. He is the author of several books, including *Thematic Origins of Scientific Thought*, *Science and Anti-Science*, *The Advancement of Science*, and *Einstein, History and Other Passions* (all from Harvard University Press). His honors include The George Sarton Medal, The Millikan Medal, election to the presidency of the History of Science Society, and the selection as Jefferson Lecturer.

Chiung-Wen Hsu received her B.S. in Industrial Management from the National Taiwan Institute of Technology in 1983 and continued graduate studies in Management Science at the National Chiao Tung University. After receiving her M.S. in 1985, she joined the Industrial Technology Research Institute's (ITRI) Offices of Planning where she worked as a researcher. While working for ITRI, she entered the Ph.D. program in Management Science at the National Chiao Tung University in 1989, and received her Ph.D. in 1994. Dr Hsu has been teaching at the National Tsing Hua University since 1994. In 2000, she transferred to the Industrial Economics & Knowledge Center, ITRI, where she currently serves as a research manager. Most of her research work has been in the area of technology transfer, project management, and industrial technology policy.

Damien Ihrig earned his B.A. in Developmental Psychology and Russian History from the University of Iowa in 1996. He worked from 1992 to 1996 as a research assistant in the Childrens' Cognitive Research Laboratory coordinating research for Dr Jodie Plumert. He then coordinated research activities for 2 years in the Traumatic Brain Injury Laboratory for Dr Jeffrey Max, a child psychiatrist at the University of Iowa Hospitals and Clinics. He earned his M.A. in Educational Measurement and Applied Statistics from the University of Iowa in May of 2000. His responsibilities at the Belin-Blank Center include administrating the Belin-Blank Exceptional Student Talent Search, a nationwide above-level testing program for academically talented youth; coordinating research activities; and directing the evaluation of Belin-Blank Center student and teacher programs.

Annamária Inzelt is a Founding Director of IKU Innovation Research Centre (1991) and Founding Co-Director of the Centre for Innovation Policy Research and Education for Central and Eastern Europe, Budapest, Hungary (1999). She is a member of several international organisations and has organised some international conferences. Dr Inzelt is author and editor of several books, chapters and articles and regularly participates in international research projects. Her main research interests are: systems of innovations, the role of internationalisation in technology upgrading and research and development, knowledge flows through mobility and migration.

Richard Joseph is an Associate Professor of Telecommunications Management in the Murdoch Business School, Murdoch University, Perth, Western Australia. He has over 20 years of experience in both government and academia, having worked as a science and technology policy advisor for various Australian government departments in Canberra. He has also worked for the OECD in Paris. He has provided advice to the United Nations Development Program and Australian and New South Wales Governments on science and technology parks.

Ralph Katz is a Professor of Management at Northeastern University's College of Business Administration and Principal Research Associate at MIT's Sloan School of Management. He has published numerous articles in leading professional journals and several books, the most recent of which is entitled *The Human Side of Managing Technological Innovation*. In 1981, Dr Katz received the National Academy of Management's 'New Concept Award' for that year's most outstanding contribution in organizational behavior. He was also the 1990 and 1991 recipient of the Academy of Management TIM Division's 'Best Paper' Awards. Dr Katz serves on several journal editorial boards and was the R&D/Innovation and Entrepreneurship Departmental Editor of *Management Science* for the past 10 years.

James C. Kaufman is Assistant Professor at California State University, San Bernardino where he is also the director of the Learning Research Institute. He received his Ph.D. from Yale, and worked as an Associate Research Scientist at Educational Testing Service for two years. He is the Associate Editor of the journal *Research in the Schools*. He is the author of *The Evolution of Intelligence* (with R. Sternberg) and *The Creativity Conundrum* (with R. Sternberg & J. Pretz), as well as the forthcoming *Gender Differences in Mathematics* (with Ann Gallagher) and *Faces of the Muse: Creativity in Different Domains* (with John Baer). Kaufman is interested in many aspects of creativity, including different domains, measurement, its relationship to mental illness, and its relationship with other constructs such as motivation and academic achievement.

Geir Kaufmann is Professor of Psychology at the Norwegian School of Economics and Business Administration. His major research interests are in the field of the psychology of problem-solving and behavioral decision-making, with a special focus on the creativity aspect of these processes. Dr Kaufmann has published a large number of scientific articles and books in these fields. Currently his main interest is in the relationship between affect and problem-solving, and he is engaged in developing and testing a model of the effects of mood on creative problem-solving.

Nigel King has a Ph.D. in psychology from the University of Sheffield, and is currently Reader in Psychology at the University of Huddersfield, U.K. He has published widely in the area of creativity, innovation and change in organisations, including the recent book *Managing Innovation and Change: A Critical Guide for Organisations* (with Neil Anderson). His research has mostly been in health care settings, and he is Director of the Primary Care Research Group at Huddersfield. He has a strong interest in the use of qualitative methods in work and organisational psychology, especially those from a phenomenological perspective.

Ronald Neil Kostoff received a Ph.D. in Aerospace and Mechanical Sciences from Princeton University in 1967. He worked at Bell Labs, Department of Energy, and Office of Naval Research. He invented and patented (1995) the Database Tomography process, a computer-based textual data mining approach that extracts relational information from large text databases. He is listed in *Who's Who in America*, 56th Edition (2002), *Who's Who in America, Science and Engineering*, 6th Edition (2002), and *2000 Outstanding Intellectuals of the 21st Century*, First Edition (2001). He has published many papers on technical, evaluation, and text mining topics, and has edited three journal special issues since 1994 (*Evaluation Review* (February, 1994), *Scientometrics* (July, 1996), *Journal of Technology Transfer* (Fall, 1997)).

Uma Kumar is a Professor of Management Science and Director of the Research Centre for Technology Management at the Eric Sprott School of Business, Carleton University. Her research is in the area of Management of Technology including forecasting and monitoring technology, efficiency in new product development through e-commerce, quality in R&D, managing R&D internationally, R&D and innovation policy, and performance metrics in e-commerce. Dr Kumar has published over 60 articles in journals and refereed proceedings. Her seven papers have won best-paper awards at prestigious conferences. Twice, she has won the Scholarly Achievement Award at Carleton University. Dr Kumar is the recipient of a number of research grants from SSHRC and NSERC.

Vinod Kumar is the Director of the Eric Sprott School of Business at Carleton University where he is also a Professor of Technology and Operations Management and the Head of the Manufacturing Systems Centre. His research is in enterprise system adoption and implementation, e-commerce technology strategy, supply chain management, improving performance of production and operation systems, manufacturing flexibility, technology transfer, quality in R&D, and innovation management in the defence and high-tech sector. Dr Kumar has published more than 75 articles in refereed journals and proceedings. He has won several Best Paper Awards in prestigious conferences. At Carleton University, Dr Kumar has also obtained the Scholarly Achievement Award twice and University Research Achievement Award twice.

Todd I. Lubart is a Professor of Psychology at the University of Paris 5, France. His research focuses on creativity, with emphases on individual differences, the development of creativity, and the role of emotion in creativity. Dr Lubart has been a visiting scholar at the Paris School of Management (ESCP). He is co-author of *Defying the Crowd: Cultivating Creativity in a Culture of Conformity* (Free Press, 1995).

Edward Major completed his Ph.D. in innovation and small firms at the Innovation and Technology Assessment Unit at Cranfield University in 2000. He is currently a Business Analyst with Bedfordshire Police Force in the U.K. and a Part-Time Lecturer in Strategy and Small Business Management at Middlesex University, also in the U.K.

Dora Marinova is the Head of the Institute for Sustainability and Technology Policy at Murdoch University, Perth, Western Australia where she has been working in the last 12 years and is currently a Senior Lecturer. She completed her Ph.D. in Bulgaria in the field of innovation theory. Dr Marinova's research interests include innovation, science and technology policy, environmental policy, Eastern Europe and technology transfer. She has published over 50 refereed journal articles and book chapters, and has been involved with more than 15 contract reports for government and industry. She is also a Fellow of the Modeling and Simulation Society of Australia and New Zealand.

Kavita Mehra has a Ph.D. in plant tissue culture from Delhi University and is a research scientist at National Institute of Science Technology and Development Studies (NISTABS), CSIR, India. Her research focuses on issues related to innovations, innovation networks and commercialization of technologies. Dr Mehra has dealt with these issues across various sectors and technologies such as plant tissue culture, floriculture, sericulture, dairy and information technology. Also worked on the concepts and methodologies of Action Research and for that, has been a visiting faculty outside India. She has published several papers in international Journals and contributed chapters in books edited by internationally recognized authors. The editorship of a special issue of an international journal '*AI & Society*' (Springer Press) on Enterprise Innovations is there to her credit.

Yukio Miyata is a Professor at the Department of Economics, Osaka Prefecture University, Osaka, Japan. After obtaining his B.A. degree in Economics from Osaka University in 1983 and B.S. degree in Materials Engineering from the University of Washington, Seattle, in 1987, he obtained an M.S. degree in Engineering and Policy in 1989 and Ph.D. in Economics in 1994, both from Washington University, St Louis. His current research interest is economic and policy analysis of technological innovations. He is particularly interested in inter-organizational collaboration for R&D. His publications include papers in *Technovation* and *Japan and the World Economy*, and three books in Japanese.

Laure Morel is a Researcher at Ecole Nationale Supérieure en Génie des Systèmes Industriels in Nancy, France. She works on constructivism knowledge management and organizational learning, and conducts research in the chemical and food industries.

Steven Muegge is completing an M.Eng. degree in Telecommunications Technology Management at Carleton University. He has eight years of telecommunications industry experience in various managerial and technical roles. His responsibilities have spanned all phases of product development, from concept definition to customer deployment, including management of hardware, software, and systems products. He has published at the International Intersociety Electronics and Photonic Packaging Conference (InterPACK) and is the inventor of a U.S. patent. Steven graduated from McMaster University with a B.Eng. in Engineering Physics, with concentrations in optical communications and solid-state devices. He is a certified Project Management Professional (PMP).

Rajiv Nag is a Doctoral Candidate in the Department of Management and Organization at the Smeal College of Business Administration, Penn State University. His current research interests focus on organizational knowledge, managerial cognition, and strategic management. Prior to commencing his doctoral studies, Rajiv worked in areas of marketing and manpower development with Fiat SpA and Leo Burnett Inc.

Thomas Nickles, Professor and Chair of the Philosophy Department at the University of Nevada, Reno, specializes in history and philosophy of science and technology. He has published numerous articles on scientific explanation, reduction of theories and of problems, problem-solving, heuristic appraisal, and innovation. He is editor of *Scientific Discovery, Logic, and Rationality* and *Scientific Discovery: Case Studies* (both Reidel, 1980) and of *Thomas Kuhn* (Cambridge University Press, 2003).

Paul Nightingale is a Senior Research Fellow in the Complex Product Systems Innovation Centre at the Science Policy Research Unit (SPRU), University of Sussex, U.K. Originally a chemist, he has industrial experience managing R&D projects for a large blue-chip firm. His Ph.D. research analyzed the introduction of computer simulation technologies in R&D processes. His research interests include IT project management in financial services, the relationship between scientific research and technical change, and innovation in instrumentation technologies. He has written on banking innovation, the introduction of genetics technologies, the role of science in society, and aerospace project management. He publishes in *Research Policy* and *Industrial and Corporate Change.*

Ikujiro Nonaka is a Professor at the Graduate School of International Corporate Strategy of Hitotsubashi University in Tokyo. He is also the Xerox Distinguished Professor in Knowledge at the Haas School of Business, University of California at Berkeley, and Visiting Dean and Professor at the Centre for Knowledge and Innovation Research of Helsinki School of Economics and Business Administration, Finland.

A. Jai Persaud is a senior policy advisor with Natural Resources Canada (NRCan), Corporate Policy and Portfolio Coordination Branch. He previously held several positions including Senior Economist and Head, Industry and Supply Analysis, with the Energy Policy Branch, Energy Sector, NRCan; the National Energy Board; the Bureau of Competition Policy; and the Alberta Energy Utilities Board. Mr Persaud has held several key assignments, including his most recent involvement with the high-profile Canada Transportation Act Review, and has undertaken significant research in the area of energy. He holds a B.A. and an M.A. in economics from the University of Calgary and an M.B.A. from the University of Ottawa. He is a Ph.D. candidate in the Sprott School of Business, Carleton University in the area of technology management.

John Phillimore has worked for the past 10 years at the Institute for Sustainability and Technology Policy at Murdoch University, Perth, Western Australia, where he is a Senior Lecturer. He has a Ph.D. from the Science Policy Research Unit at the University of Sussex, U.K. He has worked as an advisor to the Minister for Technology and the Minister for Housing, Local Government and Regional Development in the Western Australian State Government. His research interests include innovation and industry policy, research policy, vocational training policy, environmental policy, and the future of social democracy.

Evgueni Ponomarev is at the Department of Physics, University of Ottawa, Canada. His current research interests include organizational learning and development, innovation technology management, and R&D management.

Andrea Prencipe is Associate Professor of Economics and Management of Innovation, Università G. D'Annunzio, Italy and a Research Fellow at the CoPS Innovation Centre in SPRU—Science and Technology Policy Research Unit, University of Sussex, U.K. His qualifications include a Ph.D. in Technology and Corporate Strategy (SPRU), an M.A. in Innovation Management (S. Anna School of Advanced Studies, Pisa, Italy), an M.Sc. in Technology Innovation Management (SPRU), and a B.A. in Economics and Business (University G. d'Annunzio). His research interests include the co-ordination and division of innovative labour in multitechnology industries and organizational memory in project-based organisations. He has published in journals such as *Administrative Science Quarterly*, *Industrial and Corporate Change*, and *Research Policy*.

Jean E. Pretz is a doctoral candidate in the Psychology Department at Yale University. She received a B.A. in psychology and music from Wittenberg University, Springfield, Ohio, and an M.S. and M.Phil. from Yale University. She has received awards from the American Psychological Association, the American Psychological Foundation, and the American Psychological Society Student Caucus for her research on intuition and insight in problem solving. Her research interests include creativity, expertise, and transfer in problem-solving.

Sally M. Reis is a Professor and the Department Head of the Educational Psychology Department at the University of Connecticut where she also serves as Principal Investigator of the National Research Center on the Gifted and Talented. She was a teacher for 15 years, 11 of which were spent working with gifted students on the elementary, junior, and high school levels. She has authored more than 130 articles, nine books, 40 book chapters, and numerous monographs and technical reports. Her research interests are related to special populations of gifted and talented students, including: students with learning disabilities, gifted females and diverse groups of talented students.

Jean Renaud is Assistant Professor of Concurrent Engineering at Ecole Supérieure en Génie des Systèmes Industriels in Nancy, France. His research focuses on knowledge capitalization. Dr Renaud currently serves as innovation expert to French firms, and he heads a National Association on Project Management.

Joseph Renzulli is the Raymond and Lynn Neag Distinguished Professor of Educational Psychology at the University of Connecticut, where he also serves as Director of the National Research Center on the Gifted and Talented. His research has focused on the identification and development of creativity and giftedness in young people, and on organizational models and curricular strategies for differentiated learning environments and total school improvement. A focus of his work has been on applying the pedagogy of gifted education to the improvement of learning for all students.

Tudor Rickards is Professor of Creativity and Organisational Change at Manchester Business School. He is also head of the Organisation Behaviour group at MBS. His earlier career included posts in a medical school, and as a technical professional within an industrial Research at Development Laboratory. He has published extensively on creativity, innovation, and change management. He has won numerous awards for teaching in these areas, and has been Alex Osborn visiting professor for creativity at the State University of New York, Buffalo. He is founder of the Creativity and Innovation Management journal, and co-founder of the European Association for Creativity and Innovation. His latest book is *Creativity and the Management of Change*, published by Blackwell.

Maxine Robertson is a Lecturer in Organizational Behavior at Warwick Business School, University of Warwick, U.K. She has a Ph.D. in Organization Studies from Warwick Business School. She is a member of the Innovation, Knowledge and Organizational Networking (IKON) Research Centre in Warwick Business School. Her main research interests are the management of knowledge workers and knowledge intensive firms, knowledge management and innovation. She has published extensively in all of these areas.

Michele Root-Bernstein is author or co-author of three books, the most recent being *Sparks of Genius, The Thirteen Thinking Tools of the World's Most Creative People* (1999), a close look at creative imagination across the arts and sciences. As an independent scholar and writer, with a Ph.D. in History from Princeton University, she has published articles and essays in a diverse array of scholarly and literary journals and has taught in elementary, secondary and university classrooms. In addition, she has consulted on creative process and transdisciplinary schooling for educational institutions and museums across the United States and abroad.

Robert Root-Bernstein is the author or co-author of four books, including *Discovering* (1989) and, with Michele Root-Bernstein, of *Sparks of Genius, The Thirteen Thinking Tools of the World's Most Creative People* (1999), both of which explore the nature of creative thinking. He earned his Ph.D. in History of Science at Princeton

University, received a MacArthur Fellowship while doing post-doctoral research at the Salk Institute for Biological Studies in 1981, and is currently a Professor of Physiology at Michigan State University. In addition to his academic studies, he has also consulted widely for the pharmaceutical and biotech industries and has several patents himself.

Keigo Sasaki is an Assistant Professor of business administration at Yokohama City University, Japan.

Harry Scarbrough is a Professor in Warwick Business School and is Director of the ESRC's Evolution of Business Knowledge research program in the U.K. Previously he was Research Director at the University of Leicester Management Centre. Dr Scarbrough's current interests include the influence of social networks on innovation, the transfer of learning between projects, and management practices for the evaluation of human capital within firms.

Larry Scripp has conducted extensive research in art and education at the Harvard Graduate School of Education's Project Zero and has published many articles about children's musical development, computers and education, and the acquisition of music literacy skills. He has also become a nationally known educator and researcher through his work for Harvard Project Zero's Arts PROPEL project, the Leonard Bernstein Center for Education Through the Arts, and his formative role in creating The Research Center for Learning Through Music and the National Music-in-Education Consortium at the New England Conservatory. In addition he is a founding director for the Conservatory Lab Charter School in Boston and is currently chair of the new Music-in-Education Program at New England Conservatory.

Kavita L. Seeratan has a Masters Degree in Human Development and Applied Developmental Psychology, obtained from the University of Toronto. She also has an Honours Bachelors Degree with a Specialist in Psychological Research and is currently pursuing her Doctoral studies at the University of Toronto. She works as a Research Associate at the University of Toronto's Adaptive Technology and Resource Centre and is Director of Research and Development for the Learning Disabilities Resource Community Project. Kavita is also currently involved as a key member in a project development team aimed at the development, testing and application of a new Meaning Equivalence Instructional Methodology (MEIM) geared at the assessment of deep comprehension.

Larisa V. Shavinina is a Professor at the Department of Administrative Sciences in the University of Québec en Outaouais, Hull, Canada. Her research program includes such projects as Psychological Foundations of Innovation, Innovations in Silicon Valley North, Innovation Leadership, Innovation and Knowledge Management, and Innovation Education. Originally focusing on psychology of high abilities (i.e. talent, giftedness, creativity), Dr Shavinina's research has expanded to encompass innovation. She is engaged in research aimed at unifying the field of innovation, that is, to merge psychological, management, and business perspectives together. Her publications have appeared in the *Review of General Psychology, New Ideas in Psychology, Creativity Research Journal*, and others. She co-edited *CyberEducation* and *Beyond Knowledge*.

Keng Siau is Associate Professor of Management Information Systems at the University of Nebraska-Lincoln. He is Editor-in-Chief of the *Journal of Database Management* and Editor of the *Advanced Topics in Database Research* book series. He received his Ph.D. degree from the University of British Columbia where he majored in Management Information Systems and minored in Cognitive Psychology. His Master and Bachelor Degrees are in Computer and Information Sciences from the National University of Singapore. Dr Siau has edited 7 books and written 15 book chapters. In addition, he is the author of over 45 refereed journal articles and over 60 refereed conference papers. His journal papers have appeared in such journals as *Journal of Creative Behavior, Management Information Systems Quarterly, Communications of the ACM, IEEE Computer, Information Systems, DATABASE, IEEE Transactions, Journal of Computer Information Systems, Journal of Database Management, Journal of Information Technology,* and *International Journal of Human-Computer Studies*.

Dean Keith Simonton is Professor of Psychology at the University of California, Davis. He published more than 250 articles and chapters plus eight books, including *Genius, Creativity, and Leadership, Scientific Genius, Greatness, Genius and Creativity, Great Psychologists,* and *Origins of Genius*, which received the William James Book Award from the American Psychological Association. Other honors include the Sir Francis Galton Award for Outstanding Contributions to the Study of Creativity, the Rudolf Arnheim Award for Outstanding Contributions to Psychology and the Arts, and others. He served as President of the International Association of Empirical Aesthetics from 1998 to 2000. His research focuses on the cognitive, personality, developmental, and socio-cultural factors behind exceptional creativity, leadership, genius, talent, and aesthetics.

Hock-Peng Sin is currently a graduate student in industrial and organizational psychology at Pennsylvania State University. His research interests include performance appraisal and management, work group effectiveness, work motivation, as well as levels and temporal issues in organizational research.

Gerald F. Smith is a Professor in the Department of Management at the University of Northern Iowa, where he teaches managerial decision-making and problem-solving. Dr Smith studies and writes about critical thinking, problem-solving, creativity, and other topics related to higher-order thinking, especially in practical, real-world, contexts.

Vangelis Souitaris is a lecturer and research coordinator at the Entrepreneurship Centre of Imperial College, London, U.K. He is a Chemical Engineer with an M.B.A. from Cardiff University, and a Ph.D. in Technological Innovation Management from Bradford University. His research is on technology and innovation management and entrepreneurship. He has published in such journals as *Research Policy, R&D Management, Technovation, International Journal of Innovation Management,* and *British Journal of Management.*

Robert J. Sternberg is IBM Professor of Psychology and Education in the Department of Psychology at Yale University and Director of the Center for the Psychology of Abilities, Competencies, and Expertise. He is 2003 President of the American Psychological Association.

Rena F. Subotnik is Director of the Center for Gifted Education Policy funded by the American Psychological Foundation. The Center's mission is to generate public awareness, advocacy, clinical applications, and cutting-edge research ideas that will enhance the achievement and performance of children and adolescents with special gifts and talents in all domains (including the academic disciplines, the performing arts, sports, and the professions). She is author of *Genius Revisited: High IQ Children Grown Up* and co-editor of *Beyond Terman: Contemporary Longitudinal Studies of Giftedness; Talent, Remarkable Women: Perspectives on Female Talent Development,* and the 2nd edition of the *International Handbook of Research on Giftedness and Talent.* Dr Subotnik is the 2002 Distinguished Scholar of the National Association for Gifted Children.

Jon Sundbo is a Professor of Business Administration specializing in innovation and technology management at Department of Social Sciences, Roskilde University, Denmark. He is also a Director of the Center for Service Studies, coordinator of the Department's research in the area of Innovation and Change Processes in Service and Manufacturing, Director of the University's Ph.D. program Society, Business and Globalization. Dr Sundbo has been doing research on innovation, service management and the development of the service sector, human resources management and organization. He has published articles on innovation, entrepreneurship, service and management and has published several books, including *The Theory of Innovation* and *The Strategic Management of Innovation.*

Jacky Swan is Professor in Organizational Behavior at Warwick Business School, University of Warwick. She completed her Ph.D. in Psychology. She is founding member of IKON—research Centre in Innovation, Knowledge, and Organizational Networks—and conducts her research in related areas. Professor Swan's current interests are in linking innovation and networking to processes of managing knowledge. She has been responsible for a number of U.K. Research Council projects in innovation and is currently working on a study of managing knowledge in project-based environments. She has published widely, including articles in *Organization Studies, Organization, Human Relations*, and is co-author of *Managing Knowledge Work* (Palgrave, 2002).

George Swede is Chair of the Department of Psychology at Ryerson University in Toronto, Canada. He has published two books and a number of papers on the psychology of art and creativity. Furthermore, he has published 30 collections of poetry, six poetry anthologies as editor, one book of literary criticism, as well as poems, short stories and articles in over 125 different periodicals around the world. His work has also appeared on radio and television in six countries and has been translated into 15 languages.

Hung-Kei Tang obtained his B.S. and M.S. degrees in electronics engineering from Queen's University in Canada and University College in London, U.K., respectively. After practising as an engineer and manager in Canada, Hong Kong, the Netherlands and Singapore, he returned to academia. He obtained his Ph.D. in innovation management from the Nanyang Technological University in Singapore where he is currently an Associate Professor. He has more than 60 academic publications. He is the Chairman of the IEEE Engineering Management Society in Singapore.

Paul Tesluk is an Associate Professor in Management and Organization at University of Maryland, Robert H. Smith School of Business. His research interests include the implementation and utilization of technology in the workplace, design and implementation of high-involvement workplace systems, work team performance, and employee and managerial development.

Paul Trott is a Principal Lecturer, Business School, University of Portsmouth. He holds a Ph.D. from Cranfield University. He is author of many reports and publications in the area of innovation management. His book *Managing Innovation and New Product Development* published by Financial Times Management is now in its second edition.

Nancy Vail has focused her career on human resources in the United States Office of Personnel Management while earning graduate degrees at the University of California, Davis and the University of Florida, Gainesville. Her interest in aging is complemented by further studies in clinical psychology at William Paterson State University of New Jersey.

Larry Vandervert is a theoretician, writing in psychology and the neurosciences for the last two decades. He co-founded The Society for Chaos Theory and the Life Sciences in 1991. In the past two decades he has authored many articles on a brain-based theory and epistemology. Dr Vandervert proposes that the true fundamentals of science must eventually be stated in terms of brain algorithms. In recent years, Dr Vandervert has published a series of papers that describe how he believes that mathematics arises from collaborative functions of brain areas. He believes that mechanisms in the brain that generate and manipulate patterns are responsible for both the discovery of mathematics and the processes that lead to innovation.

Robert W. Veryzer is an Associate Professor in the Lally School of Management and Technology at Rensselaer Polytechnic Institute, Troy, New York. His work experience includes product planning and product management positions with *Fortune 500* firms as well as design and new product development consulting. He holds an M.B.A. from Michigan State University and a Ph.D. in consumer research and marketing from the University of Florida. His research focuses on product design, new product development, radical innovation, and various aspects of consumer behavior. Dr Veryzer's articles appear in leading professional journals, and he is co-author of *Radical Innovation: How Mature Firms Outsmart Upstarts* (Harvard Business School Press, 2000).

Robert Weisberg is Professor of Psychology at Temple University. He received a Ph.D. in psychology from Princeton University, and his areas of interest are problem-solving and creative thinking. He has published papers investigating cognitive mechanisms underlying problem-solving, and has published papers and books examining cognitive processes underlying creative thinking. His current work in problem-solving is investigating mechanisms underlying 'insight'. He is also carrying out research in creative thinking, which focuses on case studies of major creative advances, using historical 'data' to draw conclusions concerning cognitive processes involved in creative thinking.

John Wetter is Manager of e-Services Business Unit at the Xerox Corporation He is also a Doctoral Candidate at the GWU SBPM MSII Program. He leads a specialized group focused on consulting to major corporations regarding their document management needs. He developed and implemented a number of service offerings, including Digital Fulfillment, Financial Transaction Health Care and Digital Imaging Management. Prior to his current position, he managed Globalization & Localization Services for Software Development as Vice President-General Manager. Other positions held during his career were Manufacturing Manager, Materials Manager, Sales & Marketing, and Production Control Manager.

Beate E. Wilhelm, Ph.D., is Managing Director of z-link, Zurich, a non-profit institution at the interface between higher education, industry, and government in the Greater Zurich Area. Previously, she was a member of the executive board of z-link. She has many years of experience as consultant and scientist in promoting strategic innovation processes to develop regional economies. Dr Wilhelm also worked at the University of St Gallen, Switzerland, and at the Fraunhofer-Institute for Production Technology and Automation, Department of Business Organization, Germany. She is founder member and member of the board of the Swiss Association for Studies in Science, Technology, and Society. Her main research interests include knowledge production, knowledge- and technology transfer, strategic innovation processes, regional innovation policy, and self-organization.

Richard W. Woodman is the Fouraker Professor of Business and Professor of Management at Texas A&M University where he teaches organizational behavior, organizational change, and research methodology. He served as Head of the Department of Management from 1993 to 1997. His research interests focus on organizational change and organizational creativity. His current research and writing explore the role of the organization in individual change, the relationship between processes of creativity and organizational change, and reoccurring patterns in the failure of change programs. Woodman is co-editor of the JAI Press annual series, *Research in Organizational Change and Development*; his co-authored text, *Organizational Behavior*, is in its 9th edition.

Khim-Teck Yeo obtained his Ph.D. in Project Management from the University of Manchester. He has been a visiting scholar at the Sloan School of Management, MIT and visiting professor in Industrial Engineering at the University of Washington. His main academic interests are in project, systems and technology management. He has published widely in these fields with over 120 papers in international journals and conference proceedings. He serves on the Editorial Board of the *International Journal of Project Management*. He was the founding President of the Project Management Institute (PMI), Singapore Chapter. His is currently an Associate Professor in Nanyang Technological University, Singapore, and Director of the Centre for Project Management Advancement.

Maximilian von Zedtwitz is Professor of Technology Management at IMD—International Institute for Management Development, Lausanne, Switzerland. He holds Ph.D. and M.B.A. degrees from the University of St Gallen, and M.Sc. and B.Sc. degrees from ETH Zurich. He was a research associate at the Institute for Technology Management in Switzerland, and a Visiting Fellow at the Graduate School of Arts and Sciences at Harvard University in Cambridge, Massachusetts. He joined IMD in summer 2000 to teach international innovation strategy, R&D management, and technology-based incubation. He has published two books and more than 40 papers on international innovation management and R&D. He won the 1998 RADMA prize for the best paper in the *R&D Management Journal*.

Jing Zhou is Associate Professor and Mays Fellow in the Management Department at the Mays Business School at Texas A&M University. Her current research interests focus on personal and contextual factors that promote or inhibit employee creativity and innovation. She has published widely in top journals in the field, such as *Academy of Management Journal, Journal of Applied Psychology, Journal of Management*, and *Personnel Psychology*. She currently serves on the editorial boards of *Journal of Applied Psychology* and *Journal of Management*.

Preface

There is no doubt that innovations were, are, and will be extremely important for the individual and society. One way to understand the history of human culture is via its inventions and discoveries. All human cultural development builds on the amazing technological, scientific, educational, and moral achievements of the human mind. Today, people increasingly realize that innovations are even more critical than in the past. Thus, industrial competition is increasingly harsh and companies must continuously bring innovative products and services to the global market. To survive, companies need creative and inventive employees whose novel ideas are, to a certain extent, a necessity for the companies' continued existence and future success. Consequently, modern society desperately requires highly able citizens who can produce innovative solutions to current challenges and contribute new ideas that aid in the development and growth of the market for a particular product or service. People of exceptionally innovative ability thus remain as extremely important source of innovation and renewal. Hence, the new millennium is characterized by the need for innovative minds. Contemporary society, without doubt, is highly reliant on innovations. The future will be synonymous with innovation, since it will need an extremely high saturation of innovations in all areas of human endeavour. Despite the quite evident importance of innovations in the life of any societal 'organism', one should acknowledge that the phenomenon of innovation is far from well understood. Because of this, a handbook on innovation is an exceptionally timely endeavour.

The field of innovation is lively on many fronts. For the most part, it is studied today by many different disciplines as one of their components. However, contemporary demands on innovative ideas, solutions, products, and services mark an imperative of a special unified, multifaceted and multidimensional discipline: science of innovation. It was a combination of my interest in the topic of innovation with my perception of the field as needing unification that encouraged me to initiate the present project. I came to believe that the time was ripe for an *International Handbook on Innovation*—a volume that would help guide research in innovation during at least the next decade and, therefore, would advance the field.

The purpose of the handbook is multifold: (a) to pose critical questions and issues that need to be addressed by research in a given subfield of innovation; (b) to review and evaluate recent contributions in the field; (c) to present new approaches to understanding innovation; and (d) to indicate lines of inquiry that have been, and are, likely to continue to be valuable to pursue. This handbook does not provide the kind of literature reviews usually found in textbooks. The conventional understanding of handbook—as a compendium of review chapters suggesting a guide to practice—seems to be very restricted in the context of the field of innovation. The 'handbook' title suggests a guide to practice only in cases where the body of knowledge is understood to be complete and more or less unchanging. For example, 'Handbook of Mathematical Formulae', or 'Handbook of Motorcycle Repair'. However, the study of innovation is a body of knowledge under dynamic theoretical development, and so I prefer to use the 'International Handbook on Innovation' instead of the 'International Handbook of Innovation'. I hope readers will find the present chapters lively and provocative, stimulating greater interest in the science of innovation.

The handbook covers a wide range of topics in innovation. The handbook offers a broad analysis of what innovation is, how it is developed, how it is managed, how it is assessed, and how it affects individuals, groups, organizations, societies, and the world as a whole. The handbook will therefore serve as an authoritative resource on many aspects of theory, research, and practice of innovation. In short, the handbook can be considered as the first serious attempt to unify the field of innovation and, consequently, as the beginning of unified science of innovation. I hope that readers of this handbook will view it as serving that function.

The target international audience for this handbook is broad and includes a wide range of specialists—both researchers and practitioners—in the areas of psychology, management and business science, education, art, economics, technology, administration, and policy-making. Non-specialists will also be interested readers of this handbook, and it will be useful in a wide range of graduate courses as supplementary reading. Because the coverage of the handbook is so broad, it can be read as a reference or an as-needed basis for those who would like information about a particular topic, or from cover to cover either as a sourcebook or as a textbook in a course dealing with innovation. In short, anyone interested in knowing the wide range of issues regarding innovation will want to read this handbook.

The handbook hopes to accomplish at least four things for readers. First, the reader will obtain expert insight into the latest research in the field of innovation. Indeed, some of the world's leading specialists agreed to contribute to this handbook. Second, the handbook will present many facets of innovation including its nature, its development, its measurement, its management, and its social, cultural, and historical context (most books are devoted to only one of these facets). This breadth will allow the reader to acquire a comprehensive and panoramic picture of the nature of innovation within a single handbook. Third, based on this picture, the reader will develop an accurate sense of what spurs potentially creative and innovative people and companies toward their extraordinary achievements and exceptional performances. Fourth, and perhaps most importantly, the reader will be able to apply the ideas and findings in this handbook to critically consider how best to foster their own innovative abilities and innovative performance in their organization.

There are many people to thank for helping this handbook come to fruition. Most important are the authors: I thank them for their willingness to undertake the difficult and challenging task of contributing chapters. I am particularly grateful to Professor Marina A. Kholodnaya, my former Ph.D. supervisor, who to a great extent 'made' me a researcher, developing my perception of scientific problems. She continually inspires me to undertake innovative endeavors. I am also grateful to my editors at Elsevier Science—Gerhard Boomgaarten and Geraldine Billingham, Publishing Editors, Lesley Roberts, Marketing Editor, and Debbie Raven, Senior Administrative Editor—who provided just the right blend of freedom, encouragement, and guidance needed for successful completion of this project.

I also wish to acknowledge my debt of gratitude to my parents, Anna Shavinina and Vladimir Shavinin, who aroused a passionate intellectual curiosity in me. Finally, I owe my biggest debt of gratitude to my husband, Evgueni Ponomarev, and our four-year-old son, Alexander. In countless ways, Evgueni has been a true colleague, critic, and friend throughout the four years of the project. He provided the moral, financial, and technical support, and—more importantly—the time I needed to complete this project. He did so by performing a number of great tasks, from cooking and administering PC problems when I worked at nights, to assuming the lion's share (and the lioness's, too) of child care for our Alexander, who was born during the course of this project. Very simply, this is his handbook, too.

I especially wish to thank Alexander, whose entry into the world taught me more about innovation and the need for it than has any other single event in my life. He demonstrates everyday that innovation is constantly possible. I sincerely hope that today's children from around the world will grow up to be innovative individuals—in both professional and daily life.

Larisa V. Shavinina

Part I

Introduction

The International Handbook on Innovation
Edited by Larisa V. Shavinina

Understanding Innovation: Introduction to Some Important Issues

Larisa V. Shavinina

Département des Sciences Administratives, Université du Québec en Outaouais, Canada

Abstract: This introduction provides an overview of the multidisciplinary and multi-faceted research on innovation presented in the following chapters of this handbook. The main contents of each chapter are summarized, and approaches taken by chapter authors are described.

Keywords: Innovation; Creativity; Approaches to understanding innovation; Management; Business science.

Introduction

Most scholars and decision-makers will agree that innovations are necessary for individuals, groups, and society as a whole. There is also a consensus that human beings must advance in their study of innovation in greater detail. But there are a relatively small number of individuals around the world who study innovation. My goal in bringing them together in this handbook is to present a comprehensive picture of contemporary innovation research by integrating the quite diverse findings obtained by scholars from highly specialized and frequently remote disciplines, and to outline directions for further research, thus advancing the field. In choosing chapter authors, I was particularly interested in those new models, theories, and approaches, which they proposed. My deepest belief is that any handbook on any scientific topic should not only report the current findings in the field, but must also advance that field by presenting challenging new ideas. In one way or another each chapter in the handbook adds something new to our existing edifice of knowledge about innovation. This is the main merit of this handbook, which is international in scope, reflecting American, Canadian, Asian, European, and global perspectives. The chapter authors take a number of different approaches, both empirical and theoretical, reflecting a variety of possible perspectives and research methods aimed at understanding innovation. These range from case studies and autobiographical and biographical methods to experimental methods. I will briefly describe these approaches below. The handbook is divided into XV parts. The first part provides a general introduction to the work. Parts II to

XIV, consisting of 69 chapters, represent distinctive, although sometimes overlapping, approaches to understanding innovation. The final part of the handbook integrates these approaches.

Part I comprises just the present chapter, Chapter 1, which sets the stage for understanding innovation. This chapter describes the various approaches used by authors of this handbook in understanding innovation and briefly summarizes the main contents of each chapter.

Part II of the handbook describes work aimed at the understanding the multifaceted nature of innovation, its basic mechanisms and its various facets. This part presents neurophysiological, psychological, philosophical, sociological, economic, management and business science perspectives on innovation. This part comprises 15 chapters.

In Chapter 1 of Part II, *The Neurophysiological Basis of Innovation*, Larry Vandervert describes for a broad audience how the repetitive processes of working memory are modeled in the brain's cerebellum. He argues that when these models are subsequently fed back to working memory they are experienced as new, more efficient concepts and ways of doing things. As this process is repeated, the resulting degree of generalization (abstraction) increases. When multiple pairs of models are learned in working memory, they may give rise to sudden experiences of insight and intuition. To illustrate the working memory/cerebellar process of innovation, Vandervert walks the reader through three of Albert Einstein's classic subjective accounts of discovery. This is the only chapter in the handbook, which sheds light on the neuropsychological nature of innovation. The neuropsychological foundations of innovation is a promising new direction in research on innovation.

Chapter 2, *On the Nature of Individual Innovation*, by Larisa V. Shavinina and Kavita L. Seeratan, introduces a fascinating theme, which also will be discussed in other chapters of the handbook from various angles. The theme is why some individuals are exceptionally able to generate new ideas, which lead to innovation. The chapter presents a new psychological conception of individual innovation. According to the conception, individual innovation is a result of a specific organization of an individual's cognitive experience. This organization is, in turn, a result of the protracted inner process of the actualization, growth, and enrichment of one's own cognitive resources and their construction into an unrepeatable cognitive experience during accelerated mental development. The direction of this process is determined by specific forms of the organization of the individual cognitive experience (i.e. conceptual structures, knowledge base, and mental space). The unique structure of the mind, which makes possible the creative ideas leading to innovation, is being formed on the basis of this process. The uniqueness of innovators' minds expresses itself in objective representations of reality; that is, in their unique intellectual picture of the world. This means that innovators see, understand, and interpret the world around them by constructing an individual intellectual picture of events, actions, situations, ideas, problems, any aspects of reality in a way that is different from other people.

In contrast to one of the psychological understandings of innovation presented in Chapter 2, Dora Marinova and John Phillimore's *Models of Innovation*, Chapter 3, reviews a number of models developed by economists, management and business scholars, sociologists, geographers, and political scientists that are used to explain the nature of innovation. Their overview includes six generations of models: the black box model, the linear model, the interactive model, the systems model, the evolutionary models, and the innovative milieu model. The authors view innovation mainly as a process leading to generating new products. Marinova and Phillimore analyze each model, its explanatory power and related concepts, and draw further research directions. A comprehensive review of the six models of innovation provides readers with a panoramic picture of the evolution of researchers' views on the nature of innovation.

In Chapter 4, *Evolutionary Models of Innovation and the Meno Problem,* Thomas Nickles proposes a philosophical view of innovation. He presents universal Darwinism as a new approach to understand innovation. According to this approach, innovation is the product of blind variation plus selective retention (BV + SR) and is thus a kind of adaptation, that is, a selective-adaptive process. In Nickles' view, accepting

BV + SR enables human beings to recognize sources of innovation other than new ideas of creative and talented individuals. He, however, points out that the human design model is not entirely wrong, but it turns out to be based on previous applications of BV + SR.

Chapter 5, *Three-Ring Conception of Giftedness: Its Implications for Understanding the Nature of Innovation,* by Joseph S. Renzulli, is the chapter where an author was given an explicit assignment; in this case, to apply his well-known conception of giftedness toward achieving an understanding of innovation. According to the conception, giftedness that leads to innovation emerges from the interaction and overlap of three clusters of traits; high ability in a particular domain, task commitment, and creativity, and occurs in certain individuals, at certain times, under certain conditions. This is another psychological attempt to address the intriguing issue about where innovation 'comes from'. Renzulli's Three-Ring Conception of Giftedness focuses on creative/innovative productivity, which differs from academic giftedness, and is thus appropriate for understanding forces leading to the appearance of innovators.

In Chapter 6, *Innovation and Strategic Reflexivity*, Jon Sundbo presents a strategic innovation theory, which explains activities that lead to innovations in firms. His theory, in which strategic reflexivity is the core concept, is based on evolutionary theory. According to the theory, market conditions and internal resources are the drivers of the innovation process. Firms manage their innovation process and market position through the strategy. Reflexivity is considered as a process during which managerial staff and employees define their strategy. Sundbo presents case studies of innovation in service firms, which support his strategic innovation theory.

Chapter 7, *The Nature and Dynamics of Discontinuous and Disruptive Innovations from a Learning and Knowledge Management Perspective*, by Elias G. Carayannis, Edgar Gonzalez and John Wetter, discusses the nature and dynamics of innovation as a socio-technical phenomenon. Specifically, these authors analyze the evolutionary and revolutionary dimensions of innovation, that is, discontinuous and disruptive types of innovation, respectively. They claim, and provide evidence to support that claim, that the key to organizational competence for generating and leveraging discontinuous—and especially disruptive—innovations is in the individual and organizational capacity for higher-order learning and for managing the stock and flow of knowledge. This chapter is the first in the Handbook to introduce important topics of organizational learning and knowledge management and their growing role in the emergence of innovation. This role is so significant that it is generally recognized today that we live in a society, which is largely based upon a knowledge-based economy.

In Chapter 8, *Profitable Product Innovation: The Critical Success Factors*, Robert G. Cooper analyzes the critical success factors that underlie new product performance, relying upon his and other's research into hundreds of new product launches, probing the question: 'what distinguishes the best from the rest?' Ten common denominators or factors appear to drive new product success, profitability and time-to-market. The chapter outlines these ten critical success factors, and notes the management implications of each.

Chapter 9, *Types of Innovations*, by Robert J. Sternberg, Jean E. Pretz and James C. Kaufman, describes various innovative forms, each representing a different kind of creative contribution. Based on Sternberg's propulsion model of creative contributions, the authors present the following eight types of innovation: replication, redefinition, forward incrementation, advance forward incrementation, redirection, reconstruction, reinitiation, and integration. For example, a conceptual replication is a minimal innovation, simply repeating with minor variations an idea that already exists (e.g. when Mercury puts the 'Mercury' label on what is essentially an already-existing Ford car). Forward incrementations represent next steps forward in a line of progression (e.g. the 2001 version of a 2000 Ford car). Redirections represent a totally different direction for products that diverges from the existing line of progress (e.g. electric cars). The authors discuss these types of innovations and the circumstances under which they are likely to be more or less successful.

In Chapter 10, *Problem Generation and Innovation*, Robert Root-Bernstein further extends our understanding of innovation. Following Albert Einstein and many other innovators in the sciences and engineering, Root-Bernstein argues that problem generation or problem-raising is far more critical to innovation than problem solution, involving not just a thorough grasp of what is known (epistemology), but of what is not known (nepistemology). The key thesis of nepistemology states that we must know what we do not know before we can effectively solve any problem. People are creative and innovative only when they need to do something that cannot yet be done. Root-Bernstein explores strategies used by successful innovators to generate productive problems.

In Chapter 11, *The Role of Flexibility in Innovation*, Asta S. Georgsdottir, Todd I. Lubart and Isaac Getz, define flexibility as the ability to change, emphasizing that innovation encompasses different types of change. Innovation is essentially a science about change, so it is not surprising that 'innovation' is frequently considered synonymous with 'organizational change'. The authors analyze different types of flexibility, particularly concentrating on adaptive flexibility (the ability to change as a function of task requirements) and spontaneous flexibility (the tendency to change for intrinsic reasons, to try out a variety of methods). The

issue of how these types of flexibility are important at different stages in the innovation process is also considered.

In Chapter 12, *The Effect of Mood On Creativity in the Innovative Process*, Geir Kaufmann focuses on creativity aspects of innovation, discussing a recent stream of new research on the importance of mood and affect in the process of creativity. He addresses the issue of the effect of mood states on creative problem-solving as part of the process of innovation. Kaufmann criticizes the dominant opinion that there exists a positive causal link between positive mood and creativity. He analyzes research findings, which demonstrate that under certain conditions positive mood may in fact impair creativity, while negative and neutral moods may facilitate searching for creative solutions to existing problems. Finally, Kaufmann presents a new theory of mood and creative problem-solving and provides data supporting it.

Chapter 13, *Case Studies of Innovation: Ordinary Thinking, Extraordinary Outcomes*, by Robert W. Weisberg, presents another approach to understanding the nature of innovation. He challenges the existing view that innovation is the result of extraordinary thought processes, such as Wertheimer's productive (as opposed to reproductive) and Guilford's divergent (as opposed to convergent) thinking. Weisberg asserts that innovation is the result of the use of ordinary thinking process; creative thinking is simply ordinary thinking that has produced an extraordinary outcome. The author uses quasi-experimental quantitative methods to examine case studies of innovators in the arts and science to support his approach. A sampling of case studies includes Picasso's development of his painting Guernica, Edison's electric light, Mozart's compositions, and the Beatles' stylistic innovations.

In Chapter 14, *Innovation and Evolution in the Domains of Theory and Practice*, James R. Bailey and Cameron M. Ford claim that innovation appears when individuals produce novel solutions, and members of the relevant domain adopt it as a valuable variation of current practice. The authors assert that at the individual level, creative or innovative actions are adoptive responses to tensions between the person and situation. In domains such as the arts or sciences, person-situation tensions are best resolved by favoring novelty, whereas in domains such as business, the same tensions are best resolved by favoring value. Bailey and Ford employ a neo-evolutionary view of creativity to propose that these within domains tensions create intractable tensions between domains.

Chapter 15, *E-Creativity and E-Innovation*, by Keng Siau, is about developments in the field of artificial intelligence, which provide researchers another means of analyzing the creative process. The author reviews germane work and discusses the existing approaches to e-creativity and their application for understanding e-innovation. He concludes that we can build creative programs, which have the potential to shape the future of innovation.

Although many of the above-mentioned chapters consider individual differences in innovation to some extent, Part III of the Handbook, *Individual Differences in Innovative Ability*, concentrates directly on this issue. This Part consists of one chapter, *The Art of Innovation: Polymaths and Universality of the Creative Process*, by Robert Root-Bernstein, that discusses individual differences in innovative ability, which are caused by human innovative thinking. The author takes an interesting approach to the topic: he focuses on polymaths, that is, those innovative artists who have made scientific discoveries, innovative scientists who have made artistic contributions, and those who bridge both sets of disciplines without claiming allegiance to one the other. Root-Bernstein argues that examples of such scientists and artists are unexpectedly common. He concludes that it is precisely those polymaths, that is, those people who incorporate modes of thinking belonging to many cognitive domains that are those most likely to become innovators. Root-Bernstein's analysis of the most innovative polymaths leads him inevitably to the consideration of sciences–arts interactions from the viewpoint of their mutual contribution to innovation.

Part IV of the Handbook, *Development of Innovation Across the Life Span*, is aimed at understanding innovation mainly from the viewpoint of developmental psychology. It thus reflects developmental perspectives on innovation, showing how it develops in individuals from early years through late adulthood and until the end of the personal life. The developmental approaches to innovation explain many of the individual differences in innovation caused by the specificity of human development.

Chapter 1, *Young Inventors*, by Nicholas Colangelo, Susan Assouline, Laurie Croft, Clar Baldus, and Damien Ihrig, analyzes a special kind of innovation in children and adolescents, namely inventiveness. The authors briefly review the history of the study of young inventors. They further describe research on the young inventors who were part of the Invent Iowa program and who have been evaluated as meritorious inventors at local and regional invention competitions qualifying for the State of Iowa Invention Convention. This research revealed a wide range of important findings regarding perceptions of young inventors about the inventiveness process, their attitudes toward school, toward students, and an analysis of their inventions. It is interesting to note that boys and girls have equally strong interest and equal participation in inventiveness programs.

In Chapter 2, *Exceptional Creativity Across the Life Span: The Emergence and Manifestation of Creative Genius*, Dean Keith Simonton connects major innovations in the arts and sciences to the output of creative geniuses. He addresses the two main questions: "How

do great innovators appear?"; and "How does their creativity manifest itself?". Considering the first question, Simonton analyzes the early experiences that contribute to the development of extraordinary creative potential. Those factors include family background, education, and professional training. In order to address the second question, Simonton focuses on the typical career trajectory of great innovators. He discusses the ages at which geniuses tend to produce their first great work, their best work, and their last great work.

Chapter 3, *Innovations by the Frail Elderly*, by Tomas Heinzen and Nancy Vail, presents an exceptional approach to understanding innovation. The authors point out that the study of extreme, unusual, or unlikely populations represents one seldom-used yet insightful research strategy. In their opinion, nonnormative populations can provide insights about innovation that may generalize to larger populations. One such population that appears to be unlikely to demonstrate significant innovation is the frail elderly. The nine innovation principles presented in this chapter are abstracted from research, naturalistic observation, and clinical experience of the 'pre-hospice' population of frail elderly people living in nursing homes. Heinzen and Vail's principles state that 'the impetus for innovation may be external, unpleasant, and unwelcome'; 'innovation does not require a 'creative personality'; ordinary personalities in extraordinary circumstances will innovate'; 'both frustration and suffering can inspire innovation'; 'frustration will not lead to innovation unless it is sufficiently annoying to force new ways of thinking about a problem'; 'innovative behavior is both externally and internally self-rewarding'. 'changing ourselves represents one form of innovation'. Analyzing situation-driven and personality-driven kinds of innovation, Heinzen and Vail conclude that maximal probability of innovation is the product of the interaction between external stressors such as necessity, desperation, and perceived threat combined with internal capabilities such as habit, preparation, and motivation.

I would like to note that one of the aims of the handbook is to extend the existing understanding of innovation. For the most part, within the psychological science, the term 'innovation' is perceived as referring specifically to 'innovative ideas'. Within the business and management science, the concept of 'innovation' is highly associated with 'new product development' or 'innovation management'. I believe, however, that 'innovative behavior', 'innovation in daily life', and other similar phenomena should be explored, and this is another focus of the Handbook. In this light the Heinzen and Vail chapter definitely extends our understanding of innovation, pointing out a new promising direction in research on innovation.

Part V of the Handbook, *Assessment of Innovation*, comprises just one chapter about the measurement of innovation. In this chapter, Ronald E. Goldsmith and Gordon R. Foxall analyze the assessment issues from a marketing perspective. Innovativeness refers to individual differences in how people react to innovations and accounts for much of their success or failure. Their conceptualization of innovativeness includes global, consumer, and domain-specific levels of innovativeness. Goldsmith and Foxall present many of the current perspectives on innovativeness and describe a variety of measures that reflect them.

Taking into account today's ever increasing demand for innovative ideas, solutions, products, services, and the commensurate demand on the creative, gifted, and talented individuals able to produce them, Part VI of the Handbook, *Development of Innovation*, consists of chapters with clear educational implications. Their authors analyze a variety of psychological methods, creativity techniques, and educational strategies best suited for the development of human abilities to innovate. The term education is broadly interpreted in the context of this Handbook. It refers to: (a) school and university education for innovation devoted to the development of students' innovative abilities, innovative behavior, and innovative attitude to the world around them; (b) organizational learning; (c) creativity and innovation training; and (d) teaching of innovation. This Part discusses what should be done to encourage the appropriate development of innovative abilities in children and adults, and how this can be accomplished within and outside of traditional educations frameworks.

Chapter 1, *Developing High Potentials for Innovation in Young People Through the Schoolwide Enrichment Model*, by Sally M. Reis and Joseph S. Renzulli, presents the Schoolwide Enrichment Model (SEM), developed during 20 years of research in programs for gifted students, which enables each school to develop unique programs for talent development and creative productivity based upon local resources, student demographics, and school dynamics as well as upon faculty strengths and creativity. The major goal of SEM is to promote challenging and enjoyable high-end learning across a wide range of school types, levels, and demographic differences. The idea is to create a repertoire of enrichment opportunities and services that challenge all students. Any individual student does better in a school when all students appreciate creativity and innovation. SEM is one of the excellent ways to develop the creative potential of students within a traditional educational setting and inspire in them a spirit of innovation, inquiry, entrepreneurship, and a sense of power to effect change.

In Chapter 2, *Towards a Logic of Innovation*, Gerald F. Smith asserts that though it is affected by the organizational and social context, and by the individual's personality and motivation, the generation of innovative ideas depends to a considerable degree on

consciously controlled mental activities. His chapter discusses how an aspiring innovator can utilize specific mental activities and exercises in order to improve the prospects for successful idea generation. Smith points out that traces of such logic can be found in the many idea generation methods that have been proposed, but rarely tested, in the creativity literature. He also analyzes the logic of scientific discovery, another component of such logic that has been the target of philosophical and psychological development for the past half century. In Smith's view, the final and most promising stream of research is recent work that tries to develop useful prescriptions for innovation through the analysis of past cases of innovative activity. The most significant of these efforts is TRIZ, the theory of inventive problem-solving developed by Genrich Altshuller.

Chapter 3, *The Development of Innovative Ideas through Creativity Training*, by Maria M. Clapham, examines the effectiveness of various creativity training programs in enhancing innovation. In order to develop innovative thinking and to increase human ability to generate new ideas, creativity training programs of various types have became widespread in educational and business settings. Clapham provides an overview of various types of creativity training programs, which vary widely in methodology and scope, and analyzes recent research findings regarding their effectiveness for stimulating the development of innovative ideas.

In Chapter 4, *Intuitive Tools for Innovative Thinking*, Robert Root-Bernstein and Michele Root-Bernstein consider the role of intuitive thinking skills in the appearance of great innovations in the arts and sciences. The authors identified 13 non-verbal, non-mathematical, non-logical thinking tools that innovative individuals in a wide variety of disciplines say they use and discuss these tools in detail. The tools are: observing, imaging, abstracting, recognizing and forming patterns, analogizing, body thinking, empathizing, dimensional thinking, modeling, playing, transforming and synthesizing. R. Root-Bernstein and M. Root-Bernstein conclude that private, unarticulated insights generated by means of these tools are then translated in an explicitly secondary step into verbal, mathematical and other modes of public communication. Educational efforts to promote creative thinking must thus recognize and exercise intuitional thinking skills and directly address the process of translating idiosyncratic subjective thought into objectified public forms of discourse.

Chapter 5, *Stimulating Innovation*, by Ronald N. Kostoff, describes how innovation can be increased through the discovery by cross-discipline knowledge transfer. The author's approach entails two complementary components: one literature-based, the other workshop-based. The literature-based component identifies the science and technology disciplines related to the central theme of interest, the experts in these disciplines, and promising candidate concepts for innovative solutions. These outputs define the agenda and participants for the workshop-based component. An example of this approach is presented for the theme of Autonomous Flying Systems.

In Chapter 6, *Developing Innovative Ideas Through High Intellectual and Creative Educational Multimedia Technologies*, Larisa V. Shavinina and Evgueni A. Ponomarev examine how human innovative abilities can be increased by the means of contemporary educational multimedia technologies. They analyze innovations in instructional technology and their implications for developing an individual's intellectual and creative abilities. The authors present high intellectual and creative educational multimedia technologies (HICEMTs) as a special kind of psycho-educational multimedia technology, which are aimed at the development of people's innovative abilities. They argue that a real goal of education should be seen not in knowledge transfer, but in the development of the individual intellectual and creative potential.

Part VII of the Handbook, *Innovations in Different Domains*, is aimed at understanding domain-specific innovations. One of the facets of the uniqueness of the Handbook and its fresh perspectives should be seen in analyzing the specificity of innovation(s) in various areas like science, art, education, management, business, technology, finance, and so on. In other words, the chapter authors reflect a domain-specific view of innovation, considering scientific innovations, technological and industrial innovations, financial innovations, innovations in education, and so on. In this regard, the Handbook addresses the exceptionally important question: Does the uniqueness of each domain predetermine the specific mechanisms of innovation within it?

Chapter 1, *Dimensions of Scientific Innovation*, by Gerald Holton, concentrates on scientific innovation of geniuses. The author describes essential but largely hidden mechanisms of scientific innovation as they manifest themselves in works of Johannes Kepler, Henri Poincare, Enrico Fermi, and the discoverers of high-temperature superconductivity.

In Chapter 2, *Do Radical Discoveries Require Ontological Shifts?* Michelene T. H. Chi and Robert G. M. Hausmann claim that many great revolutionary discoveries in science may have occurred because the scientists have undertaken an ontological shift. That is, the scientists re-conceptualized or re-represented the problem (i.e. the phenomenon to which she/he is seeking an explanation) from the perspective of one ontology or ontological category to another ontology. Examples of the highest level of ontological categories are entities, processes, and mental states. Chi and Hausmann explain what this re-representation or shifting across ontological categories entails, and why it is unusual to undertake, therefore shedding light on

the low frequency of exceptionally revolutionary scientific discoveries. The authors discuss examples from contemporary science and the history of science, which provide evidence in favor of their radical ontological change hypothesis.

Chapter 3, *Understanding Scientific Innovation: The Case of Nobel Laureates*, by Larisa V. Shavinina, addresses an important issue in understanding scientific innovation, namely: 'Why are Nobel Laureates so capable of innovation?' The author demonstrates that scientific innovation of Nobel laureates is determined in part by specific preferences, feelings, and beliefs, which constitute a whole field of unexplored or weakly explored scientific phenomena. These phenomena constitute Nobel Laureates' extracognitive intelligence that accounts for their exceptionally developed tacit knowledge, which, in turn, significantly contributes to the emergence of scientific discoveries. Based on autobiographical and biographical accounts of Nobel laureates, Shavinina describes all of these phenomena in detail and shows that they predict scientific productivity of the highest level resulting into innovations and, as such, displaying an outstanding talent of Nobel caliber.

In Chapter 4, *Innovation in the Social Sciences: Herbert A. Simon and the Birth of a Research Tradition*, Subrata Dasgupta describes the cognitive-social-historical process of innovation in the realm of the social sciences. In his view, innovation presumes creativity but creativity does not necessarily result in innovation. The latter involves both the cognitive process of creation and the social-historical process by which the created product is assimilated into a milieu. Dasgupta analyzes how the American polymath Herbert A. Simon (1916–2001) was led to a particular model of human decision making which in turn gave birth to a radically new research tradition—the cognitive tradition or, more simply, cognitivism—first in organization theory and economics, and then in other domains of the human sciences. For his seminal role in the creation of this research tradition, Simon was awarded the 1978 Nobel Prize for Economic Science. In Dasgupta's view, the emergence of a research tradition in any science signifies a major innovation in that science.

In Chapter 5, *Poetic Innovation*, George Swede points out that every artist wants his or her work to be considered innovative, in both historical and contemporary terms. A psychologist, poet and editor of poetry periodicals and anthologies, the author discusses how and why poets strive towards these goals. His results can also be applicable to other kinds of artistic innovation. Swede's analysis is based upon the findings of psychological research on poets and their poetry, literary theory, and upon his own rich poetic and editorial experience.

Chapter 6, *Dual Directions for Innovation in Music: Integrating Conceptions of Musical Giftedness into*

General Educational Practice and Enhancing Innovation on the Part of Musically Gifted Students, by Larry Scripp and Rena F. Subotnik, discusses what should be done for systemic and sustainable innovation in public school music education. Scripp and Subotnik's analysis of innovations in music education lead them to propose a five-step course of action, potentially resulting into future innovations. These steps are: reconcile perspectives regarding the purpose of music education, establish a framework for comprehensive, interdisciplinary programs intended to benefit all children, prepare advanced college-conservatory students to contribute to the development and sustainability of innovative programs as 'artist-teacher-scholars', form networks of college-conservatory, arts organization, and public school partnerships, and, finally, explore and promote new conceptions of giftedness that result from the implementation of innovative forms of comprehensive, interdisciplinary music education.

In Chapter 7, *Determinants of Technological Innovation: Current State of the Art, Modern Research Trends, and Future Prospects*, Vangelis Souitaris reviews various methodologies used to identify the distinctive characteristics of innovative firms in the field of technology, that is, determinants of technological innovation. He discusses a wide range of issues, including the diverse nature and non-standardized definition and measurement of innovation, non-standardized measurements of the determinants, interrelated variables, different characteristics of firms, and, finally, different economic regions where the surveys take place. The author also presents a portfolio model, which synthesizes the existing research results and may be used for country- or industry-specific studies.

As follows from its title, Chapter 8, *Innovation in Financial Services Infrastructure*, by Paul Nightingale, is a thorough review of the state of research on innovations in the financial services infrastructure. Despite their evident importance to the world economy, both innovations in services in general and in financial infrastructure in particular are essentially neglected topics within the academic innovation literature. Nightingale discusses how external infrastructure technologies between institutions improve market liquidity by increasing the reach of markets. The author also analyzes internal infrastructure technologies within institutions, which are used to coordinate the profitable allocation of resources. The heavy regulation of the financial industry, the software intensity of modern infrastructure technologies, the way in which they have multiple users and their increasing complexity create extra uncertainties in their design and development. All of these cause very different patterns of innovation in financial services infrastructure than is observed in traditional consumer goods.

In Chapter 9, *Innovation in Integrated Electronics and Related Technologies: Experiences with Large-Scale Multidisciplinary Programs and Single Investigator Programs in a Research University*, Ronald J. Gutmann overviews advances in university research innovation in integrated electronics via his own many decades involvement in the field. The author distinguishes between discontinuous (or radical) and continuous (or incremental) university knowledge-based innovations. He discusses the impact of the Semiconductor Research Corporation (SRC) initiated in 1981, SEMATECH initiated in 1988 and the Microelectronics Advanced Research Corporation (MARCO) initiated in 1998 on the development of innovations in integrated electronics.

Parts VIII–XIII are about innovations in social context, broadly defined. The importance of social context in the development and implementation of innovations is widely recognized by innovation scholars. The aim of these Parts is to overview research in this direction. They comprise chapters on innovation in different group contexts, for example, companies, universities, and countries. They thus discuss innovation at the group, organizational, regional, national, and international levels. A wide range of topics such as the impact of innovation on organizational members, innovation management, innovation leadership, and national innovation systems is under consideration.

Part VIII, *Basic Approaches to the Understanding of Innovation in Social Context*, contains seven chapters. Chapter 1, *The Barriers Approach to Innovation*, by Athanasios Hadjimanolis, is a comprehensive review on the various types of barriers to innovation, including internal (i.e. strategy-related) barriers and external (e.g. market-related) barriers. The author analyzes the nature of barriers, critically examines their various classifications, discusses their impact on the innovation process, considers theoretical explanations of barriers, and overviews the existing empirical studies. He further describes the specificity of barriers in different social contexts. For instance, barriers have a particular impact in small firm innovation and in small countries. Finally, the chapter presents a set of suggestions regarding how to overcome barriers.

In Chapter 2, *Knowledge Management Processes and Work Group Innovation*, James L. Farr, Hock-Peng Sin, and Paul E. Tesluk present a dynamic model of work group innovation. The model integrates recent advances in taxonomies of work group processes and stages of the innovation process with a focus on the temporal nature of innovation. This model identifies transition and action phases with each of two major stages of innovation: a creativity stage and an innovation implementation stage. The transition phases both involve primarily planning and evaluation tasks that guide later goal accomplishment. The action phases involve primarily in acts that directly contribute to goal accomplishment. Within the creativity stage, the transition phase consists of interpretation of issues and problem identification and the action phase consists of idea generation. Within the innovation implementation stage, the transition phase consists of the evaluation of the generated ideas as possible solutions and selection of the one(s) to implement and the action phase consists of the application of the idea(s) to the problem.

Chapter 3, *Innovation and Creativity = Competitiveness? When, How, and Why*, by Elias G. Carayannis and Edgar Gonzales, describes the circumstances within which creativity and innovation occur in organizations. Using empirical findings from public and private companies, the authors discuss how and why creativity triggers innovation and vice versa, and the resulting implications for competitiveness. Carayannis and Gonzales analyze the existing literature on the topic and field interviews on the practice and implications of creativity and innovation from the perspective of competitiveness.

In Chapter 4, *Innovation Tensions: Chaos, Structure, and Managed Chaos*, Rajiv Nag, Kevin G. Corley, and Dennis A. Gioia presents a framework for understanding the tensions that underlie an organization's ability to manage innovation effectively in the face of a turbulent competition. These tensions are: (1) the fundamental tension between the desire for structure and need for creative chaos, and (2) the on-going tension between technology-push and market-pull approaches to innovation. The authors discuss the nature and boundaries of these tensions and characterize them as four distinct 'innovation contexts'. Nag, Corley, and Gioia present a case study of one high-technology organization as an example that supports their approach. The authors examine the notion of 'managed chaos', a concept that helps understand the role of innovation in the maintenance and change of an organization's identity.

Chapter 5, *Innovation and Identity*, by Nigel King, continues the theme of identity and innovation within an organizational context. The author specifically focuses on how innovation processes shape people's work-related identities. King further argues that the concept of identity is useful for considering the relationship of the person to the organization in the context of innovation. He analyzes the potential contributions of Social Identity Theory and Constructivist/Constructionist accounts of identity to the field of innovation. The author also presents findings from case studies of innovations in the British health service to demonstrate the value of an interpretive approach to innovation and identity.

In Chapter 6, *Manager's Recognition of Employees' Creative Ideas: A Social-Cognitive Model*, Jing Zhou and Richard W. Woodman concentrate on organizational creativity as a basis for the appearance of innovation. They employ a social-cognitive approach to explaining the conditions under which a manager is likely to consider rewarding and/or implementing an employee's idea. The recognition and support of

employee creative ideas is a critical facet in organizational creativity. Zhou and Woodman present a model, according to which the manager's 'creativity schema' determines recognition of creative ideas in the organizational context. In the authors' view, the creativity schema is influenced by personal traits of the manager, by aspects of the manager's relationship with the employee, and by a number of organizational influences. Finally, Zhou and Woodman describe implications of the social-cognitive approach for innovation research and business practice.

Chapter 7, *Venture Capital's Role in Innovation: Research Methods and Stakeholder Interests*, by John Callahan and Steven Muegge, is a comprehensive review of how venture capital contributes to innovation. The important role of venture capital in innovation is widely recognized by innovation research literature. Callahan and Muegge begin with the history and current state of venture capital. They describe the process of venture capital financing and how it relates to the innovation process. The authors then review the research literature related to venture capital investment decision-making, the venture capital-entrepreneur relationship, and the fostering of innovation by venture capital. Finally, using a stakeholder perspective, Callahan and Muegge emphasize the value of current research for different stakeholders and call for more qualitative, longitudinal research that contributes better stories and richer data on variable interrelationships.

In Part IX, *Innovations in Social Institutions*, chapter authors analyze innovations along a broad spectrum of social institutions, including small and medium-size firms, multitechnology companies, transnational corporations, universities, network forms of organizations, and technopoleis.

Chapter 1, *Encouraging Innovation in Small Firms through Externally Generated Knowledge*, by Edward Major and Martyn Cordey-Hayes, continues the topic of knowledge management previously addressed in the two other chapters. Specifically, Major and Cordey-Hayes analyze the conveyance of externally generated knowledge to small firms. The authors emphasize that successful innovation requires firms to draw on multiple sources of knowledge. Many small firms take little note of external sources, thus restricting their potential innovative base. Major and Cordey-Hayes present the concepts of knowledge translation and the knowledge translation gap to illustrate why so many small firms fail to access externally generated knowledge. The authors' findings from research into U.K. small firms and national innovation schemes demonstrate how intermediary organizations can be used to bridge the knowledge translation gap. Finally, they discuss implications for government innovation policy for intermediaries and for small firms.

In Chapter 2, *Linking Knowledge, Networking and Innovation Processes: A Conceptual Model*, Jacqueline Swan, Harry Scarbrough, and Maxine Robertson present a model that relates specific kinds of networking to episodes of innovation, that is, invention, diffusion, and implementation and to knowledge transformation processes. The operation of the model is illustrated through three longitudinal case studies, each focusing on a different innovation episode. The authors' thorough review of the relevant literature demonstrates that the need for innovation is frequently cited as a major reason for the emergence of network forms of organization. Through networks, it is assumed, knowledge needed for innovation is transferred more easily. However, Swan, Scarbrough, and Robertson point out that relatively little research has addressed the links between networks and the development and utilization of knowledge during processes of innovation. Research that does exist tends to focus on networks, in the structural sense, as channels for the communication of knowledge, which is seen as relatively unchanging. This research provides a useful starting point but tends to be rather static in its treatment of networks and knowledge flows during innovation. It also tends to emphasize the diffusion episode of the innovation process. In contrast, there is relatively little research on the relations between networking, as a dynamic process, and the development of innovation. The authors' model provides a more dynamic account of the links between networking and innovation processes.

Chapter 3, *Innovation in Multitechnology Firms*, by Andrea Prencipe, is about the generation of innovations in companies specializing on complex multitechnology products. He identifies two major dimensions of capabilities of such companies: synchronic systems integration and diachronic systems integration. Within each of these two dimensions, multitechnology companies maintain absorptive capabilities to monitor and identify technological opportunity from external sources and generative capabilities to introduce innovations at the architectural and component levels. Prencipe concentrates on company's generative capabilities and shows that these capabilities enable a company to frame a particular problem, enact an innovative vision, and solve the problem by developing new manufacturing processes. The author concludes that frame-enact-solve is the primary feature of a company's generative capability. Prencipe presents the case study that support his approach.

In Chapter 4, *Innovation Processes in Transnational Corporations*, Oliver Gassmann and Maximilian von Zedtwitz describe innovation in transnational business settings. The authors identify two phases in innovation process: (1) a pre-project phase fostering creativity and effectiveness, and (2) a discipline-focused phase ensuring efficiency of implementation. Such differentiation enables transnational companies to replicate and scale innovation efforts more easily in remote locations, exploiting both economies of scale and of scope. The distinctive features of these phases are different,

however: a few companies mentioned by Gassmann and von Zedtwitz have consistent and differentiated techniques to manage and lead the overall innovation effort specific to each phase.

Chapter 5, *An Analysis of Research and Innovative Activities of Universities in the United States*, by Yukio Miyata, examines how American universities contribute to innovation through their research and collaboration with industry. Universities are an important component of the national innovation system: they supply highly qualified personnel and advanced scientific and technological knowledge to public and private industry. Therefore, the state of innovative activities in universities should be an essential topic in innovation research. In light of the leading role of American science, it is especially interesting to analyze innovative efforts in the American universities. Miyata shows that American universities with a high quality of research tend to be productive in generating academic publications and research results that are close to commercialization. However, license revenue results from a small number of 'hit' inventions that are often in the field of medical research. It is difficult for universities to finance their research by license revenue, so the role of the central government is critical to maintain research quality.

In Chapter 6, *Incubating and Networking Technology Commercialization Centers among Emerging, Developing, and Mature Technopoleis Worldwide*, David Gibson and Pedro Conceição discuss the development of innovations at the regional and global levels. Technopoleis (Greek for technology and city state) refer to regions of accelerated wealth and job creation through knowledge creation and technology use. Innovation is considered as the adoption of 'new' knowledge that is perceived as new by the user. The access to knowledge and the ability to learn and put knowledge to work is essential to regional economic development and for globalization today. The authors present the conceptual framework for leveraging knowledge through Internet and web-based networks, face-to-face communication, and training programs. The aim is to accelerate regional economic development through globally linked Technology Commercialization Centers (TCCs).

Chapter 7, *Science Parks: A Triumph of Hype over Experience*? by John Phillimore and Richard Joseph, considers the role of science parks in innovation. For example, such parks can serve as technology and/or business incubators thus fostering the growth of start-ups. Phillimore and Joseph emphasize that there is a disjuncture between the critical assessment of the academic literature about most science parks and their growing number worldwide. The authors analyze the skepticism frequently expressed about science parks in the literature and how it contrasts with their continued international popularity. Phillimore and Joseph also discuss possible new directions for science parks.

Part X, *Innovation Management*, is devoted to one of the most central topics within innovation research, that is, how innovation processes should be managed. Today academic and professional journals publish more articles on innovation management than on any other innovation topic. This fact alone highlights an important role of innovation management. Four chapters included in this Part are aimed at the analysis of various aspects of this role.

Chapter 1, *Challenges in Innovation Management*, by John Bessant, reviews the question of managing innovation and particularly looks at some of the key challenges, which must be addressed if company plans to manage innovation successfully. A need to understand innovation, to build an innovation culture, and to extend participation in innovation process, as well as continuous learning are among these challenges.

In Chapter 2, *Managing Technological Innovation in Business Organizations*, Ralph Katz was given an 'assignment' to consider technological innovations and innovation management issues in technological companies. The author thus analyzes the patterns of innovation that usually take place within an industry and how such patterns affect a company's ability to manage its streams of innovative projects along technological cycles.

In Chapter 3, *Towards a Constructivist Approach of Technological Innovation Management,* Vincent Boly, Laure Morel, and Jean Renaud describe a constructivist approach to the understanding of technological innovation management. The key aspects of this approach include the development of a value-oriented strategy and a systemic vision of innovation management through its three levels: strategy, piloting, and sparking. The authors present findings from technological innovation survey in French small and medium size enterprises (SMEs), which support the constructivist approach to technological innovation management.

Chapter 4, *Promotors and Champions in Innovations: Development of a Research Paradigm*, by Jürgen Hauschildt, considers a typology of those individuals within an organizational context who enthusiastically support innovations. The success of innovations depends to a great extent on the activities of such 'champions' or 'promoters'. The author analyzes 30 years of research in this direction. Hauschildt concludes that three types of champions or promoters are necessary for successful development of innovations. First, innovations need a technical expert who acts as 'promotor by expertise'. Second, innovations need top management's sponsorship by a 'power promoter'. Third, innovations need boundary-spanning skills of a 'process promoter'. The size of a company and the diversification of its products, as well as the complexity and newness of innovation, considerably influence this 'troika' model.

Part XI, *Innovation Leadership*, comprises just one chapter, 'Innovation and Leadership', by Jean Philippe

Deschamps. This chapter is about innovation leaders, those critical senior executives which top management sees as the linchpins of its innovation process and the 'evangelists' of an innovation and entrepreneurship culture. The author begins with a well-accepted list of generic leadership imperatives as they relate to innovation. He further describes the common traits of innovation leaders in terms of personal profiles and behavioral attributes. Finally, Deschamps discusses the beliefs and management philosophy on innovation leadership adopted by some innovative companies and their CEOs and CTOs.

Part XII, *Innovation and Marketing*, presents a marketing perspective on innovation. This Part includes two chapters. In Chapter 1, *Innovation and Market Research*, Paul Trott examines the debate about the use of market research in the development of innovative products and discusses the extent to which market research is justified. The author points out that market research can provide a valuable contribution to the development of innovative products. Trott presents the conceptual framework, which should help product and brand managers to consider when and under what circumstances market research is most effective.

Chapter 2, *Marketing and the Development of Innovative New Products*, by Robert W. Veryzer, further develops the market-based view of innovation. Specifically, he analyzes innovation from the perspective of marketing concerns and challenges. The author highlights the critical value of market vision for companies, that is, the ability to bridge technological capability and market need and opportunity. Veryzer emphasizes that market vision is particularly important for high innovation products because they typically involve a significant degree of uncertainty about exactly how an emerging technology may be formulated into a usable product and what the final product application will be. He presents the conceptual framework that can help to clarify the innovation and adoption context with respect to the marketing challenge(s). Marketing thus provides a necessary and useful function in helping to shape an innovative idea into a product offering that meets the needs and desires of the people who are intended to use it.

Part XIII, *Innovation Around the World: Examples of Country Efforts, Policies, Practices and Issues*, discusses innovation practices and innovation policy issues in different countries. The 11 chapters included in this Part are essentially about national innovation systems and describe the unique paths of various states to innovation. It is beyond the scope of this Handbook to include chapters about innovations in all, or even in most, nations. The solution therefore was the following: first, to describe innovations in a few countries whose systems of innovation are unique, or which include unique and useful aspects; and second, to describe innovation systems which are representative of several or many other countries. For example, in

Chapter 1, *Innovation Process in Hungary*, Annamária Inzelt discusses innovations in Hungary, the country, which is, to a significant degree, representative of other post-socialist countries, all of which are characterized by transition economy. Inzelt based her chapter on the analysis of findings of the Hungarian innovation survey and concludes that the level of innovation in the Hungarian economy is low. For instance, Hungarian companies are aware of the importance of developing new products and accessing new markets, but in practice they mainly do this within the limited Hungarian context. The author largely attributes this low level of innovation to the financial stringency imposed by the economic transition process.

In Chapter 2, *Innovation under Constraints: The Case of Singapore*, Hung-Kei Tang and Khim-Teck Yeo examine the specificity of innovation in Singapore, a young nation that has achieved outstanding infrastructural innovations. The authors point out that the real challenge of innovation lies not only in identifying opportunities and committing to action, but also in identifying and mitigating the internal and external constraints that impede the process of innovation. Using four case studies, Tang and Yeo explore how constraints of different types can give rise to innovation as well as cause innovation efforts to fail. The Singaporean government has recognized the need for a better environment for innovation and entrepreneurship and has embarked on several strategic initiatives aimed at fostering that environment.

Chapter 3, *Continuous Innovation in Japan: The Power of Tacit Knowledge*, by Ikujiro Nonaka, Keigo Sasaki, and Mohi Ahmed, describes how Japanese companies innovate. Management practices of the knowledge-creating Japanese corporations had attracted a great deal of attention worldwide. The authors consider the issue of what the basic pattern of innovation is in the knowledge-creating business organizations. Specifically, Nonaka, Sasaki, and Ahmed analyze the basic pattern of innovation at Nippon Roche. They conclude that learning and rule-breaking alone are not enough for continuous innovation. Rather, individuals as well as organizations need to possess tacit knowledge.

In Chapter 4, *Innovation in Korea*, Sunyang Chung reviews the Korean national innovation system. Korea is an innovative and dynamic country. It has implemented a relatively competent national innovation system in three decades. Korea has invested a lot of resources in order to enhance the efficiency of its national innovation system and increase the innovation capabilities of major innovation actors. The author concludes that the dynamic development of the Korean economy within the context of global economy is connected to its efficient national innovation system.

Chapter 5, *Regional Innovations and the Economic Competitiveness in India*, by Kavita Mehra, examines the innovation process in India. The study of innovation in India is particularly interesting, given that country's dual position as both a developing nation, and one with an ancient, and rich, civilization. Mehra describes the case studies of regional innovations from diverse locations in India, which are very typical to those regions and bound to local culture, knowledge, and resources. She points out that India has a vast storehouse of knowledge in various fields, particularly in tacit form. For this reason the tacit knowledge based innovations are one of the main kinds of innovations in India. These innovations are territorially specific ones, because of their embodiments in individuals, in their social and cultural context. The author concludes that this is exactly what allows developing nations to retain their identity, and preserve local art and craft as the national heritage in the era of globalization.

In Chapter 6, *Innovation Process in Switzerland*, Beate Wilhelm analyzes the national innovation system in one of the most economically advanced countries in the world. She emphasizes that the primary role of Swiss innovation and technology policy is to foster the utilization of scientific knowledge via organized technology transfer from universities to industries. Despite the robust health of the Swiss economy, Wilhelm concludes that the national innovation system needs improvement. Specifically, the existing innovation policy must be developed in order to better coordinate science, industry, government and societal demands.

Chapter 7, *Systems of Innovation and Competence Building across Diversity: Learning from the Portuguese Path into the European context*, by Pedro Conceição and Manuel Heitor, considers the current Portuguese path towards an innovative society. It focus on Portugal within a European scene, considering a context increasingly characterized by uncertainty and diversified environments, which are particularly influenced by social and institutional factors. The authors present a conceptual framework that helps to understand the contemporary demands for being innovative. The concepts of learning society, knowledge accumulation, competence building, and systems of innovation are main components of this framework. Finally, Conceição and Heitor propose suggestions aimed at the further development of innovation policy in Portugal.

In Chapter 8, *Innovation in Taiwan*, Chiung-Wen Hsu and Hsing-Hsiung Chen examine the Taiwan innovation system, which contributes to a great extent to its accelerated technology-based industrial development. The authors consider the most representative characteristics of the Taiwan innovation system, including: (1) the Technology Development Program of the Ministry of Economic Affairs planning the industry innovation policy; (2) the research and development and technology diffusion strategy of the Industrial Technology Research Institutes; (3) the Hsinchu Science-Based Industrial Park's method of technology commercialization; and (4) the recruitment of overseas experts and cultivation of talents. Hsu and Chen show the Taiwan innovation system 'in action' on the example of the successful development of the Taiwan integrated circuit industry.

Chapter 9, *Innovation in the Upstream Oil and Gas Sector: A Strategic Sector of Canada's Economy*, by A. Jai Persaud, Vinod Kumar, and Uma Kumar, is about the Canadian innovation system and innovations in oil and gas industry. The authors discuss the main facets of Canada's innovation agenda aimed at the advanced economic development. The government of Canada is interested in making Canada a highly innovative economy. Persaud, V. Kumar, and U. Kumar consider the Canadian innovation strategy and related policy issues. The authors focus on the oil and gas sector of Canada's economy—its strategic sector—and analyze the key aspects of innovation in this sector. Innovations in the oil and gas industry can make greater contributions to sustainable development and play a crucial role in reducing greenhouse gas emissions.

Chapter 10, *The National German Innovation System: Its Development in Different Governmental and Territorial Structures*, by Hariolf Grupp, Icíar Dominguez Lacasa, and Monika Friedrich-Nishio, is about innovation system in the united Germany. Germany existed as one country from 1871 to 1945. It existed as a large array of individual states before that period and was divided into West and East from the Second World War to 1990, with extremely different innovation policies. Since then, it was reunited again. Nevertheless, there are many indications of a national innovation system through all these periods, government structures and territorial changes.

In Chapter 11, *Frankenstein Futures? German and British Biotechnology Compared*, Rebecca Harding analyzes biotechnology policies in the U.K. and Germany. She compares and contrasts the market based biotechnology policies of the U.K. with the regionally engineered biotechnology policies of Germany in the light of the national and regional systems of innovation. The author demonstrates that innovation systems generally and regional innovation systems in particular are still useful concepts in explaining the clustering of biotechnology. Harding shows that German biotechnology policy has been particularly successful in

stimulating rapid catch-up with U.K. and global levels of research.

Part XIV, *Innovations of the Future,* consists of two chapters. *Future Innovations in Science and Technology*, by Joseph F. Coates, describes future innovations in genetics, brain science, information technology, nanotechnology, materials science, space technology, energy, and transportation. A scientist and futurist, the author presents a 25-year look into the future, considering scientific developments and their practical technological applications. He concludes that a general result of future innovations will be continuing enhancement of the quality of human life, leading to unprecedented richness, on a worldwide scale.

In Chapter 2, *The Future of Innovation Research*, Tudor Rickards presents his view of future advances in the field of innovation. He identifies a few research directions in which the main developments should be expected. Rickards also describes challenges, which will be faced by innovation scholars, and discusses the multifaceted impact of innovation research on practice.

Part XV, *Conclusion*, contains a single chapter, *Research on Innovation at the Beginning of the 21st Century: What Do We Know About It?* by Larry R. Vandervert, which serves to integrate the other chapters in the Handbook. This chapter points out common as well as unique features of the various accounts of innovation and suggests directions in which future research, practice, and policy might lead us.

The chapters of this Handbook therefore demonstrate that the phenomenon of innovation is inherently multidimensional, multifaceted, interdisciplinary, personally demanding, socially consequential, cross-cultural, and frequently surprising. As a result, understanding the scientific principles that underlie innovation requires a variety of research approaches. Authors presented a wide range of approaches to understanding the nature of innovation at the individual, group, organizational, societal, and global levels. This Handbook thus provides what is perhaps the most comprehensive account available of what innovation is, how it is developed, how it is managed, how it is measured, and how it affects individuals, companies, societies, and the world as a whole.

Part II

The Nature of Innovation

Part II

The Nature of Innovation

The International Handbook on Innovation
Edited by Larisa V. Shavinina

The Neurophysiological Basis of Innovation

Larry R. Vandervert

American Nonlinear Systems, USA

Abstract: It is proposed that: (a) the recursive processes of working memory (online consciousness) are modeled in cognitive control processes of the cerebellum; and (b) when these new, more efficient control processes are subsequently fed back to working memory, they are learned in a manner that facilitates innovation. Walking through this cerebellar-working memory sequence, Einstein's experiential accounts of discovery are examined. Methods of encouraging innovation are outlined. It is concluded that innovation is a recursive neurophysiological process that constantly reduces conceptual thought to patterns, thus constantly opening new and more efficient design spaces.

Keywords: Einstein; Cerebellum; Design space; Innovation; Mathematics; Mental models; Working memory.

Introduction

Many scholars and researchers have described the close parallels between biological evolution on the one hand and the processes that govern innovation on the other (e.g. Ziman, 2000). This broad-based literature leaves little doubt that innovation is an integral part of overall biological and socio-cultural evolution. But what, precisely, is the primary evolutionary mechanism behind innovation? Why is human innovation so prolific? And why does the pace of innovation seem to be accelerating? The purpose of this chapter is to describe neurophysiological processes going on in the human brain that can address these questions. Understanding how the human brain extends itself into the production of new ideas and technologies will help build a basic science of innovation. Such a basic science will perhaps permit us to unleash more of the seemingly limitless potential of innovation.

An attempt will be made to keep the jargon of neurophysiology at a minimum. Technical terms will be used only in a manner that will likely be understood by readers from a broad variety of disciplinary backgrounds.

The Neurophysiology of Mathematical Discovery Represents a Generalized Model for All Innovation

Recently, I described a theory of how mathematical discovery arises through the collaboration of working memory and *patterns* generated in the cerebellum (Vandervert, in press).[1] In the present chapter I provide: (a) a description of how the brain processes that lead to mathematical discovery represent a general neurophy-

---

[1] Historically, mathematics has been defined first as geometry, and, later, as number. With the advent of high-speed computers, however, mathematics is seen in a new, more fundamental light. We realize now that mathematics is actually all about *patterns*:

> Mathematics is the science of patterns. The mathematician seeks patterns in number, in space, in science, in computers, and in imagination. Mathematical theories explain the *relations* (italics added) among patterns; functions and maps, operators and morphisms bind one type of pattern to another to yield lasting mathematical structures. Applications of mathematics use these patterns to 'explain' and predict natural phenomena that fit the patterns (Steen, p. 616).

See also Devlin (1994).

Today, we understand that the core of mathematics is about *relations* (functions and maps, operators and morphisms) among patterns that, ultimately, fit *all* real-world phenomena. Mathematics, as we know it, consists of patterns among phenomena in terms of scientific operations upon which there is social agreement (scientific validity). It is a thesis of this chapter that the cerebellar models of movement, perception, and thought are coded in functions, mappings, operators and morphisms pertinent to all processes taking place in the cerebral cortex. Thus it is proposed that innovation in all fields can be modeled after the brain processes that lead to mathematical innovation.

siological model for *all* innovation; and (b) a walk-through of three of Einstein's classic experiential accounts of discovery that corroborates the neuro-physiological model.

Why the Cerebellum and Working Memory?

During the course of repetitive bodily movement, a person becomes able to move more quickly and precisely. These increases in efficiency are the result of control models that are learned in the cerebellum and subsequently fed back to motor areas of the cerebral cortex. In this chapter, three interrelated arguments are presented. First, it is proposed that in the same way that cerebellar models for bodily movement are learned and fed back to the cerebral cortex, models for working memory processes are learned in the cerebellum and fed back to working memory, making its visuospatial, speech, and central executive functions significantly more efficient (see Desmond & Fiez, 1998; Doya, 1999; Ito, 1993, 1997; Leiner, Leiner & Dow, 1986, 1989 for the extension of cerebellar modeling to thought processes). Second, because the components of working memory contain the attribute of conscious awareness (Baddeley, 1992; Baddeley & Andrade, 1998; Baddeley & Hitch, 1974; Teasdale, Dritschel, Taylor, Proctor, Lloyd, Nimmo-Smith & Baddeley, 1995), it is argued that the newer, high-efficiency *patterns* of the cerebellar models learned in working memory provide the experiential basis for innovation. Finally, it is the contention of this chapter that the experience of innovative discovery is the result of the action of multiple paired cerebellar models (see Haruno, Wolpert & Kawato, 1999 for a description of such models).

Three general theoretical premises constitute the working memory/cerebellar theory of innovation.

(1) Innovation is a process of evolutionary adaptation.
(2) The selective advantages of innovation arise from efficiencies that accrue through the reciprocal learning relationship between working memory (i.e. the visuospatial sketchpad, the speech loop, and the central executive) and the perceptual-cognitive functions of the cerebellum.
(3) The working memory/cerebellar processes that lead to mathematical innovation represent the generalized model of innovation (Vandervert, in press). Thus the neurophysiological model of mathematical innovation can serve as the fundamental model for the brain processes that lead to innovation in all fields.

The Plan of this Chapter

Working Memory

First, the components of working memory will be examined. Since the concept of working memory is widely understood and resources on the topic are easily obtained, only a brief account that relates its compo-

nents to selected areas of the cerebral cortex and cerebellum will be described.

The Cognitive Functions of the Cerebellum

Next, a review of findings from newer research on the cognitive functions of the cerebellum will be presented. Because much of the research on the cognitive functions of the cerebellum tends to be relatively new and is spread over many sub-areas of investigation, a more lengthy discussion will be presented.

A Demonstration of the Generalized Model of Innovation: Einstein's Experiential Accounts of Mathematical Discovery

Finally, the interaction of working memory and the cognitive functions of the cerebellum will be illustrated within the framework of three of Albert Einstein's classic experiential accounts of mathematical discovery. While the neurophysiological 'walk through' of Einstein's subjective accounts is admittedly tentative, it is an attempt to both round out and demonstrate the generalized model of innovation that is suggested by the behavioral studies of working memory, and the neurophysiological studies of the collaborative cognitive functions of the cerebellum and working memory.

Working Memory: The Ongoing Stream of Cognitive Consciousness

According to Baddeley (1992) the term 'working memory' refers to processes in the brain that provide temporary storage and manipulation of the information necessary for such complex cognitive tasks as language comprehension, learning, and reasoning. Working memory may be characterized as our 'on-line' cognitive consciousness (e.g. Baddeley & Andrade, 1998). Working memory may be thought of as 'working consciousness'. It will be important to keep this latter conception of working memory in mind when we examine Einstein's experiential accounts of mathematical discovery.

When our attention is on a particular stream of verbal and/or visuospatial consciousness, such as when I am writing this sentence, or when it is being read, the essential conscious, cognitive framework is working memory. Working memory consists of three interacting components: a central executive and two subsidiary slave systems—a visuospatial sketchpad and a phonological (speech) loop. To maintain information in a conscious, on-line state in the central executive, the visuospatial sketchpad and the speech loop are in a continual process of repetitive rehearsal and updating. We can use the above-mentioned writing of a sentence to conveniently illustrate the on-line, repetitive interplay of the three components of working memory. First, *attentional control* of the action of writing (or reading) the sentence is carried on by working memory's central executive functions. Attentional functions of the central executive supervise, schedule and integrate information from different sources. The *visuospatial sketchpad* and the *phonological loop* are short-term memory stores/rehearsal processes for maintaining, respectively, visuospatial images and speech information that is needed for the on-line composition (or deciphering, if reading) of the sentence. Because the sketchpad and speech loop are *short-term memory* stores/rehearsal processes, they are within the ongoing grasp of consciousness, but additional and updating information continually enters working memory via the attentional functions of the central executive.

The Evolutionary Birth of Human Working Memory and its Collaboration with the Cognitive Functions of Cerebellum

The relationships among the three components of working memory are both phylogenetically and ontogenetically dynamic and interactive. Mandler (1988, 1992a, 1992b) proposed how image-schemas (conceptual primitives) drive the ontological development of language. Following in the vein of Mandler's description of the development of language from image-schemas, I proposed that image-schemas and language evolved in a yoked fashion due to the following two-part, compounding selective advantage: (a) their visuo-linguistic advantage as state variables for cognitive simulation and communication of possible future states of events in working memory; and (b) new, higher efficiency advantages in movement, thought and communication as a result of their being modeled in the cerebellum, and then being fed back to working memory (Vandervert, 1997). I believe that this is the substantial account of the evolutionary selection toward the constantly improving, reciprocal relationships among working memory's central executive, visuospatial sketchpad, and phonological (speech) loop.

There are three highly supportive lines of evidence that the activity of working memory is indeed modeled in the cerebellum. First, Ito (1993, 1997) points out that, at the neurological level, movements and thoughts are identical control objects:

> In thought, ideas and concepts are manipulated just as limbs are in movements. There would be no distinction between movement and thought once encoded in the neuronal circuitry of the brain; therefore, both movement and thought can be controlled with the same neural mechanisms [namely, those of the cerebellum] (1993, p. 449).

Thus, the control of the components of working memory in solving problems is not different from the control of hands and feet in solving problems. And, as the use of the components is repeated, the cerebellum acts to make the manipulations smoother, faster, and more efficient. Second, as will be seen below in the sections on the cognitive functions of the cerebellum,

there is abundant evidence that *patterning* representing both visual and linguistic aspects of cognition is learned in the cerebellum (see, e.g. Desmond & Fiez, 1998; Houk & Wise, 1995). Third, it has recently been found that the cerebellum continually updates its models of variations in cognitive activity (Imamizu, Miyauchi, Tamada, Sasaki, Takino, Pütz, Yoshioka & Kawato, 2000). This newer evidence is critical to the argument here because, to be of selective advantage to the rapidly changing, online nature of working memory, the infusion of parallel, rapidly updated cerebellar models is required.

Working Memory and Associated Brain Areas

Functions of working memory can be related to brain areas. Before proceeding to this description it is important to note that the distribution of working memory functions across brain areas is likely quite complex. For example, there is strong evidence that storage and rehearsal functions involve differing brain areas (Awh, Jonides, Smith, Schumacher, Koeppe & Katz, 1996). Figure 1 is a simplified illustration of key brain areas involved in working memory:

(a) The central executive functions can be tentatively thought of as spread over the SMA (supplementary motor area—long associated with 'will' or intentionality (e.g. Deecke, Kornhuber, Lang, Lang & Schreiber, 1985), Brodmann area 9, with overlappings in Broca's area (Brodmann areas 44/45), lateral and inferior parietal areas (Brodmann areas 39/40), and, I will argue below, the cerebellum (Desmond & Fiez, 1998; Fiez, Raife, Balota, Schwarz, Raichle & Peterson, 1996).

(b) The phonological loop involves at least Brodmann areas 44/45, and the cerebellum (Desmond & Fiez, 1998).

(c) The visuospatial sketchpad most likely involves, Brodmann areas 40/18/19 (Fiez, Raife, Balota, Schwarz, Raichle & Peterson, 1996).

In brief summary of this section, the brain areas described above and shown in Fig. 1 are those most fundamentally associated with working memory and, I believe, with the production of mathematical discovery and innovation. The next section describes the cognitive functions of the cerebellum in learning and feeding adaptive control models to the cognitive processes of working memory.

The New Perception of the Cerebellum: The Cognitive Functions of the Cerebellum

Research and Theory on the Cerebellum: A Fast Computational System for Patterns of Both Movement and Thought

Understandings of the cerebellum have moved far beyond the earlier, more traditional idea that its functions are limited to motor control (e.g. Kornhuber,

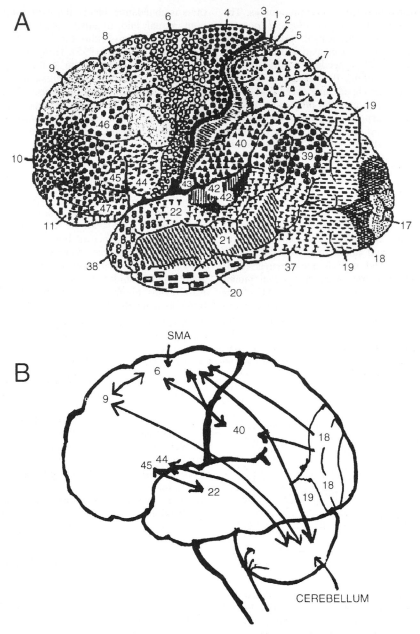

Figure 1. (A) Brodmann areas; (B) simplified connections among Brodmann areas, and between Brodmann areas and the cerebellum.

1974). A number of newer and converging lines of research and theory suggest that the cerebellum provides a fast computational system for the *patterning* of both motor and *cognitive* processes that take place in the cerebral cortex (Desmond & Fiez, 1998; Doya, 1999; Houk & Wise, 1995; Ito, 1993, 1993; Leiner & Leiner, 1997; Leiner, Leiner & Dow, 1986, 1989). This newer knowledge about the cognitive control functions of the cerebellum was the thrust of Schmahmann's (1997) volume, *The Cerebellum and Cognition.* More recently, research on the cognitive functions of the cerebellum has been increasing at an accelerating rate. Table 1 contains a collection of chapter and paper summaries of such research that is particularly relevant

Table 1. Newer research and theory supporting cognitive control functions of the cerebellum.

Source:	Methods:	Conclusions:
Akshoomoff, N., Courchesne, E. & Townsend, J., 1997; Desmond, J. & Fiez, J., 1998	Human PET and functional MRI	The cerebellum is a master computational system that adjusts responsiveness in networks including those of declarative and working memory, attention, arousal affect, language, speech, sensory modulation and motor control
Doya, K. 1999	Mathematical modeling of the respective functions of the cerebellum, basal ganglia, and cerebral cortex	
Haruno, M., Wolpert, D. & Kawato, M. 1999; Imamizu, H., Miyauchi, S., Tamada, T., Sasaki, Y., Takino, R., Pütz, B., Yoshioka, T. & Kawato, M. 2000; Ito, M. 1993, 1997; Kawato, M. 1999 ; Kawato, M. & Gomi, H. 1992; Kawato, M., Furukawa, K. & Suzuki, R. 1987 ; Pribram, K. 1971, 1991	Mathematical and Physiological modeling of cerebellar cortex	During repeated thoughts in the cerebral cortex, faster-than-real-time (FTRT) model is transferred to cerebellum to allow that model later to be fed forward back to cortex circumventing need to check outcome of thought—thus, FTRT thought. FTRT cerebellar models are short-cut circuits (abstractions) of action and thought patterns of the cerebral cortex
Ivry, R., 1997	Physiological and behavioral modeling of cerebellar timing systems	Cerebellar timing system provides a near-infinite set of interval type timers that can become attached to perceptual tasking through learning
Leiner, H. & Leiner, A., 1997; Leiner, H., Leiner, A. & Dow, R., 1986; Leiner, H., Leiner, A. & Dow, R., 1989	Theoretical computer model of cerebellar anatomy and physiology	Computing capabilities of the cerebellum have expanded with the evolution of the cerebral cortex to specify 'which', 'when', and 'where' for sequencing cognitive processing (including the linguistic manipulation of ideas that precedes planned behavior)

to the purposes of this chapter. It can be seen in the right-hand column of Table 1, that the cerebellum is now understood to be a builder of *models* that govern the rapid manipulation of both motor and cognitive activities (including language and working memory).

The Recent Evolution of the Relationship Between the Cerebellum and the Cerebral Cortex: Expansion Toward the Control of Cognitive Functions

Before proceeding to specific cognitive modeling functions of the cerebellum, it will be helpful to briefly sketch the recent evolution of the cerebellum in its relation to the cerebral cortex. Leiner, Leiner & Dow (1986, 1989) proposed that the massive three- to fourfold increase in the size of the cerebellum (and not in the basal ganglia which concurrently enlarged at a very low rate) in humans arose in conjunction with parallel increases in size and cognitive complexity of the cerebral cortex:

A detailed examination of cerebellar circuitry suggests that its phylogenetically newest parts may serve as a fast information processing adjunct of the association cortex and could assist this cortex in the performance of a variety of manipulative skills, including the skill that is characteristic of anthropoid apes and humans: the skillful manipulation of ideas (1986, p. 444).

Ito (1984, 1993, 1997) anticipated and later elaborated on Leiner, Leiner and Dow's work by proposing *how* neurophysiological processes that take place in the cerebellum can apply equally well to both movements and mental processes:

This view may be justified in view of the common features of movement and thought as control objects . . . . The cerebellum may be viewed in this way as a multipurpose learning machine which assists all kinds of neural control, autonomic, motor or mental (verbal or nonverbal) (1993, p. 449).

The cerebellum is now conceived as learning machine, as Ito put it, that facilitates the neural control of all information flows in the cerebral cortex.

Thus, the selective advantage of the greatly enlarging cerebellum was that its computing capacities were being harnessed as an 'operating system' (determining the which, when and where of information flows) for the evolving yoked complexities of movement, language, and thought—including, of course, operating system control of the visuospatial sketchpad and speech loop of working memory. Leiner & Leiner (1997) have described the computing capacity of neural connections between the cerebellum and the cerebral cortex (some 40 million nerve tracts) in some detail. The details of this enormous amount of cerebellar connectivity with the cerebral cortex are beyond the scope of this article, but see Leiner & Leiner (1997, pp. 542–547) and Leiner, Leiner & Dow (1989). It is significant to recognize here that the influence of the cerebellum in the spatial and temporal control in the *internal* world of the brain rivals the influence of the nerve tracts of the visual system on the perceived spatial and temporal dimensions of the *external* world. It will be important to recall this idea, later in this chapter, when we examine Einstein's internal world.

The Neurophysiological Basis of Innovation
In this section we will examine how the interplay between the cerebellum and the cerebral cortex leads to innovation.

Cerebellar Models
Biological *feedback* loops are relatively slow, and, therefore, rapid and coordinated movement and thought patterns cannot be smoothly or competitively executed through feedback control alone (e.g. Kawato, 1999; Kawato & Gomi, 1992). For example, the movements and thoughts of professional athletes would be disastrously handicapped if dependent only on feedback information. However, the cerebellum has evolved rapid control of such movements and thoughts by way of internal *feedforward models* that are acquired through learning (Ito, 1984, 1993, 1997; Leiner, Leiner & Dow, 1986, 1989).[2] As one learns to

---

[2] According to the *computational* scheme for *mental models* set for originally by Craik (1943) and extensively elaborated by Johnson-Laird (1983), thought processes construct mental models that are imitative, small-scale computational representations of the external world that retain the external world's relation-structure. Craik described how the preserved relation-structure of the model is computationally parallel to that which it imitates:

A calculating machine, an anti-aircraft 'predictor', and Kelvin's tidal predictor all show the same ability. In all of these latter cases, the physical process, which it is desired to predict, is *imitated* (relation-structure is preserved) by some mechanical device or model which is cheaper, quicker, or more convenient in operation (1943, p. 52).

Thus, the similarity of the relation-structure of the model captures the fundamentals of operation of that which it imitates, but makes the operations faster and cheaper, and in the case of cerebellar models, 'neurologically cheaper and simpler'. This abstractive-imitative modeling advantage, it is proposed, is precisely what takes place between working memory and the cerebellum.

play basketball or play the piano, for example, movement and thought strategies become smoother, quicker, and more efficiently executed.

More important to the premises of this chapter, cerebellar models also greatly increase the efficiency of purely *conceptual thought* (Doya, 1999; Ito, 1993, 1997; Leiner & Leiner, 1997; Leiner, Leiner & Dow, 1986, 1989). Leiner, Leiner & Dow (1989) described the processing of conceptual thought related to novel situations:

> In confronting a novel situation, the individual may need to carry out some preliminary mental processing before action can be taken, such as processing to estimate the potential consequences of the action before deciding whether to act or to refrain from acting. In such decision-generating processes, the prefrontal cortex is activated. This cortex, via its connections with the cerebellum, could use cerebellar preprogramming [e.g. learned cerebellar models that involve rapid strategies of search, planning, pattern recognition, and induction] to manipulate conceptual data rapidly. As a result, a quick decision could be made. This could be communicated to the motor areas, including the supplementary motor areas (p. 448).

The foregoing account of innovative problem-solving clearly involves the visuospatial sketchpad, speech loop and central executive functions of working memory. But how, exactly, does such 'faster-than-real-time' cerebellar preprogramming lead to innovation?

Cerebellar Models are Dynamics Models

Cerebellar feedforward models are neural representations of the dynamics of movement and thought (Ito, 1993, 1997). A dynamics model learns the *dynamics* of control objects (e.g. hands, legs, or the components of working memory) instead of a specific motor command for a movement or a specific cognitive control for the manipulation of conceptual information (Ito, 1993, 1997; Kawato, 1999; Kawato & Gomi, 1992). This means that a dynamics model is a set of control instructions that *generalizes* to a broad variety of situations that could potentially occur in the state-space in which it was originally learned. This capacity for generalization, along with the speed advantage, permits a great flexibility of movement and thought, and it is why cerebellar dynamics models have powerful selective advantage when, as described above, the individual confronts novelty. The capacity for generalization that is inherent in dynamics models of the cerebellum is the key to innovation. But this is only the beginning of the story.

Manipulation of Thought: Unconscious Generalization

In the process of learning, the cerebellum acquires two types of feedforward dynamics models, namely, dynamics models and *inverse* dynamics models. Dynamics (g) models are neural representations of the transformation from motor commands to movement and thought patterns in behavior and cognitive processing. However, the inverse dynamics (1/g, i.e. motor dynamics inversely equal to g) models are defined as neural representations of the transformation from the movement and thought patterns to the motor commands required to achieve them (see Ito, 1997, p. 481; Kawato & Gomi, 1992, pp. 445–446). In everyday language, this means that dynamics models are associated with rapid, skilled movement/thought while under conscious control, while cerebellar inverse dynamics models permit the motor cortex to be bypassed, thus allowing rapid, skilled movement/ thought to take place at an unconscious level.

Ito (1997) provided an example of the respective operation of dynamics and inverse dynamics models that is highly pertinent to the learning of models of working memory activity by the cerebellum:

> According to the psychological concept of a mental model (Johnson-Laird, 1983) [see footnote 2], thought may be viewed as a process of manipulating a mental model formed in the parietolateral association cortex (see Fig. 1) by commands from the prefrontal association cortex. A cerebellar microcomplex may be connected to neuronal circuits involved in thought [notably language processing] and may represent *a dynamics or an inverse dynamics model of a mental model* (italics added). In other words, a mental model might be transferred from the parietolateral association cortex to the cerebellar microcomplex during repetition of thought. By analogy to voluntary movement, one may speculate that formation of a dynamics model in the cerebellum would enable us to think correctly [and rapidly] in a feedforward manner, i.e. without the need to check the outcome of the thought. . . . However, an inverse dynamics model in the cerebellum would enable us to think automatically without conscious effort (p. 483).

The inverse dynamics model helps explain how generalizations can be formed outside a person's conscious awareness. This is one of the reasons that innovative insight may seem to leap out of 'nowhere'.

The Cerebellar/Working Memory Basis of Innovation: Details

In this section I describe in more detail: (1) how the resulting new cerebellar patterns of operation are learned in working memory; and (2) why innovative discovery (actually the discovery of new cerebellar

patterns in working memory) is often experienced as intuition or insight.

Detail: How Cerebellar Patterning ('Innovation') is Learned in Working Memory

What is the neurophysiological basis, in working memory, of 'seeing' (in the visuospatial sketchpad), 'hearing' (in the speech loop), and 'making decisions about' (in the central executive) innovative versions of conceptual information? Houk & Wise (1995) propose that the output from the cerebellum guides the frontal cortex by training its cortical networks toward more efficient use of problem spaces. I propose that the experience of innovation arises as cerebellar commands are executed toward broadened generalization in the conceptual thought of working memory (see also Vandervert, in press). Houk & Wise describe these cerebellum-initiated changes in the frontal cortex (and other modules of the cerebral cortex) as follows:

> On the one hand, these alterations [arriving from the cerebellum] would serve to modify the collective computations being performed by the cortical network on a moment-to-moment basis. On the other hand, the same alterations would promote changes in the weights of the Hebbian synapses on pyramidal neurons, that, in the long run, would move the network's attractors closer to the points being forced by cerebellar and basal ganglia modifications. As a consequence, the frontal cortex would become trained to perform, in a highly efficient and automatic fashion, those particular functions being forced on it by its subcortical [cerebellar and basal ganglia] inputs (1995, p. 106).

The above training of the frontal cortex may involve the input of new efficiencies and generalization from either *within* a cerebellar dynamics space or *between* cerebellar dynamics spaces (see Haruno, Wolpert & Kawato, 1999). This would account for the fact that innovations may occur either within the narrow scope of a single technology or from diverse interdisciplinary efforts.

The alteration of cortical networks toward the experience of new ideas or technology may be experienced as an innovative insight, just as is, for example, an innovative hand movement that more efficiently renders a drawing, makes a signal, or throws a Frisbee. Since such alteration is a continuous updating process in working memory, innovation is a feature of learning that occurs regularly in everyone. What makes an innovation an 'important' innovation, or a deeply experienced 'insight' is a matter of its cultural or organizational context, and its degree of generalization.

Detail: The Dynamics of Dynamics Models are Compounded: The Deepening of Generalization

There is a rather profound corollary to the foregoing principle of dynamics learning in working memory. Once a new model of the cerebellar mental model is learned in working memory and given its new, more generalized expression, the newer expression in the neural networks of the cerebral cortex, under continued recursive processing in working memory, *itself* yields even more generalized dynamics—a deeper insight. Redrafting in writing, art or model-making is an example of this compounded recursive process. This ever-new level of realized generalization is one of the reasons, perhaps the most important reason, that 'yesterday's' writing often appears to somehow be 'sub-standard'.

Within the working memory/cerebellar theory of innovation, this 'model-of-a-mental-model loop' produces an ever-deepening level of generalization or abstraction. These new levels of generalization may be accomplished either by the individual who produced the initial level of abstraction or by importing generalized information from others that could be further 'reduced' in its recipient new 'host'. I believe that this conceptually simple, selective mechanism is why the history of innovation is the story of continual refinement and an ever-deepening level of abstraction. It is an epistemological question as to how far this model-of-a-mental-model process may continue into the realm of abstraction. (See Constant (2000) for a discussion of recursive innovative processes at the socio-cultural level.)

Einstein's Experiential Accounts of Mathematical Innovation: The Generalized Case for All Innovation

In preparation for the examination of Einstein's subjective account of mathematical innovation, I will layout a summary diagram showing the hypothesized flow of information in terms of the collaborative contributions of recursive control patterns of working memory and the cerebellum. This articulated flow of brain functions provides a neurophysiological basis upon which to follow the mental steps to mathematical innovation as described by Einstein. In this way, we may assume for a moment that Einstein was perhaps describing actual, operationally specifiable milestones (within the general experimental framework of working memory developed by Baddeley and his colleagues, as it is influenced by the cognitive functions of the cerebellum) that constitute the gestation and birth, as it were, of mathematical axioms. Indeed, for Einstein, such a general assumption was the key to understanding how pure thought is directly connected to the axioms of science and mathematics (see Holton, 1979). Einstein took great pains on many occasions to tell us that the internal world of thought processes is

the epistemological nidus of axiomatic knowledge (Einstein, 1949, 1956; Holton, 1979).[3]

A General Structure for the Analysis of Einstein's Experiential Account

Figure 2 is a summary illustration of the flow of information between working memory and the cerebellum that will be used to study Einstein's experiential account.

While the mental processes that lead to the cognition of mathematical axioms are ultimately dependent upon the world of sensory experience, and such discovery must be verified in terms of tests in relation to the sensory world, the focus in Fig. 2 is upon events taking

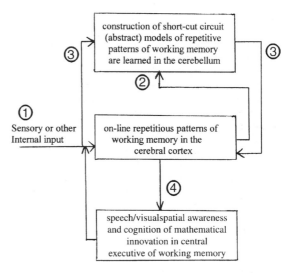

Figure 2. Flow between working memory functions of the cerebral cortex and their mapping representation in the cerebellum. Routing sequence: (1) sensory and/or other internal brain processes utilized by working memory; (2) mapping of patterns of working memory learned in the cerebellum; (3) upon elicitation by sensory, perceptual-motor, or other internal input, mappings fed forward from the cerebellum to facilitate smooth and rapid working memory processing, (4) awareness/cognition of innovation in the central executive. Since all innovations are based upon the cerebellar categorization and generation of patterns, mathematical discovery can serve at the basic model for all innovations (see footnote 1). Compare flow with simplified flows in brain areas illustrated in Fig. 1B.

[3] Similarly, it is the epistemological contention behind the theory of innovation presented in this chapter that the *patterns* that constitute mathematics (see footnote 1) arise from the reference frame of bodily movement in relation to perceptual-cognitive processing (Vandervert, 1993, 1997, 1999, 2001). This position does not deny the existence of a real world 'out there', but proposes that mathematics is based on patterns that are constructed in the brain as it adapts to its environment. See also the epistemological position on mathematics espoused by Lakoff & Núñez (1997, 2000).

place *within* the brain. Houk & Wise (1995, pp. 105–106) have presented an excellent overarching picture of brain functions that supports the processing plan proposed here. See especially their concluding section titled, 'The Search for Intelligent Behavior'. It is important to note that Fig. 2 obviously does not represent an entire account of brain areas that may have a bearing on innovation. Such a detailed, full account is beyond the purposes of this chapter.

In Fig. 2, recursive patterns of the activity of working memory are learned in the cerebellum. Shortcut mappings (generalizations or 'abstractions') of these patterns are subsequently fed back to working memory's central executive to facilitate more efficient on-line processing of conceptual thought as described in the previous section (Ito, 1993, 1997).

Bringing Cerebellar Models to Awareness Within Mental Models Employed by Working Memory: Cognizing Abstract Predictive Mappings

Baddeley (1993, 1998, especially chapters 6 and 18) has suggested that the central executive manages and comprehends available information by 'running' it within the predictive *mental model* frameworks proposed by Johnson-Laird (1983) (see footnote 2). According to Johnson-Laird, thought processes construct mental models that are *computational* representations of the external world. The selective advantage of such mental models is that they permit an organism to predict probable future states that may be critical to its survival (Vandervert, 1997). At the same time, Ito (1993, 1997) and Kawato & Gomi (1992) propose that the short-cut mappings generated in the cerebellum are also predictive mental models of the Johnson-Laird type. (See Ito (1997) on the subject of dynamics and inverse dynamics cerebellar models in the earlier section on the cognitive functions of the cerebellum.) In the brain-internal dialogue depicted in Fig. 2, then, the mental models employed by working memory and the cerebellum are computationally compatible predictive models. They would, of course, have to be compatible if cerebellar models were to act as short-cut circuits for movement and thought patterning (see Ito, 1997).

Thus within the flow depicted in Fig. 2, it is hypothesized that: (a) the *patterning* basis for mathematical innovation is contained in models generated by the cerebellum; and (b) the central executive has the capacities to bring these patterns to conscious awareness. We will now examine Einstein's experiential accounts of mathematical innovation within the foregoing perspective.

Einstein's Inner World

There are three classic sources that contribute to Einstein's subjective account of mathematical discovery: (a) Einstein's responses to Hadamard's (1945) study of invention/discovery in the mathematical field;

(b) Einstein's (1949) comments on the nature of thinking that appeared in his 'Autobiographical Notes'; and (c) his model of the construction of axioms that appeared in a 1952 letter to Maurice Solovine (Einstein, 1956). (A thorough theoretical and epistemological discussion of the letter to Maurice Solovine appears in Holton (1979).) The above sources extend over several years and provide an excellent sampling of Einstein's persistent views on the workings of his own experiential world. All three of the reports contain a great deal on how the structure of his thoughts evolved, and are therefore especially amenable to analysis in terms of the structure of processing depicted in Fig. 2.

A View of the Central Executive

Although we could begin with any one of the three sources on Einstein's experiential world, we will begin with the earliest account. This is perhaps our best chance to view the operation of the central executive of Einstein's working memory. The account was in response to a question in Hadamard's (1945) survey of the habits and mental methods of mathematicians. The specific question read as follows:

> It would be very helpful for the purpose of psychological investigation to know what internal or mental images, what kind of 'internal world' mathematicians make use of; whether they are motor, auditory, visual, or mixed, depending on the subject they are studying (Hadamard, 1945, Appendix I, p. 140).

Einstein answered this question in the following structured manner:

> (A) The words or the language, as they are written or spoken, do not seem to play any role in my mechanism of thought. The psychical entities which seem to serve as elements in thought are certain signs and more or less clear images which can be 'voluntarily' reproduced and combined.

> There is, of course, a certain connection between those elements and relevant logical concepts. It is also clear that the desire to arrive finally at logically connected concepts is the emotional basis of this rather vague play with the above mentioned elements. But taken from a psychological viewpoint, this combinatory play seems to be the essential feature in productive thought—before there is any connection with logical construction in words or other kinds of signs which can be communicated to others.

> (B) The above mention elements are, in my case, of visual and some of muscular type. Conventional words or other signs have to be sought laboriously only in a secondary stage, when the

       mentioned associative play is sufficiently established and can be reproduced at will.

(C) According to what has been said, the play with the mentioned elements is aimed to be analogous to certain logical connections one is searching for (Appendix II, pp. 142–143).

Einstein says that, in the process of mathematical discovery, he consciously ('voluntarily') plays with combinations of psychical elements, but that they are not words or language as they are written or spoken. Because the search he describes was under voluntary control, it appears that Einstein was talking about ongoing awareness in the (his) central executive. However, because conventional words and other signs had to be 'sought for laboriously only in a *secondary stage*' (italics added), it sounds as if the essential combinatory play Einstein describes is a process of 'decoding' less conventionalized sources of information (and not simply retrieval processing). What might be the origin of such less conventionalized sources of information? Within the framework of this chapter, the only source of this type of information would be cerebellar inverse dynamics models of thought patterns that Einstein had repeated many times as he mulled over problems of mathematics. That is, these are perhaps cerebellar models emerging into the consciousness of Einstein's central executive. Recall Houk & Wise's (1995) description of how cerebellar patterns are learned in the frontal cortex.

But we need to know more about the *structure* of the 'signs and more or less clear images' (psychical entities) he mentions. The next source will greatly clarify that picture.

Visualizing Automaticity in the Visuospatial Sketchpad: Decoding Cerebellar Models

In his 1949 'Autobiographical Notes' Einstein focuses on the psychical elements and processing structure of his internal world. He offers a visual record of the combinatory play and structure that brings forth an ordering element. This ordering element provides the logical connections that he mentioned 'searching for' in his above response to Hadamard. Einstein referred to the process that unlocks the ordering element of combinatory play as 'thinking':

       What, precisely, is 'thinking?' When, at the reception of sense-impressions, memory-pictures emerge, this is not yet 'thinking'. And when such pictures form series, each member of which calls forth another, this too is not yet 'thinking'. When, however, a certain picture turns up in many such series, then—precisely through such return—it becomes an *ordering element* for such series, in that it connects series which in themselves are unconnected. Such an element becomes an instrument, a concept. I think that the transition from free association or 'dreaming' to thinking is charac-

terized by the more or less dominating role which the 'concept' plays in it. It is by no means necessary that a concept must be connected with a sensorily cognizable and reproducible sign (word); but when this is the case thinking becomes by means of that fact communicable (Einstein, 1949, p. 7).

In his picture series, each picture calling forth another picture, and, as well, in the certain picture that turns up in many series, Einstein reveals the automaticity that underlies what he is referring to as axiomatic-level 'thinking'. (See Ito's (1997) comments on the automaticity of inverse dynamics models in the section on the cognitive functions of the cerebellum. See also Kihlstrom (1987) on automaticity in the cognitive unconscious.) The automatic flow of various series of memory pictures Einstein describes fits the computational architecture of *linked* multiple pairs of forward (predictive) and inverse (controller) models (Haruno, Wolpert & Kawato, 1999; Kawato, 1999). Within this architecture, a large number of separate but interconnected pairs of forward and inverse models cover a range of learning and control contexts. The paired models govern both conscious learning and unconscious control aspects related to various environment contexts, e.g. manipulation of physical objects, manipulation of language communication, and manipulation of working memory processes. A responsibility predictor function (forward model of a pair) automatically calls forth its context-appropriate inverse controller model. *In the case where working memory is mulling over several contexts, the responsibility predictor function would result in the calling forth of a series of inverse models and their visuospatial (and central executive and phonological) content (images or 'pictures')*. Thus the paired model computational architecture of the cerebellum appears to be a good candidate for the control of the automatic, each-picture-calling-forth-another portion of Einstein's description. I believe this is the selective strategy by which working memory forms new mathematical concepts (Vandervert, in press). But, we have not yet explained the crucial appearance of the ordering element that Einstein mentions.

The Axiomatic Ordering Element: The Working Memory/Cerebellar Confabulation of Intuition and Insight

Our final report from Einstein helps clarify *how* the ordering element, which would be an axiom in mathematical discovery, is 'constructed'. In a 1952 letter to his good friend Maurice Solovine (Einstein, 1956), Einstein described a complete diagrammatic model of the discovery of axioms. In the model, discovery begins with immediate sensory experience (E). Then, in the next step, Einstein shows an *intuitive leap* to axioms (A). In the letter he described the situation as follows:

A are the axioms from which we draw consequences. Psychologically the A are based upon the E. There is, however, no logical path from E to A, but only an intuitive (psychological) connection, which is always subject to 'revocation' (p. 121).

Einstein went on to say that once axioms were intuitively conceptualized, logical assertions could be deduced, and that, finally, these assertions could be tested against experience. The idea that axioms could *only* be arrived at by an intuitive leap was Einstein's most persistent epistemological belief.

Einstein suggested that we prepare ourselves for the intuitive leap to axioms (concepts or ordering principles) by focusing our attention on "certain repeatedly recurring complexes of sense impressions" and "relating to them a concept" (Einstein, 1954, p. 291). This focus of attention (on the part of the central executive functions of working memory), I believe, is the basis of the confabulation and appearance of the ordering principle or 'certain picture' mentioned above in Einstein's (1949) classic description of 'thinking'.

But why would this new, common picture that Einstein mentions begin to turn up in many series? This new, common picture seems to be the key to intuitive discovery, and I think it is an expression of the *dynamics* principle described earlier in this chapter. That is, I believe that this picture is the result of *generalization* from a newly generated level of abstraction that has originated in the cerebellum (recall the recursive model-of-a-mental-model principle described earlier).

The Cerebellar Trigger of Intuition

Why would the new picture appear *suddenly*, and *how* would it give new conceptual order to the several series? First, the appearance of the new picture would occur among many series as Einstein describes it, because the responsibility predictor functions of the several series (of multiple paired models) had simultaneously responded (via the neural computation of responsibility estimates)[4] to a newly generated level of abstraction as a *novel* control object. That is, since the new generalization had been recognized as a new control object by the series of multiple paired models, they would, depending on the coherence of their several contexts, simultaneously generalize their prediction and control *to* it. Such simultaneous generalized prediction and control would result in a new 'picture' in the visuospatial sketchpad that would appear 'suddenly' (either through simultaneous acquisition or through rapid recursion) in the many series mentioned by Einstein. Second, this new picture would be a synthesis based upon joint generalization in the

several impacted series of multiple paired models, and thus it would serve as an ordering element for otherwise unconnected series. When such simultaneous generalization among collections of multiple paired models occurs in working memory, it would be experienced as *insight* or *intuition*.[5]

Encouraging the Production of Innovation

The neurophysiological process that leads to innovation works pretty much the same way in all people, although, due to a variety of learning and genetic predispositions, the emergence of innovation is perhaps not experienced to the same degree of clarity or of productivity. The main point is that everyone is born with, and carries throughout life, the neurophysiological 'apparatus' of innovation. And, like other cognition involving the cerebellum and working memory, this apparatus is highly educable.

The most fundamental product of the interplay between working memory and the cerebellum appears to be its boundless capacity for the deepening of abstraction. And, we can conclude from the earlier discussion of the subjective reports of Albert Einstein that the abstraction of pattern is the fundamental building block of innovation. Root-Bernstein and Root-Bernstein (chap. 5, 1999) have convincingly shown that abstraction is a product of a recursive 'play' involving some problem in art, science, poetry and so on. These authors show that when abstraction is achieved in this way, it is inevitably tied to discovery and innovation. Of course, in regard to the working memory/cerebellar theory of innovation, such recursive play is exactly what leads to the transfer of working memory information to the cerebellum for pattern abstraction, and then back to working memory where it may be consciously experienced. The reason that pattern abstraction produces something new is that an abstraction moves the system (physiological or mental) toward the fundamental rules, designs, and laws that connect collections of movements and thoughts. Abstractions thereby make new connections among disciplines and among topics within disciplines. Indeed, the deepest level of abstraction, which history seems to show we are neurophysiologically destined to achieve, may eventually lead humans to a verifiable grand 'theory of everything'.

Steps That Will Encourage Abstraction and Innovation

Root-Bernstein and Root-Bernstein (chap. 5, 1999) provide several classic examples, beginning with the art of Picasso, which illustrate how pattern abstraction

---

[4] Haruno, Wolpert & Kawato (1999) provide equations for the computation of the functions of each component of the multiple paired model architecture.

[5] Einstein's example appears to involve mostly the central executive and visuospatial sketchpad of working memory. However, the same effects of multiple paired models, if arising in the other components or combinations of components of working memory, would lead to insight and intuition, although appearing in differing experiential form.

leading to innovation develops. To encourage the development of abstractions, in both children and adults, they offer the following advice:

> Abstracting... is a process beginning with reality and using some tool (a personally-chosen, repetitive technique) to pare away the excess to reveal a critical, often surprising, essence. ... Use one of the abstraction sequences discussed in this chapter—Picasso's *Bulls*, for example, or 'Observations on a Water-Worn Stone'—as a guide. Choose your subject and your abstracting tool; think about them realistically; play around with their various properties or characteristics; get at what might be most essential; then consider and reconsider your results from a distance of time and space. Say your abstraction, mime it, sing it, write it in prose, write it in poetry, extract a concept or metaphor. Practice with artwork, or, if you are scientifically inclined, practice with simple experiments or mathematical concepts. If you are a dancer, replicate the real movements of real people or of animals, then try to find the essence characterizing that personality or that species. Describe in music the distillation of birdness or windness or a carousel. Find the minimum vocabulary to convey a maximum amount of sense and sensibility (p. 90).

One can see the interplay of the components of working memory and its associative executive areas of the cerebral cortex on the one hand, and cerebellar model making on the other running all through the Root-Bernstein's approach to abstracting. Of course, their approach can also be looked upon as a guide to the improvement of the process of thinking in general. I will clarify this point in a moment.

I believe that the how-to-develop-an-abstraction plan that Root-Bernstein and Root-Bernstein have suggested is good, solid advice for systematically encouraging cerebellar/working memory processes toward generalization and innovation. And, as they say, it would be increasingly effective as it were carried out over longer periods of time. If we were to disassemble and examine Root-Bernsteins' recommendations as we did Einstein's experiential account of 'thinking' in the previous section, we would find that they both make use of the same cerebellar/working memory processes. In other words, processes very much like the abstractive processes Root-Bernstein and Root-Bernstein describe are iterated automatically in the mind of, for example, an Einstein or a Picasso. The key to highly productive innovation would seem to be an approach that keeps a person (or an organization) at tasks that are contexted in such 'play' for an extended period of time.

Discussion

Dasgupta (1996) argued that innovation is a lawful process that can be seen running across individual innovators and across innovations in different disciplines. And, he says, the study of innovation can therefore be carried out in a scientific manner.

As a part of such a science, I have described how the recursive collaboration of working memory (online consciousness) and the cognitive functions of the cerebellum constitutes the primary evolutionary mechanism behind innovation. The repetitive activities of the visuospatial loop, speech loop, and central executive control processes of working memory are constantly reduced to new and more efficient spatiotemporal patterning in the cerebellum. These control patterns are subsequently fed back to working memory where neural networks constantly learn the new and ever more efficient control instructions for conceptual thought. Human innovation is prolific and its pace has accelerated, because constantly increasing levels of generalization are produced by the reciprocal relationship between the cerebellum and working memory. This process makes its appearance as ever-new levels and combinations of design space in working memory. (See Stankiewicz (2000) for a discussion of design space as implied here.)

Highly reduced patterning involving collections of multiple paired cerebellar models becomes the basis for the discovery of mathematical knowledge when it is learned in the neural networks of working memory. This cerebellar-working memory sequence of mathematical discovery was examined by way of three of Einstein's classic experiential accounts of mathematical discovery. The neurophysiology of the discovery of mathematics constitutes the general process of innovation in all fields and accounts for their unique subjective experiences of insight and intuition.

The transparency between mathematical discovery and innovation in other fields can be seen most clearly in technological innovation. This transparency becomes increasingly evident as technology has become increasingly electronic as in computing, communication and telepresence. At this advanced level of technological development, mathematics can be literally 'reverse engineered' from technology. In other words, mathematics is embodied in technology, just as it is embodied in the neurophysiological operation of the human mind. This is the essential theoretical-epistemological basis of cybernetics, as defined by Wiener (1948): "the entire field of control and communication theory, whether in the machine or in the animal" (p. 19).

Acknowledgements

I would like to thank Marc Nagel for his help in securing the illustration of the Brodmann brain areas that appears in Fig. 1. The illustration was kindly provided by Hermann Strohmaier, Director of the Korbinian Brodmann Museum, Hauptstr. 43, 78355 Hohenfels, Germany.

References

Akshoomoff, N., Courchesne, E. & Townsend, J. (1997). Attention coordination and anticipatory control. In: J. D. Schmahmann (Ed.), *The Cerebellum and Cognition* (pp. 575–598). New York: Academic Press.

Awh, E., Jonides, J., Smith, E., Schumacher, E., Koeppe, R. & Katz, S. (1996). Dissociation of storage and rehearsal in verbal working memory: Evidence from positron emission tomography. *Psychological Science*, **7**, 25–31.

Baddelely, A. (1992, January 31). Working memory. *Science*, **255**, 556–559.

Baddeley, A. (1993). Working memory and conscious awareness. In: A. Collins, S. Gathercole, M. Conway & P. Morris (Eds), *Theories of Memory* (pp. 11–28). Hillsdale, NJ: Lawrence Erlbaum Associates.

Baddeley, A. (1996). Exploring the central executive. *The Quarterly Journal of Experimental Psychology*, **49A**, 5–28.

Baddeley, A. (1998). *Human memory: Theory and practice*. Needham Heights, MA: Allyn & Bacon.

Baddeley, A. & Andrade, J. (1998). Working memory and consciousness: An empirical approach. In: M. Conway, S. Gathercole & C. Cornoldi (Eds), *Theories of Memory* (Vol. 2, pp. 1–23). Hove, U.K.: Psychology Press.

Baddeley, A. & Hitch, G. (1974). Working memory. In: G. Bower (Ed.), *The Psychology of Learning and Motivation* (Vol. 8, pp. 47–89). New York: Academic Press.

Constant, E. (2000). Recursive practice and the evolution of technological knowledge. In: J. Ziman (Ed.), *Technological Innovation as an Evolutionary Process* (pp. 219–233). Cambridge: Cambridge University Press.

Craik, K. (1943). *The nature of explanation*. Cambridge: Cambridge University Press.

Dasgupta, S. (1996). *Technology and creativity*. New York: Oxford University Press.

Deecke, L., Kornhuber, H., Lang, W., Lang, M. & Schreiber, H. (1985). Timing functions of the frontal cortex in sequential motor and learning tasks. *Human Neurobiology*, **4**, 143–154.

Desmond, J. & Fiez, J. (1998). Neuroimaging studies of the cerebellum: Language, learning and memory. *Trends in Cognitive Sciences*, **2**, 355–362.

Devlin, K. (1994). *Mathematics: The science of patterns: The search for order in life, mind, and the universe*. New York: W. H. Freeman.

Doya, K. (1999). What are the computations of the cerebellum, the basal ganglia and the cerebral cortex? *Neural Networks*, **12**, 961–974.

Einstein, A. (1949). Autobiographical notes. In: A. Schillp (Ed.), *Albert Einstein: Philosopher-scientist* (Vol. 1, pp. 1–95). La Salle, IL: Open Court.

Einstein, A. (1954). *Ideas and opinions*. New York: Crown Publishers.

Einstein, A. (1956). *Lettres à Maurice Solovine*. Paris: Gauthier-Villars.

Fiez, J., Raife, E., Balota, D., Schwarz, J., Raichle, M. & Peterson, S. (1996). A positron emission tomography study of the short-term maintenance of verbal information. *The Journal of Neuroscience*, **16**, 808–822.

Hadamard, J. (1945). *The psychology of invention in the mathematical field*. New York: Dover.

Haruno, M., Wolpert, D. & Kawato, M. (1999). Multiple paired forward-inverse models for human motor learning and control. In: M. S. Kearns, S. A. Solla & D. A. Cohn (Eds), *Advances in Neural Information Processing Systems* (pp. 31–37). Cambridge, MA: MIT Press.

Holton, G. (1979). Constructing a theory: Einstein's model. *The American Scholar*, **48**, 309–339.

Houk, J. & Wise, S. (1995). Distributed modular architectures linking basal ganglia, cerebellum, and cerebral cortex: Their role in planning and controlling action. *Cerebral Cortex*, **2**, 95–110.

Imamizu, H., Miyauchi, S., Tamada, T., Sasaki, Y., Takino, R., Pütz, B., Yoshioka, T. & Kawato, M. (2000). Human cerebellar activity reflecting an acquired internal model of a new tool. *Nature*, **403**, 192–195.

Ito, M. (1984). Is the cerebellum really a computer? *Trends in Neurosciences*, **2**, 122–126.

Ito, M. (1993). Movement and thought: Identical control mechanisms by the cerebellum. *Trends in Neurosciences*, **16** (11), 448–450.

Ito, M. (1997). Cerebellar microcomplexes. In: J. D. Schmahmann (Ed.), *The Cerebellum and Cognition* (pp. 475–487). New York: Academic Press.

Ivry, R. (1997). Cerebellar timing systems. In: J. D. Schmahmann (Ed.), *The Cerebellum and Cognition* (pp. 555–573). New York: Academic Press.

Johnson-Laird, P. (1983). *Mental models*. New York: Cambridge University Press.

Kawato, M. (1999). Internal models for motor control and trajectory planning. *Current Opinion in Neurobiology*, **9**, 718–727.

Kawato, M. & Gomi, H. (1992). The cerebellum and VOR/OKR learning models. *Trends in Neuroscience*, **15**, 445–453.

Kihlstrom, J. (1987). The cognitive unconscious. *Science*, **237**, 1445–1452.

Kornhuber, H. (1974). Cerebral cortex, cerebellum, and basal ganglia: An introduction to their motor functions. In: F. Schmitt & F. Worden (Eds), *The Neurosciences: Third Study Program* (pp. 267–280). Cambridge, MA: MIT Press.

Lakoff, G. & Núñez, R. (1997). The metaphorical structure of mathematics: Stretching out cognitive foundations for a mind-based mathematics. In: L. English (Ed.), *Mathematical Reasoning: Analogies, Metaphors, and Images* (pp. 21–89). Mahwah, NJ: Lawrence Erlbaum.

Lakoff, G. & Núñez, R. (2000). *Where mathematics comes from: How the embodied mind brings mathematics into being*. New York: Basic Books.

Leiner, H. & Leiner, A. (1997). How fibers subserve computing capabilities: Similarities between brains and machines. In: J. D. Schmahmann (Ed.), *The Cerebellum and Cognition* (pp. 535–553). New York: Academic Press.

Leiner, H., Leiner, A. & Dow, R. (1986). Does the cerebellum contribute to mental skills? *Behavioral Neuroscience*, **100**, 443–454.

Leiner, H., Leiner, A. & Dow, R. (1989). Reappraising the cerebellum: What does the hindbrain contribute to the forebrain? *Behavioral Neuroscience*, **103**, 998–1008.

Mandler, J. (1988). How to build a baby: On the development of an accessible representational system. *Cognitive Development*, **3**, 113–136.

Mandler, J. (1992a). How to build a baby II: Conceptual primitives. *Psychological Review*, **99**, 587–604.

Mandler, J. (1992b). The foundations of conceptual thought in infancy. *Cognitive Development*, **7**, 273–285.

Pribram, K. (1971). *Languages of the brain*. Englewood Cliffs, NJ: Prentice-Hall.

Pribram, K. (1991). *Brain and perception: Holonomy and structure in figural processing*. Hillsdale, NJ: Lawrence Erlbaum.

Root-Bernstein, R. & Root-Bernstein, M. (1999). *Sparks of genius*. New York: Houghton Mifflin.

Schmahmann, J. (Ed.) (1997). *The cerebellum and cognition*. New York: Academic Press.

Stankiewicz, R. (2000). The concept of 'design space'. In: J. Ziman (Ed.), *Technological Innovation as an Evolutionary Process* (pp. 234–247). Cambridge: Cambridge University Press.

Steen, L. A. (1988). The science of patterns. *Science*, **240**, 611–616.

Teasdale, J., Dritschel, B., Taylor, M., Proctor, L., Lloyd, C., Nimmo-Smith, I. & Baddelely, A. (1995). Stimulus-independent thought depends on central executive resources. *Memory & Cognition*, **23**, 551–559.

Vandervert, L. (1993). Neurological positivism's evolution of mathematics. *The Journal of Mind and Behavior*, **14**, 277–288.

Vandervert, L. (1997). The evolution of Mandler's conceptual primitives (images schemas) as neural mechanisms for space-time simulations structures. *New Ideas in Psychology*, **15**, 105–123.

Vandervert, L. (1999). A motor theory of how consciousness within language evolution led to mathematical cognition: The origin of mathematics in the brain. *New Ideas in Psychology*, **17**, 215–235.

Vandervert, L. (2001). A provocative view of how algorithms of the human brain will embed in cybereducation. In: L. Vandervert, L. Shavinina & R. Cornell (Eds), *Cybereducation: The Future of Long Distance Learning* (pp. 41–62). New York: Mary Ann Liebert.

Vandervert, L. (in press). How working memory and cognitive modeling functions of the cerebellum contribute to discoveries in mathematics. *New Ideas in Psychology*. [Vol. 21(2), Erratum version].

Wiener, N. (1948). *Cybernetics*. New York: John Wiley.

Ziman, J. (Ed.) (2000). *Technological innovation as an evolutionary process*. Cambridge: Cambridge University Press.

The International Handbook on Innovation
Edited by Larisa V. Shavinina

On the Nature of Individual Innovation

Larisa V. Shavinina[1] and Kavita L. Seeratan[2]

[1] *Département des Sciences Administratives, Université du Québec en Outaouais, Canada*
[2] *Department of Human Development and Applied Psychology, University of Toronto, Canada*

Abstract: In this chapter a new conception of individual innovation, seeking to explain its very nature, will be presented. Individual innovation refers to innovation at the level of an individual. This conception ventures to explore an issue of exceptional importance necessary for a scientific understanding of the inner essence of innovation, namely: why innovative ideas emerge in human minds. To successfully grasp this issue, we believe that mainly developmental and cognitive mechanisms must be taken into account. The internal structure of individual innovation is presented at five levels: (1) developmental foundation of innovation; (2) its cognitive basis; (3) its intellectual manifestations; (4) its metacognitive manifestations; and (5) its extracognitive manifestations.

Keywords: Innovation; Individual innovation; Giftedness; Talent; Creativity; Cognitive experience.

Introduction

A growing body of literature suggests that innovation originates from within the individual, that is, from his or her new idea(s) (Amabile, 1988, 1996; King, 1990, this volume; Simonton, this volume). Various definitions, models, and theories of creativity and innovation include the 'generation of new ideas' as one of their components (Bailey & Ford, this volume; Clapham, this volume; Dusgapta, this volume; Lubart, 2001–2002; Runco & Pritzker, 1999; Sternberg, 1999; Sternberg et al., this volume). For instance, many innovation scholars—including us—would agree with Kanter (1983) who states that innovation is the generation, acceptance, and implementation of new ideas, processes, products, or services. In this chapter, we focus on individual innovation, that is, innovation which appears at the individual level and which is mainly responsible for the generation of new ideas. In spite of advances in innovation research, determined mostly by studies of business scholars and management science specialists (Christensen, 1997; Katz, 1997; Leonard-Barton, 1995; Tidd, Bessant & Pavitt, 1997; Tushman & O'Reilly, 1997; Van de Ven et al., 1999), we do not know for certain why it is that some individuals are exceptionally able to generate new ideas and others not. An important goal of this chapter is to try to shed some light on this important issue by presenting a new conceptual framework for understanding individual innovation.

Individual innovation begins from the generation of new ideas, which is eventually externally manifested in the extraordinary innovative achievements we see in any field of human activity. According to our conception, individual innovation is a result of a specific organization of an individual's cognitive experience which functions as a carrier of all the manifestations of individual innovation (i.e. its traits and characteristics). Cognitive experience expresses itself in a specific type of the representations of reality (i.e. how an individual sees, understands, and interprets the world around), that is, in an individual's intellectual picture of the world.

The essence of individual innovation rests in the uniqueness of the individual's intellectual picture of the world. In other words, innovators' unique view, understanding, and interpretation of what is going on in the surrounding reality are keys for the scientific understanding of individual innovation.

In our view, the internal structure of individual innovation is presented at five levels: (1) a developmental foundation for innovation; (2) the cognitive basis of innovation; (3) the level of intellectual manifestations of innovation; (4) the level of metacognitive manifestations of innovation; and (5) the level of extracognitive manifestations of innovation. Below we consider each of them in detail.

Before beginning a presentation of our conception of individual innovation, it is important to note three things. First, from our point of view, individual innovation belongs to a general construct of high abilities, which includes creativity, exceptional intelligence, giftedness, and talent. Thus, in this chapter we do not draw a distinction between individual innovation and creativity, talent, and giftedness. In our view, innovators are gifted, creative, and talented individuals.

Second, it should be emphasized that in this chapter we deal especially with individual innovation, which differs from the general understanding of innovation (e.g. Kanter's definition). Specifically, our chapter addresses the issue of *why* new ideas—from which any innovation begins—appear in the minds of only certain individuals. In this light, a range of questions arises such as: What is exceptional about the personalities of those people who are able to generate new ideas leading to innovations? Do they have extraordinary minds? Is their exceptionality a result of a combination of their unique minds and personalities? In order to successfully address these questions, one should recognize—as we have—that those individuals who are able to generate new ideas resulting into innovations (i.e. innovators) are gifted, creative, and talented.

Finally, since many chapters in this handbook provide comprehensive reviews of the existing literature on innovation appropriate in the context of our chapter (see, for example, Chi et al., this volume; Georgsdottir et al., this volume; Kaufmann, this volume; Root-Bernstein, this volume; Simonton, this volume; to mention just a few), we do not include our own literature review in this chapter.

A New Conception of Individual Innovation

Cognitive Experience as a Psychological Basis of Individual Innovation

As we previously mentioned, the nature of individual innovation is still an area of innovation research that has been left relatively unexplored. Although the traits, characteristics, features, properties, and qualities of innovation and innovative people (i.e. their external manifestations in any real activity) have been the subject of scientific research, the psychological basis (or psychological carrier) of these manifestations has not been investigated (Kholodnaya, 2002; Shavinina & Kholodnaya, 1996). Attempts to understand the nature of any psychological phenomenon based solely on listing and describing its external manifestations, including its characteristics, traits, features, qualities, and properties, are inadequate. Contradictions and crises in psychology testify to this (Vekker, 1981).

There is a need for a new direction for research that considers individual innovation as the sum of its two important aspects: its external manifestations and its

psychological basis. Consequently, there is a need to re-examine the question of the nature of individual innovation. Researchers should not simply answer the question 'What is individual innovation?' by merely listing its characteristics and traits (i.e. its external manifestations). Rather, they should answer the question: 'What is the carrier (a basis) of the characteristics and traits associated with individual innovation?' (Shavinina & Kholodnaya, 1996).

From this fundamentally changed viewpoint, scientists should study an individual's mental or cognitive experience—more precisely, the specificity of its structural organization. We assert that the individual cognitive experience is the psychological basis of individual innovation or the psychological carrier of its manifestations (Shavinina & Kholodnaya, 1996). Mental or cognitive experience is defined as a system of the available psychological mechanisms, which forms a basis for the human cognitive attitude towards the world and predetermines the specificity of his or her intellectual activity (Kholodnaya, 2002). Cognitive experience—that is the cognitive level in the structural organization of individual innovation—is formed by conceptual structures (i.e. conceptual thinking), knowledge base, and subjective mental space (Kholodnaya, 2002). These are all forms of the organization of the cognitive experience.

The importance of *conceptual structures* is determined by scientific findings, which indicate that conceptual thinking is the integrated cognitive formation, that is, a form of the integrated functioning of human intelligence (Kholodnaya, 2002). The more conceptual thinking is a form of the integrated work of intelligence, the better organization of an individual's intellectual activity will be. That means that intelligence will function perfectly (Kholodnaya, 1983; Vekker, 1981; Vygotsky, 1978). Chi & Hausmann (this volume) discuss the importance of conceptual structures in the context of their approach to understanding scientific innovation.

The *knowledge base* is the second form in the organization of the cognitive experience. Many researchers emphasize the role of the knowledge base in the development of an individual's intellectual resources as crucial (Bjorklund & Schneider, 1996; Chi & Ceci, 1987; Chi, Feltovich & Glaser, 1981; Kholodnaya, 2002; Pressley, Borkowski & Schneider, 1987; Rabinowitz & Glaser, 1985; Schneider, 1993; Shavinina & Kholodnaya, 1996; Shore & Kanevsky, 1993; Sternberg, 1985). Thus, the quantity and quality of specialized knowledge play a critical part in highly intellectual performance and in the process of acquiring new knowledge (Bjorklund & Schneider, 1996). For example, productive problem-solving cannot occur without relevant prior knowledge (Chi & Ceci, 1987). The knowledge base can facilitate the use of particular strategies, generalize strategy use to related domains, or even diminish the need for strategy activation

(Schneider, 1993). It was demonstrated that intellectually gifted people are distinguished by an adequate, well-structured, well-functioning, and elaborate knowledge base, which is easily accessible for actualization at any time (Kholodnaya, 1997; Rabinowitz & Glaser, 1985). Moreover, this rich knowledge base can sometimes compensate for overall lack of general cognitive abilities (Pressley et al., 1987; Schneider, 1993).

Conceptual structures and the knowledge base generate subjective mental space, the third form in the organization of cognitive experience. Individual differences in flexibility, differentiation, integration, and hierarchical structure of the mental space influence a persons' cognitive attitude to the world and, therefore, predetermine his or her intellectual and creative abilities, which lead to new ideas resulting into innovation. A more detailed review of the influence of flexibility on innovation can be found in the chapter by Georgsdottir et al. (this volume).

Cognitive experience expresses itself in a specific type of the objective representations of reality—that is, how an individual sees, understands, and interprets what is going on in the surrounding reality and in the world around him or her. Kholodnaya's (2002) investigations of the nature of individual intelligence have demonstrated that one of the basic phenomena (i.e. a proto-phenomenon) of an individual's intellectual life and his or her experience as a whole is his or her representations. Many psychologists have viewed representations to be important in understanding the essence of intelligence, intellectual giftedness, and innovation (Bamberger, 1986; Chi et al., this volume; Chi, Feltovich & Glaser, 1981; Klix, 1988; Oatley, 1978; Piaget, 1969). For example, Kholodnaya (1990) found that the main function of human intelligence lies in the construction of adequate representations of the world around.

In modern cognitive psychology, the expert–novice research paradigm also gives credence to the importance of the phenomenon of representations in the understanding of the nature of human intellectual abilities (Chi et al., 1981; Chi, Glaser & Rees, 1982; Kanevsky, 1990; Schneider, 1993; Shore & Kanevsky, 1993; Sternberg & Powell, 1983). For example, Chi et al. (1981) demonstrated that the main difference between experts and novices in physics has to do with their problem representations. They found that experts classified problems according to underlying principles and rules, but that novices tended to use superficial meanings of words and diagrams in their own classification of the same problems.

Therefore, cognitive experience manifests itself in the specific types of representations. This means that intelligent persons—and particularly innovators, who are intellectually gifted, creative, and talented—see, understand, and interpret the world around them by constructing an individual intellectual picture of events, actions, situations, ideas, problems, any aspects

of reality in a way that is different from other people. Because of that, their individual intellectual picture of the world (i.e. world view) is a unique one (Kholodnaya, 2002; Shavinina & Kholodnaya, 1996). One aspect of this uniqueness is connected with their objectivization of cognition, that is, they see, understand, and interpret everything in a very objective manner. Robert Root-Bernstein in his chapter, *The Art of Innovation* (this volume), emphasizes that sometimes perceiving the world differently is the key to making discoveries. Contributors to this handbook also highlight the importance of changing perspective for new ideas to appear (Chi et al., this volume; Georgsdottir et al., this volume).

It is important to note that scholars in the field of innovation point out that innovation does not necessarily imply acceptance and implementation of only objectively new ideas. Ideas can also be subjectively new ones—new only for some individuals or companies, but not for the rest of the world. As individual innovation deals mainly with the generation of new ideas, our conception of individual innovation emphasizes *objectively* new ideas, because the very essence of innovators—which, in our view, are intellectually gifted individuals—resides in their ability to see the world from an objective point of view. Kaufmann (this volume) also points out that novelty of ideas must be objective, and not only subjectively novel to its originator. Kholodnaya's (1990) understanding of human intelligence as the mechanism for structuring specific representations of reality representations connected with the reproduction of 'objective' knowledge—suggests that the degree of development of the ability for the objectivization of cognition determines one's intellectual and creative productivity. She showed that one of the distinguishing features of gifted individuals' representations of reality is their objective character. In this respect the most important conclusion is that:

> ... the significance of *intellectually gifted individuals* in society should be seen not only in that they solve problems well and create new knowledge, but mainly in the fact that they have *the ability to create an intellectual (objective) picture of the world*, i.e. they can see the world as it was, as it is, and as it will be in its reality (Kholodnaya, 1990, p. 128; italics added).

The same is true in the case of innovators. Let us consider, for example, Albert Einstein, whose scientific innovations have been undeniable for decades. A countless number of people who came along before him tried to understand the fundamental questions of time, space, and relativity. However, the fact is that only Einstein was able to propose specialized and generalized theories of relativity—the monumental achievements of the human mind of the 20th century. In our view, it was possible because Einstein saw the world—with its fundamental questions of time, space, gravity, and relativity—in its objective reality. That is—in accordance with Kholodnaya's (1990, p. 128) quotation—he saw the world as it was, as it is, and as it will be in its reality. Einstein's case as an innovator is especially compelling, because his ability to objectivize cognition was extremely pronounced (Shavinina, 1996).[1] Based on our study of Einstein, R. B. Woodward, Jean Piaget, and V. I. Vernadsky, we suggested that scientific innovations (i.e. new discoveries) are probably generated in the world of objectivization. In other words, the highly productive work of gifted, creative, and talented scientists—that is, their numerous discoveries and inventions—are made possible by their ability to objectivize cognition. Being active on a number of fronts and making important advances (i.e. formulating objectively new fundamental laws, rules, novel ideas, etc.) can be

---

[1] This is very clearly shown in his life ideal: "The *cognition of the world in its unity and its rational comprehension*" (as cited in Kuznetzov, 1979, p. 326; free translation from Russian, italics added). Being very young, Einstein wanted to go beyond various personal everyday problems to active work on *superpersonal* rational ideas, which take place in the *superpersonal world,* beyond the world of daily interests. Later, his desire to achieve objective and superpersonal cognition of the world became very strong. He was interested in general laws of the universe, nature, and existence. He wrote:

> The most important level in the development of a person like me is reached then, when the *basic interest of life goes beyond the temporary and personal, and concentrates more and more on the desire to understand mentally the nature of the world* (as cited in Kuznetzov, 1979, p. 326; free translation from Russian, italics added).

Indeed, all Einstein's scientific activity was at the highest level of abstractness. His great theories are evidence of this. No doubt, such a level of abstraction was dependent on Einstein's ability to objectivize cognition. Probably only in the world of objectivization was it possible for Einstein to go beyond the known facts, principles, laws, and so on to the new constructive theories, which could adequately explain unknown phenomena of the universe and nature and move against the prevailing concepts in theoretical physics. As a whole, Einstein's productive work on the frontier problems of physics in 1905 depended on his ability to objectivize cognition. It is not therefore surprising that he preferred "to work alone and out of the academic mainstream" (Miller, 1988, p. 171), because only in the world of objectivization can scientists interact directly with their subject (i.e. experimental findings, ideas arising from their thoughts, and so on) to the maximum degree. As a result, Einstein was able to add fresh and unusual, objectively new ideas from his thought experiments to the information available, in order to formulate his innovative view of physical theory (Shavinina, 1996).

considered a result of their objectivization of cognition. It should be noted that despite its importance to the understanding of individual innovation, the phenomenon of the objectivization of cognition with regards to innovators has been, thus far, a relatively unexplored topic. Although we have begun to investigate this topic in the case of our studies of Einstein and other innovators in science (Shavinina, 1996), further research is needed in this direction.

We suggest that the ability to objectivize cognition is important not only for innovators in science but the same is true for innovators working in business settings (we will call them 'business innovators'). Why are some companies' employees able to generate ideas resulting into new products or services and others not? In our view, business innovators are able to objectively see either hidden consumers' needs, or potential developments in these needs, or changes in technology, or something else. Of course, everyone agrees that innovation in contemporary companies is a team sport, an endeavor of many players. Modern companies have Research & Development departments, marketing channels, and so on for successful development of new ideas and their transformation into new and profitable products. This is essentially what allows one to call innovation a 'team sport'. But one cannot deny that new ideas appear in the minds of certain individuals. It may be surprising that innovation is not a very frequent event in today's companies, in spite of its quite evident importance. In our view, one main reason is that companies do not have enough intellectual and creative employees and senior executives with the ability to objectivize cognition, that is, to objectively see every aspect of their business activity. In this light the study of the objectivization of cognition of business innovators—who are responsible for initial ideas for successful new products in today's companies—is a promising research direction for innovation scholars within management schools.

The whole structural organization of an individual's cognitive experience (i.e. its conceptual structures, knowledge base, and subjective mental space) determines innovators' unique intellectual picture of the world. As our previously mentioned study of Einstein and other Nobel laureates (Shavinina, 1996) demonstrated, the objectivization of cognition is one of the important aspects of this uniqueness. Our comparative experimental study of the individual cognitive experience of the 'gifted' high school students in physics and mathematics and those high school students who were not identified as 'gifted' (i.e. 'average') allowed us to shed light on other aspects of the unique intellectual picture of the world of gifted, creative, and talented people (Shavinina & Kholodnaya, 1996). In comparison with 'average' students, 'gifted' students' representations of the reality as a whole consisted of a predominance of categorical (generalized) cognition. The 'gifted' groups' representations of the future focused on the differentiation of the 'vision' of future events. 'Gifted' students were also distinguished by more complex and rich conceptual representations (i.e. their representations are quite unfolded and a clearly articulated phenomena; Shavinina & Kholodnaya, 1996). Accordingly, the cognitive mechanisms of the construction of these representations play a key role in the organization of the experience of the gifted. These mechanisms are responsible for the construction of more categorical, differentiated, integrated and conceptually complex individual intellectual picture of the world. As innovators are gifted, creative, and talented individuals, then we can suggest, on the basis of these experimental findings, that their cognitive experience is a differentiated and integrated phenomenon. Correspondingly, their representations are generalized, categorical, conceptually rich, and complex. This allows innovators to have a unique intellectual picture of the world, which expresses itself in their exceptional performance and achievements (e.g. in their ability to generate new ideas).

Thus far, a few determinants, which constitute the uniqueness of an individual's intellectual picture of the world, have been discussed. These are: (1) innovators' objectivization of cognition; (2) their generalized, categorical, conceptually rich, and complex representations; and (3) their differentiated and integrated cognitive experience.

So far, the basis of individual innovation (i.e. an individual's cognitive experience) was considered. In other words, the psychological carrier of its various manifestations (i.e. traits, characteristics, features, and properties) was described. We now briefly look at the three levels of the manifestations of individual innovation.

Manifestations of Individual Innovation

The first level is the *level of intellectual and creative abilities* and it is composed of intellectual productivity, individual specificity of intellectual activity, and creativity (Kholodnaya, 2002; Shavinina & Kholodnaya, 1996). *Intellectual productivity* includes three types of the properties of human intelligence: *level properties, combination properties*, and *process properties*. In other words, all properties of human intelligence identified by psychological science were categorized into these three types (Kholodnaya, 1990, 2002).

Level properties characterize the achieved level of intellectual functioning—both verbal and non-verbal. These properties form a basis for such cognitive processes as rate of perception, capacity of short- and long-term memory, attention, vocabulary, and so on. Typical examples of the level properties of intelligence are those intellectual properties assessed by the Wechsler intelligence scales.

Combination properties characterize the ability to decipher various links, connections, and relations between different concepts or combination of concepts. In general, it is the ability to combine the components of experience in various ways (spatial, verbal, etc.). These intelligence properties are measured by tests of verbal analogies or the Raven Progressive Matrices Test, as well as reading comprehension tests and the so-called sorting tests.

Process properties characterize the elementary processes of information processing, as well as operations and strategies of intellectual activity. Piaget's theory describes such properties of intelligence. Standardized intelligence tests may be used to measure process properties of intelligence (e.g. the Wechsler intelligence scales, the Stanford–Binet Intelligence Test, the Raven Progressive Matrices Test, and others).

Individual specificity of intellectual activity is the second component in the level of intellectual and creative abilities. This component manifests itself in the cognitive styles of an individual. Cognitive styles provide valuable information about individual differences in the functioning of human cognitive processes. For example, originally introduced by Kagan et al. (1964), reflectivity-impulsivity cognitive style displays individual differences in the speed and accuracy with which people propose and formulate hypotheses and make decisions under conditions of uncertainty. Today's reality, in the fast-paced business world, is that a company's employees and managerial staff must be able to propose and formulate new hypotheses and ideas and make informed, critical, high-staked decisions under conditions of uncertainty. Our experimental studies showed that intellectually gifted high school students are distinguished by a reflective cognitive style (Shavinina & Kholodnaya, 1996). Results from our administration of the Kagan's Matching Familiar Figures (MFF) test also demonstrated that gifted students made fewer errors in the situation of multiple choices. From the viewpoint of basic cognitive mechanisms, it means that the gifted accurately analyzes visual space up to the moment of making decisions, that is, they are more careful in evaluating alternatives, hence making few errors. Average students, however, presumably hurry their evaluation thereby making more mistakes. The active character of visual scanning by the gifted indicates, in particular, a capacity to delay or inhibit a solution in MFF test performance containing response uncertainty, and also a capacity to differentiate unimportant and essential features of the external stimulus. When we assert that innovators are exceptionally able to generate new ideas, this also implies—in accordance with our conception—that they are able to carefully evaluate those ideas as well as the possible alternatives. This example of the reflectivity-impulsivity cognitive style demonstrates its significance for understanding individual innovation. Other cognitive styles (e.g. cognitive complexity-simplicity) are equally important, since they shed light on the manifestations of the cognitive experience, which, as we propose, is a basis of individual innovation.

Creativity refers to the originality, fluency, and flexibility of thinking, and to the ability to generate original and appropriate ideas (Georgsdottir et al., this volume; Kholodnaya, 2002; Lubart, 2001–2002; Runco & Pritzker, 1999; Shavinina & Kholodnaya, 1996). Although creativity manifestations seem to be the most appropriate in the context of our chapter, we will not consider them in detail here, because there is a vast body of literature on this topic. In contrast to what many innovation researchers assert, a real novelty of our conception resides in our emphasis that the originality, fluency, and flexibility of thinking are not the basis of individual innovation. As we discussed previously, the real basis of individual innovation is an individual's cognitive or mental experience, which serves as a psychological carrier of all manifestations of individual innovation, including the creative ones. That is, the originality, fluency, and flexibility are derivatives from the individual cognitive experience. For example, flexibility, differentiation, and integration of an individual's mental space determine one's own creative abilities. The individual mental space is one of the above-mentioned forms in the structural organization of one's own cognitive experience.

The second level of the manifestations of individual innovation is formed by *metacognitive abilities* (i.e. metacognitive awareness and regulatory processes). Metacognitive awareness refers to: (a) a system of knowledge about the basic manifestations of intellectual activity in general and about one's own individual cognitive possibilities; (b) the ability to evaluate the 'strong' and 'weak' aspects of his or her own intellectual functioning, including the ability to compensate for one's own weaknesses and rely on strengths; and (c) the ability to manage his or her mental work using various stimulation methods. Regulatory processes include planning, guiding, monitoring, and coordinating one's own cognitive processes (Kholodnaya, 1990; Shavinina & Kholodnaya, 1996).

According to many psychological accounts, metacognitive abilities are critical for the productive functioning of the human mind (Brown, 1978, 1987; Butterfield, 1986; Campione & Brown, 1978; Flavell, 1976; Sternberg, 1985). Knowledge about one's own intellectual creative abilities and the whole cognitive set-up, evaluating their efficiency, advantages and limitations, as well as planning, monitoring, and executive control are among important human abilities (Brown, 1978, 1987; Flavell, 1976; Pressley et al., 1987; Shavinina & Kholodnaya, 1996; Shore & Kanevsky, 1993; Sternberg, 1985). Moreover, research showed that less intelligent persons are characterized by a more superficial metacognitive understanding of

their own cognitive systems and of how the functioning of these systems depends upon the environment. It is also found that less intelligent people use executive processes that are not as complete and flexible for controlling their thinking (Butterfield, 1986).

Finally, the third level of the manifestations of individual innovation is presented by specific preferences, beliefs, and feelings, which can be referred to as *extracognitive abilities* (Kholodnaya, 2002; Shavinina, 1994). All of these abilities characterize the mental work of innovators. A special chapter in this volume by Shavinina analyzes this issue.

So far, we considered the second, third, fourth, and fifth levels in the internal structure of individual innovation. In the next section we are going to discuss its developmental foundation. Specifically, we will address the issue of what might happen in the individual development of certain people, which might eventually make them more open to innovations and consequently able to generate new ideas.

Developmental Foundation of Individual Innovation

Deschamps' (this volume) research on innovation leadership shows that innovation leaders are extremely fascinated and receptive to new ideas proposed by their colleagues and subordinates, which may potentially lead to new products or services. In our opinion, innovation leaders are open to innovation due to their sensitivity to everything new. Martindale (1999) noted that many creative and innovative people—in any field of human endeavor—point out that sensitivity is one of the essential characteristics of their personalities. We think that sensitivity as a personality characteristic has its roots in the specificity of the individual development of these people, particularly in their advanced development during childhood. In other words, if one wishes to know why it is that some individuals in, say, today's companies are able to produce new ideas resulting into innovative products and services or are open to supporting innovation in others (as in the case of innovation leaders), then one should look at their advanced childhood development.

We define advanced development as the development, which leads to the significant expression of an individual's potential—in the forms of innovation, giftedness, exceptional creativity, or extraordinary intelligence—and results in any socially valuable human achievement or performance (e.g. in new ideas leading to new products and/or services). According to our conception, the essence of advanced development in gifted, creative, and talented children—many of which will grow into adult innovators—can be seen in the specificity of a child's age. In the individual development—including cognitive, intellectual, emotional, personality, psychomotor, and social aspects—of a child, there are certain age periods of heightened sensitivity, which are known as *sensitive periods*. The underlying mechanism of the advanced development in the gifted, creative, and talented that actualizes their potentially high abilities can be seen in sensitive periods.

Research demonstrates that human development is not a smooth process. Instead, it contains certain stages or periods, that is, it is 'periodical' in its essence (Ananiev, 1957; Case, 1984a, 1984b; Fischer & Pipp, 1984; Flavell, 1984; Vygotsky, 1956, 1972). For example, Piagetian stages of cognitive development reflect different ways in which the child processes information and interacts with the environment and also highlights specific age ranges within which these stages are likely to occur. Investigations into sensitive periods, in developmental psychology especially, indicate a periodical nature of human development (Bateson, 1983; Bateson & Hinde, 1987; Bornstein, 1987a, 1987b; Colombo, 1982; Gottlieb, 1988; Leites, 1960, 1971, 1978; Lewis, 1988; Oyama, 1979). Psychologists working in the field of high abilities also point out the periodical essence of gifted development (Feldman, 1986a, 1986b; Leites, 1985, 1988, 1996; Morelock, 1992; Shavinina, 1997, 1999; Silverman, 1997). For example, Feldman (1986a) sees giftedness as the "movement through the stages that leads to performance superior to that of most others" (p. 302). The Columbus Group's approach to the understanding of giftedness as asynchronous development and Bamberger's (1986) findings on 'midlife crisis' also provide strong evidence for the 'periodical' nature of development of the gifted, creative, and talented.

The Concepts of Sensitive Periods and Sensitivity

Sensitive periods are defined here as special periods during human development when individuals show great openness to everything in the world around them. Specifically, *sensitivity* involves an individual's idiosyncratic, personal, and heightened responsiveness to everything going on around him or her (Leites, 1971, 1996; Shavinina, 1997, 1999; Vygotsky, 1956, 1972). Such definitions of sensitivity and sensitive periods might seem rather general; however, they appear to be expedient on the contemporary level of the study of these phenomena in innovators, where the research, specific to the gifted, creative, and talented, is restricted.

The literature provides clear indications that age sensitivity takes a certain place in the advanced development of the gifted, creative, and talented (Feldman, 1986b; Jellen & Verduin, 1986; Leites, 1960, 1971, 1996; Kholodnaya, 1993; Piechowski,

1979, 1986, 1991; Shavinina, 1997; Silverman, 1994, 1995, 1997). For instance, Feldman (1986b) incorporated unusual sensitivity into his theory of prodigy phenomenon, in the "individual psychological qualities" component. Sensitivity is also one of the important elements in Jellen & Verduin's (1986) conception of giftedness. Leites (1971) concluded that the child's sensitivity and sensitive periods are critical phenomena in the development of prodigies. Piechowski (1991) considered sensitivity to be an individual's heightened response to selective sensory or intellectual experiences—asserting that unusual sensitivity reveals the potential for high levels of development, especially for self-actualization and moral vigor (Piechowski, 1979, 1986). Sternberg (1986a) viewed "sensitivity to external feedback" as one of the metacomponents of his theory of intellectual giftedness. Kholodnaya (1993) considered child prodigies as the result of the specific development of a child during the early years.

Shavinina (1997) distinguished cognitive (i.e. sensitivity to any new information), emotional (i.e. sensitivity to one's own inner world and to the inner words of other people), and social kinds of sensitivity, which intersect with one another, forming mixed kinds of sensitivity. Leites (1971) emphasized that each child's age is characterized by one or numerous kinds of sensitivity. Vulnerability, fragility, empathy, and moral and social responsiveness are among some of the manifestations of sensitivity (Shavinina, 1999; Silverman, 1994, 1997). Cognitive sensitivity is extremely important in a child's development in general and in the advanced development of the gifted in particular. Thus, the first years of a child's life are characterized by the ease and stability of knowledge acquisition and of the development of many abilities, skills, and habits (for example, linguistic abilities; Leites, 1996). Shavinina (1999) suggested that because of cognitive sensitivity, children's knowledge acquisition is very quick; it may take place even from the very first experience.

The Characterization of Sensitive Periods in Gifted Children, Future Innovators

An individual's sensitivity is not a constant—it varies in the course of development. At some points in the personal development the child is characterized by heightened sensitivity and these special age periods are called *sensitive periods*. At other points in his or her development the child's sensitivity is low or even absent (Leites, 1971). There are exceptionally favorable inner conditions (i.e. conditions provided by the process of child development itself) and extraordinary possibilities for cognitive and intellectual development during sensitive periods. Vygotsky (1956, 1972) specifically emphasized this issue. Introducing the concept of sensitive periods, he characterized their essence in the following way:

> during these periods, certain influences have a big impact on the entire course of individual development by provoking one or another deep changes. During other periods, the same influences might have no effect or even an opposite impact on child development. Sensitive periods coincide fully with (. . .) the optimal times for learning (Vygotsky, 1956, p. 278).

Surprisingly, in light of Vygotsky's great impact on contemporary developmental theory, his view about sensitive periods appears never to have been translated into English, although it remains an important part of his developmental theory. Following Vygotsky, a general tendency in Russian psychology is to consider sensitive periods as an exceptionally productive time for learning (Leites, 1960, 1971, 1996; Shavinina, 1997, 1999) because they present extremely favorable internal conditions for mental development (Leites, 1985, 1988). In this respect, all human development—even in infancy—is characterized by individual periods of heightened sensitivity.

The years over which a child acquires language are one of the best-known examples of sensitive periods (Leites, 1996). Over a very short period of time, young children easily learn different forms and constructions of languages, but it becomes increasingly more difficult to do this in later years. It is fascinating and seems paradoxical that at a time when, for example, children are learning and speaking foreign languages with relative ease in the appropriate environment, adults who have a more developed mind—and therefore seem to be able to easily manage any linguistic difficulties—cannot do as well. Everyday life provides many examples of the difficulty with which adults learn and speak foreign languages. It seems clear that a certain age or age range—early childhood—is best suited for the specific mental activities involved in language acquisition. It appears that sensitive periods during childhood prepare and temporarily conserve favorable inner possibilities for advanced development. Elaborating further Vygotsky's ideas of sensitive periods, Zaporozhets (1964) asserted that "each period of a child's development has its age sensitivity, and because of that learning is more successful in the early years than in the elder ones" (p. 678). This is a key to the explanation of fast knowledge acquisition by gifted children that in turn leads to their advanced development. Cases of gifted, creative, and talented children at

sensitive periods support this assertion.[2] Taken together, the theoretical and empirical findings demonstrate that the changes of a child's age bring unique determinants of the advanced development: sensitive periods. Sensitive periods mean a qualitatively new strengthening of the possibilities for mental growth, which appear during the early childhood years. The strengthening of such possibilities leads to the general heightening of a child's cognitive resources.

Sensitive Periods: Developmental Losses and Individual Acquisitions

Favorable possibilities for individual development granted by these sensitive periods will weaken at a slow or fast rate—this is reality (Leites, 1971). Thus, the following question arises: can sensitive periods experienced by a child be predictors of his or her intellectually creative productivity in adulthood, which may eventually lead to innovation? We believe that the answer to this question will be 'yes' only if two important requirements are fulfilled during childhood. First, all *developmental* capacities (i.e. new abilities, habits, skills, qualities, traits, etc. acquired during a certain sensitive period) should be transformed into stable *individual* acquisitions. Second, these acquired

---

[2] For example, Alexander expressed his unusual abilities at a very early age. He started to read very well and to calculate before he was four years old. His interest in numbers probably indicated the first sensitive period. The boy continually demanded that all adults around him set up simple arithmetical tasks for him; he was hungry for them. In this period, he also liked to write various numbers. The number seven (7) was especially attractive for him: he wrote it everywhere in different forms and sizes painted in various colors. This 'digital' period eventually came to the end.

Just after his fourth birthday, Alexander entered a new sensitive period—the 'geographical' one. He read a lot about continents, countries, cities, seas, mountains, and rivers. All his questions to adults concerned only geographical issues. He asked his parents to buy geography books for him and he looked for related articles in newspapers. He watched TV programs that concerned travels around the globe and meteorological news (in this case, he could always see a map of his country). As a result, Alexander's acquired knowledge of geography was impressive. However, his new sensitive period did not consist only of the acquisition of new knowledge about geography. All of Alexander's cognitive activity was directed toward achieving one clear goal: to make a map of the world. All his time was devoted to this task. He prepared the map, conveying shapes and names of geographical objects (i.e. continents, countries, etc.) with amazing accuracy. Such an activity certainly required his artistic skills in drawing and painting, which were significantly developed during this period. In a few months, Alexander's second sensitive period—the geographical—was almost over (Shavinina, 1999).

individual capacities should, in turn, be transformed into unique cognitive experiences (Shavinina, 1999).

Although all stages of childhood can be distinguished by the heightened sensitivity (as compared to that of adults), sensitive periods have their own 'life story'. Sensitive periods emerge, exist, and even disappear during a child's development (Leites, 1971). What is important is what remains in the child at the end of sensitive period(s), as he or she grows older and favorable opportunities for advanced development weaken. It is important to note that although the favorable possibilities opened up by sensitive periods allow a child to advance significantly in his or her development by acquiring new and valuable knowledge, skills, and habits, he or she can also lose these acquisitions when a sensitive period ends. That seeming paradox is at the crux of, and is a real problem of, sensitive periods. Because of that Leites (1985, 1988, 1996) differentiates between *developmental* and *individual* aspects of sensitive periods.

If, at the end of a sensitive period a child loses almost all the exceptional capacities that he or she acquired during the given period, then one can assert that these capacities were mainly a *developmental* phenomenon (i.e. *developmental* capacities that disappear with age). This is key to understanding why so many gifted individuals who demonstrated exceptional abilities in childhood become ordinary adults who do not display extraordinary talents or outstanding creativity. Gifted children lose their unique abilities and talents in the process of their own individual development (Shavinina, 1999).

At the same time, sensitive periods are a true foundation for powerful *individual* gains. If new extraordinary capacities acquired during a certain sensitive period remain in the developing child after this period, then one can assert that these capacities have been transformed into *individual* acquisitions. Only in this case one can suppose, to a significant extent, that the child has the potential to be an intellectually creative adult, potential innovator.[3]

The Nature of Advanced Development in Today's Gifted Children, Tomorrow's Innovators

Our analysis of the gifted at sensitive periods demonstrated the following tendency: their sensitive periods usually are linked sequentially or in chains (although the periods may overlap; Shavinina, 1999). This means that gifted children are always at sensitive period(s). In other words, the gifted's sensitivity does not disappear completely. In contrast to the previously mentioned opinion that sensitive periods emerge, exist, and disappear during childhood development (Leites, 1971, 1985, 1988), the chain of sensitive periods in gifted development testifies to the lasting sensitivity of gifted children. Research supports this conclusion. Thus, Silverman (1993) pointed out that "extraordinary levels of sensitivity and compassion do not disappear with maturity. A capacity for rich, intense emotions remains in the personality throughout the lifespan" (p. 642). Probably, this depends on the kind of sensitivity (i.e. cognitive, emotional, or social). Maybe emotional sensitivity, more than any other kind, remains in the individual during his or her life, whereas cognitive sensitivity changes periodically (but certainly it does not disappear in the gifted). Such characteristics of the gifted as sensitivity to a new experience and openness of mind—which are mentioned by many contributors to this volume as essential traits of innovators—can be regarded as evidence of this tendency of cognitive sensitivity. Perhaps the availability of cognitive sensitivity throughout the life span determines the exceptional mental abilities of an individual (Shavinina, 1999). Furthermore, if sensitivity remains in gifted children for a long time, then it is quite reasonable to state that new capacities acquired during a certain sensitive period will also remain in the gifted for a long time. These capacities are fortified and

---

[3] In this context it is expedient to further consider the case of Alexander in detail. When Alexander was a five and a half, a seven-year-old girl began to live temporarily in their family. She was admitted to school and a new educational period started for Alexander. The children did exercises together, solved mathematical tasks, and learned poems by heart. Alexander started kindergarten. His teacher was impressed both by Alexander's mental development and by his ability to draw and paint. The accuracy of the presentation of even the smallest details was a distinguishing characteristic of his paintings. When he was seven, Alexander successfully passed all examinations and was admitted directly from the kindergarten into the fourth-grade class for 11-year-old pupils. All of his school grades were 'excellent'. At that time, the third

sensitive period started, which can be called an 'ornithological' period (Shavinina, 1999).

At the age of seven, Alexander read three volumes of Brem's book for university students, *The Life of Animals*. This was the beginning of his interest in zoology; birds were especially attractive to him. The essence of a new sensitive period consisted of his writing (or, more precisely, creating) a book about birds. Alexander wrote down the summary of the corresponding chapters of Brem's volumes and made many illustrations for them. He also used two lengthy articles about birds that he found. The scale of his work was impressive: the manuscript ran more than 300 pages with more than 100 drawings. The text was divided into chapters, and all chapters were interconnected with an internal integrity. Alexander had an extensive vocabulary. His linguistic abilities were manifested in the absence of mistakes in the writing of his manuscript, which included many foreign words and biological terms. Alexander continued to read a lot, preferring scientific literature. He often used an encyclopedia and a dictionary of foreign words.

developed later, and finally they are transformed into true *individual* acquisitions that have a potential to remain in the person throughout the life span. In this case one can predict to a certain extent the transition of a gifted child into a talented adult innovator who will be able to excel intellectually and creatively.[4]

Moreover, the revealed chain of sensitive periods in the development of gifted children indicates a natural *overlapping* of age sensitivity. But to demonstrate that children may be continually in sensitive period(s) does not appear to be enough to explain truly advanced development. The overlapping of sensitivities is one of the keys to discerning the inner nature of advanced development. Such an *overlapping* of age sensitivities means that a child's sensitivity originates from different (i.e. previous, current, and subsequent) childhood periods.[5] Furthermore, the overlapping of a child's sensitivity predetermines *duplication* and even *multiple strengthening* of the foundations for the rapid intellec-

---

[4] In this context the example of Alexander should be considered briefly once again. At 14 and a half, Alexander graduated from high school with excellent grades. He was equally successful at the Department of Biology at Moscow State University and became a distinguished ornithologist. He participated in many expeditions, mainly to the north, related to the investigations of various birds. As an adult, he was characterized by the ability to work very hard, an intense interest in learning, the ability to draw well (especially birds), high ability in learning foreign languages, and an extremely clear and detailed memory. He was a competent, respected, and high-achieving scientist. Alexander's colleagues wrote: "He was distinguished by a combination of abstract thinking with deep knowledge of ecology and birds that he had learned during his numerous expeditions. He had strong will, excellent memory, and was able to work hard. He was also able to read in many European languages including Scandinavian languages; he was perfect in English. Alexander was an excellent scientist, untiring, purposeful, persistent. He could make an instant draft of a bird or landscape. He had an unlimited ability to work. He created his own scientific school" (Leites, 1996, p. 156).

What is remarkable is the complete coincidence between the description of 43-year-old Alexander, the talented scientist (by his colleagues) and the description of Alexander, the exceptionally gifted child (by the psychologist) (Leites, 1960, 1996). Alexander's life shows that his *developmental* capacities (i.e. those new capacities acquired during sensitive periods in childhood) were indeed transformed into powerful *individual* abilities that remained throughout the life span. In turn, Alexander, the profoundly gifted child, evolved naturally into Alexander, the highly creative researcher.

[5] The description of Alexander's childhood summers supports this assertion. At that time, he had at least two sensitive periods: the 'bird' period and the 'butterfly' period. His interest in birds can be considered as three distinct, yet overlapping, periods: the first (his intense interest at age seven), second (his summertime interest in the observation of birds), and third (his renewed intense interest, developing as a variation of his previous interest). A new—and partially concurrent—sensitive period was that of his interest in butterflies.

tually creative growth that results into advanced development of the gifted. The phenomenon of the overlapping of age sensitivity of gifted children also testifies to the above-mentioned fact that they are always in sensitive periods. In this instance the likelihood of the transformation of all *developmental* capacities into the *individual* abilities is getting significantly high. Being permanently in sensitive periods also implies the actualization of the gifted's immense cognitive potential and acceleration of their mental development. The latter implies rapid accumulation of the gifted's cognitive resources and the construction of those resources into the unique cognitive experience that continues to enrich itself in the process of the further accelerated development governed mainly by heightened cognitive sensitivity[6] (Shavinina & Kholodnaya, 1996). Their unique cognitive experience means their unrepeatable intellectual picture of the world. All the above written concerning sensitive periods demonstrates that they are not a factor, condition, characteristic, feature, or trait in a child's development. They should be understood as an inner mechanism of advanced development of the gifted, creative, and talented.

Sensitive periods, therefore, constitute a developmental foundation of individual innovation in that that they provide a basis for an extremely accelerated mental development in childhood. This intellectual acceleration leads to fast actualization and development of one's own mental potential and its transformation into the unique cognitive experience that, in turn, forms a cognitive basis of individual innovation.

Summing Up

This chapter presented a new conception of individual innovation, which explains: (1) its developmental foundation; (2) its cognitive basis; (3) its intellectual manifestations; (4) its metacognitive manifestations; and (5) its extracognitive manifestations. The developmental foundation, the cognitive basis, and intellectual, metacognitive, and extracognitive manifestations constitute the first, second, third, fourth, and fifth levels, respectively, of the internal structure of individual innovation.

---

[6] The case of Alexander is quite appropriate again. Thus, although his 'ornithological' sensitive period was very long, his cognitive experience was different in each stage of childhood. For example, his initial drawings of birds were based on verbal descriptions from books because Alexander's first ornithological knowledge was from books, whereas his 'summer' drawings were based on his personal experience of natural observations. As mentioned above, at that time, his knowledge of birds reached a new level. Because of this, one can assert that Alexander had, in fact, a few 'ornithological' sensitive periods. These periods can be considered as previous, current, and subsequent sensitive periods sometimes simultaneously.

The developmental foundation of individual innovation is connected with the advanced childhood development of innovators that manifests itself in their accelerated mental growth, beyond which there are periods of heightened cognitive sensitivity. Specifically, during the childhood years, certain 'temporary states'—or sensitive periods—emerge at each age stage, providing significant opportunities for advanced development. Sensitive periods accelerate a child's mental development through the actualization of his or her intellectual potential and the growth of the individual's cognitive resources resulting in the appearance of a unique cognitive experience. Such accelerated development further facilitates rapid and deep knowledge acquisition, intellectual functioning, and the creation of something new and original. This leads to advanced development of future innovators. In turn, cognitive experience expresses itself in the unrepeatable intellectual picture of the world of innovators and is responsible for their exceptional achievements and performance (e.g. generation of new ideas).

In our view, individual innovation is not a 'gift', chance event, or a consequence of socialization. Individual innovation is a result of a specific organization of an individual's cognitive experience. This organization is, in turn, a result of the protracted inner process of the actualization, growth, and enrichment of one's own cognitive resources and their construction into an unrepeatable cognitive experience during accelerated mental development. The direction of this process is determined by specific forms of the organization of a person's cognitive experience (i.e. conceptual structures, knowledge base, and mental space). The unique structure of the mind, which makes possible the creative ideas leading to innovation, is being formed on the basis of this process. The uniqueness of innovators' minds expresses itself in specific, objective representations of reality—that is, in their unique intellectual picture of the world—which can manifest itself in a wide range of intellectual, metacognitive, and extracognitive manifestations.

References

Amabile, T. M. (1988). A model of creativity and innovation in organizations. *Research in Organizational Behavior*, **10**, 123–167.

Amabile, T. M. (1996). *Creativity in context*. Boulder, CO: Westview.

Ananiev, B. G. (1957). About the system of developmental psychology. *Voprosi Psichologii*, **5**, 112–126.

Bamberger, J. (1986). Cognitive issues in the development of musically gifted children. In: R. J. Sternberg & J. E. Davidson (Eds), *Conceptions of Giftedness* (pp. 388–413). Cambridge: Cambridge University Press.

Bateson, P. (1983). The interpretation of sensitive periods. In: A. Oliverio & M. Zappella (Eds), *The Behavior of Human Infants*. Plenum: New York.

Bateson, P. & Hinde, R. A. (1987). Developmental changes in sensitivity to experience. In: M. H. Bornstein (Ed.), *Sensitive Periods in Human Development* (pp. 19–34). Hillsdale, NJ: Erlbaum.

Bjorklund, D. F. & Schneider, W. (1996). The interaction of knowledge, aptitude, and strategies in children's memory development. In: H. W. Reese (Ed.), *Advances in Child Development and Behavior* (pp. 59–89). San Diego, CA: Academic Press.

Bornstein, M. H. (1987a). Sensitive periods in development: Definition, existence, utility, and meaning. In: M. H. Bornstein (Ed.), *Sensitive Periods in Human Development* (pp. 3–17). Hillsdale, NJ: Erlbaum.

Bornstein, M. H. (Ed.) (1987b). *Sensitive periods in human development*. Hillsdale, NJ: Erlbaum.

Brown, A. L. (1978). Knowing when, where, and how to remember: A problem of metacognition. In: R. Glaser (Ed.), *Advances in Instructional Psychology* (Vol. 1, pp. 77–165). Hillsdale, NJ: Erlbaum.

Brown, A. L. (1987). Metacognition, executive control, self-regulation, and other even more mysterious mechanisms. In: F. E. Weinert & R. H. Kluwe (Eds), *Metacognition, Motivation, and Understanding* (pp. 64–116). Hillsdale, NJ: Erlbaum.

Butterfield, E. C. (1986). Intelligent action, learning, and cognitive development might all be explained with the same theory. In: R. J. Sternberg & D. K. Detterman (Eds), *What is Intelligence? Contemporary Viewpoints on its Nature and Definition* (pp. 45–49). Norwood, NJ: Ablex Publishing Corporation.

Campione, J. C. & Brown, A. L. (1978). Toward a theory of intelligence: Contributions from research with retarded children. *Intelligence*, **2**, 279–304.

Case, R. (1984a). *Intellectual development: A systematic reinterpretation*. New York: Academic Press.

Case, R. (1984b). The process of stage transition: A neo-Piagetian view. In: R. J. Sternberg (Ed.), *Mechanisms of Cognitive Development* (pp. 19–44). New York: W. H. Freeman and Company.

Chi, M. T. H. & Ceci, S. J. (1987). Content knowledge: Its role, representation and restructuring in memory development. In: H. W. Reese (Ed.), *Advances in Child Development and Behavior* (Vol. 20, pp. 91–142). Orlando, FL: Academic Press.

Chi, M. T. H., Feltovich, P. J. & Glaser, R. (1981). Categorization and representation of physics problems by experts and novices. *Cognitive Science*, **5**, 121–152.

Chi, M. T. H., Glaser, R. & Rees, E. (1982). Expertise in problem-solving. In: R. J. Sternberg (Ed.), *Advances in the Psychology of Human Intelligence* (Vol. 1, pp. 7–75). Hillsdale, NJ: Erlbaum.

Chi, M. T. H. & Greeno J. G. (1987). Cognitive research relevant to education. In: J. A. Sechzer & S. M. Pfafflin (Eds), *Psychology and Educational Policy* (pp. 39–57). New York: The New York Academy of Sciences.

Christensen, C. (1997). *The innovator's dilemma*. Cambridge, MA: Harvard Business School Press.

Colombo, J. (1982). The critical period hypothesis: Research, methodology, and theoretical issues. *Psychological Bulletin*, **91**, 260–275.

Farr, J. & Ford, C. (1990). Individual innovation. In: M. West & J. Farr (Eds), *Innovation and Creativity at Work* (pp. 63–80). Chichester, U.K.: John Wiley.

Feldman, D. H. (1986a). Giftedness as a developmentalist sees it. In: R. J. Sternberg & J. E. Davidson (Eds), *Conceptions of Giftedness* (pp.285–305). Cambridge: Cambridge University Press.

Feldman, D. H. (1986b). *Nature's gambit: Child prodigies and the development of human potential.* New York: Basic Books.

Fischer, K. W. & Pipp, S. L. (1984). Processes of cognitive development: Optimal level and skill acquisition. In: R. J. Sternberg (Ed.), *Mechanisms of Cognitive Development* (pp. 45–80). New York: W. H. Freeman and Company.

Flavell, J. H. (1984). Discussion. In: R. J. Sternberg (Ed.), *Mechanisms of Cognitive Development* (pp. 187–209). New York: W. H. Freeman and Company.

Flavell, J. H. (1976). Metacognitive aspects of problem solving. In: L. B. Resnick (Ed.), *The Nature of Intelligence* (pp. 231–235). Hillside, NJ: Erlbaum.

Gottlieb, G. (1988). Sensitive periods in development: Interdisciplinary perspectives. *Contemporary Psychology,* **33** (11), 949–956.

Jellen, H. & Verduin, J. R. (1986). *Handbook for differential education of the gifted: A taxonomy of 32 key concepts.* Carbondale, IL: Southern Illinois University Press.

Kagan, J., Rosman, B., Day, D., Albert, J. & Phillips, W. (1964). Information processing in the child: Significance of analytic and reflective attitudes. *Psychological Monographs,* **78, 1** (Whole No. 578).

Kanter, R. M. (1983). *The change masters: Innovations for productivity in the American corporation.* New York: Simon & Schuster.

Kanevsky, L. S. (1990). Pursuing qualitative differences in the flexible use of a problem solving strategy by young children. *Journal for the Education of the Gifted,* **13**, 115–140.

Katz, R. (1997). *The human side of managing technological innovation.* New York: Oxford University Press.

Kholodnaya, M. A. (1990). Is there intelligence as a psychical reality? *Voprosu psihologii,* **5**, 121–128.

Kholodnaya, M. A. (1993). Psychological mechanisms of intellectual giftedness. *Voprosi psihologii,* **1**, 32–39.

Kholodnaya, M. (2002). *Psychology of intelligence* (2nd ed.). Moscow: Piter.

King, N. (1990). Innovation at work: The research literature. In: M. West & J. Farr (Eds), *Innovation and creativity at work* (pp. 15–60). Chichester, U.K.: John Wiley.

Klix, F. (1988). *Awakening of thinking.* Kiev: Vutscha schola (in Russian).

Kuznetsov B. G. (1979). *Albert Einstein: Life, death, immortality.* Moscow: Nauka (in Russian).

Leites, N. S. (1960). *Intellectual giftedness.* Moscow: APN Press.

Leites, N. S. (1971). *Intellectual abilities and age.* Moscow: Pedagogica.

Leites, N. S. (1978). Towards the problem of sensitive periods in human development. In: L. I. Anziferova (Ed.), *The Principle of Development in Psychology* (pp. 196–211). Moscow: Nauka.

Leites, N. S. (1985). The problem of the relationship of developmental and individual aspects in students' abilities. *Voprosu Psihologii,* **1**, 9–18.

Leites, N. S. (1988). Early manifestations of giftedness. *Voprosu Psihologii,* **4**, 98–107.

Leites, N. S. (Ed.) (1996). *Psychology of giftedness of children and adolescents.* Moscow: Academia.

Leonard-Barton, D. (1995). *Wellsprings of knowledge: Building and sustaining the sources of innovation.* Boston, MA.: Harvard Business School Press.

Lewis, M. (1988). Sensitive periods in development. *Human Development,* **32** (1), 60–64.

Lubart, T. I. (2001–2002). Models of the Creative Process: Past, present and future. *Creativity Research Journal,* **13**, 295–308.

Martindale, C. (1999). Biological bases of creativity. In: R. J. Sternberg (Ed.), *Handbook of creativity.* Cambridge: Cambridge University Press.

Miller A. I. (1988). Imagery and intuition in creative scientific thinking: Albert Einstein's invention of the special theory of relativity. In: D. B. Wallace & H. E. Gruber (Eds), *Creative People at Work* (pp. 171–187). New York: Oxford University Press.

Morelock, M. J. (1992). Giftedness: The view from within. *Understanding Our Gifted,* **4** (3), 1, 11–15.

Oatley, K. (1978). *Perceptions and representations.* Cambridge: Cambridge University Press.

Oyama, S. (1979). The concept of the sensitive period in developmental studies. *Merrill-Palmer Quarterly of Behavior and Development,* **25**, 83–103.

Piaget, J. (1969). *Selected psychological research.* Moscow: Nauka (in Russian).

Piechowski, M. M. (1979). Developmental potential. In: N. Colangelo & R. T. Zaffrann (Eds), *New Voices in Counseling the Gifted* (pp. 25–27). Dubuque, IA: Kendall/Hunt.

Piechowski, M. M. (1986). The concept of developmental potential. *Roeper Review,* **8** (3), 190–197.

Piechowski, M. M. (1991). Emotional development and emotional giftedness. In: N. Colangelo & G. Davis (Eds), *Handbook of Gifted Education* (pp. 285–306). Boston, MA: Allyn & Bacon.

Pressley, M., Borkowski, J. G. & Schneider, W. (1987). Cognitive Strategies: Good strategy users coordinate metacognition and knowledge. In: R. Vasta (Ed.), *Annals of Child Development* (Vol. 4, pp. 89–129). Greenwich, CT: JAI Press.

Rabinowitz, M. & Glaser, R. (1985). Cognitive structure and process in highly competent performance. In: F. D. Horowitz & M. O'Brien (Eds), *The Gifted and Talented. Developmental Perspectives* (pp. 75–97). Washington, D.C.: American Psychological Association.

Runco, M. A. & Pritzker, S. R. (Eds) (1999). *Encyclopedia of creativity* (Vol. 2, pp. 373–376). San Diego, CA: Academic Press.

Schneider, W. (1993). Acquiring expertise: Determinants of exceptional performance. In: K. A. Heller, F. J. Mönks & A. H. Passow (Eds), *International Handbook of Research and Development of Giftedness and Talent* (pp. 311–324). Oxford: Pergamon Press.

Shavinina, L. V. (1994). Specific intellectual intentions and creative giftedness. *European Journal for High Ability,* **5** (2), 145–152.

Shavinina, L. V. (1996). The objectivization of cognition and intellectual giftedness. *High Ability Studies,* **7** (1), 91–98.

Shavinina, L. V. (1997). Extremely early high abilities, sensitive periods, and the development of giftedness. *High Ability Studies,* **8** (2), 245–256.

Shavinina, L. V. (1999). The psychological essence of child prodigy phenomenon: Sensitive periods and cognitive experience. *Gifted Child Quarterly,* **43** (1), 25–38.

Shavinina, L. V. & Kholodnaya, M. A. (1996). The cognitive experience as a psychological basis of intellectual giftedness. *Journal for the Education of the Gifted*, **20** (1), 4–33.

Shore, B. M. & Kanevsky, L. S. (1993). Thinking processes: being and becoming gifted. In: K. A. Heller, F. J. Mönks & A. H. Passow (Eds), *International Handbook of Research and Development of Giftedness and Talent* (pp. 133–147). Oxford: Pergamon Press.

Silverman, L. K. (1993). Counseling needs and programs for the gifted. In: K. A. Heller, F. J. Mönks & A. H. Passow (Eds), *International Handbook of Research and Development of Giftedness and Talent* (pp. 631–647). Oxford: Pergamon Press.

Silverman, L. K. (1994). The moral sensitivity of gifted children and the evolution of society. *Roeper Review*, **17** (2), 110–116.

Silverman, L. K. (1995). The universal experience of being out-of-sync. In: L. K. Silverman (Ed.), *Advanced Development: A Collection of Works on Giftedness in Adults* (pp. 1–12). Denver, CO: Institute for the Study of Advanced Development.

Silverman, L. K. (1997). The construct of asynchronous development. *Peabody Journal of Education*, **72** (3&4), 36–58.

Sternberg, R. J. (1985). *Beyond IQ: A new theory of human intelligence*. Cambridge: Cambridge University Press.

Sternberg, R. J. (1986). A triarchic theory of intellectual giftedness. In: R. J. Sternberg & J. E. Davidson (Eds), *Conceptions of Giftedness* (pp. 223–243). New York: Cambridge University Press.

Sternberg, R. J. (1999). *Handbook of creativity*. Cambridge: Cambridge University Press.

Sternberg, R. J. & Powell, J. (1983). The development of intelligence. In: J. H. Flavell & E. M. Markham (Eds), *Handbook of Child Psychology: Cognitive Development* (Vol. 3, pp. 341–419). New York: John Wiley.

Tidd, J., Bessant, J. & Pavitt, K. (1997). *Managing Innovation*. Chichester, U.K.: John Wiley.

Tushman, M. & O'Reilly, C. (1997). *Winning Through Innovation*. Cambridge, MA: Harvard Business School Press.

Van de Ven, A. et al. (1999). *Innovation journey*. New York: Oxford University Press.

Vekker, L. M. (1981). *Psychological processes* (Vol. 3). Leningrad: Leningrad University Press.

Vygotsky, L. S. (1956). *Selected papers*. Moscow: APN Press.

Vygotsky, L. S. (1972). A problem of age periods in child development. *Voprosi psichologii*, **2**, 53–61.

West, M. A. & Farr, J. L. (Eds) (1990). *Innovation and creativity at work*. Chichester, U.K.: John Wiley.

Zaporozhets, A. V. (1964). *Child psychology*. Moscow: Pedagogica.

The International Handbook on Innovation
Edited by Larisa V. Shavinina

Models of Innovation

Dora Marinova and John Phillimore

Institute for Sustainability and Technology Policy, Murdoch University, Australia

Abstract: This chapter presents a historical examination of models used to explain innovation. It focuses on the attempts to describe innovation as a process generating new products and methods and outline the activities involved. The purpose of these models is to explain how all parties come together to generate commercially viable technologies. The overview includes six generations of models, namely black box, linear, interactive, systems, evolutionary models and innovative milieux. Each of them is explained by addressing the following issues: conceptualization background, the model itself and its elements, explanatory power, related models and concepts, and further research directions.

Keywords: Innovation; Black box; Linear: Interactive; Systems; Evolutionary models; Innovative milieux.

Introduction

Technological change and innovation have become important factors in economic and policy debates. The qualitative nature of socioeconomic changes induced by innovation (in the form of new products, technologies, outputs, activities, actors, institutions and organizations, among others) also translates into quantitative measures, such as increased company turnover, profits, market shares, exports and GDP. While many people are aware of this phenomenon, our understanding of the process of technological innovation is still quite limited.

Since the 1960s, an ever-increasing number of researchers have tried to put together pictures of the process of generation of new products and production methods and outline the activities involved in this. Efforts to explain technological change have come from a variety of disciplines, including economics, management, sociology, geography and political science. The main aim of these models of innovation is to explain how all parties come together to generate commercially viable technologies. Four decades later, we can ascertain that there are a number of process characteristics, which we have come to understand. Nevertheless, the complete phenomenon is still covered under a veil of mystery, intuition and intelligent decisions in situations of risk, uncertainty and lack of information.

The main purpose of this chapter is to review the innovation models developed so far and in doing so

outline their explanatory powers and weaknesses. The focus is on technological innovation, as distinct from social, educational or organizational innovations, for example. The models examine innovation as a creative process engaging a variety of activities, participants and interactions[1] the outcome of which is a technological product or process.

This overview includes six generations of models, namely the black box model; linear models; interactive models; system models, including networks and linkages; technology learning and evolutionary models; and innovative milieux. After a brief explanation of the historical background, each generation of models is examined separately with references to the main related theoretical developments and publications. The concluding section of this chapter draws together the progress made so far in our understanding and description of innovation. Having done this, we also recognize the challenging task of capturing a snapshot of a phenomenon, which in each specific occurrence

---

[1] The innovation models discussed in this chapter specifically refer to innovation by firms. As pointed out by Beije (1998), this modelling activity does not concern individuals or not-for-profit organizations. Also not included are models representing the impact that innovation has on the economy and society (e.g. economic growth, comparative advantages or technology gap models). Nor does the chapter differentiate innovation by sector (see Pavitt, 1984) or by 'importance' or 'spread' of technological impact (see Freeman, 1982, 1984).

involves qualitatively different features, and where the exceptions constitute the rules.

Generations of Models

In the early 1990s, Rothwell (1992) described five generations of innovation models: technology push; need pull; coupling model (with feedback loops); integrated model (with simultaneous links between R&D, prototyping and manufacturing) and systems integration/networking model (with emphasis on strategic linkages between firms). The approach taken in this overview reflects a similar chronology; however, it extends Rothwell's typology. The sections to follow cover six generations of innovation models, namely:

(1) First generation—the black box model;
(2) Second generation—linear models (including technology push and need pull);
(3) Third generation—interactive models (including coupling and integrated models);
(4) Fourth generation—systems models (including networking and national systems of innovation);
(5) Fifth generation—evolutionary models; and
(6) Sixth generation—innovative milieux.

Each of them is explained by addressing the following issues: conceptualization background, the model itself and its elements, explanatory power, related models and concepts and further research directions.

The Black Box

Background

Irrespective of the widely acknowledged importance of resource allocation in all spheres of human activity, the economics of science, research and development and innovation has remained for decades an underdeveloped field of enquiry (David, 1994). The first attempt to incorporate technological progress in the economic equation was the influential mid-1950s production function study of Solow (1957) who analysed U.S. total factor productivity during the period from 1909 to 1949. His approach was that the component of economic growth, which changes in capital and labour could not explain, is due to technological advances. He concluded that about 90% of the per capita output could be attributed to technological change. This apparent invisibility of what happens when you invest in science and technology gave rise to the so-called black box innovation model.

The Model

According to Rosenberg (1982), economists have treated technological phenomena as events transpiring inside a black box. Although they have recognized their importance, "the economics profession has adhered rather strictly to a self-imposed ordinance not to inquire too seriously into what transpires inside that box" (Rosenberg, 1982, p. vii).

The black box model, borrowed from cybernetics,[2] states that the innovation process itself is not important and that the only things that count are its inputs and outputs. For example, money invested in R&D (input into the black box) will generate, as a rule of thumb, new technological products (outputs) but economists do not need to analyse the actual mechanisms of transformation.

Explanatory Power

The model places innovation as an important economic activity for firms. Although it does not explain research and development characteristics, it draws attention to the fact that "firms and industries that spend relatively large amounts on R&D may tend to have managements that are relatively progressive and forward looking" (Mansfield, 1995, p. 259). The black box coupled with the appropriate and timely management activities makes certain firms more successful than others.

Related Models and Concepts

The black box model of innovation arose along with, and sits neatly alongside, sociological theories of science that emphasized the importance of scientific autonomy and independence as essential for the flourishing of science (Merton, 1973). The threat to scientific freedoms under Nazism and communism, coupled with the astonishing successes of scientific and technological projects in the West (such as radar, the Manhattan project and nuclear energy), led to a strong belief in the power of science to produce radical technologies, if the scientific enterprise was given sufficient resources and 'space' with which it could be free to set its own methodologies and, to a large extent, its own goals. In such a climate, the fact that the actual innovation process was effectively a 'black box' was neither surprising nor particularly of concern; rather, it could be seen as a protective cover within which scientific inquiry could flourish. 'All' that was required was innovative outcomes in return for the input of resources. While this idea of 'big science' arose in the government funding of science, it translated easily into the management of innovation in large corporations, with the establishment and growth of corporate research laboratories, many of which became internationally renowned for their innovations, but whose inner workings were only partially understood by corporate management.

Further Research

The black box model and the reluctance of economists and other researchers to address the link between science, technology and industrial development were major factors in the lack of public policy encouraging innovation. The reliance on market mechanisms to

---

[2] A black box is any apparatus whose internal design is unknown.

support technological developments demonstrated the perception of a large number of economic analysts and policy-makers that there was little need to understand how innovation actually works.

The black box model also generally refers to the research and development component of innovation and/or has a tendency to equate the two. However, non-R&D activities, such as marketing, manufacturing start-up, tooling and plant construction, are crucially important for the introduction of new products and processes. The need to open the black box and explore its interior gave rise to a number of other models which are discussed in the sections to follow.

Linear Models

Background

The 1960s and 1970s witnessed the opening of the black box of innovation with researchers becoming interested in the specific processes that generate new technologies and the learning involved in technological change. The expectations were that understanding innovation would also open the road to formulating policies, which would stimulate R&D and consequently the development of new products and processes. Innovation started to be perceived as a step-by-step process, as a sequence of activities that lead to the technologies being adopted by the markets.

The Model

The first linear description of innovation was by the so-called 'technology push' model, which was closely related to the 'science push' model of science policy advocated by Vannevar Bush in his groundbreaking Science: The Endless Frontier report. According to the report, "discoveries in basic science lead eventually to technological developments which result in a flow of new products and processes to the market place" (Rothwell & Zegveld, 1985, p. 49). The step sequence is as follows:

Basic Science $\Rightarrow$ Applied Science and Engineering

$\Rightarrow$ Manufacturing $\Rightarrow$ Marketing $\Rightarrow$ Sales.

The stages of the model may differ slightly (see, for example, Beije, 1998 or Feldman, 1994), but the focus is on technological newness as a driving force for innovation.

The 'technology push' model is also associated with the name and theoretical work of Schumpeter who studied the role of the entrepreneur as the person taking the risk and overcoming the barriers in order to extract the monopolistic benefits from the introduction of new ideas (Coombs et al., 1987).

The linear 'need pull' (or 'market-driven') model was developed not long afterwards in recognition of the importance of the marketplace and the demands of potential consumers of technology. It states that the causes of innovation are existing demands so that

the step sequence becomes as follows (Rothwell & Zegveld, 1985):

$\Rightarrow$ Market Place $\Rightarrow$ Technology Development

$\Rightarrow$ Manufacturing $\Rightarrow$ Sales.

The main exponent of demand-led innovation is considered to be Schmookler who studied patterns in patents and investments (Coombs et al., 1987; Hall, 1994). His conclusion was that fluctuations in investments can be explained better by external events (e.g. demand) than by trends in inventive activities.

Explanatory Power

The 1960s and 1970s witnessed an enormous amount of research on success factors for innovation. The technology-push/need-pull dichotomy was used to explain not only a wide range of successfully introduced new technologies but also numerous cases of failure (see the description of some projects in Coombs et al., 1987).

Policy-makers around the world have also adopted much of the simplistic linear 'technology push' model because of its clear message and economic rationale (i.e. market failure as the main justification for public investment in research and development). However, funding R&D without supporting other innovation related areas often leads to disillusionment with research and criticism that researchers do not deliver the outputs that have been promised or expected.

Related Models and Concepts

A related concept to the linear model of innovation is the so-called 'barriers to innovation' or factors which impede the adoption of new technologies (see Hadjimanolis, this volume). Similarly, a lot of research has been conducted into defining the factors for successful inovation (Cooper, this volume; Freeman, 1982), such as understanding user needs, attention to marketing and publicity, good communications and the existence of key individuals within the firm. Any deficiencies or weak links in the causal chain could cause innovations or technologies to fail in the marketplace. Both barriers and success factors can be broadly classified as belonging to the push or pull side of the innovation process.

Further Research

Although very clear and easy to understand, the linear models have always been too much of a deviation from reality. They were soon replaced by more sophisticated concepts in theory but in practice they have developed firm ground in the way research is financed and in the outcomes expected by funding bodies and corporate management. The question of what comes first—technology or need—has turned out to be a chicken and egg question, and that field of research has remained relatively quiet.

Interactive Models

Background

Both linear models were regarded as an extremely simplified picture of the generally complex interactions between science, technology and the market. There was a need for deeper understanding and a more thorough description of all the aspects and actors of the innovation process. The sequential nature of innovation has begun to be questioned and the process subdivided into separate stages, each of them interacting with the others.

The Model

According to Rothwell and Zegveld (1985, p. 50), "(t)he overall pattern of the innovation process can be thought of as a complex net of communication paths, both intra-organizational and extra-organizational, linking together the various in-house functions and linking the firm to the broader scientific and techno-logical community and to the marketplace". The stages are as follows (Rothwell, 1983):

New need   ⇔   Needs of society and the market place

⇓   ⇓   ⇓   ⇓   ⇓

| Idea Conception | Development | Manufacturing | Marketing and sales | Market |

⇑   ⇑   ⇑   ⇑   ⇑

New techno-logical capability   State-of-the-art in technology and production techniques

By permission of The British Library.

Beiji (1998) stresses that in such an interactive model, innovation is no longer the end product of a final stage of activity but can occur at various places throughout the process. It can also be circular (iterative) rather than sequential. An example of this was the 'chain-link' model suggested by Kline & Rosenberg (1986) which includes feedbacks and loops allowing potential inno-vators to seek existing inter- and intra-firm knowledge as well as carry out or commission additional research to resolve any problems arising from the market-design-production-distribution process.

Explanatory Power

The main power of the model is the explanation of the variety of interactions necessary for the success of innovation. Further studies expanded the number of boxes and provided insight into the iterative nature of innovation. According to Dodgson & Bessant (1996), the acceptance of the interactive model is now widespread. The original simple model has been extended in numerous variants to describe more players and organizations or to make it specific to a certain situation.

Related Models and Concepts

The interactive model drew the attention of researchers to the lag between new technological ideas and economic outcomes. Issues such as whether the innovation cycle is becoming shorter became a field of intensive research.

Another derivative of this model were the 'techno-logical gap' studies that explored deficiencies in firms' competences (Dodgson & Bessant, 1996) in relation to the various components and interactions required to make innovation happen.

Further Research

The interactive model was an attempt to bring together the technology-push and market-pull approaches into a comprehensive model of innovation. As a result, it provided a more complete and nuanced approach to the issue of the factors and players involved in innovation. However, it still did not explain what drives the engine of innovation and why some companies are better at doing it than others. Nor did it provide an answer as to how organizations learn or what is the role of their operational environment.

System Models

Background

The complexity of innovation requires interactions not only from a wide spectrum of agents within the firm but also from cooperation amongst firms. The well-established hierarchical mechanisms seem to break and in many cases are being replaced by new entities, which cross between organizational boundaries as well as market entities. Marceau (1992) calls this phenome-non 'permeability' of firms while Sako (1992) describes it as the existence of dynamic, industrial, strategic or innovation networks. The main focus of this approach is on innovation as a system, which includes emphasis on interactions, inter-connectedness and synergies.

The Model

The system model argues that firms that do not have large resources to develop innovation in-house, can benefit from establishing relationships with a network of other firms and organizations. Hobday (1991) summarizes the following advantages of such an organization for innovation:

- groups of small firms can maintain leading edge technologies by using the support of the other organizations within the network;
- skill accumulation and collective learning occurs within the network and benefits all participants;
- the network promotes flows of key individuals between firms;

- skills can be combined and re-combined to overcome bottlenecks;
- innovation time and costs can be reduced;
- the network provides entry into the industry for small innovative firms;
- individual firms within the network operate with high flexibility and in low cost ways, including small overheads.

The most well-known system model is the so-called national systems of innovation (e.g. Freeman, 1991; Lundvall, 1992; Nelson, 1993, 2000). It deals with the diversity in approaches to innovation in countries around the globe which differ in size, level of economic development, historical traditions or level of concern about specific policy problems (e.g. education or global warming). According to one study, this is reflected in the way the main actors in the innovation process (firms, public and private research organizations, government and other public institutions) interact and the forms, quality and intensity of these interactions (OECD, 1999, p. 22). A national system of innovation is defined as a set of institutions, which jointly and individually contribute to the development and diffusion of new technologies and provide a framework for the implementation of government policies influencing the innovation process (Metcalfe, 1995). The most important feature of this set is its interconnectedness, the way the various elements interact.

Explanatory Power

The main power of this model is in explaining the place and role of small firms in innovation and how they can survive the competition and pressures from large companies. The synergistic effect of the innovation networks explains their capacity to produce positive sum effects for all the participants (DeBresson & Amesse, 1991; Freeman, 1991). They are also flexible and can adapt more easily to the changing requirements coming from various clients and markets. They are better equipped to deal with technological risk and uncertainty. The systems facilitate communications (Tisdell, 1995), the flow of information and transfer of formal and tacit knowledge.

The national systems of innovation concept explains the differences between countries and the various role governments play. It highlights specific patterns of scientific, technological and industrial specialization, institutional profiles and structures and, most importantly, how different countries learn, "since innovation is—by definition—novelty in the capabilities and knowledges which make up technology" (Smith, 1998, p. 25). The concept is also not confined to the level of the nation state and can be applied worldwide or to regions and localities.

Related Models and Concepts

Related concepts are innovation chains (Marceau, 1992) and complexes (Gann, 1991, 2000). Chains refer to the relationships between core manufacturers and their suppliers and distributors (Dodgson, 1993). Complexes integrate not only firms but publicly funded institutions and specific industry-based research organizations (examples of this are the military or construction industry complexes).

Strategic networks (or alliances) is another related concept (Jarillo, 1988; Sako, 1992) which refers to long-term, purposeful arrangements between for-profit organizations in order to gain competitive advantage over players outside the network or alliance.

A particular case of networks is the regional network (Dodgson, 1993) in which geographic location has prime importance. In addition to physical proximity, regions share similar culture, industrial mix, economic and administrative homogeneity as well as political and governance environment, which can promote distinctive styles and modes of innovation within regions (Cooke, 1998). Regional systems of innovation are closely linked to the concept of innovative milieux, which is discussed further below.

Further Research

The potential of networks in promoting innovation compared to that of large firms is not always clearly understood. Networks are generally very dynamic, and there is not much evidence of how long they last. Their competitive advantages may be easily lost if (or when) some firms become larger, quit the network and/or take over other firms. Trust building is a crucial component in the networked innovation, and the ways to achieve and sustain this are not always clear. The mechanisms of simultaneous cooperation and competition within the network (be it in different areas or on different projects) also require further investigation and understanding.

Edquist (1997) lists nine characteristics of the systems models, which, as well as describing the approach, shed light on its weaknesses and the need for further conceptualization. They are: (1) innovations and learning are at the centre of the model; (2) it offers a holistic and interdisciplinary approach; (3) a historical perspective is natural; (4) there are differences between systems and there is no optimality; (5) it emphasizes interdependence and non-linearity; (6) it encompasses product technologies and organizational innovations; (7) it points to the central role of institutions; (8) there is a place for various kinds of ambiguities and diffusion of concepts; and (9) it provides a broad conceptual framework rather than formal theories.

Another aspect of networking which requires further investigation is the role of government, proactive policies and the regulatory environment in creating favorable conditions for such linkages and interactions. For example, a special case of facilitating networking could be public encouragement of cooperation in the development of environmental technologies.

Evolutionary Models

Background

According to Saviotti (1996, p. 29), the need for an evolutionary approach in economics was proposed on the basis of a number of failures in neoclassical economics, including its inability to deal with dynamic qualitative changes, which are internal features of technological innovation. Hodgson (1993) argues that the mechanical metaphor adopted in orthodox economic thinking has weak explanatory power, as economics and innovation are products of living creatures. Hence, the biological metaphor is more useful and parallels can be made with the Darwinian evolution of species (more explanation on this approach is provided in the reviews by Nelson, 1995; Dosi & Nelson, 1994). More recently, evolutionary studies of technological change have combined fundamentals not only from biology, but also from equilibrium thermodynamics, organizational theory and heterodox approaches in economics.

The Model

Saviotti (1996) explains the key concepts in an evolutionary approach to innovation as being the following:

- Generation of variety—innovations are seen as equivalent to mutations. They continuously generate new products, processes and forms and contribute to increased variety. Not all mutations (new technological developments) are successful, but the ones, which are often replace older products and processes consequently making them extinct.
- Selection—selection processes act together with variety-generating mechanisms. The outcome is the 'survival' (which could also be interpreted as introduction or maintenance) of some products, technologies and firms as a result of their adaptation to the environment in which they operate, and the demise of others.
- Reproduction and inheritance—firms are perceived as producing organizations and inheritance is expressed in the continuity in which organizations make decisions, develop products and generally do their business. Firms are learning entities but any developed expertise is difficult to inherit or transfer to other firms.
- Fitness and adaptation—Darwin's 'survival of the fittest' principle is represented by the propensity of an economic unit to be successful in a given environment.

- Population perspective—variation is an essential component for an evolutionary process. Hence, not only average values but also variances in the population of firms/products should be analyzed.
- Elementary interactions[3]—these include mainly competition (between products or firms) and are the most studied interaction in economics. More recently, collaboration has also become a recognized type of interaction.
- External environment—a key element in the evolutionary approach. It traditionally covers the socioeconomic (including regulatory) environment in which technologies are developed. It is determined by mechanisms such as patent regimes, market structures, standards and regulations. More recently (in the case of the green or ecologically friendly technologies) it has also started to include the link with the natural environment.

Nelson & Winter (1982) were the first to translate the conceptual evolutionary model into a computer simulation model describing business behaviour on the basis of the so-called 'routines' or regular and predictable behavioural patterns and habits of firms. This model was initially applied to data used by Solow (1957) who studied U.S. productivity. Nelson and Winter were successful in demonstrating that 'realistic' firm behaviour could account for macroeconomic outcomes at least as well as Solow's growth modelling production function. With the further advance of computer technologies, the 1980s and 1990s witnessed further interest in evolutionary modeling (e.g. Kwasnicki, 2000).

Explanatory Power

The evolutionary model challenged the central concept of economic theory, which traditionally focused on market equilibrium and complete information. This new approach explains that innovation by definition involves change, and decisions are made not merely on price. They evolve from historical context, social conventions and relationships between people and organizations. Metcalfe (1995, p. 26) concludes that in a "fundamental sense, innovations and information asymmetries are one and the same phenomenon". In other words, imperfections are necessary conditions for technical change to occur in a market economy.

The uncertainty and imperfect information inherently associated with the innovation process is a departure from the 'rational individual' concept in neoclassical economics. The evolutionary model

---

[3] In biology, elementary interactions between species include competition (for a common resource in short supply), commensalisms (one species stimulate the growth of the other) and predation (one species inhibits the growth of the other).

stresses 'bounded rationality' (Dosi & Egibi, 1991) and the value of diversity (Dowrick, 1995). It also shows how, under meaningful economic values of certain parameters, such as technological opportunities, and established decision-making rules, firms can be dynamic self-organized systems (Dosi & Orsenigo, 1994).

The selection process and the importance of the surrounding environment shed light on the processes of failure of generally fit technologies and the success of technologies which are considered inferior (e.g. MS-DOS computer operation system or VHS videorecording system). According to Tisdell (1995, p. 128), "a fit technique... may fail to be selected because its surrounding environment at the time of its occurrence is unfavourable".

According to Bryant & Wells (1998), the evolutionary school of thought is likely to become very influential in policy considerations. Since World War II, governments have consistently funded outcome oriented research. What the evolutionary model emphasizes is that the process is as (if not more) important as the results from R&D. For example, a 1996 OECD report recommends governments to increase the population of innovative firms as a main policy goal rather than correcting market failure (OECD, 1996).

The evolutionary model also points out that outcomes are to a large degree determined by the evolutionary process, be it at the level of company or country. Shedding light on how decisions are made and how the various participants interact to produce innovations, is a major explanatory feature of this model. Therefore governments should be urged to create conditions conducive to the process of innovation by shaping relationships, encouraging learning, and balancing competition with cooperation. The model has less normative power, and is less focused on the implications for innovation strategy at the level of the firm beyond the need for firms to protect diversity and a range of competencies (of people, product ranges, technologies, etc.).

Related Models and Concepts

Related to the evolutionary model are a number of generalizations, which appeared in the 1970s and 1980s. They refer to technological imperatives (Rosenberg, 1976), innovation avenues (Sahal, 1981), technological trajectories (e.g. Biondi & Galli, 1992; Pavitt et al., 1989), technological (Dosi, 1982, 1988) and technoeconomic paradigms (Freeman & Perez, 1988; Perez, 1983). The main argument is that during a particular time period we witness certain regularities in the development of technologies (represented by the nature of the applied principles and practical solutions), but they are often hindered by the delays with which institutions adapt to the new potential of these technologies.

Further Research

In addition to explanatory power, an extremely important aspect of any model is its predictive potential. If a model is a relatively accurate representation of reality, the traditional scientific approach expects it to also deliver forecasts and predictions of future parameter values. The evolutionary model in general lacks such a capacity as it describes constant change, and hence, its parameters are always in flux. Nevertheless, there is some degree of predictability if we can explain the mechanisms supporting the continuity of the old and the introduction of the new and if we can characterize the turning points in between.

Innovative Milieux

Background

Since the 1970s, a large body of literature has developed dealing with aspects of the growth of regional clusters of innovation and high technology (Feldman, 1994; Keeble & Wilkinson, 2000). The importance of geographical location for knowledge generation gave rise to the innovative milieux explanatory model. The concept is the main contribution by geographers, regional economists and urban planners to a field, which traditionally has been studied by economists and sociologists. The real-estate rule of: 'Location! Location! Location!' started to attract attention to the natural, social and built environment surrounding establishments where technologies are developed. The model includes networking and linkages but goes beyond that to emphasize the importance of quality-of-life factors.

The Model

The innovative milieu model states that "innovation stems from a creative combination of generic know-how and specific competencies" and "territorial organization is an essential component of the process of techno-economic creation" (Bramanti & Ratti, 1997, p. 5). According to Longhi & Keeble (2000, p. 27), "the innovation process is not spaceless. On the contrary, innovation seems to be an intrinsically territorial, localized phenomenon, which is highly dependent on resources which are location specific, linked to specific places and impossible to reproduce elsewhere".

An early description of innovative milieux by Camagni (1991) lists the following components:

- a productive system, e.g. innovative firm;
- active territorial relationships, e.g. inter-firm and inter-organizational interactions fostering innovation;
- different territorial socio-economic actors, e.g. local private or public institutions supporting innovation;
- a specific culture and representation process;
- dynamic local collective learning process.

Camagni & Capello (2000) emphasize that the inter-actions creating the innovative milieu are not necessarily based on market mechanisms but include movement and exchange of goods, services, informa-tion, people and ideas among others. They are not always formalized in cooperative agreements or any other contracts. Major features of such an environment are the ease of contact and trust between partners, which reduce uncertainty in the development of new technologies and prove to be a source of exchange of tacit knowledge.

In addition to the components of a productive working environment, more recently other factors have started to impact on the capacity of locations to generate innovative firms. They relate to the social, cultural and natural characteristics of the place, such as proximity to recreational sites, climate and air quality, quality of life for family members, including children, to mention a few. The high density of links is also being expressed in relation to interactions with the local community (Willoughby, 1995).

Explanatory Power

The innovation milieu concept helps explain the success of small and medium-sized enterprises, which in general lack the resources to maintain aggressive R&D strategies and operate at the cutting edge of technologies. The existing supporting network com-pensates for that and provides an operational "microcosm in which all those elements which are traditionally considered as the sources of economic development and change within the firm operate as if they were *in vitro*" (Camagni & Capello, 2000, p. 120).

The model also explains why certain localities give birth to a large number of small innovative firms, which are situated in close proximity and share a similar cultural and business ethos. It also highlights the fact that different localities have different patterns and paths in knowledge development and transfer of high technology.

Related Models and Concepts

Porter's (1990) analysis of groups of firms located in geographic proximity is often referred to as 'innovation clusters'. According to the OECD (1999), the concept of clusters is closely linked to firms networking but it goes beyond that as it captures all forms of knowledge sharing and exchange within a specific locality. Many clusters have developed over extended periods of time and have deep historical roots. Often they are also linked to the particular natural, human and other resources available in the region.

Other closely related concepts are 'the learning region' (e.g. Florida, 1995; Kirat & Lung, 1999; Macleod, 1996; Simmie, 1977) and 'collective learn-ing' (Keeble, 2000; Lawson, 2000). They stress that learning is the most important feature of any economy

and successful regions provide particular combinations of institutions and organizations to encourage knowl-edge development within the community and learning by local firms through conscious and unconscious mechanisms.

Further Research

The innovative milieux model has not so far addressed the links between innovation and ecology. It is still predominantly anthropocentric, and innovators rarely address the issues of harmony with the natural environment. Nevertheless, there is increasing evi-dence that the development of technologies is not a means of its own but a mechanism to achieve broader goals. An example of this is the Finnish environmental cluster research program (Honkasado, 2000), which covers projects encouraging cooperation between entrepreneurs who utilize the natural environment for eco-business and promotes innovative enterprises spe-cializing in environmental technologies. The interest in the locality will hopefully result in increased con-sideration of issues such as ecologically sustainable development and social justice.

Conclusion

What becomes apparent from this overview of the six generations of innovation models is that the more we study innovation, the more we realize how complex a process it is and how difficult it is to 'master' it, whether at a corporate or government policy level. Hence, while all the models discussed here stem at least in part from an interest on how innovation occurs in the firm, the explanatory factors have broadened out markedly. Innovation models have moved from a concentration on factors wholly or to a greater or lesser degree within the control of the firm (i.e. R&D management, marketing, financial resources, etc.) to factors external to the firm, so that networks of firms, government institutions and policies, and even culture and geography, have become relatively more impor-tant. The models studied here provide the basics of an explanation of the innovation phenomenon. Never-theless, they leave a lot of questions unanswered. Every model is intrinsically a simplification of reality and as a rule omits the myriad details which make every innovation case so unique in its success or failure.

References

Beije, P. (1998). *Technological change in the modern economy*, Cheltenham, U.K.: Edward Elgar.

Biondi, L. & Galli, R. (1992). Technological trajectories. *Futures*, **24**, 580–592.

Bramanti, A. & Ratti, R. (1997). The multi-faceted dimen-sions of local development. In: R. Ratti, A. Bramanti & R. Gordon (Eds), *The Dynamics of Innovative Regions: The GREMI Approach* (pp. 3–44). Aldershot, U.K.: Ashgate.

Bryant, K. & Wells, A. (1998). A new school of thought. In: Department of Industry, Science and Resources, *A New Economic Paradigm? Innovation-based Evolutionary Sys-tems* (pp. 1–3). Canberra: Commonwealth of Australia.

Camagni, R. (1991). 'Local milieu', uncertainty and innovation networks: Towards a new dynamic theory of economic space. In: R. Camagni (Ed.), *Innovation Networks: Spacial Perspectives* (pp. 121–143). London: Belhaven Press.

Camagni, R. & Capello, R. (2000). The role of inter-SME networking and links in innovative high-technology milieux. In: D. Keeble & F. Wilkinson (Eds), *High-Technology Clusters, Networking and Collective Learning in Europe* (pp. 118–155). Aldershot, U.K.: Ashgate.

Cooke, P. (1998). Introduction. In: H-J. Braczyk, P. Cooke & M. Heidenreich (Eds), *Regional Innovation Systems: The Role of Governance in a Globalised World* (pp. 2–25). London: UCL Press.

Coombs, R., Saviotti, P. & Walsh, V. (1987). *Economics and technological change*. London: MacMillan.

David, P. A. (1994). Positive feedbacks and research productivity in science: Reopening another black box. In: O. Granstrand (Ed.), *Economics of Technology* (pp. 65–89). Amsterdam: North-Holland.

DeBresson, C. & Amesse, F. (1991). Networks of innovators: A review and introduction to the issue. *Research Policy*, **20**, 363–379.

Dodgson, M. (1993). *Technological collaboration in industry: Strategy, policy and internationalisation in innovation*. London: Routledge.

Dodgson, M. & Bessant, J. (1996). *Effective innovation policy: A new approach*. London: International Thomson Business Press.

Dosi, G. (1982). Technological paradigms and technological trajectories: A suggested interpretation of the determinants and directions of technical change. *Research Policy*, **11**, 147–162.

Dosi, G. (1988). Sources, procedures, and microeconomic effects of innovation. *Journal of Economic Literature*, **26**, 1120–1171. Or in: C. Freeman (Ed.), *The Economics of Innovation* (1990, pp. 107–158). Aldershot, U.K.: Edward Elgar.

Dosi, G. & Egibi, M. (1991). Substantive and procedural uncertainty: An exploration of human behaviour in changing environments. *Journal of Evolutionary Economics*, **1**, 145–168.

Dosi, G. & Nelson, R. R. (1994) An introduction to evolutionary theories in economics. *Journal of Evolutionary Economics*, **4**, 153–172.

Dosi, G. & Orsenigo, L. (1994). Macrodynamics and microfoundations: An evolutionary perspective. In: O. Granstrand (Ed.), *Economics of Technology* (pp. 91–123). Amsterdam: North-Holland.

Dowrick, S. (Ed.) (1995). *Economic approaches to innovation*. Aldershot, U.K.: Edward Elgar.

Edquist, C. (1997). Systems of innovation approaches—their emergence and characteristics. In: C. Edquist (Ed.), *Systems of Innovation Technologies, Institutions and Organisations* (pp. 1–35). London: Pinter.

Feldman, M. (1994). *The geography of innovation*, Dordrecht: Kluwer Academic Publisher.

Florida, R. (1995). Toward the learning region. *Futures*, **27**, 527–536.

Freeman, C. (1982). *The economics of industrial innovation* (2nd ed.). London: Pinter.

Freeman, C. (Ed.) (1984). *Long waves in the world economy*, London and Dover NH: Pinter.

Freeman, C. (1988). Japan: A new national system of innovation. In: G. Dosi et al. (Eds), *Technical Change and Economic Theory* (pp. 330–348). London: Pinter.

Freeman, C. (1991). Networks of innovators: A synthesis of research issues. *Research Policy*, **20**, 499–514.

Freeman, C. & Perez, C. (1988). Structural crises of adjustment: Business cycles and investment behaviour. In: G. Dosi et al. (Eds), *Technical Change and Economic Theory* (pp. 38–66). London: Pinter.

Gann, D. (1991). Technological change and the internationalisation of construction in Europe. In: C. Freeman, M. Sharp & W. Walker (Eds), *Technology and the Future of Europe* (pp. 231–244). London: Pinter.

Gann, D. (2000). *Building innovation: Complex constructs in a changing world*. London: Thomas Telford.

Hall, P. (1994). *Innovation, economics and evolution: Theoretical perspectives on changing technology in economic systems*. New York: Harvester Wheatsheaf.

Hobday, M. (1991). *Dynamic networks, technology diffusion and complementary assets: Explaining U.S. decline in semiconductors*. DRC Discussion Papers, 78. Falmer, U.K.: Science Policy Research Unit, University of Sussex.

Hodgson, G. (1993). *Economics and evolution: Putting life back into economics*. Oxford: Polity Press.

Honkasado, A. (2000). Eco-efficiency, entrepreneurship and cooperation: The Finnish environmental cluster research programme. In: *Innovation and the Environment* (pp. 137–142). Paris: OECD.

Jarillo, J. (1988). On strategic networks, *Strategic Management Journal*, **19**, 31–41.

Keeble, D. (2000). Collective learning processes in European high-technology milieux. In: D. Keeble & F. Wilkinson (Eds), *High-Technology Clusters, Networking and Collective Learning in Europe* (pp. 199–229). Aldershot, U.K.: Ashgate.

Keeble, D. & Wilkinson, F. (2000). SMEs, regional clustering and collective learning: An overview. In: D. Keeble & F. Wilkinson (Eds), *High-Technology Clusters, Networking and Collective Learning in Europe* (pp. 1–20). Aldershot, U.K.: Ashgate.

Kirat, T. & Lung, Y. (1999). Innovation and proximity: territories as loci of collective learning processes. *European Urban and Regional Studies*, **6**, 27–38.

Kline, S. J. & Rosenberg, N. (1986). An overview of innovation. In: R. Landau & N. Rosenberg (Eds), *The Positive Sum Strategy* (pp. 275–305). Washington, D.C.: National Academy Press.

Kwasnicki, W. (2000). Monopoly and project competition: There are two sides of every coin. In: P. P. Saviotti & B. Nooteboom (Eds), *Technology and Knowledge: From the Firm to Innovation Systems* (pp. 47–79). Cheltenham, U.K.: Edward Elgar.

Lawson, C. (2000). Collective learning, system competences and epistemically significant moments. In: D. Keeble & F. Wilkinson (Eds), *High-Technology Clusters, Networking and Collective Learning in Europe* (pp. 182–198). Aldershot, U.K.: Ashgate.

Longhi, C. & Keeble, D. (2000). High-technology clusters and evolutionary trends in the 1990s. In: D. Keeble & F. Wilkinson (Eds), *High-Technology Clusters, Networking and Collective Learning in Europe* (pp. 21–56). Aldershot, U.K.: Ashgate.

Lundvall, B.-Å. (1992). *National systems of innovation: Towards a theory of innovation and interactive learning.* London: Pinter.

Macleod, G. (1996). The cult of enterprise in a networked, learning region? Governing business and skills in lowland Scotland. *Regional Studies*, **30**, 749–755.

Mansfield, E. (1972). Contribution of R&D to economic growth in the United States. *Science*, **175**, 487–494. Or in: E. Mansfield (Ed.), *Innovation, Technology and the Economy* (1995, Vol. I, pp. 255–273). Aldershot, U.K.: Edward Elgar.

Marceau, J. (1992). *Reworking the world: Organisations, technologies and cultures in comparative perspective.* Berlin: De Gruyter.

Merton, R. K. (1973). *The sociology of science: Theoretical and empirical investigations.* Chicago: N. W. Storer.

Metcalfe, J. S. (1995). Technology systems and technology policy in an evolutionary framework. *Cambridge Journal of Economics*, **19**, 25–46.

Metcalfe, S. (1995). The economic foundations of technology policy: Equilibrium and evolutionary perspective. In: P. Stoneman (Ed.), *Handbook of the Economics of Innovation and Technological Change* (pp. 409–512). London: Blackwell.

Nelson, R. R. (1993). *National innovation systems: A comparative analysis.* New York: Oxford University Press.

Nelson, R. (1995). Recent evolutionary theorising about economic change. *Journal of Economic Literature*, **33**, 48–90.

Nelson, R. (2000). National innovation systems. In: Z. Acs (Ed.), *Regional Innovation, Knowledge and Global Change* (pp. 11–26). London: Pinter.

Nelson, R. & Winter, S. (1982). *An evolutionary theory of economic change.* Cambridge, MA: Harvard University Press.

Organisation for Economic Cooperation and Development (OECD) (1996). *The OECD jobs strategy: Technology, production and job creation* (Vol. 2). Paris: OECD.

Organisation for Economic Cooperation and Development (OECD) (1999). *Managing national innovation systems.* Paris: OECD.

Pavitt, K. (1984). Sectoral patterns of technical change: Towards a taxonomy and a theory. *Research Policy*, **13** (6), 343–373. Or in: C. Freeman (Ed.), *The Economics of Innovation* (1990, pp. 249–279). Aldershot, U.K.: Edward Elgar.

Pavitt, K., Robson, M. & Townsend, J. (1989). Accumulation, diversification and organisation of technological activities in U.K. companies, 1945–83. In: M. Dodgson (Ed.), *Technology Strategy and the Firm: Management and Public Policy* (pp. 38–67). Harlow, U.K.: Longman.

Perez, C. (1983). Structural change and the assimilation of new technologies in the economic system. *Futures*, **15**, 357–375.

Porter, M. (1990). *The competitive advantage of nations.* New York: Free Press.

Rosenberg, N. (1976). *Perspectives on technology.* Cambridge: Cambridge University Press.

Rosenberg, N. (1982). *Inside the black box: Technology and economics.* Cambridge: Cambridge University Press.

Rothwell, R. (1983). Information and successful innovation. British Library R&D Report No. 5782.

Rothwell, R. (1992). Successful industrial innovation: Critical factors in the 1990s. *R&D Management*, **22**, 3.

Rothwell, R. & W. Zegveld (1985). *Reindustrialization and technology.* Harlow, U.K.: Longman.

Sahal, D. (1981). *Patterns of technological innovation.* New York: Addison-Wesley.

Sako, M. (1992). *Price, quality and trust: How Japanese and British companies manage buyer supplier relations.* Cambridge: Cambridge University Press.

Saviotti, P. P. (1996). *Technological evolution, variety and the economy.* Cheltenham, U.K.: Edward Elgar.

Simmie, J. (Ed.) (1997). *Innovation, networks and learning regions?* London: Jessica Kingsley.

Smith, K. (1998). Innovation as a systemic phenomenon: Rethinking the role of policy. In: *A New Economic Paradigm? Innovation-based Evolutionary Systems.* Discussions in Science and Innovation 4. Canberra: Department of Industry, Science and Resources.

Solow, R. M. (1957). Technical change and the aggregate production function. *Review of Economics and Statistics*, **39**, 312–320.

Tisdell, C. (1995). Evolutionary economics and research and development. In: S. Dowrick (Ed.), *Economic Approaches to Innovation* (pp. 120–144). Aldershot, U.K.: Edward Elgar.

Willoughby, K. (1995). The 'Kingdom of Camelot' and the 'Quest for the Holly Grail'. In: J. Phillimore (Ed.), *Local Matters: Perspectives on the Globalisation of Technology.* Perth, Western Australia: Murdoch University.

The International Handbook on Innovation
Edited by Larisa V. Shavinina

Evolutionary Models of Innovation and the Meno Problem

Thomas Nickles

Philosophy Department, University of Nevada, Reno, USA

Abstract: 'Universal Darwinism' is an innovation paradigm superior to the intelligent design and romantic paradigms. All innovation, including new knowledge, is the product of blind variation plus selective retention (BV + SR) and is thus a kind of adaptation. Yet many methodologists continue to regard BV + SR as a limited model of innovative problem-solving. I show how this clash can be (partially) resolved by exploring (1) the roots of the problem of innovation in Plato's Meno paradox and (2) the implications of the 'No Free Lunch' theorems. The BV + SR paradigm recognizes sources of innovation other than novel ideas of creative individuals.

Keywords: Creativity; Innovation; Universal Darwinism; Universal evolution; Evolutionary epistemology; Design; Adaptation; 'No free lunch' theorems; Meno paradox.

Introduction

Popular accounts of innovation in the sciences and the arts appeal to special powers of creative individuals. Da Vinci, Newton, Mozart, Einstein, Picasso, and perhaps also Napoleon, Thomas Edison, and Henry Ford, are said to be geniuses able to see into the future or into the deep-structure of the universe, to have special powers of intuition that transcend those achievable by mastery of one's subject area. In one way or another, these intellectual heroes possess clairvoyance, prescience, or foreknowledge. In this chapter I reject all such accounts as nonexplanatory, question-begging, or even regressive in the sense that invoking them makes the explanation problem harder rather than easier. What we are left with on a thoroughly naturalistic account is the ordinary abilities of human beings (cf. Weisberg, this volume). To be sure, we humans vary tremendously, both in our natural abilities and in the expertise that we bring to various domains; but none of us have supernatural powers. None of us are gods, and not even the brightest of innovators can foresee or control the future implications of their work with guaranteed accuracy.

In avoiding the supernatural, my account agrees with others in this Handbook. However, my emphasis will be less on especially creative individuals (that is, on 'great men and their ideas') and more on general processes that constitute sources of innovation. I do not

deny that exceptional people exist, but here I am interested in the question: What are the implications of 'going natural?' Even the more sophisticated studies of canonical innovators tend to overplay the degree to which they knew enough about what they were doing to preplan their actions in accurate detail. Building on the work of Donald Campbell, Daniel Dennett, and others, I shall try to get to the bottom of the innovation issues in cognitive and epistemological terms.[1] Although I propound nothing worth calling a general *theory* or general *method* of innovation, I claim that the subject can be interestingly addressed at a general level. This will require thinking about innovation in a rather abstract way.

If I am successful, the overall result will be a framework, with some usable constraints, for thinking about genuine innovation of any kind—namely an evolutionary framework of a broadly Darwinian type. The major implication is that blind variation plus selective retention—an evolutionary process—plays an

---

[1] See the references at the end of the chapter to Campbell and Dennett as well as to Richard Dawkins, Gerald Edelman, Henry Plotkin, Aharon Kantorovich, Gary Cziko, Dean Keith Simonton (who has a chapter in the present Handbook), and William Calvin. These authors are, of course, not responsible for the particular twist that I give to the BV + SR model of innovation.

essential role (although not the only role) in all innovative inquiry. This in turn implies that, at least to a first approximation, all innovation, all design, and, indeed, all knowledge can be construed as adaptation; and, contrariwise, in a broad sense, all adaptation can be construed as design or knowledge and, when new, as innovation. Insofar as a design is *fully* the product of intelligent planning, its production is routine, not novel, although the plan itself may of course be original and hence itself the product of prior trial and error.

Learning theories and, by extension, theories of innovation are sometimes classified into three types. *Providential* theories ultimately appeal to divine providence or some other transcendent process as the source of new knowledge or design, for example, a doctrine of innate a priori knowledge or a special capacity of clairvoyance. *Instructionist* theories claim that the world directly instructs the knowing agent by imposing its form on the receptive mind. The classical model is the signet ring impressing its pattern in wax. During the modern period we get the simple empiricist view that we can read the truth directly off nature, by passive induction, without any need for interpretive trial and error. One version of the modern view is British empiricist philosopher David Hume's psychological theory of learning or habit acquisition through mere repetition of an environmental contingency. Hume simply took for granted that the agent unproblematically *notices* the repetitive pattern.

Finally, *selectionist* theories postulate some sort of variation and selection process or iterative trial and error. In this chapter I defend the third approach, and I reject providential theories on the ground that they are question begging and viciously regressive. I shall not have space to critique instructionist theories (but see Campbell, 1974; Cziko, 1995).

At its core, Darwinian evolution consists of three mechanisms: a mechanism of blind or undirected variation, a mechanism of selection, and a mechanism of transmission or heredity, in short, BV + SR. Darwin appreciated that whenever these mechanisms operate in a reasonably stable environment with selection pressures, evolution of the species in question *will* occur. He elaborated a mechanism of natural selection but was ignorant of the precise mechanisms of variation and transmission. Although not a mathematician, Darwin thought statistically, in a qualitative sort of way, and thereby introduced populational thinking into biology more clearly than anyone had before. Moreover, his theory in effect introduced a new kind of control system into our thinking about natural processes, a kind of feedback mechanism of a subtle, indirect sort (one that is purely eliminative, with death or reproductive failure as the 'teacher'), operating over a distributed population of agents or processors responding only to their local conditions (Cziko, 2000, ch. 1 *et passim*). We should not consider evolution (and

BV + SR in general) as a mechanistic theory of a Newtonian kind.

It is no secret that evolutionary mechanisms can produce remarkably novel design. Indeed, biological evolution is surely the most innovative process that we know. What other process has produced anything as innovative and original as the biological world in all its amazing complexity and variety? This fact alone makes Darwinian evolution worthy of serious attention from students of creativity and innovation. What is it that enables such a blind- and dumb-looking process, one that completely lacks creative individuals in the human sense, to produce so much innovative design? Can we 'reverse engineer' this process and put it to human use? Have we already done that implicitly? While evolutionary processes are fascinating producers of innovation, I want to defend a stronger claim: they are the *only* truly innovative processes that we currently know!

I believe that an evolutionary process is implied by a thoroughly naturalistic treatment of innovation. At least there is no alternative account in sight at the present time, with the possible exception of self-organizing systems as a partial explanation. Yet many people interested in innovation find the selectionist model of learning and innovation highly counterintuitive. At the end of the chapter I shall reply to a few of the more frequent objections to the evolutionary model of innovation. Meanwhile, I shall maintain that, *since* it is the only innovative process that we really know and understand, we need to reconceive innovation, including learning and research. To a first approximation, we should conceive of all genuine innovation, including new knowledge, as a kind of adaptation.[2] And we should view all genuine design or fit as the product of an evolutionary process. This grand generalization of Darwinian evolution is often called 'universal Darwinism'. The label is convenient, but it misleadingly suggests too close a tie to Darwinian *biological* evolution. Accordingly, I shall often call it 'the BV + SR model of innovation' or 'Campbell's thesis'. I now motivate this generalization of the innovation problem and solution from an unexpected philosophical angle, namely, Plato's paradox of the *Meno*.

---

[2] Although this is a broadly biological conception of innovation and learning, I mean only that they are underlain by BV + SR processes, not necessarily strictly Darwinian biological processes. Still, my thesis can be viewed as an extension of the adaptationist paradigm in evolutionary theory. I do not commit myself slavishly to adaptation as the only possible origin for innovation (given such things as genetic drift and the possibility of self-organizing systems); hence my 'to a first approximation'. However, I believe that adaptation is the most important factor. And if a physical system in effect explores a phase space and hits on a stable, self-organized state or falls into an attractor basin, that is rather like blind search and selection, although it will not be iterative unless the state is self-reproducing or is shaken out of its basin with the possibility of finding an even deeper basin.

Innovation, the Meno Paradox, and Darwinian Evolution

The Meno paradox presents us with the problem of how successful inquiry is possible. The paradox denies that inquiry is possible and hence denies that we can learn anything new. By two plausible extensions, the Meno problem can also be interpreted as: (1) the problem of how human innovation is possible; and (2) the problem of how novel design anywhere can emerge, whether or not that design is accomplished by the deliberate, conscious activity of intelligent agents. Indeed, taking a cue from Plotkin (1993) and Dennett (1995) as well as Campbell, I claim that all knowledge can be treated as a form of design and, conversely, that all adaptive design can be construed as a form of knowledge. The issues are too complex to defend in detail here, but I hope to make them at least intelligible.

The paradox, formulated in Plato's dialogue *Meno*, 80d-e, has the logical form of a dilemma. In its original form the problem is to show how successful inquiry is possible, that is, how is it possible to gain new knowledge. Meno (the character in dialogue with Socrates) concludes that it is not possible. For either you already know the answer (or problem solution) that you seek or you do not. If you do, you cannot really inquire, for you already have the answer. But if you do not know the answer, then you also cannot inquire, since you would not *recognize* the answer even should you stumble upon it accidentally. Ability to recognize is the key, and you have to know an item already in order to recognize it.

To many readers, the paradox is nettlesome word-play, to others an unnecessary abstraction. There is a bit of verbal confusion, to be sure; but in my view the paradox presents us with *the* core problem of innovation; and its solution possesses practical importance as well as intellectual interest.

As usual, there are two 'horns' to the dilemma. If you already know that magnetic monopoles exist, then you cannot genuinely inquire into whether or not they exist; and so you impale yourself on the first horn of the dilemma. This is true whether you had previously learned that alleged fact by empirical investigation, by reading a text, or by direct revelation by Providence. However, if you have no idea whether monopoles exist and even what exactly they are and what would count as detecting them, then you impale yourself on the second horn, since you would not recognize one even should you 'stumble upon it' accidentally. To change the example: Had Aristotle asked what makes salt flash yellow when tossed into the fire, and a twentieth-century quantum theory textbook had fallen out of the sky at his feet, he would have been incapable of recognizing that the text provides the answer to his question (namely in terms of electron state transitions of sodium ions, which flash the familiar sodium yellow, as first proposed by Niels Bohr).

The Aristotle illustration is extreme. We all know that inquiry *is* possible, as is innovation. What, then, is the solution to the paradox? I take up this question more fully in Section 5, but to anticipate: As far as I know, the only viable solution involves an element of trial and error, that is, a process of blind variation plus selective retention (BV + SR). Of course, pure, blind or 'random' trial and error that simply discards all problem solutions that are imperfect, by some pre-established standard, is very inefficient. But things are very different if the standard of comparison is greater fitness relative to other members of an extant population, if the fitter individuals have a greater chance of selection, and if the retention mechanism involves transmission, with variation, to a new generation such that we get iterated 'cycles' of BV + SR. In that case we get gradual evolution. Darwin was the first to recognize this, and he further recognized that a BV + SR process, when it is in place, does not merely make evolution *possible*, it makes it (nomically) *necessary*, assuming a reasonably stable environment with selection pressures. As long as these conditions remain in place, there is no stopping evolution!

My claim is that human inquiry in the various sciences and the arts frequently satisfies the conditions that make evolution and hence innovation almost inevitable. So we have passed from puzzlement about whether and how innovation is possible at all to understanding why, under a variety of situations, it is inevitable—virtually necessary.

Although it is not news that human innovation is possible, the BV + SR model once again raises the question whether we can improve our innovative techniques by reverse engineering biological evolution, whether we can distill out the secret to its innovative powers, and 'bottle' it for our own use. 'Simulation of the evolutionary process is tantamount to a mechanization of the scientific method', wrote Lawrence J. Fogel, one of the founders of the field of evolutionary computing.[3]

This chapter and two companion pieces (Nickles, 2003 and forthcoming) can be viewed as an exploration of Fogel's claim. I believe that there is some truth in it. However, this does not mean that biological BV + SR is immediately applicable as a powerful scientific method (let alone as 'the' scientific method), for several reasons. One is that there are many kinds of BV + SR processes besides the biological family of evolutionary processes discovered by Darwin. Another is that the more powerful processes are highly domain-specific. It turns out that you cannot reliably or knowingly use a powerful method unless you already know a great deal

---

[3] L. J. Fogel (1999, p. 44), quoted by Jacob (2001, p. 3).

about the subject domain. So there remains this much truth in the Meno paradox. Roughly speaking, it takes knowledge to get knowledge, a claim that obviously raises 'ultimate origins' problems analogous to those faced by biological evolution. It also raises questions about how explicit that knowledge has to be and whether biological or mechanical processes can 'know' in this sense. I shall not be able to pursue such large questions in this chapter.

It was Darwin who first found the entrance to the vast space of BV + SR mechanisms and the new sort of control theory that accompanies them. In appreciating the potential of BV + SR to generate novel design, Darwin and Alfred Russel Wallace achieved the extraordinary but controversial insight that here is a process capable of producing more design (or at least apparent design) from less and hence a process that is genuinely creative. Still more controversial was Darwin's insight that, ironically, the only previously known way of explaining design—*the intelligent design model*—actually explains no truly *innovative* design at all; and that, hence, the *evolutionary model* replaces it as the only game in town! Late nineteenth-century Darwin sympathizers such as Thomas Henry Huxley and William James already realized, to some degree, that there is no reason to restrict the evolutionary process to biology and to Darwin's specific mechanisms—that bodies of knowledge claims and cultural practices may also be considered designs, designs that at one time were innovative and adaptive. Darwinian *biological* evolution is only one of many possible instances of BV + SR.

Now whether and how we can create more from less is precisely the knowledge problem of the *Meno*. Restated in our terms, the Darwinian claim is therefore that the intelligent design model does not really solve the Meno problem after all; only evolutionary models do. If this is correct, then, given the absence of viable alternatives, we can say that it is only within the last century and a half that we have been able to solve the Meno and hence learned the secret of inquiry, the key to innovation. Interestingly, even the young Darwin of 1838 had some notion of the relevance of his emerging ideas about evolution to the Meno problem. In response to a reference to Plato's doctrine of innate ideas owing to the soul's journey between lives, Darwin wrote in one of his notebooks 'read monkeys for preexistence (of the soul)'.[4]

In our time, Donald Campbell and several other investigators (see Note 1) have revived and defended a *universal* evolutionary or selection theory—the claim that, ultimately, *all* innovative design is produced by

one or another variation-plus-selection-plus-transmission process.[5] This is the aforementioned 'universal Darwinism'. Meanwhile, scientists such as Nils Jerne[6] and Macfarlane Burnet, founders of the 'clonal selection theory' of antibody formation in immunological theory have fruitfully applied generalized Darwinian models to areas of biology other than traditional evolutionary theory. More speculative applications include those of Gerald Edelman, author of an account of a semi-Darwinian brain development that he calls 'neural Darwinism', and William Calvin, with his clonal model of competing representations in the brain. Just as biological evolution can be considered a phylogenetic inquiry process or learning system, neural system development and the vertebrate immune system may be considered ontogenetic learning systems. Thus, evolutionary models are proving useful in developmental biology. Universal evolutionists extend the claim to learning theory and to all production of novel design.

Since the 1960s, founder John Holland and the many other contributors to the field of evolutionary computing have done more than anyone to extend the BV + SR methodology far beyond Darwinian evolutionary biology, to the extent that Darwinian biological evolution can now be represented as a single point or small region in a large space of BV + SR processes. In the past decade evolutionary computing has enjoyed explosive growth and is now employed in thousands of technical papers in dozens of disciplines every year.[7]

If the universal evolutionists are correct, then we have (or need) a new paradigm of innovation. For on this view *all* successful innovation, the emergence of all novel design, is the product of BV + SR. Not only is all adaptation design but also all design is a kind of evolutionary adaptation. This includes inquiry, where we need to think of problem solutions and other results as things that have evolved, grown, or emerged by a process of adaptation rather than as things that have simply been inferred by a logical process, whether deductive or inductive. In this sense the evolutionary paradigm supplants or transcends the old *logical*

---

[4] Desmond & Moore (1991, p. 263), quoted by Dennett (1995, p. 130).

[5] See the references at the end of the chapter. The BV + SR claim has other late-nineteenth-century sources that are not necessarily Darwinian, including Alexander Bain and Paul Souriau (Campbell 1960, 1974a). Also, by 'universal Darwinism', Dawkins (1976) originally referred to the Darwinian basis for life forms that may exist elsewhere in the universe.
[6] Interestingly, Jerne was led to his solution to the central problem of immunology by seeing it as similar to Plato's Meno problem.
[7] Key references for this paragraph include Campbell (1960, 1974a, 1974b), Rechenberg (1965), Holland (1992), Plotkin (1993), Dennett (1995), Cziko (1995, 2000), Edelman (1987), Schwefel (1995), and Calvin (1996).

paradigm of innovation.[8] It also replaces the *romantic paradigm*, which relies heavily upon processes that are both nonnatural and nonlogical.

In insisting that all innovation involves an element of trial and error, the BV + SR paradigm incorporates at least a modicum of luck or chance. It thereby claims that all innovation involves an element of serendipity, normally many instances of scarcely recognized micro-serendipity rather than one big, lucky break. The BV + SR paradigm is not deterministic, but neither does it depend solely upon sheer blind luck. The process is iterative and cumulative and therefore builds on its previously established platforms (Dawkins, 1986, ch. 3). Nor need it be directionless, as biological evolution is. After all, Darwin himself was much instructed by the artificial selection techniques practiced by animal and plant breeders. However, according to the new paradigm of innovation, it would take *considerable* luck to realize the goals of a research program precisely specified in advance and adhered to religiously. As Campbell (1974, 435ff) points out, serendipitous discoveries depend upon having multiple selection criteria, upon being flexible and opportunistic. Ironically, then, it is the old design paradigm (the rigid, intelligent design or rational planning model) that actually requires luck in large quantities. The irony is doubled by the fact that the need for large chunks of luck arises out of the very attempt to eliminate the need for luck through detailed rational planning.

What is Innovation?

It is impossible to define 'innovation' and 'creativity' precisely in a context-free manner. Clearly, what is innovative or creative for you may not be for your community, much less for the human race or the universe as a whole. The advanced inhabitants of Lagado or of the planet Zork may have achieved it already. Let us call 'the attribution problem' or 'the credit-assignment problem' the task of assigning credit for an innovation. Margaret Boden (1990, p. 32) distinguishes an individual's psychologically creative (P-creative) efforts, or personal firsts, from historically creative (H-creative) accomplishments, or historical firsts. (She limits the term to human history.) According to the naturalistic line that I am taking in this chapter, nothing is truly innovative if it is simply externally injected by an outside agency such as a teacher or even God. Good students are not necessarily creative students. They may 'get the point' without being able to improvise upon it. And if God gave Schrödinger and Heisenberg the laws of quantum

mechanics, then those two men are proportionately less creative, and their work is practically a form of plagiarism! (Recall the literal meaning of 'inspire'.) Similarly, simple instructive empiricism, according to which we can directly read the truth off Nature, diminishes attributions of innovation.

What about the apparently similar case in which someone or something arrives at a result by mechanically following an available recipe or method or algorithm? Isn't the result already implicitly contained in the method and so already implicit in the method-plus-user system? I would say so, but things become less clear when the methods are less specific and the algorithms complex. This latter issue somewhat resembles the question of whether you can get more out of a computer than you program into it. According to the 'Lady Lovelace objection', the computer deserves no credit for innovation.[9] I agree with Boden (1990, 6ff) and many others that you *can* get more out than you put in.

One issue is how we characterize what we put in. With simple, highly specific programs you do not get more than you put in, except, typically, more speed and accuracy. (These can make a real difference in performance, as the chess, checkers, and domino victories over human champions vividly demonstrate.) But with more general or complex programs the case is different. Here a distinction between potential and actual is necessary. From any set of statements or rules, an infinite number of logical consequences follow and are potentially recognizable by a rational cognizer. However, finite cognizers can actually recognize only a relatively small, finite set of these in any real-time situation. For this reason a purely deductive consequence can be *epistemically* novel, even though it is not *deductively* novel. Logically nonampliative inference can be epistemically ampliative. The computer, in exploring areas of consequence space that we have not explored, can produce something epistemically novel for us.

Similarly, by using a method we may be led to explore areas of search space previously unknown to us. Thus we cannot say that a method can produce nothing more than was explicitly built into it. Obviously, what its human designers built into it explicitly was already known and so cannot count as innovation. Here we encounter a tradeoff, to be further discussed below. Insofar as a method is both helpfully directive and warranted by domain knowledge, it relies upon knowledge already achieved in order to achieve a rather routine sort of problem-solving. And insofar as it is directive but unwarranted by domain knowledge, it is risky: it can lead us to major new results, but it will often fail badly. In this respect, such a method is hypothetical in character. So although I do not deny

---

[8] Logical inference often plays crucial roles in the BV + SR process. My point is that we should not essentially characterize innovation in terms of the old logical paradigm. There is a sense in which the BV + SR model also replaces the knowledge-based paradigm of artificial intelligence. See below and Koza et al. (1999).

[9] Ada Lovelace was the daughter of Lord Byron, who assisted Charles Babbage in the design of his analytical engine.

that methods can help us to innovate, the old view that methods are what produce discoveries is exactly backwards in my view. The methods routinely employed by the scientific community to solve already mastered problem domains are the final *result* of the discovery process, not the tool that produced the discoveries in the first place. A method is a final-stage streamlining of a trial-and-error process that was once full of false starts and blind alleys.

The credit-assignment problem can also be raised for the BV + SR model of innovation. Since BV + SR 'only' reorganizes and selects materials already present, with an occasional, exogenously caused chance mutation thrown in, does it really produce anything new? Isn't all the 'knowledge' already built into the system consisting of population-plus-environment? Well, yes and no. Given the power of the process to explore vast regions of design space efficiently, and given the fact that the original designs were very simple (for probabilistic reasons), there remains a lot of novel design to be 'found' or generated.[10] Still, to count as genuinely novel design, it has to be something that did not actually exist before.

In some sense, then, "genuine originality must be a form of creation *ex nihilo*" (Boden, 1990, p. 29). Boden rejects this view, on the ground that it requires a miraculous explanation. However, there is a qualified sense in which evolution does just this. New design *emerges* from something that was not design, or at least not *that* design. But, of course, not even evolution creates something from absolutely nothing. Rather, as many authors in this volume point out, it works by adapting mechanisms or designs or resources already available. New design genuinely emerges from old. Novel design is an *emergent* phenomenon. It is precisely characteristic of emergent phenomena that they manifest novel features although resulting from combinations of ingredients already present. They reduce to what was already there, except for their organizational pattern or design, and thus do not require appeal to any distinct sphere of metaphysical reality; yet they are also nonreductive in manifesting that novel design. Any satisfactory solution to the Meno problem must show how genuinely new knowl-

edge can emerge from available resources, including old knowledge.

In speaking of design, I follow Dennett's (1995) formulation of innovation as bringing new design into being. And so I have generalized the Meno problem as asking how it is possible that original design enters the world. Speaking of design does have the advantage that it embraces the arts and other human doings in addition to specifically scientific and technological innovation. However, I cannot claim more than a heuristic function for this move, since what counts as design in the relevant sense is also difficult to say. Clearly, not just any pattern counts as design in the relevant sense (as Dennett fully recognizes). Roughly speaking, to be a design, it must be adaptive to some context or purpose, it must display a degree of fitness.

Similarly, I agree with Robert Sternberg et al. (this volume) that creativity (or at least innovation) involves more than producing novelty. It must be useful novelty, in a suitably broad sense of 'useful'. To this degree, 'innovation', like 'discovery', is a success term, an achievement term. For me such success requires recognition (part of the larger-scale Meno problem!) that can only be conferred by a community of people who can authoritatively locate the innovation as a contribution to their project. However, there can be attempted innovations that fail because, although verbally recognized, they are not integrated into the practices of the community. So in this respect, 'innovation' is not as definitive an achievement term as 'discovery'.

Traditionally, credit assignment for innovation has been highly individualistic. I want to emphasize that social recognition and attribution are crucial, although they are not, of course, the whole story.[11] In my view, unlike that of many philosophers and psychologists, creativity and innovation tend to be highly distributed across communities, and this is especially true of deep innovation. At the frontier of research or of artistic production, no one can know the extent and magnitude of the ramifications (see below). A highly centralized

---

[10] Useful though it is, talk of search spaces and design spaces, as if all the possibilities already existed there waiting for us to discover them, can be misleadingly Platonic. It also threatens to reduce all discovery or recognition to the Columbus-discovers-a-new-continent model. And if we consider the knowledge to be already implicit in the population-plus-environment system, then one facet of the Meno problem becomes how to *transfer* this knowledge more explicitly to those members of the population that we call scientists (or artists or whatnot). See Section 5 for the Meno transfer and transformation problems.

[11] In short, I employ the term 'discovery' in a broad, non-realist sense that leaves room for social construction. For sophisticated accounts of the negotiation and construction of discovery attributions, see Brannigan (1981) and Simon Schaffer (1986, 1994), who point out that the attributions should not be naively accepted as accurate history of who did what. Thus Brannigan and other historical investigators have exploded the myth that Mendel was the neglected discoverer of modern genetics. More controversially, Thomas Kuhn (1978) has exploded the myth that Planck was the founder of quantum theory. I agree with the sociologists and social historians that social attribution is crucial, but I disagree insofar as they suggest that this is the whole story and that cognitive psychological accounts (for example) are beside the point.

cognitive process is not nearly efficient enough to account for the problem-formulating-and-solving successes we find in the sciences—and in the arts for that matter. Nor is innovation simply a matter of having new ideas, for inarticulate practices, institutions, social structures, and the like can also be creative. The gradual emergence of a market economy, from the Italian Renaissance on, was not the deliberate, conscious design of social planners. Not until hundreds of years later (Adam Smith in 1776) were some of the essential features even recognized.

According to the universal evolutionary paradigm, innovation does not flow directly either from logical inference alone or from the creative inspiration of a poetic individual alone, for it does not flow directly from anything. Although pieces of inquiry can be highly directed and the larger chunks directed somewhat more loosely, the process as a whole is quite indirect (but less indirect than biological evolution). No trial-and-error process in which the goals and problems themselves are initially only loosely defined can be very direct overall. And it is precisely characteristic of highly innovative work that the goals, standards, and problems themselves evolve and become sharpened during the course of the work.

So is trial and error essential to innovation or is it to be avoided? My answer in the remainder of this chapter will be Yes and No! According to universal Darwinism, yes, it is in some way essential. But no, this does not mean that all problem-solving and all productive practices are nothing more than blind, mechanical trial and error. My position is far from a simple reductivist one.

The Evolutionary Design vs. Intelligent Design Models, Human and Divine

As indicated, the leading alternative model, which has dominated the explanation of innovation until recently, is the intelligent design model, according to which deliberate, directed, pre-planned inquiry is both necessary and sufficient for innovation, aside from the occasional, improbable lucky hit. The intelligent design model has two well-known versions: the human-design model and the God-design model. The central premise, or principle, of intelligent design models is that *there can be no intelligent design without an intelligent designer; no genuine innovation, or at least no systematic innovation, without an intelligent innovator; no genuine design without a rational planner*. This idea is familiar from old, natural theological debates about the design of biological organisms and the creation of biological species. However, the premise holds equally for the extension of the theological argument from design to human innovation. After all, as William Paley's (1802) example of the watch found on the heath reminds us, the human intelligent design model itself is the source of the God-design model and of the intelligent design

model of innovation in general.[12] Paley himself considered biological organisms and specific organs such as the eye to be artifacts, artifacts constructed by the Great Artificer. The God-design model is the human-design model writ large.[13]

While many students of innovation today steer clear of divine inspiration and the like, the human-design model still underlies many accounts of discovery and innovation in science and technology as well as in the arts. According to its proponents, trial-and-error methods are typically brought in only as a last resort, when rational planning fails. Yet the human-design model leaves us with a very poor understanding of human innovation, for it is plausible only insofar as its alleged applications remain vaguely stated. We never get enough detail to see how the model could really work. Indeed, the more detail we get for a given case, the easier it is to see that the *pure* human-design model gets no purchase at all. By contrast, a progressive scientific understanding yields better-defined models and/or stronger empirical confirmation the more closely the problems are examined.

Ironically, if the evolutionists are correct, it is the BV + SR model that is fundamental and not a mere stopgap, last resort when all else fails. The claim by Campbell and a host of other evolutionists is not that rational, informed planning and directed inquiry are impossible, but rather that all of the knowledge that those activities presuppose was originally acquired by means of BV + SR processes, and that even such planning, insofar as it goes beyond already available design, must resort to trial and error. Fully planned activities are not in themselves innovative.

If we do treat *knowledge* as a kind of design (as we did in our extension of the Meno problem), then the intelligent design model is committed to the thesis that there is no epistemically successful design without a *knowledgeable* designer, no intelligent inquiry without an at least equally intelligent and informed inquirer. But if the designer, or the intelligent method employed by the designer, already contains the design, then it is difficult to see how this process can produce genuine innovation, or more knowledge from less. Stated more explicitly in terms of knowledge, the point is that if the intelligent designer must already know everything necessary, then the new knowledge is not a genuine

---

[12] Very briefly, Paley's argument was this. Suppose that in walking on the heath you come upon something that looks and works exactly like a watch. What is the probability that this object is the product of a chance coming together of the natural elements rather than the product of intelligent human design? Answer: negligible. Well, we may conceive of the human body from head to toe as a machine that is far more complex than the watch. So the probability of intelligent design is even more overwhelming in this case.

[13] My purpose is not to construct arguments against God's existence but to study the conditions that can produce genuine innovation.

innovation at all but simply a routine application of what the designer already knows. There is no genuine inquiry, for the process is impaled on the first horn of the Meno dilemma.

This point parallels the criticism leveled by evolutionists against the God hypothesis: that appeal to God cannot really explain design, for creationists are merely positing more design (infinitely more!) to explain less, rather than the other way around.[14] Theists may reply that it is possible that such a God does in fact exist and that God is self-explanatory or is beyond human understanding. I will not argue here against such a possibility but only point out that although God then provides an in-principle explanation of all novelty, we are far short of a precise model or explanation, since an infinitely powerful being can do *anything* (possible). To be able to explain everything so easily in these terms, without having to provide a precise mechanism, is really to explain nothing. We understand no more about how innovations can emerge than when we began. The theistic account replaces the mystery of creativity by a much bigger mystery.

In its strong form, then, as a complete solution to the Meno problem, the intelligent design model turns out to be no solution at all, for it holds that general human intelligence or human reason already contains all possible human knowledge, at least potentially or implicitly, a view reminiscent of Plato's own solution to the Meno problem (see Section 5). One reason why this point is little noted is that traditional epistemology (theory of knowledge) has been conservative and retrospective, focusing on the justification of knowledge claims already in hand rather than upon the process of gaining serious new knowledge candidates at the frontier.[15]

Surprisingly, common accounts of scientific method are committed to the intelligent design model. For the method allegedly contains the general principles that define scientific intelligence—principles sufficient to mine scientific discoveries from incoming data streams and that, in this sense, already contain those discoveries. But how could a powerful method of discovery be warranted in advance?

Thus we encounter an exasperating clash of views. Just when we seem to have discovered the key to all innovation, including the growth of knowledge, and hope to methodize this discovery as the only known solution to the Meno problem, that problem itself appears to deny us the ability to methodize or mechanize the secret. For the method itself would already contain the design, at least potentially, and hence would merely instantiate the old thesis of no design without a designer and, as a result, would impale itself upon the first horn of the Meno dilemma. This clash will be spelled out more fully below.

The thesis that BV + SR lies behind all genuine innovation has always met considerable resistance. Because it threatens traditional views about human nature, it has had to fight once again the old battles once faced by physicists and evolutionary biologists about the reality and explanatory power of *chance*, the possibility of genuinely statistical-probabilistic *explanation* (Lestienne, 1993). For example, we meet again the objection that appeal to chance only masks our ignorance, that if we could locate the underlying causes or (better yet) the logic or the specific inspiration, then we could discover the genuine springs of innovation. Unlike intelligent design, it is alleged, chance disappears upon closer inspection. Chance and design appear to be mutually exclusive. In my view, this objection has things exactly backwards.

Somewhat as God was invoked in the nineteenth century to account for the stability of social and physical statistics, many writers are tempted to invoke a special, non-natural intuitive power to explain how we humans hit on as many successful ideas as we do.[16] Likewise, there is a tendency to dismiss as mere bothersome noise the variation that we find in scientists' and technologists' beliefs and practices, their numerous misunderstandings[17] of one another's work, and the historical contingencies that often affect the direction of their work. This noise allegedly masks the true process of scientific innovation and is something

---

[14] See Dennett (1995, chap. 3) for a version of this objection.

[15] See Dewey (1929, chap. 7). In virtually all of his major works (e.g. 1962), Popper has featured the problem of the growth of knowledge as his central problem, while denying that there is a method of innovation. For Popper the absence of a logic of discovery implies that the search for new theories must be, to some degree, blind. Popper explicitly developed a BV + SR methodology (see Campbell, 1974). I believe that a BV + SR model is also implicit in Thomas Kuhn's account of normal and revolutionary science. In any case, the preface to Kuhn (1962) indicates that he will not employ the discovery-justification distinction invidiously and that much of his work will concern discovery. See Nickles (2003) for details.

[16] These remarks should not be misconstrued. I am not saying that all appeal to chance, including the BV + SR account of innovation, essentially depends upon an underlying physical or psychological indeterminism.

[17] These misunderstandings often result from a scientist's reading the work of other scientists whiggishly, from the standpoint of his or her *own* problem. Thus Kuhn (1978) argues that Einstein and Ehrenfest in 1905 misread Planck's 1900 works as addressing *their* problem (which Ehrenfest dubbed 'the ultraviolet catastrophe') rather than the Clausius–Boltzmann problems that Planck was actually addressing. Brannigan (1981) makes the same claim about the reception of Mendel's work. Dawkins (1986, p. 130) remarks: "Insofar as I can claim to have had any original scientific ideas, these have sometimes been misunderstandings, or misreadings, of other people's ideas". In this vein, Alfred North Whitehead supposedly quipped: "Everything has been said before, but not necessarily by someone who knew they were saying it".

to be eliminated whenever possible—and certainly to be eliminated from our methodological accounts of scientific research. But according to universal evolutionism, this is again the wrong attitude, for variation is essential to the process (Edelman, 1987). Far from being ideal, a noiseless system would be completely uncreative.

Noise also enters the system at the selection and transmission stages of the evolutionary BV + SR process, so we should not see these as strictly temporal stages in which the variants are produced entirely antecedently to stages two and three. Whether someone's work catches hold depends upon audience reception and transmission (reproduction). While scientific recognition systems canonize a few 'greats', the actual authorship of deep developments is typically more distributed and more anonymous. Here the 'author function' largely disappears and can be regarded as a post-hoc social construction (Schaffer, 1986, 1994).

Insofar as the world at bottom is genuinely stochastic, the creationist 'model' is stretched even thinner, for then not even an omniscient being can create a specific world by setting up a deterministic causal system and letting its history unfold. Rather, such a being would have to survey in advance all possible probabilistic worlds, history and all, and then choose a particular world (whatever that could mean). For in a genuinely stochastic universe, as in a deterministic universe, not even an omniscient being is allowed to interfere to keep it on track. Furthermore, a being that is thus omniscient cannot really be creative in the sense of generating novel design, since it *already* possesses all possible design. Making the world is 'simply' applying what it already knows.

In what follows I shall unpack and defend a version of Campbell's BV + SR claim that somewhat reconciles the evolutionary model and the human-design model. Much of the debate has been spoiled by a tiresome polarization between two extremum positions, neither of which is held by any responsible writer today. At one extreme we have the view that scientific discovery (say) is determined by application of 'the' scientific method and leaves no room for chance. Such a view of discovery comes close to being oxymoronic. Meanwhile, advocates of evolutionary accounts are often characterized as committed to the opposite extreme, as saying that there is nothing more to creative work than totally random or mechanical trial and error.

Campbell and his precursors already insisted that the BV + SR thesis does not at all deny the existence of rational planning or of problem-solving or learning algorithms of other kinds, only that these are not yet known or justified at the frontier of research where genuine innovation occurs. Innovation has to do with pushing back the frontier, not with the routine application of known rules. While there is a grain of truth to the claim that trial and error is employed as a 'last

ditch' effort when intelligent design runs out of intelligence, the point is that, at the frontier, it always *does* run out. Since that knowledge itself was once a product of frontier research, the grand irony is that BV + SR processes ultimately underlie the knowledge presupposed by intelligent design. In this logical and empirical respect, trial and error is a 'first ditch' effort rather than last ditch.

While I defend Campbell's BV + SR thesis,[18] I do not share his tendency to conclude that, because they are blind, BV + SR processes are the very antithesis of method; and that, therefore, there is no scientific method or method of innovation in any field. In other words, I adopt a wider conception of method than the traditional one that is linked to the intelligent design model. The BV + SR thesis may thus have positive methodological significance after all.

Campbell is correct if he only means to say that any particular BV + SR learning rule is too weak to be considered an efficient *general* method of innovation, but he is mistaken if he denies that they are methods at all. For in fact, BV + SR methods can be powerful and creative search engines when tailored appropriately to a domain. Campbell himself deeply appreciated the creative power of evolutionary processes, but he perhaps underestimated their methodological possibilities. After all, Campbell's famous paper, 'Evolutionary Epistemology', was written in praise of Karl Popper, who adamantly rejected the possibility of methods of discovery. The same paper quotes Souriau's antimethodological stance at length, with seeming approval (Campbell, 1974a, 428ff). The thesis of Souriau (1881) is 'le principe de l'invention est le hazard'. But in fact, the emerging field of evolutionary computing generalizes and methodizes the Darwinian insight.

As Campbell knew, there is also a certain vagueness in speaking so broadly of trial and error or even of BV + SR in the same breath as evolution. There are many such processes, ranging from 'random' guessing to highly directed search, that are not evolutionary in the full-blooded sense of selection from a *population* of competing variants, a selection that in turn will breed the next generation of the population. Consider a few of the many variations on the trial-and-error idea. A one-trial, all-or-nothing search clearly is not evolutionary. (E.g. you examine the first 28 characters formed by a monkey at a typewriter and 'select and retain' this string if and only if it perfectly matches 'Methinks it is like a weasel'.) Nor is an iterated version of this kind of search evolutionary (e.g. you give one monkey a thousand chances, or a thousand monkeys one chance

---

[18] Campbell himself usually wrote BVSR instead of BV + SR. I shall retain the 'plus' to distinguish my version from his.

each). However, if selection of partial matches is allowed and successful matches are allowed to accumulate,[19] then we approach the idea of an evolutionary process (albeit one based on goal-directed artificial selection rather than natural selection), even though we still have one trial at a time rather than a population of simultaneously competing individuals. For even in this simple sort of slow, serial evolution, success is allowed to 'breed true' into the next generation, while the incorrect parts are varied. Also, depending on how it is set up, a process that involves a population of competing and potentially crossbreeding individuals (trials) can amount to parallel processing.

This discussion makes clear that only a small minority of BV + SR processes are completely blind, let alone random in the above sense. Not even biological evolution is nearly random in this sense, because it involves: (a) highly constrained variation (the offspring of rabbits resemble their parents, not to mention other rabbits); (b) including nonrandom mutation and cross-over operations; (c) an iterated BV + SR process rather than a one-shot chance-arrangement of materials; and so on. In the field of evolutionary computation, specific choices of codings for individuals, choice of search operators, and of parameter values all represent departures from purely random search. In general, however, the weaker our domain knowledge, the weaker and less constrained (hence the more blind) the search procedures available to us. I shall develop this point below.

What then of the metaphor of monkeys at typewriters to describe the bare baseline of inquiry, the complete absence of method or direction? In the absence of cumulative selection, monkeys at typewriters are certainly inefficient producers of interesting novelty, but even here we could do worse. For example, a student who uses the 'copy from my neighbor' method of taking a true–false test but who gets the questions one number out of order may do much worse than chance!

Is the innovation process *convergent* or *divergent* according to the BV + SR model? The answer to this highly ambiguous old question is, Both! The production of variants is: in a sense, divergent, while the selection process could be said to be convergent. However, the question is misleading.[20] The degree of divergence and convergence will depend upon how tightly constrained the processes are. In biological

reproduction, for example, the production of variants is highly constrained: they result from such normal factors as sexual mating, noise in the system (copy errors and the like), and the occasional cosmic ray. And the selection process is convergent in weeding out the vast majority of variants that stray very far from the norm. As Campbell (1974, Section 4) points out, there is a tension between variation and selection. In effect, large variants are not recognized by the community-plus-niche system. Overall, this makes biological evolution a pretty conservative process in the short run. This tension corresponds to Thomas Kuhn's 'essential tension' between innovation and tradition as manifested in normal science. He describes normal science as a convergent process even though it is a puzzle-solving process and, as such, involves the production of variant attempts to solve nonroutine puzzles (Nickles, 2003).

Both the Meno problem, with its two horns, and the BV + SR model predict that there will always be an essential tension between innovation and tradition. Innovation implies change, but changes that are too great will not be recognized as a contribution to *that* enterprise. In the case of human inquiry, decisions as to whether to include any significant change are socially negotiated (Schaffer, 1994). Is it art? Is it good science? Is it good solid-state physics? Does it really solve the original problem or does it change the subject?

The Meno Problem Elaborated

What was Plato's own solution to the Meno paradox? Plato's Socrates concludes that learning is really only 'recollection' of knowledge already gained by souls between lives, when they supposedly enjoyed unencumbered epistemic access to reality, which knowledge was rendered subconscious by the rigors of childbirth. Thus learning, successful inquiry, is really only a matter of transforming this implicit, subconscious knowledge into explicit knowledge. That is, for Plato the only possible kind of inquiry is conversion of knowledge from one form to another. This amounts to a kind of conservation principle: the amount of knowledge (in all its forms) possessed by embodied human beings is a conserved quantity in the sense that normal human inquiry cannot increase it. (Notice that the God model of design also conserves knowledge at the cosmic level, since an omniscient God already possesses all knowledge in advance. Nothing can be gained or lost.) Plato's solution begs the Meno question, of course, for he fails to explain exactly how it is possible for souls to learn between lives. But his solution does contain grains of truth.

The problem of whether and how knowledge can be converted from one form to another is extremely important. Many researchers today are investigating the relations of various forms of knowledge representation, knowledge embodiment, and knowledge

---

[19] Dawkins (1986, ch. 3), who gave the weasel example, convincingly demonstrates the power of accumulative BV + SR.

[20] If a research result decisively favors one theory or practice over another and thus reduces the set of competitors, is this convergence? Is it still convergence if the eliminated theory is the reigning paradigm?

engineering, including habit formation,[21] tacit knowledge, skills, procedural knowledge (knowing-how) vs. declarative, propositional knowledge (knowing-that), how items in short-term memory get fixed in long-term memory, how ignorance is transformed into a good problem, and so on.[22] Let us term this the *epistemic conversion* or *epistemic transformation* form of the Meno problem.

There are also fundamental problems in understanding the 'transfer' of knowledge from person to person (e.g. teacher–pupil), person to machine, and so on. This is *the epistemic transfer problem*. After all, the central question of the *Meno* is whether virtue can be taught by one person to another. In this case the recipient of the transfer clearly learns, but the system as a whole does not. Here, for example, we face the problem of understanding the transfer of scientific, technological, and artistic knowledge from one generation to the next, for it is by means of reproducing themselves in this way that disciplinary enterprises sustain themselves.

A third important Meno problem remains: how to acquire genuinely new knowledge (previously unknown to anyone or anything who could otherwise transfer it to you). Let us call this *the problem of epistemic innovation*.[23] This tripartite division of Meno problems is rather crude, but it is sufficient for our purposes. The three problems overlap, and the processes involved in solving any of them can be extremely complex, requiring much refinement of our ordinary talk about skills, habits, and the like.[24]

Even Plato's original formulation of the Meno problem suggests (correctly, in my view) that explicit inquiry is a kind of *search* and that the Meno problem is basically a problem of *recognition* (or construc-

tion),[25] namely, how to recognize and select the (or a) correct answer, or at least a somewhat promising answer, out of a noisy background of zillions of potential alternatives.[26] Plato's text also raises the crucial issue of the relevance of old knowledge to the search for new knowledge. It is on this second problem that Plato begs the question with his epistemic transmigration story. Within a normal human life, he seems to conclude, there is no answer to the Meno. In this life, genuine inquiry, and hence genuine innovation, are indeed impossible, and knowledge is conserved.

There are other kinds of learning discussed in the educational, psychological, and artificial intelligence (AI) literature. For example, improving one's problem-solving skills certainly counts as one kind of learning even when the capacity to solve *new* kinds of problems does not increase. Much of AI has been concerned with developing problem-solving systems that operate more efficiently.[27] However, my focus here will remain on epistemic innovation as described above. In this case the Meno problem is 'How is innovation possible at all?' and not 'What is the most efficient or optimal means to achieve it?'

Many nonnatural accounts of cognition suppose that we possess a kind of divine spark in the form of human intellect or a faculty of reason. But invoking a theological or supernatural account of knowing will not help here. The limiting case is an omniscient god, but, as noted above, such a being cannot inquire at all, given the first horn of the dilemma, for it already knows everything. The God model of knowing is not helpful to theory of inquiry—or even possible. Nor, for the same reason, is the intelligent design 'theory' of genuinely innovative design a viable model.

Ditto for the historical attempts to evade the paradox by positing special forms of human knowing such as the intellectual intuition of the rationalist philosophers, clairvoyance, precognition, or some other sort of

---

[21] In effect, Aristotle solved the problem of the Meno by means of his theory of potentiality, applied to habit formation.

[22] See Karmiloff-Smith (1992) and Root-Bernstein's chapters in this Handbook for examples.

[23] I could have kept Popper's old label, "the problem of the growth of knowledge". There are other ways to parse the Meno problems. One could argue, I suppose, that even learning new things about actual natural phenomena is a kind of information-transfer problem (as simple instructionists from Aristotle to the British empiricists sometimes seem to have thought), whereas the full innovation problem concerns the creation of designs that previously were no part of the learning system (agent-plus-object) at all. Instructionist theories are clearly incapable of explaining this sort of innovation.

[24] Even simple verbal communication of an everyday sort ("I'll meet you at the clock in Grand Central Station at the same time tomorrow") is an extremely complex affair. See, Brandom (1994) for one recent account of how semantic content is constituted.

---

[25] For simplicity I shall characterize the Meno problem and the BV + SR selection problem in terms of recognition, which means that the system in question registers sufficient agreement with fitness criteria. The term 'recognition' can be contentious, for like the term 'discovery' vs. 'construction' or 'invention', it can beg questions of realism. In some contexts at least we do not discover (by recognizing) a solution just waiting there to be found; rather, we construct a solution. Moreover, whether discovered or constructed, the solution need not be optimal, as Simon (1947) emphasized in his discussion of satisficing.

[26] I do not mean that a search always must be literal, deliberate, conscious search for something predesignated, for I want to extend the idea of evolutionary biology to subconceptual human practices.

[27] See Simon & Lea (1979) and Dietterich (1986) for 'classical' discussions of these issues.

prescence.[28] Most of these solutions are as question begging as Plato's own response to the problem of innovation. Some of them, e.g. prescience, again impale themselves upon the first horn of the dilemma. Typically, the 'mechanism' by which the extraordinary faculty works remains unspecified. Notice that if 'intuition' and 'inspiration' (whatever they might be)[29] are analogous to ordinary vision, then they, too, involve an element of blind search. As Campbell has shown convincingly, even systematic visual scanning, e.g. by a creature that may be both predator and prey, is blind in the specific sense that the animal does not know in advance whether predator or prey will show up at all, or where and in what form, in the region scanned.

The more promising attempts to solve the Meno problem all 'go between the horns', thereby dismissing the argument as a false dilemma. The basic idea is that there are grades of knowledge. Our epistemic situation is not that of completely explicit, justified knowledge or else nothing at all. Rather, we possess inferior grades of knowledge or quasi-knowledge that, through inquiry, can be upgraded or converted into a more usable form. Hence we have the possibility of search that impales itself on neither horn of the dilemma. We can search because, although we do already have the answer in one respect, in another respect we do not. Moreover, we can hope to recognize the thing sought, should we stumble upon it, since we already possess some knowledge of it, explicit or tacit. Despite the fatal difficulties, there is something right in Plato's account.

The power of the iterative, 'multi-pass' BV + SR solution to the problem resides in the fact that recognition need not be an all-or-nothing affair and in the fact that BV + SR processes can be massively parallel, with many agents exploring vast regions of the search space rather than a single agent, who is likely to get stuck on a local fitness peak. (This second point does concern efficiency but more than that, since the parallel nature of the process helps BV + SR avoid traditional difficulties of hill-climbing heuristics, such as getting stuck on a local maximum.) BV + SR can parlay even the coarsest, partial recognition or 'fit' by one or more candidate solutions into a series of better

solutions. That is, low-grade recognition by one or more candidate solutions or 'variants' triggers a new generation of variants parented by those first generation variants of above-average fitness. If there is a pattern or solution to be found or constructed, it is likely that some second-generation variants will achieve still higher fitness, and merit their selection for production of the third generation of variants; and so on. Stated somewhat differently, the earlier populations of variants provide a widely dispersed set of explorers of the fitness landscape. Those found a little higher on the hills or peaks of the landscape will have a higher probability of reproducing. As the process iterates from generation to generation, the exploratory focus on the more promising regions of the search space increases exponentially. According to the clonal selection model, this is basically the manner in which the vertebrate immune system can muster its limited resources to 'recognize' any of a vast domain of invading antigens and then can proceed to sharpen its criteria of recognition.[30]

It is this ability to utilize low-grade information as a pointer that makes iterative BV + SR *heuristically* powerful. The selection criteria at each stage function as a mechanism for heuristic appraisal, that is, evaluation of the promise of pursuing a line of investigation (Nickles, 1989). (Here we humans have an advantage over nature in that we can look ahead to longer-term goals and do not always have to worry about the immediate survivability of the present model in competition with more robust but less promising ones.) Thus heuristic appraisal does not reduce to standard confirmation theory as discussed by philosophers. Scientific survivability depends upon far more than empirical track record to date.

One implication of this gradualist solution to the Meno problem of recognition is that 'micro-aha' recognition experiences will vastly outnumber the 'macro-aha' experiences widely reported in the popular romantic literature on innovation (cf. Gruber, 1974). In fact, most of the microrecognition surely takes place at the subconscious level.

The Meno problem can also be parsed into two (or more) subproblems in a different way. First, assuming that a still unanswered question or unsolved problem has already been posed (which itself requires a sophisticated sort of recognition or construction), there is *the starting problem*, the problem of knowing where and how to search for an answer. Second, there is the *stopping problem*, the problem of knowing whether and when you have found a suitable answer, or at least

---

[28] It is remarkable how many people take seriously the 'predictions' of Nostradamus, for example. There are problems even if you seem to receive direct instruction from a crystal ball or the voice of God. How do you determine whether this crystal ball is veridical? And even if it is, do you have to use it as a scanning device? How do you know what you are seeing? In the other case, how do you know that it is the voice of God and, if so, what God's message means?

[29] Here I am speaking of intuition and inspiration of a transcendental kind. I have no objection to speaking of intuition and inspiration of a down-to-earth, naturalistic sort. Indeed, I believe that culturally acquired intuition and the intuitive flow we can acquire through skilled practice are important phenomena.

---

[30] Nowadays, the clonal selection model is no longer the last word in immunology. See, for example, Tauber (1994). However, later developments do not undermine my point.

an answer worth pursuing in order to achieve a still better fit.[31]

The process of solving at least the first subproblem would seem to require a degree of blind search in the form of trial and error. For even if you already possess some knowledge of where to look for the answer, you must still search for it once you get there; and, by definition, search involves an element of blindness in Campbell's sense. For if no search at all were required, then you would already have the answer. Of course, if you know where to search, this will be highly constrained search. Your search procedure will be 'biased' (in the good sense in favorable cases) by knowledge you already possess about which small region of a potentially much larger search space you need to examine. Of course, such biases can also be unfavorable in cases where an unexpected sort of answer is superior.

The second subproblem is just the problem of recognition. A third subproblem would then be what to do with the item(s) now recognized in terms of using them to conduct a more refined search, that is, how to conduct the iteration.

Taken together, these processes amount to a blind generate and test procedure. Thus the Meno problem itself points toward its own solution, a process of BV + SR. Only resistance to the idea of blind trial and error can explain why it took so long to recognize this path to solution. Yet this is not so surprising when we recall that, until Darwin, there seemed no way other than intelligent design to explain any significant design innovation, including the epistemological design that constitutes a body of knowledge.

Search Operations: Generality and Efficiency

A look at developments in artificial intelligence (AI) over the past 50 years will reveal the importance of domain knowledge as a constraint on efficient evolutionary computation.

It has taken the AI community a long time to appreciate the need for, and power of, evolutionary adaptation as a search process.

The breakthrough results that got AI rolling were the Logic Theorist and General Problem Solver of Allen Newell and Herbert Simon, developed in the 1950s and 1960s. Their premise was the then-plausible idea that human reason or intelligence resides in relatively few general principles, namely, logical and heuristic princi-

ples that would do for problem-solving what the laws of mechanics do for physics. But despite the historical, revolutionary importance of this approach, it turned out not to have the problem-solving power expected.[32]

From the late 1960s on, prominent members of the AI community promulgated the informal truth that completely general, a priori, content-neutral, 'logical' problem-solving methods are inefficient at best and not very innovative, that inquiry can be highly focused only insofar as we already possess substantial knowledge of the domain in question.[33] It takes domain-specific knowledge to solve domain-specific problems (and thus to gain interesting knowledge) with any efficiency. As Edward Feigenbaum, a former student of Simon and one of the most enthusiastic leaders of the new movement once put it:

> There is a kind of 'law of nature' operating that relates problem-solving generality (breadth of applicability) inversely to power (solution successes, efficiency, etc.) and power directly to specificity (task-specific information).[34]

Newell and Simon already recognized, of course, that the more highly constrained a problem could be, the more readily it could be solved. As Newell (1969) remarked:

> Evidently there is an inverse relationship between the generality of a method and its power. Each added condition in the problem statement is one more item that can be exploited in finding the solution, hence in increasing the power.

But the new knowledge-based computation movement went considerably further in the direction of domain-specific knowledge by incorporating large quantities of domain-specific knowledge in the program itself and not only in the statement of the problem. The basic idea is that intelligence is a function of domain knowledge, not of general logic and heuristics alone. Thus we get a new paradigm of innovation: innovation flows from specific knowledge, not from logic or reason. It is sometimes quipped that the ratio of A to I in AI is very high. Be that as it may, it is surely an important insight that knowledge is artificial, something constructed (in various ways and to various degrees), and hence something that can, in principle, be engineered.

The lesson that it takes knowledge to get new knowledge at all efficiently was already implicit in

[31] Gigerenzer & Todd (1999) divide heuristics into search rules (what I am calling starting rules), stopping rules that determine when it is no longer cost-effective to continue searching, and decision rules that select an action decision among those options available at stopping time. My account is simplified. I also make no attempt here to extend this 'knowledge' and 'search' talk to biological processes, although I believe that it is fruitful to do so. In a sense nature solves problems that it does not pose.

[32] For their masterful summary of their previous work, see Newell & Simon (1972). For a fairly comprehensive 'insider' history of AI through the early 1990s, see Crevier (1993).

[33] Compare the rejection of traditional syllogistic logic as sterile because it is not ampliative, by Bacon (1620), Descartes (1637), and many others. Critics make the same point more generally against the rational, pre-planned design strategy of the human-design model.

[34] Feigenbaum (1969), repeated in Feigenbaum et al. (1971, p. 167).

Plato's treatment of the Meno paradox and is readily apparent in the history of any branch of science or mathematics—indeed, in the history of any major constructive endeavor, anything that results in 'design'.[35] The short history of AI merely recapitulates the history of science in this respect. Scientists themselves have recognized this characteristic of scientific inquiry. In his 1901 Presidential Address to the British Association, Lord Kelvin remarked:

> Scientific wealth tends to accumulate according to the law of compound interest. Every addition to knowledge of the properties of matter supplies the naturalist with new instrumental means for discovering and interpreting phenomena of nature, which in their own turn afford foundations for fresh generalizations.[36]

Insofar as it is correct (and of course there are exceptions, as when all the major problems in a field have been solved and sterility sets in), the claim that problem-solving power depends upon appropriate domain knowledge has major implications for the debate about scientific method. For the idea of a completely general, content-neutral scientific method becomes oxymoronic. On the one hand, such a method would be too strong—already containing potentially any number of possible scientific discoveries across many fields. On the other hand, it would be such a weak search tool as to be practically useless.[37] The traditional idea of a general scientific method harbors a greedy desire for a one-step solution to the problem of the growth of knowledge, namely: Follow my method!

Incidentally, since the mid-nineteenth century, the most widely touted general method of science is the so-called hypothetico-deductive (H-D) method. This is the method of hypothesis and test, and it can be regarded as a slow BV + SR process: the H-D method typically considers hypotheses serially and in isolation instead of in parallel, as members of a population of simultaneously competing hypotheses.[38] Its complete, content-neutral generality renders it incapable of providing guidance to hypothesis formation in any particular case. But in specific domains in which we already possess knowledge, the process can be more directed. One way in which it can be more directed is by considering variants of a previous hypothesis that showed some promise. So it is better to regard the general H-D method as a high-level schema that includes domain- or problem-specific H-D methods as instances rather than as a specific method in itself. There are various H-D methods, depending on the modes of variation, selection, and retention and, specifically, the ways of modifying previous hypotheses.

Returning now to AI: the leading implementation of knowledge-based computation was and is that of 'expert systems'. The idea is to elicit the domain knowledge of human experts and then to incorporate this content into the problem-solving program, typically in the form of rules (e.g. if–then production rules). At first the possibilities for expert systems seemed unbounded, but progress on this front, while genuine, was also ultimately disappointing. Knowledge did not compound in anything close to Kelvin's manner. One difficulty was that the more knowledge that was incorporated into expert systems, the slower they ran.[39] Another was that of transferring problem-solving expertise (domain knowledge-how as well as knowledge-that) from human expert to machine. Among other things, this obstacle suggested that explicit rules might be the wrong way to represent most expert knowledge. So here we have both the epistemic transfer and epistemic conversion forms of the Meno problem that I distinguished above.

More serious for our study of innovation is that standard expert systems do not happily solve the innovation problem either, for they turn out to be innovative only to a limited degree. They can be

---

[35] The universal evolution thesis applies to any process that produces novel design, whether or not that process is conceptual or subconceptual, conscious or subconscious or nonconscious.

[36] Kelvin (1901, p. 114). Gigerenzer (1994) adds the reverse heuristic, which he calls 'the tools-to-theories heuristic'.

[37] I like to illustrate this point with my 'crowbar' or 'pry bar' model of method. A method is a tool that we use to investigate the world. Consider a crowbar as a representative tool. Now a crowbar is useful only insofar as it fits the world at one end and the human hand at the other. If the world were tomato soup or if we were fish, a crowbar would be useless. Similarly for methods in general: they are useful only insofar as they fit the world at one 'end' and human capabilities at the other. A good method cannot be totally out of kilter with the way the world is, and it must also sufficiently fit human capacities (which may be considered another part of the world). Thus we cannot regard usable or successful methods as content-neutral anymore than hypotheses are.

[38] Another irony! Despite the historical resistance to BV + SR and to any suggestion that chance, luck, or serendipity could be an element of method, the dominant, H-D method itself turns out to be a slow BV + SR procedure. Proposals such as Chamberlain's (1897) 'method of multiple working hypotheses' and Paul Feyerabend's (1975, ch. 3; 1981, pp. 104ff and 139ff) theory proliferation methodology in effect turn the process into a fully populational one, although the populations will usually be rather small.

[39] Kevin Kelly (1994, p. 295) quotes Danny Hillis et al. as saying:

> The more knowledge you gave them, the slower computers got. Yet with a person, the more knowledge you give him, the faster he gets. So we were in this paradox that if you tried to make computers smarter, they got stupider. . . .
>
> There are only two ways we know of to make extremely complicated things. One is by engineering, and the other is evolution. And of the two, evolution will make the more complex.

immensely valuable, of course, in coordinating incoming information or queries with large databases, e.g. to give faster and more reliable medical diagnoses than human experts. But from the standpoint of genuine innovation, the popular old saw that a computer contains only that knowledge deliberately programmed into it remained too close to the truth in this case (Crevier, 1993, ch. 8). Even a perfect lateral transfer of routine problem-solving ability from human to machine would not necessarily endow the machine with significant creative powers, so there would be no overall gain in knowledge or creativity by the human–machine system. (Compare the knowledge transfer from professor to dependable but uncreative student.) Of course, once a machine with greater speed became available, that might give it a higher productivity and hence more creative 'hits' than the human investigator whom it simulates; and hit rate is one measure of creative potential (see Simonton, this volume).

In sum, knowledge-based computation of the traditional, expert-systems type runs into some practical and theoretical difficulties with the first and second Meno problem (the epistemic conversion and transfer problems) and into major difficulties with the third Meno problem (the innovation problem). Evolutionary epistemologists, including evolutionary computational theorists, believe that their approach will have more success.[40] Interestingly, some of them tout their approach as a new, general paradigm that does *not* depend heavily upon domain-specific knowledge (Koza et al., 1999).

The 'No Free Lunch' (NFL) Theorems

Recently, the above knowledge-based insights have received formal backing. David Wolpert and William Macready (1997 and elsewhere) have proved a series of seemingly fundamental theorems, commonly called the 'No Free Lunch' theorems (NFL theorems), according to which there is no one, best, general, problem-solving algorithm. The proofs are highly technical results showing, basically, that the average performance, for any relevant performance measure, is the same for any two algorithms, when averaged over all cost functions and the space of all problems, or, as we might say, all inductive learning problems, in all possible problem domains or possible worlds.

Although these results are still undergoing critical evaluation and interpretation, they are not totally surprising, certainly not to knowledge-based computation advocates and to those philosophers of science

who have noticed the domain specificity of powerful problem-solving methods and research techniques in the history of the various sciences. In fact, David Hume himself (1748, Section IV) vaguely anticipated such results when he wrote (among many similar things) that Adam, upon coming afresh into the world with full powers of reasoning but without any experience, would have no knowledge of anything in nature. In particular, Adam would be able to draw no inferences, either logical or psychological, about what would happen next. Hume's argument was that, absent empirical knowledge of what the world is like, reason is powerless to provide an inductive projection rule that is either correct, or the best possible, or better than average, or even better than some specified rule. That is, there is no reason to think that any rule is best or even better than some other rule, when averaged across all possible worlds. Contrary to the rationalists, pure reason can teach us nothing about empirical reality. Stated differently still, the point is that, under a veil of complete ignorance about which world we inhabit, or of the problem domain in which we are now working, we have no reason to believe that any particular induction rule will dominate all others, or, indeed, *any* others. Nor is there any a priori justification for projecting any perceived pattern in a given data set. In this situation, one can only proceed by adopting a (so far unjustified) learning bias and seeing how well it works. Readers of Ludwig Wittgenstein's *Tractatus* or Rudolf Carnap's *Continuum of Inductive Methods* will come away with similar intuitions in sharpened form.[41]

It is these intuitions that the NFL theorems formalize and prove. In fact, Wolpert himself quotes Hume's *Treatise* as the motto for Wolpert (1996), where he broadly characterizes the NFL theorems as follows:

(T)he implication of the NFL theorems that there is no such thing as a general-purpose learning algorithm that works optimally . . . is not too surprising. However, even if one already believed this implication, one might still have presumed that there are algorithms that usually do well and those that usually do poorly, and that one could perhaps choose two algorithms so that the first algorithm is usually

---

[40] Another approach was/is to train an inductive learning system on sets of examples rather than to try to extract rules from human experts. This method can be used with or without evolutionary algorithms. For a discussion of 'Feigenbaum's bottleneck' in philosophical context, see Gillies (1996, pp. 25ff).

[41] For an introduction to these ideas, see Carnap (1950, 1952) and Salmon (1966, pp. 70ff), on Carnap's use of state descriptions and structure descriptions in possible worlds. Assignment of equal weights to state descriptions corresponds to a world in which learning from experience is not possible. Unfortunately, Carnap's logical conception of probability required a priori choices of weights. See also Mitchell's insightful paper (1990), originally written in 1980. The Carnap sort of treatment also makes the issue simply one of inductive connectivity among phenomena, whereas it is also one of metaphysics—the nature of the underlying entities and processes that constitute a domain.

superior to the second. The NFL theorems show that this is not the case (Wolpert, 1996, p. 1362).

What implications, if any, do the NFL theorems have for our topic? Their interpretation is controversial, as some commentators deny that they seriously affect inquiry in our world. Here is the significance that I read into them, as a series of overlapping claims.

(1) 'No free lunch' suggests that something is conserved, that you can't get something for nothing, that it takes domain knowledge in order to warrant using any inductive rule with confidence, much less to establish that it is superior to another rule. This means empirical domain knowledge in the case of learning about the natural world. One way in which to state this idea is that the inductive success of any given rule will be zero when averaged over all possible learning situations. It will fail as often as it succeeds. C. Schaffer (1994) and Rao et al. (1995) provide technical discussions of conservation.

(2) We cannot determine purely a priori the relative power or efficiency of learning rules for an arbitrary domain.

(3) Efficient learning procedures are domain-specific and not general, regardless of whether or not anyone knows which rules or algorithms these are. The theorems themselves are 'objective' and not relative to the state of our knowledge.

(4) There is no completely general scientific method (across all possible problem domains or 'worlds') that is more efficient than other possible methods or learning rules, let alone a method that can be justified a priori or even by convention, as methodologists from Descartes to Popper have supposed. Specifically, there is no maximally efficient hypothetico-deductive method, let alone a maximally efficient 'logic of discovery'.

(5) Available domain knowledge can be used to select 'biased' (in the nonpejorative, engineering sense) search procedures with superior efficiency. That is, we *can* know that a rule is more (or less) efficient than the average rule or some specific rule, in a particular domain, given relevant domain knowledge. Choice of an efficient learning procedure is justified only insofar as we possess relevant domain knowledge.

(6) Conversely, the less domain knowledge we possess, the more 'blind' and general, and hence weak and inefficient, our search procedure becomes. For as long as significant domain knowledge is lacking, we have no way of knowing which of the possible rules is indeed more efficient.

(7) Thus Campbell's BV + SR thesis (the necessity of using BV + SR at the research frontier) is true.

This follows from the NFL theorems when they are *relativized* to our epistemic situation, that is, to relevant, domain knowledge that we possess rather than interpreting them (solely) as claims about objective reality.[42] Given an unknown domain, the NFL theorems do *not* imply that no rule is better than any other but rather that we cannot *know* which.[43]

(8) For in a state of ignorance we can only proceed blindly, by trying this solution or rule or that one. Once we have determined that some rules or techniques work better than others, we can then try, *post hoc*, to *explain* this fact by attributing a particular structure to the domain. Confirmation of that attribution can then warrant the use of biased rules. This basic idea was already incorporated in the traditional method of hypothesis.

(9) Thus at the frontier of knowledge, inquiry is consequentialist rather than generative, relative to the domain knowledge already available (see below).

(10) It takes knowledge to get knowledge at all efficiently, yet, contrary to Plato's version of the conservation of knowledge, we can still get more knowledge from less. This claim goes beyond the NFL theorems in assuming that the domain in question does have some structure and that a BV + SR process can find some of it. Under these conditions, creation *ex nihilo* is possible, not in the sense of innovation from absolutely nothing but in the sense of the emergence of more knowledge from less, of more design from less.

(11) However, the more innovative the knowledge (relative to our ignorance of the domain), the weaker, more inefficient our search procedures will be and the more difficult the corresponding recognition problem.

(12) Therefore no strong methods of discovery are possible in the sense of efficient methods for producing major innovations. However, weak discovery procedures are possible. And, when iterated, the ultimate result can be highly innovative.

(13) Random trial and error is not the weakest of all methods. For, averaged over all possible domains, it is as strong (or weak) as any other method.

---

[42] See Thornton (2000, pp. 141ff). Of course, neither the NFL theorems nor the data-compression techniques that Thornton proceeds to promote ultimately *solve* Hume's philosophical problem of induction. Inductive moves will always remain risky.
[43] I cannot address here the converse question of whether a sufficiently weak BV + SR procedure will work to produce learning if any rule will work.

(14) Moreover, for any particular domain, there exist any number of methods (learning rules) that are worse than random trial and error.[44]

(15) Given that there are many distinct varieties of BV + SR, not all BV + SR rules should be labeled 'mere' trial and error or 'random' trial and error.

(16) Specific BV + SR methods will be more or less efficient than each other in a given domain but equally good (or bad) when averaged across all possible domains of problems.

A few words of explanation are in order. With reference to (7) and (8), the NFL theorems do not single out BV + SR rules as objectively special in any way. Given any particular domain, there is no reason to think that a BV + SR rule will be among the objectively more efficient rules for that domain. So, as objective existence claims, the NFL theorems offer no positive support for the Campbell thesis that innovative inquiry must employ some BV + SR. It may even appear that they undermine the thesis. However, they do not, for Campbell's thesis concerns inquiry under conditions of ignorance and not the mere mathematical existence of learning rules for a given domain. The relevance of the NFL theorems to this epistemic situation is the familiar Humean one: we have no a priori reason to prefer one rule over another. In this case, BV + SR rules of one sort or another become special in the sense that we have no alternative but to use one of them. A new scientific domain may be subject to very powerful search rules. The trouble is, the investigators first opening up the domain do not know what they are. They can only proceed blindly, according to the degree of their ignorance.

Obviously, even when a powerful rule happens to exist, you cannot use it until you hit upon it, and you cannot make justified efficiency claims for it until you have shown its superiority to other rules. To be sure, you may select any arbitrary learning rule or problem-solving procedure as a hypothesis, but, a priori, it will have as much or as little chance as any other rule. This choice in itself is therefore blind. You may subsequently try another rule and yet another. You are now implementing a BV + SR procedure at the level of rules.

Once your research community achieves some domain knowledge, you may graduate to a stronger form of inquiry. Suppose that previous BV + SR has produced sufficient domain knowledge to justify more direct procedures for solving problems in this domain.

Now you may use these newly found rules to solve these classes of problems routinely, but of course this exercise is no longer seriously innovative. As I said before, discovery of efficient problem-solving methods is the *product* of innovation and not the original source and explanation of those problem solutions. The problem solutions are streamlined and methodized as the final stage of the discovery process, as it were. Meanwhile, research goes on at the frontier of the problem domain, where new problems remain unsolved. And here the conditions of the previous paragraph obtain: you can only proceed (more or less) blindly. Typically, your domain knowledge will place constraints on the candidates you generate, but, beyond that, you must grope for a promising solution. In the domain of mechanical problems, for example, it would be crazy to keep generating solution candidates that violate conservation of energy or momentum, once those principles become known. To take another sort of example, a systematic scan of a domain (e.g. a woodchuck scanning the horizon) requires domain knowledge in order for the scan to be systematic. In general, the more domain knowledge available, the more constrained or 'biased' the variation and selection mechanisms can be.

With reference to (9), a *consequentialist* methodology of innovation or discovery is one that proceeds by *post hoc* or consequential testing.[45] The old H-D method of science is an example: propose a solution and then check it by deriving testable consequences from it. Insofar as the solution goes beyond any previous knowledge and hence any epistemically justified selection criterion, it is selected solely on the basis of the success of its logical consequences. By contrast, a *generative* process proceeds from premises to novel conclusions and can, in principle, be deductive. However, if the universal evolutionists are correct, then even the process of finding deductive arguments to novel conclusions will involve an element of BV + SR. Nonetheless, this latter case shows that ultimate *justification* can be generative even when first selection or 'discovery' is not. A new deductive proof must be found by a BV + SR process, but, once found, it is, after all, a deductive proof. Finally, it is worth noting that neither a generative nor a consequentialist approach need restrict itself to deductive inferences.

The previous paragraph is misleading insofar as it suggests that inquiry is globally consequentialist, that is, consequentialist 'all the way down'. For domain knowledge that enables us to cut down the size of the search space and/or to select a more efficient learning procedure amounts to a generative contribution to inquiry. The fact that that knowledge itself is a former product of BV + SR does not prevent it from serving a generative role in the current stage of inquiry.

---

[44] Writes Wolpert (1996, p. 1355):

In other words, one can just as readily have a target for which one's algorithm has *worse than random guessing* as one in which it performs better than random. The pitfall we wish to avoid in supervised learning is not simply that our algorithm performs as poorly as random guessing, but rather that our algorithm performs worse than randomly!

[45] For further discussion of consequentialist and generative methodologies, see Laudan (1980) and Nickles (1987).

With reference to (13), Campbell always insisted that blind search is not the same as purely random search.

It is important to realize that there are zillions of actual and possible BV + SR procedures available. The class of these processes or learning rules is not at all restricted to the one or relatively few operative in biological evolution. Depending on the kinds of individuals and populations one is considering, there are any number of different possible mechanisms of variation, selection, and transmission. So here, as elsewhere, the wise choice of particular learning rule will depend upon domain knowledge.

David Fogel, one of the leading researchers in evolutionary computing, applies the NFL results to his field thus:

[The NFL theorem shows that] there is no best algorithm, whether or not that algorithm is 'evolutionary', and moreover whatever an algorithm gains in performance on one class of problems is necessarily offset by that algorithm's performance on the remaining problems.

This simple theorem has engendered a great deal of controversy in the field of evolutionary computation, and some associated misunderstanding. There has been considerable effort expended in finding the 'best' set of parameters and operators for evolutionary algorithms since at least the mid-1970s. These efforts have involved the type of recombination, the probabilities for crossover and mutation, the representation, the population's size, and so forth. Most of this research has involved empirical trials on benchmark functions. But the no free lunch theorem essentially dictates that the conclusions made on the basis of such sampling are in the strict mathematical sense limited to only those functions studied. Efforts to find the best crossover rate, the best mutation operation, and so forth, in the absence of restricting attention to a particular class of problems are pointless.

For an algorithm to perform better than even random search (which is simply another algorithm) it must reflect something about the structure of the problem it faces. By consequence, it mismatches the structure of some other problem. Note too that it is not enough to simply identify that a problem has some specific structure associated with it: that structure must be appropriate to the algorithm at hand. Moreover, the structure must be specific (Fogel, 1999, p. 56).

Fogel notes that this result contradicts much conventional practice in evolutionary computing. Numerous investigators have sought general results as to which types of representations or codings of individuals in a population and which types and specific values of variation, selection, and heritability operators are more efficient than others. While specific claims of this sort

may still hold, they are false if made completely general.

In sum, the NFL theorems sharpen the intuitive remarks of Hume, Wittgenstein, and Carnap. Insofar as we possess relevant domain knowledge, a more restricted selection of BV + SR procedures is possible; but as long as there are important, unsolved problems beyond the limits of current domain knowledge, we can only proceed blindly. Here we may apply Hume's so-called method of challenge: If you remain unconvinced that we must resort to BV + SR procedures at the frontier of research, then, pray tell, what alternative procedure is there, and how do you justify it?

Having touted knowledge-based inquiry in the last two sections, a word of caution is in order. It is possible that our perspective may change as faster machines and machines of different architectures (e.g. parallel machines and neural networks) become readily available. For example, John Koza and associates (Koza et al., 1999) are developing, with some success, genetic programs capable of solving problems that are given as high-level problem descriptions without explicit instruction as to how to go about solving them. Their aim is to realize the dream of 'getting a computer to do what needs to be done without telling it how to do it'. Such a method would be pretty general and not knowledge-intensive in the way that knowledge-based computation is. The ratio of intelligence-out to intelligence-in would be quite high, by contrast with most knowledge-based computation. In a sense the machine is self-programming. If this sort of research program succeeds, it reopens the possibility of a more general problem-solving method. (See Nickles, forthcoming, for further discussion of this idea.) For the purposes of this chapter it is important to note that Koza's approach is inspired by Darwinian evolution. Basically, he breeds populations of computer programs and tests the individuals of each generational cohort for fitness as defined by the problem to be solved. His approach is therefore highly representational, since computer programs are symbolic structures. However, like biological evolution, the artificial evolutionary computation is, in a sense, parallel and distributed. Other investigators are pursuing nonrepresentational, subconceptual sorts of computation using neural nets.

Some Objections Considered: From Impossible to (Almost) Necessary

The BV + SR thesis has a peculiar status in discussion today. Many investigators still consider it an absurdly impossible mode of innovation, or at the very least hugely inefficient compared to the techniques that scientists, artists, *et alia* actually employ in their creative work. This is the 'monkeys at typewriters' family of objections. At the other extreme, some critics consider the thesis trivial or even tautologous. Here I can respond only briefly to a few of the more frequent objections. Additional responses can be found in Nickles (forthcoming).

1. The BV + SR thesis is trivial, even tautologous, a priori true. Hence, it cannot explain anything. For it follows from the very meanings of the key words that any attempt to go beyond present knowledge must proceed blindly. How else could it possibly proceed? By your own account nonnaturalistic modes of inquiry are incoherent insofar as they claim to avoid BV + SR. And you claim that the Meno problem itself virtually requires a BV + SR solution, as do the NFL theorems under conditions of ignorance.

The thesis has surprisingly far-ranging philosophical, scientific, and policy consequences, as Campbell and Dennett, among others, have shown. So it can hardly be trivial. Nor is it tautologous or a priori true. I do claim that today broadly Darwinian selectionist processes are the only viable account we have, but the future remains open. After all, 'no design without an intelligent designer' seemed virtually tautologous until Darwin came along. It remains to be seen whether present-day and future competitors to the adaptationist program in evolutionary biology—and their analogues for wider selection theory—can seriously complement or (in some cases) replace selectionist models. The argument that, at the frontier of research, trial and error is unavoidable at least provides strong heuristic motivation for the selectionist-adaptationist program.

Tautologies do have the merit of being true, and some have the merit of being useful, even in a creative context. (After all, the Gödel theorems are just logic!) But is the BV + SR claim really tautologous? If it is, then its denial is logically false, and so are alternatives such as divine creation, Larmarckianism, and instructionism. (I do believe that these alternatives are false, but not logically false.) In short, as a thoroughly naturalistic account of design, universal Darwinism can hardly be a priori true. Moreover, not all forms of naturalism embrace universal Darwinism, so the thesis is again at risk. And even within evolutionary biology, the adaptationist thesis (according to which, roughly, all adaptive traits were selected because of their contribution to fitness) is subject to empirical challenge. So here are three levels at which universal BV + SR is at risk.[46]

2. On the contrary, the universal evolution thesis is absurdly wrong. We can easily provide an alternative model of innovation, namely intelligent design. Scien-

tists and other inquirers can often use methods far more powerful than blind trial and error. The human design model is both familiar and conveys intuitive understanding. Paley's explanation of the origin of the watch found on the heath was exactly right. In similar terms we explain the invention and development of the automobile, the airplane, television, and scientific theories. Such explanations are far more plausible and explanatory than appeals to blind chance.

This is basically the reply that creation theorists and religious intelligent design theorists give to biological evolutionists (e.g. Dembski, 1998, 2001). The reply assumes that we well understand the springs of innovation and that they lie in deliberate, intelligent design. The view I have defended holds, quite to the contrary, that we still possess a very poor understanding of creativity and that the intelligent design model (no design without an intelligent designer possessing at least as much intelligence as that incorporated in the design) is both question-begging and hopelessly unspecific. In this case, familiarity with the God-design story breeds intellectual laziness. I claim that when we look closely at any genuinely innovative design, we discover a design history consisting of step-by-step modifications and recombinations of previously extant designs. It is not for nothing that the first automobiles resembled horse buggies. Moreover, at each stage we typically find a lot of trial and error search for suitable modifications in order to solve some problem or other. That story can certainly be told for the design of Paley's watches. As Campbell suggests, the human design model suppresses its evolutionary past.

3. But the previous objection is not just a 'religious' objection. Don't scientists and engineers and other innovators often use problem-solving methods far more powerful than trial and error? Don't artists sometimes carefully design their work?

Of course they do. Once we learn how to solve a class of problems, their solution can often be reduced to routine, a process that is often automated. However, such efforts only provide streamlined ways to solve problems that we can already solve. As such, they can support further research and routinized components of larger design projects, but they themselves do not produce novel design. It is also normally the case that searches for novel solutions are *directed* to some degree. Sometimes they are highly direct or constrained. That is the whole point about the importance of relevant domain knowledge and constrained generation. Nevertheless, constrained search is still search and, to that degree, blind. Campbell (1974a) provides an elaborate discussion under the head of a 'nested hierarchy of vicarious selectors'. No one claims that inquiry in a typical scientific or artistic or

---

[46] For further discussion of the a priori issue, see Gamble (1983). For the adaptationist controversy, see Sober (2000, ch. 5). A universal Darwinist need not be an extreme adaptationist, I think, and can allow design features that are by-products of other adaptive processes; but the issues become sticky here.

administrative problem situation is so completely lacking in constraints that any and every candidate solution, however crazy, must be considered. No one denies that routine problem-solving is helpful in many design tasks, only that it does not suffice for genuinely novel design. As noted above, the BV + SR model does call attention to the importance of heuristic appraisal of variants—better than chance ways of deciding which candidates are worth pursuing. This is a topic sadly neglected by traditional confirmation theory in philosophy of science (Nickles, 1989).

4. Changing the representation of a problem can sometimes render a difficult problem easily solvable. There are also other ways of reducing a challenging problem to one that is already familiar. Are not these cases counterexamples to the BV + SR thesis?

Interesting question, but no; for we must first *search* for an alternative problem representation or a reduction technique. Again, every problem representation has its own evolutionary history. It is hitting upon a representation that makes the problem easily solvable that produces many 'aha' experiences, which are themselves Meno recognition events.

5. Some discoveries are purely deductive and thus do not involve BV + SR.

There are two points to this reply. First, the discovery of a genuinely innovative proof is not a matter of simple, routine deduction, else much of mathematics would be trivial. On the contrary, some results are totally unexpected: e.g. the (relative) consistency of non-Euclidean geometry, Gödel's theorems. Dietterich (1986, p. 12) and Cherniak (1986) rightly reject defining the knowledge of an agent in terms of the deductive closure of everything the agent knows. Simon & Lea (1979, p. 26) write:

> (F)rom a logical standpoint the processes involved in problem-solving are inductive, not deductive. To be sure, the proof of a theorem in a formal mathematical or logical system (such as Logic Theorist) is a deductive object; that is to say, the theorem stands in a deductive relation to its premises. But the problem-solving task is to discover this deduction, this proof; and the discovery process, which is the problem-solving process, is wholly inductive in nature. It is a search through a large space of logic expressions for the goal expression—the theorem.

Any program that requires genuine search, any program that includes any genuine test operations, such as generate-and-test, has an element of BV + SR

(although not necessarily evolutionary BV + SR in the full-blooded sense). Hence, contrary to appearances, Campbell's BV + SR thesis is not incompatible with Simon's work.

Second, we cannot take even deductive reasoning as an innate, God-given human ability that needs no further explanation. At the microcognitive level, various naturalistic theorists treat recognizing deductive structures of even a routine sort as a matter of pattern matching, that is, trial-and-error fitting, much of which occurs at the subconscious level (Johnson-Laird & Byrne, 1991; Margolis, 1987).

6. Nowadays, Biblical creationists as well as intelligent design theorists often grant that some evolution has occurred but insist that it is microevolution rather than macroevolution, that is, adaptation within a species rather than the emergence of entirely new species. They therefore reject Darwinian claims for the immense creative power of BV + SR processes. How do we know that evolutionary computation can get beyond microevolution, so to speak, to produce genuinely novel epistemic designs?

This is an important question. We cannot really know until more results are in. However, there are reasons for optimism. How else could we have done it in the past than by implicitly implementing an evolutionary process? I believe that evolutionary biologists have an adequate answer to the biological objection. Moreover, human cultural BV + SR processes can be much more rapid than biological ones. For example, we humans can identify and study the errors of our trial and error processes in order to learn the reasons for failure. We can learn from our mistakes more directly than biological nature can, where mistakes are simply eliminated from the population of the next generation. In nature, roughly speaking, learning or creative problem-solving becomes a matter of sex and death (Sterelny & Griffiths, 1999). In addition, our design failures are not fatal. We can make modifications that are seriously deleterious without totally losing that line of development to extinction. In terms of fitness landscapes, we are therefore able to descend from a local maximum through a valley in order to reach a higher peak. Also, we are not restricted to a branching tree structure of design, corresponding exactly to biological speciation. In effect, we can combine good ideas from different branches into a single new design. With the advent of fast, parallel computational possibilities we are also beginning to catch up with nature's ability to try out huge populations of variants at once.

Notice that even microevolution refutes the *old* model of design, according to which you cannot get more design from less.[47] So there can no longer be the same, principled objection to macroevolution.

7. The Meno problem of inquiry, as you characterize it, is so abstract and artificial that it fails to fit scientific and artistic practice. Specifically, all inquiry begins from a problem, and to have a problem at all is to have some idea of what would count as a solution. So we are never in a position of having to go from zero knowledge to some knowledge, completely blindly.

Correct, but this point does not threaten the BV + SR thesis. As long as we do not currently know how to solve a problem, we can only proceed blindly. We are at a 'personal frontier' (or at a frontier for the whole scientific community) as far as usable knowledge is concerned. A well-formulated problem already incorporates knowledge, or design, and that knowledge itself could have grown out of ignorance only on the basis of prior applications of BV + SR. The history of every well-formulated problem will *also* be an evolutionary history.

There are, of course, larger issues here that I can only mention. One is the origins problem. How did inquiry begin, so to speak? And how could BV + SR methods apply there? (Compare the problem of origins in evolutionary biology, a problem that Darwin himself sidestepped.) This brings us to a second issue. Biological nature is frequently and fruitfully considered, metaphorically, as a problem-solver. But here the variation and selection and transmission mechanisms are not, of course, 'known' to the creative agency. Rather, they are implicit in nature itself. Again, learning or creative problem-solving becomes a matter of sex and death. Much more remains to be said about bridging these different ways of considering creative problem-formulating and -solving.

8. In many cases, for problems at the furthest frontier of research or problems that have been posed but are not yet really on the research agenda, we may have no domain knowledge and hence have to use blind search as a 'method'. However, many of these problems do in fact have sufficient structure, or belong to a sufficiently structured domain, that a stronger method will be far more efficient.

Yes, but without our domain knowledge we do not know which methods these are or how to recognize and

apply them even should we hit upon them by accident. We can only search for these methods randomly at this point, namely, by learning more about the domain itself. So we are back to square one. It is important not to confuse the mathematical existence of a rule with our knowledge of it. The point of scientific inquiry, after all, is to find the strongest rules we can for a domain. But inquiry means bootstrapping our way to those rules from a condition of relative ignorance.

9. The NFL theorems imply that if an algorithm works better than average over one domain of problems, then its performance must be worse than average over another. Remarks Wolpert (1996, p. 1362), "There are just as many priors (prior probabilities) for which your favorite algorithm performs worse than pure randomness as for which it performs better". Doesn't this mean that there are domains in which no particular BV + SR procedure works well?

Yes, but this consequence does not undercut the BV + SR thesis. Campbell and other universal Darwinists do not claim that BV + SR will always work. There are possible domains in which *nothing* will work in the sense of finding humanly intelligible scientific results. Nor is it Campbell's claim that a BV + SR is always, or ever, optimal, objectively speaking. The claim is that, in a condition of ignorance, we have nothing better to go on than some BV + SR process or other.

10. In a completely 'cracked' Humean universe, learning is impossible. There it does seem plausible that no learning rule is better than any other. But we know that our universe is not like that. So why must we take seriously the NFL theorems?

Knowing that our *universe* is 'not like that' provides some empirical knowledge about our universe, but this does not automatically carry over to every domain that we happen to define. Such general knowledge should not be confused with domain-specific knowledge. As Wolpert and Fogel say, in the passages quoted in Section 7, the choice of an efficient learning rule must match the structure of that particular domain. One large-scale example is the familiar lesson that it is not a good idea to model all other sciences on physics. Early investigators assumed not only that the world is lawfully 'connected' but also that it is connected everywhere just like simple mechanics. One cannot take for granted the existence of coherent domains of phenomena or problems that call for a unified treatment. They, too, have their evolutionary histories (Shapere, 1974). Think how much effort it took to determine the state variables of physical and chemical systems.

11. In a condition of minimal domain knowledge, no learning rule is justified. Therefore, no particular version of BV + SR is justified either. So how can

---

[47] A slight qualification is in order. Paley and company did not claim that small chance combinations in nature are absolutely impossible, although they would not have considered these productive of genuine design. But for them it was overwhelmingly improbable that you could get even complex *apparent* designs by chance. In other words, complex apparent designs are almost certainly *genuine* designs and hence (by definition of 'genuine design') the work of an intelligent designer.

Campbell's BV + SR thesis be true? For choosing a non-BV + SR rule is equally justified.

To be sure, one can choose a non-BV + SR rule, but, in a state of ignorance, one is then choosing blindly at the *rules* level, the meta-level, so to speak. Thus one is still resorting to trial and error, to some form of BV + SR.

12. Artificial intelligence experts discuss the weaknesses and the limits of genetic algorithms. How is that possible if the BV + SR thesis is correct?

These experts are (or should be) discussing the comparative strength of algorithms, their optimality, dominance characteristics, and so on for specific domains. Campbell's thesis claims neither that BV + SR rules are generally stronger than others nor that any particular BV + SR rule is universally useful.

13. If BV + SR underlies all innovative design, then why so much resistance to the idea, including resistance among scientists and AI experts as well as the general public?

Ignorance, Madam! This objection had more intuitive punch a decade or two ago, before the present boom in evolutionary computing. Resistance can be largely explained in the usual ways: reluctance to give up comfortable, traditional views; failure to think critically about them; and exaggerating the degree of blindness and chance in evolutionary processes while neglecting their cumulative nature. BV + SR advocates do not say that major innovations come in one big, improbable, chance occurrence. This is the counterpart to the wrongheaded creationist objections to biological evolution—that evolution equates the appearance of organisms with an explosion in a print shop producing the works of Shakespeare, or a hurricane blowing through a junkyard and producing and Boeing 747. On the contrary, BV + SR advocates are gradualists, step-by-step evolutionists, in which later steps build upon earlier ones. Big, greedy jumps are more characteristic of the romantic model of innovation.

Conclusion

Plato's Meno paradox concludes that genuine learning and (by extension) innovation are *impossible*. Darwin and his successors, however, argue convincingly that, when the relevant conditions are satisfied (blind variation plus selective retention in a stable environment with selection pressures), then evolution is *inevitable*. Campbell and others contend, in effect, that the conditions for evolutionary adaptation *are*, in fact, satisfied in many problem contexts, including all of those in which genuine learning or innovation occurs, and that learning and innovation (the introduction of novel design) can therefore be regarded as adaptation. Conversely, all adaptations can be considered knowledge (often implicit knowledge). As Holland (1995,

p. 31) puts the crucial point, "Perpetual novelty is the hallmark of complex adaptive systems". Short of destroying the system, we cannot prevent novel developments even if we want to (e.g. mutating viruses and bacteria, and attempts to 'fix' permanently a natural language such as French).

So the Meno problem is solved by a quasi-biological process of emergence of novel variants from predecessors in the genealogy. BV + SR is the only way we know to produce more design from less, the only way we know to produce genuine innovation. Although I cannot pretend to have offered a complete defense of that claim here, it does appear that an intelligent design model, whether theological or secular (as in appeals to 'the scientific method' in place of God), cannot, by itself, explain how innovation is possible. Thus we must be very careful how we describe the work of highly creative individuals working at an investigative frontier.

Following Campbell and company, I do *not* claim that unconstrained BV + SR is the source of all innovative design. Hence, I do not claim that adaptations are solely the product of BV + SR. Physical constraints are always operative, and perhaps the physical systems are sometimes self-organizing to some degree. However, insofar as there is cumulative BV + SR at all (and there always is at the frontier of innovation), then it makes sense to speak of adaptation.

My chapter explains and defends this BV + SR thesis with some qualifications. I have tried to meet some leading objections to the thesis without pretending to survey all possible modes of innovation. In particular, I have given little attention to biological processes themselves. To extend the discussion to those areas would require modifying some of my talk about domain knowledge and ignorance, e.g. to speaking instead of variations on a given biological platform. Nor have I investigated the sources of BV + SR other than to make some casual remarks about the noise introduced by linguistic slack, misinterpretation, and historical contingencies. These sources surely include situational and social-contextual as well as individual factors. I have also assumed, without much argument, that learning and innovation amount to bringing novel design into the world.

One point with which Campbell would surely have agreed is that not all BV + SR processes are evolutionary in the full sense (although they are still fully naturalistic). He is less likely to have agreed with my claim that trial and error processes, with their corresponding elements of chance and luck, should be considered *methods* or learning rules rather than signaling the complete absence of method; but such a view is now standard in evolutionary computing. More generally, BV + SR, and full evolutionary BV + SR, when available, can be considered a methodological shell or family of shells that yield usable methods when the variation, selection, and retention mechanisms are specified in concrete contexts (Nickles, forthcoming). After all, if philosophers count the so-called hypothetico-deductive method as a method, how can they refuse the label to BV + SR?

In addition, I have argued that the NFL theorems support the BV + SR thesis, given that, at the frontier of research, we lack relevant domain knowledge to justify some non-BV + SR rule. BV + SR is the default position in the absence of domain knowledge. Typically, more than one BV + SR process will be available, but it is important to remember that the *warranted* use even of a particular BV + SR rule requires domain knowledge. The NFL theorems may be regarded as formal articulations of Humean intuitions.

And I have shown why the fact that BV + SR rules are not often, or even usually, the most efficient rules for a given domain does not undercut the BV + SR thesis. Campbell's thesis does not assert that BV + SR processes are optimal in some objective sense, and certainly not for routine problem-solving; but rather that they are the *only* procedures available at the frontier of innovation, given our ignorance of what lies beyond. Thus articles such as Baum et al. (2001), while certainly interesting, are not relevant, one way or the other, to the fate of the BV + SR thesis.

Overall, the chapter supports the claim that innovation is better considered a selective-adaptive process than a theorem-proving process or a rational planning or human design process in the traditional sense. Application of this point to human contexts places more emphasis on innovative communities, their traditions and practices, than upon individual creative 'geniuses'. The human design model is not completely wrong, of course, but it turns out to rest on previous applications of BV + SR. Hence the difficult and ironic historical intellectual reversal: when it comes to genuine innovation, BV + SR is the *first* resort rather than the last resort. Not only *can* there be novel, intelligent design without a prescient, intelligent designer, but also we have no other defensible account of it!

References

Bacon, F. (1620). *Novum Organum (The New Organon)*. Many translations.

Baum, E, Boneh, D. & Garrett, C. (2001). Where genetic algorithms excel. *Evolutionary Computation*, 9, 93–124.

Boden, M. (1991). *The creative mind: Myths and mechanisms*. New York: Basic Books.

Boden, M. (Ed.) (1994). *Dimensions of creativity*. Cambridge, MA: MIT Press.

Brandom, R. (1994). *Making it explicit: Reasoning, representing, and discursive commitment*. Cambridge, MA: Harvard University Press.

Brannigan, A. (1981). *The social basis of scientific discovery*. Cambridge: Cambridge University Press.

Calvin, W. H. (1996). *The cerebral code*. Cambridge, MA: MIT Press.

Campbell, D. T. (1960). Blind variation and selective retention in creative thought as in other knowledge processes. *Psychological Review*, 67, 380–400.

Campbell, D. T. (1974a). Evolutionary epistemology. In: P. A. Schilpp (Ed.), *The Philosophy of Karl R. Popper* (pp. 413–463). LaSalle, IL: Open Court.

Campbell, D. T. (1974b). Unjustified variation and selective retention in scientific discovery. In: F. S. Ayala & T. Dobzhansky (Eds), *Studies in the Philosophy of Biology* (pp. 139–161). London: Macmillan.

Campbell, D. T. (1997). From evolutionary epistemology via selection theory to a sociology of scientific validity. *Evolution and Cognition*, 3, 5–38.

Carnap, R. (1950). *Logical foundations of probability*. Chicago: University of Chicago Press.

Carnap, R. (1952). *The continuum of inductive methods*. Chicago: University of Chicago Press.

Chamberlain, T. C. (1897). Studies for students. *Journal of Geology*, 5, 837–848. Reprinted with the title 'The method of multiple working hypotheses'. In: C. Albritton (Ed.), *Philosophy of Geohistory, 1785–1970* (pp. 125–131). Stroudsburg, PA: Dowden, Hutchinson & Ross.

Cherniak, C. (1986). *Minimal rationality*. Cambridge, MA: MIT Press.

Crevier, D. (1993). *AI: The tumultuous history of the search for artificial intelligence*. New York: Basic Books.

Cziko, G. (1995). *Without miracles: Universal selection theory and the second Darwinian revolution*. Cambridge, MA: MIT.

Cziko, G. (2000). *The things we do: Using the lessons of Bernard and Darwin to understand the what, how, and why of our behavior*. Cambridge, MA: MIT Press.

Dawkins, R. (1976). *The selfish gene*. New York: Oxford University Press.

Dawkins, R. (1986). *The blind watchmaker*. New York: Norton.

Dembski, W. (1998). *The design inference: Eliminating chance through small probabilities*. Cambridge: Cambridge University Press.

Dembski, W. (2001). Introduction: What intelligent design is not. In: W. Dembski & J. Kushiner (Eds), *Signs of Intelligence: Understanding Intelligent Design* (pp. 7–23). Grand Rapids, MI: Brazos Press.

Dennett, D. (1995). *Darwin's dangerous idea*. New York: Simon & Schuster.

Desmond, A. & Moore, J. (1991). *Darwin*. New York: Warner.

Descartes, R. (1637). *Discourse on method*. Many later editions and translations.

Dewey, J. (1929). *The quest for certainty*. New York: Minton, Balch & Co.

Dietterich, T. (1986). Learning at the knowledge level. *Machine Learning*, 1, 287–316. In: J. Shavlik & T. Dietterich (Eds), (1990, pp. 11–25).

Edelman, G. (1987). *Neural Darwinism: The theory of neuronal group selection*. New York: Basic Books.

Feigenbaum, E. (1969). Artificial intelligence: Themes in the second decade. In: A. J. H. Morrell (Ed.), *Information processing 68: Proceedings IFIP Congress 1968* (Vol. 2, pp. 1008–1024). Amsterdam: North-Holland.

Feigenbaum, E., Buchanan, B. & Lederberg, J. (1971). On generality and problem solving: A case study using the DENDRAL program. *Machine Intelligence, 7*, 165–190.

Feyerabend, P. K. (1975). *Against method*. London: New Left Books.

Feyerabend, P. K. (1981). *Realism, rationalism & scientific method: Philosophical papers* (Vol. 1). Cambridge: Cambridge University Press.

Fogel, D. B. (1999). Some recent important foundational results in evolutionary computation. In: K. Miettinen, M. Mäkelä, P. Neittaanmaki & J. Périaux (Eds), *Evolutionary Algorithms in Engineering and Computer Science* (pp. 55–71). Chichester, U.K.: John Wiley.

Fogel, L. J. (1999). *Intelligence through simulated evolution, forty years of evolutionary programming*. New York: John Wiley.

Gamble, T. (1983). The natural selection model of knowledge generation: Campbell's dictum and its critics. *Cognition and Brain Theory, 6*, 353–363.

Gigerenzer, G. (1994). Where do new ideas come from? In: Boden (Ed.), (1994, pp. 53–74).

Gigerenzer, G., Todd, P. & The ABC Research Group (1999). *Simple heuristics that make us smart*. New York: Oxford.

Gillies, D. (1996). *Artificial intelligence and scientific method*. Oxford: Oxford University Press.

Gruber, H. (1974). *Darwin on man: A psychological study of scientific creativity*. New York: Dutton.

Holland, J. (1992). *Adaptation in natural and artificial systems*. Cambridge, MA: MIT. (Originally published 1975.)

Holland, J. (1995). *Hidden order: How adaptation builds complexity*. Reading, MA: Addison-Wesley.

Hume, D. (1748). *An enquiry concerning human understanding*. Many reprints.

Jacob, C. (2001). *Illustrating evolutionary computation with mathematica*. San Francisco: Morgan Kaufmann.

Johnson-Laird, P. & Byrne, R. (1991). *Deduction*. Hillsdale, NJ: Lawrence Erlbaum.

Kantorovich, A. (1993). *Scientific discovery: Logic and thinking*. Albany, NY: SUNY Press.

Karmiloff-Smith, A. (1992). *Beyond modularity: A developmental perspective on cognitive science*. Cambridge, MA: MIT Press.

Kelly, K. (1994). *Out of control: The new biology of machines, social systems, and the economic world*. Reading, MA: Addison Wesley.

Kelvin, Lord (William Thomson) (1901). Presidential address to the British association for the advancement of science. In: G. Basalla, W. Coleman & R. Kargon (Eds), *Victorian Science: A Self-Portrait Through the Presidential Addresses of the British Association for the Advancement of Science* (pp. 101–128). Garden City, NY: Doubleday.

Koza, J., Bennett III, F., Andre, D. & Keane, M. (1999). *Genetic programming III: Darwinian invention and problem solving*. San Francisco: Morgan Kaufmann.

Kuhn, T. S. (1978). *Black-body theory and the quantum discontinuity, 1894–1912*. Oxford: Oxford University Press.

Laudan, L. (1980). Why was the logic of discovery abandoned? In: T. Nickles (Ed.), *Scientific discovery, logic, and*

rationality (pp. 173–183). Dordrecht: Reidel. Reprinted in Laudan's *Science and hypothesis* (1981, pp. 181–189). Dordrecht: Kluwer.

Lestienne, R. (1993). *The creative power of chance*. Urbana: University of Illinois Press.

Margolis, H. (1987). *Patterns, thinking, and cognition: A theory of judgment*. Chicago: University of Chicago Press.

Mitchell, T. (1990). The need for biases in learning generalizations. In: Shavlik & Dieterich (Eds) (1960, pp. 184–191). Originally published in 1980 as a Rutgers University Technical Report.

Newell, A. (1969). Heuristic programming: Ill-structured problems. In: J. S. Arnofsky (Ed.), *Progress in Operations Research III*. New York: John Wiley.

Newell, A. & Simon, H. A. (1972). *Human problem solving*. Englewood Cliffs, NJ: Prentice-Hall.

Nickles, T. (1987). From natural philosophy to metaphilosophy of science. In: P. Achinstein & R. Kargon (Eds), *Kelvin's Baltimore Lectures and Modern Theoretical Physics: Historical and Philosophical Perspectives* (pp. 507–541). Cambridge, MA: MIT Press.

Nickles, T. (1989). Heuristic appraisal: A proposal. *Social Epistemology, 3*, 175–188.

Nickles, T. (2003). Normal science: From logic to case-based and model-based reasoning. In: T. Nickles (Ed.), *Thomas Kuhn* (Contemporary Philosophers in Focus series, pp. 142–177). Cambridge: Cambridge University Press.

Nickles, T. (forthcoming). The strange story of scientific method. In a volume edited by Joke Meheus.

Paley, W. (1802). *Natural theology*. London.

Plotkin, H. (1993). *Darwin machines and the nature of knowledge*. Cambridge, MA: Harvard University Press.

Rao, R. B., Gordon, D. & Spears, W. (1995). For every generalization action, is there really an equal and opposite reaction? Analysis of the conservation law for generalization performance. Draft of paper of the Machine Learning Conference 1995.

Rechenberg, I. (1973). *Evolutionsstrategie: Optimierung technischer systeme nach princzipien der biologischen evolution*. Stuttgart: Frommann-Holzboog.

Salmon, W. C. (1966). *Foundations of scientific inference*. Pittsburgh, PA: University of Pittsburgh Press.

Schaffer, C. (1994). A conservation law for generalization performance. In: Cohen & Hirsch (Eds), *Machine Learning: Proceedings of the Eleventh International Conference* (pp. 259–65). San Mateo, CA: Morgan Kaufmann.

Schaffer, S. (1986). Scientific discoveries and the end of natural philosophy. *Social Studies of Science, 16*, 387–420.

Schaffer, S. (1994). Making up discovery. In: Boden (1994, pp. 13–51).

Schwefel, H-P. (1995). *Evolution and optimum seeking*. New York: John Wiley.

Shapere, D. (1974). Scientific theories and their domains. In: F. Suppe (Ed.), *The Structure of Scientific Theories* (pp. 518–565). Urbana, IL: University of Illinois Press. Reprinted in Shapere (Ed.), *Reason and the Search for Knowledge* (1984, pp. 273–319). Dordrecht: Reidel.

Shavlik, J. & Dietterich, T. (Eds) (1990). *Readings in machine learning*. San Mateo, Cal.: Morgan Kaufmann.

Simon, H. A. (1947). *Administrative behavior*. New York: Macmillan.

Simon, H. A. & Lea, G. (1979). Problem solving and rule induction: A unified view. In: L. Gregg (Ed.), *Knowledge and Cognition*. Potomac, MD: Lawrence Erlbaum. Reprinted in Shavlik & Dietterich (Eds) (1990, pp. 26–43).

Souriau, P. (1881). *Theorie de l'invention*. Paris: Hatchette.

Sterelny, K. & Griffiths, P. (1999). *Sex and death: An introduction to philosophy of biology*. Chicago: University of Chicago Press.

Tauber, A. I. (1994). *The immune self: Theory or metaphor?* Cambridge: Cambridge University Press.

Thornton, C. (2001). *Truth from trash: How learning makes sense*. Cambridge, MA: MIT.

Wittgenstein, L. (1921). *Tractatus logico-philosophicus*. English edition (1961). London: Routledge and Kegan Paul.

Wolpert, D. (1996). The lack of a priori distinctions between learning algorithms. *Neural Computation*, **8**, 1341–1390.

Wolpert, D. & Macready, W. (1997). No free lunch theorems for optimization. *IEEE Transactions on Evolutionary Computation*, **1**, 67–82 (condensed version of 1995 Santa Fe Institute Technical Report, SFI TR 95–02–010, 'No Free Lunch Theorems for Search'.)

The International Handbook on Innovation
Edited by Larisa V. Shavinina

The Three-Ring Conception of Giftedness: Its Implications for Understanding the Nature of Innovation

Joseph S. Renzulli

The National Research Center on the Gifted and Talented, The University of Connecticut, USA

Abstract: The Three-Ring Conception of Giftedness focuses on creative/innovative productivity, which differs from schoolhouse or lesson-learning giftedness. Giftedness that leads to innovation emerges from the interaction and overlap of three clusters of traits—well above average ability in a particular domain, task commitment, and creativity—and occurs in certain individuals, at certain times, under certain conditions. A broader conception of giftedness will allow researchers, theorists, and educators to more fully address, answer, and respond to the most pressing questions about where innovation 'comes from'. Simultaneously, this model, when applied, will assist in the development of individuals most capable of engaging in innovation and improving the quality of life for society.

Keywords: Innovation; Gifted; Creativity; Motivation; Task commitment

Introduction

The age old issue of 'what makes giftedness' and how the contributions of gifted individuals have helped us understand the nature of innovation has been debated by scholars for decades. In the past 20 years, a renewed interest has emerged on this topic. This chapter will attempt to shed some light on this complex and controversial question by describing a broad range of theoretical issues and research studies that have been associated with the study of gifted, talented, and innovative persons. Although the information reported here draws heavily on the theoretical and research literature, it is written from the point of view of an educational practitioner who respects both theory and research, but who also has devoted a major amount of his efforts to translating these types of information into what he believes to be defensible identification and programming practices. Those in the position of offering advice to school systems that are faced with the reality of identifying and serving highly able students must also provide the types of underlying research that lend credibility to their advice. Accordingly, this chapter might be considered a theoretical and research rationale for two separate publications that describe a plan for identifying and programming

for gifted and talented students (Renzulli, 2000; Renzulli & Reis, 1997).

This chapter attempts to show a connection between the development of giftedness in young people and the ways in which gifted behaviors influence the process of innovation. Innovation is defined as original, solution-oriented *actions* that address previously unsolved problems in unique and creative ways. 'Action' is emphasized because innovation requires that the innovator will *do something* that leads to development or improvement of a product or process rather than merely thinking about or contemplating that product or process. The chapter is divided into three sections. The first section deals with several major issues that might best be described as the enduring questions and sources of controversy in a search for the meaning of giftedness and related attempts to define this concept. It is hoped that a discussion of these issues will establish common points of understanding between the writer and the reader and, at the same time, point out certain biases that are unavoidable whenever one deals with a complex and value-laden topic.

The second section describes a wide range of research studies that support the writer's 'three-ring' conception of giftedness. The section concludes with an explicit definition and a brief review of research studies that have been carried out in school programs using an identification system based on the three-ring concept. The final section examines a number of questions raised by scholars and practitioners since the time of the original publication (Renzulli, 1978) of this particular approach to a conception of giftedness.

Issues in the Study of Conceptions of Giftedness

Purposes and Criteria for a Definition of Giftedness

One of the first and most important issues that should be dealt with in a search for the meaning of giftedness is that there must be a purpose for defining this concept. The goals of science tell us that a primary purpose IS to add new knowledge to our understanding about human conditions, but in an applied field of knowledge there is also a practical purpose for defining concepts. Persons who presume to be the writers of definitions should understand the full ramifications of these purposes and recognize the practical and political uses to which their work might be applied. A definition of giftedness is a formal and explicit statement that might eventually become part of official policies or guidelines. Whether or not it is the writer's intent, such statements will undoubtedly be used to direct identification and programming practices, and therefore we must recognize the consequential nature of this purpose and the pivotal role that definitions play in structuring the entire field. Definitions are open to both scholarly and practical scrutiny, and for these reasons it is important that a definition meet the following criteria:

(1) It must be based on the best available research about the characteristics of gifted individuals rather than romanticized notions or unsupported opinions.
(2) It must provide guidance in the selection and/or development of instruments and procedures that can be used to design defensible identification systems.
(3) It must give direction and be logically related to programming practices such as the selection of materials and instructional methods, the selection and training of teachers; and the determination of procedures whereby programs can be evaluated.
(4) It must be capable of generating research studies that will verify or fail to verify the validity of the definition.

In view of the practical purposes for which a definition might be used, it is necessary to consider any definition in the larger context of overall programming for the target population we are attempting to serve. In other words, the way in which one views giftedness will be a primary factor in both constructing a plan for identification and in providing services that are relevant to the characteristics that brought certain youngsters to our attention in the first place. If, for example, one

identifies giftedness as extremely high mathematical aptitude, then it would seem nothing short of common sense to use assessment procedures that readily identify potential for superior performance in this particular area of ability. And it would be equally reasonable to assume that a program based on this definition and identification procedure should devote major emphasis to the enhancement of performance in mathematics and related areas. Similarly, a definition that emphasizes artistic abilities should point the way toward relatively specific identification and programming practices. As long as there are differences of opinion among reasonable scholars there will never be a single definition of giftedness, and this is probably the way that it should be. But one requirement for which all writers of definitions should be accountable is the necessity of showing a logical relationship between definition on the one hand and recommended identification and programming practices on the other.

Two Kinds of Giftedness

A second issue that must be dealt with is that our present efforts to define giftedness are based on a long history of previous studies dealing with human abilities. Most of these studies focused mainly on the concept of intelligence and are briefly discussed here to establish an important point about the process of defining concepts rather than any attempt to equate intelligence with giftedness. Although a detailed review of these studies is beyond the scope of the present chapter, a few of the general conclusions from earlier research are necessary to set the stage for this analysis.[1]

The first conclusion is that intelligence is not a unitary concept, but rather there are many kinds of intelligence and therefore single definitions cannot be used to explain this complicated concept. The confusion and inconclusiveness about present theories of intelligence have led Sternberg (1984) and others to develop new models for explaining this complicated concept. Sternberg's 'triarchic' theory of human intelligence consists of three subtheories: a contextual subtheory, which relates intelligence to the external world of the individual; a two-facet subtheory, which relates intelligence to both the external and internal worlds of the individual; and a componential subtheory, which relates intelligence to the internal world of the individual. The contextual subtheory defines intelligent behavior in terms of purposive adaptation to, selection of, and shaping of real-world environments relevant to one's life. The two-facet subtheory further constrains this definition by regarding as most relevant to the demonstration of intelligence contextually intelligent behavior that involves adaptation to novelty or

---

[1] Persons interested in a succinct examination of problems associated with defining intelligence are advised to review 'The Concept of Intelligence' (Neisser, 1979).

automatization of information processing, or both. The componential subtheory specifies the mental mechanisms responsible for the learning, planning, execution, and evaluation of intelligent behavior.

In view of this work and numerous earlier cautions about the dangers of trying to describe intelligence through the use of single scores, it seems safe to conclude that this practice has been and always will be questionable. At the very least, attributes of intelligent behavior must be considered within the context of cultural and situational factors. Indeed, some examinations have concluded that "(t)he concept of intelligence *cannot* be explicitly defined, not only because of the nature of intelligence but also because of the nature of concepts" (Neisser, 1979, p. 179).

A second conclusion is that there is no ideal way to measure intelligence, and therefore we must avoid the typical practice of believing that if we know a person's IQ score, we also know his or her intelligence. Even Terman warned against total reliance on tests: "We must guard against defining intelligence solely in terms of ability to pass the tests of a given intelligence scale" (1921, p. 131). E. L. Thorndike echoed Terman's concern by stating, "to assume that we have measured some general power which resides in (the person being tested) and determines his ability in every variety of intellectual task in its entirety is to fly directly in the face of all that is known about the organization of the intellect" (Thorndike, 1921, p. 126).

The reason I have cited these concerns about the historical difficulty of defining and measuring intelligence is to highlight the even larger problem of isolating a unitary definition of giftedness. At the very least we will always have several conceptions (and therefore definitions) of giftedness; but it will help in this analysis to begin by examining two broad categories that have been dealt with in the research literature. I will refer to the first category as 'schoolhouse giftedness' and to the second as 'innovative giftedness'. Before going on to describe each type, I want to emphasize that:

(1) Both types are important.
(2) There is usually an interaction between the two types.
(3) Special programs should make appropriate provisions for encouraging both types of giftedness as well as the numerous occasions when the two types interact with each other.

Schoolhouse Giftedness

Schoolhouse giftedness might also be called test-taking or lesson-learning giftedness. It is the kind most easily measured by IQ or other cognitive ability tests, and for this reason it is also the type most often used for selecting students for entrance into special programs. The abilities people display on IQ and aptitude tests are exactly the kinds of abilities most valued in traditional school learning situations. In other words, the games people play on ability tests are similar in nature to games that teachers require in most lesson-learning situations. Research tells us that students who score high on IQ tests are also likely to get high grades in school. Research also has shown that these test-taking and lesson-learning abilities generally remain stable over time. The results of this research should lead us to some very obvious conclusions about schoolhouse giftedness: it exists in varying degrees; it can be identified through standardized assessment techniques; and we should therefore do everything in our power to make appropriate modifications for students who have the ability to cover regular curricular material at advanced rates and levels of understanding. Curriculum compacting (Renzulli, Smith & Reis, 1982), a procedure used for modifying curricular content to accommodate advanced learners, and other acceleration techniques should represent an essential part of any school program that strives to respect the individual differences that are clearly evident from scores yielded by cognitive ability tests.

Although there is a generally positive correlation between IQ scores and school grades, we should not conclude that test scores are the only factors that contribute to success in school. Because IQ scores correlate only from 0.40 to 0.60 with school grades, they account for only 16%–36% of the variance in these indicators of potential. Many youngsters who are moderately below the traditional 3–5% test score cutoff levels for entrance into gifted programs clearly have shown that they can do advanced-level work. Indeed, most of the students in the nation's major universities and 4-year colleges come from the top 20% of the general population (rather than just the top 3–5%) and Jones (1982) reported that a majority of college graduates in every scientific field of study had IQs between 110 and 120. Are we 'making sense' when we exclude such students from access to special services? To deny them this opportunity would be analogous to *forbidding* a youngster from trying out for a basketball team because he or she missed a predetermined 'cutoff height' by a few inches! Basketball coaches are not foolish enough to establish *inflexible* cutoff heights because they know that such an arbitrary practice would cause them to overlook the talents of youngsters who may overcome slight limitations in inches with other abilities such as drive, speed, teamwork, ball-handling skills, and perhaps even the ability and motivation to outjump taller persons who are trying out for the team. As educators of gifted and talented youth, we can undoubtedly take a few lessons about flexibility from coaches!

Innovative Giftedness

If scores on IQ tests and other measures of cognitive ability only account for a limited proportion of the common variance with school grades, we can be equally certain that these measures do not tell the whole story when it comes to making predictions about innovative giftedness. Before defending this assertion with some research findings, let us briefly review what is meant by this second type of giftedness, the important role that it should play in programming, and, therefore, the reasons we should attempt to assess it in our identification procedures—even if such assessment causes us to look below the top 3–5% on the normal curve of IQ scores.

Innovative giftedness describes those aspects of human activity and involvement where a premium is placed on the development of original material and products that are purposefully designed to have an impact on one or more target audiences. Learning situations that are designed to promote innovative giftedness emphasize the use and application of information (content) and thinking processes in an integrated, inductive, and real-problem-oriented manner. The role of the student is transformed from that of a learner of prescribed lessons to one in which she or he uses the modus operandi of a firsthand inquirer. This approach is quite different from the development of lesson-learning giftedness that tends to emphasize deductive learning, structured training in the development of thinking processes, and the acquisition, storage, and retrieval of information. In other words, innovative giftedness is simply putting one's abilities to work on problems and areas of study that have personal relevance to the student and that can be escalated to appropriately challenging levels of investigative activity. The roles that both students and teachers should play in the pursuit of these problems have been described elsewhere (Renzulli, 1982, 1983).

Why is innovative giftedness important enough for us to question the 'tidy' and relatively easy approach that traditionally has been used to select students on the basis of test scores? Why do some people want to rock the boat by challenging a conception of giftedness that can be numerically defined by simply giving a test? The answers to these questions are simple and yet very compelling. The research reviewed in the second section of this chapter tells us that there is much more to the making of a gifted person than the abilities revealed on traditional tests of intelligence, aptitude, and achievement. Furthermore, history tells us it has been the innovative and productive people of the world, the producers rather than consumers of knowledge, the reconstructionists of thought in all areas of human endeavor, who have become recognized as 'truly gifted' individuals. History does not remember persons who merely scored well on IQ tests or those who learned their lessons well.

The Development of Potential Innovations

Implicit in any efforts to define and identify gifted youth is the assumption that we will 'do something' to provide various types of specialized learning experi-

ences that show promise of promoting the development of characteristics implicit in the definition. In other words, the *why* question supersedes the *who* and *how* questions. Although there are two generally accepted purposes for providing special education for the gifted, I believe that these two purposes in combination give rise to a third purpose that is intimately related to the definition question.

The first purpose of gifted education[2] is to provide young people with maximum opportunities for self-fulfillment through the development and expression of one or a combination of performance areas where superior potential may be present.

The second purpose is to increase society's supply of persons who will help to solve the problems of contemporary civilization by becoming innovators and producers of knowledge and art rather than mere consumers of existing information.

We value innovators, as opposed to replicators, because they are the persons who go beyond our current levels of knowledge and understanding to bring forth new ideas, questions, solutions to problems, and new products and services that did not exist prior to their application of the creative process. It is the innovators that, in effect, create thousands of jobs for replicators. One creative idea such as a book, a symphony, a new device or software program for a computer starts the wheels of production turning, thereby creating thousands of jobs in both manufacturing and a host of other areas such as finance, advertising, marketing, packaging, transportation, and sales. Thomas Edison's invention of the storage battery gave rise to an entire industry that still prospers today, and has the added benefit of providing a vehicle for the many innovative modifications and improvement that have been made on this innovative product over the years.

Although there may be some arguments for and against both of the above purposes, most people would agree that goals related to self-fulfillment and/or societal contributions are generally consistent with democratic philosophies of education. What is even more important is that the two goals are highly interactive and mutually supportive of each other. In other words, the self-satisfying work of scientists, artists, and leaders in all walks of life usually produces results that might be valuable contributions to society. Carrying this point one step farther, we might even conclude that appropriate kinds of learning experiences can and should be engineered to achieve the twofold goals described above. Keeping in mind the interaction of these two goals, and the priority status of the self-fulfillment goal, it is safe to conclude that supplementary investments of public funds and systematic

---

[2] The term *gifted education* will be used in substitution for the more technically accurate but somewhat awkward term *education of the gifted*.

effort for highly able youth should be expected to produce at least some results geared toward the public good. If, as Gowan (1978) has pointed out, the purpose of gifted programs is to increase the size of society's reservoir of potentially innovative and productive adults, then the argument for gifted education programs that focus on creative productivity (rather than lesson-learning giftedness) is a very simple one. If we agree with the goals of gifted education set forth earlier in the chapter, and if we believe that our programs should produce the next generation of leaders, problem-solvers, and persons who will make important contributions to the arts and sciences, then does it not make good sense to model our training programs after the modus operandi of these persons rather than after those of the lesson learner? This is especially true because research (as described later in the chapter) tells us that the most efficient lesson learners are not necessarily those persons who go on to make important contributions in the realm of creative productivity. And in this day and age, when knowledge is expanding at almost geometric proportions, it would seem wise to consider a model that focuses on how our most able students access and make use of information rather than merely on how they accumulate and store it.

In other words, innovation almost always takes place in context—in part because of innovation's close association with creativity, which is integrally related to context (Sternberg, Pretz & Kaufman, in press). A person is not always innovative or never innovative. Innovation occurs in certain people (not all people), at certain times (not all the time), and under certain circumstances. To be certain, there are some people who have long and continuous records of creative productivity, but there are others for whom lightening strikes only once! So while we celebrate the monumental record of more than 2000 patents by Edison, we also recognize the singular contribution of *Gone With the Wind* by Margaret Mitchell. The key issue for educators is creating the context that nurtures innovation. By providing young people with opportunities, resources, and encouragement *within their existing or developing areas of interest*, and by helping them experience the joys and rewards of creative productivity, we create a contextual environment that, for some people, will become a lifestyle. It is these people who usually gravitate toward the innovative rather than replicative roles in their respective professions—the medical research scientist rather than the medical practitioner, the fashion designer rather than the person who produces the garments. Both roles (innovator and replicator) are highly valued and necessary for progress in any field, but the trigger for any and all productivity is the one pulled by the innovator.

The Gifted and the Potentially Gifted

A further issue relates to the subtle but very important distinction that exists between the 'gifted' and the 'potentially gifted'. Most of the research reviewed later in the chapter deals with student and adult populations whose members have been judged (by one or more criteria) to be gifted. In most cases, researchers have studied those who have been identified as 'being gifted' much more intensively than they have studied persons who were not recognized or selected because of unusual accomplishments. The general approach to the study of gifted persons could easily lead the casual reader to believe that giftedness is a condition that is magically bestowed on a person in much the same way that nature endows us with blue eyes, red hair, or a dark complexion. This position is *not* supported by the research. Rather, what the research clearly and unequivocally tells us is that *giftedness can be developed* in some people if an appropriate interaction takes place between a person, his or her environment, and a particular area of human endeavor.

It should be kept in mind that when I describe, in the paragraphs that follow, a certain trait as being a component of giftedness (for example, creativity), I am in no way assuming that one is 'born with' this trait, even if one happens to possess a high IQ. Almost all human abilities can be developed, and therefore my intent is to call attention to the potentially gifted (that is to say, those who could 'make it' under the right conditions) as well as to those who have been studied because they gained some type of recognition. Implicit in this concept of the potentially gifted, then, is the idea that giftedness emerges or 'comes out' at different times and under different circumstances. Without such an approach there would be no hope whatsoever of identifying bright underachievers, students from disadvantaged backgrounds, or any other special population that is not easily identified through traditional testing procedures.

Are People 'Gifted' or Do They Display Gifted Behaviors?

A fifth and final issue underlying the search for a definition of giftedness is more nearly a bias and a hope for at least one major change in the ways we view this area of study. Except for certain functional purposes related mainly to professional focal points (i.e. research, training, legislation) and to ease of expression, I believe that a term such as *the gifted is* counterproductive to educational efforts aimed at identification and programming for certain students in the general school population. Rather, it is my hope that in the years ahead we will shift our emphasis from the present concept of 'being gifted' (or not being gifted) to a concern about developing *gifted behaviors* in those youngsters who have the highest potential for benefiting from special education services. This slight shift in terminology might appear to be an exercise in heuristic hairsplitting, but I believe that it has significant implications for the entire way we think about the concept of giftedness and the ways in which we structure the field for important research endeavors[3] and effective educational programming.

For too many years we have pretended that we can identify gifted children in an absolute and unequivocal fashion. Many people have been led to believe that certain individuals have been endowed with a golden chromosome that makes them 'gifted persons'. This belief has further led to the mistaken idea that all we need to do is find the right combination of factors that prove the existence of this chromosome. The further use of such terms as the 'truly gifted', the 'moderately gifted', and the 'borderline gifted' only serves to confound the issue and might result in further misguided searches for silver and bronze chromosomes. This misuse of the concept of giftedness has given rise to a great deal of confusion and controversy about both identification and programming, and the result has been needless squabbling among professionals in the field. Another result has been that so many mixed messages have been sent to educators and the public at large that both groups now have a justifiable skepticism about the credibility of the gifted education establishment and our ability to offer services that are qualitatively different from general education.

Most of the confusion and controversy surrounding the definition of giftedness can be placed in proper perspective by raising a series of questions that strike right at the heart of key issues related to this area of study. These questions are organized into the following clusters:

(1) Are giftedness and high IQ one and the same? And if so, how high does one's IQ need to be before one can be considered gifted? If giftedness and high IQ are not the same, what are some of the other characteristics that contribute to the expression of giftedness? Is there any justification for providing selective services for certain students who may fall below a predetermined IQ cutoff score?

(2) Is giftedness an absolute or a relative concept? That is, is a person either gifted or not gifted (the absolute view) or can varying kinds and degrees of gifted behaviors be displayed in certain people, at certain times, and under certain circumstances (the relative view)? Is gifted a static concept (i.e. you have it or you don't have it) or is it a dynamic concept (i.e. it varies both within persons and within learning-performance situations)?

(3) What causes only a minuscule number of Thomas Edisons or Langston Hugheses or Isadora Duncans to emerge as innovators, whereas millions of others

with equal 'equipment' and educational advantages (or disadvantages) never rise above mediocrity? Why do some people who have not enjoyed the advantages of special educational opportunities become innovators, whereas others who have gone through the best of educational programming opportunities fail to make an innovative contribution?

In the section that follows, a series of research studies will be reviewed in an effort to answer these questions. Taken collectively, these research studies are the most powerful argument that can be put forth to policymakers who must render important decisions about the regulations and guidelines that will dictate identification practices in their states or local school districts. An examination of this research clearly tells us that gifted behaviors can be developed in those who are not necessarily individuals who earn the highest scores on standardized tests. The two major implications of this research for identification practices are equally clear.

First, an effective identification system must take into consideration other factors in addition to test scores. Research has shown that, in spite of the multiple criterion information gathered in many screening procedures, rigid cutoff scores on IQ or achievement tests are still the main if not the only criterion given *serious* consideration in final selection (Alvino, 1981). When screening information reveals outstanding potential for gifted behaviors, it is almost always 'thrown away' if predetermined cutoff scores are not met. Respect for these other factors means that they must be given equal weight and that we can no longer merely give lip service to non-test criteria; nor can we believe that because tests yield 'numbers' they are inherently more valid and objective than other procedures. As Sternberg (1982a) has pointed out, *quantitative* does not necessarily mean *valid*. When it comes to identification that might lead to the development of innovators, it is far better to include rather than exclude those with the potential for developing innovative ideas and products. In order to create environments that produce innovators, educators could focus on opportunities for students to engage in authentic learning. This brand learning consists of *applying* relevant knowledge, thinking skills, and interpersonal skills to the solution of real problems and what happens in real-world situations. For example, students might focus on their learning activities by addressing these essential questions in any given discipline:

(1) What do people with an interest in this area do?
(2) What products do they create and/or what services do they provide?
(3) What methods do they use to carry out their work?
(4) What resources and materials are needed to produce high-quality products and services?

---

[3] For example, most of the research on the 'gifted' that has been carried out to date has used high-IQ populations. If one disagrees (even slightly) with the notion that giftedness and high IQ are synonymous, then these research studies must be reexamined. These studies may tell us a great deal about the characteristics, and so on, of high-IQ individuals, but are they necessarily studies of the gifted?

(5) How, and with whom, do they communicate the results of their work?

(6) What steps need to be taken to have an impact on intended audiences?

This type of focus would allow students to build a foundation of knowledge and pursue innovative strategies to solve a real problem. Rather than providing students with answers to these questions, the teacher organizes and guides instruction but does not dominate the exploration process. As the facilitator of situations to foster innovation, the teacher helps students select challenging projects. These activities should be student-driven, with the teacher playing an advisory role. Learning experiences that follow this model will provide for opportunities for students to take creative risks which may lead to the development of innovations.

The second research-based implication will undoubtedly be a major controversy in the field for many years, but it needs to be dealt with if we are ever going to defuse a majority of the criticism that has been justifiably directed at our field. Simply stated, we must reexamine identification procedures that result in a total pre-selection of certain students and the concomitant implication that these young people are and always will be 'gifted'. This absolute approach (i.e. you have it or you don't have it) coupled with the almost total reliance on test scores is not only inconsistent with what the research tells us, but almost arrogant in the assumption that we can use a single one-hour segment of a young person's total being to determine if he or she is 'gifted'.

The alternative to such an absolutist view is that we may have to forgo the 'tidy' and comfortable tradition of 'knowing' on the first day of school who is gifted and who is not gifted. Rather, our orientation must be redirected toward developing 'gifted behaviors' in certain students (not all students), at certain times (not all the time), and under certain circumstances. The tradeoff for tidiness and administrative expediency will result in a much more flexible approach to both identification and programming and a system that not only shows a greater respect for the research on gifted and talented people but is both fairer and more acceptable to other educators and to the general public.

Research Underlying the Three-Ring Conception of Giftedness

One way of analyzing the research underlying conceptions of giftedness is to review existing definitions along a continuum ranging from conservative to liberal. *Conservative* and *liberal* are used here not in their political connotations but rather according to the degree of restrictiveness that is used in determining who is eligible for special programs and services.

Restrictiveness can be expressed in two ways. First, a definition can limit the number of specific performance areas that are considered in determining eligibility for special programs. A conservative definition, for example, might limit eligibility to academic performance only and exclude other areas such as music, art, drama, leadership, public speaking, social service, and creative writing. Second, a definition can limit the degree or level of excellence that one must attain by establishing extremely high cutoff points.

At the conservative end of the continuum is Terman's (1926) definition of giftedness as 'the top 1% level in general intellectual ability as measured by the Stanford–Binet Intelligence Scale or a comparable instrument' (1926, p. 43).

In this definition, restrictiveness is present in terms of both the type of performance specified (i.e. how well one scores on an intelligence test) and the level of performance one must attain to be considered gifted (top 1%). At the other end of the continuum can be found more liberal definitions, such as the following one by Witty (1958):

> There are children whose outstanding potentialities in art, in writing, or in social leadership can be recognized largely by their performance. Hence, we have recommended that the definition of giftedness be expanded and that we consider any child gifted whose performance, in a potentially valuable line of human activity, is consistently remarkable (p. 62).

Although liberal definitions have the obvious advantage of expanding the conception of giftedness, they also open up two 'cans of worms' by introducing a values issue (What are the potentially valuable lines of human activity?) and the age-old problem of subjectivity in measurement.

In recent years the values issue has been largely resolved. There are very few educators who cling tenaciously to a 'straight IQ' or purely academic definition of giftedness. 'Multiple talent' and 'multiple criteria' are almost the bywords of the present-day gifted student movement, and most persons would have little difficulty in accepting a definition that includes almost every area of human activity that manifests itself in a socially useful form of expression.

The problem of subjectivity in measurement is not as easily resolved. As the definition of giftedness is extended beyond those abilities that are clearly reflected in tests of intelligence, achievement, and academic aptitude, it becomes necessary to put less emphasis on precise estimates of performance and potential and more emphasis on the opinions of qualified human judges in making decisions about admission to special programs. The crux of the issue boils down to a simple and yet very important question: How much of a trade-off are we willing to make on the objective-subjective continuum in order to allow recognition of a broader spectrum of human abilities? If

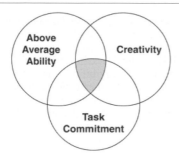

Figure 1. The three-ring conception of giftedness.

some degree of subjectivity cannot be tolerated, then our definition of giftedness and the resulting programs will logically be limited to abilities that can be measured only by objective tests.

The Three-Ring Conception of Giftedness

Research on innovative people has consistently shown that although no single criterion can be used to determine giftedness, persons who have achieved recognition because of their unique accomplishments and creative contributions possess a relatively well-defined set of three interlocking clusters of traits. These clusters consist of above-average, though not necessarily superior, ability, task commitment, and creativity (see Fig. 1). It is important to point out that no single cluster 'makes giftedness'. Rather, it is the interaction among the three clusters that research has shown to be the necessary ingredient for innovative accomplishment (Renzulli, 1978). This interaction is represented by the shaded portion of Fig. 1. It is also important to point out that each cluster plays an important role in contributing to the display of gifted behaviors. This point is emphasized because one of the major errors that continues to be made in identification procedures is to overemphasize superior abilities at the expense of the other two clusters of traits.

Well-Above-Average Ability

Well-above-average ability can be defined in two ways.

General ability consists of the capacity to process information, to integrate experiences that result in appropriate and adaptive responses in new situations, and the capacity to engage in abstract thinking. Examples of general ability are verbal and numerical reasoning, spatial relations, memory, and word fluency. These abilities are usually measured by tests of general aptitude or intelligence, and are broadly applicable to a variety of traditional learning situations.

Specific abilities consist of the capacity to acquire knowledge, skill, or the ability to perform in one or more activities of a specialized kind and within a restricted range. These abilities are defined in a manner that represents the ways in which human beings express themselves in real-life (i.e. non-test) situations.

Examples of specific abilities are chemistry, ballet, mathematics, musical composition, sculpture, and photography. Each specific ability can be further subdivided into even more specific areas (e.g. portrait photography, astrophotography, photo journalism, etc.). Specific abilities in certain areas such as mathematics and chemistry have a strong relationship with general ability, and, therefore, some indication of potential in these areas can be determined from tests of general aptitude and intelligence. They can also be measured by achievement tests and tests of specific aptitude. Many specific abilities, however, cannot be easily measured by tests, and, therefore, areas such as the arts must be evaluated through one or more performance-based assessment techniques.

Within this model the term *above-average ability will* be used to describe both general and specific abilities. *Above average* should also be interpreted to mean the upper range of potential within any given area. Although it is difficult to assign numerical values to many specific areas of ability, when I refer to 'well above average ability' I clearly have in mind persons who are capable of performance or the potential for performance that is representative of the top 15–20% of any given area of human endeavor.

Although the influence of intelligence, as traditionally measured, quite obviously varies with specific areas of performance, many researchers have found that creative accomplishment is not necessarily a function of measured intelligence. In a review of several research studies dealing with the relationship between academic aptitude tests and professional achievement, Wallach (1976) has concluded that: "Above intermediate score levels, academic skills assessments are found to show so little criterion validity as to be a questionable basis on which to make consequential decisions about students' futures. What the academic tests do predict are the results a person will obtain on other tests of the same kind" (p. 57).

Wallach goes on to point out that academic test scores at the upper ranges—precisely the score levels that are most often used for selecting persons for entrance into special programs—do not necessarily reflect the potential for innovative accomplishment. Wallach suggests that test scores be used to screen out persons who score in the lower ranges and that beyond this point, decisions should be based on other indicators of potential for superior performance.

Numerous research studies support Wallach's findings that there is a limited relationship between test scores and school grades on the one hand, and real-world accomplishments on the other (Bloom, 1963; Harmon, 1963; Helson & Crutchfield, 1979; Hudson, 1960; Mednick, 1963; Parloff et al., 1968; Richards, Holland & Lutz, 1967; Wallach & Wing, 1969). In fact, in a study dealing with the prediction of various dimensions of achievement among college students, Holland & Astin (1962) found that "getting good grades in college has little connection with more remote and more socially relevant kinds of achievement; indeed, in some colleges, the higher the student's grades, the less likely it is that he is a person with creative potential. So it seems desirable to extend our criteria of talented performance" (pp. 132–133). A study by the American College Testing Program (Munday & Davis, 1974) entitled 'Varieties of Accomplishment After College: Perspectives on the Meaning of Academic Talent', concluded that:

> the adult accomplishments were found to be uncorrelated with academic talent, including test scores, high school grades, and college grades. However, the adult accomplishments were related to comparable high school nonacademic (extra curricular) accomplishments. This suggests that there are many kinds of talents related to later success which might be identified and nurtured by educational institutions (p. 2).

The pervasiveness of this general finding is demonstrated by Hoyt (1965), who reviewed 46 studies dealing with the relationship between traditional indications of academic success and postcollege performance in the fields of business, teaching, engineering, medicine, scientific research, and other areas such as the ministry, journalism, government, and miscellaneous professions. From this extensive review, Hoyt concluded that traditional indications of academic success have no more than a very modest correlation with various indicators of success in the adult world and that "There is good reason to believe that academic achievement (knowledge) and other types of educational growth and development are relatively independent of each other" (p. 73).

The experimental studies conducted by Sternberg (1981) and Sternberg & Davidson (1982) added a new dimension to our understanding about the role that intelligence tests should play in making identification decisions. After numerous investigations into the relationship between traditionally measured intelligence and other factors such as problem-solving and insightful solutions to complex problems, Sternberg (1982b) concludes that:

> tests only work for some of the people some of the time—not for all of the people all of the time—and that some of the assumptions we make in our use of tests are, at best, correct only for a segment of the tested population, and at worst, correct for none of it. As a result we fail to identify many gifted individuals for whom the assumptions underlying our use of tests are particularly inadequate. The problem, then, is not only that tests are of limited validity for everyone but that their validity varies across individuals. For some people, tests scores may be quite informative, for others such scores may be worse than useless. Use of test score cutoffs and formulas

results in a serious problem of underidentification of gifted children (p. 157).

The studies raise some basic questions about the use of tests as a major criterion for making selection decisions. The research reported above clearly indicates that vast numbers *and* proportions of our most productive persons are *not* those who scored at the 95th percentile or above on standardized tests of intelligence, nor were they necessarily straight A students who discovered early how to play the lesson-learning game. In other words, more innovative persons came from below the 95th percentile than above it, and if such cutoff scores are needed to determine entrance into special programs, we may be guilty of actually discriminating against persons who have the greatest potential for high levels of accomplishment.

The most defensible conclusion about the use of intelligence tests that can be put forward at this time is based on research findings dealing with the 'threshold effect'. Reviews by Chambers (1969) and Stein (1968) and research by Walberg (1969, 1971) indicate that accomplishments in various fields require minimal levels of intelligence, but that beyond these levels, degrees of attainment are weakly associated with intelligence. In studies of creativity it is generally acknowledged that a fairly high though not exceptional level of intelligence is necessary for high degrees of creative achievement (Barron, 1969; Campbell, 1960; Guilford, 1964, 1967; McNemar, 1964; Vernon, 1967).

Research on the threshold effect indicates that different fields and subject matter areas require varying degrees of intelligence for high-level accomplishment. In mathematics and physics the correlation of measured intelligence with originality in problem-solving tends to be positive but quite low. Correlations between intelligence and the rated quality of work by painters, sculptors, and designers is zero or slightly negative (Barron, 1968). Although it is difficult to determine exactly how much measured intelligence is necessary for high levels of innovative accomplishment within any given field, there is a consensus among many researchers (Barron, 1969; Bloom, 1963; Cox, 1926; Harmon, 1963; Helson & Crutchfield, 1970; MacKinnon, 1962, 1965; Oden, 1968; Roe, 1952; Terman, 1954) that once the IQ is 120 or higher other variables become increasingly important. These variables are discussed in the paragraphs that follow.

Task Commitment

A second cluster of traits that consistently has been found in innovative persons is a refined or focused form of motivation known as task commitment. Whereas motivation is usually defined in terms of a general energizing process that triggers responses in organisms, task commitment represents energy brought to bear on a particular problem (task) or specific performance area. The terms that are most frequently used to describe task commitment are perseverance, endurance, hard work, dedicated practice, self-confidence, and a belief in one's ability to carry out important work. In addition to perceptiveness (Albert, 1975) and a better sense for identifying significant problems (Zuckerman, 1979), research on persons of unusual accomplishment has consistently shown that a special fascination for and involvement with the subject matter of one's chosen field "are the almost invariable precursors of original and distinctive work" (Barron, 1969, p. 3). Even in young people whom Bloom & Sosniak (1981) identified as extreme cases of talent development, early evidence of task commitment was present. Bloom & Sosniak report that "after age 12 our talented individuals spent as much time on their talent field each week as their average peer spent watching television" (p. 94).

The argument for including this nonintellective cluster of traits in a definition of giftedness is nothing short of overwhelming. From popular maxims and autobiographical accounts to hard-core research findings, one of the key ingredients that has characterized the work of gifted persons is their ability to involve themselves totally in a specific problem or area for an extended period of time.

The legacy of both Sir Francis Galton and Lewis Terman clearly indicates that task commitment is an important part of the making of a gifted person. Although Galton was a strong proponent of the hereditary basis for what he called 'natural ability', he nevertheless subscribed heavily to the belief that hard work was part and parcel of giftedness:

> By natural ability, I mean those qualities of intellect and disposition, which urge and qualify a man to perform acts that lead to reputation. I do not mean capacity without zeal, nor zeal without capacity, nor even a combination of both of them, without an adequate power of doing a great deal of very laborious work. But I mean a nature which, when left to itself, will, urged by an inherent stimulus, climb the path that leads to eminence and has strength to reach the summit—on which, if hindered or thwarted, will fret and strive until the hindrance is overcome, and it is again free to follow its laboring instinct (Galton, 1869, p. 33, as quoted in Albert, 1975, p. 142).

The monumental studies of Lewis Terman undoubtedly represent the most widely recognized and frequently quoted research on the characteristics of gifted persons. Terman's studies, however, have unintentionally left a mixed legacy because most persons have dwelt (and continue to dwell) on 'early Terman' rather than the conclusions he reached *after* several decades of intensive research. As such, it is important to consider the following conclusion that he reached as a result of 30 years of follow-up studies on his initial population:

A detailed analysis was made of the 150 most successful and 150 least successful men among the gifted subjects in an attempt to identify some of the non-intellectual factors that affect life success.... Since the less successful subjects do not differ to any extent in intelligence as measured by tests, it is clear that notable achievement calls for more than a high order of intelligence.

The results (of the follow-up) indicated that personality factors are extremely important determiners of achievement.... The four traits on which (the most and least successful groups) differed most widely were *persistence in the accomplishment of ends, integration toward goals, self-confidence, and freedom from inferiority feelings*. In the total picture the greatest contrast between the two groups was in all-round emotional and social adjustment, and in *drive to achieve* (Terman, 1959, p. 148; italics added).

Although Terman never suggested that task commitment should replace intelligence in our conception of giftedness, he did state that 'intellect and achievement are far from perfectly correlated'.

Several research studies support the findings of Galton and Terman and have shown that innovative persons are far more task-oriented and involved in their work than are people in the general population. Perhaps the best known of these studies is the work of Roe (1952) and MacKinnon (1964, 1965). Roe conducted an intensive study of the characteristics of 64 eminent scientists and found that *all* of her subjects had a high level of commitment to their work. MacKinnon pointed out traits that were important in creative accomplishments: "It is clear that creative architects more often stress their inventiveness, independence and individuality, their *enthusiasm, determination*, and *industry*" (1964, p. 365; italics added).

Extensive reviews of research carried out by Nicholls (1972) and McCurdy (1960) found patterns of characteristics that were consistently similar to the findings reported by Roe and MacKinnon. Although the studies cited thus far used different research procedures and dealt with a variety of populations, there is a striking similarity in their major conclusions. First, academic ability (as traditionally measured by tests or grade point averages) showed limited relationships to innovative accomplishment. Second, nonintellectual factors, and especially those related to task commitment, consistently played an important part in the cluster of traits that characterized highly productive people. Although this second cluster of traits is not as easily and objectively identifiable as general cognitive abilities are, these traits are nevertheless a major component of giftedness and should, therefore, be reflected in our definition.

Creativity

The third cluster of traits that characterizes gifted persons consists of factors usually lumped together under the general heading of 'creativity'. As one reviews the literature in this area, it becomes readily apparent that the words *gifted, genius*, and *eminent creators* or *highly creative persons* are used synonymously. In many of the research projects discussed above, the persons ultimately selected for intensive study were in fact recognized *because* of their creative accomplishments. In MacKinnon's (1964) study, for example, panels of qualified judges (professors of architecture and editors of major American architectural journals) were asked first to nominate and later to rate an initial pool of nominees, using the following dimensions of creativity:

(1) Originality of thinking and freshness of approaches to architectural problems.
(2) Constructive ingenuity.
(3) Ability to set aside established conventions and procedures when appropriate.
(4) A flair for devising effective and original fulfillments of the major demands of architecture, namely, technology (firmness), visual form (delight), planning (commodity), and human awareness and social purpose (p. 360).

When discussing creativity, it is important to consider the problems researchers have encountered in establishing relationships between creativity tests and other more substantial accomplishments. A major issue that has been raised by several investigators deals with whether or not tests of divergent thinking actually measure 'true' creativity. Although some validation studies have reported limited relationships between measures of divergent thinking and creative performance criteria (Dellas & Gaier, 1970; Guilford, 1967; Shapiro, 1968; Torrance, 1969) the research evidence for the predictive validity of such tests has been limited. Unfortunately, very few tests have been validated against real-life criteria of creative accomplishment; however, future longitudinal studies using these relatively new instruments might show promise of establishing higher levels of predictive validity. Thus, although divergent thinking is indeed a characteristic of highly creative persons, caution should be exercised in the use and interpretation of tests designed to measure this capacity.

Given the inherent limitations of creativity tests, a number of writers have focused attention on alternative methods for assessing creativity. Among others, Nicholls (1972) suggests that an analysis of creative products is preferable to the trait-based approach in making predictions about creative potential (p. 721), and Wallach (1976) proposes that student self-reports about creative accomplishment are sufficiently accurate to provide a usable source of data. Simonton (in press)

and Weisberg (in press) provide extensive summaries of issues related to the study of creativity and creative potential; much of the literature they review confirms the importance and complexity of the interactions among nature, the environment—including the Zeitgeist—and ability.

Although few persons would argue against the importance of including creativity in a definition of giftedness, the conclusions and recommendations discussed above raise the haunting issue of subjectivity in measurement. In view of what the research suggests about the questionable value of more objective measures of divergent thinking, perhaps the time has come for persons in all areas of endeavor to develop more careful procedures for evaluating the products of candidates for special programs.

Discussion and Generalizations

The studies reviewed in the preceding sections lend support to a small number of basic generalizations that can be used to develop an operational definition of giftedness. The first of these generalizations is that giftedness consists of an interaction among three clusters of traits—above-average but not necessarily superior general abilities, task commitment, and creativity. Any definition or set of identification procedures that does not give equal attention to all three clusters is simply ignoring the results of the best available research dealing with this topic.

Related to this generalization is the need to make a distinction between traditional indicators of academic proficiency and innovation. A sad but true fact is that special programs have favored proficient lesson learners and test takers at the expense of persons who may score somewhat lower on tests but who more than compensate for such scores by having high levels of task commitment and creativity. It is these persons whom research has shown to be those who ultimately make the most innovative contributions to their respective fields of endeavor.

A second generalization is that an operational definition should be applicable to all socially useful performance areas. The one thing that the three clusters discussed above have in common is that each can be brought to bear on a multitude of specific performance areas. As was indicated earlier, the interaction or overlap among the clusters 'makes giftedness', but giftedness does not exist in a vacuum. Our definition must, therefore, reflect yet another interaction, but in this case it is the interaction between the overlap of the clusters and any performance area to which the overlap might be applied. This interaction is represented by the large arrow in Fig. 2.

A third and final generalization concerns the types of information that should be used to identify superior performance in specific areas. Although it is a relatively easy task to include specific performance areas in a definition, developing identification procedures

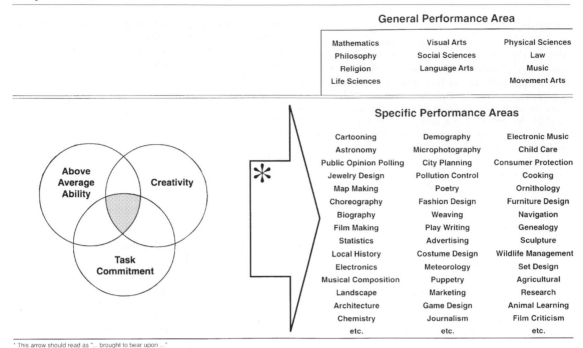

Mathematics	Visual Arts	Physical Sciences
Philosophy	Social Sciences	Law
Religion	Language Arts	Music
Life Sciences		Movement Arts

Specific Performance Areas

Cartooning	Demography	Electronic Music
Astronomy	Microphotography	Child Care
Public Opinion Polling	City Planning	Consumer Protection
Jewelry Design	Pollution Control	Cooking
Map Making	Poetry	Ornithology
Choreography	Fashion Design	Furniture Design
Biography	Weaving	Navigation
Film Making	Play Writing	Genealogy
Statistics	Advertising	Sculpture
Local History	Costume Design	Wildlife Management
Electronics	Meteorology	Set Design
Musical Composition	Puppetry	Agricultural
Landscape	Marketing	Research
Architecture	Game Design	Animal Learning
Chemistry	Journalism	Film Criticism
etc.	etc.	etc.

* This arrow should read as "... brought to bear upon ..."

Figure 2. Relationship between the three-ring conception of giftedness and contexts for innovation.

that will enable us to recognize specific areas of superior performance is a more difficult problem. Test developers have thus far devoted most of their energy to the development of measures of general ability, and this emphasis is undoubtedly why these tests are relied on so heavily in identification. However, an operational definition should give direction to needed research and development, especially in the ways that these activities relate to instruments and procedures for student selection. A defensible definition can thus become a model that will generate vast amounts of appropriate research in the years ahead.

A Definition of Gifted Behavior

Although no single statement can effectively integrate the many ramifications of the research studies I have described, the following definition of gifted behavior attempts to summarize the major conclusions and generalizations resulting from this review of research:

Gifted behavior consists of behaviors that reflect an interaction among three basic clusters of human traits—these clusters being above average general and/or specific abilities, high levels of task commitment, and high levels of creativity. Gifted and talented children are those possessing or capable of developing this composite set of traits and applying them to any potentially valuable area of human performance. Children who manifest or are capable of developing an interaction among the three clusters require a wide

variety of educational opportunities and services that are not ordinarily provided through regular instructional programs.

A graphic representation of this definition is presented in Fig. 2, and the following 'taxonomy' of behavioral manifestations of each cluster is a summary of the major concepts and conclusions emanating from the work of the theorists and researchers discussed in the preceding paragraphs:

Well-above-average ability
General ability:

(1) High levels of abstract thinking, verbal and numerical reasoning, spatial relations, memory, and word fluency.
(2) Adaptation to and the shaping of novel situations encountered in the external environment.
(3) The automatization of information processing; rapid, accurate, and selective retrieval of information.

Specific ability:

(1) The application of various combinations of the above general abilities to one or more specialized areas of knowledge or areas of human performance (e.g. the arts, leadership, administration).
(2) The capacity for acquiring and making appropriate use of advanced amounts of formal knowledge, tacit knowledge, technique, logistics, and strategy

in the pursuit of particular problems or the manifestation of specialized areas of performance.

(3) The capacity to sort out relevant and irrelevant information associated with a particular problem or area of study or performance.

Task commitment

(1) The capacity for high levels of interest, enthusiasm, fascination, and involvement in a particular problem, area of study, or form of human expression.

(2) The capacity for perseverance, endurance, determination, hard work, and dedicated practice.

(3) Self-confidence, a strong ego and a belief in one's ability to carry out important work, freedom from inferiority feelings, drive to achieve.

(4) The ability to identify significant problems within specialized areas; the ability to tune in to major channels of communication and new developments within given fields.

(5) Setting high standards for one's work; maintaining an openness to self and external criticism; developing an aesthetic sense of taste, quality, and excellence about one's own work and the work of others.

Creativity

(1) Fluency, flexibility, and originality of thought.

(2) Openness to experience; receptive to that which is new and different (even irrational) in the thoughts, actions, and products of oneself and others.

(3) Curious, speculative, adventurous, and 'mentally playful'; willing to take risks in thought and action, even to the point of being uninhibited.

(4) Sensitive to detail, aesthetic characteristics of ideas and things; willing to act on and react to external stimulation and one's own ideas and feelings.

As is always the case with lists of traits such as the above, there is an overlap among individual items, and an interaction between and among the general categories and the specific traits. It is also important to point out that not all of the traits need be present in any given individual or situation to produce a display of gifted behaviors. It is for this reason that the three-ring conception of giftedness emphasizes the interaction among the clusters rather than any single cluster. It is also for this reason that I believe gifted behaviors take place in certain people (not all people), at certain times (not all the time), and under certain circumstances (not all circumstances).

Discussion About the Three Rings

Since the original publication of the three-ring conception of giftedness (Renzulli, 1977), a number of questions have been raised about the overall model and the interrelationships between and among the three rings. In this section, I will use the most frequently

asked questions as an outline for a discussion that will, I hope, clarify some of the concerns raised by persons who have expressed interest (both positive and negative) in this particular approach to the conception of giftedness.

Are There Additional Clusters of Abilities that Should be Added to the Three-Ring Conception of Giftedness?

One of the most frequent reactions to this work has been the suggestion that the three clusters of traits portrayed in the model do not adequately account for the development of gifted behaviors. An extensive examination of the research on human abilities has led me to an interesting conclusion about this question and has resulted in a modification of the original model. This modification is represented figurally by the houndstooth background in which the three rings are now imbedded (see Fig. 1).

The major conclusion is that the interaction among the original three rings is still the most important feature leading to the display of gifted behaviors. There are, however, a host of other factors that must be taken into account in our efforts to explain what causes some persons to display gifted behaviors at certain times and under certain circumstances. I have grouped these factors into the two traditional dimensions of studies about human beings commonly referred to as personality and environment. The research[4] clearly shows that each of the factors listed in Table 1 plays varying roles in the manifestation of gifted behaviors. What is even more important is the interaction between the two categories and among the numerous factors listed in each column (In fact, a houndstooth pattern was selected over an earlier checkerboard design in an effort to convey this interaction.) When we consider the almost limitless number of combinations between and among the factors listed in Table 1, it is easy to realize why so much confusion has existed about the definition of giftedness.

Each of the factors is obviously a complex entity in and of itself and could undoubtedly be subdivided into numerous component parts. The factor of socio-economic status, for example, accounts for such things as prenatal care and nutrition, educational opportunities, and even things such as 'occupational inheritance'. Werts (1968) found, for example, that there is a clear tendency for college students to gravitate toward the occupation of their fathers. On the personality side of the ledger, MacKinnon (1965) found that in studies of highly effective individuals it was discovered time and time again that persons of the

---

[4] Literally hundreds of research studies have been carried out on the factors listed. For persons interested in an economical summary of personality and environmental influences on the development of gifted behaviors, I would recommend the following: D. K. Simonton (1978); B. T. Eiduson & L. Beckman (1973).

Table 1. Personality and environmental factors influencing giftedness.

Personality factors	Environmental factors
Perception of self	Socioeconomic status
Courage	Parental personalities
Character	Education of parents
Intuition	Stimulation of childhood interests
Charm or charisma	Family position
Need for achievement	Formal education
Ego strength	Role-model availability
Energy	Physical illness and/or well-being
Sense of destiny	Chance factors (financial inheritance, death, living near an art museum, divorce, etc.)
Personal attractiveness*	Zeitgeist

* Although personal attractiveness is undoubtedly a physical characteristic, the ways in which others react to one's physical being are quite obviously important determinants in the development of personality.

most extraordinary effectiveness had life histories marked by severe frustrations, deprivations, and traumatic experiences. Findings such as these help to highlight the complexity of the problem. The advantages of high socioeconomic status, a favorable educational background, and early life experiences that do not include hardship, frustration, or disappointment may lead to a productive career for some individuals, but for others it may very well eliminate the kinds of frustration that might become the 'trigger' to a more positive application of one's abilities.[5]

An analysis of the role that personality and environment play in the development of gifted behaviors and potentially innovative behaviors is beyond the scope of this chapter, and in many ways for school persons who are charged with the responsibilities of identifying and developing gifted and potentially innovative behaviors, they are beyond the realm of our *direct* influence. Each of the factors above shares one or a combination of two characteristics. First, most of the personality factors are long-term developmental traits or traits that in some cases are genetically determined. Although the school can play an important role in developing things like courage and need for achievement, it is highly unrealistic to believe that we can shoulder the major responsibility for overall personality formation. Second, many factors such as socioeconomic status, parental personalities, and family position are chance factors that children must take as givens when they are born and that educators must take as givens when young people walk through the schoolhouse door. We cannot tell a child to be the firstborn or to have parents who stress achievement! It is for these reasons that I have concentrated my efforts on the three sets of clusters set forth in the original model. Of course, certain aspects of the original three clusters are also

[5] I am reminded of the well-known quote by Dylan Thomas: 'There's only one thing that's worse than having an unhappy childhood, and that's having a too-happy childhood'.

chance factors, but a large amount of research clearly has shown that creativity and task commitment are in fact modifiable and can be influenced in a highly positive fashion by purposeful kinds of educational experiences (Reis & Renzulli, 1982). And although the jury is still out on the issue of how much of one's ability is influenced by heredity and how much by environment, I think it is safe to conclude that abilities (both general and specific) can be influenced to varying degrees by the best kinds of learning experiences.

Are the Three Rings Constant?

Most educators and psychologists would agree that the above-average-ability ring represents a generally stable or constant set of characteristics. In other words, if an individual shows high ability in a certain area such as mathematics, it is almost undeniable that mathematical ability was present in the months and years preceding a 'judgment day' (i.e. a day when identification procedures took place) and that these abilities will also tend to remain high in the months and years following any given identification event. In view of the types of assessment procedures most readily available and economically administered, it is easy to see why this type of giftedness has been so popular in making decisions about entrance into special programs. Educators always feel more comfortable and confident with traits that can be reliably and objectively measured, and the 'comfort' engendered by the use of such tests often causes them to ignore or only pay lip service to the other two clusters of traits.

In our identification model (Renzulli et al., 1981), we have used above-average ability as the major criterion for identifying a group of students who are referred to as the Talent Pool. This group generally consists of the top 15–20% of the general school population. Test scores, teacher ratings, and other forms of 'status information' (i.e. information that can be gathered and analyzed at a fixed point in time) are of practical value in making certain kinds of first-level decisions about accessibility to some of the general services that should be provided by a special program. This procedure guarantees admission to those students who earn the highest scores on cognitive ability tests. Primary among the services provided to Talent Pool students are procedures for making appropriate modifications in the regular curriculum in areas where advanced levels of ability can be clearly documented. It is nothing short of common sense to adjust the curriculum in those areas where high levels of proficiency are shown. Indeed, advanced coverage of traditional material and accelerated courses should be the 'regular curriculum' for youngsters with high ability in one or more school subjects.

The task commitment and creativity clusters are a different story! These traits are not present or absent in the same permanent fashion as pointed out in our mathematics example above. Equally important is the fact that we cannot assess them by the highly objective and quantifiable means that characterize test score assessment of traditional cognitive abilities. We simply cannot put a percentile on the value of a innovative idea, nor can we assign a standard score to the amount of effort and energy that a student might be willing to devote to a highly demanding task. Creativity and task commitment 'come and go' as a function of the various types of situations in which certain individuals become involved.

There are three things that we know for certain about the creativity and task commitment clusters. First, the clusters are variable rather than permanent. Although there may be a tendency for some individuals to 'hatch' more innovative ideas than others and to have greater reservoirs of energy that promote more frequent and intensive involvement in situations, a person is not either creative or not creative in the same way that one has a high ability in mathematics or musical composition. Almost all studies of highly accomplished individuals clearly indicate that their work is characterized by peaks and valleys of both creativity and task commitment. One simply cannot (and probably should not) operate at maximum levels of output in these two areas on a constant basis. Even Thomas Edison, who is still acknowledged to be the world's record holder of original patents, did not have an innovative idea for a new invention every waking moment of his life. And the most productive persons have consistently reported 'fallow' periods and even experiences of 'burnout' following long and sustained encounters with the manifestation of their talents.

The second thing we know about task commitment and creativity is that they can be developed through appropriate stimulation and training. We also know that because of variations in interest and receptivity, some people are more influenced by certain situations than others. The important point, however, is that we cannot predetermine which individuals will respond most favorably to a particular type of stimulation experience. Through general interest assessment techniques and a wide variety of stimulus variation we can, however, increase the probability of generating a greater number of innovative ideas and increased manifestations of task commitment in Talent Pool students. In our identification model, the ways in which students *react* to planned and unplanned stimulation experiences has been termed 'action information'. This type of information constitutes the second level of identification and is used to make decisions about which students might revolve into more individualized and advanced kinds of learning activities. The important distinction between status and action information is that the latter type cannot be gathered before students have been selected for entrance into a special program. Giftedness, or at least the beginnings of situations in which gifted behaviors might be displayed and developed, is in the *responses* of individuals rather than in the stimulus events. This second-level identification procedure is, therefore, part and parcel of the general enrichment experiences that are provided for Talent Pool students, and is based on the concept of *situational testing* that has been described in the theoretical literature on test and measurements (Freeman, 1962, pp. 538–554).

Finally, the third thing we know about creativity and task commitment is that these two clusters almost always stimulate each other. A person gets an innovative idea; the idea, is encouraged and reinforced by oneself and/or others. The person decides to 'do something' with the idea, and thus his or her commitment to the task begins to emerge. Similarly, a large commitment to solving a particular problem will frequently trigger the process of creative problem solving. In this latter case we have a situation that has undoubtedly given rise to the old adage 'necessity is the mother of invention.'

This final point is especially important for effective programming. Students participating in a gifted program should be patently aware of opportunities to follow through on innovative ideas and commitments that have been stimulated in areas of particular interest. Similarly, persons responsible for special programming should be knowledgeable about strategies for reinforcing, nurturing, and providing appropriate resources to students at those times when creativity and/or task commitment are displayed.

Are the Rings of Equal Size?

In the original publication of the three-ring conception of giftedness, I stated that the clusters must be viewed as 'equal partners' in contributing to the display of gifted behaviors. I would like to modify this position slightly, but will first set forth an obvious conclusion about lesson-learning giftedness. I have no doubt that the higher one's level of traditionally measured cognitive ability, the better equipped he or she will be to perform in most traditional (lesson) learning situations. As was indicated earlier, the abilities that enable persons to perform well on intelligence and achievement tests are the same kinds of thinking processes called for in most traditional learning situations, and therefore the above-average ability cluster is a predominant influence in lesson-learning giftedness.

When it comes to innovative giftedness, however, I believe that an interaction among all three clusters is necessary for high-level performance. This is not to say that all clusters must be of equal size or that the size of the clusters remains constant throughout the pursuit of innovative endeavors. For example, task commitment may be minimal or even absent at the inception of a very large and robust innovative idea; and the energy and enthusiasm for pursuing the idea may never be as large as the idea itself. Similarly, there are undoubtedly cases in which an extremely innovative idea and a large amount of task commitment will overcome somewhat lesser amounts of traditionally measured ability. Such a combination may even cause a person to increase her or his ability by gaining the technical proficiency needed to see an idea through to fruition. Because we cannot assign numerical values to the creativity and task commitment clusters, empirical verification of this interpretation of the three rings is impossible. But case studies based on the experience of innovative individuals and research that has been carried out on programs using this model (Reis, 1981) clearly indicate that larger clusters do in fact compensate for somewhat decreased size on one or both of the other two areas. The important point, however, is that all three rings must be present and interacting to some degree in order for high levels of productivity to emerge.

Summary: What Makes Innovative Giftedness?

In recent years we have seen a resurgence of interest in all aspects of the study of giftedness and related efforts to provide special educational services for this often neglected segment of our school population. A healthy aspect of this renewed interest has been the emergence of new and innovative theories to explain the concept and a greater variety of research studies that show promise of giving us better insights and more defensible approaches to both identification and programming. Conflicting theoretical explanations abound and various interpretations of research findings add an element of excitement and challenge that can only result in greater understanding of the concept in the years ahead. So long as the concept itself is viewed from the vantage points of different subcultures within the general population and differing societal values, we can be assured that there will always be a wholesome variety of answers to the age-old question: What makes innovative giftedness? These differences in interpretation are indeed a salient and positive characteristic of any field that attempts to further our understanding of the human condition.

This chapter provides a framework that draws upon the best available research about gifted, talented, and innovative individuals. There is a growing body of research offered in support of the validity of the three-ring conception of giftedness. The conception and definition presented in this chapter have been developed from a decidedly educational perspective because I believe that efforts to define this concept must be relevant to the persons who will be most influenced by this type of work. I also believe that conceptual explanations and definitions must point the way toward practices that are economical, realistic, and defensible in terms of an organized body of underlying research and follow-up validation studies. These kinds of information can be brought forward to decision-makers who raise questions about why particular identification and programming models are being suggested by persons who are interested in serving gifted youth.

The task of providing better services to our most promising young people cannot wait until theorists and researchers produce an unassailable ultimate truth, because such truths probably do not exist. Educators need to create learning environments with numerous opportunities to foster the development of innovation and support the potential innovators in our society. We can not ignore this vital potential asset.

References

Albert, R. S. (1975). Toward a behavioral definition of genius. *American Psychologist*, **30**, 140–151.

Alvino, J. (1981). National survey of identification practices in gifted and talented education. *Exceptional Children*, **48**, 124–132.

Barron, F. (1968). *Creativity and personal freedom*. New York: Van Nostrand.

Barron, F. (1969). *Creative person and creative process*. New York: Holt, Rinehart & Winston. In: B. S. Bloom (Ed.), *Taxonomy of Educational Objectives: Handbook 1. Cognitive Domain* (1956). New York: McKay.

Bloom, B. S. (1963). Report on creativity research by the examiner's office of the University of Chicago. In: C. W. Taylor & F. Barron (Eds), *Scientific Creativity: Its Recognition and Development*. New York: John Wiley.

Bloom, B. S. & Sosniak, L. A. (1981). Talent development vs. schooling. *Educational Leadership*, **38**, 86–94.

Campbell, D. T. (1960). Blind variation and selective retention in creative thought as in other knowledge processes. *Psychological Review*, **67**, 380–400.

Chambers, J. A. (1969). A multidimensional theory of creativity. *Psychological Reports*, **25**, 779–799.

Cooper, C. (1983). *Administrators' attitudes towards gifted programs based on enrichment Triad/Revolving Door Identification Model: Case studies in decision-making*. Unpublished doctoral dissertation, University of Connecticut, Storrs.

Cox, C. M. (1926). *Genetic studies of genius (Vol. 2). The early mental traits of three hundred geniuses*. Stanford, CA: Stanford University Press.

Delisle, J. R., Reis, S. M. & Gubbins, E. J. (1981). The revolving door identification model and programming model. *Exceptional Children*, **48**, 152–156.

Delisle, J. R. & Renzulli, J. S. (1982). The revolving door identification and programming model: Correlates of creative production. *Gifted Child Quarterly*, **26**, 89–95.

Dellas, M. & Gaier, E. L. (1970). Identification of creativity: The individual. *Psychological Bulletin*, **73**, 55–73.

DuBois, P. H. (1970). *A history of psychological testing.* Boston, MA: Allyn & Bacon.

Eiduson, B. T. & Beckman, L. (1973). *Science as a career choice: Theoretical and empirical studies.* New York: Russell Sage Foundation.

Freeman, F. S. (1962). *Theory and practice of psychological testing.* New York: Holt, Rinehart & Winston.

Gowan, J. C. (1978, July 25). Paper presented at the University of Connecticut, Storrs.

Gubbins, J. (1982). *Revolving door identification model: Characteristics of talent pool students.* Unpublished doctoral dissertation, University of Connecticut, Storrs.

Guilford, J. P. (1964). Some new looks at the nature of creative processes. In: M. Fredrickson & H. Gilliksen (Eds), *Contributions to Mathematical Psychology.* New York: Holt, Rinehart & Winston.

Guilford, J. P. (1967). *The nature of human intelligence.* New York: McGraw-Hill.

Harmon, L. R. (1963). The development of a criterion of scientific competence. In: C. W. Taylor & F. Barron (Eds), *Scientific Creativity: Its Recognition and Development.* New York: John Wiley.

Helson, R. & Crutchfield, R. S. (1970). Mathematicians: The creative researcher and the average Ph.D. *Journal of Consulting and Clinical Psychology*, **34**, 250–257.

Holland, J. L. & Astin, A. W. (1962). The prediction of the academic, artistic, scientific and social achievement of undergraduates of superior scholastic aptitude. *Journal of Educational Psychology*, **53**, 182–183.

Hoyt, D. P. (1965) *The relationship between college grades and adult achievement: A review of the literature* (Research Report No. 7). Iowa City, IA: American College Testing Program.

Hudson, L. (1960). Degree class and attainment in scientific research. *British Journal of Psychology*, **51**, 67–73.

Jones, J. (1982). The gifted student at university. *Gifted International*, **1**, 49–65.

MacKinnon, D. W. (1962). The nature and nurture of creative talent. *American Psychologist*, **17**, 484–495.

MacKinnon, D. W. (1964). The creativity of architects. In: C. W. Taylor (Ed.), *Widening Horizons in Creativity.* New York: John Wiley.

MacKinnon, D. W. (1965). Personality and the realization of creative potential. *American Psychologist*, **20**, 273–281.

Marland, S. P. (1972). *Education of the gifted and talented: Report to the Congress of the United States by the U.S. Commissioner of Education.* Washington, D.C.: U.S. Government Printing Office.

McCurdy, H. G. (1960). The childhood pattern of genius. *Horizon*, **2**, 33–38.

McNemar, Q. (1964). Lost: Our intelligence? Why? *American Psychologist*, **19**, 871–882.

Mednick, M. T. (1963). Research creativity in psychology graduate students. *Journal of Consulting Psychology*, **27**, 265–266.

Munday, L. A. & Davis, J. C. (1974). *Varieties of accomplishment after college: Perspectives on the meaning of academic talent.* (Research Report No. 62). Iowa City: American College Testing Program.

Neisser, U. (1979). The concept of intelligence. In: R. J. Sternberg & D. K. Detterman (Eds), *Human Intelligence* (pp. 179–189). Norwood, NJ.: Ablex.

Nicholls, J. C. (1972). Creativity in the person who will never produce anything original and useful: The concept of creativity as a normally distributed trait. *American Psychologist*, **27**, 717–727.

Oden, M. H. (1968). The fulfillment of promise: 40-year follow-up of the Terman gifted group. *Genetic Psychology Monograph*, **77**, 3–93.

Parloff, M. B., Datta, L., Kleman, M. & Handlon, J. H. (1968). Personality characteristics which differentiate creative male adolescents and adults. *Journal of Personality*, **36**, 528552.

Reis, S. M. (1981). *An analysis of the produchvity of gifted students participating in programs using the revolving door identification model.* Unpublished doctoral dissertation, University of Connecticut, Storrs.

Reis, S. M. & Cellerino, M. B. (1983). Guiding gifted students through independent study. *Teaching Exceptional Children*, **15**, 136–141.

Reis, S. M. & Hebert, T. (in press). Creating practicing professionals in gifted programs: Encouraging students to become young historians. *Instructor*.

Reis, S. M. & Renzulli, J. S. (1982). A research report on the revolving door identification model: A case for the broadened conception of giftedness. *Phi Delta Kappan*, **63**, 619–620.

Renzulli, J. S. (1977). *The enrichment triad model: A guide for developing defensible programs for the gifted and talented.* Mansfield Center, CT: Creative Learning Press.

Renzulli, J. S. (1978). What makes giftedness? Reexamining a definition. *Phi Delta Kappan*, **60**, 180–184, 261.

Renzulli, J. S. (1982). What makes a problem real: Stalking the illusive meaning of qualitative differences in gifted education. *Gifted Child Quarterly*, **26** (4), 148–156.

Renzulli, J. S. (1983). Guiding the gifted in the pursuit of real problems: The transformed role of the teacher. *The Journal of Creative Behavior*, **17** (1), 49–59.

Renzulli, J. S. (1984). *Technical report of research studies related to the revolving door identification model* (rev. ed.). Bureau of Educational Research, University of Connecticut, Storrs.

Renzulli, J. S. (2000). *Enriching Curriculum for All Students.* Arlington Heights, IL: Skylight Publishers.

Renzulli, J. S., Reis, S. M. & Smith, L. H. (1981). *The revolving door identification model.* Mansfield Center, Conn.: Creative Learning Press.

Renzulli, J. S. & Reis, S. M. (1997). *The schoolwide enrichment model: A how-to guide for educational excellence.* Mansfield Center, CT: Creative Learning Press.

Renzulli, J. S., Smith, L. H. & Reis, S. M. (1982). Curriculum compacting: An essential strategy for working with gifted students. *The Elementary School Journal*, **82**, 185–194.

Richards, J. M., Jr., Holland, J. L. & Lutz, S. W. (1967). Prediction of student accomplishment in college. *Journal of Educational Psychology*, **58**, 343–355.

Roe, A. (1952). *The making of a scientist.* New York: Dodd, Mead.

Shapiro, R. J. (1968). Creative research scientists. *Psychologia Africana* (Suppl. 4).

Simonton, D. K. (1978). History and the eminent person. *Gifted Child Quarterly*, **22**, 187–195.

Simonton, D. K. (in press). Exceptional creativity across the life span: The emergence and manifestation of creative genius. In: L. V. Shavinina (Ed.), *The International Handbook on Innovation*. Oxford: Elsevier Science.

Stein, M. 1.(1968). Creativity. In: E. Borgatta & W. W. Lambert (Eds), *Handbook of Personality Theory and Research*. Chicago: Rand McNally.

Sternberg, R. J. (1981). Intelligence and nonentrenchment. *Journal of Educational Psychology*, **73**, 1–16.

Sternberg, R. J. (1982a). Paper presented at the Annual Connecticut Update Conference, New Haven.

Sternberg, R. J. (1982b). Lies we live by: Misapplication of tests in identifying the gifted. *Gifted Child Quarterly*, **26** (4), 157–161.

Sternberg, R. J. (1984). Toward a triarchic theory of human intelligence. *Behavioral and Brain Sciences*, **7** (2), 269–316.

Sternberg, R. J. & Davidson, J. E. (1982, June). The mind of the puzzler. *Psychology Today*, **16**, 37–44.

Sternberg, R. J., Pretz, J. E. & Kaufman, J. C. (in press). The propulsion model of creative contributions applied to invention. In: L. V. Shavinina (Ed.), *The International Handbook on Innovation*. Oxford: Elsevier Science.

Terman, L. M. (1954). The discovery and encouragement of exceptional talent. *American Psychologist*, **9**, 221–230.

Terman, L. M. & Oden, M. H. (1959). *Genetic studies of genius: The gifted group at mid-life*. Stanford, CA: Stanford University Press.

Terman, L. M. et al. (1926). *Genetic studies of genius: Mental and physical traits of a thousand gifted children* (2nd ed.). Stanford, CA: Stanford University Press.

Thorndike, E. L. (1921). Intelligence and its measurement, *Journal of Educational Psychology*, **12**, 124–127.

Torrance, E. P. (1969). Prediction of adult creative achievement among high school seniors. *Gifted Child Quarterly*, **13**, 223–229.

Vernon, P. E. (1967). Psychological studies of creativity. *Journal of Child Psychology and Psychiatry*, **8**, 153–164.

Walberg, H. J. (1969). A portrait of the artist and scientist as young men. *Exceptional Children*, **35**, 5–12.

Walberg, H. J. (1971). Varieties of adolescent creativity and the high school environment. *Exceptional Children*, **38**, 111–116.

Wallach, M. A. (1976). Tests tell us little about talent. *American Scientist*, **64**, 57–63.

Wallach, M. A. & Wing, C. W., Jr. (1969). *The talented students: A validation of the creativity intelligence distinction*. New York: Holt, Rinehart & Winston.

Ward, V. (1961). *Educating the gifted: An axiomatic approach*. Westerville, OH: Merrill.

Weisberg, R. W. (in press). Case studies of innovation: Ordinary thinking, extraordinary outcomes. In: L. V. Shavinina (Ed.), *The International Handbook on Innovation*. Oxford: Elsevier Science.

Werts, C. E. (1968). Paternal influence on career choice. *Journal of Counseling Psychology*, **15**, 48–52.

Witty, P. A. (1958). Who are the gifted? In: N. B. Henry (Ed.), *Education of the Gifted. Fifty-seventh Yearbook of the National Society for the Study of Education (Part 2)*. Chicago: University of Chicago Press.

Zuckerman, H. (1979). The scientific elite: Nobel laureates' mutual influences. In: R. S. Albert (Ed.), *Genius and Eminence* (pp. 241–252). Elmsford, NY: Pergamon Press.

The International Handbook on Innovation
Edited by Larisa V. Shavinina

Innovation and Strategic Reflexivity: An Evolutionary Approach Applied to Services

Jon Sundbo

Department of Social Sciences, Roskilde University, Denmark

Abstract: Innovation in firms is discussed by using the notion of strategic reflexivity and applying a strategic innovation theory, which is within the evolutionary tradition. Market conditions and internal resources are the drivers of the innovation process. Firms manage their innovation process and market position through the strategy. Reflexivity is a social process in the firm in which managers and the employees consider the innovation process and the strategy. The social process includes organizational patterns, interaction and role patterns and organizational learning. The theory is applied to case studies of development of service concepts in service firms.

Keywords: Innovation; Strategic reflexivity; Evolutionary approach; Services; Social process.

Introduction

Aim of the Chapter

This chapter will discuss how firms develop by innovating. Innovation will be understood within the framework of the strategic innovation theory, which sees innovation as being determined by the firm's strategy (Sundbo, 1998a, 2001, 2002). Strategic reflexivity (Sundbo & Fuglsang, 2002) is an important concept within this theory. The concept refers to behavior which guides the innovation process. Strategic reflexivity leads to organizational learning. The aim of the chapter is to demonstrate how the concepts of strategic reflexivity and innovation can be used to attain a general understanding of the firm's development.

Innovation and Strategic Reflexivity

The topic of innovation belongs, in part, to economics (see for example Swedberg & Granovetter, 2001), but the approach taken here is primarily sociological. The explanation of innovation provided here takes a middle path between those who argue that innovation is determined by social structure and those who maintain that it is solely dependent on the acts of individuals. The view is that there is an element of truth in both of these approaches, which is to say that whilst social structure heavily influences the behavior of individuals, individuals' actions also determine social structure.

This view is common amongst sociologists (e.g. Bourdieu & Wacquant, 1992; Giddens, 1984).

Strategic reflexivity is a concept that can unite economic and sociological explanations of innovation. The contribution from economics is that we can understand the development of the firm in terms of a market game with an arbitrary outcome (a game in which the outcome cannot be predicted, but the outcome can be stated with a certain degree of probability). Since the outcome of market game is arbitrary, firms do not—and cannot—successfully just follow certain trajectories or just rely on being part of institutions (such as a stable network). The market changes continuously, as does the rest of the society, and the firm needs to act in new, path and institutional breaking ways to survive.

Economics has invented some theoretical approaches, such as game theory, to explain the situation with arbitrary outcome (Gintis, 2000; Owen, 1995). Such situations have also been described by cybernetics and systems theory (Beer, 1959; Boulding, 1985). However, this chapter will focus on the sociology behind the firm's understanding of possible outcomes and market strategy. The analysis emphasizes the social process of defining, maintaining and changing the way that the firm tackles a market with arbitrary outcome. The fundamental approach here is that the firm is not a living entity. A firm cannot assess probabilities and decide routes. People within (or perhaps outside) the firm assess and decide routes. They do so not as anarchistic individuals (such as the classic entrepreneur, cf. Schumpeter, 1934) or as social beings determined by structure (for example by institutions). They do assess probabilities and decide routes within a social system (namely the firm's organization) characterized by interaction, repeating patterns of acting, norms and so on as well as regularly changing patterns of behavior.

The firm formulates a strategy, which defines the goals for its future activities. Innovation is a social process. The social process is *reflexive* in the way that the employees and managers consider how the firm should develop to avoid the external threats from competitors, changes in customers preferences and political regulation, and further, how the firm could utilize the possibilities for new market positioning (e.g. marketing new products or decreasing prices through process innovations). The overall idea of how the firm should develop is encapsulated in the strategy.

This chapter thus discusses how the interaction processes within the firm can be understood as an explanation of the firm's innovation activities. The latter may only, to a certain degree, be done in theoretical terms because the interaction processes follow varied patterns. Empirical studies are necessary to investigate the different empirical forms of the interaction processes.

The strategic reflexive approach is within the tradition of evolutionary theory (e.g. Metcalfe, 1998).

Core Concepts

The core concepts that will be used in the chapter are the following:

Innovation takes place when either a new element or a new combination of old elements (cf. Schumpeter, 1934) is introduced. This is Schumpeter's classic definition of innovation which emphasizes both the act (to introduce something) as well as the result (the new product or the new organization). Schumpeter's definition will be employed here.

However, there are various types of innovation and innovative activity.

Product innovation refers to the introduction of a new product to the market.
Process innovation refers to the introduction of new production processes such as those enabled by new technology, or new work routines.
Organizational innovation designates the introduction of a new organizational form or a new management philosophy.
Market innovation denotes a firm's new market behavior such as a new strategy, new marketing, new alliances and so forth.

Innovative behavior concerns the firm's organization of the innovative activities: the production of ideas or inventions and their development into marketed products or practically implemented processes or organizational forms. Innovative behavior can be observed from outside the firm. The internal social process by which an innovation is developed and implemented is called the *innovation process*. The innovation process is an interaction between several actors (managers and employees); the constellation of these actors is called an *interaction system.*

There are many *small changes* in firms—products, processes, etc. Not any change is an innovation. To be classified as an innovation, the change must be substantial, which means that it is reproduced (cf. Gallouj, 1994; Sundbo, 1997, 1998b): if the innovation is a product (or a service), it is an innovation when it is produced in many copies. A process or an organizational innovation is one that is implemented throughout, or at least a large part of, the organization. However, even small changes which are not substantial enough to be called innovations may develop the firm. These will also be dealt with in this chapter.

Learning is the collection of experiences of what is good and what is bad. In relation to innovation and strategic reflexivity, it means the experience of which strategies and innovations were successful. It also refers to evaluating how the process of formulating strategies and organizing innovation activities can be successful. Learning can be both *individual* (each employee and manager learns) and *organizational* (the individual learning is dealt with others in the organization) (cf. Argyris & Schon, 1978).

Firm development concerns a firm developing either when it grows (in turnover, number of employees or profit) because of innovation or small changes or when it gains a competitive advantage by innovating (even though it does not grow immediately). Development can be achieved through innovation, small changes, learning, and by strategic reflexivity.

Structure of the Chapter

I will begin the chapter by discussing how the development of the firm can be understood theoretically. First, I will explain how strategic innovation theory, including the notion of strategic reflexivity, is based on evolutionary theory. Next, I will state the axioms of the strategic reflexivity explanation of innovation, which will then be outlined in the following section.

The theoretical discussion is applied to empirical cases, namely the development of service concepts in service firms.

A *service concept* is the total system of a service product (Desrumaux et al., 1998). A service concept includes the way the service product is produced and delivered (including the involvement of the customer and how the service solves the problem), the market position that the service enterprise and its products obtain, and the internal life of the firm (such as corporate culture, HRM and so on). The development of a new service concept combines product, process, organization and market innovations, as well as technological and social innovations.

The cases presented here have been selected because innovation in services normally includes many factors: new products, new processes, organizational and delivery systems, new market behavior, etc. Thus the cases can illustrate the various aspects of innovation.

The final section will sum up the theory and findings.

The Basic Position: A 'Soft' (or Non-functionalistic) Evolutionary Approach

The use of the concept of strategic reflexivity is not based on a universal interpretation of the development of the firm. The development of the firm is historical, which means that firms act in different ways in different situations. Further, the theory of innovation employed here is not a general theory of the firm. This theory only concerns how firms develop, which means how they change and innovate.

The theory of innovation presented here may be characterized as an evolutionary theory. However, it is not a 'hard' evolutionary theory such as functionalistic versions of Darwin's biological theory of 'the survival of the fittest' (cf. Smith, 1977). Such theories are based on the belief that a limited number of factors guide the selection process and the individuals or species which meet those requirements will survive. The task for the firm, according to the hard version of a social evolutionary theory, is to find out what these factors are, and then the future development of the firm can be predicted. The hard versions are based on linear determination chains.

The hard versions of evolutionary theory can also be found within economics and sociology. Hannan & Freeman (1989) have postulated a theory of organizational ecology concerning which firms will survive competition. An example of a hard version is also Nelson & Winter's (1982) attempt to find the ultimate mechanism of the development of the firm, a mechanism they called a 'routines'. Another example is the use of the concepts of knowledge and competence as explanation of the development of the firm, cf. Hodgson (1998), and Hamel & Prahalad (1994). The theories of social change in the 1950s and 1960s (Barnett, 1953; LaPierre, 1965; Moore, 1963; Ogburn, 1950; Rogers, 1995—most of which were based on Tarde, 1895) are examples of sociological theories, which used the diffusion of technology as a core concept to explain social change. A 'hard' evolutionary theory is fundamentally functionalistic: those organisms (e.g. in economics—firms) that develop the best functions in relation to the environment will survive.

Modern biology has, based on studies of ecological systems, launched a 'soft' version of evolutionary theory (Eldredge, 1999; Jørgensen, 1992; Krebs, 1988;

Prigogine & Stengers, 1986). The development of species is created by many factors and by relations between these factors. There are many determinants, and even relations may be determinants (cf. the discussion within physics between Bohr, Einstein and others about relativity, e.g. that measuring instruments distort the results). Such interdependencies mean that imbalances in the system often happen, but there are also periods of balance. The development mechanisms are extremely difficult to grasp because there are so many, and they differ from situation to situation. Future development cannot, or only with extreme difficulty, be predicted and general laws may only be stated in terms of probability. In such a system, the individual or species has better possibilities to influence the system to its benefit if this person or species can act consciously. However, the individual or species may be prepared of other individuals and species will also attempt to influence the system to own their benefit. The entrance of other actors means that not only will there be a competition between these two actors, but the competitive behavior of the actors may also affect the system, and thus change the rules of survival. This 'soft' evolutionary model can be applied to social and economic systems such as a firm, and this is what is attempted here. The 'soft' variant is not functionalistic: it claims that the firms which survive and grow are not necessarily the 'fittest'. The firms which are most successful at the moment can be the least successful in the very near future. The firm can not rely on routines (cf. Nelson & Winter, 1982) and existing capacities: it must continuously reflect and frequently develop new strategies to influence the market system and to grow. It may even change the rules of survival.

The 'soft' evolutionary system can be characterized as a 'chaotic system' as in the modern version of the mathematical chaos theory (Devaney, 1992). This theory says that the system is complex, and there are no straight lines and no simple cause–effect relationships, so one cannot predict the outcome of a certain causal factor. The system may seem to be without any order, but, some patterns are repeated throughout the system. For example, cascading running water seems to flow coincidentally, but some types of whirl can be found at several places in the whirlpool. Thus there is a certain probability of finding a particular whirl type in a certain part of the whirlpool. Chaos theory has been applied to social systems as well (Kiel & Elliott, 1996; Quinn, 1985; Stacey, 1993). An example of how it has been applied is project-organized firms such as those which are research- and development-intensive. Their organization may be without much formal structure, having no formal leaders, and every person has several roles in different projects. The organization seems to be anarchistic to a certain degree, However, one may find some patterns that are repeated. For example, in most project teams, certain roles are developed such as those of the conflict-solving person, the critical person who asks questions, and the person who presents creative, new ideas, etc.

The 'soft' evolutionary system could also be characterized by a metaphor drawing on the modern version of health, a series of physical, psychological and social factors, all of which influence our health. A complete explanation of the total system cannot be provided, but explanations of parts of the system can, however, even sub-systems can be influenced by changes in other sub-systems. If, for example, social conventions among the managers in the managerial system are changed, they may influence the informal social conventions that exist among the employees making them less creative and less interested in participating in innovative activities.

The insecurity of the evolutionary system is the reason why strategic reflexivity is so important. Strategic reflexivity is the firm's attempt to manage its competitive behavior. The firm should beat its competitors and set the rules under which the market system functions so that they are advantageous to that particular firm.

Axioms of the Innovation Theory

I will now discuss how innovation and change can be theoretically understood by using the notion of strategic reflexivity. This is a new approach and demands that some axioms be stated. I will therefore start with three statements that define the framework for the coming detailed discussion.

First Statement: Innovation Concerns Human Beings Within the Firm

In recent decades, in both economics and business administration, there has been an increasing interest in innovation, knowledge management, technological development, organizational learning and so forth. Innovation may be seen as the core notion within this group. Innovation is something that is going on in firms and in systems of firms (cf. Edquist, 1997; Nelson, 1993).

Fundamentally, the reason for the increasing interest in these topics is a desire to understand the development of the firm as an explanation for economic growth. If we are really to develop such an understanding, we need to go back to Schumpeter's work (1934) and his views concerning economic development. To Schumpeter, economic development means new solutions to society's problems and needs. One could see the process of developing new solutions in terms of social change, which a sociological discipline was called (e.g. LaPiere, 1965; Moore, 1963). However, one does not need to formulate the notion of social change in a functionalist manner within a 'hard' evolutionary approach. The process of diffusing the innovation does not need to be rational in the sense that a problem can be defined, and the new element solves

that problem. The diffusion of a new element can be caused by people just choosing another way of living, which the new elements make possible (Barnett, 1953).

Introducing the concept of strategic reflexivity re-introduces the idea of social change but in a 'soft' evolutionary way. However, here the idea is restricted to social change within firm organizations, and it is also limited to economic actions. The latter means actions which—directly or indirectly—lead to initiatives taken in the market.

The objective of the strategic reflexivity approach thus is broader than new, physical products of technology. The approach emphasizes the behavior that leads to the development of the firm, economic growth, and social change. Innovation is traditionally identified as the result of that behavior—a new product, a new production process, etc. (Coombs et al., 1987; Freeman & Soete, 1997). The results (the innovations) are of course important since they change the world. However, if we want to understand and explain the development process so that we are in a position to guide it, we need to study it. The process is therefore more important than the results. That is why innovation in this chapter has been defined as the process as well as the result (however, compared to the traditional innovation theory, the emphasis is on the process).

Innovation could be interpreted as a process of social change. New elements are invented and diffused throughout society changing it, our lives or particular sub-structures of the society (such as how we transport ourselves, how we spend our free time, how we work, etc.).

Actions, including innovation, are undertaken by persons in a social system. The approach here is to understand innovation and change as sociological processes which take place within the firm organization and between the persons within the firm and outsiders.

Second Statement: The Changes Are Neither Institutional Nor Unpredictable Anarchical

The process of change that innovation (and the small changes) brings about do not follow a constant trajectory and are not totally anarchical and unpredictable. Innovation and change is a trial-and-error process, but one which can be managed to some degree. However, the management must be ready to introduce new principles of management in each phase of the trial-and-error process, since the conditions for the innovative activity may change for each phase.

Understanding innovation in terms of the notion of strategic reflexivity differs from the institutional approach that has become popular in social sciences during the last decade (Hodgson, 1993; Nielsen & Johnson, 1998). Innovation patterns and change are not institutionalized processes and cannot be. Some institutional theories try to respond to this fact and maintain the institutional approach by saying that change itself can be institutionalized (Johnson, 1992; Nelson & Winter, 1982). A social system such as a firm can be oriented towards change, and change becomes a value and a norm. However, the point is that the conditions of change are constantly changing, and so new goals, values, norms and types of actions are regularly needed. Thus it makes no sense to call such a change system an institution. Institutions are systems where norms, action patterns and goals are relatively permanent (e.g. the family, the church). To call the firm's innovation and change system an institution within the framework outlined here would be a complete watering down of the concept 'institution', which would then just mean a social system or structure. The interpretation of the development of the firm presented in this chapter thus implies a critical approach to institutional theory's explanation of change.

The interpretation also implies that a firm's development cannot be explained by any deterministic, trajectorial approach. Theories of path dependency have been presented (e.g. Dosi, 1982; Freeman & Perez, 1988; Teece, 1986), stating that people's behavior become institutionalized. People follow well-known paths, even when they participate in a change process. The fact that they participate in a social system also leads them to follow existing paths because the change process is a social process, and people are conservative when they operate within a social system. Innovations are seen as a result of logical trajectories which exist within the system of innovation (cf. Dosi, 1982; Freeman & Soete, 1997). This is only partly true. People and even social systems are able to break the inertia and trajectory and create new paths. They can innovate rules and new ways in which the innovation process is managed and executed.

The path dependency theories come from within the technology-economic paradigm (e.g. Dosi et al., 1988; Mensch, 1979; Piatier, 1984; Rosenberg, 1972—cf. Sundbo, 1998a), which emphasizes the technology and science push. The technology and science push precondition is valid in some industries and firms (e.g. high-tech) where the scientific discoveries and technological inventions are so useful and marketable that they guarantee market success. However, this only happens in the minority of situations. In most firms, it is difficult to market an innovation and difficult to know which innovations will be successful. Therefore, the firms will act carefully and follow the way of strategic reflexivity. The rise and fall of the ICT-software-Internet industry in 1999–2000 illustrates how, in such a short time, even a high-tech industry can rely on technology push as its driving force.

Recently, there has been a tendency to treat knowledge as the key factor in explaining the development of the firm and economy (Eliasson, 1989; OECD, 1996). However, this is also insufficient. Some theories explain firms' behavior as if knowledge in itself

presents the solutions and way forward. According to these theories, knowledge is the determinant and therefore the most important factor in explaining economic growth (e.g. Boden & Miles, 2001; OECD, 1996). Within the framework adopted here, knowledge is not such a determinant. Knowledge is an important raw material, but it is people who, in the social change process, create changes by using knowledge. The people decide which knowledge to select. Firms may often follow a technological or other trajectory (e.g. a service professional trajectory; cf. Sundbo & Gallouj, 2000). However, they may often have greater success by not adapting to the most advanced steps in a trajectory, but by using low technology or technology from another trajectory, adding services to the product, or just changing the trajectory. Innovation through strategic reflection is not just knowledge but action, as the entrepreneur theory states (Binks & Vale, 1990; Schumpeter, 1934). The human will to act—even if it does not make sense in terms of existing knowledge—is the most important. This assumption also makes innovation into a social process where self-interest, firm policy and so forth play a role.

Certain patterns can be established for firms' innovative behavior. Such patterns could, for example, be that some firms follow a technological trajectory where the technological inventions determine the innovations that the firms develop. Another pattern could be that the firm establishes a strategy, picks up different technological elements and mixes them with new organizational and marketing elements (cf. the chapter by Veryzer in this book).

An individual-anarchical approach has characterized the early innovation and social change literature (e.g. Johannisson, 1987; McClelland, 1961; Menger, 1950; see also Sexton & Kasarda, 1992), inspired by the individual approach of the two classic authors, Schumpeter (1934) and Tarde (1895). The entrepreneur has been seen as the change agent and the one who creates innovation, and the entrepreneur is an individual. Since entrepreneurs are creative, the change process is anarchical and unpredictable; there is no pattern. However, single individuals alone do not create innovation within modern firms. They may play an important role (cf. Drucker, 1985; Kanter, 1984; Pinchot, 1985), but the innovation is developed within a social system in which several individuals take part in the process. The implication of these assumptions is that the innovation process in the firms follows certain—but not universal—patterns and is not completely anarchical.

Third Statement: Managers are Seeking Interpretation and Social Interaction

Managers are seeking the efficient solution for the firm's development. To find that demands interpretation of the environment and of their own possibilities (statement of the internal resources; cf. the resource based theory of the firm, e.g. Penrose, 1959; Grant, 1991). A seeking interpretation becomes the result of the attempt to efficiently develop the firm. The managers within the firm look for the best interpretation of the future way to go and the best kind of innovations. However, they are uncertain of which interpretation and which solution is the best. Therefore, they turn to other people—particularly the employees —to discuss and develop the best solution. They engage in reflexive strategic behavior, which is an interactive social process that creates innovation (as already stated; cf. also Gallouj, 2002).

The Development of the Firm: Innovation and Change by Reflexive Learning

In this section I will outline the theory of strategic reflexivity. I will first discuss the relationship between innovation and small changes, and how both these phenomena can be included in the theory. Next, the discussion of the theory starts by emphasizing its point of departure, the market and strategy, which are connected to firms' market relations. After that, I will discuss the concept of reflexivity, which emphasizes the internal processes that take place in the firm. I will sum up by discussing how the innovation and change processes result in reflexive organizational learning.

Innovation and Change

The border between innovations and what are called here small changes is fluid (cf. Sundbo, 1997). The small changes are improvements in given situations and are not reproduced, and individually they do not, therefore, lead to a significant development of the firm. How can the management benefit from the many small changes? By making them as reproducible as possible, which is achieved by creating a general learning system in the organization. The other employees in the firm ought to imitate these changes, and the organization should collect knowledge about the experiences.

Firms develop through a mixture of two types of activities. First, more formalized innovation activities, which may be both top-down and bottom-up. Second, many small changes which are made general through organizational learning, which means the imitation of best practices within the firm.

The firm develops by learning from the innovation activities and the launching of new products and other innovations, as will be explained later. This form of development can be characterized as reflexive learning based. In the reflexive organizational learning the firm unites the innovation and small change efforts into a common one, which is to develop a better awareness throughout the company of possibilities for the future development of the firm.

Innovation is a Market-Oriented Development Process

The point of departure, and the end goal, for any innovation process is the market because that is where

the firm makes its profit, which is the ultimate goal. Marketing as an explanation of the development of innovation processes can also be found in the chapters by Veryzer and Trott in this book (see also Kotler, 1983).

The firm has to interact with the market. It must 'read' market trends and must innovate either to keep up with the trends or to alter them and be in a position to create the market trend. The firm also has to follow the effects of its own market behavior to see whether it is successful or not.

This market-oriented approach emphasizes the pull side of the development process—how the market possibilities determine the development process. However, as mentioned earlier, the firm is also characterized by following some trajectories (the push side).

Strategy

I will now go on to the core discussion of the firms' development process. The notion of strategic reflexivity is the core of the development theory discussed here. Therefore, I will start by defining what I mean by strategy in this section, and by reflexivity in the next one.

Important for the firm's development is not the actual market situation, but the future one. The firm has to make an interpretation of where the future market, including the wants of the customers and the competitors, is moving. For example, whether customers' preferences in the future become more ethically determined, they want extra services attached to the products or new, innovative competitors may occur in the future. Then the firm has to choose its direction and make a strategy which defines its goals in terms of its market position and its development. Furthermore, the strategy defines (at least generally) how the firm should reach these goals. The strategy is the means that the firm uses to navigate through the uncertain waters of the market.

The strategy is based not only on market possibilities, but also on the internal resources (competencies, financial strength, customer relations, innovation capabilities, etc.) that the firm possesses (Sundbo, 2001). Further, the strategy is based on the organization and the employees (including the managers) that the firm has and their competencies. The future development of the firm must be created by that organization and the employees and managers, or by some other employees or managers that the firm hires externally. The firm may also change the organization and its way of functioning, for example by creating a new corporate culture, a new organizational structure, or by employing new people with new competencies.

When I say *the firm* interprets the future market possibilities, and decides the strategy, I mean some person(s) within the firm, namely the top management. It may very well be that the other employees are also involved in this process, or it may be that the interpretation and strategy are formulated by only some persons within the top management, or by managers outside the top management. However, the top management has the final responsibility.

The strategy is the framework for making decisions about the development of the firm, including innovation, and is also a source of inspiration for innovative ideas (since the strategy tells people where the firm wants to go). Innovation and change activities should follow the strategy and be within its limits. The top management has the task of ensuring that the firm's development is in tune with the strategy, but the firm's development is not a totally rational process where the management can guarantee the outcome. The management will attempt to carry out the process rationally and will succeed in many situations but also fail in many others. The management could fail, for example, because the interpretation of the market and competitors' moves turns out to be mistaken, or the people in the organization act in unforeseen ways, or the necessary resources cannot be procured.

The strategy is therefore continuously under consideration. If it seems to be wrong, it should be corrected. Again this is not always a smooth rational process. It is a difficult decision when and how to change the strategy, and it cannot be rationally based, although some rules of sense may be used.

Strategy has been understood in different ways (cf. Chaffee, 1985; Mintzberg, 1994; Sundbo, 1998b): rational, processual, intentional, and political. Here I present an interpretation of strategy as interpretative and processual, which means that the strategy is a result of reflections about the business environment. Further, strategy is also a result of reflections over internal resources, and an interpretation of where the business environment will move in the future.

Reflexivity

Until now I have discussed strategy. Strategy marks *that* the firm has to find its way through an uncertain but arbitrary predictable future market. The other part of the central concept employed here is reflexivity, which concerns *how* the firm finds its way, and when it should change its ways or perhaps even strategy. The concept of reflexivity belongs to theories of organization, and emphasizes the process aspect of those theories (which is another aspect of the organization apart from the structural one).

Reflexivity as a Form of Human Behavior

Reflection is based on the assumption that people seek to act rationally. They try to set up a strategy for themselves, or the social system in which they participate. However, they cannot in situations which are changing rapidly just rely on existing routines and trajectories, and consequently individuals' behavior becomes reflexive.

Reflection occurs at the level of the individual member of the organization whether he or she is a manager or an employee. Reflection means that the individual considers an unknown future with arbitrary outcomes. One must mobilize all the formal, trajectorial knowledge that one has, and use it creatively to produce new solutions. One has to break norms and patterns to develop new solutions, and get them accepted and implemented. At the same time, one has to ensure that the social system (in our case the organization) is not completely dismantled, because learning from the past, which is useful, may then end, leading to the demise of the firm. Otherwise, one has to choose to follow existing routines and patterns.

Is reflection an automatic reaction, a biological, behavioristic response to a stimulus (as Pavlov's behavioristic theory expresses? (Pavlov, 1966)) Or is reflection an intentional behavior, manifestation of which demands an act of will, and here even an act of creativity (which means that the individual or a social system has to remember to analyze the situation, and that it may be possible to do something new in a situation)? The assumption here is definitely the latter. Studies of innovation tell us that there are no systems, based on nature, which simply respond to economic stimuli.

Reflections are considerations that individuals make about innovation possibilities, and the development of the firm based on three premises:

(1) The goal for the development of the firm (which is expressed in the strategy, which is mostly, but not always, taken as the framework for reflections);
(2) The knowledge that the individuals possess;
(3) Ideas of what could be concretely done.

The concept of reflexivity has become popular within contemporary sociology (e.g. Bourdieu & Wacquant, 1994) and is one of the core concepts of Ulrik Beck's work (1992). He discusses reflexivity in relation to what he describes as the risk society—a complex and dangerous world, which is primarily due to environmental pollution, in which we now live. The citizen becomes occupied by the possibility of considering what pollution means for his or her life and what can be done about it. The citizen becomes dependent on experts for making these considerations. Another approach within sociology emphasizes that people themselves become reflexive or employ strategies towards the world in their everyday life. This latter approach is based on the assumption that contemporary citizens have become more individualistic, better educated, more career-oriented, etc. This approach has been argued for by Anthony Giddens (1984), for example. However, such ideas are not new; similar thoughts can be found in Durkheim (1933), where he expressed his worries concerning the early industrial society.

Reflexivity as an Organizational Factor

Reflection implies interaction. The individual makes his personal reflections in interaction with other people, and the firm's reflections are also based on interaction: the result of a collective, social process as mentioned before. Generally, the group will be the unit which reflects, and may often be good at doing so, but individuals may do better. Despite being potentially better, individuals often talk to others to reduce their responsibility for unsuccessful innovations.

I view that part of the organization which carries out the innovation activities as being a dual organization (cf. Daft, 1978; Sundbo, 1998b). On one level it consists of the top management, who primarily have the task of managing the firm's development, and lower line management which is responsible for day-to-day production. The top management may consist of one or more managers. Generally, one may talk about a management structure (which may differ depending on the number of managers). The top management concentrates on the firm's general lines of the firm's development, which primarily means the strategy, but also ensures that the innovation and the reflection processes are running smoothly. On the other level, the dual organization consists of the employees and the middle managers They are all placed within a formal structure, but they establish a network structure, which is informal, loosely coupled and interactive. This loosely coupled network structure is not a part of the formal structure of the organization; it cannot be found in the official schemes depicting the organization. Nevertheless, this structure exists and is created by the organization's members. It is part of their social nature, and it creates a common organizational culture. The network structure is a 'greenhouse' for innovative ideas and informal reflections because it constitutes a collective, a creative milieu. The network structure operates in relation to specific tasks such as innovations, analyses, the storing and use of knowledge and experiences, etc.

The development process (innovation and reflection) may be a top-down process where the top management takes the initiative and involves the employees. It may also be a bottom-up process where ideas, knowledge and entrepreneurship come from individual employees or from a collective process within the network structure, or managers start a process of development of innovation and reflection. Often it will be a mixture of both. Whatever, the top management will guide the process and finally decide upon the usefulness of the innovations and what consequences can be drawn from the reflections that have been made by the employees and managers in the network structure. The top management structure and network structure interact. Their mutual life is often harmonious and complementary. The top management needs the network structure to generate innovative ideas. The top managers can rarely get the ideas they need themselves and they cannot just rely on scientific or expert trajectories. They also need the network structure for reflections on whether the strategy, the innovative ideas, and the way the innovation process is organized are the right ones in a given market situation. However, if the management were not there, the firm would be broken down by individual, anarchic entrepreneur behavior. The top management keeps the firm together, the network structure cannot contribute to the firm's development without the effort of the top management. However, the relation between the two parts might also be conflictual and destructive.

Special expert departments such as an R&D department or a strategy department, which have development (innovation, strategy, analyses or reflexivity) as their formal task, are in a third position in the organization. Mostly expert departments are related to the top management because they are placed in a hierarchical structure, but they may also play their own game, which will be organized more formally than the aforementioned bottom-up process.

Does reflexivity here mean leaving the formulation and fulfillment of the business strategy to experts or to 'ordinary people', which here would be the employees and middle managers in the firms (cf. the discussion in 'Reflexivity as a Form of Human Behavior')? Traditional innovation theory (e.g. Coombs et al., 1987; Freeman & Soete. 1997) emphasizes experts as those who innovate (researchers in R&D departments, marketing experts, etc.). The assumption here is that the effort of the 'ordinary people'—the employees and middle managers—is the most important, but that they and experts work together in the innovation process. The 'ordinary people' often ensure that innovation takes place, since experts outside the R&D department may hinder innovation (cf. Trott's analysis in his chapter of this book on how marketing experts hinder innovation).

Sometimes, the expert department (for example an R&D department) undertakes the whole innovation process, but this is very rare. Since the market and the customers' needs and reactions to innovations are the crucial factors, the firm needs information and reflections concerning these factors. The employees interact with the customers and are therefore play a necessary part in the process. Besides, many firms (service firms and small firms in particular) have no R&D department or any other expert department.

The involvement of the employees in the innovation process and the reflection process may be formally organized. This organizing will often take the form of creating project teams. The empirical part of this chapter investigates the forms the reflexivity process may take under different conditions.

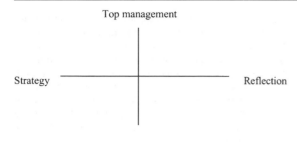

Figure 1. Model of the firm's core development factors.

The firm interacts also with several external actors and participates in formal and informal external networks to develop innovations. However, the external network is not a crucial notion in understanding the development of the firm from within the strategic reflexivity framework (as it is in the 'Uppsala school'; cf. Håkansson, 1987).

The Functional Core of Strategic Reflexivity— Organizational Learning as the Overall Process

To sum up the theoretical discussion, we can state the core of the interpretation of the firm's development function presented here and express it in Fig. 1.

The development of the firm is a dialectic process between action and analysis. The strategic part of the core notion compares to the action aspect of the model, and the reflexive part to the analytical aspect. By dialectic, I mean a situation where there are two factors, which are in opposition to each other but nevertheless only function if they interact with each other. Reflection is in conflicts with strategy because when actors reflect, they question strategy. Nevertheless, a workable strategy only exists if there are continuous reflections about its appropriateness.

The success of the firm depends on the mixture of strategy and reflection, and the management of the daily production because innovation and development are not always the primary goals. Firms develop cyclically. They will have steady periods in which they are driven by some non-innovative logic of production. If the firm is going well, there is no need to make radical changes that could destroy it. Innovation and change is risky. Successful firms may be assumed to take steps towards development only when they are aggravated by situations in the market situation (or a little before).

The other dimension in Fig. 1 is the dialectic between the top management and the informal network structure.

The whole dialectic process can be characterized as one of organizational learning, which is the overall attempt of the firm to guide the development. Organizational learning is the experience of how the four

elements of this model (cf. Fig. 1) should be combined in concrete situations. Even organizational learning can help the firm only some of the way. There are often several new factors in new situations, with the consequence that former experiences cannot be used. The learning must be combined with an entrepreneurial daring to try something new—an action perspective. I will come back to how organizational learning is carried out in 'Reflexive Feedback and Organizational Learning'.

A Case Study: The Creation of New Service Concepts and the Development of the Firm

As I have stated, the concrete way in which strategic reflexivity is developed varies according to the situation. It is assumed that certain patterns can be found, and the conditions for these patterns specified. These patterns are discovered by concrete empirical analyses, one of which we will now turn to.

Innovations in services can take many forms (cf. Gallouj, 1994). Earlier empirical analyses (Sundbo, 1998b) have demonstrated that the strategy defines the framework for the innovative activity. Here I will investigate whether strategy is developed reflexively and whether the process of developing innovations is reflexive. If the process is reflexive, it will combine considerations and corporate entrepreneurship, creativity, knowledge and learning.

I will begin this section by presenting the cases. Next I will describe how the service firms have formulated their strategies, and then move on to describe how the innovation process was carried out in the firms. Finally I will discuss learning.

The notion of strategic reflexivity and the theoretical statements established above will be the general framework for understanding the development processes. The concrete patterns that the development process follows will be investigated inductively. In the final part of the section, I will attempt to conclude whether several more general (although not universal) patterns can be identified. As far as general patterns can be identified, some models will be presented.

Empirical Date and Method

The empirical data are interviews and documentary material from case studies in different service firms. The aim is to investigate the formulation and importance of strategy, how the innovation process is undertaken, and how organizational learning is carried out. This investigation is done by studying how the service firms develop new service concepts.

Eight case studies will be used. The cases are the following: Two insurance companies, ((a) Hafnia and (b) Family Insurance), a bank ((c) Lån & Spar), a payment and credit card company ((d) The Payment Company), a chain of lawyers ((e) Ret & Råd), an engineering consultancy company ((f) COWI), a company providing cleaning and other manual services ((g) ISS) and (h) a municipal home help in one district of Copenhagen. The data have been collected throughout the 1990s in a project concerning the organization of innovation in service firms (Sundbo, 1998b) and a project on service development, internationalization and competence development (SIC, 1999).

The cases are analyzed in greater detail elsewhere (case (a), (b), (d): Sundbo, 1998b, case (c): Sundbo, 2000, case (e): Henten, 2000, case (f): Larsen et al., 2001, case (g): Sundbo, 1999; Illeris & Sundbo, 2000, case (h): Fuglsang, 2000, 2001).

The Formulation of Strategies

The firms had different ways of formulating the strategies and made different strategies. The interesting questions are: Why did they take different paths, and what made them choose different strategies?

Different Patterns of Strategy Formulation in Service Firms

I will describe briefly what the firms did—which means what the top management and other employees did. This can be placed on a scale, which is not exact, relating to the degree of control by the top management. I will start with the firm where top management's control was the greatest and work my way down to where it was least.

The insurance company Hafnia, was dominated by a top entrepreneur (who was employed as a general manger but acted like an entrepreneur. cf. Sundbo, 1998b). He decided on the strategy on the basis of his own personal purposes: insurance companies have large capital because people pay the premium before the company has to pay for damages. He wanted to use the company's fortune for investment to create industrial activities. He wanted to be a new type of capitalist. Therefore, the strategy was to create as large a premium as possible, even if the company might have high damages. The company thus selected a market segmentation strategy: they chose the private upper income market. There was a clear entrepreneurial strategy, but no reflection about whether this strategy was wise. The role of the employees was only to restrict the 'wildness' of the top entrepreneur and that was only pure reaction, not prospectively reflexive. This case demonstrates that the absence of reflexivity can be a disaster, because the company went bankrupt as the entrepreneur and some of his closest managers—who were not reflexive either—were 'too creative' in the firm's investment and accountancy policy. The company went into deficit and then went bankrupt. The social system within the firm had become a back up and follow system for this one entrepreneur and was not a reflexive one.

The home help's strategy was also decided from above, i.e. by the political system. There had been

criticisms of how the public home-help system functions for the users, absenteeism among the employees had been large, and there had been several organizational problems. A new strategy was introduced to improve client quality and job satisfaction by making the system more flexible. There had been much reflection during the political process about the purpose of the home-help organization. Further, the detailed goals and means had been discussed nationwide by professional groups within the system. However, the reflections seem to have limited effects nationwide since the same problems continued. In this particular district the strategy of the home help was adapted in a way that made it work because of the professional tradition and a particular team system that was developed (see 'Examples of Service Concepts').

The Payment Company also had a top-down strategy process. The top management formulated the strategy. However, the firm was owned by a number of banks, and it had to be very careful in its strategy not to provoke the owners, who could very easily become competitors as well. This forced the top management to take a grip of the strategy process. In return, The Payment Company had established an innovation department. Its task was to initiate the development of new service concepts among the employees and to develop a detailed strategy based on innovative ideas and the strategic arguments behind them. They should then be responsible for its implementation. The reflection thus was professionalized and the responsibility for reflection allocated to a formal department. The result was a cautious strategy of product development.

Family Insurance had a top-down strategy process as well. The choice of this form of process was due to experience. The company had had a top entrepreneur as managing director a few years before I carried out the interviews. He had got the employees and managers so occupied with innovation and change that it took up time from daily production and sale, and the company nearly went bankrupt. However, the employees and managers were still involved in innovation activities, which led to ideas concerning strategic changes, and the process was more controlled by the top management. In this case, reflections concerning the strategy were based on experience. The outcome of the process was a market segmentation strategy—the company chose the lower-income private market.

ISS is in an in-between position. The strategy process was run by an office in the headquarters, but the strategy was discussed by the managers throughout the company. The particular structure of the organization—many employees worldwide (about 250,000) and a large labor turnover—made it impossible for the firm to involve the employees and even lower managers. It was the upper managers, who were involved in the strategy process. The involvement of the upper managers was achieved by having them discuss the strategy in local groups combined with general meetings in which all involved managers were assembled. The process resulted in a strategy which emphasized productivity increase and the development of new, specialized, manual services.

In COWI the top management had the responsibility for the strategy. The market for engineering consultancies was undergoing constant change so the firm had adopted a fluent strategy. The ideas for new strategy elements mostly came from the daily product, which is solving engineering problems for customers, organized as teamwork. The company combines technical solutions with administrative, sociological and economic approaches such as the creation of a democratic administration in a developing country. Reflection was continuous, based on the experiences from projects that should solve the customers' problems. The top management combined these with speculations about what competitors—who potentially can come from other industries—will do in the future. The strategy formulation process resulted in a careful product development strategy. Old products were renewed and some new products (among those management consultancy) were introduced.

Lån & Spar has had a very broad strategy process in which all employees and managers have been involved. The bank is a small bank in an industry with tough competition, so the strategy is a matter of survival. The top management organized training courses where the employees learned what a strategy is, were obliged to discuss the bank's strategy and were requested to present proposals for a new strategy. The final strategy was a result of very comprehensive interactive reflection processes. The top management guided the process to a certain degree since the general manager decided on the final strategy with help from his nearest managers. The bank had chosen a market segmentation strategy which went for the middle and upper income private customers.

Ret & Råd had the most extreme bottom-up process. The chain was established by a group of lawyers, who could see that the market for lawyers would be dominated by large firms. They believed that if small, independent lawyers were to survive, they needed to unite in some way. The strategy thus was one of economics of scale. The strategy of Ret & Råd was also very reflexive. Since Ret & Råd was a chain of independent lawyers, all partners had to agree to any change in the strategy. This condition ensures that everybody is equally involved in the reflection and decision process (perhaps except the employees of the single lawyer), which means that they in fact do not need to be reflexive, but only need to say 'no' to hinder the formulation of a strategy. A common secretariat has been established. The leader of the secretary wanted to develop the chain and strategy further, but it had been extremely difficult to do this. Such a chain of independent professionals seems to be conservative,

and inertia and actual self-interest seem to be impediments to the long-term development of the strategy.

General Lessons from the Service Firms

The lessons from the cases are:

The service firms have generally become more strategy- and reflection-oriented during the period under study (the 1990s). In the beginning, they were mostly focused on competitors and only moved when the competitors did. The reflexive basis was limited to competitor watch. Later, they all employed a strategy and were broadly oriented towards many actors, trajectories and internal resources and competencies. Their reflection basis was broader.

The different patterns of strategy formulation can be summed up as the following:

Some of the firms (such as Hafnia) had a top-down process of strategy formulation. Others (such as Ret & Råd) had a bottom-up process. However, most of the firms had a process that was both top-down and bottom-up. These firms had a dual interaction form of strategy formulation. This means that the top management had the final responsibility and control of the strategy process, but the employees and managers were involved in the process. The employees made the reflections in interaction with other employees and actors. The top management also discussed the strategy with their closest managers and sometimes other employees. The two systems, the top-down management system and the bottom-up loosely coupled network system among the employees, interacted with each other. The interaction implies that the initiatives and ideas came sometimes from above and sometimes from below, and these two sets of initiatives and ideas met each other and sometimes melted together. They acted in accordance with the theoretical model established in 'Reflexivity as an Organizational Factor'.

The question to address here is: Which pattern is the most efficient if the firm is to develop successfully? This is difficult to determine. In the cases presented here the two most extreme (top-down and bottom-up) ones were not efficient. The pure top-down system of top entrepreneurship in Hafnia led to bankruptcy because of insufficient reflection, and at the other extreme anarchical entrepreneurship formerly executed in Family Insurance nearly did the same. The extreme egalitarian system of Ret & Råd led to action paralysis. The home help is part of a public, political system with many actors involved in the formulation of the strategy. This did not lead to an efficient strategy, since the clients and employees saw many problems which the system had difficulties in solving. This result also suggests that a decision and reflection system that is too broad but has a weak top management is not the most efficient development system. The system in the home help is not sufficiently dual.

The majority of service firms, who had a dual form of strategy process followed a pattern in which there was a more equal balance between entrepreneurship, top-down control and employee involvement. Such a pattern was the case with ISS, COWI, Lån & Spar and Family Insurance (during the period studied). These firms were all characterized by a defensive market situation with stagnant markets and hard competition. Therefore, they needed all the inputs they could get without losing control. The Payment Company had professionalized a reflexive strategy process. This company was in a growing market with a good, almost monopolized, situation, but with very hard owners. These firms were more successful in formulating a sustainable strategy.

One may, on the basis of the results, state another hypothesis that the more pressed the firm is on the market, the more they involve employees and managers in the development process. In a very pressed market situation, the firm needs a more radical development with a new strategy and many innovations, which is difficult in that situation. The less pressed they are on the market, the more the development process will be professionalized. In a less pressed situation, innovations within the existing strategic framework will be useful to the firm. However, in that situation, experimenting too much with new strategies should not disturb the favorable market situation. Risky experimentation could be the case if the formulation of the strategy is left to the employees. The above considerations also underlines a hypothesis of a cyclical movement (between pressed and non-pressed market situations) in the development and thus in strategic reflexivity.

The Innovation Process

I shall now turn to discuss the innovation process within the firms. The discussion and conclusions relating to the theory emphasize several elements of the organizational system such as interaction and roles. These organizational elements are important factors in explaining the innovation producing system in the service firms.

The analysis will deal only with innovation. Non-reproduced small changes exist in service firms. Since they are not reproduced, there is no process to describe, and no organizational system to analyze. These changes are important, but they are created by single employees. To study them would demand a comprehensive method of observation or diary writing. There have been insufficient resources in the research projects to use such methods. The process of creating small changes must therefore, unfortunately, be left out of this analysis.

Examples of Service Concepts

To begin with I briefly describe a few selected service concepts and how they were developed in the case firms, just to give an impression of what a service concept can be.

In Family Insurance a new insurance concept for bicycles was invented. The insurance policy could include or exclude validity at different places where the theft of bicycles has a high probability and the premium was correspondingly differentiated. The concept was developed as a collective process in one department.

An insurance agent (which is close to its customers) invented a particular questionnaire to investigate the customers' insurance needs in Hafnia. Somebody else in the company heard about the questionnaire, some of the managers discussed it with the insurance agent, and the questionnaire was developed and made general for use throughout company. The questionnaire was a product as well as a marketing tool, and functioned also as a process innovation (a tool in creating individually modulized insurance by combining standard elements).

ISS developed a new cleaning concept for small and medium-sized business customers. Instead of the normal cleaning assistants following a detailed routine managed by inspectors, the management of the cleaning division had developed a new concept. Flexible teams were established, and they agreed with the customers and planed their work themselves. The team leaders gained a special role. Not only did they manage the day-to-day work, but they did it in a way that involved the team members, and they created attractive social conditions in the teams. This behavior resulted in increased quality, productivity, customer satisfaction, and decreased absenteeism. The behavior was developed by a particular type of team leaders, typically 30 to 40-year-old women.

In the home help a similar type of team leader was developed. The process was managed by a group of managers in the district but again was the particular team leader type developed through the process.

The Payment Company developed a chip-based cash payment card. The card can be used in payment instead of cash. The card was more like a traditional manufacturing product and was developed in an R&D like department.

Lån & Spar had organized a large organization development project where the employees were involved in strategy development and innovation activities in teams. A couple of the results were two service concepts. One was a new credit card characterized by a low interest rate, and it only took very short time (about 15 minutes) to get the card due to a new IT-based credit assessment system. The credit card concept was developed by one team, which had been set up as a result of a training course. One employee became particularly engaged in the process and later became the leader of the group that produced and marketed the credit card. Another new concept was a customer database with information about the customers. The database was useful for marketing and advising. It was developed as a collective project within the IT department.

Ret & Råd was based on an organizational innovation, namely to unite independent lawyers in a chain and to utilize their different competencies. A division of labor was established, and the lawyers remained independent. The idea was created by a few lawyers and was diffused to, and further developed by, other lawyers, who entered the chain.

The service concepts thus are of a very different kind and are developed in different ways, but in most of these cases they were developed by involving a large part of the organization.

The Innovation Process and Strategic Framework

Earlier analyses of some of these cases have demonstrated that the strategy was the framework for the innovation activities (Sundbo, 1998b). The strategy thus generally functions as a limitation of, but also inspiration for, innovative activities. This is as more the true the more top management emphasizes the strategy, and invites the employees to develop new ideas. The middle managers have a particular role here. They are generally the most innovative people, they implement the strategy and top management's effort in the day to day work, and they often become leaders in the interaction process.

One might assume that the employees and managers would have a greater self-interest that makes them fight for their own ideas. They do, but their ideas were within the strategy. They had to be that to be accepted in the firm and this acceptance is the basis for personal success and career. New ideas were, in rare cases, implemented just by entrepreneurship and were then results not of reflections but of pure action orientation. Sometimes the entrepreneurs succeeded, which resulted in a change of the real strategy, but mostly they were squeezed out and may have established their own firm.

Thus, the process of strategic reflexivity is continuing in the innovation process. Corporate culture, which has been seen as a fundamental factor for creating innovation (e.g. Sjölander, 1985), might not be the most important factor as my empirical studies demonstrated. Culture is a long-lasting, fundamental factor and therefore not reflexive and not strategically based. Culture cannot change as fast as strategy. Corporate culture might be reflexive in the way that it is created by the management as Schein (1984) says. However, the corporate culture is then part of the strategy, and the strategy is the core key to understanding the social processes within the firm.

The case studies have demonstrated that people within the organization can be oriented towards being innovative. Such an innovation orientation demands a comprehensive effort from the management, but that may be done, as the Lån & Spar case demonstrates (Sundbo, 2000) even though there are problems in

doing this. The problems are also demonstrated in the Lån & Spar case. The employees in the bank have been more occupied by the general motivation factor—how they relate to other people and how they can personally learn from change processes—than by innovating.

Patterns of Interaction

The ideas for innovation are often developed in a certain pattern within the interaction system. These ideas may be the result of the employees' daily interaction, for example with colleagues and customers. Even when the customers are the source of ideas, the innovative idea in most cases is a result of the interaction between the employee and the customer. The customer has a problem or a want, which the employee cannot guess unless he meets the customer. The customer does not express his or her wants or problem as an idea for a new service concept. The employee translates the interaction into an idea for a service concept. That was how the insurance agent in Hafnia developed the Hafnia questionnaire. The development of the idea and the implementation of the service concept are also interaction processes in which several employees and managers from different departments participate.

In the case firms I found different interaction patterns, which can be generalized. The interaction patterns reflect the way in which ideas for innovations come to the surface and the innovations are developed. These are generalizations, and the interactions had more details, which followed different ways in the single case. Now I will outline the main patterns (a development of innovative organizational models that have been studied previously; Sundbo, 1998b). The interaction patterns also show the power system within the firm and how the organization in general is built and functions.

Four different interaction patterns can be identified. The first one is the most top-down pattern found in Family Insurance and Hafnia. The top management guided the process by signalizing which type of service concept was wanted and often picked out which persons should work with an idea. The interaction was of course between these persons, but the top management was deeply involved in almost all interactions. This pattern could be called *the management pattern*. A particular variant was Hafnia where the managing director was the entrepreneur himself. The interaction system was largely centered around commenting and developing his ideas.

The pattern in Ret & Råd and The Payment Company could be called *the professionalized pattern*. A particular department had the task of initiating new service concepts and of initiating interactions in the organization to develop the concepts. In Ret & Råd, the professional secretariat of the network had to convince the independent lawyers of the new ideas, which made the task extremely difficult. The leader of the secretariat quitted his job during the period of investigation because of the difficulties in getting changes accepted in the network.

The third pattern is *the team pattern*. Service concepts are developed through an intense interaction process within a team. The team members also interact with external individuals and the top management, but the main part of the interaction is within the team. Only in the final implementation period is the interaction pattern diffused to a larger part of the organization. This pattern was found in COWI, ISS and the home help.

Lån & Spar had the most extended interaction pattern for development of new service concepts. A very large part of the organization was involved. The top management participated in the daily interactions—although the managing director had the power to decide on all service concepts, and did that after having consulted other managers. This fourth pattern could be called *the total pattern*.

Role Patterns

In the development of new service concepts people had different roles in the interaction process. These roles have different functions in getting the innovation process running. Role is a core concept within social psychology and sociology (Maccoby et al., 1966; Parsons, 1951) and is a useful concept in this analysis. Formerly some roles in Family Insurance and The Payment Company have been described (Sundbo, 1998b). These roles were: the entrepreneur role, which is creation of ideas and struggle for implementing them (the entrepreneur is a type exposing him—or herself very much); the analyst role, which has the reflection function and analyses the possibilities and outcomes—this is a purely reflexive role; and the producer role. The first two had a function in the innovation process while the third concerns the effort of having the daily production system running. A role is something that the organization members play when he or she is at work. An individual does not need to play the role the whole time and can play several roles simultaneously.

From the other case studies referred to here, I can identify three roles which fulfil three functions in the innovation process of developing service concepts. These are the *entrepreneur* and the *analyst*, as were formerly found, plus a new role, the *interactor* (Sundbo, 2001), which has the function of having the interaction process running. Theoretically, one may state that a balance with all three roles equally represented is the optimal for innovation (here we can forget the producer role because we only are dealing with innovation). This can be expressed in the model of Fig. 2.

Interaction is what makes this system of role mixture reflexive in relation to the pure entrepreneurship system. The process where more than one individual is involved leads to more reflection due to a certain

opposition (cf. Tarde, 1897). Opposition means that there is always a certain skepticism towards new ideas.

Reflexive Feedback and Organizational Learning

Theoretically one must state that if the social system is reflexive, the innovation processes should result in learning: the experiences about how the process went and how the innovations worked in practice should be collected, analyzed and result in strategic changes. The reflexive practice should result in a feedback to the overall goal, the strategy.

The feedback mechanism concerns how suitable the actual strategy and the way in which the innovation process is carried out are. Firstly, the feedback mechanism includes how the learning of the innovation processes could result in a change of the strategy. An experience might for example be that the firm has difficulties in implementing certain innovations, which could result in a change of the strategy. Secondly, the feedback mechanism includes how learning about the innovation process could result in a change in the way in which the service concept development process is organized and carried out. It might, for example, be that some project teams function badly or there are difficulties in engaging the employees in developing new ideas.

The total feedback and learning development process can theoretically be expressed in the model shown in Fig. 3.

The process starts with reflections concerning the market and the internal resources. The reflections are interpreted by the management, which leads to a strategy. The strategy leads to innovation processes that—in this case—resulted in new service concepts. Experiences from the innovation process may lead to a change in the strategy or in the way of carrying out the innovation activities, which is another reflection. When the final outcome of the innovation process (here, the service concept) has worked for some time, further considerations may lead to the conclusion that a new market position or new internal resources are needed, which are further processes of reflection.

All these processes are experience-based learning, which leads to feedbacks to earlier factors in the model. The reflections may involve theoretical elements (found in management books or procured from consultants), but these will be considered within the framework of the concrete experiences.

The model above is of course very formalized. The model might give associations to cybernetics (e.g. Beer, 1964). The real processes are not fully rational and may have all variants of extra loops, inefficiencies, and mistakes as the empirical cases demonstrate. Therefore, the postulate here is not that the process is a smooth-running efficient process that can be described by a cybernetic model. The model is an abstract heuristic model that presents the elements in

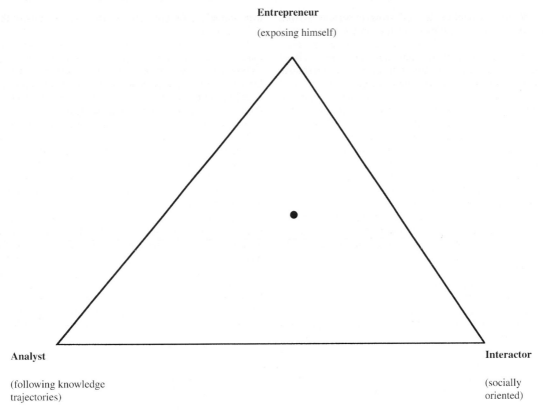

Figure 2. Model of the roles in the innovation process.

the process. The real pattern is chaotic as stated, so the real model should be extremely complex, which would be impossible to draw.

Examples of some chaotic elements of the model in real life can be found in the case studies. Organizational learning functioned generally poorly in the service firms studied. The firms wanted to collect the experiences from innovation processes but had difficulties. For example Lån & Spar had difficulties, even though the management actively attempted to create a feedback system, and the title of the total organiza-

tional development project was 'the learning organization' (Sundbo, 2000). The difficulties were caused by at least two factors. One was that the employees and managers were not interested in formally registering the general experiences. Such registering was considered an activity that counts not as a valuable competence but as a bureaucratic duty. Further, the bank had difficulties in storing the experiences. They tried to use note sheets in the IT system, but that did not work as a general learning data bank and the experiences were not systematized.

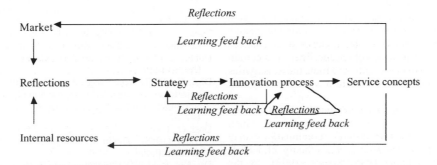

Figure 3. Model of the organizational learning process and feedback mechanisms.

COWI had difficulties in collecting experiences from the work and development in the project teams. The team members were already engaged in new projects before the first ones were completed and never did have time to systematize the experiences and communicate them to other parts of the organization.

The strategy was changed as a consequence of innovation experiences in some of the service firms. Family Insurance changed their strategy from being innovative and going for an advanced market segment to going more conservatively for the medium- and lower-income private market due to its experiences with problems of managing the internal personnel resources (whether they should be innovative or productive). The Payment Company has recently changed its strategy away from launching technical tools such as the chip-based cash payment card back to basics, credit cards, because of a failing market.

Summing Up

This chapter has presented a theoretical approach to the development of the firm, namely a strategic innovation theory (cf. Sundbo, 1998a, 2001, 2002). The theory focuses on the activities that lead to innovations. This theoretical approach is a 'soft' evolutionary one. Primarily, the firm relates to the market developments and this is decisive for its innovation activities. However, the firm also relates itself to the internal resources and their ability to contribute to the development. External trajectories, which include new knowledge and technology, also play a role, but only by presenting possibilities; they do not determine the development of the firm. The firm's relation to these factors is expressed in the strategy, which is not just a written plan, but a management policy.

The strategy defines the goal(s) for the development of the firm in broad terms and is the overall means for attempting to guide the firm's development. This statement expresses a quasi-rational approach: the top management attempts to steer the process, but that is only possible to a certain degree. The top management becomes dependent on involving the employees and middle managers in all steps of the process—from formulating the strategy over innovation activities to organizational learning.

Innovation and change can be characterized as a process of strategic reflexivity. The strategic reflexive process is a process of innovation and of small changes made by individuals or small groups. There is a certain organizational structure within which these activities are carried out, but the structure varies in different situations.

Therefore, innovation cannot be described by one causal or descriptive model, but certain patterns, which vary and are repeated in different combinations, may be observed (as is expressed in chaos theory; cf. Kiel & Elliott, 1996; Quinn, 1985; Stacey, 1993). In the chapter, such patterns were investigated in one type of firm, namely different service firms. The results of this investigation were:

The processes are taking place in a reflexive, interactive organization system as was demonstrated by an empirical case: the development of service concepts. Four interaction patterns within the innovation process could be observed. The interaction process depends on three core roles as entrepreneur, analyst and interactor.

A theoretical organizational learning system in the form of a feedback mechanism was outlined. This learning system is learning by experience and reflexive. The feedback loops may lead to a change in strategy (and thus the way the firm should go in its innovation activities) and of how the innovation activities should be carried out and organized.

Empirically, it was observed that the service firms had difficulties in establishing efficient feedback mechanisms and learning because employees and managers were not motivated to do this and because of a lack of tools for storing the experiences.

References

Argyris, C. & Schon, D. (1978). *Organizational learning: A theory of action perspective*. Reading, MA: Addison-Wesley.

Barnett, H. G. (1953). *Innovation: The basis of cultural change*. New York: McGraw-Hill.

Beck, U. (1992). *Risk society*. London: Sage.

Beer, S. (1964). *Cybernetics and management*. New York: John Wiley.

Binks, M. & Vale, P. (1990). *Entrepreneurship and economic change*. London: McGraw-Hill.

Boden, M. & Miles, I. (Eds) (2000). *Services and the knowledge-based economy*. London: Continuum.

Boulding, K. (1985). *The world as a total system*. Beverly Hills, CA: Sage.

Bourdieu, P. & Wacquant, L. (1994). *An invitation to reflexive sociology*. Oxford: Polity.

Chaffee, E. (1985). Three models of strategy, *Management Review*, **10** (1), 89–98.

Coombs, R., Saviotti, P. & Walsh, V. (1987). *Economics and technological change*. Basingstoke, U.K.: Macmillan.

Daft, R. (1978). A dual-core model of organizational innovation. *Academy of Management Journal*, **21** (2), 193–210.

Desrumaux, P., Gemmel, P. & van Ossel, G. (1998). Defining the service concept. In: B. Looy, R. van Dierdonck & P. Gemmel (Eds), *Services Management*. London: Financial Times/Pitman.

Devaney, R. (1992). *A first course in chaotic dynamic systems*. Reading, MA: Addison-Wesley.

Dobbs, I. (2000). *Managerial economics*. Oxford: Oxford University Press.

Dosi, G. (1982). Technological paradigms and technological trajectories: A suggested interpretation of the determinants and directions of technical change. *Research Policy*, **2** (3), 147–62.

Dosi, G., Freeman, C., Nelson, R., Silverberg, G. & Soete, L. (Eds) (1988). *Technical change and economic theory*. London: Pinter.

Drucker, P. (1985). *Innovation and entrepreneurship*. New York: Harper & Row.

Durkheim, E. (1933). *The division of labor in society*. New York: Macmillan.

Edquist, C. (1997). *Systems of innovation*. London: Pinter.

Eldredge, N. (1999). *The pattern of evolution*. New York: W. H. Freeman.

Eliasson, G. (Ed.) (1989). *The knowledge-based information economy*. Stockholm: The Industrial Institute for Economic and Social Research.

Freeman, C. & Perez, C. (1988). Structural crisis of adjustment: Business cycles and investment behaviour. In: G. Dosi, C. Freeman, R. Nelson, G. Silverberg & L. Soete (Eds), *Technical Change and Economic Theory*. London: Pinter.

Freeman, C. & Soete, L. (1997). *The economics of industrial innovation*. London: Pinter.

Fuglsang, L. (2000). *Menneskelige ressourcer i hjemmehjælpen: fra pelsjæger til social entreprenør (Human resources in the home help: From trapper to social entrepreneur)*, Report no. 4 from the project Service development, Internationalisation and Competencies, Centre of Service Studies, Roskilde University, Roskilde. www.ssc.ruc.dk/css/sic/published.html

Fuglsang, L. (2001). Management problems in welfare services: The role of the 'social entrepreneur' in home-help for the elderly, the Valby case. *Scandinavian Journal of Management*, **17** (4), 437–455.

Gallouj, F. (1994). *Économie de l'innovation dans les services*. Paris: Harmattan.

Gallouj, F. (2002). Interactional innovation: A neo-schumpeterian model. In: J. Sundbo & L. Fuglsang (Eds), *Innovation as Strategic Reflexivity*. London: Routledge.

Giddens, A. (1984). *The constitution of society*. Cambridge: Polity.

Gintis, H. (2000). *Game theory evolving: A problem-centered introduction to modeling strategic behaviour*. New York: Princeton University Press.

Grant, R. (1991). The Resource-based theory of competitive advantage: Implications for strategy formulation. *California Management Review*, (Spring), 114–135.

Håkansson, H. (Ed.) (1987). *Industrial technology development. A network approach*. London: Routledge.

Hamel, G. & Prahalad, C. K. (1994). *Competing for the future*. Boston: Harvard Business School Press.

Hannan, M. T. & Freeman, J. (1989). *Organizational ecology*. Cambridge, MA: Harvard University Press.

Henten, A. (2000). *Ret & Råd—organisatorisk innovation (Ret & Råd—organizational innovation)*. Report no. 6 from the project Service development, Internationalisation and Competences, Centre of Service Studies, Roskilde University, Roskilde. www.ssc.ruc.dk/css/sic/published.html

Hodgson, G. (1993). *Economics and institutions*. Cambridge: Polity Press.

Hodgson, G. (1998). Competence and contract in the theory of the firm. *Journal of economic behavior and organization*, **35**, 179–201.

Illeris, S. & Sundbo, J. (2000). *Innovation og kompetenceudvikling i rengøring. Case rapport om ISS (Innovation and competence development in cleaning. Case report on ISS)*. Report no. 2 from the project Service development, Internationalisation and Competencies, Centre of Service Studies, Roskilde University, Roskilde. www.ssc.ruc.dk/css/sic/published.html

Jørgensen, S. E. (1992). *Integration of ecosystem theories: A pattern*. Dordrecht: Kluwer.

Johannisson, B. (1987a). Anarchists and organizers—entrepreneurs in a network perspective. *International Studies of Management and Organization*, **17**(1), 49–63.

Johnson, B. (1992). Institutional learning. In: B-Å. Lundvall (Ed.), *National Systems of Innovation*. London: Pinter.

Kanter, R. M. (1984). *The change masters*. London: Unwin.

Kiel, D. & Elliott, E. (Eds) (1996). *Chaos theory in the social sciences: foundations and applications*. Ann Arbor, MI: University of Michigan Press.

Kotler, P. (1983). *Principles of marketing*. Englewood Cliffs, NJ: Prentice Hall.

Krebs, C. J. (1988). *The message of ecology*. New York: Harper & Row.

LaPiere, R. T. (1965). *Social change*, New York: McGraw-Hill.

Larsen, J. N., Illeris, S. & Pedersen, M. Kühn (2001). *COWI casen (The COWI case)*. Report no. 7 from the project Service development, Internationalisation and Competencies, Centre of Service Studies Roskilde University, Roskilde. www.ssc.ruc.dk/css/sic/published.html

Maccoby, E., Newcomb, T. & Hartley, E. (Eds), *Readings in Social Psychology* (1966). London: Methuen.

McClelland, D. (1961). *The achieving society*. Princeton, NJ: Van Norstrand.

Menger, C. (1950). *Principles of economics*. Glencoe, IL: Free Press.

Mensch, G. (1979). *Stalemate in technology*. Cambridge, MA: Harvard University Press.

Metcalfe, S. (1998). *Evolutionary economics and creative destruction*. London: Routledge.

Mintzberg, H. (1994). *The rise and fall of strategic planning*. New York: Free Press.

Moore, W. E. (1963). *Social change*. Englewood Cliffs, NJ: Prentice Hall.

Nelson, R. (1993). *National innovation systems*. Oxford: Oxford University Press.

Nelson, R. & Winter, S. G. (1982). *An evolutionary theory of economic change*. Cambridge, MA: Belknap.

Nielsen, K. & Johnson, B. (1998). *Institutions and economic change*. Cheltenham, U.K.: Elgar.

OECD (1996). *Employment and growth in the knowledge-based economy*. Paris: OECD.

Ogburn, W. F. (1950). *Social change*. New York: Viking Press.

Owen, G. (1995). *Game theory*. San Diego, CA: Academic Press.

Parsons, T. (1951). *The social system*. Glencoe, IL: Free Press.

Pavlov, I. P. (1966). *Essential works of Pavlov* (edited by M. Kaplan). New York: Bantam.

Penrose, E. T. (1959). *The theory of the growth of the firm*. New York: Blackwell.

Piatier, A. (1984). Innovation, information and long-term growth. In: C. Freeman (Ed.), *Long Waves in the World Economy*. London: Pinter.

Pinchot, G. (1985). *Intrapreneuring*. New York: Harper & Row.

Prigogine, I. & Stengers, I. (1986). *Order out of chaos: Man's new dialogue with nature*. London: Fontana.

Quinn, J. B. (1985). Managing innovation: Controlled chaos. *Harvard Business Review*, **63** (3), 73–84.

Rogers, E. M. (1995). *Diffusion of innovation*. New York: Free Press.

Rosenberg, N. (1972). *Perspectives on technology*. Cambridge, MA: Cambridge University Press.

Schein, E. (1984). *Organizational culture and leadership*. San Francisco: Jossey-Bass.

Schumpeter, J. (1934). *The theory of economic development*. Harvard, MA: Oxford University Press.

Schumpeter, J. (1943). *Capitalism, socialism and democracy*. London: Unwin.

Sexton, D. L. & Kasarda, J. (Eds) (1992). *The state of the art of entrepreneurship*. Boston, MA: PWS-Kent.

SIC (1999). *Service development, Internationalisation and competences*. Report no. 2, Danish Service Firms' Innovation Activities and Use of IT, Centre of Service Studies, Roskilde University, Roskilde.

Sjölander, S. (1985). *Early stage management of innovation*. Department of Industrial Management, Chalmers University of Technology. Göteborg.

Smith, J. M. (1977). *The theory of evolution*. Harmondsworth, U.K.: Penguin.

Stacey, R. (1993). *Managing chaos*. London: Kogan Page.

Sundbo, J. (1997). Management of innovation in services. *The Service Industries Journal*, **17** (3), 432–455.

Sundbo, J. (1998a). *The theory of innovation. Entrepreneurs, technology and strategy*. Cheltenham, U.K.: Elgar.

Sundbo, J. (1998b). *The organization of innovation in services*. Copenhagen: Roskilde University Press.

Sundbo, J. (1999). *The manual service squeeze*. Centre of Service Studies, Roskilde University, Roskilde. www.ssc.ruc.dk/css

Sundbo, J. (2000). Empowerment of employees in small and medium-sized service firms. *Employee Relations*, **21** (2), 105–127.

Sundbo, J. (2001). *Strategic management of innovation*. Cheltenham, U.K.: Elgar.

Sundbo, J. (2002). Innovation as a strategic process. In: J. Sundbo & L. Fuglsang (Eds), *Innovation as Strategic Reflexivity* (2002). London: Routledge.

Sundbo, J. & Fuglsang, L. (Eds) (2002). *Innovation as strategic reflexivity*. London: Routledge.

Sundbo, J. & Gallouj, F. (2000). Innovation as a loosely coupled system in services. *International Journal of Services Technology and Management*, **1** (1), 15–36.

Swedberg, R. & Granovetter, M. (2001). *The sociology of economic life*. Boulder, CO: Westview.

Tarde, G. (1895). *Les lois de l'imitation*. Paris: Alcan.

Tarde, G. (1897). *L'opposition universelle*. Paris: Alcan.

Teece, D. J. (1986). Profiting from technological innovation: implication for integration, collaboration, licensing and public policy. *Research Policy*, **15** (6), 285–305.

The International Handbook on Innovation
Edited by Larisa V. Shavinina

The Nature and Dynamics of Discontinuous and Disruptive Innovations from a Learning and Knowledge Management Perspective

Elias G. Carayannis[1], Edgar Gonzalez[1] and John Wetter[2]

[1] *School of Business and Public Management, The George Washington University, Washington, D.C., USA*
[2] *Xerox Corporation, USA*

Abstract: In this chapter we will discuss and profile the evolutionary and revolutionary dimensions of the nature and dynamics of innovation as a socio-technical phenomenon. We will focus in particular on the *process, content, context, and impact* of both *discontinuous* and *disruptive* innovations. We postulate that at the heart of the competence to generate and perhaps more significantly *to leverage discontinuous and in particular disruptive innovations, lies the individual and organizational capacity for higher-order learning* and for managing the stock and flow of specialized and domain-specific knowledge. We will provide both concepts and cases to illustrate our ideas and supply thematic anchors for academic and practitioner contexts.

Keywords: Disruptive innovation; Discontinuous innovation; Higher-order technological learning; Knowledge management.

Introduction

A little revolution now and then, is a good thing
Thomas Jefferson

'Innovation' is a word derived from the Latin meaning 'to introduce something new to the existing realm and order of things'. In this sense, innovation is endowed with a faculty of discontinuity and possibly disruptiveness in the form of a continuum of discontinuities reflected by a simple analogy to the way we walk. From a business perspective, an innovation is perceived as the happy ending of the commercialization journey of an invention, when that journey is indeed successful and leads to the creation of a sustainable and flourishing market niche or new market. Not all innovations are discontinuous, and not all discontinuous innovations prove to be disruptive. This is determined by the scope, timing, and impact of the innovation under consideration.

The literature on innovation, particularly regarding technological innovation, is populated by a number of taxonomies which attempt to categorize innovations by significance, similarity (and dissimilarity), technical domain, and other characteristics. As the vocabulary used to describe innovation has grown and evolved, scholars naturally generate multiple taxonomies which are at times overlapping, redundant, or divergent. A recent review of the literature on new product development found that in just 21 empirical studies, researchers have developed 15 different constructs for describing various aspects of innovation (Garcia & Calantone, 2002). Some of the distinctions produced by previous authors include *process* vs. *product* innovation (Utterback & Abernathy, 1975), *incremental vs. radical* innovation (Henderson & Clark, 1990), and *evolutionary vs. revolutionary* innovation (Utterback, 1996):

Technological innovation is defined here as a situationally new development through which people extend their control over the environment. Essentially, technology is a tool of some kind that allows an individual to do something new. A technological innovation is basically information organized in a new way. So technology transfer amounts to the communication of information, usually from one organization to another.

This chapter focuses on the recent discussion by Christensen (1997) of *disruptive* (as opposed to sustaining) technologies, and the related concept of *discontinuous* (as opposed to continuous) innovation (Anderson & Tushman, 1990). This particular type of innovation is significant, as there have been many attempts to determine the extent to which discontinuous innovations can be 'managed', and how companies can try to predict and leverage the emergence of disruptive technologies. In this chapter we will discuss and profile the evolutionary and revolutionary dimensions of the nature and dynamics of innovation as a socio-technical phenomenon and focus in particular on the process, content, context, and impact of both discontinuous as well as disruptive innovations.

We postulate that at the heart of the competence to generate and perhaps more significantly to leverage discontinuous and in particular disruptive innovations, lies the individual and organizational capacity for higher-order learning and for managing the stock and flow of specialized and domain-specific knowledge. We will provide both concepts and cases to illustrate our ideas and supply thematic anchors for academic and practitioner contexts.

The Nature and Dynamics of Innovation

Basic research is what I am doing when I do not know what I am doing.

Dr Wehrner von Braun

Before a definition of innovation can be discussed, the related term 'invention' must be understood. Florida considers invention as a breakthrough and innovation as an actualization (Florida, 1990). Hindle further clarifies invention by labeling it as the creative origin of new process and the enabler of innovation (Hindle, 1986), which has impacts on social, economic, and financial processes. Thus the emerging definition of invention may be stated as the creative process of progress while innovation is defined by the impact on societies and markets (actualization). "Innovation generally lowers the cost of responding to a change in the commercial environment" (Wallace, 1995). Thus, innovation has the connotation of market influence.

Identifying the source of innovation may assist in the definition. The pace of improvements brought about by innovation, the rate of innovation, may be determined by the technology pull or market push factors. The question of a specific source of innovation is brought about by a process of '*learning by doing*' (Rosenberg, 1976). In this is meant that innovation, through the continuous incremental effects of knowledge acquisition, has the effect of cumulatively impacting on future innovations.

Other related terms, like science and technology, should be defined in the context of innovation. Traditional epistemology defines science and scientific knowledge as the world of objective theories, objective problems and objective arguments. Further clarification is found in Kuhn (1962) defining *science* as research firmly based on one or more past achievements.

Technology is defined as that "which allows one to engage in a certain activity. . . with consistent quality of output", the "*art of science and the science of art*" (Carayannis, 2001) or '*the science of crafts*' (von Braun, 1997). Diwan adds that technological foundations are market size, standards, innovation, high motivation, and supply of capital (Diwan, 1991) The impact of innovation may be directed to multiple sectors. For example, Jonash lists product/service, process, and business innovation as the key impact areas. Product/service is the development and commercialization of hard goods, process is new ways of producing and delivering cost-time-quality advantages, and business innovation is new models of conducting business for competitive advantage (Jonash & Sommerlatte, 1999).

A fundamental challenge to the present analysis is the distinction between what is and what is not an innovation. When related to technologies, one common definition of an innovation is "an idea, practice, or object that is perceived as new by an individual or other unit of adoption" (Rogers, 1983, p. 11). Thus, a technological innovation is a new idea, practice, or object with a significant technology component.

A technical discovery or invention (the creation of something new) is not significant to a company unless that new technology can be utilized to add value to the company, through increased revenues, reduced cost, and similar improvements in financial results. This has two important consequences for the analysis of any innovation in the context of a business organization.

First, an innovation must be integrated into the operations and strategy of the organization, so that it has a distinct impact on how the organization creates value or on the type of value the organization provides in the market.

Second, an innovation is a social process, since it is only through the intervention and management of people that an organization can realize the benefits of an innovation.

The discussion of innovation clearly leads to the development of a model, to understand the evolving nature of innovation. Innovation management is concerned with the activities of the firm undertaken to yield solutions to problems of product, process, and administration. Innovation involves uncertainty and dis-equilibrium. Nelson & Winter (1982) propose that almost any change, even trivial, represents innovation. They also suggest, given the uncertainty, that innovation results in the generation of new technologies and changes in relative weighting of existing technologies (ibid). This results in the *disruptive process* of dis-equilibrium. As an innovation is adopted and diffused, existing technologies may become less useful (reduction in weight factors) or even useless (weighing equivalent to '0') and abandoned altogether. The adoption phase is where uncertainty is introduced. New technologies are not adopted automatically but rather, markets influence the adoption rate (Carayannis, 1997, 1998). Innovative technologies must propose to solve a market need such as reduced costs or increased utility or increased productivity. The markets, however, are social constructs and subject to non-innovation related criteria. For example, an invention may be promising, offering a substantial reduction on the cost of a product which normally would influence the market to accept the given innovation; but due to issues like information asymmetry (the lack of knowledge in the market concerning the invention's properties), the invention may not be readily accepted by the markets. Thus the innovation may remain an invention. If, however, the innovation is market-accepted, the results will bring about change to the existing technologies being replaced, leading to a change in the relative weighting of the existing technology. This is in effect dis-equilibrium.

Given the uncertainty and change inherent in the innovation process, management must develop skills and understanding of the process a method for managing the disruption. The problems of managing the resulting disruption are strategic in nature. The problems may be classified into three groups, *engineering, entrepreneurial, and administrative* (Drejer, 2002). This grouping correlates to the related types of innovation namely, *product, process, and administrative innovation*:

- The engineering problem is one of selecting the appropriate technologies for proper operational performance.
- The entrepreneurial problem refers to defining the product/service domain and target markets.
- Administrative problems are concerned with reducing the uncertainty and risk during the previous phases.

In much of the foregoing discussion, a recurring theme about innovation is that of *uncertainty*, leading to the conclusion that an effective model of innovation must include a multi-dimensional approach (uncertainty is defined as unknown unknowns whereas risk is defined as known knowns). One model posited as an aide to understanding is the Multidimensional Model of Innovation (MMI) (Cooper, 1998). This model attempts to define the understanding of innovation by establishing three-dimensional boundaries. The planes are defined as product-process, incremental-radical, and administrative-technical. The product-process boundary concerns itself with the end product and its relationship to the methods employed by firms to produce and distribute the product. Incremental-radical defines the degree of relative strategic change that accompanies the diffusion of an innovation. This is a measure of the disturbance or dis-equilibrium in the market. Technological-administrative boundaries refer to the relationship of innovation change to the firm's operational core. The use of technological refers to the influences on basic firm output while the administrative boundary would include innovations affecting associated factors of policy, resources, and social aspects of the firm.

A Historical and Socio-Technical Perspective on Innovation

> But in capitalist reality, . . . it is not price competition which counts but the competition from the new commodity, the new technology, the source of supply, the new type of organization, . . . competition which . . . strikes not at the margins of the existing firms but at their foundations and their very lives.

Joseph A. Schumpeter
Capitalism, Socialism and Democracy, 1942

To review the history of innovation, one must look toward the classic works of Schumpeter. Schumpeter, an economist, wrote 'The Theory of Economic Development' in 1934 as an inquiry into profit, capital, credit, interest, and the business cycles. His main contributions were: (a) the expansion of Adam Smith's economic principles of land–labor–capital into land–labor–capital–technology–entrepreneurship; and (b) the introduction of the concept of dis-equilibrium into economic discourse.

It is interesting to note that Schumpeter was a socialist and believed that the capitalist system would eventually collapse from within and be replaced by a socialist system. On this point he agreed with Marx, but his version of socialism was in many respects very different. Marx felt very strongly that the economic model employed would determine the construct of society. The cornerstone of his theoretical structure was the 'Theory of Value' (*Das Kapital*) where the value of a commodity, given perfect equilibrium and perfect competition, is proportional to the input of labor. Schumpeter disagreed with Marx on this issue offering the conclusion that both perfect equilibrium and perfect competition were problematic at best. Additional disagreements centered on the inclusion of the value of land in the equation. Another point on which Schumpeter disagreed, is Marx's contention that the capitalist system would implode (*Zusammenbruchstheorie*) as a result of its intrinsic inequities. In Schumpeter's view, the natural evolution of capitalism would destroy the foundations of capitalism from within. In fact, he believed that the economic depression of the 1930s was an indication of a paradigm shift, reinforcing his beliefs. Schumpeter viewed capitalism in much the same way as he viewed the process of innovation. Both were generally considered stable processes (under perfect conditions) from a theoretical model perspective but Schumpeter introduced the conceptual theory of dis-equilibrium as the key influential factor and this could be further expanded into the concept of continuum of punctuated dis-equilibria (Carayannis, 1994b) to capture and articulate the concept of successive Fisher–Pry curves (S-curves) with discontinuous and/or disruptive innovations causing a change of curve and/or change of 'the rules of the game' as we will see later:

> Michael Tushman and Charles O'Reilly suggest that discontinuous innovation involves breaking with the past to create new technologies, processes, and organizational 'S-curves' that result in significant leaps in the value delivered to customers. Similarly, Clay Christensen, Gary Hamel and C. K. Prahalad, and James Utterback describe discontinuous innovation as involving 'disruptive technologies', 'discontinuities', or 'radical innovations' that permit entire industries and markets to emerge, transform, or disappear (Kaplan, 1999).

Early capitalism is often referred to as 'laissez-faire' but post-WWII capitalism is much more bounded by social, political and legal norms. In following Schumpeter's principle of evolutionary capitalism, it may be that the bounded capitalism of the modern era is a logical extension of Schumpeter's theory.

The concept of innovation as a 'socio-technical' system is well established. Rogers (1995), for example, defined innovation in terms of the perceptions of the individuals or groups which adopt an innovation. Attempts to classify innovations in purely technical terms fall into the trap of portraying the result of a social process as something entirely divorced from human influence.

We propose an approach to classifying and subdividing the concepts of innovation along four fundamental dimensions:

(1) The *process* of innovation (the way in which the innovation is developed, diffused, and adopted)
(2) The *content* of innovation (the specific technical or social nature of the innovation itself)
(3) The *context* of innovation (the environment in which the innovation emerges, and the effect of that environment on the innovation)
(4) The *impact* of innovation (the social and technological change which results from the completion of the innovation process) (Carayannis, 2002).

Using these four dimensions of innovation, we can delve more deeply into the social implications of disruptive and discontinuous innovation, which in turn facilitates the integration of innovation management concepts with those of organizational learning and knowledge management. In putting these elements in

perspective, one needs to bear in mind the following key creativity and innovation drivers and qualifiers:

(1) Context: In what context do all of the above occur?
(2) Process: What is the process by which the above are realized?
(3) Content: What is the content of the above in terms of reaction on the others?
(4) Impact: What is the impact of each of the above on the others?
   • All of these attributes must be considered at all levels including the firm, industry, national and global levels;
• What you invent determines the content of the innovation;
• Commercialization is a necessary but not sufficient condition for innovation;
• Creativity and competition may be exogenous factors to competitiveness;
• Competition facilitates or suppresses competitiveness (see Fig. 1);
• Consolidation may breed complacency;
• Disruptive technologies can renew competitiveness.

However, excessive rivalry may sap competitiveness leading to the *Acceleration Trap* (von Braun, 1997) and the *Differentiation Trap* (Christensen, 1997) (see Fig. 1). These are situations of increasingly shorter and unsustainable product cycles and spiraling R&D costs with shrinking profit margins and market shares—the result of excessive competition and declining competitiveness (what we term *hyper-rivalry* in the private sector). In these situations, change takes place so fast that firms often fail to benefit fully from it (their learning curves are not steep enough) and they also end up using resources inefficiently and undermining their market position by engaging in price wars or frivolous innovation races. Then firms can find themselves 'trapped' in a vicious spiral of increasing competition and declining competitiveness and end up rendering their market niches increasingly hard to sustain.

Common Frameworks and Typologies for Characterizing Innovation

> Comforted by idols, we can lose the urge to question and thus we can willingly arrest our growth as persons: 'One must invoke tremendous counter-forces in order to cross this natural, *all too natural progressus in simile, the continual development of man toward the similar, average, herdlike common!*'
>
> Nietzsche, Thus Spake Zarathustra, 58

Innovation may be generally categorized as product, process, or administrative (Tidd, 2001). Others classify innovation by regional influences (Evangelista et al., 2001), or decision criteria (Rogers, 1995). Still others view innovation as product-process-radical-techno-

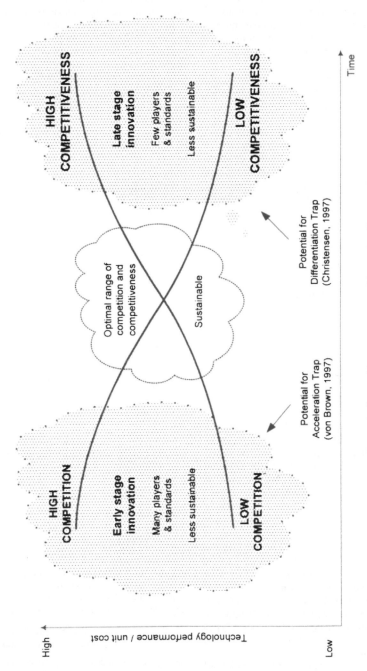

Figure 1. Competitiveness vs. competition trade-offs.

logical (Cooper, 1998). Another view of classifying types characterizes innovation by decision systems (Rogers, 1995). This method relies on the principle that adoption of innovation may be influenced by both individuals and entire social systems. There is also a distinction between sustaining and disruptive innovations (Christensen, 1997) and continuous and discontinuous innovations (Tushman, 1990):

> Discontinuities are often described as technological breakthroughs that help companies rewrite industry rules or create entirely new industries. Rarely have distinctions been made within the concept of 'discontinuity', not to mention how to identify these radical innovations. For the corporate strategist, a big question remains: how to actually structure opportunity identification so it becomes a rational process-one that yields breakthroughs reliably (vs. waiting for opportunities to arise serendipitously) (Kaplan, 1999).

Process innovation refers to change in the methods employed by a firm in delivering products or services. An example is the use of Internet technologies for supply-chain management, where the process of ordering, tracking, and billing would be Internet-based. *Product innovation* reflects change in the end product or service of the firm. An example of product innovation is the addition of a new feature such as adding a remote to a television to improve the user interaction. *Administrative innovation* refers to change in the characteristics of organizational or institutional elements. Changes in policy, organization structure, or resource allocation are examples of administrative innovations.

Using regional differences to classify innovation is a very narrow view, usually reserved to a specific technology innovation comparison. One of the drawbacks with this method is assessment of the regional nature of an innovation. For example, in the case of R&D measured by the number of patents, the region of patent invention may differ from the locale of registration—especially in the case of multinational corporations (MNC). A patent for an invention of Asian origin may be initiated in a U.S. patent filing if the headquarters is a U.S. MNC—thus the patent would be considered U.S. if measured regionally.

Integrating numerous past studies on technological innovation (especially those by Abernathy, Anderson, Clark, Henderson, Tushman & Utterback) produces a common framework distinguishing four generic types of technological innovation: *incremental, generational, radical* and *architectural*.

Incremental innovations exploit the potential of established designs, and often reinforce the dominance of established firms. They improve the existing functional capabilities of a technology by means of small scale improvements in the technology's value-adding attributes such as performance, safety, quality, and cost.

Generational or next-generation technology innovations are incremental innovations that lead to the creation of a new but not radically different system.

Radical innovations introduce new concepts that depart significantly from past practices and help create products or processes based on a different set of engineering or scientific principles and often open up entirely new markets and potential applications. They provide 'a brand-new functional capability which is a discontinuity in the then-current technological capabilities'.

Architectural innovations serve to extend the radical-incremental classification of innovation and introduce the notion of changes in the way in which the components of a product or system are linked together.

Another common distinction is the difference between *evolutionary* innovation, where technological change appears to follow a process of 'natural selection' (with technical improvements resulting from the 'survival of the fittest') and *revolutionary* innovation, where the change appears as a break or non-contiguous change in the course of the technology. These two approaches to envisioning innovation are not mutually exclusive, however.

Using the four perspectives given above, we can show how these concepts relate to one another in a more complete framework for the analysis of innovation.

Process	Content
Evolutionary innovation	Incremental innovation or Generational innovation
Revolutionary innovation	Radical innovation or Architectural innovation

The complete framework with all four dimensions provides us with a way to relate *discontinuous and disruptive* technologies to these other concepts.

Process	Content	Context	Impact
Evolutionary innovation	Incremental innovation	Continuous innovation	Non-disruptive or
	Generational innovation	Continuous innovation	Disruptive innovation
Revolutionary innovation	Radical innovation	Discontinous innovation	Non-disruptive or
	Architectural innovation	Discontinuous innovation	Disruptive innovation

Not all innovations are discontinuous, not all discontinuous innovations prove to be disruptive and not all disruptive innovations are discontinuous. This is determined by the scope, timing, and impact of the innovation under consideration and there are different strategies to deal with the challenges and opportunities arising from planned or serendipitous technological discontinuities and disruptions. Christensen (1997, p. 179) recommends three strategies for leveraging such contingencies and specifically in the case of '*technological performance over-supply*' that creates the potential for *an acceleration and/or a differentiation trap* (von Braun, 1997) (see Figs 2 and 3):

- Strategy 1 is to ascend the trajectory of sustaining technologies into ever-higher tiers of the market;
- Strategy 2 is to march in lock step with the needs of customers in a given tier of the market;
- Strategy 3 is to use marketing initiatives to steepen the slopes of the market trajectories so that customers demand the performance improvements that the technologists provide.

Kaplan (1999) discusses four strategies for leveraging such contingencies:

> Substantial growth over the long horizon requires discontinuous innovation—disruptive technologies, radical innovations and discontinuities that permit entire industries and markets to emerge. Soren Kaplan's experiences as process technology manager with Hewlett-Packard's Strategic Change Services in Palo Alto serves as a framework for all businesses dealing with the new innovation paradigms. He proposes four strategies: radical cannibalism, competitive displacement, market innovation and industry genesis. A strategy involving industry creation has a big advantage in that direct competition does not usually exist. It results in a new form of customer value with a new-to-the-world value proposition.

The Process of Innovation

> The lowest form of thinking is the bare recognition of the object.
> The highest, the comprehensive intuition of the man who sees all things as part of a system.
>
> Plato

An adequate definition of the process of innovation is inherently problematic. The field is nascent and there seems to be as many different definitions as there are researchers. However, there is sufficient information available to evoke a common understanding on many points.

The innovative process is defined by the correlation of its elements of study (Nelson, 1977). Inventions may be measured and the R&D process may be studied and defined. Science and invention may be linked, sources of innovation elaborated upon, organization factors investigated, the evolution of technology studied, diffusion of innovation measured, and the learning phenomena exposed. Invention is viewed as complementary, cumulative, and leap-frogging (Rosenberg, 1982). *Complementary invention* is the invention of a new process/product that is related to an existing technology; the invention of the mouse to support computer–human interaction is an example. *Cumulative inventions* are those that build upon, or 'tweak' an existing invention, such as a product improvement like the pouring spout on juice containers. Leap-frog invention infers a radical change away form existing technologies and echoes discontinuity in markets.

In understanding the process, one must understand the concept of innovation 'imperative' (Cooper, 1998) as a key driver. In a competitive environment, managers are driven to success, both individually and organizationally. In order to achieve organization success, the manager must do more than develop, implement, and approve innovation. They are compelled to constantly innovate in order to attain success, driving the organization to higher levels of innovation diffusion.

Most models of innovation are based on three basic ideas (Drejer, 2002). First, organization can act to create or choose their environment. Second, management's strategic choices shape the organization's structure and processes. Third, once chosen, the structure and processes constrain strategy. This is a very interesting insight into innovation models. If an organization can choose its environment, and if the choice is rational, it should be able to choose the best environment for success of its strategy. There are numerous examples of firm strategies that did not perform as expected. Is this principle negated by non-performance of strategy? It may be that exogenous factors influence the choice of environment. This is an interesting question for further study but it is not in the scope of this paper.

In the U.S., economic policy has an influence on innovation. In general, U.S. policy may be categorized as selective targeting (Nelson, 1982). Historically, U.S. policy could not necessarily be labeled as supportive of innovation. Advances have been uneven (disruptive) and slow to influence productivity and relative costs. This is evidenced by a review of Total Factor Productivity (TFP) comparisons:

> *TFP was developed by Solow in 1957 as the Growth Theory and has become the dominant approach to measuring productivity. Solow's theorem is that the Productivity Residual is uncorrelated with any variable that is uncorrelated with the rate of growth or in other words the Productivity Residual is a measure of the shift of the production function (increase in efficiency). TFP considers the traditional inputs to productivity of labor and output and adds*

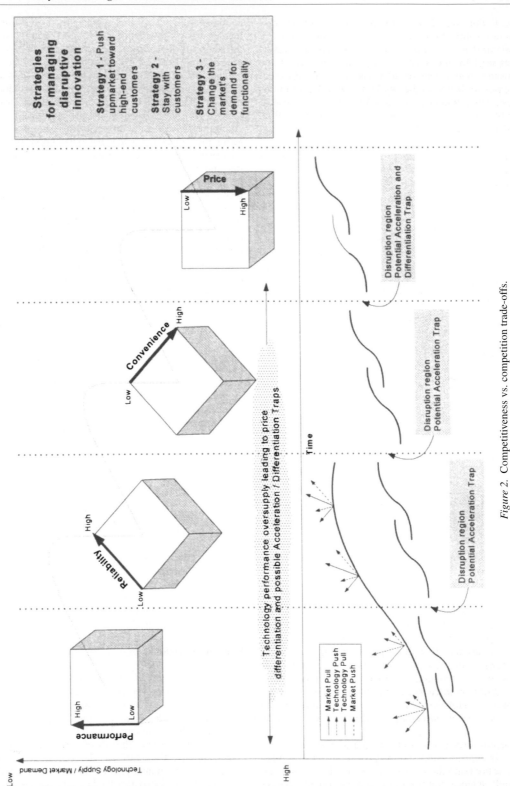

Figure 2. Competitiveness vs. competition trade-offs.

the dimension of the influence of capital. TFP is often referred to as Solow's residual. Prior to TFP, measurement of productivity was subject to factors that may incorrectly influence the outcome, like a rise in demand or a rise in price would cloud the real measurement. It is interesting to note that the TFP calculation is neutral to a rise in demand or a rise in price.

The TFP residual is considered to be an indicator of R&D performance and, as such, can be a measure of the effectiveness of innovation—at the industry or national level. Many researchers (Nelson, 1982) have concluded that TFP residual, as a measure of industry wide R&D effort, is more influential than measuring a single firm.

There are several key recurring principles of innovation. They are an integrated organizational approach, incentives for innovators, a systematic process to convert invention into innovation, team skills, communications, learning, and project management (Rolfe, 1999). These principles are instrumental in developing a innovation process. It is interesting to note the interdependencies of learning and team skills to innovation. Generally, in a team environment, individual members of a team do not possess sufficient knowledge in themselves but if collectively the team 'knowledge sum' is greater than non-team knowledge, the team will be a successful implementor of innovation. Since the common construct of teams is subject to change, the ability of the team to retain knowledge through effective learning is an important criterion for long-term success.

Identifying innovation as a process as opposed to a discrete event or outcome is generally credited to Peter Drucker (Cooper, 1998; Drejer, 2002). The control of the process of innovation is referred to as innovation management. In this context, innovation management is defined by five key activities; technological integration, the process of innovation, strategic planning, organizational change, and business development (Drejer, 2002). Technological integration refers to the relationship between technologies and the product of the firm. The process of innovation is the set of cross-functional activities that create and sustain innovation. Strategic planning involves the planning of technologies related to the innovation. Organizational change comprehends the disruptive nature of innovations on knowledge/skill requirements, new markets, new employees, etc. Business development refers to the creation of new markets for the products of innovation. It is interesting to note that innovation may be a driver of business development and also may be driven by it. This dichotomy may be explained by the fact that, in the early stages, innovation causes a disruptive change in the organization by its very nature, creating new markets for example. As the business evolves, 'technology pull' becomes evident. As competition catches up

or competitive innovations become evident, the requirement for more and more innovation to maintain market position will surface, thus causing the firm to drive innovation.

The organization is influenced by innovation in several ways. Creativity is driven by competition, change, externalities, learning, climate, communications, processes, and social interaction of individuals (Rolfe, 1999). While innovation is a purposeful act, the prime characteristic is uncertainty (Nelson, 1977). This characteristic tends to influence the set of drivers affecting the organization. In this way, as characteristics such as creativity drive innovation, the creativity itself is impacted. The impact may be positive or negative, and so the creativity may be changed and strategic plans may ineffectual. Soren Kaplan (1999) discusses the four types of discontinuities identified at Hewlett Packard and outlines a framework that could serve as a guideline for technology managers and policy makers alike:

> We have discovered four types of discontinuities through our work at HP. As a result, we have developed a framework to help leaders with discontinuous innovation opportunity identification—the process of exploring new revenue streams and identifying compelling propositions for providing heightened forms of customer value. This is the strategic intent that defines compelling new business possibilities capable of driving substantial growth. The framework takes the perspective of an organization that wishes to explore opportunities for discontinuous innovation and is founded upon three assumptions. First, we believe discontinuous innovation involves creating new forms of customer value within existing or new markets. Second, by pursuing discontinuous innovations, organizations create new competitive space or displace existing methods of delivering value to customers. Our final assumption involves the structure of the model itself. We define four discrete innovation strategies but suggest that these classifications not be regarded as mutually exclusive. Instead, these categories should focus efforts on opportunity identification by providing an understanding of 'gray areas' that all too often cloud the definition of 'discontinuity' (Kaplan, 1999).

Measures of Innovation

Truth emerges more readily from error than from confusion.

Sir Francis Bacon

How should innovation be measured, if indeed it can be measured? Research and Development (R&D) is generally the initial measurement tool utilized (Evangelista et al., 2001) but R&D itself may be measured based on different attributes. For example, as an R&D/ Intellectual Property rights (IPR) measurement, the number of patents is generally the measurement axiom. However, other attributes are frequently measured also, such as research funding budgets, number of researchers, number of significant inventions, number of new products, amount of published research, etc. (Tidd, 2001). Still, other attributes are linked in a more subtle way, such as increased productivity and growth or lower costs (Nelson & Winter, 1982). Another classification of measurable characteristics is the social impact of innovation. Examples would include the ability to measure the user benefits, lower consumer prices, user time savings, and other social enablers (Mansfield et al., 1977). A typology of measurable characteristics may help to bring together the disparate measurables (Table 1).

The main categorization is between 'hard' and 'soft' measurables. Hard measurables are those characteristics that may be directly linked to the innovation process. For example, the number of patents issued is direct outcome of the process of research and generally is not influenced by outside factors. Productivity improvements, however, may be the direct result of an innovation but the link is less clear due to other influential characteristics—productivity increases could be influenced by the mere fact of managerial increased interest surrounding the implementation of a productivity innovation. This is not to assume that the innovation was not the primary influence of productivity gains but rather the measurement process may not be sufficiently rigorous to differentiate the various influences.

R&D has a direct effect on output. In studies conducted in the manufacturing field, it was noted that *applied* research and development funding was a more powerful explanation of differences in productivity growth across manufacturing industries than total R&D funding by the entire industry (Nelson, 1977). This would indicate that R&D expenditures are a direct measure of firm productivity. Firm productivity is greater than the norm, as expressed by industry norms.

The adoption of measures of innovation may be influenced by a firm's business and technology strategy. A firm with a high profit objective, may choose to measure innovation characteristics that have a proclivity to specific goals (Nelson & Winter, 2000). This type of weighting may be more beneficial when characteristics are more directly linked, namely, hard measurables.

Managing Innovation Through Knowledge Management and Learning

Until philosophers are kings, or the kings and princes of this world have the spirit and power of philosophy, . . . cities will never have rest from their evils—no, nor the human race as I believe . . .

Plato, The Republic, Vol. 5, p. 492

The proposition that innovation can be 'managed' has been explored by numerous authors. For example, Burns & Stalker (1961) wrote their book *The Management of Innovation* based in part on an earlier study of the research and development laboratory at a local company. Whereas industrial innovation previously seemed to occur in a haphazard and disorganized manner, the post-war era brought strong interest in the idea that innovation could occur systematically, and could even be 'planned'. The merging of the field of organization studies (e.g. Cyert & March, 1963) and the study of the function of management (e.g. Barnard, 1938; Drucker, 1999a, 1999b) provided a new foundation for understanding the innovation process. Further studies on innovation form the basis for a new field of expertise and knowledge on the nature of technological and organizational innovation. Knowledge management is not always fully or properly understood by managers. Instead, in many cases, practitioners and often academics mean information and technology

Table 1. Innovation measures—hard vs. soft.

Hard Measurables		Soft Measurables	
Characteristic	Measure	Characteristic	Measure
R&D	• Patents • R&D Budget • New Products • R&D Staff • Publications • R&D Incentives • New Features • Inventions • New Markets • Product Extensions • Conferences • CRADAs • Partnerships	Impact Social	• Productivity • Growth • Lower Costs • Flexibility • Supply/Demand • Firm Size • Market Influence • User Benefits • Lower Prices • Social Enablers • Time Savers

management when they talk about knowledge management. Knowledge management is more about the art of truly understanding organizational culture dynamics and accessing, leveraging and sharing tacit know-how:

> A McKinsey survey of 40 companies in Europe, Japan, and the United States showed that many executives think that knowledge management begins and ends with building sophisticated information technology systems.
>
> Some companies go much further: they take the trouble to link all their information together and to build models that increase their profitability by improving processes, products, and customer relations. Such companies understand that true knowledge management requires them to develop ways of making workers aware of those links and goes beyond infrastructure to touch almost every aspect of a business (Hauschild et al., 2001).

The Role of Knowledge in Innovation

Since innovation is not a purely technical undertaking, the knowledge required for the successful management of innovation goes beyond science and engineering. Innovation can be subdivided into two domains: technical knowledge and knowledge transfer (Bohn, 1994), and learning about the administrative processes appropriate for managing technology (Jelinek, 1979). To facilitate the systematic development of innovations, an organization needs to have access to both types of knowledge: technical and administrative:

> To be *organizational, rather than individual, learning*, knowledge must be accessible to others rather than the discoverer, subject to both their application or use, and to their change and adaptation . . . . Organizational learning, to be learning rather than 'mere adaptation' must be generalized. It must go beyond simple replication to application, change, refinement. It must include 'rules for learning' and their change and adaptation, rather than the rote iteration of past successful actions . . . . Finally, if learning is to include innovation, it must encompass a system for governing the future as well as the present (Jelinek, 1979, pp. 162–163).

The most challenging aspect of studying the application of knowledge to innovation is to distinguish what information is most relevant and significant to the management of innovation and what is not. The features of the knowledge embedded in the innovation process can vary greatly. Some of that knowledge will be explicit, in the form of technical papers, drawings, and other documents which are codified and easily defined, and some will be tacit, integrated into organizational routines which are transferred only through socialization and collaboration. Therefore, the successful management of innovation can clearly

benefit from a systematic approach to knowledge management.

Knowledge, learning and cognition are classical terms that have been re-discovered in the context of the information technology and knowledge management revolutions. Knowledge management can be viewed as a socio-technical system of tacit and explicit business policies and practices. These are enabled by the strategic integration of information technology tools, business processes, and intellectual, human, and social capital (Conference Board, 1996). Managerial and organizational cognition can be perceived as the human and organizational capability for individual and collective reasoning, learning, emoting and envisioning. Organizational memory, intelligence, and culture are important determinants of cognition processes at both the individual and the organizational levels. We perceive managerial and organizational cognition and knowledge management as transitions across progressively higher levels of knowledge and meta-knowledge.

The Relationship Between Knowledge and Learning
Knowledge/Meta-knowledge

> The main Greek achievement was to remove explanation of the workings of the world from the realms of religion and magic, and to create a new kind of explanation—*rational explanation*—which was the subject of a new kind of enquiry.
>
> Peter Chechkland
> Systems Thinking, Systems Practice, 1981, p. 32

Beckman (1998) compiled a number of useful and relevant definitions of knowledge and organizational knowledge:

- Knowledge is organized information applicable to problem-solving (Dictionary (1));
- Knowledge is information that has been organized and analyzed to make it understandable and applicable to problem-solving or decision-making (Turban (2));
- Knowledge encompasses the implicit and explicit restrictions placed upon objects (entities), operations, and relationships along with general and specific heuristics and inference procedures involved in the situation being modeled (Sowa (3));
- Knowledge consists of truths and beliefs, perspectives and concepts, judgments and expectations, methodologies and know-how (Wiig (4));
- Knowledge is the whole set of insights, experiences, and procedures which are considered correct and true and which therefore guide the thoughts, behaviors, and communication of people (van der Spek & Spijkervet (5));
- Knowledge is reasoning about information to actively guide task execution, problem-solving, and

decision-making in order to perform, learn, and teach (Beckman (6));

- Organizational knowledge is the collective sum of human-centered assets, intellectual property assets, infrastructure assets, and market assets (Brookings (26));
- Organizational knowledge is processed information embedded in routines and processes which enable action. It is also knowledge captured by the organization's systems, processes, products, rules, and culture (Myers (27)).

Beckman (1998) proposes a five-level *Knowledge Hierarchy* in which knowledge can often be transformed from a lower level to a more valuable higher level.

A number of other authors have also proposed knowledge typologies. Nonaka & Takeuchi (7) have divided knowledge accessibility into two categories: tacit and explicit. Beckman (1998) identifies three stages of accessibility: tacit, implicit, and explicit:

(1) Tacit (human mind, organization)—accessible indirectly only with difficulty through knowledge elicitation and observation of behavior;
(2) Implicit (human mind, organization)—accessible through querying and discussion, but informal knowledge must first be located and then communicated;
(3) Explicit (document, computer)—readily accessible, as well as documented into formal knowledge sources that are often well organized.

Relationship Between Knowledge and Learning

But even though the first step along the road to a momentous invention may be the outcome of a conscious decision, here, as everywhere, the spontaneous idea—the hunch or intuition—plays an important part. In other words, the unconscious collaborates too and often makes decisive contributions. So it is not the conscious effort alone that is responsible for the result; somewhere or other the unconscious with its barely discernible goals and intentions, has its finger in the pie. . . . *Reason alone does not suffice.*

Carl Jung
The Undiscovered Self, 1958, pp. 99–100

Early research on organizational learning in the context of organization theory focused most substantially on attempting to describe learning processes in organizational settings, without necessarily assigning a normative value to learning (cf. Cyert & March, 1963; Levitt & March, 1988; March & Simon, 1958; Nelson & Winter, 1982). Learning as an organizational activity is perceived as an integration of individual efforts and group interactions. Thus, organizational learning becomes a process embedded in relationships among

individuals, through such mechanisms as information sharing, communication, and organizational culture. Some authors use the action-oriented concept of the 'learning organization' to identify paths to maximize organizational learning through a systems approach (Ciborra & Schneider, 1992; Senge, 1990). This perspective presumes that firms which are better at organizational learning will perform better than others in the market.

Other authors point out that learning can decrease organizational performance. Huber (1991) notes, "Entities can incorrectly learn, and they can correctly learn that which is incorrect". Ineffective or inappropriate learning processes can erode firm competitive advantage if they reinforce incorrect linkages between managerial activities and firm performance (Levitt & March, 1988). Even effective learning processes can be undermined by changes in market and environmental conditions which render them irrelevant, or worse, damaging to firm performance. Thus, learning activities can change from core competencies to core rigidities (Leonard-Barton, 1992). It is also possible that competence-destroying technological learning can limit firm performance in the short run but lead to superior performance in the long-term when market conditions adapt to new technologies (Christensen, 1997). Therefore, there is no linear relationship between learning and organizational performance; rather, performance improvement depends on the quality, not quantity, of organizational learning.

(1) Individual learning;
(2) Organizational learning;
(3) Inter-organizational learning.

Types of Learning

Computo, ergo sum. Particeps sum, ergo sum. Cogito, ergo sum.

Rene Descartes

We identify three levels of learning, based on previous theory-building about the impact of learning on building firm capabilities and on modifying operating modes (Carayannis, 1994a, 1994b, 1994c; Carayannis & Kassicieh, 1996). In this hierarchy, we posit three layers of technological learning:

(1) operational learning or learning and unlearning from experience;
(2) tactical learning or learning how-to-learn and unlearn from experience; and
(3) strategic learning or learning to learn-how-to-learn and unlearn from experience.

On the operational learning level, we have accumulated experience and learning by doing: we learn new things (Carayannis, 1994b). This is the short- to medium-term perspective on learning, focusing on new or improved capabilities built through the content learned by an organization. This learning contributes to the management of core organizational capabilities (Prahalad & Hamel, 1990), resource allocation (Andrews, 1965), and competitive strategy (Porter, 1991).

On the tactical learning level, we have learning of new tactics about applying the accumulating experience and the learning process (redefinition of the fundamental rules and contingencies of our short-term operating universe): we build new contingency models of decision-making by changing the rules for making decisions and/or adding new ones (Carayannis, 1994b). This is the medium- to long-term perspective on learning, resulting in a process of re-inventing and re-engineering the corporation. Tactical learning enables firms to approach new organizational opportunities in a more efficient and more effective manner, and to leverage or combining existing core capabilities in novel formations for greater competitive advantage.

On the strategic learning level, we have development and learning (internalization and institutionalization) of new views of our operating universe or Weltanschauungen (Hedberg, 1981); hence we learn new strategies of learning (Cole, 1989). Thus, we redefine our fundamentals (our rules and contingencies) for our decision-making, or we redefine the fundamentals of our operating universe. This is the very long-term perspective on learning, that focuses on re-shaping our re-inventing and re-engineering organizational 'tools' (methods and processes) (Bartunek, 1987; Bateson, 1972, 1991; Krogh & Vicari, 1993; Nielsen, 1993). The strategic learning level involves the expansion and reformulation of concepts about the limits and potential of the firm's strategic environment, where the older framework is seen as simply a 'special case' within the new, more inclusive framework (akin to the relationship between 'normal' and 'revolutionary' science developed by Kuhn, 1962).

Strategic learning serves to 'leap-frog' to a new competitive realm and "to increase the slope of the learning curve as well as the rate by which the slope per se increases by means of enhanced and innovative organizational routines" (Carayannis, 1994b, pp. 582–583). The result is what other authors refer to as "changing the rules of the game" (Brandenburger & Nalebuff, 1996; D'Aveni, 1994) or creating new "ecologies of business" (Moore, 1996). The firm pioneers a new conceptualization of its business, its market, and/or its overall competitive environment, which gives it greater strategic flexibility not only in its own course of actions but also in influencing and leading the firms around it.

Learning/Meta-Learning

The primary process by which firms change their capabilities to better fit the environment is through learning. In the case of learning like the case of most other fundamental concepts, there is little consensus as to what learning is and how it takes place. In economics, learning is perceived as tangible, quantifiable improvements in value-adding activities, in management, learning is seen as the source of "sustainable competitive efficiency" (Dodgson, 1993, p. 376), whereas in the innovation literature, learning is considered as a source of "comparative innovative efficiency" (ibid, p. 376). As noted in Doz (1996), there is a distinction in the organizational context between cognitive learning and behavioral learning. Cognitive learning occurs when members of a firm realize that some change is needed in a given situation; behavioral learning occurs when the organizational routines of that firm are actually changed (the implementation of cognitive learning). Expanding the concept of learning further, organizational learning occurs when the new behavior is replicated throughout the firm, leading to broad-based organizational change (Teece et al., 1997, p. 525).

Knowledge Management

Knowledge management is defined as: "the systematic, explicit, and deliberate building, renewal, and application of knowledge to maximize an enterprise's knowledge-related effectiveness and returns from its knowledge assets" (Wiig, 1993). Sveiby (1998) defines knowledge management as "the art of creating value from an organization's intangible assets". Moreover, Sveiby (1998) identifies two main tracks of knowledge management activities: one track focuses on knowledge management as the management of information and the other track as management of people.

Cognition/Meta-Cognition

Cognition is the human capacity to perceive, interpret, and reason about environmental and conceptual environmental or organizational stimuli and meta-cognition the capacity "to think about thinking, as meta-learning means to learn about learning" (Carayannis, 1994a, p. 8).

The knowledge creation, transfer, selection, acquisition, storage, and retrieval processes can be viewed from an information-theoretic (Shannon & Weaver, 1949) and a meta-cognitive (Halpern, 1989; Simon, 1969; Sternberg & Frensch, 1991) linguistic perspective (Chomsky, 1971, 1993), where the human problem-solver and technology manager is seen as both a technician and a craftsman (Schon, 1983), a 'lumper' and a 'splitter' (Mintzberg, 1989). Individuals, teams, and organizations rely on multi-layered technological learning and unlearning (Carayannis, 1992, 1993, 1994a, 1994b, 1994c; Dodgson, 1993) to create, maintain, and enhance the capacity of individuals, groups, and organizations to transfer and absorb embodied and disembodied (von Hippel, 1988) technology in the form of artifacts, beliefs, and evaluation routines (Garud & Rappa, 1994) and tacit and explicit knowledge (Nonaka, 1988, 1994; Polanyi, 1958, 1966). Moreover, it is critical to realize that individual and organizational learning and knowledge are mutually complementing and reinforcing entities using the medium of organizational memory:

> Moreover, this learning process must be endowed with an organizational memory that is both, accurate and precise to build, maintain, and renew continuously the firm's reservoir of skills and competencies: If an organization is to learn anything, then the distribution of its memory, the accuracy of that memory, and the conditions under which that memory is treated as a constraint become crucial characteristics of organizing. (Weick, 1979, p. 206.) (in Carayannis, 1994b, 2001).

In this context, it is important to remember that "knowledge does not grow in a linear way, through the accumulation of facts and the application of the hypothetico-deductive method, but rather resembles an upward spiral, so that each time we reevaluate a position or place we've been before, we do so from a new perspective" (in Carayannis, 1994b, p. 52). This concept sets the scene for the development of the Organizational Cognition Spiral (OCS), as part of our organizational knowledge management model. Organically relevant to these concepts is intuition, which Weick defines as 'compressed expertise' (Davenport & Prusak, 1998, p. 11) along with meta-knowledge as knowledge (awareness) about the knowledge you possess.

The OCS Model

A model for understanding the key issues involved in organizational knowledge management is proposed. The model defines different knowledge states that are a function of two dimensions—knowledge (K) and meta-knowledge (MK) as defined earlier and it consists of successive 'knowledge cycles' where an individual or organization can transition and traverse four stages of awareness or ignorance. As each cycle is traversed and then a transition to the next cycle occurs, the overall level of knowledge and meta-knowledge is rising (see Fig. 3).

Typically, but not always the transition takes place from *ignorance of ignorance* (you do not know what you do not know) to *awareness of ignorance* (you know what you do not know) to *awareness of awareness* (you know what you know: the result of search, discovery, and learning) to *ignorance of awareness* (you do not know what you know: the result of routinized as well as tacit awareness).

For the sake of simplicity, the dimensions are assumed to be at two levels representing the presence and absence of (meta) knowledge. The levels of the two dimensions are thus represented as K and K̲, and MK and M̲K̲. These two levels of the two dimensions result in a total of four knowledge states:

(1) K, M̲K̲ (Awareness of awareness);
(2) K, M̲K̲ (Ignorance of awareness);
(3) K̲, MK (Awareness of ignorance);
(4) K̲, M̲K̲ (Ignorance of ignorance).

Organizations may exist in any one of these states that include current, desired, and/or intermediate levels. The states can be represented in the following way (Fig. 3).

Knowledge management can be viewed as the process of managing the transitions across these four states.

The evolutionary knowledge transformation process is *both differential and integrative in nature* (Carayannis, 1992, 1993, 1994a, 1994b, 1996, 1997, 1998, 1999, 2001, 2002), in that it consists of *unlearning, learning and meta-learning* components, differentiating new against old experience, selecting and retaining the currently useful knowledge modules and integrating the lessons learned throughout. This process reflects the dynamic of the synthetic progression at the individual and the organizational *levels from data to information to knowledge to wisdom to intuition.* In this manner, increasingly broader and deeper levels of *organizational knowing* (Chun Wei Choo, 1998) are attained and both quantitative as well as qualitative transformations of the organizational and individual knowledge stock and flow occur.

Case in Point: Innovation at Xerox: From Unrealized Potential to Real Success

XEROX Background & History

The last decade has given rise to the terms *old economy and new economy.* The old economy, industry based, traditionally has been characterized by economies of scale while the new economy, knowledge based, is considered the economy of networks—as a collaborative (Shapiro, 1999). The shift from the old economy to the new economy can be described as a paradigm shift. In Kuhn, a paradigm is defined as "an object for further articulation and specification under new or more stringent specifications" (Kuhn, 1962). In Moore, the traditional old economy is defined as a firm going up against its competition, in a win–loose scenario (Moore, 1996). The new economy paradigm is defined as market creation or co-evolution in a win–win scenario.

Xerox has a rich history of both success and failure in innovation. The successes are evident in the office environment of today. Copiers, laser printers, and

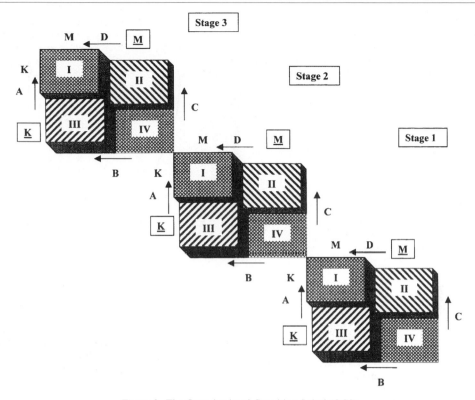

Figure 3. The Organizational Cognition Spiral (OCS).

State I: Awareness of Awareness (MK, K)—State II: Ignorance of Awareness (<u>MK</u>, K)—State III: Awareness of Ignorance (MK, <u>K</u>)—State IV: Ignorance of Ignorance (<u>MK</u>, <u>K</u>). (Possible Pathways: a. IV to II to I or b. IV to III to I). The successive stages (I, II, III) denote cumulatively higher levels of knowledge (K) and meta-knowledge (MK).

networked offices are all around us, thanks to the successful innovation of Xerox. Not only has Xerox been successful on the hardware side of the office; the service (maintenance) of copiers and supplies (toner, paper, etc.) businesses is extremely successful—as well as supporting consulting services and document processing services (solutions).

The innovations of Xerox proliferate, as evidenced by over 7000 active patents in their intellectual property portfolio. However, despite the innovation successes, there have been failures along the way.

The invention of the personal computer with a GUI, desktop, mouse, Ethernet, and the first WYSIWYG word processor never led to a Xerox innovation. The same is true of the first laser printer. In both cases Xerox invented but did not innovate. It took other firms to capture and market the inventions to reach the stage of innovation.

(1) What criteria led to the success?
(2) What criteria led to failure?

These are important questions to be asked. The answers may help to categorize the hegemony of ideas and

define success criteria, allowing the development of methodologies for creating and sustaining innovation best practices. In studying innovation it is best to look at both successes and failures as a root-cause analysis. The following is a case study of innovation at Xerox Corporation.

On October 22, 1938 in Astoria, Queens, Chester Carlson invented what later would be referred to as Xerography. In Carlson's thinking, he was revolutionizing the office process, but he would find that the world did not see his invention in the same frame of reference as he did. Carlson, born in 1906, worked as a printer's helper during his early career, including publishing a small newsletter on his own.

This early experience impressed him with the physical difficulty of putting words on paper and sharing knowledge. Later he received a degree in physics from the California Institute of Technology and began to work as a research engineer at Bell Labs. A depression-era job loss, followed by a degree in Law, led him to a second career as a patent attorney. As a patent attorney he faced the continuous problem of never having enough carbon copies.

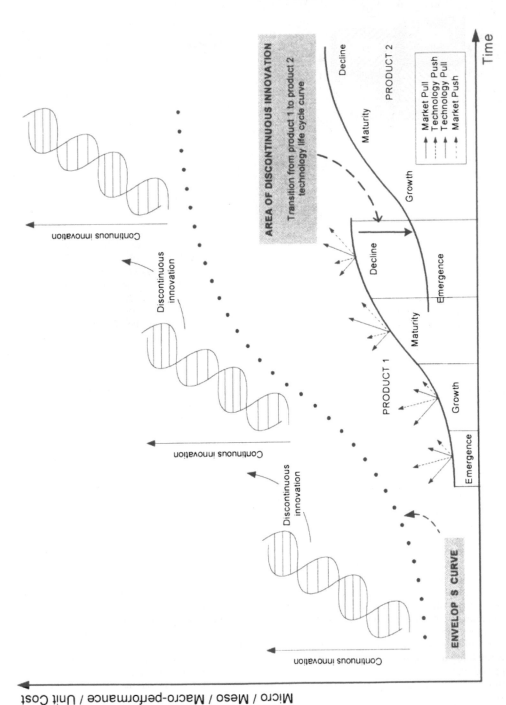

Figure 4. Continuous and discontinuous innovation: Technology performance road map.

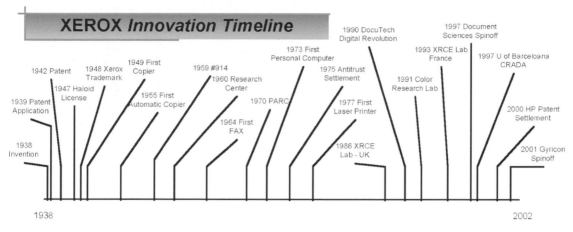

Figure 5.

The only alternatives were to use an expensive photographic process or retyping (and reproofing) lengthy patent applications. In his spare time he researched alternative technologies, eventually finding the work of Hungarian physicist Paul Selenyi on photoconductivity. He performed experiments in his kitchen, eventually replicating the image of '10–22–38 ASTORIA' on a sulfur-coated zinc plate. Innovation is not an easy process, as Carlson was to find out. He searched for a company that would be interested in financing further research in his invention. For 10 years he had no success.

The market was not ready for alternative solutions—the common thinking was that carbon copies, the current technology, were sufficient and another technology was not needed. In 1944, Battelle Memorial Institute, a non-profit research institute, became interested in assisting Carlson in further developing his invention. During the Battelle era, selenium was introduced as an improved photoconductor, and dry ink toner was developed. Finally, in 1947, the Haloid Company, a manufacturer of photo-paper, acquired a license to manufacture a machine to produce xerography. Within one year, the first Xerox copier was shipped, beginning the age of Xerography.

The early copy machine was complex to operate, but it found a niche in producing working masters for offset duplicators—one should remember that the printing technology of the period was 'letterpress', printing from individually cast metal images; a very expensive process. 'Offset' printing was in its infancy and was utilized in low-end, inexpensive printing environments.

By 1959, Haloid improved the equipment and released the no. 914 copier—the first true office copier. The no. 914 was an innovative breakthrough.

The competition of the time, the AB Dick mimeograph, 3M Thermo-Fax, and the Kodak Verifax, were eclipsed within a short period of time. The no. 914 was

so successful that it continued to be the leading technology until 1972.

Xerox continued to innovate throughout its history, albeit not always successfully. In 1973 the first desktop personal computer was developed, leading to the PC revolution. Xerox, due to marketing strategies to be discussed later, did not benefit from this development. Again, in 1977 Xerox developed laser printing but did not move quickly to capture the early market for laser printers like its competitor Hewlett-Packard.

Xerox corrected its innovation strategy when it introduced the DocuTech high-volume black & white print system in 1990, thus creating the digital revolution in putting words on paper. Xerox plans to introduce the iGen3, a color version of the DocuTech technology. It is expected that this will mark the beginning of another revolution, digital color, and bring Xerox the financial rewards of innovation.

In order to fully understand Xerox, one must view the mix of products and services and the market space that is occupied. Currently Xerox products and services may be divided into two major categories—products and services. These can be further divided from products into office, production, equipment maintenance, and the necessary supporting supplies. The services can be further divided into outsourcing, process re-engineering, Solutions (integration services), and software applications.

It is interesting to note that research into the history of products uncovered some interesting products that were once core to the Xerox portfolio—electronic typewriters, workstations, and computer systems. These are no longer in the current portfolio.

Xerox sells its products through a variety of channels including direct sales, telebusiness, resellers, agents, concessionaires, and via the Internet. These channels are managed by various organizations—see table below. The sales organization is global and divided into regional territories. The largest territory is

Table 2. Xerox innovation group research & technology centers.

Group	Location	Est.	Core Expertise	Additional Functions
Digital Imaging Technology Center	Palo Alto, CA	1994	Digital imaging 1:1 communications On-demand printing	Core competence in digital imaging
Solutions and Services Technology Center	Webster, NY	2001	Color science Image sciences and processing Systems architecture Work process studies Software development	Create, assess, acquire, apply technologies Coherence and architected set of solutions and services
Wilson Center for Research and Technology	Webster, NY	1960	Image evaluation Image processing Marking processes Media handling Microsystems Embedded controls Device controls	Color technology Image quality Cost of ownership Xerography patent basis
Xerox Research Centre of Canada	Ontario, Canada	1974	Materials research Imaging materials Inks and Toners Photoreceptors Specialty media Organic materials	Research concepts Supplies and consumables Environmental issues Materials design, Synthesis and evaluation
Xerox Research Centre Europe	Grenoble France	1993	Content analysis contextual computing Mobile computing Ubiquitous computing image processing Work practices knowledge portals E-leering, Publishing, Emerging office	License technology Document access Knowledge sharing Coordinates research, engineering, and business development activities
Palo Alto Research Center Incorporated (PARC)	Palo Alto, CA	1970	Electronic materials Micro-electro-mechanical systems (MEMS) Semiconductor devices software Engineering, Image analysis Human–computer interaction Wireless computing Knowledge representation	Ubiquitous computing Licensing, collaborative research and spin-off creation Developed first personal computer with WYSIWYG word processor, mouse, Ethernet, and GUI Developed first laser printer

the U.S. covered by the North American Solutions Group (NASG).

It was noted that almost 50% of total Xerox employees work for Xerox Services, most of which are located physically within a customer site. One area of understanding that many researchers of Xerox find difficult, is that of the Solutions business. Xerox defines Solutions as "an integrated offering that includes hardware, software and people-based services, which solves a problem, improves a work process, or creates a market or competitive advantage". Xerox has divided its Solutions delivery into four main business operations and focuses on production market (graphic arts firms), the office market, and services. The four groups are Documents Systems and Solutions Group (DSSG), Office Systems Group (OSG), Office Printing Business Group (OPBG), and Xerox Global Services (XGS).

Xerography was invented in 1938 but it was not until 1959 that the initial invention was actualized and became an innovation. The 19-year travel from invention to innovation was consumed with both finding a financial partner to further develop the idea (1938–1947) and later by an attempt to define the market (1948–1959). In the 1930s to 1950s the technology of the office was defined by carbon paper and the emerging offset printing process.

Carbon paper allowed for a real-time duplicate of a document, possibly up to eight copies, and early offset duplication was available for 8 + copies, but the cost of the 8–500 copies was somewhat prohibitive. What Chester Carlson and Haloid first found in the search for a market was that the perceived need for the innovation was non-existent. The challenge to Haloid was to develop a market.

The first machine to reproduce copies via Xerography was released in 1949. The market it captured was within the nascent offset printing technology. Specifically, the first Xerox copier was targeted toward making masters for offset duplication. The masters would then in turn be used in the duplicating process, thus making 'copies'. The Xerox master copy process for offset duplication was expensive and complex to operate, and would soon be overtaken by other, photo-based less expensive technologies. As Haloid refocused its attention back toward replacing carbon paper technology, it found a winner in the 1959 introduction of the no. 914 copier. This combination of market pull and technology push would drive revenue and profits well into the 1970s. From the early Haloid days of unstructured innovation, Xerox has developed a culture of organizing for innovation. Currently, the Xerox organization chart reflects the Innovation Group as directly reporting to the CEO. This underscores the relative importance affixed to innovation within the organization.

Innovation—Faux Pas

In 1970 Xerox formed the Palo Alto Research Center (PARC), famed as the epicenter of the computer revolution. Researchers at PARC were given freedom to conduct basic research form the beginning. This led to, among other inventions, the first personal computer in 1973 and the first laser printer in 1977.

The personal computer was advanced for its time and consisted of an operating system, a WYSIWYG word processor, and a graphical user interface with a desktop, a mouse, and Ethernet connectivity. With this advanced invention in its portfolio, Xerox should have led the computer revolution—but as history shows us, Xerox did not capitalize on its inventive resources and let others drive the birth of a new market. The question to the student of innovation is why did Xerox let this happen and what can be done to avoid this type of costly mistake in the future! In other words, what is the lesson to be learned from this?

In order to understand Xerox strategies, the authors have taken the approach of investigating the history of Xerox innovation and interviewing key players of the era. In one such interview with Mr. RT, a 30-year veteran of Xerox and an executive involved in West Coast operations for much of his career, the following information was developed.

In the 1970s, in addition to PARC, Xerox had substantial operations centered on the West Coast. One must remember that Xerox is Rochester, NY-based, where the single largest geographical concentration of employees (about 16,000) is located. The visionary of the 'office of the future' was Joe Wilson II, then Chairman of Xerox. At that time, in addition to PARC, Xerox-West consisted of Versatec (plotters), XSoft (software application development), Xerox Network Services (Ethernet, networks), Shugart (disk manufacturing), Total Recall (scan and retrieval applications), plus a substantial copier hardware manufacturing facility. This was a very advanced portfolio of technical capabilities and capacity for the time.

PARC itself was a central clearinghouse of computer information in the early 1970s. PARC developed a professional forum as a tool for motivating researchers. Each week, PARC would host a public event to allow its researchers to present their research findings. This forum was well attended by non-Xerox professionals from universities, engineers from the nascent computer industry, and others interested in the research. This early knowledge sharing aided the birth of the computer industry in the Silicon Valley area.

When the personal computer was originally developed, Xerox strategy was to market the PC as a proprietary business-to-business tool. It was more of a 'portable' computer than a 'personal' computer. The computer consisted of a 32″ wide portable unit, a hard drive basically, that could be ported or moved from location to location as required. The computer was to be inserted into a docking station to be used. The original operating system, MESA, was unique. MESA, it should be noted, eventually became the basis for artificial intelligence systems of today.

The PC was code named STAR and it was soon marketed as the 6085 workstation. Eventually, a spin-off called GlobalView was formed and the computer later became known as the GlobalView System. There were approximately 50 applications developed, like word processing, spreadsheets, graphics programs, specialized graphics (chemical and math applications), messaging, hyperlinks, browsers, etc. It contained many features like the 'Clearing House', giving the user a shared knowledge area. One application allowed the user to create custom applications (a forerunner of JAVA). All applications were proprietary and could only be used on the GlobalView system.

Around this time, Xerox began to staff its West Coast management ranks with former IBM executives, mostly from the mainframe business. It should be noted that the emerging PC market in the 1970s was influenced by the big three of the time—Xerox, IBM, and WANG. As one may know from various studies of IBM, the mainframe management mentality was not easily aligned with what eventually became the PC market. Importing IBM mainframe-experienced managers was probably one of the greatest mistakes made by Xerox. The former IBM executives did not fit well into the existing culture of Xerox and had difficulty dispersing their ideas within the Xerox management infrastructure. Xerox executives had the correct vision but they lacked proper execution.

While Xerox may have had the vision of the 'office of the future', it was not sure how to market it. Xerox was known for selling copiers and it did this very well ('selling iron'). The PC market was forming-storming-norming and Xerox management found it difficult to predict the path the industry would take. It placed its bet on business-to-business channels, ignoring the 'personal' or hobbyist market as it was known. As business-to-business was the target market, the selection of proprietary systems may have seemed the best strategy. Hindsight gives us the knowledge that the PC industry formation was driven by the hobbyist, who was the bridge between—offering insights as to the utility of the computer for both personal and business use. With the rise of Apple Computers, the market was better defined. It is interesting to note that the main attraction of the Apple computer was the graphical interface/desktop/mouse, an idea borrowed during a visit to PARC. Another issue within Xerox, limiting the market strategy, was the sales force alignment. Xerox had a well-trained and well-equipped sales force—aligned to copiers—hardware/service/supplies driven. In order for Xerox to market its new innovation, the PC, it needed a sales force that was aligned to a different conceptual product—software driven. Xerox strategy did not take into account the realignment of its

sales force, particularly the compensation plans. Xerox possessed a successful sales force primarily because its compensation plan was very liberal. In order for the existing compensation plans to benefit the individual sales person, the only end sale of computers that was attractive, from a compensation perspective, was a multi-million dollar computer sale. In the 1970s the only clients able to invest millions in computers were already mainframe customers of IBM, WANG, Digital and others—the market for PC-based business-to-business was too ill defined for major capital investment. Large corporations were not ready to move from the mainframe to the PC, even if it was networked. In essence, Xerox was compensating its sales force to sell to a market niche that did not yet exist. In a situation like this, Xerox strategy should have been to realign its compensation plan to incentivize its sales force. The consequences were obvious. Another inhibitor to success was cultural. Xerox was Rochester, NY-based and the computer revolution was West Coast-oriented. The resulting clash of cultures led to a '*not invented here*' syndrome in Rochester. Because the resources and management for supporting innovation were Rochester-based, the new inventions of West Coast origin were not readily understood. An example of this is Xerox development of web-capability technologies. The technology was developed in West Coast operations and, after development, it was ported to Rochester for further development—a clear case of cultural conflict as Rochester had little infrastructure to support the emerging web technology. Marketing decisions and funding, also Rochester-based, did not understand and strategize to align to emerging market perceptions. Copier marketing strategy was the concentration, and any marketing for the PC was aligned to copier marketing strategies. The issue of new marketing strategies for newly emerging markets was misinterpreted. The culture differences were not recognized early and not managed properly. Eventually, GlobalView was ported to non-proprietary environments like Sun workstations, the IBM 6000, and the IBM/Microsoft compatible platforms in an effort to compete but the decision was made too late in the technology life cycle for any hope of securing a place in the market. The porting itself was burdened with technological inconsistencies. For example, when it was ported to the MS platform, the nominal PC did not have sufficient memory to run GlobalView and, given that memory was expensive at the time, the initial cost of setup was prohibitive.

An attempt was made to spin-off the technology development organizations but cultural influences intervened. By the early 1990s Xerox strategy seemed to be one of closing the West Coast technology operations and merging them into Rochester-based organizations. Today, technology organizations are managed and geographically centered in Rochester. PARC in Palo Alto, California and Xerox Research

Centers in Ontario, Canada and Grenoble, France are driven, directed and managed by Rochester technology management.

Another influence on Xerox innovation strategy was a 1975 antitrust settlement. In the settlement, Xerox agreed to open its protective envelope of intellectual property confinement and agree to license what was previously considered proprietary technologies. While the settlement did not produce an instant change in culture for Xerox, it eventually influenced the innovation strategy as evidenced by the current organization of the Innovation Group.

Intellectual property eventually became a source of revenue for Xerox and not just a cost of doing business. It took almost a generation for this culture change to be fully implemented.

Conclusions

On the other side of success is a list of failures to innovate. The failures of innovation are summarized as follows:

(1) Intellectual Property Management—Patent and Trade Secret Strategies;
(2) Cultural Influences and Strategies to Mitigate;
(3) Marketing Strategy.

One cannot accept the reasons for failure without an evaluation of the lessons learned. Such a process allows for a metamorphosis of failures into the successes of the future.

In the first failure, *Intellectual Property Management—Patent and Trade Secret Strategies*, the strategy employed by PARC to recognize the output of researchers had the disastrous effect of exposing trade secrets to competitors without managing the exchange within a boundary—such as a CRADA (Cooperative Research and Development Agreement), licensing agreement, or other controlled knowledge-sharing arrangement. Diffusion of technology must be protected and a financial strategy is imperative. Capturing inventions and actualizing them into innovations is difficult and should not be hindered by uncontrolled information flows.

The second failure is one of *Cultural Influences and Strategies to Mitigate*. This is a very intricate issue as culture may not be readily evident within an organization. In the case of Xerox, there are two distinct cultural influences to be considered.

Initially, the culture of the corporation was heavily influenced by a 'home office' viewpoint. Rochester, NY is the de-facto operational hub of Xerox, with an employee concentration of over 20% of the global workforce. This made for an endemic situation. Rochester is also home to much of the historical innovation influence, since the mid 1940s. In 1970, when the innovation center (PARC) was developed on the West Coast, there was a natural resistance in the Rochester employee group to and invention emanating out of a non-Rochester-based organization. Additionally, the management of the West Coast operations mainly consisted of recent hires from IBM. The culture of IBM and the culture of Xerox were not compatible and this led to further separation from Rochester.

The third failure, *Marketing Strategy*, is intrinsically linked to the existing cultural influences. Because Rochester, home of the marketing group, did not link with the West Coast operations culturally, marketing did not understand the pith of the inventions made in PARC and other West Coast operations. This lack of understanding was pernicious to any marketing plans developed. Rochester did not understand the true nature of the inventions, which tended to exacerbate the situation. This lack of understanding led to faulty marketing plans and an underestimation of the market potential.

What are the lessons learned from this case study? Innovation may be considered a two-sided coin. On one side there is success—a rich history of invention evolving into innovation. On the other side of the coin is failure—either the lack of inventing or not transforming the invention into an innovation. Remember the definition of innovation given earlier in this paper is one of *actualization*; using the invention for improvement or taking the invention to market.

The case of Xerox Corporation gives us examples of both sides of innovation, the rich history of successes and disappointment of failure. Also, it supports our definition of innovation—actualization, as the key criteria in separating invention from innovation.

The attempt in the paper to add to the understanding of innovation is but a mere scratch on the surface of the literature. Innovation, its theory and practice, is beginning to emerge as a separate discipline of study. From what was briefly touched upon in the previous material, the influence of actors and organizational variables on innovation would be an interesting cause for further investigation and study.

References

Anderson, P. & Tushman, M. L. (1990). Technological discontinuities and dominant designs: A cyclical model of technological change. *Administrative Science Quarterly*, **35**, 604–633.

Andrews, K. E., Christensen, L. E. et al. (1965). *Business policy. Text and cases*. Homewood, IL: Richard D. Irwin Inc.

Argyris, C. & Schon, D. (1978). *Organizational learning: A theory of action perspective*. New York: Addison Wesley.

Barnard, C. (1938). *The function of the executive*. New York: New York University Press.

Bartunek, J. et al. (1987). First-order, second-order, and third-order change and organization development interventions: A cognitive approach. *The Journal of Applied Behavioral Science*, **23**, 4.

Bateson, G. (1972). *Steps to an ecology of mind*. New York: Ballantine.

Bateson, G. (1991). *A sacred unity: Further steps to an ecology of mind* (R. Donaldson, Ed.). New York: Harper Collins.

Beckman, T. (1998). *Knowledge management: A technical review*, GWU Working Paper.

Bohn, J. (1994). *Formal transformational reasoning about reactive systems in the theorem prover LAMBDA*. New York: Oxford University Press.

Brandenburger, A. M. & Nalebuff, B. J. (1996). *Co-opetition*. New York: Doubleday.

Burns, T. & Stalker, G. (1961). *The management of innovation*. Tavistock: Tavistock Institute.

Carayannis, E. (1992). An integrative framework of strategic decision making paradigms and their empirical validity: The case for strategic or active incrementalism and the import of tacit technological learning. *RPI School of Management, Working Paper Series No 131*. October, 1992.

Carayannis, E. (1993). Incrémentalisme stratégique. *Le Progrès Technique*. Paris, France.

Carayannis, E. et al. (1994a). *A multi-national, resource-based view of training and development and the strategic management of technological learning: Keys for social and corporate survival and success*. 39th International Council of Small Business Annual World Conference, Strasbourg, France, June 27–29.

Carayannis, E. (1994b). *The strategic management of technological learning: Transnational decision making frameworks and their empirical effectiveness*. Published Ph.D. dissertation, School of Management, Rensselaer Polytechnic Institute. New York: Troy.

Carayannis, E. (1994c). Gestion stratégique de l'acquisition de savoir-faire, *Le Progrès Technique*. Paris, France.

Carayannis, E. & Kassicieh, S. (1996). *The relationship between market performance and higher order technological learning in high technology industries*. Fifth International Conference on Management of Technology, Miami, FL, February 27–March 1.

Carayannis, E. (1996a) Re-engineering high risk, high complexity industries through multiple level technological learning: A case study of the world nuclear power industry. *Journal of Engineering and Technology Management, 12*, 301–318.

Carayannis, E. & Kassicieh, S. (1996b). *The relationship between market performance and higher order technological learning in high technology industries*. Fifth International Conference on Management of Technology, Miami, FL, February 27–March 1.

Carayannis, E. (1997). Data warehousing, electronic commerce, and technological learning: Successes and failures from government and private industry and lessons learned for 21st century electronic government, *Online Journal of Internet Banking and Commerce*, (March).

Carayannis, E. (1998). Higher order technological learning as determinant of market success in the multimedia arena: A success story, a failure, and a question mark: Agfa/Bayer AG, enable software, and sun microsystems. *International Journal of Technovation, 18* (10), 639–653.

Carayannis, E. (1999). Fostering synergies between information technology and managerial and organizational cognition: The role of knowledge management. *International Journal of Technovation, 19* (10), 219–231.

Carayannis, E. (2001). *The strategic management of technological learning*. Washington, D.C.: CRC Press.

Carayannis, E. (2002). Is higher order technological learning a firm core competence, how, why, and when: A longitudinal, multi-industry study of firm technological learning and market performance. *International Journal of Technovation,22*, 625–643.

Chomsky, N. et al. (1971). *Chomsky: Selected readings*. New York: Oxford University Press.

Chomsky, N. (1993). *Language and thought*. Newport: Moyer Bell.

Christensen, C. (1997). *The innovator's dilemma: When new technologies cause great firms to fail*. Cambridge, MA: HBS Press.

Chun Wei Choo (1998). *The knowing organization: How organizations use information to construct meaning, create knowledge, and make decisions*. Oxford: Oxford University Press.

Ciborra, C. U. & Schneider, L. S. (1992). Transforming the routines and contexts of management, work and technology. In: P. S. Adler (Ed.), *Technology and the Future of Work*. Cambridge, MA: MIT Press.

Cole, R. (1989). *Strategies for learning: Small group activities in American, Japanese, and Swedish industry*. Berkeley, CA: Berkeley University Press.

The Conference Board (1996). *Knowledge management in organizations*. New York: Tavistock Institute.

Cooper, J. R. (1998). A multidimensional approach to the adoption of innovation. *Management Decision, 36* (8), 493–502.

Cyert R. M. & March, J. G. (1963). *A behavioral theory of the firm*. Englewood Cliffs, NJ: Prentice-Hall.

D'Aveni, R. (1994). *Hypercompetition: Managing the dynamics of strategic maneuvering*. New York: The Free Press.

Davenport, T. & Prusak, L. (1998). *Working knowledge*. Boston, MA: Harvard University Press.

Diwan, R. K. & Chakraborty, C. (1991). *High technology and international competitiveness*. New York: Praeger.

Dodgson, M. (1993). Organizational learning: A review of some literatures. *Organization Studies, 14* (3), 375–394.

Doz, Y. L. (1996). The evolution of cooperation in strategic alliances: Initial conditions or learning processes? *Strategic Management Journal, Special Issue on Evolutionary Perspectives on Strategy, 17*, 55–83

Drejer, A. (2002). Situations for innovation management: Towards a contingency model. *European Journal of Innovation Management, 5* (1), 4–17.

Drucker, P. (1999a). *The effective executive*. London: Butterworth-Heinemann.

Drucker, P. (1999b). *Practice of management*. London: Butterworth-Heinemann.

Evangelista, R., Iammarino, S., Mastrostefano, V. & Silvani, A. (2001). Measuring the regional dimension of innovation. Lessons from the Italian innovation survey. *Technovation, 21*, 733–745.

Florida, R. L. & Kenney, M. (1990). *The breakthrough illusion: Corporate America's failure to move from innovation to mass production*. New York: Basic Books.

Garcia, R. & Calantone, R. (2002). A critical look at technological innovation typology and innovativeness terminology: a literature review. *Journal of Product Innovation Management, 19*, 110–132.

Garud, R. & Rappa, M. (1994). A socio-cognitive model of technology evolution: The case of cochlear implants. *Organization Science, 5* (3, August).

Halpern, D. (1989). *Thought and knowledge: An introduction to critical thinking*. Mahwah, NJ: Lawrence Erlbaum.

Hauschild, S., Licht, T. & Stein, W. (2001). Creating a knowledge culture. *The McKinsey Quarterly*, **1**, 74–81.

Hedberg, B. (1981). How organizations learn and unlearn. In: Nystrom & Starbuck (Eds), *Handbook of Organizational Design*. New York: Oxford University Press.

Henderson, R. & Clark, K. (1990). Architectural innovation: The reconfiguration of existing product technologies and the failure of established firms. *Administrative Science Quarterly*, **35**, 9–30.

Hindle, B. & Lubar, S. D. (1986). *Engines of change: The American industrial revolution, 1790–1860*. Washington, D.C.: Smithsonian Institution Press.

Huber, G. P. (1991). Organizational learning: The contributing processes and the literature. *Organization Science*, **2** (1), 88–115.

Jelinek, M. (1979). *Institutionalizing innovation: A study of organizational learning systems* (pp. 162–163). New York: Free Press.

Jonash, R. S. & Sommerlatte, T. (1999). *The innovation premium*. Boston, MA: Perseus Publishing.

Kaplan, S. (1999). Discontinuous innovation and the growth paradox. *Strategy and Leadership*, (March/April), 16–21.

Krogh, von G. & Vicari, S. (1993). An autopoiesis approach to experimental strategic learning. In: Lorange et al. (Ed.), *Implanting Strategic Processes: Change, Learning, and Cooperation*. London: Basil Blackwell.

Kuhn, T. S. (1962). *The structure of scientific revolutions*. Chicago, IL: University of Chicago Press.

Leonard-Barton, D. A. (1992). Core capabilities and core rigidities: a paradox in managing new product development. *Strategic Management Journal*, **13**, 111–125

Levitt, B. & March, J. G. (1988). Organizational learning. *Annual Review of Sociology*, **14**, 319–340.

Mansfield, E., Rapport, A. R., Wagner, S. & Beardsley, G. (1977). Social and private rates of return from industrial innovations. *The Quarterly Journal of Economics*, **91**, (2, May), 221–240.

March, J. G. & Simon, H. (1958). Organizations. *Administrative Science Quarterly*, **30**, 160–197.

Mintzberg, H. (1978). Patterns in strategy formation. *Management Science*, **24** (9), 934–948.

Mintzberg, H. (1989). *Mintzberg on management*. New York: The Free Press.

Mintzberg, H. (1991a). Brief case: Strategy and intuition—a conversation with Henry Mintzberg. *Long Range Planning*, **2**, 108–110.

Mintzberg, H. (1991b). Learning 1 planning 0: Reply to Igor Ansoff. *Strategic Management Journal*, **12**, 463–466.

Moore, J. F. (1996). *The death of competition*. New York: HarperCollins.

Nelson, R. & Winter, S. (1982). *An evolutionary theory of economic change*. Cambridge, MA: Harvard University Press.

Nelson, R. & Winter, S. (2000). *Evolutionary theory of economic change*. New York, NY: NYU Press.

Nelson, R. R. (1977). In search of useful theory of innovation. *New Holland Research Policy*, **6**, 37–76.

Nielsen, R. (1993). Woolman's 'I Am We' triple-loop action-learning: Origin and application in organization ethics. *The Journal of Applied Behavioral Science*, (March).

Nonaka, I. (1988). Creating organizational order out of chaos: Self-renewal in Japanese firms. *California Management Review*, (Spring).

Nonaka, I. (1994). A dynamic theory of organizational knowledge creation. *Organization Science*, (February).

Nonaka, I. & Takeuchi, H. (1995). *The knowledge-creating company: How Japanese companies create the dynamic of innovation*. New York: Oxford University Press.

Polanyi, M. (1966). *The tacit dimension*. Routledge.

Porter, M. (1991). Towards a dynamic theory of strategy. *Strategic Management Journal*, **12**, 95–117.

Prahalad, C. K. & Hamel, G. (1990). The core competence of the corporation, *Harvard Business Review*, (May–June).

Quinn, J. B. (1980). *Strategies for change: Logical incrementalism*. Chicago, IL: Richard D. Irwin.

Quinn, J. B. (1992). *The intelligent enterprise: A new paradigm*. New York: The Free Press.

Rogers, E. (1983). *The diffusion of innovations*. New York: MacMillan.

Rogers, E. M. (1995). *Diffusion of innovations* (4th ed.). New York: The Free Press.

Rolfe, I. (1999). Innovation and creativity in organizations: A review of the implications for training and development. *Journal of European Industrial Training*, (23/4/5), 224–237.

Rosenberg, N. (1976). On technological expectations. *The Economic Journal*, **86** (343, September), 523–535.

Rosenberg, N. (1982). *Inside the black box: Technology and economics*. Cambridge: Cambridge University Press.

Schon, D. (1983). *The reflective practitioner: How professionals think in action*. New York: Basic Books.

Schon, D. (1991). *The reflective turn*. New York, NY: Teachers' College Press.

Senge, P. (1990). *The fifth discipline: The art and practice of the learning organization*. New York: Doubleday.

Shannon, C. E. & Weaver, W. (1949). *The mathematical theory of communication*. Chigaco, IL: University of Illinois Press.

Shapiro, C. & Varian, H. (1999). *Innovation rules: A strategic guide to the network economy*. Boston, MA: Harvard Business School Press.

Simon, H. (1969). *The sciences of the artificial*. Cambridge, MA: MIT Press.

Sternberg, R. & Frensch, P. (1991). *Complex problem solving: Principles and mechanisms*. Mahwah, NJ: Lawrence Erlbaum.

Sveiby, K. *What is knowledge management?* http://www.sveiby.com.au

Teece, D. J., Pisano, G. & Shuen, A. (1997). Dynamic capabilities and strategic management. *Strategic Management Journal*, **18** (7), 509–533.

Tidd, J. (2001). *Innovation management in context: Environment, organizational & performance*. SPRU Science and Technology Policy Research. Working paper no. 55.

Utterback, J. M. (1996). *Mastering the dynamics of innovation*. Boston, MA: Harvard Business School Press.

Utterback, J. M. & Abernathy, W. J. (1975). A dynamic model of process and product innovation, omega. *The International Journal of Management Science*, **3** (6), 639–656.

von Braun, C. F. (1997). *The innovation war*. Englewood Cliffs, NJ: Prentice Hall.

von Hippel, E. (1988). *The sources of innovation*. Oxford: Oxford University Press.

Wallace, D. (1995). *Environmental policy and industrial innovation: Strategies in Europe, the U.S.A, and Japan. London: Royal Institute of International Affairs*. Washington, D.C.: Energy and Environmental Programme: Earthscan Publications.

Weick, K. E. (1979). *The social psychology of organizing* (2nd ed.). Reading, MA: Addison-Wesley.

Wiig, K. (1993). *Knowledge management foundation*. New York, NY: Schema Press.

The International Handbook on Innovation
Edited by Larisa V. Shavinina

Profitable Product Innovation:
The Critical Success Factors

Robert G. Cooper

Product Development Institute Inc. and McMaster University, Hamilton, Ontario, Canada

Abstract: This chapter focuses on the critical success factors that underlie new product performance. It is based on the author's and other's research into hundreds of new product launches, probing the question: What distinguishes the best from the rest? Ten common denominators or factors appears to drive new product success, profitability and time-to-market; this chapter outlines these ten critical success factors, and notes the management implications of each. The chapter concludes with a discussion of portfolio management techniques, issues and challenges.

Keywords: Product innovation; New product development; Management; Portfolio management; Success factors.

Introduction

Product innovation—the development of new and improved products—is crucial to the survival and prosperity of the modern corporation. According to a PDMA survey, new products currently account for 33% of company sales, on average (Griffin, 1997). And product life cycles are getting shorter: a 400% reduction over the last 50 years, the result of an accelerating pace of product innovation (Von Braun, 1997).

New products fail at an alarming rate, however. Approximately one in ten product concepts succeeds commercially, according to several studies (Page, 1991; Griffin, 1997), while only one in four development projects is a commercial success (Cooper, 2001). Even by the launch phase—after all the product tests and market testing are done—one-third of product launches still fail commercially. In the high-tech field, the odds are even worse: only 20% of software projects are commercial winners, according to a survey by Kleinschmidt (1999); companies scrap almost one-third of new projects for a loss of $80 billion annually (King, 1997); and "only one-quarter of IT projects are completed on time, on budget and with all the functions originally specified" (*Risk Magazine*, 1999).

The huge amounts at stake coupled with the high odds of failure make product innovation one of the riskiest endeavors of the modern corporation. Thus many managers, researchers and pundits have sought answers to the age-old question: What makes a new product a winner? And why are some businesses so much more successful at product development than the rest?

This chapter looks first at the critical success factors that underlie success in product innovation.[1] Next, these many success factors are crafted into a methodology, game plan or process for managing new product projects—a systematic new product process or Stage-Gate system. The chapter concludes with a look at portfolio management—methods for selecting the right new product projects.

The Critical Success Factors

Much research over the years has uncovered those factors that separate winning new products from less successful ones, and businesses that succeed at product innovation (Cooper, 1999b, 1999c; Cooper & Kleinschmidt, 1986, 1993, 1995a, 1996; Di Benedetto, 1999; Maidique et al., 1984; Mishra et al., 1996; Montoya-Weiss et al., 1994; Rothwell et al., 1974; Sanchez et al., 1991; Song & Parry, 1996). Here are the more important success factors based on these studies:

A Unique Superior Product

A superior and differentiated product is the number one driver of success and profitability, with success rates reported to be three to five times higher than for 'me

---

[1] This chapter is taken from a number of articles and books by the author. See Cooper (1996, 1999a, 2000, 2001).

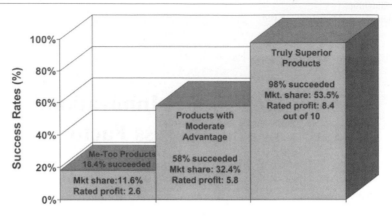

Figure 1. Impact of product superiority on success.

Source: Cooper (2001).

too', reactive products (Cooper, 1999b, 1999c; Cooper & Kleinschmidt, 1993, 1995a, 1996; Mishra et al., 1996; Song & Parry, 1996; Souder & Song, 1997). The NewProd studies[2] show the dramatic impact of product superiority: here the sample of firms was divided into the top 20% and bottom 20% in terms of product superiority: note the differences in performance in Figure 1—much higher success rates, market shares and profitabilities for unique, superior products (Cooper, 2001; Cooper & Kleinschmidt, 1986, 1996). For example, such products (the top 20% in Figure 1) when compared to those with the least degree of differentiation (the bottom 20%):

• have an exceptional commercial success rate of 98.0%, vs. only 18.4% for undifferentiated ones;
• have a market share of 53.5% of the defined target market, vs. only 11.6% for 'me too' new products;
• have a rated profitability of 8.4 out of 10 (vs. only 2.6 out of 10 for undifferentiated products—here 10 = exceptional profits, far exceeding the company's minimum hurdle); and
• meet company sales and profit objectives to a greater degree than do undifferentiated products.

The same studies show, however, that *reactive products, undifferentiated products* and *technically driven products that lack customer benefits* are the rule rather than the exception, however; and the majority fail!

What are the ingredients of these unique, superior products with significant competitive advantage?

According to research cited above, they are superior to competing products in terms of meeting users' needs, offer unique features not available on competitive products, or solve a problem the customer has with a competitive product. They provide excellent product quality, relative to competitors' products, and in terms of how the user measures quality; they feature good value for money for the customer, reduce the customer's total costs (high value-in-use), and boast excellent price/performance characteristics. And such winning products offer benefits or attributes easily perceived as useful by the customer, and benefits that are highly visible. A point of distinction: *benefits* are what customers or users value and pay money for; by contrast, *attributes* are product features, functionality and performance—the things that engineers and scientists build into products. And note that the product must be superior in the eyes of the customer, not just the eyes of the engineer, scientist or designer: often developers heavily overestimate the customer benefits and desirability of their products (Friar, 1995).

A Strong Market Orientation—Market-Driven, Customer-Focused

A thorough understanding of customers' needs and wants, the competitive situation, and the nature of the market is an essential component of new product success. This finding is supported in virtually every study of product success factors (Cooper, 2001; Di Benedetto, 1999; Mishra et al., 1996; Montoya-Weiss, 1994; Song & Montoya-Weiss, 1998; Song & Parry, 1996). Recurring success themes include:

• need recognition;
• understanding user needs;
• market need satisfaction;
• constant customer contact;
• strong market knowledge and market research;
• quality of execution of marketing activities; and

---

[2] The *NewProd studies*: a series of investigations by the author and co-workers between 1975 and 2000 that identified the success factors in new product development—the factors that differentiated between winners and losers. The typcial methodology was a paired comparison of sucessful vs. unsuccesful new product projects. NewProd™ is a legally registered tradename of R. G. Cooper & Associates Consultants Inc.

• more spending on the up-front marketing activities.

Even in the case of technology-driven new products (where the idea comes from a technical or laboratory source), the likelihood of success is greatly enhanced if customer and marketplace inputs are built into the project soon after its inception.

Conversely, a failure to adopt a strong market orientation in product innovation, an unwillingness to undertake the needed market assessments and to build in the voice of the customer, and leaving the customer out of product development spells disaster. Poor market research, inadequate market analysis, weak market studies, test markets, and market launch, and inadequate resources devoted to marketing activities are common weaknesses found in almost every study of why new products fail (de Brentani, 1989; Calantone et al., 1979; Griffin & Page, 1993).

A strong market orientation is missing in the majority of firms' new product projects, however. Detailed market studies are frequently omitted—in more than 75% of projects, according to several investigations (Cooper, 2001; Cooper & Kleinschmidt, 1986). Further, marketing activities are the weakest rated activities of the entire new product process, with relatively few resources and little money spent on the marketing actions (less than 20% of total project expenditures).

A strong market orientation must be built into every stage of the new product process in order to achieve success (Veryzer, 1998):

• Idea generation: focusing on the customer, and using the customer as a source of ideas;
• Product design: employing market research as an input to product design, not just an after-the-fact check;
• During development via constant customer contact and feedback (e.g. continuous customer testing of facets of the product);
• After development: undertaking customer trials, preference tests and test markets to verify market acceptance and launch plan;
• Launch: employing a well-designed, carefully targeted, properly resourced, guided by a well-conceived marketing plan, based on solid market information.

3. The World Product—An International Orientation in Product Design, Development, and Target Marketing

International products targeted at world and nearest neighbor export markets are the top performers (Cooper, 2001; Kleinschmidt, 1988). By contrast, products designed with only the domestic market in mind, and sold to domestic and nearest neighbor export markets, fare more poorly. The magnitude of the differences between these international and exported new products vs. domestic products is striking: differences of two- or three-to-one on various performance gauges.

The comfortable strategy of "design the product for domestic requirements, capture the home market, and then export a modified version of the product sometime in the future" is myopic. It leads to inferior results today; and with increasing globalization of markets, it will certainly lead to even poorer results in the years ahead. The threat is that one's domestic market has become someone else's international market: to define the new product market as 'domestic' and perhaps a few other 'convenient countries' severely limits market opportunities. For maximum success in product innovation, the objective must be to design for the world and market to the world. Sadly, this international dimension is an often overlooked facet of new product game plans or one which, if included, is handled late in the process, or as a side issue.

An international orientation means defining the market as an international one, and designing products to meet international requirements, not just domestic. The result is either a *global* product (one version for the entire world) or a *glocal* product (one development effort, one product concept, but perhaps *several product variants* to satisfy different international markets). An international orientation also means undertaking market research, concept testing and product testing in multiple countries, rather than just the home country; and it means relying on an international or global project team (about one NPD project team in five is now reported to be a global development team (McDonough et al., 2001)).

Pre-development Work—The Up-front Homework

The steps that precede the design and development of the product make the difference between winning and losing (Cooper, 1999b, 1999c; Cooper & Kleinschmidt, 1993, 1995a, 1996; Mishra et al., 1996; Montoya-Weiss et al., 1994; Song & Parry, 1996; Thomke & Fujimoto, 2000). New product projects that feature a high quality of execution of activities that precede the development phase—the *fuzzy front end*—are more successful:

• a success rate of 75.0% (vs. only 31.3% for projects where the predevelopment activities are found lacking);
• a higher rated profitability (7.2 out of 10 vs. only 3.7 for projects where pre-development activities are poorly undertaken); and
• a market share of 45.7% (vs. 20.8%) (Cooper, 2001).

Successful businesses spend about twice as much time and money on these vital up-front or pre-development activities, such as initial screening, preliminary market and technical assessments, detailed market studies or marketing research (success factor 2 above) and business and financial analysis just before the decision to 'Go to Development' (Cooper, 2001).

Most businesses confess to serious weaknesses in the up-front or pre-development steps of their new product process. Small amounts of time and money are devoted to these critical steps: only about 7% of the expenditure and 16% of the effort (Cooper, 2001). Far from adding extra time to the project, research reveals that homework pays for itself in reduced development times, the result of shaper and more stable product definition, and fewer surprises (and time wasters) later in the project (Thomke & Fujimoto, 2000).

Sharp and Early Product Definition

Successful products have *much sharper definition prior to development* (Cooper, 1999b, 1999c; Cooper & Kleinschmidt, 1993, 1995a, 1996; Mishra et al., 1999; Rauscher & Smith, 1995; Wilson, 1991). Projects that have these sharp definitions are 3.3 times as likely to be successful; have higher market shares (by 38 share points on average); are rated 7.6 out of 10 in terms of profitability (vs. 3.1 out of 10 for poorly defined products); and do better at meeting company sales and profit objectives (see Figure 2).

This definition includes:

- specification of the target market: exactly who the intended users are;
- description of the product concept and the benefits to be delivered;
- delineation of the positioning strategy (including the target price);
- and a list of the product's features, attributes, requirements and specifications.

Unless the these four items are clearly defined, written down and agreed to by all parties prior to entering the development stage, the odds of failure increase by a factor of three.

Projects that have sharp product and project definition prior to development are considerably more successful. The reasons are as follows (Cooper, 2001):

(1) Building a definition step into the new product process forces more attention to the up-front or pre-development activities;
(2) The definition serves as a communication tool and guide. All-party agreement or 'buy in' means that each functional area involved in the project has a clear and consistent definition of what the product and project are—and is committed to it;
(3) This definition also provides a clear set of objectives for the development phase of the project, and the development team members. With clear product objectives, development typically proceeds more efficiently and quickly: no moving goalposts and no fuzzy targets!

6. A Well-conceived, Properly Executed Launch Backed by a Solid Marketing Plan

The best products in the world will not sell themselves! A strong marketing and selling effort, a well-targeted selling approach, and effective after-sales service are central to the successful launch of the new product (Cooper & Kleinschmidt, 1986; Di Benedetto, 1999; Mishra et al., 1996; Song & Montoya-Weiss, 1998; Song & Parry, 1996). But a well-integrated and properly targeted launch does not occur by accident; it is the result of a *fine-tuned marketing plan*, properly backed and resourced, and proficiently executed.

There are four requirements for an effective market launch plan:

(1) The development of the market launch plan is an *integral part of the new product process*: it is as central to the new product process as is the development of the physical product;

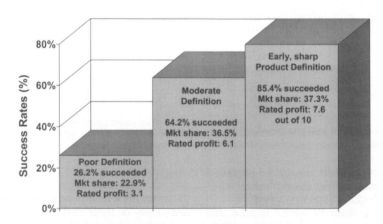

Figure 2. Impact of early, sharp product definition.

Source: Cooper (2001).

(2) The development of the market launch plan *must begin early* in the new product project. It should not be left as an afterthought to be undertaken as the product nears commercialization;

(3) A market launch plan is only as good as the *market intelligence* upon which it is based. Market studies designed to yield information crucial to marketing planning must be built into the new product game plan;

(4) These who will execute the launch—the sales force, technical support people, other front line personnel—must be engaged in the development of the market launch plan. This ensures valuable input and insight into the design of the launch effort, availability of resources when needed, and buy-in by those who must execute to the product and its launch—elements so critical to a successful launch (Hultink, E. J. & Atuahene-Gima, 2000).

7. The Right Organizational Structure, Design and Climate

Product innovation is not a one-department show. It is very much a multidisciplinary, cross-functional effort. Investigations into new product success consistently cite interfaces between R&D and marketing, coordination among key internal groups, multidisciplinary inputs to the new product project, and the role of project teams and the team leader or champion (Cooper, 2001; Markham & Griffin, 1998; Song et al., 1997; Song & Parry, 1996). Successful new product projects feature a balanced process consisting of critical activities that fall into many different functional areas within the firm: marketing and marketing research, engineering, R&D, production, purchasing, and finance (Cooper, 1998; Cooper & Kleinschmidt, 1995b, 1995c, 1996). The Stanford Innovation Project study of new product launches in high-technology firms reveals that a critical distinguishing factor between success and failure is the "simultaneous involvement of the create, make, and market functions" (Maidique et al., 1984). The NewProd studies show that projects undertaken by empowered multifunctional teams are more successful (Cooper, 1999a, 1999c; Cooper & Kleinschmidt, 1993, 1995b, 1996). Similarly, analyses of Japanese successes emphasize their attention to manufacturability from the start of development efforts, the location in one place of engineers, designers, and manufacturers, and the conception of management unconstrained by traditional American functionalism (Peters, 1986, p. 261). Finally Griffin's review of multiple benchmarking studies, along with her own PDMA best practices study, reveals that effective cross-functional teams are fundamental to success (Griffin, 1997).

Product development must be run as a multidisciplinary, cross-functional effort. Good organizational design means projects that are:

- organized as a *cross-functional team* with members from R&D, Engineering, Marketing & Sales, Operations and so on (as opposed to each function doing its own part independently);
- where the team is dedicated and focused (i.e. devotes a large percentage of their time to this project, as opposed to spread over many projects or other work);
- where the team members are in constant contact with each other, via frequent but short meetings, interactions, project updates and even co-location;
- where the team is accountable for the entire project from beginning to end (as opposed to accountability for only one stage of the project); and
- where there is a strong project leader or champion who leads and drives the project (Barczak, 1995; Markham & Griffin, 1998).

While the ingredients of good organizational design should be familiar, surprisingly many firms have yet to get the message (Cooper & Kleinschmidt, 1996).

Peters argues strongly in favor of project teams: "the single most important reason for delays in development activities is the absence of multifunction (and outsider) representation on development projects from the start" (Peters, 1986). Peters continues: "The answer is to co-mingle members of all key functions, co-opt each function's traditional feudal authority, and use teams".

A second organizational success ingredient is climate and culture. A positive climate is one which supports and encourages intrapreneurs and risk-taking behavior; where new product successes are rewarded and recognized (and failures not punished); and where team efforts are recognized, rather than individuals. For example, the Best companies in the PDMA best practices study emphasize employee recognition for new product performance (Griffin, 1997). A positive climate also means senior managers refraining from 'micro-managing' projects and second-guessing the team members; and resources and time being made available for creative people to work on their own 'unofficial projects'. Idea-submission schemes (where employees are encouraged to submit new product ideas), and open project review meetings (where the entire project team participates), are other facets of a positive climate.

Top Management Support

Top management support is a necessary ingredient for product innovation. But it must be the right kind of support. The Stanford project, a Hewlett-Packard study, and other investigations have found top management support to be directly linked to new product success (Cooper, 2001; Maidique & Zirger, 1984; Song & Parry, 1996; Wilson, 1991).

Top management's role in product development is as a facilitator—to set the stage—and not be an actor front and center. One important role of senior management is

143

to articulate a new product strategy for the business, something that is often missing. An effective new product strategy means defined new product goals (e.g. percentage of the business's sales to be derived from new products); delineated arenas of focus (e.g. product-types, markets, technologies and technology platforms where the business unit intends to concentrate its development efforts); and strategies that have a longer-term orientation and are visible to everyone in the business. Management must also deploy the necessary product development resources, and keep the commitment. And it must commit to a disciplined process to drive products to market. These three factors—an articulated new product strategy, adequate resources, and a disciplined new product process—are the strongest drivers of new product performance at the business unit level, according to a major benchmarking study (Cooper, 1998; Cooper & Kleinschmidt, 1995b, 1995c, 1996). Finally, senior management must empower project teams and support committed champions by acting as mentors, facilitators, 'godfathers', or executive sponsors to project leaders and teams—acting as 'executive champions', as Peters calls them (Peters, 1986, p. 302).

Leveraging Core Competencies

Leverage and synergy is the common thread that binds the new business to the old. When translated into product innovation, the ability to leverage existing and in-house strengths, competencies, resources, and capabilities increases the odds of success of the new product project. By contrast, 'step-out' projects take the firm into territory that lies beyond the experience and resource base of the company, and increase the odds of failure (Cooper, 1999a, 1999c; Cooper & Kleinschmidt, 1995a, 1995b, 1995c, 1996; Mishra et al., 1996; Song & Parry, 1996).

The reasons for the impact of leverage are clear:

(1) Resources are available and at marginal cost;
(2) Operating within one's field of expertise—either familiar markets or familiar technologies—provides considerable 'domain knowledge', which is available to the project team;
(3) The more often one does something, the better one becomes at doing it—the experience factor. If new product projects are closely related to (leveraged from) current businesses, the chances are that there has been considerable experience with such projects in the past, hence lower costs of execution and fewer misfires.

Two types of leverage are important to product innovation:

(1) *Technological leverage*: the project's ability to build on in-house development technology, utilize inside engineering skills, and use existing manufacturing or operations resources and competencies.
(2) *Marketing leverage*: the project/company fit in terms of customer base, sales force, distribution channels, customer service resources, advertising and promotion, and market-intelligence skills, knowledge and resources.

These two dimensions of leverage—technological and marketing, and their ingredients—become obvious checklist items in a scoring or rating model to help prioritize new product projects. And if leverage is low, yet the project is attractive for other reasons, then steps must be taken to bolster the internal resources and competencies. Low leverage scores signal the need for outside resources—partnering or out-sourcing (Bonaccorsi & Hayashi, 1994). But neither solution is a panacea—there are risks and costs to both routes to securing the needed resources and competencies (Campbell & Cooper, 1999).

Market Attractiveness

Market attractiveness is an important strategic variable. Porter's 'five forces' model considers various elements of market attractiveness as a determinant of industry profitability (Porter, 1985). Similarly, various strategic planning models—for example, models used to allocate resources among various existing business units—employ market attractiveness as a key dimension in the two-dimensional map or portfolio grid (Day, 1986).

In the case of new products, market attractiveness is also important: products targeted at more attractive markets are more successful (Cooper, 1999a, 1999c; Cooper & Kleinschmidt, 1995a, 1995b, 1995c, 1996; Maidique et al., 1985; Mishra et al., 1996; Song & Parry, 1996). There are two dimensions of market attractiveness:

(1) *Market potential*: positive market environments, namely large and growing markets—markets where a strong customer need exists for such products, and where the purchase is an important one for the customer. Products aimed at such markets are more successful.
(2) *Competitive situation*: negative markets characterized by intense competition, competition on the basis of price, high quality, and strong competitive products; and competitors whose sales force, channel system, and support service are strongly rated. Products aimed at such negative markets are less successful, according to the studies cited above.

The message is this: both elements of market attractiveness—market potential and competitive situation—impact the new product's fortunes; and both should be considered as criteria in any model or scoring scheme for project selection and prioritization.

Tough Go/Kill Decision Points and Better Focus

Most companies suffer from too many projects and not enough resources to mount an effective or timely effort on each—a lack of focus. This stems from inadequate project evaluation and poor project prioritization. Project evaluations are consistently cited as weakly handled or nonexistent: decisions involve the wrong people from the wrong functions (no functional alignment); no consistent criteria are used to screen or rank projects; or there is simply no will to kill projects at all (Barczak, 1995; Cooper et al., 1998, 1999; Cooper, 1998; Cooper & Kleinschmidt, 1986, 1995a, 1995b, 1995c, 1996).

The desire to weed out bad projects coupled with the need to focus limited resources on the best projects means that tough Go or Kill and prioritization decisions must be made. Some companies have built *funnels* into their new product process via decision-points in the form of gates. At gate reviews, senior management rigorously scrutinizes projects, and makes Go or Kill and prioritization decisions based on visible Go/Kill criteria (see Figure 5 for sample Go/Kill criteria). Progressive businesses are also moving to portfolio management, which attempts to select the right set of new product projects in order to maximize the value of the portfolio, achieve the right balance of projects, and yield a portfolio of projects that supports the business's strategy (Cooper et al., 1997a, 1997b; Roussel et al., 1991).

Quality of Execution

Certain key activities—how well they are executed, and whether they are done at all—are strongly tied to profitability and reduction in time-to-market. Particularly pivotal activities include: the vital homework actions outlined above; and market-related activities. But proficiency of most activities in the new product process impacts on outcomes, with successful project teams consistently doing a better quality job across many tasks (Cooper, 1999a, 1999c; Cooper & Kleinschmidt, 1986, 1995d, 1996; Mishra et al., 1996; Song & Montoya-Weiss, 1998; Song & Parry, 1996).

There is a *quality crisis* in product innovation, however. Investigations reveal that the typical new product project is characterized by serious errors of omission and commission:

- Pivotal activities, often cited as central to success, are omitted altogether. For example, more than half of all projects typically leave out detailed market studies and a test market (trial sell);
- Quality of execution ratings of important activities are also typically low. In postmortems on projects, teams typically rate themselves as 'mediocre' in terms of how good a job they did on these vital activities.

New product success is thus very much within the hands of the men and women leading and working on projects. To improve quality of execution, the solution that leading firms in a variety of industries, such as DuPont, Procter & Gamble, Exxon, Bayer, Lego, ITT, International Paper and Pillsbury-General Mills, have adopted is to treat *product innovation as a process*. They have adopted a formal stage-and-gate product delivery process; they build in quality-assurance approaches, such as check points and metrics that focus on quality of execution; and they design quality in by making mandatory certain vital actions that are often omitted, yet are central to success.

The Necessary Resources

Too many projects simply suffer from a lack of time and money commitment. The results are predictable: much higher failure rates (Cooper, 1998; Cooper & Kleinschmidt, 1995b, 1995c, 1996). Some facts:

- A strong market orientation is missing in the typical new product project. And much of this deficiency is directly linked to a lack of marketing resources available for the project;
- Another serious pitfall is that the homework does not get done. Again much of this deficiency can be directly attributed to a lack of resources: simply not enough money, people, and time to do the work.

The reason: as the competitive situation has toughened, companies have responded with restructuring and doing more with less, and so resources are limited or cut back (Cooper & Edgett, 2002). This short-term focus takes its toll. Certain vital activities, such as market-oriented actions and predevelopment homework are highly under-resourced, particularly in the case of product failures.

Speed—But Not at the Expense of Quality of Execution

Speed yields: competitive advantage (the first on the market) (Song et al., 2000; Song & Montoya-Weiss, 1998); less likelihood that the market situation has changed; and a quicker realization of profits (Ali et al., 1995). So the goal of reducing the development cycle time is admirable. A word of caution here: speed is only an interim objective; the ultimate goal is profitability. While studies reveal that speed and profitability are connected, the relationship is anything but one-to-one (Cooper, 1995; Lynn et al., 1999). Further, often the methods used to reduce development time yield precisely the opposite effect, and in many cases are very costly. The objective remains successful products, not a series of fast failures! Additionally, an over-emphasis on speed has led to trivialization of product development in some businesses—too many product modifications and line extensions, and not enough real new products and new platforms (Cooper & Edgett, 2002; Crawford, 1992).

Some of the ways which project teams have reduced time-to-market have been highlighted above: a true cross-functional team; solid up-front homework; sharp, early product definition (Rauscher & Smith, 1995); and better focus—doing fewer projects (Cooper, 1995, 2001; Cooper & Kleinschmidt, 1994). Other methods include:

- *Parallel processing*: activities are undertaken in parallel (rather than sequentially) with the team members constantly interacting with each other. New product rugby with time compression is the result;
- *Flowcharting*: the team maps out its entire project from beginning to end, and focuses on reducing the time of each element or task in the process;
- *A time-line and discipline*: project teams use computer software to plan their projects in a critical path or Gantt chart format. The rules are simple: practice disciple; the time-line is sacred; and resources can be added but deadlines never relaxed.

A Multistage, Disciplined New Product Process

A systematic new product process—a *Stage-Gate™ process*[3]—is the solution that many companies have turned to in order to overcome the deficiencies which plague their new product programs (Cooper, 1998, 2001; Lynn et al., 1999; Menke, 1997). *Stage-Gate* processes are simply road maps or 'play books' for driving new products from idea to launch, successfully and efficiently. About 68% of U.S. product developers have adopted Stage-Gate processes, according to the PDMA best practices study (Griffin, 1997). And the payoffs of such processes have been frequently reported: improved teamwork; less recycling and rework; improved success rates; earlier detection of failures; a better launch; and even shorter cycle times (by about 30%).

Doing Projects Right: A Stage-Gate™ Process or Road Map from Idea to Launch

A Stage-Gate process is a *conceptual and operational roadmap* for moving a new product project from idea to launch—a blueprint for managing the new product process to improve effectiveness and efficiency. The goals of a Stage-Gate process include:

- Doing projects right—such processes lay out pre-scribed actions and best practices as a guide to the project team;
- Building in the critical success factors—as was seen above, most project teams miss the mark on success factors 1–14. And so Stage-Gate processes attempt to build these in to every project by design.

Stage-Gate approaches break the innovation process into a predetermined set of stages, with each stage

---

[3] The term 'Stage-Gate' was coined by the author and first appeared in print in Cooper (1988). Stage-Gate is a legal tradename of the author. For more on Stage-Gate™ methods, see Cooper (2001); also the Stage-Gate web-page: www.prod-dev.com

consisting of a set of prescribed, cross-functional and parallel activities (see diagram below). The entrance to each stage is a gate, which serves as the quality control and Go/Kill check points in the process.

The Stages

Stages are where the action occurs. The players on the project team undertake key tasks in order to gather information needed to advance the project to the next gate or decision point. Stages are cross-functional: There is no R&D or Marketing stage. Rather, each stage consists of a set of parallel activities undertaken by people from different functional areas within the firm, working together as a team and led by a project team leader. In order to manage risk via a Stage-Gate method, the parallel activities in a certain stage must be designed to gather vital information—technical, market, financial, operations—in order to drive down the technical and business risks. Each stage costs more than the preceding one, so that the game plan is based on incremental commitments. As uncertainties decrease, expenditures are allowed to mount: risk is managed.

A Typical Stage-Gate™ Process

The general flow of the typical or *generic* stage-gate process is shown in Figure 3 (Cooper 2001). Here, following idea generation or Discovery, there are five key stages:

Stage 1. Scoping

A quick investigation and sculpting of the project. This first and inexpensive homework stage has the objective of determining the project's technical and marketplace merits. Stage 1 involves desk research or detective work—little or no primary research is done here. Prescribed activities include preliminary market, technical and business assessments.

Stage 2. Build the Business Case

The detailed homework and up-front investigation work leading to a *business case*—a defined product, a business justification and a detailed plan of action for the next stages. This second homework stage includes actions such as a detailed market analysis, user needs and wants studies to build in voice of customer, concept testing, detailed technical assessment, source of supply assessment, and a detailed financial and business analysis.

Stage 3. Development

The actual design and development of the new product. Stage 3 witnesses the implementation of the Development Plan and the physical development of the product. Lab tests, in-house tests or alpha tests ensure that the product meets requirements under controlled conditions. The 'deliverable' at the end of Stage 3 is a lab-tested prototype of the product. While the emphasis in Stage 3 is on technical work, marketing and

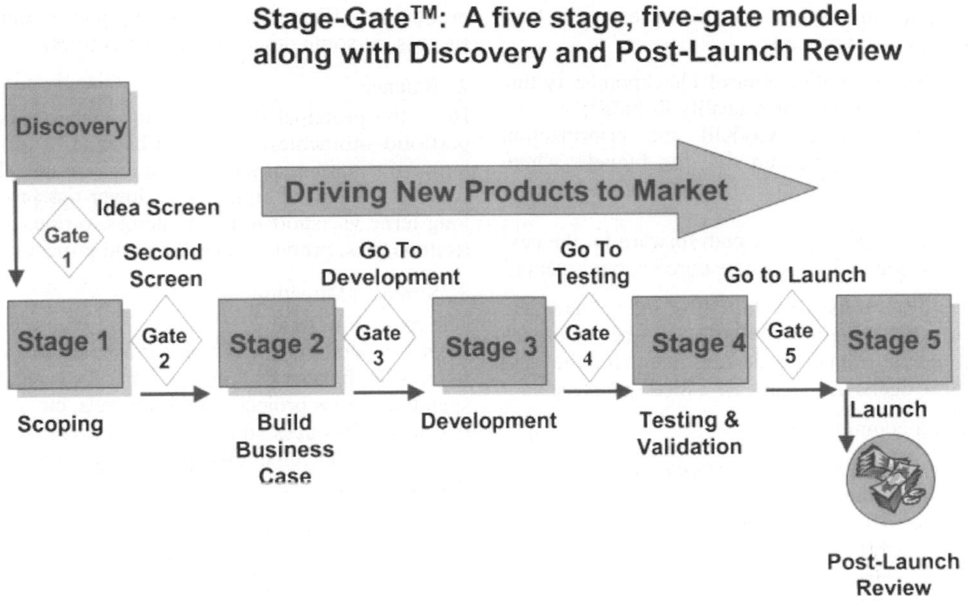

Stage-Gate™: A five stage, five-gate model along with Discovery and Post-Launch Review

Figure 3. A typical Stage-Gate™ new product process.

Stage-Gate™ is a legal tradename of RG Cooper & Associates Consultants Inc.
Source: Cooper (2001).

operations activities also proceed in parallel. For example, market-analysis and customer-feedback efforts continue concurrently with the technical development, with constant customer opinion sought on the product as it takes shape during development. Additionally, the manufacturing (or operations) process is mapped out, the marketing launch and operating plans are developed, and the test plans for the next stage are defined. An updated financial analysis is prepared, while regulatory, legal, and patent issues are resolved.

Stage 4. Testing & Validation
The verification and validation of the proposed new product, its marketing and production. This stage tests and validates the entire viability of the project: the product itself, the production process, customer acceptance, and the economics of the project. A number of activities are undertaken at Stage 4, including in-house product tests (extended lab tests or alpha tests to check on product quality and product performance); user or field trials of the product (to verify that the product functions under actual use conditions, and to establish purchase intent); trial, limited, or pilot production (to test, debug, and prove the production process); pre-test market, test market, or trial sell (to gauge customer reaction and measure the effectiveness of the launch plan); and revised business and financial analysis.

Stage 5. Launch
Full commercialization of the product—the beginning of full production and commercial launch and selling.

Here the market roll-out or launch plan is implemented. Final production equipment installation and commissioning occurs, along with full production start-up. Finally, the post-launch plan—monitoring and fixing—is implemented, along with early elements of the life cycle plan (new variants and releases, continuous improvements).

Some 12–18 months after launch, the Post Launch Review occurs: the performance of the project vs. expectations is assessed along with reasons why and lessons learned; the team is disbanded but recognized or rewarded; and the project is terminated.

At first glance, this overview portrays the stages as relatively simple steps in a logical process. But what you see above is only a high-level view of a generic process. In an operational process in a real company, drilling down into the details of each stage reveals a much more sophisticated and complex set of activities: a detailed list of activities within each stage, the how to's of each activity, best practices that the team ought to consider, and even the required deliverables from each activity in that stage (for example in the format of templates). In short, the drill-down provides a *detailed and operational road map* for the project team—everything they need to know and do to successfully complete that stage of the process and project.

The Gates
Preceding each stage is an entry gate or a Go/Kill decision point, shown as diamonds in Figure 3.

147

Effective gates are central to the success of a fast-paced, new product process:

• Gates serve as quality control checkpoints: is this project being executed in a quality fashion?;
• Gates also serve as Go/Kill and prioritization decisions points: gates provide the funnels, where mediocre projects are culled out at each successive gate;
• Finally, gates are where the path forward for the next stage is decided, along with resource commitments.

Gate meetings are usually staffed by senior managers from different functions, who own the resources required by the project leader and team for the next stage. These decision-makers are called 'gatekeepers'.

Gates have a common format:

• *Deliverables*: these are the *inputs* into the gate review—what the project leader and team deliver to the meeting. They are the results of the actions of the previous stage, and are based on a standard menu of deliverables for each gate;
• *Criteria*: these are questions or metrics on which the project is judged in order to make the Go/Kill and prioritization decision;
• *Outputs*: these are the results of the gate review—a decision (Go/Kill/Hold/Recycle); an action plan approved; and the date and deliverables for the next gate agreed upon.

Doing the Right Projects: Project Selection and Portfolio Management

An effective new product process, such as *Stage-Gate*™ in Figure 3, improves the odds of success. And it leads to the second way to win at new products, namely doing the right projects or and effective portfolio management.[4] Portfolio management deals with the vital question: How should the business most effectively invest its R&D and new product resources (Roussel et al., 1991)? Much like a stock market portfolio manager, those senior executives who optimize their R&D investments—select winning new product projects and achieve the ideal balance of projects—will win in the long run.

Three Goals in Portfolio Management

There are three macro or *high-level goals* in portfolio management (Cooper et al., 2002a, 2002b):

1. Value Maximization

Here, one selects projects so as to maximize sum of the values or *commercial worths* of all active projects in terms of some business objective (such as long-term

profitability, EVA, return-on-investment, likelihood of success, or some other strategic objectives).

2. Balance

Here, the principal concern is to develop a balanced portfolio—to achieve a desired balance of projects in terms of a number of parameters; for example, the right balance in terms of high-risk vs. lower-risk projects; or long-term vs. short-term; or across various markets, technologies, product categories, and project types.

3. Strategic Direction

The main goal here is to ensure that, regardless of all other considerations, the final portfolio of projects truly reflects the business's strategy—that the breakdown of spending across projects, areas, markets, etc. is directly tied to the business strategy (e.g. to areas of strategic focus that management has previously delineated); and that all projects are 'on strategy'.

Goal 1: Maximizing the Value of the Portfolio

A variety of methods can be used to achieve this goal, ranging from financial models through to scoring models (Cooper et al., 1997a, 1997b, 2000). Each has its strengths and weaknesses. The end result of each method is a rank-ordered or prioritized list of 'Go' and 'Hold' projects, with the projects at the top of the list scoring highest in terms of achieving the desired objectives: the value in terms of that objective is thus maximized.

Net Present Value (NPV)

The simplest approach is merely to calculate the NPV of each project on a spreadsheet; and then rank all projects according to their NPVs. The Go projects are at the top of the list. . . one continues to add projects down the list until out of resources. Logically this method should maximize the NPV of the portfolio.

Fine in theory! But the NPV method ignores probabilities and risk; it assumes that financial projections for new products are accurate (they usually are not!); it assumes that only financial goals are important; and it fails to deal with constrained resources—the desire to maximize the value for a limited resource commitment. A final objection is more subtle: the fact that NPV assumes an all-or-nothing investment decision, whereas in new product projects, the decision process is an incremental one—more like buying a series of options on a project (Deaves & Krinsky, 1997; Faulkner, 1996). In spite of their apparent rigor, financial approaches—NPV as well as some listed below—yield the worst performing portfolios, and were poorly rated by mangers in a major portfolio study of methods in practice (Cooper et al., 1998, 1999, 2001a).

Expected Commercial Value (ECV)

This method seeks to maximize the value or *commercial worth* of the portfolio, subject to certain budget

---

[4] Much of this section on Portfolio Management is based on a book by the author and co-workers (Cooper et al., 2002a) and a chapter in the PDMA handbook (Cooper et al., 2002b).

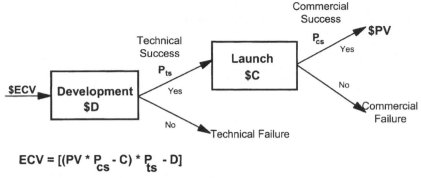

$$ECV = [(PV * P_{cs} - C) * P_{ts} - D]$$

ECV = Expected Commercial Value of the project
P_{ts} = Probability of Technical Success
P_{cs} = Probability of Commercial Success (given technical success)
D = Development Costs remaining in the project
C = Commercialization (Launch) Costs
PV = Present Value of project's future earnings (discounted to today)

Figure 4. Determination of expected commercial value of a project.

Source: Cooper, Edgett & Kleinschmidt (2002a).

constraints, and introduces the notion of risks and probabilities. The ECV method determines the risk-adjusted value or commercial worth of each project to the corporation, namely its *expected commercial value*. The calculation of the ECV is based on a decision tree analysis, and considers the future stream of earnings from the project, the probabilities of both commercial success and technical success, along with both commercialization costs and development costs (see Figure 4 for the calculation and definition of terms). A variety of techniques are available for estimating probabilities of success, including scoring models, data tables and behavioral approaches such as Delphi (Cooper et al., 2002a; Davis et al., 2001).

In order to arrive at a prioritized list of projects, the ECV of each project is first determined. Then, one considers what resources are scarce or limiting—for example, R&D or marketing people or dollars or work-months, or even capital funds. Thus, the ratio of what one is trying to maximize (namely the ECV) *divided by the constraining resource* (for example, the R&D costs per project) is determined for each project. Projects are rank-ordered according to this ECV/R&D Cost ratio until the budget limit is reached. Those projects at the top of the list are Go, while those at the bottom (beyond the budget limits) are placed on Hold. The method thus ensures the greatest 'bang for buck': that is, the ECV is maximized, for a given R&D budget.

This ECV model has a number of attractive features: it recognizes that the Go/Kill decision process is an incremental one (the notion of purchasing options). Indeed some experts indicate that this ECV decision tree method comes very close to approaching *a real options approach to valuing projects* (Deaves & Krinsky, 1997; Faulkner, 1996). Finally, the ECV method deals with the issue of constrained resources, and attempts to maximize the value of the portfolio in light of this constraint.

The major weakness of the method is the *dependency on extensive financial and other quantitative data*. Accurate estimates must be available for *all* projects' future stream of earnings and their costs and also for probabilities of success—estimates that are often unreliable or, at best, simply not available early in the life of a project. A second weakness is that the method *does not look at the balance* of the portfolio—at whether the portfolio has the right balance between high- and low-risk projects, or across markets and technologies. A third weakness is that the method considers only *a single financial criterion* for maximization.

Scoring Models as Portfolio Tools

Scoring models have long been used for making Go/Kill decisions at gates. But they also have applicability for project prioritization and portfolio management. Projects are scored on each of a number of criteria by management. (A sample list of project prioritization criteria, from which a scorecard is constructed, is shown in Figure 5). The Project Attractiveness Score is

149

1. Strategic:
    *   degree to which project aligns with BU's strategy
    *   strategic importance
2. Product Advantage:
    *   unique benefits
    *   meets customer needs better
    *   value for money
3. Market Attractiveness:
    *   market size
    *   market growth
    *   competitive situation
4. Synergies (Leverages Core Competencies):
    *   marketing synergies
    *   technological synergies
    *   manufacturing / operations synergies
5. Technical Feasibility:
    *   technical gap
    *   complexity
    *   technical uncertainty
6. Risk Vs. Return:
    *   expected profitability (magnitude; e.g. NPV)
    *   return (e.g. IRR)
    *   payback period
    *   certainty of return/profit estimates

✓ **The six Factors (in bold) are scored (e.g. 1-5 or 0-10) on a scorecard**
✓ **Factor scores must clear minimum hurdles**
✓ **Also added (weighted/unweighted) to yield the Project Attractiveness Score, which is used to make Go/Kill decisions & also to prioritize projects**

Figure 5. Best practice go-to-development decision criteria.

Copyrighted: Cooper, Edgett & Kleinschmidt (2002a).

the weighted addition of the factor ratings, and becomes the basis for developing a rank-ordered list of projects. Projects are ranked until there are no more resources.

Scoring models generally are praised in spite of their limited popularity. Research into project selection methods reveals that scoring models produce a strategically aligned portfolio and one that reflects the business's spending priorities; and they yield effective and efficient decisions, and result in a portfolio of high value projects (Cooper et al., 2002a).

Goal 2: A Balanced Portfolio
The second major goal is a balanced portfolio—a balanced set of development projects in terms of a number of key parameters. The analogy is that of an investment fund, where the fund manager seeks balance in terms of high risk vs. blue chip stocks; and balance across industries, in order to arrive at an optimum investment portfolio.

Visual charts are favored in order to display balance in new product project portfolios. These visual representations include portfolio maps or bubble diagrams (Figure 6).

Risk–Reward Bubble Diagrams
The most popular bubble diagram is a variant of the *risk/return chart* (see Figure 6). Here one axis is some measure of the *reward* to the company, and the other is a *success probability* (Cooper et al., 2002a; Evans 1996; Roussel et al., 1991). A sample bubble diagram is shown in Figure 6 for a business unit of a major chemical company. Note that the size of each bubble shows the annual resources spent on each project.

Traditional Charts for Portfolio Management
There are numerous parameters, dimensions or variables across which one might wish to seek a balance of projects. As a result, there are an endless variety of histograms and pie charts which help to portray portfolio balance. Some examples:

Project types is of vital concern. What is the spending on genuine new products vs. product renewals (improvements and replacements), or product extensions, or product maintenance, or cost reductions and process improvements? And what should it be? Pie charts effectively capture the spending split across project types—actual vs. desired splits (Figure 7).

150

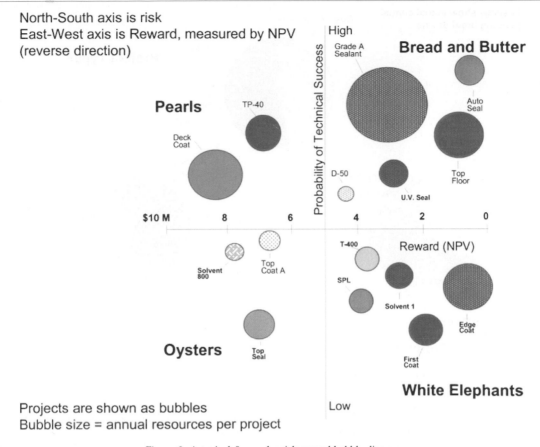

North-South axis is risk
East-West axis is Reward, measured by NPV
(reverse direction)

Projects are shown as bubbles
Bubble size = annual resources per project

Figure 6. A typical & popular risk-reward bubble diagram.

Source: Cooper (2001).

Markets, products and technologies provide another set of dimensions across which managers seek balance. Pie charts are again appropriate for capturing and displaying this type of data (again, Figure 7).

Goal 3: Building Strategy into the Portfolio

Strategy and new product resource allocation must be intimately connected. *Strategy becomes real when one starts spending money!* The mission, vision and strategy of the business are made operational through the decisions management makes on where to spend money. For example, if a business's strategic mission is to 'grow via leading edge product development', then this must be reflected in the mix of new product projects underway—projects that will lead to growth (rather than simply to defend) and products that really are innovative.

Linking Strategy to the Portfolio: Approaches

There are three main ways to build in the goal of strategic alignment:

(1) *Bottom up—building strategic criteria into project selection tools*: here strategic fit is achieved by including numerous strategic criteria into the Go/Kill and prioritization tools;
(2) *Top-down*—Strategic Buckets method: this begins with the business's strategy and then moves to setting aside funds—envelopes or buckets of money—destined for different types of projects;
(3) *Top-down*—Product Roadmap: here the business's strategy defines what major initiatives or platform developments to undertake.

Bottom Up—Strategic Criteria Built into Project Selection Tools

Not only are scoring models effective ways to maximize the value of the portfolio, but they can also be used to ensure strategic fit. One of the multiple objectives considered in a scoring model, along with profitability or likelihood of success, can be to *maximize strategic fit*, simply by building into the scoring model a number of strategic questions. In the

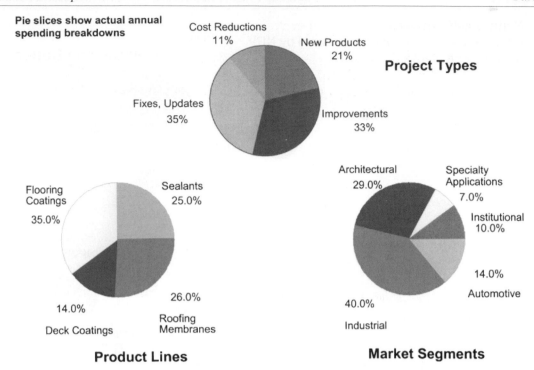

Figure 7. Breakdown of resources by project types, product lines & markets.

scoring model displayed in Figure 5, for example, three major factors out of six are strategic; and low scores on these will all but kill the project. Thus, projects that fit the business's strategy, offer competitive advantage, and leverage the firms' core competencies likely to rise to the top of the list. Indeed, it is inconceivable how any 'off strategy' projects could make the active project list at all: this scoring model naturally weeds them out.

Top-down Strategic Approach—Strategic Buckets Model

While strategic fit can be achieved via a scoring model, a top-down approach is the only method designed to ensure that where the money is spent mirrors the business's strategy (Cooper et al., 2002a). The Strategic Buckets model operates from the simple principle that *implementing strategy equates to spending money on specific projects*. Thus, setting portfolio requirements really means 'setting spending targets'.

The method begins with the business's strategy, and requires the senior management of the business to make forced choices along each of several dimensions—choices about how they wish to allocate their scarce money resources. This enables the creation of 'envelopes of money' or 'buckets'. Existing projects are categorized into buckets; then one determines whether actual spending is consistent with desired spending for each bucket. Finally projects are priori-

tized within buckets to arrive at the ultimate portfolio of projects—one that mirrors management's strategy for the business.

In practice, senior management first develops the vision and strategy for the business. Next, they make forced choices across key strategic dimensions; that is, based on this strategy, development resources are split across categories on each dimension. Some common dimensions are:

*   *Strategic goals*: Management splits resources across the specified strategic goals. For example, what percentage should be spent on Defending the Base? On Diversifying? On Extending the Base? and so on;
*   *Product lines*: Resources are split across product lines: e.g. how much should be spent on Product Line A? On Product Line B? On C? A plot of product line locations on the product life cycle curve is used to help determine this split;
*   *Project type*: What percentage of resources should go to new product developments? To maintenance-type projects? To process improvements? To fundamental research? etc;
*   *Familiarity Matrix*:[5] What should be the split of resources to different types of markets and to

---

[5] The familiarity matrix was first proposed as a strategic tool by Ed Roberts (Roberts & Berry, 1983).

152

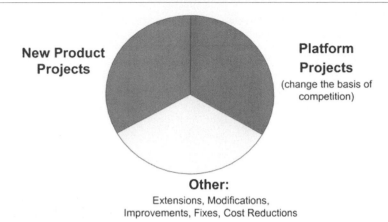

New Product Projects

Platform Projects
(change the basis of competition)

Other:
Extensions, Modifications,
Improvements, Fixes, Cost Reductions

The business's strategy dictates the split of resources into buckets; projects are rank ordered within buckets, but using different criteria in each bucket.

Figure 8. Strategic buckets method of resource allocation.

Source: AlliedSignal-Honeywell; reported in: Cooper, Edgett & Kleinschmidt (2002a).

different technology types in terms of their *familiarity to the business*? (Roberts & Berry, 1983);

• *Geography*: What proportion of resources should be spent on projects aimed largely at North America? At Latin America? At Europe? At the Pacific? Or at global?

Now, management develops *strategic buckets*. Here the various strategic dimensions (above) are collapsed into a convenient handful of buckets. Then the desired spending by bucket is determined: the *'what should be'*. This involves a consolidation of desired spending splits from the strategic allocation exercise above. Next comes a gap analysis. Existing projects are categorized by bucket and the total current spending by bucket is added up (the *'what is'*). Spending gaps are then identified between the *'what should be'* and *'what is'* for each bucket.

Finally, projects within each bucket are rank-ordered, using either a scoring model or financial criteria to do this ranking within buckets. Portfolio adjustments are then made, either via immediate pruning of projects, or by adjusting the approval process for future projects.

A somewhat simpler example is shown in Figure 8. The leadership team of the business begins with the business's strategy, and uses the Mercedes-Benz emblem (the three-point star) to help divide up the resources. There are three buckets: fundamental research and platform development projects which promise to yield major breakthroughs and new technology platforms; new product developments; and maintenance—technical support, product improvements and enhancements, etc. Management divides the

R&D funds into these three buckets, and then rates and ranks projects against each other within each bucket. In effect, three separate portfolios of projects are created and managed. And the spending breakdowns across projects mirror strategic priorities.

The major strength of the Strategic Buckets Model is that it firmly links spending to the business's strategy; further, it recognizes that all development projects that compete for the same resources should be considered in the portfolio approach. Finally, different criteria can be used for different types of projects. That is, one is not faced with comparing and ranking very different types of projects against each other.

Top Down Strategic Approach—Product Roadmap

A product roadmap[6] is an effective way to map out a series of assaults in an attack plan. What is a roadmap? It is simply a management group's view of how to get where they want to go or to achieve their desired objective (Albright, 2001; Myer & Lehnerd, 1997). The roadmap is a useful tool that helps the group make sure that the capabilities to achieve their objective are in place when needed. Two useful types of roadmaps are the product roadmap and the technology roadmap:

(1) The *product roadmap* defines the product and product releases along a timeline—how the product line will evolve over time, and what the next generations will be. It answers the question: what products? An example is shown in Figure 9. Here

---

[6] This section is based on Lucent Technologies. See R. E. Albright (2001). For more on platforms and roadmaps, see Myer & Lehnerd (1997).

The road map shows the major initiatives and developments: new platforms, platform extensions, and major new product lines – depicted over time

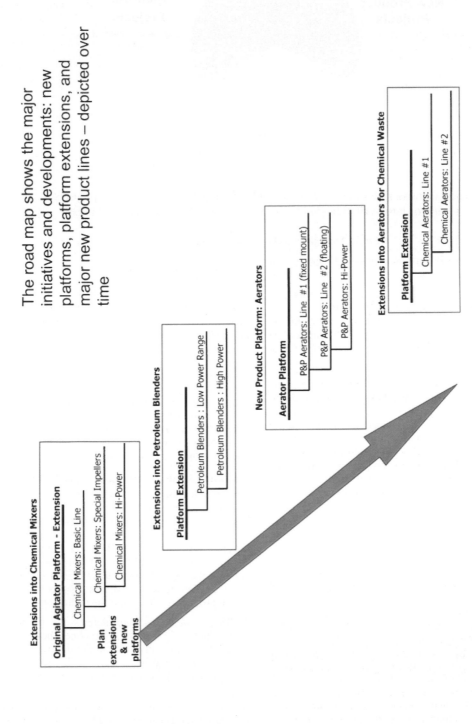

Figure 9. A product roadmap for a process equipment manufacturer.

Source: Cooper (2001).

the product roadmap for a process equipment manufacturer not only defines the platforms and platform extensions needed but also maps out the various product releases, specifies features or functionality that will be built into each new release, and indicates timing (launch date);

(2) The *technology roadmap* is derived from the product roadmap, but also specifies how you will get there. That is, it lays out the technologies and technological competencies that are needed in order to implement (develop and source) the products in the product roadmap. The technology roadmap is a logical extension of the product roadmap and is closely linked to it.

Conclusion

Product innovation is a vital task for the modern corporation. The company's success at new product conception, development and launch may well decide the fate of the entire business. Many companies miss the mark when it comes to the 15 critical success factors outlined earlier in this chapter, and the results are predictable—a mediocre new product effort. Further, if the company lacks a systematic new product process, or if portfolio management for new products is a foreign concept, then best practices are missing. The effort to implement a robust Stage-Gate™ process coupled with effective portfolio management is significant: it may take a year or so of hard work. But those firms that have successfully implemented Stage-Gate and portfolio management methods for product innovation are reaping the benefits (Griffin, 1997; Menke, 1997).

References

Albright, R. E. (2001). Roadmaps and roadmapping: linking business strategy and technology planning. *Proceedings, Portfolio Management for New Product Development*. Ft. Lauderdale, FL: Institute for International Research (IIR) and Product Development & Management Association (PDMA).

Ali, A., Krapfel, R. & LaBahn, D. (1995). Product innovativeness and entry strategy: impact on cycle time and breakeven time. *Journal of Product Innovation Management*, **12** (1), 54–69.

Barczak, G. (1995). New product strategy, structure process and performance in the telecommunications industry. *Journal of Product Innovation Management*, **12** (3), 224–234.

Bonaccorsi, A. & Hayashi, A. (1994). Strategic partnerships in new product development: An Italian case study. *Journal of Product Innovation Management*, **11** (2), 134–145.

Brentani, U. de (1989). Success and failure in new industrial services. *Journal of Product Innovation Management*, **6**, 239–258.

Calantone, R. J. & Cooper, R. G. (1979). A discriminant model for identifying scenarios of industrial new product failure. *Journal of the Academy of Marketing Science*, **7**, 163–83.

Campbell, A. J. & Cooper, R. G. (1999). Do customer partnerships improve success rates? *Industrial Marketing Management*, **28** (5), 507–519.

Cooper, R. G. (1988). The new product process: a decision guide for managers. *Journal of Marketing Management*, **3** (3), 238–255.

Cooper, R. G. (1995). Developing new products on time, in time. *Research-Technology Management*, **38** (5), 49–57.

Cooper, R. G. (1996). New products: what separates the winners from the losers. In: M. D. Rosenau, Jr (Ed.), *PDMA Handbook for New Product Development*. New York: John Wiley.

Cooper, R. G. (1998). Benchmarking new product performance: results of the best practices study. *European Management Journal*, **16** (1), 1–7.

Cooper, R. G. (1999a). New Product Development. In: M. J. Baker (Ed.), *International Encyclopedia of Business & Management: Encyclopedia of Marketing* (1st ed., pp. 342–355). London: International Thomson Business Press.

Cooper, R. G. (1999b). The invisible success factors in product innovation. *Journal of Product Innovation Management*, **16** (2), 115–133.

Cooper, R. G. (1999c). New product leadership: building in the success factors. *New Product Development & Innovation Management*, **1** (2), 125–140.

Cooper, R. G. (2000). Doing it right—winning with new products. *Ivey Business Journal*, 54–60.

Cooper. R. G. (2001). *Winning at new products: Accelerating the process from idea to launch* (3rd ed.). Reading, MA: Perseus Books.

Cooper, R. G. & Edgett, S. J. (2002). The dark side of time and time metrics in product innovation. *Visions*, **XXVI** (22), 14–16.

Cooper, R. G., Edgett, S. J. & Kleinschmidt, E. J. (1997a). Portfolio management in new product development: lessons from the leaders—Part I. *Research Technology Management*, **40** (5), 16–28

Cooper, R. G., Edgett, S. J. & Kleinschmidt, E. J. (1997b). Portfolio management in new product development: lessons from the leaders Part II. *Research-Technology Management*, **40** (6), 43–52.

Cooper, R. G., Edgett, S. J. & Kleinschmidt, E. J. (1998). Best practices for managing R&D portfolios. *Research-Technology Management*, **41** (4), 20–33.

Cooper, R. G., Edgett, S. J. & Kleinschmidt, E. J. (1999). New product portfolio management: practices and performance. *Journal of Product Innovation Management*, **16** (4), 333–351.

Cooper, R. G., Edgett, S. J. & Kleinschmidt, E. J. (2000). New problems, new solutions: making portfolio management more effective. *Research-Technology Management*, **43** (2), 18–33.

Cooper, R. G., Edgett, S. J. & Kleinschmidt, E. J. (2001a). Portfolio management for new product development: Results of an industry practices study. *R&D Management*, **31** (4), 361–380.

Cooper, R. G., Edgett, S. J. & Kleinschmidt, E. J. (2002a). *Portfolio Management for New Products* (2nd ed.). Reading, MA: Perseus Books.

Cooper, R. G., Edgett, S. J. & Kleinschmidt, E. J. (2002b). Portfolio management: fundamental to new product success. In: P. Beliveau, A. Griffin & S. Sommermeyer (Eds),

The PDMA Toolbox for New Product Development (pp. 331–364). New York: John Wiley.

Cooper, R. G. & Kleinschmidt, E. J. (1986). An investigation into the new product process: steps, deficiencies and impact. *Journal of Product Innovation Management*, **3** (2), 71–85.

Cooper, R. G. & Kleinschmidt, E. J. (1993). Major new products: What distinguishes the winners in the chemical industry. *Journal of Product Innovation Management*, **10** (2), 90–111.

Cooper, R. G. & Kleinschmidt, E. J. (Nov. 1994). Determinants of timeliness in new product development. *Journal of Product Innovation Management*, **11** (5), 381–396.

Cooper, R. G. & Kleinschmidt, E. J. (1995a). New product performance: Keys to success, profitability and cycle time reduction. *Journal of Marketing Management*, **11**, 315–337.

Cooper, R. G. & Kleinschmidt, E. J. (1995b). Benchmarking the firm's critical success factors in new product development. *Journal of Product Innovation Management*, **12** (5), 374–391.

Cooper, R. G. & Kleinschmidt, E. J. (1995c). Benchmarking firms' new product performance and practices. *Engineering Management Review*, **23** (3), 112–120.

Cooper, R. G. & Kleinschmidt, E. J. (1995d). Performance typologies of new product projects. *Industrial Marketing Management*, **24**, 439–456.

Cooper, R. G. & Kleinschmidt, E. J. (1996). Winning businesses in product development: Critical success factors. *Research-Technology Management*, **39** (4), 18–29.

Crawford C. M. (1992). The hidden costs of accelerated product development. *Journal of Product Innovation Management*, **9** (3), 188–199.

Davis J., Fusfeld A., Scriven E. & Trittle G. (2001). Determining a project's probability of success. *Engineering Mgmt Review*, **29** (4), 56–61.

Day, G. (1986). *Analysis for strategic marketing decisions*. St. Paul, MN: West Publishing.

Deaves R. & Krinsky I. (1997). *New tools for investment decision-making: Real options analysis*. Working paper, McMaster University, Hamilton ON Canada.

Di Benedetto C. A. (1999). Identifying the key success factors in new product launch. *Journal of Product Innovation Management*, **16** (6), 530–544.

Evans, P. (1996). Streamlining formal portfolio management. *Scrip Magazine*.

Faulkner, T. (1996). Applying 'options thinking' to R&D valuation. *Research-Technology Management*, 50–57.

Friar J. H. (1995). Competitive advantage through product performance innovation in a competitive market. *Journal of Product Innovation Management*, *12* (1), 33–42.

Griffin, A. (1997). *Drivers of NPD success: The 1997 PDMA Report*. Chicago: Product Development & Management Association.

Griffin, A. & Page, A. L. (1993). An interim report on measuring product development success and failure. *Journal of Product Innovation Management*, **9** (1), 291–308.

Hultink, E. J. & Atuahene-Gima, K. (2000). The effect of sales force adoption on new product selling performance. *Journal of Product Innovation Management*, **17** (6), 435–450.

King, J. (1997). Poor planning kills projects, pushes costs up. *Computerworld*.

Kleinschmidt, E. J. (1999). *Unpublished research results*. McMaster University, Hamilton, ON, Canada.

Kleinschmidt, E. J. & Cooper, R. G. (1988). The performance impact on an international orientation of product innovation. *European Journal of Marketing*, **22**, 56–71.

Lynn, G. S., Skov, R. B. & Abel, K. D. (1999). Practices that support team learning and their impact on speed to market and new product success. *Journal of Product Innovation Management*, **16** (5), 439–454.

Maidique, M. A. & Zirger, B. J. (1984). A study of success and failure in product innovation: The case of the U.S. electronics industry. *IEEE Trans. Engineering Management*, (EM–31), 192–203.

Markham, S. K. & Griffin, A. (1998). The breakfast of champions: association between champions and product development environments, practices and performance. *Journal of Product Innovation Management*, **15** (5), 436–454.

McDonough, E. F. III, Kahn, K. B. & Barczak, G. (2001). An investigation of the use of global, virtual and collocated new product development teams. *Journal of Product Innovation Management*, **18** (2), 110–121.

Menke, M. M. (1997). Essentials of R&D strategic excellence. *Research-Technology Management*, **40** (5), 42–47.

Mishra, S., Kim, D. & Lee, D. H. (1996). Factors affecting new product success: cross country comparisons. *Journal of Product Innovation Management*, **13** (6), 530–550.

Montoya-Weiss, M. M. & Calantone, R. (1997). Determinants of new product performance: a review and meta-analysis. *Journal of Product Innovation Management*, **11** (5), 397–417.

Myer, M. H. & Lehnerd, A. P. (1997). *The power of product platforms*. New York: Free Press.

Page, A. L. (1991). *PDMA new product development survey: Performance and best practices*. Paper presented at PDMA Conference, Chicago.

Peters, T. (1988). *Thriving on chaos*. New York: Harper & Row.

Porter, M. E. (1985). *Competitive advantage: Creating and sustaining superior performance*. New York: Free Press.

Rauscher, T. G. & Smith, P. G. (1995). Time-driven development of software in manufactured goods. *Journal of Product Innovation Management*, **12** (3), 186–199.

Risk Magazine (Oct., 1999). No cited author. Doubts linger over clearing solution. *Risk*, 44 (October).

Roberts, E. B. & Berry, C. A. (1983). Entering new businesses: selecting strategies for success. *Sloan Management Review*, 3–17 (Spring, 1983).

Rothwell, R., Freeman, C., Horseley, A., Jervis, V. T. P., Robertson, A. B. & Townsend, J. (1974). SAPPHO updated—Project SAPPHO Phase II. *Research Policy*, **3**, 258–291.

Roussel, P., Saad, K. N. & Erickson, T. J. (1991). *Third generation R&D, managing the link to corporate strategy*. Boston, MA: Harvard Business School Press & Arthur D. Little Inc.

Sanchez, A. M. & Elola, L. N. (1991). Product innovation management in Spain. *Journal of Product Innovation Management*, **8** (1), 49–56.

Souder W. E. & Song X. M. (1997). Contingent product design & marketing strategies influencing new product success & failure in U.S. & Japanese electronics firms.

Journal of Product Innovation Management, **14** (1), 21–34.

Song, X. M., Di Benedetto, C. A. & Song, L. Z. (2000). Pioneering advantages in new service development: a multi country study of managerial perceptions. *Journal of Product Innovation Management*, **17** (5), 378–392.

Song, X. M. & Montoya-Weiss, M. M. (1998). Critical development activities for really new vs. incremental products. *Journal of Product Innovation Management*, **15** (2), 124–135.

Song, X. M., Montoya-Weiss, M. M. & Schmidt J. B. (1997). Antecedents & consequences of cross-functional cooperation: A comparison of R&D, manufacturing & marketing perspectives. *Journal of Product Innovation Management*, **14** (1), 35–47.

Song, X. M. & Parry, M. E. (1996). What separates Japanese new product winners from losers. *Journal of Product Innovation Management*, **13** (5), 422–439.

Thomke, S. & Fujimoto, T. (2000). The effect of 'front-loading' problem solving on product development performance. *Journal of Product Innovation Management*, **17** (2), 128–142.

Veryzer, R. W. (1998). Key factors affecting customer evaluation of discontinuous new products. *Journal of Product Innovation Management*, **15** (2), 136–150.

Von Braun, C. F. (1997). *The innovation war*. Upper Saddle River, NJ: Prentice Hall.

Wilson, E. (1991). *Product development process, product definition guide, release 1.0*. Internal Hewlett-Packard document, Palo Alto, CA.

The International Handbook on Innovation
Edited by Larisa V. Shavinina

Types of Innovations

Robert J. Sternberg[1], Jean E. Pretz[1] and James C. Kaufman[2]

[1] Department of Psychology, Yale University, USA
[2] Department of Psychology, California State University, USA

Abstract: Innovations can be of eight different types. Each represents a different kind of contribution. For example, a conceptual replication is a minimal innovation, repeating with minor variations an idea that already exists (e.g. Mercury's putting the 'Mercury' label on what is essentially an already-existing Ford car). Forward incrementations represent next steps forward along existing lines of progression (e.g. the 2001 version of a 2000 car). Redirections represent a totally different direction for products that diverge from the existing line of progress (e.g. electric cars). We will discuss the types of innovations and the circumstances leading to success.

Keywords: Innovation; Creativity; Incrementation; Propulsion; Redirection; Reinitiation; Replication.

Introduction

In the erstwhile television series, *All in the Family*, there was an episode in which a character created a novel invention: a remote-control doorbell ringer. One could ring someone's doorbell remotely, so that one actually did not have to go up to the door of the person in order to ring the bell. It quickly became obvious that, although the invention was novel, it somehow was not creative. What it lacked was a second property of creativity beyond novelty: usefulness. All inventions and, indeed, innovations of any kind, start with some kind of creative enterprise, and the enterprise must produce work that is not just novel, but useful.

Creativity is the ability flexibly to produce work that is novel (i.e. original, unexpected), high in quality, and useful, in that it meets task constraints (Lubart, 1994, this volume; Ochse, 1990; Sternberg, 1988, 1999b; Sternberg & Lubart, 1995, 1996). The remote-control doorbell ringer was novel, but not useful for any particular task anyone could think up—at least on the show. Creativity is a topic of wide scope that is important at both the individual and societal levels for a wide range of task domains. At an individual level, creativity is relevant, for example, when solving problems on the job and in daily life. At a societal level, creativity can lead to new scientific findings, new movements in art, new inventions, and new social programs. The economic importance of creativity is clear because new products or services create jobs. Furthermore, individuals, organizations, and societies must adapt existing resources to changing task demands to remain competitive. *Innovation* is the channeling of creativity so as to produce a creative idea and/or product that people can and wish to use.

Creativity may be viewed as taking place in the interaction between a person and the person's environment (Amabile, 1996; Csikszentmihalyi, 1996, 1999; Feldman, 1999; Feldman, Csikszentmihalyi & Gardner, 1994; Sternberg, Kaufman & Pretz, 2002; Sternberg & Lubart, 1995). According to this view, the essence of creativity cannot be captured just as an intrapersonal variable. Thus, we can characterize a person's cognitive processes as more or less creative (Finke, Ward & Smith, 1992; Rubenson & Runco, 1992; Weisberg, 1986, this volume) or the person as having a more or less creative personality (Barron, 1988; Feist, 1999). We further can describe the person as having a motivational pattern that is more or less typical of creative individuals (Hennessey & Amabile, 1988), or even as having background variables that more or less dispose that person to think creatively (Simonton, 1984, 1994, this volume). But we cannot fully judge that person's creativity independent of the context in which the person works.

For example, a contemporary artist might have thought processes, personality, motivation, and even background variables similar to those of Monet, but that artist, painting today in the style of Monet, probably would not be judged to be creative in the way

Monet was. He or she was born too late. Artists, including Monet, have experimented with impressionism, and unless the contemporary artist introduces some new twist, he or she will likely be viewed as imitative rather than creative. To a large extent, creative innovation starts with a decision to be creative and to think intuitively and 'out of the box' (see Root-Bernstein & Root-Bernstein, this volume).

The Propulsion Model of Creative Contributions

Sternberg (1999c) has presented what he refers to as a propulsion model of creative contributions (see also Sternberg, Kaufman & Pretz, 2001, 2002). The idea is that creative contributions 'propel' a field forward in some way—they are the result of creative leadership on the part of their creators. The propulsion model is a descriptive taxonomy of eight types of creative contributions. Although the eight types of contributions may differ in the extent of creative contribution they make, there is no a priori way of evaluating *amount* of creativity on the basis of the *type* of creative contribution. Certain types of creative contributions probably tend, on average, to be greater in amounts of novelty than are others. For example, replications tend, on average, not to be highly novel. But creativity also involves quality of work, and the type of creative contribution a work makes does not necessarily predict the quality of that work.

The eight types of creative contributions are:

(1) *Replication*. The creative contribution represents an effort to show that a given field is where it should be. The propulsion is intended to keep the field where it is rather than moving it;

(2) *Redefinition*. The creative contribution represents an effort to redefine where the field currently is. The current status of the field thus is seen from a new point of view;

(3) *Forward Incrementation*. The creative contribution represents an attempt to move the field forward in the direction in which it already is moving, and the contribution takes the field to a point to which others are ready to go;

(4) *Advance Forward Incrementation*. The creative contribution represents an attempt to move the field forward in the direction it is already going, but the contribution moves beyond where others are ready for the field to go;

(5) *Redirection*. The creative contribution represents an attempt to move the field from where it is currently headed toward a new and different direction;

(6) *Reconstruction/Redirection*. The creative contribution represents an attempt to move the field back to where it once was (a reconstruction of the past) so that the field may move onward from that point, but in a direction different from the one it took in the past;

(7) *Reinitiation*. The creative contribution represents an attempt to move the field to a different and as yet not reached starting point and then to move the field in a new direction from that point;

(8) *Integration*. The creative contribution represents an attempt to move the field by putting together aspects of two or more past kinds of contributions that formerly were viewed as distinct or even opposed. This type of contribution shows particularly well the potentially dialectical nature of creative contributions, in that it merges into a new Hegelian type of synthesis two ideas that formerly may have been seen as opposed (Sternberg, 1999a).

The eight types of creative contributions described above are viewed as qualitatively distinct. However, within each type, there can be quantitative differences. For example, a forward incrementation may represent a fairly small step forward for a given field, or it may represent a substantial leap. A reinitiation may restart an entire field or just a small area of that field. Moreover, a given contribution may overlap categories. For example, a forward incrementation may be the result of an integration of somewhat closely related concepts in the field.

Thus, when people are creative, they can be creative in different ways. The exact nature of these ways depends upon the theory of types of creative contributions one accepts. What is a creative contribution and why do we need a taxonomy of types of creative contributions? A consensual definition of a creative contribution is of something that is: (a) relatively original; and (b) high in quality vis-à-vis some purpose (see Sternberg & Davidson, 1994; Sternberg & Lubart, 1995, 1996). Starting with creative contributions rather than creative contributors can have several advantages. First, a given contributor may make a variety of different types of contributions. The contributor thus is not limited to any one of the eight types of creative contributions described in this book. Contributions may be primarily of a given type, but creators may not be. For example, much, but certainly not all of Picasso's work, set off in strikingly original and bold directions. Through the proposed theory, one can evaluate individual works of a creator and not just the 'average' or typical type of work the creator produced. Second, even if contributors tend to have a type of contribution they make most often, observing differences in the types of contributions individuals make can help elucidate differences in the types of creativity the contributors typically tend to show. Third, the emphasis on contributions rather than creators underscores the point that people can modify the kinds of contributions they make. Someone early in a career may be afraid to depart too much from accepted ways of doing things, for fear that such a departure will put his or her career at risk of being derailed. Later in a career, however, for example, after attaining tenure or, at least, financial security, the creator may be willing to take risks and make contributions that earlier he or she would not have felt comfortable making earlier. The creator's goals and purposes may change. Even from the days when composers had to compose music or artists had to paint works in order to please their royal patrons, creative individuals have always operated under societal constraints. Given the importance of purpose, creative contributions must always be defined in some context. If the creativity of an individual is always judged in a context, then it is important to understand how the context interacts with how people are judged. In particular, what are the types of creative contributions a person can make within a given context? Most theories of creativity concentrate on attributes of the individual (see Sternberg, 1988, 1999b). But to the extent that creativity is in the interaction of person with context, we would need as well to concentrate on the attributes of the individual and the individual's work relative to the environmental context.

How can one apply the propulsion model to invention? We consider this question next, reviewing inventions that represent each of the eight types of creative contributions listed above.

Examples of Each of the Types of Creative Propulsion

Replication

In replication, inventors essentially duplicate an existing product, sometimes, with improvements in pricing or in quality. Many people are familiar with Altoids™, which are a particularly strong kind of breath mint. When these mints first came onto the market, they were an immediate hit. People quickly started buying them up. As so often happens when a product is successful, other manufacturers see money to be made by imitating the original product. One of the authors recently went to a drug store to buy a tin of these mints, and was only a bit surprised to see that, where before the one brand had had this particular section of the drugstore counter to itself, now it was sharing space with three competitors. As far as the author could tell, the three competitors basically had replicated the Altoids™ formula for success, with minor variations.

Such copies are commonplace in industry. The success in IBM personal computers rapidly gave rise to large numbers of replications, which were so similar to the original IBMs that they actually were (and still are) referred to as 'clones'. Kleenex™ tissue paper gave rise to large numbers of imitators, as did bran cereals with raisins. Sometimes the replicators ultimately achieve a larger market share than did the originators, typically when the new products are either notably better in quality or notably less expensive than the original product.

Redefinition

Video games have become an integral part of American culture. Just as the invention of the television in the 1950s redefined entertainment and indirectly influenced many arenas of American life, the introduction of the video game in the 1970s has initiated a change in the nature of childhood play throughout society.

Where did video games come from? Who first had the idea to make television interactive, to give people the power to control the action on the screen? The origins can be found in one of the first video games, Pong. Video games actually stem from a very simple concept invented in the 1950s by an unsuspecting physicist working at Brookhaven National Laboratories in Upton, New York, Willy Higinbotham. Higinbotham's original 'video game' was designed to add some spice to a public tour of his instrumentation laboratory. Because Brookhaven was involved in nuclear research, these tours were organized to show visitors that the research was entirely peaceful and posed no risk to residents. Higinbotham knew that the tour of the facility was not very exciting, so he tooled around with an oscilloscope and a simple computer, built a few controls with buttons and knobs, and created what can be considered the first video game. The game was conceptually very similar to Pong: one player presses a button to launch a ball from the left side of the screen, adjusts a knob so the ball clears a barrier in the middle of the screen, and a second player hits the ball back over the 'net'. With these modest beginnings, the video game industry was born.

Little did Higinbotham know that not only would his simple tennis game relieve boredom on the tour of his laboratory, but that players would form lines, eager to try their hand at the novel challenge. Before Higinbotham understood the potential of the demand he had created, marketers seized hold of the idea. In fact, Pong is simply a redefinition of Higinbotham's basic tennis game. The originators of Pong did not seriously revamp the design, but they did bring to it a new perspective. Realizing the game's potential for fascination (and remuneration), they grabbed their opportunity to latch onto the public's interest. As almost every American child can attest, their fascination has not waned, and the video game industry continues to profit (Flatow, 1992). Redefinitions are not limited, of course, to video games.

A split-second decision can also represent an attempt to redefine a field. Jeno Paulucci was a businessman who had just purchased Chun King, a canned food company. Paulucci wanted to use this company to sell his own food inventions, Cantonese vegetables with Italian spices. Paulucci had a meeting with the main buyer for a top food chain, and he was trying to sell his Chun King cans to be distributed by this chain. If the businessman was able to make the sale, his company would surely take off and become profitable; if not, it would be a difficult task to maintain financial solvency.

Demonstrating his product, Paulucci opened up a can of chop suey vegetables—and saw a cooked grasshopper sitting right in the middle of the can. Paulucci had one brief moment in which the grasshopper was hidden from the buyer by the open can lid. What could he do? If the buyer saw the grasshopper, the sale would surely not go through—and Chun King's reputation would likely suffer. Think about what you might do in such a situation.

In those few seconds, Paulucci redefined his role as salesman. Not only was Paulucci responsible for presenting his product to a potential buyer, but his chances of making a sale would be increased if he could convince the customer that he was a consumer himself. A typical businessman might have either tried to surreptitiously remove the grasshopper from the can (and run a high risk of being caught) or tried to explain the many legitimate reasons why the dead grasshopper was not indicative of his company's product. Paulucci decided to attempt neither of these two strategies. Instead, he looked the buyer in the eye and said that the vegetables looked so good that he wanted to have the first bite himself. Paulucci took a large forkful of vegetables—including the fateful grasshopper—and ate the large bite with a big smile. He got the sale (Hay, 1988).

Forward Incrementation

In forward incrementation, a new product is invented that moves an existing product line to the next step consumers are ready to take. The invention of the incandescent lamp is a good example of the forward incrementation typical of normal science (Kuhn, 1970). Although Americans strongly associate the name of Thomas Edison (1847–1931) with the invention of the light bulb, Joseph Swan (1828–1914) of England deserves equal scientific credit for his work on the problem, as do many others who also contributed small steps along the way toward the successful product.

The first attempts at creating a light bulb can be traced to 1838, when Jobart "sealed a carbon rod inside a vacuum and watched it glow as a current was passed through it" (Clark, 1985, p. 220). By 1847, progress had reached a point where the concept of an evacuated bulb with a carbon filament had been solidified. But specific problems remained. Incandescent lamps in the mid-1800s burned for a very short period of time due to a combination of two factors. First, the filaments were not durable, and second, there was too much air inside the bulb (Yarwood, 1983).

As the technology was not yet available to create a vacuum inside the bulb, experiments focused on perfecting the filament. Swan experimented with a carbon filament treated in various ways, whereas Edison believed that a platinum filament was the answer. Earlier, he had worked with carbon unsuccess-

fully. In 1865, Hermann Sprengel introduced the mercury vacuum pump, and 14 years later, Edison devised and his group devised a vacuum pump combining the Sprengel and Geissler pumps. He finally met success with a carbon filament (Clark, 1985).

Step by step, scientists settled on the carbon filament and began to produce light bulbs in quantity by 1882. At that point, the only impediment to more widespread use of the bulbs was the lack of electrical wiring in homes and businesses. The next step of the path was cleared as Edison and his colleagues set out to create a system of electricity to bring power to cities. Even after Thomas Edison and Joseph Swan joined forces to create United Electric Company, experiments on metal filaments continued (Yarwood, 1983). Eventually, the tungsten filament became preferred because of its high melting point, and it remains the standard in the light bulbs of today.

Whereas some discoveries are characterized by distinct moments of insight that create a discontinuous path of progress, the light bulb is better viewed as the product of a scientific evolution directed by careful experimentation and simple trial and error. Such forward steps or incrementations are key to the progress of every scientific field.

Another example of a next step in science is Pasteur's development of vaccine therapy following his germ theory of disease. After the germ theory had described how airborne bacteria are the mechanism for the cause of disease, the next step was to extend the theory's implications for disease prevention, namely, vaccine therapy. The concept of vaccination did not originate with Pasteur (nor did pre-germ theory advances in sanitary practices in medical settings). For example, Edward Jenner had earlier discovered the effectiveness of using small amounts of cowpox to vaccinate humans against small pox.

Pasteur reasoned that the administration of similarly attenuated germs might prevent the development of bacterially caused diseases. His first attempt, using fowl cholera, proved a success. Although he at first did not realize that the attenuation was caused by prolonged exposure to warm air, later experiments with anthrax in cattle confirmed that time and heat did indeed sufficiently weaken the bacteria. Pasteur's great triumph of vaccine therapy was in a dire case of a young boy attacked by a rabid dog. After administering a culture developed from the brains of dogs and rabbits, the boy's survival was celebrated and the miraculous effectiveness of vaccination assured (Meadows, 1987).

Vaccine therapy was not a new idea to Pasteur, in that Jenner had already used it in the treatment of smallpox. In fact, ancient Chinese culture had already recognized this technique in the 10th century. Alchemists in China administered a smallpox vaccine by inserting slightly infected plugs of cotton into the noses of healthy individuals (James & Thorpe, 1994). But Pasteur extended this method to include new diseases caused by germs he had identified, the logical next step, an increment forward on a path well chosen.

Advance Forward Incrementation

Advance forward incrementations can be found in products that, in essence, are ahead of their time. Consider, for example, the domain of computers. "If anyone was born out of his time it was Babbage; his ideas about the computer were only to find their rightful place by the middle years of the twentieth century" (Cardwell, 1994, p. 420).

In 1834, a British mathematician named Charles Babbage (1791–1871) envisioned the universal computer. Just over a century before the first modern computers were created, Babbage wrote up plans for a powerful analytical engine that had the capacity to perform diverse and complex functions. Babbage's analytical engine was a mechanical computer, complete with a memory store and the ability to operate conditional logic. Input and output were recorded in binary code on familiar punch cards. Although many engineers and scientists in the 19th century knew of Babbage's work, they did not pursue the potential of his ideas. They simply were not ready for them. Thus, Babbage's creative contribution can be viewed as an advanced forward incrementation.

The uniqueness of this example is not merely in its precocity, but also in its original emphasis on pure science. Whereas Babbage had hoped to create a universal computer, the demand for computers in businesses and the military did not require the flexibility of a universal engine, but instead was focused on creating specialized machines. The first computers that emerged around the turn of the twentieth century thus were business machines. One well-known early computer was Hermann Hollerith's business machine, which was designed to aid in calculating the census of 1890. During the wars, great progress was made in creating specialized computers that could make pointed calculations related to battle. Such computers were powerful, but suffered from severely limited memories. For example, in order for such a computer to change its specialty, it had to be completely reprogrammed, a process that required some parts of the computer to be modified and the machine rebuilt.

The concept of a universal computer, however, was not lost among scientists whose orientations were more basic. While computers were being developed to estimate ballistics during World War II, German engineer Konrad Zuse was aiming to create a universal computer, a scientific machine with great flexibility and less specificity. By 1941, he had built his Z3 computer. It was a mechanical universal machine that performed various tasks at the command of computer programs. Around the same time, from 1939 to 1944, International Business Machines (IBM) built a scientific universal computer according to specifications set out by Howard Aitken of Harvard University. This machine was entirely mechanical, stretching 50 feet in length and run by a 4-horsepower motor. Although the first IBM computer did not have the capability of using conditional logic, its creators could be proud of their success in carrying out Babbage's vision of an analytical engine whose purpose was purely scientific. In fact, British mathematician Alan Turing took Babbage's concept of a universal computer to a hypothetical extreme with his invention of what has become known as the Turing machine. This machine exists only in an abstract sense, but it was instrumental in Turing's essay 'On computable numbers, with an application to the *Entscheidungsproblem*', proving that numbers existed whose precise value could not be computed.

Computers have had a revolutionary impact on many areas of modern life, from the first business machines to today's miniature day-planner computers. Charles Babbage's notion of the universal machine came a century before the first machines were built, a prototypical example of an idea ahead of its time. Perhaps the demands of the day dictated the path of progress on computer development, but Babbage's universal analytical engine was bound to return to the fore once the conditions were right. Incidentally, the revolution that has followed the advent of the computer age would not have been a surprise to the one who was there at its conception. Babbage has been quoted as predicting that "as soon as an analytical engine exists, it will necessarily guide the future course of science" (Cardwell, 1994, p. 483).

In the early 19th century, scientists were fascinated with understanding the relationship between magnetism and electricity. Michael Faraday (1791–1867) was a British chemist who had also become intrigued with the relationship between these two physical phenomena. In 1821, Faraday demonstrated that not only can electric current act like a magnet, but magnets can induce electrical current. Faraday's contributions to this line of investigation included the invention of several instruments, such as the electric motor, the electrical transformer, and the dynamo, devices that demonstrated the conversion of magnetism to electricity and of electricity to mechanical motion (Meadows, 1987).

Common theorizing about magnetic forces considered magnetism to be similar to gravity. Just as two bodies are attracted to one another, more or less depending on their distance from each other, two magnetic bodies also possess an amount of attraction. The strength of the attraction between two objects was thought to be a direct function of the distance between them. However, Faraday focused precisely on this distance, proposing that the space between two magnetic or electric objects was not uniform, but rather composed of 'lines of force'. These are the lines observed in the textbook example showing the position of iron filings in the vicinity of two magnets. Faraday's field theory held that magnetic forces acted as fields between objects rather than as forces that lay within the objects themselves (Meadows, 1987).

Acceptance of this view of magnetism would necessitate a fundamental shift in scientific understanding at the time. Unfortunately, the import of Faraday's field theory was not appreciated by his contemporaries. These contemporaries were not satisfied with the theory and awaited a mathematical proof of it. The scientific world waited for this proof until 1873, when Scottish mathematician James Clerk Maxwell (1831–1879) published his *Treatise on Electricity and Magnetism*. This work not only proved the validity of Faraday's field theory, but also confirmed another prediction of Faraday's, which stated that light is actually a kind of electromagnetic wave (Meadows, 1987). Finally, the foresight of a great chemist was vindicated.

Some things that we take for granted today represented advance forward incrementations when they began. In the 1930s, with America mired in the Great Depression, Michael Cullen had a variety of ideas that changed the way that most people do their shopping. He saw a need for lower food prices. At that time, the predominant way that people bought food was through mom-and-pop grocery stores. These stores, however, were both slow (clerks completed a customer's order, resulting in long lines) and somewhat expensive (the quantities of food that the mom-and-pop stores could purchase would not be in large enough quantities to give them discounts). Cullen also noted that there were several other factors in place that made a new idea possible; most notable of these factors was the automobile, which enabled people to be able to travel farther to purchase food (Panati, 1991).

What Cullen did was to advance the field of food purchasing—and, by extension, shopping—drastically: He opened up the first supermarket. Cullen opened up huge stores that stocked enormous quantities of food. He chose low-rent locations that were not in the center of town. Yet they were locations that were suddenly accessible to people because of the automobile. Cullen even set aside paved lots for people to park their cars, an innovation by itself. He invented the art of balancing prices—by making a profit on one product, he could sell another product at cost. These bargains went a long way in attracting a large audience for the new supermarket. He introduced an early version of the shopping cart (an improvement over baskets). In addition, Cullen borrowed (from a different chain of stores) the idea of self-service, which helped cut down on his overhead (Panati, 1991).

Unlike many others who introduced advanced forward incrementation, Cullen's creation was immensely—and immediately—successful. One reason for this success was that while the field was not ready for his creation, the public was. Mom-and-pop

grocers tried to convince the government to make Cullen's price-cutting illegal. They also tried to get newspapers to reject advertisements for supermarkets. But the Great Depression had created a public that was eager to save money in any way possible, and the objections of the other grocers meant little. Indeed, many other supermarkets soon opened up in competition with Cullen's stores (Panati, 1991).

Redirection

Redirections are represented by products that are different in kind, in some way, from products that have existed before. Consider the concept of a 'custom-made' product. Almost any product ordered 'custom made' today will take longer to be produced and will likely be more expensive than off-the-shelf products. For the custom-made price, you are allowed to designate options according to your personal preferences and special needs. Americans living in the 19th century enjoyed the luxury (whether they wanted it or not) of custom-made everything, from kitchen cabinets to household machines. But with the introduction of mass production and assembly lines, the standards in industry were about to change, and with them, the standard of living of the average American citizen.

Eli Whitney in the U.S. as well as others in Europe had introduced mass-production techniques to the arms industry. However, a strong promoter of this industrial movement was the assembly line. The first notable assembly line in use during the 19th century was actually a *disassembly* line. Employees at a Midwest meat-packing company harvested cuts of meat from carcasses hanging from a trolley overhead. When the work at one station was done, the pieces were easily transported to the next point on the line (*History of the organization of work*, 2000).

It was in this context that Henry Ford first launched his constant motion assembly line for the manufacture of the Model T Ford in 1913. Whereas large machines had previously been mass-produced from a single location to which large stocks of components were hauled, Ford's innovation was to install a moving assembly line in which identical parts could be added to the car as it passed down the line (Cardwell, 1994). On the assembly line, one of Ford's Model T's could be produced in 93 minutes. This reduction in time led to a commensurate reduction in cost to the consumer. In particular, this unprecedented cost-effectiveness forced Ford's competitors to join him and play the new production game (*History*, 2000).

While his application of this technique to the automobile industry is no great feat of creative thinking, its impact on the field was strongly redirective. Ford's promotion of the assembly line and of mass-production techniques was an impetus with wide-reaching impact. Whereas others had begun using these methods, Ford's introduction of the assembly line and mass production to the production of cars initiated a change in manufacturing whose impact reached beyond the automobile industry to include a broad range of industrial domains.

The increased use of mass production has influenced the nature of the industrial workforce and the economies of countries through the world, and has led to related techniques such as automatization. For example, factories rely primarily on unskilled or semi-skilled labor while machines have taken over the technical difficulties of the job. As mass-production facilities have become more ambitious, more elaborate supervisory hierarchies have become necessary, and positions for management specialists as well as distributors and salespeople have been created. Although Ford himself did not conceive of this evolution of manufacturing industry, his introduction of the assembly line and the technique of mass production may be considered the crucial step which led to a complete redirection in the field of manufacturing.

An invention that had redirective effects on the field of communication as well as on the wider social and intellectual world is the printing press with movable type. The first printed book dates to 9th century China although the printing press did not appear in Europe until the mid-15th century. Whether these developments were independent or related is of less concern than is the impact of the printed word on Western society at the time of its introduction.

Prior to the invention of movable type and the printing press, books were laboriously copied by hand. This process resulted in books' being rare and precious. When Johannes Gutenberg (1394/99–1467) introduced the concept of movable type in 1448 in Mainz, Germany, a revolution followed. Especially in Europe, the 26-letter alphabet was particularly well suited to the use of movable type. Any book could be reproduced by mixing and matching many multiples of these 26 basic prototypes. (This was not the case in China. Given the large numbers of characters used in the Chinese language, movable type was simply not as practical in China as it turned out to be in Europe (Cardwell, 1994)). It has been estimated that there were more books published in Europe between 1450 and 1500 than had been published in the previous 1,000 years (Yarwood, 1983).

The immediate and lasting effect of Gutenberg's contribution can be summarized as the grand facilitation of the dissemination of ideas. Gutenberg's process affected many aspects of culture, including the dissemination of information about religion. Perhaps the best-known printed volume associated with Gutenberg's name is his version of the Latin Bible. As a case in point, greater public access directly to scripture had revolutionary (or, at least, reformatory) effects on the culture at the time. As Bibles became more available, lay people who were not proficient in Latin wished to read the text in their native language. As Bibles became available in the vernacular, the theology of the Church was also in a parallel transition. Martin Luther (1483–1546) advocated direct access to the scripture by lay people, preaching a new Reformation theology that argued for a direct and personal relationship with God (Rubenstein, 2000).

One of the most popular products of early presses were indulgences, certificates that could be purchased from the Catholic Church in exchange for an absolution of sins. Appropriately, the presses also helped disseminate the famous 95 theses against the sale of indulgences first posted by Martin Luther in 1517. Although such heretical theological ideas would normally have been discussed among Luther's colleagues at the university in Wittenberg, the printed versions of his theses no doubt spread his controversial ideas to others, gaining subtle support for his call for reform.

The invention of the printing press greatly facilitated the spread of ideas in an age of great progress. From the careful molding and casting of movable metal type to the printing and binding of the illustrative Gutenberg Bibles, the printing press has been pivotal in redirecting the history of European culture.

Another redirection occurred in the field of telecommunications with the invention of the telephone in 1876. During the 19th century, people communicated with one another over long distances using the revolutionary telegraph network. However, as the technique became more popular, supply could no longer meet demand. The problem was that only one message could travel over a given wire at any given time, and there simply were not enough wires to carry all the messages that people wanted to send. In 1872, the race was on to find a way to send more than one message simultaneously over the same wire (Flatow, 1992).

Alexander Graham Bell (1847–1922), a teacher who enjoyed tinkering, believed he might have a solution to this problem. Based on his experience with tuning forks, Bell reasoned that it might be possible to send two messages over the same wire. He experimented with tuning-fork transmitters set at different pitches and receivers in the form of reeds set to vibrate at the same frequency as the tuning forks. Once Bell was able to tune the reeds to resonate with the forks, he was able to show that his hypothesis was correct: each fork caused a response in its respective reed, and the two frequencies did not interfere with one another! This was Bell's harmonic telegraph (Flatow, 1992).

Follow-up experiments with his new musical instrument revealed that when the tuning fork was plucked, its tone was transmitted to the receiver. Bell's suspicions that speech could be transmitted similarly over wire were beginning to become plausible. Bell knew that he only needed a transmitter that could convert speech into electrical signals, and the telephone then could become a reality. Meanwhile, Elisha Gray, a successful inventor, had been working on the same telegraph problem, and had obtained similar results. However, Gray was discouraged by businessmen and others who believed that the telephone would never be profitable. Unfortunately, Gray listened to the short-sighted businessmen, and his design of the telephone was set aside (Flatow, 1992).

This is the point where Bell's daring to defy the crowd paid off. Bell's recent breakthrough with the harmonic telegraph propelled him forward. Flatow (1992) writes, "Bell could no longer ignore (opportunity's) knock. He would give up on the telegraph and concentrate on the telephone" (p. 81). Rather than follow the path of normal science, working to perfect the telegraph system, Bell saw an opportunity to make a new contribution, one that could redirect the field toward a communication system based not on dits and dahs, but rather on the authentic human voice.

The rest of the story is well known. Bell patented his telephone in 1876, just hours before Gray's application arrived at the patent office (Flatow, 1992). A few years later, the invention of the microphone by David Hughes (1831–1900) greatly improved the quality of telephone communication and spurred its more widespread use. The first telephone lines were installed shortly thereafter in 1878 in New Haven, Connecticut. By the 1920s telephones had become a standard mode of communication (Williams, 1987). Although the telegraph did not immediately fall into disuse, the addition of the telephone pushed our communicative repertoire in a new direction.

Reconstruction/Redirection

One relatively recent invention that is based on an old idea is the amateur photographer's one-use camera. The small unit is purchased pre-loaded with film, and the film is developed simply by dropping off the whole camera at a camera center. The film processors remove the film from the camera, develop it, and return the prints to the customer. This convenient service saves consumers the trouble of carrying a fragile, expensive camera with them on vacation. It also gives a second chance to those vacationers who forgot their personal cameras at home. And it even allows newlyweds to collect candid shots of their friends and relatives at their wedding reception without having to hire a professional. You think this is a wonderful example of an invention for modern convenience? Think again.

Actually, the one-use camera was invented in 1888, one of the original Kodaks to be marketed to amateurs (Flatow, 1992). The first cameras for personal use were marketed under the motto, "You press the button, we do the rest", a slogan which would certainly apply to many of today's automatic cameras (Flatow, 1992, p. 49). Whereas early photography was characterized by its reliance on cumbersome equipment and dangerous chemicals, this new Kodak was improved for amateur use. Kodak's first personal camera was small and lightweight. Its film was stored in a compact roll

that could be unwound with a key, and the operation of the camera required no special chemicals or setup. Consumers bought a camera that was preloaded with a roll of film of 100 exposures. When the roll was shot, the entire camera was shipped to Kodak for processing and returned fully reloaded and ready to go. By 1900, Eastman Kodak had perfected their concept of the personal-use camera with the introduction of the Brownie camera. The six-exposure Brownie was cheap and extremely simple to use, and caught the imaginations of many Americans (Flatow, 1992).

The reinvention of the original Kodak as a modern one-use, disposable camera is an example of a creative reconstruction/redirection. The Kodak company recognized the potential of this old idea to move the field in a new direction. Whereas the original camera first enabled non-professionals to try their hand at photography, the modern reconstruction contributes in a different way. In the field of modern photography, it is no new concept that amateurs have access to quality cameras, so the creative contribution of the new Kodak original is its convenience. Now there is an option for people who seek an inexpensive, temporary solution to their photography needs. Wouldn't they be surprised to learn that a similar convenience existed at the turn of the century?

Reinitiation

Reinitiations are represented by inventions that take off in an imaginative and wholly new way. An especially creative reinitiation lies in the field of written language. Imagine you were raised to communicate in a language that was only spoken and that had no written form. Suppose you wanted to create a system to represent the sounds and meanings of your native tongue. Where would you start? How many symbols would you need to sufficiently record the multitude of nuances in your oral tradition?

This is exactly the challenge that Cherokee Native American Sequoyah took upon himself in 1809. Sequoyah grew up in Alabama and spent time fighting the white settlers in the early 1800s. In 1809, when Sequoyah was approximately 35 years old, he realized that the Cherokee language might suffer under the effects of the social upheaval taking place at the time, so he decided he would do what he could to preserve the oral tradition with which he and his people had been raised. Unfamiliar with any particular written language, English or otherwise, Sequoyah began to create symbols to represent the sounds of Cherokee.

As he noted the many sounds that were part of his spoken words, he took symbols found in printed materials, including some from the English, Greek, and Arabic alphabets to represent these sounds. In 1821, after 12 years of hard work, Sequoyah had created a list of 85 syllables, the Cherokee syllabary. Based on this syllabary, Sequoyah began to read and write and to teach others to do the same. By 1828, the *Cherokee*

Phoenix became the first Native American newspaper to be published. The fact that this written language was created single-handedly and in such rapid time is a feat of genius and perseverance (Zahoor, 1999).

The contribution of Sequoyah to the written language may be viewed as a reinitiation through the alphabets he employed in creating it. Sequoyah took letters from English, Greek, and Arabic and assigned to them entirely new Cherokee phonemes without any regard to their sounds in the original languages. Although there is no reason to believe that Sequoyah did this with the intent to reject the source alphabets, his new syllabary is effectively a reinitiation of the alphabets he drew on in creating his Cherokee syllabary.

Reinitiative contributions are often bold and daring gestures. Such contributions are often found in inventive performance and artistic production. One prime example can be found in sculpture, with Marcel Duchamp's 1917 Fountain. Duchamp's Dada piece is simply a urinal turned on its back. The very act of entering such a piece in an art show is a statement about art—Duchamp's sculpture made art-making focus on the definition of exactly what art is and what art can be. Duchamp's urinal became a piece of art, and he and his fellow Dada creators set the stage for other modern art that exists, in part, to challenge our ideas of what 'art' encompasses (Hartt, 1993).

Another radical reinitiator is one of Duchamp's friends, the composer John Cage. He often employed unconventional sound materials and spent a period in which his compositional process (and often performance) was determined entirely by chance. The philosophy that led Cage to compose in this unorthodox manner can be considered essentially a rejection of some basic tenets of the Western musical tradition, including the definition of music itself. Cage declared music to be all sound, including the whispers and heartbeats we perceive while silent. Cage's affinity for Eastern philosophy caused him to focus on the importance of awareness in the human experience, and he used his music to foster awareness in his listeners.

An illustration of this point is his piece *4' 33"*. The performance of this piece consists of four minutes and 33 seconds of 'silence', or rather, in Cage's terminology, 'unintentional sound'. In performance, the instrumentalist approaches her instrument, prepares to play, and proceeds to sit, motionless and without sound, for four minutes and 33 seconds. The only pauses are those indicated to signal the change of movement. The music, therefore, is that sound which exists in the environment during that period of time. Cage's statement is that there is music being played around us all the time; we must reject the notion of music as organized melody, harmony and rhythm to include all intentional sound, even the rush of traffic beyond the door and the buzzing of the fluorescent lights above our heads (Cage, 1961; Hamm, 1980).

A reinitiation in science was Lavoisier's invention of a revolutionary new chemistry. This reinitiation had an immediate and lasting impact on the field. In the 18th century, the predominant view of combustion was that of the German chemist Georg Stahl (1660–1734). Stahl's model of combustion was founded on the premise that combustible matter was composed of water and a substance called phlogiston. According to this early view, burning a metal resulted in the loss of phlogiston. Support for this theory was demonstrated by combining the oxide of a metal with a material containing phlogiston (for example, charcoal). This experiment would yield a pure sample of the metal, consistent with the theory that the addition of phlogiston could restore the initial substance (Meadows, 1987).

However, it became clear that chemical reactions of combustion yielded slightly heavier products, despite the claim that the process of combustion required the loss of matter, namely, phlogiston. This apparent paradox led many chemists of the day to attempt to explain the result in the context of phlogiston theory. However, Antoine Lavoisier (1743–1794) was skeptical enough of the vague and sparse evidence for the actual existence of this mysterious substance called phlogiston that he dared to discard the concept altogether. Lavoisier proceeded to explain how combustion could take place in its absence. Based on careful analysis, Lavoisier confirmed through observation that metals actually did gain weight during combustion. This result was attributed to the presence of oxygen during combustion. When a metal burned, oxygen in the air was consumed and was captured in the resulting compound, an oxide (Cohen, 1985).

The key failure of phlogiston theory is its lack of an understanding of the gaseous state. Early chemists did not consider air to play a role in combustion reactions. At most, the air was a mere waste bin for the phlogiston lost in the reaction. Lavoisier discovered that elements could exist in solid, liquid, and gaseous states. It was this realization that became the foundation of Lavoisier's chemical revolution. Lavoisier pointed out that air is composed of many different substances, one of which is oxygen.

In contrast to many reinitiative contributions, Lavoisier's revolution in the field of chemistry made an impact almost immediately. Through the introduction of a standardized chemical language, a new periodical in 1788, a textbook in 1789, and a host of disciples, the new chemistry became respected by all but a few skeptical scientists (notably, Joseph Priestly) within a few decades (Meadows, 1987). This revolution is known as the triumph of the Antiphlogistians (Cohen, 1985).

Lavoisier's opposition to phlogiston theory was not wholly a radical leap from the normal science of his time. There was no dearth of results in studies of combustion that did not fully support phlogiston theory. However, Lavoisier's confident declaration of his rejection of this long-held theory and his suggestion of a plausible alternative view reinitiated the field of chemistry. Especially notable are Lavoisier's subsequent contributions to the field in terms of developing a standardized chemical language and his laying out a textbook describing the new chemistry (Meadows, 1987). In this way, the contribution of a new chemical theory was not merely a next logical step, but a reinitiation from a new starting point that rejected what had been crucial assumptions of the field.

While a reinitiation can certainly be the result of a great and world-class innovator, unemployed advertising men can have reinitiative ideas as well, if on a different scale. In 1975, Gary Dahl listened to his friends talk about how expensive and time-consuming it was to own a pet. Dogs and cats are not a simple responsibility, and even lower-maintenance pets like fish or birds require some time, effort, and money. It was at this point that Dahl had the idea that a pet did not have to be a living creature. And, in fact, he proceeded along in that direction to create an entirely new idea: a pet rock (Panati, 1991).

While primarily (of course) a parody of pet owners, the pet rock does represent an entirely new way of looking at what it means to own a pet. Dahl took the idea and expanded on it. He wrote an owner's manual (which included instructions on how to train the pet rock to roll over or play dead). He packaged it in a box with air holes, and he also sold pet food, in the form of rock salt (Panati, 1991). The pet rock, as many people know, became an instant success; it is one of the few fads that is still remembered years later. Dahl may not have even been intending to make a creative statement, but he did. He conceptualized the field of pet ownership as being in a different place, and then he showed a new way for that field to go. And, indeed, perhaps the current feelings of possessiveness and ownership that some people feel for their cars is not that far removed from the odd tenderness one might feel for a pet rock.

Integration

An invention that is an integration is one that combines ideas from two distinct domains of invention. An example of an historical integration is the complex evolution of the Japanese written language. Originally, Japanese was strictly a spoken language, but around AD500, Japanese speakers began to use Chinese kanji (pictographs) to represent Japanese words. While written Japanese looked identical to the Chinese, any one character was pronounced in two different ways, one of which approximated the Chinese pronunciation, and the other of which corresponded to the Japanese word bearing the same meaning. Since 1945, close to 2,000 kanji have been designated as Kanji for Daily

Use in the Japanese language. These are the kanji that children learn in school today (Miyagawa, 2000).

By AD1000, however, the Japanese needed a supplemental system to represent Japanese words that had no Chinese counterpart. To meet this purpose, syllabaries called kana were derived from Chinese characters. Rather than representing the meaning of a whole word, a kana represents a syllable or sound. There are two systems of kana in modern Japanese. The first, called hiragana, originated with Buddhist priests. While they were translating Chinese works into Japanese, they began inserting syllables next to characters to designate a particular Japanese inflection or alternate meaning from the Chinese (Vogler, 1998). The second system of kana, katakana, is primarily used to represent foreign words.

The original use of Chinese characters to represent Japanese words may be viewed as a redefinition of sorts. That is, the meaning remained the same while the label, or way of reaching that meaning, changed. However, the entire Japanese language is best seen as an integration (and subsequent forward incrementation) of symbol systems. Specifically, the introduction of kana to modify kanji is a particularly good example of melding two representational systems into one entirely new system. The meanings of words with kanji and kana are not fully decipherable based solely on the kanji, evidence that a true integration has taken place.

An example of integration in science is Isaac Newton's (1642–1727) formulation of the universal law of gravity. In essence, this discovery was a synthesis of ideas proposed by thinkers such as Ptolemy, Copernicus, Kepler, Galileo, and Hooke. Newton succeeded in explaining how a singular concept of gravity could explain a variety of observations made by scientists past and contemporary. While Copernicus had dared to oppose the Ptolemaic notion that planets encircle the sun and Kepler had demonstrated that the orbits were elliptical, no one had been able to explain why such elliptical orbits would emerge. Newton discerned the mathematical rule by which elliptical orbits would result, thus explaining the relationship between planets and the sun. In his 1687 'The Mathematical Principles of Natural Philosophy', Newton described basic principles that could mathematically account for the diverse predictions made by Copernicus and Kepler.

This contribution is a grand synthesis. Not only did Newton bring together a variety of ideas to explain planetary motion. He also recognized the relationship of this phenomenon to another and essentially integrated the two problems by offering a single solution to both. Newton showed that his principles governing the movement of bodies in the sky could also account for a second, formerly unrelated, problem: that of the motion of objects on the earth. Essentially, Newton's universal law of gravity declared that matter attracts matter. That is, the sun attracts the planets just as the

Earth's gravity pulls on objects near its surface. By achieving this insight, Newton explained two major physical questions of his time with one universal law, an integrative contribution that has remained the basis of physics ever since.

Conclusion

When we think about inventions, we tend to think about their differing primarily in terms of their novelty, their usefulness, or perhaps their profitability. The propulsion model of creative contributions may lead us to think about them in terms of the type of innovation they represent. How is the propulsion model useful in understanding invention?

First, at a theoretical level, the model may be useful in understanding how inventions differ from one another, not just in their level of creativity, but in the type of creativity they demonstrate in the first place. Using the propulsion model, it may be possible even to trace historically how inventions of different kinds have influenced the development of various kinds of technologies.

Second, it may help us understand why certain types of inventions are less readily accepted than others. Advance forward incrementations, for example, may be beyond people's ready grasp. Reinitiations as well may be difficult for people to understand or accept. Forward incrementations may have the best shot for a ready market, on average. But they have to move far enough. For example, new editions of existing software that seem to be only minor variants and small improvements do not tend to sell well, unless manufacturers rig other software so that it will work only with the newer versions (and thereby risk consumer resentment and possibly government intervention!). At the same time, software that is beyond people's ready capabilities for use may be equally unappealing, simply because it is too complex for most people to handle.

Third, at a practical level, the model may be of use to organizations that specialize in invention. In order to maximize their profitability, they may wish purposely to distribute the kinds of inventions they produce across the various types in order to maximize both short-term and long-term investments in inventions.

In conclusion, we offer the propulsion model of creative contributions as a means for understanding different kinds of creative contributions, in general, and inventions, in particular. We believe the model may help all of us understand not only what the kinds of inventions are, but why some inventions are more or less successful than others, often independent of their quality or level of novelty.

Acknowledgments

Preparation of this chapter was supported by Grant REC–9979843 from the National Science Foundation and by a government grant under the Javits Act Program (Grant No. R206R000001) as administered by the Office of Educational Research and Improvement, U.S. Department of Education. Grantees undertaking such projects are encouraged to express freely their professional judgment. This chapter, therefore, does not necessarily represent the positions or the policies of the U.S. government, and no official endorsement should be inferred.

References

Amabile, T. M. (1996). *Creativity in context*. Boulder, CO: Westview.

Barron, F. (1988). Putting creativity to work. In: R. J. Sternberg (Ed.), *The Nature of Creativity* (pp. 76–98). New York: Cambridge University Press.

Cage, J. (1961). *Silence*. Middletown, CT: Wesleyan University Press.

Cardwell, D. (1994). *The Fontana history of science*. Fontana Press: London.

Clark, R. W. (1985). *Works of man*. New York: Viking Press.

Cohen, I. B. (1985). *Revolution in science*. Cambridge, MA: Belknap/Harvard.

Csikszentmihalyi, M. (1996). *Creativity*. New York: HarperCollins.

Csikszentmihalyi, M. (1999). Implications of a systems perspective for the study of creativity. In: R. J. Sternberg (Ed.), *Handbook of Creativity* (pp. 313–335). New York: Cambridge University Press.

Feist, G. J. (1999). The influence of personality on artistic and scientific creativity. In: R. J. Sternberg (Ed.), *Handbook of Creativity* (pp. 273–296). New York: Cambridge University Press.

Feldman, D. H. (1999). The development of creativity. In: R. J. Sternberg (Ed.), *Handbook of Creativity* (pp. 169–186). New York: Cambridge University Press.

Feldman, D. H., Csikszentmihalyi, M. & Gardner, H. (1994). *Changing the world: A framework for the study of creativity*. Westport, CT: Praeger.

Finke, R. A., Ward, T. B. & Smith, S. M. (1992). *Creative cognition: Theory, research, and applications*. Cambridge, MA: MIT Press.

Flatow, I. (1992). *They all laughed . . .* New York: Harper Collins.

Hamm, C. (1980). John Cage. In: *The new Grove dictionary of music and musicians* (Vol. 3, pp. 597–603). London: Macmillan.

Hartt, F. (1993). *Art: A history of painting, sculpture, architecture* (4th ed.). Englewood Cliffs, NJ: Prentice Hall.

Hay, P. (1988). *The book of business anecdotes*. New York: Facts on File Publications.

Hennessey, B. A. & Amabile, T. M. (1988). The conditions of creativity. In: R. J. Sternberg (Ed.), *The Nature of Creativity* (pp. 11–38). New York: Cambridge University Press.

History of the organization of work (2000). In: *Encyclopaedia Britannica* (on-line). Available: http://www.britannica.com/bcom/eb/article/printable/1/0.5722.115711.00.html

James, P. & Thorpe, N. (1994). *Ancient inventions*. New York: Ballantine Books.

Kuhn, T. S. (1970). *The structure of scientific revolutions* (2nd ed.). Chicago: University of Chicago Press.

Lubart, T. I. (1994). Creativity. In: R. J. Sternberg (Ed.), *Thinking and Problem Solving* (pp. 290–332). San Diego, CA: Academic Press.

Meadows, J. (Ed.) (1987). *The history of scientific discovery*. Phaidon: Oxford.

Miyagawa, S. (2000). *The Japanese Language* (on-line). Available: http://www-japan.mit.edu/articles/Japanese Language.html

Ochse, R. (1990). *Before the gates of excellence*. New York: Cambridge University Press.

Panati, C. (1991). *Panati's parade of fads, follies, and manias*. New York: HarperPerennial.

Rubenson, D. L. & Runco, M. A. (1992). The psychoeconomic approach to creativity. *New Ideas in Psychology*, **10**, 131–147.

Rubenstein, G. (2000). *Gutenberg and the historical movement in Western Europe*. http://www.digitalcentury.com/encyclo/update/print.html

Simonton, D. K. (1984). *Genius, creativity, and leadership*. Cambridge, MA: Harvard University Press.

Simonton, D. K. (1994). *Greatness: Who makes history and why?* New York: Guilford.

Sternberg, R. J. (Ed.) (1988). *The nature of creativity: Contemporary psychological perspectives*. New York: Cambridge University Press.

Sternberg, R. J. (1999a). A dialectical basis for understanding the study of cognition. In: R. J. Sternberg (Ed.), *The Nature of Cognition* (pp. 51–78). Cambridge, MA: MIT Press.

Sternberg, R. J. (Ed.) (1999b). *Handbook of creativity*. New York: Cambridge University Press.

Sternberg, R. J. (1999c). A propulsion model of types of creative contributions. *Review of General Psychology*, **3**, 83–100.

Sternberg, R. J. & Davidson, J. E. (Eds) (1994). *The nature of insight*. Cambridge, MA: MIT Press.

Sternberg, R. J., Kaufman, J. C. & Pretz, J. E. (2001). The propulsion model of creative contributions applied to the arts and letters. *Journal of Creative Behavior*, **35**, 75–101.

Sternberg, R. J., Kaufman, J. C. & Pretz, J. E. (2002). *The creativity conundrum*. Philadelphia, PA: Psychology Press.

Sternberg, R. J. & Lubart, T. I. (1995). *Defying the crowd: Cultivating creativity in a culture of conformity*. New York: Free Press.

Sternberg, R. J. & Lubart, T. I. (1996). Investing in creativity. *American Psychologist*, **51**, 677–688.

Vogler, D. (1998). *An overview of the history of the Japanese language* (on-line). Available: http://humanities.byu.edu/classes/ling450ch/reports/japanese.htm

Weisberg, R. W. (1986). *Creativity, genius and other myths*. New York: Freeman.

Williams, T. I. (1987). *The history of invention*. New York: Facts on File.

Yarwood, D. (1983). *Five hundred years of technology in the home*. London: B. T. Batsford.

Zahoor, A. (1999). *Sequoyah and Cherokee syllabary* (on-line). Available: http://salam.muslimsonline.com/~azahoor/sequoyah1.htm.

The International Handbook on Innovation
Edited by Larisa V. Shavinina

Problem Generation and Innovation

Robert Root-Bernstein

Department of Physiology, Michigan State University, USA

Abstract: Most of the literature on innovation concerns fostering better solutions to existing problems. Many innovators in the sciences and engineering argue, however, that problem generation is far more critical to innovation than problem solution, involving not just a thorough grasp of what is known (epistemology), but of what is not known (nepistemology). The proper definition of a problem gets an innovator more than half way to its solution; poorly posed questions divert energy, resources, and ideas. This chapter explores how problem definition and evaluation act as catalysts for insight and examines strategies used by successful innovators to generate productive problems.

Keywords: Nepistemology; Ignorance; Problem generation; Problem evaluation; Types of innovation.

Introduction: The Nature of Ignorance

Most people believe that creativity and innovation, especially in the sciences and technology, are forms of effective problem-solving. I, however, believe, that creativity and innovation consist of effective problem-raising. People are creative only when they need to do something that cannot yet be done. Identifying, structuring, and evaluating problems in ways that allow their solution are therefore as important—arguably *more* important—than finding solutions. We must know what we do not know before we can effectively solve any problem.

One could, of course, argue that all of the important questions have already been asked, that ignorance is finite. One might believe that the number of questions that can be asked is limited so that ignorance decreases in direct correspondence to the increase in the volume of knowledge. One might, however, assert the opposite: that ignorance is infinite. The greater the volume of knowledge we accumulate, the greater the sphere of ignorance we can recognize around us. Every question breeds more questions without end. I favor the latter view. Every time someone in history has proclaimed that a field such as physics, medicine, philosophy or art has finally reached the end of its possible progress, a revolution has already been under way in that field, opening up unexpected vistas for exploration (Root-Bernstein, 1989, p. 45). For example, even as Lord Kelvin was preaching that physics was a closed book

with no new surprises to yield, Einstein was inventing relativity theory and Planck was creating the quantum revolution. Indeed, Einstein wrote that the very idea that physics could ever become a closed field was repugnant to a physicist in the twentieth century:

> It would frighten him to think that the great adventure of research could be so soon finished, and an unexciting if infallible picture of the universe established for all time (Einstein et al., 1938, p. 58).

On the contrary, Einstein argued, in the struggle for every new solution,

> new and deeper problems have been created. Our knowledge is now wider and more profound than that of the physicist of the nineteenth century, but so are our doubts and difficulties (Einstein et al., 1938, p. 126).

This is the situation in every fecund field. New explanations and new techniques always create unforeseen sets of new problems. In consequence, what drives progress is not the search for ultimate knowledge, but the search for ever more wondrous questions.

Creative people in every discipline recognize the importance of generating or discovering new problems. That which we cannot yet do impels us to invention. "Recognizing the need is the primary consideration for design", said Charles Eames (Anon, 2000, p. 4). He and his wife Ray were the first designers to utilize

170

molded plywood, fiberglass, wire-mesh, and cast aluminum to make practical everyday objects such as furniture. Novelist Dick Francis says that whether you are trying to solve a mystery or write book, "You have to ask the right questions" (Francis, 1992, p. 288). Filmmaker Godfrey Reggio agrees. Commenting on his film 'Koyaanisquatsi' he cautioned viewers that it is not a solution to the world's problems, but an unfolding question:

> You know, the question is really more important than the answer. I can frame the question, but I don't know the answer (Kostelanetz et al., 1997, p. 251).

In fact, the more important the question, the more important the ideas to which it will give rise. Sir Peter Medawar, whose research made possible medical transplant technologies, has written that scientists interested in important discoveries require important problems. "Dull or piffling problems", he says, "yield dull or piffling answers" (Medawar, 1979, p. 13). And Werner Heisenberg, of uncertainty principle fame, stated another oft-repeated principle known to all scientists: "Asking the right question is frequently more than halfway to the solution of the problem" (Heisenberg, 1958, p. 35). One of the keys to creativity according to all of these people lies in the study of problem recognition and generation. It is not what we know, but what we do not know that drives inquiry in every discipline. Learning how to question deeply and well is therefore, as Socrates made clear 2,500 years ago, one of the most important keys for unlocking hidden knowledge and one that opens doors in every field of endeavor.

Nepistemology: The Types and Origins of Ignorance

The philosophical study of what we do not know falls under the heading of *nepistemology*. Epistemology, as many people will be aware, is the study of how knowledge comes into being. Its complement, nepistemology, is the study of how ignorance becomes manifest. Despite the extraordinary importance of nepistemology, the field has little literature and even fewer practitioners. For some unfathomable reason, the existence of problems is taken for granted and their nature and origins generally ignored. In a world in which education, business, and government are focused on problem-*solving*, it is worth remembering that problems must be invented just as surely as their solutions. Because good solutions can be generated only to well-defined problems, problem recognition and problem generation is arguably a critical step. Failure to ask questions results in the stagnation of knowledge. Asking the wrong question yields irrelevant information. Asking a poorly framed question yields confusion. Asking trivial questions yields trivial results. However, learning to recognize what we do not

know, developing the skills to transform ignorance into well-defined questions, and being able to identify the important problems among them, sparks innovation.

Few programs in history have been designed specifically to teach problem generation. Socrates, of course, asked questions but it is not clear that he taught his students how to ask their own. The Bauhaus, a design movement based in Germany during the 1920s and early 1930s, certainly incorporated problem generation as a fundamental part of its Foundation Course. When assigned to design a table or paint a painting students were challenged to begin not by seeking answers, but by seeking questions: "what is a table? . . . what is a painting?" (Lionni, 1997, p. 166) Only by questioning the assumptions they brought to their work, they were taught, was it possible to produce innovative answers. Similarly, English sculptors such as Henry Moore, Jacob Epstein, and Barbara Hepworth reinvented sculpture by "asking themselves the question: 'What do we mean by sculpture?' " (Harrison, 1981, p. 217). By returning to so-called primitive techniques of direct carving and structural rather than realistic forms, they discovered new questions about the nature of art and thereby new answers. Learning to ask the right question always opens up new possibilities (Browne et al., 1986).

One of the very few contemporary questioning programs is 'The Curriculum on Medical Ignorance' founded by surgeons Marlys and Charles Witte and philosopher Anne Kerwin at the University of Arizona Medical School in Tucson, Arizona. The purpose of the program is to teach medical students how to recognize their own ignorance and that of other doctors in order to better define what still remains mysterious and take a first step towards enlightenment. The Wittes and Kerwin have identified six basic types of ignorance. (Witte et al., 1988, 1989). First, there are things we know we do not know (known unknowns, or explicit ignorance). A second type of ignorance consists of things we do not know we do not know (unknown unknowns, or hidden ignorance). The next kind of ignorance is that consisting of things we think we know but do not (misknowns, or ignorance masquerading as knowledge). Fourth, ignorance may be found among the things we think we do not know, but we really do (unknown knowns, or knowledge masquerading as ignorance). Next, some ignorance persists due to social conventions against asking certain types of questions in the first place (taboos, or off-limits ignorance). Ignorance may also persist due to refusal to look at some types of answers to perfectly legitimate questions (blinkers, or persistent ignorance).

Two essential points must be understood about this typology of ignorance. First, each distinct type of ignorance requires a different set of techniques in order to reveal itself clearly. Second, problem recognition and invention are active processes. Proficient questioners are those who can draw upon a wide range of problem-finding tools and apply them appropriately to diverse questioning contexts. If nothing else, the Curriculum on Medical Ignorance reveals the poverty of simply asking 'why' or 'how' or 'what' for every observation one makes.

Explicit ignorance, the things we know we do not know, consists of questions and problems that experts in a field formally recognize because current explorations repeatedly fail to provide useful answers. Explicit problems from diverse fields include questions such as the following: How does the immune system differentiate 'self' from 'nonself?' Should animals have the same sorts of rights accorded to human beings? Does art progress or merely change to reflect the values of each new generation?

Explicit ignorance can be discovered using a wide range of proven approaches. One is to question the assumptions underlying the question. Some questions turn out to be unanswerable because they contain implicit concepts that are not valid. For example, what do we mean by 'progress' and how can we apply that term to art? What is a 'right?' Do 'rights' necessarily entail responsibilities? Can an animal that cannot understand or implement its obligations under the law be accorded 'rights' or must such rights reside in a legal guardian, as is the case for mentally incompetent human beings? Any problem, like these, that exists for any period of time demonstrates that the experts not only do not have the answer, but do not have the question. Persistence of questions implies a lack of proper question formulation. Questioning the question itself can often yield insight.

A second way to identify explicit ignorance is to focus on paradoxes. Paradoxes exist only where two or more well-formulated sets of knowledge come into conflict and thus represent the locus of the unknown. Such loci should be embraced, for where there are paradoxes, there is, as the physicist Neils Bohr repeatedly pointed out, hope of progress (Moore, 1966, p. 196).

A third method for recognizing explicit ignorance is to learn how to perceive the sublimity of the mundane. 'Sublimity of the mundane' refers to the inordinate beauty manifested in everyday things. Unfortunately, as practitioners of every discipline are all too well aware, that which is commonplace is easily overlooked or, in the terms of artist Jasper Johns, is recognized without being seen (Root-Bernstein et al., 1999, p. 32). Whether a person sets out to perceive what makes a flag flag-like, as Johns did in his art, or to ask why a banana turns brown but an orange does not, as biochemist Albert Szent-Gyorgyi did, such everyday problems, precisely because they are so commonplace, will also turn out to be fundamental. Johns's art has provided new insights into how we see, while Szent-Gyorgyi's researches on banana browning resulted in the discovery of vitamin C, for which he earned a Nobel Prize.

A final method for *revealing* explicit ignorance is to err. As Francis Bacon once wrote, truth comes out of error more rapidly than out of confusion. A clear mistake does far more good to any investigation by identifying the specific nature of the problem being addressed than any other method. Hence the trite (but true) aphorism that we learn best from our mistakes. People who never err not only never succeed, but have no mistakes (and hence generate no problems) from which to learn.

Hidden ignorance, the things we do not know we do not know, is perhaps the most difficult form of ignorance to discover precisely because it is not explicit. One of the most important characteristics of hidden ignorance stems from the fact that human beings tend to account for any phenomenon in any field in terms of what they already know. Thus, many forms of ignorance remain hidden simply because people are too proud to say the simple sentence, "I do not know". Instead, we tend to make up stories to hide our ignorance. For example, prior to the discovery of electricity, every culture attributed lightning to the wrath of the 'gods' rather than simply admitting ignorance. Similarly, prior to the discovery of germs, infectious diseases were attributed to bad air, bad water, and bad habits. No medical expert in any culture seems to have had the courage to simply admit that they did not really know how diseases spread.

To discover hidden ignorance requires very different methods than to address explicit ignorance. Hidden ignorance must be surprised. One way to surprise ignorance according to Sir Peter Medawar is to design research to challenge expectations. (Medawar, 1979, passim) His advice is predicated on the philosophy of science espoused by Sir Karl Popper, who maintains that theories can never be proven (because the observation you have not yet made may invalidate your position), but theories can always be *disproven*. The object of any research project in any discipline should therefore be to find the evidence that will disprove one's expectations. Physicians, for example, often call for tests to 'rule out' a possible but unlikely diagnosis. The specific way in which expectation is disproven will then reveal the nature of one's unsuspected ignorance.

Another method related to challenging expectation is to look for information where no reasonable person would look. There are many ways to implement this strategy but each entails to a greater or lesser degree the risk of appearing to be foolish or even crazy. The most acceptable method is to run many highly varied and even outrageous controls with any experiment. Winston Brill, a member of the National Academy of Sciences (USA), for example, recounts that his most important breakthrough occurred when he was designing a pesticide based on a particular theory of how its structure should look. He synthesized several versions of his compound in what he predicted to be active and inactive forms and also took a few chemicals off his shelves at random to use as controls. None of the chemicals he had designed as a pesticide worked well, but one of his random choices was a stunning success, representing a new class of pesticides that no one could have predicted in advance (personal communication). Only by doing the unexpected can the unforeseen reveal surprises.

Another approach to revealing unknown unknowns is purposefully to turn things on their heads to yield a new perspective. The artist Wassily Kandinsky, for example, unthinkingly turned one of his early representational paintings upside-down and left it for several weeks. Re-entering his studio one evening, having forgotten about the painting, he suddenly became aware of it, illuminated by the setting sun, but did not recognize it or its content. All he saw was form and color. Suddenly he realized that it might be possible to create totally non-representational art, something that even the abstract painters of that time had not yet considered. By viewing his own work literally upside down he discovered that there was a whole realm of art that no one had realized was there, let alone begun to explore (Herbert, 1964, 32ff)

Playing contradictions (sometimes known as playing the devil's advocate) is another fecund source of unknown unknowns. Nothing reveals how little we know than to take a proposition such as 'the world is spherical' and assert instead that 'the world is flat'. As the novelist H. G. Wells pointed out a century ago, very few are the supposedly educated members of society who are able to produce evidence or experiments (especially ones that do not resort to the data gathered from space craft) that will clearly distinguish one of these possibilities from the other. (Wells, 1975, pp. 32–25) That which is obvious often becomes inobvious when its opposite is asserted, thereby raising questions and possibilities that would never otherwise come to light.

Finally, one can reveal hidden ignorance by drawing out the most absurd implications of any idea. Many children excel at this strategy, taking anything an adult says and applying it literally—too literally!—to any situation they meet. Much satirical literature ('Candide', Voltaire's send-up of rosy philosophical optimism as espoused by Leibnitz and Pope, is a good example) uses this strategy to reveal the limitations inherent in popular philosophies. What extrapolation of this sort reveals are the boundaries beyond which an otherwise quite reasonable proposition or method fails. The point at which it fails is often the point at which our ignorance begins.

Ignorance masquerading as knowledge, or misknowns, presents yet another class of problems. A classic example from medicine was the belief, rampant among physicians for most of the twentieth century, that stomach ulcers are due to mental stress or eating spicy foods or alcohol and can be treated only by a bland diet and relaxation techniques. Stomach ulcers are, in fact, caused by the bacterium *Helicobacter pylori* and can be cured by a proper regimen of antibiotics. Stress, alcohol and spicy foods exacerbate the consequences of the bacterial infection. Thus, secondary factors were confused with primary causes.

Several methods exist for discovering that the king is wearing no clothes. One is to doubt all correlations. Everyone is taught that correlation is not causation, but in practice this warning is all too often ignored. In the case of ulcers, because spicy foods and alcohol *exacerbated* the symptoms, they appeared to be so highly correlated with the onset of ulcers that they were mistakenly identified as causative agents. Evidence that many people eat spicy foods and drink lots of alcohol without getting ulcers was ignored.

A second method for identifying misknowns is to question habit. For centuries, painters habitually propped their canvases up on easels in order to work on a surface perpendicular to their view. It took the courage of Jackson Pollock to take the canvas off of the easel and place it on the ground to experiment with the artistic possibilities of working in a new way.

Two additional methods also exist for identifying misknowns. Perhaps the most important is to doubt most those results or findings that accord best with our preconceptions. Those results or observations that best fit our preconceptions and expectations are the least interesting and yield the fewest questions and insights. Ignore them. They do not yield problems. The things that are the most interesting in terms of revealing ignorance are those that are the most disturbing or which conflict most clearly with strongly held beliefs or practices. Therefore, pay attention to the heretics, revolutionaries, and people stepping to their own drummer. Many of their ideas will be wrong, but many of the problems they reveal will be valid.

A related method for identifying misknowns is to collect anomalies. Anomalies are phenomena and observations that do not fit within the theoretical or explanatory structure of a field. They are the things that experts ignore as inconvenient nuisances. As philosopher Thomas Kuhn pointed out in his book *The Structure of Scientific Revolution* (1959) every great discovery has resulted from an individual paying attention to the anomalous results that most practitioners of a field have refused to countenance or have actively ignored because anomalies generate problems.

Finally, try doing the impossible. There are two kinds of impossible things: those that nature will not allow us to do (e.g. perpetual motion machines) and those that our predecessors have found beyond their means. The history of medicine is full of procedures that were declared to be 'impossible', such as blood transfusions and transplants, that are now part and parcel of everyday clinical practice. Indeed, the inventor of the instant camera and its Polaroid film, Edwin Land, advised innovators not to "do anything that someone else can do. Don't undertake a project unless it is manifestly important and nearly impossible" (Root-Bernstein, 1989, p. 415).

Next we must deal with things we do not know that we know, or hidden knowledge. There is a story that one of the great eighteenth-century mathematicians sat down one day to try to calculate the perfect dimensions of a container for storing and shipping alcoholic beverages such as beer or wine. After taking into account the available materials, limitations imposed by the fact that the containers had to be carried by men from one place to another, and so forth, he generated his solution—only to find that he had described the wooden barrels already in use. Similarly, a recent study of medical treatments has found that many of the most 'innovative' are based on folk remedies that have been around for millennia. Just for example, honey pastes have been used since at least Babylonian times to treat severe wounds and burns and have recently been introduced in many hospitals to treat infections and ulcerations that are beyond the scope of current antiseptic, antibiotic, and surgical treatments. As both these cases show, the basic problem underlying hidden knowledge is that it exists either among people who are not considered to be valid sources of knowledge by disciplinary 'experts' or resides in historical documents bypassed by subsequent developments in the name of 'progress' (Root-Bernstein et al., 1997, Chapters 3 and 12).

The causes of hidden knowledge suggest solutions. The most obvious is to go outside the boundaries of acceptable disciplinary sources for information. Charles Darwin broke all of the standards of science in his day (not to mention social conventions) by writing to farmers, pigeon fanciers, horse breeders, and other non-scientists in order to discover the accumulated wisdom about artificial selection that existed outside of the academic community. The physicians who rediscovered the modern therapeutic uses of honey paid attention to non-traditional practitioners of their art such as family folk medical traditions or local traditional healers. Similarly, many innovators, such as the cubist artists Brach and Picasso and the modern composers Richard Reich and David Glass, having mastered formal Western techniques during their early training, then turned to non-academic sources of inspiration such as traditional African and Hindu arts.

Albert Szent-Gyorgyi, the Nobel laureate who discovered vitamin C, has suggested that another general way to reveal problems of unknown knowns is to go back and re-do classic experiments from 50 or 100 years ago using modern techniques (Root-Bernstein, 1989, p. 412). These re-creations, he says, almost always yield novel insights because the investigator brings to them different problems and theories that change their interpretation. Artists such as Henry Moore have similarly used ancient techniques such as direct carving, invented by Egyptian and Greek

sculptors, to revolutionize modern art by applying it to new materials.

Finally we reach persistent unknowns (taboos) and unacceptable or blinkered knowns. Both classes of ignorance are caused by the refusal to ask certain types of questions or to entertain certain types of answers. These classes of ignorance often result from social taboos or customs and sometimes from economic or political considerations as well. For example, many women living in underdeveloped nations (especially those dominated by Muslim sects) may not ask "How can I control my fertility?" Artists in the United States have met similar constraints on their ability to ask questions such as whether man has been formed in God's image or vice versa. And biomedical researchers studying AIDS have found it so unacceptable to ask whether the human immunodeficiency virus explains everything about AIDS that many have lost their funding and positions. The characteristic of taboo questions is that one risks one's career or even one's life by asking them.

Certain types of answers may also be avoided just as adamantly as certain kinds of questions. For example, alternative medical therapies are used by more than one third of European and American adults, but they are rarely subjected to the kind of double-blind controlled clinical trials that are required of pharmaceutical drugs. Both taboos and blinkered knowns help explain the situation. In the first place, it is not in the interest of purveyors of alternative medicines to have to meet the standards of either efficacy or purity set for pharmaceutical agents. In no case would meeting such standards increase their market share or profit margin and in many cases the standards might eliminate their product from the market. Thus, it is in the interest of these purveyors to foster the persistence of ignorance about their products. However, it is too expensive for Food and Drug Administrations or pharmaceutical companies to seek answers to issues of efficacy and safety. The only motive that these institutions have to test alternative medicines is to accredit them as prescription drugs. It costs, however, approximately one quarter to one half a billion U.S. dollars to obtain Food and Drug Administration approval for a new drug. In order to make such an investment worthwhile, a pharmaceutical company must be able to obtain patents on the drug and its manufacture so that other companies cannot take advantage of the work they have put into proving safety and efficacy. Unfortunately, alternative medicines are, by their nature, non-novel agents already in the public domain, and hence unpatentable. Thus, there is no business incentive to find answers to the problems posed by alternative medicine usage (Root-Bernstein, 1995, passim). We therefore persist in our ignorance about such therapies and will do so until other incentives make it worthwhile for us to seek answers.

Additional examples of blinkered answers can be found in various approaches to AIDS adopted around the world. In Africa, for example, it is against the social mores of many men to use condoms, although they know that condoms are the most effective method of protecting themselves against AIDS. In some Muslim countries, AIDS-prevention counselors are not allowed to mention the fact that AIDS is spread most commonly by homosexual men and female prostitutes because neither group may be mentioned in conversation or print. Thus, programs that have proven to be very effective in controlling AIDS in some countries cannot be used in others because of such taboos.

Because the roots of taboo questions and blinkered answers are social, the methods for addressing it are also social—or, more accurately, anti-social. One of the most noteworthy aspects of many creative acts, whether in the sciences, humanities, social sciences or the arts, is that the people who carry them out are labeled heretics, provocateurs, misfits and worse. The greatest questioner of all time, Socrates, was put to death for the anti-social implications of his questions. Galileo was charged with heresy by the Catholic Church for daring to question the Ptolemaic view of the universe that underpinned Church doctrine. Darwin became a pariah among fundamentalists of many religions for questioning whether God had indeed created man in His image. Margaret Sanger and Marie Stopes, the revolutionaries who gave Western women contraceptive knowledge, each went to jail more than once. So did the many suffragettes who worked to give women the vote, simply for asking why women should not be able to do the things men do. One must have the courage of such people in order to break the taboos that prevent most of us from asking certain types of questions or facing the consequences of certain types of answers. The more dangerous the questions are to people in power, the greater the courage needed to ask them.

Problem Evaluation and Classification

Now, once problems have been recognized or generated, they must be evaluated. Just as there are different kinds of ignorance, they yield a great diversity of kinds of problems. Problems can be classified according to *type*, *class*, and *order*. Type refers to the technique necessary to *solve* a problem. Class refers to the *degree* to which the problem is solvable. And *order* refers to the placement of a problem within the context of the universe of other related problems. While each of these concepts may initially sound very abstract and academic, each is, in fact, extremely useful in practice. Classifying problems allows each to be linked to an appropriate method for achieving a solution and allows links between problems to be explored explicitly.

Just as it is true that not all forms of ignorance can be recognized or discovered in the same way, not all problems can be solved using the same methods. There are, in fact, at least ten types of problems, each distinguished by the manner in which it must be addressed in order to achieve a solution. The ten basic problem types are: (1) problems of definition; (2) problems of theory; (3) problems of data; (4) problems of technique; (5) problems of evaluation; (6) problems of integration; (7) problems of extension; (8) problems of comparison; (9) problems of application; and (10) artifactual problems (Root-Bernstein, 1982, *passim*; Root-Bernstein, 1989, p. 61). Readers will immediately recognize a significant overlap in the terminology and concepts embedded in this typology of problems with the typology of innovation types proposed by Sternberg et al. in this volume. The type of problem an innovator chooses to address therefore determines the type of innovation she or he produces.

Problems of definition concern the purity of the language and its meaning as used by any given discipline. Representative issues include 'what is velocity' or 'what is a legal right' or 'what is beauty'. Such questions cannot be addressed by experiment or observation or any form of research. Such definitions may be determined axiomatically in the same way that we accept definitions of basic concepts such as the numbers one, two, three, etc. or when we state that velocity is distance traveled divided by the time of travel. Definitions may also be addressed by means of a process, such as the legal system that determines whether something is a right.

Problems of theory or explanation involve any attempt to find or make sense of the relationships in existing data or observations that are currently incomprehensible or anomalous. Theory problems require the recognition or invention of a pattern that makes coherent sense of available information.

Problems of data involve the collection of experiences or information relevant to addressing some particular kind of ignorance. Relevant techniques include observation, experiment, and any kind of playing with materials that yields novel data.

Technique problems are those that concern the manner in which novel data, observations or effects are to be achieved. Such problems generally require the invention of instruments, methods of analysis or display, or techniques that allow new phenomena to be observed or created.

Evaluation problems arise when it is necessary to determine how adequate a definition, theory, observation or technique may be for any given application or situation. For example, is any particular observation an anomaly or an artifact of the way in which the observation was made? Evaluation problems require the invention of criteria and methods for evaluation. One might, for example, consider the entire field of statistics to be a set of solutions to evaluation problems.

Problems of integration often involve contradictory or paradoxical situations in which it is not obvious whether two theories, data sets, methods, or styles of research are compatible. For example, is there any benefit to teaching art to science students and science to art students? If one were to do so, how would one go about it? Integration problems generally require the rethinking of existing definitions and theories to determine what hidden assumptions may provide bridges between apparently disparate concepts.

Extension problems concern the range of possible uses to which any method, technique, theory, or definition can be put. Ray and Charles Eames, for example, experimented widely with plywood to find out what its possibilities and limitations were as a design material. Such extensions require extrapolations from existing practices and ideas to unknown ones in the form of predictions, play, and testing. The object in addressing extension problems is to discover the boundaries limiting the valid use of any particular solution methodology.

Problems of comparison arise when more than one possible solution exists to a problem and one needs to determine which is the best. Explicit criteria must be generated to make such comparisons. Such criteria are embodied in logic and in the use of analogies. While it may be relatively obvious how to compare the relative merits of two types of glue, for example, it may be very difficult to determine whether glue, staples, screws, nails, or hooks are the best means for attaching any two objects together. Unless one has a good understanding of the nature of the problem to be solved, criteria for generating such comparisons cannot be made.

Problems of application attend any attempt to extend a solution from one problem instance to another. Chemical engineers, for example, are highly sensitive to the fact that reactions that occur quickly and nearly completely in a test tube may be disasters when scaled up to hundreds of gallons. Similarly, while the principles of flight are the same for a model airplane and a jumbo jet, the application of those principles differs in obvious ways. Thus, problems of application often involve modeling or scaling-up issues, or the transfer of a solution from one discipline to another.

The final type of problem is the artifactual problem. Artifactual problems arise from misunderstandings. Such misunderstandings can arise from inaccurate or misleading assumptions, as in the classic lawyer's question: "When did you stop beating your wife?" Or such misunderstandings can arise from ignorance. For example, I was once asked by a reviewer of a paper I had written to provide a correlation coefficient for a trait that had no variance: all of the members of one group had the trait; none of the members of the other group had it. By definition, no correlation coefficient can be calculated in such a case and I had to point out that the problem was not valid.

Just as there are many types of problems, there are many classes of problems defined by the degree to which a complete answer can be achieved. Problems can be classified by whether they are: (1) unsolvable; (2) solvable only by approximation; (3) exactly solvable; (4) solvable as a class; or (5) solvable only for particular cases. Knowing the degree to which a problem can be solved is essential not only for evaluating the degree of precision that one can expect in an answer, but also for determining how important a problem may turn out to be.

Although practitioners of every field will recognize that some problems are amenable to more precise or general solutions than others, the only professionals who explicitly classify their problems according to such criteria are mathematicians (Wilf, 1986, pp. 178–221). In mathematics, problems are described as 'P', 'NP complete' and 'NP incomplete'. 'P' problems are those for which it can be demonstrated that the entire class of such problems can be solved using a common algorithm, or general problem-solving technique. An example of such a problem is whether any two numbers, x and y, are divisible by a common factor. It is possible to prove that any such problem can be solved. Thus, mathematicians can prove that 'P' problems are solvable even before addressing any particular manifestation of such a problem and before generating any particular solution. 'NP' problems, in contrast, are those for which it is impossible to solve the class of such problems, but for which individual solutions may be possible. One can always prove that any solution to an 'NP' problem is valid, but not whether such a solution exists in advance of actually finding it. An example is the classic traveling salesman problem in which a salesman must visit a large number of cities in a country and wants to do so in less than a certain number of days. There may or may not be a solution that satisfies the salesman's needs. Worse, there is no algorithm that will allow him to find out. He must generate possible solutions to the problem and hope that one satisfies his criteria. Worse yet, for most such traveling salesman problems, the number of possible solutions is too large to explore in any reasonable way, so that one can never prove that one has achieved the optimal solution to the problem, even if one finds a viable solution. The best one can do is to generate many solutions and compare them, looking for the one that is best of the batch of solutions on hand. Most real-life problems are of this NP type and require the generation of many possible solutions that are compared for their effectiveness. (See Root-Bernstein, 'The Art of Innovation', in this volume.)

Once an 'NP' problem has been solved, it may or may not yield general solutions to other such problems. In the case of the traveling salesman problem, each case must be solved individually within the entire class of such problems. Mathematicians call such 'NP' problems 'NP incomplete'. The class of problems that can be solved by the same algorithm as an already solved 'NP' problem is called the class of 'NP complete' problems. Such NP complete problems are not P problems because it is cannot be shown in advance that they have a solution—they must be found

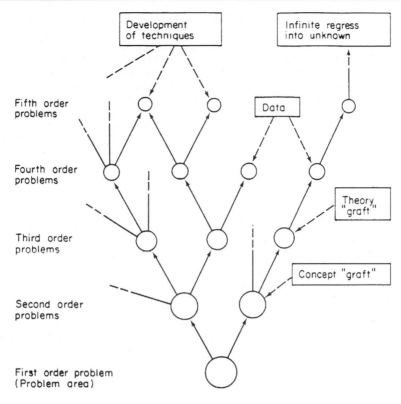

Figure 1. **Caption copy not supplied**.

one case at a time by trial and error. Moreover, as with NP problems in general, no one can ever prove that the optimal solution has been achieved for any member of an NP complete class. A better mouse trap might still be possible.

Mathematicians also recognize that there are problems that are neither 'P' nor 'NP', but are demonstrably unsolvable. This is the basis of Goedel's proof, which demonstrates that it is impossible to devise a mathematical system that has no unprovable assumptions. It has also been proven that it is impossible to trisect an angle using a compass and ruler. One might call such problems 'I', for impossible.

The existence of 'I' problems such as trisecting an angle raises another important point concerning the nature of problems. I have been stressing the principle that poorly defined problems yield confusing answers, whereas well-defined problems yield clear answers. One of the clearest possible answers to a well-defined problem is that it is not solvable at all. As astrophysicist Gregory Benford has cautioned, "The existence of a well-defined problem does not imply the existence of a solution" (Benford, 1989, p. 155). For example, many people have desired to create perpetual motion machines. The criteria defining the problem are

extremely well defined: such a machine must be capable of creating more energy than it uses. Stated as a question, the problem becomes: how does one create energy *de novo*? Anyone who has physics knowledge knows that, while this is structurally and logically a well-formulated question, it is not a reasonable question. To create energy *de novo* would violate the laws of thermodynamics. Thus, despite the fact that the problem can be stated exactly, it can also be shown that the problem has no solution. The question of whether God performs miracles is of the same class because miracles are, by definition, metaphysical or supernatural events beyond human comprehension, thereby placing any evidence beyond our ability to validate or replicate it. Such questions are therefore beyond rational discourse and belong, properly so, to questions of faith.

Finally, problems must be evaluated in terms of their order, which is to say with regard to their relationship to other problems. The most graphic and useful way of exploring the relationship between problems is to generate a hierarchical tree displaying their logical and methodological connections (Fig. 1). The logical problem tree was invented by biologist James Danielli to illustrate the fact that only very rarely is any problem addressed in isolation (Danielli, 1966, *passim*; Root-

Bernstein, 1989, p. 63). A problem of theory may require the invention of a new method for gathering relevant data and need to be solved in order to determine the feasibility of some practical application. An ordered problem tree allows investigators and inventors to determine which problems need to be solved in what order, using what methods. More importantly, a problem tree illustrates the degree to which any specific problem is more or less general and more or less connected to other problems of various types and classes.

Ordered problem trees are particularly useful for evaluating claims of significance. It is common for investors in new technologies and for funding agencies offering grants to require inventors to provide a justification of the utility of the work they propose to carry out. Thus, a cell biologist may justify the utility of his or her studies of cell division by the fact that cancer cells divide abnormally and therefore research on cell division may yield a cancer cure. Such claims may or may not be accurate, since the number of problems that need to be solved to link basic bench research with a clinical application may be so large as to be beyond reason. The degree of ignorance that must be addressed can only be evaluated when an ordered problem tree has been constructed to demonstrate the types, classes, and number of related problems that define the general problem area.

Problem Evaluation

An understanding of how problems are generated and the various types, classes, and orders into which they can be categorized provides useful information about how important or trivial any given problem is likely to be. Once again, there is virtually no formal research on this subject, but it is probably fair to say that what are called 'important' problems are those that are completely solvable for a large class of cases, and that this property of solvability is directly related to the ability to construct a very strongly connected problem tree that links many types of problems robustly. But the fact that problem evaluation requires a detailed understanding of where any given problem exists within a tree of other linked problems should warn us that there is nothing intrinsic to any given problem that makes it important or not. A seemingly trivial problem may provide the key to solving an entire problem tree of great significance; or a seemingly important problem may provide only a trivial answer because it can yield only a specific answer that connects with and informs no other problem. Many problems that may appear to be intrinsically interesting or worthy of attention (e.g. world peace) may be dependent on so many sub-problems of such an intractable nature (local economic conditions and resources, cultural habits, education, etc.) that there is no practical way to address them. Working directly on sub-problems may be a more practical goal.

However, the irresolvable nature of general or even specific problems should never be a barrier to problem-generating activities. The most important problems are always those that need to be solved and cannot be. Simply perceiving the detailed nature of such nested problems can be of practical value in and of itself. One of the strategies employed by creative people in every field is to construct a problem tree in order to identify the critical problems that cannot yet be addressed and, having identified and characterized these gaps in our knowledge, wait until answers are supplied by someone else. When these sub-problems are finally addressed, the problem tree becomes complete, and the importance of what may have appeared to be very trivial, individual problems suddenly takes on vast significance. One of the characteristics of the most innovative people is to discover the world in a grain of sand.

Another characteristic of great questions is that they are bold or even dangerous and it takes at least a modicum of bravery or bravado to pose them. Solving daring problems is of course important, but daring to pose them in the first place is still the key to making breakthroughs in any field. Many people are capable of solving well-defined problems, since solutions, as Einstein and his collaborator Leopold Infeld wrote,

> may be merely a matter of mathematical or experimental skill. To raise new questions, new possibilities, to regard old problems from a new angle, requires creative imagination and works real advance in science (Einstein et al., 1938, p. 5).

Thus, learning how to pose the most insightful, provocative, and challenging questions is surely an art as much in need of training and practice as those arts devoted to problem-solving.

But in the end, the creative urge to generate problems comes down to motivation. What makes an explorer in any discipline is their attraction to the mysterious, the incomprehensible, the paradoxical, the unknown. As Albert Szent-Gyorgyi, a Nobel laureate in Medicine and Biology wrote,

> A scientific researcher has to be attracted to these (blank) spots on the map of human knowledge, and if need be, be willing to give up his life for filling them in (Root-Bernstein, 1989, p. 407).

Fellow Nobel laureate and physicist I. I. Rabi agreed: "The only interesting fields of science are the ones where you still don't know what you're talking about" (Root-Bernstein, 1989, p. 407). Thus, the most creative people are always explorers quite literally in the way that Magellan, Columbus, or Lewis and Clark were explorers, aiming for the regions of our most manifest ignorance.

References

Anonymous (2000). *The work of Charles and Ray Eames: A legacy of invention*. At the Museum (LACMA Calendar), June, 4.

Benford, G. (1989). *Tides of light*. New York: Bantam.

Browne, M. N. & Keeley, S. M. (1986). *Asking the right questions. A guide to critical thinking* (2nd ed.). New York: Prentice-Hall.

Danielli, J. F. (1966). What special units should be developed for dealing with the life sciences... In: *The Future of Biology* (pp. 90–98). A Symposium Sponsored by the Rockefeller University and SUNY November 26 and 27, 1965. New York: SUNY Press.

Einstein, A. & Infeld, L. (1938). *The evolution of physics*. New York: Simon & Schuster.

Francis, D. (1992). *Driving force*. New York: Fawcett Crest.

Harrison, C. (1981). *English art and modernism 1900–1939*. London: Allen Lane; Bloomington, IL: Indiana University Press.

Heisenberg, W. (1958). *Physics and philosophy*. New York: Harper.

Herbert, R. L. (1964). *Modern artists on art. Ten unabridged essays*. Englewood Cliffs, NJ: Prentice-Hall.

Kostelanetz, R. & Flemming, R. (Eds) (1997). *Writings on glass: Essays, interviews, criticism*. New York: Schirmer Books.

Kuhn, T. (1959). *The structure of scientific revolution*. Chicago, IL: Chicago University Press.

Lionni, L. (1997). *Between worlds. The autobiography of Leo Lionni*. New York: Knopf.

Medawar, P. (1979). *Advice to a young scientist*. New York: Harper & Row.

Moore, R. (1966). *Niels Bohr: The man, his science, and the world they changed*. New York: Knopf.

Root-Bernstein, R. S. (1982). The problem of problems. *Journal of Theoretical Biology*, **99**, 193–201.

Root-Bernstein, R. S. (1989). *Discovering, inventing and solving problems at the frontiers of knowledge*. Cambridge, MA: Harvard University Press (reprinted by Replica Books, 1997).

Root-Bernstein, R. S. (1995). The development and dissemination of non-patentable therapies (NPTs). *Perspectives in Biology and Medicine*, **39**, 110–117.

Root-Bernstein, R. S. & Root-Bernstein, M. M. (1997). *Honey, mud, maggots and other medical marvels*. Boston, MA: Houghton Mifflin.

Root-Bernstein, R. S. & Root-Bernstein, M. M. (1999). *Sparks of genius, the thirteen thinking tools of the world's most creative people*. Boston, MA: Houghton Mifflin.

Taton, R. (1957). *Reason and chance in scientific discovery* (trans. A. J. Pomerans). New York: Philosophical Library.

Wells, H. G. (1975). *Early writings in science and science fiction* (R. Philmus & D. Hughes, Eds). Berkeley, CA: University of California Press.

Wilf, H. S. (1986). *Algorithms and complexity*. Englewood Cliffs, NJ: Prentice-Hall.

Witte, M. H., Kerwin, A. & Witte, C. L. (1988). Communications: Seminars, clinics and laboratories on medical ignorance. *Journal of Medical Education*, **63**, 793–795.

Witte, M. H., Kerwin, A., Witte, C. L. & Scadron, A. (1989). A curriculum on medical ignorance. *Medical Education*, **23**, 24–29.

The International Handbook on Innovation
Edited by Larisa V. Shavinina

The Role of Flexibility in Innovation

Asta S. Georgsdottir[1], Todd I. Lubart[1] and Isaac Getz[2]

[1] *Université René Descartes—Paris V, Laboratoire Cognition et Développement CNRS (UMR 8605), France*
[2] *ESCP-EAP-European School of Management, France*

Abstract: Flexibility is the ability to change. Innovation involves different types of change. In this chapter we will examine the importance of flexibility for different aspects of innovation. Different types of flexibility will be considered throughout the chapter, such as adaptive flexibility (the ability to change as a function of task requirements) and spontaneous flexibility (the tendency to change for intrinsic reasons, to try out a variety of methods). Finally, we will discuss how different types of flexibility can be important at different stages in the innovation process.

Keywords: Innovation; Flexibility; Creativity.

Introduction

Have you ever wondered why the dinosaurs did not make it? According to recent evidence, dinosaurs' living conditions changed brutally after a comet hit the earth, putting a great deal of dust and debris in the atmosphere. Being unable to adapt to the new, darker and colder climatic conditions, they became extinct. Sometimes survival depends on flexibility. And what about the earliest human beings who made it through the Ice Age, populated the earth from Africa to Iceland and became a dominant species? Often prosperity also depends on flexibility.

Flexibility can be defined as the capacity to change (Thurston & Runco, 1999). This may be either a change in how one approaches a situation in the external environment (adaptive flexibility) or a natural tendency to change for intrinsic reasons (spontaneous flexibility).

Flexibility may vary from one species to another, or from one individual to another; some species are more flexible than others in general. Within each species, some individuals will be able to adapt to even the most difficult circumstances whereas others will not. This is true for organizations as well. Some organizations are like dinosaurs, slow to react to a changing environment, whereas others are, like our ancestors, able to adapt. Even Microsoft, one of the largest world corporations faced a threat to its survival when at one point, its management found out that a rival Netscape had 700 people working on Internet applications, while it had only four; Microsoft managed to rebound in this

case in six months. This ability to adapt quickly and effectively to the environment is particularly important nowadays given the speed of technological evolution and globalization. All this puts pressure on individuals, companies and societies to make changes and to be adaptable if they are to survive and prosper (Mumford & Simonton, 1997).

As important as flexibility is for survival and prosperity under pressure, it does not directly cause it. There are mediating behaviors involved. One candidate is innovation. Innovation is typically defined as the generation, acceptance and implementation of new ideas, processes, products or services (Kanter, 1983). It is through the implemented novel productions (economic, social or technological) that an individual or an organization can survive and prosper when the environment changes, through productions that are appropriate to the new environment (economic, social, or technological). In this chapter we suggest that flexibility influences innovation in several important ways, and research supporting this assertion will be presented. Three approaches to creativity and innovation will be presented successively: the creative person, the environmental press and the creative process. We start by examining flexibility as a characteristic of the innovator him- or herself, that is, the cognitive and conative factors that relate to flexible thinking and the creation of new ideas, new productions. (It is important to note that though we speak of the innovator, the discussion can refer to either a single person or a group of people working together.) Then we will consider flexibility as a characteristic of the environment, the audience for innovations, the ways that individuals, embedded in cultural contexts, differ in their readiness to accept new ideas and innovations. Finally we will discuss how different types of flexibility (adaptive and spontaneous) can be important at different stages in the innovation process. We conclude by discussing several issues concerning the implications of the study of flexibility for individual and group innovation, but also for the organizational culture that can either facilitate or block innovation.

Before beginning an examination of flexibility and innovation, it is important to note that creativity, defined as an ability to produce work that is novel and appropriate (Amabile, 1996; Lubart, 1994), is a phenomenon closely related to innovation, and sometimes the terms are used interchangeably (West & Richards, 1999). Thus, in this chapter we do not draw a distinction between creativity and innovation.

Flexibility as a Characteristic of the Innovator

Cognitive Aspects

Flexibility is widely regarded as a cognitive ability important to creativity and innovation (Chi, 1997; Jausovec, 1991, 1994; Runco & Charles, 1997; Runco & Okuda, 1991; Thurston & Runco, 1999; Torrance, 1974). For example, for Thurston & Runco (1999), flexible cognition can facilitate creativity in several ways, such as helping to change strategies to solve a problem more effectively or to see a problem from a new perspective. It can also help one to switch easily between conceptual categories thus facilitating the production of diversified ideas. A flexible person can avoid getting stuck on one part of a problem, or can let go of the problem for a while when stuck and come back to it later. In discussing flexibility as a cognitive ability, research distinguishes adaptive flexibility and spontaneous flexibility, each facilitating creativity in a different way.

Adaptive flexibility involves changing perspectives on the problem, redefining it, or changing one's strategy to solve it when old perspectives or strategies have proven unsuccessful. For example, the candle task (Duncker, 1945), requires the participant to fix a candle on a wall without the wax dropping on the floor using only a candle, a box of matches and a box of tacks. Many people cannot find the solution to this problem because they try to attach the candle itself on the wall and consider the box as simply a recipient that holds the matches and/or the tacks provided. Once they change their perspective on the box, it becomes apparent that fixing a box on the wall with a few tacks and putting the candle in or on the box solves the problem perfectly. The change in perspective involves redefining the box as a support rather than a container, which then leads to an adaptive solution. Inflexibility in

one's conception of a common object, such as a box, is called 'functional fixedness' in the literature.

A number of authors have drawn attention to how adaptive flexibility in response to environmental constraints can lead to creative and innovative outcomes. For Barron (1988) an important ingredient of creativity is "seeing in new ways" (p. 78). According to Runco (1999) changes in perspective are related to both artistic and scientific creativity. Lipshitz & Waingortin (1995) call these changes in one's point of view on a problem 'reframing', which when applied to real-world problems, can lead to innovative solutions. For example, a hotel manager received frequent complaints about the elevators being too slow. He could follow the (very expensive) suggestion from an engineer to replace the elevators with a faster model, but decided to get a second opinion from an industrial psychologist. For the psychologist the problem was "not slow elevators but bored people" (Lipshitz & Waingortin, p. 153). He suggested installing large mirrors in front of the elevators so that people could pass the time watching themselves or others. The manager opted for this (much cheaper) solution and heard no more complaints about slow elevators. Thus, taking a different perspective on the problem led to an innovative and low-cost solution. The iMac from Macintosh is another example. In their design conception and publicity campaign, the computer is seen from a different point of view—as a decorative object rather than merely as an information-processing machine.

Flexible strategy use has also been found to characterize creative problem-solving as shown by gifted persons (Jausovec, 1991, 1994). Jausovec (1991) studied the relationship between flexible strategic thinking and problem-solving performance in two studies. In the first study, he compared the responses of gifted students to those of average and poor students on a number of tasks designed to provoke two types of rigid answers: response set and perceptual set. Response set is the inability to break a habit and use a different strategy to solve the task when there is a better way of solving it, and perceptual set is the tendency to see a problem just from one perspective. Gifted students showed fewer rigid answers than both average and poor students, who did not differ among themselves.

In the second study, the gifted and average students verbalized their strategies for solving both open and closed problems. Closed problems are stated in clear terms so that there is no need to redefine them, and they also have a single correct solution. Open problems are often less clearly stated, so clarifying what the goals are is part of the solution, and they do not usually have a single correct solution, but rather several solutions that work (Jausovec, 1994). Convergent thinking and logic are efficient strategies for solving closed problems, whereas divergent thinking and redefinition of the problem are more helpful for solving open problems (Jausovec, 1994). Jausovec (1991) identified the following strategies from students' verbalizations: subgoals (breaking the problem into smaller problems), working backward, modeling (making a simplified representation of a complex information), inference, trial and error, goal discovery, memory recall (recalling relevant knowledge from memory), generating analogies, and finally intuition and insight. The results showed that gifted students used more varied strategies than average students when solving the problems, and they also used them more selectively, applying different strategies to different problems depending on the requirements of the task. Thus, gifted students were more flexible in their strategy use.

Kaizer & Shore (1995) found also a tendency for competence in mathematical problem-solving to be associated with flexible strategy use. When comparing the performance of more and less competent mathematics students on mathematical word problems, the more competent students alternated between appropriate strategies (verbal or visual) to solve the problems, whereas the less competent were as likely to resort to trial and error as to use one of the appropriate strategies. Taken together, these studies suggest that adaptive flexibility may play a key role in problem-solving because it facilitates the use of a variety of strategies in response to the requirements of the task.

Spontaneous flexibility is the capacity to find a variety of solutions to a problem when there is no external pressure to be flexible. For example, in divergent thinking tasks, a participant may be requested to find as many ideas as possible concerning an object or a situation. However, whether the ideas generated come from one or many different categories of use depends on the individuals' natural, spontaneous flexibility of thought. The tendency to change conceptual categories easily and to cover many different categories in the responses indicates spontaneous flexibility. It is integrated in the scoring system of Torrances' (1974) Test of Creative Thinking (TTCT).

For example, in one version of the 'unusual uses' task, participants are asked to indicate uses for a cardboard box. The (spontaneous) flexibility score corresponds to the number of conceptual categories from which the responses are drawn. A flexible answer to the box task would be, for example, three ideas (make a toy house, play kick the box and store old clothing) all belonging to different categories (the categories of construction, box as toy and box as container, respectively). A rigid answer would be three ideas (put shoes in a box, put pencils in a box, put toys in a box), all belonging to a single category (box as container).

One important question explored in research is how flexibility facilitates the generation of new ideas. Several authors have drawn attention to the flexibility of existing knowledge structures as a source of new ideas, and the rigidity of it as an obstacle to creativity (Chi, 1997; Mumford, Baughman, Maher, Costanza & Supinski, 1997; Mumford & Gustafson, 1988). Knowledge is structured into conceptual categories—in groups of entities (concepts or images) that people believe belong together. The facility by which people work with different categories is thus an indicator of the flexibility of their knowledge structures.

For Chi (1997), the essence of creativity is the flexibility with which people cross category boundaries. She represents knowledge as stored on distinct associative trees. These trees can be a barrier to creativity because it is difficult for many people to move from one tree to another—from one category to another. When a concept on one knowledge tree is re-represented in the context of another tree, the result can be a new idea. For example, when the sound of footsteps in the snow is re-represented in the context of pop music, the result can be an original song (Björk, 2001, see Fig. 1).

Similarly, for Perkins (1988), crossing significant boundaries is central to creativity. For example, the impressionists crossed paradigm boundaries in the arts

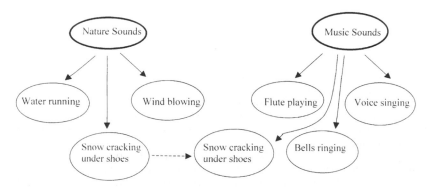

Figure 1. Example of two conceptual trees: nature sounds and music sounds. Flexibility involves considering (either spontaneously or purposefully) a knowledge element in a non-habitual conceptual tree, which may in turn lead to creative, innovative ideas.

when trying to capture the fleeting nature of light in their paintings, and Einstein crossed them in science when going against the traditional principles of electrodynamics. Operating on or across boundaries demands a profound change in perspective, as it involves questioning the limits of existing knowledge structures. By directly paying attention to boundaries, one can start operating explicitly on these boundaries, fiddling with them to experiment, or intentionally crossing them if one has reached an impasse.

According to Mumford et al. (1997), who also explored the flexibility of knowledge structures, combination and reorganization of existing knowledge play an important role in creating new ideas, but the processes underlying the combinations and reorganizations are not the same when working with similar concepts or distant ones. In their study, subjects combined and reorganized items from different unrelated categories in order to create a new concept that could account for all the items. They found that performance on this category combination task was related to the creativity of solutions of novel and ill-defined problems. Moreover, they found that instructions to use different methods for making connections between items (feature mapping or metaphors) contributed differently to performance for close categories and distant categories; when people were working with close (similar) categories, instructions to use feature mapping helped in finding connections between items, whereas when working with distant categories, instructions to use metaphors led to better performance.

In a similar vein, Mumford & Gustafson (1988) believe that the processes underlying major and minor creative contributions differ. According to them, integration and reorganization of cognitive (knowledge) structures is likely to underlie major creative contributions, but application of existing cognitive structures is likely to underlie minor contributions. These authors suggest that cognitive flexibility, whether it is due to outside requirements or to intrinsic reasons, is an important aspect of creative thought.

In order to implement a creative idea, however, flexible cognition is not enough. A number of recent models of creativity consider that in addition to flexibility, a combination of various cognitive factors, personality and motivational traits, emotional factors and environmental conditions is necessary in order to lead to creative, innovative productions (Getz & Lubart, 2000; Getz, Lubart & Biele, 1997; Lubart, 1994, 1999; Lubart & Sternberg, 1995; Mumford & Gustafson, 1988). In the next section we will examine the conative aspects of flexibility—personality traits and cognitive style—associated with innovation.

Conative Aspects

In this section we treat flexibility as a personality trait, a relatively stable, preferred way of behaving, rather than as a cognitive ability discussed in the previous section. In the literature, the personality of highly creative, innovative individuals as compared to their less creative, less innovative counterparts has received a great deal of attention, and a number of personality traits have been identified as related to high creative achievement. We focus on the work that relates to flexibility.

According to Gough (1995), *flexibility as a personality trait* refers to having a preference for change and novelty. The trait of flexibility has been traditionally measured with self-report measures, questions concerning a person's reactions in various situations, whereas flexibility as a cognitive ability has been evaluated through performance on specific problems that tend to involve mental blocks and set effects.

The personality trait of flexibility has been associated with creative performance in several studies (Dellas & Gaier, 1970; Feist, 1998, 1999; McKinnon, 1962). For example, McKinnon (1962) found creative architects to be more flexible than their less creative peers. In a meta-analysis of the literature on personality and creative achievement, Feist (1998) found creative scientists to be more flexible than less creative scientists and that artists were more flexible than non-artists as measured by the California Psychological Inventory Flexibility scale.

In addition, other personality dimensions and traits, also associated with creativity, seem to have an important aspect of flexibility built into them. The Openness to experience factor in the Five-Factor Model of personality (Costa & McCrae, 1985b) relates to flexibility. The positive pole of Openness to experience is anchored with terms such as, 'flexible', 'need for change', 'open-minded' and 'wide interests', whereas the negative pole concerns rigidity of beliefs, opinions and behaviors. For example, in Feist's (1998) meta-analysis, creative scientists, compared to less creative scientists, scored higher on openness to experience. Artists, compared to non-artists, also were more open to experience. Furthermore, McCrae (1987) found that openness to experience correlated positively with the number of ideas generated on divergent thinking tests.

Another sign of conative flexibility linked with creativity is the capacity to integrate apparently conflicting or opposite personality traits. Some authors have suggested that creative persons often have a combination of personality traits that seem contradictory. Traits that normally do not go together, such as sensitivity and coldness (Feist, 1998) or introversion and dominance (Eysenck, 1997), co-occur in the creative individual. Creative people, it seems, are flexible enough to integrate very dissimilar traits in their personality without losing their integrity (Eysenck, 1997), and, in fact, complexity seems to be a characteristic of creative people (Barron & Harrington, 1981; Eysenck, 1997).

Cognitive styles, at the interface between personality and cognition, refer to preferred ways of using one's cognitive abilities to approach a task. Lubart & Sternberg (1995) proposed that certain styles facilitate creativity and innovation more than others. For example, the legislative stylist prefers to invent his or her own rules and methods of doing things rather than to focus on implementation of these rules (executive style). The progressive stylist has a preference for novelty, welcoming change and innovation, whereas the conservative stylist prefers to do things the way they have always been done. Finally, the global stylist prefers working with a task in its large context, specially important for problem-finding and problem definition, whereas local stylists prefer to work with the details, on a narrower level.

Although specific styles have been associated with creativity, flexibility in shifting between different styles has received less attention (Martinsen, 1997). For example, the ability to shift between global and local levels of processing could be important for innovation, permitting the innovator to apply a different kind of treatment to different tasks or to different phases in the innovation process. It might be better to apply global processing in the beginning of a task, in order to identify the need for a new product in the marketplace. In contrast, during the elaboration of details of the particular product, local processing could be more helpful. Thus, in this view, flexibility may be due to an absence of a dominant cognitive style because a strong style would lead to preferential but sometimes non-optimal behavior in certain parts of a task or in certain tasks in general.

We have seen how flexibility contributes to creativity and innovation when viewed as a characteristic of individuals (both cognitive and conative). However, a possible downside of high flexibility could be that it could cause difficulties for the creative person to stick to a project and not to go from one thing to another. At this point it is important to distinguish between the *invention* or generation of an original idea and *innovation* or the project of converting it into a useful production (Roberts, 1988). In order to invent new ideas, flexibility is needed to see problems from a different perspective and come up with alternative ideas whereas perseverance is needed in order to stick to the project in spite of obstacles on the road of implementation. In this way other characteristics that play a role in creativity may balance any possible negative effects of flexibility.

However, in order to fully understand creativity, studying the individual is not enough (Csikszentmihalyi, 1988). No matter how creative the innovator, if the environment is not ready for creative, innovative ideas then original thinking will not flourish. In the next part of this chapter we will discuss the importance of the

social environment's willingness to accept innovations, and ways in which innovation can be facilitated through increased flexibility of the audience.

Flexibility as a Characteristic of the Audience for Innovations

Creativity takes place in a context, in a certain time and place that influence if the outcome will be seen as creative or not. For Csikszentmihalyi (1988), creativity is the result of an interaction between the environment (the cultural domain and the field) and the individual. Without a culturally defined domain and a group of peers to evaluate whether a proposal is original and useful, it is not possible to specify what is creative or not. Research on consumer behavior indicates that some people accept innovations more readily than others (Goldsmith & Foxall, this volume; Goldsmith & Hofacker, 1991; Goldsmith, d'Hauteville & Leisa, 1998). Some people will quickly try new products whereas others will continue to use well-known products well after innovative ones have been widely adopted in the marketplace. This phenomenon, called 'consumer innovativeness', has been shown to be related negatively to the age of consumers perhaps due to well-worn habits or risk aversiveness and is positively related to the trait of sensation-seeking. Thus some segments of the audience may be more rigid than others when considering potential innovations and adopting them. Beyond the level of individuals, organizational culture is a particular type of cultural environment that has been shown to play an important role in the destiny of new ideas. Consider now how one of its aspects—organizational flexibility—may influence innovation.

Flexibility in Organizational Climate and Culture

Organizational creativity is "deliberately changing procedures to make new, superior levels of quantity, quality, cost, and customer satisfaction possible" (Basadur & Hausdorf, 1996, p. 21). To achieve this goal, creative and innovative management is needed (Fernald, 1989).

A corporate climate refers to a coherent set of observable behaviors, attitudes and feelings in an organization (Ekvall, 1991). It is a manifestation of the deeper culture of the organization. According to Schein (1990), an organizational culture is a pattern of basic assumptions of a group, working to cope with problems of external adaptation and internal integration. These assumptions are considered valid, as they have worked in the past, and are therefore passed on to newcomers. Ekvall (1991) considers organizational climate as an intervening variable that affects the outcomes of the organization's actions through its effect on organizational and psychological processes like communication, decision-making and employee motivation. Together with the available resources, the organizational climate will have effects on outcomes such as productivity, quality of products, innovation and job satisfaction. These outcomes in turn influence both the resources and the climate.

A number of authors have suggested that one of the hallmarks of corporate cultures that support creativity and innovation is their flexibility or, in other words, their lack of rigidity. For example, for Hisrich (1990), flexibility is one of the common characteristics of companies that have an excellent reputation, as they show the ability to adapt to changes in the marketplace. According to this author, traditional corporate climates emphasize cautiousness and avoiding risks; they favor conservative decisions, and are rigid in the sense that they are hierarchical with established lines of authority, instructions and working hours. Furthermore, short-term gains are preferred over long-term potential as each manager protects his or her own job. This emphasis on not making any mistakes hinders creativity and reduces flexibility. A climate that nurtures innovation tends to have a relatively flat (non-hierarchical) organization with teamwork, sponsors and mentors that facilitate communication, trust and cooperation. In addition, it has a long-term vision, goals and action plans to follow them up. Actions are rewarded, suggestions and experimentation are welcome, and creation and development in all areas is supported.

The willingness of managers and other employees to change their ways of doing things and thereby to take some risk is important in order for new ideas to be translated into action. For example, in interviews with research and development scientists focusing on qualities of the environment that either facilitate or inhibit creativity, Amabile (1988) found that an overemphasis on maintaining the status quo—the unwillingness to change anything—was a frequently mentioned obstacle to organizational creativity.

In two empirical studies, Ekvall (1991) identified features of a creative or innovative climate in organizations and in organizational departments. The first study compared the creative and innovative climate in 27 Swedish organizations, evaluated (by product innovation criteria) as being either innovative or stagnated. The second study compared the creative climate of organizational departments in an American industrial company that had been evaluated by independent judges as being either innovative or stagnant. The result of both studies went in the same direction. The innovative organizations and the innovative departments were rated higher on all aspects of creative climate measured (challenge, freedom, idea-support, trust, dynamism, playfulness, debates, risk-taking and idea-time) and had less conflict than the stagnated organizations and departments. In terms of organizational culture, three types of profiles were identified. These profiles reflected value orientations that were focused differentially on either structure, people or change. Of the three, the change-oriented cultures were the most flexible. Change-oriented cultures were

characterized by entrepreneurial and transformational management, in which the manager is more a leader who sets goals and makes tasks meaningful rather than one who gives instructions and makes decisions.

Furthermore, in a meta-analysis of organizational innovation, Damanpour (1991) found several potential determinants of innovation that relate to flexibility. The first was functional differentiation, or the existence of differentiated units of professionals, who work explicitly on elaborating, introducing, and influencing change. The second was lack of formalization with a low emphasis on strict rules. The third determinant was a positive managerial attitude toward change. A manager who is favorable toward change creates a good climate for innovation, especially useful during the implementation stage, when conflict resolution and coordination of efforts are important. Of these three determinants, a significant positive relationship to innovation was observed empirically for functional differentiation and positive managerial attitude towards change.

Large organizations in particular have rigidity problems that can be obstacles to innovation. According to Rosenfeld & Servo (1991) increasing size is related to depersonalization and decreased communication. Both invite the risk of becoming rigid as each individual has less impact and there is less interaction between divisions. Remaining open to change while maintaining organizational integration becomes more difficult with increased size. Innovation should be possible however, despite size, if an organization uses its human resources in a flexible manner. One approach to achieve this involves managing properly innovative individuals inside organizations, as well as other key players in the process of innovation, such as technological gatekeepers, sponsors and champions. Technological gatekeepers are experts on technology that will give their opinion on whether an innovation is cutting-edge technology or not. Champions are the advocates of the innovator when facing the organization, supporting the new idea (see also Hausschildt, this volume), and sponsors are senior, higher-status members of the organization, who dispose of material and human resources and use them to develop an idea. Flexibility is also a desirable characteristic for these different actors in the innovation process. For example, Howell & Shea (2001) found that the tendency to seek information from diverse sources was positively related to champion behavior. Breath of interests or spontaneous preference for variety led people to scan diverse domains in search for information that could become critical to identify a promising innovation opportunity. Champions, thus, did not stick to subjects relevant to their own job, but showed interest in a variety of other topics. For organizations to identify people capable of facilitating innovation, it is important for them to have a flexible attitude towards the employee—to be able to set a formal job description aside for a while and see a person from a different perspective. Finally, some authors (e.g. Hammer, 2001) point out that innovation that depends on exceptional persons identified as champions is *not* well managed. Rather, everyone in the company has to have the flexibility to adopt the role of innovation generator, champion and implementor within a well-described innovation process such as product development (see also a later discussion of the system for managing ideas). In a similar vein, observers of the creative and innovative practices in business (e.g. Hamel, 2000; Peters, 1999) point out that companies that have a large flexibility in terms of job contents independent of one's title or function are often the most innovative. Such organizational flexibility allows them to assemble teams for innovative projects based on employees' intrinsic motivation and not on the organizational rigidified role.

We have seen different ways in which flexibility and the willingness to change are important for innovations, but how can we help organizations to be more flexible, more open to change? We will now examine some of the efforts that have been made to stimulate innovation in organizations, in particular those pertaining to being flexible.

How Can We Facilitate Innovation Through Flexibility?

As important as it is for organizations to have creative individuals on board, going against the stream in a rigid, hierarchical and authoritarian structure can be difficult. In order to facilitate the task for innovators, several authors have suggested how flexibility could be increased, both at the organizational and at the individual level.

Burnside (1990) presents a model of how to improve organizational climates for creativity. Central to this model is goal clarity, both in the long term and in the short term, in order to channel employee's creativity. A clear long-term goal is a specific idea of where the organization is attempting to go in the future, whereas clear short-term goals are smaller concrete objectives in order to get there. After goal-setting, three important components in motivating an employee to be creative are challenge, freedom and encouragement. A challenging task is meaningful and interesting to the employee. Challenge suggests that the task will require the employee to go beyond established knowledge or known skills, moving the individual into his/her 'zone of proximal development'. The employee has freedom when he or she is given maximal autonomy to decide how to do the task, and encouragement is provided by the manager who helps the employee to build courage to come out with his or her ideas and to take risks. Part of encouragement is accepting mistakes when an employee is willing to learn a lesson from an error.

To help create these conditions in large organizations, some have adopted a decentralized structure called an 'office of innovation', first developed by the Eastman Kodak Company. The office of innovation is not attached to a particular department, and also works independently from the human resource department, allowing for input from all departments and the flow of ideas between them. In the office of innovation, facilitators receive and look for individuals who play different roles in the innovation process (technological gatekeepers, sponsors and champions) in all departments of the organization (Rosenfeld & Servo, 1991). This decentralization increases flexibility in the organizational structure as it goes beyond department barriers in the conscious effort to innovate.

Fernald (1989) describes other examples of how different organizations have made an effort to stimulate creativity and innovation. At Intel, engineers are encouraged to meet with their engineer-peers to practice 'constructive confrontation' (p. 211), in order to eliminate bad ideas and make good ideas better. In these meetings, ideas are examined in a critical light, thus putting them in a different perspective. In this way, the engineers practice deliberately engaging their cognitive flexibility by changing their point of view on a problem in a group context.

Saturn Corporation illustrates an organization (General Motors) that considers flexibility and change necessary to be productive and innovative. In order to produce cars capable of competing on the world market, Saturn departs considerably from the traditional, hierarchical corporate structure. First, it considers its employees as creative and innovative people, capable of taking well-considered risks. Second, it keeps the bureaucratic structure as lightweight and flat as possible. For example, all employees contribute to the decision-making process, all are salaried personnel with a part of their earnings tied to the company's profit, and job classifications are limited and simple (Fernald, 1989). In 2001, however, General Motors made a decision to bring its Saturn division closer to its other divisions which lead many industry observers to question if its culture could survive in the long run.

A more global approach to achieve innovation through both specific organizational processes and flexible management of human resources is proposed in the *System for Managing Ideas* (SMI, Getz & Robinson, in preparation). SMI—a set of idea management processes and of human resource management practices aimed to encourage, act upon and recognize employees' ideas—is increasingly found in excellent companies known for the creativity and innovation of their employees. The essential belief on which SMI is founded is that if employees are not creative, it is not their fault (lack of necessary personal traits and abilities), but the company's which does not manage its employees' creative potential properly. According to the SMI, every employee is viewed as potentially creative, each to a different degree. Thus, SMI's set of practices aims to make every one's potential a reality—

through encouraging employees' ideas, providing all the necessary support for their implementation, and recognizing their innovative productions appropriately. In this sense, the SMI approach is contrary to the earlier mentioned approach of flexibly managing particular individuals—innovators, gatekeepers, sponsors and champions. Although doing the latter is better than doing nothing, relying on particular extraordinary individuals in any organizational process, be it product development or innovation, is generally viewed as a sign of non-systematic management. For innovation, this approach may be less reliable than the systematic idea management that involves flexibility in managing all company employees' creative potential.

Manager's attitudes towards creativity and innovation are important for the future of new ideas, as they must often introduce potential changes, at least on the middle and lower levels of organizations (Damanpour, 1991). According to Basadur & Hausdorf (1996), manager's attitudes towards creativity can be influenced through experience. In their study, managers who received training in using flexibly different modes of thought (alternatively using divergent and convergent thinking) in order to be more creative had more positive attitudes towards creativity after the training than before it. A manager who has positive attitudes towards creativity is more likely to be creative him or herself and encourage creativity in employees.

Thus, flexibility of the audience and, more generally, the environment is important for both stimulating innovation and appreciating it. In the next section, we will look at the process of innovation and how different types of flexibility can become important at different moments in the process.

Different Types of Flexibility at Different Stages in the Innovation Process

According to creativity research, the creative process is not viewed as a homogeneous phenomenon. It involves different stages or sub-processes that probably call for different approaches. We will start by briefly reviewing the different stages and sub-processes of the creative process before turning to how different types of flexibility are relevant for the different parts of the innovative process.

The creative process is frequently regarded as a special case of problem-solving, with the problem-solving process described in sequential stages (Lubart, 2000–2001). The best-known stage model was proposed by Wallas (1926), and includes four stages. The first is preparation, followed by incubation, illumination and finally verification. During the preparation stage, relevant knowledge and skills are used to analyze and define the problem. Next the problem is set aside for awhile, at least consciously, allowing different associations to the problem to be formed unconsciously, and impasses to be rejected. Illumination is a sudden insight into the solution of the problem that

then needs to be tested during the verification stage. In addition, some authors consider problem-finding, the realization that there exists a problem to be solved, as a separate stage, preceding the preparation stage (Amabile, 1996; Dudec & Côté, 1994; Runco & Chand, 1994).

In a similar vein, Amabile (1988, 1996) proposed a stage model for the creative and innovative process in organizations including the following stages: (1) identifying the problem to be solved; (2) preparing by assembling the relevant information; (3) generating the response after searching possible solutions; (4) (critically) evaluating and communicating the response; and finally (5) making a decision to go through with the idea or not. At any point in this process the person or group may need to return to a previous phase to rework problems that become apparent as the work progresses.

Flexibility could play a role at different points in the creative process. Both adaptive and spontaneous flexibility are important, but the two do not function in the same way. Adaptive flexibility is applied when there is a pressure from the outside to change, whereas spontaneous flexibility reflects change for intrinsic reasons. For example, in the preparation phase, spontaneous flexibility may be useful for gathering information from diverse domains that might be relevant to solve the problem, avoiding a limited search in a few domains that seem directly relevant. Also in the preparation stage, defining the problem may call for adaptive flexibility, to see the problem from a different perspective, which may result in an insightful approach. For example, in a study by Getzels & Csikszentmihalyi (1976), a group of art students were observed while making still-life drawings, and the originality of these drawings was then evaluated by expert judges. They observed a range of behaviors during the preparation phase. Some students examined only a few objects before moving on to drawing them, whereas others examined many different objects in detail, and sometimes changed the original arrangement they had made, after having started the drawing. These latter students demonstrated flexibility (changing their original approach to the problem and adopting a new one). This attitude was also positively related to the rated originality of the final drawings. Amabile (1996) separates problem-finding from the other stages in her model. Problem-finding calls for flexibility in order to identify a problem others did not see. In working through the process, flexibility can also facilitate moving with ease from one stage to another, depending on the sub-problems that may arise in the process.

The fact that incubation is postulated to take place mostly on an unconscious level limits what we can say about flexibility at that stage. Nevertheless, the ability to set the problem aside for a while can be regarded as an act of flexibility in itself instead of stubbornly struggling with it when the present approach is clearly not effective. A different act of flexibility in the incubation stage can occur in preconscious emotional processing. A series of studies by Getz & Lubart (1998, 2000) and Lubart & Getz (1997) have shown that when individuals rely on emotional processing of their experiential memory representations they produce more creative associations than individuals relying on the cognitive processing of their conceptual memory representations. Thus spontaneous flexibility could be involved during incubation if an individual switches from cognitive to emotional processing. These findings are also corroborated in the well-known organizational practice of fostering the production of creative ideas by going and experiencing (feeling) directly the problems. Indeed, within the Japanese Kaizen (continuous improvement) approach, recognizing both the importance of 'feeling' the problem and the cultural reluctance (rigidity) to do so encourages a problem-solver to 'become' flexible and to go and experience the problem before trying to look for its solutions (Robinson, 1991).

The illumination stage is the moment where the different elements of the problem suddenly 'click' into place and the solution becomes clear. Adaptive flexibility could be essential at this stage, allowing one to change one's approach to the problem or to rearrange its different elements in a new way, which can result in a new insight. The verification stage involves testing the solution produced—determining if it solves the problem or not. This is perhaps the stage where convergent thinking is the most important—verifying if a particular solution is right for the particular problem. The facility to switch from divergent thinking to convergent thinking involves adaptive flexibility, as each mode is best suited to different portions of the problem-solving process.

The idea of a sequential stage-based process has been recently questioned, as some authors find it too simplistic to describe what really takes place during creative problem-solving (Lubart, 2000–2001). Many sub-processes that play a role in creativity have been studied independently of stage models of creativity, some of which involve flexibility, such as definition of the problem, divergent thinking, reorganization of information and remote association. For example, remote association is a process that is hypothesized to involve spreading activation in memory (Mednick, 1962). A spontaneously flexible person, able to search in many varied categories, should have more chance of finding varied associations in response to a given problem and thus a greater capacity for remote association. Lubart & Mouchiroud (in press) propose an alternative, dynamic view in which the creative process is made up of different sub-processes that are mutually reinforcing. This approach does not postulate a fixed sequential order of the different processes but rather an interaction of different processes, or their

simultaneous activation. In this type of approach to the creative process, the importance of flexibility to jump back and forth between different modes of thought or sub-processes, or to apply them at the same time, is evident.

Conclusion

In this chapter we have seen how flexibility is important for innovation. The innovator needs it in order to identify new problems and to get away from the traditional ways of solving them, and the audience needs it in order to give new and unusual ideas a chance to exist and show their value. Finally, the process of innovating involves flexibility in numerous ways.

These have several implications, both for the functioning of organizations and for future research on creativity and innovation. For example, in modern organizations, innovation often depends on a group of individuals being able to work as a team. In the beginning of the chapter we specified that our discussion of individual flexibility could also apply to a group of people working together. In a team, having a flexible personality is probably important for the individual members to work well together, but additionally, having a group of individuals with different perspectives on the problem could also contribute to a creative and innovative solution. This would be especially true if team members voluntarily confront their different point of views in order to solve the problem better.

Moreover, innovative individuals and teams need a flexible environment if their ideas are to become real products. Organizational creativity involves deliberately changing established procedures, which implies that managers at all levels in organizations can contribute to flexibility. It is important that managers be flexible enough to give the employees space to be innovative, for example by being willing to take well-considered risks, having low emphasis on rules and offering the possibility of flexible working hours. While all these things are important to create a more flexible corporate culture, the System for Managing Ideas (Getz & Robinson, in preparation) offers an alternative approach. Organizational innovation is placed in a new perspective by redefining the problem. In the System for Managing Ideas, innovation is not about exceptional individuals innovating against all odds under traditional management, but about exceptional management practices that enable the creative potential of ordinary individuals. This approach could encourage interest for the employees' efforts to be innovative, and tolerance towards these employees who just cannot seem to fit nicely into their job descriptions, instead of interpreting their behavior as a simple defiance against authority.

Organizations today need to be ready to change without much notice and to prepare as well as they can for the unpredictable hazards of the organizational environment: economic crises, smart competitors, changes in consumer behavior and a workforce that is evolving as well. Awareness of why and how flexibility is important for organizations can help them take action towards more adaptability and thereby be less vulnerable if the environment suddenly changes. As for the innovative process itself, we have seen how the process of creating and innovating can be seen as an essentially flexible one. For example, in the sequential stage-based approach, flexibility seems to be important at each stage. However, with more recent, non-sequential approaches to the creative process, new venues open up for research on flexibility. In this non-sequential and interactive view of the creative process, the role of flexibility to synchronize the use of different sub-processes becomes a central issue.

References

Amabile, T. M. (1988). A model of creativity and innovation in organizations. *Research in Organizational Behavior*, **10**, 123–167.

Amabile, T. M. (1996). *Creativity in context*. Boulder, CO: Westview.

Barron, F. (1988). Putting creativity to work. In: R. J. Sternberg (Ed.), *The Nature of Creativity: Contemporary Psychological Perspectives* (pp. 77–98). Cambridge: Cambridge University Press.

Barron, F. & Harrington, D. M. (1981). Creativity, intelligence, and personality. *Annual Review of Psychology*, **32**, 439–476.

Basadur, M. & Hausdorf, P. A. (1996). Measuring divergent thinking attitudes related to creative problem solving and innovation management. *Creativity Research Journal*, **9**, 23–32.

Björk (2001). *Aurora*. On *Vespertine (CD)*. London: One Little Indian Records.

Burnside, R. M. (1990). Improving corporate climates for creativity. In: M. A. West & J. L. Farr (Eds), *Innovation and Creativity at Work* (pp. 265–284). Chichester, U.K.: John Wiley.

Chi, M. T. H. (1997). Creativity: Shifting across ontological categories flexibly. In: T. B. Ward, S. M. Smith & J. Vaid (Eds), *Creative Thought: An Investigation of Conceptual Structures and Processes* (pp. 209–234). Washington, D.C.: American Psychological Association.

Costa, P. T. Jr. & McCrae, R. R. (1985b). *The NEO personality inventory manual*. Odessa, FL: Psychological Assessment Resources.

Csikszentmihalyi, M. (1988). Society, culture and person: A system view of creativity. In: R. J. Sternberg (Ed.), *The Nature of Creativity: Contemporary Psychological Perspectives* (pp. 325–339). Cambridge: Cambridge University Press.

Damanpour, F. (1991). Organizational innovation: a meta-analysis of effects of determinants and moderators. *Academy of Management Journal*, **34**, 555–590.

Dellas, M. & Gaier, E. I. (1970). Identification of creativity: The individual. *Psychological Bulletin*, **73**, 55–73.

Dudec, S. Z. & Côté, R. (1994). Problem finding revisited. In: M. A. Runco (Ed.), *Problem Finding, Problem Solving and Creativity* (pp. 130–150). Norwood, NJ: Ablex.

Duncker, K. (1945). On problem solving. *Psychological Monographs*, **58** (Whole No. 270).

Ekvall, G. (1991). The organizational culture of idea-management: A creative climate for the management of ideas. In: J. Henry & D. Walker (Eds), *Managing Innovation* (pp. 73–79). London: Sage.

Eysenck, H. J. (1997). Creativity and personality. In: M. A. Runco (Ed.), *The Creativity Research Handbook* (Vol. 1, pp. 41–66). Cresskill, NJ: Hampton Press.

Feist, G. J. (1998). A meta-analysis of personality in scientific and artistic creativity. *Personality and Social Psychology Review*, **2**, 290–309.

Feist, G. J. (1999). The influence of personality on artistic and scientific creativity. In: R. J. Sternberg (Ed.), *Handbook of Creativity* (pp. 273–296). New York: Cambridge University Press.

Fernald, L. W. (1989). A new trend: Creative and innovative corporate environments. *Journal of Creative Behavior*, **23**, 208–213.

Getz, I. & Lubart, T. I. (1998). The emotional resonance model of creativity: Theoretical and practical extensions. In: S. W. Russ (Ed.), *Affect, Creative Experience, and Psychological Adjustment* (pp. 41–56). Philadelphia, PA: Bruner/Mazel.

Getz, I. & Lubart, T. I. (2000). An emotional-experiential perspective on creative symbolic-metaphorical processes. *Consciousness and Emotion*, **1**, 89–118.

Getz, I., Lubart, T. I. & Biele, G. (1997). L'apprentissage du changement dans l'organisation: La perspective de la créativité. (The acquisition of change in the organization: the perspective of creativity). In: P. Besson (Ed.), *Dedans— dehors: Les nouvelles frontières de l'organisation. (Inside—outside: The organizations' new frontiers)* (pp. 205–217). Paris: Vuibert.

Getz, I. & Robinson, A. G. R. (in preparation) *Une formidable force: Comment les idées de tous transforment l'entreprise et l'économie. (A tremendous strength: How everyone's' ideas can transform the firm and the economy)*.

Getzels, J. & Csikszentmihalyi, M. (1976). *The creative vision: A longitudinal study of problem-finding in art*. New York: Wiley-Interscience.

Goldsmith, R. E., d'Hauteville, F. & Leisa, R. (1998). Theory and measurement of consumer innovativeness: A transactional evaluation. *European Journal of Marketing*, **32**, 340–353.

Goldsmith, R. E. & Foxall, G. R. (this volume). The measurement of innovativeness. In: L. V. Shavinina (Ed.), *International Handbook on Innovation*. Oxford: Elsevier Science.

Goldsmith, R. E. & Hofacker, C. F. (1991). Measuring consumer innovativeness. *Journal of the Academy of Marketing Science*, **19**, 209–221.

Gough (1995). *Guide pratique d'interprétation du CPI. (Practical guide for interpretation of the CPI)*. Paris: ECPA.

Hamel, G. (2000). *Leading the revolution*. Boston, MA: Harvard Business School Press.

Hammer, M. (2001). *The agenda: What every business must do to dominate the decade*. New York: Crown Business.

Hauschildt, J. (this volume). The measurement of innovativeness. In: L. V. Shavinina (Ed.), *International Handbook on Innovation*. Oxford: Elsevier Science.

Hisrich, R. D. (1990). Entrepreneurship/Intrapreneurship. *American Psychologist*, **45**, 209–222.

Howell, J. M. & Shea, C. M. (2001). Individual differences, environmental scanning, innovation framing, and champion behavior: key predictors of project performance. *The Journal of Product Innovation Management*, **18**, 15–27.

Jausovec, N. (1991). Flexible strategy use: A characteristic of gifted problem solving. *Creativity Research Journal*, **4**, 349–366.

Jausovec, N. (1994). *Flexible thinking: An explanation for individual differences in ability*. Cresskill, NJ: Hampton Press.

Kaizer, C. & Shore, B. (1995). Strategy flexibility in more and less competent students on mathematical word problems. *Creativity Research Journal*, **8**, 77–82.

Kanter, R. M. (1983). *The change masters: Innovations for productivity in the American corporation*. New York: Simon & Schuster.

Lipshitz, R. & Waingortin, M. (1995). Getting out of ruts: A laboratory study of a cognitive model of reframing. *Journal of Creative Behavior*, **23**, 151–171.

Lubart, T. I. (1994). Creativity. In: R. J. Sternberg (Ed.), *Thinking and Problem Solving* (pp. 289–332). New York: Academic Press.

Lubart, T. I. (1999). Creativity across cultures. In: R. J. Sternberg (Ed.), *Handbook of Creativity* (pp. 339–350). New York: Cambridge University Press.

Lubart, T. I. (2001–2002). Models of the Creative process: Past, present and future. *Creativity Research Journal*, **13**, 295–308.

Lubart, T. I. & Getz, I. (1997). Emotion, metaphor and the creative process. *Creativity Research Journal*, **10**, 285–301.

Lubart, T. I. & Mouchiroud, C. (2003). Creativity: A source of difficulty in problem solving. In: J. Davidson & R. J. Sternberg (Eds), *The Psychology of Problem Solving* (pp. 127–148). New York, NY: Cambridge University Press.

Lubart, T. I. & Sternberg, R. J. (1995). An investment approach to creativity: Theory and data. In: S. M. Smith & T. B. Ward et al. (Eds), *The Creative Cognition Approach* (pp. 271–302). Cambridge, MA: the MIT Press.

Martinsen, O. (1997). The construct of cognitive style and its implications for creativity. *High Ability Studies*, **8**, 135–158.

McCrae, R. R. (1987). Creativity, divergent thinking and openness to experience. *Journal of Personality and Social Psychology*, **54**, 1258–1265.

McKinnon, D. W. (1962). The nature and nurture of creative talent. *American Psychologist*, **17**, 484–495.

Mednick, S. A. (1962). The associative basis of the creative process. *Psychological Review*, **69**, 220–232.

Mumford, M. D., Baughman, W. A., Maher, M. A. & Supinski, E. P. (1997). Process-based measures of creative problem-solving skills: IV. Category combination. *Creativity Research Journal*, **10**, 59–71.

Mumford, M. D. & Gustafson, S. B. (1988). Creativity syndrome: Integration, application and innovation. *Psychological Bulletin*, **103**, 27–43.

Mumford, M. D. & Simonton, D. K. (1997). Creativity in the workplace: People, problems and structures. *Journal of Creative Behavior*, **31**, 1–6.

Perkins, D. N. (1988). The possibility of invention. In: R. J. Sternberg (Ed.), *The Nature of Creativity: Contemporary Psychological Perspectives* (pp. 362–385). Cambridge: Cambridge University Press.

Peters, T. (1999). *The circle of innovation*. New York: Vintage Books.

Roberts, E. B. (1988). Managing invention and innovation. *Research-Technology Management*, (Jan.–Feb.), 11–29.

Robinson, A. G. (Ed.) (1991). *Continuous improvement in operations*. Portland, OR: Productivity Press.

Rosenfeld, R. & Servo, J. C. (1991). Facilitating innovation in large organizations. In: J. Henry & D. Walker (Eds), *Managing Innovation* (pp. 28–39). London: Sage.

Runco, M. A. (1999). Perspectives. In: M. A. Runco & S. R. Pritzker (Eds), *Encyclopedia of Creativity* (Vol. 2, pp. 373–376). San Diego, CA: Academic Press.

Runco, M. A. & Chand, I. (1994). Problem finding, Evaluative thinking and creativity. In: M. A. Runco (Ed.), *Problem Finding, Problem Solving and Creativity* (pp. 40–76). Norwood, NJ: Ablex.

Runco, M. A. & Charles, R. E. (1997). Developmental trends in creative potential and creative performance. In: M. A. Runco (Ed.), *Creativity Research Handbook* (Vol. 1, pp. 115–152). Cresskill, NJ: Hampton Press.

Runco, M. A. & Okuda, S. M. (1991). The instructional enhancement of the flexibility and originality scores of divergent thinking tests. *Applied Cognitive Psychology*, **5**, 435–441.

Schein, E. H. (1990). Organizational culture. *American Psychologist*, **45**, 109–119.

Thurston, B. J. & Runco M. A. (1999). Flexibility. In: M. A. Runco & S. R. Pritzker (Eds), *Encyclopedia of Creativity* (Vol. 1, pp. 729–732). San Diego, CA: Academic Press.

Torrance, E. P. (1974). *Tests de pensée créative. (Tests of creative thinking). Manuel*. Paris: ECPA.

Wallas, G. (1926). *The art of thought*. New York: Harcourt, Brace.

West, M. A. & Richards, T. (1999). Innovation. In: M. A. Runco & S. R. Pritzker (Eds), *Encyclopedia of Creativity* (Vol. 2, pp. 45–55). San Diego, CA: Academic Press.

The International Handbook on Innovation
Edited by Larisa V. Shavinina
© 2003 Published by Elsevier Science Ltd.

The Effect of Mood On Creativity in the Innovative Process

Geir Kaufmann

Norwegian School of Economics and Business Administration, Bergen, Norway

Abstract: The division of labor between the concepts of creativity and innovation is discussed. The effect of mood states on creative problem-solving as part of the process of innovation is then addressed. The prevailing notion that there exists an unconditional positive causal link between positive mood and creativity is criticized. Evidence is reviewed that shows that under certain, important conditions, positive mood may actually impair creativity, while negative and neutral moods may facilitate finding insightful and highly creative solutions to problems. A new theory of mood and creative problem-solving is developed.

Keywords: Innovation; Innovative process; Creativity; Mood; Creative problem-solving.

The Conceptual Terrain of Creativity, Innovation and Mood

We have previously argued that there is a string of concepts in the innovation domain that, put together, form a fairly tight and coherent conceptual equation (Kaufmann, 1993). At the base, there is *originality* that is charaterized by novelty of ideas, which is a necessary, but not sufficient condition for *creativity*, which, in addition requires the concept of value or usefulness. Creativity, in turn, is a necessary, but not sufficient condition for *invention*, which in turn requires *incremental* novelty, in the sense that the novelty in question must be objective, and not only subjectively novel to its originator. In the last piece of the conceptual equation, invention is a necessary but not sufficient condition for *innovation*, which in addition requires that the condition of realizability be fullfilled.

In the present discussion we will concentrate mainly on the creativity space of the innovation domain, and discuss the status and implications of a recent stream of new research that has sparked considerable attention and controversy. In this new line of research, the functional significance of mood and affect in the process of creativity has been addressed.

Affect and Mood—Figure and Ground

According to Morris (1989), we can distingusih between the 'figure' and 'ground' of mood. As ground, mood is the backdrop against which a person experiences and evaluates events; as figure, mood comes into the forefront of attention, which may result in attempts to explain why a particular mood is present and, if it is negative, attempts to 'repair' or replace the current mood state. Moods are seen to be pervasive and global and have the capability to influence a broad range of thought processes and behaviors.

Vosburg & Kaufmann (1999) argued that a more precise conceptual outlook would be to regard mood as the background state, and affect as its corresponding, overt manifestation. Thus positive mood may be manifested in the affective state of elation, and negative mood, in the overt state of depression or anger. By definition, then, mood is a background state not consciously articulated by the individual. This is a matter of practical importance, in the sense that, at least on some theories, mood effects are only obtainable when not consciously experienced and articulated in an explicit way by the individual (cf. Forgas, 2000a).

Roadmap of the Chapter

In the present chapter we will review and critically discuss the current status of research and theories on mood and creativity. We will argue that a premature closure on a certain bottom line conclusion has been taken, and plead for a more nuanced perspective on the complex interrelationships between mood states and different aspects of the process of creativity, and present our perspective in the form of a new theory of mood effects on creative problem-solving. Last, but not least, we will place the discussion of mood effects on creativity in the larger context of process of innovation.

From 'Cold' to 'Hot' Cognition

In psychology, there has always been a tendency to drive a wedge between cognition and emotion, in the sense that they are seen as two distinct, separate systems, that may run exclusively on their own wheels, but sometimes may interact in different ways (e.g. Forgas, 2000a).

The 'Cold' Approach

This general position has, to a large degree, dominated traditional cognitive approaches to the study of creativity, and cognitive theories of creativity are consequently largely focused on elucidating 'cold' cognitive mechanisms and strategies involved in creative thinking. In Newell, Shaw and Simon's classical work on creativity conceptualized from a cognitive information processing perspective, there is a passing mention of the importance of 'motivation' and 'persistence' as one of the important defining criteria. However, the remainder of the treatment of the subject is almost entirely concentrated on identifying the most relevant heuristic strategies that are used by problem-solvers in handling the creativity aspect of problem solving (Newell, Shaw & Simon, 1979). Along similar lines, Weisberg (this volume), argues uncompromisingly that the traditional, well-known mechanics of ordinary problem-solving is the only road that takes us to the peak levels of creativity, excluding any privileged role of affect in the process of creativity that other theorists would like to grant (e.g. Getz & Lubart, 1999; Getz & Lubart, 2000; see also Root-Bernstein & Root-Bernstein, this volume).

Also in the most recent, sophisticated 'creative cognition' approach launched by Finke, Ward & Smith (1992) and Smith, Ward & Finke (1995), where the concepts, theories and methods of cognitive science are applied as a platform for a rigorous study of creativity, the focus is exclusively on identifying the kind of cognitive mechanisms that are at work when people are engaged in creative endeavours. No place in their theories has been given to the 'hot' dimension of cognition, that is fueled by desires, affect and mood.

Indeed, as Forgas (2000a) points out, in the classical cognitive tradition there has generally been a tendency to think of affect as a 'subtractive' factor that is 'dangerous' in the sense of having the potential to subvert efficient cognitive processing. More recently, we see this perspective expressed in theories espoused by Ellis & Ashbrook (1988) and Mackie & Worth (1991), where both positive and negative affect are regarded as states that take up scarce processing capacity.

Cognition is Heating Up

During the last 20 years, however, we also have seen an increasing interest in embracing the concepts of emotion, affect and mood within more integrated cognitive theories, like the associative network theory (Bower, 1981). Here affective content is also granted a potentially positive function, in providing extra cues in memory. On this perspective, openness to feelings may become a useful adjunct to the machinery of effective, rational thinking. This perspective is also adopted specifically with regard to creativity by Getz & Lubart (1999), in their elegant Emotional Resonance Model of Creativity.

Of particular relevance to the area of creativity is also the view put forward in one of the most prominent theories in the field advocated by Forgas (1995, 2000b). In his Affect Infusion Model (AIM), a core thesis is that tasks that require elaborate and constructive processing—exactly what is likely to be needed in the open and unstructured problem spaces that require creative thinking—there will be more reliance on affectively primed information, and greater influence of affect and mood states on performance. Such effects are held to be both detrimental and facilitative of performance, partly due to the kind of information processing that is triggered by the particular affective state in question with regard to the task at hand.

Following Forgas (2000a) we may regard the division of labor between the different emotion concepts in the way that affect is a general, superordinate concept, where mood refers to long-term, relatively stable affective states, and emotions constitute more short-lived, specific affective episodes.

Research on affect and creativity has been focused largely on the effect of positive and negative mood, compared to a neutral state on a large variety of creative task performances (Russ, 1993, 1999).

Mood and Creativity: A Straight or a Crooked Relationship?

We may start out by asking if the research done so far gives us a fairly clear and consistent picture of the impact of mood and creativity, or if moderating contingencies is the rule rather than the exception.

In spite of many differences regarding the theoretical interpretations of mood effects on cognitive performance in general (Forgas, 2000a; Hirt, 1999; Hirt, McDonald & Melton, 1996; Martin & Stoner, 1996), a specific position has frequently been cited as a valid general conclusion on the relationship between mood and creative problem-solving. This is the conclusion that *positive mood facilitates creative problem-solving* (e.g. Benjafield, 1996; Hirt, 1999; Hirt, McDonald & Melton, 1996; Isen, 1993, 1999; Isen & Baron, 1991; Shapiro & Weisberg, 1999; Shapiro, Weisberg & Alloy, 2000).

Hirt (1999) is a representative voice of these views when he claims that "Individuals in positive mood states have been reliably shown to be more creative on a range of tasks than are individuals in other mood states" (p. 242). Not only that, but "the effects of (positive) mood on creativity appear to be remarkably robust both in terms of the mood induction procedure used and the range of possible creativity tasks that have been measured" (p. 242). Isen (1999) agrees, and claims that ". . . positive affect is associated with greater cognitive flexibility and improved creative problem-solving across a broad range of settings" (p. 3).

The Supporting Evidence

It is true that findings from a large number of studies seem to support this general conclusion (Hirt, 1999; Russ, 1993, 1999). In a classical, and frequently cited study, Isen and her coworkers have shown positive mood to facilitate performance on standard indicators of creative problem-solving ability, such as insight problems (Isen, Daubman & Nowicki, 1987; see also Greene & Noice, 1988), and the Remote Associates Test (RAT) (Isen, Daubman & Nowicki, 1987; cf. Estrada, Isen & Young, 1994 for a replication and extension). Unfortunately, a serious flaw exists in all these studies, in the form of lack of discriminant validation of the findings. No control tasks, *not* presumed to be affected by mood, or to be affected to a lesser degree than creativity tasks, were included. Thus, we do not know if these effects are constrained to creative problem-solving specifically, or if they reflect a general, positive performance effect of positive mood, for instance through reducing test anxiety. However, in further studies it has also been shown that positive mood promotes more original responses to a word association task, and encourages broader and more inclusive categorization performance (Isen & Daubman, 1984; Isen, Johnson, Mertz & Robinson, 1985; cf. Greene & Noice, 1988). Such findings are taken to mean that positive mood selectively facilitates the kind of basic cognitive, associative mechanisms that may lie behind creative performances. These are interesting findings, but a direct demonstration that the kinds of associations primed by positive mood do indeed significantly impact on creative problem-solving performance would have been desirable. Thus, we are left with interesting hints as to a facilitative effect of positive mood on creative problem-solving, but by no means with conclusive evidence.

Consistent with the positive mood-creativity hypothesis, Bowden (1994), in a review of psychometric studies of cognitive characteristics associated with elevated positive mood, reported that this state of mind was associated with tendencies toward over-inclusion and loose conceptual boundaries. Similar results were reported by Jamison (1993) when highly creative individuals were observed by monitoring their mood systematically and recording reports of creativity, a mild hypomanic state was observed to significantly coincide with high levels of ideational fluency, speed of association, combinatorial thinking (including incongruent combinations and metaphors) and 'loose' processing involving irrelevant intrusions in thought (cf. Schuldberg, 1990, 1999; Shapiro & Weisberg, 1999; Shapiro, Weisberg & Alloy, 2000). In a positive sense, this cluster of cognitive characteristics may tend to facilitate originality and creativity in problem-solving.

Recent experimental research adds further support to these findings (Hirt, McDonald & Melton, 1996; Murray, Sujan, Hirt & Sujan, 1990; Showers & Cantor, 1985). In these experiments, positive mood subjects, compared to controls, were found to be better able to generate and use broader categories when identifying similarities, and also more able to shift to narrower categories when focusing on differences. Such findings have been linked to creative problem-solving by the argument that subjects in whom positive mood had been induced gained access to more unusual and diverse information, and were also more flexible when categorizing (Hirt, McDonald & Melton, 1996; Isen & Baron, 1991).

These findings are consistent with results from other studies showing positive mood to increase fluency in divergent thinking tasks. Abele (1992a) induced positive, negative and neutral mood by way of autobiographical recall, and demonstrated that positive mood resulted in superior performance on ideational fluency tasks. Vosburg (1998a) recorded mood at arrival through an adjective checklist immediately prior to task performance and found that positive mood facilitated and negative mood inhibited fluency of idea production.

Related to these findings are results showing facilitative effects of positive mood on heuristic problem-solving tasks, like the Means End Problem Solving Test (Mitchell & Madigan, 1984; cf. Abele, 1992b). Similarly, Carnevale & Isen (1986) examined the effect of mood on the quality of problem-solving in negotiation behavior. They focused on the ability to find a creative integration from which both parties gain. Again, positive mood was found to yield superior performance. Staw & Barsade (1993) examined the relationship between positive affectivity and performance in a complex managerial In-Basket task. Positive affectivity was found to be significantly positively related to all important problem-solving attributes, including flexibility and originality.

The Explaining Models

The conclusion that positive mood facilitates creative problem-solving is reached from different theoretical premises. Isen (1984, 1987, 1993, 1999) posits that positive, compared to negative and neutral, material is more extensively connected and better integrated in memory. This is held to promote spreading activation and increase the likelihood of making remote associations conducive to creative thought. However, no direct and independent evidence in support of this general principle has ever been offered. The idea that positive mood promotes spreading activation is largely gratuitous and used in a *post hoc* manner to explain observed, empirical findings.

According to the cognitive tuning theory originally proposed by Schwarz (1990), and further developed by Schwarz & Bless (1991) and Clore, Schwarz & Conway (1994) (see also Schwarz, 2000), the essential function of emotional states is to inform the individual about the state of the current task environment (cf. Frijda, 1988). Negative mood indicates a problematic situation, whereas positive mood signals a satisfactory state of affairs. Consequently, individuals in negative mood will be tuned to an analytic, 'tight' mode of processing (cf. Fiedler, 1988, 2000), where the situation is treated carefully and systematically. Positive mood individuals relax on the processing requirements, and are more prone to use simplifying heuristics and 'loose' processing (cf. Fiedler, 1988, 2000). As such, they may be more willing to explore novel procedures and possibilities that could increase the likelihood of finding creative solutions (cf. Russ, 1993).

Much evidence does exist, in fact, that lends support to the general thesis that positive mood promotes a heuristic style of problem-solving (e.g. Abele, 1992b; Forgas, 2000a, 2000b).

However, the idea that positive mood tends to promote a 'no problem' attitude to the task at hand, whereas negative mood leads to perceiving the (same) task as problematic, and in need of careful scrutiny, may just as well be used as an argument in favor of the notion that negative mood may put the individual on higher alert. This may facilitate new interpretations and reframing of a task. Conversely, positive mood may put people's cognition on automatic and may lead them to overlook important new features in a task that is superficially disguised as a traditional, well-handled one. This is an interesting type of problem that has tended to be overlooked in the problem-solving literature. Such tasks are described and classified as 'deceptive problems' by Kaufmann (1988).

In a similar theoretical approach, which is gaining in popularity, Fiedler (2000) and Bless (2000) link the effects of positive and negative mood, respectively, to the processes of assimilation and accommodation. In a strange twist, they argue, however, that somehow assimilation promotes creativity and accommodation does not. But accommodation involves reorganizations of established schema into new, more adaptible ones, and is, in fact, the mechanism that normally is seen as most closely linked to creative thinking in the classical theory of Piaget (e.g. Furth, 1969). How do we reconcile the view specifically adopted by Bless (2000) that a happy mood tends to make people rely on stereotypes and prior judgment in the face of the frequently held definitional criterion of creativity to the effect that one of its essensials involves rejecting or seriously modifying conventional perspectives and solutions? (cf. Kaufmann, 1988; Newell, Shaw & Simon, 1979; Sternberg & Lubart, 1995.)

All in all, theories supposed to explain the positive-mood creativity link are either ad hoc rationalizations, or problematic as clearcut interpretations of the posited relationship, and may even be internally inconsistent.

Some Cause for Alarm

Several studies suggest that positive mood may also significantly impair problem-solving performance. Mackie & Worth (1989, 1991) argued that mood states will activate task-irrelevant material in memory and promote cognitive overload. They demonstrated that positive mood subjects showed reduced processing when shown stimuli for the same amount of time as neutral mood subjects. Increased exposure times were needed for positive mood subjects to systematically process information to the same level as negative mood subjects. It is unclear, however, why this reduction of capacity in processing is particularly marked in the case of positive mood, since the argument made by Mackie and Worth makes no provisions for an asymmetrical decrement, and should, in principle, also be obtained under negative mood. In line with the observation that positive mood may lead to reduced processing efforts, however, Martin, Ward, Achee & Wyer (1993) demonstrated that subjects stopped searching for task-relevant information sooner under positive as compared with negative mood conditions when asked to stop when they thought they had enough information.

Other findings also are consistent with the hypothesis that positive mood leads to less effort being spent in problem-solving. Sinclair & Mark (1995) found happy subjects to be less accurate than neutral and sad subjects in judging the magnitude and direction of correlation coefficients associated with each of nine scatterplots. Melton (1995) found subjects in a positive mood to be more prone than negative mood subjects to select unqualified solutions in solving linear syllogisms, and also to be more influenced by atmosphere effects in their problem-solving performance. In this study, Melton also threw doubts on the robustness of some key findings in the mood and creativity field. Positive mood subjects were not more adept than negative mood subjects in solving RAT tasks. As noted above, several studies also indicate that positive mood leaves people more open to biases in thinking and judgments. In contrast, negative (mildly depressed) mood seems to encourage more realistic perceptions and judgments, and decreases the tendency to be subject to biases (Alloy, 1986; Alloy & Abramson, 1979; Alloy, Abramson & Viscuti, 1981; Forgas, 1998, 2000a, 2000b; Tabachnik, Crocker & Alloy, 1989).

This is particularly alarming, since lack of creativity is often linked to what we may term 'strategy bias', where the subject carries on using a procedure that is no longer apt for solving the problem. This is the case in the classical Einstellung experiments reported by Luchins (Luchins & Luchins, 1950; cf. Kaufmann, 1979, 1988). Here the experimental procedures were varied in such a way that a previously successful procedure was no longer adequate (or possible) for solving subsequent tasks with some new twists.

It may, however, be argued that most of the studies demonstrating processing decrements for positive mood are not specifically carried out in the context of creative problem-solving tasks, and that somehow what is bad for systematic, logical thinking may be good for creativity.

Some Reasons Why Negative May Be Positive

Researchers in the creativity field (e.g. Boden, 1991; Mumford & Gustafson, 1988; Mumford, in press; Rothenberg, 1990; Weisberg, 1986, 1993) often emphasize the need for rational and 'tight' processing modes in high-level creative problem-solving. There are also good theoretical reasons to expect facilitating effects of negative mood on creative problem-solving. Virtually by definition, creative problem-solving entails a *modification* or *rejection* of conventional solutions (Boden, 1991; Kaufmann, 1993; Newell, Shaw & Simon, 1979)—processes that might well be prompted by negative mood. Runco (1994, 1999a) has argued convincingly that 'tension' and 'dissatisfaction' appear to be important prerequisites for creative problem-solving. In accordance with such a view, Mraz and Runco (1994) reported that an indicator of strongly negative mood (frequency of suicidal thoughts) was, indeed, significantly *positively* related to problem finding ability, indicative of an ability to imagine new and interesting problems. More recently, Zhou & George (2001) have demonstrated that workers who reported the strongest job dissatisfaction were rated as the most creative. This relationship was contingent on a high continuance commitment to the organization. It is true that we lack direct evidence on the mediating role of negative mood here. Yet the results may be reasonably interpreted as consistent with a theory like the one proposed by Runco above, to the effect that negative mood may promote a 'creative tension' and may lead people more easily to question the status quo. 'Defying the crowd' is, as Sternberg & Lubart (1995) argue, close to the essence of creativity. At least it may be a necessary prerequisite for creative problem-solving.

In line with such arguments, we find that problem *finding* has been seen as a key element in creativity, in the sense that it is often a necessary antecedent to creative problem-solving (cf. Getzels & Csikszentmihalyi, 1976; Runco, 1994; see also Root-Bernstein, this volume). Thus, there is a need for expanding studies in the field to also include tasks of problem finding as well as problem-solving. If we take seriously the idea that creative thinking is more than a happy-go-lucky idea excursion, then the processing decrements demonstrated for positive mood above do seem to present a problem for the general assumption of a straightforward positive causal link between positive mood and creativity.

Inconsistent Evidence on the Positive Mood–Creativity Link

Even if we keep strictly within the specific task domain of creative problem-solving, however, several findings anomalous to the positive mood-enhance-creativity theory have been reported. Jausovec (1989), comparing the effects of positive, negative and neutral mood on analogical transfer in insight problems, reported a complex set of findings. In one task, a facilitating effect of positive mood on an analogical transfer task was obtained. In another task, positive mood was detrimental to performance, whereas in a third task, no significant differences between the different moods appeared. It is interesting to note that the task where positive mood was found to be detrimental is a typical insight problem (the Radiation Problem; Duncker, 1945), held in the classical literature to be a representative indicator of core creative processes in creative problem-solving (e.g. restructuring; cf. Wertheimer, 1958). Weisberg (1994) pointed to an important distinction between productivity and creativity. With reference to the evidence indicating a significant relationship between hypomania and creativity cited above, Weisberg conducted an interesting case study of Schumann's mood bipolarity. The results suggested that the relationship demonstrated between elevated positive mood and creativity may reflect increased productivity, in the sense of quantity of products, but it did *not* generalize to a higher *quality* of creativity. Vosburg (1998b) made similar observations. In her study, positive mood significantly enhanced ideational fluency, compared to negative mood, but no significant differences between positive and negative mood were obtained on scores of originality and usefulness of ideas.

Evidence to the Contrary

Even more serious evidence against the hypothesis of a general positive effect of positive mood on creative problem-solving has recently been reported by Kaufmann & Vosburg (1997). They demonstrated a *reversal* of the standard finding that positive mood facilitates creative problem-solving. In the first study, subjects were given the task of solving two insight problems (the Hatrack and the Two String problem), as well as two control tasks (standard reasoning problems; Sentence Completion and Analogies). Mood was measured immediately prior to task performance by way of Russell's Adjective Check List (Russell, 1979). The results showed that positive mood was significantly *negatively* related to task performance. In a second study, subjects were given the same insight tasks. Positive and negative mood, as well as arousal, were measured by way of the Russell scale, as in the first study. In addition, positive, negative and neutral mood were experimentally induced by way of short video-clips. Both error rates and solution latencies were recorded as measures of performance. Natural positive mood measured by the adjective checklist was again found to be significantly negatively related to task performance. Moreover, experimentally induced positive mood subjects were found to be the poorest problem solvers compared to control, neutral and negative mood subjects, in that particular rank order. These findings argue strongly against any simple and unconditional proposition stating that positive mood facilitates creative problem-solving. Interestingly, no mood effects were observed on the control tasks, indicating that the mood selectively operated in the creative problem-solving tasks, in general line with Affect Infusion Theory (Forgas, 1995, 2000a).

Closely related, in a series of experiments, Gasper (in press) was able to empirically cash in on the hypothesis originally proposed by Kaufmann & Vosburg (1997), to the effect that induced positive mood, as compared to induced negative mood, promotes the Einstellung effect in problem-solving in selected tasks in the Luchins tradition.

A key task in the new 'creative cognition' tradition explored by Finke & Slayton (1988), Finke, Ward & Smith (1993) and Smith, Ward & Finke (1995), is the so called 'creative mental synthesis' task. Here the subject is presented with randomly generated shapes and alphanumeric figures, such as a line, a square, or the capital letters L, D, and X. The task is to combine a given number of such elements into a recognizable pattern. An example of a very easy task is giving the subjects a J and a D, and asking them to combine them into a meaningful figure. No doubt, you have already constructed the umbrella that most people easily can visualize. This task can be made very difficult by adding the number of elements to three and five, and can be scored on both correspondence (how well integrated the different constituent elements are), and creativity (how original and ingenious the configuration of the elements are).

Anderson, Arlett & Tarrant (1995) directly investigated the effect of mood on performance in this task by way of comparing induced positive, negative and neutral mood to various indicators of problem-solving performance. The Velten (1968) mood induction procedure was employed, where the subjects were given 60 statements, happy or sad, and asked to experience each statement fully. The results showed that positive mood had a significantly negative effect on performance, in particular compared to the neutral mood condition. Anderson, Arlett & Tarrant (1995) claim that the results

show positive mood to have a detrimental effect on the constraining of elements into a whole pattern, by settling for solutions with poor correspondence and lower creativity.

In line with such findings, in a recent experiment, Kaufmann & Vosburg (2002) found a crossed inter-action between mood and early vs. late idea production. Positive mood subjects scored significantly higher in early production, whereas negative and neutral mood subjects significantly outperformed the positive mood subjects in late production. Indeed, the positive mood condition seemed to produce a steep response gradient held by Mednick (1962) to be characteristic of non-creative individuals (cf. Martindale, 1990; Fasko, Jr., 1999), whereas the negative/neutral conditions were closer to the flat association gradient typical of creatives.

Recently, Szymanski & Repetto (2000) also reported that induced negative mood facilitated creative prob-lem-solving, measured by a story-telling task. In a fresh study conducted in a field situation, George & Zhou (2001), in close line with our arguments above, make the observation that "people in positive moods . . . may evaluate the status quo positively as well as their own ideas which may cause them not to put forth high levels of effort to make suggestions for significant improvements that are, in fact useful" (p. 3). They base their study primarily on the contextual Mood-as-Input Model advocated by Martin & Stoner (1996, 2000). From this vantage point they suggest that under important contextual conditions conducive to creativity and conducive to facilitating the effects of mood states, such as perceived recognition and rewards for crea-tivity as well as clarity of feelings, negative mood will tend to facilitate, whereas positive mood will be detrimental to, creativity. Creativity was measured by supervisory ratings. The negative mood hypothesis was supported but only weakly. However, there was a clearly significant detrimental effect for positive mood on creativity, in support of the theoretically derived hypothesis. Such findings, at the very least, clearly warn against any general and straightforward conclu-sion to be drawn from such findings as reported in the frequently cited study by Isen, Daubman & Nowicki (1987).

Among the results referred above, the Kaufmann and Vosburg findings seem particularly difficult to recon-cile with the classical findings reported by Isen, Daubman & Nowicki (1987), where performance on a task strongly similar to the ones used in Kaufmann and Vosburg experiments was facilitated by positive mood. There is, however, one potentially critical difference between the Isen et al. and Kaufmann and Vosburg study that may account for the discrepant findings. In the Isen et al. experiments, the task was solved in a practical setting that allowed the subjects continuous feedback on their solution attempts. In the Kaufmann and Vosburg studies, the tasks were performed in writing with no external feedback on solution ideas. Thus, the subjects were left to set their own subjective standards for acceptable solutions. Kaufmann & Vosburg (1997) and Vosburg & Kaufmann (1999) suggested that the existing theories are insufficient to account for all of the available evidence on the effect of mood on creative problem-solving performance. Specifically, they argued that there is a need to take account of the task-solution requirements in predicting mood effects. We will return to this important point as a means of reconciling seemingly contradictory findings when we present our general theory of mood effects in creative problem-solving in the last section of this chapter.

Moving Away from the Singular Positive Mood-Enhance-Creativity Myth

We can now clearly see that we have come far away from the simple positive mood-enhance-creativity hypothesis. The functional relationship between mood and creativity seems to be as complex as the phenomenon of creativity itself. We should not be surprised. The fabric of creativity is made up from a large number of constituent psychological and contextual components, and is not likely to be captured by simple formulas, and neither is its relationship to such broad variables as mood. We have seen that there is good reason to think that some important aspect of creativity, such as an easy flow of ideas, may indeed be facilitated by positive mood, and inhibited by negative mood. This is, indeed, a very important part of the total creative problem-solving process, particularly in the early 'lift off' stage of creative thinking. However, other critically important aspects of creativity, such as reframing, and striving hard to find insightful and original solutions, may, indeed be prevented by positive mood and facilitated by negative mood. It seems clear that we need a more complex theory to account for the large number of seemingly inconsistent and even paradoxical findings in this area. In the last section, we will sketch the essential elements and structure of a general theory of mood and creative problem-solving that is based on independently verified empirical premises and may do the job of integrating the bewildering array of findings in the mood and creativity area.

A Theory of Mood Effects on Creative Problem Solving

A theory of mood effects on creative problem-solving needs to be anchored in a more general theoretical perspective on the functions of emotions, particularly with respect to their role in cognition.

In general, we agree with Frijda's functionalist account of emotions (Frijda, 1986), where emotions are seen as basically having a signal, or heuristic cue function (cf. Kaufmann, 1996; Vosburg & Kaufmann,

1999). More specifically, we agree with the basic premise behind the cognitive tuning theory developed by Clore, Schwarz & Conway (1994) (see also Schwarz (1990, 2000)). Here a positive emotional state signals a satisfactory state in the task environment, whereas a negative emotional state signals that the task environment is problematic. Thus, mood may induce a 'frame of mind' (cf. Morris, 1989) that serves as a mental backdrop for choosing a particular strategy to deal with the task at hand.

We do, however, believe that this is only *half* the story, and that the theory needs to be supplemented by a second premise to the effect that mood also signals a state in the 'internal environment' of the information processor. According to Bandura (1997), self-efficacy may be defined as the "belief in one's capabilities to organize and execute the courses of action required to produce given attainments" (p. 3). Thus, we postulate that positive mood will lead to enhanced confidence in one's own ability to solve the problem at hand, and thus promote a higher level of self-efficacy. Conversely, negative mood is thought to lower confidence, promote pessimism, increase susceptibility to fear of failure and thus decrease the individual's level of self-efficacy. Such a relationship between mood and self-efficacy has in fact previously been empirically established in a study by Kavanagh & Bower (1985).

Why this second premise? Without it, we believe it will be difficult to reconcile some of the seemingly discrepant findings reported in the literature, such as the diametrically opposed findings on the solution of insight problems discussed above. We will return to this important point after we have presented our theory in more detail.

Mood and Its Influence on Different Aspects of Problem Solving

The principles stated above may now be applied more specifically to different aspects of problem-solving, and testable hypotheses may be derived. Different levels of abstraction may be chosen for the development of a theory in this field. For the present purpose, we will focus on four general dimensions of problem-solving. These are: (1) *Problem perception*, i.e. how the problem is represented to the individual in general terms; (2) *Solution requirements*, i.e. what are the criteria for an adequate solution to the problem; (3) *Process*, i.e. what type of processing of problem information is most adequate to deal with the problem at hand; and (4) *Strategy*, i.e. what kind of general method or tactic of solving the problem is required.

We do not claim that these dimensions are exhaustive. There are additional dimensions that might be considered, such as problem ownership, importance of the problem, reversibility/irreversibility of solutions, etc. (cf. Mitchell & Larson, 1988). But we do think that the four dimensions chosen here are all of primary importance whereas creative problem-solving is concerned, and that a limitation in variables to be dealt with at this stage of the inquiry is needed. This caveat also applies to the further examination of variables within the four dimensions.

Mood and Problem Perception

The issue of problem perception (or problem representation) is thought to be highly important in the study of problem-solving. What happens at this stage may to a strong degree determine the further processing of the problem and, ultimately, the success of solving the problem (cf. Simon, 1979), particularly with regard to *creative* problem-solving (Getzels & Csikszentmihalyi, 1976; Hayes, 1978; Kaufmann, 1988; Unsworth, 2001). There are many important sub-issues in this regard (cf. Runco, 1994), but here we will concentrate on the more general level of perception of the problem. A highly important aspect of problem representation that may be intimately linked to affect and mood states is the so-called 'valence' of the problem, in the sense of classifying the problem as an 'opportunity' or as a 'threat' (e.g. Jackson & Dutton, 1988; cf. Ginsberg & Venkatraman, 1992; Thomas et al., 1993).

A problem perceived as an opportunity is seen as positive, controllable and involving a potential gain, whereas a problem perceived as a threat is seen as negative, uncontrollable and involving a potential loss. A number of studies have demonstrated predictable differences in information processing as a consequence of defining the problem as threats or opportunities (e.g. Thomas et al., 1993). Seeing the problem as an opportunity results in more open information processing and stronger belief in one's own ability to solve the problem. In contrast, a threat representation leads to more restricted information search and to a more pessimistic outlook on one's ability to solve the problem (Jackson & Dutton, 1988).

From the theory presented above, it follows that positive mood should promote a frame of mind that is conducive to perceiving a problem as an opportunity, whereas negative mood should increase the likelihood of perceiving the problem as a threat. Jackson & Dutton (1988) make this assumption on a more intuitive basis. In a recent study by Mittal & Ross (1998), this posited link between mood states and the valence in categorizing the problem has been directly tested.

Mittal & Ross used business scenarios as problem situations, where the demographics of the customer base of a company was changing significantly. Mood was manipulated by presenting half of the subjects with a sad story and the other half with a happy story. Among the many interesting findings observed in this study, significant differences in issue interpretation in a strategic decision context were observed in the two mood groups. Subjects in a positive mood were more

likely to interpret the strategic issue as an opportunity compared to negative mood subjects.

Mood and Solution Requirements

Again it should be emphasized that there are many different issues that are important in this category. A cardinal consideration may, however, be said to reside in the distinction between optimizing vs. satisficing criteria originally proposed by Simon (e.g. 1956) and further developed by Simon (1977) and March (1994). Rather then viewing these criteria as distinct categories, we have previously argued that it is more convenient to regard them as opposite, extreme points on a continuum (Kaufmann, 1996; Kaufmann & Vosburg, 1997, 2002; Vosburg & Kaufmann, 1999). At the satisficing end of this continuum, the individual is held to construct a very simplified mental model of the solution space and accept the first solution that meets the corresponding aspiration level for an adequate solution. At the optimizing end of the continuum, the individual will attempt to perform an exhaustive search and rational evaluation of the expected utilities of all alternative solutions available. Kaufmann & Vosburg (1997) and Vosburg & Kaufmann (1999) posited that positive mood will promote a satisficing strategy of solution finding, whereas negative mood will be more appropriate under optimizing conditions. In their experiments, the tasks employed ranked extremely high in solution requirements (highly ingenious and unconventional solutions were required to fulfill all the solution requirements). They explained their findings that positive mood was detrimental and negative mood facilitative of finding the ingenious, 'ideal' solutions with reference to the hypothesis that positive mood promotes a satisficing and negative mood an optimizing solution strategy.

One important determinant of solution criteria along this continuum is the degree to which the task environment is perceived as satisfactory or problematical. Carrying this argument a step further, we may also argue that satisficing vs. optimizing criteria will be applied in the course of solving a problem, and that different individuals may settle for solutions of highly different quality. From the theory presented above, it follows that being in a positive mood would lead to perceiving a task as less requiring of high-quality solutions than when being in a negative mood, where a stricter criterion for an acceptable solution is more likely to be upheld. We argued above that this mechanism is the likely explanation for the 'paradoxical' mood effects observed in the Kaufmann and Vosburg study cited above. Also, we have seen above that Anderson, Arlett & Tarrant (1995) made similar observations and interpretations of a significant negative effect of positive mood on performance in the creative mental synthesis task.

We may now posit the general principle that positive mood will promote a satisficing strategy of solution finding, whereas negative mood will be more appropriate under optimizing conditions.

We would, however, like to add that, concerning our theory, availability of *feedback* on solution attempts may significantly change the postulated pattern. It is expected that positive mood subjects, when continuously informed of their success, and lack of success at the task, may be more persistent in pursuing the tasks, and may, under such circumstances, possibly also be more flexible in shifting strategies, than negative mood subjects. The latter may be expected to have their lower confidence in eventually being able to solve the task confirmed, and their performance may more easily 'freeze' and deteriorate correspondingly. These important qualifications follow from the postulated effect of mood on the 'inner environment', where differences in self-efficacy are held to be the crucial mediating variable. We have seen that making this distinction in mood effects may serve to reconcile the findings of Isen et al. (1987) and Kaufmann & Vosburg (1997). With no feedback on a self-set criterion for solution, positive mood subjects will more likely settle for solutions that have a lower quality, or, indeed, do not qualify as solutions at all. With negative performance on solution attempts, our theory leads us to expect that the situation may change significantly. Here the positive effect of self-efficacy may kick in among the positive mood subjects, and lead to more persistent and vigilant information processing in reaching out for the ideal solution of the task (cf. Phillips & Gully, 1997).

Such important contingencies are in line with the thrust of the argument in the Mood-as-Input Model developed by Martin & Stoner (1996) and Martin (2000), also advocated in the context of creativity by George & Zhou (2001).

Mood and Type of Process

In cognitive information processing theory, the kind of process is often distinguished in terms of the level of processing and breadth of processing (Anderson, 1990).

The level of processing refers to the question of whether processing occurs at a surface level, such as the sensory level, or at a deeper level, such as the semantic level, where information is further processed in terms of meaning and organizational structure in memory. Breadth of processing refers to the distance between informational units that are related during processing.

As we have seen above, the breadth dimension has been particularly targeted by mood theories (e.g. Isen & Daubman, 1984; Isen, Johnson, Mertz & Robinson, 1985). One possible explanation is that positive material is more richly interconnected than negative material, and that positive mood may provide a retrieval cue for linking information at a broader level, as suggested by Isen & Daubman (1984). This hypothesis also entails, however, that positive material

is better organized than negative material. We should then expect that positive mood promotes both a higher level and a broader information processing than negative mood. In our theory, however, positive mood is linked to broader information processing on the premise that positive mood promotes a less problematic perception of the task, and possibly also to overconfidence in ability to handle the task in question. Thus, positive mood may lead to a less cautious approach to the task than negative mood, promoting a broader but also a more superficial processing. More concisely, positive mood promotes broad and shallow processing, whereas negative mood leads to more constricted but deeper processing. This may be an advantage in a task that requires the generation of new ideas, but may be detrimental when the task requires careful considerations and deeper processing. In a recent experimental study, Elsbach & Barr (1999) observed the effect of induced positive and negative mood on the cognitive processing of information in a structured decision protocol. Here the subjects were given the task of assessing and weighing information in a complex case involving the core issue of whether a racing team should participate in an upcoming race. Negative mood subjects made much more careful and elaborate considerations than the positive mood subjects that were more prone to thrust superficial, 'holistic' judgment without going into the depth and details of the available information.

Mood and Strategy

This is probably the dimension that has received the most attention in discussion of the effects of mood on problem-solving behavior. As stated above, several theorists claim that positive mood increases the likelihood of employing heuristic-intuitive strategies whereas negative mood promotes the use of controlled, systematic and analytic information processing methods (e.g. Abele, 1992a).

A hallmark of a heuristic strategy is, of course, simplification. Analytic strategies are costly in terms of cognitive economy, and often the problem-solver has to resort to cruder and more general strategies in dealing with a fairly complex task. The choice of strategy may, however, also be determined by the general assessment of the structure of the task, and the individual's confidence in his or her ability to solve the problem. It follows, then, that positive mood would increase the likelihood of employing heuristic, short-cut strategies, whereas negative mood should lead to more cautious, analytic and systematic methods of dealing with the task at hand. As argued above, the available evidence tends to support this contention (cf. Forgas, 2000a, 2000b).

Task or Mood?

A highly pertinent problem in mood research is the question of how much of an effect mood really has on

TASK CONSTRAINTS

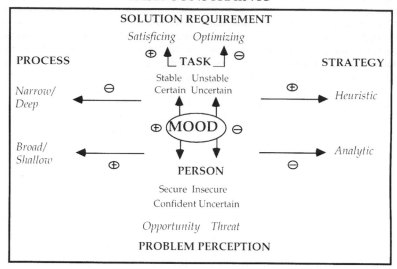

Figure 1. A theory of mood effects on creative problem solving.

performance, compared to the requirements of the task. Is mood a very significant factor with pervasive and persistent effects on task performance, or is mood a much more ephemeral factor that fairly easily yields to the requirements of the task? Is positive mood more fleeting than is negative mood, or vice versa? Are there individual differences in the reliance of mood as a heuristic cue to information processing? Is it possible, for example, that people with an 'intuitive' cognitive style rely more on mood in their choice of strategy and process, than individuals with an 'analytic' cognitive style, who are barely affected by mood as a heuristic cue for choice of mode of information processing? Indeed, in a recent study by Kuvaas, Kaufmann & Helstrup (2002), the typical, and normally robust mood dependency effect in memory and judgment was obtained only for subjects with little need for cognition.

These are important questions to consider, and very little is known about what the answers may be. So far, we have seen that mood effects seem to have their strongest effect in ill structured tasks, where an open and constructive information processing is required. Much more research is, however, needed to validate this general claim. A possible limitation of this principle may be that the conditions postulated by Forgas (2000a) may selectively open the door for positive mood effects to operate, whereas structured problem spaces, like that employed by Elsbach & Barr (1999), may be selectively favoring the operation of negative mood effects.

To take account of these considerations in our model, we have added task constraint as an external

boundary variable, which may significantly moderate the effect of mood as shown in Fig. 1, where the full model is graphically illustrated.

How wide or narrow this boundary is, if it varies for tasks and/or individuals, are important questions to consider to further our understanding of the functional importance of mood not only in creativity in the process of innovation, but also more generally in information processing and cognitive performances.

Concluding Comments on the Effect of Mood on Creativity in the Process of Innovation

Contrary to some bold statements made from adherents of the mainstream position to the effect that there is a straightforward, unconditional facilitative effect of positive mood on creativity, we have seen that the road from mood to creativity is far more crooked and full of twists and turns, often where we least expect them. The picture is one of complexity, not simplicity as the mainstream view would like to present it. Yet we have seen that a broad, general model enables us to view the differential effects of positive and negative mood as forming a meaningful and coherent pattern.

The complexity is apparent when we consider the relationship between mood and creativity per se. When we expand the equation to include creativity in the process of innovation, we may expect the picture to become even more complicated. As argued above, creativity may be seen as a necessary, but not sufficient, condition for innovation. Conceptually, innovation also requires that a creative product may be exploitable and possible to implement and diffuse in its market place, broadly conceived. Often this process is more difficult than the process of invention in the first place. There are important and careful considerations to be made before a decision is made to launch the effort and resources that go with the attempts at trying to capture a productive niche for a novel product, procedure or policies. From what we have learned so far, a positive mood would not be facilitative of the cognitive processes behind such deliberations and decisions. Quite to the contrary, we have seen consistently that the kind of nuanced, fine tuned and critical thinking that will enter into this stage of thinking in the process of innovation, would more likely be disrupted by positive mood, and facilitated by negative mood. Or we may even find that, in the total, complex process of innovation, there is a place for an ahedonic, cool mind too!

References

Abele, A. (1992a). Positive and negative mood influences on creativity: Evidence for asymmetrical effects. *Polish Psychological Bulletin*, **23**, 203–221.

Abele, A. (1992b). Positive vs. negative mood influences on problem solving: A review. *Polish Psychological Bulletin*, **23**, 187–202.

Alloy, L. B. (1986). *Cognitive processes and depression*. New York: Guilford.

Alloy, L. B. & Abramson, L. Y. (1979). Judgments of contingency in depressed and non-depressed students: Sadder but wiser? *Journal of Experimental Psychology: General*, **108**, 441–485.

Alloy, L. B., Abramson, L. Y. & Viscusi, D. (1981). Induced mood and illusion of control. *Journal of Personality and Social Psychology*, **41**, 1129–1140.

Anderson, R. E., Arlett, C. & Tarrant, L. (1995). Effect of instructions and mood on creative mental synthesis. In: G. Kaufmann, T. Helstrup & K. H. Teigen (Eds), *Problem Solving and Cognitive Processes* (pp. 183–197). Bergen: Fagbokforlaget.

Bandura, A.(1997). *Self-efficacy: The exercise of control*. New York: Freeman.

Benjafield, J. G. (1996). *Cognition*. Englewood Cliffs, NJ: Prentice-Hall.

Bless, H. (2000). The interplay of affect and cognition: The mediating role of general knowledge structures. In: J. P. Forgas (Ed.), *Feeling and Thinking: The Role of Affect in Social Cognition* (pp. 201–222). New York: Cambridge University Press.

Boden, M. A. (1991). *The creative mind: Myths and mechanisms*. New York: Basic Books.

Bowden, C. L. (1994). Bipolar disorder and creativity. In: M. P. Shaw & M. A. Runco (Eds), *Creativity and Affect* (pp. 201–222). Norwood, NJ: Ablex.

Bower, G. H. (1981). Mood and memory. *American Psychologist*, **36**, 129–148.

Carnevale, P. J. D. & Isen, A. M. (1986). The influence of positive affect and visual access on the discovery of integrative solutions in bilateral negotiation. *Organizational Behavior and Human Decision Processes*, **37**, 1–13.

Clore, G. L., Schwarz, N. & Conway, M. (1994). Affective causes and consequences of social information processing. In: R. S. Wyer & T. K. Srull (Eds), *Handbook of Social Cognition* (Vol 1). *Basic Processes* (2nd ed., pp. 323–417). Hillsdale, NJ: Lawrence Erlbaum.

Duncker, K. (1945). On problem solving. *Psychological Monographs*, **5** (Whole No. 270).

Ellis, H. C. & Ashbrook, P. W. (1988). Resource allocation model of the effects of depressed mood states on memory. In: K. Fiedler & J. P. Forgas (Eds), *Affect, Cognition and Social Behavior* (pp. 25–43). Toronto: Hogrefe.

Elsbach, K. E. & Barr, P. S. (1999). The effect of mood on individuals' use of structured decision processes. *Organization Science*, **10**, 181–189.

Estrada, C. A., Isen, A. M. & Young, M. J. (1994). Positive affect improves creative problem solving and influences reported source of practice satisfaction in physicians. *Motivation and Emotion*, **18**, 285–299.

Fasko, D. Jr. (1999). Associative theory. In: M. A. Runco & S. R. Pritzker (Eds), *Encyclopedia of Creativity* (Vol. 1, pp. 135–146). New York: Academic Press.

Fiedler, K. (1988). Emotional mood, cognitive style, and behavior regulation. In: K. Fiedler & J. P. Forgas (Eds), *Affect, Cognition and Social Behavior* (pp. 101–119). Toronto: C. J. Hogrefe.

Fiedler, K. (2000). Toward an account of affect and cognition phenomena using the BIAS computer algorithm. In: J. P. Forgas (Ed.), *Feeling and Thinking: The Role of Affect in Social Cognition* (pp. 223–252). Paris: Cambridge University Press.

Finke, R. A. & Slayton, K. (1988). Explorations of creative visual synthesis in mental imagery. *Memory & Cognition*, **21**, 283–293.

Finke, R. A., Ward, T. B. & Smith, S. M. (1992). *Creative cognition: Theory, research, and application*. Cambridge, MA: The MIT Press.

Forgas, J. P. (1995). Mood and judgment: The affect infusion model. *Psychological Bulletin*, **117**, 39–66.

Forgas, J. P. (2000a). Introduction: The role of affect in social cognition. In: J. P. Forgas (Ed.), *Feeling and Thinking: The Role of Affect in Social Cognition* (pp. 1–28). New York: Cambridge University Press.

Forgas, J. P. (2000b). Affect and information processing strategies. An interactive relationship. In: J. P. Forgas (Ed.), *Feeling and Thinking: The Role of Affect in Social Cognition* (pp. 253–280). New York: Cambridge University Press.

Frijda, N. H. (1986). The laws of emotion. *American Psychologist*, **43**, 349—358.

Furth, H. G. (1969). *Piaget and knowledge: Theoretical foundations*. London: Prentice-Hall.

Gasper, K. (in press). When necessity is the mother of invention: Mood and problem solving. *Journal of Experimental Social Psychology*.

Ginsberg, A. & Venkatraman, N. (1992). Investing in new information technology: The role of competitive posture and issue diagnosis. *Strategic Management Journal*, **13**, 37–53.

George, J. M. & Zhou, J. (2001). *Understanding when bad moods foster creativity and good ones don't: The role of clarity of feelings*. Washington, D.C.: The Academy of Management Conference.

Getz, I. & Lubart, T. I. (1999). The Emotional Resonance Model for generating associations. In: S. W. Russ (Ed.), *Affect, Experience, and Psychological AdjustMent* (pp. 41–65). Philadelphia, PA: Brunner/Mazel.

Getz, I. & Lubart, T. I. (2000). An emotional-experiential perspective on creative symbolic-metaphorical processes. *Consciousness and Emotion*, **1**, 89–118.

Getzels, J. W. & Csikszentmihalyi, M. (1976). Problem finding and creativity, *The creative vision: A longitudinal study of problem finding in art* (pp. 236—251). New York: Wiley-Interscience.

Greene, T. R. & Noice, H. (1988). Influence of positive affect upon creative thinking and problem solving in children. *Psychological Reports*, **63**, 895–898.

Hayes, J. R. (1978). *Cognitive psychology: Thinking and creating*. Homewood, IL: Dorsey Press.

Hirt, E. R. (1999). Mood. In: M. A. Runco & S. R. Pritzker (Eds), *Encyclopedia of Creativity* (Vol. 2, pp. 241–250). New York: Academic Press.

Hirt, E. R., McDonald, H. E. & Melton, R. J. (1996). Processing goals and the affect–performance link: Mood as main effect or mood as input? In: L. L. Martin & A. Tesser (Eds), *Striving and Feeling: Interactions Among Goals, Affect, and Self-regulation* (pp. 303–328). Mahwah, NJ: Erlbaum.

Isen, A. M. (1984). Toward understanding the role of affect in cognition. In: R. Wyer & T. Srull (Eds), *Handbook of Social Cognition* (Vol. 3, pp. 179–236). Hillsdale, NJ: Erlbaum.

Isen, A. M. (1987). Positive affect, cognitive processes, and social behavior. In: L. Berkowitz (Ed.), *Advances in Experimental Social Psychology* (Vol. 20, pp. 203–253). New York: Academic Press.

Isen, A. M. (1993). Positive affect and decision making. In: M. Lewis & J. Haviland (Eds), *Handbook of Emotions* (pp. 261–277). New York: Guilford.

Isen, A. M. (1999). On the relationship between affect and creative problem solving. In: S. W. Russ (Ed.), *Affect, Creative Experience and Psychological Adjustment* (pp. 3—17). Philadelphia, PA: Brunner/Mazel.

Isen, A. M. & Baron, R. A. (1991). Positive affect as a factor in organizational behavior. *Research in Organizational Behavior*, **13**, 1–53.

Isen, A. M. & Daubman, K. A. (1984). The influence of affect on categorization. *Journal of Personality and Social Psychology*, **47**, 1206–1217.

Isen, A. M., Daubman, K. A. & Nowicki, G. P. (1987). Positive affect facilitates creative problem solving. *Journal of Personality and Social Psychology*, **52**, 1122–1131.

Isen, A. M., Johnson, M. M. S., Mertz, E. & Robinson, G. F. (1985). The influence of positive affect on the unusualness of word associations. *Journal of Personality and Social Psychology*, **48** (6), 1413–1426.

Jackson. S. E. & Dutton, J. E. (1988), Discerning threats and opportunities. *Administrative Science Quarterly*, **33**, 370–387.

Jamison, K. R. (1993). *Touched with fire: Manic depressive illness and the artistic temperament*. New York: The Free Press.

Jausovec, N. (1989). Affect in analogical transfer. *Creativity Research Journal*, **2**, 255–266.

Kaufmann, G. (1979). The Explorer and the Assimilator: A cognitive style distinction and its potential for innovative problem solving. *Scandinavian Journal of Educational Research*, **23**, 101–108.

Kaufmann, G. (1988). Problem solving and creativity. In: K. Gronhaug & G. Kaufmann (Eds), *Innovation: A cross-disciplinary approach* (pp. 87–137). Oslo: Norwegian University Press/Oxford University Press.

Kaufmann, G. (1993). The content and logical structure of creativity concepts: An inquiry into the conceptual foundations of creativity research. In: S. G. Isaksen, M. C. Murdock, R. L. Firestien & D. J. Treffinger (Eds), *Understanding and Recognizing Creativity: The Emergence of a Discipline* (pp. 141–157). Norwood, NJ: Ablex.

Kaufmann, G. (1996). *The mood-and-creativity puzzle: Toward a theory of mood effects on problem solving*. Paper presented at the Convention for the American Psychological Association.

Kaufmann, G. & Vosburg, S. K. (1997). 'Paradoxical' mood effects on creative problem solving. *Cognition and Emotion*, **11**, 151–170.

Kaufmann, G. & Vosburg, S. K. (2002). Mood in early and late idea production. *Creativity Research Journal*, **14**, 317–333.

Kavanagh, D. J. & Bower, G. H. (1985). Mood and self-efficacy: Impact of joy and sadness on perceived capabilities. *Cognitive Therapy and Research*, **9**, 507–525.

Kuvaas, B., Kaufmann, G. & Helstrup, T. (2002). *Need for cognition and mood dependency in memory and judgment*. Paper presented at the Academy of Management Conference.

Luchins, A. S. & Luchins, E. H. (1950). New experimental attempts at preventing mechanization in problem solving. *Journal of General Psychology*, **42**, 279–297.

Mackie, D. M. & Worth, L. T. (1989). Processing deficits and the mediation of positive affect in persuasion. *Journal of Personality and Social Psychology*, **57**, 27–40.

Mackie, D. M. & Worth, L. T. (1991). Feeling good, but not thinking straight: The impact of positive mood on persuasion. In: J. Forgas (Ed.), *Emotion and Social Judgments* (pp. 201—219). London: Pergamon.

March, J. G. (1994). *A primer of decision making*. New York: Simon & Schuster.

Martin, L. L. (2000). Moods do not convey information: Moods in context do. In: J. P. Forgas (Ed.), *Feeling and Thinking: The Role of Affect in Social Cognition* (pp. 153–177). New York: Cambridge University Press.

Martin, L. L. & Stoner, P. (1996). Mood as input: What we think about how we feel determines how we think. In: L. L. Martin & A. Tesser (Eds), *Striving and Feeling: Interactions Among Goals, Affect, and Self-regulation* (pp. 279–301). Mahwah, NJ: Erlbaum.

Martin, L. L., Ward, D. W., Achee, J. W. & Wyer, R. S. (1993). Mood as input: People have to interpret the motivational implications of their moods. *Journal of Personality and Social Psychology*, **64**, 317–326.

Martindale, C. (1990). *Cognitive psychology: A neural network approach*. New York: Wadsworth.

Mednick, S. A. (1962). The associative basis of the creative process. *Psychological Review*, **69**, 220–232.

Melton, R. J. (1995). The role of positive affect in syllogism performance. *Personality and Social Psychology Bulletin*, **21**, 788—794.

Milgram, R. M. & Milgram, N. (1978). Quality and quantity of creative thinking in children and adolescents. *Child Development*, **49**, 385–388.

Mitchell, J. E. & Madigan, R. J. (1984). The effects of induced elation and depression on interpersonal problem solving. *Cognitive Therapy and Research*, **8**, 277–285.

Mitchell, T. R. & Larson, R. J. (1988). *People in organizations*. New York: McGraw-Hill.

Mittal, V. & Ross, W. T. (1998). The impact of positive and negative affect and issue framing on issue interpretation and risk taking. *Organizational Behavior and Human Decision Processes*, **76**, 289–324.

Morris, W. N. (1989). *Mood: The frame of mind*. New York: Springer-Verlag.

Mraz, W. & Runco, M. A. (1994). Suicide ideation and creative problem solving. *Suicide and Life-Threatening Behavior*, **24** (1), 38—47.

Mumford, M. D. (in press). Where have we been, where are we going to? Taking stock in creativity research. *Creativity Research Journal*.

Mumford, M. D. & Gustafson, S. B. (1988). Creativity syndrome: Integration, application, and innovation. *Psychological Bulletin*, **103**, 27–43.

Murray, N., Sujan, H., Hirt, E. R. & Sujan, M. (1990). The influence of mood on categorization: A cognitive flexibility interpretation. *Journal of Personality and Social Psychology*, **59**, 411–425.

Newell, A., Shaw, J, C. & Simon, H. A. (1979). The processes of creative thinking. In: H. A. Simon (Ed.), *Models of Thought* (Vol. 1, pp. 144–174). New Haven, CT: Yale University Press.

Philips, J. M. & Gully, S. M. (1997). Role of goal orientation, need for achievement, and locus of control in self-efficacy and goal setting process. *Journal of Applied Psychology*, **82**, 793–802.

Rothenberg, A. (1990). *Creativity and madness: New findings and old stereotypes*. Baltimore, MD: Johns Hopkins University Press.

Runco, M. A. (1994). Creativity and its discontents. In: M. P. Shaw & M. A. Runco (Eds), *Creativity and Affect* (pp. 102–126). Norwood, NJ: Ablex

Runco, M. A. (1999a). Tension, adaptability, and creativity. In: S. W. Russ (1999). *Affect, Creative Experience, and PsychoLogical Adjustment* (pp. 165–194). Philadelphia, PA· Brunner/Mazel.

Runco, M. A. (1999b). Time. In: M. A. Runco & S. R. Pritzker (Eds), *Encyclopedia of Creativity* (Vol. 1, pp. 659–663). New York: Academic Press.

Runco, M. A. & Charles, R. E. (1993). Judgments of originality and appropriateness as predictors of creativity. *Personality and Individual Differences*, **15**, 537–546.

Russ, S. W. (1993). *Affect and creativity: The role of affect and play in the creative process*. Hillsdale, NJ: Lawrence Erlbaum.

Russ, S. W. (Ed.) (1999). *Affect, creative experience, and psychological adjustment*. Philadelphia, PA: Brunner/Mazel.

Russell, J. A. (1979). Affective space is bipolar. *Journal of Personality and Social Psychology*, **37** (3), 345–356.

Schuldberg, D. (1990). Schizotypal and hypomanic traits, creativity, and psychological health. *Creativity Research Journal*, **3**, 218–230.

Schuldberg, D. (1999). Creativity, bipolarity, and the dynamics of style. In: S. W. Russ (Ed.), *Affect, Creative Experience, and Psychological Adjustment* (pp. 221–237). Philadelphia, PA: Brunner/Mazel.

Schwarz, N. (1990). Feelings as information: Informational and motivational functions of affective states. In: E. T. Higgins & R. Sorrentino (Eds), *Handbook of Motivation and Cognition: Foundations of Social Behavior* (Vol. 2, pp. 527–561). New York: Guilford.

Schwarz, N. (2000). Emotion, cognition and decision making. *Cognition and Emotion*, **14**, 433–440.

Schwarz, N. & Bless, H. (1991). Happy and mindless, but sad and smart? The impact of affective states on analytic reasoning. In: J. P. Forgas (Ed.), *Emotion and Social Judgments* (pp. 55–71). New York: Pergamon Press.

Shapiro, P. J. & Weisberg, R. W. (1999). Creativity and bipolar diathesis: Common behavioral and cognitive components. *Cognition and Emotion*, **13**, 741–762.

Shapiro, P. J. Weisberg, R. W. & Alloy, L. B. (2000). *Creativity and bipolarity: Affective patterns predict trait creativity*. Paper presented at the Convention of the American Psychological Society.

Simon, H. A. (1977). *The new science of management decision*. Englewood Cliffs, NJ: Prentice-Hall.

Simon, H. A. (1979). *Models of thought*. New Haven, CT: Yale University Press.

Sinclair, R. C. & Mark, M. M. (1995). The effects of mood state on judgmental accuracy: Processing strategy as a mechanism. *Cognition and Emotion*, **9**, 417 438.

Smith, S. M., Ward, T. B. & Finke, R. A. (1995). *The creative cognition approach*. Cambridge, MA: The MIT Press.

Staw, B. M. & Barsade, S. G. (1993). Affect and managerial performance: A test of the Sadder-but-Wiser vs. Happier-and-Smarter Hypotheses. *Administrative Science Quarterly*, **38**, 304–331.

Sternberg, R. J. & Lubart, T. I. (1995). *Defying the crowd*. New York: The Free Press.

Szymanski, K. & Repetto, S. E. (2000, August). *Does negative mood increase creativity?* Paper presented at the Convention for the American Psychological Association.

Tabachnik, N., Crocker, J. & Alloy, L. B. (1989). Depression, social comparison, and the false consensus effect. *Journal of Personality and Social Psychology*, **45**, 688–699.

Thomas, J. B., Clark, S. M. & Gioia, D. A. (1993). Strategic sensemaking and organizational performance: Linkages among scanning, interpretation, action, and outcomes. *Academy of Management Journal*, **36**, 239–270.

Unsworth, K. (2001). Unpacking creativity. *Academy of Management Review*, **26**, 289–297.

Velten, W. (1968). A laboratory task for induction of mood states. *Behavior Research & Therapy*, **6**, 473–482.

Vosburg, S. K. (1998a). The effects of positive and negative mood on divergent thinking performance. *Creativity Research Journal*, **11**, 165–172.

Vosburg, S. K. (1998b). Mood and the quantity and quality of ideas. *Creativity Research Journal*, **11**, 315–324.

Vosburg, S. K. & Kaufmann, G. (1999). Mood an creativity. A conceptual organizing perspective. In: S. W. Russ (Ed.), *Affect, Creative Experience and Psychological Adjustments* (pp. 19–39). Philadelphia, PA: Brunner/Mazel.

Weisberg, R. W. (1986). *Creativity: Genius and other myths*. New York: W. H. Freeman.

Weisberg, R. W. (1993). *Creativity: Beyond the genius*. New York: W. H. Freeman.

Weisberg, R. W. (1994). Genius and madness? A quasi-experimental test of the hypothesis that manic-depression increases creativity. *Psychological Science*, **5**, 361–367.

Wertheimer, M. (1958). *Productive thinking*. New York: Harper & Row.

Zhou, J. & George, J. M. (2001). When job dissatisfaction leads to creativity: Encouraging the expression of voice. *The Academy of Management Journal*, **44**, 682–696.

The International Handbook on Innovation
Edited by Larisa V. Shavinina

Case Studies of Innovation: Ordinary Thinking, Extraordinary Outcomes

Robert W. Weisberg

Department of Psychology, Temple University, USA

Abstract: Researchers studying the creative process have been faced with the issue of how to study the thought processes of individuals at the highest levels of accomplishment. Much research in this area uses qualitative case studies. The present chapter uses what can be called quasi-experimental quantitative methods to examine case studies, which make possible rigorous testing of hypotheses concerning the creative process. Examples discussed include Picasso's development of his painting *Guernica* and Edison's electric light. Results from the case studies serve as the basis for conclusions concerning the thought processes underlying innovation at the highest levels.

Keywords: Innovation; Creativity; Thought processes; Thinking; Case studies; Creative process.

Introduction

In my work, I use the term 'creative' to refer to the production of goal-directed novelty (Weisberg, 1993, chap. 8). Thus, the creative process consists of the cognitive processes that result in production of goal-directed novelty, a creative individual is one who intentionally produces such, and a creative work is a novel product produced intentionally by some individual. In this chapter I will use *innovation* as synonymous with *creative product*. This definition differs from that used by most researchers who study creativity, who usually include as a criterion for calling something creative that it be *of value* (see chapters in this volume). That is, a proposed solution to a problem must actually solve the problem in order to be called creative; an invention must carry out the task for which it was designed; a work of art must find an audience. For reasons discussed elsewhere (Weisberg, 1993, chap. 8), I believe that including *value* in the definition of creative and related concepts significantly clouds several important issues. However, in the present context the specifics of the definition are irrelevant, since all the innovations to be discussed in this chapter are undoubtedly of the highest value. Therefore, no one would argue that the conclusions drawn are irrelevant because they are drawn from analyses of products which are not creative.

Methods for Studying Creative Thinking

Laboratory Investigations and Their Limitations

Researchers who study creative thinking usually have approached the subject matter from one of two directions. On the one hand, investigators have carried out experimental studies, typically centering on the study of undergraduates working on laboratory problems (e.g. Gick & Holyoak, 1980; Glucksberg & Weisberg, 1966; Weisberg, 1995, 1999; Weisberg & Alba, 1981; Weisberg & Suls, 1973). Successful problem solving, *per se*, requires creative thinking, so such investigations have face validity as studies of creative processes. However, one can raise a question concerning their external validity: To what population can one generalize conclusions drawn from such studies? On the basis of controlled investigations of undergraduates working on small-scale laboratory problems, one may not be able to make inferences about such individuals as Picasso, Mozart, Edison, and Einstein, who are also part of the population of creative thinkers one wishes to understand. Indeed, it could be argued that those latter individuals, and others like them, are the most important component of that population. If we wish to be able to claim at some point that we truly understand creative thinking, we will have to be able to explain how innovations at the highest level are produced. Examining undergraduates working on laboratory exercises may not provide an understanding of the thought processes underlying the achievements of the most eminent among us. Of course, one could simply assume that laboratory investigations can illuminate aspects of the creative process that are relevant to all creative thinkers. However, such an assumption, if made without empirical support, seems to beg a basic question which motivates the study of creative thinking: What are the thought processes which bring about the greatest advances?

Studying High-Level Creativity

Thus, there is a potential limitation in our understanding of the creative process if we rely solely on results from experimental studies of undergraduates. Researchers have therefore also attempted to investigate directly the creative process in individuals of the first rank. First, researchers have observed such individuals at work (e.g. Dunbar, 2001). Second, researchers have collected autobiographical reports concerning how individuals of great accomplishment carried out their work (e.g. Ghiselin, 1952). A third method has been to carry out biographical studies of eminent creators, hoping thereby to illuminate the creative process (e.g. Gardner, 1993). Fourth, researchers have used archival information to try to reconstruct the processes underlying significant creative advances (e.g. Gruber, 1981). Finally, researchers have used historiometric methods to carry out quantitative analysis of historical data (e.g. Martindale, 1990; Simonton, this volume). After a brief review of these various methods, I will describe another method, which I call the 'quantitative historical case study'. This method centers on quantitative analysis of the work of individuals, based on archival records, and in some circumstances allows us to carry out rigorous tests of predictions concerning the creative process in eminent individuals for whom laboratory data are not available (Ramey & Weisberg, 2002; Weisberg, 1994, 1999, 2002; Weisberg & Buonanno, 2002).

Studying High-Level Creativity in Real Time: '*In Vivo*' Investigations

One way to overcome the limitations inherent in using results from studies of undergraduates to draw conclusions concerning creative thinking in people of the highest rank is to try to observe such people directly. Dunbar (e.g. 1995, 2001) over several years observed the ongoing activities in four high-level research laboratories in molecular biology. He was given complete access to the laboratory activities by the directors, who in each case was a scientist of high repute, with numerous publications and grants, and who had been awarded prizes based on the quality of the research carried out in his or her laboratory. Dunbar regularly attended and recorded laboratory meetings, discussed ongoing work with the scientists involved,

and was given copies of research papers in various stages of completion. He became so integral a part of those various laboratories that he was informed several times when discoveries were imminent, so that he could be present.

These observations enabled Dunbar to make several discoveries concerning the processes underlying creative work in those laboratories (2001). As one example, he found that scientists' use of analogies in theorizing is relatively limited, and that a scientist may not move far afield when trying to think of a situation analogous to some poorly understood phenomenon that he or she is facing. Any analogies produced will usually come from domains closely related to the one in which the scientist is working. This finding indicates that creative scientists do not rely on far-ranging leaps of thought, based on remote analogies, when trying to understand a recalcitrant phenomenon. Also, Dunbar (1995) has found that a scientist's conception of his or her own work is sometimes changed radically as the result of input from colleagues during laboratory meetings in which data and analyses are discussed. The scientist alone is less likely to try to deal with recalcitrant data. Thus, the scientist may not experience a sudden insight into some phenomenon, based on a concomitant restructuring of understanding, without help from people who may be reluctant to accept the scientist's perspective on the phenomenon. Dunbar has also found that, perhaps surprisingly, senior scientists are more likely than their junior colleagues to change their analysis of some phenomenon in light of disconfirming data. One might have thought that senior scientists, being more set in their ways, would be more likely to cling to old ways of looking at things, but this was not the case in the laboratories studied by Dunbar. It seems that senior scientists, because of experience with being mistaken, are less likely to hold strongly to a hypothesis when it is contradicted by data.

In sum, Dunbar's research has shown how scientific research can be studied *in vivo*, and has produced a number of important results. However, although the laboratories studied by Dunbar are directed by scientists of strong reputation, those individuals and their research groups are as yet nowhere near the significance of Einstein, Darwin, and the like, and so there is still the issue of how the conclusions from Dunbar's research would illuminate the latter. In addition, taking Dunbar's tack leaves us without a way of gaining an understanding of historically significant figures who are not available to provide access to their laboratories. Thus we are led to other options, which deal with historically significant individuals *per se*.

Subjective Reports of the Creative Process

A number of investigators have used personal reports of individuals of extraordinary accomplishment to provide entrée into the creative process (e.g. Ghiselin, 1952). These reports, coming from individuals from across the spectrum of creative fields, including poetry, literature, music, visual arts, and the sciences, are in the form of letters, addresses before scientific societies or other groups, responses to questionnaires, and in-depth interviews. Ghiselin made clear the motivation for use of such reports as the basis for theorizing concerning creativity.

> (A) large amount of comment and description of individual processes and insights has accumulated, most of it fragmentary, some of it not perfectly reliable. Among these materials the most illuminating and entertaining are the more full and systematic descriptions of invention and the reflections upon it made by the men and women most in position to observe and understand, the thinkers and artists themselves (1952, p. 11).

Ghiselin's enthusiasm notwithstanding, first-hand reports concerning the creative process suffer from several shortcomings. They are usually made long after the fact, which raises questions about their accuracy (Perkins, 1981; Weisberg, 1986, 1993). For example, the well-known report by Poincaré (excerpted in Ghiselin, 1952, pp. 33–42), concerning how several of his most important mathematical discoveries came about, was made many years after the fact (Miller, 1996). Similarly Kekulé's often-cited report of his 'dream' of a snake grasping its tail and dancing before his eyes, which served as the basis for his formulation of the circular structure of the benzene ring, was presented some 25 years after it occurred, as part of an address he gave at a Festschrift commemorating his discovery (Rocke, 1985). Long lags between an event and the report of that event raise questions about the accuracy of the report. Dunbar (2001) has found that the investigators in the laboratories that he studied were likely to forget how advances in their thinking came about. Because Dunbar had taped laboratory meetings, he was able to check the accuracy of the investigators' recollections of events in the meeting that had resulted in changes in conceptions of some phenomenon. Even after as short a period as 6 months, people were able to remember the outcome of the lab meeting—e.g. a new way of conceptualizing some phenomenon—but were not able to remember how that change had been brought about. Thus, subjective reports given years after the fact are of questionable value as data for theories of creative thinking.

Even if the subjective report were given very soon after the events in question, which might reduce potential memory problems, in most cases we have no way of verifying the accuracy of the report, because usually there is no objective evidence to support it. Indeed, given the subject matter of many of these autobiographical reports—they are presented to give us insight into what the isolated person was thinking when he or she made some creative advance—it is not clear that there could be information that could serve to verify them. The difficulty verifying subjective reports is particularly troubling in the face of evidence that some often-cited published reports are not true, including 'Mozart's letter' (in Ghiselin, 1952, pp. 44–45) concerning his process of musical composition, and Coleridge's description of how the poem 'Kubla Kahn' came to be (in Ghiselin, 1952, pp. 84–85; see Weisberg, 1986, for further discussion).

Furthermore, the individuals providing those reports, although of undeniable eminence in their fields, have no training as behavioral scientists, which may limit their ability to provide valuable data, even if such are available. That is, contrary to Ghiselin's belief quoted above, the creative individual—painter, poet, or physicist—may *not* be in the best position to observe and understand the processes underlying his or her achievements. The position taken in this chapter is that the cognitive scientist, equipped with tools to analyze objective data, is the individual most likely to make valid observations about the creative process. Finally, subjective reports describe autobiographical events and as such provide no information directly amenable to quantitative analysis. This limits their usefulness as the basis for development of scientific theories, the goal of which is to understand creative thinking.

Biographical Case Studies

In a move away from the problems inherent in using autobiographical reports as the basis for theories of creative thinking, Gardner (1993) carried out biographical studies of seven of the most eminent creative individuals of the 20th century, Freud, Einstein, Picasso, Stravinsky, Martha Graham, T. S. Eliot, and Gandhi, each of whom exemplifies one of Gardner's proposed multiple intelligences. He used these biographies to derive a number of conclusions concerning how each individual brought about his or her groundbreaking work. For example, Gardner emphasizes the role of a support group in providing a sympathetic arena in which the individual can introduce radical ideas. Gardner's approach is a significant step toward an objective basis for a theory of creative thinking, because it relies on historical data. However, a quantitatively oriented psychologist is left unsatisfied with biographies as the basis for an analysis of creative thinking, because there is very little in the way of a unique psychological contribution to the analysis. As one example of this lack, biographies, although undoubtedly informative, provide little in the way of quantitative data to serve as the basis for scientific theorizing; e.g. there are no data tables or graphs in the 400-plus pages of Gardner's book.

Historical Case Studies: Archival Data and Reconstruction of Process

A number of investigators have examined individual cases of the highest level of creative achievement, such as Gruber's (1981) analysis of Darwin's development of the theory of evolution through natural selection (see also Holmes (1980) and Tweney (1989)). Gruber's ground-breaking study was based on archival data, i.e. Darwin's notebooks, and his work stimulated interest in case studies of creative thinking (e.g. Perkins, 1981; Weisberg, 1986). One difference between Gruber's case study of Darwin and Gardner's (1993) biographical case studies is Gruber's concentration on Darwin's development of his theory. That is, one could contrast Gruber's and Gardner's perspectives by saying that Gruber presented a biography, not of Darwin, but of the theory of natural selection. In the case-study perspective, the emphasis is less on the creator than on the work. Inferences about the creative process are made on the basis of changes that occurred in the work, as evidenced by objective data in notebooks, laboratory records, and the like.

Gruber (1981) has proposed on the basis of his analysis of Darwin that the creative process may be unique in each individual, and that there may be no generalizations to be made about the creative process or the creative person, because each creative accomplishment is carried out by a unique individual in a unique set of circumstances. Such a conclusion, however intriguing it may be, must be supported or rejected through the analysis of other case studies. Furthermore, Gruber's method of analysis may make it difficult to discover generalizations about the creative process, even the generalization that no generalizations can be made about the creative process, because the qualitative and descriptive slant of his method provides little in the way of data to be analyzed in the search for generalizations. As in Gardner's (1993) biographical studies, there are no data tables in Gruber's study of Darwin. Gruber may be correct in his claim that no strong generalizations will come out of case studies of creative thinking, but the only way to know is to carry out more of them, in a form which allows latent generalizations to become manifest.

A sub-genre of case studies has centered on computer modeling of the creative processes that led to significant innovations (e.g. Kulkarni & Simon, 1988; Langley, Simon, Bradshaw & Zytkow, 1987). In these studies, computer programs are given data comparable to those presumed available to the investigator whose case study is being analyzed. The computer program is usually minimally structured, so that the only methods available to it are general-purpose heuristics, and the question of main interest is whether a program so structured will produce the discovery that the researcher produced. It has been concluded by a number of investigators that such programs have been successful, which supports the claim that a number of innovations in science have been brought about using very general methods (Klahr & Simon, 1999; Langley et al., 1987). The case studies to be discussed in the present chapter can be seen as being influenced by the overall conception behind the computer-simulation studies, that is, the idea that it is possible to analyze in relatively straightforward terms the cognitive processes that produced innovations of the highest order. One basic difference between the present chapter and computer simulation studies is that I have made no attempt to model any case studies using computers.

Historiometric Methods

Simonton (e.g. this volume) and Martindale (e.g. 1990) have applied quantitative methods to historical data in order to formulate and test hypotheses concerning creative thinking. For example, Simonton has investigated the influence of war and other social upheaval on creativity, by breaking the last two millennia into 20-year 'epochs', and determining for each the frequency of social unrest and of creative accomplishment, based on such measures as years in which active war was carried out, and numbers of creative individuals who flourished. Using methods such as time-lagged coefficients of correlation, Simonton has attempted to distill causal relations from historical data, and has concluded, for example, that occurrence of war involving a nation results in a decrease in creative accomplishment in that nation in the following epoch. Also, the occurrence of a significant number of individuals of high levels of accomplishment during one epoch is positively related to the level of accomplishment in the next generation, which Simonton takes as supporting the idea that one generation serves as role models for the next. These methods have thus produced sometimes striking findings, but they have usually been applied to the analysis of groups of individuals, over relatively long periods of time. This method does not preclude the study of the creative process in individuals, however, and the studies I will discuss can be looked upon as based on the philosophy behind these historiometric methods.

Quantitative Historical Case Studies

Methods used heretofore to attempt to come to grips with the production of innovation have not for the most part provided quantifiable data concerning the creative process in individuals of great renown. In an attempt to apply quantitative analyses to data from individual cases of creativity at the highest level, I and my students have recently carried out a number of what could be called quantitative historical case studies (Ramey & Weisberg, 2002; Weisberg, 1994, 1999, 2002; Weisberg & Buonanno, 2002). The specific antecedent for those studies was the work of Hayes (1981, 1989), who examined the role of experience— what Hayes called 'preparation'—in the production of creative masterpieces. Hayes measured the amount of time that elapsed between an individual's beginning a career in the fields of musical composition, painting, and poetry, and the production of that individual's first 'masterpiece'. Hayes defined a masterpiece in objective terms as, e.g. a musical composition that had been

207

recorded at least five times, a painting that was cited in reference works, or a poem that was included in compendia. There was a regular relationship between time in a career and production of the first masterpiece: all of the individuals in all of the domains required significant periods of time—approximately 10 years—before production of their first masterpiece. Thus, Hayes provided quantitative evidence for the claim that immersion in a discipline is necessary before an individual can produce world-class work (see Weisberg (1999) for further discussion).

Several of the case studies that I will discuss in this chapter have been carried out using methods similar to those of Hayes. Although the 'data' from case studies of innovations are not always amenable to rigorous quantitative analysis, we will see that bringing a quantitative orientation to case studies sometimes results in discoveries that would not have been apparent using only qualitative presentations of historical information. I will now examine the question of whether ordinary processes could serve as the basis for innovation, and whenever possible, case studies will be presented as evidence. Before discussing ordinary processes in innovative thinking, it is first necessary to briefly examine the opposite view, which has a long history in psychology, i.e. the idea that great innovative advances are the result of extraordinary thought processes.

Innovation and Extraordinary Thinking

Because creative thinking brings about innovations that most of us are not capable of producing, many researchers who study creativity have assumed that there is something extraordinary about the creative process. That is, extraordinary outcomes must be the result of extraordinary processes. Two examples of such processes are the Gestalt psychologists' notion of productive—as opposed to reproductive—thinking (e.g. Scheerer, 1963; Wertheimer, 1982) and Guilford's notion of divergent—as opposed to convergent—thinking (e.g. Guilford, 1950).

Extraordinary Processes

According to the Gestalt psychologists, true creative advances require that the person use productive thinking to go beyond what had been done before (staying with what had been done before was mere reproductive thinking). Furthermore, if one relied on the past and 'mechanically' reproduced habitual responses, one would not be able to deal with the particular demands of the novel situation that one faced, and would therefore be doomed to failure. Similarly, Guilford (1950) reasoned that the first step in the creative process must be a breaking away from the past, which is the function of divergent thinking. As the name implies, this type of thinking *diverged* from the old, and produced numerous novel ideas, which could serve as the basis for a creative product. The highly

creative individual is assumed to be high in divergent-thinking ability (e.g. Csikszentmihalyi, 1996). Once divergent thinking has served to produce multiple new ideas, then convergent thinking can be used to narrow down the alternatives to the best one. Current theoretical analyses of creative thinking sometimes assume that it is based on divergent thinking, even if that term is not explicitly used. As an example, Simonton (1999, p. 26) proposes that the creative process must begin with "the production of many diverse ideational variants". These multiple and varied ideas, produced presumably as a result of divergent thinking, provide the basis for the thinker's ability to deal with the new situation that he or she is facing.

Although there are differences between the concepts of productive/reproductive thinking and divergent/convergent thinking, there is a basic underlying similarity: the first step in the creative process is breaking away from the past, and there is a special kind of thought process that brings this about (see Weisberg (1999) for further discussion). This idea has become part of our culture, as evidenced in the often-heard comment about the need for 'out-of-the-box' thinking in situations demanding creativity. This instruction is a direct extension of the Gestalt idea that true creativity requires productive thinking, and the related idea that reliance on experience can interfere with dealing with the demands of the present. The well-known work by Luchins and Luchins (e.g. 1959) on problem-solving set is often brought forth as a graphic example of the dangers that can arise from a too-strong (and unthinking) reliance on the past.

Innovation as the Expression of Ordinary Processes

The view to be presented in this chapter, in contrast to those just discussed, assumes that innovation comes about through the use of ordinary thinking processes; creative thinking is simply ordinary thinking that has produced an extraordinary outcome—ordinary thinking writ large (Weisburg, 1986, 1993, 1999). Variations of this view have been presented by Simon and colleagues (e.g. Simon, 1976, pp. 144–174; Simon, 1986) and Perkins (1981). When one says of someone that he or she is 'thinking creatively', one is commenting on the *outcome* of the process, not on the process itself. Although the impact of innovations can be profound—as when a scientific theory changes the way we conceive the world, or when an invention produces changes in the way we live, or when an artistic product can arouse a strong emotional response in all of us—the mechanism though which the innovation comes about, as I will show, can be very ordinary.

The view that ordinary thought processes serve in creative endeavors has several facets. The most striking aspect of ordinary thinking is that it is based strongly on experience: ordinary thinking is based on *continuity* with the past. We build on the old to produce the new. Ordinary thinking is also *structured*: we are usually able to understand why one thought follows another in our consciousness. Third, ordinary thinking consists of a family of related activities (Perkins, 1981). We use the term *thinking* to refer to a number of specific activities, such as remembering, imagining, reasoning, and so forth. Finally, ordinary thinking is sensitive to environmental events. All these components of ordinary thinking are important in innovative thinking as well.

Continuity in Thinking

Consider the processes involved in an activity that would *not* be labeled creative, such as preparing dinner from a recipe. Perhaps the most striking aspect of this activity is that the more we carry it out, the better we get; we take it for granted that we will show learning. If you watched the first time I followed that recipe, and then came back and watched me do it a tenth time, presumably I would be better at it. I would not have to check the directions in the cookbook, I could do it from memory; I would carry out each of the components faster; and there would be less wasted effort. Thus, the basic component of thinking as expressed in our day-to-day activities is that it builds on what has come before; it is based on *continuity* with the past.

If creative thinking is ordinary thinking, then it too should demonstrate continuity with the past. Several different sorts of evidence would demonstrate continuity in creative thinking. First, there should be a learning curve in creative disciplines: people who work in creative disciplines should require a significant amount of time to become good at what they do. Specific data that would demonstrate the development of skill within a discipline would be increasing productivity and increasing quality of a person's work. We have already seen some evidence to support this general perspective from Hayes's (1989) study of preparation in creative fields—10 years are needed before a masterwork is produced—and additional evidence will be adduced shortly. A second aspect of continuity with the past is that we build on our experiences. If creative work also builds on the past, then it should be possible to discern connections between innovations and what came before. That is, there should be antecedents for creative works (Weisberg, 1986, 1993). Another way of expressing this aspect of continuity—antecedents for creative works—is to say that we move incrementally away from the past as we go on to something new; we do not reject the past, we use it. If innovations are based on continuity with the past, rather than rejection of it, we should see incremental movement beyond what has been done before.

Structure in Thinking

As has been argued for millennia, at least from Aristotle, our ordinary thinking is structured: one thought follows another in a comprehensible manner as we carry out our ordinary activities. There are several components to the structure of our ordinary thinking activities. Sometimes our thoughts are linked through associative bonds, which are the result of links between events in our past. This is expressed in Hobbes's familiar statement concerning the stream of associations in ordinary thinking (reprinted in Humphrey, 1963, p. 2).

> The *cause* of the *coherence* or consequence of one conception to another is their first *coherence* or consequence at that *time* when they are produced by sense: as for example, from St. Andrew the mind runneth to St. Peter, because their names are read together; from St. Peter to a *stone* for the same cause; from *stone* to *foundation*, because we see them together; and for the same cause from *foundation* to *church*, from church to *people*, and from people to *tumult:* and according to this example the mind may run from almost anything to anything (emphases in original).

A second basis for one thought following another is *similarity*; as Aristotle noted, common content will tend to make one thought call forth another (Humphrey, 1963, pp. 3–4). So, for example, thinking about your team's last game may bring to mind earlier games you attended. Similarly, an environmental event can remind one of a similar event from one's past.

Ordinary thinking also possesses structure because sometimes we use reasoning processes of various sorts in our ordinary activities, and this provides a basis for moving from one thought to the next. As an example, a friend says to you: 'If it rains tomorrow morning, I am not coming'. The next morning, you look out the window, see that it is raining, and think 'She is not coming'. That thought arises from the logic of what your friend said. If innovative thinking depends on ordinary thinking, it too should possess structure of these sorts; we should be able to understand the succession of thoughts as a creator brings a product into existence.

Thinking is a Family of Activities

Ordinary thinking is made up of a family of cognitive activities. The phrase 'I'm thinking', can refer to any of a large group of activities, some of which are the following: trying to remember something; imagining some event that you witnessed; planning how to carry out some activity before doing it; anticipating the outcome of some action; judging whether the outcome of an anticipated action will be acceptable; deciding between two alternative plans of action; determining the consequences of some events that have occurred (through deductive reasoning); perceiving a general pattern in a set of specific experiences (through inductive reasoning); comprehending a message; determining if two statements are contradictory. Creative thinking should also be constructed from these sorts of basic cognitive activities (Perkins, 1981).

Creative Thinking and Environmental Events

Ordinary thinking is sensitive to environmental events, which often change the direction of thought and action. For example, you might change your beliefs about someone by witnessing their behavior in some situation. You might observe someone, who you thought was kind, behaving cruelly toward someone without any reason that you can see. That might change your opinion of the person's character. Creative thinking also should be sensitive to external events.

We now turn to an examination of evidence supporting the hypothesis that each of these facets of ordinary thinking is also seen in creative thinking. In each of these areas, much of the evidence will be drawn from case studies.

Continuity in Creative Thinking

Several sorts of evidence would support the idea that creative thinking depends on continuity with the past. First, if individuals must learn to be creative, there should be a significant amount of time before an individual puts a distinctive stamp on his or her work. Second, during this time period, one should see an increase in an individual's productivity, as he or she learns the skills relevant to the domain; and in the quality of that individual's work, as he or she begins to put a distinctive stamp on the work. Third, one should be able to find antecedents for creative works, which is another way of saying that creative works should be the result of small-scale, or incremental, advances beyond the past, rather than being the result of wholesale rejection of the past.

Learning to be Creative: The 10-Year Rule

The work of Hayes (1981, chap. 10; 1989), already briefly reviewed, provides evidence that creative work requires a significant period of development. If production of a masterpiece in Hayes's terms can be taken as evidence for having mastered the discipline, then a significant period of time is required before this level of accomplishment is reached. Hayes's work was based on earlier research on what has become known as the '10-Year Rule' in the development of expertise in problem solving. The 10-Year Rule was postulated by Chase and Simon (1973) on the basis of their study of chess-masters' skills. In an extension of work by de Groot (1965), Chase and Simon examined, among other things, the ability of chess masters to recall almost perfectly, after an exposure of only 5 seconds, the positions of the 25–30 pieces from the middle of a typical master-level game. They concluded that the basis for this skill was a database of familiar groups of pieces, or 'chunks', acquired through years of intensive study and play, which are applied to any new board position which the master faces. The master uses this database to perceive the board position upon initial exposure, choose the next move from that position, and remember the position at a later time. Thus, the memory for the board position was based on the same information that served in the master's choice of the next move in the game, which can be considered a component in a creative act at the highest level (i.e. playing master-level chess). In estimating how many chunks had to be available in order for the master to be able to recall 25–30 pieces from a brief exposure, Chase and Simon concluded that approximately 50,000 chunks of 3–6 pieces each were needed, and that 10 years or so of intense study of chess was required to store these chunks in memory (hence, the '10-Year Rule').

As we have already seen, Hayes (1981, Chapter 10; 1989) extended Chase & Simon's analysis to creative production in the domains of musical composition, painting, and poetry, and provided quantitative evidence to support the need for immersion in the discipline as a necessary condition for the production of world-class work. In addition, this relationship was seen for even the most extraordinarily precocious individuals, such as Mozart. Mozart began his musical career at age six, under the tutelage of his father, a professional musician of some renown. His first masterpiece by Hayes's criterion was the Piano Concerto #9 (K. 271), written in 1786, when Mozart was 21, some 15 years into his career. Gardner's (1993) biographical studies of seven eminent 20th-century individuals also supported the 10-Year Rule: each of those individuals needed a significant amount of time before carrying out his or her first major accomplishment. Furthermore, those of Gardner's subjects who produced more than one major work required significant amounts of time before each of them.

In conclusion, research on the 10-Year Rule supports the idea that even great individuals need a significant period of time to gradually perfect their skills (Weisberg, 1999). However, the research discussed so far does not tell us in any detail what happens during the period of time when the individual is developing. Based on the notion of continuity in thinking, one would expect that there would be increases in an individual's productivity and in the quality of his or her work during that time.

Learning to Write Great Music. 1: Mozart

Productivity

As we just saw, Mozart's first masterwork was composed some 15 years into his career. There were, of course, compositions produced during Mozart's earlier years, as shown in Fig. 1A, and those data support the hypothesis that Mozart's production would increase with his experience as a composer. Over the first few years of his career, production gradually increased, until after about 12 years it more or less leveled off,

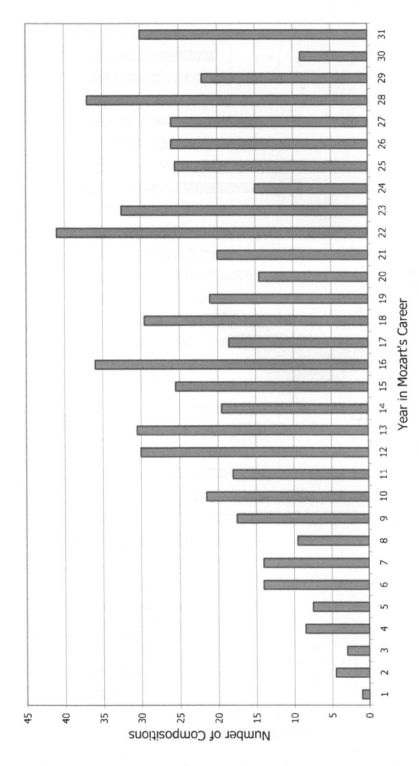

Figure 1A. Number of compositions per year of Mozart's career.

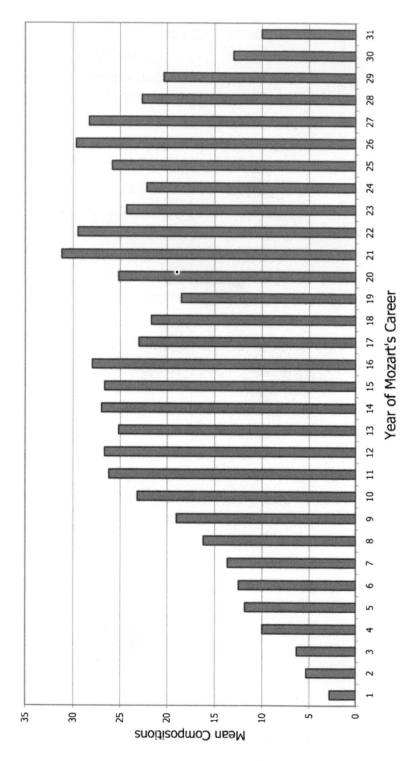

Figure 1B. Mozart's productivity—three-year running averages.

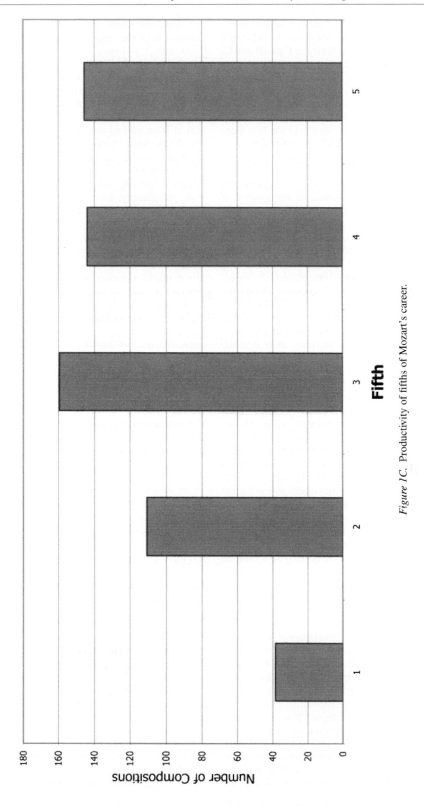

Figure 1C. Productivity of fifths of Mozart's career.

although there is considerable variability in productivity from year to year. The increase in Mozart's productivity can be seen more clearly in Figs 1B and C, which present, respectively, the average numbers of his compositions grouped over three-year segments (a three-year running average of his productivity), and grouped by fifths of his career. Here we see relatively clearly that the number of compositions increased over the first 12 years or so. The increase over the first three fifths of Mozart's career in Fig. 1C is significant.

Quality

Based on the notion of continuity in creative thinking, one would also expect an increase in the quality of Mozart's compositions over time. Possible changes in the quality of Mozart's early compositions can be seen if we use the number of recordings of each of his compositions as a continuous measure of quality, which allows one to include 'pre-masterworks' in the analysis. This measure differs from Hayes's 5-recording criterion for masterworks, which is an all-or-none measure. Figure 2A presents the average number of recordings for each of the compositions for each year of Mozart's career, and three interesting points emerge. First, similarly to Mozart's productivity, there is much variability in quality from year to year. Second, consistent with the notion of continuity, there is an overall increase in the quality of compositions over time; compositions later in Mozart's career are recorded on average more than early compositions. However, in contrast with productivity (Fig. 1), the increase in the quality of Mozart's compositions as indexed by the average number of recordings is much less gradual. This is seen more clearly in Figs 2B and C, which present three-year running averages and average numbers of recordings per composition for the fifths of Mozart's career, respectively, and the contrast with Mozart's productivity is striking. The data in Fig. 2 indicate that Mozart seems to have languished in what one could call mediocrity until he moved up to a new level of quality, approximately 10 years into his career. The overall difference in numbers of recordings for the fifths of Mozart's career shown in Fig. 2C is significant, with the first and second fifths being significantly lower than the others. The first and second fifths do not differ, nor do the last three fifths.

This analysis of Mozart's development as a composer has provided support for two aspects of the notion of continuity in creative thinking: we have seen an increase in Mozart's productivity and in the quality of his compositions over the first portion of his career. However, we have also seen possible differences in the pattern of development of these two aspects of Mozart's production, with quantity increasing gradually over time, and quality increasing more suddenly. One interesting point to be taken away from the quantitative analysis in Figs 1 and 2 is that one may become aware of patterns in a creator's work that are

not apparent when one simply presents a qualitative discussion of the same phenomenon.

The Question of Originality

The finding that Mozart's earliest compositions are of lower quality than his later ones leads to the question of what differences between early vs. late compositions might account for the differences in quality. Assume that at least part of the process of becoming a 'world-class' composer involves developing a distinctive voice or style—producing music that leads listeners to say 'That is by Mozart'. In general terms, reaching world-class level in any artistic field requires development of some originality, i.e. one's own style, so that one's work is recognizable. One might extend this view to creative work in the sciences and technology, where one also must establish a 'distinctive voice' before one's work can be considered innovative. There are at least two possibilities concerning how this distinctive voice might develop. On the one hand, the young person might produce distinctive work from the beginning of his or her career, but the quality of those early innovations might be low, so they make no impression, or even a negative impression, and are not acknowledged as masterworks. On the other hand, the young artist might not be particularly innovative in his or her early works, and that is why modern-day audiences find them of less interest than later works. Coming back to Mozart, his earliest works might not be 'Mozartian'.

Evidence to support the latter view comes from examination of some of Mozart's earliest compositions, which turn out to be little more than reworkings of compositions of other composers (Weisberg, 1999). Table 1A summarizes Mozart's first seven works in the genre of the piano concerto, and one sees that those works contain almost nothing original: the music was composed by others. Mozart arranged it for piano and other instruments. So Mozart seems to have begun learning his skill through study and small-scale modification of the works of others; Mozart's earliest works, which perhaps should not even be labeled as being by Mozart, contained little that was original and have been of little lasting musical interest.

Even when Mozart began to write music of his own, those pieces seem to have been based relatively closely on works by other composers, as can be seen in his production of symphonies (Zaslaw, 1989; see Table 1B). Mozart's final three symphonies (#39–41), usually acknowledged as his greatest, are those with which listeners are most familiar. They comprise four large-scale movements, and require approximately 30–40 minutes to perform. In addition, each contains innovative elements, which places it on a high plane of artistic achievement in the history of music. As one example, the final movement of Mozart's last symphony (K. 551, the *Jupiter*) is structured to a degree of complexity never before seen in a symphonic work. Mozart's early

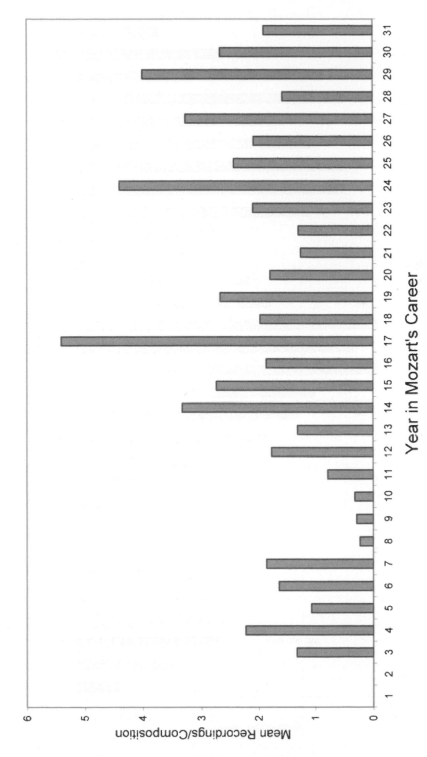

Figure 2A. Mozart's development—mean recordings/composition over the years of Mozart's career.

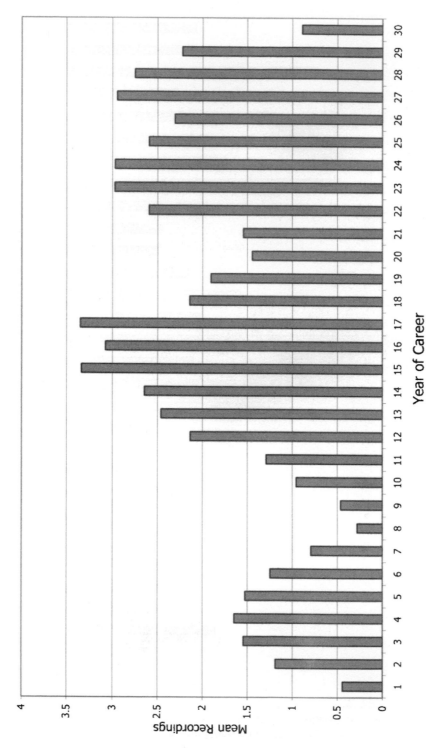

Figure 2B. Mean recordings per composition per year—three-year running average.

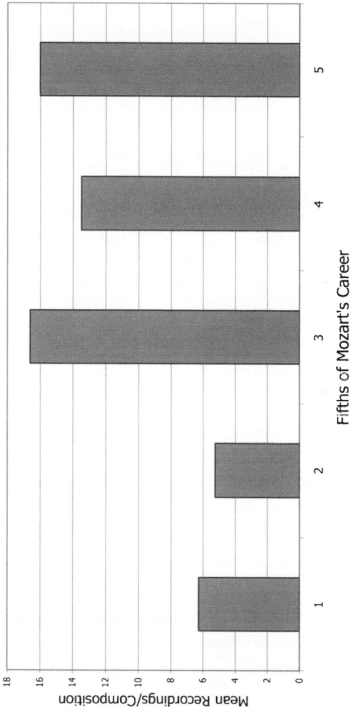

Figure 2C. Mean recordings per composition by fifths of Mozart's career.

Table 1. Mozart's early compositions.

A. Piano Concertos

# (K.#)	Sources
– (K. 107, 1–3)	J. C. Bach Op. 5 #2, 3, 4
1 (K.37)	H. F. Raupach, [composer not known], L. Honauer
2 (K.39)	H. F. Raupach, J. Schobert, H. F. Raupach
3 (K.40)	L. Honauer, J. G. Eckard, C. P. E. Bach
4 (K.41)	L. Honauer, H. F. Raupach, L. Honauer

B. Symphonies

	#	Date, Location	Movements	Tempos	Performance duration
J. C. Bach	Op. 3, #1	1765, London	3	Fast, Slow, Fast	10'
	Op. 3, #2	1765, London	3	F, S, F	10'
	Op. 3, #3	1765, London	3	F, S, F	10'
	Op. 3, #4	1765, London	3	F, S, Menuetto	15'
	Op. 3, #5	1765, London	3	F, S, F	10'
	Op. 3, #6	1765, London	3	F, S, F	9'
Abel	Op. 7, #1	1764, London	3	F, S, F	10'
	Op. 7, #2	1764, London	3	F, S, F	8'
	Op. 7, #3	1764, London	3	F, S, F	11'
	Op. 7, #4	1764, London	3	F, S, Menuetto	9'
	Op. 7, #5	1764, London	3	F, S, F	8'
	Op. 7, #6	1764, London	3	F, S, F	14'
Mozart	1 (K16)	1764, London	3	F, S, F	10'
	– (K19a)	1765, London	3	F, S, F	12'
	4 (K19)	1765/6, Holland	3	F, S, F	12'
	5 (K22)	1765, Holland	3	F, S, F	7'
	– (K45a)	1765/6, Holland	3	F, S, F	11'
	39 (K543)	1788, Vienna	4	(S intro) F, S, Men., F	33'
	40 (K550)	1788, Vienna	4	F, S, Men., F	33'
	41 (K551)	1788, Vienna	4	F, S, Men, F	38'

symphonies are very different from the later ones in structure, emotional scale, and innovation. His first symphony (K. 16) was produced in London in 1764, when he was eight; several others were produced in the next two years. All those very early symphonies were composed when Mozart and his father were visiting London and Holland, as part of a 'grand tour' of Europe, which afforded the young composer opportunities to meet important musical figures and members of the nobility.

While the Mozarts were in London, young Wolfgang became close to Johann Christian Bach (youngest son of J. S. Bach), who had established himself in England as a composer. Bach had composed a number of symphonies for use in concerts which he promoted in London in partnership with W. F. Abel, another German composer who had also written symphonies for their concerts. Those early symphonies almost always were composed of three short movements, usually in tempos of fast, slow, fast, and were built with

a simple harmonic structure. Mozart's first symphonies closely parallel those of Bach and Abel in structure and substance (see Table 1B). They consist of three short movements, usually in tempos of fast, slow, fast. In addition, the harmonic structure of Mozart's early symphonies does not go beyond the structure of those models. Thus, Mozart's earliest works are not very original, which supports the hypothesis that the ten years of 'apprenticeship' provide the basis for later originality.

Mozart's Development: Conclusions

The analysis of Mozart's career development supports the notion of continuity in creative innovation, as we found the developmental sequence expected on the basis of continuity. A significant amount of time passed before a unique contribution was made, and there was evidence for the development of Mozart's skill as a composer during this time: his productivity increased with experience, as did the quality of his compositions.

In addition, there was evidence that his early works were based relatively directly on those of others.

Learning to Write Great Music. 2: The Beatles

Further evidence for this developmental sequence— immersion in the works of others before producing innovations of one's own, with early works being of lower quality than later ones—can be seen in the career of the Beatles, specifically, the Lennon–McCartney songwriting team (Weisberg, 1999). I analyzed the development of the Lennon–McCartney songwriting team to examine the hypotheses that: (1) Lennon and McCartney developed their skill as songwriters through long-term immersion in the works of others; and (2) their skill as songwriters improved over their years together. The first step in the analysis was to demonstrate a long period of immersion in the works of others before the production of significant works of their own. I used records of the Beatles' performances (Lewisohn, 1992) to determine how often they played together before they became world-famous in 1964. As shown in Fig. 3, before becoming well known, the Beatles had spent more than five years performing, and were on stage on average more than once a day during those years. The results in Fig. 3 are evidence for a long period of development, but we also have to consider the content of those performances, in order to show that they involved immersion in the works of others. As expected, during those early performances, the Beatles were mainly playing songs written by others. Table 2A presents, for each year of the Beatles' performing career, the proportion of new songs in their performance repertoire that were Lennon–McCartney songs. Only in the last three years of their career did they add significant numbers of their own works to their repertoire. Thus, Lennon and McCartney began their careers as composers with a long apprenticeship of immersion in the works of others; at the very least, they were playing the works of others much more than they were playing their own.

The fact that the Beatles were playing few Lennon–McCartney songs early in their careers says nothing directly about the relative quality of those early songs. Those early Lennon–McCartney works, in contrast to Mozart's early works, might have been high in quality. Table 2B displays the proportion of early and late Beatles' compositions from their performance repertoire that were released on records at any time during their time together (which extended beyond their performing career). Assuming again that probability of recording is a measure of a song's quality, then it is clear that early Beatles' songs were of a lower quality, because many of them were not released while the Beatles were together. Many of those early songs were released on recordings in significant numbers only relatively recently, when numerous anthologies of early Beatles' material have been produced. Similar conclu-

sions can be drawn concerning the development of originality in the music of Lennon and McCartney. Critics are in agreement that their originality became manifest most strongly in such works as *Rubber Soul, Revolver*, and *Sergeant Pepper's Lonely Hearts Club Band* (e.g. Kozinn, 1995; Lewisohn, 1992). These works were produced several years into their career.

Development of Musical Creativity: Conclusions

Analysis of the development of musical creativity in two cases from very different domains—18th-century classical music and 20th-century popular music— supports the same conclusions: composers begin by getting to know well the works of other composers. It is only after a significant period of immersion in others' music that the composer's own identity begins to emerge. At the beginning of a composer's career, the music of others seems to serve relatively directly as the basis for new compositions. Over a period of time, the composer's distinctive voice develops. Later works are of higher quality than early works according to objective measures, such as the number of recordings made of each composition; they are more original as well, according to the opinions of critics. We can now examine the generality of this view, looking beyond the domain of music to two examples of scientific innovation.

Continuity in Scientific Innovation

One aspect of continuity of thought that is particularly clear in scientific innovation is the presence of antecedents to many ground-breaking works. Antecedents to a seminal scientific innovation can be seen in the development of Darwin's theory of evolution (Gruber, 1981). One can analyze Darwin's work into two stages: his theory based on natural selection with which we are all familiar is the second of those stages. The first stage in Darwin's theorizing was based very directly on Lamarckian evolution (i.e. the idea that evolution of species depended on the transmission of acquired characteristics from one generation to the next) and on the concept of the monad (the idea that each species sprang from a single living form—the monad—that appeared spontaneously from nonliving matter and then evolved in response to environmental conditions). These two ideas had been advocated earlier by other theorists, one of whom was Darwin's grandfather, Erasmus Darwin, who had, independently of Lamarck, proposed a theory of evolution based on acquired characteristics. Darwin went on to reject those earlier views and develop the theory which revolutionized scientific thinking about the development of species, but he began with his feet firmly planted in the work that came before him. Furthermore, it is interesting to note that at least part of the basis for

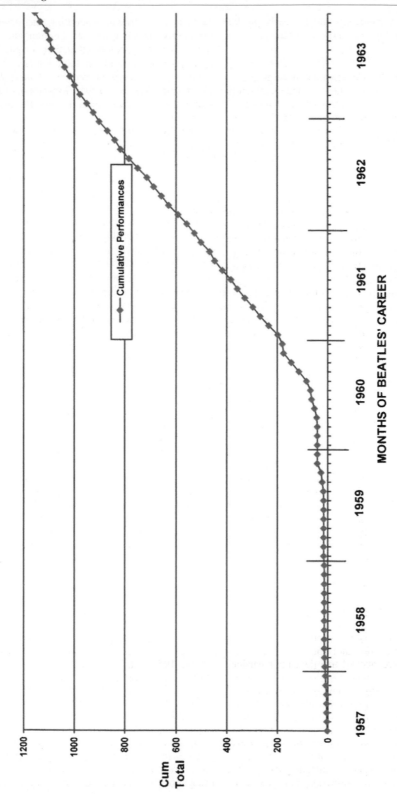

Figure 3. Beatles' cumulative performances until the advent of Beatlemania (1964). From Weisburg, R. W. (1999). Creativity and knowledge: A challenge to theories. In: R. J. Sternberg (Ed.), *Handbook of Creativity* (pp. 226–250). Cambridge: Cambridge University Press. Reprinted with permission.

Table 2. Development of Lennon–McCartney songwriting team.

A. Proportion of Lennon & McCartney Songs in Beatles' Performance Repertoire

	Year								
	1957	1958	1959	1960	1961	1962	1963	1964	1965/1966
Number of new songs in repertoire	53	20	20	98	50	46	15	11	10
Proportion of Lennon & McCartney songs	0.23	0.10	0.05	0.02	0.06	0.17	0.67	1.00	0.80

B. Proportion of Lennon & McCartney Songs Released on Records During Performance Career

	First Half of Career (1957–1961)	Second Half of Career (1962–1966)
Number of songs written	20	37
Proportion recorded	0.16	0.92

C. Proportion of Lennon & McCartney Songs Released on Records During Performance Career

	Thirds of Performance Career		
	1957–1959	1960–1962	1963–1966
Number of songs written	15	13	29
Proportion recorded	0.13	0.54	1.00

Darwin's rejection of previous work was his acquisition of new pieces of information, which raised problems for his first theory. This sensitivity to environmental events, already noted as a characteristic of ordinary thinking, will be discussed in more detail later.

Another example of antecedents in scientific innovation is seen in the development of Watson and Crick's double-helix model of DNA. Watson and Crick's work can also be divided into two phases, with the double helix being the second phase. Their initial work centered on the possibility that DNA was a three-stranded or triple helix. Their general orientation, that is, the working assumption that DNA was probably helical in shape, was based on then-recent work by Pauling, who had proposed a helical model for the structure of the protein alpha-keratin (Olby, 1994; Watson, 1968; Weisberg, 1993). As with Darwin, Watson and Crick's final product went considerably beyond the earlier work, but that work was crucial in setting the direction that ultimately proved successful. Watson and Crick were pushed to reject their earlier work and to go beyond it at least in part by new pieces of information that became available to them, in a further example of sensitivity to environmental events in innovative thinking.

We see here examples from science of two aspects of continuity in innovation, specifically, the use of the past as the foundation for the development of the new, and the role of environmental events in pushing the thinker to go beyond the past. In neither of these cases did the innovation come about through a leap away from the past.

Hidden Continuities in Apparent Discontinuities

When examining case studies of innovation, one sometimes finds radical breaks—what seem to be *discontinuities*—in thinking; i.e. a shift in thinking leads to the development of a new line of work. A central thesis of the present viewpoint is that such radical breaks are more apparent than real, and that they are still based on the processes that underlie ordinary thinking. Sometimes what seems to an observer to be a discontinuity occurs because the individual's database is not known in detail. Therefore, the individual in actuality produces an innovation that is built on the past, but the ignorance of the observer results in the mistaken belief that the innovation has come out of nothing. An example of such a 'hidden antecedent' can be seen in the development of Jackson Pollock's 'poured paintings', which were critical in making New York City the capital of the art world around 1950 (Landau, 1989). Pollock's works, with their swirls of paint and a complete lack of recognizable objects, were totally different from anything he or anyone else in American painting had produced until that time (see Weisberg, 1993). Perhaps the singular revolutionary aspect of Pollock's technique was his complete rejection of traditional methods of applying paint to canvas. Rather than using a brush or palette-knife to apply paint, Pollock poured the paint directly from a can, or used a stick to fling the paint on the canvas, which was laid out on the floor, rather than hung on an easel.

Although many people today are still struck by the seeming radicalness of Pollock's innovative technique and subject matter, if we examine Pollock's background, we can find a likely source for his developments (Landau, 1989, pp. 94–96). In 1935, as part of the WPA, a number of artist's workshops were organized in New York. One of these workshops was directed by David Alfaro Siqueiros, an avant-garde painter from Mexico who, along with his compatriots Diego Rivera and José Clemente Orozco, had established a presence in the New York art scene. These painters had as part of their agenda a bringing down of art from what they saw as its elitist position in society, and making it more accessible to the masses. One aspect of that agenda was incorporating modern materials and techniques into painting, including industrial paints, available in cans, in place of traditional tubes of oil paint. There was also experimentation with new ways of applying paint, such as airbrushing, rather than the traditional brush. In the Siqueiros workshop, there also was experimentation with such techniques as throwing paint directly on the canvas, thereby avoiding use of the brush, which one of Pollock's friends who attended the workshop derided as "the stick with hair on its end". In 1936, Siqueiros himself created a painting, *Collective Suicide,* for which he poured paint directly from a can on the painting, which was flat on the floor, and he also flicked paint on it with a stick. Although Siqueiros did not create that painting at the workshop, it is reasonable to assume that similar techniques were demonstrated there. The attendees of the workshop collaborated on projects involving these new materials and methods, and one of those attendees was Pollock, who worked on a project with two other young artists involving applying paint using pouring and spilling, among other methods. In addition, there is evidence that in the early 1940s, several years before his 'breakthrough' works, Pollock had produced at least one painting in which paint is used in a nonrepresentational manner. Interestingly, in the context of the earlier discussion of the 10-Year Rule, the structural complexity of that painting and of Pollock's technique are very primitive compared to his mature works.

Thus, Pollock's advances, when placed in context, are seen to have come out of his experiences. We (and artists and critics in the late 1940s) were surprised by Pollock's work because we (and they) were not familiar with his history. This is not to say that Pollock's work was not highly innovative and of singular importance in the history of mid-20th century art; rather, it is simply to point out that even what seem to be the most radical innovations in any domain may have direct antecedents within the experience of the creator. Simply because some innovation appears as if it came out of nowhere, one must not assume that that

is how it came about. Looking below the surface can sometimes reveal direct sources for what seem to be the most radical of innovations.

Continuity in Innovative Thinking: Conclusions

We have examined evidence that supports several expectations arising from the idea that creative thinking is based on continuity. First, there is a learning curve in creative disciplines: people who reach the highest levels of accomplishment in the arts and sciences do so after a long period of immersion in their discipline (the 10-Year Rule). During this period of time, there is an increase in the individual's productivity and also in the quality of his or her work. This increase in quality of work is accompanied by an increase in originality as well. In the cases discussed, it has been possible to point out antecedents for innovations, and to show that innovations build on the past rather than reject it. We now turn to an examination of the structure of the thought processes that bring forth innovations. Specifically, we will examine the claim that creative thinking is structured in much the same way as ordinary thinking.

Structure in the Creative Process: Picasso's Development of *Guernica*

I have recently completed a quantitative case study of Picasso's creation of his great painting *Guernica* (Weisberg, 2002; see Fig. 4), which provides information concerning several aspects of Picasso's thinking. First, the study examined the preliminary sketches that Picasso produced in the process of creating *Guernica*, which allowed inferences concerning structure in Picasso's thought processes. Second, the study also examined the relationship between *Guernica* and what came before, which provided further evidence concerning antecedents in artistic innovation, and evidence concerning factors that link ideas in creative thinking.

Guernica, a landmark of 20th century art, was painted in response to the bombing, on April 26, 1937, of the Basque town of Guernica, in northern Spain, by the German air force (Chipp, 1988, chap. 3). The destruction of the town and killing of innocent people horrified the world, and Picasso's painting quickly became a great anti-war document. *Guernica* is massive in scale, measuring approximately $12' \times 26'$, and is painted in monochrome: black, white, and shades of gray. In the left-hand portion of the painting, a bull stands over a mother whose head is thrown back in an open-mouthed scream, holding a dead baby whose head lolls backward. Below them, a broken statue of a warrior holds a broken sword and a flower. Next to the bull, a bird flies up toward a light. In the center of the painting, a horse, stabbed by a lance, raises its head in a scream of agony. In the upper center, a woman leans out of the window of a burning building, holding a light to illuminate the scene. Beneath the light-bearing woman, another woman with

Figure 4. Guernica. © 2002 Estate of Pablo Picasso/Artists Rights Society (ARS), New York.

bared breasts hurriedly enters the scene. At the far right, a burning woman falls from a burning building.

The Preliminary Sketches

At the time of the bombing of Guernica, Picasso had been working on a painting of an artist's studio that was to be displayed in the pavilion of the Spanish Loyalist government in a World's Fair opening in Paris in June, 1937 (see Fig. 5). When news of the bombing reached Paris, Picasso dropped work on the studio painting and began work on *Guernica*, producing his first sketches on May 1; the last sketch is dated June 4. He began painting on approximately May 11; the completed work was put on display early in June. Picasso dated and numbered all the preliminary works for *Guernica*, a total of 45 sketches. There are several different types of preliminary works. Seven *composition studies* present overviews of the whole painting; the remaining sketches, *character studies*, examine characters individually or in small groups. Samples of preliminary sketches are shown in Fig. 6.

The preliminary sketches provide us the opportunity to get inside Picasso's creative process, and can serve to answer several questions:

(1) We can determine through analysis of the sketches whether or not Picasso systematically worked out the structure of *Guernica*. Do we see a pattern in how Picasso worked, for example, concentrating first on composition sketches and then on character sketches, or did he simply jump from one type of sketch to another, without any logic or pattern? If we find that Picasso worked first on composition sketches, say, before spending much time on individual characters, it would be evidence that he worked out the structure of the painting—the overall idea—before fleshing it out by considering the individual characters.

(2) After determining whether or not Picasso was systematic in working out the overall structure of the painting, we can investigate the specifics of how Picasso decided on that structure. Did he experiment with several different possible structures for the painting, or did he have one structure in mind when he began to work? If we find that there was a common core to all the composition sketches, it would be evidence for the hypothesis that Picasso had a basic idea in mind in response to the news of the bombing, and this idea structured the process whereby the painting was produced.

(3) The next question we can examine is where the structure for *Guernica* came from. That is, can we relate the overall structure of *Guernica*, as seen in the composition sketches and in the painting itself, to any antecedents, in Picasso's work or in the work of other artists?

(4) We can also use the sketches to examine the question of whether or not Picasso's thought

process was structured when he worked on individual characters. For example, if we look at all the sketches that contained one particular character, say, the horse, do we find that aspects of that character are randomly varied from one sketch to the next, or is Picasso systematic in his explorations of the characters?

(5) Finally, parallel to the question of possible antecedents for the overall structure of *Guernica*, we can examine the question of antecedents for the individual characters in the painting. Was *Guernica* constructed out of components that we can trace to other works?

Systematicity in the Sketches?

Picasso worked on the sketches for *Guernica* over a period of a little more than a month; for ease of exposition, this period can be summarized into three phases of work: the first two days (May 1–2); an additional six days, commencing about a week later (May 8–13); and a final two weeks of work, which began about a week later (May 20–June 4). As can be seen in Table 3A, there was structure in Picasso's thinking as exemplified in the sketches: the first two days resulted in composition studies and studies of the horse, arguably the central character in the painting. In the second phase, the composition studies are fewer, and other characters are examined. In the last phase, there are no composition studies, and peripheral characters (e.g. the falling person) are seen for the first time. This pattern can be made clearer by combining categories of sketches, as shown in Tables 3B and 3C. These results support the conclusion that Picasso spent the bulk of his early time working on the overall structure of the painting and on the main character, and then moved on to other aspects of the painting. There is a significant decrease over the three periods in the frequency of composition sketches. Thus, analysis of the temporal pattern in the whole set of sketches has provided an affirmative answer to the first question outlined above: Picasso was systematic in working out the structure of *Guernica*.

Deciding on an Idea: Analysis of the Composition Studies

The next question focuses on how Picasso decided on the final structure of the painting. Examination of the content of the composition studies can show us the specific path through over which he traveled. As can be seen in Table 4, the structure of the painting is apparent in the composition studies produced on the first day of work. In seven of the eight composition studies, including the very first one, the light-bearing woman is in the center, overlooking the horse. In addition, each of the central characters (horse, bull, light-bearing woman) is present in almost all of the composition sketches, with other characters appearing less frequently. This pattern supports the view that Picasso had

Figure 5. Artist's studio. © 2002 Estate of Pablo Picasso/Artists Rights Society (ARS), New York.

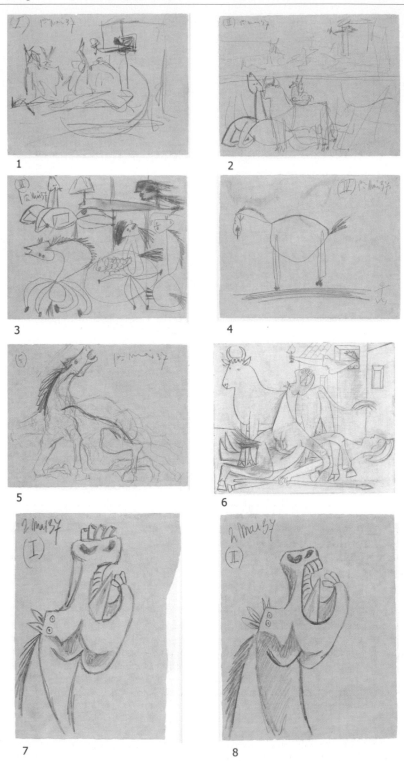

Figure 6. Examples of preliminary sketches. © 2002 Estate of Pablo Picasso/Artists Rights Society (ARS), New York.

Table 3. Summary of Picasso's preliminary sketches for Guernica. *(From Weisburg, in press. Copyright Baywood Publishing Co. Reprinted with permission).*

A. All Preliminary Works Tabulated by Three Periods of Work

Period	Comp.	Horse	Bull	Mother & Child	Woman	Hand	Falling Person	Man	Total
1 (May 1–2)	6	5	0	0	0	0	0	0	11
2 (May 8–13)	2	4	2	5	1	1	0	0	15
3 (May 20–June 4)	0	2	2	2	8	1	3	1	19

B. Composition Sketches vs. All Others

Period	Composition	All Others	Total
1 (May 1–2)	6	5	11
2 (May 8–13)	2	13	15
3 (May 20–June 4)	0	19	19
Total	8	37	45

C. Composition Sketches + Horse + Bull vs. All Others

Period	Composition + Horse + Bull	All Others	Total
1 (May 1–2)	11	0	11
2 (May 8–13)	8	7	15
3 (May 20–June 4)	4	15	19
Total	23	22	45

at least the 'skeleton' or 'kernel' of *Guernica* in mind when he began to work: *Guernica* is the result of Picasso's working out of this kernel idea. We see here further evidence for structure in Picasso's thought process: news of the bombing stimulated him to paint, and his thought process was relatively constrained from the beginning. Indeed, if the sketches can be taken at face value as the record of Picasso's thought processes concerning the painting, then from the very beginning there was only one idea that he considered.

Antecedents to the Structure of Guernica?

Analysis of *Guernica* as arising from a kernel idea that Picasso had available from the beginning of his work immediately raises another question: Whence did the kernel idea arise? Chipp (1988), in his extensive analysis of *Guernica*, was struck by the quick gestation of the painting. Based on the discussion so far, one might expect that the structure of *Guernica* was based on works that came before, and this could help account

for the speed with which the painting took shape. When Picasso painted *Guernica*, he was in his mid-50s and had been an artist for most of that time. Therefore, he had available a history of his own to draw on, and that history played a significant role in the creation of *Guernica*, which is closely related to many of Picasso's works from the 1930s. One striking example of a work that presages *Guernica* is *Minotauromachy*, an etching made by Picasso in 1935 (see Fig. 7). In this composition, a dead woman in a matador's costume, holding a sword in one hand, is draped over the back of a rearing horse. A minotaur (the mythological half-man half-bull) raises a hand in front of his eyes to shield them from the light from a candle held by a young woman who is observing the scene. Two other women observe the scene from a window above, where two birds also stand. On the far left, a man is climbing a ladder.

Table 5 summarizes a number of correspondences between *Guernica* and *Minotauromachy*, and indicates that *Minotauromachy* contains the same kernel idea,

Table 4. Guernica Composition studies: presence of characters. (*From Weisburg, in press*).

Sketch #	Date	Final Structure?	Horse	Bull	Light-Woman	Mother Child	Mother Adult	Fleeing Woman	Fallen Woman	Fallen Warrior	Flying Animal	Wheel	Upraised Arm
1	May 1	Yes	X(?)	X	X						X		
2a		Yes		X	X								
2b		No	X	X							X		
3		Yes	X	X	X								
6		Yes	X	X	X					X	X		
10	May 2	Yes	X	X		X			X	X			
12	May 8	Yes	X	X		X				X			
15	May 9	Yes	X	X	X		X					X	X
Summary			0.88	0.88	0.75	0.25	0.13	0	0.13	0.38	0.38	0.13	0.13

Table 5. Corresponding elements in Guernica *and* Minotauromachy. *(From Weisburg, in press).*

Minotauromachy	Guernica
Bull (Minotaur)	Bull
Horse—head raised	Horse—head raised (stabbed—dying)
Dead person	Dead person (broken statue)
Sword (broken—in statue's hand)	Sword (in Minotaur's hand)
Flowers (in girl's hand)	Flower (in statue's hand)
Two women above observing woman on ground holding light	Woman above observing + holding light
Birds (standing in window above)	Bird (flying up toward light)
Vertical person (Man fleeing)	Vertical person (Burning woman falling)
Sailboat	
	Electric light
	Mother & Child
	Woman running in

Table 6. Comparison of elements in Artist's Studio and Guernica. *(From Weisburg, in press).*

Artist's Studio	Guernica
	Bull
	Horse—head raised (stabbed—dying)
	Broken statue
	Sword
	Flower (in statue's hand)
	Woman above observing + holding light
	Bird flying up toward light
	Burning woman falling
Electric light (above and spotlight below)	Electric light above
	Mother & Child
	Woman running in
Reclining model	
Artist	
Male spectator	
Easel	
Window	Window

Figure 7. Minotauromachy. © 2002 Estate of Pablo Picasso/Artists Rights Society (ARS), New York.

and may have served as a source for *Guernica*. In order to demonstrate that those correspondences in Table 5 reflect more than chance, however, one needs a 'control' painting to compare with *Guernica*. As indicated earlier, just before the bombing of Guernica, Picasso was working on a painting of an artist's studio. He never got beyond sketches for that work, one of which was shown in Fig. 5. If one compares that work with *Guernica*, as shown in Table 6, one finds very little overlap in subject matter, especially as regards major characters and structure.

One can carry out a statistical test of the degree of correspondence between *Guernica* and the *Artist's Studio* vs. *Minotauromachy*. One can count the number of rows in Tables 5 and 6, and use each of those numbers as the denominator of a fraction, the numerator of which is the number of lines which correspond in the respective tables. Based on this measure, the degree of correspondence of *Guernica* and the *Artist's Studio* is 0.13 (2/16), and the degree of correspondence of *Guernica* and *Minotauromachy* is 0.67 (8/12). The proportion of correspondence is significantly higher for the latter than the former. Furthermore, the strong

correspondence between *Minotauromachy* and *Guernica* shown in Table 5 is actually an underestimation of the true correspondence. *Minotauromachy*, an etching, was printed from a drawing made by Picasso on a printing plate, so the scene Picasso drew on the plate was actually reversed from left to right in comparison with the print shown in Fig. 7. The 'vertical person' was drawn on the right, and the bull was on the far left. The light-bearing female also faces in the same direction as the corresponding character in *Guernica*. In conclusion, not only does *Minotauromachy* contain many characters similar to those in *Guernica*, but the absolute spatial organization of the two works is also similar. One might raise questions about the specific entries in Tables 5 and 6, but it seems clear that there is a much higher degree of similarity between *Guernica* and *Minotauromachy* than between *Guernica* and the *Artist's Studio*.

This analysis of the composition studies for *Guernica* has supported the claim that the kernel idea for the painting was in Picasso's mind from the beginning. Further, examination of similarities between *Guernica* and *Minotauromachy* indicates that that idea was one which can be found in earlier works in Picasso's career. Thus, at the time of its creation, *Guernica* was the most recent variation on a theme present in Picasso's work. So at least part of the reason why Picasso was able to create *Guernica* so quickly was that it was related to ideas that he had used before.

The Link Between *Monotauromachy* and *Guernica*

Assume for the sake of discussion that when Picasso began to paint *Guernica*, he had *Minotauromachy* in mind and used it as a model for the new work. This raises the question of why the bombing of Guernica caused Picasso to think of *Minotauromachy*. There are several links that can be traced between the bombing and *Minotauromachy* and *Guernica* which can help us understand why Picasso's thinking might have taken the direction that it did. First, the bombing took place in Spain, Picasso's native land, and *Minotauromachy* is a representation of a bullfight, which obviously has deep connections to Spain and to Picasso, who painted bullfight scenes from his very earliest years (Chipp, 1988). In addition, the emotionality of the bombing might have provided a further link to the bullfight, an event of great emotional significance for a Spaniard. *Guernica* also contains the skeleton of a bullfight: a bull and horse, a person with a sword (the statue), and 'spectators' overlooking the scene. It may also be of potential importance that when Picasso was growing up in Spain, the bull was not the only victim in a typical bullfight. The horse that carried the *picador* (the lance-carrier, whose task is to drive the lance into the shoulders of the charging bull) was not padded, and was often an innocent victim of the bull's charge. Based on this reasoning, the horse in the center of *Guernica*, whose head is raised in a scream of agony, can be seen as a representation of an innocent victim, and one can understand how that symbolization might have arisen in Picasso's mind. Thus, *Guernica* and *Minotauromachy* are linked by a web of interrelationships, and it is not hard to understand why the bombing might have stimulated Picasso to think of *Minoaturomachy*, which then played a role in directing his further thinking. It is also notable that this web of relations does not seem different than those discussed centuries ago by philosophers interested in the structure of ordinary thinking.

Structure in Development of Individual Characters

We can also use the sketches of individual characters to examine the question of structure in Picasso's thought process. First, did he tend to concentrate on only one character at any given time, and second, did he examine systematically the characteristics of each character? As with the development of the overall structure of the painting, Picasso was systematic in his development of the individual characters.

Attention to One Character at a Time

In order to determine whether Picasso tended to concentrate on a given character at a given point in time, we can analyze the sequential pattern over all the sketches (see Table 3A). As can be seen, the sketches for the horse were concentrated in the first two periods, the mother and child were most frequent in the second period, and the isolated woman and the falling person were most frequent in the third. Thus, Picasso seems to have been systematic in his working on the individual characters over time.

Development of Individual Characters

If we consider the development of individual characters over the series of sketches, we can also find evidence for structure in Picasso's thought. For several of the individual characters, one can focus on elements which Picasso varied separately. As one example, in the sketches of the horse, Picasso varied the position of the head: up vs. down. Another example is in the sketches of the woman: whether her eyes are dry or tearing. A third is whether the woman is alone or with another individual (usually the baby). We can examine each of those components, in order to uncover structure in Picasso's thinking as he worked on each character.

Including composition studies, the horse was sketched a total of 19 times in the first two periods of Picasso's work. The position of the head of the horse in those periods is summarized in Table 7A, and a clear differentiation is seen: in the earlier sketches, the head is predominantly down, which changes in the later sketches. A similar pattern is seen in the sketches containing women. Tables 7B–7D summarize all 20 sketches, composition sketches and character studies, in which there was at least one woman participating in the action (the light-bearing woman was ignored, as were any dead women). Once again there is a pattern in the presence of the various elements of the women over the two periods of work in which women appeared in sketches. In the early sketches, the woman usually is holding a dead person, whereas in the later sketches she is usually alone (Table 7B). Similarly, in the early sketches, she is screaming without tears; in the later sketches, tears are almost always present (Table 7C). Finally, Table 7D summarizes for all the sketches the relationship between the facial expression of the woman and whether she is presented alone or with a dead person. When she is alone, she is almost always weeping; when she is holding the dead person, she sheds no tears. In conclusion, analysis of the character sketches supports the conclusions drawn from analysis of the composition studies: Picasso was systematic in his working out of the elements of *Guernica*.

Antecedents to Characters in Guernica

The final question to be examined is whether there were antecedents to specific characters in *Guernica*, and one can find what seem to be specific connections

Table 7. *Summaries of presentation of the horse and of women in the sketches for* Guernica. *(From Weisburg, in press).*

A. Position of Head of Horse Summarized Over Periods 1 and 2

Period	Head Up	Head Down
1 (May 1–2)	8	1
2 (May 8–20)	2	7

B. Types of Women ín Periods 2 vs. 3

Period	Mother & Child	Solitary Woman
2 (May 8–13)	8	1
3 (May 20–June 3)	2	10

C. Expressions of Women in Periods 2 vs. 3

Period	Open-mouthed	Weeping
2 (May 8–13)	9	0
3 (May 20–June 3)	2	10

D. Relationship Between Social Environment and Emotional Expressions of Women

Type of Woman	Open-mouthed	Weeping
Mother & Child	8	2
Solitary Woman	3	8

to characters in *Guernica* from works of other artists. Two examples will serve to make the point that Picasso's thought process was structured in various ways by art with which he was familiar and which was relevant to the theme of Guernica. One particularly distinctive character in the sketches is the falling man in sketch 35 (see Fig. 8A). Picasso included no men among the actors in Guernica (the only male is the broken statue), so the content of this sketch is intriguing. This distinctive individual, with sharply drawn profile, striking facial expression, facial hair, and placement of eyes, as well as the falling posture with outstretched arms, bears striking resemblance to the man shown on the right in Fig. 8B. The latter drawing is from an etching in a series by Goya (1746–1828), called *Disasters of War*, which was created more than 100 years before *Guernica*. Picasso had great respect for and knowledge of Goya's works (Chipp, 1988), and it would be not surprising if the events that stimulated Picasso's painting of Guernica also resulted in his recollection and use of Goya's work as the basis for his own, especially given the commonality of theme. Picasso changed the man into a woman in the painting, but the falling woman in *Guernica*

bears residue of the Goya etching from which she began: her profile is similar to that of Goya's man, and her outstretched hands with exaggeratedly splayed fingers echo those of the Goya.

A second example of a correspondence between one of Picasso's characters and another work from Goya's *Disasters of War* is shown in Figs 8C and D. Picasso's sketch 14 contains a mother and child, with the woman distinctive in her sharply profiled head thrown back; her pose, with her outstretched left leg producing a distinctive overall triangular shape; and her skirt folding between her legs. The woman in Goya's etching is similar in facial profile and expression, and in her posture, with an outstretched left leg producing an overall triangular shape, and her skirt folded between her legs.

Structure in Creative Thinking—Conclusions

Examination of the content of the sketches for *Guernica* indicated that Picasso's thought process was structured in several ways during his creation of this great work. First, he considered a relatively narrow range of subject matter when he began to work; the creation of *Guernica* can be looked upon as the

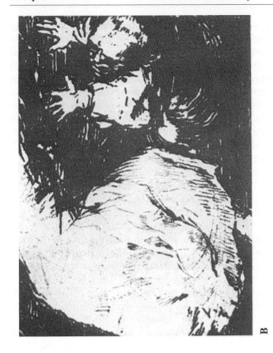

Figure 8. (A) Picasso's falling man. (B) Goya's falling man. (C) Picasso's mother and child from sketches. (D) Goya's woman from *Disasters of War*. A and C © 2002 Estate of Pablo Picasso/Artists Rights Society (ARS), New York.

elaboration of a kernel idea, which is seen clearly at the very beginning of Picasso's work on the painting. Furthermore, this idea was a straightforward extension of earlier work done by Picasso, perhaps in conjunction with his knowledge of Goya's *Disasters of War.* Second, he worked systematically over time, beginning work with the overall structure of the painting, and then spending time on the specific representations of the individual characters. Third, the development of the individual characters showed a similar pattern. For the representation of the horse, as well as of the woman, Picasso examined a very limited set of possibilities, and those limited possibilities were examined systematically. Further evidence for structure in the creative process will be presented in a later section, as part of the discussion of Edison's invention of the light bulb.

This case study of *Guernica* also has provided further evidence for continuity in thinking; specifically, we have seen much evidence for antecedents for Picasso's great painting, both in his own work an in that of others, although I have concentrated only on the work of Goya. (For further analysis, see Chipp (1988).) This analysis has provided evidence for what one could call layers of antecedents to *Guernica.* The overall structure, based on the kernel idea seen in *Minoaturomachy,* is one level of structure. Within that structure or framework, specific characters are based on other antecedents, meaning that one can trace antecedents nested within antecedents in *Guernica.* If the present analysis is accepted as valid, it means that it has been possible to trace the origins of some microscopic aspects of *Guernica,* for example, the facial expressions and postures of some characters, as well as the appearances of the characters' hands. Thus, the present analysis supports the proposal that creative works may be closely linked to previous works, although how often this occurs and how close the links are remain to be answered, through the analysis of other works, in painting as well as in other domains (for further discussion, see Weisberg (1993)). We now turn to consideration of the question of the cognitive components which comprise innovative thinking (Perkins, 1981).

Cognitive Components of the Creative Process

Ordinary thinking is built out of a family of cognitive processes, which includes deductive reasoning, both formal and informal, as well as induction; processes of comprehension; evaluation; imagination; and so forth. We now examine the notion that the creative process is built out of those same sorts of cognitive components.

Ordinary Thinking in Creative Problem-Solving

On the basis of laboratory research examining problem-solving, Perkins (1981) proposed that we could understand creative leaps, or Aha! experiences in problem-solving, as being the result of ordinary cognitive processes, such as recognizing or realizing

that something is true or false, without assuming the operation of extraordinary thought processes. He presented the Fake-Coin problem (see Table 8A), and when the participant solved it, he or she was to report immediately on the thought processes that had led up to solution. Perkins took precautions to ensure that those reports would be as accurate as possible: his participants were given some training and practice in producing verbal reports, and the reports were collected immediately after solving the problem, which kept potential difficulties to a minimum.

Two of Perkins's reports are presented in Table 8B. The two people solved the problem differently, with one (whom Perkins called Abbott) reporting that the solution 'just snapped' together in a leap of insight; the other (Binet) worked out the solution in a logical series of steps. When Perkins examined the reports further, however, he concluded that the thought processes carried out by Abbott and Binet were in actuality very similar. First, both Abbott and Binet focused on, or *recognized*, the date as the crucial piece of information. Second, Abbott's 'leap' turns out to have required only a couple of steps of reasoning on Binet's part; that is, the insight process turns out not to have done much in the way of cognitive work. Most importantly, the 'leap of insight' could be explained by our understanding of ordinary cognitive processes: what was required was that the thinker *realize* the contradiction in the coinmaker's knowing that Christ would be born at some later date. Perkins pointed out that we often experience such realizations in our ordinary cognitive activities. As one example, we constantly fill in information when we comprehend sentences as part of our ordinary language activities, and in so doing we can come to realize that two statements are contradictory. Thus, one can understand Abbott's 'leap of insight' as an example of the ordinary process of realizing that something was impossible. Perkins concluded that it was not necessary to assume that leaps of insight are brought about by anything in the way of extraordinary mental processes. Rather, sometimes we use *reasoning* in order to work out the consequences of some state of affairs, and other times we can realize the consequences directly, without reasoning anything out. As a parallel situation, Perkins points to our understanding of jokes: sometimes we can 'get' a joke directly, as we hear the punch-line, while other times we have to have the logic of the joke explained to us. Getting a joke as we hear it involves realization of the same sort that plays a role in leaps of insight.

In conclusion, Perkins showed that the 'insights' involved in solving some sorts of problems might consist of very small steps, rather than large leaps, and that whatever leaps occurred might be brought about by ordinary cognitive processes. One might argue that Perkins's results depended on the specific problem he studied, i.e. the Fake Coin, which is not particularly complex. Perhaps really important advances, such as

Table 8. Perkins's study of insight in problem solving.

A. Fake Coin Problem.

A museum curator is approached by man offering to sell him an ancient coin. The coin is made of metal of the typical sort, and on one side it has an engraving of an ancient emperor with the date 44 B.C. barely readable. The curator has had dealings with the man before, but this time he immediately calls the police and accuses him of fraud. Why?

B. Perkins's two protocols on Fake Coin Problem (Perkins, 1981)

Abbott
1. Couldn't figure out what was wrong after reading through once.
2. Decided to read problem over again
3. Asked himself, do architects dig up coins? Decided yes.
4. Asked himself, could the problem have something to do with bronze? Decided no.
5. Saw the word 'marked'. This was suspicious. Marked could mean many different things.
6. Decided to see what followed in the text.
7. Saw 544 B.C. (Imagined grungy coin ion the dirt; had an impression of ancient times.)
8. Immediately realized—"it snapped"—that B.C. was the flaw.

Binet
1. Thought perhaps they didn't mark coins with the date then.
2. Thought they didn't date at all—too early for calendar. (Image of backwards man hammering 544 on each little bronze coin.)
3. Focused on 544 B.C.
4. Looked at B.C.
5. Realized 'B.C.—that means Before Christ'.
6. Rationalized that it couldn't be before Christ since Christ wasn't born yet.
7. Saw no possible way to anticipate when Christ was going to be born.
8. Concluded 'Fake!'

those involved in high-level innovations, might involve larger leaps and extraordinary thought processes. In response to this objection, we now turn to Edison's invention of the light bulb, an innovation at the highest level from a domain very different from those discussed so far, and here too we will see numerous examples of the role of ordinary cognitive processes in furthering innovative work. In addition, development of the light bulb will provide us with further evidence for structure in innovative thinking, the role of antecedents in innovation, and the incremental nature of innovative advances.

Cognitive Components in High-Level Innovation: Edison's Invention of the Light Bulb

Edison invented the light bulb in October, 1879. The invention looks very similar to today's light bulbs. Electric current is passed through a thin filament of carbon ('the burner'), which is enclosed inside a glass bulb, in a vacuum. The current flowing through the carbon causes it to heat to the point of glowing or 'incandescence', thereby producing light.

The Myth

Edison's invention of the electric light is part of American mythology. Here is a description from the *New York Herald* of the process through which the light bulb was invented, published around the time of Edison's demonstration of the successful light bulb.

Sitting one night in his laboratory reflecting on some of the unfinished details, Edison began abstractly rolling between his fingers a piece of compressed lampblack (carbon obtained from the soot deposited on the glass flues of gas lamps) until it had become a slender filament. Happening to glance at it, the idea occurred to him that it might give good results as a burner if made incandescent. A few minutes later the experiment was tried, and to the inventor's gratification, satisfactory, although not surprising results were obtained (*New York Herald*, December, 21, 1879; quoted in Friedel & Israel, 1986, p. 94).

The invention of the light bulb is a story which many of us think we know. Edison, who is famous for saying that genius is 1% inspiration and 99% perspiration (Freidel, 1986), is legendary for working through innumerable possibilities before finding just the right material that would serve in his bulb. And if the *Herald* report is accurate, he just stumbled across that material. From that story, one gets the impression that Edison had not thought about the usefulness of carbon before 'absently' rolling the piece of lampblack in his fingers. If all Edison did was try one material after another, until he stumbled on one that worked, we have little reason to look more deeply into his accomplishment. In reality, however, the story is different (Friedel, 1986; Friedel et al., 1986; see also Jehl, 1937; Weisberg & Buonanno, 2002), and examination of Edison's accomplishment will enable us to consider in a different context the cognitive components of innovative thinking, as well as the role of antecedents in innovation. I will present this case historically, to provide a context for understanding Edison's achievement and analyzing the thought processes involved. Sternberg, Pretz and Kaufman (this volume) discuss Edison's light bulb in the context of their system of classification of innovations. Their overall conclusion is supported by the present study.

Antecedents to Edison

As one might expect, based on the discussion thus far, there had been numerous attempts to produce a working incandescent electric light bulb in the years before Edison began to tackle the problem, and he was aware of what had been done before (Freidel & Israel, 1986). Table 9 summarizes some of these earlier attempts, almost all of which used either carbon or platinum as the burner in the bulb. However, there were

difficulties with each of those elements, which earlier investigators had been unable to overcome. At the heat needed to produce light, carbon would quickly oxidize (burn up), rendering the bulb useless. In order to eliminate oxidation, it was necessary to remove the carbon burner from the presence of oxygen, and most early attempts placed the carbon either in a vacuum or surrounded by an inert gas. However, the vacuum pumps available before Edison's work could not produce a sufficient vacuum to save the burner. Platinum presented a different problem: when used as a burner, its temperature had to be controlled very carefully, because if it got too hot, it would melt and crack, thereby rendering the bulb useless. Thus, when Edison began his work, perfection of the vacuum was crucial to the carbon-burner light bulb; control of the temperature was crucial in utilizing platinum.

In analyzing Edison's innovation, we shall examine what we can call the phases of Edison's work (Weisberg & Buonanno, 2002). For each phase, we

Table 9. Summary of History of Work on the Electric Light (Adapted from Freidel & Israel (1986), p. 115).

Who invented the incandescent lamp?

Edison was by no means the only, or the first, hopeful inventor to try to make an incandescent electric light. The following list, adapted from Arthur A. Bright's *The Electric Lamp Industry*, contains over 20 predecessors or contemporaries. What was different about Edison's lamp that enabled it to outstrip all the others?

Date	Inventor	Nationality	Element	Atmosphere
1838	Jobard	Belgian	Carbon	Vacuum
1840	Grove	English	Platinum	Air
1841	De Moleyns	English	Carbon	Vacuum
1845	Starr	American	Platinum	Air
			Carbon	Vacuum
1848	Staite	English	Platinum/iridium	Air
1849	Petrie	American	Carbon	Vacuum
1850	Shepard	American	Iridium	Air
1852	Roberts	English	Carbon	Vacuum
1856	de Changy	French	Platinum	Air
			Carbon	Vacuum
1858	Gardiner & Blossom	American	Platinum	Vacuum
1859	Farmer	American	Platinum	Air
1860	Swan	English	Carbon	Vacuum
1865	Adams	American	Carbon	Vacuum
1872	Lodyguine	Russian	Carbon	Vacuum
			Carbon	Nitrogen
1875	Kosloff	Russian	Carbon	Nitrogen
1876	Bouliguine	Russian	Carbon	Vacuum
1878	Fontaine	French	Carbon	Vacuum
1878	Lane-Fox	English	Platinum/iridium	Nitrogen
			Platinum/iridium	Air
			Asbestos/carbon	Nitrogen
1878	Sawyer	American	Carbon	Nitrogen
1878	Maxim	American	Carbon	Hydrocarbon
1878	Farmer	American	Carbon	Nitrogen
1879	Farmer	American	Carbon	Vacuum
1879	Swan	English	Carbon	Vacuum
1879	Edison	American	Carbon	Vacuum

shall consider any new developments and how they came about. The question to be asked about each phase is whether it is necessary to assume anything beyond ordinary thought processes in attempting to understand how the new developments came about.

Edison's Early Phases: Beginning with the Past

Edison began the first phase of his electric-light work in 1877, with a bulb comprising a burner of carbon in a vacuum (see the left-hand column in Table 10). This work, built directly on the past, was, like earlier work with carbon burners, not successful: the burner oxidized. Since, at the time, there was no way available to improve the vacuum, due to limitations in technology, Edison dropped the carbon burner. About a year later, he carried out a second phase of work on the light bulb, in which platinum served as the burner (see Table 10).

Here too the work was built directly on what had been done in the past. In order to try to stop the platinum from melting, the platinum bulbs contained 'regulators', devices like thermostats in modern heating systems, designed to regulate the temperature of the platinum, and thereby keep it from melting. Edison and his staff in his laboratory in Menlo Park, New Jersey developed regulators of many different types, but it proved impossible to control the temperature of the platinum burner. Thus, Edison's first two phases of work were based directly on work from the past with which he was familiar. Furthermore, the initial attempts using carbon, the rejection of carbon and the switch to platinum, and the incorporation of regulators in the platinum bulbs, can all be understood on the basis of ordinary thought processes. Carbon was rejected because it kept burning up; platinum was used because

Table 10. Summary of Phases of Edison's Invention of the Electric Light. (From Weisburg & Buonanno, 2002).

Phase and Activity	Basis for New Developments
A. Phase 1 (early 1877; Autumn, 1877)	
A1. Carbon burner	A1. Continuity: previous work by others
A2. Vacuum	A2. Continuity: previous work by others
B. Phase 2 (August–early October, 1878)	
B1. Platinum burner	B1. Failure of carbon; continuity
B2. Regulators	B2. Need to control burner temperature
C. Phase 3 (October, 1878–February, 1879)	
C1. Analysis of broken platinum burners	C1. Consistent failures of platinum
C2. Escaping hydrogen ⇒ need for vacuum	C2. Deduced from observation of burners
D. Phase 4 (early spring, 1879)	
D1. Platinum in vacuum	D1. Deduced from analysis of platinum's reaction to heating
Analysis of Requirements of Electrical System	
E. Phase 5 (Spring, 1879)	
E1. High-resistance platinum	E1. Deduced from system requirements
E2. Search for insulating material	E2. Problems with long platinum spirals
F. Phase 6 (July–August, 1879)	
F1. Improved vacuum pumps	F1. Failures with platinum ⇒ need better pumps; hiring of Boehm; article by de la Rue and Muller
G. Phase 7 (September–early October, 1879)	
G1. Platinum in high vacuum	G1. New pumps available
H. Phase 8 (mid-October, 1879)	
H1. Return to carbon, now in high vacuum	H1. Failure of platinum; new pumps overcame problems with carbon

it did not burn (and because it had been tried in the past). The regulators in the platinum bulbs were necessary in order to control the temperature of the platinum burner.

Systematic Analysis of Platinum

After the consistent failures with platinum, Edison looked carefully at why the platinum burners failed (Phase 3 in Table 10). He observed the broken burners under a microscope, and he and his staff thought they found evidence that the melting and cracking were caused by escaping hydrogen gas, which platinum under normal conditions had absorbed from the atmosphere. The hydrogen escaped when the platinum was heated, causing holes to form, which facilitated melting and cracking of the burner. Edison reasoned that the platinum might be stopped from cracking if the hydrogen could be removed slowly. In order to bring this about, he reasoned further that the platinum would first have to be heated slowly in a vacuum, which would allow the hydrogen to escape without destruction of the platinum. These Phase 3 advances came about as a result of information gained from the observations of the failed burners under the microscope, and Edison's reasoning from those observations. Edison turned to the systematic analysis of platinum because of the consistent failures with platinum-burner lamps in Phase 2.

Platinum in a Vacuum

Based on the new Phase 3 information and the deductions from it, Phase 4 of Edison's work involved a platinum burner being heated slowly in a vacuum and then sealed. The removal of the hydrogen from the platinum burner made it last longer and burn brighter, but it did not eliminate the basic problem: the burners still overheated and melted (Friedel & Israel, 1986, pp. 56–57, 78). In the spring of 1879, Edison and the staff at his laboratory attempted to extend the life of the platinum burner through the use of more efficient vacuum pumps and through a variety of coating techniques. Edison was able to obtain a state-of-the-art Geissler vacuum pump in late March (Friedel & Israel, 1986, p. 78), but even with this device, it was still very difficult to achieve near-complete vacuums that lasted for the time needed to heat the platinum to remove the hydrogen. Phase 4 of Edison's work, arising directly from the new information in Phase 3, brought some improvement, but overall the work was still a failure. As far as cognitive processes are concerned, nothing extraordinary is seen in this phase.

The Light Bulb in Context: Edison's Analysis of the Electrical System and Its Needs

So far we have been analyzing Edison's work on the light bulb in isolation from any other activities. However, Edison had strong entrepreneurial interests, and he was developing the electric light bulb as part of a *system* of electric lighting, which he hoped would replace the gas lighting then in use (Israel, 1998). Consideration of the light within the context of a lighting system put certain constraints on its characteristics, and by the late fall of 1878, Edison began to realize that the light would need to be highly resistant to the passage of the electrical current if it were to operate efficiently as part of a larger system (Friedel & Israel, 1986, p. 56; see Weisberg & Buonanno (2002) for further discussion).

Electrical resistance (measured in ohms) refers to how easily an electrical current passes through some material. Some substances, such as many metals, are low in resistance, so that electrical current passes easily through them. In contrast, materials such as glass and rubber are high in resistance, and thus are poor electrical conductors. The actual resistance of any given material when it is placed in an electrical circuit depends upon its physical configuration, e.g. for a wire, its diameter and length. Resistance of an electric wire decreases as the diameter of the wire increases. That is, all other things equal, a wire of small diameter is more resistant than one of larger diameter. In addition, the resistance of any material increases as its length increases: a 2-foot length of wire is twice as resistant as a 1-foot length of the same wire.

In the system as envisioned by Edison, there were to be many individual lamps, each receiving electricity through copper feeder wires from a central generating station. In order for each lamp to receive enough electrical energy, there would have to be either a large current or a high voltage, because energy = (voltage) × (current) (Friedel, 1986). Sending large amounts of current through the feeder wires raised another problem: much of the energy would be given off in the form of heat, as the current flowed to the individual user. This would reduce the energy available to light the bulb. Because the amount of heat given off by a flowing electrical current depends on the resistance of the material through which it flows, in theory this problem could be solved by decreasing the resistance of the feeder lines. This would mean increasing their diameter, which would result in spending an enormous amount of money on copper lines; this additional expense would have made electrical lighting economically unfeasible. Alternatively, one could increase voltage and reduce the current in order to ensure that each lamp received enough energy. Because voltage = (current) × (resistance), Edison reasoned that if he were to use a low current in his system, then it would be necessary for each lamp to have a high-resistance burner. Thus, Edison concluded that the use of high-resistance lamps was necessary for electrical lighting to become a commercially viable alternative to gas lighting. This conclusion, which had important ramifications on Edison's work on the light bulb, came about through logical/mathematical analysis of the requirements of the lighting system.

High-Resistance Platinum and Insulation

In Phase 5 of his work, Edison was concerned with the dimensions of the burner, in response to the requirement that the lamp be of high resistance. He concluded that there would have to be a very thin and long platinum wire, wound into a spiral, packed inside the evacuated glass bulb (Friedel & Israel, 1986, p. 79). However, as a result of the tight packing of the burner, the turns in the wire would often come into contact with one another, resulting in a short-circuiting of the lamp. To solve this problem, Edison tried to coat the platinum spirals with an insulating material that would prevent them from coming into direct contact with one another. During the spring of 1879, Edison and the staff devoted much of their time to finding an insulating material that would adhere to the platinum wire under high-temperature conditions (Friedel & Israel, 1986, p. 79), but by the end of April 1879, such material had yet to be found. The work in Phase 5 arose from the requirements of the electrical system and the resulting problems with packing the high-resistance platinum burner inside the bulb.

Improved Vacuum Pumps

Work on the electric lamp did not resume until the summer of 1879. Edison at this time made substantial progress in designing a more efficient vacuum pump. In August, he hired Ludwig Boehm, a full-time glassblower who had apprenticed under Heinrich Geissler (the creator of the Geissler vacuum pump that had been used in Edison's earlier work). The hiring of Boehm was an indication of Edison's belief that a more efficient vacuum pump would be an indispensable component of a successful platinum lamp (Friedel & Israel, 1986, p. 82). Edison and his staff spent much of August attempting to develop more efficient pumps. They eventually produced a vacuum pump that was a combination of two pumps—a Geissler pump and a Sprengel pump, another type of advanced vacuum pump. The idea of combining two vacuum pumps was first presented in an article by de la Rue and Muller (Friedel & Israel, 1986, pp. 61–62). Using this new pump, Edison was able to reduce the pressure inside the bulb to one-hundred-thousandth of normal atmospheric pressure, which was the most nearly complete vacuum then in existence (Fridel & Israel, 1986, pp. 62, 82). Further work in the autumn of 1879 produced pumps that were even more efficient. Edison's work in Phase 6 is summarized in Table 10; the advances in vacuum pump technology benefited from the new expertise brought by Boehm, as well as information from the article by de la Rue and Muller.

Platinum in a High Vacuum

Edison used the new vacuum pump with a platinum-burner bulb, and in early October 1879, Edison's light consisted of a highly resistant filament of tightly wound platinum inside a glass bulb that was evacuated to approximately one 1-millionth of normal atmospheric pressure (Friedel & Israel, 1986, pp. 87–88). However, this advance had not solved the basic problems: even after being heated in the vacuum, the platinum filaments would last for only a few hours and would tolerate only a minimal amount of electrical current, making it very difficult to generate light of useful brightness. Also, the platinum burners had a resistance that was far too low to be of any use in the system Edison had envisioned. Edison had calculated that it would be desirable to obtain lights that had a resistance of 100 ohms or greater, but the lights constructed in early October at Menlo Park had a resistance of only 3 or 4 ohms. Phase 7 of Edison's work is summarized in Table 10, and the work in that phase was a straightforward application of the advances in Phase 6.

Return to Carbon and Success

In early October, 1879, Edison began to experiment once again with carbon as an incandescent substance. This goes against the *Herald* story quoted above, which makes no mention of the fact that Edison's successful use of carbon was his second attempt with that element. The return to carbon follows directly from the situation at the end of Phase 7: (1) the platinum bulb was still not successful; (2) an improved vacuum pump was available; and (3) Edison knew that earlier attempts with carbon had failed due to problems with the vacuum. Experiments with carbon continued through October. On October 19th, Edison's assistant Francis Upton experimented with raising a stick of carbon to incandescence in a vacuum. The results were encouraging, producing a light that was the equivalent of 40 candles. On October 21st, Upton recorded that he had raised a half-inch stick of carbon that had a diameter of 0.020 inches to 'very good light'; however, the stick had a resistance of 2.3 ohms when heated, much too low to be useful (Friedel & Israel, 1986, p. 100). On October 22nd, another assistant, Charles Batchelor, conducted experiments using a 'carbonized' piece of cotton thread placed inside of an evacuated bulb. The thread had been baked in an oven until it turned into pure carbon. This filament initially had a resistance of 113 ohms, which increased to 140 ohms as it was being heated. Although this was encouraging, the filament yielded a weak light that was the equivalent of only half a candle. Batchelor continued to experiment with a variety of carbon materials throughout the day, and at 1:30 a.m. the next morning he attempted once again to raise a carbonized cotton thread to incandescence (Friedel & Israel, 1986, p. 104). This light burned for a total of 14.5 hours, with an intensity of 30 candles.

With hindsight, we can see that Edison had produced a successful light bulb, and the date of October 22, 1879 is generally given as the date of the invention of the light bulb. However, from Edison's viewpoint, there was no sudden point when it was obvious that everything was in place. The cotton-thread experiments were an undeniable success, but Edison still did not consider the lamp to be a completed invention (Friedel & Israel, 1986, p. 103). One problem was that the cotton filament was extremely fragile. The staff spent much of late October, 1879 continuing to search for another carbon material. (Here may be part of the basis for the *Herald* story.) Carbonized cardboard filaments in the shape of a horseshoe seemed to produce especially promising results. By early November, 1879, Edison felt sufficiently confident in the success of the carbon lamp that he filed for an electric light patent with the U.S. patent office. On New Year's Eve, 1879, Edison opened the lab, illuminated by interior electric lights and electric streetlights, to the public, and crowds came to view and marvel at what the wizardry of Edison had accomplished (Friedel & Israel, 1986, p. 112).

Invention of the Light Bulb: Summary and Conclusions

This analysis of Edison's seminal invention is consistent with the other case studies presented: his work provides further evidence for continuity and structure in innovative thinking, as well as providing evidence that innovative thinking is made up of a family of components. There has been no need to assume anything beyond ordinary thinking in order to understand Edison's accomplishment. Edison's first attempts were based directly on what had been done in the past. He moved beyond past work first by trying to determine the reasons for his consistent failures. In addition, his logical/mathematical analysis of the requirements of the electrical system that he hoped to build led to the conclusion that the lamp had to be of high resistance. Improvements in vacuum-pump technology and the need for a high-resistance lamp in turn led back to carbon and ultimate success. All these developments can be understood as being the outgrowth of ordinary thought processes, and we can see how innovation comes about as an orchestration of a family of cognitive processes.

Critical Analysis and Discontinuity in Thinking

The invention of the light bulb has provided an example of a discontinuity in Edison's thinking—use of a platinum burner in a vacuum. Edison began his work believing that a platinum burner could be exposed to air, but later the vacuum became a crucial component of the platinum-burner bulb. This discontinuity, which was critical in Edison's success, was brought about by ordinary cognitive processes: in Phase 3 Edison examined under the microscope the condition of the broken platinum burners, which led him to conclude that escaping hydrogen had contributed to their destruction. (At least one earlier investigator had

placed the platinum burner in a vacuum (see Table 10). I am assuming that Edison was not aware of that attempt, although if he was, it would mean that platinum in a vacuum would be a further example of continuity.) In considering how to remove the hydrogen without destroying the burner and how to keep the burner isolated from hydrogen, placing it in a vacuum was a logical conclusion. Edison's critical analysis in Phase 3 of the difficulties that arose in Phase 2 led to the realization of the need for the vacuum: nothing new is needed for us to understand Edison's thinking.

A discontinuity arising in a similar way can be seen in James Watt's development of the steam engine (Basalla, 1988; Weisberg, 1993, pp. 126–130). Although Watt is usually given credit for inventing *the* steam engine, Watt's invention was in fact a modification of an extant engine—the Newcomen engine, a steam-powered engine invented by Newcomen a generation earlier. Watt was employed as a laboratory technician at the University of Glasgow, and he was preparing a small-scale Newcomen engine for a lecture demonstration. He was struck by the difficulty he had in getting it to run; it would carry out a few cycles and then would stop, due to lack of steam because all the coal had been quickly consumed. Watt then modified the Newcomen engine so that it ran more efficiently. Thus, Watt's innovation was driven by his critical analysis of the poor performance of the Newcomen engine; again, ordinary cognitive processes are central in this innovation. Discontinuity based on critical analysis of one's own earlier work can also be seen in Watson and Crick's development of the double-helix model of DNA. They originally proposed a triple-helix model of the structure (based in part on Pauling's alpha-helix, as noted above). When this model was found to be inadequate on empirical and logical grounds (Olby, 1994; Watson, 1968; Weisberg, 1993), they modified it, producing the double helix. It is also important to note that some of the inadequacies in the triple-helix model of DNA were pointed out by other researchers, with whom Watson and Crick discussed their model. This parallels Dunbar's (1995) finding, cited earlier, of the importance of input from others in changing a scientist's analysis of a phenomenon.

Cognitive Processes in Innovation: Conclusions

This section has centered on analysis of Edison's invention of the light bulb to provide support for the claim that creative thinking is constructed out of the same family of cognitive processes that are seen in ordinary thinking (Perkins, 1981). It was concluded that it was possible to understand each phase in Edison's thinking without introducing processes beyond those found in ordinary thinking (see Table 10). In addition, several other cases, presented more briefly, also support the idea that radical innovation can be stimulated by such mundane processes as critically examining deficiencies in an extant product, either one's own or someone else's. This section also provided further evidence for continuity in innovation, specifically the role of antecedents in the development of new products.

Sensitivity to External Events

Ordinary thinking is sensitive to outside events, which can serve to trigger new directions in thinking. This is a second way in which discontinuity in thinking can occur: an external stimulus may suddenly change the direction of a person's work. An example of a triggering event can be seen in Alexander Calder's development of mobiles in the 1930s in Paris (Calder, 1966, p. 113). These abstract moving sculptures, which depended on the wind to provide motion, were different than anything Calder had produced before (see Fig. 9A). For the present discussion, we will examine only the abstract or non-representational nature of these sculptures, which was in itself a significant breakthrough (for further discussion, see Weisberg, 1993). Figure 9B reproduces an exhibit from 1931 in Paris of Calder's then-recent work, and one can see graphic evidence that his work took a radical shift in subject matter around 1930, moving relatively suddenly away from representation. In this exhibit, one sees representational and abstract works together.

In examining Calder's work before this point (e.g. the wire sculptures hanging near the ceiling in the exhibit in Fig. 9B), one finds little or nothing in the way of non-representation. Calder himself (1966) provides the clue to why his style changed relatively suddenly. In the late 1920s, Calder had made a name in avant-garde art circles in Paris, through his performances of his *Circus*, a miniature big-top, complete with multiple rings, a high-wire act, animals, acrobats, jugglers, clowns, music, a chariot race, a sword-swallower, etc., and grandstands for the spectators. Many of the most well-known of the modern artists in Paris witnessed Calder's circus performances, which resulted in his establishing connections to many artists. One of the artists who attended a circus performance was Piet Mondrian, whose work at that time involved highly abstract geometrical paintings, built out of blocks of primary colors. Calder reported that in 1930 he visited Mondrian's studio, which was painted white, with blocks of primary colors on the walls and examples of his latest work also present (see Fig. 9C). Very soon thereafter, Calder began to paint in a style much like Mondrian's, but he then adapted that style to sculpture, with which he felt more comfortable. One can see that at least one of the abstract sculptures in Fig. 9B has a base painted in the style of Mondrian. So the radical change in Calder's style came about at least in part as a result of his adoption of the style of Mondrian, which means that the radical change within Calder's style was not radical at all in the modern-art milieu of Paris around 1930. The change seems radical only in the context of Calder's work until that point.

Another example of a discontinuity triggered by an external event occurred during Picasso's production, in 1906–1907, of his seminal painting *Les Demoiselles d'Avignon*, arguably the most important painting of the 20th century (Rubin, 1984). The style of this painting was radically different from anything that he or any other artist had produced before, and this radical stylistic shift has been attributed to Picasso's exposure to non-Western art (Rubin, 1984; Weisberg, 1993, pp. 197–198).

External events can also play a role in the development of innovation in domains beyond art. A similar process may have occurred in the Wright brothers' development of the 'wing-warping' control system for their early glider, which became a crucial component of their successful powered airplane. This system, which is the direct ancestor of the ailerons seen on modern aircraft, was designed so that the tips of the trailing edges of the two wings on the Wrights' machine could be made to lift or drop by the operator. The mechanism was constructed so that the two sets of wing tips moved in opposition; i.e. when the left wing tips were lifted, the right ones dropped, and vice versa. This caused the plane to turn in one direction or the other. This control system, which differed radically from those developed by other would-be airplane inventors, may have been stimulated, at least in part, by Wilbur Wright's observations of birds' movements of their wing tips when gliding (Weisberg, 1993, pp. 139–140).

It is also interesting to note that one difference between the Wright brothers and other researchers working on the problem of flight was that the Wrights were from the beginning very concerned with how they would control their flying machine once they got it into the air. Most of the other would-be inventors assumed that the machine could be made inherently stable with a few simple passive design components, such as a V-shaped wing (Crouch, 1992). It was assumed that such components would serve to automatically keep the plane stable in flight. The Wrights, in contrast, assumed from the beginning that the human operator would have to play an active role in controlling the machine in the air. A possible reason for the Wrights' concern about control (besides the fact that more than one person had been killed investigating possibilities of flight) is that they came to the problem from a background of designing and building bicycles, which have problems of stability remarkably like those seen with flying machines. Most of the other people working on the problem of flight approached it with the belief that movement in the air would be like a boat's movement on the surface of a body of water, which turned out to be an incorrect analogy. The Wrights' adopting this overall perspective is further support for the notion of continuity in thinking: the general orientation of their work on controlling the flying machine was based on their experience with bicycles.

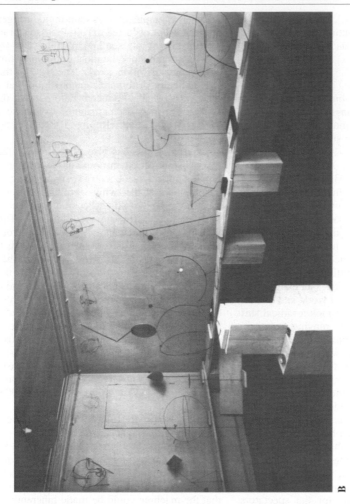

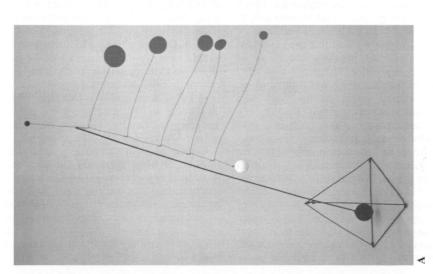

Figure 9. (A) A Calder mobile from around 1930. (B) Exhibition at Gallery Percier. © 2002 Estate of Alexander Calder/Artists Rights Society (ARS), New York.

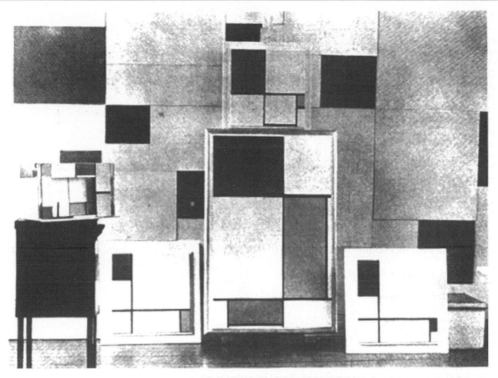

Figure 9. (C) Mondrian's studio with works. 1929; © 2002 Mondrian/Holtzman Trust c/o Beeldrecht/Artists Rights Society (ARS), New York.

External triggers have also played important roles in innovation in science. Darwin's final synthesis of the theory of evolution based on natural selection may have been stimulated by his reading a passage in Malthus's well-known essay *On Population*, which discussed the competition among organisms for limited supplies of food (Gruber, 1981; for further discussion see Weisberg, 1993). In this example, the external stimulus provided the capstone on a process that had gone on for more than 20 years for Darwin. External events can also serve to start the process of innovation. As an example, the collaboration between Watson and Crick may have come about in part as the result of an external trigger. Watson's initial desire to work on the problem of the structure of DNA at the Cavendish Laboratory at Cambridge University—where he met Crick—was stimulated in part by his exposure to an X-ray photograph of crystalline DNA which he saw at a conference in Naples (Watson, 1968). This external event played a role in initiating the whole process. In addition, comparable to Malthus's influence on Darwin, Watson and Crick's thinking at several places was pushed in a critical direction as a result of their exposure to new experimental results which allowed them to draw conclusions about important features of the DNA structure (Olby, 1994; for further discussion see Weisberg, 1993).

Innovation and External Events—Conclusions

Examples from a broad range of domains—the arts, invention, and science—support the hypothesis that new directions in innovative thinking are sometimes the result of external events. External events can play a role at every point in the creative process, from initiation of interest in some phenomenon to the final piece of information that results in things coming together. These examples are further evidence that innovation comes about through processes no different than those underlying our day-to-day activities.

The Ordinary Basis for Creative Thinking: Conclusions and a Remaining Question

Conclusions

At the beginning of the chapter, several facets of ordinary thinking were outlined, and the discussion has presented support for the hypothesis that each of those facets is also seen in innovative thinking. We can summarize the discussion in the chapter by examining those facets once again (see Table 11).

Continuity in Thinking

Continuity in innovative thinking is supported first by the learning curve that has been found in creative disciplines: people who work in creative disciplines

243

Table 11. Summary of case studies.

Continuity in Thinking
  Learning to be Creative
    10-Year Rule (Hayes, 1981)
      Gardener's (1993) biographical studies
    Mozart
      10-Year Rule
      Increasing productivity
      Increasing quality of works
      Increasing originality
    Lennon & McCartney
      10-Year Rule (although fewer years)
      Increasing quality of works
      Increasing originality
  Antecedents to Innovation
    Mozart
      Early piano concertos constructed out of those of others
      Early symphonies modeled on those of others
    Lennon & McCartney
      Immersion in early rock & roll
    Darwin
      First stage: Lamarckian evolution and the monad
    Watson & Crick
      Pauling and helical structures
    Picasso
      Guernica and *Minotauromachy*
      Characters based on Goya
    Pollock
      Stylistic innovation: Siquieros workshop
    Edison
      First attempts: Carbon and platinum burners; based on earlier work
    Wright brothers
      Controlling bicycle to controlling a flying machine
Structure in Innovative Thinking
  Picasso's *Guernica*
    Antecedents related in content
    Worked on overall composition first, then characters
    Systematic working out of overall composition
    Systematic working out of individual characters
  Edison
    Analysis of problems with platinum burners; based on failures
    Analysis of needs of electrical system
Cognitive Components in Innovative Thinking
  Perkins—experimental studies of insight in problem-solving
  Edison
    Logical/mathematical thinking
      Analysis of failed platinum burners
      Analysis of requirements of electrical system
      Need for high-resistance burner
    Evaluation of consequences
  Watt
    Critical analysis of problems with Newcomen engine
  Watson & Crick
    Critical analysis in rejection of triple-helix model
External Triggers to Innovation
  Watson & Crick
    Criticism by others of the triple-helix model of DNA
    Information from others' research that pointed the way to the double helix
  Calder
    Mondrian's studio and abstraction
  Picasso
    'Primitive' art and *Les Demoiselles d'Avignon*
  Wright brothers
    Bird flight and 'wing warping'

require a significant amount of time to get good at what they do, as we have seen from studies of the 10-Year Rule (Hayes, 1989). Gardener's (1993) biographical studies supported the same conclusion. There was also evidence from case studies in music for increasing productivity and increasing quality of work during this learning period. A further aspect of continuity, that in producing innovations we build on our experiences, was supported by the finding of antecedents for innovations in a wide variety of areas (Weisberg, 1986, 1993). Similarly, based on a consideration of the relationship between creators' early works and works produced by those around them, we saw incremental movement beyond what had been done before, rather than wholesale rejection of the past.

Structure in Thinking

Ordinary thinking is structured in several ways: (1) our thoughts are linked through associative bonds, which are the result of links between events in our past, and on the basis of similarity of content; (2) we use reasoning processes of various sorts, which provide a basis for moving from one thought to the next. The analysis of Picasso's creation of *Guernica* and Edison's invention of the light bulb provided evidence for the structured nature of innovative thinking.

Thinking is a Family of Activities

Ordinary thinking is made up of a large family of cognitive activities, some of which are: planning how to carry out some activity before doing it; anticipating the outcome of some action; judging whether the outcome of an anticipated action will be acceptable; deciding between two alternative plans of action; determining the consequences of some events that have occurred (through deductive reasoning); perceiving a general pattern in a set of specific experiences (through inductive reasoning). Several sorts of evidence support the claim that creative thinking is also constructed from those sorts of basic cognitive activities (Perkins, 1981). Analysis of the thought processes underlying Edison's invention of the light bulb indicated that the phase-to-phase transitions cold be understood as being the result of ordinary processes. For example, Edison's decision in Phase 4 to put the platinum burner in a vacuum came about as a consequence of his analysis in Phase 3 of the failed platinum burners. Similarly, his decision to return to carbon in Phase 8 was the result of the consistent failures with platinum, the still-remaining need for a burner of high resistance, and the availability of the newly developed high-efficiency vacuum pump.

Thinking is Responsive to Environmental Events

Ordinary thinking and innovative thinking are both sensitive to environmental events: radical changes in thinking are sometimes based on the innovator's exposure to what could be called external triggers. Support for this conclusion come from the arts (Calder's radical style shift in his sculpture around 1930, Picasso's radical style shift in painting *Les Demoiselles d'Avignon*); science (Darwin's reading of Malthus; Watson & Crick's collaboration); and invention (the Wright brothers' control system based on bird flight).

A Remaining Question: What then is the Basis for Innovation?

If one assumes for the sake of discussion that innovation is indeed the result of ordinary thinking processes, we are still left with one important question: What are the differences between the great innovators and the rest of us? If innovation at all levels, including the highest, is brought about through thought processes which are available to essentially all of us, then why do very few of us produce seminal advances? In response to this question, one should note that, even if we all possess the same basic set of cognitive processes, there are still ways in which we could be different which might bring about the differences among us in innovative achievement. Let us take the hypothetical case of two individuals who are researchers in the same field—why might one produce ground-breaking work while the other does little or nothing in the way of innovation? One crucial element of difference could be disparities among individuals in information-processing capacities. As one example, it was noted earlier that Mozart's last symphony was different in complexity than earlier symphonies, those by Mozart and by others. That advance leads us to the question of how Mozart was able to produce a greater degree of complexity than others had (and than even he had previously produced). One possibility is that Mozart might have had a larger working-memory capacity than others (Baddeley, 1990). That is, Mozart might have been better able to keep track of multiple strands of music, which might have contributed to more complexity in his compositions. However, it should also be noted that an individual's ability to process and remember information is based on the person's expertise in the domain. Therefore, Mozart's 'larger working-memory span' might not have been some innate characteristic of his information-processing system. Rather, it might be a secondary result of his long immersion in music. In addition, on this view, the fact that Mozart's last symphony is the most complex of all of his works in that genre is because his ever-increasing expertise might have paved the way for works of greater and greater complexity.

This leads to a second possible difference between people who produce notable innovations and those who do not: the different databases that different individuals bring to their work. If one individual has broader and deeper knowledge than the other, such a difference could affect the chances of one individual rather than the other producing innovation. Related to the issue of the size of the database is how it is acquired—all other things equal, a larger database would be the result of more intense and/or longer-lasting immersion in the discipline. This difference in immersion might in turn be the result of differences in the motivation with which individuals approach their work. Motivation might be crucial in another way as well. It has been proposed that one crucial element in producing innovative work at the highest level ('genius-level' work) is the motivation to persist in an activity even when it seems unlikely that success will be obtained (see Howe, in press). Although innovation may ultimately come about from the application of one's knowledge to the situation that one is facing, and this application may come about as the result of structured processes of thinking that we all can carry out, having the information to work with only comes through years of immersion in the discipline. Not everyone is equipped with the motivation to stay the course.

References

Baddeley, A. (1990). *Human memory. Theory and practice.* Boston, MA: Allyn & Bacon.

Basalla, G. (1988). *The evolution of technology.* New York: Cambridge University Press.

Calder, A. (1966). *Calder. An autobiography with pictures.* New York: Pantheon.

Chase, W. G. & Simon, H. A. (1973). Perception in chess. *Cognitive Psychology, 4*, 55–81.

Chipp, H. B. (1988). *Picasso's 'Guernica'.* Berkeley, CA: University of California Press.

Crouch, T. (1992). Why Wilbur and Orville? In: R. J. Weber & D. J. Perkins (Eds), *Inventive Minds. Creativity in Technology* (pp. 80–92). New York, NY: Oxford.

Csikszentmihalyi, M. (1996). *Creativity. Flow and the psychology of discovery and invention.* New York: Harper Collins.

de Groot, A. (1965). *Thought and choice in chess.* The Hague: Mouton.

Dunbar, K. (1995). How scientists really reason: Scientific reasoning in real-world laboratories. In: R. J. Sternberg & J. E. Davidson (Eds), *The nature of insight* (pp. 365–395). Cambridge, MA: MIT Press.

Dunbar, K. (2001). The analogical paradox: Why analogy is so easy in naturalistic settings, yet so difficult in the psychological laboratory. In: D. Gentner, K. J. Holyoak & B. Kokinonv (Eds), *Analogy: Perspectives from cognitive science* (pp. 313–334). Cambridge, MA: MIT Press.

Friedel, R. (1986). New light on Edison's electric light. *American Heritage of Invention and Technology, 1*, 22–27.

Friedel, R. & Israel, P. (1986). *Edison's electric light. Biography of an invention.* New Brunswick, NJ: Rutgers University Press.

Gardner, H. (1993). *Creating minds. An anatomy of creativity seen through the lives of Freud, Einstein, Picasso, Stravinsky, Eliot, Graham, and Gandhi.* New York: Basic.

Ghiselin, B. (1952). *The creative process. A symposium.* New York: Mentor.

Gick, M. L. & Holyoak, K. J. (1980). Analogical problem solving. *Cognitive Psychology, 12*, 306–355.

Glucksberg, S. & Weisberg, R. (1966). Verbal behavior and problem solving: Some effects of labeling in a functional fixedness task. *Journal of Experimental Psychology, 71*, 659–664.

Gruber, H. E. (1981). *Darwin on Man. A psychological study of scientific creativity* (2nd ed.). Chicago, IL: University of Chicago Press.

Guilford, J. P. (1950). Creativity. *American Psychologist, 5*, 444–454.

Hayes J. R. (1981). *The complete problem solver.* Philadelphia, PA: Franklin Institute Press.

Hayes, J. R. (1989). Cognitive processes in creativity. In: J. A. Glover, R. R. Ronning & C. R. Reynolds (Eds), *Handbook of Creativity* (pp. 135–145). New York: Plenum.

Holmes, F. L. (1980). Hans Krebs and the discovery of the ornithine cycle. *Federation Proceedings, 39*, 216–225.

Howe, M. J. A. (in press). Some insights of geniuses into the causes of exceptional achievements. In: L. V. Shavinina & M. Ferrari (Eds), *Beyond knowledge: Extracognitive facets in developing high ability.* Mahwah, NJ: Erlbaum.

Humphrey, G. (1963). *Thinking: An introduction to its experimental psychology.* New York: John Wiley.

Israel, P. (1998). *Edison. A life of invention.* New York: John Wiley.

Jehl, F. (1937). *Menlo Park reminiscences.* Dearborn, MI: The Edison Institute.

Klahr, D. & Simon, H. A. (1999). Studies of scientific discovery: Complementary approaches and convergent findings. *Psychological Bulletin, 125*, 524–543.

Kozinn, A. (1995). *The Beatles.* London: Phaidon Press.

Kulkarni, D. & Simon, H. A. (1988). The processes of scientific discovery: The strategy of experimentation. *Cognitive Science, 12*, 139–175.

Landau, E. (1989). *Jackson Pollock.* New York: Abrams.

Langley, P., Simon, H. A., Bradshaw, G. L. & Zytkow, J. M. (1987). *Scientific discovery. Computational explorations of the creative process.* Cambridge, MA: MIT Press.

Lewisohn, M. (1992). *The complete Beatles chronicle.* New York: Harmony.

Luchins, A. S. & Luchins, E. H. (1959). *Rigidity of behavior.* Eugene, OR: University of Oregon Press.

Martindale, C. (1990). *The clockwork muse. The predictability of artistic change.* New York: Basic.

Miller, A. I. (1996). *Insights of genius. Imagery and creativity in science and art.* New York: Copernicus, an Imprint of Springer-Verlag.

Olby, R. (1994). *The path to the double helix. The discovery of DNA.* New York: Dover.

Perkins, D. N. (1981). *The mind's best work.* Cambridge, MA: Harvard.

Ramey, C. H. & Weisberg, R. W. (2002). *The 'poetical activity' of Emily Dickinson: A further test of the hypothesis that affective disorders foster creativity.* Unpublished manuscript, Temple University.

Rocke, A. J. (1985). Hypothesis and experiment in the early development of Kekulé's benzene theory. *Annals of Science, 42*, 355–381.

Rubin, W. (1984). Picasso. In: W. Rubin (Ed.), *'Primitivism' in 20th century art. Affinity of the tribal and modern* (pp. 240–343). New York: Museum of Modern Art.

Scheerer, M. (1963). On problem-solving. *Scientific American, 208*, 118–128.

Simon, H. A. (1979). *Models of thought.* New Haven, CT: Yale University Press.

Simon, H. A. (1986). The information-processing explanation of Gestalt phenomena. *Computers in Human Behavior*, **2**, 241–255.

Simonton, D. D. (1999). *Origins of genius. Darwinian perspectives on creativity*. New York: Oxford.

Tweney, R. D. (1989). Fields of enterprise: On Michael Faraday's thought. In: D. B. Wallace & H. E. Gruber (Eds), *Creative People at Work. Twelve Cognitive Case Studies* (pp. 91–106). New York: Oxford University Press.

Watson, J. D. (1968). *The double helix: A personal account of the discovery of the structure of DNA*. New York: New American Library.

Weisberg, R. W. (1986). *Creativity: Genius and other myths*. New York: Freeman.

Weisberg, R. W. (1993). *Creativity: Beyond the myth of genius*. New York: Freeman.

Weisberg, R. W. (1994). Genius and madness? A quasi-experimental test of the hypothesis that manic-depression increases creativity. *Psychological Science*, **5**, 361–367.

Weisberg, R. W. (1995). Prolegomena to theories of insight in problem solving: Definition of terms and a taxonomy of problems. In: R. J. Sternberg & J. E. Davidson (Eds). *The*

Nature of Insight (pp. 157–196). Cambridge, MA: MIT Press.

Weisberg, R. W. (1999). Creativity and knowledge: A challenge to theories. In: R. J. Sternberg (Ed.), *Handbook of Creativity* (pp. 226–250). Cambridge: Cambridge University Press.

Weisberg, R. W. (n press). *On structure in the creative process: A quantitative case study of the creation of Picasso's* Guernica. *Empirical Studies in the Arts..*

Weisberg, R. W. & Alba, J. W. (1981). An examination of the alleged role of 'fixation' in the solution of several 'insight' problems. *Journal of Experimental Psychology: General*, **110**, 169–192.

Weisberg, R. W. & Buonanno, J. (2002). *Edison and the Electric Light: A Case Study in Technological Creativity*. Unpublished manuscript. Temple University.

Weisberg, R., & Suls, J. M. (1973). An information-processing model of Duncker's candle problem. *Cognitive Psychology*, **4**, 255–276.

Wertheimer, M. (1982). *Productive thinking* (Enlarged Edition). Chicago, IL: University of Chicago Press.

Zaslaw, N. (1989). *Mozart's symphonies. Context, performance practice, reputation*. Oxford: Oxford.

The International Handbook on Innovation
Edited by Larisa V. Shavinina

Innovation and Evolution: Managing Tensions Within and Between the Domains of Theory and Practice

James R. Bailey[1] and Cameron M. Ford[2]

[1] *Department of Management & Technology, School of Business and Public Management, George Washington University, USA*
[2] *Management Department, College of Business Administration, University of Central Florida, USA*

Abstract: Innovation occurs when individuals produce *novel* solutions, and members of the relevant domain adopt it as a *valuable* variation of current practice. At the individual level, creative or innovative actions are adoptive responses to tensions between the *person* and *situation*. In domains such as the arts or sciences, person–situation tensions are best resolved by favoring novelty, whereas in domains such as business, the same tensions are best resolved by favoring value. We employ a neo-evolutionary view of creativity to propose that these *within* domains tensions create intractable tensions *between* domains.

Keywords: Innovation; Creativity; Evolution; Domain; Tensions.

Introduction

That progress is the progeny of diverging and converging forces is true in innovation as in all endeavors. Such operative forces, or *tensions*, are not new as explanatory frameworks. At a planetary level, geological history is best understood as long-grinding terrestrial and atmospheric pressures that shaped modern topography. Even the origin of a species can be viewed as friction between which variation is best suited to environmental demands—a logic to which we will soon return. And in human affairs, societies advance when social equilibrium is upset by countervailing movements as diverse as immigration and technology.

The common theme in all these tensions is *change* and *stability*—a contrast inherent in virtually all discussions of creativity and innovation. Other parallel and analogous polar anchors include subjective vs. objective, emotion vs. logic, conceiving vs. organizing and, in our minds the one most indicative of creativity and innovation, *novelty* vs. *value*. A truly innovative idea, process, service or product achieves both, and can thereby be said to have successfully resolved the tension. But how, exactly, is that accomplished?

In the pages that follow we make two interrelated arguments. The first is that innovation begins when an individual presents a novel proposal. But because proposed variations are evaluated in specific contexts or domains—replete with idiosyncratic expectations and reinforcements—novel proposals will be more compatible with and frequently adopted in some domains rather than others. Hence, explaining creative action in a given context requires appeal to the classic *person–situation* debate. By conceiving individuals as goal-oriented, future thinking and feedback-seeking entities, we advocate a neo-evolutionary approach wherein individual creativity introduces variations that are subsequently selected or rejected by the demands and predilections of the domain environment. In domains where novelty holds intrinsic value (e.g. the arts, science), creative behavior is an adaptive response that mitigates person–situation tensions. Alternatively, in domains where 'new' doesn't necessarily mean 'improved' (business management—cf. Ford & Gioia, 2000), person–situation tensions are best addressed by adherence to established practices.

Our second argument is trans-application in nature, based on the premise that innovations achieve their greatest impact when they are adapted across inter-dependent domains. Reciprocal adaptation is especially valuable in the worlds of *theory* and *practice*. We hold that in theoretical domains such as organizational studies, selection processes favor novelty, whereas in practical domains such as business, selection processes favor demonstrated value. Hence, over time theoretical structures have become increasingly complex, jargon-laden, pluralistic and unproven, whereas application concentrates on easily apprehended, broadly diffused 'best practices' whose value has been benchmarked.

In an ironic twist, the very evolutionary forces that engender innovation within these domains drive a divide between them. That is, a consequence of individuals' striving to resolve the person–situation tensions imposed by selection processes *within* their domains is exacerbated misalignment *between* them. Hope for maximally integrating theory and practice, then, lies in a recognizing that innovation reflects both novelty *and* value, and thus deploying intermediaries and designing institutional resource dependencies capable of sustaining divergent and convergent variations.

A Fork in the Road: Person–Situation Tensions

If tensions are critical in all endeavors, what is the one most representative of innovation? On the one hand lies the stark realization that innovation, in the final and fundamental analysis, is the spawn of individual imagination. To be sure, findings from academic journals build to novel theses and companies bring new products to market. But in every case, if enough layers are peeled or the shrouds of history are lifted, some clever, brave, or just plain lucky individual stands at the center. On the other hand, though, innovation does not unfold in a vacuum. The germination of an idea in one's mind stems from thousands of diverse threads, and the execution of that idea is the product of many minds and hands. In these ways, innovation is social in nature. Furthermore, there is ample evidence that throughout history some societies have been more progressive than others, and some firms have managed to sustain inventive cultures better than others. In these ways, innovation is institutional in nature. The oft-repeated aphorism that 'necessity is the mother of invention' is testimony to the creative potential of collective will.

All of this brings us to the much traveled *person–situation* debate, which is itself a variant of the grand nature–nurture debate, writ small. This is the tension most characteristic of innovation, garnering enormous examination over the last half century. The literature is too vast to review here, but the polar structure is easy to grasp.

At one end of the continuum is the hardcore 'individual differences' camp that dates back to Allport (1937) and draws heavily from research into disposi-tions, personality, traits, values, needs and attitudes. This approach assumes that individuals possess meas-urably stable and enduring qualities that determine behavior across a wide variety of situations. Although seldom explicitly stated, the intellectual commitment is that too many variables comprise a given situation at a given moment, making them a fragile foundation for useful prediction. Situations, then, are a proverbial goose chase. And although people do change, they still constitute the most stable factor in the equation, and hence should be focus of analysis.

The other end of the continuum is anchored, predictably, by the 'situationalists'. Chief among them is Walter Mischel, whose 1968 tome, *Personality and Assessment*, built the case that situations vary in strength, depending on extant mores, norms, reward structures, supervision, tightly scripted roles and the like. In strong situations, the behavioral patterns imbedded in personality would be trumped by con-textual cues. In this way, Mischel showed the Behaviorist tendency to assume that the conditions of one's life determine one's behavior. Change the conditions (via public policy and programs) and the behavior will follow. Behavior, then, is the flotsam and jetsam of the situation's ebb and flow. (For an instructive, recent exchange on the manifestation of this debate in organizational studies, see Davis-Blake & Pfeffer, 1989 and House, 1996.)

Of course, any reasonable person—including those that staked claim in this debate—recognizes that behavior is the function of both the person and the situation. For this reason, a strong contingent of 'interactionists' has emerged who examine motiva-tional and creativity processes associated with innovation (Amabile, 1996; Ford, 1996; Woodman, Sawyer & Griffin, 1993). To encapsulate, the person–situation tension is enacted when individuals' choose from a range of alternative behaviors in a particular circumstance. That choice is a coping behavior that either facilitates or constrains personally appealing courses of action.

With this as a background, we will distinguish between two broad options for coping with person–situation tensions—*novel behavior* and *routine behavior*—that together represent a 'fork in the road' (Ford, 1996). Whether consciously or not, in any given situation individuals are presented with a choice between new and routine behavioral options that instigate change or reinforce the status quo respec-tively. To support this view, we focus on three interactionist theories that treat individuals as goal-oriented, future-thinking and feedback-seeking entities attempting to resolve the person–situation tensions they face on a day-to-day basis.

Goals Model

This model assumes that individuals are goal-directed, summarized nicely with its founder's own words:

> (B)ehavior is the expression of the interplay among many goals, and there is a complex relation between

goals and behaviors, such that the same behavior may express different goals and the same goal may lead to different behaviors, according to the demands of the situation (Pervin, 1989, p. 355).

Pervin is asserting that individuals' action is organized to advance the achievement of a variety of personally desired outcomes, the saliency of which varies depending on the availability of particular outcomes in particular situations. Thus, individuals 'read' a situation to determine its operative goals, its reward structure, and their own ability to behave according to those demands. Such internal (i.e. goals) and external (i.e. situations) canvassing becomes the basis for choosing among alternative behavioral options. Elected behaviors address the person–situation tension by attaining desired outcomes or avoiding unpleasant ones.

Social Cognitive Model

This model focuses on individual adaptability to situations as a function of social intelligence (Cantor & Kihlstrom, 1987). Social intelligence is not necessarily that typically associated with the raw problem-solving ability reflected in IQ. One can have a high IQ and be woefully socially inept. Rather, it refers to having readily accessible memory structures of the concepts, rules, dynamics and skills that were operative in previous situations. This is a 'social-cognitive' model because individuals draw on the knowledge of what did or did not work in previous situations to formulate strategies for handling new ones. This *feed-forward* process creates the expectancies and instrumentalities underlying individual motivation, as well as specific performance strategies designed to attain personally desired outcomes. Adaptability, then, becomes a competency that turns on seamless rehearsal of relevant past experience and careful assessment of current circumstances to organically integrate behavior. Evidence shows that: (a) strategies can be considered effective only to the extent that the person using them, and the situation in which they are used, are considered; (b) effective strategies do not necessarily generalize across contexts; and (c) individuals who are inflexible in strategic deployment run into a whole host of problems.

Self-Appraisal Model

This school of thought is based on the notion that individuals strive to reduce uncertainty about their abilities in relation to environmental demands (Bailey, Strube, Yost & Merbaum, 1994). What distinguishes it from the two previous models is the notion that self-knowledge is predicated by one's willingness to seek out feedback. People can be arranged on a continuum from those low in self-appraisal motivation (meaning they tend to shy away from diagnostic feedback, favoring instead that which is skewed toward enhanc-

ing their self-image) to those high in self-appraisal (meaning a keen desire for accurate and complete feedback, regardless of its self-esteem implications).

This is a theory of ego operation that examines how the very human and hedonistic urge to feel good about oneself is manifested. Those low in self-appraisal do so by choosing tasks and eliciting feedback that reflects positively on them. In contrast, those high in self-appraisal take the lumps that negative feedback might inflict on their ego in order to understand better what they do well and what they do poorly. This, subsequently, puts them in a better position to select or create environments where their abilities are well matched to the demands, which ultimately would lead to positive feedback via successful performance, and hence serve the ego well. The Self-Appraisal model goes beyond the others by emphasizing the possibility of enacting or creating environments to discover the 'fitness' of specific behaviors—that is, the extent to which they accomplished desired outcomes.

These three models provide a robust foundation for understanding the person–situation tension as it relates to creativity and innovation. The Goals Model highlights how objectives are made salient and pursued depending on extant circumstances, and thus involves *direction*. The Social Cognitive Model looks at how future contingencies could help or hinder alternative course of action. Imagining possible consequences is an anticipatory *feedforward* activity that forms and shapes expectations. The Self-Appraisal Model poses that the ego strength to face *feedback* regarding abilities in relation to demands is the path to self-regard and, although threatening, could lead ultimately to placing oneself in a complimentary environment. These three elements are illustrated in Fig. 1 as a framework for illustrating multifaceted person–situation tensions.

A Neo-Evolutionary Resolution of Person-Situation Tensions

The evolutionary metaphor of the person–situation tension would be that between *variation* and *selection*. At a surface level, variation represents individual creative action and selection the environment (referred to interchangeably as the domain) in which that action takes place. Although accurate so far as it goes, as with all metaphors, these parallels break down when extended. Standard evolution theory holds that the cycles of production—that which produces variation—are causally distinct from cycles of maintenance—the environmental demands which select or reject variation. So, for example, the environment did not cause the giraffe's neck to be long. Rather, the giraffe's neck is long because of tendencies toward spontaneous genetic variation. The environment did, however, select for that neck inasmuch as it was a 'fit' adaptation that increased the likelihood of procreation (see Bailey, 1994, for a review).

Person   **Situation**

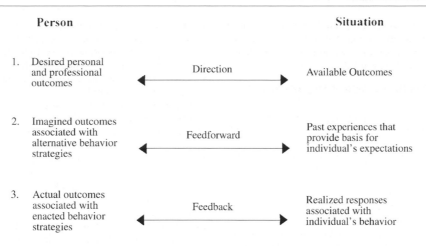

Figure 1. Managing person-situation tensions.

Casual separation of variation and selection may hold true in biology, but it is untenable in the social psychology of creativity and innovation. Individual creative action is not merely an accidental, spontaneous variation, and the domain in which individuals live and work is not merely a removed, disinterested, passive environment that selects after the fact. Rather, as our previous discussion demonstrates, the person and the situation are reciprocal and interdependent. Thus, we appeal instead to neo-evolutionary theory as an optimal framework to describe person–situation resolution. Originally described at the individual level by Campbell (1960) and later elaborated on and organized by Weick (1979), this perspective contends that each level of analysis has three sequentially ordered process categories, *enactment, selection,* and *retention* (E-S-R).

An E-S-R approach is especially useful for understanding how creativity and innovation unfold in social settings because it explicitly depicts the tension between divergence and convergence, change and stability, novelty and value, and creating and organizing. An 'enactment' refers to an individual choice that is, in our parlance, either new or routine and sensitive to the demands of the domain. That is, individuals' operative goals and expectations are conditioned by past and current environmental features such as resources dependencies that acknowledge, allocate and reward. Assuming that individuals seek to achieve desired outcomes and avoid undesired ones, it makes perfect logical and hedonistic sense that they would choose enactments that are likely to be well received in their domains. In this fashion, either new *or* routine behaviors can be labeled adaptive.

A neo-evolutionary approach also recognizes that domains change as a result of enactments that are selected and retained as a legitimate practice. That is, once an idea, process or product is accepted, it is now

available for further adoption within the domain. Individual creative actions represent novel enactments that may instigate a commensurate E-S-R cycle at the next highest level (e.g. a team or department). This kind of analytic scheme better depicts how individual dispositions and interpretations are expressed in organizational settings, and how patterns of social interaction at various levels of analysis impact on creative outcomes (Weick, 1979). Just as the introduction of a new species alters the floral and faunal equilibrium of an ecosystem, so it is with innovation.

With this background, creativity and innovation are defined from the perspective of those working within a particular domain. Thus, an idea, process or product is considered 'creative' to the extent that informed judgments by knowledgable players in a given domain deem it to have merit. In this way, selection processes are determined by those individuals who enjoy influence, authority or expertise. Creativity is a domain-specific, subjective judgment of the novelty and value of a particular action (Ford, 1996).

The E-S-R approach illustrates the person–situation interaction that leads individual actors to choose new or routing behavioral options. It also helps to surface the priorities that drive selection processes between domains. The next sections show how innovating (i.e. enactments, variations) and organizing (i.e. selection, retention) are differentially prized in theory and practice. Equally apparent, however, is that these processes are reciprocal; the former cannot proceed in the absence of the knowledge and purpose that results from latter, and if the latter becomes too narrowly focused and rigid, it will stagnate creating. Innovating and organizing are like rival siblings who constantly squabble but who are nevertheless intimately linked. The following section will employ this neo-evolutionary perspective to describe how individuals' efforts to reduce person–situation tensions within domains can increase tensions between related domains with different methods of ascribing value. We focus specifically on the tension between theory and practice created by the processes illustrated in Fig. 2.

Tensions within the Domain of Organizational Studies

The primary tension facing tenure-track faculty in organizational studies is simple yet profound: publish or parish. The old axiom is overstated perhaps, but few would disagree with its essence. Bedeian (1996) described the road to professional success in organizational studies by saying:

> In economic analogy, publications are the major currency of the realm. Whereas there may be diversity in academic reward structures at the institutional level (e.g. teaching, research, service), the reward structure at the national and international level is monolithic rather than plural. Thus, whereas scholars may draw their paychecks locally, academic recognition and the rewards that follow (e.g. editorial appointments, professional board memberships, fellow designations) are conferred elsewhere as a consequence of judgments made by the larger academic community . . . Publications mean visibility, esteem, and career mobility (p. 6).

Recent research supports this characterization. Gomez-Mejia & Balkin's (1992) analysis of determinants of faculty pay in organizational studies found that the correlation between the number of top-tier publications (defined as the top 21 organizational studies journals as judged by a sample of department chairs) and 9-month faculty pay was 0.61. The authors of this study suggest that given the nonprogrammable nature of faculty's work, incentive systems are used in lieu of monitoring as a means of directing behavior (i.e. selecting desired, and rejecting undesired, behavior). Of course, this environmental demand is a source of person–situation tension only to the extent that it is poorly aligned with individual aspirations. If one wants to do nothing but publish, academe provides an exceptional fit. But such is not the case. Gomez-Majia and Balkin found strong evidence that most faculty not only are resigned to a singular and narrow path toward career success, but do not like it:

> Of the respondents to our survey, 11% were 'strongly satisfied' and 58% were either 'dissatisfied' or 'strongly dissatisfied' with the statement that 'when it come down to it, publishing is the overwhelming criterion for faculty pay decisions'; in contrast, 80% agreed that this statement was true (p. 948).

Thus, the adaptive strategy for organizational scholars who seek tenure and high pay is, like it or not, to publish research in top-tier journals. In order to thrive, one must become savvy about the criterion used by journal editors and reviewers to evaluate submissions. From the neo-evolutionary perspective we employ to describe person–situation interactions related to creativity, this is akin to understanding the selection processes that favor one type of variation over another, and enacting accordingly.

Recent empirical research sheds further light on the selection processes employed by top-tier journals. Beyer, Chanove & Fox (1995) found that submitted papers were likely to be viewed as more significant to the field, and therefore more likely to be published, if they made explicit claims regarding novelty. Mone & McKinley (1993) found evidence that a 'uniqueness value' is held by organizational scholars that influences their conception and execution of research programs. Specifically, they found that the dialogue between researchers submitting papers and editors deciding the papers' fate was defined by explorations and identification of novelty. The key question reviewers ask to gauge the worth of a submitted paper is 'what's new here?' Locke & Golden-Biddle (1997) went so far as to analyze specific gambits that researchers use to convince reviewers of the novelty of their work because, ". . . what counts as a contribution is that which is perceived as unique or novel in light of the extant literature" (p. 1024). The rhetorical devices they identified include specifying gaps in current theory and knowledge, noting oversights or neglected issues and advocating alternative theoretical approaches.

Taken together, our neo-evolutionary view suggests that resolving the person–situation tension facing those in the domain of organizational studies involves publishing novel theoretical proposals in top-tier journals. This strategy leads to job security, high pay, recognition and mobility, and therefore constitutes an adaptive response that advances personally desired outcomes. Given the selection processes imposed by the domain and the nature of adaptive responses most likely to be retained, what are the implications for theory in organization studies?

We believe that the description offered by Astley (1985) remains valid, perhaps more so now than when originally presented more than 15 years ago. Astley argued that organizational studies value iconoclasm over truth, and that creative theorists were much more likely to be successful than those who sought to verify the validity of existing theories, catalog the success rates of current practices or the like. Because novelty is assessed relative to existing theory rather than current practice (Locke & Golden-Biddle, 1997), most research tends to be literature-driven rather than problem-driven. Consequently, organizational studies continue to evolve as a fragmented discipline addressing loosely related topics. Astley's conclusion that "New theoretical advances do not seem to build cumulatively on previous findings; instead they add to the bewildering variety of perspectives within the field" (p. 504) still rings true.

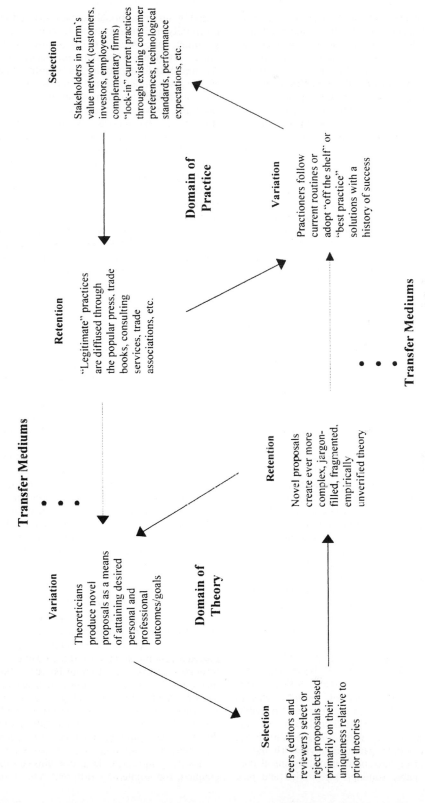

Figure 2. A neo-evolutionary view of tensions within and between the domains of theory and practice.

Our description leads to the disheartening conclusion that theoretical novelty is more important than practical value to those seeking to advance their careers in the discipline. The resulting 'management theory jungle' (cf. Koontz, 1961) should come as no surprise. Our stance is that optimism with respect to theoretical pluralism lies in changing the selection processes in a manner that more closely aligns the values facing scholars with those affecting management practitioners. But first we examine the tensions faced by practitioners.

Tensions Within the Domain of Management Practice

Despite wide recognition of its importance, few organizations can claim to have mastered the challenges associated with managing creativity. Why is this so? We maintain that the primary characteristic of firms is the presence of normative cues, emanating from various sources, that describe and reinforce adherence to routine or habitual behaviors (Ford, 1996). Organizations primarily exist to align the collective efforts of diverse individuals toward common goals. Formal aspects such as structural alignment, operational procedures and rules, stated objectives and compensation systems, as well as informal aspects such as roles, norms and culture, all serve to induce conformity in the thoughts and actions of organization members. Industrial and professional identifications carry equally as defined notions of appropriate and inappropriate behavior. The comportment expected of a financial officer in a bank, say, can be as constricted as an actor in a play.

These processes are the functional equivalent of environmental pressures that form the dynamics of selection processes. All are very important and even indispensable means of aligning collective behavior and empowering individuals to act on behalf of organizational interests. Shared interpretive frameworks held by participants encourage reliance on well-understood, previously tested solutions, but they also discourage the generation or consideration of new possibilities. As a consequence, the benefits of reliability and efficiency associated with normative processes come at the cost of low creativity. This is especially problematic in large organizations where innovation may even be considered 'illegitimate' (Dougherty & Heller, 1994).

On the whole, there is every reason to believe that the domain of practice privileges value over novelty, organization over innovation. March & Simon (1958) described firms as collections of routines wherein innovation resulted only when the repertoire of habitual responses failed. Burns & Stalker (1961) saw firms as ranging between routine-laden and mechanistic vs. innovation-oriented and organic archetypes. Thompson (1967) partitioned organizations into a technical core that maintained stable routines surrounded and protected by an institutional domain that addressed novel, potentially threatening demands and circumstances. In a similar vein, Weick (1979) illustrated an organizational antimony whereby firms address conflicting requirements for flexibility and stability. March (1991) argued that organizations comprise incommensurate activities related to either exploiting current routines or exploring novel alternatives. Christensen's (1997) research on 'the innovator's dilemma' presents compelling evidence that the operational configurations necessary to dominate existing technologies or markets make it impossible for firms to participate in highly creative departures from the status quo, and vice versa. Decades of research suggest that organizing is the antithesis of innovating, where the former promotes routines and the latter disrupts them.

As previously argued, the person–situation tension means that individuals must choose between new and routine behavioral options. Unfortunately, the paradox between conforming and creating looks rather one-sided to most employees. Especially in 'strong' organizational settings—characterized by detailed job descriptions, rigid reporting relationships, close supervision and persuasive cultures—powerful normative processes furnish an environment where thought and behavior become routine-driven. Therefore, firms become unreceptive to creative enactments because they run counter to potent forces toward convergence. Given a choice between time-tested routines that produce relatively certain, near-term consequences, and creative alternatives that are typically disruptive, uncertain, and prone to failure, most individuals opt for the safe haven of sanctioned routines. This is true at the organizational level as well; corporations tend to adopt industry 'best practices', avoiding risks associated with 're-inventing the wheel', but promoting look-alike strategies and operations that doom an industry to competition based solely on price (Porter, 1996). It is hardly surprising, then, that corporations have trouble harnessing employees' creative potential. Evolutionary tensions between enactments and selection commonly follow a trajectory that favors known value over novelty.

Tension Between Theory and Practice

The connection between theory and practice—in this case, organization studies and management practice—is readily apparent and much lauded. Indeed, academic research can and should serve the business community through careful examination of processes and practices that improve implementation. Rigorous research can identify trends, refine procedures and instigate change. In every way, successful enterprise management is inextricably linked to the scholarship that informs and directs it (see Bailey & Eastman, 1996, for a detailed treatment of the interdependent benefits).

However, in reality these domains are only loosely coupled, represented by different sets of stakeholders with different preferences with respect to novelty and value. As we have shown, novelty is intrinsically valuable in academe, whereas it is eschewed in practice, which gravitates toward demonstrated value instead. Consequently, academics and practitioners are motivated by quite dissimilar and even competing interests embedded in their domains. As basic testimony to this tensions, knowledge that defines theoretical domains is well documented, archived and easily accessible, whereas knowledge that defines practice is often proprietary, and tends to be documented in a haphazard fashion (even within firms).

Not surprisingly, then, attempts to enhance and manage innovation often encounter a paradox between the respective benefits of aligning collective effort—which promotes individual routine choices that emphasize organization—and empowering creativity—which embraces new choices that emphasize innovation. Conformity to current practices, values and norms can indeed direct the collective energies of employees toward realizing desired organizational outcomes, and thus is a critical component of success. However, individual creativity allows companies to resolve intransigent and strategically critical problems and develop new market offerings and technologies. This is especially important in the long term because creative actions contribute to learning and competitive differentiation; the heart and soul of sustained vitality. Ultimately, business success requires capitalizing on advantages associated with both creating and organizing. The squabbling children must be reconciled.

Resolving Between Domain Tensions

How, then, can these inherent tensions between the highly interdependent domains of theory and practice be resolved? We offer two broad strategies, arranged around intermediaries and aligning interests.

Intermediary Roles

Fortunately, there are several avenues through which innovative theories are communicated to practitioners, and best practices are noticed by professors. The extraordinary growth of the 'guru business' demonstrates the economic value provided by those who fulfill intermediary roles between professors and business people. The central problem with such boundary spanners is that they do not enjoy full acceptance in either domain. Professors often shun gurus for their simplistic treatment of complex issues. Practitioners just as often dismiss them as 'witch doctors' who have renamed and reframed last year's fashion, angling for personal profit over firm performance. Despite this, popular management books are perennial bestsellers, testimony to their influence.

Although there is unquestionable value in the guru, there are other ways to structure an intermediary role. We believe there is enormous value in business schools appointing 'practitioner' or 'professional' professors,

drawn from the ranks of business people. Such individuals do not need to possess doctorates, and their role could be defined specifically to facilitate interaction between faculty and practitioners, novelty and value. That is, in addition to teaching elective courses, these professional faculty could be paired with more traditionally academic faculty to explore ways in which novel ideas could be framed and communicated to a practitioner audience. This might include arranging meetings and presentations with corporate counterparts or pursing publication in trade publications as well as highly respected outlets like the *Harvard Business Review*. Further on-site action research opportunities could be brokered by such professional professors. Their general role would be to help connect the world of theory to the world of practice.

An inverse strategy for developing intermediaries is the 'professional sabbatical'. In this scenario, professors are encouraged, and perhaps even required, to spend a percentage of their sabbaticals as 'professors in residence' of a firm. Finance professors could reside in banks, entrepreneurship professors in small business, organizational behavior professors in consulting firms, and so on. Because they are paid by the school, this option is low-cost for firms, involving only an office and minimum support. The idea here is for professors to work side by side with those who apply, in some way, shape or form, academic principles. Their role would be to encourage further the value of novelty, to push practitioners to question assumptions of best practices, the processes that are erected around institutional innovation and the championing of new ideas. That is, by bringing fresh perspectives and recent research to bear, professors could add value to the apprehension and resolution of real organizational issues.

Aligning Interests

The primary means available for resolving the domain tensions between theory and practice is to alter the selection process of the domains themselves. Specifically, the world of theory would have to encourage, reward and embrace practical value as much as it does novelty. Conversely, the world of practice would have to encourage, reward and embrace novelty as much as it does value.

How could this ever be accomplished? The answer is deceptively simple in conception but extremely difficult in implementation. Frankly stated, it involved introducing resource dependencies that reward academics for value and practitioners for novelty. That is, universities should alter their current reward system to encourage publications in trade magazines and popular managerial outlets. Right now, the correlation between faculty salary and publication records in top-tier academic journals is skewed and should be balanced. Further, activities such as designing practicum and supervising internships, all of which foster a keener sense of appreciation for the challenges faced by businesses, should be explicitly acknowledged and rewarded.

For firms, the answer lies less in promoting creativity than in not punishing it. Organizations will always reward those who control costs and oversee smooth operations. Such is to be expected whenever large-scale strategic directions require careful administrative support. The problem is that those who do take risks or recommend novel processes are discouraged by red-tape and territorial defenses, and can even be shunned or marginalized. As a result, a kind of organizational level 'groupthink' occurs where those who have truly creative ideas do not even bother advancing them. This kind of creativity malaise is not uncommon; how many executives tell tales of shelving great ideas because of the resistance they engendered? Firms need to offer mechanisms by which such individuals can advance their ideas in a non-threatening forum and, if adopted, enable their quick and fluid implementation.

Conclusion

The same processes that alter the selection mechanisms of a domain can also create a different path, one where creative action is central to the interests and identity of employees and the organization. Neo-evolutionary processes related to creativity have been described by both Ford's (1996) and Drazin, Glynn & Kazanjian's (1999) theories of organizational creativity. These theories adhere to the requirements of *co-evolution* research by describing multilevel processes, multidirectional causality, nonlinear processes, positive feedback processes, and path and history dependence (Lewin & Volberda, 1999).

Thinking of creative action in this light raises an interesting 'chicken and egg' problem. Which comes first: a creativity-friendly environment that nurtures the talents of would-be creators, or creative actions that test the boundaries of legitimate behavior and reveal the true character of the environment? Co-evolution theory suggests that one must consider recursive influences that are path-dependent and prone to rapid escalation. In the context of creativity research, one must simultaneously consider how individual behavior shapes proximal and distal features of the work environment, and how the environment shapes individual behavior, and how these reciprocal influences evolve over time.

The final hope for resolving between domain tensions so as to harmonize the interdependent advantages of theory and practice, novelty and value, lies, paradoxically, in macro-managing the environmental selection pressures. We say paradoxically because we built this paper around the premise that innovation begins with an act of individual creativity. However, we also argued that that creativity does not occur in a vacuum, but rather is intimately linked with the context in which it takes place. The creation of that context, and thus the triggering of the co-evolutionary processes described above, is a matter of organizational structure, policy, reward systems and culture. Unlike in the biological realm, organizational environments can be consciously managed. And they must be, if innovation is the objective.

References

Allport, G. (1937). *Personality: A psychological interpretation*. New York: Holt.

Amabile, T. (1996). *Creativity in context*. Boulder, CO: Westview Press.

Astley, W. G. (1985). Administrative science as socially constructed truth. *Administrative Science Quarterly*, **30**, 497–513.

Bailey, J. R. (1994). Great individuals and their environments revisited: William James and contemporary leadership theory. *Journal of Leadership Studies*, **1** (4), 28–36.

Bailey, J. R. & Eastman, W. N. (Eds) (1996). Science and service in organizational scholarship. *Journal of Applied Behavioral Science*, **32**, 350–464.

Bailey, J. R., Strube, M. J., Yost, J. H. & Merbaum, M. (1994). Proactive self-appraisal in the organization. In: R. H. Kilmann, I. Kilmann & Associates (Eds), *Managing Ego Energy: The Transformation of Personal Meaning Into Organizational Success* (pp. 128–148). San Francisco, CA: Jossey-Bass.

Bedeian, A. G. (1996). Lessons I learned along the way. In: P. J. Frost & S. M. Taylor (Eds), *Rhythms of Academic Life: Personal Accounts of Careers in Academia*. Thousand Oaks, CA: Sage.

Beyer, J. M., Chanove, R. G. & Fox, W. B. (1995). The review process and the fates of manuscripts submitted to AMJ. *Academy of Management Journal*, **38** (5), 1219–1260.

Burns, T. & Stalker, G. M. (1961). *The management of innovation*. London: Tavistock.

Campbell, D. T. (1960). Blind variation and selective retention in creative thought as in other knowledge processes. *Psychological Review*, **67**, 380–400.

Cantor, N. & Kihlstrom, J. F. (1987). *Personality and social intelligence*. Englewood Cliffs, NJ: Prentice-Hall.

Christensen, C. M. (1997). *The innovator's dilemma: When new technologies cause great firms to fail*. Harvard Business School Press, Boston, MA.

Davis-Blake, A. & Pfeffer, J. (1989). Just a mirage: The search for dispositional effects in organizational research. *Academy of Management Review*, **14**, 385–400.

Dougherty, D. & Heller, T. (1994). The illegitimacy of successful product innovations in established firms. *Organization Science*, **5**, 200–218.

Drazin, R., Glynn, M. A. & Kazanjian, R. K. (1999). Multilevel theorizing about creativity in organizations: A sensemaking perspective. *Academy of Management Review*, **24**, 286–307.

Ford, C. M. (1996). A theory of individual creative action in multiple social domains. *Academy of Management Review*, **21**, 1112–1142.

Ford, C. M. & Gioia, D. A. (2000). Factors influencing creativity in the domain of managerial decision making. *Journal of Management*, **26**, 705–732.

Gomez-Mejia, L. R., Balkin, D. B. (1992). Determinants of faculty pay: An agency theory perspective. *Academy of Management Journal*, **35** (5), 921–955.

House, R. J., Shane, S. A. & Herold, D. M. (1996). Rumors of the death of dispositional research are vastly exaggerated. *Academy of Management Review*, **21**, 203–224.

Koontz, H. (1961). The management theory jungle. *Academy of Management Journal*, **4**, 174–188.

Lewin, A. Y., Long, C. P. & Carroll, T. N. (1999). The coevolution of new organizational forms. *Organization Science*, **10**, 535–550.

Locke, K. & Golden-Biddle, K. (1997). Constructing opportunities for contribution: Structuring intertextual coherence and 'problematizing' in organizational studies. *Academy of Management Journal*, **40** (5), 1023–1062.

March, J. G. (1991). Exploration and exploitation in organizational learning. *Organization Science*, **2**, 71–87.

March, J. & Simon, H. (1958). *Organizations*. New York: John Wiley.

Mone, M. A. & McKinley, W. (1993). The uniqueness value and its consequences for organizational studies. *Journal of Management Inquiry*, **2**, 284–296.

Mischel, W. (1968). *Personality and assessment*. New York: John Wiley.

Pervin, L. (1989). Persons, situations, interactions: The history of a controversy and a discussion of theoretical models. *Academy of Management Review*, **14**, 350–360.

Porter, M. E. (1996). What is strategy? *Harvard Business Review*, **74** (6), 61–78.

Thompson, J. (1967). *Organizations in action*. New York: McGraw Hill.

Weick, K. (1979). *The social psychology of organizing*. Reading, MA: Addison-Wesley.

Woodman, R. W., Sawyer, J. E. & Griffin, R. W. (1993). Toward a theory of organizational creativity. *Academy of Management Review*, **18**, 293–321.

The International Handbook on Innovation
Edited by Larisa V. Shavinina

E-Creativity and E-Innovation

Keng Siau

College of Business Administration, University of Nebraska-Lincoln, USA

Abstract: Creativity, most people tend to assume, is inborn, mysterious, unanalyzable, and unteachable. Psychologists have been trying to understand this phenomenon for decades. The literature is overwhelmed with inconsistent theories and hypotheses. Lately, development in artificial intelligence has provided researchers with another means of analyzing the creative process. In this chapter, we review some of the work in this area and discuss some approaches for e-creativity.

Keywords: Creativity; Innovation; E-Creativity; E-Innovation.

Introduction

May 11, 1997 is a milestone in computer evolution. On that day, Garry Kasparov, who had never lost a chess match, conceded defeat for the first time ever to a chess-playing computer—IBM's Deep Blue computer. After losing the first game in the match, the Deep Blue IBM computer won in game two, and managed draws in games three, four, and five. In game six, a game lasting only an hour, Deep Blue won the game by dominating on the board and capturing Kasparov's queen. The final score was 3.5 points for Deep Blue and 2.5 for Kasparov. The match drew worldwide attention, not only because 34-year-old Kasparov was widely considered the greatest chess player ever, but also because of its compelling man-vs.-machine theme.

With the win by Deep Blue, a few age-old questions are popping up again. Are computers good enough to substitute for human beings? Can computers be creative? In this chapter, we look at the quest for electronic-creativity, e-creativity or e-innovation. E-creativity relies on Artificial Intelligence (AI), which emphasizes defined operations that can yield the same sorts of ideas that are produced by creative human beings. Some of the e-creativity efforts and projects are reviewed in this chapter, and some criteria that are necessity for e-creativity are discussed.

What is Creativity?

Creativity is an important, but elusive, phenomenon (Siau, 1995, 1996, 1999, 2000). Different writers have different expressions for this moment of insight and for the process leading to it. Gestalt psychologists named it the 'Aha' experience. Pearce (1971) called it 'the crack in the cosmic egg'. Bruner initially referred to it as the 'empty category' (1956) and later as 'thinking with the left brain' (1962). Guilford (1959) and de Bono (1970) have used the broader terms 'divergent thinking' and 'lateral thinking', respectively, to refer to the process by which new ideas are generated.

Since the beginning of civilization, people have been amazed by one of our most precious abilities—creativity. Greek philosophers, such as Plato, Socrates, and Aristotle, occasionally discussed creativity in their philosophical account. For instance, Plato valued creative creation as a result of inspiration, which can only be activated by divine power. Plato saw creative activity as a topic that cannot be subjected to rational analysis.

The word 'creativity' has long been associated with mysticism. Koestler (1975) wrote about creativity:

The moment of truth, the sudden emergence of a new insight, is an act of intuition. Such intuitions give the appearance of miraculous flashes, or short-circuit of reasoning. In fact they may be likened to an immersed chain, of which only the beginning and the end are visible above the surface of consciousness.

Creativity is also defined in many other different ways. For example, Torrance (1970) defined creativity as the process of sensing gaps or missing elements, and forming ideas or hypotheses concerning them. Boden (1996) identified two senses of being 'creative'. One is psychological creative (P-creative) and the other is historical creative (H-creative). The psychological sense refers to ideas that are fundamentally novel with respect to the individual that has the idea. For example, if John comes up with the idea of the wheel for the first time (he has not heard or seen a wheel before), his idea is P-creative—no matter how many people had the same idea before him. The historical sense applies to ideas that are fundamentally novel with respect to the whole human history. In this case, John's invention is H-creative only if no one has ever had that idea before him. By definition, all H-creative ideas are P-creative too.

Boden (1996) also defined two types of creativity— the *improbabilist* and the *impossibilist*—and further summarized how creativity works along this dimension. The *improbabilist* creativity involves combinations of familiar ideas to give a 'statistically surprising', or improbable, new idea. Several AI models that fit into this framework of *improbabilist creativity* are the analogy-based model, the associative-based model, and the induction-based model. *Impossibilist* creativity, however, centers on the formation of new ideas that could not have emerged in the current state of the domain. According to Boden, impossibilist creativity involves the transformation of conceptual spaces (paradigm-shifting creativity).

The Birth of E-Creativity

Cognitive psychologists did not start to explore the internal realm of mind and the operations involved in creative thinking until the invention of the von Neumann machine. Researchers in artificial intelligence (AI), especially Herbert Simon and Allan Newell, pushed forward these explorations. Newell (1990) proposed *the unified theory of cognition*. Langley et al. (1987) used the process-tracing technique to access the thought process of individuals involved in creative thinking to formulate hypotheses and to write computer programs to mimic the related process for problem-solving in various domains. Boden (1989, p. 2) described the computational psychologists as people who:

adopt a functionalist approach to the mind ... conceive of the mind as a representational system, and see psychology as the study of the various computational processes whereby mental representation are constructed, organized, interpreted, and transformed ... think about neuroscience ... in a broadly computational way, asking what sorts of logical operations or functional relations might be embodied in neural networks ...

The computational concepts developed by AI-researchers are now influencing more traditional disciplines such as psychology. The next section introduces AI and some of its approaches.

AI—An Enabling Technology

AI studies how the computer can be programmed to achieve intelligent attributes of human beings. With a goal to explain what the mind is and how it works, AI is closely related to cognitive psychology. It has created new approaches for creativity studies, and its computational concepts have shifted the fundamental focus from the external measurements of creativity to the internal description of processes involved in creative activities. Three of the popular approaches in AI are: classical/symbol-system AI, neural networks, and genetic algorithms.

Classical/Symbol-System AI

The central assumption that leads to classical AI is *the physical symbol system hypothesis* proposed by Newell and Simon (1976, p. 116):

A physical system consists ... symbols ... contains a collection of processes that operate on expressions to produce other expressions ... A physical symbol system is a machine that produces through time an evolving collection of symbol structures ... has the necessary and sufficient means for general intelligent action.

Classical AI has a diverse research scope and application domains. They vary from planning, game-playing (e.g. chess), general or expert problem-solving to theorem-proving, perception, and natural language understanding. Because knowledge is essential for intelligent reasoning, and processes are required to manipulate the knowledge, knowledge representation and processes are the two issues in classical AI systems. A number of techniques for knowledge representation are available in AI. They include semantic networks, scripts, frames, rule-based systems, and several others.

The most successful and widely used classical AI architecture is the rule-based production system. Most of the existing e-creativity programs are based on this technique. A production rule typically takes the form of:

> IF conditions are true
> THEN some actions
> ELSE some other actions

An inference engine is used to realize the reasoning of the production-rule systems. It employs two approaches in the reasoning process: backward chaining and forward chaining. In backward chaining, a specified goal is used to select rules in the knowledge base, which will in turn call for additional information and eventually work backward to information that is available to the system. Forward chaining, however, is an inference method in which known facts are used to select rules, which will provide additional facts and thus, eventually, the solution to the problem.

Neural Networks

Neural networks, or connectionist systems, are parallel processing systems with computational properties that are modeled after the human brain. They are based on a theory of association (pattern recognition) among cognitive elements (Baer, 1993). To mimic the human cognitive processes, connectionist systems are not programmed, but are designed to learn from experience by 'self-organization'. The connectionist paradigm maintains that what appears to be lawful cognitive behavior may in fact be produced by a mechanism in which symbolic representations are unnecessary—no rules are "written in explicit form anywhere in the mechanism" (Rumelhart & McClelland, 1986). An example of simple connectionist architecture is shown in Fig. 1.

The key to understanding connectionist architecture is the 'hidden units'—units that are neither input nor output. Unlike the intermediate states of computation in symbolic accounts, these units represent uninterpretable states of the system. That is, the patterns of connections and interactions between the hidden units cannot be expressed in the form of rules or propositions. In other words, the 'meaning' of these units and their various combinations cannot be stated in words.

Connectionism can be approached in two ways. First, it can be considered simply as a way to implement production systems. The learning that takes place in a connectionist system is an induction of rules, which, although not stated formally as rules, operates on inputs in the same way that rules in a production system operate on symbols. The set of weights that the system derived, in this 'weak' connectionist scenario, would simply be the program's coding of the 'explicit inaccessible rules' (Rumelhart & McClelland, 1986).

Second, the connectionist model may be used to replace existing symbolic-processing models as the explanations of cognitive processing. This is an extreme claim made by many connectionists, and it is this claim ('strong' connectionism) that caused the most interest and controversy (Churchland, 1988, 1989; Pinker & Prince, 1988; Smolensky, 1988).

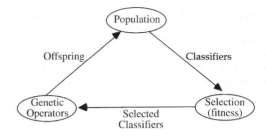

Figure 2. The genetic algorithm cycle.

Pinker and Prince termed this strong connectionist claim 'eliminative connectionism'.

Genetic Algorithms

Genetic algorithms are general-purpose search algorithms that use principles inspired by natural population genetics to evolve solutions to problems. The basic idea is that over time, evolution will select the 'fittest species'. In this sense, genetic algorithms emulate biological evolutionary theories to solve optimization problems. The central loop of the genetic algorithm is depicted in Fig. 2.

(a) Select pairs from the set of classifiers according to their strengths—the stronger a classifier is, the more likely it is to be selected.

(b) Apply genetic operators to the pairs, creating 'children' or 'offspring'. These genetic operators perform the role of variation. Chief among the genetic operators is crossover, which simply exchanges a randomly selected segment between the pairs. The other operators, such as mutation and inversion, have lesser roles, mainly providing 'insurance' against overemphasis on any given kind of schema.

(c) Replace the weakest classifiers with new offspring.

Attempts at E-Creativity

Several approaches have been attempted to implement e-creativity. Some of them are discussed below.

(a) *Generative grammars:* Generative grammar systems use a predefined grammar to specify the way the output is to be generated. The grammar is specified as a set of production rules. For example, Rumelhart (Rowe & Patridge, 1993; Rumelhart, 1975) used the following for a story grammar:

> Story → setting, episode
> Setting → time, place, characters
> Episode → event, reaction

Then the story is built in a top-down fashion by replacing story with setting and episode, replacing setting with time, place, and characters until everything becomes irreplaceable, or terminals (in

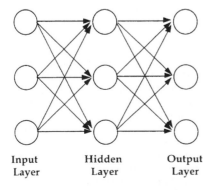

Input   **Hidden**   **Output**
Layer   **Layer**   **Layer**

Figure 1. A simple connectionist architecture.

this case, words and phrases). The system can use random selection for the replacement to produce different stories each time the system is executed. The weakness of this system, according to Rowe & Patridge (1993), is the rigidity of the grammar. For example, the stories generated using the grammar described above always have a fixed structure of story line. Other examples of generative-grammar systems are Rumelhart's story-writer (Rumelhart, 1975), TALE-SPIN (Meehan, 1981), Johnson-Laird's jazz improviser (Johnson-Laird, 1987), AARON (McCorduck, 1991), MINSTREL (Turner, 1992), and EMI (Cope, 1991).

(b) *Analogy-based systems:* Koestler (1964) asserted that most of the so-called creative scientific progress sometimes is a discovery of novel analogy with another unrelated domain. Gentner's Structure Mapping Theory (SMT) (1989) proposed that an analogy could be formed by mapping objects/concepts and their relations to one another. A computer model, called the Structure-Mapping Engine (SME) (Falkenhainer et al., 1989) was constructed to prove Gentner's theory. Other examples of analogical problem-solver include processes of induction (PI) (Holyoak & Thagard, 1989a, 1989b, 1994), and Copycat (Mitchell, 1993).

(c) *Meta-rules:* Meta-rules are 'rules to control rules'. In other words, they are used to create, delete, or modify rules in a system. The reason for using meta-rules is that the rules in a rule-based system are often so rigid that they constrain the behavior of the system. Meta-rules can be used to provide variations to the governing rules of the system so that new behavior can emerge. For example, the grammar rules in a story-generator limit the possible story lines that the generator can produce. Meta-rules can be introduced to modify some of these rules to provide new variations of storylines. For example, Meta-DENDRAL is an add-on module to an expert system DENDRAL (Lindsay et al., 1993). DENDRAL is an expert system used to assist the chemists in molecule decomposition analysis. The purpose of Meta-DENDRAL is to find possible decomposable molecules. Although Meta-DENDRAL has been very successful in its domain (i.e. it H-created some new rules in molecule decomposition), Boden (1996) argued that its success comes from the highly sophisticated chemical knowledge of DENDRAL. Other systems that make use of meta-rules are MU.S.CA-DET (Paestre, 1989), and DAY-DREAMER (Muller, 1987).

(d) *Discovery systems:* A discovery system works to discover new information from a given input. Examples of such systems are BACON (Langley et al., 1987) and AM (Davis & Lenat, 1982). BACON is able to find a theory (formula) to explain a set of data. Given a set of measurements of several properties, BACON can find relationships of these properties and produce a formula that fulfills the measurements. For example, Kepler's laws are rediscovered when measurements of planetary co-ordinates and trajectories are given to BACON as input. Other laws rediscovered by BACON are Boyle's law and Ohm's law. Rowe & Patridge (1993) described this as an "ad hoc curve-fitting exercise" because BACON never uses any semantic constructs like models or structures to achieve its goal. Langley's (1987) team developed GALUDER, STAHL, and DALTON based on BACON concepts.

AM works to form new mathematical concepts given a set of initial mathematical concepts and heuristic rules to discover and judge the importance of the concepts. In AM, a mathematical concept is represented as a frame with slots such as definition, examples, domain, and range. A concept can be another concept's example. Some slots in the initial concepts are purposely left blank for AM to fill in appropriate items. AM will occasionally create some new concepts in the process of filling these slots. New concepts are then added into its concepts pool and the process goes on until AM finds 'nothing interesting'. With the initial set of concepts like sets, lists, ordered pairs, intersection, union, and canonization, AM discovers natural numbers, addition, multiplication, primes, prime factorization, and Goldbach's conjecture.

(e) *Decentralized systems:* A decentralized system is generally defined as a system without a central control mechanism. It might contain several independent (homogeneous or heterogeneous) modules/agents interacting with each other to achieve global-level behavior (emergent behavior). An example of such a system is Copycat (Mitchell, 1993). Copycat generates an analogy for strings of alphabets. For example, given an input such as:

If abc → bca, then def → ??

Copycat can produce strings that will fulfill the analogy. The internal mechanism of Copycat is quite complicated. It uses a network of permanent concepts (slipnet), such as leftmost, rightmost, middle, same, group, alphabetic successor, and alphabetic predecessor to create a relationship map of the input structure. Weights are assigned to the links of the network of permanent concepts and used to determine the possible solutions. Then parallel processing among structuring agents (codelets) determines the final solution(s). Another example of this approach is Letter Spirit (Hofstadter & McGraw, 1993).

(f) *Induction-based systems:* Induction includes inferential processes that expand knowledge in the face of uncertainty. The study of induction-based machines in computational creativity assumes that the key to creativity lies in the ability to learn (i.e. see Shank & Childers, 1988). Thus, induction-based systems are intimately associated with learning and self-improving machines. Examples of such systems are PI (Holyoak & Thagard, 1989a, 1989b, 1994) and Copycat (Mitchell, 1993).

Criteria for E-Creativity

The formal study of creativity as a science started under the name of cognitive psychology. Guildford's theory of divergent thinking (1950) was probably one of the most influential theories on creativity. Guildford claimed that there is no such thing as general intelligence that enables 'general creativity' in any domain. He proposed four criteria that are considered important to divergent thinking: fluency, flexibility, originality, and elaboration.

(a) Fluency—the ability to produce a large number of ideas
(b) Flexibility—the ability to produce a wide variety of ideas
(c) Originality—the ability to produce unusual ideas
(d) Elaboration—the ability to develop/embellish ideas, and to produce detail

In his quest for a general, domain-transcending theory that would account equally well for creativity irrespective of the task domain, Mednick (1962) proposed five factors that are critical for creativity to occur:

(a) Domain-specific knowledge
(b) Numbers of associations generated, not generation speed
(c) Differences in cognitive or personality styles
(d) Selection of creative combinations
(e) Associative hierarchy—how associations are organized

In their attempts to move from general creativity to computational creativity, Rowe & Partridge (1993) surveyed psychological theories of human creative behavior and proposed five characteristics required for e-creativity to occur:

(a) Organize knowledge in such a way that the number of possible associations (the creative potential) is maximized.
(b) Allow for ambiguity in representations. The programs should, under appropriate circumstances, accept the generation of seemingly incorrect associations in order to build connections between concepts.
(c) Enable multiple representations. A single concept should not just be appropriate in one situation. Rather, a concept should be indexed to many situations. This would help to avoid the problem of functional fixity.
(d) Evaluate new combinations. New combinations should be examined and their usefulness assessed.
(e) Elaborate new combinations to elicit their consequences.

Discussion

Based on prior studies and a literature review, we categorized the criteria for e-creativity into knowledge representation, randomness, and learning.

(a) *Knowledge representation:* Knowledge representation refers to the way knowledge is represented so that it can be processed efficiently and effectively by the functioning system. Most of the researchers in computational creativity agree that knowledge must be represented in the most flexible way for the functioning system to manipulate it efficiently. Boden (1990) argued that a creative system needs a multilevel internal map of its 'conceptual space' (knowledge schema), so that the system can explore the map in a flexible and efficient manner or even transform the 'conceptual space' to enable *impossibilist* creativity. Minsky (1990), however, suggested using several different representations in a system to compensate for deficiencies in each of the knowledge representation schemes.
(b) *Randomness:* Randomness can be used to produce ambiguous behaviors and to provide a certain level of novelty in the output of the system. Racter is an example of such a system. Racter constructs sentences, poems or stories using 'syntax directives' and random selections from words and phrases. Such systems, however, might surprise the reader by its creation only for the first time. The 'creative' impression of Racter does not last long. Rowe & Patridge (1993) argued that randomness is not *sufficient* for creative behavior. Yazdani (1989) also reported that pure randomness will only produce interesting output by chance, with all other output being useless. He claimed that randomness should be used by artists and scientists as a starting point for generating a large decision space. Boden (1990) reported that if a random process happens to produce creative output, it can only result in 'first-time curiosity'. Other examples of systems that use randomness are Johnson-Laird's jazz improviser (Johnson-Laird, 1987) and TALE-SPIN (Meehan, 1981). In short, randomness helps to 'think outside the box' but is not sufficient, by itself, to produce sustain creative behavior.
(c) *Learning:* Learning is viewed as an important criterion in computational creativity. Systems that are capable of making changes in and by themselves over time, with the goal of improving their performance on the tasks in a particular environment, are more interesting than systems that are

not capable of such changes. Neural networks and genetic algorithms are two popular approaches for achieving learning. Most of the existing learning systems, however, based their performance on factors such as minimizing costs and maximizing profits. For computational creativity, we are more interested in generating new and unpredictable outcomes.

Conclusion

To attribute creativity to divine inspiration, or to some unanalyzable power of intuition, is to suffer from a paucity of ideas. Even to describe it as the work of the unconscious or subconscious mind, or as a combination of old concepts in new ways, does not get us very far. But, surprising as it may seem, we can build creative programs and study their characteristics. E-creativity has the potential to shape the future of innovation.

References

Baer, J. (1993). *Creativity and divergent thinking: A task-specific approach*. Mahwah, NJ: Lawrence Erlbaum.

Boden, M. A. (1989). Introduction. In: M. A. Boden (Ed.), *Artificial Intelligence in Psychology* (pp. 1–24), Cambridge, MA: The MIT Press.

Boden, M. A. (1990). Understanding Creativity. *Journal of Creative Behavior*, **26** (3), 213–216.

Boden, M. A. (1996). Creativity. In: M. A. Boden (Ed.), *Artificial Intelligence* (pp. 267–289). San Diego, CA: Academic Press.

Bruner, J., Goodnow, J. & Austin, G. (1956). *A Study of Thinking*. New York: John Wiley.

Bruner, J. (1962). *On knowing: Essays for the left hand*. Cambridge, MA: Belknap Press.

Churchland, P. M. (1988). *Matter and consciousness*. Cambridge, MA: MIT Press.

Churchland, P. S. (1989). From Descartes to neural networks. *Scientific American*, **261** (1), 118.

Cope, D. (1991). *Computers and musical style*. Oxford: Oxford University Press.

Davis, R. & Lenat, D. (1982). *Knowledge based systems in artificial intelligence*. New York: McGraw-Hill.

De Bono, E. (1970). *Lateral thinking*. Penguin Books.

Falkenhainer, B., Forbus, K. D. & Gentner, D. (1989) The structure-mapping engine: Algorithm and examples. *AI Journal*, **41**, 1–63.

Gentner, H. (1989). *The mechanisms of analogical reasoning*. Cambridge: Cambridge University Press.

Guildford, J. P. (1950). Creativity. *American Psychologist*, 205–208.

Guilford, J. (1959). Three faces of intellect. *American Psychologist*, **14**, 369–379.

Hofstadter, D. R. & McGraw, G. (1993). Letter spirit: An emergent model of the perception and creation of alphabetic style (CRCC Tech. Rep. No. 68. Bloomington: Indiana University), Department of Computer Science.

Holyoak, K. J. & Thagard, P. R. (1989a). Analogical mapping by constraint satisfaction. *Cognitive Science*, **13**, 295–356.

Holyoak, K. J. & Thagard, P. R. (1989b). A computational model of analogical problem solving. In: S. Vosniadou & A. Ortony (Eds), *Similarity and Analogical Reasoning* (pp. 242–264), Cambridge: Cambridge University Press.

Holyoak, K. J. & Thagard, P. R. (1994). *Mental leaps: Analogy in creative thought*. Cambridge, MA: MIT Press.

Johnson-Laird, P. N. (1987). Reasoning, imagining and creating. *Bulletin of the British Psychological Society*, **40**, 121–129.

Koestler, A. (1964), *The act of creation*. London: Hutchinson.

Koestler, A. (1975). *The act of creation*. London: Picador.

Langley, S. P., Bradshaw, H. G. L. & Zytkow, J. M. (1987). *Scientific discovery*. Cambridge, MA: MIT Press.

Lindsay, R., Buchanan, B. G., Feigenbaum, E. A. & Lederberg, J. (1993). DENDRAL: A case study of the first expert system for scientific hypothesis formation. *Artificial Intelligence*, **61**, 209–262.

McCorduck, P. (1991). *Aaron's code*. San Francisco, CA: Freeman.

Mednick, S, A, (1962). The associative basis of the creative process. *Psychological Review*, **69**, 220–232.

Meehan, J. (1981). TALE-SPIN. In: R. C. Schank & C. J. Riesbeck (Eds), *Inside Computer Understanding: Five Programs Plus Miniatures* (pp. 197–226). Hillsdale, NJ: Erlbaum.

Minsky, M. (1990). Logical vs. analogical or symbolic vs. connectionist or neat vs. scruffy. In: P. H. Winston & S. A. Shellard (Eds), *Artificial Intelligence at MIT, Expanding Frontiers* (Vol. 1, pp. 218–243). Cambridge, MA: MIT Press.

Mitchell, M. (1993). *Analogy-making as perception*. Cambridge, MA: MIT Press.

Muller, E. T. (1987). *Day-dreaming and computation: A computer model of everyday creativity, learning and emotions in the human stream of thought*. Unpublished doctoral dissertation, University of California, Los Angeles.

Newell, A. (1990). *Unified theories of cognition*. Cambridge, MA: Harvard University Press.

Newell, A. & Simon, H. A. (1976). Computer Science as empirical inquiry: Symbol and search. *Communications of the ACM*, **19** (3), 113–126.

Paestre, D. (1989). MUSCADET: An automatic theorem proving system using knowledge and metaknowledge in mathematics. *Artificial Intelligence*, **38** (3), 257–318.

Pearce, J. (1971). *The crack in the cosmic egg*. New York: Julian Press.

Pinker, S. & Prince, A. (1988). *On language and connectionism: Analysis of a parallel distributed processing model of language acquisition*. Cambridge, MA: Center for Cognitive Science.

Rowe, J. & Partridge, D. (1993). Creativity: A survey of AI approaches. *Artificial Intelligence Review*, **7**, 43–70.

Rumelhart, D. E. (1975). Notes on a schema for stories. In: D. G. Bobrow & A. Collins (Eds), *Representation and Understanding: Studies in Cognitive Science* (pp. 211–236). New York: Academic Press.

Rumelhart, D. E. & McClelland, J. L. (1986). PDP models and general issues in cognitive science. In: D. E. Rumelhart, J. L. McClelland & The PDP Research Group (Eds), *Parallel Distributed Processing: Explorations in the Micro-*

structure of Cognition (Vol. 1) *Foundations* (pp. 110–146). Cambridge, MA: MIT Press.

Shank, R. C. & Childers, P. (1988). *The creative attitude: Learning to ask and answer the right questions.* New York: Macmillan.

Siau, K. (1995). Group creativity and technology. *The Journal of Creative Behavior*, **29** (3), 201–216.

Siau, K. (1996). Electronic creativity techniques for organizational innovation. *The Journal of Creative Behavior*, **30** (4), 283–293.

Siau, K. (1999). Internet, world wide web, and creativity. *Journal of Creative Behavior*, **33** (3), 191–201.

Siau, K. (2000). Knowledge discovery as an aid to creativity. *Journal of Creative Behavior*, **34** (4), 248–258.

Smolensky, P. (1988). On the proper treatment of connectionism. *Behavioral and Brain Sciences*, **11**, 1–23.

Torrance, E. P. (1970). Inference of dyadic interaction on creative functioning. *Psychological Reports*, **25**, 319–394.

Turner, S. (1992). *MINSTREL: A model of story-telling and creativity (Tech. Note UCLA-AI-17–92).* Los Angeles: University of California, AI Laboratory.

Yazdani, M. (1989). A computational model of creativity. In: R. Forsyth (Ed.), *Machine Learning: Principles and Techniques* (pp. 171–183). London: Chapman & Hall.

Part III

Individual Differences in
Innovative Ability

The International Handbook on Innovation
Edited by Larisa V. Shavinina

The Art of Innovation: Polymaths and Universality of the Creative Process

Robert Root-Bernstein

Department of Physiology, Michigan State University, USA

Abstract: Many people view arts and sciences as being different because sciences yield objective answers to problems whereas arts produce subjective experiences I argue that art and science are on a continuum in which artists work with *possible* worlds whereas scientists are constrained to working in *this* world. But sometimes perceiving this world differently is the key to making discoveries. Thus, arts and sciences are on a continuum in which artistic thinking produces possibilities that scientists can evaluate for efficacy here and now. Not surprisingly, then, many of the most innovative scientists have had avocations in the arts, and some of the most innovative artists have had avocations in the sciences. These polymaths have often written or spoken about how their arts involvments have benefitted their scientific creativity and may provide a model for fostering a more innovative education.

Keywords: Innovation; Polymathy; Artscience; Creative process; Avocations; Hobbies.

Introduction: The Universality of Creative Thinking

Innovation in science and engineering is often portrayed as if it were distinct from that in the fine arts, perhaps because most definitions of innovation center on the idea of *effective problem-solving*. Science and engineering are supposed to be objective, intellectual, analytical, and reproducible so that it is clear when an effective solution has been achieved to a problem. The arts, literature, and music, by contrast, are portrayed as being subjective, sensual, empathic, and unique, so that it is often unclear whether a specific problem is being addressed, let alone whether a solution is achieved. It therefore comes as a considerable surprise to find that many scientists and engineers employ the arts as scientific tools, and that various artistic insights have actually preceded and made possible subsequent scientific discoveries and their practical applications. These trans-disciplinary interactions must cause us to reconsider how we think about innovation.

There are four important ways in which innovative ideas flow between the professions. One is through problem generation (see Root-Bernstein, 'Problem Generation and Innovation' in this volume). The arts often invent or discover phenomena and observations unknown to the sciences. A sceond role that the arts

play is to provide scientists and engineers with non-traditional physical and mental tools, analogies and models that can be used to solve problems. More on this below. Third, the arts often provide scientists with the tools necessary to communicate their results—with the considered use of words, images, and modeling techniques necessary to reify ideas as theories and explanations. The aesthetic portrayal of results in the sciences is just as important as in the arts and relies upon the same tools. And finally, the arts contribute to scientific innovation through *fantasy*, that is to say, through *the generation of possible worlds* that scientists can test according to the constraints of what is known about the real world. This fourth type of sciences–arts interactions is the one most often overlooked and of the widest application to understanding innovative thinking in general.

All real-world innovation is a process that involves the elaboration through fantasy (sometimes called imagination) of many possible solutions to any given problem, and the use of the widest range of mental and physical tools to constrain and evaluate which of these possibilities is most adequate to any given need. In this sense, all innovation is a process of survival of the fittest in which multiple variations of ideas are selected by social, economic, cultural and other factors

(Nickles, this volume; Root-Bernstein, 1989). What the arts provide the sciences is the ability to imagine possibilities—possible problems, possible tools, possible solutions—through synthetic and sensual forms of thinking to which analytical and logical forms of thinking can later be applied as part of the selection process. In short, I maintain that effective solutions to problems (i.e. innovation or creativity itself) can only be achieved if a *range* of possible solutions have been elaborated that can be examined for their relative effectiveness. Arts foster sciences and technology by elaborating possible worlds that can be evaluated for the insights they provide to the real world.

Note at the outset that I am not claiming that art is science or science, art. A painting is not usually a scientific diagram; each has distinct purposes and goals. A short story is not usually a scientific paper and no intelligent person could confuse them under most circumstances. A sculpture is not usually equivalent to a molecular model, nor do they have the same functions. But sometimes they do.

Sometimes what an artist imagines might be possible turns out to be what actually is. Unexpectedly, a painting can sometimes be a way to generate scientific ideas; fiction can explore and even propose new scientific theories; and sculptures can be scientific or mathematical models. The purpose of this essay is to argue that there exist fundamental connections between sciences and arts that provide non-trivial windows onto the gardens of the mind where innovative thinking is cultivated.

The key to understanding my approach to innovative thinking in the arts and sciences is to distinguish clearly between disciplinary products and trans-disciplinary processes. I believe, along with Koestler (1964), Bronowski (1967), and many others including Weisberg in this volume, that no distinction exists between the arts and sciences at the level of the creative process itself (Root-Bernstein, 1984, 1989). The ways in which artists and scientists discover and invent problems, experiment with ways to come to grips with them, and generate and test possible solutions is universal. In fact, most of the greatest innovators in every discipline have been polymaths—Renaissance people like Leonardo da Vinci— who demonstrated their creative abilities in several fields of endeavor. Thus, I propose that creative people are generally creative, and their general creative ability comes from mastering a common set of thinking tools (see Root-Bernstein & Root-Bernstein, 'Intuitive Tools for Innovative Thinking' in this volume) and a creative process that is similar across all disciplines. This point can be made particularly effectively through examples of artists who have made scientific discoveries, scientists who have made artistic contributions, and those who bridge both sets of disciplines without claiming allegiance to one or another. As rare as such examples may be (I will argue that they are unexpectedly

common), the very fact that polymaths make trans-disciplinary discoveries should warn us against making too-easy distinctions between disciplines or cognitive domains. On the contrary, I will argue that those people who incorporate (in a quite literally visceral way) modes of thinking belonging to many cognitive domains are those most likely to become innovators.

Polymathy as a Predictor of Success

Santiago Ramon y Cajal, one of the first Nobel Prize Winners in Physiology or Medicine, is remembered today for his ground-breaking work on neuroanatomy. He was also a talented artist and photographer who attributed much of his scientific success to these avocations. Not only did he proclaim that, "it is not without reason that all great observers are skillful in sketching", but he also avowed that only through neuranatomical studies were his strong "aesthetic instincts" satisfied by the "incomparable artistic emotions" he experienced. Ramon y Cajal recognized that his polymathic tendencies were unusual, but asserted in his autobiography that the most successful scientists were, like him, "endowed with an abundance of restless imagination spend(ing) their energy in the pursuit of literature, art, philosophy, and all the recreations of mind and body. To him who observes them from afar, it appears as though they are scattering and dissipating their energies, while in reality they are channeling and strengthening them" (Ramon y Cajal, 1951, p. 171). To be creative, Ramon y Cajal said, one had to have wide experience with the process of creating.

A surprising number of his contemporaries agreed. Charles Richet, another Nobel laureate and a celebrated playwright, wrote: "Generally those who later become illustrious (in science) have shown from the first, by their aptitude for history, science, literature, languages, that they were superior to their contemporaries" (Richet, 1927, p. 128). Similarly, J. H. van't Hoff, the first Nobel laureate in Chemistry (1901) and himself a musician and poet, proposed that the development of the scientific imagination requires the development of artistic, musical, and poetic talents. In a famous address on 'Imagination in Science' (van't Hoff, 1878), he listed dozens of examples of eminent scientists who were multiply talented, including such notables as Kepler (a musician), Galileo (an artist), Davy (a poet), and Pasteur (another artist). He concluded his address by arguing that often the poetic vision outstrips the scientific showing the latter the way to the truth. This is a point to which I shall return below.

Many of van't Hoff's contemporaries saw the same connections between artistic proclivities and scientific success that he did. Van't Hoff's friend and colleague Wilhelm Ostwald produced a large body of work on scientific genius that validated the polymathy hypothesis (Ostwald, 1909). The English polymath, Francis Galton found that polymathy was unusually common among members of the British Royal Society (Galton, 1874). Botanist P. J. Moebius, the grandson of the famous mathematician, and the Frenchman Henri Fehr both noted independently the unusually high incidence of artistic and musical proclivities among two large groups of mathematicians (Fehr, 1912; Moebius, 1900). Jacques Hadamard confirmed these findings several decades later in his classic, *The Psychology of Invention in the Mathamatical Field* (Hadamard, 1945). All of these studies were summarized by the Nobel laureate and pianist Max Planck when he suggested that scientific success depends upon the use of an '*artistic* imagination' (his emphasis; Planck, 1949, p. 14).

When so many successful scientists all say that artistic imagination fosters good science, one must wonder whether it might not be true. In fact, a large body of psychological literature supports the hypothesis that polymathy is correlated with career success. For example, Stanford psychologist Lewis Terman, the father of the Stanford Binet IQ test and one of the foremost investigators of high achievers, wrote in 1941 that: "While the versatility of geniuses has long been stressed by biographers (e.g. of Da Vinci), the less spectacular cases are usually overlooked. People prefer to believe that the genius, as a rule, is no better than the rest of us except in one particular. The facts are very different . . . . there are few persons who achieved great eminence in one field without displaying more than average ability in one or more other fields" (Seagoe, 1975, p. 221). His conclusion was based upon decades of study of high achievers followed from their school days well into their careers, much of it published by his collaborator Catherine Cox. He also drew on studies of historical figures carried out by his colleague R. K. White. Analyzing hundreds of historical figures, White had found that "the typical genius surpasses the typical college graduate in range of interests and . . . he surpasses him in range of ability" (White, 1931, p. 489). Similarly, historian of science Paul Cranefield found that there was a direct correlation between the eminence a scientist achieved and his range of activities. The number of avocations practiced by a scientist correlated with the number of different scientific areas in which he worked. The number of different areas in which each scientist worked correlated with the number of significant discoveries they made. And the range and nature of the subjects that a scientist addressed in their research correlated with their cultural and philosophical avocations (Cranefield, 1966).

Subsequent cognitive studies have tended to validate the notion that the versatility of genius provides useful mental skills. For example, studies by Rauscher et al. and Gardiner et al. have suggested that direct relationships may exist between art and musical skills and improved spatial and mathematical reasoning in children (Graziana, Peterson & Shaw, 1999; Gardiner et al., 1996, p. 284; Rauscher, Shaw & Ky, 1997). Similar results exist for adults. An 11-year follow-up of participants in Project TALENT by Humphreys et al. (1993) has shown that in college students neither high grades nor high scores on verbal and mathematics tests are predictive of future participation or success in engineering and physical sciences. High scores on spatial and mechanical comprehension tests were, however, predictive, especially when combined with high mathematics scores. Schaer et al. (1985), Woody Flowers at MIT, and various other investigators (Stewart, 1985) have shown that it is possible to train scientists and engineers through explicit visualization and drawing exercises to improve their imaging and modeling skills, thus bringing the arts–sciences connection full circle.

Avocations can be as useful in training the creative mind as can formal classwork. Students who develop artistic skills through natural inclination tend to have much improved success in all careers according to a huge, long-term study of Israeli professions carried out by Milgram et al. (1993). They found that high IQ, standardized test scores, and high school grades were not good predictors of career success either independently or as aggregate measures. The best single predictor of success in any field, including the sciences, was participation as an adolescent in what Milgram calls "challenging leisure-time activities", i.e. avocations that require significant intellectual and time commitments. These include music performance and composition, painting, drawing, photography, chess, electronics, programming as a hobby, and creative writing, among others. These activities appear to be surrogate measures not only of intellectual ability, but energy, self-motivation, task commitment, cognitive breadth, and other attributes that strongly influence success.

Continued participation in the arts as an adult is also highly predictive of success as a scientist. A convenience sample of 40 scientists recruited by Bernice Eiduson in 1955 for a psychological study was analyzed by Root-Bernstein et al. in 1988 (Root-Bernstein, Bernstein & Garnier, 1993). By then, the group included four Nobel laureates, 11 members of the National Academy of Sciences (USA), many typical university professors, as well as several individuals who left academia for industry. Two measures of success were employed: impact (the ratio of citations to publications) and citation cluster (people with one or more papers having more than 100 citations over a 15 year period; people with one or more papers having between 10 and 100 citations in a single year; people with one or more papers having between 10 and 100 citations over 15 years; and people who met none of the previous criteria). The Nobel laureates were all in the top impact group and citation cluster, as were most of the National Academy members. Statistically significant correlations were found between success as measured by impact ratio and participation as an adult in painting, drawing, and sculpting. Significant correlations were found between success as measured by citation cluster and painting, collecting art, writing poetry, photography, crafts, singing, and most strongly with the sum of all hobbies. In short, these correlations validated the anecdotal evidence collected by Ramon y Cajal, Richet, van't Hoff and so many other Nobel laureates (Root-Bernstein et al., 1995).

Root-Bernstein et al. (1995) also found that statistically significant correlations existed between various hobbies and the modes of thinking that the scientists reported using during their scientific work. Artistic scientists tended to be visual and imageless-nonverbal thinkers whereas musicians were predominately visual thinkers. The link found between music training and improved visualization by Graziano et al. (1999), Gardiner et al. (1996) and Rauscher et al. (1997), therefore seems to extend into adulthood. Sculptor-scientists were mainly imageless-nonverbal and kinesthetic thinkers. Writer-scientists, not surprisingly, were mainly verbal thinkers; and those who participated in electronics-related hobbies tended to use the widest range of modes of thinking. How we think about problems therefore seems to be a function of what mental skills we develop through practice. Moreover, modes of thinking were independently correlated with measures of scientific success. Success as measured by impact ratio correlated significantly with visual thinking, use of verbal/auditory patterns, kinesthetic thinking, and other unusual forms of thinking such as use of word images, acoustic images, and talking to oneself. Success as measured by citation clusters, however, correlated most strongly with use of visual images, and visualized symbols and words.

These results do not distinguish between three non-exclusive possibilities. First, the most successful scientists may simply be very bright people who could succeed at anything. Second, innate talents are expressed in both scientific style and avocations. And third, practice using certain modes of thinking leads to skill development that is trans-disciplinary. Interviews with the scientists lead me to favor the latter hypotheses. The most successful scientists all percieved the sciences and arts to be complementary or compatible, as opposed to the less successful, who saw them in conflict. Some of the most successful went on to describe, like Ramon y Cajal, ways in which their avocations stimulated their scientific acumen.

I am not claiming that to become a successful scientist one *must* cultivate the arts, music, or literature. There are many ways of learning how to think visually, kinesthetically, and by analogies, including the practice of science itself. In fact, until recently, both free-hand and mechanical drawing, model building, and writing were integral parts of many science and engineering curricula. But due to larger classes, fewer labs, and monetary restrictions, arts and crafts programs now retain almost exclusive hegemony over such skills. Still, their continued necessity in science and engineering training has been urged in books as varied as Eugene Ferguson's *Engineering and the Mind's Eye* (Ferguson, 1992), Cyril Stanley Smith's *A Search for Structure* (Smith, 1981), Geri Berg's *The Visual Arts and Medical Education* (Berg, 1983), Phillip Ritterbush's *The Art of Organic Forms* (Ritterbush, 1968), and my own *Discovering* (Root-Bernstein, 1989). All of these studies tell us that artistic scientists and engineers have more image-ination, musically talented ones duet (do it) better, and the verbally inclined have the skills to become pundits. Seriously. Being cultured is still a prerequisite to being educated, and education is still a requirement for being successful.

The Uses of Arts by Scientists

The fact is that scientists and inventors not only explore the arts as avocations, but use them in their professional work (Root-Bernstein, 1985, 1987, 1990, 1997, 2001). Astrophysicist and novelist Gregory Benford points out, "Many believe that science fiction (SF) writers get their ideas from science and often this is so. Fewer recall that ideas have also flowed the opposite way" (Benford, 2001, p. 1). The Russian rocket pioneer Konstantin Tsiolkovsky was inspired to begin inventing by the works of Jules Verne; Leo Szilard took out the first patent on nuclear reactors after reading a short story by H. G. Wells about the potential for unlocking the energy stored within the nucleus of atoms; and Benford himself explores in his fiction the ideas that he cannot carry out as a scientist.

The use of fiction to explore novel scientific ideas is hardly rare. Robert L. Forward, an astrophysicist and inventor who is also an award-winning science fiction writer, says of his work: "Those of (my) far-out ideas which can be accomplished using present technology (I) do as research projects. Those that are too far out (I) write about in speculative science articles or develop in (my) short stories and novels" (Forward, 1985, book-jacket). Forward insists that his science and his fiction are on a continuum, his fiction simply consisting of the ideas that are currently beyond his means to implement. That does not mean that these ideas are any less valuable or insightful. As science fiction writer Jeff Hecht notes, "Fictional inventions take real skill and some prove truly prescient" (Hecht, 1999, p. 59). Recognition that important innovations are foreseen by scientific novelists has now become so explicit that the European Space Agency recently announced that they are scouring science fiction for new space propulsion technologies (Anon., 2000, p. 41). Similarly, corporations such as Global Business Network hire scientist-novelists such as computor innovator Vernor Vinge because they provide "an unbelievably fertile perspective from which to look back at and reunderstand the present. It's that ability to conceptualize whole new ways of framing issues .... He has contributed to the turnarounds of at least two well-known technology companies" (Hafner, 2001, D9).

Poetic scientists find similar connections between their poetry and their science. Words are scientific tools just as much as they are artistic ones, and therefore the exercise of words can be as enlightening to a scientist as to a writer. Thus, Nobel-prize-winning chemist Roald Hoffmann says that, "I begin with a vision of unity of creative work in science and in the humanities and arts . . . . I have no problem doing (or trying to do) both science and poetry. Both emerge from my attempt to understand the universe around me, from my own personal affection for communicating, teaching what I've learned, and from my infatuation with language . . . . It seems obvious to me to use words as best I can in teaching myself and my coworkers. Some call that research . . . . The words are important in science" (Hoffmann, 1988, p. 10). Similarly, physician and poet Jack Coulehan maintains that writing poetry makes him a better physician: "Both disciplines require careful observation. Both focus on the concrete—an illness, an event, a feeling—over the abstract. In fact, William Carlos Williams' famous aphorism about poetry, 'No ideas but in things', is also a good prescription for medical practice . . . . Moreover, a physician's ability to empathize with a paitient is a creative act analogous to the poet's act of exploring the subject of a poem . . . . (And), like poetry, medicine achieves much of its powerful effect through the use of symbol and metaphor" (Coulehan, 1993, p. 57). To learn how to manipulate images and feelings through the words of a poem is therefore, according to Hoffmann and Coulehan, to learn how to manipulate the images and feelings that are expressed in scientific symbols as well. Equally important, learning to manipulate language teaches verbal imagination without which great science is not possible.

Ethologist and Surrealist painter Desmond Morris has said that he uses his art for the same reasons that scientific fiction writers and poetic scientists practice theirs. "Being a biologist and a student of evolution, I attempted to evolve my own world of biomorphic shapes, influenced by but not directly related to the flora and fauna of our planet. From canvas to canvas I have tried to let them grow and develop in a natural way, without ever crudely borrowing specific elements from known animals or plants" (Morris, 1987, p. 17). He says that his biomorphic shapes undergo the same evolutionary stages that real organisms do so that his art becomes a mental laboratory for exploring the nature of evolutionary processes in the abstract. He realizes explicitly that his "paintings are very biomorphic, very preoccupied with biological shapes, and that my biological writings are largely concerned with visual patterns of behaviour. I have never resisted that kind of leakage . . ." (Remy, 1991, p. 18). In fact, he has made that leakage the source of fertile studies of the biological origins of art and of the evolution of primitive arts among humans, an interest he shared with another evolutionary pioneer, Mary Leakey, who began her scientific career as an illustrator for other anthropologists (Leakey, 1984, pp. 39–43).

Many evolutionary and ethological studies have been influenced by artistic biologists. Thomas Henry Huxley, Alfred Russel Wallace (Nelson, 2001, p. 1260), and Ernst Haeckel are only a few of the pioneers of evolutionary theory who were also artists. Haeckel, whose prints (Haeckel, 1905) had a significant impact on the artists of his day (Kockerbeck, 1986), was the first to fully realize Darwin's concept of an evolutionary tree as an articulated image. Huxley for his part claimed that accurate scientific observation was impossible without a facility for drawing and argued that art classes should be required for all scientists: "I should make it absolutely necessary for everybody, for a longer or shorter period, to learn to draw . . . . I do not think its value can be exaggerated, because it gives you the means of training the young in attention and accuracy, which are the two things in which all mankind are more deficient than in any other mental quality whatever" (T. H. Huxley, 1900, vol III, pp. 183–184).

Contemporary artistic scientists include ethologist Jonathan Kingdon (Gautier-Hion et al., 1988) and paleontologist-novelist Robert Bakker (Bakker, 1995), both of whom draw their own specimens and create three-dimensional models from their sketches in order to better understand with their material. Bert Holldobler, a professor of entomology and an amateur painter, also draws the illustrations for all of his publications. He says that, "It is my urge that I make every paper I write as crystal clear as I can in words and illustrations. A scientific publication should be a piece of art" (personal communication). His success is proven by the fact that he won a Pulitzer Prize for his collaboration with E. O. Wilon on their book, *The Ants*, of which Holldobler says, "It was our first intention to write a scholarly book, but we were also driven to show the beauty of the life of our subjects, both in our writing and in the illustrations of the book" (personal communication). For these men, an illustration is therefore both art and science.

Another example of art having a scientific component and science having an intrinsic artistic element can be found in the subject of moire patterns. Moire patterns are created when figures with periodic patterns are made to overlap so that both patterns are still visible. The word 'moiré' comes from the French word for 'watered' and has classically been used to describe the particular type of silk fabric that has the shimmer of water waves. This shimmer results from folding the silk onto itself at a slight bias and then pressing it under high pressure and heat to imprint the slightly offset pattern of one fold of the fabric onto the other. Similar patterns can be created by overlapping two or more screens or meshes, or by drawing overlapping patterns on paper. Physicist Gerald Oster demonstrated that moire patterns have many interesting mathematical properties, including the potential to perform analogue computations, thereby calculating by visual means (Oster & Nishjima, 1963). Oster not only published widely on the scientific uses of moire patterns, but also exhibited them as art in several well-known galleries (Mueller, 1967, pp. 197–198).

In a similar way, music has played its role in scientific developments. Musical analogies are so commonly used by scientists that historian of science Jamie Kassler has written a book describing their many problem-solving functions (Kassler, 2001). Not only did Johannes Kepler derive many of his astronomical discoveries from attempts to optimize the 'harmony of the spheres' (which he wrote out literally as music!) but physicist-musician Louis de Broglie earned himself a Nobel prize by comparing atoms to tiny stringed instruments and then showing that they would have the same kinds of harmonics and overtones, but with frequencies proportionally smaller. In these cases, the links between music and science are apparent and often describable as mathematics. Sometimes scientists therefore take advantage of the music–mathematics equation to convert quantitative results into music! Many university professors and corporations, including Bell Labs, Exxon Research and Engineering Corporation, and Xerox have experimented with transforming complex quantitative data into sound. They have discovered that the ear has powers of pattern recognition that extend far beyond the discriminatory abilities of the eye, and therefore that turning data into music rather than into graphs or charts often aids analysis. Applications of this technique have included exploring the patterns within the human genome, chemical analyses, complex physiological data, and multivariate economic indicators (Peterson, 1985a).

Sometimes scientists use music less analytically and more inspirationally, raising interesting issues about what stimulates creative thinking. Cardiology researcher Richard Bing, for example, is also a composer. He says that, "Writing music enriches me to look at science in a different way. It helps me emotionally to feel more about science. You see, I am a romanticist. I perceive science as an emotional exercise of searching the unknown" (Root-Bernstein, 1987, p. 2). Bing believes, as did Einstein, that "both (music and science) are born of the same source and complement each other through the satisfaction they bestow" (Clark, 1981, p. 106). This may explain why scientists from Darwin to Einstein and inventors such as Charles Martin Hall have all turned to music when faced by an apparently unresolvable problem. Somehow music frees the analytical mind allowing intuition to yield its fruits (Root-Bernstein, 2001).

Even dance has placed its footprints in scientific innovation. Berkely physicist Marvin Cohen has collaborated with choreographer David Wood and his

dancers to explore forms of dynamic symmetry that inform the theory of super-conductivity. Cohen describes the dance not only as a novel way to communicate his otherwise inaccessible mathematical theories, but also as a form of 'research'. He hopes that the dancers will invent forms of dynamic interactions that physicists have yet to have considered as possible models for electron interactions (NOVA, 1988). Dance notations are playing a similar role in elucidating the visualization problems that plague neurology and other areas of medicine. Describing and accurately communicating the nature of the motor impairments that characterize particular neurological lesions and various genetic diseases has always been extremely difficult. How does one explain to someone that multiple sclerosis is typified by this type of gait, but amyotrophic lateral sclerosis presents a different set of motor impairments that look quite different? That problem was first solved about 20 years ago by neurologists Ilan Golani and Philip Teitelbaum, when they applied Eshkol–Wachman movement notation, a well-known tool for recording dance movements, to the description of laboratory animal and human movement diseases (Golani & Teitelbaum, 1979). Since that time, Labanotation and Benesh Movement Notation have also been applied to the recording and analysis of physical actions. Someday, they may inform our design and control of robots as well.

The arts, in short, often supply scientists with ways of looking at the world that complement purely logical and analytic modes. This cross-fertilization is common enough, and the insights yielded by applying artistic methods to scientific problems useful enough, that an increasing number of investigators are suggesting that for exploring the human dimensions and implications of science and technology, artistic methods may even be superior to scientific ones. Chemist Carl Djerassi, for instance, the inventor of the birth control pill, has begun writing 'science-in-fiction' in order to explore the ethical and social implications of the latest biomedical innovations in meaningful ways impossible through purely analytical discourse (Djerassi, 1990, p. 16). For the same reasons, Nobel-prize-winning economist John Kenneth Galbraith said of his novel *The Triumph* that it "is a story I have tried to tell before in articles and lectures. But it has occurred to me that maybe there are truths that best emerge from fiction" (Galbraith, 1968, p. 7). Physicist/musician Victor Weisskopf has suggested that, "Especially in human relations, a piece of art or a well-written novel could be much more revealing than any scientific study. In many respects, *Madame Bovary* is a piece of sociology—in fact, better sociology than much of what is done by aping the techniques and language of the natural sciences" (Weisskopf, 1977, p. 410). And chemist/poet Roald Hoffmann has warned scientists to beware of chavinism in their dealings with the arts: "One thing is certainly not true: that scientists have some greater insight into the workings of nature than poets" (Hoffmann, 1988, p. 10). In sum, these men are arguing that the analytical methods of the sciences are not the only possible methods for revealing truth, and that particularly in the human sciences, but perhaps in all sciences, the artist's approach may be more insightful than that of the mathematician. This is an interesting and potentially revolutionary message that bears serious consideration, especially when it comes from such accomplished scientists.

Arts Advance the Sciences

The thesis that the arts may provide insights beyond the methodological capabilities of the sciences comes from the fact that the artistic innovations often precede and make possible subsequent scientific ones. Examples are legion.

Alexander Graham Bell's first invention, multiplex telegraphy, resulted from his avocation as a pianist. He was playing a well-known parlor game of striking an 'A' and showing his party that only the harmonics of 'A' resonated; and similarly for 'B' and 'C', and so on, when it suddenly struck him that one could design a telegraph along similar lines. One could send multiple messages along a single telegraph wire if one sent one metaphorically in the key of 'A', another in the key of 'B', and so forth. His work on multiplex telegraphy was what led him to the telephone (Mackay, 1997, 9ff).

In an intellectually related innovation, professional musician George Antheil and actress Hedy Lamarr collaborated to create one of the most influential inventions of the last century: frequency hopping. Lamarr was married to a major munitions manufacturer in Germany before divorcing him and moving to the United States. She was aware that munitions such as torpedos were usually radio-controlled, but that it was relatively easy to jam the radio frequency by which control was maintained. She discussed this problem with Antheil, a polymath who not only wrote some of the most revolutionary music of the twentieth century (e.g. 'Ballet mecanique'), but also wrote about endocrinology, crime, and war. They realized that radio signals are like music: if one changes the radio frequency at which one sends a message, one keeps the same 'notes' within a message, just as one retains the musical message when one alters the key. But if one changes the 'key' (radio frequency) on a random basis, then the person trying to intercept or jam the message will have a very hard time doing so. Their invention is the basis for almost all secure communications, control systems, and anti-jamming devices currently in use worldwide (Braun, 1997).

A musical device also inspired one of the first heart pacemakers. Engineer Earl Bakken was looking for a way to create miniaturized regulators for the electrical output of apparatuses being designed to steady heart rhythms. Bakken found the answer in an already existing device, the electronic metronome that musicians had been using for years. Instead of generating a sound, as musical metronomes do, Bakken reconfigured his device to generate regular pulses of electricity to stimulate the heart. His company, Medtronics, became a leader in the development of the implantable pacemakers that are used today (Jeffrey, 1997).

Visual arts have also stimulated many scientific innovations. Hermann Rorschach, the Swiss psychiatrist who invented the well-known Rorschach ink blot tests, was also an amateur artist. The idea for ink blot tests came to him from a popular party game based on decalcomania, an artistic technique in which an image is transferred from one piece of paper to another, or within a piece of paper by folding it. For entertainment, people would drip a variety of colored inks or paints onto paper that was then folded and pressed. When the paper was unfolded, unexpected images were revealed. Rorschach noticed not only that people had a tendency to comment on what they saw in the ink blots, but that different people saw very different things that were suggestive of their personalities and problems (Larson, 1958). Thus, Rorschach's innovation resulted from taking advantage of a technique invented and popularized by professional and amateur artists.

Perceptual psychologists also owe many debts to artists. Those studying motion perception, for example, often rely upon the rotoreliefs of Marcel Duchamp. Rotoreliefs are a form of artwork based upon tops. If a spiral is painted on a top, it appears to move continuously, even though it actually has a finite length. A similar effect can be observed if a spiral is painted on a round platen and rotated on a record player. Duchamp invented very large platens that he painted with extraordinary patterns that create effects that still puzzle perceptual psychologists and are therefore useful in their research. One of the most intriguing is a double spiral that appears to spiral both in and out simultaneously (Sekuler & Levinson, 1977, p. 61). Another set of perceptual phenomena that have become a major focus of psychological reasearch are the size and space illusions invented by Adelbert Ames, a lawyer and artist. The most famous of Ames' innovations is the so-called 'Ames room' that creates the illusion that a person standing at one side of the room is a dwarf and a person standing at the other side of the room is a giant, when in fact they are the same size (Behrens, 1994).

Sometimes artists and scientists collaborate in an intricate dance of images. One example of such a dance was initiated by M. C. Escher, the famous graphic artist. Escher specialized in drawing impossible things, such as hands emerging out of a flat piece of paper drawing themselves. Roger Penrose, a mathematical physicist and amateur artist, visited an exhibition of Escher's work in 1954 and was stimulated to invent his own impossible objects. The result of his experimentation is the famous 'impossible tribar' a two-dimensional rendering of a three-dimensional triangle that appears to twist both forwards and backwards simultaneously. Roger Penrose showed his tribar to his father L. S. Penrose, a biologist who also dabbled in art, and L. S. Penrose soon invented the 'impossible staircase', in which the stairs appear to spiral both up and down simultaneously. The Penroses' published their 'discoveries' in 1956 the *British Journal of Psychology* and sent Escher a copy of the publication as thanks. Escher then developed the artistic possibilities of the impossible tribar and staircase in ways that have since become famous not only in artistic circles, but as fodder for psychological experimentation (Ernst, 1992, pp. 71–80).

The impossible tribar was only one of Roger Penrose's artistic forays. Another was his invention of aperiodic tilings, which also owe a debt to Escher. A tiling is a pattern made out of repeating visual or design elements that covers a surface such as a plane or a sphere. A periodic tiling is one that does so using a regularly repeating pattern. Much of Escher's graphic work involved periodic tilings interpreted as birds, fish, and other animals. Penrose spent many years mastering the techniques of making periodic tilings like those popularized by Escher. But he tired of the regularity of the patterns and began searching for new possibilities (personal communication). One of these was aperiodic tiling. Aperiodic tilings are composed of a limited number of invariant shapes that nonetheless result in a pattern that never repeats. Penrose was one of the pioneers of this field. He says he created them mainly as an artistic puzzle. He later investigated the mathematical properties of them. Even more recently, after his artistic renderings were popularized by Martin Gardner in his mathematics column in *Scientific American* (Gardner, 1977), crystallographers realized that many metal alloys have structures that are described as aperiodic tilings (Peterson, 1985b, p. 188). Thus, an artistic hobby resulted in the discovery of novel structures that subsequently shed light on the nature of previously mysterious properties of metals.

Artists often invent novel structures that subsequently reveal unexpected properties of nature. R. Buckminster Fuller, working as an artist and architect, laid out an entire theory of structural stability based on tetrahedral forms that eventually led to his invention of the geodesic dome. Fuller claimed that this structure combined the least material to encapsulate the greatest space with the greatest stability. As such, he suggested that it would turn out to be a fundamental construct of nature. In fact, when virologists began to study the structures of spherical viruses, they turned to Fuller for help in solving the possible structures and soon discovered that most spherical viruses are geodesic domes, as Fuller had predicted (Fuller, 1965, p. 72). More recently, chemists have discovered that extremely stable carbon compounds, appropriately called 'buckminsterfullerenes', can be made that have geodesic structures, and similar structures are beginning to be discovered among biological macromolecules.

Almost as well known as Fuller's geodesic concept is his student Kenneth Snelson's concept of tensegrity, in which stable structures result from a juxtaposition of rigid pieces and elastic tensions to yield structures with great structural integrity. Snelson's tensegrity sculptures exist in major collections (e.g. the Smithsonian) all over the world and attracted a great deal of attention during the 1970s. Among those who became enamoured enough with tensegrity sculptures to build some of their own were biologists Steve Hiedemann and Donald Ingber (personal communications), who more than a decade later were to realize that the structures within cells that maintain their shapes have many similarities to tensegrity structures (Brookes, 1999, p. 43). The tensegrity theory of cell structure has captured the cover of *Scientific American* (Ingber, 1998, p. 48), and the new mathematics of tensegrity captured the cover of *American Scientist* (Connelly & Back, 1998, p. 142), once again demonstrating that artistic inventiveness can provide basic scientific insights.

Artist Wallace Walker has similarly galvanized solid geometry. While a student at the Cranbrook Academy in the 1960s, he was asked to make a three-dimensional object out of a sheet of paper only by folding and gluing it. The result was a complex donut that could be folded through its center hole to take on a kaleidoscopic variety of shapes. Walker's invention not only earned him a patent but also attracted the attention of Doris Schattschneider, a mathematician specializing in geometric objects. Schattschneider determined that Walker's paper sculpture was the first of a novel class of geometric objects, now called kaleidocycles (Schattschneider & Walker, 1977).

Another version of paper folding has also become the source of major mathematical innovations in recent years, and that is the ancient oriental art of origami. Mathematicians have recently discovered that the rules of origami embody (literally) a set of mathematical algorithms that determine whether an object can be created by folding, or whether it must be cut. This observation has opened an entirely new field of research into what is being called 'plication' that is now yielding unexpected benefits (Hayes, 1995). A set of engineers at the Lawrence Livermore National Laboratory in Berkeley, California have recently invented a huge, foldable lens, many times larger than any lens heretofore created, for use in space laboratories (Anon., 2001). Other scientists have discovered links between origami and logic that may transform the way computers are designed and programmed (Cipra, 1998).

Novel structures are not the only ways in which artists contribute to scientific innovation. Some contributions stem from artistic observations. Since artists are trained to observe what other people overlook, they sometimes think about what other people never see. Leonardo da Vinci, for example, was the first to observe that a cross-section through the branches of a tree from the twigs down to the trunk will always conserve the same total area of wood. 'Leonardo's principle' as this observation is now known, is still a viable area of research in botany and engineering. The invention of the concept of camouflage provides another, more modern example. Ever since people began hunting, they undoubtedly noticed that animals tend to blend into their environments, but it was not until the 1880s, when a professional painter of portraits and angels named Abbott Thayer looked carefully at nature through the eyes of an amateur evolutionist that anyone thought to question how this blending came to be. After more than a decade of visual experimentation, Thayer described the entire range of possible mechanisms by which camouflage might be expressed in nature (Maryman, 1999, 116ff).

Artistic techniques have also been the basis for scientific developments. One such technique is anamorphosis, meaning 'shape change', which derived historically from the discovery of perspective. Perspective drawing involves the mapping of a three-dimensional object onto a flat surface. Renaissance artists quickly realized that two-dimensional objects could also be mapped onto three-dimensional surfaces, including spheres, cones, and rods (Gardner, 1975). Such transformations became central to D'Arcy Thompson's *On Growth and Form* (Thompson, 1930, Ch. 17). and Julian Huxley's *Problems of Relative Growth* (J. Huxley, 1932, Ch. 4), both of which describe evolutionary and embryological processes as anamorphic distortions. Anamorphosis also underlies Wilder Penfield's and neurologist-artist C. N. Woolsey's studies of the motor and sensory mappings of primates onto the cortex of their brains, which yield the familiar 'homunculi' with their huge lips, hands and feet, and tiny bodies (Woolsey, 1978).

Another striking example of how artistic techniques inform science is the reification of logic in modern computer chips. The logic embedded in computer chips is actually a pattern directing electron flow. This pattern is drawn using techniques as old as drafting architectural plans and then shrunk using photographic techniques to produce a tiny template. This template is then used to transfer the pattern to silica wafers using methods adapted directly from silk screening and etching. Thus, the physical embodiment of logic as a functional image on a chip contains within it hundreds of years of artistic experience (Root-Bernstein, 2000).

Finally, the arts can foster scientific advances through the development of new aesthetics (Root-Bernstein, 1996). The use of pixels, false coloring, and abstractions provide three cases. The process of breaking a picture into discrete areas of color (pixels) was invented by pointillist painters such as Seurat. The technique of falsely coloring objects was invented by Fauvist painters. And abstract art, in which a single element of a complex phenomenon, such as its pattern, structure, or color, is chosen for selective description, was pioneered by Picasso, Braque and Kandinsky during the 1920s. Examine any scientific illustration carefully and you are likely to find one or more of these artistic techniques being employed to focus attention on one particular aspect of data. Without an excellent sense of history, it is too easy to overlook the artistic origins of many of the scientific tools we use for analyzing our results (Root-Bernstein, 2001).

Artscience and Innovation: The Future of Polymathy

Fortunately, there is growing understanding that art fosters science. Mitchell Feigenbaum, one of the pioneers of chaos theory, believes that understanding how artists paint will provide the cognitive insights necessary to do better science: "It's abundantly obvious that one doesn't know the world about us in detail. What artists have accomplished is realizing there's only a small amount of this stuff that's important, and then seeing what it was. So they can do some of my research for me" (Gleick, 1984, p. 71). Similarly, C. S. Smith, of MIT spent a lifetime studying oriental arts and crafts for the insight they gave him into metallurgy: "I have slowly come to realize that the analytic, quantitative approach I had been taught to regard as the only respectable one for a scientist is insufficient . . . . The richest aspects of any large and complicated system arise from factors that cannot be measured easily, if at all. For these, the artist's approach, uncertain though it inevitably is, seems to find and convey more meaning" (Smith, 1978, p. 9).

In fact, some scientists are formally inviting artists to help them perceive new realities. Milton Halem, chief of the Space Data and Computing Division at NASA's Goddard Space Flight Center in the early 1990s, invited Sara Tweedie, a design instructor in the Corcoran School of Art, to help his engineers invent new ways to visualize the huge amounts of data being generated from satellite sources. "Visualizing that data, coloring that data, enables the mind to more quickly assimilate the information and the image", he noted. He says he needs artists such as Tweedie to push "modeling a step beyond where it's gone in the past" (Mercier, 1990, p. 28). The National Supercomputing Facility and Bell Laboratories (now Lucent Technologies) have hired artists for similar reasons.

But the real future of sciences–arts interactions must be within the minds of individuals. Artists and scientists too often speak different languages and use different tools. In order for them to collaborate effectively, to perceive in each others' problems and methods opportunities for insight, we must have a large cadre of artist-scientists and scientist-artists. Some of these people already exist. Desmond Morris for his part refuses to be labeled as a scientist or an artist: "If my paintings do nothing else, they will serve to demonstrate that such titles are misleading. In reality, people today are not scientists or artists . . . they are explorers or non-explorers, and the context of their explorations is of secondary importance. Painting is no longer merely a craft, it is a form of personal research . . . . So, in the end, I do not think of myself as being part scientist and part artist, but simply as being an explorer, part objective and part subjective" (Morris, 1987, p. 27). He adds that, "perhaps the time will come when we will give up the folly of separating sub-adults into the imaginative and the analytical—artists and scientists—and encourage them to be both at once" (Remy, 1991, p. 18).

This melding of the artistic and scientific mind within a single individual may even have benefits for both since, as the art critic Kenneth Clark has suggested, art and science emerge from the same imaginative sources: "Art and science . . . are not, as used to be supposed, two contrary activities, but in fact draw on many of the same capacities of the human mind. In the last resort, each depends on the imagination. Artist and scientist alike are both trying to give concrete form to dimly apprehended ideas" (Clark, 1981, p. 24). Rather than forcing individuals to choose a scientific approach to problems or an artistic one, thereby devaluing the other, Clark admonishes us, to "wait patiently for our faculties to be reunited" (Clark, 1981, p. 29).

Clearly, the arts and sciences are as capable of full integration today as they were in the Renaissance, and there is every reason to expect their union to be as fruitful. But to derive the fruits of their union, we must foster the connections and the people who can make them. Richard Mueller, an MIT-trained engineer, novelist and artist agrees with Clark. He writes in his stimulating book, *The Science of Art*, "In many ways, I think, the scientist is delaying his own understanding and development in science by discouraging, not only the artistically inclined members of his clan, but also the artistic urges within himself . . . . Art may be a necessary condition for constructing the new consciousness from which future science gets its structural realities to match nature, in which case it is more important than we generally admit . . ." (Mueller, 1967, p. 320).

Thus, we need a new kind of education that fosters interactions between disciplines rather than divisions between them and which trains people who can bridge C. P. Snow's 'Two-Cultures' divide. We need such curricula and people not only because of the fragmentation of knowledge that must result in their absence, but more importantly as a stimulus to the highest forms of innovation. For specialization can never suffice. As General Electric engineer, Charles Steinmetz, pointed out nearly a century ago, "technical training alone is not enough to fit a man for an interesting and useful life" (Seymour, 1966, p. 119). He urged his students to study languages, literature, philosophy, art, music and history, arguing that if an engineer failed to produce a workable invention, it was his own fault for not understanding the greater needs of society and the factors that control its manufacturing and economic functions. Similarly, composer-architect-engineer Iannis Xenakis argues today that "the artist-conceptor will have to be knowledgeable and inventive in such varied domains as mathematics, logic, physics, chemistry, biology, genetics, paleontology . . . the human sciences and history; in short, a sort of universality, but one based upon, guided by and oriented towards forms and architecture" (Xenakis, 1985, p. 3). All this is necessary so that we can address the truly important problems of the world, added embryologist and art historian C. H. Waddington: "The acute problems of the world can be solved only by whole men (and women), not by people who refuse to be, publicly, anything more than a technologist, or a pure scientist, or an artist. In the world of today, you have got to be everything or you are going to be nothing" (Waddington, 1972, p. 360). Buckminster Fuller agreed: "Overspecialization leads to extinction. We need the philosopher-scientist-artist—the compherensivist, not merely more deluxe quality technician mechanics" (Fuller, 1979, p. 104).

To invent and to create requires an understanding that incorporates all that is known sensually and abstractly, subjectively and objectively, imaginatively and concretely. And because of their wide disciplinary training in the imaginative skills, handicrafts and expressive languages, only polymaths will have the tools necessary to do so. Thus, the future of innovation will reside, as it always has resided, in the minds of mulitply talented people who transcend disciplinary boundaries and methods. We can recognize this phenomenon by fostering artscience—a term promoted by artist-inventor-psychologist Todd Siler (Siler, 1990)—or we can retard it by creating educational and workplace systems that prevent arts and sciences from meeting. As Siler has pointed out, artscience is both the past and future of innovation because innovators cannot help drawing upon any form of thinking that will spur their imagination. We ignore this profound truth at our peril.

References

Anonymous (2000). Random samples. *Science*, **289**, 41.
Anonymous (2001). Origami: Folded space. *The Economist* (31 March), 74–75.
Bakker, R. T. (1995). *Raptor red*. London: Transworld.
Behrens, R. R. (1994). Adelbert Ames and the cockeyed room. *Print*, **XLVIII** (II), 92–97.
Benford, G. (2001). *Imagining the future*. Unpublished manuscript.

Berg, G. (Ed.) (1983). *The visual arts and medical education.* Carbondale, IL: Southern Illinois University Press.

Braun, H-J. (1997). Advanced weaponry of the stars. *Invention & Technology*, **12**, 10–17.

Bronowski, J. (1967). *Scientific genius and creativity.* New York: W. H. Freeman.

Brookes, M. (1999). Hard cell, soft cell. *New Scientist*, **164**, 42–47.

Cipra, B. (1998). Proving a link between logic and origami. *Science*, **279**, 804–806.

Clark, K. (1981). *Moments of vision.* London: John Murray.

Connelly, R. & Back, A. (1998). Mathematics and tensegrity. *American Scientist*, **86**, 142–151.

Coulehan, J. (1993). Physician as poet, poem as patient. *Poets & Writers Magazine.* (March/April), pp. 57–59.

Cranefield, P. (1966). The philosophical and cultural interests of the biophysics movement of 1847. *Journal of the History of Medicine*, **21**, 1–7.

Djerassi, C. (1990). Illuminating scientific facts through fiction. *The Scientist*, 23 July, 16.

Ernst, B. (1992). *Optical illusions.* Koln: Taschen.

Fehr, H. (1912). *Enquete de l'enseignment mathematique sur la methode de travail des mathematiciens.* Paris: Gauthier-Villars; Geneva; George et Cie.

Ferguson, E. S. (1992). *Engineering and the mind's eye.* Cambridge, MA: MIT Press.

Forward, R. L. (1965). *Starquake.* New York: Ballantine.

Fuller, R. B. (1965). Conceptuality of fundamental structures. In: G. Kepes (Ed.), *Structure in Art and in Science* (pp. 66–88). New York: George Braziller.

Fuller, R. B. (1979). In: P. H. Wagschall & R. D. Kahn (Eds), *R. Buckminster Fuller on Education.* Amherst, MA: University of Massachusetts Press.

Galbraith, J. K. (1968). *The triumph.* Boston, MA: Houghton Mifflin.

Galton, F. (1874). *English men of science.* London: Macmillan.

Gardiner, M. F., Fox, A., Knowles, F. & Jeffrey, D. (1996). Learning improved by arts training. *Nature*, **381**, 284.

Gardner, M. (1975). The curious magic of anamorphic art. *Scientific American*, **232**, 110–116.

Gardner, M. (1977). Extraordinary nonperiod tiling that enriches the theory of tiles. *Scientific American*, **234**, 110–122.

Gautier-Hion, A., Bourliere, F., Gauteir, J-P. & Kingdon, J. (Eds) (1988). *A primate radiation.* Cambridge: Cambridge University Press.

Gleick, J. (1984). Solving the mathematical riddle of chaos (Interview with Mitchell Feigenbaum). *New York Times Magazine*, (10 June), 69–73.

Golani, I. & Teitelbaum, P. (1979). A proposed natural geometry of recovery from akinesia in the lateral hypothalamic rat. *Brain Research*, **164**, 237–267.

Graziano, A. B., Peterson, M. & Shaw, G. L. (1999). Enhanced learning of proportional math through music training and spatial-temporal training. *Neurological Research*, **21**, 139–152.

Hadamard, J. (1945). *The psychology of invention in the mathematical field.* Princeton, NJ: Princeton University Press.

Haeckel, E. (1905). *Ernst Haekcls Wanderbilder nach eigenen aquarellen und olgemalden.* Berlin: Koehler.

Hafner, K. (2001). A scientist's art: computer fiction. *New York Times* (2 Aug), D1 & D9.

Hayes, B. (1995). Pleasures of plication. *American Scientist*, **83**, 504–510.

Hecht, J. (1999). Patent right. *New Scientist* (23 Oct), 59.

Hoffmann, R. (1988). How I work as poet and scientist. *The Scientist* (21 March), 10.

Humphreys, L. G., Lubinski, D. & Yao, G. (1993). Utility of predicting group membership and the role of spatial visualization in becoming an engineer, physical scientist, or artist. *Journal of Applied Psychology*, **78**, 250–261.

Huxley, J. (1932). *Problems of relative growth.* London: Dial Press.

Huxley, T. H. (1900). *Collected essays.* 10 volumes. New York: Macmillan.

Ingber, D. E. (1998). The architecture of life. *Scientific American*, **278**, 48–59.

Kassler, J. C. (2001). *Music, science, philosophy: Models in the universe of thought.* London: Ashgate.

Kirk, J. (1997). Many paths to the pacemaker. *Invention & Technology*, **12**, 28–39.

Kockerbeck, C. (1986). *Ernst Haeckels 'kunstformen der nature' und ihr einfluss auf die deutsche bildende kunst der jahrundertwende.* Frankfurt: Peter Lang.

Koestler, A. (1964). *The act of creation.* London: Hutchinson.

Larson, C. A. (1958). Hermann Rorschach and the ink-blot test. *Science Digest* (Oct.), 44.

Leakey, M. (1984). *Disclosing the past.* Garden City, NY: Doubleday.

Mackay, J. (1997). *Sounds out of silence. A life of Alexander Graham Bell.* Edingburgh: Mainstream Publishing.

Maryman, R. (1999). A painter of angels became the father of camouflage. *Smithsonian*, **30**, 116–128.

Mercier, A. M. (1990). NASA's Halem illustrates need for info visualization. *Federal Computer Week.* (8 Oct), 28.

Milgram, R. & Hong, E. (1993). Creative thinking and creative performance in adolescents as predictors of creative attainments in adults: A follow-up study after 18 years. In: R. Subotnik & K. Arnold (Eds), *Beyond Terman: Longitudinal Studies in Contemporary Gifted Education.* Norwood, NJ: Ablex.

Moebius, P. J. (1900). *Ueber die anlage zur mathetmatik.* Leipzig: Barth.

Morris, D. (1987). *The secret surrealist. The paintings of Desmond Morris.* Oxford: Phaidon.

Mueller, R. E. (1967). *The science of art.* New York: John Day.

Nelson, G. (2001). Wallace at center stage. *Science*, **293**, 1260–1261.

NOVA (1988). *Race for the superconductor.* Boston, MA: WNET for PBS video.

Oster, G. & Nishijima, Y. (1963). Moire patterns. *Scientific American*, **208**, 54–63.

Ostwald, W. (1909). *Grosse maenner.* Leipzig: Akademische Verlagsgesselshaft.

Peterson, I. (1985a). The sound of data. *Science News*, **127**, 348–350.

Peterson, I. (1985b). The five-fold way for crystals. *Science News*, **127**, 188–189.

Planck, M. (1949). *Scientific autobiography and other papers.* F. Gaynor (Trans.). New York: Philosophical Library.

Ramon y Cajal, S. (1951). *Precepts and counsels on scientific investigation: Stimulants of the sprit.* J. M. Sanchez-Perez (Trans.). Mountain View, CA: Pacific Press Publishing Association.

Rauscher, F. H., Shaw, G. L. & Ky, K. N. (1997). Music training causes long-term enhancement of preschool children's spatial-temporal reasoning. *Neurological Research*, **19**, 2–8.

Remy, M. (1991). *The surrealist world of Desmond Morris*. L. Sagaru (Trans.). London: Jonathan Cape.

Richet, C. (1927). *The natural history of a savant*. O. Lodge (Trans.). London: J. M. Dent.

Ritterbush, P. (1968). *The art of organic forms*. Washington, D.C.: Smithsonian Press.

Root-Bernstein, R. S. (1984). Creative process as a unifying theme of human cultures. *Daedalus*, **113**, 197–219.

Root-Bernstein, R. S. (1985). Visual thinking: the art of imagining reality. *Transactions of the American Philosophical Society*, **75**, 50–67.

Root-Bernstein, R. S. (1987). Harmony and beauty in biomedical research. *Journal of Molecular and Cellular Cardiology*, **19**, 1–9.

Root-Bernstein, R. S. (1989). *Discovering*. Cambridge, MA: Harvard University Press.

Root-Bernstein, R. S. (1990). Sensual education. *The Sciences* (Sep/Oct), 12–14.

Root-Bernstein, R. S. (1996). The sciences and arts share a common creative aesthetic. In: A. I. Tauber (Ed.), *The Elusive Synthesis: Aesthetics and Science* (pp. 49–82). Netherlands: Kluwer.

Root-Bernstein, R. S. (1997). Art, imagination and the scientist. *American Scientist*, **85**, 6–9.

Root-Bernstein, R. S. (2000). Art advances science. *Nature*, **407**, 134.

Root-Bernstein, R. S. (2001). Music, science, and creativity. *Leonardo*, **34**, 63–68.

Root-Bernstein, R. S., Bernstein, M. & Garnier, H. (1993). Identification of scientists making long-term high-impact contributions, with notes on their methods of working. *Creativity Research Journal*, **6**, 329–343.

Root-Bernstein, R. S., Bernstein, M. & Garnier, H. (1995). Correlations between avocations, ascientific style, work habits, and professional impact of scientists. *Creativity Research Journal*, **8**, 115–137.

Schaer, B., Trenthan, L., Miller E. & Isom, S. (1985). Logical development levels and visual perception: Relationships in undergraduate engineering graphic communications. Paper presented at the Mid-South Educational Research Association, Biloxi, Mississippi, 6 Nov.

Schattschneider, D. & Walker, W. (1977). *M. C. Escher kaleidocycles*. Corte Madera, CA: Pomegranate Artbooks, Inc.

Seagoe, M. (1975). *Terman and the gifted*. Los Altos, CA: W. Kaufmann.

Sekuler, R. & Levinson, E. (1977). The perception of moving targets. *Scientific American*, **236**, 60–73.

Seymour, A. (1966). *Charles Steinmetz*. Chicago, IL: Follett.

Siler, T. (1990) *Breaking the mind barrier*. New York: Touchstone.

Smith, C. S. (1978). Structural hierarchy in science, art and history. In: J. Wechsler (Ed.), *On Aesthetics in Science* (pp. 9–53). Cambridge, MA: MIT Press.

Smith, C. S. (1981). *A search for structure*. Cambridge, MA: MIT Press.

Stewart, D. (1985). Teachers aim at turning loose the mind's eyes. *Smithsonian* (August), 44–55.

Thompson, D'A. (1930). *Growth and form*. Cambridge: Cambridge University Press.

Van't Hoff, J. H. (1878). Imagination in science. Amsterdam. G. F. Springer (Trans. 1967), *Molecular Biology, Biochemistry and Biophysics*, **1**, 1–18.

Waddington, C. H. (1972). *Biology and the history of the future*. Edinburgh: Edinburgh University Press.

Weisskopf, V. (1977). The frontiers and limits of science. *American Scientist*, **65**, 405–411.

White, R. K. (1931). The versatility of genius. *Journal of Social Psychology*, **2**, 482.

Woolsey, T. A. (1978). C. N. Woolsey—scientist and artist. *Brain, Behavior and Evolution*, **15**, 307–324.

Xenakis, I. (1985). *Arts/sciences: Alloys*. New York: Pendragon Press.

Part IV

Development of Innovation
Across the Life Span

Part IV

Development of Innovative
Areas in Life Span

The International Handbook on Innovation
Edited by Larisa V. Shavinina

Young Inventors

Nicholas Colangelo, Susan Assouline, Laurie Croft, Clar Baldus
and Damien Ihrig

*The Connie Belin & Jacqueline N. Blank International Center for Gifted Education and Talent
Development, The University of Iowa, USA*

Abstract: This chapter focuses on young inventors—children and adolescents. A brief review of
the history and documentation of young inventors is presented. Research is presented on the
young inventors who were part of the Invent Iowa program and who have been evaluated as
meritorious inventors at local and regional invention competitions qualifying for the State of Iowa
Invention Convention. This research includes perceptions of young inventors about the
inventiveness process, their attitudes toward school and toward students, and an analysis of their
inventions. Boys and girls have equally strong interest and equal participation in inventiveness
programs.

Keywords: Innovation; Inventiveness; Flexibility; Originality; Invent Iowa; Iowa Inventiveness
Inventory.

Introduction

When we think of inventors, we typically form the
image of a Thomas Edison or a Henry Ford: adults who
made fame and fortune through their inventions. We
also have more ambiguous images of adults who had
clever ideas but for one reason or another achieved
neither fame nor fortune. These adults may get the
moniker of tinkerer, oddball, or 'ahead of his time'.
Regardless of whether the adult was successful or not,
the constant in the image is the adult (typically a male
adult). It is rare that our image of the inventor is a child.
Yet children and adolescents do invent. We know little
about the young inventor in terms of both the person
and the invention. The focus of this chapter is to 'look'
at young inventors, their inventions, and their attitudes
and perceptions.

Early in the 1980s, the United States Patent and
Trade Office committed to a comprehensive effort to
introduce thinking at all levels of school curricula.
Project XL was initiated in 1985 as a national effort to
create and disseminate new programs and materials to
promote critical and creative thinking and problem-
solving skills fundamental to innovation and invention
(United States Patent and Trade Office, 1997). The
program reflected the emerging belief among leaders in
business and science and technology, as well as in
education, that schools could nurture the creative and

innovative spirit essential for future leaders (Treffinger,
1989). In spite of widespread support for the cultivation of inventiveness among children, however, little
comprehensive research explores the behaviors,
thoughts, and attitudes of school-age inventors. The
literature on inventiveness and invention programs for
children emphasizes thinking skills and creativity and
this makes inventiveness 'invisible'.

Westberg (1996) called inventiveness a "manifestation of creativity" (p. 265), and she suggested
(Westberg, 1998), for example, "to stimulate children's
creative thinking abilities, consider teaching them how
to develop inventions" (p. 18). Rossbach (1999) suggested that through the invention process, "students of
all levels work toward the common goal of realizing
their creative potential" (p. 8). In addition to creative
thinking, other specific skills associated with teaching
children about the inventive process include critical
thinking, problem-finding, problem-solving, decision-
making, and the development of enhanced research
skills (Perkins, 1986; Rossbach, 1999; Treffinger,
1989). Gorman and Plucker (n.d.) found that a
structured inventiveness course enables secondary
students to develop their abilities to collaborate with
team members and to better understand the collaboration, politics, patience and persistence required in the
invention and patent processes.

281

Invention programs for children share attributes that parallel recommendations for stimulating children's creative thinking. Perkins (1986) suggested that educators focus on intrinsic motivation, minimizing evaluative feedback in favor of more informative feedback. Treffinger (1989) emphasized establishing challenging but appropriate expectations and standards, as well as delineating criteria to help students identify their strengths and weaknesses. Treffinger also explored the invention-evaluation process, encouraging formative evaluations that were respectful of the diversity of student processes as well as outcomes.

Inventiveness programs for school-age children share organizational features. Young inventors are typically introduced to:

- the definitions, goals, and purpose related to inventing;
- the societal contributions made by inventions and inventors, frequently including examples of school-age inventors;
- experiences creating Rube-Goldberg kinds of inventions with random available materials;
- the importance of journals or logs;
- activities to help them identify needs or wants;
- skills to research relevant content areas and to evaluate proposed solutions;
- ways to design prototypes or models;
- general strategies for naming inventions;
- audiences for sharing their inventions (Gorman & Plucker, n.d.; Plucker & Gorman, 1995; Shlesinger, Jr., 1982; Treffinger, 1989; Westberg, 1996, 1998).

McCormick (1984) and Gorman and Plucker (n.d.) structured the exploration of inventiveness around historical immersion in the lives of inventors and their inventions. Gorman and Plucker developed online interactive models (Gorman, 1994), enabling students to simulate the 19th century competition between Alexander Graham Bell and Elisha Gray as they pursued different strategies to develop the telegraph. Structured explorations of the invention process allowed children to develop significantly greater numbers of inventions (Westberg, 1996). Also, inventiveness activities have often been evaluated for outcomes related to academic subjects rather than enhanced inventiveness. Shlesinger, Jr. (1982) reported that teaching an inventiveness process enhanced students' general interest in social science and history. McCormick (1984) also found that teaching inventiveness improved student attitudes toward science as well as inspiring student flexibility and originality.

As implied earlier, inventiveness programs for children are submerged under the more familiar school concepts of critical thinking and creativity. Inventiveness (and invention curriculum), while viewed positively in schools, has not become a major part of school and school curriculum. It is viewed as a special or extra-curricular unit. Where it does appear, it seems to be more 'justified' in how it positively affects more typically used school concepts such as creativity. Inventiveness is not well researched among children and adolescents because it is not part of the tradition of school. The State of Iowa has a unique situation whereby its statewide Invent Iowa program and the Invent Iowa curriculum offered to schools allow an opportunity to better understand young inventors and the inventiveness process with this age group. (Note: while some other states have State invention programs, Iowa's is extensive and has been in existence over 15 years.)

The remainder of the chapter provides a brief description of the Invent Iowa program and the research done with young inventors in this program. The research focuses on attitudes about school, and the types of inventions they produce.

Invent Iowa

Invent Iowa is a comprehensive, statewide program developed to assist Iowa's educators in promoting the invention process as part of their regular kindergarten through high school curriculum. The program was initiated in 1987 through the support of state political, business, and educational leaders in response to the future of rapidly expanding technology and the decline in American inventiveness in relation to other nations. Since 1989, Invent Iowa has been administrated through The Connie Belin and Jacqueline N. Blank International Center for Gifted Education and Talent Development at The University of Iowa (Belin-Blank Center).

Invent Iowa supports educators in teaching creative-thinking and problem-solving skills associated with the invention process. As a means of teaching these skills, students are encouraged to identify real-world problems and develop their own inventions that successfully solve these problems. This multidisciplinary process enables students to use several academic skills in combination: reading, library and field research, science and technology, creative and critical thinking, writing, art, and persuasive speaking. Students who participate in the program have the opportunity to display their inventions at their local, regional, and state invention conventions. Each year an estimated 30,000 students in grades K-12 participate in the Invent Iowa program. Invent Iowa sponsors a series of 'invention conventions' at the local, regional, and state levels. At each level, inventions are evaluated, and students are recognized for their achievements. The *Invent Iowa State Invention Convention*, organized by the Belin-Blank Center, is the pinnacle of inventiveness competitions in Iowa. Over 300 students qualify and are invited to the *State Invention Convention* each spring.

Invent Iowa is not limited by gender, race, class, academic standing, or any other classification. Invent Iowa helps students gain the sense of accomplishment and purpose vital to the development of a creative, competent, and involved workforce.

The Invent Iowa Evaluation Rubric

Students participating in Invent Iowa must submit their invention and a detailed and structured journal of this invention. Additionally, a student must be able to articulate the purpose and process employed in completing his or her invention. The inventor's journal is also signed by a school representative and a family member, confirming that the invention is the idea and work of the student.

Students are evaluated at the local, regional, and state level according to the criteria of the Invent Iowa Evaluation Rubric (Rubric) (See Appendix A). The Rubric is used by the judges to evaluate the actual invention in terms of the dimensions of appeal, usefulness, and novelty (criteria for invention).

Each of the criteria dimensions is evaluated according to four levels of mastery, from Expert Inventor (highest rating) to Beginner Inventor (lowest rating). The other part of the Rubric (criteria for Entry) guides judges in evaluating the visual, written and oral presentation by a student of his or her invention.

The Rubric is designed to provide subscores, as well as a total score. The subscores are particularly useful in providing the young inventor specific feedback on the various aspects of the invention and the presentation. Judges provide oral and written feedback to the inventor. The total score is used to determine the local and regional qualifiers for State, and at the State Invention Convention to determine the special Merit Awards.

Research with Young Inventors

Much of what we know about young inventors has to do with whether or not they 'win' an invention contest. While success in such competitions tells us something useful about the skill of the young inventor, it does not inform us about the psychological profile of the inventor including his or her attitudes, feelings, interest, motivation, etc. The research reported in this study informs on some of the perceptions and attitudes of young students who have demonstrated their inventiveness in major competitions.

A unique feature of this chapter is its focus on young inventors and research with young inventors. The research is based on those students, grades 3–8, who qualified for the Iowa State Invention Convention in the year 2001. Each of these students had to receive merit recognition at the local and regional level in order to be selected for State. Also, all competitors were part of the Invent Iowa program. In the year 2001, approximately 30,000 students in K-8 participated in local and regional conventions. Of this group 364 were selected for the State Convention in grades 3–8. (Note: Students in grades K-2 can only participate in local and regional conventions. Students are not eligible for state com-

petition until they reach 3rd grade. Students in grades 9–12 did not participate in the research.)

For the remainder of this chapter the term 'young inventors' will refer to those students from grades 3–8 who were selected for the Iowa State Invention Convention. They met the criterion for 'inventor' in that they had to compete against a large number of students in order for their inventions to qualify for State.

The Invent Iowa Survey—2001

The students in grades 3–8 who qualified for the Iowa State Invention Convention in 2001 were asked to complete the Invent Iowa Survey (IIS). The IIS is a self-report instrument developed at the Belin-Blank Center, which assesses the perceptions of young inventors in the following:

(a) perceptions about the inventiveness process;
(b) attitudes about school;
(c) attitudes about status of being a young inventor.

Of the 364 students who qualified for the Iowa State Invention Convention, 294 (81%) completed the IIS. Table 1 indicates by grade and gender those who qualified for the State Convention and who completed the IIS.

Stereotypically, inventing and entering invention contests have been associated more with boys than with girls. Of the 364 students who qualified for the State Convention ($N = 207$ girls; $N = 157$ boys), girls are very much part of the Invent Iowa program and have been very successful in making it to the State Convention. (Note: The minority population of the State of Iowa is about 6% of the general population. Of those who participated in 2001, 6% minorities qualified for the State Convention. Since the numbers were small, we did not do any separate analyses by ethnicity, only by gender.)

Table 1. Young inventors who completed the Invent Iowa Survey in 2001.

Grade	Males (Response Rate)	Females (Response Rate)	Total (Response Rate)
3	14 (93%)	14 (78%)	28 (85%)
4	26 (76%)	41 (75%)	67 (75%)
5	45 (78%)	43 (78%)	88 (78%)
6	21 (84%)	45 (83%)	66 (84%)
7	13 (100%)	16 (89%)	29 (94%)
8	11 (92%)	5 (71%)	16 (84%)
Total	130 (83%)	164 (79%)	294 (81%)

The Inventiveness Process

In 1992, Colangelo, Kerr, Huesman, Hallowell and Gaeth developed the Iowa Inventiveness Inventory (III), which measured attitudes and perceptions of adult inventors (males only) who held one or more patents. The III included items that tapped perceptions of the inventiveness process among adult inventors. These items (1–19) were included in the Invent Iowa Survey (IIS) and were used to elicit attitudes of young inventors regarding the inventiveness process. For each of the items (1–19) a student selected from one of five choices:

(1) Strongly Disagree
(2) Disagree
(3) Neutral
(4) Agree
(5) Strongly Agree

A multivariate analysis of variance (MANOVA) was performed to look for significant global differences between males and females in their mean responses to

items 1–19. In other words, when one examines all of the items at the same time, is there a difference between the responses of males and females? A MANOVA procedure was significant at the $p < 0.05$ level ($F(19,251) = 2.58$), indicating a difference between the answers of males and females on items 1–19. The eta-squared value is a measure of effect size, or 'practical significance'. It indicates that, although the global test was significant, only a small proportion (16%) of the variance in all the dependent variables is accounted for by the differences between the two groups. Thus, although there was a difference between males' and females' mean responses, the practical significance of the difference may be questioned.

A univariate analysis of variance was performed as a follow-up to the significant MANOVA (Table 2). This was done to identify the individual items males and females were responding to differently. The analysis

Table 2. Univariate analysis of variance for items 1–19 on the 2001 Invent Iowa Survey.

Item	Males (Mean)	Females (Mean)	Type I Sum of Squares	df	Mean Square	F	Sig.	Eta-Squared
1. I like to hang around places where people are making or fixing things	3.76*	3.48	4.90	1	4.90	5.67	0.018*	0.0206
2. I can figure things out that others have trouble understanding	4.02	3.98	0.03	1	0.03	0.05	0.830	0.0002
3. I need to get a lot of ideas in order to get a really good idea	3.09*	3.39	5.59	1	5.59	4.19	0.042*	0.0153
4. When I start working on a project, I just won't quit	3.82	3.88	0.09	1	0.09	0.10	0.757	0.0004
5. I want to invent only things which can help people	3.40	3.40	0.19	1	0.19	0.14	0.708	0.0005
6. Inventiveness is something you can be taught. (Reversed)	3.55	3.65	0.48	1	0.48	0.38	0.539	0.0014
7. It seems like too much school could cause you to lose your creativity	2.39	2.25	1.11	1	1.11	0.78	0.378	0.0029
8. I like to be alone when I am thinking of new ideas	3.03	3.13	0.56	1	0.56	0.42	0.520	0.0015
9. I sometimes have a lot of projects going at once	3.61	3.80	1.83	1	1.83	1.33	0.249	0.0049
10. I like to have a place to make things	3.75	3.61	1.42	1	1.42	1.35	0.246	0.0050
11. When I think of a new thing or machine, I can 'see' the complete idea in my mind before I start working on it	4.05	3.76	4.00	1	4.00	3.77	0.053	0.0138
12. My best friend and I build things together	3.09	3.04	0.03	1	0.03	0.02	0.880	0.0001
13. Whenever I look at a machine, I can see how to change it	3.09	2.84	3.47	1	3.47	3.68	0.056	0.0135
14. I don't really consider myself a good student, but I do have the ability to solve problems	2.40	2.16	5.12	1	5.12	3.56	0.060	0.0130
15. When I decide to solve a problem, my mind is full of that problem	3.60	3.45	0.69	1	0.69	0.65	0.422	0.0024
16. You can't explain inventiveness	3.27	3.17	0.34	1	0.34	0.24	0.628	0.0009
17. You don't have to have good grades in school to be a good inventor	4.00	3.83	1.68	1	1.68	1.14	0.286	0.0042
18. I am a person who likes to take things apart	4.07*	3.21	45.00	1	45.00	29.89	0.0000001*	0.1000
19. I am more interested in how a toy or object works than in playing with it	3.25*	2.70	16.28	1	16.28	12.64	0.0004*	0.0449

* $p < 0.05$.

Table 3. Univariate analysis of variance for items 20–22 on the 2001 Invent Iowa Survey.

Dependent Variable	Males (Mean)	Females (Mean)	Type I Sum of Squares	df	Mean Square	F	Sig.	Eta-Squared
20. It is important to keep notes about my invention	3.61*	3.97	9.95	1	9.95	8.31	0.004*	0.028
21. I like to draw my invention before I start building it	3.70	3.70	0.02	1	0.02	0.02	0.903	0.0001
22. I like to make a smaller model of my invention before I start building it	2.52	2.61	0.61	1	0.61	0.47	0.493	0.002

* Significant difference between male and female responses $p < 0.05$.

indicates mean responses by males and females on items (1) 'I like to hang around places where people are making or fixing things', (3) 'I need to get a lot of ideas in order to get a really good idea', (18) 'I am a person who likes to take things apart', and (19) 'I am more interested in how a toy or object works than in playing with it' were significantly different. Males tended to indicate stronger agreement with items 1, 18, and 19 than females. Females tended to indicate stronger agreement with item 3. The measures of effect size (eta-squared) for the responses to items 1 (2%), 3 (1.5%), 18 (10%), and 19 (5%) were small, however.

From the research of Colangelo et al. (1992) and Colangelo, Assouline, Kerr, Huesman and Johnson (1993) the higher the mean on each item, the more the response is consistent with the responses of adult inventors with patents. Two items elicited a different response from young inventors verses adult inventors: item 7 'It seems like too much school could cause you to lose your creativity', and item No. 14 'I don't really consider myself a good student, but I do have the ability to solve problems'. Adult inventors were very negative about the role of school (see Colangelo et al., 1992, 1993) and did not see themselves as good academic students, compared to young inventors. Perhaps the school environment has been changing and is currently more conducive to creative and inventive endeavors.

On all items except four, boys and girls were similar. On item 3 girls were more like adult inventors than boys and on item 7, 18 and 19, boys were more like adult inventors than are girls. But the important finding here is that young inventors are similar in perceptions about the inventiveness process regardless of gender.

Item 20, 21, 22 also come from Colangelo et al. (1992) and focus on *initiating* the inventiveness process. Table 3 indicates that girls are more likely to keep notes (as do adult inventors), and both boys and girls are likely to want to draw their concepts before building. Unlike adult inventors, young inventors are not inclined to want to make a model before building.

A multivariate analysis (MANOVA) was performed to look for significant global differences between males and females in their mean responses to items 20–22.

Again, when one examines all of the items at the same time, there is a difference between the responses of males and females. A MANOVA procedure was significant at the $p < 0.05$ level ($F(3,283) = 3.24$), indicating a difference between the answers of males and females on items 20–22. The eta-squared value indicates that, although the global test was significant, only a small proportion (3%) of the variance in all the dependent variables is accounted for by the differences between the two groups.

A univariate analysis of variance was performed as a follow-up to the significant MANOVA. It can be seen from Table 3 that the global significant result can be attributed to item 20 'It is important to keep notes about my invention'. Females tended to indicate more agreement with this statement then males. Again, the proportion of variance (about 3%) accounted for by the difference in responses between males and females is small.

Attitudes About School

The fact of the matter is that young inventors are going to be in school and they will be in school for the vast majority of their young life. Thus, their feelings about school are important. Colangelo, Assouline, Kerr, Hussman and Johnson (1993) reported that adult inventors did not have positive memories and attitudes

Table 4. Young inventors' attitudes about school: by gender.

How do you feel about school in general?	Gender*			
	Males		Females	
	N	%	N	%
A. I love it	27	21.3	49	30.4
B. I like it	89	70.1	105	65.2
C. I don't like it	9	7.1	6	3.7
D. I don't like it at all	2	1.6	1	0.6
Total	127	100	161	100

* Percentages may not add to 100 because of rounding.

Table 5. Young inventors' attitudes about school: by grade.

How do you feel about school in general?	Grade*											
	3		4		5		6		7		8	
	N	%	N	%	N	%	N	%	N	%	N	%
A. I love it	11	40.7	19	28.8	26	30.2	15	23.1	5	17.9	0	0.0
B. I like it	13	48.1	42	63.6	57	66.3	48	73.8	18	64.3	16	100.0
C. I don't like it	2	7.4	5	7.6	3	3.5	2	3.1	3	10.7	0	0.0
D. I don't like it at all	1	3.7	0	0.0	0	0.0	0	0.0	2	7.1	0	0.0
Total	27	100	66	100	86	100	65	100	28	100	16	100

* Percentages may not add to 100 because of rounding.

about school. Most of their recollections were either neutral or negative about school and its importance to their inventiveness.

In responding to an item on the IIS regarding attitude about school, young inventors were overwhelmingly positive. Table 4 reports the general attitude about school by gender. While girls are a bit more positive, especially in 'loving school', the pattern is consistent for boys and girls. The general attitudes about school are also presented by grade (see Table 5) and indicate no unique patterns other than highly positive at each grade.

Attitudes About Types of Students

One of our interests was to understand how young inventors viewed the hierarchy of types of students in their school. Putting it in basic terms, we wanted to know who the young inventors thought were the 'cool' students in school.

The item represented in Table 6 asked young inventors to rank the types of students that seem to be most attractive to the kids in school. Athletes were ranked highest (most attractive) by both males and females, and Hard Working (surprisingly) was ranked 2nd. Class Comedian and Popular were next. Inventor was ranked low on the list. Interestingly, however, the girls ranked it higher. We also did an analysis by grade (see Table 7), but no strong patterns emerged. Looking at the category of Inventor, this was rated less attractive by the 7th and 8th graders and more attractive by elementary school students. There may be a trend regarding the attractiveness of being an inventor by age and grade. This trend would need to be further investigated since the numbers in this study were too small to warrant definitive statements.

So, what do young inventors invent? We thought it would be revealing to do a systematic analysis of the types of inventions that were displayed at the State Invention Convention in 2001. There were 288 inventions by 364 young inventors at State in 2001. (The reason there are more inventors than inventions is that

students can team up on an invention (maximum of 2 inventors on a team) and we had several teams.)

The 288 inventions were classified using the general categories developed by the United States Patent Office to categorize patents. The detailed definitions of each category can be found by going to the U.S. Patent Office Website at www.USPTO.gov. The descriptions are much too extensive to include in a chapter. However, we have provided a very brief description of each category (see Appendix B) to give a sense of what type of inventions would be included in each category. Each invention was analyzed by a team of three raters in order to determine which category it fit best. Table 8 provides the results of that analysis by gender and grade.

The most popular categories of inventions were Tools, Kitchen/Bath, and Organization, which accounted for 141 of 288 inventions (49%). The

Table 6. Perceptions of attractiveness of types of students: by gender.

What do you think kids your age see as the 'coolest' thing to be in school?	Gender*			
	Males		Females	
	N	%	N	%
Artist	23	6.3	20	4.4
Athlete	94	25.9	119	26.0
Class Comedian	51	14.0	52	11.4
Class Leader	21	5.8	33	7.2
Good Writer	5	1.4	8	1.7
Hard Worker	82	22.6	111	24.2
Inventor	11	3.0	29	6.3
Math Wiz	18	5.0	8	1.7
Musician	15	4.1	14	3.1
Really Smart Student	11	3.0	11	2.4
Popular	32	8.8	53	11.6
Total	363	100	458	100

* Percentages may not add to 100 because of rounding.

Table 7. Perceptions of attractiveness of types of students: by grade.

What do you think kids your age see as the 'coolest' thing to be in school?	Grade*											
	3		4		5		6		7		8	
	N	%	N	%	N	%	N	%	N	%	N	%
Artist	9	11.1	8	4.0	13	5.5	11	6.0	0	0.0	2	4.4
Athlete	22	27.2	53	26.8	53	22.5	49	26.6	23	29.9	13	28.9
Class Comedian	7	8.6	15	7.6	30	12.7	29	15.8	16	20.8	6	13.3
Class Leader	5	6.2	16	8.1	16	6.8	9	4.9	4	5.2	4	8.9
Good Writer	1	1.2	4	2.0	4	1.7	3	1.6	1	1.3	0	0.0
Hard Worker	17	21.0	44	22.2	51	21.6	51	27.7	21	27.3	9	20.0
Inventor	3	3.7	12	6.1	18	7.6	6	3.3	0	0.0	1	2.2
Math Wiz	2	2.5	8	4.0	7	3.0	5	2.7	0	0.0	4	8.9
Musician	2	2.5	10	5.1	10	4.2	2	1.1	3	3.9	2	4.4
Really Smart Student	2	2.5	2	1.0	7	3.0	9	4.9	0	0.0	2	4.4
Popular	11	13.6	26	13.1	27	11.4	10	5.4	9	11.7	2	4.4
Total	81	100	198	100	236	100	184	100	77	100	45	100

* Percentages may not add to 100 because of rounding.

Table 8. 2001 state invention convention: categories of inventions by gender.

Category	2001 Invention Convention Categories by Gender					
	Gender*					
	Male		Female		Total	
	N	%	N	%	N	%
Tools	24	18.6	26	16.4	50	17.4
Kitchen/Bath	19	14.7	30	18.9	49	17.0
Organization	14	10.9	28	17.6	42	14.6
Clothes/Accessories	11	8.5	14	8.8	25	8.7
Safety/Protection/Rescue	11	8.5	12	7.6	23	8.0
Farm	12	9.3	10	6.3	22	7.6
Amusement	12	9.3	7	4.4	19	6.6
Pets	9	7.0	10	6.3	19	6.6
Automotive	4	3.1	5	3.2	9	3.1
Furniture	2	1.6	6	3.8	8	2.8
Medical	4	3.1	4	2.5	8	2.8
Cleaning	2	1.6	4	2.5	6	2.1
Electronic	4	3.1	1	0.6	5	1.7
Disabled	1	0.8	2	1.3	3	1.0
Total	129	100	159	100	**288**	**100**

categories of Clothes/Accessories, Safety/Protection/ Rescue, and Farm accounted for 70 inventions (24%). Categorized inventions by gender were fairly comparable except in the categories of Kitchen/Bath and Organization where the girls outnumbered boys in these categories by slightly less than 2:1. Proportionately, inventions in the category of Tools more often were submitted by boys.

Discussion/Summary

From our research there are some issues that emerge:

(1) It is a stereotype that 'inventing' is a boy thing. Boys and girls are comparably inventive and interested in participating in invention contests;

(2) Inventive boys are similar to inventive girls in their attitudes about school and inventiveness;

(3) In contrast to well-established adult inventors, the young inventors in our study had a positive attitude about school and their academic abilities;

(4) The inventions of young inventors run a gamut of the classifications of the U.S. Patent Office. There are some gender differences in the classifications of types of inventions generated by young inventors.

The inventiveness spirit is quite alive and well in children and adolescents. Programs like Invent Iowa give expression and outlet to this spirit. It is important to promote inventiveness programs because inventiveness *is a talent* not usually identified and nourished in the traditional curriculum of schools. Historically, we have envisioned ourselves as a nation of 'doers', of innovative and practical people. It seems the accuracy of this is reflected in our young inventors.

References

Colangelo, N., Assouline, S. G., Kerr, B. A., Huesman, R. & Johnson, D. (1993). Mechanical inventiveness: A three-phase study. In: G. R. Bock & K. A. Krill (Eds), *The Origins and Development of High Ability* (pp. 160–174). Chichester, U.K.: John Wiley.

Colangelo, N., Kerr, B. A., Huesman, R., Hallowell, K. & Gaeth, J. (1992). The Iowa Inventiveness Inventory: Toward a measure of mechanical inventiveness. *Creativity Research Journal*, **5** (2), 157–163.

Gorman, M. (1994). The invention of the telephone: An active learning module. Retrieved August 4, 1999, from http://jefferson.villiage.virginia.edu/~meg3c/id/TCC315/intro.html

Gorman, M. & Plucker, J. (n.d.). Teaching invention as critical creative processes: A course on Technoscientific creativity. Manuscript submitted for publication.

McCormick A. J. (1984). Teaching inventiveness. *Childhood Education*, **60**, 249–255.

Perkins, D. N. (1986). *Knowledge as design*. Hillsdale, NJ: Lawrence Erlbaum.

Plucker, J. & Gorman, M. (1995). Group interaction during a summer course on invention and design for high ability secondary students. *The Journal of Secondary Gifted Education*, 258–271.

Rossbach, J. (1999). Inventive differentiation. *NRC/GT Newsletter* (Electronic version). Retrieved from http://www.sp.uconn.edu/~nrcgt/news/fall99/fall993.html

Shlesinger, Jr., B. E. (1982). An untapped resource of inventors: Gifted and talented children. *The Elementary School Journal*, **82** (3), 215–220.

Treffinger, D. J. (1989). *Student invention evaluation kit—field test edition*. Honeoye, NY: Center for Creative Learning.

United States Patent and Trade Office, Special Projects Assistant (1997). *The inventive thinking curriculum project: An outreach program of the United States Patent and Trademark Office* (Electronic version). Washington, D.C.: Author. Retrieved from http://www.ed.gov/inits/teachers/Eisenhower/

Westberg, K. L. (1996). The effects of teaching students how to invent. *Journal of Creative Behavior*, **30** (4), 249–267.

Westberg, K. L. (1998). Stimulating children's creative thinking with the invention process. *Parenting for High Potential*, **18–20**, September, 25.

Appendix A: Invent Iowa Evaluation Rubric

CRITERIA FOR INVENTION — LEVELS OF MASTERY

CRITERIA FOR INVENTION	Expert INVENTOR 5	Skillful INVENTOR 4	Amateur INVENTOR 3	Beginner INVENTOR 2	POINTS
Novelty — A. **Does the invention involve a *novel* idea?**	A significant level of difference between this invention and prior products — 5	A substantial level of difference between this invention and prior products — 4	**Some simple differences** between this invention and prior products — 3	**Very similar** to prior Products — 2	Novelty A ___ +
Novelty — B. **Is the invention a *fresh or unexpected* idea?**	Unique and exciting product idea — 5	**Very interesting** product idea — 4	**Attractive but predictable** idea — 3	**Traditional** product idea — 2	B ☐ =
Usefulness — C. **Is the invention *workable*?**	**Clear and convincing** evidence that this invention will work effectively — 5	Sufficient evidence that this invention will work effectively — 4	**Minimal evidence** that this invention will work effectively — 3	**Little evidence** that this invention will work effectively — 2	Useful C ___ +
Usefulness — D. **Is the invention *appropriate* for the stated need or idea?**	A clear and convincing connection between the problem or idea and the invention — 5	Sufficient evidence of a connection between the problem or ideas and the invention — 4	**Minimal evidence** of a connection between the problem or idea and the invention — 3	**Little evidence** of a connection between the problem or idea and the invention — 2	D ☐ =
Appeal — E. **Does the invention have strong *interest* and *appeal* to intended audience?**	Professional in appearance and well-suited to the intended audiences — 5	Attractive and **useful** for the intended audiences — 4	**Useful to some people** in the intended audience — 3	Idea might have **potential** interest and appeal — 2	Appeal E ___ +
Appeal — F. **Is the invention *well-crafted and complete*?**	**Clear and convincing** evidence that materials and construction are the best for the invention — 5	Sufficient evidence that materials and construction can be used for the invention — 4	**Minimal evidence** that materials and construction can be used for the invention — 3	**Little evidence** that materials and construction can be used for the invention — 2	F ☐ =
Invention Point Subtotals	___	___	___	___	

Appendix A—Continued

CRITERIA FOR ENTRY

	LEVELS OF MASTERY				POINTS
	Expert INVENTOR 5	*Skillful* INVENTOR 4	*Amateur* INVENTOR 3	*Beginner* INVENTOR 2	
G. Is the invention entry *diagram* presented professionally?	**Elaborate and attractive** diagram with all parts clearly labeled and explained 5	An **attractive** diagram with all parts labeled 4	Diagram with **most** parts labeled 3	A **simple** drawing of the invention 2	**Invention Presentation** G ____ +
H. (1) Is the invention entry *model* a clear idea representation?	A **highly-detailed** and **comprehensive** representation of the Invention 5	A **comprehensive** representation of the invention 4	An **adequate** representation of the invention 3	A **simplified** representation of the invention 2	H (1) ____ +
OR					**OR**
H. (2) Is the invention entry *prototype* an exact replica?	A **highly-detailed** and **comprehensive** working replica of the invention 5	A **comprehensive** working replica of the invention 4	An **adequate** working replica of the invention 3	A **simplified** working replica of the invention 2	H (2) ____ +
I. Is the Inventor's *Log* thorough and complete?	A **well-developed** description of the invention process 5	A **description** containing the **highlights** of the invention process 4	**Brief statement** of the invention idea and : solution 3	A **partial description** of the invention 2	I ____ +
J. Is the Inventor's *oral presentation* thorough and complete?	Communicates a **high level** of knowledge and understanding of the process leading to this invention 5	Communicates **some knowledge** and understanding of the process leading to this invention 4	**Describes** the invention idea and the solution 3	**Briefly describes** the invention 2	J ____ □ =
Entry Point Subtotals	____	____	____	____	

Appendix A—Continued

Subtotals	Evidence of Expertise	Evidence of Skillfulness	Evidence of Amateur Status	Evidence of Beginner Status	POINTS
Invention Subtotals from Page 1	☐	☐	☐	☐	☐
Entry Subtotals from page 2	☐	☐	☐	☐	☐
TOTAL SCORE	☐	☐	☐	☐	☐

Comments:

Appendix B: Brief Definitions of U.S. Patent and Trade Office Patent Classifications

(1) Tools:

Inventions categorized as *Tools* included hand tools and presses.

(2) Kitchen and Bath:

Inventions categorized as *Kitchen and Bath* included devices or methods used in baths, closets, sinks, and spittoons, cutlery, refrigeration, food and beverage preparation, treating, and preservation, and power-driven conveyors.

(3) Organization:

Inventions categorized as *Organization* included packages, flexible/portable closures, partitions, or panels, special receptacles or packages, and support racks.

(4) Clothes and Accessories:

Inventions categorized as *Clothes and Accessories* included apparel, boots and shoes, buckles, and jewelry.

(5) Safety, Protection, and Rescue:

Inventions categorized as *Safety, Protection, and Rescue* included equipment used for body restraint or protective covering, rescue, or as a fire extinguisher or casing.

(6) Farm:

Inventions categorized as *Farm* included methods or devices used in animal husbandry, methods and structures used to raise and care for bees, and devices or methods for crop threshing or separating.

(7) Amusement:

Inventions categorized as *Amusement* included sporting goods, toys, games, and devices or methods related to music.

(8) Pets:

Inventions categorized as *Pets* included harnesses, fluid handling, farriery, and dispensing of solids.

(9) Automotive:

Inventions categorized as *Automotive* included devices and methods used by or for motor vehicles.

(10) Furniture:

Inventions categorized as *Furniture* included devices for supporting the weight of a person in a seated position.

(11) Medical:

Inventions categorized as *Medical* included devices used in a variety of medical situations.

(12) Cleaning:

Inventions categorized as *Cleaning* included devices or chemicals used for the removal of foreign material.

(13) Electronic:

Inventions categorized as *Electronic* included devices for producing light or related to television functioning.

(14) For the Disabled:

Inventions categorized as *For the Disabled* included devices or methods used to accommodate people with physical disability.

The International Handbook on Innovation
Edited by Larisa V. Shavinina

Exceptional Creativity Across the Life Span: The Emergence and Manifestation of Creative Genius

Dean Keith Simonton

Department of Psychology, University of California, Davis, USA

Abstract: Major innovations in the arts and sciences can be largely attributed to the output of creative geniuses. But how do such great innovators emerge? And how does their creativity manifest itself? The first question shall be addressed by examining the early experiences that contribute to the development of extraordinary creative potential. The factors include family background, education, and professional training. The response to the second question concentrates on the typical career trajectory of illustrious creators. Features of this trajectory include the ages at which geniuses tend to produce their first great work, their best work, and their last great work.

Keywords: Age; Creativity; Genius; Creative potential; Career trajectories; Life-span development.

Introduction

The terms 'creativity' and 'innovation' are sometimes used interchangeably, and other times are considered to represent quite distinct phenomena. However, creativity involves the capacity to produce some idea or product that is both original and functional. By the same token, innovation involves the act of introducing something new. Hence, creative individuals are necessarily innovators, and their ideas or products can be considered innovations. However, sometimes researchers prefer to distinguish between the origination of a new idea and the dissemination or adoption of that idea by others. This distinction is especially useful when discussing technological change. It is one thing for an inventor to devise and patent a 'better mousetrap', but quite another for that mousetrap to become widely adopted in households or businesses. A person who adopts the new mousetrap is then called an 'innovator' even when he or she had absolutely nothing to do with its creation. However, this distinction between creators and innovators becomes much less tenable when we examine other domains of achievement in the arts and sciences. For example, when Albert Einstein applied Max Planck's new quantum theory to explain the photoelectric effect, Einstein was acting as an innova-

tor in the sense that he was adopting and disseminating a new theory. Yet Einstein's innovation only has significance because it took the form of a creative product—a novel and successful treatment of a critical phenomenon. Indeed, so important was Einstein's application that it earned him a Nobel Prize, just as Planck had received one for the original theory. Speaking more generally, innovation usually takes place when one creative product becomes the basis for another creative product. Hence, in this chapter I shall use creativity and innovation as essentially equivalent terms.

Creativity or innovation can be studied from several different perspectives. Some researchers investigate the phenomenon from the standpoint of the psychological processes that underlie the origination of a creative product or innovation (e.g. Kaufmann, 2003; Root-Bernstein & Root-Bernstein, 2003; Weisberg, 2003). Other investigators examine the characteristics of the products that emerge from these processes (e.g. Simonton, 1980c, 1986c; Sternberg, Pretz & Kaufman, 2003). Yet other researchers concentrate on the attributes of the person that enable him or her to engage those processes or generate those products. It is this person-approach that I will adopt in the present chapter. In particular, I wish to scrutinize those individuals who seem to exhibit the highest levels of this creative capacity, namely, those who display genius-level creativity.

The Creative Genius

Ever since Galton's (1869) pioneering work, psychologists have become accustomed to look at the distribution of human capacities in terms of the normal 'bell-shaped' curve. Most people are of average ability, and persons with higher-than-average or lower-than-average abilities become more rare to the degree that they depart from the population mean. Although the capacity for creativity might be viewed in the same fashion (Nicholls, 1972), such a conception has no empirical or theoretical support (Simonton, 1997b, 1999c). If creative ability is gauged according to the production of major innovations—of outright creative products in a particular domain of achievement—then the distribution of that ability is far from normal (Dennis, 1954a, 1954c, 1955; H. Simon, 1955). On the contrary, the cross-sectional distribution of total output is highly skewed right, so that a small proportion of the creators is credited with a lion's share of the innovations produced. In concrete terms, the top 10% of the innovators in a particular domain are typically responsible for about half of everything produced, whereas the bottom 50% can usually only claim about 15% of the contributions. Indeed, the mode of the distribution is almost invariably a single innovation only. Most inventors have only one patent; most poets publish only one poem. Moreover, the most prolific contributors to a domain are at least a 100 times more productive than their least productive colleagues. As Cesare Lombroso (1891) once affirmed in his classic *The Man of Genius*, "the appearance of a single great genius is more than equivalent to the birth of a hundred mediocrities" (p. 120). Creative productivity clearly displays a highly elitist distribution.

In line with Lombroso's remark, we can refer to those individuals who make up the productive elite as the *creative geniuses* of their chosen achievement domain. A prototypical example is Thomas Edison, whose output of inventions was so prodigious that he still holds the record for patents at the United States Patent Office. Moreover, among his more than a thousand inventions are several that radically transformed modern civilization, such as the microphone, the phonograph, the motion picture, and the incandescent lamp. Edison's status as an inventive genius is unquestionable.

The goal of this chapter is to examine creative genius from the standpoint of lifespan developmental psychology. From this perspective, the life of outstanding creators consists of two major phases (Simonton, 1975b, 1997b). The first phase represents the period in which the individual acquires the necessary capacity for creativity—what may be styled 'creative potential'. The second phase encompasses the period in which this acquired potential becomes actualized in the form of creative products, whether those products be discoveries, inventions, treatises, novels, plays, poems, paintings, or musical compositions. This chapter then concludes with a discussion of the factors involved in the final termination of this second phase—the creator's death and the works that close the creator's career.

The Development of Creative Potential

"Genius must be born, and never can be taught", claimed John Dryden (1693/1885, p. 60). When the English dramatist said this he was expressing an idea already centuries old. However, the first scientist to subject this belief to empirical test was Francis Galton in his 1869 *Hereditary Genius*. Galton accomplished this by examining the family pedigrees of eminent achievers in a diversity of domains, including outstanding creators in science, literature, music, and painting. According to the results of his systematic statistical analysis, creative geniuses are highly likely to come from family lines that contain other eminent individuals, very often in the same domain of accomplishment. Moreover, the likelihood of such a distinguished pedigree is far greater than would be expected according to any reasonable baseline. Although this finding has been replicated many times (e.g. Bramwell, 1948), the theoretical significance of these results has been often contested. In fact, the first major attack on Galton's conclusions came only a few years later, when Candolle (1873) published an empirical study of the environmental factors that contribute to the emergence of creative genius in the sciences. Candolle's investigation inspired Galton (1874) to conduct his own inquiry into the origins of scientific creativity. The resulting book was called *English Men of Science: Their Nature and Nurture*, the subtitle suggesting that Galton had backed off a little from his extreme genetic determinism. At the same time, Galton's 1874 study introduced the 'nature–nurture issue' into the behavioral sciences (cf. Teigen, 1984). Henceforth, investigators would need to determine the relative impact of genetic and environmental influences in the development of creative potential.

Below I review some of the key findings regarding this problem. I begin by discussing the role of the environment and then end by discussing the impact of genetic endowment.

Environmental Factors

Since the time of Galton (1874) and Candolle (1873), researchers have unearthed an impressive inventory of circumstances and conditions that appear to nurture the acquisition of creative potential (Simonton, 1987a, 1994). These diverse influences may be roughly grouped into three categories: family, school, and society.

Family Background

Galton's (1874) pioneer investigation devoted a considerable amount of scrutiny to the home environments of his highly eminent subjects. This emphasis was well placed. The home dominates infancy and childhood, and continues to exert some influence all the way through adolescence. One of the most critical of these long-term factors appears to be the socioeconomic status of the parents. Specifically, creative individuals tend to come from professional or entrepreneurial homes in which the parents place a high value on learning and education (e.g. Chambers, 1964; Cox, 1926; Ellis, 1926; Galton, 1874; Roe, 1953). This value almost invariably takes the form of a home replete with opportunities for intellectual stimulation, such as a large and varied library, ample and diverse magazine subscriptions, and family outings to museums and galleries (see also Schaefer & Anastasi, 1968; Terman, 1925). Not surprisingly, the children raised in such homes exhibit deep and varied interests, with a particularly strong inclination toward omnivorous reading, which makes a powerful contribution to creative development (Goertzel, Goertzel & Goertzel, 1978; McCurdy, 1960; Simonton, 1986b).

At the same time, families of religious or ethnic minorities produce eminent creators out of proportion to their representation in the population (Galton, 1869; Hayes, 1989; Helson & Crutchfield, 1970). A case in point is the conspicuous representation of Jews among the eminent creators of European civilization (Arieti, 1976; Hayes, 1989). However, this asset only appears when minorities enjoy some of the same basic rights as the majority (Hayes, 1989; Simonton, 1998a).

Interestingly, some family background variables assume a major role in determining the domain in which creative potential develops. One example is birth order. Galton (1874) indicated that his eminent scientists were more likely to be firstborns, but subsequent work suggests that the relation between ordinal position and creativity is more complicated than that (Clark & Rice, 1982; Ellis, 1926; Sulloway, 1996). Firstborns appear to gravitate to those domains that are high in prestige, power, and respect for authority, whereas laterborns tend to go into those areas that encourage more independent, individualistic, high-risk, and iconoclastic forms of creative achievement (Simonton, 1999b; Sulloway, 1996). Hence, scientific creators are more likely to be firstborns (Eiduson, 1962; Galton, 1874; Roe, 1953; Terry, 1989), whereas artistic creators are more likely to be laterborns (Bliss, 1970; see also Eisenman, 1964). If a scientist is a laterborn, he or she is more likely to become a revolutionary rather than a defender of the received paradigms (Sulloway, 1996). Likewise, more conservative forms of artistic creativity, such as classical music, tend to attract firstborns

more than laterborns (Schubert, Wagner & Schubert, 1977).

Another interesting example concerns the impact of traumatic experiences. Surprisingly, exceptional creativity does not always emerge from the most favorable home environments (e.g. Eisenstadt, 1978; Goertzel, Goertzel & Goertzel, 1978; Walberg, Rasher & Parkerson, 1980). On the contrary, creative potential seems to require a certain amount of exposure to: (a) diversifying experiences that help weaken the constraints imposed by conventional socialization; and (b) challenging experiences that help strengthen a person's capacity to persevere in the face of obstacles (Simonton, 1994). Yet these developmental inputs seem especially important for artistic forms of creative behavior (Berry, 1981; Brown, 1968; Post, 1994; Raskin, 1936; Simonton, 1986b). Not only is artistic creativity more unrestrained than scientific creativity, but in addition the career paths for artists often require more struggle than the typical career path in science.

Education and Training

Research on talent development has pointed to the supreme importance of training (Howe, Davidson & Sloboda, 1998). That is, one does not acquire world-class competence in fields like sports, chess, or music performance without first devoting about a decade to extensive and deliberate practice (Ericsson, 1996). To a certain extent, the same '10-year rule' applies to creative genius (Gardner, 1993; Hayes, 1989; Simonton, 1991b). Creative individuals do not produce new ideas *de novo*, but rather those ideas must arise from a large set of well-developed skills and a rich body of domain-relevant knowledge (Csikszentmihaly, 1990). At the same time, research on creative development suggests that this process is far more complicated than first appears (cf. Weisberg, 2003). To begin with, creators vary greatly in the amount of preparation they need, the greater the creativity manifested in adulthood the less time was needed to attain mastery (Clemente, 1973; Simonton, 1991b, 1992b; Zuckerman, 1977). So the creative genius can somehow acquire the requisite knowledge and skill more quickly than the norm. Furthermore, it is not necessarily the case that the greater creators have attained higher levels of formal training or education in their chosen domain. On the contrary, frequently the creative genius has acquired somewhat less than normally expected (Goertzel, Goertzel & Goertzel, 1978; Simonton, 1976a, 1987a). In fact, there is evidence that overtraining or overspecialization can actually harm rather than enhance the growth of creative potential (Simonton, 1976a, 2000). Sometimes highly creative individuals will have received considerable training, but in another field that is marginal to the domain in which they later attain eminence (Hudson & Jacot, 1986; Simonton, 1984e).

Complicating this picture all the more is the fact that education and training operate somewhat differently depending on whether the creative potential is directed toward the arts or the sciences (Simonton, 1999b). For the most part, formal instruction and practice play a bigger role in scientific creativity than in artistic creativity (Hudson, 1966; Schaefer & Anastasi, 1968; Simontion, 1984d). As a consequence, the scientific genius will tend to display higher levels of scholastic performance and training than will the artistic genius.

Sociocultural Context

To make the case against Galton's (1869) extreme biological determinism, Candolle (1873) presented data proving that creative genius was very much contingent on the economic, political, cultural, and social milieu. So persuasive is this dependence that some sociologists and anthropologists have argued that creativity is mostly if not entirely a sociocultural phenomenon (e.g. Kroeber, 1944). However, it is possible to accept the influence of these forces while still maintaining that creativity represents a psychological phenomenon (Simonton, 1984d). This possibility ensues from the fact that the sociocultural environment, to a very large extent, influences the course of creative development. To see how this can be so, let us turn to the following three sets of findings:

(1) If Galton's (1869) genetic determinism were correct, the number of creative geniuses would not drastically alter from one generation to the next. Such rapid fluctuations cannot happen because the gene pool for any given human population cannot change rapidly. Yet as Kroeber (1944) pointed out, creative geniuses tend to cluster over the course of history, forming dramatic 'Golden Ages' separated by dismal 'Dark Ages' (see also Schneider, 1937; Sorokin & Merton, 1935). Furthermore, Kroeber suggested that these clusters or 'cultural configurations' likely reflect the operation of a social learning process—what he styled 'emulation'. Each generation provides role models and mentors for the next generation. Using generational time-series analysis, this conjecture has been confirmed for European, Chinese, and Japanese civilizations (Simonton, 1975b, 1988b, 1992a). That is, for any given domain of creative achievement, the number of eminent creators in generation g is a positive linear function of the number of eminent creators in generation $g-1$ (Simonton, 1984c). This inter-generational effect occurs because the productive period of those in a given generation overlaps the developmental period of those in the following generation (see also Simonton, 1977b, 1984a, 1996).

(2) Certain political environments also operate during the developmental stages of an individual's life, either encouraging or discouraging the acquisition of creative potential. Thus, on the one hand, growing up in times of anarchy, when the political world is plagued by assassinations, coups d'état, and military mutinies, tends to be antithetical to creative development (Simonton, 1975b, 1976c). On the other hand, growing up when a civilization is fragmented in a large number of peacefully coexisting independent states tends to be conducive to the development of creative potential (Simonton, 1975b). In fact, nationalistic revolts against the oppressive rule of empire states tend to have positive consequences for the amount of creativity in the following generations (Kroeber, 1944; Simonton, 1975b; Sorokin, 1947/1969). Many nations have experienced Golden Ages after winning independence from foreign domination, ancient Greece providing a classic example.

(3) The rationale for the last-mentioned consequence seems to be that nationalistic rebellion encourages cultural heterogeneity rather than homogeneity (Simonton, 1994). Rather than everyone having to speak the same language, read the same books, follow the same laws, and so on, individuals are left with more options. This suggests that cultural diversity may facilitate creative development, and there is evidence that this is the case. Creative activity in a civilization tends to increase after it has opened itself to extensive alien influences, whether through immigration, travel abroad, or studying under foreign teachers (Simonton, 1997c). By enriching the cultural environment, the ground may be laid for new creative syntheses. This finding is consistent with what was noted earlier about the developmental asset of ethnic marginality. The result also falls in line with research indicating the creativity-augmenting effects of exposure to linguistic, ideological, and behavioral diversity (Campbell, 1960; Lambert, Tucker & d'Anglejan, 1973; Nemeth & Kwan, 1987).

The above sets of findings taken together show how the Zeitgeist will often affect genius during the critical phase in which an individual acquires creative potential.

Genetic Factors

The preceding section did not by any means exhaust the list of environmental conditions and events that affect creative development. Some psychologists have concluded from such inventories that creative genius is 100% nurture and 0% nature (e.g. Howe, 1999). In other words, Dryden's quote got everything backwards. Genius is not born, but made, and Galton's (1869) original thesis was way off the mark. Yet such an inference would be very much mistaken. Sufficient evidence has already accumulated to suggest that to a certain, even if limited, extent, the genes inherited at the moment of conception influence the individual's capacity for creativity. Consider, for example, the following four items:

(1) A distinctive set of personality and cognitive traits are associated with the magnitude of creativity displayed (Martindale, 1989; Simonton, 1999a). That is, certain traits distinguish creative from non-creative people, and creative geniuses from those who are less notably creative (Feist, 1998). Yet, significantly, almost all of the characteristics associated with creativity have nontrivial heritability coefficients (Bouchard, 1994; Eysenck, 1995). By nontrivial I mean traits in which 30–50% of the variance can be ascribed to genetic inheritance. Examples include intelligence, introversion, energy level, and psychoticism. If creativity is linked with a specific cluster of traits, and if many of those traits have significant genetic components, then it is logically impossible to infer that creative genius is entirely an environmental product.

(2) It is become increasingly recognized that the genetic basis of creativity may not operate according to a simplistic additive model (Waller, Bouchard, Lykken, Tellegen & Blacker, 1993). Instead, the numerous genetic components of genius may function according to some kind of multiplicative function (Simonton, 1999c; cf. Burt, 1943; Sternberg & Lubart, 1995). This multiplicative inheritance has been called 'emergenesis' (Lykken, 1998), and it has several critical implications (Simonton, 1999c). These include: (a) a highly skewed cross-sectional distribution for innate creative ability (precisely like that observed for individual differences in productive output); (b) low probabilities of familial inheritance (contrary to Galton's, 1869, belief); and (c) attenuated validity coefficients for predicting creativity using an additive model (the inevitable procedure).

(3) Modern behavioral genetics has shown that the influence of inherited characteristics does not appear all at once, but rather often unfolds gradually during the course of personal development (e.g. Bouchard, 1995; Plomin, Fulker, Corley & DeFries, 1997). It is for this reason that identical (monozygotic) twins reared apart in separate homes actually become more similar as they age. This epigenetic development, when combined with the aforementioned emergenic inheritance, produces additional complexities in the acquisition of creative potential (Simonton, 1999c). For instance, a combined emergenic–epigenetic process supports the emergence of 'late bloomers' whose creative abilities might not have been apparent in childhood. The reverse can happen as well, once promising talents may fail to realize their creative potential because of unfavorable epigenetic trajectories.

(4) Behavior geneticists have devised powerful statistical methods for partitioning the variance in any individual characteristic into three sources: the genetic, the shared environment, and the nonshared environment (Bouchard, 1994). The shared environment includes those things that all siblings have in common, such as the socioeconomic background and child-rearing practices of their parents, whereas the nonshared environment encompasses those things that are unique to each sibling, such as birth order or peer relationships (Harris, 1995; Sulloway, 1995). The fascinating finding is that for a majority of human traits, the nonshared environment accounts for much more variance than does the shared environment (Bouchard, 1994). This would seem to contradict some of the findings regarding the impact of family background on the development of creative potential. If shared environment plays such a small part, why is creative genius more likely to emerge in specific home conditions? The answer is that many of these so-called environmental factors are actually genetic factors operating incognito (Scarr & McCartney, 1983). For example, parents who value education and who have professional occupations are likely to have above-average intelligence, and the latter asset is passed down to their offspring. The fact that the parents fill their homes with intellectually stimulating materials may only be an outward sign of the intellectual superiority of their children rather than a cause of that superiority. Many other supposed effects of the family environment may similarly reflect underlying genotypic traits shared by both parent and child (Simonton, 1994).

The foregoing four sets of findings suggest that the nature–nurture debate has not ended with an outright victory for the proponents of nurture. Creative development involves some combination of genetic and environmental factors, albeit the latter probably play the larger role. Creative genius is both born and made.

The Realization of Creative Potential

Once nature and nurture combine to produce an individual with high creative potential, how is that potential creativity converted into actual creative productivity? From a lifespan perspective, this issue can be broken down to several subsidiary questions: When does creative output begin? When does the creative genius reach his or her career peak? When does the creative career effectively end? In other words, what is the trajectory of productivity across the course of a creator's career? Remarkably, the oldest empirical study in the behavioral sciences was specifically devoted to this very issue. In 1835 Quételet published the classic *A Treatise on Man and the Development of His Faculties*. Here he pioneered the use of statistics to study individual differences and longitudinal changes. Among the studies in the latter category was an inquiry into the relation between age and creative productivity, with a specific focus on the careers of English and French playwrights. Perhaps Quételet was ahead of his time, for this particular investigation inspired no follow-up investigations. For instance, when Beard (1874) turned to the same subject nearly 40 years later, he did so using qualitative methods that Quételet (1835/1968) had already rendered obsolete. About a century after Quételet, however, a quantitative approach was revived by Harvey C. Lehman, culminating in his 1953 book *Age and Achievement* (see also Lehman, 1962). Since that time, quantitative methods have been substantially improved so that a great deal is now known about the career trajectories for creative genius (Simonton, 1988a, 1997b). The key findings of this research literature can be grouped under two major headings, namely, those that concern endogenous factors and those that regard exogenous factors.

Endogenous Career Development

To investigate the age-productivity curve, Quételet's (1835/1968) tallied the number of plays produced in consecutive 5-year age periods. The resulting aggregate tabulation yielded a curvilinear function with a single peak—or what may be described as an inverted-backward J-curve (Simonton, 1980c, 1988a). Subsequent investigators, and especially Lehman (1953), have replicated this curve over and over, perhaps making it the best replicated finding in developmental psychology (see also Lehman, 1962). Across a great diversity of creative achievements, the same longitudinal pattern emerges. Creative productivity usually begins sometime in the middle 20s. The output rate per unit of time then increases rapidly until the peak productive age is reached sometime in the 30s or 40s. Thereafter creativity output slowly declines until, by the final decades of life, works are being produced at about half the rate witnessed at the career acme. Hence, there appears to be an optimal age for making a major contribution to human civilization.

Despite the robustness of these results, the findings were often attacked on methodological grounds (e.g. Cole, 1979; Dennis, 1954d, 1956; Zuckerman & Merton, 1972). These criticisms largely focused on the post-peak decrement, arguing that it was some kind of artifact. Of the various criticisms, the following three were the most important (Simonton, 1988a):

(1) The empirical age curves were obtained by aggregating output across large numbers of notable creators. Although this seems like a reasonable procedure, it opens the way for the introduction of spurious results. There is no guarantee that what holds at the aggregate level will in fact be descriptive of what holds at the level of all the individuals composing the aggregate. That is, there may intrude some form of aggregation error. The most obvious instance are the consequences of the fact that not all creators live to the same ripe old age (Dennis, 1956). Because no creative genius can ever continue output after death, the number of works produced by those in their 80s will be necessarily less than those produced by those in their 40s simply because more people live to 50 than live to 90. Hence, the aggregate tabulations will display an exaggerated age decrement. Fortunately, there are several methods for removing this artifact (Simonton, 1988a, 1991a). For instance, Quételet (1835/1968) calculated creative output per age period after making adjustment for the number of creators still alive each age period. Dennis (1966), however, confined his samples to creators who had lived to become octogenarians. Whatever the procedure adopted, the outcome is clear: the age decrement is not the result of some aggregation error (Simonton, 1988a).

(2) For most creative domains, the number of creators making contributions to those domains has increased over historical time (Lehman, 1947; Simonton, 1975b, 1988b, 1992a). This increase is due to both the exponential growth of the human population in general (Taagepera & Colby, 1979) and the explosive growth of activity in certain domains, such as those in the sciences (Price, 1963). What this means is that as creators get older, they have to endure more competition from colleagues than they had to face at the beginning of their career (Dennis, 1954d). That increased competition may imply increased difficulties getting papers published, patent application approved, paintings exhibited, or films distributed, depending on the creative domain. One problem with this criticism, of course, is that it implicitly assumes that the vehicles or outlets for disseminating creative products have not increased in the same proportion as the number of active creators. In the sciences, for instance, the number of professional journals has also grown explosively. More important, however, is the fact that the age decrement in creative output cannot be dismissed as a spurious consequence of increased competition (Simonton, 1988a). The post-peak decline still appears even after introducing controls for the number of competitors active in each age period (e.g. Simonton, 1977a).

(3) Many of the early studies, such as those conducted by Lehman (1953), did not tabulate total output, but rather only included high-impact or highly acclaimed works in the tabulations. It has been argued that this practice exaggerates the observed age decrement (e.g. Dennis, 1955d). The hypothesized distortion arises from the supposition that the age curves for total output has a more or less flat

post-peak age trend (see, e.g. Dennis, 1966). However, ever since Quételet's (1835/1968) pioneering investigation, this assumption has been shown to be plain wrong (see also Simonton, 1977a, 1985, 1997b). When lifetime output is split into 'hits' and 'misses' and then tabulated into separate time series, the resulting age curves are virtually identical. Those age periods with the most hits are also those with the most misses. Hence, across the career course, quality of output is closely associated with total quantity. Even more astonishing, the quality ratio or hit rate—the number of hits per age period divided by the total number of attempts (hits plus misses)—does not systematically increase or decrease over time, nor does it exhibit some curvilinear form (Oromaner, 1977; Over, 1989; Simonton, 1977a, 1985, 1997b; Weisberg, 1994). Instead, the quality ratio fluctuates unpredictably over the course of the career, hovering around some average hit rate. This longitudinal constancy in the proportion of hits to total attempts has been called the *equal-odds rule* (Simonton, 1997b; cf. Simonton, 1988a).

In light of the foregoing, the conclusion is inevitable that an age decrement does indeed tend to appear in the latter part of the career. Even so, there do exist several factors that govern the specific shape and location of the age curve, including the magnitude of the post-peak decline. The more crucial of these moderating variables are discussed below.

Career Age Versus Chronological Age

The original research plotted output as a function of the creator's chronological age (Dennis, 1966; Lehman, 1953; Quételet, 1835/1968). However, this is not the only longitudinal function possible. A significant alternative is to consider creative output as a function of how long the creator has been producing original ideas in a given domain. This alternative has been called 'career' or 'professional' age (Bayer & Dutton, 1977; Lyons, 1968; Simonton, 1998a). For instance, the productivity of a scientist might be gauged from the age in which he or she received the Ph.D. in the chosen specialty (e.g. Lyons, 1968). Admittedly, in many instances the chronological age and career age will correlate very highly across large samples of individual creators (Bayer & Dutton, 1977). This correlation results from the fact that the age at career onset will often be very similar—usually sometime in the middle or late 20s (Raskin, 1936; Simonton, 1991a). Even so, the age at which the career begins is by no means fixed, permitting some appreciable discrepancies to take place (Simonton, 1998b). Given this possible variation, the question then arises: Which longitudinal function is most descriptive, one based on chronological age or one based on career age? Recent empirical research points to the second definition as the more appropriate

(Simonton, 1988a, 1991a, 1997b). The longitudinal changes in creative output are best described in terms of career age. If a creator launches the career earlier than normal, the peak will appear earlier in terms of chronological age but in the same place in terms of career age. Likewise, a late start in the career will shift the whole career trajectory over to the later years of life.

The dependence of creative output on career age has two valuable implications:

(1) Nothing prevents a creative genius from attaining a career peak at a chronological age when the vast majority of creators are well into the age-decrement portion of the age curve (i.e. 'over the hill'). A concrete example of such an exceptional 'late bloomer' is the Austrian composer Anton Bruckner. Not discovering his mission as a symphonist until he was in his late 30s, his first undoubted masterpiece did not appear until he was 50, and his last great work was left incomplete when he died in his early 70s. His best works appeared over a 20-year interval, which is normal, but the whole career was atypically shifted over by more than 10 years. Another fascinating instance is the career of the French scientist Michel Eugène Chevreul. Having already had an extremely productive career as a chemist, in his 90s he began to notice the effects of aging. This observation inspired him to switch fields and thereby become a pioneer in gerontological research—and simultaneously resuscitate his creative potential. His last publication appeared when he was 102, just one year before his death at age 103.

(2) If the longitudinal function involves career age rather than chronological age, then creative productivity across the life span must be the consequence of something intrinsic to the creative process itself. In fact, I have offered a mathematical model that explains this phenomenon in terms of a two-step cognitive process (Simonton, 1984b, 1991a, 1997b). In the first step, creative potential is gradually converted to various ideas that make up 'works-in-progress' (the ideation stage); in the second step, these ideas are transformed into finished products (the elaboration stage). However, attempts to explain the career trajectory according to extrinsic factors are doomed to fail. For instance, insofar as the aging process is contingent on chronological age, then the age decrement cannot be attributed to declines in cognitive or physical functioning (cf. Lehman, 1953; McCrae, Arenberg & Costa, 1987). Indeed, as the Bruckner case illustrates, nothing prevents someone from reaching a career peak at an age when the aging process should already have been well advanced. This is not tantamount to saying that these various extrinsic factors are totally irrelevant. For example,

it could very well be that a career peak that occurs in the 30s or 40s might display a higher quantity of output than a career peak that occurs in the 50s or 60s. Nevertheless, the shape of the age curve is decided by how long the creative genius has been active in the domain, not by how old he or she happens to be.

This last point will receive additional support in the next section.

Domain of Creative Achievement

The two-step cognitive model mentioned above predicts that the specific shape of the career trajectory will necessarily vary according to the rates of ideation (in step one) and elaboration (in step two). Moreover, the ideation and elaboration rates will depend on the nature of the ideas that make up a given domain of creativity (i.e. their abstractness, diversity, number, and complexity). Hence, as an immediate consequence, the model predicts that the expected trajectories should vary across distinct domains (Simonton, 1984b, 1991a). This prediction has been abundantly confirmed (Dennis, 1966; Lehman, 1953; Simonton, 1991a). For instance, lyric poets, in comparison to novelists, tend to reach their peaks earlier and to exhibit much steeper post-peak declines (Lehman, 1953). In fact, the characteristic career trajectories can be used to estimate the ideation and elaboration rates for various creative domains (Simonton, 1984b, 1989a). Needless to say, because the ideation and elaboration rates are intrinsic properties of the creative process, these results again confirm that longitudinal changes in creative output are determined more by career age than by chronological age (Simonton, 1989a, 1991a, 1997b). After all, the aging process is probably the same for all creators, whether they be poets or novelists, mathematicians or geologists, and thus it would be difficult to explain, except on a *post hoc* basis, the distinctive career peaks and age decrements.

The domain contrasts in career trajectories have important consequences regarding the longitudinal location of a creator's *career landmarks* (Simonton, 1991a, 1997b; cf. Raskin, 1936). These landmarks indicate three events: the age of the creator's first great work, the age of the creator's single best work, and the age of the creator's last great work. In other words, the landmarks pinpoint the onset, acme, and termination of that period of the career in which high-impact works are being produced. According to the equal-odds rule mentioned earlier, quality is a function of quantity, and therefore those periods that feature the greatest total output will tend to be those that contain the most influential work. Because the single best work will be found among the best works, the age for this career landmark must differ appreciably across creative domains, closely tracking the expected age for the

maximum output rate. That prediction has been amply confirmed for domains in both the arts and sciences (Simonton, 1988a, 1997a). For instance, lyric poets produce the best work at a younger age than do novelists, a difference that is transhistorically and cross-culturally invariant (Simonton, 1975a). An analogous set of arguments apply to the other two career landmarks. For example, those fields that display the most severe post-peak age decrements in total output will tend to be those in which the last influential work will appear earlier in the career (Simonton, 1991a). Conversely, those domains in which the decline is negligible will tend to see great contributions appear very late in life. This tends to be the case for such fields as history and philosophy (Dennis, 1966; Lehman, 1953; Simonton, 1989a).

Nonetheless, the next moderating factor is even more powerful in determining the longitudinal location of the first and last major work.

Individual Differences in Creative Potential

At this chapter's outset I noted how much creators can vary in terms of total lifetime output. The most outstanding creators in a given domain will often be hundreds of times more productive than are their least prolific colleagues (viz., those who made only one contribution each). According to the two-step cognitive model, these individual differences in lifetime output can be credited to an underlying latent variable called *creative potential* (Simonton, 1991a, 1997b). The higher the level of creative potential an individual enjoys, the more ideas can be generated in a given unit of time. In other words, high creative potential manifests itself as a higher mean output rate throughout the career course, which necessarily includes a higher than normal maximum output rate. However, someone with low creative potential will exhibit a low mean output rate and a low maximum output rate. Otherwise, the career trajectories will be unchanged. The following pair of implications result:

(1) Those who are highly prolific in their 40s will have been highly prolific in their 20s and 30s and will continue to be so in their 50s, 60s, and subsequent decades, as their life spans permit. In contrast, those who are less prolific in their 40s will have been correspondingly less prolific in the decades either before or after. So creative output per unit of time must exhibit considerable longitudinal stability (Cole, 1979; Dennis, 1954b; Helmreich, Spence & Thorbecke, 1981). In particular, the correlations among the output levels for various pairs of decades will be uniformly high. Even more critically, the correlations will all be approximately the same size (Simonton, 1997b). Hence, the correlation between output in the 30s and output in the 50s will be about the same magnitude as the

correlation between output in the 30s and output in the 40s. This turns out to be an especially important consequence of there being a single latent variable, creative potential, underlying output in each period (Simonton, 1997b). It means that the intercorrelations follow what is known as a 'simplex structure', a pattern that completely contradicts alternative theories of longitudinal stability (cf. Simonton, 1991c). According to the theory of cumulative advantage (Allison, Long & Krauze, 1982; Allison & Stewart, 1974), for example, output in adjacent decades should be more highly correlated than those in nonadjacent decades, yielding a correlation typical of an autoregressive process (Simonton, 1997b). But that prediction is strictly disproven.

(2) Individual differences in creative potential also have critical repercussions for the location of the three career landmarks (Simonton, 1991a, 1997b). To keep the argument simple, imagine two creators working within the same discipline and who started their careers at the same age, but who differ greatly in creative potential. Then the career trajectories for the two creators will look identical, except that the curve for the high-potential creator will have a higher amplitude. That is, high creative potential is associated with more productivity throughout the career course. In the case of the middle career landmark—the age at the best work—this difference in amplitude will make no difference whatsoever. Because the best work will be located around the age of maximum total output, both high- and low-potential creators will produce their most influential product at the same age. Nonetheless, something very different occurs for the first and last career landmarks. Because the creator with the highest potential is accumulating output at a faster rate in the first decade of the career, the chances increase that he or she will get their first hit sooner. Similarly, because the high-potential creator is producing at a higher rate in the final decade of the career, he or she will claim the last hit later in life. Hence, the higher the creative potential, the earlier appears the first landmark, the later appears the last career landmark, but the location of the middle career landmark remains unchanged (Simonton, 1991a, 1997b). Whether creative potential is gauged by total lifetime output, maximum output rate, mean output rate, or achieved eminence, these expectations have been amply confirmed in empirical research (Raskin, 1936; Simonton, 1991a, 1991b, 1992b, 1997b).

Significantly, cross-sectional variation in creative output is far greater than longitudinal variation (Levin & Stephan, 1989, 1991; Over, 1982a, 1982b; Simonton, 1997b). As a consequence, if one is trying to predict a creator's productivity in a particular age period, it is far more crucial to know who the person is than how old that person is. Take the typical case of a domain with an age decrement of 50% between the peak in the 40s and the decade of the 70s. Someone with a low creative potential who produces a total of two products at their career acme would be expected to produce but one product in the 70s. In contrast, someone with a high creative potential might generate 100 products in the 40s, and thus anticipate an output of 50 products in the 70s. Hence, the highly creative septuagenarian would be 25 times more productive than the low-creative individual at his or her career peak!

Career Flexibility and Openness

Judging from what has been said thus far, creative potential may seem to operate like a retirement account. During the developmental period of a future creator's life, deposits are made into this account, so that creative potential slowly builds up to a sizable 'nest egg'. Then during the productive period of the active creator's career, withdrawals are made from this account to 'purchase' creative products to offer the world. Some creators may have big accounts that enable them to buy a large inventory, whereas others only have a meager account that barely maintains some semblance of productivity. But in either case, the account is gradually used up, only those with the most impressive savings at the outset having any hope of having the supply last until death renders the funds unnecessary anyway.

This conclusion is not entirely unjustified. The amount of time and energy that a person can devote to creative development is far greater in childhood and adolescence than in adulthood, when a myriad of responsibilities and distractions may interfere. For instance, future great scientists usually devoted about 50 hours per week in graduate school to attaining mastery of their chosen discipline (Chambers, 1964). It is very difficult for a fulltime professor to maintain this level of continuing education. Course preparation and grading, committee meetings, and other professorial tasks compete for on-the-job time, while ever growing family responsibilities and activities vie for what time remains. Even so, studies of eminent scientists suggest that they manage to adopt strategies that serve to resuscitate creative potential at least to some degree. The following three strategies seem especially valuable:

(1) Omnivorous reading was previously indicated as playing a major role in creative development. Such reading habits continue to preserve creative potential throughout the careers of highly successful scientists (Blackburn, Behymer & Hall, 1978; Dennis & Girden, 1954). The latter not only read extensively in their own specialty, but also try to keep up on the latest advances in disciplines adjacent to their own (R. Simon, 1974).

(2) High-impact scientists tend to have extensive contacts with other scientists (Crane, 1972; Simonton, 1992c). The notion of the 'lone genius' is thus pure myth. These contacts consist of colleagues, collaborators, and competitors who, in one way or another, help stimulate and thus maintain the capacity to generate new and fruitful ideas. Those scientists who have these exchanges to an exceptional degree accordingly tend to have longer and more productive careers (Simonton, 1992c).

(3) Great scientists do not work on one idea until it is exhausted, and then switch to another idea to push to the hilt (Garvey & Tomita, 1972). Instead, they tend to have several works in progress going on simultaneously, each at various degrees of development, and each having variable odds of having a successful outcome (Dunbar, 1995; Hargens, 1978; Root-Bernstein, Bernstein & Garnier, 1993; R. Simon, 1974). Not only does this provide them with a backup project when another project encounters unforeseen obstacles, but also crosstalk will often occur among the various projects so that the solution to one problem may lead to the solution to another, even seemingly unrelated problem (Tweney, 1990; see, e.g. Poincaré, 1921). The latter possibility is reinforced by the fact that prolific scientists tend to engage in a 'network of enterprise', that is, a collection of diverse but nonetheless interrelated ideas (Gruber, 1989; see also Feist, 1993; Simonton, 1992b). The ideational interaction among these projects helps the scientist get more mileage out of his or her creative potential.

Although the above three points concern scientific creativity, it is likely that similar strategies facilitate the maintenance of creative potential in artistic creativity. Thus, great artists, in contrast to their lesser colleagues, also tend to participate in abundant interrelationships with fellow artists (Simonton, 1984a). Likewise, the greatest composers appear to sustain their creativity by switching back and forth between different genre (Simonton, 2000) and by engaging in various stylistic shifts (Martindale, 1990; Simonton, 1977b). Hence, it is probably a general phenomenon. If a creative genius wants to avoid 'drying up' or 'running out of steam', it behooves him or her to be flexible and open throughout the career course (see also Georgsdottir, Lubart & Getz, 2003).

Exogenous Career Influences

The creative genius does not generate ideas in isolation from the outside world. On the contrary, many external events impinge on creators during the course of their careers. These extraneous events can deflect the career trajectory from what would be anticipated on the basis of what has been discussed so far. Such external events are of two kinds, the impersonal and the personal.

Impersonal

Earlier in this chapter I listed some of the political events that contribute to creative development, such as civil disturbances. Some political events, however, have little or no effect on the acquisition of creative potential but instead have an impact on the realization of that potential. The best example is war. In general, war has a negative repercussion on creativity, both the quantity and quality of creative products declining when conceived under wartime conditions (Fernberger, 1946; Price, 1978; Simonton, 1976b, 1980a, 1986c). To be more precise, major wars between nation states are associated with an overall decline in creativity in the participating countries. In contrast, wars fought far away from the homeland, like European imperial or colonial wars in Africa, Asia, or the Americas, had no consequences whatsoever, whether positive or negative (Simonton, 1980a). Thus, the two World Wars were detrimental to creative output in European nations, but not the Boer War in South Africa, the Spanish Conquest of Mexico or Peru, or the Opium War in China. In short, the war had to be of the sort that would have a direct impact on the creator. It is also essential to note that the negative repercussion of military conflict tends to be short-lived or transient (Price, 1978; Simonton, 1975b). When peacetime conditions return, creativity recovers quickly, at least at the individual level. The latter stipulation is necessary because sometimes the impact of war is so devastating on a particular country that many of its most creative denizens end up seeking their fortunes in nations less harmed by the events. A case in point is the episode of the numerous and illustrious creators who enriched the United States by the end of the Second World War.

Incidentally, dramatic political events and circumstances like war affect creative output in yet another manner—the very nature of the piece. For instance, the thematic content or style of a creative product tends to betray its conception under wartime conditions (Cerulo, 1984; Martindale, 1975; Simonton, 1986a, 1986c). That is, to a certain extent the product reflects the external conditions in which it was created. This even holds for scientific creativity, including creative products in psychology (McCann & Stewin, 1984; Padgett & Jorgenson, 1982; Sales, 1972).

Personal

No matter what the magnitude of achievement, the creative genius remains a human being, with a personal life that can sometimes interfere with work. One obvious source of interference is any serious bout with physical illness (Lehman, 1953; Simonton, 1977a). Perhaps more subtle, but still a harmful intrusion is having a family life. Francis Bacon (1597/1942) put the problem this way:

> He that hath wife and children hath given hostages to fortune; for they are impediments to great enter-

prises, either of virtue or mischief. Certainly the best works, and of greatest merit for the public, have proceeded from the unmarried or childless men, which, both in affection and means, have married and endowed the public. (p. 29)

In support of this observation, Havelock Ellis (1926) noted from his scrutiny of British geniuses that there was "a greater tendency to celibacy among persons of ability than among the ordinary population" (p. xiv). Not counting priests, the rate was nearly 1 out of 5. Another investigation into the lives of a more elite sample of historic figures found that 55% never married (McCurdy, 1960). Marriage often tends to abbreviate or depress the creative career (Kanazawa, 2000), an adverse consequence that may mostly result from the cares and responsibilities imposed by parenting children (Hargens, McCann & Reskin, 1978; Kyvik, 1990).

It goes without saying that these negative repercussions of marriage can be more severe for female creators than for their male colleagues. Thus, women who win an entry in *Who's Who* are four times more likely than similarly illustrious men to be unmarried (Hayes, 1989). In addition, those who do get married tend to do so at a later age than is the norm for their social class (Ellis, 1926), and these successful women who somehow fit marriage in their lives are three times more likely to be childless in comparison to comparably accomplished married men (Hayes, 1989). In fact, between 1948 and 1976 in the United States, the proportion of doctorates that were granted to women correlated −0.94 with the average cohort fertility, a very remarkable aggregate-level correlation (McDowell, 1982).

I mentioned that impersonal factors may shape a creative product's content or style. Personal factors can also have this consequence. For instance, research on the musical compositions making up the classical repertoire has shown that the originality of the melodic material is positively associated with the intensity of stress that the composer was experiencing at the time of a work's creation (Simonton, 1980b, 1987b). In a sense, the unpredictability of the musical themes reflects the unpredictability of the creator's life.

The Termination of a Creative Life and Career

This chapter began at the moment of conception—when creators-to-be inherit a distinctive collection of genes from their parents. From that initial impetus we traced the acquisition of creative potential, including family experiences and conditions, education and training, and even the epigenetic growth of the individual's inborn potential. We next switched gears to examine how that potential was realized during the course of a creative career. This examination included endogenous factors like age at career onset, domain of creativity, and magnitude of creative potential as well

as exogenous factors like war, family, and health. The end result of these diverse factors is a body of work on which all creators must stake their respective reputations (Galton, 1869; Simonton, 1991c). Even so, a somewhat pessimistic saying laments that 'all good things must come to an end'. Even the most outstanding genius must eventually die, an event that must once and for all cut off transformation of potential creativity to actual creative products.

Curiously, although the creative life must be ended by death, the very timing of the moment of death is partly contingent on the nature of that creative life. In particular, life expectancies vary according to the specific domain in which a genius attains eminence (Cox, 1926; Simonton, 1997a). Not only do writers tend to die younger than creators in other fields do (Kaun, 1991), but among writers, poets tend to have shorter life spans than do the rest (Simonton, 1975a). Although scientists as a group tend to enjoy long lives (Cassandro, 1998; Cox, 1926), eminent mathematicians constitute a notable exception (Simonton, 1991a). According to these data, anyone who aspires to exceptional longevity should avoid becoming either a mathematician or a poet!

Exactly why these life-expectancy contrasts appear is more problematic. One explanation is predicated on the fact that life span is negatively correlated with creative precocity (Simonton, 1977b; Zhao & Jiang, 1986). That is, those who achieve their first career landmark at a younger age tend to die younger as well. Because the various domains of creativity differ in the age at which the first contribution tends to appear, it is conceivable that the life expectancies merely track the contrasts in precocity (Simonton, 1988a). Hence, poets and mathematicians may die younger because they are prone to create influential products at younger ages than creators in other artistic or scientific domains. Yet this answer somewhat begs the question. Why do precocious creators die younger in the first place? Conceivably this might merely represent a sampling artifact. Because eminence is impossible without producing at least one high-impact work, those who accomplish this achievement at younger ages can still die young and claim some place in the history of their discipline.

Although this explanation seems plausible, empirical studies show that it cannot completely account for the domain contrasts in life expectancy (Cassandro, 1998; McCann, 2001). Other factors must be operating as well. One such variable is psychopathology, for the rates of mental disorder vary across disciplines (Ludwig, 1995; Post, 1994). For instance, poets are more disposed towards psychopathology than are other creators (Ludwig, 1995). In addition, such psychological difficulties are indicative of a shortened lifespan. The latter effect may result from outright suicide (Lester, 1991) or from unhealthy behaviors, such as alcoholism (Davis, 1986; Lester, 1991; Post,

1996). Both suicide and alcoholism are in fact very common among poets as well. Nonetheless, this explanation raises yet another question: Why is psychopathology unevenly distributed among the alternative domains of creative achievement? Is it because young creative talents choose the domain that best fits their own personality disposition? Or is it because achievement domains differ greatly in the stresses and strains that creators must confront to achieve distinction? Apropos of the latter alternative, there is evidence that the stresses of creative achievement can increase psychopathological behaviors (Schaller, 1997) and that different achievement domains vary in the amount of stress they impose (Simonton, 1997a). Yet this explanation cannot completely solve the puzzle. For example, highly versatile scientific creators—those who make contributions to more than one domain—have shorter life spans than do colleagues who confine themselves to a single field, but the same does not hold for artistic creators (Cassandro, 1998). How can this be? In all likelihood the connection between life span and the creative domain is the joint function of a host of factors operating in complex interactions.

Whatever the explanation, creative geniuses are seldom blind to the approaching termination of their careers. Except in cases of lethal accidents or homicides, creators will often see the end coming. The decline in mental functioning and physical health will become ever more apparent, and the rate at which death takes away a colleague, friend, or family member will gradually accelerate. Under such circumstances, creators may experience some kind of 'life review' (Butler, 1963). Past goals, ambitions, plans, and hopes will undergo reminiscence and reassessment. This end-life process may then influence the creative ideas that emerge from their minds.

Now there are data showing that creative individuals can experience significant psychological transformations as death approaches near. For instance, they are more likely to display some decline in conceptual complexity (Suedfeld, 1985; Suedfeld & Piedrahita, 1984). It is significant that this tendency toward cognitive simplification is not seen in those who died unexpectedly, and that the declines are a function of proximity to death, not of age per se. More to the point, research has indicated that changes do indeed take place in the creative products that emerge toward the end of life. For example, artists frequently evolve a distinctive 'old-age style' (Lindauer, 1993; see also Munsterberg, 1983). A frequently cited case is that of Titian, the Italian painter whose late *Christ Crowned with Thorns*, a product of his 90s, departs dramatically from the style on which he based his fame. Naturally, some might be inclined to dismiss these late-life creations as sad illustrations of the deteriorating powers of a once-great genius. But such a judgment would be incorrect. In fact, the works that emerge during those closing years can surpass anything that the creator has done for years. It is almost as if the creator is pouring into these final thoughts every remaining ounce of their creative potential. Accordingly, these products can constitute 'last artistic testaments' that provide fitting capstones to the creator's entire career. This more favorable interpretation is supported by the *swan-song phenomenon* (Simonton, 1989b).

Music critics and historians have often suggested that classical composers can feature creative transformations in their final years (e.g. Einstein, 1956). Instances include the *Four Serious Songs* of the 63-year-old Johannes Brahms or the *Four Last Songs* of the 85-year-old Richard Strauss. Of course, these are both cases in which the composers were very old. Yet last-work effects have been attributed to works of much younger composers, such as Franz Schubert's appropriately titled *Schwanengesang*, published posthumously after the composer died only 31 years old. Supporting these conjectures is a quantitative study of the works created by 172 classical composers, including those just noted (Simonton, 1989b). Each composition was scored on several variables, such as melodic originality, duration, aesthetic quality, and popularity in the repertoire. The analysis then determined how these variables changed as a function of the work's proximity to the composer's death year. To avoid confounding the closeness of death with the composer's age, the latter was statistically partialed out (along with numerous other potential sources of artifact). The outcome was clear. Regardless of age, compositions created within or close to the year of death were highly distinctive. To be specific, they tended to be shorter and to contain less original melodies. At the same time, late-life works tended to score high in aesthetic significance and to secure a prominent place in the classical repertoire. These were not inferior creations by any means, but masterpieces that managed somehow to say more with less, to encapsulate a career in an elegant and forthright affirmation. Even if outstanding creators cannot usually decide *when* their creativity is going to end, they retain capacity to pronounce *how* their creativity is going to end.

References

Allison, P. D., Long, J. S. & Krauze, T. K. (1982). Cumulative advantage and inequality in science. *American Sociological Review*, **47**, 615–625.

Allison, P. D. & Stewart, J. A. (1974). Productivity differences among scientists: Evidence for accumulative advantage. *American Sociological Review*, **39**, 596–606.

Arieti, S. (1976). *Creativity: The magic synthesis*. New York: Basic Books.

Bacon, F. (1942). *Essays and the New Atlantis*. Roslyn, NY: Black. (Original works published 1597 and 1620).

Bayer, A. E. & Dutton, J. E. (1977). Career age and research—professional activities of academic scientists: Tests of alternative non-linear models and some implications for higher education faculty policies. *Journal of Higher Education*, **48**, 259–282.

Beard, G. M. (1874). *Legal responsibility in old age*. New York: Russell.

Berry, C. (1981). The Nobel scientists and the origins of scientific achievement. *British Journal of Sociology*, **32**, 381–391.

Blackburn, R. T., Behymer, C. E. & Hall, D. E. (1978). Correlates of faculty publications. *Sociology of Education*, **51**, 132–141.

Bliss, W. D. (1970). Birth order of creative writers. *Journal of Individual Psychology*, **26**, 200–202.

Bouchard, T. J., Jr. (1994). Genes, environment, and personality. *Science*, **264**, 1700–1701.

Bouchard, T. J., Jr. (1995). Longitudinal studies of personality and intelligence: A behavior genetic and evolutionary psychology perspective. In: D. H. Saklofske & M. Zeidner (Eds), *International Handbook of Personality and Intelligence* (pp. 81–106). New York: Plenum.

Bramwell, B. S. (1948). Galton's 'Hereditary' and the three following generations since 1869. *Eugenics Review*, **39**, 146–153.

Burt, C. (1943). Ability and income. *British Journal of Educational Psychology*, **12**, 83–98.

Butler, R. N. (1963). The life review: An interpretation of reminiscences in the aged. *Psychiatry*, **26**, 65–76.

Campbell, D. T. (1960). Blind variation and selective retention in creative thought as in other knowledge processes. *Psychological Review*, **67**, 380–400.

Candolle, A. de (1873). *Histoire des sciences et des savants depuis deux siècles*. Geneva: Georg.

Cassandro, V. J. (1998). Explaining premature mortality across fields of creative endeavor. *Journal of Personality*, **66**, 805–833.

Cerulo, K. A. (1984). Social disruption and its effects on music: An empirical analysis. *Social Forces*, **62**, 885–904.

Chambers, J. A. (1964). Relating personality and biographical factors to scientific creativity. *Psychological Monographs: General and Applied*, **78** (7, Whole No. 584).

Clark, R. D. & Rice, G. A. (1982). Family constellations and eminence: The birth orders of Nobel Prize winners. *Journal of Psychology*, **110**, 281–287.

Clemente, F. (1973). Early career determinants of research productivity. *American Journal of Sociology*, **79**, 409–419.

Cole, S. (1979). Age and scientific performance. *American Journal of Sociology*, **84**, 958–977.

Cox, C. (1926). *The early mental traits of three hundred geniuses*. Stanford, CA: Stanford University Press.

Crane, D. (1972). *Invisible colleges*. Chicago, IL: University of Chicago Press.

Csikszentmihaly, M. (1990). The domain of creativity. In: M. A. Runco & R. S. Albert (Eds), *Theories of Creativity* (pp. 190–212). Newbury Park, CA: Sage.

Davis, W. M. (1986). Premature mortality among prominent American authors noted for alcohol abuse. *Drug and Alcohol Dependence*, **18**, 133–138.

Dennis, W. (1954a). Bibliographies of eminent scientists. *Scientific Monthly*, **79**, September, 180–183.

Dennis, W. (1954b). Predicting scientific productivity in later maturity from records of earlier decades. *Journal of Gerontology*, **9**, 465–467.

Dennis, W. (1954c). Productivity among American psychologists. *American Psychologist*, **9**, 191–194.

Dennis, W. (1954d). Review of *Age and achievement*. *Psychological Bulletin*, **51**, 306–308.

Dennis, W. (1955, April). Variations in productivity among creative workers. *Scientific Monthly*, **80**, 277–278.

Dennis, W. (1956). *Age and achievement*: A critique. *Journal of Gerontology*, **9**, 465–467.

Dennis, W. (1966). Creative productivity between the ages of 20 and 80 years. *Journal of Gerontology*, **21**, 1–8.

Dennis, W. & Girden, E. (1954). Current scientific activities of psychologists as a function of age. *Journal of Gerontology*, **9**, 175–178.

Dryden, J. (1885). Epistle to Congreve. In: W. Scott & G. Saintsbury (Eds), *The Works of John Dryden* (Vol. 11, pp. 57–60). Edinburgh: Paterson. (Original work published 1693)

Dunbar, K. (1995). How scientists really reason: Scientific reasoning in real-world laboratories. In: R. J. Sternberg & J. E. Davidson (Eds), *The Nature of Insight* (pp. 365–396). Cambridge, MA: MIT Press.

Eiduson, B. T. (1962). *Scientists: Their psychological world*. New York: Basic Books.

Einstein, A. (1956). *Essays on music*. New York: Norton.

Eisenman, R. (1964). Birth order and artistic creativity. *Journal of Individual Psychology*, **20**, 183–185.

Eisenstadt, J. M. (1978). Parental loss and genius. *American Psychologist*, **33**, 211–223.

Ellis, H. (1926). *A study of British genius* (rev. ed.). Boston, MA: Houghton Mifflin.

Ericsson, K. A. (Ed.) (1996). *The road to expert performance: Empirical evidence from the arts and sciences, sports, and games*. Mahwah, NJ: Erlbaum.

Eysenck, H. J. (1995). *Genius: The natural history of creativity*. Cambridge: Cambridge University Press.

Feist, G. J. (1993). A structural model of scientific eminence. *Psychological Science*, **4**, 366–371.

Feist, G. J. (1998). A meta-analysis of personality in scientific and artistic creativity. *Personality and Social Psychology Review*, **2**, 290–309.

Fernberger, S. W. (1946). Scientific publication as affected by war and politics. *Science*, **104**, August 23, 175–177.

Galton, F. (1869). *Hereditary genius: An inquiry into its laws and consequences*. London: Macmillan.

Galton, F. (1874). *English men of science: Their nature and nurture*. London: Macmillan.

Gardner, H. (1993). *Creating minds: An anatomy of creativity seen through the lives of Freud, Einstein, Picasso, Stravinsky, Eliot, Graham, and Gandhi*. New York: Basic Books.

Garvey, W. D. & Tomita, K. (1972). Continuity of productivity by scientists in the years 1968–1971. *Science Studies*, **2**, 379–383.

Georgsdottir, A. S., Lubart, T. I. & Getz, I. (2003). The role of flexibility in innovation. In: L. V. Shavinina (Ed.), *International Handbook on Innovation*. Oxford: Elsevier Science.

Goertzel, M. G., Goertzel, V. & Goertzel, T. G. (1978). *300 eminent personalities: A psychosocial analysis of the famous*. San Francisco, CA: Jossey-Bass.

Gruber, H. E. (1989). The evolving systems approach to creative work. In: D. B. Wallace & H. E. Gruber (Eds), *Creative People at Work: Twelve Cognitive Case Studies* (pp. 3–24). New York: Oxford University Press.

Hargens, L. L. (1978). Relations between work habits, research technologies, and eminence in science. *Sociology of Work and Occupations*, **5**, 97–112.

Hargens, L. L., McCann, J. C. & Reskin, B. F. (1978). Productivity and reproductivity: Fertility and professional achievement among research scientists. *Social Forces*, **57**, 154–163.

Harris, J. R. (1995). Where is the child's environment: A group socialization theory of development. *Psychological Review*, **102**, 458–489.

Hayes, J. R. (1989). *The complete problem solver* (2nd ed.). Hillsdale, NJ: Erlbaum.

Helmreich, R. L., Spence, J. T. & Thorbecke, W. L. (1981). On the stability of productivity and recognition. *Personality and Social Psychology Bulletin*, **7**, 516–522.

Helson, R. & Crutchfield, R. S. (1970). Mathematicians: The creative researcher and the average Ph.D. *Journal of Consulting and Clinical Psychology*, **34**, 250–257.

Howe, M. J. A. (1999). *The psychology of high abilities*. Washington Square, NY: New York University Press.

Howe, M. J. A., Davidson, J. W. & Sloboda, J. A. (1998). Innate talents: Reality or myth? *Behavioral and Brain Sciences*, **21**, 399–442.

Hudson, L. (1966). *Contrary imaginations*. Baltimore, MD: Penguin.

Hudson, L. & Jacot, B. (1986). The outsider in science. In: C. Bagley & G. K. Verma (Eds), *Personality, Cognition and Values* (pp. 3–23). London: Macmillan.

Kanazawa, S. (2000). Scientific discoveries as cultural displays: A further test of Miller's courtship model. *Evolution and Human Behavior*, **21**, 317–321.

Kaufmann, G. (2003). The effect of mood on creativity in the innovative process. In: L. V. Shavinina (Ed.), *International Handbook on Innovation*. Oxford: Elsevier Science.

Kaun, D. E. (1991). Writers die young: The impact of work and leisure on longevity. *Journal of Economic Psychology*, **12**, 381–399.

Kroeber, A. L. (1944). *Configurations of culture growth*. Berkeley, CA: University of California Press.

Kyvik, S. (1990). Motherhood and scientific productivity. *Social Studies of Science*, **20**, 149–160.

Lambert, W. E., Tucker, G. R. & d'Anglejan, A. (1973). Cognitive and attitudinal consequences of bilingual schooling: The St. Lambert project through grade five. *Journal of Educational Psychology*, **65**, 141–159.

Lehman, H. C. (1947). The exponential increase of man's cultural output. *Social Forces*, **25**, 281–290.

Lehman, H. C. (1953). *Age and achievement*. Princeton, NJ: Princeton University Press.

Lehman, H. C. (1962). More about age and achievement. *Gerontologist*, **2**, 141–148.

Lester, D. (1991). Premature mortality associated with alcoholism and suicide in American writers. *Perceptual and Motor Skills*, **73**, 162.

Levin, S. G. & Stephan, P. E. (1989). Age and research productivity of academic scientists. *Research in Higher Education*, **30**, 531–549.

Levin, S. G. & Stephan, P. E. (1991). Research productivity over the life cycle: Evidence for academic scientists. *American Economic Review*, **81**, 114–132.

Lindauer, M. S. (1993). The old-age style and its artists. *Empirical Studies of the Arts*, **11**, 135–146.

Lombroso, C. (1891). *The man of genius*. London: Scott.

Ludwig, A. M. (1995). *The price of greatness: Resolving the creativity and madness controversy*. New York: Guilford Press.

Lyons, J. (1968). Chronological age, professional age, and eminence in psychology. *American Psychologist*, **23**, 371–374.

Lykken, D. T. (1998). The genetics of genius. In: A. Steptoe (Ed.), *Genius and the Mind: Studies of Creativity and Temperament in the Historical Record* (pp. 15–37). New York: Oxford University Press.

Martindale, C. (1975). *Romantic progression: The psychology of literary history*. Washington, D.C.: Hemisphere.

Martindale, C. (1989). Personality, situation, and creativity. In: J. A. Glover, R. R. Ronning & C. R. Reynolds (Eds), *Handbook of Creativity* (pp. 211–232). New York: Plenum Press.

Martindale, C. (1990). *The clockwork muse: The predictability of artistic styles*. New York: Basic Books.

McCann, S. J. H. (2001). The precocity-longevity hypothesis: Earlier peaks in career achievement predict shorter lives. *Personality and Social Psychology Bulletin*, **27**, 1429–1439.

McCann, S. J. H. & Stewin, L. L. (1984). Environmental threat and parapsychological contributions to the psychological literature. *Journal of Social Psychology*, **122**, 227–235.

McCrae, R. R., Arenberg, D. & Costa, P. T. (1987). Declines in divergent thinking with age: Cross-sectional, longitudinal, and cross-sequential analyses. *Psychology and Aging*, **2**, 130–136.

McCurdy, H. G. (1960). The childhood pattern of genius. *Horizon*, **2**, 33–38.

McDowell, J. M. (1982). Obsolescence of knowledge and career publication profiles: Some evidence of differences among fields in costs of interrupted careers. *American Economic Review*, **72**, 752–768.

Munsterberg, H. (1983). *The crown of life: Artistic creativity in old age*. San Diego, CA: Harcourt-Brace-Jovanovich.

Nemeth, C. J. & Kwan, J. (1987). Minority influence, divergent thinking and detection of correct solutions. *Journal of Applied Social Psychology*, **17**, 788–799.

Nicholls, J. G. (1972). Creativity in the person who will never produce anything original and useful: The concept of creativity as a normally distributed trait. *American Psychologist*, **27**, 717–727.

Oromaner, M. (1977). Professional age and the reception of sociological publications: A test of the Zuckerman–Merton hypothesis. *Social Studies of Science*, *7*, 381–388.

Over, R. (1982a). Does research productivity decline with age? *Higher Education*, **11**, 511–520.

Over, R. (1982b). Is age a good predictor of research productivity? *Australian Psychologist*, **17**, 129–139.

Over, R. (1989). Age and scholarly impact. *Psychology and Aging*, **4**, 222–225.

Padgett, V. & Jorgenson, D. O. (1982). Superstition and economic threat: Germany 1918–1940. *Personality and Social Psychology Bulletin*, **8**, 736–741.

Plomin, R., Fulker, D. W., Corley, D. W. & DeFries, J. C. (1997). Nature, nurture, and cognitive development from 1 to 16 years: A parent–offspring adoption study. *Psychological Science*, **8**, 442–447.

Poincaré, H. (1921). *The foundations of science: Science and hypothesis, the value of science, science and method* (G. B. Halstead, Trans.). New York: Science Press.

Post, F. (1994). Creativity and psychopathology: A study of 291 world-famous men. *British Journal of Psychiatry*, **165**, 22–34.

Post, F. (1996). Verbal creativity, depression and alcoholism: An investigation of one hundred American and British writers. *British Journal of Psychiatry*, **168**, 545–555.

Price, D. (1963). *Little science, big science*. New York: Columbia University Press.

Price, D. (1978). Ups and downs in the pulse of science and technology. In: J. Gaston (Ed.), *The Sociology of Science* (pp. 162–171). San Francisco, CA: Jossey-Bass.

Quételet, A. (1968). *A treatise on man and the development of his faculties*. New York: Franklin. (Reprint of 1842 Edinburgh translation of 1835 French original.)

Raskin, E. A. (1936). Comparison of scientific and literary ability: A biographical study of eminent scientists and men of letters of the nineteenth century. *Journal of Abnormal and Social Psychology*, **31**, 20–35.

Roe, A. (1953). *The making of a scientist*. New York: Dodd, Mead.

Root-Bernstein, R. & Root-Bernstein, M. (2003). Tools for thinking intuitively. In: L. V. Shavinina (Ed.), *International Handbook on Innovation*. Oxford: Elsevier Science.

Root-Bernstein, R. S., Bernstein, M. & Garnier, H. (1993). Identification of scientists making long-term, high-impact contributions, with notes on their methods of working. *Creativity Research Journal*, **6**, 329–343.

Sales, S. M. (1972). Economic threat as a determinant of conversion rates in authoritarian and non-authoritarian churches. *Journal of Personality and Social Psychology*, **23**, 420–428.

Scarr, S. & McCartney, K. (1983). How people make their own environments: A theory of genotype → environmental effects. *Child Development*, **54**, 424–435.

Schaefer, C. E. & Anastasi, A. (1968). A biographical inventory for identifying creativity in adolescent boys. *Journal of Applied Psychology*, **58**, 42–48.

Schaller, M. (1997). The psychological consequences of fame: Three tests of the self-consciousness hypothesis. *Journal of Personality*, **65**, 291–309.

Schneider, J. (1937). The cultural situation as a condition for the achievement of fame. *American Sociological Review*, **2**, 480–491.

Schubert, D. S. P., Wagner, M. E. & Schubert, H. J. P. (1977). Family constellation and creativity: Firstborn predominance among classical music composers. *Journal of Psychology*, **95**, 147–149.

Simon, H. A. (1955). On a class of skew distribution functions. *Biometrika*, **42**, 425–440.

Simon, R. J. (1974). The work habits of eminent scientists. *Sociology of Work and Occupations*, **1**, 327–335.

Simonton, D. K. (1975a). Age and literary creativity: A cross-cultural and transhistorical survey. *Journal of Cross-Cultural Psychology*, **6**, 259–277.

Simonton, D. K. (1975b). Sociocultural context of individual creativity: A transhistorical time-series analysis. *Journal of Personality and Social Psychology*, **32**, 1119–1133.

Simonton, D. K. (1976a). Biographical determinants of achieved eminence: A multivariate approach to the Cox data. *Journal of Personality and Social Psychology*, **33**, 218–226.

Simonton, D. K. (1976b). Interdisciplinary and military determinants of scientific productivity: A cross-lagged correlation analysis. *Journal of Vocational Behavior*, **9**, 53–62.

Simonton, D. K. (1976c). Philosophical eminence, beliefs, and zeitgeist: An individual-generational analysis. *Journal of Personality and Social Psychology*, **34**, 630–640.

Simonton, D. K. (1977a). Creative productivity, age, and stress: A biographical time-series analysis of 10 classical composers. *Journal of Personality and Social Psychology*, **35**, 791–804.

Simonton, D. K. (1977b). Eminence, creativity, and geographic marginality: A recursive structural equation model. *Journal of Personality and Social Psychology*, **35**, 805–816.

Simonton, D. K. (1980a). Techno-scientific activity and war: A yearly time-series analysis, 1500–1903 A.D. *Scientometrics*, **2**, 251–255.

Simonton, D. K. (1980b). Thematic fame and melodic originality in classical music: A multivariate computer-content analysis. *Journal of Personality*, **48**, 206–219.

Simonton, D. K. (1980c). Thematic fame, melodic originality, and musical zeitgeist: A biographical and transhistorical content analysis. *Journal of Personality and Social Psychology*, **38**, 972–983.

Simonton, D. K. (1984a). Artistic creativity and interpersonal relationships across and within generations. *Journal of Personality and Social Psychology*, **46**, 1273–1286.

Simonton, D. K. (1984b). Creative productivity and age: A mathematical model based on a two-step cognitive process. *Developmental Review*, **4**, 77–111.

Simonton, D. K. (1984c). Generational time-series analysis: A paradigm for studying sociocultural influences. In: K. Gergen & M. Gergen (Eds), *Historical Social Psychology* (pp. 141–155). Hillsdale, NJ: Lawrence Erlbaum.

Simonton, D. K. (1984d). *Genius, creativity, and leadership: Historiometric inquiries*. Cambridge, MA: Harvard University Press.

Simonton, D. K. (1984e). Is the marginality effect all that marginal? *Social Studies of Science*, **14**, 621–622.

Simonton, D. K. (1985). Quality, quantity, and age: The careers of 10 distinguished psychologists. *International Journal of Aging and Human Development*, **21**, 241–254.

Simonton, D. K. (1986a). Aesthetic success in classical music: A computer analysis of 1935 compositions. *Empirical Studies of the Arts*, **4**, 1–17.

Simonton, D. K. (1986b). Biographical typicality, eminence, and achievement style. *Journal of Creative Behavior*, **20**, 14–22.

Simonton, D. K. (1986c). Popularity, content, and context in 37 Shakespeare plays. *Poetics*, **15**, 493–510.

Simonton, D. K. (1987a). Developmental antecedents of achieved eminence. *Annals of Child Development*, **5**, 131–169.

Simonton, D. K. (1987b). Musical aesthetics and creativity in Beethoven: A computer analysis of 105 compositions. *Empirical Studies of the Arts*, **5**, 87–104.

Simonton, D. K. (1988a). Age and outstanding achievement: What do we know after a century of research? *Psychological Bulletin*, **104**, 251–267.

Simonton, D. K. (1988b). Galtonian genius, Kroeberian configurations, and emulation: A generational time-series analysis of Chinese civilization. *Journal of Personality and Social Psychology*, **55**, 230–238.

Simonton, D. K. (1989a). Age and creative productivity: Nonlinear estimation of an information-processing model.

International Journal of Aging and Human Development, **29**, 23–37.

Simonton, D. K. (1989b). The swan-song phenomenon: Last-works effects for 172 classical composers. *Psychology and Aging,* **4**, 42–47.

Simonton, D. K. (1991a). Career landmarks in science: Individual differences and interdisciplinary contrasts. *Developmental Psychology,* **27**, 119–130.

Simonton, D. K. (1991b). Emergence and realization of genius: The lives and works of 120 classical composers. *Journal of Personality and Social Psychology,* **61**, 829–840.

Simonton, D. K. (1991c). Latent-variable models of posthumous reputation: A quest for Galton's *G. Journal of Personality and Social Psychology,* **60**, 607–619.

Simonton, D. K. (1992a). Gender and genius in Japan: Feminine eminence in masculine culture. *Sex Roles,* **27**, 101–119.

Simonton, D. K. (1992b). Leaders of American psychology, 1879–1967: Career development, creative output, and professional achievement. *Journal of Personality and Social Psychology,* **62**, 5–17.

Simonton, D. K. (1992c). The social context of career success and course for 2,026 scientists and inventors. *Personality and Social Psychology Bulletin,* **18**, 452–463.

Simonton, D. K. (1994). *Greatness: Who makes history and why.* New York: Guilford Press.

Simonton, D. K. (1996). Individual genius and cultural configurations: The case of Japanese civilization. *Journal of Cross-Cultural Psychology,* **27**, 354–375.

Simonton, D. K. (1997a). Achievement domain and life expectancies in Japanese civilization. *International Journal of Aging and Human Development,* **44**, 103–114.

Simonton, D. K. (1997b). Creative productivity: A predictive and explanatory model of career trajectories and landmarks. *Psychological Review,* **104**, 66–89.

Simonton, D. K. (1997c). Foreign influence and national achievement: The impact of open milieus on Japanese civilization. *Journal of Personality and Social Psychology,* **72**, 86–94.

Simonton, D. K. (1998a). Achieved eminence in minority and majority cultures: Convergence vs. divergence in the assessments of 294 African Americans. *Journal of Personality and Social Psychology,* **74**, 804–817.

Simonton, D. K. (1998b). Political leadership across the life span: Chronological vs. career age in the British monarchy. *Leadership Quarterly,* **9**, 195–206.

Simonton, D. K. (1999a). Creativity and genius. In: L. A. Pervin & O. John (Eds), *Handbook of Personality Theory and Research* (2nd ed.). New York: Guilford Press.

Simonton, D. K. (1999b). *Origins of genius: Darwinian perspectives on creativity.* New York: Oxford University Press.

Simonton, D. K. (1999c). Talent and its development: An emergenic and epigenetic model. *Psychological Review,* **106**, 435–457

Simonton, D. K. (2000). Creative development as acquired expertise: Theoretical issues and an empirical test. *Developmental Review,* **20**, 283–318.

Sorokin, P. A. (1969). *Society, culture, and personality.* New York: Cooper Square. (Original work published 1947)

Sorokin, P. A. & Merton, R. K. (1935). The course of Arabian intellectual development, 700–1300 A.D. *Isis,* **22**, 516–524.

Sternberg, R. J. & Lubart, T. I. (1995). *Defying the crowd: Cultivating creativity in a culture of conformity.* New York: Free Press.

Sternberg, R. J., Pretz, J. E. & Kaufman, J. C. (2003). The propulsion model of creative contributions applied to invention. In: L. V. Shavinina (Ed.), *International Handbook on Innovation.* Oxford: Elsevier.

Suedfeld, P. (1985). APA presidential addresses: The relation of integrative complexity to historical, professional, and personal factors. *Journal of Personality and Social Psychology,* **47**, 848–852.

Suedfeld, P. & Piedrahita, L. E. (1984). Intimations of mortality: Integrative simplification as a predictor of death. *Journal of Personality and Social Psychology,* **47**, 848–852.

Sulloway, F. J. (1996). *Born to rebel: Birth order, family dynamics, and creative lives.* New York: Pantheon.

Taagepera, R. & Colby, B. N. (1979). Growth of Western civilization: Epicyclical or exponential? *American Anthropologist,* **81**, 907–912.

Teigen, K. H. (1984). A note on the origin of the term 'nature and nurture': Not Shakespeare and Galton, but Mulcaster. *Journal of the History of the Behavioral Sciences,* **20**, 363–364.

Terman, L. M. (1925). *Mental and physical traits of a thousand gifted children.* Stanford, CA: Stanford University Press.

Terry, W. S. (1989). Birth order and prominence in the history of psychology. *Psychological Record,* **39**, 333–337.

Tweney, R. D. (1990). Five questions for computationalists. In: J. Shrager & P. Langley (Eds), *Computational Models of Scientific Discovery and Theory Information* (pp. 471–484). San Mateo, CA: Kaufmann.

Walberg, H. J., Rasher, S. P. & Parkerson, J. (1980). Childhood and eminence. *Journal of Creative Behavior,* **13**, 225–231.

Waller, N. G., Bouchard, T. J., Jr., Lykken, D. T., Tellegen, A. & Blacker, D. M. (1993). Creativity, heritability, familiality: Which word does not belong? *Psychological Inquiry,* **4**, 235–237.

Weisberg, R. W. (1994). Genius and madness? A quasi-experimental test of the hypothesis that manic-depression increases creativity. *Psychological Science,* **5**, 361–367.

Weisberg, R. W. (2003). Case studies of innovation: Ordinary thinking, extraordinary outcomes. In: L. V. Shavinina (Ed.), *International Handbook on Innovation.* Oxford: Elsevier Science.

Zhao, H. & Jiang, G. (1986). Life-span and precocity of scientists. *Scientometrics,* **9**, 27–36.

Zuckerman, H. (1977). *Scientific elite.* New York: Free Press.

Zuckerman, H. & Merton, R. K. (1972). Age, aging, and age structure in science. In: M. W. Riley, M. Johnson & A. Foner (Eds), *Aging and Society: Vol 3. A Sociology of Age Stratification* (pp. 292–356). New York: Russell Sage Foundation.

The International Handbook on Innovation
Edited by Larisa V. Shavinina
© 2003 Published by Elsevier Science Ltd.

Innovations by the Frail Elderly

Thomas E. Heinzen and Nancy Vail

Department of Psychology, William Paterson University of New Jersey, USA

Abstract: Discovering the principles that underlie innovation requires a variety of research strategies. One counter-intuitive strategy is to study populations whose extreme circumstances make innovation unlikely. The principles abstracted here are derived from case studies of frail elderly people living in nursing homes. Varied innovations are described and interpreted as a consequence of the interaction between external stressors and internal characteristics, despite significant environmental and personal limitations. The article examines the validity of the fit between these observations and Amabile's description of necessary and sufficient components for innovative behavior.

Keywords: Innovation; Frail elderly; Creativity; Aging; Circumstances; Counter-intuitive; Frustration; Motivation; Stressors.

Introduction

Understanding the principles that reliably predict innovation will require all of the customary tools in our collective research tool kit. Even then, it is likely that we will have to develop still more techniques because the nature of the phenomenon we are seeking to understand is not only pervasive but also inherently surprising and complex. Consequently, the present chapter begins by offering a brief rationale for the mildly unusual approach presented here: naturalistic/clinical observations of a population *un*likely to practice innovation, specifically, the frail elderly living in nursing homes.

Interspersed throughout the chapter are three case study reports that progressively describe: (a) a problem requiring innovation; (b) failed or marginally successful solutions to that problem that force the innovating individuals to clarify the problem and search for alternative answers; and (c) eventual innovations. The chapter abstracts principles that emerge from the combination of relevant literature and these case observations.

The only modifications to these case studies are those details necessary to honor the customary constraints of confidentiality and informed consent.

The Complex Causes of Innovation

Like the proverbial swan gliding elegantly on the water but paddling furiously beneath the surface, the product of innovation may appear to be a single, elegant act, but the process that produces that innovation is a multi-factored, emotionally complex process (Shaw, 1994). A variety of theoretical models with varying degrees of empirical support agree on this fundamental observation about the complex nature of innovation (Runco & Shaw, 1994). One such model, useful because it is directive (Albert & Runco, 1990) that also proposes creativity-enhancing interventions (Amabile, 1990), is Amabile's (1983) componential framework, used to understand the innovations described here. Those components include domain-relevant skills, creativity-relevant skills, and task motivation.

At the broadest definitional level, Amabile (1990, p. 66) recognizes that there is a fairly high consensus that both creativity and innovation are accurately defined as consisting of both novelty (or originality) and whatever works (effectiveness). At a more specific level, the components articulated by Amabile continue to bear a fairly high level of face validity as well as allowing for more discriminating predictions and applications. For example, her componential definition worked well in a study of a particular technological evaluation (Heinzen, 1990) revealing that one form of distance education (one-way video, two-way audio) met or exceeded participant expectations within each domain except for that related to creativity-relevant skills, particularly creative thinking. Participants were disappointed in the degree to which this technology did not facilitate creative thinking. The twin elements of novelty and effectiveness were embedded in this more detailed analysis of a particular technology. The model also appears to offer some measure of construct validity, as suggested by the way it clarified the distinctive roles played by mid-level managers in state government documented as producing innovation in spite of the bureaucratic restraints (Heinzen, 1990, 1994). In this example, novelty and effectiveness represented distinctive roles played by particular managers that, when functioning as a team, led to innovation with state government. The model also more pragmatically articulated unplayed roles that tended to block innovation while contributing to a particular kind of organizational analysis.

Each of the components is complex. For example, the affective component of the innovation process appears to include a range of cognitively driven self-assessment activities at various points in the process (Shaw, 1994). Similarly, the pragmatic skills which support innovation within particular domains (such as medical research, non-fiction literature, cooking, or public administration) are themselves unique to the domain and even more specific to the particular tasks within those sub-domains (Heinzen, 1994b). For example, innovations within public administration involve fairly mundane skills such as being able to run a good meeting to a detailed knowledge of the causes of public transportation successes and failures in diverse communities, as well as the differential causes of those outcomes (Heinzen, 1994a). But the component most difficult to understand appears to be the affectively related component of task motivation.

The motivational component of innovation is the subject of intense, thoughtful theoretical and empirical debate. The debate focuses on both the measurement and validity of the distinction between intrinsic and extrinsic motivation and their varying consequences (Eisenberger & Cameron, 1996). This slow but important level of empirical debate is beneficial to our understanding of innovation as it forces us to clarify our terms and intended meanings. But for the time being, we can use this brief review of innovation literature to assert a fairly obvious but nevertheless important recognition: *the process that produces innovation is complex.*

Case Study 1: The Difficulties of a Smoking Habit in a Nursing Home

B. V. is an 84-year-old, long-term cigarette smoker who was compelled, by a stroke, financial circumstances, and limited family support, to move to a nursing home following a stroke. The stroke left B. V. confined to a wheelchair with limited use of her left side in both the upper and lower extremities. In addition to the stroke, B. V. was suffering from severe, long-term depression related to the shock and adjustment requirements of moving away from her

previously supportive but aging community support network to a nursing home much closer to her nearest relative. In these circumstances, B. V. declared that the chief satisfaction of her life was "to smoke cigarettes, and I don't care at all if it kills me". She especially wished to start her day with an early morning cigarette.

The nursing home was understandably not eager to support a smoking habit within a facility where many residents were immobile and maintaining respiratory function through oxygen support. In addition to concerns about B. V.'s health, the nursing home also worked hard to remove all fire hazards and was particularly sensitive to threats posed by long-term smokers. Consequently, B. V. reluctantly agreed to a 'contract' limiting her cigarette smoking to one cigarette in the morning and another in the evening, outside the facility, and only while supervised. However, B. V. also was free to smoke more if she could persuade others to help her achieve these safety requirements. Regardless of whatever ethical issues are engaged by this decision, the resulting situation was that B. V. was a highly motivated smoker. She was limited by severe mobility deficits (barely strong enough to wheel her own wheelchair), by living on the third floor of a nursing home, and by needing an elevator to take her to the ground floor where she might be able to go outside and smoke. It is within the context of her severe limitations as well as her highly motivated (i.e. addictive) desire to fulfill this goal of smoking as much as possible that we can view and come to understand her innovations.

Innovation and Unusual Methodologies

Although there is frequently significant reluctance to give credence to innovative scientific methods and their sometimes disturbing conclusions, the history of science suggests that our degree of openness to new ways of addressing scientific issues significantly contributes to the energy that drives scientific revolutions (Kuhn, 1962). Scientific discoveries in psychology certainly are no exception. For example, a significant amount of research on attraction depended on a simple experimental paradigm developed by Byrne (1971). As another example, research on self-recognition among both infants and infants depended upon 'dreaming up' the idea of placing colored dots on foreheads and observing their behavior before a mirror (Gallup, 1977, 1995). That idea was adapted by Cameron (1988) to observe infants 'playing' with their shadows. This methodological innovation clarified (and lowered) the age at which we can demonstrate infant self-recognition.

The slow progression of knowledge in psychology is the consequence of these innovative methodologies as surely as other sciences have depended upon improved capabilities to view previously unseen phenomena by progressively more inventive microscopes. Moreover, as a subject of investigation, innovation is by definition both surprising (novel) and wide (pervasive), yet nonetheless deep (complex). The principles that predict and support innovation are diverse because innovation occurs across 'disciplines', drives evolution (Heinzen, 1994b), yet produces adaptive responses that individuals report as personally transforming. So a second common-sense assertion about innovation, also drawn from the literature, is that *understanding innovation will require a wide range of methodological approaches*. The study of unlikely populations is one such methodological innovation capable of producing insights into innovation that may generalize to other populations.

Case Study 2: How to Repair a Velcro Strap

A. S. is a 94-year-old male who is in generally good health. However, he fatigues quickly and has difficulty using his fingers due to arthritis. Nevertheless, he retains a sense of humor and enthusiasm for problem-solving, reminiscing, and eating even though his taste sensitivity has been severely limited. A. S. lives in a nursing home, and his source of discretionary spending is limited to fifty dollars per month. He is obligated by the practice of this particular home to spend most of that money on a clothing allowance and toiletries in addition to those already supplied, and any gifts, cards, or 'extras' that he may desire.

His immediate problem was that he spilled broccoli soup on the strap of his sneakers. The straps use velcro to secure the shoes to his feet. The soup proved to be difficult to clean and the consequent scrubbing damaged the velcro so that the velcro strap no longer 'grabbed' efficiently; although the straps fastened initially, they came apart when A. S. walked. A. S. was reluctant to ask the aide for further assistance because he felt embarrassed at being needy in any form; he also enjoyed the challenge of trying to solve problems himself. This was a fairly urgent problem because A. S. could not walk confidently or very far unless his shoes were securely fastened to his feet.

Insights From Non-Representative Samples

Studying unlikely or inappropriate populations is among the counter-intuitive approaches and techniques that appear to have provided an unreasonable level of scientific advance. For example, the particular and very small proportion of the HIV-positive population who do not develop full-blown AIDS is now serving as one promising model for how to understand, treat, and maintain immune function and thus manage the disease (Kolata, 2001). In this case, the non-representative exception to the rule may prove to be instructive.

The effectiveness of a different kind of 'error' was demonstrated by Piaget. Piaget broke several fundamental rules of population sampling by his attentive (perhaps we should dare to say 'loving' or at least 'fascinated') observations of his own children. Although details of his theories have been modified, it was Piaget's own inventiveness in documenting his observations that led to their subsequent refinement. Moreover, his core observation of cognitive development, as well as the methods he used to document those observations, has led to thousands of studies, theoretical variants, and distinctive interventions.

The example of the child development theories proposed by Piaget is particularly applicable to the present chapter because these case studies document unusual, innovative cognitive activity among a different yet distinctive age-population: the frail elderly. What would Piaget have made of these observations?

Subsequent research regarding Piaget's many ideas, inspired mostly by Piaget's methods, has led to several modifications. Specifically, Berk (1988) noted that Piaget probably underestimated the competencies of infants and children and that he inaccurately doubted that training could influence development. In addition, improved methods suggest that cognitive stages are more gradual than discrete and that they probably continue much longer into the lifespan. However, Piaget's initial observations of his children, the methods he employed to validate those observations, and the observation of continuing cognitive growth have retained their validity. They have inspired ever more innovative research, and suggest that non-normative populations can lead to productive, creative thinking that holds up under more careful, empirical tests. It is not unusual to learn from extreme or unlikely populations; indeed, they may compel more creative thinking that our customary theories generally allow. The observations documented here represent one way to rethink our attitudes and expectations regarding the capabilities of the frail elderly, as well as the factors that facilitate or impede innovation.

Consequently, even though the abstraction of principles from such unlikely samples would not appear to generalize to larger populations, they appear to provide an opportunity for insight that might otherwise be lost or relatively inaccessible. So a third assertion of this chapter is that *non-normative populations can provide insights that may generalize to larger populations*.

Case Study 3: How to Accept the Unacceptable

L. N. is a 96-year-old female living in a nursing home. She is diagnosed with Parkinson's Disease, mild dementia, arthritis, macular degeneration, and depression. L. N. is a lifelong religiously oriented individual who carried a personal 'secret' for approximately 80 years. Raised in a small, rural, conservative and very religious community outside the United States, she became pregnant when she was 15 years old. L. N. was quietly moved by her family to a larger city in the United States, in order to avoid the social embarrassment of an unwanted pregnancy. The resulting infant was returned to the home community where it was raised by relatives of the father while L. N. herself continued her life by moving and working in the New York metropolitan area. L. N. eventually married, produced several children, grandchildren, and great-grandchildren.

During our first interview, L. N. agreed to further visits but only under the condition that "there are some things that I will never tell you". A few months later, L. N. received a letter from her first child, now 81 years of age. This daughter's daughter had 'found' her maternal grandmother through a variety of church records and genealogies published on the Internet. L. N. did not want to meet with this 'new' family because she wished to maintain her lifelong secret. However, the letter seeking information about her had become known throughout her family during the genealogical search, and L. N. was obliged to meet the daughter she had given up for adoption 81 years earlier. Against her will, L. N.'s secret had been revealed.

Learning About Innovation From the Frail Elderly

One population that appears to be unlikely to demonstrate significant innovation is the frail elderly. The factors that would seem to oppose innovation by individuals living in nursing homes are many. First, that population is pre-selected as physically needy, usually the result of a major health crisis such as a debilitating stroke, Alzheimer's disease, vascular dementia, or functional limb loss such as a hip replacement due to a fall, often related to osteoporosis. Moreover, many psychological disorders are co-morbid with these physical difficulties. The proportion of the elderly who carry the risk factors associated with depression is estimated at 80% with significant co-morbidity accompanied by other innovation-suppressing factors such as "disabling illness, chronic illness, financial strain, lack of social support, prior history of depression, and family history of depression" (Tangalos et al., 2001). Innovation seems unlikely to flourish in such a setting.

However, this population of frail elderly does offer something that other more conventional populations do not. They are living under conditions of chronic stress. These exceptional levels of stress are highly motivating personal circumstances and support reactive creativity and consequent innovation (Heinzen, 1994). Verbal innovations seem particularly likely to flourish since this is the one remaining domain-relevant skill that is common to most, but not all, of this particular population (Heinzen, 1996). This suggests a fourth, more tentative assertion about innovation: *The impetus for innovation may be external, unpleasant, and unwelcome.*

Case Study 1: The Real Problem is Keeping the Elevator Open

From the moment she awoke, B. V.'s primary purpose was to smoke cigarettes as much and as often as possible. However, there were several obstacles to her goal, each occurring at three distinct stages. First, she had to persuade staff to care for her before others so she could be in her wheelchair as soon as possible. Second, she had to make her way to the elevator, reach the call button, enter the elevator, direct it to the proper floor, and exit the elevator. Third, B. V. needed to find the individuals who had access to her cigarettes and matches and persuade someone downstairs to take her outside and sit with her while she smoked. Each of these steps presented very real physical and emotional difficulties that B. V. persistently sought to overcome throughout all her waking hours.

The first need for cooperation from nursing aides and nurses who administer medications was set within the context of many, complex personalities, limited resources, and a high level of demand from fellow residents. In addition, the personalities holding some measure of responsibility and consequent authority over a nursing home resident are many, an unavoidable feature of a 24-hour care facility that includes among fellow residents, nurses, nurses' aides, physicians, and a variety of therapists (physical, occupational, speech, and psychological).

For example, each nurse's aide is responsible for as many as seven to ten patients. In addition to the intrinsic variability of their personal needs, each patient is acutely aware of the importance of their own needs and understandably less aware of the severity of the needs of others they cannot see (but may hear). Furthermore, the aides do not work seven days per week and occasionally take vacations. Combined with an every-three-month employee rotation pattern across the floor, a particular individual has at least three aides to cover their 24-hour needs plus another three aides to accommodate various shift changes. Each three-month rotation of nurse's aides (usually a contractual feature of the workplace) introduced a new set of 'helping' personalities to the frail, elderly residents. Consequently, B. V. was obliged to discover ways in which each aide and nurse might be more or less effectively persuaded to prioritize her morning care so that she could satisfy her strong desire for cigarettes.

Once she arrived downstairs, B. V. faced a similarly daunting set of inter-personal challenges to find a sympathetic, available helper so that she could achieve her goal of maximal cigarette smoking. Stages one and three of B. V.'s problem-solving were difficult, constantly shifting, inter-personal problems. To a moderate degree, B. V. solved these problems through flattery, loud complaining, accusations, praise, and threats of poor behavior. In short, B. V. used the customary currency of human relations to cajole individuals to help her fulfill the demands of her cigarette addiction. However, the focus of the present observation is on B. V.'s far more tangible, intermediate goal of manipulating the elevator.

There were several obstructions confronting B. V. in her attempt to achieve what appears to be a simple goal. Although, the elevator was approximately 75 feet from the entrance to her room, the staff was keenly aware that B. V. represented a potential fire hazard with her willingness to break the facility rules for cigarette smoking. Consequently, they were not generally disposed to be helpful when B. V. sought assistance to get to the elevator, believing also that greater help could be offered to B. V. by helping her to stop cigarette smoking entirely. Second, B. V.'s stroke as well as the long-term effects of cigarette smoking made the act of wheeling herself to the elevator exhausting. Third, the buttons for working the elevator were hidden behind a spring-loaded cover that was placed at a height of about four feet. This height represented a significant reach upward for a slightly built individual with osteoporosis and confined to a wheel chair.

In addition to these constraints, B. V.'s stroke limited her to having only one arm available to position the wheelchair, manipulate the cover, reach the button, and then wheel herself into the elevator after the door opened. Finally, and perhaps most daunting to B. V., was that after achieving all of this, the door to the elevator remained open only for a limited period of time. If B. V. did not cross the electric eye with her wheelchair quickly enough, the door would close and B. V. would have to reposition her wheelchair and repeat the entire, laborious process once again.

Situation-Driven Versus Personality-Inspired Innovation

Taken together, these three accounts demonstrate a 'source of motivation continuum' that describes the range of motivations to innovate. At one end of the continuum is situation-driven innovation, as demonstrated in these case study accounts of innovation by the frail elderly and noted as an impetus to innovation in diverse other settings (Heinzen, 1994b). Individual circumstances sometimes require innovation in order for some goal to be realized. In our case studies, these included cigarette smoking, fastening damaged velcro, or discovering a way to emotionally accommodate oneself when a life-long secret has been revealed. This situation-driven innovation would not occur unless the situation provided no other alternative.

At the other end of the continuum is personality-driven innovation. This is creative and self-initiating innovation. We tend to attribute 'creativity' to those personalities that chronically explore, change, adjust, invent, and seem disposed to do so whether or not circumstances are pressuring them. Moreover, it seems likely that such creative personalities are likely to innovate more readily when faced with challenging circumstances. However, such 'creative types' seem to be relatively rare, in keeping with the low frequency of genius and the corresponding low probability of any individual being strongly equipped with all three of Amabile's necessary and sufficient components for creativity.

Consequently, we can offer this fifth assertion that complements the observations of our case study: *Innovation does not require a 'creative personality'; ordinary personalities in extraordinary circumstances will attempt to innovate.*

Case Study 2: An Unsatisfactory Solution to the Velcro Problem

A. S. was highly motivated to walk. The velcro that continually 'came apart' when he tried to walk represented a severe threat to his independence and to his sense of himself as a competent individual. His inability to afford new sneakers, his reluctance to ask for help, and his determination to continue walking motivated A. S. to search for new ways to solve his velcro problem. His first attempt was to wrap rubber bands around his sneakers. This was a relatively difficult task to achieve, given his level of arthritis, limited movement, and the necessity of applying the rubber bands while his sneakers were on his feet.

However, A. S. did achieve some limited success using his rubber band solution: the velcro stayed in place, but only if it was not very tight. In addition, the rubber bands tended to move as A. S. shuffled across the floor. The rubber bands proved to be only a temporary, relatively unsatisfactory solution.

Frustration and Innovation

These three case studies all have in common the experience of significant frustration, even suffering. This commonality begins to tie these observations to the literature describing the relationship between frustration and innovation. A. S. was frustrated in his attempt to fix the velcro on his sneakers. While a trivial inconvenience for most of the population, this frustration represented a significant loss of mobility, some social embarrassment, and a personal humiliation. Similarly, B. V. was deeply frustrated in her desire to smoke cigarettes. She applied a lifetime's worth of human relationship skills to persuade different individuals to assist her to smoke yet remained frustrated by her inability to manipulate the elevator doors. L. N. also experienced frustration but it was about a less tangible goal. She wished to preserve a personal secret that she had successfully shielded from various family and friends for more than 80 years. A. S. and B. V. were frustrated in active pursuit of a goal; L. N. was frustrated when an outside force imposed frustration on the goal that she already had achieved. All three individuals, in their own particular way, were frustrated.

Frustration, we now recognize, leads to many consequences. The most well-known, frequently studied consequence of frustration is aggression (Dollard et al., 1939). But this does not exhaust the network of possible consequences of frustration. Other candidates for the consequences of frustration include the kind of depression known as learned helplessness, renewed perseverance, a variety of intra-psychic defense mechanisms, resetting one's goal, and innovation (Heinzen, 1994a). It is likely that each of the participants in these three case studies experienced each of these reactions, in varying degrees.

The psychological literature has addressed the relationship between frustration and innovation in a variety of ways. Psychodynamic authors identified 'sublimation' as a defense mechanism whose purpose was to transform privately psychic frustrations into socially acceptable works of creativity (Freud, 1933; May, 1975). Using a more straightforward, social perspective, Lewin (1935) expressed a similar relationship between frustration and creativity but in a far more pragmatic manner. Lewin articulated frustration as an obstruction to a goal that provoked a variety of responses, including circumvention and consequent innovation as a means of goal attainment. Heinzen (1994b) proposed a relatively coherent model of responses to frustration in his description of the frustration-response network. Consequently, a variety of literatures, both theoretical and empirical, affirm the experiences identified in these case studies: *Frustration inspires innovation.*

Case Study 3: Recognizing the Lack of Alternatives

L. N.'s initial reaction to being 'discovered' by the child she gave up for adoption 81 years earlier was extremely negative. She reported disappointment in her failure to maintain what she considered to be an embarrassing secret. She experienced both sadness and guilt that her grandchildren and great-grandchildren would be burdened with a sense of shame and social embarrassment. She expressed annoyance that she was still alive. She became 'mad at God' for allowing this to take place. She felt fear at having to meet and face the possible judgment of her 'new' daughter. She re-experienced the social disapproval of particular family members and a private sense of shame, not so much for bearing a child 'without benefit of clergy', but for singling herself out in such a public way.

L. N. also experienced increasing anxiety as the details of flights, family meetings, and related arrangements systematically counted down towards the actual day of the meeting, aggravated by the general sense of excitement that family members conveyed to her. L. N. had successfully 'buried' her secret and experienced very real suffering as her secret was systematically revealed against her will. However, L. N. also recognized the inevitability of the event, was unwilling to consider taking her own life as a way to avoid the situation, and slowly started to reconcile herself to facing the situation she had spent 80 years avoiding.

Suffering Also Can Lead to Innovation

The term 'frustration' does not accurately convey the quality of experience that L. N. reported when she was first contacted by her 81-year-old daughter. A more appropriate term may be 'emotional suffering' and there is additional evidence that this deeper, longer lasting, more profound experience also is capable of producing innovation and creativity. In a book appropriately titled *Hope Never Dies*, holocaust survivor Sarah Wahrman (1999) describes the forced labor required in Nazi concentration camps, the depth of suffering imposed upon those victims, and the startling and inventive coping that this suffering inspired.

Specifically, Wahrman describes working the night shift at Guben, Germany, processing fine wire and metal spools in order to support the German war industry. Prompted by hunger pains and the need to barter for food, this survivor created wire flowers using the colors provided in the fine wire (green, bronze, and silver). She fashioned these onto heavier wires and created flowered pins, the shape of a rabbit, the appearance of a French poodle, and a reindeer with horns. She bartered these with the camp cook for extra food that was then divided among a larger group of 'inmates'. This kind of invention in the face of shared disaster suggests that innovation is more than just a clever ornament that periodically improves daily existence—innovation is a fundamental survival tool, a product of human evolution focused first on survival rather than amusement, decoration, or improvement.

Similarly, during the more recent genocide in Rwanda, the terror experienced by children was represented artistically in an innovative attempt to help children recover from the severe trauma of that more recent holocaust. In only 100 days during the year 1994, more than 100,000 people were murdered in Rwanda as "an integral part of the Hutu government's plan to exterminate the Tutsi population . . . This holocaust claimed 800,000 lives and created 300,000 orphans". A portion of those children were cared for by relief agencies seeking to help them deal with their private traumas and it was within this context that Salem (2000) was able to document their innovation.

These children indicated the range of response anticipated by the frustration-response network. They reacted "with aggression and rage; others live in

constant fear. More than 60% of the children interviewed in the aftermath of the Rwandan genocide said
they did not care if they grew up" (Clinton, 2000).
Relief workers and trauma counselors, desperate to
find some way to alleviate the stress experienced by
some of these 300,000 children, noticed that the
children "spontaneously created" drawings "some
three to four years after the genocide ended" (Salem,
2000, p. 7). Their drawings and comments documented
their private memories of that atrocity and appear to
have contributed in some small measure to their
personal adjustment. It is difficult to imagine the effect
on a small child of witnessing not only the murder of
one's parents, but the systematic destruction of all the
individuals in your home village.

These same authors report that one desperate mother
placed her well-dressed child on the road and directly
in the path of oncoming soldiers, hoping they would
spare him. But when she observed other children being
slaughtered, she created a disruption, drawing the
soldiers off the road after herself and away from her
child. Although the mother was killed, her child was
rescued by field medics (p. 42), some of whom later
reported the story. Suffering may not merely lead to
innovation, it may require innovation and creativity as
a coping mechanism in the cause of survival and
adjustment to a more peaceful world.

We can tentatively assert from these descriptions that
*suffering threatens survival, clarifies goals, and
thereby facilitates innovation.* It is worth mentioning
two related applications of these principles, one
cautionary and the other challenging. First, since
innovation is complex and requires many components,
there is no guarantee that suffering will lead to
innovation and no consequent need to perversely
impose suffering in order to produce innovation.
Second, it is nonetheless still helpful to recognize that
not all suffering is inherently detrimental. Necessary
surgical interventions are often painful. Similarly, most
parents agree that shaping the attitudes and behavior of
children in a positive way sometimes requires uncomfortable levels of constraint and discipline. Personal
growth may involve, or even require, some level of
suffering. Regardless of whatever particular psychological mechanism may be at work, our general
conclusion is this: *Suffering can nurture innovation.*

Case Study 1: Adapting a Tool to Solve the Elevator Problem

B. V. was moderately successful at negotiating the
complex personality issues related to persuading a
wide variety of individuals to provide minimal
assistance in order to help her smoke. They were
willing to 'get her up' in time for an early cigarette.
This is a remarkable success, partly because B. V.'s
desire to smoke early in the morning was especially
inconvenient. The morning is an especially demanding time of the day for the staff. From

7.00 a.m.–8.30 a.m., the staff was busy delivering
meals, administering medications, and attending to a
wide ranger of personal needs among the residents
of the facility.

However, after arriving at the elevator doors, B. V.
repeatedly found herself in an unproductive, repetitive pattern of behavior. With her one good arm,
B. V. laboriously positioned herself near the button
calling for the elevator. She then reached up to the
metal door protecting the buttons, opened it, pressed
the button, and then failed to move past the electric
eye that would keep the door open long enough for
her to enter the elevator. B. V.'s own frustration
appeared to be increasing as staff refused to assist
her. On the day that I was unobtrusively observing
her, B. V. repeated this cycle six times, with rest
pauses imposed as the elevator doors closed and the
elevator traveled to other floors.

After the sixth failure to get to the elevator's
electric eye quickly enough to keep the doors open,
B. V. stared at the closed elevator door for a lengthy
period of time and then retreated to her room.
Although I assumed that she was discouraged and
had temporarily 'given up', B. V. reappeared in the
hallway some minutes later. Wedged in her wheelchair was a two foot long 'grab bar', a device
designed to assist individuals with limited mobility
to reach further, fasten on light objects, and retrieve
or place them as needed. It is customarily used for
light clothing, sections of a newspaper, or other
light-weight objects.

Although she was now extremely tired by her
efforts, B. V. first attempted to use the grab bar to
manipulate the door to the call buttons from a
distance, but the grab bar was not equal to the
dexterity required for this maneuver. However, after
calling for the elevator in her customary way and
positioning herself as close to the doors as possible,
B. V. frantically waved the grab bar in front of the
electric eye after the door opened. She succeeded in
keeping the elevator door open until she could enter
the elevator. She then used the grab bar to press the
appropriate button and move her further along the
path to her goal.

Levels of Frustration and Cognitive Clarity

Are we justified in referring to B. V.'s behavior as
innovative? Clearly reminiscent of tool use by primates, B. V.'s slight modification of the use of the grab
bar to manipulate the electric eye of the elevator was an
innovation for her. Her action met the twin definitional
requirement of being both original and effective and
helped her achieve a goal that was important to her. It
is interesting to note that B. V. appeared to come up
with the idea after experiencing increasing levels of
frustration and then calming herself sufficiently to
focus on the details of the particular problem she was
trying to solve. When we interviewed her later about

how she came up with the solution to this problem, her only clarification was, "I just thought of it, that's all". B. V.'s increasing frustration and her eventual self-calming response so that she could think up a solution suggests an idea indicated by other empirical literature (Heinzen, 1989): *Frustration will not lead to innovation unless it is sufficiently annoying to force new ways of thinking about a problem.*

Case Study 2: Using a Common Object in a New Way

A. S. continued to have difficulty with the velcro strap of his sneakers and remained frustrated by his relative inability to walk. The combination of limited discretionary money, his private insistence on not asking for help, and his painful arthritis made his temporary solution of applying rubber bands around his shoes inadequate. It also was difficult to maintain since the rubber bands frayed and broke quickly and were somewhat painful to his feet. They were also mildly unattractive and produced some negative comments from others. The rubber band solution seemed to broadcast his disability and A. S. found this public display of his disability and his limited means embarrassing. Consequently, A. S. remained highly motivated to discover a better solution.

The solution came while A. S. was sitting on a couch and observing a secretary stapling papers together. When the secretary left for a brief period, A. S. furtively borrowed her stapler, took it back to his couch, removed his sneakers, opened the stapler so that the bottom would not close the ends of the staple, and painfully pressed one staple through the velcro strap. He tested it, noted that it would grip, and applied about a half dozen staples to substitute for the damaged velcro. This innovation 'held' the strap together. Although it was somewhat more difficult to remove the strap, that inconvenience was preferable to several re-fastenings per day.

Why Innovation Survives, Even in a Nursing Home

The innovation that A. S. achieved also appears to fit the general definition of innovative behavior. A. S. certainly succeeded in adapting an existing device to solve a new problem. Moreover, A. S.'s efforts strike me as a more distinctive application of a stapler than B. V.'s adaptation of a grab bar to solve her particular problem. But both innovations were adaptations to the peculiarities of their private situation. When we interviewed A. S. about his innovative behavior, he proudly reported, "I'm pretty damn clever, aren't I? Well, you have to be in a place like this".

The additional observation that may generalize from this particular observation is that A. S. appeared to be aware of his own habits of innovation, attributed them somewhat to the constraints of living in a nursing home, and enjoyed thinking and behaving in an innovative fashion. This association between innova-

tive behavior and positive affect has been explored in a variety of ways. Isen et al. (1987), for example, demonstrated how small acts that made people feel good appear to 'prime' particular affectively sensitive cognitive structures and thereby increase positive affect and facilitate problem-solving. Isen et al. (1978) suggest that such priming facilitates access to cognitive structures, thus engaging greater complexity and consequent creativity. Given the affective complexity of innovation (Shaw, 1994), both suffering (as a clarifying impetus to solve problems) and positive affect (as an intrinsically satisfying activity) are acknowledged parts of the process that can lead to innovation. Not only does innovation solve a practical problem, but the process of moving from frustration or even suffering to problem-solving and innovation is self-rewarding, both pragmatically and emotionally. The underlying principle suggested here about innovation is that *innovative behavior is both externally and internally self-rewarding.*

Case Study 3: When Changing Yourself is the Only Alternative

L. N. was very clear within herself that she did not want her secret to be made public. For more than 80 years, she had maintained this lifelong secret of giving birth and then giving up a child when she was a 15-year-old mother. However, the manner in which she was 'discovered' made it impossible to shield this news from any of the individuals within her family that she had been trying to protect. Furthermore, L. N. recognized that there was no person appropriate to 'blame' for this revealing of her private secret. L. N. knew that it was her grand-daughter, not her actual daughter, who had sought her out. Her motivation was a genealogical search rather than an anticipation that she still was alive. In short, the only acceptable solution to this state of affairs was for L. N. to change herself. The best solution was for L. N. to accept, at the age of 96, what had been unacceptable to her for more than 80 years. Over time, L. N. even succeeded at discovering a way to enjoy this surprising event, even though the process of arriving at that emotional state was uncomfortable.

We interviewed L. N. several times after the meeting took place between herself, her daughter, her granddaughter, and various members of her more immediate family. There were additional surprises for L. N. as a result of this meeting, not the least of which was that she discovered that she was a great-great-grandmother. It took several weeks for L. N. to experience and to clarify for herself how she felt about this meeting.

Over time L. N. reported that her greatest and most pleasant surprise was the lack of shame or embarrassment conveyed to her by her immediate family, especially the grandchildren. She was also startled by the striking physical resemblance between herself and her newly discovered daughter. She reported, "It was like looking in a mirror". She indicated that, "it was a lot for an old woman to go through". L. N. reported that the general stress of learning this news, anticipating the meeting, experiencing the new relationships, and coping with family reactions as well as the emotional ups and downs after the meeting had left her feeling fatigued. She seemed to conclude her own perceptions of this event with the words, "I'm glad that it happened; after all, I don't really have any other choice than to be happy about it, do I?"

Changing Ourselves as the Innovation of Last Resort

The innovation suggested in this last of the three case studies does not have a tangible product or technique to which we can point. When L. N. adjusted herself to her new circumstances, it did not feature B. V.'s novel use of a grab bar to get in the elevator or A. S.'s adaptation of a stapler to hold together the velcro straps of his sneakers. However, we think L. N. qualifies as a clinical innovation because L. N. created something new within herself. Specifically, she created a novel and effective attitude for herself that allowed her to accept and even enjoy what had been, for more than 80 years, unacceptable. Consequently, we can suggest another principle of innovation that may rank somewhat higher on the evolutionary scale of human development: *Changing ourselves represents one form of innovation.*

The principles of innovation, proposed from these case studies and related literature, are summarized below.

(1) The process that produces innovation is complex;
(2) Understanding innovation will require a wide range of methodological approaches;
(3) Non-normative populations can provide insights about innovation that may generalize to larger populations;
(4) The impetus for innovation may be external, unpleasant, and unwelcome;
(5) Innovation does not require a 'creative personality'; ordinary personalities in extraordinary circumstances will innovate;
(6) Both frustration and suffering can inspire innovation;
(7) Frustration will not lead to innovation unless it is sufficiently annoying to force new ways of thinking about a problem;
(8) Innovative behavior is both externally and internally self-rewarding;
(9) Changing ourselves represents one form of innovation.

These principles of innovation have emerged from people living in notoriously noxious circumstances: the coerced institutionalization of frail, elderly individuals due to a variety of physical and mental disabilities. Consequently, these principles are biased in the direction of insights shaded by these distressing circumstances. Nevertheless, as we asserted in our opening three principles, there is reason to believe that they may prove to be enduring principles that will help us in the daunting task we have assigned ourselves: understanding innovation.

Comparing the Case Studies to the Componential Model of Creativity

Amabile (1983) proposed that novel (i.e. innovative) behavior required three necessary and sufficient components: domain-relevant skills, creativity-relevant skills, and task motivation. She subsequently developed the principle of intrinsic motivation, which in turn generated considerable debate (Eisenberger & Cameron, 1996). The case studies presented here may not clarify that debate, but they do suggest that sufficient external frustration, even suffering, can lead to innovation. Whether that experience qualifies as intrinsic or extrinsic motivation is probably best resolved by acknowledging the complex, fluctuating nature of affect in the creative process (Shaw, 1994).

These case studies also suggest that Amabile is right to focus on the motivational component of innovation, in spite of its inherent definitional difficulties. These individuals enjoyed minimal domain-relevant and creativity-relevant capabilities. Nevertheless, the motivational pressure to innovate spurred them to maximize what capabilities they still possessed to solve everyday problems in remarkable ways.

References

Albert, R. S. & Runco, M. A. (1990). Observations, conclusions, and gaps. In: M. A. Runco & R. S. Albert (Eds), *Theories of Creativity* (pp. 263–269). Newbury Park, CA: Sage.

Amabile, T. M. (1983). *The social psychology of creativity.* New York: Springer-Verlag.

Amabile, T. M. (1990). Within you, without you: The social psychology of creativity, and beyond. In: M. A. Runco & R. S. Albert (Eds), *Theories of Creativity* (pp. 61–91). Newbury Park, CA: Sage.

Berk, L. E. (1998). *Development through the lifespan.* Boston, MA: Allyn & Bacon.

Byrne, D. (1971). *The attraction paradigm.* New York: Academic Press.

Cameron, P. A. & Gallup, G. G. (1988). Shadow recognition in human infants. *Infant Behavior & Development,* **11** (4), 465–471.

Clinton, H. (2000). In: *Witness to Genocide: The Children of Rwanda.* New York: Friendship Press.

Dollard, J., Doob, L. W., Miller, N. D., Mowrer, O. H. & Sears, R. R. (1939). *Frustration and aggression.* New Haven, CT: Yale University Press.

Eisenberger, R. & Cameron, J. (1996). Detrimental effects of reward: Reality or myth? *American Psychologist,* **51** (11), 1153–1166.

Freud, S. (1933). *New introductory lectures on psycho-analysis.* New York: Norton.

Gallup, G. (1977). Self-recognition in primates. A comparative approach to the bidirectional properties of consciousness. *American Psychologist*, **32**, 329–338.

Gallup, G. G. (1995). Mirrors, minds, and cetaceans. *Consciousness & Cognition: An International Journal*, **4** (2), 226–228.

Heinzen, T. E. (1989). Moderate challenge increases creative ideation. *Creativity Research Journal*, **2**, 3–6.

Heinzen, T. E. (1990). Creating creativity in New York State government. *Public Productivity & Management Review*, **14** (1), 91–98.

Heinzen, T. E. (1994a). *Everyday frustration and creativity in government*. Norwood, NJ: Ablex.

Heinzen, T. E. (1994b). Proactive and reactive creativity and situational affect. In: M. P. Shaw & M. A. Runco (Eds), *Affect and Creativity* (pp. 127–146). Norwood, NJ: Ablex.

Heinzen, T. E. & Alberico, S. A. (1990). Using a creativity paradigm to evaluate teleconferencing. *The American Journal of Distance Education*, **4** (3), 3–12.

Isen, A., Daubman, K. & Nowicki, G. (1987). Positive affect facilitates creative problem solving. *Journal of Personality and Social Psychology*, **52** (6) 1122–1131.

Isen, A., Shalker, T. E., Clark, M. & Karp, L. (1978). Affect and accessibility of material in memory, and behavior: A cognitive loop? *Journal of Personality and Social Psychology*, **36**, 1–12.

Kolata, G. (2001). On research frontier, basic questions. *The New York Times*, June 5, F1, F9.

Kuhn, T. S. (1962). *The structure of scientific revolutions* (2nd ed.). Chicago, IL: University of Chicago Press.

Lewin, K. (1935). *A dynamic theory of personality*. New York: McGraw-Hill.

May, R. (1975). *The courage to create*. New York: Norton.

Piaget, J. (1952). *The origins of intelligence in children*. New York: Norton.

Poizner, H. (1987). *What the hands reveal about the brain*. Cambridge, MA: MIT Press.

Runco, M. A. & Shaw, M. P. (1994). Conclusions concerning creativity and affect. In: M. P. Shaw & M. A. Runco (Eds), *Affect and Creativity* (pp. 261–270).

Salem, R. A. (2000). *Witness to genocide: The children of Rwanda*. New York: Friendship Press.

Shaw, M. P. (1994). Affective components of scientific creativity. In: M. P. Shaw & M. A. Runco (Eds), *Affect and Creativity* (pp. 3–43). Norwood, NJ: Ablex.

Tangalos, E. G., Maguire, G. A., Siegal, A. P. & DeVane, C. L. (2001). Behavioral manifestations of psychosis and depression in long-term care settings: From clinical trials to bedside therapy. *Annals of Long-Term Care*, **9** (5) (Suppl.), 1–4.

Wahrman, S. (1999). *Hope never dies*. Brooklyn, NY: Mesorah Publications.

Part V

Assessment of Innovation

Modeling of Separation

The Measurement of Innovativeness

Ronald E. Goldsmith[1] and Gordon R. Foxall[2]

[1] *College of Business, Florida State University, USA*
[2] *Cardiff Business School, Cardiff University, Wales, U.K.*

Abstract: Diffusion of innovations is a theory that describes the spread of new things through social systems as they are adopted or rejected by individuals. Innovativeness refers to interindividual differences in how people react to these new things and accounts for much of their success or failure. Innovators may welcome them; the majority may gradually adopt them; laggards either slowly or never adopt them. Thus, the measurement of innovativeness is an important activity for both theory testing and practical purposes. This chapter presents many of the current theoretical perspectives on innovativeness and describes the measures that reflect them.

Keywords: Innovativeness; Measurement of innovativeness; Interindividual differences; Diffusion of innovations.

Introduction

The way in which innovativeness is measured depends upon the intentions of the researcher and the conception of innovativeness that is driving his or her work. Only when these matters have been resolved can the measurement of innovativeness itself be considered. This chapter discusses each of these in turn. First, the reasons researchers would have for measuring innovativeness are addressed. These range from intellectual curiosity—the desire simply to find out—to practitioner concerns—the expectation that behavior can be modified by the application of knowledge. Second, the researcher's understanding of the nature of 'newness', 'change', and 'innovation' is addressed. The role of discontinuity in the definition of innovativeness is discussed, and the points are illustrated with material from research on innovativeness in industry, marketing, and consumer choice. The conceptualization of innovativeness as a trait or latent process as opposed to a behavior-based understanding of innovativeness is also raised. Third, the intellectually driven approach will be illustrated by the role of psychometrics in assessing innovativeness. Discussion follows on the relationship of innovativeness to: (a) personality; and (b) situations, and the measurement issues arising from each are examined. Global vs. situation-specific measures of innovativeness are compared. Both qualitative and quantitative approaches to measurement are con-

sidered, though issues of quantification are treated in greater detail.

Diffusion of Innovations

Diffusion theory describes how new things (such as new products) spread through a social system (Rogers, 1995). The diffusion of innovations has been and remains an important topic in marketing management and consumer behavior owing to the importance of new products to the health of many companies. As consumers make their individual *adoption* decisions, these aggregate to produce the timing and pattern of *diffusion*. Thus, adoption is an individual or micro decision process, while diffusion is a social or macro process. In diffusion theory, when the time of adoption of a new thing since its introduction is plotted as a frequency distribution, the result is a normal or bell-shaped curve. A cumulative plot of adoptions yields an S-shaped curve describing the spread of the new thing through the social system as increasing numbers (followed by decreasing numbers) of individuals adopt. Most discussions of adoption and diffusion are based on these basic principles.

The importance of diffusion theory was recognized by some of the first academics to describe consumer behavior as a topic of systematic study (e.g. Zaltman, 1965), and diffusion was prominently featured in some of the first consumer behavior texts (Engel, Kollat &

Blackwell, 1968; Howard & Sheth, 1969; Wasson & McConaughy, 1968). Diffusion continues to be an important topic requiring space in recent texts in both marketing management (e.g. Kotler, 2000) and consumer behavior (e.g. Hawkins, Best & Coney, 1998). A key element in diffusion theory is the concept of 'innovativeness'. While there are several definitions, they all incorporate the notion that people differ in their reaction to the novel, ranging from quick acceptance to outright rejection. These concepts form the subject of this chapter.

What is an Innovation?

The term 'innovation' has many meanings. It can refer to the inventive process by which new things, ideas, and practices are created; it can mean the new thing, idea, or practice itself; or it can describe the "process whereby an existing innovation becomes a part of an adopter's cognitive state and behavioral repertoire" (Zaltman, Duncan & Holbek, 1973, pp. 7–8). The first meaning is the domain of the New Product Development (NPD) process, an important aspect of marketing management (Thomas, 1993, 1995). Our concern is with the second and third of these meanings. We understand innovations to be things, ideas, or practices that are perceived to be new to the audience to which they are introduced (Rogers, 1995). How consumers react to them, whether they are adopted or not, and how rapidly they spread if adopted are the results of a decision-making process by individual adopters who are subject to a variety of influences both internal and external. The measurement of 'innovativeness' is important to this process.

Definitions of 'innovation' abound. This is because the concept of innovation appears in many different fields of study and social theories. In the area of high technology, for instance, we are told that "Innovation is the use of new knowledge to offer a new product or service that customers want. It is invention + commercialization" (Afuah, 1998, p. 13; see also Foxall, 1984). In the context of educational research (Evans, 1967, pp. 15–16) an innovation is described as having two subcomponents: "First, there is the idea or item which is novel to a particular individual or group and, second, there is the change which results from the adoption of the object or idea". Evans adds an important qualification. "We would also include among innovations those items or ideas which represent a recombination of previously accepted ideas". In high-technology businesses innovations can be totally new devices or ways of using devices, or they can be modifications of existing machines or processes (Foxall, 1988; von Hippel, 1988). Examples of medical innovations are a new drug, apparatus, or treatment (Coleman, 1966). Agricultural innovations are new seeds, fertilizers, equipment, or farm practices (Lionberger, 1960).

In the marketing context, new products are very important to the long-run success of a business. We can observe several different types of new products:

- Modifications—these are the venerable 'new and improved' versions of brands. They may be relatively minor changes, and they replace the existing version, which vanishes;
- Line extensions—These are different varieties of the product. They are new sizes, flavors, formulations, and packages that supplement the product line and sit side by side with the existing versions;
- Brand extensions—Companies execute brand extension strategies by putting a brand name onto products in a different category than the first;
- New Brands—In this instance, the product may be common, but the brand name is new;
- Innovations—We want to reserve the term 'innovation' to refer to a new-to-the-world product. It may replace an old version or sit beside it, but the product itself is unlike anything that has existed before.

In the marketing context, any of the first four types of new products may involve the fifth concept, but they do not have to in order to qualify as new products. That is, all innovations can be new products, but not all new products are innovations.

Thomas S. Robertson (1971, p. 7) proposed that new products could be arrayed along a continuum of "newness in terms of consumption effects" describing how continuous or discontinuous their effects were on established consumption patterns. This scheme described three categories:

(1) A *continuous* innovation has the least disrupting influence on established consumption patterns. Alteration of a product is usually involved, rather than the creation of a new product. An example would be a new microwaveable snack food.
(2) A *dynamically continuous* innovation has more disrupting effects than a continuous innovation. Although it still generally does not involve new consumption patterns, it involves the creation of a new product or the alternation of an existing one. An electric car is arguably 'dynamically continuous' owing to the relatively minor changes owners would make in fueling and driving patterns.
(3) A *discontinuous* innovation involves the establishment of new consumption patterns and the creation of previously unknown products. The fax machine was discontinuous, as was downloading music as MP3 files.

Thus, the concept of 'innovation' is broad enough to cover new things in many domains and in many forms. But what does 'new' mean? By explication of this concept we can get a better idea of what innovations are all about.

One way of viewing this issue is to argue that newness implies at least three different qualities. The first is *recency*. Things are new when they are encountered or acquired recently (Richins & Bloch, 1986). Thus, consumers speak of the 'new car' just as if they describe the 'new baby'. A second aspect of newness is *originality*. Things are judged as new when they are unfamiliar because they are so original. 'That's new to me', we might say, upon viewing unfamiliar art or an unfamiliar product. Finally, newness is a function of *similarity* (Barnett, 1953). How similar or different a thing is from exiting things of the same type lead to perceptions of newness. Fashionable clothing, for example, might be considered innovative in so far as it has only recently appeared in the marketplace, it is creative and original, and it is stylistically different from existing clothing.

Since newness is often in the eye of the beholder, it should be possible to measure perceived newness of innovations by measuring perceived recency of acquaintance, perceived originality, or perceived similarity to existing things. That is, the perceived recency, originality, or similarity of a stimulus may be assessed in much the same way, via semantic differential scales for instance, as researchers assess the perceived qualities of any other stimulus. It may be as important to measure 'perceived newness' as it is to measure innovativeness since these two constructs may interact as the adoption decision process proceeds.

Why Measure Innovativeness?

The measure of innovativeness depends in part on the motives of those doing the measurement and on the contexts of their measurement. In other words, investigators may want to measure innovativeness for different purposes. Thus, researchers may want an interval level measure of innovativeness as an individual difference variable, much like any other personality or intelligence measure; or they may want simply to categorize people into an 'adopter category'. This latter notion comes from Rogers (1995, p. 262) who used the bell-shaped distribution of adoption partitioned into its standard deviations from average time of adoption to form five categories of adopters:

(1) innovators (the first 2.5% to adopt);
(2) early adopters (the next 13.5% of adopters);
(3) the early majority (34%);
(4) the late majority (34%);
(5) laggards (the last 16% to adopt).

Some researchers study diffusion theory, seeking to understand this social phenomenon in greater detail. They may, for instance, be trying to identify the influence that innovativeness has on other market-related phenomena, such as search for and processing of information, decision-making, or brand loyalty (Klink & Smith, 2001). Innovativeness may be incor-

porated into models as an independent variable, a moderator or covariate, or as a dependent variable. Multiple-item interval level measures facilitate the incorporation of innovativeness into structural equation models. Theory-oriented research emphasizes innovativeness measures that correspond closely to the way innovativeness is conceptualized for a specific theoretical context (Foxall, 1988). They will most likely want an interval level measure that possesses high levels of content and construct validity, and emphasize reliability, generalizability, or uniqueness. Ease of use and convenience may be sacrificed to attain these psychometric characteristics.

Other researchers have applied commercial reasons for studying innovativeness. They are trying to solve specific problems involving the spread of new things, especially new products. If members of the different adopter categories can be identified, even crudely as a tripartite classification of 'early adopters', 'majority consumers', and 'laggards', this information may help those responsible for developing strategies to promote or to retard innovation spread. While desiring the psychometric rigor of theory-oriented measures, they will likely put more emphasis on speed, simplicity, and easy of use. McDonald & Alpert (1999) describe six practical reasons for identifying innovators:

(1) *To enlist the cooperation of innovators in refining and improving new products.* The idea of enlisting innovative lead users for new product ideas and evaluations is well established in industrial marketing (von Hippel, 1988) and in the marketing of consumer goods (e.g. McCarthy, 1998). We can suggest that a related reason for identifying innovators is that they may be able to identify new products that are destined to fail with cutting-edge buyers.

(2) *To enhance the speed of new product diffusion in order to generate cash flow.* Thus, the innovators can be targeted for concentrated marketing effort designed to persuade them to adopt quickly. Marketing spending for a relatively small number of early buyers should be more productive than for the larger mass markets further along the diffusion curve. And because innovative consumers are relatively price-insensitive (Goldsmith & Newell, 1997) sales should yield greater gross margins.

(3) *Early adopters promote new products to other buyers.* The diffusion process is driven largely by social communication or word-of-mouth (Gatignon & Robertson, 1991). Innovators play a key role in initiating this process as they provide positive role models and recommendations to later adopters. Finding them and cultivating their good will and good opinions will enhance the acceptance of the new product in the rest of the market.

(4) *Early adopters are often heavy users of the product category.* While making up a minority of buyers,

heavy users account for a disproportionate share of sales and profits in most product fields (Hallberg, 1995), and the overlap between innovativeness and heavy usage is well documented (Goldsmith, 2000a; Taylor, 1977). Innovators are experienced consumers who have a level of product knowledge and expertise in consumption plus a degree of wealth that allows them to make earlier adoption decisions and to act on them. Their capacity to make adoption decisions relatively earlier and relatively quickly stems in large part from their being heavy users of the product category (Foxall, 1993).

(5) *Early adopters help create a 'market leader' image.* Firms may want to be seen as market or product leaders, the source of breakthrough products that are the choice of the innovative consumer in their product field: "Product leaders have to prepare markets and educate potential customers to accept product that never before existed" (Treacy & Wiersema, 1995, p. 87). Identifying and understanding lead users (innovators) may be a key to success in this strategy.

(6) *Some may want to stop the diffusion of an undesirable innovation.* "Not all innovations are good, and some are socially undesirable or destructive" (McDonald & Alpert, 1999). Agencies that want to inhibit the spread of undesirable innovations might start with identifying those innovative souls who would be the first to adopt them and target them with counter-adoptive strategies.

Companies or agencies wishing to measure innovativeness for one or more of these reasons are chiefly interested in identifying innovators (and membership in the other adopter categories). They wish to categorize consumers and may likely be willing to sacrifice content and construct validity for criterion-related (predictive) validity and easy of use and flexibility (adaptable to a variety of data-collection contexts, such as personal and telephone interviews). They may be able to classify innovators based on records of prior behavior and not need to conduct primary data gathering activities. Of course, innovativeness measures developed and used by one group of researchers might also be used by others; the distinction is not mutually exclusive.

Three Concepts of Innovativeness

The concept of innovativeness refers to interindividual differences that characterize people's responses to new things. There are at least three approaches to the conceptualization of innovativeness, each of which carries its own implications for the measurement of the construct: behavioral, global trait, and domain-specific activity. Each makes its own contribution to the purposes of the investigator and requires its own interpretation of the results it produces.

Behavioral

The behavioral perspective on innovativeness identifies the concept with the act of adoption. Consumers are thus designated as innovators or not depending on whether they adopt a new product or not. Moreover, the degree of innovativeness they possess depends on how quickly they adopt after encountering the innovation. This simple, time-based approach to the conceptualization of innovations has given rise to a more sophisticated behavioral approach to the diffusion of innovations, which emphasizes the external rewards available to consumers at each successive stage of the product-market life cycle (Foxall, 1993, 1994b). Conceived within a broader behavioral perspective approach to consumer behavior (Foxall, 1990), this depicts the behavior of the earliest adopters of new products (*consumer initiators*) as determined by the high levels of both utilitarian (functional, technical, economic) and symbolic (social, psychological) rewards available to the consumer at this initial phase of the life cycle. Only consumers with the appropriate learning history of innovative behavior are likely to purchase at this stage. Subsequently, *earlier* and *later imitators* are induced to adopt by patterns of reward that emphasize first the utilitarian and then the symbolic benefits of purchasing at that time. Finally, the *last adopters* venture into the market place when the benefits of adopting the 'new' item are obvious to all, and the alternative products these consumers have hung on to for so long have themselves disappeared from the market (Foxall, 1996).

Global Personality Trait

The global trait view argues that innovativeness is a type of personality trait. Personality traits are thought to be relatively enduring patterns of behavior or cognition that differentiate people. Innovativeness describes reactions to the new and different. These reactions range from a very positive attitude toward change to a very negative attitude. Across the population, these attitudes are hypothesized to follow a bell-shaped normal distribution (Rogers, 1995). In Jackson's (1976) personality theory, innovation exists along side other personality traits such as conformity, risk taking, or tolerance as one of a battery of traits that describe "a variety of interpersonal, cognitive, and value orientations likely to have important implications for a person's functioning" (Jackson, 1976, p. 9). Another example can be found in the Five Factor Model of Personality, which contains a trait called 'openness to experience' that has been described as "how willing people are to make adjustments in notions and activities in accordance with new ideas or situations" (Popkins, 1998). Costa & McCrae (1992) characterize openness as curiosity and a motivation toward learning. Hurt, Joseph & Cook (1977) describe innovativeness as a willingness to try new things

(1) Global Innovativeness: A personality trait 'willingness to try new things'.
Related to other personality traits: risk taking, openness to experience. Of the demographic variables, only youth may be related. Only weakly related to any specific overt behaviors.

(2) Consumer Innovativeness: Describes consumers who want to be the first to buy new products.
Related to other consumer characteristics: marketplace knowledge, opinion leadership, price insensitivity. Higher levels of income and gregariousness (cosmopolitanism) may characterize consumer innovators.

(3) Domain Specific Innovativeness: Describes consumers in specific product fields who wish to be cutting edge, the owners of the newest products in the field.
Related to consumer characteristics: product-category knowledge, domain-specific opinion leadership, involvement in the product category, a heavy user of the product. May be characterized by specific demographic characteristics, such as age and gender, and these will vary by product category.

Figure 1. Conceptual levels of innovativeness.

(Goldsmith, 1991). This global trait can be compared to other traits at a similar level of specificity (e.g. Goldsmith, 1987).

Domain-Specific Personality Trait

An alternative to the global view of innovativeness suggests that while it is true that people can be differentiated in this way, for the purposes of prediction and explanation in marketing, it is useful to think also of innovativeness as a domain-specific characteristic. That is, consumers are thought of as being more or less innovative within specific product categories, such as a fashion enthusiast, a wine connoisseur, or a movie buff. Innovativeness does not overlap across product categories unless these are closely related (Goldsmith & Goldsmith, 1996). For example, the wine connoisseur may have no interest in new movies but may be enthusiastic about new restaurants; the movie buff may keep up with the latest in popular music but have no interest in new wines.

In this view, innovativeness may manifest itself at different levels of generality/specificity or abstraction/breadth (see Clark & Watson, 1995). A consumer may be characterized by an overall level of global innovativeness, and also by different levels of domain-specific innovativeness. Global innovativeness may be linked to domain-specific innovativeness (Goldsmith, Freiden & Eastman, 1995), but not necessarily (Foxall & Szmigin, 1999). Possessing higher levels of global innovativeness predisposes individuals to seek the new and different across many facets of life, and this influence may extend to consumption. The desire for the new and different may manifest itself in specific product domains where the individual is innovative, seeking the new and different products as they appear, but perhaps shunning newness in other product categories. Conceivably, a given product innovator may be generally conservative when it comes to new things in general.

Figure 1 summarizes the three concepts of global, consumer, and domain-specific innovativeness. Each is related to the lower level construct, but relationships

with overt behaviors grow stronger as one proceeds down the scale of specificity.

Three Ways to Measure Innovativeness

Behavioral

The behavioral perspective of innovativeness identifies the concept with: (1) adoption or non-adoption of an innovation; and (2) the time of adoption. Consumers are classified as innovators or non-innovators depending on their purchase (consumption) of a new product. They may alternatively be graded on their innovativeness by marking their time of adoption as measured from the instant the new product appears in the marketplace. This latter approach, termed the "temporal conception" of innovativeness (Midgley & Dowling, 1978), was for many years the most commonly used way to measure innovativeness. Unfortunately, trying to measure innovativeness by the time-of-adoption method has many flaws.

Theoretically, it confuses the behavioral phenomenon to be explained and predicted with one of the chief concepts employed to explain and predict it. That is, what social scientists often want to explain and predict is time of adoption, and one of the most theoretically relevant and powerful concepts used to explain and predict time of adoption is 'innovativeness'. Innovativeness is a hypothetical or latent construct. It refers to interindividual differences in personality. The relative recency with which a new product is adopted is a behavior that is partially explained by innovativeness: higher levels of innovativeness, *ceteris paribus*, lead to faster adoption. One would not use 'years married' as a measure of spousal affection, although the latter may explain and predict the former.

Methodologically, time of adoption is most appropriate as an indicator of precisely what it refers to: how much time passed since the new product was introduced before it was adopted. Even as an indicator of behavior, however, time of adoption is a fallible measure. The exact time when a new product is introduced into a specific market may be unknown.

There may be no record of purchase to show exact time of adoption. Moreover, consumers may not be able to accurately recall when they did adopt, and memory shortcomings are a well-documented flaw in many measures used in marketing research. There may be a difference between when the new product is introduced to the market and when the consumer learns about it. Although innovators are earlier to adopt than later consumers, there will be differences in the time between the appearance of a new product and when it is perceived. Finally, innovators may reject new products; non-innovative consumers may buy them. Not buying a new product because it is of poor quality, inappropriate for one's needs, or because one cannot afford it does not make an innovative consumer less innovative. A non-innovative consumer, for instance, may buy a new product as a gift for an innovator.

An alternative measurement technique proposed to overcome the shortcomings of time-of-adoption is termed the 'cross-sectional' method (Midgley & Dowling, 1978). This approach provides a list of new products, and level of innovativeness is indicated by how many of these the consumer has purchased. Not only does this measurement approach have many of the problems of the time-of-adoption method, but also it would seem to be indicative of the more global concept of innovativeness than of innovativeness in any specific product category, unless one limited the list of new products to a single category. (But what if there were few new products in the category, or they were widely spaced in time?) Determining which products are new and whether consumers were exposed to all of them (not all new products may be available in a specific market area) contribute to the methodological difficulties of measuring innovativeness by the cross-sectional method.

It can be argued that these behavioral measures are best thought of as measures of what they are simply said to be: behavior. Time of adoption is a measure of when a consumer purchased a new product after its appearance in the market. This may be an important dependent variable in a larger study of new product adoption but is not suitable as a measure of innovativeness as a latent construct. The cross-sectional method denotes ownership of a variety of new products, again, possibly a key dependent variable in a study of ownership of new products (e.g. *America's Tastemakers*, 1959). Both measures, it should be kept in mind, are subject to measurement errors.

Global Trait

Innovativeness as a global trait is best measured by means of a standardized self-report. Such scales are normally the products of rigorous validation procedures, so that they can be used with some confidence that they meet conventional standards of reliability and validity. Their psychometric shortcomings should be well described by their scale developers (in contrast to the unknown and likely unknowable errors in the behavioral measures).

Four scales have been described in literature that seem to do a reasonable job measuring global innovativeness: Jackson, Kirton, NEO, Hurt et al. (1977).

The Jackson Personality Inventory (Revised) contains 300 True/False items comprising 15 scales that are organized in terms of five higher-order dimensions. One of these 20-item scales is termed 'innovation' and measures the global personality dimension as described above.

M. J. Kirton (1989) developed a pencil-and-paper self-report, the Kirton Adaption-Innovation Inventory or KAI, to measure individual differences in decision-making and problem-solving. Innovators seek to change the context in which problems are imbedded, while adaptors try to disturb these frameworks as little as possible to solve the problem. Adaptors try to do things better, while innovators seek to do things differently. The KAI consists of 32 items organized into three subscales. High scores identify innovators and low scores adaptors, but no especial value is accorded to either tendency, as both cognitive styles are important. The bipolar nature of this scale has important implications for the measurement and conceptualization of innovativeness. Expecting consumer innovators to show the characteristics of Kirton's *innovators*, Foxall (1994c) undertook a series of studies of early adopters of new food products and brands, which identified the heaviest purchasers of these items as *adaptors*. Further investigation revealed that adaptors who were highly involved with the product category bought most new products or brands, followed by innovators (whether high- or low-involved) and finally less-involved adaptors (Foxall, 1995; Foxall et al., 1998). These results are of considerable practical relevance since they explain: (a) why marketing research has consistently found only weak relationships between innovator personality characteristics and early adoption; and (b) why so many new consumer products, aimed promotionally at innovators (in Kirton's sense), fail at the point of consumer acceptance. Clearly, marketing campaigns need to take adaptors into greater consideration. Similar results have been found for consumers' adoption of new financial products, use of credit cards, use innovativeness with respect to computer software, and organizational computer utilization (Foxall, 1999).

The NEO Personality Inventory (Costa & McCrae, 1992) contains the Openness to Experience subscale. The sense of this scale is that synonyms are "original, imaginative, broad interests, and daring" (McCrae & Costa, 1987, p. 87).

Hurt et al. (1977) describe a 20-item scale to measure innovativeness as a global personality trait characterized by a "willingness-to-change" or a "willingness-to-try new things". Their original scale actually contained items measuring this concept as well as items identifying a "creative and original person". The scale has been evaluated for its psychometric characteristics, and there is good evidence that the Innovation Subscale is a valid measure of the global trait (Goldsmith, 1991; Pallister & Foxall, 1998).

Consumer Innovativeness

While global personality traits are important concepts in the explanation of behavior, they have proved to be only weakly associated with specific consumer behaviors (see Foxall & Goldsmith, 1988). For this reason, efforts have been made to conceptualize 'consumer innovativeness' as the tendency to buy new products soon after they appear in the marketplace (Foxall, Goldsmith & Brown, 1998, pp. 40–45). Thus, consumer innovativeness is a more restricted or less general concept than global innovativeness.

The consumer innovator has long been of interest to marketers and to advertisers (How Consumers Take to Newness, 1955; Who are the Marketing Leaders?, 1958). Most of the earliest studies in marketing of consumer innovativeness used time of adoption as a means of identifying innovators. Some thought, however, was given to developing other ways of measuring innovativeness. An early study of American consumers (*America's Tastemakers*, 1959) identified a leading-edge group of consumers termed the 'high mobiles' who disproportionally adopted many of the current innovations. Membership in this group was derived from their geographical, social, and economic mobility.

In the 1960s advertising researchers originated the concept of lifestyle. To measure lifestyle (psychographics) they developed large batteries containing items designed to tap a variety of lifestyle dimensions. In one of the Activities, Interests, and Opinions (AIO) batteries developed in advertising to measure lifestyles can be found in sets of items labeled in ways that suggest a measure of innovativeness. Wells & Tigert (1971) present three items in a large AIO scale to designate the 'New Brand Tryer'. These items would form a Likert scale, self-report measure of consumer innovativeness. Leavitt & Walton (1975) attempted to develop a self-report measure of general consumer innovativeness. Their Open Processing scale showed some promise as a measure of consumer innovativeness, but little effort was made toward a thorough psychometric evaluation. One study did show that four of these innovativeness scales (Jackson, KAI & Hurt et al., Open Processing) generally exhibited convergent validity indicating that they are measuring either the same or highly related constructs (Goldsmith, 1986).

Within the framework of measuring consumer innovativeness, some scholars have expanded and enriched this concept by conceptualizing and measuring varieties or dimensions of consumer innovativeness. Venkatraman & Price (1990) distinguish 'cognitive' from 'sensory' innovativeness. The former refers to individuals who prefer to engage in activities that stimulate the mind, while the latter seek sensory stimulation. Their 16-item Likert scale contains two eight-item subscales to measure the two types of innovativeness. Similarly, Manning, Bearden & Madden (1995) distinguish two aspects of consumer innovativeness: consumer independent judgement-making and consumer novelty-seeking. The former is defined as the degree to which an individual makes innovation decisions independently of the communicated experience of others (Midgley & Dowling, 1978) while the latter is defined as the desire to seek out new product information (Hirschman, 1980). Finally, Price & Ridgway (1983) formulated the concept of 'use innovativeness'. They defined this concept as the use of previously adopted products in novel ways (Hirschman, 1980) and developed a scale to tap the five dimensions of the concept: creativity/curiosity, risk preference, voluntary simplicity, creative reuse, and multiple use potential.

Domain-Specific Innovativeness

Although conceptualizing innovativeness at this level of generality/specificity may be an improvement over the global personality approach for marketing purposes, it still leaves something to be desired for researchers interested in the new buyers for a specific product. Because consumer innovativeness is largely domain-specific, measures of general consumer innovativeness will identify the effects of that construct on behavior, but will not be good measures of consumer innovativeness within a specific product category; consumer innovativeness tends to manifest itself within specific categories with little overlap to other categories (Goldsmith & Goldsmith, 1996).

Consequently, researchers who wish to study consumer innovativeness will likely focus on a single product category to discover how this individual difference variable functions within a network of other variables where it can operate as an independent variable, a dependent variable, a mediating variable, or a moderating variable. Applied researchers would like to be able to identify innovators within the product category of their firm so that they can target innovators with unique marketing strategies, seek the input of innovators in the NPD process, or test how different consumers react to various marketing and advertising strategies. With these concerns in mind, Goldsmith & Hofacker (1991) developed the Domain Specific Innovativeness Scale (DSI). This is a balanced Likert scale with three positively and three negatively worded items (Bearden & Netemeyer, 1999, pp. 86–87).

The DSI has been evaluated for its psychometric characteristics (e.g. Goldsmith & Flynn, 1995). It has been shown to be internally consistent and free from both social desirability and acquiescent response bias. There is evidence for its convergent, discriminant, and nomological validity as well as its predictive and known-group validity (Goldsmith, 2000b, 2000c; Goldsmith, d'Hauteville & Flynn, 1998). Using the DSI yields a distribution of scores with a theoretical range of 6–30 with a mean of 18. The items can be used to assign a single score to each consumer studied, or the items can be used as multiple measures of the construct in a structural equations model.

Conclusion

This chapter has discussed the measurement of innovativeness in the context of research into innovation and has, therefore, taken concepts of innovativeness and the purposes of the investigator into central consideration. It has shown how researchers' expectations of a simple relationship between the behavior of the earliest adopters of new products and their personality traits and types gave way to a more sophisticated treatment of both the concept of 'newness' and the measures that could be brought to bear on the empirical identification of innovativeness and its impact on behavior. In this conclusion, we draw attention once again to the interconnectedness of concepts of innovativeness and the ways in which this construct is measured.

We employed the idea that innovation lies, ultimately, in the 'eye of the beholder'. This remains a convenient but unanalyzed term: as behavioral scientists we seek the underlying psychological characteristics that may make such terminology empirically intelligible. Foxall (1994a) draws attention to the confusion inherent in much usage of 'innovation', 'innovator', and 'innovativeness'. Such confusion arises in part from the same term being used to describe distinct conceptual levels from the hypothetical and abstract notions of 'innate' and 'inherent' innovativeness, to the concrete and observable 'actualized' innovativeness. To use the same term in each case, while claiming that the former provides an explanatory basis for the latter, is to prejudge the issue of whether innovative behavior is attributable to an underlying personality system. Between these extreme concepts there are a plethora of terms that refer to measurable intervening variables: sensory innovativeness, cognitive innovativeness, hedonic innovativeness, adaption innovation, and so on. Finally, at the level of consumption rather than purchasing, the term 'use innovativeness' has been proposed to refer to the deployment of a product in a novel application (Hirschman, 1980).

One means of addressing this problem is to adopt terms that distinguish the earliest adopters of new products (currently called 'innovators') from the underlying trait that is hypothesized to account for their behavior ('innovativeness'). Foxall (1994a) proposed 'consumer initiators' for the earliest adopters, a term that emphasizes the role of these purchasers in the initiation of markets. A synonym is 'initial adopters', which similarly depicts an observable level of analysis

which may be related on the basis of further empirical and conceptual work to an underlying trait or to the environmental consequences (the pattern of utilitarian and symbolic rewards) that induce adoption at each stage of the diffusion process, or to both. This approach leaves open the question of what causes innovative behavior and, in addition, encourages the investigation of a diversity of underlying traits (e.g. adaption as well as innovation, to use Kirton's terms) which may form the basis of empirical research. This approach also readily embraces Robertson's continuum of innovations (from the most continuous to the most discontinuous) and the idea of use innovativeness (perhaps 'use initiation' would be a more consistent term).

This is one source of solution, of course, but one that has been incorporated in both the consumer behavior and marketing research literatures as a means of both encouraging the resolution of terminological confusion and emphasizing that the measurement of innovativeness cannot be divorced from the meanings we attach to the term.

References
Afuah, A. (1998). *Innovation management*. New York: Oxford University Press.
America's Tastemakers (1959). *America's tastemakers no. 1 and no. 2*. Princeton, NJ: Opinion Research Corporation.
Barnett, H. G. (1953). *Innovation: The basis of cultural change*. New York: McGraw-Hill.
Bearden, W. O. & Netemeyer, R. G. (1999). *Handbook of marketing scales* (2nd ed.). Thousand Oaks, CA: Sage.
Clark, L. A. & Watson, D. (1995). Constructing validity: Basic issues in objective scale development *Psychological Assessment, 7* (3), 309–319.
Coleman, J. (1966). *Medical innovation: A diffusion study*. New York: Bobbs-Merrill.
Costa, P. T. & McCrae, R. R. (1992). *Revised NEO personality inventory and NEO five-factor inventory professional manual*. Odessa, FL: Psychological Assessment Resources.
Engel, J. F., Kollat, D. T. & Blackwell, R. D. (1968). *Consumer behavior*. New York: Holt, Rinehart and Winston.
Evans, R. I. (1967). *Resistance to innovation in higher education*. San Francisco, CA: Jossey-Bass.
Foxall, G. R. (1984). *Corporate innovation: Marketing and strategy*. London: Croom Helm; New York: John Wiley.
Foxall, G. R. (1988). Marketing new technology: Markets, hierarchies, and user-initiated innovation. *Managerial and Decision Economics, 9*, 237–250.
Foxall, G. R. (1990). *Consumer psychology in behavioural perspective*. London: Routledge.
Foxall, G. R. (1993). Consumer behaviour as an evolutionary process. *European Journal of Marketing, 27* (8), 46–57.
Foxall, G. R. (1994a). Consumer initiators: *Both* adaptors *and* innovators. In: M. J. Kirton (Ed.), *Adaptors and Innovators: Styles of Creativity and Problem Solving* (pp. 114–136). London: Routledge.
Foxall, G. R. (1994b). Behaviour analysis and consumer psychology. *Journal of Economic Psychology, 15*, 5–91.

Foxall, G. R. (1994c). Consumer initiators: Adaptors and innovators. *British Journal of Management*, **5**, S3–S12.

Foxall, G. R. (1995). Cognitive styles of consumer initiators. *Technovation*, **15**, 269–288.

Foxall, G. R. (1996). *Consumers in context: The BPM research program*. London: Routledge.

Foxall, G. R. (1999). Consumer decision making: Process, involvement and style. In: M. J. Baker (Ed.), *The Marketing Book* (4th ed., pp. 109–130). London: Butterworth-Heinemann.

Foxall, G. R. & Goldsmith, R. E. (1988). Personality and consumer research: Another look. *Journal of the Market Research Society*, **30** (2), 111–125.

Foxall, G. R., Goldsmith, R. E. & Brown, S. (1998). *Consumer psychology for marketing*. London: International Thomson Business Press.

Foxall, G. R. & Szmigin, I. (1999). Adaption-innovation and domain specific innovativeness. *Psychological Reports*, **84**, 1029–1030.

Gatignon, H. & Robertson, T. S. (1991). Innovative decision processes. In: T. S. Robertson & H. H. Kassarjian (Eds), *Handbook of Consumer Behavior* (pp. 316–348). Englewood Cliffs, NJ: Prentice-Hall.

Goldsmith, R. E. (1986). Convergent validity of four innovativeness scales. *Educational and Psychological Measurement*, **46**, 81–87.

Goldsmith, R. E. (1987). Self-monitoring and innovativeness. *Psychological Reports*, **60**, 1017–1018.

Goldsmith, R. E. (1991). The validity of a scale to measure global innovativeness. *Journal of Applied Business Research*, **7**, 89–97.

Goldsmith, R. E. (2000a). Characteristics of the heavy user of fashionable clothing. *Journal of Marketing Theory and Practice*, **8** (4), 1–9.

Goldsmith, R. E. (2000b). How innovativeness distinguishes online buyers. *Quarterly Journal of Electronic Commerce*, **1** (4), 323–333.

Goldsmith, R. E. (2000c). Identifying wine innovators: A test of the domain specific innovativeness scale using known groups. *International Journal of Wine Marketing*, **12** (2), 37–46.

Goldsmith, R. E., d'Hauteville, F. & Flynn, L. R. (1998). Theory and measurement of consumer innovativeness. *European Journal of Marketing*, **32** (3/4), 340–353.

Goldsmith, R. E. & Flynn, L. R. (1995). The domain specific innovativeness scale: Theoretical and practical dimensions. In: *Association of Marketing Theory and Practice Proceedings* (pp. 177–182).

Goldsmith, R. E., Freiden, J. B. & Eastman, J. K. (1995). The generality/specificity issue in consumer innovativeness research. *Technovation*, **15** (10), 601–612.

Goldsmith, R. E. & Goldsmith, E. B. (1996). An empirical study of overlap of innovativeness. *Psychological Reports*, **79**, 1113–1114.

Goldsmith, R. E. & Hofacker, C. F. (1991). Measuring consumer innovativeness. *Journal of the Academy of Marketing Science*, **19** (3), 209–221.

Goldsmith, R. E. & Newell, S. J. (1997). Innovativeness and price sensitivity: Managerial, theoretical, and methodological issues. *Journal of Product and Brand Management*, **6** (3), 163–174.

Hallberg, G. (1995). *All customers are not created equal*. New York: John Wiley.

Hawkins, D. I., Best, R. & Coney, K. A. (1998). *Consumer behavior: Building marketing strategy*. Boston, MA: McGraw-Hill.

Hirschman, E. C. (1980). Innovativeness, novelty seeking, and consumer creativity. *Journal of Consumer Research*, **7** (3), 283–395.

'How Consumers Take to Newness' (1955). *Business Week*, (September 24), 41–45.

Howard, J. A. & Sheth, J. N. (1969). *The theory of buyer behavior*. New York: John Wiley.

Hurt, H. T., Joseph, K. & Cook, C. D. (1977). Scales for the measurement of innovativeness. *Human Communications Research*, **4** (1), 58–65.

Jackson, D. N. (1976). *Jackson personality inventory manual*. Goshen: Research Psychologists Press.

Kirton, M. J. (1989). *Adaptors and innovators: Styles of creativity and problem-solving*. London: Routledge.

Klink, R. R. & Smith, D. C. (2001). Threats to the external validity of brand extension research. *Journal of Marketing Research*, **38** (August), 326–335.

Kotler, P. (2000). *Marketing management: The millennium edition*. Upper Saddle River, NJ: Prentice Hall.

Leavitt, C. & Walton, J. R. (1975). Development of a scale for innovativeness. In: M. J. Schlinger (Ed.), *Association for Consumer Research (Proceedings)* (Vol. 2, pp. 545–552). Association for Consumer Research.

Lionberger, H. F. (1960). *Adoption of new ideas and practice*. Ames, IA: The Iowa State University Press.

McCarthy, M. J. (1998). Stalking the elusive teenage trendsetter. *Wall Street Journal*, (November 19), B1.

McCrae, R. R. & Costa, P. T. (1987). Validation of the five-factor model of personality across instruments and observers. *Journal of Personality and Social Psychology*, **52** (1), 81–90.

McDonald, H. & Alpert, F. (1999). Are innovators worth identifying? In: *Australia–New Zealand Marketing Academy Conference Proceedings*.

Manning, K. C., Bearden, W. O. & Madden, T. J. (1995). Consumer innovativeness and the adoption process. *Journal of Consumer Psychology*, **4** (4), 329–345.

Midgley, D. F. & Dowling, G. R. (1978). Innovativeness: The concept and its measurement. *Journal of Consumer Research*, **4** (March), 229–242.

Pallister, J. G. & Foxall, G. R. (1998). Psychometric properties of the Hurt–Joseph–Cook scales for the measurement of innovativeness. *Technovation*, **18** (11), 663–675.

Popkins, N. C. (1998). *The five-factor model: Emergence of a taxonomic model for personality psychology*. http://www.personalityresearch.org

Price, L. L. & Ridgway, N. M. (1983). Development of a scale to measure use innovativeness. In: R. P. Bagozzi & A. M. Tybout (Eds), *Advances in Consumer Research* (Vol. 10, pp. 679–684). Association for Consumer Research.

Richins, M. L. & Bloch, P. H. (1986). After the new wears off: The temporal context of product involvement. *Journal of Consumer Research*, **13** (September), 280–285.

Robertson, T. S. (1971). *Innovative behavior and communication*. New York: Holt, Rinehart and Winston.

Rogers, E. M. (1995). *Diffusion of innovations* (4th ed.). New York: The Free Press.

Taylor, J. W. (1977). A striking characteristic of innovators. *Journal of Marketing Research*, **14** (February), 104–107.

Thomas, R. J. (1993). *New product development*. New York: John Wiley.

Thomas, R. J. (1995). *New product success stories*. New York: John Wiley.

Treacy, M. & Wiersema, F. (1995). *The discipline of market leaders*. Reading, MA: Addison-Wesley.

Venkatraman, M. P. & Price, L. L. (1990). Differentiating between cognitive and sensory innovativeness: Concepts, measurement, and implications. *Journal of Business Research*, **20** (4), 293–315.

von Hippel, E. (1988). *The sources of innovation*. New York: Oxford University Press.

Wasson, C. R. & McConaughy, D. H. (1968). *Buying behavior and marketing decisions*. New York: Appleton-Century-Crofts.

Wells, W. D. & Tigert, D. J. (1971). Activities, interests, and opinions. *Journal of Advertising Research*, **11** (August), 27–35.

'Who are the Marketing Leaders?' (1958). *Tide*, **32**, 53–57.

Zaltman, G. (1965). *Marketing: Contributions from the behavioral sciences*. New York: Harcourt, Brace, and World.

Zaltman, G., Duncan, R. & Holbek, J. (1973). *Innovations and organizations*. New York: John Wiley.

Part VI

Development of Innovation

The International Handbook on Innovation
Edited by Larisa V. Shavinina

Developing High Potentials for Innovation in Young People Through the Schoolwide Enrichment Model

Sally M. Reis and Joseph S. Renzulli

The National Research Center on the Gifted and Talented, The University of Connecticut, USA

Abstract: The Schoolwide Enrichment Model (SEM): (1) focuses on a pedagogy that brings innovative productive persons to our attention; and (2) organizes programs around a continuum of services model that accommodates a wide variety of potentials across academic domains as well as other areas of human endeavor. The SEM also applies the knowledge gained about how we can help students become more innovative to the process of school change and improvement. Based on the belief that 'a rising tide lifts all ships', the SEM includes specific vehicles for providing *all* students with opportunities for 'high end' learning, creativity, and innovative thinking.

Keywords: School-based enrichment; Programming; Opportunities; Innovation; Creative productivity; Gifted students; SEM.

Introduction

How can we develop the creative potential of young people and inspire in them a spirit of innovation, inquiry, entrepreneurship, and a sense of power to change things? How can educators and parents help children learn to think creatively, and to value opportunities for creative work of their choice? The Schoolwide Enrichment Model (SEM), developed during 20 years of research in programs for talented and gifted students enables each faculty the flexibility to develop unique programs for talent development and creative productivity based on local resources, student demographics, and school dynamics as well as faculty strengths and creativity. The major goal of SEM is to promote both challenging and enjoyable high-end learning across a wide range of school types, levels and demographic differences. The idea is to create a repertoire of enrichment opportunities and services that challenge all students. Any individual student does better in a school when all students appreciate creativity and innovation; and therefore the underlying theme of schoolwide enrichment is 'a rising tide lifts all ships'. This approach allows schools to develop a collaborative school culture that takes advantage of resources and appropriate decision-making opportuni-

tics to create meaningful, high-level and potentially creative opportunities for students to develop their talents. In this chapter, innovation is used to describe these creative opportunities and their innovative outcomes.

The SEM suggests that educators examine ways to make schools more inviting, friendly, and enjoyable places that encourage talent development instead of regarding students as repositories for information that will be assessed with the next round of standardized tests. Not only has this model been successful in addressing the problem of high potential in all students who have been under-challenged, but it also provides additional learning paths for gifted, talented, and creative students who find success in more traditional learning environments.

At the heart of the SEM is the Enrichment Triad Model (Renzulli, 1977) developed in the mid-1970s and initially implemented by school districts primarily in Connecticut in the United States. The model, which was originally field-tested in several districts, proved to be quite popular, and requests were received from all over the United States for visitations to schools using the model and for information about how to implement the model increased. A book about the Enrichment

Triad Model (Renzulli, 1977) was published, and more and more districts began asking for help in implementing this approach. It was at this point that a clear need was established for research about the effectiveness of the model and for practical procedures that could provide technical assistance for interested educators to help develop programs in their schools. We became fascinated by the wide range of Triad programs developed by different types of teachers in different school districts, some urban, rural and suburban. In some programs, for example, teachers consistently elicited high levels of creative productivity in students while others had *few* students who engaged in this type of work. In some districts, many enrichment opportunities were regularly offered to students not formally identified for the program, while in other districts only identified gifted students had access to enrichment experiences.

In the more than two decades since the Enrichment Triad Model has been used as the basis for many educational programs for gifted, talented, and creative students, an unusually large number of examples of creative productivity have occurred on the parts of young people whose educational experiences have been guided by this programming approach. Perhaps, like others involved in the development of theories and generalizations, we did not fully understand at the onset of our work the full implications of the model for encouraging and developing creative and innovative productivity in young people. These implications relate most directly to teacher training, resource procurement and management, product evaluation, and other theoretical concerns (e.g. motivation, task commitment, self-efficacy) that probably would have gone unexamined, undeveloped, and unrefined without the favorable results that were reported to us by early implementers of the model. We became increasingly interested in how and why the model was working and how we could further expand the theoretical rationale underlying our work, and the population to which services could be provided. Thus, several years of conceptual analysis, practical experience, and an examination of the work of other theorists brought us to the point of tying together the material in this chapter, which represents approximately 23 years of field-testing, research, evolution and dissemination.

In another chapter in this handbook, Renzulli (2003) has provided a description of the conception of giftedness upon which this model is based. This chapter presents a description of the Enrichment Triad Model and a chronology of how the model has expanded and changed. Space does not permit a review of the numerous research studies that have been carried out on the model over the years, but interested readers are invited to visit the Research and Evaluation section of the folder on The Schoolwide Enrichment Model at our website (www.gifted.uconn.edu).

An Overview of the Enrichment Triad Model

The Enrichment Triad Model was designed to encourage creative productivity on the part of young people by exposing them to various topics, areas of interest, and fields of study, and to further train them to *apply* advanced content, process-training skills, and methodology training to self-selected areas of interest (see Root-Bernstein, this volume). Accordingly, three types of enrichment are included in the Triad Model (see Fig. 1).

Type I enrichment is designed to expose students to a wide variety of disciplines, topics, occupations, hobbies, persons, places, and events that would not ordinarily be covered in the regular curriculum. In schools that use this model, an enrichment team consisting of parents, teachers, and students often organizes and plans Type I experiences by contacting speakers, arranging minicourses, demonstrations, or performances, or by ordering and distributing films, slides, videotapes, or other print or non-print media.

Type II enrichment consists of materials and methods designed to promote the development of thinking and feeling processes (see Root-Bernstein & Root-Bernstein, this volume). Some Type II training is general, and is usually carried out both in classrooms and in enrichment programs. Training activities include the development of: (1) creative thinking and problem-solving, critical thinking, and affective processes; (2) a wide variety of specific learning how-to-learn skills; (3) skills in the appropriate use of advanced-level reference materials; and (4) written, oral, and visual communication skills. Other Type II enrichment is specific, as it cannot be planned in advance and usually involves advanced methodological instruction in an interest area selected by the student. For example, students who become interested in botany after a Type I experience might pursue additional training in this area by doing advanced reading in botany; compiling, planning and carrying out plant experiments; and seeking more advanced methods training if they want to go further.

Type III enrichment involves students who become interested in pursuing a self-selected area of personal interest or that they consider a problem (see Root-Bernstein, problem generation, this volume) and are willing to commit the time necessary for advanced content acquisition and process training in which they assume the role of a first-hand inquirer. The goals of Type III enrichment include:

- providing opportunities for applying interests, knowledge, creative ideas and task commitment to a self-selected problem or area of study;
- acquiring advanced level understanding of the knowledge (content) and methodology (process) that are used within particular disciplines, artistic areas of expression and interdisciplinary studies;

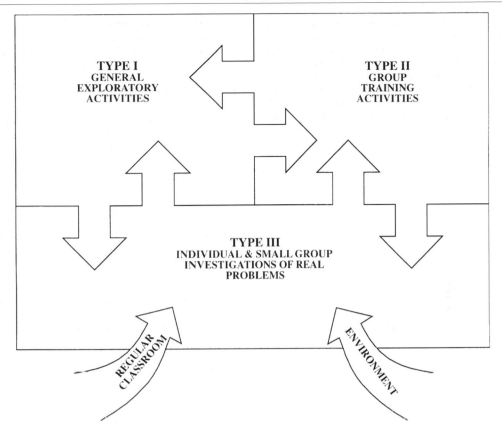

Figure 1. Enrichment triad model.

- developing authentic products that are primarily directed toward bringing about a desired impact upon a specified audience;
- developing self-directed learning skills in the areas of planning, organization, resource utilization, time management, decision-making and self-evaluation;
- developing task commitment, self-confidence, and feelings of creative accomplishment.

The Revolving Door Identification Model

As our experience with Triad Programs grew, our concern about the students who were being identified to participate and those who were not being included in these programs also grew. We became increasingly concerned about students who were not able to participate in enrichment programs because they did not score in the top 1–3% of the population in achievement or intelligence tests.

Research conducted by Torrance (1962, 1974) had demonstrated that students who were rated highly on creativity measures do well in school and on achievement tests but are often not selected for gifted programs because their scores are often below the cutoff for admission. Some of our own research (Reis & Renzulli, 1982) indicated that when a broader population of students (15–20% of the general population called the 'talent pool') were able to participate in Types I and II enrichment experiences, they produced equally good Type III products as the traditional 'gifted' students (the top 3–5%). This research produced the rationale for the Revolving Door Identification Model (RDIM) (Renzulli, Reis & Smith, 1981) in which a talent pool of students receives regular enrichment experiences and the opportunity to 'revolve into' Type III creative productive or innovative experiences. In RDIM, we recommend that students be selected for participation in the talent pool on the basis of multiple criteria that include indices of creativity, because we believe that one of the major purposes of gifted education is to develop creative thinking and creative productivity in students. Once identified and placed in the talent pool through the use of test scores, teacher, parent, or self-nomination, and examples of creative potential or productivity, students are observed in classrooms and enrichment experiences for signs of advanced interests, creativity, or task commitment. We have called this part of the process 'action information' and have found it to be an instrumental part of the identification process in assessing students' interest and motivation to become involved in Type III creative or innovative productivity. Further support for expanding identification procedures through the use of these approaches has recently been offered by Kirschenbaum (1983) and Kirschenbaum & Siegle (1993) who demonstrated that students who are rated or test high on measures of creativity tend to do well in school and on measures of achievement. The development of the RDIM led to the need for a guide dealing with how all of the components of the previous Triad and the new RDIM could be implemented, and the resulting work was entitled *The Schoolwide Enrichment Model* (SEM) (Renzulli & Reis, 1985, 1997).

The Schoolwide Enrichment Model (SEM)

In the SEM, a talent pool of 10–15% of above-average-ability/high-potential students is identified through a variety of measures including: achievement tests, teacher nominations, assessment of potential for creativity and task commitment, as well as alternative pathways of entrance (self-nomination, parent nomination, etc.). High achievement test and IQ test scores automatically include a student in the talent pool, enabling those students who are underachieving in their academic schoolwork to be included.

Once students are identified for the talent pool, they are eligible for several kinds of services; first, interest and learning styles assessments are used with talent pool students. Informal and formal methods are used to create or identify students' interests and to encourage students to further develop and pursue these interests in various ways. Learning-style preferences which are assessed include: projects, independent study, teaching games, simulations, peer teaching, programmed instruction, lecture, drill and recitation, and discussion. Second, curriculum compacting is provided to all eligible students for whom the regular curriculum is modified by eliminating portions of previously mastered content. This elimination or streamlining of curriculum, enables above-average students to avoid repetition of previously mastered work and guarantees mastery while simultaneously finding time for more appropriately challenging activities (Reis, Burns & Renzulli, 1992; Renzulli, Smith & Reis, 1982). A form, entitled The Compactor (Renzulli & Smith, 1978), is used to document which content areas have been compacted and what alternative work has been substituted. Third, the Enrichment Triad Model offers three types of enrichment experiences. Type I, II, and III Enrichment are offered to all students; however, Type III enrichment is usually more appropriate for students with higher levels of ability, interest, and task commitment.

Separate studies on the SEM demonstrated its effectiveness in schools with widely differing socio-economic levels and program organization patterns

(Olenchak, 1988; Olenchak & Renzulli, 1989). The SEM has been implemented in thousands of school districts across the country (Burns, 1998), and interest in this approach continues to grow.

Newest Directions for the Schoolwide Enrichment Model

The present reform initiatives in general education have created a more receptive atmosphere for more flexible and innovative approaches to challenge all students, and accordingly, the SEM has been expanded to address three major goals that we believe will accommodate the needs of gifted students and, at the same time, provide challenging learning experiences for all students. These goals are:

- To maintain and expand a continuum of special services that will challenge students with demonstrated superior performance or the potential for superior performance in any and all aspects of the school and extracurricular program;
- To infuse into the general education program a broad range of activities for high-end learning that will: (a) challenge all students to perform at advanced levels; and (b) allow teachers to determine which students should be given extended opportunities, resources, and encouragement in particular areas where superior interest and performance are demonstrated;
- To preserve and protect the positions of gifted education specialists and any other specialized personnel necessary for carrying out the first two goals.

School Structures

The Regular Curriculum

The regular curriculum consists of everything that is a part of the predetermined goals, schedules, learning outcomes, and delivery systems of the school. The regular curriculum might be traditional, innovative, or in the process of transition, but its predominant feature is that authoritative forces (i.e. policy-makers, school councils, textbook adoption committees, state regulators) have determined that the regular curriculum should be the 'centerpiece' of student learning. Application of the SEM influences the regular curriculum in three ways. First, the challenge level of required material is differentiated through processes such as curriculum compacting and textbook content modification procedures. Second, systematic content intensification procedures should be used to replace eliminated content with selected, in-depth learning experiences. Third, the types of enrichment recommended in the Enrichment Triad Model (Renzulli, 1977) are integrated selectively into regular curriculum activities. Although our goal in the SEM is to influence rather than replace the regular curriculum, application of certain SEM components and related staff develop-

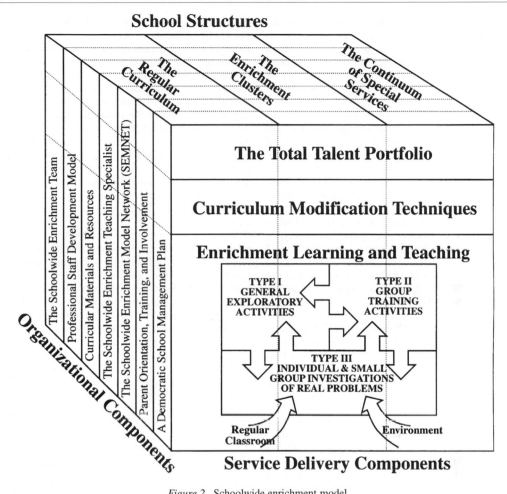

School Structures

Figure 2. Schoolwide enrichment model.

ment activities has resulted in substantial changes in both the content and instructional processes of the entire regular curriculum.

The Enrichment Clusters

The enrichment clusters, a second component of the Schoolwide Enrichment Model, are designed to celebrate creativity and innovation. They are non-graded groups of students who share common interests, and who come together during specially designated time blocks during school to work with an adult who shares their interests and who has some degree of advanced knowledge and expertise in the area. The enrichment clusters usually meet for a block of time weekly during a semester. All students complete an interest inventory developed to assess their interests, and an enrichment team of parents and teachers tally all of the major families of interests. Adults from the faculty, staff, parents, and community are recruited to facilitate enrichment clusters based on these interests, such as

creative writing, drawing, sculpting, archeology and other areas. Training is provided to the facilitators who agree to offer the clusters, and a brochure is developed and sent to all parents and students highlighting the role of student interests and select choices of enrichment clusters. A title and description that appeared in a brochure of clusters in a school using the SEM follows:

Invention Convention

Are you an inventive thinker? Would you like to be? Brainstorm a problem, try to identify many solutions, and design an invention to solve the problem, as an inventor might give birth to a real invention. Create your invention individually or with a partner under the guidance of Bob Erikson and his students, who work at the Connecticut Science Fair. You may decide to share your final product at the Young Inventors' Fair on March 25th, a statewide daylong celebration of creativity and innovative.

337

Students select their top three choices for the clusters, and scheduling is completed to place all children into their first or, in some cases, second choice. Like extracurricular activities and programs such as 4-H and Junior Achievement, the main rationale for participation in one or more clusters is that *students and teachers want to be there*. All teachers (including music, art, physical education, etc.) are involved in teaching the clusters; and their involvement in any particular cluster is based on the same type of interest assessment that is used for students in selecting clusters of choice.

The model for learning used with enrichment clusters is based on an inductive approach to solving real-world problems through the development of authentic innovative products and services. Unlike traditional, didactic modes of teaching, this approach, known as enrichment learning and teaching (described fully in a later section), uses the Enrichment Triad Model to create a learning situation that involves the use of methodology, develops higher-order thinking skills, and authentically applies these skills in creative and situations. Enrichment clusters promote cooperativeness within the context of real-world problem solving, and they also provide superlative opportunities for promoting self-concept. "A major assumption underlying the use of enrichment clusters is that *every child is special if we create conditions in which that child can be a specialist within a specialty group*" (Renzulli, 1994, p. 70).

Enrichment clusters are organized around various characteristics of differentiated programming for gifted students on which the Enrichment Triad Model (Renzulli, 1977) was originally based, including the use of major disciplines, interdisciplinary themes, or cross-disciplinary topics (e.g. a theatrical/television production group that includes actors, writers, technical specialists, costume designers). The clusters are modeled after the ways in which knowledge utilization, thinking skills, and interpersonal relations take place in the real world. Thus, all work is directed toward the production of a product or service. A detailed set of lesson plans or unit plans is not prepared in advance by the cluster facilitator; rather, direction is provided by three key questions addressed in the cluster by the facilitator and the students:

(1) What do people with an interest in this area (e.g. film making) do?
(2) What knowledge, materials, and other resources do they need to do it in an excellent and authentic way?
(3) In what ways can the product or service be used to have an impact on an intended audience?

Enrichment clusters incorporate the use of advanced content, providing students with information about particular fields of knowledge, such as the structure of a field as well as the basic principles and the functional concepts in a field (Ward, 1960). Ward defined functional concepts as the intellectual instruments or tools with which a subject specialist works, such as the vocabulary of a field and the vehicles by which persons within the field communicate with one another. The methodology used within a field is also considered advanced content by Renzulli (1988a), involving the use of knowledge of the structures and tools of fields, as well as knowledge about the methodology of particular fields. This knowledge about the methodologies of fields exists both for the sake of increased knowledge acquisition and for the utility of that know-how as applied to the development of products, even when such products are considered advanced in a relative sense (i.e. age, grade, and background considerations).

The enrichment clusters are not intended to be the total program for talent development in a school, or to replace existing programs for talented and creative youth. Rather, they are one vehicle for stimulating interests and developing talent potentials across the entire school population. They are also vehicles for staff development in that they provide teachers an opportunity to participate in enrichment teaching, and subsequently to analyze and compare this type of teaching with traditional methods of instruction. In this regard the model promotes a spill-over effect by encouraging teachers to become better talent scouts and talent developers, and to apply enrichment techniques to regular classroom situations.

The Continuum of Special Services

A broad range of special services is the third school structure targeted by the model; a diagram representing these services is presented in Fig. 3. Although the enrichment clusters and the SEM-based modifications of the regular curriculum provide a broad range of services to meet individual needs, a program for total talent development still requires supplementary services that challenge our most academically talented young people who are capable of working at the highest levels of their special interest and ability areas. These services, which cannot ordinarily be provided in enrichment clusters or the regular curriculum, typically include: individual or small group counseling, various types of acceleration, direct assistance in facilitating advanced-level work, arranging for mentorship with faculty members or community persons, and making other types of connections between students, their families, and out-of-school persons, resources, and agencies.

Direct assistance also involves setting up and promoting student, faculty and parental involvement in special programs such as Future Problem Solving, Odyssey of the Mind, the Model United Nations program, and state and national essay competitions, mathematics, art, and history contests. Another type of direct assistance consists of arranging out-of-school

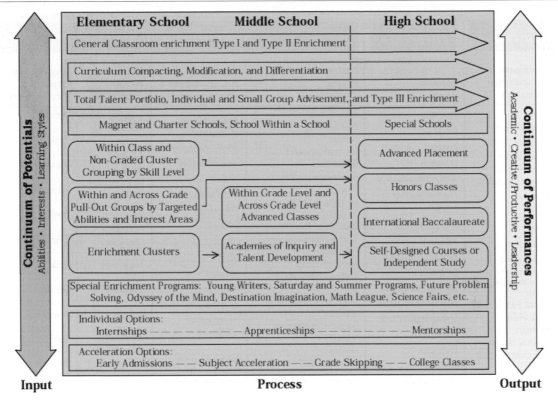

Figure 3. The integrated continuum of special services.

involvement for individual students in summer programs, on-campus courses, special schools, theatrical groups, scientific expeditions, and apprenticeships at places where advanced-level learning opportunities are available. Provision of these services is one of the responsibilities of the schoolwide enrichment teaching specialist or an enrichment team of teachers and parents who work together to provide options for advanced learning. Most schoolwide enrichment teaching specialists spend two days a week in a resource capacity to the faculty and three days providing direct services to students.

Service Delivery Components

The Total Talent Portfolio

The Schoolwide Enrichment Model targets specific learning characteristics that can serve as a basis for talent development. Our approach to targeting learning characteristics uses both traditional and performance-based assessment to compile information about three dimensions of the learner—abilities, interests, and learning styles. This information, which focuses on strengths rather than deficits, is compiled in a management form called the 'Total Talent Portfolio' (see

Fig. 4) which is used to make decisions about talent development opportunities in regular classes, enrichment clusters, and in the continuum of special services. The major purposes of the Total Talent Portfolio are to *collect* several different types of information that portray a student's strength areas and to regularly update this information. This information is periodically *reviewed and analyzed* to make purposeful decisions about providing opportunities for enrichment experiences in the regular classroom, the enrichment clusters, and the continuum of special services. Various acceleration and enrichment learning options and opportunities are then provided for students through participation in a shared decision-making process with their teachers and parents. This information is used for educational, personal, and career counseling about talent development opportunities and their child's involvement in them.

This expanded approach to identifying talent potentials is essential if we are to make genuine efforts to include more under-represented students in a plan for *total* talent development and the development of innovation. This approach is also consistent with the more flexible conception of *developing* gifts and talents

339

Joseph S. Renzulli

Abilities	**Interests**	**Style Preferences**			
Maximum Performance Indicators	*Interest Areas*	*Instructional Styles Preferences*	*Learning Environment Preferences*	*Thinking Styles Preferences*	*Expression Style Preferences*
Tests	Fine Arts	Recitation & Drill		Analytic	Written
•Standardized	Crafts	Peer Tutoring	*Inter/Intra Personal*	(School Smart)	
•Teacher-Made	Literary	Lecture			Oral
Course Grades	Historical	Lecture/Discussion	•Self-Oriented		
Teacher Ratings	Mathematical/Logical	Discussion	•Peer-Oriented	Synthetic/ Creative	Manipulative
Product Evaluation	Physical Sciences	Guided Independent Study *	•Adult-Oriented	(Creative, Inventive)	
•Written	Life Sciences	Learning /Interest Center	•Combined		Discussion
•Oral	Political/Judicial	Simulation, Role Playing, Dramatization, Guided Fantasy	*Physical*	Practical/ Contextual	Display
•Visual	Athletic/Recreation		•Sound	(Street Smart)	
•Musical	Marketing/Business		•Heat		Dramatization
•Constructed	Drama/Dance	Learning Games	•Light		
(Note differences between assigned and self-selected products)	Musical Performance	Replicative Reports or Projects*	•Design	Legislative	Artistic
	Musical Composition	Investigative Reports or Projects*	•Mobility	Executive	Graphic
Level of Participation in Learning Activities	Managerial/Business	Unguided Independent Study*	•Time of Day		Commercial
	Photography		•Food Intake	Judicial	
Degree of Interaction With Others	Film/Video	Internship*	•Seating		Service
	Computers	Apprenticeship*		Ref: Sternberg, 1984, 1988, 1990	
Ref: General Tests and Measurements Literature	Other (Specify)	*With or without a mentor	Ref: Amabile, 1983; Dunn, Dunn, & Price, 1977; Gardner, 1983		Ref: Kettle, Renzulli, & Rizza, 1998; Renzulli & Reis, 1985
	Ref: Renzulli, 1997				
		Ref: Renzulli & Smith, 1978			

Figure 4. The total talent portfolio.

that has been a cornerstone of our work and our concerns for promoting more equity in special programs.

Curriculum Modification and Differentiation Techniques

The second service delivery component of the SEM is a series of curriculum modification techniques designed to: (1) adjust levels of required learning so that all students are challenged; (2) increase the number of in-depth learning experiences; and (3) introduce various types of enrichment into regular curricular experiences. The procedures used to carry out curriculum modification are curriculum compacting, textbook analysis and removal of repetitious material from textbooks, and a planned approach for introducing greater depth into regular curricular material. Due to space restrictions, curriculum compacting is described in depth here, and other modification techniques are described in detail in other publications (see, for example, Reis et al., 1993; Renzulli, 1994).

How to Use the Compacting Process

Defining goals and outcomes. The first of three phases of the compacting process consists of defining the goals and outcomes of a given unit or segment of instruction. A major goal of this phase of the compacting process is to help teachers make individual programming decisions; a larger professional development goal is to help teachers be better analysts of the material they are teaching and better consumers of textbooks and prescribed curricular material.

Identifying students for compacting. The second phase of curriculum compacting is identifying students who have already mastered the objectives or outcomes of a unit or segment of instruction that is about to be taught. The first step of this phase involves estimating which students have the potential to master new material at a faster than normal pace; knowing one's students is, of course, the best way to begin the assessment process. Scores on previous tests, completed assignments, standardized achievement tests, and classroom participation are the best ways of identifying highly likely candidates for compacting.

Being a candidate for compacting does not necessarily mean that a student knows the material under consideration. Therefore, the second step of identifying candidates involves finding or developing appropriate tests or other assessment techniques that can be used to evaluate specific learning outcomes. Unit pretests, or end-of-unit tests that can be administered as pretests are appropriate for this task, especially when it comes to the assessment of basic skills. An analysis of pretest results enables the teacher to document proficiency in specific skills, and to select instructional activities or practice material necessary to bring the student up to a high level on any skill that may need some additional reinforcement.

The process is slightly modified for compacting content areas that are not as easily assessed as basic skills, and for students who have not mastered the material, but are judged to be candidates for more rapid coverage. First, students should have a thorough understanding of the goals and procedures of compacting, including the nature of the replacement process. The amount of time for completion of the unit should be specified, and procedures such as periodic progress reports or log entries for teacher review should be agreed upon. Of course, an examination of potential acceleration and/or enrichment replacement activities should be a part of this discussion.

Another alternative is to assess or pretest all students in a class when a new unit or topic is introduced; although this may seem like more work for the teacher, it provides the opportunity for all students to demonstrate their strengths or previous mastery in a given area. Using a matrix of learning objectives, teachers can fill in test results and establish small, flexible, and temporary groups for skill instruction and replacement activities.

Providing acceleration and enrichment options. The final phase of the compacting process can be one of the most exciting aspects of teaching because it is based on cooperative decision-making, innovation, and creativity on the parts of both teachers and students. Efforts can be made to gather enrichment materials from classroom teachers, librarians, media specialists, and content area or gifted education specialists. These materials may include self-directed learning activities, instructional materials that focus on particular thinking skills, and a variety of individual and group project oriented activities that are designed to promote hands-on research and investigative skills. The time made available through compacting provides opportunities for exciting learning experiences such as small-group, special-topic seminars that might be directed by students or community resource persons, community-based apprenticeships or opportunities to work with a mentor, peer tutoring situations, involvement in community-service activities, and opportunities to rotate through a series of self-selected mini-courses. The time saved through curriculum compacting can be used by the teacher to provide a variety of enrichment or innovative opportunities for the student.

Enrichment strategies might include a variety of Type I, II, or III or a number of options included on the continuum of services. Acceleration might include the use of material from the next unit or chapter, the use of the next chronological grade level textbook or the completion of even more advanced work. Alternative activities should reflect an appropriate level of challenge and rigor that is commensurate with the student's abilities and interests.

Decisions about which replacement activities to use are always guided by factors such as time, space, and the availability of resource persons and materials. Although practical concerns must be considered, the ultimate criteria for replacement activities should be the degree to which they increase academic challenge and the extent to which they meet individual needs. Great care should be taken to select activities and experiences that represent individual strengths and interests rather than the assignment of more-of-the-same worksheets or randomly selected kits, games, and puzzles! This aspect of the compacting process should also be viewed as a creative opportunity for an entire faculty to work cooperatively to organize and institute a broad array of enrichment experiences. A favorite mini-course that a faculty member has always wanted to teach or serving as a mentor to one or two students who are extremely invested in a teacher's beloved topic are just some of the ways that replacement activities can add excitement to the teachers' part in this process as well as the obvious benefits for students. We have also observed another interesting occurrence that has resulted from the availability of curriculum compacting. When some previously bright but underachieving students realized that they could both economize on regularly assigned material and 'earn time' to pursue self-selected interests, their motivation to complete regular assignments increased; as one student put it, 'Everyone understands a good deal!'

The best way to get an overview of the curriculum compacting process is to examine an actual example of how the management form that guides this process is used. This form, 'The Compactor', presented in Fig. 5, serves as both an organizational and record-keeping tool. Teachers should fill out one form per student, or one form for a group of students with similar curricular strengths. The Compactor is divided into three sections:

(1) The first column should include information on learning objectives and student strengths in those areas. Teachers should list the objectives for a particular unit of study, followed by data on students' proficiency in those objectives, including test scores, behavioral profiles and past academic records;

INDIVIDUAL EDUCATIONAL PROGRAMMING GUIDE

The Compactor

Prepared by Joseph S. Renzulli
Linda M. Smith

NAME_____ AGE_____ Individual Conference Dates And Persons
Participating in Planning Of IEP

SCHOOL_____ TEACHER(S)_____

GRADE_____ PARENT(S)_____

CURRICULUM AREAS TO BE CONSIDERED FOR COMPACTING Provide a brief description of basic material to be covered during this marking period and the assessment information or evidence that suggests the need for compacting.	PROCEDURES FOR COMPACTING BASIC MATERIAL Describe activities that will be used to guarantee proficiency in basic curricular areas.	ACCELERATION AND/OR ENRICHMENT ACTIVITIES Describe activities that will be used to provide advanced level learning experiences in each area of the regular curriculum.

☐ Check here if additional information is recorded on the reverse side.

Figure 5. The compactor.

(2) In the second column, teachers should detail the pretest vehicles they select, along with test results. The pretest instruments can be formal measures, such as pencil and paper tests, or informal measures, such as performance assessments based on observations of class participation and written assignments;

(3) The third column is used to record information about acceleration or enrichment options; in determining these options, teachers must be fully aware of students' individual interests and learning styles. We should never replace compacted regular curriculum work with harder, more advanced material that is solely determined by the teacher; instead, students' interests should be taken into account. If, for example, a student loves working on science fair projects, that option may be used to replace material that has been compacted from the regular curriculum. We should also be careful to help monitor the challenge level of the material that is being substituted. We want students to understand the nature of effort and challenge, and we should ensure that students are not simply replacing the compacted material with basic reading or work that is not advanced.

Rosa: A Case Study in Curriculum Compacting

Rosa is a fifth grader in a self-contained heterogeneous classroom; her school is located in a lower socio-economic urban school district. While Rosa's reading and language scores range between four or five years above grade level, most of her 29 classmates are reading one to two years below grade level. This presented Rosa's teacher with a common problem: What was the best way to instruct Rosa? He agreed to compact her curriculum. Taking the easiest approach possible, he administered all of the appropriate unit tests for the grade level in the Basal Language Arts program and excused Rosa from completing the activities and worksheets in the units where she showed proficiency (80% and above). When Rosa missed one or two questions, the teacher checked for trends in those items and provided instruction and practice materials to ensure concept mastery.

Rosa usually took part in language arts lessons one or two days a week; she spent the balance of the time with alternative projects, some of which she selected. This strategy spared Rosa up to six or eight hours a week with language arts skills that were simply beneath her level. She joined the class instruction only when her pretests indicated she had not fully acquired the skills or to take part in a discussion that her teacher thought she would enjoy. In the time saved through compacting, Rosa engaged in a number of enrichment activities. First, she spent as many as five hours a week in a resource room for high-ability students. This time was usually scheduled during her language arts class,

benefiting both Rosa and her teacher, since he did not have to search for all of the enrichment options himself. The best part of the process for Rosa was that she did not have to make up regular classroom assignments because she was not missing essential work.

Rosa also visited a regional science center with other students who had expressed a high interest and aptitude for science. Science was a second strength area for Rosa, and based on the results of her *Interest-A-Lyzer*, a decision was made for Rosa to proceed with a science fair project on growing plants under various conditions. Rosa's Compactor, which covered an entire semester, was updated in January. Her teacher remarked that compacting her curriculum had actually saved him time—time he would have spent correcting papers needlessly assigned! The value of compacting for Rosa convinced him that he should continue the process. The Compactor was also used as a vehicle for explaining to Rosa's parents how specific modifications were being made to accommodate her advanced language arts achievement level and her interest in science. A copy of The Compactor was also passed on to Rosa's sixth-grade teacher, and a conference between the fifth and sixth grade teachers and the resource teacher helped to ensure continuity in dealing with Rosa's special needs.

The many changes that are taking place in our schools require all educators to examine a broad range of techniques for providing equitably for *all* students. Curriculum compacting is one such process that has demonstrated that many positive benefits can result from this process for both students and teachers.

Enrichment Learning and Teaching

The third service delivery component of the SEM, which is based on the Enrichment Triad Model, is enrichment learning and teaching that has roots in the ideas of a small but influential number of philosophers, theorists, and researchers such as Jean Piaget (1975), Jerome Bruner (1960, 1966), and John Dewey (1913, 1916). The work of these theorists coupled with our own research and program development activities, has given rise to the concept we call enrichment learning and teaching. The best way to define this concept is in terms of the following four principles:

(1) Each learner is unique, and therefore, all learning experiences must be examined in ways that take into account the abilities, interests, and learning styles of the individual;

(2) Learning is more effective when students enjoy what they are doing, and therefore, learning experiences should be constructed and assessed with as much concern for enjoyment as for other goals;

(3) Learning is more meaningful and enjoyable when content (i.e. knowledge) and process (i.e. thinking

skills, methods of inquiry) are learned within the context of a real and present problem; and therefore, attention should be given to opportunities to personalize student choice in problem selection, the relevance of the problem for individual students at the time the problem is being addressed, and authentic strategies for addressing the problem;

(4) Some formal instruction may be used in enrichment learning and teaching, but a major goal of this approach to learning is to enhance knowledge and thinking skill acquisition that is gained through formal instruction with applications of knowledge and skills that result from students' own construction of meaning (Renzulli, 1994, p. 204).

The ultimate goal of learning that is guided by these principles is to replace dependent and passive learning with innovative, engaged learning. Although all but the most conservative educators will agree with these principles, much controversy exists about how these (or similar) principles might be applied in everyday school situations. Numerous research studies and field tests in schools with widely varying demographics have been carried out (Renzulli & Reis, 1994). These studies and field tests provided opportunities for the development of large amounts of practical know-how that are readily available for schools that would like to implement the SEM. They also have shown that the SEM can be implemented in a wide variety of settings and used with various populations of students including high-ability students with learning disabilities and high-ability students who underachieve in school.

Concluding Thoughts

The many changes taking place in general education have resulted in some unusual reactions to the SEM that might best be described as the good news/bad news phenomenon. The good news is that many schools are expanding their conception of giftedness, and they are more willing than ever to extend a broader continuum of services to larger proportions of the school population. The bad news is that with increasing attention paid to raising test scores, it is even more challenging to continue arguing for opportunities for creativity and innovation in our schools. We are pleased to conclude this chapter with some non-negotiables about the SEM. Before continuing, however, it is important to say that the material that follows is based on the assumption that the reader is familiar with the model and has reviewed the work in this chapter that summarizes our most recent book on this approach, entitled *The Schoolwide Enrichment Model: A Comprehensive Plan for Educational Excellence* (Renzulli & Reis, 1997).

First, although we have advocated a larger talent pool than traditionally has been the practice in gifted education, and a talent pool that includes students who

gain entrance on *both* test and non-test criteria (Renzulli, 1978, 1982, 1986, 1988b, 2002), we firmly maintain that the concentration of services necessary for the development of high level potentials cannot take place without targeting and documenting individual student talents and abilities. Targeting and documenting does not mean that we will simply play the same old game of classifying students as 'gifted' or 'not gifted', and let it go at that. Rather, targeting and documenting are part of an ongoing process that produces a comprehensive and always evolving 'Total Talent Portfolio' about student abilities, interests, and learning styles. The most important thing to keep in mind about this approach is that *all information should be used to make individual programming decisions about present and future activities, and about ways in which we can enhance and build upon documented strengths.* Documented information: (1) will enable us to recommend enrollment in advanced courses or special programs (e.g. summer programs, college courses, etc.); and (2) will provide direction in taking extraordinary steps to develop specific interests and resulting projects within topics or subject matter areas of advanced learning potential.

Enrichment specialists must devote a *majority* of their time to working directly with talent-pool students, and this time should mainly be devoted to facilitating individual and small group investigations (i.e. Type IIIs). Some of their time with talent-pool students can be devoted to stimulating interest in potential Type IIIs through *advanced* Type I experiences and *advanced* Type II training that focuses on learning research skills necessary to carry out investigations in various disciplines. To do this, we must encourage more classroom teachers to become involved in talent development through both enrichment opportunities and curriculum modification and differentiation within their classrooms.

A second non-negotiable is that SEM programs must have specialized personnel to work directly with talent-pool students, to teach advanced courses and to coordinate enrichment services in cooperation with a schoolwide enrichment team. The old cliché, 'Something that is the responsibility of everyone ends up being the responsibility of no one', has never been more applicable than when it comes to enrichment specialists. The demands made upon regular classroom teachers, especially during these times of mainstreaming and heterogeneous grouping, leave precious little time to challenge our most able learners and to accommodate interests that clearly are above and beyond the regular curriculum. In a study recently completed at The National Research Center on the Gifted and Talented (Westberg, 1991), researchers found that in 84% of regular classroom activities, *no differentiation was provided for identified high-ability students.* Accordingly, time spent in enrichment programs with specialized teachers who understand how to encourage student creativity and innovation is even more important for high-potential students. Related to this non-negotiable are the issues of teacher selection and training, and the scheduling of special program teachers. Providing unusually high levels of creative challenge requires advanced training in the discipline(s) that one is teaching, in the application of process skills, and in the management and facilitation of individual and small group investigations. It is these characteristics of enrichment specialists rather than the mere grouping of students that have resulted in achievement gains and high levels of creative productivity on the parts of special program students.

Summary

The SEM creates a repertoire of services that can be integrated in such a way to create 'a rising tide lifts all ships' approach. The model includes a continuum of services, enrichment opportunities and three distinct services: curriculum modification and differentiation, enrichment opportunities of various types, and opportunities for the development of individual portfolio including interests, learning styles, product styles and other information about student strengths. Not only has this model been successful in addressing the problem of high-potential students who have been under-challenged but it also provides additional important learning paths for creative students who achieve academic success in more traditional learning environments but long for opportunities for innovation in school.

The absence of opportunities to develop creativity in all young people, and especially in talented students, is troubling. In the SEM, students are encouraged to become responsible partners in their own education and to develop a passion and joy for learning. As students pursue creative enrichment opportunities, they learn to acquire communication skills and to enjoy creative challenges. The SEM provides the opportunity for students to develop their gifts and talents and to begin the process of lifelong learning, culminating, we hope, in higher levels of creative and innovative work of their own selection as adults.

References

Amabile, T. (1983). *The social psychology of creativity*. New York. Springer-Verlag.

Bruner, J. S. (1960). *The process of education*. Cambridge, MA: Harvard University Press.

Bruner, J. S. (1966). *Toward a theory of instruction*. Cambridge, MA: Harvard University Press.

Burns, D. E. (1998). *SEM network directory*. Storrs, CT: University of Connecticut, Neag Center for Gifted Education and Talent Development.

Dewey, J. (1913). *Interest and effort in education*. New York: Houghton Mifflin.

Dewey, J. (1916). *Democracy and education*. New York: Macmillan.

Dunn, R., Dunn, K. & Price, G. E. (1977). Diagnosing learning styles: Avoiding malpractice suits against school systems. *Phi Delta Kappan*, **58** (5), 418–420.

Gardner, H. (1983). *Frames of mind*. New York: Basic Books.

Kettle, K., Renzulli, J. S. & Rizza, M. G. (1998). Products of mind: Exploring student preferences for product development using 'My Way . . . An Expression Style Instrument'. *Gifted Child Quarterly*, **42** (1), 49–60.

Kirschenbaum, R. J. (1983). Let's cut out the cut-off score in the identification of the gifted. *Roeper Review: A Journal on Gifted Education*, **5**, 6–10.

Kirschenbaum, R. J. & Siegle, D. (1993). *Predicting creative performance in an enrichment program*. Paper presented at the Association for the Education of Gifted Underachieving Students 6th Annual Conference, Portland, OR.

Olenchak, F. R. (1988). The schoolwide enrichment model in the elementary schools: A study of implementation stages and effects on educational excellence. In: J. S. Renzulli (Ed.), *Technical Report on Research Studies Relating to the Revolving Door Identification Model* (2nd ed., pp. 201–247). Storrs, CT: University of Connecticut, Bureau of Educational Research.

Olenchak, F. R. & Renzulli, J. S. (1989). The effectiveness of the schoolwide enrichment model on selected aspects of elementary school change. *Gifted Child Quarterly*, **32**, 44–57.

Piaget, J. (1975). *The development of thought: Equilibration on of cognitive structures*. New York: Viking.

Reis, S. M., Burns, D. E. & Renzulli, J. S. (1992). *Curriculum compacting: The complete guide to modifying the regular curriculum for high ability students*. Mansfield Center, CT: Creative Learning Press.

Reis, S. M., Westberg, K. L., Kulikowich, J., Caillard, F., Hébert, T. P., Plucker, J. A., Purcell, J. H., Rogers, J. & Smist, J. (1993). *Why not let high ability students start school in January? The curriculum compacting study* (Research Monograph 93106). Storrs, CT: University of Connecticut, The National Research Center on the Gifted and Talented.

Renzulli, J. S. (1976). The enrichment triad model: A guide for developing defensible programs for the gifted and talented. *Gifted Child Quarterly*, **20**, 303–326.

Renzulli, J. S. (1977). *The enrichment triad model: A guide for developing defensible programs for the gifted and talented*. Mansfield Center, CT: Creative Learning Press.

Renzulli, J. S. (1978). What makes giftedness? Re-examining a definition. *Phi Delta Kappan*, **60**, 180–184, 261.

Renzulli, J. S. (1982). What makes a problem real: Stalking the illusive meaning of qualitative differences in gifted education. *Gifted Child Quarterly*, **26**, 147–156.

Renzulli, J. S. (1986). The three ring conception of giftedness: A developmental model for creative productivity. In: R. J. Sternberg & J. E. Davidson (Eds), *Conceptions of giftedness* (pp. 53–92). New York: Cambridge University Press.

Renzulli, J. S. (1988a). The multiple menu model for developing differentiated curriculum for the gifted and talented. *Gifted Child Quarterly*, **32**, 298–309.

Renzulli, J. S. (Ed.). (1988b). *Technical report of research studies related to the enrichment triad/revolving door model* (3rd ed.). Storrs, CT: University of Connecticut, Teaching The Talented Program.

Renzulli, J. S. (1994). *Schools for talent development: A practical plan for total school improvement.* Mansfield Center, CT: Creative Learning Press.

Renzulli, J. S. (2002). The three-ring conception of giftedness: Its implications for understanding the nature of innovation. In: L. V. Shavinina (Ed.), *International Handbook on Innovation.* Oxford: Elsevier Science.

Renzulli, J. S. & Reis, S. M. (1985). *The schoolwide enrichment model: A comprehensive plan for educational excellence.* Mansfield Center, CT: Creative Learning Press.

Renzulli, J. S. & Reis, S. M. (1994). Research related to the Schoolwide Enrichment Model. *Gifted Child Quarterly,* **38,** 2–14.

Renzulli, J. S. & Reis, S. M. (1997). *The schoolwide enrichment model: A how-to guide for educational excellence.* Mansfield Center, CT: Creative Learning Press.

Renzulli, J. S., Reis, S. M. & Smith, L. H. (1981). *The revolving door identification model.* Mansfield Center, CT: Creative Learning Press.

Renzulli, J. S. & Smith, L. H. (1978). *The compactor.* Mansfield Center, CT: Creative Learning Press.

Renzulli, J. S., Smith, L. H. & Reis, S. M. (1982). Curriculum compacting: An essential strategy for working with gifted students. *The Elementary School Journal,* **82,** 185–194.

Sternberg, R. J. (1984). Toward a triarchic theory of human intelligence. *Behavioral and Brain Sciences,* **7,** 269–287.

Sternberg, R. J. (1988). Three facet model of creativity. In: R. J. Sternberg (Ed.), *The Nature of Creativity* (pp. 125–147). Boston: Cambridge University Press.

Sternberg, R. J. (1990). Thinking styles: Keys to understanding student performance. *Phi Delta Kappan,* **71**(5), 366–371.

Torrance, E. P. (1962). *Guiding creative talent.* Englewood Cliffs, NJ: Prentice-Hall.

Torrance, E. P. (1974). *Norms-Technical manual: Torrance tests of creative thinking.* Bensenville, IL: Scholastic Testing Service.

Ward, V. S. (1960). Systematic intensification and extensification of the school curriculum. *Exceptional Children,* **28,** 67–71, 77.

The International Handbook on Innovation
Edited by Larisa V. Shavinina

Towards a Logic of Innovation

Gerald F. Smith

*Department of Management, College of Business Administration, University of Northern Iowa,
USA*

Abstract: This chapter identifies the most promising avenues for promoting and improving idea generation, and assesses the possibilities and limitations of such an endeavor. The chapter will discuss research streams and programs, identifying specific heuristics and strategies that can be employed to generate innovative ideas. It will also consider fundamental questions regarding the extent to which idea generation can be 'rationalized' and the ways in which a useful logic of innovation might be developed. Traces of such a logic can be found in the many idea-generation methods that have been proposed, but rarely tested, in the creativity literature.

Keywords: Innovation; Creativity; Idea generation, Scientific discovery; TRIZ; Inventive problem-solving.

Thomas Edison's famous analysis of genius—that it is '1% inspiration and 99% perspiration'—pertains to innovation and other endeavors that depend on the generation of creative ideas. His epigram juxtaposes two contrasting views of the mental activities involved in innovation. Per the inspiration account, innovative ideas result from dazzling mental leaps and 'Aha' experiences. This view, part of the popular image of creative genius, has attracted the attention of scientists intent on explicating the creative process (Smith, Ward & Finke, 1995). By contrast, the perspiration account sees innovative breakthroughs as resulting from more effortful and mundane activities (Weisberg, 1993, 2003). Innovators often employ a brute force, trial-and-error strategy in which many possibilities are successively tried until an acceptable solution is found. Less appealing to empirical investigators, this view was favored by Edison.

Their value notwithstanding, each account has inadequacies. Inspiration is unreliable, incapable of being enacted at will, so would-be innovators derive few benefits from this perspective. The perspiration account, however, fails to explain how people generate insightful, non-obvious possibilities that result in major innovative breakthroughs. Thus, the poetic qualities that make Edison's epigram memorable also make it misleading. Recognizing the most divergent aspects of innovative thought, it overlooks elements that are more central, notably the importance of domain knowledge and intelligent deliberation. Reliance on knowledge

and reflective thought were the hallmarks of Edison's own approach to invention; witness his conscious use of analogies (Weisberg, 1993) and employment of expert scientists (Friedel, 1992).

These resources also predominate in more pedestrian idea-generation efforts, as when a writer devises the opening paragraphs of a chapter like this. Rather than waiting for inspiration or relying on simple-minded trial and error, most writers (the present author included) are tacitly aware, from their reading and writing experiences, of strategies for latching onto a reader's attention: ask a provocative question; narrate an interesting vignette; make an outrageous claim; or explore the implications of an epigram. These strategies suggest possibilities that can be shaped, by reflection, to fit the situation at hand.

In this chapter I consider whether innovation exhibits or can be equipped with a 'logic', that term being construed broadly to encompass any consciously controlled mental activities that enhance one's prospects for generating valuable ideas. Such a logic would lie in the considerable space between inspiration and perspiration, providing mental tools that innovators can deploy in their search for solutions. My interest in this issue reflects a belief that idea generation is a critical part of the innovation process and that much can be done to improve it.

The chapter's next section frames the issue. This is followed by a discussion of the search for a logic of scientific discovery, a parallel endeavor of direct relevance to the chapter's topic. The chapter's longest section reports and analyzes various research programs devised to support the generation of innovative ideas. The most promising extract lessons from past incidents of creative thought. The contents of these sections inform the chapter's concluding discussion of the revealed nature of the logic of innovation.

Can Idea Generation Be Empowered?

What is innovation? Most simply, it is the introduction of or change to something new. When innovating, one makes progressive changes that result in novel states of affairs. So understood, innovations can be characterized in several respects. They have an object or target, that which is being changed. This can be a product, a process, or such things as an individual's lifestyle, an organization's strategy, and a society's culture. Innovations vary in extent or magnitude, the degree to which one deviates from the past. Though there is 'nothing new under the sun', some innovations diverge more substantially than others from known precedents. A final characteristic is the impact of the change, the significance or range of its effects. Boden's (1991) distinction between psychological and historical creativity—having ideas that are novel for oneself vs. novel for humankind—is apposite. Some innovations

change the trajectory of history; others change the way a CD collection is organized.

Organizing the Conceptual Landscape

The innovation process begins with a felt need for change and culminates in its successful implementation. Key process stages include the generation of ideas, development of the most promising, and their acceptance by relevant parties. A process perspective highlights several concepts related to innovation, most being concerned with the generation of innovative ideas, this chapter's focus. These concepts and their relationships are depicted in Fig. 1.

Innovation is closely related to *problem-solving*, as suggested by similarities between models of the two processes (Schroeder, Van de Ven, Scudder & Polley, 2000). A problem is a difficult or challenging situation that bears improvement; problem-solving is the thinking done to improve things (Smith, 1998b). Problem-solving is not always innovative—one can adopt tried-and-true solutions—but innovation invariably includes problem-solving since the generation and implementation of ideas for change never transpire without difficulty.

The early stages of the innovation process, concerned with the generation and development of novel ideas, are called *invention*. This term is especially

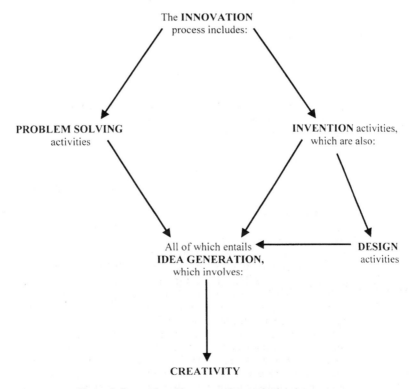

Figure 1. Innovation, idea generation, and related concepts.

applicable if the ideas could be patented. Because it is a challenging mental task, invention involves problem-solving activity. Much of this thinking/problem-solving is *design* work. Design is the development or devising of complex things. Not all design is invention, the latter having a higher standard for originality (Dasgupta, 1996). Architects design but do not invent buildings. However, an architect might invent by devising a novel solution to a design problem.

Idea generation, the mental production of possibilities or alternatives, must be performed at various points in most problem-solving episodes. Lying at the heart of both invention and design, it is widely acknowledged to be a key part of the innovation process (Van de Ven, Angle & Poole, 2000; Weber & Perkins, 1992). One's idea-generation ability and performance are usually attributed to the person's *creativity*. Creativity can be regarded as a fixed trait or as an acquired repertoire of mental skills, habits, and attitudes.

Innovation Research

Progressive change, the core meaning of innovation, is a matter of much importance and widespread applicability. This is reflected in innovation research: there is a huge amount of it spread across many fields (cf., Downs & Mohr, 1976; Rogers, 1983; Van de Ven, Angle & Poole, 1989). No academic disciplines exist exclusively for the study of innovation, and few journals are dedicated to the topic. Technological innovation has attracted the most attention because of its role in the rise of Western civilization (Mowery & Rosenberg, 1998). There are substantial streams of research that study innovation in different fields (e.g. education, healthcare); it has also been investigated from various disciplinary perspectives, notably economics and sociology (cf., Mansfield, 1968; Steil, Victor & Nelson, 2002).

Since the topic's practical importance is responsible for the great volume of innovation research, it follows that this research should be relevant to practice: it should contribute to increases in the amount and effectiveness of innovative activities. Most scientists would accept this need for practical relevance. At the same time, however, they would address it indirectly, following Lewin's dictum that "nothing is as practical as a good theory" (1945, p. 129). Scientists develop a deep theoretical understanding of empirical phenomena, holding that such understanding will enable prediction and control of practical outcomes. This theory-based strategy has been enormously successful in the physical and biological sciences, where scientific theories underlie technological development of practical artifacts and interventions. However, the strategy has rarely been effective in the human sciences. This may be because few human phenomena are susceptible to being deeply explained by scientific theories (Dupre, 1993; Rosenberg, 1994).

Innovation is, of course, a quintessentially human phenomenon. Within this vast area of research, two large streams would seem to have considerable potential for practical application. One is psychological research on creativity, which could presumably help people generate innovative ideas. The other, research on the management of innovation, tries to promote effective innovation practices in organizations. Have these two bodies of research delivered the practical goods? Reasonable people could disagree in their assessments, but in this author's judgment, neither of these substantial research streams has had a significant impact on innovation practice. Granting this claim, if only for the sake of argument, why might it be true? What has kept scientific research from having a larger effect on innovation practice?

With regard to research on the management of innovation, it is likely that the phenomenon in question—innovation activity in organizations—is too diverse, complex, and contingent to be explained by a scientific theory. Recent descriptive studies have concluded that prior academic assumptions about organizational innovation were simplistic, and that the process may be the result of a "nonlinear dynamic system" (Van de Ven, Polley, Garud & Venkataraman, 1999, p. 5). This characterization offers no added leverage for process improvement. Proposed theories of organizational innovation processes (e.g. Poole & Van de Ven, 2000) do not provide the explanatory insights required for prescriptive support. Indeed, the most useful research in this area is indifferent to theory, focusing instead on the identification and validation of 'best practices', what innovative companies do that makes their change efforts successful (Kanter, Kao & Wiersema, 1997; Zairi, 1999). Even here, there are prescriptive limitations: innovation begins with and depends on the generation of ideas, an individual-level mental activity that organizations can enable but not cause.

This limitation does not apply to psychological research on creativity, which is expressly concerned with idea generation. Moreover, psychology is among the most highly developed of the human sciences, having produced knowledge of considerable value in many practical applications—child development, education, and marketing, among others. Why has psychological research not been as useful with regard to innovation? Several reasons come to mind:

- First and foremost, creativity is difficult to understand. Psychology has been hard-pressed to develop adequate theories of higher-order thinking processes like judgment and reasoning, and creativity is assuredly the most elusive of these:
- Second, much research treats creativity as a trait that is fixed for each person (Torrance, 1988). This approach offers no support for individual idea generation performance;

- Third, a substantial portion of creativity research studies the influence of factors—personality and culture, for instance—that are not susceptible to control (Csikszentmihalyi, 1988);
- Fourth, the cognitive research strategy that is most promising, from a prescriptive standpoint, is relatively new and undeveloped (Smith, Ward & Finke, 1995). But even cognitive accounts run into impasses, capacities and processes like intuition and insight that are resistant to conscious direction and improvement.

Surprisingly, elements of the idea generation process that are subject to individual control—for instance, the selection and use of creativity techniques—have not attracted much research attention (Smith, 1998a). Indeed, setting aside a few noteworthy exceptions (e.g. Perkins, 1981), it might seem that creativity researchers are not especially concerned with practical applications of their work

Is Prescription Possible?

Perhaps creativity research has had so little effect on practice because there is little effect it could have. The generation of innovative ideas may be an activity that simply cannot be improved by outside interventions. Such a view follows from Campbell's (1960) claim that creativity is a blind variation and selective retention process. It is suggested by trait theories of creativity, which regard idea-generation performance as determined primarily by an individual's innate endowment of creative capacity (Torrance, 1988). And it follows from cognitive theories that emphasize uncontrollable mental processes like intuition (Bowers, Farvolden & Mermigis, 1995). These concerns raise doubts about the prospects for improving idea generation. Are there grounds for believing the opposite, for claiming that people can be helped to generate innovative ideas?

There has been a widespread tendency—among researchers and lay people—to regard idea generation as unfettered, free-wheeling thought (Adams, 1979; Buzan, 1983). Artistic imagination is taken as its model, creativity being seen as the unconstrained exploration of possibilities that satisfy few real world demands. Indeed, the wilder the idea, the better. This conception overlooks a critical distinction: whereas many artists regard art's role as one of disturbing mainstream beliefs, this is not the case for other forms of innovation. Outside the realm of art, innovation pursues the goal of progressive change, making things better, more adequate to human needs. For this to happen, innovative ideas must present possibilities that are feasible, capable of being implemented within relevant contexts. An idea's craziness per se is not a virtue; the wildly improbable must still be doable. Because implementation contexts are dense, posing many constraints and demands, the generation of innovative ideas is a knowledge-intensive process.

Losing some of its similarity to artistic imagination, idea generation takes on qualities associated with practical problem-solving (Sternberg et al., 2000).

Thinking that is more constrained, as by feasibility considerations, is more knowledge-intensive. Knowledge of relevant constraints and demands must be taken into account for such thinking to reach acceptable conclusions. Thinking that is more knowledge-intensive is more susceptible to individual monitoring and control. Inspiration gives way to perspiration; instead of relying on free-flowing imagination, it becomes more productive to mentally work one's way through to a solution (Weisberg, 2003). Such thinking is also susceptible to assistance and support from techniques, heuristics, and other interventions (Root-Bernstein, 2003). These tools can embody relevant knowledge or bring it to mind. Idea generation becomes more reasoned, reflective, analytical, and logical as one tries to deal with pertinent constraints and demands, avoiding the dead ends they impose and exploiting the possibilities they present. Thinking also becomes more experience-based, past instances providing examples of what works and what does not. This enables support by interventions that bring experiential knowledge to bear on idea generation tasks.

The role of analytical and knowledge-based thought in idea generation is evidenced by many creativity techniques (Smith, 1998a). A method like morphological analysis (Allen, 1962), which decomposes a problem into parameters and components, searching for combinations that have promising solution implications, demonstrates how logical analytical thought can lead to innovative ideas. The checklists used by product developers to generate new product ideas (Higgins, 1994)—"Vary the size, shape, or color of an existing product"—summarize knowledge compiled from successful past innovations.

Thus, the knowledge-intensive nature of innovation tasks creates the potential for idea generation to be intelligently disciplined and directed. Though divergent imaginative thought has a role, convergent reasoning and analysis are also productive. Experiential knowledge is important, and there may be general principles of innovation to identify and exploit. In short, there are good reasons for believing that a logic of innovation exists or can be developed.

The Logic of Scientific Discovery

The attempt to develop a logic of innovation can profit from an investigation of scientific discovery. Discovery and innovation are closely related activities and achievements. Commonly conceived as relying on creative thought processes that are immune to conscious control, each is often viewed as being unamenable to prescription. We can create propitious environments for scientific discovery and innovation, and we can insure that motivated, creative people are employed to perform such tasks. But thereafter, it is argued, little can be done: scientists, inventors, and the rest of society can only hope that insights, inspirations, and breakthroughs occur.

Research on scientific discovery has challenged this portrayal. Studies from various disciplinary perspectives have identified the considerable role of reflective thought in the discovery process (Dasgupta, 2003; Giere, 1992; Kantorovich, 1993; Nickles, 1980b). There is strong evidence of a logic of scientific discovery, noting again that 'logic' is to be taken, not as deductive, algorithmic procedures, but as thinking that is susceptible to conscious control and improvement. These findings lend encouragement to the search for a logic of innovation. They offer insights as to what a logic of innovation might consist of and provide content that can be directly incorporated into that logic.

Task Analysis

Discovery is often contrasted with invention, both being ways of coming up with something new: one can either find it or make it. Though everyday uses of the terms are more elaborate, these root meanings provide a valuable contrast. Innovation, as noted earlier, is a broad notion that includes acts of invention. It also takes in some acts of discovery. For instance, Fleming's discovery of penicillin led to innovations in the treatment of bacterial infections.

However, scientific discovery is more diverse and complex, encompassing achievements that are not really discoveries in the simple sense of finding something new. Indeed, the most significant scientific discoveries are more like inventions than findings. Newton formulated his laws of motion and conceived that objects were attracted by gravity, rather than finding these 'things' fully realized during his studies. Darwin devised an evolutionary account of life on earth, and Einstein developed his theories of relativity. Even Lavoisier's discovery of oxygen has an element of invention, the construction of a plausible account of reality. The reason scientific discovery has this inventive character is that scientists rarely look for things that can simply be seen. Rather, they search for evidence of theoretical entities, things like atoms and genes that are invisible to human eyes. Most of their searches are not for things at all, but are instead for scientific explanations, deep theoretical accounts of empirical phenomena. Explanations are created, not found, constructed through extended mental efforts rather than being directly perceived. Simon's development of the cognitive approach to individual and organizational decision-making exemplifies this kind of effortful achievement (Dasgupta, 2003).

The dual nature of scientific discovery is reflected in a distinction by Kantorovich (1993). In 'discovery by exposure', the scientist becomes aware of something that is already there—for instance, Galileo's discovery of the moons of Jupiter. This process is usually perception-driven. With 'discovery by generation', however, "the object of discovery . . . is in a sense a product of the discovery process" (Kantorovich, 1993, pp. 32–33). The scientist invents a theory of certain phenomena, creating an explanation that did not previously exist. Even something as eternally real as gravity was, in a significant sense, created through its discovery: though it had always been there, gravity became 'real' for people as a result of Newton's theory.

Several influential accounts of science (Laudan, 1977; Nickles, 1980b; Simon, 1977) view it as a form of problem-solving. Each type of scientific discovery cited by Kantorovich matches up with a major conceptual component of the problem-solving process. Discovery by exposure results from search activity. Search, the systematic exploration of a set of possibilities, is, per Newell & Simon (1972), the fundamental problem-solving process. Search is guided by knowledge that directs the problem-solver's attention in directions most likely to yield solutions. Being susceptible to prescriptive support, knowledge-driven heuristic search provides grounds for believing that discovery by exposure is similarly susceptible.

The other kind of discovery, by generation, is a form of design, another important problem-solving activity. As noted earlier, design is the development of complex things, the creation of artifacts serving certain purposes. A scientific theory is such an artifact. An artifact's design is subject to constraints, this being the case with theories (Nickles, 1980a). Design includes search activity—for instance, an architect's search for possible building entryways. This is replicated in the design of theories, when scientists search for variables and relationships that provide explanatory power (Bechtel & Richardson, 1993). Being shaped by past experience and prevailing practices, design is knowledge-intensive. Viewed as design, scientific discovery offers opportunities for the deployment of strategies, methods, heuristics, and other forms of procedural knowledge.

Thus, a task analysis of scientific discovery suggests substantial room for development of a supporting logic. It clarifies the close relationship between scientific discovery, innovation, and other concepts featured in Fig. 1. This provides grounds for believing that studies of the logic of scientific discovery will have implications for innovation.

Elements of Logic

Interest in the logic of scientific discovery originated with Reichenbach's (1938) distinction between the discovery and justification of scientific claims. Philosophers of science concluded that while justification had a logic that scientists employ and philosophers could improve, such was not the case for discovery. In his book *The Logic of Scientific Discovery*, Popper denied that there was any such thing, saying that "the act of conceiving or inventing a theory, seems to me neither to call for logical analysis nor to be susceptible of it" (1959, p. 31). Discovery occurs through inspiration or intuition, making it a matter for psychological investigation, rather than rational reconstruction by philosophers. Note, however, that these philosophers employed a narrow, deduction-centered, notion of logic and that they wrote at a time when psychologists had virtually no understanding of higher-order mental processes.

The first significant argument on behalf of a logic of discovery was made by Hanson (1965), who contended that scientific discovery evidences the use of abductive reasoning, inference to the best explanation. Observing certain surprising phenomena, a scientist infers the hypothesis that most plausibly explains them, this hypothesis becoming the object of further investigation. This account has serious inadequacies—for instance, it says why a hypothesis is selected, but not how it came to be identified in the first place. Hanson's work stirred up interest in scientific discovery, leading to a program of research by philosophers informally known as the 'friends of discovery'. Following the example of Kuhn (1962), scholars analyzed historical cases of discovery, searching for elements of method and logic (cf. Nickles, 1978).

Since the 1970s, this program has received support from the newly prominent field of cognitive science. Cognitivism revived the psychological study of thinking, developing scientifically rigorous ways of studying unobservable mental activity. As their interests expanded, cognitive researchers became interested in the thinking of scientists. Viewing discovery as problem-solving, Simon and others developed computational models that generate theoretical concepts and educe scientific laws (Langley, Simon, Bradshaw & Zytkow, 1987; Simon, Langley & Bradshaw, 1981; Thagard, 1988, 1992). For instance, Langley's BACON program used heuristic processes to derive Kepler's laws from its analyses of relevant data sets (Langley et al., 1987). Though critics allege that these 'discoveries' were assisted by structuring devices unavailable to working scientists, computational studies demonstrate the role and utility of heuristic strategies in scientific discovery. Thus, by the end of the twentieth century, a broad-based research effort had identified many aspects and elements of a logic of scientific discovery.

Scientific thinking, including discovery, is influenced in the most general way by fundamental presuppositions, methodological rules and principles, and core concepts, all of which Holton (1973, 1978, 2003) subsumed under the notion of 'theme'. These would include a belief in causality and forms of conservation, and commitments to norms of deductive and inductive reasoning. The principle of parsimony is a theme, as is the assumption that the universe is governed by laws that can be represented mathemat-

ically. Scientists can benefit from an awareness of this intellectual milieu. For instance, when trying to explain certain phenomena, it is helpful to be aware of the kinds of explanation used in one's field—for instance, functionalist accounts, stage theories, and mathematical models. Since scientific discovery is given direction and made more intelligible by thematic resources and constraints, they are part of its logic.

At the other extreme in terms of specificity, and much closer to traditional notions of logic, are relatively formal techniques used for discovery purposes. Most of these are statistical methods used to identify patterns in data. Factor analysis, curve-fitting methods, and exploratory data analysis are among the most popular. These clearly belong to the logic of discovery, being analytical tools devised for the performance of discovery-related tasks.

Certain general mental skills and practices are widely employed as part of discovery. These include normal mental capacities—mental imagery, for instance—that scientists use to generate theories and hypotheses. There is ample evidence that scientists conduct thought experiments as a way of pre-testing hypotheses (Gooding, 1992). They manipulate mental images to achieve insights into phenomena (Nersessian, 1992). They also reason analogically, their thinking being affected by the technical analogues—clockworks or computers, for instance—favored at the time (Bechtel & Richardson, 1993). Some of these mental practices can be improved through prescriptive support (Root-Bernstein & Root-Bernstein, 2003). For instance, there are rules of analogical rigor—maintain structural consistency, avoid mixed analogies, among others—that scientists can learn (Gentner & Jeziorski, 1989). The informed deployment of these mental capacities is a source of rationality in the discovery process. It is likely to be a source of logic in innovation as well.

The most numerous instances of discovery logic are the many strategies and heuristics used by scientists. They have been identified in various ways. Dunbar (1995) studied working scientists in molecular biology labs, noting how they used negative evidence, explored analogies, followed up on surprising results, and were influenced by their social context. In contrast, Tweney (1989) examined historical documents from the work of Michael Faraday, the 19th century English physicist, finding that Faraday employed confirmatory–disconfirmatory heuristics proposed by researchers: scientists should seek confirming evidence early in the life of a hypothesis, but shift toward seeking disconfirming evidence as things progress. Gigerenzer (1994) looked at various research programs in proposing his tools-to-theories heuristic: the tools scientists use to justify their theories provide concepts and metaphors that inspire new theories.

Adopting a perspectivist approach to research, McGuire (1989) proposed heuristics for use during the formative stages of a research project: explore the meaning of key variables through conceptual analysis; conjecture the limits of hypothesized relationships; and hypothesize the contrary relationship, trying to explain it. An extensive set of heuristic advice was reported by Root-Bernstein (1989). He collected ideas from an array of sources, mostly famous scientists, compiling them into a 'Manual of Strategies for Discovery'. The following are among the most interesting: Delbruck's principle: "Be sloppy enough that something unexpected happens but not so sloppy that you can't tell what happened" (p. 412); Macfarlane's law: "When a number of conflicting theories coexist, any point on which they all agree is the one most likely to be wrong" (p. 413); George's strategy: "Vary the conditions over the widest possible range" (p. 416); and Burnet's advice: "Do as large a proportion as possible of your experiments with your own hands" (p. 418). Many of these suggestions apply to innovation tasks.

Some methods and heuristics are domain-specific; others only apply to certain discovery tasks, though these could appear in many disciplines. Bechtel & Richardson (1993) discussed the task of developing a mechanistic explanation, explaining a complex system's behavior in terms of the functions performed by its parts and relationships among these parts. Such explanations often appear in biology, psychology, and the social sciences. Per these authors, scientists who develop mechanistic explanations rely on two heuristics: decomposition, breaking the system down into components; and localization, identifying system activities with the behavior of specific components. Various strategies—analysis of system breakdowns, inhibitory or deficit studies, and excitatory studies—can generate information and insights regarding system functioning. These methods may pertain to innovation, prompting analyses of existing or analogous systems that suggest ideas for improvement.

A final form of prescriptive support derives from research on the 'illogic' of scientific discovery, thinking mistakes that scientists, like the rest of us, are prone to commit. It has long been recognized that scientists are not immune from such mistakes as over-valuing results from small samples, being biased by vivid information, and maintaining beliefs that have been discredited by new evidence (Kahneman, Slovic & Tversky, 1982; Nisbett & Ross, 1980). Making scientists more aware of common mental mistakes enhances their capacity for metacognitive self-monitoring. Thus, it is another aspect of the logic of scientific discovery.

Implications for Innovation

The foregoing constitutes a substantial array of mental tools that scientists consciously deploy for discovery purposes. The value of these resources notwithstanding, they do not exhaust the discovery process; assuredly, many important activities transpire beyond their reach. Relevant knowledge may or may not come

to mind; key patterns may or may not be seen; insights and intuitions may or may not occur. Scientific discovery involves mental activities that are not susceptible to prescriptive support. Nonetheless, it is heartening to find so many opportunities for aiding and improving the process.

Notably absent from this account is anything like a formal logic or algorithmic procedure. The statistical techniques employed for discovery purposes only suggest promising areas for investigation; computational models fall far short of automating discovery in real world contexts. Studies of discovery consistently demonstrate that, because of its context sensitivity and knowledge dependence, this ill-structured generative process cannot be fully proceduralized. Rather than a few powerful procedures, scientific discovery involves the use of many informal methods, strategies, and heuristics. These have varying degrees of generality and, following the well-known generality-power trade-off, the more general tend to be less effective. One would expect this to be true for innovation as well: effective idea generation practice is likely to involve a multitude of informal heuristic methods.

Generating Innovative Ideas

Where would one look to find a logic of innovation, consciously deployable means of generating innovative ideas? Some sources mirror those prominent in the logic of scientific discovery. Others are unique to innovation, reflecting its differences from discovery. This section discusses four sources of innovation logic. The first, process studies, investigates mental activities of innovators, seeking effective practices that can be employed by others. Another source derives lessons from analyses of inventions and other innovation products. The work of Genrich Altshuller (1984, 1996) typifies and motivates most of these efforts. A third source addresses process and other non-product targets of innovative activity. Finally, there is overtly prescriptive material—creativity techniques—that directly addresses the idea-generation task. Each of these sources will be discussed with the intent of identifying elements of innovation logic.

Process Studies

Several streams of research contribute to our understanding of the process by which innovative ideas are generated. Studies of inventors have identified mental capacities and activities that drive their achievements (Weber & Perkins, 1992). Studies of designers offer insights into their creative performances (Lawson, 1980; Rowe, 1987). Cognitive research uses experimental settings to explore facets of creativity (Smith, Ward & Finke, 1995). Research in these areas tries to demystify the creative process, solving Perkins' (1988) *ex nihilo* problem by explaining how people can create something that is genuinely new. Process studies have identified four kinds of resources that can be brought to bear for this purpose: the innovation task or problem; mental capacities; informal heuristics, methods, and strategies; and prior knowledge and experience.

Innovation Tasks/Problems

Innovative thought should be adapted to or constrained by the problem one is trying to solve (Isaak & Just, 1995). Effective innovators exploit problem characteristics, shaping their mental activities to the task. For instance, since there are many acceptable solutions to innovation problems, a search usually proceeds forward in light of one's objectives, rather than working backwards from a predefined goal. Since innovation is often aimed at developing a physical artifact, visual imagery, graphic representations, and physical models are powerful tools for generating, evaluating, and developing solution alternatives.

Innovation problems start out as general objectives—say, a cure for cancer. Poorly specified at the outset, goals are elaborated as candidate solutions emerge and develop. Goals are a kind of constraint, the broader set of requirements an innovation must satisfy (Smith & Browne, 1993). Some constraints originate externally—building codes, for instance. Internal constraints emerge when solution alternatives are developed; the design of one part of a product limits the way other parts can be constructed. Constraints can be soft or hard, more or less amenable to violations. Innovators surface and assess self-imposed constraints, their preconceptions about the problem's solution (Lawson, 1980). They challenge boundaries implicit in prevailing artifactual forms. The proliferation of internal constraints leads innovators to employ a 'least commitments' design strategy (Goel & Pirolli, 1989): Solutions to sub-problems should be minimally constraining on subsequent design work. Solutions to some problems are organized as 'homing spaces' (Perkins, 1992), allowing one to know if progress is being made towards the goal. The Wright Brothers demonstrated another way of exploiting problem structure: unlike their competitors who searched design space for a machine that would fly, they decomposed the problem of flight into functional components—lift and lateral control, for instance. Searches of these function spaces yielded solutions that were combined in the first airplane (Bradshaw, 1992).

Mental Capacities

Like scientific discovery, innovative thinking employs mental capacities that are susceptible to improvement. The most fundamental is our ability to form and manipulate mental representations. Gorman (1997) included mental models—dynamic representations that can be operated in the imagination—in his cognitive framework for understanding innovation. Research on creative cognition focuses on 'pre-inventive structures', mental representations of promising ideas that might be developed into creative products (Ward, Smith &

Finke, 1999). Design theorists note the importance of a 'primary generator', an appealing idea that narrows the search space while being elaborated into a detailed solution (Lawson, 1980).

Studies of creative cognition have identified processes people use to generate and evaluate mental structures (Ward, Smith & Finke, 1999). These processes rely on existing knowledge, as when established concepts are stretched to suggest new possibilities. Our capacity for mental imagery figures prominently in innovative thought. Finke's laboratory studies demonstrated how visual forms suggest new functions (1990); Crouch claimed that the Wright Brothers were set apart by their "genius for visualizing the abstract" (1992, p. 84). Mental images can be made explicit in pictures, diagrams, and other graphic representations, which appear prominently in historical studies of invention (cf. Carlson & Gorman, 1990; Gorman & Carlson, 1990) and in the work of designers (Smith & Browne, 1993).

Innovative thinking involves the use of analogy as a vehicle for carrying ideas from one domain to another. Engineers recognize the value of natural analogies, how nature solves a problem, looking, for instance, to clam shells for hinge ideas. Architects use iconic analogies, designing buildings that resemble an appropriate object—praying hands for Wright's Unitarian church and sails for Utzon's Sydney Opera House (Rowe, 1987).

These capacities are employed in a process that is fundamentally conjectural: promising ideas are generated and developed as possible solutions, conjectures that will be discarded if insurmountable barriers or more appealing possibilities arise. The generation of innovative ideas can be aided by interventions that facilitate their visual depiction, or which highlight stimulating concepts, relationships, and analogies (Root-Bernstein & Root-Bernstein, 2003). The evaluation and development of innovative ideas involve reflective thought that is highly susceptible to knowledge-based support.

Heuristics, Methods, and Strategies

Consciously employed mental practices that innovators can acquire through experience or training, heuristics, methods, and strategies are by nature part of the logic of innovation. Recent interest in heuristics can be traced to Polya (1957), who offered suggestions for solving mathematical problems. Some of these—look for related problems, draw a figure, separate the parts of the condition—are applicable to innovation. Almost as venerable are 'weak methods'—means–ends analysis, working backwards, and hill-climbing, among others—identified by cognitive research on problem-solving (Lesgold, 1988). The draghunt device favored by Edison (Gorman & Carlson, 1990) is a generate-and-test method that systematically tries many alternatives.

Though many heuristics are equally applicable to scientific discovery, others only apply to innovation. Thus, the development of the electron microscope was motivated by a realization that optical microscopes had reached their upper performance limit, and the heuristic of exploring different solution principles when that is the case (Weber & Perkins, 1992). Informal strategies and pieces of advice often seem like common sense—deliberately move away from old paths; try to integrate the desirable properties of many things; if something works, use it again (Perkins & Weber, 1992)—but they can have a huge impact on thinking.

Decomposition—breaking large problems down into many simple ones—is the most important strategy for invention. Its power was demonstrated by the Wright Brothers, who combined solutions to lift, control, and power sub-problems into a successful airplane. Decomposition is the prototypical means of addressing design problems (Simon, 1973). Design affords decomposition "because systems can be understood as hierarchies of components at different levels" (Smith & Browne, 1993, p. 1215), levels being defined around the system's structural parts or intended functions. Design problems are recursive, decomposing into smaller design problems that can be solved or further broken down. While an innovation might begin as a vague idea, its development entails extensive decompositional elaboration, a process that tests the idea's feasibility.

Prior Knowledge and Experience

Though innovation, by definition, involves a change to something new, process studies demonstrate that changes inevitably use and partially preserve the old and established (Weisberg, 1993, 2003). Innovations embody elements of the past; nothing is totally new or unprecedented. The generation of innovative ideas relies on knowledge of existing artifacts and practices (Ward, Smith & Finke, 1999). The use of analogies is typical of a process in which ideas, rather than being generated *ex nihilo*, are borrowed from various sources and shaped to fit the situation at hand. Studies have identified precedents supporting development of the telephone (Gorman & Carlson, 1990), moving pictures (Carlson & Gorman, 1990), ultrasonic imaging (Wild, 1992), and other inventions. Carlson and Gorman use the term 'mechanical representations' to refer to "the specific, working components an inventor uses to construct physical models of his invention" (1990, p. 392). Mechanical representations include the "levers, screws, wedges, wheels" and other artifacts that are "basic building blocks of invention" (p. 393). Some inventors have favorite devices that reappear in their creations; Edison was partial to the cylinder and stylus, double-action pawl, and polar relay (Jenkins, 1984). But all innovators draw on a culture's stock of artifacts, its repertoire of tried-and-true means of achieving certain ends. Studies of design note the same phenomenon, using the term 'design types' or 'prototypes' to refer to standard solutions to design problems—for instance, ways of constructing watertight roofs (Smith & Browne, 1993).

The importance of standard solutions explains the salience of historical material in design education. Rather than learning powerful idea generation methods, architectural students learn how their forbearers solved certain problems, so those solutions will be mentally available when practical situations require. Similarly for engineers and other inventors: education is largely a matter of becoming aware of means that inventors past have made available for the solution of current problems. This part of an innovator's education and idea generation practice can be assisted by technology, for instance, case-based reasoning systems that provide designers with past cases pertaining to their current problem-solving needs (Kolodner, 1993).

In addition to these resources for the generation of innovative ideas, process studies have identified common mental mistakes made by innovators. Lawson (1980) warns against the 'category trap'—categorizing a problem in solution terms—and other mental pitfalls. This is another element for our emerging logic of innovation. Thus far, that logic closely resembles the logic of scientific discovery. However, we now consider an approach to innovation that has no parallel in science.

Product Analysis

The attempt to develop 'logical' creativity methods is as old as the study of creativity itself. During the past several decades, this endeavor has been fueled by a powerful new strategy that has delivered impressive results. The strategy is product-analytic: the results of innovative activities are analyzed in the hopes of identifying patterns that can be used to direct idea generation. This strategy is based on a structuralist assumption that the infinite variety of things making up our world is shaped by a smaller set of principles. Once discovered, these can be represented in rules, methods, and heuristics for the generation of innovative ideas.

The first, most highly developed, and most influential program of product analysis was undertaken by Genrich Altshuller. Altshuller is a Russian inventor and patent inspector who began, in 1946, to develop an "algorithm for the solution of inventive problems". The approach he used was to study patents issued in the Soviet Union for technical inventions. Since then, he has analyzed over 40,000 patents, doing much of this work during six years spent in a Siberian prison camp—his criticisms of Soviet innovation policy rubbed Stalin the wrong way. Altshuller identified useful regularities of various kinds. These were organized and developed into a 'Theory of Inventive Problem Solving', commonly known by its Russian acronym, TRIZ. During the 1960s, TRIZ achieved prominence in the Soviet Union, even being taught to high school students. Altshuller's work (1984, 1996) began to appear, in translation, to Western audiences in the 1980s. During the 1990s, Altshuller achieved the ultimate apotheosis: his work was picked up by American consulting firms which market related products and services as breakthrough boosts to creative thinking.

Like many others, Altshuller conceives the idea-generation problem as one of reducing the size of the search space. He is skeptical of psychological studies of the creative process, which do not address the 'primary and objective side of creativity' (1984, p. 8). He is also unimpressed by brainstorming and traditional creativity methods, which rarely solve difficult problems. The 'objective side of creativity' has to do with overcoming contradictions that are at the core of most problems. Technical contradictions exist insofar as improvements in part of a system create troubles in other parts. These derive from underlying physical contradictions.

By analyzing thousands of patents, Altshuller developed knowledge of different kinds of contradictions and means of overcoming them. Most of this knowledge existed initially as particularized insights and pieces of advice, mostly having to do with physical effects and phenomena that could be exploited for problem-solving purposes. Thus, an electrical effect called the 'Corona Discharge' can be used to measure the gas pressure inside a light bulb (Altshuller, 1996). More abstract insights were also identified and confirmed through repetition in multiple cases—for instance, the strategy of separating contradictory properties in space or time, or the principle of preparatory action (do it in advance). Much of TRIZ reflects Altshuller's attempt to encompass myriads of insights in a 'theory', a compelling conceptual framework that integrates these parts into a meaningful whole. His theoretical endeavors tend to be 'scientistic', using language—S-field analysis, contradiction, algorithm, law—which implies a level of rigor that may not be achievable with such subject matters.

More modestly conceived, Altshuller's project can be viewed as producing three important outputs. First, and at the lowest conceptual level, TRIZ includes a substantial set of physical effects and devices that inventors can use to achieve particular purposes. Thus, phase transitions can be used to stabilize temperatures; movement can be controlled through the Toms, Bernoulli, or Weissenberg effects; and dimensions of objects can be measured by applying and reading magnetic and electrical markers. These means are summarized in a table (Altshuller, 1984), a compendium of stock solutions or raw materials for innovations involving physical phenomena.

Second, at a middle level of abstraction, Altshuller identified a wealth of heuristics that innovators can learn and apply. Some of these—change the state of the physical property; introduce a second substance, for instance—are tied to the kinds of physical inventions

Altshuller studied. Others are more general. Do it inversely; do a little less; fragmentation/consolidation; Ideal Final Result; and Model with Miniature Dwarfs (Altshuller, 1996, p. 168): all can be applied with socio-technical systems and other problems that are not solved at the chemistry-and-physics level. Some of these heuristics have been identified in other fields. Forcing an object to serve multiple functions (Altshuller, 1996, p. 159) is a standard design strategy; as will be seen, Weber (1992) independently recognized the value of incorporating multiple objects into one system (Altshuller, 1996, p. 29). This aspect of TRIZ, which resembles Polya's (1957) work on mathematical discovery, is a significant contribution to the logic of innovation.

The most abstract and intendedly general element of Altshuller's program is his 'algorithm for solving inventive problems', initially called ASIP, later ARIZ. This procedure includes typical problem-solving steps—selection of the problem, constructing a model of the problem, and so forth—fleshed out with pertinent sub-tasks and heuristics. ARIZ includes S-field analysis, a method Altshuller devised for representing problematic systems in ways that suggest solutions. These structured procedural elements of TRIZ may be useful, but they are unlikely to be as powerful as their maker suggests. Indeed, their effectiveness is likely to derive from the lower-level heuristics they include. Arguably, Altshuller's basic conceptual scheme, focused on overcoming contradictions, lacks the depth needed to drive a problem-solving method that is both powerful and general.

In summary, Altshuller's work is significant, even though his theories, algorithms, and laws do not satisfy criteria normally associated with those terms. If he has not developed an idea-generation method of unprecedented power, his work has yielded a wealth of heuristic knowledge and advice from which innovators in many fields can profit. Of special importance for current purposes is Altshuller's methodology—the analysis of innovative products—a research strategy he has developed further than anyone else.

Recognizing the importance of Altshuller's work, others have tried to extend and capitalize on it. One line of development involves software implementations; the 'Invention Machine Lab' of Invention Machine Corp. and 'Innovation Tools' by Ideation International Inc. are representative products. Another approach uses Altshuller's ideas to create new methods for solving innovation problems. Mohrle & Pannenbacker (1997) combined Altshuller's content with Five-Field-Analysis—a problem-solving scheme that looks at the current state, resources, goals, the intended state, and transformations—creating a technique they call 'Problem-Driven Inventing'. Helfman (1992) presented 'Analytic Inventive Thinking' as a practical model for the support of innovation. His model has nine stages that organize Altshullerian devices—a focus on contradictions, the use of 'tools' and 'elves', consideration of ideal states—in a functional analysis of problematic systems. As was the case with Altshuller's own work, while these techniques may be useful, their power is likely to derive from lower-level heuristics, not from the procedural frameworks constructed around those elements.

A more significant offspring of Altshuller's program is the work of Goldenberg, Mazursky & Solomon (1999a, 1999b, 1999c). Rather than replacing Altshuller's content, these scholars borrowed his product-analytic methodology, using it to study things other than patent applications. Their analyses of new products identified 'templates', common patterns in a product's development. For instance, Attribute Dependency occurs when two previously independent product variables are related, as when Domino's Pizza connected price and delivery by offering a discount for late deliveries (Goldenberg & Mazursky, 2000). Subsequent experimental research has shown that the use of a template-driven approach improves idea-generation performance over traditional creativity techniques (Goldenberg et al., 1999c). Employing the same methodology, these researchers studied product advertisements, identifying six 'creativity templates'—for instance, pictorial analogy and extreme situations—commonly employed in high-quality ads (Goldenberg et al., 1999b). Again, empirical research indicates that utilization of the templates enhances idea-generation performance. Goldenberg and his associates have demonstrated that Altshuller's product analytic methodology can be applied to different kinds of innovation tasks.

A final program of research, undertaken by Robert Weber (1992), resembles Altshuller's work, though it was conducted independently. Weber analyzed inventions, mostly everyday artifacts like hand tools, to discover principles underlying their success. He concluded that heuristics are "the engine that drives inventive variation" (1992, p. 83), identifying a substantial set of these. For instance, many inventions incorporate an inverse function, a means of undoing their primary action: pencils come with erasers; hammers include nail-pulling claws. The Swiss Army Knife exemplifies the complement heuristic—combining tools used in the same context—and the shared-property heuristic—join common parts of tools to eliminate redundancies. Weber's work is more modest in its claims than Altshuller's—he does not propose laws, theories, or formal methods for generating innovative ideas—but his findings are a comparably significant contribution to the logic of innovation.

Summing up, product-analytic research has generated a wealth of heuristics and other insights that support idea generation. These can be incorporated into formal techniques, although this structuring adds little value, over and beyond that carried by the heuristics themselves. The product-analytic approach is effective because it taps into regularities that underlie and are embodied in observed artifacts. These regularities may not be laws, as claimed by Altshuller, but they express principles that can inform idea-generation practice. And, as the work of Goldenberg and his associates demonstrates, the product-analytic approach can be applied in almost any field where innovative product ideas must be generated.

Process Analysis

Innovation efforts have always been directed primarily at the development of products, especially new inventions and consumer products. During the 1980s, however, organizational processes became a prominent target of innovation activity (Davenport, 1993). This prominence has not diminished; many companies maintain teams and departments with ongoing responsibility for improving organizational processes. Our interest is in whether these process-innovation efforts exhibit a logic, intelligible reliable practices that can be used to generate new process ideas.

Before the 1980s, organizational processes were an almost invisible aspect of corporate life, like a building's electrical system, part of the infrastructure that attracts attention only when something goes wrong. Industrial engineers rationalized production activities—say, the movements of product assemblers—but no one paid attention to business processes—the handling of vendor shipments and customer orders, the processing of insurance claims. The quality movement changed this. Quality was the dominant business buzzword of the 1980s, and the pursuit of product quality led to a recognition that related processes—of product design, production, and customer service, among others—had to be improved. When Total Quality Management (TQM) made quality improvement an organization-wide imperative, managers realized that all organizational products and processes, internal and external, needed attention (Juran, 1992). This put process innovation on the agenda of every manager.

Though valuable, TQM's approach to process improvement was conservative. Existing organizational processes were analyzed with an eye towards reducing errors and creating economies, usually by eliminating unnecessary activities. Rather than being totally replaced, processes were revamped, sometimes to great effect. This approach was challenged, during the early 1990s, by reengineering, a more aggressive program of process innovation (Davenport, 1993; Hammer & Champy, 1993). TQM's successor in management buzzword history, reengineering called for the wholesale abandonment of existing organizational processes and their replacement by radically new ones, designed from the ground up (Hammer, 1990). Reengineering's start-from-scratch, clean-slate approach to process innovation was motivated by and relied heavily on developments in information technology (IT). Advances in computing and communications had created resources that existing organizational processes were not using and that new processes could be designed to exploit. In a sense, IT was reengineering's equivalent of Altshuller's (1984) table of physical effects and phenomena: an important resource that innovators had not been adequately aware of. Michael Hammer's 1990 article in the *Harvard Business Review*, which initiated the reengineering revolution, showed what could be done. Hammer described how the Ford Motor Company used IT in a redesigned accounts-payable process to reduce the headcount of its 500 person operation by 75%.

If IT is a process-innovation resource, there remains the question of how innovative ideas for improvement are generated. Much of the reengineering literature begs this question, offering homilies about critical and creative thinking. Thus, Davenport admits that "there is less to say about the design phase of process innovation", describing design activity as "largely a matter of having a group of intelligent, creative people review the information collected . . . and synthesize it into a new process" (1993, p. 153). Hammer's work is more useful. His 1990 article identified principles of reengineering—for instance, link parallel activities instead of integrating their results. Hammer & Champy (1993) identified themes commonly encountered in reengineered processes (e.g. workers make decisions), traditional rules that can be broken by modern IT (e.g. field personnel can send and receive information wherever they are), and symptoms of reengineering opportunities (e.g. complexity, exceptions, and special cases reflect accretion onto a simple base process). These insights can be incorporated into a process-innovation logic. Beyond the use of these heuristics, however, it is likely that reengineering experts get their ideas the old-fashioned way: by adapting known solutions to process design problems to meet current situational demands. Like every other form of innovation, reengineering draws heavily on the past.

Benchmarking, another process-improvement methodology, is clear about the origins of its ideas: they come from other organizations, especially world-class performers in an area one is trying to improve. Thus, in the study that made benchmarking famous, when Xerox was trying to improve the order-handling performance of its distribution units, it examined the order-fulfillment process at L. L. Bean, identifying cost-saving methods that could be implemented in Xerox facilities (Camp, 1989). As this case demonstrates, benchmarking reaches beyond industry boundaries, searching for 'best practices' wherever they can be found. Nonetheless, its copy-cat strategy has been derided by reengineering proponents who claim that "benchmarking is just a tool for catching up, not for jumping way ahead" (Hammer & Champy,

1993, p. 132). This criticism is unjustified since there is no innovation method or approach, reengineering included, that consistently advances the state-of-the-art. As with creative performances (Boden, 1991), most innovations are locally, rather than globally, new. This is especially true for organizational processes, things that are not patented or publicized. Whether one is borrowing from other organizations or from the past, it is still innovation if progressive changes are made.

Indeed, benchmarking's borrowing strategy is at the heart of a broader 'best practices' approach that constitutes an important new perspective on management. Rather than looking to academic theories or the proclamations of gurus for advice, managers are advised to study practices used in world-class organizations. Osborne & Gaebler's (1992) book, *Reinventing Government*, demonstrated what public organizations could learn from their peers. Since then, a wealth of literature (e.g. Hiebeler, Kelly & Ketteman, 1998) has identified business processes, structures, strategies, and other characteristics of high-performing entities that can be adapted to serve the needs of other organizations. This is a major source of ideas for innovation, a source based on a simple but powerful precept: borrow intelligently from others.

During the mid-1990s, I engaged in a project aimed at quality and process improvement. The project employed a case-analytic strategy similar to that used by Altshuller, though at the time I was unaware of Altshuller's work. The quality boom had created a large literature of case histories of problem-solving efforts, written by people in organizations who wanted to share their success stories. I collected over 700 cases of this kind from books and journals, analyzing each in search of lessons having a useful degree of generality. My methodological approach is described in Smith (1994); the project's results are reported in *Quality Problem Solving* (Smith, 1998b).

Case analyses yielded many insights. Some were included in chapters that discussed key problem-solving tasks, like problem definition and diagnosis (Smith, 1998b). Most appeared in chapters devoted to particular types of quality problems—efficiency and process design, among others (Smith, 1998b). Thus, I identified 15 kinds of unnecessary activities that could be eliminated to improve process efficiency. For instance: discontinue activities that create unused outputs; eliminate unnecessary controls by empowering employees and using periodic audits; discontinue tests and inspections that rarely find exceptions; develop templates for activities with reusable outputs; discontinue activities initiated in response to one-time needs; unless the consequences of an occurrence are severe, discontinue activities that protect against rare events; and add storage capacity to reduce the frequency of materials handling (Smith, 1998b, p. 222). Other case-derived prescriptions addressed such process design issues as process flow and layout, input screening and control, exception handling, setup, coordination and consolidation of activities, and process triggers.

This project applied Altshuller's case-analytic strategy to organizational problem-solving episodes. It yielded lessons applicable to problem situations, mostly process-related, that often appear in organizations. The lessons serve as heuristic advice for making improvements or innovations in organizational processes. Process innovations are rarely new in a universalistic sense. But they are new for that organization or that process, and their performance effects can be substantial. Since they offer useful guidance to would-be innovators, these case-based heuristics, business best practices, and reengineering principles comprise a process-innovation logic.

Idea-Generation Methods

Creativity or idea-generation methods might be regarded as existence proofs of the logic of innovation. They are, after all, mental tools that people use to generate innovative ideas. Barring widespread irrationality, their popularity indicates that there are consciously deployable means of improving idea-generation performance. The huge variety of creativity methods invites consideration as to the nature of their functioning. How do idea-generation methods work, and do the devices they employ comprise part of the logic of innovation?

This line of reasoning would fail if creativity methods do not really drive the generation of ideas, but only evoke or channel mental activities, largely uncontrolled, that do the actual work. On this view, a method only initiates a creative process that proceeds without conscious control, so the method's value-added is marginal. Indeed, there are idea-generation methods for which this may be the case. Free association techniques or methods like Idea Tracking (Van Gundy, 1981), which sets the problem aside so unconscious mental processes can take over, are examples. Thus, some idea-generation methods depend on means that are not part of innovation logic, on account of being non-operational, insusceptible to conscious control.

At the other extreme lies a contrasting possibility. Perhaps the generation of innovative ideas can be proceduralized, reduced to an algorithm, a strong method that reliably produces creative outputs. If this were true, the logic of innovation would become a logic in the narrower, traditional sense of deduction. All the informal strategies and heuristics that we have uncovered would be trumped, marked as inferior, second-rate tools that must necessarily give way to more powerful procedures. Over the years, more or less extreme versions of this claim have been proposed (Crovitz, 1970; Zwicky, 1969). A number of people have suggested that creativity can be captured in an algorithm. The argument runs as follows: all possible solutions to an innovation problem possess certain attributes, each of which can take on various values. An algorithmic creativity method will generate all relevant attributes, all possible values for each, and all possible attribute–value combinations. From this exhaustive set of alternatives, the most promising solutions can be identified. Zwicky's (1969) 'morphological box' pursued this approach. The basic combinatorial strategy can be seen in Crovitz's (1970) method of 'relational algorithms', which uses a set of relational words to generate combinations. Tauber's (1972) 'heuristic ideation' technique is a relatively pure implementation of the strategy, intended for use in the generation of new product ideas.

The fact that these methods have not driven their less proceduralized cousins into extinction indicates that something is awry. It is possible that such techniques are done in by combinatorics, that they generate more possibilities than an army of evaluators could sift through. More likely, the approach is mistaken in its assumption that one can identify all the attributes, and values of attributes, around which creative alternatives can be defined. Arguably, the most innovative solutions to problems are marked by attributes and values that had not previously been recognized as problem-relevant. So, for the time being at least, our emerging informal logic of innovation does not seem to be threatened by a more powerful, formalized competitor.

Several years ago I conducted research on idea-generation methods, the results of which were published in the *Journal of Creative Behavior* (Smith, 1998a). The project was motivated by a belief that the many creativity methods in use drew on a limited set of 'active ingredients', devices that plausibly promote idea generation. The research tried to specify those active ingredients. To this end, I identified 172 idea-generation methods and used an iterative analytical procedure to determine how they worked, how the method might promote idea generation. This analysis resulted in the identification of 50 idea-generation devices. This set is reasonably complete: 46 of these 50 devices were initially identified in the first 90 (out of 172) idea-generation methods that were analyzed, and no new devices were discovered in the last 32 methods that I analyzed.

Three kinds of idea-generation devices were differentiated. Strategies are active means for generating ideas. The most numerous and significant type, most strategies refer to identifiable mental activities. The following are typical: Fantasy, conceiving of states in which reality constraints have been dropped; Relationship Search, looking for relationships among two or more things; Boundary Stretching, exploring extreme values of variables in the situation; and Combination, combining elements, attributes, and other aspects of the problem. Tactics, the least common type of device, are stimulatory tools that support strategies. Among the tactics are: Elaboration, enriching the problem situation to provide idea-generation material; Changing

Environment, mentally or physically leaving one's normal thinking environment; and Concrete Stimuli, using physical things or pictures as stimuli during idea-generation sessions. Finally, enablers are passive means of promoting idea generation. Enablers create conditions under which ideas are more likely to occur. Goal Setting, establishing quotas or time deadlines for idea generation, is an enabler, as is Non-Disclosure, not stating the problem at the outset of sessions, and Anonymity, insuring that ideas cannot be traced to their originators. The research identified 31 idea-generation strategies, seven tactics, and 12 enablers.

Many devices are familiar from work discussed in this chapter. Thus, the strategies of Analogy, Decomposition, Abstraction, and Mental Simulation are widely employed by innovators. So too with Enhancement, the strategy of modifying ideas to make them more feasible and effective. Perkins and Weber noted that sometimes invention involves "searching for a purpose to fit a thing" (1992, p. 320), an approach reflected in the Circumstance strategy: think of circumstances in which an idea might be effective. The Checklists strategy, using an established set of ideation prompts to generate alternatives, suggests the templates identified by Goldenberg et al. (1999a, 1999b, 1999c) and other outputs of the product-analytic approach.

However, many idea-generation devices do not have clear precedents. This is true of interpersonal strategies like Group Interaction—verbalizing thoughts in a group so one person's ideas prompt others—and the Nominal Group strategy—generate and share ideas silently within a group. Rearrangement, changing the structure of a situation by rearranging its parts, might be used informally by inventors and designers. More outlandish idea-generation strategies—for instance, Identification, imaginatively becoming a non-human part of the problem, and Negation, adopting counter-assumptions for problem-relevant beliefs—are less likely to be useful in innovation tasks where most reality constraints must be honored.

Some devices apply beyond idea generation, to other mental tasks. Challenge Assumptions, questioning beliefs associated with the problem, pertains to most high-level cognitive activities (Root-Bernstein, 2003), as does Change of Perspective, thinking about the problem from the viewpoints of different agents. Dialectic, conducting a debate between opposing sides on an issue, applies more to evaluation than idea generation, and Structure, organizing information to reveal relationships, is used by many diagram-based problem-definition methods. Two enablers—Block Removal, removing mental barriers that inhibit idea generation, and Incubation, setting the problem aside to escape mental ruts—are widely applicable means of clearing one's mind.

Their prominence in creativity techniques notwithstanding, some devices seem weak, unlikely to be effective. Thus, Association, a strategy of mentally following associative links among ideas in memory, lacks the direction found in more powerful tools. Change of Attitude, adopting different attitudes toward the situation, is not promising. Remote Stimuli, providing stimuli unrelated to the task, is a widely used tactic that has no compelling rationale (Perkins, 1983), the same being true of Force Fit, an enabler in which ideas are forced together to create breakthroughs. De Bono (1992) endorsed Provocation, drawing attention to the problem or task, but this enabler is unlikely to provide much mental leverage.

Many of these tools are valuable additions to the logic of innovation. This is true of strategies like Bootstrapping, analyzing known alternatives to generate new families of possibilities, and Integration, combining alternatives to make better solutions, and of many devices already discussed. Even enablers can be productive: both Competition, arranging idea generation contests between groups or individuals, and Deferred Evaluation, withholding evaluation of ideas so as not to inhibit generation, can have positive effects on idea-generation performance.

Not surprisingly, then, creativity techniques embody effective idea-generation devices, and their prescribed mental behaviors are ones that successful innovators often employ on their own. Some devices fall short on the operationality requirement for innovation logic: their performance is not sufficiently subject to conscious control. None exhibits the degree of procedural structure and power associated with algorithms. But this study of the 'active ingredients' employed in idea-generation methods did yield noteworthy additions to the logic of innovation.

In Conclusion

This chapter has demonstrated that the process of generating innovative ideas can be directed intelligently. Though unbridled imagination has a role, the idea-generation process can benefit significantly from deliberative, reflective thought that uses knowledge to achieve its goals. As support for this conclusion, the chapter identified elements of the logic of innovation, pieces of declarative and procedural knowledge that promote idea generation. This logic is heuristic, not formal. No strong idea-generation procedures have been uncovered. Rather, innovation logic encompasses a variety of informal methods, strategies, and pieces of advice.

Though innovation logic can be viewed as consisting of concepts, principles, methods, heuristics, and other such components, another kind of characterization may be illuminating. Some of the logic consists of recommendations that innovators use certain innate mental capacities. Thus, they are told to fantasize and free associate; to construct mental models and images; to conduct thought experiments and conceive bold conjectures and hypotheses. Little guidance is provided concerning how to do these things; it is assumed that the capacities exist and can readily be employed. Advice of this kind is warranted by evidence that successful innovators think in these ways. Another kind of element warns innovators against mental errors or shortcomings. In addition to the design traps discussed by Lawson (1980), there are more general errors, for instance, judgmental biases (Kahneman et al., 1982). Other shortcomings are addressed by the block removal devices used in creativity techniques (Adams, 1979), and the incubation enabler that helps people get out of mental ruts (Smith, 1998a). This advice responds to predictable inadequacies in innovative thought, warning against avoidable mistakes as a form of prevention, and offering assistance for overcoming common weaknesses.

Both kinds of elements are purely psychological. They suggest that idea generation will be improved if innovators think in some ways and not others, these ways being definable in process terms, without reference to the content one thinks about. This psychological approach to improving innovation is exemplified by research on creative cognition (Ward, Smith & Finke, 1999), work that tries to lay bare the mental activities leading to creative outcomes.

There are psychological elements in the logic of innovation that have a different character. Innovators are encouraged to challenge assumptions and to look at things from different perspectives. Advice of this kind prompts one to question dominant ways of thinking in a field, to think about its content. The recommendation to think more abstractly about the innovation task aims at inducing a deeper understanding of the problem, one that grasps its fundamental nature. This advice also fosters engagement with content. The use of analogies, a common motif in the chapter, reflects an assumption that similarity is a useful basis for inference. It is, in this respect, content-driven: the thinking practice only makes sense if reality has the character assumed. These examples suggest that the logic of innovation is as much about the world and innovation tasks as it is about cognitive processes.

The content or declarative knowledge side of innovation logic is clearly represented in Altshuller's (1984) table of physical effects and the knowledge of information technology possessed by reengineering experts. This knowledge of how the world works and of the means available to achieve certain ends is an essential ingredient in the idea-generation process. Innovation can be promoted by making such knowledge more readily available.

Even more important for innovation purposes is knowledge derived from past experience. Such knowledge appears repeatedly in the chapter, in various forms: as mechanical representations cited by Carlson & Gorman (1990) and prototypes used by designers; as templates that Goldenberg, Mazursky, & Solomon (1999a, 1999b, 1999c) identified through case analysis; as lessons learned from benchmarking studies and

research on quality problem-solving. The relevant principle is expressed in Polya's (1957) advice to look for related problems. Experiential knowledge is important because innovation tasks always have precedents; there are solutions to problems in other times and organizations that are usefully pertinent to the current situation. So knowledge of precedents is a valuable source of ideas, and a key part of innovation logic.

The most substantial part of this logic consists of activities, mostly mental, that are infused with or devised in light of knowledge about innovation tasks and the world. Innovators are advised to think about certain things in certain ways because these ways of thinking are consonant with reality; they reflect and exploit its underlying structure. Decomposition is an effective strategy because many things can be decomposed into nearly independent parts, and our mental limitations make it easier for us to solve many small problems than one large one. Some innovation problems exist in homing spaces (Perkins, 1992), so a hill-climbing strategy may be viable and effective. Problem difficulties often result from tradeoffs—strength vs. weight of materials, or speed vs. accuracy of a performance, for instance—so Altshuller's focus on contradictions and the means of overcoming them offers insights. Tools should be designed with an inverse function because one may need to undo a previous action. Indeed, experiential knowledge is valuable for the same reason: previously solved problems are structurally similar to the one currently being addressed, so what worked then might work now.

Rather than being an abstract formal or purely psychological procedure, the logic of innovation is infused with content. Recommended thinking practices embody knowledge of innovation tasks and the nature of reality. In this respect, innovation logic conforms to the psychological dictum that behavior is an adaptation to reality and, like the meanderings of Simon's (1981) ant, can only be understood in light of the task environment. Borrowing from Gibson (1979), innovation logic exploits environmental 'affordances'; it uses opportunities provided by the idea-generation task. Many opportunities derive from the underlying structure of reality, its regularities, expressed in principles of widespread applicability. Innovation logic is evidence for the structuralist claim that a more fundamental reality underlies the phenomenal world of experience. Effective thinking about innovation tasks taps into this reality, relevant principles being expressed in innovation logic.

This suggests the possibility that higher-order thinking in general is infused with content. Effective thinking practices of whatever kind—inferential, judgmental, and so forth—are effective by virtue of being adapted to the realities they address, as a result of being informed by knowledge of those realities. Accordingly, learning how to think is largely a matter of acquiring insights into reality and thinking tasks that humankind has accumulated during its history. One's ability to identify causes is improved as a result of learning the distinction between causes and conditions, realizing that causes can be proximate or remote, and knowing the types of causes recognized by Aristotle. The 'hows' of thinking tasks are filled in, not with refined cognitive operations, but with declarative knowledge content and content-informed mental activities.

The content dependence of thought might lead some to conclude that thinking is domain-dependent, that effective thinking behaviors vary substantially across academic disciplines and fields of practice. Baer (1998), for instance, argued that creativity is domain-specific. There are, to be sure, domain-specific elements of thought, as can be seen with the logic of innovation. The templates for product innovations and effective advertisements identified by Goldenberg and his associates are specific to those artifacts and the marketing domain to which they belong. However, content dependence of thought does not imply domain dependence when content itself is general. Thinking that is driven by content will be common across domains to the extent that the driving content is common. Causal thinking has considerable generality for this reason. Decomposition and most other heuristics in the logic of innovation also possess a substantial degree of generality. Thinking that is adapted to underlying structures shares the generality of those structures.

Thus, the logic of innovation applies to innovation tasks in most, if not all, fields of practice. There is a need to develop both the general and domain-specific elements of this logic. General elements can be included in thinking skills instruction made available to all students and adults. Domain-specific elements can be included in courses and training programs intended for particular professions (e.g. chemical engineers), with more focused versions being developed for professionals working in specific organizations or on particular innovation tasks. But the bottom-line message of this chapter should be clear: there is a way between the horns of Edison's inspiration–perspiration dilemma. The way—logic of innovation—uses consciously controlled mental activities informed by content knowledge to generate promising ideas and alternatives.

References

Adams, J. L. (1979). *Conceptual blockbusting: A guide to better ideas* (2nd ed.). New York: Norton.

Allen, M. S. (1962). *Morphological creativity*. Englewood Cliffs, NJ: Prentice-Hall.

Altshuller, G. S. (1984). *Creativity as an exact science: The theory of the solution of inventive problems*. New York: Gordon & Breach.

Altshuller, G. S. (1996). *And suddenly the inventory appeared: TRIZ, the theory of inventive problem solving*. Worcester, MA: Technical Innovation Center.

Baer, J. (1998). The case for domain specificity of creativity. *Creativity Research Journal*, **11** (2), 173–177.

Bechtel, W. & Richardson, R. C. (1993). *Discovering complexity: Decomposition and localization as strategies in scientific research*. Princeton, NJ: Princeton University Press.

Boden, M. (1991). *The creative mind*. New York: Basic Books.

Bowers, K. S., Farvolden, P. & Mermigis, L. (1995). Intuitive antecedents of insight. In: S. M. Smith, T. B. Ward & R. A. Finke (Eds), *The Creative Cognition Approach* (pp. 27–51). Cambridge, MA: MIT Press.

Bradshaw, G. (1992). The airplane and the logic of invention. In: R. N. Giere (Ed.), *Minnesota Studies in the Philosophy of Science* (Vol. 15) *Cognitive Models of Science* (pp. 239–250). Minneapolis, MN: University of Minnesota Press.

Buzan, T. (1983). *Use both sides of your brain*. New York: Dutton.

Camp, R. (1989). *Benchmarking: The search for industry best practices that lead to superior performance*. White Plains, NY: Quality Resources.

Campbell, D. T. (1960). Blind variation and selective retention in creative thought as in other knowledge processes. *Psychological Review*, **67** (6), 380–400.

Carlson, W. B. & Gorman, M. E. (1990). Understanding invention as a cognitive process: The case of Thomas Edison and early motion pictures, 1888–91. *Social Studies of Science*, **20**, 387–430.

Crouch, T. D. (1992). Why Wilbur and Orville? Some thoughts on the Wright Brothers and the process of invention. In: R. J. Weber & D. N. Perkins (Eds), *Inventive Minds: Creativity in Technology* (pp. 80–92). New York: Oxford University Press.

Crovitz, H. F. (1970). *Galton's walk*. New York: Harper & Row.

Csikszentmihalyi, M. (1988). Society, culture, and person: A systems view of creativity. In: R. J. Sternberg (Ed.), *The Nature of Creativity: Contemporary Psychological Perspectives* (pp. 325–339). Cambridge: Cambridge University Press.

Dasgupta, S. (1996). *Technology and creativity*. New York: Oxford University Press.

Dasgupta, S. (2003). Innovation in the social sciences: Herbert A. Simon and the birth of a research tradition. In: L. V. Shavinina (Ed.), *International Handbook on Innovation*. Oxford: Elsevier.

Davenport, T. H. (1993). *Process innovation: Reengineering work through information technology*. Boston, MA: Harvard Business School Press.

De Bono, E. (1992). *Serious creativity*. New York: Harper-Collins.

Downs, G. W. & Mohr, L. B. (1976). Conceptual issues in the study of innovation. *Administrative Science Quarterly*, **21**, 700–714.

Dunbar, K. (1995). How scientists really reason: Scientific reasoning in real-world laboratories. In: R. J. Sternberg & J. E. Davidson (Eds), *The Nature of Insight* (pp. 365–395). Cambridge, MA: MIT Press.

Dupre, J. (1993). *The disorder of things*. Cambridge, MA: Harvard University Press.

Finke, R. A. (1990). *Creative imagery: Discoveries and inventions in visualization*. Hillsdale, NJ: Lawrence Erlbaum Associates.

Friedel, R. (1992). Perspiration in perspective: Changing perceptions of genius and expertise in American invention. In: R. J. Weber & D. N. Perkins (Eds), *Inventive Minds: Creativity in Technology* (pp. 11–26). New York: Oxford University Press.

Gentner, D. & Jeziorski, M. (1989). Historical shifts in the use of analogy in science. In: B. Gholson, W. R. Shadish, R. A. Neimeyer & A. C. Houts (Eds), *Psychology of Science: Contributions to Metascience* (pp. 296–325). Cambridge: Cambridge University Press.

Gibson, J. J. (1979). *The ecological approach to visual perception*. Boston, MA: Houghton Mifflin.

Giere, R. N. (Ed.) (1992). *Minnesota studies in the philosophy of science* (Vol. 15) *Cognitive models of science*. Minneapolis, MN: University of Minnesota Press.

Gigerenzer, G. (1994). Where do new ideas come from? In: M. A. Boden (Ed.), *Dimensions of Creativity* (pp. 53–74). Cambridge, MA: MIT Press.

Goel, V. & Pirolli, P. (1989). Motivating the notion of generic design within information-processing theory: The design problem space. *AI Magazine*, **10**, 18–36.

Goldenberg, J. & Mazursky, D. (2000). First we throw dust in the air, then we claim we can't see: Navigating in the creativity storm. *Creativity and Innovation Management*, **9** (2), 131–143.

Goldenberg, J., Mazursky, D. & Solomon, S. (1999a). Creative sparks. *Science*, **285**, 1495–1496.

Goldenberg, J., Mazursky, D. & Solomon, S. (1999b). The fundamental templates of quality ads. *Marketing Science*, **18** (3), 333–351.

Goldenberg, J., Mazursky, D. & Solomon, S. (1999c). Toward identifying the inventive templates of new products: A channeled ideation approach. *Journal of Marketing Research*, **36** (2), 200 210.

Gooding, D. (1992). The procedural turn; or, why do thought experiments work? In: R. N. Giere (Ed.), *Minnesota Studies in the Philosophy of Science* (Vol. 15) *Cognitive Models of Science* (pp. 45–76). Minneapolis, MN: University of Minnesota Press.

Gorman, M. E. (1997). Mind in the world: Cognition and practice in the invention of the telephone. *Social Studies of Science*, **27**, 583–624.

Gorman, M. E. & Carlson, W. B. (1990). Interpreting invention as a cognitive process: The case of Alexander Graham Bell, Thomas Edison, and the telephone. *Science, Technology & Human Values*, **15** (2), 131–164.

Hammer, M. (1990). Reengineering work: Don't automate, obliterate. *Harvard Business Review*, (July–August), 104–112.

Hammer, M. & Champy, J. (1993). *Reengineering the corporation: A manifesto for business revolution*. New York: Harper-Collins.

Hanson, N. R. (1965). *Patterns of discovery*. Cambridge: Cambridge University Press.

Helfman, J. (1992). The analytic inventive thinking model. In: R. J. Weber & D. N. Perkins (Eds), *Inventive Minds: Creativity in Technology* (pp. 251–270). New York: Oxford University Press.

Hiebeler, R., Kelly, T. B. & Ketteman, C. (1998). *Best practices: Building your business with customer-focused solutions*. New York: Simon & Schuster.

Higgins, J. M. (1994). *101 creative problem solving techniques*. Winter Park, FL: New Management Publishing Company.

Holton, G. (1973). *Thematic origins of scientific thought: Kepler to Einstein*. Cambridge, MA: Harvard University Press.

Holton, G. (1978). *The scientific imagination: Case studies*. Cambridge: Cambridge University Press.

Holton, G. (2003). Dimensions of scientific innovation. In: L. V. Shavinina (Ed.), *International Handbook on Innovation*. Oxford: Elsevier.

Isaak, M. I. & Just, M. A. (1995). Constraints on thinking in insight and invention. In: R. J. Sternberg & J. E. Davidson (Eds), *The Nature of Insight* (pp. 281–325). Cambridge, MA: MIT Press.

Jenkins, R. V. (1984). Elements of style: Continuities in Edison's thinking. *Annals of the New York Academy of Sciences*, **424**, 149–162.

Juran, J. M. (1992). *Juran on quality by design*. New York: Free Press.

Kahneman, D., Slovic, P. & Tversky, A. (Eds) (1982). *Judgment under uncertainty: Heuristics and biases*. Cambridge: Cambridge University Press.

Kanter, R. M., Kao, J. & Wiersema, F. (1997). *Innovation: Breakthrough thinking at 3M, DuPont, GE, Pfizer, and Rubbermaid*. New York: Harper-Business.

Kantorovich, A. (1993). *Scientific discovery: Logic and tinkering*. Albany, NY: State University of New York Press.

Kolodner, J. (1993). *Case-based reasoning*. San Mateo, CA: Morgan Kaufmann.

Kuhn, T. S. (1962). *The structure of scientific revolutions*. Chicago, IL: University of Chicago Press.

Langley, P., Simon, H. A., Bradshaw, G. L. & Zytkow, J. M. (1987). *Scientific discovery: Computational explorations of the creative processes*. Cambridge, MA: MIT Press.

Laudan, L. (1977). *Progress and its problems*. Berkeley, CA: University of California Press.

Lawson, B. (1980). *How designers think*. London: Architectural Press.

Lesgold, A. (1988). Problem solving. In: R. J. Sternberg & E. E. Smith (Eds), *The Psychology of Human Thought* (pp. 188–213). Cambridge: Cambridge University Press.

Lewin, K. (1945). The research center for group dynamics at Massachusetts Institute of Technology. *Sociometry*, **8** (2), 126–136.

Mansfield. E. (1968). *The economics of technological change*. New York: W. W. Norton.

McGuire, W. J. (1989). A perspectivist approach to the strategic planning of programmatic scientific research. In: B. Gholson, W. R. Shadish, R. A. Neimeyer & A. C. Houts (Eds), *Psychology of Science: Contributions to Metascience* (pp. 214–245). Cambridge: Cambridge University Press.

Mohrle, M. G. & Pannenbacker, T. (1997). Problem-driven inventing: A concept for strong solutions to inventive tasks. *Creativity and Innovation Management*, **6** (4), 234–248.

Mowery, D. C. & Rosenberg, N. (1998). *Paths of innovation: Technological change in 20th-century America*. Cambridge: Cambridge University Press.

Nersessian, N. J. (1992). How do scientists think? Capturing the dynamics of conceptual change in science. In: R. N. Giere (Ed.), *Minnesota Studies in the Philosophy of Science* (Vol. 15) *Cognitive Models of Science* (pp. 3–44). Minneapolis, MN: University of Minnesota Press.

Newell, A. & Simon, H. A. (1972). *Human problem solving*. Englewood Cliffs, NJ: Prentice- Hall.

Nickles, T. (1978). Scientific problems and constraints. In: I. Hacking & P. Asquith (Eds), *PSA 1978* (Vol. 1, pp. 134–148). East Lansing, MI: Philosophy of Science Association.

Nickles, T. (1980a). Introductory essay: Scientific discovery and the future of philosophy of science. In: T. Nickles (Ed.), *Scientific Discovery, Logic, and Rationality* (pp. 1–59). Dordrecht, Holland: D. Reidel Publishing.

Nickles, T. (Ed.) (1980b). *Scientific discovery: Case studies.* Dordrecht, Holland: D. Reidel Publishing.

Nisbett, R. & Ross, L. (1980). *Human inference: Strategies and shortcomings of social judgment.* Englewood Cliffs, NJ: Prentice-Hall.

Osborne, D. & Gaebler, T. (1992). *Reinventing government.* New York: Penguin.

Perkins, D. N. (1981). *The mind's best work.* Cambridge, MA: Harvard University Press.

Perkins, D. N. (1983). Novel remote analogies seldom contribute to discovery. *Journal of Creative Behavior,* **17,** 223–239.

Perkins, D. N. (1988). The possibility of invention. In: R. J. Sternberg (Ed.), *The Nature of Creativity: Contemporary Psychological Perspectives* (pp. 362–385). Cambridge: Cambridge University Press.

Perkins, D. N. (1992). The topography of invention. In: R. J. Weber & D. N. Perkins (Eds), *Inventive Minds: Creativity in Technology* (pp. 238–250). New York: Oxford University Press.

Perkins, D. N. & Weber, R. J. (1992). Conclusion: Effable invention. In: R. J. Weber & D. N. Perkins (Eds), *Inventive Minds: Creativity in Technology* (pp. 317–336). New York: Oxford University Press.

Polya, G. (1957). *How to solve it.* Princeton, NJ: Princeton University Press.

Poole, M. S. & Van de Ven, A. H. (2000). Toward a general theory of innovation processes. In: A. H. Van de Ven, H. L. Angle & M. S. Poole (Eds), *Research on the Management of Innovation: The Minnesota Studies* (pp. 637–662). Oxford: Oxford University Press.

Popper, K. R. (1959). *The logic of scientific discovery.* New York: Harper & Row.

Reichenbach, H. (1938). *Experience and prediction.* Chicago, IL: University of Chicago Press.

Rogers, E. M. (1983). *Diffusion of innovations* (3rd ed.). New York: Free Press.

Root-Bernstein, R. S. (1989). *Discovering.* Cambridge, MA: Harvard University Press.

Root-Bernstein, R. S. (2003). Problem generation and innovation. In: L. V. Shavinina (Ed.), *International Handbook on Innovation.* Oxford: Elsevier.

Root-Bernstein, R. S. & Root-Bernstein, M. (2003). Intuitive tools for innovative thinking. In: L. V. Shavinina (Ed.), *International Handbook on Innovation.* Oxford: Elsevier.

Rosenberg, A. (1994). *Instrumental biology or the disunity of science.* Chicago, IL: University of Chicago Press.

Rowe, P. G. (1987). *Design thinking.* Cambridge, MA: MIT Press.

Schroeder, R. G., Van de Ven, A. H., Scudder, G. D. & Polley, D. (2000). The development of innovation ideas. In: A. H. Van de Ven, H. L. Angle & M. S. Poole (Eds), *Research on the Management of Innovation: The Minnesota Studies* (pp. 107–134). Oxford: Oxford University Press.

Simon, H. A. (1973). The structure of ill structured problems. *Artificial Intelligence,* **4,** 181–201.

Simon, H. A. (1977). *Models of discovery and other topics in the methods of science.* Dordrecht, Holland: D. Reidel Publishing.

Simon, H. A. (1981). *The sciences of the artificial* (2nd ed.). Cambridge, MA: MIT Press.

Simon, H. A., Langley, P. W. & Bradshaw, G. L. (1981). Scientific discovery as problem solving. *Synthese,* **47,** 1–27.

Smith, G. F. (1994). Quality problem solving: Scope and prospects. *Quality Management Journal,* **2** (1), 25–40.

Smith, G. F. (1998a). Idea-generation techniques: A formulary of active ingredients. *Journal of Creative Behavior,* **32** (2), 107–133.

Smith, G. F. (1998b). *Quality problem solving.* Milwaukee, WI: ASQ Quality Press.

Smith, G. F. & Browne, G. J. (1993). Conceptual foundations of design problem solving. *IEEE Transactions on Systems, Man & Cybernetics,* **23** (5), 1209–1219.

Smith, S. M., Ward, T. B. & Finke, R. A. (Eds) (1995). *The creative cognition approach.* Cambridge, MA: MIT Press.

Steil, B., Victor, D. G. & Nelson, R. R. (Eds) (2002). *Technological innovation and economic performance.* Princeton, NJ: Princeton University Press.

Sternberg, R. J., Forsythe, G. B., Hedlund, J., Horvath, J. A., Wagner, R. K., Williams, W. M., Snook, S. A. & Grigorenko, E. L. (2000). *Practical intelligence in everyday life.* New York: Cambridge University Press.

Tauber, E. M. (1972). HIT: Heuristic ideation technique—A systematic procedure for new product search. *Journal of Marketing,* **36,** 58–70.

Thagard, P. (1988). *Computational philosophy of science.* Cambridge, MA: MIT Press.

Thagard, P. (1992). *Conceptual revolutions.* Princeton, NJ: Princeton University Press.

Torrance, E. P. (1988). The nature of creativity as manifest in its testing. In: R. J. Sternberg (Ed.), *The Nature of Creativity: Contemporary Psychological Perspectives* (pp. 43–75). Cambridge: Cambridge University Press.

Tweney, R. D. (1989). A framework for the cognitive psychology of science. In: B. Gholson, W. R. Shadish, R. A. Neimeyer & A. C. Houts (Eds), *Psychology of Science: Contributions to Metascience* (pp. 342–366). Cambridge: Cambridge University Press.

Van de Ven, A. H., Angle, H. & Poole, M. S. (1989). *Research on the management of innovation.* Cambridge, MA: Ballinger.

Van de Ven, A. H., Angle, H. & Poole, M. S. (Eds) (2000). *Research on the management of innovation: The Minnesota studies.* Oxford: Oxford University Press.

Van de Ven, A. H., Polley, D. E., Garud, R. & Venkataraman, S. (1999). *The innovation journey.* New York: Oxford University Press.

VanGundy, A. B. (1981). *Techniques of structured problem solving.* New York: Van Nostrand Reinhold.

Ward, T. B., Smith, S. M. & Finke, R. A. (1999). Creative cognition. In: R. J. Sternberg (Ed.), *Handbook of Creativity* (pp. 189–212). Cambridge: Cambridge University Press.

Weber, R. J. (1992). *Forks, phonographs, and hot air balloons: A field guide to inventive thinking.* New York: Oxford University Press.

Weber, R. J. & Perkins, D. N. (Eds) (1992). *Inventive minds: Creativity in technology.* New York: Oxford University Press.

Weisberg, R. W. (1993). *Creativity: Beyond the myth of genius*. New York: W. H. Freeman.

Weisberg, R. W. (2003). Case studies of innovation: Ordinary thinking, extraordinary outcomes. In: L. V. Shavinina (Ed.), *International Handbook on Innovation*. Oxford: Elsevier.

Wild, J. J. (1992). The origin of soft-tissue ultrasonic echoing and early instrumental application to clinical medicine. In:

R. J. Weber & D. N. Perkins (Eds), *Inventive Minds: Creativity in Technology* (pp. 115–141). New York: Oxford University Press.

Zairi, M. (Ed.) (1999). *Best practice: Process innovation management*. Oxford: Butterworth- Heinemann.

Zwicky, F. (1969). *Discovery, invention, research through the morphological approach*. New York: Macmillan.

The International Handbook on Innovation
Edited by Larisa V. Shavinina

The Development of Innovative Ideas Through Creativity Training

Maria M. Clapham

Drake University, USA

Abstract: Creativity training became popular in the 1950s with programs such as Osborn's brainstorming approach to problem-solving. Half a century later, in an effort to enhance innovative thinking, creativity training programs of various types have proliferated in educational institutions and business environments. These programs vary in methods and scope. With such popularity and diversity of programs, it is appropriate to examine their effectiveness in enhancing innovation. This chapter will provide an overview of various types of creativity training programs and will examine recent research findings regarding their effectiveness for stimulating the development of innovative ideas.

Keywords: Innovation; Creativity; Innovative ideas; Creativity training; Innovative thinking; Creative problem-solving.

The word 'innovate' comes from the Latin word 'innovare', which means 'to renew, to make new'. Innovation is a form of problem-solving that begins, according to Smith (2003), with the feeling that change is needed and ends with the successful implementation of an idea. A critical component of innovation is idea generation, or ideation. One must first develop a new idea before it can be introduced. Divergent thinking, a cognitive process that focuses on developing multiple possibilities rather than finding a single solution, results in greater ideation. Ideation is important during several phases of innovative problem-solving, including the development of ideas about problems to solve and the development of solutions to those problems (Doolittle, 1995). The term 'creativity' has been associated with innovation in various ways. Sometimes it has been used to refer exclusively to the process of ideation and at other times it has been used synonymously with innovation to refer to both the development and implementation of new ideas (Unsworth, Brown & McGuire, 2000). In either case it is clear that creativity is closely linked to the process of innovation. This chapter will focus on the ideational component of the innovation process.

In the world of business, innovation is a key to success. Rapid changes and advances characterize today's business environment, and in order to remain competitive in the global marketplace, companies must develop and implement new ideas. Business organizations, more than ever before, recognize that they need employees who think creatively in order to maintain their competitive edge. Many in the past decade have turned to creativity training as a means of enhancing innovative thinking in their employees. The belief underlying this movement is that most employees are capable of making creative contributions in their work (Farr, 1990; Weisberg, 1986). If they are not doing so, it is because they do not have the skills or motivation to think creatively (Steinmetz, 1968). Popular magazines have touted the benefits of creativity training (Campbell, 1993; Higgins, 1994; Wise, 1991; Zelinski, 1989), and many anecdotal reports exist of its effectiveness (Gundry, Kickul & Prather, 1994). Does creativity training really work? Can it improve organizational innovation? If so, what forms of creativity training are most effective? These are important questions to address.

In the past, creativity tended to be viewed as a fixed inborn trait. Currently, while exceptions exist, experts generally conceptualize creativity as a multifaceted construct that is affected by both nature and nurturing processes. A widely held view of creativity is that we are born with a range of creative potential, and environmental factors will influence the extent to which our creativity develops to its maximum capacity (Plucker & Runco, 1999). According to the popular model proposed by Amabile (1990), creative performance in any domain requires domain-relevant skills, creativity-relevant skills, and task motivation. Ripple (1999) similarly states that creativity consists of a combination of abilities, skills, motivation, and attitudes. Skills, motivation and attitudes are modifiable characteristics. If these indeed form part of the creative process, then some aspects of creativity are malleable. But are these malleable aspects of creativity subject to individual control? A widely held belief is that creativity is too intuitive a process to be regulated by an individual, that one cannot control how and when 'insight' will come (Smith, 2003). Skills, however, are learned proficiencies, so it seems logical to conclude that creativity could be enhanced through deliberate training and education. Furthermore, it is possible that training and education could affect the motivational element of creative performance. Many educators support the view that creativity can be developed through educational techniques (Baer, 1987; Christensen, 1988; Doolittle, 1995; Fasko, 2000; Ghosh, 1993; Sternberg & Williams, 1996; Torrance, 1972) and believe that promoting innovative thinking should begin in childhood (Carter, 1984; McCormack, 1984). According to Smith (1998), we are handicapped in our practical application of these techniques by the fact that limited research effort has been devoted to examining creative processes that are subject to individual control.

Creativity Training Techniques and Programs

What are some of the techniques proposed to increase innovation? A number of recent reviews of methods used to stimulate creativity exist, including those of Parnes, Plucker and Runco, Ripple, and Runco in *The Encyclopedia of Creativity* (1999). This section provides a summary of general approaches to enhancing ideation, as well as a brief description of the most frequently used and researched instructional programs.

The development of innovative ideas occurs, according to Parnes (1999), when new associations are made between already existing pieces of information. He uses an analogy to describe the necessary ingredients for this process to take place: creativity requires fuel to make it run and the removal of brakes to allow it to run. The fuel consists of sensory impressions from any source including books, environment and experience. The brakes consist of any constraints, internal or external, that limit our mental exploration. Using this model, we can loosely categorize techniques used in creativity training programs as being designed to either 'add something' or 'subtract something'. Training may teach individuals tactics for scanning the environment to come up with new associations. Such tactics include checklist, forced relationship, attribute listing, analogies, scanning the environment, and imagery. Training may also focus on tactics or strategies for removing

blocks. These may include deferring judgment, engaging in relaxation, enhancing self-confidence or self-efficacy, increasing appreciation for creativity, allowing oneself time and space to let ideas flow, or providing the necessary resources to facilitate the flow of ideas. Most creativity training programs use a combination of both 'adding' and 'subtracting' techniques.

Many creativity training programs exist. They may focus on one or several stages of the innovation process, and can be taught by themselves or in combination with other cognitive processes. Thus, creativity training programs may simply try to teach individuals ways of coming up with ideas, or they may teach people how to work through the various stages involved in solving problems, from the problem-finding stage to the solution implementation stage. They may also teach people how to manage their cognitive processes to effectively alternate between the generation of ideas and the evaluation of ideas. Furthermore, training programs may vary depending upon the target population. Some programs are designed for children, while others are focused on adults in educational or business settings. Some techniques focus on enhancing individual ideation while others teach people to work as groups, such as the popular techniques of ideawriting, delphi, and nominal groups (Moore, 1987). A brief description of instructional programs follows:

Brainstorming

Introduced by Alex Osborn, this approach to enhancing ideation is based on the premise that quantity of ideas breeds quality that production of many different ideas increases the likelihood of coming up with a high-quality idea. In this approach, idea generation is separated from idea evaluation. Proponents believe that early evaluation of ideas restricts the process of idea generation, so participants are taught to defer judgment until the idea-generation stage has concluded. In addition, participants practice various idea-generating tactics. Research has found that initial ideas are frequently not chosen as the most preferred (Basadur & Thompson, 1986), supporting the contention that effort extended toward developing more ideas is beneficial. Brainstorming is a technique for idea generation that has been incorporated into many more extensive creativity training programs.

Creative Problem-Solving (CPS)

The Osborn–Parnes CPS process structures problems into five stages: fact-finding, problem-finding, idea-finding, solution-finding, and acceptance-finding. Working through these stages requires the appropriate application of idea generation and idea evaluation. Participants learn a variety of techniques for managing these cognitive processes at each stage of creative problem-solving through practice. Principles of brainstorming are central to idea generation in this approach. According to Parnes (1987, 1999), while the basic elements of CPS remain the same, procedures used in training have evolved over time. Currently, participants are using more imaginative exercises in problem-finding and developing plans for action, the beginning and end stages of the creative problem-solving process. In addition, participants are focusing more on visions than on goals, a process that Parnes calls 'visionizing'.

Synectics

Developed by Gordon & Prince, synectics means the joining together of seemingly unrelated elements. The program is similar to CPS in that it addresses all stages of the creative problem-solving process, and emphasizes differentiation between idea generation and idea evaluation. The strategies used in training, however, vary. Synectics relies heavily on metaphors to 'make the strange familiar and the familiar strange'. Proponents of the use of metaphors in creativity training argue that, by examining the similarity between apparently dissimilar objects, people will be able to look at objects differently. Synectics teaches the use of three metaphoric forms: the personal analogy emphasizes empathic involvement by having subjects try to identify with the object of the analogy; the direct analogy focuses on making connections between the object of the analogy and external facts/knowledge; the symbolic analogy is a two-word description of the object of the analogy in which the words appear to contradict each other (Griffith, 1987). Synectics distinguishes between three roles: the facilitator is in charge of managing the process, but does not direct content; the client is the problem owner; the participants are others that contribute their ideas about the problem at hand. Because participants do not own the problem, it is believed that they can provide the unique perspectives necessary for finding an innovative solution. Kostoff (2003) also emphasizes the importance of obtaining fresh perspectives through cross-disciplinary access and experience.

Lateral and Vertical Thinking

Edward de Bono contends that there are two forms of thinking: vertical thinking involves the implementation and utilization of already existent ideas ("digging the same hole deeper") whereas lateral thinking involves developing new ideas ("digging a whole somewhere else") (cited in Parnes, 1999). His program focuses on cognitive strategies to increase the development of new ideas. According to de Bono, two processes necessary to stimulate lateral thinking are 'escape' and 'provocation'. Escape consists of rejecting assumptions and pre-formed concepts by shifting perspectives, and provocation consists primarily of suspending judgment (Murray, 1992). De Bono stresses the importance of

positive emotions for lateral thinking, and thus uses strategies like humor, fantasy and play extensively (Sikka, 1991).

Hemisphericity

Hemispheric approaches to creativity training, influenced by Ned Herrmann, are based on the notion that the two hemispheres of our brain are specialized for handling different types of tasks. According to this view, the left hemisphere is more effective at performing tasks requiring sequential processing of information. The right hemisphere, however, is more effective at performing tasks involving simultaneous processing of information and is therefore better able to make associations between remote elements. Hemispheric training approaches encourage the use of the right hemisphere through the practice of information-processing tasks that are thought to be dominated by the right hemisphere or that require more balanced usage of both hemispheres. They rely heavily on imagery techniques, relaxation, art and music. They also use physical and sensory exercises such as heterolateral walking, a form of walking in which the opposite arm and leg are forward, and upside down drawing (Carter, 1983).

Khatena Training Method

Khatena's method consists of five main strategies for enhancing ideation. Breaking away from the obvious and commonplace involves viewing the environment in a different way. Transposing ideas means transferring an idea into a different mode of expression. Exploring analogies forces participants to examine unexpected similarities between elements. Restructuring involves reorganizing the various components of a structure. Synthesizing ideas consists of incorporating new ideas into the existing structure (Sikka, 1991). In applying these five strategies, Khatena relies heavily on guided imagery and relaxation (Vaught, 1983).

Packaged Educational Programs

A number of programs intended for educational settings have been packaged into a format of several lessons. The Purdue Creative Thinking Program consists of numerous lessons in which divergent thinking is emphasized. Stories about famous creative people are presented to stimulate interest in creativity, techniques to enhance creativity are suggested, and then practice exercises are conducted. The Productive Thinking Program is aimed at developing creative problem-solving skills, and improving attitude and self-confidence toward such problem-solving. Program materials include mysteries or detective problems that require use of both convergent and divergent thinking to be solved (Sikka, 1991). The Torrance Ideabooks are workbooks for practicing perceptual and cognitive skills that are the basis for divergent thinking.

Psychogenics and Psychosynthesis

These methods take a 'let it happen' approach to developing creativity. Wenger's psychogenics is based on Wallas's notion of incubation, and emphasizes the use of meditation and imagery. Assagioli's psychosynthesis is based on the view that the unconscious plays a vital role in creativity, and emphasizes reliance on one's intuition (Du Pont de Bie, 1985). Markley (1988) supports the use of intuition for strategic innovation.

Technology-Based Programs

A number of programs rely heavily on the use of computers. Such programs may consist of interactive computer games (Doolittle, 1995), software that guides individuals through brainstorming or the stages of creative problem-solving (Markas & Elam, 1997; Rickards, 1987; Small, 1992), or programs that generate unusual word associations (Brown, 1997).

Reviews of Effectiveness of Creativity Training

Several important reviews of creativity training research have been conducted. Torrance (1972) reviewed 142 studies involving evidence about the effectiveness of teaching children to think more creatively. He classified the methods used to teach creativity in these studies into nine categories: Osborn–Parnes CPS techniques; packaged programs such as the Productive Thinking Program or the Purdue Creativity Program; other disciplined approaches such as semantics or creativity research; creative arts; media and reading programs; curricular and administrative arrangements; teacher-classroom variables such as fostering a creative classroom environment; direct motivational techniques such as rewards; and testing conditions. This review was conceptualized very broadly as it includes teaching approaches that directly focus on developing skills for self-regulating creativity as well as approaches that involve the manipulation of environmental factors that influence creativity. The most frequent criteria for effectiveness in these studies were scores on divergent thinking tests. In the great majority of the studies included in the review (103 out of 142), the criteria consisted of scores on the *Torrance Tests of Creative Thinking* (TTCT). This is a set of divergent thinking tests that provides scores in fluency (the number of ideas produced), flexibility (the number of different types of ideas produced) and originality (the uniqueness of the ideas) in both verbal and figural form. Other criteria consisted of Guilford-based divergent thinking measures, creative writing, research projects, problem-solving, art projects, question-asking, and more. Torrance found that the most frequently examined approaches to teaching creativity were Osborn–Parnes programs, complex packages, and teacher-classroom variables, while the most successful approaches, as indicated by the extent to which the approach achieved its specified criteria for success, were Osborn–Parnes CPS programs and other disciplined approaches. Other fairly successful approaches included complex packages, creative arts, and media and reading programs. Torrance concluded that creativity training is effective for enhancing ideation and that the most successful programs emphasize both cognitive and emotional components.

According to Mansfield, Busse & Krepelka (1978), Torrance's conclusions are optimistic. These authors conducted a review of studies involving only multiple-session training programs. The primary programs reviewed were the Productive Thinking Program, the Purdue Creative Thinking Program, the Osborn–Parnes CPS Program, the Myers–Torrance Workbooks, and the Khatena training method. The authors found mixed support for the effectiveness of these programs; studies of sound design that showed the highest rate of success were those using criteria similar to the tasks used during training. Of the programs evaluated, the CPS program showed the highest rate of success for improving idea generation. Overall, the authors concluded that creativity training can improve idea generation, but that results may not transfer easily to 'real-life' innovation. Because of the combination of elements required for creativity at the professional level, the authors suggest that creativity training may not be very useful for improving organizational innovation. Furthermore, they stated that it is unclear why the programs are effective for increasing idea generation. Are gains in scores the result of an improvement in skills, enhanced motivation, a clearer understanding of the desired responses, or experimental demands placed on the subjects? If either the skills do not transfer to other tasks, or the motivation to perform is limited to the treatment situation, then transfer of training will be limited. Feldhusen & Clinkenbeard (1986) concurred with Mansfield et al. (1978) in expressing reservations about creativity training's effectiveness. They argued that, because of conceptual and methodological problems in research studies, conclusions about the effectiveness of such training are premature.

In an effort to better quantify the effectiveness of creativity training, meta-analytic techniques have been applied. Rose & Lin (1984) used several criteria to select studies for their meta-analysis. First, similarly to Mansfield et al. (1978), only studies that examined the effect of several lessons/training treatments were included. Second, to facilitate comparability of studies, only studies that used the TTCT as criteria were included. Third, studies needed to provide sufficient data to be included in the meta-analysis. A total of 46 studies met the criteria. Of these, 22 were doctoral dissertations, 13 were articles in periodicals, and 11 were unpublished studies. Training programs used in these studies were classified into six categories: Osborn–Parnes-type programs, the Productive Thinking Program, Purdue Creative Thinking Program, other programs combining several elements of creativity, classroom arrangements, and special techniques such as dramatics or meditation. The authors found that creativity training accounted for 36% of the variance in verbal creativity and 14% of the variance in figural creativity. This difference in effectiveness for verbal and figural creativity is not surprising: the majority of training programs rely on verbal activities that may transfer more easily to performance on verbal tests. When comparing scores on originality, fluency and flexibility, the greatest impact of training in both verbal and figural creativity was found on originality scores. Training accounted for 25% of the variance in verbal and figural originality scores. The authors concluded that, while further research is needed to understand its processes, creativity training can have a positive impact on ideation. Cohen (1984) conducted another meta-analysis of creativity training research. In her examination of 106 dissertations and article publications, she found greater effectiveness in studies that used creativity tests as criteria and when tasks employed during training were similar to the tasks used as the performance criteria. Both of these meta-analyses indicate that creativity training can improve scores on creativity tests.

Past qualitative and quantitative reviews of creativity training studies have resulted in similar conclusions. They find that creativity training can be effective for enhancing idea generation as indicated by performance on divergent thinking tests. They also indicate that training is most likely to show positive effects when post-training tasks are similar to training experiences. Most of the research on creativity training incorporated in these reviews was conducted in educational settings with children, teenagers or college students. These issues lead to concerns about transfer of training to other types of tasks and environments and lend support to the view that creativity is domain-specific (Baer, 1994). Arguing that studies of creativity may lack generalizability to real world situations, Roweton (1989) stated that "we drifted away from recognizible forms of creativity" (p. 250) in our efforts to gain experimental control. Can creativity training really improve organizational innovation? If so, which forms of training are most effective? Perhaps recent research can address these questions.

Recent Research on Effectiveness of Creativity Training

What have we learned about creativity training since the reviews conducted in the 1970s and 1980s? To address this question, a search was conducted in primary databases of psychology (PsychInfo), education (ERIC), and business (Business Source Primer) for studies published from 1982 to 2002 that examine creativity training. The 1982 date was selected to ensure coverage of studies that may not have been included in the 1984 meta-analyses due to publication lag time. A decision was made to only report studies that used experimental or quasi-experimental designs because of questions about causal effects in other studies. Of the studies found published within this period, the great majority met the design criteria. The search resulted in 11 dissertations, and 29 publications in periodicals. This review, while not inclusive of all studies of creativity training, is representative of the body of research in the field. A summary description of the characteristics of these studies, organized by target population, follows:

Creativity Training for Children and Teenagers

The majority of recent studies examining creativity training have focused on training children or teenagers in educational settings. Creativity training for children is based on two assumptions: it is inherently beneficial for children, and it can increase the pool of innovative individuals entering the work force in the future. Of the studies reviewed here, seven involved training children, eleven trained adolescents/teenagers, and two studied both children and adolescents/teenagers. Using an approach that also has the ultimate goal of improving student creativity, three studies examined how training teachers affected teacher attitudes and behaviors in the classroom.

Studies conducted with children generally used the TTCT as criteria. Markewitz (1982) found that 12 divergent thinking sessions had a significant effect on kindergartener's TTCT flexibility scores. Jaben (1983, 1986) found that a 12-week Purdue Creativity Training Program had a significant positive effect on TTCT verbal fluency, flexibility and originality scores of behaviorally disordered and learning disabled children. Meador (1994) found that training in synectics twice a week for 12 weeks had a significantly positive effect on TTCT figural scores and verbal skills of both gifted and regular kindergarten students. In Glover (1982), 5th grade teachers underwent two days of training in applied behavior analysis emphasizing the reinforcement of fluency, flexibility, and originality. Their students obtained higher scores than a comparison group on both figural and verbal TTCT 12 weeks after the training. Furze, Tyler & McReynolds (1984) showed significant effects of a 14-week training program involving artist-educators in the classroom on a different criterion, a test that the authors argue measures originality called the *Obscure Figures Test*. Taken together, these studies indicate that extended programs emphasizing ideation can have a significant positive impact on children's divergent thinking.

One study involving children used expert evaluations of creative performance as criteria rather than divergent-thinking scores. Baer (1994) found that 2nd graders who experienced 16 one-hour sessions of divergent-thinking training that emphasized brainstorming showed significant improvements in the creativity of various verbal tasks such as writing stories, telling stories and writing poems, but not in collage-making. These results, according to Baer, support the contention that creativity training effects do not generalize well to tasks that differ from those practiced in training.

Studies conducted with adolescents and teenagers from both gifted and non-gifted programs show mixed results. The studies involving brainstorming or the CPS model generally showed positive results. Baer (1988) used the Osborn–Parnes CPS approach in combination with other techniques such as Synectics in a three-day training program, and found significant positive effects 6 months later on creative problem-solving scores. Westberg (1996) found that teaching the process of invention had positive effects on verbal divergent-thinking scores and number of invention ideas produced. While the study did not show positive effects on the quality of ideas produced, the author expressed concerns about the psychometric characteristics of the measure used to assess quality revealed in her study. Kovac (1998), using brainstorming and directed imagination training over 10 months, found significantly higher flexibility scores in the training group than the control group. Russell & Meikamp (1994) showed that training in brainstorming had a significantly positive effect on the metacognitive strategy of concept mapping in gifted, regular education, and learning-disabled students.

Other creativity training techniques have been less successful in enhancing teenagers' verbal or figural divergent thinking. In studies of imagery training across 10 sessions (Vaught, 1983), imagery training and the Khatena technique in four 45-minute sessions (Sikka, 1991), the Scamper ideation technique (Mijares-Colmenares, Masten & Underwood, 1988; Mijares-Colmenares et al., 1993), and a hemispheric approach with short duration (Carter, 1983; Masten, 1988, Masten, Khatena & Draper, 1988), the training groups did not show significantly higher scores in divergent thinking than the comparison groups.

Two studies have examined the effects of training gifted students in several grade levels over an extended period. LeRose (1987) conducted a 12-year longitudinal study on students starting in kindergarten. Students in the experimental groups learned various divergent thinking strategies, and showed higher TTCT flexibility scores in 1st and 9th grade than a comparison group of gifted students. Heiberger (1983) examined the effects of workbook activity sheets several hours a week during most of an academic year in 2nd–7th grade, and found that such activity increased scores on the figural TTCT.

Studies examining the effects of providing teachers with creativity training have also found positive training effects. McConnell & LeCapitaine (1988) found that student ratings of teacher acceptance and openness to new ideas improved in teachers who participated in 40 hours of Synectics training. Participation in a combination Osborn–Parnes CPS/hemi-sphericity program (Murray, 1992), and an Osborn–Parnes CPS program (Mammuraci, 1989) resulted in more positive teacher attitudes toward creativity and higher external ratings of the creativity of teachers' lessons.

The recent studies with children and teenagers confirm the findings of previous reviews: creativity training can improve divergent-thinking performance and, of the training approaches tested to date, the most effective tend to focus on ideational strategies such as brainstorming or Osborn–Parnes-type techniques. These studies also suggest that such training can be effective for various types of students, including gifted, regular and disabled students. They do little, however, to verify that training effects generalize to applied problems, or to explain the mechanisms through which creativity training works.

Creativity Training Applied to College Students

The ten studies of college students reviewed in this chapter generally used scores on divergent-thinking tests as criteria. Clapham & Schuster (1992) tested the effects of a two-hour training session designed to improve divergent thinking with undergraduate engineering students. Students who participated in this training showed increased scores one week later on the *Structure of the Intellect Learning Abilities* test and the *Owens Creativity Test* compared to students who participated in interview training. The significant effects of general ideational training on the *Owens*, a mechanical ideation test, shows that training effects can generalize to content applications not covered in the training. Harkins & Macrosson (1990) examined the effects of a 10-week program that primarily emphasizes right-hemisphere utilization through drawing exercises on the verbal and figural TTCT of undergraduate business students. The comparison group received only study time. Results showed higher scores of the training group on figural fluency and flexibility, but not on any verbal scores, supporting the contention that creativity is domain-specific. Griffith (1987) examined how 15 sessions of Synectics and lateral thinking training affected the performance of undergraduate management students on the figural and verbal TTCT, and the *Remote Associates Test* (RAT). He found that the training group performed significantly higher than a control group on overall scores of the verbal and figural TTCT as well as subscores of verbal originality and figural fluency and originality. No significant effects were found for the RAT. While these studies show positive effects of creativity training on divergent thinking scores, the results are mixed regarding the generalizability of scores to content different from that encountered in training.

In another study, Kabanoff & Bottger (1991) examined the effects of a program modeled after the Osborn–Parnes CPS program, taught over 10 weeks with two 80-minute sessions per week, on the verbal

TTCT scores of MBA students. The comparison group consisted of MBA students who received no training but were given incentives to show improvement in creativity. Both the training and the comparison groups were told that the number and originality of ideas affected scores. Both groups improved in fluency, while only the training group improved in originality, suggesting that motivational, practice or demand characteristics may have affected fluency and flexibility whereas the training itself affected originality. Rose and Lin in their meta-analysis also found that creativity training had the greatest impact on originality. In their examination of personality factors affecting self-selection into the course, the authors found that individuals with a high preference for achievement and dominance and low deference were less likely to take the creativity course. These results contribute to our understanding of the self-selection process into these courses.

Several comparative studies examined what components or types of creativity training are most effective. Clapham (1997) compared the effects of a 30-minute training seminar, incorporating both motivational and skills components, to a training session of 10 minutes involving exclusively ideational tactics. She found that both types of training were effective in increasing scores on the figural TTCT compared to a control group who received word-processing training. Blissett & McGrath (1996) compared the effects of two 5-hour training programs, the Khatena Training Method and interpersonal problem-solving training, on verbal TTCT scores and interpersonal problem-solving ability of undergraduate students. Results showed that the creativity training had a significant effect on ideation but not on interpersonal problem-solving, suggesting that these are independent processes. Furthermore, neither training had a significant effect on self-perceptions of innovative style, as measured by the *Kirton Adaption-Innovation Inventory.* This is in accordance with two other studies: Murdock, Isaksen & Lauer (1993) found that undergraduates in a college creativity course based on the Osborn–Parnes CPS program showed no overall change in scores on the *Kirton Adaption-Innovation Inventory* compared to a control group of marketing students. In another study, Daniels, Heath & Enns (1985) found that two creativity training sessions designed to enhance ideation did not improve the self perceptions of creativity in undergraduate women as measured by the *Something About Myself (SAM) Creative Perception Inventory.* Thus, while the training appears to improve divergent thinking, it does not positively affect self-ratings of creative style.

Two studies reported real-world creative performance indicators. Parrott (1986) compared the effects of imagery utilization training to divergent thinking training using Parne's Creative Behavior Workbook in first-year mechanical engineering students. Both types

of training took place across eight weeks, one hour per week. Results showed that both types of training resulted in a higher self-reported vividness of visual imagery. Furthermore, while not tested directly, some evidence of transfer was found in that self-ratings of use of imagery in a design project were positively associated with performance in engineering examinations. Kellstrom (1985) examined how a 50-minute training program involving a combination of imagery and relaxation affected the creativity of projects developed by undergraduate instructional design students. The projects were instructional design materials rated on creativity and effectiveness by trained experts who were blind to the treatment conditions of participants. Results showed that the imagery training groups had higher ratings on both creativity and effectiveness of their materials than an alternate training group. Both of these studies suggest that creativity training may transfer to real-life projects.

The studies of college students do more than confirm the ideational effects found with children and teenagers. Some of the studies, in comparing effectiveness of different types of programs or program components, address the question of what makes this training work. Other studies, in examining the effects of training on project performance, address the issue of transfer of training. While these studies are not yet conclusive, they are leading us toward a better understanding of creativity training, and suggest that training may enhance project innovation.

Creativity Training in Business Settings

The research on creativity training in business setting is quite limited. A total of seven studies were found. Min Basadur and his colleagues have been at the forefront of this work. Arguing that most of the research on creativity training has focused on the ideational stage, they have conducted a number of studies examining the stages of creative problem-solving. In Basadur, Graen & Green (1982), engineers, engineering managers and technicians underwent a total of three days of training and assessment divided. The experimental group underwent an Osborn–Parnes-type training program in creative problems solving by working through the ideation and evaluation components in each of the stages of problem-finding, problem-solving and solution implementation in an experiential way. A placebo group received alternate training, and a matched control group received no training. Results showed that the creativity training resulted in higher preference of ideation in problem-solving, practice of ideation in problem-finding, and problem-finding performance immediately after the training. Furthermore, the results showed some transfer of training in that the experimental group showed higher scores in practice of ideation in problem-solving two weeks after the training as reported by self and by coworkers. Basadur, Graen & Scandura (1986) reported that a three-day

CPS training program had a positive impact on the attitudes toward ideation in problem-solving of manufacturing engineers as measured by self-report and reports of superiors both immediately after the training and five weeks later.

Other studies have focused on specific components of the CPS model. Fontenot (1993) examined the creative problem-finding components of the model in an eight-hour training program with business people. Results showed that the training was effective in enhancing group performance on fluency in data-finding and problem-finding, flexibility in problem-finding, but not the complexity of final problem statements presented when examining business cases immediately after training. Basadur & Finkbeiner (1985) distinguished between two components of ideation: active divergence consists of aggressively generating many ideas and premature convergence consists of judging ideas in their early stages. They found that a combination of training in high active divergence and low premature convergence resulted in greater fluency and higher-quality ideas. Basadur, Wakabayashi & Takai (1992) found that training in creative problem-solving had a positive impact on the attitudes toward active divergence and premature convergence in Japanese managers.

Yet other studies have examined how individual preferences and style mediate the effectiveness of creativity training. Basadur, Wakabayashi & Graen (1990) found that the CPS program significantly reduced attitudes toward premature convergence in managers and nonmanagers, and increased attitudes toward active divergence in managers. The effects varied, however, depending upon the style of the trainee. Four styles were identified: optimizers prefer thinking and evaluation, generators prefer experience and ideation, conceptualizers prefer thinking and ideation, and implementors prefer experience and evaluation. In the overall sample, the training had the greatest impact on attitudes toward active divergence in optimizers. In contrast, the training had a positive impact on attitudes toward premature convergence in all four styles. In another study, Bush (1998) examined the relative effectiveness of two types of training for enhancing the creativity of engineers' concept designs. Engineers in this study were involved in a one-day program in which they were pretested for creativity, personality and style preferences, and then trained for two and one half hours in either creative attitude blocks or de Bono's Provocation and Movement process. Participants were then given a company problem to solve independently for three hours. Their proposed solutions were then rated for creative value by three judges. Results showed that engineers who tended to have a preference for small incremental changes and detailed work produced more creative concepts with the Provocation and Movement training, whereas those who preferred larger-scale changes and high-level

organizing work performed better with the attitude-oriented training. These studies provide initial evidence that individual differences interact with training effectiveness.

The studies of creativity training in business environments, while limited in number, have greatly expanded our understanding of creativity training effectiveness. Their results indicate that creativity training, and in particular the Osborn–Parnes CPS model for training, can have a positive impact on the creative problem-solving capacity of professionals by improving performance at several stages of creative problem-solving. Furthermore, this research suggests that training effects interact with personal variables.

Conclusions

Creativity training can improve ideational innovation. We see this in studies that examine creativity training in children, college students and professionals. Furthermore, creativity training has positive effects on both divergent thinking test scores and project performance. However, many questions remain about creativity training.

We do not yet know how creativity training works. Are effects due to an increase in skills, a change in attitude, or something else? Is the process different for children than for adults? Perhaps, as Runco (1999) suggests, understanding the mechanisms by which it works is not as important as knowing that it does work. I would argue that a greater understanding of the mechanisms through which it works has practical as well as theoretical benefits. It can aid us in the development of more effective and efficient training programs, and contribute to our understanding of the construct of creativity.

We do not have a thorough understanding of conditions that facilitate the transfer of training to real-world creativity. As indicated at the beginning of this chapter, because of the importance ascribed to organizational innovation, creativity training has become very popular in business settings, yet surprisingly little research has been published on the effectiveness of creativity training in business organizations. While exceptions exist, the majority of studies continue to use divergent thinking test scores as criteria. Furthermore, most studies in business settings apply the CPS training model. Is this the most effective approach for businesses? Little research exists comparing its effects to other approaches for enhancing organizational innovation. We also do not fully understand how domain specificity affects transfer of training. The extent to which domain specificity limits generalizability of training effects will have implications for how creativity training is conducted.

The organizational climate will also impact transfer of training. Because organizational innovation depends upon several factors beyond ideational skills, it is likely that transfer of training is limited in practice. Many researchers argue that creativity is a process that occurs in a social context (Amabile, 1990, 1996) and if this is the case, then the organizational context must be taken into consideration to maximize transfer of training to professional behavior. It is unfortunate if we do not maximize the use of creative skills in organizations by providing a climate that promotes ideation. A combination of creativity training and creative climate would be optimal for organizational innovation.

Other questions about creativity training need further examination. What is the association between learning styles and effectiveness of training? What are the long-term effects of creativity training? What training approaches are most appropriate for different types of jobs? Basadur (1995) has initiated work in this area. What other benefits might be obtained from creativity training? Researchers suggest that benefits may include improved mental health and increased learning (Conti, Amabile, & Pollack, 1995; Cropley, 1990; Hickson & Housley, 1997). While questions about creativity training remain, it is clear that research is providing more information about the benefits of such training for innovation.

References

Amabile, T. M. (1990). Within you, without you: The social psychology of creativity, and beyond. In: M. A. Runco & R. S. Albert (Eds), *Theories of Creativity* (pp. 61–91). London: Sage.

Amabile, T. M. (1996). *Creativity in context: Update to the social psychology of creativity.* Boulder, CO: Westview Press.

Baer, J. (1994). Divergent thinking is not a general trait: A multidomain training experiment. *Creativity Research Journal*, 7, 35–46.

Baer, J. M. (1988). Long-term effects of creativity training with middle school students. *Journal of Early Adolescence*, 8, 183–193.

Baer, S. (1987). Teaching for creativity, teaching for conformity. *Teaching English in the Two Year College*, 14 (3), 195–204.

Basadur, M. (1995). Optimal ideation–evaluation ratios. *Creativity Research Journal*, 8 (1), 63–75.

Basadur, M. S. & Finkbeiner, C. T. (1985). Measuring preference for ideation in creative problem-solving training. *Journal of Applied Behavioral Science*, 21, 37–49.

Basadur, M., Graen, G. B. & Green, S. G. (1982). Training in creative problem solving: Effects on ideation and problem finding and solving in an industrial research organization. *Organizational Behavior and Human Performance*, 30, 41–70.

Basadur, M. S., Graen, G. B. & Scandura, T. A. (1986). Training effects on attitudes toward divergent thinking amond manufacturing engineers. *Journal of Applied Psychology*, 71, 612–617.

Basadur, M. & Thompson, R. (1986). Usefulness of the ideation principle of extended effort in real world professional and managerial creative problem solving. *Journal of Creative Behavior*, 20, 23–34.

Basadur, M., Wakabayashi, M. & Graen, G. B. (1990). Individual problem-solving styles and attitudes toward divergent thinking before and after training. *Creativity Research Journal*, 3 (1), 22–32.

Basadur, M., Wakabayashi, M. & Takai, J. (1992). Training effects on the divergent thinking attitudes of Japanese managers. *International Journal of Intercultural Relations*, 16, 129–345.

Blissett, S. E. & McGrath, R. E. (1996). The relationship between creativity and interpersonal problem-solving skills in adults. *Journal of Creative Behavior*, 30 (3), 173–182.

Brown, A. S. (1997). Computers that create: No hallucination. *Aerospace America*, 35, 26–27.

Bush, D. H. (1998). Creativity in real world engineering concept design (Doctoral dissertation, University of Minnesota, 1998). *Dissertation Abstracts International*, 58 (11-B), 6071.

Campbell, P. G. (1993). Creativity training requires discipline. *Folio: The Magazine for Magazione Management*, 22, 33–34.

Carter, L. K. (1983). The effects of multimodal creativity training on the creativity of twelfth graders (Doctoral dissertation, Kent State University, 1983). *Dissertation Abstracts International*, 44 (7-A), 2091.

Christensen, J. J. (1988). Reflections on teaching creativity. *Chemical Engineering Education*, 170–176.

Clapham, M. M. (1997). Ideational skills training: A key element in creativity training programs. *Creativity Research Journal*, 10, 33–44.

Clapham, M. M. & Schuster, D. H. (1992). Can engineering students be trained to think more creatively? *Journal of Creative Behavior*, 26, 156–162.

Cohen, C. M. G. (1984). Creativity training effectiveness: A research synthesis (Doctoral dissertation, Arizona State University, 1984). *Dissertation Abstracts International*, 45, 8-A, 2501.

Conti, R., Amabile, T. M. & Pollack, S. (1995). The positive impact of creative activity: Effects of creative task engagement and motivational focus on college students' learning. *Personality and Social Psychology Bulletin*, 21 (10), 1107–1116.

Cropley, A. J. (1990). Creativity and mental health in everyday life. *Creativity Research Journal*, 3(3), 167–178.

Daniels, R. R., Heath, R. G. & Enns, K. S. (1985). Fostering creative behavior among university women. *Roeper Review*, 7 (3), 164–166.

Doolittle, J. H. (1995). Using riddles and interactive computer games to teach problem-solving skills. *Teaching of Psychology*, 22 (1), 33–36.

Du Pont de Bie, A. (1985). Teaching creativity—creatively with psychosynthesis. *Gifted Education International*, 3 (1), 43–46.

Farr, J. L. (1990). Facilitating individual role innovation. In: M. A. West & J. L. Farr (Eds), *Innovation and Creativity at Work* (pp. 207–230). Chichester, U.K.: John Wiley.

Fasko, D. Jr. (2000). Education and creativity. *Creativity Research Journal*, 13, 317–327.

Feldhusen, J. F. & Clinkenbeard, P. R. (1986). Creativity instructional materials: A review of research. *Journal of Creative Behavior*, 20, 153–182.

Fontenot, N. A. (1993). Effects of training in creativity and creative problem finding upon business people. *The Journal of Social Psychology*, 133 (1), 11–22.

Furze, C. T., Tyler, J. G. & McReynolds, P. (1984). Training creativity of children assessed by the obscure figures test. *Perceptual and Motor Skills*, 58, 231–234.

Ghosh, S. (1993). An exercise in inducing creativity in undergraduate engineering students through challenging examinations and open-ended design problems. *IEEE Transactions on Education*, **36**, 113–119.

Glover, J. A. (1982). Implementing creativity training of students through teacher inservice training. *Educational Research Quarterly*, **6**, 13–18.

Griffith, T. J. (1987). An exploration of creativity training for management students (Doctoral dissertation, Boston University, 1987). *Dissertation Abstracts International*, **49** (3-A), 411.

Gundry, L. K., Kickul, J. R. & Prather, C. W. (1994). Building the creative organization. *Organizational Dynamics*, **22**, 22–37.

Harkins, J. D. & Macrosson, W. D. K. (1990). Creativity training: An assessment of a novel approach. *Journal of Business and Psychology*, **5** (1), 143–148.

Heiberger, M. A. (1983). A study of the effects of two creativity-training programs upon the creativity and achievement of young, intellectually-gifted students (Doctoral dissertation, University of Tulsa, 1983). *Dissertation Abstracts International*, **44** (11-A), 3223–3224.

Hickson, J. & Housley, W. (1997). Creativity in later life. *Educational Gerontology*, **23**, 539–547.

Higgins, J. M. (1994). Creating creativity. *Training & Development*, **48**, 11–15.

Jaben, T. H. (1983). The effects of creativity training on learning disabled students' creative written expression. *Journal of Learning Disabilities*, **16**, 264–265.

Jaben, T. H. (1986). Effects of creativity training on behaviorally disordered students' creative written expression. *Canadian Journal for Exceptional Children*, **3** (2), 48–50.

Kabanoff, B. & Bottger, P. (1991). Effectiveness of creativity training and its relation to selected personality factors. *Journal of Organizational Behavior*, **12**, 235–248.

Kellstrom, M. R. (1985). Mental imagery as an efficacious adjunct to instructional design (Doctoral dissertation, Kansas State University, 1985). *Dissertation Abstracts International*, **47** (1-A), 156.

Kostoff, R. N. (2003). Stimulation innovation. In: L. V. Shavinina (Ed.), *International Handbook on Innovation*. Oxford: Elsevier Science.

Kovac, T. (1998). Effects of creativity training in young soccer talents. *Studia Psychologica*, **40**, 211–218.

LeRose, B. H. (1987). An investigation of the effects of a creativity training program on measures of creative thinking and achievement for gifted students: A longitudinal experimental study (Doctoral dissertation, University of Wisconsin-Madison, 1987). *Dissertation Abstracts International*, **48** (3-A), 551.

Mammucari, D. R. (1989). A study to determine the impact of teacher training programs in creativity on improving observable traits of creative teaching (Doctoral dissertation, Temple University, 1989). *Dissertation Abstracts International*, **50** (7-A), 2022.

Mansfield, R. S., Busse, T. V. & Krepelka, E. J. (1978). The effectiveness of creativity training. *Review of Educational Research*, **48**, 517–536.

Marakas, G. M. & Elam, J. J. (1997). Creativity enhancement in problem solving: through software or process? *Management Science*, **43**, 1136–1146.

Markewitz, D. A. (1982). The influence of creativity intervention training on the adjustment potential of kindergarten children (Doctoral dissertation, University of Santo Tomas, 1982). *Dissertation Abstracts International*, **44** (12-A), 3638.

Markley, O. W. (1988). Using depth intuition in creative problem solving and strategic innovation. *Journal of Creative Behavior*, **22** (2), 85–100.

Masten, W. G. (1988). Effects of training in creative imagination with adaptors and innovators. *Educational and Psychological Reports*, **8**, 331–345.

Masten, W. G., Khatena, J. & Draper, B. R. (1988). Hemispheric learning style and stimulation of creativity in intellectually superior students. *Educational and Psychological Research*, **8**, 83–92.

McConnell, D. M. & LeCapitaine, J. E. (1988). The effects of group creativity training on teachers' empathy and interactions with students. *Reading Improvement*, **25** (4), 269–275.

McCormack, A. J. (1984). Teaching inventiveness. *Childhood Education*, **60**, 249–255.

Meador, K. S. (1994). The effects of synectics training on gifted and nongifted kindergarten students. *Journal of the Education of the Gifted*, **18** (1), 55–73.

Mijares-Colmenares, B. E., Masten, W. G. & Underwood, J. R. (1988). Effects of the scamper technique on anxiety and creative thinking of intellectually gifted students. *Psychological Reports*, **63**, 495–500.

Mijares-Colmenares, B. E., Masten, W. G. & Underwood, J. R. (1993). Effects of trait anxiety and the scamper technique on creative thinking of intellectually gifted students. *Psychological Reports*, **72**, 907–912.

Moore, C. M. (1987). *Group techniques of idea building*. Newbury Park: Sage.

Murdock, M. C., Isaksen, S. G. & Lauer, K. J. (1993). Creativity training and the stability and internal consistency of the Kirton adaption-innovation inventory. *Psychological Reports*, **72**, 1123–1130.

Murray, A. M. (1992). Training teachers to foster creativity using the 4MAT model (Doctoral dissertation, University of Massachusetts, 1992). *Dissertation Abstracts International*, **53** (6-A), 1873.

Parnes, S. J. (1987). Visioneering—State of the art. *Journal of Creative Behavior*, **21** (4), 283–299.

Parnes, S. J. (1999). Programs and courses in creativity. In: M. A. Runco & S. R. Pritzker (Eds), *Encyclopedia of Creativity* (Vol. 2, pp. 465–477). San Diego, CA: Academic Press.

Parrott, C. A. (1986). Visual imagery training: Stimulating utilization of imaginal processes. *Journal of Mental Imagery*, **10**, 47–64.

Plucker, J. A. & Runco, M. A. (1999). Enhancement of creativity. In: M. A. Runco & S. R. Pritzker (Eds), *Encyclopedia of Creativity* (Vol. 1, pp. 669–675). San Diego, CA: Academic Press.

Rickards, T. (1987). Can computers help stimulate creativity? Training implications from a postgraduate MBA experience. *Management Education and Development*, **18** (2), 129–139.

Ripple, R. E. (1999). Teaching creativity. In: M. A. Runco & S. R. Pritzker (Eds), *Encyclopedia of Creativity* (Vol. 2, pp. 629–638). San Diego, CA: Academic Press.

Rose, L. H. & Lin, H. (1984). A meta-analysis of long-term creativity training programs. *Journal of Creative Behavior*, **18**, 11–22.

Roweton, W. E. (1989). Enhancing individual creativity in American business and education. *Journal of Creative Behavior*, **23**, 248–257.

Runco, M. A. (1999). Tactics and strategies for creativity. In: M. A. Runco & S. R. Pritzker (Eds), *Encyclopedia of Creativity* (Vol. 2, pp. 611–615). San Diego, CA: Academic Press.

Russell, R. S. & Meikamp, J. (1994). Creativity training—A practical teaching strategy. In: D. Montgomery (Ed.), *Rural Partnerships: Proceedings of the Annual National Conference of the American Council on Rural Special Education* (ERIC Document Reproduction Services No. ED369621).

Sikka, A. (1991). The effect of creativity training methods on the creativity of fourth, fifth, and sixth grade minority students (Doctoral dissertation, Mississippi State University, 1991). *Dissertation Abstracts International*, **53** (12-A), 4260.

Small, C. H. (1992). Innovation software stimulates engineering creativity. *EDN-Technology Update*, **37**, 59–65.

Smith, G. F. (1998). Idea-generation techniques: A formulary of active ingredients. *Journal of Creative Behavior*, **32** (2), 107–133.

Smith, G. F. (2003). Towards a logic of innovation. In: L. V. Shavinina (Ed.), *International Handbook on Innovation*. Oxford: Elsevier Science.

Steinmetz, C. S. (1968). Creativity training: a testing program that became a sales training program. *Journal of Creative Behavior*, **2** (3), 179–186.

Sternberg, R. J. & Williams, W. M. (1996). *How to develop student creativity*. Alexandria, VA: Association for Supervision and Curriculum Development.

Torrance, E. P. (1972). Can we teach children to think creatively? *Journal of Creative Behavior*, **6** (2), 114–143.

Unsworth, K. L., Brown, H. & McGuire, L. (2000). *Employee innovation: The roles of idea generation and idea implementation*. Paper presented at the meeting of the Society for Industrial and Organizational Psychology, New Orleans, LA.

Vaught, L. D. (1983). The effects of evocative imagery training and receptive imagery training on measures of creativity, imagery vividness, and imagery control in high school students (Doctoral dissertation, Oklahoma State University, 1983). *Dissertation Abstracts International*, **44** (7-A), 2100.

Weisberg, R. W. (1986). *Creativity: Genius and other myths*. New York: W. H. Freeman.

Westberg, K. L. (1996). The effects of teaching students how to invent. *Journal of Creative Behavior*, **30** (4), 249–267.

Wise, R. (1991). The boom in creativity training. *Across the Board*, **28**, 38–42.

Zelinski, E. J. (1989). Creativity training for business. *Canadian Manager*, **14**, 24–25.

The International Handbook on Innovation
Edited by Larisa V. Shavinina

Intuitive Tools for Innovative Thinking

Robert Root-Bernstein[1] and Michele Root-Bernstein[2]

[1] *Department of Physiology, Michigan State University, USA*
[2] *Independent Scholar, East Lansing, MI, USA*

Abstract: In this chapter we examine the fundamental role of intuitive thinking skills in creative endeavor across the arts and sciences. The imagination manifests itself in a set of 13 non-verbal, non-mathematical, non-logical thinking tools that innovative individuals in all disciplines say they use: observing, imaging, abstracting, recognizing and forming patterns, analogizing, body thinking, empathizing, dimensional thinking, modeling, playing, transforming and synthesizing. Private, unarticulated insights generated by means of these tools are then translated in an explicitly secondary step into verbal, mathematical and other modes of public communication. Any educational effort to promote creative thinking must therefore recognize and exercise intuitional thinking skills and directly address the process of translating idiosyncratic subjetive thought into objectified public forms of discourse.

Keywords: Innovation; Intuition; Imagination; Insight; Synesthesia; Observation; Visualization; Pattern; Thinking tool.

Creative Process and 'Tools for Thinking'

Creative thinking is inseparable from intuition and aesthetic experience. While asserting such a basis for artistic activity may not seem odd, it may be surprising to find that even in the sciences and technology, ideas emerge as insights that cannot at first be communicated to other people because they exist as emotional and imaginative formulations that have no formal language. Indeed, practitioners of disciplines across the arts and sciences, including physics and mathematics, have commented that all creative thinking begins in private, sensual feelings that reveal unexpected problems (see Root-Bernstein, 'Problem Generation and Innovation', this volume) and unforeseen opportunities. Once a person feels the existence of a problem or a possibility, he or she must then work with attendant emotions and sensations to translate them, in an explicitly secondary step, into forms that can be communicated. Thus it is necessary, in any description of the creative process, to distinguish between intuitive 'tools for thinking' (Root-Bernstein & Root-Bernstein, 1999) that yield those personal insights and the translation skills necessary to turn insights into verbal, logical-mathematical, visual, kinesthetic and other public modes of communication (what Howard Gardner has called 'intelligences'). It is also necessary to reassert the fundamental role of the private and sensual in creative thinking, so often overlooked. Indeed, understanding the non-verbal, non-logical basis for imaginative thought is essential for stimulating creativity and innovation. Exercising the 'tools for thinking' that comprise this pre-linguistic form of intuitional cognition is as necessary to education as formal training in the languages and logic of public communication.

Thinking With Feeling

It is very difficult to find any major figure in any art or science who has said that creative work is done using words, mathematics, logic, or any of the other higher order forms of thinking that are supposed to characterize intelligence. Even the most verbal poets and mathematical scientists maintain that their creative work emerges from feelings, emotions, and sensual images. Consider the case of T. S. Eliot, who has been characterized by Howard Gardner in his book *Creating Minds* (1993) as a prototypical 'verbal thinker'. Eliot himself wrote that 'the germ of a poem' emerges from a musical "feeling for syllable and rhythm ... (that) bring to birth the idea and the image" (Eliot, 1975, pp. 113–114). The object of poetry is to "find words for the inarticulate ... to capture those feelings which people hardly even feel, because they have no words for them" (Lu, 1966, p. 134). All this occurs, Eliot wrote in *The Music of Poetry*, "before it (the poem) reaches expression in words" (Eliot, 1975, p. 114). The words of a poem, Eliot wrote, are only a translation: "With a poem you can say, 'I got my feeling into words for myself. I now have the equivalent in words for that much of what I have felt'" (Eliot, 1963, p. 97). What makes poets and novelists writers is not that they think in words, but that they express themselves preferentially in words.

Eliot's description of his creative thinking and the difficulty of translating pre-verbal thoughts into words is typical of other writers, as we have demonstrated in our book *Sparks of Genius* (1999). Most find that they can write only after they feel, see, and hear their material in their imagination. For Robert Frost, "a poem . . . begins as a lump in the throat, a sense of wrong, a homesickness, a love sickness. It is never a thought to begin with" (Plimpton, 1989, p. 68). Similarly, E. E. Cummings said that, 'The artist is not a man who describes but a man who FEELS' (*). Thus we find that for poet Gary Synder, writing comes from a process of visualizing situations that give rise to feelings:

I'll replay the whole experience again in my mind. I'll forget all about what's on the page and get in contact with the preverbal level behind it, and then by an effort of reexperiencing, recall, visualization, revisiualization, I'll live through the whole thing again and try to see it more clearly.

As the emotional images become clearer to Snyder, they give rise to the same sort of musical rhythms experienced by Eliot:

The first step is the rhythmic measure, the second step is a set of preverbal visual images which move to the rhythmic measure, and the third step is embodying it in words.

The notion that writing is a translation process occurs in autobiographical accounts of many other writers, too. Stephen Spender insisted that the challenge of writing is to find words for emotional images that have no words:

Can I think out the logic of images? How easy it is to explain here the poem that I would have liked to write! How difficult it would be to write it. For writing it would imply living my way through the imaged experience of all those ideas, which here are mere abstractions.

Novelist Dorothy Canfield Fisher also found that words would come only after "intense visualizations of scenes . . .". Novelist Isabel Allende, however, relies upon gut feelings to bring forth words:

Books don't happen in my mind, they happen somewhere in my belly . . . . I don't know what I'm going to write about because it has not yet made the trip from the belly to the mind . . . . It is something that I've been feeling but which has no shape, no name, no tone, no voice.

Novelist and composer William Goyen characterizes the process of writing as "the business of taking it from the flesh state into the spiritual, the letter, the Word".

The same distinction between creative thinking and modes of expression can be used to characterize scientists. Consider Albert Einstein, the man Howard Gardner characterizes in *Creating Minds* (1993) as his prototypical "logical-mathematical thinker". Just as Eliot said he did not think in words, Einstein said that "No scientist thinks in formulae". The "essential feature in productive thought", he wrote, is an associative play of images and feelings:

The words of the language, as they are written or spoken, do not seem to play any role in my mechanism of thought. The psychical entities which seem to serve as elements in thought are certain signs and more or less clear images which can be 'voluntarily' reproduced and combined . . . . The above mentioned elements are, in my case, of visual and some of muscular type.

Einstein went on to describe thinking as the ability to associate images and muscular feelings in a repeatable way with problems upon which he was working, adding that ". . . Conventional words or other signs have to be sought for laboriously only in a secondary stage, when the associative play already referred to is sufficiently established and can be reproduced at will".

Again, Einstein's non-verbal, non-mathematical thought is typical of scientists. Fellow Nobel laureate Richard Feynman described his problem-solving as kinesthetic, acoustic, and visual:

It's all inspired picturing . . . . In certain problems that I have done, it was necessary to continue the development of the picture as the method, before the mathematics could really be done.

Harvard astrophysicist Margaret Geller recounts a similar approach:

I have to have a visual model or a geometric model or else I can't do it (physics). Problems that don't lend themselves to that I don't do.

Barbara McClintock, yet another Nobelist, also described a non-verbal approach:

When you suddenly see the problem, something happens that you have the answer—before you are able to put it into words. It is all done subconsciously . . . . You work with so-called scientific methods to put it into their frame after you know.

Logic and mathematics, in other words, are the translations that scientists use to communicate their insights, just as writers use words.

Let there be no mistake: the thought processes that these scientists describe are a form of intuition. Einstein made the point explicitly:

> Only intuition, resting on sympathetic understanding, can lead to it (insight); . . . the daily effort comes from no deliberate intention or program, but straight from the heart.

His colleague Henri Poincaré, perhaps the greatest mathematician of the early twentieth century, agreed: "It is by logic that we prove, but by intuition that we discover". Mathematicians Edward Kasner and James Newman write similarly that, "Mathematical induction is . . . an inherent, intuitive, and almost instinctive property of mind" (Kasner & Newman, 1940, p. 35). That which is important must be deeply felt, as mathematical physicist Wolfgang Pauli made clear. During the initial phases of problem solving, "the place of clear concepts is taken by images of powerful emotional content". Indeed, according to botanist Agnes Arber, without this emotional content, creative scientific thought is stymied:

> New hypotheses come into the mind most freely when discursive reasoning (including its visual component) has been raised by intense effort to a level at which it finds itself united indissolubly with feeling and emotion. When reason and intuition attain this collaboration, the unity into which they merge appears to possess a creative power which is denied to either singly.

Thinking and feeling are, in short, just as inseparable to a scientist as to a writer or artist.

We want to emphasize the point that, despite the very real differences between the products created by artists, writers, and scientists, people in all fields use a similar set of pre-verbal, pre-logical forms of creative thinking. Pauli says that scientists must FEEL just as deeply as poet E. E. Cummings. Feynman's development of the picture as a method could just as easily be Spender's 'logic of the images'. The physical concepts emerging from Einstein's muscles could just as well be novels regurgitated from Allende's belly. The important point is that each of these creative individuals knew something sensually and somatically before they were able to describe it formally to anyone else. Until we are able to access, practice and use such pre-linguistic, somatic thinking explicitly, we are cut off from our most innovative sources of thought.

Tools for Thinking

The emotional, intuitional, pre-verbal nature of creative thinking does not place it beyond comprehension. Just as logic and language build upon skills that can be learned and practiced, so does intuition. Hundreds of autobiographical and archival sources, interviews, and formal psychological studies reveal that every creative person uses some subset of a common imaginative 'tool kit'. This tool kit consists of a baker's dozen of pre-logical, pre-verbal skills:

(1) *observing*;
(2) *imaging*;
(3) *abstracting*;
(4) *pattern recognizing*;
(5) *pattern forming*;
(6) *analogizing*;
(7) *bodily kinesthetic thinking*;
(8) *empathizing*;
(9) *dimensional thinking*;
(10) *modeling*;
(11) *playing*;
(12) *transforming*; and
(13) *synthesizing*.

We emphasize that this tool kit consists of the imaginative skills common to all creative people, and the labels are those terms they use to describe their own thinking. Artist Brent Collins, who transforms mathematical equations into stunning wood sculptures, provides an apt example. In one brief passage describing his artistic process, he refers to the relationships between logic and image, aesthetics and intuition, and to his use of physical and mental tools:

> I made (two-dimensional) templates exactly to scale . . . . The entire mathematical logic of the sculpture is inherently readable from the template. There are, however, many aesthetic choices . . . The template serves as a guide for a spatial logic I somehow intuitively know how to follow. Using common woodworking tools and proceeding kinesthetically, I am able to gradually feel and envision its visual implications . . . The linear patterns issue as abstractions (Collins, np).

While few innovators are as succinct in their description of the tools for thinking that underlie their creative work, reference to *Sparks of Genius* will show that many are just as explicit. It is therefore worth considering what mental operations each tool for thinking represents and the many ways in which each can be used.

Observing is perhaps the first and most basic of thinking tools. As human beings we are all equipped to sense the world, but observing is a skill that requires additional patience, concentration and curiosity. The American painter Georgia O'Keeffe looked carefully at things, and forces us to do so, too, in her very large paintings of flowers. "Still—in a way—" she said, "nobody sees a flower—really—it is so small—we haven't the time—and to see takes time, like to have a friend takes time". Observing is paying close attention to what is seen, but also what is heard, touched, smelled, tasted and felt within the body. In dense jungles, biologists such as Jared Diamond observe and identify birds by sound; in the absence of sight, the blind biologist Geermat Vermeij observes seashells with his hands, by touch; bacteriologists and doctors observe bacteria by smell; chemists and doctors have—historically at least—observed sugar in the urine by taste. Inventors and engineers, and the mechanics they rely on, similarly observe kinesthetically by cultivating hands-on experience with tools and machines—they know how tightly the nut is screwed onto the bolt by the feel of it.

Imaging, also a primary thinking tool, depends upon our ability to recall or imagine the sensations and feelings we observe in the absence of external stimulation. We can image visually and also aurally, and with smells, tastes, tactile and muscular feelings as well. If you can close your eyes and see a thing, or imagine the taste, touch, smell, or sound of it when it is not present, then you are imaging. For example, those of us who are already good at visualizing can close our eyes and see a triangle—and if we're practiced, we can make it change color and dimension, rotate it, etc. And if we're really good at visualizing, we can imagine an object with a triangular profile from all sides—or the much more complex object Charles Steinmetz, inventor of electrical generators, was asked to envision. A group of colleagues at General Electric once approached him with a problem they could not solve: "If you take a rod two inches in diameter and cut it (in half) by drilling a two-inch hole through it, what is the cubic content of the metal that's removed?" Steinmetz was able to answer the question quickly, first by visualizing the removed core, then by applying equations that calculated its volume. Such visualizing, Eugene Ferguson argues in *Engineering and the Mind's Eye*, plays a central role in engineering and invention. Without it, the engineer cannot foresee the invention he wishes to make. By the same token, the chef cannot foretaste the delicacy she wishes to create in the absence of imaging; the musician cannot forehear the symphony she wishes to write down.

Abstracting is yet another important thinking tool. Because sense experience and sense imagery are so rich and complex, creative people in all disciplines use abstracting to concentrate their attention. Abstracting means focusing on a single property of a thing or process in order to simplify it and grasp its essence. Scientists and engineers work with abstractions all the time, for instance stripping a physical situation of all extraneous characteristics such as shape, size, color, texture, etc. and zeroing in on point mass, spring and distance. "I'll tell you what you need to be a great scientist . . ." says physicist Mitchell Wilson. "You have to be able to see what looks like the most complicated thing in the world and . . . find the underlying simplicity". Similarly, in the arts, abstracting means choosing which simplicity captures the

essence of some concrete reality. Pablo Picasso tells us how:

> To arrive at abstraction, it is always necessary to begin with a concrete reality . . . . You must always start with something. Afterward you can remove all traces of reality . . . .

And he does just that in a series of etchings called 'The Bull'. Searching for the essence of bull, its minimal suggestion, he finally finds it in the simple linear description of its tellingly distorted shape, the tiny head surmounted by enormous horns, the massive body balanced by a short, hanging tail.

Abstracting often works in tandem with *patterning*, a tool with two parts. We organize what we see, hear, or feel by grouping things all the time. Sometimes we do so visually, as in a quilt or a graph, but of course, we can group things with all our senses. *Recognizing patterns* means perceiving a (repetitive) form or plan in apparently random sets of things and processes, whether in the natural world or in our man-made world. While the ability to recognize faces, and patterns that look like faces, seems to be ingrained in every normal human being, recognizing patterns is often influenced by culture. Westerners are inclined to hunt for a linear, back and forth, or up and down arrangement of information and our tables, graphs, books, and even architecture mirrors this predilection. Thus, although spirals are a common natural form (snails, sea shells, tornadoes, pinecones, whorls of hair on head), Westerners seldom use this pattern to design buildings, graphs or tables. Culture therefore plays a major role in what patterns we recognize and expect to perceive.

Recognizing patterns is also the first step toward creating new ones. Novel *pattern forming* always begins by combining two or more elements or operations in some consistent way that produces a (repetitive) form. For instance, the pattern found in 'watered' silk is created by folding the fabric at a slight bias and then pressing it under high heat and steam with great force. This process imprints the rectilinear pattern of the warp and woof of each fold of the fabric onto the opposing material at a slight offset. The result is what is known as a Moire pattern. Such Moire patterns can be produced by overlapping almost any regular grid over another, as when we look through two window screens or two sections of link fencing. The creation of novel Moire patterns is limited only by the imagination of the individual choosing what regular patterns to overlay.

Pattern forming is also at work when engineers design complex machines. There are only a very small number of basic machines—levers, wheels, screws, cogs and so forth—from which every mechanical device is constructed. Technological invention is the process of forming new patterns with simpler components by combining elements and operations in novel patterns. The same can be said of pattern forming in language and the language arts, since a finite number of words, grammars and narrative structures can be potentially combined and recombined to myriad, innovative effect (J. Gardner, 1983, pp. 52–53).

Recognizing and forming patterns leads directly to *analogizing*, that is, recognizing a functional likeness between two or more otherwise unlike things. We use analogies all the time to broaden our understanding of things. For instance, biologists often describe different bird beaks as if they work like human tools. A nutcracker and a particular bird beak may not look the same, but they function similarly and therefore are analogous. Analogy also has an important place in engineering and invention. Velcro, as no doubt everyone knows, was developed by analogy to the grasping properties of the common bur. Biomimicry, the use of nature as source of ideas, has in fact, become a well-recognized method of innovation. One of the more striking, recent examples of bio-analogy in architecture and engineering is the Gateshead Millennium Bridge. Chris Wilkinson Architects in Great Britain took the human eyelid for its analogical model and designed a drawbridge that works like the eyelid. When the 'lid' is closed, the bridge is down and people can move across. When a ship approaches, the lid is raised and ships can pass under the resulting arch.

While reading the above description of the Gateshead Bridge, you may have paid unusual attention to the way your eyelid functions and feels. This is an example of *body* or *kinesthetic thinking*. Body thinking means just that: thinking with the body. It is based upon sensations of muscle, sinew and skin—sensations of body movement, body tensions, body balance, or, to use the scientific term, proprioception. For instance, if you can imagine how it feels in your hand to set various gears in motion, if you can imagine in your muscles how they feel in motion, you are thinking with your body. Charles 'Boss' Kettering, director of research at General Motors for many decades, is said to have chided his engineers when they became overly analytical and mathematical. Always remember, he told them, "what it feels like to be a piston in an engine". Cyril Stanley Smith, the chief metallurgist for the Manhattan Project, clearly understood his creative debt to body thinking:

> In the long gone days when I was developing alloys, I certainly came to have a very strong feeling of natural understanding, a feeling of how I would behave if I were a certain alloy, a sense of hardness and softness and conductivity and fusibility and deformability and brittleness—all in a curiously internal and quite literally sensual way.

The same kinesthetic and tactile imagination is at work, too, in what is often considered the abstract reasoning of mathematics. The mathematician Stanis-law Ulam said he calculated "not by numbers and symbols, but by almost tactile feelings . . .". While at work on the atomic bomb at Los Alamos he imagined the movements of atomic particles visually and proprioceptively, feeling their relationships with his whole body well before he was able to express the quantum equations in numbers. This same muscular sense for the body in motion may also provide insight into engineering and architecture. At Princeton University one architecture student recently combined a dance production called 'The Body and the Machine' with a senior thesis, explaining that "exploring conceptual issues (in architecture) kinetically helps me understand them" (Moseley, 18).

Empathizing, our next tool, is related to body thinking, for this imaginative skill involves putting yourself in another's place, getting under their skin, standing in their shoes, integrating 'I' and 'it', feeling the objective world subjectively. Empathizing with other people, with animals, with characters on stage or in a book is standard fare for novelists, actors, and even physicians. But artists and scientists also empathize with nonhuman, even non-animal things and processes. Isamu Noguchi reified this sort of empathy in his sculpture, 'Core', a piece in basalt with carved holes. "Go ahead", he told visitors to his studio. "Put your head into it. Then you will know what the inside of a stone feels like". By putting her head 'in there', focusing her attention at the level of the corn chromosomes she studied, Nobel laureate Barbara McClintock was able to develop a 'feeling for the organism' so complete that she described herself as being down inside her preparations, and their genes became her 'friends'. And astrophysicist Jacob Shaham talked of 'reading' his equations like scripts for a play in which the 'actors'—energy, mass, light and so on—have intents and motives that he could physically act out.

Yet another tool that we most often learn unconsciously is *dimensional thinking*, rooted in our experience of space and time. Creative individuals think dimensionally when they alter the scale of things, as artists Claes Oldenburg and Coosje van Bruggen did in their *Batcolumn* in Chicago. Their ten-story-high rendition of a baseball bat strikes us very differently than the three-foot version. As any architect knows, size and mass can be altered to convey anything from flowery delicacy to dominating power. Moreover, the engineering of scale changes can be complex: different structural designs and different materials are almost certainly required as artist-engineers work dimensionally with properties such as strength and durability. Inventive individuals also think dimensionally when they map things that exist in three dimensions onto two dimensions, for instance in maps or blueprints. Indeed, this kind of dimensional thinking is at the heart of drawing in perspective. Artists, scientists and engineers also think dimensionally when they try to reconstruct three-dimensional phenomena from information recorded in two dimensions. Construction engineers interpret and build three-dimensional structures from two-dimensional instructions. In fact, how we orient ourselves in space has implications for the patterns we form in two and three dimensions. Cartesian coordinates assume a world of right angles; polar coordinates map a spherical universe. Buckminster Fuller rejected both in favor of a tetrahedral coordinate system and, based upon that system, invented his geodesic dome. Each coordinate system permits us to recognize and solve a different set of problems.

The tools for thinking briefly sketched up to this point are what might be called primary tools. They can be learned and practiced somewhat independently, though they are always interacting. Body thinking is a kind of imaging; observing feeds into abstracting and patterning; patterning in turn merges with analogizing and so forth. The last four tools for thinking, however, are clearly tools that rely upon the acquisition of primary tools and integrate them into composite tools.

The first of these composite tools is *modeling*, that is, plastically representing a thing or a process in abstract, analogical and/or dimensionally altered terms. The point of modeling is to depict something real or imagined in actual or hypothetical terms in order to study its structure or function. Artists make and use models all the time by preparing maquettes, smaller conceptualizations of pieces in planning. Scientists and engineers also create simplified models of objects and processes. In the case of flight simulators, engineers model the hands-on experience of flying planes for educational purposes by imitating the reality of that experience in space and time. Molecules that can never actually be seen or touched are built millions of times their actual size out of plastic or wood. Stars, which are beyond our ability to comprehend in any realistic sense, become a series of equations describing their actions over time frames beyond the entire experience of humanity. Modeling, as many practitioners have said, is like playing god, toying with reality in order to discover its unexpected properties.

Playing, of course, is itself another integrative tool that builds upon the other primary skills. We play when we do something for the fun of it, when we break or bend the rules of serious activity and elaborate new ones. Play is the exercise of our minds, bodies, knowledge, and skills for the pure emotional joy of using them. Unlike work, play has no set, serious goal; yet by encouraging fun, play is useful, for when creative individuals play with techniques and ideas they very often open up new areas of understanding through serendipitous discovery.

Among the greatest of players was the sculptor Alexander Calder, whose early training was in engineering. One manifestation of his play was a lifelong habit of designing toys for children (and for himself, too) out of wire and wood. In fact, Calder's first true success in the art world was as a result of having built himself a working model of a circus, complete with animals, props, entertainers with movable parts, a trapeze with a net and a tent. He actually played circus, too, inviting friends and acquaintances in the Parisian intelligentsia to watch him enact sights, sounds and stories under the big top. He was just having fun, yet his toys have been called a 'laboratory' for his subsequent, ground-breaking work. From movable toy figures he graduated to kinetic sculptures— hand-driven, then motor-driven—and finally to free-floating mobiles. In keeping with his playful spirit, however, he always refused to call his sculpture 'art', deeming the word too serious for his intentions.

Even the most serious innovations often have their origins in play. Alexander Fleming's discovery of penicillin has been traced to his hobby of collecting colored microbes for the 'palette' with which he created microbial 'paintings' on nutrient agar. Charles 'Fay' Taylor, the MIT engineer who made major strides in automotive engine design, explored mechanical objects by playing with kinetic sculptures. And Nobel laureate Richard Feynman said that his Nobel-winning work in quantum mechanics began when he started playing with the rotation of plates thrown in the air.

Play teaches us that how one learns something has no bearing on the importance of the lesson learned. What counts is the practice gained in extending the abilities and experience of one's mind and body. What counts is the practice gained in the use of more than one thinking tool at a time. Playing thus feeds into yet another imaginative tool, *transforming*, the serial or simultaneous use of multiple imaginative tools in such a way that one tool or set of tools acts upon another. To play is to transform, for one takes an object, observes it, abstracts essential characteristics from it, dimensionally alters the scale, and then, using body skills, creates a physical or mental representation of the object with which one can play. Take a look at any creative endeavor and you'll find such combinations of thinking tools being used to transform ideas and insights into one or more expressive languages.

In order to invent strobe photography, for example, engineer Harold Edgerton of MIT first transformed his mental image for a strobe light for ultrafast flash photography into a visual diagram, and then transformed the diagram into a working model. He played around with different versions of the strobe until he achieved one that matched his mental picture. Then, using his prototype, he played with setup conditions, different kinds of subjects and motions until, finally, he transformed all these components—film, camera, strobe, subject—into the results he wanted: a photograph that was both a scientific experiment and a work of art. In retrospect we can see that Edgerton made use of several imaginative tools: visualizing, modeling, playing, and something more, too, for without the ability to translate his ideas into words, diagrams, strobe and photograph his imaginative invention of ultrafast flash photography would have come to naught. Indeed, such transformations are typical even of data, as Edward Tufte has beautifully demonstrated in his books on visual information. Every table or graph or illustrated set of instructions for assembling something is a transformation of one kind of knowledge into another.

The necessary consequence of transformational thinking is our final mental tool: *synthesizing*, the combining of many ways of thinking into a synthetic knowing. When one truly understands something, emotions, feelings, sensations, knowledge and experience all combine in a multimodal, unified sense of comprehension. One feels that one knows and knows what one feels. Einstein, for example, claimed that when he sailed he felt the equations of physics playing out through the interactions of the boat, the wind, and the water. He became a little piece of nature. Similarly, artists and writers describe the creative process as a melding of sight, sound, taste, touch, smell, and emotion in which all become interwoven in an experience so powerful that they lose their sense of self. Feeling and thinking become one in a process that is often described as 'synesthetic'.

Synesthesia is a neurological term that refers to the experience that some people have of seeing colors when they hear certain sounds, or perceiving tactile feelings when tasting various foods. Artists and musicians, many of whom have some form of neurological synesthesia, often describe the ultimate aesthetic experience as being one in which a performer or observer of an art experiences all possible sensations simultaneously. A picture or a symphony may, for example, generate visual, acoustic, and tactile sensations along with definite emotions and even tastes, smells, and movements in the observer. One way to judge art is the degree to which it provokes such a multi-modal experience.

If we refer back to the descriptions of scientific thinking given by Einstein, Feynman, McClintock, Arber and other scientists in the opening of this chapter, then it is clear that scientists, too, experience a form of synesthesia. Ideas are inseparable from the emotions, the visual and tactile images and other sensations that accompany their genesis. Since the result of such sensory and somatic integration is not just an aesthetic experience, but also an intellectual one, we have suggested that it be called 'synosia', from a combination of 'syn', meaning together, and 'osia' from 'gnosis', the Greek word for knowledge. Synosia, in short, is the combination of knowledge and emotion, objective and subjective understanding into a synthetic whole.

The fact is that true understanding (by which we mean the ability to act upon the world), as opposed to knowledge (which is the merely passive acquisition of facts, often without the skills to use them), is always

synthetic. Immanuel Kant wrote many years ago that "The intellect can intuit nothing, the senses can think nothing. Only through their union can knowledge arise". He understood that we recognize that which is important by its emotional impact on us and use our senses to explore how to respond. Thus, we can now understand why Einstein, Poincaré, and so many other innovators have claimed that intuition rather than reason is the basis of creative thought. To feel is to think, just as to think is to feel. Only when the two are integrated is innovation possible.

Training Intuition

Since intuition develops from the kinds of non-verbal, non-mathematical tools for thinking that we have just outlined, it can be exercised. The use of mental tools is no different than the use of physical tools: both require training and practice. Fortunately, many of the innovative people who have discussed how they have used observing, imaging, patterning, analogizing, and all the rest of the tools, have also described how they acquired skill in using these tools. The one thing they all say is that intuition results from doing things, not passively learning about them. One builds up a sense of how things should work by having experienced how they actually do (or do not) work. Thus, more than one innovator has stated that an expert is an individual who has made all the mistakes in the field.

Observing and imaging, for example, are often learned together through the practice of fine and applied arts and hobbies of all sorts. Collecting anything from stamps or coins to butterflies or buttons teaches an individual visual discrimination and memory. These talents are raised to a higher level by the practice of applied and fine arts. The artist-writer Leo Lionni's first drawing teacher was his architect uncle who gave him lessons as a small boy. Similarly, the writer Vladimir Nabokov also learned as a child to make detailed drawings both from life and from memory of objects that he examined over and over again. By his own admission, he used his observing and imaging skills equally in his research on butterflies at the Harvard Museum of Comparative Zoology and in his literary undertakings. Many Nobel laureates in the sciences have echoed Santiago Ramon y Cajal's statement that "that which has not been drawn has not been seen". And the same lessons have applied to observing well in sound, smell, taste, and touch and recalling the images derived from these senses. Pioneering composer Charles Ives was taught by his musician-father to hear the 'music' in a thunderstorm or the tone of a pane of glass when it is tapped—things that most of us overlook, or more accurately overhear. Chemical ecologist Thomas Eisner was taught by his father, a perfumer, how to use his nose to identify the composition of substances. Eisner now uses that faculty to study the ways in which insects use odors to communicate with one another.

Abstracting can also be learned and practiced by observing how other people have performed the process and by copying them. Even the expert artists have to learn the abstracting process of eliminating all the unnecessary clutter to reveal some basic property of an object. This process is beautifully illustrated in Randy Rosen's extraordinary book, *Prints* (1978). He shows how Pablo Picasso and Roy Lichtenstein both eliminated various features of a bull, step by step over many months, to yield very different and yet very evocative abstractions of 'bullness'. Guides to good writing, such as *The Elements of Style* by Strunk and White, recommend that writers revise by cutting out words, sentences, paragraphs that are unnecessary—in other words, they advise writers to abstract, to jettison all but what is essential to the work. No better example of written abstracting exists than the one-line plot descriptions given in the TV Guide. Trying to duplicate such one-line descriptions is excellent training in discovering the essence of things.

Patterning can be learned by similar experience. Richard Feynman recounted that his first formal introduction to patterns was as a very young child. His father gave him a set of small ceramic tiles, some blue and some white, and then had him create simple patterns: all blue; all white; alternating blue and white; two blue and one white; one blue and two white; etc. Simply learning that patterns have permutations was the beginning of one of Feynman's greatest ideas, which is that nature always employs every possible path to achieve any given end. There is a lesson here for creative thinking, too. The greater the number of patterns one knows, the greater one's understanding of possibilities. Many forms of pattern recognition require formal training in music, poetry, and symmetry, and books about these subjects abound. Far better, however, is active participation in composing music, poetry, and artwork, since doing always teaches more than reading. For the same reasons, much can be learned about patterns by playing word games, building puzzles, learning to dance, becoming a chess master, or doing recreational mathematics. When one can recombine what one knows to invent new chess puzzles, choreograph a new dance, or invent new mathematical problems or poetic forms, then one has graduated to pattern forming, which brings creative joys unmatched by any passive hobby.

Artist-inventor-psychologist Todd Siler has written extensively on how to generate patterns connecting like and unlike using a process he calls 'metaphorming'. To metaphorm, one uses any and all forms of connection-making—including visual analogy, metaphoric figures of speech, narrative cause and effect and rational hypothesis—to explore the meaning inherent in the comparison of two or more things. Take any given object, he advises, and ask yourself what else is this like, what does it remind me of? And why? Articulate the connection as metaphor, as hypothesis, a symbol, as pun. To metaphorm the mind with garden means to assert that the mind is a garden, that there are gardens of the mind. Thoughts germinate like flowers. The imagination is the soil in which they grow. The mind, layered like an onion, requires cultivation and nourishment. Ideas root themselves and become difficult to dislodge. Dangerous ideas create 'mind fields'. Taken literally, of course, there are 'mind fields', which can be studied by means of functional magnetic resonance imaging and other neurological techniques. Are we on the verge, as mind-gardeners, of intervening physically to enhance or otherwise influence the growth of a mind-plant? Metaphorming ideas in as many ways as possible is good practice in making the structural connections and functional analogies that animate art, science and technology. Similar pattern forming techniques have been adapted for elementary and secondary classroom use, for instance in *The Private Eye, Looking/Thinking by Analogy* guide for learning in art, writing, science, math and social studies (Ruef, 1992). Analogizing, that particular search for similarity of function, especially involves looking at things and processes in order to discover not simply how they work, but how they might work outside their given context. Young people exposed to such training acquire the active habits of mind necessary to the intuitive generation of novel ideas.

Body thinking is another tool best developed through active participation with the world. This may seem self-evident, but in an age when people spend increasing amounts of time in front of computers, simple body skills among students are declining dramatically. Children—and adults, too—spend less and less time handwriting, drawing, running, jumping and playing physical games and sports of all kinds. But the truth is, they cannot learn to ride a bike simply by reading about it. Nor can they really understand structural forms such as buildings or bridges without experience of thinking about the muscular supports of their own body; they cannot really understand physical processes such as the molecular behavior of solids, liquids and gases, without incorporating notions of speed and vector within themselves. All kinds of physical activity, including organized arts and athletics, work to develop body-thinking skills. Sports and dancing build gross body-thinking; finer body thinking skills result from making music, art, and building things. For added bonus, body thinking can be reviewed and practiced mentally. For instance, the pianist can see and feel herself playing a piece of music, remembering every detail without so much as moving a finger. The downhill skier can imagine each moment on a race course without leaving his room. Studies of people in every discipline from sports to music, engineering to design, show that imaging how it will feel to perform a particular set of actions can actually improve subsequent performance.

Dimensional thinking must also be learned by doing. One must learn how to translate a three-dimensional object into two dimensions by drawing or photographing it. One must learn how to transform the information given on a two-dimensional blueprint or assembly diagram into the three-dimensional object. Such skills can be acquired through formal classes in drafting and modeling, or through informal experience building furniture, knitting, sewing, or doing any other craft. Perhaps most challenging is learning how to transform a linear set of mathematical symbols into a graph or physical model of the equation—an exercise that was once common in geometry and algebra classes and which should be re-instituted universally.

Playing, and the modeling that it so often entails, is especially important to the exercise and training of intuition. Most innovators build models of sorts, play with a wide variety of games and tools, and generally have extensive experience with making things of all sorts. Carl G. Jung, the famous psychologist, recalled untold hours building models of castles as a teenager. He then took up painting, through which he discovered the function of mandalas (world images) as models for the psychological lives of his patients. Einstein, of course, spent his most creative years in a patent office, daily analyzing and playing with models of inventions. Many artists and writers, including Claes Oldenberg and H. G. Wells, created entire imaginary civilizations with which they played as children and teenagers and from which they subsequently drew novel ideas for their arts as adults. Alexander Calder, as mentioned above, modeled a circus and derived from his experience not only contacts with the art world, but specific ideas about how to design moving sculptures. Such experiences are common among imaginative people. Indeed, one of the few good correlations that exists to predict which individuals will be creative reveals that they have, often from childhood, *made* things with hands and mind.

Just as all roads lead to Rome, all the experiences gained from the exercise of imaginative thinking tools lead towards synthesizing, that ability to pull together all one imagines with all one knows, that drive to meld sensual knowledge with received wisdom into a unified knowing that we have called synosia. We are often most aware of the 'rational' or sense-making character of synthetic breakthroughs in human thought—for instance, the explanatory power of Alberti's drawings in perspective or Einstein's theory of relativity, but non-rational feelings and perceptions play an equally important role in the generation of synthesis. There is that deeply troubling sick feeling in the pit of one's stomach when one looks at a situation and knows that something is wrong; or the unmatchable 'high' that accompanies the 'Aha!' of an unexpected insight. For mathematician-philosophers Bertrand Russell and Norbert Wiener, creative work almost always began with feelings of physical discomfort evoked by certain unsolved problems in mathematics (Hutchinson, 1959, p. 19; Wiener, 1956, pp. 85–86; Wiener, 1953, pp. 213–214). Equally physical, orgasmic feelings of relief and achievement attended the solution of those problems. Nobel laureate Sabrumanyam Chandrasekhar has called this "shuddering before beauty" (Curtin, 1982, p. 7).

Ultimately, all thinking tools, but especially modeling, playing, transforming, and synthesizing, give birth to the inarticulate sense-making called intuition. Intuition involves non-explicit expectations of what should happen when something is tweaked, of how a system will behave when it is twisted, of what kind of response a person will give in a particular situation. We build vague models of how things work and people behave based on our experiences. These models often owe a great deal to playing with tools, games, people, and systems to find out how they respond to various stimuli. We develop a 'feel' for what should happen, But because we have not analyzed our experience in any formal way, we cannot explain the resulting 'intuitions'. They remain what philosopher and physicist Michael Polanyi has described as 'personal knowledge'—pre-verbal understanding that yields insight before it yields the means to explain insight. Though personal knowledge is just that, personal and unspeakable, it is nonetheless valid and useful. In fact, Neils Bohr used to chide his students with the comment, "You're not thinking; you're just being logical!" (Frisch, 1979, p. 95). His colleague Enrico Fermi was known to dismiss mathematical 'proofs' of concepts with the comment that his 'intuition' told him they were wrong. Because Fermi had so much experience actually doing physics, building, making and inventing things, most of his colleagues trusted his intuitions, which were often right (Wilson, 1972). Learning to pay attention to that which moves us—to an accumulation of unarticulated but felt experience that forms our intuition—is key to creative work (Root-Bernstein, 2002).

Intuition and the Future of Innovative Education

Having placed intuition on a comprehensible footing, and outlined its role in the comprehensive, creative knowing that is synosia, we can now think about the educational implications such recognition must imply. Education in every discipline rightly emphasizes analytical, logical, technical, objective, descriptive aspects of each field. These inform the nature of public discourse between practitioners and their formal communication of disciplinary knowledge. But, as must by now be evident, the subjective, emotional, intuitive, synthetic, sensual aspects that make up the private human face of all creative inquiry deserve equal educational recognition. It is this human face, after all, that fuels desire to discover, to invent, to know. Without it, creative work has no motivation, no driving force. This is not to argue that practice of imaginative thinking and the exercise of intuition is of greater import than mastery of the logical, analytical, technical aspects of any discipline. Far from it. Innovation is possible only when individuals emotionally engage in a subject and intuit novel ideas *and also* evaluate ideas and results logically and translate them into forms appropriate for communication and analysis by other people. Synosia cannot do one without the other, nor can an education that truly seeks to prepare students for innovation and invention.

Unfortunately, not only does our education system generally ignore the emotional and subjective aspects of creativity, so do the cognitive sciences. This is too bad, for theories in cognitive psychology do not mirror so much as they inform educational practice. And as mathematician Seymour Papert of MIT makes clear, the enormous impact cognitive theories often have on educational practice can be to the detriment of innovation. He writes:

> Popular views of mathematics, including the one that informs mathematical education in our schools, exaggerate its logical face and devalue all connections with everything else in human experience. By doing so, they fail to recognize the resonances between mathematics and the total human being which are responsible for mathematical pleasure and beauty . . . (Papert, 1978, p. 104).

Papert finds grounds in this oversight to question the validity of cognitive theories as they inform education:

> Implicit in the confrontation of these views of mathematics is a broader question about the legitimacy of theories of psychology, often called cognitive, which seek to understand thinking in isolation from considerations of affect and aesthetics (Papert, 1978, p. 104).

Papert has a point. The separation of cognition from somatic sensation and aesthetic feeling is both inaccurate and inappropriate: inaccurate, for if, as Einstein and McClintock both said, one must become a piece of nature in order to discover the hidden mysteries of nature, then the oversight of imaginative and intuitive thinking undermines our understanding of creative endeavor; inappropriate, because the same dualistic divorce of mind and body, emotion and reason, has had a deleterious effect on education. Psychologist Jeanne Bamberger has documented just how harmful. She studied a group of Boston teachers and some of their students who were considered bright but who performed poorly in school. Teachers and students were brought to Bamberger's Laboratory for Making Things in Cambridge, Massachusetts where they were asked to build mobiles. Most of the children had no difficulty building mobiles, but when asked how they did it and what physical principle they used, they were unable to answer. As one young man said, he "just knew I

had a feeling of it, like on a teeter totter" (Bamberger, 1991, p. 38). The teachers, however, had learned the principle underlying mobile construction, which is the same as balancing two weights on a lever: 'weight times distance must be equal on both sides of the fulcrum'. They, however, were mostly unable to implement this principle in practice, and few built a functional mobile (Bamberger, 1991, p. 44). There is something obviously wrong with an educational system that can produce students unable to explain how they do what they do and teachers unable to do what they can explain.

The crux of the matter with education lies in the dissociation of mind from body and thus sciences from arts. For most of the twentieth century, psychology was dominated by an over-simplified use of Lewis Terman's theory of intelligence, which relied solely upon verbal and mathematical measures of problem-solving ability (Seagoe, 1975). Practitioners overlooked the fact that Terman himself had actually found that for very creative people, but not for average people, verbal and mathematical scores were sufficient to predict high achievement on visual, analogical, mechanical, physical and other tests. Initially, at least, communication skills with words and numbers were understood as predictors/indicators for some of our most important imaginative thinking skills: visualizing or imaging, analogizing, modeling and body thinking. As Terman's work affected the field, however, the communication skills that predicted creative intelligence took precedence over the imaginative substance of that intelligence—as evidenced by the heavy testing of verbal and mathematical skills at all levels of schooling.

Unfortunately recent multiple intelligences theories such as Howard Gardner's (1983) threaten to exacerbate the problem by focusing much-needed attention on a broader set of communications skills such as kinesthetic, musical, verbal, visual, and inter-personal abilities without simultaneously distinguishing them from creative thinking skills. The fact that people can be highly verbal, extraordinarily artistic, or wonderfully musical and at the same time have little or no creative ability seems generally to have been overlooked or ignored. Most creative people are, in fact, polymathic and utilize their skills in multiple domains (see Root-Bernstein, 'The Art of Innovation', this volume).

Bamberger, Papert and others point the way towards a more balanced view of innovative thinking by forcing us to look at mind as part of the body. Neurologists such as Antonio Damasio (1994) remind us that people who, for reasons of disease or accident, lose emotional affect also lose their ability to act reasonably. Rational decision-making, he argues, cannot be divorced from emotional affect. The anecdotal reports of so many of the world's most creative people are finally finding an analytical basis.

The implications of these findings for cognitive sciences and education cannot be underestimated. What they tell us is that any theory of mind that claims to account for creative thinking must describe the sensual, emotional, and somatic manifestations of thought as well as their analytical, objective, and communicable formulations. Moreover, the transformational process by which ideas are translated from their personal, bodily forms into formal languages for communication must be made explicit. Educationally, each of these points has equivalent importance. Words and numbers are not sufficient to produce innovative people, nor are the tools for thinking that we have outlined here. Tools for thinking are necessary to develop the sensual, emotional, bodily forms of thinking from which new ideas emerge, but tools for thinking are not sufficient for communicating these ideas to other people. In order to provide a complete education, tools for thinking need to be taught in an integrated fashion with a variety of expressive skills— verbal and mathematical, to be sure, but also bodily-kinesthetic, visual-spatial and others that pertain to Gardner's multiple domains. Translating and transforming skills that link imaginative tools to expressive modes and expressive modes one to another are equally necessary. Only when mind and body, synthesis and analysis, personal thought and public communication skills are all part and parcel of cognitive studies and educational practice will an enhanced capacity for innovation become available to everyone.

References

* All quoted material with the exception of individual items that are followed by a reference may be found in Root-Bernstein, R. S., and Root-Bernstein, M. M. (1999). *Sparks of genius*. Boston, MA: Houghton Mifflin.

Bamberger, J. (1991). The Laboratory for Making Things. In: D. Schone (Ed.), *The Reflective Turn: Case Studies in and on Educational Practice* (pp. 38–44). New York: Teachers College Press.

Collins, B. (1991). *Wood sculpture and topological allegories. Exhibit brochure*. AAAS Art of Science and Technology Program, Washington, D.C., 9 April—7 June.

Curtin, D. (Ed.). (1982). *The aesthetic dimension of science*. The Sixteenth Nobel Conference, 1980. New York: Philosophical Library.

Damasio, A. R. (1994). *Descartes's error: Emotion, reason, and the human brain*. New York: G. P. Putnam's Sons.

Eliot, T. S. (1963). T. S. Eliot interview. *In Writers at Work, The Paris Review Interviews* (2nd series, pp. 95–110). New York: Viking Press.

Eliot, T. S. (1975). Selected prose of T. S. Eliot (F. Kermode, Ed.). New York: Harcourt Brace Jovanovich/Farrar, Straus and Giroux.

Frisch, O. R. (1979). *What Little I Remember*. Cambridge: Cambridge University Press.

Gardner, H. (1983). *Frames of mind. The theory of multiple intelligences*. New York: Basic Books.

Gardner, H. (1993). *Creating minds*. New York: Basic Books.

Gardner, J. (1983). *The art of fiction*. New York: Vintage Books.

Hutchinson, E. D. (1959). *How to think creatively*. New York: Abington-Cokesbury Press.

Kasner, E. & Newman, J. (1940). *Mathematics and the imagination*. New York: Simon & Schuster.

Lu, F-P. (1966). *T. S. Eliot, The dialectical structure of his theory of poetry*. Chicago, IL: University of Chicago Press.

Moseley, C. (1994). Mind, body spirit. On the boards with choreographer Ze'eva Cohen. *Princeton Alumni Weekly*, **94**, 16–22.

Papert, S. (1978). The mathematical unconscious. In: J. Wechsler (Ed.), *On Aesthetics in Science*. Cambridge, MA: MIT Press.

Plimpton, G. (Ed.) (1989). *The writer's chapbook*. New York: Viking.

Root-Bernstein, R. S. (2002). Aesthetic cognition. *International Studies in the Philosophy of Science*, **16**, 61–77.

Root-Bernstein, R. S. & Root-Bernstein, M. M. (1999). *Sparks of genius*. Boston, MA: Houghton Mifflin.

Rosen, R. (1978). *Prints. The facts and fun of collecting*. New York: E. P. Dutton.

Ruef, K. (1992). *The private eye, looking/thinking by analogy*. Seattle, WA: The Private Eye Project.

Seagoe, M. (1975). *Terman and the gifted*. Los Altos, CA: W. Kaufmann.

Siler, T. (1996). *Think like a genius*. Denver, CO: ArtScience Publications.

Wiener, N. (1953). *Ex-prodigy: My childhood and youth*. New York: Simon & Schuster.

Wiener, N. (1956). *I am a mathematician*. London: Gollancz.

Wilson, M. (1972). *A passion to know*. Garden City, NY: Doubleday.

The International Handbook on Innovation
Edited by Larisa V. Shavinina
© 2003 Published by Elsevier Science Ltd.

Stimulating Innovation

Ronald N. Kostoff

Office of Naval Research, USA

Abstract: Innovation is critical for maintaining competitive advantage in a high-tech global economy, especially for organizations or nations that do not possess low-cost labor forces. Many studies on innovation attempt to identify endogenous and exogenous variables that impact innovation (Kostoff, 1997a) in order to better understand the environment that promotes innovation. The author's recent efforts have focused on developing processes for enhancing innovation that exploit the transference of information and insights among seemingly disparate disciplines.

Keywords: Innovation; Discovery; Cross-discipline knowledge transfer; Literature-based discovery; Text mining.

The objective of this chapter is to describe how innovation can be promoted through the enhancement of discovery by cross-discipline knowledge transfer. The approach developed entails two complementary components—one literature-based, the other workshop-based. The literature-based component identifies the science and technology disciplines related to the central theme of interest, the experts in these disciplines, and promising candidate concepts for innovative solutions. These outputs define the agenda and participants for the workshop-based component. An example of this combined approach is presented for the theme of Autonomous Flying Systems. The hybrid approach appears to be an excellent vehicle for generating discovery and enabling innovation. However, it requires substantial time and effort in both phases.

Introduction

Innovation reflects the metamorphosis from present practice to some new, hopefully 'better' practice. It can be based on existing non-implemented knowledge, discovery of previously unknown information, discovery and synthesis of publicly available knowledge whose independent segments have never been combined, and/or invention. In turn, the invention could derive from logical exploitation of a knowledge base, and/or from spontaneous creativity (e.g. Edisonian discoveries from trial and error).

The process of innovation is of immense social interest and impact. Classical studies by Mansfield (1980, 1991), Griliches (1958, 1979, 1994), and Terleckyj (1977, 1985) focused on the relationship between innovation and micro- or macro-economics. Studies by Wenger (1999) on combined visualization/ brainstorming techniques, Patton (2002) and Taggar (2001) on the impact of group stimulation to creativity, Chen (1998) and Siau (1996) on contributions of electronic technology to creativity, and books by Boden (1991) and DeBono (1992) on mental processes in creativity, focused on the process of creativity and its contributions to innovation. Large-scale studies by the Department of Defense (DoD, 1969), Illinois Institute of Technology Research Institute (IITRI, 1968), Battelle (Battelle, 1973), and the Institute for Defense Analysis (IDA, 1990, 1991a, 1991b) focused on identifying the environmental and management conditions most conducive to innovation. Recent symposia have focused on the relation of innovation to: technology policy (Conceicao, 1998, 2001); technology forecasting (Arciszewski, 2000; Grupp & Linstone, 1999), competitive advantage (Hitt et al., 2000); and economic growth and impact (Archibugi & Michie, 1995; Spender & Grant, 1996; Van de Klundert et al., 1998). Yet both the process and impacts of innovation remain poorly understood.

One of the least studied components of innovation is the discovery and synthesis of publicly available knowledge whose independent segments have never been combined; i.e. the transfer of information and understanding developed in one or more disciplines to other, perhaps very disparate, disciplines. With the explosion in availability of information, the number of opportunities to synthesize knowledge and enhance discovery from disparate disciplines increases non-linearly. Conversely, with accelerating production of information, scientists and technologists find it increasingly difficult to remain aware of advances within their own discipline(s), much less advances in other seemingly unrelated ones. Paradoxically, the growth in science has led to the balkanization of science!

As science and technology become more specialized, the incentives for interdisciplinary research and development are reduced, and this cross-discipline transfer of information becomes more difficult. The author's observation, from examination of many science and technology sponsoring agencies and performing organizations, supplemented by a wide body of literature (Bauer, 1990; Bruhn, 1995; Butler, 1998; Metzger, 1999; Naiman, 1999), is that strong cross-disciplinary disincentives exist at all phases of program/project evolution, including selection, management and execution, review, and publication. To overcome cross-discipline transmission barriers, and thereby enhance innovation, systematic methods are required to heighten awareness of experts in one discipline to advances in other disciplines. Most desirable are methods that *incorporate/require cross-disciplinary access as an organic component*.

This chapter presents two different, yet complementary, approaches to increase cross-discipline knowledge transfer and provide the framework for enhancing innovation. One is literature-based, and the other is workshop-based. Each approach individually represents a major advance in enabling discovery and subsequent innovation, and the hybrid of the two approaches provides a synergy that multiplies their combined benefits.

The literature-based approach is summarized first, followed by the workshop-based approach. The advantages of combining the two approaches are then presented. The details of each approach are presented in the appendices.

Accessing Linked Literatures for Enhancing Innovation-Summary

The first approach searches for relationships between linked, overlapping literatures, and discovers relationships or promising opportunities not obtainable from reading each literature separately. The general theory behind this approach, applied to two separate literatures, is based upon the following considerations (Swanson, 1986).

Assume that two literatures with disjoint components can be generated, the first literature AB having a central theme 'a' and sub-themes 'b', and the second literature BC having a central theme(s) 'b' and sub-themes 'c'. From these combinations, linkages can be generated through the 'b' themes that connect both literatures (e.g. AB → BC). Those linkages that connect the disjoint components of the two literatures (e.g. the components of AB and BC whose intersection is zero) are candidates for discovery, since the disjoint themes 'c' identified in literature BC could not have been obtained from reading literature AB alone.

Some initial applications of the first approach have been published in the medical literature (Swanson, 1986). One interesting discovery was that dietary eicosapentaenoic acid (theme 'a' from literature AB) can decrease blood viscosity (theme 'b' from both literatures AB and literatures BC) and alleviate symptoms of Raynaud's disease (theme 'c' from literature BC). There was no mention of eicosapentaenoic acid in the Raynaud's disease literature, but the acid was linked to the disease through the blood viscosity themes in both literatures. Subsequent medical experiments confirmed the validity of this literature-based discovery (Gordon & Lindsay, 1996). (A website (Swanson & Smalheiser, 1998b) overviews the process used to generate this discovery, and contains software that allows the user to experiment with the technique. Finn (1998) outlines perceptions of different knowledgeable individuals on Swanson and Smalheiser's general technique.)

This literature-based discovery approach is in its infancy. Public and private financial support for this technology are minimal. *It is a research area of unlimited potential that seems to have fallen through the cracks.* There is essentially one group that is publishing results of literature-based innovation and discovery in the credible peer-reviewed literature (Smalheiser, 1994, 1998a, 1998b; Swanson, 1986, 1997, 1999), two groups that have published concept papers (Hearst, 1999; Kostoff, 1999a), and a few other groups that have replicated Swanson's initial results (Gordon & Lindsay, 1996; Weeber et al., 2001). Presently, the approach is not automatic. It requires much thought, expertise, and effort. The author's group is examining different approaches to make the process more systematic, while reducing the manual labor intensity. Given the potential benefits of the literature-based approach for stimulating innovation, it is truly a technology whose time has come. Appendix 1 generalizes and expands upon the literature-based approach, using the Database Tomography techniques and experience developed by the author since 1991 (Kostoff, 1993, 1994, 1998, 1999b, 2000a, 2000b). It outlines the theory of the expanded approach, the implementation details, and overviews the range of applications possible with this technique.

Interdisciplinary Workshops for Enhancing Innovation-Summary

The second approach consists of convening workshop(s) of experts from different disciplines focused on specific central themes. The purpose of such a workshop is to achieve multi-discipline synergies and cross-discipline transfers to generate promising research directions for these central themes. The theory behind this approach is described in Appendix 2. To test this theory, a workshop on Autonomous Flying Systems was convened in December 1997. Its implementation mechanics and results are described in detail in Appendix 2.

The total workshop process consisted of three phases:

(1) A two-month pre-meeting e-mail phase in which each participant provided descriptions of advanced capabilities and promising research opportunities from his/her discipline to all other participants;
(2) A two-day meeting at the Office of Naval Research during which the promising opportunities identified beforehand were discussed, crystallized, and enhanced; and
(3) A post-meeting e-mail phase in which each participant provided additional or embellished opportunities.

A number of important lessons were extracted from the conduct of this workshop, and they can be summarized as follows:

(a) The workshop approach broke new ground toward stimulating innovative thought. It was not easy, simple, or effortless, and required substantial planning and work in order to be effective. One should not throw people from 15 different disciplines together in a room for two days and hope to get new ideas synthesized. There needs to be a common generic thread woven through the different disciplines represented to spark the innovative thought process.

   Interdisciplinary workshops, when performed correctly, are the wave of the future in defining new research (and technology) areas and approaches. Because of the intensity and effort involved throughout the process, they are most appropriate for large-scale 'grand challenges' in full-blown workshop form, but appropriate as well for smaller scale issues.

(b) Representatives from diverse technical disciplines, organizations, and development categories attended the workshop. There was substantial value in having a balance of discipline, category, and organization diversity at the same meeting. The different perspectives presented benefited all participants.

   The use of modern information technology can expand the degree of diversity dramatically. Some of the concepts and group software proposed for network-centric peer review (Kostoff, 2001) can be easily adapted for use in innovation workshops. This would allow many more people, disciplines, and organizations to be represented, further enhancing the potential for cross-discipline information transfer and resultant innovation and discovery.

(c) Problem selection is crucial. The problem should be sufficiently general that many diverse disciplines can link to it. Given the choice of equally relevant problems, there is more potential for impact in selecting problem areas for which a large interdisciplinary community is not yet obvious.

(d) It is important to select participants by the most objective processes available. A combination of expert recommendation and strategic topical maps based on computational linguistics, publications, and citations was used for the selection process, and this approach produced highly knowledgeable individuals. Incorporation of the full literature-based approach to innovation in the discipline or participant selection process could further enhance confidence that the most appropriate mix of disciplines and experts has been chosen.

(e) It is extremely important that individuals selected for participation be world-class experts in their particular areas. There are relatively very few individuals producing the seminal works in any field (Kostoff, 1998, 1999b), and it is these people who should be central to any truly innovative workshops. However, in addition to these established experts, highly competent individuals new to the field should also be selected. One benefit of transcending selection of known experts is that fresh faces new to established communities appear. They can sometimes challenge established paradigms and offer concepts typically not advanced through panels based solely upon well-known, over-used panelists.

(f) The e-mail component of the workshop is crucial. The gestation period between the input of promising ideas and their actual discussion at the workshop allows consideration of many different approaches and syntheses. It also saves substantial time at the workshop by clarifying confusing issues beforehand. However, in the first experience reported here, the stimulation of dialogue in the e-mail phase among most of the participants did not occur. The only participant to raise questions was the author, and this occurred only a few times. Nonetheless, in these instances, the dialogue was extremely valuable in clarifying issues and surfacing points of contention. In future workshops, it is strongly recommended that a few individuals representing different disciplines be asked to assume a role of facilitator, with the task of stimulating dialogue and raising questions during the workshop build-up phase.

(g) All the attendees at the workshop were required to participate; there were no pure observers. This meant that they had to submit accomplishments and opportunities statements by e-mail. They also had to be prepared to lead discussions at the workshop. This participation requirement was valuable in that each attendee obtained a sense of ownership in the workshop and its outcome. His/her contribution tended to be more substantive and creative than is typically the case at standard workshops. Those who contributed more in the e-mail phase tended to contribute more in the workshop phase. In addition, there was a sense of equality among participants when all were required to contribute, as opposed to an audience/performer environment with passive onlookers. The requirement that each attendee be an active participant translates directly into a limitation on audience size. However, it was concluded that the participation of a limited number of motivated and active individuals contributed more to the innovation process than the standard workshop of few active participants and many observers.

(h) In general, there needs to be some incentive to motivate participation of world-class experts in these workshops. Unless they are able to envision some type of substantive impact resulting from their participation, either on larger science and technology issues or in their individual disciplines, they could be reluctant to invest the substantial amount of time required for serious participation. This, however, did not turn out to be a problem for the Autonomous Flying Systems workshop, apparently because of the limited size of the field and the interest of the participants in the type of workshop conducted.

In addition, during the workshop, participants did not appear to have reluctance in sharing new concepts. This is in stark contrast to some workshops the author has attended where novel ideas were held very closely. In the Autonomous Flying Systems workshop, there was a spirit of cameraderie and cooperation that pervaded the proceedings, and helped overcome the barriers to sharing. This spirit was fostered in the pre-meeting e-mail dialogue phase, and further nurtured during the meeting by having all attendees participate in the proceedings as equal partners.

Finally, interdisciplinary workshops are a powerful potential source of radically innovative ideas if conducted properly. There are three central requirements for success:

(1) A problem of significant interest to the sponsoring organization must be selected;
(2) An optimal mix of world-class experts appropriate to the problem must be chosen;
(3) Conditions must be created which will motivate the participants to share their novel concepts.

The Autonomous Flying Systems workshop addressed these three requirements to a significant degree. A preliminary concept proposal emerged, and a copy of this proposal is available from the author.

Need for Literature/Workshop Synergy

Most organizations use some variant of a workshop/group dynamics approach for brainstorming or other proxies for stimulating innovation. The most current information is available, and real-time information exchange is unmatched. The attendees and participants in these groups tend to be focused subject experts representing a small fraction of the relevant technical community; there is rarely any complementary sophisticated literature analysis performed, and there are rarely experts present from strongly divergent disciplines. The outputs and discussion are highly subjective. The workshop techniques tend not to make full use of many of the information-technology advances of recent years. Probably most importantly, there are strong disincentives for the participants to reveal the latest innovations. What many workshops produce in practice are forums for 'selling' completed or near-completed research efforts.

A few performers, individuals or small groups of individuals, pursue the literature-based computer-assisted approach. This literature approach tends to be more sophisticated and technologically advanced than the workshop approach, and is more objective. It is more comprehensive, since it encompasses science and technology beyond the scope of any individual, or group of individuals, and can access data from many technical disciplines and many global sources. The source data are not as current as the workshop approach, due to the documentation time lag. However, with the advent of extensive on-line documentation, this time lag has been reduced considerably. One intrinsic limitation is that only a relatively modest amount of science and technology performed globally is documented and readily accessible to the wider user community (Kostoff, 2000c); obviously, any science and technology not documented cannot be accessed. The literature-based approach has not received widespread attention and may fall short of the interpretive and analytical strengths of the workshop approach. As a result, the literature approach is not widely used (e.g. Finn, 1998).

While either the workshop approach or the literature approach can be done independently to help stimulate discovery, they should be done in tandem to maximize the benefit provided by each. There is nothing on record to indicate that this joint approach to innovation has been implemented, or even considered. The Autonomous Flying Systems workshop described in this chapter has some elements of the combined approach. Some of the Database Tomography proximity analysis tools were used to identify the scope of related literatures, and the prolific individuals in these literatures. These individuals were then invited to the workshop. However, time constraints precluded using the full capabilities that the literature-based approach can offer.

In a joint workshop–literature effort, the literature approach would be included in the background pre-meeting phase of the workshop approach (as developed in Appendix 2). Accordingly, the literature study would provide:

(1) Background reading for the workshop participants in related yet disparate science and technology areas;

(2) Strategic maps of the broader science and technology literature as outlined in the DT papers referenced above;

(3) Promising opportunities for innovation and discovery; and

(4) The disparate science and technology disciplines from which the experts for the workshop could be drawn.

The hybrid literature–workshop approach would eliminate the limitations of each approach done separately. The right people from the right combination of disciplines could be identified by the literature-based approach and invited to the workshop. The literature-based analysis could structure the technical relationships, and provide an objective starting point for discussion. Network-centric peer review would allow linking, and fusing information from, large numbers of reviewers to incorporate more representative opinion sampling from the larger technical community. The only limitation not overcome is the disincentive for the participants, or document authors, to reveal their latest science and technology advancements.

There is extra time and cost involved with two approaches, and if responses were required with severe time limitations, then only one approach might prove feasible. For organizations that are serious about stimulating discovery and subsequent innovation, the additional time should not be a factor, given the potential high marginal benefits. Government could probably draw upon a more eclectic group than industry. Because of the competitive aspects, industry would probably rely more upon internal participants and contracted consultants, whereas government would draw upon individuals from many organizations.

Conclusions

The advent of large databases, and the parallel advances in computer hardware and software, provide the opportunity to augment and amplify traditional approaches of human creativity in generating discovery and subsequent innovation. This chapter has shown that multi-discipline structured workshops can enhance the science and technology discovery and subsequent innovation processes, and has shown that multi-discipline literature-based analyses can enhance the science and technology discovery process. The document has shown conceptually that the combination of computer-enhanced literature-based analyses and multi-discipline structured workshops has the synergistic potential to dramatically improve the discovery and subsequent innovation process relative to the already strong capabilities available from each process separately. This literature–workshop synergy represents a potential major breakthrough for systematically identifying: (1) the most promising disciplines to be used in the workshop; (2) specific experts from these different disciplines; (3) candidate promising concepts that form the basis for discussion.

(The views expressed in this chapter are those of the author and do not represent the views of the Department of the Navy.)

References

Archibugi, D. & Michie, J. (Eds) (1995). Technology and innovation. *Cambridge Journal Of Economics*, **19** (1), 1–4 (February, Special Issue).

Arciszewski, T. (Ed.) (2000). Innovation: The key to progress in technology and society. *Technological Forecasting and Social Change*, **64** (2–3, 119–120 (June–July, Special Issue).

Battelle (1973). *Interactions of science and technology in the innovative process: some case studies*. Final Report. Prepared for the National Science Foundation. Contract NSF-C 667. Battelle Columbus Laboratories. March 19.

Bauer, H. H. (1990). Barriers against interdisciplinarity—implications for studies of science, technology, and society (sts). *Science, Technology, and Human Values*, **15** (1), 105–119 (Winter).

Boden, M. (1991). *The creative mind*. New York: Basic Books.

Bruhn, J. G. (1995). Beyond discipline: Creating a culture for interdisciplinary research. *Integrative Physiological and Behavioral Science*, **30** (4), 331–341 (September–December).

Butler, D. (1998). Interdisciplinary research 'being stifled'. *Nature*, **396**, 202 (19 November).

Chen, Z. (1998). Toward a better understanding of idea processors. *Information and Software Technology*, **40** (10), 541–553 (15 October).

Conceicao, P., Heitor, M. V., Gibson, D. V. & Shariq, S. S. (Eds) (1998). The emerging importance of knowledge for development: Implications for technology policy and innovation. *Technological Forecasting and Social Change*, **58** (3), 181–202 (July, Special Issue).

Conceicao, P., Gibson D. V., Heitor, M. V. & Sirilli, G. (Eds) (2001). Beyond the digital economy: A perspective on innovation for the learning society. *Technological Forecasting and Social Change*, **67** (2–3), 115–142 (June–July, Special Issue).

DeBono, E. (1992). *Serious creativity*. New York: Harper Collins.

DOD (1969). *Project hindsight*. Office of the Director of Defense Research and Engineering. Washington, D.C. DTIC No. AD495905. October.

Finn, R. (1998). Program uncovers hidden connections in the literature. *The Scientist*, **12** (10), 12–13 (11 May).

Gordon, M. D. & Lindsay, R. K. (1996). Toward discovery support systems: a replication, re-examination, and extension of Swanson's work on literature-based discovery of a connection between Raynaud's disease and fish oil. *Journal* of the *American Society for Information Science*, **47** (2), 116–128.

Griliches, Z. (1958). Research costs and social returns: hybrid corn and related innovations. *Journal of Political Economy*, **66** (5), 419–431.

Griliches, Z. (1979). Issues in assessing the contribution of research and development to productivity growth. *The Bell Journal of Economics*, **10** (1), 92–116 (Spring).

Griliches, Z. (1994). Productivity, R&D, and the data constraint. *The American Economic Review*, **84** (1), 1–23 (March).

Grupp, H. & Linstone, H. A. (Eds) (1999). National technology foresight activities around the globe—Resurrection and new paradigms. *Technological Forecasting and Social Change*, **60** (1), 85–94 (January, Special Issue).

Hearst, M. A. (1999). Untangling Text Data Mining. *Proceedings of ACL 99, the 37th Annual Meeting of the Association for Computational Linguistics*. University of Maryland. June 20–26.

Hitt, M. A., Ireland, R. D. & Lee, H. U. (Eds) (2000). Technological learning, knowledge management, firm growth and performance. *Journal Of Engineering and Technology Management*, **17** (3–4), 231–246 (September–December, Special Issue).

IDA. *DARPA technical accomplishments*. Volume I. IDA Paper P-2192. February 1990; Volume II. IDA Paper P-2429. April 1991; Volume III. IDA Paper P-2538. July 1991. Institute for Defense Analysis.

IITRI (1968). *Technology in retrospect and critical events in science*, Illinois Institute of Technology Research Institute Report. December.

Kostoff, R. N. (1993). Database tomography for technical intelligence. *Competitive Intelligence Review*, **4** (1), 38 43.

Kostoff, R. N. (1994). Database tomography: origins and applications. *Competitive Intelligence Review*, **5** (1), 48–55 (Special Issue on Technology).

Kostoff, R. N. (1997a). *The handbook of research impact assessment* (7th ed.). DTIC Report Number ADA296021. Summer. Also located at www.dtic.mil/dtic/kostoff/index.html

Kostoff, R. N. (1997b). Database tomography for information retrieval. *Journal of Information Science*, **23** (4), 301–311.

Kostoff, R. N. (1998). Database tomography for technical intelligence: a roadmap of the near-earth space science and technology literature. *Information Processing and Management*, **34** (1), 69–85.

Kostoff, R. N. (1999a). Science and technology innovation. *Technovation*, **19** (10, October), 593–604.

Kostoff, R. N. (1999b). Hypersonic and supersonic flow roadmaps using bibliometrics and database tomography. *Journal of the American Society for Information Science*, **50** (5), 427–447 (15 April).

Kostoff, R. N. (2000c). The underpublishing of science and technology results. *The Scientist*, **14** (9), 6–6 (1 May).

Kostoff, R. N. (2001). Advanced technology development peer review—A case study. *R&D Management*, **31** (3), 287–298 (July).

Kostoff, R. N., Braun, T., Schubert, A., Toothman, D. R. & Humenik, J. (2000a). Fullerene roadmaps using bibliometrics and database tomography. *Journal of Chemical Information and Computer Science*, **40** (1), 19–39 (January–February).

Kostoff, R. N., Green, K. A., Toothman, D. R. & Humenik, J. (2000b). Database tomography applied to an aircraft science and technology investment strategy. *Journal of Aircraft*, **37** (4, July–August), 727–730. Also, see Kostoff, R. N., Green, K. A., Toothman, D. R. & Humenik, J. A. *Database tomography applied to an aircraft science and technology investment strategy*. TR NAWCAD PAX/RTR-2000/84, Naval Air Warfare Center. Aircraft Division. Patuxent River, MD.

Mansfield, E. (1980). Basic research and productivity increase in manufacturing. *The American Economic Review*, **70** (5), 863–873 (December).

Mansfield, E. (1991). Academic research and industrial innovation. *Research Policy*, **20** (1), 1–12.

Metzger, N. & Zare, R. N. (1999). Interdisciplinary research: from belief to reality. *Science*, **283**, 642–643 (29 January).

Naiman, R. J. (1999). A perspective on interdisciplinary science. *Ecosystems*, **2**, 292–295.

Patton, J. D. (2002). The role of problem pioneers in creative innovation. *Creativity Research Journal*, **14** (1), 111–126.

Siau, K. (1996). Electronic creativity techniques for organizational innovation. *Journal Of Creative Behavior*, **30** (4), 283–293.

Smalheiser, N. R. & Swanson, D. R. (1994). Assessing a gap in the biomedical literature—magnesium—deficiency and neurologic disease, *Neuroscience Research Communications*, **15** (1), 1–9.

Smalheiser, N. R. & Swanson, D. R. (1998a). Calcium-independent phospholipase a (2) and schizophrenia. *Archives General Psychiatry*, **55** (8), 752–753.

Smalheiser, N. R. & Swanson, D. R. (1998b). Using ARROWSMITH: a computer assisted approach to formulating and assessing scientific hypotheses. *Computational Methods and Progress in Biology*, **57** (3), 149–153.

Spender, J. C. & Grant, R. M. (Eds) (1996). Knowledge and the firm. *Strategic Management Journal*, **17** (5–9, Special Issue).

Swanson, D. R. (1986). Fish oil, Raynauds syndrome, and undiscovered public knowledge. *Perspectives in Bioogy and Medicine*, **30** (1), 7–18.

Swanson, D. R. (1999). Computer-assisted search for novel implicit connections in text databases. *Abstracts of Papers of the American Chemical Society*, **217**.

Swanson, D. R. & Smalheiser, N. R. (1997). An interactive system for finding complementary literatures: a stimulus to scientific discovery. *Artificial Intelligence*, **91** (2), 183–203.

Taggar, S. (2001). Group composition, creative synergy, and group performance. *Journal Of Creative Behavior*, **35** (4), 261–286.

Terleckyj, N. (1977). *State of science and research: some new indicators*. Boulder, CO: Westview Press.

Terleckyj, N. (1985) Measuring economic effects of federal R&D expenditures: recent history with special emphasis on federal R&D performed in industry. Presented at NAS Workshop on 'The Federal Role in Research and Development'. November.

Van de Klundert, T. C. M. J. & Palm, F. C. (Eds) (1998). Market dynamics and innovation. *Economist*, **146** (3), 387–390 (October, Special issue).

Weeber, M., Klein, H., de Jong-van den Berg, L. T. W. & Vos, R. (2001). Using concepts in literature-based discovery: Simulating Swanson's Raynaud-fish oil and migraine-magnesium discoveries. *Journal of the American Society for Information Science and Technology*, **52** (7), 548–557 (May).

Wenger, W. (1999). *Discovering the obvious: techniques of original, inspired scientific discovery, technical invention, and innovation*. Gaithersburg, MD: Psychegenics Press.

Appendix 1—Literature Approach

Overview

The theoretical basis of the literature approach mirrors the scientific process in many ways. Information from diverse literatures, with relevant interfaces, is examined. All information is first analyzed and then synthesized to produce discovery and innovation. Initial work (Gordon, 1996; Swanson, 1986) examined three variable classes or themes (c, b, a) in two literature categories (C and B) using two different approaches (start with 'c', determine 'b', then determine 'a'; start with 'c' and 'a', then determine 'b').

The principal thematic variables determine a thematic literature. From the previous example, if Raynaud's disease is the thematic variable specified initially, then the corresponding thematic literature might be all the papers in a given database that contain the phrase Raynaud's disease. The remaining thematic variables and literatures are determined by applying different algorithms to the initial thematic literature and subsequent derived literatures. Again, from the previous example, an algorithm would be applied to the Raynaud's disease thematic literature to determine the thematic variable blood viscosity, and a derived literature could then be determined as all the papers in a given database that contain the phrase 'blood viscosity'.

The first approach in the initial reported work (Gordon, 1996; Swanson, 1986) could be viewed as addressing the question: What variables 'a' could influence variable 'c' through mechanisms 'b', or, in the example described above, What treatment factors 'a' could influence Raynaud's disease 'c' through the different mechanisms 'b'? This approach started with thematic variable 'c' (e.g. Raynaud's disease), and used this variable to develop thematic literature C. Algorithms were applied to this thematic literature database to identify thematic variable 'b' values (b1, b2, etc., representing characteristics such as blood viscosity, blood flow, blood platelets, poor circulation, and others) closely linked to thematic variable 'c'. Each value or theme of variable 'b' (b1, b2, etc.) was used to develop a thematic literature B1, B2, etc. Algorithms were applied to each of the thematic B literatures to identify thematic variable 'a' values (a1, a2, etc. representing characteristics such as fish oil, eicosapentaenoic acid, and others) closely linked to the specific thematic variable 'b' of each thematic B literature. Values of the thematic 'a' variables in each of the thematic B literatures not found in thematic literature C defined a subset of the thematic B

literatures that was disjoint from thematic literature C (e.g. the term 'fish oil' was not found in the Raynaud's disease literature). These disjoint thematic 'a' variables and their associated thematic B literature subsets became candidates for discovery and innovation.

The other approach reported could be viewed as addressing the question: What are the mechanisms 'b' through which variable 'a' could impact variable 'c'? This approach started with variables 'c' and 'a', and their associated literatures C and A, and identified variables 'b' that were linked to both variables 'c' and 'a'. The same types of algorithms as in the first approach were used to identify closely linked variables, and the requirement for disjointness between literatures C and A was used as a basis for discovery.

From the experience of these two approaches, it becomes clear that the independent and dependent variables chosen, and the algorithmic approach selected, depend on the question being asked. Further examination shows that other approaches beyond these two are possible to answer other questions. The present chapter examines seven approaches to generate innovation and discovery that are structured to answer seven different questions, and shows how the algorithms and techniques developed in Database Tomography are used in these approaches.

Specific Approaches

The following discussion will be limited to scenarios of three variables 'a', 'b', 'c', and two literatures. In future studies, more complex cases could be candidates for analysis and experimentation.

For the simple two literature/three variable case, seven separate generic cases are possible, where the variables specified can be viewed as 'independent' and the variables determined can be viewed as 'dependent':

(1) specify 'a', determine 'b' and 'c';
(2) specify 'c', determine 'a' and 'b';
(3) specify 'b', determine 'a' and 'c';
(4) specify 'a' and 'c', determine 'b';
(5) specify 'a' and 'b', determine 'c';
(6) specify 'b' and 'c', determine 'a';
(7) specify 'a' and 'b' and 'c', validate linkage existence.

Cases (1), (2), and (3) are the most open-ended and least constrained. In each case, one variable is specified, and the other two are determined using the DT algorithms, the condition of disjointness and, most importantly, expert judgement. Cases (4), (5), and (6) are more constrained, since two variables are specified, and the third is determined using similar processes to the above. Case (7) is fully constrained, and its purpose is to ascertain literature support for validation of a hypothetical relation between specified values of the three variables. Cases (4) and (5) are subsets of case (1); cases (4) and (6) are subsets of case (2); cases (5) and (6) are subsets of case (3); case (7) is a subset of cases (1) through (6). The solution mechanics for each of these seven cases will now be outlined.

Opportunity Driven

This first case addresses the question: What are the potential variable 'c' impacts that could result from variable 'a', and what are the variable 'b' mechanisms through which these impacts occur? One specific variant of this question is of particular interest and importance to the science and technology community: What are the potential impacts on research, development, systems, and operations that could result from research on a given topic?

If the generic question of this first case is applied to the above example for the case where variable 'a' is 'fish oil' only, it could be phrased as: What are the potential impacts or benefits (positive or negative) resulting from fish oil that would not be obvious from examining the fish oil literature alone? This is an open-ended question, and places no restrictions on the mechanisms 'b' or the types of impact 'c'. The first case is represented schematically as:

$$a \rightarrow b \rightarrow c.$$

Here, 'a' is the independent variable, and 'b' and 'c' are the dependent variables that result from the solution process. The operational sequence is to start with the variable 'a' and generate a literature A. Again following the above example and using the abbreviations FO (fish oil), BV (blood viscosity), and RD (Raynaud's disease), this means that the process would start by identifying the FO literature (call this A1). Many approaches could be used to define this literature; the approach recommended here is the one used in recent Database Tomography studies (Kostoff, 2000a, 2000b) for defining literatures. As an example of one literature definition approach, the iterative Simulated Nucleation method (Kostoff, 1997b) would be used to identify all the papers in the Science Citation Index which contained FO (and other related terms in the query) in the title, keywords, and abstract fields. This collection of papers would constitute the FO literature.

The next step in the process is to identify the variables 'b' (b1, b2, . . .) linked closely to variable 'a1', and then identify the literatures B associated with variable 'b' (B1, B2, . . . the BV literatures). For this step, the proximity analysis method used in the recent Database Tomography studies (or other co-occurrence techniques) would be employed. For a journal-based database, this method conceptually identifies phrases in paper titles or abstracts or main texts physically located near the term of interest. As an example, if the term of interest in a given database is Raynaud's disease, then the proximity analysis method would provide a list of all phrases in close physical proximity to the term Raynaud's disease for all occurrences of this term in the text. The proximity analysis approach of Database Tomography is based on the experimental findings that phrases within a semantic boundary (same sentence, paragraph, etc.) located physically close to the term of interest are contextually and conceptually close to the term of interest. Continuing the above example, this step uses the proximity analysis of Database Tomography to identify phrases in the FO literature physically close to the term FO, such as 'b1', 'b2', etc.

For each of these identified phrases 'b1', 'b2', etc., a literature (B1, B2, . . .) is established by querying the SCI. The next step is, for each of these B literatures, to identify the linked variables 'c' (c1, c2, . . .) The process used to identify the variables 'b1', 'b2', etc. linked to variable 'a1' is repeated to obtain the variables 'c1', 'c2', etc. linked to each value of variable 'b'. The subsets of the B literatures which are disjoint from literature A1 (e.g. the B literatures which do not contain the term FO) must then be identified, and the variables 'c' (and their associated linking mechanisms 'b' to variable 'a1') within these disjoint B literature subsets then become the candidates for discovery and innovation.

It is obvious that the process can easily mushroom out of control unless stringent limiting constraints are placed on the number of B literatures and 'c' variables selected. For example, suppose that three 'b' variables 'b1', 'b2', 'b3' (and their associated three B literatures (B1, B2, B3) are identified as closely linked to FO. Suppose also that each of these three 'b' variables is closely linked to five 'c' variables. Then four literature searches are required (A1, B1, B2, B3), and 15 abc linked pathways must be examined for disjointness and discovery, according to the following:

$$a1 \rightarrow b1 \rightarrow c11; \ a1 \rightarrow b1 \rightarrow c12; \ a1 \rightarrow b1 \rightarrow c13;$$
$$a1 \rightarrow b1 \rightarrow c14; \ a1 \rightarrow b1 \rightarrow c15;$$
$$a1 \rightarrow b2 \rightarrow c21; \ a1 \rightarrow b2 \rightarrow c22; \ a1 \rightarrow b2 \rightarrow c23;$$
$$a1 \rightarrow b2 \rightarrow c24; \ a1 \rightarrow b2 \rightarrow c25;$$
$$a1 \rightarrow b3 \rightarrow c31; \ a1 \rightarrow b3 \rightarrow c32; \ a1 \rightarrow b3 \rightarrow c33;$$
$$a1 \rightarrow b3 \rightarrow c34; \ a1 \rightarrow b3 \rightarrow c35$$

In reality, there will be hundreds, if not thousands, of candidate 'b' and 'c' variables. However, there are different ways by which the 'b' and 'c' variables can be sharply limited in number. First, the analysts performing the study would eliminate all non-technical content phrases that passed through the trivial word filter in the Database Tomography algorithm. Second, the numerical indices for each phrase generated by the Database Tomography proximity algorithm would be used as one figure of merit for pre-selection of key phrases. Third, those 'c' variables that reappear in different abc pathways would have a higher priority for selection. Fourth, analyst judgement would be applied to weight the potential value of the different abc pathways in computing figures of merit.

The literature searches and proximity analyses are fairly straightforward, and have been refined in the Database Tomography process. The main intellectual efforts must be focused on prioritizing and reducing the number of linked variables or literatures to be examined, and interpreting the relationships among the final disjoint literatures to generate potential discovery relationships.

Requirements Driven

This second case addresses the question: What are the variables 'a' that could impact variable 'c', and what are the variable 'b' mechanisms by which these impacts are produced? Applied to the above example for the case where 'c' is Raynaud's disease only, it could be phrased as: What are the factors and their associated mechanisms that could impact the course of Raynaud's disease that would not be obvious from examining the Raynaud's disease literature alone? This second case is represented schematically as:

$$a \leftarrow b \leftarrow c$$

Here, 'c' is the independent variable, and 'b' and 'a' become the dependent variables. The operational sequence is to start with variable 'c', and generate a literature C. Again following the above example, this means that the process would start by identifying the RD literature (call this C1). The same literature definition process as in the first case would be used. The next step would be to identify the linked variables 'b' (b1, b2, etc.) to variable 'c1', and then their associated literatures B (B1, B2, the BV literatures). For this step, the proximity analysis method used in the recent DT studies would be employed again as in the first case. Continuing the above example, this step uses the proximity analysis of DT to identify phrases in the RD literature physically close to the term RD, such as 'b1', 'b2', etc.

For each of these identified phrases b1, b2, etc. a literature (B1, B2, etc.) is established by querying the SCI. The next step is, for each of these B literatures, to identify the variables 'a' (a1, a2, etc.) linked to variable 'b'. The process used to identify the variables 'b1', 'b2', etc. linked to variable 'c1' is repeated to obtain the variables 'a1', 'a2', etc. linked to each value of variable 'b'. The subsets of the B literatures that are disjoint from literature C1 (e.g. the B literatures which do not contain the term RD) must then be identified, and the variables 'a' within these disjoint B literature subsets (and their associated linking mechanisms 'b' to variable 'c1') then become candidates for discovery and subsequent innovation. The same stringent limits on variables and literatures used in the first case are applicable here.

Mechanism Driven

The third case addresses the question: For a given mechanism 'b', what are the variables 'a' that could impact the variables 'c'? Applied to the above example

for the case where 'b' is blood viscosity, it could be phrased as: What combinations of variables that could effect a change in the blood viscosity mechanism and could be impacted by a change in the blood viscosity mechanism are candidates for discovery that were not obvious from examining only the blood viscosity literature? The third case is represented schematically as:

$$a \leftarrow b \rightarrow c$$

Here, 'b' is the independent variable, and 'a' and 'c' are dependent variables. The operational sequence starts with variable 'b' and generates a literature B. Again following the above example, this means that the process would start by identifying and generating the BV literature (call this B1). The same literature definition and generation process as in the first case would be used. The next step would be to identify the variables 'a' (a1, a2, etc.) and 'c' (c1, c2, etc.) linked to variable 'b1', and then their associated literatures A (A1, A2, the FO literatures) and C (C1, C2, the RD literatures). For this step, the proximity analysis method used in the first two cases would be employed for the BV literature (B1). Continuing the above example, this step uses the proximity analysis of DT to identify phrases in the BV literature physically close to the term BV, such as 'a1', 'a2', etc. (FO literature) and 'c1', 'c2', etc. (RD literature). However, an arbitrary step is required at this point, since the proximity analysis only provides the aggregate of the linked variables 'a' and 'c'. The analyst is required to divide the aggregate linked variables obtained from the proximity analysis into two groups: 'a' variables and 'c' variables. In the above example, the proximity analysis would generate the linked variables such as fish oil and Raynaud's disease. The analyst would be required to specify two categorizations for these variables, such as 'dietary factors' for the 'a' variables and 'diseases' for the 'c' variables. This step will depend heavily on the analyst's expertise in the technical area and ability to create taxonomies.

The next step is to identify/generate A and C literatures using the approach described above. The final step is to identify the subsets of A literatures and C literatures that are disjoint. Each group of articles from the A literature and the C literature that contains a 'b1' variable is considered to be a linked group. The subsets of these literatures that are linked through the common 'b1' variable and that are disjoint (i.e. the C literature does not contain the 'a' variable and the A literature does not contain the 'c' variable) must then be identified. The variables 'a' and 'c' within these disjoint A and C literature subsets linked through the 'b1' variable then become the candidates for discovery and subsequent innovation. The same stringent limits on variables and literatures used in the first approach are applicable here.

Opportunity-Requirements Driven

This fourth case addresses the question: What are the mechanisms 'b' through which variable 'a' could impact variable 'c'? Applied to the above example for the case where 'c' is Raynaud's disease only, and 'a' is fish oil only, it could be phrased as: What are the mechanisms through which fish oil could impact Raynaud's disease that would not be obvious from examining only the Raynaud's disease literature or the fish oil literature? The fourth case is represented schematically as:

$$a \rightarrow b \leftarrow c$$

Here, variables 'a' and 'c' are independent, and variable 'b' is the dependent variable. The operational sequence is to start with the variable 'c', and generate a literature C, and with variable 'a', and generate a literature A. Again following the above example, this means that the process would start by generating the RD literature (call this C1) and the FO literature (call this A1). The same literature definition and generation process as in the first case would be used. The next step would be to identify the linked variables 'b', and then their associated literatures B for both the A1 literature and the C1 literature. For this step, the proximity analysis method used in the first two approaches would be employed, for the FO literature (A1) and the RD literature (C1). Continuing the above example, this step uses the proximity analysis of DT to identify phrases in the RD literature physically close to the term RD, such as 'b1', 'b2', etc. and to identify phrases in the FO literature physically close to the term FO, such as b51, b52, etc. The next step is to identify the subsets of the A1 literature and C1 literature that are linked. Each group of articles from the A1 literature and the C1 literature that contains a 'b' variable is considered to be a linked group. The subsets of these literatures linked through the common 'b' variables that are disjoint (i.e. the C1 sub-literature that does not contain the 'a1' variable and the A1 sub-literature that does not contain the 'c1' variable) must then be identified, and the variables 'b' within these disjoint A1 and C1 literature subsets then become the candidates for discovery and subsequent innovation. The same stringent limits on variables and literatures used in the first case are applicable here.

Opportunity-Mechanism Driven

The fifth case addresses the question: What are the variables 'c' which could be impacted by variable 'a' through mechanism(s) 'b'? While the schematic shown for this case is identical to that of case 1, the two schematics should be interpreted differently. In case 1, the intermediate mechanism(s) 'b' are not specified beforehand, but are a result of the solution process. In the present case, these 'b' mechanism(s) are specified beforehand. Applied to the above example for the case where 'b' is blood viscosity only, and 'a' is fish oil

only, the question in this case could be phrased as: What abnormalities could be influenced from the impact of fish oil on blood viscosity that would not be obvious from examining only the abnormality's literature or the fish oil literature? The fifth case is represented schematically as:

$$a \rightarrow b \rightarrow c$$

Here, 'a' and 'b' are the independent variables, and 'c' is the dependent variable. The operational sequence is to start with the variable 'a', and generate a literature A, and with variable 'b', generate a literature B. Again following the above example, this means that the process would start by generating the FO literature (A1) and the BV literature (B1). The same literature definition and generation process as in the first case would be used. The next step would be to identify the linked variables 'c', and then their associated literatures C (the collection of RD literatures) for the B1 literature. For this step, the proximity analysis method used in the previous cases would be employed for the B1 literature only. Continuing as before, this step uses the proximity analysis of DT to identify phrases in the BV literature physically close to the term BV, such as 'c1', 'c2', etc. The resulting C literatures are automatically linked to the A1 literature through the linking variable 'b1'. The 'c' variables which are disjoint to the A1 literature (i.e. the C sub-literature that does not contain the 'a1' variable and the A1 literature does not contain the 'c' variables) must be identified, and become the candidates for discovery and subsequent innovation. The same stringent limits on variables and literatures used in the first case are applicable here.

Requirements-Mechanism Driven

The sixth case addresses the question: What are the variables 'a' that could impact variable 'c' through mechanism 'b'? Applied to the above example for the case where 'b' is blood viscosity only, and 'a' is fish oil only, it could be phrased as: What factors could impact Raynaud's disease by impacting blood viscosity that would not be obvious from examining only the factors' literature or the Raynaud's disease literature? The sixth approach is represented schematically as:

$$a \leftarrow b \leftarrow c$$

Here, 'b' and 'c' are the independent variables, and 'a' is the dependent variable. The operational sequence is to start with the variable 'c', and generate a literature C, and with variable 'b', and generate a literature B. Again, this means that the process would start by identifying and generating the RD literature (C1) and the BV literature (B1). The same literature definition and generation process as in the first case would be used. The next step would be to identify the linked row of variables 'a' (a1, a2, etc.), and then their associated literatures A (the FO literatures) for the B1 literature. For this step, the proximity analysis method used in the previous cases would be employed, for the B1 literature only. Continuing as before, this step uses the proximity analysis of DT to identify phrases in the BV literature physically close to the term BV, such as 'a1', 'a2', etc. The resulting A literatures are automatically linked to the C1 literature through the linking variable 'b1'. The 'a' variables which are disjoint to the C1 literature (i.e. the A sub-literature does not contain the 'c1' variable and the C1 literature does not contain the 'a' variables) must be identified, and become the candidates for discovery and subsequent innovation. The same stringent limits on variables and literatures used in the first case are applicable here.

Opportunity-Mechanism-Requirements Validation

The seventh case addresses the question: Does the literature support the possibility that variable 'a' could impact variable 'c' through mechanism 'b'? Applied to the above example for the case where 'a' is fish oil only, 'b' is blood viscosity only, and 'c' is Raynaud's disease only, it could be phrased as: Does the literature support the possibility that fish oil could impact Raynaud's Disease by altering blood viscosity in a way that would not be obvious from examining only the fish oil literature or the Raynaud's disease literature? The seventh approach is represented schematically as:

$$a \leftrightarrow b \leftrightarrow c$$

Here, 'a' and 'b' and 'c' are independent variables. The operational sequence could start with either 'a' or 'b' or 'c'. For the present discussion, the operational sequence starts with the variable 'b', and generates literature B. Again following the above example, this means that the process would start by identifying and generating the BV literature (B1). The same literature generation process as in the first approach would be used. The next step would be to extract the B1 sub-literatures which contain the variables 'a1' (literature A1) and 'c1' (literature C1).

The final step is to validate the existence of disjoint A1 and C1 sub-literatures (i.e. A1 sub-literature that does not contain the 'c1' variable and a C1 literature that does not contain the 'a1' variable). The 'a1'–'b1'–'c1' sequence then becomes a candidate for discovery and subsequent innovation. The same stringent limits on variables and literatures used in the first approach are applicable here.

Appendix 2: Crossing the Bridge: Interdisciplinary Workshops for Innovation

Background

The Office of Naval Research established a series of workshops in 1997 aimed at promoting innovation while also enhancing organization, category, and discipline diversity components. The focus of the first novel workshop founded on this plan was 'Autonomous Flying Systems', an area of perceived long-term interest not only to the Navy and Department of Defense, but also to the National Aeronautics and Space Administration and other governmental and industrial organizations. The process employed was designed, starting with a clean slate, and was intended for application to very significant technical challenges. The present appendix further describes the process that was used to identify the technical theme of the workshop, select the participants, and conduct all three phases of the total workshop.

Workshop Theme Identification

It was decided that the initial workshop theme should: (1) focus on problems related to the main science and technology emphasis area of the author's home organization, Strike Technology; and (2) help establish the most supportive environment for innovation. The problem selected should be focused and understandable, and it should have a generic technical base amenable to soliciting people from many different disciplines. The topic finally selected was autonomous control of unmanned air vehicles, including takeoff and landing from limited areas on smaller Navy ships. It was apparent that the underlying science and technology permeated many different disciplines, including aerodynamics, controls, structures, communications, guidance, navigation, propulsion, sensing, and systems integration. Also, the naval applications for some aspects of this problem were sufficiently unique that probably not a great deal of work had been done in this area. Subsequent literature analyses validated this assumption.

Present naval air systems are either manned (most aircraft) or tele-operated, semi-autonomous (weapons and some aircraft). The weapons are a mix ranging from 'dumb' bombs and shells to 'smart' missiles. The future trend is toward 'smart' autonomous or semi-autonomous aircraft and weapons. Since a major role of the Office of Naval Research is to proactively address the technology that will influence future naval forces, it seemed natural to examine science and technology roadblocks on the path to unmanned autonomous 'smart' flight systems. Consequently, the focus of the initial workshop was defined as identification of the fundamental operational principles of autonomous flying systems over a fairly wide range of flight environments. In particular, the workshop was aimed at examining what had been learned about autonomous or semiautonomous operation from the animal (mainly flying) kingdom and from other unmanned autonomous/semiautonomous tele-operated systems such as autonomous underwater vehicles and locomoted robots. Animals are now being studied as integrated systems by scientists on the forefront of biological research. The issues of aerodynamics, flight mechanics, dynamic reconfiguration, materials, control, neuro-sciences, and locomotion are not being studied as separate disciplines by these scientists, but rather are being studied in parallel in the same animal system and in their relation to the function and mission of the animal system. While this integrative biological research is in its infancy, and results are only starting to emerge, the time seemed appropriate for assembling these diverse groups and exploiting their synergy. Not only could there be benefit to the Navy from such cross-discipline interaction, but benefit could be possible for each of the contributing disciplines as well.

A major thrust of the workshop was projected to be identification of the autonomous operational principles for each unique system and the relation of these principles to mission and function, then extraction of the generic operational principles that underlay all the systems, both biological and man-made. It was hoped that the cross-fertilization of disciplines would be able to further elucidate and clarify the more important generic concepts, and then provide insight that could be utilized to enhance the autonomous operation of naval flying systems.

Participant Selection

Once the theme of the workshop was established, a sub-theme taxonomy was developed to focus the agenda and to identify workshop participants. A dual approach was followed to generate the taxonomy.

Discussions were held with agency experts on the generic theme concerning the taxonomy structure. In parallel, the Science Citation Index was queried for papers related to the generic theme. Both bibliometric and computational linguistics analyses of these papers were performed to provide strategic maps of the topical area, identifying key performers, journals, institutions, and their relations to the technical themes and sub-themes of the workshop. A taxonomy was constructed based on these strategic maps. (For a description of how the bibliometric and computational analyses are combined to generate strategic maps, see Kostoff (1998, 1999)).

Both of these taxonomy sources, in-house experts and the Science Citation Index, then provided initial candidates for participation in the workshop. These candidates were contacted, and asked to suggest additional candidates. This procedure continued until a large pool of potential candidates was established. Three main selection criteria for workshop participants were established:

(1) Multiple recommendations;
(2) Significant publications is the field; and
(3) Literature citations.

These three criteria were tempered with judgment to insure that bright young individuals, who had not yet established a track record, were not excluded from the pool, and that the panel as a whole had the correct level of discipline, category, and organization balance. In addition, a guideline was established that all workshop attendees would be active participants, so the number of attendees was limited to facilitate discussion and interactions.

All these constraints, guidelines, and selection criteria were used to arrive at the final panel size and structure. The result was a panel of slightly more than 20 people representing a mix of disciplines that included biologists (experts in bird, bat, frog, fish, or insect studies), robotics, artificial intelligence, controls, autonomous aircraft, fluid dynamics, sensors, neuroscience, cognitive science, autonomous underwater vehicles, aerodynamics, propulsion, and avionics.

Overview of Workshop Process Steps

Workshop Buildup

The buildup period for the workshop in question started about two months before the meeting. Specific guidance for the conduct of the workshop was sent to the participants by e-mail, including a statement of the naval technical problems to be addressed. The technical component of the buildup phase was then conducted by e-mail.

The main purpose of this buildup phase technical component was to have each participant generate new ideas from his/her discipline for all other participants to consider. The other participants could then dialogue by e-mail to clarify/modify/embellish these ideas. At a minimum, even if no dialogue resulted, there would be a gestation period of about two months for each participant to absorb these concepts from other disciplines. Specifically, each participant was requested to:

- Submit half a dozen leading-edge capabilities or accomplishments in his/her discipline(s) that could potentially impact the naval technical problems;
- Identify several leading-edge capabilities or accomplishments projected in his/her discipline(s) over the next decade that could potentially influence the naval technical problems; and
- Submit a few leading-edge capabilities or accomplishments in his/her discipline(s) whose impact on the naval technical problems was not obvious to him/her, but might be obvious to someone else.

The participants were free to comment on potential relations among any of the capabilities, accomplishments, or combinations of capabilities and accomplishments, and any of the naval technical problems, or combinations of problems. All of the comments received were then sent to all the participants. This exercise helped stimulate the thinking of the participants, and provided a documented record of the process. One of the functions of the participants from the author's organization was to facilitate and stimulate dialogue by raising questions and issues on the submitted information.

If any of the participants saw a capability or accomplishment from another participant that could impact a problem in his/her discipline, but not impact a naval technical problem, then the two participants were free to dialogue together without informing all the participants. However, these two participants engaged in independent dialogue were requested to keep a record of their exchange for possible inclusion in the final workshop report as potential discovery. This would cover the real possibility of discovery occurring in topics other than the one targeted.

Workshop Meeting

As a result of the ideas presented during the buildup phase, it appeared that the seeds existed for a new science and technology program on Autonomous Flying Systems. Therefore, an agenda was sent to the participants with further guidance to address promising science and technology opportunities at the workshop, that would serve as the foundation of such a program. Specifically, the participants were asked to address the following issues at the workshop:

• What are the present leading-edge capabilities in your discipline?
• What are the desired future capabilities in your discipline?
• What are the leading research opportunities in your discipline and what additional capabilities could they provide if successful?
• What is the level of risk of these opportunities successfully achieving their targets?
• How would these potentially enhanced capabilities contribute to, or translate into, improved understanding and/or operation of autonomous flying systems?

The meeting occurred on 10–11 December 1997 at ONR. Since some of the leading-edge capabilities and potential accomplishments appeared to have applicability to naval technical problems (identified during the e-mail buildup period), the proponent for the capability or accomplishment item took the lead in fleshing out his/her ideas and leading the discussion at the meeting. As a result, the workshop meeting tended to evolve into full panel discussions on each of these potential capabilities.

There were two rounds of discussion at the workshop. The first round consisted of presentations and discussions by each proponent. The second round of the workshop involved each participant identifying his/her leading promising research opportunities.

Workshop Cleanup

The participants were requested to provide any additional narrative information that added to or modified their ideas as a result of the workshop experience. The outcomes of the workshop included both the tangible and intangible.

Three immediate tangible outcomes were projected:

(1) A concept proposal for a science and technology program focused on Autonomous Flying Systems would be generated;
(2) Technical papers may be submitted to leading science journals based on innovations identified; and
(3) One or more papers on the complete workshop experience might be submitted to leading science journals.

In addition to developing specific topics, it was anticipated that new, un-exploited ideas in interdisciplinary research and development might surface during contact between panelists. These novel subjects might form the basis of additional workshops. In addition, extensive lessons were learned as a result of the workshop process. These lessons were summarized in Section 1b.

The International Handbook on Innovation
Edited by Larisa V. Shavinina

Developing Innovative Ideas Through High Intellectual and Creative Educational Multimedia Technologies

Larisa V. Shavinina[1] and Evgueni A. Ponomarev[2]

[1] *Département des Sciences Administratives, Université du Québec en Outaouais, Canada*
[2] *Department of Physics, University of Ottawa, Canada*

Abstract: This chapter presents high intellectual and creative educational multimedia technologies (HICEMTs) as one of the possible methods to facilitate the development of innovative ideas in people. HICEMTs will constitute one of the innovative breakthroughs in science and technology of the 21st century and will lead to a new wave of innovations in psychology, education, and organizational learning. HICEMTs appear at the intersection of many subdisciplines of psychology, including general, cognitive, developmental, educational, personality, media, cyber, and applied psychology, education, and multimedia. The general and specific nature of HICEMTs is considered. The importance of HICEMTs is discussed from a technological, economic, societal, educational, and psychological perspective.

Keywords: Innovation; Educational multimedia; Technology; HICEMTs; Innovative ideas.

Introduction

Innovation begins with new ideas. Creative ideas are the foundation on which innovation is built (Amabile, 1988, 1996). Reis & Renzulli (this volume), Clapham (this volume), Kostoff (this volume), R. Root-Bernstein & M. Root-Bernstein (this volume), Smith (this volume) discuss a wide range of methods for enhancing an individual's ability to generate new ideas. Our chapter explores the role of contemporary educational multimedia technologies, especially high intellectual and creative educational multimedia technologies (HICEMTs), in the development of human abilities to produce new ideas.

Psychoeducational multimedia technologies (PMTs) and HICEMTs were first introduced in 1997 (Shavinina, 1997a). PMTs are multimedia technologies that base their five-part educational essence (discussed later in this chapter) on fundamental psychological processes and phenomena (Shavinina, 1998a). HICEMTs constitute a special type of PMT whose general content is elaborated in accordance with underlying psychological mechanisms and states and whose special content is developed, structured, presented, and delivered according to the key principles of an individual's intellectual functioning and creative performance (Shavinina, 2000). The term *high* in HICEMTs refers to a significant saturation of the special content of these technologies through educational materials directed toward the actualization and development of human intellectual potential and creative abilities.

HICEMTs emerge at the crossroads of many subfields of psychology (e.g. general, cognitive, developmental, educational, personality, media, cyber, and applied), education, and multimedia technology. In describing HICEMTs, this article proceeds as follows. First, it seems expedient to discuss briefly the impact of information technology on human beings that sets in motion the emergence of HICEMTs. Second, today's educational multimedia applications[1] available on the global educational multimedia market are analyzed from a psychological viewpoint to demonstrate how they contribute to the appearance of HICEMTs. Third, the importance of HICEMTs for contemporary people is described. Finally, the general and specific nature of HICEMTs is considered.

Psychological View of an Information-Based Society

Extraordinarily rapid information development is a key feature of the contemporary era. Extremely fast growth of information technology has an enormous impact on people, resulting in significant changes in everyday life and determining its qualitatively new level. In the professional, educational, public, and home settings, we are faced with the strong presence of new information and communication technologies. No one is surprised by individuals' daily use of the Internet or compact discs (CDs). Cyberspace is becoming an integral part of individuals' inner, psychological world.

---

[1] Educational multimedia applications refer to on-line (e.g. Internet) and off-line (e.g. CD-ROMs) multimedia products and services devoted to learning, teaching, and training. In turn, 'multimedia' refers to a new generation of communication tools that can draw on a full range of audio-visual resources, ranging from text and data to sound and pictures, and that store and process all of these diverse data in a single integrated delivery system. A general multimedia application thus delivers an integrated presentation that combines at least three of the following: (a) text (including notes, captions, subtitles, and resources such as tables of contents, indexes, dictionaries, and help facilities); (b) data (such as tables, charts, graphs, spreadsheets, statistics, and raw data of various kinds); (c) audio (including speech, music, atmospheric background noise, and sound effects); (d) graphics (often ranging from traditional media such as drawings, prints, maps and posters to images processed or created entirely within a computer); (e) photographic images from negatives, slides, prints, or even digital cameras (which record photographic images directly as computer graphics); (f) animation (whether recorded on film or video or created with a computer); and (g) moving pictures (specifically, digital video either converted from analogue film and video or created entirely within a computer).

Educational multimedia applications are almost completely unexplored from a psychological point of view, because multimedia tools in their modern forms (i.e. CD-ROMs and Internet-based programs) have appeared only during the last 5–6 years. It should be emphasized that this chapter deals only with multimedia, not computer-assisted instruction (CAI) or computer-based training (CBT). Much has been written concerning CAI and CBT over the past few years, provoked by the active entry of computers into educational and professional settings. Today, it is multimedia technology that is attracting attention, and it holds great promise in education in general and in organizational learning in particular. Of course, CAI–CBT and contemporary educational multimedia have some common properties (Honour & Evans, 1997).

Society's saturation by an unparalleled quantity of hardware, software, and applications has reached its threshold level, leading people to realize that we are actually living in an information society. Today's society means cybersociety. Everyday news about tremendous discoveries and amazing inventions in the area of information and communication technologies leaves no doubt concerning a novel reality of our life: We are living in an information era (Shavinina, 1998a).

The present state of information technology and its even more promising future (Lan & Gemmill, 2000) lead to an understanding of the unique challenges it presents. For example, global economic competition is increasingly harsh, and companies must rapidly bring innovative products and services to the global market. To survive and prosper, companies desperately need intellectual and creative employees whose novel ideas are, to a certain extent, a guarantee of companies' existence and success. Consequently, modern society desperately requires highly able citizens who can bring innovative solutions to its current challenges and at the same time produce new ideas for its ongoing advancement. The distinguishing characteristic of the new millennium is, consequently, a need for creative and intellectual people for further social progress (Shavinina, 1997a, 1997b).

Of course, there are courses on creative thinking and intelligence training, and creativity consultants actively travel around the world, teaching strategies for increasing people's abilities to overcome the everyday problems they face in their work and daily life. However, creativity and intelligence training are still available only to a small percentage of people,[2] and current textbooks and instruction manuals do not change this situation. This is especially problematic for

---

[2] Creativity and intelligence training are currently available only to a small percentage of people, mainly for economic reasons. The services of creativity consultants conducting such training are expensive, and, for the most part, only affluent companies can use these consultants. The available textbooks do not change this situation at all, mainly for psychological reasons. They do not provide a sufficient level of intrinsic motivation to follow their content. This is particularly important for children, who often need a teacher or parent to guide and stimulate their learning through textbooks. Only multimedia technology allows people to develop their intellectual and creative abilities while being guided and motivated by means of educational multimedia technology itself. This is possible because of the accessibility of any website practically from any place on Earth and at any time. Distinguishing features of multimedia technology, presenting a combination of text, sound, graphics, and animation, provide a higher level of stimulation than traditional textbooks. Conventional textbooks and learning manuals are low in cost, and this is their advantage. However, as more and more of today's schools, universities, and companies are connecting to the Web, many Internet-based programs are becoming free for students and employees.

children, who often need someone (i.e. parents, teachers, or other caregivers) to monitor and encourage them as they proceed through learning manuals. The appearance of multimedia technology gives people an exceptional opportunity to develop their intellectual and creative potential through being guided and stimulated by the means of educational multimedia technology itself (i.e. through specific multimedia effects and educational contents). A significant problem is how to elaborate the most successful educational multimedia technologies that could productively develop human abilities. This chapter represents one such attempt.

Certainly, the idea of developing an individual's abilities through computer technology is not a new one, having been discussed in the psychological literature (Bowen, Shore & Cartwright, 1993; Olson, 1986; Salomon, Perkins & Globerson, 1991). The novelty resides in the suggestion of developing an individual's intellectual and creative resources through multimedia technology. However, in adopting from educational content alone, multimedia technology, with its multiple effects and advantages, provides almost nothing for the development of human creative and intellectual potential. What is particularly needed is a well-developed variety of educational contents that, together with multimedia, form 'educational multimedia technology' (Shavinina, 1998a).

Furthermore, if today intelligence and creativity guarantee personal and economic success, at least to some extent, then the contemporary educational system across all of its levels should provide productive teaching and learning to increase people's creative and intellectual resources in direct accordance with the requirements of an information society. This is particularly important for children, whose adulthood will take place in a cyber environment entirely different from that of the present. Children must be prepared for this environment of the future. This means that, even today, they should be taught through radically new educational programs that will correspond to the needs of the 21st century information era (Olson, 1986; Salomon et al., 1991).

It is interesting to note that national governments realize this need for more productive teaching and learning models (Shavinina, 1997b). This expresses itself, for instance, in the rising attention of national governments to the development of citizens' intellectual and creative potential. For example, the following demonstrate the concern of the European Union about the future life of Europeans, which is predetermined by the human ability to innovate: (a) a white paper published on teaching and learning in a cognitive society ('Teaching and Learning', 1995); (b) the 'Education in the Information Society' initiative (1996); and (c) a resolution on educational multimedia prepared by the Counsel of the Ministers of Education of the European countries ('Educational Multimedia', 1996). The common view across these documents was

expressed by Olli-Pekka Heinonen, Finland's minister of education: "We want creative citizens, who can take responsibilities and who are able to solve problems that do not yet exist today" (Heinonen, 1996). The next step of the European Union in the face of growing challenges of cybersociety was establishment of its Educational Multimedia Task Force. The aim of the task force is to encourage and stimulate the development of high-quality learning, teaching, and training multimedia resources and to raise awareness of the educational potential of multimedia applications in teaching and learning through financing the most promising projects. Canada, in attempting to increase its citizens' ability to innovate, has established such foundations and initiatives as the Canadian Innovation Foundation (federal initiative) and the Innovators Alliance (Ontario provincial government initiative).

Indeed, it seems that today's educational multimedia technology holds the most promise for the development of human intellectual and creative resources. But multimedia opportunities that will considerably increase people's abilities are not yet well understood. More than 16 years ago, David Olson (1986) underlined that "the potential impact of computers . . . lies, in fact, on coupling the resources of the mind with the resources of the computer" (p. 355). This is generally true, but the idea is far from fully realized. Today multimedia applications—through the simultaneous use of audio, text, multicolor images, graphics, movies, and so on—provide an excellent opportunity for the development of educational technologies going far beyond computer-assisted instruction or computer-based training, existing educational software, and traditional curricula. For example, productive training of creative thinking cannot be formulated and delivered only in verbal–written form; it also requires visual imagery and oral expressions, feedback and user-friendly means, and so on. Such full-scale delivery can be obtained through multimedia technology. With the appearance and advanced development of such technology, people are in a unique position to elaborate and 'place' almost anything into a computer format. As a result, multimedia technology may significantly influence the development of an individual's mind. But the contemporary educational multimedia technologies existing on the global market are far from reaching this goal. An understanding of this issue requires an understanding of the current state of educational multimedia products. However, before proceeding with this topic, we consider existing approaches to instructional technology.

Instructional Technology Innovations

It seems appropriate to distinguish the following approaches in the large and rapidly growing body of literature on instructional technology innovations and their pedagogical outcomes: computer-mediated communication (CMC), virtual classrooms, simulation training, and intelligent tutoring systems. One of the current tendencies in the field is to combine some of these approaches in developing educational multimedia technologies. For example, Oren, Nachmias, Mioduser & Lahav (2000) used both CMC and virtual classrooms in their Learnet model. Scardamalia & Bereiter (1994, 1996) did the same in developing computer-supported intentional learning environments (CSILE) technology (articulated below). In part, such intersections of instructional technology approaches can be explained by the endless novelty introduced by modern information and communication technologies. As a result, another distinctive tendency of the field is a lack of consensus about terms, notions, definitions, and conceptual approaches regarding technology-based education.

CMC, a central characteristic of cyberspace, refers to the Internet, the World Wide Web, local area networks, bulletin board systems, electronic mail, and computer conferencing systems (Allen & Otto, 1996; Hiltz, 1986; Jonassen, 1996; S. Jones, 1995; Koschmann, 1996a, 1996b; McAteer, Tolmie, Duffy & Corbett, 1997; Rapaport, 1991; Romiszowski, 1992; Romiszowski & Mason, 1996). CMC links students and employees with one another, with their teachers and managers, respectively, as well as with experts in their fields of inquiry and with the community at large (Sherry, 2000). Such methods of asynchronous (time-independent) communication as e-mail, listservs, and electronic conferences gather participants in on-line discussions together, transmit content, and provide a forum for group discussions. With regard to education, CMC refers to on-line learning and teaching. The fundamental idea behind on-line learning is that learning is a collaborative activity. There are two schools of psychological thought that relate to CMC education: activity theory and situated learning (Tolmie & Boyle, 2000).

Within activity theory, CMC is considered as a mediating tool in an activity system. Drawing on the investigations of Russian psychologists (mainly Leont'ev, 1981; Tikhomirov, 1981), Engestrom (1996) defined the framework of an activity system as consisting of six interrelated components: (a) the subject (e.g. a student, teacher, or expert who is carrying out an activity); (b) the object of activity (e.g. a product or message posted on the Internet); (c) the mediating tools of the activity (e.g. multimedia tools); (d) the community of learners (students, teachers, and all people who are connected electronically by the network and are concerned with the problems and issues discussed on it); (e) the division of labor (the responsibilities commonly associated with the roles of student, teacher, expert, and so on); and (f) the rules and norms regarding appropriate social actions (posting, moderating a discussion, seeding a conference, and so on). Consequently, when exploring CMC as a way of enhancing teaching and learning, one should

consider its effects on the entire sociocultural system into which it is introduced, including the people who compose it, the tools they use, the products and performances they create, the norms and conventions of tool use, the roles and responsibilities of individual group members, and the meanings they share as a cultural group (Engestrom, 1996).

Understanding of on-line learning as a collaborative activity led to the appearance of computer-supported collaborative learning (CSCL), an emerging paradigm of education that emphasizes a delicate balance between the individual mind and socially shared representations developed through ongoing discourse and joint activities that take place within a learning community (Koschmann, 1996b). CSCL uses CMC in both its synchronous (real-time) and asynchronous forms to develop shared knowledge bases and to promote common understandings.

The philosophical foundations of CSCL are based on situated learning (Lave & Wenger, 1991; Suchman, 1987), communities of learners (Brown, 1994; Brown & Palincsar, 1989), and cognitive apprenticeships (Collins, Brown & Newman, 1989). Within these closely related approaches, scientists are attempting to expand investigations of human cognition and conceptual change beyond the individual mind to include learning that is built through mediated conversations among members of peer groups and learning communities. CSCL shifts the focus of education from learning as acquisition of knowledge and skills to learning as entry, enculturation, and valued activities situated within a community of practice (Sherry, 2000). Used as a mediating tool, CMC enables students, teachers, experts, and other members of a learning community to share distributed representations (Allen & Otto, 1996) and to use distributed cognition (Norman, 1993) to overcome the limitations of the individual human mind. For instance, Bereiter (1994) pointed out that an important feature of on-line discourse is that "understandings are being generated that are new to the local participants and that the participants recognize as superior to their previous understanding" (p. 4). Hence, participants in on-line communication suspend their individual thinking and begin to share collectively, thus creating commonly shared meaning and constructing a shared purpose that leads to expansion of ways of knowing.

An example of successful on-line learning is the National Geographic Society's Kids' Network. In a study of 36 schools that participated in the network, scientists found that participating students outperformed students in usual classrooms in terms of their grasp of certain scientific concepts and outperformed control group students on issues unrelated to their unit of study (Sherry, 2000).

Oren et al.'s (2000) Learnet is another example of on-line learning. Learnet is a virtual community hosted by an appropriate virtual environment and embodying advanced pedagogical ideas. CyberSchool (http://www.cyberschool.4j.lane.edu), Willoway Cyber-school, and Science Learning Network (http://www.sln.org; a consortium of museums that defines itself as an on-line community of educators, students, schools, and science museums) can be considered examples of the Learnet model. Therefore, through using CMC as a mediating tool and: (a) sharing information; (b) fostering multiple viewpoints; (c) suggesting promising strategies; and (d) negotiating shared meaning among students, teachers, and experts, on-line learning enhances and expands the traditional types of teaching and provides new opportunities for learning.

The virtual classroom (laboratory, university, or company) is another innovation in today's instructional technology (Hiltz, 1990; Javid, 2001; Jonassen, 1996; Lan & Gemmill, 2000; Oren et al., 2000; Westera & Sloep, 2001). CMC generates a new learning environment, a virtual learning environment, where communities of learners are involved in collaborative learning. The philosophical, theoretical, and methodological foundations of the virtual classroom approach are essentially the same as in the CMC approach (Bereiter, 1994; Brown, 1994; Collins et al., 1989; Lave & Wenger, 1991), because the successful implementation and use of CMC inevitably lead to the development of virtual learning environments.

The term *virtual learning environment* refers to any educational site on the Internet that includes information, learning activities, or educational assignments or projects (Oren et al., 2000). 'Virtual' is typically used in a general form, such as the possibility of accessing a site from any place at any time, thus eliminating some of the physical constraints of the real world. These websites offer a range of instructional modes, from the retrieval of curricular resources to be integrated in regular classroom activities to complete educational units that include information resources, pedagogical approaches, and technological tools; exist only on the World Wide Web; and serve on-line distance learning (Oren et al., 2000). In other words, it is possible to distinguish two types of virtual learning communities: geographically bounded and interest-bounded ones. The first are based on existing communities (e.g. schools and universities), and they use the capability of on-line technology to support these communities. The educational advantages of on-line technology are connected to the support and enrichment it provides to real classrooms. Edmonds and Kamiak Cyberschools (Javid, 2001), the Maryland Virtual High School of Science and Mathematics (Verona, 2001), and Roosevelt Middle School (http://www.flinet.com/~rms/) are examples of this type of virtual learning community. Interest-bounded learning communities are established through CMC and use of the Web as a 'meeting place'. The communities are totally dependent on the virtual environment in which they exist and function in a

manner similar to distance learning courses (Schuler, 1996). A successful example of this type of virtual learning community is the CSILE Knowledge forum project (considered in detail later).

Another successful example of the implementation of a virtual classroom approach is an innovative educational model called the Virtual Learning Company (Westera & Sloep, 2001). This model was developed and implemented at the Open University of the Netherlands. The Virtual Learning Company is a distributed, virtual learning environment that embodies the functional structures of real companies; it offers students a rich and meaningful context that resembles the context of real work in many respects. Although it makes extensive use of advanced information and communication technologies, the true innovative power of the Virtual Learning Company resides in the underlying, new educational framework. Students in the Virtual Learning Company assume professional roles and run the business; that is, they deliver knowledge-centered products and services to authentic, external customers. The approach differs from regular role-playing games, simulations, and various forms of apprenticeship learning in that its processes do not reflect predefined scenarios and outcomes. In the Virtual Learning Company, students are in charge of a business system that freely interacts with the outside world. The educator's role reflects that of an in-company training coordinator: It is restricted to facilitating, monitoring, and supporting the growth of the 'employees'. The Virtual Learning Company strives to bring together the contexts of education and work. It attempts to offer a concrete and meaningful environment that closely resembles students' future workplaces (Westera & Sloep, 2001).

The Virtual Learning Company is built on two entirely different but strongly interdependent processes. First, in the educational process, novice students are transformed into competent students; second, in the business process, the students act on orders of external customers and turn them into knowledge-centered products and services. This duality is not unique to the Virtual Learning Company; it strongly matches modern ideas on knowledge-centered businesses, the importance of human capital, human-resource management, performance improvement and personnel development, and work organized through virtual business teams. The pedagogical principles underlying the Virtual Learning Company include: (a) competence learning rather than reproduction of codified knowledge; (b) custom-made education rather than boring uniformity; (c) student control instead of teacher control; and (d) closing the gap between learning and working. Various studies demonstrate that virtual learning companies are a powerful and promising tool to meet today's educational needs, anticipating a worldwide shift to cybereducation (Westera & Sloep, 2001).

Computer-based simulations represent a special type of innovation in contemporary instructional technology (Bliss & Ogborn, 1989; Ferrari, Taylor & VanLehn, 1999; Gredler, 1990, 1992, 1996; Harper, Squires & McDougall, 2000; Jonassen, 1993, 1996). A *simulation* can be defined as a "classroom experience, typically using a computer, which gives students information analogous to that obtained by working in a specific workplace, requires them to make decisions similar to those that the workplace demands, and experience the results of those actions" (Ferrari et al., 1999, p. 26). For example, flight simulators have been developed for both the aviation and entertainment industries. Businesses and business schools routinely use simulations that place students in the role of a CEO or other high-level decision-maker. Medical schools often allow students to practice diagnosis and treatment with computer-simulated patients. Ferrari et al.'s (1999) detailed analysis of 39 computer-based simulations of workplaces indicated that simulations fell into three main types: role-playing, strategy, and skill.

Role-playing simulations involve a high-fidelity portrayal of the information and activities that a real jobholder would experience in a typical day on the job. Examples are Parkside Hotel (manager of a hotel), Starr Medical (nurse at a large hospital), SWAT (rookie police officer on a SWAT team), and HBM (human resources manager for a large manufacturing company). Another good example is Court Square Community Bank, in which students are placed in a virtual office and play the role of bank vice-president.

In strategy simulations, the information and actions of the student are more abstract and powerful than those of the real occupation so that the student can implement long-range strategies and see their results. Examples are Sim City 2000 (managing the growth of a city), Transport Tycoon (developing a national transportation business), C.E.O. (managing a large conglomerate), and Entrepreneur (developing a start-up company). Another example is Capitalism. The goal of this simulation is to set up a profitable corporation by outwitting the competition and gaining a greater market share while overcoming a number of realistic obstacles.

In skill simulations, students practice a specific skill or task that is one component of an occupation. Examples are TRACON (controlling air traffic from an airport tower), Microsoft Flight Simulator (flying a small plane), Train Dispatcher (routing trains between Washington, D.C. and Boston), and M1A2 Abrams Tank (driving an army tank). Another good illustration of a skill simulation is the flight simulation Apache. Throughout this simulation, the student is in the cockpit of an Apache helicopter and uses the joystick to steer the helicopter while using the keyboard to control the weapons, radar, and other systems.

Simulations can be used not only for training purposes but also in traditional educational settings, to assist students in acquiring basic academic skills such as math or science. The constructivist educational perspective is especially applicable for the development of educational simulations (Jonassen, 1993). Many writers have expressed the hope that constructivism will lead to better educational software and better learning (J. S. Brown, Collins & Duguid, 1989; Papert, 1993). They stress the need for open-ended exploratory, authentic learning environments in which learners can develop personally meaningful and transferable knowledge and understanding. These writers have proposed guidelines and criteria for the development of constructivist software (Grabinger & Dunlap, 1995; Hannafin & Land, 1997; Honebein, Duffy & Fishman, 1993; Rieber, 1992; Savery & Duffy, 1995; Squires, 1996). A main thesis of these guidelines is that learning should be authentic. Harper et al.'s (2000) review of the literature indicates three important concepts originating from the notion of authenticity: credibility, complexity, and ownership.

If students are to perceive that an environment offers realistic learning, they need to be able to explore the behavior of systems or environments; one way to do this is through working with simulations. The environment should provide the student with intrinsic feedback, which represents the effects of the student's action on the system or environment. Students should be able to express personal ideas and opinions, experiment with them, and try out different solutions to problems (Ainsworth, Bibby & Wood, 1997).

Grabinger & Dunlap (1995) emphasized that learners should be presented with complex environments that represent interesting and motivating tasks rather than contrived, sterile problems. Only in complex, rich environments learners will have the opportunity to construct and reconstruct concepts in idiosyncratic and personally meaningful ways. Learners may need help in coping with complexity. Strategies to help learners include scaffolding, anchoring, and problem-based environments (Cognition and Technology Group at Vanderbilt, 1990).

A sense of ownership should be a prominent feature of learning. The established idea of locus of control (Blease, 1988; Goforth, 1994; Wellington, 1985) is relevant in this context. Working in software environments, which provide high levels of user control, will help students perceive that they are instrumental in determining the process of the learning experience. Metacognition, in which learners reflect on their own cognition to improve their learning, is also appropriate here (Scardamalia, Bereiter, McLean, Swallow & Woodruff, 1989). The point is that, through a conscious, personal appraisal of cognitive processes, an individual can improve his or her capacity to learn. If this is to be effective, however, the learner must feel a sense of ownership of the learning.

The use of simulations as learning environments has a long history. Initial claims for the educational benefits of using simulations tended to emphasize pragmatic solutions to classroom problems. Both lengthy processes (e.g. population growth or genetic change) and short processes (e.g. changes in impulsive force during a collision) are possibilities for simulation. Difficult, dangerous, or expensive processes are also candidates for simulation (e.g. experiments with radioactive materials).

Gredler (1996) categorized simulations as either symbolic or experiential. Symbolic simulations are dynamic representations of the behavior of a system or set of processes or phenomena. The behavior that is simulated is usually the interaction of two or more variables over time, and the learner can manipulate these variables to discover scientific relationships, explain or predict events, or confront misconceptions. A simulation of a laboratory experiment is a classic example of a symbolic simulation.

Experiential simulations aim to establish a particular psychological reality and place learners in defined roles within that reality. The main elements of an experiential simulation are as follows: (a) a scenario of a complex task or problem that unfolds in part in response to learner actions; (b) a serious role taken by the learner in which she or he executes the responsibilities of the position; (c) multiple plausible paths through the experience; and (d) learner control of decision making (Gredler, 1996, p. 523). Role-playing simulations of environmental planning are classic examples of experiential simulations. For instance, the CD-ROM, *Investigating Lake Iluka*, an experiential simulation, is based on the concept of an information landscape that incorporates the biological, chemical, and physical components of a range of ecosystems that make up a coastal lake environment. Users are given problem-solving strategies to investigate this information in a variety of ways using the range of physical tools provided (Harper et al., 2000).

Therefore, the possibilities afforded by new multimedia technology, combined with contemporary ideas about learning, have opened up new perspectives for educational simulations. In particular, the use of sophisticated multimedia environments has led to the design of experiential simulations in which the learner plays an authentic role, carrying out complex tasks.

Intelligent tutoring systems (ITS) approaches also provide new ways of teaching and learning (Anderson, Corbett, Koedinger & Pelletier, 1995; Costa, 1992; Frasson & Gauthier, 1990; Merrill, Reiser, Ranney & Trafton, 1992; Park, 1996; Psotka, Massey & Mutter, 1988; Schofield & Evans-Rhodes, 1989; Shute & Psotka, 1996; VanLehn, 1999). ITSs are adaptive instructional systems developed with the application of artificial intelligence methods and techniques (Park, 1996). ITSs are developed to resemble what actually occurs when student and teacher sit down one on one and attempt to teach and learn together (Anderson et al., 1995; Shute & Psotka, 1996). ITSs have compo-

nents representing content to be taught, the inherent teaching or instructional strategy, and mechanisms for understanding what the student does and does not know (Park, 1996; Park & Seidel, 1989). These components are referred to as the problem-solving or expertise module, the student-modeling module, and the tutoring module. The expertise module evaluates the student's performance and generates instructional content during the instructional process. The student-modeling module assesses the student's current knowledge and makes hypotheses about the conceptions and reasoning strategies he or she used to achieve his or her current state of knowledge. The tutorial module usually consists of a set of specifications for the selection of instructional materials the system should present and how and when they should be presented (Seidel & Park, 1994; Shute & Psotka, 1996). Artificial intelligence methods for the representation of knowledge (e.g. production rules, semantic networks, and script frames) make it possible for ITSs to generate knowledge to present to the student on the basis of his or her performance on the task rather than selecting the presentation according to predetermined branching rules (Anderson et al., 1995; Park, 1996; Psotka et al., 1988; Schofield & Evans-Rhodes, 1989).

There are many types of ITSs (VanLehn, 1999). One common type is the coached practice environment. The tutor coaches the student as the student solves a multistep problem (e.g. solving a complex algebra word problem or discovering the laws governing a simulated economy). The student works with software tools, such as spreadsheets, graphs, and calculators. The tools are often designed especially for the task; an example is a kind of scratch paper that facilitates entering of the types of equations or other notations that the task demands. To solve the problem, the student must complete many user interface actions. After each action, the coach remains silent or may make comments. These comments represent one main form of instruction. When the problem is finished, the coach may review certain key steps in the solution, a process called reflective follow-up. This is another important form of instruction (VanLehn, 1999).

Examples of ITSs include the LISP tutor (e.g. Anderson et al., 1995), which provides instruction in LISP programming skills; Smithtown (Shute & Glaser, 1990), a discovery world that teaches scientific inquiry skills in the context of microeconomics; Bridge (Shute, 1991), which teaches Pascal programming skills; the Geometry Tutor (Anderson, Boyle & Reiser, 1985), which provides an environment in which students can prove geometry theorems; and WITS (Whole-course Intelligent Tutoring System; Callear, 1999), which provides an environment to teach a course on solid-state electronics. In general, results from evaluation studies demonstrate that these tutors do, in fact, accelerate learning with, at the very least, no degradation in outcome performance relative to appropriate control groups (Shute & Psotka, 1996). However, there are criticisms that ITS developers have failed to incorporate many valuable learning principles and instructional strategies developed by instructional researchers and educators (Park, Perez & Seidel, 1987). Cooperative efforts among experts in different domains, including learning—instruction sciences and artificial intelligence, are required to develop more powerful ITSs (Park & Seidel, 1989; Seidel, Park & Perez, 1988).

Therefore, even a brief analysis of contemporary innovations in the field of instructional technology demonstrates that CMC, virtual classroom, simulation training, and intelligent tutoring systems approaches involve new methods of teaching and learning, expanding the resources of the human mind and broadening the conventional boundaries of education. However, none of these approaches incorporate the specificity of HICEMTs described subsequently.

The First Generation of Educational Multimedia Technologies

Today, many people are accustomed to the advanced educational tools included in existing educational multimedia products. Strong development of multimedia technology leads to rapid emergence of a great number of educational off-line and on-line applications. Because, for example, they are easily accessible, educational multimedia products and services are becoming international. The first question that should be addressed is: What is really out there on the global educational multimedia market?

We have come to the conclusion that the current educational multimedia products constitute, in fact, the first generation of educational multimedia (Shavinina, 1997a, 1997b, 1998a; Shavinina & Loarer, 1999). Five types of educational multimedia products within this first generation can be identified. The first is learning manuals used in the framework of traditional school or college curricula and higher education (i.e. history, chemistry, physics, mathematics, biology, and languages). An example would be "history in the university: from the ancient times to the beginning of the 20th century". It is necessary to note that the great number and diversity of language multimedia products lead some authors to place them in a separate category of educational multimedia applications. This seems to be inappropriate, because the main purpose of these educational multimedia titles is the learning of native or foreign languages.

The second type is multimedia products focusing on general knowledge in various domains (e.g. automobiles). The third type is reference sources (i.e. various encyclopedias, dictionaries, and atlases). An example would be an encyclopedia of the human body. The fourth type is edutainment (i.e. different games with educational aspects, multimedia versions of famous tales, and interesting stories). The final type is cultural cognition (i.e. on-line and off-line multimedia applications on arts and culture). An example would be 'the best museums of the world'.

Analyzing today's educational multimedia applications from the viewpoint of their content, we found that they are 'domain-specific' products and that they now dominate on the international market (Shavinina, 1997a, 1997b). Domain-specific educational multimedia products are directed toward knowledge acquisition and skill development in a particular domain (e.g. language, arts, history, physics, literature, biology, and so on) and, very often, one narrow topic (Shavinina, 2000). As argued earlier (Shavinina, 1998a), these domain-specific educational multimedia applications are, in reality, computer versions of printed content (i.e. textbooks and learning manuals). In other words, they are simply 'placed' on CDs or the Internet, or both. Of course, contemporary multimedia technologies change the mode of presentation of traditional manuals and textbooks. However, the nature of their content has not changed. The lack of such transformation can be considered the main shortcoming of today's educational multimedia technologies and the biggest problem for their future. It should be emphasized that, in their current form, domain-specific educational multimedia products do not have realistic chances for further, more advanced development. Content continues to be a problematic issue; in many cases, so-called 'educational multimedia' products are nothing more than simple applications of pure multimedia technology in education. A solution can be found in the development of PMTs and HICEMTs (Shavinina, 1997a, 1997b, 1998a).

Analyzing current educational multimedia products from the standpoint of multimedia technology (i.e. specific features of multimedia), we concluded that contemporary educational multimedia applications simply use multimedia effects without any attempt to present or perhaps modify these effects in accordance with users' psychological organization (Shavinina, 1997b, 2000). It seems that companies develop educational multimedia applications simply because they have the required multimedia technology. This is also not a promising direction for the development of educational multimedia technologies.

The first generation of educational multimedia will undoubtedly be replaced by the second generation: PMTs and HICEMTs (Shavinina, 1997a, 1998a). The future belongs to HICEMTs, which will be distinguished by new principles in the creation of their content. The content of educational multimedia products and services is exceptionally important for the advancement of the field, because the most innovative breakthroughs can be accomplished in this area. From this standpoint, it is not surprising that executives of multimedia companies see successful developments in their firms in terms of substantial improvements in the content of their products. For example, the president of

Quebecor Multimedia, Monique Lefebvre (1997), has insisted that "we must stop putting the emphasis on electronic might and more on content" (p. 20).

Taking into account the foregoing ideas and arguments, it seems that the need for a new, second generation of educational multimedia is self-evident. However, before proceeding with a description of HICEMTs, it is important to discuss more precisely their importance or, in other words, why people need them.

The Importance of HICEMTs

The value of HICEMTs can be discussed from various multifaceted perspectives, but five closely related perspectives—technological, economic, societal, educational, and psychological—are particularly important. The *technological* perspective refers to the earlier-mentioned realities of the information age and its challenges for people. Thus, one of today's realities is extremely fast development in terms of high technology; the challenges range from a huge change in individuals' lives in business and private settings to a strong need to know and use appropriately the extraordinary quantity of developed software and applications. New breakthroughs in high technology encourage companies to rapidly introduce technological innovations related to business practices. For example, all of Bill Gates's 12 rules for succeeding in the cyber era (Gates, 1999) are related to new technologies. But companies forget that it is, first of all, people who use today's information and communication technologies. Unfortunately, the human mind does not improve as rapidly as new high technologies! Consequently, current realities of cybersociety reveal an exceptional need for the elaboration of special technologies whose primary goal is the development of human mental abilities (Shavinina, 2000). Therefore, there is a large and growing impetus for the development of HICEMTs.

The value of HICEMTs is clearly revealed from an *economic* standpoint. As mentioned earlier, today's industrial competition is increasingly harsh, and firms must bring new products and services to the global market with unprecedented speed. Nevertheless, the speed of human thinking lags behind the speed of the international market. There is thus an increasing need to develop special technologies for the mind—and especially HICEMTs—so as to accelerate the speed of human intelligence and creativity and generate new ideas and solutions in accordance with the demands of the world market. Such development is one of the key purposes of HICEMTs.

The importance of HICEMTs is also evident from a *societal* perspective. Everyone knows that modern society has many unsolved problems, including demographic, medical, environmental, political, economic, moral, and social problems. Accordingly, contemporary society can be characterized by a strong need for

highly able minds that can productively solve the numerous social problems and make appropriate social decisions.

As mentioned earlier, today's cybersociety produces many challenges to citizens, who must be exceptionally able if they are to bring innovative solutions to these challenges and, as a result, contribute to further societal progress. In short, intellectually creative citizens are guarantees of political stability, economic growth, industrial innovations, scientific and cultural enrichment, psychological health, and the general prosperity of any society in the 21st century.

HICEMTs also have value from an *educational* perspective. As discussed elsewhere (Shavinina, 1997a, 1997b), the real and the most important goal of education should be seen not as knowledge transfer but as development of people's intellectual and creative abilities. Today's children and adults must have at their disposal a set of productive educational and training technologies to reach this goal. The main purpose of HICEMTs is developing human mental resources, which allow people to both successfully apply existing knowledge and produce new knowledge.

At the same time, education is—by its nature—a psychological process based on underlying psychological mechanisms. The essence of HICEMTs coincides with the goals and nature of education and with the basic functioning of users' minds. Only in the case of a maximal matching between HICEMTs and the goal of education, on the one hand, and between HICEMTs and fundamental mechanisms of the functioning of human intelligence and creativity, on the other, one can really make assertions about productive learning and training. Any society that wishes to improve the quality of education should focus attention on the actualization and development of its citizens' intellectual and creative resources (Shavinina, 1998a).[3]

Finally, among many possible *psychological* arguments in favor of HICEMTs, it is necessary to mention

---

[3] Certainly, some opponents can point out that current realities of the African continent—where large segments of the population live at extreme levels of poverty—prevent local government officials and parents from investing in educational technology. However, this is not entirely true. The success of the WorLD program testifies to this. In almost all 15 WorLD countries (e.g. Mauritania, Senegal, and Uganda), parents are so excited about the opportunities their children are gaining through the program that they have set aside parts of their meager incomes to help the school pay its monthly connectivity costs. The same is true for officials in these countries. For example, in Mauritania, when the minister of education, a former biology teacher, was introduced to various websites for use in biology, chemistry, and other sciences, he was impressed by the depth and breadth of information available. Shortly thereafter, at his request, the ministry of finance agreed to finance leased-line Internet connectivity for all secondary schools in Mauritania (Carlson & Hawkins, 2001).

(Continued overleaf)

409

at least two. First, the human mind is a unique phenomenon in nature, and its development must be realized by individualized means designed to be appropriate for the advanced development of mental resources. Certainly, educational multimedia applications in general provide a good opportunity for education when, where, and how users wish. However, HICEMTs allow people to advance in learning and training according to the unique characteristics of their minds (i.e. personal abilities, speed, and time). HICEMTs provide an exceptional opportunity for individualized, personalized, differentiated, and flexible learning and training because the underlying idea of these novel technologies is the following: many differing people can find in HICEMTs the individual means (i.e. the individual approach) for actualizing their own mental potential and the subsequent development of their intellectual and creative abilities (Shavinina, 2000). Second, the earlier-mentioned conclusions about the current state of educational multimedia show that there is at present no educational multimedia technology that can satisfy all the discussed features of these five perspectives as they manifest themselves in their contemporary requirements of the human mind in general and intellectual and creative resources in particular.

When the Medium is 'Mental'

McLuhan (1964) argued that all of a culture's technologies are extensions of the human body and nervous system. He suggested that, as culture advances, the new technology (the new medium) used to frame and convey information creates a new environment that is more influential than the content of the information it conveys. In his now famous phrase, McLuhan described this superordinance of the medium over its content as "the medium is the message". Vandervert (1999, 2001) argued that, in the case of contemporary cybereducation in general and in the case of HICEMTs in particular, for the first time in human history "the medium (i.e. the new technology) is mental". Current levels of advancement in multimedia technology and psychological science form a strong foundation for the development of such mental technologies.

---

[3] (Continued)

   Likewise, critics can note that in America 20% of all children are born in poverty and about the same percentage drop out of high school. There is also the sad problem of school violence. As a result, American society's educational focus is on broader social and economic issues other than the development of new instructional technologies. However, this is not the whole story. For example, the National Science Foundation (NSF) and other government and private funding agencies provide many grants for the development of new educational technologies, which supports our assertion. For instance, in 1997 NSF provided grants to 52 of 120 universities involved in the development of Internet 2, the next-generation Internet (Lan & Gemmill, 2000).

New Scientist magazine's November 7, 1998, issue was devoted to Silicon Valley's phenomenal capacity to generate and nurture ideas leading to the birth of dozens of the world's most successful high-technology companies. This issue emphasized that Silicon Valley's companies (e.g. Hewlett-Packard) profoundly understand the need to invest in ideas. Moreover, "now, even private entrepreneurs create companies which exist solely to produce ideas with money-making potential" ('Editorial Introduction', 1998, p. 30). HICEMTs can be viewed as instruments that will enable individuals to generate such ideas. It is clear today that great ideas—resulting in innovative discoveries and inventions—are beginning to rule the global economy. Energy-information calculations show that the full information-processing capacity of each person's brain is approximately 22 legacies (Vandervert, 2001). Thus, the human brain is characterized by huge creative potential. The primary purpose of HICEMTs is the actualization and development of this potential. These new technologies will help people produce innovative ideas leading to significant achievements and exceptional performance in all areas of human endeavor.

HICEMTs: General Characteristics

The nature of HICEMTs can be understood through a set of their general and specific characteristics. The general characteristics of HICEMTs are as follows: (a) general psychological basis; (b) actualization of fundamental cognitive mechanisms; (c) new targets of educational and developmental influences; (d) better adaptation to individuals' psychological organization;[4] and (e) 'psycho-edutainment' as an overall framework. Each of these five characteristics actually represents clusters of characteristics, and these within-cluster characteristics are highly interrelated.

General Psychological Basis

Their general psychological basis implies that HICEMTs are based on psychological processes and phenomena and, especially, on the mechanisms of human intellectual and creative functioning. General psychological processes and phenomena include attention, perception, short-term and long-term memory, visual thinking, knowledge base, mental space, concept formation, analytical reasoning, metacognitive abilities, cognitive and learning styles, critical thinking, motivation, and many other psychological mechanisms. One of the strong arguments to place psychological foundations at the heart of HICEMTs is the fact that the educational process is always a

---

[4] Working on the development of a series of new textbooks on mathematics, which are largely based on her theory of human intelligence, Kholodnaya (1997) introduced the principles of new targets of educational influences and a need to better adapt learning and teaching materials to children's psychological organization.

psychological process. As mentioned earlier, learning, teaching, and training are based on fundamental psychological mechanisms.

For example, a person's knowledge base is an important psychological foundation for the process of knowledge transfer, which is traditionally considered a goal of education and of any educational tool. Psychologists have demonstrated a significant role for knowledge bases in the successful functioning of human intelligence and creativity (Bjorklund & Schneider, 1996; Chi & Greeno, 1987; Chi & Koeske, 1983; Kholodnaya, 1997; Runco, in press; Runco & Albert, 1990; Schneider, 1993; Shavinina & Kholodnaya, 1996; Sternberg, 1985, 1990). The quantity and quality of specialized knowledge play an essential role in highly intellectual performance and in the process of acquiring new knowledge. For example, productive problem-solving cannot occur in the absence of relevant prior knowledge (Chi & Greeno, 1987). The knowledge base can facilitate the use of particular learning strategies, generalize strategy use to related domains, or even diminish the need for strategy activation. Moreover, a rich knowledge base can sometimes compensate for an overall lack of general cognitive abilities (Bjorklund & Schneider, 1996; Schneider, 1993). However, current educational multimedia products do not take into account this specificity of the psychological nature of the human knowledge base, particularly with respect to its optimum functioning, which distinguishes exceptional intellectual and creative performance (Rabinowitz & Glaser, 1985; Shavinina & Kholodnaya, 1996; Shore & Kanevsky, 1993; Sternberg & Lubart, 1995). Other examples of psychological processes and phenomena and their appropriateness for the development of HICEMTs have been described elsewhere (Shavinina, 1997a, 1997b, 1998a).

Fundamental Cognitive Mechanisms

HICEMTs are directed toward actualization of fundamental cognitive mechanisms, which play a significant role in an individual's intellectual and creative functioning. This is the second key characteristic of HICEMTs. The following cognitive processes and phenomena are viewed as important in contemporary accounts of human intelligence and creativity (Brown, 1978, 1984; Flavell, 1976, 1979; Kholodnaya, 1997; Mumford, in press; Runco & Albert, 1990; Shavinina & Kholodnaya, 1996; Sternberg, 1984, 1985, 1988a, 1988b, 1990): (a) conceptual structures (Case, 1995; Kholodnaya, 1983, 1997); (b) knowledge base (Chi & Greeno, 1987; Chi & Koeske, 1983); (c) mental space (Kholodnaya, 1997; Shavinina & Kholodnaya, 1996); (d) cognitive strategies (Bjorklund & Schneider, 1996; Pressley, Borkowski & Schneider, 1987); (e) meta-cognitive processes (Borkowski, 1992; Borkowski & Peck, 1986; Brown, 1978, 1984; Campione & Brown, 1978; Flavell, 1976, 1979; Shore & Dover, 1987;

Sternberg, 1985, 1990); (f) specific intellectual intentions (Shavinina, 1996b; Shavinina & Ferrari, in press (a), in press (b)); (g) objectivization of cognition (Kholodnya, 1997; Shavinina, 1996a); and (h) intellectual and cognitive styles (A. E. Jones, 1997; Kholodnaya, 1997; Martinsen, 1997; Riding, 1997; Sternberg, 1985, 1987). Other processes and phenomena should also be embedded in HICEMTs. These psychological mechanisms provide a necessary foundation for the further successful development of human creative and intellectual abilities, because enhancement of an individual's mental potential must be based on already-actualized cognitive resources (Shavinina, 1998a).

The great French scientist Blaise Pascal asserted that "chance favors the prepared Mind". To rephrase Pascal's assertion with regard to HICEMTs, the prepared mind will gain much more advantage from specially elaborated psychoeducational multimedia technologies designed for the development of mental abilities. Through repeated exposure to the fundamental processes and phenomena of the human cognitive system, HICEMTs become 'know-how' learning and training multimedia technologies, and they thus provide an underlying educational basis for the subsequent development of an individual's mind.

New Targets of Educational and Developmental Influences

HICEMTs bring new targets of educational and developmental influences. The conventional wisdom of companies and developers of current educational multimedia technologies is to address their products to the abstract 'user' as a whole. Developers and experts in this field are mostly unconcerned with where their products and services are directed. There is a general user (i.e. children, adolescents, or adults), and what is needed is to specify to what age category a given CD-ROM or Internet-based program is addressed (Shavinina, 1997b). For example, in analyzing the contemporary state of educational multimedia available on the Internet, Roberts (1996) did not even raise the question as to where these learning programs are directed. Instead, the goal is seen in another issue: "How do we redesign and adapt our classroom teaching and learning approaches in ways that are effective and appropriate to distance education?" This shows that 'players' in the contemporary arena of educational multimedia do not go beyond the simple 'placing' of the content of printed learning manuals on the Internet or CDs.

From a psychological standpoint, it is unproductive to direct educational multimedia technologies to users in general, because people represent in themselves complex psychological systems with many hierarchical components, multidimensional variables, multifaceted parameters, and structural interrelations. And to direct any educational influence to such complex systems as a whole is to decrease immediately the quality of education. As a result, the exact targets of the existing educational multimedia applications are missing, although they could be responsible for the higher productivity of the educational process. For instance, in analyzing the current educational multimedia technologies, we can always ask ourselves: "To what exactly—in the structure of users' intelligence or personality—is the given educational multimedia technology directed?" However, it is not easy to answer this question from a psychological point of view (Shavinina, 1997a).

The main goal of current educational multimedia is 'knowledge transfer'. But today such transfer is not fully realized, because successful knowledge transfer is a derivative of fundamental psychological mechanisms. Consequently, it is the basic mental processes and phenomena that should be viewed as the real targets of learning, teaching, and training. Therefore, the development of HICEMTs leads to a change in the traditional audience (i.e. targets) of available educational multimedia products from 'users as a whole' to the underlying mechanisms of human intellectual and creative functioning. The primary objective of education has also changed: from realizing simple knowledge transfer to developing intellectual and creative abilities. Such changes will result in significant increases in the quality of learning and will contribute to the advancement of educational multimedia products and services (Shavinina, 1998a, 2000).

Better Adaptation to Individuals' Psychological Organization

The foregoing features of HICEMTs allow them to be better adapted to an individual's psychological organization than current educational multimedia. HICEMTs can take into account numerous psychological characteristics of users, such as behavioral, developmental, emotional, motivational, personality, and social features. HICEMTs are directly built on the psychological specificity of users (Shavinina, 2000).

For example, any two CDs or Internet-based courses for learning of a foreign language are certainly different for children and adults. It is clear that this difference is mainly connected to content. And developers of educational multimedia, for the most part, are limited by content issues. Their way of thinking is the following: "Children cannot understand some educational material, so we need to present them with more simple information". But this is not enough. Another difference between CDs for children and for adults concerns the psychological organization of the two groups of users. Adults have an internal motivation to study foreign languages, whereas children should have a strong, ongoing external motivation in addition to the internal one (or even instead of it; Shavinina, 1998a).

HICEMTs have the potential to generate the necessary conditions for the appearance and maintenance on the appropriate level of a child's motivation to learn, cognitive behavior, emotional involvement, and personal satisfaction with her or his gradual progress through the educational content (Shavinina, 1997a, 1997b). It does not matter by what means developers of educational multimedia can reach this goal (e.g. exciting scenarios, specific multimedia effects, user-friendly means, innovative learning methods, and so on; certainly, their combination would be desirable). The principal thesis is that educational multimedia technologies should fit the internal psychological structures of human intellectual, creative, behavioral, cognitive, developmental, emotional, motivational, and other systems. HICEMTs are developed to accomplish this goal, and contemporary achievements of information and communication technologies allow today's developers of educational multimedia to reach it.

Psycho-Edutainment

As we have argued and predicted elsewhere (Shavinina, 1998a), the appearance of 'psycho-edutainment'—a new area of the global educational multimedia market—is an inevitable event. This innovative multidisciplinary multimedia field that emerges at the crossroads of existing multimedia fields (i.e. education, entertainment, and edutainment) and a new area—psychology—is the only scientifically viable framework for the development of HICEMTs. Taking into consideration that: (a) education is, in its essence, a psychological process; (b) entertainment involves its own psychological mechanisms related to games; and (c) play is a preferable and leading form of children's activity, one can conclude that HICEMTs cannot be developed other than through the synthesized regrouping of contemporary fields of multimedia (Shavinina, 2000). The nature of HICEMTs cannot be associated with one particular multimedia field; they may be created only in the space of 'psycho-edutainment'.

The five features just described provide the general characterization of HICEMTs. The other considerable portion of the features of HICEMTs relates to their more specialized characterization.

HICEMTs: Specific Characteristics

Among special characteristics of HICEMTs, the following can be distinguished: (a) 'intellectual' content; (b) 'creative' content; and (c) 'intellectually creative edutainment'. The first two characteristics deal with the specific content of HICEMTs, and they demonstrate '*what*' should be included in these technologies. Intellectually creative edutainment, however, represents a substantial portion of the important characteristics related to the mode of presentation of this content. Consequently, intellectually creative edutainment describes '*how*' specific content might be embedded in HICEMTs.

Intellectual and creative contents cover many different, multidimensional, and interrelated aspects.

However, it is not possible to consider all of them here, because they vary significantly and depend on developers. At the current development level of psychology, there are many different theories of human intelligence and creativity (Detterman, 1994; Kholodnaya, 1997; Miller, 1996; Runco, in press; Runco & Albert, 1990; Shavinina, 1998b; Simonton, 1988; Sternberg, 1982, 1985, 1988b, 1990; Sternberg & Lubart, 1995) that predetermine developers' conceptions of creative and intellectual contents. A variety of the psychological approaches in the field of individual intelligence and creativity will strongly influence differences in the content of HICEMTs through developers' conceptions. These different approaches predetermine what is intellectual and creative in HICEMTs.

From our point of view, one of the possible approaches to developing the content of HICEMTs can be based on Kholodnaya's (1997) theory of individual intelligence. This theory is one of many theories that might underlie the development of HICEMTs. Sternberg's triarchic theory of human intelligence and Gardner's multiple intelligences theory are well-known examples of other theories that could also serve as foundations for the development of HICEMTs. Our goal is not to consider all possible theoretical approaches that may underlie HICEMTs. It is beyond the scope of one chapter to analyze the advantages and limitations of the existing theories of intelligence and creativity in their application to the design of HICEMTs. This is a topic for another chapter or even a book.

The joining of entertainment and education has led to the appearance of 'edutainment', a rapidly developing multimedia field. And, as mentioned earlier, the combining of psychology and edutainment has led to the emergence of 'psycho-edutainment', a promising new multimedia area. Equally, the combination of intelligence and creativity with edutainment leads to 'intellectually creative edutainment', a framework for the development of HICEMTs. HICEMTs will significantly transform 'edutainment'. Therefore, modern multimedia technology, educational games, and entertainment—built on the fundamental psychological processes and basic principles of human intellectual functioning and creative performance—form a type of 'intellectually creative edutainment' that provides a real opportunity to develop HICEMTs.

How Many HICEMTs?

Of course, an inevitable question arises: How many HICEMTs can be developed? The answer is quite clear: An unlimited number can be developed. There are at least three scientific reasons for this. First, as mentioned earlier, the variety of psychological approaches and theories in the areas of human intelligence and creativity—which provide a foundation for developers' conceptions of creativity and intelligence—exclude, in principle, a limited number

of HICEMTs. Second, the complex, multidimensional nature of intellectual and creative abilities (toward the actualization and development of which HICEMTs are directed) excludes only a single educational multimedia technology that would be more productive than other technologies, because the development of mental abilities can be achieved through various psychoeducational methods. Third, the transdisciplinary nature of HICEMTs (i.e. in which, in contrast to traditional interdisciplinary and multidisciplinary approaches, "the complete fundamental-level merging of 'disciplinary' sources of knowledge is the focus"; Vandervert, 2001) provides a basis for the development of a variety of new psychoeducational multimedia technologies (Vandervert, Shavinina & Cornell, 2001). Finally, technological and economic factors also result in the exclusion of a limited quantity of HICEMTs.

General and Specific HICEMTs

HICEMTs are not a simple 'placing' on CD-ROM or the Internet of already-existing creativity and intelligence training such as Osborn's brainstorming, Parnes's creative problem-solving, Feldhusen's critical thinking, Torrance's futuristic creative problem-solving, or Sternberg's (1986) intelligence program. This is a rather trivial task. More precisely, such products should be referred to as 'computer versions' of printed matters (i.e. books, manuals, tests, and so on), but not as HICEMTs, because they do not use the multiple possibilities of multimedia. HICEMTs can be developed in two main directions: (a) as derivatives of special creativity and intelligence training, but essentially changed, enriched, and presented in multimedia form; and (b) as derivatives of any knowledge domain (i.e. chemistry, biology, mathematics, history, and so on). In other words, one can distinguish general and specific HICEMTs. In the case of the latter, the aforementioned principles of HICEMTs will provide a foundation on which a special content of any domain can be built. These principles will serve as a basis for the development of an immense quantity of HICEMTs across various content domains.

HICEMTs in Action: A Practical Illustration

One of the possible examples of today's educational multimedia products that addresses many of the general characteristics of HICEMTs is Knowledge forum, a second-generation CSILE product.[5] This Web-based educational technology can be considered, to a certain extent, as a partial practical illustration of

---

[5] For more information on CSILE research, development, team members, and applications, visit the CSILE website (http://www.csile.oise.on.ca). For CSILE product information, contact Learning in Motion, 500 Seabright Avenue, Suite 105, Santa Cruz, California 95062 (mcappo@learn.motion.com).

HICEMTs on-line. Its authors assert that they are developing a miniature version of a knowledge-building society using CSILE software (Scardamalia & Bereiter, 1996).

Created by Marlene Scardamalia, Carl Bereiter, and the staff of educational and psychological researchers and computer scientists at the Ontario Institute for Studies in Education of the University of Toronto, Knowledge forum was born over a decade of research on 'expert learning' (Bereiter & Scardamalia, 1993; Scardamalia & Bereiter, 1994). Expert learning refers to learning that leads to the gradual acquisition of expert knowledge through continual reinvestment of students' mental resources in addressing problems at higher levels (Bereiter & Scardamalia, 1993). Scardamalia and Bereiter's cognitive research on the nature of expertise and expert learning has resulted in a knowledge society model for an educational network that engages students in constructing, using, and improving knowledge (Scardamalia & Bereiter, 1996).

The aim of Knowledge forum is to enhance students' knowledge building and understanding, with a particular emphasis on kindergarten through Grade 12. When students analyze and evaluate information and revise and reshape ideas, real knowledge building occurs. In other words, Knowledge forum supports knowledge construction. In brief, this educational technology proceeds as follows. Students initially represent and develop their ideas and questions through the production of graphics and text notes. However, as research on expertise demonstrates, often students' initial ideas and theories are based on misconceptions. As students reach a deeper understanding, their initial theories change. Students then reference other work (and can track back to the original source with a simple mouse click), identify gaps in knowledge, and ask for collaboration to increase their understanding. Referencing and 'building-on' facilities, therefore, help students create a multidimensional knowledge-building framework for their ideas. Sharing of comments helps them revise their ideas and initial theories. As students structure and connect their ideas beyond the usual 'topic titles', they improve their ability to understand and apply new concepts. Continual improvement of ideas is supported by features of CSILE software that encourage explanation-seeking, theory-building, problem-solving, argumentation, publication, and other socio-cognitive operations that help advance understanding. Through a variety of linking, searching, commenting, and visiting activities, the network encourages consideration of various ideas from different perspectives. The purpose is to enable participants to gain knowledge and understanding. In short, Knowledge forum provides an environment in which knowledge can be examined and evaluated for gaps, added to, revised, and reformulated.

Bereiter (1994) introduced the concept of *progressive discourse*, defining it as a set of innumerable local discourses that consist of clarifications and resolutions of doubts and that generate ideas advancing the larger discourse. He asserted that meaning making and new conceptual structures arise through a dialectic process in which members of a learning community negotiate contradictions and begin to synthesize opposing standpoints into a more encompassing scheme. Progressive discourse is another label for what most educators call inquiry learning. It is especially useful in problem-based learning, wherein the field of inquiry is ill defined and there are no simple or straightforward answers. The concept of progressive discourse is also closely related to the concept of *engaged learning*, which integrates authentic tasks, interactive instruction, collaborative knowledge building, heterogeneous grouping, and co-exploration by students and teachers (Sherry, 2000).

Evaluation research shows that students using CSILE, the first generation of Knowledge forum, outperformed control groups on both standardized tests and knowledge tests (Scardamalia & Bereiter, 1996). Students working in Knowledge forum demonstrated impressive results in terms of textual and graphical literacy, theory improvement, implicit theories of learning, and comprehension of difficult concepts. The depth of their explanation and understanding was higher than that of control groups. Scardamalia and Bereiter (1996) also found that pupils working in Knowledge forum became actively involved in building and richly linking databases, pointing out discrepant information, contributing new information and ideas, considering ideas from different viewpoints, and forming important new working relationships and study groups.

As can be seen, Knowledge forum is built on general psychological processes and fundamental cognitive mechanisms associated with the successful growth of an individual's knowledge base, leading to the acquisition of expertise. Knowledge forum also brings new targets of educational influences. In other words, Knowledge forum incorporates general psychological characteristics of HICEMTs.[6]

---

[6] Other possible examples of today's educational multimedia technologies that incorporate some of the general characteristics of HICEMTs are the Web-based Integrated Science Environment (WISE) and the Learning by Design (LBD) project. Developed by Marcia Linn and her laboratory in the Department of Education, University of California at Berkeley, WISE is a simple yet powerful learning environment where students examine real-world evidence and analyze current scientific controversies. WISE is a powerful tool that combines advanced educational approaches and cutting-edge multimedia technology. WISE Version 2 is a more powerful and versatile generation of the WISE software, with increased capabilities for expansion, internationalization, and customization–localization. See http://www.wise.berkeley.edu for more details.

Summary

The economic, technological, scientific, societal, and cultural future of human beings will apparently be synonymous with a permanent need for intellectual and creative minds, because all areas of endeavor will require an extremely high level of innovation. And only exceptionally able people can produce new ideas of such quality that lead to real innovations. Despite the evident importance of individuals with creative and intellectual abilities in the life of any societal 'organism', one should acknowledge that their development through psychoeducational methods is far from well understood. Thus, the development of HICEMTs seems to be a very timely and needed endeavor.

Therefore, by rethinking the role of education in the information era, using contemporary multimedia technology, using findings of psychological science in general and knowledge on human intelligence and creativity in particular, and elaborating 'intellectually creative entertainment', an entirely new wave of technological innovations in psychology—HICEMTs—can be developed. HICEMTs will dominate the global educational multimedia market of the 21st century.

References

Ainsworth, S. E., Bibby, P. A. & Wood, D. J. (1997). Information technology and multiple representations. *Journal of Information Technology for Teacher Education*, **6**, 93–104.

Allen, B. S. & Otto, R. G. (1996). Media as lived environments. In: D. H. Jonassen (Ed.), *Handbook of Research for Educational Communications and Technology* (pp. 199–225). New York: Macmillan.

Amabile, T. M. (1988). A model of creativity and innovation in organizations. *Research in Organizational Behavior*, **10**, 123–167.

Amabile, T. M. (1996). *Creativity in context*. Boulder, CO: Westview.

Anderson, J. R., Boyle, C. & Reiser, B. (1985). Intelligent tutoring systems. *Science*, **228**, 456–462.

Anderson, J. R., Corbett, A. T., Koedinger, K. R. & Pelletier, R. (1995). Cognitive tutors: Lessons learned. *Journal of the Learning Sciences*, **4**, 167–207.

Bereiter, C. (1994). Implications of postmodernism for science, or, science as progressive discourse. *Educational Psychologist*, **29**, 3–12.

Bereiter, C. & Scardamalia, M. (1993). *Surpassing ourselves*. Chicago, IL: Open Court.

Bjorklund, D. F. & Schneider, W. (1996). The interaction of knowledge, aptitude, and strategies in children's memory development. In: H. W. Reese (Ed.), *Advances in Child Development and Behavior* (pp. 59–89). San Diego, CA: Academic Press.

Blease, D. (1988). *Evaluating educational software*. London: Croom Helm.

Bliss, J. & Ogborn, J. (1989). Tools for exploratory learning. *Journal of Computer Assisted Learning*, **5**, 37–50.

Borkowski, J. G. (1992). Metacognitive theory: A framework for teaching literacy, writing, and math skills. *Journal of Learning Disabilities*, **25**, 253–257.

Borkowski, J. G. & Peck, V. A. (1986). Causes and consequences of metamemory in gifted children. In: R. J. Sternberg & J. E. Davidson (Eds), *Conceptions of Giftedness* (pp. 182–200). Cambridge: Cambridge University Press.

Bowen, S., Shore, B. M. & Cartwright, G. F. (1993). Do gifted children use computers differently? *Gifted Education International*, **8**, 151–154.

Brown, A. L. (1978). Knowing when, where, and how to remember: A problem of metacognition. In: R. Glaser (Ed.), *Advances in Instructional Psychology* (Vol. 1, pp. 77–165). Hillsdale, NJ: Erlbaum.

Brown, A. L. (1984). Metacognition, executive control, self-regulation, and other even more mysterious mechanisms. In: F. E. Weinert & R. H. Kluwe (Eds), *Metacognition, Motivation, and Learning* (pp. 60–108). Stuttgart, Germany: Kuhlhammer.

Brown, A. L. (1994). The advancement of learning. *Educational Researcher*, **23**, 4–12.

Brown, A. L. & Palincsar, A. S. (1989). Guided, cooperative learning and individual knowledge acquisition. In: L. B. Resnick (Ed.), *Knowing, Learning, and Instruction* (pp. 76–108). Hillsdale, NJ: Erlbaum.

Brown, J. S., Collins, A. & Duguid, P. (1989). Situated cognition and the culture of learning. *Educational Researcher*, **18**, 32–41.

Callear, D. (1999). Intelligent tutoring environments as teacher substitutes: Use and feasibility. *Educational Technology*, 6–11.

Campione, J. C. & Brown, A. L. (1978). Toward a theory of intelligence: Contributions from research with retarded children. *Intelligence*, **2**, 279–304.

Carlson, S. & Firpo, J. (2001). Integrating computers into teaching: Findings from a 3-year program in 15 developing countries. In: L. R. Vandervert, L. V. Shavinina & R. Cornell (Eds), *CyberEducation: The Future of Long Distance Learning* (pp. 85–114). Larchmont, NY: Liebert.

Carlson, S. & Hawkins, R. (1998). Linking students around the world: The World Bank's new educational technology program. *Educational Technology*, 57–60.

Case, R. (1995). The development of conceptual structures. In: W. Damon, D. Kuhn & R. Siegler (Eds), *Handbook of Child Psychology* (pp. 745–800). New York: John Wiley.

Chi, M. T. H. & Greeno, J. G. (1987). Cognitive research relevant to education. In: J. A. Sechzer & S. M. Pfafflin (Eds), *Psychology and Educational Policy* (pp. 39–57). New York: New York Academy of Sciences.

Chi, M. T. H. & Koeske, R. D. (1983). Network representation of a child's dinosaur knowledge. *Developmental Psychology*, **19**, 29–39.

---

[6] (Continued)

Created by Janet Kolodner and her colleagues at the EduTech Institute, Georgia Institute of Technology, the LBD project has as its goal developing a design-based curriculum for middle school science and software to go with it. LBD takes its cues from case-based reasoning, according to which one can best learn from experience if one does a good job of interpreting one's experiences and anticipating when the lessons they teach might be useful. See http://www.cc.gatech.edu/edutech for more details.

Cognition and Technology Group at Vanderbilt (1990). Anchored instruction and its relationship to situated cognition. *Educational Researcher*, **19**, 2–10.

Collins, A., Brown, J. S. & Newman, S. E. (1989). Cognitive apprenticeship. In: L. B. Resnick (Ed.), *Knowing, Learning, and Instruction* (pp. 154–185). Hillsdale, NJ: Erlbaum.

Costa, E. (Ed.) (1992). *New directions for intelligent tutoring systems*. Heidelberg, Germany: Springer-Verlag.

Detterman, D. K. (Ed.) (1994). *Current topics in human intelligence* (Vol. 4). *Theories of intelligence*. Norwood, NJ: Ablex.

Editorial introduction to the special issue on Silicon Valley (1998). *New Scientist*, **168**, 30.

Education in the information society (an initiative of the European Commission) (1996). Brussels: Office for Official Publications of the European Community.

Educational multimedia in Europe (a resolution on educational multimedia prepared by the Counsel of the Ministers of Education of the European countries) (1996). Brussels: Office for Official Publications of the European Community.

Engestrom, Y. (1996). Interobjectivity, ideality, and dialectics. *Mind, Culture, and Activity*, **3**, 259–265.

Ferrari, M., Taylor, R. & VanLehn, K. (1999). Adapting work simulations for school: Preparing students for tomorrow's workplace. *Journal of Educational Computing Research*, **21**, 25–53.

Flavell, J. H. (1976). Metacognitive aspects of problem solving. In: L. B. Resnick (Ed.), *The Nature of Intelligence* (pp. 231–235). Hillside, NJ: Erlbaum.

Flavell, J. H. (1979). Metacognition and cognitive monitoring: A new area of cognitive-development inquiry. *American Psychologist*, **34**, 906–911.

Frasson, C. & Gauthier, C. (Eds) (1990). *Intelligent tutoring systems*. Norwood, NJ: Ablex.

Gates, W. H. (1999). Bill Gates' new rules. *Time*, **153**, 30–35.

Goforth, D. (1994). Learner control = decision making + information: A model and meta-analysis. *Journal of Educational Computing Research*, **11**, 1–26.

Grabinger, R. S. & Dunlap, J. C. (1995). Rich environments for active learning. *Association for Learning Technology Journal*, **2**, 5–34.

Gredler, M. E. (1990). Analyzing deep structure in games and simulations. *Simulations/Games for Learning*, **20**, 329–334.

Gredler, M. E. (1992). *Designing and evaluating games and simulations: A process approach*. London: Kogan Page.

Gredler, M. E. (1996). Educational games and simulations: A technology in search of a (research) paradigm. In: D. H. Jonassen (Ed.), *Handbook of Research for Educational Communications and Technology* (pp. 521–540). New York: Macmillan.

Hannafin, M. J. & Land, S. M. (1997). The foundations and assumptions of technology-enhanced student centres learning environments. *Instructional Science*, **25**, 167–202.

Harper, B., Squires, D. & McDougall, A. (2000). Constructivist simulations: A new design paradigm. *Journal of Educational Multimedia and Hypermedia*, **9**, 115–130.

Heinonen, D. P. (1996). Citoyens de l'Europe. (European Citizens). *Le Monde*, 3.

Hiltz, S. R. (1986). The 'virtual classroom': Using computer-mediated communication for university teaching. *Journal of Communication*, **36**, 95–104.

Hiltz, S. R. (1990). Evaluating the virtual classroom. In: L. M. Harasim (Ed.), *Online Education: Perspectives on a New Environment* (pp. 133–183). New York: Praeger.

Honebein, P. C., Duffy, T. M. & Fishman, B. J. (1993). Constructivism and the design of authentic learning environments. In: T. M. Duffy, J. Lowyck & D. H. Jonassen (Eds), *Designing Environments for Constructive Learning* (pp. 87–108). Berlin: Springer-Verlag.

Honour, L. & Evans, B. (1997). Shifting the culture towards electronic learning media: Learning as an interactive experience. In: *Open Classroom II Conference: Papers and Presentations* (pp. 285–294). Athens: Kastaniotis.

Javid, M. (2001). Edmonds and Kamiak cyberschools: Two innovative emerging models for cybereducation. In: L. R. Vandervert, L. V. Shavinina & R. Cornell (Eds), *Cyber-Education: The Future of Long Distance Learning* (pp. 185–218). Larchmont, NY: Liebert.

Jonassen, D. H. (1993). A manifesto to a constructivist approach to the use of technology in higher education. In: T. M. Duffy (Ed.), *Designing Environments for Constructive Learning* (pp. 231–247). New York: Springer-Verlag.

Jonassen, D. H. (Ed.) (1996). *Handbook of research for educational communications and technology*. New York: Macmillan.

Jones, A. E. (1997). Reflection–impulsivity and wholist-analytic: Two fledglings? ... or is R–I a cuckoo? *Educational Psychology*, **17**, 65–77.

Jones, S. (Ed.) (1995). *Cybersociety: Computer-mediated communication and community*. Thousand Oaks, CA: Sage.

Kholodnaya, M. A. (1983). *The integrated structures of conceptual thinking*. Tomsk, Russia: Tomsk University Press.

Kholodnaya, M. A. (1997). *The psychology of intelligence*. Moscow: APN Press.

Koschmann, T. (Ed.) (1996a). *CSCL: Theory and practice of an emerging paradigm*. Mahwah, NJ: Erlbaum.

Koschmann, T. (1996b). Paradigm shifts and instructional technology. In: T. Koschmann (Ed.), *CSCL: Theory and Practice of an Emerging Paradigm* (pp. 1–23). Mahwah, NJ: Erlbaum.

Lan, J. & Gemmill, J. (2000). The networking revolution for the new millennium: Internet 2 and its educational implications. International *Journal of Educational Telecommunications*, **6**, 179–198.

Lave, J. & Wenger, E. (1991). *Situated learning*. New York: Cambridge University Press.

Lefebvre, M. (1997). For better or for worse: The multimedia reign. *Focus*, **3**, 20–22.

Leont'ev, A. N. (1981). The problem of activity in psychology. In: J. V. Wertsch (Ed.), *The concept of activity in Soviet psychology* (pp. 37–71). Armonk, NY: M. E. Sharpe.

Martinsen, O. (1997). The construct of cognitive style and its implications for creativity. *High Ability Studies*, **8**, 135–158.

McAteer, E., Tolmie, A., Duffy, C. & Corbett, J. (1997). Computer-mediated communication as a learning resource. *Journal of Computer Assisted Learning*, **13**, 219–227.

McLuhan, M. (1964). *Understanding media: The extensions of man*. New York: McGraw-Hill.

Merrill, D. C., Reiser, B. J., Ranney, M. & Trafton, J. G. (1992). Effective tutoring techniques: A comparison of human tutors and intelligent tutoring systems. *Journal of the Learning Sciences*, **2**, 277–306.

Miller, A. (1996). *Insights of genius: Visual imagery and creativity in science and art*. New York: Springer-Verlag.

Mumford, M. (in press). Cognitive approaches to creativity. In: M. A. Runco (Ed.), *Handbook of Creativity* (Vol. 2). Norwood, NJ: Ablex.

Norman, D. A. (1993). *Things that make us smart*. Reading, MA: Addison-Wesley.

Olson, D. R. (1986). Intelligence and literacy: The relationships between intelligence and the technologies of representation and communication. In: R. J. Sternberg & R. K. Wagner (Eds), *Practical Intelligence* (pp. 338–360). Cambridge: Cambridge University Press.

Oren, A., Nachmias, R., Mioduser, D. & Lahav, O. (2000). Learnet-A model for virtual learning communities in the World Wide Web. *International Journal of Educational Telecommunications*, **6**, 141–157.

Papert, S. (1993). *The children's machine: Rethinking school in the age of the computer*. New York: Basic Books.

Park, O. (1996). Adaptive instructional systems. In: D. H. Jonassen (Ed.), *Handbook of research for educational communications and technology* (pp. 634–664). New York: Macmillan.

Park, O., Perez, R. S. & Seidel, R. J. (1987). Intelligent CAI: Old wine in new bottles or a new vintage? In: G. Kearsley (Ed.), *Artificial Intelligence and Instruction: Applications and Methods* (pp. 11–45). Reading, MA: Addison-Wesley.

Park, O. & Seidel, R. J. (1989). A multidisciplinary model for development of intelligent computer-assisted instruction. *Educational Technology Research and Development*, **37**, 78–80.

Pressley, M., Borkowski, J. G. & Schneider, W. (1987). Cognitive strategies: Good strategy users coordinate metacognition and knowledge. In: R. Vasta (Ed.), *Annals of Child Development* (Vol. 4, pp. 89–129). Greenwich, CT: JAI Press.

Psotka, J., Massey, L. D. & Mutter, S. A. (Eds.) (1988). *Intelligent tutoring systems: Lessons learned*. Hillsdale, NJ: Erlbaum.

Rabinowitz, M. & Glaser, R. (1985). Cognitive structure and process in highly competent performance. In: F. D. Horowitz & M. O'Brien (Eds), *The Gifted and Talented: Developmental Perspectives* (pp. 75–97). Washington, D.C.: American Psychological Association.

Rapaport, M. (1991). *Computer-mediated communications*. New York: John Wiley.

Riding, R. (1997). On the nature of cognitive style. *Educational Psychology*, **17**, 29–49.

Rieber, L. P. (1992). Computer-based microworlds: A bridge between constructivism and direct instruction. *Educational Technology Research and Development*, **40**, 93–106.

Roberts, J. M. (1996). The story of distance education: A practitioner's perspective. *Journal of the American Society for Information Science*, **47**, 811–816.

Romiszowski, A. J. (1992). *Computer-mediated communications*. Englewood Cliffs, NJ: Educational Technology.

Romiszowski, A. J. & Mason, R. (1996). Computer-mediated communication. In: D. H. Jonassen (Ed.), *Handbook of Research for Educational Communications and Technology* (pp. 438–456). New York: Macmillan.

Runco, M. A. (Ed.) (in press). *Handbook of creativity* (Vols 1 and 2). Norwood, NJ: Ablex.

Runco, M. A. & Albert, R. S. (Eds) (1990). *Theories of creativity*. Newbury Park, CA: Sage.

Salomon, G., Perkins, D. & Globerson, T. (1991). Partners in cognition: Extending human intelligence with intelligent technologies. *Educational Researcher*, **20**, 2–9.

Savery, J. R. & Duffy, T. M. (1995). Problem based learning: An instructional model and its constructivist framework. *Educational Technology*, **35**, 31–38.

Scardamalia, M. & Bereiter, C. (1994). Computer support for knowledge-building communities. *Journal of the Learning Sciences*, **3**, 265–283.

Scardamalia, M. & Bereiter, C. (1996). Engaging students in a knowledge society. *Educational Leadership*, **11**, 6–10.

Scardamalia, M., Bereiter, C., McLean, R., Swallow, J. & Woodruff, E. (1989). Computer supported intentional learning environments. *Journal of Educational Computing Research*, **5**, 51–68.

Schneider, W. (1993). Domain-specific knowledge and memory performance in children. *Educational Psychology Review*, **5**, 257–273.

Schofield, J. W. & Evans-Rhodes, D. (1989). Artificial intelligence in the classroom. In: D. Bierman, J. Greuker & J. Sandberg (Eds), *Artificial Intelligence and Education* (pp. 238–243). Springfield, VA: IOS.

Schuler, J. (1996). *Cyberspace as dream world*. (Retrieved January 18, 1998, from the World Wide Web: http://www.rider.edu/~suler/psycyber/cybdream.htm)

Seidel, R. J. & Park, O. (1994). An historical perspective and a model for evaluation of ITS. *Journal of Educational Computing Research*, **10**, 103–128.

Seidel, R. J., Park, O. & Perez, R. S. (1988). Expertise of ICAI: Development requirements. *Computers in Human Behavior*, **1**, 235–256.

Shavinina, L. V. (1996a). The objectivization of cognition and intellectual giftedness. *High Ability Studies*, **7**, 91–98.

Shavinina, L. V. (1996b). Specific intellectual intentions and creative giftedness. In: A. J. Cropley & D. Dehn (Eds), *Fostering the Growth of High Ability: European perspectives* (pp. 373–381). Norwood, NJ: Ablex.

Shavinina, L. V. (1997a). *Educational multimedia of 'tomorrow': High intellectual and creative psychoeducational technologies*. (Paper presented at the European Congress of Psychology, Dublin, Ireland.)

Shavinina, L. V. (1997b, September). *High intellectual and creative technologies as an educational multimedia of the 21st century*. (Paper presented at the European Open Classroom II Conference: School Education in the Information Society, Sisi, Crete, Greece).

Shavinina, L. (1998a). Interdisciplinary innovation: Psychoeducational multimedia technologies. *New Ideas in Psychology*, **16**, 189–204.

Shavinina, L. V. (1998b). On Miller's insights of genius: What do we know about it? *Creativity Research Journal*, **12**, 183–185.

Shavinina, L. V. (2000). High intellectual and creative educational multimedia technologies. *CyberPsychology & Behaviour*, **3**, 1–8.

Shavinina, L. V. & Ferrari, M. (in press (a)). Non-cognitive mechanisms in high ability. In: L. V. Shavinina & M. Ferrari (Eds), *Beyond Knowledge: Extracognitive Facets in Developing High Ability*. Mahwah, NJ: Erlbaum.

Shavinina, L. V. & Ferrari, M. (Eds) (in press (b)). *Beyond knowledge: Extracognitive facets in developing high ability.* Mahwah, NJ: Erlbaum.

Shavinina, L. V. & Kholodnaya, M. A. (1996). The cognitive experience as a psychological basis of intellectual giftedness. *Journal for the Education of the Gifted*, **20**, 3–35.

Shavinina, L. V. & Loarer, E. (1999). Psychological evaluation of educational multimedia. *European Psychologist*, **4**, 33–44.

Shavinina, L. V. & Sheeratan, K. L. (in press). Extracognitive phenomena in the intellectual functioning of gifted, creative, and talented individuals. In: L. V. Shavinina & M. Ferrari (Eds), *Beyond Knowledge: Extracognitive Facets in Developing High Ability*. Mahwah, NJ: Erlbaum.

Sherry, L. (2000). The nature and purpose of online discourse. *International Journal of Educational Telecommunications*, **6**, 19–51.

Shore, B. M. & Dover, A. C. (1987). Metacognition, intelligence and giftedness. *Gifted Child Quarterly*, **31**, 37–39.

Shore, B. M. & Kanevsky, L. S. (1993). Thinking processes: Being and becoming gifted. In: K. A. Heller, F. J. Mönks & A. H. Passow (Eds), *International Handbook of Research and Development of Giftedness and Talent* (pp. 133–147). Oxford: Pergamon Press.

Shute, V. J. (1991). Who is likely to acquire programming skills? *Journal of Educational Computing Research*, **7**, 1–24.

Shute, V. J. & Glaser, R. (1990). A large-scale evaluation of an intelligent discovery world: Smithtown. *Interactive Learning Environments*, **1**, 51–76.

Shute, V. J. & Psotka, J. (1996). Intelligent tutoring systems: Past, present, and future. In: D. H. Jonassen (Ed.), *Handbook of Research for Educational Communications and Technology* (pp. 514–563). New York: Macmillan.

Simonton, D. K. (1988). *Scientific genius: A psychology of science.* New York: Cambridge University Press.

Squires, D. (1996). Can multimedia support constructivist learning? *Teaching Review*, **4**, 10–17.

Sternberg, R. J. (Ed.) (1982). *Handbook of intelligence.* New York: Cambridge University Press.

Sternberg, R. J. (Ed.) (1984). *Mechanisms of cognitive development.* New York: Freeman.

Sternberg, R. J. (1985). *Beyond IQ: A triarchic theory of human intelligence.* New York: Cambridge University Press.

Sternberg, R. J. (1986). *Intelligence applied.* San Antonio, TX: Harcourt Brace.

Sternberg, R. J. (1987). Intelligence and cognitive style. In: R. E. Snow & M. J. Farr (Eds), *Aptitude, Learning and Instruction* (Vol. 3). *Conative and Affective Analyses* (pp. 77–97). Hillsdale, NJ: Erlbaum.

Sternberg, R. J. (Ed.) (1988a). *The nature of creativity.* New York: Cambridge University Press.

Sternberg, R. J. (1988b). A three-facet model of creativity. In: R. J. Sternberg (Ed.), *The Nature of Creativity* (pp. 125–147). New York: Cambridge University Press.

Sternberg, R. J. (1990). *Metaphors of mind: Conceptions of the nature of intelligence.* New York: Cambridge University Press.

Sternberg, R. J. & Kaufman, J. C. (1997). Innovation and intelligence testing. *European Journal of Psychological Assessment*, **12**, 175–182.

Sternberg, R. J. & Lubart, T. (1995). *Defying the crowd: Cultivating creativity in a culture of conformity.* New York: Free Press.

Suchman, L. (1987). *Plans and situated actions: The problem of human/machine communication.* New York: Cambridge University Press.

Teaching and learning—Towards cognitive society (a white paper of the European Commission) (1995). Brussels: Office for Official Publications of the European Community.

Tikhomirov, O. K. (1981). The psychological consequences of computerization. In: J. V. Wertsch (Ed.), *The Concept of Activity in Soviet Psychology* (pp. 256–278). Armonk, NY: M. E. Sharpe.

Tolmie, A. & Boyle, J. (2000). Factors influencing the success of computer mediated communication (CMC) environments in university teaching. *Computers & Education*, **34**, 119–140.

Vandervert, L. R. (1999). Maximizing consciousness across the disciplines: Mechanisms of information growth in general education. In: J. S. Jordan (Ed.), *Modeling Consciousness Across the Disciplines* (pp. 3–25). New York: University Press of America.

Vandervert, L. R. (2001). A provocative view of how algorithms of the human brain will embed in cybereducation. In: L. R. Vandervert, L. V. Shavinina & R. Cornell (Eds), *CyberEducation: The Future of Long Distance Learning* (pp. 41–62). Larchmont, NY: Liebert.

Vandervert, L. R., Shavinina, L. V. & Cornell, R. (Eds) (2001). *CyberEducation: The Future of Long Distance Learning.* Larchmont, NY: Liebert.

VanLehn, K. (1999). *Intelligent tutoring systems.* (Retrieved September 12, 2000, from the World Wide Web: http://www.pitt.edu/~vanlehn/ITSseminar9)

Verona, M. E. (2001). WebSims-Creating an online science lab. In: L. R. Vandervert, L. V. Shavinina & R. Cornell (Eds), *CyberEducation: The Future of Long Distance Learning* (pp. 237–256). Larchmont, NY: Liebert.

Wellington, J. J. (1985). *Children, computers and the curriculum.* New York: Harper & Row.

Westera, W. & Sloep, P. (2001). The future of competence learning in cyberreality: The virtual learning company and beyond. In: L. R. Vandervert, L. V. Shavinina & R. Cornell (Eds), *CyberEducation: The Future of Long Distance Learning* (pp. 115–136). Larchmont, NY: Liebert.

Part VII

Innovations in Different Domains

The International Handbook on Innovation
Edited by Larisa V. Shavinina

Dimensions of Scientific Innovation

Gerald Holton

Jefferson Physical Laboratory, Harvard University, USA

Abstract: Research in the history of scientific innovations reveals that apart from the usual textbook presentations of how science advances, there are—especially in the early, private stages of individual scientists' work—a great number of different procedures in actual use of which the final published papers rarely give even a hint. Examples of these often essential but largely hidden mechanisms of scientific innovation are given from the study of works by Johannes Kepler, Henri Poincaré, Enrico Fermi, and the discoverers of high-temperature superconductivity.

Keywords: Innovation; Scientific innovation; Discovery; Creativity; Kepler; Poincaré; Fermi.

Introduction

One of the last physicist-philosophers, P. W. Bridgman, defined the scientific method as "doing your damnedest, no holds barred". Bridgman, a Nobel Prize winner (1946) for establishing the field of high-pressure physics, and also the father of the movement in philosophy of science known as operationalism, was venting his exasperation with the simplified, schoolbook view of how scientists do their work, especially when they face a really difficult problem. For he knew, as all science practitioners do, that during those early, 'private' stages of research, when euphoria is about to give way to despair, enlightenment may come from the most unexpected, even 'nonscientific' direction. In modern times, scientists have been taught to keep this fact secret; the publications of the final research results are silent about it. The great biologist Peter Medawar once confessed, "It is no use looking to scientific 'papers', for they not merely conceal but actively misrepresent the reasoning that goes into the work they describe".

So let me pull away the curtain that usually covers the human drama, and discuss a few examples of the difficult struggle—not to demean but rather to celebrate it. And it makes a good story to boot.

Pursuing Cosmic Harmony, but Finding a New Law

The first example of a pioneer breaking through the boundaries of current knowledge in an unexpected way is Johannes Kepler. Since antiquity, splendid scientific ideas and results had been accumulating. But it is not unfair to characterize natural philosophy which Kepler faced at the start of the 17th century by the work of a contemporary of his, Robert Fludd of Oxford, whose writings have been called a mixture of the "astrological, alchemical, magical, cabalistic, theosophic, mock-mystic, and pseudo-prophetic". Kepler's main findings were quite different, and are useful to this day. He discovered the three basic laws of solar system astronomy that determine how all planets move. The most difficult to uncover was his third law, which states that for every planet, the square of its period of revolution (T) around the sun, when divided by the third power of its mean distance (R) from the sun, yields the same number. In short, T^2/R^3 is a constant, the same for all planets of our solar system.

How dull that result sounds now! But this simple law revealed to Kepler and to the entire world the commonality that binds together all the planets into one harmonic system, despite their vastly different individual orbits, speeds, and sizes. And that commonality also made more plausible Copernicus's heliocentric theory, which in Kepler's time was still widely resisted. Equally important, the third law, together with Galileo's mechanics, helped Newton to discover the dynamics of the solar system—the high point of the 17th-century Scientific Revolution.

In short, Kepler's equation was a crucial breakthrough, away from the stagnant prison of the narrow Aristotelian–Ptolemaic astronomy and into the infinitely open landscape of the modern worldview. We know the date of the climatic moment of Kepler's discovery: May 15, 1618. After years of agonizing exploration, while living in misery as a servant to city governors and the mad Emperor Rudolph II, he now

421

could express the feelings of an exhilarated iconoclast. He wrote: "What 25 years ago I already had a vague premonition of; what I determined 16 years ago as the whole aim of research; what I devoted the best part of my life to—*that* I have at last brought into the light . . . . I yield freely to the sacred frenzy . . . . The die is cast, I am writing the book . . . . It can wait a century for a reader, as God himself has waited six thousand years for a witness".

But how did Kepler do it? With what motivation and what help did he batter his way through the wall that had stood for millennia? On first opening Kepler's voluminous published works, we see little that seems promising. His three golden laws are scattered like rare jewels in a mountain of confusing and mystical material. To some degree the cultural soil on which he stood was the same as that of Robert Fludd. Early in Kepler's first book, the *Mysterium Cosmographicum* of 1596, he searched for the causes for the numbers, dimensions, and motions of the orbits of the planets. Surely the Creator must have had in mind some secret, harmonious order when He placed the planets' orbits in what looks a seemingly accidental manner, with huge, empty spaces between the orbs. Kepler first thought he found the answer in his famous fantasy that God had arranged a set of the five regular Platonic bodies, one nestled inside the other, to define and separate the paths of the six then-known planets. The secret harmony of the solar system might thus be revealed by geometry. And it almost worked. The fit with existing observations was close, but not quite good enough, and Kepler knew he had to look elsewhere to find the secret.

So when we open his *Harmonices Mundi* of 1619 we find at last, in its fifth section, Kepler's powerful third law, which establishes that all the orbits have a simple common mathematical property. But to our astonishment, the law appears in just one sentence, in the middle of a book that is mostly concerned with harmony in music. Though earlier Kepler had been motivated by a neo-Platonic philosophy, in which he had looked to geometry for the celestial harmony, now his search turned to associating the motions of the planets to relations among musical notes.

That sounds far-fetched now. But Kepler's university training had centered on the quadrivium: the four fields of arithmetic, geometry, music, and astronomy. In addition, Kepler had rigorously studied music theory, which was then much debated among intellectuals. Kepler had read not only Ptolemy's *Harmony*, but also a book by one of the period's most interesting innovators of music theory, namely Galileo's father, Vincenzio Galilei, the author of the *Dialogo della musica antica e moderna*. In Kepler's mind, God, in the act of Creation, had made use of musical consonances in designing the world system, with the resulting harmony radiating down from heaven, to resonate within the soul of each human being.

Specifically, what Kepler attempted was to make concrete and quantitative an old dream about the harmony of the spheres that went back to Ancient Greece. He discovered that when he associated with each planet in the solar system musical notes which represented their speeds at various moments, there were 48 occasions, repeated through eternity, when the speeds of the planets happened to be equivalent to musical chords that we would regard as harmonious. For example, comparing the speed of Saturn at its two extreme points along its ellipse, the ratio of these two speeds comes to 106 divided by 135, which is close to 4 divided by 5—only a churlish person would deny that! This ratio, 4/5, corresponds musically to a major third—pleasant to the ear and to the mind.

Kepler applied such reasoning to all the planets. In this way, he found that during their orbital motions, Earth and Venus from time to time sing out a minor and a major sixth. Kepler even found a heavenly six-part chord, generated by the orbital velocities of all six planets known at the time. Now it all worked—the harmony of the world system was revealed—although to more rational scientists, including Galileo, it must have seemed like the ravings of a madman. But that real, permanent breakthrough, Kepler's third law, was simply a byproduct which Kepler had stumbled on while carefully comparing the parameters of the planetary orbits in his search for those celestial harmonies.

It is regrettable that in our schoolbooks it now all comes down to merely learning the formula of Kepler's law, and that the cultural and historic context is neglected. But that this separation is possible carries an important lesson, too. For contrary to today's fashionable relativists and constructivists of scientific history, it is precisely a strength of science that, in the long run, it can throw off the various cultural scaffoldings which, in the nascent phase, helped some individual in building part of the ever-unfinished Temple of Isis. The concepts and theories of mature science, as used in daily life and publications, work well without any residue of their various individual human origins. It is left only to the historian to record the cries of agony and ecstasy of the pioneers.

The best metaphor for the scientist working, like Kepler, at the extreme intellectual limit seems to me that of a person trapped in a narrow prison, down in a shaft where the high walls consist of what is *already known*, the knowledge that satisfies the ordinary mortal but which oppresses the extraordinary one, as with claustrophobia. He claws desperately against those walls, as Kepler did for 25 years, trying to climb up or break out, to escape from the incompleteness of the known into the world beyond, of which he has only a haunting premonition. Robert Oppenheimer once compared working on very difficult problems with trying to crawl upward inside a mountain, not knowing when and where one will find the escape.

The Relativism of Impotency

At this point, I should recognize that a few scientists and philosophers do speak of having found epistemological limits in the science of the real world. The usual examples mentioned include Heisenberg's so-called principle of uncertainty (or, as he initially referred to it, the principle of inaccuracy (*Ungenauigkeit*)); that the speed of light is an upper limit for transmitting information; that no perpetual motion machine can exist. Fifty years ago, Sir Edmund Whittaker proposed the charming terminology that these and similar cases reveal the existence of the *Principle of Impotency*. But the feeling that these are really epistemological limits rests on a misunderstanding. The examples cited, and others like them, are only discoveries about the behavior of nature that, relative to a previous stage of science, were unexpected. Thus earlier forms of science for centuries gave false hope that a perpetuum mobile can be built. The fact that it cannot is an important, positive finding about nature, expressed in a law of thermodynamics. Similarly, the Newtonian style of science before Einstein and Heisenberg, before relativity theory and quantum mechanics, allowed one to think that positions and momenta of moving particles can both be found simultaneously to arbitrary degrees of accuracy, or that there exist Absolute Space, Absolute Time, Absolute Simultaneity, and instant propagation of information. These hopes within the old framework turned out to be false, revealing instead a more accurate picture of nature's operations.

All such cases are analogous to Galileo's discovery, by means of his telescope, that the moon has mountains like the earth, instead of being perfectly round and smooth as previous science had maintained. The impotency to find a smooth moon is an impotency only in the context of Aristotelian science. In the context of the new Galilean science, it became an unexpected fact of nature, and an important discovery.

'An Anticipatory Consonance with Nature'

The next case represents another point on the wide spectrum of those inexhaustible ingenuities and variety of means by which scientists try to transcend the limits of contemporary understanding. This one reveals a trick or tool that some scientists turn to, at least for a while. The scientist-philosopher Michael Polanyi coined for it the term "tacit knowing". Partly in rebellion against positivism, and in direct confrontation with the famous ending of Wittgenstein's *Tractatus*, Polanyi defined his concept in one sentence: "I shall start from the fact that *we can know more than we can tell*".

The concept is perfectly familiar from everyday experience (recognizing a face; or the skill displayed by an athlete, a musician, a craftsman). At first glance it seems out of place in science, but it is not far from what Einstein called "*Fingerspitzengefühl*", a feeling at the tips of your fingers where the way out of the difficulties lies.

When struggling with a problem that cannot be solved by standard procedures such as induction from good data, some scientists, often unconsciously, seem capable of using personal intuition. To return to our earlier metaphor, they can, so to speak, look over the top of the high wall before them. It is a precious gift, for which the physicist-philosopher Hans Christian Oersted has provided the happy term, an "anticipatory consonance with Nature", not too far from Leibniz's concept of resonating with the Pre-established Harmony.

A striking example of tacit knowledge is the moment when Enrico Fermi's hand moved in a way he did not understand, but eventually helped him to unravel a mystery which led him effectively to launch the nuclear age. It happened one morning in October 1934, in the old physics laboratory at the University of Rome. Fermi, a brilliant leader of a group of scientists almost as young as he, was puzzled: fast neutrons, when sprayed at a sample of silver sitting on a wooden table, copiously produced artificial radioactivity in the silver. However, they failed miserably to do so when the experiment was repeated on a marble table in another part of the laboratory. As was realized later, on the first table some of the neutrons that reached the silver had first hit the table top and then been scattered up from its wood, having been slowed down by collisions with the nuclei of the hydrogen atoms in the wood. However, the heavy atoms in the marble could not produce this slowing effect on neutrons—and contrary to accepted theory at the time, it was *slow* neutrons, not fast ones, which strongly induced radioactivity in silver. As one would say today, the hydrogen nuclei had acted as 'moderator', a concept that later became essential for constructing nuclear reactors.

Fermi described the crucial moment where the tacit dimension of scientific understanding asserted itself in this example:

> I will tell you how I came to make the discovery which I suppose is the most important one I have made. We were working very hard on the neutron-induced radioactivity and the results we were obtaining made no sense. One day, as I came to the laboratory, it occurred to me that I should examine the effect of placing a piece of lead before the incident neutrons. [I interject here that this was not going to work because lead, with its heavy atoms, would not slow down neutrons significantly] . . . (but) I was clearly dissatisfied with something: I tried every 'excuse' to postpone putting the piece of lead in its place. (Finally) . . . I said to myself: 'No, I do not want this piece of lead here; what I want is a piece of paraffin'. It was just like that: with no

advanced warning, no conscious, prior reasoning. I immediately took some odd piece of paraffin . . . and placed it where the piece of lead was to have been.

At once, the induced radioactivity presented itself abundantly, even for silver sitting on the marble table. Fermi's former student and co-worker, Emilio Segrè, called it "the miraculous effects of filtration". Within hours, Fermi understood what was happening: the hydrogen nuclei in the piece of paraffin slowed down during collisions; the material in the wooden table had done the same before. Fermi's "anticipatory consonance with Nature", long before the purely rational part of science could produce the correct theory of nucleonics, turned the research project of that whole group into a powerhouse of innovation. On coming across such events, where tacit knowing asserts itself to help leap over a barrier, I am reminded of Aristotle's observation: "The intellect approaches the object of its research while that is still opaque and obscure: but the intellect circles around it, and suddenly, thanks to its noetic light, the matter becomes visible and comprehensible".

Invoking the Personal Thematic Presupposition

Turning to more recent times, we may point to a study that shows again how the most unlikely-appearing tool, in the hands of the right persons, can be used to cut through a well-established border at the edge of known science. It is the story of the discovery of superconductivity at high temperatures. Superconductivity is the loss of electrical resistance below a critical temperature (T_c). The phenomenon of ordinary superconductivity at very low temperatures had been known since 1911, when Heike Kamerlingh Onnes discerned it in mercury cooled down to just a few degrees above absolute zero (to about 4 Kelvin). Many investigators took up the search for electrical conductors with higher critical temperatures, where it would be easier and cheaper to keep a substance resistance-free, and where the practical applications could be fabulous.

Here and there, small progress was made. But by 1973, over six decades after the original discovery, the best efforts in many research laboratories working with a great variety of materials had brought the critical temperature T_c only up to about 23 Kelvin. The general frustration of failing to break through this limit, year after year, was crystallized in a remark by Bernd Matthias, a highly respected physicist at the Bell Laboratories: in the absence of successful theories to guide one, it would be best to give up the search, for otherwise "all that is left in this field will be these scientific opium addicts, dreaming and reading one another's absurdities in a blue haze".

All this changed practically overnight, when superconductivity existing at about 30 Kelvin was discovered in 1986 by a two-man team, Karl Alex Müller and his former student Georg Bednorz, both working at the IBM Zurich Research Laboratory. Their announcement became an instant sensation, not just because of those extra 7 degrees, but mainly because the materials used by the Swiss team were distinctly different from all previously known superconductors, and this alone opened up the promise of reaching much higher ceilings. Also, the compound they used was relatively easy to prepare and to modify. No new technology was involved; the particular ceramic they used had been available for decades but had never been considered a candidate for high-temperature superconductivity research.

The news of the discovery caused a so-called "Woodstock of Physics", an impromptu session at the meeting of the American Physical Society in 1987; over 3,500 scientists crowded into the meetings rooms to hear about the new achievement. The Nobel Prize committee awarded the 1987 Prize for Physics to Müller and Bednorz with maximum possible speed. All over the world, researchers could now look in new directions for ever-higher temperatures T_c. The present record stands at over 160 Kelvin. For some physicists, the hope has been revived that eventually a superconductor will be found that works even near room temperature. The practical applications, already substantial, would be immense.

Now, the most interesting question for me was: *What were the crucial elements* that made the scientific discovery of Müller and Bednorz possible in the first place? Their success depended not only on their own skills, but on two other factors. The first was of course the work of other scientists who had preceded them, and on whose findings or inventions they built—as is true in every advance. But the other factor was a crucially important leap of the imagination, a typical characteristic of the barrier smashers throughout the history of science. The most intriguing addition here, as so often, is the private dimension of scientific discovery, the motivation.

Because of the tradition of formality in science writing today, the personal aspect of the discovery was of course absent from the published record. But here I was lucky. Dr Müller and Dr Bednorz kindly allowed me to interview them, and what I found was revealing indeed. The two men had first spent two fruitless years trying to get to a higher T_c with the types of material that had been used by all the others. But earlier they had worked with success on quite different problems in physics, using a class of highly symmetrical (cubic) crystals. As Dr Müller said, since his student days, they had always "worked for him". Nobody had so far tried to use these perovskite-type crystals for studying high-temperature superconductivity, and there were no convincing reasons from existing theory why anyone should have. Nevertheless, the Swiss team decided to try its luck on a perovskite-type material, a copper-containing oxide, fundamentally different from the materials that had been ransacked by the pioneers in the field. When other physicists heard about it, they called it a 'crazy' idea. But it worked.

Why was it not a crazy idea for Müller and Bednorz? I asked Dr Müller to share his motivation with me. It turned out that his unlikely choice of a perovskite in that search was guided not just by the force of (well rewarded) habit. As he put it: "I was always dragged back to this symbol", the cubic crystal. He first became fascinated with the cubic crystal structure in 1952, when he was working on his doctorate. It so happened that Wolfgang Pauli was one of Müller's professors at the Swiss Technical Institute (ETH). Just at that time, Pauli had published an essay entitled 'The Influence of Archetypal Ideas on the Scientific Theories of Kepler', in a book with Carl Jung. Much impressed by that essay, Müller started to read Kepler avidly, and so he encountered Kepler's commitment to the use of three-dimensional structures of high symmetry—the cube and the other four Platonic solids—in his early work on planetary motion.

Dr Müller continued as follows: "If you are familiar with Carl Jung's terminology, the (cubical) perovskite structure was for me, and still is, a symbol of it—it's a bit high-fetched, but—of holiness. It's a Mandala, a self-centric symbol which determined me . . . . I dreamt about this perovskite symbol while getting my doctor's degree. And more interesting about this, is also that (in this dream) the perovskite was not just sitting on a table, but was held in the hand of Wolfgang Pauli, my teacher". At the time, Müller had divulged this extraordinary aspect of his inspiration, which encouraged him eventually to try a perovskite for superconductivity, only to friends and to Pauli's last assistant. He has since discussed it in an introspective essay, illustrated with the Dharmaraja Mandala.

To the historian this is familiar ground. Some scientists, from Kepler to Francis Crick and James Watson, were guided in the early stages of their research by a visually powerful, highly symmetric geometrical design. So in faithfulness to Müller's self-report, his personal thematic presupposition has to be added to the resources he used and was as important as any of the others. With that inclusion, we recognize the inheritance of a line of empowering conceptions, reaching in this case back first to Wolfgang Pauli and then to the works of Kepler, who gloried in his thematic preference for high symmetry four centuries earlier.

Thematic Influences in Scientific Discovery

In the cases already addressed, we have seen how preexisting commitments in culture, philosophy, and society have shaped the very process and direction of discovery. In the final, most extensive case, we turn to a scholar who was doubly influential as a scientist and as a culture carrier, directing further exploration across academic and disciplinary boundaries.

Henri Poincaré (1854–1912)—mathematician, physicist, and all-round polymath—was a very symbol of the establishment, his five professorships held simultaneously, and a member of the *Académie des Sciences* as well as the *Académie Française*—a rare combination, making him in effect doubly immortal. Nor did it hurt his standing in society that his cousin was Raymond Poincaré, the Prime Minister and eventually President of the French Republic. By temperament, Henri Poincaré was conservative, but also immensely productive, with a torrent of nearly 500 published papers on mathematics—many of them fundamental discoveries, from arithmetic to topology and probability, as well as in several frontier fields of mathematical physics, including launching in his book on celestial mechanics what is now called chaos theory. But regarding himself justly as a culture carrier, and perhaps sensitized by a then fashionable turn-of-the-century movement brandishing the slogan 'the Bankruptcy of Science', Poincaré also wrote with translucent rationality on the history, psychology, and philosophy of science, including his philosophy of conventionalism, in such semi-popular books as *Science and Hypothesis* of 1902. For cultured persons all over the world, these were considered required reading. And all his work was done with intense focus and at great speed, one finished right after another.

Poincaré was a mathematical genius of the order of Carl Friedrich Gauss. Before he was 30, Poincaré became world-famous for his discovery, through his ingenious use of non-Euclidian geometry, of what he called the fuchsian functions (named after the German mathematician Immanuel Lazarus Fuchs, who was publishing in the 1860s to 1880s). I shall come back to the circumstances of that discovery in a while.

Although Poincaré generally kept himself well informed of new ideas in mathematics, on this particular point his biographer, Jean Dieudonné, allows himself the schoolmasterly remark that "Poincaré's ignorance of the mathematical literature, when he started his researches, is almost unbelievable . . . . He certainly had never read Riemann" – referring to Bernard Riemann, the student of Gauss and the person first to define the *n*-dimensional manifold in a lecture in June 1854. That lecture, later published, and the eventual opus of Riemann's three-volume-long series of papers, are generally recognized to be monuments in the history of mathematics, inaugurating one type of non-Euclidean geometry.

For over 2,000 years, Euclid's *Elements of Geometry* had reigned supreme, showing that from a few axioms the properties of the most complicated figures in a plane or in three-dimensional space followed by deduction. Euclid has been the bane of most pupils in their early lessons, but to certain minds his *Geometry* was and is representative of the height of human accomplishment. To Albert Einstein, who received what he called his "holy" book of Euclidean geometry at age 12, it was a veritable "wonder". He wrote: "Here were assertions ... which—though by no means evident—could nevertheless be proved with such certainty that any doubt appeared to be out of the question. This lucidity and certainty made an indescribable impression upon me ...".

Galileo, too, was stunned by his first encounter with Euclid. It is said that as a youngster, destined to become a physician, he happened to enter a room in which Euclidean geometry was being explained. It transfixed him, and set him on his path to find the mathematical underpinnings of natural phenomena. To Immanuel Kant, of course, Euclidean geometry was such an obvious necessity for thinking about mathematics and nature that he proposed it as an exemplar of the synthetic *a priori* that constituted the supporting girder of his philosophy.

But in the early part of the nineteenth century, a long-simmering rebellion came to a boil against the hegemony of Euclidean geometry, and especially its so-called fifth axiom—the one which implies, as our schoolbooks state clumsily, that through a point next to a straight line only one line can be drawn which is parallel to it, both of them intersecting only at infinity. Here, the main rebels were Riemann, the Hungarian János Bolyai, and the Russian Nicolai Ivanovich Lobachevsky. As to Poincaré's initial ignorance of that particular literature, ignorance can be bliss, not merely because reading Riemann's turgid prose is no fun, but chiefly because Poincaré's approach was original with him.

When I look at the work of scientists, I try to be alert to the role the thematic presuppositions may have motivated their discoveries. In the case of Poincaré, the most prominent thematic element, common both to his mathematics, his physics, and his philosophy, is his embrace, for better or worse, of the thema of *continuity*. Thus, as his biographer summarizes it, the "continuity methods dear to Poincaré led him to adopt continuous parameters and groups in a field that would have benefited from greater flexibility". Poincaré's adopted "Hermite's method of 'continuous reduction' in the arithmetic theory of forms", had a policy for looking for 'invariance under transformations', and proved the "existence of a whole continuum of solutions" in the three-body problem. It was Poincaré who had to inform H. A. Lorentz that Lorentz's transformation equations in the early version of relativity theory formed a group that left invariant the quadratic expression $x^2 + y^2 + z^2 - c^2 t^2$. In sum, "The main leitmotiv of Poincaré's mathematical work is clearly the idea of continuity".

That had a parallel predilection in Poincaré's physics. Indeed, Poincaré may well have achieved a true special theory of relativity before Einstein, had he not clung obstinately to the old concept of that continuum in space, the ether, which Einstein was to dismiss in 1905 in one casual phrase. In fact, Poincaré

never made his peace with Einstein's ether-free relativity. His beautiful and seductive prose often hid a tentative method of argument just where he was getting most into trouble. Thus he toyed with the idea of a uniform time, a Newtonian sensorium of God; yet on rejecting it as untestable, he put nothing else in its place. In 1912, long after Einstein's and Minkowski's achievements in relativity had become acknowledged by more and more excellent physicists despite the initial shock, Poincaré bared his conventionalist philosophy by writing on those radical innovations:

> What shall be our position in view of these new conceptions? Shall we be obliged to modify our conclusions? Certainly not; we had adopted a convention because it seemed convenient and we had said that nothing could constrain us to abandon it. Today some physicists want to adopt a new convention. It is not that they are constrained to do so; they consider this new convention more convenient; that is all. And those who are not of this opinion can legitimately retain the old one in order not to disturb their own habits. I believe, just between us, that this is what they should do for a long time to come.

Worse still, Poincaré's long researches in electromagnetism, including the Maxwellian theories of light which he had introduced in France, caused him to be repelled by another of Einstein's other great discoveries of 1905, that of the quantum of light, later called the photon. The confrontation between these two different minds came to a head in 1911 at the so-called Solvay Conference in Brussels. A large fraction of the world's best physicists were gathered for a meeting to unpuzzle the strange new quantum theory, which contradicted the centuries-long Newtonian idea that nature does not move by jumps, and instead proposed the introduction of discreteness and discontinuity into nature's phenomena.

As the proceedings of that conference show, Poincaré and Einstein appeared to be on opposite sides, in part because Poincaré, surprisingly, had come to the conference without having seriously studied the new quantum physics publications. The British physicist J. H. Jeans came away from the Conference converted to the new physics, writing simply: "The keynote of the old mechanics was continuity, *natura non facit saltus* (as Newton had put it). The keynote of the new mechanics is discontinuity". But Poincaré, writing in 1912 after the summit, and shortly before his unexpected death, concluded, "The old theories, which seemed until recently able to account for all known phenomena, have recently met with an unforeseen check .... An hypothesis (of quantization of energy) has been suggested ... but so strange an hypothesis that every possible means was thought for escaping it. The search has revealed no escape so far .... Is discontinuity destined to reign over the physical universe, and will its triumph be final?" It was a cry from the heart, the more so as the antithema of discontinuity was invading classical science from many directions, for example, in genetics ('mutations', entering in the 1890s), in radioactivity ('transmutations', a few years later), and in the experimental findings in C. T. R. Wilson's cloud chamber that showed individual collisions between an alpha-particle and an atom.

On the evidence given so far, it would be easy to judge Poincaré to be, and to regard himself as, a continuist to the core. But we have learned that especially those innovators, in both science and the arts, who are far above our own competences and sensibilities, are not so easily pinned down. They tend to have what seem to us to be contradictory elements, but which somehow nourish their creativity.

In the case of Poincaré, his readers can encounter also a *discontinuist* side. Having warned, at the very end of the Solvay Conference, that quantum physics contradicted the very basis of classical triumphs, namely the use of differential equations, he nevertheless threw himself at once into a serious study of the consequences of quantum ideas, declaring them to be potentially without doubt the greatest and most profound revolution that natural philosophy has undergone since Newton. The other main evidence for Poincaré's willingness to face the force of sudden changes emerged strikingly in his ideas about the psychology of invention and discovery.

The reference here may be widely familiar, but is so startling that it deserves nevertheless to be mentioned. I refer to the famous lecture on "L'invention mathématique", which he gave in 1908 at the *Société de Psychologie* in Paris. The fine mathematician Jacques Hadamard, in his book *The Psychology of Invention in the Mathematical Field* (New York: Dover, 1945), remarked on that lecture that it "throw(s) a resplendent light on relations between the conscious and the unconscious, between the logical and the fortuitous, which lie at the base of the problem (of invention in the mathematical field)" (p. 12). In fact, Poincaré was telling the story about his first great discovery, the theory of fuchsian functions and fuchsian groups. Poincaré had attacked the subject for two weeks with a strategy (typical in mathematics) of trying to show that there could not be any such functions. Poincaré reported in his lecture, "One evening, contrary to my custom, I drank black coffee and could not sleep. Ideas rose in crowds; I felt them collide until pairs interlocked, so to speak, making a stable combination" (p. 14). During that sleepless night he found that he could in fact build up one class of those functions, though he did not yet know how to express them in suitable mathematical form.

Poincaré explained in more detail: "Just at this time, I left Caen, where I was living, to go to a geological excursion . . . . The incidence of the travel made me forget my mathematical work. Having reached Coutance, we entered an omnibus to go someplace or other. At the moment when I put my foot on the step, the idea came to me, without anything in my former thoughts seeming to have paved the way for it, that the transformations I had used to define the fuchsian functions were identical with those of non-Euclidean geometry. I did not verify the idea; I should not have had time, as, upon taking my seat in the omnibus, I went on with a conversation already commenced, but I felt a perfect certainty. On my return to Caen, for conscience's sake, I verified the results at my leisure" (p. 13).

Perfecting his idea came unexpectedly sometime later, when he was engaged in entirely different work: "One morning, walking on the bluff (at a seaside resort), the idea came to me with just the same characteristics of brevity, suddenness and immediate certainty, that the arithmetic transformations of indefinite ternary quadratic forms were identical with those of non-Euclidian geometry". Poincaré analyzed those two events in these terms: "Most striking at first is this appearance of sudden illumination, a manifest sign of long, unconscious prior work. The role of this unconscious work in mathematical invention appears to me incontestable". "It seems, in such cases, that one is present at one's own unconscious work, made particularly perceptible to the overexcited consciousness . . ." (pp. 14–15).

Hadamard collected a number of similar reports, where out of the continuous, subterranean incubation in the subconscious, there appeared, in a discontinuous way, a rupture of startling intensity, the conscious solution. He mentioned Gauss himself who spoke of such a rupture as "a sudden flash of lightning", and similar observations by Hermann Helmholtz, Wilhelm Ostwald, and Paul Langevin—not to forget Mozart, who spoke memorably about the source of his musical thoughts as follows: "Whence and how do they come? I do not know, and I have nothing to do with it".

Poincaré himself also expressed puzzlement about the source of his sudden ideas. In the full text of Poincaré's talk of 1908, soon widely read in chapter 3 of his popular book *Science and Method* (1908), he confesses, "I am absolutely incapable even of adding without mistakes, (and) in the same way would be but a poor chess player" (p. 49). But he reported to having "the feeling, the intuition, so to speak, of this order (in which the elements of reasoning are to be placed), so as to perceive at a glance the reasoning as a whole". He celebrated "this intuition of mathematical order that makes us divine hidden harmonies and relations" (pp. 49–50). To be sure, after that intuition comes labor: "Invention is discernment, choice". But for that, priority must be granted to aesthetic sensibility in privileging unconscious phenomena, 'beauty', and 'elegance'. How congenial this must have sounded to the artists among his readers!

Poincaré had discussed the nature of discovery also earlier, especially in his book, *Science and Hypothesis*. A point he made that particularly struck home at the time was the notion that concepts and hypotheses are not given to us uniquely by nature itself, but are to a large degree conventions chosen by the specific investigator for reason of convenience, and guided by 'predilections' (p. 167). As to the principles of geometry, he announces his belief that they are only 'conventions' (p. 50). But such conventions are not arbitrary. Eventually they have to result in a mathematics that sufficiently agrees with what we can compare and measure by means of our senses. And in the specific case of physics, any convenient presupposition still would eventually have to pass the test of serving in the explanation of observable phenomena.

A significant part of *Science and Hypothesis* was Part II, consisting of three chapters, devoted to non-Euclidean and multi-dimensional geometries. In those pages there are neither equations nor illustrations; but there are great feats of trying to clarify things by analogy. To give only one widely noted example: in attempting to make the complex space of higher geometry plausible, Poincaré introduced a difference between geometric space and conceptual or 'representative' space. The latter has three manifestations: visual space, tactile space, and motor space. The last of these is the space in which we carry on our movements, leading him to write, in italics, *"Motor space would have as many dimensions as we have muscles"* (p. 55).

These and Poincaré's other attempts to popularize the higher geometries must have left the lay reader excited, but, to tell the truth, not really adequately informed. Happily, at just about that time help came from a few popularizations by others which further enhanced the glamour of higher mathematics in the imagination of the young artists in Paris. One of these aids was a book that plays an important role in our story. In 1903, a year after *Science and Hypothesis*, there was published in Paris a volume entitled *Elementary Treatise on the Geometry of Four Dimensions: An Introduction to the Geometry of n-Dimensions*. In it, Poincaré's ideas and publications are repeatedly invoked, from the second page on. The author was a now almost forgotten figure, E. Jouffret (the letter E. hiding his wonderful first names, Esprit Pascal). He identified himself on the title page as a retired Lieutenant Colonel in the Artillery, former student at the Ecole Polytechnique, officer of la Légion d'honneur, officer of public instruction, and member of the Mathematical Society of France. In 1906 he added a more stringent treatment, in his volume *Mélanges de Géométrie a Quatre Dimensions*.

There is also a good deal of evidence that a friend of the circle of artists in Paris, an insurance actuary named Maurice Princet, was well informed about the new mathematics, and acted as an intermediary between the painters and such books as Jouffret's.

Of course non-Euclidean geometry had been around for many decades, Lobachevsky writing in 1829 and Bolyai in 1832. But both were little read even by mathematicians until the 1860s. Then, for two decades to either side of 1900, a growing flood of literature, professional and popular, fanned the enthusiasm about that unseen fourth spatial dimension.

This cultural phenomenon can have a variety of possible explanations. First, one must not forget that this branch of mathematics was still at the forefront of lively debate among mathematicians themselves. There were about 50 significant professionals in Europe and in the United States working in that area. Second, to lay persons, the new geometry could be a liberating concept, by hinting at an imaginative sphere of thought not necessarily connected to the materialistic physical world that had been presented by 19th-century science—which in any case was itself in upheaval thanks to a stunning series of new discoveries. And above all, the new geometries lent themselves to wonderful and even mystical excesses of the imagination, not least by literary and figurative artists and musicians. They included Dostoyevsky; H. G. Wells; the science fiction writer Gaston de Pawlowski; Alfred Jarry of pataphysics; Marcel Proust; the poet Paul Valery; Gertrude Stein; Edgar Varèse; George Antheil; the influential Cubists Albert Gleizes and Jean Metzinger, and so on—not to speak of Ouspensky and the Theosophists. Some artists were explicit about their interest. For example, Marcel Duchamp, in his private notes, wrote about his innovating encounter with Jouffret's book, and explained to friends that part of his masterwork, the "Large Glass", incorporated representation of four-dimensional objects. Kazimir Malevich gave the subtitle "Color masses in the Fourth Dimension" to a work in 1915; and as late as 1947, the surrealist Max Ernst produced a painting he called "L'homme intrigué par le vol d'une mouche non-euclidienne".

***

The scientific imagination at times avails itself of the most unexpected, 'extra-scientific' conceptions, while 'doing their damnedest', be it through music, tacit knowledge, thematic presuppositions, aesthetic (or philosophical) preferences. We now know of the influence of *Naturphilosophie* and the philosophy of Immanuel Kant on Hans Christian Oersted's discovery of the magnetic field around electric currents; of Niels Bohr's debt to the psychology of William James, in proposing the Complementarity Principle; and of the encouragement Einstein found for his whole unification program, from the special theory of relativity on, in the literary works of Johann Wolfgang von Goethe, which Einstein studied from his school days on. The examples are endless.

But here one must avoid a careless extrapolation. The work of major composers, artists and authors bears forever their sacred individual authority, as well as the stamp of the cultural context of their time. But the products of research which eventually become absorbed into the *scientific* canon, into the corpus of mature and rational science, have left behind those extra-scientific, often 'non-rational' influences which, in the nascent phases, may have led individual scientists to their major discoveries. No matter what tools they used, or what the momentary social context of their labors was, the lasting results are only revelations of Nature's own regularities; and those are in principle open to anyone—as is proved by the fact that not infrequently the same law, such as the Conservation of Energy, was discovered more or less simultaneously by several researchers in different countries, approaching it in their unique ways. When the building is finished, the scaffolding falls away.

Albert Einstein put it this way: "Science as something coming-into-being, as a goal, is just as subjectively, psychologically conditioned as all other human endeavors". But at the same time he warned: "Science as something already in existence, already completed, is the most objective, impersonal thing that we humans know".

The second lesson from our case histories is that the imagined epistemological limits of natural science have turned out to be porous or illusory. I have seen no solid evidence that it will ever be otherwise. On the contrary, nature has constantly revealed ever more startling surprises—and mankind's comprehensibility to deal with them has been ever more agile.

This leaves me less with satisfaction than with a sense of wonderment, of awe. Another insight of Einstein contains, for me, the final word on this topic, especially because it shows in his last sentence where human beings do encounter a real, insuperable mystery—although not one of reason but one of the spirit.

In a letter of 1952 to his old friend Maurice Solovine, Einstein wrote: "You find it strange that I consider the comprehensibility of the world ... as a wonder or as an eternal secret. Well, *a priori* one should expect a chaotic world which cannot be grasped by the mind in any way. One *should* expect the world

to be subject to law only to the extent that we order it through our intelligence. Ordering of this kind would be like the alphabetical ordering of the words of a language. By contrast, the kind of order created by Newton's theory of gravitation, for instance, is of a wholly different character. Even if the axioms of the theory are proposed by a human being, the success of such a project (finding order in nature through science) presupposes that a high degree of ordering exists in the objective world; and this could not be expected *a priori*. That is the 'wonder'—which is being constantly reinforced, precisely as our knowledge expands".

Further Reading

Gerald Holton, *Einstein, history, and other passions* (Cambridge, MA: Harvard University Press, 2000), Chapter 3 ('"Doing one's damnedest": The evolution of trust in scientific findings'), and Chapter 4 ('Imagination in Science').

Max Caspar, *Kepler* (New York: Dover, 1993).

Edmund Whittaker, *From Euclid to Eddington: A study of conceptions of the external world* (New York: Dover, 1958).

Michael Polanyi, *Personal knowledge: Towards a post-critical philosophy* (London: Routledge, 1998).

Emilio Segrè, *Enrico Fermi, physicist* (Chicago, IL: University of Chicago Press, 1970).

Gerald Holton, *The scientific imagination* (Cambridge, MA: Harvard University Press, 1998), 'Introduction: How a scientific discovery is made: The case of high-temperature superconductivity'.

Peter Galison, Stephen R. Graubard & Everett Mendelsohn (Eds), *Science in culture* (New Brunswick, NJ: Transaction Press, 2001), Chapter 1, 'Einstein and the cultural roots of modern science'.

Henry Poincaré, *Science and hypothesis* (New York: Dover, 1952 (originally 1902)), and *Science and method* (New York: Dover, 1952 (originally 1908)).

Jean Dieudonné, 'Henri Poincaré'. In: C. Gillispie (Ed.), *Dictionary of scientific biography* (Vol. 11) (New York: Charles Scribner's Sons, 1973).

Linda Dalrymple Henderson, *The fourth dimension and non-Euclidean geometry in modern art* (Princeton, NJ: Princeton University Press, 1983).

Gerald Holton, 'Henri Poincaré, Marcel Duchamp, and Innovation in Science and Art', *Leonardo*, vol. 34, no. 2 (2001).

The International Handbook on Innovation
Edited by Larisa V. Shavinina

Do Radical Discoveries Require Ontological Shifts?

Michelene T. H. Chi and Robert G. M. Hausmann

Learning Research and Development Center, University of Pittsburgh, USA

Abstract: The theoretical stance explicated in this chapter assumes that scientific discoveries often require that the problem-solver (either the scientist or the inventor) re-conceptualizes the problem in a way that crosses ontological categories. Examples of the highest level of ontological categories are ENTITIES, PROCESSES, and MENTAL STATES. Discoveries might be explained as the outcome of the process of switching the problem representation to a different ontological category. Examples from contemporary and the history of science will be presented to support this radical ontological change hypothesis.

Keywords: Scientific discovery; Ontological shift; Emergent systems; Contemporary science; History of science; Representation.

Introduction

This chapter entertains a simple claim. The claim is that many great revolutionary discoveries in science may have occurred because the scientists have undertaken an ontological shift. That is, the scientists had re-conceptualized or re-represented the problem (i.e. the phenomenon to which she/he is seeking an explanation) from the perspective of one ontology or ontological category to another ontology. The remaining chapter explains what this re-representation or shifting across ontological categories entails, and why it is unusual to undertake, thereby explaining the low frequency of revolutionary scientific discoveries. Examples from contemporary and the history of science are cited to exemplify this claim.

The Nature of Ontological Categories

All concepts and ideas (or concepts within ideas and theories) belong to a category. Psychologists define concepts in the context of their category membership. The term concept is used here broadly to refer to a category instance (such as a cat, as an instance of the category Animal, a thunderstorm as an instance of a Process category, and an idea as an instance of a MENTAL STATES category). To assume that all concepts belong to a category is quite standard. For instance, White (1975) also assumed that a concept signifies a way of classifying something. The advan-

tage of categorization is that it allows people to assign the same label to a new instance of the category, and to make inductive and deductive inferences about a new category member (Chi, Hutchinson & Robin, 1989; Collins & Quillian, 1969; Medin & Smith, 1984). Thus, there is clearly a cognitive advantage of having a categorical representation.

What structure does a categorical representation take? Psychologists have by and large addressed the hierarchical (subset–superset) nature of categorical representation. Questions of interests have been: Within such a hierarchy, which categories are the basic level (Rosch, Mervis, Gray, Johnson & Boyes-Braem, 1976)? How are super-ordinate categories coalesced and subordinate categories differentiated (Smith, Carey & Wiser, 1985)? But most of the attention has been paid to the nature of the features that coheres category members, such as whether they are defining features, characteristic features, core features, explanation-based or theory-based features. In our work, we are concerned primarily with lateral (rather than hierarchical) relationship among the categories. We define the lateral relationship to be an ontological one (Chi, 1997).

Ontology refers to the categorical structure of reality. It has long been assumed, since the times of Aristotle, that things belong to *fundamentally different* ontological categories. What is in dispute, to some extent, is the structure of the ontological categories. Categorical structure is typically hierarchical, with 'being' as the topmost category, subsuming everything that exists (Lowe, 1995). Deciding on the structure of ontological categories is a metaphysical question. Our concern, however, is to propose that *psychologically*, people represent categories with ontological boundaries as well. We are not focusing, at the moment, on the exact structure of such psychological ontological categories, but merely to propose the influence of such ontological categories on discoveries.

We can begin by assuming a few major ontological categories as psychologically real and distinct, as our working definition. Consistent with theories provided by Sommers (1971) and Keil (1979), there is likely no dispute that ENTITIES (or material kinds), PROCESES, and MENTAL STATES are three basic ontological categories, both metaphysically and psychologically. These major categories (or sometimes we will refer to them as trees)[1], with subcategories that they subsume, are shown in Fig. 1. Thus, Fig. 1 shows a plausible ontological categorical structure. Other major ontological categories that are not shown may be TIME, SPACE, and so forth. Another issue that we are

not dealing with is whether metaphysical ontological categories map directly onto psychological ontological categories. We suspect the mapping is not perfect; however, we will sidestep this issue for now.

The psychological reality of ontological categories can be determined by their ontological attributes, which is a set of constraints or properties governing the behavior of members of a given ontological category. For example, objects in the ENTITIES category must have a certain set of constraints that dictate their behavior and the kinds of properties they can have. ENTITIES (such as sand, paint, or men) can have ontological attributes such as being containable, storable, have volume and mass, can be colored, and so forth. In contrast, Events such as war (belonging to the PROCESS ontological tree) do not have these ontological attributes and obey a different set of constraints. Therefore, members of two ontological categories that occupy different trees have mutually exclusive ontological attributes.

Notice that the definition of an ontological attribute is quite different from all the other kinds of categorical features that psychologists have studied. It is not a defining feature, which is an essential feature that a member of a category must have; nor is it a

---

[1] Trees will be referred to in capital letters, and ontological categories will have the first letter capitalized.

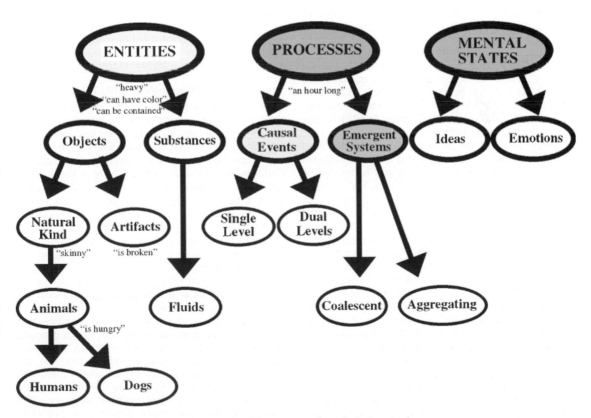

Figure 1. A plausible structure of ontological categories.

characteristic feature, in which the members of a category frequently or typically have (Chi, 1997). Instead, an ontological attribute is defined by Sommers (1971) as a feature that an entity has the potential to have, such as being colored, even though it does not necessarily have it. War or pain, for instance, does not have the ontological attribute of 'being red', whereas squirrels can have such an ontological attribute even though squirrels are commonly not red. Thus, ontological attributes are distinct from other kinds of categorical features in that they are constraints that govern a member of an ontological category, without that member necessarily having that feature.

The psychological reality of ontological categories can be determined by a linguistic test commonly used by philosophers—the predicate sensibility test. As illustrated in Fig. 1, each of the major trees (ENTITIES, PROCESSES and MENTAL STATES) generates a hierarchy of subcategories. A predicate (indicated in quotes), which modifies one concept, will sensibly modify all other concepts below it on the same (branch of a) tree (commonly known as 'dominates', e.g. Sommers, 1963; or 'spans' Wall, 1972), even if it is false. For example, 'hungry' can be applied to the category of Animals and all of its subcategories, such as 'humans' and 'dogs'. Therefore, Animals and Humans are not ontologically distinct. Thus, a bee, an Animal on the ENTITIES tree, has the potential to be 'heavy' even though it is false, whereas a bee cannot be 'an hour long', a predicate of the PROCESS tree. Conversely, an event can be 'an hour long' but not 'skinny'. The point is that predicates on the same tree can modify concepts below it sensibly even if it is false because it is plausible that a specific concept possesses that attribute, and moreover, the truth or falsity of the sentence can be checked.

When an attribute of one category cannot span members of another category, then the two categories are ontologically distinct, and they will be referred to here as 'lateral' or 'parallel' categories. For example, 'is broken' (an attribute of the Artifact category) cannot be used sensibly to modify an entity from a parallel category such as a dog (a member of the Animal category). That is, 'dogs' and 'Artifacts' do not involve a direct hierarchal relationship. Similarly, it makes no sense to say, 'the thunderstorm is broken'. Such statements are called category mistakes. 'The thunderstorm is broken' is not merely a falsehood, for otherwise 'the thunderstorm is unbroken' would be true. 'The thunderstorm is broken' is a category mistake because 'broken' is a predicate used to modify physical objects made by humans, whereas a thunderstorm is a type of PROCESS of the Emergent Systems kind. By this definition then, categories on parallel branches of the same tree (such as Natural Kind and Artifacts, see Fig. 1), as well as on branches of different trees (such as Natural Kinds in the ENTITIES tree and Events in the PROCESS tree), are onto-

logically distinct. However, parallel categories on the same tree do share some attributes. For example, both an Artifact such as a teapot and a Natural Kind, such as a dog, can have color. Nevertheless, they will be considered to be ontologically distinct, as the evidence to be cited below shows, because they have mutually exclusive sets of ontological attributes. In contrast, the attributes of the trees are totally, mutually exclusive.

Gelman (1988) and Keil (1979, 1989) tested the psychological reality of a few ontological categories depicted in Fig. 1. Besides the sensibility judgment task, Keil (1989) has also used a physical transformation task such as surgical operations to show that even young 5-year-old children deny the possibility that a toy bird could ever be made into a real bird. Thus, they adamantly honor the distinction between Natural Kinds and Artifacts even at such a young age.

In general, categories are ontologically distinct if one is not super-ordinate of the other. Thus, branches on the same tree can presumably form distinct ontological categories as well, as in the case of the distinction between Natural Kinds and Artifacts. For purposes of our discussion, we will consider the categories on the three major ontological trees in Fig. 1 to be ontologically distinct, as well as 'branches' on the same tree that do not occupy a subordinate/super-ordinate relationship. Whether or not there are degrees of ontological distinctiveness (e.g. are humans and dogs less ontologically distinct than Natural Kinds and Artifacts?) remains an epistemological and psychological issue that is not addressed here.

Are Processes of Ontological Shifts Difficult?

What is an ontological shift? By our definition, an ontological shift is merely the re-assignment or re-categorizing of an instance from one ontological category to another. Suppose we acknowledge that Fish and Mammals are distinct ontological categories. Children often mistake 'whales' to be a kind of Fish rather than Mammals. However, when children are told that a whale is a mammal, they do not seem to have difficulty reassigning whales to the Mammals category. So it seems that the process of shifting itself may be straightforward, and perhaps it is analogous to a routine learning process of linking or integrating new ideas with old, provided that the old ideas already exist. That is, children already know about the category of Mammals, so that transferring the category membership of whales is not difficult.

The apparent ease of shifting across ontological categories, as demonstrated with the whales example, can be further seen in the ease with which adults shift across ontological boundaries when they use and understand metaphors. That is, metaphors often take the form of combining a predicate from one ontology to modify an entity from another ontology. Take Lakoff's example of anger, which is a MENTAL STATE. In metaphors, anger is conceptualized as a mass Substance, and takes the grammar of mass nouns, as opposed to count nouns. Thus, one would say *How much anger has he got in him?* but not *How many angers does he have in him?* Because anger is treated as a mass Substance, it can occupy space, can be contained, and can come out, such as *He 'let out' his anger* or *I could barely 'contain' my rage*. This anger Substance is sometimes conceived of as a kind of heated Fluid, such as *She got all 'steamed up'* or *His pent-up anger 'welled up' inside him*. The quoted terms can all be understood as predicates from the Substance category.

Thus, it seems as if shifting across ontological categories is fluid and easily undertaken. However, there is a class of situations in which an ontological shift is necessary, but people find it extremely difficult to do. This is the case when students misunderstand concepts such as heat transfer, electricity, natural selection (Chi, 1997; Ferrari & Chi, 1998). Thousands of studies have documented the misconceptions students hold and the failure of innovative instruction to remove their misconceptions. This difficulty in correcting their misconception may have arisen, we proposed, from the possibility that correcting their misconceptions requires an ontological shift (Chi, in press; Chi & Roscoe, 2002). For example, students tend to conceive of concepts such as heat transfer as a kind of a non-emergent Causal Event, whereas they are in fact a kind of an Emergent System (see Fig. 1). Causal Events are processes with activities that occur with temporal and spatial contiguity, often with an identifiable agent that directly or indirectly produces an effect. Emergent Systems, however, are similar to complex dynamic systems (Casti, 1994; Holland, 1998), in which numerous individual agents or elements independently interact in some uniform way to give rise to complex behavior at the macro level that is not exhibited at the level of the individual elements. For example, each element can follow a simple rule simultaneously and independently along with all the other elements, without any explicit goal to achieve. Yet, a macro pattern will emerge that looks as if the system is intending to achieve a certain goal, and that the process of achieving this goal is orchestrated by some director/ agent in some sequential way.

Besides this special case of misconceptions in the context of learning and understanding, shifting ontologically may be difficult (if not directly told to do so) for three basic reasons. The first is a lack of the alternative category. The second is the lack of awareness of the need to shift. Finally, the third is that it is resource-intensive. We discuss the first reason briefly since it does not apply to scientific discoveries and elaborate on the latter two reasons more extensively.

Lack of the Alternative Categories

This reason may not be a relevant one for discovery, but in learning science concepts, students often fail to successfully re-conceptualize a phenomenon or concept with respect to the right ontological category because they may lack knowledge about the target ontological category. In particular, as mentioned above, we have identified a class of science concepts that entail an Emergent System kind rather than a non-emergent Causal Event kind of process (see Fig. 1 under PROCESSES). Students are unfamiliar with this Emergent kind of PROCESSES; therefore, one could assume that they cannot shift their representation of processes from a Causal to an Emergent kind when that category does not yet exist in their knowledge (Chi, in press).

Lack of Awareness

Even if people do have the alternative ontological categories, they may still not shift spontaneously because they may not be aware that they have to shift their representation of an entity or phenomenon from one ontological category to another. This lack of awareness may arise from four sources. First, it is a low-frequency phenomenon. That is, we do not routinely need to re-classify a phenomenon from being of one kind to another kind. We generally categorize an entity correctly on the basis of its outward appearance. How we represent an everyday entity or phenomenon usually corresponding to its true identity. For instance, when we identify an animal that is furry with a wagging tail as a Dog, our categorization is usually accurate. Thus, it is seldom the case that the core features of an entity or phenomenon are in fact different from its surface features, so that we have misclassified and have to shift. There are a few exceptions. For example, children often classify whales as a kind of Fish because it has many of the characteristic features of a fish, such as swimming in water, shaped like a fish and so on. This is one of the few instances in which the surface features belie the true core feature. Occasionally, we must shift in the context of reading stories and watching films. For example, in the popular children's novel *Indian in the Cupboard*, the central theme of the book is that a toy Indian comes alive. A great deal of suspense usually surrounds such a kind of conversion and the main character is clearly surprised. A similar ploy is used in *Velveteen Rabbit*. Similarly, in the film *The Crying Game*, a male character is disguised as a female. The viewers are surprised when the disguise is revealed. Notice that these kinds of revelations in stories and movies are much more dramatic than mistaken identity. Misidentifying one person (e.g. an undercover agent) for another (a preacher), as often occurs in mysteries, is suspenseful, but not as dramatic because no ontological categories had to be crossed: both an undercover agent and a preacher are Males, whereas discovering that a character is really a Male rather than a Female requires shifting that person's categorical membership laterally. The drama, in the case of crossing ontological categories, arises because the viewers or readers must revisit and revise all the previous assumptions and implications about the character when he was a male because a new set of properties and features has to be inherited when an instance changes membership. Thus, since we are not accustomed to re-represent or re-classify a phenomenon or entity from one ontological category to another, we are not aware that such a shift is sometimes necessary.

However, mistaken classification based on the perceptual features is precisely the source of misconceptions in science (Chi & Roscoe, 2002; Chi, Slotta & de Leeuw, 1994). For example, we perceive and feel heat flow from one location to another. Therefore, when heat flow is conceptualized as a Causal Event, it appears as if the hot molecules move from one location to another location, when in fact the flow sensation derives from the collective random bombardment of all the moving molecules moving at different speeds (Chi, in press).

A second reason for a lack of awareness for the need of ontological shift is that our mis-representation is often adequate for everyday functioning. This is especially true in the case of our science misconceptions. It is perfectly adequate for us to close the window in order not to let the heat escape, thereby treating heat as a kind of Substance that can flow from one space to another space, when in fact heat is a kind of Emergent Systems PROCESS. Thus, our mistaken classification provides an adequate explanation for the phenomena that we encounter everyday, so that we are not aware of our misconceptions.

Finally, we have a tendency to pursue endeavors in a rigid and routine manner, analogous to functional fixedness (Ohlsson, 1992). For example, consider Duncker's famous candle problem. The participant is presented with three objects on a table: a book of matches, a box of tacks, and a candle. The goal is to attach the candle to the wall so it will burn normally. The solution requires the solver to reclassify the box, not as a *container*, but as a *platform*. Subjects solve the problem correctly 86–100% of the time when the problem is presented with the box of tacks empty, vs. a solution rate of 41–43% when the box is full (Weisberg & Suls, 1973). That is, there is no reason for us to think about re-classifying the box when it is full since we can continue to think of it as a container. Thus, even reclassifying within the same ontology is uncommon.

Resource Intensiveness

Although we have assumed in a preceding section that shifting itself may not be difficult in the context of using and understanding metaphors, shifting across ontological categories may be resource-intensive in other contexts. That is, the shifting itself may not be effortful, but once undertaken, efforts and cognitive resources have to be allocated for the process of re-inheriting a new set of features. Although there is no direct evidence on this point, there is some suggestive evidence that even shifting perspectives takes extra effort. Black, Turner & Bower (1979) have found, for example, that people prefer to maintain the same point of view when processing a narrative text. (This implies that changing perspective or one's representation may require extra cognitive resources, even though changing perspectives is not the same as shifting across ontological categories. See discussion below.)

In sum, our assumption is that shifting per se is not difficult. The ease of shifting per se can be supported by two kinds of anecdotal evidence mentioned above. In the first case, the ease of shifting can be seen in children when they are told that whales are a kind of mammal. Similarly, in novels and movies, children and adults have no difficulty in shifting their representation of a stuffed toy to a live person, or from a man to a woman, and so forth. A second example of ease of crossing ontological categories comes from adults' comfortable use and understanding of metaphors. However, in both of these cases, shifting occurs when we are explicitly told to shift, or we are presented with a metaphorical usage that includes a combination of two ontological categories. Shifting on our own, without being told, may be more uncommon, largely because we are not aware of the necessity to shift. Finally, the shifting itself may not be the bottleneck (assuming one knows about the category to which one is shifting), what is difficult may be the resource intensiveness of re-inheriting all the attributes of the alternative category. For example, it is difficult for some cognitive psychologists to consider cognition not as knowledge (a Substance) in the head but as a PROCESS of interaction; the difficulty arises not because we cannot undertake such a shift, but because we cannot easily understand all the ramifications for our views on memory, learning, and problem-solving (see the debate between Anderson, Reder & Simon, 1996 & Clancey, 1996).

Re-representation and Ontological Shifts

An ontological shift is a kind of re-representation. However, it is distinctly different from other kinds of re-representation. We discuss five other kinds of re-representation. But before discussing them, a few remarks should be made about general issues regarding representation. First, there is no dispute that certain representations of a problem may make a problem easier or harder to solve. For example, in order to give commands to the programming tool 'turtle graphics' (Papert, 1980) for the turtle to draw lines across the graphics screen, some problems are easier to solve by taking the turtle's point of view, and other problems are easier to solve by taking a more global point of view. Similarly, Hayes & Simon (1977) found differences in solution times for isomorphic representations of the classic Tower of Hanoi problem. Solution times were twice as fast if the problem was represented as

Monsters exchanging Globes, rather than Monsters changing the size of the Globes. One of the reasons that a problem can be solved more easily in one representation than another has to do with the constraints of the problem. Sometimes the constraints become more obvious in one representation vs. another. At other times the constraints become an integral part of the representation (being built in in some sense). And yet at other times the constraints can be released. Thus, whether a specific representation is more or less beneficial to arriving at the problem solution is a different issue from the issue that concerns us here: namely, whether or not shifting between representations (or re-representing) is beneficial, and if so, why it is not undertaken more frequently, as may be necessary in scientific discoveries.

Changing Perspectives

The most often discussed notion of changing representation is in the context of development. Piaget & Inhelder (1956), for example, discuss the way younger children cannot (but older children can) see things from another physical perspective besides their own. Similarly, younger children cannot (but older children can) acknowledge the listener's point of view. For example, Shatz & Gelman (1973) showed that young 2-year-olds did not adjust their speech to the age (and knowledge) of the listener, whereas 4-year-olds did adjust their speech depending on whether they were speaking to another peer or an adult. Thus these developmental studies implicate that children are capable of shifting their perspectives as they get older.

In the adult literature, besides Papert's (1980) work cited above, Hutchins & Levine (1981) have shown that problem-solvers do change perspectives as they solve the Missionaries and Cannibals problem. They used deictic verbs such as 'come', 'go', 'take', 'send', 'bring', and place adverbs such as 'here', 'there', 'across' to determine the solver's point of view, such as viewing the river that the Missionaries and Cannibals have to cross, from either the left bank or the right bank. One of their interesting findings was that when solvers were 'blocked' in their solution, in the sense that they have made two non-progressive moves out of a problem-solving state (in the problem space), then they were successful in becoming unblocked when they changed their point of view. So clearly, being able to shift one's point of view is beneficial for problem-solving, and it is a form of re-representation.

Re-representing at Different Levels

Instead of changing perspectives, representational shifts can also be considered in the context of levels. There are four ways to think about levels.

Shallow and Deep Levels

One way to think about levels is in the context of shallow surface features (such as the entities mentioned in a problem statement) and deeper semantic features (such as the principles that govern the solution procedure). So for example, Chi, Feltovich & Glaser (1981) showed that expert physicists represented routine physics problems at a conceptually deeper level (in terms of the principles that guide the solution), whereas physics novices (those who have taken one course in college with an A grade), tended to represent the same problems according to their literal components (such as the pulleys and the inclined planes). Obviously the underlying principles of the problems determine their problem solution and not the literal components. The implication of this work is that as one acquires expertise, one's representation changes so that it is organized according to the principles of physics rather than the physical components.

Subset–superset Hierarchical Levels

Besides re-representation between levels in the sense of from a shallow more surface-oriented level to a deeper more conceptually oriented level, a second form of re-representation occurs between hierarchical (subset–superset) levels of categories. Suppose we ask students to solve the following two 'insight' problems:

(1) A man who lived in a small town married 20 different women in that town. All are still living and he never divorced a single one of them. Yet, he broke no laws. How can you explain this?
(2) Two strings hang from a ceiling. They are hung far enough apart that a person cannot reach both strings at the same time. The goal is to tie the strings together. Lying on a nearby table are a hammer and a saw.

These problems are typically considered to be difficult (thereby called 'insight' problems). To solve them requires that one re-represents an entity within the problem. In the first problem, the solution is to re-represent the man not as a 'bachelor', but as a 'clergyman' (a specific type of 'bachelor'). Similarly, in order to solve the second problem, one must re-represent hammer not as a 'tool', but as a 'heavy tool'. This heavy tool can then act as a weight to be tied to one of the strings to create a pendulum. One simply swings the pendulum, grabs the first string, and then catches the other string on its return. In both cases, the entities in the problems are re-represented as an instance of a subcategory. This is an example of re-representation within hierarchical levels. Hierarchical levels satisfy the relationship of 'kind of'. That is, a clergyman is a kind of 'bachelor', and a 'hammer' is a kind of 'heavy tool'.

Component Levels

A third way to think about levels is in terms of decomposition. For example, if we were asked to explain how the human circulatory system works in terms of its function of delivering oxygen to body parts, we would explain it by appealing to the components of the circulatory system, such as the heart, the lungs, blood, and blood vessels, and that it is the contraction of the heart that sends blood to different parts of the body. One can then further ask how the heart contracts. To answer this question, we would have to discuss the components of the heart, such as the role of the rise of ventricular pressure and so on. Basically, each question and its accompanying explanation must reduce each component into its finer and finer constituent parts. Miyake (1986) collected protocol data that illustrated representational shifts in terms of a reduction-decomposition approach to levels. She showed that dyads, in attempting to understand how a sewing machine works, would move to progressively lower levels when they recognized that they had not understood the mechanism. For example, in figuring out how a stitch is made, one can understand it by explaining that the needle pushes a loop of the upper thread through the material to the underside, so that the upper thread loops entirely around the lower thread. However, in order to understand how this looping mechanism works, one has to explain the mechanism at a lower level, of how the bottom thread is able to go through the loop of the upper thread.

Component levels satisfy a 'part of' relationship, rather than a 'kind of' relationship. In these examples, ventricular pressure is part of the cardiovascular system, and looping is part of the mechanism of making a stitch.

Emergent Levels

A fourth way to think about levels is one in which the relationship between them is an 'emergent' one (Chi, in press; Resnick & Wilensky, 1998). (See the discussion about Emergent Systems above.) In this conception of levels, the (often) observable 'macro' level behaves independently of the 'micro' level objects. Moreover, the macro observable level arises from local interactions of the micro level individuals. The most commonsensical example is a traffic jam. A traffic jam is a gridlock of cars, in which cars can no longer move at the same speed before the jam. This is the macro-level phenomenon. The behavior of the individual cars in a jam is independent of the jam. Each individual car may be following the same simple rule, which is to accelerate if there was no car in front within a certain distance, and slow down when a car comes within a given distance. The independence of the macro and micro levels can be seen in that the jam itself can move backward even though the individual cars move forward. However, the jam arises or emerges from the interaction of the cars.

For another example, changes in a moth's pigmentation, due to the smoky industrialization in England in the middle of the 19th century, can be understood in terms of emergent levels as well. That is, the emergent pattern, that moths were getting darker

over generations, occurred because individual moths were independently being eaten or not eaten by birds. It so happened that lighter moths, which perched on tree trunks that were getting increasingly sooty, became more apparent and visible to hungry birds. Thus, it was more likely that the lighter moths were more visible and thereby were eaten while darker moths tended to survive. Even though one cannot specify on an absolute basis whether a given dark or light moth (on a relative scale) would survive or not (since it depended on a number of local conditions, such as whether a moth happens to land on a darkly sooted tree trunk, and whether a hungry bird happens to be nearby and can see the moth), nevertheless, over generations, the probability is such that the darker ones tended to survive and reproduce, thus producing the changing pattern of moths getting darker over generations. Thus, this overall pattern is an emergent one. It is not caused by any specific actions on the part of any specific moth or groups of moths. This is not a case of re-representation from one level to another lower level, as in the example of the sewing machine. Rather, this is a re-representation from a macro level to a representation of the relationship between the micro and the macro level, in order to understand the macro level pattern. Representing the inter-level emergent relationship is extremely difficult for students (Chi, in press).

In sum, two different kinds of re-representation are described above, one involving changing spatial perspectives and one involving four kinds of changing levels. Among these five examples of re-representation, only the last one constitutes an ontological shift. In that case, the shift is between the ontologies of 'Causal Events' and 'Emergent Systems' (see Fig. 1 again). The fact that only one kind of re-representation involves an ontological shift (or sometimes referred to as radical conceptual change, Chi, 1992) may explain why it occurs infrequently. In the next section, we revisit whether or not we have evidence that ontological shifts are possible.

Evidence of Ontological Shifts

Besides the anecdotal evidence that people can shift across ontological categories readily, as in the case of children re-representing *whale as a mammal* or the case of understanding stories about an *Indian in the Cupboard*, is there evidence that ontological shifts actually occur? There are two kinds of evidence. In this section, we suggest evidence of ontological shifts between novices and experts. In one study, we compared the explanations of ninth grade students with those of graduate physics students on simple conceptual problems, such as *Explain which coffee cup is a better insulator, a styrofoam cup or a ceramic cup?* We coded their explanations not for accuracy (which would not be fair to the high school students), but for the kind of predicates they used in their explanations. The predicates they use would indicate which ontological category they represent the concept/phenomenon. Ninth graders (the novices) would explain their choice of the coffee cup with justifications such as *The coffee in the ceramic mug is hotter because the heat in the styrofoam cup is 'gonna escape'*. In contrast, the graduate students' (the experts) answers might say something to the effect that *The energy loss in the ceramic cup is through transfer of heat from a hotter source to a cooler source, due to the movement or 'motion of electrons'*. The basic difference in their explanations is that novices use Substance-based predicates. That is, viewing heat as a kind of Substance that can be contained, stored, or escaped; whereas experts use PROCESS-based predicates such as motion of electrons (Slotta, Chi & Joram, 1995). Thus, experts appeared to have undertaken an ontological shift whereas novices have not.

A better example that is not contaminated by knowledge of the domain jargon comes from our study with expert swimming coaches. Two experts with a minimum of 12 years as full-time head coach and who have produced from 20 to 100 top national caliber swimmers were compared with two novice coaches with a maximum of two years' of full-time coaching experience. Their task was to diagnose underwater films of four swimmers performing the freestyle stroke. The general diagnoses rendered by the novices focused on specific body parts, such as *the elbow was bent on extension*, or *the right arm was not underneath the body*. In contrast, experts tended to diagnosis more holistically, referring to processes, such as *Little and unequal body roll*, or *stroke unbalanced*. Basically, the more novice coaches diagnosed swimming deficiencies by attending to individual body parts (Concrete Entities), whereas the experts focused on aspects of the swimming PROCESSES, such as body roll (Leas & Chi, 1993). According to the aforementioned studies (Leas & Chi, 1993; Slotta et al., 1995), one undergoes an ontological shift in the process of acquiring expertise.

Does Ontological Shift Underlie Discoveries?

As in the case of resistance to ontological shifts in learning about many science concepts (i.e. the robustness of misconceptions), it may be the case that scientists are also resistant to change and choose to persist in generating explanations or hypotheses within the same ontology. The thesis of this chapter is to explore the claim that major scientific discoveries may have arisen because the scientist underwent an ontological shift, in terms of the nature of the novel, revolutionary explanations. Few empirical studies directly support this claim; however, Hutchins & Levine (1981) found that subjects who were blocked in their problem-solving became unblocked when they shifted their perspective.[2] However, as we said earlier,

shifting perspective is not exactly the same thing as shifting across ontological categories. Below, we review two domains of scientific discovery and/or scientific revolution: one in contemporary (20th century) science, and the other in the history of science, to show that these discoveries did require an ontological shift. By scientific revolution, we mean changes that are more extensive than at the level of individual theories, but rather changes that involve the research tradition or paradigm under which specific theories are constructed. We use the term 'paradigm' in the same sense as Kuhn (1996), Lakatos (1970), and Laudan (1977), who considered research paradigms as worldviews that encompass a set of theories with similar assumptions and similar concepts. 'Similar' can be conceptualized, in our opinion, as those belonging to the same ontology. So, for instance, 'evolutionary theory' really refers to a family of doctrines that all assume that organic species have common lines of descent. All variants of evolutionary theories would implicitly make that assumption. Thus, theories within the same paradigm basically are modifications and extensions of one another, since they adopt the same assumptions and concepts, scrutinize the same set of problems, and use the same sort of methodologies and instruments to do their science.

Some Examples of Ontological Shifts in Contemporary 20th Century Theories

In this section, we briefly describe what may be considered major breakthroughs in the 20th century, drawing on examples from the treatment of diseases as well as theories for natural or scientific phenomena. This treatment is very superficial, and captures merely the highlights of the changes in the scientific explanations, as rendered by other scientists.

Peptic Ulcers

Thagard (1998a, 1998b) detailed the historical shifts in the causal explanation for peptic ulcers. The prevailing hypothesis prior to 1979 for the cause of stomach ulcers was an excess acidity in the stomach. The increased production of gastric acid was thought to be caused by elevated levels of stress experienced by an individual. The causal sequence of events that eventually lead to an ulcerated stomach lining was: increase in stress levels, which lead to the increased production of stomach acidity, which then leads to erosion of the stomach lining. There was a general agreement and satisfaction with the stress leading to excess acidity model, although treatment was only successful at reducing acidity (the symptom), not stress (the cause).

Evidence against the prevailing explanation came from a source not initially designed to explain the etiology of ulcers. In 1979, while Warren (a pathologist) was performing an autopsy on a patient, who was diagnosed with nonulcer dyspepsia and gastritis, he

---

[2] See also (Andersen, 2002) for historical evidence.

observed an unusually high count of bacteria in his stomach. Warren's observation went against conventional wisdom because the medical community believed the stomach was too hostile an environment for the survival of microorganisms. Warren began collaborating with Marshall (a gastroenterologist), and they noted a correlation between gastritis and bacterial infection, as well as duodenal ulcer and bacteria. To strengthen their hypothesis, Warren and Marshall successfully treated 90% of their patients, who were diagnosed with duodenal ulcers, with antibiotics. After several different types of experimental studies, Warren and Marshall finally proposed the hypothesis that the bacteria, *Helicobacter pylori*, were responsible for peptic ulcers.

Today, the bacteria hypothesis remains the most viable explanation for peptic ulcers because it succeeds in treating patients at the level of the hypothesized causal mechanism, instead of merely controlling the symptoms. Instead of using antacids to control the increased stomach acidity, physicians now also prescribe a regimen of antibiotics to eradicate the infection (Graham, 1993). Notice that a breakthrough in finding a better causal explanation of peptic ulcer consisted of shifting the ontological categories of the explanations, from one of MENTAL STATES (stress) or Substance (acid) to one of Animate Concrete Entities (bacteria).

Dinosaur Extinction

Clearly, one of the most intriguing scientific issues of the modern era has been to provide a coherent explanation for the massive extinction of dinosaurs, nearly 65 million years ago. Several interesting ideas, ranging from infertility due to rising climatic temperature, ill reactions to newly evolved flowering plants, cataclysmic events (such as massive meteors or massive volcanic activity), attempted to explain why dinosaurs became extinct (Gould, 1985). However, several of these theories are not viable because they are not consistent with other empirical observations. For example, because the extinction was on a global scale, an acceptable theory must also explain why oceanic life (such as plankton) also died at the same time, and it must be consistent with the fossil record. Therefore, the first two explanations (infertility and increased toxicity in flowering plants) must be ruled out because neither hypothesis can explain why plant and oceanic life also died out (Gould, 1985).

The prevailing dogma, which finds its intellectual roots in the gradualist ideas put forth by geologist Charles Lyell, was that the Earth was modified over millions of years (W. Alvarez, 1997). The Earth has undergone long periods of climatic changes (the 'ice age' of the Quaternary Period, for example), as well as changes in sea level. To explain dinosaur extinction, gradualists assumed that, like every other species, dinosaurs became extinct because they could not adapt to the climatic changes. Scientists of the day argued

that an incomplete fossil record is evidence that dinosaurs died out over a protracted period (W. Alvarez, 1997). At this point, iridium did not play a role in the development of a theory of dinosaur extinction. According to the gradualist perspective, if iridium existed on Earth, then it would have had an Earth-bound origin (volcanic activity, for example).

Although the gradualist explanation seemed congruent with the available evidence, Walter & Luis Alvarez proposed a different theory (L. W. Alvarez, Alvarez, Asaro & Michel, 1980). They proposed that an enormous meteor struck the earth, which created enough dust to block the sun around the planet. They reasoned that the obstruction of sunlight would cause the process of photosynthesis to stop, thereby causing the entire food chain to suffer. The impact hypothesis satisfies the two aforementioned constraints. First, the impact hypothesis explains why there is global extinction (not just animals, but plants and oceanic life as well). Second, the hypothesis is consistent with the fossil record because the amount of iridium embedded within the Crustaceous–Tertiary boundary is several orders of magnitude higher than the amount found on Earth during any other time period. High concentrations of iridium are mainly found on asteroids or comets, as well as the core of the Earth. To rule out the alternative hypothesis that volcanic activity caused the extinction of the dinosaurs, the Alvarez team turned to a different strand of evidence. They demonstrated that a specific type of rock, called 'shock quartz', was also observed at the same time period. Shock quartz is formed by a violent strike to a layer of quartz. Because of the overlapping evidence, the meteor hypothesis gained popularity.

These two hypotheses have maintained the scientists' interest perhaps because they are ontologically distinct. The impact theory is based on the notion of a one-time discrete Causal Event, involving a Concrete Entity such as a meteorite, whereas the gradualist explanation can be considered a continuous PROCESS of climatic and sea-level changes, perhaps likened to an Emergent System.

Coronary Heart Disease

Discovering the cause for coronary heart disease (such as acute myocardial infarction) represents one of the most significant challenges to the medical community because an estimated 7,500,000 Americans die each year from myocardial infarction (American Heart Association, 2002). The direct explanation for the cause of heart attacks has been atherosclerosis, which is the thickening of the coronary arteries, thereby restricting the flow of blood. The question is what caused such thickening to occur. The prevailing explanation is that arteries are thickened by cholesterol deposits into the artery walls and production of atherosclerotic plaques. The plaques cause a narrowing of the vessels. In the last two decades, Americans have been obsessed with diets that can reduce their cholesterol level to an acceptable healthy level.

More recently, however, two alternative theories have been proposed. One is the hypothesis that, instead of cholesterol, it is iron overload that causes heart attacks. This hypothesis is consistent with a number of pieces of evidence. For instance, men who regularly donate blood have a lower risk of heart disease, and pre-menopausal women who regularly lose blood, and thus iron, also have a lower risk of heart disease. Tuomainen, Kari, Nyyssonen and Salonen (1998), using a case-control study of Finnish men, found that individuals with high iron levels were more likely to have suffered an acute myocardial infarction in the past 6–7 years. Presumably, iron can increase 'oxidative stress' on the lining of the blood vessels. That is, iron can somehow interfere with nitric oxide, a chemical that relaxes blood vessel walls, allowing the blood to flow more freely.

A second promising new hypothesis is currently being proposed, that coronary heart disease is caused by inflammation in the bloodstream. This hypothesis is compatible with the finding that many people (in fact, over 50%) with no known risk factors (such as high cholesterol levels) nevertheless do have heart attacks. Moreover, an enzyme called myeloperoxidase (MPO) and a substance called interleukin 6 were both elevated among people who had heart attacks and narrowed coronary arteries (Zhang et al., 2001). Both substances are associated with inflammation. For example, MPO is normally found in infection-fighting white blood cells, so that their elevation indicates the body's attempt to fight inflammation. Using a cross-sectional survey, Meier, Derby, Jick, Vasilakis and Jick (1999) found that adults who had taken tetracyclines and quinolones (antibiotics) showed a reduced likelihood of experiencing an acute myocardial infarction. The exact mechanism that mediates atherosclerosis through inflammation is not completely clear. One possibility is that bacteria and other infections (such as gum infection) can cause clot-forming cells (or platelets) to clump together, thereby causing arteriole blockage.

It seems that there is a rapidly growing body of evidence that inflammation in the bloodstream can cause heart attacks. Moreover, there is a growing consensus among scientists that other disorders, such as colon cancer, Alzheimer's disease, may also be caused by chronic inflammation. Hence, inflammation seems to be the hypothesis that is favored at the moment (in the sense that an increasing number of studies are testing this hypothesis and trying to understand its mechanism). The excitement, we believe, arises from the fact that a hypothesis based on inflammation (a PROCESS) is ontologically distinct from a hypothesis based on both cholesterol deposits and iron overload (both cholesterol and iron are a kind of Substance). Thus, an ontological shift in hypothesiz-

ing implies that the new hypothesis has inherited many other ontological attributes that require scientists to further investigate its ramifications.

Some Examples from the History of Science

In this section, a few examples of major scientific discoveries are cited that may be thought of as having undergone ontological shifts.

Epilepsy and Other Diseases

Between 430 and 330 BC, the approach to diseases was a religious one. For example, epilepsy was caused by divine visitation (Thagard, 1996). Divine beliefs fit the ontology of a kind of MENTAL STATES. Consequently, treatment of diseases was consistent with the explanation at the time, such as by appealing to the gods for mercy or other magic.

After 330 BC, Hippocrates rejected the religious approach by postulating that epilepsy is caused by an excess of phlegm, one of the four fluids (or humors) that were thought to constitute the human body. Healthy bodies supposedly had the correct proportion of each of the four fluids (blood, phlegm, yellow bile, and black bile). Imbalances in these fluids were thought to produce diseases, and different kinds of imbalances produced different diseases. Treatment of course, was again compatible with the prevailing explanation. To redress humeral imbalance, one would change one's diet, or rid the body of excess phlegm by inducing vomiting or evacuation of the bowels, and/or letting blood from the veins.

One can say that Hippocrates' humeral theory is radically different from the divine visitation theory because the explanation of epilepsy had undergone a major ontological shift, from a kind of a MENTAL STATE to a kind of Substance. In the 19th century, another major conceptual shift occurred, this time to postulate that diseases were caused by bacteria and other microorganisms. This breakthrough occurred from, presumably, the analogy that Pasteur made in the late 19th century between fermentation and disease. Since fermentation was caused by yeast and bacteria, so too can diseases be caused by microorganisms. Thus, another ontological shift had occurred, from one of Inanimate Substance (imbalance of phlegm) to Animate ENTITIES (micro-organisms).

Notice that before Pasteur, Fracastoro had posed another theory, the notion of 'contagion', which can be thought of as 'seeds' that can be passed from one person to another. Treatment would therefore consist of expelling and destroying these seeds, rather than restoring humeral imbalance. Although the contagion theory and the humeral theory appear to be rather different, they were not ontologically distinct. They are both a kind of Substance. Thus, Fracastoro's theory was not a breakthrough and had little influence after 1650 (Thagard, 1996).

Theory of Evolution Through Natural Selection

In the 18th century, Lamarck postulated that environmental forces modified an organism's morphology. The classic example is the giraffe's long neck. To explain why the giraffe has a long neck, Lamarck hypothesized that reaching for leaves on tall trees gave the giraffe a longer neck. These changes in animal physiology could then be passed onto the organism's offspring. The central idea in Lamarck's theory of evolution was that traits could be acquired from intentional use. If there were a change in the animal's environment, the animal would then be able to respond (i.e. adapt) by changing its traits. Thus, the Lamarckian theory proposed that evolution is a kind of a Causal Event, in that the agent (the giraffe in this case) can directly cause and manipulate (by using and stretching his neck, for example) his own traits to suit the environment.

In the 19th century, Darwin rejected the idea that animals have the capacity to actively change their heritable traits (Gould, 1985). Instead, he postulated a radically different mechanism: natural selection. Darwin's theory can be stated in the following way. Basically, each giraffe is different from every other giraffe, from variability in the inherited genes. So some giraffes have longer necks and others shorter necks. Those with longer necks can reach their food source, so they are more likely to survive and reproduce, although some of the shorter neck giraffes may also survive if some short trees happen to be around. But in general, the taller giraffes have an advantage, given that trees tend to be tall. Because offspring inherit such a trait (of long necks), then the taller giraffes, having had more opportunities to survive and reproduce, tend to produce offspring with longer necks. Over generations, the entire population of giraffes tends to have increasingly longer necks.

The shift from intentional modification of heritable traits to probabilistic survival and reproduction due to random variation in a trait represents a radical ontological shift. Evolution via intentional adaptation might be categorized as a kind of Causal Event, whereas Darwin's theory might be best understood as a kind of Emergent System. (See Chi, in press, for more details.)

Electricity

Another ontological shift can be found in the theoretical development surrounding the topic of electricity. Among the very first documented cases of static electricity (i.e. ancient Greece) included the observation that amber attracted small pieces of straw after it was rubbed (Meyer, 1971). These early experiments are now considered demonstrations in electrostatics. The explanations generated by early theorists and experimenters were targeted at explaining why two objects are attracted to one-another after being rubbed with a cloth or fur. The first explanation for the electromotive forces began with a type of Material Substance called 'effluvium'. In 1600, William Gilbert proposed that the effluvium stretched out across space to hold two objects together. The larger the space, the thinner the effluvium became, and the weaker the felt attraction. The experimental data confirming the existence of the effluvium as a Substance was the observation that the interposition of any object would break the attraction between two objects (Home, 1981). This was taken as evidence as a Material Substance because one property of material objects is that they are unable to occupy the same space. Home (1981) provides a description of the antiquated construct:

> The effluvia are described at one point as 'fine', at another as 'subtle'; however, it is emphasized that they are material, and as such are subject to ordinary physical laws. The presence of moisture and other impurities in the atmosphere can hinder the action for the effluvia in two ways: the moisture will condense on the surface of the glass or other substance to be electrified, and in doing so may block the pores from which the effluvia are normally emitted; alternatively, the mere presence of impurities in the atmosphere should increase its density and hence the resistance it offers to the motion of the effluvia (p. 5).

Although initially successful, the effluvium theory was unable to account for a few enigmatic observations. For example, the theory required some medium for the action at a distance. It was hypothesized that the effluvium stretched out across the air, from one object to the other. However, experiments conducted after the introduction of the vacuum demonstrated that a medium was unnecessary to observe the same results. Another difficulty was the observation that some charges attract while others repel. Therefore, it became increasing apparent that the effluvium theory was incomplete.

The effluvium theory was eventually replaced by Benjamin Franklin's single fluid theory of electricity (Meyer, 1971). Objects were considered positive if they contained more fluid, and negative if they had less fluid. The flow of electricity, according to Franklin, was from positive (objects with more fluid) to negative (objects with less fluid). Franklin also made the assertion that the electrostatic demonstrations in the lab also applied to natural phenomena, such as lightning (Holyoak & Thagard, 1995). Franklin did, however, express some doubt as to the fluid theory. He stated, "The electrical matter consists of particles extremely subtle since it can permeate common matter, even the densest, with such freedom and ease as not to receive any appreciable resistance" (quoted in Hecht, 1994). It seems reasonable for Franklin and his contemporaries to conceptualize electricity as a fluid, given the popularity of 'fluid-like' explanations for other physical phenomena (e.g. the caloric theory of heat and the impetus theory of motion).

Electricity, however, did not remain in the flow of a fluid. First, in a series of experiments, Faraday provided empirical support for the claim that different 'forms' of electricity were actually manifestations of a common origin. That is, electricity generated by a rotating magnet was derived from the same cause as a voltaic pile or frictional (i.e. static) electricity (Harre, 1981). Second, Faraday went on to propose a field theory of electricity, which stated that electricity was induced in a current-carrying wire when moved relative to a magnetic field. The movement of the wire thereby 'cut' the lines of forces surrounding the magnet, causing the induction of an electric current (Nersessian, 1992; Tweney, 1989). James Clerk Maxwell, a contemporary of Faraday, went on to formalize these ideas with his famous field equations.

A unified, mechanistic explanation for the various forms of electricity was provided by the discovery of the electron in 1899 (Harre, 1981). Sir Joseph John Thomson, who was credited with the discovery, attempted to measure both the mass and charge on the electron (Hecht, 1994). Although not everyone at the time subscribed to the atomic theory of matter, Thomson made the argument that there were subatomic particles, which could be observed experimentally. Given Thomson's discovery, electricity could now be understood as net movement of free electrons through a closed circuit. Thus, the flow of electricity is a kind of Emergent System.

The history of electricity thus demonstrates a non-radical shift from the progression of the effluvium to the fluid theory, in that both theories involved substances moving from one place to another, as a kind of Causal Event. The breakthrough did not occur until Thomson's ideas were established, that current flow is the net movement of subatomic particles. This revolutionary idea required an ontological shift by considering electrical current to be a kind of an Emergent System rather than a kind of a Causal Event.

The Caloric Theory of Heat

Fluid-type models of physical phenomena were used to explain not only electricity, but also heat transfer. The caloric theory of heat, proposed largely by Lavoisier, was the dominant theory of the day (Spielberg & Anderson, 1995). The theory held two basic tenets. First, heat transfer was the result of two bodies, of unequal temperature, coming into contact with one another. The caloric was understood to flow from the relatively 'hot' object to the 'cold' object. The second tenet stated that the caloric was conserved (i.e. it can neither be created, nor destroyed). Although the theory was useful in understanding new developments, such as the steam engine, it became apparent that the theory was not totally correct.

One anomaly, which could not be explained by the two tenets, was the explanation of heat generated during friction (Spielberg & Anderson, 1995). Because the caloric was always conserved, it was hard to explain why two objects, which start in thermo-equilibrium, could produce heat when rubbed together. This observation violated the second tenet. To investigate this anomaly more systematically, Joule designed a method for precisely measuring the heat generated from friction. His apparatus included a container, with several compartments. The compartments were filled with fluid, and a paddle agitated the fluid contained within. Joule measured the temperature of the liquid with a thermometer. He found that the temperature of the liquid increased slightly when the paddle was turned. Because the system was in thermo-equilibrium at the beginning of the experiment, the only source of the heat generated was the agitated fluid. Therefore, Joule concluded that the caloric theory of heat must be incorrect, and the motion of the fluid led him to postulate a new theory of heat. The kinetic theory of heat dispensed with the concept of a separate fluid; instead, it treated heat as a property of matter. Joule reframed heat as the motion of the molecules that made up the fluid in his apparatus.

The switch from the caloric theory to the kinetic theory of heat is hailed as "among the greatest intellectual achievements of the nineteenth century" (Hecht, 1994, p. 565). The kinetic theory can be considered a major ontological shift in two ways. First, the caloric was considered a material Substance, which can 'flow' from one object to the next. However, the kinetic theory conceived of temperature as the average speed with which molecules move, so that heat energy is not a kind of Substance, rather it refers to a PROCESS, such as speed of molecules. Moreover, heat energy is transferred from one object to the next by changing the velocity of the molecules that comprise the object. Thus, heat transfer is a kind of Emergent Process.

Alternative Mechanisms Leading to Scientific Discoveries and Theory Change

What this chapter proposes is a mechanism for inducing a radically new hypothesis or theory, to explain a set of phenomena or data. The idea is that it may be the case that novel theories are often noticed, focused upon, explored, and sometimes succeeded in explaining the phenomena better, when the novel theories are ontologically distinct from the former theories. We are not suggesting that a novel theory, in order to be successful, must be ontologically distinct. Rather, we are suggesting that when a given theory, and its related family of theories, are inadequate for explaining a phenomenon or a disease for many years, perhaps an ontological shift in hypothesizing can allow a major breakthrough to occur. We illustrated this possibility with a few examples.

What other mechanisms have been proposed for scientific discoveries? The crux of understanding scientific discovery is to understand how a new hypothesis is induced, especially a radically new one. The new hypothesis presumably can explain a pattern of observations that other existing hypotheses have failed to explain successfully. Psychological research attempting to understand how new discoveries are made often focuses on the generation of experiments to test the hypothesis, as well as evaluation of the resulting evidence. Far less work has been done on the processes of inducing a new hypothesis. However, Klahr & Dunbar's (1988) research does address the issue of inducing new hypotheses. Their task involved discovering the function of a mystery key, labeled 'RPT', on a robotic toy. The participants tested their hypotheses by writing small programs that included the mystery key. Most participants used their semantic knowledge of 'repeat' to induce their initial hypothesis. As they received feedback from the device, it became evident that the repeat function had violated some participants' initial hypotheses. However, a new hypothesis, in this task, is not ontologically distinct from earlier hypothesis. So the findings learned from this task cannot generalize to revolutionary discoveries that did involve ontological shifts.

Similarly, computational research on the process of inducing a new hypothesis took a few directions that also differed from those involving ontological shifts. Langley, Simon, Bradshaw & Zytkow's (1987) work, for example, addressed the problem of discovering laws given a set of empirical observations. For instance, BACON.5 can look for a linear relation among a set of variables by using a standard linear regression technique. If the variables and their values fit a monotonically increasing relation, then the ratio between the two terms is found, given the signs of the values of the two terms are the same. If a monotonically decreasing trend occurs, then a product is defined, and so on. BACON.5's discovery is restricted to finding an appropriate fit of a mathematical expression to the empirical values that correspond to a pre-specified set of variables. This process corresponds to the formulation of a new theory or principle (in the form of a mathematical expression) on the basis of experimental results. This is essentially a curve-fitting process, which seems to be quite different from the process of inducing a radically new hypothesis.

Another computational approach was Thagard's (1992) model, ECHO. ECHO basically describes the mechanism by which a specific hypothesis is accepted or rejected, on the basis of the number of pieces of evidence that support or contradict the hypothesis. In this sense, ECHO is not a model of hypothesis generation per se, but rather, it is a model that evaluates which of several hypotheses better fit the evidence.

Thagard (1992) mentioned two other mechanisms that characterize scientific discoveries. One is replacement of an old theory by a new theory. Replacement does accurately portray the *result* of ontological shift,

but it does not refer to the ontological shift per se. Thagard (1992) and others (e.g. Hampton, 1997) also proposed conceptual combination, in arriving at new ideas and concepts. However, it is difficult to imagine how two concepts from different ontologies can be combined, or how combining two concepts from the same ontology can derive a radically (i.e. ontologically) different concept. A third mechanism that is often proposed is analogical reasoning (Gentner et al., 1997; Holyoak & Thagard, 1995). However, by our assumption, analogy can only produce similar accounts since, by definition, analogies have similar structures. For example, it was mentioned that diseases caused by bacteria were discovered by analogizing to the process of fermentation caused by yeast. Thus, yeast and bacteria are analogous. The ontological shift (from phlegm to bacteria) occurred when the scientist noticed the similarity between yeast and bacteria. That was the insight. The analogical mapping between yeast causing fermentation and bacteria causing diseases was straightforward. Thus, we would argue that the analogical mechanism per se, mapping yeast to fermentation and bacteria to disease, did not create the discovery. Perhaps some prior ontological shift (from phlegm to bacteria) was the mechanism that caused the scientist to notice the similarity between yeast and bacteria in the first place.

Hence, in general, scientific discoveries, in the form of generating or inducing a radically novel hypothesis that explains the observed pattern of findings, have not been explored extensively. What has been explored extensively is scientific thinking more broadly, such as skills of weighing all the available evidence, ways of experimenting that are systematic, such as holding some variables constant while varying others (Klahr & Dunbar, 1988; D. Kuhn, Amsel & O'Loughlin, 1988). However, few of these research findings bear directly on the mechanism of inducing a radically novel hypothesis. This chapter proposes one such mechanism: the possible role of ontological shifts in representation as required for the generation of a truly new hypothesis, one that can be considered a major scientific discovery.

Acknowledgments

The authors are grateful for funding from the Spencer Foundation Grant No. 200100305. We would also like to thank Ryan D. Tweney for his insightful comments on an earlier version of our chapter.

References

Alvarez, L. W., Alvarez, W., Asaro, F. & Michel, H. V. (1980). Extraterrestrial cause for the Cretaceous–Tertiary extinction. *Science*, **208** (4448), 1095–1108.

Alvarez, W. (1997). *T. rex and the crater of doom*. Princeton, NJ: Princeton University Press.

American Heart Association (2002). *2002 Heart and stroke statistical update*. Dallas, TX: American Heart Association.

Andersen, H. (2002). The development of scientific taxonomies. In: L. Magnani & N. J. Nersessian (Eds), *Model-based Reasoning: Science, Technology, Values* (pp. 95–112). New York: Kluwer Academic.

Andersen, J. R., Reder, L. M. & Simon, H. A. (1996). Situated learning and education. *Educational Researcher, 25* (4), 5–11.

Black, J. B., Turner, T. J. & Bower, G. H. (1979). Point of view in narrative comprehension, memory, and production. *Journal of Verbal Learning and Verbal Behavior, 18* (2), 187–198.

Casti, J. L. (1994). *Complexification: Explaining a paradoxical world through the science of surprise.* New York: Harper Perennial.

Chi, M. T. H. (1992). Conceptual change within and across ontological categories: Examples from learning and discovery in science. In: R. N. Giere (Ed.), *Cognitive Models of Science: Minnesota Studies in the Philosophy of Science* (Vol. XV, pp. 129–186). Minneapolis, MN: University of Minnesota Press.

Chi, M. T. H. (1997). Creativity: Shifting across ontological categories flexibly. In: T. B. Ward, S. M. Smith & J. Vaid (Eds), *Creative Thought: An Investigation of Conceptual Structures and Processes* (pp. 209–234). Washington, D.C.: American Psychological Association.

Chi, M. T. H. (in press). Emergent versus non-emergent causal processes: How misconceptions in science arise and how they can be overcome. *Journal of the Learning Sciences.*

Chi, M. T. H., Feltovich, P. & Glaser, R. (1981). Categorization and representation of physics problems by experts and novices. *Cognitive Science, 5,* 121–152.

Chi, M. T. H., Hutchinson, J. E. & Robin, A. F. (1989). How inferences about novel domain-related concepts can be constrained by structured knowledge. *Merrill-Palmer Quarterly, 35* (1), 27–62.

Chi, M. T. H. & Roscoe, R. D. (2002). The process and challenges of conceptual change. In: M. Limon & L. Mason (Eds), *Reconsidering Issues in Conceptual Change: Theory and Practice.* Dordrecht: Kluwer Academic.

Chi, M. T. H., Slotta, J. D. & de Leeuw, N. (1994). From things to processes: A theory of conceptual change for learning science concepts. *Learning and Instruction, 4,* 27–43.

Clancey, W. J. (1996). Conceptual coordination: Abstraction without description. *International Journal of Educational Research, 27* (1), 5–19.

Collins, A. M. & Quillian, M. R. (1969). Retrieval time from semantic memory. *Journal of Verbal Learning and Verbal Behavior, 8* (2), 240–247.

Ferrari, M. & Chi, M. T. H. (1998). Naive evolutionary explanations and radical conceptual change. *International Journal of Science Education, 20* (10), 1231–1256.

Gelman, S. A. (1988). The development of induction within natural kind and artifact categories. *Cognitive Psychology, 20* (1), 65–95.

Gentner, D., Brem, S., Ferguson, R. W., Markman, A. B., Levidow, B., Wolff, P. et al. (1997). Analogical reasoning and conceptual change: A case study of Johannes Kepler. *The Journal of the Learning Sciences, 6* (1), 3–40.

Gould, S. J. (1985). *The flamingo's smile: Reflections in natural history.* New York: W. W. Norton & Company.

Graham, D. Y. (1993). Treatment of peptic ulcers caused by *Helicobacter pylori. New England Journal of Medicine,* **328** (5), 349–350.

Hampton, J. A. (1997). Emergent attributes in combined concepts. In: T. B. Ward, S. M. Smith & J. Vaid (Eds), *Creative Thought: An Investigation of Conceptual Structures and Processes* (pp. 83–110). Washington, D.C.: American Psychological Association Press.

Harre, R. (1981). *Great experiments in science.* Oxford: Phaidon.

Hayes, J. R. & Simon, H. A. (1977). Psychological differences among problem isomorphs. In: N. J. Castellan, D. B. Pisoni & G. R. Potts (Eds), *Cognitive Theory* (Vol. 2, pp. 21–41). New York: Lawrence Erlbaum.

Hecht, E. (1994). *Physics.* Pacific Grove, CA: Brooks/Cole.

Holland, J. H. (1998). *Emergence: From chaos to order.* Reading, MA: Addison-Wesley.

Holyoak, K. J. & Thagard, P. (1995). *Mental leaps.* Cambridge, MA: The MIT Press.

Home, R. W. (1981). *The effluvial theory of electricity.* New York: Arno Press.

Hutchins, E. & Levine, J. (1981). *Point of view in problem solving* (Technical Report 105). La Jolla, CA: Center for Human Information Processing.

Keil, F. C. (1979). *Semantic and conceptual development: An ontological perspective.* Cambridge, MA: Harvard University Press.

Keil, F. C. (1989). *Concepts, kinds, and cognitive development.* Cambridge, MA: The MIT Press.

Klahr, D. & Dunbar, K. (1988). Dual space search during scientific reasoning. *Cognitive Science, 12* (1), 1–48.

Kuhn, D., Amsel, E. & O'Loughlin, M. (1988). *The development of scientific thinking skills.* Orlando, FL: Academic Press.

Kuhn, T. S. (1996). *The structure of scientific revolutions* (3rd ed.). Chicago, IL: The University of Chicago Press.

Lakatos, I. (1970). Falsification and the methodology of scientific research programmes. In: I. Lakatos & A. Musgrave (Eds), *Criticism and the Growth of Knowledge* (pp. 91–195). Cambridge: Cambridge University Press.

Langley, P., Simon, H. A., Bradshaw, G. L. & Zytkow, J. M. (1987). *Scientific discovery: Computational explorations of the creative processes.* Cambridge, MA: The MIT Press.

Laudan, L. (1977). *Progress and its problems.* Berkeley, CA: University of California Press.

Leas, R. R. & Chi, M. T. H. (1993). Analyzing diagnostic expertise of competitive swimming coaches. In: J. L. Starkes & F. Allard (Eds), *Advances in Psychology: Cognitive Issues in Motor Expertise* (Vol. 102, pp. 75–94). Amsterdam: Elsevier Science.

Lowe, E. J. (1995). Ontology. In: T. Honderich (Ed.), *The Oxford Companion to Philosophy* (pp. 634–635). Oxford: Oxford University Press.

Medin, D. L. & Smith, E. E. (1984). Concepts and concept formation. *Annual Review of Psychology, 35,* 113–138.

Meier, C. R., Derby, L. E., Jick, S. S., Vasilakis, C. & Jick, H. (1999). Antibiotics and risk of subsequent first-time acute myocardial infarction. *Journal of the American Medical Association, 281* (5), 427–431.

Meyer, H. W. (1971). *A history of electricity and magnetism.* Cambridge, MA: MIT Press.

Miyake, N. (1986). Constructive interaction and the iterative process of understanding. *Cognitive Science, 10,* 151–177.

Nersessian, N. J. (1992). How do scientists think? Capturing the dynamics of conceptual change in science. In: R. N. Giere (Ed.), *Cognitive Models of Science: Minnesota Studies in the Philosophy of Science* (Vol. XV, pp. 3–44). Minneapolis, MN: University of Minnesota Press.

Ohlsson, S. (1992). Information-processing explanations of insight and related phenomena. In: M. T. Keane & K. J. Gilhooly (Eds), *Advances in the Psychology of Thinking* (Vol. 1, pp. 1–44). London: Harvester-Wheatsheaf.

Papert, S. (1980). *Mindstorms: Children, computers, and powerful ideals*. New York: Basic Books.

Piaget, J. & Inhelder, B. (1956). *The child's conception of space*. London: Routledge & K. Paul.

Resnick, M. & Wilensky, U. (1998). Diving into complexity: Developing probabilistic decentralized thinking through role-playing activities. *Journal of the Learning Sciences*, **7** (2), 153–172.

Rosch, E. R., Mervis, C. B., Gray, W. D., Johnson, D. M. & Boyes-Braem, P. (1976). Basic objects in natural categories. *Cognitive Psychology*, **8**, 382–439.

Shatz, M. & Gelman, S. A. (1973). The development of communication skills: Modifications in the speech of young children as a function of listeners. *Monographs of the Society for Research in Child Development*, **38** (5, Serial No. 152).

Slotta, J. D., Chi, M. T. H. & Joram, E. (1995). Assessing students' misclassifications of physics concepts: An ontological basis for conceptual change. *Cognition and Instruction*, **13** (3), 373–400.

Smith, C., Carey, S. & Wiser, M. (1985). On differentiation: A case study of the development of the concepts of size, weight, and density. *Cognition*, **21**, 177–237.

Sommers, F. (1963). Types and ontology. *Philosophical Review*, **72** (3), 327–363.

Sommers, F. (1971). Structural ontology. *Philosophia*, **1**, 21–42.

Spielberg, N. & Anderson, B. D. (1995). *Seven ideas that shook the universe* (2nd ed.). New York: John Wiley.

Thagard, P. (1992). *Conceptual revolutions*. Princeton, NJ: Princeton University Press.

Thagard, P. (1996). The concept of disease: Structure and change. *Communication and Cognition*, **29**, 445–478.

Thagard, P. (1998a). Ulcers and bacteria I: Discovery and acceptance. *Studies in History and Philosophy of Science. Part C: Studies in History and Philosophy of Biological and Biomedical Sciences*, **29**, 107–136.

Thagard, P. (1998b). Ulcers and bacteria II: Instruments, experiments, and social interactions. *Studies in History and Philosophy of Science. Part C: Studies in History and Philosophy of Biological and Biomedical Sciences*, **29**, 317–342.

Tuomainen, T-P. P., Kari, Nyyssonen, K. & Salonen, J. T. (1998). Association between body iron stores and the risk of acute myocardial infarction in men. *Circulation*, **97** (15), 1461–1466.

Tweney, R. D. (1989). A framework for the cognitive psychology of science. In: B. Gholson & W. R. Shadish Jr. (Eds), *Psychology of Science: Contributions to Metascience* (pp. 342–366). Cambridge: Cambridge University Press.

Wall, R. E. (1972). *Introduction to mathematical linguistics*. Englewood Cliffs, NJ: Prentice-Hall.

Weisberg, R. & Suls, J. M. (1973). An information-processing model of Duncker's Candle Problem. *Cognitive Psychology*, **4** (2), 255–276.

White, A. R. (1975). Conceptual analysis. In: C. J. Bontempor & S. J. Odell (Eds), *The Owl of Minerva* (pp. 103–117). New York: McGraw-Hill.

Zhang, R. M. D. P., Brennan, M-L. P., Fu, X. M. S., Aviles, R. J. M. D., Pearce, G. L. M. S. & Penn, M. S. M. D. P. et al. (2001). Association between myeloperoxidase levels and risk of coronary artery disease. *Journal of the American Medical Association*, **17**, 2136–2142.

The International Handbook on Innovation
Edited by Larisa V. Shavinina

Understanding Scientific Innovation:
The Case of Nobel Laureates

Larisa V. Shavinina

Département des Sciences Administratives, Université du Québec en Outaouais, Canada

Abstract: Why are Nobel laureates so innovative? Although the capacity of Nobel laureates for scientific innovation has attracted researchers' attention for decades, the phenomenon is far from understood. This chapter argues that scientific innovation of Nobel laureates is determined in part by specific preferences, feelings, beliefs, and intuitive processes, which constitute a whole field of unexplored or poorly understood scientific phenomena. Based on autobiographical and biographical accounts of Nobel laureates, the chapter explores these phenomena, demonstrating that they predict scientific productivity of the highest level that results in innovations and, consequently, outstanding scientific talent of Nobel caliber.

Keywords: Innovation; Scientific innovation; Creativity; Nobel laureates; Preferences; Feelings; Intuition; Giftedness.

Introduction

The capacity of Nobel laureates for scientific innovation has been an attractive research topic for decades. However, it is not known why Nobel laureates are so capable of innovation. This chapter discusses some hidden psychological mechanisms, which significantly contribute to the innovative capacity of Nobel laureates.

Autobiographical and biographical findings on distinguished scientific geniuses demonstrate that their mental functioning is determined in part by *specific feelings, preferences*, *beliefs* and other similar phenomena. But researchers rarely study these phenomena, which can be referred to as 'extracognitive phenomena'. One of the greatest minds of the 20th century, Albert Einstein, in discussions with Max Wertheimer (1959) about the development of the theory of relativity and the way of thinking, which led to it, emphasized that:

> ... during all those years there was the *feeling of direction*, of going straight toward something concrete. It is, of course, very hard to express that feeling in words; but it was decidedly the case, and clearly to be distinguished from later considerations about the rational form of the solution (p. 228; italics added).

Similarly, Jacques Hadamard (1954) in his study on the psychology of invention in the mathematical field cited Henri Poincaré, a famous French scientist, who asserted that:

> it may be surprising to see emotional sensibility invoked à propos of mathematical demonstrations which, it would seem, can interest only the intellect. This would be to forget the *feeling of mathematical beauty*, of the *harmony of numbers and forms*, of *geometric elegance*. This is a true esthetic feeling that all real mathematicians know, and surely it belongs to emotional sensibility (p. 31; italics added).

Poincaré (1913) also insisted that "pure logic would never lead us to anything but tautologies. It is by logic that we prove. It is by *intuition* that we discover" (p. 208; italics added). In a similar way, Rosenblueth and Wiener (1945) emphasized that:

> An *intuitive flair* for what will turn out to be the most important general question gives a basis for selecting some of the significant among the indefinite number of trivial experiments which could be carried out at that stage. Quite vague and tacit generalizations thus influence the selection of data at the start (p. 317).

Even such a brief account shows the extremely significant role which 'feeling of direction', 'sense of beauty', those processes usually referred to as 'intuition', and other similar phenomena played and play in the appearance of some of the most celebrated creative scientific discoveries.

In the context of this chapter, the phenomenon of the extracognitive refers to four interrelated—and at the same time obviously different—components. These are:

(1) specific intellectual feelings (e.g. feelings of direction, harmony, beauty, and style);
(2) specific intellectual beliefs (e.g. belief in elevated standards of performance);
(3) specific preferences and intellectual values (e.g. the 'inevitable' choice of the field of endeavor by certain geniuses and internally developed standards of intellectual working); and
(4) intuitive processes.

The word 'specific' embodies the uniqueness of these components in the intellectual functioning of gifted, creative, and talented individuals. It is interesting to note that other psychologists also use the word 'specific' in their accounts of the intellectually creative processes at the highest level. For example, Marton, Fensham & Chaiklin (1994) wrote about a 'specific form' of the feeling of being right and made reference to other 'specifics' in their study of scientific intuition in Nobel Prize winners.

The chapter consists of four parts. The first part deals with methodological issues related to the use of biographical and autobiographical literature as well as case-study approaches as a means of investigating the psychological underpinnings of Nobel laureates. The second part of this chapter presents findings regarding the extracognitive phenomena of Nobel laureates. The third part, a discussion, considers the relationship between the components of the extracognitive, its link to metacognition, and its functions. Finally, the fourth part concludes that the extracognitive should be viewed as the highest level of the manifestation of the intellectual and creative potentials of an individual— that is exactly the case of Nobel laureates—and, consequently, as an important predictor of scientific innovation.

Methodological Issues

Two main data sources were used in my studies: (1) biographical and autobiographical accounts on Nobel laureates; and (2) the analysis and amalgamation of research literature presenting the psychological investigations as well as other related studies pertaining to Nobel laureates. The use of the biographical and autobiographical accounts and the case-study method can be, and has been to a certain extent, a controversial matter. However, any attempt at a comprehensive

review of the research on Nobel laureates and other distinguishing individuals is unfeasible without some reliance on such accounts. These accounts and the case-study method are perfectly suited for capturing the special characteristics of highly creative and intellectual individuals and their discoveries or inventions. Autobiographical and biographical literature is essential for the research and description of persons or events distinguished by their rarity as is the case with Nobel laureates. Using this literature, researchers can describe the idiosyncratic features of gifted, creative, and talented individuals including a focus on the characteristics of their extracognitive phenomena. Often biographical and autobiographical accounts provide a holistic view of the subject (Foster, 1986; Frey, 1978) allowing researchers to develop and validate theories grounded in a more direct 'observation' of the individuals (Gross, 1994; Merriam, 1988). The analysis of subjective phenomena such as the individual's feelings, preferences, beliefs, and views enables a more comprehensive 'picture' of the individual than that which is possible with only objective methods.

The use of autobiographical and biographical literature for the study of outstanding individuals—for example, Nobel scientists—presents, however, certain limitations, such as:

(1) The possible subjectivity of biographers resulting from their individual interpretations of events, thoughts, and states. These interpretations may be influenced by their personal attitudes towards the person about whom they write, an attitude potentially swayed in part by whether the latter is living or not. Autobiographers can also be very subjective and contradictory in their accounts of their own thinking processes, psychological states, and the surrounding events, which lead up to and follow their particular innovative breakthroughs in science;
(2) Time of writing the biography or autobiography, normally after an individual has already become a brilliant personality—often relies on vague memories of one's thinking processes which may have likely been weakened or altered over time. Hence, the conclusions and reports of autobiographers and especially of biographers are not always very reliable. It certainly raises the issue surrounding the validity and reliability of subjective reports when they are used as data. This problem has been addressed extensively in the literature (Brown, 1978, 1987; Ericsson & Simon, 1980). However, despite such critiques, scholars in the field of creativity view the use of self-reports as an "effective means of learning about scientific thinking" (John-Steiner, 1985, p. 181).

It should be noted that Nobel laureates and other distinguishing personalities—who are themselves the subjects of biographies and autobiographies—grasped the above-mentioned problems very well. For example, Einstein himself realized this clearly when he wrote at the beginning of his *Autobiographical Notes*:

> The exposition of that which is worthy of communication does nonetheless not come easy; today's person of 67 is by no means the same as was the one of 50, of 30, or of 20. Every reminiscence is colored by today's being what it is, and therefore by a deceptive point of view . . . (Einstein, 1949, p. 47).

Holton (1973) further pointed out that it is not only growth or change that colours one's interpretations—it is also the difference between experience lived and experience reported.

> In this case it is well possible that such an individual in retrospect sees a uniformly systematic development, whereas the actual experience takes place in kaleidoscopic particular situations (Einstein, 1949, p. 47).

However, even with these limitations in mind, the use of biographical and autobiographical literature is probably the single best source available for the investigation of the extracognitive phenomenon in Nobel laureates since such accounts are enriched with information pertaining to an individual's specific feelings, preferences, beliefs, and intuitive processes.

The Extracognitive Phenomena in Nobel Laureates

As it is well known from the history of science, any new field or topic—being at the initial stages of its development—starts with a descriptive stage. Since the research on Nobel laureates is not a well-developed topic and the study of the extracognitive is a relatively new one, the findings presented below are descriptive in nature. That is, they describe the different components of the extracognitive phenomenon in Nobel laureates and, as such, act to provide an introduction to this new concept.

The purpose of this section is to draw at least an approximate picture of the extracognitive in Nobel laureates. In order to do so, I primarily examine autobiographical and biographical accounts of Nobel laureates and existing psychological studies on these outstanding scientists. However, taking into account the fact that psychological publications on this topic are not numerous, I also refer to appropriate sociological investigations—for example, a famous study of Nobel laureates in the United States of America carried out by Harriet Zuckerman (1977). Below I present the findings regarding each of the four above-mentioned components of the extracognitive phenomenon.

Specific Intellectual Feelings

Specific Scientific Taste, Including Sense of 'Important Problems', 'Good' Ideas, 'Correct' Theories, and Elegant Solutions, and Feeling of Being Right

Zuckerman (1977, 1983) showed that a *specific scientific taste* is an extremely important virtue of Nobel Prize winners. Its primary criteria include a *sense* for distinguishing the 'important problem' and an appreciation of stylish solutions. For them, "deep problems and *elegant solutions* distinguish excellent science from the merely competent or commonplace" (Zuckerman, 1983, p. 249; italics added). For example, discussing his own feelings in the process of scientific creativity, Stanley Cohen, Nobel laureate in medicine, 1986, noted:

> . . . to me it is *a feeling of* . . . . 'Well, I really don't believe this result', or 'This is a trivial result', and 'This is an important result', and 'Let us follow this path'. I am not always right, but I do have feelings about what is an important observation and what is probably trivial (quoted in Marton et al., 1994, p. 463; italics added).

Analyzing Nobel laureates' replies to the question "In the absence of rational, logical support for scientific intuitions, what makes the scientist follow them?", Marton et al. (1994) found that nine out of 72 laureates said that it was a 'feeling' that made them persevere. Marton et al. (1994) pointed out that the feeling could be "a *feeling of being right, being wrong*, or *having come across something important* . . . . The feeling is often an immediate, intense feeling of certitude, of being right, especially when an answer to a problem one has been struggling with appears suddenly, without any preceding steps whatsoever" (p. 463). It is interesting to note that Marton et al. (1994) perceived a *'feeling of great certitude'* in Nobel laureates.

Marton et al. (1994) also found that the "feeling of being right often seems to originate from artistic and/or—in the metaphoric sense—sensory or quasi-sensory qualities. You sense, see, recognize, feel in your fingers or produce certain qualities" (p. 463). As Paul Berg, Nobel laureate in chemistry, 1980, highlighted:

> There is another aspect that I would add to it, and that is, I think, *taste. Taste in almost the artistic sense.* Certain individuals see art in some undefinable way, can put together something which has a *certain style*, or a certain class to it. A certain rightness to it (quoted in Marton et al., 1994, p. 463; italics added).

Csikszentmihalyi's (1996) study of ten Nobel Prize-winners in science, among another 91 exceptionally creative persons, provided additional evidence for the existence of a *sense* for an 'important problem'. He pointed out that Nobel Laureates seem to have an ability to distinguish between 'good' and 'bad' ideas. For example, Manfred Eigen, Nobel laureate in chemistry, 1967, was one of several scientists in Csikszentmihalyi's study who asserted that the only difference between them and their less creative colleagues was that they could tell whether a problem was soluble or not, and this saved enormous amounts of time in addition to many false starts. Similarly, at Linus Pauling's (Nobel laureate in chemistry, 1954) 60th birthday celebration, a student asked him, "Dr Pauling, how does one go about having good ideas?" He replied, "You have a lot of ideas and throw away the bad ones" (Csikszentmihalyi, 1996, p. 67). Likewise, George Stigler, Nobel laureate in economics, 1982, claimed:

> I consider that I have *good intuition and good judgment on what problems are worth pursuing and what lines of work are worth doing*. I used to say (and I think this was bragging) that whereas most scholars have ideas which do not pan out more than, say, 4% of the time, mine come through maybe 80% of the time (quoted in Csikszentmihalyi, 1996, p. 61; italics added).

Arthur L. Schwalow, Nobel laureate in physics, 1981, says the following about a similar feeling accompanying his scientific creativity:

> . . . you store in your mind a *feeling for a magnitude of things*, how big things really are, so you'll get a *feeling whether something will go, or not*, if you try to put two ideas together (quoted in Marton et al., 1994, p. 466; italics added).

Similarly, Niels Bohr's *feeling for correct and incorrect theories* was a legendary one. For example, after hearing Wolfgang Pauli's presentation to a professional audience a new theory of elementary particles, Bohr summarized the subsequent discussion by saying that "we are all agreed that your theory is crazy. The question which divides us is whether it is crazy enough to have a chance of being correct. *My own feeling is that it is not crazy enough*" (quoted in Cropper, 1970, p. 57).

In an overview of Robert Burns Woodward's accomplishments, who was one of the great organic chemists of all time and Nobel laureate in chemistry, 1965, his friend and colleague at Harvard University, Frank Westheimer stated:

> Even scientists who mastered his methods could not match his *style*. For there is an *elegance* about Woodward's work—his chemistry, his lectures, his publications—that was natural to him, and as unique as the product itself (quoted in C. E. Woodward, 1989, p. 229; italics added).

Therefore, the personal accounts of Nobel laureates demonstrate that specific scientific taste is a very

critical facet in scientific creativity of an exceptionally high level. As was documented by various examples, this taste expresses itself in a variety of feelings, senses, and styles.

Feeling of Beauty

Describing his discovery of the positron, Paul Adrienne Maurice Dirac (1977) emphasized that:

> It was sort . . . of faith . . . that any equations which describe fundamental laws of Nature must have *great mathematical beauty* in them (e.g.) . . . the *beauty of relativity* . . . . (I was working on) the physical interpretation and transformation theory of quantum mechanics . . . (p. 136; italics added).

Also, upon receiving the Nobel Prize for physics, Dirac (1963) remarked:

> It seems that if one is working from the point of view of *getting beauty* in one's equations and if one has really sound insight, one is on a sure line of progress (p. 47; italics added).

Likewise, Dirac (1963), discussing why Ervin Schrödinger failed to publish a relativistic wave equation, asserted that "it is more important to *have beauty in one's equations* than to have them fit experiment" (p. 47; italics added). Similarly, physicist Allan Cormack, Nobel laureate in medicine, 1979, said in an interview with Rothenberg (1996):

> The abstractions (I do in mathematics) are just as *beautiful* (as in art) and I find them more satisfactory . . . . I think there's a great deal of satisfaction in seeing ideas put together or related. And there is a structural thing there just as much as in sculpture or painting or anything of that sort—form and economy of means . . . . Very often in biology you say, 'If such-and-such went that way, will this go that way?' Very often the reason you ask why is because you found the previous thing to be *attractive* somehow (p. 212; italics added).

Werner Heisenberg (1971) wrote the following about his own feeling of beauty and its role in scientific creativity:

> You may object that *by speaking of simplicity and beauty I am introducing aesthetic criteria of truth*, and I frankly admit that *I am strongly attracted by the simplicity and beauty of the mathematical schemes* which nature presents us. You must have felt this too: the almost frightening simplicity and wholeness of the relationship, which nature suddenly spreads out before us . . . (p. 68).

Similarly, describing his investigations in the area of X-ray crystallography, Robert Huber, Nobel laureate in chemistry, 1988, asserted in an interview with Rothenberg (1996):

> It's not just joy. This first stage (in X-ray crystallography) of seeing a crystal is not only a *beauty*, there's also so much promise behind it, so many hopes. And, then, very much later, when analysis has succeeded, ah, to see the molecule for the first time, that's similar (p. 212; italics added).

In the same way, recalling his research on the discovery of the instructed mixture paradigm, Jean-Marie Lehn, Nobel laureate in chemistry, 1987, pointed out in an interview with Rothenberg (1996): "Once you have recognized this *harmony* which exists in self-recognition, then you should say, 'But what is the self/non-self?'" (p. 229, italics added). Ochse (1990) pointed out that "Max Born hailed the advent of relativity as making the universe of science not only grander but also more *beautiful*" (p. 123).

Likewise, R. B. Woodward's daughter wrote about her famous father's "*feeling for art*" in the synthesis of organic compounds (C. E. Woodward, 1989, p. 235, italics added). In the Cope talk of 1973, in which he reviewed the background of his orbital symmetry work in chemistry, Robert Burns Woodward asserted:

> For almost 50 years now, I have been involved in an affair with chemistry. It has been throughout a richly rewarding involvement, with numerous episodes of high drama and intense engagement, with the joys of enlightenment and achievement, with the special pleasures which come from the *perception of order and beauty in Nature*—and with much humor (quoted in C. E. Woodward, 1989, p. 230; italics added).

In general, R. B. Woodward perceived an amazing sense of beauty in chemistry. This sense of beauty manifested itself in every facet of his work. For example, he claimed "I love crystals, the *beauty of their forms*—and their formation . . ." (quoted in C. E. Woodward, 1989, p. 237; italics added). His daughter wrote, "the *aspects of art and beauty in Woodward's work* are contained not only in the forms found in or built into the fixed structures of molecules. They have perhaps more to do with the way in which he manipulated the molecules, in his design of the synthetic steps, in a process that was not tortuous but *harmonious*, that *felt right* and was *elegant*" (C. E. Woodward, 1989, p. 234; italics added).

Ethologist, artist, writer, and Nobel laureate in medicine Konrad Lorenz (1952) expressed a general attitude of a man to the beauty of nature when he wrote that:

> He who has once seen the intimate *beauty of nature* cannot tear himself away from it again. He must become either a poet or a naturalist and, if his eyes are good and his powers of observation sharp enough, he may well become both (p. 12).

It should be acknowledged that other researchers have also discussed and highlighted the importance of the feeling of beauty so strongly expressed by Nobel laureates. Ochse (1990) considered the feeling of beauty in the context of the analysis of aesthetic sensitivity; Rothenberg (1996) in the context of aesthetic motivation; Kuhn (1970) in the context of aesthetic considerations; and McMorris (1970), Miller (1981, 1992, 1996), and Wechsler (1978) in the context of aesthetics in science as a whole. Furthermore, Kuhn (1970) argued that application of aesthetic sensitivity—that scientists mainly express as a feeling or sense of beauty—was indeed essential to the progress of science:

> Something must make at least a few scientists *feel they are on the right track*, and sometimes it is only *personal and inarticulate aesthetic considerations* that can do that. Men have been converted by them at times when most of the articulate technical arguments pointed the other way . . . even today Einstein's general theory attracts men principally on *aesthetic grounds*, an appeal that few people outside of mathematics have been *able to feel* (Kuhn, 1970, p. 158; italics added).

Ochse (1990) suggests that behind every famous scientist's expressions about their sense of beauty is "the suggestion that underlying scientific creativity is an intellectual motivation that is fuelled by a positive evaluation of learning and achievement, and guided by aesthetic sensitivity—which may relate to a need for emotional satisfaction" (p. 124).

Feeling of Direction and Similar Feelings (in one's own scientific activity, in search of mentors and of one's own unique domain in science, and so on)

As it follows from Albert Einstein's (Nobel laureate in physics, 1921) citation at the beginning of this chapter, the feeling of direction played an important role in his work on the development of the Theory of Relativity. Michael S. Brown, Nobel laureate in medicine, 1985, described a similar feeling in a round-table discussion—called 'Science and Man'—with the Nobel laureates in physics, chemistry, and medicine:

> And so . . . as we did our work, I think, *we almost felt at times that there was almost a hand guiding us*. Because we would go from one step to the next, and somehow we would *know which was the right way to go*. And I really can't tell how we knew that, how we knew that it was necessary to move ahead (quoted in Marton et al., 1994, pp. 461–462).

It can be said that intellectually creative individuals have very specific manifestations of cognitive direction in the study of scientific problems and of the world as a whole.

Zuckerman (1977, 1983) found that the Nobel laureates, "in their comparative youth, sometimes went to great lengths to *make sure* that they would be working with those they considered the best in their field" (Zuckerman, 1983, pp. 241–242; italics added). This belief (i.e. self-confidence, which is very often explained as good fortune, luck, fate or chance, etc.) led them to the masters of their craft—to the scientists of Nobel caliber. It is thus possible to confirm the existence of a feeling of direction in the scientific elite. For instance, the biochemist Hans Krebs, Nobel laureate in chemistry, (1967) noted:

> If I ask myself how it came about that one day I found myself in Stockholm, I have not the slightest doubt that I owe this *good fortune* to the circumstance that I had an outstanding teacher at the critical stage in my scientific career . . . (p. 1444; italics added).

The 'good fortune' that Hans Krebs so humbly refers to is nothing else but the deep intellectual *feeling of direction* in the fulfillment of his own scientific career.

Likewise, recalling his work on the creation of monoclonal antibodies, Georges Kohler, Nobel laureate in medicine, 1984, said in an interview with Rothenberg (1996):

> I was one of the first Ph.D. students in an institution (Basel Institute) which turned out, later on, to be a very important one. I studied genetic diversity of antibodies and I thought that (the somatic mutation hypothesis) was a very clever idea: to have one gene, and from that somatically you can have many variants. I knew there were variants because I started on that in my Ph.D. And variants were made by Cesar Milstein in the lab, so I said, *"Okay, I'm going to study how these variants are going to be made. I am going to Cesar Milstein . . ."*. I knew that the Cesar Milstein group had done experiments with cell fusion—another field in which I was reading a book and was interested. And I remember that when I was about to go to Cambridge, I was talking to somebody in the library and saying, *"I'm going to Cesar Milstein and make a lot of fusions"* (p. 227; italics added).

These accounts bring to light the exceptional role of one's 'feeling of direction' and similar feelings—which are in fact one's more latent manifestations of the feeling of direction—to the intellectual activity of Nobel laureates. Their scientific creativity is determined by such feelings, which guide Nobel laureates in their work.

Specific Intellectual Beliefs

Belief in Specific Standards of Performance

Kholodnaya (1991) pointed out that a belief in the existence of some principles and specific standards, by which the nature of scientific research is determined, and a priori confidence in the truth of a certain vision of things are among important beliefs of outstanding scientists. Thus, C. E. Woodward (1989) highlighted R. B. Woodward's *'feeling for experimentation'* that was based on his conviction that theory without experimental proof was worthless. "Ideas and theory could have an aesthetic aspect but their beauty and elegance were always tied to a concrete relationship with a physical reality" (C. E. Woodward, 1989, p. 237). Similarly, Zuckerman (1977) found that *elevated standards of performance* (the methods and quality of first rate research) are essential for Nobel laureates.

Specific Intellectual Preferences (e.g. choice of 'difficult' scientific problems or problems at the leading edge of science)

Specific intellectual preferences of Nobel laureates manifest themselves in a variety of ways. Some have a preference for 'difficult' scientific problems, which were and had remained unsolved for many decades. Richard Feynman is a fine example of this pursuit to solve the unsolved. Feynman, in his early twenties at Los Alamos, was an *"enfant terrible*, bubbling with quick brilliance on the theoretical problems of bomb building that came his way" (Wilson, 1972, p. 10). Feynman's account is as follows:

> It was a succession of successes—but easy successes. After the war, I moved over to the kind of problems (like the self-energy of the electron) that men spend years thinking about. On that level there are no easy successes; and the satisfaction you get when you're proved right is so great that even if it occurs only twice in a lifetime, everything else is worth it! (quoted in Wilson, 1972, pp. 10–11).

Nobel laureates' preference for problems at the very frontiers of science is another way their specific intellectual preferences are manifested. This desire to decipher, understand and explore the relatively unknown areas of a particular field is demonstrated as we follow the life trajectory of Nobel Prize winner, Enrico Fermi. In the 1920s, he was:

> one of the leaders among the young European physicists who were developing the quantum mechanical wave theories of atomic structure. Then in 1929 Fermi made a sharp decision . . . Fermi felt that . . . the only remaining area of physics to attack where all was still unknown at that time was the heart of the atomic structure—nuclear physics. So in 1929 Fermi moved away from research in atomic theory, where he had made a great reputation, and went into the unknown of neutron physics, where he worked with such ingenuity on the interaction between neutrons and atomic nuclei that within seven years he won a Nobel prize. Fifteen years later, at the end of World War II in 1945, Fermi changed

again. By that time, he probably had learned more about the neutron than anyone else in the world, but now it was nuclear physics that seemed to have been all cleaned up for him. So once again he moved on— this time into the uncharted domain of high-energy particles. To Fermi, it was always necessary to be at work where the big mystery was (Wilson, 1972, p. 16).

Such examples allowed Wilson (1972) to conclude that "when there is no longer enough mystery in a subject to attract" scientists, "they move on to new fields" (p. 16). In other words, they always follow their inner, specific intellectual preferences.

Highly Developed Intuitive Processes

As it was emphasized in the psychological literature, certain individuals have an intuitive feeling, as they begin their intellectual activity, about what their final product will be like (Gardner & Nemirovsky, 1991; Gruber, 1974; Hadamard, 1954; Ochse, 1990; Policastro, 1995; Simonton, 1980). This is especially true in the case of Nobel laureates. It is widely recognized by these great scientists that intuition is an essential component of creative thinking that leads to innovative discoveries. For example, Max Planck asserted that the pioneer scientist working at the frontier of science "must have a vivid *intuitive* imagination, for new ideas are not generated by deduction, but by an artistically creative imagination" (Planck, 1950, p. 109). At the same time, he recognized that intuition alone is not sufficient. For instance, in his autobiographical account on the discovery of the constant and the quantum of action, Max Planck wrote:

> So long as it (the radiation formula) had merely the standing of a law disclosed by lucky *intuition*, it could not be expected to possess more than a formal significance. For this reason, on the very day that I formulated this law, I began to devote myself to the task of investing it with true physical meaning (Planck, 1950; p. 41; italics added).

Similarly, Einstein highly appreciated intuition in creative processes. Thus, he wrote about "Bohr's *unique instinct*, which enabled him to discover the major laws of spectral lines and the electron shells of the atoms", (Einstein, 1949; cited in John-Steiner, 1985, p. 194). Clearly, 'Bohr's *unique instinct*', as referred to by Einstein, is nothing else but his 'unique intuition'. Wilson (1972) exemplified the value placed on intuition in this personal observation about the famous Enrico Fermi:

> Years ago, as a graduate student, I was present at a three-way argument between Rabi, Szilard, and Fermi. Szilard took a position and mathematically

stated it on the blackboard. Rabi disagreed and rearranged the equations to the form he would accept. All the while Fermi was shaking his head. 'You're both wrong', he said. They demanded proof. Smiling a little he shrugged his shoulders as if proof weren't needed. '*My intuition tells me so*', he said. I had never heard a scientist refer to his intuition, and I expected Rabi and Szilard to laugh. They didn't. The man of science, I soon found, works with the procedures of logic so much more than anyone else that he, more than anyone else, is aware of logic's limitations. *Beyond logic there is intuition . . .* (pp. 13–14; italics added).

Another example where the importance of intuition becomes evident came from the work on the discovery of DNA's structure, for which James Watson and Francis Crick were awarded the Nobel Prize. Thus, following her conversation with Sir Francis Crick, John-Steiner (1985) pointed out that one of Crick's contributions to the team's efforts—that led to the discovery of DNA's structure—was his "*intuition* and his ability to work with a minimum number of assumptions while approaching a problem" (p. 187).

Marton et al. (1994) studied intuition of Nobel Prize winners analyzing interviews conducted between 1970 and 1986 with laureates in physics, chemistry, and medicine by a Swedish Broadcasting Corporation. Practically all the laureates regard scientific intuition as a "phenomenon distinctively different from drawing logical conclusions, step by step" (Marton et al., 1994, p. 468). Eighteen out of 72 subjects in this study emphasized that intuition *feels* different from logical reasoning and cannot be explained in logical terms. Marton et al. (1994) concluded that Nobel laureates consider scientific intuition as "an alternative to normal step-by-step logical reasoning" (p. 468). Nobel laureates in their scientific activity:

> do something or something happens to them without their being aware of the reasons or the antecedents. The acts or the events are, however, guided or accompanied by feelings which sometimes spring from a quasi-sensory experience. Intuition is closely associated with a sense of direction, it is more often about finding a path than arriving at an answer or reaching a goal. The ascent of intuition is rooted in extended, varied experience of the object of research: although it may *feel* as if it comes out of the blue, it does *not* come out of the blue (Marton et al., 1994, p. 468).

To date, Marton et al.'s (1994) study is the only one that conducts a systematic investigation of intuition in Nobel laureates.

So far, we emphasized the important positive role of intuitive processes in scientific creativity of an extremely high level by quoting exceptionally accomplished scientists. However, it is also pertinent to mention the case when the absence of scientific intuition resulted in quite mediocre work. For example,

> Robert Oppenheimer was a brilliant interpreter of other men's work, and a judge who could make piercing evaluations of other men's work. But when it came time—figuratively speaking—to write his own poetry in science, his work was sparse, angular, and limited, particularly when judged by the standards he himself set for everyone else. He knew the major problems of his time; he attacked them with style; but *he* apparently *lacked that intuition*—that faculty beyond logic—*which logic needs in order to make great advances*. If one were speaking not of science but of religion, one could say that Oppenheimer's religiosity was the kind that could make him a bishop but never a saint (Wilson, 1972, p. 13; italics added).

Although the above-mentioned personal accounts of Nobel laureates demonstrate that 'intuition' plays a significant role in highly creative and intellectual processes, we do not know for sure its psychological nature and origin. The common wisdom is to use the term 'intuition' in association with the term 'insight' which does not follow logically from available information, and is interpreted as inexplicable (so-called 'instinctive' insights) (Ochse, 1990). This kind of insight is often considered to be an innate quality explained by superior functioning of the right hemisphere of the brain. Hadamard (1954) described intuition as something 'felt', in contrast to something 'known', emphasizing that it involves emotional empathy.

Ochse (1990) attributed intuitive thinking to the operation of automatic mental routines, "unconsciously triggered by configurations of exogenous and/or endogenous stimuli. More specifically, intuition may be viewed as unconsciously triggered *automatic integration of relevant elements of information*, and an 'intuitive feeling' may be seen as part of the experiential outcome of such processes—somewhat equivalent to a feeling of recognition" (p. 243). His standpoint fits perfectly to Bruner's (1960) idea that "intuitive thinking characteristically does not advance in careful well-planned steps. It tends to involve manoeuvres based on an implicit perception of the total problem. The thinker arrives at an answer which may be right or wrong, with little if any awareness of the process by which he reached it" (pp. 57–58).

Ochse (1990) also suggested that creators develop "well-established bases for intuition because they are constantly involved with their subject of interest, and this practice would lead to the establishment of routines that enable them to integrate relevant actions and items of information. Moreover, creators work

independently rather than following prescribed curricula and instructions, which favors the acquisition of a relatively wide repertoire of *generalizable routines*" (p. 244).

Policastro (1995) distinguishes between phenomenological and technical definitions of creative intuition. According to the first, intuition is defined as "a vague anticipatory perception that orients creative work in a promising direction" (p. 99). This definition is phenomenological in that it points to the subject's experience: How does it feel to have a creative intuition? What is that like? The study by Marton et al. (1994) mentioned above analyzed mainly this kind of intuition in Nobel laureates.

According to the technical definition, intuition is "a tacit form of knowledge that broadly constrains the creative search by setting its preliminary scope" (Policastro, 1995, p. 100). This implies that intuition is based on cognitive foundations in the sense that it arises from knowledge and experience. Similarly, Simonton (1980) stressed that intuition involves a form of information processing that might be more implicit than explicit, but which is not at all irrational. Policastro (1995) emphasized that both definitions (phenomenological and technical) are important because they complement each other in fundamental ways.

Bowers, Regher, Balthazard, & Parker (1990) defined *intuition* as "a preliminary perception of coherence (pattern, meaning, structure) that is at first not consciously represented, but which nevertheless guides thought and inquiry toward a hunch or hypothesis about the nature of the coherence in question" (p. 74). Bowers et al. conducted experimental studies of intuition, two of which revealed that "people could respond discriminatively to coherences that they could not identify" (p. 72). A third experiment showed that this tacit perception of coherence "guided people gradually to an explicit representation of it in the form of a hunch or hypothesis" (p. 72).

In spite of the differences in the psychological interpretation of the nature of intuition in general and scientific intuition in particular, it is clear however that intuitive processes are extremely important for the productive functioning of human mind. The truth is that any successful scientist—and first of all Nobel laureates—relies on his or her intuition. It is important to note that other psychologists also include intuitive processes in their conceptualizations of higher psychological functions. For example, Sternberg et al. (2000) consider intuition as one of the prototypical forms of developed practical intelligence.

Findings from autobiographical and biographical accounts of Nobel laureates as well as related data from the research literature regarding their extracognitive phenomena were presented in this section. The main conclusion is that the above-described specific scientific taste, intellectual feeling of beauty, feeling of direction, specific beliefs, intellectual preferences, scientific intuition, and their variants are exceptionally important characteristics in extraordinary creative achievements of the world's best scientists, Nobel laureates. The integration of the autobiographical and biographical findings on Nobel laureates with the findings of the psychological studies on these distinguishing scientists demonstrates that their mental functioning is determined—at least in part—by their extracognitive abilities.

Discussion

A range of intellectual phenomena, which describe and comprise the four components of the extracognitive, was presented above. Now I would like to address the issue of the relationship between these components taking as an example the relationship between a feeling of direction and intuitive processes. To start the discussion, it seems appropriate to raise the following question: What is behind the feeling of direction? Marton et al. (1994) suggest that this is one's own intuition. They found that although famous examples of intuition associated with names such as Archimedes, Kekulé and Poincaré emphasize arriving at answers or the solutions to problems, other Nobel laureates in their sample stressed the outcome of intuition as their starting rather than end points. In other words, if intuition is considered by Nobel laureates as outcome or result (i.e. intuition denotes an idea, a thought, an answer, or a feeling), then there are two main alternative ways of experiencing what intuition yields. The first has to do "with direction when moving *from* a certain point *towards* something as yet unknown" (Marton et al., 1994; p. 462; italics added). It concerns "finding, choosing, *following a direction, a path*" (Marton et al., 1994; p. 461; italics added). For instance, 37 out of 72 Nobel laureates see the result of scientific intuition as "finding or following a path" (Marton et al., 1994; p. 462). It seems safe to assert that such an understanding of intuitive processes is behind the 'feeling of direction'. The other main way of experiencing the outcome of intuition by Nobel Prize winners deals with *coming* to a certain point, *arriving at* an answer, that "which one was moving towards is illuminated and seen clearly" (i.e. end points; Marton et al., 1994; p. 462). Marton et al. (1994) concluded that "the most fundamental aspect of scientific intuition is that the scientists choose directions or find solutions for which they do not have sufficient data in the computational sense" (p. 468). In describing the close relationship between intuition and sense of direction, Marton et al. (1994) came close to the introduction of the concept of the extracognitive in intellectual functioning of gifted individuals, identifying some of its main components.

It seems that there are reciprocal relationships between intuition and various feelings of Nobel laureates described above. I submit that not only do intuitive processes form a basis for the actualization and subsequent development of these feelings, but also the latter may influence the growth of intuition. As a result of their study of the experience of intuition in Nobel laureates, Marton et al. (1994) described a 'cumulative structure' of intuition consisting of its higher and lower levels. On the lower level subjects simply indicated that "the experience of scientific intuition differs from the experience of making explicit, logical conclusions based on available information" (p. 463). On the higher level a crucial component is "a feeling, a feeling of something being right (or, occasionally, something being wrong)" (Marton et al., 1994; p. 464). On the next level again a specific form of this feeling emerges: "the sudden appearance of an idea or of an answer accompanied by a feeling of great certitude" (Marton et al., 1994; p. 464). Consequently, as one can see on the example of the feeling of direction and intuitive processes, the components of the extracognitive are closely related to each other.

The Extracognitive Phenomena and Metacognition

It appears that there is a reciprocal relationship between the extracognitive phenomena and the metacognitive processes of an individual. In other words, if an individual possesses any component of the extracognitive—say, feeling of direction or feeling of beauty (but the findings show that any Nobel laureate has all the above-described components of the extracognitive)—then he or she is also expected to possess highly developed metacognitive abilities.

Metacognition is broadly defined as "any knowledge or cognitive activity that takes as its object, or regulates, any aspect of any cognitive enterprise" (Flavell, 1992, p. 114). The essence of metacognition already implies some psychological phenomena, which exist at the intersection of metacognition and the extracognitive. For example, psychologists studying metacognition use and investigate such concepts as awareness and self-awareness (Ferrari & Sternberg, 1996) and feeling of knowing (Brown, 1978), just to mention a few. The most basic form of self-awareness is the "realization that there is a problem of knowing what you know and what you do not know" (Brown, 1978, p. 82). Very often an individual unconsciously comes to such a realization. Probably, intuition is behind any unconscious understanding of any object of cognition including his or her own cognitive apparatus.

Moreover, the well-known title of Brown's (1978) famous article, *Knowing When, Where, and How to Remember: A Problem of Metacognition*, which can be considered synonymously with the definition of metacognition, indicates that feeling of direction, specific scientific taste, and feeling of beauty also play a role in metacognitive functioning. Thus, the feeling of direction corresponds to 'where' (i.e. *guiding function* of the extracognitive). Specific scientific taste and the feeling of beauty relate to 'how' (i.e. *function of evaluation and judgement* of the extracognitive). Altogether—including intuition—they correspond to 'when'. Therefore, the phenomenon of the extracognitive contributes to the development of a person's metacognitive abilities. I would even add that the extracognitive lies somewhere in the heart of metacognition and, consequently, allow psychologists to better understand its anatomy and, hence, its nature. In its turn, metacognition leads to the further development of the extracognitive, strengthening and crystallizing its components in an individual's intellectual functioning. For example, the individual with developed metacognitive abilities will be more open to his or her own feeling of direction, feeling of beauty, and intuitive processes.

Empirical findings support our view about the existence of a direct link between the extracognitive and metacognition. Thus, Marton et al. (1994) concluded that scientific intuition of Nobel laureates can be interpreted in terms of awareness, namely as an "initially global grasp of the solution, a kind of metaphorical 'seeing' of the phenomenon being searched for, an *anticipatory perception* of its 'shape' or its gross structure" (p. 468). For example, the choice of a direction in scientific work might possibly be understood as "reflecting a marginal (not fully conscious) awareness of the nature of a phenomenon, a metaphorical 'seeing' of the phenomenon as a whole without knowing its parts, a seeing 'through a glass, darkly' " (Marton et al., 1994, pp. 468–469).

The Functions of the Extracognitive in Scientific Creativity

The goal of this section is to discuss the roles that the extracognitive phenomenon plays in intellectual creative activity at exceptionally high levels, as that of Nobel laureates. It appears reasonable to distinguish the following functions of the extracognitive: cognitive function, guiding function, function of evaluation and judgement, criterion function, aesthetic/emotional function, motivational function, quality/ethical function, and advancing function.

Cognitive function implies that intuitive processes, specific intellectual feelings, beliefs, and preferences provide a mode of acquiring new scientific knowledge—in the form of novel ideas, solutions, models, theories, and approaches. As it was mentioned above, intuition can be, for example, interpreted as 'intuitive understanding' (Marton et al., 1994) or 'intuitive thinking' (Bruner, 1960). The extracognitive is, therefore, a particular cognitive mode of human thinking that appears in advance of any logical, conscious accounts of an individual's intelligence.

Guiding function of the extracognitive implies that specific feelings, beliefs, preferences, and intuition lead scientists in the process of their creative endeavors towards right theories, approaches, and models. The extracognitive phenomenon guides scientists' "sense of 'this is how it has to be', their sense of rightness" (Wechsler, 1978, p. 1) or scientific truth. The quotations of Albert Einstein, Michael S. Brown, Hans Krebs, and Georges Kohler highlighted in this chapter clearly demonstrate the guiding function of the extracognitive in their work. Analyzing the Nobel laureates' views of scientific intuition, Marton et al. (1994) concluded that there appears to be an "experience of the phenomena the scientists are dealing with which is of a quasi-sensory nature and which may encompass all relevant previous experiences of the phenomena. When this experience yields an *intensely felt verdict on the direction to go, the step to take or on the nature of the solution to the problem* the scientist is engaged in what we speak of 'scientific intuition' " (p. 471; italics added).

The *function of evaluation and judgement* means that the phenomenon of the extracognitive plays a key role in the appreciation and, consequently, acceptance or disapproval and, hence, rejection of any new idea, theory, model or approach. The above considered accounts of Niels Bohr, Stanley Cohen, Paul Berg, George Stigler, Paul Dirac, Werner Heisenberg, and other Nobel laureates underline this function of specific feelings, developed intellectual beliefs, preferences, and intuitive processes. These are just "*feelings* which enable scientists to follow their intuitions in the absence of rational, logical support" (Marton et al., 1994, p. 467).

Probably, the scientists' very first evaluation of everything new in science is unconscious in its nature. At the same time, as one can see from the Nobel laureates' accounts, it is not, however, entirely unconscious. The phenomenon of the extracognitive—for example, feelings of direction and of beauty, specific preferences, beliefs, and intuition—provide certain criteria for creative work of scientific minds. This is what the *criterion function* is all about. It should be noted that other functions of the extracognitive also execute to some extent the criterion function (e.g. the aesthetic/emotional function).

The *aesthetic/emotional function* highlights the important role of feeling of beauty and specific scientific taste in scientific creativity. Firstly, passionately pursuing their intellectual quest, scientists feel intense *aesthetic pleasure* (Wilson, 1972) in discovering new laws of nature. Secondly, the phenomenon of the extracognitive helps to generate "*aesthetic criteria of truth*" (Heisenberg, 1971) which aesthetic experience presents to scientists. This is well illustrated by Heisenberg's (1971) quotation mentioned above which definitely attests to the importance of the feeling of beauty in his work. In this respect the aesthetic/emotional function of the extracognitive is closely related to its function of evaluation and judgement: the feeling of beauty and specific scientific taste delineate the very initial basis for such evaluation and judgement.

Feeling pride, enjoyment, and delight in discovering new ideas and theories, in finding new problems and fresh perspectives, in penetrating new puzzles of nature, and in solving very old and previously unsolved problems, Nobel laureates experience great pleasure and personal happiness, feelings which motivate them to go further and further in their intellectual search. This is what the *motivational function* of the extracognitive is all about. As Wilson (1972) pointed out, scientists' "*success stimulates* themselves and their colleagues to still more exacting studies of the phenomena of nature" (p. 16; italics added).

Quality/ethical function of the phenomenon of the extracognitive is closely related to scientists' belief in high and far-reaching standards of scientific activity. Having elevated inner standards of performance (Kholodnaya, 1991; Zuckerman, 1977, 1983), outstanding scientists assure an exceptionally high quality of first-rate research choosing or inventing the methods, approaches, and techniques necessary to do so. Because of that, they very often cannot accept a work of poor quality. Due to their own superior standards of work and certain principles, by which the nature of scientific research is determined, Nobel laureates set new, or transcend old, ethical rules in science. This can concern the behavior of scientists, their important role in the public arena, their social and moral responsibility (Gruber, 1986, 1989), and so on. Very often such scientists act as role models for future generations of scientists.

Advancing function refers to the important role of the phenomenon of the extracognitive in advancing science in general and the intellectually creative activity of individual researchers in particular. It is commonly accepted that in order to advance any field of science, scientists must build on it. Not only must they have a great deal of comprehensive and complex knowledge in the particular discipline, but at the same time, they should be able to step outside of it consequently expanding it (Csikszentmihalyi, 1996; Marton et al., 1994). Considering this position of scientists as a paradoxical one, Marton et al. (1994) assert that intuition resolves this paradox providing a sudden shift in the structure of a scientists' awareness. "Scientific intuition is just a special case of intuitive understanding. It is a sudden shift from a simultaneous awareness of all that it takes for that understanding to come about, to a highly singular, focused awareness of that which the understanding is an understanding of" (Marton et al., 1994, p. 469).

Therefore, the phenomenon of the extracognitive plays a critical role in exceptional scientific creativity of Nobel laureates performing the cognitive, guiding, aesthetic/emotional, motivational, and advancing functions, as well as the function of evaluation and judgement, criterion function, and quality/ethical func-

tion. The exceptional creativity results in unique contributions to science, that is, scientific innovations.

Conclusions

The findings presented in this chapter demonstrate that the mental functioning of Nobel laureates is determined in part by subjective, internally developed feelings, beliefs, preferences, standards and orientations, as well as intuitive processes, which constitute the different components of the extracognitive phenomenon. In this light, their negative reaction to any attempts to impose external standards on intellectually creative behavior is not surprising.

The extracognitive phenomenon guides Nobel laureates in their understanding of the nature providing intuitive aesthetic/emotional criteria for the appropriate evaluations and judgements leading to quality work and high ethical standards. This phenomenon also motivates Nobel laureates to go beyond the limits of their intellectual pursuits and advance scientific knowledge about the world.

There are at least a few reasons to assert that the phenomenon of the extracognitive should be considered as the highest level of the manifestation of the intellectual and creative potentials of an individual. First, the fact that so many Nobel laureates—whose extraordinary intelligence and creativity are unquestionable—(a) expressed almost all components of the extracognitive; (b) were very attentive to its manifestations in their own work; and (c) stressed its important role in scientific search testifies to the exceptionally high status of the extracognitive in the structure of Nobel laureates giftedness (Shavinina & Kholodnaya, 1996). Second, taking into account the direct link between metacognition and the extracognitive phenomenon, it becomes apparent that developed metacognition is strongly associated with highly developed intelligence (Sternberg, 1985, 1988). From this, one can suggest that a person displaying feelings of direction and of beauty, specific scientific taste, preferences, and intuition is also distinguished by exceptional intellectual abilities. Third, the phenomenon of the extracognitive carries out such versatile and multidimensional functions in the functioning of the human mind during the process of scientific creativity, which probably no other psychological processes could do. One can, consequently, suggest that if an individual exhibits his or her extracognitive in his or her own activity, then it means that this individual has already reached an integrated, well-balanced, and advanced level in his or her intellectually creative development. Altogether, these reasons allow us to conclude that the phenomenon of the extracognitive is probably the highest level of the manifestation of the intellectual and creative resources of a personality and, therefore, an important criterion of intellectually creative giftedness and a determinant of innovators in science of Nobel caliber.

This conclusion has essential educational implications. For example, this criterion should be taken into consideration in the process of the identification of gifted and talented children and adolescents for special educational options. Also, gifted education programs—both enrichment and acceleration classes—must include elements, which would direct to the development of pupil's extracognitive abilities. At the same time, intelligence and creativity tests should be developed or modified in ways that would allow us to examine an individual's extracognitive abilities.

To conclude, this chapter presented rich findings regarding the phenomenon of the extracognitive that covers a whole field of unexplored or weakly explored scientific phenomena (i.e. specific feelings, beliefs, preferences, and intuition). The above-considered findings indicate that the phenomenon of the extracognitive predicts scientific productivity of the highest level resulting in significant discoveries and, as such, showing an outstanding talent of Nobel caliber.

Research on the extracognitive phenomenon in exceptional individuals is a new enterprise. This chapter does not attempt to account for all possible facets of the extracognitive in Nobel laureates, and it is sometimes both vague and speculative in its formulations. However, it nevertheless provides a useful attempt to understand and conceptualize the valuable psychological phenomena in the successful functioning of Nobel laureates' minds leading to scientific innovations. Future investigations will help to unravel many more unknown components and manifestations of this phenomenon.

References

Bowers, K. S., Regher, G., Balthazard, C. & Parker, K. (1990). Intuition in the context of discovery. *Cognitive Psychology*, **22**, 72–110.

Brown, A. L. (1978). Knowing when, where, and how to remember: A problem of metacognition. In: R. Glaser (Ed.), *Advances in Instructional Psychology* (Vol. 1, pp.77–165). Hillsdale, NJ: Erlbaum.

Brown, A. L. (1987). Metacognition, executive control, self-regulation, and other even more mysterious mechanisms. In: F. E. Weinert & R. H. Kluwe (Eds), *Metacognition, Motivation, and Understanding* (pp. 64–116). Hillsdale, NJ: Erlbaum.

Bruner, J. S. (1960). *The process of education*. Boston, MA: Harvard University Press.

Cropper, W. H. (1970). *The quantum physicists*. New York: Oxford University Press.

Csikszentmihalyi, M. (1996). *Creativity*. New York: Harper-Perennial.

Dirac, P. A. M. (1963). The evolution of the physicist's picture of nature. *Scientific American*, (May).

Dirac, P. A. M. (1977). Recollections of an exciting era. *Varenna Physics School*, **57**, 109–146.

Einstein, A. (1949). Autobiographical notes. In: P. A. Schlipp (Ed.), *Albert Einstein: Philosopher and Scientist* (pp. 3–49). New York: The Library of Living Philosophers.

Ericsson, K. A. & Simon, H. A. (1980). Verbal reports as data. *Psychological Review*, **87**, 215–251.

Ferrari, M. J. & Sternberg, R. J. (1996). *Self-awareness*. New York: Guilford Press.

Flavell, J. H. (1992). Perspectives on perspective taking. In: H. Beilin & P. Pufall (Eds), *Piaget's Theory: Prospects and Possibilities* (pp. 107–139). Hillsdale, NJ: Erlbaum.

Foster, W. (1986). The application of single subject research methods to the study of exceptional ability and extraordinary achievement. *Gifted Child Quarterly*, **30** (1), 333–337.

Frey, D. (1978). Science and the single case in counselling research. *The Personnel and Guidance Journal*, **56**, 263–268.

Hadamard, J. (1954). *The psychology of invention in the mathematical field*. New York: Dover.

Heisenberg, W. (1971). *Physics and beyond*. New York: Harper & Row.

Holton, G. (1973). *Thematic origins of scientific thought: Kepler to Einstein*. Cambridge, MA: Harvard University Press.

Gardner, H. & Nemirovsky, R. (1991). From private intuitions to public symbol systems. *Creativity Research Journal*, **4** (1), 3–21.

Gross, M. U. M. (1994). The early development of three profoundly gifted children of IQ 200. In: A. Tannenbaum (Ed.), *Early Signs of Giftedness* (pp. 94–138). Norwood, NJ: Ablex.

Gruber, H. E. (1974). *Darwin on man*. Chicago, IL: University of Chicago Press.

Gruber, H. E. (1986). The self-construction of the extraordinary. In: R. J. Sternberg & J. E. Davidson (Eds), *Conceptions of giftedness* (pp. 247–263). Cambridge: Cambridge University Press.

Gruber, H. E. (1989). The evolving systems approach to creative work. In: D. B. Wallace & H. E. Gruber (Eds), *Creative People at Work* (pp. 3–23). New York: Oxford University Press.

John-Steiner, V. (1985). *Notebooks of the mind: Explorations of thinking*. Albuquerque, NM: University of New Mexico.

Kholodnaya, M. A. (1991). Psychological mechanisms of intellectual giftedness. *Voprosu psichologii*, **1**, 32–39.

Krebs, H. (1967). The making of a scientist. *Nature*, **215**, 1441–1445.

Kuhn, T. (1970). *The structure of scientific revolutions*. Chicago, IL: University of Chicago Press.

Lorenz, K. (1952). *King Solomon's ring*. New York: Crowell.

Marton, F., Fensham, P. & Chaiklin, S. (1994). A Nobel's eye view of scientific intuition: Discussions with the Nobel prize-winners in physics, chemistry and medicine (1970–1986). *International Journal of Science Education*, **16** (4), 457–473.

McMorris, M. N. (1970). Aesthetic elements in scientific theories. *Main Currents*, **26**, 82–91.

Merriam, S. B. (1988). *Case study research in education: A qualitative approach*. San Francisco, CA: Jossey Bass.

Miller, A. I. (1981). *Albert Einstein's special theory of relativity: Emergence (1905) and early interpretation (1905–1911)*. Reading, MA: Addison-Wesley.

Miller, A. I. (1992). Scientific creativity: A comparative study of Henri Poincaré and Albert Einstein. *Creativity Research Journal*, **5**, 385–418.

Miller, A. I. (1996). *Insights of genius: Visual imagery and creativity in science and art*. New York: Springer Verlag.

Ochse, R. (1990). *Before the gates of excellence: The determinants of creative genius*. Cambridge: Cambridge University Press.

Planck, M. (1950). *Scientific autobiography and other papers*. London: Williams & Norgate.

Poincaré, H. (1913). *Mathematics and science* (New York: Dover, 1963). Originally published by Flammarion in 1913.

Poincaré, H. (1952). *Science and hypothesis*. New York: Dover.

Policastro, E. (1995). Creative intuition: An integrative review. *Creativity Research Journal*, **8** (2), 99–113.

Rosenblueth, A. & Wiener, N. (1945). Roles of models in science. *Philosophy of Science*, **XX**, 317.

Rothenberg, A. (1996). The Janusian process in scientific creativity. *Creativity Research Journal*, **9** (2 & 3), 207–231.

Shavinina, L. V. & Kholodnaya, M. A. (1996). The cognitive experience as a psychological basis of intellectual giftedness. *Journal for the Education of the Gifted*, **20** (1), 3 35.

Simonton, D. K. (1980). Intuition and analysis: A predictive and explanatory model. *Genetic Psychology Monographs*, **102**, 3–60.

Sternberg, R. J. (1985). *Beyond IQ: A triarchic theory of human intelligence*. New York: Cambridge University Press.

Sternberg, R. J. (1988). *The triarchic mind: A new theory of human intelligence*. New York: Viking.

Sternberg, R. J., Forsythe, G. B., Hedlund, J., Horvath, J. A., Wagner, R. K., Williams, W. M., Snook, S. A. & Grigorenko, E. L. (2000). *Practical intelligence in everyday life*. New York: Cambridge University Press.

Wechsler, J. (1978). *On aesthetics in science*. Cambridge, MA: MIT Press.

Wertheimer, M. (1959). *Productive thinking*. West Port, CT: Greenwood Press.

Wilson, M. (1972). *Passion to know*. Garden City, NY: Doubleday.

Woodward, C. E. (1989). Art and elegance in the synthesis of organic compounds: Robert Burns Woodward. In: D. B. Wallace & H. E. Gruber (Eds), *Creative People at Work* (pp. 227–253). New York: Oxford University Press.

Zuckerman, H. (1977). *Scientific elite*. New York: Free Press.

Zuckerman, H. (1983). The scientific elite: Nobel Laureates' mutual influences. In: R. S. Albert (Ed.), *Genius and Eminence*. Oxford: Pergamon Press.

The International Handbook on Innovation
Edited by Larisa V. Shavinina

Innovation in the Social Sciences: Herbert A. Simon and the Birth of a Research Tradition

Subrata Dasgupta

Institute of Cognitive Science, University of Louisiana at Lafayette, USA

Abstract: Innovation presumes creativity, but creativity does not necessarily entail innovation. The latter involves both the cognitive process of creation and the social-historical process by which the created product is assimilated into a milieu. My concern in this essay is with innovation in the *social sciences*. In particular, I examine how the American polymath Herbert Simon was led to a model of human decision-making that gave birth to a new *research tradition* in the human sciences. For his role in the creation of this research tradition, Simon received the 1978 Nobel Prize in Economics.

Keywords: Innovation; Cognitivism; Research tradition; Simon; Behavioral economics; Nobel Prize; Creativity.

Introduction

Innovation presumes creativity, but creativity does not necessarily entail innovation. Creativity is, fundamentally, a cognitive process, the product of which is a particular concept, idea, scientific discovery, technological invention, new design, or distinctive literary, musical or artistic work. However, we normally reserve the word innovation to mean a change of some sort in a social, economic, cultural or intellectual milieu, brought about by a product of the creative process. Innovation, thus, involves both a cognitive process of (usually individual) creativity, and a social—historical process by which the created product is accepted by, influences, or is assimilated into, a particular milieu. Innovation occurs in every domain of human endeavor. My concern here is with innovation in the *social sciences*. As we all recognize, in the domain of science, a scientist practices his or her craft within a particular framework, which the philosopher of science Larry Laudan (1977) called a *research tradition*. According to Laudan, a research tradition has certain traits including, most fundamentally, the presence of one or more empirical (that is, descriptive) theories, and one or more normative (or prescriptive) ontological and methodological commitments. The emergence of a research tradition in any science signifies, unequivocally, a major innovation in that science. In this essay, I will discuss the cognitive-social-historical process of innovation in the realm of the social sciences. In

particular, I will examine how the American polymath Herbert A. Simon (1916–2001) was led to a particular model of human decision-making which in turn gave birth to a radically new research tradition, first in organization theory and economics, and then in other domains of the human sciences. This new tradition I call the *cognitive tradition* or, more simply, *cognitivism* For his seminal role in the creation of this research tradition, Simon was awarded the 1978 Nobel Prize for Economic Science.

This essay will also reveal how and why a major innovation in the social sciences faced significant resistance—both before and after the Nobel—to the extent that in economics, cognitivism has neither replaced the dominant (neoclassical) research tradition nor been assimilated into the latter. Rather, cognitivism became (and remains) an alternative research tradition with its own adherents and practitioners. This historical fact gives empirical credence to Laudan's thesis that a given science may be populated by more than one research tradition at any given time—in contradiction to Thomas Kuhn's more celebrated theory, in *The Structure of Scientific Revolutions* (1970), that a science is governed, at any one time, by just one paradigm.

A 'Puzzling' Nobel Prize

When the Royal Academy of Sciences in Sweden awarded the 1978 Nobel Memorial Prize in Economic

Sciences to Herbert Simon, the award occasioned considerable bemusement in the community of professional economists—a reaction that the popular press was quick to sense (Cairncross, 1978; Rowe, 1978; Williams, 1978).

The official Nobel Foundation citation tells us that the prize was awarded for Simon's "pioneering research into the decision-making process in economic organization" (Carlson, 1979). In fact, despite the apparent puzzlement amongst practicing economists in 1978, Simon had come to be regarded, by the early 1960s, as one of the founders of the "behavioral approach to economic analysis" largely because of his emphasis on the process of decision-making in administrative and economic contexts (Cyert & March, 1963; Machlup, 1967; Marris, 1964). A relatively recent history of economics has attributed the origin of the *"behavioural approach* to the theory of the firm" to Simon (Screpanti & Zamagni, 1993, p. 378).

Unfortunately, the words 'behavior' and 'behavioral' in the context of the social sciences are used in somewhat confusing and contradictory ways—and this has obfuscated, to some extent, the precise nature of the 'behavioralism' which led Simon to the accolade of a Nobel Prize. Simon himself made a clear distinction between 'behavioralism' and 'behaviorism'. For him, the former term refers generally to the study of human behavior in social processes and institutions, and includes the processes by which behavior is manifested, while the latter term is restricted to the particular school of psychology associated originally with John B. Watson, and which limits itself to the study of those aspects of human behavior that can only be observed.[1] However, in the economics literature, the two concepts are often conflated. For instance, Machlup (1967, p. 4), in the same breath, spoke of "the program of behaviorism" that relies "only on observations of overt behavior" (i.e. Watsonian behaviorism) and yet is required to observe "by what processes (businessmen) . . . reach decisions", mindless of the fact that recognition of such processes, being mental, are anathema to behaviorist psychologists; Julian Margolis (1958) used 'behaviorism' in the title of a paper to refer to what in the text is clearly *not* behaviorism (in the behaviorist psychologist's sense). Likewise, in discussing the very work for which Simon would later receive the Nobel, Robin Marris (1964) claimed to be writing about 'behaviorist arguments' and, accordingly, presented his 'general view about behaviorism in economics'.

To add to the confusion, when the economist Amartya Sen (1979) speaks of the "behavioral foundations of economic theory" and Simon (1955) of a "behavioral model of rational choice", they are referring to two quite different things. Sen's reference is to

the consumer's *preference* amongst alternative bundles of goods as 'revealed' by the choices they are observed to make; Simon is writing of the *psychological process* by which the economic agent makes a decision.

Clearly, the concept of 'behavioralism' is both too broad and too ambiguous as an accurate descriptor of Simon's contribution to economic thought. And it certainly does not explain *why* Simon's thinking was so much outside the mainstream of economic thought, *circa* 1978, that the Nobel award for that year occasioned so much puzzlement in economic circles.

As we will see, the explanation lies in the novelty and unfamiliarity of the research tradition that Simon brought into being.

Research Traditions

Every scientist practices his or her craft within a particular framework—broadly, an integrated network of theories, facts, methods, assumptions and values. For Thomas Kuhn (1970), this framework constituted a 'paradigm' or 'disciplinary matrix'; Imre Lakatos (1978) called it a 'research programme'; and for Larry Laudan (1977), the framework constituted a 'research tradition'. While there are common features amongst these three proposals, there are also important differences. For our purpose here, the nature of the phenomenon I wish to explore falls most appropriately within the Laudanian concept of the research tradition.

A *research tradition* (RT), according to Laudan (1977, p. 89 et seq.), has the following fundamental traits: (a) It contains one or more *specific theories* some of which are temporary successors of earlier ones while others are 'current' or coexisting at some given time. By 'theory', Laudan meant assertions, propositions, hypotheses or principles that can lead to specific, empirically testable predictions, or can provide detailed explanations of phenomena. Theories are fundamentally descriptive; furthermore, the theories within a RT need not all be mutually consistent; (b) It is characterized by certain *ontological* and *methodological commitments* which collectively distinguish it from other RTs; (c) Because of possible inconsistencies among its constituent theories, a RT may (in order to resolve these inconsistencies) *evolve* through several versions. Generally, then, a RT itself is neither explanatory nor predictive. For one thing, it is too general for specific explanations or predictions to emanate from it; for another, since it is grounded in certain ontological and methodological commitments, a RT has a strong prescriptive role which, by its very nature, makes the RT unsuitable for detailed explanation or prediction; (d) The success of a RT depends on its *problem-solving effectiveness*—that is, its effectiveness, by way of its constituent theories, in adequately solving an increasing range of problems; (e) A RT and its theories interact in a number of ways. First, the RT plays a role in determining the *problems* with which its

---

[1] H. A. Simon, Personal communication, e-mail, Nov. 14, 1999.

constituent theories must contend. Second, the RT has a 'heuristic role': it can provide 'vital clues' for theory construction, indeed, it should provide important guidelines as to how its theories can be modified so as to enhance the RT's 'problem-solving capacity'. Third, the RT has a 'justificatory role': it may help to rationalize or justify theories; (f) An important and distinct aspect of the Laudanian RT is that it explicitly recognizes that problems may be *empirical* or *conceptual*. Broadly speaking, an empirical problem— Laudan (1977, pp. 14–15) also called this a 'first order' problem—is posed by the observation or detection of some phenomenon about the basic objects that constitute the domain of a given science. In contrast, a conceptual—or 'higher-order'—problem arises in the context of a previously established or proposed theory, where the latter is perceived to be unsatisfactory on logical, philosophical or even aesthetic grounds.

More generally, then, a RT is "a set of . . . assumptions about the entities and processes in a domain of study and about the appropriate methods to be used for investigating the problems and constructing the theories in that domain" (Laudan, 1977, p. 813), within the specific constraints described in (a)–(f) above.

Simon's Model as a Research Tradition

Simon himself never referred to his work in economics (or any of the other domains in which he worked) as a 'research tradition'. His preferred, and more modest, term was 'model' by which he meant a theory that is precise enough to be empirically testable.[2] For the present, then, I will refer to the work for which he received the Nobel Prize simply as *Simon's model*.

What exactly was Simon's model, and how can we demonstrate that it constituted a RT? That is, how can we show that it was not just an invention but the birth of a genuine innovation? My demonstration will appeal to the historical process itself. I will describe how Simon's model came into existence over a period that began in the late 1930s, but properly took shape between 1952 and 1957. We will see that this model emerged in response to *Simon's dissatisfaction with the dominant research tradition in microeconomics*. We will further see that the model was not simply a corrective or adjustment to, or a new constitutive theory of, the dominant RT: it neither enriched nor enhanced the problem-solving effectiveness of the dominant RT, nor did the latter provide a 'heuristic role' in the construction of Simon's model. Rather, the model emerged as an *alternative* and, thus, as a *competitor* to the dominant RT in microeconomics. We

will then see that *structurally* Simon's model possesses all the characteristics of a RT. I will then argue that Simon's model is fundamentally a *cognitive* model— that is, it is a model that coheres with the way in which cognitive scientists regard 'cognition'. Finally, we will examine the evidence to argue that Simon's cognitive model does not just structurally conform to a RT but that it was *historically* the beginning of a new RT, the *cognitive tradition*, not only in economics but more generally in the social sciences.

The Dominant Research Tradition in Economics

The *neoclassical* tradition, the dominant RT in economics, emerged in about the last quarter of the 19th century. As a research tradition, it has evolved considerably in the course of the 20th century, and though it is by no means the only framework guiding economic thought, it is fair to say that in important areas of economic theory and practice, it became, and remains, the dominant tradition.

For the neoclassicist, the social entity of interest is the market and its two principal constituents, the producer and the consumer. The most significant characteristic of this social entity is that producers and consumers are assumed to behave so as to maximize their respective satisfactions—profit in the case of producers, and 'utility' in the case of consumers. Neoclassical *Homo economicus* is a 'maximizer' of personal satisfaction and, in this precise sense, is a 'perfectly rational' being.

Two other characteristics of neoclassicism are worth noting. First, the economic universe is *ahistorical* and *acultural*: the behavior of the market and it constituent actors are taken to be independent of cultural differences and historical time. Second, the economic universe is a *formal world*, amendable to exact mathematical treatment.

Simon's intellectual acquaintance with the neoclassical tradition reached back to his undergraduate days in the 1930s at the University of Chicago where, he confessed 60 years later, he had had "about as much economics as if . . . I had majored in it"[3]—though, in fact, his formal baccalaureate and doctoral degrees were in political science. In the 1940s, while teaching at the Illinois Institute of Technology, he had begun to publish papers in such 'mainstream' economics journals as the *Quarterly Journal of Economics*, and had started a long and close association with the Cowles Commission for Research in Economics, whereby he was in intellectual exchange with some of the foremost American exponents of the neoclassical tradition.

---

[2] H. A. Simon, Personal communication, e-mail, Jan.1, 2001.

[3] H. A. Simon, Personal communication, e-mail, Nov. 14, 1999.

Thus, *socially*, Simon was well ensconced in the neoclassical domain.

An Empiricism/Formalism Polarity in Simon's Scientific Style

Yet, steeped though he was in the formalism inherent in the neoclassical tradition, Simon was rooted still more deeply in the empiricism that had guided him in his most substantial work on administrative theory, first as a doctoral student between 1938 and 1942, and then as a young faculty member in the Illinois Institute of Technology and (from 1949) at the Carnegie Institute of Technology. In fact, Simon's long engagement with the social sciences reveals a continuing polarity between empiricism and formalism. On the one hand, there was his 'one long argument' for an empirical science of administration, a science that appealed to the way in which people in organizations actually function and make decisions. Simon's empiricism in administrative theory drew upon case studies reported in the organizational literature, in the observations of psychologists and sociologists, and in his own field studies. Most markedly, perhaps, his empiricism was manifested in his commitment to operational principles, and in the very nature of the specific principles described in his first two books, the second jointly authored with two colleagues (Simon, 1947; Simon, Smithburg & Thompson, 1950).

Coexisting with this empiricism was a strong formal or mathematical strain which, from the very onset of his doctoral work, longed to lend 'precision and clarity' to all the social sciences, including sociology and political science. In the realm of organization theory, Simon's formalism was most evident in his series of papers on group behavior.

Despite the presence of this polarity in what we might call Simon's personal scientific style, there was, yet, a *tension* between these two elements of his style. He was, he explained in 1945, a social scientist not a mathematician, and he was not willing to let "formal theory . . . lead the facts around the nose".[4] Nor was he prepared to be seduced by the pure mathematician's desire for rigor for the sake of rigor. "Mathematics is a language" which scientists should be able to read, speak and write, he wrote in 1953. But they should not be turned into 'grammarians'.[5] Furthermore, "mathematical theorizing" must originate in the "field about which the theorizing is to be done" (Simon, 1954).

The Conflict Between Simon's 'Administrative Actor' and Neoclassical *Homo Economicus*

In his own perception, then, Simon was a social scientist not a mathematician. Theory must originate in the 'field'. Here lay the problem with neoclassical economics: its assumptions did not originate in the 'field'; it was, rather, a case where formal theory *did* 'lead the facts around the nose'. At the very least, the assumptions surrounding the behavior of neoclassical *H. economicus* were inconsistent with the model of the *administrative actor* which, following Chester Barnard (1938), Simon had helped shape. The economic actor, according to the neoclassicist, was a maximizer of satisfaction, a perfectly rational, omniscient being with complete knowledge at his disposal and possessed of the cognitive capability to enforce his maximization objective. Simon's administrative actor was a purposive or goal-oriented being: she desired to take action or make decisions that were conductive to the attainment of that goal. For Simon, purposive behavior is rational if the choice of means leads to the attainment of the goal. However, such purposive behavior faces several oppositions. There are limits to the actor's innate cognitive capability; limits to one's knowledge about the relevant factors that need to be taken into account in order to arrive at a 'rational' choice; limits to the actor's ability to cope with the complexity of the environment in which decisions have to be made, and with the alternative courses of action (choices) and their consequences. Ultimately, all these constraints suggest that *it is exceedingly difficult for an individual to achieve fully rational behavior.* Simon (1947) termed such behavior 'subjective rationality'; exactly a decade later, he reframed this notion and retermed it *bounded rationality* (Simon, 1957).

The contrast between neoclassical *Homo economicus* and Simonian administrative actor was stark. The former exuded a confidence the latter lacked. Both economic theory and Simonian administrative theory concerned themselves with *behavior*. Yet the behavior the actor indulges in according to the neoclassical tradition makes no appeal to the cognitive and other psychological factors that prompted and constrained the behavior of the Simonian administrative actor.

Simon's 'New Problem'

This *inconsistency* between his model of the boundedly rational administrative actor and the perfectly rational neoclassical economic actor became the source of a *problem* for Simon. By the middle of 1950, he was embarking on a new scientific activity, the objective of which was to create a *H. economicus* whose behavior manifested the kind of empirical plausibility—social and cognitive—that had shaped the Simonian administrative decision-maker (Simon, 1950a). His goal, he wrote in 1951, was to construct a bridge between

(T)he economist with his theories of the firm and of factor allocation, and the administrator with his theories of organizations—a bridge wide enough to permit some free trade of ideas between two intellectual domains that have hitherto been quite effectively isolated from each other (Simon, 1951).

However, for Simon, this was to be a one-way bridge, leading *from* administrative theory *to* economic theory. He wished to investigate, he wrote to the mathematical economist Tjalling Koopmans in 1952, "how the processes of rational decision making are influenced by the fact that decisions are made in an administrative context".[6]

Furthermore, this bridge was to be paved with psychological matter, for

(A)s my work has progressed, it has carried me further into the psychology of decision making processes. I have on several previous occasions expressed my conviction that a real understanding of 'rational choice' will require further investigation of these psychological problems.

The *Zeitgeist* Concerning Behavioral Economics, *Circa* Early 1950s

In effect, Simon wished to espouse an alternative to the neoclassical tradition, one we might tentatively call the 'behavioral' tradition. Taken narrowly, this takes into account cognitive, motivational and other psychological factors that enter into economic decision making. More broadly, it extends beyond psychology: as Simon would put it several decades later, the behavioral tradition emphasizes the "factual complexities of our world" (Simon, 1999).

In this broader sense, the behavioral approach reaches back to such thinkers as Thorsten Veblen and John R. Commons. There is no evidence that Simon was influenced by Veblen, but he had read Commons (Simon, 1991) and acknowledged the latter's influence in *Administrative Behavior* (Simon, 1947, p. 136).

But behavioral economics had more modern practitioners. In particular, there was George Katona. Katona was both an economist and a psychologist (Campbell, 1980), and for him, the psychology of economic behavior went well beyond *behaviorism*. Stimulus and response were not sufficient to comprehend behavior; there were other, internal, 'intervening variables' that mediate between stimulus and response. Katona (1951, p. 32) cited as examples, 'organization, habit, motive, attitude'. Katona was a cognitivist in that he espoused gestalt psychology; he had been influenced by, and had worked with, the gestaltist Max Wertheimer (Campbell, 1980, p. 4). A human being confronted with a set of external situations and factors—a 'geographical environment'—*perceives* this environment according

to the 'organization of his perception'. And 'motives, attitudes and frames of reference' affect the organization of the person's perception and how he or she responds to the environment.

In the realm of economic rationality, Katona questioned the assumption of *H. economicus* as a striver of maximum satisfaction (Katona, 1951, pp. 32, 36, 70–71); that the economic actor is entirely egocentric; that the consumer is driven by just one motive. It was obvious to him that psychological research could shed light on these issues.

A particular instance when the psychologist was useful, Katona noted, had to do with the 'attitude' toward income and expenditure. Psychologists had studied the *level of aspiration* humans identify in various goal-directed behaviors. A person may wish to hold to some ideal level but such a level has 'no psychological reality', since one may not be able to state what the ideal is. Rather, she establishes an aspiration level for which she *can* strive. Aspiration levels *do* have 'psychological reality'—since one can raise or lower it depending on the success or failure with a prior level (Katona, 1951, pp. 91–92).

Katona was not alone in wishing to draw upon psychology to understand economic behavior. There was already a considerable knowledge base concerning the relevance of psychology to economic decision-making, as Ward Edwards' (1954) lengthy and critical review article attests to.

Edwards' article and Katona's book provide independent evidence of a certain *Zeitgeist*, a certain dissatisfaction with the neoclassical view of economic rationality. However, neither Katona nor Edwards makes any reference to Simon's theory of bounded ('subjective') rationality, originally articulated in *Administrative Behavior* and included in the 1950 text on *Public Administration*.

The Emergence of Simonian *H. economicus*: I. Tentative Steps

Even before Simon confided his intention to Koopmans, he had already made tentative inroads toward the attainment of his goal—tentative in that he *straddled* both the neoclassical RT in economics and the behavioralism or psychologism of his model of administrative man. In a paper published in 1951, Simon addressed a special instance of his 'problem'. This concerned the question of how an employer/boss and an employee/worker would enter into an employment contract. The neoclassical version of this situation was devoid, Simon claimed, of the reality of how such contracts are made as suggested by administrative theory (Simon, 1951). Simon's approach was to draw an *analogy* between the boss–worker relation in economic theory and the superior–subordinate relation in administrative theory; he then transferred some crucial concepts from the latter to the former. Thus, he said, a worker (W) enters into an employment contract

---

[6] H. A. Simon to T. C. Koopmans, Sept. 29, 1952. HASP.

with a boss (B) when W agrees to accept B's authority to select a particular set of tasks ('behavior') for W to perform. Simon contrasted this with the contractual model assumed in traditional (neoclassical) price theory. There, B promises to pay a sum of money, and W promises in return a specific quantity of labor.

But under what conditions is W willing to give B authority over W's behavior? Appealing to arguments put forth in *Administrative Behavior*, Simon proposes that he will do so if the behavior falls within an 'area of acceptance' or a 'zone of indifference' (Simon, 1947, p. 133). There are other conditions also that will determine how boss and worker, employer and employee, will enter into a contract. For example, if B *does not know* for certain at the time the contract is made what behavior he or she wants for W, then B may be willing to additionally compensate W for the 'privilege of postponing' the choice of behavior until some time after the contract is entered into.

This, Simon suggested, is how people actually behave in a contractual situation; this is what each party is actually prepared to accept of the other's behavior. Here is a glimpse of the effect of imperfectness of information, incompleteness of one's knowledge, on the decision-maker's rationality.

However, *maximization of satisfaction* so precious to the neoclassical tradition remained: Simon assumed that both boss and worker wish to maximize their respective well-being, and that their well-being could be formalized mathematically by a 'satisfaction function'. Here, then, was the maximizing neoclassical *H. economicus* dosed with a sprinkling of the administrative actor's behavioral characteristic.

This same characteristic appeared in another paper published in 1952, in which Simon explicity compared the microeconomic theory of the firm ('F-theory') with administrative/organization theory ('O-theory') (Simon, 1952a). The decision-maker, according to F-theory (basically the entrepreneur), is an optimizer; the decision-maker according to O-theory (the participant within an organization) is an *adaptive* actor. The firm, according to F-theory, operates to maximize the entrepreneur's satisfaction; the firm, according to O-theory, seeks to ensure its own survival. But Simon also appealed to neoclassically grounded welfare economic theory to introduce a criterion of optimality into O-theory: a viable decision is optimal if no further increase can be made in the net satisfaction of any one participant without decreasing the satisfaction of at least one other participant. Welfare economists refer to this condition as 'Pareto optimality' (Sen, 1979, p. 86). Thus optimality was not eliminated on this O-theoretic view of the firm. Vestiges of neoclassicism were still evident in the behavioralistic O-theory.

The notion of the adaptive decision-maker was not by any means new in Simon's thinking. He had discussed it in *Administrative Behavior*. Just as in the natural world an organism's survival is achieved by its adapting to new environmental situations, so also in the artificial world of organizations, the latter's survivability is achieved by the adaptive behavior of its members. Such adaptation arises because the organization's objectives change in response to the values, interests and needs of the organization's members as well as to the changing demands of customers. Adaptive behavior is the means by which both organisms and organizations cope with an uncertain future and their inability to predict the future with any degree of accuracy. This very uncertainty, and the paucity of knowledge about the future, places severe limits on the decision-maker's ability to make completely rational decisions. O-theory is predicated on the recognition of this fact; hence the decision-maker adapts from moment to moment to the situation at hand.

Also in 1952, taking up another special problem from the economic domain, Simon further explored the nature of adaptive decision-making. The context was the industrial management of production and inventory (Simon, 1952b). The issue was to control the rate of production of an item so as to minimize the cost of manufacture over a period of time, where 'cost' is a function of the variation in the manufacturing rate and the inventory of finished goods. The latter, in turn, is affected by the number of customer orders per unit time.

Simon once more resorted to *analogy*—this time, between the production/inventory control situation and *feedback mechanisms*. He was thoroughly familiar with the principle of feedback in servomechanisms. As far back as 1935, he had read Alfred Lotka's classic work (Lotka, 1956) in which he had detected "(t)he basic idea of goal orientation and feedback".[7] He had read the paper 'Behavior, Purpose and Teleology' by Rosenbluth, Wiener & Bigelow (1943) soon after its appearance.[8] Wiener's *Cybernetics* had also been published recently (Wiener, 1948), and by 1950, he was contemplating the "study of organizations and other social and human servos" (Simon, 1950b).

For our purposes here, two particular aspects of Simon's 1952 study of production control are noteworthy. First, he was able to explore, in some detail, an *economic* decision system that circumvented uncertainty or lack of knowledge about the future by acting adaptively. Like Lotka's adaptive organism, the (human) production controller receives information about the state of the 'environment'. The relevant 'state' consists of the 'optimum'—meaning desired— inventory and customer orders. The controller computes the actual inventory as a function of the production rate and customer orders; the computed value becomes the 'feedback' information which is used to calculate the difference ('error') between

---

[7] H. A. Simon, Personal communication, e-mail, Feb. 25, 2000.
[8] Ibid.

optimum and actual inventories. This error value prompts the controller to take action—that is, increase or decrease production. Thus, there is no need to predict the future in order to take action 'now'. Rather, action is taken 'now' based on knowledge of the *immediate past*.

The second and more compelling aspect of this exploration was what it revealed of adaptive behavior compared to optimal behavior. The general criterion of optimality for the production controller was the minimization of production cost. This objective was recast into a goal to adjust the production rate as a function of the error separating actual from desired inventory. Here was a more everyday-sense view of optimality. The criterion of optimality was not to minimize or maximize something; rather, it was to adjust some factor depending on the deviation of actual performance from desired performance. *The neoclassical sense of optimization as a characteristic of rational behavior was noticeably absent from this analysis.*

The Emergence of Simonian *H. economicus*: II. The 'Model'

As we have seen, in the three papers published in 1951–1952, Simon had taken several steps toward achieving his original goal—to comprehend how rational decision-making in the microeconomic setting is influenced by administrative/organizational contexts. In essence, this meant drawing upon his work on administrative behavior in which he had already worked out a theory of rationality. Thus, he already had a model of rational behavior; but, apparently this model had not infiltrated the world of economic theory.

In a series of publications appearing between 1955 and 1957, Simon completed the task he had tentatively begun (Simon, 1955, 1956, 1957). However, his solution was not a case of simply adapting economic behavior to his model of administrative behavior; rather, it *entailed creating a new model of H. economicus* that: (a) manifested relevant characteristics reflecting the administrative/organizational context in which *H. economicus* resides; and (b) introduced additional features of the decision-maker that were not present in Simon's administrative actor. In this process, the *language of discourse for describing rational decision-making was revised*—in both the economic and administrative domains.

What were these 'additional factors' that Simon introduced? First, he made a distinction between the constraints on which the decision-maker *has no control* and his *actual actions*. In *Administrative Behavior*, the former had been identified in the specific context of administration: they were cognitive as well as social (i.e. organizational) constraints. Now, the constraints collectively constituted the decision maker's *environment*. One part of this environment lay inside the 'skin' of the economic actor and pertained to innate or (unconsciously) acquired characteristics that the actor possessed: perceptive, cognitive, computational and motor capacities; the other part lay outside the actor's 'skin'—external variables beyond the actor's control (Simon, 1955, p. 101).

Second, Simon adopted the psychological notion of *aspiration level* which Katona (and other psychologists before him) had discussed. The economic actor establishes a level of aspiration as the goal to achieve.

Third, we see the presence in Simon's model of the idea of *search*. The action or behavior that the decision-maker undertakes entails sequential search for a behavior that meets the aspiration level. In *Administrative Behavior*, Simon had articulated the notions of plans and strategies, drawing on von Neumann and Morgenstern (1946), but the concept of search did not appear there or in *Public Administration*.

Simon's formulation of search was rather precise and operational, and was intimately linked with strategies employed in chess. In fact, it is difficult to state whether the strategies of the chess player were, collectively, the *source* of his idea of search, or whether the latter came independently, and he noticed a congruence in chess. His interest in chess reached back to boyhood, he recalls; in high school, he had studied the game seriously to the extent that he had mastered the basic decision-making issues in chess and tic-tac-toe.[9]

In the early 1950s, he was aware of some of the first ideas on computer chess published at the time by Claude Shannon (1950) and Alan Turing (1953) and had read, in the original Dutch, the pioneering work by Adrian deGroot (1946) on the psychology of chess.[10]

Regardless of the precise temporal and causal link between his knowledge of the psychology of (both human and machine) chess, what we *can* say with confidence is that in 1955, he drew an analogy between the chess player's rationality in choosing a move and the economic actor's rationality in making an economic decision (Simon, 1955).

A significant notion that Simon's analogy brought forth was that the chess player undertakes a *mental* search, indeed a mental *simulation* of the game-in-progress. The concept of mental simulation was not original to Simon. It was discussed, for example, by Kenneth Craik (1943). Simon appeared to be ignorant of Craik's work. The search stops when an outcome is 'discovered' that meets the chess player's aspiration level, namely, a 'win' state. Furthermore, since the 'space' of possible strategies the chess player would have to explore before choosing a move would be potentially beyond the cognitive capacity of the player—bounded rationality in action—the player abstracts from this complexity by employing various

---

[9] H. A. Simon, Personal communication, e-mail, Nov. 4, 2000.
[10] Ibid.

heuristics to reduce the search process. The economic decision-maker is thus, in Simon's model, *a heuristic or selective searcher.*

Fourth, we see the emergence of *computation* as a metaphor for the process of human thinking. This notion was influenced in part by the ideas advanced by Shannon (1950) and von Neumann on the problem of designing a chess computer,[11] and in part by his interest in the application of computer technology to decision-making and organization theory—an interest he had expressed in 1950 (Simon, 1950b). Certainly, by 1955 (and probably earlier, in 1953), the computational metaphor had entered Simon's vocabulary. He could freely write about an organism's 'computational capacities' in making rational choice, invoke a model of chess playing inspired by the idea of a chess playing program, and compare the 'I.Q. of a computer with that of a human being'.

The interpenetration of computation with human thinking and rationality was, perhaps, most telling in Simon's fleeting comment that a study of the various 'definitions of rationality' might lead to the design of computers that might achieve "reasonably good scores on some of the factors of intelligence in which present computers are moronic" (Simon, 1955).

Fifth (and finally), the economic actor, administrator, chess player, 'simple-minded' animals and mechanical beings (automata) were all subsumed into a *universal organism:* an organism that compensates for its bounded rationality by adapting its behavior according to its goals and the information received from the environment, and who seeks to *satisfice;* that is, perform, act or behave so as to achieve an aspiration level that may not be optimal (that is, not the very best) but one that is satisfactory or 'good enough'. In Simon's model then, the economic decision-maker, the administrative decision-maker, the exploratory 'simple-minded' animal and the mechanical animal (such as *Machina speculatrix* described by W. Grey Walter (1961) with which Simon was familiar) are all instances of a *universal, boundedly rational, satisficing, adaptive organism.* And with this organism, Simon had finally shed the last vestiges of neoclassical *H. economicus.*

The Coherence of Simon's Model to a Laudanian Research Tradition

So what was the outcome of Simon's endeavor? It was, in fact, a model of the human decision-maker; and shorn of its historical development, we see that it consisted of three major components: (a) a *prescriptive* postulate about the fundamental nature of the model's domain; (b) a set of *descriptive* propositions about

decision-making; and (c) a set of *prescriptive* postulates concerning theories about decision-makers.

The postulate about the nature of the model's domain is that *organizational behavior is a network of decision processes* (Simon, 1947, p. 220).

The descriptive propositions are fourfold: (a) *The Principle of Bounded Rationality (PBR)*: "the capacity of the human mind for formulating and solving complex problems is very small compared with the size of the problems whose solution is required for objective rational behavior in the real world" (Simon, 1957, p. 198); (b) *The Principle of Satisficing (PS)*: the decision-maker establishes 'satisficing' goals as aspirations, and seeks decisions (or choices) that meet such aspirations; (c) *The Principle of Heuristic Search (PHS)*: the decision-maker seeks a satisficing decision by sequentially searching the space of possible choices and selecting the first that meets the satisficing aspiration or goal. The search process employs both the aspiration and the structure and properties of the environment to manage the complexity of search; and (d) *The Principle of Adaptive Behavior (PAB)*: organisms and organizations cope with the uncertainty of the future and their inability to predict the future with any degree of accuracy by means of adaptive behavior—by continually adjusting actions or behavior to the changing environment so as to meet given goals according to information received from the environment.

Finally, there are three prescriptive postulates about decision-making: (a) *Operationalism*: any proposition, hypothesis or theory concerning the behavior of decision-makers must rest on operational concepts, so that one can carry out procedures to confirm, corroborate or falsify the proposition; (b) *Empiricism*: a social science (such as economics or organization theory) must be grounded in the ways in which people actually function, operate and make decisions. That is, the propositions of a social science must appeal to actual human behavior; (c) *Formalism-as-a-Language:* formalism, especially mathematics, is to serve as a language to lend precision and clarity to the propositions of a social science. It should not override operationalism or empiricism.

Stated thus, Simon's model *manifests the structural characteristics of a Laudanian research tradition.*[12] The four descriptive propositions (PBR, PS, PHS, and PAB) comprise the theories of a RT; the three prescriptive postulates (operationalism, empiricism and formalism-as-a-language) constitute methodological commitments; and the prescriptive postulate about the nature of the domain is an ontological commitment.

Simon's Model is a *Cognitive* Model of Human Decision-Making

As summarized above, Simon's model has no explicit link with economics. It is, rather, a more general model

---

[11] Simon had attended a lecture by von Neumann in 1952 on computer chess and was sufficiently stimulated by it to actually engage, by mid-1953, with the design of chess playing programs. H. A. Simon to J. von Neumann. June 24, 1953. HASP.

[12] See section on 'Research Traditions' above.

of decision-making. There is, yet, another vital characteristic of his model. We noted (in the 'Introduction') that Simon has come to be closely associated with the 'behavioral approach' to economics and administrative theory. We further noted the confusion amongst social scientists about 'behavioralism' and its occasional conflation with the behaviorist school of psychology.

When we examine the structural features of Simon's model, especially its four descriptive principles, both the confusion and the ambiguity surrounding 'behavioralism', *as it applies to Simon's model*, disappear. This is because Simon's model is something more radical and specific than is suggested by the ambiguous term 'behavioral'. It is not just that decision-making *behavior* is being modeled; in arguing that such behavior entails cognitive limitations causing bounds on rationality, setting goals, drawing upon information from the environment, mentally searching, and continually adapting action and behavior based on feedback from the environment, the decision-making organism is viewed as a 'cognizer'. In other words, *Simon's model is a cognitive model of human decision-making, and therein lies its originality (and Simon's creativity) in the social sciences.*

How are we justified in claiming that Simon's model of human decision-making is a cognitive model? This depends on what we mean by 'cognition'. We might appeal to one of several elaborate theories of cognition (von Eckardt, 1998), but I think we can capture the essence of the concept in the following terms (Dasgupta, 2000).

On the one hand, we have behavior: observable and overt characteristics of humans and animals, manifested as interactions with the external world. On the other hand, we have the brain: a corpus of physical matter that obeys the laws of physics and chemistry, the various activities of which give rise to behavior.

The problem is, there is a gap between overt behavior and brain matter sufficiently large as to make it difficult to explain behavior directly in terms of neurophysiological events. Furthermore, the latter are subject to natural (i.e. physico-chemical) laws, and such laws are nonpurposive, whereas behavior is purposive. This gives rise to the added problem of explaining how nonpurposive, physico-chemical events and processes can give rise to purposive behavior.

What scientists very often do in order to bridge large conceptual gaps between two types of events that are (believed to be) causally related is to propose or search for one or more *intermediate levels* of abstraction, explanation and description. Effectively, what is created is a *hierarchy* of abstraction levels, the assumption being that scientists are better off attempting to bridge the narrower conceptual gaps thus created between the adjacent levels in such a hierarchy (Pattee, 1973; Whyte, Wilson & Wilson, 1969).

The essence of cognitivism is the proposition that there exists one or more intermediate levels ('cognitive

levels') of organization and explanation between the extremeties of overt, purposive behavior and the non-purposive physico-chemical activities of brain matter (Newell, 1990, p. 111 et seq.). The nature of these cognitive levels and how they affect the mapping of behavior onto brain matter is the domain of *cognitive science*. And, while there remains considerable disagreement as to the 'architecture' of cognition, the following assumptions are more or less accepted by most practitioners of cognitive science: cognition: (a) refers to certain *functional entities* that are physically realized by brain matter and that mediate between behavior and brain matter; and (b) explanations of these functional entities entail the generation, organization, manipulation and processing of *representations*.

Returning to our topic, the four propositions PBR, PS, PHS and PAB constituting the empirical theories in Simon's model of decision-making *refer precisely to such functional entities*. It is because of this that it is not sufficient to call Simon's model a behavioral model; rather, a stronger claim can be made that it is a cognitive model.

Simon's Cognitive Model as the Origin of a Research Tradition

My argument thus far has led to the thesis that Simon has created a distinctly cognitive model of decision-making, and that this model has the characteristic structure of a Laudanian RT. But this does not mean that Simon's model did *in fact* create a research tradition. Paraphrasing Eric Hobsbawm (1983), a tradition, at the very least, must be a significant *practice* that has been in existence for some period of time. How can we demonstrate that Simon's cognitive model of decision-making not only possesses the structural characteristics of a RT but is also a genuine tradition (in Hobsbawm's sense)?

My claim is that this cognitive model achieved the status of a genuine RT because its core theories, viz., PBR, PS, PHS and PAB, *entered into the very languages of discourse of a range of disciplines often called the human sciences.*

Economics, political science, and organization theory were the first of the human sciences in which we see the explicit presence of Simonian cognitivism (Banfield, 1957; Cyert & March, 1963; Leibenstein, 1960; Machlup, 1967; March & Simon, 1958; Margolis, 1958; Marris, 1964; Storing, 1962). But these were not the only ones. Not surprisingly, the language of psychology itself became rapidly infused by both the methodological commitments and the core theories of the Simonian model, after appropriate reformulations in information processing terms. The new cognitive psychology to which the Simonian model contributed became one of the ingredients of the cognitive science that came into being in the mid-1970s (Gardner, 1985; Miller, Galanter & Pribram, 1960; Newell, Shaw & Simon, 1958). And by its influence on cognitive psychology, the Simonian model entered, from as early as the mid-1960s and expanding over the next three decades, the consciousness of those who practiced and wrote on the design disciplines, in particular, in the realms of architecture, industrial design and computing systems design (Alexander, 1964; Dasgupta, 1991; March, 1976; Steadman, 1979), and what the sociologist Donald Schon (1983) called 'professional practice'.

Cognitivism as an 'Alternative' Research Tradition in Economics

Thus, there is sufficient evidence that by the late 1970s, when Simon was awarded the Nobel Prize, the cognitive model he had created had become the basis and essence of a RT that infused several of the human sciences. This makes it all the more perplexing the issue raised at the start of the paper: Why did the cognitive tradition fail to be assimilated into the theoretical mainstream of economics? As one recent commentator, sympathetic to Simonian cognitivism, has pointed out, college textbooks on microeconomics continue to present the neoclassical version (Earl, 1995, p. 28–29).

The resistance to cognitivism has come from several directions, and the nature of these objections gives us some clues as to the essence of the resistance. But it is worth noting that even before such objections were summoned forth against Simon in particular, other voices had been raised in defense of the neoclassical tradition. In particular, there was Milton Friedman's influential essay of 1953 in which he argued that whether or not the assumptions underlying a theory are realistic is irrelevant; what is important is that the theory yields accurate or 'good enough' predictions about the economic universe, or that the theory yields predictions that are superior to those yielded by alternative theories (Friedman, 1953). From this perspective, it is sufficient, Friedman asserted, to assume that firms and businessmen behave *as if* they are capable of making completely rational decisions; that is, they are capable of making optimal choices. (It is ironic that Friedman himself advanced what was essentially a satisficing argument in support of the neoclassical, omniscient, perfectly rational *H. economicus*!)

At about the same time, the statistician Leonard Savage (1954) admitted that the assumption that a decision-maker assesses all alternatives before making a decision—Savage called this a 'look before you leap' strategy—might appear 'preposterous' when taken literally. Nonetheless, he held that 'look before you leap' was the 'proper subject' of any theory of decision-making.

In an early criticism of Simonian cognitivism—his 'program of behaviorism'—Fritz Machlup objected that the kind of 'real world complexity' that Simon's economic actor must allegedly deal with does not, in fact, exist. "The *Homo economicus* I have encountered in the literature was not such a perfectionist; and, indeed, did not need to be so because *H. economicus* does not have to deal with the 'entire environment', not even with the 'relevant aspects of the entire environment', but only with the relevant changes in environmental conditions". The axioms of maximizing behavior in the neoclassical tradition are postulated only to predict or analyze such changes. Thus, for instance, "The theory of prices and allocation viewed as a theory of adjustment to change does not call for impossible (i.e. omniscient) performances". One can, perfectly well, postulate maximizing behavior in such situations (Machlup, 1967, pp. 25, 25n).

Machlup raised a further objection. He considered a particular scenario in which the government placed a certain surcharge on import duties. Neoclassical theory, with its profit maximizing assumption, 'will without hesitation' predict that imports will fall. And he asked: "What will satisficing theory predict?" In partial answer, Machlup offered a quote from Simon in which the latter described the *thought process* that would guide decision-making according to his model. Thus, Machlup speculated, satisficing theory does not provide any real "quantitative . . . (or) qualitative predictions" (Machlup, 1967, p. 26n).

The point is, Machlup argued, Simon's 'behavioral' economics applies to a concept of the firm that is relevant to issues in organization theory and management science, but not to the firm or its behavior as relevant to such microeconomic issues as competitive prices and allocation, innovation and growth, and welfare theory. There are no doubt, he admitted, problems to which 'behavioral theory' applies. But those problems are not of concern to economists (Machlup, 1967, pp. 30–31).

For the mathematical economist William Cooper (who, with Abraham Charnes, made important contributions to the theory and application of linear programming, one of the major post-World War II optimization techniques that enriched neoclassical economics), and a friend and associate of Simon, the 'real world' complexity to which Simon appealed, to argue the case for satisficing, did not merit abandonment of the optimization assumption of neoclassical economics. The very complexity of the real world, Cooper wrote to Simon, may prompt scientists to 'synthesize artificial problems' which are easier to deal with and yet have "certain properties in common with the 'real' problem".[13]

Like Machlup, Cooper was suggesting that real world complexity is not necessarily relevant in the realm of effective economic problem-solving.

In 1957, a second edition of *Administrative Behavior* was published. In a review of this edition by Edward Banfield, we detect a resonance with Machlup's and

---

[13] W. W. Cooper to H. A. Simon, April 17, 1961. HASP.

Cooper's reservations, but this time in the realm of organization theory (Banfield, 1957). For Banfield, the principle of omniscient, optimal behavior which Simon's satisficing principle was intended to replace, served a purpose the new principle did not: the former was a prescriptive device for a 'practical science of administration'; Simon's model, in contrast, was 'sociological'—meaning descriptive. There remained a lacuna between the replaced and replacing principles. Simon, Banfield (1957, p. 284) declared,

> has destroyed the rationale of the old conceptual schema without offering any new one . . . so far as 'good' administration is concerned, he has no basis for judging what criteria are relevant and what are not.

Like Machlup, Banfield doubted the utility of the satisficing principle as a prescriptive device. Like both Machlup and Cooper, Banfield thought that real world complexity is not always relevant to actual problematic situations.

What exactly was it about the nature of the cognitive research tradition that brought about these various reservations? I suggest that at the very core is the following: the new tradition—its constituent principles in particular—was concerned with cognitive *procedures* (or *processes*); consequently, it demanded a radical *switch in perspective* (what Kuhn might have called a 'paradigm shift') from that underlying the neoclassical tradition.

This change in perspective becomes clearer when we examine one of the seminal texts belonging to the new cognitive RT: Richard Cyert's and James March's *A Behavioral Theory of the Firm* (1963). Cyert and March—both close associates of Simon—began their book by stating a set of propositions which, they claimed, captured the essence of the neoclassical model of the firm. These propositions—part of the constituent theories of the neoclassical RT—are precisely of the predictive form that Machlup had approvingly alluded to.

They then proceeded to present their 'behavioral' theory. Woven into this theory were several ideas distilled from Simon's writings on administrative theory and microeconomics: satisficing, aspiration levels, sequential search, the influence of the order of search on decision-making; the focus of recent organization theory on the 'decision to belong' within an organization, and how positions within the organization impact decision-making, individual goals, and perceptions; the role of plans in organizations; the idea that a firm viewed as an organization is a coalition of participants with conflicting goals that have to be reconciled; the theory of human problem-solving; and the way executives are compensated.

Ultimately, the Cyert–March 'behavioral theory' is presented in terms of a set of economic and organizational concepts. Some of these describe the variables (factors) that enter into a firm's behavior (e.g. variables having to do with the organization's goals and expectations); others are concerned with the relationship amongst variables (e.g. relationships having to do with the resolution amongst conflicting goals, or with the avoidance of uncertainty).

However, these variables and relational concepts by themselves do not constitute the theory. We get an early glimpse of what their 'behavioral' theory is going to be when they stress that there is, in the theory, "an explicit emphasis on the actual *process* of organizational decision-making as its basic research component" (Cyert & March, 1963, p. 125). It is the explication of *process* which draws upon the variables and relationship between variables as the building blocks that completes the 'behavioral' theory.

How does one describe or understand this process? The 'natural theoretical language' for specifying this process, according to Cyert and March, is "the language of a computer program"; and the "general structure" of the process can be "conveniently represented" in the form of a flowchart (Cyert & March, 1963, p. 125). The computational *metaphor* Simon had employed in 1955 had metamorphosed into the language of *actual* computation in Cyert's and March's theory.

The new theory was, thus, embedded not in the form of the familiar *declarative* form of knowledge in which scientific propositions were habitually stated but in the form of *procedural* knowledge—flow charts, algorithms, and computer programs. *This* was the change in perspective—and a change in the language of discourse—that the cognitive tradition engendered. I suggest that it was because this *procedural worldview* at the heart of the cognitive tradition was (and continues to be) so much at odds with the essentially *declarative* world view underlying the neoclassical tradition that cognitivism failed to be assimilated into the mainstream of neoclassicism, and that as a result, cognitivism became (and remains) an *alternative research tradition* in the realm of microeconomics.

Conclusion

I began this paper with Simon's Nobel Prize, and it seems appropriate to return to this occasion in my conclusion. The main body of the two-page Nobel Prize citation juxtaposes the Simonian model of the firm with the neoclassical model; it compares, in some detail, one with the other. Thus, although it did not speak of 'research traditions' or 'paradigms', or of 'switches in world views' or ' paradigm shifts', the citation leaves the reader in little doubt that the award was, in fact, in recognition of Simon's contribution to the genesis of a radically distinct research tradition in economic science.

In fact—and this is particularly noteworthy—the citation went further: it recognized explicitly that the work for which the Prize was being awarded had implications for "a number of research fields with similar problems, both in economics and in other disciplines" (Carlson, 1979). Thus, the citation alludes to what I think is the most fundamental feature of the cognitive research tradition which Simon was so instrumental in bringing into being: the fact that it was a *transdisciplinary* research tradition which encompassed not just economics, or organization theory, but a range of the human sciences including (as noted earlier) architecture, computer programming, and other design disciplines, professional practices (such as in medicine), and, since the late 1980s, the history and philosophy of science and technology (see, e.g. Alexander, 1964; Blum, 1996; Cross, 1984; Dasgupta, 1991, 1996; Giere, 1988; Langley et al., 1987; Magnani, Nersessian & Thagard, 1999; March, 1976; Nersessiain, 1995; Rowe, 1987; Steadman, 1979; Thagard, 1988; Weber & Perkins, 1994). Ultimately, Simon's creativity and innovation lay in the transdisciplinarity and, one might claim, universality of the cognitive tradition.

Acknowledgments

This essay has drawn extensively on unpublished papers and correspondence located in the Herbert A. Simon Papers Collection in the archives of the Carnegie-Mellon University Library. I also owe a debt to the late Professor Herbert Simon for his ready, prompt and patient responses to my queries, in both interviews and through email correspondence. I also thank Professor Amiya K. Bagchi for his comments on an earlier version of this essay. This work was supported by the Computer Science Eminent Scholar Trust Fund of the University of Louisiana at Lafayette Foundation.

References

Alexander, C. (1964). *Notes on the synthesis of form.* Cambridge, MA: Harvard University Press.

Banfield, E. C. (1957). The decision making schema. *Pub. Administrative Review,* **17** (4), 278–282.

Barnard, C. I. (1938). *The functions of the executive.* Cambridge, MA: Harvard University Press.

Blum, B. (1996). *Beyond programming: To a new era of design.* New York: Oxford University Press.

Cairncross, F. (1978). Professor Simon says .... *The Guardian,* (October 24).

Campbell, A. (1980). An introduction to George Katona. In: H. A. Simon et al. (Eds), *The 1979 Founders Symposium, The Institute of Social Research Honoring George Katona.* Ann Arbor, MI: University of Michigan.

Carlson, S. (1979). The prize for economic science. In Memory of Alfred Nobel, *Les Prix Nobel 1978,* Stockholm: The Nobel Foundation.

Craik, K. (1943). *The nature of explanation.* Cambridge: Cambridge University Press.

Cross, N. (Ed.) (1984). *Developments in design methodology.* Chichester, U.K.: John Wiley.

Cyert, R. & March, J. G. (1963). *A behavioral theory of the firm.* Englewood- Cliffs, NJ: Prentice Hall.

Dasgupta, S. (1991). *Design theory and computer science.* Cambridge: Cambridge University Press.

Dasgupta, S. (1994). *Creativity in invention and design.* New York: Cambridge University Press.

Dasgupta, S. (1996). *Technology and creativity.* New York: Oxford University Press.

Dasgupta, S. (2000). The origins and nature of the cognitive paradigm: An overview. *Indian Journal of Physiology & Pharmacology,* **44** (4), 379–391.

deGroot, A. (1946). *Het denken van den Schaker.* Amsterdam: North Holland.

Earl, P. E. (1995). *Microeconomics for business and marketing.* Aldershot, U.K.: Edward Elgar.

Edwards, W. (1954). The theory of decision making. *Psychology Bulletin,* **51** (4), 380–417.

Friedman, M. (1953). The methodology of positive economics. In: *Essays in Positive Economics.* Chicago, IL: University of Chicago Press.

Gardner, H. (1985). *The mind's new science.* New York: Basic Books.

Giere, R. (1988). *Explaining science: A cognitive approach.* Chicago, IL: University of Chicago Press.

Hobsbawm, E. (1983). Inventing tradition. In: E. Hobsbawm & T Ranger (Eds), *The Invention of Tradition* (pp. 1–14). Cambridge: Cambridge University Press.

Katona, G. (1951). *Psychological analysis of economic behavior.* New York: McGraw-Hill.

Kuhn, T. S. (1970). *The structure of scientific revolutions.* Chicago, IL: University of Chicago Press.

Lakatos, I. (1978). *The methodology of scientific research programmes.* Cambridge: Cambridge University Press.

Langley, P., Simon, H. A., Bradshaw, G. L. & Zytkow, J. (1987). *Scientific discovery.* Cambridge, MA: MIT Press.

Laudan, L. (1977). *Progress and its problems.* Berkeley, CA: University of California Press.

Leibenstein, H. (1960). *Economic theory and organizational analysis.* New York: Harper & Brothers.

Lotka, A. J. (1956). *Elements of mathematical biology.* New York: Dover.

Machlup, F. (1967). Theories of the firm: Marginalist, behavioral, managerial. *American Economic Review,* **57** (1), 1–33.

Magnani, L., Nersessian, N. & Thagard, P. (Ed.) (1999). *Model-based reasoning in scientific discovery.* New York: Kluwer Academic.

March, J. G. & Simon, H. A. (1958). *Organization.* New York: John Wiley.

March, L. (1976). The logic of design and the question of values. In: L. March (Ed.), *The Architecture of Form.* (pp. 1–60). Cambridge: Cambridge University Press.

Margolis, J. (1958). The analysis of the firm: Rationalism, conventionalism and behaviorism. *Journal of Business,* (July), 187–199.

Marris, R. (1964). *The economic theory of 'managerial' capitalism.* New York: The Free Press of Glencoe.

Miller, G. A., Galanter, E. & Pribram, K. H. (1960). *Plans and the structure of behavior.* New York: Holt, Rinehart & Winston.

Nersessian, N. (1995). Opening the black box: Cognitive science and history of science. *Osiris,* **10**, 196–215.

Newell, A. (1990). *Unified theories of cognition.* Cambridge, MA: Harvard University Press.

Newell, A., Shaw, C. J. & Simon, H. A. (1958). Elements of a theory of human problem solving. *Psychological Review*, **65** (3) 151–166.

Pattee, H. H. (Ed.) (1973). *Hierarchy theory.* New York: George Braziller.

Rosenbluth, A., Wiener, N. & Bigelow, J. (1943). Behavior, purpose and teleology. *Philosophy of Science*, **10**, 18–24.

Rowe, J. L. (1978). Outspokenness costs Robinson a Nobel. *The Washington Post*, (October 22).

Rowe, P. G. (1987). *Design thinking.* Cambridge, MA: MIT Press.

Savage, L. (1954). *The foundations of statistics.* New York: John Wiley & Sons.

Schon, D. (1983). *The reflective practitioner.* New York: Basic Books.

Screpanti, E. & Zamagni, S. (1993). *An outline of the history of economic thought.* Oxford: Clarendon Press.

Sen, A. K. (1979). Rational fools: A critique of the behavioral foundations of economic theory. In: H. Harris (Ed.), *Scientific Models and Man.* Oxford: Clarendon Press.

Shannon, C. E. (1950). Programming a digital computer for playing chess. *Philosophical Magazine*, **41**, 356–375.

Simon, H. A. (1947). *Administrative behavior* (3rd ed., 1976). New York: Macmillan Press, The Free Press.

Simon, H. A. (1950a). *Administrative aspects of allocative efficiency.* Cowles Comm. Disc. Paper Econ. No. 281. April 25.

Simon, H. A. (1950b). Modern organization theories. *Advances in Management*, (October 2–4).

Simon, H. A. (1951). A formal theory of the employment relationship. *Econometrica*, **19**, 293–305.

Simon, H. A. (1952a). A comparison of organization theories. *Review of Economic Studies*, **20** (1), 1–19.

Simon, H. A. (1952b). Applications of servomechanism theory to production control. *Econometrica*, 20.

Simon, H. A. (1954). Some strategic considerations in the construction of social science models. In: P. Lazarsfeld (Ed.), *Mathematical Thinking in the Social Sciences* (pp. 383–415). Glencoe, IL: The Free Press.

Simon, H. A. (1955). A behavioral model of rational choice. *Quarterly Journal of Economics*, **69**, 99–118.

Simon, H. A. (1956). Rational choice and the structure of the environment. *Psychological Review*, **63** (2), 129–138.

Simon, H. A. (1957). Rationality in administrative decision making. In: *Models of Man* (pp. 196–206). New York: John Wiley.

Simon, H. A. (1986). Preface to handbook of behavioral economics. In: B. Gilhead & S. Kaish (Eds), *Handbook of Behavioral Economics.* Greenwich, CT: JAI Press. Reprinted in (Simon, 1999, pp. 275–276).

Simon, H. A. (1991). *Models of my life.* New York: Basic Books.

Simon, H. A. (1999). *Models of bounded rationality* (Vol. 3) *Empirically grounded economic reasoning.* Cambridge, MA: MIT Press.

Simon, H. A., Smithburg, D. & Thompson, V. A. (1950). *Public administration.* New York: Alfred A. Knopf.

Steadman, P. (1979). *The evolution of designs.* Cambridge: Cambridge University Press.

Storing, H. J. (1962). The science of administration: Herbert A. Simon. In: H. J. Storing (Ed.), *Essays on the Scientific Study of Politics* (pp. 63–100). New York: Holt, Rinehart & Winston.

Thagard, P. R. (1988). *Computational philosophy of science.* Cambridge, MA: MIT Press.

Turing, A. M. (1953). Digital computers applied to games. In: B. V. Bowden (Ed.), *Faster than Thought.* London: Pitman.

von Eckardt, B. (1998). *What is cognitive science?* Cambridge, MA: MIT Press.

von Neumann, J. & Morgenstern, O. (1944). *Theory of games and economic behavior.* Princeton, NJ: Princeton University Press.

Walter, W. G. (1961). *The living brain* (1st ed., 1953). Harmondsworth: Penguin Books.

Weber, R. J. & Perkins, D. N. (Ed.) (1994). *Inventive minds: Creativity in technology.* New York: Oxford University Press.

Whyte, L. L., Wilson, A. G. & Wilson, D. (Ed.) (1973). *Hierarchical structures.* New York: Elsevier.

Wiener, N. (1948). *Cybernetics.* Cambridge, MA: MIT Press.

Williams, J. (1978). A life spent on one problem. *The New York Times*, (November 26).

The International Handbook on Innovation
Edited by Larisa V. Shavinina

Poetic Innovation

George Swede

Department of Psychology, Ryerson University, Canada

Abstract: Every artist wants his or her work to be considered innovative, in both historical and contemporary terms. In this paper, I focus on how and why poets strive towards these goals, but the results generalize readily to other kinds of artistic innovation. My discussion is based on the results of psychological research on poets and their poetry, some literary theory, as well as my long experience as a poet and editor of poetry periodicals and anthologies.

Keywords: Creative; Innovative; Novelty; Personal one creativity; Personal two creativity; Universal creativity; Universal two creativity.

Introduction

I have previously defined creativity in terms of a fourfold typology: universal one and two and personal one and two (Swede, 1993). Creative acts are universal when they stand out as unique and valuable in the history of the human race. Such accomplishments are the ones that get the attention of scholars and sometimes the general public. Most creative behavior, however, exists on another continuum, that of individual life spans. All of us do many things that are new and meaningful for ourselves, but already have been done by others, often for generations, such as falling in love or driving a car for the first time. Personal creativity is the stuff of character and gets noticed by counselors and novelists.

The numbers associated with each type indicate whether one individual or more than one is involved. If only one person is responsible for the creative act, it is called universal one or personal one; if two or more individuals collaborated, it is universal two or personal two. The majority of creative behaviors, whether universal or personal, involve collaborations among two or more individuals, such as the making of a motion picture or winning a local softball championship. Research on innovation tends to focus on universal one creators because individuals are easier to study than groups.

My typology does not distinguish, however, among major and minor achievements within any of the four categories. A throwaway whodunit novel and Dostoevsky's masterpiece *Crime and Punishment* are both considered as universal one. So also are the inventions of the Frisbee and the telephone, although they are obviously of different social significance (Swede, 1993). Various attempts have been made to develop rating scales for creative acts, but so far little that is definite has emerged (Besemer & Treffinger, 1981). Rating achievements in one field, such as inventions or novels, is hard enough, but the difficulty rises exponentially when comparing across fields, for instance, the telephone vs. *Crime and Punishment*.

The term innovative is almost identical in meaning to creative; therefore, I will use them interchangeably. Thus, universal one and two innovators and creators are the same, as are the personal one and two counterparts. The word creativity, however, has no equivalent among the words related to innovation. It refers to an ability presumed to exist within someone who shows creative or innovative behavior.

For a poem to be considered as universally creative or innovative, it must possess the two criteria mentioned earlier, i.e. be both original and meaningful. These two general characteristics are paramount. No matter how well written, a poem will not be published in a respected journal unless it meets both these criteria. A poem can be original without being meaningful, such as when 20 randomly selected words are strung together. Or, a poem can be meaningful without being original as when a poet imitates a well-known work, such as Edgar Allen Poe's 'The Raven' or Sylvia Plath's 'Daddy'. According to my scheme, such a poem is personally creative (as is the poet responsible). Obviously, it will not pass editorial screening. Poems by eminent poets, however, exist on a different level. They stand out as meaningful and original on the continuum of world history. They reveal universal one

471

innovation or what Barsalou and Prinz (1997) call exceptional, as opposed to mundane, creativity. The focus of this chapter will be on the poets who write these great poems.

The study of universal innovators in poetry might help us to understand the relationship between language, cognition and imagination. Oral poems were likely the first artistic acts performed by our ancient ancestors. They required no tools except those with which they were born: ears, mouth, tongue, words (even if they were grunts) and some sort of imaginative twist in thinking. What we learn about the production of poetry might help us to understand other realms of human innovation since the mechanics of creative thinking seem to be roughly the same no matter what the activity (see Kaufmann, R. Root-Bernstein & M. Root Bernstein, Simonton, Sternberg et al., Vandervert, Weisberg, this volume). Of course, any findings will be especially relevant to those endeavors that involve the imaginative use of words, such as the translation of novels and poems as well as scientific and technological discoveries from one language into another, not to mention the generation of new languages to run the super computers of the future.

The Search for Novelty

Most of the work most of us do is routine. Our job definitions certainly do not specify that we be universally creative all the time. Those working in the arts, however, do have this pressure. Artists are expected to do something new or novel *every* time they start a new project. Thus, the pressure on a painter, film-maker, novelist, sculptor, ad writer and poet to be universally creative or innovative is omnipresent. Martindale (1973), who has extensively examined the processes by which poets make their poems, believes that the search for novelty is their main driving force:

> The value on novelty is the one value that cannot be violated by the poet since it is the thing that differentiates his role from those concerned with recitation or reproduction (p. 319).

For a poem to be judged as successful, it must stimulate a number of responses in the reader: deep emotions, vivid images, different levels of meaning as well as a sense of truth. But, above all else, the reader must experience the poem as universally new or special. To create such a poem, the beginning poet must develop abilities specific to the poetic craft as well as the capacity to think in an original manner, or what Amabile (1983) calls 'domain relevant skills' and 'creativity-relevant skills' (p. 363). Once the emerging poet is able to make poems editors will want to publish, he or she will begin a lifetime struggle to stay original, i.e. to avoid imitating others as well as oneself.

For some poets I know, the search for novelty becomes such a powerful concern that they limit their reading of poems by contemporaries to avoid uncon-

scious duplication. Harold Bloom (1973), the noted literary scholar, gives the reason for this as being 'the anxiety of influence':

> Poets, by the time they have grown strong, do not read the poetry of X, for really strong poets can read only themselves. For them, to be judicious, is to be weak, and to compare, exactly and fairly, is to be not elect (p. 19).

Such an attitude seems to be vital to maintaining the relentless pursuit of the original. Mexican poet Octavio Paz, who won the Nobel Prize in 1990, points out that the endless search for novelty became dominant during the early 19th century, with the rise of Romanticism:

> From the Romantic era onward, a work of art had to be unique and inimitable. The history of art and literature has since assumed the form of a series of antagonistic movements: Romanticism, Realism, Naturalism, Symbolism. Tradition is no longer a continuity, but a series of sharp breaks. The modern tradition is the tradition of revolt (p. 17).

This outlook often leads to behavior that others, especially those outside the arts, perceive as egotistical or self-serving (Amabile, 1996; Barron, 1969; Ferris, 1978; Middlebrook, 1991). I suspect the reason is to maintain the belief that poetry is important despite the fact that the 21st-century world considers it almost irrelevant. Billy Collins, former U.S. poet laureate, states that his country needs to have a National Poetry Month to remind the populace that this form of expression still matters:

> We don't have National Television Month. We don't have National Go to the Movies Month. So to have a month for something is to admit that there is a neglect taking place, that you should pay attention (Pushing poetry, p. 58).

Czeslaw Milosz, a Polish poet who won the Nobel Prize in 1980, describes this dearth of interest more poetically:

> Poets in the twentieth century are by nature isolated, deprived of a public, 'unrecognized', while the great soul of the people is asleep, unaware of itself and learns of itself only in the poetry of the past (p. 30).

A quick perusal of the entertainment segments on TV and radio, in magazines, newspapers and the Internet reveals an utter fascination, bordering on idolization, of movie actors, film-makers, pop, jazz (and even opera) singers, orchestra conductors, various kinds of musicians and pop novelists. Poets and their collections are rarely, if ever, featured.

An advantage accrues to such anonymity, however: poets are freer to be original because they are spared the judgments of, and subsequent control by, the marketplace. Martindale (1973, 1975, 1990) has shown via an ingenious series of studies involving major

English poets (from 1700 to 1840) and French poets (from 1800 to 1940) that the more autonomy poets have been granted by their period in history, the more innovative they become.

Poets have another advantage over other artists, especially novelists. Given the right circumstances, they can complete a poem in minutes. Canadian poet Gwendolyn MacEwen (1980) succinctly describes the experience:

> A poem is a much more rewarding thing to write because very often in the space of half an hour or sometimes even five or ten minutes, you can see the whole thing mapped out before you, at least in rough form. And sometimes the first draft can be the final version of the poem, and it's wonderful that relatively little time expires between the original version and the result . . . . So I certainly am more immediately satisfied when I'm writing a poem than when I'm writing prose (p. 65).

My own experience, as well as those of other writers with whom I have discussed this particular issue, confirm MacEwan's observations. Thus poets, while almost invisible compared to prose writers, do have more occasions to privately feel a sense of accomplishment. This must be one of the chief reasons to continue composing poems in the absence of more tangible rewards.

Foreground and Background Factors

All universally creative outcomes, whether artistic or scientific, involve the union or fusion of previously unrelated ideas. Koestler (1964) refers to such processes as 'bisociative thought' (p. 121), Rothenberg (1979) calls them 'Janusian thought' (p. 55) and Ward, Smith & Vaid (1997) depict them as 'conceptual combination and expansion' (p. 10). These terms effectively describe the necessary central associations; for instance, Albert Einstein's theory of relativity fuses energy and matter, or, on a simpler level, this one-line poem by Yukio Mishima (1959/1968), "My solitude grows more and more obese like a pig" (p. 1081), unexpectedly connects pig and solitude.

However, such conceptions are misleading because they imply that creative thinking has to connect only two different planes of thought. While two may be central, they are never enough to fully explain what happened. Einstein had to integrate not only energy and matter, but also time, space, motion and light. Mishima had to include the idea of obesity and the considerations of word choice and word order. Without them, the relationship of pig to solitude would never have worked. The union of two thoughts always involves other ideas, considerations or conditions, such as the routines people establish for doing their work (Swede, 1993). The right setting can facilitate creative work, the wrong one inhibit it. Certain routines are well known, such as those of the popular horror-fiction writer

Stephen King. He gets up each day at 9 a.m., swallows a multiple vitamin pill with a glass of water, turns on the radio and types until 5 p.m. with only a few breaks (Kanfer, 1986). The novelist Marcel Proust could only write in a cork-lined room, and the poet Friedrich Schiller had to sniff rotten apple cores before starting on a poem (Gardner, 1982).

Several poets I know do most of their initial drafts and subsequent revisions in busy cafés and restaurants, often sitting at a table for hours sipping cups of coffee accompanied by the occasional doughnut or piece of pie. Perhaps such circumstances enable a poet to reduce the sense of isolation and therefore better focus on writing. However, these settings might actually *increase* feelings of alienation insofar as the poet's sense of being an outsider is heightened by seeing others absorbed in the more ordinary activities of living. There is evidence to suggest that this hypothesis is more correct, i.e. negative mood seems to be a spur for creativity (see Kaufmann, this volume).

As these examples illustrate, creative thinking involves many factors, not just the central ideas in the finished product. Without a coming together of the right thoughts at a suitable time and place, the creative process flounders, and efficiency falls. An individual in a physically threatening environment—extreme heat or cold and with low water and food supplies—is unlikely to muster the cognitive wherewithal to coordinate the variables needed to make a poem, particularly a great one. For this to happen, the poet must feel safe. However, as Martindale's (1975) and Simonton's (1984) historiometric studies have revealed, personal security is not enough. Unless the culture in which one lives stresses achievement and encourages experimentation, creativity will falter, no matter how safe a person feels. Research shows that universal or exceptional creativity requires that certain background conditions, sometimes surprising ones, such as *not* being married (see Simonton, this volume), exist in addition to the core or foreground ideas.

My interest in the importance of background factors began during graduate school when I published a study showing how surrounding colors and shapes became a part of learning tasks without subject awareness (Swede & McNulty, 1967). This finding altered my views on the extent to which a poet should be considered the sole creator of a poem. I could no longer accept the idea that a poet was completely in charge of the creative process and take all the credit for what he or she produces. Not surprisingly, other poets considered my views misguided, if not downright crazy. My position seemed less exotic when Skinner (1972) published an article in *Saturday Review* based on a talk he had given at the New York Poetry Center. Skinner presents a convincing case that a poet can take very little credit for a poem, that it is mainly the outcome of the poet's genetic and environmental histories over which he or she has little control:

The poet often knows that some part of his history is contributing to the poem he is writing. He may, for example, reject a phrase because he sees that he has borrowed it from something he has read. But it is quite impossible for him to be aware of all his history, and it is in this sense that he does not know where his behavior comes from. Having a poem, like having a baby, is in large part a matter of exploration and discovery, and both the poet and mother are often surprised by what they produce (p. 35).

I wish I had been at the Poetry Center when Skinner gave his talk. The question period must have crackled with the sparks generated by disgruntled poets.

In Skinner's view, then, poetic (or any other kind of) originality is either rewarded or not rewarded by the environment. If it is rewarded, then a person is likely to keep writing poems, provided his or her genes predispose to verbal skills. Amabile's (1983, 1996) extensive research into the social factors involved in creativity chiefly confirm what Skinner says, although she cautions that premature and excessive reinforcement, especially in childhood, might undermine the process.

In what follows, I will discuss what is known about how a person becomes and then stays a poet, and by poet I mean, a *published* poet. By published, I mean that the poet's work has gone through a typical editorial process and has been judged as unique and valuable (and therefore universally creative/innovative) by an editor or team of editors working for a recognized periodical or book publisher, traditionally based or on the Internet. Patrick (1935) established a similar set of criteria (without the option of the Internet, of course) in the selection of the 55 poets for her pioneering study of the steps involved in the composition process.

Becoming a Poet

Chukovsky (1925/1971), a Russian writer who studied the word play of thousands of preschoolers, believed that the child from two to five is a 'linguistic genius' (p. 7). Children at this age, in their attempts to master their first language, invent words, and experiment with rhyme, rhythm and metaphor. They see this learning process not as work, but as play. Chukovsky provides numerous examples, for instance this verse by a four-year-old boy:

The raven looked at the moon—oon—oon
And saw in the sky a yellow balloon
With eyes, nose, and mouth in a round face,
Swimming with clouds at a slow pace (p. 76).

Such complete poems are less typical of preschoolers than word inventions or short rhymes. Here are two charming examples of rhymes by Chukosky's son at four years. The first involved him "mounted on a broomstick, shouting like one possessed":

I'm a big, big rider,
You're smaller than a spider (p. 64).

The second occurred moments later when sitting down to dinner and "scanning with his spoon, he declaimed":

Give me, give me, before I die,
Lots and lots of potato pie! (p. 65).

With Chukovsky as my inspiration, I kept notes on interesting things said by my two sons during their preschool and early school years. Here are two that startle with their originality:

16/7/1972, Andris, age 3 years, 10 months (in anger to his brother Juris): "I'm going to pull your bones out and swim in you". 19/1/1973, Juris, age: 5 years four months: "I had a bad dream last night, a nightmirror" (Swede, 1976).

Juris's 'nightmirror' could be construed as a mistake, i.e. he really meant to say 'nightmare'. But the aptness of the substitution suggests poetic imagination at play. If instead Juris had said 'nightmayor', then the likelihood that he simply made a homophonic error would be greater. However, the statement by Andris is clearly an intentional use of metaphor, worthy of Stephen King.

My examples and the thousands of statements recorded by Chukovsky as well as by Schwartz and Weir (as cited in Winner, 1982), indicate that metaphorical or poetic thinking is an integral part of first language learning during the preschool years. In fact, suggests Chukovsky (1925/1971), this poetic proclivity has much in common with the adult poet. Then, by five or six, when we have mastered the basics of our first language, such as being able to understand and utter complex sentences, our interest in verbal play diminishes as we pursue new challenges offered by school: how to read and write, do arithmetic, understand science and history. For most students, poetry loses its central importance; it becomes just another topic in the curriculum. If this did not happen, Chukovsky (1925/1971) claims that a child by "the age of ten (would) eclipse any of us with his suppleness and brilliance of speech" (p. 7).

More recent research suggests that the school-age child's capacity for linguistic creativity does not decline as much as Chukovsky and others believed:

When children are asked to explain and produce figurative expressions or to create new ones, they exploit the productive nature of language and are able not only to explain the semantic structure, namely whether certain types of idioms mean what they mean . . . but also to produce idioms and create new figurative expressions for concrete events and abstract mental states (Cacciari, Levorato & Cicogna, 1997, p. 173).

Cacciari et al. (1997) also point out the special nature of verbal creativity compared to other kinds of expression, such as drawing from the imagination nonexistent creatures:

> Whereas language allows individuals to recombine meanings and referents in an almost infinite number of ways (with the important limitation of the balance between novelty and comprehensibility), the ways in which one might imagine (draw) nonexistent creatures are more limited . . . they are more likely to be predictable from what we know about everyday categorization processes (p. 174).

In other words, drawing imagined creatures will be bound to the limited number of existing schemas for human and/or animal shapes. However, developing verbal metaphors for such things can make use of a virtually unlimited number of word combinations taken from both the natural and human-made worlds. Poets have a definite advantage over visual artists.

What, then, determines that only a few students will go on to have careers as poets? Is there a 'poetry gene'? Very likely not, since complex behaviors involve the interaction of both heredity and environment (Simonton, 1999, this volume). Furthermore, as Lykken (1998) points out, the hereditary component is likely to have a polygenic basis which determines not only a specific ability, but also such traits as persistence, concentration, curiosity and certain physical attributes required by the creative activity, such as hand size for virtuoso pianists (Simonton, 1999). Such characteristics will then interact with the background factors of parental upbringing, peer experiences and type of formal education.

Perhaps the answer lies in adult behavior that can then be traced back to similar actions in childhood. The most noticeable characteristic of exceptionally creative people is their remarkable drive to be original (see Weisberg, this volume). They possess in the extreme what Amabile terms 'task motivation' (1983, p. 393). For example, a well-designed study of outstanding social, biological and physical scientists concluded:

> The one thing all of these sixty-four scientists have in common is their driving absorption in their work. They have worked long hours for many years, frequently with no vacations to speak of because they would rather be doing their work than anything else (Roe, 1952, p. 51).

Artists are no different. I have already referred to Stephen King's long hours of work, but Walt Disney perhaps was even more driven. When taking his wife out for dinner Disney would often suggest dropping by the studio for a few minutes, only to end up working until the early morning while his wife slept on the couch (Schickel, 1985, p. 167). This workaholic attitude is likely sown during the childhood and adolescent years, as was the case with the Welsh poet Dylan

Thomas who published his first poem at the age of 11 in his grammar school magazine and at 13 sold his first poem to a Cardiff newspaper (Ferris, 1978). Likewise, Gwendolyn MacEwen, at age 15, published her first poem in a national magazine and left school at 18 to pursue writing prose as well as poetry full-time (Pearce, 1980).

Such an intense drive to compose poetry is partly the result of reinforcement for original verbal utterances or writings by parents, teachers or friends as well as others who capture the young person's imagination—family, teachers, friends who are very interested in poetry or are themselves published poets (Amabile, 1996; Skinner, 1972). Some of these influential persons will likewise serve as models for the child to imitate (Bandura, 1997). The general cultural milieu also will have a strong influence. One that values poetry will stimulate more parents to be supportive of children who show an interest in composing poems. Simonton's (1984, 1997, 1999, this volume) extensive historiometric and archival studies have shown that such influences are especially crucial during the developmental periods of childhood, adolescence and early adulthood.

Of course, children are not merely passive recipients of environmental influences and will interpret the feedback they get from their attempts to create poems and act accordingly to formulate realistic goals. Bandura (1997) describes the origins of the feeling of self-efficacy that is required for achievements of any kind:

> Self-efficacy beliefs are constructed from four principal sources of information: enactive mastery experiences that serve as indicators of capability; vicarious experiences that alter efficacy beliefs through transmission of competencies and comparison with the attainments others; verbal persuasion and allied types of social influences that one possesses certain capabilities; and physiological and affective states from which people partly judge their capableness, strength, and vulnerability to dysfunction. Any given influence, depending on its form, may operate through one or more of these sources of efficacy information (p. 79).

Thus, accomplishment is more than the result of environmental reward and punishment. It also involves how we interpret such experiences and parlay our understanding into behavior that is effective in getting to our goals, in this case, the writing and publishing of poems.

In his autobiography, the influential American poet William Carlos Williams (1958) provides evidence of all these forces at work. He became seriously interested in poetry at age 16 when forced to give up baseball and running because of a heart murmur, "I was forced back on myself. I had to think about myself, look into myself. And I began to read" (p. 1). Someone else might have taken up stamp collecting, but Williams from early childhood had been raised in a family that valued language and literature:

> My father was an Englishman who never got over being an Englishman. He had a love of the written word. Shakespeare meant everything to him. He read the plays to my mother and my brother and myself. He read well. I was deeply impressed (p. 2).

Williams took the sense of self-efficacy formed by success in baseball and track and redirected it toward other areas for which he was also well prepared—reading and, eventually, writing poems—pursuits which continued even during the rigors of medical school. As mentioned earlier, Dylan Thomas was even more precocious due in large part to a father who was an English teacher and who encouraged Thomas' 'natural passion for words' (Ferris, 1978, p. 29) from the early pre-school years, thus developing even earlier a sense of self-efficacy for poem-making. A look at the biographies and autobiographies of a dozen other poets on my bookshelves revealed similar pre-adult experiences.

However, students shaped by such circumstances to become serious poets face daunting tasks: how to become a voice distinct from contemporaries as well as poets from the past and how to maintain this uniqueness over the course of a career. As mentioned earlier, such a focus on novelty or newness is most acute in the arts, but some other pursuits, such as inventor, research scientist, and explorer, have similar pressures.

Staying Productive and Innovative—Behavioral Traits

An Intense Devotion to Work

In the preceding discussion, I have cited controlled research as well as anecdotal evidence which shows that an intense devotion to work is characteristic of universally creative individuals in all walks of life, not just in the arts. However, much speculation exists about why such driven behavior occurs and how it is maintained. Woodman (1981) examined the explanations of major personality theorists and concludes that none provide the entire answer. Donald MacKinnon (1978) suggests the possible reasons why:

> Creative behavior is not different from any other form of behavior in being almost certainly the expression of not only one, but many motives. We need not choose among the different motivational theories of creative behavior. We can find confirming cases for each of them. Our need is to seek for still other and as yet unnoticed motivational factors leading to creative striving, and demonstrate, not how each of them is a given factor in any given bit of creative behavior for a given person, but how they act in concert in the creative striving of persons (p. 197).

MacKinnon's motivational criteria are largely met by Bandura's (1997) theory of self-efficacy which explains the complex interaction of experiences necessary for the high drive needed for great achievement: a sense of mastery from successful achievement (a finished poem); learning from and comparing oneself to significant others (admired, eminent poets); getting the approval from significant others (valued teachers, critics, or eminent poets); and finally, the positive affective and physiological states accompanying these experiences (mental and physical well-being). Self-efficacy theory is thus an effective blend of well-established results from conditioning, social-learning, cognitive science and the physiology of stress and coping.

Quality and Quantity of Output

Of the variables, other than hard work, that seem to be necessary for exceptional creativity, two of the most relevant are high quality and high quantity of output. David Perkins (1981) found that poets judged as superior outperformed less talented colleagues on both measures:

> They could get more quantity for the same quality, or more quality for the same quantity (p. 142).

The better poets were simply more efficient in achieving quality, that is, they simply did more good work. Studies of other kinds of universal creators have shown similar results: American scientists who have won the Nobel Prize published an average of 3.24 papers per year compared to 1.48 by a matched comparison sample (Zukerman's study as cited in Simonton, 1984, pp. 84–85); the top 10% of individuals in seven different fields produced 50% of the work (Dennis' study as cited by Simonton, 1984, p. 79).

Perhaps a cause-and-effect relationship exists: quantity begets quality. The more one produces, the more likely something worthwhile will occur, i.e. the more streams a prospector pans for gold, the more nuggets he or she will find. However, this suggests a purely mechanical relationship between effort and success which does not explain individuals concerned chiefly with quantity of output. For instance, during the 1980s, an American artist received media attention for oil-painting thousands of landscapes per year, yet not one of them reached more than a minimally acceptable standard. Around the same time, a Canadian writer gained much publicity for publishing more novels (several hundred) than any other writer in history, but none of his books earned critical acclaim, and likely never will. Obviously, a large capacity for work and a huge output are not enough for greatness.

Then there are those universal innovators whose work reveals a high quality but who have a low output. Gregor Mendel, the father of modern genetics, published only seven papers in his lifetime, and Bernard

Reiman, an influential mathematician, totaled a mere 19 (Simonton, 1984). Compared to others of his stature and longevity, Walt Whitman wrote a small number of poems, all of which ended up in his only collection, *Leaves of Grass*. The first edition of this masterpiece, self-published in 1855, contained only 12 poems, while the ninth and last edition, done in 1892, contained 400 pieces (Whitman, 1969). By comparison, both eminent and average universal poets frequently publish two or three times as many, either during their lifetime, posthumously, or both.

But simply counting the number of poems is a process full of pitfalls. Some poets, like Whitman, write mainly long poems (more than a page with lengthy individual lines, while others, like Emily Dickinson, write chiefly short ones, usually less than a page and with brief individual lines (Allen et al., 1965). To date, Dickinson has had close to 600 of her poems published, mainly posthumously (Dickinson, 2002). This is 50% more than the output of Whitman, but in terms of the number of actual words in print, perhaps Whitman has more. In spite of such technical details, the overall evidence clearly shows that work of high quality is necessary and sufficient for perceived greatness—high output by itself is not.

However, we must beware of the dizzying effects of circular reasoning in relation to quality and greatness. How do we know someone is a genius? Because others say so. And, how do we know the work is of high quality? Because the creator is a genius. This kind of logic is very difficult, if not impossible, to avoid to some degree.

Further complicating matters is the fact that historical assessments of the quality of an artist's work can change dramatically with passing time, i.e. artists' reputations are not constant. Except for literary scholars, who now remembers these poets, celebrated as much as pop music stars in their time: Richard Brautigan, Kahlil Gibran and Rod McKuen? Then there are those poets who were unappreciated while they lived, yet acquired legendary status after their deaths: Emily Dickinson and Walt Whitman.

The Possession of High Standards and the Need to Prove Oneself

What motivates one universal creator to seek high quality and another to hold quantity more dear? Barron (1969) reports on a series of controlled studies at the Institute of Personality Assessment and Research (IPAR) at Berkeley which compared outstanding writers, architects, mathematicians to those with only average success in these fields on a number of personality measures. The former possessed significantly higher ideals for what they produced. The elite group seemed to have higher standards (p. 71), ones that go beyond the norms for professional competence. A typical example of the search for such excellence is Anne Sexton who "would willingly push a poem

through twenty or more drafts" (Maxine Kumin as quoted by Middlebrook, 1991, p. 94).

Another finding reported by Barron has a clinical undertone:

> Creative individuals are very much concerned about their personal adequacy, and one of their strongest motivations is to prove themselves (1969, p. 70).

Barron's interpretation suggests that the need to prove oneself is a characteristic different from the possession of high standards. More likely, they are merely different expressions of a strong sense of self-efficacy. Someone who has written outstanding poems in the past will want to maintain his or her high standards in the future. Naturally, some anxiety about possible failure will occur, and this might be what the IPAR research was tapping through their testing and interviewing procedures. Such anxiety is perfectly normal, not pathological, as hinted at by Barron's interpretation.

A problem with the criterion of high standards is that it does not remain constant in the lives of eminent creators. Most of the work they produce is not of the highest quality. Of the two or three thousand poems an outstanding poet typically publishes, only a handful, usually 50 or 60, get reprinted in various anthologies of best poems. The rest are deemed merely very good, but minor works. The same can be said for all other enterprises, scientific as well as artistic (Dennis, 1966; Simonton, this volume; Weisberg, this volume).

A Strong Sense of Reality and Intrinsic Motivation

According to the IPAR studies, a fourth feature of exceptional creators seems to be their strong sense of reality, or what Barron (1969) refers to as 'ego strength' (p. 73). Using the results of interviews and personality tests, such as the Minnesota Multiphasic Personality Inventory and the California Psychological Inventory, Barron reports that the outstanding architects, mathematicians and writers scored consistently higher than representative samples in the same occupation who, in turn, scored higher than the general population. Such findings make considerable sense because writing a poem involves having to make unusual or unexpected connections among disparate objects or events. Only someone with a strong sense of how things actually are would feel secure enough to do this. But again we must be cautious. Mackinnon (1978) points out that no one motive is sufficient. Among the individuals in the general population studied by the IPAR must have been some individuals with a very strong sense of reality but with no desire to become eminent achievers.

Another characteristic, often commented upon, is the independence or intrinsic motivation shown by exceptional creators. The studies reported by Barron (1969) confirm this, and so do many other observations, informal and formal (Amabile, 1996; Kenner, 1971;

477

Middlebrook, 1991; Simonton, 1984; Winner, 1982). As Amabile (1996) points out, however, not everyone falls into a neat formula. The self-motivation of some creators, especially writers, is vulnerable to external forces. For instance Amabile analyses Sylvia Plath's spiral to self-destruction as the result of over-dependency on the approval of others:

> The greatest burden that impeded Plath's writing during her postcollege years is an extrinsic constraint that, perhaps more than any other specific social factor, appears to undermine the creativity of outstanding individuals: the expectation of external evaluation, and the attendant concern with external recognition (p. 11).

While Plath became increasingly reliant on approval from others, her contemporary, Anne Sexton, became more independent with age. She accomplished this despite serious problems with alcohol and drugs and critics who could not abide a woman speaking frankly about "menstruation, abortion, masturbation, incest, adultery and drug addiction at a time (the 1960s) when the proprieties embraced none of these as proper topics for poetry" (Middlebrook, 1991, pp. 143–144).

How do we explain the difference between Plath and Sexton? Amabile (1996) suggests that in childhood Sexton learned how to deal with external pressure and stay true to her beliefs, although not without a price, as indicated above. In Bandura's (1997) terms, she developed the necessary beliefs of self-efficacy, at least in relation to her poetry. Most elite poets with whom I am familiar seem to be more like Sexton than Plath.

Correlation Is Not Causation

So far, I have discussed six traits associated with outstanding universal achievers: hard-working, high output, high standards, a need to prove oneself, fierce independence and a strong sense of reality. But a problem exists: we know of these characteristics only from correlational studies and not from ones designed to discover cause-and-effect relationships. Weisberg (1986) describes the problem succinctly:

> If creative artists are more autonomous . . . it is then concluded that the autonomy contributed to their creativity. Such a conclusion is unjustified, however—all that has been demonstrated is a correlation between autonomy and creativity, not that the former was a cause of the latter . . . the exact opposite might be true—being creative might make one autonomous (p. 78).

The eminent universal creators might have become more productive and formed a stronger sense of reality *after* they became successful. This is quite likely if we consider that positive reinforcement creates a sense of self-efficacy which then leads to more rewards and more feelings of accomplishment and ultimately a strong belief in one's own worthiness (Bandura, 1997).

Creativity and Madness

I have left discussion of the following trait to the last because it has received the most attention and is widely believed to be true: that universally creative artists, writers and scientists are more likely to be crazy than the average person. The following 20th-century writers and their circumstances immediately come to mind: Virginia Woolf, Sylvia Plath, Ernest Hemingway, and Yukio Mishima committed suicide; Dylan Thomas and Jack Kerouac drank themselves to death; Ezra Pound, Theodore Roethke and Robert Lowell spent considerable time in mental institutions. These examples are well known to me because of my involvement with literature. Readers with interests in the visual arts, the sciences or leadership could produce similar lists just as easily.

Recent articles in well-respected journals have reinforced the idea that high achievement and madness are allied. For instance, one of the most widely quoted by the media has been 'Manic-Depressive Illness and Creativity' by Kay Redfield Jamison which appeared in a 1995 issue of *Scientific American*. Jamison marshals evidence to suggest that:

> Highly creative individuals experience major mood disorders more often than do other groups in the general population (p. 66)

Her catalogue of elite creative persons with presumed manic-depressive psychosis is impressive. Among the 18th- and 19th-century poets she lists are William Blake, Lord Byron, Edgar Allen Poe, Alfred Lord Tennyson, while the poets from the 20th-century include John Berryman, Randall Jarrell, Robert Lowell, Sylvia Plath, Theodore Roethke, Delmore Schwartz and Anne Sexton. Among other kinds of writers she mentions are Ernest Hemingway, Herman Hesse, Mark Twain, Tennessee Williams and Virginia Woolf. Non-writers include Vincent van Gogh, Gustav Mahler, Charles Mingus, Georgia Okeeffe, Mark Rothko and Robert Schumann. Jamison's main argument is that the manic phase helps spur universal creativity:

> The common features of hypomania [mild mania] seem highly conducive to original thinking; the diagnostic criteria for this phase of the disorder include 'sharpened and unusually creative thinking and increased productivity' . . . the cognitive styles associated with hypomania (namely, expansive thought and grandiose moods) can lead to increased fluency and frequency of thoughts (p. 66).

478

Nevertheless, Jamison cautions against leaving this illness untreated:

> Manic-depressive illness often worsens over time—and no one is creative when severely depressed, psychotic or dead (p. 67).

Basically, Jamison points out the dual nature of a particular illness. In its early stages, manic-depressive illness can help *some* persons, not all, to be more creative. But in its later stages, if left untreated, it is more debilitating than beneficial.

In apparent agreement, Barron (1969) reports that exceptional writers have more psychopathology than average writers, and both have more than the individuals at large. Thus, we have a paradox. The highly creative are both more in touch with reality (see above) and also less in touch. However, Barron provides as good an explanation as anyone has to date:

> If one is to take these test results seriously, creative individuals appear to be both sicker and healthier psychologically than people in general. Or, to put it another way, they are much more troubled psychologically, but they also have far greater resources with which to deal with their troubles. This jibes rather well with their social behavior . . . They are clearly effective people who handle themselves with pride and distinctiveness, but the face they turn to the world is sometimes one of pain, often one of protest, sometimes of distance and withdrawal; and certainly they are emotional (p. 75).

Of course, this does not mean that a person who has these characteristics will necessarily become exceptionally creative. All we can say for sure is that the traits uncovered at the IPAR co-exist with the capacity for high achievement. We cannot say that they are necessary for creative success, for surely there are persons who are both sicker and healthier who do not accomplish anything, nor have the desire to do so.

Studies which came after those conducted at the IPAR, and which have compared the frequency of psychiatric problems in the general population with those in the sub-group of the exceptionally creative, show that these two populations do *not* differ, i.e. both have the same degree of psychological pathology, around 10% (Simonton, 1984). Yet the belief that madness is married to creativity persists. Perhaps exceptional creators behave in ways that, while not psychotic, are strange or eccentric. And maybe observers call this behavior crazy or mad when they really mean to say the behavior is unusual or beyond the norm. Schubert & Biondi (1977) add some more details to this view:

> Some individuals find it difficult to respond creatively while in the mainstream of daily chores. They require . . . unprogrammed activity time for the more

fulfilling release of their creative potential. Withdrawal may provide them with the needed time (p. 192).

The striving for creative goals sometimes requires unusual coping strategies which are misinterpreted by others as abnormal. They are abnormal (read unusual), but they are not crazy in the clinical sense, just eccentric.

Another variable at play here is that people *expect* the exceptionally creative to display strange behavior. And, we know that when others expect us to do something, we tend to oblige. Bandura (1997) succinctly terms this interaction 'reciprocal determinism' (p. 6). The behavior might not really be so strange in itself but be unusual given the circumstances. For instance, a poet in the fever of composition might be late for a wedding at which he is the best man. Continuing with something while inspired is perfectly normal, but not when others are counting on you to perform a duty at a certain time. The groom, however, is likely to excuse his poet-friend precisely because he is a poet, i.e. poets are unreliable. Meanwhile, the poet in question might actually have made a conscious decision to be late because he knows that his friend believes poets tend to be late.

Still one more factor complicates matters: historical variations in cultural expectations for the kinds of thought process which the poet or artist should employ. Quite likely, poets who lived during times when their culture demanded realistic images and a particular form (French Romantic and English Neo-Classical), had different personalities to those who lived during eras which required unusual images and breaks with form (French Symbolist and English Metaphysical). Colin Martindale (1990) found evidence for more psychopathology among the latter group of poets (Baudelaire, Rimbaud and Donne, Marvell) than among the former (Hugo, Chateaubriand and Pope, Dryden). Of course, such finding leads to the classic chicken and egg argument: Do people with psychopathological tendencies become poets when distant associations and a break with convention are the poetic styles or do the demands for unusual imagery and rule-breaking create pathology in poets who started out as normal? Today, no clear cultural expectations exist, and thus poets have more freedom to follow their inclinations. In my experience, persons who show unusual thought processes in their daily interactions also tend to prefer writing poetry that uses primordial thinking and regressive imagery, as found with the Symbolists and Metaphysicals. In other words, the individual appears to select the style best suited to his or her personality.

Staying Productive and Innovative—Background Influences

I have dealt with behaviors common to elite universal creators, focusing on the poets among them. I have also referred to a number of background influences, such as genetics, family, age and education, historical period and culture. Now I would like to elaborate on four of these—intelligence, higher education, gender and age.

Intelligence, Creativity and Poetic Creativity

Over the years, evidence has mounted that creativity is an ability distinct from intelligence, at least as defined by IQ tests (Sternberg, 1988; see also Sternberg et al., this volume). Nevertheless, creativity and intelligence are positively related. Before persons can be creative in fields such as mathematics or architecture, they need above-average intelligence, that is, an IQ higher than 100. Otherwise, they will be unable to master the body of knowledge required (Simonton, 1984).

But mathematicians, architects, or writers with IQs above those of their colleagues will not necessarily be more creative. In fact, Barron (1969) reports findings at the IPAR which reveal that outstanding and average creators in all fields studied have almost identical IQs. The most eminent writers have IQs no higher than their less illustrious colleagues. Simonton (1984) summarizes the relationship of IQ to creativity in the following way:

> There is a positive relationship between intelligence and creativity . . . but it tends to vanish in the upper reaches of intelligence. Beyond an IQ of around 120, further gains in IQ do not increase the likelihood of creative achievement. An IQ of 120 is not a very selective cut-off point: it marks the average intelligence of college students . . . and about 10% of the general population has an IQ of 120 or higher (p. 45).

The implications are important for the education of our children because, for generations, educators have valued intelligence more than creativity. This can be seen in the use of the term 'gifted'. Educators almost always apply it to students with high IQs. Whether the students also have high creativity or not has little bearing.

In North America, an elementary school student who obtains extraordinarily high marks will be given an IQ test, or some other kind of similar test with definite right or wrong answers, to determine whether the child is truly very intelligent or is an overachiever. If the student scores high on the independent test(s), he or she will be deemed gifted. Creativity tests, in which there are no right or wrong answers, are rarely given in such circumstances.

During visits as a poet to hundreds of schools, I have found that the natural urge for poetic creativity will spring to life in even the most depressed children, ones about whom anxious teachers have warned me beforehand. In a free-spirited, non-threatening workshop environment, such children soon will be not only reading and enjoying all kinds of poetry, but writing

their own as well. In such sessions, the so-called 'at risk' children often reveal heretofore hidden above-average capacity for what Root-Bernstein and Root-Bernstein call 'thinking with feeling' (see this volume). Dozens of teachers have privately told me how surprised they were that some of their worst students wrote the best poems. A number of other writers who have conducted workshops in schools report similar experiences (Swede, 1990).

What this shows is the importance of inserting creativity sessions (painting, dance, music as well as poetry) into the school curriculum, ones in which there are no grades and, thus, no failures. Given such an opportunity, the poorly performing students, who have practically given up in their regular classes, might discover above-average creative abilities. The subsequent improvement in feelings of self-efficacy would likely result in better school marks.

Higher Education and Poetic Creativity

According to Billy Collins, a recent U.S. poet laureate, over 200 master of fine arts programs offer an education in how to write poetry ('Pushing Poetry', 2001). Seemingly, higher education is necessary for creating poems that people will want to read. Ironically, this might be the case for modest, but not for exceptional, universal creators. Simonton (1984) has done historiometric studies on the greatest achievers in science, the arts, humanities and leadership which suggest that the relationship between education and success is not at all straightforward. After examining the educational backgrounds of a large variety and range of universal creators who lived between 1450 and 1850, he found that three to four years of college seemed the optimum for the greatest achievements. Or, put another way, persons with *fewer* than three or four years of college as well as those with *more* than three or four years were universal creators of lesser rank. In fact, the lowest-rated creators were those with doctorates. Simonton also examined the educational levels of 125 20th-century exceptional individuals in the arts and humanities and found an almost identical pattern. But his analysis of 20 modern scientists revealed a slight change from the earlier sample. Most went on to do graduate work, but without getting a degree.

A possible explanation for these findings is that a few years of university stimulate creativity and that further training causes too much dependence on traditional perspectives. Such reasoning is dangerously close to being circular, however, for it goes little beyond what the data reveal in the first place. Simonton offers a better interpretation:

> Perhaps the most securely directed and confident geniuses go to school only until they obtain the required knowledge and technical ability, and then quit. Creators of the highest rank tend to split off from conventional perspectives, and they may

discover that higher education does not contribute to goals that lie outside the mainstream. Their less illustrious colleagues, who attain fame not so much by changing the course of history as by advancing it, find that formal education improves their opportunity for achievement. It is important to note that . . . graduate education is (not) detrimental in any direct way to the development of the highest creative potential, but only that it may be irrelevant (p. 73).

This interpretation only has a partial fit for poets. An exceptional physicist will gain the knowledge he or she needs and then quit higher education and go to work as a physicist. But poetry's low popularity makes it necessary for elite universal poets to find other ways to support themselves. None can make a living through sales of books, tapes, and videos, or from fees for public appearances. Therefore, many elite universal poets, like Billy Collins, stay in school in order to qualify for employment in departments of English at universities where they teach literature and usually creative writing as well. Others continue with studies not related to poetry in order to support themselves. William Carlos Williams became a medical doctor to sustain his poetry career. Wallace Stephens became a lawyer. In other cases, a poet will find someone else to support them, as with Anne Sexton who married a prosperous businessman. Sometimes a poet, such as Gwendolyn MacEwen, will do other kinds of writing to supplement her income.

Gender and Poetic Creativity

Most psychological studies of universal innovation continue to be based on men (see this volume: Root-Bernstein & Root-Bernstein, Simonton, Sternberg et al., Vandervert, Weisberg). The chief reason is that exceptional creators have been, and continue to be, almost exclusively male in the domains of science, technology, business, politics and many, but not all, of the arts. Of course, this does not mean that women are uncreative. No evidence exists to show that women are less curious than men, that they have less desire for novelty and change. But there is much data to show how powerful social and political pressures have restricted women's opportunities chiefly to personal one and two kinds of creativity, ones that according to traditional views are unacknowledged as creative: child-rearing, teaching, entertaining, sewing, cooking, gardening and charity work (Swede, 1993)

One has only to look at the history of women's rights to see how recently political circumstances have improved. Less than 100 years ago, in 1907, Norway granted women the right to vote and then was followed by other Western countries over the next decade or so (Swede, 1993). While such legislation changed women's legal rights, social pressures continue to limit the opportunities of women in business, science and technology (Bandura, 1997; Gornick, 1983). Most

women still learn during childhood and adolescence that some occupations are more suitable for men. They internalize such distinctions in spite of wide-ranging efforts by educators, business and governments to encourage girls to consider suitable careers in science and technology (Bandura, 1997; Gornick, 1983). If women do decide to become scientists, their salaries, however, will be below those of men with a similar education and job title (Gornick, 1983).

In the arts, women have fared better, at least according to what we see in the popular media. Female actors, singers and musicians seem to be as successful as their male counterparts, if not more so. This is quite a leap forward considering that Western women were not allowed on stage until the 17th century (Montagu, 1968). But apart from these impressions, no comprehensive data exist on either male or female artists. One reason seems to be that most poets, novelists, painters, and musicians, whether women or men, do not earn enough from their craft and must make money doing other things, with the result that their artistic earnings become difficult to locate in government files on income (Swede, 1993). Another difficulty is the lack of specific criteria for the term 'artist'. How do we classify the sound person on a movie set—as a scientist or artist? What about newscasters: are they actors or not? And are the writers of the memorable jingles for retail products lyricists/composers or business people?

To get a sense of how women are doing in poetry, I tabulated the number of males and females who are in the *League of Canadian Poets-Membership List, 2001/2002*. The full membership criteria, as described in *The League By-Law and Guide For Members*, fit well with my working definition of an average universal poet described earlier insofar as eligibility is ultimately based on 'sufficient poetic achievement' (p. 5). Of the 304 full members, roughly 55% were women. Clearly, in Canada at least, women have taken giant strides in the domain of poetry, their numbers approximating those in the general population (Census, 2002; Statistics Canada, 2001).

The focus of this chapter is, however, on eminent universal poets, ones who go beyond merely being sufficient achievers. Getting data about such persons is a daunting task because one cannot find agreement about whom to include on this list, for no by-laws or membership criteria exist. I finally chose to focus on the Poet Laureates of the United States and England because the selection process is very formal, involving numerous checks and balances. In England, the office was created in 1591 and became official in 1668 and has always been awarded for life. At the time of the Laureate's death, the prime minister nominates a number of possible successors from which the reigning monarch chooses one (Poets Laureate of England, 2002). The U.S. criteria have undergone a number of changes since the office originated in 1937. Today, the rules governing the position are as follows:

The Poet Laureate is appointed annually by the Librarian of Congress and serves from October to May. In making the appointment, the Librarian consults with former appointees, the current Laureate, and distinguished poetry critics (Poet Laureate Consultant in Poetry, 2002, p. 3).

Of the 23 English Laureates, official and unofficial, *none* have been women (Poet Laureates of England, 2002) and of the 41 U.S. Laureates, only eight have been female (Poet Laureate Consultant in Poetry, 2002). These data reveal that gender equality does not extend to the upper levels of achievement. Such a finding is likely due to the different things boys and girls learn are important in life:

> The male instrumental role, with its goal and achievement orientation, helps males succeed in work, while the female expressive role, with its emphasis on nurturing and dependency, may not. As a result, women may lack some of the skills, values, and attitudes toward the self that would lead them to the highest levels of occupational achievement (Lemme, 2002, p. 323).

In the literary arts, however, male dominance of the highest rungs of accomplishment is somewhat surprising in light of the long-standing evidence of a slight female superiority in verbal skills (Joseph, 2000). What this demonstrates is the pervasive influence of gender role socialization.

Age and Poetic Creativity

At what age are the exceptionally creative most likely to do their best work? The answer will depend, to some extent, on the kind of activity. Eminent achievement in dance surely reaches an apex before that in more sedentary activities, such as star-gazing or writing poetry. While age and achievement have yet to be studied in many fields—dance and astronomy included—enough data exist to draw some tentative, general conclusions. Creativity, in both quality and quantity, usually peaks around the age of 40, give or take five years (Dennis, 1966; Lehman, 1953; Simonton, 1984). This usually follows a pattern: creativity rapidly rises to this peak and then declines gradually, with the consequence that about one-half of all significant contributions to culture are made by persons 40 years or older (Dennis, 1966; Lehman, 1953; Simonton, 1984).

Another aspect is the relationship of quality to quantity. According to Dennis (1966), the ratio of major to minor achievements remains roughly the same throughout adulthood, and therefore the periods of greatest output also contain the greatest number of significant works. Recent data gathered by Weisberg (this volume) suggest a more complicated relationship, i.e. early major works might be of lesser quality than

those which arise when the universal creator has mastered the required skills more fully.

Dennis (1966) also found differences among the areas of scholarship, science and the arts. Productivity in the arts peaks, as well as declines, earlier than in the sciences and scholarship. Artists tend to be more productive in their twenties than scientists and scholars and less productive in their seventies. In addition, the peak periods in some disciplines within scholarship (history, philosophy) and the sciences (botany, invention, mathematics) occur during the decades of the fifties and sixties.

Dennis provides widely accepted explanations for these differences. He believes that the early peak in the arts is due to less demanding educational requirements. Individuals in the sciences and scholarship must digest vast amounts of formal knowledge before they can conduct original experiments or write a book, whereas those in the arts do not need such extensive background information before putting brush to canvas, foot to stage or pen to paper.

The faster decline in the arts, Dennis feels, can be explained by different working styles. Many artists—painters, novelists, poets, composers—are universal one creators in that they typically work alone. Most scientists and scholars are universal two creators because they get the assistance of graduate students, or they collaborate with colleagues who have the same goals. Working in such teams likely conserves energy and time and thus prolongs creative life.

Explanations, of course, are never that simple. Dennis found that poets peak in the forties and novelists in the fifties. Simonton (1975), in a more precise comparison, found that poets reach their peak at an earlier age than novelists, 39 vs. 43, and provides an explanation for why a difference of four years is significant:

> It is large enough to help explain why the reputations of poets may survive even if the poets die at younger ages than do novelists . . . . Twice as much of a poet's lifetime output comes from the twenties as is the case for novelists (p. 101).

Both Dennis and Simonton suggest that the reason for this difference is that a novelist requires more life experience to write a novel than a poet does to construct a poem.

A further complication is the fact that many of the arts are as collaborative as the sciences (or even more so). In fact, the most popular of the arts—music and film—always involve universal two creativity in the production phase. A Broadway musical or Hollywood film might start in the mind of an individual composer or writer, but in the end they involve a large number of individuals who are also being creative: the producers and directors, the musicians, the singers, the actors and the various kinds of stage and movie technicians. To date, no universal two creators have been the subjects of study. Perhaps, because they engage in collaboration, directors, actors and musicians might have longer creative lives than painters, poets and novelists, thus limiting the conclusions of Dennis.

Also, what happens to creativity when a person changes fields? Does a poet who becomes a novelist lengthen his or her creative life? Does a novelist who becomes a poet shorten it? Simonton (see this volume) suggests that both will start with fresh creative potential in their new careers and follow the typical productivity pattern for those endeavors described earlier. Personal observation appears to confirm this view. For instance, Margaret Atwood began as a poet, publishing a pamphlet in 1961, and a book-length collection in 1966, which won Canada's prestigious Governor General's Award. Eventually, she switched most of her energy and time to writing novels and other prose for which she first received worldwide acclaim in the 1980s (Benson & Toye, 1997). But the changeover has not been complete for, in 1995, she published a collection of poetry. Simonton (this volume) believes that such back-and-forth flexibility sustains creativity.

Concluding Remarks

This chapter has looked at what we know about how someone becomes an eminent universal poet. Was information conveyed that would be helpful to students of poetry, to aspiring poets and to literary scholars? My guess is both yes and no. Apart from those who believe that great poetry comes from some sort of divine inspiration, most will not be surprised to learn that to become exceptional, a poet must have experienced early nurturing of verbal skills as well as reinforcement for subsequent poetic displays. Students, poets and scholars will appreciate, however, the attempts to explain the details of how this happens. In particular, they should find Bandura's (1997) efficacy theory helpful in explaining why someone will continue to write poems in a culture that does not value poets as much as it values other kinds of artists, such as novelists, film-makers and musicians.

Examined too were the things a poet must do to remain outstanding. Here again some findings merely confirm the obvious while others do illuminate. That the elite universal poet is driven to work hard and to have an output of high quality and, typically, also of high quantity, will not be stunning revelations to fellow poets and literary critics. Neither will the conclusions that the exceptional poet possesses high standards, a need to prove oneself, a strong sense of reality and fierce independence. However, the suggestion that these characteristics develop as a result of success, rather than being causes of it, should be newsworthy. Not so enlightening will be the finding that personality might determine the kind of poetry someone chooses to write rather than the reverse. But, especially interesting should be the revelation that madness is no more prevalent among super universal creators than it is among the general population.

This chapter also points out the importance of background influences. The finding that creativity and intelligence are related and yet ultimately distinct definitely will be of interest. That poets do not need to finish a university degree in literature or creative writing in order to produce poems will not come as a surprise, nor will the scarcity of women among eminent poets. The relationship of age to creativity might raise some eyebrows in poetry circles for it is not generally known that poets bloom earlier and fade sooner than scientists and scholars (see Simonton, this volume).

Of course, a few questions remain unanswered. They are mainly about the precise nature of the creative process—how poetically meaningful associations are formed, step by step. As a result of some ingenious studies involving practicing poets, Perkins (1981) concludes that the making of poems is mainly a trial-and-error process and occurs on the conscious rather than unconscious level—findings many poets will not greet kindly. Ultimately, we need to know what happens at the level of brain regions and neuronal circuitry. Data from such research are imminent (see Vandervert, this volume) and will provide a stringent test for Perkins' conclusions.

References

Allen, G. W., Rideout, W. B. & Robinson, J. K. (1965). *American poetry*. New York: Harper & Row.

Amabile, T. (1983). The social psychology of creativity: A componential conceptualization. *Journal of Personality and Social Psychology*, **45** (2), 357–376.

Amabile, T. (1996). *Creativity in context*. Boulder, CO: Westview Press.

Bandura, A. (1997). *Self-efficacy: The exercise of control*. New York: W. H. Freeman and Company.

Barron, F. (1969). *Creative person and creative process*. New York: Holt, Rinehart and Winston.

Barsalou, L. W., & Prinz, J. J. (1997). Mundane creativity in perceptual symbol systems. In: T. B. Ward, S. M. Smith & J. Vaid (Eds), *Creative Thought: An Investigation of Conceptual Structures and Processes* (pp. 267–307). Washington, D.C.: American Psychological Association.

Benson, E. & Toye, W. (Eds) (1997). *The Oxford companion to Canadian literature* (2nd ed.). Toronto: Oxford University Press.

Bloom, H. (1973). *The anxiety of influence: A theory of poetry*. New York: Oxford University Press.

Cacciari, C., Levorato, M. C. & Cicogna, P. (1997). Imagination at work: Conceptual and linguistic creativity in children. In: T. B. Ward, S. M. Smith & J. Vaid (Eds), *Creative Thought: An Investigation of Conceptual Structures and Processes* (pp. 145–177). Washington, D.C.: American Psychological Association.

Chukovsky, K. (1971). *From two to five* (M. Morton, Trans.). Berkeley, CA: University of California Press. (Original work published in 1925).

Cox, C. (1926). *The early mental traits of three hundred geniuses*. Stanford, CA: Stanford University Press.

Dennis, W. (1966). Creative productivity between the ages of 20 and 80 years. *Journal of Gerontology*, **21**, 1–8.

Dickinson, E. (2000). *The complete poem of Emily Dickinson (with an introduction by Martha Dickinson Bianchi).* (Online). Available: http://www.bartleby.com/br/113.html (originally published in 1924 by Little, Brown, and Company).

Ferris, P. (1978). *Dylan Thomas*. Harmondsworth, U.K.: Penguin Books.

Gornick, V. (1983). *Women in science*. New York: Simon & Schuster.

Holyoak, K. J. & Thagard, P. (1999). *Mental leaps*. Cambridge, MA: The MIT Press.

Houston, J. P. & Mednick, S. A. (1963). Creativity and the need for novelty. *Journal of Abnormal and Social Psychology*, **66**, 137–141.

Jamison, K. R. (1995, February). Manic-depressive illness and creativity. *Scientific American*, **272**, 62–67.

Joseph, R. (2000). The evolution of sex differences in language, sexuality, and visual-spatial skills. *Archives of Sexual Behavior*, **29** (1), 35–66.

Kanfer, S. (1986). King of horror. *Time*.

Kenner, H. (1971). *The pound era*. Berkeley, CA: University of California Press.

Koestler, A. (1964). *The act of creation*. London: Pan Books.

League of Canadian Poets—Membership list, 2001/02 (2002). Toronto: League of Canadian Poets.

Lemme, B. H. (2002). *Development in adulthood* (3rd ed.). Boston, MA: Allyn & Bacon.

Lykken, D. T. (1998). The genetics of genius. In: A. Steptoe (Ed.), *Mind: Studies of Creativity and Temperament* (pp. 15–37). New York: Oxford University Press. Abstract from: APA/PsycINFO Accession Number 1999-02186-001.

MacEwen, G. (1980). The fourth dimension. In: J. Pearce (Ed.), *Twelve Voices* (pp. 63–73). Ottawa: Borealis Press.

MacKinnon, D. W. (1978). *In search of human effectiveness*. Buffalo, NY: Creative Education Foundation.

Martindale, C. (1973). An experimental simulation of literary change. *Journal of Personality and Social Psychology*, **25** (3), 319–326.

Martindale, C. (1975). *Romantic progression: The psychology of literary history*. Washington, D.C.: Hemisphere.

Martindale, C. (1990). *The clockwork muse*. New York: Basic Books.

Middlebrook, D. W. (1991). *Anne Sexton: A biography*. Boston, MA: Houghton Mifflin.

Milosz, C. (1983). *The witness of poetry*. Cambridge, MA: Harvard University Press.

Mishima, Y. (1959/1968). In: J. Bartlett (Ed.), *Bartlett's familiar quotations* (p. 1081). Boston, MA: Little, Brown.

Montagu, A. (1968). *The natural superiority of women*. London: Collier-Macmillan.

Patrick, C. (1935). Creative thought in poets. *Archives of Psychology*, (178).

Paz, O. (1974). *Alternating current*. London: Wildwood House.

Perkins, D. N. (1981). *The mind's best work*. Cambridge, MA: Harvard University Press.

Poet Laureate Consultant in Poetry (2002). (Online). Available: http://lcweb.loc.gov/poetry/laureate.html

Poets Laureate of England, 1591–2002 (2002). (Online). Available: http://www.hycyber.com/CLASS/laureate_uk.html

Pushing poetry to lighten up—and brighten up (2001). *Newsweek*, 58.

Roe, A. (1952/1982). A psychologist examines sixty-four eminent scientists. In: P. E. Vernon (Ed.), *Creativity* (pp. 43–51). New York: Penguin Books.

Schickel, R. (1985). *The Disney version*. New York: Simon & Schuster.

Schubert, D. S. P. & Biondi, A. M. (1977). Creativity and mental health: Part III—Creativity and adjustment. *The Journal of Creative Behavior*, **II** (3), 186–197.

Simonton, D. K. (1975). Age and literary creativity: A cross-cultural and transhistorical survey. *Journal of Cross-Cultural Psychology*, **6**, 259–277. Abstract from: ERIC File EJ125293.

Simonton, D. K. (1984). *Genius, creativity and leadership*. Cambridge, MA: Harvard University Press.

Simonton, D. K. (1991). Personality correlates of exceptional personal influence: A note on Thorndike's (1950) creators and leaders. *Creativity Research Journal*, **4** (1), 67–78.

Simonton, D. K. (1999). Talent and its development: An emergenic and epigenetic model. *Psychological Review*, **106** (3), 435–457.

Skinner, B. F. (1972). On 'having' a poem. *Saturday Review*, pp. 32, 34, 35.

Statistics Canada 2001 Census (2002). (Online). Available: http://www12.statcan.ca/english/census01

Swede, G. (1976). (Metaphoric utterances by my two sons prior to the age of seven). Unpublished raw data.

Swede, G. (Ed.) (1990). *The universe is one poem*. Toronto: Simon & Pierre.

Swede, G. (1993). *Creativity: A new psychology*. Toronto: Wall & Emerson.

The League by-law and a guide for members (1988). Toronto: The League of Canadian Poets.

Ward, T. B., Smith, S. M. & Vaid, J. (1997). *Creative thought*. Washington, D.C.: American Psychological Association.

Weisberg, R. W. (1986). *Creativity: Genius and other myths*. New York: W. H. Freeman.

Whitman, W. (1969). *Leaves of grass: Selections form a great poetic work* (Selections and supplementary material by Francis Griffith). New York: Avon Books. (Original work published in 1892.)

Williams, W. C. (1967). *I wanted to write a poem: The autobiography of the works of a poet* (Reported and edited by Edith Heal). Boston, MA: Beacon Press.

Winner, E. (1982). *Invented worlds: The psychology of the arts*. Cambridge, MA: Harvard University Press.

Woodman, R. W. (1981). Creativity as a construct in personality theory. *Journal of Creative Behavior*, **15** (1), 43–66.

The International Handbook on Innovation
Edited by Larisa V. Shavinina

Directions for Innovation in Music Education: Integrating Conceptions of Musical Giftedness into General Educational Practice and Enhancing Innovation on the Part of Musically Gifted Students

Larry Scripp[1] and Rena F. Subotnik[2]

[1] *New England Conservatory and Conservatory Lab Charter School, USA*
[2] *Center For Gifted Education Policy, American Psychological Foundation, USA*

Abstract: Five strategic steps posited here suggest a course of action that can lead to systemic and sustainable innovation in public-school music education: (1) reconcile perspectives regarding the purpose of music education; (2) establish a framework for comprehensive, interdisciplinary programs that are intended to benefit all children; (3) prepare advanced college-conservatory students to contribute to the development and sustainability of innovative programs as 'artist-teacher-scholars'; (4) form networks of college-conservatory, arts organization, and public school partnerships; and (5) explore and promote new conceptions of giftedness that result from the implementation of innovative forms of music education practices in public schools.

Keywords: Innovation; Music education; Innovative programs; Giftedness; College-conservatory networks; Learning through Music; Music in Education; Artist–Teacher–Scholar; Conservatory Lab Charter School.

Introduction

Only an inventor knows how to borrow, and every person is or should be an inventor (Ralph Waldo Emerson).

Research conducted during the last several decades suggests new frameworks for understanding the relationship between conceptions of giftedness in music and innovative music education practices in public schools. Two path-breaking approaches to music education have demonstrated that ordinary young children can develop musical performance ability at a level formerly associated with giftedness in music. The two approaches include: (1) a musical 'language acquisition' system (such as Suzuki's (1981, 1969) violin study method[1] that builds on the cumulative memorization of progressively more difficult repertoire); and (2) a 'language literacy' system (such as

---

[1] See Suzuki's *Ability Development from Age Zero* (1981) and *Nurtured by Love* (1969) for descriptions of his 'mother tongue' approach to learning the violin and violin music, an approach which Suzuki claims makes musical skill development analogous to language acquisition. Hence the characterization of his teaching as 'talent development' allows him to claim that prior musical talent has relatively little to do with the success of his violin method. Despite the worldwide acclaim for unprecedented success in training young musicians to memorize violin repertoire at an early age, Suzuki's approach is thought to be too narrowly focused on instrumental skills and does not lend itself well to general musical abilities, including the ability to read, compose, or analyze music.

485

Kodály's (1974) choral method[2] based on the ability to use syllables to sight-sing progressively more difficult materials derived from folk songs). Taken together, these practices have greatly expanded educators' and parents' notions of every child's musical potential (Davidson & Scripp, 1989).[3]

For those unaware of recent innovations in violin instruction, witnessing groups of young children performing violin concertos together remains astonishing. However, parents who support their children in Suzuki violin classes for a few years can take for granted that their children will acquire a level of violin performance repertoire previously demonstrated only by the most advanced children. Similarly, all parents who enroll their children in Kodály classes now can expect their children to develop basic literacy skills that most adults make no pretense of mastering. Although the ability to read books is expected of every child, most adults still believe that only a select number of musically gifted children can sight-sing choral pieces without prior rehearsal or accompaniment (a cappella). In short, measures of basic intelligence expected in other modes of education—the ability to read and recite familiar literature—have widely been considered as exclusively within the purview of education for musically gifted children.

Innovations in pedagogy have made the study of violin playing and music reading accessible to all young children who engage seriously with these pedagogical approaches, yet neighborhood schools remain remiss in providing these opportunities. The lingering impression that musical skill exists as a measure of a special talent keeps music in the box of special education for the gifted. Parents who can afford violin lessons for their children know that all children can benefit enormously from this experience, yet little impetus exists for making high-quality musical education available to all children.

Failure to incorporate proven innovations in music education also blinds us to the contributions that learning in and through music can make to both the ordinary and the extraordinary public-school student. Environmental 'conditions of support' that can be provided by virtually any family are known now to predict children's musical development (Davidson & Scripp, 1994).[4] Yet the gap in attitudes toward musical education is widening among parents. One group of parents assumes that young children who do not have access to, or who do not seem to benefit from, formal lessons in violin or music reading early on cannot acquire significant musical ability at a later time. These parents often see no need for their children to pursue any form of music education as part of their public school education. Another group assumes that young children who do profit from early music lessons are by definition musically gifted, confusing normal musical ability with extraordinary talent. A third group of parents understands that music should be an essential element of education, regardless of early indications of musical ability. This group welcomes the innovations of the past century that affirm the view that any child can develop musical abilities and that music education provides important opportunities for cognitive, aesthetic, social-emotional and neurological development far beyond the relatively narrow range of musical performance abilities previously associated with musical talent.

If we mistake early violin performance or music-reading skill as signs of giftedness, we will continue to marginalize music's role in education, and also misidentify those with truly exceptional abilities. Until schools incorporate the music education innovations of the past century into their curriculum and assessment practices, new forms and assessments of learning in the field of education in music will not evolve at the pace of other domains such as language, science, or mathematics.

Breaking the Mold of Music's Role in Public Education

The purpose of this chapter is to demonstrate that the current era of school reform provides a new context for changes in music education. Emerging examples for innovation have resulted from the need to meet three different challenges:

(1) *To develop a more comprehensive view of music as having value for its own sake.* Only by exploiting and assessing a wider variety of learning processes

---

[2] See *The Selected Writings of Zoltán Kodály* (1974) for descriptions of the training of early literacy skills as the basis for a nationwide choral program in Hungary. The success of his sight-singing methods led to its dissemination in the United States in the 1970s which resulted in modest scale studies of the effect of music reading through sight-singing on academic learning that have been replicated in the 1990s (Gardiner, 2000; Gardiner, Fox, Knowles & Jeffrey, 1996). The Kodaly approach remains a desired professional certification, yet whole school or district implementation in the United States remains elusive, partly because of its focus entirely on choral repertoire for young children and sight-singing skills based on a strict and somewhat inflexible progression of folk tunes.

[3] In Davidson & Scripp's 'Education and development from a cognitive perspective' (1989) innovations in music pedagogy are assessed in relation to developmental frameworks for musical development that emerged from their work at Harvard Project Zero.

---

[4] In this article Davidson and Scripp argue that social and environmental conditions that stimulate musical cognitive development in early childhood play a predominant role in the development of giftedness in music.

in traditional music classes[5] (Gardner, 1989b; Winner, Davidson & Scripp, 1992; Wolf & Pistone, 1991) will music gain status in the public-school curriculum equal to other core subject areas;

(2) *To incorporate musical content and processes as resources for interdisciplinary learning that is authentic both to music and other subject areas.* Music integrated with other subject areas transcends the traditional boundaries of music as a separate field of study and provides important occasions for 'teaching for transfer' across disciplines (Catterall, 2002; Perkins, 1992; Salomon & Perkins, 1989; Tunks, 1992)[6] based on fundamental concepts and learning processes shared between music and other subject areas (Bamberger, 2000; Burnaford, Aprill & Weiss, 2001; Gardiner, 2000; Perkins, 1989; Scripp, 2002).[7] Only when learning in music is integrated with other subject areas will public schools be able to afford comprehensive and authentic music programs for every child;

(3) *To establish new benchmarks for giftedness in the domain of music.* Determination of musical giftedness will have to keep pace with innovative practices now available to public schools. These practices are based on a deeper and wider view of

learning in music and its application to learning in other subject areas. New expectations for excellence in music in public schools will challenge educators to re-conceptualize conditions and indications of musical giftedness (Davidson & Scripp, 1993) that are both discipline-specific and interdisciplinary in nature.

The impact of innovation also must be evaluated by its degree of application in public schools. For example, if music educators continue to offer the teaching practices envisioned by Suzuki and Kodály only as after-school programs or as an option for interested students once or twice a week, neither the essence nor the scope of music's role as part of the core public-school curriculum will have been realized. Although these innovative practices of the past influenced a whole generation of music educators—both Suzuki's & Kodály's work and writings spawned the growth of professional societies of teachers, professional development strategies, conferences, publications, and instruments—public schools were never able to justify the inclusion of both of these programs into the core public-school curriculum.

However, circumstances emerging from this era of school reform have created new opportunities for change in music education. Recent national and state standards for music education have challenged school communities to consider music both as a basic intellectual capacity to be learned for its own sake (Gardner, 1983)[8] and as a tool for positive enhancement of academic and social-emotional development (Butzlaff, 2001; Hetland, 2000a,b; Scripp, 2002; Standley, 1996; Vaughn, 2000).[9] Consequently, innovations in music education most likely will stem from the need to create music programs that feature a broad synthesis of musical skills and processes employed both for the sake of the musical-cognitive development

---

[5] The Arts Propel Project, a collaboration among Harvard Project Zero, the Educational Testing Service, and the Pittsburgh Public Schools, stimulated the growth of alternative assessment processes (portfolio assessment, performance assessment) based on evaluating perceptual skills, learning products, and documentation of reflective thinking authentic to the musical, visual, and language arts. These alternative assessment models have become common in progressive public schools' curricula.

[6] In *Smart Schools*, (1992, pp. 99–130) Perkins argues that learning transfer does not often occur in schools without a conceptual framework and a consistent method for 'teaching for transfer' that draw on earlier seminal essay on transfer with Salomon (1989). Earlier research on music learning and transfer is summarized by Tunks (1992). James Catterall's article The Arts and the Transfer of Learning (2002) provides an up-to-date view of the research that supports a growing acceptance of expectations of learning transfer between the arts and other subject areas.

[7] Scripp (2002), Bamberger (2000), and Gardiner (2002) review research and present curricular design work which suggest that meaningful transfer that occurs between music and other subject areas occurs through the simultaneous employment of fundamental concepts and learning strategies shared among these disciplines. Perkins' (1989) essay is a particular view of standards of coherency necessary for discussing the arts as resources for a deeper understanding of scientific, literary, or historical concepts. Burnaford, Aprill & Weiss (2001) provide design perspectives from a group of arts educations and action researchers who develop curricula from the point of view of teaching for transfer through the arts.

[8] Gardner's theory of multiple intelligences (1983) is rooted in the premise of separate intelligences that constitute a particular profile of learners and argues for the teaching of disciplines for their own individual contribution to cognitive development; since the emergence of this theory, music has been seen increasingly as 'an early organizer of intelligences' that have particular impact on cognitive and emotional development outside of the domain of music.

[9] The new publication *Critical Links: Learning in the Arts and Student Academic and Social Development* (2002) contains the latest summary of a very large body of research on the impact of music and other aspects of learning and behavior. Scripp (2002) provides the summarizing essay, Hetland (2000a,b) provides meta-analyses of research on the 'Mozart Effect' and the relationship between musical training and spatial-temporal intelligence, Vaughn (2000) and Butzlaff (2000) summarize research on the impact of music on mathematical and language skills respectively, and Standley (1996) offers a meta-analysis of the impact of music on behavior, cognitive, and social-emotional development.

of every child and for the sake of music's contribution to the whole school curriculum.

Laboratories for Innovation

Innovations in music education are now emerging in laboratory school programs where music teachers, classroom teachers, curriculum specialists, and researchers work together—and in collaboration with consultants from schools of music—to design comprehensive, interdisciplinary music studies for the benefit of all children. These laboratory programs—not to be confused with magnet school programs that accept only musically gifted students—offer music to all students, regardless of ability or degree of initial support from parents. One such example is the Conservatory Lab Charter School, a school created in partnership with the New England Conservatory and its Research Center for Learning Through Music[10] (Street & Scripp, 2002). The mission of this public elementary school is to create a laboratory program that uses the intensive and interdisciplinary study of music to enhance cognitive, physical (neurological), and social-emotional development in all children. The mission of the New England Conservatory—in addition to the training of world-class musicians—is to focus on the effect of the synthesis of authentic and interdisciplinary-based musical studies on school environments.

The Lab School employs methods that draw on the innovations of Suzuki and Kodály, and music is awarded equal status with all core subjects of the school curriculum. That is, throughout the entire six years of elementary school, *all students* are required to:

• take weekly violin lessons and classes and perform in recitals several times a year;
• participate in a comprehensive music curriculum that features lessons based on musical works, music reading, composition, and improvisation;
• participate in special programs or events such as creating opera in schools and both music;

and academic teachers are required to:

• provide instruction in mathematics, science, social studies, and language arts that include 'Academic Enhancement Lessons',[11] both in the music class and in the academic classroom. These lessons result in the study of fundamental concepts and processes shared between music and other disciplines;
• design music listening programs that provide an array of listening experiences for the purpose of developing perceptual skills, priming students cognitively and emotionally for daily activities, finding new ways to make transitions in the school day emotionally satisfying, and serving as a resource for exercises in social or emotional development (Standley, 1996);[12]
• monitor musical development and music-integrated learning across the curriculum and provide evidence, by way of tests and portfolio assessment, of learning both in and through music.

As Lab School parents marvel at their children's ability to play the violin, read music at sight, or simply learn to match pitch, they discover, as Kodály and Suzuki suggested years ago, that all children can develop a significant degree of these abilities in the course of an enriched public-school education. They also witness students learning how to use their understanding of language and musical skills by setting words to music, or investigating the spatial-temporal relationships in musical composition and performance. In addition, these parents see teachers using music to prime and reinforce good behavior and listening skills. Children in violin ensembles begin to exhibit higher levels of self-control and social empathy; by choosing their own music at certain times of the school day, children reflect on the relationship between music they prefer and their own emotional well-being.

Assessment of a comprehensive array of laboratory school practices, not surprisingly, reflects a wider range of learning in and through music than do conventional music education programs. Test results reflect significant, positive correlations among tests of basic musical and academic skills (Gardiner, 1993, 2000; Gardiner & Scripp, 2002), suggesting that music, for very young children, is a tool for developing symbolic processing skills as well as exploring other concepts shared between music, mathematics, and language. As program development continues, the entire school

---

[10] *Learning Through Music* represents a research-based field of inquiry into the wide range of learning that occurs in the context of comprehensive study and experience of music that includes, yet extends beyond the boundaries of musical skill or perception into areas of cognitive or social–emotional development normally associated with other subject areas or contexts; frameworks for comprehensive, interdisciplinary *Learning Through Music* programs were first designed by faculty at New England Conservatory's Research Center and later used as the basis for program implementation and research with public-school partnerships principally including the Conservatory Lab Charter School, a public school founded for the purpose of developing and disseminating a replicable model of Learning Through Music practices to school children selected by lottery and not screened for musical ability.

---

[11] 'Academic Enhancement Lessons' are an essential ingredient of the Learning Through Music Program designed by the New England Conservatory's Research Center and implemented in collaboration with staff at the Conservatory Lab Charter School.

[12] Standley's (1996) meta-analysis of the impact of music on social-emotional development provides a basis for Music Listening Programs designed by New England Conservatory and implemented and evaluated in partnership with the Conservatory Lab Charter School.

Photo 1. This photograph was taken on the first day children received their violin at the Conservatory Lab Charter School. Violin instruction for all children in this public school was part of the Learning Through Music Program developed in partnership with New England Conservatory.

community understands better how the knowledge of music and a command of music symbol systems provide new perspectives of linguistic, musical, mathematical, and behavioral development of young children.

Five Strategies for Innovation in Music Education

The work of laboratory school programs such as the New England Conservatory/Conservatory Lab Charter School Partnership[13] suggests a strategic framework for innovation.

As inventive as one laboratory program may be, the evolution and large-scale dissemination of new approaches to music education depend on a shared set of principles and responsibilities to which music teachers, classroom teachers, and administrators willingly subscribe. The steps presented below and discussed throughout the rest of this chapter suggest the conditions by which the next phase of music education must evolve beyond musical innovations of the previous century. Already in process in several

communities, innovation in music education can be achieved by partnerships among schools of music, arts organizations, and public-school communities through a series of five strategic steps:

- Strategy 1: Reconcile the false dichotomy between the 'essentialist' and 'instrumentalist' perspectives regarding the purpose of music in public-school education;
- Strategy 2: Establish a framework for innovative forms of music education based on comprehensive, interdisciplinary programs that are intended to benefit all children in public schools;
- Strategy 3: Create new frameworks for preparing advanced college-conservatory students to contribute to innovative programs as 'artist-teacher-scholars';
- Strategy 4: Form a network of college-conservatory, arts organization, and public-school partnerships that are committed to designing and implementing innovative, research-based forms of 'music in education' practices, and to influencing policy concerning the essential role of music in the public schools;
- Strategy 5: Explore and promote new conceptions of giftedness that result from the implementation of innovative forms of comprehensive, interdisciplinary music education practices in public schools.

---

[13] Similar laboratory programs such as Georgia State University School of Music's 'Sound Learning' Program in Atlanta and Fulton County Public Schools are under way in several communities.

The Lesson Plan in Action   Logical-Quantitative Entry Point

Sample Student Work:

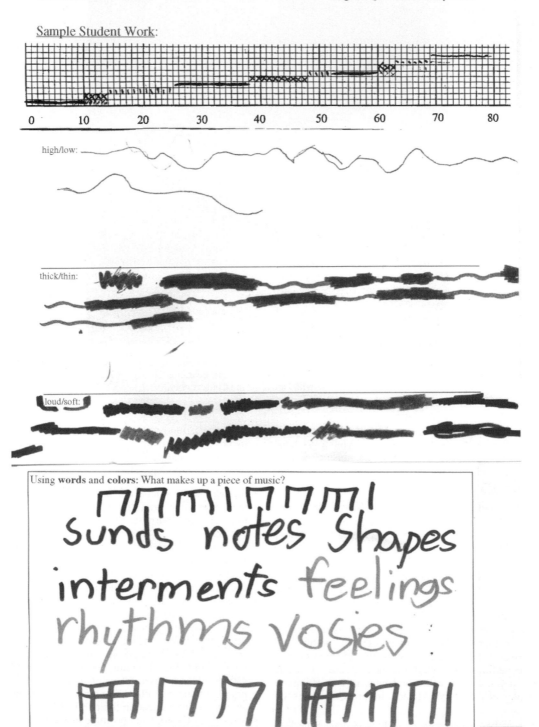

Photo 2. At the Conservatory Lab Charter School, Mona Rashad, a New England Conservatory violin major and Music-in-Education guided intern, documented children's 'multiple representations' of their response to music-listening tasks.

Strategy 1: Reconcile the false dichotomy between the 'essentialist' and 'instrumentalist' perspectives regarding the purpose of music in public-school education.

> What we must first seek to answer is whether music is to be placed in education or not, and what power it has ... whether as education, play or pastime (Aristotle).

A major assumption of this chapter is that innovation in the field of music education will take place only when musicians and educators are able to reconcile two philosophical strands that stand in each other's way. That is, music education must evolve out of the 'false dichotomy' between the 'essentialist' position of music education 'for its own sake' and the 'instrumentalist' position of 'music for education's sake' (Burnaford, Aprill & Weiss, 2001; Hope, 2000).[14]

Perhaps one reason policy-makers have been reluctant to support music as a core subject in public schools is that educators, administrators, artists, and parents seem to be divided in their advocacy. On one side, the 'essentialists' argue that music should be taught for its own sake. Essentialists maintain that, while evidence exists for several kinds of ancillary benefits from music instruction, music teachers should focus on music's own set of skills and literature and not be responsible for drawing out 'extra-musical benefits' from this instruction. By contrast, those sympathetic to the 'instrumentalist' point of view believe that music does not exist in a vacuum, that it is connected intrinsically to other subject areas and art forms, and that learning in music inevitably draws on and engages learning processes and fundamental concepts shared across many subject areas—often simultaneously.

As research emerges that establishes stronger relationships between music and learning in other areas of the curriculum,[15] advocates from both camps are caught in a complicated bind. Schools are under tremendous pressure to demonstrate increased academic achievement. Music had been viewed as dispensable even before the arrival of the current climate of school accountability. Clearly, if we ignore aspects of learning transfer between music and other subject areas, music education will remain outside of the mainstream of public education and will survive only as an educational elective for the talented or highly motivated few.

From the school-reform perspective, music educators are expected to cast music as an enhancement to learning in all subject areas, and as a support for social-emotional goals of the school curriculum. School administrators express *instrumentalist* objectives: music must support the learning goals of other areas of the school curriculum in the same way that mathematics enhances science, and reading is key to the social-studies curriculum.

Kodály and Suzuki teachers have shown how traditional music training can be disseminated effectively through deployment in schools, yet these music educators failed to convince a sufficient number of school communities that 'music for its own sake' is a reasonable and sustainable aspiration for all children. To the extent that the 'instrumentalist's' view prevailed as a rationale for innovative programs of the past, the deployment of these programs for the sole purpose of enhancing math or reading test scores became a symbol of how music's particular value in education could be trivialized. To the extent that these programs were seen from the 'essentialist' perspective only, school communities could ignore music's potential for positive interaction with learning in other domains.

Reconciliation of these opposing views will be a necessary condition for future innovation in education. Although music study takes place in isolation from mathematics and language learning in schools today, research cited above suggests that music functions as a catalyst for cognitive skills and social-emotional development across disciplines. These effects are most evident when conditions for transfer are optimized. Thus, effective music lessons depend on a general education background and they profit by analogous understanding of fundamental concepts of language (theme, character, phrasing, phonemic and phonological skills, decoding skills, etc.), mathematics (quantity, proportion, pattern, sequence, distribution, etc.), and higher-order cognitive skills (systems thinking, analysis, inquiry, invention, discovery, explanation, demonstration, interpretation, hypothesis-testing, reflection, etc.).

Reconciliation of the seemingly opposite purposes of music education will allow the public to recognize a new wave of innovation that transforms the practices and scope of music in public education.

A Taxonomy of Terms Emerging from the Reconciliation of the Essentialist and Instrumentalist Views of Music Education

- The term *music education* is defined for the purposes of this chapter as the extension of pedagogical and training practices based on traditional goals of music education 'for its own sake';
- The term *music-in-education* is offered here as a label for building on innovations in traditional music teaching, and, at the same time, committing music as an appropriate tool that enhances general education;
- The concept of *learning through music* signifies music education's effect on learning outside of the

---

[14] See Sam Hope's essay, 'Integrity in integrative programs: challenge to learning through music' in the *Journal for Learning Through Music* and Burnaford, Aprill & Weiss's *Renaissance in the Classroom* (2001), for further discussion of the ongoing tension about the role of music and the arts in education.

[15] Such as the *Critical Links* (2002) publication cited previously.

narrow parameters of musical performance, analysis, or aesthetic experience;

• *Comprehensive, interdisciplinary music program* is a phrase intended to capture music educators' intent to reconcile the essentialist and instrumentalist perspectives. Innovation takes the form of programs that are wide ranging in scope and, at the same time, wedded to serving the individual school's core curriculum. The comprehensive study of music is designed to benefit both 'music for its own sake' and music for the sake of learning across the curriculum.

The path of innovation in music education described earlier can be summarized by the matrix below (Fig. 1). This figure plots the evolution of music education from extra-curricular, performance-based programs (top left) to the far more inclusive forms of comprehensive, interdisciplinary programs (bottom right) that may emerge in the future as a result of research-based laboratory school programs. As we shall see later, these innovative programs will depend also on coherent conceptual frameworks for learning, teacher preparation, and cross-institutional collaborations with schools of music.

Strategy 2: Establish a framework for innovative forms of music education based on comprehensive, interdisciplinary programs that are intended to benefit all children in public schools.

> My father taught me that the symphony was an edifice of sound. And I learned pretty soon that it was built by the same kind of mind and the same kind of way that a building is built (from *Frank Lloyd Wright*, a film by Ken Burns and Lynn Novick, written by Geoffrey Ward, 1998, the American Lives Film Project, Inc. PBS Home Video).

Frank Lloyd Wright, America's best-known architect of the 20th century, learned to play the piano as a child and developed a view of music as a creative process directly related to his career work. Not only did Wright's son report hearing him play Chopin late at night, but his grandson recalls Wright's insistence on piping music into the fields and workrooms at Taliesin as his staff and apprentices did their work. The concerts, poetry readings, and plays performed every Saturday night at Taliesen, were, Wright believed, "as important as working in the drafting room".

Music educators may aspire to this level of influence, yet no public-school music programs have been designed to expect such a result. First, music educators will recognize that Wright's ability to play piano music by Chopin indicates that his musical education extended far beyond what is commonly expected of an American public-school student today. Second, educators in various academic subjects will recognize that Wright's strong assertion that musical composition and architecture are created by 'the same kind of mind' and

that the professions 'are quite similar', indicates a level of interdisciplinary understanding that is rarely required or assessed in American public schools. Comprehensive, interdisciplinary music programs promise more than innovative approaches to either goal: they offer a progressive view of music as a coherent curriculum that is taught to support learning transfer across the curriculum.

In order for an *interdisciplinary* music curriculum to gain a status equal with other core curriculum subjects, it must be deeply coherent from an intra-disciplinary perspective. In the manner of a rich language arts curriculum that features both the mechanics of reading and writing as well as the study and composition of various literary forms, a *comprehensive* music curriculum should include:

• training in various forms and traditions of performance including improvisation and standard repertoire using the voice and on more than one instrument;
• experience with computer-assisted music composing, drill and practice, and keyboard practice;
• exposure to a wide range of music literature and aesthetic experience through listening skills, concert events, artist-in-residence programs;
• expectations for fluent music reading skill; and
• time to investigate open-ended musical problems.

Although music educators, parents and administrators may endorse a more *comprehensive* music program for those with musical talent, schools have yet to invest in comprehensive programs for all children, let alone comprehensive, fully integrated interdisciplinary approaches to music education that serve the entire school community.

In a comprehensive, interdisciplinary program, music educators can claim that the study of music enhances understanding of basic mathematical concepts in much the same way that the study of mathematics makes for a clearer understanding of music. Singing words increases phonological awareness of language, as the ability to read enhances the study of musical repertoire. In this context, the study of opera provides a natural context for studying language (text setting), history (narrative), social studies (character), and mathematics (geometry of set design or the spatial-temporal aspects of organizing any performance event) (Wolf, 1999).[16]

Faced with ever-expanding lists of standards mandated by national and state boards of education, a comprehensive, interdisciplinary music program emerges as a much-needed innovation for both music training itself and for integrated learning in public schools. The argument for investing in innovative, comprehensive, interdisciplinary music programs is

---

[16] See Wolf's (1999) essay on the value of studying opera in terms of interdisciplinary skill development in the context of an interdisciplinary musical medium.

Changing Contexts for Innovation in Music Education

	Music Education 'for its own sake' (as a separate subject or intellectual domain)	Music in Education 'for the sake of learning in and across the curriculum' (both as a separate subject and as an interdisciplinary study linked with multiple forms of representation, concepts, and skills that are shared across other subject areas or intellectual domains)
Music Learning primarily assessed as a performance skill and offered as an extra-curricular option (The study of music as an isolated discipline or subject that is taught as a non-compulsory subject or provided as intensive performance training for the talented few)	Conventional Role Of Music Education In Public School Practices Of The Last Century (Innovations in instrumental, vocal, and music reading instruction in early music education created standards of musical performance for ordinary children formally associated with giftedness in music, yet most public schools ignored these innovations and continued to offer conventional or reduced music programs)	Experimental Interdisciplinary Music Programs Over The Last Century (Interdisciplinary music programs, offered in some private schools for all children (e.g., Waldorf Schools) were virtually absent in public school music programs of the previous century)
Music Education taught and assessed from the point of view of a comprehensive curriculum for all children (The comprehensive study of music as part of the compulsory core curriculum and supported by artist in schools programs or interns from schools of music that supports a diverse music education for every child and provides new opportunities for excellence for both ordinary and gifted children)	Innovation In Music Education That Supports A More Comprehensive View Of The Music Curriculum Available To All Students (Schools are now finding ways to adopt innovations in music technology and cross-cultural studies and, as a result, design and assess music programs based on more inclusive and comprehensive music curricula that include multiple-instrument instruction, voice and sight-singing classes, opportunities for computer-assisted composition and ear training-music reading drills, and a range of cross-cultural approaches to musical perception, performance, improvisation, composition, and reflective understanding of music)	The Evolution of Research-Based Comprehensive, Interdisciplinary Music Programs that Suggests a New Role and Purpose for Music in Public Education (Research that demonstrated the impact of high quality music programs on academic achievement and experimental laboratory school programs that have created innovations in early music education suggest that innovation in music education will take the form of "music in education" programs that require authentic forms of musical performance and reading skill, and, at the same time, provide comprehensive, interdisciplinary programs for learning in and through music for the benefit of every public school student)

Figure 1. A four-way comparison of the focus and scope of the music programs that suggests a path of innovation in music education from performance-based, non-compulsory curricula for the benefit of the talented few toward comprehensive, interdisciplinary programs designed for the benefit of every child in the school community.

made more feasible by charter-school legislation and the willingness of experts in the field of music education to design charter schools as laboratories for innovation.

Yet, in the school-reform world—where many claim they are working creatively on issues of curriculum and pedagogy—school leaders must proceed responsibly. If a school receives a charter based on the fundamental idea that music can be used as a powerful *medium* and *model* for teaching and learning across the curriculum, the whole school community must work together to dedicate itself as a laboratory for innovation.

The *comprehensive, interdisciplinary* program described below originated through the efforts of the

New England Conservatory faculty and a founding coalition that petitioned for, and received funding for, the Conservatory Lab Charter School, an elementary school dedicated to the development and dissemination of innovative *Learning Through Music* practices (Street & Scripp, 2002). The New England Conservatory created a Research Center to support innovations in music education while the laboratory school provided a platform for implementation in a public-school setting.[17] The basis for innovation in this context is best represented by a set of premises, each of which is necessary, yet not sufficient in itself, for the development of comprehensive, interdisciplinary music programs:

The Premises of the Conservatory Lab Charter School 'Learning Through Music' Program

(1) Music serves as both a medium and a model for learning in public schools.

- As a *medium*, musical literature and music-making skills serve as essential components of every child's elementary education;
- As a *model*, music represents a configuration of teaching and learning processes fundamental not only to music but to other disciplines by way of listening (perception, observation), questioning (investigating, analyzing), creating (inventing, transforming), performing (demonstrating, interpreting), and reflecting (connecting, personalizing, evaluating).

(2) The comprehensive and intensive study of music should be provided at all grade levels both for its own sake and as a field of study of equal importance with other academic areas.

- As part of the core public school curriculum, the study of music and the development of musical skill becomes part of the child's compulsory education. The child is evaluated according to standards of musical literacy, skill development, and knowledge of literature at all grade levels authentic to a conservatory preparatory education.

(3) Music should be used as a model and resource for interdisciplinary learning.

- Comprehensive, interdisciplinary music studies should include art forms intrinsically related to music, such as opera or ballet;
- Interdisciplinary music studies should address aspects of academic disciplines that share fundamental concepts with music. Some examples include the study of proportion in music and math, the study of themes and character in music and language arts, the study of historical periods from the point of view of music and social studies, and the study of wavelength or resonance from the perspectives of music and physics.

(4) Music should be employed as a medium and model for social-emotional development.

- Music listening should be used for exploring concepts of preference, choice, style, and cultural identity among students;
- The experience of music performance or composing (composing and performing a cappella vocal works, for example) should be used to model social-emotional skills such as empathy, perspective-taking, or cooperative behavior;
- The systemic home practice of music should serve as a model for self-discipline and the ability to set and evaluate personal goals.

(5) Music should be used as a model for curriculum development, teaching, assessment, professional development, and engagement of the whole-school community.

- Music becomes a model for interdisciplinary learning skills and interdisciplinary content throughout the curriculum;
- Music becomes a model for teaching that includes a rich array of teaching processes and modalities of learning;
- Assessing musical development can parallel assessing for such skills as phonemic awareness, diction, calculation, and problem-solving;
- *Learning Through Music* becomes a new form of teacher professional development as a way to re-evaluate and revitalize teaching in all subject areas;
- Music-making brings communities together in positive and constructive ways.

The *Learning Through Music* partnership with New England Conservatory stems from the need to create and disseminate a research-based, replicable approach to innovative curriculum development and assessment practices. At the Conservatory Lab Charter School (CLCS), the premise of *Learning Through Music* informs virtually every aspect of the school's development. One of the most pressing challenges is in the realm of professional development. At CLCS, music and classroom teachers must adapt to strategies and

---

[17] Larry Scripp, Chair of Music Education at New England Conservatory, created the founding coalition for the Conservatory Lab Charter School at the same time proposing a new Research Center for Learning Through Music at New England Conservatory. From its inception, the Lab School was proposed as a research-generating public school program where New England Conservatory research faculty (cited extensively in this article) serve as consultants and staff members, and conservatory 'Music-in-Education' students serve as guided interns.

Learning Through Music *Academic Enhancement* LESSON PLAN
(independently designed and implemented by either academic or music teacher)

Grade Level__1_____Subject Area(s) Music___Math___Language___Science

Lesson: The Dinosaur Music Book

List Goals (lesson objectives relevant to Literacy Focus Abstracts):
- Word segmentation
- Vocabulary (for Science Project)

List Chosen Fundamental Concept Shared Between Music and Math and/or Language Literacy
- Text Segmentation matched to word/melodic segmentation (of a well known song)
- Performance of created words/melody and verification of parallel structure of both

List Resources Need (materials, media, worksheets, etc.):
- Paper, pens, rhythm sticks, source materials for theme of the text
- Journal writing materials

List Sequence of Events:
- Read books on dinosaurs and discuss and categorize their various perceptions and reflections of their physical features
- Draw features on paper
- Compose lyrics relevant to the features of the dinosaurs that match the word or syllable segmentation of the words and music of "Mary Had a Little Lamb" onto text-setting sheets
- Have children perform each other's work and investigate whether the words are matching the melody and the content appropriate to the physical description of the dinosaur.
- Collect all the individual dinosaur pages and assemble them into a book
- Report on Journal writing on the process, goals, and personal meaning of the project

List Chosen Learning Artifact (journal writings, worksheets, work in progress, final products, demonstrations, performances, etc.):
- Journals
- Drawings
- Text Setting Sheets
- Class books

Describe Assessment Process and Expected Results (scoring rubric, student reflections test, teacher observation, etc.):
- Journal writing assessed for detail of perception, richness and specificity of reflective writing
- Text setting is assessed for vocabulary and spelling match between the observations and their incorporation into a song (syllable segmentation, accent, syntactical sense of the music. etc.)
- Pre-post test on dinosaur vocabulary, word segmentation, and melody reading for units, contour and accent

Figure 2. A 'Learning Through Music' Academic Enhancement Lesson Planning Sheet used at the Conservatory Lab Charter School. Unlike conventional lesson plans, Academic Enhancement Lessons can be implemented by both the academic teacher and the music teacher. The form requires the teacher to specify concepts shared between music, language and science that will be integrated throughout the lesson sequence. In this case the shared concept is limited to reinforcing academic content through musical song. Later examples illustrate how fundamental concepts such as measurement, proportion, and time can serve as shared "cognates" between math, music, and science.

Brontosaurus had a long
neck,
A very long neck, a very long
neck.
Brontosaurus had a long neck
That stretched to the tops of
trees.

Photo 3. A page from the 'Learning Through Music' Dinosaur Project as described in the Academic Lesson Plan above. First grade students remember facts about the Brontosaurus by composing new lyrics to the tune, 'Mary Had a Little Lamb' that illustrate the physical features of dinosaurs. See following examples for more sophisticated versions of Learning through Music "Academic Enhancement Lessons".

approaches that have not been a part of their previous teacher preparation and practice.

Strategy 3: Create new frameworks for preparing advanced college-conservatory students to contribute to the development and sustainability of innovative programs as 'artist-teacher-scholars'.

The mission of New England Conservatory is to become an advocate for music's role in our society. This mission is purposefully broad. It includes two primary aspects: first, to provide the best possible training of professional musicians to contribute to the highest standards of music in our culture, and second, to provide ways for musicians to become better communicators, teachers, and, in addition, effective advocates for the contribution music can make to enhance learning in other subject areas in our public schools (Daniel Steiner, President, New England Conservatory of Music).

Innovative music-education practices in schools will not survive without parallel innovation in teacher preparation programs. New England Conservatory, the

longest-running conservatory of music in the United States, is an example of how changes in teacher preparation for the most gifted music majors are needed to foster, evaluate, and sustain innovation in music education.

Until recently, music education has been treated as an area of narrow specialization in higher education. At most music schools, a sharp division has existed between those who will perform and those who will teach. Students who excel in musical performance have been encouraged not to 'dilute' their performance training by taking courses designed for music teachers or scholars; conversely, those with less potential in the performance arena are often relegated to the music education degree program. Because conventional music education offerings (which feature intensive preparation for band, choral, or orchestra directors) are not perceived as applicable to the goals of the aspiring performer, the most advanced performance majors rarely sacrifice precious rehearsal time to take music-education courses.

However, both the young musicians who develop successful careers in music and the performance majors who become interested primarily in teaching eventually will be expected to bring a high degree of artistic and scholarly sophistication to bear on their roles as educators. Some of the skills commonly required of today's music teachers and professional musicians engaged in education outreach programs, yet not sufficiently developed in conventional music education programs, include: (1) providing a comprehensive array of educational experiences (voice and instrumental instruction, improvisation, computer-assisted composing, cross-cultural approaches to music, etc.) that extend beyond conventional general music and performance ensemble classes; (2) designing and implementing interdisciplinary music programs that connect with school-wide curricula and serve all students at every level of musical skill; (3) campaigning effectively for providing educational experiences in collaboration with community arts organizations (e.g. artist-in-residence programs); and (4) educating policy-makers on the ways in which music enhances other types of learning by identifying relevant research and presenting findings to support the existence of music education. These skills require training more broadly based in *artistry* and *scholarship*—in addition to teaching—than what is typically offered in conventional music-education programs.

The gulf between the training of gifted musicians as performers and the need to prepare conservatory graduates as educators is revealed in a survey of recent graduates of the New England Conservatory. The results of this survey revealed an astonishing statistic: 85% of performance majors teach in some capacity, yet fewer than 5% had received pedagogical training.[18] At least four explanations for these results seem possible: (1) there is little opportunity for musicians to sustain their careers purely as performers, yet there is a terrific demand for music teachers with conservatory training; (2) successful performance careers often demand teaching responsibilities; (3) training as a performance major provides good preparation for teaching; and (4) teaching is a desirable and productive way to sustain development as a performer.

At the New England Conservatory, all four explanations appear relevant and, as a result, faculty—with support from the senior administration—decided to pursue a new framework for curricular reform in music education. They resolved to:

(1) regard the development of the Artist/Teacher/Scholar (ATS) as part of the core mission of the school;
(2) create a Music-in-Education Concentration as an institution-wide service to all performance majors who wish to prepare themselves sooner, rather than

later, for the integration of teaching and scholarship relevant to their career as musical artists.

The figure below suggests how the Artist/Teacher/Scholar framework integrates seemingly isolated strands of any musician's education or professional career. Young musicians invariably begin their musical studies strictly as performers. At first, students only view their private instructors as gifted artists who can also demonstrate how young performers can play their instrument better. The most gifted students know that schools of music determine the admission of students almost entirely on the basis of their performance audition. Although subsequent professional training includes required studies in music theory, history, and liberal arts, students at first experience these subjects as ancillary, if not irrelevant, to their career as a performing artist. By the third year of undergraduate training, however, it becomes increasingly clear to music students that the integration of these subject areas can contribute significantly to their aspirations for a career in music. With the added option of pedagogical or music-education internships, young musicians then begin to adopt a new mode of development. That is, they begin to understand that teaching experiences force them to reflect on and communicate what one knows which, in turn, enhances their development as complete musicians. Typically, New England Conservatory students who graduate with an undergraduate degree in performance with a music-in-education concentration reflect on their development as performing musicians in terms of a developing synthesis of learning and teaching. As one voice major states in an exit interview:

> The great joy and intellectual enrichment comes from being deeply challenged and excited by my work in the music-in-education program. Teaching has become a source of inspiration, fueling my own journey of music learning. My desire to become a better musician is now inseparable from my desire to become a better teacher.[19]

The most gifted music students invariably teach. Whether in outreach programs in symphony or opera programs, or as faculty members of college or community school programs, conservatory graduates typically thread their career in music with teaching. Although conservatory faculty may specialize in teaching performance, composition, theory, or music history as separate subjects, virtually all faculty members report that they later integrate the three facets of the framework as inseparable from their continuing evolvement as musical artists, educators, and think-

---

[18] Similar statistics are obtained in other major schools of music, such as the Juilliard School or the Manhattan School of Music.

[19] Interview data and other reports concerning the development of the Music-in-Education Program at New England Conservatory are available in an internal FIPSE report from the NEC's Research Center.

	Musician As Artist	Musician As Teacher	Musician As Scholar
Artistry	**The Practice, Habits of Mind, and Products of the Musical Artist**	(The musician who develops the artistry of teaching music)	(The musician who produces scholarship focused on music, music history, creative process, etc.)
Teaching	(The musician who teaches or guides the development of the musical artist)	**The Practice, Habits of Mind, and Products of the Music Educator**	(The musician who teaches or guides scholarship in music and musical processes)
Scholarship	(The musician who focuses on the scholarship of music and musical processes)	(The musician who focuses on the scholarship of teaching and learning music)	**The Practice, Habits of Mind, and Products of the Music Scholar**

The Artist/Teacher/Scholar (ATS) Framework is based on the belief that a successful musical education must foster the development and integration of all three aspects of musicianship: artistry, teaching, and scholarship. The Framework demands high standards of every musician in the areas of:

- *Artistry* [1] (the ability and desire to engage and command artistic processes resulting in high-quality musical performance, composition, listening and reflective thinking skills, as well as knowledge of musical works, creative processes, and learning skills that can be applied to a wide range of disciplines);
- *Teaching* [ii] (the ability and desire to provide effective instruction, coaching, and personal mentoring in a range of musical settings as well as abilities as a public presenter or liaison at presentations of diverse musical works);
- *Scholarship* [iii] (the ability and desire to generate inquiry, research, and reflection on one's personal artistic and educational work while taking into account other perspectives, including a wide range of historical, psychological, social, and artistic sources).

---

[1] See Music and the Mind by Anthony Storr or Artistry: The Work of Artists by Vernon Howard for philosophical frameworks for defining musical artistry.

[ii] See Artistry In Teaching by Louis J. Rubin, Margaret Gullette's The Art and Craft of Teaching, or Allan Pearson's The Teacher as references on teaching in conjunction with artistry.

[iii] See We Scholars, for David Damrosh's criticism over specialization and the lack of socialized or collaborative scholarship and Ernest Boyer's Scholarship Reconsidered for a discussion of the four new aspects of scholarship now being reassessed in universities—the scholarship of research, applied scholarship, integrative-interdisciplinary scholarship, and the scholarship of teaching.

Figure 3. The Artist/Teacher/Scholar (ATS) Framework.

---

ers.[20] Thus, for a successful faculty member, the 'teaching of musical artistry' is as important as their concerns for 'artistry of their teaching' (or the ability to communicate or impart knowledge as artists). Likewise, the 'teaching of scholarship' (e.g. conducting historical research) is as important as the 'scholarship of teaching' (e.g. educational research or personal reflections on teaching). The developmental track of the Artist/Teacher/Scholar most often begins with private lessons, then studio teaching, and listening to recordings of the works they are studying. The path of Artist/Teacher/Scholar is a course where each area is increasingly synthesized into the whole.

---

[20] Interview data are analyzed as part of an internal report on the development of the Artist/Teacher/Scholar Committee and Music-in-Education Program at New England Conservatory sponsored by the United States Department of Education's FIPSE (Funds for the Improvement of Post Secondary Education) Program.

On the surface, the mission statement cited above may not suggest a framework for innovation in music education. However, the conception of the Artist/ Teacher/Scholar reflects the need for schools of music to integrate three *complementary* responsibilities: to provide every student with the means of becoming a 'complete' musician, to enable students to handle the diverse challenges of a life in music, and to ensure that music will play a vital role in the overall education and development of school children. Innovation in music education for gifted performers at major schools of music can support the highest caliber of professional training for musicians and concurrently nurture innovation in public schools through the comprehensive, interdisciplinary programs described above.

Educating gifted artist-teacher-scholars requires intensive curriculum development. The United States Department of Education's Funds for the Improvement of Post Secondary Education grant (FIPSE) supported New England Conservatory in its efforts to create a Music-in-Education Concentration program, a curriculum open to all graduate and undergraduate performance and composition majors.

Music-in-Education guided internships provide five entry points for conservatory students to engage with public schools: private lessons, music classes, performance ensemble classes, music-based interdisciplinary classes or projects (e.g. the Creating Original Opera program designed by the Metropolitan Opera Guild), and artist in residence programs. Taken together, these internships—when guided by conservatory faculty— promote the success and sustainability of authentic, comprehensive, and interdisciplinary school learning programs.

With new courses and internship opportunities available to students in all majors, music education has moved from the periphery to the center of NEC's curriculum structure. Prior to the establishment of the Music-in-Education Concentration in 1998, 3% of all NEC students enrolled in music education courses. Three years later, that figure has increased to 18%. By the end of the 2000–2001 academic year, 64 students were engaged in internships with partnering schools and arts organizations where they gained valuable, hands-on experience in dealing with the challenges that all musicians—including performers, administrators, teachers, and scholars—face in society today.

Music students are not the only beneficiaries of Music-in-Education activities. An unanticipated boon created by the internship program is the extent to which supervision of internships, and associated involvement in public schools, has constituted valuable professional development for conservatory faculty. For the first time, conservatory faculty members invest their time developing classroom materials, collaborating with public school teachers on project design and implementation, observing projects in action, providing feedback on student and teacher work, and engaging in assessment.

In the context of these new curricular structures, gifted performers become not only artists but also emissaries for music's role in education—to the benefit of both. Participating conservatory students who were interviewed provide evidence that a Music-in-Education program can transform their view of the role and impact of the performing musician in public schools. A brass performance major who led an honors ensemble at New England Conservatory and continues to work in educational outreach programs reports:

> Since leaving NEC to become a faculty member at the University of Maryland, I have discovered that NEC is taking a leadership role in the field of music education programs both by training musicians for outreach and preparing performers to present *Learning Through Music* principles in concert master-classes and interdisciplinary activities.

Similarly, participation in the Music-in-Education program expands the level of engagement of faculty members who choose to serve as mentors for students involved in a wide variety of educational settings. Conservatory faculty members report that gifted young performers, in the midst of intensive professional training, benefit greatly from contact with the outside world of public education while it stimulates faculty to discover new contexts for mentoring and developing aspiring Artist-Teacher-Scholars. Faculty and students both report that guided internships benefit, often in tandem, a musician's performance skills, audience or classroom communication skills, the ability to reflect on their own education as artists, and their desire to contribute positively to the needs of the local community.

Exit interviews reveal that conservatory students who participate in education internships at laboratory school settings, for example, become better prepared for teaching opportunities in school districts that hold music programs responsible for contributing to the overall goals of the school district. Mona Rashad, a conservatory-trained violin major, states that the work she did as a teaching intern confirmed the value of teaching as a musical performer who can advocate for music's essential role in the public-school curriculum:

> The chance to work at the Lab school has really given me the confidence and experience I need. My interview for a teaching position in a string program ended with an immediate offer. I had previously thought that the Music-in-Education program was way ahead of its time. After talking with school administrators and teachers, I realize now that the Conservatory's program is exactly where educators and administrators are going.

Ms. Rashad contrasted her experience with a traditional music education experience by saying that it

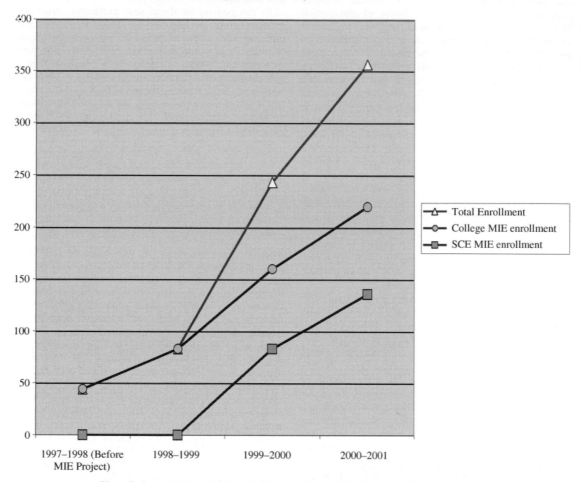

Impact of MIE Program on College and School of Continuing Education
Enrollment at New England Conservatory

Years Before and After the Music-in-Education Program Development at New England Conservatory.

Figure 4. Graph demonstrating the effect of the Music-in-Education Programs on student performance majors and continuing education teacher professional development enrollment.

"would never have prepared me to be so ready to meet the expectations of music education as it is now being practiced". She continues:

> ... after the interview was over, the superintendent told me that the team had come to talk to the state Department of Education about the same issues we discussed in my interview. He said that our discussion about interdisciplinary aspects of teaching music was far better than the one they had with the state people! That really opened my eyes.

The shift from music education degree programs to the creation of Music-in-Education concentrations for performance majors led to a new consensus on the value of education in the training of gifted musicians committed to a conservatory education. This innovation—the first systemic curricular change at New England Conservatory in 40 years—has had a significant impact on the quality and diversity of the institution's music education experience. Student responses such as the one below continue to inspire many faculty and staff as they work on this project with the hopes that a flexible, more inclusive, and more socially committed Music-in-Education program will benefit gifted music students as much as the institutions and communities who employ their talents in the future:

> Teaching has become a source of inspiration, fueling my own journey of music learning. My desire to

become a better musician is now inseparable from my desire to become a better teacher (voice major).

Strategy 4: Form a network of college-conservatory, arts organizations, and public-school partnerships that are committed to designing and implementing innovative, research-based forms of 'Music-in-Education' practices, and to influencing policy concerning the essential role of music in public schools.

The secret of success is constancy of purpose (Benjamin Disraeli).

Another strategy for successful innovation in music education concerns the dissemination of new practices at music schools who partner with public schools. In September 2000, a conference at New England Conservatory entitled 'Making Music Work in Public Education: Innovative Practices and Research from a National Perspective'[21] brought together researchers and administrators from arts organizations and college/

conservatories to discuss the integration of two approaches to music education. After an initial presentation a principal question emerged: How can arts organizations, institutions of higher education, and public schools network together to establish high standards for Music-in-Education practices informed by 'best practices' in the fields of both music-integrated interdisciplinary instruction and authentic music curricula in the context of school reform?

The discussions and presentations at this conference led to the formation of a National Consortium for Music-in-Education, now funded by a U.S. Department of Education grant and the National Endowment for the Arts. The principal consortium members are nationally known arts and education organizations that have expanded their initiatives to include a wider array of perspectives. The mission of the consortium is to develop, evaluate, and disseminate nationally:

(1) a shared practice-based theory of music in education that will reveal the effectiveness of music programs in enriching and sustaining academic and artistic excellence;

---

[21] Sponsored by the Spencer Foundation.

Photo 4. The Borromeo Quartet, the quartet in residence at the New England Conservatory, provides a Learning Through Music presentation for children at the Conservatory Lab Charter School.

(2) cross-institutional professional development of teachers and artists who promote national standards of comprehensive, interdisciplinary music education goals in the context of arts in education programs;

(3) shared standards for learning, teaching, and community support for music in the curriculum of public school education;

(4) training for young, talented musicians as role models for music in education in public school settings.

The principal organizations of the consortium complement each other by design. New England Conservatory, along with Northwestern University, Georgia State University and the Mannes College of Music have created several different models for Music-in-Education programs in laboratory-school partnership settings. The Metropolitan Opera Guild, Boston Symphony Orchestra, and the Ravinia Festival Orchestra Program in Chicago are models of cultural organizations dedicated to transforming school curricula through opera, orchestral music, and the participation of local performing arts groups in educational programs. The Kenan Institute for the Arts and the Chicago Arts Partnerships in Education offer extensive cultural and arts education initiatives through local partnerships supported by city and state educational policy. *From The Top*, a nationally syndicated radio show, collaborates with New England Conservatory to promote the role of young artists as peer role models and cultural leaders in schools.

The consortium's work will be guided by key conceptual and organizational frameworks for innovation in music education, including the Artist/Teacher/Scholar Framework. From each consortium member's perspective, this framework will be used to expand the focus of music from the more traditional model of 'music education' as a separate, often ancillary curriculum in schools, toward several Music-in-Education program models which support performing artists as guided interns in a variety of educational settings. Finally, the consortium will serve as a model for the 'triangulation' of community cultural and educational partnerships focused on improving the quality of learning in and through music for children and youths.

The structure of the consortium represents an opportunity for influencing changes in the way music education is defined and applied in public schools, and its work is intended to support the emerging field of Music-in-Education. Some examples include:

(1) *A Music-in-Education Professional Development Exchange Program* for training musical artists as teachers and as 'action researchers' (a form of scholarship of teaching), based on the exchange of data and practice models between the researchers and artist educators within each organization. Evaluation of this program will focus on the

effectiveness of professional development programs that increase the capacity of musicians and music students to serve as 'Artist/Teacher/Scholars' trained to work with children and youths. The evaluation should provide evidence for the effectiveness of the consortium's professional development practices, and it will be referenced to standards set by the National Board for Professional Teaching Standards.

(2) *The Publication of Music-Integrated Learning Curricular Units* with explicit learning objectives based on authentic, comprehensive, and interdisciplinary Music-in-Education practices, will represent new forms of collaboration between higher education faculty, teaching artists and classroom teachers in the curriculum design and implementation process. This collaboration works also to evaluate and prepare curricula for publication in a consortium-approved Music-in-Education 'handbook of proven and promising practices'. Program evaluation will focus on the impact of the consortium's curricular units toward increased knowledge, skills, and understanding of music, musical processes, diverse musical literature, and learning related to other subject areas through music by public-school children and youths. The evaluation of the data (student interview responses, academic and musical performance with respect to curricular units, expert evaluation of student portfolio work, and the evaluation of academic achievement results with respect to learning in and through music) will provide evidence of the effectiveness of the consortium's curricular units and will be referenced to national and state standards within and across disciplines relevant to the consortium institutions.

(3) *Models for Sustainable Practices* based on collaborations among educators, artists, and K-12 teachers who stress the musical arts as a model for community involvement with schools (parent classes, parent–teacher performing groups, parent roles which affect school cultures, parent–child engagement, and other opportunities for the entire school community to support Music-in-Education practices in their schools). The evaluation of the Consortium's work in each city will focus on the capacity of each participating organization to influence local policy and sustainable community practices that support Music-in-Education programs. The evaluation of the data (interviews with stakeholders of each organization, expert analyses of the quality of research conducted throughout the consortium, and expert analyses on the consortium as a model for increasing local community collaboration and resources that support arts learning) will provide evidence of the effectiveness of the consortium's programs and assess the group's future as an organization capable of setting stan-

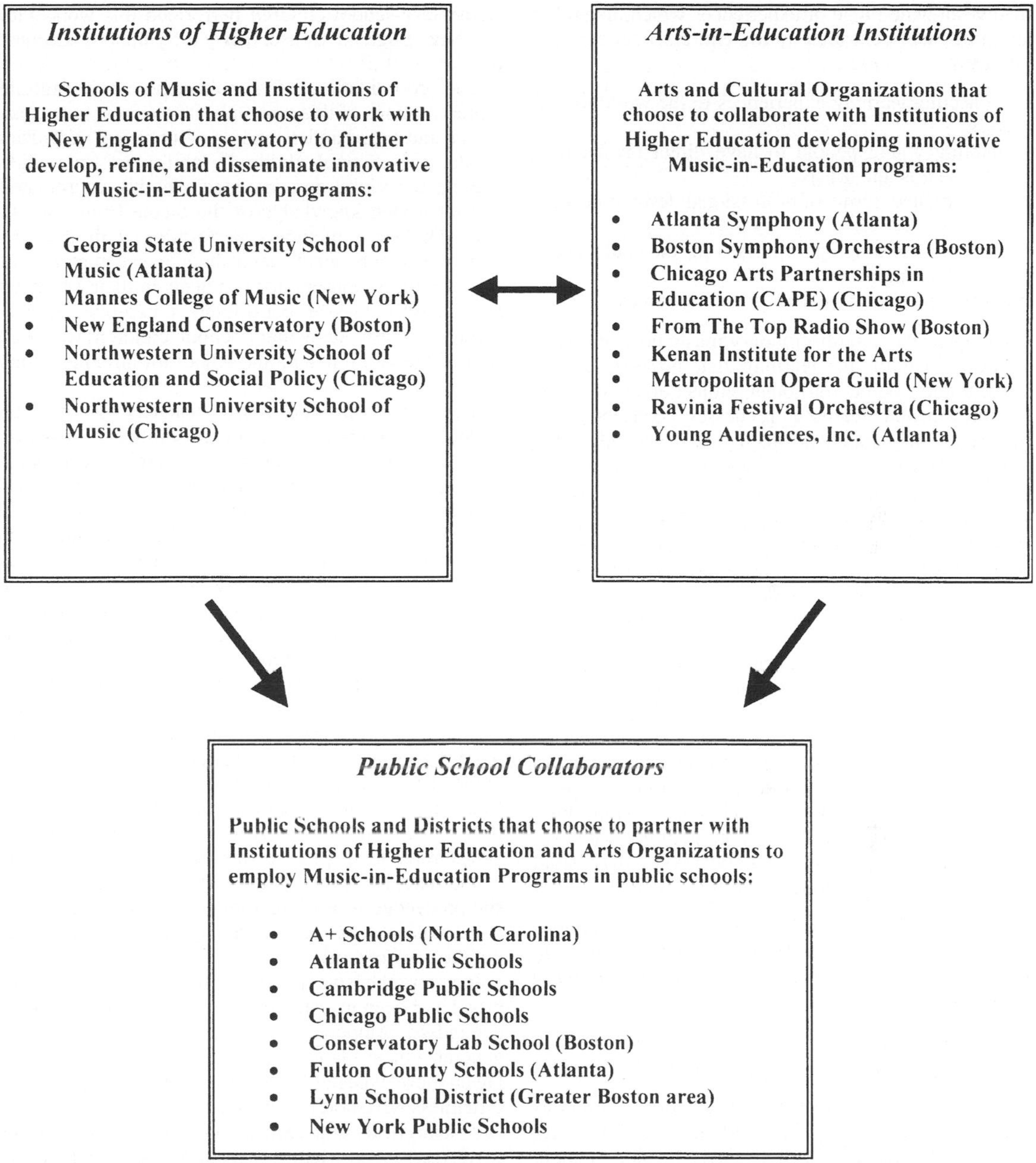

Institutions of Higher Education

Schools of Music and Institutions of Higher Education that choose to work with New England Conservatory to further develop, refine, and disseminate innovative Music-in-Education programs:

- Georgia State University School of Music (Atlanta)
- Mannes College of Music (New York)
- New England Conservatory (Boston)
- Northwestern University School of Education and Social Policy (Chicago)
- Northwestern University School of Music (Chicago)

Arts-in-Education Institutions

Arts and Cultural Organizations that choose to collaborate with Institutions of Higher Education developing innovative Music-in-Education programs:

- Atlanta Symphony (Atlanta)
- Boston Symphony Orchestra (Boston)
- Chicago Arts Partnerships in Education (CAPE) (Chicago)
- From The Top Radio Show (Boston)
- Kenan Institute for the Arts
- Metropolitan Opera Guild (New York)
- Ravinia Festival Orchestra (Chicago)
- Young Audiences, Inc. (Atlanta)

Public School Collaborators

Public Schools and Districts that choose to partner with Institutions of Higher Education and Arts Organizations to employ Music-in-Education Programs in public schools:

- A+ Schools (North Carolina)
- Atlanta Public Schools
- Cambridge Public Schools
- Chicago Public Schools
- Conservatory Lab School (Boston)
- Fulton County Schools (Atlanta)
- Lynn School District (Greater Boston area)
- New York Public Schools

Figure 5. The Music-in-Education National Consortium (MIENC) is a project funded in part by a U.S. Department of Education FIPSE (Funds for the Improvement of Post Secondary Education) grant and the National Endowment for the Arts.

dards for, and providing leadership on behalf of, the field of music in education.

In order for the form of innovation in music education described here to take root in standard educational practice, the soil must be prepared for acceptance by a larger constituency of stakeholders than was typical in the traditional music education organizations of the last century. Research cited above suggests that comprehensive, interdisciplinary music programs will succeed in transforming schools not just because of the uniqueness of musical studies, but because music involves a wide range of experiences which, like other art forms, will contribute to the entire school community by:

- connecting learning experiences to the world of real work;
- enabling young people to have direct involvement with the arts and artists;
- requiring new forms of professional development for teachers;
- providing learning opportunities for the adults in the lives of young people;
- engaging community leaders and resources.

Comprehensive, interdisciplinary music programs constitute a medium for learning that provides unique musical entry points into other subject areas by virtue of fundamental learning processes, concepts, and representations shared between music and other disciplines. As the research cited above suggests, positive conditions for the appropriate interaction of music among several disciplines rely on high-quality instruction in each discipline. Only in schools where mathematics, music, and language are all taught effectively and with an equal degree of emphasis can we assume that music will promote learning transfer across disciplines.

The work of such a consortium will more likely raise public consciousness of the essential role of music in education than would each of the consortium's participating organizations in isolation. As research reported in popular media continues to expand its coverage to include evidence that connects music to many aspects of human potential, the National Music-in-Education consortium can provide practical guidelines for innovative program development and education policy based on a responsible interpretation of this research.

Strategy 5: Explore and promote new conceptions of giftedness that result from the implementation of innovative forms of comprehensive, interdisciplinary music education practices in public schools.

Man's mind stretched to a new idea never goes back to its original dimension (Oliver Wendell Holmes).

Once comprehensive, interdisciplinary music programs become available to every child in a school, larger numbers of students will more likely be able to attain higher levels of musical development. As a result, the bar for what parents or music educators would deem evidence of giftedness in music will be raised.

Consider this example of what is commonly expected at the Conservatory Lab Charter School: the ability to play the violin and sing at the same time. That is, every student learned to sing the names of notes and the German text of Beethoven's 'Ode to Joy', at the same time they performed the melody or various harmonic underpinnings of the tune on the violin (see Fig. 6 below). This skill was demonstrated when these

elementary-school children performed this work with the New England Conservatory's Symphony Orchestra.

The coordination of simultaneous and multiple representations of this tune indicates to most musicians a deep and broad understanding of a musical skill. That is, the Lab School children's ability to sing while playing the violin demonstrated a highly coordinated and integrated knowledge of the theme from 'Ode to Joy'. Orchestra members performing with the Lab School students were astonished to see that a skill required of all conservatory students in their advanced musicianship classes constituted a baseline requirement for these elementary-school children who had studied the violin in a public school for little more than a year.

The classroom and music teachers in the Lab School's comprehensive, interdisciplinary program have observed other aspects of academic and musical development that might have been associated with giftedness in the past. The ability to read and pronounce phonetic German spelling, the ability to start with any word in the text and sing the song from that point (at actual pitch), and the ability to sing the song with scale degree numbers (sometimes backwards and forwards) all indicate a deep operational and relational knowledge of the song.

Another example of musical ability that might have been considered an indication of giftedness is young Maya's ability to represent her knowledge of her violin music. The figure below illustrates how this second-grader used a matrix to represent the first phase of a folk song she played on the violin. She later created a mirror image of the melody on a second sheet of paper. When asked to sing the forward and backward versions of the tune, she exclaimed, "I like this both ways!" as she proceeded to explore symmetrical relationships in other musical tunes and original compositions throughout the project work.

In comprehensive, interdisciplinary music programs, broad and deep investigations of musical tunes become a standard component for all children in their violin, general music, language, and mathematics classes. Given these expectations for all children, indications of giftedness will more likely be characterized by creative and analytical engagement in multiple modes of performance knowledge, a symbolic grasp of music, composition, and a reflective understanding of musical design and meaning through interdisciplinary-based investigation.

As comprehensive, interdisciplinary music programs become a standard for public schools, music educators will undoubtedly discover new forms of musical giftedness based on:

- *Music as a medium for the integration of skills.* Rather than looking first toward accelerated performance skills based on imitation of conventional

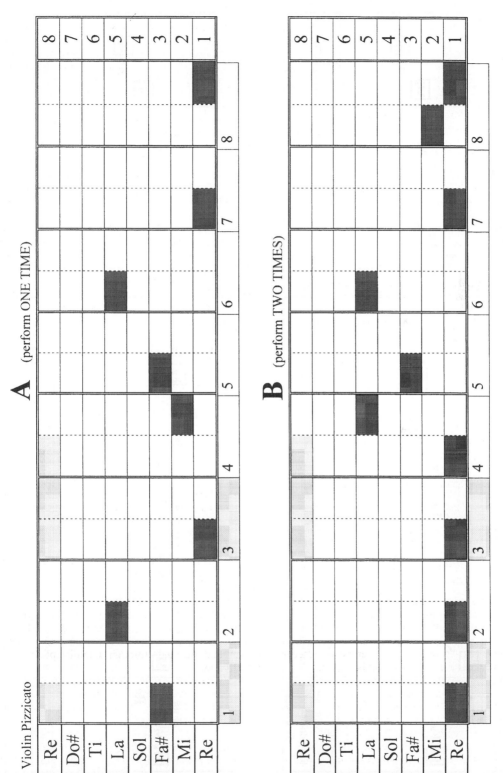

Figure 6. All students at the Conservatory Lab Charter School prepare to read music in many different, and simultaneously coordinated, ways. Here students read an accompanying part to Beethoven's 'Ode to Joy' as a mathematical matrix to be performed with syllable names and scale degree numbers.

Sing with SOLFEGE/violin pizzicato (first time)/PHONETIC GERMAN (second time)

A (perform ONE TIME)

Fa Fa Sol La Mi	Re Re Mi Fa	Fa Mi Mi	Fa Fa Sol La	La Sol Fa Mi	Re Re Mi Re Re
Freu-de Schö-ner	Göt-ter- -fun-ken Toch-ter	aus E- -ly-si-um	Wir be- -tre-ton	Feu-er trun-ken Himm-li -sche dein	Hei-lig-tum!
Froy-duh Shurn-nuh	Gurt-tuh- -foon-kin Toch-tuh- Ows Eh-	-lee-zee-oom	Feer beh- -treh-ton	Foy-uh troon-ken Himm-li -schuh Dine	Hie-lig-toom!

B (perform TWO TIMES)

Fa# Mi Mi Re	Mi Mi Sol Fa Fa Re	Sol Fa Fa Mi Re La	Fa Fa Sol La Mi	La Sol Fa Mi	Re Re Mi Re Re
Dei-ne Zau-ber Bin-den Wie-der	Was die Mo-de Streng ge-teilt	Al-le Men-schen	Wer-den Brü-der Wo Dein Sanf-ter	Flü-gel Welt.	
Die-nuh Zow-buh Bin-den Vee-duh	Vass Dee Mo-duh Shtreng guh- tielt	All-luh Men-shen	Ver-den Brew-duh Vo Dine Zanf-tuh	Flew-gull Vialt.	

Figure 7. Lab School Students also learn to perform 'Ode to Joy' as a play- and sing-exercise, reading phonetic German or note name syllables while they pluck the selected notes on the violin.

practices, we will view musical talent as the ability to create or improvise diverse forms music, analyze systematically diverse styles and genres of music, and translate music into mathematical representations and vice versa;

- *Music as an indication of the ability to work with multiple representations.* Rather than thinking of the ability to read music as a mechanical, stimulus–response mechanism that a gifted student may learn more quickly, we would view the ability to switch performance representations flexibly (from playing to singing, from silent imagination to external performance). In addition, symbolic representations (numbers, notes, words, gestures, etc.) would provide a more useful framework for evaluating and nurturing musical talent;

- *Understanding musical concepts and skills through investigation of other subject areas.* Seen through the lens of interdisciplinary music programs, giftedness will more likely be evaluated by a student's ability to translate information from one field to another, or explain concepts intrinsic to music by borrowing

fundamental concepts shared with other disciplines (such as probability in mathematics, or sonic analysis related to physics);

- *Music as a model for artistic process.* Giftedness in music will more easily be judged by how individuals relate music to other art forms (such as theater, opera, dance, or film) in terms of subject matter, social context, emotional effect, and structural content;

- *Music as a medium and model for understanding interpersonal relationships.* Music is predominately a social art form and relies ultimately on working with fellow performers, teachers, and audiences. Interpersonal giftedness in music can be discerned in individuals who teach or coach others, who present music effectively to audiences, and who learn from others in different performing and studio ensemble contexts;

- *Music as a medium and model for intra-personal giftedness.* Musical development is predicated also on the ability of an individual to sustain personal practices consistently and profitably over long peri-

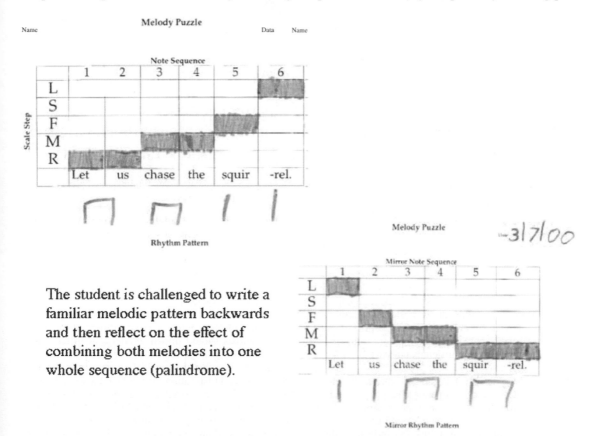

The student is challenged to write a familiar melodic pattern backwards and then reflect on the effect of combining both melodies into one whole sequence (palindrome).

Photo 5. In this exercise, children at the Conservatory Lab Charter School are encouraged to explore different versions of the same musical phrase. In this case, both the frontward and backward versions of the tune 'Let Us Chase The Squirrel' are explored for their symmetrical qualities and its aesthetic effect.

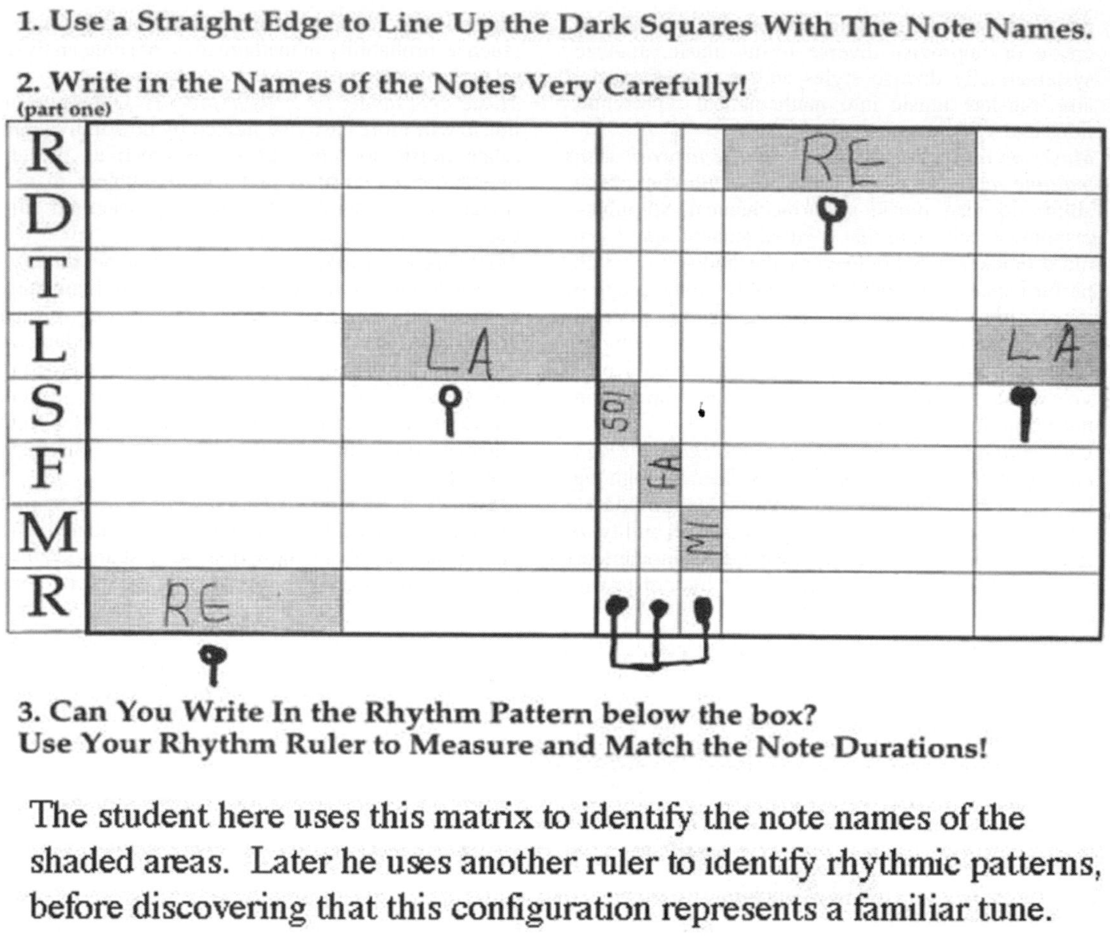

1. Use a Straight Edge to Line Up the Dark Squares With The Note Names.

2. Write in the Names of the Notes Very Carefully!
(part one)

R					RE	
D					♩	
T						
L		LA				LA
S		♩	SO/	·		♩
F			FA			
M			MI			
R	RE		♩♩♩			

♩

3. Can You Write In the Rhythm Pattern below the box?
Use Your Rhythm Ruler to Measure and Match the Note Durations!

The student here uses this matrix to identify the note names of the shaded areas. Later he uses another ruler to identify rhythmic patterns, before discovering that this configuration represents a familiar tune.

Photo 6. Later on, children learn to identify sophisticated music through multiple representations explored previously.

ods of time. Giftedness in music can be seen in terms of the ability to set goals and priorities, and in the ability to derive satisfaction from personal practice and constructive self-assessment;
• *Music as a form of self-education for social good.* Giftedness in music should be measured by more than the ability to follow instructions or to adhere to a set of prescribed procedures. Giftedness in this area is characterized by awareness of the worth of music both to one's personal development and to the community to which musicians belong.

From the perspective of higher education, redefining giftedness in music can stimulate changes in the entire process of a college education or professional training. Gifted music students can be admitted on the basis of diverse forms of musical ability (as indicated above) and their willingness to embrace a broader view of the place of music in society. In this context, the Artist/Teacher/Scholar framework serves to further a musical

giftedness by way of artistic development integrated with teaching experiences and rigorous scholarship. The result may lead to rethinking what form musical concerts can take, what roles musicians can play in music education, or what contribution the education of gifted musicians can make toward creating better citizens in our society.

It is up to today's educators to decide whether the testimony of last century's most influential musicians and scientists should inform models for giftedness in contemporary public education. If education policy-makers can see that the musical process is an essential ingredient and inspiration for public education, then we all will make better sense of learners who link artistic production conjoined with the higher-order cognitive skills as indications of giftedness in both the arts and academics. Perhaps only then will educators in the arts and academics use the words of master artists such as Pablo Picasso as a means for others to understand that creative processes are virtually indistinguishable from critical thinking skills associated with scientific method or academic excellence:

> paintings are but research and experiment. I never do a painting as a work of art. All of them are researches. I search constantly and there is a logical sequence in all this search (Pablo Picasso, cited in Gardner, *The Arts and Human Development: A Psychological Study of the Artistic Process*, 1973).

Comprehensive, interdisciplinary programs in music education that raise expectations for literacy and performance skills for all children not only will suggest new criteria for the determination of gifted musicians but also will enlarge the range of responsibilities and roles that these musicians will be expected to play in furthering our society's understanding of the relationship between music and education.

Implications for Innovation in Music Education

> Creative thinking may simply mean the realization that there is no particular virtue in doing things the way they have always been done (Rudolph Flesch, educator).

Music has reached an uncertain plateau in American public school education. Bolstered by innovations of the past, new standards exist for instruction in music-reading and instrumental skill development, yet these resources are not made available in public schools. Without adopting a new set of strategies for building toward a more progressive view of music programs that serve all children in public schools, educators and parents will continue to see musical ability as a special talent assessed primarily from the relatively parochial view of early emergence of musical performance skill.

At this point, however, there appears no particular virtue in maintaining current music-education practices. A summary of research conducted over the past two decades suggests four important themes that inform the basis for invention of new programs in music education (Scripp, 2002):

(1) Theme 1: Meta-analyses on large bodies of research over the last few decades reveal consistently strong, positive relationships between music and learning in other subject areas. This unambiguous finding should be studied carefully so as to maximize its application to public-school Music-in-Education practices.

(2) Theme 2: Generative neurological and cognitive frameworks for learning transfer have emerged from research on music and learning; studies that suggest learning transfer through music, especially in—but not limited to—the field of spatial-temporal skills, should continue to be tested in school settings for their application to, and impact on,

learning in comprehensive, interdisciplinary school programs.

(3) Theme 3: Although the strong relationship between music and learning in other subject areas now is recognized, there is an underlying tension between the 'one-way cause and effect' and 'two-way interaction' models of research on music and learning. There is a need to interpret and design new research studies to help school communities understand the complexities of learning both in music and in the academics, and to better teach the principles shared between these subject areas.

(4) Theme 4: The use of music as a tool for social-emotional development and behavior modification in schools has been well established in recent meta-analyses; the systemic employment of music-listening and music-making needs to be employed for goals outside of the traditional performance arena or aesthetic education and made available to classroom and music teachers as a tool for measurable outcomes in the areas of inter-personal and intra-personal skills.

Based on these findings and the preliminary work of innovative schools such as the Conservatory Lab Charter School and their partnerships with schools of music and other arts organizations, it is time to implement new conceptions of music education built on the innovations in music instruction of the past. Yet now we also must include a wider range of entry points, assessments, and interdisciplinary connections.

Researchers studying the effect of arts-integrated instruction on academic achievement warn, however, that to oversimplify expectations for learning transfer between music and the academics may limit our ability to assess the impact of innovative music programs on the whole school community (Adkins & McKenney, 2001; Catterall, 1998; Corbett, McKenney, Noblit & Wilson, 2001; Nelson, 2001)[22]. Furthermore, researchers at the Center for Arts Education Research at Columbia University report, for example, that significant instances of learning transfer across domains occur primarily because of the inherent interdependency of the subject areas, and not because one subject area 'causes' learning in another.

> Arts learning ... calls upon a constellation of capacities and dispositions that are layered and unified in the construction of forms we call paintings, poems, musical compositions, and dances. ... What is critical is not that the capacities and

---

[22] Reports from the A+ Schools Project in North Carolina (Corbett, McKenney, Noblit & Wilson, 2001; Nelson, 2001 in particular) show that the evaluation of the impact of arts programs in schools should not be limited to test score results as evidence that suggests 'causal links' between A+ arts programs and academic achievement.

Photo 7. All students at the Conservatory Lab Charter School performed with the New England Conservatory Symphony Orchestra their rendition of 'Ode to Joy'.

dispositions transfer from the arts to other subject areas, as has often been argued, but that they are exercised broadly across different knowledge domains. Given this interpretation, no subject has prior rights over any other subject, for to diminish one is to diminish the possibility and promise of them all. If the arts help define our path to the future, they need to become curriculum partners with other subject disciplines in ways that will allow them to contribute their own distinctive richness and complexity to the learning process as a whole (from Burton, Horowitz & Abeles, 1999).

In order for systemic change to occur in public schools, innovation also must be supported by a wider and more cohesive array of resources and institutional partnerships. In order for Music-in-Education programs to be successful they will need to rely on institutional partnerships with colleges of music, local arts organizations, and publications that serve the professional needs of teachers. The National Consortium for Music in Education has been formed for the specific purpose of sharing resources and research needed to support changes in educational policy with regard to the role of music in public school education. The Artist/Teacher/Scholar conception of the conservatory-trained musician, the evolving role of gifted musicians as resources for public schools, and the scope of Music-in-Education programs now serve as guiding principles for sharing innovative practices and research for a new national consortium of schools of music, arts organizations, and their partnering public-school communities.

Furthermore, the dissemination and sustainability of innovation in music education face many challenges and obstacles that only new forms of teacher preparation, ongoing professional development, curriculum development, and evaluation methods may solve. Comprehensive, interdisciplinary music curricula will require music teachers who demonstrate a deep and diverse understanding of music and academics as is necessary to understand and facilitate connections across disciplines for all students. One cannot teach effectively the relationship between the concept of proportion in relation to musical rhythm without significant background musical skill; similarly, one cannot incorporate the concept of rhythm to the teaching of proportion without a deep understanding of mathematical problem-solving skills. It will be the shared understanding of the interactions between musical and academic learning processes that will enable entire school communities to accept music as an invaluable subject for its own sake and as a resource for enhancing learning throughout the school curriculum.

Should the five strategies for innovation in music education discussed in this chapter become established in the next decade, innovations resulting from these strategies will challenge the lingering perception that learning in music is primarily a rare talent that is first and foremost measured by performance ability and has no direct bearing on the benefits of a musical education for all children across the school curriculum. Giftedness redefined as a synthesis of learning in and through music and the education of professionally trained musicians as artist-teacher-scholars both promise to serve our musical culture and education system equally well.

References

Adkins, A. & McKenney, M. (2001). Placing A + in a national context: A comparison to promising practices for comprehensive school reform. *Report No. 6 in a Series of Seven Policy Reports Summarizing the Four-Year Pilot of A + Schools in North Carolina*. Winston-Salem, NC: Kenan Institute of the Arts.

Bamberger, J. (2000). Music, math and science: Towards an integrated curriculum. *Journal for Learning Through Music*, **1**, 32–35.

Boyer, E. (1990). *Scholarship reconsidered: Priorities of the professoriate*. Princeton, NJ: Carnegie Foundation.

Burnaford, G., Aprill, A. & Weiss, C. (Eds) (2001). *Renaissance in the classroom: Arts integration and meaningful learning*. Mahwah, NJ: Lawrence Erlbaum.

Burton, J., Horowitz, R. & Abeles, H. (2000). Learning in and through the arts: the question of transfer. In: E. Fiske (Ed.), *Champions of Change: The Impact of the Arts on Learning* (pp. 228–257). Washington, D.C.: The Arts Education Partnership and the President's Committee on the Arts and the Humanities.

Butzlaff, R. (2000). Can music be used to teach reading? *The Journal of Aesthetic Education*, **34** (3), 167–178.

Catterall, J. (1998). Involvement in the arts and success in secondary school. *Americans for the Arts Monographs*, **1** (9).

Catterall, J. (2002). The arts and the transfer of learning. In: D. Deasy (Ed.), *Critical Links: Learning in the Arts and Student Academic and Social Development*. Washington, D.C.: Arts Education Partnership.

Catterall, J., Chapleau, R. & Ivanaga, J. (1999). Involvement in the arts and human development: Extending an analysis of general associations and introducing the special cases of intensive involvement in music and in theater arts. Champions. In: E. Fiske (Ed.), *Champions of Change: The Impact of the Arts on Learning, Arts Education Partnership*. Washington, D.C.

Corbett, D., McKenney, M., Noblit, G. & Wilson, B. (2001). The A + schools program: School, community, teacher, and student effects. *Report No. 6 in a Series of Seven Policy Reports Summarizing the Four-Year Pilot of A + Schools in North Carolina*. Winston-Salem, NC: Kenan Institute of the Arts.

Damrosch, D. (1995). *We scholars: Changing the culture of the university*. Cambridge: Harvard University Press.

Davidson, L. & Scripp, L. (1989). Education and development in music from a cognitive perspective. In: D. J. Hargreaves (Ed.), *Children and the Arts: The Psychology of Creative Development*. Philadelphia, PA: Open University Press.

Davidson, L. & Scripp, L. (1994). Conditions of giftedness: Musical development in the preschool and early elementary years. In: R. Subotnik & K. Arnold (Eds), *Beyond Terman: Contemporary Longitudinal Studies of Giftedness and Talent* (155–185).Norwood, NJ: Ablex.

Deasy, D. (Ed.) (2002). *Critical Links: Learning in the Arts and Student Academic and Social Development*. Washington, D.C.: Arts Education Partnership (http://www.aep-arts.org).

Gardiner, M. (2000). Music, learning, and behavior: A case for mental stretching. *Journal for Learning Through Music*, **1**, 72–93.

Gardiner, M., Fox, A., Knowles, F. & Jeffrey, D. (1996). Learning improved by arts training. *Nature*, **381**, p. 284.

Gardiner, M. & Scripp, L. (in press). Patterns of correlation between musical and academic progress: Interrelationships between early musical and early literacy, and numeracy skill development at the Conservatory Lab Charter School. *Journal for Learning Through Music*, **2**.

Gardner, H. (1973). *The arts and human development: A psychological study of the artistic process*. New York: John Wiley.

Gardner, H. (1983). *Frames of mind: The theory of multiple intelligences*. New York: Basic Books.

Gardner, H. (1989a). Project zero: An introduction to arts PROPEL. *Journal of Art and Design Education*, **8** (2), 167–182.

Gardner, H. (1989b). Zero-based arts education: An introduction to arts PROPEL. *Studies in Arts Education*, **30** (2), 71–83.

Graziano, A., Peterson, M. & Shaw, G. L. (1999). Enhanced learning of proportional math through music training and spatial-temporal training. *Neurological Research*, **21**, 139–152.

Gullette, M. (Ed.) (1982). *The art and craft of teaching*. Cambridge, MA: Harvard University Free Press.

Hetland, L. (2000a). Learning to make music enhances spatial reasoning. *The Journal of Aesthetic Education*, **34** (3–4), 179–238.

Hetland, L. (2000b). Listening to music enhances spatial-temporal reasoning: evidence for the 'Mozart Effect'. *The Journal of Aesthetic Education*, **34** (3–4), 105–148.

Hope, S. (2000). Integrity in integrative programs: a challenge to learning through music. *Journal for Learning Through Music*, **1**, 12–15.

Howard, V. (1982). *Artistry: The work of artists*. Cambridge, MA: Hackett.

Kodály, Z. (1974). *The selected writings of Zoltán Kodály*. London: Boosey & Hawkes.

Moga, E. Burger, K. & Winner, E. (2000). Does studying the arts engender creative thinking? Evidence for near but not far transfer. *The Journal of Aesthetic Education*, **34** (3–4), 91–104.

Nelson, C. A. (2001). The arts and education reform: Lessons from a four-year evaluation of the A + schools program, 1995–1999. *Executive Summary of the Series of Seven Policy Reports Summarizing the Four-Year Pilot of A +*

Schools in North Carolina. Winston-Salem, NC: Kenan Institute of the Arts.

Pearson, A. (1989). *The teacher*. New York: Routledge, Chapman & Hall.

Perkins, D. (1989). Art as understanding. *Journal of Aesthetic Education*, **22** (1), 111–131.

Perkins, D. (1992). *Smart schools: From training memories to educating minds*. New York: Free Press.

Rubin, L. (1985). *Artistry in teaching*. New York: McGraw-Hill.

Salomon, G. & Perkins, D. (1989). Rocky roads to transfer: Rethinking mechanisms of a neglected phenomenon. *Educational Psychologist*, **24** (2), 113–142.

Scripp, L. (2002). An overview of research on music and learning. In: Deasy, D. (Ed.), *Critical Links: Learning in the Arts and Student Academic and Social Development*. Washington, D.C: Arts Education Partnership.

Scripp, L. & Davidson, L. (1994). Giftedness and professional training: The impact of music reading skills on musical development of conservatory students. In: R. Subotnik & K. Arnold (Eds), *Beyond Terman: Contemporary Longitudinal Studies of Giftedness and Talent* (pp. 186–211). Norwood, NJ: Ablex.

Scripp, L., Davidson, L. & Keppel, P. (Eds) (2000). *Journal for Learning Through Music*. A publication of New England Conservatory of Music (nec-musicined.org).

Standley, J. M. (1996). A meta-analysis on the effects of music as reinforcement for education/therapy objectives. *Journal of Research in Music Education*, **44** (2), 105–133.

Storr, A. (1985). *Music and the mind*. New York: The Free Press.

Street & Scripp, J. M. (2002). Starting with music: the challenges of innovation in elementary school education. *Perspectives*, March, 4–6.

Suzuki, S. (1969). *Nurtured by love*. New York: Exposition Press.

Suzuki, S (1981). *Ability development from age zero*. Athens, OH: Ability Development Associates.

Tunks, T. (1992). The transfer of music learning. In: R. Colwell (Ed.), *Handbook of Research on Music Teaching and Learning* (pp. 437–447). New York: Schirmer Books.

Vaughn, K. (2000). Music and mathematics: Modest support for the oft-claimed relationship. *The Journal of Aesthetic Education*, **34** (3–4), 149–166.

Winner, E., Davidson, L. & Scripp, L. (Eds) (1992). *Arts PROPEL: A handbook for music*. Cambridge, MA: Harvard Project Zero.

Wolf, D. & Pistone, N. (1991). *Taking full measure: Rethinking assessment through the arts*. New York: College Entrance Examination Board.

Wolf, D. P. (1999). Why the arts matter in education or just what do children learn when they create an opera? In: E. Fiske (Ed.), *Champions of Change: The Impact of the Arts on Learning* (pp. 92–98). Washington, D.C.: The Arts Education Partnership and the President's Committee on the Arts and the Humanities.

The International Handbook on Innovation
Edited by Larisa V. Shavinina

Determinants of Technological Innovation: Current Research Trends and Future Prospects

Vangelis Souitaris

The Entrepreneurship Centre, Imperial College, London, U.K.

Abstract: This chapter is a review of several methodologies, which have been used to identify the distinctive characteristics of innovative firms (determinants of technological innovation). Some of the problems affecting this research field are the diverse nature and non-standardised definition and measurement of innovation itself, non-standardised measurements of the determinants, interrelated variables, different characteristics of firms targeted and finally different economic regions where the surveys take place. The chapter presents a portfolio model, which synthesises previous research results and may be used for country or industry specific studies.

Keywords: Innovation; Technological innovation; Determinants of technological innovation; Portfolio model.

Introduction

The evidence from the literature strongly supports the view that technological innovation is a major influence on industrial competitiveness and national development (Tidd, 2001; Zaltman et al., 1973). Some firms are more technologically innovative than others, and the factors affecting their ability to innovate are important to management scholars, practising managers, consultants, and technology policy-makers.

The author adopted the OECD definitions of technology and technological innovation. Technology can be interpreted broadly as the whole complex of knowledge, skills, routines, competence, equipment and engineering practice which are necessary to produce a product or service. A new product requires a change in this underlying technology. Technological innovation occurs when a new or changed product is introduced to the market, or when a new or changed process is used in commercial production. The innovation process is the combination of activities—such as design, research, market investigation, tooling up and management—which are necessary to develop an innovative product or production process (OECD, 1992).

The factors that affect a firm's innovation[1] rate are called 'determinants of innovation'. They derive from a wide range of aspects of the company such as internal and external communications, managerial beliefs, the financial situation, size, structure, quality of personnel, R&D effort, technical capabilities and market conditions (Souitaris, 1999).

This chapter examines the methodologies and tools researchers have used in order to identify the determinants of technological innovation, within organisations. The results of a large number of studies are summarised and a framework is extracted. Afterwards, some current research trends are presented. Finally a view of what should be done in the future is proposed, in order to expand upon our current knowledge.

On the Methodology of Studies on Determinants of Innovation

From as early as the late 1950s, much literature has been published on the determinants of technological innovation. To illustrate the amount of academic research in this area, Rogers (1983) refers to 3,085

---

[1] Wherever the word 'innovation' is used in the text, it always refers to technological innovation.

publications about the diffusion of innovation, of which 2,297 are empirical research reports. He also adds that the number of publications was almost eight times more in 1983 than in 1962 (Rogers, 1983, p. xv).

In order to present better the objectives and the methodology of these studies, we can use the following categorisations:

(1) Categorisation according to the approach.

   (a) Studies researching at the project level, looking for the determinants of success or failure of innovative projects.

   The main characteristic of these studies is that the sample comprises new technological projects. The objective is to correlate the success rate of the projects to a number of predefined possible determinants. This kind of research is known as the 'innovation or decision design' (Downs & Mohr, 1976) or the 'object approach' (Archibugi et al., 1994). Examples of the object approach are the following research works: Rubenstein et al. (1976), Rothwell (1977, 1992), Maidique & Zinger (1984) and Cooper (1979, 1999, 2002).

   (b) Studies researching at the firm level, looking for the determinants of the firms' ability to innovate.

   The unit of analysis in these studies is the firm, and this kind of research is known as the 'multiple innovation research' (Downs & Mohr, 1976) or the 'subject approach' (Archibugi et al., 1994). There are two possibilities in this approach: the variables can determine either the firm's rate of innovation, or its ability to succeed and to benefit from its innovative technology. The interest in the rate of innovation as a dependent variable stems from the implicit hypothesis that firms introducing innovation regularly are more likely to sustain a large number of successful innovative products and processes (even if some of the projects fail). Examples of the subjective approach are the following research works: Mohr (1969), Miller (1983), Ettlie et al. (1984), Khan & Manopichetwattana (1989a) and Hajihoseini & de la Mare (1995).

(2) Categorisation according to the number of tested determinants of innovation.

   (a) Studies testing a large set of factors. These studies try to identify important determinants of innovation, testing integrated models with a wide range of variables. They usually have the intention, using regression equations with the determinants as independent variables, of predicting the highest possible proportion of the variation of the dependent variable (innovation rate). Examples of research works of this kind are: Duchesneau et al. (1979), Miller & Friesen (1984), and Swan & Newell (1994).

   (b) Studies that test one or a few specific factors like, for example, participation in professional associations, or formalisation of structure. Usually those studies use more sophisticated and detailed measures of the variable(s), intended to identify a possible correlation between the tested variable(s) and the dependent variable (innovation). However, they are only able to explain a portion of the variance in the rate of innovation. Examples of research works of this kind are: Mansfield (1963) on size and structure, Sapolsky (1967), and Hage & Dewar (1973) also on structure, Kets de Vries (1977) on personality of the entrepreneur, Tushman & Scandel (1981) on technology gatekeepers, Miller et al. (1982) on top executive locus of control, Chon & Turin (1984) on structure and decision-making procedures, Newell & Swan (1995) and Swan & Newell (1995) on membership in professional associations (for detailed lists of studies see Chiesa et al. (1996) and Brown & Eisenhardt (1995)).

The Problem of the Inconsistency of the Results

To date, the research carried out has been unable to conclude on the relevant variables or their exact impact on innovation. Although similar variables have been tested by different researchers, the results have shown differing degrees of impact on innovation. Most often, from the tested set of determinants, different ones were found to be significantly correlated to the innovation rate in each empirical survey. In some cases there was even disagreement as to whether a factor actually correlated positively or negatively to the rate of innovation (Downs & Mohr, 1976; Wolfe, 1994). For example, firm size is a highly disputed variable (Khan, 1990). The instability of the determinants from case to case frustrates integrated theory-building efforts. (Downs & Mohr, 1976; Tidd, 2001; Wolfe, 1994).

Duchesneau et al. (1979) demonstrated this inconsistency of results, duplicating a large number of previous studies. They deliberately used the same measures of determinants, but the sample was from one specific industrial sector (footwear industry). Their results were different from the original studies, mainly concerning the relative extent to which different variables correlate to innovation.[2]

---

[2] We should mention here Damanpour's (1991) objection to the various assertions about instability of results. His view is based on a meta-analysis of previous results and is worth reading.

Having reviewed a large number of studies, the author has identified a number of possible reasons for the inconsistency of these results (Souitaris, 1999). Determinants of innovation, and in particular their degree of correlation to the rate of technological innovation, are dependent upon the following factors:

(1) Nature, definition and measurement of innovation itself.

The important determinants can be differentiated by the nature of innovation, for example high-cost vs. low-cost innovation, simple vs. complex innovation and incremental vs. radical innovation (e.g. Dewar & Dutton, 1986; Tornatzky & Klein, 1982). Determinants of high-cost innovation appear very different to those of low-cost innovation. Downs & Mohr (1976) found that a wealth of resources would predict the former very differently to the latter. Ettlie et al. (1984) found that, while 'incremental' innovation may be enhanced by a decentralised structure, 'radical' innovation requires a more centralised structure with particular emphasis on decision-making and a higher level of support and involvement from top management.

More recent innovation typologies are produced by Clark & Wheelwright (1992) (research or advanced development; breakthrough development; platform or generational; derivative or incremental) and Christensen (1997) ('sustaining' innovation vs. 'disruptive' innovation).

An additional problem is the lack of a standard definition of technological innovation (Garcia & Calantone, 2002). What is included in, or excluded from, the definition of technological innovation is an important issue which needs to be addressed. Should aesthetic improvements in the matters of style, design or re-packaging be included as technological innovation? What degree of change is required for a product or process to be considered as technological innovation? There is a difficulty in interpreting the terms used in definitions such as 'significant' or 'considerable' (Smith, 1988 outlines the variations in the definition of innovation). In addition, should the definition distinguish between product and process innovations or between the development of completely new products and the incremental modification of existing products in a systematic way? The different definitions and interpretations of technological innovation have led to variations in the identified determinants—hence the ongoing research interest into this subject (for good and current discussions of this problem see Garcia & Calantone, 2002; Souitaris, 1999; Tidd, 2001).

Also, there is no standard measurement of technological innovation. There are two levels of innovation measurement referred to in literature

(Duchesneau et al., 1979). Firstly, at the micro-level—where the adoption of a number of industry specific innovations is measured—innovations are selected as being representative, by a group of industrial experts or by the researchers reading industry specific magazines. Second, there is the aggregated level where the rate of innovation of a firm is measured in various ways such as the number of new products and processes, the percentage of sales due to new products or the number of patents filed. Whatever decision is taken on the measurement of innovation to be used in a particular study can influence the results on the innovation determinants.

(2) Measurement of the determinants of innovation.

In the literature, there are two types of determinants of innovation. The first type includes variables measuring facts such as the size of the firm, number of graduates, size of innovation budget, etc. The above are straightforward and easy-to-measure variables, with highly reliable measurements. They are also easily transferable over different studies. For instance, the number (or the logarithm) of personnel is a generally standardised measure of a firm's size (see, for example, Kimberly & Evanisko, 1981).

The second type of variables includes perceptions and attitudes of the respondents (such as perceptions of the intensity of competition or attitudes towards risk-taking), as well as general and usually subjective concepts (like centralisation of power,[3] complexity of knowledge[4] and awareness of strategy). Many of those variables are then broken down into a number of items and are measured using scales (see section 'The Portfolio Model of Starting Variables' for references on individual constructs and measures).

This second type of variables is no less important a predictor of innovation performance than the hard measures of the first type. However, the definitions of the soft variables are subjective and depend upon the author's perception. There are no consistent definitions and measures of concepts as general as the 'scanning of the environment' or 'environmental heterogeneity' (Miller & Friesen, 1984), or 'formalisation'[5] and 'centralisation' (Hage & Aiken, 1970). The concepts may be similar but the way in which they are actually broken down into variables and measured using scales differs. The different definitions and

---

[3] Degree to which power and control in a system are concentrated in the hands of relatively few individuals.
[4] Degree to which an organisations' members possess a relatively high level of knowledge and expertise.
[5] Degree to which an organisation emphasises following rules and procedures in the role performance of its members.

measurement of determinants referring to similar concepts makes the comparison of results more difficult.

For example, it is difficult to compare a general variable called "scanning of the environment" (Miller, 1983)—which is measured using a large number of scale items including many forms of communication with the external environment— with specialised variables like "the number of contacts made each year with representatives of machinery suppliers" and "the number of trade journals read or scanned by innovation decision makers" (Webster, 1970). In both cases the variables refer to the same concept of collecting information but they are not directly comparable, due to the different scope of the definitions.

Another problem for someone trying to compare previous results is that not all studies define the measurements of each variable clearly. Many of them (especially journal papers) give just lists of determinants, underestimating the importance of their actual measurement (for example, in the seminal work of Miller (1983), the measurement of the variables is not clear).

In addition to this, it is worth mentioning that most of the variables are interrelated, and this creates problems in the interpretation of the results. For instance, size is probably a surrogate measure of several dimensions that lead to innovation such as total resources, slack resources and organisational structure (Rogers, 1983). These relationships between variables and their effect on the final results are not easy to understand clearly due to the complexity of the issue.

(3) Effect of different stages of innovation process on innovation rate.

Inconsistent results and low correlations of organisational structure variables with innovation can also be caused by some of the variables being related to innovation in one direction during initiation of innovation, and in the opposite direction during implementation of innovation. It has been argued that low centralisation, high complexity and low formalisation can facilitate initiation in the innovation process and that the same structural characteristics can also hinder the implementation of an innovation within an organisation (Sapolsky, 1967; Zaltman et al., 1973).

(4) Different kind of firms used as sample.

Miller (1983), Khan & Manopichetwattana (1989b) and Damanpour (1991), among others, found that different types of firms show different determinants of technological innovation. For example, Rothwell, (1974, 1977) showed that factors associated with innovation were significantly different, or at least showed a different order

of importance, in different industrial sectors. For example, within the chemical industry, technical factors were most important, while in the scientific instruments industry, market factors dominated (Rothwell, 1974). Mohr (1969) has also referred to a moderating effect of the size of the firm on the relative importance of its determinants of innovation. For example, top management characteristics and attitudes were found to be more important innovation determinants for small firms, due to the more active involvement of top managers in the innovation process (also see Carrier, 1994; Lefebvre et al., 1997).

These and similar findings indicated that studies which only look at the industry as a whole cannot be generalised. They cannot be directly compared to studies, which deal only with one industrial sector, or one particular type of organisation (e.g. multinational). The fact that different results are achieved, according to the type of firm, illustrates the problem involved in trying to achieve a unified theory of determinants of innovation, which can be applied to all situations.

(5) Different geographical regions in which the empirical surveys take place.

Much of the available literature in this area is biased towards investigating determinants of innovation in the U.S. or other industrialised Western countries. Often the importance of the region in the interpretation of the results is overlooked (Boyacigiller & Adler, 1991; Drazin & Schoonhoven, 1996).

White (1988) and Souitaris (1999), among others, indicated that the characteristics of innovative firms are strongly influenced by economic development and the management culture in the region.

In conclusion, the fact that there seemed to be no unified theory concerning the determinants of innovation has reduced the amount of published studies and the effort devoted to the subject after the late 1970s. Rogers (1983) argued that after several hundred studies of organisational innovativeness were completed in the 1950s, 1960s and 1970s, this approach to innovation in organisations became passé. However, the problem was not resolved, and the crucial question of what the determinants of innovation are still remains open.

The Portfolio Model of Starting Variables

Forrest (1991) and Tidd et al. (1997), among others, argued that there is no one best way of managing the innovation process as it depends on firm specific circumstances. Nelson & Winter (1977) introduced the concept of 'routines', which are particular ways of behaviour which emerge as a result of repeated experiments and experience around what appears to be good practice. Different firms use different routines with various degrees of success. There are general recipes from which general suggestions for effective routines can be derived, but they must be customised to particular organisations and related to particular technologies and products (Tidd et al., 1997).

It is difficult to produce a universally applicable model of the determinants of technological innovation. Differences in the industrial sectors and geographical regions all have an effect, which is very hard to quantify or exclude. Taking this into account, the author has developed a working 'portfolio model' of potential determining variables (Souitaris, 1999, 2002). The full list of determinants in the model is not always applicable—there are different sets of important determinants, depending upon certain environmental dimensions that underlie the analysis (such as economic development and managerial culture). The study's model is intended to be a starting point for empirical research, in order to explore the contingencies.

The routines associated with innovation are extensive, and their strength of association is specific to particular conditions for reasons explored in the previous section. However, collectively the determinants of innovation tend to cluster around key themes (Tidd et al., 1997) presented in Table 1. The table demonstrates a comparative presentation of models in the literature that attempt to integrate the determinants of innovation. Common classes of factors appear throughout the different models focusing on 'context' (external environment and firm's profile), 'strategy',

'scanning external information' and 'organisational structure'.

Innovation textbooks (such as Ettlie, 2000; Tidd et al., 2001; Trott, 1998) and papers with practical orientation (Bessant, 2003; Cooper, 1999, 2002) advise students and practitioners drawing from this generally acceptable body of knowledge. However, despite the apparent similarity of integrative models of determinants of innovation at the aggregated level, there is more variety when it comes to operationalisation and empirical testing. The literature includes a large number of individual indicators falling into the above general variable categories;[6] The variables in our portfolio model were categorised in four classes, in line with the integrative models of determinants of innovation reviewed previously (see Table 1 and Fig. 1).

The presentation of the portfolio model which follows covers two types of sources: (1) conceptual works that introduced the general themes and proposed their relationship with firm innovation; and (2) studies (mainly empirical) that associated innovation rate with specific indicators, within the general themes.

(1) Contextual Variables.

Organisations are viewed by several theoretical perspectives as adaptive systems, and this suggests that contextual variables may have a causal influence on strategy and structure. Examples of such theoretical perspectives are the contingency

---

[6] Chiesa et al. (1996) and Souitaris (1999) offered detailed literature-based frameworks of operational indicators.

Table 1. *A comparison of 'integrated' models of determinants of innovation.*

Miller & Friesen (1984)	Khan & Manopichet- wattana (1989)	R. Miller & Blais (1992)	Rothwell (1992)	Tidd, Bessant & Pavitt (1997)	Souitaris (1999)	This Portfolio
Environment	Competitive environment	Context			Economic variables	Context
	Firm's profile		Corporate conditions			
Decision- making	Strategy	Strategy		Strategy*	Strategic variables	Strategy
	Entrepreneurial attitudes					
	Functions	Process		Implementation mechanisms		
Information- processing			Tactical variables	External communications	External communications	External communications
Structure	Structure	Structure		Organisational context	Internal capabilities	Organisational context

* Contextual variables included in the 'strategy' theme.

Contextual variables

Firm's Profile
- years of operation
- growth rate of size
- growth rate of sales
- growth rate of profits
- earnings from exports

Competitive environment
- perception of rate of changing customer needs
- perception of intensity of competition

External communications

Communication with the stakeholders
- consultation with customers in person
- consultation with panels of customers
- feedback from customers through post or phone
- consultation of suppliers of raw materials
- consultation of suppliers of equipment
- use of market research

Networking - Scanning external information
- acquiring information from public agencies
- exchange information with other domestic firms
- international contacts
- membership of professional associations
- subscription to scientific and trade journals
- attendance of trade fairs
- access and use of the world wide web
- use databases to search for new technology
- existence of technology gatekeeper
- monitoring the competitors

Co-operation with external organisations
- universities and research institutions
- public and private technology consultants
- Other firms (joint R&D projects)
- Licensing
- financial institutions (borrowing for R&D)
- absorption of public technology funds

 Rate of technological innovation

Strategic variables

Innovation budget
- existence of an innovation budget
-consistency of the innovation budget

Business strategy
- definition of the business strategy
- communication of the business strategy
- inclusion of new technology plans
  in the strategy
- horizon of the strategy

Management attitudes
- locus of control
- attitude towards risk
- attitude towards new technology pay-back period
- perception about performance gap

CEO's profile
- age
- owner vs. Appointed

Organisational Competencies

Technical competencies
- intensity of R&D
- intensity of quality control

Market competencies
- strength in marketing
- breadth of distribution system

Education of personnel
- proportion of staff with a university degree
- proportion of staff with engineering or science
  degree

Breadth of experience of personnel
- proportion of staff performing a managerial role
- proportion of engineers-scientists-managers
  with experience in another company
- proportion of engineers-scientists-managers
  with experience abroad

Training of personnel
- intensity of training of engineers and managers
- intensity of training of production employees

Internal Processes
- degree of formalisation
- slack (thinking) time of engineers and managers
- interdepartmental teams working on innovation
  projects
- existence of project champion
- extent of formal internal communication
- incentives to employees to encourage new ideas

Figure 1. The portfolio model of determinants of innovation.

Source: Souitaris (2002a).

theory (Burns & Stalker, 1961; Donaldson, 1996), institutional theory (Parsons, 1966), resource dependence (Aldrich, 1979; Barney, 2001), population ecology (Hannan & Freeman, 1977), and industrial economics (Freeman, 1982). The literature also includes interesting discussions on the impact of environmental variables (Miller & Blais, 1992). There are those who consider the impact of the environment on the firm's strategy and behaviour to be highly important (Weber, 1947), and others who claim that it is the organisations which select and even structure their environment (Miller, 1989). For an excellent review and critique of all the above main theories in the 'environmental school' of the management literature see Mintzberg et al. (1998).

The current study has used two types of contextual variables in the portfolio model:

(a) Firm's profile: Literature in this area connects innovation with factors such as the age of the firm (Nejad, 1997), growth rate (Smith, 1974), profitability (Mansfield, 1971) and earnings from exports (Calvert et al., 1996);

(b) Competitive Environment: Evidence in the literature points to the fact that the high rate of change of customer needs and intense competition are closely associated with a high innovation rate (Khan & Manopichetwattana, 1989a; Miller & Friesen, 1984).

(2) Strategy-Related Variables.

A firm's strategy can be viewed as a network of decisions, which need to be made in order to position the firm within its environment and to create the organisational structure and processes. Since the 1960s, when the idea of corporate strategy was first noted, there has been much debate between the two main schools of thought: the 'rationalist' school (Ansoff, 1965) and the 'incrementalist' school (Mintzberg, 1987). Porter (1980) explicitly linked technology to 'five forces' which drive competition within the industry (bargaining power of suppliers, threat of new entrants, bargaining power of buyers, threat of substitutes and intensity of rivalry). Porter's 'rationalist' approach suggests that managers need to analyse the external environment and, based on this analysis, they must define a course of action. However, the 'incrementalists' Teece & Pisano (1994) suggested a different approach to corporate strategy, that of 'dynamic capabilities' underlining the importance of dynamic change and corporate learning.

Cooper (1984) was one of the first of the empirical scholars to identify an association between corporate strategy and innovation performance of firms (see also Cooper's chapter in

this volume). Our model incorporates four subsets of strategy-related indicators:

(a) *Innovation budget.* Literature showed that where there is a budget for innovation and in particular when this budget is consistent over time, the rate of innovation will be increased (Khan, 1990; Twiss, 1992);

(b) *Business strategy.* In firms with a well-defined business strategy, including plans for new technology, the rate of innovation was found to be higher (Rothwell, 1992; Swan & Newell, 1995). Moreover, those firms, which had a strategy with a long-term horizon and could communicate it to their employees, showed a higher rate of innovation (Khan & Manopichetwattana, 1989a);

(c) *Management attitude.* Literature also indicates that top managers of the more innovative companies have an internal 'locus of control'. They consider that the performance of their firm depends on manageable practices rather than the influence of external environmental factors which they cannot control. (Miller et al., 1982). In addition, the top managers of the most innovative firms appear to have less fear of risk-taking (Khan & Manopichetwattana, 1989b) and recognise that in a shorter-than-expected time scale, the new technology costs can be recovered (Eurostat, 1996). Finally, these managers consider that there is a 'performance gap' between how the firm currently performs and how it could perform in an ideal situation (Duchesneau et al., 1979);

(d) *CEO's profile.* This particularly relates to the age and status of the CEO—i.e. whether he/she is also the owner or an appointed executive. The literature implies that a younger CEO who is also the owner will be more receptive to innovation (Khan & Manopichetwattana, 1989b).

(3) External Communications.

Another positive influence on the rate of innovation identified in the literature is the acquisition and scanning of information (Tidd et al., 1997). Therefore, three subsets of innovation-related communications variables have been incorporated in the model.

The first subset comprises the factors related to *communication with the firms' stakeholders*. These are:

(a) Customers: personal meetings (Chiesa et al., 1996; Maidique & Zinger, 1984; Rochford & Rudelius, 1992), panel discussions (Chiesa et al., 1996), postal or telephone feedback (Chiesa et al., 1996), or quantitative market research for a broader customer profile (Khan & Manopichetwattana, 1989b);

(b) Suppliers of machinery and equipment: (Duchesneau et al., 1979; Rothwell, 1992).

The second subset incorporates the factors related to the *collection and scanning of information* which can be from sources such as public agencies (Carrara & Duhamel, 1995) or other firms (Alter & Hage, 1993; Bidault & Fiscer 1994; Trott, 2003). The membership of professional associations, (Swan & Newell, 1995), subscription to scientific and trade journals (Khan & Manopichetwattana, 1989b), attendance at trade fairs (Duchesneau et al., 1979), access to and use of the Internet, and use of electronic patent and research databases to search for new technology are other ways of collecting information on innovation, albeit less direct. A 'technology gatekeeper'—i.e. someone whose role is specifically to search for information on new technology—is another determining variable according to some literature (Allen, 1986; Rothwell, 1992). Finally, simply by monitoring one's competitor's activities, a great deal of useful and critical information can be discovered (Chiesa et al., 1996).

The third subset refers to the *co-operation of the firm with third parties* such as universities and research institutions (Bonaccorsi & Piccaluga, 1994; Lopez-Martinez et al., 1994); public and private consultants (Bessant & Rush, 1995; Pilogret, 1993); other firms in the form of joint ventures (Alter & Hage, 1993; Rothwell, 1992; Swan & Newell, 1995) or licensing (Lowe & Crawford, 1984); and financial institutions as a source of venture capital (EUROSTAT, 1996). The absorption of public technology funds, where they are available, can be another determinant of innovation. (Smith & Vidvei, 1992).

(4) Variables Related to the 'Organisational Context'.

Bureaucracy theory (Weber, 1947), classical management (Gulick & Ulrick, 1938) and organisational sociology (Blay & Schoenherr, 1971) all emphasise the dominant influence of the structural attributes of an organisation on its behaviour. However, this appears to work both ways—while predefined structural factors may either hinder or encourage innovation, yet others insist that structure can be modified as a function of strategy to enhance the innovative potential of firms (Miller & Blais, 1992).

The organisational competencies incorporated in the portfolio model are classified into six subsets and are all based on the empirical literature:

(a) *Technical competencies.* Both the intensity of R&D (Ducheneau et al., 1979; Ettlie et al., 1984) and the intensity of quality control (Rothwell, 1992; Zairi, 1996) are associated with innovation;

(b) *Market competencies.* Cooper (1984) and Maidique & Zinger (1984) and Veryzer (2003) associated an effective marketing programme and a broad distribution system with innovation.

(c) *Education of personnel.* In firms which had a higher number of educated and technically qualified staff, there appeared to be a more responsive attitude to innovation (Carter & Williams, 1957; Nejad, 1997). Miller & Friesen (1984) suggested that 'technocrats' came up with more than average innovative ideas.

(d) *Breadth of experience of personnel.* The broader the base of employees within a firm who had managerial responsibilities, the higher the rate of adoption of innovations (Becker & Stafford, 1967). Organisations in which the staff have more varied backgrounds, for example working experience in other companies and/or abroad will generally have a more positive attitude towards innovation (Carroll, 1967). Such employees can often suggest and implement ideas for innovation.

(e) *Training.* Hage & Aiken (1970) and Dewar & Dutton (1986) associated innovation with 'knowledge depth', measured by the level of professional training. On-the-job training has also been linked to the rate of innovation by more recent authors (Nejad, 1997; Swan & Newell, 1995)—this training refers to both professional training for engineers and managers and technical training offered to the production employees.

(f) *Internal 'process' variables.* Innovative companies are less formalised than non-innovative ones (Cohn & Turin, 1984). The business innovative performance can be also enhanced by introducing thinking (or 'slack') time for engineers and management (EUROSTAT, 1994) and by using cross-functional interdisciplinary teams (Clark & Fujimoto, 1991; Cooper, 1990; Hise et al., 1990).

Another critical factor influencing innovation is the existence of a 'project champion' (Cooper, 1979; Hauschildt, 2003; Rothwell, 1992). The 'project champion' is an individual who dedicates herself to an innovation project and will give a personal commitment to fulfilling that project (Scon, 1973). Burns & Stalker (1961), Rogers & Shoemaker (1971) and Rothwell (1992), have identified an association between internal communication and technological innovation. Finally, authors such as Felberg & DeMarco (1992), Twiss (1992) and Chiesa et al. (1996) made a case that a firm's innovation potential can be enhanced by allowing employees to generate their own new ideas, by encouraging the circulation and communication of such ideas, and by offering incentives of some form to the employees.

It is worth stressing again that the model of this study was not intended to be exhaustive. The factors that can be related to innovation are numerous and possibly change over time as management practice is a dynamic process.

More Recent Issues and Considerations

Measuring Innovation—Using Portfolios of Indicators

One of the major problems facing innovation research is the absence of common ground or definition. The most commonly used indicators at the 'aggregated level' are technology-based ones, including capital expenditure, expenditure on research and development and patent activity (OECD, 1982; Tidd et al., 1996) and these have been used for the longest time. The strengths and weaknesses of technological indicators have long been recognised (see Pavitt & Patel, 1988; Smith, 1992). Although the definitions of these indicators are relatively consistent and data are collected on a routine basis, it can be argued that they measure innovation input (effort towards innovation) rather than innovation output (actual results from the innovation effort).

More recently, there has been a tendency for those undertaking innovation surveys to use innovation output or 'market' indicators, such as the number of new products and new processes adopted during a specific time period (for good reviews of innovation surveys see Archibugi et al., 1994 and Smith, 1992). These 'innovation-count' indicators have the drawback that products and processes are not directly comparable across different industries. Neither can they account for the economic significance of the innovations (Smith, 1992).

As a response to these disadvantages, some researchers have used 'impact' indicators, which attempt to collect data on the proportion of sales directly related to new products over a particular time period (see for example Meyer-Krahmer, 1984). These indicators show the rate at which a firm changes its product lines and vary across different industries and probably over time. However, impact indicators are good measures of both technological newness and economic significance (Smith, 1992).

Empirical literature seems to have suffered from inconsistent results over the years because of the difficulty in capturing the complexity of innovation with a simple, accurate measure (Duchesneau et al., 1979). Saviotti & Metcalfe (1984) and Tidd (2001) suggested that multi-indicators of innovation can offer a more complete picture of innovation performance, since the issue could be investigated from several different points of view and the problem of incomplete-

ness of each one of the individual measures could be minimised. Hence, I propose a portfolio of seven widely used innovation indicators:

(1) Number of incrementally innovative products introduced in the past 3 years;
(2) Number of radically innovative products introduced in the past 3 years;
(3) Number of innovative manufacturing processes introduced in the past 3 years;
(4) Percentage of current sales due to incrementally innovative products introduced in the past 3 years;
(5) Percentage of current sales due to radically innovative products introduced in the past 3 years;
(6) Expenditure for innovation in the past 3 years over current sales. This includes R&D expenditure as well as a wide set of other expenditures related to innovation, such as the acquisition of technology and know-how, tooling up, industrial engineering, industrial design, production start-up, training linked to innovation activities and marketing of new products;
(7) Number of patents acquired in the past 3 years.

Three types of indicators are used in the above portfolio:

(1) 'Input' measures (variables 6 and 7) indicating the effort made towards innovation;
(2) 'output' measures (variables 1, 2 and 3) capturing the rate of implementation of innovation; and
(3) 'impact' measures (variables 4 and 5) indicating the impact of the company's innovative products.

Each type of measure is in itself incomplete (see Hansen, 1992; Smith, 1992; Souitaris, 1999), but collectively they can be used to measure innovation activity. They have now been accepted by the OECD as standardised tools for future innovation surveys (OECD, 1992). One of the limitations of this indicator-portfolio model is its inability to capture innovation failure and therefore to reveal the project success vs. failure ratio (Smith, 1989). This is a limitation which has to be acknowledged, because the level of analysis is the firm as a whole and not the individual project.

Two more 'composite' indicators that future readers might want to consult before selecting their innovation measures are presented by Hollenstein (1996) and Coombs et al. (1996).

Narrowing the Scope—Taxonomies of Firms with Similar Determinants of Innovation

As a response to the inconsistency of the innovation determinants, the contingency school of thought emerged (see Burns & Stalker, 1961; Downs & Mohr, 1976; Tidd et al., 1997, 2001), suggesting that there is no universal 'best' way to manage innovation as the phenomenon is context-specific. In order to make the results more meaningful and comparable, Wolfe (1994) urged future researchers to define clearly the contextual settings of their surveys (i.e. the stage of the innovation process, the innovation attributes and the organisational context).

Many researchers have realised that innovation determinants can vary in different contexts and have narrowed down the scope of their work. Some have decided to concentrate on a narrow range of firms—by selecting similar firm size (for example small and medium-sized enterprises) or firms of the same industry. For instance, Khan & Manopichetwattana (1989a) and Rothwell (1978) focused only on small firms and Duchesneau et al. (1979) on the footwear industry.

Some authors proposed taxonomies of firms with different determinants of technological innovation. The first taxonomy was proposed by Burns & Stalker (1961). They distinguished between 'mechanistic' and 'organic' organisations. Mechanistic organisations have a lower complexity, higher formalisation and centralisation, and lower internal and external communication than organic organisations. In the 1980s, Miller & Friesen (1984) identified two types of firm configurations[7] with different innovation determinants. These were 'conservative' firms with positive and significant correlation of innovation with information-processing, decision-making and structural variables, and 'entrepreneurial' firms with negative correlation of innovation with information-processing, decision-making and structural integration variables. The structure of the firm took a back seat while the goals and strategies of the company were viewed as the more important driver for innovation. Khan & Manopichetwattana (1989b) developed five clusters of small firms with different strategy, structure and managerial attitudes. Each of these was shown to have its own specific factors determining innovation.

These taxonomies (for good discussions on innovation taxonomies see Souitaris (2002a) and Tidd (2001) had an unquestionable and novel value in the innovation management literature, accepting that the characteristics of highly innovative firms are specific to particular conditions and trying to identify clusters of firms with common important determinants of innovation. However, the proposed classifications were weighted towards perceptual criteria (such as the risk-taking, proactiveness, entrepreneurial strength and belief in luck) rather than the factual measures such as size, industrial sector and common innovation type, hence the ongoing requirement for development of more precise and factual taxonomies. These could help to clarify conflicting research results on determinants of innovation (see section 'The Problem of the Inconsistency of the Results').

---

[7] Readers that would like to know more about the configuration school in strategic management (supporting the idea of taxonomies), should refer to Miller (1986 and 1996) and Dess et al. (1993).

The author of this chapter tested the applicability of Pavitt's (1984) taxonomy (which derived from the economic school of thought) as an effective factual classification that could benefit the management literature searching for the determinants of innovation. Pavitt suggested that industrial sectors differ greatly in the sources of technology they adopt, the users of the technology they develop, and the methods used by successful innovators to appropriate the benefits of their activities. He produced a simple and practical classification with four categories of firms:

(1) *'Supplier-dominated firms'*. These firms are usually small, they do not place much emphasis on R&D, and they have lower engineering capabilities. They take the majority of their innovative ideas from firms, which supply them with equipment or materials;

(2) *'Large-scale producers'*. These firms tend to be much bigger and instigate their own process technologies. They concentrate their resources in this area, and usually diversify vertically into technological equipment, which is related to their own technology. As a result they contribute to a large extent to innovation in all sectors of their activity;

(3) *'Specialised suppliers'*. These firms tend to be smaller, perhaps mechanical or instrumental engineering firms. They also produce a high proportion of their own process technologies but focus more of their innovative activities on new products for use in other sectors. There is little diversification of technology and a relatively small contribution to innovations produced in their principal sector of activity. Their end users and other firms outside the sector make a more significant contribution;

(4) *'Science-based firms'*. These companies are usually firms in the chemical, pharmaceutical and electrical and electronic engineering sectors, whose main source of technology is internal R&D. They produce a relatively high proportion of their own process technology and of product innovations used in other sectors of the industry. Usually large, most of their technological diversification is within the corporation, and they produce a relatively high proportion of all the innovations made in their principal sector of activity.

Pavitt's taxonomy was selected for the test because it produced firm classes with a similar size, industrial sector and innovation type (three important moderators causing result instability in the management literature). The author expected that a simultaneous 'control' of all the three moderators would reduce the variation of the innovation determinants within classes and increase the variation across classes.

An empirical test in a sample of 105 Greek companies showed that firms in different trajectories (categories of firms) of Pavitt's taxonomy showed differences in the rate of technological innovation (for the detailed results of the study, see Souitaris (2002a)). Innovation for 'supplier-dominated' firms was related to the competitive environment, acquisition of information, technology strategy, risk attitude and internal co-ordination. For 'scale-intensive' firms the important determinants were related to the ability to raise funding as well as the education and experience of personnel. For 'specialised suppliers' innovation was associated with high growth rate and exporting, as well as training and incentives offered to the employees to contribute towards innovation. 'Science-based' firms depended upon technology-related variables, education and experience of personnel, growth in profitability and panel discussions with lead customers.

Using Pavitt's taxonomy management scholars can simplify the problem of multi-dimensional moderation. Size, industrial sector and type of innovation are combined into a single dimension: the 'technological trajectory'.

On the basis of my own research described above, I propose a 'two-step' methodology to identifying the distinguishing characteristics of innovative firms:

(1) A classification of firms according to 'industrial-level' moderators. Pavitt's taxonomy has a high practical value at this level. It conveniently aggregates 'industrial-level' factors, producing four sectoral firm classes, rather than a long list of sectors;
(2) Identification of a set of management-related determinants of innovation specific to each sectoral class. In practice, this method offers the opportunity to customise innovation questionnaires and measure the right 'type' of variables according to the firm's class.

The International Dimension

Most of the empirical research on the determinants of innovation has been carried out in industrialised developed countries. Recently, there has also been some interest in the particular conditions in Asian Newly Industrialised Countries (NICs) (see Hobday, 1995; Kim et al., 1993), in developing countries such as Iran (Nejad, 1997) and in transition economies in Eastern Europe (Inzelt, 2003). Several authors suggested that using the findings of innovation studies in technologically advanced countries to explain the innovative behaviour in countries with a less developed technological base is likely to be inappropriate (Drazin & Schoonhover, 1996; Mishra et al., 1996; Nejad, 1997; Souitaris, 1999).

A number of research paradigms have attempted to explain the international differences in technological development and innovation at a conceptual level. Neo-classical economic theorists have placed emphasis on the importance of a local supply of skills, specific local demands, openness of communication, pressure from competition and market structure (Nabseth & Ray, 1974; Porter, 1990). The 'national innovation systems' paradigm underlined the important role of deliberate intangible investment in technological learning activities (involving institutions such as other firms, universities and governments and the links among them). Innovation systems theorists also stressed the national incentive structures of temporary monopoly profit from innovation and the firm-specific competencies (Lundvall, 1998; Patel & Pavitt, 1994). The neo-contingency school of thought put forward the case for the way in which the diffusion and utilisation of innovation in different countries could be affected by systematic differences in business strategies, organisational forms and specific social processes, all of which are mutually dependent (Slappendel, 1996; Sorge, 1991). The neo-institutional theorists placed more importance on the prevailing national institutional frameworks and networks (e.g. professional associations). These could create standards of best practice which would encourage some technologies to be diffused more widely than others (Di Maggio & Powell, 1983; Swan et al., 1999).

In spite of all the research into the national differences in the patterns of technological innovation, there is a need for more empirical research in order to fully understand the complexity of the issue (Moenaert et al., 1994; Patel & Pavitt, 1994; Swan et al., 1999). Moenaert et al. (1994) proposed an operational framework for future empirical research, which combines the elements of most of the conceptual paradigms. According to Moenaert et al. the innovation process in different countries depends upon four 'socio-economic' dimensions: technological heritage, administrative heritage, market structure and regional entrepreneurship with additional influence of the national 'cultural context'.

I have attempted to use this framework in order to empirically identify the determinants of innovation in Greece (an example of a European newly industrialised nation with a less developed technological base).[8] The 'Greek studies' (Souitaris, 2001a, 2001b, 2002b) are based on a sample of 105 manufacturing companies in Greece, and the results are briefly summarised below.

Major-importance 'organisational competencies' determining innovation were found to be the intensity of R&D, strength in marketing, proportion of university graduates and engineers in the staff, proportion of staff with managerial responsibility, proportion of professional staff with previous experience in another company and incentives offered to the employees to contribute to innovation (Souitaris, 2002b). Regarding

---

[8] The average GDP per person is $11,739 per annum, which indicates a medium-level development compared for instance to $23,478 per annum for a large Western European country like the U.K. and $1,352 per annum for a developing country like Iran (Economist, 1998).

'strategic variables' important determinants of innovation included incorporation of technology plans in the business strategy, managerial attitude towards risk, perceived intensity of competition and rate of change of customer needs and finally status of the CEO (owner-CEOs were associated with a higher innovation rate than appointed CEOs). In general, top-management characteristics proved to be more important 'strategic' determinants of innovation for the Greek firms than corporate practices (Souitaris, 2001b). Regarding 'external communications' the empirical results supported two hypotheses for industrialising countries (proposed by Souitaris, 2001a): (1) searching for product-specific information is more important for innovation than scanning more general market and technological information; (2) the co-operation with partnering organisations (such as investing firms and joint venture partners) is more important for innovation than the co-operation with assisting organisations (such as universities, consultants or government agencies) (Souitaris, 2001a).

A common observation which emerged from the 'Greek studies' is that the 'major importance' determinants were generally scarce in the country's context. For example, the Greek national culture is generally risk-averse (Hofstede, 1991), but the attitude towards risk was a highly important variable (Souitaris, 2001b). In Greece there is a low indigenous production and supply of technology (Giannitsis & Mavri, 1993), but the R&D intensity and the incorporation of technology planning into the business strategy were important predictors of a high innovation rate (Souitaris, 2001b, 2002b). The Greek market has a traditionally low level of competition because of protectionism measures, but the perception of intense competition and demanding customers was strongly associated with a high innovation rate (Souitaris, 2001b). Despite the fact that Greece suffers from an outdated educational system which does not consider the needs of the industry (Tsipouri, 1991), education-related variables proved important determinants of innovation (Souitaris, 2002).

The findings of the Greek studies put forward the hypothesis that the most important determinants of innovation in newly industrialised countries are those which are generally absent in the country-specific institutional market and social context. In other words, the most innovative companies are those which can overcome the traditional rigidities of the context of their countries and incorporate rare attitudes and practices for the local business environment. This hypothesis requires further testing in innovation research.

Where are We Going from Now?

As time passes by and management styles evolve, new determining variables appear, and the relative importance of the old ones changes. Hence, it is recommended that holistic empirical surveys be carried out periodically, to act as yardsticks of our current knowledge. Qualitative methodologies such as observation and case studies would be useful from time to time in order to explore the perceptions of practising managers, to capture emerging determinants and to identify new lines of thinking for further quantitative research.

The fact that the results show different patterns depending on the region and/or the sectoral class, should be accepted and lived with. Hence, instead of devoting time and resources to the search for a unified theory of innovation, we can use portfolio models such as that presented in this chapter as a starting point and then identify the determining variables with the highest predictive power for the particular context. Using the set of important determinants as a base, auditing systems can then be developed putting the research results into practice.

In my view, the most fruitful direction for further research would be to untangle the 'black box' of the contingency theory. Contingency theory has been accused of having rather abstract and vague dimensions of the environment (Mintzberg et al., 1998). We need to map what determinants work under what exact environmental circumstances. Despite the fact that this is a highly complex problem due to the number of intervening variables, I propose work in two directions.

(1) Empirical research on the important determinants of innovation in countries and regions with different managerial cultures and stages of economic development. International surveys carried out under exactly the same conditions (same industries and same measurements for innovation and its determinants) would be particularly useful;
(2) Empirical research in order to confirm and establish the use of taxonomies, such as Pavitt's 'technological trajectories'. The creation of taxonomies of firms is encouraged in theory development, as it allows large amounts of complex information to be collapsed into more convenient categories, which are easier to comprehend (Carper & Snizek, 1980).

We always have to keep in mind that research on the determinants of innovation can have immediate usable and practical outcomes. The results of these studies will be valuable for: (1) company managers and consultants who want to identify the keys to high rate of innovation and (2) public policy-makers, who can see the impact of general 'infrastructure' variables like education, training, venture capital and information on the company's innovation potential.

Acknowledgments

The author would like to thank Dr. R. F. de la Mare for his contribution to these ideas and particularly Deborah

Salmon for her valuable language-related editing of the text.

References

Aldrich, H. E. (1979). *Organisations and environments*. Englewood Cliffs, NJ: Prentice-Hall.

Allen, T. J. (1986). *Managing the flow of technology*. Cambridge, MA: MIT Press.

Alter, C. & Hage, J. (1993). *Organizations working together*. Newbury Park, CA: Sage.

Ansoff, I. (1965). The firm of the future. *Harvard Business Review*, Sept–Oct., 162–178.

Archibugi, D., Cohendet, P., Kristensen, A. & Schaffer, K. (1994). *Evaluation of the community innovation survey*. EIMS Project Mr. 93/40, European Commission DG XIII D.

Barney, J. B. (2001). Is the resource-based view a useful perspective for strategic management research? Yes. *Academy of Management Review*, **26** (1), 41–57.

Becker, S. W. & Stafford F. W. (1967). Some determinants of organisational success. *Journal of Business*, **40**, 511–518.

Bessant, J. (2003). Challenges in innovation management. In: L. V. Shavinina (Ed.), *International Handbook on Innovation*. Oxford: Elsevier Science.

Bessant, J. & Rush, H. (1995). Building bridges for innovation: the role of consultants in technology transfer. *Research Policy*, **24**, 97–114.

Bidault, F. & Fischer, W. (1994). Technology transactions. Networks over markets. *R&D Management*, **24** (4), 373–386.

Blay, P. M. & Schoenherr, R. A. (1971). *The structure of organisations*. New York: Basic Books.

Bonaccorsi, A. & Piccaluga, A. (1994). A theoretical framework for the evaluation of university–industry relationships. *R&D Management*, **24** (4), 229–247.

Boyacigiller, N. A. & Adler, N. J. (1991). The parochial dinosaur: Organisational science in a global context. *Academy of Management Review*, **16** (2), 262–290.

Brown, S. & Eisenhardt, K. M. (1995). Product development: Past research, present findings and future directions. *Academy of Management Review*, **20** (2), 343–378.

Burns, T. & Stalker, M. (1961). *The management of innovation*. London: Tavistock.

Calvert, J., Ibarra, C., Patel, P. & Pavitt, K. (1996). *Innovation outputs in European industry. Proceedings of the EU conference Innovation measurement and policies*. Luxembourg.

Carper, W. B. & Snizek, W. E. (1980). The nature and types of organisational taxonomies: An overview. *Academy of Management Review*, **5** (1), 65–75.

Carrara, J. L. & Duhamel, M. (1995). *Technology brokers in Europe*. Brussels: European Commission, EIMS publication.

Carrier, C. (1994). Intrapreneurship in large firms and SMEs: A comparative study. *International Small Business Journal*, **12** (3), 54–60.

Carroll, J. (1967). A note on departmental autonomy and innovation in medical schools. *Journal of Business*, **40**, 531–534.

Carter, C. F. & Williams, B. R. (1957). *Industry and technical progress: Factors governing the speed of application in science*. London: Oxford University Press.

Chiesa, V., Coughlan, P. & Voss, C. A. (1996). Development of a technical innovation audit. *Journal of Product Innovation Management*, **13**, 105–135.

Chon, S. F. & Turin, R. M. (1984). Organisational structure, decision making procedures and the adoption of innovations. *IEEE Transactions of Engineering Management*, **31**, 154–161.

Christensen, C. (1997). *The innovators dilemma*. Boston, MA: Harvard Business School Press.

Clark, K. B. & Fujimoto, T. (1991). *Product development performance*. Boston, MA: HBS Press.

Clark, K. B. & Wheelwright, S. C. (1992). *Managing new product and process development: Text and cases*. New York: Free Press.

Coombs, R., Narandren, P. & Richards, A. (1996). A literature-based innovation output indicator. *Research Policy*, **25**, 403–413.

Cooper, R. G. (1979). The dimensions of industrial new product success and failure. *Journal of Marketing*, **43**, 93–103.

Cooper, R. G. (1984). The strategy–performance link in product innovation. *R&D Management*, **14** (4), 247–259.

Cooper, R. G. (1990). New products: What distinguishes the winners? *Research and Technology Management*, **33** (6), 27–31.

Cooper, R. G. (1999). From experience. The invisible success factors in product development. *Journal of Product Innovation Management*, **16**, 115–133.

Cooper, R. G. (2003). Profitable product innovation. In: L. V. Shavinina (Ed.), *International Handbook on Innovation*. Oxford: Elsevier Science.

Damanpour, F. (1991). Organisational innovation: A meta-analysis of effects of determinants and moderators. *Academy of Management Journal*, **34** (3), 555–590.

Dess, G., Newport, S. & Rasheed, A. M. A. (1993). Configuration research in strategic management: Key issues and suggestions. *Journal of Management*, **19** (4), 775–795.

Dewar, R. & Dutton, J. E. (1986). The adoption of radical and incremental innovations: An empirical analysis. *Management science*, **32**, 1422–1433.

Di Maggio, P. J. & Powell, W. W. (1983). The iron cage revisited: institutional isomorphism and collective rationality in organisational fields. *American Sociological Review*, **48**, 147–160.

Donaldson, L. (1996). *For positivist organisational theory*. London: Sage.

Downs, G. W. & Mohr, L. B. Jr. (1976). Conceptual issues in the study of innovation. *Administrative Science Quarterly*, **21**, 700–714.

Drazin, R. & Schoonhoven, C. B. (1996). Community, population and organisation effects on innovation: A multilevel perspective. *Academy of Management Journal*, **39** (5), 1065–1083.

Duchesneau, D., Cohn, S. F. & Dutton, J. E. (1979). *A study of innovation in manufacturing: Determinants, processes and methodological issues*. Social Science Research Foundation, University of Maine at Orono.

The Economist (1998) The world in 1999. The Economist Publications.

Ettlie, J., Bridges, W. P. & O'keefe, R. D. (1984). Organization strategy and structural differences for radical vs. incremental innovation. *Management Science*, **30** (6), 682–695.

Ettlie, J. E. (2000). *Managing technological innovation*. New York: Wiley.

EUROSTAT (1994). *The community innovation survey. Status and perspectives*. Brussels: European Commission.

EUROSTAT (1996). Innovation in European Union. *Statistics in focus: Research and Development*, Vol. 2. Brussels: European Commission.

Felberg, J. D., DeMarco, D. E. (1992). New idea enhancement at Amoco Chemical: An early report from a new system. *Journal of Product Innovation Management*, 9, 278–286.

Forrest J. E. (1991). Models of the process of technological innovation. *Technology Analysis & Strategic Management*, 3 (4), 439–453.

Freeman, C. (1982). *The economics of industrial innovation*. Cambridge, MA: MIT Press.

Garcia, R. & Calantone, R. (2002). A critical look at technological innovation typology and innovativeness terminology: a literature review. *Journal of Product Innovation Management*, 19, 110–132.

Giannitsis, T. & Mavri, D. (1993). *Technology structures and technology transfer in the the Greek industry* (in Greek). Athens: Gutenberg.

Gulick, L. H. & Ulrick, L. F. (1937–1938). Papers on the science of administration. New York: Columbia University Press.

Hage, J. & Aiken, M. (1970). *Social change in complex organisations*. New York: Random House.

Hage, J. & Dewar, R. (1973). Elite vs. organizational structure in predicting innovation. *Administrative Science quarterly*, 18, 279–290.

Hajihoseini, H. & de la Mare, R. F. (1995). The impact of indigenous technological capability on the technological development process in Iran. *Proceedings of the R&D Management conference 'Knowledge, Technology and Innovative Organisations'*, Piza (Italy).

Hannan, M. & Freeman, J. (1977). The population ecology of organisations. *American Journal of Sociology*, 83, 929–964.

Hansen, J. A. (1992). *New indicators of industrial innovation in six countries. A comparative analysis*. U.S. National Science Foundation, State University of New York, College of Fredonia.

Hauschildt, J. (2003). Promoters and champions in innovation—development of a research paradigm. In: L. V. Shavinina (Ed.), *International Handbook on Innovation*. Oxford: Elsevier Science.

Hise, R. T., O'Neal, L., Parasuraman, A. & McNeal, J. U. (1990). Marketing/R&D interaction in new product development: Implications for new product success rates. *Journal of Product Innovation Management*, 7 (2), 142–155.

Hobday, M. (1995). *Innovation in East Asia: The challenge to Japan*. Guilford, U.K.: Edgar.

Hofstede, G. (1991). *Cultures and organisations*. London: McGraw-Hill.

Hollenstein, H. (1996). A composite indicator of a firm's innovativeness. An empirical analysis based on survey data for Swiss manufacturing. *Research Policy*, 25, 633–645.

Inzelt, A. (2003). Innovation process in Hungary. In: L. V. Shavinina (Ed.), *International Handbook on Innovation*. Oxford: Elsevier Science.

Kets De Vries, M. F. R. (1977). The entrepreneurial personality: A person at the crossroads. *Journal of management studies*, 14, 34–57.

Khan, A. M. (1990) Innovation in small manufacturing firms. In: J. Allesch (Ed.), *Consulting in Innovation*. Oxford: Elsevier Science.

Khan, A. M. & Manopichetwattana, V. (1989a). Models for innovative and non innovative small firms. *Journal of Business Venturing*, 4, 187–196.

Khan, A. M. & Manopichetwattana, V. (1989b). Innovative and non-innovative small firms: types and characteristics. *Management Science*, 35, 597–606.

Kim, Y., Kwangsun, S. & Jinjoo, L. (1993). Determinants of technological innovation in the small firms of Korea. *R&D Management*, 23 (3), 215–226.

Kimberly, J. R. & Evanisko, M. J. (1981). Organisational innovation: The influence of individual, organisational and contextual factors of hospital adoption of technological and administrative innovations. *Academy of Management Journal*, ??, 689–713.

Lefebvre, L. A., Mason, R. & Lefebvre, E. (1997). The influence prism in SMEs: The power of CEOs' perceptions on the technology policy and its organisational impacts. *Management Science*, 43 (6), 856–878.

Lopez-Martinez, R. E., Medellin, E., Scanlon, A. P. & Solleiro, J. L. (1994). Motivations and obstacles to university industry co-operation (UIC): A Mexican case. *R&D Management*, 24 (1), 17–31.

Lowe, J. & Crawford, N. (1984). *Technology licensing and the small firm*. London: Gower.

Lundvall, B. (1998). Why study systems and national styles of innovation? *Technology Analysis and Strategic Management*, 10 (4), 407–421.

Maidique, M. A. & Zinger, B. J. (1984). A case study of success and failure in product innovation: The case of the U.S. electronics industry. *IEEE Transactions on Engineering management*, 31 (4), 192–203.

Mansfield, E. (1963). Size of firm, structure, and innovation. *Journal of Political Economy*, 71, 556–576.

Mansfield, E. (1971). *Research and innovation in the modern corporation*. New York: W. W. Norton.

Meyer-Krahmer, F. (1984). Recent results in measuring innovation output. *Research policy*, 13, 175–182.

Miller, D. (1983). The correlates of entrepreneurship in three types of firms. *Management Science*, 29 (7), 770–791.

Miller, D. (1986). Configurations of strategy and structure: Towards a synthesis. *Strategic Management Journal*, 7, 223–249.

Miller, D. (1996). Configurations revisited. *Strategic Management Journal*, 17, 505–512.

Miller, D. & Friesen, P. H. (1984). *Organizations: A quantum view*. Englewood Cliffs, NJ: Prentice-Hall.

Miller, D., Kets de Vries, M. F. R. & Toulouse, J. M. (1982). Top executive locus of control and its relationship to strategy, environment and structure. *Academy of management Journal*, 25, 237–253.

Miller, R. (1989). *New location dynamics*. Cambridge, MA: MIT Centre for Technology and Policy.

Miller, R. & Blais, R. (1992). Configurations of innovation: Predictable and maverick modes. *Technology Analysis and Strategic Management*, 4 (4), 363–386.

Mintzberg, H. (1987). Crafting strategy. *Harvard Business Review*, (July–August), 66–75.

Mintzberg, H., Ahlstrand. B. & Lampel, J. (1998). *Strategy Safari*. London: Prentice Hall.

Mishra, S., Kim, D. & Dae Hoon, L. (1996). Factors affecting new product success: Cross country comparisons. *Journal of Product Innovation Management*, **13**, 530–550.

Moenaert, R. D., de Meyer, A. & Clarysse, B. J. (1994). Cultural differences in new technology management. In: W. M. E. Souder & J. D. Sherman (Eds), *Managing New Technology Development* (pp. 267–314). New York: McGraw-Hill.

Mohr, L. B. (1969). Determinants of innovation in organizations. *American Political Science Review*, **63**, 111–126.

Nabseth, L. & Ray, G. F. (Eds) (1974). *The diffusion of new industrial processes: An international study*. Cambridge: Cambridge University Press.

Nejad, J. B. (1997). Technological innovation in developing countries: Special reference to Iran. Unpublished doctoral dissertation, University of Bradford, U.K.

Nelson, R. & Winter, S. (1977). In search of a useful theory of innovation. *Research Policy*, **5**, 36–76.

Newell, S. & Swan, J. (1995). Professional associations as important mediators of innovation process. *Science Communication*, **16** (4), 371–387.

OECD (1981) *OECD the measurement of scientific and technical activities. Frascati Manual*. Paris: OECD.

OECD (1992) *OECD proposed guidelines for collecting and interpreting technological innovation data: Oslo Manual*. Paris: OECD.

Parsons, C. (1966). *Societies: evolutionary and comparative perspectives*. Englewood Cliffs, NJ: Prentice-Hall.

Patel, P. & Pavitt, K. (1994). National Innovation Systems: Why they are important, and how might they be measured and compared? *Economics of Innovation and New Technology*, **3**, 75–95.

Pavitt, K. (1984). Sectoral patterns of technical change: Towards a taxonomy and a theory. *Research Policy*, **13**, 343–373.

Pavitt, K. & Patel, P. (1988). The international distribution and determinants of technological activities. *Oxford Review of Economic Policy*, **4** (4).

Pilogret, L. (1993). Innovation consultancy services in the European Community. *International Journal of Technology Management*. Special Issue on Industry—University—Government co-operation, **8** (6/7/8), 685–696.

Porter, M. (1980). *Competitive strategy*. New York: Free Press.

Porter, M. E. (1990). *The competitive advantage of nations*. London: Free Press.

Rochford, L. & Rudelius, W. (1992). How involving more functional areas within a firm affects the new product process. *Journal of Product Innovation Management*, **9** (4), 287–299.

Rogers, E. M. (1983). *Diffusion of innovations* (3rd ed.). New York: The Free Press.

Rogers, E. M. & Shoemaker, F. (1971). *Communication of innovations: A cross-cultural approach*. New York: Free Press.

Rothwell, R. (1974). SAPPHO Updated: Project SAPPHO phase II. *Research Policy*, **3** (3), 258–291.

Rothwell, R. (1977). The characteristics of successful innovators and technically progressive firms (with some comments on innovation research). *R&D Management*, **7** (3), 191–206.

Rothwell, R. (1978). Small and medium sized manufacturing firms and technological innovation. *Management Decision*, **16** (6), 362–370.

Rothwell, R. (1992). Successful industrial innovation: critical factors for the 1990s. *R&D Management*, **22** (3), 221–239.

Rubenstein, A. H., Chakrabarti, R. D., O'Keefe, W. E., Souder, W. E. & Young, H. C. (1976). Factors influencing innovation success at the project level. *Research Management*, (May), 15–20.

Sapolsky, H. (1967). Organizational structure and innovation. *Journal of Business*, **40**, 497–510.

Saviotti, P. P. and Metcalfe, J. S. (1984). A theoretical approach to the construction of technological output indicators. *Research Policy*, **13**, 141–151.

Slappendel, C. (1996). Perspectives on innovation in organisations. *Organisation Studies*, **17** (1), 107–129.

Smith, K. (1988). Survey-based technology output indicators and innovation policy analysis. In: A. F. J. van Raan, A. J. Nederhof and H. M. Moed (Eds), *Selected Proceedings of the First International Workshop on Science and Technology Indicators*. Leiden: European Commission.

Smith, K. (1989). Survey-based technology output indicators and innovation policy analysis. In: A. F. J van Raan, A. J. Nederhof & H. F. Moed (Eds), *Science and Technology Indicators*. Leiden: DSWO Press, University of Leiden.

Smith, K. (1992). Technological innovation indicators: experience and prospects. *Science and Public Policy*, **19** (6), 383–392.

Smith, K. & Vidvei, T. (1992). Innovation activity and innovation outputs in Norwegian industry. *STI Review*, 11.

Smith, R. F. (1974). Shuttleless Looms. In: L. Nabseth & G. F. Ray (Eds), *The Diffusion of New Industrial Processes: An International Study*. Cambridge: Cambridge University Press.

Sorge, A. (1991). Strategic fit and the social effect: Interpreting cross-national comparisons of technology, organisation and human resources. *Organisation Studies*, **12** (2), 161–190.

Souitaris, V. (1999). Research on the determinants of technological innovation. A contingency approach. *International Journal of Innovation Management*, **3** (3), 287–305.

Souitaris, V. (2001a). External communication determinants of innovation in the context of a newly industrialised country: A comparison of objective and perceptual results from Greece. *Technovation*, **21**, 25–34.

Souitaris, V. (2001b). Strategic influences of technological innovation in Greece. *British Journal of Management*, **12** (2), 131–147.

Souitaris, V. (2002a). Technological trajectories as moderators of firm-level determinants of innovation. *Research Policy*, **31** (6), 877–898.

Souitaris, V. (2002b). Firm-specific competencies determining innovation. A survey in Greece. *R&D Management*, **32** (1), 61–77.

Swan, J. A. & Newell, S. (1994). Managers beliefs about factors affecting the adoption of technological innovation. A Study using cognitive maps. *Journal of Managerial Psychology*, **9** (2), 3–11.

Swan, J. A. & Newell, S. (1995). The role of professional associations in technology diffusion. *Organisation studies*, **16**, 846–873.

Swan, J. A., Newell, S., Robertson, M. (1999). Central agencies in the diffusion and design of technology: A comparison of the U.K. and Sweden. *Organisation Studies*, **20** (6), 905–931.

Teece, D. & Pisano, G. (1994). The dynamic capabilities of the firms: an introduction. *Industrial and Corporate Change*, **3**, 537–556.

Tidd, J. (2001). Innovation Management in context: Environment, organisation and performance. *International Journal of Management Reviews*, **3** (3), 169–184.

Tidd, J., Bessant, J. & Pavitt, K. (1997 & 2001). *Managing innovation. Integrating technological, market and organisational change*. Chichester, U.K.: John Wiley.

Tidd, J., Driver, C. & Saunders, P. (1996). Linking technological, market and financial indicators of innovation. *Economics of Innovation and New Technology*, **4**, 155–172.

Tornatzky, L. G. & Klein, K. J. (1982). Innovation characteristics and innovation adoption—implementation. *IEEE Transactions in Engineering Management*, **29**, 28–45.

Trott, P. (1998). *Innovation management and new product development*. London: Pitman Publishing.

Trott, P. (2003). Innovation and market research. In: L. V. Shavinina (Ed.), *International Handbook on Innovation*. Oxford: Elsevier Science.

Tsipouri, L. J. (1991). The transfer of technology revisited: some evidence from Greece. *Entrepreneurship and Regional Development*, **3**, 145–157.

Tushman, M. & Scandal, T. (1981). Boundary spanning individuals. Their role in information transfer end their antecedents. *Academy of Management Journal*, **24**, 289–305.

Twiss, B. (1992). *Managing technological innovation*. London: Pitman.

Veryzer, R. (2003). Marketing and the development of innovative new products. In: L. V. Shavinina (Ed.), *International Handbook on Innovation*. Oxford: Elsevier Science.

Weber, M. (1947). *The theory of social and economic organisation*. New York: Oxford University Press.

Webster, F. E. Jr. (1970). Informal communication in industrial markets. *Journal of Marketing Research*, 7.

White, M. (1988). *Small firm's innovation: Why regions differ*. London: Policy Studies Institute Report.

Wolfe, R. A. (1994). Organisational innovation: Review, critique and suggested research directions. *Journal of Management Studies*, **31** (3), 405–431.

Zairi, M. (1996). *Benchmarking for best practice*. London: Butterworth-Heinemann.

Zaltman, G., Duncan, R. & Holbek, J. (1973). *Innovations and organizations*. New York: John Wiley.

The International Handbook on Innovation
Edited by Larisa V. Shavinina

Innovation in Financial Services Infrastructure

Paul Nightingale

Complex Product System Innovation Centre, Science Policy Research Unit (SPRU), University of Sussex, U.K.

Abstract: Financial infrastructure is essential to the world economy yet is largely neglected within the academic innovation literature. This chapter provides an overview of innovation in financial infrastructure. It shows how external infrastructure technologies between institutions improve market liquidity by increasing the reach of markets. Internal infrastructure technologies within institutions are used to co-ordinate the profitable allocation of resources. The heavy regulation of the industry, the software intensity of modern infrastructure technologies, the way in which they have multiple users and their increasing complexity create extra uncertainties in their design and development. As a consequence, they have very different patterns of innovation from traditional consumer goods.

Keywords: Innovation; Capital goods; Infrastructure; Technology; Financial services.

There have been, since the world began three great inventions. . . . The first is the invention of writing, which alone gives human nature the power of transmitting, . . . its laws, . . . contracts, . . . annals, and . . . discoveries. The second is the invention of money, which binds together . . . civilised societies. The third is the Oeconomical Table . . . which completes (the other two) . . . by perfecting their object: (This is) the great discovery of our age, but of which our posterity will reap the benefit.

Adam Smith, *The Wealth of Nations*, **IV** (ix), 38

Aims

This chapter aims to give an overview of how innovation takes place in financial infrastructure, and uses a contingency theory approach to show how it can be conceptualised using established models of innovation in capital goods.

Introduction

Within the academic innovation literature, services have traditionally been considered non-innovative.[1]

---

[1] Important exceptions can be found in Bell (1973) and Fuchs (1968).

This perception is reflected in the disproportionate bias towards research into mass-production manufacturing technologies. This may partly be attributed to a relative lack of visibility compared to consumer goods, and partly to the fact that service innovation is often intangible or dependent on innovation in bespoke infrastructures, and is therefore poorly reflected by traditional innovation indicators, such as patents. However, it is also because the nature of innovation in services is different from innovation in the mass production consumer goods traditionally studied in the literature.

This academic neglect is unfortunate given the importance of service innovation to GDP growth, employment, the economy, social change and the geography of global cities such as London or New York (Barras, 1986, 1990; Freeman & Perez, 1988; Soete & Miozzo, 1989). In the OECD, services account for approximately two-thirds of employment and economic value added (Bryson & Daniels, 1988; Gallouj & Weinstein, 1997; cf. Eurostat, 1998). The fact that the extent of employment of scientists and engineers in services overtook manufacturing in the USA in 1990 suggests that services are not as low-tech as their profile in the literature might suggest.

If one looks at financial services in OECD countries, they typically represent some 5–10% of GDP and a similar proportion of employment. For example, 5.4 million people are employed in the U.S. banking sector, which is more than twice the combined number of employees in automobiles, computers, pharmaceuticals, steel and clothing (Frei et al., 1998, p. 1). Banking is also extremely high-tech: global banks often spend over $1 billion a year on IT (the cost of technology is second only to wages for most financial services companies) and, as Barras (1986, 1990) has argued, it is a vanguard sector that has been one of the first to adopt and diffuse new information technology.

This chapter will use a contingency theory approach to achieve three aims: firstly, to identify and explore the key features of innovation in financial services; secondly, to relate these specific features to various empirical studies on innovation; and thirdly, to try and explain how the contingent features of innovation in financial services infrastructure—in particular, complexity, software intensity, its multi-user nature and heavy regulation—can explain the various aspects of their patterns of innovation. It aims to provide an overview of how innovation takes place within financial infrastructure projects in order to draw out the important features.

The chapter explores the history of innovation in financial services and provides definitions of some of the terms used in this chapter; compares innovation in infrastructure and complex capital goods with innovation in mass production consumer goods; and explores the specific problems associated with innovation in embedded software. The final section draws conclusions.

Financial Services—Definitions and History

Early research on services tended to stress their intangible nature, heterogeneity, and the importance of time constraints on service delivery (Lovelock, 1983). Time is important because the output of services is often a 'performance rather than an object' (Lovelock, 1983; Lovelock & Yip, 1996; McDermott et al., 2001, p. 333). Storage is impossible when production and consumption are simultaneous, making the reliable and timely co-ordination and control of resources highly important. More recent academic literature has tended to move away from clear-cut distinctions between intangible services and tangible manufacturing and now understands them as a continuum (Brady & Davies, 2000; Miles, 1996; Uchupalanan, 2000). The intangible aspect is dominant in certain areas such as teaching, and the tangible aspect in areas such as detergent production. There is also a large middle ground where they overlap, such as fast food. In this chapter, the difference between services and manufacturing along this continuum is understood in terms of the tangibility of functions, so that service firms are paid to *perform a function*, while manufacturing companies produce objects that are bought to *provide a function* (Nightingale & Poll, 2000b).

The financial system performs the overall function of moving money between savers and borrowers. This allows people to *transfer their ability to use money to transform the world through time and space*. This links the present to the future, allowing borrowers and savers to switch between current income and future spending.[2] Prior to the development of money, a barter system was used for trade but was limited to the 'here and now'. Gradually, precious metals began to be used as *commodity* money to *measure* and *store value* through time and space as a medium of exchange. Coins, for example, were in use by the eighth century BC Later on, the risk and cost of carrying around large amounts of precious commodities saw the development of *contract* money that could be exchanged for deposited gold and silver. Contract money gradually developed into *fiat* or *fiduciary* money that maintains its value even though it is not backed by gold. This development allowed a more extensive allocation of resources through time and space, but its geographical locus was limited by its basis in personal rather than institutional trust. Financial institutions perform various functions within the financial system, including using technological infrastructure to profitably and reliably deliver services, and so extend the institutional trust upon which the financial system depends.

Financial institutions are typically used to mediate the relationship between savers and borrowers because they have specialised technical capabilities (see Merton, 1975a, 1995; Merton & Brodie, 1995; Nightingale & Poll, 2000a). Firstly, they have extensive specialised knowledge of the risks involved. Secondly, there are generally large differences in the amounts of money that borrowers and lenders have or need. Thirdly, the liquidity requirements of lenders and borrowers are different. Liquidity refers to the ability to turn assets into cash quickly and cheaply. Lenders generally want quick access to their cash, while borrowers want longer-term, more stable funding. Fourthly, financial institutions allow the pooling of savings and risk which increases the liquidity of long-term debt. Lastly, financial institutions can take advantage of economies of scale by spreading the fixed costs of investments in infrastructure over a large number of contracts.

The development of specialised financial institutions and a sophisticated division of labour came about through an expansion of the market for financial services following the emergence of the fiscal-military state in the 16th and 17th centuries. As early-Modern wars were fought by attrition, the ability to allocate funds through time and space effectively determined military power and enabled smaller countries like

---

[2] Money and financial contracts in general are only means to other ends—they allow one to inter-convert goods and labour through time and space: for example, allowing the young to borrow to buy a house and the old to save productively.

Holland and England to take on France and Spain.[3] This 'financial revolution' started when the provincial States of Holland accepted collective responsibility for war loans by securing them on future tax returns (Dickson, 1967, p. 63). The Bank of Amsterdam was set up in 1609, with the English waiting until the cost of a war with France in 1689 forced them to form the Bank of England in 1694.[4]

Improved commercial finance quickly followed to fund the activities of the newly emerging joint stock companies. The first joint-stock company, the Muscovy Company, was set up in London in 1553. As trade expanded, merchant banks emerged and acted as *accepting houses* that charged interest and commission on the bills of exchange that were used to finance trade. A virtuous cycle emerged in which expanded trade provided capital for financial institutions, who in turn provided capital for expanded trade and new firms. As joint stock companies became more important, the buying and selling of shares became more formalised, and exchanges were set up. The New York Stock Exchange was started under a buttonwood tree in 1792. The institution that would develop into the London Stock Exchange was started in 1760, renamed in 1773 and officially regulated in 1809.

Distinct institutions emerged within this heavily regulated financial system. While there is huge diversity within the sector, the three main forms of financial institutions are *banks, exchanges* and *investment institutions*. The institutional differences are determined by functional differences in how firms move money between savers and borrowers. Each method used, in turn, defines the technological trajectories (Dosi, 1982) that the firms follow, and influences the kind of technologies and infrastructure they use (Buzzacchi et al., 1995; Penning & Harianton, 1992).

Banks are traditionally divided into two kinds, depending on how they move surplus funds from investors to borrowers (Berger et al., 1995). Commercial or retail banks rely on deposits drawn from individual savers that they re-lend at a profit; the savers are typically paid a nominal amount of interest. Investment banks, however, make their money through fees charged from arranging complex financial deals (Eccles & Crance, 1988). Unlike commercial banks that load the assets they hold, investment banks help allocate surplus savings by underwriting securities that are sold to other investors. Securities have the advantage of being liquid, and consequently allowing investors to make rapid changes to their portfolios.

Exchanges are institutionalised markets for the trading of financial contracts. The two main types of contracts sold on exchanges are company shares (which are sold in equity markets), and government and company debt (which are sold in bond markets).[5] Markets have an advantage over institutional investment mechanisms in that they publicise information about the price of resources, allowing financial actors to improve their resource allocation. Exchanges are normally closed institutions with their own rules and regulations, which allow well-established financial processes to be developed and used in a more 'trustworthy' environment, improving the liquidity of trading. Since the value of assets generally increases with their liquidity (though there are exceptions), the ability to easily trade a contract has important economic implications.

Investment institutions such as pension funds, mutual funds (called unit trusts in the U.K.) and life-insurance companies are a fairly recent development. Up until the late 19th century, private individuals were the main investors, but after that date, investment trusts became increasingly important, followed by unit trusts in the 1930s (Golding, 2000). In the 1960s, pension funds took off, and the life-insurance industry became a major force in financial investment. Investment institutions now hold about two-fifths of U.S. household financial assets, and the largest five fund managers place assets larger than the combined GDPs of France and the U.K. They bundle together the savings and insurance contributions of individuals to invest in a range of assets on a long-term basis. Strictly speaking, investment institutions rely on fund managers for their investment management, but in practice most investment institutions are directly involved in investing in quoted companies. During the 1980s and 1990s, investment institutions grew rapidly, until they now control approximately US$26 trillion of funds, US$13 trillion of which are in the U.S.—approximately three times the GDP. Each type of investment institution has different investment preferences. Pension funds have predictable long-term liabilities, while insurance funds require more liquid assets in case they should have to pay out for a disaster.

Because financial services are so important to the wider global economy, they have traditionally been heavily regulated (Berger et al., 1995). Understanding regulation and regulatory loopholes is essential for understanding the development of the financial services industry (Calomiris & White, 1994; Hall, 1990; Merton, 1995b). In the USA regulations limited the ability of financial institutions to compete in a range of product and geographic markets. In Europe and Japan the operations of financial institutions are similarly regulated, which in turn influences what they do, and the structural possibilities of the technologies they use

---

[3] Philip II had been defeated in 1575 not because of military superiority but because the cost of his army had bankrupted him, and he could no longer pay his troops.

[4] London did not emerge as a financial centre until the Napoleonic wars eliminated Amsterdam.

---

[5] Equities, unlike bonds, are 'real assets' that can protect the owner from inflation.

(Channon, 1998). Because financial infrastructure technologies influence firm behaviour, their development will almost inevitably involve national and possibly international regulators. These regulators may require changes to legal frameworks, which complicates and lengthens the design and development processes. While this is true of other sectors, it does not normal occur to the same extent.

The shift towards more market-based regulation in the late 20th century is commonly referred to as 'deregulation', but is perhaps better thought of as a process of re-regulation. Specific restrictions have been removed, but government and international regulators still play a vital role in the functioning of financial markets.[6] The ongoing changes in national and regional regulations have led to an increased internationalisation of markets and capital flows. This has, in turn, produced a corresponding increase in the geographical scope of infrastructure and allowed firms to exploit new economies of scale, generating market growth and increased concentration.

The Importance of Liquidity

The increasing concentration of the financial services industry over the 1980s and 1990s has seen the emergence of investment institutions with extremely large capital funds. The size of these funds has intensified concern about liquidity and the corresponding ability to cheaply adjust portfolios when markets move. For example, a fund of US$500 million will not be able to make significant changes to its performance without trading units of about US$10 million (Golding, 2000, p. 67). Buying or selling such an amount of shares in an 'illiquid' firm that trades only US$250,000 of shares a day is going to be both time-consuming and expensive, and is likely to alert other players in the market, pushing the price up or down.

As a consequence, institutional investors have a preference for firms with large market capitalisation and very liquid shares. The size of modern financial institutions means that they have a major influence on market behaviour (and that the economist's notion of large numbers of small independent investors is becoming unrealistic).[7] Since the 1980s, there has also been a trend towards the securitisation of a range of assets. This means that financial assets such as mortgages and credit card liabilities have been pooled and resold as contracts in markets where they can be traded. These bonds, backed by securitised assets, tend

to be substantially more liquid:[8] not only are there fewer bonds than shares to choose from, but they are also more frequently traded. The New York Stock Exchange, for example, trades about US$350 billion of bonds a day, compared to only US$28 billion a day in equities.

Given that the liquidity of assets is determined in part by the number of potential buyers, infrastructure technologies that can bring a larger, more diverse set of buyers into contact with sellers will increase liquidity and the value of assets. Financial institutions rely on infrastructure technologies to do this. These technologies need not, however, be owned by the same people who operate the markets, and the divergent technological trajectories of stock exchanges and telecommunications networks has produced a shift towards the outsourcing of telecommunications.

Financial Institutions as Socio-Technical Systems

Financial institutions provide, monitor, and maintain the processes whereby funds are pooled, matched to borrower's requirements, and then allocated. Financial institutions take financial contracts (funds, bonds, etc.) as their inputs, and then process them, before reengineering them into new forms of contract that are then sold to customers. In doing so, they use sophisticated processes comprising people, knowledge and technology, to match the financial requirements of borrowers to those of savers. As such, they can be conceptualised as socio-technical systems (Hughes, 1983, 1987), where technology is used to improve the processes involved in matching financial contracts to customers by replacing person-based (and often market-based) mechanisms with an organised technology-based mechanism. This allows financial institutions to provide better services, develop new products, and exploit improved economies of scale, scope and speed (Barras, 1986, 1990; Ingham & Thompson, 1993; McMahon, 1996; Nightingale & Poll, 2000b).

The relationship between innovation in infrastructure technologies and performance improvement is fairly well understood in large manufacturing firms following the work of Alfred Chandler (1990). He showed the way in which firms invest in high fixed-cost infrastructures that improve the capacity and speed of production processes. The effective use of technology could then generate the fast, high-volume flows that would turn low-cost inputs into high-value outputs. In this way, the high costs of the infrastructure are then spread over a large volume of output to keep unit costs low.

If the volume of production is too low, then high fixed costs cannot be adequately spread and unit costs

---

[6] The links between internationalisation and regional regulation have led Howells (1996) to question the usefulness of the concept of globalisation.

[7] It is now not uncommon for large British companies to have 80–90% institutional ownership (Golding, 2000, p. 31).

[8] Bonds backed by regular mortgage payments can be sold instantaneously and very cheaply, while repossessing and selling several thousand homes is extremely costly and difficult.

will rise (Chandler, 1990, p. 24). As a result, manufacturing firms organise and co-ordinate the resources required to fully utilise the capacity of production processes and increase the average speed of 'flows'.[9] This is done in two ways. Firstly, firms developed sophisticated managerial techniques for controlling processes and, secondly, they exploited external and internal infrastructure technologies such as the telegraph and railway systems to ensure that production was uninterrupted.

Within the financial services sector things are slightly different. Profitability is linked to the efficient contextualisation and processing of information rather than the utilisation of capital machinery (Nightingale & Poll, 2000a). This means that profits are far less constrained by infrastructure technology than they are in manufacturing. For example, the profitability contribution of a worker on a production line or in a steel mill is largely determined by the technology itself, and a worker would be hard pressed to increase profits substantially. By contrast, a trader in a bank who was smarter, faster or had superior analysis than others in the market could very easily make significantly more profits with very few restrictions from technology. The amount of profit made on a trade of US$100 million will be more than on a trade of US$1 million, even though the same telecommunications system might be used. This is called *leverage*.

The flip side of high leverage is that while profits are almost unlimited, the same is true of losses, creating an incentive to understand and manage the extent and likelihood of these losses occurring, namely *risk*. As infrastructure technology has allowed larger and more complex contracts to be produced and traded, understanding risk has increased in importance- and risk-management technologies are key aspects of modern financial service infrastructure.

These differences underpin Barras' (1986, 1990) concept of the reverse product life cycle, whereby innovation in financial services is process- rather than product-driven (1986, 1990). Following Vernon (1966), Utterback & Abernathy (1975) argued that innovation in manufacturing follows a cycle that initially concentrates on product innovation until an established design is formulated, after which time competition is based on process innovation. Barras (1986, 1990), by contrast, argues that financial services follow a pattern whereby innovations in processes (and infrastructure technologies) allow new products to be introduced (1986). Both Utterback & Abernathy (1975) and Barras' (1986, 1990) ideas have been heavily criticised (Pavitt & Rothwell, 1987; Uchupalanan, 2000). In financial services, where the process is often the product, the sharp analytical distinction between product and process innovation is questionable (cf. Easingwood & Storey, 1996).

Another way to look at innovation in financial services is to look at how, on the one hand, technology is used to increase the scale and liquidity of financial transactions by increasing 'reach' and bringing more buyers and sellers into contact with one another and, on the other hand, at how technology can be used to better match savers and lenders. In this way the manufacturing categories of process and product innovation are subsumed into the overlapping categories of service provision and control. This has the advantage of allowing us to see how innovation in financial service infrastructure co-evolves with innovation in infrastructure technologies outside the financial system. For example, the infrastructure of the railway network co-evolved with the telegraph system and the development of new financial products and services, which allowed the transportation of goods to be profitably co-ordinated.

It also allows one to conceptualise the direction of innovative activity. Infrastructure technologies that allow increases in the scale and liquidity of financial services tend to do so by expanding the range of savers and borrowers with whom firms interact because the scale and liquidity of financial transactions are closely related (Peffers & Tuunainen, 2001). For example, prior to the introduction of the telegraph, the New York Stock Exchange only managed to sell 31 shares over a single day in March 1830 (DuBoff, 1983, p. 261). However, infrastructure technologies that improve the internal allocation and control of service provision tend to focus on improving the accuracy, scope, speed and reliability of control processes (Nightingale & Poll, 2000b).

Technologies of communication and transportation are therefore particularly important for financial services, as delays in communication lead to deviations in market prices and opportunities for arbitrage. Consequently, the evolution of the financial services sector has been influenced by developments in information and transportation infrastructure technologies. For example, when communication between London and Amsterdam was dependent on sailing boats, it took three days for information to travel between the markets. This inefficiency in the communication of market information created opportunities for arbitrage. Similarly, when communication within Britain was undeveloped and irregular local exchanges flourished, but with the advent of regular mail coaches, leaving from Lombard Street in the City of London (starting in 1784), there was a pull towards the larger market in London (Michie, 1997, p. 306).

The development of the telegraph was a significant advance in infrastructure technology. Starting in the United States in 1844, by 1860 there were 5 million messages being transmitted annually along 56,000 miles of wire and 32,000 miles of telegraph poles

---

[9] One way to maintain capacity utilisation (and therefore lower unit costs) is to exploit unused production capacity in new product lines, thereby generating economies of scope.

(DuBoff, 1983, p. 255). In 1851 a link was introduced between London and Paris which transcended the previous 12-hour communication times and allowed real-time price communication. In 1866 the first link between London and New York was set up, by 1872 the telegraph had linked London and Melbourne, and by 1898 there were 15 undersea cables (only nine were working) under the Atlantic (Michie, 1997, p. 310). By 1890, information could travel the 400 miles from Glasgow to London in 2.5 minutes (ibid.).

The telegraph had important implications for commodity exchanges as, together with improved infrastructure of railroads and storage, it allowed contracts to be linked to the point of production. As a consequence 'to arrive' contracts started to replace advanced payment based on 'certified' samples (DuBoff, 1983). This produced greater price stability, as large amounts of commodities were not being dumped in the illiquid markets of commercial centres. With the advent of the telegraph a far wider range of buyers and suppliers could be searched enabling exchanges and market makers to better match supply and demand. Infrastructure technologies such as the telegraph therefore increased liquidity and allowed the centralisation of exchanges (DuBoff, 1983).

The telegraph technology was very limited until the 1860s, with transmission rates of about 15 words per minute. With the invention of the stock telegraph in 1867, this went up to 500 words per minute, which again improved market performance (DuBoff, 1983, p. 263). Middlemen could be cut out of transactions, buyers and sellers did not need to travel, time lags were reduced, and there was a substantial reduction in the risks involved. As DuBoff notes:

> The expected savings from a given market search will be higher the greater the dispersion of prices, the greater the number of production stages, and the greater the expenditure on the resources or service. For example, the only way to know all the prices which various buyers and sellers are quoting at a given moment would be to bring about a complete centralisation of the market, only then will costs of canvassing, or search, be at a minimum. Conversely, with infinite decentralisation these costs will reach a maximum. To lower them it pays to centralise—to reduce spatial dispersion and the number of independent decision makers (1983, p. 266).

The temporal and geographic reach of buyers and sellers was further improved by the introduction of the telephone. The geographic scope was increased again when the first transatlantic telephone cable was laid in 1956. The previous radio-based telephone infrastructure could only deal with 20 people, but by 1994 submarine cables could handle some 600,000 calls at a time (Michie, 1997, p. 318).

While much of the early innovation in infrastructure involved financial services firms 'piggybacking' on established technologies like the telegraph, by the late 1980s firms were investing heavily in their own communications networks that could link local branches together. In doing so they relied heavily on external suppliers such as telecommunications companies and specialised financial information suppliers like Reuters, Dow Jones/Telerate, Knight Ridder and Bloomberg. By linking more institutions and people together with technical infrastructure, financial institutions have been able to save time in allocating financial contracts and increase the scope and number of buyers and sellers. This has made markets more liquid and allowed customers to reduce their inventories and the financial resources needed to maintain them. In doing so, the process of buying and selling has become disintermediated, with middlemen who traditionally matched buyers and sellers being replaced. This process of centralisation allows larger institutions to exercise improved control and secrecy, which in turn allows them to exploit new economies of scale (cf. Berger et al., 1995; DuBoff, 1983).

Internal Infrastructure

Once telecommunications systems had increased the reach of markets, and railroads had made improved national transportation of goods possible, market liquidity and the scale of transactions improved, creating new innovation challenges. Firms making large numbers of transactions needed to work out how best to allocate resources between different customers. Until recently, this had been done using unsophisticated technologies, largely subjective assessments of risk, and a 'my word is my bond' trust-based attitude towards risk control.

Since the 1980s the development of sophisticated theoretical tools in financial engineering has allowed products to be better priced and controlled (Marshall & Bansal, 1992). The initial developments came in 1952 when Harry Markovitz developed the fundamentals of portfolio theory (1952). William Sharpe helped develop the Capital Asset Pricing Model (CAPM) in the 1960s, and in the 1970s, Merton, Scholes & Black developed their option-pricing model (Black & Scholes, 1973). These theoretical developments have co-evolved with developments in internal IT infrastructure technologies that can integrate data and perform complex pricing and risk calculations (Bansal et al., 1993; Nightingale & Poll, 2000a).

The ability to value options has transformed finance because it has allowed risk to be approximated mathematically (Berstein, 1996). As a result, internal risk-management processes that previously relied on individual bankers' subjective assumptions, can now be modelled. This allows risk management to be more accurately related to the real risk exposures that financial institutions face. The development of these new tools and new internal software-intensive, IT infrastructure technologies increased the ease and

Table 1. Qualitative changes in risk management infrastructure.

Feature	Craft	Mass Standardisation	Complex
Number of servers	1–100	100–1000	1000+
Maintenance	Craft-based	Standardised	Very complex
Architecture	Decentralised	Centralised	Centralised
Key factor	Automation	Change function	Risk & reliability
Problems	Few	Predictable	One in a million
Importance	Limited	Business cost	Business critical
Risk analysis	Limited	End of day	Quicker and wider

Adapted from Nightingale & Poll (2000a).

accuracy of analysis and product pricing, and has allowed a range of new products and services to be developed (Nightingale & Poll, 2000a).

The shift towards a more theoretical basis for financial engineering was dependent on a transformation in the relationship between front and back office functions. Previously, the back office function in financial institutions had been automated to reduce costs. Risk exposures were typically computed on large mainframe technologies that would provide analysis of positions at the end of the day. With the development of sophisticated computer workstations and analytical software packages on traders' desks, many back-office functions were brought to the front office where they could be carried out closer to the customer. Risk analysis is typically now performed at various organisational levels within financial institutions, and traders, for example, will have a degree of sophisticated analysis available at their desks. This analysis can be performed in close to real time, allowing improved control over positions and exposures, but also imposing a large cost on organisations (Brady & Target, 1996).

This shift in the architecture of control has been dependent on the development of increasingly powerful software-intensive technologies and systems that are able to perform complex calculations quickly on large data sets. This has required a number of the larger financial institutions to develop technological capabilities in information technology, and these capabilities are beginning to be detected within the patent statistics (Pavitt, 1996). Financial service firms are developing their own capabilities in IT and in the mathematical algorithms needed to derive solutions to their risk and pricing calculations.

Nightingale & Poll (2000a) described how an investment bank developed the internal infrastructure technology needed to produce an increasingly sophisticated range of financial products. They showed that the ability to control the pricing and risk of financial contracts is dependent on the scope of the data that are accessed, the accuracy of the models, the speed of calculation and the reliability of the system.

The scope of control is important because calculated risk exposures will come closer to real risk exposures if the infrastructure allows the scope of risk analysis to extend and include more trades that the bank is party to. For example, if the bank's offices in London, New York and Tokyo are all exposed to the same position, each office may be within its local risk limits but the bank as a whole may be over-exposed. This exposure can be reduced if a wider scope of trades is included in the analysis.

Similarly, the ability to calculate exposures and prices is dependent on the accuracy of the models used and the power of the computer systems. If the models are inaccurate, or it takes a long time to calculate a position, the quality of the approximation will be reduced as inaccuracies build up or markets move. These two factors, and the drive towards more scope, lead to increases in the size and power of these IT systems. As Table 1 shows, increases in size and complexity lead to qualitative changes in the technology. Problems that might occur very rarely on individual workstations, for example, may occur every day in IT systems comprising several thousand globally linked servers, making maintenance a very complex and increasingly centralised process. Consequently, modern financial institutions are among the most innovative and demanding customers for the telecommunications and IT sectors (Barras, 1986, 1990).

Financial Infrastructure

Financial institutions depend on innovation in infrastructure technologies in two main ways. Firstly, they depend on *external* infrastructure technologies to underpin their trading and bring buyers and sellers together. Typically these infrastructure technologies, like the old railroads and telegraphs or modern IT and telecommunications networks, link wide geographic areas and allow financial firms to co-ordinate an increasingly diverse selection of customers. This increases the liquidity of markets, reduces middlemen and allows centralised firms to exploit new economies of scale and scope. Often these technologies are produced and maintained by external suppliers, but many firms possess sophisticated technological capabilities and may operate their own Virtual Private Networks with telecommunications suppliers.

Secondly, once financial institutions have reached a certain scale they can exploit *internal* infrastructure technologies to calculate and control how their internal resources should be allocated. This use of internal infrastructure technology is dependent on firms building their internal technological capabilities to develop and use IT. Even when firms outsource their IT and telecommunications, they must still have sufficient technical capabilities to be intelligent customers and operate and use the infrastructure reliably.

This first half of this chapter has explored why infrastructure is important to the financial services industry, showing the importance of regulations and regulatory compliance to their development, and how the technologies comprise a range of complex, software-intensive capital goods that form part of, and are linked to, wider Large Technical Systems, such as global telecommunications networks. Despite their economic importance, there is little academic literature on the specific problems of financial infrastructure innovation, compared to other industries such as pharmaceuticals, for example. Consequently, the next section looks at the wider innovation literature, and examines what it can illuminate about the features of infrastructure outlined here. It will explore how infrastructure is similar to, and differs from, the mass production consumer goods traditionally studied in the innovation literature, and how the particularities of financial services influence the relevance of this wider research.

Innovation in Infrastructure

The Nature of Innovation

Innovation is generally characterised by high levels of uncertainty and sector specificity (Dosi, 1988; Freeman, 1982; Pavitt, 1984, 1989). Technological novelty and complexity contribute towards uncertainty, and mean that attempts to develop and use innovations often run into difficulties. These uncertainties can, however, be managed (Tidd et al., 1997) and a series of empirical studies have attempted to establish the factors that contribute to success. Project 'SAPPHO' was one of the first attempts to systematically identify what differentiates pairs of successful and unsuccessful innovations (Rothwell et al., 1974; SPRU, 1972). Although the original study was based on process innovations in the chemical industry and product innovations in the scientific instruments sector, three major findings came through clearly and have been supported by subsequent studies (Bacon et al., 1994; Clark et al., 1989; Cooper, 1983; Cooper & de Brentiani, 1991; Cooper & Kleinschmidt, 1990, 1993;

Prencipe, 2002; Shenhar et al., 2002; Tidd et al., 1997). These are: the importance of understanding user needs, the importance of good internal knowledge co-ordination, and the importance of having strong internal technical capabilities in order to access and incorporate external sources of knowledge.

While these findings remain important for understanding innovation in financial infrastructure, care must be taken to recognise the sector specificity of technical change (Freeman & Soete, 1998). Pavitt (1984) has produced a taxonomy of sectoral patterns of innovation that divides innovations into three types with different characteristics. These are: supplier-dominated, production-intensive and science-based. More recently the taxonomy was updated to include information-intensive software and services (Pavitt, 1990, cf. 1996). While the original paper is highly cited, many people miss that it added an additional "fourth category . . . to cover purchases by governments and utilities of expensive capital goods, related to defence, energy, communications and transport" (1984, p. 276).

The notion that innovation in complex capital goods is different from innovation in consumer goods was developed in the military-technology literature (Walker et al., 1988). Walker et al. (1988) formulated a hierarchy of military technologies that extends from very-low-cost materials and components (such as nuts and bolts), to high-cost components (such as jet fighter engines), and on to entire military systems costing billions of dollars (such as missile defence systems). They noted that:

> as the hierarchical chain is climbed products become more complex, few in number, large in scale, and systemic in character. In parallel, design and production techniques tend to move from those associated with mass-production through series- and batch-production to unit production. Towards the top of the hierarchy, production involves the integration of disparate technologies, usually entailing large-scale project management and extensive national and international co-operation between enterprises. Thus, the pyramid is also one of increasing organisational and managerial complexity (Walker et al., 1998, pp. 19–20).

During the 1980s and 1990s a growing body of research analysed these highly engineered, bespoke capital goods. The research found that they tend to be produced by temporary networks of systems integration firms and their suppliers and regulators (Burton, 1992; Hobday, 1998; Hughes, 1987, 1983; Miller et al., 1995; Prencipe et al., 2002; Shenhar, 1998; Walker et al., 1988). Research on the management of their development has stressed the importance of good project management, good risk management, and effective control over the various suppliers involved in

Table 2. Contrast between innovation in complex capital goods & commodity products.

Feature	Complex Capital Goods	Commodity Products
Product Characteristics	Very high cost	Low cost
	Multi-functional upstream capital goods	Single function downstream consumer goods
	Complex components and interfaces	Simple components and interfaces
	Many bespoke components	Small number of standardised components
	Hierarchical and systemic	Simple architectures
	High degree of embedded software	Little software
Production Characteristics	One-off projects or small batch production	High volume
	Highly uncertain	Well understood
	Systems Integration	Efficient production
	Scale-intensive mass production not relevant	Incremental process cost improvements
Innovation Process	User-producer-driven	Supplier-driven
	Highly flexible craft-based	Formalised
	Innovation and diffusion collapsed	Innovation and diffusion separate
	Innovation path agreed *ex ante* among suppliers, customers and regulators	Innovation path mediated by market selection
Industrial Co-ordination	Elaborate temporary network	Structured around large firms and their supply chains
	Project-based, multi-firm alliances	Mass production by single firm
Market Characteristics	Duopolisitic structure or internal provision	Many competing buyers and sellers
	Few large transactions	Large number of transactions
	Business to business	Business to consumer
	Administered markets	Regular market mechanism
	Heavily regulated and often politicised	Limited regulation

project development (Dvir et al., 1998; Hobday, 1998; Lindkvist et al., 1998; Might & Fischer, 1985; Miller et al., 1995; Morris, 1990, 1994; Pinto et al., 1993, 1998; Shenhar, 1998; Shenhar et al., 2002; Shenhar & Dvir, 1996; Tatikonda & Rosenthal, 2000; Williams, 1995; Williams et al., 1995 for a review of recent research). Insightful work has also revealed the particular problems associated with large-scale software development (Boehm, 1991; Brady 1997; Brooks, 1995; Gibbs, 1994; Parnas, 1985; Walz et al., 1993). In an important study of 110 Israeli defence projects Dvir et al. (1998) found that the balance of success factors was far from universal, with software projects being very different from hardware projects, and risk management and budget control less important for small-scale projects but vital for large ones.

In two seminal papers Hobday explored innovation in complex products and systems (CoPS) and noted that they are characterised by their business-to-business, capital good nature, their batch production processes, high-cost, inherent uncertainty, and the degree of embedded software. They tend to have production process that are based on temporary, negotiated and bureaucratically driven project-based firms (PBFs) and organisations (involving webs of users, producers and regulators) (cf. Gann & Salter, 1998; Lemley, 1992; Marquis & Straight, 1965; Might

& Fischer, 1985). This is particularly true of financial infrastructure, like the CREST or TAURUS systems, which are used and developed by many firms. Further research has highlighted the importance of early user involvement and the heavily regulated, bureaucratically administered nature of the market and development processes (Burton, 1993; Morgan et al., 1995; Morris, 1990; Sapolski, 1972; Walker et al., 1988). These features can be contrasted with mass-production goods that are low-cost, well understood, contain little embedded software and are sold in largely unregulated markets. This contrast can be seen in Table 2 (derived from Hobday, 1998, p. 699).

The Process of Infrastructure Development

The combination of high technological complexity and uncertainty, complicated user needs, long development times, high costs and high risk makes infrastructure development extremely difficult. There tends to be substantially more stress on risk management, project co-ordination, and uncertainty management than in traditional innovation processes. Despite these attempts at dealing with uncertainty, infrastructure projects suffer from a range of innovation problems and frequent project failures (Flowers, 1996; Hobday, 1998; Morris, 1990; Nightingale, 2000; Sauer & Waller, 1993; Tatikonda & Rosenthal, 2000).

One way of dealing with the uncertainty and high risk is to reverse the traditional method of *ex post* selection of consumer products in markets. Consumer goods innovation typically starts with R&D and moves into development, then production, then marketing and finally product launch and the selection of products by market mechanisms (see, for example, Utterback & Abernathy, 1975, cf. Barras, 1990).

By contrast, in complex infrastructure technologies the high risks and costs involved mean that customers, suppliers, regulators and government bodies negotiate contracts, product designs and production methods before development is begun (Hobday, 1998; Peck & Scherer, 1962; Walker et al., 1988; Woodward, 1958). This is intended to reduce risk and ensure that the end product matches the various stakeholders' requirements. As a consequence, infrastructure innovation processes will typically start with marketing and sales, and only after an outline design and production process is specified will the contracts be signed and development started.

The importance of understanding user needs throughout the development process is well recognised within the innovation literature. The SAPPHO project found it to be the major determinant of success (Rothwell et al., 1974; SPRU, 1972). This has been supported by subsequent research (Cooper, 1986; Grieve & Ball, 1992; Keil & Carmel, 1995; Lundvall, 1988; Mansfield, 1977; Rothwell, 1976; Teubal et al., 1977). In particular, von Hippel (1976) has shown the importance of users as sources of innovation, and the case studies of the SAPPHO project illustrated the importance of users making adjustments to technologies after they had been acquired (Rothwell, 1977). Bacon et al. (1994) found that using development engineers rather than marketing staff to liase with customers, and using prototypes to aid customer feedback, produced superior results.

This initial stage of infrastructure development involves understanding user requirements, dealing with technical uncertainties, finding solutions that are acceptable to the various customers and users and then obtaining commitment from a whole network of stakeholders to an uncertain and potentially very risky venture. Simply getting the various users and customers, as well as national and international regulators, to define the infrastructure's function in any detail can be extremely difficult. Infrastructure technologies' initial cost estimates and overarching function are generally vague enough to interest a range of parties, but once one moves to the more specific architectural layout of the technology, the engineering trade-offs become increasingly politicised.

Choices about technologies become politicised because the implementation of new infrastructures often causes disruptions to established practices within institutions. The changes potentially impact a range of actors who may be resistant to the proposed project if they feel that their interests will suffer. The 'Big Bang' in the City of London, which moved the City towards electronic trading, for example, required a large push from the U.K. government and met resistance from established groups within the City. The introduction of any major technology will have positive and negative effects for different groups, and those who are negatively affected may have very rational reasons to be against the project. The very late shift towards electronic trading in the New York Stock Exchange, for example, is indicative of the power of entrenched interests. Even with a largely positive group of users, it is not necessarily the case that they will have the same requirements or demands, and substantial negotiation is needed to define acceptable solutions (cf. Moynihan, 2002). These negotiations will necessarily involve trade-offs and the final proposals may not necessarily match anyone's specific requirements, a feature that applies as much within organisations as it does between them (Barki & Hartwick, 2001).

The problems involved in specifying the functions of a major infrastructure technology are complicated because many firms and institutions lack the technology capabilities to be 'intelligent customers'. Infrastructure technologies, after all, are not produced by banks everyday. As a consequence, financial institutions are susceptible to being misled by consultants and contractors into producing overly complex technologies that do not match their needs (Collins, 1997). Despite the substantial in-house capabilities that financial institutions have for developing and using information technology, many still lack the expertise needed to fully comprehend the complexities and potential difficulties they face in developing and implementing new infrastructure. Software firms are generally reluctant to be open about these potential problems, as they have an incentive to downplay them in order to secure contracts (Flowers, 1996). Even using external consultants to assess the bids may not overcome this problem, as their independence from the software industry is often questionable (Collins, 1997; Flowers, 1996).

The costs, complexities and risks involved in developing infrastructure mean that development projects require commitment from a range of actors. This typically comes in the form of written legally enforceable contracts, organisational commitment, and political endorsement. Unfortunately, these can easily lock counter-parties into a particular direction of technical change that may involve using inferior technologies and architectures (Collingridge, 1983; Walker, 2000). Similar problems emerge within organisations and make changes to heavily committed decisions difficult to undertake (Flowers, 1996). The process of securing commitment to a technology can make 'pulling the plug' on failing projects very difficult. This is a common problem in IT-intensive infrastructure. Keil et al. (2000a) found that between

30% and 40% of all the information systems in their sample exhibited some degree of "escalation of commitment to failing courses of action", where the projects in question spiraled out of control (cf. Keil, 1995; Keil et al., 2000b; Keil & Montealegre, 2000; Smith et al., 2001). As with infrastructure projects in developing countries, larger projects require greater commitment, which makes changes in the light of potential failure more difficult.

Project Uncertainty and Process Flexibility

Once the initial specifications are defined and the customers are committed, the process moves to the next stage where engineers and technologists propose, test and modify solutions. As Vincenti (1990, p. 9) notes, the design process:

> for devices that constitute complex systems is multi level and hierarchical . . . The levels run more or less as follows from the top on down:
>
> (1) Project Definition—translation of some usually ill defined military or commercial requirement into a concrete technical problem for level 2;
> (2) Overall Design—layout of arrangements and properties . . . to meet the project definition;
> (3) Major Component Design—division of project . . . ;
> (4) Subdivision of areas of component design from level 3 according to engineering discipline required . . . ;
> (5) Further Division of categories in level 4 into highly specific problems.

Thus, for complex infrastructure technologies the design process will involve specifying components and architecture in increasing detail. This process is complicated for infrastructure technologies because many sub-components are systemically related. As Nelson points out:

> A particular problem in R&D on multi-component systems arises if the appropriate design of one component is sensitive to the other components. Such interdependencies mitigate against trying to redesign a number of components at once, unless there is strong knowledge that enables viable design for each of these to be well predicted *ex ante* or that there exists reliable tests of cheap models of the new system (1982, p. 463).

This systemic complexity means that the effects of incorrect design are magnified as design modifications spread to other systemically related subsystems.

The inherent uncertainties involved mean that initial solutions will rarely work correctly (Petroski, 1986). Instead, an iterative process of trial and error design is used to bring proposed solutions closer to the desired function. Each design iteration will cause the innova-tion process to *feed back* to earlier stages *adding to its cost and schedule* (Nightingale, 2000; Williams et al., 1995). With simple technologies this process is rarely problematic, but with complex infrastructure technologies there are a larger number of possible feedback loops and potential failures within and between components.

If the process of updating design specifications takes time, the design changes are extensive, or the components are related to a large number of 'sensitive' components, the amount of redesign work can be very large. There is consequently a danger of 'redesign chain reactions' that spread through the different systems with disastrous effects (Nightingale, 2000). As the next section will show, this is a particular problem with software-intensive systems which tend not to scale in a linear way, so that small, local design modifications have the potential to grow into funda-mental design changes at the project level.

The complexity of infrastructure projects means that even without these redesign loops, unforeseeable emergent problems can develop during the production process. As a consequence, production and develop-ment overlap to a far greater extent than in traditional innovation processes, where development problems are typically ironed out before production begins. The overlap between development and production is also required because the long timescales involved in developing infrastructure technologies (often years) mean that some component technologies can undergo radical technical changes during the lifetime of the project. This is especially true of rapidly changing technologies such as IT. Consequently, the project's designs and processes need to be flexible and able to incorporate new developments as the project proceeds (Hobday, 2000).

The fluidity of design specifications and the high risk and cost of infrastructure make good internal, cross-functional knowledge co-ordination important for project success. Research on other sectors has high-lighted the performance differences between firms that integrate functional disciplines and those that have a sequential innovation process (Bowen et al., 1995; Clark & Fujimoto, 1991; Leonard-Barton, 1995; Iansiti, 1995; Rothwell, 1992, 1993; Womack et al., 1991). The importance of organisational structure for knowledge integration has been highlighted by Gal-braith (1973), Wheelwright & Clark (1993) and Clark & Fujimoto (1991), building on the original insights of Burns & Stalker (1961). The importance of different organisational forms in the development of large, complex technologies has been well recorded in the literature (Burns & Stalker, 1961; Larson & Gobeli, 1987, 1989; Miles & Snow, 1986). Larson and Gobeli, for example, show how the success of projects is dependent on appropriate organisational structures and relate that to the complexity, technological novelty, managerial capabilities and functional definition of projects (1989). Rothwell (1993, 1992) has shown how information technology can be used to improve the integration of knowledge (cf. Bacon et al., 1994).

The Importance of Embedded Software

The problems associated with financial infrastructure innovation are exacerbated by the increasing importance of embedded software. This software is used to control how financial information is routed and processed and can radically improve infrastructure performance (Nightingale & Poll, 2000a). However, the incorporation of software adds to the complexity and uncertainty of development (Nightingale, 2000, p. 5) and turns what were previously straightforward engineering tasks into high-risk development projects (Brooks, 1986; Hobday, 1998).

Software development is problematic because of its vulnerability to errors and fragility (Brooks, 1995; cf. Boehm, 1981, 1991; Mills, 1971; Parnas, 1985; Ropponen & Lyytinen, 2000; Royce, 1970; Willcocks & Grithiths, 1994). Software is more vulnerable to design errors than other technologies because the abstract construction of interlocking concepts, data sets, relationships, algorithms and function invocations that make up a piece of software must all work perfectly if the software is to function properly. Unfortunately, the potential problems inherent in a string of code are not easy to find, making debugging and testing embedded software extremely difficult and time-consuming, and as a result, systems are often launched without being properly tested (Flowers, 1996).

The fragility of embedded software refers to the difficulties involved in modifying software as compared to many other technologies. As software is vulnerable to 'bugs', must work almost perfectly, and typically scales in non-linear ways, minor changes in code can necessitate extensive redesigns. These, in turn, can produce feedback loops within the software development process that can snowball into resource-intensive redesign processes and lead to increasingly fragile and low-quality products.

Software also adds to the development problems associated with the more tangible parts of the infrastructure because it may increase the degree of systemic interactions between physical sub-systems. While this can improve system performance, it can also reduce the ability to break complex innovation problems into sub-problems and modularise development.

The problems associated with producing large software systems are well recognised in the literature (Collins, 1997; Flowers, 1996; Fortune & Peters, 1995). This literature is mainly drawn from large projects that developed software from scratch, but financial service infrastructure often involves building on older legacy systems. Such systems may have inappropriate architectures and data structures that were designed for an older generation of infrastructure.

Similarly, the code may be written in a language that is no longer in common use, and the system experts may no longer be with the organisation, or, if they are still within the organisation, they may be committed to the older system and highly resistant to change.

The intangible nature of software further complicates the testing process. With a physical piece of infrastructure such as a road, it is possible to reliably evaluate how much further work is needed, but with software this is often extremely difficult. Software engineers have a saying that software is '90% finished 90% of the time'. This difficulty in evaluating the extent of further work creates extra uncertainties that can delay the cancellation of a failing project well beyond the point at which it would have been stopped had the full extent of the required work been known (Block, 1983; Flowers, 1996; Staw & Ross, 1987).

The problems involved in developing software-intensive infrastructure can be gauged by Gibbs' (1994, p. 72) point that "for every six new systems that are put into operation, two others are cancelled. The average software development project overshoots its schedule by half; larger projects generally do worse and some three quarters of all large scale systems are 'operating failures' that either do not function as intended or are not used at all". Similarly, the General Accounting Office has reviewed large U.S. government IT projects and noted:

> During the last 6 years, agencies have obligated over $145 billion building up and maintaining their information technology infrastructure. The benefits of this vast expenditure, however, have frequently been disappointing. GAO reports and congressional hearings have chronicled numerous system development efforts that suffered from multi-million dollar cost overruns, schedule slippages measured in years, and dismal mission-related results (GAO, 1997, p. 6).

Since financial services are significant users of IT infrastructure, they are particularly vulnerable to these large-scale IT failures. There have been a number of failed projects within the sector, ranging from the high-profile TAURUS system in the London Stock Exchange to a whole host of lower-profile failures within other institutions. These smaller failures are often covered up as financial institutions attempt to maintain their reputations for reliability, but interviews suggest that they are extremely common. Even more difficult to analyse are the numerous operational failures that produce significant losses for the institutions involved. Typically a software glitch or design error may cause a financial institution to miss-sell a series of trades. Whether these operational failures are due to the technology, training and operations, or auditing and management, may be impossible to tell. Whatever their cause, there are numerous cases of financial institutions losing hundreds of millions of dollars.

Implementation and Failure

In addition to the difficulties described above, the implementation of financial infrastructure projects can also cause a successful design project to ultimately be an operational failure. To a large extent, successful implementation is dependent on effective project management (Pinto & Slevin, 1997, cf. Currie, 1994; Pinto & Covin, 1989; Slevin & Pinto, 1987). This includes, first and foremost, realistic planning in terms of both resources and time (ibid.) and recognising the importance of 'soft' human resource issues (Corbato, 1992; Levasseur, 1993). Within the U.K. financial-services sector, contractors work on a rule of thumb that implementation will be twice as expensive and take twice as long as the best initial estimates.

As infrastructure technologies are often business-critical, and financial services are time-dependent, the introduction of new systems can be very risky (Nightingale & Poll, 2000a). The business-critical nature of these technologies means that back-up systems must be in place and maintained to a very high standard. It is not uncommon for system failures to result in significant financial loss and damage to reputation. One interviewee likened the implementation process to "changing the foundations of a house with the house still standing" and added that "you also, at any given time, must be able to return the foundations back to their original state without anyone noticing you are there". The business-critical nature of these technologies often means that multiple back-up systems are put in place (many of which are used for simulation-based training).

One particular area of risk involves data conversion and migration from old to new systems. In some instances this may be so difficult that it is judged quicker and easier to re-key all the data. This problem is gradually receding as more and more products are built on similar database platforms and are released with Other Data-Base Connectivity (ODBC) drivers. This highlights the point that effective design involves considering implementation issues at an early stage, in particular customers' end-to-end business systems and requirements. Unfortunately, the changes that new infrastructure make to these processes are extremely difficult to foresee before implementation begins.

The risks and uncertainties involved in system implementation mean that it is common to run parallel systems, often for months, in order to iron out last-minute design problems. These systems are also used for staff training before implementation. Typically, within the City of London at least, and where it is technically feasible, the new systems will be run in a minimum of three environments; a live system, a QA (or quality assurance) system, and a test system. The test system often contains scrambled data (allowing

wider access than confidential customer information would permit) and is used for training and experimental late-stage design. The QA system is used for more sophisticated and rigorous testing of design changes, before components are implemented into the live system. There are limitations to the use of parallel runs, particularly in larger, more complicated, real-time, environments and where systems are being installed that are fundamentally different from the earlier technology.

The problems associated with the design and implementation of financial service infrastructure, highlighted in this chapter, mean that project failures are extremely common. No reliable figures are available, but anecdotal evidence suggests that at least a third of these projects are terminated or are operationally compromised. A classic, and well documented, example of financial infrastructure failure was the City of London's TAURUS system (cf. Currie, 1997; Flowers, 1996).

The TAURUS System

The TAURUS (Transfer and Automated Registration of Uncertified Stock) system was an infrastructure project within the City of London that was intended to create a paperless share-trading environment. Legal requirements in the U.K. meant that registers of all share trades had to be held by all publicly listed companies. Consequently, even small trades involved a very inefficient process whereby at least three pieces of paper were physically moved around the City. Unsurprisingly, an international report had criticised the settlements system, and it was clear that the system would be unable to cope with the mass share ownership that would follow the privatisation of U.K. public utilities. By August 1987, for example, a backlog of nearly 650,000 unsettled deals had built up (Flowers, 1996, p. 101).

Moving to a computerised system would, it was hoped, reduce time and cut costs by removing the need to physically transfer paper during trading, and the middlemen who controlled the process. The London Stock Exchange (LSE) had previously installed a less ambitious Talisman system in 1979 as part of a computerised share settling system, and its success led to the proposal in 1981 to automate the entire market and end paper trading. Consequently, work started on the system that was to become TAURUS in 1983, and the LSE as an institution committed itself very publicly to develop a world-class system that would ensure its dominance over other European exchanges.

The original architecture of the project involved replacing the registrars (who recorded information on trading) with a single large database administered by the LSE. Unfortunately, the project had many stakeholders who had vested interests in maintaining their positions. Instead of re-engineering from scratch, and then automating, the project recreated the highly inefficient organisation of the exchange. The initial design was pushed forward before the various parties had reached agreement about what it should do, before legislation and regulations had been changed and before the processes had been simplified. The registrars quickly saw that the technology would deprive them of their lucrative (but inefficient) livelihoods and became an entrenched group of actors hostile to the new technology (Drummond, 1996a, 1996b).

The LSE carried on with the project from 1983 until 1988, but the share registrars' opposition produced an independent technical review. This showed that the proposed project would be extremely costly (£60 million) and very technically complex, requiring two IBM 3090 mainframes and 560 disk drives to cover the transactions in the trillion or so shares in issue by the LSE (Flowers, 1996, p. 102). Taurus 1 was consequently cancelled, and the Bank of England (the U.K. financial regulator at the time) became involved and set up the Security Industry Steering Committee on Taurus (SISCOT Committee).

In the spring of 1989, the SISCOT Committee suggested the far cheaper and less risky option of extending the Talisman system from the original 32 Market Makers to around 1,000 other financial institutions (ibid.) Shares would be held in accounts on behalf of clients by TACs (Taurus Account Controllers) who would maintain the records. However, TACs could pass the maintenance of records on to registrars, which had the unfortunate effect of replicating the previous paper-based system in parallel with the electronic TACs. This design, however, made it extremely difficult for companies to know who owned their shares, and consequently, if they were being prepared for a hostile takeover. Together with stockbrokers, who were concerned about costs, the large firms forced through a series of design changes.

In March 1990 a detailed outline of the project was published, and in October 1990 the technical specifications were published. These showed that rather than build a system from scratch, the project would involve buying and modifying a U.S.-based system. The development work was however based on split locations, with 25, out of a team of 40, based at Vista Concepts New York office (Drummond, 1996a, 1996b; Flowers, 1996). By December 1991, as costs escalated and schedules slipped, the new legal framework for the operation of TAURUS was produced. By this time the press were becoming hostile as rumours emerged about software-development problems. By February 1993, a technology review predicted that it would take another three years to build the system and that its costs would double. By March the project was cancelled at a cost of about £75 million. The total costs to the City were probably in the order of £400 million.

The Taurus project illustrates the range of problems involved in producing financial infrastructure. In particular, it shows the major innovation problems in

financial infrastructure concerning the 'soft' issues involved in co-ordinating a wide range of users who have divergent needs and concerns about the technology. The main cause of the failure was the inability of the various agents involved to restructure the inefficient internal processes before the development of the technology was begun. This meant that the project re-created vested interests who were hostile to changes. The slow-moving regulation process, the decision to undertake substantial redesign of a packaged system, and confusion about ultimate control over the project further complicated matters.

The complexities inherent in the design of such a substantial system were made worse by constant design changes. The software-intensive nature of the project meant that it was very difficult to know the extent of further redesign work. The lack of organisational co-ordination within the development process, and in particular the use of two main development sites on different sides of the Atlantic further complicated the process. The story does, however, have a happy ending. After the Taurus fiasco, the Bank of England project managed the development of the CREST system which appeared on schedule and to budget in the mid-1990s.

Conclusion

This chapter has given a brief overview of an extremely complex field. It has shown both the diversity of different forms of financial service infrastructure, and the heterogeneity of the various firms and organisations that use it. Financial services are extremely high-tech in many areas, and the traditional view of services as un-innovative clearly needs to be revised. While there are encouraging signs that this is taking place, our understanding of innovation in services, and financial services in particular, is a long way behind our understanding of manufacturing innovation.

Financial services differ from manufacturing because they involve firms performing functions for customers rather than providing goods that perform functions. These functions are typically consumed as they are produced, making their provision time-dependent in a way that manufactured products which can be stored are not. Firms rely on a range of infrastructure technologies for this temporal control of their internal processes. The internal processes themselves are also typically dependent on infrastructure.

Financial services are also different in that they depend heavily on the liquidity of markets. This liquidity can be improved by using external infrastructure technologies to bring together larger numbers of diverse buyers and sellers. As they are based on the contextualisation and processing of information, rather than physical materials, financial services have the potential to leverage technology to produce more profits than is possible in manufacturing. Unfortunately, the ability to generate large profits goes hand in hand with the ability to suffer substantial losses. This, in turn, creates a much larger incentive for managing risk than is common in manufacturing.

In this chapter, we have divided infrastructure technologies into internal and external. Infrastructure that is external to financial institutions is often provided by third parties and is used to increase the reach of markets in time and space. This makes markets more liquid and improves the financial value of assets, and makes trading more efficient. Typically, these technologies will involve telecommunications systems that have historically ranged from the earliest telegraph systems to today's high-powered settlement systems and global networks.

Internal infrastructure technologies, however, are used within firms to allocate resources to customers. They involve technologies such as customer-focused information systems, ATM machines and their networks, and internal risk-management systems. In both cases, there is a tendency for the infrastructures to increase in size and complexity. External infrastructure increases in size and complexity because this allows the liquidity of markets to improve. Internal infrastructure is driven towards increasing coverage, computing power, speed and reliability, because this improves the ability to profitably control transactions.

The tendency towards increases in the size and complexity of these infrastructure technologies has consequences for innovation. As projects increase in size and complexity they become more uncertain and more risky. This increased uncertainty comes from unpredictable emergent phenomena caused by components interacting in new ways, from technical changes to sub-systems as the time taken for the projects increases, changes in regulations, and changes in business practice. All of these factors make defining the systems' requirements and freezing them extremely difficult. The complexity has become even more problematic in the last two decades with the introduction of embedded software, and the need to work with legacy systems.

The complexity and uncertainty of innovation mean that when problems emerge, the design process has to feedback to earlier stages in the innovation process. This adds to the cost and schedule of the project. Within software the problem is exacerbated as repeated redesign can lead to an increasingly unreliable and fragile product.

However, the really important issue in innovation in financial infrastructure is not technical at all. The big problems concern 'soft' issues about dealing with multiple users, dealing with regulators and coping with the politics of the organisational and institutional changes that the introduction of new infrastructure brings (Murray, 1989). These factors probably go further towards explaining why the financial services industry is riddled with failed infrastructure projects that have not delivered what they were intended to provide.

Despite the constant stream of failed projects and, in particular, IT-based failures, the financial services industry does manage to produce infrastructure that is reliable and performs its task well. Failure may be common, but successes abound. Given the complexities of the innovative tasks involved, this success in both development and operation is a major achievement. Given the importance of these successes to the modern financial system, and by extension their impact on the global economy, understanding the nature of their innovation is important and worthy of further research.

References

Bacon, G., Beckman, S., Mowery, D. & Wilson, E. (1994). Managing product definition in high-technology industries: A pilot study. *California Management Review*, **2**, 32–56.

Bansal, A., Kauffman, R. J., Mark, R. M. & Peters, E. (1993). Financial risk and financial risk management technology (RMT)—issues and advances, *Information and Management*, **24** (5), 267–281.

Barki, H. & Hartwick, J. (2001). Interpersonal conflict and its management in informational systems development. *MIS Quarterly*, **25** (2), 195–228.

Barras, R. (1986). Towards a theory of innovation in services: The vanguard of the services revolution. *Research Policy*, **15** (3), 161–173.

Barras, R. (1990). Interactive innovation in financial and business services: The vanguard of the services revolution, *Research Policy*, **19** (3), 215–237.

Bell, D. (1973). *The coming of post-industrial society*. London: Heinemann.

Berger, A. N., Kashyap, A. K. & Scalise, J. M. (1995). The transformation of the U.S. banking industry what a long strange trip its been. *Brookings Papers on Economic Activity*, **2**, 55–218.

Berstein, P. L. (1996). *Against the gods: The remarkable story of risk*. London: John Wiley.

Black, F. & Scholes, M. (1973). The pricing of options and corporate liabilities. *Journal of Political Economy*, **81** (3), 637–659.

Block, R. (1983). *The politics of projects*. New York. Yourdon Press.

Boehm B. W. (1981). *Software engineering economics*. Englewood Cliffs, NJ: Prentice Hall.

Boehm, B. (1991). Software risk management principles and practices. *IEEE Software*, **1**, 32–41.

Bowen, H. K., Clark, C., Holloway, C. & Wheelwright, S. (Eds) (1995). *The perpetual enterprise machine: High performance product development in the 1990s*. New York: Oxford University Press.

Brady, T. (1997). *Software make or buy decisions in the first forty years of business computing*. Unpublished D. Phil thesis, SPRU, University of Sussex, U.K.

Brady, T. & Target, D. (1996). Strategic information systems in the banking sector—holy grail or poison chalice. *Technology and Strategic Management*, **7** (4), 387–406.

Brooks, F. P. (1986). No silver bullet—Essence and accidents of software engineering. In: H. J. Kugler (Ed.), *Information Processing 86* (pp. 1069–1076). Amsterdam: Elsevier Science (North Holland).

Brooks, F. P. (1995). *The mythical man month, essays on software engineering* (25th Anniversary Edition). New York: Addison-Wesley.

Bryson, J. R. & Daniels, P. W. (Eds) (1988). *Service industries in the global economy* (2 vols). Cheltenham, U.K.: Edward Elgar.

Burns, T. & Stalker, G. M. (1961). *The management of innovation*. London: Tavistock.

Burton, J. G. (1993). *The pentagon wars*. Annapolis, MD: Naval Institute Press.

Buzzacchi, L., Colombo, M. G. & Mariotti, S. (1995). Technological regimes and innovation in services: the case of the Italian banking industry. *Research Policy*, **24** (2), 151–168.

Calomiris, C. W. & White, E. N. (1994). The Origins of Federal Deposit Insurance. In: C. Goldin & G. Libecap (Eds), *The regulated economy: A historical approach to political economy* (pp. 145–188). Chicago, IL: University of Chicago Press.

Chandler, A. D. Jr. (1990). *Scale and scope: The dynamics of industrial capitalism*. Cambridge, MA: Belknap Press.

Channon, D. F. (1998). The strategic impact of IT on the retail financial services industry. *Journal of Strategic Information Systems*, **7** (3), 183–197.

Clark, K. B. & Fujimoto, T. (1991). *Product development performance: Strategy, organisation, and management in the world auto industry*. Boston, MA: Harvard Business School Press.

Clark, K. B., Hayes, R. S. & Wheelwright, C. (1989). *Dynamic manufacturing*. New York. Free Press.

Collingridge, D. (1983). *Technology in the policy process*. London: Frances Pinter.

Collins, T. (1997). *Crash, ten easy ways to avoid a computer disaster*. London: Simon & Schuster.

Cooper, R. G. (1983). A process model for industrial new product development. *IEEE Transactions on Engineering Management*, **EM-30**.

Cooper, R. G. (1986). *Winning at new products*. Reading MA: Addison Wesley.

Cooper, R. G. & de Brentiani, U. (1991). New industrial financial services: what distinguishes the winners. *Journal of Product Innovation Management*, **8**, 75–90.

Cooper, R. G. & Kleinschmidt, E. J. (1990). *New products: the key factors in success*. Chicago, IL: American Marketing Association.

Cooper, R. G. & Kleinschmidt, E. J. (1993). Major new products: What distinguishes the winners in the chemical industry. *Journal of Product Innovation Management*, **10**, 90–111.

Corbato, F. (1992). On building systems that will fail. *Communications of the ACM*, **343** (9), 79.

Currie, W. (1994). The strategic management of a large scale IT project in the financial services sector. *New Technology, Work and Employment*, **9** (1), 51–66.

Currie, W. (1997). Computerising the stock exchange: A comparison of two information systems. *New Technology, Work and Employment*, **12** (2), 75–83.

Davies, A. & Brady, T. (2000). Organisational capabilities and learning in complex product systems: towards repeatable solutions. *Research Policy*, **29** (7–8), 931–953.

Dickson, P. G. M. (1969). *The financial revolution in England: A study in the development of public credit, 1688–1756*. London: Macmillan.

Dosi, G. (1982). Technological paradigms and technological trajectories: a suggested interpretation of the determinants and directions of technical change. *Research Policy*, **11** (3), 147–162.

Dosi, G. (1988). The nature of the innovation process. In: G. Dosi, C. Freeman, R. R. Nelson, G. Silverberg & L. Soete (Eds). *Technical Change and Economic Theory* (chap. 10). Brighton, U.K.: Wheatsheaf.

Drummond, H. (1996a). *Escalation in decision-making: The tragedy of Taurus*. New York: Oxford University Press.

Drummond, H. (1996b). The politics of risk: Trials and tribulations of the Taurus Project. *Journal of Information Technology*, **11** (4), 347–357.

DuBoff, R. B. (1983). The telegraph and the structure of markets in the United States 1845–1890. *Research in Economic History*, **8**. 253–277.

Dvir, D., Lipovetsky, S., Shenhar A. J. & Tisher, A. (1998). In search of project classification: A non-universal approach to project success factors. *Research Policy*, **27** (9), 915–935.

Easingwood, C. J. & Storey, C. (1996). The value of multi-channel distribution systems in the financial services sector. *Service Industries Journal*, **16** (2), 223–241.

Eccles, R. G. & Crance, D. D. (1988). *Doing deals: Investment banks at work*. Reveling, MA: Harvard Business School Press.

Eurostat (1998). *Statistics in focus, distributive trade, services & transport* (No. 5/98). Services in Europe—Key Figures Luxembourg, Eurostat.

Flowers, S. (1996). *Software failure management failure*. London: John Wiley.

Fortune, J. & Peter, G. (1995). *Learning from failure: The systems approach*. John Wiley.

Freeman, C. (1982). *The economics of industrial innovation* (2nd ed.). London: Pinter.

Freeman, C. & Perez, C. (1988). Structural crises of adjustment, business cycles and investment behaviour. In: G. Dosi, C. Freeman, R. R. Nelson, G. Silverberg & L. Soete (Eds), *Technical Change and Economic Theory*. Brighton, U.K.: Wheatsheaf.

Freeman, C. & Soete, L. (1998). *The economics of industrial innovation* (3rd ed.). London: Pinter.

Frei, F. X., Harker, P. T. & Hunter, L. W. (1997). *Innovation in retail banking*. Wharton Financial Institutions Centre Working paper 97-48-B.

Fuchs, V. (1968). *The service economy*. New York: National Bureau of Economic Research.

Galbraith, J. R. (1973). *Designing complex organisations*. Reading, MA: Addison-Wesley.

Gallouj, F. & Weinstein, O. (1997). Innovation in services. *Research Policy*, **26** (4–5), 537–556.

Gann, D. & Salter, A. (1998). Learning and innovation management in project-based, service-enhanced firms. *International Journal of Innovation Management*, **2** (4), 431–543.

G.A.O. (1997). *High risk series—1997*. Washington, D.C.: General Accounting Office U.S. HR 97-9.

Gibbs, W. W. (1994). Software's chronic crisis. *Scientific American*, **9**, 72–81.

Golding, T. (2000). *The city: Inside the great expectation machine: Myth and reality in institutional investment and the stock market*. Financial Times, Prentice Hall.

Grieve, A. & Ball, D. F. (1992). The role of process plant contractors in transferring technology. *R&D Management*, **22** (2), 183–192.

Hall, M. (1990). The bank for international settlements capital adequacy 'rules': Implications for banks operating in the U.K. *The Service Industries Journal*, **10** (1), 147–171.

Hobday, M. (1998). Product complexity, innovation and industrial organisation. *Research Policy*, **26**, 689–710.

Hobday, M. (2000). The project based organisation: An ideal form for innovation in CoPS? *Research Policy*, **29** (7–8), 871–893.

Howells, J. (1996). Technology and globalization: The European payments system as a case of non-globalisation. *Technology Analysis and Strategic Management*, **8** (4), 455–466.

Hughes, T. (1983). *Networks of power: Electrification in western society, 1880–1930*. Baltimore, MD: John Hopkins University Press.

Hughes, T. P. (1987). The evolution of large technical systems. In: W. E. Bijker, T. P. Hughes & T. Pinch (Eds), *The Social Construction of Technology: New Directions in the Sociology and History of Technology*. Cambridge, MA: MIT Press.

Iansiti, M. (1995). Technology integration: Managing technological evolution in a complex environment. *Research Policy*, **24** (6), 521–542.

Ingham, H. & Thompson, S. (1993). The adoption of new technology in financial services: The case of building societies. *Economic Innovation and New Technology*, **2**, 263–274.

Keil, M. (1995). Pulling the plug: Software project management and the problem of project escalation. *MIS Quarterly*, **19** (4), 421–447.

Keil, M. & Carmel, E. (1995). Customer developer links in software development. *COMMUN ACM*, **38** (5), 33–44.

Keil, M., Mann, J. & Rai, A. (2000a). Why software projects escalate: An empirical analysis and test of four theoretical models. *MIS Quarterly*, **24** (4), 631–664.

Keil, M. & Montcalcgrc, R. (2000). Cutting your losses: Extricating your organisation when a big project goes awry. *Sloan Management Review*, 5–68.

Keil, M., Tan, B. C. Y. & Wei, K. K. (2000b). A cross cultural study on escalation of commitment behaviour in software projects. *MIS Quarterly*, **24** (2), 299–325.

Larson, E. W. & Gobeli, D. H. (1987). Matrix management: Contradictions and insights. *Californian Management Review*, **29** (4), 126–138.

Larson, E. W. & Gobeli, D. H. (1989). Significance of project-management structure on development success. *IEEE Transactions on Engineering Management*, **36** (2), 119–125.

Lemley, J. K. (1992). The channel tunnel: Creating a modern wonder-of-the-world. *PMNetwork. The Professional Magazine of the Project Management Institute*, **7**, (14–22 July).

Leonard-Barton, D. (1995). *Wellsprings of knowledge: Building and sustaining the sources of innovation*. Cambridge, MA: Harvard Business School Press.

Levasseur, R. E. (1993). People skills—how to improve the odds of a successful project implementation. *Interfaces*, **23** (4), 85–87.

Lindkvist, L., Soderlund, J. & Tell, F. (1998). Management product development projects: On the significance of fountains and deadlines. *Organisational Science*, **19** (6), 931–951.

Lovelock, C. H. (1983). Classifying services to gain strategic marketing insights. *Journal of Marketing*, **47** (3), 9–20.

Lovelock, C. H. & Yip, G. S. (1996). Developing global strategies for service businesses. *California Management Review*, **38** (2), 64–74.

Lundvall, B. (1988). Innovation as an interactive process: From user-producer interaction to the national system of innovation. In: G. Dosi, C. Freeman, R. R. Nelson, G. Silverberg & L. Soete (Eds), *Technical Change and Economic Theory*. London: Francis Pinter.

Mansfield, E. (1977). *The production and application of new industrial technology*. New York: Norton.

Markowitz, H. (1952). Portfolio selection. *Journal of Finance*, **7** (1), 77–91.

Marquis, D. G. & Straight, D. M. Jr. (1965). *Organisational factors in project performance*. Paper presented to the Second Conference on Research Program Effectiveness, July 25, 1965. Reprinted in NASA NsG 235. Washington, D.C.: Office of Navel Research.

Marshall, J. F. & Bansal, V. K. (1992). *Financial engineering: A complete guide to financial innovation*. New York: NYIF.

McDermot, C. M. K., H. & Walsh, S. (2001). A framework for technology management in services. *IEEE Transactions on Engineering Management*, **48** (3), 333–341.

McMahon, L. (1996). Dominance and survival in retail financial services: use of electronic delivery channels to optimise distribution strategy. *Journal of Financial Services Marketing*, **1** (1), 35–47.

Merton, R. C. (1975). The theory of finance from the perspective of continuous time. *Journal of Finance*, **10** (4), 659–674.

Merton, R. C. (1995a). A functional perspective on financial intermediation. *Financial Management*, **24** (2), 23–41.

Merton, R. C. (1995b). Financial innovation and the management and regulation of financial institutions. *Journal of Banking and Finance*, **19** (3–4), 461–481.

Merton, R. & Brodie, Z. (1995). A conceptual framework for analysing the financial environment. In: *The Global Financial System: A Functional Perspective*. Cambridge, MA: Harvard Business School Press.

Michie, R. C. (1997). Friend or foe? Information technology and the London Stock Exchange since 1700. *Journal of Historical Geography*, **23** (3), 304–326.

Might, R. J. & Fischer, W. A. (1985). The role of structural factors in determining project-management success. *IEEE Transactions in Engineering Management*, **32** (2), 71–77.

Miles, I. (1996). *Innovation in services: Services in innovation*. Manchester, U.K.: Manchester Statistical Society.

Miles, R. E. & Snow, C. C. (1986). Organisations: New concepts for new norms. *Californian Management Review*, **28** (3).

Miller, R., Hobday, M., Lewroux-Demer, T. & Olleros, X. (1995). Innovation in complex systems industries the case of flight simulation. *Industrial and Corporate Change*, **4** (2), 363–400.

Mills, H. D. (1971). Top-down programming in large systems. In: R. Rustin (Ed.), *Debugging Techniques in Large Systems*. Englewood Cliffs, NJ: Prentice Hall.

Morgan, R., Cronin, E. & Severn, M. (1995). Innovation in banking: New structures and systems. *Long Range Planning*, **28** (3), 91–100.

Morris, P. W. G. (1990). The Strategic Management of Projects. *Technology in Society*, **12**, 197–215.

Morris, P. W. G. (1994). *The management of projects.* London: Thomas Telford.

Moynihan, T. (2002). Coping with client-based 'people problems': Theories of action of experienced IS/software project managers. *Technology Communications,* **49** (19), 61–80.

Murray, F. (1989). The organisational politics of information technology: Studies from the U.K. financial services industry. *Technology Analysis and Strategic Management,* **1** (30), 285–297.

Nelson, R. R. (1982). The role of knowledge in R&D Efficiency. *Quarterly Journal of Economics,* **97**, 453–470.

Nightingale, P. (2000). The product-process-organisation relationship in complex development projects. *Research Policy,* **29** (7–8), 913–930.

Nightingale, P. & Poll, R. (2000a). Innovation in investment banking: The dynamics of control systems in the chandlerian firm. *Industrial and Corporate Change,* **9**, 113–141.

Nightingale, P. & Poll, R. (2000b). Innovation in services: The dynamics of control systems in investment banking. In: S. Metcalfe & I. Miles (Eds), *Innovation Systems and the Service Economy.* Amsterdam: Kluwer Academic Press.

Parnas, D. L. (1985). Software aspects of strategic defence systems. *Communications of the ACM,* **28** (12), 1326–1335.

Pavitt, K. (1984). Sectoral patterns of technological change: Towards a taxonomy and a theory. *Research Policy,* **13**, 343–374.

Pavitt, K. (1990). What we know about the strategic management of technology. *California Management Review,* **32** (3), 17–26.

Pavitt, K. (1996). National policies for technical change: Where are the increasing returns to economic research? *Proceedings of the National Academy of Science USA,* **93** (23), 12693–12700.

Pavitt, K. & Rothwell, R. (1976). A comment on 'A dynamic model of process and product innovation'. *OMEGA,* **4** (4), 375–376.

Peck, J. & Scherer, F. M. (1962). *The weapons acquisitions process.* Cambridge, MA: Harvard University Press.

Peffers, K. & Tuunainen, V. K. (2001). Leveraging geographic and information technology scope for superior performance: An exploratory steely in international banking. *Journal of Strategic Information Systems,* **10** (3), 175–200.

Penning, J. M. & Harianton, P. (1992). The diffusion of technological innovation in the commercial banking industry. *Strategic Management Journal,* **13**, 29–46.

Petroski, H. (1986). *To engineer is human: The role of failure in successful design.* London: Macmillan.

Pinto, J. K. & Covin, J. G. (1989). Critical factors in project implementation: A comparison of construction and R&D projects. *Technovation,* **9**, 49–62.

Pinto, J. K. & Slevin, D. P. (1987). Critical factors in successful project implementation. *IEEE Engineering Management,* **34** (1), 22–27.

Pinto, M. B., Pinto, J. K. & Prescott, J. E. (1993). Antecedents and consequences of project team cross functional co-operation. *Management Science,* **39** (10), 1281–1297.

Prencipe, A. (2002). *Strategy, systems and scope: Managing systems integration in complex products.* London: Sage Publications.

Prencipe, A., Davies, A. & Hobday, M. (Eds) (2002). *The business of systems integration.* Oxford: Oxford University Press.

Ropponen, J. & Lyytinen, K. (2000). Components of software development risk: How to address them? A project manager survey. *IEEE Transactions on Software Engineering,* **26** (2), 98–112.

Rothwell, R. (1976). *Innovation in textile machinery: Some significant factors in success and failure.* Brighton: University of Sussex. SPRU Occasional Paper no. 2.

Rothwell, R. (1977). The characteristics of successful innovators and technically progressive firms. *R&D Management,* **7**, 191–206.

Rothwell, R. (1992). *Successful industrial innovation crucial factors for the 1990s.* SPRU 25th Anniversary, Brighton: University of Sussex. Reprinted in *R&D Management,* **22** (2), 221–239.

Rothwell, R. (1993). *Towards the fifth generation innovation process,* Brighton: SPRU Working Paper.

Rothwell, R., Freeman, C., Horley, A., Jervis, V. I. P., Robertson, Z. B. and Townsend, J. (1974). SAPPHO updated. Project SAPPHO Phase II. *Research Policy,* **3**, 258–291.

Royce, W. W. (1970). Managing the development of large software systems: Concepts and Techniques. *Proceedings WESCON.*

Sapolski, H. M. (1972). *The polaris system development: Bureaucratic and programmatic success in government.* Cambridge, MA: Harvard University Press.

Sauer, C. & Waller, A. (1993). *Why information systems fail: A case study approach.* Henley on Thames, U.K.: Alfred Waller.

Shenhar, A. J. (1998). From theory to practice: Towards a typology of project-management styles. *IEEE I Engineering Management,* **45** (1), 33–48.

Shenhar, A. J. & Dvir, D. (1996). Towards a typology of project-management, *Research Policy,* **25** (4), 607–632.

Shenhar, A. J., Tischer, A. & Dvir, D. (2002). Refining the search for project success factors: A multivariate, typological approach. *R&D Management,* **32** (2), 111–126.

Slevin, D. P. & Pinto, J. K. (1987). Balancing strategy and tactics in project implementation. *Sloan Management Review,* **29** (1), 33–41.

Smith, A. (1776/2000). *An inquiry into the nature and causes of the wealth of nations.* New York: Liberty Fund.

Smith, H. J., Kiel, M. & Depledge, G. (2001). Keeping mum as the project goes under, towards an explanatory model. *Journal of Management Information Systems,* **18** (2), 189–227.

Soete, L. & Miozzo, M. (1989). *Trade and development in services: a technological perspective.* Working Paper No. 89-031, MERIT, Maastricht.

SPRU (Science Policy Research Unit) (1972). *Success and failure in industrial innovation.* Brighton, U.K.: Centre for the Study of Industrial Innovation.

Staw, B. & Ross, J. (1987). Behaviour in escalation situations: antecedents, prototypes and solutions. In: L. Cummings & B. Staw (Eds), *Research in Organisational Behaviour.* Greenwich, CT: JAI Press.

Tatikonda, M. V. & Rosenthal, S. R. (2000). Technological novelty, project complexity and product development execution success: A deeper look at task uncertainty in product innovation. *IEEE Transactions on Engineering Management*, **47** (1), 74–87.

Teubal, M., Amon, N. & Trachtenberg, M. (1976). Performance in innovation in the Israeli electronic industry. *Research Policy*, **5** (4), 354–379.

Tidd, J., Bessant, J. & Pavitt, K. (1997). *Managing innovation: Integrating technology, market and organisational change*. London: John Wiley.

Uchupalanan, K. (2000). Competition and IT-based innovation in banking services. *International Journal of Innovation Management*, **4** (4), 455–489.

Utterback, J. & Abernathy, W. J. (1975). A dynamic model of process and product innovation. *International Journal of Management Science*, **3** (6), 424–441.

Vernon, R. (1966). International investment and international trade in the product cycle. *Quarterly Journal of Economics*, **80**, 190–207.

Vincenti, W. G. (1990). *What engineers know and how they know it*. Baltimore, MD: Johns Hopkins University Press.

Von Hippel, E. (1976). The dominant role of users in the scientific instrument innovation process. *Research Policy*, **5**, 212–239.

Walker, W. (2000). Entrapment in large technology systems: Institutional commitment and power relations. *Research Policy*, **29** (7), 833–846.

Walker, W., Graham, M. & Harbor, B. (1988). From components to integrated systems: Technological diversity and integration between the military and civilian sectors. In: P. Gummett & J. Reppy (Eds), *The Relations Between Defence and Civilian Technology*. Amsterdam: Kluwer Academic.

Walz, D. B., Elam, J. J. & Curtis, B. (1993). Inside a software design team—knowledge acquisition, sharing and integration. *Communications of the ACM*, **36** (10), 63–77.

Wheelwright, S. & Clark, K. B. (1993). *Revolutionising product development: Quantum leaps in speed, efficiency and quality*. New York: Free Press.

Willcocks, L. & Grithiths, C. (1994). Predicting the risk and failure in major information technology projects. *Technical Forecasting and Social Change*, **47** (2), 205–228.

Williams, T. (1995). A classified bibliography of recent research relating to project risk management. *European Journal of Operational Research*, **85** (1), 18–38.

Williams, T., Eden, C. & Ackermann, F. (1995). The effect of design changes and delays on project costs. *Journal or the Operational Research Society*, **46** (7), 809–818.

Womack, J., Jones, D. & Roos, D. (1991). *The machine that changed the world*. Cambridge, MA: MIT Press.

Woodward, J. (1958). *Management and technology*. London: HMSO.

The International Handbook on Innovation
Edited by Larisa V. Shavinina

Innovation in Integrated Electronics and Related Technologies: Experiences with Industrial-Sponsored Large-Scale Multidisciplinary Programs and Single Investigator Programs in a Research University

Ronald J. Gutmann

Center for Integrated Electronics, Rensselaer Polytechnic Institute, USA

Abstract: University research innovations in integrated electronics are presented with a focus on the impact of industry support. University knowledge-based innovations are divided into discontinuous (or radical) and continuous (or incremental), with the relative contributions affected by program funding guidelines, program review methodology, industrial mentoring, and in-house industrial research and development in such a rapidly evolving industry. The impact of the Semiconductor Research Corporation (SRC) initiated in 1981, SEMATECH initiated in 1988 and the Microelectronics Advanced Research Corporation (MARCO) initiated in 1998 is highlighted.

Keywords: Integrated electronics; Discontinuous/radical innovations; Continuous/incremental innovations.

Introduction

The semiconductor/integrated circuit (IC) industry has made significant continuous and discontinuous innovations in the past four decades, thereby fueling the information-technology revolution that has transformed our society. During the past two decades this industry has financially supported and guided research universities in the required disciplines, both to obtain a source of research and development personnel and to obtain leading-edge research; the latter can be divided into continuous (or incremental) innovations, often involving contributions to the scientific knowledge base, and discontinuous (or radical) innovations, namely high-risk, high-payoff ideas which are not embedded in critical paths of the industry roadmap. This paper presents a single investigator perspective of the research university role in both types of innovation,

derived from three decades of experience of involvement.

The main tenets of the chapter include the following: the semiconductor/IC industry has been very astute in dealing with research universities; the research universities have benefited tremendously from the financial support and guidance; continuous or incremental innovations are more easily accomplished and more easily measured, often involving insight into the underlying knowledge base of industry practice; discontinuous innovation in this field is dominantly achieved by industrial research laboratories, with research universities as fast-followers; large-scale multi-investigator programs are very effective for continuous innovations and for providing multidisciplinary educational and research experiences, but have not established a strong record of discontinuous innovation; programs work best with a strong personal commitment for research success in a professional and enjoyable environment.

Innovation by the semiconductor/IC industry has fueled the information-technology (IT) revolution that has transformed our society. The technology base was established mostly by industrial organizations, both vertically integrated high-tech companies like AT&T (now Lucent) Bell Laboratories and IBM as well as Silicon Valley companies that grew with the semiconductor/IC industry explosive growth like Intel. The cost reductions and performance enhancements of this industry as succinctly summarized by Moore's Law are well known; less well known is the role of the semiconductor/IC industry in supporting, and working with, research universities—in many ways as unique a characteristic of this industry as IC performance advances.

This chapter is a personal reflection of an individual participant in the university research and education enterprise. The author has served as Program Director at the U.S. National Science Foundation (NSF), where he was on the first Technical Activities Board (TAB) of the Semiconductor Research Corporation (SRC)[1] in a government liaison role. He has also served two terms on the University Advisory Committee of the SRC and on the SEMATECH[1] University Advisory Board. He has served as co-Director of a SEMATECH Center of Excellence (SCOE) on Multilevel Interconnects at Rensselaer and participated in other SRC, NSF Engineering Research Center (ERC), Defense Advanced Research Program Agency (DARPA), and industry-sponsored multi-disciplinary programs—and many smaller research programs sponsored by U.S. government agencies and national/international companies. For five years, the author served as Director of the Rensselaer Center for Integrated Electronics.

The chapter emphasizes the research university role in innovation for the semiconductor/IC industry and highlights the role of the SRC (and related industry organizations); the role of U.S. government support from NSF, DARPA and other mission agencies is only briefly mentioned in highlighting innovative issues, as this role is similar in all disciplines. Innovation is divided into two categories: continuous (or incremental) innovation and discontinuous (or radical) innovation. One main tenet of the paper is that the SRC-based funding has been extremely successful in continuous innovation (as well as in providing funding for graduate student education and training) but has been less successful in achieving discontinuous innovations.

The semiconductor/IC industry has recently moved to more discontinuous innovation-focused funding through the Microelectronics Advanced Research Corporation (MARCO). MARCO is a wholly owned

subsidiary of SRC, enabling utilization of SRC management functions (such as contracting agreements) as appropriate. MARCO was established because of both the decline in long-range research funding in major industrial laboratories and the increasing impediments to scaling as projected by Moore's Law and delineated in industry-generated roadmaps (the National Technology Roadmap for Semiconductors (NTRS) and the more recent International TRS and ITRS).

Normally radical and incremental innovation refers to business practices or products in a business environment. In the context of this chapter referring to university research, discontinuous (or radical) and continuous (or incremental) innovation refers to the effect of the research results. University knowledge-based research often provides the scientific underpinnings of current technological practice, thereby enabling better engineering design. However, university research often establishes new technological approaches that lead to significantly improved performance, lower-cost and/or higher-reliability technology. When successful, these higher-risk endeavors are referred to as discontinuous (or radical) innovations.

The remainder of this chapter is organized as follows. First, a brief history of the industry funding of universities is presented, focusing on the major role of the SRC, but also including SEMATECH and MARCO. Second, the major tenets of the paper are developed, using the technology area of most involvement by the author (IC on-chip interconnects). Third, these tenets are discussed using the recently initiated MARCO Focus Centers as a benchmark on these tenets. Finally, the author's perspectives are summarized.

Semiconductor/IC Industry Funding of Research Universities

The semiconductor/IC industry initiated a collaborative method of funding research universities in 1982, when the Semiconductor Industry Association (SIA) established the Semiconductor Research Corporation (SRC). Originally three center-type programs were established (single-university multiple-investigator programs) with many smaller single-investigator programs. The emphasis was entirely on silicon-based research, as the U.S. government-sponsored university research programs were dominantly in compound semiconductors. The SIA clearly established the program both to establish a source of pre-competitive silicon-based research and to increase the source of graduates with education and training in silicon IC technology and design.

In the first decade the SRC funding clearly impacted the U.S. research programs in semiconductors/IC technology by funding both center and single-investigator research programs at major research universities, and by establishing educational programs

[1] The SRC and SEMATECH are described more fully in the next section of this chapter.

(both funding selected curriculum developments and graduate fellowships for U.S. citizens and permanent residents). The SRC is a research management organization, with various industry boards to provide input into funding decisions and program evaluations. The SRC-sponsored core programs have always been focused to industry concerns and more micro reviewed than government funded university research, with the anticipated pluses and minuses of such a research management process. However, the SRC funding and program mentoring clearly was positive for the university research community and for the graduate students involved, although the micro-review process has often been a source of friction. Originally considered by many university researchers as a minor supplement to government-funded research, the SRC evolved into an important complementary component of university research within a decade. The industrial leveraging of government-funded research established a broad-based three-way partnership (industry, government and research university).

In the late 1980s the SIA established SEMATECH as a research and development cooperative to help revitalize the U.S. semiconductor industry. As part of the post-site selection process in which Austin, Texas was selected as the venue for SEMATECH, various SEMATECH Center of Excellence (SCOE) research programs were established in 1989. The SCOEs focused on different research areas of semiconductor manufacturing, received approximately $1,000,000 ($1M) per year for five years, and were managed by the SRC with its already established infrastructure for managing and reviewing university research programs. Generally the SCOEs were focused on shorter-range research than the SRC programs, although the time line is difficult to establish firmly in such a rapidly evolving area of research. Some conflict arose between the more focused view of SEMATECH with a well-defined research and development agenda and the SRC with a longer research mission and a longer time line for research results (although shorter range than many government-sponsored research programs). After the initial five years, the research funding for the SCOEs was incorporated into the SRC program to a large extent, although SEMATECH continues to fund smaller, nearer-term more-focused university research programs directly.

The second decade of the SRC was marked by full incorporation into the fabric of U.S. research universities, relative stability of a mode of operation, absorption of SCOE program emphasis and research funds (as described above), establishing a close working relationship with U.S. government funding agencies (particularly NSF and DARPA) including joint support of center programs, establishing a new SIA-funded SRC-managed multi-university research center funding organization (MARCO) and the internationalization of both the SRC and SEMATECH. The SRC is fully established as a major university funding source for semiconductor/IC materials, processing, devices, design and manufacturing research, with an agenda varying from short-term to long-term focus (although still not as long range as government-sponsored research except for the newer MARCO programs). Moreover, the past five to eight years has seen significant collaboration with U.S. government funding agencies in areas of mutual interest, including formal agreements and collaborative funding.

The two more recent modifications have resulted from the internationalization of the SRC and the establishment of the MARCO Focus Centers. While the impact of the former is small at this time, the impact could be significant in the next five years as international companies (particularly European and Pacific Rim) participate in collaborative university research and the SRC funds more research at international universities. The company overhead required to fully take advantage of the SRC program (including participation in program reviews, funding authorization/allocation meetings, and long-range policy meetings) may inhibit a rapid evolution to internationalization of the SRC; full internationalization will be difficult to achieve.

The second recent modification has clearly been significant, namely the establishment of MARCO Focus Centers for multi-university collaboration in long-range research necessary to maintain the rate of advancements of price and performance predicted by Moore's Law in the presence of upcoming limitations to CMOS IC scaling as delineated in the NTRS and ITRS reports. The SIA, stimulated by Dr. Craig Barrett of Intel, established MARCO to lead a new university research thrust with a long-range focus in such technology needs. The first two multi-university centers were established in 1998 (Interconnect as well as Design and Test), with two additional centers established in 2000 (Materials, Structures and Devices as well as Circuits, Systems and Software) and two more anticipated in 2002–2003. Steady-state funding of each center is projected to be $10 million/year.

These MARCO centers are impacting the research university landscape in a manner similar to the NSF Engineering Research Center (ERC) program begun in the mid-80s and the earlier DARPA (now NSF) Material Research Laboratory (MRL) program begun in the 60s. The long range impact of MARCO centers compared to the long-standing MRL and ERC programs is difficult to project, but the impact could become as significant if the industry continues a commitment to such large-scale, long-term university research throughout its cyclic business environment. Since the MARCO program is jointly supported by SIA Board of Director companies (50%), equipment/materials/software suppliers (25%) and DARPA (25%), funding stability and broad support have been established in a relatively short time frame.

Development of Major Tenets

In this section the author combines his experience and background outlined in the Introduction with the Semiconductor/IC Industry research funding/impact outlined in the previous section to develop the major tenets of this paper. The section is split into three parts: first, a technology summary of IC on-chip interconnects so that the experience and perspective described can be focused more specifically; second, personal experiences from major interconnects research programs such as the SCOE on Multilevel Interconnects, the SRC Center on Advanced Interconnect Science and Technology (CAIST) and the MARCO Interconnect Focus Center (IFC) for Giga-scale Integration, complemented by related single investigator programs; third, an innovation perspective abstracted from these experiences.

IC On-Chip Interconnects: A Technology Summary

In the late 1980s CMOS technology scaling indicated that conventional IC on-chip interconnects would limit microprocessor speed within the next decade. On-chip interconnects had gradually evolved in the number of metal interconnect levels, but the technology had not changed significantly—aluminum lines (or trenches), oxide interlevel dielectric (ILD), patterning of metal lines followed by ILD deposition with gap fill and planarization constraints, and a change from aluminum to tungsten vias for vertical interconnection between aluminum lines. As the minimum feature size continually decreased, CMOS devices become faster, but the interconnect delay does not scale similarly. As a result the interconnect delay continually increases relative to the device delay, projecting that microprocessors would become interconnect-limited.

In the search for reduced interconnect delay, metals with a higher electrical conductivity and dielectrics with a lower dielectric constant (low-k ILD materials) were explored in the 1990s. The Research Division of IBM led the development, both in copper and alternative high conductivity conductors and in polymers and other low-k dielectrics. While IBM had done appreciable research in the mid-1980s and beyond, other companies and the research universities did not become appreciably involved until the 1990s when the interconnect bottleneck became more widely realized and the limitations of aluminum and oxide as the key on-chip interconnect materials became clear.

Since the choice of conductors is relatively limited (copper, silver and gold) and the choice of low-k ILDs is relatively large, the industry settled initially on copper as the interconnect conductor of choice. However, since copper cannot be patterned by reactive ion etching (RIE) at room temperature (easily done with aluminum), an alternative patterning strategy was established for copper. In the so-called Damascene technique, the ILD is patterned for trenches (or lines) and vias, followed by copper and appropriate liner deposition and then chemical-mechanical planarization (CMP) to eliminate copper and liner between the trenches and vias. The CMP process was originally introduced to planarize the ILD and/or vias in interconnect structures to allow an increasing number of on-chip interconnect levels, but the use in a Damascene patterning strategy was a key development for on-chip copper interconnects.

The research and development in this field was highlighted by the IBM announcement in the Fall of 1997 when copper was announced as going into manufacturing, with products introduced within a year. Three years later IBM announced the first low-k ILD to go into manufacturing, a spin-on polymer from Dow Chemical called SiLK (for silicon low-k technology). Numerous IC manufacturers, both large vertically integrated companies and manufacturing foundries, have introduced copper interconnects in manufacturing, with some low-k ILDs (mostly organosilicate glass rather than SiLK).

In summary, the 1990s was the interconnect technology decade, with four major advances being introduced into IC manufacturing:

(1) CMP enabling six-to-eight metal levels of on-chip metallization with high yield during IC manufacturing;

(2) copper metallization for lines and vias to replace aluminum lines and tungsten vias for improved electrical conductivity and electromigration capability;

(3) dual Damascene patterning to replace metal reactive ion etching (RIE) and dielectric gap fill to improve line definition and to lower manufacturing costs;

(4) low-k ILDs to replace oxide for lower line and coupling capacitance and, when combined with copper trenches, lower interconnect delay.

SCOE, CAIST, MARCO and Related Research

The recognition of the growing importance of on-chip interconnect technology in the late 1980s coincided with the establishment of SEMATECH and the funding of the SCOEs. The New York SCOE was established at Rensselaer in 1988–1989 in Multilevel Interconnects, with research activities to extend the knowledge base in more conventional technology as well as in new areas of copper metallization and CMP. The award of this program initiated a multi-disciplinary research program in IC interconnects at Rensselaer and has led to ensuing center-based programs such as the SRC CAIST launched in 1996 (centered at Rensselaer) and strong participation in the MARCO Interconnect Focus Center (IFC) led by Georgia Tech and launched in 1999. Faculty at Rensselaer who have had key leadership roles in this decade-plus thrust in IC interconnects include Tim Cale, Dave Duquette, Bill Gill, Ron Gutmann (author), Toh-Ming Lu, Jim Meindl, Shyam Murarka and Arjun Saxena, full professors from four different academic departments.

In parallel with this semiconductor/IC industry consortium support, this support was highly leveraged with other research support, both cost-sharing funds from Rensselaer and New York State and additional research support. The latter includes both U.S. government support and single company support for specific research. While most of these programs were relatively small, a large-scale program that was funded by IBM in parallel with the SCOE emphasized polymer materials for low-k ILDs, an area not funded in the SCOE at the time. This interactive program with IBM allowed Rensselaer to establish a useful base in low-k materials, processing and characterization, and completed an advanced interconnect technology portfolio.

The Rensselaer programs in copper CMP and copper metallization technology (liners, alloys and Damascene patterning strategies) became particularly well known, with Rensselaer established as a major research university in IC interconnects as a result of the SCOE and CAIST programs (and related leveraged research support). However, the role of IBM Research in initiating and leading the development of copper interconnect research is clear, with Rensselaer and other research universities providing an underlying science base in many areas and in providing a database for other company investments and directions. The differences between the industrial contributions and the research university contributions are best compared by tracking not only refereed journal articles and conference papers, but also the patent literature.

The author believes that the SRC program review, a relatively micro-look at deliverables and annual results, is effective in educating graduate students for industrial opportunities and in providing a desirable scientific knowledge base for industrial practice (i.e. continuous innovation). However, the process does not encourage long-term research focused on discontinuous innovations, as the programs are reevaluated and redirected with modified budgets on an annual basis. Most importantly, the industry participants are asked to evaluate individual tasks based upon the impact of the research to their company rather than to the industry at large. While this approach is desirable to keep individual companies pleased with their SRC investment, the impact on long-term research of a high-risk high-payoff nature with discontinuous innovation results can be (and has been in the author's opinion) negative.

The SIA has recognized the need for different mode of research management to better encourage 'out-of-the-box' thinking and discontinuous innovation in establishing MARCO as a new subsidiary of the SRC. National and international roadmaps by the industry (NTRS and ITRS) indicate many future needs where

no known solutions exist, resulting in a consensus that increased funding of consortium-managed university research was essential. The resulting major differences between MARCO and SRC core research programs have been the individual program size, the emphasis on long-range research and the increased emphasis on discontinuous innovation research results, as well as the mode of research management. The SIA requires more high-risk high-opportunity research in all the MARCO programs, and anticipates many discontinuous innovations to emerge.

As an example, the Rensselaer IFC program (part of the New York State program led by the University at Albany) emphasizes wafer-scale three-dimensional (3D) ICs, optical interconnects for chip-to-chip and on-chip broadband interconnect, terahertz technology for interconnects and characterization, carbon nanotubes for interconnects, nano-metrology techniques and multiscale materials and process modeling. The 3D program uses the on-chip interconnect technology, fundamental understanding and research experience from the 1990s to establish a new approach to monolithic wafer-scale 3D ICs, using dielectric adhesives to bond two fully processed IC wafers and copper Damascene patterning to form inter-wafer interconnects. Such a program would not be possible without the research expertise established by the SCOE, CAIST and related interconnect programs, companion research expertise in IC design and packaging technology and the new funding paradigm established by the MARCO Focus Centers.

Innovation Perspective

Based upon this experience and perspective, the author believes that the SRC-funded core research programs are effective at continuous innovation, with the university research community providing the scientific underpinnings for recently developed industry innovations. In addition, the research universities can be effective fast followers when stimulated and mentored by leading industrial research laboratories, thereby contributing to truly discontinuous innovations affecting the semiconductor/IC industry. The more recent MARCO program has been initiated to provide incentives and freedom to pursue truly discontinuous innovations by the research university community. The results of this initiative will take five years or more to evaluate fully.

Another perspective is the generally inhibiting role of large university centers on truly discontinuous innovation. While in some disciplines centralized shared facilities may require larger multi-investigator programs, the relatively small single (or an interactive few) investigator(s) may be best for truly discontinuous innovations, where new approaches may be investigated over some extended duration in relative privacy without specific deliverables and milestones. Large multi-disciplinary programs have many advantages and many attractive results, but the author does not believe that discontinuous innovations are best achieved in such an environment, particularly with traditional means of SRC program management and review. Effective management of technological innovation is difficult enough within an industrial organization, as described by both J. Bessant and R. Katz in this handbook.

In fact, a main purpose of this chapter is to put forth a personal perspective (rather than fully annotated with references) to stimulate further discussion, not only in the semiconductor/IC research community, which has been extremely innovative in dealing with universities and where these issues have been examined, but also in other industrial sectors. The SRC core and traditional center programs have clearly been successful; complementary programs like the SCOEs have had an important role. The MARCO program may be another complementary program with a relatively short life-time, but the MARCO program is a great opportunity for the university research community to extend its role in the discontinuous innovation arena. Hopefully, MARCO will become a long-term important ingredient of the SRC research portfolio, with many discontinuous innovation accomplishments. Challenges in innovation management as described by J. Bessant in this handbook are magnified with such industrial consortia. However success can be achieved with commitment and vision.

This unique industry support of integrated electronics has been in parallel with government support. Research and innovation contributions of such programs have not been included here, but the integrated electronics field has a similar history as presented by Y. Miyata elsewhere in this handbook. However, the impact of the U.S. Semiconductor industry support of universities described here is more positive than the more general evaluation by Miyata (2003). Perhaps this research field is unique in ways the author has been unable to abstract.

Discussion of Specific Tenets

Main tenets of the chapter include the following: the semiconductor/IC industry has been very astute in dealing with research universities; the research universities have benefited tremendously from the financial support and guidance; continuous or incremental innovations are more easily accomplished and more easily measured, often involving insight into the underlying knowledge base of industry practice; discontinuous innovation in this field is dominantly achieved by industrial research laboratories, with research universities as fast-followers; large-scale multi-investigator programs are very effective for continuous innovations and for providing multidisciplinary educational and research experiences but have not established a strong record of discontinuous innovation; programs work best with a strong personal

commitment throughout. In this section these major tenets are presented with a brief discussion of each.

Semiconductor/IC Industry Astute in Dealing with Research Universities

The SIA in launching the SRC in 1982, the SCOE program with SEMATECH in 1988–1989 and the MARCO Focus Research Centers in 1998 has demonstrated an ability to understand the operation of research universities and to establish a mode of operation which accommodates professional independence and creativity for the industry advantage. Like any effective enterprise, the SRC has established a basic operating mode in becoming a permanent fixture in research university support.

The research university community has benefited significantly from SRC programmatic interactions, in the funding support that is provided (direct costs and 'full' government-equivalent indirect costs) and in the mentoring of research program directions and graduate students. While the SRC mode of operation is different than government funding agencies or single company support/interaction, the complementary nature of the time frame of the research focus and the review process offers a complementary breadth to more classical sponsor interactions.

Continuous Innovations More Easily Accomplished and Mentored

The SRC core programs, both center-based and individual investigator-focused, are more amenable to continuous innovations with major contributions to the scientific underpinnings of present or near-future industry practice or relatively small advancements to the field. The proposal and program review practices are the major factors, as the research activities are reviewed in detail annually from the perspective of the individual company sponsors. This process mandates continuous contributions, which leads to short-time horizon research.

Discontinuous Innovations Led by Industrial Research Laboratories

Key discontinuous innovations in the semiconductor/IC industry have resulted from major industrial research laboratories. While this statement may be debatable in isolated areas, the author believes that the tenet is widely true and has presented the IC on-chip interconnect paradigm shift in the 1990s as an example. Universities can be fast-followers and make major contributions and may occasionally introduce discontinuous innovations from SRC and government funded programs (e.g. Rensselaer in interconnects in the 1990s), but the overall record is not outstanding. The new MARCO Focus Centers are an excellent opportunity for research universities to become a stronger contributor in discontinuous innovation, at a time when major industrial research laboratories are often shrinking their time horizon and when discontinuous innovation is needed to maintain progress according to Moore's Law as CMOS shrinking becomes more difficult and more expensive. The MARCO research objectives, funding level and research management approach provide such an opportunity.

Large Center Program Perspective on Innovation

Large center programs have a few tremendous advantages, both those within a research university and those with multiple research university participation. They encourage/require meaningful multi-disciplinary interactions, sharing of expertise and facilities, and effective interactions among faculty, research staff, graduate students and industrial researchers and managers. The experience for all stakeholders is worthwhile, particularly for graduate students. However, as described, the author feels strongly that such programs tend to focus on continuous innovations rather than discontinuous innovation where a greater risk of failure is involved. The MARCO program removes most of the impediments; five years will be needed before the impact can be evaluated.

Strong Personal Commitment Necessary for Success

All such programs work best where there is a true commitment to the objectives of the program, principally the participating faculty and the industrial mentors. Clear and honest communication, including respecting the intellectual property (IP) of the participants, is a necessary ingredient for achieving the most effective research results in a professional and enjoyable environment.

Summary

The author's perspectives and major tenets of this chapter are presented in the previous section. While the perspectives presented are a personal viewpoint, the author believes that many in the SRC research community hold similar views, both the university researchers and industrial mentors. Whatever the view toward continuous and discontinuous innovation, the SIA has clearly changed the scope and focus of research universities in the field of semiconductors/ICs, both in the past two decades and for the foreseeable future. The initiative with the MARCO Focus Centers indicates that the SIA continues to evolve different funding mechanisms in the future, but probably not until the MARCO Focus Centers become fully established (5-year funding ramp anticipated) and can be fully evaluated (2008 time frame). Other industry sectors could well benefit from a careful review of the two-decade experience of the semiconductor/IC industry in forging new relationships between industry, government and research universities. The research university role in the contributing to the semiconductor/IC industry has clearly been enhanced by these SIA investments as managed by the SRC.

Acknowledgments

The author gratefully acknowledges his faculty colleagues for their technical contributions to the Rensselaer program and for shaping the author's perspectives presented here, namely T. S. Cale, T. P. Chow, D. J. Duquette, P. S. Dutta, W. N. Gill, T. M. Lu, J. F. McDonald, J. D.Meindl, S. P. Murarka, K. Rose, E. J. Rymaszewski and A. N. Saxena. He has also benefited over this period from supervising many talented doctoral students, including S. Banerjee, C. Borst, K. Chatty, T. Chuang, M. Heimlich, C. Hitchcock, Y. Hu, N. Jain, J. Jere, S. Khan, V. Khemka, B. C. Lee, T. Letavic, P. Lossee, K. Matocha, D. McGrath, J. Neirynck, D. Price, R. Saxena, P. Singh, S. Saroop, B. Wang, C. Wong, Y. Xiao, and A. Zeng. The author's perspective on discontinuous innovation has been influenced by interactions with faculty in the Rensselaer Lally School of Management, namely R. Leifer, C. McDermott, J. Morone, G. O'Connor, L. Peters, M. Rice and R. Veryzer.

Discussions over these issues with many industrial research leaders have helped shape the author's perspectives, particularly J. Carruthers (Intel), G. DiPiazza (AMP), D. Fraser (Intel), D. Havemann (TI), W. Holton (TI/SRC), R. Isaac (IBM), J. Ryan (IBM), R. Schinella (LSI Logic), W. Siegle (AMD), L. Sumney (SRC) and B. Weitzman (Motorola). Discussions with colleagues from other research universities have also been very beneficial, particularly J. Gibbons (Stanford), G. Haddad (University of Michigan), D. Hodges (University of California, Berkeley), A. Kaloyeros (University at Albany), N. Masnari (North Carolina State University), T. McGill (Cal Tech), R. Reif (MIT), A. Tasch (University of Texas, Austin), and E. Wolf (Cornell).

The funding of the SIA through the SRC, SEMATECH and MARCO, many semiconductor/IC companies and government agencies over this period is gratefully acknowledged and is much appreciated.

References

Semiconductor Industry Association (SIA) Roadmaps

—National Technology Roadmap for Semiconductors, (NTRS), 1992, 1994, and 1997
—International Technology Roadmap for Semiconductors (ITRS), 1999 and 2001

all available from the:

Semiconductor Industry Association
4300 Stevens Creek Boulevard
San Jose, CA 95129

Semiconductor Research Corporation (SRC)
P.O. Box 12053
Research Triangle Park, NC 27709

SEMATECH, Inc.
2706 Montopolis Drive
Austin, TX 78741

Semiconductor/Integrated Circuits (ICs) Technology

—S. K. Ghandhi, 'VLSI Fabrication Principles', Wiley Interscience, 2nd edition, 1994.
—Y. Nishi and R. Doering, 'Handbook of Semiconductor Manufacturing Technology', Marcel Dekker, 2000 (Chapter 37 by G. D. Hutcheson entitled 'Economics of Semiconductor Manufacturing' contains an excellent description of Moore's Law).

IC Interconnect Technology

—IBM Journal of Research and Development special issue entitled 'On-Chip Interconnection Technology', Vol. 39, July 1995, pp. 369–520 (excellent description of pre-copper interconnects).
—S. P. Murarka, I. V. Verner and R. J. Gutmann, 'Copper-Fundamental Mechanisms for Microelectronic Applications', Wiley Interscience, 2000.

—J. M. Steigerwald, S. P. Murarka and R. J. Gutmann, 'Chemical Mechanical Planarization of Microelectronic Materials', Wiley Interscience, 1997.

Innovation Literature

—R. Leifer, C. M. McDermott, G. Colarelli, O'Connor, L. S. Peters, M. Rice and R. W. Veryzer, 'Radical Innovation—How Mature Companies Can Outsmart Upstarts', Harvard Business School Press, 2000 (this book uses 'radical' and 'incremental' innovation in a similar manner as 'discontinuous' and 'continuous' innovation in this chapter, with a focus on commercial products rather than the underlying knowledge base emphasized here).
—Y. Miyata, 'An Analysis and Innovation Activities of U.S. Universities', in L. V. Shavinina, editor, 'International Handbook on Innovation', 2003.
—J. Bessant, 'Challenges in Innovation Management', in L. V. Shavinina, editor, 'International Handbook on Innovation', 2003.
—R. Katz, 'Managing Technological Innovation in Business Organizations', in L. V. Shavinina, editor, 'International Handbook on Innovation', 2003.

Note: Since writing this chapter, the MARCO centers have matured with the promise of discontinuous innovation being realized in various areas.

RJG, April 2003

Part VIII

Basic Approaches to the Understanding of Innovation in Social Context

Part VII

The International Handbook on Innovation
Edited by Larisa V. Shavinina

The Barriers Approach to Innovation

Athanasios Hadjimanolis

Department of Business Administration, Intercollege, Cyprus

Abstract: The nature of barriers is first clarified and their effect on innovation is broadly outlined. The various taxonomies of barriers are presented and critically evaluated. Their impact and mechanisms of action are then developed. The pattern of barriers in different contexts is considered and various aspects of a theoretical explanation of barriers are discussed. Since barriers are especially important in small firm innovation and in difficult environments, e.g. in small countries, these special cases are studied in some depth. Finally, the empirical studies on barriers are reviewed. The chapter ends with suggestions to overcome barriers and a conclusions section.

Keywords: Innovation; Barriers.

Introduction

The importance of innovation for the competitiveness of firms and as an engine of growth at a regional or country level is widely recognized (Hitt et al., 1993; Tidd et al., 1997). At the same time it is believed that there is an 'innovation problem', in the sense that the majority of organizations are not doing enough to introduce or adopt innovations (Storey, 2000). One of the approaches to examining the reasons for inadequate innovation is the study of constraints or factors inhibiting innovation—that is the 'barriers to innovation' approach (Piatier, 1984). There exists a large amount of literature on innovation barriers (Bitzer, 1990; Piatier, 1984; Witte, 1973). Despite the extensive empirical research on barriers it seems that there is, however, no conceptual framework that would integrate the factors acting as barriers and would permit an explanation of their combined effect.

What follows is a relatively selective review that attempts to present a reasonably comprehensive account of the theory and the existent research. The study of barriers provides an insight into the dynamics of innovation, while it is also a first step in the process of overcoming them. Bannon & Grundin (1990) argue that the existence of barriers in innovation is the rule rather than the exception and that:

"In most cases organizational and business procedures work against both successful development and use of innovative products" (Bannon & Grundin, 1990, p. 1).

The *aim* of this chapter is therefore to relate the review of innovation barriers to a practical understanding of the innovation process and to action for its facilitation with the elimination of barriers.

Innovation as a complex phenomenon needs a multilevel model of analysis (Drazin & Schoonhoven, 1996). Barriers can then be studied at various levels starting from the individual and moving up to the firm, the sector or community, and the country level. While all levels are considered in the following sections, the main emphasis is on barriers to innovation at the level of the firm. The other levels are also considered mainly in relation to the firm. For example, innovative action at the individual level is considered from the point of view of managers or employees, rather than consumers or members of social groups. Innovation is initially viewed in a very broad sense; later the focus is on technological innovation in the context of the private firm. The following broad definition is used:

"Innovation is the search for and the discovery, development, improvement, adoption and commercialization of new processes, new products and new organizational structures and procedures" (Jorde & Teece, 1990, p. 76).

Technological innovation mainly focuses on new processes and new products.

In the following sections the nature of barriers is first clarified, and various ways of classifying barriers are considered in detail. The broad classification into internal and external barriers is used for a descriptive exposition of the main barriers. The next section discusses the role of barriers within the innovation process, their points of impact, as well as their effects on innovation. Since barriers may act at various points of the innovation process, this process is briefly discussed and the various models for innovation are mentioned. The static view of considering barriers as antecedents of innovation and predictors of outcome is expanded in this section into a dynamic analysis of their evolution, and interaction, during the various phases of the innovation process.

Next, the theoretical explanations for the existence of barriers are considered. Barriers—especially internal ones—may emerge as symptoms, and their deeper causes and underlying factors have to be accounted for. Since barriers are especially critical in the case of innovative small firms, as well as in difficult environments, as is the case with small and developing countries or countries in transition, these special cases are studied in some depth. The section also includes a short overview of some of the existing empirical data on barriers in both industrial products and services innovation in small firms. Then the methodological limitations of such empirical studies are discussed. Some measures and ways to overcome barriers by the firm itself and by regional/national authorities are proposed. The chapter ends with conclusions and suggestions for further research.

Nature and Classification of Barriers

Nature of Barriers

A barrier to innovation is any factor that influences negatively the innovation process (Piatier, 1984). The factors with a positive influence are called facilitators. Barriers to, and facilitators of, innovation are, however, related. On the one hand, facilitators may turn to barriers, or vice versa, as the firm evolves throughout its life cycle stages or as external conditions change (Koberg et al., 1996). However, many barriers are actually due to lack of facilitators. It is only then, for analytical convenience, that barriers are studied separately from facilitators of innovation and for a complete picture the study of both is necessary. Barriers are also known as obstacles, constraints, and inhibitors. Although there may be subtle differences in the meaning of these terms, they are used as synonyms here.

It is important at this point to consider some assumptions in the barriers approach.

- An implied assumption is that innovation is inherently a good thing and any resistance to it by employees or managers, which could be interpreted as a barrier, is unwelcome (Frost & Egri, 1991). This is not always true, and resistance or skepticism may

be actually well founded and a positive action for the good of the firm (King, 1990).

- It is also frequently assumed that removal of barriers will somehow restore the natural flow of innovation. This is far from true, because innovation is a relatively unnatural phenomenon in the sense that it needs motivation, extraordinary effort, tolerance to risk and coordination of the activities of many actors (Hadjimanolis, 1999; Tidd et al., 1997). It seems that the removal of barriers is a necessary—but not a sufficient condition—for innovation to take place.

- A third assumption is that existence of barriers is by itself a bad thing, and all efforts should be made to remove them. While this is generally the case, barriers may occasionally turn into positive factors stimulating innovation or providing valuable for the future learning experience for the firm (Tang & Yeo, 2003). For example learning to live with barriers or their gradual elimination at a local level may be a necessary first step in the internationalization process of innovative firms. Internationalization is sometimes vital for the long-term survival and success of firms having small national markets (Fontes, 1997).

- A dubious assumption is that focusing on innovation barriers is more important than focusing on reinforcing positive factors for innovation. Perhaps both are equally necessary and complementary.

Classification of Barriers

Due to the multitude of barriers, a classification scheme would be useful in their study. Barriers can be classified in a number of ways, and there are several typologies. They are usually based on the origin or source of barriers. A useful classification is in distinguishing between internal to the firm and external to the firm barriers (or endogenous and exogenous respectively (Piatier, 1984)). Similarly other types of barriers, e.g. export barriers, are classified into internal and external (Leonidou, 1995). External barriers have their origin in the external environment of the firm and cannot be influenced by it, while the firm can influence internal barriers. Barriers can further be classified into direct/indirect according to their impact on the innovation process and into general /relative. General barriers are barriers affecting all firms, while relative barriers selectively affect some of them (e.g. in specific sectors). Barriers could also be classified as tangible or objective and cognitive or perceptual. The latter are not 'real' barriers, but are subjective and perceived by the firm. This distinction is further considered in the next sections, although it should be noted that the existence and significance of all barriers is related to the perceptions of the firm's managers and employees.

As mentioned above, obstacles can also be considered at various levels starting from the micro-level and ending up at the macro-level. These are the individual, group, firm level, inter-organizational level, and regional/national level (King, 1990). The first three

Table 1. External and internal barriers to innovation.

External	Internal
1. Market related	1. People related
2. Government related	2. Structure related
3. Other	3. Strategy related

can be considered as internal barriers, and the last three as external. Barriers can refer to the presence or absence of some factors. The most common classification of external and internal barriers is used here for a discussion of barriers. Table 1 illustrates the classification with some sub-categories of each major type. *External* barriers can be subdivided into market related, government related and other.

External Barriers

(i) The market-related barriers refer to various types of market failure and other market induced innovation-hampering factors. One type of market failure refers to insufficient appropriability (i.e. ability of the innovating firm to capture rents or profits created through innovation (Teece, 1986)). Other types include market risk, inadequate size of R&D that is undertaken by private firms, and externalities (Cohen & Noll, 1991; Sanz-Menendez, 1995). The public good character of innovation may lead to know-how leakage and other spillovers, which impair innovation incentives and may act as barriers (Jorde & Teece, 1990). Supply and demand deficiencies may also present barriers. For example lack of skilled employees in the market or lack of innovative users. The nature and intensity of competition within the market affect the profitability and strategy of firms and are indirect causes of barriers.

Another market-related barrier is what has been called the 'short-termism' problem (Storey, 2000). It is an effect of pressure, e.g. from the stock exchange market on public quoted firms, to show profits in the short term. Investments with a long-term payback period, as many innovation projects tend to be, are then neglected in firms with a short-term horizon. Such projects are however necessary for the success and eventual survival of the firm, although they may have an adverse short-term impact on profits.

Under the market-related barriers the most frequently mentioned type is that of financial barriers (Piatier, 1984). These barriers may result from the reluctance of lenders, e.g. commercial banks, to share—the perceived as high-risk of innovation projects. Information asymmetry between lenders and borrowers is especially high in the case of innovation, and outside capital providers have difficulty in the financial assessment of innovative projects (Pol et al., 1999). This fact aggravates the risk and uncertainty factor. Innovators are also frequently unable to provide the collateral for loans as a security for the bank.

Financial barriers are especially important for small firms and for start-ups (Storey, 1994). The lack of venture capital for innovative high technology start-ups is a frequent complaint as further discussed in a later section.

(ii) Government and its policies and regulations are a frequent source of barriers to innovation (Piatier, 1984; Pol et al., 1999). Many policies directly or indirectly related to innovation are designed to correct market failure. Problems may arise, however, due to unintended consequences of such policies and side effects of regulations. Standards imposed by government or by supra-national organizations, such as the European Union, may also act as obstacles to innovation. Bureaucratic procedures in getting licenses or grants and in other contacts with governmental organizations are also a frequent cause of barriers. Problems in policy communication may induce discrimination against some firms, e.g. micro-firms and small firms, preventing them from getting the support they are entitled to.

Laws and regulations may give rise to barriers due to either their side effects or inadequacies in implementation. Firms have to comply with regulations at the local, regional, national, and even supra-national level (for example European Union directives). Regulations may discourage innovative activities and hinder firms from entering new markets, by increasing uncertainty and risk. They may also prevent some firms from undertaking promising projects, because they increase their time frame, cost and risk (Preissl, 1998). In other instances, they may impose unnecessary limitations on the operations of the firm. It is important to note that the same regulations may be beneficial for innovation in some industrial sectors and detrimental in others.

Examples of legal constraints include labor and consumer protection legislation, environmental regulation, and anti-trust legislation (Jorde & Teece, 1990). The legal frame for the protection of intellectual property has an even more direct impact. A weak intellectual property regime, for example, allows the easy copying of innovations and acts as a disincentive and inhibitor for firms to undertake costly innovation that could easily, and at a fraction of cost, be exploited by their competitors (Chesbrough, 1999). The tax system is a potential source of indirect barriers by reducing incentives to innovate. Trade barriers, for example, the so-called non-tariff barriers, may prevent foreign market entry and reduce the commercial success of an innovative new product.

Regulations, standards, and rules mentioned above are examples of institutions. Institutions related to innovation include the science and technology infrastructure and the physical infrastructure. Many institutions are therefore under the direct or indirect control of central or regional government. The term 'institutions' is however used here in a very broad sense to include all political, social and cultural institutions—formal and informal—and also all related rules and procedures. These institutions characterize a society and cannot easily change without the cooperation of wider social forces, e.g. firms and their associations, labor and government (Sanz-Menendez, 1995).

Lack of suitable institutions, inadequate performance of existing ones, and what has been called institutional inertia or rigidity, i.e. resistance of old institutions to change, may lead to innovation barriers (Freeman, 1994). Institutional structures may have adverse effects on transaction costs, making innovation more costly or hardly affordable. Institutional factors also affect the extent of cooperation, trust and mutual consideration of firms and the formation of alliances and other forms of cooperation.

(iii) The 'other' category includes *technical, societal and inter-organizational barriers.* Technical barriers may originate from predominant standards, e.g. in telecommunications, or arise due to changes in technology (Freeman, 1994). Risk of technology obsolescence, destruction of a firm's competences with change of technology, and dangers from picking the wrong technology, are major considerations in some fields of high technology (Starbuck, 1996). Other technical obstacles are due to the scale of capital requirements for entering a particular new technology field and scale of experience effects (technological entry barriers).

Societal factors may form important innovation barriers (Shane, 1995). Norms and values of a society and attitudes towards science, socio-economic change and entrepreneurship determine the innovation climate (Piatier, 1984). The latter, if it is negative, has an adverse effect on innovation efforts and on the willingness of Government to assist innovation.

External barriers may also arise at the *inter-organizational* level when firms have to cooperate at a regional, national or international level (Tidd et al., 1997). For example, barriers to innovation occur during cooperation along the supply chain, when customers discourage product changes or access to distribution channels is problematic for a new firm. Although the latter example refers to vertical cooperation, there are similar problems in horizontal cooperation between firms of the same sector when there is no tradition of such cooperation or there is lack of trust.

Inter-firm networks are frequently seen as facilitators of innovation by being sources of ideas, information and resources (Swan et al., 2003). They can, however, act as obstacles to innovative change due to technical, knowledge, social and administrative dependencies. There are barriers to exit from a network arising from investments made by the company itself and by other network members (Hakansson, 1990). While inter-organizational barriers are treated here as external, there is also an internal dimension in the sense that firms should have special competences in order to develop, maintain and take advantage from inter-organizational relationships.

Internal Barriers

Internal barriers relate to the characteristics of organizational members, the characteristics of the organization, and the management of innovation as a change process. They can be conveniently classified into people related, and structure- and strategy-related.

(i) People related. They can be studied at the individual and the group level and, if necessary, separately for managers and employees. They are due to perceptions, including biases and lack of motivation, deficits in skills, but also to vested interests and personal goals differing from organizational ones. For example, innovation may affect the status and privileges of experts by making their expertise obsolete. Such experts then resist innovation and change. To overcome this natural resistance, so-called 'innovation champions' are needed, and their absence may prove a major barrier to innovation (Gemuenden, 1988; Hauschildt, 2003). An innovation champion is an individual recognizing the potential in a new technology or a market opportunity, adopting the relevant project as his/her own, committing to it, generating support from others and advocating vigorously on behalf of the project (Markham & Aiman-Smith, 2001). The role of champions is further discussed in a later section.

Management may be preoccupied with the current operations and have a conservative attitude, which may lead to perceiving innovation as being risky and difficult. Lack of commitment of top management to innovation, as indicated by not rewarding risk taking and lack of toleration of failure, is mentioned as a major innovation barrier (Hendry, 1989). The decision-making process of managers, constrained by their bounded rationality, and its organization regarding search procedures, information sources, and evaluation rules, is also a source of barriers (Schoemaker & Marais, 1996).

Witte (1973) classifies people-related barriers into two categories, i.e. those due to lack of will, and those due to lack of competence. Barriers of the first category refer to the attraction of the status quo and fear of the unknown, but also fear of failure and being blamed for it (Bitzer, 1990). Factors causing will-related barriers include the effects of specific personality traits and feelings of managers and employees—acting as individuals or as members of teams. For example, perceived favoritism, jealousies and resentments have detrimental effects on innovation (Webb, 1992). Causes of people-related barriers are considered in more detail later.

Competence barriers are due to lack of creativity and specific new knowledge required by the innovation (Tang & Yeo, 2003).

Inhibiting factors or blocks to individual creativity, such as lack of training, autonomy, and extrinsic motivation are closely related to innovation barriers and have been extensively studied, but their detailed examination is beyond the scope of this chapter (Amabile, 1997). The lack of skills, as a competence barrier, has several dimensions. For example, Yap & Souder (1994) refer to the lack of both breadth and depth of personnel (i.e. number and variety of specialists) as an innovation barrier. Similarly, Staudt (1994) refers to the lack of suitably qualified managerial personnel as a barrier to innovation, but also to incumbent managers having competences in fields becoming obsolete, rather than competences in emerging fields.

(ii) Structural. Structure affects the behavior of organizational members during the innovation process and determines the problem-solving capacity of the firm. Structural obstacles include inadequate communication flows, inappropriate incentive systems, and obstruction problems by some departments (Hauschildt, 2001). The latter is also referred to as lack of inter-functional integration (Hitt et al., 1993). Collaboration between marketing and R&D for example is vital, especially for product innovation. Problems in this collaboration, due to different values, motivations, and goals, have an adverse effect on innovation (Hendry, 1989).

Centralization of power in an organization affects negatively innovation in older firms (while being positively correlated with innovation in new ventures (Koberg et al., 1996)). Mechanistic structure (i.e. a rigid hierarchical structure without many participation possibilities for employees) in a turbulent environment has been mentioned as a barrier to innovation in early studies on innovation (Burns & Stalker, 1961). Schoemaker & Marais (1996) refer to firm inertia and formalized procedures as obstacles to process innovations. Webb (1992) mentions the contradiction between formal organization and actual management practices as a problem leading to defensiveness and distrust on the part of employees, with detrimental effects on innovation. Lack of time is a frequently mentioned internal barrier (Hadjimanolis, 1999). While time can be seen as a resource, it is also clearly related to structural issues like organization of the work, delegation of tasks, and specialization.

Structural inertia may be accompanied by cultural inertia and internal politics games (Maute & Locander, 1994; Starbuck, 1996). Culture refers to the shared norms, values, and beliefs of the firm (Armenakis & Bedeian, 1999). Cultural barriers are then due to the existing beliefs and values of the firm that are not supportive of change. A culture of blame and fear of responsibility, for example, obstructs experimentation, change, and innovation. Cultural barriers are related to motivation and reward and punishment systems, and are intertwined with the people-related barriers mentioned above. The work environment has a direct impact on the intrinsic motivation for creativity and innovation, and an adverse environment may stifle innovation efforts (Amabile, 1997).

Problems related to *systems* may also be included here. Inadequate search and information acquisition systems from external sources (Madanmohan, 2000) and problematic internal dissemination mechanisms may hamper innovation (Sheen, 1992). Other examples are out-of-date accountancy systems (Rush & Bessant, 1992) and lack of planning systems.

(iii) Strategy related. Many internal barriers are related to strategy. Lee (2000), for example, mentions the failure of strategy in British firms to connect the introduction of flexible manufacturing systems with the long-term aims of the firm, e.g. its competitive position. Technical people may also be unaware of strategy and objectives, and cannot therefore persuade senior managers of the benefits and necessity of new technology, while senior managers—being technologically ignorant—cannot see these benefits themselves. Other barriers may be goal-related in the sense that senior managers may fail to appreciate the necessity for innovation or are too risk-averse to attempt to innovate. Markides (1998) mentions complacency, satisfaction with the status quo, and reluctance to abandon a certain present (adequately profitable) for an uncertain future, as potential innovation barriers. Fear of cannibalizing sales of existing products may be a more specific excuse for avoiding innovation.

Strategy is today related to the development of core capabilities and resources that are difficult for competitors to imitate (Peteraf, 1993). Some key capabilities related to innovation are technological ones, such as the capacity to produce ideas and develop them to products. Other capabilities are, marketing and service skills and legal skills to protect the firm's intellectual property. Also ability to network, form alliances and span inter-firm boundaries (Rosenkopf et al., 2001). The lack of the above capabilities, or their inadequate level, may form major internal barriers to innovation. Core capabilities may however turn to 'core rigidities' with environmental, e.g. technological change and develop into traps and barriers when they stop offering a competitive advantage (Leonard-Barton, 1995). Rigidities are also related to the 'sunk cost' fallacy and commitment to existing technologies (Schoemaker & Marais, 1996).

Resource-related barriers include lack of internal funds (e.g. from cash flow), and lack of machinery, testing or other technical equipment (Bitzer, 1990). Important barriers may arise from the lack of an own R&D department, a low percentage of organizational resources dedicated to development work, and technical problems due to inadequate experience or knowledge. Resource-related are also what Teece

(1986) has called appropriability constraints, i.e. lack of complementary assets or capabilities to take full advantage of an innovation that a firm has developed or adopted. Slack resources are considered important for innovation, but there is disagreement over whether their lack, forms a barrier to it. Nohria & Gulati (1996) claim that there is a U-shaped relation between the degree of slack and innovation. Initially innovation increases, with an increasing degree of slack, up to an inflection point; then negative effects set in and innovation decreases.

Impact of Barriers on the Innovation Process

Innovation Process Models

We have briefly considered above the nature of barriers. This section describes their effect or impact on the innovation process and also the frequency of their occurrence and their intensity. The relevant actor, whose action is inhibited by a barrier, could be an individual, a group, a firm, etc. The emphasis here is on private firms as actors. Barriers may act on one or more stages of innovation, and their impact may be different at their various points of action. Of particular interest is the role of barriers during the initial stages of the innovation process, since inability of the firm to overcome them leads to a passive attitude and avoidance of innovation. In order to facilitate the discussion of the impact of barriers, we have first to describe briefly the innovation process.

The traditional linear model of innovation conceived innovation as a linear sequence of events from research to development, production and commercialization. It has since been recognized that the innovation process is a much more complex phenomenon with many players and feedback and feed-forward loops between the various stages. The simultaneous model (Kline & Rosenberg, 1986) corrected some of the deficiencies of the linear model by recognizing the existence of tight linkages and feedback mechanisms between the stages of the innovation process.

The current interaction models of innovation (for example the system integration and networking model (SIN) of Rothwell (1992)) emphasize the internal interaction among the various departments of the firm and the external interaction with suppliers, customers, technology providers, and governmental institutions— even competitors.

The interaction model is an outcome of a systemic approach to innovation, which recognizes that firms do not innovate in isolation, but as a part of a much broader system. The innovation system can be conceived at a local or regional level, at a national level or even at an international level. The regional dimension of innovation will be considered in the next section, while the national innovation system (NIS) concept is discussed in more detail in the section entitled 'Barriers in Special Cases'.

The interaction model gives an indication of the complexity of the innovation process, the many actors involved, and the multiplicity of interaction. It implies, therefore, the fluidity of the process. Another dimension of this fluidity is the fact that goals are not fixed at the beginning of the innovation process. They change along the decision process as new alternatives appear and interact with the problem-solving activities (Gemuenden, 1988). Barriers have, therefore, a dynamic nature due to the characteristics of the innovation process itself. They should not then be considered as pre-existing and given (i.e. as antecedents of innovation), but rather as fluid and evolving. This fact adds further difficulties to their study and evaluation.

While the identification of barriers, the estimation of the frequency of their appearance, and the ranking of their importance, is not without problems, the evaluation of their impact is much more difficult due to a number of reasons presented below. By 'impact' we mean the exact final effect on innovation, i.e. its partial or complete inhibition. In other words the barrier may stop innovation completely, delay innovation or increase its cost. Apart from these negative effects, positive effects may also arise, for example an increased sensitivity and awareness of barriers and a valuable learning experience for future innovation efforts. Evaluation of the impact of barriers includes the determination of the point or stage of the innovation process at which they act and the mechanism of action.

The main difficulties of this evaluation are as follows:

- barriers may have a dynamic nature, i.e. a systematic variation according to the stage of innovation. Their evolutionary character increases the difficulty of assessing their exact impact;
- barriers may not act in isolation, but may interact mutually, reinforcing their action and leading to a vicious circle. Mohnen & Rosa (2000) refer to the complementarity of barriers and suggest a systemic approach for their elimination or reduction;
- barriers may not act directly on the firm, but may have their effect during one or more stages of the innovation process through intermediates, such as banks, customers or competitors (Piatier, 1984). This implies that barriers tend to act on the interfaces of the firm with other actors within the innovation system.

Patterns of Barriers in Different Contexts

We have so far discussed barriers to innovation in general. Most studies, however, challenge the universality of barriers and tend to suggest that their nature, frequency and impact probably vary in accordance with the context of innovation (Pol et al., 1999). There are therefore different patterns of constraints for different contexts. Some of these contexts include the type of innovation, the type of innovator, the size of the firm, the sector, the location, and probably even the business cycle. They are briefly considered below.

(i) Type of innovation. Incremental or radical innovation types probably involve different types of barriers. In other words, the degree of novelty of the innovation is related to the level of difficulty to innovate. What matters, as an innovation barrier, is the *perceived* degree of novelty by the innovation actor (Tidd et al., 1997). The type (product, process, or social innovation) and characteristics of innovation and its complexity determine to some extent the difficulties that the firm finds producing or adapting innovation to its needs. With increasing complexity of innovation, for example, problems of communication and process management become more severe (Hauschildt, 2001).

(ii) Type of innovator. We can distinguish here between non-innovators, new venture innovators (start-up firms), first time innovators (but already established firms) and frequent innovators, i.e. firms that are experienced innovators and innovate on a continuous basis. The type of innovator is therefore both related to the firm's life cycle, from start-up phase through maturity to decline, and the firm's experience in innovation. Barriers tend to be higher during the transition stages from one phase to another. Mohnen & Roeller (2001) suggest that innovation intensity in established innovators and the probability of becoming an innovator are two different innovation processes that are subject to different sets of constraints. There are also differences between producers of innovation (that is those developing their own innovations internally) and adopters of innovation developed elsewhere. In the case of adoption, barriers to diffusion and technology transfer have to be considered (Godkin, 1988), as well as, the problem of internal resistance to a foreign innovation, also known as NIH (Not Invented Here) syndrome (Katz & Allen, 1982).

(iii) Size of the firm. Barriers may vary by the size of the firm, i.e. for small and large firms (Mohnen & Rosa, 2000; Piatier, 1984). The size of the firm probably determines not only the nature, but also the importance of barriers, with small firms perceiving their impact as more severe. It is widely believed that the main innovation barriers in large firms are largely the internal ones, while such firms have the resources and the know-how to overcome any existing external barriers (Vossen, 1998). Internal barriers arise from their complexity and are due to lack of motivation, problems of communication and coordination and possibly lack of incentives. According to Quinn (1985) the main bureaucratic barriers to innovation in large firms include top management isolation from produc-

tion and markets, intolerance of entrepreneurial fanatics, short time horizons, the accounting practices, excessive rationalism and bureaucracy, and inappropriate incentives. In the case of small firms external barriers are very important, while resource-related, internal ones may also be critical. Barriers in small firms, due to their importance, are further considered in the section entitled 'Barriers in Special Cases'.

(iv) Sector. Barriers probably vary by sector (Preissl, 1998). Some barriers may be industry-specific in the sense that they are typical for the firms of one sector, but not for firms of other sectors. Similarly, the perceived importance of export barriers is reported to vary across industries due to industry-specific factors (Leonidou, 1995). The inter-sectoral differences in barriers arise from business demographic differences, i.e. firm size distributions by sector, but also from the context for innovation and the level of innovation in each sector. Innovation is expected to be much higher and continuous in high technology sectors with a high percentage of R&D such as information technology and biotechnology, against low technology sectors as glass or woodworking. Barriers may also be related to the industry life cycle, since opportunities for innovation and resources may differ between these stages. While traditionally there are many studies on barriers in various sectors of the manufacturing industry, studies on innovation barriers in services have only recently received attention (Mohnen & Rosa, 2000; Preissl, 1998).

(v) Business cycle. Barriers may also vary during the different phases of the business cycle of the economy, i.e. recession and growth, due to the differentiated availability of resources and the investment climate. Their variation may also be attributed to the different extent of government interference in the economy in each of these phases.

(vi) Location. We refer here to the effects of the specific location of the firm, any regional resource deficits and problems of the national innovation system of the country. Country specific institutions, regulations, and other conditions create country specific barriers. National cultural factors may also affect differentially the perception of barriers (Shane, 1995). The size of the country and its level of industrial development are major factors in innovation and the existence of barriers. Country effects are further elaborated later. The regional dimension of innovation has received considerable attention in recent years (Love & Roper, 2001). A number of tangible factors, such as the local industrial structure, local institutions and regional policies and regulations may affect innovation. There are also several intangible factors, such as culture, social capital, and the extent of local networking that may be relevant.

Theoretical Explanations of Barriers

Barriers have been studied by various disciplines such as economics, sociology, psychology and management. We discuss briefly each of these approaches.

Economists tend to concentrate on external barriers, i.e. those resulting from market failures, governmental policy implementation, institutional inadequacies or rigidities and supply/demand deficiencies. They concentrate on externalities, imperfect and asymmetric information available to the various actors involved in innovation and imitation, as well as their effect as incentives and disincentives for innovation. Economists also study the effect of various factors on costs, including opportunity costs, and how they impact on the perceptions of risk and uncertainty. The transaction costs perspective (Williamson, 1985) is a useful approach in such studies. When they study internal barriers, economists focus on incentives and resources. The aim of the economic approach is frequently to connect the problems with national policy aspects and suggest policy changes.

The resource-based view (Conner & Prahalad, 1996; Peteraf, 1993) is an interesting perspective in the economics and strategic management literatures, which could illuminate some aspects of the appearance of barriers. This view concentrates on the unique resources and capabilities owned by the firm and contributing to the development of its competitive advantage. The acquisition of resources and capabilities is a long-term, evolutionary, and cumulative process. Innovation depends on the availability of technological resources and knowledge and demands a number of capabilities. The lack of these resources and capabilities can be manifested as internal barriers to innovation. Obtaining these resources from the environment is costly and difficult. Cohen & Levinthal (1990) have suggested that the firm must have an 'absorptive capacity' in order to be able to obtain, to adapt, and use externally available technological information and knowledge.

Even economists recognize the importance of perceptions. Competitive intensity, for example, is not just a matter of market concentration, but also an issue of perception by the firm itself. There is now a branch of evolutionary economics studying aspects of perceptions, mental models, and learning processes, their impact on innovation, and other economic issues (Howells, 1995). The study of perceptions is however mainly the realm of social psychologists. The latter give emphasis to the internal barriers and study the role of attitudes of managers and employees at an individual and a group level. Psychologists also examine perceptions of mainly internal, but also of external barriers. We shall briefly consider various aspects of the social psychology approach to the understanding of barriers.

Innovation implies change; it usually represents a major change for the organization involved. The study of the more general phenomenon of organizational change can then provide a useful framework for the study of barriers and their effects (Armenakis & Bedeian, 1999). Barriers to innovation can then be considered as a subset of barriers to change and have analogies and similarities to barriers identified in other organizational changes. As examples of such changes we can cite barriers to growth of small firms (Chell, 2001; Storey, 1994), barriers to export and internationalization (Leonidou, 1995), barriers to learning (Kessler et al., 2000; Steiner, 1998), as well as barriers to structural reorganization (Armenakis & Bedeian, 1999).

In order to understand barriers to change, we have to study the perceptions, assumptions, interpretations and cognitions of managers and employees. Particular emphasis is given to managers (especially in the case of small firms), since managers determine organizational priorities and make resource allocation decisions (Storey, 2000). Their perceptions are therefore vital for change and innovation. Organizational members interpret environmental signals and the external reality in a sense-making process (Weick, 1995). In this process they form mental representations and models and develop their cognitions. Carl Weick (1979) has introduced the concept of the enacted environment, which is shaped by managerial interpretation and strategic choice in contrast to the 'objective' reality of an external environment. Mental models affect the readiness for change (Swan, 1995). For example, deeply held assumptions about the market may affect the process of identifying or creating opportunities and may therefore act as barriers to adoption of new ways of thinking in technological innovation. Perceptions filter information and therefore affect the assessment of benefits and costs of innovation to be developed internally or adopted from external sources (Tidd et al., 1997). Different perceptions of risk within the organization may prevent managers from reaching an internal consensus on the need to innovate.

A more specific model that could serve as a basis for the understanding of barriers, is the cognitive infrastructure model of the intent to innovate (Krueger, 1997), which uses the Ajzen-Fishbein framework based on their theory of planned behavior. The model maintains that attitudes, beliefs, and social norms affect intentions, which subsequently influence the readiness for change and innovation. The intentions themselves are also affected by perceived competencies and other factors. The model is useful because it incorporates the main influences on the intention to innovate. It could be criticized as giving a partial picture of the phenomenon of barriers since many barriers operate at its second stage. These barriers appear between the formed intentions and the realization of innovation, where the model does not give much information on factors—

other than good intentions—that affect the particular innovation.

Innovation, both as development of new products or processes within the firm and as technology adoption, can be seen as a learning process (Dodgson, 1993). At the same time perceptions and mental models of managers and employees are formulated and reformulated during the processes of change (and innovation is such a process of change) by learning and interaction between members of an organization. Organizational learning links, then, cognitions and innovation action. Thus a powerful perspective, which can be used in the comprehension of innovation barriers, is that of learning. While learning at an individual level has long been recognized as an essential process for the adaptation of human beings to the changing environmental conditions, more recently the organizational learning process has received attention by social psychologists, organizational and management scientists (Steiner, 1998).

Adaptive learning processes in organizations frequently have a collaborative and interpersonal nature, which leads naturally to the consideration of these interpersonal relations in a network perspective (Tidd et al., 1997). We consider, then, the role of such learning networks in innovation. Learning can be classified as internal, when the firm creates new knowledge leading to innovation internally, or as external learning when knowledge is obtained from external sources. Internal and external networks are involved, respectively, in the learning process. Different sets of barriers operate in these two cases (Kessler et al., 2000). For organizational learning to take place, information and knowledge have to be transferred throughout the organization and shared across many different groups. Then information should be stored, retrieved, processed and finally utilized. Impediments to internal learning may originate from the culture of the organization if values such as risk-taking, communication openness and appreciation of teamwork are absent. Even if formally espoused, but not properly rewarded, they may act as barriers.

Barriers to external learning, i.e. obstacles to the transfer of knowledge from external sources such as universities, research centers, or other firms, may come from a lack of boundary-spanning individuals or from political resistance to externally generated ideas. This resistance is known as the NIH (Not Invented Here) syndrome (Katz & Allen, 1982). It happens especially if the ideas appear revolutionary and threatening to the current status of managers or other employees, and there is no innovation champion or promotor (Hauschildt, 2001). The innovation champion may be the same person as the boundary-spanner or another individual. External learning faces more barriers than internal learning. Learning has, however, sometimes to be preceded by 'unlearning' of ineffective technologies and practices (Starbuck, 1996).

Sociologists and organizational scientists concentrate on political processes, power structures and status within firms, as well as their role in change processes as, for example, innovation (Frost & Egri, 1991). Innovation may alter the status quo and balance of power within the firm, leading to intense political activity of the affected individuals or groups (Gemuenden, 1988). Political games and use of power to access resources and to protect vested interests or improve career prospects may present powerful barriers to innovation (Burns & Stalker, 1961). Collaboration between rival interest groups is frequently problematic (Hendry, 1989). The concept of organizational inertia is useful in visualizing the role of barriers. Organizations tend to resist new ideas, adhering to old routines and maintaining hierarchies and power structures, and innovate only when a particular force pushes them to overcome this inertia (Aldrich & Auster, 1986). Such forces, acting as catalysts, include innovation champions and environmental threats.

Management theorists try to integrate the concepts of the various disciplines mentioned above. Their models for the explanation of barriers include many personal (e.g. attitudes and leadership), structural, and systems variables. They recognize the interaction of these variables and their dynamic nature (Tidd et al., 1997). Management scientists suggest that barriers are frequently due to the failure of the firm strategists to evaluate the competitive and long-term implications of investing (or not investing) in new technology and developing new products in time. The risk of not innovating is not immediately visible. Management writers also emphasize the bounded rationality limitation in strategic decisions (i.e. that they are not necessarily rational or top-down, but a result of political and cognitively-biased processes (Schoemaker & Marais, 1996)).

Pol et al. (1999) view barriers as a component of the innovation climate of a country in line with a systemic approach to innovation. Other components or parts of the innovation climate include the national innovation system, incentives to innovation, and international linkages. This is an interesting conceptual framework for external barriers, but it could be criticized in that the innovation climate itself, incentives and barriers to innovation, can also be seen as components of the national innovation system. An extension of this model would consider the internal innovation climate within the firm and the interaction of internal and external 'climates'. The concept of internal innovation climate emphasizes the interplay of structural, cultural and political factors.

Barriers in Special Cases

Disadvantages Facing Small Firms in Innovation

While there may be barriers to innovation in large firms, due to the complexity of their organization as already mentioned, it is widely recognized that small firms, because of their size and limited resources, face particular challenges and barriers to innovation (Vossen, 1998). We follow the definition of the European Union for small firms, i.e. those with up to 50 employees. Medium-size firms, i.e. those with up to 250 employees, frequently face similar problems, perhaps to a lesser degree, and are often included in this discussion.

The liability of smallness is a well-established concept in the literature (Aldrich & Auster, 1986). Small firms face resource gaps in terms of time, staff, and money (Garsombke & Garsombke, 1989) and tend to depend on the external infrastructure for techno-logical and other services. They have, however, a weak ability to interface with the infrastructure (for example universities and other technological centers), even if this is available for their support (Major & Cordey-Hayes, 2003). This weakness is due to lack of time, and inadequate managerial resources, knowledge and experience. Small firms lack economies of scale and scope in production, and R&D, and have a low ability to influence evolving technical standards (Yap & Souder, 1994).

Small firms are not usually able to match the wage rates, career opportunities and job security of larger firms and cannot easily recruit skilled labor and managerial talent. These recruitment barriers may prove major innovation barriers, although they are not insurmountable. Sometimes there are advantages in working in small firms, such as recognition and a better work environment. Small firms may also face a bureaucratic burden, i.e. a disproportionate cost of dealing with government agencies (Levy, 1993). Barriers may also arise in intellectual property protection, i.e. in searching for existing patents, filing for patents or defending them in case of infringement, due to the high cost and lack of suitable personnel.

Small firms depend on the capacity of their owner to receive/interpret signals from the market environment and identify opportunities for innovation. In contrast, large firms have a formal mechanism for such activities, including teams of scientists and managers. The innovation strategies of small firms focus on flexibility and exploitation of market niches (Keogh & Evans, 1999; Vossen, 1998). Small firms have some advantages in that they have simple structures, direct face-to-face communication, and a friendly internal climate, which tend to eliminate many internal barriers (Rothwell, 1984).

We briefly present here the results of empirical research on innovation barriers in small firms. According to Piatier (1984) the three major barriers to innovation in small firms are: 'education and training', 'finance', and 'product standards'. The top five external barriers in another empirical study (Hadjimanolis, 1999) include 'innovation too easy to copy', 'government bureaucracy', 'lack of governmental assistance',

'shortage of skilled labor', and 'bank policies on credit'. In the same study, the importance of barriers as perceived by the owners/managers of firms was not found statistically correlated to innovativeness. One possible explanation offered was that innovative small firms find ways to overcome barriers, while less innovative firms are not adequately aware of them. The top ranking barriers according to a survey of Garsombke & Garsombke (1989) include lack of capital, staff, time, and knowledge of available technology. A recent ADAPT project study (Schemman, 2000) found finance and lack of skilled workers to be major barriers in small European firms. Similarly Keogh & Evans (1999) have found 'lack of cash and finance' as the top barrier. A thorough review of the empirical studies is beyond the scope of this chapter. There are, however, remarkable similarities in the nature of barriers in the empirical studies, although importance ranks may vary.

Small firms form an extremely heterogeneous group, and some categorization regarding their innovation features would be useful. According to Tidd et al. (1997) they can be distinguished into supplier-dominated, specialized supplier firms, and new technology-based firms (NTBFs). Supplier-dominated firms only need competencies to adopt and assimilate technology developed by others, usually their suppliers. Specialized supplier firms have little R&D, but significant design and production skills. New technology-based firms are start-up firms in electronics, software or biotechnology. The latter deserve some further discussion in terms of innovation barriers.

Start-up firms attempting to innovate, face more—and to some extent different—barriers than established firms. Due to complexity and riskiness, they have problems in obtaining finance in the first stages of their development, especially in countries where venture capital finance is underdeveloped (Storey, 1994). New innovative firms entering an industry may face sig-nificant entry barriers, which can be overcome if their technology is clearly superior to that of the incumbent firms (Love & Roper, 2001).

The Case of Small and Less Developed Countries

The size of a country has a direct impact on the local supply and demand of technology and affects in other ways the national innovation system (NIS). A national innovation system is defined as:

> "The network of agents and a set of policies and institutions that affect the introduction of technology that is new to the economy" (Dahlman & Frischtak, 1993, p. 414).

Small national innovation systems are usually also 'weak' systems. The smallness of the local market and in the case of small peripheral countries, isolation and distance from major foreign markets limit the opportunities for technological innovation (Hadjimanolis & Dickson, 2001).

Nations differ in their institutional environments (Chesbrough, 1999). This is especially true when small countries are compared with large ones. The national institutional setting in small countries is frequently a source of problems. Vital institutions may be totally lacking, for example institutions providing technological services and research results, or their number is limited, such as business incubators and technology parks. Institutions include not only formal organizations, but also informal rules and procedures. Institutional rigidities and resistance to change is another problem (Souitaris, 2003). In addition, national institutions constrain the state capacity to define and implement an innovation policy and shape its outcome (Sanz-Menendez, 1995).

A usual problem faced by small national innovation systems is difficulties in flows within the system—for example human flows of researchers between universities and private firms—or financial flows towards innovating firms. Linkages between institutions and firms, necessary for such flows, may be limited or non-existent (Argenti et al., 1990). All the above factors raise external barriers to innovation and require extra efforts from the firms to innovate. The majority of these firms are small or very small and have, in addition, the liability of smallness as mentioned above. Opportunities for innovation in small firms are much more influenced by the national innovation system than those for large firms, since they depend more on others for innovation, e.g. their suppliers (Tidd et al., 1997). In the case of small countries, barriers to international technology transfer are also quite significant, since small countries depend on such transfer for a large part of their technological needs.

Less developed countries face similar problems to the smaller industrialized countries regarding their national innovation systems, but these problems are usually much more acute in their case. Countries in transition such as those in Eastern Europe are undergoing wide-ranging institutional and societal change, and face unique innovation barriers. The discussion of such barriers (Staudt, 1994) is beyond the scope of this chapter.

Methodological Difficulties of Empirical Studies on Innovation Barriers

Some indicative empirical studies on innovation barriers have already been summarized in the previous section. The present section focuses on the methodological difficulties in researching barriers.

Researching barriers is a difficult undertaking. Most empirical studies are based on surveys using lists of barriers derived from the literature. They investigate the opinions and beliefs of managers (very rarely those of employees) and their perceptions of barriers. The fact that the studies are not based on a sound theoretical

Table 2. Main difficulties in studies on barriers.

1.	Failed and disappeared innovators not counted
2.	Overemphasis on external barriers by managers
3.	Sensitivity of innovator and obstacles interlinked
4.	Inadequate methodology and analytical techniques
5.	Dynamic nature of barriers

framework leads to difficulties of interpretation of the results and to heterogeneity. It also creates problems of comparison with previous research and of generalization beyond the current research context.

Research based on case studies permits a deeper understanding of the particular context, although the above-mentioned problems are not avoided. There are additional difficulties in studies on barriers, whether surveys or case studies, which are summarized in Table 2 and then briefly analyzed.

Enterprises that failed to innovate and subsequently disappeared are not counted in the research, although barriers are most operative in such cases (Piatier, 1984). In addition, managers tend to attribute major importance to external barriers. This shifts attention away from internal barriers and shifts responsibility away from managers to some external and uncontrollable sources. This *ex post* rationalization by managers pointing to external constraints, when in fact internal constraints were the problem, is frequently mentioned in research (Barkham et al., 1996). Another major difficulty is that the sensitivity of the innovator, i.e. the level of awareness of barriers, and the problems and obstacles that are *actually* encountered are inextricably linked.

In the analysis of results, researchers frequently use factor and cluster analysis in order to bundle barriers together, in order to investigate underlying factors or classify firms into groups according to the barriers they face (Hadjimanolis, 1999). These techniques illuminate interrelations of barriers, but do not go far enough toward an explanation of why barriers occur and which factors affect their appearance. The dynamic nature of barriers is also a major problem. Barriers may change when environmental forces change; even the perception of barriers by firms is affected by environmental change.

Overcoming Barriers

While concentration on opportunity thinking, in contrast to obstacle thinking, is a preferable thought pattern (Neck et al., 1999), the realistic awareness of barriers is justifiably necessary. Identification of barriers is then essential in order to deal with them and attempt to eliminate them, thus increasing the innovation performance of the firm. The usual types of firms' response to barriers are either the avoidance of dealing with them or the *ad hoc* approach. A systematic approach to overcoming barriers is more rarely observed. Piatier (1984) proposes a *perception cycle* for barriers. Initially the problem is not appreciated; its existence is even denied or underestimated. At a second step the problem is exaggerated (seen as bigger than it actually is). Then it is properly assessed, and in a final step it is effectively surmounted.

Research into barriers was referred to earlier as an external academic or consultancy approach, having as an aim to increase knowledge in this field and help firms and public policy makers in the identification and prioritization of barriers, as well as, the evaluation of their adverse effects on innovation. We can also visualize an internal process of continuously identifying and overcoming barriers to innovation. Staudt (1994) refers to the useful analogy of developing a radar system to navigate through a sea with icebergs.

This internal process should start with a realization of the importance of identifying barriers and a systematic look into their sources, both externally and within the firm. The organization has usually control over the internal barriers, but their elimination cannot be successful without a detailed plan of action. While raising awareness and attention is a first step, their classification by importance and devising ways to overcome barriers are equally important steps in the process of eliminating barriers. It is particularly important to eliminate or minimize barriers at the early stages of innovation. Methods to reduce barriers probably have to be specific to each stage of the innovation process (Bitzer, 1990). Mohnen & Roeller (2001)—in a study, which used European data on obstacles to innovation—have found that:

> The lack of internal human capital (skilled personnel) is complementary with all the other obstacles in almost all industries (Mohnen & Roeller, 2001, p. 15).

Measures against barriers at both the firm and the innovation policy level should then concentrate on improving human capital as a first priority. Measures for such improvement include training, but also a suitable motivation system with rewards (monetary, promotion, etc) and sanctions.

The role of champions or promoters in overcoming resistance to innovation as emphasized in the literature (Hauschildt, 2003) has been mentioned above. They are frequently necessary in addition to any other structural and system-related arrangements (King, 1990). To summarize, we would say that barriers need a specific strategy and resources, within the frame of a systematic course of action, to be overcome successfully. The overall aim should be to create a favourable innovation climate within the firm, which encourages idea generation, circulation, experimentation and risk-taking. To this effect, suitable organizational mechanisms and systems are needed, e.g. for developing new ideas and rewarding creative work (Amabile, 1997). The internal organizational process of overcoming barriers can be helped by benchmarking, i.e.

observing innovation management best practice in successful innovating firms. What such firms do to avoid and overcome the inevitable barriers, provides a useful yardstick for action.

Although most discussion in the existing literature is on ways to overcome barriers, a proactive approach to *prevent* barriers before they occur is apparently a more rational strategy. This implies proactive management of the innovation change process and periodic measurement of the innovation potential of the organization, in order to determine gaps and take the necessary action.

While firms should do their part, regional authorities and national governments must also act against external barriers by trying to control and remove them. This is usually done in the context of a national innovation policy that addresses barriers according to their importance. It is essential to emphasize that the nature and importance of barriers should be established for the particular context through research. Measures should then be based on the results of such research, rather than subjective perceptions of policy makers of what could possibly constitute barriers. Different sets of measures are probably needed to increase the innovation intensity in established innovators on the one hand, and to stimulate entry to innovation among non-innovators on the other (Mohnen & Roeller, 2001). An important consideration is that barriers frequently arise as side effects of support policies with good intentions. The 'ex ante' and 'ex post' evaluation (i.e. before and after implementation) of innovation support policies should, then, consider the occurrence of side effects and a possible modification of the original policy to remove them.

Conclusions

The barriers approach to innovation focuses on the main problems that may occur during the complex and delicate process of innovation. The innovation process is fraught with difficulties, as it demands the close cooperation—over an extended period of time—of many people. The actions of all these people have to be coordinated, and their talent combined with internal and external resources for a successful outcome. The difficulties of motivation, coordination, development of capabilities, and acquisition of resources emerge as barriers to innovation. The available theories provide only a partial comprehension of the underlying mechanisms that generate these barriers. Based on that knowledge, a process of identifying and eventually overcoming these barriers can be set up within the firm. At the same time regional authorities and national governments have a significant role to play in the elimination of external barriers.

Much research is still needed to illuminate the joint action of barriers as a system, and their dynamic nature. The usual top-down approach in research that is focused on top managers and their views on barriers should be complemented with more 'bottom-up' research, i.e. with investigation of the views of employees. There is also need for longitudinal research to follow the development and interaction of barriers during the innovation process and to find causal links between innovation barriers and innovation performance.

More comparative research (inter-country and inter-sectoral) is required to illustrate the patterns of barriers in different contexts, their similarities and differences and lessons that can be drawn from them. Similarly Drazin & Schoonhoven (1996) call for more comparative studies of alternative forms of organizing for innovation on a global scale. There is a particular need to focus on studying barriers in the case of non-innovators, i.e. use the latter not as a reference group during studies of barriers for innovators, but as the main group for study. Such an approach was already followed for non-exporters in the case of export barriers (Leonidou, 1995).

References

Aldrich, H. & Auster, E. (1986). Even dwarfs started small: Liabilities of age and size and their strategic implications. In: L. Cummings & B. Staw (Eds), *Research in Organizational Behavior*, **8** (pp. 165–198). San Francisco, CA: JAI Press.

Amabile, T. (1997). Motivating creativity in organizations: On doing what you love and loving what you doing. *California Management Review*, **40** (1), 39–58.

Argenti, G., Filgueira, C. & Sutz, J. (1990). From standardization to relevance and back again: Science and technology indicators in small peripheral countries. *World Development*, **18** (11), 1555–1567.

Armenakis, A. & Bedeian, A. (1999). Organizational change: A review of theory and research in the 1990s. *Journal of Management*, **25** (3), 293–315.

Bannon, L. & Grundin, J. (1990). Organizational barriers to innovation in IT development and use. In: R. Hellman, M. Ruohonen & P. Sorgaard (Ed.), *Proceedings of 13th IRIS Conference*, Vol. 1 (pp. 49–60). Turku, Finland.

Barkham, R., Gudgin, G., Hart, M. & Hanvey, E. (1996). *The determinants of small firm growth*. London: Jessica Kingsley Publishers.

Bitzer, B. (1990). *Innovationshemmnisse in unternehmen*. Wiesbaden: Deutsche Universitaets Verlag.

Burns, T. & Stalker, G. M. (1961). *The management of innovation*. London: Tavistock Publications, pp. 96–125.

Chell, E. (2001). *Entrepreneurship: Globalization, innovation and development*. London: Thomson Learning.

Chesbrough, H. (1999). The organizational impact of technological change: A comparative theory of national institutional factors. *Industrial and Corporate Change*, **8** (3), 447–485.

Cohen, L. & Noll, R. (1991). New technology and national economic policy. In: L. Cohen & R. Noll (with J. Banks, S. Edelman & W. Pegram) (Ed.), *The Technology Pork Barrel* (pp. 1–16). Washington, D.C.: The Brookings Institution.

Cohen, W. & Levinthal, D. (1990). Absorptive capacity: A new perspective on learning and innovation. *Administrative Science Quarterly*, **35**, 128–152.

Conner, K. & Prahalad, C. K. (1996). A resource-based theory of the firm: Knowledge vs. opportunism. *Organization Science*, **7** (5), 477–501.

Dahlman, C. & Frischtak, C. (1993). National systems supporting technical advance in industry: The Brazilian experience. In: R. Nelson (Ed.), *National Innovation Systems—A Comparative Analysis* (pp. 414–450). New York: Oxford University Press.

Dodgson, M. (1993). On organizational learning: A review of some literatures. *Organization Studies*, **14**, 375–394.

Drazin, R. & Schoonhoven, C. (1996). Community, population, and organizational effects on innovation: A multilevel perspective. *Academy of Management Journal*, **39** (5), 1065–1083.

Fontes, M. & Coombs, R. (1997). The coincidence of technology and market objectives in the internationalization of new technology-based firms. *International Small Business Journal*, **15** (4), 16–35.

Freeman, C. (1994). The economics of technical change. *Cambridge Journal of Economics*, **18**, 463–514.

Frost, P. & Egri, C. (1991). The political process of innovation. *Research in Organizational Behavior*, **13**, 229–295.

Garsombke, T. & Garsombke, D. (1989). Strategic implications facing small manufacturers: The linkage between robotization, computerization, automation and performance. *Journal of Small Business Management*, October, 34–44.

Gemuenden, H. (1988). Promotors: Key persons for the development and marketing of innovative industrial products. In: K. Gronhaug & G. Kaufmann (Eds), *Innovation: A Cross-Disciplinary Perspective* (pp. 347–374). Oslo: Norwegian University Press.

Godkin, L. (1988). Problems and practicalities of technology transfer. *International Journal of Technology Management*, **3** (5), 587–603.

Hadjimanolis, A. (1999). Barriers to innovation for SMEs in a small less developed country (Cyprus). *Technovation*, **19** (9), 561–570.

Hadjimanolis, A. & Dickson, K. (2001). Development of national innovation policy in small developing countries: The case of Cyprus. *Research Policy*, **30** (5), 805–817.

Hakansson, H. (1990). Product development in networks. In: D. Ford (Ed.), *Understanding Business Markets* (pp. 487–507). London: Academic Press Ltd.

Hauschildt, J. (2001). Teamwork for innovation—The 'Troika' of Promoters. *R&D Management*, **31** (1), 41–49.

Hauschildt, J. (2003). Promotors and champions in innovations—development of a research paradigm. In: L. V. Shavinina (Ed.), *International Handbook on Innovation*. Oxford: Elsevier Science.

Hendry, J. (1989). Technology, marketing, and culture: The politics of new product development. In: R. Mansfield (Ed.), *Frontiers of Management* (pp. 96–108). London: Routledge.

Hitt, M., Hoskisson, R. & Nixon, R. (1993). A mid-range theory of interfunctional integration, its antecedents and outcomes. *Journal of Engineering and Technology Management*, **10**, 161–185.

Howells, J. (1995) A socio-cognitive approach to innovation. *Research Policy*, **24** (6), 883–894.

Jorde T. & Teece, D. (1990). Innovation and cooperation: Implications for competition and antitrust. *Journal of Economic Perspectives*, **4** (3), 75–96.

Katz, R. & Allen, T. J. (1982). Investigating the Not Invented Here (NIH) syndrome: A look at the performance, tenure and communication patterns of 50 R&D project groups. *R&D Management*, **12**, 7–19.

Keogh, W. & Evans, G. (1999). Strategies for growth and barriers faced by new technology-based SMEs. *Journal of Small Business and Enterprise Development*, **5** (4), 337–350.

Kessler, E., Bierly, P. & Gopalakrishnan, S. (2000). Internal vs. external learning in new product development: Effects on speed, costs and competitive advantage. *R&D Management*, **30** (3), 213–223.

King, N. (1990). Innovation at work: The research literature. In: M. West & J. Farr (Eds), *Innovation and Creativity at Work* (pp. 15–59). Chichester, U.K.: John Wiley.

Kline, T. & Rosenberg, N. (1986). An overview of innovation. In: N. Rosenberg & R. Landau (Eds), *The Positive Sum Strategy* (pp. 275–305). Washington, D.C.: National Academy Press.

Koberg, C., Uhlenbruck, N. & Saranson, Y. (1996). Facilitators of organizational innovation: The role of the life-cycle stage. *Journal of Business Venturing*, **11** (2), 133–149.

Krueger, N. (1997). Organizational inhibitions: Perceptual barriers to opportunity emergence. ICS conference proceedings. http//www.sbaer.uca.edu/docs/proceedingsII/97ics007.text.

Lee, B. (2000). Separating the wheat from the chaff: FMS, flexibility and socio-organizational constraints. *Technology Analysis and Strategic Management*, **12** (2), 213–228.

Leonard-Barton, D. (1995). *Wellsprings of knowledge: Building and sustaining the sources of innovation*. Boston, MA: Harvard Business School Press.

Leonidou, L. (1995). Export barriers: Non-exporter's perceptions. *International Marketing Review*, **12** (1), 4–25.

Levy, B. (1993). Obstacles to developing indigenous small and medium enterprises: An empirical assessment. *The World Bank Economic Review*, **7** (1), 65–83.

Love, J. & Roper, S. (2001). Location and network effects on innovation success: Evidence for U.K., German, and Irish manufacturing plants. *Research Policy*, **30** (4), 643–661.

Madanmohan, T. (2000). Failures and coping strategies in indigenous technology capability process. *Technology Analysis and Strategic Management*, **12** (2), 179–192.

Major, E. & Cordey-Hayes, M. (2003). Encouraging innovation in small firms through externally generated knowledge. In: L. V. Shavinina (Ed.), *International Handbook on Innovation*. Oxford: Elsevier Science.

Markham, S. & Aiman-Smith, L. (2001). Product champions: Truths, myths, and management. *Research-Technology Management*, May-June, 44–50.

Markides, C. (1998). Strategic innovation in established companies. *Sloan Management Review*, **39** (3), 31–42.

Maute, M. & Locander, W. (1994). Innovation as a socio-political process: An empirical analysis of influence behavior among new product managers. *Journal of Business Research*, **30**, 161–174.

Mohnen, P. & Rosa, J. (2000). Les obstacles a l'innovation dans les industries de service au Canada. Working paper 2000 s-14. Scientific series, CIRANO, Montreal, Canada.

Mohnen, P. & Roeller, L. H. (2001). Complementarities in innovation policy. Working paper 2001s-28 Scientific series, CIRANO, Montreal, Canada.

Neck, P., Neck, H., Manz, C. & Godwin, J. (1999). 'I think I can; I think I can' A self-leadership perspective toward enhancing entrepreneur thought patterns, self-efficacy, and performance. *Journal of Managerial Psychology*, **14** (6), 477–501.

Nohria, N. & Gulati, R. (1996). Is slack good or bad for innovation? *Academy of Management Journal*, **39**, 1245–1264.

Peteraf, M. A. (1993). The cornerstones of competitive advantage: A resource based view. *Strategic Management Journal*, **14**, 179–191.

Piatier, A. (1984). *Barriers to innovation*. London: Frances Pinter Publishers Ltd.

Pol, E., Crinnion, P. & Turpin, T. (1999). Innovation barriers in Australia: What the available data say and what they do not say. Working Paper University of Wollongong, Australia.

Preissl, B. (1998). Barriers to innovation in services. SI4S topical paper STEP Group.

Quinn, J. B. (1985) Managing innovation: controlled chaos. *Harvard Business Review*, May/June, 73–84.

Rosenkopf, L. & Nerkar, A. (2001). Beyond local search: Boundary spanning, exploration, and impact in the optical disk industry. *Strategic Management Journal*, **22** (4), 287–306.

Rothwell, R. (1984). The role of small firms in the emergence of new technologies. *Omega*, **12** (1), 19–27.

Rothwell, R. (1992). Successful industrial innovation: Critical factors for the 1990s. *R&D Management*, **22** (3), 221–239.

Rush, H. & Bessant, J. (1992) Revolution in three quarter time: Lessons from the diffusion of advanced manufacturing technologies. *Technology Analysis and Strategic Management*, **4** (1), 3–19.

Sanz-Menendez, L. (1995). Policy choices, institutional constraints and policy learning: The Spanish science and technology policy in the eighties. *International Journal of Technology Management*, **10** (4/5/6), 622–641.

Shane, S. (1995). Uncertainty avoidance and the preference or innovation championing roles. *Journal of International Business Studies*, 1st quarter, 47–68.

Sheen, M. (1992). Barriers to scientific and technical knowledge acquisition in industrial R&D. *R&D Management*, **22** (2), 135–143.

Schemman, R. (2000). Innovation in SMEs: Concepts, experiences and recommendations ADAPT project transnational report, Web version (http://www.ifer-sic.de/report/index.htm).

Schoemaker, P. & Marais, L. (1996). Technological innovation and firm inertia. In: G. Dosi & F. Malerba (Eds), *Organization and Strategy in the Evolution of the Enterprise* (pp. 179–205). Hampshire, U.K.: MacMillan Press Ltd.

Souitaris, V. (2003). Determinants of technological innovation: Current state of the art, modern research trends and future prospects. In: L. V. Shavinina (Ed.), *International Handbook on Innovation*. Oxford: Elsevier Science.

Starbuck, W. (1996). Unlearning ineffective or obsolete technology. *International Journal of Technology Management*, **11** (7/8), 725–737.

Staudt, E. (1994). Innovation barriers on the way from the planned to the market economy: The management of non-routine processes. *International Journal of Technology Management*, **9** (8), 799–817.

Steiner, L. (1998). Organizational dilemmas as barriers to learning. *The Learning Organization*, **5** (4), 193–201.

Storey, D. (1994). *Understanding the small business sector.* London: Routledge.

Storey, J. (2000). The management of innovation problem. *International Journal of Innovation Management*, **4** (3), 347–369.

Swan, J. (1995). Exploring knowledge and cognitions in decisions about technological innovation: Mapping managerial cognitions. *Human Relations*, **48** (11), 1241–1270.

Swan, J., Robertson, M. & Scarbrough, H. (2003). Knowledge, networking and innovation: Development and understanding of process. In: L. V. Shavinina (Ed.), *International Handbook on Innovation*. Oxford: Elsevier Science.

Tang, H. K. & Yeo, K. T. (2003). Innovation under constraints: The case of Singapore. In: L. V. Shavinina (Ed.), *International Handbook of Innovation*. Oxford: Elsevier Science.

Teece, D. (1986). Profiting from technological innovation. *Research Policy*, **15**, 285–306.

Tidd, J., Bessant, J. & Pavitt, K. (1997). *Managing innovation*. Chichester, U.K.: John Wiley.

Vossen, R. (1998). Relative strengths and weaknesses of small firms in innovation. *International Small Business Journal*, **16** (3), 88–94.

Webb, J. (1992). The mismanagement of innovation. *Sociology-The Journal of the British Sociological Association*, **26** (3), 471–492.

Weick, K. (1979). *The social psychology of organizing.* Reading, MA: Addison-Wesley.

Weick, K. (1995). *Sensemaking in organizations.* Thousand Oaks, C.A: Sage Publications.

Williamson, O. E. (1985). *The economic institutions of capitalism.* New York: Free Press.

Witte, E. (1973). *Organization fuer innovationsentscheidungen: Das promotoren modell.* Goettingen: Ottto Schwarz & Co.

Yap, C. M. & Souder, W. (1994). Factors influencing new product success and failure in small entrepreneurial high technology electronic firms. *Journal of Product Innovation Management*, **11**, 418–432.

The International Handbook on Innovation
Edited by Larisa V. Shavinina

Knowledge Management Processes and Work Group Innovation

James L. Farr[1], Hock-Peng Sin[1] and Paul E. Tesluk[2]

[1] *Department of Psychology, Pennsylvania State University, USA*
[2] *Department of Management & Organization, University of Maryland, USA*

Abstract: Following a selective review of theoretical models and empirical research on work group effectiveness and innovation, we present a dynamic model of work group innovation. Our model integrates recent advances in taxonomies of work group processes and stages of the innovation process with a focus on the temporal nature of innovation. We also provide a discussion of the specific inputs, group processes, emergent states, and outcomes that appear to be most relevant for each of the various stages of work group innovation.

Keywords: Work group innovation; Group processes; I-P-O models.

Introduction

Our focus is to bring together three streams of theory and research that have separately addressed factors that are thought to influence organizational effectiveness: group-based work, knowledge management, and innovation. Although each of these domains has received considerable attention from organizational scholars in the past decade, little empirical research or conceptual development has occurred that might integrate them. We argue here that knowledge management processes related to how work groups seek, share, store, and retrieve information relevant to their performance (both taskwork and teamwork) are important factors associated with group effectiveness, including group innovation. We present an original theoretical framework based on several existing Input-Process-Outcome (I-P-O) models of group effectiveness that provides our initial understanding of how work groups' knowledge management processes influence their capacity for generating and implementing innovative solutions to problems. Our focus is primarily on the group level of analysis, but include relevant factors at the individual (e.g. knowledge and skill) that influence group-level knowledge management processes and innovation in important ways.

We present brief reviews of the extant literature on work group effectiveness, work group innovation, and knowledge management, focusing on recent theoretical frameworks in each domain. These reviews are followed by our model.

Work Group Effectiveness

Work group effectiveness is most often studied through a lens that uses an (I-P-O) perspective (Guzzo & Shea, 1992; McGrath, 1991). Because it is the dominant perspective for studying work groups, the I-P-O framework can be used to examine different approaches to innovation in work groups from a common vantage point. One of the advantages of the I-P-O paradigm is that it is inherently temporal (Marks, Mathieu & Zaccaro, 2001) and, because innovation involves consideration of criteria that can be viewed from both short-term (e.g. idea generation) and long-term (e.g. implementation, learning) perspectives, the I-P-O framework is useful for incorporating this time-based perspective.

Following the I-P-O lens, *inputs* refer to factors such as group composition, design, leadership, and organizational context conditions (rewards, training, information systems) that influence the processes by which group members engage with each other and their environment as they work toward their objectives. *Outcomes* are often considered to be in three forms:

(1) the productive output of the team as defined by those who evaluate it;
(2) the satisfaction of team members; and
(3) the capabilities of the members of the team and their willingness to continue working together over time (Hackman, 1987).

Processes have been defined in a number of different ways but typically are considered to be interactions among members of a team and/or with other groups or individuals outside the team that serve to transform inputs (e.g. members' skills) and resources (e.g. materials, information) into meaningful outcomes (Cohen & Bailey, 1997; Gladstein, 1984; Marks et al., 2001). They are, therefore, typically considered as mediating the relationship between input factors and group outcomes (McGrath, 1991), and in most models of team effectiveness, processes are depicted as serving a central role (e.g. Gist, Locke & Taylor, 1987; Gladstein, 1984; Guzzo & Shea, 1992; Hackman, 1987). Indeed, an accumulating amount of research in different team settings has shown that processes such as coordination and communication facilitate team performance in terms of outcomes such as productivity and manager ratings of group effectiveness (e.g. Campion, Medsker & Higgs, 1993; Hackman & Morris, 1975; Hyatt & Ruddy, 1996). Yet, while there is clear evidence of their importance to team effectiveness, group process remains very much a 'black box' when it comes to looking past short-term or immediate outcomes to understanding innovation in the team context. One of our objectives is to suggest new directions for research to better understand how team processes impact innovation in team settings.

Group processes from a temporal perspective. A clear reality and challenge for many teams is how to manage multiple tasks and objectives simultaneously and adjust to shifting priorities. To understand the role of how processes enable teams to do this requires incorporating a temporal perspective. While demonstrating the importance of team processes, existing research tells us little about how teams balance the interests of competing outcomes from internal/external and short-term/long-term perspectives. Because this requires understanding how teams manage multiple goals and sets of activities over time (McGrath, 1991), recent theoretical work taking a temporal perspective to team processes is particular promising for understanding the functioning of healthy teams.

One perspective is offered by Kozlowski, Gully, Nason & Smith (2001), who identify the critical phases in team development, how they build upon and transition from one to another, and the primary processes that take place at individual, dyadic, and team levels as teams progress in their development. According to their framework, after their initial formation, teams focus on establishing an interpersonal foundation and shared understanding of the team's purpose and goals through socialization and orientation processes. Then, attention shifts to members' development of task competencies and self-regulation through individual skill acquisition experiences. This, in turn, enables team members to immediately develop an understanding of their own and their teammates' roles and responsibilities. Finally, once team members have

a good understanding of each other, their own individual task requirements, and their teammates' roles and responsibilities, the team then focuses on how they can best manage interdependencies both in routine and novel situations by attending to the network of relationships that connect team members to each other.

What the Kozlowski et al. (2001) perspective makes clear is that adaptable teams are those that are able to successfully manage the different challenges presented in each phase of their development. Their model also suggests that, because the phases are progressive and yield enable team capabilities that build on each other and are necessary for meeting the requirements for the next set of developmental challenges, teams that fail to effectively engage in critical sets of processes at earlier stages of development and achieve requisite cognitive, affective, and behavioral outcomes will be at a distinct disadvantage at later developmental stages. For instance, even mature teams with a substantial shared history will be limited in their adaptability and learning capabilities if team members do not develop a high level of familiarity with their team members and a strong team orientation through team socialization experiences.

Marks and colleagues (2001), who also use a temporal approach to describe team processes, offer another complementary theoretical perspective. Although they do not take a developmental view of how teams build progressively complex capabilities per se, they do focus on team performance as a recurring set of I-P-O action and transition episodes where outcomes from initial episodes serve as inputs for the next cycle. Thus, consistent with Kozlowski et al. (2001), team outcomes at earlier stages influence team functioning in subsequent phases. Furthermore, Marks et al. (2001) present a taxonomy that organizes different forms of team processes depending on the phase of task accomplishment. Sometimes groups are actively engaged in working toward goal accomplishment and so *action processes* are dominant (e.g. coordination, communication, team and systems monitoring). At other times, groups are either planning for upcoming activities or reflecting on past performances and so members are involved with *transition processes* (e.g. planning, performance analysis, goal specification). Finally, other interactions that involve *interpersonal processes*, such as managing conflict, affect, and teammates' motivation, occur during both transition and action phases of teamwork. Together, these different types of processes capture unique forms of member interaction, but merge together as a sequenced series of team member activities and interactions as a team goes about performing its work.

This distinction between different types of processes based on when they occur during a team's set of performance episodes is helpful both in terms of theory and practice. For instance, interactions supporting a climate that encourages openly discussing problems, mistakes, and errors—a necessary condition for innovation through learning to occur in teams (Edmondson, 1996)—depends on interpersonal processes such as affect and conflict management in order to help raise alternative perspectives and dissenting viewpoints and keep disagreements focused on the task rather than becoming personally directed at team members (Jehn, 1995). These types of interpersonal processes are necessary both when a team is actively working toward task accomplishment as well as during transition periods such as, during routine pre-/post-shift meetings. Effective interpersonal processes are important not only for contributing to a team's ability to recognize and interpret errors and utilize its experiences to derive learnings that enhance capabilities to meet future demands, but also for maintaining the long-term viability of the team by supporting cohesion and members' commitment to the team by constructively managing disagreements, conflict and contributing to individual team members' skill development and team longevity.

In contrast, if we are interested in studying how teams prioritize being innovative, attention should be focused on transition processes such as whether, and how, teams specifically identify, discuss, and emphasize innovation as an explicit goal during transitional phases in a team's life cycle. From an intervention standpoint, this might mean training teams on how to more effectively utilize opportunities such as pre-shift meetings or 'down-time' periods to work on prioritizing process improvement goals and reviewing their performance and conditions influencing the development of new ideas. To help teams more effectively weight potentially competing goals such as achieving high levels of productivity and generating novel ideas, interventions could also be designed to help teams explicitly identify, prioritize, and try to balance different, and perhaps competing, team objectives.

Work Group Innovation: I-P-O-Based Models

While most researchers have adopted the I-P-O model of work group effectiveness, it has been noted that the conceptual development of the outcome domain has been the least developed (Brodbeck, 1996). Most studies have relied on the group's productive output (i.e. task performance) as the default criterion measure for its effectiveness. Another common criterion measure for group effectiveness has been team viability, or the extent to which members want to continue their team or group involvement. In one sense these two output variables correspond with Hackman's (1987) suggestion that a comprehensive assessment of success in ongoing groups must capture current effectiveness (i.e. task performance) and future effectiveness (i.e. team viability). However, Brodbeck (1996) pointed out that two other important effectiveness dimensions for

work group research are individual well-being and group innovation.

An emphasis on work group innovation is consistent with contemporary writings that note, organizations must continually innovate in order to increase or even maintain competitiveness (see e.g. Banbury & Mitchell, 1995; Hamel & Prahalad, 1994; Wolfe, 1994). Innovation can occur at the organizational level, the work group level, and the individual level, but the group level has been less researched than either of the other two levels (Anderson & King, 1993), despite the increasing use of work groups as the basis for accomplishing work tasks in many organizations. Recently, however, West et al. (1998) have proposed a group-level model of innovation.

West, Borrill & Unsworth (1998) Model of Work Group Effectiveness

Drawing on Brodbeck's (1996) work, West et al. (1998) included team innovation as one of the outcome variables in their review and synthesis of the research related to IPO models of work group effectiveness. They suggested three possible input variables that might influence team innovation. First, heterogeneity in group composition is considered important as it has been found to be related to group innovation (e.g. McGrath, 1984; Jackson, 1996). For instance, although not directly examining this issue, Ancona & Caldwell (1992) found that when a new member from a certain functional area joined an existing team, communication increased dramatically in that functional area. This, in turn, might favor innovation through the introduction of additional and different ideas and models (Agrell & Gustafson, 1996). More direct evidence has come from Wiersema & Bantel (1992), who studied a sample of top management teams in the Fortune 500 companies and found that the top managers' cognitive perspectives, as reflected in the team's demographic characteristics, were linked to the team's strategic management initiatives. Specifically, teams with higher educational specialization heterogeneity were more likely to undergo changes in corporate strategy. Bantel & Jackson (1989) reported that more innovative banks were managed by more educated teams who were heterogeneous in terms of their functional areas of expertise. Similarly, a more recent study (Drach-Zahavy & Somech, 2001) has found that team heterogeneity, as defined by differences in organizational roles, was positively related to team innovation. Therefore, heterogeneity might affect team innovation because team members possess different skills and expertise and, hence, have broader informational resources and knowledge.

Second, West et al. (1998) noted that team tenure might have a negative effect on team innovation. For example, Katz (1982) suggested that as team tenure increases, team members are less likely to communicate internally within the group or externally with key information sources. In addition, there is some evidence that team tenure might be related to team homogeneity, which, in turn, may have negative effects on team innovation (e.g. Jackson, 1996). On the other hand, Bantel & Jackson (1989) did not find a direct effect of average team tenure on team innovation. West & Anderson (1996) also reported no correlation between average team tenure and overall level of innovation among a sample of top management teams, but did find a positive relationship between average team tenure and impact of innovation on staff well-being.

Third, West et al. (1998) suggested that group composition with respect to personality or dispositional characteristics of the group members might be related to group innovation. For example, West & Anderson (1996) reported that the proportion of innovative members within the group predicted the introduction of radical innovation. This is consistent with the assumption that idea generation is a cognitive process residing in individual group members, although the translation from ideas to actions still requires a variety of situational attributes (Mumford & Gustafson, 1988). In addition, McDonough & Barczak (1992) found that when the technology was familiar, an innovative style on the part of the team as a whole led to faster product development, but that the leader's style was not a significant predictor.

Drawing on previous reviews (e.g. Agrell & Gustafson, 1996; Guzzo & Shea, 1992), West et al. (1998) also identified four possible group process characteristics that might influence group innovation. First, the existence and clarity of group goals or objectives was found to be positively related to group effectiveness in general (e.g. Guzzo & Shea, 1992; Pritchard, Jones, Roth, Stuebing & Ekeberg, 1988). Pinto & Prescott (1988) studied the life cycle of over 400 project teams and found that a clearly stated mission was the only factor that predicted success at all the four stages of the innovation process. They suggested that clarity of goals aided the innovation process because it enabled focused development of new ideas.

Given that innovation is not solely stimulated through a burst of creativity by a talented individual, but is influenced by an interactive process among the team members (Agrell & Gustafson, 1996; Mumford & Gustafson, 1988; West, 1990), West et al. (1998) argued that high levels of participation of group members in decision making should result in more innovation. According to Agrell & Gustafson (1996), although the innovation process begins with the production of ideas from individuals, often the potential innovation may be abandoned or defeated if these ideas are not properly discussed in a dialogue that involves the whole team.

Task-related team conflict, which implies divergent thinking and perspectives, appears to serve as an important process that contributes to successful innova-tion or generation of ideas (Mumford & Gustafson, 1988). Tjosvold and colleagues have provided some interesting empirical evidence that constructive controversy leads to improved quality of decision-making and, therefore, innovation. For example, Tjosvold (1982) reported that supervisors who used a cooperative-controversy style explored, understood, accepted, and combined workers' arguments with their own to make a decision. It was concluded that task-related controversy within a cooperative context can result in curiosity, understanding, incorporation, and an integrated decision. In another study, Tjosvold, Wedley & Field (1986) asked 58 managers to describe successful and unsuccessful decision-making experiences by indicating the extent to which those involved in making the decision experienced constructive controversy. Results indicated that constructive controversy was significantly related to successful decision-making. De Dreu & West (2001) found that minority dissent (the public opposition by a minority of a group to the beliefs, ideas, or procedures of the group majority) stimulated divergent thinking and creativity, and that such dissent, when combined with high levels of participation in decision making among group members, led to more frequent group innovations.

Fourth, innovation is more likely to occur when the organizational and/or group contexts are supportive of innovation (Amabile, 1983). Burningham & West (1995) examined the contribution of individual innovativeness and team climate factors to the rated innovativeness of work groups in a study of 59 members of 13 teams in an oil company. Support for innovation was found to be the most consistent predictor for predicting externally rated group innovativeness. In addition, West & Anderson (1996) found that support for innovation was positively correlated with their measures of overall amount of team innovation, the number of innovations introduced, and the rated novelty of the innovations.

West's (2002) Model of Work Group Innovation and Creativity

Drawing on the earlier model of work group effectiveness (West et al., 1998) that we have just described, West (2002a) has recently proposed a model focused on the work group innovation process. This model suggests that four sets of factors are the primary determinants of work group innovation. They include characteristics of the group task, group knowledge diversity and skills, external demands on the group, and integrating group processes. The effects of group task characteristics and group knowledge diversity and skills on innovation are both hypothesized to be fully mediated by integrating group processes, whereas external demands are hypothesized to have both a direct effect on innovation and an indirect effect, also mediated by group processes. In addition, West argues that the innovation process includes two stages, creativity and innovation implementation. Creativity is concerned with idea generation and development, whereas innovation implementation is the application of those ideas to produce innovative products, services, and procedures (West, 2002). West further proposes that external demands on groups can have very different effects on creativity and innovation implementation, although he argues that the effects of task characteristics, knowledge diversity and skills, and integrating group processes on creativity and implementation are similar.

An examination of the West (2002) model's specific content of the four determinants of creativity and innovation implementation and of the forms of the effects they are hypothesized to have both clarifies how West believes that the innovation process unfolds and raises some questions about the model. First we present more details of his model and then we note some concerns.

Group Task Characteristics include factors taken from sociotechnical systems (e.g. Cooper & Foster, 1971) and Job Characteristics Theory (Hackman & Oldham, 1980), such as autonomy, task significance, task identity or completeness, varied demands, opportunities for social interaction, opportunities for learning, and opportunities for development. The model predicts that higher levels of these factors will be correlated with higher levels of both creativity and innovation implementation.

West contends that *requisite diversity* of knowledge and skills among group members is needed for creativity and implementation innovation, that is, an 'optimal' level of knowledge diversity exists for a given task that will encourage creativity and innovation through enhanced task performance capabilities, varieties of perspectives and approaches to problems, and constructive conflict. Too little diversity leads to conformity and common approaches to problems. However, too much diversity (or insufficient overlap in knowledge and skills) among group members may result in disparate mental models and poor levels of coordination and communication that, in turn, negatively affect the innovation process. Thus, diversity is likely to have a curvilinear relationship with group processes that mediate its link with creativity and innovation implementation.

Certain work group processes, termed *integrating group processes* by West (2002a), are affected by group task characteristics and knowledge and skill diversity and mediate the impact of the task and knowledge factors on the innovation process. Integrating group processes allow group members to work collaboratively to capitalize on their diverse knowledge and skills (West, 2002a). Integrating group processes include clarifying and ensuring commitment to group objectives, participation in decision making, managing conflict effectively, minority influence, supporting innovation, developing intra-group safety, reflexivity, and developing group members' integration skills.

West (2002a) argues that the external context of the work group affects directly the group's innovation process and also the group's integrating processes. The external context, whether internal or external to the organization of which the group is a part, makes demands that can have differential and complex effects. Demands can motivate, but they can sometimes be perceived as threats to the group. In general, West predicts that external demands inhibit the creativity stage of the innovation process, but facilitate the implementation of innovation, although excessive demand levels may make effective implementation be seen as impossible and lessen group members' motivation to implement the proposed innovation. External demands may also force the development of more effective group integrating processes by serving as a driver of change in the ways that group members work together.

In summary, the model of the innovation process proposed by West (2002a) makes important contributions by clearly delineating two major stages of the process, a creativity stage and an innovation implementation stage; by specifying the roles of group tasks characteristics, the diversity of group members' knowledge and skills, and integrating role processes in the innovation process; and by noting the differential effects that external demands and threats have on the creativity and implementation stages. However, we believe that West's model does not fully address very recent advances in thinking about I-P-O models of group effectiveness (e.g. Marks et al., 2001). In addition, we believe that how work groups manage their individual and team knowledge requires more extended elaboration than it has received in existing models of group innovation. Consistent with this belief, we now consider recent research and theory concerning knowledge management. Following that, we present our conceptual framework for work group innovation.

Knowledge Management

Increasingly, organizations that have effective means of creating, storing, and transferring knowledge have been seen by organizational scholars as having competitive advantage over those competitors who do not (Nahapiet & Ghoshal, 1998; Thomas, Sussman & Henderson, 2001). We will use *knowledge management* (see e.g. Hedlund, 1994) to refer to the way in which an organization and its units acquire, store, retrieve, share, and transfer information both across organizational units and among members of a single unit.

Because we are primarily interested in work group-level phenomena, most of our attention and theoretical development related to knowledge management is devoted to this level. Hedlund (1994) has noted that the work group is the level at which much knowledge transfer and learning takes place in organizations, especially with regard to innovation and product development. Hinsz, Tindale & Vollrath (1997) in a review of research and theory on groups as information processors note that a distinction can be made in terms of the contributions (including knowledge) that individual members bring to the group interaction and the processes involved in the way these individual contributions are combined (aggregated, pooled, or transformed) during group interaction to produce group outcomes. Thus, the knowledge of individual group members is an individual-level input and the combination of the members' knowledge is a group process in an I-P-O approach to group effectiveness.

Following Mohammed & Dumville (2001), we organize our discussion of knowledge management within the context of team mental models. Team mental models are shared understandings and knowledge structures that team members have regarding the key elements of the group's environment, including the group task, equipment, working relationships, and situations (Mohammed & Dumville, 2001). We suggest that knowledge management within a work group that is attempting to create and implement an innovation is ultimately concerned with the development of: (a) a shared mental model of the desired end-product (the innovative product, service, or system) and the means by which it can be achieved; and (b) an accurate inventory of the task-related and teamwork-related knowledge and skills possessed by group members. The shared mental model defines the requisite knowledge and skills perceived to be needed for an effective innovation. A comparison of the requisite needs and the inventory of existing knowledge and skills defines the necessary knowledge and skill acquisition that the group must attain by learning or other means.

Several specific aspects of group-level knowledge management have interesting implications for work group innovation, including transactive memory, cognitive consensus, knowledge distribution and information sharing, imputation of own and others' knowledge, and group learning behavior. We briefly discuss these below.

The concept of *transactive memory systems* was introduced by Wegner (1987) and has been applied to groups (e.g. Moreland & Myaskovsky, 2000). It refers to systems of memory aids that groups may use to help ensure that important information is recalled. Such memory is a social phenomenon and individuals within groups may use each other as a form of external memory aid to augment personal memory. For transactive memory to be effective, group members must have a shared awareness of what knowledge is known by what members of the group. Furthermore, new knowledge that comes to the group should be stored with the member who is the group's expert for that knowledge domain, resulting in increasing specialization among individual members' memories

(Mohammed & Dumville, 2001). We suggest that the type of information may moderate whether such specialization is desired: specialization of task-related information memory may reduce cognitive load and be useful, whereas teamwork-relevant knowledge may require that it be stored in the memory of all group members. Most existing research in transactive memory has emphasized task-related knowledge (Mohammed & Dumville, 2001), so there may be limits to generalizing to team mental models that include other forms of information.

The members of work groups that are formed to represent a wide range of perspectives and constituencies (e.g. cross-functional teams and task forces) and charged with developing an innovative solution to a complex organizational problem frequently have very different perspectives and interpretations of the issues involved (Mohammed & Dumville, 2001). Such groups need to reach *cognitive consensus* on the interpretation of the issues before they can develop effective and mutually acceptable decisions about courses of action to take (Mohammed & Ringseis, 2001). While cognitive diversity has been suggested as predictive of idea generation and creativity, cognitive consensus may be required in order to develop an acceptable group mission or set of goals.

Research on knowledge distribution and information sharing in groups (e.g. Stasser, Taylor & Hanna, 1989) suggests that group members tend to discuss what they believe to be shared information known to all group members. This finding implies, that groups in which knowledge is distributed may not effectively use their diverse knowledge unless group processes explicitly elicit unique knowledge from all group members (Mohammed & Dumville, 2001). Stasser (1991) has suggested that group members be told that they each may hold unique information and that others may also have unique information.

While one might assume that most individuals in work groups would understand that information is so distributed among members, Nickerson (1999) in a review of the research on *imputing what other people know* concluded that individuals often impute their own knowledge to others, i.e. assume that others know what they know, especially when they have little direct experience with the other people. While the tendency to overimpute one's knowledge to others decreases as one learns more about others, there is a danger in newly formed work groups for errors to occur that may hinder the development of shared mental models. To the extent that groups do not develop transactive memory systems and fail to discuss their unique knowledge, however, such learning about others may not occur effectively and erroneous knowledge imputation may continue.

Edmondson (1999) suggested that work groups would be more effective to the extent that they engage in *team learning behaviors*, such as seeking feedback, sharing information, asking for help, talking about errors, and experimenting. However, some of these learning behaviors are potentially costly because an individual exhibiting such behaviors may appear to be incompetent and, thus, threaten his or her image (Brown, 1990). Despite the potential benefit of learning behaviors for the team or organization, research has shown that people are generally reluctant to disclose their errors and are often unwilling to ask for help (Lee, 1997). Therefore, when faced with situations that are potentially embarrassing, people tend to behave in ways that inhibit learning (Argyris, 1982). Such behaviors would be detrimental for successful group innovation.

That said, Edmondson (1996) found that people are more likely, and willing, to ask for help, admit errors, seek feedback, and discuss problems when they perceive the interpersonal threats in the situations as sufficiently low. Drawing on these insights, Edmondson (1999) proposed that teams differ in their levels of team psychological safety, which is defined as the "shared belief that the team is safe for interpersonal risk taking" (p. 354). Psychological safety is a team climate characterized by interpersonal trust and mutual respect such that people are comfortable and willing to engage in learning behaviors that make them vulnerable to threats and ridicules (Robinson, 1996). For example, Edmondson (1999) found that: (a) team psychological safety was positively related to team learning behavior; (b) team learning behavior was positively related with team performance; and (c) team learning behavior mediated that relationship between team psychological safety and team performance. Therefore, teams with higher levels of team psychological safety engaged in more learning behavior, which in turn led to higher levels of team performance. Equally important, Edmondson also found that team psychological safety mediated the effects of team leader coaching and contextual support on team learning behavior. In other words, coaching and context support promotes team psychological safety, and team psychological safety promotes team learning.

Integration of Current Research and Theory on Group Effectiveness and Innovation

As we have noted earlier, West and colleagues (West et al., 1998; West, 2002) have provided valuable insights concerning work group effectiveness and innovation. We seek to elaborate West's (2002) model of work group innovation in three ways. First, although West emphasized the important difference between a creative or an idea generation stage and an innovation implementation stage in the overall innovative process, we draw from Marks et al. (2001) and suggest that the creativity stage and the implementation stage each contain both a transition phase and an action phase.

Second, again following Marks et al. (2001), we emphasize the distinction between team processes and team emergent states. Specifically, team processes are interdependent acts members take to yield collective outcomes. However, emergent states refer to "constructs that characterize properties of the team that are typically dynamic in nature and vary as a function of team context, inputs, processes, and outcome" (Marks et al., 2001, p. 357). We contend that some of the processes (e.g. intragroup safety) presented in West's (2002) model should be regarded as team emergent states. The distinction between group processes and emergent states is an important one. Team processes depict the nature of team member interactions (e.g. participation, conflict management), whereas emergent states depicts the cognitive (e.g. team efficacy; shared mental model), affective (e.g. team cohesion and team psychological safety), and motivational (e.g. team's intrinsic motivation) states of the teams. In addition, emergent states can be considered as both inputs to the team's current phase of the innovative process and as proximal outcomes that then become inputs for the next innovative phase.

Third, West (2002) proposed in his model that most influences on the innovative process (i.e. task characteristics, diversity in knowledge and skills, and integrating group processes, but not external demands) have identical relationships with both the creativity and innovation implementation stages. Instead, we propose that our elaborated temporal perspective of work group innovation suggests that different input and process variables are relevant and important in predicting the respective outcomes in each phase of the innovation process. Concomitantly, we attempt to include a more comprehensive and inclusive set of variables that are relevant for the investigation of work group innovation. In particular, we elaborate on the knowledge management processes and systems that groups use during innovation.

We now present our model in more detail.

Dynamic Model of Work Group Innovation

Figure 1 depicts the temporal sequence of various phases in the process of work group innovation, drawing on both West (2002) and Marks et al. (2001). (Although it is clear that the innovation process is not linear (West, 2002), we have portrayed it in Fig. 1 as if it were for sake of parsimony in our graphic depiction of the process.) First, there are two distinct stages in the innovation process: creativity and innovation implementation. Within both the creativity and implementation stages are transition and action phases. Nested within each phase are various input and process variables that influence the interim outcomes for that phase. Table 1 presents our initial thoughts about specific inputs, processes, and outcomes that are most relevant for each of the four phases of the innovation

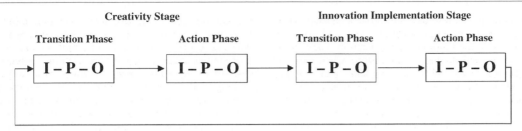

Figure 1. Dynamic model of work group innovation.

process. Below we briefly discuss the major components of our model.

Outcomes

We believe it is helpful to first discuss the outcomes because they help to demarcate the change from one phase of the innovation process to another. Referring to Table 1, there are clearly identifiable outcomes that punctuate the phases. In the transition phase of the creativity stage, the team interprets the relevant issue(s) and identifies the problem the team or organization is facing. Once issues are interpreted and the problems

Table 1. Inputs, processes, task-related outcomes, and emergent states in work group innovation.

	Creativity Stage		Innovation Implementation Stage	
	Transition Phase	Action Phase	Transition Phase	Action Phase
Inputs:				
• Task characteristics	Autonomy, Completeness, Significance (*Intrinsic motivation*)			Autonomy, Completeness, Significance (*Intrinsic motivation*)
• Individual	Expertise	Goal orientations Openness to experience	Expertise Agreeableness	Goal orientation Conscientiousness
• Group	Leader behavior (*Psychological safety*)	Requisite diversity in knowledge and skills Demographic diversity	Requisite diversity in knowledge and skills	Diversity in personality and attitudes Social network
• External demands	Competition Uncertainty	Goal orientation (*Group efficacy*) Time constraints Threats	Problem importance	Goal orientation (*Group efficacy*) Time constraint
Processes:				
• Transition	Mission analysis (*Shared mental model*)		Goal specification Strategy formulation (*Shared mental model*)	
• Action		Monitoring progress toward goals		Monitoring progress toward goals Team monitoring Coordination System monitoring
• Interpersonal	Conflict management Affect management (*Group cohesion*)	Conflict management Affect management (*Group cohesion*)	Conflict management Affect management (*Group cohesion*)	Motivation and confidence building (*Group efficacy*)
Task-related outcomes	Interpretation of issues Problem identification and recognition	Generation of creative ideas/solutions	Evaluation and selection of ideas/solutions	Application of ideas/solutions to problem
Group outcomes: Emergent states	Psychological safety; Group efficacy; Shared mental models; Group cohesion; Group affect			

identified, the team moves into an action phase where the goal is to generate creative ideas and solutions pertaining to the problem. The transition phase of the innovation implementation stage starts when the team begins to evaluate and assess the ideas and solutions generated earlier. Finally, the action phase of the implementation stage is the team's concerted effort to apply the idea or solution to the problem. Note that there is also a 'feedback loop' from the end of implementation stage to the beginning of the creativity stage (see Fig. 1). This happens when the team identifies related problems or fails to implement the initial innovation successfully.

Input Variables

Task Characteristics
Based on socio-technical systems theory (e.g. Cooper & Foster, 1971) and job characteristic theory (Hackman & Oldham, 1975), West (2002) proposed that certain task characteristics help to evoke 'task orientation' or intrinsic motivation in the team which will in turn facilitate innovation. In other words, when teams experience autonomy, perceive the team as responsible for completing a whole task, and consider the task to be important, a state of high intrinsic motivation is more likely to emerge. According to Amabile and colleagues, a high level of intrinsic motivation is fundamental to creativity and innovation (Amabile, 1983; Amabile & Conti, 1999). We reason that high levels of intrinsic motivation are most crucial when teams face uncertainties or difficulties requiring novel approaches. Hence, ensuring high levels of intrinsic motivation is beneficial during the transition phase of the creativity stage, where the team is still trying to assess and make sense of the problem at hand. Similarly, intrinsic motivation is crucial during the action phase of the implementation stage because the team is likely to encounter significant roadblocks and obstacles.

Individual Variables
Some individual level variables that are relevant for understanding innovation are level of expertise, personality, and goal orientation. First, level of expertise is considered important because knowledge is the basic requirement before one can create something new within a particular domain (Amabile & Conti, 1999). However, Sternberg (1999) has noted that extremely high levels of expertise can also be a hindrance to creativity and innovation. (Diversity in the types of expertise within a team can help to reduce this problem as we discuss later.) We believe that expertise is most relevant during the transition phase of the creativity and implementation stages because problem identification and evaluation of ideas link directly to domain knowledge.

Second, research at the individual level has found some relationships between personality and innovation.

For example, Barrick & Mount (1991) reported in their meta-analytic finding that openness to experience is related to various learning and training criteria, which suggests that individuals who are high on openness to experience are more likely to adopt new ways of thinking and to embrace changes. We reason that openness to experience is most relevant in the action phase of the creativity stage because that is when new or even seemingly 'wild' ideas must be entertained and encouraged. Next, agreeableness has been found to be negatively related to creative achievement but not with creative thinking (King, Walker & Broyles, 1996). In other words, individuals who are agreeable may suggest more ideas but also often find it difficult to evaluate or find fault with others' ideas. Hence, we propose that agreeableness is relevant during the transition phase of the implementation stage when lower levels of agreeableness may result in better evaluation and selection of ideas. Patterson (2002) reasoned that conscientiousness might be negatively associated with innovation because individuals high on conscientiousness tend to comply with rules and organizational norms (e.g. Hogan & Ones, 1997). We would agree for the creativity action phase. However, we propose that conscientiousness might be positively related to innovation implementation (i.e. action phase of the implementation stage) because a high level of persistence is needed when there are resistance and obstacles to change (e.g. Amabile, 1983).

Third, we propose that goal orientation is relevant during both action phases of the innovation process. Dweck (1986) has demonstrated that the way in which individuals approach a task can influence their behavior in a number of respects. When approaching a task from a learning goal orientation, the individual's main objective is to increase his or her level of competence on a given task. Alternatively, when approaching a task from a performance goal orientation, individuals are primarily concerned with demonstrating their competency to others via task performance. Farr, Hofmann & Ringenbach (1993) reviewed the goal orientation research and noted ways that it could be applied to work behavior, including learning. A number of these suggested research directions have been recently pursued. For example, Colquitt & Simmering (1998) found learning orientation to be positively related to motivation to learn both initially and after performance feedback had been given, whereas performance goal orientation was negatively related to motivation to learn. VandeWalle & Cummings (1997) found in two studies that learning goal orientation and performance goal orientation were positively and negatively related to feedback seeking, respectively. Hence, during the action phase of the creativity stage, we suggest that individuals with strong learning goal orientations will be more willing to participate in idea generation without fear of appearing incompetent. In addition, individuals with strong learning goal orientations may be more persistent when encountering problems and obstacles during the action phase of the implementation stage.

Group Variables

Some group level variables that are relevant for understanding innovation are diversity, social networks, and leadership role behavior. After reviewing the extant literature, Milliken & Martins (1996) suggested that various types of work group diversity (observable traits, such as demographic characteristics; unobservable traits, such as personality and values; and functional characteristics, such as knowledge, skills and organizational experience) may be differentially related to group processes and outcomes. We propose that all three types of diversity (i.e. demography, personality and attitudes, and knowledge and skills) are important predictors of work group innovation but are differentially important in different phases of the innovation process. To summarize our predictions, we believe that demographic diversity is most relevant for idea generation, diversity in knowledge and skills is relevant for both idea generation and evaluation, and that diversity in personality and attitudes is more relevant for application and implementation of the ideas. Since very high levels of diversity can result in dysfunctional conflict within a team (Milliken & Martins, 1996), we use West's (2002) concept of requisite diversity to suggest that there are optimal levels of diversity that lead to effective creativity and innovation implementation.

Second, research has shown that the team's social network within the organization is related to the action phase of the implementation stage (e.g. Tsai, 2001; Tsai & Ghosal, 1998). Tsai & Ghoshal (1998) found that social interaction and trust among organizational teams were related to the extent of resource exchange among teams and product innovation within the company. Drawing on a network perspective on organizational learning, Tsai (2001) found that organizational groups can produce more innovations and enjoy better performance if they occupy central network positions that provide access to new knowledge developed by other units but only if units have the ability to successfully replicate the new knowledge. Thus, empirical evidence from network research suggests that teams that are connected and embedded within the social network of the organization can more readily garner support and resources that result in successful implementation of their innovative ideas.

Third, as noted earlier, Edmondson (1999) found that team leaders' coaching on learning behaviors promotes psychological safety in teams, an emergent state which we believe is crucial for efficient communication and interactions among team members, especially in the idea generation and idea evaluation phases of the innovative process.

External Demands

According West (2002), competition and uncertainty in the external environment will facilitate innovation. Hence, they are relevant in the transition phase of the creativity stage in that they set off the whole innovation process. West (2002) also noted that having time constraints pushes the teams to actively implement the innovation, but that external demands that are perceived as threats or constraints on the team are likely to result in fewer ideas of lower novelty being generated by a work group.

Next, while goal orientation is often conceptualized as a dispositional characteristic, its theoretical foundation also recognizes that it can be activated by a variety of situational factors (Dweck, 1986; Farr et al., 1993). For example, Martocchio (1994) found that trainees in a situation that elicited a learning goal orientation experienced a significant decrease in computer anxiety between pre- and post-training assessments, but not those trainees in a situation that led to a performance goal orientation. Also, trainees in the learning goal orientation condition experienced a significant increase in computer efficacy beliefs, while trainees in the performance goal orientation condition experienced a significant decrease in computer efficacy between the pre- and post-training assessments. Relatively few studies have looked at goal orientation in group or team research aside from considering fit between individual and team goal orientations (Kristof-Brown & Stevens, 2001). However, given that goal orientation is closely related to learning behaviors (Colquitt & Simmering, 1998; VandeWalle & Cummings, 1997), it is potentially a useful variable to be considered in team research, especially in the area of group innovation. For example, the proportion of members with learning or performance goal orientation within the group might predict the level of learning behaviors engaged by the team, which lead to amount and/or quality of group innovation. It might also be possible to enhance the emergent state of group efficacy by manipulating and inducing a learning goal orientation in the work environment (Martocchio, 1994; Winters & Latham, 1996).

Process Variables

As described earlier, Marks et al. (2001) provide an excellent taxonomy of team processes that are relevant for different phases of task accomplishment. In the following section, we briefly describe each process and how they might be relevant for various phases of the innovation process.

Transition Processes

Three activities included as transition processes are, mission analysis, goal specification, and strategy formulation. For mission analysis, the team's major objective is to identify the main tasks at hand, which is most crucial for the innovation process during the transition phase of the creative stage, where the team has to identify the target problem and interpret the relevant issues. Goal specification refers to "the identification and prioritization of goals and subgoals for mission accomplishment" and strategy formulation refers to "the development of alternative courses of action for mission accomplishment" (Marks et al., 2001, p. 365). We propose that both goal specification and strategy formulation are particularly relevant during the transition phase of the implementation stage, where the team has to develop concrete plans of actions to implement the ideas. Note that all the three transition processes help to cultivate the emergent state of shared cognitions or team mental model in the team. It is likely that cognitive consensus (Mohammed & Ringseis, 2001) is important for both of the transition

Action Processes

Action processes include activities such as system monitoring, goal monitoring, team monitoring, and coordination. Groups monitor progress toward their goals by assessing the discrepancy or gaps between the goals and the current situation, which is needed both when the team is generating ideas as well as when implementing the solutions. Team monitoring involves feedback, coaching, or assistance to other group members in relation to task accomplishment. Coordination refers to actions targeted at managing the interdependent actions of team members toward task accomplishment. System monitoring refers to activities that track the team's resources as well as the environmental conditions. Note that all four action processes are important when the team is engaging in the innovation implementation, whereas we propose that goal monitoring is the primary action process needed in the creativity stage.

Interpersonal Processes

Conflict management, affect management, and motivation/confidence building are processes that work groups engage in to manage their interpersonal relationships. Conflict management can include both preventive and reactive approaches to managing interpersonal disagreements or disputes. Note that effective conflict management is not a focus on conflict avoidance, because task-or idea-focused conflict can be constructive to the innovation process (West, 2002). Affect management involves monitoring and regulating the team members' level of emotional arousal. Both conflict and affect management are especially important during idea generation and evaluation given the higher probability of disagreement and conflict. Note that effective conflict and affect management will enhance the team's emergent state of group cohesion. Motivation and confidence building involves activities that preserve or enhance the team's sense of efficacy beliefs. Hence it will lead to the emergent state of group efficacy, which is crucial during the implementa-

tion stage since the team is likely to encounter high level of resistance and obstacles toward change.

Summary

We have provided an I-P-O model of the work group innovation process that identifies transition and action phases with each of two major stages of innovation: a creativity stage and an innovation implementation stage. The transition phases both involve primarily planning and evaluation tasks that guide later goal accomplishment. The action phases are both involved primarily in acts that directly contribute to goal accomplishment. Within the creativity stage, the transition phase consists of interpretation of issues and problem identification and the action phase consists of idea generation. Within the innovation implementation stage, the transition phase consists of the evaluation of the generated ideas as possible solutions and selection of the one(s) to implement and the action phase consists of the application of the idea(s) to the problem. We also provide our initial thoughts on the specific inputs, group processes, and emergent states that are important for each phase of our model. We trust that our model will generate innovative research examining and extending it in novel ways.

Acknoweledgment

Paul Tesluk would like to acknowledge support from the National Science Foundation (Grant #0115147) for his work on this chapter.

References

Agrell, A. & Gustafson, R. (1996). Innovation and creativity in work groups. In: M. A. West (Ed.), *Handbook of Work Group Psychology* (pp. 317–344). Chichester, U.K.: John Wiley.

Amabile, T. M. (1983). The social psychology of creativity: A componential conceptualization. *Journal of Personality and Social Psychology*, **45**, 357–376.

Amabile, T. M. & Conti, R. (1999). Changes in the work environment for creativity during downsizing. *Academy of Management Journal*, **42**, 630–640.

Ancona, D. G. & Caldwell, D. F. (1992). Bridging the boundary: External activity and performance in organizational teams. *Administrative Science Quarterly*, **37**, 634–665.

Anderson, N. & King, N. (1993). Innovation in organizations. In: C. L. Cooper & I. T. Robertson (Eds), *International Review of Industrial and Organizational Psychology*, Vol. 8, (pp. 1–34). Chichester, U.K.: John Wiley.

Argyris, C. (1982). *Reasoning, learning and action: Individual and organizational.* San Francisco: Jossey Bass.

Banbury, C. & Mitchell, W. (1995). The effect of introducing important incremental innovations on market share and business survival. *Strategic Management Journal*, **16**, 161–182.

Bantel, K. A. & Jackson, S. E. (1989). Top management and innovations in banking: Does the demography of the top team make a difference? *Strategic Management Journal*, **10**, 107–124.

Barrick, M. R. & Mount, M. K. (1991). The Big Five personality dimensions and job performance: A meta-analysis. *Personnel Psychology*, **44**, 1–26.

Brodbeck, F. (1996). Work group performance and effectiveness: Conceptual and measurement issues. In: M. A. West (Ed.), *The Handbook of Work Group Psychology* (pp. 285–316). Chichester, U.K.: John Wiley.

Brown, R. (1990). 'Politeness theory: Exemplar and exemplary'. In: I. Rock (Ed.), *The Legacy of Solomon Asch: Essays in Cognitive and Social Psychology: 23–37*. Hillsdale, NJ: Erlbaum.

Burningham, C. & West, M. A. (1995). Individual, climate and group interaction processes as predictors of work team innovation. *Small Group Research*, **26**, 106–117.

Campion, M. A., Medsker, G. J. & Higgs, A. C. (1993). Relations between work group characteristics and effectiveness: Implications for designing effective work groups. *Personnel Psychology*, **46**, 823–850.

Cohen, S. G. & Bailey, D. E. (1997). What makes teams work: Group effectiveness research from the shop floor to the executive suite. *Journal of Management*, **23**, 239–290.

Colquitt, J. A. & Simmering, M. J. (1998). Conscientiousness, goal orientation and motivation to learn during the learning process: A longitudinal study. *Journal of Applied Psychology*, **83**, 654–665.

Cooper, R. & Foster, M. (1971). Sociotechnical systems. *American Psychologist*, **26**, 467–474.

De Dreu, C. K. W. & West, M. A. (2001). Minority dissent and team innovation: The importance of participation in decision making. *Journal of Applied Psychology*, **86**, 1191–1201.

Drach-Zahavy, A. & Somech, A. (2001). Understanding team innovation: The role of team processes and structures. *Group Dynamics*, **5**, 111–123.

Dweck, C. S. (1986). Motivational processes affecting learning. *American Psychologist*, **41**, 1040–1048.

Edmondson, A. (1996). Learning from mistakes is easier said than done: Group and organizational influences on the detection and correction of human error'. *Journal of Applied Behavioral Science*, **32**, 5–32.

Edmondson, A. (1999). Psychological safety and learning behavior in work teams. *Administrative Science Quarterly*, **44**, 350–383.

Farr, J. L., Hofmann, D. A. & Ringenbach, K. L. (1993). Goal orientation and action control theory: Implications for industrial and organizational psychology. In: C. I. Cooper & I. T. Robertson (Eds), *International Review of Industrial and Organizational Psychology* (Vol. 8, pp. 193–232). Chichester, U.K.: John Wiley.

Gist, M. E., Locke, E. A. & Taylor, M. S. (1987). Organizational behavior: Group structure, processes, and effectiveness. *Journal of Management*, **13**, 237–257.

Gladstein, D. (1984). Groups in context: A model of task group effectiveness. *Administrative Science Quarterly*, **29**, 499–517.

Guzzo, R. A. & Shea, G. P. (1992). Group performance and intergroup relations in organizations. In: M. D.Dunnette & L. M. Hough (Eds), *Handbook of Industrial and Organizational Psychology*, (Vol. 3, pp. 269–313). Palo Alto, CA: Consulting Psychologists Press.

Hackman, J. R. (1987). The design of work teams. In: J. W. Lorsch (Ed.), *Handbook of Organizational Behavior* (pp. 315–342). Englewood Cliffs, NJ: Prentice-Hall.

Hackman, J. R. & Morris, C. G. (1975). Group tasks, group interaction processes, and group performance effectiveness: A review and proposed integration. In: L. Berkowitz (Ed.), *Advances in Experimental Social Psychology* (Vol. 8, pp. 45–99). New York: Academic Press.

Hackman, J. R. & Oldham, G. (1975). Development of the job diagnostic survey. *Journal of Applied Psychology*, **60**, 159–170.

Hamel, G. & Prahalad, C. K. (1994). Competing for the future. *Harvard Business Review*, **72** (4), 122–128.

Harrison, D. A., Price, K. H. & Bell, M. P. (1998). Beyond relational demography: Time and the effects of surface- and deep-level diversity on work group cohesion. *Academy of Management Journal*, **41**, 96–107.

Harrison, D. A., Price, K. H., Gavin, J. H. & Florey, A. T. (2002). Time, teams, and task performance: Changing effects of surface- and deep-level diversity on group functioning. *Academy of Management Journal*, **45**, 1029–1045.

Hedlund, G. (1994). A model of knowledge management and the N-form corporation. *Strategic Management Journal*, **15**, 73–90.

Hinsz, V. B., Tindale, R. S. & Vollrath, D. A. (1997). The emerging conceptualization of groups as information processors. *Psychological Bulletin*, **121**, 43–64.

Hogan, R. & Ones, D. (1997). Conscientiousness and integrity at work. In: R. Hogan, J. Johnson & S. Briggs (Eds), *Handbook of Personality Psychology* (pp. 873–886). New York: Academic Press.

Hyatt, D. E. & Ruddy, T. M. 1997. An examination of the relationship between work group characteristics and performance: Once more into the breech. *Personnel Psychology*, **50**, 553–585.

Jackson, S. E. (1996). The consequences of diversity in multidisciplinary work teams. In: M. A. West (Ed.), *Handbook of Work Group Psychology* (pp. 53–76). Chichester, U.K.: John Wiley.

Jehn, K. A. (1995). A multimethod examination of the benefits and detriments of intragroup conflict. *Administrative Science Quarterly*, **40**, 256–282.

Katz, R. (1982). The effects of group longevity on project communication and performance. *Administrative Science Quarterly*, **27**, 81–104.

King, L. A., Walker, L. & Broyles, S. J. (1996). Creativity and the five factor model. *Journal of Research in Personality*, **30**, 189–203.

Kozlowski, S. W. J., Gully, S. M., Nason, E. & Smith, E. M. (2001). Developing adapting teams: A theory of compilation and performance across levels and time. In: D. R. Ilgen & E. D. Pulakos (Eds), *The Changing Nature of Performance: Implications for Staffing, Motivation, and Development* (pp. 240–292). San Francisco, CA: Jossey-Bass.

Kristof-Brown, A. & Stevens, C. K. (2001). Goal orientation in project teams: Does the fit between members' personal mastery and performance goals matter? *Journal of Applied Psychology*, **86**, 1083–1095.

Lee, F. (1997). 'When the going gets tough, do the tough ask for help?' Help seeking and power motivation in organizations'. *Organizational Behavior and Decision Processes*, **72**, 336–363.

Marks, M. A., Mathieu, J. E. & Zaccaro, S. J. (2001). A temporally based framework and taxonomy of team processes. *Academy of Management Review*, **26**, 356–376.

Martocchio, J. J. (1994). Effects of conceptions of ability, and academic achievement. *Journal of Applied Psychology*, **79**, 819–825.

McDonough, E. F., III & Barczak, G. (1992). The effects of cognitive problem-solving orientation and technological familiarity on faster new product development. *Journal of Product Innovation Management*, **9**, 44–52.

McGrath, J. E. (1984). Groups: *Interaction and Performance*. Englewood Cliffs, NJ: Prentice Hall.

McGrath, J. E. (1991). Time, interaction, and performance (TIP): A theory of groups: *Small Group Research*, **22**, 147–174.

Milliken, F. & Martins, L. (1996). Searching for common threads: Understanding the multiple effects of diversity in organizational groups. *Academy of Management Review*, **21**, 402–433.

Mohammed, S. & Dumville, B. C. (2001). Team mental models in a team knowledge framework: Expanding theory and measurement across disciplinary boundaries. *Journal of Organizational Behavior*, **22**, 89–106.

Mohammed, S. & Ringseis, E. (2001). Cognitive diversity and consensus in group decision making: The role of inputs, processes, and outcomes. *Organizational Behavior and Human Decision Processes*, **85**, 310–335.

Moreland, R. L. & Myaskovsky, L. (2000). Exploring the performance benefits of group training: Transactive memory or improved communication? *Organizational Behavior and Human Decision Processes*, **82**, 117–133.

Mumford, M. D. & Gustafson S. B. (1988). Creativity syndrome. Integration, application and innovation. *Psychological Bulletin*, **103**, 27–43.

Nahapiet, J. & Ghoshal, S. (1998). Social capital, intellectual capital, and the organizational advantage. *Academy of Management Review*, **23**, 242–266.

Nickerson, R. S. (1999). How we know—and sometimes misjudge—what others know: Imputing one's own knowledge to others. *Psychological Bulletin*, **125**, 737–759.

Patterson, F. (2002). Great minds don't think alike? Person-level predictors of innovation at work. In: C. I. Cooper & I. T. Robertson (Eds), *International Review of Industrial and Organizational Psychology* (Vol. 17, pp. 115–144). Chichester, U.K.: John Wiley.

Pinto, J. K. & Prescott, J. E. (1988). Variations in critical success factors over the stages in the project life cycle. *Journal of Management*, **14**, 5–18.

Pritchard, R. D., Jones, S. D., Roth, P. L., Stuebing, K. K. & Ekeberg, S. E. (1988). Effects of group feedback, goal setting, and incentives on organizational productivity. *Journal of Applied Psychology*, **73**, 337–358.

Robinson, S. L. (1996). 'Trust and breach of psychological contract'. *Administrative Science Quarterly*, **41**, 574–599.

Stasser, G. (1991). Pooling of unshared information during group discussion. In: S. Worchel & W. Wood (Eds), *Group Process and Productivity*. Newbury Park, CA: Sage.

Stasser, G., Taylor, L. A. & Hanna, C. (1989). Information sampling in structured and unstructured discussions of three- and six-person groups. *Journal of Personality and Social Psychology*, **57**, 67–78.

Sternberg, R. J. (Ed.) (1999). *Handbook of Creativity*. New York: Cambridge University Press.

Thomas, J. B., Sussman, S. W. & Henderson, J. C. (2001). Understanding 'strategic learning': Linking organizational learning, knowledge management, and sensemaking. *Organization Science*, **12**, 331–345.

Tjosvold, D. (1982). Effects of approach to controversy on superior's incorporation of subordinates' information in decision making. *Journal of Applied Psychology*, **67**, 189–193.

Tjosvold, D., Wedley, W. C. & Field, R. H. G. (1986). Constructive controversy, the Vroom-Yetton model, and managerial decision making. *Journal of Occupational Behavior*, **7**, 125–138.

Tsai, W. (2001). Knowledge transfer in intraorganizational networks: Effects of network position and absorptive capacity on business unit innovation and performance. *Academy of Management Journal*, **44**, 996–1004.

Tsai, W. & Ghoshal, S. (1998). Social capital and value creation: The role of intrafirm networks. *Academy of Management Journal*, **41**, 464–476.

VandeWalle, D. & Cummings, L. L. (1997). A test of the influence of goal orientation on the feedback seeking process. *Journal of Applied Psychology*, **82**, 390–400.

Wegner, D. M. (1987). Transactive memory: A contemporary analysis of the group mind. In: B. Mullen & G. R. Goethals (Eds), *Theories of Group Behavior* (pp. 185–205). New York: Springer-Verlag.

West, M. A. (1990). The social psychology of innovation in groups. In: M. A. West & J. L. Farr (Eds), *Innovation and Creativity at Work* (pp. 309–333). Chichester, U.K.: John Wiley.

West, M. A. (2002). Sparkling fountains or stagnant ponds: An integrative model of creativity and innovation implementation in work groups. *Applied Psychology: An International Review*, **51**, 355–424.

West, M. A. & Anderson, N. R. (1996). Innovation in top management teams. *Journal of Applied Psychology*, **81** (6), 680–693.

West, M. A., Borrill, C. S. & Unsworth, K. L. (1998). Team effectiveness in organizations. In: C. I. Cooper & I. T. Robertson (Eds), *International Review of Industrial and Organizational Psychology* (Vol. 13, pp. 1–48). Chichester, U.K.: John Wiley.

Wiersema, M. F. & Bantel, K. A. (1992). Top management team demography and corporate strategic change. *Academy of Management Journal*, **35**, 91–121.

Winters, D. & Latham, G. P. (1996). The effect of learning vs. outcome goals on a simple vs. complex task. *Group and Organization Management*, **21**, 236–250.

Wolfe, B. (1994). Organizational innovation: Review, critique and suggested research directions. *Journal of Management Studies*, **31**, 405–431.

The International Handbook on Innovation
Edited by Larisa V. Shavinina

Creativity and Innovation = Competitiveness?
When, How, and Why

Elias G. Carayannis and Edgar Gonzalez

School of Business and Public Management, The George Washington University, USA

Abstract: In this chapter, we propose to look at both for-profit and not-for-profit entities to examine:

(a) when, how, and why creativity and innovation occur;
(b) how and why creativity triggers innovation and vice versa; and
(c) what are the connections and implications for competitiveness of the presence or absence of creativity and innovation using empirical findings from both the public and private sectors.

We combine literature sources (including those of the authors) as well as field interviews on the practice and implications of creativity and innovation from the perspective of competitiveness.

Keywords: Innovation; Creativity; Competitiveness; Institutional learning; Entrepreneurial learning; Public-private sector partnerships.

Introduction

Zum sehen geboren; Zum schauen bestellt:
{Born to see; Meant to look}

(Faust, Goethe)

In Greek, the word for creator also denoting God, is the word *poet*. This underlines the dynamic underlying creativity, in that it encompasses both a structured, disciplined, scientific as well as an artistic element: one could say that creativity emerges from the interplay and interfacing of science and art—in a way, it could be conceived of as being *the art of science and the science of art* (Carayannis, 1998b).

The relationship between creativity, innovation and competitiveness at a very basic level appears readily apparent: creativity is a necessary (but not sufficient) factor enabling innovation, and innovation of different types can improve national economic competitiveness. This relationship is extremely significant, however, in that it links three levels of analysis: creativity (mostly at *the individual level or micro level*), innovation (mostly at *the organizational or meso level*), and competitiveness (mostly at *the national or macro level*) (see Figs 1, 3 and 8). Understanding the specific links and dynamics contained within this relationship may provide significant insight into the ability of nations to build and sustain conditions of competitiveness.

In this chapter, we explore what creativity and innovation are and what their significance and role is for people and organizations (and by extension potentially nations) as ingredients, catalysts, and possibly inhibitors of competitiveness. We look at both *for-profit and not-for-profit* entities to examine:

(a) when, how, and why creativity and innovation occur;
(b) how and why creativity triggers innovation and vice versa; and
(c) what are the connections and implications for competitiveness are of the presence or absence of creativity and innovation.

Creativity

Management is, all things considered, the most creative of all arts.
It is the art of arts, because it is the organizer of talent.

(Jean-Jacques Servan-Schreiber)

Starting at the individual level, **creativity** may be defined as the capacity to '*think out of the box*', to think laterally, to perceive, conceive, and construct ideas, models, and constructs that exceed or supersede established items and ways of thinking and perceiving.

Creativity is related to the capacity to imagine, since it requires the creator to perceive future potentials that are not obvious based on current conditions. From a cognitive perspective, creativity is the ability to perceive new connections among objects and concepts—in effect, reordering reality by using a novel framework for organizing perceptions.

Creative types such as artists, scientists, and entrepreneurs often exhibit attributes of *obsessed maniacs* and *clairvoyant oracles* (Carayannis, 1998–2002, George Washington University Lectures on Entrepreneurship) as well as the capacity and even propensity for creative destruction that is how Joseph Schumpeter qualified innovation. Albert Scentzgeorgi, a Nobel Prize laureate, defined creativity as *'seeing what everyone sees and thinking what no one has thought before'*.

Innovation

Discovery consists of looking at the same thing as everyone else and thinking something different.
Albert Szent-Gyorgyi—Nobel Prize Winner

Innovation is a word derived from the Latin, meaning to introduce something new to the existing realm and order of things or to change the yield of resources as stated by J. B. Say quoted in Drucker (Drucker, 1985).

In addition, innovation is often linked with creating a sustainable market around the introduction of new and superior product or process. Specifically, in the literature on the management of technology, technological innovation is characterized as the introduction of a new technology-based product into the market:

Technological innovation is defined here as a situationally new development through which people extend their control over the environment. Essentially, technology is a tool of some kind that allows an individual to do something new. A technological innovation is basically information organized in a new way. So technology transfer amounts to the communication of information, usually from one organization to another (Tornazky & Fleischer, 1990).

The broader interpretation of the term 'innovation' refers to an innovation as an "idea, practice or material artifact" (Rogers & Shoemaker, 1971, p. 19) adopted by a person or organization, where that artifact is "perceived to be new by the relevant unit of adoption" (Zaltman et al., 1973). Therefore, innovation tends to change perceptions and relationships at the organizational level, but its impact is not limited there. Innovation in its broader socio-technical, economic, and political context, can also substantially impact, shape, and evolve ways and means people live their lives, businesses form, compete, succeed and fail, and nations prosper or decline (see Fig. 1).

Specifically, Fig. 1 attempts to illustrate the nature and dynamics of an emerging globalization framework in which creativity and innovation—as enabler of technological effort in manufacturing and as an engine of industrial development—can lead to improved competitiveness and sustained development. On the other hand, lack of creativity and innovation constitutes a factor for failure in manufacturing performance and, as a result, is a factor for failure in economic performance, too. For those countries in which creativity and innovation is applied effectively, globalization can be an engine of beneficial and sustainable economic integration. However, globalization can be a powerful force for deprivation, inequality, marginalization and economical disruption in those non-competitive countries.

Government or market success or failure is determined by how they take advantage of the four major elements that shape the setting for creativity, innovation and competitiveness in the globalized world:

(1) the coordination and synergy in the relationship between governments, enterprises, research laboratories and other specialized bodies, universities and support agencies for small and medium enterprises (SMEs);
(2) the power of information and communication technology;
(3) the efficiency that managerial and organizational systems can bring to production and commerce; and
(4) the international agreements, rules and regulations.

All the four elements of this framework will impact on creativity and innovation at the micro level (firm level) as well as on innovation and competitiveness at the macro level (industry, national, global).

Competitiveness

Competitiveness is the capacity of people, organizations, and nations to achieve superior outputs and especially outcomes, and in particular, *to add value*, while using the same or lower amounts of inputs (see Fig. 2).

Moreover, entrepreneurial value-adding and entrepreneurial learning by doing, learning by analogy, and learning by failing, does not belong to the realm of for-profit entities only, but also in the domain of not-for-profit entities. This is shown in Fig. 2 with the overlapping circles connecting creativity and innovation activities across for-profits and not-for-profits.

The standard for judging whether these results are 'superior' can encompass both prior capabilities of a particular organization or nation and a comparison with other organizations or nations. The critical assumption of competitiveness, then, is that it is accomplished through a process of organizational improvement, where the institutions in an economy leverage people,

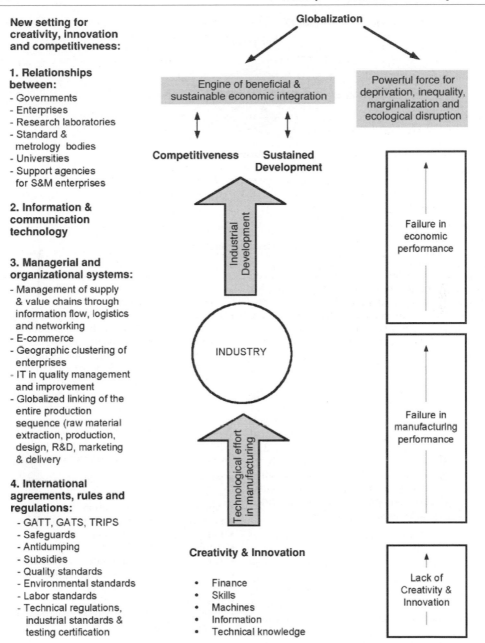

Figure 1. The CIC value chain: Global and local perspectives.

knowledge and technologies to rearrange relationships and enable higher states of production.

When, Why and How Creativity Arises

Imagination is more important than knowledge. To raise new questions, new possibilities, to regard old problems from a new angle, requires creative imagination and marks real advance in science.

(Albert Einstein)

The problem with 'creativity' is that it is an intangible. While we generally know when something is creative, we often don't know why. It seems difficult to articulate a precise definition of the topic.

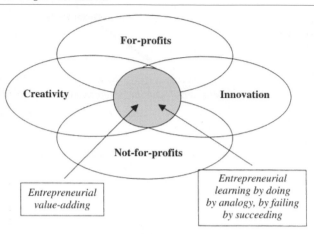

Figure 2. CIC: Value-adding and learning topology.

Aristotle, for example, suggested that inspiration involved a form of madness whereby great insights began as a result of a person's own thoughts progressing through a series of associations (Dacey & Lennon, 1998, p. 17). This view of the creative individual as mad, or potentially so, continued through the nineteenth century.

Freud believed creative ability was a personality trait that tends to become fixed by experiences in the first five years of life (Dacey & Lennon, 1998, p. 36). He maintained that creative expression was a means of expressing inner conflicts that otherwise would result in neuroses. Creativity was a sort of emotional purgative that kept men sane (Kneller, 1965, p. 21). During the first half of the twentieth century, B. F. Skinner and other behaviorists considered creative production to be strictly the result of 'random mutation' and a product of appropriate reinforcers provided by society (Dacey & Lennon, 1998, p. 138).

Cognitive View of Creativity (Personal Creativity)
Kneller (1965, p. 3) suggested that definitions of creativity seem to fall into four categories. Creativity is considered from the standpoint of the person who creates, in terms of mental processes, in terms of its products, or focuses on environmental and cultural influences. He states that "an act or an idea is creative not only because it is novel, but also because it achieves something that is appropriate to a given situation" (1965, p. 6). We create when we discover and express something that is new to us. The operative phrase is 'new to us'; even if another person has discovered something, it is still creativity if we have re-discovered it for ourselves.

Amabile (1996, p. 33) appears to provide the most complete definition available to date. She suggests a two-part definition of creativity:

(1) that a product or response is creative to the extent that appropriate observers independently agree it is

creative. Appropriate observers are those familiar with the domain in which the product or the response articulated (p. 33); and

(2) that a product or response ₤ ill be judged as creative to the extent that it is both a novel and appropriate task at hand, and the task is heuristic rather than algorithmic. She defines algorithmic tasks as those for which the path to the solution is clear and straightforward; heuristic tasks are those for which algorithms must be developed. She calls these tasks 'problem discovery' (p. 35).

Amabile (1996, p. 90) also lists personality traits that appear repeatedly in summaries of empirical work on the characteristics of creative persons:

- High degree of self-discipline in matters concerning work.
- Ability to delay gratification.
- Perseverance in the face of frustration.
- Independence of judgment.
- A tolerance for ambiguity.
- A high degree of autonomy.
- An absence of sex role stereotyping.
- An internal locus of control.
- A willingness to take risks.
- A high level of self-initiated, task-oriented striving for excellence.

Of their nine principal traits, it may be helpful to further define three: stimulus freedom, functional freedom, and flexibility. Stimulus freedom (Getzels, Taylor, Torrance, cited by Dacey & Lennon, 1998, p. 100) occurs when people are likely to bend the rules to meet their needs, if the stated rules of a situation interfere with their creative ideas. Functional freedom is the ability to use items for other creative, or unique uses. Dacey and Lennon contend that the more education a person has, the more rigid his or her perception of function is likely to become. Also, because education tends to encourage complexity of thought, this may produce a convoluted thinking style which works against producing simple ideas—the ones that comprise many of the world's greatest solutions. Flexibility is the capacity to see the whole of a situation, rather than just a group of uncoordinated details.

Gestalt psychologists believed that creative problem-solving is similar in important ways to perception. They argued that it is primarily a reconstruction of gestalts, or patterns, that are structurally deficient. Creative thinking begins with a problematic situation that is incomplete in some way. The thinker grasps this problem as a whole. The dynamics of the problem itself and the forces and tensions within it, set up similar lines of stress within his or her mind. By following these lines of stress, the thinker arrives at a solution that restores harmony of the whole (Kneller, 1965, p. 27). Restructuring and productive thinking often do not occur because problem-solvers tend to become fixated on attempting to apply past experience to the problem, and thus do not deal with the problem on its own terms (Weisberg, 1992, p. 51).

Creativity in an Organizational Context

Culture is the invisible force behind the tangibles and observables in any organization, a social energy that moves people to act. Culture is to the organization what personality is to the individual —a hidden, yet unifying theme that provides meaning, direction, and mobilization.
(Killman, R., *Gaining Control of the Corporate Culture*, 1985)

In the business context, creativity now is championed by certain authors as the critical element enabling change in organizations. Kao (1996, xvii) defines creativity as:

the entire process by which ideas are generated, developed and transformed into value. It encompasses what people commonly mean by innovation and entrepreneurship. In our lexicon, it connotes both the art of giving birth to new ideas and the discipline of shaping and developing those ideas to the stage of realized value.

Kao views creativity as the "result of interplay among the person, the task, and the organizational context" (cited in Gundry et al., 1994). Drazin et al. (1999) agree with this assertion. They conclude that creativity is both an individual and group level process. Complex, creative projects found within large organizations require the engagement of many individuals, rather than just a few. It is often difficult to assign credit to any one individual in a creative effort (Sutton & Hargadon, cited in Drazin et al., 1999). Creativity, they believe, is an iterative process whereby individuals develop ideas, interact with the group, work out issues

in solitude, and then return to the group to further modify and enhance their ideas. Their sense making perspective of creativity illustrates the notion that individuals are influenced in their creative efforts by such factors as conflict, political influence, and negotiated order at the group level.

Environmental Effects on Creativity

When I am, as it were, completely myself, entirely alone, and of good cheer ... it is on such occasions that my ideas flow best and most abundantly. *Whence and how* they come, I know not; nor can I force them. Those ideas that please me I retain in memory.
(W. A. Mozart, quoted in Brewster Ghiselin, 1952, p. 34)

Woodman & Schoenfeldt (1990, p. 18) stress the importance of social environment. They state: "it is clear that individual differences in creativity are a function of the extent to which the social and contextual factors nurture the creative process. Research on creativity has led to a recognition of the fact that the kind of environment most likely to produce a well-adjusted person is not the same as the kind of environment most likely to produce a creative person". Because of the dearth of research in this area, we will briefly examine the factors through an ever-widening circle of social influences—from family to culture.

Amabile (1996, p. 179) reports that there appear to be three social factors that are important for creative behavior:

- Social facilitation (or social inhibition), brought about by the presence of others: She reports that the presence of others can impair performance on poorly learned or complex tasks, but enhance performance on well-learned or simple tasks (p. 181). In addition, there is much evidence that subjects perform more poorly on idea-production tests when they work together than when they work alone.
- *Modeling, or the imitation of observed behavior*: Research suggests that a large number of creative models in one generation will stimulate general creative production in the next generation (Simonton, cited on p. 189). At the individual level, the pattern of influence seems to be complex. At the highest levels of creative eminence, modeling may be relatively unimportant. In addition, although exposure to creative models may stimulate early high-level productivity, it may be important at some point to go beyond the examples set by one's mentors.
- *Motivational orientation, or an individual's intrinsic or extrinsic approach to work*: Studies suggest that intrinsic orientation leads to a preference for challenging and enjoyable tasks, whereas an extrinsic orientation leads to a preference for simple, predictable tasks (p. 192).

591

There is some evidence that cultures may promote or inhibit creativity. Arieti (1976, p. 303) explored cultural influences on creativity and suggests that the potentiality for creativity is deemed much more frequent than its occurrence. Some cultures promote creativity more than others and he labeled these cultures as 'creativogenic'. He held that people become creative (or to use his term, 'genius') because of the juxtaposition of three factors:

(1) The culture is right. He uses the example that the airplane would not have been invented if gasoline had not been invented.
(2) The genes are right. The person's intelligence, which is known to be genetic, must be high. Creativity, which may or may not be genetic, must also be high.
(3) The interactions are right. He offers the example of Freud, Jung, and Adler. If Jung and Adler had not had Freud to compete over, and against, it is questionable whether either Jung or Adler would hold such a high position in psychology today.

Hofstede (1980, p. 43), in a study of the culture of 40 independent nations, found four criteria by which their cultures differed: power distance, uncertainty avoidance, individualism-collectivism and masculinity-feminity. These dimensions appear to have a powerful influence on the 'collective mental programming of the people in an environment'. They are also grounded in our collective cultural history. Americans, for example, tend to exhibit high individualism, small power distance, and weak uncertainty avoidance. That they show these tendencies reflects American history which has placed high value on equality, independence, and willingness to take risks.

This cultural influence is qualitatively different than the social influences mentioned in previous creativity models. For want of a better term, we call it 'cultural embeddedness', because it implies more than a society's norms, values, and mores. It is what defines our reality. In light of this additional component, we are proposing a new model of creativity which not only illustrates the components of creativity, but the creative process as well. In this model, personality and cognitive factors interact with the individual and vice versa. The social environment interacts with the three factors and vice versa; the individual initiates and participates in the creative process. Cultural embeddedness influences not only all of the creative factors but all steps of the creative process.

When, Why and How Innovation Arises

Innovation is creative destruction.

The success of everything depends upon *intuition*, the capacity of seeing things in a way which afterwards proves to be true.

(Joseph Schumpeter)

If creativity can be seen as a process and a product or event, the use of the term innovation in terms of creativity seems to muddy the waters. If one consults the business literature, 'innovation' and 'creativity' appear to be used interchangeably.

Innovation is seen as the panacea for competing successfully in today's global marketplace, but in much of the literature the concept is a vague one. Managers are told they must promote innovation, but they are not given the specifics of how this is to be accomplished. The articles often cite one or two examples of companies that are profiting from 'innovation', and the reader is left to grapple with the mechanics of extrapolating useful information that is transferable to his or her own situation.

In an extensive review of popular and academic business literature, we found that the information provided ranged from the esoteric notion of promoting an innovative organizational climate to concrete steps in creative problem-solving. Given the target audience, some of the information provided is aimed at practical applications. Apart from the confusion over terminology, it appears that the literature may be divided into the following broad categories: the nature of innovation, the individual and creativity/innovation, how to promote innovation within organizations, the political nature of innovation, and enhancing creativity.

Drucker also links innovation with entrepreneurship by noting that entrepreneurs appear to not have a certain kind of personality in common, but a commitment to the systematic practice of innovation. He describes innovation as 'the means by which the entrepreneur either creates new wealth-producing resources or endows existing resources with enhanced potential for creating wealth' (Drucker, 1998, p. 149).

In her study of the longevity of successful pharmaceutical companies, Henderson (1994) found that the companies constantly challenged conventional wisdom and stimulated a dynamic exchange of ideas. 'They focused on continuously refurbishing the innovative capabilities of the organization. They actively managed their companies' knowledge and resources' (Henderson, 1994, p. 102). Gundry et al. (1994) further defines organizational innovation by stating that: "Organizations that encourage employee creativity share certain characteristics: They capitalize on employee attributes, enhance employees' conceptual skills, and cultivate an organizational culture that fosters experimentation and stimulates creative behavior".

Chesbrough & Teece (1996) attack the problem of definition from a different viewpoint. They state that there are two types of innovation. Autonomous innovations are those that can be pursued independently from other innovations. They use the example of a new turbocharger to increase horsepower in an engine, which can be developed without the complete redesign of the engine or the rest of the car. Systemic innovations, on the other hand, are those which must be accomplished along with related complementary innovations. Redesigning a workflow process in a factory would be an example of a systemic innovation, because it requires changes in supplier management, personnel, and information technology.

The Relationship Between Creativity and Innovation

> *Learning* is formulating intelligence. *Deciding* is implementing intelligence.
> *Entrepreneurship* is organized abandonment.
>
> (J. B. Say)

To promote and support innovation within organizations, a corporate culture must be adopted that will accept and defend the vagaries of individual and group creativity. Leaders must ensure that employees believe the expectation that innovation is part of their jobs.

They do this by providing a safe environment, where there is freedom to fail. A high tolerance for failure allows for trial and error, experimentation, and continuous learning (Ahmed et al., 1999).

People, culture, and technology serve as the institutional, market, and socio-economic 'glue' that binds, catalyzes, and accelerates interactions and manifestations between creativity and innovation as shown in Fig. 3, along with public-private partnerships, international R&D consortia, technical/business/legal standards such as intellectual property rights as well as human nature and the 'creative demon'. The relationship is highly non-linear, complex and dynamic, evolving over time and driven by both external and internal stimuli and factors such as firm strategy, structure, and performance as well as top-down policies and bottom-up initiatives that act as enablers,

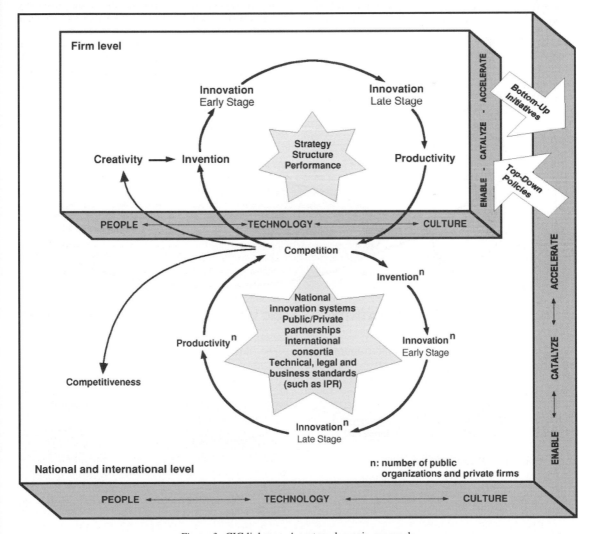

Figure 3. CIC linkages: A system dynamic approach.

catalysts, and accelerators for creativity and innovation that leads to competitiveness.

C.III. Creativity, Innovation and Competitiveness: Concepts and Empirical Findings from the Public and Private Sectors

> Leaders in learning organizations have the ability to conceptualize strategic insights, so that they can become public knowledge open to public debate and improvement
>
> (Peter Senge, *The Fifth Discipline*, 1990, p. 356)

Creativity is the result of inspiration and cognition, the liberation of talent in a nurturing and even provocative context and it is mostly an intensely private and individualistic process—it operates at the *micro* (individual) level (see Fig. 4). Innovation is a team effort and takes place at the *meso* (group/organizational) level, as it needs to combine the blessings of creativity

with the fruits of invention and the propitiousness of the market conditions—timing, selection, and sequencing are important as well as 'divine providence', obsession and clairvoyance. Competitiveness is the edifice resting on the pillars of creativity, invention, and innovation and it materializes at the *macro* (industry/market/national/regional) level (see Fig. 4).

We are inspired from the Nobel-prize-winning discovery of *the double helix* as nature's fundamental scaffold and evolutionary competence, to elucidate and articulate the nature and dynamics of the inter-relationship between and among creativity, innovation and competitiveness and their evolutionary pathways. We attempt to do that by means of the *Creativity, Innovation and Competitiveness Double Helix (CIC2Helix)* (Fig. 4) in which one strand represents the flow and record of creativity and the other that of competitiveness. At any point along their evolution, these two strands are linked by the value-adding chain

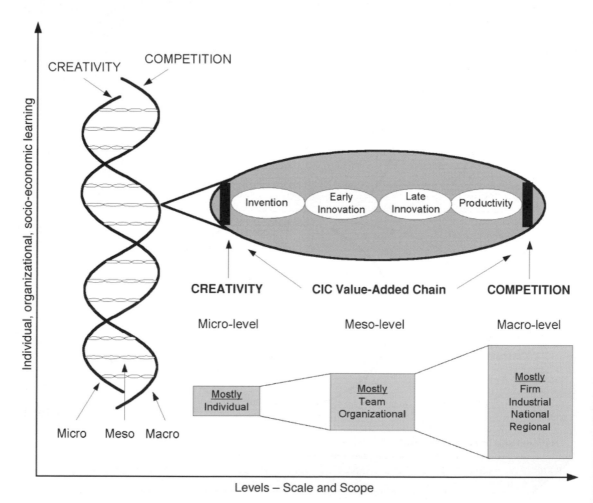

Figure 4. The CIC spiral and value-added chain.

of creativity, invention, early-stage innovation, late-stage innovation, productivity, competitiveness (*CI3PC value-adding chain*). This chain serves as the catalyst and accelerator of *social, economic, organizational and individual learning* and *meta-learning* which allows for the CI2C Helix to continue evolving by enhancing and advancing the effectiveness of generated knowledge and the efficiency of knowledge transmission and absorption. In so doing, *knowledge economies of scale and scope* are attained at increasingly higher levels, allowing for more to be accomplished with less, faster, cheaper and better. These gains are manifested in diverse ways at the micro, meso and macro levels, namely higher standards of living, more competitive firms, more robust economies, and accelerated and more sustainable development trends.

We further attempt to shape and corroborate our perspectives with field research questionnaires that were responded to by public and private sector managers from a number of countries around the world (see Appendix), dealing with issues related to drivers, critical success and failure factors and measures of creativity, innovation and competitiveness.

Our overall findings from talking to practitioners from the public, and the private sector, point to the following main challenges that encompass as well potential opportunities for growth and development if innovation and creativity is allowed to blossom:

- Failure of imagination (where the policy makers and managers fail to envision the future and confront the present).
- Failure of courage (where the decision makers are too afraid to confront real challenges and as a result, they shy away from critical reality checks).
- Fear to succeed (where the decision makers and other stakeholders are individually or institutionally hesitant to embrace the potential and changes of success—either consciously or subconsciously and thus undermine or undercut their own efforts).
- Fear to fail (where the policy makers and managers are so concerned with failure and fail to realize that one can not avoid risk but that one can only manage it as best possible; as a result, they end up mismanaging risk and engendering processes and trends that lead to failure even when not necessary).
- Too short term focus on earnings (the 'tyranny of the marketplace' often precipitates decisions that are poor from a mid- to long-term perspective and only serve as short-term expediencies). In the case of the public sector, the equivalent case is that of politicians that only care about or are forced to focus only on winning the next elections.
- Strategic versus Tactical choices and actions (as a result of all the above-mentioned 'pathologies' of decision-making, tactical choices trigger actions that

often preempt or impede strategic choices and actions).

We further organize and present our findings by key themes:

Theme 1: Key dimensions of Innovation and Creativity.

Theme 2: Drivers of Innovation—Catalysts and Inhibitors.

Theme 3: A glimpse of the current situation in several countries—Challenges and Opportunities.

- **Theme 1: Key Dimensions of Innovation and Creativity**

In the public and private sectors, innovation can be understood as a way of rethinking and reshaping government, repositioning public service/public organizations, managing/leading the change process, restructuring programs and service delivery, redesigning and improving service delivery for citizens, redesigning accountability frameworks and performance measures and revitalizing public service providers and private firms (see Table 1).

Based on the responses we received from public and private sector practitioners from a number of countries, *innovation* is seen as encompassing the following attributes with the most important ones listed first:

- Inventing something new.
- Seeing something from a different perspective.
- Introducing changes.
- Improving something that already exists.
- Spreading new ideas.
- Performing an existing task in a new way.
- Generating new ideas only.
- Following the market leader.
- Adopting something that has been successfully tried elsewhere.
- Attracting innovative people.

- **Theme 2: Drivers of Innovation**

As we can see from Figs 1, 5 and 6, there are a host of internal and external enablers, catalysts, and accelerators of innovation since it is a complex, non-linear, and interactive process with human, technological and cultural underpinnings. Our empirical findings are reflected in these figures and are further discussed below.

According to the empirical findings, there are factors that act as catalysts or inhibitors for creativity, innovation and competitiveness in the public and in the private sector:

Catalysts

(1) Leadership, vision, strategic plan (with the right goals). Relative organizational autonomy and certain degree of authority to innovate.
(2) Innovation/creativity rewards system in place.

(3) Protection of intellectual property rights (IPR).

(4) Propitious organizational environment for converting tacit ideas and knowledge into explicit proposals for improvement: open and frequent communication and dialogue, strategic rotation between different functions and technologies and access to information (Nonaka & Takeuchi, 1995).

(5) The right mix of people and esprit de corps manifested in well-functioning teams.

(6) Sense of urgency (if you feel you're up against it, as happens frequently in the private sector, you become very innovative and competitive).

(7) One innovates when in need or 'necessity is the mother of invention': review of innovation experiences in governments during the last 25 years shows that innovation occurred in almost all of the cases in an environment of financial scarcity and even crisis (Glor, 2001).

(8) Willingness of governments to innovate—for that, motivation is required in both, central and front line government officials and management.

(9) In the private sector, supportive management willing to take risks and encourage fresh thinking. Public officials and private managers with enough time to formulate and implement innovation initiatives.

(10) Government support for R&D and incentives for investment in R&D such as R&D tax credit.

(11) Availability of risk capital including angel investment and venture capital.

(12) Compromise between the political and economic power and existence of social control.

(13) Innovation networks and clusters such as: existence of educational institutions of higher learning, think tanks, training programs and technical teams, existence of institutions that act as conveners of networking events, collaboration among different countries and institutions in international collaborative R&D projects.

(14) Diversity of people and free flow of ideas (generation of widely divergent views, challenging of assumptions, testing of hypotheses,

Table 1. Competitiveness, productivity and innovation measures.

	Competitiveness	Productivity	Innovation
National	• Standards of Living • Gross Domestic Product (GDP) • Expenditures • Gross National Product (GNP) • World Economic Forum 8 Factors • Unemployment • Exchange Rate • Purchasing Power Parity • Equity Markets • Bond Markets • Interest Rates • LIBOR and Money Rates • Dow Jones Global Indexes	• GDP/worker • BW Production Index • Total Factor Productivity (TFP) • Compensation/Hour • Tornqvist and Fisher Indexes	• Research & Development (R&D) as % GDP • R&D • National Labs • Nobel Prizes
Industry	• Sales • Market Share • Dow Jones U.S. • Dow Jones Global • Inventories • Profitability	• Output/worker • Profitability • Industry Groups • Compensation/Hour • Tornqvist Sector Output • Federal Reserve Board Index • Bureau of Labor Statistics KLEMS	• R&D as % GDP • Patents • Scientists • R&D Expenditure • R&D Personnel • R&D % of Profit
Firm	• Sales • Market Share • Equity • Profitability	• Output/worker • Profitability • Output/hour • Standard Costs	• R&D as % Sales • R&D Expenditure • Patents • Scientists • R&D Personnel • National Labs

compensation for cultural or intellectual myopia).

Inhibitors

(1) Resistance comes from the elites in that elites are inclined to screen out innovations whose consequences threaten to disturb the status quo, for such disruption may lead to a loss of position for them (Rogers, 1995).

(2) Much innovation fails due to resistance to change: failure of courage and failure of imagination can prove to be formidable deterrents of innovation.

(3) Sense of 'comfort': why should I push myself and disturb convenient routines; conservatism in its multiples forms, e.g. 'don't change the established path' syndrome, 'no risk policy', 'no non-proven alternatives', 'don't skip the line'.

(4) Lack of courage in government officials to face opposition, fear of losing support from the electorate and accompanying lack of long-term vision (focus only on short term gains), instability and high turnover of public officials in their positions.

(5) Lack of courage by Chief Executive Officers and the Board of Directors in the private sector, to embrace change or take a long-term view, pressure from stakeholders to increase earnings per share in the short term, fear of losing support from stakeholders if they do not respond to the short-term pressure for results, frequent turnover and little time to formulate and implement long-range growth initiatives (Perel, 2002).

(6) The way innovation is introduced is an important determinant and predictor of its likelihood of success: more incremental innovations have a lower impact on hierarchical power relationships and as a result, are confronted with smaller resistance and inertia than more radical ones. The 'dangerous innovations' that are resisted the most, are those of a disruptive and restructuring nature, rather than those that will only affect fine-tune the functioning of the system (Christensen, 1997).

(7) Rigidity of hierarchical structures and lack of management for results; instability in the rules of the game (arbitrariness, favoritism), corruption and lack of transparency; poverty and political

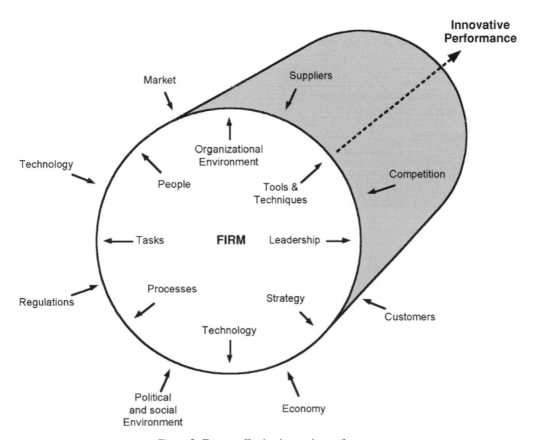

Figure 5. Factors affecting innovative performance.

struggle; centralized bureaucracies, top-down policies and government control for the sake of control.

- **Theme 3: A Glimpse of the Current Situation in Several Countries—Challenges and Opportunities**

  Our findings show that there are several major challenges and potential opportunities associated with innovation and creativity-supporting initiatives and policies. Intrinsic in this is the role as well as the successes and failures of the Multilateral Development Banks (MDBs). There are also challenges and opportunities facing the private sector, emanating from the high and increasing rates of technological change, globalization, intensity of rivalry, and dilution of national franchises:

  (1) There is great potential for creativity, innovation and competitiveness at the individual level, but there is a lack of public policies to foster and take advantage of this capacity.

  (2) In some countries, government policies, while providing no direct financial assistance, were invaluable in providing a market for product development and subsequent sales opportunities.

  (3) Usually in developing countries, policies are generated top-down without any consensus, debate or agreement with civil society. The main emphasis is control. There are no channels for participation and formulation of a bottom-up policy. In this regard, the Congress is not even a representative body of the civil society. Also, there is no transparency, nor accountability.

  (4) In some European countries there is an urgent need for stronger links of education and research with the real economy, and integration into the European research and innovation system.

  (5) In many countries the private sector is regarded as more competent in terms of creativity, innovation and competitiveness than the public sector. The public sector seems to be paralyzed because of norms and regulations that block development. The political indecisiveness affects all the economic activities and makes economy sluggish.

  (6) Universities are excellent, shining examples of public sector environments that fuel creativity and innovation.

  (7) Many countries have not fully tackled in a way that policy and practice merge into a single unified outcome. Most of the examples cited in relation with this type of cooperation has occurred in pockets throughout the public sector with the initiative being taken by individ-

uals, rather than through organizational planning.

  (8) The major challenge is for less developed countries because of lack of adequate capacity and infrastructure necessary to transform vision into action and lack of a continuous and stable dynamic that foster creativity, innovation and competitiveness.

  (9) Most developing countries still think in state-centered terms. Public sector too dominant, too prone to doing wrong things for political or even personal reasons. Private sectors are disorganized; companies have poor management and strategic planning skills. Public and private sectors do not collaborate enough; they often play the blame game. Governments policies often help the politically powerful rather than expose firms to competitive pressure in order to build strength to compete on a global basis.

  (10) Overall, in developed as well as in developing countries, no specific, explicit and systematic measurements for creativity, innovation and competitiveness seem to be undertaken. Usually, social and economic performance indicators are considered as proxies to measure performance in this field.

  (11) Multilateral Development Banks (MDBs), although traditionally dominated by paradigms and ideological postures that resist debate and change, and therefore, creativity and innovation, have independently and also in collaboration, launched numerous initiatives to further promote competitiveness and higher levels of development in their borrowing member countries.

Lately, the advances of Internet technology and high speed connectivity have allowed member countries to exchange vital information to participants in various programs and projects, promoting distance learning in a modern and cost effective manner. However, more work is needed in MDBs to measure their effectiveness in terms of fostering creativity, innovation and competitiveness at the national and regional levels.

Recommendations for Action based on Field Study

To learn one must be humble.

(James Joyce, Ulysses)

We compile and categorize thematically below, the responses of the public and private sector practitioners we collected via our field research to provide the foundation for recommendations to practitioners and a roadmap for future research.

The Role of the Public Sector in Promoting Creativity, Innovation and Competitiveness (CIC)

The role of public sector innovation is decisive as catalyst and accelerator of social and economic development. Public sector CIC promotion can lead to better, more efficient and cost-effective ways of managing public sector operations and social welfare functions; it can as well help improving market functioning in competitive environments.

The opportunity cost for the public sector not promoting the creation of a more competitive environment is enormous, since the social and economic costs associated with obsolete or outmoded ways of conducting and managing business are substantial, especially in less developed economies.

The role of the government in a developing country is much more critical, as often the private sector has no ways and means to raise venture capital. Therefore unless the government steps into the vacuum there is very little chance that the innovations will ever see the light of day.

Usually, less developed countries lack adequate capacity and infrastructure necessary to transform vision to action; whereas in industrial countries these conditions are met to a greater extent. The public sector can act as a catalyst to CIC in developing countries for it can promote/assume works in areas where the private sector finds inadequate profit incentives for engagement.

The *modus operandi* of the public sector can be a function of the degree of market development, its functions and the demand for public/private sector services. A well-developed public infrastructure should ideally allow for CIC promotion from both public and private sector entities.

Government active support for innovations is essential in developing countries due to the political and economic system inadequacies that might disturbed innovative technology initiatives in public or private sector.

The challenge is to have reasonable ways for the science developed in the public sector to migrate to commercialization when warranted, without the public sector becoming effectively competitors with the private sector. The public sector should not be operating as competitor to private sector entities, crowding out entrepreneurial initiative.

The public sector requires more aggressive policies to foster creativity in its own public management, promote the innovation within the government and raise its own competitiveness in the process.

- **The Public Sector could promote CIC in several ways**

  (1) Creating an environment which supports CIC. It includes issuing policies, norms and regulations that enable CIC; giving awards and incentives like tax breaks, insurance and other favorable conditions to take advantage of international experience; providing effective stimuli to research and scientific development through investing resources adequately.

  (2) Using the purchase power of the government (around 30% of Gross National Product in Latin American countries) to foster competitiveness (besides efficiency and transparency).

  (3) Building social safety nets for those that fail when seeking to invent, as well as support mechanisms for those that need additional support to invent/innovate.

  (4) Acting over those market failures, where the private sector can not act alone because of lack/asymmetry of information or problems of scale. In this regard, the promotion of non-traditional exports or subsidies to technological innovation in small and medium enterprises could be cited as examples.

  (5) Trying to commercialize research generated by the public sector, e.g. from National Aeronautic Space Administration, federal laboratories, or Department of Defense.

  (6) Building an efficient innovation system, the main areas of focus are related to research and innovation networks, innovation and technological transfer programs (scientific and technological parks), innovative SMEs and improved capacity of economy to absorb R&D achievements.

  (7) Making available resources for fundamental research. These monies do create an environment that is less inclined to be driven by economics and be very focused on 'pure' science.

Public-Private Sector Partnerships to Promote CIC

In developed countries where markets are functioning more efficiently private sector participation in ICC is more pronounced. Besides, partnership arrangements between public/private sector are as such as to allow a higher degree of private sector participation in certain operations/areas despite the relatively higher level of competitiveness and/or lower profit margins (in these countries, private firms often have an incentive to engage in partnerships with the government for other reasons such as to acquire a larger market share, or just for marketing, advertisement and promotion). Moreover, in developing countries with less stable socio-economic and/or political structures concerns over government's credibility often act as a deterrent for private sector participation

Some areas in the private sector require support from the public sector to improve creativity, innovation and competitiveness. However, it seems that what the private sector needs, more than public/private sector partnerships, is a more competitive and transparent public procurement process (to fight corruption) and incentives for university/industry as well as domestic/foreign, public/private sector partnerships.

Support new projects and initiatives in the private sector (i.e. research in applied sciences or manufacturing technological development, etc.). In the last case, there is the possibility of partnerships between public and private sector but, only in those areas in which the private sector is weak and in those areas really innovative.

Support for new projects and initiatives in the private sector (i.e. research in applied sciences or manufacturing technological development, etc.) through public/private partnerships is desired mainly in those areas in which the private sector is lacking innovative capacity and creative competence and especially for those areas where discontinuous innovations (potentially disruptive in nature) are the main pursuit.

Description of criteria to identify potential initiatives/projects to conform partnerships between public and private sector, criteria and process for selecting the candidates in the private sector, and conditions in which public venture capital should be involved:

Criteria for identifying potential initiatives/partners

(1) Government priority areas.
(2) Potential practical application and benefit for society and/or economy.
(3) Low cost or profitable project.
(4) Long term project.
(5) Candidates with probity and enough financial means.
(6) Willingness to change.
(7) Competitive spirit.
(8) Futuristic vision.
(9) Complementary experience and resources.
(10) A good record of accomplishments.
(11) Their contribution to innovation.
(12) Expertise and market access.

Process
(1) This is a very important and yet paradoxical process, to pick a winner there is almost an element of gambling that comes into play, yet the public sector has a responsibility to the community not to gamble with public funds.
(2) The process would need to involve the community and the business sector in order to ensure that it was transparent. The rules and procedures would also have to be extremely clear as would the involvement of any public official. This would in particular be extremely critical should the product progress to the next stage where venture capital would be involved.
(3) An 'Experts Committee' (EC) could be established for selection, oversight and evaluation. A 'Quality Assurance' Committee/Group could complement the functions of the 'EC' to review the quality/

progress at a later stage once relevant works have been initiated. Financing arrangements are always particular to the specific operations in question.

The Role of the MDBs in Promoting CIC
The role of MBDs in supporting public and private sectors innovation is decisive. MDBs could:

(1) promote the formulation of national policies and action plans for CIC with short and long term goals, with challenging but feasible objectives; train and stimulate the staff responsible of implementing projects and activities in private and public organizations; allocate the needed resources and tools for performing the appropriated actions, and provide continuous technical assistance and supervision;
(2) create the conditions to foment a climate where innovations in developing nations could be brought to fruition;
(3) help developing countries get the underlying economic and regulatory policies right so that conditions to encourage new ventures and innovation can work within the private sector;
(4) share global best practices with developing nations. Strengthen the private sector institutions so that business can make a greater contribution to policy decisions;
(5) contribute to eliminate trade barriers that affect the developing countries and destroy the possibility of any innovative development;
(6) disseminate information, knowledge and successful experiences on CIC among member countries;
(7) promote CIC through agreements and treaties and incorporate them in the policies for development.

To be able to contribute much more effectively to promote CIC, MDBs should:

(1) incorporate CIC in its own management. It is necessary that MDBs are co-responsible (with the countries) for the results of its cooperation. This responsibility should be measured in terms of national development and citizens' wellbeing in the target countries;
(2) be organizations more creative and more innovative, less rigid and less bureaucratic;
(3) support the governmental projects in which the government holds strong leadership and ownership of the plan. In other words, no financial support should be provided to the government that has no independent ownership of development ideas or sustainable leadership;
(4) give more flexibility to the countries in finding their better path to development (no predefined magic formulas);
(5) being more accountable for their achievements in terms of national productivity and competitiveness.

Conclusions and Future Research

The empires of the future are the empires of the mind.

(Sir Winston Spencer Churchill)

In this chapter, we have attempted to address the following issues concerning for-profit and not-for-profit entities:

(a) when, how, and why creativity and innovation occur;
(b) how and why creativity triggers innovation and vice versa; and
(c) what are the connections and implications for competitiveness of the presence or absence of creativity and innovation using empirical findings from both the public and private sectors.

We combine literature sources (including those of the authors) as well as field interviews to enrich our chapter with current academic and practitioner insights on the practice and implications of creativity and innovation from the perspective of competitiveness.

We believe that *competitiveness is a product and a function of creativity and innovation stocks and flows* that are determined and modified through diverse types, ways and means of learning (top-down, bottom-up, by doing, by analogy, by succeeding, by failing, trans-national, domestic, via skills exchanges, technology-leveraging, partnerships, etc.) as well as individual cognition and inventiveness (*the 'when', 'how', and 'why' of creativity and innovation*) (see Figs 6 and 7).

In Fig. 6, we show the participatory and synergistic interaction between the public sector, the private sector, and key institutions for collaboration like universities, research institutions and Non-Governmental Organizations (NGOs) in building strategic alliances towards the objective of higher levels of competitiveness in developing countries. In this context:

- governments are accountable for establishing a stable and predictable political and macroeconomic environment, issuing transparent policies and enforcing legal and property rights, facilitating cluster development, creating a business environment with low transaction costs, and supporting and giving incentives to creativity and innovation;
- enterprises have to mount competitive strategies, develop networks and clusters for achieving efficiency (social capital), increase the intensity of their technological effort (more resources for R&D), build new capabilities and skills (human and intellectual capital) and develop a modern infrastructure. Suppliers of physical and service inputs and infrastructure have to meet international standards of costs, quality and delivery;
- universities and research institutions have to align curricula to business needs and craft public and private partnerships to develop new capacities and

skills in public and private sectors. NGOs should serve as enablers, catalysts and accelerators of public and private partnerships.

The *top-down institutional learning* complementing *bottom-up entrepreneurial learning* act as enabler, catalyst, and accelerator of economic development and convergence across developed and developing countries as well as cross-pollination and transfer of technology and best practices across developed and developing countries as well as public and private sectors, universities and research institutions and NGOs (see Figs 3 and 6).

We find from our field research that innovation and creativity are becoming increasingly important for public sector reforms as well as private sector survival and prosperity, given the current challenges and opportunities facing public and private sectors around the world. Some of the challenges and opportunities facing the public sector with serious implications for the private sector as well, encompass the following:

(a) shrinking budgets and shifting demographics with ageing populations;
(b) the higher-rent-yielding tax base of knowledge workers attaining increased mobility;
(c) increased pressures for accountability and transparency driven by privatization, globalization, and an increasingly informed and sophisticated voter base;
(d) increased pressures on and from the private sector to become more competitive, demanding in return a more competitive public sector;
(e) and last but not least the role of the Multilateral Development Banks and especially the International Monetary Fund which are relentless in demanding gains in efficiency and transparency in public sector policies, structures and practices.

Competitiveness is also a way of *looking back* as it reflects the past achievements of creative genius and innovative energy as well as shaping the future by providing the foundation and skeleton for emerging private and public sector endeavors. Moreover, at the present, the manifestations of creative and innovative endeavor serve as auguries of the *emerging horizons* of socio-economic and institutional development and learning which further interact with individual learning and creativity in an never-ending spiral process (see Figs 5, 6, and 7).

We also find that creativity and innovation do not always lead to enhanced competitiveness (at least in the short- to medium-term) (Carayannis, 2001a, 2001b, 2002).

All in all, our foray into the domain of creativity and innovation as they relate and impact competitiveness has identified many areas of interest that warrant further focused research to better understand and more

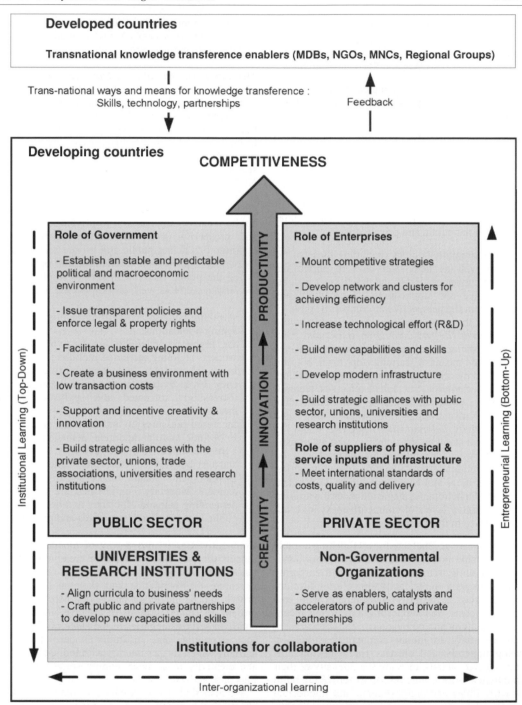

Developed countries

Transnational knowledge transference enablers (MDBs, NGOs, MNCs, Regional Groups)

Trans-national ways and means for knowledge transference :
Skills, technology, partnerships

Feedback

Developing countries

COMPETITIVENESS

PRODUCTIVITY — INNOVATION — CREATIVITY

Institutional Learning (Top-Down)

Entrepreneurial Learning (Bottom-Up)

Role of Government

- Establish an stable and predictable political and macroeconomic environment

- Issue transparent policies and enforce legal & property rights

- Facilitate cluster development

- Create a business environment with low transaction costs

- Support and incentive creativity & innovation

- Build strategic alliances with the private sector, unions, trade associations, universities and research institutions

PUBLIC SECTOR

Role of Enterprises

- Mount competitive strategies

- Develop network and clusters for achieving efficiency

- Increase technological effort (R&D)

- Build new capabilities and skills

- Develop modern infrastructure

- Build strategic alliances with public sector, unions, universities and research institutions

Role of suppliers of physical & service inputs and infrastructure
- Meet international standards of costs, quality and delivery

PRIVATE SECTOR

UNIVERSITIES & RESEARCH INSTITUTIONS

- Align curricula to business' needs
- Craft public and private partnerships to develop new capacities and skills

Non-Governmental Organizations

- Serve as enablers, catalysts and accelerators of public and private partnerships

Institutions for collaboration

Inter-organizational learning

Figure 6. CIC learning and institutional linkages.

clearly exemplify and articulate the key drivers and dimensions (content, process, context, impact) of creativity and innovation and the role of learning in this process (Carayannis, 2002). In particular, we feel that it would be useful to empirically map the nature and dynamics of learning and meta-learning along the CIC2Helix (Fig. 5) regarding both the public and the private sectors.

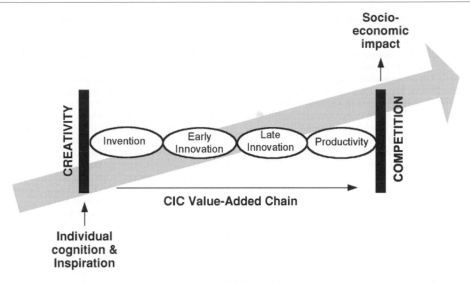

Figure 7. CIC value-added chain.

A better understanding of such processes could result in enhancing and advancing the effectiveness of generated knowledge and the efficiency of knowledge transmission and absorption. In so doing, *knowledge economies of scale and scope* could be attained at increasingly higher levels, allowing for more to be accomplished with less, faster, cheaper and better (Carayannis, 1998a, 1998b, 1999a, 1999b, 2001a, 2001b, 2002).

These gains can be manifested in diverse ways at *the micro, meso and macro levels* (see Fig. 5), namely higher standards of living, more competitive firms, more robust economies, and accelerated and more sustainable development trends (see Figs 5, 6 and 7). Indeed, one could identify both a challenge and an opportunity for private and especially public sector policy makers and managers with regards to the risks and possibilities inherent in the three figures we refer to. Specifically, Figs 5, 6 and 7 capture and reflect upon a combination of dynamic, complex and powerful forces at play: human cognition, individual creativity, organizational productivity, and national competitiveness as well as institutional inertia, political short-sightedness, and market as well as government failures. Understanding how to better manipulate and leverage those forces can result in spectacular results (to some extent, success stories in socio-economic development such as Singapore and South Korea may be cases in point), while attempting to work against those forces and trying to suppress them can lead to dangerous and unsustainable regimes of poverty and autocracy. Providing rewards and incentives to foster creativity and innovation and even rewarding failure may be a strong enabler, catalyst and accelerator for creativity and competitiveness.

In our other chapter on discontinuous and disruptive innovations, we continue our study and analysis of the nature and dynamics of the relationship between creativity and innovation and when, how and why quantum leaps in creative destruction occur following the Schumpeterian line of reasoning (Schumpeter, 1934, 1942).

References

Ahmed, P. K., Loh, A. Y. E. & Zairi, M. (1999). Cultures for continuous improvement and learning. *Total Quality Management,* July.

Amabile, T. (1996). *Creativity in context: Update to the social psychology of creativity.* Boulder, CO: Westview Press.

Arieti, S. (1976). *Creativity: The magic synthesis.* New York: Basic Books.

Carayannis, E. (2002), Is higher order technological learning a firm core competence, how, why, and when: A longitudinal, multi-industry study of firm technological learning and market performance. *International Journal of Technovation,* **22**, 625–643.

Carayannis, E. (2001a), Learning more, better, and faster: A multi-industry, longitudinal, empirical validation of technological learning as the key source of sustainable competitive advantage in high-technology firms. *International Journal of Technovation,* May.

Carayannis, E., Hagedoorn, J. & Alexander, J. (2001b). Strange bedfellows in the personal computer industry: Technology alliances between IBM and Apple. *Research Policy,* **30** (6).

Carayannis, E. & Roy, S. (1999a). Davids vs. Goliaths in the small satellite industry: The role of technological innovation dynamics in firm competitiveness. *International Journal of Technovation,* May.

Carayannis, E. & Alexander, J. (1999b). Winning by co-opting in knowledge-driven, complex environments: The formation of strategic technology Government-University-Industry (GUI) partnerships. *Journal of Technology*

Transfer, **24** (2/3), 1999 Lang-Rosen Award for Best Paper by the Technology Transfer Society.

Carayannis, E. (1998a), Fostering synergies between information technology and managerial and organizational cognition: The role of knowledge management. *International Journal of Technovation*, May.

Carayannis, E. (1998b), Higher order technological learning as determinant of market success in the multimedia arena; A success story, a failure, and a question mark: Agfa/Bayer AG, Enable Software, and Sun Microsystems. *International Journal of Technovation*, **18** (10), 639–653.

Chesbrough, H. W. & Teece, D. J. (1996). When is virtual virtuous? Organizing for innovation. *Harvard Business Review*, **74** (1), 65–73.

Christensen, C. (1997). *The innovator's dillemma: When new technologies cause great firms to fail.* HBS Press.

Dacey, J. S. & Lennon, K. H. (1998). *Understanding creativity: The interplay of biological, psychological, and social factors.* San Francisco, CA: Jossey-Bass.

Drazin, R., Glynn, M. A. & Kazanjian, R. K. (1999). Multilevel theorizing about creativity in organizations: A sensemaking perspective. *The Academy of Management Review*, **42** (2), 125–145.

Drucker, P. F. (1998). The discipline of innovation. *Harvard Business Review*, **76** (6), 149–157 (originally published in May-June 1985).

Eleanor, D. G. (2001). Innovation patterns. *The Innovation Journal*, **2**, 130–135.

Gundry, L. K., Prather, C. W. & Kickul, J. R. (1994). Building the creative organization. *Organizational Dynamics*, Spring.

Henderson, R. (1994). Managing innovation in the information age. *Harvard Business Review*, **72** (1), 100–105.

Hofstede, G. (1980). Motivation, leadership, and organization: Do American theories apply abroad? *Organizational Dynamics*, **9**, 42–63.

Kao, J. (1996). *Jamming: The art and discipline of business creativity.* New York: HarperCollins.

Kneller, G. F. (1965). *The art and science of creativity.* New York: Holt, Rinehart & Winston, Inc.

Nonaka, I. & Takeuchi, H. (1995). *The knowledge-creating company: How Japanese companies create the dynamic of innovation.* New York: Oxford University Press.

Perel, M. (2002). Corporate courage: Breaking the barrier to innovation, *Research Technology Management*.

Rogers, E. M. & Shoemaker, F. F. (1971). *Communication of innovations.* New York: The Free Press.

Rogers, E. M. (1995). *Diffusion of innovations* (4th ed.). New York: The Free Press.

Schumpeter, J. A. (1934). *The theory of economic development.* Oxford: Oxford University Press.

Schumpeter, J. A. (1942). *Capitalism, socialism, and democracy.* New York: Harper & Brothers Publishers.

von Braun, C. F. (1997). *The innovation war.* Englewoord Cliffs, NJ: Prentice Hall.

Weisberg, R. W. (1992). *Creativity: Beyond the myth of genius.* New York: W. H. Freeman & Co.

Woodman, R. W. & Schoenfeldt, L. F. (1990). An interactionist model of creative behavior. *Journal of Creative Behavior*, **24**, 10–20.

Zaltman, G., Duncan, R. & Holbek, J. (1973). *Innovations and organizations.* New York: John Wiley.

Appendix I: Field Research Questionnaire
Questions: Please respond by return email within ten days of receipt and email your responses to: *caraye@gwu.edu*

(1)   Do you think that there is a role for public sector in fostering creativity, innovation and competitiveness?

(2)   According with your experience, what are the main factors that act as catalysts or inhibitors for creativity, innovation and competitiveness in the public and in the private sector?

(3)   What is your definition of 'innovation?"

- Inventing something new   [ ]
- Generating new ideas only   [ ]
- Improving something that already exists   [ ]
- Spreading new ideas   [ ]
- Performing an existing task in a new way   [ ]
- Following the market leader   [ ]
- Adopting something that has been successfully tried elsewhere   [ ]
- Introducing changes   [ ]
- Attracting innovative people   [ ]
- Seeing something from a different perspective   [ ]
- Other (please, define)   [ ]

(4)   What kind of benefits are expected when public sector plays an active role in promoting creativity, innovation and competitiveness?
- Benefits to government/public sector   [ ]
- Benefits to society   [ ]
- Profit

(5)   How could the public sector measure the benefits of fostering creativity, innovation and competitiveness?

(6)   To foster creativity, innovation and competitiveness, should the public sector play a rather passive role (formulating policy, enabling legislation and regulations) or should the public sector play a more active role in new ventures that foster creativity, innovation and competitiveness?

(7)   Should the public sector foster creativity, innovation and competitiveness from outside (by partnering and/ or supporting private sector in new ventures) and/or from inside (by developing its own initiatives)?

(8)   How the public sector should select the winners in the private sector for partnering or receiving support in projects and programs promoting creativity, innovation and competitiveness? What should be the criteria? How should be the process? In what stage of the process a venture capital should be involved?

(9)   How does the current policy in your country impact in creativity, innovation and competitiveness?

(10)   Do you see any difference in the role of government/public sector with respect of creativity, innovation and competitiveness in developed VS developing countries?

(11)   Do you measure and benchmark creativity and/or innovation in your organization and, if yes, how?

(12)   Do you benchmark creativity and/or innovation in your organization to that other organizations and are they public, private or both?

(13)   Would you consider the public or the private sector more competent in terms of creativity and innovation, and why?

(14)   What examples could you mention to describe the dynamic of creativity, innovation and competitiveness in your organization/country?

(15)   What should be the role of the Multilateral Development Banks in supporting creativity, innovation and competitiveness in developing and developed countries?

Appendix II: Field Research Respondents & Affiliations
Countries

Argentina, Australia, Colombia, El Salvador, Greece, Jordan, Mexico, Nicaragua, Peru, Japan, U.S.

Contributors from Field Study and Affiliations

Partner, venture capital firm (3)
Partner, consulting firm (2)
Coordinator, E-Government Program—Office of the President (1)
Consultant, Industry & Technology Ministry (1)
Secretary, Office of the President (1)
Secretary, of Government (1)
Director, Chamber of Commerce (1)
Procurement official—Technical Secretary—Office of the President (1)
Principal, Private Company (1)
Consultant, International Development Agency (1)
International development consultancy (1)
Official, Private Bank (1)
Official, Multilateral Development Bank (2)
Official, University (1)
Official, European Commission (2)
CEO, High Tech Firm (1)

The International Handbook on Innovation
Edited by Larisa V. Shavinina

Innovation Tensions: Chaos, Structure, and Managed Chaos

Rajiv Nag[1], Kevin G. Corley[2] and Dennis A. Gioia[1]

[1] *Smeal College of Business, Penn State University, USA*
[2] *College of Business Administration, University of Illinois, USA*

Abstract: This chapter presents a framework for understanding the tensions that underlie an organization's ability to manage innovation effectively in the face of a turbulent competitive environment: (1) the fundamental tension between the desire for *structure* and need for creative *chaos*, and (2) the on-going tension between *technology-push* and *market-pull* approaches to innovation. We explore the nature and boundaries of these tensions and characterize them as four distinct 'innovation contexts'. Using one high-technology organization's struggle as an example, we discuss the notion of 'managed chaos', a concept that helps understand the role of innovation in the maintenance and change of an organization's identity.

Keywords: Innovation; Chaos; Structure; Organizational identity.

Introduction

Over the course of the last generation, innovation has progressed from being a 'nice to have' to an 'ought to have' to a 'must have' in many organizations. Quantity and quality of innovation have gone from mainly being performance yardsticks in R&D departments and new product development teams to becoming the *raison d'etre* for many of the organizations in which R&D departments and NPD teams are housed. The rapid advance of technology is the root cause of this shift, of course, and the need for increased follow-on technological innovation has been the widespread response. We now see a trend in which more and more organizations are treating innovation as a 'critical to have'—i.e. as a survival imperative.

This trend is perhaps most evident in those industries most violently buffeted by technological change and technological competition (where staying on the bleeding edge is itself a business imperative). Yet, the writing on the wall suggests that this heretofore localized trend is quickly spreading to a much wider domain. Business executives in such contexts can no longer assume that they are primarily competent managers of the status quo or shepherds of incremental change. Instead, they need to assume a considerably more complicated role that places a premium on maintaining the dynamic balance between efficiently managing resources to increase shareholder value and

effectively innovating to remain competitive and insure survival.

A theoretical focus on this dynamic balancing act highlights several significant internal tensions associated with innovation, especially in high-tech organizations and industries. Perhaps most evident is the key tension between the creation of stable organizational mechanisms to exploit a particular business model and the subsequent destruction of those same mechanisms and models to cope with the ever-changing requirements of a highly competitive environment (March, 1991). Given the acceleration of the latter process over the last decade, understanding these innovation tensions becomes paramount for understanding the survival and growth of high-tech companies in the modern age.

Innovation most often has been studied by examining the implementation of creative ideas within an organization (Amabile, 1988)—usually the implementation of ideas that are not necessarily new to a given domain, but are new or innovative to the focal organization itself (Ford, 1995). In our view, however, this approach does not adequately account for the most dynamic aspects of organizational innovation. In particular, it does not account for the internal tensions arising from a turbulent competitive environment. The successful exploitation of known technologies requires the creation of stable organizational mechanisms that enhance the level of coordination. Such mechanisms require much lower levels of flexibility and risk-taking than the creative exploration, development, and implementation of new technologies (which also tend to engender a great deal of turmoil within an organization). When exploitation and exploration processes are necessarily juxtaposed, as they now so frequently are, a predictable tension arises.

We characterize this juxtaposition as a tension between *structure* and *chaos* (a set of provocative first-order labels used by informants in one of our research projects, but which arguably have application to a much wider domain in the study of innovation). Structure/chaos tension makes the nature of organizational innovation different for exploitative approaches and explorative approaches to organizational well-being and survival. Furthermore, and importantly, this is not an either-or choice for most organizations, but rather a continuous dialectic that is closely linked to changes in the organization's environments.

Exploring an essential feature of this structure/chaos tension reveals a different sort of innovation tension, one that emerges as the organization attempts to balance the tendency to become a more inwardly oriented *'technology push'* organization versus the need to become a more externally responsive *'market pull'* organization. A pure technology push approach is characterized by internal creative processes that are focused on designing and developing cutting-edge technology as a basis for advancement and achieve-

ment, with the organization pushing the technology into the marketplace without first identifying a market need. A pure market pull perspective, on the other hand, focuses less on the cutting-edge advancement provided by a technology and more on the technology's ability to fulfill a need in the marketplace that will result in sales and revenue as technology creation and development is pulled by market demand. We believe that this technology push/market pull dichotomy captures a common tension in many contexts involving innovation, perhaps especially in R&D groups and high-tech organizations devoted to R&D. Furthermore, we have found that this technology push/market pull tension implicates an organization's very identity and affects strategic issues such as positioning within the marketplace and relationships with key stakeholders.

In this chapter, we extend traditional notions of organizational innovation in high-tech companies and industries by examining the implications of juxtaposing these two tensions. We argue that organizational structure/chaos and technology push/market pull tensions are a defining part of current organizational practice and, therefore, need to be more closely examined to further our understanding of modern innovation processes. When these two tensions are arrayed in a 2×2 matrix (see Fig. 1), each of the resulting four cells represents a distinct context in which organizations can attempt to nourish innovative practices (Structure-Push, Structure-Pull, Chaos-Push, and Chaos-Pull). We explore the specifics of each of

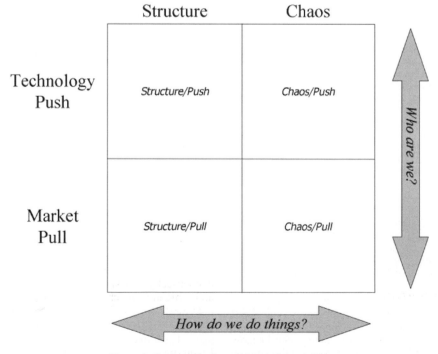

Figure 1. Juxtaposing the two innovation tensions.

these 'innovation contexts' and discuss managerial practices necessary to facilitate innovation within each cell.

Additionally, to account for the dynamic environments with which organizations must cope, we examine the issues and implications involved in movement among the different innovation contexts. It is no longer feasible to think of an organization sustaining its innovative practices within a static world. Consequently, organizations must assume that their innovative contexts are in flux and, therefore, must conduct their innovation efforts under changing circumstances. We examine this new state of affairs by considering what it takes to manage these tensions in a way that transitions from one cell to another are not only possible, but also effective. To demonstrate the relevance and usefulness of this framework, we present a case example of one organization's attempt to manage these tensions and its movement across cells. Our analysis highlights the issues faced by managers, as well as the implications of these tensions for organizational survival and growth.

We conclude our discussion of these innovation tensions with the explication of a model that articulates the main challenges faced by organizations driven to make innovation a vital part of their strategic efforts. Dealing with these challenges inevitably raises fundamental questions about an organization's identity. The key feature of this model is the notion of 'managed chaos', a concept that suggests a way to manage these tensions by surfacing and questioning basic assumptions about the role of innovation in the maintenance or alteration of organizational identity. We posit that managed chaos is exercised not just by destabilizing existing organizational structures and routines, but also by reconsidering innovation's *raison d'etre* in many high-tech companies. We suggest that by working towards a state of managed chaos, an organization can maximize the potential found in the intersection of the structure/chaos and technology push/market pull tensions of organizational innovation.

The Fundamental Tensions of Organizational Innovation

Before delving into the complexities involved with understanding the dynamic aspects of organizational innovation, it is first necessary to understand each tension better and to explore the details of each of the four 'ideal contexts' that arise from their juxtaposition. We begin by examining the Structure-Chaos tension in more depth, followed by the Push-Pull tension.

Structure-Chaos Tension

Organizations are composed of individuals who bring together divergent skills and resources that need to be integrated to achieve a common goal. This working premise is as complex as it is simplistic and obvious. Lawrence & Lorsch's (1967) dimensions of differ-entiation and integration contain an elemental dialectic that challenges the very act of organizing. The complexities of any modern organizational environment require functional differentiation and division of labor. Each function/department specializes in managing its own part of the environment. To do so, each function/department needs the autonomy to frame the nature of its sub-environment and develop the necessary organizing mechanisms and structure to deal with it.

Uncontrolled differentiation, however, can result in an organization's sub-units pulling away from each other by pursuing their own independent agendas. To reign in the different sub-units, an organization needs to develop integrating mechanisms that impose a common sense of purpose and some standard operating procedures on the sub-units. Traditionally, the top management of an organization plays the integrative role by setting up control systems and creating the context for the development of a common organizational culture. These integrative mechanisms, however, can result in excessive centralization of decision-making and loss of local flexibility. This problem is exacerbated in turbulent environments, where the need for fast and frequent differentiation to face changing external complexities must co-exist with the integrative need to work together with the rest of the organization (c.f. Brown, Corley & Gioia, 2001).

Another dialectic that is fundamental to organizing is the need to achieve a workable balance between current efficiency and future effectiveness. In other words, organizations face two divergent modes of existence—one that is built on the successful exploitation of existing technologies, markets, and products, and the other that is built on a continuous exploration of new technologies and market opportunities. Successful exploitation of a specific technology requires an organization to focus on the selection, execution, and implementation of a limited set of technologies and/or business models to maintain a stable equilibrium (March, 1991). Thus, a successful exploitative strategy would necessitate the creation of stable organizational structures and routines that persist over an extended period of time (Nelson & Winter, 1982). Organizational resources and energies are focused on maintaining stability and predictability in internal and external environments. Although new ideas might emerge, these are most likely aimed at improving the existing technology and enhancing the efficiency of the current organizational structure and processes.

An organization engaged in exploration follows a diametrically opposing approach, however. Its strategy and structure are characterized by variation, experimentation, uncertainty, and risk (March, 1991). Organizational resources and energies are focused on bringing new data into the organization, turning it into useful information, and putting it to use with other organizational knowledge for the creation of new structures and processes. Change is treated as a natural part of the organization's culture, while routines (to the extent that they exist) are often focused on facilitating such change. Explorative organizations often can take the form of 'organized anarchies' (Cohen, March & Olsen, 1972) that have unclear technologies and unstructured decision-making processes characterized by solutions looking for problems.

These two elemental rivalries—'differentiation versus integration' (Lawrence & Lorsch, 1967) and 'exploitation versus exploration' (March, 1991)—most affect the design of an organization, especially where innovation is at issue. An organization that focuses on stable technologies and predictable routines and mechanisms to reconcile or resolve these rivalries represents, in our terms, an emphasis on '*structure*'. An organization that focuses on the proactive search for new technologies and employs inventive mechanisms and decision-making styles to reconcile these rivalries represents an emphasis on '*chaos*'. Although this structure-chaos dichotomy might be somewhat of an oversimplification, it provides a useful conceptual basis for examining different approaches to organizing for innovation.

Push-Pull Tension

Although structure-chaos tensions are vital for understanding the organizing process, they alone cannot provide an explanation for the most common difficulties organizations face in trying to become innovative. In practice, the tension between 'technology push' and 'market pull' orientations tends to be more fundamental and to present the more difficult challenge for effectively managing organizational innovation— mainly because the technology push, market pull issue more strongly implicates organizational identity. Although the structure-chaos tension could be described as raising questions about 'how' things are done in an organization, the push-pull tension might be best characterized as provoking questions about 'who' the organization thinks it is—which is obviously a more deeply-rooted and central issue.

In high-tech companies, tension emerges as organizations attempt to balance the tendency to become an inwardly oriented 'technology push' organization (one that values creative and innovative technological invention over practical relevance and marketability) against the need to become an externally responsive 'market pull' organization (one that values developing technological inventions to meet market needs). The notion of technology push presumes the precedence of technological innovation for its own sake, often on the assumption that new technologies can create a new market if one does not exist. If an organization sees itself as a 'technology company' (and many high-tech companies use this label as part of their self description), the tendency to focus on basic research or innovative product development can be overwhelm-

ingly strong ('It's who we are as a company!'). Then, if the organization wants to profit from the technology it develops from this strong internal (and identity-consistent) focus, it must actively push the technology out into the marketplace and try to discover or develop a market need.

Alternatively, the notion of market pull is characterized by technological design and development focused on serving an identified need or filling a niche in the market (thus the idea of the market 'pulling' the technology out of the organization). In this case, the organization brings its technical resources and capabilities to bear on a market problem to be solved. Innovative capacity is directed toward the creative solution of externally identified needs. The organization does not need to spend as much time or as many resources selling the technology, but also most likely has less discretion in designing and developing the specific aspects of the technology itself.

The Organizational Identity Issue

As noted, the difficulty for the organization in adopting a push versus pull orientation is found in the strong ties this tension has to the organization's identity. Organizational identity can be thought of as the shared theory that members of an organization have about who they are as a collective (Stimpert, Gustafson & Sarason, 1998), or as the perception of the organization held by insiders that helps to answer the question 'Who are we as an organization?' (Gioia, Schultz & Corley, 2000). If key members of the organization see their identity as being a collection of top-tier scientists dedicated to the production or advancement of scientific knowledge, then the notion of market pull, or 'science in the name of profits', is a tough sell and difficult to implement. Likewise, if the members of the organization are at heart business-oriented and see profit-generation as the driver of R&D decisions, the notion of 'science for science's sake' seems not only foreign, but downright dangerous for the organization's survival. A further difficulty might arise because often the organization is split in its desires, with top echelons of the organization being more market pull oriented and the lower echelons being more technology push oriented (e.g. managers versus scientists/engineers)—a case where an organization might be said to have 'multiple identities' (Pratt & Foreman, 2000).

In principle, neither pure state appears particularly healthy, especially over time. Pure technology push might lead an organization to be renowned for the technological advancement of the field, but it can endanger the company's chances at survival (especially if there is inadequate internal marketing expertise). Pure market pull, on the other hand, can result in better business standing and more profits, but might easily relegate the company to the status of a 'technology reactor', always behind the curve in terms of advancing technologies that can capture the markets' interest (especially when the timeline for patents is considered). Thus, some sort of middle range approach would seem to be most effective, because an organization able to balance this tension would be able to take advantage of the benefits on both aspects of the tension. Because technology push/market pull tension is so closely linked with organizational identity, however, the inertial pressures pushing an organization away from the center and toward one end of the dichotomy can be great.

Juxtaposing the Tensions

Figure 1 provides a graphical representation of four prototypical innovation contexts that arise by arraying these two tensions in a 2×2 matrix. The horizontal dimension represents the tension between *Structure* and *Chaos*, or how an organization answers the question 'How do we do things?' The vertical dimension represents the tension between *Technology Push* and *Market Pull*, or how the organization answers the question, 'Who are we?' Of course, Fig. 1 represents an oversimplified visualization of the tensions underlying organizational innovation. The bifurcations

between Structure/Chaos and Technology Push/Market Pull represent prototypes, whereas many organizations might find themselves straddling these extremities. To emphasize the details of these underlying tensions, however, we present a graphical description of the pure states (details in Fig. 2), along with a brief description of each innovation context.

Chaos, Structure, and Technology Push

As noted earlier, technology push involves an organization developing technologies and products without expressed consideration of immediate or specific market requirements. The tensions involved in managing structure/chaos tradeoffs within a technology push orientation is one that has implications for the breadth and flexibility of the R&D agenda and ensuing organizational mechanisms. A technology push R&D environment favoring a chaos approach is characterized by apparently haphazard endeavors focused on solving a loosely-articulated scientific problem. There is usually a high level of ambiguity as to what research questions need to be asked and how the likely solutions are to be generated and pursued. The search for

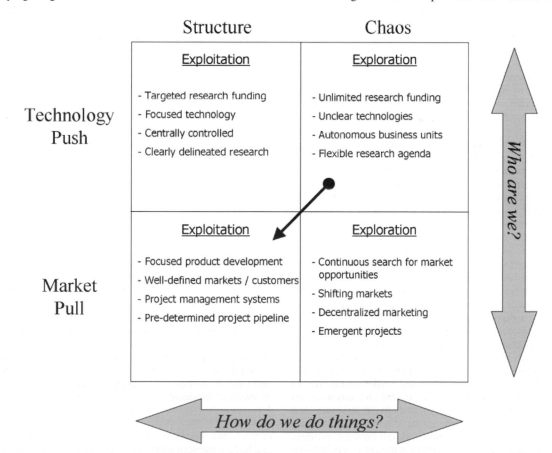

Figure 2. Inchoate's underlying assumptions and change path.

scientific breakthroughs is often serendipitous and is aided by an enhanced employee autonomy and decentralization of decision-making. Typically, adequate research funding is available to pursue a wide range of technological approaches. Politics, however, can be fierce, as autonomous units vie for the funding and attention from the best technology people.

A technology-driven organization that favors a structured approach is one that usually channels its funding and efforts into specific technologies. This focused technology push can arise for historical reasons ('this what we are good at' or 'this is what we do here'), for external reputation reasons ('this is what we are known for'), or even for prestige reasons ('this is what will bring us visibility'). The focal technologies, however, often involve long term projects where centrally controlled processes and procedures ensure that resources are applied to the right efforts at the right time. Close controls and monitoring are usually in place to guarantee that projects maintain efficiencies, possibly to the detriment of the groundbreaking discovery that a less focused process might produce.

Chaos, Structure, and Market Pull

The other aspect of the vertical dimension in Fig. 2 represents the market pull orientation. A market-driven organization is one that seeks to get closer to its customer's needs, forecast changes in the market, and stay ahead of the competition by pre-empting the moves of other players in the market (Day, 1999). In short, the organization is externally aligned to the changing requirements of the market. As depicted in Fig. 1, a market pull orientation can be accompanied either by chaos or structure, but with different outcomes.

In a turbulent external environment, a market driven organization that favors a chaos approach engages in a continuous search for new market opportunities. The internal organization is characterized by a continuous emergence of new commercial ideas that vie with each other for top management support. The marketing orientation is most often decentralized across the organization and the senior management encourages independent initiatives. The organization typically does not adhere to specific markets for a long term but engages in a continuous search for new and unexplored markets that might require completely different business approaches and technological expertise.

A market-driven organization that favors a structured approach is one that successfully positions itself in one or more markets and follows a leadership strategy. The emphasis on structure is meant to derive efficiencies from cross-pollination of creative ideas and commercial opportunities. The emphasis on structure also provides a certain amount of direction and predictability to the organization's developmental projects. Such an organization creates integrative mechanisms (Lawrence & Lorsch, 1967) in the form of well-defined

and widely-disseminated reporting systems and project appraisal mechanisms. These systems in turn provide an opportunity for the transfer of insights and learning from one part of the organization to the other. Structure enables a market-driven organization to successfully exploit product ideas and to maintain a healthy stream of creative initiatives contemporaneously. It thus able attempts to capitalize on the efficiency benefits of exploitation and the creative advantages of exploration.

Changing Innovation Contexts: A Case Example of Structure/Chaos, Push/Pull Innovation Tensions

Although examining the four 'prototypical' innovation contexts provides some insight into better understanding organizational innovation in high-tech companies, this static representation does not go far enough. Because modern organizations face rapidly changing competitive environments and fast paced technological shifts, different innovation contexts are often required to help cope with these changes. Therefore, in addition to understanding the contexts themselves, it also is necessary to understand some of the issues and implications of organizations moving among these innovation contexts. To help highlight these issues and implications, we employ a demonstrative case analysis of one organization's attempts to move within the innovation tension grid.

Innovation and Change at Inchoate Inc.

In this section, we examine the ways in which Inchoate Inc. (a pseudonym), effectively managed the innovation tension between structure and chaos, but struggled in trying to manage the innovation tension between technology push and market pull. It was Inchoate's difficulties in managing this more fundamental tension that characterized its problems as it attempted to adapt to changes in its competitive environments. We acquired access to the organization as a group of researchers interested in understanding the links between organizational identity, change, and innovation. During a span of six months, we carried out extensive one to two hour interviews with top executives and senior managers, and analyzed various secondary data sources such as the organization's intranet, internal corporate communications, and reports in the external media. Before examining the details of Inchoate's change experiences, it is helpful first to present an overview of its history and some of the more significant changes it underwent leading up to our study, because tracing Inchoate's historical trajectory is important to understanding its current innovation tensions.

Historical Roots

Inchoate Inc. began as a dedicated research laboratory of a large American electronics firm over a half century ago, flush with funds to do basic scientific research. It maintained that status until the mid-1980s, when its parent company was acquired by another large technology conglomerate that already had an R&D division, whereupon Inchoate was reconstituted as a free-standing R&D corporation wholly responsible for its own strategic direction and performance. While studying the organization, we found that the structure/chaos tensions faced by Inchoate Inc. were evident early in its history, even if they might be more prominent today. What has changed most dramatically for Inchoate is the nature and scope of its business and the level of turbulence in the external environment that is in direct conflict with its institutional heritage. During its time as the R&D division of its parent company, Inchoate's key human resources were composed almost solely of talented scientists and engineers. In those early years, they were focused on carrying out basic scientific research and developing innovative technologies in the fields where its parent company competed. During this initial period, the emphasis on the type of research alternated between basic research and applied research. Despite these fluctuations, though, the research arm continued to be inwardly oriented, with its research agenda defined either by the application requirements of the parent company or by the top management's visions of the future of science.

The next distinct period in the company's history, spanning the 1950s and 1960s was a time when the emphasis on basic scientific research gained clear ascendance over applied research. During this period, the research arm ventured into uncharted scientific areas that broadened its knowledge base and allowed its scientists to create new fields through their discoveries. The subsequent period in the early 1970s, however, saw a return to the application of technologies for solving problems for the manufacturing divisions of the parent corporation. In the late 1980s, when the parent corporation was acquired by a conglomerate that already had its own R&D division, Inchoate was spun off as an independent entity. The spinoff agreement stipulated that Inchoate would receive five years of (progressively decreasing) funding to help establish itself as an independent technology company. At the end of this period, Inchoate's scientists and engineers would lose access to the former parent's funds for basic research, as well as the closed, well-defined environment fostered by a parent corporation. As the weaning period wound down, Inchoate Inc. had to redefine itself as a freestanding organization, open to market and environmental forces, and responsible for seeking its own revenues and funds for further research.

The newly found independent status was a watershed in Inchoate's evolution. Suddenly, the same scientists and engineers whose main job for years had been to work on projects that excited their intellectual desires and aspirations were now forced to search for external clients and be sensitive to the needs of the

market. Funding for projects could now come only from Inchoate's own success at developing products and ventures. Only if a proposed project was deemed to have potential for market success would its own top management approve explorative funding. Science-for-science's-sake was no longer a criterion for funding. Scientists were asked not only to think of the technological and scientific implications of their work, but also the marketing, development, and distribution implications of any ideas they were pursuing. The organization began hiring market-savvy people with business development backgrounds to join project teams and help ensure that the scientists were focusing their efforts on technologies that could be turned into marketable products. Top management pointedly characterized this shift as one that moved Inchoate from being a 'technology push' to a 'market pull' organization.

This change, and the tensions it produced, plainly tapped a much more fundamental dialectic than the historical tension between structure and chaos. Up until the late 1980s, the challenge for Inchoate's precursor (the research arm) was to manage its irregular changes among basic research (an activity that demanded an unfettered and chaotic internal environment to facilitate serendipitous explorations in science and technology) and applied research (which required a periodic focus on more structured and orderly processes). From the early 1990s, however, this 'traditional' dialectic was compounded by the need to transform from a new technology-generating orientation into a market-responsive orientation—a transformation that represented a profound shift in the organization's identity.

Identity change at Inchoate

It is important to emphasize that most of Inchoate's key employees (including most members of the top management team) were scientists and engineers trained to carry out trailblazing research without previously having to think about revenue streams and cost efficiencies. Their only major constraint had been an occasional emphasis on focused applications of technology for meeting specific requirements of the parent corporation's manufacturing division. We believe that this state of affairs is, in principle, common to many high-tech firms, who often tend to hire mainly for technical expertise and then find themselves confronting environments that demand more business acumen.

During our involvement with Inchoate, we found that it had had great difficulty giving up this orientation, despite the demands of sometimes merciless markets, and despite the very real possibility that the organization might not survive *unless* a market pull orientation could be adopted. This difficulty in adapting is directly traceable to the legacy of the organization and the people who constitute it. The research scientists, as well as the top management members (who rose from the scientific research ranks), are predominantly university-trained researchers, whose main reason to work in an organization like Inchoate is to engage in creative research that satisfies their intellectual curiosities and justifies their superior training. To this day there is still greater value in the company for scientific brilliance over commercial viability (and thus also a strong value for a more chaotic approach to discovery and innovation).

Inchoate engendered and exacerbated this chaos in a different way by creating many independent business units based on loosely defined technology areas. High autonomy granted to the business units meant they could chart their own research agendas and hire their own research staff, although the central organization had broad control over profitability. This approach led to the creation of what senior executives described as 'stovepipes' out of the business units; i.e. there was no evident integration of resources and even little evident sharing of knowledge across the units. The only distinct prior effort to structure the technology push orientation in Inchoate that we found was a focus on licensing of technologies to the government and other companies. The licensing process required the development of organizational mechanisms that would enable the protection of Inchoate's intellectual property and allow its appropriation. We also found no evidence that the licensing department of Inchoate, which is the organization's intellectual property repository, has played a significant role in the creation and dissemination of knowledge from the intellectual property to the business units. For all these reasons, we classify the initial state of Inchoate (at the inception of their intended transformation) as being in the Chaos-Technology Push quadrant of Fig. 2.

The removal of virtually unlimited research funds as part of their (not necessarily welcome) independence led Inchoate into an era where it had to seek out new revenue sources that would fund future research agendas. To do that, Inchoate faced a new reality of listening to customers in the market who were not necessarily looking for brilliant technology but solutions to their specific problems. The change from 'creators of technologies' to 'marketers of solutions' has been a tough one for Inchoate's scientists and engineers. It has demanded that they change their long-held views on how research should be carried out. More importantly, it has brought into their laboratories an unpredictable and dynamic element—the forces and vagaries of the market.

Inchoate's top management, in its efforts to transform the organization into a market-driven organization, tried to impose a 'structure' approach to displace the existing 'chaos' approach. This distinction is especially important when viewed in the light of the starting point of this transformation (technology push/ chaos) to its intended outcome (market pull/structure). Inchoate instigated this intended transformation by

bringing in a new group of senior professionals who were not scientists and engineers, but rather people who knew market trends and who had business development, rather than technological skills. These senior managers were brought in to design and implement systems that would bring a high level of predictability to the research and developmental projects and ensure that they had commercial viability and revenue potential. Inchoate also adopted a novel structural initiative that employed two person teams made up of a technological specialist and a business development professional. These teams would champion research projects that had market potential and satisfied specific needs of customers.

Innovation, Change, and Inchoate's Problems

The bold arrow in Fig. 2 depicts Inchoate's path of intended transformation. The attempted change was not just in 'how' activities were carried out in the organization (structure/chaos) but also 'who' the organizational members thought they were (technology push/market pull). Throughout its history, Inchoate and the earlier research arm managed to shift between basic and applied research while still maintaining the essential character of the organization as based in scientific and technological excellence without much regard to the commercial utility. The change from technology push to market pull, however, required a redefinition of the fundamental character of the organization and its people. This is the elemental intersection of tensions that Inchoate's top management did not succeed in managing, even though they were able to recognize it.

Inchoate's lack of success in their transformation attempt can be ascribed, at least in part, to the adoption of a change in innovation strategy that tried to take a direct leap from chaos/technology push to structure/ market pull. Inchoate's problems with the hoped-for change were hindered because the change effort implicated two tensions simultaneously (both the 'how' and 'who' dimensions). The end result of the change effort was that although Inchoate was able to adopt a number of structural mechanisms to enhance the commercial viability of its research projects, it could not transform its fundamental organizational identity from technology push to market pull, thus rendering most of the structural changes ineffective.

Innovation as Managed Chaos—The Reverse 'Z'

The model presented in Fig. 3 highlights the key challenges faced by organizations that are driven by the simultaneous need to innovate and to alter their identities to accommodate dynamic innovation contexts. These challenges, as previously noted, are exacerbated in high turbulence environments where the bases for competitive leadership and survival constantly shift (a condition that now arguably applies to many R&D firms and high-tech industries). In this

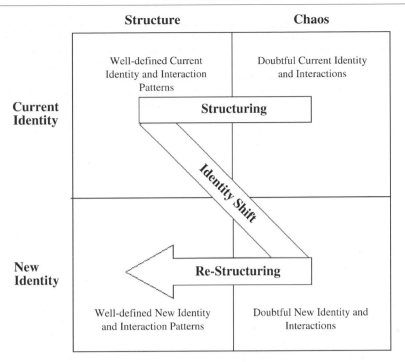

Figure 3. Managing the innovation tensions via managed chaos.

final section we attempt to enlarge our conceptual understanding of the link between organizational structure and organizational identity in the changing contexts of innovation by employing a 'structurational' lens (Giddens, 1984). The underpinnings of structuration theory provide a useful general framework for analyzing issues arising from shifts within the innovation tension grid because of its focus on how social actors continuously and reflexively create and recreate social structures. The key insight from structuration theory is that the interactions among actors serve the dual roles of forming the basis of social structures while also creating the means for changing those structures.

In brief outline, the process occurs as follows: Initially there is ambiguity among organization members concerning appropriate modes of thought, action, and interaction. Over a period of time, however, actors' interactions coalesce into structured patterns governed by rules that in turn both reflexively constrain and yet also enable further action. Structuration, or *structuring* (to highlight its inherent dynamism), can be viewed as a process by which patterns of interaction between members of a collective or an organization attain a rule-like 'objectivity', and these rules (which form the basis of the organization's structure) in turn delimit the behavior of the very organizational members who created them. Yet, rules and structured patterns can be reconstituted on an on-going basis via thought and

action to create revised structures (Nag, 2001). Thus, organizational structures are, by nature, fluid and continuously being shaped by organizational members, even as those structures act to constrain and guide organizational behavior.

The idea of structuration, therefore, provides a link between our previously discussed notions of chaos and structure. Most organizations start with ill-defined activities and nebulous goals and, over a period of time, develop relatively stable and temporally persistent coordinating mechanisms that can be discerned as their structures. However, these structures can be disturbed by environmental jolts (such as the one faced by Inchoate Inc. and many other modern high-tech companies), bringing about a state where the organization's activities and goals are again infused with ambiguity and are better characterized as chaotic. In highly turbulent environments, then, a shift can occur, from a somewhat stable conception of 'organizational structure' to a more dynamic view of 'organizational structuring' to better capture the interplay between present chaos and prior structure. In Fig. 3, this notion is captured in the two horizontal ends of the reverse 'Z' that represent the dynamic process of structuring/ restructuring, linking the states of structure and chaos.

Although the concept of structuring provides a link between chaos and structure, it still leaves us searching for a similarly useful fulcrum to track the second tension between technology push and market pull.

As noted before, we view this tension as more fundamental because it relates more directly to the organization's identity. Scholars of organizational identity typically define identity as organizational members' collective understanding of features presumed to be central, enduring, and distinctive for the organization (Albert & Whetten, 1985). However, recent theoretical conceptualizations of organizational identity view it as less enduring and more flexible in character than previously presumed (Corley, 2002; Gioia, Schultz & Corley, 2000). These views highlight the mutable character of an organization's identity as it continuously interacts with perceptions of the organization held by entities external to the organization. Although the labels used to describe organizational identity may remain stable during normal periods of incremental change, the meanings underlying those labels nonetheless tend to change over time as the organization attempts to adapt to changes in its external environments. In this manner, organization's enjoy the benefits arising from a sense of stability around answers to the question 'who are we?', while still finding it possible to adapt to fluctuating environments.

Extending this notion of an adaptable organizational identity to our innovation grid, we contend that an organization's identity develops and evolves contemporaneously with its structure. The process of structuring, which leads to the institutionalization of interaction patterns, can also lead to a crystallization of organizational identity attributes, especially in the labels used to describe the organization's core attributes. Although structure and identity might develop out of interactions between organizational members, however, the latter exists at a more fundamental and tacit level. Therefore, it is likely that an organization will respond to external changes, even dramatic environmental jolts, merely by changing its structures, without realizing a corresponding need to change its identity. That is, a horizontal movement in the innovation grid (illustrating a change in structure) might seem like an appropriate response to competitive pressures in one's industry, but might not be successful without a corresponding vertical movement (illustrating an identity change), as well. Thus, we posit that the structure-chaos tension and the more fundamental push-pull tension implicating an identity change need to be seen as mutually interlinked and recursive processes that should be carefully identified and managed by organizational leaders.

For us, the notion of '*managed chaos*' captures this sense of simultaneously managing necessary changes in organizational structure and organizational identity to cope with changing innovation contexts. Some scholars have focused on the importance of viewing chaos as a reality of organizations in highly dynamic industries and have prescribed continuously evolving and flexible organizational structures and systems to maintain an inherent instability (Brown & Eisenhardt, 1998; Nonaka, 1988; Schoonhoven & Jelinek, 1990). The notion of managed chaos extends this thinking by recognizing that it is necessary to surface and then challenge the present organizational assumptions underlying how the organization sees itself before structural changes can have a lasting impact.

The reverse 'Z' symbol encapsulates a plausible prescription for attaining managed chaos in the face of turbulent environments that force organizations to manage the interlinked tensions between chaos, structure, and organizational identity. In illustrating this notion of managed chaos in Fig. 3, we have replaced the first-order concepts of 'technology push' and 'market pull' with the labels 'current identity' and 'new identity' to help generalize beyond those organizations dealing with the push-pull tension. Other identity-related tensions exist beyond the push-pull tension and it is important to illustrate how the managed chaos model also applies to them.

After analyzing Inchoate's experiences, we concluded that its problems stemmed fundamentally from top management's attempts to change it from technology push to market pull through structural changes only (for example, bringing in a new team of business development professionals). That is, Inchoate failed to realize that the change it was undertaking involved more than just changing its internal structures; it also required a corresponding change in its identity. As depicted in Fig. 2, the new business development team tried to take Inchoate directly from a technology push/chaos mindset to a market pull/structure mindset. The technology push/chaos mindset of Inchoate was emblematic of the organization's longstanding inward focus on pure technology research and lack of formal structures to develop a specific set of technological ideas for commercial success. The new business development team tried to implement project evaluation processes that were meant to ensure the market relevance and commercial viability of the research projects carried out by Inchoate's scientists. This change, however, remained only superficial in nature. There was strong evidence for this phenomenon in the organization members' comments during the process of change. We found that in spite of having made structural changes towards the espoused market-pull transformation, the organizational members nonetheless continued to use identity adjectives that were reminiscent of their technology-push character.

Gioia et al. (2000) have argued that despite using the same identity labels in the face of incremental change, nonetheless, there can be subtle changes in the meanings and actions underlying those labels, which can create an adaptive instability in organizational identity that facilitates organizational change. Certainly, under conditions of incremental change the subtleties involved in maintaining familiar labels while underlying meanings and actions change can be quite useful in managing non-disruptive change. This line of thinking, however, can be modified and extended to account for more radical change. Inchoate's experiences suggest that deliberate changes in actions and structures (in the face of severe environmental jolts) might augur for a more pronounced change in identity labels themselves in order to execute transformative change. Changes in actions, structure, meanings *and* labels are likely to be necessary to produce managed chaos under such conditions.

Inchoate failed to unearth, and change, the basic assumptions accompanying the technology push paradigm. It failed to realize that to become a market pull company, Inchoate and its members had to change important aspects of their identity and fundamental assumptions underlying the nature of their business. Thus, the fact that they attempted a direct leap across the grid instead of the more steadying movement found in the reverse-Z model of managed chaos ultimately stymied the change effort, to the dismay of top management and the change champions.

Conclusion

Overall, on the basis of our recent research, we have concluded that to better understand effective innovation practices we need to expand the conceptualization of the dynamics involved in organizational innovation. We have identified two dimensions that apply to many high-tech firms and industries. Those dimensions focus on the tensions and tradeoffs involved in: (1) how an organization pursues innovation (ranging from very structured to relatively chaotic processes); and (2) how an organization conceives its identity (in terms of technology push versus market pull orientations). The intersection of these dimensions suggests four prototypical contexts, each with different requirements for managing innovation processes. Furthermore, the framework developed in Fig. 3 acknowledges that environmental shifts can and do demand changing innovation assumptions and practices. Changes in the modern environment have produced an inordinately complex innovation-management challenge—one that might be addressed by adopting a 'managed chaos' approach.

The concept of managed chaos suggests that organizations facing turbulent environments need to exhibit flexibilities in both structure and identity. The linkage between these two basic notions is an important fulcrum for understanding the tensions of managing organizational innovation in contemporary organizations. The reverse 'Z' model of managed chaos presented here beckons managers to proactively evaluate the organizational change initiatives in terms of their implications for both the emergent organizational structure and the organization's identity. A corresponding change in an organization's identity is essential for the new structure to make sense to the organizational members. If structure is the tool to achieve organizational objectives then identity is the lens by which organizational members look at themselves and the world outside. This hitherto underdeveloped relationship between 'How we do things' and 'Who we are' is an important dialectic that forms the crux of managing innovation in the face of extreme turbulence.

References

Albert, S. & Whetten, D. (1985). Organizational identity. In: L. L. Cummings & B. M. Staw (Eds), *Research in Organizational Behavior* (Vol. 7). Greenwich, CT: JAI Press.

Amabile, T. M. (1988). A model of creativity and innovation in organizations. In: B. M. Staw & L. L. Cummings (Eds), *Research in Organizational Behavior* (Vol. 10). Greenwich, CT: JAI Press.

Brown, M., Corley, K. G. & Gioia, D. A. (2001). Growing pains: The precarious relationship between off-line parents and on-line offspring. In: N. Pal & J. Ray (Eds), *Pushing the Digital Frontier* (pp. 117–134). New York: AMACON.

Brown, S. L. & Eisenhardt, K. M. (1998). *Competing on the edge: Strategy as structured chaos.* Boston, MA: Harvard Business School Press.

Cohen, M. D., March, J. G. & Olsen, J. P. (1972). A garbage can model of organizational choice. *Administrative Science Quarterly*, **17**, 1–25.

Corley, K. G. (2002). *Breaking away: An empirical examination of how organizational identity changes during a corporate spin-off.* Unpublished dissertation. The Pennsylvania State University.

Day, G. S. (1999). Creating a market driven organization. *Sloan Management Review.*

Ford, C. M. (1995). Creativity is a mystery: Clues from the investigators' notebooks. In: C. M. Ford & D. A. Gioia (Eds), *Creative Action in Organizations: Ivory Tower Visions and Real World Voices* (pp. 12–49). Newbury Park: CA: Sage.

Giddens, A. (1984). *The constitution of society: Outline of the theory of structuration.* Polity Press.

Gioia, D. A., Schultz, M. & Corley, K. G. (2000). Organizational identity, image and adaptive instability. *Academy of Management Review*, **25** (1), 63–81.

Lawrence, P. R. & Lorsch, J. W. (1967). *Organization and environment.* Boston, MA: Harvard Business School.

March, J. G. (1991). Exploration and exploitation in organizational learning. *Organization Science*, **2** (1), 71–87.

Nag, R. (2001). *Toward group learning as group learning.* Proceedings. 4th Organizational Learning & Knowledge Management Conference, University of Western Ontario, Canada, 415–424.

Nelson, R. & Winter, S. (1982). *An evolutionary theory of economic change.* MA: Harvard University Press.

Nonaka, I. (1988). Creating organizational order out of chaos: Self-renewal in. *California Management Review.*

Pratt, M. G. & Foreman, P. O. (2000). Classifying managerial responses to multiple organizational identities. *Academy of Management Review*, **25** (1), 18–42.

Schoonhoven, C. B. & Jelinek, M. (1990). Dynamic tension in innovative, high technology firms: Managing rapid technological change through organizational structure. In: M. A. Glinow & Mohrman (Eds), *Managing Complexity in*

High Technology Organizations. Oxford, New York: Oxford University Press.

Stimpert, J. L., Gustafson, L. T. & Saranson, Y. (1998). Organizational identity within the strategic management conversation: Contributions and assumptions. In: D. Whetten & P. Godfrey (Eds), *Identity in Organizations: Developing Theory Through Conversations* (pp. 83–98). Thousand Oaks, CA: Sage.

The International Handbook on Innovation
Edited by Larisa V. Shavinina

Involvement in Innovation: The Role of Identity

Nigel King

Department of Behavioural Sciences, University of Huddersfield, U.K.

Abstract: The impact of innovation on organizational members has been examined in various ways. However, relatively few studies have addressed how innovation processes shape people's work-related identities (and vice versa). This chapter argues that the concept of identity is useful for considering the relationship of the person to the organization in the context of innovation. It evaluates the potential contributions of Social Identity Theory and Constructivist/ Constructionist accounts of identity to this area. Finally, it presents data from case studies of innovations in the British health service to illustrate the value of an interpretive approach to identity and innovation.

Keywords: Innovation; Identity; Functionalism; Postmodernism; Interpretivism.

Introduction

Innovation has been a prominent research topic in work and organizational (w/o) psychology for over two decades, and as such can claim many successes. It has, for example, shown how organizational structures can impede or facilitate innovation (Miles & Snow, 1978); it has highlighted the problems that often occur through autocratic leadership and the refusal to devolve decision-making power (Bass, 1985, 1999a, 1999b; Kanter, 1983); it has demonstrated the key role that climate and culture can play in shaping the innovative performance of teams and organizations (Nystrom, 1990; West & Anderson, 1992). This is not to say that we now have definitive answers to the key questions in such areas—on the contrary, they remain the foci of debate, as new and competing ways of explaining the phenomena are advocated. What we *do* have is a sense of a vigorous, constructive discussion capable of clarifying and refining our understanding, and offering managers and practitioners potentially valuable insights into the what is happening in their own organizations. In some areas, though, there has been a paucity of theoretical and empirical research and as a result a failure to properly address important real-world issues. For example, there has been a much greater focus on the processes by which new ideas are generated and adopted in organizations, and much less on the implementation and routinization of change (Kimberly, 1981; King & Anderson, 2002). I will argue in this chapter that

another major area of neglect has been the involvement in the innovation process of organizational members *other* than those with direct managerial responsibility for it. After a critical overview of the relevant literature, considering the likely causes of this neglect, I will propose that occupational identity is a useful concept to utilize in developing our understanding of involvement in innovation. I will illustrate this point with examples of current and ongoing research into innovation in primary health care settings in the United Kingdom. Finally, I will conclude by drawing some implications for future research and practice.

Defining Innovation

Before turning to the main theme of this chapter, it may be useful to consider how innovation should be defined. The difficulty of defining innovation has long been recognized, and while there is consensus that novelty is a central feature of innovation, a range of positions are proposed regarding the degree of novelty (absolute or relative?), the scale of changes which 'count' as innovations, and the way innovation should be distinguished from related concepts such as creativity and organizational change (Kimberly, 1981; King & Anderson, 2002; West & Farr, 1990; Zaltman et al., 1973). Nicholson (1990) argues that any attempt to define innovation objectively is undermined by the fact that the phenomenon is intrinsically relative and subjective—how innovative a new product, process or procedure is depends on the perspective of the observer. He likens this to an audience at a jazz concert; what may sound like novel and creative improvisation to one member, may be recognized as derivative and unoriginal by another. This position is congruent with the central arguments of the present chapter, in terms of its advocacy of non-positivist approaches to the study of innovation. I would therefore claim that the question of 'what is innovation?' is not one to be answered in the abstract prior to commencing research, but should itself be a key focus of innovation research.

Involvement in the Innovation Process: Overview of the Literature

In identifying it as a neglected topic within innovation research, I am not suggesting that writers have thought organizational member involvement to be of no great importance. On the contrary, many of the common recommendations from innovation research are based on assumptions about organizational member responses to, and participation in, change; from Lewin's (1951) early work on democratic leadership styles to the literature championing transformational leadership (Howell & Higgins, 1990). The rationale for promoting participative leadership styles, for example, is that these will result in members feeling a greater sense of ownership in innovations, reducing the likelihood of resistance (e.g. Kanter, 1983). Similarly, arguments for risk-tolerant climates and cultures are based on a recognition that fear of the consequences of failure can inhibit individual and team propensity to innovate (e.g. West, 1990). The problem is that most of this literature has either treated organizational member involvement as a black box (between the 'inputs' of structure, leadership, resources etc. and the 'output' of innovation), or has reduced its complexities to the single issue of 'resistance'. Exceptions include a relatively small number of studies that have sought to understand innovation involvement as a meaning-making process, related to values, relationships and inter-group dynamics (e.g. Meston & King, 1996; Symon, 2000).

Resistance to Change

The literature in this area has long been driven by a strong practical concern to offer reliable advice to organizations on how to overcome resistance (e.g. Kotter & Schlesinger, 1979; Lawrence, 1969). While this applied focus is in many ways commendable, too often it is associated with a set of unacknowledged assumptions that can result in a very blinkered view of the processes involved. First, work on resistance is often infused with the kind of 'pro-innovation bias' discussed by Rogers (1983), which portrays innovation in general as a 'good thing' and thus resistance as 'bad'. Second, the research agenda is commonly dominated by the concerns of senior management, who

control the researcher's access to the organization, and in some cases may have directly funded the research. Finally, there is also a bias in much of the literature towards pluralistic models of change (as Glendon, 1992, notes) which recognize that different groups in an organization will have different perspectives on innovations, but assume that if the process is managed effectively, all parties can benefit from positive outcomes.

The result of this combination of biases is to make questions such as 'how can organizations best overcome resistance to innovation?' appear neutral and non-contentious. I would argue on the contrary, that applied to any particular example this question is almost inevitably controversial and politically-charged. As with the concept of innovation itself (e.g. Nicholson, 1990), what counts as resistance depends on one's perspective (Aydin & Rice, 1991; Burgoyne, 1994). Thus a group of staff perceived by managers to be willfully resisting change may see themselves as struggling to do the best they can with increased demands, or attempting to make sense of unclear aims and expectations. Managers may believe that an innovation is clearly for the general good, and thus interpret opposition as irrational intransigence; staff may view it as inevitably producing winners and losers in terms of power, status and/or material rewards. The very notion that resistance *should* be overcome becomes dangerous if one accepts that some innovations may have a socially malign impact. If an administrative innovation in a hospital was putting patient safety or confidentiality at risk, would we feel comfortable in advising managers on how to overcome the resistance of doctors and nurses?

It is not my intention here to dismiss as irrelevant the bulk of previous research for exhibiting the biases I have noted above. There is much valuable work showing, for example, the contingencies that managers should consider when adopting a change-management strategy (Kotter & Schlesinger, 1979), or the key role of two-way communication in minimizing resistance (Plant, 1987). My point is rather that by failing to recognize the unspoken assumptions behind their work, researchers provide a very partial view of the way organizational members respond to, and are involved in, innovation. This needs to be complimented by research taking different perspectives which move away from the 'overcoming resistance' framework in their conceptualizing of involvement in the innovation process.

Meaning-Making in the Innovation Process

Rather than locating organizational members' reactions to innovation on a dimension of resistance—acceptance, a small but growing number of researchers have focused on the ways in which they construct their responses, and the purposes such constructions serve (e.g. Bouwen & Fry, 1996; King, Anderson & West,

1991; Whetten & Godfrey, 1999). Although this research comes from a variety of theoretical perspectives, much of the empirical work uses a qualitative methodology, because of its suitability for examining meaning-making in specific contexts. One example of this kind of research is a study I carried out with Carolyn Meston (Meston & King, 1996), looking at the introduction of a staff training innovation (National Vocational Qualifications, or 'NVQs') in a residential care home for older people. We used a grounded theory methodology which seeks to build a theoretical account of a particular phenomenon through detailed qualitative analysis (Strauss & Corbin, 1990). This 'bottom-up' approach to the development of theory is in contrast to the traditional 'top-down' perspective of positivist social science. Data were gathered by participant observation over a six month period, supplemented by semi-structured interviews with key informants. The grounded theory model developed from our analysis is shown in Fig. 1. As can be seen, personal values in the workplace, and the extent to which individuals felt a need for peer group approval, were important factors shaping the way members of staff evaluated the innovation and responded to it in specific situations. To take one example; Michael (a Care Assistant) strongly values his social relationships at work, but also gains great satisfaction from a sense of doing his job well. These values pull him in two directions in his responses to NVQs. He perceives the peers whose company and esteem he values as largely hostile to the innovation, but at the same time he personally can see the benefits of the new training for quality of care. The result is that the responses he exhibits to NVQs vary markedly depending on which of his colleagues he is working alongside on a particular shift.

A different perspective on meaning-making processes in relation to innovation comes from scholars drawing on ideas from social constructionism. This tradition has developed most strongly in European Social Psychology, and is centrally concerned with the way our social reality is constructed through the everyday use of language. It eschews any attempt to 'explain' social phenomena in terms of such internal psychological factors as 'beliefs', 'attitudes', 'personality traits' and so on. In terms of empirical research, it utilizes a range of methods—most prominently discourse analysis (e.g. Burman & Parker, 1993)—which seek to show in fine detail how language operates in particular texts to construct particular versions of social reality. (See Burr, 1995, for a good overview of the differing strands of social constructionist theory). René Bouwen and colleagues have argued that organizational innovation can be seen as 'a joint conversational event where new configurations of meaning are constructed' (Steyaert, Bouwen & Van Looy, 1996, p. 67). They use the term 'logics' to refer to the different stances organizational members may take

towards an innovation, as encapsulated in their talk about it (Bouwen, De Visch & Steyaert, 1992). Using interviews from case studies of organizational innovations, they show how old and new logics compete within participants' accounts; for instance, in one Flemish high-tech company, an established 'academic', research-oriented logic is challenged by a 'commercial' customer relations-based logic in the wake of innovations introduced by a new, entrepreneurial management (Steyaert et al., 1996).

A different social constructionist approach is taken by Symon (2000), who examines the use of rhetoric in the construction and justification of responses to innovation. The case she considers is the introduction of a networked PC system in a British public sector organization. The professional staff (referred to as 'inspectors') utilize a range of rhetorical devices to justify their opposition to the innovation. They construct an identity for themselves as technically expert, and as the staff group who best understand the business of the organization (they use the metaphor of working 'at the coal face'). This provides legitimacy for their criticisms of the position of IT specialists who are championing the innovation. They also turn the main argument of proponents of the change on its head—that far from saving resources, it will waste them, in the form of the valuable time of professional staff who will be expected to do their own typing as one of the consequences of the new system.

Qualitative research focusing on meaning-making in responses to innovation has begun to make apparent the complex ways in which organizational members may be involved in innovation processes (Symon, Cassell & Dickson, 2000). When we look closely at individual cases it becomes clear that traditional quantitative methodologies such as attitude surveys present a static and rather superficial view of the way people are affected by and involved in innovation attempts. As such, they can only provide a partial view of why organizational members respond as they do, and offer little insight into how their responses are formed. For critics of qualitative approaches (e.g. Morgan, 1996), their principle limitation is that through their insistence on the over-riding importance of context, they are unable to draw generalized conclusions about organizational innovation, which may serve as the basis for recommendations for practice. I will return to this issue in the concluding section of the chapter.

Innovation and Identity

As the previous section shows, researchers have begun to move towards a more rounded and context-sensitive view of involvement in organizational innovation (Bouwen & Fry, 1996; Carrero, Peiró & Salanova, 2000; Meston & King, 1996; Symon, 2000). Longitudinal research examining the development of particular innovations over time has also helped highlight the complexities of the process, underlining the

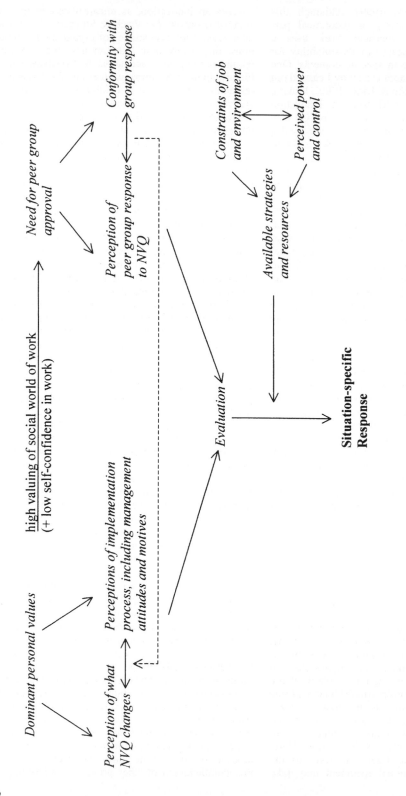

Figure 1. A grounded theoretical model of responses to the introduction of a training innovation at 'Hazel Hill' nursing home.

Reproduced by kind permission of the authors and *Psychology Press*, from Meston, C. M. & King, N. (1996). Making sense of 'resistance': Responses to organizational change in a private nursing home for the elderly. *European Journal of Work and Organizational Psychology*, **5**, 91–102.

inadequacies of simple, uni-dimensional understandings of member responses; the Minnesota Innovation Research Program (MIRP) is the most important example of this kind of work (Van de Ven, Polley, Garud & Venkataraman, 1999). I want to argue here that to further develop our understanding in this area, a focus on issues of identity construction and change should play a central role.

Relevance to Innovation Research

Identity is a key concept in Social Psychology and Sociology, and one which is utilized in a variety of ways to address a range of issues in those disciplines. Du Gay, Evans & Redman (2000) argue that 'the term 'identity' often provides only simple cover for a plethora of very particular and perhaps non-transferable debates' (p. 2); it is therefore likely that any attempt to provide a universally-acceptable definition will be in vain. For the purposes of this chapter I will propose a broad working definition, taking *identity* to refer to *a person's relatively enduring sense of who they are*. Two points of elaboration are required here. Firstly, I assume identity to be intrinsically social; one can only have a sense of oneself through a sense of how one relates to others—be they specific individuals and groups (friends, family, workmates and so on), or wider social categories (such as gender, class or ethnic group). Secondly, in saying that identity is relatively enduring, I am not implying that there is necessarily anything essential about it. Whether there are essential and universal aspects to identity is a question of vigorous debate amongst different theoretical traditions, such as psychoanalysis (Frosh, 1989), social constructionism (Burr, 1995), symbolic interactionism (Blumer, 1969), and many others (Du Gay et al., 2000).

Despite its prominence in the social sciences generally, the concept of identity has received relatively little attention in work and organizational (w/o) psychology. This is beginning to change now, for example through the recognition that studying diversity at work inevitably means studying identities (Nkomo, 1995). I would contend that the field of organizational innovation and change—and especially the topic of organizational member involvement in innovation processes—is a particularly fruitful one in which to employ it. This is principally due to the fact that innovation and change commonly have a strong impact on work-related identities. The nature of this impact is highly varied. Because innovations often result in changes to work roles and practices, people may feel their occupational identities to be under threat, and thus may adopt a resistant stance—as was the case for the 'inspectors' in Symon's (2000) study, cited above. Innovations may also surface tensions between different aspects of identity; Michael's ambiguous response to NVQs in Meston & King's study (1996) (*op cit*) may be interpreted as resulting from a conflict between his peer group identity and his occupational identity as a caring, competent Care Assistant. Finally, people may reconstruct their identity in response to organizational innovation, a process exemplified in Steyaert et al's (1996) (*op cit*) depiction of the shift from old to new 'logics' in the course of change implementation.

Approaches to Identity and Innovation

In the recent edited volume 'Identity in Organizations' (Whetton & Godfrey, 1998), Gioia (1998) distinguishes three main types of theoretical approach to the area, which he describes as 'lenses for understanding organizational identity'; functionalist, postmodern and interpretivist. I will use the same categories to review the main theories which have been, or could be, applied to issues of involvement in innovation processes.

1. Functionalism: Social Identity Theory

One of the most influential theories in Social Psychology over the last two decades has been Social Identity Theory, as devised by Henri Tajfel (1978) and developed by writers such as Turner (Turner, Hogg, Oakes, Reicher & Wetherell, 1987), Hogg & Abrams (1988), and Brewer (1991). The theory states that in order to make manageable the huge amount of social information available to us, we rely on a cognitive process of categorization to simplify it. Because we are motivated to view ourselves in a positive light, we seek ways of comparing the groups we identify with ('in-groups') favourably with those we do not identify with ('out-groups'). Note that the theory sees identity as plural, but with different identifications being salient in different circumstances. For example, a person's identification as a supporter of a particular soccer team may be unimportant at work, and therefore not serve as a basis for in-group/out-group comparisons. In contrast, when attending a match, the supporter identity will be highly salient, and 'superiority' over the opposing team's supporters is likely to be symbolized in colourful verbal exchanges between the groups.

Social Identity Theory has had its greatest impact in w/o psychology in the area of diversity (see Jackson & Ruderman, 1995, for several examples). It is equally applicable to the study of involvement in and responses to innovation. There are many situations in which the nature of group identifications is likely to influence how people respond to innovations. The effectiveness of change agents, for example, may depend on the extent to which they are seen as representing a disliked or distrusted out-group by staff in a position to effectively undermine an innovation attempt. Innovations may make salient inter-group distinctions which had previously been unimportant, by seeming to benefit some groups at the expense of others, raising the likelihood of inter-group conflict focused on the innovation. Equally, innovations may stimulate people to change their identifications, in order to maintain a positive evaluation of the in-group. Social Identity

Theory can thus provide a well-developed framework for studying involvement in innovation, allowing predictions to be made and tested about organizational member responses.

Despite these strengths, some important criticisms of Social Identity Theory have been made, which are pertinent to its application to the area of organizational innovation. Hartley (1996) has pointed out the substantial differences between the kinds of groups used in most experimental studies and the groups that exist in work organizations. Experimental groups are, for example, lacking in history, untroubled by internal power and status issues, expect no long-term consequences as a result of their actions and decisions, and are usually composed of schoolchildren or students. She argues that the effects on identification produced by the most trivial manipulations of categorization in such experimental groups may well not occur in work groups where history, power, status, and anticipations of future consequences are likely to be highly salient to members. She also suggests that there is a need to distinguish between 'group identification' which may be relatively transient and open to change, and 'social identity' which is more enduring and resistant to change.

2. Postmodern Critique: Social Constructionist Perspectives on Social Identity

Hartley's warning against assuming that effects produced in experimental groups will be found in real-world ones raises the issue of the extent to which group context shapes social identity processes. Because Social Identity Theory proposes a universal social-cognitive mechanism of categorization as underlying social identity formation, it is open to the criticism that it underplays the extent to which the context of a group influences the nature of social identifications within it. Social constructionist critics emphasize that identities are constructed (and reconstructed) through everyday interactions, drawing on the discourses of the person's society and culture to achieve particular ends in particular contexts. One consequence of this is that identities are more shifting and unstable than Social Identity Theory suggests, as people use them for differing purposes—even in the course of a single interaction (Potter & Wetherell, 1987). To give a hypothetical example relevant to the topic of this volume; a middle-manager might in the course of a conversation with her superior identify herself with the traditions of the company at one point, and as an enthusiast for innovation at another. The apparent conflict between 'traditionalist' and 'innovator' identities does not imply she is being intentionally deceitful; it simply reflects the multiplicity of identities, which are available to this individual.

A further social constructionist critique of Social Identity Theory focuses on the role of argument and persuasion in identity construction. Michael Billig

(1987) contends that Social Identity Theory presents a model of identification processes which is rather mechanistic, operating as a kind of 'bureaucratic filing system' in which a particular ready-formed identity is retrieved or disposed of according to the prevailing social contingencies. In contrast, Billig (1987) claims that identity formation and change occurs chiefly in the context of dialogue and argument; we discuss, challenge, debate each others' identifications in a two-way process through which both sides' identities may shift.

Social constructionist approaches to identity have played an important part in drawing attention to the role of everyday interaction in the forming and reforming of identities. This has been recognized even by some influential adherents of Social Identity Theory such as Abrams & Hogg (1992), though they insist that there is still a need to understand the identities apparent in talk in terms of underlying cognitive mechanisms. Other critics who are more sympathetic to social constructionism's rejection of universalism and cognitivism are troubled by its tendency to 'lose the person' in its accounts (e.g. Crossley, 2000) and to deny the possibility of personal agency.

3. Interpretivism: Symbolic Interactionist and Phenomenological Approaches

Symbolic interactionism (Denzin, 1989) and phenomenology (especially phenomenological psychology; Giorgi, 1970; Moustakas, 1994), though distinct and broad traditions, share common features in the way they theorize identity. Like social constructionism, they argue for identity to be understood in context, and emphasize its location in everyday interactions. They are, however, less exclusively focused on the minutiae of language use and stress the importance of understanding the personal and collective projects that people engage in. Looking at an occupational group such as lawyers or doctors, for example, interactionists have examined how they have organized themselves in order to achieve the status of 'profession', and at how individual members of the group take on values and develop their careers in a way that enables them to become recognized as successful 'professionals' (Macdonald, 1995). Although interactionist and phenomenological writers do not see identity as in any sense fixed, they tend to represent it as less fluid and shifting than do social constructionists. Biography (individual experience) and history (collective experience) constrain the ways in which identity can be formed just as the immediate social context does (Denzin, 1989; Moran, 1999).

Within this tradition, Butt (1996, 1998, 2001) has developed an approach to identity which has much to offer to the study of organizational member involvement in innovation. He presents an interpretation of Kelly's (1955) Personal Construct Psychology (PCP) as a phenomenological theory, also drawing strong parallels with G. H. Mead's interactionism (Butt,

2001). PCP states that each person has their own unique set of constructs, representing the common ways they perceive themselves and their world. Constructs are usually seen as cognitive entities located within the individual (e.g. Mancuso, 1996). In contrast, Butt (2001) emphasizes that construing should be seen as a form of social action, and constructs therefore located principally in our interactions with others. Furthermore, he supports the view of the phenomenological philosopher Merleau-Ponty (1962) that we generally do not stop to reflect before acting; rather, that most of our engagement with the world is 'pre-reflective'.

Organizational innovations commonly disrupt the patterns of social interaction through which, according to Butt (2001), our identities are constructed. As a result, organizational members may be drawn to deliberate upon aspects of identity, which are usually taken for granted and not the subject of reflection (i.e. normally remain pre-reflective). Precisely how any one member construes the implications of organizational change for their identity at work is neither the result of purely individual cognitions nor is it determined by social structural forces. Instead, it is mediated by their personal construct system, which—though unique to each person—will inevitably reflect the constraints of a particular organizational and professional/occupational context. Butt (2001) uses the example of fashion to illustrate how context moulds and limits personal agency;

> Our personal constructs do not arise in a vacuum, but in the context of the social constructs which surround us. Just as it is difficult to choose clothes which are not manufactured and on offer, so we cannot easily be a particular man or woman that is not sketched out in our culture (p. 90).

Burr & Butt (1997) have used this approach to examine personal change at work following role change from shopfloor worker to Supervisor, a situation which has some parallels with that of innovations that result in role change. In the next section of this chapter I will provide an outline of research from this perspective looking directly at how organizational members respond to innovation and change.

Innovation and Identity in a Primary Health Care Setting

In the British National Health Service (NHS), most patients' initial access to healthcare is through the primary care system. This consists principally of community-based General Medical Practitioners (GPs) and the teams of other health professionals employed by or attached to their Practices, including Community Nurses and a range of therapy services. Most incidences of ill-health are dealt with within the primary care system, with only a minority being referred on to specialist secondary care in hospitals. Primary care is an especially fruitful area for the study of innovation and identity for two main reasons. Firstly, over the last decade or so it has been subject to an unprecedented series of radical policy, organizational and practice changes, instigated by successive Conservative and Labour governments (e.g. Department of Health, 1989, 1998, 2000). Secondly, the work force consists of several professional groups with strong—and in the case of medicine and nursing, long-established—identities, which are likely to be challenged by the recent and current macro-level changes in the sector. Particularly challenging is the increased emphasis on multi-disciplinary collaborative working, both amongst different health care professions and between the health and social care sectors (Burch & Borland, 2001; Poxton, 1999). (Note that in the U.K., social care is the responsibility of Local Authority Social Services departments).

The Primary Care Research Group at the University of Huddersfield has been involved in evaluating a number of multi-disciplinary service innovations (e.g. King, Roche & Frost, 2000; King & Ross, in press). I would like to draw on data from two of these ongoing projects to illustrate how identity (defined from an interpretivist position) can be a valuable concept in understanding organizational members' experiences of and involvement in innovation processes. Both innovations were concerned with the provision of care outside of the normal working daytime hours, and were led by Community Nursing services; however, to function effectively both required good collaboration with social services staff, with other primary health care professions, and with the secondary care sector. I will provide some brief background to each innovation, before focusing on some specific points relating to identity. Pseudonyms for places and people will be used throughout, to protect confidentiality.

Case One: Fast Response Service, Buckton

Buckton is a large city in the north of England, with a population of around half a million which includes a substantial ethnic minority population (especially from the Indian sub-continent, but also Afro-Caribbean and Eastern European). It encompasses several significant areas of high deprivation. The Fast Response Service was set up in 1998 to provide short periods (up to two weeks) of support for patients in their own homes during episodes of acute illness. A major aim was to prevent admission to hospital in cases where this was not medically imperative, but without a service like the FRS would be necessary because of a lack of support in the home. The lead role in the service was played by District Nurses (community nurses with a specialist post-registration qualification); they oversaw the provision of care in the home (for up to 24 hours per day) by Health Care Assistants (who are not professionally trained). Many of the patients receiving the service were already receiving support from social services;

those who were not at the time FRS went in often needed to do so once discharged. Because of these issues of co-ordination and hand-over between services, social service managers were closely involved in the development of FRS. It should be noted that the boundaries of community nursing and social services teams within the city were not co-terminous at the time of the study.

Case Two: Out of Hours District Nursing Service, Spilsdale

Spilsdale is a borough in the north west of England, covering a mixture of urban, suburban and rural areas. Almost half of its 200,000 population live in the main town, Wigglesworth, which has a sizeable ethnic minority (like Buckton, predominantly from the Indian sub-continent) and some areas of high deprivation. The more rural parts of the borough comprise a number of small towns and villages, some relatively isolated and all with a strong sense of local identity. The Out of Hours District Nursing Service (OHDNS) was set up to address the lack of night-time and weekend District Nursing cover. As with the FRS, the introduction of the new service had important implications for co-ordination with social services, and close liaison between the two was planned, as well as with medical out of hours services. Unlike Buckton, Spilsdale's community nursing and social service boundaries were co-terminous.

Method

In both cases a qualitative methodology was used to examine the experiences of staff involved in the service innovations. The principal data collection method used was focus groups, supplemented by individual interviews. While a wide range of primary care, secondary care and social services staff were included, for the purposes of this chapter we will concentrate on the two main professional groups concerned; District Nurses and Social Workers. Summary details of participants from these professions are given in Table 1, below. All interviews were transcribed and analysed thematically, using the 'template' approach (King, 1998). For fuller details of the methodology see King & Ross (in press).

Identity Issues in the Two Service Innovations

As noted earlier, interpretivist views see identity as constructed in the interactions we undertake in our everyday lives, though inevitably also shaped by wider historical and cultural forces. The interactions of professional nurses and social workers are strongly related to the roles associated with the two groups. In our case studies the innovations impacted on professional identities because they led to perceptions or anticipations of changes to professional roles and the relationships with colleagues and patients/clients associated with them. I want to highlight below three ways in which this was manifest: in experiences of role uncertainty, in perceptions of role erosion, and in perceptions of role extension.

1. Role Uncertainty

In the two cases, some staff from both professions complained about uncertainty and confusion surrounding their roles in the wake of the innovations. Very often it was not the new service *per se* that caused difficulty; rather it was a whole range of changes, both local and national, of which the FRS and OHDNS were particular instances. This is typified by a comment from Abbie, a social services team leader in Spilsdale;

> It's very confusing. I know there are a lot of things coming up, new initiatives. Everybody is left with their heads spinning and thinking 'where does that fit into that?' I think we need to be sure who's doing what, and how we access it, and who provides what!

Role uncertainty was not inevitably seen as a threat to valued aspects of professional identities (though in some instances it could be), but it was often associated with perceptions that identities were in some degree of flux. The following two sections show how this mutability could be interpreted in quite different ways, even by members of the same professional group in the same innovation case.

2. Role Erosion

A number of participants described perceptions of 'role erosion'; that valued aspects of their role were being lost because of national and local changes in primary

Table 1. Participants in the two primary care innovation studies.

Innovation example	Management of service	Community Nursing staff	Social Services staff
Buckton City Fast Response Service	1 focus group (n = 6) 1 individual interview	2 focus groups (n = 5, n = 7)	3 focus groups (n = 3, n = 3, n = 4)
Spilsdale Out-of-hours District Nursing service	1 focus group (n = 2)	5 focus groups (n = 7, n = 3, n = 5, n = 5, n = 9)	4 focus groups (n = 3, n = 12, n = 7, n = 5)

health care and/or social care. These perceptions could colour their responses to the particular changes we were focusing on, leading to a suspicious stance even where the innovation itself did not appear to offer any real threat to their role. Often, the 'other' professional group were blamed for this erosion, potentially making collaboration in new services more difficult;

> Don't get me on about Social Services! I just feel totally out of my role. They have skimmed off the top, I feel left with the odd jobs (Julie, District Nurse: Spilsdale).

This perception of role erosion was shared by some of the community nursing managers. Sheri, Co-ordinator for the Out-of-hours scheme in Spilsdale, and a qualified District Nurse herself, expressed anxiety about the way further developments in the service might impact on her staff;

> I do worry—it does concern me the way they've always been devalued really and undermined umm and its like, you know, to give you (an example) to do with the out of hours or 24-hour District Nursing—at the moment the Trust are looking at developing a sort of crisis intervention team in addition to the 24-hour District Nursing service— but that is going to be led from the hospital . . . . and so you are going to have staff who don't have the District Nursing qualification assessing patients in the community to decide whether or not its appropriate for their care to be maintained at home, or whether they need to go into hospital, or into the intermediate care bed or discharged form hospital out into the community—well, you know in my opinion that should be a role for—that is for a trained District Nurse really.

Perceptions of role erosion related to service changes were probably more prominent amongst District Nurses than Social Workers. This may reflect the fact that nursing is a more developed profession than Social Work, with a generally more positive public and media image; they thus may have more to lose if new collaborative arrangements alter the boundaries between the two professions.

3. Role Extension

In direct contrast to those participants who felt their professional identity was threatened by role erosion were those who construed organizational changes as enhancing or extending their roles. In these cases, staff showed a willingness to redefine what it means to be a District Nurse or a Social Worker;

> We do things now that we wouldn't have done 20 years ago, it's a progression—things are still changing—it's exciting really (Tina, District Nurse: Spilsdale).

Note that the above quote is from the same professional group in the same case as the previous example of perceived role erosion. This illustrates an important point in interpretive approaches, although the construction of identities is strongly shaped by immediate and wider social contexts, it is also personal—reflecting the particular pattern of relationships and experiences of an individual professional. Divergences in interpretations of what organizational innovations imply for identities are therefore to be expected. However, in defining identity as a personal and social construction, we must be wary of slipping back towards an individualistic account which reduces social context to a set of 'external variables' having a secondary influence on essentially individual processes of identity formation (Social Identity Theory's emphasis on cognitive mechanisms of categorization is an instance of this). The person does not and cannot exist independent of their social context, even though she is not simply determined by it. In relation to our innovation examples, this was evident in the way that public perceptions of the two professions limited the scope for identity change; for example, Social Workers pointed out that many service users expected certain care tasks to be carried out by a uniformed nurse, and were uncomfortable with any redefinition of traditional roles and boundaries.

Conclusion

I have argued in this chapter that the involvement of organizational members (other than senior managers and other key decision-makers) in innovation processes is a neglected research area. There is a considerable amount of work on resistance to change, but this presents a rather partial and restricted view of organizational members' experiences. In responding to this neglect, I have proposed that the concept of identity is a particularly useful one to employ. I have outlined some of the main theoretical approaches to identity and illustrated my favored interpretivist position with examples from current research in a British Primary Healthcare setting. I want to conclude by suggesting future directions for research and practice in the area of identity and involvement in innovation.

Implications for Research

The particular interpretivist approach to identity an innovation that I have advocated, has two main strengths as a basis for further research. First, it emphasizes the need to examine organizational member involvement in the context of specific innovations and specific organizations. This is in keeping with other important developments in related areas of organizational research, such as the attention to the detail of innovation processes in the Minnesota Innovation Research Program (Van de Ven et al., 1999), or the arguments for sensitivity to context in research on work group diversity (Triandis, 1995). Second, it recognizes the essentially social nature of human beings, but also accepts that within bounds the individual organizational member has scope for a distinctive personal construal of an innovation. This approach therefore avoids the individualism of Social Identity Theory, and the denial of personal agency inherent in Social Constructionism.

In terms of an agenda for empirical work, the immediate need is for more interpretivist case studies of innovations from a widening range of types of organization. As more such material is published, fruitful areas for attention in future studies will become apparent. In the work on collaboration between health and social services in progress at Huddersfield, for example, our findings so far have led us towards a more fine-grained examination of the dynamics of personal relationships in the context of organizational change. From this we hope to gain a deeper understanding of the way professional identity is constructed in the everyday working lives of our participants, resulting in the kind of widely-varying perceptions of innovation and change that we noted above ('role erosion' versus 'role extension').

So far I have discussed identity in terms of the way organizational members perceive themselves. The concept can, however, be applied at a different level, to examine the way the organization as a whole sees and presents itself. As Fiol, Hatch and Golden-Biddle (1998) state;

> An organization's identity is the aspect of culturally embedded sensemaking that is self-focused. It defines who we are in relation to the larger social system to which we belong (p. 56).

Identity understood at this level is also relevant to organizational member involvement in innovation. Significant organizational innovations very commonly involve cultural change (whether intended or not), impacting on the identity of the organization. Member responses to innovation will be shaped by the nature of the relationship between personal identities and the changing organizational identity. For instance, where there is perceived to be an incongruity between the organizational identity and the professional/occupational identities of members, the result could be hostility, obstructiveness, or feelings of disempowerment and disengagement. Future research could usefully begin to examine these kinds of relationships between different levels of organizational identity.

Implications for Practice

Interpretive research into identity and innovation does not seek to produce general theories, which explain experience and behavior regardless of context. For some critics, this makes such work of limited value for informing practice. I would contend, however, that interpretive studies can make a strong contribution to practice, in several ways. Most directly, they can provide powerful insights for the organizations in which the actual research is carried out; insights not available to less contextually-informed approaches. They can also achieve transferability to other settings, through 'naturalistic generalization' (Stake, 1995), as I have argued elsewhere;

> The transferability of findings is based on the recognition of parallels between the research setting and other contexts, and must be on the basis of the reader's own understanding of other cases. This process must be facilitated by the researcher describing the setting, methodology and findings in sufficient depth to give the reader a strong grasp of the nature of the research context—what is commonly referred to as 'thick description' (Geertz, 1973) (King, 2000, p. 595).

Rather than seeing the reluctance to make generalizations as a weakness, I would view it as a useful corrective to the tendency in the innovation and change literature towards over-generalized prescriptions for managers (e.g. Kanter, 1984; Peters & Waterman, 1983). The one piece of general advice I am comfortable in giving is that innovation leaders should take into account the identities of organizational members when planning and implementing change. Interpretive research cannot tell them how to resolve specific issues in specific circumstances—that must be based on their own knowledge of their organization and its members. What it can do is sensitize them to areas, which may be of significance and warrant careful scrutiny.

To conclude, it is not my purpose here to claim that only the interpretive approach can advance our knowledge of organizational member involvement in innovation—although I do see it as having considerable strengths, as I hope I have shown. I would like to see growing emphasis on this area from a range of theoretical perspectives, because it is through dialogue and debate between different positions that we sharpen our concepts and enrich our understanding. Identity is, of course, not the only concept of relevance to this area, but it is one which is centrally important and which up to now has been largely overlooked by w/o psychologists investigating innovation.

References

Abrams, D. & Hogg, M. A. (1990). The context of discourse: Let's not throw out the baby with the bathwater. *Philosophical Psychology*, **3**, 219–225.

Aydin, C. E. & Rice, R. E. (1991). Social worlds, individual differences and implementation: Predicting attitudes towards a medical information system. *Information and Management*, **20**, 119–136.

Bass, B. M. (1985). *Leadership and performance beyond expectations*. New York: Free Press.

Bass, B. M. (1999a). Two decades of research and development in transformational leadership. *European Journal of Work and Organizational Psychology*, **8** (1), 9–32.

Bass, B. M. (1999b). Current developments in transformational leadership: Research and applications. *Psychologist-Manager Journal*, **3** (1), 5–21.

Billig, M. (1987). *Arguing and thinking: A rhetorical approach to social psychology*. Cambridge: Cambridge University Press.

Blumer, H. (1969). *Symbolic interactionism: Perspective and method*. Englewood Cliffs, NJ: Prentice Hall.

Bouwen, R., De Visch, J. & Steyaert, C. (1992). Innovation projects in organizations: Complimenting the dominant logic by organizational learning. In: D. M. Hosking & N. R. Anderson (Eds), *Organizational Change and Innovation: Psychological Perspectives and Practices in Europe*. London: Routledge.

Bouwen, R. & Fry, R. (1996). Facilitating group development: Interventions for a relational and contextual construction. In West, M. A. (Ed.), *Handbook of work group psychology*. Chichester, U.K.: John Wiley.

Brewer, M. B. (1991). The social self. On being the same and different at the same time. *Personality and Social Psychology Bulletin*, **86**, 307–324.

Burch, S. & Borland, C. (2001). Collaboration, facilities and communities in day care services for older people. *Health and Social Care in the Community*, **9** (1), 19–30.

Burgoyne, J. (1994). Stakeholder analysis. In: C. Cassell & G. Symon (Eds), *Qualitative Methods in Organizational Research*. London: Sage.

Burman, E. & Parker, I. (1993). *Discourse analytic research*. Routledge: London.

Burr, V. (1995). *An introduction to social constructionism*. London: Routledge.

Burr, V. & Butt, T. (1997). Interview methodology and PCP. In: P. Denicolo & M. Pope (Eds), *Sharing Understanding and Practice*. Farnborough: ECPA Press.

Butt, T. (1996). PCP: Cognitive or social psychology? In: J.Scheer & A.Catina (Eds), *Empirical Constructivism in Europe*. Giessen: Psychosozial Verlag.

Butt, T. (1998). Sociality, role and embodiment. *Journal of Constructivist Psychology*, **11**, 105–116.

Butt, T. (2001). Social action and personal constructs. *Theory and Psychology*, **11** (10), 75–95.

Carrero, V., Peiró, J. & Salanova, M. (2000). Studying radical organizational innovation through grounded theory. *European Journal of Work and Organizational Psychology*, **9** (4), 489–514.

Crossley, M. L. (2000). *Introducing narrative psychology: Self, trauma and the construction of meaning*. Buckingham, U.K.: Open University Press.

Denzin, N. K. (1989). *Interpretive interactionism*. Newbury Park, CA: Sage.

Department of Health (1989). *Caring for people: Community care in the next decade and beyond*. London: HMSO.

Department of Health (1998). *The new NHS: Modern, dependable*. London: HMSO.

Department of Health (2000). *The NHS Plan*. London: HMSO.

Du Gay, P., Evans, J. & Redman, P. (Eds) (2000). *Identity: A reader*. London: Sage.

Fiol, C. M., Hatch, M. J. & Golden-Biddle, K. (1998). Organizational culture and identity: What's the difference anyway? In: D. A. Whetten & P. C. Godfrey (Eds), *Identity in Organizations: Building Theory Through Conversations*. Thousand Oaks: Sage.

Frosh, S. (1989). *Psychoanalysis and Psychology*. London: Macmillan.

Geertz, C. (1973). *The interpretation of culture*. New York: Basic Books.

Gioia, D. (1998). From individual to organizational identity. In: D. A. Whetten & P. C. Godfrey (Eds), *Identity in Organizations: Building Theory Through Conversations*. Thousand Oaks: Sage.

Giorgi, A. (1970). *Psychology as a human science*. New York: Harper & Row.

Glendon, I. (1992). Radical change within a British university. In: D. M. Hosking & N. R. Anderson (Eds), *Organizational Change and Innovation: Psychological Perspectives and Practices in Europe*. Routledge: London.

Hartley, J. F. (1996). Intergroup relations in organizations. In: M. A. West (Ed.), *Handbook of Work Group Psychology*, Chichester, U.K.: John Wiley.

Hogg, M. & Abrams, D. (1988). *Social identifications: A social psychology of intergroup relations and group processes*. London: Routledge.

Howell, J. M. & Higgins, C. A. (1990). Champions of technological innovation. *Administrative Science Quarterly*, **35** (2), 317–341.

Jackson, S. E. & Ruderman, M. N. (Eds) (1995). *Diversity in work teams: Research paradigms for a changing workplace*. Washington, D.C.: American Psychological Association.

Kanter, R. M. (1983). *The change masters*. New York: Simon and Schuster.

Kimberly, J. R. (1981). Managerial innovation. In: P. C. Nystrom & W. H. Starbuck (Eds), *Handbook of Organizational Design*. Oxford: Oxford University Press.

King, N. (1998). Template analysis. In: G. Symon & C. Cassell (Eds), *Qualitative Methods and Analysis in Organizational Research*. London: Sage.

King, N. (2000). Making ourselves heard: The challenges facing advocates of qualitative research in work and organisational psychology. *European Journal of Work and Organizational Psychology*, **9** (4), 589–596.

King, N. & Anderson, N. R. (2002). *Managing innovation and change: A critical guide for organizations*. London: Thompson.

King, N., Anderson, N. R. & West, M. A. (1991). Organizational innovation in the U.K.: A case study of perceptions and processes. *Work and Stress*, **5**, 331–339.

King, N., Roche, T. & Frost, C. D. (2000). Diverse identities, common purpose. In: *Proceedings of the BPS Annual Occupational Psychology Conference*, Brighton, January, 199–204.

King, N. & Ross, A. (in press). Professional identities and interprofessional relations: Evaluation of collaborative community schemes. *Social Work in Health Care*.

Kotter, J. P. & Schlesinger, L. A. (1979). Choosing strategies for change. *Harvard Business Review*, March-April, 106–114.

Lawrence, P. R. (1969). How to deal with resistance to change. *Harvard Business Review*, January-February, 115–122.

Lewin, K. (1951). *Field theory in social science*. New York: Harper & Row.

Macdonald, K. M. (1995). *The sociology of the professions*. London: Sage.

629

Mancuso, J. (1996). Constructionism, personal construct psychology and narrative psychology. *Theory and Psychology*, **6**, 47–70.

Merleau-Ponty, M. (1962). *Phenomenology of perception.* London: Routledge.

Meston, C. M. & King, N. (1996). Making sense of 'resistance': Responses to organizational change in a private nursing home for the elderly. *European Journal of Work and Organizational Psychology*, **5**, 91–102.

Miles, R. E. & Snow, C. C. (1978). *Organizational strategy, structure and process.* New York: McGraw-Hill.

Morgan, M. (1996). Qualitative research: A package deal? *The Psychologist*, **9** (1), 31–32.

Moran, D. (1999). *An introduction to phenomenology.* London: Routledge.

Moustakas, C. (1994). *Phenomenological research methods.* Thousand Oaks, CA: Sage.

Nicholson, N. (1990). Organizational innovation in context: culture, interpretation and application. In: M. A. West & J. L. Farr (Eds), *Innovation and Creativity at Work: Psychological and Organizational Strategies.* Chichester, U.K.: John Wiley.

Nkomo, S. (1996). Identities and the complexities of diversity. In: S. E. Jackson & M. N. Ruderman (Eds), *Diversity in Work Teams: Research Paradigms for a Changing Workplace.* Washington, D.C.: American Psychological Association.

Nystrom, H. (1990). Organizational innovation. In: M. A. West & J. L. Farr (Eds), *Innovation and Creativity at Work: Psychological and Organizational Strategies.* Chichester, U.K.: John Wiley.

Peters, T. & Waterman, R. H. (1982). *In search of excellence: Lessons from America's best-run companies.* New York: Harper & Row.

Plant, R. (1987). *Managing change and making it stick.* Aldershot, U.K.: Gower.

Potter, J. & Wetherell, M. (1987). *Discourse and social psychology: Beyond attitudes and behaviour.* London: Sage.

Poxton, R. (1999). *Partnerships in primary and social care: Integrating services for vulnerable people.* London: Kings Fund Publishing.

Rogers, E. M. (1983). *Diffusion of innovations* (3rd ed.). New York: Free Press.

Stake, R. (1995). *The art of case-study research.* Thousand Oaks, CA: Sage.

Steyaert, C., Bouwen, R. & Van Looy (1996). Conversational construction of new meaning configurations in organizational innovation: A generative approach. *European Journal of Work and Organizational Psychology*, **5**, 67–88.

Strauss, A. & Corbin, J. (1990). *Basics of qualitative research.* Newbury Park: Sage.

Symon, G. (2000). Everyday rhetoric: Argument and persuasion in everyday life. *European Journal of Work and Organizational Psychology*, **9** (4), 477–488.

Symon, G., Cassell, C. & Dickson, R. (Eds) (2000). Special issue: Qualitative methods in organizational research and practice. *European Journal of Work and Organizational Psychology*, **9** (4).

Tajfel, H. (1978). The psychological structure of intergroup relations. In: H. Tajfel (Ed.), *Differentiation Between Social Groups: Studies in the Social Psychology of Intergroup Relations.* London: Academic Press.

Triandis, H. C. (1996). The importance of contexts in studies of diversity. In: S. E. Jackson & M. N. Ruderman (Eds), *Diversity in Work Teams: Research Paradigms for a Changing Workplace.* Washington, D.C.: American Psychological Association.

Turner, J. C., Hogg, M., Oakes, P., Reicher, S. & Wetherell, M. (1987). *Rediscovering the social group: A self-categorization theory.* Oxford, U.K.: Basil Blackwell.

Van de Ven, A. H., Polley, D. E., Garud, R. & Venkataraman, S. (1999). *The innovation journey.* New York: Oxford University Press.

West, M. A. (1990). The social psychology of innovation in groups. In: M. A. West & J. L. Farr (Eds), *Innovation and Creativity at Work: Psychological and Organizational Strategies.* Chichester, U.K.: John Wiley.

West, M. A. & Anderson, N. R. (1992). Innovation, cultural values and the management of change in British hospitals. *Work and Stress*, **6**, 293–310.

West, M. A. & Farr, J. L. (Eds) (1990). *Innovation and creativity at work: Psychological and organizational strategies.* Chichester, U.K.: John Wiley.

Whetten, D. A. & Godfrey, P. C. (Eds) (1998). *Identity in organizations: Building theory through conversations.* Thousand Oaks: Sage.

Zaltman, G., Duncan, R. & Holbek, J. (1973) *Innovations and organizations.* New York: John Wiley.

The International Handbook on Innovation
Edited by Larisa V. Shavinina

Managers' Recognition of Employees' Creative Ideas: A Social-Cognitive Model*

Jing Zhou and Richard W. Woodman

Department of Management, Mays Business School, Texas A&M University, USA

Abstract: The recognition and support of employee creative ideas is a crucial component in organizational creativity. In this paper, we explore a social-cognitive approach to explaining the conditions under which a manager is likely to consider an employee idea as creative. Our model posits that the manager's 'creativity schema' dictates recognition of creative ideas in the work setting. This creativity schema is influenced by personal characteristics of the manager, by aspects of the manager's relationship with the employee, and by a number of organizational influences. Implications of this approach for research and for practice are discussed.

Keywords: Creativity; Creativity schema; Organizational creativity; Creative ideas.

Introduction

Employee creativity plays an important role in the survival and growth of organizations (Amabile, 1988; Staw, 1984; Woodman, Sawyer & Griffin, 1993). Recognition of this dynamic has led to an increasing research interest in understanding what contextual or organizational factors facilitate employee creative performance (e.g. Amabile, Conti, Coon, Lazenby & Herron, 1996; George & Zhou, 2002; Oldham & Cummings, 1996; Shalley, 1995; Shalley & Oldham, 1997; Tierney, Farmer & Graen, 1999; Zhou, forthcoming (a); Zhou & Oldham, 2001).

Management support and encouragement have consistently been shown to have main or interactive effects in promoting employee creativity (Madjar, Oldham & Pratt, 2002; Oldham & Cummings, 1996; Scott & Bruce, 1994; Tierney, Farmer & Graen, 1999). However, we know relatively little about what factors influence when and why an idea, event, behavior, or outcome is considered creative from a manager's perspective. To fully understand the nature and dynam-

ics of management support for creativity, we first need to understand the factors that determine the recognition (or non-recognition) of employee creativity.

In this chapter, we will develop a conceptual model designed to explain the recognition of creative ideas in the work setting. We posit that a manager's creativity schema provides a useful heuristic to explain the recognition of behavior, events, ideas, and outcomes as creative. We develop a social-cognitive model to identify and describe the personal characteristics, work relationship factors, and organizational factors associated with the formation and use of such schema.

Conceptual Background

Employee creativity may be defined as the generation of novel and useful ideas (Amabile, 1988). We embed this individual creativity within the broader issue of organizational creativity which may be defined as the creation of a valuable, useful new product, service, idea, procedure, or process by individuals working together in a complex social system (cf. Woodman et al., 1993). Within the context of organizational creativity, three assumptions make research on managers' recognition of employee creative ideas particularly meaningful. First, employees seldom produce creative ideas in social isolation. Creative performance is as much of a social process as a cognitive process (Amabile, 1983). Second, employees typically do not generate creative ideas overnight. Instead, there is a

* Earlier versions of this paper were presented at 'The 21st Century Change Imperative: Evolving Organizations & Emerging Networks' Conference, Center for the Study of Organizational Change, University of Missouri-Columbia, June 12–14, 1998, and the Academy of Management Annual Meetings, Chicago, 1999.

process by which a novel and useful idea gets developed and refined (Basadur, Graen & Green, 1982; Rogers, 1983; Wallas, 1926). This process is likely to involve interaction, communication with others, and is often subject to evaluation and feedback from others (Amabile, 1996; Shalley & Perry-Smith, 2001; Zhou, forthcoming (b)). To the extent that managers are important elements of the social environment in organizations, their reactions may have substantial influence on the creative idea generation process. Third, whether an idea or event is creative is not completely objective. Individuals attach meanings and interpretations to ideas or events. Thus, the degree to which an idea, event, or outcome is creative is, to a very real extent, subjective. What one person sees as a creative idea another person may or may not agree with.

There has always been a subjective, social dimension to the judgment of creativity. As defined above, creativity implies, at a minimum, an assessment of two dimensions; novelty and usefulness. These dimensions are sometimes further broken down into component sub-dimensions (e.g. MacCrimmon & Wagner, 1994), but these are the two most commonly accepted dimensions of creativity and will suffice for our purposes here. In terms of measurement, novelty has traditionally posed less of a problem than usefulness. Novelty or originality is a (sometimes) simple matter of counting and comparing. (Although lacking the knowledge base to accurately recognize novelty could, of course, be a constraint.) The dimension of usefulness, however, has proven to pose the most tricky construct validity issues. At some level of abstraction, it seems impossible to assess the value or usefulness of a product, idea, outcome, and so on without a *judgment* about such value or usefulness. We see this most clearly in the world of art. What one person regards as an attractive, valuable painting for example, another individual may view as without merit. At first, it might seem that the usefulness criterion would be less daunting in judging creativity in the organization. A newly invented product is useful if people buy it; a newly created process is useful if it is more efficient than the process it replaces, and so on. However, at the 'idea' stage, usefulness depends very much upon a judgment about a future state of value. Also, an idea judged as creative (both novel and potentially useful) by one manager may leave another unmoved. Indeed, Epstein is so vexed by the problems plaguing the usefulness criterion that he avoids altogether the language of creativity when dealing with 'generative phenomena' (his preferred term). 'Behavior called creative by one group might be harshly judged by another' (Epstein, 1990, p. 139). Epstein even argues that, due largely to these judgmental disagreements, the creative product may ultimately provide a poor index for measuring or understanding the creative process. Nevertheless, we wish to focus on a particular type of

creative product in this chapter; specifically, the creative idea and the factors that influence managerial recognition of the idea as creative.

Csikszentmihalyi goes much further than most observers when he argued that "... creativity is not an attribute of individuals but of social systems making judgments about individuals" (1990, p. 198). In his systems view of creativity, Csikszentmihalyi conceptualizes creativity as the result of the interaction among three sub-systems: a domain, a person, and a field (Csikszenthihaly, 1990, 1996). It is within the 'field' sub-system that judgments concerning creative ideas, behavior, and outcomes are made. Within the work setting, the manager would appear to be a crucial component of Csikszentmihalyi's 'field'. In sum, we view the manager as a key actor in the social system that 'makes judgments' about the creativity of individual employees.

Recognition of Creative Ideas

What is the most useful way to understand or to explain managerial recognition of creative ideas? In this chapter, we will argue that creativity schema dictate managers' recognition of a creative idea. In this section, we will: (a) describe the nature and potential function of a creativity schema; (b) identify personal characteristics that are associated with the formation of a creativity schema; (c) identify aspects of the manager's relationship with the employee that are related to the specific dimensions of the manager's creativity schema; and (d) identify the organizational influence factors that are related to the specific dimensions of the schema. Because we are still at a very preliminary stage of developing a social-cognitive perspective of managers' recognition of creative ideas, we emphasize presenting new research ideas instead of empirical support. In addition, in identifying the variables that affect the formation and utilization of a creativity schema, we shall attempt to be illustrative rather than exhaustive.

Figure 1 displays the hypothesized linkages between creativity schema and the personal, relationship, and organizational influences factors that are associated with the schema. An important aspect of the model shown in Fig. 1 should be noted at this point. We see a critical difference among these three categories (i.e. personal, relationship, and organizational) in terms of their relationship to the schema. Specifically, we posit that personal characteristics of the manager are related to, or help to explain, the *formation* of creativity schema. In contrast, the dyadic relationship and organizational influence variables help to explain how the schema is *utilized* in judging the creativity of the employee's idea. In each category, however, we will propose relationships to specific dimensions of the schema. In developing our propositions, we will specify the nature of the association (i.e. a positive or a negative relationship) whenever possible. In some

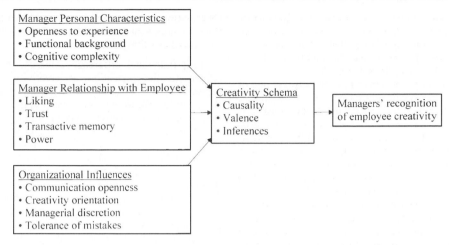

Figure 1. Managers' Recognition of Employee Creativity: A Social-Cognitive Model.

cases, however, the nature of the relationship between the variables is difficult to predict given the current state of our knowledge. Also, it is possible in a few instances that the directionality of the relationship might vary across situations. (Potential moderators are not explored in this paper.) In these cases, we choose to make a more general statement concerning the relationship.

Creativity Schema

Fiske & Taylor define a schema as a "cognitive structure that represents organized knowledge about a given concept or type of stimulus. A schema contains both the attributes of the concept and the relationships among the attributes" (1984, p. 140). In essence, a schema is a mental representation of an external target that helps the individual to make sense of the target in a simplified and organized way.

Previous theory and research in social cognitive psychology and organizational change suggests that a schema concerning a specific target can be meaningfully thought of as including three components; causality, valence, and inferences (e.g. Lau & Woodman, 1995; Markus & Zajonc, 1985; Taylor & Crocker, 1981). Causality refers to the aspect of a schema that maps the sequential relations between events. It enables individuals to make causal attributions and to understand the connections between causes and effects. For a creativity schema, causality provides managers with some explanations concerning where the creative idea has come from, why it has been produced, and how it fits in the particular context within which the employee and manager work. Valence refers to the significance of the target idea or event. For a creativity schema, valence allows the manager to decide whether the idea produced by the employee is significant and meaningful. Finally, inferences allow the individual to

predict what is going to happen in the future, and how likely these events or outcomes are to take place. This component in a creativity schema allows the manager to make inferences about whether the focal idea can be successfully implemented, and what the consequences of implementation are for the organization and for the manager.

Although little research has been conducted to examine the role of schema on managers' recognition of creative ideas generated by employees, previous research in organizational change sheds light on the construct validity of this concept. For example, Lau & Woodman (1995) investigated the content and development of change schema. Using both qualitative and quantitative data-collection methods across three different samples, they found that the measurement of change schema had satisfactory construct validity. Also, empirical work has successfully employed the change schema construct to explore changes in organizational culture and further demonstrated the construct validity of change schema as well as the utility of a cognitive approach to measuring culture change (Lau, Kilbourne & Woodman, 2003). In addition, theory and research, in general, has supported the notion that a schema has the function of directing an individual to make sense of an external target or event. Further, a schema can be useful for explaining or understanding an individual's possible attitudinal and behavioral reactions to the target or event (e.g. Markus & Zajonc, 1985; Schank & Abelson, 1977; Taylor & Crocker, 1981).

Formation of Creativity Schema: Relationships to Personal Characteristics

As shown in Fig. 1, we posit that three variables concerning the focal manager's personal characteristics are related to the formation of a creativity schema.

While we are prepared to argue for the likely importance of these three personal characteristics, we certainly are not suggesting that these individual differences are the only ones associated with the components of a creativity schema.

The first of these, openness to experience, is one of the Big Five personality dimensions (e.g. McCrae & Costa, 1985, 1999). It captures intellectual curiosity, imagination, aesthetic sensitivity, and wide interests, among other traits (John & Srivastava, 1999). Individuals scoring highly on the openness to experience dimension tend to seek and enjoy varied experiences for their own sake. Scratchley & Hakstian (2000–2001) found that openness to experience (operationalized as openness to change, openness to risk, and openness to ambiguity) was positively correlated with divergent thinking and could be used to successfully predict managerial creativity (operationalized as devising new ideas, work methods, and modes of operation useful to the organization). George & Zhou (2001) found that individuals high on openness to experience were the most creative when provided with positive feedback and when performing heuristic (as opposed to algorithmic) tasks. They concluded that openness to experience may support creative behavior when the situation allows for the manifestation of the possible influence of the trait. In addition to the production of new ideas, by extension openness to experience could be related to the recognition of new ideas as well. Thus, it appears that managers who are more open to experience would be more likely to perceive an idea to have high valence. Thus, we posit:

Proposition 1: Openness to experience is positively related to the valence component of creativity schema.

A second variable that might be related to creativity schema is the manager's functional background. This background is an indicator of the type of knowledge he or she possesses with regard to the domain of the proposed creative idea. Amabile (1988) identified 'domain-relevant skills' as being important for creativity. By extension, knowledge and skills relevant to the domain of some proposed idea could also be a factor in the recognition and appreciation of an idea as being novel and valuable. Functional background also results in a manager's selective perceptions of ambient stimuli (e.g. Dearborn & Simon, 1958; Waller, Huber & Glick, 1995). That is, the cognitive map derived from experience in a particular functional background directs a manager's attention to certain events in the surrounding environment, and to certain attributes of the creative idea. Functional background would appear to be associated with how the manager makes sense of why an idea might be developed, how it fits into the employee's work, whether the idea is significant in this context, and whether or not it is likely to be successfully implemented. Therefore, we suggest:

Proposition 2: Functional background is related to all three components—causality, valence, and inferences—of creativity schema.

A third individual difference construct that may be related to the formation of creativity schema is captured by the broad notion of cognitive complexity. Research has identified a number of cognitive abilities that are related to creativity (Hayes, 1989; Woodman et al., 1993). For example, the cognitive ability of divergent production, consisting of processes of fluency, flexibility, originality, and elaboration, has long been considered as a cognitive key to creativity (Guilford, 1984; Mumford, 2000–2001). Interestingly, with regard to the *recognition* as opposed to the production of creative ideas, the cognitive ability of convergent production or thinking may be more critical. Whereas divergent thinking might allow the individual to find or produce numerous original ideas, it is convergent thinking that allows the individual to select from among the ideas those that might be most useful or valuable. Runco (1999) explores the role of critical thinking in creativity and his conclusions suggest that this aspect of cognitive functioning might be crucial for the recognition of the value of an idea, particularly when the idea is one among many. In a similar vein, Hogarth (1987) suggested that the ability to reason and to understand the causality of events was a crucial component of creativity. In general, we would expect managers with higher cognitive complexity (with the attendant cognitive abilities implied by that construct) to be better able to process the complex information that might be needed for judging an idea as creative. Specifically, we expect:

Proposition 3: Cognitive complexity is positively related to the causality and inferences components of creativity schema.

Utilization of Creativity Schema: The Role of Relationships

Four variables in our model concern the dyadic relationship between the focal manager and the employee who has generated the creative idea (see Fig. 1). The first of these, liking, refers to the extent to which the manager has positive affect toward the employee (e.g. Judge & Ferris, 1993; Tsui & O'Reilly, 1989). Previous research showed that when a manager liked an employee, the manager tended to respond more positively to him or her, for example by giving higher performance ratings (e.g. Cardy & Dobbins, 1986; Judge & Ferris, 1993). Although we know of no research directly focused on the relationship between liking an employee and the response to ideas produced by the employee, related evidence concerning the relationship between liking and manager assessments and responses in general suggests that liking might be related to the creativity schema components of valence and inferences. When a manager likes an employee, he or she is more likely to think positively about the employee's ideas, believe such ideas are meaningful and significant, and also judge that the idea is likely to be successfully implemented. Therefore, we propose:

Proposition 4: Liking is positively related to the valence and inferences components of creativity schema.

Trust is another important variable that defines the quality of the relationship between a manager and an employee. Following Mayer, Davis & Schoorman (1995), trust is defined as "the willingness of a party to be vulnerable to the actions of another party based on the expectation that the other will perform a particular action important to the trustor, irrespective of the ability to monitor or control that other party" (Mayer et al., 1995, p. 712). Because any potentially creative idea is accompanied by some degree of uncertainty, and is, as previously discussed, somewhat subjective, whether or not the idea is congruent with other aspects of the job and work setting, as well as its significance and consequences are subjected to the manager's personal interpretations. If the manager trusts the employee, he or she should be more likely to favorably interpret questions and issues surrounding why the employee has produced the idea, the meaning and significance of the idea, and the consequences that may stem from implementation of the idea. In sum:

Proposition 5: Trust is related to the causality, valence, and inferences components of creativity schema.

Wegner and his colleagues (e.g. Wegner, 1986; Wegner, Erber & Raymond, 1991) maintain that transactive memory is a system of encoding, storing, and retrieving information that is shared by more than one person. This is an intriguing concept with potential implications for understanding the utilization of creativity schema from the perspective of the dyadic relationship between a manager and her subordinate. When a manager and an employee have developed this memory sharing system, they have knowledge about each other's memories, and have shared responsibility that could enable one person to remember something the other might not (Wegner et al., 1991). Due to the potential utility of transactive memory, a manager having such a relationship with an employee should have a good understanding of the causes, nature, and consequences of the ideas produced by the employee. Therefore, we posit:

Proposition 6: Transactive memory is related to the causality, valence, and inferences components of creativity schema.

A final aspect of the dyadic relationship between the manager and the employee that could impact managerial recognition of employee creative ideas is captured by the notion of relative power in the relationship (Hellriegel, Slocum & Woodman, 2001; Kramer &

Neale, 1998). The ability of one actor in this dyad to influence the other could range from the manager having all of the power, with the employee being virtually powerless to a relationship where power is more balanced between the actors, to, at the other extreme, the employee having strong ability to influence the manager while the manager has little ability to influence the subordinate. This latter scenario seems the least likely of the three. One might speculate that a balanced relationship (regardless of the source or reasons for that balance) might be the one where the manager is most likely to focus on the 'merits' of the creative idea in judging it. It seems conceivable that the circumstance where the manager's power is dramatically stronger than the employee's power might result in a tendency for the manager to be less willing to acknowledge the potential value of any employee's idea, creative or not. In any event, we expect that the relative power positions characterizing this dyadic relationship may influence managerial recognition of employee creative ideas. Thus:

Proposition 7: The relative power of the manager in relation to the employee is related to all three components of creativity schema.

Utilization of Creativity Schema: The Role of Organizational Influences

Finally, as shown in Fig. 1, we hypothesize that four organizational influences are related to creativity schema. The first of these factors, communication openness, refers to the extent to which there are open channels of communication in the organization. A significant body of research on organizational innovation supports the notion that the availability of information is a crucial variable in the creative process (Damanpour, 1991; Kanter, 1988; Payne, 1990). This work further suggests that constraints on information and communication have a negative impact on creativity. When a manager can openly and freely exchange ideas and acquire information, he or she is more likely to understand how an idea produced by an employee is related to other aspects of the job that the employee is doing, and how the idea might be related to the work performed by other employees across various jobs and organizational functions. As such, the manager may also have a better understanding of the meaning and significance of the proposed idea. Finally, the manager is also likely to have a better and richer understanding of whether the idea can be implemented, and the likely consequences of this implementation. Thus, we propose:

Proposition 8: Communication openness is related to all three components of creativity schema.

The second variable in the organizational influences category is creativity orientation. In general, the 'climate for creativity' has long been considered a crucial component in fostering creative behaviors and outcomes (cf. Amabile, 1983, 1996). It would seem to be intuitive that when an organization is oriented toward finding, promoting, and supporting employee creativity, a manager is more likely to see the significance of the idea generated by the employee, and is more likely as well to believe that the idea has a good chance to be successfully implemented. Therefore:

Proposition 9: Creativity orientation is related to the valence and inferences components of creativity schema.

The third variable in this category is managerial discretion. This discretion reflects the extent to which the focal manager has control over many operational and production related issues such as, choices of projects, work procedures, outcome criteria, and time-lines. Greater autonomy, in general, allows for an exploration of alternative methods for completing work. An increase in experimentation, in turn, appears to be related to more creative outcomes (Amabile, 1996; Shalley, 1991; Zhou, 1998). By extension, it seems logical that the greater the discretion a manager has, the more likely that he or she is to believe that the implementation of the idea and its consequences are predictable and manageable. Thus:

Proposition 10: Management discretion is positively related to the inferences component of creativity schema.

Finally, tolerance of mistakes refers to the extent to which the organization's reward policy, practices, and culture encourage risk-taking and forgive innocent mistakes committed in the course of trying to accomplish the organization's tasks. In such an organizational culture, producing and implementing new and useful ideas would typically be well regarded and valued. Evidence supports the notion that creative behavior and outcomes are enhanced when risk-taking is both encouraged and supported, particularly by an absence of punishment (Amabile, 1988; Burnside, 1990; Nystrom, 1990). When there is a tolerance for mistakes made in the pursuit of organizational goals, the focal manager may be more likely to feel confident and comfortable about new ideas proposed by employees. In other words, organizational tolerance of mistakes may be related to the component of the manager's creativity schema related to the consequences of the idea. That is:

Proposition 11: Tolerance of mistakes is related to the inferences component of creativity schema.

Discussion and Suggestions

In summary, the focus of the model shown in Fig. 1 and developed here has been to explore the key role of the manager in the judgment and recognition of employee creativity. A social-cognitive approach to understand-ing the manager's role in the creative process in organizations suggests several implications for organizational research and management practice.

Implications for Research

In terms of our model in Fig. 1, there are a number of implications for needed research as well as further theory development. While the dimensionality of schema (causality, valence, and inferences) has been supported by previous research (e.g. Lau & Woodman, 1995), it will be important to develop and demonstrate the construct validity of a measure of *creativity* schema. Further, we have posited only a limited number of personal characteristics of the manager that might be related to the formation of creativity schema and hence to the recognition of employee creativity. It seems reasonable to suppose a much larger list of possibilities in this regard. In addition, it could well be that greater explanatory power could be added to the socio-cognitive model shown in Fig. 1 by including specific characteristics and behavior of the employee generating the creative idea. For example, an obvious aspect of the employee's behavior that could impact the recognition of creative ideas would be past employee performance. That is, an idea generated by an employee with a history of exemplary performance might be likely to be judged as creative. At the other extreme, an employee with a checkered past in terms of job performance might more readily have his ideas dismissed or ignored by his manager or supervisor.

Finally, in terms of theory development, we have proposed a model in the most straightforward manner possible, i.e. that manager personal characteristics, characteristics of the dyadic relationship between manager and employee, and organizational influences are viewed as 'antecedent conditions' to managerial recognition of creativity. Research may show that a more complex configuration provides greater explanatory power. For example, some contextual variables (e.g. organizational influences) may well moderate the relationship between personal characteristics and managerial recognition.

There is some recent research on organizational creativity that can inform theoretical development of the model represented by Fig. 1. In addition, our perspective may contribute to the further development of several on-going lines of inquiry. For example, Shalley, Gibson & Blum (2000) investigated the degree to which work environments were structured to complement or be congruent with the 'creativity requirements' (i.e. level of creative behavior) required by a particular job. Among other notions, they posited that the level of creativity required in a job was positively associated with organizational support. While they showed only partial support for this hypothesis across a series of potential relationships, the hypothesis would seem to capture a crucial dynamic. In terms of our posited relationships, it would be valuable to know to what extent managerial recognition of employee creativity represents an important component of 'organizational support'. Further, the creative requirements of a specific job represent another explanatory variable that could be added to our model. It is logical that managerial recognition could be more important in situations where the jobs have a relatively high 'creativity requirement' than in situations where it is less important for employees to engage in creative behavior.

Organizational support for creativity was posited by Zhou & George (2001) to interact with job dissatisfaction and continuance commitment to predict creativity. Continuance commitment refers to employees being committed to their organizations not because of affective attachment or identification with the organizations' values and goals, but because of necessity (e.g. not being able to find jobs elsewhere) (Allen & Meyer, 1996). They found that, employees with high levels of job dissatisfaction and continuance commitment were more likely to exhibit creativity when perceived organizational support for creativity was high. In terms of our model, the findings from the Zhou & George study are potentially related to both the manager-employee set of variables as well as the organizational influences antecedents. Their work suggests that our straightforward diagramming of these explanatory variables, as we discussed earlier, may be far too simple. The potential for interactions among these variables across levels of analysis is very real.

Another line of inquiry with implications for our social-cognitive model is represented by the investigation performed by McGrath (2001). In a study of 56 new business development projects, she found that organizational learning was more effective when the projects were operated with high degrees of autonomy. Learning effectiveness, as operationalized in this study, is related to exploration behaviors leading to creativity and innovation. This line of inquiry suggests that the managerial discretion variable, which we include as an important component of organizational influences on managerial recognition, may be a particularly important explanatory variable. Research is needed to isolate the potentially crucial role of autonomy and discretion in the perception of employee creativity. Here again, possible interactions between autonomy and variables in our model seem likely.

Finally, a theoretical perspective advanced by Unsworth (2001) suggests yet another potentially useful avenue in developing research to explore our model. Unsworth recently suggested that treating creativity as a unitary concept has hampered development of a richer understanding of organizational creativity. Unsworth advanced a matrix of 'creativity types' consisting of the following categories; expected creativity (required solution to discovered problem), proactive creativity (volunteered solution to discovered problem), responsive creativity (required solution to

specified problem), and contributory creativity (volunteered solution to specified problem). These types differ along dimensions of driver (why engage in the creative process?) and problem (what is the initial state triggering the need for creativity?). Unsworth (2001) posits that there may be important and interesting differences in factors contributing to creativity and the creative process itself across these categories. In terms of the model in Fig. 1, an interesting research question concerns whether the classes of antecedents would vary with the type of creativity involved (e.g. would the important organizational influences vary across types?). Further, might managers in general, regardless of the constellation of antecedents, more readily recognize some types of creativity than others? Of course, the construct validity of Unsworth's creativity categories has yet to be established.

Implications for Practice

The major approach taken organizationally in terms of utilizing knowledge about creativity in the work setting has probably been through training programs. While results are somewhat mixed, in general a number of studies have shown positive results from attempting to 'train' employees to be more creative. As mentioned earlier, divergent thinking has long been considered a cognitive key to creativity. Thus, training programs have frequently targeted divergent thinking when attempting to improve the creative performance of employees and managers (e.g. Basadur, Graen & Scandura, 1986; Basadur, Wakabayashi & Graen, 1990). The widely-used training programs designed to improve the ability of groups and teams to brainstorm creative solutions to problems and/or teaching individual decision makers to engage in 'lateral thinking' (i.e. looking for alternative ways to define and understand problems) also fall into the divergent thinking arena conceptually (Ripple, 1999). A second approach has been to target creative performance related to a specific domain, for example strategic planning (cf. Wheatley, Anthony & Maddox, 1991). Based on the logic developed in our paper, a potentially fruitful arena for creativity training might be to focus this activity on enhancing the ability of key managers and decision makers to recognize creativity when they see it. Such creativity training would have the advantage of a relatively specific focus, might have wide-spread applicability across a large number of individuals of varying cognitive abilities, personalities, attitudes, and values and thus could be instrumental in helping to foster the supportive climate and acceptance considered important for organizational creativity.

Harrington (1999) suggests the crucial importance of the environmental context for creative behavior. There is a tradition of research in the psychological sciences that has focused on the effect of the environment on the creativity of gifted individuals. In general, opportunities to learn and create, ready access to needed information, the presence of a supportive social system including appropriate rewards for creative behavior, and the like are considered to be instrumental for creativity. Research on the 'climate for creativity' has been successfully extended into organizations (e.g. Amabile, 1996; Isaksen, Lauer, Ekvall & Britz, 2000–2001; Tesluk, Farr & Klein, 1997). As one example of this body of work, Tesluk et al. (1997) identified a supportive organizational culture, the utilization of appropriate goal-setting and rewards for creativity, a host of organizational characteristics (e.g. design features, human resource practices and policies), and socioemotional support as being particularly crucial in fostering individual creativity in the workplace. In sum, we have reasonable insight into many of the contextual factors (both at the organizational level and the group level) that can enhance organizational creativity and, conversely, inhibit it (cf. Woodman, et al., 1993). Based on the approach represented by the model of Fig. 1, and consistent with the systems view of creativity advanced by Csikszentmihalyi (1996), we argue that managerial recognition of creative ideas should be considered a crucial component of the necessary organizational climate, socioemotional support, appropriate reward structure, and so on needed for organizational creativity. Woodman (1995) has long argued that the high-payoff strategy for managers in terms of impacting creative behavior and outcomes in the work setting is to learn how to design, and to manage, the context affecting creativity, rather than focusing on attempting to directly manage either creative persons or the creative process. To this argument we would now add the notion that the manager needs to learn how to manage herself in terms of the ability to identify, nurture, and reward creative ideas when they appear.

Concluding Comments

We have argued that the managerial recognition of employee creative ideas can be meaningfully explained by an examination of the manager's creativity schema. This schema is influenced by personal characteristics of the manager, by the relationship that exists between the manager and the employee who creates the idea that will be judged by the manager, and by certain organizational or contextual influences that characterize the particular work setting. We have proposed a number of relationships between these sources of explanatory variation and the causality, valence, and inferences components of the manager's creativity schema.

Certainly, our proposed social-cognitive model has a number of limitations. We have made no attempt to address the exact process through which a manager's creativity schema is developed. Further, we have made no attempt to specify how much time might be needed for the formation and development of the schema. Pettigrew, Woodman & Cameron (2001) have argued

strongly that notions of time need to be incorporated into theory development concerned with *any* change processes in organizations. The model developed here would fit that broad context and will eventually need to include the dimension of time. An additional limitation concerns the explanatory variation included in Fig. 1 or, more to the point, the variables that have been omitted. The model could be expanded in several ways as was noted above.

Despite the obvious limitations at this point of theory development, the proposed model has the potential to make important contributions to organizational creativity research. To the extent that management support is crucial in fostering employee creativity, then understanding the initial conditions for management support—how and why managers might recognize an idea as creative—would seem to be a valuable line of inquiry. This approach has the potential to broaden our understanding of how to foster employee creativity. Creative idea generation is an ongoing process in organizations, and managers' recognition of and responses to potentially creative ideas put forth by employees may affect whether these ideas will be accepted, nurtured, and implemented.

References

Allen, N. J. & Meyer, J. P. (1996). Affective, continuance, and normative commitment to the organization: An examination of construct validity. *Journal of Vocational Behavior*, **49**, 252–276.

Amabile, T. M. (1988). A model of creativity and innovation in organizations. In: B. M. Staw & L. L. Cummings (Eds), *Research in Organizational Behavior* (Vol. 10, pp. 123–167). Greenwich, CN: JAI Press.

Amabile, T. M. (1996). *Creativity in context*. Boulder, CO: Westview Press.

Amabile, T. M. (1983). *The social psychology of creativity*. New York: Springer-Verlag.

Amabile, T. M., Conti, R., Coon, H., Lazenby, J. & Herron, M. (1996). Assessing the work environment for creativity. *Academy of Management Journal*, **39**, 1154–1184.

Basadur, M., Graen, G. B. & Green, S. G. (1982). Training in creative problem-solving: Effects on ideation and problem finding and solving in an industrial research organization. *Organizational Behavior and Human Performance*, **30**, 41–70.

Basadur, M., Graen, G. B. & Scandura, T. A. (1986). Training effects on attitudes toward divergent-thinking among manufacturing engineers. *Journal of Applied Psychology*, **71**, 612–617.

Basadur, M., Wakabayashi, M. & Graen, G. B. (1990). Individual problem-solving styles and attitudes toward divergent thinking before and after training. *Creativity Research Journal*, **3**, 22–32.

Burnside, R. M. (1990). Improving corporate climates for creativity. In: M. A. West & J. L. Farr (Eds), *Innovation and Creativity at Work* (pp. 265–284). Chichester, U.K.: John Wiley.

Cardy, R. L. & Dobbins, G. H. (1986). Affect and appraisal accuracy: Liking as an integral dimension in evaluating performance. *Journal of Applied Psychology*, **71**, 672–678.

Csikszentmihalyi, M. (1996). *Creativity: Flow and the psychology of discovery and invention*. New York: Harper-Collins.

Csikszentmihalyi, M. (1990). The domain of creativity. In: M. A. Runco & R. S. Albert (Eds), *Theories of Creativity* (pp. 190–212). Newbury Park, CA: Sage.

Damanpour, F. (1991). Organizational innovation: A meta-analysis of effects of determinants and moderators. *Academy of Management Journal*, **34**, 555–590.

Dearborn, D. & Simon, H. A. (1958). Selective perception: A note on the departmental identifications of executives. *Sociometry*, **21**, 140–144.

Fiske, S. T. & Taylor, S. E. (1984). *Social cognition*. Reading, MA: Addison-Wesley.

George, J. M. & Zhou, J. (2002). Understanding when bad moods foster creativity and good ones don't: The role of context and clarity of feelings. *Journal of AppliedPsychology*, **87**, 687–697.

George, J. M. & Zhou, J. (2001). When openness to experience and conscientiousness are related to creative behavior: An interactional approach. *Journal of Applied Psychology*, **86**, 513–524.

Guilford, J. P. (1984). Varieties of divergent production. *Journal of Creative Behavior*, **18**, 1–10.

Harrington, D. M. (1999). Conditions and settings/environment. In: M. A. Runco & S. R. Pritzker (Eds), *Encyclopedia of Creativity* (Vol. 1, pp. 323–340). San Diego, CA: Academic Press.

Hayes, J. R. (1989). Cognitive processes in creativity. In: J. A. Glover, R. R. Ronning & C. R. Reynolds (Eds), *Handbook of Creativity* (pp. 135–145). New York: Plenum Press.

Hellriegel, D., Slocum, J. W. & Woodman, R. W. (2001). Power and political behavior. In: *Organizational Behavior* (9th ed., pp. 264–291). Cincinnati, OH: South-Western Publishing Co.

Hogarth, R. M. (1987). *Judgment and choice: The psychology of decision*. New York: John Wiley.

Isaksen, S. G., Lauer, K. J., Ekvall, G. & Britz, A. (2000–2001). Perceptions of the best and worst climates for creativity: Preliminary validation evidence for the situational outlook questionnaire. *Creativity Research Journal*, **13**, 171–184.

John, O. P. & Srivastava, S. (1999). The big five trait taxonomy: History, measurement, and theoretical perspectives. In: L. A. Pervin & O. P. John (Eds), *Handbook of Personality: Theory and Research* (2nd ed., pp. 102–138). New York: The Guilford Press.

Judge, T. A. & Ferris, G. R. (1993). Social context of performance evaluation decisions. *Academy of Management Journal*, **36**, 80–105.

Kanter, R. M. (1988). When a thousand flowers bloom: Structural, collective, and social conditions for innovation in organizations. In: B. M. Staw & L. L. Cummings (Eds), *Research in Organizational Behavior* (Vol. 10, pp. 169–211). Greenwich, CT: JAI Press.

Kramer, R. M. & Neale, M. A. (Eds) (1998). *Power and influence in organizations*. Thousand Oaks, CA: Sage.

Lau, C. M. & Woodman, R. W. (1995). Understanding organizational change: A schematic perspective. *Academy of Management Journal*, **38**, 537–554.

Lau, C. M., Kilbourne, L. M. & Woodman, R. W. (2003). A shared schema approach to understanding organizational culture change. In: W. A. Pasmore & R. W. Woodman (Eds), *Research in Organizational Change and Develop-*

ment (Vol. 14, pp. 225–256). Oxford, U.K.: Elsevier Science.

Madjar, N., Oldham, G. R. & Pratt, M. G. (2002). There's no place like home? The contributions of work and non-work creativity support to employees' creative performance. *Academy of Management Journal*, **45**, 757–767.

Markus, H. & Zajonc, R. B. (1985). The cognitive perspective in social psychology. In: G. Lindzey & E. Aronson (Eds), *The Handbook of Social Psychology* (Vol. 1, pp. 137–230). New York: Random House.

Mayer, R. C., Davis, J. H. & Schoorman, F. D. (1995). An integrative model of organizational trust. *Academy of Management Review*, **20**, 709–734.

McCrae, R. R. & Costa, P. T. (1999). A five-factor theory of personality. In: L. A. Pervin & O. P. John (Eds), *Handbook of Personality: Theory and Research* (2nd ed., pp. 139–153). New York: The Guilford Press.

McCrae, R. R. & Costa, P. T. (1985). Openness to experience. In: R. Hogan & W. H. Jones (Eds), *Perspectives in Personality* (Vol. 1, pp. 145–172). Greenwich, CT: JAI Press.

McGrath, R. G. (2001). Exploratory learning, innovative capacity, and managerial oversight. *Academy of Management Journal*, **44**, 118–131.

Mumford, M. D. (2000–2001). Something old, something new: Revisiting Guilford's conception of creative problem solving. *Creativity Research Journal*, **13**, 267–276.

Nystrom, H. (1990). Organizational innovation. In: M. A. West & J. L. Farr (Eds), *Innovation and Creativity at Work* (pp.143–161). Chichester, U.K.: John Wiley.

Oldham, G. R. & Cummings, A. (1996). Employee creativity: Personal and contextual factors at work. *Academy of Management Journal*, **39**, 607–634.

Payne, R. (1990). The effectiveness of research teams: A review. In: M. A. West & J. L. Farr (Eds), *Innovation and Creativity at Work* (pp. 101–122). Chichester, U.K.: John Wiley.

Pettigrew, A. M., Woodman, R. W. & Cameron, K. S. (2001). Studying organizational change and development: Challenges for future research. *Academy of Management Journal*, **44**, 697–713.

Ripple, R. E. (1999). Teaching creativity. In: M. A. Runco & S. R. Pritzker (Eds), *Encyclopedia of Creativity* (Vol. 2, pp. 629–638). San Diego, CA: Academic Press.

Rogers, E. M. (1983). *Diffusion of innovations* (3rd ed.). New York: Free Press.

Runco, M. A. (1999). Critical thinking. In: M. A. Runco & S. R. Pritzker (Eds), *Encyclopedia of creativity* (Vol. 1, pp. 449–452). San Diego, CA: Academic Press.

Scott, S. G. & Bruce, R. A. (1994). Determinants of innovative behavior: A path model of individual innovation in the workplace. *Academy of Management Journal*, **37**, 580–607.

Scratchley, L. S. & Hakstian, A. R. (2000–2001). The measurement and prediction of managerial creativity. *Creativity Research Journal*, **13**, 367–384.

Shalley, C. E. (1991). Effects of productivity goals, creativity goals, and personal discretion on individual creativity. *Journal of Applied Psychology*, **76**, 179–185.

Shalley, C. E. (1995). Effects of coaction, expected evaluation, and goal setting on creativity and productivity. *Academy of Management Journal*, **38**, 483–503.

Shalley, C. E., Gilson, L. L. & Blum, T. C. (2000). Matching creativity requirements and the work environment: Effects on satisfaction and intentions to leave. *Academy of Management Journal*, **43**, 215–223.

Shalley, C. E. & Oldham, G. R. (1997). Competition and creative performance: Effects of competitor presence and visibility. *Creativity Research Journal*, **10**, 337–345.

Shalley, C. E. & Perry-Smith, J. E. (2001). Effects of social-psychological factors on creative performance: The role of informational and controlling expected evaluation and modeling experience. *Organizational Behavior and Human Decision Processes*, **84**, 1–22.

Shank, R. C. & Abelson, R. (1977). *Scripts, plans, goals and understanding*. Hillsdale, NJ: Erlbaum.

Staw, B. M. (1984). Organizational behavior: A review and reformulation of the field's outcome variables. In: M. R. Rosenzweig & L. W. Porter (Eds), *Annual Review of Psychology* (Vol. 35, pp. 627–666). Palo Alto, CA: Annual Reviews.

Taylor, S. E. & Crocker, J. (1981). Schematic basis of social information processing. In: E. T. Higgins, C. P. Herman & M. P. Zanna (Eds), *Social Cognition: The Ontario Symposium* (Vol. 1, pp. 89–134). Hillsdale, NJ: Erlbaum.

Tesluk, P. E., Farr, J. L. & Klein, S. R. (1997). Influences of organizational culture and climate on individual creativity. *Journal of Creative Behavior*, **31**, 27–41.

Tierney, P., Farmer, S. M. & Graen, G. B. (1999). An examination of leadership and employee creativity: The relevance of traits and relationships. *Personnel Psychology*, **52**, 591–620.

Tsui, A. S. & O'Reilly, C. A. (1989). Beyond simple demographic effect: The importance of relational demography in supervisor-subordinate dyads. *Academy of Management Journal*, **32**, 402–423.

Unsworth, K. (2001). Unpacking creativity. *Academy of Management Review*, **26**, 289–297.

Wallas, G. (1926). *The art of thought*. New York: Harcourt, Brace.

Waller, M. J., Huber, G. P. & Glick, W. H. (1995). Functional background as a determinant of executive's selective perception. *Academy of Management Journal*, **38**, 943–974.

Wegner, D. M. (1986). Transactive memory: A contemporary analysis of the group mind. In: B. Mullen & G. R. Goethals (Eds), *Theories of Group Behavior* (pp. 185–208). New York: Springer-Verlag.

Wegner, D. M., Erber, R. & Raymond, P. (1991). Transactive memory in close relationships. *Journal of Personality and Social Psychology*, **61**, 923–929.

Wheatley, W. J., Anthony, W. P. & Maddox, E. N. (1991). Selecting and training strategic planners with imagination and creativity. *Journal of Creative Behavior*, **25**, 52–60.

Woodman, R. W. (1995). Managing creativity. In: C. M. Ford & D. A. Gioia (Eds), *Creative Actions in Organizations* (pp. 60–64). Thousand Oaks, CA: Sage.

Woodman, R. W., Sawyer, J. E. & Griffin, R. W. (1993). Toward a theory of organizational creativity. *Academy of Management Review*, **18**, 293–321.

Zhou, J. (1998). Feedback valence, feedback style, task autonomy, and achievement orientation: Interactive effects on creative performance. *Journal of Applied Psychology*, **83**, 261–276.

Zhou, J. (forthcoming-a). When the presence of creative coworkers is related to creativity: Role of supervisor close monitoring, developmental feedback, and creative personality. *Journal of Applied Psychology*.

Zhou, J. (forthcoming-b). Job-related feedback and creative performance. In: C. Ford (Ed.), *Handbook of Organizational Creativity*. Hillsdale, NJ: Lawrence Erlbaum.

Zhou, J. & George, J. M. (2001). When job dissatisfaction leads to creativity: Encouraging the expression of voice. *Academy of Management Journal*, **44**, 682–696.

Zhou, J. & Oldham, G. R. (2001). Enhancing creative performance: Effects of expected developmental assessment strategies and creative personality. *Journal of Creative Behavior*, **35**, 151–167.

The International Handbook on Innovation
Edited by Larisa V. Shavinina

Venture Capital's Role in Innovation: Issues, Research and Stakeholder Interests

John Callahan and Steven Muegge

Department of Systems and Computer Engineering and Eric Sprott School of Business, Carleton University, Canada

Abstract: The purpose of this chapter is to review the role of venture capital in innovation. The chapter begins with the history and current state of venture capital. We also describe the process of venture capital financing and how it relates to the innovation process. Then, we review the research literature related to venture capital investment decision-making, the venture capital-entrepreneur relationship, and the fostering of innovation by venture capital. Finally, using a stakeholder perspective, we outline the usefulness of current research for different stakeholders and call for more qualitative, longitudinal research that contributes better stories and richer data on variable interrelationships.

Keywords: Innovation; Venture capital; Research.

Introduction

The purpose of this chapter is to provide a review of the role of venture capital in the innovation process. Excellent reviews of the investment issues of venture capital already exist (Gompers & Lerner, 2001a, 2001b). There are no reviews, however, that focus on venture capital's role in innovation. The present chapter aims to fill this gap.

Innovation is an ancient activity in human history. The pace of innovation, however, has accelerated significantly in the last 50 years (Agarwal & Gort, 2001). Innovation is now commonly regarded as the basis for competitive advantage between enterprises and between whole communities (Porter, 1990). Venture capitalists make high-risk equity investments in new entrepreneurial ventures. Innovation by venture capital financed start-ups is felt by many to contribute significantly to modern economic development. The fall of 2002, as this chapter is being written, is actually a good time to ask about venture capital's role in innovation and economic development. The last 15 years provide a complete up-and-down cycle.

There are many natural sources of conflict between venture capitalists and entrepreneurs. As equity investors, venture capitalists want the companies in which they invest to be successful. There are many versions of 'success', however, in any situation as complex as building a new company. For the founding team of

entrepreneurs, successful innovation can be the creation of a company of which they can be proud, that provides a good living, and may provide real equity value at some time in the future. This process might take 10, 15, even 20 years and still be successful. Entrepreneurs are normally not diversified—their entire fortunes will be tied up in their companies. On the other hand, a venture capitalist will have a reasonably diversified portfolio of a dozen or more investments. Moreover, for the venture capitalist, success is very specific and clear cut. A VC invests only with the prospect of realizing real equity value through a liquidity event like acquisition or an initial public offering—generally within a period of five to seven years (Lerner, 1994). The conflicts that naturally exist in this relationship are captured by a quote from an article in an online engineering journal (Tredennick, 2001):

> VCs know how to deal with engineers, but engineers don't know how to deal with VCs. VCs take advantage of this situation to maximize the return for the venture fund's investors. Engineers are getting short-changed.

In reviewing the role of venture capital in innovation, we cover the management issues and the research to date. Our focus is on independent venture capital firms but we also review corporate venturing for comparison.

641

The chapter begins with an overview of venture capital—its history and current state. We then describe the process of venture capital financing and how it relates to the innovation process. Next we review the research literature related to venture capital investment decision-making and the venture capital-entrepreneur relationship. We also ask the question again using the research literature—does venture capital foster innovation? We find that the jury is out on this question. Finally, using a stakeholder perspective, we outline the usefulness of current research for different stakeholders and call for more qualitative, longitudinal research that contributes better stories and richer data on variable interrelationships.

Overview of Venture Capital

Venture capital (VC) is a specialized form of financing, available to a minority of entrepreneurs in attractive industries. Many venture capital success stories have become household names—Amazon, Cisco, Compaq, eBay, Federal Express, Intel, Lotus, Netscape, Sun Microsystems, and Yahoo all received VC funding. Venture capital is not exclusive, however, to the technology sector. The growth of Staples, Starbucks, and TCBY—all 'brick and mortar' retailers with innovative business models—was also fueled by venture capital investment. In the words of VC researchers Paul Gompers & Josh Lerner (2001b, p. 83):

No matter how we look at the numbers, venture capital clearly serves as an important source for economic development, wealth and job creation, and innovation. This unique form of investing brightens entrepreneurial companies' prospects by relieving all-too-common capital constraints. Venture-backed firms grow more quickly and create far more value than nonventure-backed firms. Similarly, venture capital generates a tremendous number of jobs and boosts corporate profits, earnings, and workforce quality. Finally, venture capital exerts a powerful effect on innovation.

In addition to funding, venture capital investors (venture capitalists, or VCs) can provide specialized knowledge of a particular industry, experience successfully growing a business from start-up to publicly traded company, and access to a network of contacts that may include seasoned managers, partners, and customers. The venture capitalist brings terms, controls, expertise, and financial strength that helps form a well-managed and well-financed company that is more likely to succeed. In exchange, the venture capitalist demands a preferred equity share of the new venture, along with favorable upside and downside investment

protections.[1] The founding entrepreneurs relinquish equity and agree to contractual restrictions intended to protect the venture investment. In doing so, the founders give up exclusive ownership of the whole pie for the possibility of owning a small slice of a much larger pie, when the firm is taken public or acquired.

Venture capitalists are able to effectively exit their investments only at a liquidity event—an initial public offering of stock (IPO) on a public stock exchange, acquisition of the firm by another firm, or bankruptcy. The IPO is the most lucrative result for all investors (Cumming & MacIntosh, 2002), so in principle, the interests of the founders and venture capital investors align in this regard.

Joseph Schumpeter (1934) first proposed that small entrepreneurial firms are most likely to be the source of most innovation. Modern research supports the notion that large established firms have great difficulty managing innovations that fall outside of their previous experience, including *architectural innovations* (Henderson & Clark, 1990), *competency-destroying innovations* (Tushman & Anderson, 1986), and *disruptive technology* that changes the basis for competition in an industry (Christensen, 1997). Established firms may partially overcome these limitations through *ambidextrous organizational structures* (Tushman & O'Reilly, 1997), *radical innovation hubs* (Leifer et al., 2000), and *corporate venturing* programs that emulate venture capital (Chesbrough, 2000). Nonetheless, new firms would appear to have some natural advantages at realizing some innovations.

Innovation in small firms is difficult to finance because of four fundamental problems (Gompers & Lerner, 2001b):

(1) high uncertainty;
(2) information asymmetry;
(3) intangible soft assets;
(4) sensitivity to volatile market conditions.

High uncertainty is a fundamental trait of innovation that no amount of study or due diligence can entirely eliminate. The future is not only unknown, it is unknowable (Christensen, 1997). *Information asymmetry* refers to the large information gaps possible between innovators and investors. Because of their particular specialized expertise, innovators are likely to have a superior understanding of their innovation,

---

[1] According to Zider (1998) and Kaplan & Stromberg (2000a), these restrictions may include preferred and convertible securities to ensure that VCs are paid first if the firm is liquidated, anti-dilution constraints, prevention of early liquidation by entrepreneurs, mandatory redemption rights to force liquidation by VCs, restrictions on the sale of assets, restrictions of sales of stock that would alter ownership, non-compete and vesting provisions that make it expensive for the entrepreneur to leave the firm, and loss of control rights if the firm performs poorly.

while investors are likely to have a superior understanding of financing. *Intangible soft assets* include patents and trademarks, human capital, and future opportunities. The real value of these assets is difficult to measure; they may have great value to a particular owner, but negligible value to others. The value and liquidity of innovative firms is highly *sensitive to volatile market conditions*. During an economic boom, it may be relatively easy and lucrative to complete an IPO of a promising firm on the public stock markets; in a depressed market, it may be impossible.

These four fundamental problems make it difficult for many entrepreneurs to raise high levels of funds through traditional debt financing.[2] Venture capital fills this void by providing high levels of funding to opportunities with high uncertainty and large information asymmetries—in other words, ventures that may not otherwise have been funded.

Venture capital is neither available nor necessarily desirable to all entrepreneurs. Most start-ups do not employ venture capital, nor would they be attractive candidates for venture funding.[3] The vast majority of entrepreneurial start-ups are sole proprietorships in the service industry with limited opportunity for growth (Bhidé, 2000, p. 13). Venture capitalists do not fund laundries, family-run restaurants, or hair salons.

Some founding entrepreneurs that would qualify for venture funding may prefer to *bootstrap*—self-finance from personal savings, debt, and re-invested revenue. A number of significant Fortune 500 firms, including such technology notables as Hewlett-Packard,[4] Microsoft,[5] and Dell,[6] have grown to dominate their industries without early venture capital funding. In each example, the original founders retained significant ownership and control of their innovation.

Venture capital emerged in the United States in the years following World War II (Gompers & Lerner, 2001a). In 1946, founders from the Massachusetts Institute of Technology and Harvard Business School partnered with local businesses leaders to establish American Research and Development (ARD), the first true venture capital firm. ARD invested in emerging companies seeking to commercialize wartime technologies. ARD was a publicly traded closed-end mutual fund.[7]

Many early venture capital organizations were organized as closed-end funds or Small Business Investment Companies (SBICs).[8] In 1958, the first venture capital limited partnership was formed (Draper, Gaither & Anderson). Limited partnerships became more prevalent throughout the 1970s and 1980s, and are now the most common venture capital structure. Unlike mutual funds, limited partnerships are exempt from American Securities Exchange Commission (SEC) regulations, including exacting investment disclosure requirements.

Until 1979, investment in limited partnership venture capital funds was restricted to a limited number of institutions and wealthy individuals. In that year, changes to U.S. Department of Labor regulations opened up venture fund investment to pension funds, a rich new source of capital to fuel new growth.

Venture capital is not equally available to entrepreneurs in all countries. The U.S. venture capital pool remains the largest in the world by either absolute size or relative comparison to other economic data. In 1995, the ratio of the venture capital pool to the size of the economy was 8.7 times higher in the United States than in Asia, and 8.0 times higher in the United States than in continental Europe (Gompers & Lerner, 1999a, p. 326). In 2001, 62% of global private equity[9] was invested in North America, 21% in Western Europe, 12% in Asia Pacific, 2% in the Middle East and Africa,

---

[2] Qualification for a bank loan, for example, may require tangible collateral and agreement to a fixed repayment plan. As the perceived risk of the investment rises, the terms of the financing would become more expensive and restrictive.

[3] Bhidé (2000) reports that only 5% of 1989 *Inc. 500* companies start with VC funding, while 80% bootstrap with modest funds. The *Inc. 500* is a compilation of the fastest growing privately held companies in the United States.

[4] HP was founded in 1938 and taken public in 1957—an interval of 19 years. It achieved Fortune 500 status in 1962. In contrast, the typical VC-backed company that went public between 1984 and 1994 did so in just five years (Venture Capital Journal, February 1995, p. 45).

[5] Microsoft was founded in 1975 and taken public in 1986. In 1975, personal computers were restricted to a small number of hobbyists; neither founder had management experience, and both had dropped out of college. Microsoft did accept some late-stage funding prior to IPO, although Bhide (2000, p. 164) suggests that this decision was motivated by the desire to improve the legitimacy of the IPO to institutional investors rather than a need to raise capital.

[6] Dell was founded in 1984, taken public in 1988, and achieved Fortune 500 status in 1992, all without venture capital financing.

---

[7] A closed-end fund is a mutual fund whose shares are issued initially and subsequently trade on an exchange much like common shares. Because of their liquidity, securities regulation did not preclude them being marketed to average investors—this fact lead to brokers too commonly selling them to investors not really suited to their high risk (Gompers & Lerner, 2001a).

[8] The SBIC program was set up by the American federal government to encourage the development of venture capital for innovation after the shock of the Soviet launch of the Sputnik satellite in 1957. The program, however, was badly designed and the organizational form is no longer significant (Gompers & Lerner, 2001a).

[9] Statistics on global private equity are more widely available and better standardized than international venture capital statistics. Private equity includes venture capital, buyout funds, mezzanine debt funds, and special situation funds. Venture capital is a substantial component of private equity. In the United States in 2001, the $59.7B pool of private equity included $41.9B of venture capital.

and less than 1% in Central and Eastern Europe (PricewaterhouseCoopers, 2002). Table 1 ranks the top twenty countries for disbursements of private equity investment.

Table 1. Global private equity investment (2001).*

Rank	Country	Investment (US$B)
1	USA	59.7
2	United Kingdom	6.2
3	Germany	4.0
4	Canada	3.2
5	France	3.0
6	Japan	2.1
7	Italy	2.0
8	Sweden	1.8
9	Korea	1.8
10	Hong Kong	1.8
11	China	1.8
12	Netherlands	1.7
13	Israel	1.6
14	Australia	1.3
15	India	1.1
16	Singapore	1.1
17	Spain	1.1
18	Taiwan	0.8
19	Belgium	0.4
20	Denmark	0.3

* All dollar values in billions of U.S. dollars.
Source: PricewaterhouseCoopers (2002).

Timing is significant—the supply of venture capital money and the willingness of venture capitalists to invest are strongly dependent on the state of the equity markets and other market forces (Sahlman, 1990b). Over the long run, the pool of venture capital, and venture disbursements to portfolio firms, have grown significantly (see Fig. 1). Disbursements from U.S. venture funds have grown from just over US$1B in 1981 to nearly US$42B in 2001, a compound annual growth rate of nearly 19%. The cyclical fluctuations can be very large—disbursements in 2000, the peak year for venture capital, exceeded US$100B. While U.S. 2001 venture capital investment declined sharply from 2000 levels, 2001 was still the third-highest disbursement year in the history of the industry, trailing only the two exceptional preceding years. Internationally, the recent declines were less precipitous, with Canadian disbursements declining 27% in 2001 compared to the 65% decline in the United States. Periods with a rapid increase in capital commitments favor the entrepreneur, with less restrictive partnership agreements, larger and more frequent investments in portfolio firms, and higher valuations for investments (Gompers & Lerner, 1999a, p. 326). Periods of decline reduce the supply of VC money and favor the venture capitalist.

At the time of this writing, the technology industry is in a downturn following an exceptional period of record fundraising, IPOs, and acquisitions. Technology spending by businesses, the target customer base of many technology ventures, is in decline. Market

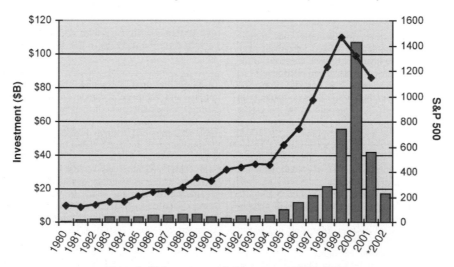

Figure 1. U.S. venture capital investment (1980–2002).

* Data from 2002 is incomplete, accounting only for the first nine months of the year.
Source: PricewaterhouseCoopers/Venture Economics/National Venture Capital Association MoneyTree Survey, Q3 2002 Quarterly Statistics.

evaluations of technology firms are low. Based on data from the first three quarters, U.S. disbursements in 2002 are expected to further decline from 2001 levels. Such 'boom and bust' cycles are not new. VC disbursements previously declined in the late 1970s, and again in the mid-1980s. Figure 2 expands the ten-year period between 1982 and 1991 to illustrate the depth and duration of the previous disbursement decline.

Despite impressive long-term growth, venture capital remains a very small fraction of the total equity markets. Gompers & Lerner (1999a) estimate that in the United States, there are a hundred dollars of publicly traded equity for every dollar of venture capital.

Venture capitalists strongly favor particular high-growth, technology industries. Tables 2 and 3 show the distribution of U.S. venture capital investment across industry classification.[10] In 2001, nearly three-quarters of total venture capital disbursement dollars went to firms in only six of the seventeen industries classifications: *software* (20%), *telecommunications* (15%), *networking and equipment* (14%), *retailing and distribution* (10%), *biotechnology* (8%), and *Information Technology services* (7%). The retailing and distribution category includes traditional 'brick and mortar' retailing as well as Internet businesses; the other five categories are exclusively technology industries. In the

most recent data available at the time of this writing, 55% of U.S. venture capital investment was awarded to Internet-related businesses, including E-commerce, Internet software, services and tools, hardware, and infrastructure.[11]

These tables also demonstrate the short-term trends common in venture capital investment. As an example, consider the retailing and distribution category. During the height of the dot-com boom of 1999 and early 2000, the fraction of venture money invested in retailing rose from typical levels of 5–10%, to 23% and 18% respectively, as venture firms invested heavily in e-commerce. In 2001, the fraction of venture investment in retailing had returned to 10%.

According to Zider (1998):

> The myth is that VCs invest in good people and good ideas. The reality is that they invest in good industries. Regardless of the talent or charisma of individual entrepreneurs, they rarely receive backing from a VC if their businesses are in low-growth market segments.

Tredennick (2001) summarizes this differently:

> VCs either all fund something or none of them will. If you ride the crest of a fad, you've got a good chance of getting funded. If you have an idea that's too new and different, you will struggle for funding.

---

[10] The seventeen industry classifications are defined by the PricewaterhouseCoopers/Venture Economics/National Venture Capital Association MoneyTree Survey.

[11] From the MoneyTree survey, Third Quarter 2002, $2464.1M of 4475.9M total VC investment in Q3 2002 was disbursed to Internet-related businesses.

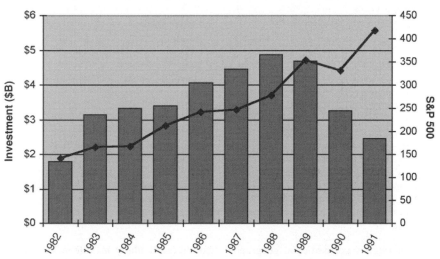

U.S. Venture Capital Disbursements (1982–1991)

Figure 2. U.S. venture capital investment (1982–1991).

Source: PricewaterhouseCoopers/Venture Economics/National Venture Capital Association MoneyTree Survey.

Table 2. U.S. venture capital investment by industry classification (1995–2001).*

	1995	1996	1997	1998	1999	2000	2001
Biotechnology	854	1,249	2,136	1,556	2,222	4,300	3,236
Business Products and Services	171	321	269	405	1,347	2,005	496
Computers and Peripherals	421	464	527	521	1,185	2,954	1,139
Consumer Products and Services	601	450	566	578	630	1,053	468
Electronics/Instrumentation	154	273	383	300	352	907	412
Financial Services	177	287	358	622	790	741	492
Healthcare Services	387	672	1,187	818	666	593	437
Industrial/Energy	652	614	942	1,387	1,500	2,256	1,336
IT Services	187	463	671	1,247	4,216	9,120	2,994
Media and Entertainment	382	939	985	1,613	5,428	8,808	2,235
Medical Devices and Equipment	705	652	987	1,200	1,438	2,543	2,047
Networking and Equipment	346	626	1,013	1,511	4,367	11,122	5,716
Other	29	11	56	128	178	249	172
Retailing/Distribution	360	827	852	2,155	12,572	19,420	4,164
Semiconductors	203	218	483	631	1,222	3,298	1,809
Software	1,081	2,308	3,256	4,228	9,348	20,402	8,545
Telecommunications	1,007	1,313	1,684	2,729	8,076	17,536	6,241
Total	7,717	11,687	16,356	21,630	55,537	107,306	41,940

* All dollar values in millions of U.S. dollars.
Source: PricewaterhouseCoopers/Venture Economics/National Venture Capital Association MoneyTree Survey, Q3 2002 Quarterly Statistics.

Table 3. U.S. relative venture capital investment by industry classification (1995–2001).

	1995 %	1996 %	1997 %	1998 %	1999 %	2000 %	2001 %
Biotechnology	11	11	13	7	4	4	8
Business Products and Services	2	3	2	2	2	2	1
Computers and Peripherals	5	4	3	2	2	3	3
Consumer Products and Services	8	4	3	3	1	1	1
Electronics/Instrumentation	2	2	2	1	1	1	1
Financial Services	2	2	2	3	1	1	1
Healthcare Services	5	6	7	4	1	1	1
Industrial/Energy	8	5	6	6	3	2	3
IT Services	2	4	4	6	8	8	7
Media and Entertainment	5	8	6	7	10	8	5
Medical Devices and Equipment	9	6	6	6	3	2	5
Networking and Equipment	4	5	6	7	8	10	14
Other	0	0	0	1	0	0	0
Retailing/Distribution	5	7	5	10	23	18	10
Semiconductors	3	2	3	3	2	3	4
Software	14	20	20	20	17	19	20
Telecommunications	13	11	10	13	15	16	15

Source: PricewaterhouseCoopers/Venture Economics/National Venture Capital Association MoneyTree Survey, Q3 2002 Quarterly Statistics.

Venture capitalists strongly favor particular geographical regions. Figure 3 shows the distribution of U.S. venture capital by state. In the United States, venture capital investment is strongly concentrated in California (particularly in Silicon Valley, Orange County, and San Diego) and Massachusetts (particularly near Route 128 that circles Boston). In the third quarter of 2002, these two states together represented over half of

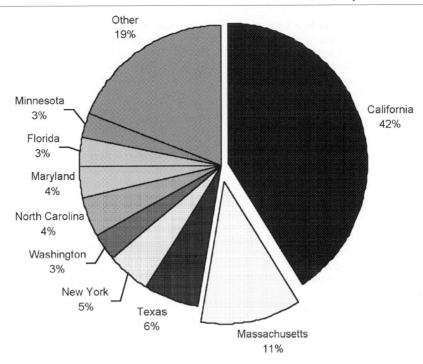

Figure 3. Venture capital investment by region (U.S. only).

Source: PricewaterhouseCoopers/Venture Economics/National Venture Capital Association MoneyTree Survey, Q3 2002 Quarterly Statistics.

all U.S. venture capital investment.[12] There appear to be two closely related factors to explain geographical clustering.

First, venture capitalists tend to invest close to home. In the United States, venture capital firms are highly clustered in California and Massachusetts.[13] Lerner (1995) reports that over half the venture-backed firms in a biotechnology sample had a venture director with an office within 60 miles of their headquarters. Powell et al. (2002) reports that more than half of all U.S. biotech firms received locally based venture funding between 1988 and 1999.

Second, regions with large venture capital activity develop *agglomeration economies* that further favor

venture capital through a virtuous circle of improved process efficiency. Intermediaries familiar with the workings of the venture process, particularly lawyers, accountants, and real estate brokers, reduce the transaction costs associated with forming and financing new firms (Gompers & Lerner, 1999b).

Each venture capital firm has a style and character unique to itself (Nesheim, 2000, p. 187). Firms differ from one another by reputation, age, experience of the general partners, preference for lead or follow-on investment, and their record of past success. Many choose to specialize in a particular industry (such as telecommunications or biotechnology), or a particular funding stage (seed, early or late-stage investments). A few firms can boast of significantly higher performance. According to Nesheim (2000, p. 180), while approximately 60% of funded firms go bankrupt, several VC partners claim to have one-third to one-half fewer bankruptcies per portfolio company. The IPOs of firms backed by VCs with strong reputations attract higher quality underwriters and are more widely held by institutional investors (Megginson & Weiss, 1991).

The Venture Capital Investment Cycle

In the dominant limited partnership form, a venture capital firm sets up one or more separate investment funds as limited partnerships. The firm becomes the

---

[12] From the birth of the VC industry, venture capital has favored California and Massachusetts. In the third quarter of 2002, California and Massachusetts respectively captured 41% and 11% of U.S. venture capital investment. In the late 1960s, Silicon Valley and Route 128 had similar levels of both high-tech employment and venture capital investment; today, Silicon Valley has significantly outpaced Route 128. For a history and analysis of regional advantage, see Saxenian (1994).

[13] Nearly 40% of U.S. venture capital firms qualifying for the 2002 MoneyTree Survey had head offices in either California or Massachusetts (251 of 668 VC investment firms). Ten more firms had offices in the New England area close to Massachusetts.

general partner in these funds and then sells units of interest in these funds to limited partners (wealthy individuals, pension funds and corporate investors).[14] The VC firm manages these funds as a general partner. A condition of limited partnership is that the limited partners play no role in managing the funds. When opening a fund, a VC firm will specify both a subscription target and an investment policy for the fund.

Limited partners pay VCs annual 'carrying' or management fees generally between 1% and 3% of the their investment. Once a fund is terminated (usually within ten years), the general partner receives 'carried interest' of around 20% (Gompers & Lerner, 1999b) of the capital gains realized by the fund over its lifetime with the limited partners receiving the rest. It is only at fund termination that the limited partners realize liquidity on their investment. A VC firm opens and terminates different funds, some appealing to retail investors and others to institutional investors, on a regular basis.

Note that VCs do not participate directly in losses although losses certainly lead to a loss of future business. Bhidé (2000, p. 144) argues that this asymmetry leads VCs to take excessive risk in their investments.

Insiders in the companies in which VCs invest, have tacit knowledge of their opportunities that is very hard to make available to outsiders. Because of information asymmetries and the related lack of 'efficient pricing', venture capital investing is very labor intensive. VC firms do not handle the volume of invested funds regularly handled by fund managers of liquid established stocks. $100 million is large for a VC firm, whereas funds over a $1 billion are common for liquid investments. As a result the fees charged to investors by VCs are correspondingly higher (Lerner, 1995; Sahlman, 1990a, p. 508).

Once the venture capital firm has received money from subscription to a fund, it sets about investing the funds. This process of raising money, and then placing it, creates a time lag that has given venture capital firms significant difficulties in recent years. Firms raised funds during good years for investing, and then when markets turned down in 2000 and 2001 they did not have good opportunities in which to invest. As a result, many firms had significant 'overhang' during this period. Some even returned funds to investors—a very costly proposition.

The investor returns on a venture capital fund are typically generated by a small fraction of their investments. One study of venture capital portfolios reported that about 7% of investments accounted for more than 60% of the profits, while fully one-third resulted in a partial or total loss (Bhidé, 2000, p. 145). Such skewed returns across the portfolio have been an attribute of venture capital investing throughout the history of the industry. ARD, the first professionally managed VC firm, generated over half of its annualized rate of return from a single $70K investment in Digital Equipment Corporation (out of total investments of $48M).[15]

Venture Capital's Role in New Venture Financing

When thinking about venture capital's role in innovation, keep in mind the relatively small percentage of innovative start-up companies that use venture capital during their development.

The Financing Sequence

For those start-up companies that do have business models that require significant up-front expenditures on product/service development and business infrastructure creation, the normal sequence of equity financing is as follows:

(i) Personal funds of the entrepreneurs

The entrepreneurs who start a company are the first to invest in the company. This may be a significant amount in the case of a company started by entrepreneurs successful from previous ventures. Normally, the amounts raised this way will be tens of thousands of dollars. This equity will likely include personal debt raised by these individuals that is invested in the start-up as equity. It may include 'sweat equity' in the form of under-compensated work. This type of initial investment can extend to employees as well. The proposition becomes: 'If you think that a job here is attractive, you should want to invest in the opportunity'.

(ii) Friends and family funds

A new venture will seldom be able to proceed to raise equity investments from organized sources if it is not able to raise equity from the friends and family of the founding entrepreneurs. The ability to go to friends and family and convince them to invest is regarded as a sign of commitment by the founding entrepreneurs to a real, quality opportunity. The family and friends round is again likely to be in the tens of thousands of dollars range.

(iii) Angel investors

Angels are wealthy individuals who invest their own money (Fenn & Liang, 1998). They are often entrepreneurs who have been successful in the

---

[14] A typical distribution of limited partners includes pension funds (50%–60%), endowments and foundations (20%–30%), other financial institutions (6%), and high net worth individuals (4%). Source: Venture Economics.

[15] The twenty-five year annualized rate of return from 1946 to 1971 was 15.8%. Excluding the DEC investment, the annualized rate of return would have been 7.4% (Bhidé, 2000, p. 162).

same area of business as that in which they invest. Angels often keep a low profile in their communities, not wanting to be pestered by start-ups looking for money, but preferring to find investment opportunities through their personal business networks. They usually invest between $100K and $500K. Although seldom organized, there are exceptions including the Band of Angels in Silicon Valley, Zero Stage Capital in New England and Purple Angel in Ottawa, Canada.[16] Many more firms receive funding from angels than from venture capitalists, but the level of funding is much lower (Freear & Wetzel, 1990).

Angel investment is important to a start-up for more than the risk capital that angels provide. They often have deep knowledge of the industry and of the entrepreneurs that drive them. As a result, they bring credibility and contacts with their investments. Start-ups that have been financed by angels have a much greater success rate in attracting subsequent venture capital. In a recent questionnaire survey study, Madill et al. (2002) found that "57% of the firms that had received private investor financing also received financing from institutional venture capitalists; only 10% of firms that had not received angel financing obtained venture capital".

(iv) Venture capital

The minimum amount invested in a venture by organized venture capital companies is generally over a million dollars. On the high side, VC investments up to $100 million are possible.

VC investments are very commonly *syndicated*—there will be a lead VC that organizes a group of VC firms to invest in a start-up (Lerner, 1994). For example, when the computer security company, Zero Knowledge, went to the venture capital market for financing in 1999, they had serious discussions with 10 venture capital firms in both Canada and the United States. In the end, they raised $12 million in equity from three American firms: Platinum, Aragon and Strategic Acquisitions.

VC investments are also commonly *staged*, so that multiple rounds of venture capital investment may be required to take an early-stage firm to liquidity (Gompers, 1995). Each funding round is negotiated at the current valuation of the firm, and dilutes the ownership of existing investors.[17]

Staging is a control mechanism that allows VCs to monitor the progress of firms and maintain the option to abandon under performing projects.

Venture capital firms supply many other things to a new venture in addition to financing. Very commonly they bring a deep knowledge of the technologies and markets, and as a result can add significant value in terms of business model and marketing strategy. Some VCs have large networks of contacts—with other investors, customers, potential partners, and managers. These contacts can be of great value to a new venture. Investment in a start-up by a prestigious VC also brings credibility in both the financial and product markets.

(v) Merchant bank financing

As a startup grows and proves its business model, investment risk can decrease. At this point the need for capital can increase substantially. Under these circumstances, a start-up can look to institutional investors called 'merchant banks' for financing. Investments at this stage are called late stage venture capital or mezzanine financing. Merchant banks have large amounts of funds available to them, and the lower risk and likely shorter horizon until liquidity of late stage venture financing can be attractive to them. They generally invest in the form of debt, sometimes convertible to equity. As debt investors, one of their principal concerns is that the company has the cash flow to service the debt.

(vi) Liquidity

Venture capitalists look to a liquidity event like divestiture (i.e. acquisition by another company) or an initial public offering (IPO) to cash out. As a result, a venture capital backed new venture must plan and work towards such a liquidity event from the start if they wish to raise venture capital.

In some cases, venture capitalists may exercise control rights to force bankruptcy of an under performing venture. This may allow the VC to recoup some investment through ownership of preferred shares that are paid out before common shares.

The investor view of this financing sequence is shown in Fig. 4, which is an adaptation of the funnel model commonly used in new product innovation management (Wheelwright & Clark, 1992, 111–132).

Opportunities arise out of ideas at the 'fuzzy front end' of innovation. Many opportunities enter the funnel; few exit to sustained profitability. The timing in the diagram is meant to be descriptive of common patterns. The line segment marking the timing of venture capital investment is dotted at both ends to indicate the variable entry points of venture capital

---

[16] The Purple Angel partners are former executives of Nortel Networks. Purple was the corporate color of Bell-Northern Research, the Nortel research and development subsidiary, until the mid-1990s.

[17] Venture capital investors commonly demand anti-dilution protections on their investments, shifting this burden to founders and other investors.

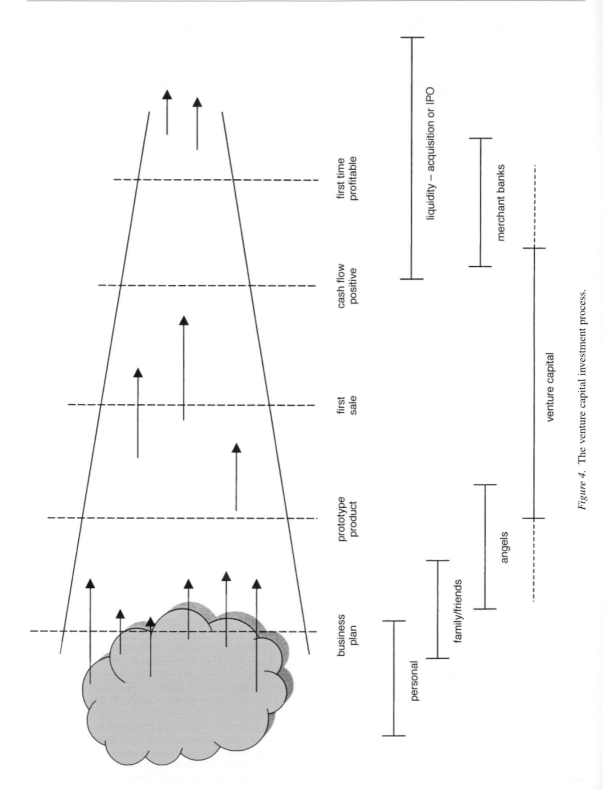

Figure 4. The venture capital investment process.

during different investment eras: for example, very early in 1998 and 1999, and much later in 2001 and 2002.

New venture opportunities emerge from the 'fuzzy front end' propelled by the drive, and personal funds, of the entrepreneurial team. As they enter the funnel, opportunities proceed through standard milestones such as having a business plan, having a prototype of their product or service, making a first sale, becoming cash flow positive, and reaching profitability. At each stage, a portion of these opportunities fail—the narrowing of the funnel represents the attrition of opportunities. Early stage investors regard success as having an opportunity exit the funnel through a liquidity event, at which time they can cash out. Very few startup ventures ever reach a VC satisfying exit like divestiture or IPO.

The Venture Capital Investment Process

Table 4 outlines the main stages of a model of the venture capital investment process (Tyebjee & Bruno, 1984). The model was developed on the basis of a questionnaire survey of 87 U.S. VCs and comments received from managers of seven of them. Tyebjee and Bruno make an important point about their results:

> The diversity of the responses, both in content and style, demonstrates the heterogeneity in practices of different venture capital firms. This heterogeneity cautions against too rigid a specification in any model describing venture capital management.

Tyebjee & Bruno's model was corroborated by Sweeting (1991) for U.K. venture firms.

Venture capital firms are interested in learning early about potential investments, and use their personal networks to locate such opportunities. In rare instances, a venture capitalist may become involved in the development of a new venture before it is ready for investments of the size and type appropriate for VCs. More commonly, however, the deals seek out VCs, who often maintain a high profile in their investment community—spending significant amounts of time at business events and conferences. The timing of VC financial entry into an opportunity can depend greatly on the supply of and demand for good opportunities by VCs. During the bubble years of 1998 and 1999, very early entry—before real sales—was the norm. Since the bubble burst in 2000, many VCs have been investing more conservatively and later in the opportunity development cycle.

VCs refer to 'deal flow' to describe the flow of investment opportunities that they see. Deal flow is the lifeblood of a VC firm. Because they normally see so many business plans, they have tough filters to control their workload. Of the business plans that they see, they finance only a very small percentage (Nesheim, 2000). Just reading a business plan can take hours, and VCs can receive hundreds per month. Some VCs do not accept any unsolicited business plans. They do take seriously, business plans brought to them by personal contacts and individuals that they know and trust (Shane & Stuart, 2002). This is one of the reasons that angel investment can be so important for a new venture intent on raising venture capital. A well-connected angel can personally introduce the founding entrepreneurs and their business opportunity to potential VC investors (Fenn & Liang, 1998).

Table 4. A processual model of U.S. venture capital fund activity.

Stage	Features
I. Deal origination	• Most deals are referred by third parties. • Referrals by other VCs are often invitations to join syndicates. • VCs are rarely proactive in searching out deals.
II. Deal screening	• Most frequently used screening criteria are: technology and/or market; stage of financing.

III. Deal evaluation	• Decision to invest based upon expected return compared with level of risk. Factors considered include:
	    Market attractiveness
	    Product differentiation
	    Management team capabilities
	    Protection of business from uncontrollable factors, e.g. competition, product obsolescence.

IV. Deal structuring	• VC funds use a wide range of approaches. An aim can be to help motivate managers to perform. • Price can be determined by: quality of opportunity; past experience with similar deals and so on.
V. Post-investment activities	• Venture funds provide management guidance and business contacts. • Representatives of venture funds normally sit on boards of operating businesses; they assist with development of business strategy. • Venture fund representatives can act as 'sounding boards' for operating business management.

Source: Sweeting (1991, p. 603).

VCs screen deals initially based on such factors as investment stage, investment size, industry sector and geography. If a deal gets through this screening, the first questions asked of the entrepreneurs driving an opportunity are of the form, "So what? Who cares? Why you?"[18] In other words: What is the core of the opportunity and why is it important? Who are the customers and what value is the start-up going to provide to them? And, what competitive advantage does the start-up bring to the table that will ensure that they can make money with the opportunity? Subsequent discussions elaborate on these themes. The decision criteria that VenGrowth uses in evaluating opportunities are the following: people, market, customer traction, competition, product idea, technology and timing. Other VC firms may have somewhat different criteria, but the core elements—experienced managers,[19] proprietary products, minimum investment thresholds, and extensive due diligence—are fairly uniform across VC firms (Bhidé, 2000). The research on venture capital deal evaluation is surveyed in a later section.

If a VC is still interested in investing after reviewing the company's business plan and talking with the principles, the VC will issue a term sheet to the company. This term sheet outlines what the VC sees as the basis for a financing deal. If the company accepts the term sheet, then *due diligence* by the VC begins in earnest on the company, the entrepreneurial team, and the opportunity. During this period of due diligence, the company is normally restricted from 'shopping the deal around' to other investors—in a sense, acceptance of the term sheet gives the VC an option to invest. This due diligence period can last several months, and is always a period of high stress and high cost in terms of management attention for the company.

Valuation of a startup, required as part of any deal, is a complex task (Timmons, 2001, Ch. 14). Quantitative models are used—multiples of sales, discounted multiples of future earnings, comparison with previous and concurrent deals, previous valuations at angel seed rounds—but many of the factors are qualitative. Qualitative factors focus on the match between what is required to be successful and the strength of the core management team, and of future market and technology trends.

Structuring the deal is the last stage before closing the investment. A good deal structure is one in which the goals of the VCs and of the entrepreneurs are aligned to the greatest extent possible. Important considerations include the equity share allocated to each party, the investment instruments used, and the staging of disbursements to the company.

The investment instruments used in VC deals have changed over the last few years. In the past, it was usual for VCs to purchase common shares of the companies in which they invested. They became investors on the same level as the founding entrepreneurs, family and friends, and angels. In the last few years, VCs have taken to insisting on convertible preferred shares and the senior liquidation rights that come with them (Kaplan & Stromberg, 2000a). These shares generally have minimum conversion values of two to three times the original sums invested. This means that when a liquidity event occurs, the VCs get paid before the common shareholders at a minimum payout that is a multiple of their initial investment. These are very tough terms.

When a deal has been signed, the start-up firm gets a check for the initial 'tranche' of the VC funds to be invested. It is rare for the full deal amount to be paid in one lump sum.[20] As part of the contract, the start-up must meet defined milestones to get successive tranches of the deal. These milestones take a variety of forms such as product development events, hiring key personnel, and meeting sales targets.

VCs are very active investors. Commonly, they participate as active members of the board; recruiting management and key technical personnel; developing business strategies; monitoring the company's performance; and facilitating subsequent financing rounds (Kaplan & Stromberg, 2000b). VC firms have even been know to function much like the chief financial officers of their client companies if these companies do not yet have adequate internal financial controls and competencies. This is usually short lived, however, and a VC will actively aid in recruiting such competencies for a company. VC-financed firms are more likely and faster to professionalize by adopting stock option plans and hiring external business executives, such as a vice-president of sales, or an external CEO (Hellman & Puri, 2000b).

As stated earlier, VCs will only invest in an opportunity if there is a good likelihood of some liquidity event within their five to seven year investment horizon.

Corporate Venturing

Corporations have also experimented with funding innovation directly through corporate venturing programs that seek to emulate the venture capital industry. The popularity of corporate venturing appears to rise and fall in approximately ten-year cycles with the

---

[18] These questions have actually been copyrighted by an Ottawa consulting company, Reid-Eddison.

[19] There is a saying in the venture capital community that 'the three important things about a deal are people, people and people'. A variation on this is that 'the five most important things about a deal are people, people, people, market and product'. Good people will find good opportunities, and more importantly, be able to execute on them.

[20] This was not the case during the Internet and dot.com 'bubble' when VCs commonly paid out the full amounts of an investment stage in one check.

venture capital industry and the broader equity markets (Block & MacMillan, 1995). Activity peaked in the late 1960s (Fast, 1978), mid-1980s (Yost, 1994) and late 1990s, declining again each time at the next market downturn (Chesbrough, 2000).

Corporate venturing includes 'intrapreneurship' programs to incubate and spin-off new entrepreneurial firms from within the corporation, as well as corporate venture capital funds (CVC) that invest corporate money directly in external start-ups in exchange for equity and control rights.

Examples of corporate venturing programs include *Xerox Technology Ventures* (1989–1995) documented by Hunt & Lerner (1995), Chesbrough & Smith (2000), and Chesbrough & Rosenbloom (2002); the *Lucent New Ventures Group* documented by Chesbrough & Socolof (2000); and the *Nortel Networks New Business Ventures* program (1997–1999), documented by O'Connor & Maslyn (2002), Leifer et al. (2001) and Hyland (2002).

Corporate venturing can provide favorable returns when compared to the returns from independent venture funds. During its eight-year lifetime, the $30M XTV fund invested in over twelve ventures, delivering capital gains of $219M. Hunt & Lerner (1995) estimate that $175M returned to Xerox, suggesting a 56% internal rate of return compared to a mean net return of 13.7% by independent VC funds over the same time period. Nonetheless, the program was discontinued,[21] underscoring the significant challenges of fostering entrepreneurship within large corporations.

Chesbrough's (2000) survey of the corporate venturing literature identifies several specific challenges that these initiatives face, including *adverse selection*, *resource allocation conflicts*, *conflicts of interest* between the new venture and parent sponsor, and potential *conflict of objectives* between financial and objectives.

Von Hippel (1977) identified the problem of adverse selection. Over time, the best performing ventures either spin-off or migrate to other divisions, leaving the corporate venturing organization with the under performing ventures. Fast (1978) noted that managers of established businesses can view successful corporate ventures as threats which compete for scarce resources. Rind (1981) explored possible conflicts of interest within new venture organizations between the success of the parent sponsor and the success of the new venture. The sponsor may constrain the marketing options of the new venture in order to prevent competition with existing businesses. Siegel, Siegel & MacMillan (1988) explored the potential conflict between two frequently cited rationales for new venture businesses. *Strategic* investments seek to

exploit the potential for additional growth latent in the parent sponsor—in other words, improve the performance of existing businesses. *Financial* investments aim to create additional revenue and profit in the new venture itself. According to Siegel et al., parental intervention to align the venture with strategic interests reduces the autonomy of the new venture, and likely reduces financial performance.

Chesbrough (2000) proposes that corporate venture structures and venture capital structures have some significant differences. Compared to venture capital, corporate venturing provides weaker incentives for success, weaker financial discipline on the downside (i.e. slower to terminate under performing ventures), internal (rather than external) monitoring, and constraints on the discovery of alternative business models (Chesbrough & Rosenbloom, 2000). Potential advantages include longer investment time horizons (unconstrained by the fixed lifetime of a VC fund), larger scale of capital investment, management of strategic complementarities, and the retention of group learning. Chesbrough argues that for corporate venturing to succeed and persist through the less exuberant market cycles, it must leverage these potential advantages to deliver strategic benefits to the sponsoring firm.

Von Hippel (1977) showed that corporate ventures were more likely to succeed when the parent firm had significant prior experience in the target market. Experience with the technology, however, did not correlate to increased likelihood of success. Athey & Stern (1997) introduced *complementarity*—the notion that corporations can benefit from closely related activities. Research suggests that intrapreneurship and CVC programs are both more effective when investing in businesses that are closely related to the core competencies of parent. In a comparison of VC and CVC investments, Gompers & Lerner (1999c) found that corporations may be able to select better ventures using information from their related businesses and provide greater value to those firms once the investments are made. CVC programs without a well-defined strategic focus have less investment success and less stability than well-defined programs. Likewise, the successful investments of the Xerox Technology Ventures program were concentrated in industries closely related to corporate parent's business (Hunt & Lerner, 1995).

Venture Capital Decision Criteria

Early venture capital studies established that venture capitalists make investment decisions based on analysis of financial fundamentals rather than intuition or 'gut feel' (Pence, 1982). Subsequent work has made progress towards elucidating the details of this process.

MacMillan et al. (1985) conducted a frequently referenced study on the decision criteria of U.S.

---

[21] For an analysis of the motivations behind this decision, see Hunt & Lerner (1995).

Table 5. Opportunity evaluation criteria.

Criteria	Percent
Capable of sustained effort	64
Thoroughly familiar with market	62
At least 10× return in 5–10 years	50
Demonstrated leadership in the past	50
Evaluates and reacts well to risk	48
Investments can be made liquid	44
Significant market growth	43
Track record relevant to venture	37
Articulates venture well	31
Proprietary technology	29

Source: MacMillan et al. (1985, p. 123).

venture capital firms. The criteria most frequently rated are shown in Table 5.

The quality of the people is the most important factor; six of the top ten criteria in Table 5 relate to characteristics of the founding entrepreneurs. Experienced founders significantly increase the attractiveness of a venture (Bhidé, 2000). Having people on the team that 'have done it before'—that is, who have previously built a start-up opportunity to create shareholder value—is regarded very positively by VC investors.

Other considerations include the size and accessibility of the total available market, the growth rate of this market, customer traction, and the technology of the product. Customer traction is highly valued and paying customers are best. VCs typically bring in resident or contract technical experts to evaluate the product design and technology. They also interview prospective customers.

Kaplan & Stomberg (2000b) analyzed investment memoranda and subsequent status reports from ten VC partnerships for 58 investments in 42 portfolio companies. Their results confirm that VCs expend a great deal of time and resources evaluating and screening transactions. VCs explicitly consider the attractiveness of the opportunity (market size, strategy, technology, customer adoption, and competition) and risk. Management risk is cited in 60% of sample investments, most often related to a need to complete the team with seasoned executives. Management risk is correlated with contract restrictions (particularly voting and board seats) and cash flow restrictions based on performance. The early appraisal of the management team was related to subsequent performance; in particular, portfolio companies with strong management teams were more likely to go public.

Shane & Stuart (2002) studied the social capital of company founders. New ventures with founders having direct and indirect relationships with venture investors were more likely to receive venture funding and less likely to fail. They conclude that founder social capital represents an important endowment for early-stage organizations. Either the venture capital decision

criteria are partially subjective or they objectively place value on established relationships beyond that observed in previous studies.

Shepherd & Zacharakis (2002) suggest that research into decision aids can potentially improve the venture capital decision process, decision accuracy, and speed up the acquisition of expertise. This is a promising avenue for future research.

The Venture Capitalist—Entrepreneur Relationship

There has been a significant amount of research done on the venture capitalists entrepreneur relationship.[22] Frequent calls are heard for more since the area is one of great importance (Sapienza & Korsgaard, 1996; Steier & Greenwood, 1995). The theoretical frameworks used for most of this research have come from economics—for example, agency theory and incomplete contracts.[23] These economic theories focus on the natural conflicts that exist between investors and entrepreneurs, and remedies based on factors such as governance structure, restrictive covenants, stage disbursement of funds and investor oversight (Giudici & Paleari, 2000).

The usual application of agency theory to the venture capitalists—entrepreneur relationship models the venture capitalist as the principle and the entrepreneur as his agent. Once the venture capitalist invests in a new venture, he or she is interested in financial success that results in an early liquidity event such as acquisition or IPO. The entrepreneur shares this objective to a certain extent but has other interests as well such as compensation, management perks, career, survival of the business, and building the business beyond liquidity.

Just differences in risk exposure can lead to significant agency type conflicts. Venture capitalists are relatively well diversified as investors; entrepreneurs are not. Based on financial theory, this means that venture capitalists are concerned with systematic risk related to the market as a whole, whereas entrepreneurs are concerned with the total risk of their investment in their venture. Callahan & Sharp (1985) show analytically that this difference in risk exposure leads to growth objectives that differ between entrepreneurs and venture capitalists.[24] As part of their model,

[22] (Bhidé, 2002; Baker & Gompers, 1999; Bratton, 2002; Gifford, 1997; Gompers, 1999; Gompers, 1995; Gompers & Lerner, 1996; Jog et al., 1991; Kaplan & Stromberg, 2001, 2002; Lerner, 1995; MacMillan et al., 1989; Steier & Greenwood, 1995; Sahlman, 1990; Sapienza & Korsgaard, 1996; Sweeting, 1991; Sweeting & Wong, 1997.)

[23] The first important article on agency theory was Jensen & Meckling (1976). See Hart (2001) for recent review of the agency theory literature. See Hart & Moore (1990) for an introduction to incomplete contracts.

[24] Callahan & Sharp's simple analytic argument is excerpted and shown in Appendix A.

Callahan & Sharp also show that conflicts between entrepreneurs and venture capitalists are likely to arise over such issues as managerial compensation and perquisite consumption. Jog et al. (1991) showed these conflicts to exist empirically using a questionnaire survey of Canadian VC partners.

The book, *The Venture Capital Cycle,* by Gompers & Lerner (1999) contains excellent examples of research on the venture capitalist—entrepreneur relationship based on agency theory. In Chapter 7, Why are Investments Staged?, they conclude that VCs stage their investments in new ventures because of a concern that entrepreneurs with inside information will continue spending investor money even when faced with losing prospects because they stand personally to lose salary, perks, and reputation. In Chapter 8, How Do Venture Capitalists Oversee Firms?, they examine the role of VCs as directors of their portfolio companies. They conclude that the representation of VCs, unlike outside board members, increases around the time of CEO turnover.[25] They also find that geographic proximity is important for VC board members. In Chapter 9, Why Do Venture Capitalists Syndicate Investments?, they find that established VCs syndicate with each other in first round financing. In subsequent rounds, they involve less-established VCs. They attribute these results to the uncertainty in the first round—having another established VC also willing to invest is an important decision factor. They also find support for the contention that syndication is a way around the unfair information advantage that would accrue to the lead VC in subsequent rounds, that will probably demand syndication because of amounts of money required, if they go in alone at first.

Procedural justice theory provides another theoretical framework for Sapienza & Korsgaard (1996) to examine entrepreneur—investor relations.[26] They carried out two studies: a simulation study using students and a survey questionnaire study of VC partners. Their principle conclusion is that 'timely feedback promoted positive relations between entrepreneurs and investors'. They suggest that entrepreneurs yield a level of control and share information so that investors will eschew monitoring, and trust and support the entrepreneurs. This suggestion is congruent with advice that comes from agency theoretic analyses of the venture capitalist—entrepreneur relationship.

Game theory also provides a different perspective on the venture capitalist—entrepreneur relationship. Cable & Shane (1997) use a prisoner's dilemma

approach to develop a number of testable hypotheses that emphasize the cooperative alternatives to mutual value creation by venture capitalists and entrepreneurs. This perspective builds on Timmons & Bygrave's (1986) finding that "an ongoing cooperative relationship between entrepreneur's and venture capitalists is more important to the performance of the venture than the provision of venture capital itself" (Cable & Shane, 1997, p. 143). Cable & Shane (1997, p. 168) maintain that:

> Modeling venture capital relationships as a principal-agent problem appears unduly restrictive given the potential for opportunistic, non-cooperative actions by venture capitalists as well as entrepreneurs. For example, while the agency approach focuses on venture capitalists' adverse selection problem when evaluating entrepreneurs, an adverse selection problem also exists for entrepreneurs since they must locate venture capitalists who can provide complimentary managerial experience, access to relevant networks, and legitimacy.

Two recent empirical papers (Schefczyk & Gerpott, 2001a; Schefczyk & Gerpott, 2001a) have referenced Cable and Shane's paper but do not build specifically on their hypothesis structure.

Shepherd & Zacharakis (2001) emphasize the necessity of a balance between trust and control in the venture capital—entrepreneur relationship. They propose that the entrepreneur can

> build trust with the VC (and vice versa) by signaling commitment and consistency, being fair and just, obtaining a good fit with one's partner, and with frequent and open communication.

They regard their study as a counter weight to economic approaches like agency theory in which control is emphasized.

Venture Capital and Innovation

Does venture capital foster innovation? There are three popular arguments:

(1) venture capital unleashes innovation. VCs free innovative firms from capital constraints and add genuine value that helps them become successful;
(2) venture capital is neutral to innovation. VCs identify the best new ventures, and are the intermediary gatekeepers for funding;
(3) venture capital stifles innovation. VCs back only conventional ideas. Unconventional innovative ventures are screened out as too risky, and never receive funding.

The research to date is inconclusive. The jury is still out on this very important question.

The venture capital research is clear in one regard— VC-backed firms are more successful than non-VC backed firms, both before and after IPO. Venture-

---

[25] A similar conclusion is drawn by Gabrielsson & Huse (2002).

[26] Procedural justice theory (Lind & Tyler, 1988) examines the impact of the *process* of decision-making on the quality of exchange relationships. See De Clercq & Sapienza (2001) for a theoretical application of both agency and procedural justice theories to venture capitalist—entrepreneur relationships.

backed firms bring product to market faster (Hellman & Puri, 2000b), 'professionalize' earlier by introducing stock option plans and hiring external business managers (Hellman & Puri, 2000a), time IPOs more effectively to the market (Lerner, 1994), and have higher valuations at least five years after IPO (Gompers & Brav, 1997). Venture-backed IPOs pay lower fees and are less under priced. (Megginson & Weiss, 1991).

Causation, however, is more difficult to establish. Do venture capitalists add value that makes it more likely for their portfolio firms to succeed, or are they simply good at picking winners?

Research suggests that VCs do have some on impact their portfolio firms. Hsu (2000) compares a group of VC-backed start-ups with a control group of start-ups that obtained government funding through the U.S. Small Business Innovative Research (SBIR) program, which does not impact ownership or governance. The study concludes that venture capital changes the path of funded projects, by altering the commercialization strategy and making the firm more sensitive to the business environment.

Other studies imply that there are limitations to the value added by VC influence. Ruhnka, Feldman & Dean (1992) investigated the strategies employed by 80 venture capital firms to deal with the 'living dead' investments in their portfolios—ventures that were self-sustaining but failed to achieve levels of growth or profitability necessary for attractive exits such as IPO or acquisition. Venture managers were able to achieve a successful turnaround or exit in 55.9% of living dead situations, regardless of the age of the VC firms, their size, or the relative availability of investor personnel for monitoring investees. From the invariance of this result, the authors argue that that causal factors are outside VC control.

Some promising recent work supports the notion of a causal link between VC and innovation. Kortum & Lerner (2000) investigated trends in patent rates as a measure of innovation. Statistical analysis showed that the rate of U.S. patent filing was correlated with early-stage venture capital disbursements, when controlling for corporate research and development expenditures. In particular, the rate of patent applications declined during the 1970s and early 1980s while corporate research and development spending increased steadily. The rate of patent applications steadily increased after 1985, following the rapid rise of early-stage venture capital disbursements in the late 1970s and early 1980s.

Some anecdotal accounts, however, present a different picture. According to Tredennick (2001), venture capitalists and their technical experts actually favor very conventional and proven ideas: "If you step too far from tradition, (the VC) will not understand or appreciate your approach . . .. Just as Hollywood would rather make a sequel than produce an original movie,

VCs look for a formula that has brought success". According to Bhidé (2000):

> VC-backed entrepreneurs face extensive scrutiny of their plans and ongoing monitoring of their performance by their capital providers. These distinctive initial conditions lead them to pursue opportunities with greater investment and less uncertainty, rely more on anticipation and planning and less on improvisation and adaptation, use different strategies for securing resources, and face different requirements for success.

Both accounts suggest that the venture capital process may actually screen out the most significant innovations in favor of minor variations of what has come before.

Other research suggests that venture capitalists frequently engage in 'herding'—making investments that are very similar to those of other firms—or what Tredennick (2001) calls 'riding the crest of a fad'. Devenow & Welch (1996) show that a variety of factors can lead to investors obtaining poor performance. Social welfare may suffer because value-creating investments in less popular technology areas may have been ignored.

During the Internet and dot-com 'bubble' of the late 1990s and early 2000, many startup ventures received large disbursements of very early venture capital funding. Since the collapse of the bubble, anecdotes have emerged describing the destructive effects of such large amounts of early money. The business model of a startup venture is like an untested hypothesis—the real test is making a profit from paying customers. Availability of early money can hide problems in a business by delaying such a test. Some very early stage startups redefined success in terms of financing—achieving the first (or the next) venture capital investment round. Bootstrapping, the creation of a significant business without significant outside financing, is again becoming popular because of the relatively limited supply of venture capital money—and it may not be a bad development. Important bootstrapped success stories include household names like Dell, Gillette, Heinz, HP, Mattel, Nike, Oracle, UPS, and Walt Disney.

The differences between bootstrapped and 'big money' startups are summarized in Table 6. Bootstrapping forces focus on cash flow and the immediate needs of customers in niche addressable markets. Freed of cash flow constraints, big money startups can try for highly engineered product 'home runs' with a view of striking it rich and cashing out. Big money allows for significant compensation packages, so the personal sacrifice of principals can be very low. When one reads about big money startups, the news all too often centers on their financing progress rather than success with real customers.

Table 6. Differences between bootstrap and big money start-ups.

	Bootstrap	Big Money
Cash	earn it	other people's
Initial focus	customers	exit
Product	incremental	fully featured
Markets	niche	$1B
Org. Structure	fluid	rigid
Time Horizon	near term	long term
Media Profile	low	high
Personal Sacrifice	high	low

Source: Presentation by Ken Charbonneau, Partner, KPMG, Ottawa, Carleton University, *Magic from a Hat* entrepreneurship lecture series, November 11, 2002.

In conclusion, venture capital would appear to at least help bring innovation to market. However, the selection process may not always identify and fund the most significant innovations, and especially in times of abundant supply, there may be disadvantages to 'big money'. Over all, venture capital may be a positive force to drive innovation—but the jury is still out.

The substantial body of financial and econometrics research offers few insights on innovation. More work is needed to specifically isolate and disentangle the influence of venture capital from that of other market forces, and relate that influence to innovation and the public good. Anecdotal accounts from seasoned practitioners and observers on possible limitations and drawbacks of venture financing remain untested with accepted research methods.

As Zider (1998) states:

> The (venture capital system) works well for the players it serves: entrepreneurs, institutional investors, investment bankers, and the venture capitalists themselves. It also serves the supporting cast of lawyers, advisers, and accountants. Whether it meets the needs of the investing public is still an open question.

Stakeholders and Research

Research on the role of VCs in innovation is carried out because there is a market for it. Researchers are producers. Consumers—the stakeholders in the output—are varied. They include entrepreneurs, venture capitalists, investors, policy makers, and researchers themselves.

Entrepreneurs need guidance on how to:

- approach VCs so as to maximize their chances of getting a good deal;
- appraise and judge VCs one from another;
- negotiate and structure equity investment deals;
- structure the participation of VCs on their boards and manage this participation subsequently;

- maintain control of their companies as they accept outside ownership;
- negotiate and work with VC set milestones;
- get through a liquidity event like an acquisition or an IPO effectively.

VCs are interested in how to:

- find good opportunities;
- screen and appraise good opportunities;
- negotiate and structure equity investment deals;
- support and control client companies;
- terminate client company relationships.

Investors need help in how to:

- judge the investment records of VC firms;
- appraise VC competencies in specific fields of investment;
- appraise the risk/return potential of VC investments;
- negotiate investment terms and covenants.

Policy makers include government officials and participants in quasi-government bodies such as the U.S. Federal Reserve Bank. They need guidance in how to:

- set tax policy;
- regulate investor access to VC funds;
- support VC activity that increases entrepreneurial value creation.

Researchers and educators, mainly in universities, want:

- research tools;
- interesting testable hypotheses;
- theories for creating new testable hypotheses;
- theories for explaining and teaching.

We can see that the interests of these consumer stakeholder groups are not the same.

Most research relevant to the role of venture capital in innovation follows the normal cycle of positivist scientific inquiry developed in the natural sciences: description of some phenomenon; theorizing that results in interesting, testable hypotheses; data gathering based on the hypotheses; hypothesis tests; and then renewed theorizing. Another approach to theory building is qualitative methods (Bailyn, 1977; Schall, 1983). Researchers using this methodology gather large amounts of text data, often interview transcripts, and then analyze the data for generalizable wisdom. They do not build theory in terms of testable hypotheses. Qualitative theory building and theory testing are closely interrelated—rather than being rigidly distinct as in positive theory building and testing. A variant of qualitative method based on case studies (Yin & Campbell, 2002) is used in the initial, descriptive stage of positive theory building.

Most large sample hypothesis tests are cross-sectional. They gather and analyze data on cases at a point in time. As a result there is little evidence of interrelationships between variables in each case. Such interrelationships are investigated across cases using multivariate statistical methods like regression. Qualitative methods can be used more easily to gather longitudinal data on a case over time. Then variable interactions within a single case can be investigated. Qualitative longitudinal data gathering, however, is very resource intensive. Processual research (Dawson, 1997; Hinings, 1997; Pettigrew, 1997; Woiceshyn, 1997) is one specific form of qualitative analysis.

The total effort of a research study is usually constrained by the availability of some resource like time or money. Given such a constraint, there is a natural trade-off between the number of cases in a sample and the amount of information gathered for each case. When testing hypotheses, it is normal to gather small amounts of very specific data on a large number of cases. Hundreds, even thousands, of cases about which little is known. When using case studies to develop theory rather than test it, large amounts of information are gathered on a very small number of cases. Research studies with one case can be published (Steier & Greenwood, 1995). Nested studies of several cases are more rigorous (Yin & Campbell, 2002) but are rare since they are very resource intensive. Qualitative studies tend to lie between these two extremes—tens of cases with significant amounts of data on each one.

These different research approaches can be complementary, but they appeal to stakeholders in different ways. Investors, policy makers and researchers can be satisfied with averages because they are interested either with long run effects or with large numbers of situations. The covariance-based techniques of large sample statistics can satisfy them. This is not true of entrepreneurs and venture capitalists. A typical entrepreneur would only be involved in a few startup ventures in his or her entire lifetime. Entrepreneurs deal with particular, specific, negotiated situations. They create new ventures, and are seldom interested in averages. In fact, they quite explicitly do not reference their situations to the average. Even venture capitalists deal with few enough investments that they do not really trust in averages or the law of large numbers to assist them. Both entrepreneurs and venture capitalists, as is common for business decision makers, test theories against their experience and intuition rather than using large scale statistical methods. Research must work to assist these decision makers to improve their intuition.

There is a striking lack of research on the role of VCs in innovation that provides rich insights into specific situations, and that can form the basis for effective theorizing as a result. Steier & Greenwood (1995) provide a notable exception. They document in detail the experiences of a single entrepreneurial new venture in the deal structuring and post-investment stages of venture capital involvement. Another is the

Zaplet Inc. case study by Leonard (2001). Leonard describes how the lead VC, Vinod Khosla, played an atypically active—even dominant—role in reformulating the business strategy and management structure of a very early start-up. There is a need for the equivalent of the studies by Burns & Stalker (1994, originally published in 1961) and Poole et al. (2000) that have been carried in innovation. Consider even the book, *Startup*, by Kaplan (1994). The book is a breathless, first person account of the story of GO Corporation, a start-up that tried and failed to develop and commercialize a hand held computer operated with a pen instead of a keyboard in the early 1990s. The book is not research—but the chapter on financing (Kaplan, 1994, pp. 59–81) contains more information useful to an entrepreneur than does the MacMillan et al. (1985) VC decision investment model referred to earlier.

We do not pretend that this call for more qualitative, processual research, and with it better stories, is new.[27] It does seem particularly important in the area of venture capital's role in innovation.

Financial Data Sources

The financial and economic data for this chapter was taken from the following sources: global private equity from PricewaterhouseCoopers (2002); venture capital investment in the United States from the *PricewaterhouseCoopers/Venture Economics/National Venture Capital Association MoneyTree Survey*, available online (http://www.pwcmoneytree.com), *Venture Economics* (http://www.ventureeconomics.com), and the *National Venture Capital Association* (NVCA) (http://www.nvca.com); venture capital investment in Canada from *Macdonald and Associates* (http://www.canadavc.com); venture capital investment in Europe from the *European Venture Capital Association* (http://www.evca.com); venture capital investment in Asia from the *Asian Venture Capital Journal* (http://www.asiaventure.com).

References

Agarwal R. & Gort, M. (2001). First-mover advantage and the speed of competitive entry, 1887–1986. *Journal of Law and Economics*, **44** (1), April, 161–177.

Amit, R., Glosten. G. & Muller, E. (1990). Does venture capital foster the most promising entrepreneurial firms? *California Management Review*, Spring, 102–111.

Athey, S. & Stern, S. (1997). An empirical framework for testing theories about complementarity in organizational design. Working paper (http://www.stanford.edu/~athey/testcomp0498.pdf).

Bailyn, L. (1977). Research as cognitive process: Implications for data analysis. *Quality and Quantity*, **11**, 97–117.

Baker, M. & Gompers, P. A. (1999). Executive ownership and control in newly public firms: The role of venture

---

[27] See, for examples, the dialogue between Dyer, Gibb & Wilkins (1991) and Eisenhardt (1989a, 1989b), and the special issue of the *Journal of Business Venturing* edited by Gartner & Burley (2002).

capitalists. Harvard Business School, working paper, November (http://www.people.hbs.edu/mbaker/cv/papers/Ownership.pdf).

Barry, C., Muscarella, C., Peavy, J. & Vetsuypens, M. (1990). The role of venture capital in the creation of public companies: Evidence from the going public process. *Journal of Financial Economics*, **27**, 447–471.

Bhidé, A. V. (2002). Taking care: How mechanisms to control mistakes affect investment decisions. Columbia Business School. Working paper, August (http://www.gsb.columbia.edu/faculty/abhide/bhide_taking_care.pdf).

Bhidé, A. V. (2000). *The origins and evolution of new businesses*. New York: Oxford University Press.

Block, Z. & MacMillan, I. (1995). *Corporate venturing: Creating new businesses within the firm*. Cambridge, MA: Harvard Business School Press.

Bratton, W. W. (2002). Venture capital on the downside: Preferred stock and corporate control. *Michigan Law Review*, **100** (5), March, 891–945.

Burns, T. & Stalker, G. M. (1994). *The management of innovation*. Oxford: Oxford University Press.

Cable, D. M. & Shane, S. (1997). A prisoner's dilemma approach to entrepreneur-venture capitalist relationships. *Academy of Management Review*, **22** (1), 142–176.

Callahan, J. & Sharp, J. (1985). Entrepreneurs and venture capitalists: differences in growth objectives. Proceedings of the 30th Annual World Conference of the International Council for Small Business, Montreal, June 16–19.

Chesbrough, H. (2002). Making sense of corporate venture capital. *Harvard Business Review*, March, 2002.

Chesbrough, H. (2000). Designing corporate ventures in the shadow of private venture capital. *California Management Review*, Spring, 31–49.

Chesbrough, H. & Rosenbloom, R. (2002). The role of the business model in capturing value innovation: Evidence from Xerox Corporation's technology spinoff companies. *Industrial and Corporate Change*, **11** (3), June, 529–555.

Chesbrough, H. & Smith, E. (2000). Chasing economics of scope: Xerox's management of its technology spinoff organizations. Working paper (http://www.people.hbs.edu/hchesbrough/spinoff.pdf)

Chesbrough, H. & Socolof, S. (2000). Commercializing new ventures from Bell Labs Technology: The design and experience of Lucent's New Ventures Group. *Research-Technology Management*, March, 1–11.

Christensen, C. (1997). *The innovator's dilemma*. Cambridge, MA: Harvard Business School Press.

Cumming, D. & MacIntosh, J. (2002). Venture capital exits in Canada and the United States. Presented at the 2002 Babson Conference on Entrepreneurship.

Davis, K. S. (1999). Decision criteria in the evaluation of potential intrapreneurs. *Journal of Engineering and Technology Management*, **16** (3–4), September-December, 295–327.

Dawson, P. (1997). In at the deep end: Conducting processual research on organisational change. *Scandinavian Journal of Management*, **13** (4), 389–405.

De Clercq, D. & Sapienza, H. J. (2001). The creation of relational rents in venture capitalist-entrepreneur dyads. *Venture Capital: An International Journal of Entrepreneurial Finance*, **3** (2) April, 107–127.

Devenow, A. & Welch, I. (1996). Rational herding in financial economics. *European Economic Review*, **40**, 603–615.

Dyer, W. G. Jr. & Wilkins, A. L. (1991). Better stories, not better constructs, to generate better theory: A rejoinder to Eisenhardt. *Academy of Management Review*, **16** (3), 613–619.

Eisenhardt, K. M. (1989a). Better stories and better constructs: The case for rigor and comparative logic. *Academy of Management Review*, **16** (3), 620–627.

Eisenhardt, K. M. (1989b). Building theories from case study research. *Academy of Management Review*, **14**, 532–550.

Fast, N. (1978). *The rise and fall of corporate new venture divisions*. Ann Arbor, MI: UMI Research Press.

Fenn, G. W. & Liang, N. (1998). New resources and new ideas: Private equity for small businesses. *Journal of Banking and Finance*, **22** (6–8), August, 1077–1084.

Forlani, D. & Mullins, J. W. (2000). Perceived risks and choices in entrepreneurs' new venture decisions. *Journal of Business Venturing*, **15** (4), July, 305–322.

Freear, J., Sohl, J. E. & Wetzel, W. E. (1995). Angels: personal investors in the venture capital market. *Entrepreneurship and Regional Development*, **7**, 85–94.

Gabrielsson, J. & Huse, M. (2002). The venture capitalist and the board of directors in SMEs: Roles and processes. *Venture Capital: An International Journal of Entrepreneurial Finance*, **4** (2), April, 125–146.

Gans, J. & Stern, S. (1999). When does funding small firms bear fruit? Evidence from the SBIR Program. MIT Sloan School of Management working paper.

Gartner, W. B. & Birley, S. (2002). Introduction to special issue on qualitative methods in entrepreneurship research. *Journal of Business Venturing*, **17** (5), 387–395.

Gifford, S. (1997). Limited attention and the role of the venture capitalist. *Journal of Business Venturing*, **12**, 459–482.

Giudici, G. & Palcari, S. (2000). The optimal staging of venture capital financing when entrepreneurs extract private benefits from their firms. *Enterprise and Innovation Management Studies*, **1** (2), May, 153–174.

Gompers, P. A. (1999). Ownership and control in entrepreneurial firms: An examination of convertible securities in venture capital investment. Harvard Business School working paper.

Gompers, P. A. (1995). Optimal investment, monitoring, and the staging of venture capital. *Journal of Finance*, **50**, December, 1461–1490. (This article forms the basis for chapter 7 of Gompers and Lerner's *The Venture Capital Cycle*)

Gompers, P. A. & Brav, A. (1997). 'Myth or reality?' The long-run underperformance of initial public offerings: Evidence from venture and nonventure capital-backed companies. *Journal of Finance*, December, 1791–1821.

Gompers, P. & Lerner, J. (2001a). The venture capital revolution. *Journal of Economic Perspectives*, **15**, 145–168.

Gompers, P. & Lerner, J. (2001b). *The money of invention: How venture capital creates new wealth*. Cambridge, MA: Harvard Business School Press.

Gompers, P. & Lerner, J. (1999a). *The venture capital cycle*. Boston, MA: MIT Press.

Gompers, P. A. & Lerner, J. (1999b). An analysis of compensation in the U.S. venture capital partnership. *Journal of Financial*, **51**, January, 3–44. (This article forms the basis for chapter 4 of Gompers and Lerner's *The Venture Capital Cycle*)

Gompers, P. A. & Lerner, J. (1999c). The determinants of corporate venture capital success: Organizational structure, incentives, and complementarities. In: R. Morck (Ed.), *Concentrated Corporate Ownership*. University of Chicago Press.

Gompers, P. A. & Lerner, J. (1996). The use of covenants: An analysis of venture partnership agreements. *Journal of Law and Economics*, **39**, October, 463–498.

Gorman, P. & Sahlman, W. A. (1989). What do venture capitalists do? *Journal of Business Venturing*, **4**, 231–248.

Hall, J. & Hofer, C. W. (1993) Venture capitalists decision criteria in new venture evaluation. *Journal of Business Venturing*, **8** (1), January, 25–42.

Hart, O. (2001). Financial contracting. *Journal of Economic Literature*, **39** (4), December, 1079–1100.

Hart, O. & Moore, J. (1990). Property rights and the nature of the firm. *Journal of Political Economy*, **98**, 1119–1158.

Hellman, T. & Puri, M. (2000a). The interaction between product market and financial strategy: The role of venture capital. *Review of Financial Studies*, Winter, 959–984.

Hellman, T. & Puri, M. (2000b).Venture capital and the professionalization of start-up firms: Empirical evidence. Stanford Business School working paper.

Henderson, R. & Clark, K. (1990). Architectural innovation: The reconfiguration of existing product technologies and the failure of established firms. *Administrative Science Quarterly*, **35**, 9–30.

Hinings, C. R. (1997). Reflections on processual research. *Scandinavian Journal of Management*, **13** (4), 493–503.

Hsu, D. (2000). Do venture capitalists affect the commercialization strategies at start-ups? MIT Industrial Performance Center working paper, June (http://globalization.mit.edu/globalization 00–006.pdf).

Hunt, B. & Lerner, J. (1995). Xerox technology ventures: March 1995. Harvard Business School case no. 9-295-127 (and teaching note no. 9-298-152).

Hyland, J. (2001). Using VC experience to create business value. In: E. J. Kelly (Ed.), *From the Trenches: Strategies from Industry Leaders on the New E-conomy*. Chichester, U.K.: John Wiley.

Jensen, M. C. & Meckling, W. H. (1976). Theory of the firm: Managerial behavior, agency costs and ownership structure. *Journal of Financial Economics*, **3**, 305–360.

Jog, V. M., Lawson, W. M. & Riding, A. L. (1991). The venture capitalist—entrepreneur interface: Expectations, conflicts and contracts. *Journal of Small Business and Entrepreneurship*, **8**.

Kaplan, J. (1994). *Startup: A silicon valley adventure*. Penguin Press.

Kaplan, S. & Stromberg, P. (2002). Financial contracting theory meets the real world: An empirical analysis of venture capital contracts. University of Chicago Graduate School of Business, working paper.

Kaplan, S. & Stromberg, P. (2001). Venture capitalists as principles: Contracting, screening, and monitoring. *American Economic Review*, **91** (2), 426–430.

Kaplan, S. & Stromberg, P. (2000). How do venture capitalists choose and manage their investments? University of Chicago working paper.

Kortum, S. & Lerner, J. (2000). Assessing the contribution of venture capital to innovation. *Rand Journal of Economics*, **31**, 674–692.

Leifer, R., McDermott, C., O'Connor, G., Peters, L., Rice, M. & Veryzer, R. (2001). *Radical innovation: How mature companies can outsmart upstarts*. Cambridge, MA: Harvard Business School Press.

Lerner, J. (1995). Venture capitalists and the oversight of private firms. *Journal of Finance*, **50**, March, 301–318. (This article forms the basis for chapter 8 of Gompers and Lerner's *The Venture Capital Cycle*).

Lerner, J. (1994). The syndication of venture capital investments. *Financial Management*, **23**, Autumn, 16–27. (This article forms the basis for chapter 9 of Gompers and Lerner's *The Venture Capital Cycle*).

Lind, E. A. & Tyler, T. (1988). *The social psychology of procedural justice*. New York: Plenum.

MacMillan, I., Kulow, D. & Khoylian, R. (1989). Venture capitalist involvement in their investments: Extent and performance. *Journal of Business Venturing*, **4**, 27–47.

MacMillan, I. C., Siegel, R. & Narasimha, P. N. S. (1985). Criteria used by venture capitalists to evaluate new venture proposals. *Journal of Business Venturing*, **1** (1), 119–128.

Madill, J. J., Haines, G. Jr., Orser, B. J. & Riding, A. L. (2002). Managing high technology SMEs to obtain institutional venture capital: A role for angels. Eric Sprott School of Business, Carleton University working paper, presented at the Babson Entrepreneurship Research Conference, June.

Megginson, W. & Weiss, K. (1991). Venture capital certification in initial public offerings. *Journal of Finance*, **46**, 879–903.

Nesheim, J. L. (2000). *High tech start up*. New York: The Free Press.

O'Connor, G. C. & Maslyn, W. T. (2002). Nortel Networks' Business Ventures Group. Babson College Case Study 102-C02.

Pence, C. C. (1982). *How venture capitalists make investment decisions*. UMI Research Press.

Pettigrew, A. M. (1997). What is processual analysis? *Scandinavian Journal of Management*, **13** (4), 337–348.

Poole, M. S., Van De Ven, A. H., Dooley, K. & Holmes, M. E. (2000). *Organizational change and innovation processes: Theory and methods for research*. Oxford: Oxford University Press.

Porter, M. E. (1990). The competitive advantage of nations. *Harvard Business Review*, March.

Powell, W. W., Koput, K. W., Bowie, J. I. & Smith-Doerr, L. (2002). The spatial clustering of science and capital: Accounting for biotech firm-venture capital relationships. *Regional Studies*, **36** (3), 291–305.

PricewaterhouseCoopers (2002). *Global private equity 2002: A review of the global private equity and venture capital markets*. PricewaterhouseCoopers.

Rind, K. (1981). The role of venture capital in corporate development. *Strategic Management Journal*, **2**, 169–180.

Ruhnka, J. C., Feldman, H. D. & Dean, T. J. (1992). The living dead phenomenon in venture capital investments. *Journal of Business Venturing*, **7** (2), 137–155.

Sahlman, W. A., Stevenson, H. H., Roberts, M. J. & Bhidé, A. (Eds) (1999). *The entrepreneurial venture* (2nd ed.). Cambridge, MA: Harvard Business School Press.

Sahlman, W. A. (1990a). The structure and governance of venture capital organizations. *Journal of Financial Economics*, **27**, 473–521.

Sahlman, W. A. (1990b). The horse race between capital and opportunity. *The Venture Capital Review*, Spring, 17–25. (Reprinted in W. A. Sahlman et al., *The Entrepreneurial Venture*.)

Sapienza, H. & Korsgard, M. (1996). The role of procedural justice in entrepreneur-venture capitalist relations. *Academy of Management Journal*, **39**, 544–574.

Sapienza, H. J. (1992). When do venture capitalists add value? *Journal of Business Venturing*, **7**, 9–27.

Saxenian, A. (1994). *Regional advantage: Culture and competition in Silicon Valley and Route 128*. Cambridge, MA: Harvard Business School Press.

Schall, M. (1983). A communication rules approach to organizational culture. *Administrative Science Quarterly*, **28**, 557–581.

Schefczyk, M. & Gerpott, T. J. (2001a). Management support for portfolio companies of venture capital firms: An empirical study of German venture capital investments. *British Journal of Management*, **12** (3), September, 201–216.

Schefczyk, M. & Gerpott, T. J. (2001b). Qualifications and turnover of managers and venture capital-financed firm performance: An empirical study of German venture capital-investments. *Journal of Business Venturing*, **16** (2), March, 145–163.

Schumpeter, J. (1934). *The theory of economic development*. Cambridge, MA: Harvard University Press.

Shane, S. & Stuart, T. (2002). Organizational endowments and the performance of university start-ups. *Management Science*, **48** (1), January, 154–170.

Shepherd, D. A. & Zacharakis, A. (2001). The venture capitalist-entrepreneur relationship: Control, trust and confidence in co-operative behaviour. *Venture Capital: An International Journal of Entrepreneurial Finance*, **3** (2), April, 129–149.

Shepherd, D. A. & Zacharakis, A. (2002). Venture capitalists' expertise—A call for research into decision aids and cognitive feedback. *Journal of Business Venturing*, **17** (1), January, 1–20.

Shepherd D. A., Ettenson, R. & Crouch, A. (2000). New venture strategy and profitability: A venture capitalist's assessment. *Journal of Business Venturing*, **15** (5–6), September-November, 449–467.

Siegel, R., Siegel, E. & MacMillan, I. (1988). Corporate venture capitalists: Autonomy obstacles and performance. *Journal of Business Venturing*, **3** (3), 233–247.

Steier, L. & Greenwood, R. (1995). Venture capitalist relationships in the deal structuring and post-investment stages of new firm creation. *Journal of Management Studies*, **32** (3), May, 337–357.

Sweeting, R. C. (1991). U.K. venture capital funds and the funding of new technology-based businesses: Process and relationships. *Journal of Management Studies*, **28** (6), November, 601–622.

Sweeting, R. C. & Wong, C. F. (1997). A U.K. 'hands-off' venture capital firm and the handling of post-investment investee-investor relationships. *Journal of Management Studies*, **34** (1), January, 125–152.

Timmons, J. A. (2001). *New venture creation*. New York: McGraw-Hill/Irwin.

Timmons, J. & Bygrave, W. (1986). Venture capital's role in financing innovation for economic growth. *Journal of Business Venturing*, **1**, 161–176.

Tredennick, N. (2001). An engineer's view of venture capitalists. *IEEE Spectrum Online*, September (http://www.spectrum.ieee.org/WEBONLY/resource/sep01/speak.html).

Tushman, M. & Anderson, P. (1986). Technological discontinuities and organization environments. *Administrative Science Quarterly*, **31**, 439–465.

Tushman, M. & O'Reilly, C. III (1997). *Winning through innovation: A practical guide to leading organization change and renewal*. Cambridge, MA: Harvard Business School Press.

Tyebjee, T. T. & Bruno, A. V. (1984). A model of venture capitalist investment activity. *Management Science*, **30** (9), 1051–1066.

Utset, M. A. (2002). Reciprocal fairness, strategic behavior & venture survival: A theory of venture capital-financed firms. *Wisconsin Law Review*, (1), 45–168.

von Burg, U. & Kenny, M. (2000). Venture capital and the birth of the local area networking industry. *Research Policy*, **29** (9), December, 1135–1155.

von Hippel, E. (1977). Successful and failing internal corporate ventures: An empirical analysis. *Industrial Marketing Management*, **6**, 163–174.

Wheelwright, S. C. & Clark, K. B. (1992). *Revolutionizing product development*. New York: The Free Press.

Woiceshyn, J. (1997). Literary analysis as a metaphor in processual research: A story of technological change. *Scandinavian Journal of Management*, **13** (4), 457–471.

Yin, R. K. & Campbell, D. T. (2002). *Case study research: Design and methods*. Sage Publications Inc.

Yost, M. (1994). The state of corporate venturing: The number of active programs levels off as corporations complete shifts back to core businesses. *Corporate Venturing*, June 1994.

Zacharakis, A. L. & Meyer, G. D. (2000). The potential of actuarial decision models: Can they improve the venture capital investment decision? *Journal of Business Venturing*, **15** (4), July, 323–346.

Zacharakis, A. L. & Meyer, G. D. (1998). A lack of insight: Do venture capitalists really understand their own decision process? *Journal of Business Venturing*, **13** (1), January, 57–76.

Zider, B. (1998). How venture capital works. *Harvard Business Review*, November/December, 131–139.

Appendix
Entrepreneurs and Venture Capitalists: Differences in Growth Objectives
Excerpted from Callahan & Sharp (1985)

The Problem

The typical entrepreneur is not diversified; as far as financial risk is concerned he has most of his personal fortune and human capital in the new venture. As an active manager, he receives a salary and fringe benefits from the company during its start-up and initial stages.

The typical venture capitalist is well diversified financially. Unlike the entrepreneur, he has a financial, rather than a personal relationship with the company. The venture capitalist is interested in the overall performance of a portfolio of investments. He is not an active manager, nor does he want to be involved in day-to-day management. He exerts control by constraint through protective covenants and membership on the board of directors. The typical venture capitalist receives no immediate cash flow from an investment but rather is interested only in capital gains.

The points of conflict between the entrepreneur and the venture capitalists are twofold:

(i) the amount of salary and fringe benefits consumed by the manager/entrepreneur; and
(ii) the growth rate of the company, resulting from strategy decisions that affect growth such as products and markets, pricing and the relative focus or diversification of the company.

We outline a model that illustrates how these two conflicting points of interest arise in a natural and logical way given the differences in diversification of the entrepreneur and the venture capitalist, and the fact that the entrepreneur receives salary and fringe benefits as a manager.

The Model

The complete model comprises three sections:

The Basic Valuation Model

The basic valuation model used is an adaptation of the capital asset pricing model of Stapleton (1971). The basic idea behind Stapleton's model is that the present value of an uncertain cash flow can be considered as the certainty equivalent of the value of the cash flow discounted to the present value at the risk free rate. Stapleton further argues that in a capital asset pricing model world this certainly equivalent is a linear function of the expected value and the standard deviation of cash flow.

In Stapleton's notation, DV is the discounted value operation using the risk free rate. Thus if C is some uncertain cash flow anticipated at time n and I is the risk free rate

$$DVC = C/(1+i)^n \qquad (1)$$

The present value of C for a diversified investor is

$$E(DVC) - S \cdot R_{CM} \cdot \sigma(DVC) \qquad (2)$$

where R_{CM} is the correlation between C and the return on the market (as defined in the capital asset pricing model) and S is a market determined risk aversion factor. In a sense, $R_{CM} \cdot \sigma(DVC)$ is the relevant risk measure for a diversified investor since he is concerned only with systematic risk, being able to eliminate unsystematic risk through diversification. For an undiversified investor though, the relevant risk factor is $\sigma(DVC)$ and the present value of C is

$$E(DVC) - S\sigma(DVC) \qquad (3)$$

We shall assume that the company goes public after n periods and that both entrepreneur and venture capitalist can (but do not have to) wind up their positions at this time. We also assume that the net value of the company at this time, V, increase with the average growth rate, g, of the company. This is a weak

assumption implicit in most conceivable company valuation models.

The Entrepreneur Model

We assume that the manager/entrepreneur receives a periodic wage, w, that is not random. It is possible to consider a random wage, or one that is performance related, but that would add little to the analysis. We further assume that the entrepreneur owns a portion α of the equity of the company and that α remains constant. Then the present value of the entrepreneur's claims on the company, V_1, is a function of g and w given by

$$V_1 = V_1(g,w) \qquad (4)$$

$$= w \cdot \{[(1+i)^n - 1]/i\} + \alpha \cdot \{E/(1+i)^n$$
$$- S \cdot \sigma/(1+i)^n\} \qquad (5)$$

$$= w \cdot \{[(1+i)^n - 1]/i\} + \{\alpha/(1+i)^n\}$$
$$\cdot \{E - S \cdot \sigma\} \qquad (6)$$

where E and σ are the mean and standard deviation of V respectively and the relevant risk for the entrepreneur as an undiversified investor is the total risk.

The Venture Capitalist Model

The present value of the venture capitalist's claim on the company, V_2, is given by

$$V_2 = V_2(g,w) \qquad (7)$$

$$= \{(1-\alpha)/(1+i)^n\} \cdot \{E - S \cdot R_{VM} \cdot \sigma\} \qquad (8)$$

where R_{VM} is the correlation between V and the return on the market as a whole given that the venture capitalist is a diversified investor.

The Analysis

Taking partial derivatives we have

$$\partial V_1/\partial w = \{[(1+i)^n - 1]/i\} + \{\alpha/(1+i)^n\}$$
$$\cdot \{1 - S \cdot \partial\sigma/\partial E\} \cdot \partial E/\partial w \qquad (9)$$

$$\partial V_1/\partial g = \{\alpha/(1+i)^n\} \cdot \{1 - S \cdot \partial\sigma/\partial E\} \cdot \partial E/\partial g \qquad (10)$$

$$\partial V_2/\partial w = \{(1-\alpha)/(1+i)^n\}$$
$$\cdot \{1 - S \cdot R_{VM} \cdot \partial\sigma/\partial E\} \cdot \partial E/\partial w \qquad (11)$$

$$\partial V_2/\partial g = \{(1-\alpha)/(1+i)^n\}$$
$$\cdot \{1 - S \cdot R_{VM} \cdot \partial\sigma/\partial E\} \cdot \partial E/\partial g \qquad (12)$$

We assume that

$$\partial E/\partial w < 0 \qquad (13)$$

i.e. the higher the wages paid to the entrepreneur the lower the expected terminal net value of the company. By virtue of our assumption that a higher growth rate increases the expected terminal value of the company, we also have

$$\partial E/\partial g > 0 \qquad (14)$$

We shall examine the extent to which the objectives of the venture capitalist and entrepreneur are congruent by considering whether a wage (w) and growth rate (g) exist that simultaneously optimize V_1 and V_2. Because the power of the venture capitalist at the stage of venture capital financing of the company is usually greater than that of the entrepreneur, we start with the venture capitalist's point of view.

Now the venture capitalist wants values of w and g such that

$$\partial V_2/\partial w \geq 0 \text{ and } \partial V_2/\partial g \geq 0$$

i.e. such that from (11) and (12)

$$\partial E/\partial w \leq 0, \ \{1 - S \cdot R_{VM} \cdot \partial \sigma/\partial E\} \leq 0 \quad \text{and}$$

$$\{1 - S \cdot R_{VM} \cdot \partial \sigma/\partial E\} \geq 0$$

or

$$\partial \sigma/\partial E = 1/\{S \cdot R_{VM}\} \tag{15}$$

In all probability, R_{VM} will be small so that the optimal value of E for the venture capitalist will be large.

For the entrepreneur, if we set

$$\partial \sigma/\partial E = 1/\{S \cdot R_{VM}\}$$

we have from (9)

$$\partial V_1/\partial w = \{[(1 + i)^n - 1]/i\} + \{\alpha/(1 + i)^n\}$$

$$\cdot \{1 - 1/R_{VM}\} \cdot \partial E/\partial w > 0 \tag{16}$$

and from (10)

$$\partial V_1/\partial g = \{\alpha/(1 + i)^n\} \cdot \{1 - 1/R_{VM}\} \cdot \partial E/\partial g < 0 \tag{17}$$

In other words, when the entrepreneur's wage and growth rate of the company are set at values that are optimal for the venture capitalist, the entrepreneur wants a higher wage and a lower growth rate for the company.

As can be seen there is no prospect of simultaneous optimization of V_1 and V_2. The values of the growth rate and wage that optimize the venture capitalist's value, V_2, certainly do not optimize the entrepreneur's value, V_1.

Equations (16) and (17) present the entrepreneur's view of the dilemma. Equation (16) merely confirms the intuition that the entrepreneur is better off with a higher wage. Equation (17) conceals a little more. Venture capital is most likely to be used as a method of financing for companies in new and innovative industries. There is a case for arguing that the earnings of such industries, unrelated as they are to the mature industries whose earnings represent the bulk of the market index, will show an unusually low correlation with the market index. It is plausible then that the term $1/R_{VM}$ is much greater than 1 so that the term $\{1 - 1/R_{VM}\}$ is considerably less than zero. In other words the optimum growth rate as seen by the venture capitalist is way above that perceived by the entrepreneur as ideal. At a growth rate that appears attractive to the venture capitalist Eq. (17) shows that the value to the entrepreneur is decreasing rapidly.

Reference

Stapleton, R. C. (1971). Portfolio analysis, stock valuation and capital budgeting decision rules for risky projects. *The Journal of Finance*, **26** (1), March, 95–117.

Part IX

Innovations in Social Institutions

The International Handbook on Innovation
Edited by Larisa V. Shavinina

Encouraging Innovation in Small Firms Through Externally Generated Knowledge

Edward Major[1] and Martyn Cordey-Hayes[2]

[1] *Bedfordshire Police Force and Middlesex University, U.K.*
[2] *Cranfield University, U.K.*

Abstract: This chapter examines the conveyance to small firms of externally generated knowledge. Successful innovation requires firms to draw on multiple sources of knowledge. Many small firms take little note of external sources, thus restricting their potential innovative base. We develop the concepts of *knowledge translation* and the *knowledge translation gap* to illustrate why so many small firms fail to access externally generated knowledge. Findings from research into U.K. small firms and national innovation schemes show how intermediary organisations can be used to bridge the knowledge translation gap. Implications follow for government innovation schemes, for intermediaries and for small firms.

Keywords: Innovation; Small firms; External knowledge; Knowledge translation; Intermediaries.

Introduction

The rise to prominence of the field of knowledge management (see Farr, Sin & Tesluk, 2003) gives witness to the centrality of knowledge in fostering innovation. Knowledge underlies innovation. It is the precursor, in many ways the lubricant, of innovation. Firms need knowledge to build innovative potential.

Innovation is 'the introduction of novelties' or 'the alteration of what is established by the introduction of new elements or forms'.[1] It thus concerns introduction and progression rather than invention. Innovation has been adjudged the main determinant by which commercially successful organisations derive competitive advantage (Tidd, Bessant & Pavitt, 1997). Successful innovation must involve several parties. A knowledge source is needed to support the innovating party (i.e. to provide the new invention/idea). Innovation within a firm might be based on knowledge held by that firm. Even so, the source of the knowledge contributing towards an innovation may be separated from the actual innovating party. Alternatively, the innovating firm might need to go to external sources for

the knowledge it needs. In this case the distance between knowledge source and innovating party is increased. For successful innovation, knowledge needs to be successfully conveyed. The conveyance, or transfer, of knowledge forms a third element in the process, between the knowledge source and the innovating party. This chapter considers the conveyance into small firms of externally generated knowledge. It thus compliments Hadjimanolis's study (2003) of the internal and external barriers to innovation that firms face. Understanding this conveyance process has implications both for the presentation of government innovation schemes and policy initiatives and for the small firms themselves.

The chapter begins with a brief discussion of the barriers to innovation in small firms (the innovating party). Consideration then turns to the knowledge sources contributing to innovation. Sources of externally generated knowledge are emphasized, including government innovation schemes and policy initiatives. The concept and models of knowledge transfer provide the basis for examination of the conveyance of external knowledge. The chapter develops knowledge transfer into a model of knowledge *translation* that enables a clear understanding of the conveyance process to be built. Findings are then reported from research into

---

[1] As defined in the Oxford English Dictionary, Oxford University Press, 1933, reprinted 1978.

small firms and innovation schemes in the U.K. Applying the knowledge translation model to this research reveals how intermediary organizations can be used to reach different *types* of small firms. Implications are drawn for the conveyance process itself, for government innovation schemes and policy initiatives, for intermediaries and for small firms. Recommendations are made which are aimed at enhancing the flow of externally generated knowledge, therefore encouraging successful small firm innovation.

Barriers to Innovation in Small Firms

Statistics show that in 1999, 99.8% of all U.K. businesses employed less than 250 people. These small and medium sized enterprises (SMEs) accounted for 55% of national employment and 51% of national turnover. The 0.2% of businesses rated as *large* contributed less than half of the nation's employment and turnover (Department of Trade and Industry, 2001). Other developed economies are similar. Storey (1994) records that businesses with less than 500 employees account for between 62% and 91% of national employment in all member states of the European Community. Figures for Japan and the United States are of the same order (Small Business Administration, 2000; Storey, 1994). Whichever way they are looked at, small firms make an immense contribution to wealth and employment.

Innovation is favored by the small firms' advantages of flexibility and reduced bureaucracy. But their innovative potential is hampered by inherent problems not faced by large firms. Rothwell & Zegvelt (1982) identify four weaknesses particular to small firms. First, limited manpower restricts their ability to perform competitive R&D. Small firms are likely to have a much lower proportion of their personnel, possibly none, concerned exclusively with R&D than large firms (Storey, 1994). Many small firm personnel have more than one role. Second, small firms have limited time and resources to devote to external communications. This limits the information base from which decisions can be made. Low in-house employment of specialists restricts communication and network formation with outside sources of expertise (Rothwell, 1991). Third, is excessive management influence. Small firms are much more prone to domination by a single manager or team that may use inappropriate skills or strategies. Fourth, small firms can have difficulties raising finance. Financial barriers to innovation in small firms are variously reflected as a lack of capital (Garsombke & Garsombke, 1989; Keogh & Evans, 1999) and unfavorable bank policies on credit (Hadjimanolis, 1999). Small firms of course are varied and different, as will be shown below. Their inherent weaknesses (and advantages) will be present to varying degrees.

Sources of Externally Generated Knowledge

The traditional way for a small firm to build its knowledge base is to generate knowledge internally. Firms use formal and informal research and development to grow their internal knowledge. In this internal process, research may be considered the knowledge source and development the means of conveyance to the firms' innovation centre (its decision-makers). Internally generated knowledge has the advantage of close proximity to the firm's decision-makers. By definition it lies within the firm's boundaries. Of course, close proximity does not ensure that internal knowledge will have any impact. It also suffers from Rothwell & Zegvelt's (1982) first weakness, limited manpower to devote to internal R&D.

By concentrating on internal R&D many firms neglect the role of externally generated knowledge. Small firms, with their limited resources are particularly prone to neglecting external knowledge. They are disadvantaged when compared to larger competitors who may be able to dedicate specific resources to examining external knowledge sources. Woolgar, Vaux, Gomes, Ezingeard & Grieve (1998) present a framework showing the environment surrounding a small firm. Their 'SME-centric universe' (Fig. 1) gives the basis for a categorization of a firm's external knowledge sources.

The Immediate Business Environment

Small firms interact most often and most closely with their immediate business environment: their customers and suppliers, and to a lesser extent their competitors. Research has shown that customers and suppliers, the *supply chain*, are always the most extensive and important external contacts that a firm has (Moore, 1996; Woolgar et al., 1998). Customer feedback is a constant source of external information, and therefore potentially usable knowledge. Suppliers can be a major source of component and production knowledge. Competitors, through their behavior, can be a further significant source of knowledge.

Intermediaries

Beyond the immediate business environment, a small firm will have communications with trade associations, colleges and schools, TECs (training and enterprise councils) and consultants, will read the trade press and may attend exhibitions. Other organizations within the business support community can be added: Chambers of Commerce, innovation and technology centres, small business agencies such as the U.K.'s Business Links and USA's Small Business Development Centers, professional institutes and research associations. Here, all of these organizations are grouped under the collective term *intermediaries*. Often a firm's intermediary contacts will be very cursory, where they can provide only a minor source of external knowledge. As

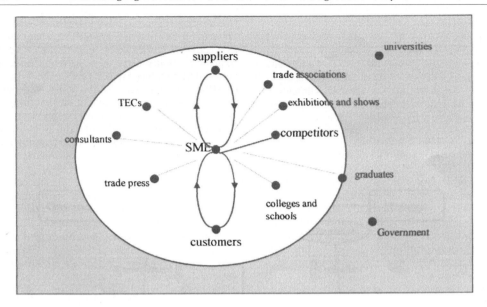

Figure 1. Woolgar et al.'s SME-centric universe.

Source: Woolgar et al. (1998), page 578.

will be seen intermediaries have a significant role to play in conveying externally generated knowledge.

Universities and Outside Research

In Woolgar et al.'s (1998) framework, universities fall well outside of a small firm's focus of attention. Universities are repositories of knowledge and expertise, but partly due to the problems considered above, many small firms have no significant contact with them. Evidencing the differences between small firms, Woolgar et al. (1998) found that those that did have well-developed links with universities had a greater appreciation of external knowledge sources than had those lacking such links.

Governments; Policy Initiatives and Innovation Schemes

Governments, like universities, generally fall beyond a small firm's normal sphere of attention. Governments try to influence small firms through business services, innovation schemes and policy initiatives. The U.K. Government's Small Business Service (SBS) is organized: to help small businesses realize their potential, to enhance small business performance through world class business support, to promote small. business enterprise across society and to provide high standard, value for money service (Department of Trade and Industry, 2001). In the USA, the Small Business Administration (SBA) provides financial, technical and management assistance to help Americans start, run and grow their small businesses.

As an example of a national policy initiative, recent years have seen many governments running national *Foresight* programs. These are structured programmes taking a forward look to identify and prepare for emerging trends in markets and technologies in the medium to long term. Foresight activities aim to:

(a) identify events and seek opinions in order to prioritize future events;
(b) contribute to the development of a well-informed support environment for resource allocation and funding prioritization; and
(c) promote cooperation between actors from different fields so as to incorporate a variety of viewpoints (Cabello et al., 1996).

The U.K. Government's national Foresight program was initiated in 1993 to identify emerging technological trends and market opportunities. When applied to individual firms, Foresight is about generating and growing the capabilities to envision and look forward, enhancing the firms' future innovative potential. Government schemes and initiatives have a poor implementation record. For example despite substantial communicative effort, the U.K. Foresight program has had minimal effect on U.K. firms (Major & Cordey-Hayes, 2000a). The small firms' inherent problems discussed above compound the difficulties of distance. Government initiatives and innovation schemes lie outside of the firms' normal sphere of attention in much the same way as the government itself. Many small firms are simply not aware of support services and initiatives.

External Networking

The final external knowledge source is a method rather than a location. An inquisitive manager will search the external knowledge sources, be they in the immediate business environment, intermediaries, universities or government schemes and initiatives. Through this external networking he or she will be better positioned to access new externally generated knowledge.

Conveying External Knowledge

Hadjimanolis (2003) notes that information and knowledge must be transferred throughout a firm, for the firm to learn from it. But externally generated knowledge lacks the advantage of close proximity to the firm's decision-makers. Unlike internally generated knowledge it must first be brought into the firm's internal sphere before it can be used. The growing subject of *knowledge transfer* informs our understanding of how externally generated knowledge is conveyed to the small firm.

The Concept of Knowledge Transfer

The concept of knowledge transfer derives from the field of innovation (Gilbert & Cordey-Hayes, 1996). Knowledge transfer is the conveyance of knowledge from one place, person, ownership, etc. to another (Major & Cordey-Hayes, 2000b).

Any transfer must involve more than one party. There has to be a source (the original holder of the knowledge) and a destination (where the knowledge is transferred to). When used to describe movement of knowledge, the term *transfer* is perhaps inappropriate. As defined, it implies that for the transferred item to be gained by (conveyed to) the destination it must be lost by (conveyed from) the source. But as an intangible asset, knowledge does not necessarily have to be given up by one party to be gained by the other. The framework developed below proposes alternative terminology that more accurately describes the process.

Models of Knowledge Transfer

Research into knowledge transfer has focused on separating the overall transfer from source to destination into comprehensible sub-processes. Trott, Seaton & Cordey-Hayes (1996) develop an interactive model of technology transfer describing four stages in the knowledge transfer process: for inward technology transfer (knowledge transfer) to be successful, an organization must be able to:

(1) search and scan for information which is new to the organization (*awareness*);
(2) recognize the potential benefit of this information by associating it with internal organizational needs and capabilities (*association*);
(3) communicate these to and assimilate them within the organization (*assimilation*); and
(4) apply them for competitive advantage (*application*)

Non-routine scanning, prior knowledge, internal communication and internal knowledge accumulation are key activities affecting an organization's knowledge acquisition ability (Trott et al., 1996). These findings correspond with those underlying Cohen & Levinthal's (1990) *absorptive capacity*, the ability of an organization to recognize the value of new information, to assimilate it and to commercially apply it.

In addition to the works of Trott et al. (1996) and Cohen & Levinthal (1990), the knowledge transfer literature includes frameworks and models by Cooley (1987), Slaughter (1995) and Horton (1997, 1999). Two streams of models can be distinguished: Node models describe nodes, discrete steps that are each gone through. Process models describe knowledge transfer by separate processes that are each undertaken. Node models are presented by Cooley (1987), Slaughter (1995) and Horton (1997). Cooley discusses *information systems* in the context of the information society:

> Most of such systems I encounter could be better described as *data* systems. It is true that *data* suitably organized and acted upon may become *information*. *Information* absorbed, understood and applied by people may become *knowledge*. *Knowledge* frequently applied in a domain may become *wisdom*, and *wisdom* the basis for positive *action* (Cooley, 1987, p. 11).

Cooley thus raises a progressive sequence of five nodes. He conceptualizes his model as a noise-to-signal ratio. The signal being transmitted (i.e. conveyed) is subject to noise, but as the information system moves from data towards wisdom, noise is reduced and the signal increases. Knowledge acquiring organizations will find low-noise wisdom more useful than high-noise data.

Slaughter (1995) presents a four-node *hierarchy of knowledge*, which he uses to describe a *wise culture*. He uses the same first four terms as Cooley (1987):

(1) *Data*; raw factual material;
(2) *Information*; categorized data, useful and otherwise;
(3) *Knowledge*; information with human significance;
(4) *Wisdom*; higher-order meanings and purposes.

Slaughter's descriptions indicate how stages further up the hierarchy are subject to less noise, and are more useful to knowledge acquiring organizations. Data, information and knowledge are stages on the path to wisdom.

While Cooley (1987) and Slaughter (1995) set their models in a macro-context, Horton (1997) focuses on industry on a micro-scale. In the specific context of the printing industry, but with wider applicability, she describes an *information value progression*. This progression again starts with data and finishes with wisdom, with information and knowledge as intermediate steps. Cooley (1987) describes data as objective, calculative and subject to noise. Horton (1997) agrees; data is hard, objective and low order, as well as being low value, voluminous and contained. At the other end, wisdom is subjective, judgmental and less subject to noise (Cooley, 1987). By Horton (1997) it is subjective, soft and high order, as well as high value, low volume and contextualized.

Similarities are seen in all three models described, despite their contextual differences. Horton (1997) however adds another intermediate step. This is the node of *understanding*, between knowledge and wisdom. Understanding:

> is the result of realising the significance of relationships between one set of knowledge and another . . . It is possible to view the ability to benefit from understanding as being wisdom (Horton, 1997, p. 3).

Understanding thus results from knowledge but precedes wisdom. Table 1 compares the three node models. With wisdom a result of understanding and the basis for positive action, the models are mutually supportive. A six-node scheme of knowledge transfer is formed.

Process models are presented by Cohen & Levinthal (1990), Trott et al. (1996) and Horton (1999). Trott et al.'s (1996) four processes of inward technology transfer, awareness, association, assimilation and application, are described above. Horton (1999) presents a three-phase process model of successful foresight. Phase one is a *collection, collation* and *summarization* phase. Information is collected from a wide range of sources, collated to give it structure, and summarized into a manageable form. Knowledge generated from stage one needs *translation* and *interpretation*, the elements of phase two. Translation puts the information in an understandable language. Interpretation is organization specific. It is about determining the meanings and implications of the translated information. Translation and interpretation is the most crucial step in the process, where most of the value is added. But it is also poorly understood, having few theoretical techniques. Phase three comprises *assimilation* and *commitment*. Understanding generated in phase two needs to be assimilated by decision-makers. If changes are to result commitment to act is needed. It is only at this point that the value of the whole three-phase process can be realized and judged.

Trott et al. (1996) and Horton (1999) both refer to an assimilation process; the means by which ideas are communicated within the organization (Trott et al., 1996), or by which understanding is embedded in the organization's decision-makers (Horton, 1999). Assimilation then is a process of internal communication. Horton's assimilation process is in the same phase as commitment. By separating these out, Horton's commitment becomes correlated with Trott et al.'s subsequent process of application. Commitment to action is equivalent to the process of application for competitive advantage. By similarly separating Horton's phase one, collection becomes analogous to Trott et al.'s first process of awareness. The collection of data is a central means for an organization to become aware of new opportunities. Similarly, Horton's collation and summarization can together be compared to Trott et al.'s association. Collection and summarization of information is equivalent to the process of associating the value of the information to the organization. This leaves Horton's second phase, translation and interpretation, with no direct analogy. The comparison suggests analogy with a process not explicitly stated by Trott et al., the process of *acquisition*. This may be described as the process by which an organization draws knowledge into itself, which it can then assimilate. Viewed from within the organization this is a process of acquisition into itself. Viewed from outside it is a process of translation of knowledge from one place to another. It should be noted that the term *acquisition* as used here refers to a discrete part of the overall process of drawing knowledge into the knowledge acquiring organizations. The term *knowledge* similarly has an overall meaning (as in the phrase knowledge transfer) as well as referring to a discrete node in the overall knowledge transfer process.

Trott et al. (1996) and Horton (1999) can now be related to Cohen & Levinthal (1990). Cohen and Levinthal describe *absorptive capacity* as:

> the ability of a firm to *recognize* the value of new, external information, *assimilate* it, and *apply* it for commercial ends (Cohen & Levinthal, 1990, p. 128).

These recognition, assimilation and application steps correspond to Horton's (1999) collation, assimilation and commitment. Again there is no explicit provision for the translation and interpretation process. Neither is

Table 1. Node schemes of knowledge transfer.

Model	node 1	node 2	node 3	node 4	node 5	node 6
Cooley	data	information	knowledge	N/A	wisdom	action
Slaughter	data	information	knowledge	N/A	wisdom	N/A
Horton	data	information	knowledge	understanding	wisdom	N/A

Table 2. Process schemes of knowledge transfer.

Model	process 1	process 2	process 3	process 4	process 5
Horton	collection	collation summarization	translation interpretation	assimilation	commitment
Trott et al	awareness	association	N/A	assimilation	application
Cohen and Levinthal	N/A	recognition	N/A	assimilation	application

there an initial collection process. However the positions of these can be filled without upsetting the basic model. Horton (1999) presents the only model to clearly distinguish the translation and interpretation phase, supporting her contention that this phase is poorly understood. Table 2 compares the three process models. A five-process scheme of knowledge transfer is formed.

Knowledge Translation

A Framework for Knowledge Translation

From showing correspondences between models within the two streams, it becomes possible to build an integrated framework, a *framework for knowledge translation*, combining both nodes and processes. The term *knowledge translation* brings a subtly different perspective than the term *knowledge transfer*. The word translation has a dual meaning. It can refer to movement from one place to another place, much the same as the word transfer. It can also refer to putting something into an understandable form, as in the second phase of Horton's (1999) process model. Both meanings are appropriate to the context here: Knowledge translation is both the movement (or transfer) of knowledge from one place to another, and the altering of that knowledge into an understandable form.

Horton's work (1997, 1999) provides the key to integrating the knowledge transfer models. Her guide to successful foresight (process model; Horton, 1999) makes use of her information value progression (node model; Horton, 1997):

> Each phase (in the process model) creates greater value than the previous one as the outputs move up the information value chain from information through knowledge to understanding (Horton, 1999, p. 5).

Thus Horton's four nodes are connected by her three phases. Looking simultaneously at the expanded node and process schemes in Tables 1 and 2, the five processes fall neatly between the six nodes. Figure 2 illustrates this, combining the contents of Tables 1 and 2 into a single combined framework for knowledge translation. The initial node is raw data. The first process is to collect this into information. The second process is collation and summarization of this information into knowledge. Knowledge is the end point of

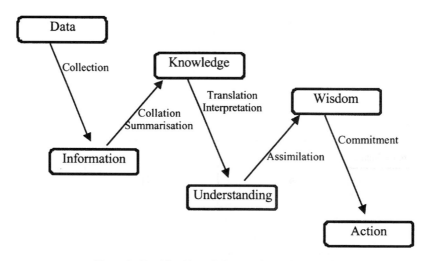

Figure 2. Combined knowledge translation framework.

Source: Expanded from Horton (1997, 1999).

Horton's (1999) phase one. The third process exactly follows Horton's (1999) phase two; the translation and interpretation of knowledge into understanding. Understanding is then assimilated within the organization into wisdom, from which a commitment to positive action can follow. The processes in this integrated framework are the ways of reaching the successive nodes.

Figure 2 presents a complete system of knowledge translation. This completeness may be misrepresentative of extant knowledge transfer or translation schemes. For example, the U.K. national Foresight program has been incomplete in its intent to encourage a forward thinking culture within industry (Major & Cordey-Hayes, 2000a). It is possible that not every process in the scheme will always be fully conducted, thus restricting the overall flow of new ideas that can lead to innovation. The latter processes of assimilation and commitment take place mainly within the destination organization (Cooley, 1987). Early stages might be carried out by an outside body, perhaps the source of the data or information. Collection of data and collation of information can be performed on a macro-scale by government innovation schemes and policy initiatives (as was done by the U.K. Foresight program). This leaves the middle process less clear, and possibly incomplete. This is the notion of *the knowledge translation gap*. For a complete and successful knowledge translation process this gap must be bridged. This implies a role for parties other than the source and destination organizations; i.e. for some outside intermediary organization. Attention is thus turned to the intermediaries, introduced as an external knowledge source in Section 4, above. The knowledge translation gap conceptualizes the barriers to external

learning identified and discussed by Hadjimanolis (2003).

Characteristics of Knowledge and the Knowledge Translation Framework

Before turning in earnest to the role of the intermediaries, some consideration of the characteristics of knowledge is needed to complete the groundwork. Major & Cordey-Hayes (2000a, 2000b) show how knowledge can be characterized along two perception-dependant dimensions. Whether internally or externally generated, firms want knowledge in a discrete and concrete form. To make decisions they want something tangible that can give them operational knowledge. Distant external sources such as universities and governments, as well as government innovation schemes and policy initiatives, are perceived to be intangible and strategic. Distinctions are being made according to the tangibility and the temporal nature of knowledge. Along one dimension, knowledge can be either *concrete* (i.e. tangible) or *abstract* (i.e. intangible, or tacit). A product can be characterized as exhibiting tangible knowledge, while a process exhibits intangible, or tacit, knowledge. Along the second dimension, knowledge can be either *strategic* (i.e. long-term, for overall direction setting) or *operational* (i.e. short-term, decision-making knowledge, for route finding within the overall direction).

These characteristics of knowledge can be illustrated by a two dimensional framework. Figure 3 shows the abstract/concrete distinction along a vertical axis and the strategic/operational distinction along a horizontal axis. Four regions are defined within this framework. The firm's decision-makers want knowledge in Region A; concrete information to help them make decisions.

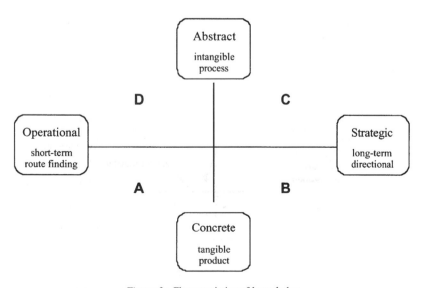

Figure 3. Characteristics of knowledge.

They want a tangible product giving them operational knowledge. External knowledge sources such as universities and government innovation schemes are perceived to be in Region C; their distance makes them intangible; their subject matter outside of the firm's operational needs and therefore strategic. Where specific outputs are produced, such as the reports of the U.K. Foresight program, a scheme or initiative might be regarded as being in Region B; it is still long-term, but has a perceived product in the form of reports that can be read. Region D represents an area of tacit knowledge, unwritten rules, processes and procedures that contribute to a firm's short-term decision-making culture.

The two-dimensional framework in Fig. 3 gives another context in which to view the framework of knowledge translation developed above. Research shows that firms view internally generated knowledge in concrete and operational terms (Major & Cordey-Hayes, 2000a, 2000b). They want discrete and tangible knowledge to enable them to take immediate actions. In the terminology of the knowledge translation framework they are using their internally generated knowledge as wisdom, from which they take actions. Where they seek externally generated knowledge, most firms do so to add to this stock of short-term, tangible knowledge. Certain firms take a much longer-term view. Forward-looking, future orientated firms have a fuller perspective of knowledge acquisition. Their

internal *processes* (the term processes itself signifying their greater appreciation of the abstract dimension) for acquiring external knowledge show them to be reaching much *farther back* through the nodes and processes of the knowledge translation framework. They seek knowledge that they can translate into understanding within the company, which gives an informed base for the wisdom underlying their actions. Even farther back they may seek the basic uncollated information that underlies knowledge.

Action is concrete, and, almost by definition, short-term. Wisdom, in the sense of firms' requirements, is also concrete and operational (though less short-term than action). Action and wisdom, the final two nodes of the knowledge translation framework fall within Region A in Fig. 3. It follows that the remaining nodes and processes of the knowledge translation framework can be located on the two-dimensional characterization of knowledge. The nodes and processes from the combined knowledge translation framework (Fig. 2) can be superimposed onto the dimensions of knowledge (Fig. 3). Figure 4 shows the resulting integrated conceptual framework for knowledge translation.

Action and wisdom are concrete and operational. At the other end, data is also concrete. It is a tangible product (hard and objective according to Cooley (1987) and Horton (1997)). However, because it lacks codification or an operational context it is strategic. As data is collected and collated into information and knowl-

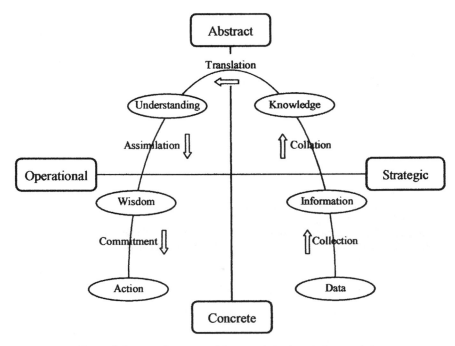

Figure 4. Integrated conceptual framework for knowledge translation.

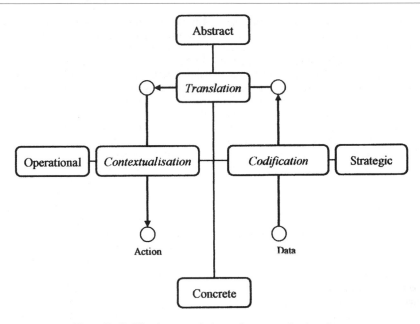

Figure 5. Codification, translation and contextualization phases.

edge, its immediate tangibility is reduced. Information is softer, less tangible, than data. Knowledge is less tangible still. While information may be talked about in product terms, knowledge is regarded more as an abstract process. The translation and interpretation phase brings the system into the operational frame of an individual firm (as the knowledge receiver), reaching a level of internal understanding. Abstract understanding, given a context through assimilation becomes tangible wisdom, the basis for commitment to action. The sequence of elements in the combined knowledge translation framework moves through the quadrants defined by the two dimensions of knowledge in the order BCDA. Concrete and strategic data must be taken and transformed through abstract elements before they can emerge as concrete and operational wisdom.

The sequence can be summarized into movements between the quadrants. Moving from quadrant B to quadrant C (strategic-concrete to strategic-abstract) is a process of *codification*. Combining the collection and collation processes, this extracts abstract meanings from the data. Quadrant C to quadrant D (strategic-abstract to operational-abstract) remains the *translation* process. This process brings externally generated knowledge into the organization's sphere, generating internal understanding. Finally, a *contextualization* process, moving from quadrant D to quadrant A (operational-abstract to operational-concrete), extracts wisdom and promotes action. This is the result of putting the understanding into the organization-specific context. These three summary *phases* equate to Hor-

ton's (1999) three phases of successful foresight. They are shown in Fig. 5.

The Role of Intermediaries in Knowledge Translation

Bridging the Knowledge Translation Gap

The previous section introduced the knowledge translation gap to describe how the overall knowledge translation process can be incomplete. In terms of Fig. 5, the codification of externally generated knowledge can be carried out by an outside body, perhaps the knowledge source itself (e.g. government as the source of a national innovation scheme). Contextualization occurs within the knowledge destination (the small firm in the present context). As suggested certain (forward-looking, future orientated) firms reach back through the knowledge translation framework to carry out the translation phase (and possibly also the codification phase) themselves. Most firms though face the possibility of a gap in the translation phase, and therefore a barrier in their access to externally generated knowledge. Some other body is needed to bridge the gap between the firm and the knowledge source. An organization fulfilling an intermediary role can translate and interpret external knowledge into a form in which receiving organizations can then contextualize it.

Intermediaries were introduced in Section 3. In broad terms, an intermediary organization is any body coming between the knowledge source and the destination (fulfilling the role of an agent in principle-agent theory). In these terms, organizations within the business support community are intermediaries. Business support agencies such as Chambers of Commerce and small business support organizations might therefore be able to fulfil this role. Industry based bodies; trade and research associations and professional institutes, are potentially well positioned to supply an intermediary service to their industries. Even universities, not previously considered as intermediaries, might bridge the gap through their specialist knowledge. Intermediaries provide a ready, extant system for connecting knowledge sources to knowledge receivers.

Categorizing Intermediaries

The roles of intermediaries are poorly understood. Previous studies expressed the somewhat muddled and amorphous perception of the business support community (Center for Exploitation of Science and Technology, 1997; Woolgar et al., 1998). Major & Cordey-Hayes (2000a) distinguish three categories of intermediary, according to the prime function they provide for their industrial clients. First are *signposters*. These are first-point-of-contact organizations. They point to sources of advice, guidance and expertise, rather than answer problems themselves. In the U.K. Training and Enterprise Councils (TECs), Business Links and Regional Technology Centers (RTCs) all signpost firms to a body appropriate to the query. Small Business Development Centers (SBDCs) and the Office of Technology perform a similar function in the U.S. Second are *facilitators*. These give advice and guidance to client firms to help them to help themselves. Trade associations' central information provision role is a facilitating function. Some industrial research associations (RAs) have a membership body of industry clients much as trade associations. These membership-based RAs (MRAs) play a similar information provision, facilitating role. Chambers of Commerce facilitate regionally, by encouraging networking between firms in a common locality. Third are *contractors*. These are specialist sources of expertise offering direct help on specific issues. University research, non-membership RAs (NMRAs) and professional institutes (through personal memberships) are prime sources of such expertise. When viewed as a continuum, patterns emerge within these three intermediary functions. On moving from signposters, through facilitators to contractors there are longer-term intermediary-client relationships, increasing involvement with client firms' operations, an increasing large firm focus and a corresponding increasing difficulty of small firm involvement.

Intermediaries then have a central role in bridging the knowledge translation gap and thus in the conveyance to small firms of externally generated knowledge. The next section shows how this bridging role can be exploited to reach the small firms.

Intermediaries and Conveyance to Small Firms of Externally Generated Knowledge

Categorizing Small Firms

Major & Cordey-Hayes's (2000a) research leads to a categorization of small firms. They propose that small firms be ordered according to the *futures orientation* of their culture and attitude. Firms with a strong futures orientation have the ability to reach back to the translation and codification processes of the knowledge translation framework (Fig. 5) to acquire externally generated knowledge, which can then contribute to their innovative potential. Four themes describing the boundary activities and network orientation of a sample of U.K. small firms were studied (Major, Asch & Cordey-Hayes, 2001; Major & Cordey-Hayes, 2000a) to identify the attributes that contribute to small firms' futures orientation. The key theme was awareness and perception of the U.K. Foresight program and the concepts it is promoting, and the use of such concepts as a regular part of their business. High awareness, accurate perception and regular use suggest a greater appreciation of long-term issues (i.e. a strong futures orientation). Such firms are able to reach out to obtain the type of knowledge that Foresight provides. Remaining themes were: (1) firms' relationships with intermediaries; (2) the importance of external networking between firm personnel and people in outside organizations; and (3) firms' reliance on their supply chains as a knowledge source.

Three distinct groups of firms emerged, 10% of the small firms sampled had a strong futures orientation. These are *strategic* firms, with a highly *involved* approach to their future. They are characterized by high and ongoing awareness and accurate perception of Foresight, and regular use of Foresight concepts predating their knowledge of the U.K. Foresight program. Their futures orientation puts them ahead of what the Foresight program can provide, and some have been giving the program the benefit of their own experience. These firms have many contacts with intermediaries, but only a select few, typically with trade and research associations and universities, are deep and ongoing. These few intermediaries are important sources of external knowledge. Outwardly focused personal attitudes underpin these firms' foresight knowledge and futures orientation. Correspondingly, external interpersonal networking was found to be highly important. However, reliance on supply chains was not important. Customers and suppliers (the supply chain) are always a firm's most extensive contacts (Moore, 1996; Woolgar et al., 1998), but for the involved, strategic firms they are not a significant source of externally generated knowledge. 70% of the small firms sampled had a weak futures orientation. Their attitude is *reactive* or *uninvolved*. These firms have little or no awareness of Foresight or its concepts. Their intermediary contacts are all at a level where they can have little impact as external knowledge sources. External networking has a low importance, driven out by concentration on the present. Completing the contrast with the *involved* firms, supply chains are their most important external knowledge source. But reliance on these tangible, short-term operational management relationships tends to drive out use of longer-term external knowledge sources. Between the two extremes, 20% of the small firms sampled were distinguished with intermediate characteristics. These firms know about Foresight and are aware of the importance of their future, but lack prior involvement with Foresight concepts. Their intermediary contacts are more selective and more useful external knowledge sources than in uninvolved firms, but lack the depth and selectivity of involved firms. The importance of external networking and reliance on the supply chain also consistently rate between involved and uninvolved firms. These are *open*, or *responsive* firms. They are open to the future, but need a stimulus to generate response and action. Compared to the involved firms open firms know what is needed, but only the involved firms can actually do it.

Targeting Small Firms—Small Firm-Intermediary Interactions

Dissemination of the U.K. Foresight program took no account of the differences between small firms. Communicating Foresight to *involved* firms, who already have a futures-orientated culture, is like *preaching to the converted*. For *uninvolved* firms futures-orientated policy initiatives are so far outside of their normal sphere of attention as to be essentially unreachable. The present, they perceive, requires their full attention. Communicating with these firms would require great expenditure of effort simply to be heard, let alone be listened to. It is proposed that the most effective audience for futures oriented initiatives like Foresight is the *open* firms. Openness means that they will listen, lack of prior involvement means that significant impacts may result. Combining management or organizational changes with a well-targeted innovation scheme or policy initiative could be the stimulus an open firm needs to become an involved firm.

Major & Cordey-Hayes's (2000a, 2000b) research shows that only a small portion of an innovation scheme or policy initiative's potential audience will respond so as to fulfil its aims. The most important outcome of the research in terms of conveyance of government schemes and initiatives was a model developed to show how intermediary organizations can provide the method to target this priority audience. Intermediaries interact with small firms as part of their regular operation. The research revealed strong patterns in how small firms use intermediaries to build up their internal knowledge and to enhance their awareness and acquisition of externally generated knowledge

(Major & Cordey-Hayes, 2000a). *Involved* firms were found to get this support from *contractor* organizations, *open* firms from *facilitators*, and *uninvolved* firms (at a minimal level) from *signposters*.

Figure 6 illustrates these relationships between small firms and intermediaries. The vertical axis ranks small firms by their futures orientation. *Involved* firms, with a strong futures orientation rank higher than *open* and *uninvolved* firms with moderate and weak futures orientation respectively. The horizontal axis ranks the intermediary organizations by the signposter-facilitator-contractor categorization. In describing the small firms-intermediaries relationships, the model shows that a move up the small firms' futures orientation scale is accompanied by a move along the intermediaries functions scale. *Involved* firms use their extensive contacts at universities, and to a lesser extent NMRAs and professional institutes, for their specific research expertise. Knowing where to look, they do not need to be signposted. Already having access to important external knowledge sources, they do not need a facilitator. *Open* firms are open to the desire for futures involvement but lack the *facility*. Links with trade associations and MRAs and links with chambers of commerce give them access to externally generated knowledge on industry and regional bases respectively. With this access, these firms too are in no need of signposting. *Uninvolved* firms use trade associations and chambers of commerce for their present-orientated representational roles, but have little desire to access external knowledge that can lead to innovation. Lacking the intermediation of the facilitators they require the basic guidance of signposters, Business Links and RTCs in the U.K., SBDCs and the SBA Office of Technology in the U.S., to point them towards external knowledge sources.

Implications and Recommendations

Implications for Innovation Schemes and Policy Initiatives

Figure 6 gives the basis for engaging intermediaries to target small firms. Though generated primarily from study of the Foresight program, the mechanism that the model describes is generic; it describes relationships, rather than a single policy initiative. Foresight is simply a *package of knowledge*, the conveyance, or translation of which the model is guiding. It plays no part in the conveyance, or translation mechanism itself. The mechanism can therefore be applied to translation of other schemes and initiatives of a similar nature to the Foresight program.

Engaging intermediaries brings schemes and initiatives closer to the small firms, overcoming some of their inherent resource problems. Increasing the distance between policy-makers and firms, reduces the effects of firms' mistrust of government and government's limited understanding of the firms' situations.

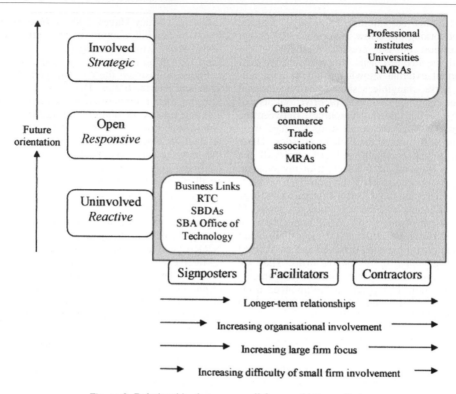

Figure 6. Relationships between small firms and intermediaries.

Compared with policy-makers, intermediaries hold greater understanding of, and are more trusted by, small firms, and have more dedicated resources to deal with them. In principle, all that is required is for the policy-maker to select the organizations that deal appropriately with the specified target audience. The previous section suggested that for Foresight, the priority target audience should be *open* firms. But whichever target group a policy initiative deems the priority, the generic nature of the model indicates which intermediaries to engage. The U.K. Foresight program has started to recognize the value of inter-mediaries. The *Foresight associate program* engaging intermediaries to bring the benefits of Foresight to their spheres of influence, is a welcome development, but the predominance of professional institutes among the associates (Office of Science and Technology, 2000) does not fully reflect the recommended targeting of open, responsive firms.

Implications for Small Firms

The theory and models developed in this chapter should encourage small firms to strengthen their links with intermediaries. Stronger links could increase the small firms' exposure to externally generated knowl-edge, be it from the intermediary itself or through

greater access to innovation schemes and policy initiatives, with consequent benefits to the firms' innovative potential.

The previous section linked small firms' futures orientation (with its effect on future innovation) to the type of intermediary they have dealings with. Involved firms' futures-orientation accompanies their deep links with contractors. Encouraging open firms to deepen their contractor links may stimulate them towards an involved attitude. Similarly, encouraging greater use of facilitators in uninvolved firms could stimulate a more open attitude.

This chapter has made recommendations for the conveyance of externally generated knowledge to small firms. The processes within the firms whereby that knowledge is manipulated to generate innovation and competitive advantage fall beyond the scope of the present chapter. This theme is taken up in the literature on organizational learning (see Gilbert & Cordey-Hayes, 1996; Vickers & Cordey-Hayes, 1999).

Implications for Intermediaries

Intermediaries have been shown to have a central role in the conveyance to small firms of externally gen-erated knowledge. They should thus be encouraged to perform this role, working with government innovation schemes and other knowledge sources, bridging the knowledge translation gap to increase the innovative potential of their small firm contacts.

Conclusions

This chapter has considered the conveyance into small firms of externally generated knowledge. It started by presenting the barriers to innovation in small firms and the sources of internally and externally generated knowledge potentially available. From the concept of knowledge transfer a framework of knowledge translation has been developed. The derived notion of the knowledge translation gap explicitly illustrates why so many small firms fail to reach external knowledge sources. The chapter has proposed that intermediaries can be the bridge that is needed to cover the knowledge translation gap. Understanding how intermediaries can be used to convey externally generated knowledge to small firms brings implications and recommendations for government innovation schemes and policy initiatives, for intermediaries and for the small firms, all of which will encourage and enhance successful small firm innovation.

References

Cabello, C., Scapolo, F., Sørup, P. & Weber, M. (1996). Foresight and innovation: The rise of initiatives at European level. *The IPTS Report*, **7**, 33–39. Seville: The Institute for Prospective International Studies.

Centre for Exploitation of Science and Technology (1997). *SMEs and access to technology*. London: Centre for Exploitation of Science and Technology.

Cohen, W. M. & Levinthal, D. A. (1990). Absorptive capacity: A new perspective on learning and innovation. *Administrative Science Quarterly*, **35** (March), 128–152.

Cooley, M. (1987). *Architect or bee: The human price of technology*. London: The Hogarth Press.

Department of Trade and Industry (2001). *Small business service*. Website: www.sbs.gov.uk

Farr, J. L., Sin, H. P. & Tesluk, P. E. (2003). Knowledge management processes and work group innovation. In: L. V. Shavinina (Ed.), *International Handbook on Innovation*. Oxford: Elsevier Science.

Garsombke, T. & Garsombke, D. (1989). Strategic implications facing small manufacturers: The linkage between robotisation, computerisation, automation and performance. *Journal of Small Business Management*, (October), 34–44.

Gilbert, M. & Cordey-Hayes, M. (1996). Understanding the process of knowledge transfer to achieve successful technological innovation. *Technovation*, **16** (6), 301–312.

Hadjimanolis, A. (1999). Barriers to innovation for SMEs in a small less developed country (Cyprus). *Technovation*, **19** (9), 561–570.

Hadjimanolis, A. (2003). The barriers approach to innovation. In: L. V. Shavinina (Ed.), *International Handbook on Innovation*. Oxford: Elsevier Science.

Horton, A. M. (1997). *Conceptual model for understanding the future role of print in the information economy*. Unpublished paper. Alpha to Omega and Beyond, London.

Horton, A. M. (1999). A simple guide to successful foresight. *Foresight*, **1** (1), 5–9.

Keogh, W. & Evans, G. (1999). Strategies for growth and barriers faced by new technology-based SMEs. *Journal of Small Business and Enterprise Development*, **5** (4), 337–350.

Major, E. J., Asch, D. & Cordey-Hayes, M. (2001). Foresight as a core competence. *Futures*, **33** (2), 91–107.

Major, E. J. & Cordey-Hayes, M. (2000a). Engaging the business support network to give SMEs the benefit of Foresight. *Technovation*, **20** (11), 589–602.

Major, E. J. & Cordey-Hayes, M. (2000b). Knowledge translation: A new perspective on knowledge transfer and foresight. *Foresight*, **2** (4), 411–423.

Moore, B. (1996). Sources of innovation, technology transfer and diffusion. In: A. Cosh & A. Hughes (Eds), *The Changing State of British Enterprise*. Cambridge, U.K.: Economic and Social Research Council.

Office of Science and Technology (2000). *Foresight*. Website: www.foresight.gov.uk

Rothwell, R. (1991). The role of small firms in technological innovation. In: J. Curran, J. Stanworth & D. Watkins (Eds), *The Survival of the Small Firm* (Vol 2). Aldershot, U.K.: Gower.

Rothwell, R. & Zegvelt, W. (1982). *Innovation and the small and medium sized firm*. London: Frances Pinter.

Slaughter, R. A. (1995). *The foresight principle: Cultural recovery in the 21st century*. London: Adamantine Press Limited.

Small Business Administration (2000). *Small business administration Home Page*. Website: www.sba.gov

Storey, D. J. (1994). *Understanding the small business sector*. London: Routledge.

Tidd, J., Bessant, J. & Pavitt, K. (1997). *Managing innovation: Integrating technological, market and organisational change*. Chichester, U.K.: John Wiley.

Trott, P., Cordey-Hayes, M. & Seaton, R. A. F. (1995). Inward technology transfer as an interactive process. *Technovation*, **15** (1), 25–43.

Woolgar, S., Vaux, J., Gomes, P., Ezingeard, J-N. & Grieve, R. (1998). Abilities and competencies required, particularly by small firms, to identify and acquire new technology. *Technovation*, **18** (9), 575–584.

Vickers, I. & Cordey-Hayes, M. (1999). Cleaner production and organizational learning. *Technology Analysis and Strategic Management*, **11** (1), 75–94.

The International Handbook on Innovation
Edited by Larisa V. Shavinina

Linking Knowledge, Networking and Innovation Processes: A Conceptual Model

Jacqueline Swan, Harry Scarbrough and Maxine Robertson

IKON (Innovation, Knowledge and Organizational Networks) Research Centre, Warwick Business School, University of Warwick, U.K.

Abstract: Innovation is frequently cited as a major reason for the emergence of network structures. However, although network structures have been studied extensively, relatively little research has examined the diverse roles played by networking processes in innovation. This chapter develops a conceptual model that relates specific kinds of networking, to particular episodes of innovation (invention, diffusion, implementation) and to processes of knowledge transformation. The operation of the model is illustrated through three case examples, each focusing on a different innovation episode. These are used to illustrate interactions among processes of networking, innovation and knowledge transformation.

Keywords: Innovation; Knowledge; Networking; Networks; Invention; Diffusion; Implementation; Knowledge transformation.

Introduction

Innovation may be defined as: "the development and implementation of new ideas by people who over time engage in transactions with others in an institutional context" (Van de Ven, 1986, p. 591). Networking—as a social communication process—is thus recognized as playing a central role in innovation. Freeman (1991), for example, observes that many studies since the 1950s have noted 'the importance of both formal and informal networks, even if the expression network was less frequently used' (p. 500). Encouraging innovation is frequently cited as the major reason for developing network forms of organization. Despite this observation, relatively little research has focused explicitly on the links between networks and innovation (Oliver & Ebers, 1998). Moreover, research that does address links between networks and innovation tends to look at the impact of network structures—social processes of networking receive relatively little attention (Pettigrew & Fenton, 2000). This chapter looks, therefore, at the roles and implications of networks for innovation focusing, in particular, on processes of networking.

The chapter begins by using existing theory and research to develop a conceptual model linking networking with innovation processes. Innovation is presented here as an episodic *process*, encompassing the design and development (invention), spread (diffusion) and implementation of ideas that are new to the adopting unit (Clark, 1987; Van de Ven, 1986). Recognising the limits of stage models of innovation, these episodes are seen as iterative, recursive and ultimately conflated, not as linear and sequential (Clark & Staunton, 1989; Ettlie, 1980). For example, pivotal modifications built into the design of the innovation during implementation may feed back into its diffusion (Fleck, 1994). Previous studies have tended to separate these different aspects of innovation and to focus on discrete episodes (i.e. either invention or diffusion or implementation—Wolfe, 1994). However, this chapter attempts to outline a more holistic model of the links between networking and innovation by tracking the roles of networks and social networking activities across the entire innovation process, and by noting how these roles may vary across different episodes.

This model is developed and illustrated by contrasting the role of inter-organizational networking (the case of a professional association) in *diffusing* technological innovation with that of intra-organizational networking (the case of a consultancy firm) in *inventing* scientific innovation for clients. It then illustrates how these different kinds of networks coalesce when *implementing* operations management technology (the case of a manufacturing firm). The aim of the chapter is to use these examples to explore the roles played by different kinds of network and social networking activities at different points across the whole innovation process.

This focus on the links between networking and innovation processes highlights an area, which is comparatively under-researched and under-theorized in the literature (exceptions being the work of Alter & Hage, 1993; Oliver & Leibeskind, 1998; Rogers, 1983). Although explanations of the emergence of network structures frequently cite, as reasons for these structures, the industrial change and market turbulence associated with product and process innovations, relatively few studies go on to address the performative role of networks in developing or promoting such innovations (Oliver & Ebers, 1998). In contrast, our study, not only highlights that role, but also suggests that innovation is closely, reciprocally and systematically intertwined with the creation and maintenance of networks (see also Gibson & Conceicao, 2003; Major & Cordey-Hayes, 2003). The argument for this view is based, firstly, on the development of a theoretical model which draws on research in this area, and, secondly, on the empirical study of network effects in widely differing innovation processes. In concise terms, the development of the argument is based on two key propositions. Namely, that a processual view of networks is as, or more, relevant to innovation studies than the conventional structural view (Pettigrew & Fenton, 2000; Wolfe, 1994) and second, that innovation is better characterized not as the production of physical artefacts but as flows and combinations of knowledge and information (Major & Cordey-Hayes, 2003; Nonaka et al., 2003; Tidd, 1997). These propositions are elaborated further below.

Network Structures and Networking Processes

Networks have been analyzed in a variety of ways and through different theoretical lenses (Alter & Hage, 1993; Ebers & Jarrillo, 1997; Grandori & Soda, 1995; Oliver & Ebers, 1998). However, in many discussions a distinction is made between *structural* characteristics or forms of networks and the *processes* involved in developing and sustaining networking relations (e.g. Alter & Hage, 1993). Existing work (cf. Ahuja, 2000; Powell et al., 1996) linking networks with innovation tends to focus on the former. Networks are viewed principally in functional terms as structured channels through which information is communicated and knowledge is transferred. Hansen (1999), for example, develops a contingency model linking network structures (in terms of the strength of network ties) to forms of knowledge transfer (in terms of relatively complex/tacit or simple/explicit forms). From a detailed empirical study of product innovation in a large electronics company he concludes that networks characterized by strong ties are most effective for the

transfer of tacit knowledge and weak ties for transfer of explicit knowledge.

Structural perspectives tend to see networks as an intermediate organizational form which lies somewhere between markets and hierarchies. A distinction is also made between inter and intra-organizational networks with much of the networking literature focusing on the latter (e.g. Alter & Hage, 1993; Ebers, 1997; Grandori & Soda, 1995; Jarrilo, 1993). Grandori & Soda (1995, p. 184), for example, define networks as 'modes of organizing economic activities through inter-firm co-ordination'. Such modes of organizing—i.e. network structures—are, it is argued, important for innovation because they allow for more open and/or extensive exchange and transfer of knowledge and information across firms.

This emphasis on the relatively formal and persistent relationships between firms usefully highlights the structural implications of *networks* in innovation. However, it tends to downplay social processes of *networking*, including the creative role of actors and agents and the importance of interpersonal and informal relationships in developing, creating and sustaining networks (Ebers, 1997). The emphasis on networks as structures for knowledge and information exchange, then, tends to overshadow the social processes through which these structures emerge and develop and the intentions of the actors involved. It also tends to downplay the performative role of networks in actually creating, defining and shaping the knowledge and information that is exchanged.

In contrast, our chapter builds from this earlier work but aims to develop a framework that will address the role of networking processes, not just structures. This underscores the importance of both structure *and* agency in the innovation process. This is not to downplay the importance of structural accounts but merely to give more serious attention to social processes of networking in innovation. In keeping with recent work that has highlighted the importance of 'social capital' (Nahapiet & Ghoshal, 1998), this chapter reveals how the development of the innovation process is intertwined with the creation and maintenance of social networks over time.

Networking involves the active search and development of knowledge and information through the creation and articulation of informal relationships within a context of more formal intra- and inter-organizational structural arrangements. Networking processes are critical for innovation precisely because they span structural forms rather than being contained within them. For example, they occur not just within network structures but also within hierarchies and markets. Networking is also self-sustaining and self-energising—one contact leads to, or precludes, the development of other contacts. Unlike the search routines highlighted in classical organization theory (Cyert & March, 1963), networking is wayward and emergent, being driven more by interest and opportunity, or by chance 'accidental encounters', than by the rational needs of a particular decision-making process (Kreiner & Schultz, 1993). It is also important for innovation because it involves the liberal sharing of knowledge and information and an open-ended outlook on collaboration (Kreiner & Schultz, 1993).

This processual view of networks has important implications for the analysis of innovation. First, it highlights the need to set aside static, institutional explanations of network development in order to recognize their emergent, formative qualities (Ebers, 1997). Where a structural view refers to persistence, stability and established relationships, a processual view denotes the need to examine the role, in innovation, of the sometimes fragile and exploratory activities based on embryonic contacts and half-formed relationships. Second, this approach suggests that the conventional separation between inter- and intra-organizational networks overstates the effect of more-or-less settled organizational boundaries and under-estimates the networking activities that are directly subversive of such boundaries. For example, in innovation projects networking among actors from different organizations (such as that between project managers and consultants) may be as close or closer than networking among actors from within the same organization. Thus inter and intra-organizational boundaries may become blurred as those involved may come to identify more with the innovation process and their role in it than with their own organization.

The pervasiveness and importance of social networking processes in creating and defining the role of networks in innovation is underlined in this chapter by three case examples. This extends the relevance of the network concept beyond the question of structural form to the wider issue of the 'organizing methods' or the 'social practices of organizing' (Knights et al., 1993) which are applied to socio-economic activity.

Innovation and Networking

Structural perspectives conceptualize networks as diffusion channels through which new ideas are spread from innovators to adopters (e.g. Abrahamsson, 1991; Ebers & Jarrillo, 1997). In contrast, the processual view presented here not only addresses diffusion, but also seeks to analyze ways in which networking processes exert shaping effects on the character and design of the knowledge and innovations diffused. The role of intermediaries (business support agencies, trade associations, professional institutes, universities etc.) in the translation of knowledge relating to innovation is also highlighted by Major & Cordey-Hayes (2003). During the diffusion of innovation, certain features technologies may be highlighted, and others downplayed, depending of the kinds of networking activity

involved and the vested interests of those concerned (Swan et al., 1999a). For example, in the mid-1980s Manufacturing Resources Planning (MRP2) was heavily pushed by technology suppliers in the USA via their engagement in a variety of networks, including professional associations. This became the dominant technology design for production management, even though other options existed at the time (e.g. Just In Time—Newell & Clark, 1993). This links the diffusion of innovation more closely to its design and implementation and suggests a much closer examination of the interactions between networking activities and knowledge flows than is usually the case with more structural accounts. In helping to understand these interactions two existing processual accounts are worthy of closer attention—those of Kreiner & Schultz (1993) and Ring & Van De Ven (1994).

Kreiner & Schultz's (1993) study of informal university-industry networks in the R&D environment proposes a multi-stage analysis of network formation. The first stage of this process is 'discovering opportunities' which is activated by the accidental encounters and exploratory trust-building of a 'barter economy'. This stage leads on to 'exploring possibilities' where initial ideas are tested and validated and projects materialize. The final stage—'consummating collaboration'—is where projects are enacted through a 'crystallized network of collaboration'. The Kreiner and Schultz model contains some important insights into the links between networking and innovation. For example, they note that in the initial stages of idea generation, the concept of knowledge exchange or 'know-how trading' (von Hippel, 1988) fails to explain the promiscuous sharing of ideas seen amongst their R&D actors. This leads them to question the assumption that the act of sharing knowledge diminishes its value to the owner. Rather, when knowledge consists of 'loose ideas and inspirations' (p. 197), the immediate value of the knowledge being shared is low, but the potential gain from combining it with the knowledge of others is high.

A further implication of the Kreiner & Schultz model is that the process of networking has a joint outcome. While networking creates knowledge (in this case R&D knowledge), it also crystallizes new network relationships that then act as a 'centre of gravity' for further networking and research. In short, networking creates path dependencies in the production and diffusion of knowledge by providing space for new network relationships. It is important to note, however, that such path dependencies do not impose rigid search behaviours on network participants. Kreiner & Schultz (1993) suggest that innovation derives from the *'blending of ideas, knowledge, competencies, experience and individuals'* (p. 200), but that this blending usually happens in unplanned and emergent ways. The importance of 'accidental encounters' in the Kreiner & Schultz (1993) study underlines the central importance to innovation, particularly in the nascent phases, of informal, opportunistic inter-personal relations in network formation. This need to recognize the role of personal and informal dimensions of networks and networking in innovation is further echoed and amplified in the work of Ring & Van De Ven (1994) who seek to develop a socio-psychological account of networking processes.

The Ring & Van De Ven (1994) model is presented as a counterpoint to conventional analyzes of networks, contrasting, for instance, the importance of perceived trust and equity with the usual emphasis on transactional efficiency. This approach leads them to propose a cyclical model of network formation encompassing four distinctive socio-psychological activities—negotiations, commitments, executions, and assessments. This model shows some important similarities with the Kreiner & Schultz (1993) analysis. First, it is process-based—network structures are seen as an outcome of networking processes. Second, it is recursive—networking interactions closely resemble the path-dependencies noted by Kreiner & Schultz (1993). Third, it highlights the importance of exploratory trust-building interactions as the catalyst to network formation. Fourth, Ring & Van de Ven (1994) also highlight the tensions between formal organizational structures, networking activities, and innovation, focusing in particular on the tensions created between organizational roles and personal interactions, with narrowly defined roles limiting the scope for networking and innovation.

Innovation and Knowledge

The important role played by networking highlighted in these studies, underscores the need to revise our understanding of innovation. Traditional views have tended to emphasize the creation and distribution of physical artefacts—ideas are invented, distributed as physical artefacts, and then implemented in firms (Clark, 1995; Rogers, 1983, 1995). This artefact-based model is increasingly challenged, however, by the growth of the service sector and the rise of knowledge-based products and processes (reflected in the growth of the consultancy industry). In this context, innovation is better conceptualized not as a materially-constituted entity but as a particular transformation of knowledge and information (Macdonald & Williams, 1992; Nonaka et al., 2003). This 'knowledge-based' view of innovation moves us beyond the linear assumptions of the artefact-based model and highlights the complex and recursive interactions that filter and shape the innovation from inception through to end-use.

One implication of this 'knowledge-based' view of innovation is the significance attached to networks as the means through which knowledge is elicited, translated and (re)combined to produce innovation. However, an emphasis on networking suggests that such networks are not passive channels for the transfer of knowledge but are also implicated in its active (re)production and appropriation (Alter & Hage, 1993; Clark, 2000). This further differentiates the processual view from the so-called 'entitive approach' of the artefact-centred model (Hosking & Morley, 1990). Entitive perspectives treat innovation as an object or thing which is invented and diffused, more or less unchanged, from one adopter to another. In contrast, the knowledge-based view, not only suggests that innovation involves complex interactions between different groups and constituencies, but also that these interactions have a shaping and filtering effect upon the innovation itself. In short, during innovation, knowledge and information is both communicated and transformed by the network of social actors.

Taking this further, the different 'episodes' within the innovation process (invention, diffusion and implementation) can be characterized as involving distinctive shifts in the ways knowledge is constituted (see also Major & Cordey-Hayes, 2003). The episode of invention, for example, may be characterized in terms of the (social) *construction* of knowledge (Bijker et al., 1987). Here, loose ideas shared through interpersonal networks may crystallize into new forms of work practices or products. In contrast, the diffusion episode is associated more closely with the *commodification* and *communication* of knowledge (Rogers, 1983). This episode requires that knowledge be made more explicit and codified in order to be translated to a wider social constituency. Diffusion involves, then, the progressive *objectification* or 'black boxing' of knowledge, such that its communication and distribution ceases to be dependent on the particular tacit understandings and social context of its creators, but is transformed into more explicit, generic and therefore, more widely portable forms (Scarbrough, 1995). In contrast, again, the episode of implementation relies on the *appropriation* of knowledge within firms (Clark, 1987). Here generic, objectified ideas about new work practices, technologies or products need to be applied to the specific context of the adopter by customizing and adapting them to local requirements. This involves unpacking knowledge from its objectified state and fusing it with local and often tacit knowledge of the organization (Fleck, 1994).

A Model to Link Innovation, Knowledge and Networking

The discussion so far has outlined the importance of networking in innovation by conceptualizing innovation as a process that involves flows and combinations of knowledge and information occurring through networks and networking. This suggests that a conceptual model that is able to link networking processes and innovation could be useful. However, the different knowledge requirements of different episodes of the innovation process also suggests that any model to link innovation and networks needs to be able to relate specific types of networking activities to particular episodes within the innovation process.

A number of different perspectives have been applied to this issue. As noted, many writers adopt a structural approach, which sees the creation of networks as a consequence of the failure of market and hierarchical forms to adequately regulate certain kinds of transactions (Casson & Cox, 1997; Grandori & Soda, 1995; Poire & Sabel, 1984; Powell et al., 1996). However, there are important limitations to this analysis. First, it emphasizes the functional characteristics of organizational and institutional structures over the agency of groups and individuals. Yet, by focusing on agency, it is clear that networking relationships may occur, for example, within markets and hierarchies and so these may be complementary, rather than competing, forms (Holland & Lockett, 1997). Second, the emphasis on individual transactions is a poor characterization of the rich social interactions involved in the transfer and exchange of knowledge. For example, Kreiner & Schultz (1993) note the inappropriateness of an exchange perspective to the R&D environment. This suggests that the potential creation of new economic benefits from the innovation—as opposed to the distribution of existing benefits—may help to lower some of the transactional barriers to the communication and exchange of knowledge (Lazonick, 1991). Third, the economists' concerns with the problems of costing and exchanging knowledge tend to gloss over social concerns to do with the legitimation of knowledge (Casson & Cox, 1997; Williamson, 1985). Before knowledge can be effectively diffused and implemented (for example, as a new form of 'best practice'), it has to be accepted as legitimate by the relevant social group. Therefore, social and institutional mechanisms of validation are also critical.

The social validation of knowledge involves establishing the credibility, the essential 'rightness' of what is proposed. In a technological context a variety of institutional networks may perform this function including, importantly, professional associations, trade associations, policy making bodies and so forth (Major & Cordey-Hayes, 2003; Swan et al., 1999a). Hence, the networks which facilitate the communication and exchange of knowledge often, simultaneously, provide the means of its validation. Persuasion and co-optation through networking activities, rather than exchange seem, then, to be key (Callon, 1980). Thus, even objectified forms of knowledge—emerging technologies, for example—will depend for their acceptance on the underpinning validation supplied by industry standards, relevant professions, and adoption of 'best' practice by 'leading' firms (King et al., 1994). Inter-organizational networks that help to communicate such standards will be crucial.

Taken together, these points suggest that networking activities are particularly suited to the dynamic and unstructured flows of knowledge associated with innovation processes. Networking operates within and across markets and hierarchies. It helps to address the transactional 'stickiness' of knowledge by promoting trust and stimulating value creation through innovation. At the same time, it also serves as means of validating knowledge by enrolling network partners. Thus, networks serve both to disseminate and to transform knowledge. Taking these arguments together suggests that a model of the role of networks in innovation should incorporate the following features:

(1) It should recognize that the roles of networks vary across episodes involved in the innovation process.
(2) The importance of knowledge transformation (involving construction, communication and exchange) should be highlighted.
(3) Both structural *and* processual dimensions of networks should be incorporated.
(4) It should recognize the changing role of personal sense-making and trust-building activities in the evolution of networks through networking activities.
(5) Path-dependency of both the innovation process and network formation should be included.
(6) The implications of different kinds of knowledge for the roles played by networks should be addressed.

A model that encompasses the theoretical points above and relates different episodes of the innovation process to networking activity is presented in Fig. 1. This focuses on the interplay between the three critical dimensions identified in the discussion above—networking activity, knowledge attributes, and the episodic innovation process. However, unlike the existing models outlined above (e.g. Kreiner & Schultz, 1993), Figure 1 attempts not just to describe the interplay between these different elements, but also to compare different innovation episodes in terms of the different kinds of networking involved.

Mapping these complex interactions in this schematic way requires a number of theoretical caveats and qualifications. For example, the relationships between the different episodes of the innovation process must be seen as operating in a non-linear, recursive fashion (Clark & Staunton, 1989; Fleck, 1994). At the same time, while the character of networking is broadly correlated with the relative codification of knowledge (global, inter-organizational networks serving to disseminate more codified knowledge, for example), it is also important to acknowledge the continuous interplay between the ways in which knowledge is created and exchanged. Although Fig. 1 suggests that the dominant forms of knowledge transfer may vary from one episode to another, the overall scope and direction of the innovation process will reflect these co-dependencies. It follows from this discussion that

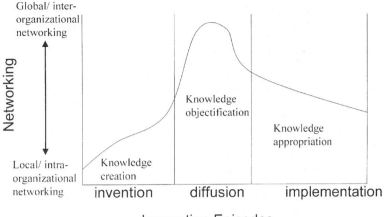

Figure 1. A model linking networking, knowledge and innovation.

the curve outlined in Fig. 1 defines the interaction between networking and innovation, not only in terms of the distinctive characteristics of networking in different episodes (e.g. inter- or intra-organizational), but also in terms of its extent or *scope*. This is denoted in Fig. 1 by the area encompassed within the curve during the different episodes. The constituent elements and assumptions of this model are characterized as follows:

Invention

During this innovation episode, the focus is on the social construction and creation of new knowledge through an exploratory and highly personalized process of networking among a fairly narrowly defined social group who have relevant tacit knowledge and interests. This is referred to in Fig. 1 as 'local intra-organizational' networking in order to reinforce the point that networking activities may cut across organizational boundaries. However, the advantages of physical proximity for interpersonal networking, means that networking within the firm would be likely to be especially crucial.

During this episode informal, interpersonal networking seeks to identify potential network participants who possess information and expertise that could be relevant to the development of new products or services. This networking will typically be wayward and emergent, with initial lose contacts quickly generating stronger ties as individuals come to realize that they have some common interest. 'Accidental encounters' may be important here (Kreiner & Schultz, 1993). Formal or informal coalitions (e.g. project teams) are assembled on the basis of (uncertain) expectations of reciprocity and trust and a willingness to share knowledge (Ring & Van de Ven, 1994). These expectations are generated through repeated social

interaction but may also be signalled by contextual cues. For example, membership of a particular organization or profession may imbue incumbents with expected expertise (Meyerson et al., 1996). A key role of social coalitions and teams is in the testing and interpretation of knowledge—what Ring & Van De Ven (1994) term 'sense-making'. The emphasis here, then, is on the sharing and creation of knowledge, rather than on the exchange of information or artefacts. Because outcomes are uncertain, economic considerations will be secondary to interpersonal, trust-based interactions.

Diffusion

During the diffusion episode, the emphasis shifts to the objectification and communication of knowledge through more global, inter-organizational networks. Ideas, now crystallized as new technical artefacts, products or services, become commodities that can be exchanged. Here the primary role of networking is to broadcast knowledge to legitimize particular inventions or new ideas (or old ideas repackaged) so that they become accepted and adopted by the wider community. For example, particular templates for technology design may be promoted to the wider community as new forms of 'best practice'.

Diffusion thus involves a social process of formal and informal information exchange among members of a social system (Rogers, 1983, 1995). This process is unequal and may be conflictual. For example, different groups of 'change agents' (such as salespeople, consultants, firms) may aggressively and opportunistically promote the adoption of particular ideas or artefacts where it is in their interests to do so (Swan & Newell, 1995). New ideas may also be diffused through the 'weak ties' linking different social groups (Granovetter, 1973; Hansen, 1999). 'Boundary spanning' individuals (e.g. key consultants) play a role in this diffusion process, involving themselves in a wide range of broader inter-organizational networks and translating ideas developed through this networking into locally applicable or organizationally useful solutions (Tushman & Scanlan, 1981). Thus, boundary spanning individuals act as 'knowledge brokers', helping to overcome social and organizational barriers (Aldrich, 1999).

Particular inter-organizational networks may also play a brokerage role in providing opportunities for members from different social communities to meet and interact. Professional associations, for example, allow links to develop between practitioners working in industry, academics, consultants and technical specialists, thereby promoting knowledge flows within a particular professional knowledge domain (Aldrich & von Glinow, 1992). At the same time, market structures create distinctive incentives towards the commodification of knowledge. Some networks (e.g. with technology suppliers) will act as distribution channels for pre-packaged knowledge and ideas (Scarbrough, 1995).

The distributed nature of networks in this episode may enforce a reliance on surrogate indices of the validity and legitimacy of the knowledge being diffused. One such surrogate may be found in the professional ethos and credentials of the networks through which knowledge is communicated. For example, professional groupings such as professional institutes and trade associations may be seen as more 'impartial' communicators of the 'state of the art' than organizations with more naked commercial interests. As we will see, this assumption is not always justified—professional networks may well be colonized by actors with commercial interests—but it may help to explain the importance of professionalized networks in communicating knowledge.

The usefulness of inter-organizational networks in legitimizing and diffusing knowledge is not without consequences for innovating firms. Such networks also exercise a shaping effect upon the range of technological options available to such firms. For example, DiMaggio & Powell (1983) argue that the greater the interconnections among firms within a particular community, the greater the tendency for 'isomorphism' to occur—for example, through mimetic, normative or coercive processes—that leads firms to resemble one another. Hence networks for diffusion may paradoxically allow new ideas to diffuse more widely but, at the same time, place tighter constraints around the particular ideas that will be considered legitimate within the community. Again, social agency is important here—by engaging in diffusion networks, particular constituents (or agents) can influence further processes of invention and implementation. For example, Rogers (1983) describes how the Dvorak design keyboard—a more efficient alternative to the existing Qwerty—was effectively precluded by core con-

stituents (e.g. manufacturers and teachers) with vested interests in maintaining the status quo.

Implementation

This episode refers to the local appropriation of new ideas as organizationally-specific solutions. Often innovations cannot be adopted by organizations as 'off the shelf' packages. Rather, they represent multifaceted 'bundles' of knowledge that require modification and reconfiguration to adapt them to specific technical and organizational contexts (Clark, 2000). In implementation episodes, the deconstruction and re-construction of knowledge comes to the fore. Generic, objectified ideas floated through global, inter-organizational networks need to be appropriated by blending with ideas about specific organizational problems and context (Clark, 1987). The customization, during implementation, of standardized software packages diffused by technology suppliers through global inter-organizational networks so that they meet local user requirements, is a good example of the knowledge appropriation involved in implementation (e.g. Robertson et al., 1996).

Here networking is more purposeful, as those with interests in implementing specific solutions use networking to mobilize the information and resources (including political and social resources) that will be required. With many kinds of implementation, local intra-organizational networking again becomes important as new generic ideas are interpreted and blended with existing local, often tacit, knowledge and as project owners attempt to generate the commitment needed from the relevant social groups (Fulk, 1993). Thus inter- and intra-organizational networks may converge during implementation—weak inter-organizational ties for information search combine with strong intra-organizational ties required for the formation of project teams.

Figure 1 depicts these episodes of invention, diffusion and implementation as unfolding recursively over time through the medium of networking. Local, intra-organizational networking clusters around the intensive episodes of invention and implementation, where tightly integrated knowledge and information flows are required. New relationships are forged on the basis of reciprocity and the development of trust—the kind of 'barter economy' described by Kreiner & Schultz (1993). Then, in diffusion, networking is extensively rather than intensively oriented and emerges around the dissemination of more explicitly objectified knowledge flows. Relationships are more structured than emergent, and are more explicitly transactional or market-based (Scarbrough, 1995). Issues about the value of what is being diffused are resolved through more institutionalized forms of legitimation and validation.

The remainder of this chapter operationalizes and further develops this model using case examples of invention, diffusion and implementation. These brief

Table 1. A summary of networks, networking and knowledge transformation during episodes of the innovation process.

Innovation Episode	Dominant networks	Features of the networking process	Role of networks	Knowledge transformation
Invention: *Universal Consulting Case Example*	Local intra-organizational —strong ties within consultants' firm Local inter-organizational —strong ties between consultants and clients Mainly interpersonal networking	Wayward and emergent: formation of many strong ties and some weak ties	Coalition building around new ideas and clients projects	Knowledge created, socially constructed from loosely structured, ambiguous and novel ideas into new products/services
Diffusion: *Professional Association Case Example*	Global inter-organizational —many weak ties across firms via professional association —relatively strong ties between MRP2 consultants and professional association Mainly informational networking plus interpersonal for an active minority	Opportunistic and conflictual: primarily through many weak/ indirect ties and few strong ties	Brokerage, broadcasting and legitimation of new technology as 'best' practice Selective promotion of knowledge (e.g. in the form of technology design templates)	Knowledge objectified, commodified and communicated using material artefacts and 'best practice' methodologies
Implementation: *LiveCo Case Example*	Local intra-organizational —strong ties among project team Local inter-organizational —formal relationship between team and IT consultant Global inter-organizational —weak ties with other firms, professional association, IT vendors. Intra/inter-organizational boundaries blurred Mainly interpersonal networking plus formal, client/ consultant relationship	Purposeful, intentional: primarily using weak ties for information search coupled with strong ties in project team	Exchange of required information and resources (including political and social)	Knowledge appropriated (unpacked and reconfigured) into workable applications in specific context by combining objectified knowledge with local (often tacit) knowledge

case 'vignettes' are taken from our earlier empirical studies of innovation and networking. The first focuses on the role of local, intra-organizational networking within a scientific consultancy generating inventions for clients. The second case examines the ways in which global inter-organizational networking made possible by a professional association can serve as a vehicle for the diffusion of innovation in the area of operations management. The third case examines how the knowledge diffused via global networks is appropriated *in situ* with that constructed via local networks during the implementation of operations management technology in a particular manufacturing firm. The links between networks, knowledge and innovation episodes identified in these cases are summarized in Table 1. The comparison between them also allows

further analyzes of the relative value of different forms of networking in different episodes of the innovation process.

Case Example 1: Universal Consultancy—The Role of Networks in Invention

This case example is based on longitudinal research at the 'Universal Consultancy' organization (see Robertson & Swan, 1998 for details). Universal was founded in 1986 by a charismatic, highly successful individual and, over time, had grown from a handful of consultants, to an organization employing 150 consultants at its main headquarters and a further 110 on an associate basis in the U.S., Japan and Europe. The organization is a laboratory-based, business and technology, consulting and investing company specializing in the invention of novel scientific products and services, which are sold to clients in the form of Intellectual Property Rights (IPR).

All consultants in Universal were educated to Ph.D. level within their respective scientific subjects and many were considered 'world' experts. The emphasis was on inventing new products and services by consultants combining their different areas of scientific knowledge with a keen commercial awareness (i.e. characteristic of the 'symbolic analyst'—Reich, 1991). The overall ethos of the firm (vehemently defended by the founder) was that sustainable competitive advantage was to be achieved through innovation and that this depended on the effectiveness of the skill base. Universal's competitive advantage was based, then, on its ability to *rapidly* develop inventions in line with client requirements. This capability was maintained and supported by the method of working that relied heavily on the local, intra-organizational networking activities of consultants.

Consultants were allocated across seven divisions reflecting broad specialisms (e.g. applied science, IT, engineering) but these existed purely for administrative purposes. In general, formal structures were quite deliberately opposed—the overriding emphasis being, instead, on maintaining a non-hierarchical (with the exception of the founder), egalitarian approach to organizing. Divisions were created, merged and disbanded over time in a reactive manner, premised on market opportunities and consultants were reallocated accordingly. Consultants would come together in self-forming teams for the duration of projects, regrouping as new projects and personal interests demanded. The approach to organizing was typical of the 'adhocracy'—an organizational form considered to stimulate innovation (Mintzberg, 1983)—and work organization was described as 'fluid'. Everyone, including the Founder, was actively involved in project working. Project working was to be largely unconstrained by organizational (hierarchical or divisional) boundaries. The 'modus operandum' was intensive, local, intra-organizational networking, mobilized around the inventions being worked on at any point in time. Interdisciplinary working across divisions was encouraged and valued—indeed many consultants had chosen to work at Universal because of its emphasis on cross-disciplinary working and informal work practices.

As the organization grew it became necessary to consider mechanisms that would sustain networking across divisions. One mechanism introduced was based on individual performance targets set within a loose financial control system. Personal revenue targets (PRTs) were established by the Board yearly and performance was monitored monthly. Given the emphasis on egalitarianism, the same monthly PRT applied to all consultants. This meant that Divisional targets were directly related to the number of consultants within each division. Personal (and therefore

Divisional) revenue was generated by consultants getting themselves involved in a project and receiving a share of the project revenue (decided through negotiation with the project leader). To get involved in new projects—and so meet Divisional targets—consultants were expected, and encouraged, to actively engage in intra-organizational networking. This intra-organizational networking allowed them to market their own expertise so it could be spotted and exploited. Although this local networking was mainly intra-organizational, it also extended to the client base. Over time then, as the organization grew, local inter-organizational networking expanded as consultants created and built their own client base. Market opportunities were discovered through a dual process of trust-building and exploring possibilities for invention with clients and consultants in other divisions, in a manner reminiscent of the university-industry R&D networks described by Kriener & Schulz (1993).

Project leaders and project team members were not formally allocated at Universal. Rather, they emerged during the proposal stage of a project. Typically the project leader was the consultant who was seen to have generated the market opportunity via his or her local networking. At the project proposal stage, the potential gain of combining knowledge with others was high—'know-how' trading (von Hippel, 1988) and the promiscuous sharing of ideas within the intra-organizational network was common (Kreiner & Schulz, 1993). Project leaders were also expected to be fair in their assessment of individual consultants' likely contribution. This helped to generate trust-based relations and perceptions of equity and so sustained and strengthened the intra-organizational network (Ring & Van de Ven, 1994).

In short, a micro-economy for knowledge existed at Universal—the PRT system demanded that individuals' knowledge be made 'visible' and traded within the organization. This micro-economy facilitated informal and interpersonal intra-organizational networking necessary to sustain invention. This networking, in turn, encouraged the social construction and creation of knowledge in projects and also enhanced the individual consultants' own knowledge base. This expanding knowledge base then further advanced the internal 'marketability' of the individual consultant. Given the absence of structurally-defined roles and responsibilities, there was an overriding emphasis on social relations—these being governed by a psychological, rather than a legal or employment-based, contract. These are defining characteristics of co-operative networking relationships (Ring & Van de Ven, 1994).

When project work began it was characterized by the exchange of tacit and explicit knowledge (Nonaka & Takeuchi, 1995) among consultants. Some client input was necessary, particularly at the early stages, but because of the consultants' high levels of scientific expertise, knowledge creation relied more heavily on intra-organizational networking activity. This intra-organizational networking was based mostly on interpersonal contact, for example, through face-to-face informal 'brainstorming' sessions. Many consultants also spent a considerable amount of time travelling so e-mail was also used intensively. However, the limitations of electronic communication, and the advantages of face-to-face interaction for creating knowledge were recognized (and referred to, often) by consultants. The successful conclusion of project work occurred when the knowledge that had been created through this intra-organizational networking activity became objectified and communicated to the client, for example in the form of a new IPR.

The Universal case highlights the ways in which the dynamics of local and intra-organizational networking, driven by an internal 'market for knowledge', could promote coalitions around new ideas and projects, and collaborative efforts to innovate, even amongst a group of workers that were (as in this case) highly individualistic. This intra-organizational networking was mostly informal and interpersonal and emerged around the inventions themselves. It was supported by the social context, in particular a strong egalitarian culture and an efficient information (e-mail) system. These kind of intensive informal interpersonal networking activities generated social mechanisms (e.g. the formation of trust, establishing reputation, negotiation of responsibilities, collaborative working and so forth) that were important for further invention. Given the nature of the work (and workers) involved, these mechanisms would have been much more difficult, if not impossible, to achieve through more formal means.

Case Example 2: A Professional Association—The Role of Networks in Diffusion

This section examines the relationship between the networking engendered through a particular professional association, and its role in the diffusion of technologies for operations management (for details of the empirical work see, Newell & Clark, 1993; Swan & Newell, 1995; Swan et al., 1999a). The association is the Institute of Operations Management in the U.K. The IOM, like its American Counterpart—the American Production and Inventory Control Society (APICS)—comprised members from different occupational sectors (e.g. manufacturing, consultancy, software and hardware suppliers, academics) and market sectors (e.g. pharmaceuticals, automotive, food and drink). This association therefore played a 'brokerage' role, operating as an extensive (or 'global') network that generated a large number of 'weak ties' across individuals from different organizations (Aldrich & von Glinow, 1992). Weak ties have been identified as important for the diffusion of innovation because they allow ideas to spread across social communities and allow firms to encounter ideas that

are not bounded by the usual norms of their particular sector (Granovetter, 1973; Rogers, 1983).

The primary aims of the IOM were: first, to keep members who work in industry up to date with new developments in operations management; and second, to enhance the professional profile of careers in operations management (these have traditionally afforded relatively low status in the U.K.). To achieve these aims, the IOM organized a formalized program of events (e.g. journals, conferences, seminars, and company presentations) aimed at broadcasting information about new technological solutions to members. It also organized educational qualifications in an attempt to provide a clearer career path for those working in operations management.

Members who actually attended formal meetings also had opportunities to meet informally and discuss new ideas. The professional association thus created an arena in which knowledge and ideas relating to innovation could be exchanged both through formal and informal networking. However, although opportunities for informal, interpersonal networking existed, only around 20% of members actively exploited these by attending events on a regular basis. The passive majority just read, or scanned, the information that was transmitted through the association's journals—their networking was informational (contact with information) rather than interpersonal (contact with other members). The role of the IOM, then, was largely one of broadcasting—it acted as a diffusion network whereby knowledge that had already been created and articulated in explicit forms by the more active members (e.g. in the form of written articles) was broadcast to the relatively passive community.

Because professional associations rely heavily on volunteers, the shape of their activities depends on the interests of those particular social groups who get involved. For example, the IOM depended heavily on volunteers to organize events, write articles for journals, teach on courses and so forth. Technology suppliers (software vendors and consultants) were particularly active in the IOM. They got involved because they saw the professional association as an important global network for marketing their particular technologies. Thus, although they comprised a minority of members (23% as compared to 70% who work as practitioners in manufacturing), technology suppliers were extremely active in articulating the information that the IOM then disseminated to members in industry. For example, the bulk of the information disseminated by two of its key activities—journals and conferences—was written or presented by the much smaller group of technology suppliers (Newell et al., 1997).

One of the most widely known technological developments for operations management is known as 'Manufacturing Resources Planning' (MRP2). The concept of MRP2 diffused widely during the mid-

1980s to mid-1990s, mainly from the U.S. to Europe (Clark, 1987). The early diffusion process was driven by aggressive marketing among suppliers of software and hardware (notably IBM) and also by consultants selling training and education to accompany the introduction of the MRP2 'philosophy' (notably the Oliver Wight Consultancy in the U.S.—Wilson et al., 1994). Market arenas, bolstered by the many problems that firms had experienced in implementing technologies to date, developed and provided incentives and opportunities for the commodification of knowledge associated with MRP2. Technology suppliers played a key role in this—effectively objectifying knowledge about MRP2 in the form of material artefacts (software packages) and tightly prescribed methodologies that could, it was claimed, be used in anywhere. For example, Oliver Wight developed a step-by-step 'Proven Path' to successful MRP2 implementation (Wight, 1984). MRP2 was promoted by many as *the* definitive 'best practice' for operations management, even though alternative technologies were known at that time (Swan & Newell, 1995). The diffusion process, then, was driven by the commodification and objectification of knowledge. Whilst this was a useful marketing strategy for technology suppliers, it caused problems for users because the objectified MRP2 solutions presented simplified the technology and de-emphasized the need for organizational appropriation (as discussed below—Clark & Staunton, 1989).

In the U.S. and U.K. the professional associations played a critical role in the diffusion of this objectified knowledge about MRP2 in two key respects. First, they acted as global inter-organizational networks for the broadcasting of information about MRP2. As seen, technology suppliers play an active role in shaping the information that these networks disseminate. In this case, a few major suppliers of MRP2 systems (e.g. IBM) took the opportunity to form strong ties with the professional association organizers (e.g. APICS). These resulted in the APICS network being enlisted to help in an 'MRP2 Campaign' to disseminate knowledge about the new MRP2 technology to its members (Vollman & Berry, 1985). Further, a global networking arrangement between the IOM and APICS was significant in providing a channel through which best practice ideas originating in the USA could be packaged and diffused to a practitioner community in the U.K. (Clark & Newell, 1993).

Second, the professional association networks played an important legitimizing role—MRP2 became accepted by firms in industry as *the* latest 'best practice' in part because communication about it was being broadcast via the professional associations. Whilst information encountered through professional association networks reaches only a subset of the relevant community, it is afforded a very high level of legitimacy and validity. Where close interpersonal trust among members of a network is not present or difficult

to develop, then problems surrounding competing claims to knowledge are solved by trusting your source (Meyerson et al., 1996). If the source is a professional association then credibility is likely to be greater than if the source is a more direct link with a technology supplier. Thus, ideas diffused via the professional association network are likely to be seen by potential adopters as impartial, even though (as seen above) they may have originated in the supply side.

This case illustrates how professional networks in the U.K. and U.S. played both a broadcasting and legitimating role in the promotion of objectified knowledge about MRP2 as a new 'best practice' technology design. This role is particularly salient because of the extensive weak ties generated by such network structures. However, because much of the networking activity of members was to do with transmitting information, rather than with developing interpersonal relationships, and because much of this information was shaped by an active supply side with interests in selling technologies, only positive features of the technology that would encourage it to be adopted were communicated. In contrast, the difficulties and complexities associated with the technology and its implementation were heavily downplayed. Thus whilst this inter-organizational networking encouraged rapid diffusion, it also generated potential problems for implementation. These are illustrated in the case that follows.

Case Example 3: A Manufacturing Firm—The Role of Networks in Implementation

This section presents the case of a manufacturing firm, which successfully implemented a new MRP2 system (for details of the empirical work see Robertson et al., 1996). It illustrates how the coalition of global and local networks played a key role in the implementation process.

The case study firm is referred to herein as LiveCo. LiveCo is a large vehicle manufacturer in the U.K. operating to a make-to-order profile. The implementation process in LiveCo began in the late 1980s when a decision was taken by members of the Board to invest in and implement an MRP2 system. At this time the company was facing a financial crisis—sales were declining in all markets due, it was claimed, to LiveCo's outdated product range. The firm decided to consolidate its manufacturing operations from 14 geographically distributed different sites into one. This major organizational change clearly had a profound destabilizing effect on local networking activity. This to some extent made it easier to develop new local networks specifically in relation to the MRP2 project.

The philosophy behind MRP2 technology is to integrate information used for different aspects of the manufacturing process with wider capacity planning and sales forecasts, so that materials are available when needed without holding unnecessarily high levels of inventory. Due to this demand for integration, the implementation of these technologies often requires considerable change in both organization and technical practice (Clark & Staunton, 1989). The implementation of MRP2 technologies, then, depends heavily on the context into which they are introduced and require a blending of both technical and organizational knowledge. The notion of a single, generic, 'best' practice with regards to MRP2, promoted (as seen) through diffusion networks, is actually quite misleading when it comes to implementation. Rather, MRP2 technologies need to be (re)configured according to the unique context in which they operate (Fleck, 1994; Swan et al., 1999b). Some researchers refer to this as a process of knowledge *appropriation* (e.g. Clark & Staunton, 1989). The need for organizational integration, in particular, has posed many problems for user firms attempting to implement MRP2 technologies, with examples of failure or partial failure littering the research on implementation (Waterlow & Monniot, 1986; Wilson et al., 1994). LiveCo was perhaps unusual in managing to implement MRP2 technology successfully, achieving both high levels of integration and appropriation of the technology within a relatively short timescale.

LiveCo was structured along traditional hierarchical and functional lines. Because of this, intra-organizational networking might have been expected to be difficult. However, because of the uncertainty surrounding reorganization, formal routines were introduced that demanded that senior managers from all functions would have an input into all major policy decisions, including those concerning new technology. In line with this policy, LiveCo developed a formal cross-functional senior project team to handle implementation (comprising operations management, manufacturing engineering, manufacturing systems, sales and marketing, and logistics). This team met regularly and over an extended period of time, mostly on a face-to-face basis. A crucial feature of implementation in this case, then, was the development of local intra-organizational network, comprising powerful individuals who were engaged in interpersonal networking that transcended functional boundaries. Because individuals in this team were more or less equal in terms of their formal status, regular negotiation took place over project commitments, directions, roles and responsibilities. Thus, the formation of this network comprised many of the social processes highlighted by Ring & van de Ven (1994) as important for inter-organizational networking. This local networking also extended beyond the LiveCo organization to include information systems support from a specialist IT consultancy. This had a long history of working with LiveCo and provided, among other things, a consultant to be a permanent member of the project team. Thus inter and intra-organizational boundaries became blurred during implementation.

Project team members developed an awareness of MRP2 'best practice' via their involvement in a number of inter-organizational networks. For example, members of the project team had heard about MRP2 from reading trade and professional association journals and from software vendors and their publicity materials. As the logistics manager commented '*it was difficult to read or speak to anyone back then without MRP2 being mentioned as the answer to all our problems*'. This manager was also an active member of the IOM and had attended IOM courses and seminars where MRP2 had been advocated. The IT consultant in the team also advocated the use of MRP2 and arranged for the Oliver Wight consultancy to present the MRP2 concept to the board. Thus inter-organizational diffusion networks played a significant role in alerting project managers to generic notions of MRP2 'best practice'.

However, armed with a good understanding of local manufacturing operations developed through their local networks, the project team rejected supplier prescriptions regarding MRP2 'best practice' implementation. They were aware that these prescriptions did not sit comfortably with their particular manufacturing profile. Instead, team members explicitly set out to use their own informal interpersonal contacts in other manufacturing firms (e.g. friends and ex-colleagues) to arrange factory visits to other 'like' companies and to see for themselves what they were doing. These site visits allowed the team to develop an understanding of a broader range of technological design templates for operations management than MRP2. During implementation then, some of these ideas were blended with the MRP2 'best practice' template, with the result being that MRP2 was implemented in a limited capacity for high level planning, whilst detailed shopfloor planning and control was achieved with a combination of in-house software and a Just-in-Time 'Kanban' system. Initial education and training for a broader group of senior managers was then provided by a consultancy specializing in MRP2. However, because project team members were aware that this only offered, as they put it a '*single-point*' solution, they also developed in-house training to show users how knowledge about MRP2 concepts was to be appropriated within the particular operational context of LiveCo.

This case illustrates how knowledge cultivated via intra- and inter-organizational networks may be blended during implementation. A major factor in LiveCo's ability to successfully develop and implement their MRP2 system was that they recognized the limitations of the tightly prescribed and commodified knowledge regarding MRP2. They were thus able to unpack the knowledge gained via inter-organizational diffusion networks and blend this with that gained via local networking activity to implement a system that was appropriate for their specific context. The fact that implementation team members were from different functions extended local networking activity to cover a broader range of expertise.

Conclusion

If the three cases (summarized in Table 1) are compared in terms of the model outlined in Fig. 1, an insight can be gained into the differing roles played by networks and networking within the innovation process. Thus, the shallow curve that we see in the typical project in Universal reflects a concern with the social construction and creation of knowledge to produce customized client solutions. Although local inter-organizational networks with clients were important for identifying market opportunities, intra-organizational networks among consultants, mobilized by the development of informal, interpersonal, and trust-based relationships, were more important for knowledge creation. This has certain advantages for the invention process. Universal is able to mobilize an extensive knowledge community in which knowledge sharing is relatively open and more or less unaffected by the problems of opportunism seen in market relations. This certainly enhances the speed and responsiveness of the invention process.

Universal Consultancy represents an organization that has managed to create a powerful local networking environment that encourages the sharing of knowledge and information. The ability of organizations to promote such an environment should not be overestimated. Despite Williamson's (1985) claim that hierarchies are generally more conducive to the sharing of knowledge than markets, it is clear that many organizations find it difficult to promote open sharing of knowledge amongst their employees. This is partly due to internal organizational boundaries (e.g. functional departments) but it also reflects the instrumental attitudes that may be fostered by the employment relationship. Put simply, even within a hierarchy, groups and individuals may view knowledge as private property to be hoarded and only grudgingly or calculatedly shared. In this case, the financial performance system developed by the consultancy, as well as the prevailing egalitarian culture, plays a powerful role in fostering the kind of organic, loosely structured sharing of knowledge that is critical in the generation of invention. At the same time, the effective use of email reinforces this by making it possible for knowledge sharing to transcend face-to-face contact.

In contrast, the steeper curve of MRP2 diffusion reflects a much greater concern with broadcasting and legitimation of knowledge and information occurring through the objectification of knowledge. Whereas the intra-organizational networking at Universal permitted a rich, highly interactive and collaborative process of knowledge creation that then generated further innovation, the inter-organizational networking that took place within the structure of the professional associa-

tion network had a more constraining effect on innovation. Certainly, both the objectification of knowledge, as well as the legitimation of selected technologies as 'best', encouraged more rapid diffusion. However, it is also clear that the validating role of the professional association network structure was, to an extent, subverted by the networking activity of technology suppliers. The latter created a pro-innovation bias and biased communication processes in favour of a particular innovation—MRP2—which actually proved much harder to implement in many organizational contexts than it appeared (Clark & Staunton, 1989). During implementation, for example in LiveCo, local inter- and intra-organizational networks needed, then, to be mobilized in order to unpack and reconfigure generic solutions into locally applicable ones.

The model in Fig. 1 allows the ways in which knowledge is organized during the innovation process to be compared across different episodes. This gives it some analytical value that goes beyond purely descriptive accounts of process. The model also attempts—perhaps ambitiously—to weave together innovation, networks and processes of knowledge transformation, and so adding to an understanding of process. However, the model—indeed any model—is schematic. The real complexity and inherent 'messiness' of networking activity during innovation is also captured within the case examples. Finally, comparing these cases has some important implications for further work in this area. In particular, the model that is outlined here is a stylized representation of particular episodes of innovation. It does not address certain important issues, notably the interaction between episodes. If innovation episodes are indeed iterative and recursive as claimed (Wolfe, 1994), then it is also important to understand the nature of these iterations and their relation to networking.

Further research into these areas is certainly called for, especially insofar as it develops a network perspective on the innovation process. In addition, it is also worth commenting on the roles of consultancy firms in the transformation of knowledge and innovation that emerges from these cases. It seems clear, then, from contrasting the case of Universal with the case of the diffusion process, that consultants may play a wide variety of roles, and that their networking activities can act both to expand and constrain innovation. When considering this final point, a major theme of this chapter is highlighted, that is, the need to recognize the diverse roles played by networks and networking alike in the processes of innovation systems.

Acknowledgments

The authors would like to acknowledge Professor Sue Newell, Bentley College, Boston, for her contribution to the empirical work and the Economic and Social

Research Council for supporting the research that informs this analysis.

References

Abrahamson, E. (1996). Management fashion. *Academy of Management Review*, **21**, 254–285.

Ahuja, G. (2000). Collaboration networks, structural holes and innovation: A longitudinal study, *Administrative Science Quarterly*, **45** (3), 425–455.

Aldrich, H. (1999). *Organizations evolving*. Thousand Oaks, CA: Sage.

Aldrich, H. & von Glinow, M. A. (1992). Personal networks and infrastructure development. In: D. V. Gibson, G. Kozmetsky & R. W. Smilor (Eds), *The Technopolis Phenomenon: Smart Cities, Fast Systems, Global Networks*. New York: Rowman and Littlefield.

Alter, C. & Hage, J.(1993). *Organizations working together*. Newbury, PA: Sage.

Bijker, W. E., Hughes, T. & Pinch, T. J. (Eds) (1987). *The Social Construction of Technological Systems*. London: MIT Press.

Callon, M. (1980). The state and technical innovation: A case study of the electrical vehicle in France, *Research Policy*, **9**, 358–376.

Casson, M. & Cox, H. (1997). An economic model of inter-firm networks. In: M. Ebers (Ed.), *The Formation of Inter-Organizational Networks* (pp. 174–196). Oxford: Oxford University Press.

Clark, P. (2000). *Organizations in action*. London: Sage.

Clark, P. A. (1987). *Anglo-American innovation*. New York: De Gruyter.

Clark, P. & Newell, S. (1993). Societal embedding of production and inventory control systems: American and Japanese influences on adaptive implementation in Britain. *International Journal of Human Factors in Manufacturing*, **3**, 69–80.

Clark, P. & Staunton, N. (1989). *Innovation in technology and organization*. London: Routledge.

Cyert, R. M. & March, J. G. (1963). *A behavioral theory of the firm*. Englewood Cliffs, N.J.: Prentice Hall.

DiMaggio, P. J. & Powell, W. W. (1983). The iron cage revisited: Institutional isomorphism and collective rationality in organizational fields. *American Sociological Review*, **48**, 147–160.

Ebers, M. & Jarillo, J. C. (1997). The construction, forms, and consequences of industry networks. *International Studies of Management & Organization*, **27**, 3–21.

Ebers, M. (Ed.) (1997). *The formation of inter-organizational networks*. Oxford: Oxford University Press.

Ettlie, J. (1980). Adequacy of stage models for decisions on adoption of innovation. *Psychological Reports*, **46**, 991–995.

Fleck, J. (1994). Learning by trying: the implementation of configurational technology. *Research Policy*, **23**, 637–652.

Freeman, C. (1991). Networks of innovators: A synthesis of research issues. *Research Policy*, **20(5)**, 499–514.

Fulk, J. (1993). Social construction of communication technology. *Academy of Management Journal*, **36** (5), 921–950.

Gibson, D. & Conceicao, P. (2003). Incubating and networking technology commercialization centres among emerging, developing and mature technopolies worldwide. In: L. V. Shavinina (Ed.), *International Handbook on Innovation*. Oxford: Elsevier Science.

Grandori, A. & Soda, G. (1995). Inter-firm networks: Antecedents, mechanisms and forms. *Organization Studies*, **16**, 184–214.

Granovetter, M. S. (1973). The strength of weak ties. *American Journal of Sociology*, **78**, 1360–1380.

Hansen, M. T. (1999). The search transfer problem: The role of weak ties in sharing knowledge across organizational sub-units. *Administrative Science Quarterly*, **44**, 82–111.

Holland, C. & Lockett, G. (1997). Mixed mode operation of electronic markets and hierarchies. In: M. Ebers (Ed.), *The Formation of Inter-Organizational Networks* (pp. 238–262). Oxford: Oxford University Press.

Hosking, D. & Morley, I. (1992). *A social psychology of organizing*. London: Harvester Wheatsheaf.

Jarillo, J. (1993). *Strategic networks: Creating the borderless organization*. Oxford: Butterworth Heineman.

King, J. L., Gurbaxani, V., McFarlan, F. W., Raman, K. S. & Yap, C. S. (1994). Institutional factors in Information Technology innovation. *Information Systems Research*, **5** (2), 136–169.

Knights, D., Murray, F. & Willmott, H. (1993). Networking as knowledge work: A study of inter-organizational development in the financial services sector. *Journal of Management Studies*, **30**, 975–996.

Kreiner, K. & Schultz, M. (1993). Informal collaboration in R&D. The formation of networks across organizations. *Organization Studies*, **14** (2), 189–209.

Lazonick, W. (1991). *Business organization and the myth of the market economy*. Cambridge: Cambridge University Press.

Macdonald, S. & Williams, C. (1992). The informal information network in an age of advanced telecommunications. *Human Systems Management*, **11**, 177–188.

Meyerson, D., Weick, K. & Kramer, R. M. (1996). Swift trust and temporary groups. In: R. M. Kramer & T. R. Tyler (Eds), *Trust in Organizations: Frontiers of Theory and Research*. New York: Sage.

Major, E. & Cordey-Hayes, M. (2002). Encouraging innovation in small firms through externally generated knowledge. In: L. V. Shavinina (Ed.), *International Handbook on Innovation*. Oxford: Elsevier Science.

Mintzberg, H. (1983). *Structures in fives, Designing Effective Organizations*. Englewood Cliffs: Prentice-Hall.

Nahapiet, J. & Ghoshal, S. (1998). Social capital, intellectual capital and the organizational advantage. *Academy of Management Review*, **23** (2), 242–266.

Newell, S. & Clark, P. (1990). The importance of extra-organizational networks in the diffusion and appropriation of new technologies. *Knowledge: Creation, Diffusion, Utilisation*, **12**, 199–212.

Newell, S., Swan, J. A. & Robertson, M. (1997). Inter-organizational networks and diffusion of information technology: Developing a framework. In: T. J. Larsen & G. McGuire (Eds), *Information Systems and Technology Innovation and Diffusion*. Idea Publishing Group

Nonaka, I. & Takeuchi, H. (1995). *The knowledge creating company*. New York: Oxford University Press.

Nonaka, I., Sasaki, K. & Ahmed, M. (2003). Continuous innovation: The power of tacit knowledge. In: L. V. Shavinina (Ed.), *International Handbook on Innovation*. Oxford: Elsevier Science.

Oliver, A. & Ebers, M. (1998). Networking network studies: An analysis of conceptual configurations in the study of inter-organizational relationships. *Organization Studies*, **19**, 549–583.

Oliver, A. L. & Liebeskind, J. P. (1998). Three levels of networking for sourcing intellectual capital in biotechnology. *International Studies of Management and Organization*, **27** (4), 76–103.

Pettigrew, A. M. & Fenton, E. (Eds) (2000). *The innovating organization*. London: Sage.

Poire, M. & Sabel, C. F. (1984). *The second industrial divide: Possibilities for prosperity*. New York: Basic Books.

Powell, W., Koput, K. & Smith-Doerr, L. (1996). Inter-organizational collaboration and the locus of innovation: Networks of learning in biotechnology. *Administrative Science Quarterly*, **41** (1), 116–139.

Reich, R. (1991). *The work of nations: Preparing ourselves for 21st century capitalism*. London: Simon and Schuster.

Ring, P. S. & Van De Ven, A. H. (1994). Developmental processes of cooperative inter-organizational relationships. *Academy of Management Review*, **19** (1), 90–118.

Robertson, M. & Swan, J. (1998). Modes of organizing in and expert consultancy: Power, knowledge and egos. *Organization*, **5**, 543–564.

Robertson, M., Swan, J. & Newell, S. (1996). The role of networks in the diffusion of technological innovation. *Journal of Management Studies*, **33**, 333–359.

Rogers, E. M. (1983). *Diffusion of Innovations* (3rd ed.; 1995, 4th ed.). New York: Free Press.

Scarbrough, H. (1995). Blackboxes, hostages and prisoners. *Organization Studies*, **16**, 991–1020.

Swan, J. & Newell, S. (1995). The role of professional associations in technology diffusion. *Organization Studies*, **61** (5), 847–874.

Swan, J., Newell, S. & Robertson, M. (1999a). The diffusion and design of technologies for operations management: a comparison of central diffusion agencies in the U.K. and Sweden. *Organization Studies*, **20** (6), 905–932.

Swan, J. A., Newell, S. & Robertson, M. (1999b). The illusion of best practice in information systems for operations management. *European Journal of Information Systems*, **8**, 284–293.

Tidd, J. (1997). Complexity, Networks and Learning: Integrative themes for research on innovation management. *International Journal of Innovation Management*, **1** (1), 1–21.

Tushman, M. & Scanlan, T. (1981). Boundary spanning individuals: Their role in information transfer and their antecedents. *Academy of Management Journal*, **24**, 289–305.

Van de Ven, A. H. (1986). Central problems in the management of innovation. *Management Science*, **32**, 590–607.

Vollman, A. & Berry, W. (1985). *Manufacturing control*. Illinois: Dow Jones-Irwin.

von Hippel, E. (1988). *The sources of innovation*. Oxford: Oxford University Press.

Waterlow, G. & Monniot, J. (1986). *A study of the state of the art in computer-aided production management*. Report for ACME Research Directorate, Science and Engineering Research Council, Swindon, U.K.

Wight, O. (1984). *Manufacturing resource planning: MRP2*. Vermont: Oliver Wight Publications.

Williamson, O. E. (1985). *The economic institutions of capitalism*. New York: Free Press.

Wilson, F., Desmond, J. & Roberts, H. (1994). Success and failure of MRP2 implementation. *British Journal of Management*, **5**, 221–240.

Wolfe, R. (1994). Organizational innovation: Review, critique and suggested research directions. *Journal of Management Studies*, **31**, 405–431.

The International Handbook on Innovation
Edited by Larisa V. Shavinina

Managing Innovation in Multitechnology Firms

Andrea Prencipe

Complex Product Systems Innovation Centre, SPRU, University of Sussex, U.K. and Faculty of Economics, University G. D'Annunzio, Pescara, Italy

Abstract: This chapter identifies two major dimensions of capabilities of firms developing multitechnology products: *synchronic systems integration* and *diachronic systems integration*. Within each of these two dimensions, multitechnology firms maintain *absorptive capabilities* to monitor and identify technological opportunity from external sources and *generative capabilities* to introduce innovations at the architectural and component levels. The chapter focuses on a firm's *generative* capabilities and illustrates that a firm's generative capabilities enables it to *frame* a particular problem, *enact* an innovative vision, and *solve* the problem by developing new manufacturing techniques. The triad *frame-enact-solve* is argued to be the primary feature of a firm's *generative capability.*

Keywords: Innovation; Multitechnology products; Systems integration capabilities; Generative capabilities.

Introduction

Early research on the management of technological innovation underlined that innovation is a complex multi-actor phenomenon (Rothwell et al., 1974). Innovation is understood as the processes thereby new ideas are commercialized (Freeman, 1972, 1984). Not only do successful industrial innovations require the involvement and co-ordination of all firm's business functions, from R&D, through engineering and manufacturing, to marketing, but also the involvement and co-ordination of external organizations to the firm, such as suppliers (Clark & Fujimoto, 1991; Freeman, 1991; Nonaka et al., 2003; Rothwell, 1992; Von Hippel, 1988). Organizing and managing the innovation process, therefore, span both intra- and inter-firm boundaries (Bessant, 2003; Gassmann & von Zedtwitz, 2003; Katz, 2003; Swan et al., 2003).

The literature on technology strategy has highlighted that several industries are increasingly characterized by multitechnology multicomponent products (Granstrand et al., 1997). Multitechnology multicomponent products have important managerial implications since they intensify the co-ordination efforts for firms developing them. The number of technologies and components is in fact too large to be managed within the firm's

organizational boundaries so that the co-ordination of external sources of components and technologies becomes paramount for the successful development of new products and processes. In other words, the multi-actor nature of the innovation process is exacerbated in firms that develop multitechnology multicomponent products because of the increasing number and relevance of external organizations, such as suppliers, customers, and universities.

Building upon the emerging literature on multitechnology corporations (Brusoni et al., 2001; Granstrand, 1998; Granstrand et al., 1997; Patel & Pavitt, 1997), this chapter identifies the different types of capabilities that firms developing multitechnology products are required to develop and maintain. It proposes a taxonomy that categorizes these capabilities into *synchronic systems integration* and *diachronic systems integration*. Within each category, firms monitor external technological developments (*absorptive capabilities*) and introduce innovative solutions at both the component and architectural levels (*generative capabilities*). The chapter focuses on a firm's *generative* capabilities and argues that *generative* capabilities enables firm to *frame* a particular problem, *enact* an innovative vision, and *solve* the problem. The triad *framing-enacting-solving* constitutes the primary feature of a firm's *generative capability*.

The chapter is organized as follows. Based on the theoretical and empirical literature on multitechnology firms and products, the next section introduces two dimensions of capabilities of firm's developing multitechnology products. This is followed by a section that focuses on the firm's *generative* capabilities and attempts to disentangle its primary feature. The final section presents the conclusions.

Multitechnology Firms and Multitechnology Products: The Multiple Roles of Firms' Capabilities

Empirical and theoretical studies on firms' capabilities, although paradoxically in their infancy (given that Penrose pioneered the resource-based approach in 1959), have provided invaluable insights to understand their nature (Dosi et al., 2000) and their role as source of a firm's competitive advantage (Grant, 1998). This section relies on the resource-based research tradition and its more recent evolution known as the capability-based approach, to single out and discuss the role of a firm's capabilities in multitechnology settings. In so doing, it extends the theoretical and empirical research on *multitechnology corporations* (Granstrand & Sjö-lander, 1990). Granstrand & Sjölander (1990) observed that technological diversification was an increasing and prevailing phenomenon among large firms in Europe, Japan and the U.S. and put forward the concept of the *multitechnology corporation*. A *multitechnology corporation* is a firm that has expanded its technology base into several technologies. Following this line of research, Patel & Pavitt (1997) showed that products, and firms developing them, are becoming increasingly multitechnology. Firms rely on a growing number of specialized bodies of scientific and technological knowledge to develop products.

The concept of a multitechnology corporation rests on the fundamental distinction between products and technologies. A product is a physical artefact made up of components that carry out specific functions and rely on specific yet different technologies. A technology is understood here as the body of knowledge underlying the design, development, and manufacture of the product (Brusoni et al., 2001). In this way, the concept of a multitechnology firm is different from that of multiproduct firm, since 'the development, production, and use of a product usually involve several technologies and each technology can usually be applied in several products. Thus the technology-product connection is not 'one-to-one' (Granstrand & Sjölander, 1990, p. 36). Also as discussed in Grant & Baden-Fuller (1995) and Pavitt (1998), the distinction between product and its underlying technologies is fundamental for theoretical interpretations of the firm and in particular for the definition of its boundaries.

The multitechnology nature of products has significant managerial implications for the firms producing them in terms of the technological capabilities that are required to be developed, maintained, and nurtured over time. In particular, 'make or buy' decisions are critical issues since firms do not and cannot develop in-house all the technologies relevant for product design and manufacturing. Multitechnology firms must increasingly make use of external sources of components and technological knowledge, such as suppliers, through the use of the market or through collaborative agreements, such as joint ventures.

In order to take full advantage of collaborative relations, firms need to be equipped with an adequate and independent set of in-house technological capabilities (Mowery, 1983). Granstrand et al. (1997) found that large firms develop capabilities over a wider number of technological fields than those in which they actually produce, and this number is increasing over time. In other words, firms retain technological capabilities about components whose production is fully outsourced. Specifically, Granstrand et al. (1997) drew some conclusions on outsourcing decisions in multitechnology firms. They distinguished two sets of factors that affect corporate outsourcing decisions: (a) the degree to which the innovation is autonomous or systemic; and (b) the number of independent suppliers outside the firm. On these grounds, they proposed a two-by-two matrix that identifies four cells. Each cell is associated with a different case calling for a particular degree of internal technological capabilities. Granstrand et al. identified four intermediate corporate positions between full integration and full disintegration, where each position is characterized by a different type of technological capabilities, namely exploratory research capability, applied research capability, systems integration capability, and full design capability. For instance, when the number of external sources is low and the innovation is systemic, then companies should maintain a wide range of in-house capabilities, from exploratory and applied research down to production engineering.

Besides the factors identified by Granstrand et al., the type of capabilities that multitechnology firms should develop may depend on the role and the ensuing importance that each component plays within a product. The importance of components within the economics of product, and therefore, of a firm varies greatly according to a number of dimensions, such as their technical features, performances, and costs. Firms conceive *components' hierarchies* in order to identify which are the peripheral and the key components and consequently adjust their technological capabilities (Prencipe, 2003b).

An interesting approach to analyzing the hierarchical role of components in a product has been put forward by Maïsseu (1995). Considering three variables, namely the impact of the component's cost on the cost of the overall system, its influence on the quality of the system, and the technological maturity of the component, he proposed a taxonomy, which identifies four categories. According to this approach, it is possible to determine the relative weight of each component. Thus, components with low impact in terms of quality and cost of the end product and whose underlying technologies are mature are to be considered to be *trivial*. Then, there are *basic components* whose cost is relatively high, while their technologies have reached the maturity stage. *Key components* are those whose characteristics heavily influence the quality of the end product and whose technologies are at the initial stage, but which do not affect the cost of the system to a great extent. Finally, there are the *critical components*. Their influence in terms of cost and quality is relatively high and their underlying technologies are at the initial stage.

It is worth stressing that this approach heavily underlines the issue that components may evolve across the hierarchy over time. Technological change occurring at different levels of the systems may shift the relative hierarchy of components and system-level critical problems. As hierarchies are usually constituted according to a series of 'rules' valid at a given point in time, they can provide predictions as long as the 'rules' remain unchanged. Therefore, a hierarchical taxonomy concerning products made up of many components may be undermined by changes in the underlying component technologies. Evolution may be endogenous in that changes can occur within the system itself, stemming from existing as well as exogenous technological trajectories, that is to say existing technologies can be replaced by new ones or new technologies can be added (Prencipe, 1997).

This may well inform firms' outsourcing decisions. For instance, Pavitt (1998) argued that a critical issue that companies take (or should take) into account when outsourcing components is the rate of change of the underlying technologies and the ensuing *technological imbalances* (Rosenberg, 1976). When technologies advance at different rates then companies should be able to keep pace with them and incorporate changes in their product and their *components' hierarchies*.

Identifying Capabilities in Multitechnology Firms

Drawing on and extending Granstrand et al. (1997), the aim of this section is to single out the diverse roles of firms' capabilities in multitechnology settings. Firms producing multitechnology products develop capabilities to generate new products and processes as well as to integrate externally produced components and co-ordinate the development of new technologies. In multitechnology settings, therefore, a firm's capabilities are not monolithic entities, and do not perform 'one role only', rather they are multifaceted and multipurpose. This is the reason why multitechnology

corporations are important and interesting empirical settings in which to study the roles of a firm's capabilities. Besides R&D, design, and manufacturing capabilities, therefore, we argue that firms producing multitechnology products must develop two main types of capabilities to compete successfully over time (Prencipe, 2003).

Synchronic systems integration refers to the capabilities to define the requirements, specify and source equipment, materials, and components, which can be designed and manufactured either internally or externally, and integrate them into the architectures of existing products. These capabilities are developed and maintained through a deliberate strategy labelled *intelligent customership* that enables firms to gain a better understanding of the underlying technologies of outsourced components in order to control and integrate changes and improvements (Prencipe, 1997). Therefore, *synchronic systems integration* relates to the capabilities to manage evolutionary changes in products and their underlying technologies through the introduction of component-level innovations.

Diachronic systems integration refers to the capabilities to co-ordinate the development of new and emerging bodies of technological knowledge across organizational boundaries in order to introduce new product architectures. Different bodies of technological knowledge relevant to the production of a multitechnology product may be characterized by uneven rates of advance. Firms that develop multitechnology products must keep pace with and, more importantly, co-ordinate uneven technological developments to incorporate them into new products and processes (Prencipe, 2004). Also, firms developing multitechnology products cannot encompass in-house, all the relevant scientific and technological fields. The management of the relationships with and co-ordination of external sources of technologies, such as universities, research laboratories, and suppliers, becomes, therefore, a central task for multitechnology firms (Lorenzoni & Lipparini, 1999). Therefore, *diachronic systems integration* relates to the capabilities required to master revolutionary changes in products and technologies.

Multitechnology firms are required to develop both *synchronic* and *diachronic systems integration capabilities* to pursue both incremental and discontinuous innovations and changes in order to compete successfully. *Synchronic* and *diachronic systems integration* may well constitute the capabilities of the *ambidextrous organization* as identified by Tushman & O'Reilly (1996). Ambidextrous organizations are those capable of competing in mature environments through incremental innovations and in new environments through discontinuous innovations.

Besides integrating and co-ordinating, multitechnology firms monitor external technological developments (*absorptive capabilities*) and introduce innovative solutions at both the component and architectural levels (*generative capabilities*). *Absorptive capabilities* are those required to monitor, identify, and evaluate new opportunities emerging from general advances in science and technology. This is close to the concept of absorptive capacity as put forward by Cohen & Levinthal (1990). *Generative capabilities* are the capabilities to innovate both at the component level and the architecture level (i.e. new paths of product configuration) also independently of external sources. While component-level innovations mostly relate to the synchronic dimension of systems integration, architectural-level innovations refer to the diachronic systems integration. Exploratory research programs play a fundamental role in the introduction of new component technologies as well as new product architectures. *Absorptive* and *generative capabilities* permeate both the synchronic and the diachronic dimension of systems integration (Prencipe, 2004).

The discussion above should be interpreted as a preliminary attempt to categorize the role of capabilities of firms developing multitechnology products. The intention is not to defend the boundaries of a specific category, particularly because there are other dimensions according to which firms' capabilities can be categorized (see, e.g. Granstrand & Sjölander, 1990; Granstrand et al., 1997). The use of the different categories is instead designed to draw attention to the multiple roles that capabilities have in the economics of the development of multitechnology products. The different roles of capabilities of multitechnology firms are discussed at length in previous works. For instance, Brusoni & Prencipe (2001) discussed the impact of modular design strategy on firm's capabilities operating in multitechnology settings. The synchronic and diachronic dimensions of systems integration and their relationships with a firm's corporate strategy are discussed in Prencipe (2003). Systems integration is scrutinized in relation to typologies of products (e.g. monotechnology versus multitechnology) and rate and stage of development of the underlying technologies in Brusoni et al. (2001). Dosi et al. (2003) provided an evolutionary economics interpretations of a firm's systems integration capabilities. Paoli & Prencipe (1999) discussed the role of a firm's knowledge base in multitechnology settings. Building on these previous works, the following section focuses on generative capabilities and attempts to identify its primary features.

Generative Capabilities: Primary Features

In the previous section, *generative capabilities* have been defined as the capabilities to introduce innovative technological solutions both at the component level and the architectural level also independently of external sources. To detail the primary features of *generative capabilities* we rely on the contributions of Dierickx & Cool (1989) on the cumulative nature of a

firm's capabilities and Dosi & Marengo (1993) on the role of a firm's *frame of reference*.

Dierickx & Cool (1989) put emphasis on the building process that affects the accumulation of a firm's capabilities. Although they talked about strategic assets, we argue that their argument holds also for a firm's capabilities. They argued that the common feature of a firm's capabilities is 'the cumulative result of adhering to a set of consistent policies over a period of time. Put differently, strategic asset *stocks* are *accumulated* by choosing appropriate time paths of *flows* over a period of time ... *while flows can be adjusted instantly, stocks cannot*. It takes a consistent pattern of resource flows to accumulate a desired change in strategic asset stocks' (1989, p. 1506, original emphasis).

Dierickx & Cool argued that the process of accumulation of *stocks* is characterized by the interplay of the following properties:

(a) *time compression diseconomies*, ('crash' R&D programmes are often characterized by low effectiveness);

(b) *asset mass efficiencies* ('success breeds success');

(c) *interconnectedness of assets stocks* (assets stock accumulation is influenced by the stock of other assets);

(d) *asset erosion* (all asset stocks decay and need to be maintained); and

(e) *causal ambiguity* (the process is not deterministic but it is characterized by stochastic elements).

This distinction between *stocks* and *flows* and the features of the building process of firms' capabilities as proposed by Dierickx & Cool (1989) underlines that a firm's capabilities must be painstakingly accumulated over time. Also, the distinction between *stocks* and *flows* underlines the relevance of the accumulated technological capabilities both as a basis for further development and the need for them to be continuously cultivated over time via dedicated investments in experimentation and personnel. Based on Dierickx & Cool, we argue that a firm's capabilities need to be built, cumulated, nurtured, and refined over time. Although the *stock-flow* dynamics captures the relevant features of each type of capability proposed in the taxonomy, as discussed below it is particularly useful to better understand the primary features of a firm's *generative* capabilities.

Dosi & Marengo (1993) argued that a firm's learning processes could not be reduced to mere information gathering and processing. Unlike Bayesan learning processes, where new information is employed to update the probability distribution within a fixed and unchanging *frame of reference*, in the learning processes Dosi & Marengo referred to, the *frame of reference* is continuously updated, constructed, evaluated, and eventually modified. Dosi & Marengo maintained that "There are fundamental elements of learning and innovation that concern much more the *representation* of the environment in which individuals operate and *problem solving* rather than simple information gathering and processing" (1993, p. 160, original emphasis).

Based on this line of reasoning, we argue that a firm's frame of reference, and more importantly its continuous renewal, constitute the distinctive base of its learning processes. A firm's frame of reference is based on its cumulated knowledge and its updating and modification are the result of continuous learning investments. What is fundamental is the *dynamic* that characterizes the frame of reference and its continuous renewal. This dynamic is well captured by the *stock-flow* binomial á *la* Dierickx & Cool. We propose that a *generative capability* hinges on a firm's frame of reference (i.e. its capability to frame and identify a problem and allocate resources to its solution) and problem solving capability (i.e. its capability to develop a solution to a problem). Also, we contend that the primary features of a firm's *generative capabilities* are *problem framing*, *vision enacting*, and *problem solving*. Firms build, update, and renew their frames of reference within which they enact an innovative vision to solve problems in turn identified by their frame of reference (their view of the world). The triad *framing-enacting-solving* is clearly inspired by the work of Weick (1969, 1985) on enactment and sense-making.

Framing, Enacting, and Solving in Context: An Example of a Firm's Generative Capabilities

The development of the first- and second-generation wide chord fan blade by Rolls-Royce Aero Engines discussed in Prencipe (2001) constitutes a good case to explain the deployment and enhancement of a firm's *generative capabilities*. The in-house technological capabilities accumulated over time by Rolls-Royce constituted the base of its learning processes and gave impetus to virtuous cycles of *framing, enacting*, and *solving*. Notwithstanding the failure of the first attempt (the all-composite wide chord fan blade), Rolls-Royce's conviction about the enormous advantage of the wide chord fan blade supported new investments aimed at developing the radically new technology. Due to the technological knowledge developed over time, Rolls-Royce was able to *frame* the problem (low performing, narrow blade) and *enact* an innovative technological vision (wide chord fan blade).

Borrowing the terms of Dierickx & Cool (1989), the knowledge garnered during the development of the first-generation wide chord fan blade represented Rolls-Royce's *stock* of cumulated technological knowledge, which was refined and advanced through dedicated in-house investments (*flows*) that led to the development of the second-generation fan blade. While the stock of in-house technological knowledge cumulated during the development of the first generation fan

blade formed the platform for Rolls-Royce's new technological solution, the dedicated investments in experimentation and personnel enhanced the company's *generative capabilities*.

The interrelations between first- and second-generation fan blades also provide empirical support to the insights of Dierickx & Cool on the features of the building process of a firm's capabilities. The success of the second generation built heavily on the first-generation's success (*asset mass efficiency*). As Dierickx & Cool argued "firms who already have an important stock of R&D know-how are often in a better position to make further breakthroughs and add to their existing stock of knowledge than firms who have low initial levels of know-how" (Dierickx & Cool, 1989, p. 1508).

The second-generation fan blade was not a mere *point extrapolation* of the first-generation, however. It contained several innovative technological features both at the product level and at the process level. The first-generation wide chord fan blade was innovative at the time of its introduction. It was, however, both complex and labour intensive in terms of engineering and manufacturing activities. This called for a fundamental change in design, analysis, and manufacturing processes. Therefore, although the second-generation fan blade development relied heavily on the *stock* of previously accumulated technological capabilities, the company's *generative capabilities* from the first generation fan blade were not only deployed, but also were nurtured and enhanced via dedicated investments in experimentation and personnel. These investments gave rise to a new virtuous cycle *frame-enact-solve*. The cycle started with the *reframing* of the complex and labour intensive engineering and manufacturing activities of the first-generation blade which led to new design concepts of hollow blades that in turn prompted to laboratory programmes on bonding and forming fabrication processes. Eventually this led to a better understanding of the manufacturing processes (Prencipe, 2001).

Concluding Remarks

This chapter has identified two main dimensions of the capabilities firms developing multitechnology products. *Synchronic systems integration* relates to the capabilities required to specify, buy, and integrate externally-designed and produced components. Firms developing multitechnology products are also required to develop *diachronic systems integration*, that is, the capabilities required for the co-ordination of change across different bodies of technological knowledge as well as across organizational boundaries. *Synchronic systems integration* and *diachronic systems integration* refer to the capabilities of the *ambidextrous organization* as identified by Tushman & O'Reilly (1996). The chapter also argued that firms are characterized by what Cohen & Levinthal (1990) labelled *absorptive* capabilities related to monitoring, identifying, and evaluating new technologies.

The chapter has then focused on *generative capabilities* needed to innovate both at the component level and the architectural level, also independently of external sources. It proposed that the primary features of a firm's *generative capabilities* are *problem framing*, *vision enacting*, and *problem-solving*. Firms build, update, and renew the frames of reference within which they enact an innovative vision to solve problems that, in turn, are identified by their frame of reference. Following Dosi & Marengo (1993), we argued that *problem framing* is the distinctive basis for organizational learning processes.

This chapter is an attempt towards a better understanding of a firm's capabilities. By considering the multitechnology empirical setting, the chapter proposed a taxonomy to categorize a firm's capabilities according to their roles. The capabilities needed to strategically manage the links with a network of suppliers are paramount for a firm's competitiveness in a multitechnology setting (Brusoni et al., 2001; Lorenzoni & Lipparini, 1999). This chapter has extended this argument and argued that in-house capabilities play an equally important role in building such competitive advantage in a high-technology dynamic environment. The firm's *generative capabilities*, discussed in-depth here, constitute an important dimension of dynamic capability (Teece & Pisano, 1994) that enables a firm to grow (Penrose, 1959).

References

Brusoni, S. & Prencipe, A. (2001). Unpacking the black box of modularity: Technologies, products, organisations. *Industrial and Corporate Change*, **10**, 179–205.

Brusoni, S., Prencipe, A. & Pavitt, K. (2001). Knowledge specialisation, organisational coupling, and the boundaries of the firm: Why do firms know more than they make? *Administrative Science Quarterly*, **26**, 4.

Clark, K. & Fujimoto, T. (1991). *Product development performance*. Boston, MA: Harvard Business School Press.

Cohen, W. A. & Levinthal, D. A. (1990). Absorptive capacity: A new perspective on learning and innovation. *Administrative Science Quarterly*, **35**, 128–152.

Dierickx, I. & Cool, K. (1989). Asset stock accumulation and sustainability of competitive advantage. *Management Science*, **35** (12), 1504–1510.

Dosi, G. & Marengo, L. (1993). Some elements of an evolutionary theory of organisational competencies. In: R. W. England (Ed.), *Evolutionary Concepts in Contemporary Economics* (pp. 157–178). Ann Arbor, MI: University of Michigan Press.

Dosi, G., Hobday, M., Marengo, L. & Prencipe, A. (2003). The Economics of Systems Integration: Toward an Evolutionary Interpretation. In: A. Prencipe, A. Davies & M. Hobday (Eds), *The Business of Systems Integration* (forthcoming). Oxford: Oxford University Press.

Dosi, G., Nelson, R. R. & Winter, S. (Eds) (2000). *The nature and dynamics of organisational capabilities*. Oxford: Oxford University Press.

Freeman, C. (1974). *The economics of industrial innovation*. London: Penguin Modern Economics Texts.

Freeman, C. (1982). *the Economics of Industrial Innovation*. London: Frances Pinter Publishers.

Freeman, C. (1991). Networks of innovators: A synthesis of research issues. *Research Policy*, **20** (5), 499–514.

Gassmann, O. & von Zedtwitz, M. (2003). Innovation processes in transnational corporations. In: L. V. Shavinina (Ed.), *International Handbook on Innovation*. Oxford: Elsevier Science.

Granstrand, O. (1998). Towards a Theory of the Technology-Based Firm. *Research Policy*, **27**, 465–489.

Granstrand, O. & Sjölander, S. (1990). Managing innovation in multitechnology corporations. *Research Policy*, **19**, 35–60.

Granstrand, O., Patel, P. & Pavitt, K. (1997). Multi-technology corporations: why they have 'distributed' rather than 'distinctive core' competencies. *California Management Review*, **39** (4), 8–25.

Grant, R. (1998). *Contemporary strategy analysis*. Malden, MA: Blackwell.

Grant, R. & Baden-Fuller, C. (1995). A knowledge-based theory of inter-firm collaboration. *Academy of Management Best Paper Proceedings*, 17–21.

Katz, R. (2003). Managing technological innovation in business organizations. In: L. V. Shavinina (Ed.), *International Handbook on Innovation*. Oxford: Elsevier Science.

Lorenzoni, G. & Lipparini, A. (1999). The leveraging of inter-firm relationships as a distinctive organisational capability: A longitudinal study. *Strategic Management Journal*, **20**, 317–338.

Maïsseu, A. P. (1995). Managing technological flows into corporate strategy, *International Journal of Technology Management*, **1** (1), 3–20.

Mowery, D. (1983). The relationship between intra-firm and contractual forms of industrial research in American manufacturing, 1900–1940. *Exploration in Economic History*, **20**, 351–374.

Nonaka, I., Sasaki, K. & Ahmed, M. (2003). Continuous innovation: The power of tacit knowledge. In: L. V. Shavinina (Ed.), *Handbook of Innovation*. Oxford: Elsevier Science.

Patel, P. & Pavitt, K. (1997). The technological competencies of the World's largest firms: Complex and path-dependent, but not much variety. *Research Policy*, **26**, 141–156.

Pavitt, K. (1998). Technologies, products and organisations in the innovating firm: What Adam Smith tells us and Joseph Schumpeter doesn't. *Industrial and Corporate Change*, **7**, 433–452.

Penrose, E. (1959). *The theory of the growth of the firm*. London: Basil Blackwell.

Paoli, M. & Prencipe, A. (1999). The role of knowledge bases in complex product systems: Empirical evidence from the aero engine industry. *Journal of Management and Governance*, **3**, 2.

Prencipe, A. (1997). Technological competencies and product's evolutionary dynamics: A case study from the aero-engine industry, *Research Policy*, **25**, 1261–1276.

Prencipe, A. (2001). Exploiting and nurturing in-house technological capabilities: Lessons from the aerospace industry. *International Journal of Innovation Management*, **5** (3), 299–321.

Prencipe, A. (2003). Corporate strategy and systems integration capabilities. In: A. Prencipe, A. Davies & M. Hobday (Ed.), *The Business of Systems Integration* (forthcoming). Oxford: Oxford University Press.

Prencipe, A. (2004). *Strategy, systems, and scope: Managing systems integration in complex products*. London: Sage Publications (forthcoming).

Rosenberg, N. (1976). *Perspectives on technology*. Cambridge: Cambridge Univerity Press.

Rothwell, R. (1992). Successful industrial innovation: Critical factors for the 1990s. *R&D Management*, **22**, 3.

Rothwell, R. Freeman, C., Horsely, A., Jervis, V. T. P., Robertson, A. & Townsend, J. (1974). SAPPHO updated: Project Sappho Phase II. *Research Policy*, **3**, 258–291.

Swan, J., Scarbrough, H. & Robertson, M. (2003). Linking knowledge, networking and innovation processes: A conceptual model. In: L. V. Shavinina (Ed.), *International Handbook on Innovation*. Oxford: Elsevier Science.

Teece, D. J. & Pisano, G. P. (1994). The dynamic capabilities of firms: An introduction. *Industrial and Corporate Change*, **3**, 537–556.

Tushman, M. & O'Reilly III, C. A. (1996). Ambidextrous organisations: Managing evolutionary and revolutionary change. *California Management Review*, **38** (4), 8–30.

Von Hippel, E. (1988). *The sources of innovation*. Oxford University Press: Oxford.

Weick, K. (1969). *The social psychology of organising*. New York: McGraw-Hill.

Weick, K. (1985). *Sensemaking in organisation*. London: Sage Publications.

The International Handbook on Innovation
Edited by Larisa V. Shavinina

Innovation Processes in Transnational Corporations

Oliver Gassmann[1] and Maximilian von Zedtwitz[2]

[1] *Institute for Technology Management, University of St. Gallen, Switzerland*
[2] *IMD-International Institute for Management Development, Switzerland*

Abstract: If innovation is considered as a process, then the differentiation of the innovation process into two phases creates several benefits. These two phases are, firstly, a pre-project phase fostering creativity and effectiveness, and a secondly a discipline-focused phase to ensure efficiency of implementation. This differentiation enables transnational companies to replicate and scale innovation efforts more easily in remote locations, exploiting both economies of scale and scope. Although the characteristics of these phases are quite distinct, few companies have consistent and differentiated techniques to manage and lead the overall innovation effort specific to each phase.

Keywords: Innovation; International project management; R&D; Creativity; Phase model.

Introduction

If innovation is considered as a process, then the differentiation of the innovation process into two phases creates several benefits. These two phases are, first, a pre-project 'cloudy' phase fostering creativity and effectiveness and, second, a discipline-focused 'component' phase to ensure efficiency of implementation. This differentiation enables transnational companies to replicate and scale innovation efforts more easily in remote locations, exploiting both economies of scale and scope. Although the characteristics of these phases are quite distinct, few companies have consistent and differentiated techniques to manage and lead the overall innovation effort specific to each phase.

Our overarching goal is to show that dividing the overall innovation process into the cloudy and component phases is a simple and easily implementable way to overcome typical communication and managerial problems in international innovation. In this chapter we first summarize earlier phase models of innovation. In this context, we refer to innovation as a company's efforts in instituting new means of production and/or bringing new products or services to market. Next we describe several innovation processes. An innovation process is a cumulative sequence of defined stages and activities leading to an innovation. Recent research, most notably work done in the Minnesota Innovation

Research Program (see Van de Ven, Angle & Poole, 1989; Van de Ven, Polley, Garud & Venkataraman, 1999), shows that innovation is usually unpredictable and difficult to manage with tight controls. The two-phase model we outline in this chapter allows chaotic and random innovation processes to occur in the early innovation phase, and argues for a narrowing and redirection of the creative energy in the first phase during execution and implementation of the initial ideas in the second phase. We then describe tools and systems that help to channel creativity into implementation suited to transnational innovation. Creativity here is understood to be the ability to produce both novel and original ideas appropriate for the task at hand. Transnational innovation is innovation that includes participants of the innovation process from geographically distributed locations, usually in other countries or time zones. We conclude with some implications for managing innovation processes in transnational settings.

Linear and Non-Linear Models of Innovation

Innovation is an inherently complex and unpredictable task. Companies that are driven by meeting financial and operational targets as well as strategic objectives have invented numerous techniques to capture the uncertainty of innovation into a measurable and hence manageable framework.

Pioneering work done by Schumpeter (1911, 1939) and Bush (1945) helped to explain the origin of organized technology development. Building on the notion of *science push*, they described innovation as a linear process from basic research, applied research, and development, through design and manufacturing, to marketing and sales (see Marinova & Phillimore, this volume). Similar models based on a linear logic (e.g. the value-added chain, Porter (1985)) reinforced the concept of a sequential innovation process. Science and technology programs in many Western countries are still based on this pipeline model and are often used to justify the financing of public research and science. This ideology implicitly assumes a causal correlation of research input and innovation output: higher investments in basic research will lead to more innovation and more advanced products.

The linear model has worked well in fields where immediate applicability and practicality was not a determining driver. A well-known example of science-driven innovation is laser technology: the theoretical foundation of this technology was built between 1900 and 1920 by famous scientists such as Max Planck (quantum effect), Niels Bohr (atomic model) and Albert Einstein (conventional sources of light emit a spontaneous photon radiation). Scientific research on laser technology itself took place in the 1950s, and the first successful laser device was constructed by Theodore Maiman in 1960. Today's applications are widespread: cutting, drilling and welding of materials; distance and gas measuring; telecommunications; and medical technology.

In the 1960s, a new paradigm emerged based on the empirical work of Schmoockler (1966) in patent statistics. Innovation was found to be determined more by *market pull* than by the classical science push. This model also assumes a linear causal innovation sequence, but in this case a market demand is what triggers innovative activity in the preceding functions. This model enhanced the position of marketing: R&D (research and development) departments and new product development teams were assigned reactive roles to develop products according to given specifications.

The market-driven model has given companies a tool that aligns internal processes according to measurable output, thus greatly increasing the role of R&D as a strategic element in achieving, building and expanding market dominance. For instance, AT&T strongly pushed transistor development at Bell Labs because telephone companies demanded smaller and more convenient switching technologies. Hippel's (1988) lead user concept, further underscored the importance of capable users, and customers, as a source of innovation. Lead users are technologically savvy customers with an urgent need for improved products who could serve as trendsetters in an emerging market. Hilti, a leading construction technology supplier, applied the lead user concept and easily halved R&D costs and time-to-market (Hippel & Herstatt, 1991).

Both the science push and the market pull approaches are linear sequential models. Only the initial source of innovation is different. However, several studies since the late 1970s have shown that innovation processes are seldom linear processes triggered by a single source—either scientific potential or market need—but rather *random processes* that are more complex and uncertain than the linear model assumes (e.g. Cohen, March & Olsen, 1972; Tushman & Anderson, 1986).

Van de Ven and colleagues (1989, 1999) have undertaken several longitudinal studies consisting of thousands of single observations, and assembled ample and rich evidence that innovation is inherently a chaotic process. The description of 3M's Cochlear Hearing Implant innovation journey (detailing more than one thousand events of the 3M innovation effort, see Van de Ven et al., 1999) is well documented and underlines how difficult it is to predict consequences from decisions to actions to outcomes.

Managers and organizations, however, assume a certain degree of predictability and cause-and-effect relationships in innovation and often introduce structured management schemes to increase the stability and coherence of their efforts. Linear models were improved by integrating science push and market pull into non-sequential *feedback models*. Regardless of the trigger of innovation, several complete feedback loops ensure that both science and market inputs are recognized and implemented. For instance, Roy (1986) described innovation as a cyclic process in which technological opportunities, invention, knowledge production and market demands were linked together. Later, Kline (1985) and Kline & Rosenberg (1986) introduced the *chain-linked model*. This model describes five paths of innovations. Some of these paths are linear and follow the invention to development to production to marketing sequence, while other paths are based on several feedback loops, i.e. reiteration to early-stage innovative activity. A major implication of their model is that the market remains a significant driver of innovation and that science-driven innovation is relatively rare, yet should not be totally neglected.

In the 1980s, several integrative approaches to R&D management pioneered by Japanese companies (e.g. the 'rugby'-approach, which advocates a team rather than a relay approach to product development) became popular under the umbrella of simultaneous or concurrent engineering. These approaches focused on overlapping innovation sub-phases, mainly in product development and manufacturing (see e.g. Liker, Kamath, Wasti & Nagamachi, 1995; Nishiguchi, 1996). Based on these interlaced models of innovation, i.e. innovation processes with overlapping sub-phases, *interaction models* were developed that emphasized the principle of interaction itself as an important source of innovation (see e.g. Durand, 1992; Schmoch, Hinze, Jäckel, Kirsch, Meyer-Krahmer & Münt, 1995). They explain innovation as the result of intense, continuous interaction of both individual and institutional protagonists, and communication becomes a key factor in innovation processes. Nonaka & Takeuchi (1995), who introduced the rugby-approach of R&D management noted above, focus on knowledge creation and sharing as the central determinants of corporate success, and consider innovation almost as a byproduct of knowledge management (see also Nonaka & Ahmed, 2003).

Over the past decades, compliance with ISO requirements has led to highly disintegrated and ineffective R&D phase concepts and the illusion that all critical innovation factors can be measured and structured. At the same time, engineers and developers have requested more creative freedom and fewer administrative chores, particularly in the early phases of innovation. Although it is now accepted that innovation processes are non-linear, managers need normative models that reflect the need for clear, unified processes throughout the organization. In the following section we compare process models that attempt to combine both linear and non-linear approaches to innovation.

Normative Models of Innovation Processes

In R&D management practice it is very rare to find clearly distinguishable and predetermined project phases executed exactly according to a predefined schedule. Although systems engineering offers some help in structuring the R&D process into linear sequential project phases, R&D managers are generally not successful in implementing these methods in the innovation process. However, there are some frameworks that guide the design of innovation processes in a company.

Classical phase segmentation and process orientation (as in modern management theory) have been combined in the *stage-gate process* (see Cooper, this volume; Cooper & Kleinschmidt, 1991; O'Connor, 1994; for transnational innovation: Gassmann, 1997). Every step—or 'stage'—necessary to complete a particular project task is linked to the next by a 'gate' at which decisions for the continuation of the project are made. Unlike milestones, gates are more flexible in terms of time, date and content. Gates allow a deliberate parallelization of phases as well as their recombination or adaptation to new requirements. At each gate the R&D project is analyzed and reviewed in its entirety, often including some competitor intelligence (i.e. evaluation of similar R&D projects by competing firms), as well as external market and technology developments. The number of stages and gates needs to be adapted to industry and project requirements. Ex-ante agreements serve as guidelines for the collaboration of project participants.

The *loose-tight concept* also plays a central role in the design of R&D processes (e.g. Albers & Eggers, 1991; Wilson, 1966). According to this concept, the success of the project depends on the degree of organization during the R&D process. In the early stage of a project, the organization should be designed loosely; towards the conclusion of the project, it should become more and more rigid and tight. The varying degrees of R&D project organization are imposed by constraints in time: Although creativity and idea generation are highly important in the early stage, the management concern shifts to efficiency and project implementation on schedule in the later stage.

The stage-gate process is successful in areas and industries dominated by market-pull innovation. Further indications for applying the stage-gate process are innovations in existing markets (e.g. transfer of product development competence); new applications of existing products and services (e.g. relaunch of an adapted product in a new market); high costs of product development and market introduction (e.g. initial product releases); and limited uncertainty in terms of expected innovation (e.g. incremental innovation). Most of today's industries and well-managed R&D processes rely on the stage-gate process to some extent.

In industries or projects where the science or technology push is the dominant driver of innovation, stage-gate processes are too rigid and slow. Innovations that are triggered by a technological invention with unknown market potential need different processes and techniques to succeed. Under these circumstances, the *probe-and-learn process* is more appropriate. This process has been described from a number of angles, including marketing and discontinuous innovations (e.g. Lynn, Morone & Paulsen, 1996), product development and experimentation (e.g. Thomke, 1995; Wheelwright & Clark, 1992) and technology strategy (e.g. Iansiti, 1998).

Traditional market research methods are based on the 'law of big numbers'—the more customers who want a new feature or product, the more valuable it is. These methods often do not work in technology-driven innovation, as target markets do not yet exist. More anticipatory and exploratory market research methods are needed, such as scenario techniques (pioneered by Royal Dutch/Shell, see Shoemaker, 1995), Delphi studies (see Best, 1974; Dalkey & Helmer, 1963), Beta-customer test groups (see Kottler, 2000) and lead user workshops (Hippel, 1986, 1988). Successful examples of this kind of innovation are 3M's Post-it, Corning's optical fiber technology, Netscape's Navigator and Schindler's LiftLoc system.

Many innovation projects in New Economy companies (i.e. those with mostly Internet-based products and services) have been characterized by a high degree of uncertainty in terms of market fit (see e.g. Trott & Veryzer, 2003). An example of an e-innovation that followed the probe-and-learn process is ICQ (pro-nounced 'I seek you'). The term was coined by four avid young computer users—who established Mir-abilis, a new Internet company, in July 1996 in Tel Aviv, Israel—to describe a new way of communicating over the Internet. Although millions of people were connected to one huge worldwide network—the Inter-net—they recognized that these people were not truly interconnected. The missing link was a technology that would enable Internet users to locate each other online on the Internet and create peer-to-peer communication channels in an easy and simple manner.

They developed a crude prototype first, and offered it free of charge on the Internet in November 1996. Still a very sketchy product, it was full of flaws and lacked important functionalities. However, based on online feedback from users and rapid prototyping techniques, they continued to fine-tune the first version. Three months after the launch, ICQ customer base reached 350,000 users; after an additional three months this number stood at 850,000. Even at this stage the product was continually refined and adapted to new user needs (e.g. the introduction of a 'I am busy' state in order to prevent communication bottlenecks). However, there was no clear product strategy.

Fourteen months after first introducing the product, Mirabilis had 8 million subscribers and handled 1.3 million users a day. Although the company operated with heavy losses, the market value increased in expectation of even higher subscription numbers. Eventually, AOL acquired Mirabilis in mid-1998 for $287 million in cash and renamed it ICQ. By November 2001, ICQ had 120 million users.

ICQ is perhaps a forerunner of product development in an increasingly international and online-based world. Two observations can be made:

Even established industries develop products with more and more online content. In addition, simulation, virtual reality and communication tools based on broadband information and communication technolo-gies (ICT) allow research, development and design to integrate users, scientists and engineers in virtual teams around the world. Greater density of information and greater geographical dispersion are two important factors in the design of modern innovation processes.

Generally, there is only a vague idea about the eventual product design at the start of a development project. Project members differ greatly in their under-standing of project objectives and methodologies. By communicating their ideas during the conception of the project, project members create shared knowledge and understanding. In the early phases of a project, tacit or implicit knowledge is transformed into explicit knowl-edge. But designing and generating product design drafts and specification lists must be done and decided on as a team. Knowledge sharing and know-how transfer is hampered not only by geographical separa-tion but also by epistemological and cultural barriers.

Stage-gate, loose-tight and probe-and-learn proc-esses were developed when most R&D was carried out in one location by one team. However, the typical development team at the beginning of the 21st century is becoming transnational in nature (Boutellier, Gass-mann & Zedtwitz, 2000; Gassmann & Zedtwitz, 1997). Probe-and-learn processes also appear to work well in web-based settings and can thus be transposed to dispersed team settings. Nevertheless, distance—and thus problems of different time zones—and culture impose barriers and further imperfections on the innovation process (see e.g. Hadjimanoli, 2003). Would it be possible to combine some of these process models and adapt the result to a truly transnational innovation process framework?

Cloudy-to-Component Process for Modular Innovations in Multinational Companies

What is the Cloudy-to-Component Process?

Companies that undertake more than just application engineering and engage in fully fledged R&D will need to split their innovation process into two phases: the 'cloudy' phase and the 'component' phase. More differentiated phase concepts are commonly accepted and applied in industrial R&D, but they suffer from the strictly sequential execution of project phases and are therefore often impractical in transnational innovation projects. The highly structured stage-gate process can easily become bogged down in bureaucracy and rigidity; the probe-and-learn process can lead to unplanned trial-and-error development and unpro-ductive chaos.

Our concept of the cloudy-to-component process (C-to-C process, see Gassmann, 1997) is especially appropriate for innovation processes in transnational companies. Remember that existing innovation pro-cesses have been fine-tuned for collocated innovation. Due to the increased internationalization of R&D and knowledge creation, it has become more difficult and rewarding at the same time to optimize global product development and integrate distributed competencies (Gassmann & Zedtwitz, 1999a).

Our C-to-C process does not imply that projects are carried out without reviews or milestones. Such projects tend to be managed both ineffectively and inefficiently. The solution lies in placing the appro-priate focus on what is to be achieved in the two phases. Too many projects are slowed down or canceled because faulty designs have to be corrected late in a project, and too many projects have not achieved their full potential because project managers have pushed for cost efficiency and short-term solu-tions too early. In transnational innovation projects, there is less slack to compensate for these management errors. At the same time, they offer great potential for 'doing it right the first time'. This separation into two phases must be well planned beforehand and must be

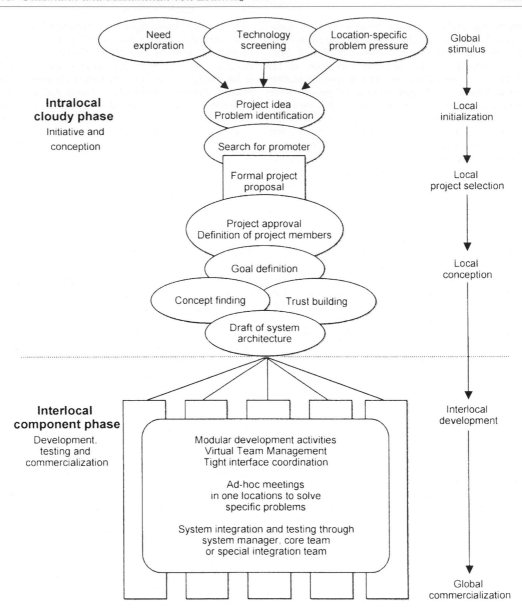

Figure 1. The separation of the R&D process into two phases allows a focus on effectiveness and efficiency in the appropriate stages.

communicated to, and accepted by, all involved parties. The 'cloudy' phase is thus reserved for wild, inventive creativity, and every idea is given a chance. Only when the project proposal is finally approved does the cost-intensive component phase set in with structured engineering methods (Fig. 1).

Inputs to the Cloudy Phase

The first phase—the 'cloudy' phase—is the domain of creative idea generation, research and advanced devel-

opment. Freedom of thought and an open playing field for engineers should be ensured. During the cloudy phase the principal product features are conceived, the main system characteristics defined, and the project is initiated. This early phase is based on market and technology research, as well as on internal problem pressures.

Market exploration in this phase is based on traditional market research tools such as panel research, focus group interviews, sales and distribution questionnaires, scenario techniques, lead market analysis, etc. More recent techniques include cooperative forms of R&D such as 'lead users' and 'anthropological expeditions', both of which help to tap into implicit user knowledge (see Leonard-Barton, 1995). Technology screening and assessment also take a variety of forms. Technology listening posts, leading-edge innovation centers, technology intelligence and technology forecasts, expert interviews, patent database research, and reverse engineering of competitor products are typical techniques and sources used here. Exploration of markets and technologies has to be conducted on a global scale, as the sources of technical knowledge are less and less limited to a few regions of innovation and markets are becoming increasingly international in nature.

Needs exploration and technology screening are the two primary sources of good project ideas. Ideally, project ideas result from a balance between market pull and technology push. Dominance of technology-focused engineers would lead to over-engineered products that would not be accepted by the customer. Conversely, short-term profit considerations of sales and marketing people with no technological vision would reduce the long-term innovation capacity of the company (striking the right balance is one of the fundamental problems of innovation management). In the early cloudy phase of a project, it is essential to allow creative input to come from all possible directions.

As well as being influenced by technology and market determinants, the generation of project ideas can be highly affected by location-specific problems and pressures such as low capacity utilization, financial difficulties and fashion trends. Low capacity utilization in a particular site (e.g. due to relocation of manufacturing to another site) will urge local management to search for new businesses. Units with negative financial results and low cash flows are under more pressure to change than units with profitable products. We also found above average creativity and a propensity to initiate projects in R&D units that were in danger of being rationalized due to global efficiency enhancement measures. If management does not succeed in communicating a clear framework and a common vision, the imminent crisis is worsened by the growing paralysis of the work force. For instance, the significant departure of qualified personnel at DASA (a German aerospace company) in 1995 was related to the long uncertainty about goals as well as changes in the leadership of MTU (a German jet engine and power plant company and close business partner of DASA). Unlike ABB, which experienced a 'creative crisis' in its GT-24/26 radical innovation project (see e.g. Imwinkelried, 1995; Zedtwitz & Gassmann, 2002), and IBM, which experienced such a crisis in its VSE (virtual storage extended) development project (see e.g. Gassmann, 1997), MTU was characterized by paralysis

that resulted in a reduction of idea generation and innovation.

Fashion trends often seduce managers into enlarging their product spectrum with the latest and most refined products in the market. Many R&D project are thus initiated not because of a clear market need or technological potential but rather to improve the image or reputation of a particular business unit. Besides company-external market pull (e.g. request for a new product) and technology push (e.g. exploitation of a technological capability), a major driver of new project ideas is therefore a company-internal problem push (e.g. justification of previous market investments and product commitments). The two external drivers prevail in a global environment, whereas the internal driver is local.

Examples of the C-to-C Process in Industry

The innovation process in the chemical and pharmaceutical industries is two stage and models very much like our C-to-C process. BASF underscores the distinction between cloudy and component phases by speaking of 'R&D activities' in the early R&D stages and of 'R&D projects' in later stages. For Bayer, milestones and project review meetings only start once the preclinical phase has been reached, when the project is started formally (Fig. 2).

General Motors calls this early phase of innovation the 'bubble-up process'. This process is driven by an interdisciplinary team, representing advanced development, strategic purchasing and advanced marketing. Most of the activities focus on strategy development and exploration of markets, brands and technologies. At Schindler, the owner of the cloudy phase is a unit called the Technology Management Area whose technological experts, representatives of innovation marketing and lead users jointly develop so-called concept elevators. These functional prototypes show technical feasibility and market acceptance; they also define the principal product architecture and technology to be used. The individual components will later be developed and fully documented by the development center.

The distinction between cloudy and component phases is therefore not an academic one, but a very real one. Companies succeeding at transnational innovation manage each phase differently and optimize the deployment and utilization of specific organizational and management techniques.

Intralocal Versus Interlocal Execution in the C-to-C Phases

There is substantial research indicating that innovation is spurred by geographical proximity between R&D and other R&D units, suppliers and customers (e.g. Allen, 1971; Hippel, 1988; Tushman, 1979). As Tushman (1979) noted, the patterns and intensity of

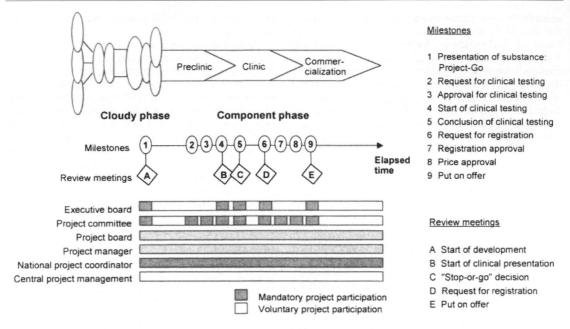

Figure 2. Bayer: The first milestones start at preclinical development.

communication differ remarkably between early creative R&D and later-stage development work.

R&D in the early cloudy phase is a contact sport. Many tools based on modern communication technology and advances in virtual engineering may allow people to collaborate more productively at a distance, but the trip to the coffee corner or across the hallway to a trusted colleague is still the most reliable and effective way to review and revise a new idea. Moreover, the internal sociopolitical game of finding and convincing an idea champion as well as building a core team is based on frequent and face-to-face encounters. The early cloudy phase is therefore heavily intralocal.

However, the stimuli for new ideas and projects— customer needs, technology potentials and performance pressures—can have very global origins. Scientists maneuver in an international scientific community in which the locus of the individual is irrelevant. Gatekeepers are active listeners and transferors of outside ideas to the internal R&D organization (Allen & Cohen, 1969). The potential of outside transparency is of course curtailed by the effects of the not-invented-here syndrome (see e.g. Katz & Allen, 1982), which still governs many contemporary R&D organizations.

Once the product or system architecture has been (locally) conceived, and most of the interdependencies between different parts of the final product have been defined and described, the actual R&D work can be separated and assigned to specialized and better prepared R&D units. Some research may still have to

be carried out with respect to the underlying properties and improvements of individual system components, but these should not affect the system as a whole. Coordination and communication about the system is now the task of the overall project management team, which controls and directs the innovation effort through interface coordination, travel and regular project reviews. The integration of local customers in the innovation process, and the restricted availability of critical engineering and testing resources, require the dispersion of project activities, making the component phase part of interlocal innovation (Gassmann & Zedtwitz, 1999b).

Building Blocks for Improving the 'Cloudy' Phase in Transnational Innovation

Intensive Idea Flow and Workflow Systems

Although creativity flourishes over shorter distances, recent advances in collaborative workflow systems (see e.g. Carmel, 1999) allow the idea generation for a single project to take place on a global scale. This was demonstrated by ABB's workflow system PIPE (Project Idea, Planning & Execution), a Lotus Notes-based workflow system designed to transmit and distribute ideas, problems, commentaries and solutions by means of modern information and communication technologies. The generator of an idea also selected the group of persons who could access his contributions. His initial idea, along with his evaluation of commercial potential and supplementary comments, was then refined and complemented with the ideas and suggestions of other PIPE participants.

If an idea received enough support, a formal project was proposed, for which detailed information about objectives, risk, possible problems and available resources was required. Upon approval of the project proposal, the program manager transferred this information into the PIPE Planning Application. The project idea was then integrated into the overall project plan. A project manager was assigned and a decision was taken on what the participating sites would contribute. Local group managers proposed local project schedules defining sub-goals, costs and means of funding. The program manager, local corporate research managers and business unit representatives then evaluated the consolidated project plan, contributing their priorities by e-mail.

PIPE also supported project execution. Simple and formalized project reports concerning costs, schedules and results served as easy-to-distribute project information. A report archive logged the project history, thus facilitating exchange of experiences across several projects.

Interestingly, after some years of experience with PIPE, ABB decided to restrict the freedom of idea generation and commenting with this workflow system. This decision was motivated by the frequent uncertainties over ownership of shared ideas and inventions. As long as reward and compensation systems are tied to the extent of measurable technological contribution, trust and confidence remain significant determinants of effective transnational idea generation (see De Meyer & Mizushima, 1989).

Good Ideas Require Good Promotion

Although stimulated by global determinants, identifying a problem usually starts with a single person or a collocated group of people. Looking for support for their ideas, they try to convince influential people in their organization about the significance of their insights (see e.g. Hauschildt, 2003; Roberts, 1968; Witte, 1973). The influence of these idea champions is based on their hierarchical position (power promoters), their knowledge (functional promoters) or their communication abilities (process promoters).

With the omnipresence of e-mail and global communication networks, one may be tempted to look for appropriate promoters regardless of their location. Experience shows, though, that personal relations are tremendously important in winning over decision-makers to new ideas. These personal relations are difficult to establish just for the purpose of championing a project idea, particularly at international distances. Decision-makers are influenced by project opponents, who bring in technological and economical arguments against a new project idea. Internal political arguments play an important role, since opponents fear

that new projects may mean a reduction in resources for their own activities.

The better the idea generator is able to communicate his intentions and visions, the more likely he is to succeed in finding top-management support. In order to find a power promoter, a project idea must be fresh and presented very soon after conception. Commercialization potential and project vision are often more important than technical decision criteria. For as long as it remains difficult to inspire people just by means of e-mail and shared workflow systems and convince them to support an idea, the quest for promoters will remain a matter of face-to-face contact.

Bottleneck Product Profitability Calculation

If a project idea finds enough support and passes preliminary evaluations, a formal project proposal demonstrating technical realization and commercialization potentials is made. Product profitability calculations are widely used as they are robust, but they are not always appropriate for project evaluation. For instance, if a market has not yet developed a dominant design of product architecture, project evaluation can lead to unrealistic market forecasts (Fig. 3).

The profitability calculation is based on forecasted market returns discounted to net present value. Before the emergence of a dominant design (e.g. mobile communication), a market is characterized by intense market dynamics, making sales forecasts highly unreliable. Numerical values with many decimal places provide an illusion of only hypocritical exactness. Future project proposals should be complemented with qualitative data and evaluated in light of technological vision.

More recently, some companies have started experimenting with real-option analysis to evaluate R&D projects (see e.g. the special issue of the *R&D Management Journal*, **31** (2), 2001, on real-options in R&D). For R&D projects that allow multi-period investment decisions, real-option analysis is better than the popular net present value approach. NPV undervalues potential projects by as much as several hundred% because it ignores the value of flexibility. Real options include the flexibility to expand, contract, extend or defer R&D projects in response to unforeseen events during the innovation phase. Managers often overrule NPV results by accepting projects with low or negative NPV for 'strategic reasons'. In essence, they are using their intuition to account for the flexibility of a project's real options (see Copeland, 2001). Real-options reasoning offers a way to capture the value of project portfolios, research programs, technological and innovation competence, and technology and product vision.

Project Approval: Rational Criteria Versus Politics

After what the idea promoters consider sufficient conceptualization and refinement, the project proposal

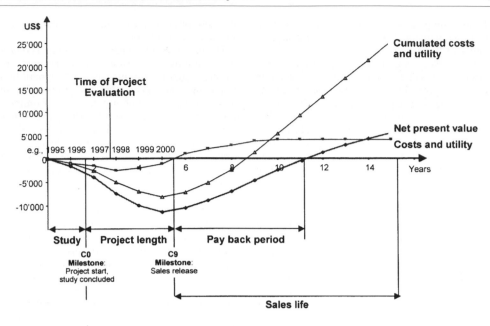

Figure 3. Product profitability and return-on-investment are inappropriate when the market is dynamic and unpredictable.

Source: Boutellier et al. (2000, p. 176).

is presented to a steering committee. This committee is not necessarily located in the same R&D unit where the project will eventually be carried out. For instance, the 'Investment Review Board' of the IBM S/390 system architecture has been meeting in New York to decide about major project activities in the IBM Germany site at Böblingen. Transnational R&D projects require particularly large budgets, which must be approved by the highest authorities of business areas. Project selection always takes place at the location of the decision-maker.

Project approval also includes a decision about key project members and participating locations. It is at this stage that it is determined whether an R&D project will be carried out transnationally or in only one place. In our experience, this decision is rarely based on a structured top-down evaluation, during which project requirements would be systematically combined with competencies, know-how bases, and available capacities of potential R&D units. As the IBM VSE development example demonstrated, project participation was determined in a political agreement finding

process (Fig. 4). Often enough, political considerations outweigh rational criteria.

Profit-center thinking usually gains the upper hand, despite the fact that resource and competency-based decisions would be economically more reasonable from a corporate perspective. Each R&D site strives for full capacity utilization, and projects funded by headquarters or central R&D unit are particularly attractive. Examples of such centrally funded strategic projects are 'Top Projects' at Bosch, 'Golden Badge Special Projects' at Sharp, 'Core R&D' and 'Strategic Business Projects' at Hitachi, and 'Core Projects' at Siemens and NEC.

System Architecture as a Critical Success Factor

Concept finding and definition, which determine the architecture of the system to be developed, partly coincide with the initial goal finding process. Especially during the subsequent interlocal component development phase, an accepted system architecture is one of the critical success factors of the entire project. Interfaces between modules and components must be

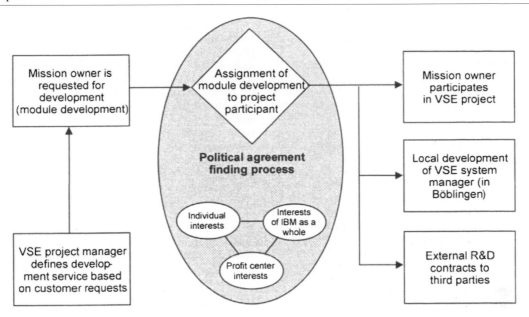

Figure 4. Rational criteria and balancing of interest determine who will eventually participate in a project.

Source: Boutellier et al. (2000, p. 177).

clearly defined. Changes in one module may not affect other modules. The stability achieved through such a system architecture reduces the number of design changes in later stages and consequently the intensity of interaction between decentralized teams (see e.g. Morelli, Eppinger & Gulati, 1995). This stability also makes standardized reporting possible within the same project.

The system architecture defines not only the success of the current product but also the success of future products and perhaps an entire product line. Only if interfaces have been designed clearly and with a wide range of applications in sight can this architecture serve as the basis of a product platform in future modular product development. A reduction in variant complexity (which is an essential part of cost reduction programs in all industries) is supported by a clear conception and designed stability in system architectures during the development phase.

System Management to Speed Up the Component Phase in Transnational Innovation

Goal Conflicts in Teams Jeopardize Project Success

The assignment of teams and locations to the R&D project is the linking step between the cloudy and component phases of innovation. Although much of the knowledge and goals of the cloudy phase may have been tacit and not necessarily well articulated, distributed teams will only be able to work with explicit and easily transferable information.

Although a system concept may have been defined and approved, these local teams may differ significantly in their interpretation and realization of the overall objective. In transnational innovation, the process of goal alignment is highly complex, since differing ideas about goals are not easily resolved because of geographical distances and cultural differences. Goal conflicts may occur between R&D (technologically advanced products), production (manufacturable products), marketing (customer-oriented products) and logistics (storable and transportable products).

In addition to these classical goal conflicts, regional perspectives may complicate the situation. Representatives of product business areas favor global standardization, whereas regional managers request country-specific product variants. Each location tries to justify the importance of its respective product variant with forecasts about how much this variant would contribute to overall product turnover.

Furthermore, every location strives for development and manufacturing share, partly to utilize as much of its R&D and production capacity as possible, and partly to develop new competencies in interesting technology areas. The various expectations of participating interest groups lead to minimal consent in the goal definition, and the participants eventually agree only to a rudimentary core concept (also known as 'concept peeling'). The overall project management team must possess excellent moderation and negotiation skills to align and focus all teams on the optimal

711

integrated solution, and to motivate and inspire the various participants for a commonly shared goal.

Know-How Redundancy and the Need for Generalists

The stability of a system architecture is made possible through the input of all members of the core team in consultation with invited experts. Much of the underlying knowledge is tacit or implicit, i.e. it resides in the heads of key project members and is not easily accessible or transferable (see e.g. Nonaka et al., 2003). During the project conception, socialization processes are therefore highly important to exchange crucial but otherwise unattainable knowledge. Constructing a common knowledge base goes in parallel with establishing redundant knowledge within project teams to improve project-internal communication.

It is possible to support both a generalist and a specialist focus. The turbine manufacturer MTU Munich introduced ABC teams and thus separated individuals who preferred to avoid working in teams from team-enthusiastic project members. The A-team is organized in the management level and defines the strategic framework for project teams (program decisions and reviews of critical milestones). B-teams carry out much of the component and parts development, while C-teams consist of highly qualified specialists (e.g. for rotor blade materials). C-team members set functional guidelines and are consulted by B-teams when specific problems occur.

Defining technical interfaces is also characterized by a socio-psychological effect. Each team manager tries to enforce high tolerance levels in the interfaces of his module, since this increases the likelihood of successful development of his module. This behavior triggers a cascading effect of tolerance-determined loss of operational range for components and products. High tolerances imply higher costs at lower effectiveness. The overall project manager thus has to ensure that safety thinking and risk aversion do not lead to excessively high tolerances.

Structured Engineering in the Component Phase

The project manager must be a competent system architect himself to implement the highly modular approach to the project successfully. He groups the functional elements of the system into components, defines clear interface standards and protocols, and assigns development tasks to specialist teams. When the system is being divided into individual work tasks, particularly high-risk tasks should not be distributed among a large number of distinct modules but are better concentrated in one component. This critical module is then tackled by a highly qualified R&D unit, preferably with superior management capacity. Risk concentration is not easy to achieve, since risk distribution is an integral part of portfolio thinking and team managers will attempt to shift risk to other locations.

Most of the development, prototyping, testing and eventually commercialization of the product takes place in the component phase. These activities are more detailed, structured and less interdependent than activities in the cloudy phase. Whereas the emphasis in the cloudy phase is on goal-adequate concept generation, the focus in the component phase shifts to efficient concept realization. Since the cost-intensive component phase consumes most of the project resources, it requires resource- and time-conscious project management. Capacity planning and multi-project management become important.

In order to ensure access to critical resources during the component phase, it is necessary for the project manager to report to a steering committee at the highest level, e.g. the executive board or directly below. Typical members of this committee are directors of business areas, R&D, marketing, manufacturing and regional areas. Such a heavyweight steering committee increases the likelihood of successful commercialization of the project outcome.

At ABB, the strategically important 'Common Technology' projects were managed directly by the business area director for power transmission. Only when the project had got off to a good start and was well established would this director hand the project management over to a lower-ranked manager. Since the steering committee was not capable of controlling the entire project because it lacked the required specialist knowledge, component-specific sub-committees were formed to evaluate the project activities. Typically, expert knowledge serves as a foundation for sound decisions in steering committees as well as in project management.

Conclusions: Success Factors for Managing Transnational Innovation Processes

Modern transnational innovation has come a long way from the unmanaged, almost haphazard, exploitation of research in the early decades of the 20th century. There are no illusions about the (un-)predictability of sequential logic in linear R&D models.

By dividing the innovation process into two phases, R&D processes can be adequately designed and managed in each transnational innovation phase. This separation improves process transparency considerably and reduces the cost-intensive development phase significantly. Although the characteristics of the two phases are different, too few companies approach them with distinctive management methods. The creative cloudy phase requires soft management methods, ensuring freedom, flexibility and inventiveness of scientists and engineers. In this phase, tacit knowledge is transformed into explicit knowledge and communicated to other members of the team. In order to keep up intrinsic motivation, it is important to create team spirit, a common project culture and a shared understanding of project goals and the underlying system architecture.

In the component phase of the project, the focus shifts to efficient implementation of these goals. Costs and milestones are used to determine the progress of the project. Compared with the cloudy phase, new coordination and control mechanisms are used to complete the project successfully. The component phase then leads into another, often globally executed, market introduction.

These R&D project phases allow a different degree of integration of international contributors to and participants in the innovation process. It is important to find the right form of organization for each phase and project. Critical success factors must be considered well in advance to ensure that upcoming problems are taken care of as they occur.

The interlocal component phase can be executed more successfully if enough emphasis has been placed on an effective creative cloudy phase, since it is here that the basis for the subsequent cost-intensive development stages is defined. Only in the component phase should the focus of innovation shift from effectiveness to measurable efficiency.

Idea generation in the early phase should be supported by modern computer technologies and software products. New tools and software packages are introduced regularly, and it is the responsibility of good innovation managers to back up the idea-finding stages of their engineers with the latest support tools. 'Computers' and 'creativity' are not a contradiction.

Good ideas must be communicated, quickly evaluated and promoted. Potential promoters must be enlisted for new ideas early in the R&D project.

In order for a project to pass from the cloudy phase to the component phase, traditional product profitability calculations must be complemented by alternative assessment models and qualitative criteria such as competence establishment and product visions.

Political power struggles should not be allowed to affect operative project work. A strong steering committee clears the way for project managers.

The actual innovators in an innovation project are rarely team-eschewing specialists, although their input is critical for project success. Separating project members into different teams (ABC-teams) can neutralize this conflict.

During the component phase, highly structured engineering is required. Measurement criteria such as on-time-delivery or first-pass-yields used in manufacturing could be helpful.

References

Albers, S. & Eggers, S. (1991). Organisatorische Gestaltungen von Produktinnovations-prozessen. Führt der Wechsel des Organisationsgrades zu Innovationserfolg? *Zeitschrift fuer Betriebswirtschaftliche Forschung (zfbf)*, **43** (1), 44–64.

Allen, T. (1971). Communication networks in R&D labs. *R&D Management*, **1**, 14–21.

Allen, T. & Cohen, S. (1969). Information flow in two R&D laboratories. *Administrative Science Quarterly*, **14**, 12–19.

Best, R. J. (1974). An experiment in delphi estimation in marketing decision making. *Journal of Marketing Research*, (November), 447–452.

Boutellier, R. & Gassmann, O. (1996). *F + E-Projektbewertung führt zu erfolgreichen Produkten—Weichenstellung in der Entwicklung für raschen Marktzutritt. Technische Rundschau*, **15**, 10–14.

Boutellier, R., Gassmann, O. & Zedtwitz, M. von (2000). *Managing global innovation, uncovering the secrets of future competitiveness* (2nd ed.). Berlin, Tokyo, New York: Springer.

Bush, V. (1945). *Science, the endless frontier*. Washington, D.C.: Office of Scientific Research and Development.

Carmel, E. (1999). *Global software teams*. Upper Saddle River: Prentice Hall.

Cohen, M., March, J. & Olsen, J. (1972). A garbage can model of organizational choice. *Administrative Science Quarterly*, **17**, 1–25.

Cooper, R. & Kleinschmidt, E. (1991). New product processes at leading industrial firms. *Industrial Marketing Management*, **20**, 137–147.

Copeland, T. (2001). The real-options approach to capital allocation. *Strategic Finance*, (October), 33–37.

Dalkey, N. & Helmer, O. (1963). An experimental application of the delphi method to the use of experts. *Management Science*, (April), 458–467.

De Meyer, A. & Mizushima, A. (1989). Global R&D management. *R&D-Management*, **19** (2), 135–146.

Durand, T. (1992). Dual technological trees: Assessing the intensity and strategic significance of technological change. *Research Policy*, **21**, 361–380.

Gassmann, O. (1997). *Internationales F&E-Management*. Munich: Oldenbourg.

Gassmann, O. & Zedtwitz, M. von (1998). Organization of industrial R&D on a global scale. *R&D Management*, **28** (3), 147–161.

Gassmann, O. & Zedtwitz, M. von (1999a). New concepts and trends in international R&D organization. *Research Policy*, **28**, 231–250.

Gassmann, O. & Zedtwitz, M. von (1999b). Organizing virtual R&D teams: Towards a contingency approach. In: D. F. Kocaoglu (Ed.), *Technology and Innovation Management—Setting the Pace for the Third Millennium*. Proceedings of the Portland International Conference on Management of Engineering and Technology (PICMET), Portland, July 25–29.

Hippel, E. von (1986). Lead users: A source of novel product concepts. *Management Science*, (July), 791–805.

Hippel, E. von (1988). *The sources of innovation*. New York: Oxford University Press.

Hippel, E. von & Herstatt, C. (1991). From experience: Developing new product concepts via the lead user method: A case study in a 'low tech' field. *Journal of Product Innovation Management*, **9** (3), 213–222.

Iansiti, M. (1998). *Technology integration: Making critical choices in a dynamic world*. Boston, MA: Harvard Business Press.

Imwinkelried, B. (1996). Internationales F&E-Projektmanagement bei ABB am Beispiel der Gasturbine GT24/26.

In: O. Gassmann & M. von Zedtwitz (Eds), *Internationales Innovationsmanagement* (pp. 83–104). Munich: Vahlen.

Katz, R. & Allen, T. (1982). Investigating the not invented here (NIH) syndrome: A look at the performance, tenure, and communication patterns of 50 R&D project groups. *R&D Management*, **12** (1), 7–19.

Kline, S. J. (1985). Innovation is not a linear process. *Research Management*, **28**, 36–45.

Kline, S. J. & Rosenberg, N. (1986). An overview of innovation. In: R. Landau & N. Rosenberg (Eds), *The Positive Sum Strategy—Harnessing Technology for Economic Growth* (pp. 275–306). Washington, D.C.: National Academy Press.

Kottler, P. (2000). *Marketing management*. Upper Saddle River: Prentice Hall.

Leonard-Barton, D. (1995). *Wellsprings of knowledge—building and sustaining the sources of innovation*. Boston, MA: Harvard Business School Press.

Liker, J., Kamath, R., Wasti, S. N. & Nagamachi, M. (1995). Integrating Suppliers into Fast-Cycle Product Development. In: J. Liker, J. Ettlie & J. Campbell (Eds), *Engineered in Japan* (pp. 152–191). New York: Oxford University Press.

Lynn, G. S., Morone, J. G. & Paulsen, A. S. (1996). Marketing and discontinuous innovation: The probe and learn process. *California Management Review*, **38** (3), 8–37.

Morelli, M., Eppinger, S. & Gulati, R. (1995). Predicting technical communication in product development organizations. *IEEE Transactions on Engineering Management*, **42** (3), 215–222.

Nest, R. J. (1974). An experiment in delphi estimation in marketing decision making. *Journal of Marketing Research*, (November), 447–452.

Nishiguchi, T. (Ed.) (1996). *Managing product development*. New York: Oxford University Press.

Nonaka, I. & Takeuchi, H. (1995). *The knowledge-creating company. How Japanese companies create the dynamics of innovation*. New York: Oxford University Press.

O'Connor, P. (1994). Implementing a stage-gate process: A multi-company perspective. *Journal of Product Innovation Management*, **11**, 183–200.

Porter, M. (1985). *Competitive advantage: Creating and sustaining superior performance*. New York: Free Press.

Roberts, E. (1968). Entrepreneurship and technology. *Research Management*, **11** (4), 249–266.

Roy, R. (1983). *Bicycles: Invention and innovation*. London: The Open University Press.

Schmoch, U., Hinze, S., Jäckel, G., Kirsch, N., Meyer-Krahmer, F. & Münt, G. (1995). The role of the scientific community in the generation of technology. In: G. Reger, & U. Schmoch (Eds), *Organisation of Science and Technology at the Watershed—The Academic and Industrial Perspective* (pp. 1–138). Heidelberg: Physica.

Schmoockler, J. (1966). *Invention and economic growth*. Cambridge: Harvard University Press.

Schumpeter, J. A. (1911). *Theorie der wirtschaftlichen Entwicklung*. Leipzig: Duncker and Humblot.

Schumpeter, J. A. (1939). *Business cycle*. New York: McGraw-Hill.

Shoemaker, P. J. H. (1995). Scenario planning: A tool for strategic thinking. *Sloan Management Review*, (Winter), 25–40.

Thomke, S. (1995). *The economics of experimentation in the design of new products and processes*. Ph.D. dissertation, MIT Sloan School of Management.

Tushman, M. (1979). Managing communication networks in R&D laboratories. *Sloan Management Review*, (Winter), 37–49.

Tushman, M. & Anderson, A. (1986). Technological discontinuities and organizational environment. *Administrative Science Quarterly*, **31**, 436–465.

Van de Ven, A., Angle, H. & Poole, M. (Eds) (1989). *Research on the management of innovation: The Minnesota studies*. New York: Oxford University Press.

Van de Ven, A., Polley, D. E., Garud, R. & Venkataraman, S. (1999). *The innovation journey*. New York: Oxford University Press.

Wheelwright, S. & Clark, K. (1992). *Revolutionizing product development*. New York: The Free Press.

Wilson, J. Q. (1966). Innovations in organizations: Notes toward a theory. In: J. D. Thompson (Ed.), *Approaches to Organizational Design* (pp. 193–218). Boston.

Witte, E. (1973). *Organisation für Innovationsentscheidungen—das Promotorenmodell*. Vandenhoeck and Ruprecht. Göttingen: Schwartz and Co.

Yin, R. (1994). *Case study research: Design and methods*. Thousand Oaks: Sage.

Zedtwitz, M. von & Gassmann, O. (2002a). Market vs. technology drive in R&D internationalization: Four different patterns of managing research and development. *Research Policy*, **31** (4), 569–588.

Zedtwitz, M. von & Gassmann, O. (2002b). Managing customer-oriented research. *International Journal of Technology Management*, **24** (2/3), 165–193.

The International Handbook on Innovation
Edited by Larisa V. Shavinina

An Analysis of Research and Innovative Activities of Universities in the United States

Yukio Miyata

Department of Economics, Osaka Prefecture University, Japan

Abstract: University-Industry collaboration in the United States has been revitalized in the last two decades, but research at universities is still primarily federally funded basic research—a pattern established after World War II. While universities actively obtaining patents, they cannot finance their research budget through licensing revenues alone. This chapter argues that high quality of research has to be supported by the federal government, and that the fruits of university research on regional economy takes a long time to be realized.

Keywords: Innovation; Innovative activities; Research; University-industry collaboration; United States; Technological change.

Introduction

Universities have been involved in promoting scientific knowledge (seeking the truth) and educating people. This chapter analyzes research and innovative activities of universities in the United States, focusing on the university-industry collaboration in research. According to Schmpeter (1934, p. 66), innovations are: (1) an introduction of new products (or products with improved quality); (2) new method of production; (3) new markets and distributing channels; (4) new sources of supply of inputs; and (5) new organizations of an industry. Although the role of business faculty of universities in helping to create new markets, new distribution channels, or new organizations should not be denied, this chapter focuses on how science and engineering knowledge, as well as personnel of universities contributes to creating new products and processes.

As innovations do not stop with application or acquisition of patents but include a necessary commercialization process, private businesses are the agents of innovation. Universities, however, which supply research personnel and scientific/technological knowledge, are an important component of the national innovation system. Understanding the way in which U.S. universities and industry relate is an important step in promoting innovation. This collaboration is now the envy of the world, and an analysis of its actual practice has important policy implications.

The collaboration between university and industry occurs in many forms: contracted research from industry to universities, cooperative research between university and industry personnel, licensing of university-owned patents to industry, informal information exchange between university and industry personnel, consultation by university personnel, and establishing start-up firms by faculty members or graduates of universities in order to commercialize their research results. In this chapter, all of these activities are referred to as 'university-industry collaboration'.

The chapter is organized as follows. The next section provides theoretical background of university-industry collaboration. Then in the next section, after briefly reviewing the history of U.S. university research, the current status of university-industry collaboration is analyzed. The section after that discusses potential problems in university-industry collaboration, followed by a section of conclusions and policy implications.

Theoretical Background

According to Kline & Rosenberg (1986), innovations are traditionally considered to be based on a 'linear model' in which basic research, applied research, development, production, and marketing occurs in succession (see Fig. 1). The fruits of basic research are the fundamental understanding of nature, and are expressed in the form of academic publication. People who read these publications develop theories into

Figure 1. Linear model.

Source: Drawn by the author from Kline & Rosenberg (1986).

commercial products; therefore, the benefits of basic research often cannot be collected by sponsors, normally being utilized by others who do not pay the cost of research. Because of this 'spillover effect', private businesses are not willing to invest in basic research, so that, under the market mechanism, basic research would have less investment than that at the socially optimal level. Thus, governments provide universities with funds to conduct basic research. According to the linear model, as long as governments support basic research at universities, firms that find potential commercial benefits in the basic research results will continue on with applied research, development, and manufacturing, thus coming up with the innovations.

However, this simple smooth process of innovation, stated by the liner model, has been criticized as 'unrealistic' or 'too optimistic', and replaced by the 'chain-linked model'. According to the chain-linked model by Klein & Rosenberg (1986), innovations result from the process which consists of the recognition of potential markets, invention and analytical design, detailed design and testing, redesign and production, and distribution, sales, and marketing (see Fig. 2). However, more importantly, among these stages, there are feedback loops. According to this model, it is rare that newly generated knowledge from research leads to innovations. Innovations often utilize and rely on existing knowledge. Scientific and technological knowledge is related to each stage of Fig. 2. Whenever problems occur, one consults with existing scientific and technological knowledge. However, if this existing scientific and technological knowledge cannot solve the problem, new research will be initiated. The universities' role is to fill the pool of scientific and technological knowledge so that firms can utilize it whenever they need.

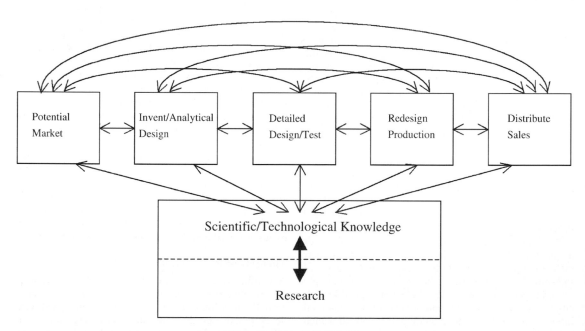

Figure 2. Chain-linked model.

Source: Drawn by the author from Kline & Rosenberg (1986).

University-industry collaboration can be understood as an application of the chain-linked model. Firms have recognized that it is too optimistic to expect investing in basic research at a central laboratory automatically leading to innovations. They must shift their research emphasis to short-term applied research and rely on universities to do basic research. Moreover, connecting scientific and technological knowledge of universities to industry needs is now important on the national level. In the 1950s and 1960s, U.S. firms technologically were far ahead of foreign competitors. When universities supplied scientific and technological knowledge through academic publication, only U.S. firms could utilize such information. However, today's foreign firms can utilize academic research results as soon as they are published in academic journals. So, it is important to build a more direct bridge between universities and industry in the U.S. for exploiting the excellent research capability of U.S. universities, which is shown by the fact that they dominate in the number of Nobel laureates and attract students from all over the world.

However, the university-industry collaboration may not generate the expected benefits for the following reasons. First, university research personnel are good at basic research but may be poor at considering marketability. Thus, firms should not expect universities to generate research results that are ready for manufacturing. Instead of completely contracting out research to universities, firms should keep in touch with university research personnel so that a feedback loop is built and a synergetic effect is generated.[1] In addition, it is not proper for firms to cut in-house research budgets by relying on universities for basic research. Even though basic research has the spillover problem mentioned above, firms should invest in basic research for accumulating scientific and technological knowledge on their own. Otherwise they will not be able to understand recent research trends, making it difficult for them to discover new research topics and partners from university research personnel. Moreover, patents are far from innovations. Technological expertise of licensees is necessary to transform licensed technology into commercial products, and technological expertise is accumulated through in-house research (Rosenberg, 1990).

Second, university research personnel may avoid firms' requests. While several universities clearly consider the contribution of faculty to regional industries through cooperative research, licensing, or consulting as 'criteria for promotion', however, promotion of faculty depends on the quantity and quality of academic research papers. Accepting money from

industry is beneficial for their research purpose, but university research personnel are neither employees nor sub-contractors of sponsoring firms. To prevent shirking, firms should again keep in touch with research progress at universities.

Third, from a social perspective, the university-industry collaboration can be criticized as an exploitation of university research capabilities that should be available to the general public. The government has been supporting university research. When a firm utilizing university research results is able to commercialize it through new products, consumers have to pay twice (tax to support university research and the price of the product). Also, research personnel should conduct research for public interest, such as assessments of pollution damage or risks of new technology, which sometimes conflicts with industrial interests.

Fourth, university-industry collaboration may deteriorate the research capabilities of universities. During university-industry collaboration, firms often ask university research personnel to withhold announcement of research progress or research results even to colleagues at the university. University research has been developed with the free exchange of information. If one knows what other research personnel are doing, they can exchange critical opinions with each other in early stages of their research, making corrections quickly and avoiding duplication of research efforts (Cohen et al., 1998). The secrecy hinders information exchange among research personnel in the academic community and is detrimental to the quality of research conducted by universities. In addition, because graduate education is often based on research experience, deteriorating research quality leads to a decline in its quality, and is costly to industry in the long-term.

Analysis of University-Industry Collaboration

Universities in the U.S. began as private institutes established by churches, followed by state universities. The expansion of state universities was owed to the Morrill Act of 1862, which allowed states to raise funds for their state universities by selling federal lands. The federal government and Congress played a role in this act, but besides that, the federal role in university research was minimal until World War II. According to Table 1, federal assistance for university research was limited to agriculture through agricultural experiment stations. State governments were bigger sponsors of university research than the federal government. Money from industry accounted for 12%, which is greater than the current level (mentioned later). Moreover, money the universities themselves used to fund research came from the gifts or donation from wealthy businesspeople. Hence, the relationship between industry and universities was strong before World War II and

---

[1] Completely contracting out basic research to universities is a strategy based on the linear model rather than the chain-linked model.

Table 1. Estimated university research funding sources in 1935.

Sources	Shares
State appropriations for agricultural experiment stations	14
Other state appropriations spent for research	14
Federal grants to universities for agricultural experiment stations	10
Non-profit foundations	16
Industry	12
University	34

Source: Sommer (1995, p. 7), Mowery & Rosenberg (1989, p. 93).

universities often contributed to the development of regional industry (Rosenberg & Nelson, 1994).

During World War II, many university research personnel worked on military related research projects including 'The Manhattan Project'. The role of university research was highly recognized by the federal government: the advancement of scientific knowledge generated by university research was expected to solve economic, social, and national security problems the nation faced. Furthermore, 'The Sputnik Shock' of 1957, which lead to the National Defense Education Act of 1958, increased federal basic research money to universities. Figure 3 indicates that R&D money that universities used, increased rapidly in the 1960s, so did federal research money to universities. Also as Fig. 4 indicates, the federal government has been the largest sponsor of university research, accounting for more than 60% of university research spending, while the share of industry or local governments has been less than 10%.

Table 2 indicates which federal agency supports university research. The National Institutes of Health (NIH) has been the largest federal sponsor of university research. The National Science Foundation (NSF), whose function is to support university research, is competing with the Department of Defense (DoD) for second place for federal sponsorship. The NSF is far from a centralized agency that administrates university research. U.S. policy supporting university research can be characterized as decentralized, where each agency supports 'directed basic research' so that research results would contribute to the needs of that agency.

As shown later in detail, federal research money to universities is heavily directed toward basic research. Although, compared with industry, universities play a minor role in performing research and development, their share as a performer of basic research increased to above 50% during the 1960s and has since remained at a high level (see Fig. 5). If we include the government-owned laboratories, which are contracted out to universities to operate, the percentage reaches about 60%.[2]

U.S. universities continued to remain in contact with industry but the relationship between universities and industry weakened in the 1960s due to an increase in federal funding. Research personnel at universities were interested in basic research. However, at the end of 1970s, federal funding stagnated and universities turned to industry once again as an important financial source. Industry also began seeing universities as a source of scientific and technological expertise. Firms no longer rely on 'the linear model' and reduced basic research at their central laboratories. As firms started utilizing outside information sources, they became interested in cooperative R&D with other firms as well as collaboration with universities. Moreover, in the fields of biotechnology and computer software, university research results were expected to directly result in commercial products.[3] University personnel started establishing start-up firms to commercialize their own research results, which was necessary for emerging high-tech industries. The interests of industry and university, therefore, coincided. In addition, policy-makers expressed concerns that, although U.S. universities were excellent in research, they did not help U.S. firms boost their competitiveness. Congress passed the Patent and Trademark Amendments of 1980 (Bayh-Dole Act), which allowed universities to own patents resulting from federal research money and to license them to firms. Then, the relationship between industry and universities was strengthened in the 1980s once again.

However, even after 1980, as Fig. 4 shows, the major source of university research money remained the federal government. Because it is difficult to see the trend, Fig. 6 focuses on the ratio of research money from industry to universities, to the amount of research

---

[2] Shapley & Roy (1985) points out that U.S. university research, including second-tier universities, had been heavily inclined towards basic research and neglected engineering, causing a decline in competitiveness of U.S. manufacturing. However, Rosenberg & Nelson (1994) states that Shapley and Roy overstated the lack of applied research in the U.S. university research. For example, in life science, the funding for medical research has been greater than for biological research. Although, in some cases, university research personnel have conducted research to satisfy their intellectual curiosity, in many others, they have had only vague ideas of the potential application of their research results.

[3] The emergence of biotechnology does not mean the revival of the linear model. While biotechnology seems to directly result from laboratory research, a great amount of clinical testing is necessary. Also, because firms conduct in-house research secretly, if one only looks at commercialized results, they are seemingly breakthrough innovations. However, in actuality, in-house research is incremental, and feedback loops among marketing, invention, designing, and manufacturing exist (McKelvy, 2000, p. 273).

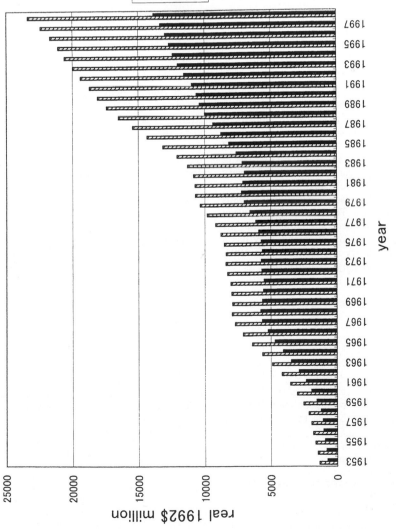
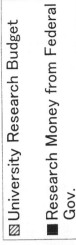

Figure 3. University research spending.

Source: USNSF (2000).

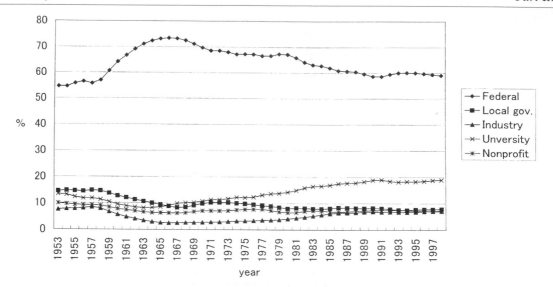

Figure 4. Sources of university research.

Source: USNSF (2000).

money that universities use. The ratio declined in the 1960s and recovered in the 1980s but was still smaller than 10%. However, research money from industry is often used in university-industry-government cooperative research projects, so industry money is estimated to affect 20%–25% of university research (Behrens & Gray, 2001). The ratio of research money from industry to universities to the research money that industry spends is shown in Fig. 7. It also recovered in the 1980s but still only to about 1.3%. The relationship between university and industry has been strengthened, but the magnitude is still not so strong. Biotechnology has a stronger tie between industry and university. In 1994, research money from industry to universities is 1.5 billion dollars, which accounted for 11.7% of life science research funds that universities obtained from external sources such as federal government or nonprofit organizations, and for 13.5% of life science research funded by food-tobacco and pharmaceutical industries (Blumenthal et al., 1996; USNSF 2000). However, even in biotechnology, research money from industry does not dominate the research budget at either, the university, or industry level.

The source of research funding is not so different between private and public universities as Table 3 shows. Private universities rely more on money from the federal government than public universities do, but money from university and non-federal government

Table 2. Shares of federal agencies in research funding to universities.

Year	NIH	NSF	DoD	NASA	DoE	DoA	Others
1970	35.1	15.4	14.7	8.9	6.8	4.4	14.7
1975	47.8	18.0	8.4	4.5	5.5	4.5	11.3
1980	47.2	16.1	11.6	3.7	6.7	5.1	9.7
1985	49.8	15.8	14.8	3.7	5.6	4.6	5.6
1990	52.3	14.5	13.3	5.2	5.5	3.8	5.5
1995	52.6	14.5	13.3	5.9	5.0	3.6	5.0
1998	56.6	14.4	10.5	5.4	4.4	3.4	5.3

NIH: National Institutes of Health
NSF: National Science Foundation
DoD: Departrnent of Defense
NASA: National Administration of Space and Aeronautics
DoE: Department of Energy
DoA: Department of Agriculture.

Source: USNSF (2000).

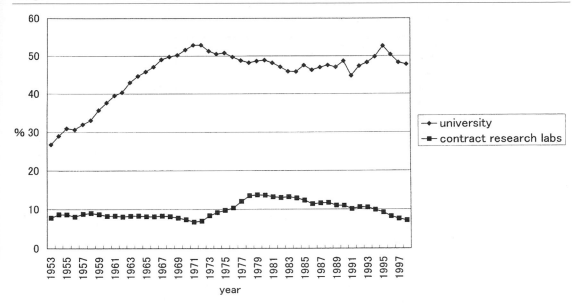

Figure 5. The share of university as performer of basic research.

Source: USNSF (2000).

sources account for a higher percentage for public universities than for private ones. However, the reliance on money from industry is the same between public universities and private universities. The percentage increased from 1977 to 1987, but it did not change between 1987 and 1997.

Table 4 shows the ratio of basic research, applied research, and development in university research. Even since university-industry collaboration was revitalized in the 1980s, two-thirds of university research funds have been used for conducting basic research. Development accounts for less than 10%. University personnel who are actually involved in collaboration with industry admit that their research has been moved away from basic research toward applied research and development (Cohen et al., 1994). However, this is not true of all university research.

A possible reason for the stable ratio of research direction at universities is that research money from industry to universities is still relatively small compared with the total university research budget. Moreover, it is interesting to point out that, as Table 5 indicates, basic research accounts for 60% of all research money from industry to universities. It would seem that industry does not think much of the universities' abilities to conduct development projects in which industry has expertise. Industry wants access to university basic research that it cannot adequately

Table 3. Comparison of research funds source between private and public universities.

	Federal Government	Non-Federal Government	Industry	University	Others*
1977					
Private	77.3%	2.3%	3.9%	6.4%	10.0%
Public	61.3	13.0	3.1	16.1	6.5
1987					
Private	74.4	2.3	7.0	8.6	7.8
Public	52.9	11.7	6.3	22.8	6.3
1997					
Private	72.3	2.1	7.0	10.1	8.5
Public	53.4	10.4	7.1	22.8	6.3

* 'Others' includes non-profit organizations.

Source: USNSF (2000).

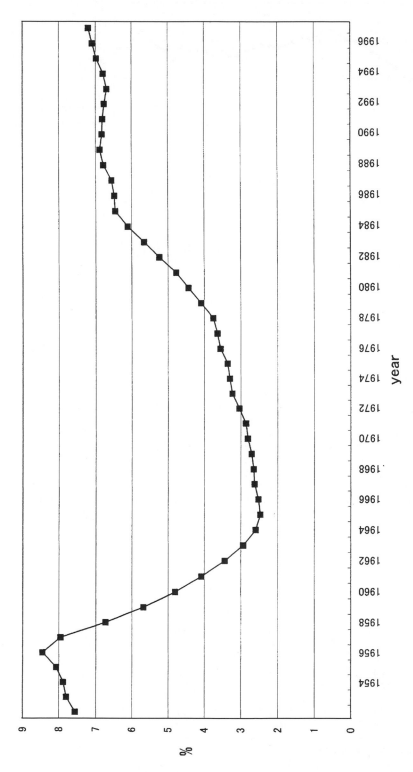

Figure 6. The ratio of industry–university money to university research spending.

Source: USNSF (2000).

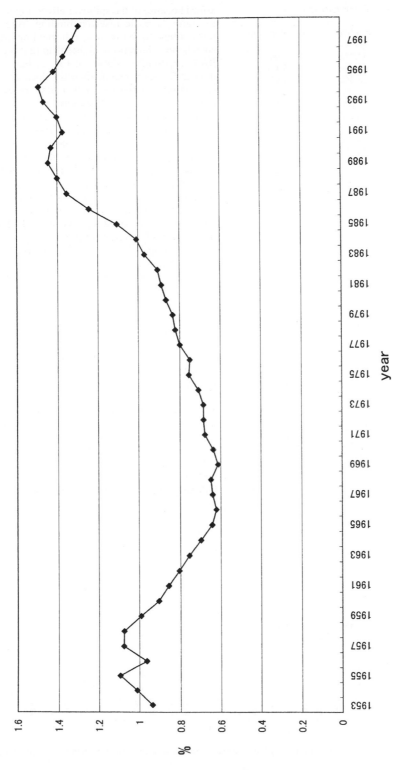

Figure 7. The ratio of industry–university money to industry research spending.

Source: USNSF (2000).

Table 4. *Share of directions of university research.*

Year	Basic	Applied	Development
1953	45.2%	49.0%	5.9%
1955	52.5%	41.4%	6.1%
1960	68.8%	26.3%	4.9%
1965	76.5%	19.0%	4.4%
1970	76.7%	18.6%	4.6%
1975	69.5%	26.2%	4.4%
1980	66.8%	25.1%	8.0%
1985	68.1%	24.5%	7.3%
1990	65.7%	26.0%	8.3%
1995	67.3%	24.8%	7.9%
1998	68.7%	24.1%	7.2%

Source: USNSF (2000).

Table 5. *Share of directions of research money from industry to university.*

Year	Basic	Applied	Development
1953	63.4%	31.4%	4.9%
1955	63.1%	31.5%	5.4%
1960	61.0%	32.6%	6.4%
1965	63.9%	31.3%	4.8%
1970	65.6%	26.5%	7.4%
1975	60.7%	32.5%	6.8%
1980	59.0%	33.4%	7.3%
1985	61.7%	31.4%	6.9%
1990	60.5%	32.4%	7.1%
1995	61.4%	31.7%	7.0%
1998	63.6%	29.9%	6.5%

Source: USNSF (2000).

afford because of the spillover effect. Figure 8 indicates the ratio of basic research funding from industry to universities to basic research funding that industry spends on. The ratio is higher than that regarding total R&D expenditure shown in Fig. 7, but it is still less than 20%. As a result, industry does not completely rely on universities to conduct basic research through university-industry collaboration.

Another reason for the large share of basic research at universities even after university-industry collaboration was revitalized in the 1980s is that development work is conducted off-campus through start-up firms which were established by faculty members or graduates to exploit their research results. So, on-campus research remains basic research (Cohen et al., 1998).

Table 6 shows the directions of research money from the federal government to universities. Federal research money actually consists of money from several departments and agencies such as the National Institutes of Health, National Science Foundation, Department of Defense, and National Aeronautics and Space Administration. The share of basic research remains the largest.

Table 7 shows that the composition of non-federal (mainly state) government money had been oriented toward basic research in the 1960s, but since the early 1970s, the percentage of basic research declined. It should be noted that, today, research money from state governments to universities is similarly proportioned for basic research, applied research, and development as research money from industry is. While both state governments and industry like to utilize the basic research strengths of universities, they also like to avoid any spillover from their basic research. State governments do not want the fruits of research funded with their money to diffuse beyond state borders. The federal government does not care much about spillover,

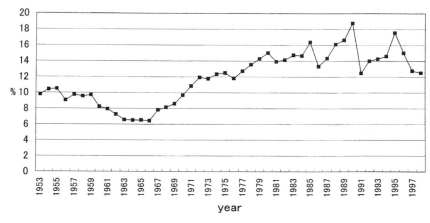

Figure 8. The ratio of industry–university basic research money to basic research spending by industry.

Source: USNSF (2000).

Table 6. Share of directions of research money from federal government to university.

Year	Basic	Applied	Development
1953	54.7%	39.6%	5.7%
1955	61.0%	33.0%	6.0%
1960	75.2%	20.6%	4.2%
1965	80.9%	15.0%	4.1%
1970	78.5%	16.6%	5.0%
1975	73.7%	22.9%	3.4%
1980	70.6%	21.0%	8.4%
1985	72.1%	20.3%	7.6%
1990	69.3%	21.5%	9.2%
1995	71.2%	20.3%	8.6%
1998	72.3%	20.1%	7.6%

Source: USNSF (2000).

Table 7. Share of directions of research money from non-federal government to university.

Year	Basic	Applied	Development
1953	16.4%	76.4%	7.2%
1955	27.8%	64.7%	7.5%
1960	49.9%	42.9%	7.0%
1965	63.1%	31.1%	5.8%
1970	75.6%	21.6%	2.7%
1975	61.1%	32.5%	6.4%
1980	59.3%	33.5%	7.3%
1985	61.7%	31.4%	6.9%
1990	60.5%	32.4%	7.1%
1995	61.4%	31.7%	7.0%
1998	63.5%	29.9%	6.6%

Source: USNSF (2000).

though they recently begin worrying about spillover of federally funded research results to countries who are good at commercializing research results published by the research personnel of U.S. universities.

Research funding allocation to each university is basically decided through a peer review in which prominent scientists can receive research funding based on their ability, excluding political consideration or 'pork barrel politics'. As a result, top ranked universities that have many able researchers, tend to obtain research money from the federal government and non-profit organizations. Universities that do not receive adequate research money from those sources are more willing to accept money from industry.

Figure 9 plots the ratio of research money from industry to universities to total university research budget (INDRAT) against the total university research budget, indicating no clear trend. If a regression analysis of FY 1997 is done for 61 universities between INDRAT and total university research budget, the coefficient is not statistically significant even at two-sided 10%.

$$INDRAT = 0.059 + 0.724 \text{ Total Research Budget}$$

$$(t = 1.27, R^2 = 0.027)$$

If the regression analysis is done between INDRAT and research money from the federal government to university, the relationship becomes weaker.

$$INDRAT = 0.067 + 0.491 \text{ Research Money from} \\ \text{Federal Government}$$

$$(t = 0.58, R^2 = 0.006)$$

Highly ranked universities receive a large amount of research money from the federal government, so the reliance on money from industry becomes weaker.

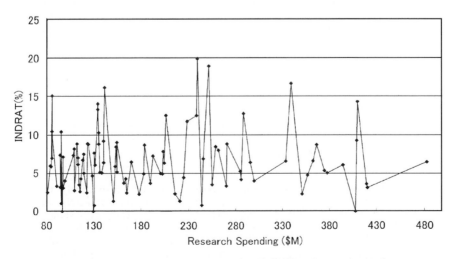

Figure 9. Industry–university money ratio (INDRAT) and research spending.

Source: USNSF (2000).

In addition, Fig. 10 shows that the number of patents issued to universities has increased since particularly the mid-1980s, several years after the Bayh-Dole Act of 1980 allowed universities to own patents resulting from federal research funding. Although the direction in university research did not move toward development so much, the surge in patents implies that university research personnel became more willing to apply for patents.[4] Another reason is that patents resulting from basic research, such as in biotechnology, have been increasing. Since 1981, three life-science/drug-related patent classes have been top-3 patents granted to universities. The share of them was 18.81% in 1981 but increased to 25.33% in 1991, then reached 40.97% in 1998 (USPTO 2000).

In biotechnology, cooperative research between university and industry is productive for generating patents. According to Blumenthal et al. (1996), amongst the 'Fortune 500' pharmaceutical firms, the number of patents issued per $10 million research spending is 1.7 for cooperative research with university, and 1.2 for 'elsewhere'.[5] For firms that are not in the 'Fortune 500', the number of issued patents per $10 million research spending is 6.7 for cooperative research with university and 3.5 for 'elsewhere'.

Mansfield (1998) investigates the percentage of innovations that could not have been produced without academic research and those which were significantly supplemented by academic research. The survey considered the contribution of academic research conducted between 1986–1994, to innovations generated in 1994. Mansfield previously did a similar survey investigating the contribution of academic research conducted between 1975–1985, to innovations generated in 1985, the results are listed together in Table 8. The survey covered 77 major firms that do not include any start-up (spin-off) firms of universities. Drug/Medical products and Instruments (including medical devices) have high scores, but the figures are 30% at most. In other industries, figures are not so high. The contribution of academic based innovations to total sales or to cost reduction was not so large. Moreover, the last row of Table 8 indicates that it takes a long time (6–7 years) to commercialize academic research results, while the time-span became a bit shorter in the 1990s compared with the 1980s. Large firms tend to keep a proper distance from university research.

While both are research intensive industries, compared with pharmaceuticals, in the semiconductor industry, industrial research has been ahead of university research. Universities' role had been limited to research personnel and entrepreneurs. The semiconductor industry organized Semiconductor Research Corporation (SRC) in 1982, which pooled research money from firms and supported research at universities. Since then, SRC has been effective in educating graduate students for industrial opportunities and in providing a desirable scientific knowledge base for industrial practice, although drastic innovations remain to be created at industrial research laboratories (See Gutmann, this handbook).

According to AUTM (1998), in FY 1997, there are 58 universities, which have at least 10 licenses, and total number of license agreements is 2,098. Among them, 10.1% went to start-up firms that were established under license from universities, 49.3% to non-start-up small firms (less than 500 employees), and 40.6% to large firms. When the Bayh-Dole Act was enacted in 1980, exclusive licenses from universities to large firms would be invalid after eight years of licensing or five years of successful commercialization, whichever comes first, so that large firms would not dominant license acquisitions. Since the dominance of large firms did not happen and the restrictions significantly prevented large pharmaceutical firms from commercializing university research results, the restriction was lifted in 1984. Even today, licenses actually go to small firms. Because the number of small firms are many, it may be natural for small firms to account for a greater share of the licensees by universities, though a large individual firm may be able to purchase many licenses from universities.

Inter-organizational technology transfer is difficult unless the recipient has technological expertise. The fact that there are many small firm licensees implies that these small firms have expertise to utilize university research. AUTM (1999) points out that more than 90% of licenses to start-up firms are exclusive, while about half of all licenses to non-start-up small firms or large firms are exclusive. Large firms tend to keep a proper distance from universities, emphasizing informal information exchange between university and industry research personnel. In contrast, start-up firms want a more direct connection to university research (Etzkowitz, 1999).

Tables 9 and 10 give the results of a survey conduced by Lee (2000) that asked university personnel and industry personnel respectively to rank the actual benefits of cooperative research from 1 (lowest) to 5 (highest) in scale. The survey shown in Table 9 covers 422 faculty members of 40 universities randomly chosen from the top 100 in terms of their total research budgets. University personnel were likely to answer that, through cooperative research with industry, they obtained money and ideas for their own research. The

---

[4] For universities such as Stanford University or University of California-Berkeley that have been active in collaborating with industry, the biotechnology revolution of the 1970s increased their patenting activities prior to the Bayh-Dole Act. However, for many other universities, the enactment of Bayh-Dole Act in 1980 was an important opportunity (Mowery, Nelson, Sampat & Ziedonis, 1999).

[5] Research carried out 'elsewhere' is mainly in-house R&D but also includes cooperative R&D with other firms or with national laboratories.

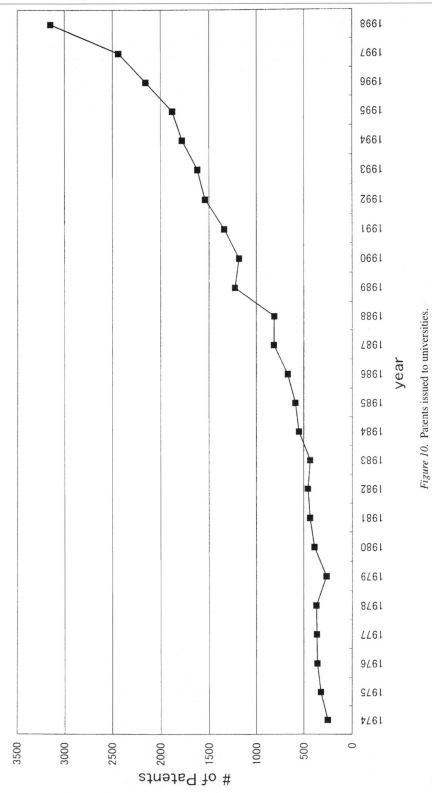

Figure 10. Patents issued to universities.

Source: USNSF (2000).

Table 8. *Innovations based on academic research.*

	Innovations that could not have been developed without recent academic research results (%)		Innovations that were developed with very substantial aid from recent academic research results (%)	
	1986–1994	1975–1985	1986–1994	1975–1985
Products				
Drug/Medical Products	31	27	13	17
Information Processing	19	11	14	17
Chemical	9	41	11	4
Electrical	5	6	3	3
Instruments	22	16	5	5
Machinery	8	N.A.	8	N.A.
Metals	8	13	4	9
Mean1	15	13	8	9
Contribution to Sales*	5.1	3.0	3.8	2.1
Processes				
Drug/Medical Products	11	29	6	8
Information Processing	16	11	11	16
Chemical	8	2	11	4
Electrical	3	3	2	4
Instruments	20	2	4	1
Machinery	5	N.A.	3	N.A.
Metals	15	12	11	9
Mean	11	10	7	7
Contribution to Cost Reduction†	2.0	1.0	1.5	1.6
Years needed to commercialize academic research results	6.2	7.0	5.1	6.7

* The percentage of new products sales to total sales.
† The percentage of cost reduction by new processes.
The survey of 1975–1985 did not cover the machinery industry.

Source: Reprinted of Table 1 of Mansfield (1998) with permission from Elsevier Science.

Table 9. *Faculty benefits experienced with industry-sponsored R&D.*

Faculty benefits	Rating
Acquired funds for research assistant and lab equipment	3.87
Gained insights into one's own academic research	3.82
Supplemented funds for one's own academic research	3.55
Field-tested one's own theory and research	3.50
Acquired practical knowledge useful for teaching	3.04
Created student internships and job placement opportunities	2.97
Led to patentable inventions	2.55
Created business opportunities	2.14

Source: Lee (2000).

Table 10. *Industry benefits derived from collaboration with university.*

Industry benefits	Rating
Gaining access to new research	4.01
Developing new product/process	3.74
Maintaining relationship with the university	3.61
Developing new patents	3.37
Solving technical problems	3.15
Improving product quality	2.38
Reorienting R&D agenda	2.34
Recruiting students	1.75

Source: Lee (2000).

enhancement of business opportunities was not highly ranked. Lee (2000) also points out that university personnel tend to obtain what they wanted. Even though lowly ranked, university personnel who sought to expand their business opportunities through cooperative research with industry achieved such goals.

The survey shown in Table 10 covers 140 firms of AUTM (Association of University Technology Mangers)[6] members including start-up firms as well as large firms. Industry personnel highly ranked 'access to new research' and 'develop new product/process.' Solving short-term problems such as 'technical problems' or 'quality improvement' was not highly ranked. The survey also indicates that, compared with cooperative research between universities and large firms, the one between universities and start-up firms tend to focus on business opportunities and does not contribute much to the enhancement of research and education capabilities of university. The distance between universities and large firms is greater than that between universities and small ones.

Table 11 indicates the correlation among the number of invention disclosures, the number of patent applications, the number of licensing agreements, and the amount of license revenues based on each university's data of FY1999 (AUTM 2000). The relationship between invention disclosures and patent applications is strong, but the amount of license revenue is not related to any other variable. It is very uncertain which invention will result in huge license revenue. USGAO (1998) states that a small number of highly successful patents or 'blockbuster' patents, which are often in the field of medical/life science, account for a large portion of the license revenue of a university. And those blockbuster patents often resulted from federally funded research. Table 12 indicates how much license

Table 11. Correlation among outputs of university research (FY 1999).

	Invention Disclosure	Patent Application	License Agreement	License Revenue
Invention Disclosure	1			
Patent Application	0.87	1		
License Agreement	0.76	0.61	1	
License Revenue	0.33	0.29	0.39	1

Source: Calculated by the author from AUTM (2000).

[6] AUTM is an organization whose purpose is to promote technology transfer from universities to industry. Firms that have membership in AUTM are strongly interested in collaborating with universities.

Table 12. Share of license revenue resulted from federally funded research (FY 1996).

University	Share
Johns Hopkins Univ. (Medical)	60.9%
Johns Hopkins Univ. (Applied Phys Lab)	25.7%
Univ. of Washington	16.9%
Stanford Univ.	92.7%
Univ. of Michigan	21.5%
Univ. of Wisconsin-Madison	98.9%
Harvard Univ.	70.0%
Columbia Univ.	95.6%
Michigan State Univ.	97.7%
Univ. of California*	51.8%

* The entire University of California campuses.

Source: USGAO (1998).

revenue of the elite universities come from the results of federally funded research results. In some universities, the percentage is above 90. It is unlikely that cooperative research between industry and university will lead to a blockbuster patent from scratch. A university's research quality has been built on money provided by the federal government for a long time, and nowadays, industry utilizes the research capabilities of universities by supporting only 10% of the research budget of the universities.

License revenue is not only uncertain but also inadequate to support university research. Table 13 compares license revenue and research spending by universities in FY1997. The ratio of license revenue to total university research budget is very small. Most of the universities have a value of less than 5%. The ratio of license revenue to research money from industry to university is also small. Only seven universities have values of 100% and the majority of universities have values of less than 25%. The license revenue accounts for a tiny portion of the total research budget. It is difficult for universities to finance their research budget with license revenue.

However, license revenue represents only small portion (usually 2%–5%) of sales of new products resulted from licensed technology. According to ATUM (2000) which assumes that 83% of the license income is associated with product sales, the royalty rate is 2% of product sales, and $151,000 is necessary to support an employee, in FY 1999, licensed technology generated $35.8 billion of sales and 237,100 jobs. Moreover when a firm receives a license, it increases investment to commercialize that technology. Licenses from universities induced $5.1 billion investment, supporting 33,800 jobs. As a result, the total impact of universities on North American economy is $40.9 billion and 270,900 jobs. This is a tiny compared with Gross Domestic Product of the U.S. ($9000 billion in 1999), but greater than the amount of research spending by universities ($26 billion in 1998). While

Table 13. Relative magnitude of license revenue of universities (FY 1997).

The Ratio to Total Research Budget		The Ratio to Research Money from Industry	
Range	No. of Universities	Range	No. of Universities
$5\% \leqq X$	10	$100\% \leqq X$	7
$1\% \leqq X < 5\%$	25	$50\% \leqq X < 100\%$	9
$0.5\% \leqq X < 1\%$	15	$25\% \leqq X < 50\%$	13
$0.2\% \leqq X < 0.5\%$	15	$10\% \leqq X < 25\%$	18
$0.1\% \leqq X < 0.2\%$	4	$5\% \leqq X < 10\%$	9
$X < 0.1\%$	4	$X < 5\%$	17
Total	73	Total	73

Source: Calculated from AUTM (1998).

universities cannot earn license revenues to finance their research spending, the impact of their innovations is greater than the research spending, implying the justification of government support for university research.

Table 14 shows the inputs and outputs of research of the elite universities with a high quality of research. As mentioned before, the elite universities receive a large amount of research money from industry, but money from the federal government is much larger, so money from industry is proportionally smaller in total of a research budget. The University of Michigan has not generated a blockbuster patent yet, so its license revenue is small compared with the other three universities. Table 14 also indicates that even elite research universities that generate many patents and licenses and earn license revenue cannot finance their research budget from it alone. Their high quality research has to be federally financed.

In the same way as university-industry collaboration was revitalized in the late 1970s, the university's role in promoting regional development has been recently re-emphasized. Stanford University and the adjacent Stanford Research Park have successfully attracted many high-tech firms, and have become a nucleus of development for the greater Silicon Valley area. Also, the state of North Carolina has created a research park called Research Triangle Park, through three prominent universities in the state; University of North Carolina, Duke University, and North Carolina State University. These two cases are the envy of the world as well as other state governments.

What is the relationship between university research and innovation in a region? Despite being somewhat dated, the database created in 1982 by the U.S. Small Business Administration (SBA) is excellent because it identifies the actual location of where those innovations occurred, rather than the location of headquarters of innovators. As well as Feldman (1994) and Audretcsh & Feldman (1996) who used this database to support spillover effects of university research on regional innovations, Varga (1998) has conducted a

Table 14. Inputs and outputs of major university research (FY 1997).

	Stanford	MIT	Harvard	Michigan
Research Money from Industry (thousand dollar)	24,000	59,000	12,000	31,000
Research Money from Federal Govermnent (thousand dollar)	332,000	311,000	223,000	296,000
Total Research Budget (thousand dollar)	395,000	411,000	300,000	483,000
No. of Invention Disclosures	248	360	119	168
No. of New Patent Application	128	200	61	80
No. of Issued Patents	64	134	39	52
No. of License Agreements	122	75	67	47
License Revenue (thousand dollar)	51,760	21,210	16,490	1,780
No. of Start-up Firms	15	17	1	6

Source: AUTM (1998), USNSF (2000).

regression analysis on state and metropolitan area levels. In both cases, the number of innovations depends on the university research budget, the research budget of other firms, and the distance between the firm and the university. As the distance increases, the impact of research spending from other organizations declines. The rate of decline is faster in the case of spillover from other firms' research than that from university research. University research is more characteristic of the public good (spillover effect) than corporate research does. However, the distance between innovator and university is still important.

Furthermore, Varga has calculated how much a university's research budget has to increase in order to generate one more innovation of a region, which he called Marginal University Research Expenditure, MURE. And he categorizes regions into four groups according to calculated MURE values. In the lowest MURE group, which is the most innovative, in order to increase the number of innovations by one, a university's research budget must increase by 5%. In the second group, the university's research budget must increase by 33%. In the third group, the university's research budget must be three-fold. In the forth group, the university's research budget must be 50 times the current level, which is impossible to implement. Regions where university research has already led to innovations can increase the number of innovations easily by increasing university research budget. However, for regions in which university research has not successfully resulted in innovations, it is very difficult, if not impossible, to increase the number of innovations by increasing university research budget. Thus, a newly built research park is not always successful.

Although the SBA database, on which the above research works were based, is excellent in identifying the actual location of innovations, it is old and was created before the onset of the Internet. The Internet makes complex information exchange possible. The extent to which the Internet will be able to substitute for face-to-face interaction in the future will affect the importance of concentrating research facilities adjacent to universities and thus the policy in building research parks. Further empirical research is necessary in this field.

AUTM (1998) states that while, in FY1997, 333 firms were established under licenses from universities, 83% of them were located in the same state as the licenser universities. It is true that research universities generate and diffuse the scientific/technological knowledge necessary for innovations, and supply and attract entrepreneurs. However, it seems these research parks are only successful in the long term. Stanford Research Park and North Carolina Research Triangle took more than twenty years to successfully develop. Patient support is necessary. Moreover, when these two were developed in the 1960s, competitors were rare. Since the 1980s, many local governments have been inter-

ested in developing research parks as part of their high-tech industrial policy, so competition is rigorous and offers no guarantee that every research park will be successful.

Another important point is that, even if a research park is successful in attracting research facilities, it does not necessarily lead to an increase in employment in the manufacturing industry. State governments do not build research parks near universities for the prestige—they do so to increase employment. Research results of a research park are, however, not necessarily manufactured in the region. Lugar & Goldstein (1991) compares research park regions against adjacent regions that have a similar size population without them. Of 45 research parks, 19 have generated less employment than the adjacent regions. Moreover, they have found that the success factors of research parks for regional economic development are: having time-honored tradition, having good research universities, and being located in regions of 0.5 to 1 million people which offers both a pool of labor forces and a market for manufactured products.

Potential Problems of University-Industry Collaboration

As mentioned previously, university-industry collaboration has potential negative aspects. Money from industry may induce university research personnel to distort research results, called a 'financial conflicts-of-interest' problem. The problem is that the public does not trust research results of university personnel who have financial ties with industry, even though they do not actually forge research results. When university research personnel spend too much time in cooperation with industry, other tasks of the university such as education and research for public interest are neglected, called a 'conflicts-of-commitment' problem. A sponsoring firm might ask university research personnel to postpone the announcement or publication of research results until the firm obtains patents or becomes ready for commercialization, called a 'secrecy' problem. These problems existed before and were discussed when the Bayh-Dole Act was enacted in 1980.[7] The atmosphere in 1980 was, however, that 'you cannot make an omelet without breaking eggs'

---

[7] It is often difficult to distinguish these three problems. For example, the secrecy problem is a neglect of duty to contribute to public interest on the part of university research personnel, so it is also the conflicts-of-commitment problem. Also, it can be viewed as a distortion of behavior by financial interests with the sponsoring firm, making it somewhat like the financial conflicts-of-interest problem. Furthermore, how much university professors should be allowed to use their students for collaboration with firms has both aspects of financial conflicts-of-interest and conflicts-of-commitment problems. Therefore, sometimes all these problems are called simply as 'conflicts-of-interest' without being categorized further.

(one should tolerant some negative aspects in order to exploit the benefit of university-industry collaboration) (White, 2000, p. 96).

An example of a negative aspect of university-industry collaboration was cited by Blumenstyk (1998), referring to that, in 1996, an article pointed out the danger of appetite suppressants. In the same edition of the journal, two other academic physicians made a commentary that minimized the study's conclusions. What was not announced to the public was that these two physicians were paid consultants to companies that made or distributed similar drugs. This is a typical financial conflicts of interest problem. Among 789 life science articles published in 1992, 33.8% of them had at least one chief author who had financial interests with firms whose activities were related to the field of the published research. Financial interests include the research personnel owning stock in a firm, working as a consultant or a director for it, or receiving research money form it (Krimsky et al., 1996). It is difficult to identify whether or not these financial interests actually cause university research personnel to alter research results in favor of the sponsor, but it is important not to make the public suspicious of any such activity.

In 1995, NIH and NSF requested universities that were receiving funds from them to establish a (financial) conflict-of-interest policy. Today, many universities have their own such policy. According to Cho et al. (2000) who surveyed 89 universities of the top 100 NIH fund recipients, it is uncommon (only 19%) for a university to specify prohibited activities. Many universities simply require research personnel to report financial interests as the first step to mitigate a financial conflicts-of-interest problem. Selection of licensees and cooperative research partners should be conducted fairly so that firms which do not have any financial relationship with research personnel would not be disadvantaged. Hence, research personnel who have financial interests with firms should not be a member of any committee that decides licensees or research partners.

In the conflicts-of-commitment problem, university personnel have been allowed to work as a consultant once a week. Although consulting is often a paid job, it is regarded as a community service or diffusion of knowledge, which is the university faculty's duty in addition to research and education. Because inter-organizational technology transfer is difficult, a university researcher often works as a consultant of a licensee firm to provide technical advice for the commercialization of licensed technology. The elite universities tend to set a maximum number of days for off-campus work, which is usually equivalent level to the traditional 'once a week' rate.

About the secrecy problem, Cohen et al. (1994) has conducted a survey regarding restrictions on communication which are placed on the faculty members who participate in industry–university cooperative research.

The survey covered 479 UIRCs (University Industry Research Centers) in which several firms participated in cooperative research programs with several departments of one university and which received financial support from both the federal and local governments. The state government is particularly interested in supporting UIRC so that it would be developed into a nucleus of research park. UIRCs have an aspect of public policy and the UIRC staff state that the purpose is to gain scientific/technological knowledge rather than to improve existing products or to create new jobs/business, but restrictions are found to be rather tight. The survey indicates that 56.6% of UIRCs have some restrictions. 21.3% of UIRCs set restrictions on communications with faculty members of the same university if they do not participate in the same cooperative research project. 28.6% of them set restrictions against the faculty members at other universities, 39.9% of them set restrictions against companies that are members of UIRCs but do not participate in the same project, and 41.5% of them set restrictions against general public.

Blumenthal et al. (1997) has surveyed 3,394 life science faculty members of the top 50 universities of NIH recipients. Among 2,167 responses, 19.8% of them answered that they had postponed publication more than six months. As NIH states that a delay of 60 days is reasonable for sponsoring firms to prepare patent application, six months is rather long. However, the delay of publication is due not only to the sponsoring firms' request but also to the intentions of the research personnel. When a researcher publishes a paper, other researchers may follow the research and publish more and better papers. A researcher wants to withhold publication until many papers are ready to be published. However, in the survey, 27.2% of researchers who participate in cooperative research with industry have experienced publication delays of more than six months, but the percentage is 11.1% for researchers who do not participate in cooperative research with industry. Hence, collaboration with industry may cause publication delays. On the other hand, in the aforementioned survey about conflicts-of-interest policy by Cho et al. (2000), only 12% of the universities set a time limit for how long a publication can be delayed.

Problems of conflicts of interest, conflicts of commitment, and secrecy have been occasionally reported. There is no time-serious data to check whether problems have recently become worse or not. Moreover, it is inconclusive whether or not the negative aspects of university-industry collaboration outweigh the benefits. However, as Table 12 indicates, universities earn a significant portion of license revenue from federally funded research. There are several policymakers who express concerns that federally funded research is being utilized for specific firms which only pay for the last stage in the completion of research, rather than for the general public interest (Campbell, 1998). The elite research universities of high research quality have a strong negotiating position with sponsoring firms because these universities obtain adequate federal research money and need not rely on money from firms. Moreover, firms definitely want access to the research capabilities of these universities. However, second-tier universities may easily accept firms' demands to attract research money from industry. Firms can threaten even the elite universities by contracting with others that are more responsive to their needs. It may be time for the entire university community to set up rules on the freedom of publication (Nelsen, 1999).

Another problem, which is often called 'institutional conflicts of interest' or 'conflicts of mission,' is that universities have become very interested in earning license revenue. Recently, universities are increasingly taking a too restrictive approach to licensing and putting too high a value on their intellectual property contributions. Firms seek second-tier universities and foreign universities for collaboration when they think that elite universities are tough to deal with (Government-University-Industry Research Roundtable, 1998). Firms complain that university TLO (Technology Licensing Organization) is too oriented toward legal staff, and should have more staff with engineering and marketing backgrounds (Siegel, Waldman & Link, 1999). Moreover, in 1999, NIH issued the guidelines regarding research tools such as cell lines, animal models, reagents, clones, or database, which resulted from NIH funding. NIH thinks that these research tools should not be treated as intellectual property right but shared by the entire research community.

Finally, if government promotes university-industry collaboration, the existence of foreign firms and foreign students would be a problem. If foreign firms can participate in university-industry collaboration, federally funded research results would help them compete against U.S. firms. The Bayh-Dole Act requires that when firms commercialize a product made with the exclusively licensed technology from universities, it should be 'substantially' manufactured in the United States.

USGAO (1992) has found that in 1991, among 197 patents resulting from funding of NIH or NSF, only 18 patents are licensed to foreign firms and 11 patents are licensed to foreign subsidiaries in the U.S. The Bayh-Dole Act does not prohibit licensing to foreign firms, but it does requires that manufacturing occurs in the U.S. However, universities do not follow where domestic licensees actually manufacture the product made with the licensed technology (USGAO, 1998).

While a similar concern was discussed to a lesser extent when the Bayh-Dole Act was enacted, access by foreign firms to U.S. universities research was hotly debated by Congress in the late 1980s when economic nationalism peaked. In particular, MIT was criticized.

About half of the members of the MIT Liaison Program, which will be discussed in detail later, were foreign firms, and MIT had a Liaison Office in Tokyo. However, in 1990, 15% of MIT research funds were supported by industry, with 20% of those funds coming from foreign firms; therefore, only 3% of all research money was from foreign firms. 86% of the licenses went to U.S. firms, and 77% of the firms for which faculty worked as consultants were U.S. firms. Among the 215 chairs endowed by firms, only 30 were supported by foreign firms. While some media and politicians expressed concern that elite universities such as MIT helped foreign competitors, the actual involvement of foreign firms was minor.[8] It was an appropriate policy decision not to impose any restriction on university-industry collaboration in terms of their nationality (U.S. Congressional Hearing, 1993).

Another concern is the existence of foreign students. As university research was increasingly oriented toward industry needs, foreign students may understand what kind of frontier research U.S. firms are interested in. And, when foreign students go home after obtaining a Ph.D. and work for firms in their home country, they compete against U.S. firms.

Figure 11 shows that the percentage of U.S. citizen recipients of science and engineering doctoral degrees[9] was declining between 1986 and 1994, while the a little increase was observed in the late 1990s. The percentage would be increasing if it included foreign-born permanent residents; however, a significant portion of doctoral degrees are offered to foreign students. The scientific/engineering manpower has to rely on foreign nationals or foreign-born residents. Foreign students could not participate in several research projects funded by the Department of Defense due to national security concerns. There was an opinion that a similar restriction might be necessary for economic competitiveness reasons. In fact, as Korea and Taiwan developed their own high-tech industries, students from those countries increasingly tended to return to their home countries. But students from India and China supplemented this decline in the 1990s.[10]

Throughout the 1990s, the percentage of foreign (including permanent residents) doctoral degree recipients in science and engineering fields who plan to stay in the U.S. has been increasing, to above 70% in 1999. The percentage of those who could actually find jobs

---

[8] U.S. firms can have an informal relationship with university research personnel, so they do not have to be members of the Liaison Program. As a result, foreign membership must be high.
[9] Science includes social and behavioral psychology sciences, in which the shares of U.S. citizens and permanent residents are relatively high.
[10] However, Chinese government recently tries to bring back Chinese research personnel by offering financial incentives.

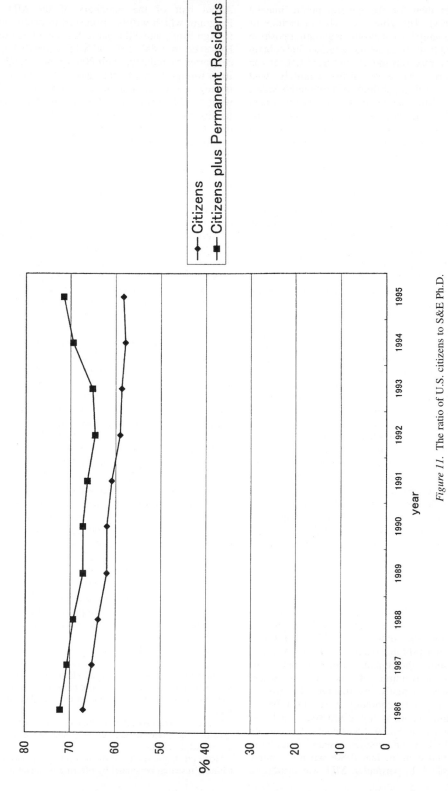

Figure 11. The ratio of U.S. citizens to S&E Ph.D.

*Source:*USNSF (1998b).

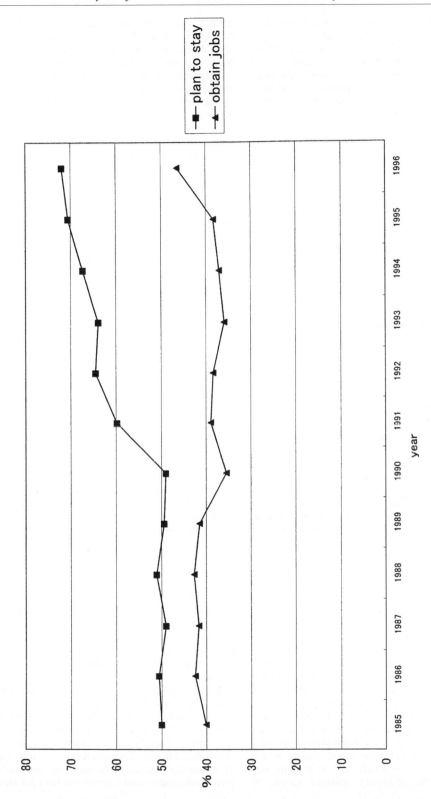

Figure 12. Foreign S&E Ph.D. recipients staying in the USA.

Source: USNSF (1998b).

and stay in the U.S. has also been increasing, as high as 50% in 1999 (USNSF 2002, Chap. 2). Foreign scientists and engineers have already become important labor force in the high-tech industries and universities Between 1995 and 1998, 29% of the start-up firms in Silicon Valley were established by Indian or Chinese people. As they keep a relationship with their home country, foreign start-up firms tend to be good at exporting (U.S. Congress Committee Report, 1999). Moreover, even if foreign students from the elite universities such as MIT and Stanford return home after graduation, they are likely to become business/political/academic leaders there, and are expected to contribute to enhanced relations with the U.S. and favor of long-term U.S. interest (U.S. Congressional Hearing, 1993). Because foreign graduate students actually conduct a significant portion of research, excluding them may deteriorate the quality of research carried out at U.S. universities (Burtless & Noll, 1998). It was the correct decision for policymakers not to impose any restrictions on enrollment of foreign students at U.S. universities.

Conclusions and Policy Implications

U.S. universities contributed to regional economies in the second half of 19th century and the early 20th century. However, since World War II, particularly after the Sputnik Shock of 1957, the federal government has been strongly supporting basic research conducted by universities. In the late 1970s, university-industry collaboration was revitalized, thanks to the Bayh-Dole Act, the biotechnology revolution, the financial needs of universities, and a revised of R&D strategy within industry based the linear model.

However, even since the 1980s, university research has still emphasized basic research funded by the federal government. The elite universities obtain a large amount of research money from the federal government as well as industry, so their reliance on industry is small. The elite universities generate both academic research results and research results close to commercialization. However, license revenue results from a few 'blockbuster' patents, often in the field of medical/life science. There is a major mechanism in university-industry collaboration in the U.S.: federal government generously supports medical/life research at universities and its research results are easily protected by patents, resulting in licenses to firms some of which generate huge license revenue for universities. However, license revenue is small compared to the research budget of universities. It is difficult for universities to maintain a high quality of research by earning money themselves. Universities cannot become 'for-profit research organizations', and tax-payer support is still critical for maintaining university research. Because federal government money is still important for university research, how to draw the line between two of the following opinions should be

further discussed: 'university research results should be openly available to the public' and 'industry should be allowed to promote the commercialization of research results that eventually benefit consumers'.

Research universities actually generate innovation by diffusing knowledge and supplying entrepreneurs. Governments of states try to build research parks that concentrate research facilities around universities, hoping to create their own 'Stanford Research Park' or 'North Carolina Research Triangle Park.' However, these two research parks took more than twenty years to become successful, and for research parks to succeed there must exist a high quality research university nurtured by long-term federal research funds. It is not easy for local governments to quickly improve the quality of research done by universities. Moreover, even if a research park successfully attracts research facilities, it is uncertain if they will significantly contribute to an increase in high wage manufacturing employment of the region.

The contribution of universities to industry should not be limited to visible results such as inventions, patents, or licenses. As Cohen et al. (1998) points out, informal linkage such as communication between research personnel is important for industry to transfer university expertise to firms. In this aspect, the role of ILP (Industrial Liaison Program) is important. Firms that become members of the ILP are introduced to proper research personnel at the university to obtain advice. TLO (Technology Licensing Organization) is an organization that licenses patented technology resulted from university research. By promoting the flow of informal information, ILP covers the primitive stage of university-industry collaboration and TLO covers the mature stage of collaboration. In the regions or nations in which university-industry collaboration is not yet so active and where universities do not have many patents to license, ILP rather than TLO should be established first.

An important point is to provide research personnel at universities with incentives for informal cooperation with industry. Visible results, such as the number of publications or patents, can be easily used to evaluate university research personnel, but assistance to regional industry through informal communication is not. The effort a university faculty member spends answering questions from industrial personnel should also be evaluated since it is important for industrial innovations, even if it does not necessarily lead to patents.

Besides contributing to industrial innovations in formal (patents, inventions, licensing, or activity of TLO) and informal ways (advising, communication, or activity of ILP), education remains an important role for universities. Basic research educates research personnel. Middle class engineers are also generated by university education. In addition, universities generate mathematics and science teachers for elementary and secondary educational institutions. Of course, graduate education has been effectively developed through a mixture of education and research, so good research universities are often strong in generating research personnel, academic research results, and inventions. However, universities can contribute to generating engineers and schoolteachers even if their research is not top-ranked. Local governments should not insist on universities generating innovations, but provide them with financial resources for research and scholarships. Good engineers and scientists have tacit knowledge by which they learn new things not written in books. Therefore, if a region has this kind of human resources, firms can realize a high rate of return on their R&D or manufacturing investment.

In conclusion, there are several ways in which universities contribute to innovations and regional development: they generate inventions, patents, licenses, informal communication with regional firms, spin-off firms, and human resources including scientists, engineers and schoolteachers. One university does not have to attempt all of these functions. The entire university system of a nation or a region should cover them so that the division of labor among their universities is desirable. Needless to say, all of these tasks take a long time to achieve. It is not proper for policymakers to expect quick results with an increase in funding to university research. Funding university research is not the way to solve cyclical recessions. In addition, firms and universities should not expect easy returns from collaboration, remembering the following statement by Doctor Lewis Branscomb, whose brilliant career spans government (National Bureau of Standards, now National Institute of Standards and Technology), private sector (IBM), and academia (Harvard University):

> If the universities value the partnership as a means of exposing faculty and students to leading-edge technical issues that are driving innovations of benefits to society, and are not basing their expectations primarily on revenues from patents, a stable, productive relationship may endure. If the firms see universities as sources of new ideas and as windows on the world of science, informing their own technical strategies, rather than viewing students as a low-cost, productive source of near term problem-solving for the firm, they too will be rewarded (U.S. Congress Committee Report 1998, p. 22).

References

Association of University Technology Managers (AUTM) (1997). *AUTM licensing survey: FY 1996*. Norwalk, Connecticut: AUTM Inc.

Association of University Technology Managers (AUTM) (1998). *AUTM licensing survey: FY 1997*. Norwalk, Connecticut: AUTM Inc.

Association of University Technology Managers (AUTM) (1999). *AUTM licensing survey: FY 1998*. Norwalk, Connecticut: AUTM Inc.

Association of University Technology Managers (AUTM) (2000). *AUTM licensing survey: FY 1999*. Norwalk, Connecticut: AUTM Inc.

Audretcsh, D. B. & Feldman, M. P. (1996). R&D spillovers and the geography of innovation and production. *American Economic Review*. **86** (3), 630–640.

Behrens, T. R. & Gray, D. O. (2001). Unintended consequences of cooperative research: Impact of industry sponsorship on climate for academic freedom and other graduate student outcome. *Research Policy*, **30**, 179–199.

Blumenstyk, G. (1998). Conflict-of-interest fears rise as university chase industry support. *The Chronicle of Higher Education*, (May 22).

Blumenthal, D., Campbell, E. G., Causino, N. & Louis, K. S. (1996). Relationships between academic institutions and industry in the life sciences—an industry survey. *New England Journal of Medicine*, **334** (6), 368–373.

Blumenthal, D., Cambell, E. G., Anderson, M. S., Causino, N. & Louis, K. S. (1997). Withholding research results in academic life science. *The Journal of the American Medical Association*, **277**, 1224–1228.

Burtless, G. & Noll, R. G. (1998). Students and research universities. In: R. G. Noll (Ed.) *Challenges to Research Universities*. Washington, D.C.: Brookings Institution Press.

Campbell, P. W. (1998). Pacts between universities and companies worry federal officials. *The Chronicle of Higher Education*, (May 15).

Cho, M. K., Shohara, R. & Schissel, A. (2000). Policies on faculty conflicts of interest at U.S. universities. *Journal of American Medical Association*, **284** (17), 2203–2208.

Cohen, W. M., Florida, R. & Goe, W. R. (1994). *University-industry research centers in the United States*. Center for Economic Development, John Heinz III School of Public Policy and Management, Carnegie Mellon University.

Cohen, W. M., Florida, R., Randazzese, L. & Walsh, J. (1998). Industry and the academy: Uneasy partners in the cause of technological advance. In: R. G. Noll (Ed.), *Challenges to Research Universities*. Washington, D.C.: Brookings Institution Press.

Etzkowitz, H., (1999). Bridging the gap: The evolution of industry-university links in the United States. In: L. M. Branscomb, F. Kodama & R. Florida (Eds), *Industrializing Knowledge*. Cambridge, MA: The MIT Press.

Feldman, M. P. (1994). *The geography of innovation*. Dordrecht, Netherlands: Kluwer Academic Publishers.

Government-University-Industry-Research Roundtable (1998). *Overcoming barriers to collaborative research*, <www.nap.edu/html/>

Kline, S. J. & Rosenberg, N. (1986). An overview of innovation. In: R. Landau & N. Rosenberg (Ed.), *The Positive Sum Strategy*. Washington, D.C.: National Academy Press.

Krimsky, S., Rothenberg, L. S., Stott, P. & Kyle, G. (1996). Financial interests of authors in scientific journals: A pilot study of 14 publications. *Science and Engineering Ethics*, **2**, 395–410.

Lee, Y. S. (2000). The sustainability of university-industry research collaboration: An empirical assessment. *Journal of Technology Transfer*, **25**, 111–133.

Lugar, M. I. & Goldstein, H. A. (1991). *Technology in the garden*. Chapel Hill: The University of North Carolina Press.

Mansfield, E. (1998). Academic research and industrial innovation: An update of empirical findings. *Research Policy*, **26**, 773–776.

McKelvey, M. (1996). *Evolutionary innovations*. Oxford: Oxford University Press.

Mowery, D. C., Nelson, R. R., Sampat, B. N. & Ziedonis, A. A. (1999). The effects of the Bayh-Dole Act on U.S. university research and technology transfer. In: L. M. Branscomb, F. Kodama & R. Florida (Eds), *Industrializing Knowledge*. Cambridge, MA: The MIT Press.

Mowery, D. C. & Rosenberg, N. (1989). *Technology and the pursuit of economic growth*. Cambridge: Cambridge University Press.

Nelsen, L. (1999). *Remarks at secrecy in science: Exploring university, industry, and government relationships* (March 29, 1999). At Massachusetts Institute of Technology. < http://www.aaas.org/spp/secrecy/Prestns/ne1sen.htm >

Noll, R. G. (1998). The future of research universities. In: R. G. Noll (Ed.), *Challenges to Research Universities*. Washington, D.C.: Brookings Institution Press.

Rosenberg, N. (1990). Why do firms do basic research with their own money? *Research Policy*, **19**, 165–174.

Rosenberg, N. & Nelson, R. (1994). American universities and technical advance in industry. *Research Policy*, **23**, 323–348.

Schumpeter, J. A. (1934). *The theory of economic development* (English ed.). Cambridge, MA: Harvard University Press.

Shapley, D. & Roy, R. (1985). *Lost at the frontier*. Philadelphia: ISI Press.

Siegel, D., Waldman, D. & Link, A. (1999). *Assessing the impact of organizational practices on the productivity of university technology transfer offices: an exploratory study*. National Bureau of Economic Research. Working Paper 7256.

Sommer, J. W. (1995). Introduction. In: J. W. Sommer (Ed.), *The Academy in Crisis*. New Brunswick: Transaction Publishers.

U.S. Congress Committee Report. (1998). *Unlocking our future; Toward a new national science policy*. House

Committee on Science (September 24, 1998). < http://www.house.gov/science_policy_report.htm >

U.S. Congress Committee Report (1999). *American leadership in the innovation economy*. Staff of Connie Meck, Chairman of the Joint Economic Committee, U.S. Congress, < http://www.senate.gov/ ~ jec/summitreport.htm >

U.S. Congressional Hearing (1993). *Access by foreign companies to U.S. universities*. Subcommittee on Science, Committee on Science, Space, and Technology, U.S. House of Representatives, 103rd Congress, First Session, October 28. Washington, D.C.: U.S. Government Printing Office.

USGAO (General Accounting Office) (1992). *University research: Controlling inappropriate access to federally funded research results*. GAO/RCED-92–104. Springfield, Virginia: National Technical Information Service, U.S. Department of Commerce.

USGAO (General Accounting Office) (1998). *Technology transfer: Administration of the Bayh-Dole Act by research universities*. GAO/RCED-98–126. < http://www.gao.gov. >

USNSF (National Science Foundation) (2000). *Science and engineering indicators, 2000*. Washington, D.C.: U.S. Government Printing Office.

USNSF (National Science Foundation) (2002). *Science and engineering indicators, 2002*. Washington, D.C.: U.S. Government Printing Office.

USOTA(Office of Technology Assessment) (1984). *Technology, innovation, and regional development: Encouraging high-technology development, background paper #2*. Washington, D.C.: U.S. Government Printing Office.

USPTO(Patent and Trademark Office). (2000). *U.S. colleges and universities: Utility patent grants, 1969–1999*. Wasihngton, D.C.: U.S. Patent and Trademark Office.

Varga, A. (1998). *University research and regional innovation*. Boston: Kluwer Academic Publishers.

White, G. D. (Ed.) (2000). *Campus, Inc.: Corporate power in the Ivory Tower*. Amherst, New York: Tower Prometheus Books.

Zucker, L. G., Darby M. R. & Armstrong, J. (1998). Geographically localized knowledge: Spillovers or markets? *Economic Inquiry*, **XXXVI** (January), 65–86.

The International Handbook on Innovation
Edited by Larisa V. Shavinina

Incubating and Networking Technology Commercialization Centers among Emerging, Developing, and Mature Technopoleis[1] Worldwide

David V. Gibson and Pedro Conceição

IC² (Innovation, Creativity, Capital) Institute, The University of Texas at Austin, USA

Abstract: The ability and desire to access knowledge and to be able to learn and put knowledge to work is central to regional economic development and for globalization to be a force for drawing the world together. This chapter presents the logic, conceptual framework, and key elements for leveraging codified knowledge and tacit know-how through Internet and web-based networks and face-to-face communication and training programs. The objective is to accelerate regional economic development and shared prosperity through globally linked and leveraged Technology Commercialization Centers (TCCs) and to enhance the competitiveness and accelerate the growth of select regionally-based SMEs.

Keywords: Incubation; Innovation; Commercialization; Entrepreneurship; Knowledge creation, diffusion, and adoption; Networking.

Introduction

Technology continues to shrink the world. There is no choice other than to participate in the global community. Science and technology is too precious a resource to be restricted from drawing the world together. That is what the 21st century is all about. (Dr. George Kozmetsky, Chairman of the Board, IC² Institute, The University of Texas at Austin)

This chapter is focused in the belief that technology can be a force for drawing the world together and that the ability to learn, and put knowledge to work is central to regional economic development. Building on these two beliefs the authors present the logic, conceptual framework, and elements of a strategy to leverage codified knowledge and tacit know-how through Internet and web-based networks and face-to-

face training programs and thereby accelerate technology-based economic development through globally linked and leveraged Technology Commercialization Centers (TCCs) that will facilitate the growth of select regionally-based small and medium sized enterprises (SMEs).

The suggested plan brings together business entrepreneurs, academia, and regional government in targeted emerging, developing, and mature technology regions or Technopoleis worldwide. Each TCC will be viewed as an 'experiential learning laboratory' where lessons learned will be used in world-class research, education, and training programs. The activities described in this chapter are designed to function as an integrated program which, over time, will contribute to and leverage local and global initiatives for knowledge transfer, accumulation, use, and diffusion to accelerate sustained economic growth and shared prosperity worldwide.

The ongoing 'communications revolution' is making the tasks of globally linking public/private collaboration and knowledge acquisition, transfer, and adoption at least feasible. Indeed, there are two important

---

[1] Technopoleis (Greek for technology and city state) are regions of accelerated wealth and job creation through knowledge/technology creation and use. Innovation is the adoption of 'new' knowledge—knowledge that is perceived as new by the user.

advantages for today's SMEs that were not available to entrepreneurs in the mid- and late-1900s:

(1) The distance canceling power of Internet and web-based communication; and consequently.
(2) The possibility of global access to talent, technology, capital, and know-how by leveraging and partnering through modern information and communication technologies (ICT).

This 'death of distance' as Frances Cairncross (1997) puts it, reduces the inherent economies of knowledge clusters and opens up the field to new entrants. These developments appear especially promising for the developing world, potentially enabling them to economically tap into informational and technical sources hitherto available only in mature technopoleis. The proposal to accelerate the development of globally linked TCCs is based on the realities that:

(1) few regions in the developing world can hope to match, at least in the short-term, the physical and smart infrastructure of established technopoleis such as Silicon Valley, California and Austin, Texas, where there is an overwhelming agglomeration of technology, talent, capital, and know-how; and
(2) regional, national, and global computer-based networks in the knowledge age allow for, if not encourage, the development of non-geographic bound or virtual technopoleis.

A key question for the 21st century, therefore, is how necessary and sufficient is the regional development of 'smart' infrastructure in all its aspects (i.e. talent, technology, capital, and know-how) or physical infrastructure (i.e. science parks, incubators, and high-tech corridors) in the emerging internet-based economy where the movement of knowledge is increasingly through ICT? And it may be asked, which sectors or components of this infrastructure must be physically co-located or digitally networked at different stages of firms becoming globally competitive?

As a starting point we focus on the importance of fostering entrepreneurship at the grassroots level. To do this we target small and mid-sized technology-based enterprises (SMEs) that might be considered relatively successful at the local level but are in need of assistance (e.g. talent, technology, capital, and know-how) to achieve accelerated growth and global market penetration. At the moment that these virtual networks are created around the SME, it becomes a Learning & Innovation Pole (LIP).

A LIP is operationalzed at the most basic level as an SME that is linked to other SMEs in an Internet and web-supported global network and also has access to a range of support activities such as training programs, workshops, and mentoring activities. Rather than relying on a well-defined geographic area to provide all of the networks and services required for success in knowledge-based economic efforts, LIPs rely on regional and global cooperative and collaborative networks and training programs to provide service and assistance on a real-time, as-needed basis. LIPs will be able to shift and grow to take advantage of emerging opportunities and market needs.

While physical proximity is becoming less and less important to regionally-based economic development because of the pervasiveness of advanced ICT, successfully fostering the growth and global competitiveness of select SMEs in targeted regions will take more than computer-based networks and web connections. There needs to be a sense of community and relationship building. There needs to be local visionaries, champions, and implementers. And to sustain the regional and global partnerships and alliances there needs to be a meaningful flow of know-how and resources and win-win partnerships among all members of the network.

Conceptual Background

Traditionally, wealth creation in developed and developing nations, has emphasized physical assets. The capital stock of a nation was thought to be a measure of national prosperity, and the attraction of foreign direct investment became a prime strategy of less developed regions. As the world moves into the 21st century, however, the emphasis is on knowledge transfer, accumulation, adoption, and diffusion as being critical to economic development. As the World Bank noted in its 1998 *World Development Report*:

> It appears that well-developed capabilities to learn— the abilities to put knowledge to work—are responsible for rapid catch-up ... The basic elements (to develop these learning abilities) appear to be skilled people, knowledge institutions, knowledge networks, and information and communications infrastructure.

This section provides a summary of the theory of knowledge transfer, accumulation, and use to create a better understanding of the process through which knowledge and learning can be leveraged for regional economic development. Knowledge transfer, adoption, accumulation, and diffusion are key to sustainable economic prosperity in the emerging global economy of the 21st century. As stated by Abramovitz & David in a 1996 OECD report, "The expansion of the knowledge base ... (has) progressed to the stage of fundamentally altering the form and structure of economic growth." Rapid advances in information and communication technologies and declining costs of producing, processing and diffusing knowledge are transforming social and economic activities worldwide (The World Bank, 1998).

While the current knowledge revolution is resulting in many positive outcomes, there is concern that it is accelerating the polarization of the 'haves' and 'have nots'. Scientific and technical advances have increased the economic welfare, health, education, and general living standards of only a relatively small fraction of humankind to unmatched levels. The unevenness of such development among, and within, both developing and developed regions has increased significantly. Two hundred and fifty years ago, for example, the difference in income per capita between the richest and poorest countries in the world was 5 to 1. Today, the difference is approaching 400 to 1 (Landes, 1998). The underlying reasons for these inequalities are complex and, according to most analyses, are to be found in the outcomes of the social and economic revolutions that pre-date the current knowledge revolution. While the industrial revolution lowered the costs of manufacturing and distribution, over time this economic and social revolution also tended to divide the world into industrialized and non-industrialized nations and fostered bi-modal societies of wealthy and poor.

The current knowledge revolution is critically different from the past industrial revolution. It is based upon a shift of wealth creating assets from physical things to intangible resources based on knowledge (Stevens, 1996). Knowledge-based economic regions tend to be located near leading universities and research centers in the most advanced regions of the world (Quandt, 1998; Smilor, Gibson & Kozmetsky, 1986). Lucas (1988) argues that people with high levels of human capital tend to migrate to locations where there is an abundance of other people with high levels of human capital. Indeed, the importance of the physical proximity of talent, technology, capital, and know-how or 'smart infrastructure' has been argued to be crucial to fostering regional wealth and job creation (Audretsch & Feldman, 1996; Audretsch & Stephan, 1996; Gibson, Smilor & Kozmetsky, 1991; Rogers & Larsen, 1982).

Despite the strong arguments for the importance of physical proximity or agglomeration of 'smart infrastructure', advances in telecommunications and information technologies are transforming our perceptions of geography (Cairncross, 1998). Advances in ICT are key to explaining the shift from the industrial age—coal, steel, and material items—to a global knowledge-based age—information, human capital, and ideas. While it is still difficult to realize what William Mitchell (1995) calls 'cities of bits'—where the majority of the world's people are connected through telephones, televisions, faxes, and computers to a world-wide web—key influencers in business, academia, and government are increasingly realizing opportunities to use the special characteristics of knowledge and ICT to foster regional development through cooperation, collaboration and competition.

A better understanding of the process through which knowledge and learning can contribute to economic development in developing as well as developed regions is urgently required. In this regard, it is important to define knowledge and to realize how it differs from physical things (Dosi, 1996). Here we follow the analysis of Conceição et al. (1998), who build on Nelson & Romer's (1996) differentiation between ideas and skills, or software and wetware.

- Software ('ideas'): Knowledge that can be codified and stored outside the human brain, for example in books, CDs, records, and computer files. Software (as defined here) is referred to as the 'structural capital' of private and public organizations and includes intellectual property that is codified (Edvinson & Malone, 1997). When employees leave their place of work the software remains.
- Wetware ('skills'): Knowledge that cannot be dissociated from individuals, is stored in each individual's brain, and includes convictions, abilities, talent, and know-how. Wetware is referred to as the 'human capital' of private and public organizations and is the know-how or intangible resources that provide key added-value for enterprise development and accelerated growth (Edvinson & Malone, 1997). When employees leave their place of work the wetware leaves with them.

These two kinds of knowledge differ: (1) in the way they are produced, diffused, and used; and (2) in the level of codification. While ideas correspond to knowledge that can be articulated (in words, symbols, or other means of expression), skills correspond to knowledge that cannot be formalized or codified. This apparently simple difference has very important consequences in terms of the way knowledge is produced, diffused and used.

The classification of knowledge in this manner is very significant in the context of this paper which aims to use Internet and web-based links to accelerate growth and job creation. The transmission of software, or codified knowledge, is not much affected by geographic distance, especially in this age of high-bandwidth and near-zero transmission cost (Swann & Prevezer, 1998). However, the transmission of wetware, or tacit knowledge, cannot be easily accomplished without face-to-face contact (ibid.).

Not only does this indicate the importance of including face-to-face contact in the proposed LIP Program, it also indicates the types of industries that will most benefit from an Internet and web-based network. That is, from a technology transfer point of view, leading edge technologies that are highly dependent on wetware skills are unlikely to benefit from only the access that LIP networks provide. But, more standard technologies that are further along their life-cycle and therefore require more codified knowledge would benefit greatly from the financing and marketing links provided through the LIP network. This is because new technologies spur the development of skills required to use them. However, as these technologies become more sophisticated, the required skill levels tend to decrease and the ability to codify the required knowledge increases. As a result, selection of SMEs in the targeted regions will not necessarily focus on advanced technology (e.g. new materials, semi-conductors, biotechnology), but will emphasize the use of appropriate ICT and business processes in the regionally based enterprises. It is clear that modest technology and innovative management processes can produce substantial wealth and job creation for a region (Conceição, Heitor & Oliveira, 1998).[2] The business and networking focus of the Learning & Innovation Poles (LIPs) therefore, will be based on the assessment of the technology and infrastructure strengths, weaknesses, opportunities, and threats (SWOT) of each selected region.

Today, the really substantial gains in wealth are to be found in the use and diffusion of knowledge. However, without skills, ideas may be irrelevant. Similarly without ideas, there may be no need for new and better skills. In short, it is important to stress that the accumulation of knowledge leads to the creation of wealth only if the knowledge is effectively transferred, adopted, and diffused.[3]

In the proposed LIP/TCC Program, personal networks and partnering programs (e.g. education and training, conferences, etc.) linked via ICT will be used to facilitate the collaboration of regionally-based

---

[2] For example, in Austin, Texas a Fortune 500 company, DELL Computers was started out of a university dorm room in 1982 by one entrepreneurial student at the University of Texas at Austin. The idea was to build customer designed computers using off-the-shelf technology and direct marketing, initially over the phone and increasingly over the Internet. Based on this modest start-up, over 7,000 people are employed in the Austin area with additional manufacturing and sales operations in Asia, Europe, and Latin America.

[3] History is full of examples where the producers of an innovative technology—by not using and diffusing it—were surpassed by others who did. Two examples serve as illustrations: One at the grand scale of the history of civilization; The other at the much smaller scale of contemporary corporate warfare. China developed what was, after the invention of writing, one of the most important ideas for the progress of humankind-the movable type printing press. This technology dramatically increased the possibilities of codifying knowledge. However, Imperial China restricted the use and diffusion of this technology to the affairs of the Emperor and his court. As a result it was Europe that benefited most from this invention by promoting its widespread use and diffusion (Landes, 1998). A more contemporary example is provided by Xerox PARC, a state-of-the-art R&D facility located in Sunnyvale, California. In the 1970s, housing some of the world's most brilliant researchers, PARC discovered many of the fundamental computer and software concepts and technologies that have become the basis of today's computer industry. Apple Computer, at the time a Silicon Valley start-up, used PARC developed knowledge and technologies in its innovative and successful Macintosh computer generating considerable wealth and jobs. In the 1980s it was Seattle-based Microsoft that benefited from the software technologies developed years earlier at PARC.

Learning & Innovation Poles as members of a global learning and innovation network. This global learning and innovation network will facilitate the transfer and use of existing knowledge and the creation of new knowledge for regional economic development. To foster equitable knowledge transfer, accumulation, diffusion and use of both *software* and *wetware*, this project holds to three principles of operation.

Principle No. 1: When establishing Learning & Innovation Poles, we must deal with social as well as physical constructs that link participating people and institutions in networks of knowledge production, sharing, adoption, and diffusion that lead to self-reinforcing learning cycles (Fig. 1).

In competitive marketplace economies, business or financial global networks often do not operate to the benefit of less developed regions, indeed such networks often contribute to unequal development. A key question is whether such networks, linking the 'haves' with the 'have nots', go beyond awareness to the actual development of capabilities for knowledge accumulation and application in the less developed sites.

Principle No. 2: To foster networks in which the interaction leads to increased learning capability in all network nodes, but in which the rate of learning is higher in the less developed nodes (Fig. 2).

This project also strives to encourage and facilitate local ownership of activities and results. To be truly sustainable, the processes of innovation must occur within, and be 'owned' by the champions in each node. Therefore, it is critical that regional champions or businessmen/women feel that they 'own' LIPs. This idea leads to **Principle No. 3**, which is to foster the regional 'ownership' of the activities and the results or return-on-investment of the network. A sense of ownership can be fostered by shared decision-making structures in which the ultimate choice and responsibilities lie with local participants rather than external facilitators.

Technology Commercialization Centers

The objective is to foster the global linking of regional champions and enterprises in view of the realities and challenges of the international marketplace. The networks that we propose building via the LIP/TCC project are centered on identifying select SMEs and regional champions for long-term partnerships. These networks will be sustained by being task focused for short-term success as well as for longer-term vision. Figure 3 depicts how LIPs know-how networks will assist targeted companies in targeted regions to cross the knowledge transfer and application gap to market applications, leading to firm diversification and expansion and new firm formation.

Technology Commercialization Centers through Learning & Innovation Pole Networks will strive to

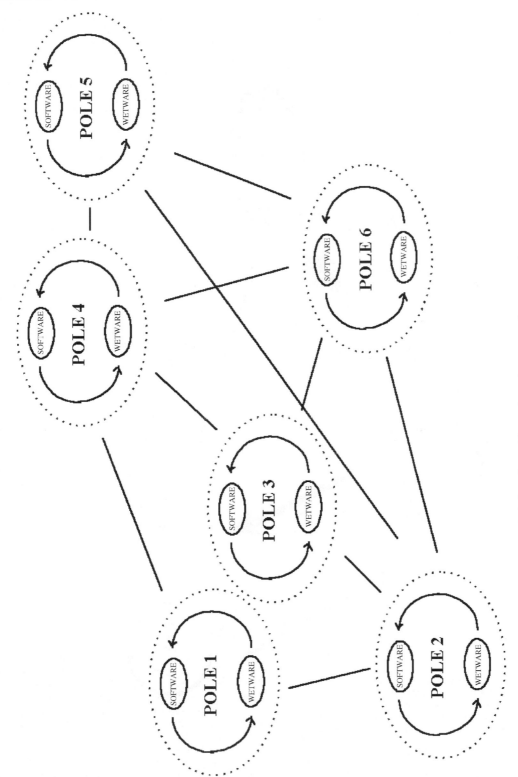

Figure 1. Learning and innovation poles as networks among regions.

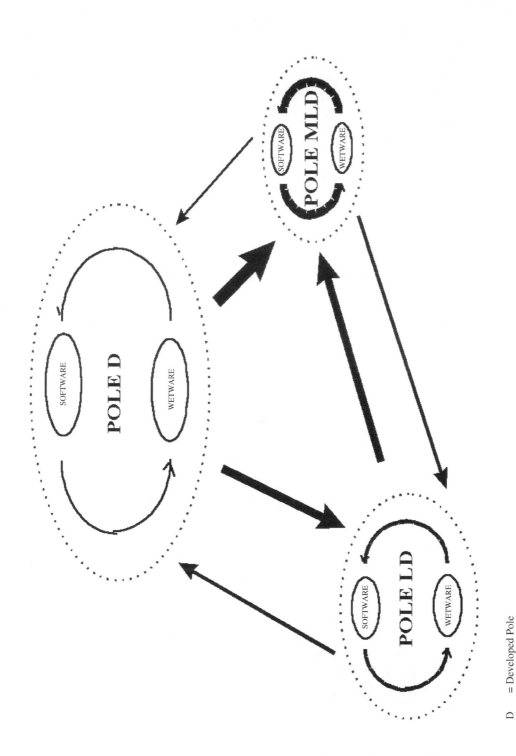

Figure 2. A learning network based on proportional reciprocity.

D = Developed Pole
LD = Less Developed Pole
MLD = Much Less Developed Pole
Arrow density implies the richness of knowledge transfer and application

Regional/National/Global/ Knowledge/Technology		Leveraging for Regional Firm:
• Government • Academic • Business: large, mid-sized, and small		• Diversification • Expansion • New Formation

Knowledge Transfer and Application with...

- Market Research know-how
- Financial know-how
- Legal know-how
- Production know-how
- Distribution, sales, and service know-how

Figure 3. Crossing the knowledge transfer and application gap with know-how.

shorten product development cycles by broadening entrepreneurs' global know-how in such areas as market research, finance, advertising, quality issues, management, sales, and service (see Figs 4 and 5).

Common challenges to having targeted SMEs think and act globally are:

(1) success in home markets and a home-country bias
(2) limited personnel
(3) limited resources
(4) limited time
(5) limited tolerance for extra problems and challenges of going global (e.g. the fear of losing control of one's intellectual property)
(6) ignorance of critical success factors in foreign markets
(7) legal, trade, and governmental constraints

These challenges need to be balanced against the benefits of a SME being part of the LIP/TCC Program. Such benefits include:

(1) access to needed and often critical knowledge
(2) global market access and niche market opportunities
(3) access to needed talent, technology, capital, and know-how
(4) minimizing mistakes and misspent resources
(5) maximizing speed to the market and the commercial potential of a venture
(6) Being aware of 'your' firm's global strengths, weaknesses, opportunities, and threats

The objective of the LIP/TCC program is to foster the global linking of regional champions and enterprises with venture financing, managerial and marketing know-how, and supporting services.

Partnerships with local champions (e.g. business leaders, professors, and students from local universities/colleges, and influencers from the local chambers of commerce) will be formed. Existing institutional data will be used, as much as possible, to conduct a benchmark/scorecard of each targeted

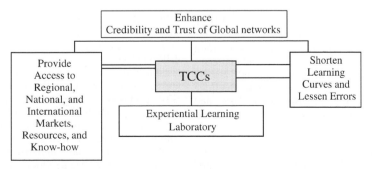

Figure 4. Technology commercialization centers as learning and innovation poles to foster venture success and accelerated growth.

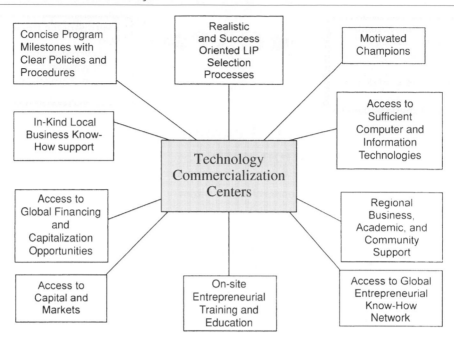

Figure 5. Ten success factors for Learning and Innovation Pole (LIP) development.

region. This benchmark's focus will be to identify and leverage existing capabilities:

(1) for nurture Technology Commercialization Centers in each target region
(2) to network these poles nationally and globally through the use of ICT and personal networks

Regional assessments will focus on:

- R&D and technical expertise related to current and future businesses activities in the region—the talent and technology base for growth of existing and new firms and industries
- core strengths and assets, competitive advantages in terms of regional, national, and global markets
- assessment of existing regional innovation systems
- scorecard (employment, founding date, spin-offs, growth, etc.) of existing small, mid-sized, and large firms in the region
- assessment of emerging clusters of activity, location of multi-nationals, branch plants, and headquarters
- interviews and survey of key regional leaders (academic, business, and government) to facilitate regional understanding by the researchers and regional support and ownership by key local champions

A number of data collection activities are recommended to provide local advisory boards and stakeholders the information necessary for them to make informed decision regarding the challenges in

facilitating technology-based economic development in selected LIPs. These aspects may include, but are not restricted to:

- specific and measurable objectives for the regional SME
- formation of business partnerships
- location and size of specific components of the regional SME
- technological profile; market and competitive orientation of the SME
- options for expansion
- financing needs and sources
- time schedule
- networks strengths and needs
- supply of services and consulting
- structure of LIP oversight

Building Networks

ICT personal networks and partnering programs will link TCCs into important resources for sustaining each LIP. Internet and other ICT links (e.g. video) will be established between each of the TCC sites with an emphasis on fostering collaboration among the LIPs. The proposed networks will attempt to identify regional champions for long-term partnerships. These networks will be sustained by being task focused for short-term success as well as longer-term vision.

While there is an appreciation for the national, institutional, and organizational contexts of each TCC, the focus is on the individual level of analysis and action (Fig. 6). The initial objective is to have the

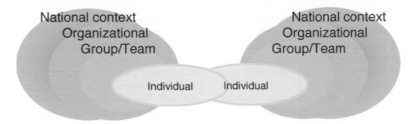

Figure 6. Focus on individual level of analysis and action.

selected TCCs benefit from the codified knowledge and tacit know-how of the LIP network and training activities that focus on access to talent, technology, capital, know-how, and market access. The LIP network will strive to shorten product development cycles and time-to-market by broadening local firms' global knowledge in such areas as finance, market research, advertising, quality control, sales, and after-sales support and service.

The LIP networks will focus on six functional objectives:

(1) *Networking for Markets*: The identification of markets and successful execution of marketing strategies is a determining factor in the success and sustainability of small and medium technology-based firms.

(2) *Networking for Capital*: Access to adequate financing is one of the most critical factors for the success of technology-based firms. Broad public/private/NGO partnerships could be established to offer integrated access to services such as financial planning, support for obtaining grants, and opportunities for access to venture, development, and seed capital.

(3) *Networking for Inter-Firm Linkages*: A networked approach is ideal for maximizing the impact of programs and projects, such as partnerships, alliances, and linkages to outside suppliers. Most clusters in developing countries tend to rely heavily on the local supplier base, which may become insufficient for their rapidly growing needs. Careful coordination is required to ensure that the local suppliers are able to match increases in demand so that jobs may be retained and created and that other substitute supply streams can be brought online as required.

(4) *Networking For Technological Support*: Electronic networks are extremely useful tools to diffuse the benefits of technological support, providing services such as technology assessment and forecasting, technology gateway (assistance on technological choices and on marketing assessment of innovative projects), and access to outside technical information. These services could also be concentrated in a few centers provided by public

agencies, private consultants and business associations.

(5) *Networking to Expand Access to Technology Transfer Opportunities*: The use of electronic networks for technology transfer is already being established in several places to stimulate investment in S&T, R&D, technology transfer, development of commercial potential of R&D, spin-offs. Networks are necessary tools to facilitate access to technology transfer opportunities worldwide.

(6) *Networking for Talent and Know-How*: SMEs often do not have and cannot afford the entire range of technical and business talents and know-how required for success in local and global markets. The process of identifying and hiring such talent and know-how on a short-term as-needed basis is also difficult for smaller enterprises. Networks of talent and know-how would be a great asset that would allow SMEs access to the experts at affordable rates and opportune moments.

While established SME firms will be the initial focus of the LIP Program, there is a critical need to develop the infrastructure and resources of the region to promote accelerated development of knowledge based firms from the bottom-up as a longer term strategy. The basic need is to improve the process of knowledge transfer, acquisition, absorption, and diffusion. The issues involved include education, physical infrastructure construction, and improved policy environments. For example, if knowledge acquisition—whether imported from abroad or created at home—is to lead to economic development, it must be absorbed and applied. This requires universal basic education as well as opportunities for lifelong learning (World Bank, 1998). Also, the extent and economy of modern ICT greatly expand the potential for both the acquisition and the absorption of knowledge, but this can only happen after a basic level of telecommunications infrastructure is acquired.

The following are potential areas for action (Quandt, 1998):

• *Creating and Strengthening Local Technopolis Management Structures*: The first step is the creation of an organizational and functional structure for the local cluster, preferably leveraging existing groups and associations. This would involve both private and public sector participants. The establishment of linkages between technopolis managers will enable a better understanding of stakeholder needs and markets and will improve organization methods. The creation of a permanent, dedicated business and technology information network would make communications more continuous and interactive, rather than sporadic exchanges that normally occur only at periodic meetings.

- *Determining Educational Needs and Offering Training*: Based on regional descriptive profiles and targeted interviews with local stakeholders, education requirements for the LIPs and targeted companies can be ascertained. Courses could then be offered through local workshops as well as via the Internet to help improve the skills of local trainers. For example, the IC^2 Institute is currently working with several global partners to offer long-distance educational programs—degree and certificate—focusing on technology commercialization and forming virtual teams of 'students' to evaluate the commercial potential of innovative technologies (www.IC2.org).

- *Fostering Personnel Exchanges*: Visits of key personnel among regions in the network would greatly facilitate knowledge, technology and know-how transfer. For example, one of IC^2 Institute's global partners, the Instituto Superior Técnico has established an IMPACT Program for leading Portuguese entrepreneurs to build markets in the United States. Exchanges of students, faculty, and entrepreneurs facilitate these processes.

- *Building Local Skills and Training IT Specialists*: A skilled workforce is one of the most important localization factors for technology-based companies, and a major constraint to the development of technopoleis in many less developed regions. This characteristic is essentially place-based, yet virtual technologies may boost the development of human resources in more remote locations through training centers, distance education, career planning, virtual job markets, and also support business development through the establishment of virtual entrepreneur schools providing all kinds of training—technical, managerial, marketing, etc.

- *Optimizing and Sharing Facilities*: For each region, the required facilities for a viable technopolis could be kept to a minimum, provided they are integrated into a shared system. The operational support infrastructure could be optimized and many facilities could be shared over the network, including prototype centers, pilot plants, online libraries, test laboratories, and online conferencing facilities.

Technology Commercialization Centers will focus on select entrepreneurs and business enterprises in targeted regions. These sites will be studied over time to provide data and case examples for research as well as for job and skills training. It is important to involve a range of regions with a variety of characteristics, challenges, and opportunities for wealth and job creation. Metrics for success will be regionally-focused and will be identified and followed over time. In general these metrics will include:

Targeted to Specific SMEs
- Global technology and business assessment
- Market assessment: regional and global, existing and emerging
- Intellectual property rights and protection
- Capital access
- Increased profit
- Accelerated growth
- Access to new technology and business processes
- Shorter time to market
- Management and employee development

Government Oriented
- Job creation
- Space utilization
- Capital creation
- Incremental revenues (i.e. taxes, services, etc.)
- Development of regional 'smart infrastructure'
- Increased global awareness and competitiveness

Academic Oriented
- Training of faculty and students
- New curriculum development
- Successful placement of students
- Experiential, on-the-job learning
- Enhanced favorable relations with community
- Research and publications
- Revenue generation (i.e. royalty, license fees, etc.)

Conclusions

Establishing a network of LIPs composed of meaningful and sustainable Technology Commercialization Centers depends upon the effective use of ICT—as well as regional and global partnerships and partnering programs that facilitate the diffusion and adoption of tacit knowledge. While the role of business, academic, and government sectors in building regionally-based technology centers (i.e. technopoleis) has been observed and experimented with for over 20 years, it remains unclear what are the key resources and conditions for accelerating the growth of virtual technopoleis and how these criteria might change depending on geographic location. The LIP Program presented here seeks to provide the experimental laboratory to address the following questions:

(a) How does one best accelerate entrepreneurial wealth and job creation in SMEs through Internet and web-based access to talent, technology, capital, know-how, and markets?

(b) What 'smart' infrastructure must be physically present at SME locations and what can be virtually linked regionally, nationally, and globally and how does this change over firm growth and maturation?

(c) What are the critical components of regionally-based technopolis development in the emerging Internet and web-based knowledge economy of the 21st century?

(d) Can a developing region 'leap-frog' the 20–40 years it has traditionally taken to build technopoleis without first building world-class research facilities and state-of-the-art science parks, and the agglomeration of 'smart' infrastructure (i.e. finance, legal, marketing, manufacturing, sales and distribution, global expertise, etc.)?

References

Abramovitz, M. & David, P. (1996). Technological change and the rise of intangible investments: The U.S. economy's growth-path in the twentieth century. *Employment and Growth in the Knowledge-based Economy*. Paris: OECD.

Audretsch, D. B. & Feldman, M. P. (1996). R&D spillovers and the geography of innovation and production. *American Economic Review*, **86** (3), 630–640.

Audretsch, D. B. & Stephan, P. R. (1996). Company-scientist location links: The case of biotechnology. *American Economic Review*, **86** (3), 641–652.

Botkin, J. (1986). Route 128: Its History and Destiny. In: *Creating the Technopolis: Linking Technology Commercialization and Economic Development* (1988). Cambridge, MA: Ballinger.

Brett, A. M., Gibson, D. V. & Smilor, R. W. (Eds) (1991). *University spin-off companies*. Savage, MD: Rowman & Littlefield Publishers, Inc.

Bruton, H. J. (1998). A reconsideration of import substitution. *The Journal of Economic Literature*, **XXXVI** (June), 903–936.

Caircross, F. (1997). *The death of distance*. Harvard, MA: Harvard Business Press.

Conceição, P. & Heitor, M. V. (1998). A knowledge-centered model of economic development: New roles for educational, science, and technology policies. *Proceedings of the 2nd International Conference on Technology Policy and Innovation*, Lisbon, 3–5 August.

Conceição, P., Gibson, D., Heitor, M. & Shariq, S. (1997). Towards a research agenda for knowledge policies and management. *Journal of Knowledge Management*, **1** (2), December, 129–141.

Gibson, D. V. & E. M. Rogers (1994). *R&D collaboration on trial: The microelectronics and computer technology corporation*. Boston, MA: Harvard Business School Press.

Gibson, D. V., Kozmetsky, G. & Smilor, R. W. (1992). *The technopolis phenomenon: Smart cities, fast systems, global networks*. Lanham, MD: Rowman & Littlefield Publishers, Inc.

Gibson, D. V. & Smilor, R. (1991). Key variables in technology transfer: A field-study based empirical analysis. *Journal of Engineering and Technology Management*, **8**, December, 287–312.

Kelly, K. (1998). *New rules for the new economy*. New York, NY: Viking.

Landes, D. (1998). *The wealth and poverty of nations: Why some are so rich and some so poor*. New York: W. W. Norton & Company.

Lucas, R. E. (1988).On the mechanics of economic development. *Journal of Monetary Economics*, **22**, 3–42.

Lugar, M. I. & Goldstein, H. A. *Technology in the garden: Research parks and regional economic development*. Chapel Hill, NC: The University of North Carolina Press.

Malecki, E. J. (1991). *Technology and economic development: The dynamics of local, regional and national change*. Essex, U.K.: Longman Group.

Massey, D., Quintas, P. & Wield, D. (1992). *High-tech fantasies: Science parks in society, science and space*. London: Routledge.

Muller, E., Gundrum, U. & Koschatzky, K. (1997). Methods for ascertaining firms' needs for innovation services. In: *Technology-Based Firms in the Innovation Process: Management, Financing and Regional Networks*. Heidelberg, Germany: Physica-Verlag.

Nelson, R. R. & Romer, P. (1996). Science, economic growth, and public policy. In: B. L. R. Smith & C. E. Barfield, *Technology, R&D, and the Economy*. Brookings, Washington, D.C.

Quandt, C. (1998). Developing innovation networks for technology-based clusters: The role of information and communication technologies. Paper prepared for the workshop on 'Tech-regioes: Ciencia, tecnologia e desenvolvimento—passado, presente e futuro, Rio de Janeiro, December 6.

Rogers, E. M. & Larsen, J. K. (1984). *Silicon Valley fever: Growth of high-tech culture*. New York, NY: Basic Books.

Saxenian. A. (1994). *Regional advantage: Culture and competition in Silicon Valley and Route 128*. Cambridge, MA: Harvard Business School Press.

Scott, A. J. (1993). *Technopolis: High-technology industry and regional development in Southern California*. Berkeley, CA: University of California Press.

Shapiro, C. & Varian, H. R. (1999). *Information rules: A strategic guide to the network economy*. Boston, MA: Harvard Business School Press.

Smilor, R. W., Kozmetsky, G. & Gibson, D. V. (1988). *Creating the technopolis: Linking technology commercialization and economic development*. Cambridge, MA: Ballinger Publishing Company.

Swann, P., Pervezer, M. & Stout, D. (1998). *The dynamics of industrial clustering*. London: Oxford University Press.

Tornatzky, L. G. & Fleischer, M. (1990). *The process of technological innovation*. Lexington, MA: Lexington Books.

World Bank (1998). *World development report 1998: Knowledge for development*. Washington, D.C.: The World Bank.

The International Handbook on Innovation
Edited by Larisa V. Shavinina

Science Parks: A Triumph of Hype over Experience?

John Phillimore[1] and Richard Joseph[2]

[1] Institute for Sustainability and Technology Policy, Murdoch University, Australia
[2] Murdoch Business School, Murdoch University, Australia

Abstract: This chapter argues that there is a disjuncture between the critical assessment of the academic literature about most science parks and their growing number worldwide. After briefly detailing the spread of science parks, the chapter explores their nomenclature as part of a broader categorization of the type of parks found today. Several rationales for parks are examined. The skepticism commonly expressed about science parks in the literature is analyzed and contrasted with their continued popularity. The chapter concludes with an assessment of possible new directions for science parks.

Keywords: Innovation; Science parks; Industrial parks; Business incubators; Research park.

Introduction

Depending on the definition used (an issue discussed below), there are now more than 400 science parks spread throughout the world. Fifty years ago, there were just two—the Stanford Research Park next to Stanford University in California and the Research Triangle Park in North Carolina. According to the International Association of Science Parks (IASP), every Park has an average of 3,900 employees, indicating total employment of around 1.5 million employees worldwide (IASP, 2000, p. 4).

Moreover, the number of parks continues to grow. The two major international associations, the European-based IASP and the U.S.-based Association of University Research Parks, each have around 230 members (with little overlap). The IASP has witnessed a doubling of member parks in less than a decade. The first parks appeared in the USA, with the U.K. and France following suit in the 1960s and 1970s. Australia and Canada experienced their main science park growth from the mid-1980s, with continental Europe getting on board in the 1990s. China and other countries in the Asia Pacific have now become the major growth area for science parks, although their size (average employees of almost 10,000 compared to just over 2,000 in Europe and North America) indicates that Asian science parks may be combining elements of more traditional industry parks with science parks proper.

Yet despite this apparently inexorable growth, there has always been a disjuncture between the promise and the reality of the science park model. The great majority of academic evaluations of science parks have invariably failed to show any particularly significant or distinctive benefits arising from science parks in terms of technology transfer between universities and industry, technology and economic development more generally, or local or regional urban renewal in particular.

The purpose of this chapter is to explore in more detail the apparent conflict between the critical assessment of most academic literature about the (lack of) achievements of most science parks and their continued popularity. The chapter begins by outlining the nomenclature of science parks and related concepts (technopoles, technology parks, research parks, etc) as part of a broader categorization of the type of parks commonly found today. Several rationales for parks are then examined in 'Definitions': technology development; regional development; the encouragement of 'synergies' (in particular technology transfer from universities to industry); real estate development; political prestige. The skepticism commonly expressed about science parks in the academic literature on innovation is then analyzed and contrasted with the continued popularity of the concept in 'Rationales for Science Parks'. The chapter concludes with an assessment of possible new directions for science parks.

Definitions

Our focus in this section is to present a nomenclature of science parks and related concepts as a framework for understanding the broad range of developments in these areas that we see today. Specific developments need not conform to a specific category and indeed 'hybrid' developments are more common. What unifies many developments is their underlying rationale. However, we will leave this discussion of rationale for the next section. What we aim for in this section is categorization with an emphasis on functional description as opposed to rationale. This section paraphrases work we have done in looking at definitional issues (Joseph, 1986, 1994; Macdonald & Joseph, 2001; Phillimore, 1999).

The concept of a science park is something that cannot be captured by a static definition. It is inherently dynamic and centered around geography, changing location patterns of industry and technological change. We will first provide a brief historical context and follow this with definitions of the spectrum of categories in use today: small business incubators; technology business incubators; science and technology parks; and business parks and high quality industrial estates.

Historical Context

The growth of residential and industrial districts was once centered around ports, waterways and railways. The development of these districts during the first half of the twentieth century has encouraged alternative forms of location patterns for industry. For example, with the growth of air transport, location near an airport became important. As patterns of industrial location changed, developments in industrial parks followed. Hence, it is now popular to cite such factors as proximity to an airport or university, together with good infrastructure (e.g. telecommunications), as important attributes of a modern industrial park.

The world's first planned industrial park was launched in Manchester (U.K.) in 1896, but it was not until after World War II in the USA that industrial parks gained widespread popularity. Operational characteristics have predominated in the design of these parks (viz. comprehensive planning, zoning, entry control of tenants, development strategy, profitability, etc.). In short, "whatever the name, the presumption is that the industrial park is a project which has been planned and developed as an optimal environment for industrial occupants" (Barr, 1983).

Since the 1970s, there has been growing integration of commercial functions with the industrial activities traditionally associated with industrial parks. One factor driving this integration is technological change—new firms and existing firms introducing new technology are demanding new environments and services (Joseph, 1989a). For example, 'new' industrial activities, such as research and information handling, require conditions similar to those in a residential environment. Commercial property developers have attempted to reflect these changes in the way industrial parks are marketed and planned. National, regional and state governments, together with universities and research institutions, have also taken a greater interest in the shaping of industrial parks, however defined (Congress of the United States, 1984).

Industrial parks are now functionally diverse in order to meet the needs of new companies and changing industries, changing technologies and a range of policy demands from governments. Because this functional diversity has manifested itself in a variety of activities, definitional problems have arisen (Macdonald, 1987; OECD, 1987). Confusion is compounded by the fact that there are national differences in the use of terms. In some countries, the term 'S&T park' has come to encapsulate all forms of government support for new firms and technology development.

While there are historical reasons explaining the diversification of functions, it is possible to define a range of activities arising from the general industrial park concept. All of them reflect diverging aspects of the central problem of meshing a changing industrial and technological environment to the notion of industry location and increased government involvement in innovation. Despite the fact that terms differ between nations and even regions, categorization is possible and can assist policy development. We have categorized the main terms as:

- small business incubators;
- technology business incubators;
- science and technology parks;
- business parks and high quality industrial estates.

Small Business Incubators and Technology Business Incubators

The terms 'small business incubator', 'enterprise center', 'business technology center', 'technology business incubator' and 'innovation center' are often used interchangeably. As a result, problems have arisen in distinguishing between incubators that support new firms in general and the more specialized incubators that deal with technology-related problems associated with the start-up of technology-oriented new firms. For instance, 'innovation center' is often used when the incubators are close to, or on, a university campus. Incubators may be located on an S&T park and it is presumed that the new start-up firms will have technology development as a core part of their business plan. In general, small business incubators have a number of characteristics:

- they are a method of increasing the efficiency of enterprise development by grouping a number of businesses in one facility under the guidance of an experienced business person;

- the facility provides low-cost, expandable space, frequently in an old building with minimum standards;
- they must have access to a business-consulting network providing low-cost services, such as product development, marketing, finance and services; and
- they provide a forum in which the people in the various businesses can assist each other (Australian Department of Industry, Technology & Commerce, 1989a).

While the characteristics of small business incubators are appropriate to all sorts of new businesses, we suggest that for incubators which are specifically accommodating companies where technology development is a central component of the business plan, the term 'technology business incubator' be used. In practice, however, depending upon local circumstances, small business incubators may also be able to accommodate some technology-based companies. With these qualifications in mind, we suggest the following definitions as a guide for policy.

Small business incubators are converted or purpose-built industrial buildings which offer accommodation and a supportive, growth-oriented environment for newly-formed companies.

Technology business incubators (TBIs) are converted or purpose-built industrial buildings which offer accommodation and a supportive, growth oriented environment for newly-formed companies which have technology development as a core component of their business plan.

Some of the services a TBI can provide, in addition to those provided by a small business incubator, are:

- programs for improving the scientific and engineering training of entrepreneurs;
- an information network incorporating skilled workers and advisers from local universities, schools and business;
- technical and management training;
- evaluation, consulting and referral services;
- an information gathering and dissemination service (e.g. guest speakers and site visits);
- specialist advice on financing technology development; and
- patenting and commercialization advice (Congress of the United States, 1984).

Science and Technology Parks

While small business incubators and TBIs focus on new enterprise development, S&T parks aim to establish concentrations of firms or industries in a particular area. S&T parks are also associated with technology transfer objectives. The terms 'research park', 'science park' and 'technology park' are frequently used interchangeably and distinctions easily become

blurred. The term 'research park' may reflect the original concept behind science parks.

A '**research park**' implies the following:

- a high quality, low density physical environment in a park-like setting;
- interaction between academics, researchers and commercial organizations and entrepreneurs; and
- an environment for research and product development, with conventional production and office activities excluded.

A '**science park**' can be defined as a property-based initiative which:

- has formal operational links with one or more universities, research centers, or other institutions of higher education;
- is designed to encourage the formation and growth of knowledge-based industries and other organizations normally resident on site; and
- has a management function which is actively engaged in the transfer of technology and business skills to tenant organizations (Australian Department of Industry, Technology & Commerce, 1989b, p. 7).

A '**technology park**' can be defined broadly as a collection of high technology industrial companies concerned with both research and manufacturing, located in attractive, well-landscaped surroundings, and situated within a reasonable catchment area of a scientific university or a major research institute (Australian Department of Industry, Technology & Commerce, 1989b, p. 17).

Key differences between the above definitions rest on: (i) what activities are permitted in the park (e.g. research only or some light manufacturing); (ii) the strength of the link with the university (e.g. close operational links or just in the vicinity); and (iii) the extent to which interaction with the university is promoted or expected to occur (e.g. active technology transfer programs in place). In short, a definition of S&T park must reflect a spectrum of activities (Joseph, 1992).

We suggest that a S&T park can be defined as a property-based initiative which:

- has a high quality, low density physical environment in a park-like setting;
- is located within a reasonable distance of a university or research institute; and
- emphasizes activities which encourage the formation and growth of a range of research, new technology or knowledge-based enterprises.

S&T parks may or may not have a TBI associated with them. It is usually the case, however, that for larger developments, a TBI is located on the park, be it a research park, science park or S&T park.

Business Parks and High Quality Industrial Estates

While S&T parks highlight a technology transfer role, there is a class of development which has little direct focus on technology transfer, but which sometimes makes claims in that direction. This class of development—business parks and high quality industrial estates—is defined below (Joseph, 1986):

Business parks provide a high quality prestigious environment, suitable for a wide range of activities, including manufacturing, assembly, sales and other office-based activities. These parks do not require close proximity to academic institutions.

High quality industrial estates are property-based developments which can legitimately claim some features attractive to certain forms of modern light industry. Often a 'high-tech' center is planned for a later stage of development of the industrial estate.

The potential for science parks to evolve into business parks and industrial estates and to lose their research and technology focus is discussed further in the next section.

Rationales for Science Parks

Massey et al. (1992, p. 21) list three main classes of science park objectives: economic development, local benefits, and transfer of technology. Castells & Hall (1994, p. 224) amend these slightly in listing three motivations for the establishment of technopoles (their generic term for science and technology parks): reindustrialization, regional development, and the creation of synergies.

The first two of these objectives are fairly straightforward, and could alternatively be described as technology development and urban or regional renewal respectively. The third aim is less clear. At its narrowest, it could be seen simply as the promotion of technology transfer from universities to companies. However, Castells & Hall describe it in more detail as 'the generation of new and valuable information through human intervention' to the extent that an 'innovative milieu', which generates constant innovation, is created and sustained (Castells & Hall, 1994, p. 224).

In the U.K., with the onset of urban decline in many cities in the 1980s, the second of these three motivations was predominant. Local authorities and universities established science parks as a form of urban redevelopment. In other countries, decentralization of economic and technological activities was an important aim. In both Japan and Korea, for example, the national government's ambition to shift activities away from Tokyo and Seoul respectively was a strong factor in determining the direction (and subsequent location) of science park developments in those countries. This involved a policy that required particular government research agencies to move from the capital and become core tenants in regional science parks (Castells & Hall, 1994; Shin, 2001).

In Australia by contrast, the first motivation was more prominent, with technology park development in the 1980s being an integral aspect of government (high) technology policy in many States. This was partly due to the relative absence of the urban problems facing Britain, but also because of the greater role undertaken by State governments in developing technology parks in Australia. These governments had a responsibility for, and interest in, industry development.

In all countries, however, technology transfer has been an avowed aim of science park development—in particular, the encouragement of knowledge transfer from universities and government research institutions to the commercial sector. However, Castells & Hall's more comprehensive notion of knowledge and information networking was, until relatively recently at least, less evident. The language and rationale behind most parks tended instead to reflect a uni-directional flow of research emanating from universities and being commercialized by park-based companies. Measuring the extent of technology transfer from university and research centers to park-based companies has formed the core of most academic assessments of science park performance—and of most of the criticism of this performance.

A further rationale for the establishment of science parks has been their real estate potential. As property-based ventures, the real estate component of science parks is ever-present, especially in privately owned and managed parks where generating income through rent and property sales is crucial to the park's (and the manager's) commercial future. This may necessitate a relaxation of 'selection' criteria for park tenants (i.e. allowing tenants without a commitment to R&D or an interest in accessing nearby universities or research centers) (Westhead, 1997, p. 57). Indeed, it has been argued that science parks can too easily become glorified business parks, attracting firms for reasons of prestige and image rather than any substantive benefits in terms of linkages with university research (Shearmur & Doloreux, 2000, p. 1067). Westhead (1997, p. 57) considers that 'property-based initiatives which link with a local HEI (higher education institution) for solely promotional reasons do the Science Park movement a disservice', and that science park associations should be stricter in applying conditions to those parks wishing to use the 'science park' label. But such moves will clash with pressures to improve the commercial success of science parks.

Image and prestige are not only of concern to companies considering locating on a science park. They are also very important to park owners and sponsors, especially governments, who are keen to demonstrate their commitment to modern technology development through showcase developments such as science parks. From this perspective, science parks can be seen as a 'place marketing' tool, promoting the virtues of a certain location in competition with other localities trying to entice inward investment to the park. The competition may be intra- or inter-metropolitan, and may or may not result in a net increase in investment for the state or country in which the science park is located. As a consequence, it has been argued that 'the popularity of science parks may be more fruitfully analyzed from this local political perspective than from a more aggregate perspective' (Shearmur & Doloreux, 2000, p. 1079).

In assessing the effectiveness of science parks and the extent to which they are meeting their objectives, it is important to keep in mind these various possible rationales behind their establishment.

Assessments

The question of making an assessment of science parks is not easy. There are various complicating factors. First, in 'Definitions', we saw that a multitude of definitions exist and as a result it is difficult to identify exactly what sort of development is being assessed, especially since 'hybrid' activities exist on many parks (eg. a TBI is frequently associated with a science park). Second, in 'Rationales for Science Parks', we observed that the rationale for parks varies and teasing out these strands is not easy either. Third, there is politics. Many park developments have received substantial support from the state or are solely public initiatives and form part of public policy support for industry or technology development. As a result, while it may be possible to subject privately funded ventures to purely commercial criteria, those with public investment usually have broader long-term policy objectives and this makes assessment based on commercial criteria alone difficult.

Furthermore, compounding this theme is the policy analysis process itself (Joseph, 1994). Since the objectives of parks are often changing, pinning down a clear cut link between identifying an objective and the successful attainment of that objective often evades policy analysts, especially if vested interests are involved. Finally, trenchant academic criticism of science parks (particularly in the U.K. over the past 10 years), has questioned whether science parks work (Grayson, 1993). It is primarily to this last point that we will direct our attention in this section.

How accurate is it to say that hype has triumphed over experience? Asking the question 'has hype triumphed over experience' is slightly different to asking the question 'do science parks work?' There are numerous studies of the latter. Many are reviewed in Grayson (1993), and there are more recent assessments of parks in Sweden (Lofsten & Lindelof, 2001), Canada (Shearmur & Doloreux, 2000), Korea (Shin, 2001), Greece (Bakouros et al., 2002) and Australia (Phillimore, 1999). None of these studies is definitive

in being able to resolve this latter question. Proponents and opponents of science parks often take rather entrenched positions.

The aim of this section is to examine why such entrenched positions have evolved and the underlying reasons for this situation. We will explore this theme below by analyzing different positions taken in the debate by proponents, the U.K. Science Parks Association (UKSPA), and opponents, Massey, Wield and Quintas (1992), who wrote a very critical book entitled *High Tech Fantasies*. Grayson (1993) reports on these two positions.

Grayson's (1993, p. 104) review of the science park literature addresses the question 'do science parks work?'

> The answer to this question may seem self evident given the rapid growth of science parks developments in both the U.K. and overseas, and the low failure rate of the companies which locate on them . . . However, there are doubts: about whether the science park is a necessary vehicle for the development of small high tech companies, whether the concept of technology transfer it embodies is a fruitful one, whether it can succeed as a focus for local or regional regeneration, and (on a broader level) whether the concept of high tech, small firm-led growth itself is reasonable.

The doubts about science parks noted in the above quotation relate to criticisms by Massey et al. (1992) and others (e.g. Macdonald, 1987) that British science parks fail to meet the criteria laid down by the UKSPA definition. As noted in 'Definitions', this definition covers the following areas:

- a science park has formal links with a university or other higher educational or research institution;
- a science park is designed to encourage the formation and growth of knowledge-based and other organizations normally resident on site; and
- a science park has a management function which is actively engaged in the transfer of technology and business skills to the organizations on site.

On all counts, Massey et al. (1992) argue that British science parks fall short.

Grayson (1993, pp. 119–133) summarizes the UKSPA's response to this criticism below.

> UKSPA argues that, far from being based on an outdated linear notion of innovation, the science park plays a more important role in the more complex interactive model by providing the conditions for feedback between academics and the market, and by contributing to technology diffusion throughout industry as a whole. It claims that much of the critical academic analysis is based on misunderstandings derived in part from the unrealis-

tic claims made by some science park proponents, but also from a misinterpretation of the evidence.

In addition to the above claims, the UKSPA strongly suggests that since so many countries have chosen the science park route, if its potential was so limited these countries would have hardly adopted these policies:

> ... the science park concept has been employed by economies and societies as diverse as France, Germany, Spain, the former Soviet Union, Taiwan, Japan and Korea; successful economies are spending huge resources—much more than Britain—on this kind of systematic conversion of science into technology and that is difficult to see how, otherwise, new industries are going to be born (Grayson, 1993, p. 123).

The above interchange from the U.K. debate allows us to generalize as to why hype and experience are not easy bedfellows in evaluating science parks. There are three reasons why hype might be able to triumph over experience, and these could be applicable in any country.

First, there is no clear-cut agreement between proponents and opponents of science parks about the model of innovation that applies. This is an area where there will perhaps never be agreement but the phrase used by the UKSPA—'systematic conversion of science into technology'—indicates that they have a vested interest in protecting a linear way of looking at the innovation process as opposed to other ways. Yet, in the same breath, the UKSPA, is comfortable in claiming that science parks are about 'more complex interactive models'. Here we see the phenomenon of 'the shifting goal posts' in policy analysis. That is, policy objectives are subtly changed so that definitive analysis based on agreed criteria become difficult. Critics have difficulty in making a criticism stick since science park proponents are able to redefine their objectives all too readily. This is particularly acute for science parks since they are long-term ventures, with success periods measured by decades, not just a few years. Little wonder, then, that there is deep-seated disagreement.

Second, the UKSPA blames academics for misunderstanding the role of science parks and partly attributes this to 'unrealistic claims made by some science park proponents'. This point glosses over and simplifies a very significant attribute of science parks—they have often been portrayed as symbolic of high technology development policy and as such, politicians have a vested interest in seeing them succeed. In our view, it is not so simple as interpreting this as a few enthusiastic land and property developers having over-stepped the mark. Rather, some science parks have become 'flagships' for political agendas, either at the national, state or regional level. From this aspect, it is very easy to see that science parks will be defended at all costs. Two defences that the UKSPA use are the 'correct' interpretation of 'reliable' statistics and the time factor over which science parks are expected to deliver benefits:

> The U.K. Science Park Association has for several years collected statistical information on the national science park movement and this provides a broad basis for the assessment of progress ... Nevertheless, behind the statistics lie factors which must be frankly recognized and taken into account when the numbers are interpreted. It is unrealistic to expect that science parks in themselves and in the short to medium term will have a major effect on the national level of employment or economic activity (Grayson, 1993, p. 120).

Once again, 'shifting goal posts' can act as a vehicle for allowing hype to triumph over experience. However, what qualifies as 'experience' is contestable since there is no agreed view on what statistics are allowed as authentic. Moreover, once political reputations and public money are involved, it is very difficult for specific science parks to be labeled as an outright failure. For example, commercial failure could be disguised by changing the entry criteria for the park. Policy failure could be disguised by obfuscating the grounds for policy analysis.

Finally, the third area where hype may triumph over experience relates to the UKSPA's view that since so many countries have chosen the science park route, if its potential was so limited these countries would have hardly adopted these policies. This too is an interesting observation in that it suggests that the copying of the science park model from one country to another implies that the model itself is successful. We claim that this argument is flawed since science parks themselves were conceived with a logic of trying to emulate Silicon Valley (Joseph, 1989b; Macdonald, 1983). The Silicon Valley model is a narrative about how the Californian high technology mecca, Silicon Valley, came about. The narrative, which has no doubt influenced and been the part justification of many policies for high technology (including science parks), includes some or all of the following features: a faith in entrepreneurialism; a vital role for venture capital; a critical role played by research universities; a healthy supply of highly qualified researchers; and benefits for firms co-locating (agglomeration economies). Apart from the process of modeling itself having its own complications (Joseph, 1997), Macdonald (1998, pp. 160–188) points out that the science park route of trying to copy or model the Silicon Valley phenomenon is flawed in a number of ways:

- policy makers who have modeled Silicon Valley have consistently misunderstood the crucial role that information plays in Silicon Valley itself;

- the Silicon Valley model provides for a mis-interpretation of the sort of information that is important to high technology; and
- the Silicon Valley model has given rise to a view that the Stanford Industrial Park, established by Stanford University, somehow caused the growth of Silicon Valley, which it has not.

More recently, Cook & Joseph (2001) have added state and cultural factors to the above in criticizing the Silicon Valley model, demonstrating just how difficult it is to transfer policy ideas such as science parks into a different cultural context (eg. southeast Asia).

What then are our conclusions about the triumph of hype over experience? Macdonald summarizes our view above:

> Policy portrayed high technology as something magical: apply public money with the right incanta-tion and wondrous things would happen. When they failed to happen, there was no incentive for policy makers to explain what had gone wrong. The good performer hides his embarrassment and quickly moves on to the next trick. The audience will never be told why the high-technology miracle never happened. But behind the scene, there is the opportunity for policy makers and their political masters to learn from what went wrong (Macdonald, 1998, p. 183).

However, it would be irresponsible of us to dismiss science parks out of hand. They form a permanent and significant element of the technology policy landscape and they are continually being copied and established in many countries. They obviously fill a need, espe-cially if tenants are willing to locate in them. However, such a need has more to do with property development than technology, industry or regional development policy.

This section has attempted to address why hype may have triumphed over experience and we have demon-strated that for some very strong reasons, criticisms of science park hype are likely to be ignored by their proponents. What this means in practice is that science parks will have a role but that role will need to reassessed in the light of fundamental criticisms of their rationale and the continuing experience of their contribution to policy objectives. Because hype may triumph over experience (or logic for that matter), there is the risk that the science park phenomenon will prove to be a pale imitation of the sort of benefits that many countries are now seeking from high technology development policy. There is an opportunity to learn and this must be grasped. It is not even a case of replacing a 'good policy measure' with a 'better policy'. Rather, science parks are an opportunity to learn and there are many obstacles preventing that lesson from being learnt. It is to this that we turn our attention in the next section.

Conclusion

Science parks have been in existence now for over fifty years, have been a regular and conscious tool of government technology and regional development policies for at least twenty years in the USA, U.K., Canada, Australia and parts of Europe, and are growing rapidly in number in Asia and parts of Latin America. The high hopes held by many government, academic and business proponents of science parks have called forth a worldwide science park 'movement'. Other observers, however, consider them to be 'high tech fantasies' based more on hype than substance and a relative waste of public money compared to how those funds might otherwise have been spent to achieve the same goals.

As the discussion in this chapter shows, the jury is still out on whether science parks can best be seen as the last hurrah of old-style, place-based industrial development, or as the harbinger of new-style, infor-mation and network-based technology development in the information economy.

While these opposing views are effectively irrecon-cilable, it is still possible for both sides to learn from the various assessments that have been made of science parks. For whatever one's views on science parks, the fact remains: they exist. And so the question also remains: where now for science parks?

In our view, the way forward comprises several possible elements:

- enlisting more pro-active and supportive science park management aimed at encouraging more frequent and more dense linkages (both formal and informal) between park companies, universities and research institutes (Westhead, 1997, p. 59);
- using the existing and expanding international sci-ence park network as a source of 'value added' for park tenants in the form of information provision, marketing expertise and contacts, technical support and business links (Phillimore, 1999, p. 679);
- encouraging science parks to integrate more closely with their civic, urban and regional environments so that they actively use their spatial characteristics to promote interaction and networking possibilities, rather than cutting themselves off from their sur-rounding innovative milieu through maintaining an 'elitist' image. Some parks in Australia, for example, are now evolving into technology precincts with the explicit aim of linking more closely to the wider community (Phillimore, 1999, p. 679);
- creating new science parks with a more specific focus, such as particular technologies (e.g. bio-technology) or environmental objectives. This enables more distinctive park developments to emerge and should make it easier to encourage closer links between park tenants themselves as well as between the tenants and relevant university and government researchers; and

- Moving beyond a linear view of innovation towards a network view which emphasizes 'the creation, acquisition and handling of information' as a core function of science parks (Joseph, 1994, p. 54).

While these initiatives should assist science parks in their quest for relevance and performance, no-one should consider that they are likely to answer the question contained in the title of this chapter. That is as much a political as it is an economic question, the answer to which is unlikely to be settled in the near future.

References

Australian Department of Industry, Technology and Commerce (1989a). *Business incubators.* Canberra: Australian Government Publishing Service.

Australian Department of Industry, Technology and Commerce (1989b). *Technology parks in Australia.* Canberra: Australian Government Publishing Service.

Bakouros, Y. L., Mardas, D. C. & Varsakelis, N. C. (2002). Science park, a high tech fantasy?: An analysis of the science parks of Greece. *Technovation*, **22**, 123–128.

Barr, B. M. (1983). Industrial parks as locational environments: a research challenge. In: F. Hamilton & G. J. R. Linge (Eds), *Regional Economics and Industrial Systems* (pp. 423–440). Brisbane: John Wiley.

Castells, M. & Hall, P. (1994). *Technopoles of the World: The making of the 21st century industrial complexes.* London: Routledge.

Congress of the United States, Office of Technology Assessment (1984). *Encouraging high-technology development.* Technology, Innovation and Regional Economic Development Background Paper No. 2, Washington: USGPO.

Cook, I. & Joseph, R. (2001). Rethinking Silicon Valley: New perspectives on regional development. *Prometheus*, **19** (4), 377–393.

Grayson, L. (1993). *Science Parks: An experiment in high technology transfer.* London: The British Library.

International Association of Science Parks (2000). IASP statistics. *IASP News No. 2*, **4**.

Joseph, R. A. (1986). *Technology parks: A study of their relevance to industrial and technological development in New South Wales.* Report to the NSW Department of Industrial Development and Decentralisation (DIDD), Sydney: DIDD.

Joseph, R. A. (1989a). Technology parks and their contribution to the development of technology-oriented complexes in Australia. *Environment and Planning C: Government and Policy*, **7** (2), 173–192.

Joseph, R. A. (1989b). Silicon Valley myth and the origins of technology parks in Australia. *Science and Public Policy*, **16** (6), 353–366.

Joseph, R. A. (1992). Technology parks: An overview. *NSW Department of State Development, State Development Occasional Paper No. 2.* Sydney: Government of New South Wales.

Joseph, R. A. (1994). New ways to make technology parks more relevant. *Prometheus*, **12** (1), 46–61.

Joseph, R. A. (1997). Political myth, high technology and the information superhighway: An Australian perspective. *Telematics and Informatics*, **14** (3), 289–301.

Lofsten, H. & Lindelof, P. (2001). Science parks in Sweden—industrial renewal and development? *R&D Management*, **31** (3), 309–322.

Macdonald, S. (1983). High technology policy and the Silicon Valley model: An Australian perspective. *Prometheus*, **1** (2), 330–349.

Macdonald, S. (1987). British science parks: reflections on the politics of high technology. *R&D Management*, **17** (1), 25–37.

Macdonald, S. (1998). *Information for innovation: Managing change from an information perspective.* Oxford: Oxford University Press.

Macdonald, S. & Joseph, R. (2001). Technology transfer or incubation? Technology business incubators and science and technology parks in the Philippines. *Science and Public Policy*, **28** (5), 330–344.

Massey, D., Quintas, P. & Wield, D. (1992). *High tech fantasies: Science parks in society, science and space.* London: Routledge.

Organization for Economic Cooperation and Development (OECD) (1987). *Science parks and technology complexes in relation to regional development.* Paris: OECD.

Phillimore, J. (1999). Beyond the linear view of innovation in science park evaluation: An analysis of Western Australian Technology Park. *Technovation*, **19**, 673–680.

Shearmur, R. & Doloreux, D. (2000). Science parks: actors or reactors? Canadian science parks in their urban context. *Environment and Planning A*, **32**, 1065–1082.

Shin, D-H. (2001). An alternative approach to developing science parks: A case study from Korea. *Papers in Regional Science*, **80**, 103–111.

Westhead, P. (1997). R&D 'inputs' and 'outputs' of technology-based firms located on and off Science Parks. *R&D Management*, **27** (1), 45–62.

Part X

Innovation Management

The International Handbook on Innovation
Edited by Larisa V. Shavinina

Challenges in Innovation Management

John Bessant

School of Management, Cranfield University, U.K.

Abstract: Innovation represents the core renewal process in any organization. Unless a business is prepared to work continuously at renewing what it offers and how it creates and delivers that offering, there is a good chance that it won't survive in today's turbulent environment . . . So managing innovation becomes one of the key strategic tasks facing organizations of all shapes, sizes and sectors. This chapter reviews the question of managing innovation and particularly looks at some of the key challenges, which must be addressed.

Keywords: Innovation; Innovation management; Organization; Business; Innovation culture; Learning.

Introduction

Innovation represents the core renewal process in any organization. And unless a business is prepared to work continuously at renewing what it offers and how it creates and delivers that offering, there is a good chance that it won't survive in today's turbulent environment. It's a sobering thought that only one firm out of the Dow Jones 100 index actually made it through from the beginning to the end of the 20th century—even the biggest enterprises have no guarantee of survival and the mortality rate for smaller firms is very high (de Geus, 1996). Nor is this only a problem for individual firms; as Utterback's study indicates, whole industries can be undermined and disappear as a result of radical innovation which rewrites the technical and economic rules of the game (Utterback, 1994). Two worrying conclusions emerge from his work; first, that most innovations which destroy the existing order originate from newcomers and outsiders to a particular industry, and second, that few of the original players survive such transformations.

So the question is not one of whether or not to innovate but rather of *how* to do so successfully. Managing innovation becomes one of the key strategic tasks facing organizations of all shapes, sizes and sectors. This chapter reviews the question of managing innovation and particularly looks at some of the key challenges which must be dealt with.

What Has to be Managed?

One of the difficulties in managing innovation is that it is something of a 'Humpty-Dumpty' word. In 'Alice in Wonderland' this character used words to mean 'whatever I want them to mean'—and in similar fashion people talking about innovation often use the term in widely different ways! The word originally comes from the Latin *innovare* meaning 'to make something new', but it helps to think of it in terms not only of 'invention'—creating something new—but also of its development and take up in practice. A useful definition is that offered by Freeman; '. . . the technical, design, manufacturing, management and commercial activities involved in the marketing of a new (or improved) product or the first commercial use of a new (or improved) process or equipment' (Freeman, 1982). Rothwell reminds us that it is not always about radical change; '. . . *innovation does not necessarily imply the commercialization of only a major advance in the technological state-of-the-art . . . . But it includes also the utilization of even small-scale changes in technological know-how . . .*' (Rothwell, 1992). Perhaps the most succinct definition is offered by the Innovation Unit of the U.K. Department of Trade and Industry who see it simply as 'the successful exploitation of new ideas'.

Getting this process right once is possible through luck—the skill comes in being able to repeat the trick. Whether the organisation is concerned with bricks, bread, banking or baby care, the underlying challenge is still the same. How to obtain a competitive edge through innovation—and through this, survive and grow? (This is as much a challenge for non-profit organizations—in police work, in health care, in education the competition is still there, and the role of innovation still one of getting a better edge to dealing with problems of crime, illness or illiteracy).

Essentially organizations have to manage four different phases in the process of turning ideas into successful reality (Tidd, Bessant et al., 2001). They have to:

- scan and search their environments (internal and external) to pick up and process signals about potential innovation;
- strategically select from this set of potential triggers for innovation those things which the organization will commit resources to doing;
- having chosen an option, organizations need to resource it—providing (either by creating through R&D or acquiring through technology transfer) the resources to exploit it;
- finally organizations have to implement the innovation, growing it from an idea through various stages of development to final launch—as a new product or service in the external market place or a new process or method within the organization.

One other phase is relevant, albeit optional. Organizations need to reflect upon the previous phases and review experience of success and failure—in order to learn about how manage the process better, and to capture relevant knowledge from the experience.

Although this process is generic, there are countless variations in how organizations actually carry it out. But innovation management is about learning to find the most appropriate solution to the problem of consistently managing this process, and doing so in the ways best suited to the particular circumstances in which the organization finds itself.

The trouble is that the innovation puzzle which organizations are trying to solve is constantly changing and mutating. This chapter looks at some of the current challenges which are involved in trying to manage innovation, but we should recognize that there is no 'one best way' to do this. There will always be a need to develop new approaches to meet new and emerging challenges.

Challenge 1: Why Change?

Innovation matters. In an uncertain world the only certainty is that simply sitting still carries with it a high risk. History teaches us that innovation is not a luxury item on the strategic agenda but a survival imperative—unless organizations are prepared to change what they offer and how they create and deliver that offering they may simply not be around in the long term (Braun & Macdonald, 1980; de Geus, 1996; Tushman & Anderson, 1987).

Innovation is what agile firms do—they constantly re-invent themselves in terms of their solutions to the puzzle posed by the threats and opportunities in their environment. This may mean adoption of new technology, or generation of their own. It may involve

reconfiguring products, processes or markets. But in each case it involves learning and unlearning, and it requires strategic direction to focus this process (Tidd, Bessant et al., 2001).

Firms often follow a 'resource-based' strategy of accumulating technological assets—essentially the building of technological competence. But evidence suggests that simply accumulating a large knowledge base may not be sufficient—as the cases of well-established firms like IBM, GM, Kodak and others demonstrate. The firms which demonstrate sustained competitive advantage exhibit *'timely responsiveness and rapid product innovation, coupled with the management capability to effectively co-ordinate and redeploy internal and external competencies'* (Teece & Pisano, 1994).

Whilst it is easy to pay lip service to this kind of challenge the historical evidence suggests that firms are not good at learning and that seeing—and responding to—the need for continuous innovation is not a widely distributed capability (Christenson, 1997; Hamel, 2000; Womack, Jones et al., 1991).

Challenge 2: What to Change?

Even if firms recognize and accept the need for continuous innovation they may find difficulties in framing an appropriate innovation agenda. With limited resources they may risk putting scarce eggs into too few or the wrong baskets. Innovation can take many forms—from simple, incremental development of what is already there to radical development of totally new options. It can range form changes in what is offered—product or service—through to the ways in which that offering is created and delivered (process innovation). It can reflect the positioning of a particular offering—for example, putting a well-established product into a new market represents a powerful source of innovation. And it can involve rethinking the underlying mental models associated with a particular product or service (Francis, 2001).

The challenge here is for firms to be aware of the extensive space within which innovation possibilities exist and to try and develop a strategic portfolio which covers this territory effectively, balancing risks and resources. Table 1 maps out some options.

Challenge 3: Understanding Innovation

Part of the problem in managing innovation is the way people think about it. Whilst the term is in common usage, the meaning people attach to it—and hence the way in which they behave—can vary widely (Dodgson & Bessant, 1996; Tidd, Bessant et al., 2001). For example, there is often considerable confusion between 'invention' and 'innovation'. The former is essentially about the moment of creative insight which first opens up a new possibility—the discovery of a new compound, the observation of a new phenomenon, the recognition of an unmet market need. But whilst this is

Table 1. The innovation agenda.

	Do what we do better—improvement innovation	Do different—radical innovation
Product or service offering	Improved versions of existing products, Mk2, and later releases, etc.	Radical new product or service concepts
Process for creating and delivering that offering	Improvements to established processes—lean approaches to driving out waste, quality improvements, etc.	Radical alternative routes to achieving the process goal—for example, float glass as an alternative way of making flat glass to traditional grinding and polishing
Positioning in market context	Repositioning or relaunching established products or services in different context—e.g. opening up new market niches	Radical re-framing of what is being offered—for example, many utility businesses are now being seen as commodities which can be traded in dealing rooms—as opposed to service businesses. Christensen's work on disk drives, mini-mills and other industries highlights the potential for reframing to create new markets.
'Paradigm' - underlying mental models	New business models	New ways of conceptualizing the problems—for example, mass production vs. craft production

essential to start the process off, invention is not enough. Taking that brilliant idea through, on an often painful journey to become something which is widely used involves many more steps and a lot of resources and problem-solving on the way. As Edison is reputed to have said, "it's 1% inspiration, 99% perspiration!"

History is littered with forgotten names which bear testament to the danger of confusing the two. Spengler invented the vacuum sweeper but Hoover brought it through to commercial reality. Howe developed the first sewing machine but Singer is the name we associate with the product because he took it from invention to widespread acceptance (Bryson, 1994).

The problem isn't just a confusion of invention and innovation. Other limits to our mental models include the view that innovation is all about science and technology creating new opportunities—what is sometimes called the 'technology push' model (Coombs, Saviotti et al., 1985; Rothwell, 1992). It has elements of truth about it but on its own it is a weak basis for managing innovation—plenty of great technological possibilities fail to find markets and never make it as innovations. Similarly, the view that 'necessity is the Mother of invention' may sound persuasive—but a totally marketing led approach to innovation may miss some important tricks. The emergence of the Walkman family of products within Sony took place despite strong marketing input to suggest there was no demand for this kind of product (Nayak & Ketteringham, 1986).

Table 2 lists some common misperceptions and partial views of innovation. The challenge to managing the process well is to ensure a broader and integrated view to underpin the structures and procedures which firms put in place to make it happen.

Challenge 4: Building an Innovation Culture

In effect, innovation poses a constantly mutating puzzle, or set of puzzles. Firms cannot afford not to play the game but the rules are anything but clear. There is not the comfort of a single 'right' answer because the question keeps changing. And even if one firm does come up with a solution which works for them it does not follow that everyone else adopting it will meet with the same degree of success. Copying may simply make the problem worse.

So what can an organization do? Research suggests that the task of managing innovation is about creating firm specific routines—repeated, reinforced patterns of behavior—which define its particular approach to the problem (Cohen, Burkhart et al., 1996; Nelson & Winter, 1982; Pavitt, 2002; Tidd, Bessant et al., 2001). 'Routine' in this sense does not mean robotic but it does mean an established pattern—'the way we do things around here'—which represents the approach a particular organization takes to dealing with the innovation challenge.

Routines of this kind are not mindless patterns; as Giddens points out '. . . *the routinized character of most social activity is something that has to be 'worked at' continually by those who sustain it in their day-to-day conduct . . .*' (Giddens, 1984). It is rather the case that they have become internalized to the point of being unconscious or autonomous. They become part of 'the way we do things round here'—in other words, part of the dominant culture of the organization.

Table 2. Common problems associated with partial views of innovation (Tidd, Bessant et al., 2001).

If innovation is only seen as . . .	. . . the result can be
Strong R&D capability	Technology which fails to meet user needs and may not be accepted
The province of specialists in white coats in the R&D laboratory	Lack of involvement of others, and a lack of key knowledge and experience input from other perspectives
Meeting customer needs	Lack of technical progression, leading to inability to gain competitive edge
Technology advances	Producing products which the market does not want or designing processes which do not meet the needs of the user and which are opposed
The province only of large firms	Weak small firms with too high a dependence on large customers
Only about 'breakthrough' changes	Neglect of the potential of incremental innovation. Also an inability to secure and reinforce the gains from radical change because the incremental performance ratchet is not working well
Only associated with key individuals	Failure to utilize the creativity of the remainder of employees, and to secure their inputs and perspectives to improve innovation
Only internally generated	The 'not invented here' effect where good ideas from outside are resisted or rejected (Katz & Allen 1982)
Only externally generated	Innovation becomes simply a matter of filling a shopping list of needs from outside and there is little internal learning or development of technological competence

Innovation routines are increasingly recognized as contributing to competitive advantage and one important feature is that such routines cannot be simply copied from one context to another; they have to be learned and practiced over a sustained period of time. (Pavitt, 2000) Thus, for example, the Toyota Production System with its high levels of participation took over 40 years to evolve and become embedded in the culture. (Monden, 1983) Whilst it is easy for Toyota executives to demonstrate this to others, it is not easy to replicate it; the Ford and General Motors needed to go through their own learning processes and come up with their own firm-specific versions of the idea. (Adler, 1992; Wickens, 1987).

Routines can begin by the chance recognition of something that worked or as the result of trying a new and different approach. But if they work repeatedly, they gradually get established and eventually formalized into structures and procedures—until finally they are part of the organization's personality. A useful analogy can be made with learning to drive a car; in the early stages the problem is one of unfamiliarity with the whole experience. Learning is concentrated on trying to understand and master the individual low level skills associated with things like steering, the clutch, the brakes, the accelerator and so on. Gradually facility with these is developed but then the challenge comes of linking the individual behaviors together in complex sequences—for example, changing gear or making a hill start. Eventually, and after extensive trial and error and practice the point is reached where you

qualify for a driving licence—but this only indicates an expression of basic competence; the real challenges of becoming a good driver, able to cope with different cars, weather conditions, road types, etc. still lie ahead. Much, much later, after years of practice in different vehicles and under different conditions, the point is reached where the conscious effort required is minimal and driving can take place simultaneously with listening to the radio, talking to other passengers or even daydreaming!

Some examples of routines—behavior patterns—which are needed for innovation are given in Table 3.

Challenge 5: Continuous Learning

The innovation agenda—as we have seen—is constantly shifting and firms needs to develop routines to deal with the key challenges, which emerge from their environment. There is plenty of scope for learning and adapting of these routines—not everything has to be engineered from scratch. The distinction is between blind copying and adopting and developing a good practice, which someone else uses.

So, for example, in the area of new product development there is now an accepted 'good practice' model of how an organization can create and deliver a stream of new products and services (Cooper, 1993, 2000, 2003; Smith & Reinertsen, 1991; Wheelwright & Clark, 1992). (It is important to stress the repeatability here—anyone can get lucky once but only those organizations, which have in place relevant routines, are in a position to repeat the trick. 'The way we do

Table 3. Key routines needed for basic innovation capability (based on Tidd, Bessant et al., 2001).

Basic capability	Contributing capabilities
Recognizing	Searching the environment for technical and economic clues to trigger the process of change
Aligning	Ensuring a good fit between the overall business strategy and the proposed change—not innovating because it's fashionable or as a knee-jerk response to a competitor
Acquiring	Recognizing the limitations of the company's own technology base and being able to connect to external sources of knowledge, information, equipment, etc. Transferring technology from various outside sources and connecting it to the relevant internal points in the organization
Generating	Having the ability to create some aspects of technology in-house—through R&D, internal engineering groups, etc.
Choosing	Exploring and selecting the most suitable response to the environmental triggers which fit the strategy and the internal resource base/ external technology network
Executing	Managing development projects for new products or processes from initial idea through to final launch. Monitoring and controlling such projects
Implementing	Managing the introduction of change—technical and otherwise—in the organization to ensure acceptance and effective use of innovation
Learning	Having the ability to evaluate and reflect upon the innovation process and identify lessons for improvement in the management routines
Developing the organization	Putting those new routines in place—in structures, processes, underlying behaviors, etc.

Table 4. Key features of emerging 'good practice' model in NPD (Bessant & Francis, 1997).

Theme	Key characteristics
Systematic process for progressing new products	Stage-gate model (Cooper, 1994) Close monitoring and evaluation at each stage (Bruce & Bessant, 2001)
Early involvement of all relevant functions	Bringing key perspectives into the process early enough to influence design and prepare for downstream problems (Wheelwright & Clark, 1992) Early detection of problems leads to less rework
Overlapping/parallel working	Concurrent or simultaneous engineering to aid faster development whilst retaining cross-functional involvement (Bowen, Clark et al., 1994)
Appropriate project management structures	Choice of structure—e.g. matrix/line/project/heavyweight project management—to suit conditions and task (Bruce & Cooper, 1997)
Cross-functional team working	Involvement of different perspectives, use of team-building approaches to ensure effective team working and develop capabilities in flexible problem-solving (Francis & Young, 1988; Shillito, 1994)
Advanced support tools	Use of tools—such as CAD, rapid prototyping, computer-supported co-operative work aids—to assist with quality and speed of development
Learning and continuous improvement	Carrying forward lessons learned—via post-project audits, etc. (Rush, Brady et al., 1997) Development of continuous improvement culture (Caffyn, 1998)

Table 5. Examples of company routines for successful innovation (based on 'How 3M keeps the new products coming'. R. Mitchell in Henry & Walker, 1990).

Company	Routines
3M	• keep divisions small so that each division manager knows the names of all staff. When divisions get too big, split them up • tolerate failure and encourage experimentation and risk-taking. Divisions must derive 25% of sales from products introduced during the past five years • motivate the champions—when someone comes up with a new product idea, he/she can recruit an 'action team' to develop it. Salaries and promotion are directly tied to the product's progress; if successful the champion ends up running his/her own division • stay close to the customer, through regular interaction between research, marketing and customers • encourage experiment—staff can spend up to 15% of their time on ideas of their own to try and prove they are workable. For those needing seed money an internal venture fund—Genesis—awards grants of up to $50,000 for development
Hewlett-Packard	• researchers are encouraged to spend up to 10% of time on their own pet projects, and have 24 hour access to laboratories and equipment • keeps divisions small to focus team efforts
Merck	• gives researchers time and resources to pursue high-risk, high payoff products
Johnson and Johnson	• freedom to fail is a key value • extensive use of autonomous small operating units for development
General Electric	• jointly develops products with customers—for example, developed the first thermoplastic body panels for cars through joint work with BMW
Rubbermaid	• 30% of sales must come from products developed in the last five years
Dow Corning	• forms research partnerships with customers

things round here' has to include structures and procedures to make this happen.) These basics of the 'good practice' model are given in Table 4.

But making the model work in a particular organization requires taking these ideas and shaping them to suit a particular and highly firm-specific context. Table 5 shows how some firms have evolved their own variants on these routines to help with the problem of product innovation.

The Moving Frontier

The trouble with innovation, as we have already seen, is that it is not a static problem to which we might try and find a particular solution, which fits. Instead, it is a mutating and shifting set of puzzles—and firms have to try and find their own particular ways of solving them and continuing to be able to do so as the puzzles shift and change. The routines identified in Table 3 are a good starting point but firms will need to adopt, adapt and create new routines as the puzzle shifts. In the next section we look at some areas and challenges, which represent the frontier of innovation management around which there are few proven routines and much room for experiment and learning.

Challenge 6: High Involvement Innovation

Traditionally innovation has been the province of the specialist who often works apart from the mainstream of the organization's operations—for example, in R&D

or IT functions. The roots of this division of labour can be traced back at least to the late 19th century and the emergent mass production models based on 'scientific management' (Boer & Berger et al., 1999; Schroeder & Robinson, 1991). Here emphasis was placed on a belief in 'one best way'—in the way work was organized, the products and services offered, etc.—and any interference with the designs produced by innovation specialists was seen as disruptive. Such separation of 'head' and 'hand' became institutionalized in the functional and hierarchical modes of organization, which became the dominant blueprint for much of the 20th century (Best, 1990; Kaplinsky, 1994).

The limitations of such an exclusive model of innovation quickly become clear. Innovation is fundamentally about creative problem-solving and as environments become more turbulent and uncertain, so the requirement for this capability increases. With uncertain markets, rapidly changing technological threats and opportunities, increasing regulatory pressures, shifting customer and competitive requirements, and a host of other variables to deal with the likelihood of getting the 'right' innovative response is low. Organizations need to increase their innovative capacity, and one powerful mechanism for doing so is to extend participation in the process to a much wider population.

This simple point has been recognized in a number of different fields, all of which converge around the view that higher levels of participation in innovation represents a competitive advantage (Caulkin, 2001; Huselid, 1995; Pfeffer & Veiga, 1999). For example, in the field of quality management it has become clear that major advantages accrue from better and more consistent quality in products and services (Brown, Bessant et al., 2000; Garvin, 1988). But the underlying recipe for achieving 'total quality' is an old one, originally articulated in the early part of the 20th century and making extensive play on the contribution which participation in the process of finding and solving quality problems could make (Deming, 1986; Juran, 1951).

Similarly, the concept of 'lean manufacturing' emerged from detailed studies of assembly plants in the car industry and has since diffused widely around the world and across business sectors. Central to this alternative model was an emphasis on team working and participation in innovation—for example, the average number of suggestions offered by workers in Japanese 'lean' plants was approximately 1 per week; in contrast the European average was around half a suggestion, per worker, per year! (Womack, Jones et al., 1991).

Studies of high performance organizations, especially those, which achieve significant productivity improvements through their workforces, place considerable emphasis on involvement in innovation (Bessant, 2003, forthcoming; Guest, Michie et al., 2000; Pfeffer, 1998). Characteristic of such cases is a blurring of the lines of responsibility for the innovation process, moving away from specialists and towards higher levels of participation by others in incremental innovation as a complement to specialist activity (DTI, 1997; Pfeffer & Veiga, 1999).

Much discussion has focused on the concept of 'learning organizations', seeing knowledge as the basis for competition in the 21st century (Garvin, 1993; Senge, 1990; Pedler & Boydell et al., 1991). Mobilizing and managing knowledge becomes a primary task and many of the recipes offered for achieving this depend upon mobilizing a much higher level of participation in innovative problem-solving and on building such routines into the fabric of organizational life (Leonard-Barton, 1995; Nonaka, 1991).

There is nothing new nor difficult in this concept; indeed, it would be hard to disagree with the premises that we need as much creative problem-solving as possible and that everyone has the basic wherewithal to do it. It is also a theme which recurs in the literature on innovation; many studies report on the importance of involvement and participation in sustained incremental improvement (Figuereido, 2001; Hollander, 1965; Tremblay, 1994). Such high involvement innovation lies at the heart of the 'learning curve' theory which has had such a strong impact on strategic thinking; learning curves only work when there is the commitment and enabling structure for participative problem-solving (Garvin, 1993).

The difficulty comes not in the concept but its implementation. Mobilizing high levels of participation in the innovation process is unfamiliar and, for many organizations, relatively untested and apparently risky. The challenge is thus one of building routines—establishing and reinforcing behavior patterns. Table 6 lists some examples of the kinds of routines which are needed to begin to build high involvement innovation. This information is drawn from an extensive European research program looking at the development of high levels of participation in continuous improvement (CI) innovation (Bessant, Caffyn et al., 2001).

Whilst the generic routines can be specified in terms of particular new behaviors which must be learned and reinforced—for example, systematic problem solving through some form of learning cycle or monitoring and measuring to drive improvement—the particular ways in which different organizations actually achieve this will vary widely. Thus routines for CI are essentially firm-specific; this is one reason why simply imitating what was done successfully within Japanese firms has proved to be such a poor recipe for many Western firms. There is no short cut in the process; CI behaviors have to be learned, reinforced and built upon to develop capability.

Challenge 7: Dealing with Discontinuity

Much innovation can be seen as a 'steady state' activity. Of course it is about change, but it takes place within a framework which is relatively consistent. Most change happens as incremental developments of what is already there—'doing what we do better'. For example, theories of innovation dynamics suggest that when a new product concept emerges there is an initial period of uncertainty during which there is considerable experiment around different configurations (Laurila, 1998; Tushman & Anderson, 1987; Utterback, 1994). But then a 'dominant design' or 'technological trajectory' becomes established and the emphasis shifts to incremental improvements and variations on this basic theme. The same can be said of process innovation—the introduction of a radically new process is followed by a long period of refinement and improvement, stretching and developing the performance of that process, driving out waste, eliminating bugs, etc.

This pattern of innovation is sometimes called 'punctuated equilibrium', borrowing a term from the field of evolutionary biology which explores how species emerge and develop. Its implication is that the way in which we organize for innovation will be around keeping up a steady stream of incremental developments within an envelope established by the original product or process concept (Tushman & Anderson, 1987). Such work might persist for decades—for example, the electric light bulb with which we are all familiar has been around since the late 19th century when work by Swann & Edison created the

Table 6. Key routines in developing high involvement in continuous improvement (CI) innovation (Based on Bessant, Caffyn et al., 2001).

Ability	Constituent behaviors
'Understanding CI'—the ability to articulate the basic values of CI	• people at all levels demonstrate a shared belief in the value of small steps and that everyone can contribute, by themselves being actively involved in making and recognizing incremental improvements • when something goes wrong the natural reaction of people at all levels is to look for reasons why etc. rather than to blame individual(s)
'Getting the CI habit'—the ability to generate sustained involvement in CI	• people make use of some formal problem-finding and solving cycle • people use appropriate tools and techniques to support CI • people use measurement to shape the improvement process • people (as individuals and/or groups) initiate and carry through CI activities—they participate in the process • closing the loop—ideas are responded to in a clearly defined and timely fashion—either implemented or otherwise dealt with
'Focusing CI'—the ability to link CI activities to the strategic goals of the company	• individuals and groups use the organization's strategic goals and objectives to focus and prioritize improvements • everyone understands (i.e. is able to explain) what the company's or department's strategy, goals and objectives are • individuals and groups (e.g. departments, CI teams) assess their proposed changes (before embarking on initial investigation and before implementing a solution) against departmental or company objectives to ensure they are consistent with them • individuals and groups monitor/measure the results of their improvement activity and the impact it has on strategic or departmental objectives • CI activities are an integral part of the individual or groups work, not a parallel activity
'Leading CI'—the ability to lead, direct and support the creation and sustaining of CI behaviors	• managers support the CI process through allocation of time, money, space and other resources • managers recognize in formal (but not necessarily financial) ways the contribution of employees to CI • managers lead by example, becoming actively involved in design and implementation of CI • managers support experiment by not punishing mistakes but by encouraging learning from them
'Aligning CI'—the ability to create consistency between CI values and behavior and the organizational context (structures, procedures, etc.)	• ongoing assessment ensures that the organization's structure and infrastructure and the CI system consistently support and reinforce each other • the individual/group responsible for designing the CI system design it to fit within the current structure and infrastructure • individuals with responsibility for particular company processes/systems hold ongoing reviews to assess whether these processes/systems and the CI system remain compatible • people with responsibility for the CI system ensure that when a major organizational change is planned its potential impact on the CI system is assessed and adjustments are made as necessary.
'Shared problem-solving'—the ability to move CI activity across organizational boundaries	• people co-operate across internal divisions (e.g. cross-functional groups) in CI as well as working in their own areas • people understand and share an holistic view (process understanding and ownership) • people are oriented towards internal and external customers in their CI activity • specific CI projects with outside agencies—customers, suppliers, etc.—are taking place • relevant CI activities involve representatives from different organizational levels

Table 6. Continued.

Ability	Constituent behaviors
'Continuous improvement of continuous improvement'—the ability to strategically manage the development of CI	• the CI system is continually monitored and developed; a designated individual or group monitors the CI system and measures the incidence (i.e. frequency and location) of CI activity and the results of CI activity • there is a cyclical planning process whereby the CI system is regularly reviewed and, if necessary, amended (single-loop learning) • there is periodic review of the CI system in relation to the organization as a whole which may lead to a major regeneration (double-loop learning) • senior management make available sufficient resources (time, money, personnel) to support the ongoing development of the CI system
'The learning organization'—generating the ability to enable learning to take place and be captured at all levels	• people learn from their experiences, both positive and negative • individuals seek out opportunities for learning/personal development (e.g. actively experiment, set their own learning objectives) • individuals and groups at all levels share (make available) their learning from all work experiences • the organization articulates and consolidates (captures and shares) the learning of individuals and groups • managers accept and, where necessary, act on all the learning that takes place • people and teams ensure that their learning is captured by making use of the mechanisms provided for doing so • designated individual(s) use organizational mechanisms to deploy the learning that is captured across the organization

dominant design for the incandescent filament bulb. Whilst there has been extensive incremental innovation within that envelope—for example, in new materials, in more efficient manufacturing processes, in new markets for light bulbs, etc.—the basic pattern for over a century has been one of incremental development within the broad framework established in the 1880s. (For more on this particular innovation see the chapters by Weisberg and by Sternberg and colleagues in this book (Sternberg & Pretz et al., 2003; Weisberg, 2003)).

But there are points at which the rules change. It may well be as a result of scientific progress creating new possibilities, or it may be a result of dramatic shifts in the demand side for innovation—for example, radical restructuring of markets through deregulation or the opening up of trade barriers. At such points the old rules may no longer apply—for example, the technical knowledge associated with a particular product or process may become redundant as it is replaced by a new one. (The shift from valve electronics in the period up to 1947 to the era of solid state and integrated circuits, which was ushered in by the invention of the transistor, is a good example of this (Braun & Macdonald, 1980).) (The chapter by Ronald Guttmann in this volume also provides a detailed discussion of this theme (Guttmann, 2003).)

This kind of transition poses very big management challenges. Historical evidence suggests strongly that when such discontinuous changes take place the old

incumbents do not usually do well and it is at this point that new entrants become key players. (See, for example, the case of the ice industry (Utterback, 1994; Weightman, 2002), the automobile industry (Altschuler, Roos et al., 1984), the cement industry (Tushman & Anderson, 1987), and the mini-mill case in steel-processing (Christenson, 1997). In part this is because of the high level of investment committed to the older generation of technology which established players have but there is also much to suggest that organizational barriers are also a problem. (In this volume Hadjimanolis discusses this 'barriers' approach to innovation (Hadjimanolis, 2003)). Established players are not always quick to pick up on the signals about change or to make sense of them, they may react too slowly and in the wrong directions—or they may simply try to deny the importance or magnitude of the change affecting their business (Henderson & Clark, 1990).

There is an inherent conflict underpinning this pattern. The routines which well-managed firms build up to sustain and develop their innovations in product and process within a particular 'envelope' are not the same as those they will need to create innovations outside that space. Radical, 'out of the box' thinking and the high-risk project management approaches which accompany completely new directions in innovation do not sit well within existing and relatively highly structured frameworks. For this reason many firms set up 'skunk works' or other mechanisms to create more and different space within which people can work (Christenson, 1997; Leifer & McDermott et al., 2000). It is also here that the role of innovation champions and promoters becomes significant—a point addressed in detail elsewhere in this volume in the chapter by Hauschildt (Hauschildt, 2003).

The challenge here is to develop what some writers have called 'the ambidextrous organization' (Tushman & O'Reilly, 1996). That is, to manage under one roof to operate routines for 'doing what we do better' innovation (within the envelope) and simultaneously to allow space for another set of routines for 'doing differently'—moving beyond the envelope into new and uncharted territory. The risk, if they cannot develop these two sets of routines, is that people will migrate out of the organization to set up their own operation (Tushman & O'Reilly, 1996). The chapter by Katz, (this volume), deals with this issue in some detail (Katz, 2003).

Challenge 8: Managing Connections

Characteristic of the routines developed and shared around 'good practice' in innovation management is the focus on the individual firm (Ettlie, 1999; Tidd, Bessant et al., 2001; Van de Ven & Angle et al., 1989). Whilst there are gaps in this model it is, as we have seen, possible to build up a useful model on which to build and develop innovation management capability. But increasingly the challenge to business organizations is to operate not in 'splendid isolation' but in relationships with others. Whether this takes place in the context of supply chains or as part of a network of small firms sharing resources, or in a network of firms sharing knowledge resources to develop a new product or service, the emphasis is shifting towards managing the inter-firm dynamics of innovation. And this calls for a whole new set of routines.

Such inter-organizational networking is becoming an issue of considerable interest amongst researchers, policy-makers and practitioners (Best, 2001; Gereffi, 1995; Porter, 1990; Nadvi & Schmitz, 1994). (The chapter by Swan (this volume) takes up this theme in detail (Swan, 2003)). In part this reflects the perception of advantages of networking over traditional transactional models of organization, in which there is often a trade-off between modes of interaction (Williamson, 1975) and in part it acknowledges the impact of technological and market changes which have blurred the boundaries between enterprises and opened up the arena in which new forms might emerge. 'Virtual enterprises', 'boundary-less organizations' and 'networked companies' are typical examples of the thinking and experimentation which is going on to try and establish different approaches to the problem of inter-organizational relationships (De Meyer, 1992; Dell, 1999; Harland, 1995; Magretta, 1998; Meade, Liles et al., 1997).

There are numerous examples of inter-firm activity where it is clear that some forms of emergent 'good practice' can be seen to operate. For example, in the area of buyer/supplier relationships there is growing recognition that the traditional arms-length bargaining type of relationships may not always be appropriate (Carlisle & Parker, 1989). In particular as firms need to work more closely in strategic areas and where their interdependence increases (as, for example, with just-in-time delivery systems), so there is a need to build more cooperative forms of relationship. The prescriptions for such cooperative relationships are relatively easy to write—the difficulty is in implementing them successfully. Effectively new routines are needed to deal with issues like trust, risk, gain-sharing, etc. (Lamming, 1993).

Similarly, there has been much discussion about the merits of technological collaboration, especially in the context of complex product systems development (Dodgson, 1993; Hobday, 1994; Marceau, 1994). Innovation networks of this kind offer significant advantages in terms of assembling different knowledge sets and reducing the time and costs of development—but are again often difficult to implement (Oliver & Blakeborough, 1998; Tidd, Bessant et al., 2001).

Studies of 'collective efficiency' have explored the phenomenon of clustering in a number of different contexts (Humphrey & Schmitz, 1996; Nadvi & Schmitz, 1994; Piore & Sabel, 1982). From this work, it is clear that the model is widespread—not just confined to parts of Italy, Spain and Germany but diffused around the world—and under certain conditions, extremely effective. For example, one town (Sialkot) in Pakistan plays a dominant role in the world market for specialist surgical instruments made of stainless steel. From a core group of 300 small firms, supported by 1,500 even smaller suppliers, 90% of production (1996) was exported and took a 20% share of the world market, second only to Germany (Schmitz, 1998). In another case the Sinos valley in Brazil contains around 500 small firm manufacturers of specialist high quality leather shoes. Between 1970 and 1990 their share of the world market rose from 0.3% to 12.5% and they now export some 70% of total production (Nadvi, 1997). Both of these examples show gains resulting from close interdependence in a cooperative network.

In all of these cases it appears that networks form for particular purposes but then offer the possibility of additional activity based on the core cooperative framework. So, for example, the clusters of middle Italy may have originally formed as an economic response, providing a way of resolving the basic difficulties of resource access for small firms (Nadvi & Schmitz, 1994; Piccaluga, 1996). But having established a core framework, which allowed for resource sharing and collective efficiency, these networks began to grow, adding a dimension of technological learning

Table 7. Core processes in inter-organizational networking.

Process	Underlying questions
Network creation	How the membership of the network is defined and maintained
Decision-making	How (where, when, who, etc.) decisions get taken
Conflict resolution	How (and if) conflicts are resolved
Information processing	How information flows and is managed
Knowledge capture	How knowledge is articulated and captured to be available for the whole network
Motivation/commitment	How members are motivated to join/ remain in the network—e.g. through active facilitation, shared concerns for development, etc.
Risk/benefit sharing	How the risks and benefits are shared
Integration	How relationships are built and maintained between individual representatives in the network

to them. The case of CITER—the Centro Informazione Tessile di Emilia Romagna in Italy is a good example here (Cooke & Morgan, 1991; Murray, 1993). The predominantly small-scale of firm operations in the fashion textile sector meant that none could afford the design technology or undertake research into process development in areas like dyeing. A cooperative research centre—CITER—was established and funded by members of the *consorzia* and chartered with work on technological problems related to the direct needs of the members. Over time, this has evolved into a world-class research institute but its roots are still in the local network of textile firms. It has become a powerful mechanism for innovation and technology transfer and has helped to upgrade the overall knowledge base of this sector (Rush, Hobday et al., 1996).

In essence, what we are seeing is a process of learning and experimentation around routines for inter-firm working. Whilst these are not yet clearly established it is possible to see a pattern in which the whole—the network—can perform at a level which is greater than the sum of its parts. It appears that firms need to explore and develop routines in at least eight critical areas which are shown in Table 7.

Arguably, inter-organizational networks will be more or less effective in the ways in which they handle these processes. For example, a network with no clear routines for resolving conflicts is likely to be less effective than one, which has a clear and accepted set of norms—a 'network culture'—which can handle the inevitable conflicts, which emerge.

Building and operating networks can be facilitated by a variety of enabling inputs—for example, the use of advanced information and communications technologies may have a marked impact on the effectiveness with which information processing takes place. In a number of cases independent facilitation appears to have a strong influence on many of the behavioral dimensions. For example, the U.K. Society of Motor Manufacturers and Traders established a shared learning network called the 'Industry Forum' in order to

help develop performance improvement amongst small and medium-sized component makers. This program makes use of a number of trained engineers who work with firms individually and collectively, facilitating learning and development (DTI, 2000). Similar programs have now been rolled out with government support to eight other sectors including chemicals, ceramics and aerospace (DTI, 2000; DTI/CBI, 2000). Similar work is going on in the food industry in Australia and in the automobile and furniture sectors in South Africa (AFFA, 2000; Barnes & Bessant et al., 2001). Developing learning networks of this kind can be carried out around a number of groupings including regional, sectoral, supply chains and those organized around a key innovation theme or topic (Bessant & Francis, 1999; Bessant & Kaplinsky et al., 2003 forthcoming; Bessant & Tsekouras, 2001; Dent, 2001).

Conclusions

We have explored some—but by no means all—of the challenges confronting firms in their quest to manage innovation successfully. In dealing with these challenges the key management task lies in creating and reinforcing patterns of behavior—building 'the way we do things around here'—through identifying and establishing innovation *routines*. Some routines are a basic prerequisite—if the organization has no mechanisms for picking up signals about the need for change in its environment it may not survive for long. Equally, if its mechanisms for managing innovation as a process are based on models of invention being the key activity, then it will be a powerhouse for new ideas but probably an unsuccessful business.

But beyond the basic routines, which firms need to learn and maintain, there are others which become increasingly important to graft on. Once the firm has mastered the basic 'skills' of innovation, it needs to look at how well it can involve the full range of its staff, how well it can manage to operate in networks rather than on its own, how well it can handle continuous and discontinuous changes, etc.

This makes the management task not simply one of building and sustaining routines for innovation but also—and most importantly—one of creating the underlying learning routines, which enable the organization to do so.

References

Adler, P. (1992). The learning bureaucracy: NUMMI. In: B. Staw & L. Cummings (Eds), *Research in Organizational Behaviour*. Greenwich CT.: JAI Press.

AFFA (2000). *Supply chain learning: Chain reversal and shared learning for global competitiveness*. Canberra: Department of Agriculture, Fisheries and Forestry—Australia (AFFA).

Altschuler, D., Roos, D., et al. (1984). *The future of the automobile*. Cambridge, MA: MIT Press.

Barnes, J., Bessant, J., et al. (2001). Developing manufacturing competitiveness in South Africa. *Technovation*, **21** (5).

Bessant, J. (2003 forthcoming). *High involvement innovation*. Chichester, U.K.: John Wiley.

Bessant, J., Caffyn, S., et al. (2001). An evolutionary model of continuous improvement behaviour. *Technovation*, **21** (3), 67–77.

Bessant, J. & Francis, D. (1997). Implementing the new product development process. *Technovation*, **17** (4), 189–197.

Bessant, J. & Francis, D. (1999). Using learning networks to help improve manufacturing competitiveness. *Technovation*, **19** (6/7), 373–381.

Bessant, J., Kaplinsky, R., et al. (2003 forthcoming). Putting supply chain learning into practice. *International Journal of Operations and Production Management*.

Bessant, J. & Tsekouras, G. (2001). Developing learning networks. *A.I. and Society*, **15** (2), 82–98.

Best, M. (1990). *The new competition*. Oxford: Polity Press.

Best, M. (2001). *The new competitive advantage*. Oxford: Oxford University Press.

Boer, H., Berger, A., et al. (1999). *CI changes: From suggestion box to the learning organisation*. Aldershot, U.K.: Ashgate.

Bowen, H., Clark, K., et al. (1994). Regaining the lead in manufacturing. *Harvard Business Review*, September/October, 108–144.

Braun, E. & Macdonald, S. (1980). *Revolution in miniature*. Cambridge: Cambridge University Press.

Brown, S., Bessant, J., et al. (2000). *Strategic operations management*. Oxford, U.K.: Butterworth Heinemann.

Bruce, M. & Bessant, J. (Eds) (2001). *Design in business*. London: Pearson Education.

Bruce, M. & Cooper, R. (1997). *Marketing and design management*. London: International Thomson Business Press.

Bryson, B. (1994). *Made in America*. London: Minerva.

Caffyn, S. (1998). *Continuous improvement in the new product development process*. Brighton: Centre for Research in Innovation Management, University of Brighton.

Carlisle, J. & Parker, A. (1989). *Beyond negotiation*. Chichester, U.K.: John Wiley.

Caulkin, S. (2001). *Performance through people*. London: Chartered Institute of Personnel and Development.

Christenson, C. (1997). *The innovator's dilemma*. Cambridge, MA: Harvard Business School Press.

Cohen, M., Burkhart, R., et al. (1996). Routines and other recurring patterns of organization. *Industrial and Corporate Change*, **5** (3).

Cooke, P. & Morgan, K. (1991). *The intelligent region: Industrial and institutional innovation in Emilia-Romagna*. Cardiff: University of Cardiff.

Coombs, R., Saviotti, P., et al. (1985). *Economics and technological change*. London: Macmillan.

Cooper, R. (1993). *Winning at new products*. London: Kogan Page.

Cooper, R. (1994). Third-generation new product processes. *Journal of Product Innovation Management*, **11** (1), 3–14.

Cooper, R. (2000). *Product leadership*. New York: Perseus Press.

Cooper, R. (2003). Profitable product innovation. In: L. V. Shavinina (Ed.), *International Handbook on Innovation*. Oxford: Elsevier Science.

de Geus, A. (1996). *The living company*. Boston, Mass: Harvard Business School Press.

De Meyer, A. (1992). *Creating the virtual factory*. INSEAD.

Dell, M. (1999). *Direct from Dell*. New York: Harper Collins.

Deming, W. E. (1986). *Out of the crisis*. Cambridge, MA: MIT Press.

Dent, R. (2001). *Collective knowledge development, organisational learning and learning networks: an integrated framework*. Swindon: Economic and Social Research Council.

Dodgson, M. (1993). *Technological collaboration in industry*. London: Routledge.

Dodgson, M. & Bessant, J. (1996). *Effective innovation policy*. London: International Thomson Business Press.

DTI (1997). *Competitiveness through partnerships with people*. London: Department of Trade and Industry.

DTI (2000). *Learning across business networks*. London: Department of Trade and Industry.

DTI/CBI (2000). *Industry in partnership*. London: Department of Trade and Industry/Confederation of British Industry.

Ettlie, J. (1999). *Managing innovation*. New York: John Wiley.

Figueredo, P. (2001). *Technological learning and competitive performance*. Cheltenham: Edward Elgar.

Francis, D. (2001). *Developing innovative capability*. Brighton: University of Brighton.

Francis, D. & Young, D. (1988). *Top team building*. Aldershot, U.K.: Gower.

Freeman, C. (1982). *The economics of industrial innovation*. London: Frances Pinter.

Garvin, D. (1988). *Managing quality*. New York: Free Press.

Garvin, D. (1993). Building a learning organisation. *Harvard Business Review*, July/August, 78–91.

Gereffi, G. (1995). International trade and industrial upgrading in the apparel commodity chain. *Journal of International Economics*, **48**, 37–70.

Giddens, A. (1984). *The constitution of society*. Berkeley, CA: University of California Press.

Guest, D., Michie, J., et al. (2000). Employment relations, HRM and business performance: an anlysis of the 1998 Workplace Employee Relations Survey. London, CIPD.

Guttman, R. (2003). Innovation in integrated electronics and related technologies: Experiences with industrial-sponsored large-scale multidisciplinary programs and single investigator programs in a research university. In: L. V. Shavinina (Ed.), *International Handbook on Innovation*. Oxford: Elsevier Science.

Hadjimanolis, A. (2003). The barriers approach to innovation. In: L. V. Shavinina (Ed.), *International Handbook on Innovation*. Oxford: Elsevier Science.

Hamel, G. (2000). *Leading the revolution*. Boston. MA: Harvard Business School Press.

Harland, C. (1995). Networks and globalisation. Engineering and Physical Sciences Research Council.

Hauschildt, J. (2003). Promotors and champions in innovations – development of a research paradigm. In: L. V. Shavinina (Ed.), *International Handbook on Innovation*. Oxford: Elsevier Science.

Henderson, R. & Clark, K. (1990). Architectural innovation: The reconfiguration of existing product technologies and the failure of established firms. *Administrative Science Quarterly*, **35**, 9–30.

Henry, J. & Walker, D. (1990). *Managing innovation*. London: Sage.

Hobday, M. (1994). Complex product systems, Science Policy Research Unit, University of Sussex.

Hollander, S. (1965). *The sources of increased efficiency: A study of Dupont rayon plants*. Cambridge, MA: MIT Press.

Humphrey, J. & Schmitz, H. (1996). The Triple C approach to local industrial policy. *World Development*, **24** (12), 1859–1877.

Husclid, M. (1995). The impact of human resource management practices on turnover, productivity and corporate financial performance. *Academy of Management Journal*, **38**, 647–656.

Juran, J. (1951). *Quality control handbook*. New York: McGraw-Hill.

Kaplinsky, R. (1994). *Easternization: The spread of Japanese management techniques to developing countries*. London: Frank Cass.

Katz, R. (2003). Managing technological innovation in business organizations. In: L. V. Shavinina (Ed.), *International Handbook on Innovation*. Oxford: Elsevier Science.

Katz, R. & Allen, T. (1982). Investigating the not invented here (NIH) syndrome. *R&D Management*, **12** (1), 7–19.

Lamming, R. (1993). *Beyond partnership*. London: Prentice-Hall.

Laurila, J. (1998). *Managing technological discontinuities*. London: Routledge.

Leifer, R., McDermott, C., et al. (2000). *Radical innovation*. Boston, MA: Harvard Business School Press.

Leonard-Barton, D. (1995). *Wellsprings of knowledge: Building and sustaining the sources of innovation*. Boston, MA: Harvard Business School Press.

Magretta, J. (1998). The power of virtual integration. *Harvard Business Review*, March-April.

Marceau, J. (1994). Clusters, chains and complexes: Three approaches to innovation with a public policy perspective. In: R. Rothwell & M. Dodgson (Eds), *The Handbook of Industrial Innovation*. Aldershot, U.K.: Edward Elgar.

Meade, L., Liles, D., et al. (1997). Justifying strategic alliances and partnering: a prerequisite for virtual enterprising. *Omega*, **25** (1).

Monden, Y. (1983). *The Toyota production system*. Cambridge, MA: Productivity Press.

Murray, R. (1993). CITER. In: H. Rush et al. (Eds), *Background/Benchmark Study for Venezuelan Institute of Engineering*. Brighton: CENTRIM, University of Brighton.

Nadvi, K. (1997). *The cutting edge: Collective efficiency and international competitiveness in Pakistan*. Institute of Development Studies.

Nadvi, K. & Schmitz, H. (1994). *Industrial clusters in less developed countries: Review of experiences and research agenda*. Brighton: Institute of Development Studies.

Nayak, P. & Ketteringham, J. (1986). *Breakthroughs: How leadership and drive create commercial innovations that sweep the world*. London: Mercury.

Nelson, R. & Winter, S. (1982). *An evolutionary theory of economic change*. Cambridge, MA: Harvard University Press.

Nonaka, I. (1991). The knowledge creating company. *Harvard Business Review*, November-December, 96–104.

Oliver, N. & Blakeborough, M. (1998). Innovation networks: The view from the inside. In: J. Grieve Smith & J. Michie (Eds), *Innovation, co-operation and growth*. Oxford: Oxford University Press.

Pavitt, K. (2000). *Technology, management and systems of innovation*. London: Edward Elgar.

Pavitt, K. (2002). Innovating routines in the business firm: what corporate tasks should they be accomplishing? *Industrial and Corporate Change*, **11** (1), 117–133.

Pedler, M., Boydell, T., et al. (1991). *The learning company: A strategy for sustainable development*. Maidenhead: McGraw-Hill.

Pfeffer, J. (1998). *The human equation: Building profits by putting people first*. Boston, MA: Harvard Business School Press.

Pfeffer, J. & Veiga, J. (1999). Putting people first for organizational success. *Academy of Management Executive*, **13** (2), 37–48.

Piccaluga, A. (1996). *Impresa e sistema dell'innovazione tecnologica*. Milan: Guerini Scientifica.

Piore, M. & Sabel, C. (1982). *The second industrial divide*. New York: Basic Books.

Porter, M. (1990). *The competitive advantage of nations*. New York: Free Press.

Rothwell, R. (1992). Successful industrial innovation: Critical success factors for the 1990s. *R&D Management*, **22** (3), 221–239.

Rush, H., Brady, T., et al. (1997). *Learning between projects in complex systems*. Centre for the study of Complex Systems.

Rush, H., Hobday, M., et al. (1996). *Technology institutes: Strategies for best practice*. London: International Thomson Business Press.

Schmitz, H. (1998). Collective efficiency and increasing returns. *Cambridge Journal of Economics*.

Schroeder, D. & Robinson, A. (1991). America's most successful export to Japan – continuous improvement programmes. *Sloan Management Review*, **32** (3), 67–81.

Senge, P. (1990). *The fifth discipline*. New York: Doubleday.

Shillito, M. (1994). *Advanced QFD: Linking technology to market and company needs*. New York: John Wiley.

Smith, P. & Reinertsen, D. (1991). *Developing products in half the time*. New York: Van Nostrand Reinhold.

Sternberg, R., Pretz, J., et al. (2003). Types of innovation. In: L. V. Shavinina (Ed.), *International Handbook on Innovation*. Oxford: Elsevier Science.

Swan, J., et al. (2003). Linking knowledge, networking and innovation processes: A conceptual model. In: L. V. Shavinina (Ed.), *International Handbook on Innovation*. Oxford: Elsevier Science.

Teece, D. & Pisano, G. (1994). The dynamic capabilities of firms: an introduction. *Industrial and Corporate Change*, **3** (3), 537–555.

Tidd, J., Bessant, J., et al. (2001). *Managing innovation* (2nd ed.). Chichester, U.K.: John Wiley.

Tremblay, P. (1994). *Comparative analysis of technological capability and productivity growth in the pulp and paper industry in industrialised and industrialising countries*. University of Sussex.

Tushman, M. & Anderson, P. (1987). Technological discontinuities and organizational environments. *Administrative Science Quarterly*, **31** (3), 439–465.

Tushman, M. & O'Reilly, C. (1996). *Winning through innovation*. Boston, MA: Harvard Business School Press.

Utterback, J. (1994). *Mastering the dynamics of innovation*. Boston, MA.: Harvard Business School Press.

Van de Ven, A., Angle, H., et al. (1989). *Research on the management of innovation*. New York: Harper and Row.

Weightman, G. (2002). *The frozen water trade*. London: Harper Collins.

Weisberg, R. (2003). Case studies of innovation: Ordinary thinking, extraordinary outcomes. In: L. V. Shavinina (Ed.), *International Handbook on Innovation*. Oxford: Elsevier Science.

Wheelwright, S. & Clark, K. (1992). *Revolutionising product development*. New York: Free Press.

Wickens, P. (1987). *The road to Nissan: Flexibility, quality, teamwork*. London: Macmillan.

Williamson, O. (1975). *Markets and hierachies*. New York: Free Press.

Womack, J., Jones, D., et al. (1991). *The machine that changed the world*. New York: Rawson Associates.

The International Handbook on Innovation
Edited by Larisa V. Shavinina

Managing Technological Innovation in Business Organizations

Ralph Katz

College of Business, Northeastern University and Sloan School of Management, MIT, USA

Abstract: Organizations in today's hypercompetitive world face the paradoxical challenges of 'dualism', that is, functioning efficiently today while innovating effectively for tomorrow. Corporations, no matter how they are structured, must manage both sets of concerns simultaneously. They must build those seemingly contradictory structures, competencies, and cultures that foster not only more efficient and reliable processes but also the more risky explorations needed to recreate the future. This chapter describes the patterns of innovation that typically take place within an industry and how such patterns affect an organization's ability to manage its streams of innovative projects along technological cycles.

Keywords: Innovation; Innovation management; Technological innovation; Technology cycles; S-curve; Dominant design.

Introduction

Innovation is often broadly defined as the introduction of something new while Luecke (2002) defines technical innovation as the embodiment, combination, or synthesis of knowledge into new products, processes, or services. More recently, Drucker has argued that innovation is change that creates a new dimension of performance (Hesselbein & Johnston, 2002). Regardless of the particular definitions used, a plethora of research studies has convincingly demonstrated over the past decades a very consistent, albeit somewhat disturbing, pattern of results with respect to the management of innovation (see Adizes, 1999; Christensen & Bower, 1996; Cooper & Smith, 1992; Grove, 1996) for some recent examples). In almost every industry studied, a set of leading firms when faced with a period of discontinuous change, fails to maintain its industry's market leadership in the new technological era. Tushman & O'Reilly (1997) nicely emphasize this point in their research when they describe how Demming, probably the individual most responsible for jump-starting the quality revolution in today's products, would highlight this recurring theme in his lectures by showing a very long list of diverse industries in which the most admired firms rapidly lost their coveted market positions.[1] It is indeed ironic that so many of the most dramatically successful organizations become so prone to failure (Miller, 1994).

This pathological trend, described by many as the *tyranny of success* (Christensen, 1997), in which winners often become losers—in which firms lose their innovative edge—is not only an American phenomenon, but a worldwide dilemma, exemplified by the recent struggles of firms such as Xerox in the U.S., Michelin in France, Philips in Holland, Siemens in Germany, EMI in England, and Nissan in Japan (see Charan & Tichy, 2000; Collins, 2001; Tushman & O'Reilly, 1997). The Xerox brand, for example, has penetrated the American vernacular so strongly that it has the distinction of being used both as a noun and as a verb though the company is probably now hoping that to be 'Xeroxed' won't eventually take on a different meaning. The histories of these and so many other outstanding companies demonstrate this time and time again. It seems that the very factors that lead to a firm's success can also play a significant role in its demise (Henderson & Clark, 1990). The leadership, vision,

---

[1] Industries such as watches, automobiles, cameras, stereo equipment, radial tires, hand tools, machine tools, optical equipment, airlines, color televisions, etc.

strategic focus, valued competencies, structures, policies, rewards, and corporate culture that were all so critical in building the company's growth and competitive advantage during one period can become its Achilles heel as technological and market conditions change over time. In the mid-1990s, an outside team of experts worked for more than three years with senior management on a worldwide cycle time reduction initiative at Motorola Corporation, arguably one of the world's most admired organizations at that time (Willyard & McClees, 1987). In just about every workshop program they conducted across the different businesses, the consultants strongly reminded all of those managing and developing new products and services of this alarming demise of previously successful companies by referring to a notable 1963 public presentation made by Thomas J. Watson, Jr., IBM's Chairman and CEO. According to Watson:

> Successful organizations face considerable difficulty in maintaining their strength and might. Of the 25 largest companies in 1900, only two have remained in that select company. The rest have failed, been merged out of existence, or simply fallen in size. Figures like these help to remind us that corporations are expendable and that success—at best—is an impermanent achievement which can always slip out of hand (Loomis, 1993).

Dualism and Conflicting Organizational Pressures

It is important to recognize, however, that this pattern of success followed by failure—of innovation followed by inertia and complacency—is not deterministic. It does not have to happen! Success need not be paralyzing. To overcome this tendency, especially in today's rapidly changing world, organizations more than ever before are faced with the apparent conflicting challenges of *dualism*, that is, functioning efficiently today while innovating effectively for the future. Not only must business organizations be concerned with the financial success and market penetration of their current mix of products and services, but they must also focus on their long-term capabilities to develop and incorporate what will emerge as the most customer-valued technical advancements into future offerings in a very quick, timely, and responsive manner. Corporations today, no matter how they are structured and organized, must find ways to internalize and manage both sets of concerns simultaneously. In essence, they must build internally those contradictory and inconsistent structures, competencies, and cultures that not only foster more efficient and reliable processes but that will also encourage the kinds of experiments and explorations needed to re-create the future even though such innovative activities are all too often seen by those running the organization as a threat to its current priorities, practices, and basis of success (Collins, 2001; Foster & Kaplan, 2001). Fortunately,

there is more recent evidence, as reported in the research studies conducted by Tushman & O'Reilly (1997), of a few successful firms, such as GE Medical System, Alcoa, and Ciba Vision, that have managed to do this very thing—to continue to capture the benefits of their existing business advantages even as they build their organizational capabilities for long term strategic renewal. They are somehow able to transform themselves through proactive innovation and strategic change, moving from today's strength to tomorrow's strength by setting the pace of innovation in their businesses.

While it is easy to say that organizations should internalize both sets of concerns in order to transform themselves into the future, it is a very difficult thing to do. The reality is that there is usually much disagreement within a company operating in a very pressured and competitive marketplace as to how to carry out this dualism (Roberts, 1990). Amidst the demands of everyday requirements, decision-makers representing different parts of the organization rarely agree on the relative merits of allocating resources and management attention among the range of competing projects and technical activities; that is, those that directly benefit the organization's more salient and immediate needs versus those that might possibly be of import sometime in the future.

Witness for example the experiences of Procter and Gamble (P&G) over the past five or more years. In the beginning, the analysts claimed that P&G was doing a very good job at managing its existing businesses but unfortunately was not growing the company fast enough through the commercialization of new product categories (Berner, 2001). Over the last couple of years, P&G impressively introduced a number of very successful new products (swiffer, whitestrips, thermacare, and febreze—just to name a few) that are collectively bringing in considerably more than a billion dollars in added revenue per year. The analysts, however, now claim that while P&G has managed to introduce some very exciting new products; in doing so, it took its eye off the existing brands and lost important market share to very aggressive competitors (Brooker, 2002). It is not particularly surprising that these same analysts now want P&G to de-emphasize its new venture strategies and investments in order to concentrate on protecting and strengthening its bedrock major brands. The pendulum just seems to keep on swinging.

While much has been written about how important it is to invest in future innovative activities—'to innovate or die'—as management guru Tom Peters so picturesquely portrays it (Peters, 1997), there is no single coherent set of well-defined management principles on how to structure, staff, and lead organizations to accommodate effectively these two sets of conflicting challenges (See John Bessant's *Challenges in Innovation Management* chapter in this volume). Classical management theories are primarily concerned with the efficient utilization, production, and distribution of today's goods and services (Katz, 1997). Principles such as high task specialization and division of labor; the equality of authority and responsibility; and the unities of command and direction (implying that employees should have one and *only* one boss and that information should *only* flow through the formal chain of command) are all concentrating on problems of structuring work and information flows in routine, efficient ways to reduce uncertainty and facilitate control and predictability through formal lines of authority and job standardization (Mintzberg, 1992). What is needed, therefore, are some comparable models or theories to help explain how to organize both incremental and disruptive innovative activities[2] within a functioning organization such that creative, developmental efforts not only take place but are also used to keep the organization competitive and up-to-date within its industry over generations of technological change.

While empirically tested frameworks and theories for achieving dualism may be sparse, there is no shortage of advice (Micklethwait & Wooldridge, 1998). In one of the more recent prominent books on the subject, for example, Hamel (2000) proposes the following ten rules for reinventing one's corporation:

(1) *Set Unreasonable Expectations*: Organizations must establish bold, nonconformist strategies to avoid the typical bland, unexciting goals that most organizations use especially if they hope to avoid behaving and thinking like managers of mature businesses or like mature mangers trying to lead mature businesses. Only nonlinear innovation, according to Hamel, will drive long-term success, or as Collins & Porras (1997) describe it—organizations that are 'built to last' have *BHAG*s (i.e. Big Hairy Audacious Goals).

(2) *Stretch Your Business Definition*: Companies need to define themselves by what they know (their core competencies) and what they own (their strategic assets) rather than by what they do. Senior managers must look continuously for opportunities outside of the business they manage, redefining their markets and challenging their assumptions in ways that allow them to challenge conventional wisdom and outgrow their competitors. One company with which I worked, for example, required all of its business units to periodically redefine their markets in larger contexts. As a consequence, the associated market

shares would be smaller but the opportunities for both extension and growth would now be more salient and enticing. General Electric's CEO, Jack Welch, adopted this approach whenever his division managers '*gamed*' his demands that their businesses be 'number 1 or 2' by narrowing the definition of their markets. Welch made them redefine their markets globally so that they had no more than 10% and had to grow (Welch & Byrne, 2001).

(3) *Create a Cause, Not a Business*: Revolutionary activity must have a transcendent purpose or there will be a lack of courage to commit and persist. Employees within the organization have to feel that they are contributing to something that will make a genuine difference in the lives of their customers or in society.

(4) *Listen to New Voices*: If a company truly wants to remain at the forefront of its business, it must refrain from listening only to the old guard that is more likely to preserve its old routines and comfort levels. Instead, management must give disproportionate attention to three often underrepresented constituencies; namely, those with a youthful perspective, those working at the organization's periphery, and newcomers who bring with them fresh ideas and less preconceived notions.

(5) *Design an Open Market for Ideas*: What is required to keep an organization innovative is not the vision per se but the way the vision is implemented through at least three interwoven markets: a market for ideas, a market for capital, and a market for talent. Employees within the organization must believe that their pursuit of new frame-breaking ideas are the welcomed means by which the organization hopes to sustain its success and they must be able to give their new possibilities 'air time' without rigid bureaucratic constraints and interference.

(6) *Offer an Open Market for Capital*: Rather than designing control and budgeting processes that weed out all but the most comfortable and risk averse ideas, the organization needs to think more like a venture capitalist and permit investments in experiments and unproven markets. Individuals or teams experimenting with small investments and unconventional ideas should not have to pass the same screens and hurdles that exist within the large established business. The goal is to make sure there are enough discretionary resources for winners to emerge—not to make sure there are no losers.

(7) *Open Up the Market for Talent*: Organizations should create an internal auction for talent across the different businesses and opportunities. The organization's professional and leadership talent cannot feel that they are locked inside moribund,

---

[2] Many other terms have been used to denote this continuum of activities, including continuous versus discontinuous; pacing versus radical; competence-enhancing versus competence-destroying; and sustaining versus disruptive innovation (see Foster & Kaplan, 2001, for the most recent publication in this area).

mature businesses. Instead, they must have the chance to try out something new, to experiment with an idea, or to proceed with their imaginations. With more fluid boundaries inside the organization, people can search out and follow through on the most promising new ventures and ideas by voting with their feet.

(8) *Lower the Risks of Experimentation*: Neither caution nor brash risk taking is likely to help an organization maintain its innovative vitality. Successful revolutionaries, according to Hamel, are both prudent and bold, careful and quick. They prefer fast, low-cost experimentation, and learning from customers to gambling with vast sums of money in uncertain environments. And just like experienced venture capitalists, the organization needs to invest in a portfolio rather than in any single project.

(9) *Make Like a Cell—Divide and Divide*: Using cell division as a metaphor, Hamel claims that innovation dies and growth slows when companies stop dividing and differentiating. Division and differentiation free the leadership, the professional talent, and the capital from the constraints and myopic evaluations of any single large business model. And just as important, it keeps the business units small, focused, and more responsive to their customers.

(10) *Pay Your Innovators Well—Really Well*: To maintain entrepreneurial enthusiasm, companies need to reward those individuals taking the risks in ways that really demonstrate that they can have a piece of the action. The upside has to look sufficiently attractive for them to stick around—to create something out of nothing. Innovators should have more than a stake in the company, they should have a stake in their ideas. If you treat people like owners, they will behave like owners.

Implicit in all of this well-intentioned exhortation, as well as in so many other similar examples of advice (Handy, 2002; Sutton, 2001) is the need to learn how to build parallel structures and activities that would not only permit the two opposing forces of today and tomorrow to coexist but would also balance them in some integrative and meaningful way. Within a technological environment, the operating organization can best be described as an '*output-oriented*' or '*downstream*' set of forces directed towards the technical support of problems within the business's current products and services in addition to getting new products out of development and into the marketplace through manufacturing and/or distribution (Katz & Allen, 1997). For the most part, these kinds of pressures are controlled through formal job assignments to business and project managers who are then held accountable for the successful completion of product outputs within established schedules and budget constraints, that is, for making their quarterly forecasts.

At the same time, there must be an '*upstream*' set of forces that are less concerned with the specific architectures, functionalities, and characteristics of today's products and services but are more concerned with all of the possible core technologies that could underlie the industry or business environment sometime in the future (Allen, 1986). They are essentially responsible for the technical health and vitality of the corporation, keeping the company up-to-speed in what could become the dominant and most valued technical solutions within the industry. And as previously discussed, these two sets of forces, downstream and upstream, are constantly competing with one another for recognition and resources which can either be harmful or beneficial depending on how the organization's leadership resolves the conflicts and mediates the priority differences.

If the product-output or downstream set of forces becomes dominant, then there is the likelihood that sacrifices in using the latest technical advancements may be made in order to meet more immediate schedules, market demands, and financial projections. Strong arguments are successfully put forth that strip the organization of its exploratory and learning research activities. Longer-term, forward-looking technological investigations and projects are de-emphasized in order to meet shorter-term goals, thereby mortgaging future technical expertise. Under these conditions, important technological changes in the marketplace are either dismissed as faddish or niche applications or go undetected for too long a period. To illustrate, Walter Robb, the general manager of GE's medical business at the time EMI was just starting to commercialize CT scanners, readily admits that GE medical was so focused on running, improving, and protecting its existing X-ray business that it could only see CT scanners as nothing more than a minor niche market application that would never seriously affect GE's medical business. The dilemma in these instances of course is that the next generation of new product developments either disappears or begins to exceed the organization's in-house technological capability. Only two of the leading producers of vacuum tubes in 1955, for example, were able to make the leap to transistors by 1975; similarly, almost none of the major 35 mm camera producers were the pioneer marketers of digital photography (see Anderson & Tushman, 1991; Foster, 1986; Iansiti, 2000; Tripsas, 2001 for additional examples and discussion).

At the other extreme, if the research or upstream technology component of the business is allowed to dominate development work within R&D, then the danger is that products may include not only more sophisticated and more costly technologies but also perhaps less proven, more risky, and less marketable technologies (Pate-Cornell & Dillon, 2001; Scigliano,

2002). This desire to be technologically aggressive—to develop and use the most attractive, most advanced, most clever, state-of-the-art technology—must be countered by forces that are more sensitive to the operational environment and the infrastructure and patterns of use. Customers hate to be forced to adapt to new technologies unless they feel some pressing need or see some real added value. Technology is not an autonomous system that determines its own priorities and sets its own standards of performance and benchmarks. To the contrary, market, social, and economic considerations, as so poignantly pointed out by Steele (1989), eventually determine priorities as well as the dimensions and levels of performance and price necessary for successful applications and customer purchases. Unwanted technology has little value and technical performance that is better than what customers want nearly always incurs a cost penalty and is rarely viable. Such effects are vividly portrayed by the painful experiences of all three satellite consortia—Motorola's, Qualcom's, and Microsoft's—as each consortia group still tries to penetrate the cellular and wireless tele-markets with their respective Iridium, Globalstar, and Telstar systems (Crockett, 2001; Morgan et al., 2001). To put it simply, few customers will pay for a racing car when what they really want is a family sedan.

A Model of Innovation Dynamics in Industry

In a previous but still seminal piece of research, Abernathy & Utterback (1978) put forth a model to capture some interesting dynamics of innovation within an industry (see Utterback, 1994, for the most thorough discussion of these issues and their implications). The model, as shown in Fig. 1, suggests that the rates of major innovations for both products and processes within a given industry often follow fairly predictable patterns over time; and perhaps more importantly, that product and process innovation interact with each other. In essence, they share an important tradeoff relationship.

The rate of product innovation across competing organizations in an industry or product class, as shown in Fig. 1, is very high during the industry's early formative years of growth. It is during this period of time, labeled the *fluid* phase by Abernathy & Utterback (1978) in their model, that a great deal of experimentation with product design and technological functionality takes place among competitors entering the industry. During this embryonic stage, no single firm has a '*lock*' on the market. It seems that once a pioneering product has demonstrated the feasibility of an innovative concept, rival products gradually appear. As long as barriers to entry are not too high, these new competitors are inspired to enter this new emerging market with their own product variations of design choices and features. No one's product is completely perfected and no organization has as yet mastered the

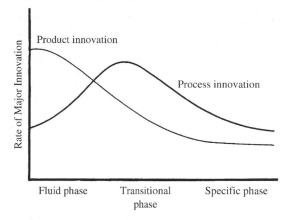

Figure 1. The dynamics of innovation (Reprinted by permission of Harvard Business School Press. From *Mastering the Dynamics of Innovation* by James Utterback, Boston, MA 1994, p. xvii).

manufacturing process or monopolized all of the means of distribution and sales. Furthermore, customers do not as yet have a firm conception of what an ideal product design would look like nor enough experience from which to indicate what they would want or even what they would be willing to pay for in terms of features, functions, and attributes in this new product class. In a sense, the market and the industry are in a *fluid* stage of development with all of the involved participants learning as they move along together.

One of the more illustrative examples offered by Abernathy & Utterback (1978) phase lies in the early years of the automobile industry in which a bewildering variety of *horseless carriages* were designed and sold to the public. Not only were internal combustion machines being commercialized as the power source at this time but a whole host of other electric and steam-driven cars were emerging from the workshops of dozens of other producers as viable competitive alternatives. Each manufacturer hoped to capture the allegiance of the public with its own novel new design and driver amenities. Electric or battery powered cars, in fact, were produced and sold to customers from the earliest years of the industry and were discontinued primarily because Henry Ford eschewed the battery option (influenced in part on advice he received from Thomas Edison) and selected the internal combustion engine as the basis of his Model T design. In addition to relatively poor acceleration and the inconvenience of finding outlets for re-charging, batteries became an unpopular choice simply because they were much too heavy for the mostly unpaved muddy roads that existed at that time.

According to the model, in many product class situations, this period of fluidity usually gives way to a *transitional* phase in which the rate of major innovations in the product's design decreases substantially but the rate of major innovations within the process begins to increase. It is within this period that the wide variety of different designs within the product class begins to disappear mainly because some aspects of a design standard have emerged. Most commonly, certain elements within the overall design have proven themselves in the marketplace as most desirable for satisfying the customers' needs or because certain design standards have been established either by formal agreements among major producers or by legal and regulatory constraints. As the basic form of the product becomes more predictable and consistent, the industry begins to concentrate more on improvements in the way the product is produced and how costs in the overall production and operation can be reduced.

In discussing their automobile example, Abernathy & Utterback (1978) describe how early imaginative designs of the auto age (three, four, and five-wheel auto designs, for example) gave way to a set of fairly standardized designs among the many competitors. As a result of this convergence, the features and basic form of the automobile achieved a reasonable degree of uniformity. In a sense, customers had developed a pretty clear understanding about what cars should look like and how they should be driven. The automobile companies developed a set of technologies and the driving public developed a set of expectations that mutually adjusted to one another over time to essentially define the basic format of the automobile. At the same time, however, substantial progress was being made in the overall ability of the firms to manufacture large quantities of cars at lower costs. By 1909, the rate of product design innovation had diminished so much that an article in the prestigious journal of *Scientific American* proudly proclaimed:

> That the automobile has practically reached the limits of its development is suggested by the fact that during the past year no improvements of a radical nature have been introduced (*Scientific American*, 1909, p. 145).

Undoubtedly, it was this kind of standardization and general stability in the overall design that probably allowed Henry Ford to devote the numerous years it took to establish successfully his assembly line manufacturing process for the Model T.

If the market for the product class continues to grow, the industry gradually passes into what Abernathy & Utterback (1978) call the *specific* phase as shown in Fig. 1. The researchers deliberately chose the term '*specific*' to emphasize the notion that during this period, the business unit strives to manufacture very specific assembled products at the highest possible rates of efficiency. During this progression, the products themselves become increasingly well-defined with product similarities greatly outpacing product differences. Customers, moreover, have grown considerably

more comfortable and experienced with the product mix; and consequently, they have become more demanding and price sensitive. With this increase in familiarity, customers are now more capable of assessing what they need and value and what they are willing to pay. What typically results from all this movement is the gradual commoditization of the marketplace in which the value ratio of perceived functional quality to price (i.e. the business unit's cost) essentially becomes the basis of competition. Furthermore, the interrelationships between product and process are so intertwined that changes in either one of them require corresponding changes in the other. Even small changes can therefore be difficult and expensive.

Dominant Designs

The critical supposition underlying this gradual shift from a *fluid* phase to the *transitional* and *specific* phases is the emergence of a dominant design within the industry. Retrospective studies of innovation within a product class typically reveal that at some point in time there emerges a general acceptance of how the principal components comprising the product's overall architecture interface with one another (e.g. Anderson & Tushman, 1990; Meyer & Lehnerd, 1997; Utterback, 1994). A dominant design does not necessarily incorporate the best technologies nor is it necessarily the most optimal. It is established and defined experientially by the marketplace when a particular design architecture becomes the one most preferred by the overwhelming majority of the product's purchasers. Ford's Model T; the Douglas DC-3 aircraft; Boeing's 700-series; the QWERTY keyboard; IBM's 360 mainframes and PCs; JVC's VHS format; the Sony Walkman; WINTEL-based PC's; Powerpoint, Word, and Excel software programs; and CDMA or GSM-type cell phones, for example, all accounted for upwards of at least 70% to 80% of their respective markets at the height of their popularity (Cusumano & Yoffie, 1998; Teece & Pisano, 1994).

What the emergence of a dominant design does seem to lead to, however, is a shift in concentration of an industry's innovative resources and energies away from coming up with significant design alternatives to focusing on process improvements in what has become the ensconced product architecture. To put it bluntly, it would be a very risky bet for any organization to make a significant and irretrievable capital investment in a manufacturing process dedicated to a specific architectural configuration while there is still major flux among competing designs within the industry. In creating a convergent set of engineering conventions and design choices, a dominant design also provides equipment and component suppliers within the industry a clearer and more well-defined product and technological context within which their engineers can work to improve their pieces of the overall manufacturing system (Fine, 1999; Reinertsen, 1997).

Once a dominant design emerges, the basis of competition within the marketplace changes substantially. The industry changes from one characterized by many firms with many unique designs to one that will eventually contain many fewer firms competing with very similar product designs (Valery, 1999). A dominant design, therefore, is not a particular product or machine but a set of engineering standards and options representing an accepted pattern of design choices within the product's scope. Customers today, for example, rarely have to ask how to start, drive, or steer their cars; use their lights, windshield wipers and brakes; check their tires; or fill their gas tanks. When children's battery-operated toys are purchased, customers expect to use either the AA or AAA-standard size batteries that can be purchased in just about every retail outlet—the initial A and B-size batteries having fallen by the wayside in most of today's product designs.[3]

More recently, scholars have begun studying the survival rate of organizations in industries as a function of their entry with respect to the dominant design. Research by Suarez & Utterback (1995) and by Christensen et al. (1998), for example, revealed that firms attempting to enter and compete in industries after dominant designs had become established faced much lower chances of success and survival. According to these scholars, the existence of a dominant design significantly restricts the engineers' freedom to differentiate their products through innovative design. As a result, there are fewer opportunities for small or entering firms to find beachheads in markets operating in a post-dominant design era. Their data on company survival further suggests that there might be a bounded 'window of opportunity' for entry in fast-moving industries. In their industry samples, not only did the firms that entered after the dominant design had emerged have much lower probabilities of survival but the firms that entered too early also had reduced chances of survival. Perhaps many of these early firms had exerted efforts that built proprietary or specialized technological and design capabilities—which in the high turbulence of the fluid phase may have been appropriate and therefore successful—but which are now no longer sought by markets enveloped by a dominant architecture in which other factors and capabilities are more relevant and highly valued (Agarwal & Bayus, 2002).

The emergence of a dominant design is a watershed event in that it delimits the *fluid* phase[4] in which a rich

mixture of design experimentation and competition had been taking place within the product class. It is important to understand that a dominant design is not predetermined. Nor does it come about through some rational, optimal, or natural selection process. Instead, it emerges over time through the interplay of technological possibilities and market choices that are influenced and pushed by the many individual and allied competitors, regulators, suppliers, and sales channels, all of whom have their own technological, political, social, and economic agendas.

Exogenous Factors Influencing the Dominant Design

Complementary Assets

By understanding the many factors that can influence the establishment and persistence of a dominant design within a given industry, managers can try to enhance the long-term success of their products by making sure they are actively involved in this process rather than dismissing, resisting, or functioning apart from it (Tripsis, 2001; Utterback, 1994). In one of the more notable articles on how companies capture value from technological innovation, Teece (1987) describes just how powerful complementary assets can be in shaping the dominant design that eventually wins in the marketplace. Complementary assets are those assets necessary to translate an innovation into commercial success. Innovative products usually embody the technical knowledge about how to do something better than the existing state of the art. However, to generate attractive financial returns and profits, this technical know-how must be linked with other capabilities or assets in order for them to be sold, used, and serviced in the market. These assets span a wide variety of non-technical properties, including brand name and company image, supply chain logistics, distribution and sales channels, customer service and support, specialized manufacturing capabilities, deep financial pockets, peripheral products, switching costs, political, regulatory, and customer knowledge, critical real estate or institutional associations, and control over raw materials or key components (Lynn & Reilly, 2002; Rosenbloom & Christensen, 1994).

The more a firm possesses control over these kinds of complementary assets, the more advantage it has over its competitors in establishing its product as the dominant design. While numerous PCs were available in the market long before IBM introduced its first personal computer in 1981, the IBM PC quickly became the dominant design, primarily because of its exceptional brand name, service reputation, and the emergence of so much other peripheral and applications software (Scott, 2001). In controlling access to homes through their established infrastructures, the 'baby bell' companies had an enormous advantage over new DSL and other telecom start-up competitors most

---

[3] According to Tony Mazzola, a technical marketing manager at Eveready, the A battery is still used in some power tools and camcorders and the B battery is used for bicycles in Europe. Also, there are still AAAA, F, J, and N-size batteries but the E and G-size have become obsolete (Hartlove, 1998).

[4] In their descriptive model of these innovation patterns, Tushman & O'Reilly (1997) call this fluid period an '*era of ferment*' to denote the intense *agitation* that transpires among alternative designs and features within the product class.

of whom, to their dismay, realized much too late just how precarious a situation they were in even though they may have had attractive cost-performance capabilities and governmental support as a result of industry deregulation (Rosenbush & Elstrom, 2001). Not surprisingly, established telephone and cable companies, accustomed to operating as near monopolies, were not particularly eager to open their systems and infrastructure to rivals bent on offering similar services. The Verizons and Time Warners of the world threw up numerous legal and technological roadblocks to the upstart service providers (Trudel, 2002).

In sharp contrast, Christensen (2001) recently concluded from his historical investigations of disruptive technologies that one of the reasons the new entrepreneurial transistor-type companies were able to successfully displace the leading vacuum tube producers was that they were, in fact, able to bypass the critical distribution channels that had been controlled, or at least heavily influenced, by the premier vacuum tube companies. The new transistor firms did not have to rely solely on the legion of existing appliance store outlets that had grown all too comfortable with the opportunistic profits they were deriving from the servicing and replacement of vacuum tube components and products. They were, instead, able to sell their '*revolutionary*' transistorized products through the many large department store chains that were also rapidly growing at that time throughout the country.

External Regulations and Standards

Government requirements and regulations can also play a significant role in defining a dominant design especially when they impose a particular standard within an industry (Bagley, 1995). The FCC's (Federal Communications Commission) approval of RCA's television broadcast standard, for example, provided RCA a tremendous advantage not only in establishing its receivers as the dominant design within the industry but also in favoring its black and white TV strategy over the color TV strategies that were being developed and lobbied by very worthy competitors at that time.[5] Many governments around the world, either individually or collectively, try to determine standards for emerging technologies within their industries to facili-

tate easier and quicker product developments and to encourage increased compatibility among system components within the infrastructure of use (Caves, 1996).

The standards that a government establishes for package labeling, high definition television, telecommunications, automobile safety, or even for the content definitions of certain foods, such as ice cream, orange juice, or peanut butter, can either favor or undermine the interests and strategies of particular competitors within an industry. Governmental requirements and regulations can also be used to enhance the attractiveness of domestic producers over foreign competitors. By knowing that the U.S. government was going to require synthetic detergent producers to eliminate phosphates in order to reduce environmental pollution, Whirlpool Corporation realized that suppliers would have to augment their synthetic detergents in the future with more potent chemical additives to compensate for the banned phosphates (Knud-Hansen, 1994). The company also understood that over time these more forceful chemical compounds would probably be substantially more corrosive to many of the washing machine's internal parts, including the inside pumps and drums. As a result, Whirlpool gained a distinct advantage in the industry when it quickly developed new washing machine appliance models that could be marketed to withstand the reformulated, albeit more corrosive, detergents. Whirlpool's products were offered with a complete three-year warranty on its newly designed machines. Most of its major competitors, in comparison, found themselves back-pedaling when they had to lower their equivalent warranties to only a couple of month. Obviously, such companies subsequently rushed to recover with stronger model designs that could resist the caustic characteristics of the new detergents' chemical compositions. An alternative scenario could easily unfold in 2007 when the Federal Government, in order to promote greater water conservation, has purportedly mandated that all washing machines sold in the U.S. will be '*front*' rather than '*top loaders*' (Van Mullekom, 2002). Front load machines are exactly the kind of design that has been made and used extensively for many years throughout all of Europe while American manufacturers have concentrated predominantly on top load designs.

Organizational Strategies

The business strategy pursued by a firm relative to its competitors can also significantly influence the design that becomes dominant. The extent to which a company enters into alliances, agreements, and partnerships with other companies within its industry can substantially effect its ability to impose a dominant design. One very notable example of this in recent times is the success of JVC's VHS system over Sony's Betamax system in the VCR (Video Cassette Recorder) industry. By establishing formidable alliances first in

---

[5] RCA's political and VHF technical strategies, masterminded by the renowned David Sarnoff for the direct benefit of its NBC network, were pitted directly against the UHF color strategy of CBS, led by the relatively young unknown Peter Goldmark, who was hired for the expressed purpose of getting CBS into television and 'the urge to beat RCA and its ruler, David Sarnoff'. When the FCC unexpectedly affirmed the monochrome standards of RCA on March 18, 1947, CBS scrambled to buy, at what became a quickly inflated price, the VHF licenses it had previously abandoned. Interestingly enough, the U.S. Government is now seriously reconsidering UHF as the possible standard for its broadband (Fisher & Fisher, 1997).

Japan, then in Europe and finally the United States, JVC was able to overcome Sony's initial market success even though it is generally acknowledged that Sony's Betamax was better technologically (Cusumano et al., 1992). To its regret, Sony deliberately chose to go it alone, relying for the most part on its own strong brand name, reputation, and movie recordings. Sony avoided the kind of technical alliances and market partnerships that would ultimately make JVC's VHS-product so much more attractive and successful than Betamax. Microsoft and Intel pursued similar relationships when they embarked on a strategy to make their '*WINTEL*' machine and its associated application programs more dominant than many alternative products (Gawer & Cusumano, 2002). At the time, many of these competitive products were seen as technically superior and most had already been successfully commercialized, including Lotus's 1-2-3 spreadsheet, Wordperfect's word processor, Harvard's graphics package, the Unix and OS/2 operating systems, or the Macintosh PC (Cusumano & Selby, 1995).

It is also quite common for companies planning to develop products in an emerging industry to send qualified participants to industrial meetings and/or professional conferences that have been convened explicitly to reach agreements on specific technical and interface standards for the common interest of all, including product developers, suppliers, and users. A great many of the technical protocols and specifications underpinning today's Internet architecture, for example, come from such organized meetings and resultant agreements. The World Wide Web, for example, is currently coordinated through a global consortia of hundreds of companies organized and led by MIT's Laboratory of Computer Science, even though Tim Berners-Lee developed the World Wide Web while he was working in Europe at CERN, the world's largest particle physics center.[6]

However, companies often disagree about the comparative strengths and weaknesses of particular technical approaches or solutions. As a result, those with similar views often band together to form coalitions that will both use and promote their technical preferences within their product class (See Gawer & Cusumano, 2002; Roberts & Liu, 2002 for some recent frameworks and examples). Alternative technical camps can subsequently evolve into rival company consortia. Direct competition within the industry

becomes centered around these different but entrenched innovative design choices. In replacing cables with wireless technology, for example, many companies have decided to design their new innovative products around the blue tooth wireless set of specifications originally developed by Ericsson (Ferguson, 2000). Contrastingly, many other companies, including Hewlett Packard, Philips, Samsung, and Sharp, have chosen to base their wireless products on the newest IEEE 802.15.3 set of professional standards (Reinhardt, 2002). The marketplace has yet to decide which if any of these will emerge as the dominant design.[7]

For a business unit to be successful over time with its innovative products, it needs to know much more than whether the innovation creates value for the targeted customers. The firm also needs to consider at least two other important factors. First, it needs to know whether it can '*appropriate*' the technology, that is, to what extent can it control the know-how within the innovation and prevent others from copying, using, or developing their own versions of it (von Hippel, 1998). And secondly, the firm needs to know whether it has or can secure the necessary complementary assets to commercialize the innovation in a timely and effective manner within the marketplace (Leifer et al., 2001).

It is critical for senior management teams to realize that their businesses should be building their strategies based on the combined answers to these two important questions. As shown in Fig. 2, economists discovered long ago that profitability within an industry can be significantly affected by the interplay of answers to these two questions (Porter, 1990; Teece, 1987). In general, the more a company can protect its intellectual knowledge and capability from competitors, whether it be through patents, license agreements, technical secrecy, or some other means, the more likely the company can derive profits from its product innovations as long as the complementary assets are freely available. However, if the complementary assets (e.g. distribution channels, raw materials, specialized machinery or key components) are tightly controlled, then there is likely to be '*tugs of war*' or alliances between the '*keepers*' of the technology and the '*keepers*' of the complementary assets. For example, although Procter and Gamble and Kimberly Clark might dominate the U.S. development and production

---

[6] Neil Calder, who was the Director of Public Relations at CERN, sadly tells of the day that Tim Berners-Lee walked into his office wearing jeans and a T-shirt to tell him of his progress and latest developments. After listening in complete bafflement, Calder thanked Tim for coming and to 'keep in touch'. It was only much later that Calder realized that Berners-Lee had just described to him the creation of the World Wide Web. 'It was the biggest opportunity I let slip by', says Calder (2002) who is now Director of Communications of the Stanford Linear Accelerator Center.

[7] It is by no means clear which wireless standard will emerge as the winner or even if a single winner will emerge. The professional IEEE 802.11specifications have split into four separate 802.15 standards project groups with differing views on low-complexity and low-power consumption solutions. The older 802.11a standard, for example, has a throughput of 54 Mbits/s while 802.11b has a throughput of only 11 Mbits/s. Additional company coalitions have formed to develop products based on other possible wireless standards including the standards known as WI-FI and IPv6. Clearly, wireless products are firmly in the *fluid* stage of Abernathy & Utterback's (1978) model.

Complementary Assets

	Freely Available	Tightly Controlled
Freely Available	Profitability Is Difficult in a Commodity Business	Holders of Complementary Assets Are Profitable
Tightly Controlled	Owners of Intellectual Property Are Profitable	Profitability Based on Power Alliances

(Row group label at left: **Intellectual Property**)

Figure 2. Profitability winners.

of new disposable diaper products technologically, the companies have to work within the constraints and pressures that Walmart imposes on them if they want to sell their products to consumers through Walmart's vast array of store outlets, i.e. its powerful complementary assets (Olsen, 2001). If the intellectual property, however, cannot be appropriated, that is, the technical knowledge and capability are widely available to all competitors, then profits are likely to be made by those firms that control the complementary assets. If both complementary assets and the intellectual property are freely available, then it will be hard for any of the major businesses operating in such an industry to secure and maintain high profitability.

Corresponding Changes Across the Product/Process Model

Organizational Changes

As a business unit forms around an initial new product category and becomes highly successful over time, it goes through the same transformation that any entrepreneurial enterprise experiences as it grows in size and scope (Morgan et al., 2001; Roberts, 1990). During the pioneering period of the new product industry, i.e. the *fluid* stage, the processes used to produce the new products are relatively crude and inefficient (Abernathy & Utterback, 1978). As the rates of product and process innovation inversely shift, however, the organization has worked to take more and more costs and inefficiencies out of its processes, making production increasingly specialized, rigid, and capital intensive. Through the early 1920s, major auto-makers could assemble a car in four to five hours but it still required three to eight weeks to paint.[8] With ever increasing

capital improvements and innovations in the painting process, especially Dupont's development of Duco lacquer, the painting time had been cut by 1930 from more than 25 separate operations requiring many weeks to a more continuous spraying process applied over a few days. Painting had become another unskilled task and the strong, independent painters' union collapsed (Funderburg, 2002).

In recounting the rich example of incandescent lighting, Utterback (1994) points out that Edison's initial lighting products were made by a laborious process involving no specialized tools, machines, or craft laborers. At the time, getting the product to work was far more exciting and important to the innovators than creating an efficient volume production process. However, as companies built capacity to meet the growing market demand for incandescent lighting more cost effectively than their competitors, vast improvements in the process were gradually developed and introduced by both manufacturers and suppliers, including the use of specialized glass-blowing equipment and molds in addition to high-capacity vacuum pumps (Graham & Shuldiner, 2001). As a result of all these process improvements, especially the development of successive generations of glass-blowing machines, the number of manufacturing steps, according to Utterback (1994), fell dramatically from 200 steps in 1880 to 30 steps in 1920. The manufacturing process had evolved over this period into an almost fully automated continuous process. Amazingly enough, the glass-blowing Ribbon Machine which was introduced in 1926 is essentially the same device that is used today, and it still remains the most cost effective way to produce light bulb blanks (Graham & Shuldiner, 2001).

As organizations make these kinds of shifts, that is, from focusing on exploratory product designs to concentrating on larger-scale operational efficiencies and the production of more standardized offerings, many other important parallel changes also take place

---

[8] Part of Ford's rationale for offering the Model T in 'any color as long as it was black' was that black absorbed more heat than lighter colors and therefore dried significantly faster. It also lasted considerably longer as black varnish was most resistant to ultraviolet sunlight (Funderburg, 2002).

within the industry (Gawer & Cusumano, 2002; Roberts & Liu, 2002; Tushman & O'Reilly, 1997; Utterback, 1994). Not only do changes in products and processes occur in the previously described inverse systematic pattern, but organizational controls, structures, and requirements also change to adjust to this pattern. In the *fluid* period of high market and technical uncertainties, for example, business leaders are more willing to take the risks required for the commercialization of radical new ideas and innovations. Individuals and cross-functional areas are able to function interdependently, almost seamlessly, in order to enhance their chances of success (Golder & Tellis, 1997; Klepper, 1996). Formal structures and task assignments are flatter, more permeable, and more flexible—emphasizing rapid response, development, and adjustment to new information and unexpected events. The company is *organic* in that it is more concerned with the processing of information and the effective utilization of knowledge than the more rigid *mechanistic* following of formal rules, bureaucratic procedures, and hierarchical positions (Mintzberg, 1992).

As the industry and its product class evolve, the informal networks, information exchanges, and fast-paced entrepreneurial spirit that at one time had seemed so natural and valued within the organic-type firms slowly give way to those leaders increasingly skilled at and experienced in the coordination and control of large, established businesses (Roberts, 1990). As operations expand, the focus of problem-solving discussions, goals, and rewards shifts away from those concerned with the introduction of new, radical innovations to those who are capable of meeting the more pronounced and complex market, production, and financial pressures facing the organization (Miller, 1990; Tushman & O'Reilly, 1997). Such demands, moreover, become progressively more immediate and interconnected as the products become more standardized, the business more successful, and the environment more predictable—all of which combine to make it even more difficult and costly to incorporate disruptive kinds of innovations. Rather than embracing potentially disruptive innovations and changes—encouraging and sponsoring explorations and new ways of doing new things—the organization seeks to reduce costs and maximize the efficiencies in its on-going tasks and routines through more elaborate rules, procedures, and formal structures (Katz & Allen, 1997).

The real dilemma in all of this is that major changes in the environment get responded to in old ways. The organization myopically assumes that the basis by which it has been successful in the past will be the same basis by which it will be successful in the future. As the technical and market environments become increasingly stable, the growth of the enterprise relies to a greater extent on stretching its existing products

and processes (Tushman & O'Reilly, 1997; Valery, 1999). The organization encourages and praises its managers and leaders for achieving consistent, steady results that predictably build on past investments and sustainable improvements. More often than not, ideas that threaten to disturb the comfortable stability of existing behavioral patterns and competencies will be seriously discouraged, both consciously and unconsciously. Contrastingly, ideas that build on the historical nature of the business's success, including its products, markets, and technical know-how, are more likely to be positively received and encouraged (Katz & Allen, 1989). A large body of research shows that when comfortable, well-run organizations are threatened, they tend to increase their commitments to their status quo—to the practices and problem-solving methodologies that made them successful in the past; not necessarily to what's needed or possible for the future (e.g. Jellinek & Schoonhoven, 1990; Hamel & Prahalad, 1994; Miller, 1990; Weick, 2000). They tend to cultivate the networks and information sources that affirm their thinking and commitments rather than diligently search for information and/or alternatives that might disagree with their inveterate patterns. Without outside intervention, they become increasingly homogeneous and inward-looking, demanding increased loyalty and conformity. They hire and attract individuals who '*fit in*' rather than those who might think and behave in significantly new and different ways (Hambrick & D'Aveni, 1988; Katz, 1982).

Industry Changes

During an industry's *fluid* stage, there is usually sufficient flux in both the technology and the marketplace that large numbers of competitors are able to enter the industry with their own product variations. In their detailed studies of the cement and minicomputer industries, for example, Tushman & Anderson (1986) show just how fragmented industries can be in this stage as each industry had more than a hundred separate firms introducing new products during the early formative years of the product's initial *fluid* cycle. Utterback (1994) draws the same conclusion when he quotes Klein's (1977) assessment of the auto industry's competitive landscape. According to Klein (1977), there were so many initial competitors jockeying for market share leadership positions in the early 1900s that it would have been impossible to predict the top ten '*winners*' over the next 15 or so years. Even today, we tend to forget that many of the initially successful PC competitors included firms that were at one time very well known but which have now disappeared, firms such as Sinclair, Osborne, and Commodore Computer.

After the emergence of a dominant design, the industry moves towards a more commodity-type product space with many fewer surviving competitors. One would expect rapid market feedback and consumer value to be based more on features and functionality than on cost per se during the industry's *fluid* period (Utterback, 1994). As markets become more stable with fewer dominant players, however, the basis of competition also shifts. Market feedback tends to be slower and direct contact with customers tends to decrease, although there are increases in the availability of industry information and statistical analyses (Tellis & Golder, 2001). While incremental changes in products may stimulate market share gains during the *transitional* and *specific* stages of the product-process lifecycle, they can usually be copied quickly and introduced by competitors, thereby resulting over time in much greater parity among the price, performance, features, and service characteristics of competitors' comparable products (Christensen, 1997; Tushman & O'Reilly, 1997). At the same time, however, there is much greater emphasis within the industry on process innovation. Each organization tries to make the kinds of changes and improvements that benefit its particular manufacturing process, changes that are not easily copied or transferable between rival production lines that have been uniquely modified and fine-tuned. In short, price and quality become relatively stronger elements within the overall competitive equation (Fine, 1999; Hamner & Champy, 1993).

As the longevity of a particular dominant design persists within the industry, the basis of competition resides more in making refinements in product features, reliabilities, packaging, and cost. In this kind of environment, the number of competing firms declines and a more stable set of efficient producers emerges. Those companies with greater engineering and technical skills in process innovation and process integration have the advantage during this period. Those that do not will be unable to compete, and ultimately, will either fail, ally together, or be absorbed by a stronger firm (Roberts & Liu, 2002). It is perhaps for these reasons that Mueller & Tilton (1969); Tushman et al. (1986); Tyre & Orlikowski (1993); Utterback (1994); and many others contend that large corporations seldom provide their people with real opportunities or incentives to introduce developments of radical impact during this stage. As a result, changes tend to be introduced either by veteran firms or new entrants that do not have established stakes in a particular product market segment. Radical innovations and changes are usually introduced by disruptive players in small niche market segments (see Christensen, 1997) for as technological progress slows down and process innovation increases, barriers to entry in the large established markets become more formidable. As process integration proceeds, firms with large market shares, strong distribution networks, dedicated suppliers, and/or protective patent positions are also the organizations that benefit the most from product extensions and incremental improvements (Utterback, 1994). It should not be particularly surprising, therefore, that in their study

of photolithography, Henderson & Clark (1990) discovered that every new emerging dominant design in process innovation in their study's sample was introduced by a different firm than the one whose design it displaced. Radical, disruptive innovations that start in small market segments are too easily viewed as distractions by the more dominant firms trying to satisfy the cost and quality demands of their large customer base. These same innovations, however, are more easily seen as opportunities by both smaller firms and new entrants.

Innovation Streams and Ambidextrous Organizations

Because industry forces and corresponding organizational priorities operate so differently between product and process innovation phases, Tushman & O'Reilly (1997) argue that a business unit must redirect its leadership attention away from a particular stage of innovation and move towards managing a series of contrasting innovations if it is to survive through recurring cycles of product-process innovations. An organization has to function in all three phases simultaneously, producing streams of innovation over time that allow it to succeed over multiple cycles of technological products. Innovation streams emphasize the importance of maintaining control over core product subsystems and proactively shaping dominant designs while also generating incremental innovations, profiting from architectural innovation introductions that reconfigure existing technologies, and most importantly, by initiating its own radical product substitutes.

To demonstrate this stream of innovations capability, Tushman & O'Reilly (1997) refer to the enviable success of the Sony Walkman. Having selected the WM-20 platform for its Walkman, Sony proceeded to generate more than 30 incremental versions within the WM-20 family. More importantly, it commercialized four successive product families over a 10-year period that encompassed more than 160 incremental versions across the four families. This continuous stream of both generating incremental innovation while also introducing technological discontinuities (e.g. flat motor and miniature battery at the subsystem level) enabled Sony to control industry standards and outperform all competitors within this product class.

The obvious lesson, according to the two authors, is that organizations can sustain their competitive advantage by operating in multiple modes of innovation simultaneously. In organizing for streams of innovation, managers build on maturing technologies that also provide the base from which new technologies can emerge. These managers emphasize discipline and control for achieving short-term efficiencies while also taking the risks of experimenting with and learning from the '*practice products*' of the future. Organizations that operate in these multiple modes are called 'ambidextrous organizations' by Tushman & O'Reilly

(1997). Such firms host multiple, internally inconsistent patterns of structures, architectures, competencies, and cultures—one pattern for managing the business with efficiency, consistency, and reliability and a distinctly different pattern for challenging the business with new experimentation and thinking.

Each of these innovation patterns requires distinctly different kinds of organizational configurations.[9] For incremental, sustained, and competence-enhancing types of changes and innovations, the business can be managed with more centralized and formalized roles, responsibilities, procedures, structures, and efficient-minded culture. Strong financial, supply chain, sales, and marketing capabilities coupled with more experienced senior leadership teams are also beneficial. In sharp contrast, the part of the organization focusing on more novel, discontinuous innovations that have to be introduced in a more fluid-type stage requires a wholly different kind of configuration—a kind of entrepreneurial, skunk-works, or start-up spirit and mentality. Such organizational entities are relatively small with more informal, decentralized, and fluid sets of roles, responsibilities, networks, and work processes. The employees themselves are usually less seasoned and disciplined—they are, however, eager to '*push the state of the art*' and to test '*conventional wisdoms*' (Tushman & O'Reilly, 1997).

What all too often becomes very problematic is that the more mature, efficient, and profitable business sees its entrepreneurial unit counterpart as inefficient and out of control—a '*renegade, maverick*' group that violates established norms and traditions. The contradictions inherent in these two contrasting organizational configurations can easily lead to powerful clashes between the traditional, more mature organizational unit that's trying to run the established business vs. the part that's trying to recreate the future. If ambidextrous organizations are to be given a real chance to succeed, the management teams need to keep the larger, more powerful business unit from trampling and grabbing the resources of the entrepreneurial entity. While the company can protect the entrepreneurial unit by keeping the two configurations physically, structurally, and culturally apart, the company also has to decipher how it wants to integrate the strategies, accomplishments, and markets of the two differentiated units. Once the entrepreneurial unit has been separated, it becomes all too convenient for the firm to either sell or kill the smaller activity. One well-known European electronics organization, for example, sadly discovered from its own retrospective investigation that the vast majority of its numerous internal skunk-works for new products in the 1990s were eventually commercialized by competitors. Even more worrisome was the finding that none of the skunk-work products commercialized by the firm *itself* became successful businesses. Clearly, the need to differentiate is critically important to foster both the incremental and radical types of innovation; but at the same time, the organization has to figure out how it is going to eventually integrate the so-called '*upstart*' unit into the larger organization so that the potential of the ambidextrous organization is not lost. The ability of senior executives and their teams to integrate effectively across highly differentiated organizational configurations and innovative activities remains the true challenge of today's successful enterprise.

References

Abernathy, W. & Utterback, J. (1978). Patterns of industrial innovation. *Technology Review*, **80**, 40–47.

Adizes, I. (1999). *Managing corporate lifecycles*. Paramus, New Jersey: Prentice Hall Press.

Allen, T. J. (1986). Organizational structure, information technology, and r&d productivity. *IEEE Transactions on Engineering Management*, **33**, 212–217.

Anderson, P. & Tushman, M. (1990). Technological discontinuities and dominant designs: A cyclical model of technological change. *Administrative Science Quarterly*, **35**, 604–633.

Agarwal, R. & Bayus, B. (2002). The market evolution and sales takeoff of product innovations. *Management Science*, **48**, 1024–1041.

Bagley, C. E. (1995). *Managers and the legal environment: Strategies for the 21st century* (2nd ed.). St. Paul, MN: West Publishing.

Berner, R. (2001). Can Procter and Gamble clean up its act? *Business Week*, March 12, 80–83.

Brooker, K. (2002). The Un-CEO. *Fortune*, September 12, 88–96.

Calder, N. (2002). New trends in internet and web technologies. Presentation at School of Engineering, MIT.

Caves, R. E. (1996). *Multinational enterprise and economic analysis*. Cambridge: Cambridge University Press.

Charan, R. & Tichy, N. (2000). *Every business is a growth business*. New York: Times Books.

Christensen, C. (1997). *The innovator's dilemma*. Cambridge, MA: Harvard Business School Press.

Christensen, C. (2001). Some new research findings for the innovator's dilemma. Presentation at Sloan School of Management, MIT.

Christensen, C. & Bower, J. (1996). Customer power, strategic investment, and the failure of leading firms. *Strategic Management Journal*, **17** (1), 97–218.

Christensen, C., Suarez., F. & Utterback, J. (1998). Strategies for survival in fast-changing industries. *Management Science*, **44**, 1620–1628.

Collins, J. & Porras, J. (1994). *Built to last*. New York: Harper Business.

Collins, J. (2001). *Good to great: Why some companies make the leap . . . and others don't*. New York: Harper Collins.

Cooper, A. & Smith, C. (1992). How established firms respond to threatening technologies. *The Academy of Management Executive*, **6**, 55–70.

Crockett, R. (2001). Motorola. *Business Week*, July 16, 34–37.

---

[9] See Tushman & O'Reilly (1997) for their original and more complete discussion of organizational congruence, culture, innovation streams, and ambidextrous organizations.

Cusumano, M. A., Mylonadis, Y. & Rosenbloom, R. (1992). Strategic maneuvering and mass market dynamics: The triumph of VHS over Beta. *Business History Review,* **Fall,** 51–93.

Cusumano, M. A. & Selby, R. (1995). *Microsoft's secrets.* New York: Free Press.

Cusumano, M. A. & Yoffie, D. B. (1998). *Competing on internet time.* New York: Free Press.

Ferguson, C. (2000). The wireless future. *Frontier,* December 4, 56–58.

Fine, C. (1999). *Clockspeed: Winning industry control in the age of temporary advantage.* New York: Perseus Publishing.

Fisher, D. E. & Fisher, M. J. (1997). The color war. *American Heritage of Invention and Technology,* **12,** 8–18.

Foster, R. (1986). *Innovation: The attacker's advantage.* New York: Summit Books.

Foster, R. & Kaplan, S. (2001). *Creative destruction: Why companies that are built to last underperform the market—and how to successfully transform them.* New York: Doubleday Press.

Funderburg, A. (2002). Paint without pain. *American Heritage of Invention and Technology,* **17,** 4–57.

Gawer, A. & Cusumano, M. (2002). *Platform leadership.* Cambridge, MA: Harvard Business School Press.

Golder, P. N. & Tellis, P. G. (1997). Will it ever fly? Modeling the takeoff of really new consumer durables. *Marketing Science,* **16,** 256–270.

Graham, M. & Shuldiner, A. (2001). *Corning and the craft of innovation.* New York: Oxford University Press.

Grove, A. (1996). *Only the paranoid survive.* New York: Doubleday.

Hamel, G. (2000). *Leading the revolution.* Cambridge, MA: Harvard Business School Press.

Hamel, G. & Prahalad, C. K. (1994). *Competing for the future.* Cambridge, MA: Harvard Business School Press.

Hambrick, D. & D'Aveni, R. (1988). Large corporate failures as downward spirals. *Administrative Science Quarterly,* **33,** 1–28.

Hamner, M. & Champy, J. (1993). *Reengineering the corporation.* New York: Harper Business.

Handy, C. B. (2002). *The elephant and the flea.* Cambridge, MA: Harvard Business School Press.

Hartlove, C. (1998). Plastic batteries. *Technology Review,* **101,** 61–66.

Henderson, R. & Clark, K. (1990). Architectural innovation: The reconfiguration of existing product technologies and the failure of existing firms. *Administrative Science Quarterly,* **35,** 9–30.

Hesselbein, F. & Johnston, R. (2002). *On creativity, innovation, and renewal.* New York: John Wiley.

Iansiti, M. (2000). How the incumbent can win. *Management Science,* **46,** 169–185.

Jellinek, M. & Schoonhoven, C. (1990). *The innovation marathon.* New York: Blackwell.

Katz, R. (1982). The effects of group longevity on project communication and performance. *Administrative Science Quarterly,* **27,** 81–104.

Katz, R. (1997). Organizational issues in the introduction of new technologies. In: R. Katz (Ed.), *The Human Side of Managing Technological Innovation* (pp. 384–397). New York: Oxford University Press.

Katz, R. & Allen, T. (1989). Investigating the Not Invented Here (NIH) syndrome: A look at the performance, tenure, and communication patterns of 60 R&D project groups. *R&D Management,* **19,** 7–19.

Katz, R. & Allen, T. J. (1997). How project performance is influenced by the locus of power in the R&D matrix. In: R. Katz (Ed.), *The Human Side of Managing Technological Innovation* (pp. 187–200). New York: Oxford University Press.

Klein, B. (1977). *Dynamic economics.* Cambridge, MA.: Harvard University Press.

Klepper, S. (1996). Entry, exit, growth, and innovation over the product life cycle. *American Economic Review,* **86,** 562–583.

Knud-Hansen, C. (1994). Historical perspective of the phosphate detergent conflict. *Conflict Resolution Consortium,* University of Colorado at Boulder.

Leifer, R., Colarelli, G. & Rice, M. (2001). Implementing radical innovation in mature firms: The role of hubs. *Academy of Management Executive,* **15,** 102–113.

Loomis, C. (1993). *Fortune,* May 3, 75.

Luecke, R. (2002) *Essentials of managing creativity and innovation.* Cambridge, MA: Harvard Business School Press.

Lynn, G. & Reilly, R. (2002). *Blockbusters: How new product development teams create them—And how your company can too.* New York: Harper Business.

Micklethwait, J. & Wooldridge, A. (1998). *The witch doctors: Making sense of the management gurus.* New York: Times Books.

Miller, D. (1994). What happens after success. *Journal of Management Studies,* **31,** 325–358.

Miller, D. (1990). *The Icarus paradox: How exceptional companies bring about their own downfall.* New York: Harper.

Mintzberg, H. (1992). *Structuring in fives: Designing effective organizations.* New Jersey: Prentice-Hall Press.

Morgan, H. L., Kallianpur, A. & Lodish, L. (2001). *Entrepreneurial marketing.* New York: John Wiley.

Mueller, D. & Tilton, J. (1969). R&D costs as a barrier to entry. *Canadian Journal of Economics,* **2,** 570–579.

Meyer, M. & Lehnerd, A. (1997). *The power of product platforms.* New York: Free Press.

Olsen, S. (2001). Wal-Mart, Amazon considering partnership. CNET News, March 5.

Pate-Cornell, M. E. & Dillon, R. L. (2001). Success factors and failure challenges in the management of faster-better-cheaper projects. *IEEE Transactions on Engineering Management,* **48,** 25–35.

Peters, T. (1997). *The circle of innovation.* New York: Vintage Books Press.

Porter, M. (1990). *The competitive advantage of nations.* New York: Free Press.

Reinertsen, D. (1997). *Managing the design factory: The product developer's toolkit.* New York: Free Press.

Reinhardt, A. (2002). Kiss those TV, DVD, and stereo wires goodbye. *Business Week,* September 16, 60.

Roberts, E. (1990). *Entrepreneurs in high tech.* New York: Oxford University Press.

Roberts, E. & Liu, W. (2002). Ally or acquire? How technology leaders decide. *Sloan Management Review,* **41,** 26–34.

Rosenbloom, D. & Christensen, C. (1994). Technological discontinuities, organizational capabilities, and strategic commitments. *Industry and Corporate Change*, **3**, 655–686.

Rosenbush, S. & Elstrom, P. (2001). Eight lessons from the Telecom mess. *Business Week*, August 13, 20–26.

Scigliano, E. (2002). Ten technology disasters. *Technology Review*, **105**, 48–52.

Scott, J. (2001). The secret history. *Business Week Online*, August 14.

Steele, L. (1989). *Managing technology: The strategic view*. New York: McGraw-Hill.

Suarez, F. & Utterback, J. (1995). Dominant designs and the survival of firms. *Strategic Management Journal*, **16**, 415–430.

Sutton, R. (2001). *Weird ideas that work: Practices for promoting, managing, and sustaining innovation*. New York: Free Press.

Teece, D. (1987). Capturing value from technological innovation: Integration, strategic partnering, and licensing decisions. In: B. Guile & H. Brooks (Eds), *Technology and Global Industry* (pp. 245–265). Washington D.C.: National Academy Press.

Teece, D. & Pisano, G. (1994). Dynamic capabilities of firms. *Industry and Corporate Change*, **3**, 537–556.

Tellis, G. J. & Golder, P. N. (2001). *Will and vision: How latecomers grow to dominate markets*. New York: McGraw-Hill.

Tripsis, M. (2001). Unraveling the process of creative destruction: Complementary assets and incumbent survival in the typewriter industry. *Strategic Management Journal*, **22**, 111–124.

Trudel, J. (2002). Innovation in sight. *IEEE Engineering Management Review*, **30**, 106–108.

Tushman, M. & Anderson, P. (1986). Technological discontinuities and organizational environments. *Administrative Science Quarterly*, **31**, 439–465.

Tushman, M., Newman, W. & Romanelli, E. (1986). Convergence and upheaval: Managing the unsteady pace of organizational evolution. *California Management Review*, **29**, 29–44.

Tushman, M. & O'Reilly, C. (1997). *Winning through innovation*. Cambridge, MA.: Harvard Business School Press.

Tyre, M. & Orlikowski, W. (1993). Exploiting opportunities for technological improvements in organizations. *Sloan Management Review*, **32**, 13–26.

Utterback, J. (1994). *Mastering the dynamics of innovation*. Cambridge, MA.: Harvard Business School Press.

Valery, N. (1999). Innovation in industry. *The Economist*, February 20, 5–28.

Van Mullekom, K. (2002). Consumer guide—Washing machines. *Consumer Reports*, February 10, 78–81.

von Hippel, E. (1998). Economics of product development by users. *Management Science*, **44**, 629–644.

Weick, K. E. (2000). *Making sense of the organization*. New York: Blackwell Publishers.

Welch, J. & Byrne, J. (2001). *Straight from the gut*. New York: Warner Books.

Willyard, C. & McClees, C. (1987). Motorola's technology roadmap process. *Research Technology Management*, **30**, 13–19.

The International Handbook on Innovation
Edited by Larisa V. Shavinina

Towards a Constructivist Approach of Technological Innovation Management

Vincent Boly, Laure Morel and Jean Renaud

Laboratoire de Recherche en Génie des Systèmes Industriels, France

Abstract: This chapter describes a constructivist approach to the understanding of technological innovation management within French SME's. Firstly, it presents findings from technological innovation surveys in French SME's and analyzes the emerging trends (e.g. the non-existence of a single ideal new product development process, a high degree of variability of technological evolution, and the importance of cognitive aspects). Secondly it describes a constructivist approach and its key aspects (e.g. development of a value-oriented strategy; a systemic vision of innovation management through its three levels: strategy, piloting, and sparking), as well as limitations of current practices in different kind of French SME's.

Keywords: Innovation; SME's; Constructivism; Strategy; Piloting; Sparking; Value; Knowledge.

Introduction

Companies having less than 500 employees represent around 90% of the total number of French enterprises. As a result, their impact on the country's global economic performance is highly significant. Our attention is directed toward innovation, in SME's (Small and Medium Enterprises). One can define innovation as:

- the steps processing an idea into a new product or service, launched on the market and creating new value;
- the development process of a new industrial activity within an existing company or in the context of a company launching project (Garcia et al., 2002);
- a cognitive approach within an industrial system, characterized by a paradigm change (out of the box thinking, rules breaking). It begins with creation tasks and ended with standardization (Swan et al., this volume);
- the technological reaction of an adaptative industrial system facing an internal signal or an external evolution (this includes changes in the boarder between the company and its environment in order to invest opportunity spaces).

Note that a technology is considered as a sum of scientific knowledge and connected know-how (knowledge aiming at the industrialization of the first one).

In its last survey, the SESSI, Statistic Service of the French Ministry of Industry (SESSI, 1999), states that the percentage of innovative SMEs is improving. Accordlingly, investment devoted to R&D is rising (Table 1): the number of small companies developing a formal research activity has doubled during the last 15 years. Nevertheless, research remains mainly the domain of international industrial groups (60.9% of them carry out formal research).

Table 1. *Percentage of companies having a formal research program in 1996.*

Number of employees	% in all French companies
20 to 49	3.7
50 to 99	9.3
100 to 249	20.7
Small Entr.	7.6
250 to 499	38.1
Total SME	9.2
Big companies	60.9
Total*	11.0

Table 2. Entreprises with technological innovation in French industry between 1994 and 1996 (% of the total number of companies).

	20 to 49 employees	50 to 99 employees	100 to 249 employees	250 to 499 employees	SME
Innovation in products or process	33.3	43.8	53.4	66.3	40.0
Innovation in products	26.6	37.3	46.3	59.5	33.3
(including new products on a market)	15.8	19.8	29.6	36.4	19.7
Innovation in process	23.1	30.9	39.0	50.5	28.4
Entreprises having on-going projects in 1997	26.7	37.1	46.4	61.3	33.4
Total innovative companies (including on-going projects)	38.4	48.9	58.1	70.6	45.0

Source: SESSI, 1999.

This analysis of the 'formal research' criteria contains several limitations and other surveys suggest a greater number of innovative SMEs (Table 2). Among these limitations one can note: some SME's do not consider as researchers part time technology developers, and on the other hand some company managers integrate any collaboration with academic structures (training period of students for example) as research activities. Moreover many French SME's have 'informal research': experiments and design activities realized off the official working time.

This difference between innovative companies and those having a formal research activity can be explained by the high degree of variability in the innovation process (Boly et al., 2000). Increased attention is directed by SME's top management toward the creation of a continuous adjustment between products and production practices, and their external technological environment. As a result, research activities do not systematically represent the first step of the innovation process. New customer requirements, quality problem solving, supplier propositions and internal ideas also induce technological evolution. Moreover, few small companies are involved in a formal innovation process. Finally, innovation often emerges from action and not from study organized by top management.

It should be noted that in new company creation, the percentage of business working in innovative fields is still improving: 5% in 1999 and 6.5% in 2000 (DIGITIP, 2001). This represents about 11,000 new innovative enterprises in 2000.

As in many other parts of the world, the strategies of French SMEs depend heavily upon the related industrial sector (3.6% more employees in innovative companies compared to 1999). Innovation investments remain weak in sectors requiring high capitalization, or high promotion costs. These include: energy and raw materials production.

In fact, most surveys on radical innovation highlight a U trend between the size of the enterprises and their performance (Bernard, 1994). Many small companies valorizing high technological and/or specialized know-how are involved in innovation as well as major companies investing in research.

In short, the number of French SMEs trying to strengthen their competitive advantage through innovation is increasing. Funds mobilized to support innovative projects tripled in 2000 compared to 1999: this includes stock offerings as well as venture capital (Ministère de l'Economie, 2001).

As far as SMEs are characterized by weak resources, high reactivity and their degree of information integration, their innovation piloting approaches constitute an interesting research area.

This section emphasizes some of the phenomena observed in innovative French SMES and the resulting new innovation piloting concepts.

Technological Innovation: Emerging Concepts

Observation of Some Basic Phenomena in French SMEs

In order to get a better understanding of the innovation process within SMEs, our team generally adopts an anthropological type approach: data gathering includes observation, listening, conversations, questions and answers and internal document reading. This approach suits our special interest in reporting qualitative and quantitative data. As a matter of fact, decision and action in the field of innovation stem not only from logical thought and a scientific approach, but also from intuition or non linear thinking (Buckler et al., 1996). Our clinical methodology implies that the researchers both participate in (strategy definition, technical design and project reviews) and study organizational change with the aim of contributing both to the advancement of knowledge and to practical questions that organizations may have. Via this involvement, it is possible to gain access to a wealth of data which is denied to other approaches (Karlsson et al., 1997).

At the end of the inquiry stage (direct observation and analysis of documents), information is gathered through clinical data tables (Fig. 1).

Information analysis is based on two systemic models: 'the result flowsheet' and the 'activity Gantt chart'. In each firm, and at the end of each experimental period, individual and collective productions

Internal and external customers of Mr/Mrs X	Results: what did Mr/Mrs X give to them	Related activities: what does Mr/Mrs X do	Resources used by Mr/Mrs X	Internal and external suppliers of Mr/Mrs X

Figure 1. Clinical data tables: description of individual or collective involvement in the NPDP.

were gathered into a series of results (Fig. 2). Thus modelling consists of the immaterial (information, advice, . . .) and material (prototype, plan, . . .) result sequence. This model is consistent with the focus on studying the dynamic evolution of innovative projects. It is indeed meaningful to describe decision making and to understand project management practices and mental perceptions.

First, information about individual or collective activities from clinical data tables is assembled into a 'Gantt chart of the activities': a diagram indicating the following activities carried out by people involved in each innovation task (Fig. 3). Activities are chronologically ordered and additional items included: who was in charge of the activity, who was concerned by this activity. This model is meaningful to determine actual responsibilities. It captures a broad spectrum of responsibility sharing: technical studies, validation of other activities and final decision-making.

Further analysis of the two models includes common activity sequence identification within the tasks of panel firms. These standard activities determine general project management practices. Second, comparison between the thirteen 'result flowsheets' is used at three different stages: initial idea, industrial report

(documents include technical, commercial and financial statements) and final innovative activity launched.

What Kind of Concepts are Emerging?
The Non Existence of a Single Ideal New Product Development Process
Two main observations are in conflict with the idea of a meta-innovation process. First, the variable position of certain activities and results within a global New Product Development Process (NPDP). Second, the different ways to undertake certain given tasks. As a consequence, it is not possible to assemble the 13 flowsheets into one model.

In the technical field as well as in that of trade, strategy and finance, the variable position of results and activities is confirmed. For example, prototype creation is generally planned after initial drawings and technical surveys. But, in some cases (Firm G, M), even before validating the start of the project, technical feasibility tests are carried out.

Moreover, some technical steps are managed in very variable ways. For example, routing definition sometimes represents a single task. But in some cases (Firm B, C, D), routing is analyzed as forecast routing before the final product plan design, and is defined as final

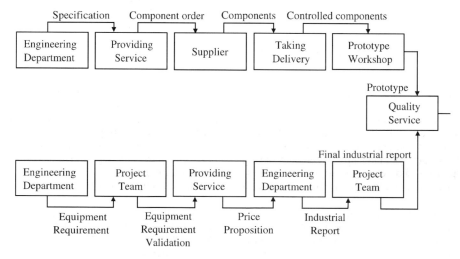

Figure 2. Modelling of NPDP through the 'result flowsheet'.

ACTIVITY	Project Leader	Top Management	Engineering	Production	Sales Force	Quality
Final Product Plan Drawing			X			
Plan Review With The Customer	X					
Agreement On Final Product Plan	X					
Internal Product Plan Circulation			X			
Final Routing Definition			X			
Final Routing Validation						X

Figure 3. Modelling of NPDP through the 'activity Gantt chart' (X means 'made by' and a grey cell means 'concerned with').

routing in a later step. Engineers argue that forecast routing allows anticipation with regard to the correlation between the new product requirements and the current production equipment.

Market analysis attests to the same variability. A 'result flowsheet' study indicates that some firms begin market analysis after prototype production (Firm D, G, I). However, companies undertake customer surveys before development (Firm B, E, F, H). Moreover, separate studies of different target markets are undertaken throughout NPDP in Firm A.

Diversity also exists in the field of partnership. Top management sometimes tends to contract with partners before plan drawing, while others wait until technical tests have been carried out on the product. In fact, partnership as a result of an external offer could occur at any moment.

It should be noted that certain activities and result transfers are lacking even when comparing all NPDPs. Market analysis, contact with potential partners and legal studies are not systematically undertaken during successful projects. For example, some firms investigate direct market confrontation and experienced retroaction (Firm M).

As a result, the project manager's capacity to select and order these activities influences success rates. Lists of activities, or intermediary results are at the manager's disposal in order to define the appropriate innovation process. Then, key factors for success

include: a correct NPDP and its on going improvement. Fundamentally, innovation induces more adaptation approaches than attempts at optimization.

The High Degree of Variability in Technological Evolution

Authors such as Song et al. (1996), state that the way an idea is processed into an innovative activity depends heavily upon factors including: strategy (i.e. whether top management investigates partnerships or non targeted markets (i.e. single principal or large size market) and production process (i.e. single know-how or combinations of technologies). NPDP results from these internal decisions but also from external events (Ducharme et al., 1995) and is made up of highly variable ordered standard activities. Hence, purposely or in reaction to external constraints, personnel involved in the project do or do not undertake certain activities. Consequently, competitive advantages emerge from activity position modification within NPDP. One particular aspect is described by Swan et al. (this volume): the role of networks and the way they are managed, vary across the innovation process.

Our outcomes highlight that NPDP is not the only source of instability. Within our panel, we observe that various strategies were applied by top management during the projects. Although the same people participated throughout the project, initial ideas always differed from the final activities launched (we had only one exception: Firm D). The improvement of technological knowledge via R&D or feasibility studies may partly explain this evolution. However other factors influence not only the product but also the global targeted industrial structure, such as contact with suppliers of funds, sales opportunities during development stages or new norms. Innovation emerges from a phenomenon of confrontation.

Variability may also be considered through the impact of some of the technical decision taken during the project. For example, the innovative ergonomic manholes produced by Firm K, were based on design decisions requiring little evolution in production know-how whereas success induced important changes in the field of organization. However, some innovations are only made possible by adding new abilities (at the design, development or production launching stages).

Within some enterprises and at the same time, different organizational structures are developed to manage innovative projects: multifunctional project teams, single functional departments or individual missions (Griffin, 1997). And yet, the concepts of innovation and project are not strictly correlated. Informal structures may lead to innovation. We had the example of a company that innovated by applying the proposition of a student at the end of a training period.

Basically, innovation relies, in part, on this variability. Indeed, people facing a context of instability have to be reactive and creative. These are two basic abilities promoting innovation.

The Process of Technology Transformation and its Impact on Success

The success of a new product depends on its characteristics, on external variables (such as customer acceptability or political events) but also on its development process. The McKinsey survey on the financial margin brought by innovative projects offers some thoughts regarding the impact of the technology transformation process. Delayed launching, real production costs higher than the objective and (but with a slighter impact) under-estimated research costs constitute major obstacles when aiming at maximum value. This outcome suggests that:

- the use of project management tools (including planning and cost control) (Firm B, C, D, L, K);
- human resource management (including time and individual assignment to the project), (Firm A, H, C, J, M);
- information processing approaches (such as objective cost design) (Firm J, C);

influence the success of an innovation.

One other dimension of the impact of the innovation process on the success of the product is the partnership strategy. Mitchell et al. (1996) states that codesigned products have a better survival rate: 57% for innovations managed by two or more companies together against 39% for products developed by a single enterprise. The existence of confrontation or negotiation steps between the partners within the innovation process has an impact. In our panel, we observed that project duration was reduced in the case of partnership. Handling the agenda seems easier when two or more top managers are involved.

Other considerations rely on the sociological dimension of the innovation process. While participating in development tasks, people try to forecast their own place in the future activity. Hopes, desires and fears influence these representations. As a result, at the individual level, decisions are not only based on objective reasoning approaches. Interrelation management and individual variables may change decisions about the product, its production organization and, as a result, the corresponding global activity.

The Necessary Integration of Complexity, Risk and Uncertainly

Pascale (1999) considers innovative companies as complex adaptative systems. As long as innovation is 'technological activity in a constant state of flux' an anthroposocial dimension is to be taken into account. As a consequence, describing and understanding innovation phenomena relies on the concept of evolution and its complex related principles.

The multivariable structure of the innovation process (Iansiti et al., 1997) and the multiplicity of links between these variables, represents the first phenomenological aspect of complexity (Benghozi et al., 2001). Any evolution in the production process for example, may have an impact on other processes such as distribution, maintenance and quality control (Firm B, G). Moreover, all industrial activities require financial resources and innovation often implies a modification of investment allocation principles. Even the relations between the industrial system and its environment move to a new temporary balance: innovation success may depend on new supplier selection or other responsibility sharing with traditional suppliers (Firm J). The environment itself changes following the development of the innovative projects of the company. Firm G had to adapt to European norms in the field of acoustics for its new activity, whereas, there were legal constraints relating to mechanics for its previous product.

In summary, multivariable aspects and analyzing *in situ* observation in French companies, we can state that innovation influences (Boly et al., 2000):

- the results, being the common outcome of several internal processes (an efficient prototype is the result of quality and design processes);
- the transverse activities (forecasting the new product cost is an activity concerning people in charge of design, finance and production);

- the resources to be shared between different processes (including investment);
- the cross impact between process performance: stock control and production performance for example.

Complexity also arises from the contradiction in the objectives of the company departments. Innovation aims at improving *value* in the company. But global value evolution is not the simple sum of local value improvements. Transforming value improvement strategy into local objectives remains difficult: safety and productivity, ecological impact and yield, quality and financial results. As far as innovation creates local disturbances (evolution of quality control procedures, logistic evolution, responsibility modifications in the purchasing department in the case of Firm C) another coherence level has to be defined to further innovation, taking into account that the responsability area of an individual is no longer adapted. Moreover, the impact of design decisions between global and local levels has to be managed.

The concept of discontinuance also leads to complexity. Considering innovation as a radical phenomenon upsetting the equilibrium between an industrial system and its environment, we observed an improvement in the susceptibility of the company to changes. As a consequence, innovation modifies the necessary reactions of the company to external disturbances. Extra costs and negotiations with the customer are the consequences of a temporary production delay (Firm E of our panel). But, not setting the agenda to produce the final prototype of a new product to be presented to the annual trade fair may cause definitive project failure. Complexity arises from these evolutions in daily manager behavior necessitated by innovation.

Uncertainty is the second basic notion to integrate while aiming at a better understanding of innovation. In many cases, negative outcomes from feasibility studies are not consistent with the following success of an innovation. Thus, Firm J believed there would be an improvement in the global market for road verge cutting machines following the introduction of its innovation. They established a profitable business plan although first analysis of the traditional sales volume was negative. In fact, success is often based on antinomical decisions. Moreover, any member of a team involved in the innovation process faces hazards (events they foresee but without knowing their final form: currency levels for example) and uncertain (unforeseeable) events (bankruptcy of a major customer for example). As a result, top management gains useful insight by defining a possible risk level: what is the maximum gap between the final result described by the project team and the real future one. A risk reduction policy may represent a limitation to innovation.

More precisely, links between innovation and uncertainty can be detailed as follows:

- considering innovation as a decision process, we can state that uncertainty relies on the quality of the main resource of this process: information. For example, precision in the customer needs description remains fundamental. As time to market reduces, that quality and volume of information available may be critical;
- uncertainty is positively correlated to the novelty degree (Beaudoin, 1984): people in companies are inexperienced with the new know how linked with innovation. As a consequence, their understanding and control of new technical aspects are not optimum;
- complexity is associated with uncertainty. Thus, the exhaustive prediction of the impact of an innovation on the previous industrial system or its environment, is generally impossible for the human brain. Due to this limitation, problems remain unstudied at the end of a project;
- uncertainty depends on the strategy: uncertainty is often greater when aiming to be first at launching a new product than when deciding to be a follower. Moreover, innovation decisions imply an evolution in the global strategy of the firm;
- uncertainty emerges from the constructivist approach in innovation management. New information modifies the initial representation of the project and as a consequence the 'future product' is characterized by on going evolution. Moreover the project team itself may change;
- innovation considered as breaking the rules in the company (technical concepts, production process, organizational structures, and so on) is source of uncertainty as long as people lack references;
- uncertainty emerges as innovation requires top managers ability to identify opportunities within their environment. Innovation success then partly depends on unstable variables like curiosity, perspicacity, intuition and culture;
- uncertainty is related to the company's size. In particular, financial resources may be critical in the case of small businesses to achieve complete feasibility studies and experimental tasks;
- basically, each technology is highly adapted to the particular characteristic of the industrial system. Uncertainty arises from the abilities of people to adjust general knowledge to their specific context (machine, organization, or behavior);
- the sociological dimension of innovation is also a source of uncertainty: relations between people involved in the project, the emergence of 'champions' as project leaders, individual strategy influences the outcomes of the innovation process;
- is there a chaotic phenomenon in project management? Does the repetitive use of some methodological tools lead to results that are clearly different from expectations (many functional analyses may lead to a lack of formalization)? In this case,

uncertainty would arise from the systematization of management practices.

In summary, innovation requires a complex approach to problems. It is not approximation thinking nor disorder management (The Minessota Studies, 2000). Integrating complexity deals with creating methodologies and developing reasoning modes, in order to analyse links between variables, to articulate different concepts and to organize the reflection between local and global approaches.

Local Performance Optimization Versus Creation of New Value

Because of the increasing complexity of industrial practices, the general evolution of economic demand and the non linearity of innovation processes, the notion of local performance evolves toward a more global notion of value. Indeed, firms require the development of new abilities to be able to cope with complex problems in all their dimensions, and to initiate and direct new development actions. To transform the constraints into opportunities of development represents one of these abilities.

Optimization of industrial unit processes (such as production monitoring, product/process design and project monitoring) is a major concern in companies focusing on performance control and an intangible objective. As a consequence, finding solutions to the problems by adopting a short-term emergency mentality remains the priority. The result is often a technical solution with a better financial return. But, only visible indicators are considered (facts or objects) such as product quality or better productivity.

Obviously this type of management presents limitations. Indeed, this logic does not take into account individual and collective behavior whereas proposing innovative solutions requires the mobilization and implication of all people involved. Furthermore, this management, based on a sequential approach to each operational problem avoids any feedback effects. So,

diagnosis seems completely disconnected from action and can lead to aberrations (for example, within our panel, some firms driving an environmental approach, were completely disrupted by the fact that indicators could give contradictory results depending on the point of view chosen: quality or environment).

As a result, innovation managers gain useful insight by developing an integrated and simultaneous approach to the technological, human and organizational dimensions (as a whole). This, so called 'logic of value creation', guarantees both the improvement of the technological system as well as the integration of individual behavior. It should be noted that value is an evolving concept according to specific criteria (see 2.2).

Value development rests mainly on the principle of integration, highlighting the dimension or dimensions, which may be used as a lever when attempting to solve a global problem. Each action or decision has to be made or taken considering its impact on the technical variables of the project, the organization of the company, the individual behaviors and the present reasoning modes of the employees.

As a consequence, innovation appears as a new principle of action based on novelty, transforming a set of constraints into an opportunity of new value development and taking into account both local optimization (development of tools and methodologies) and global development (study of the links and interactions existing between the dimensions described in Fig. 4). Finally, this challenge also relies on the firms' abilities to create a synergy between the social environment, technology and organizational processes to effectively contribute to the satisfaction of its customers.

The Importance of Cognitive Aspects

The complex aspect of innovation ends in the creation of a new status for knowledge and thus action.

First, innovation suggests a confrontation of individual representations. The purpose of managers is to

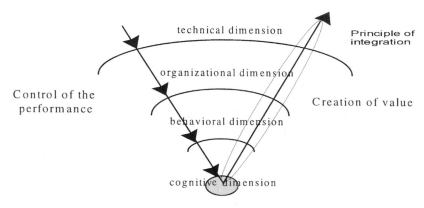

Figure 4. Control of the performance to the creation of the value.

build gradually a *common vision* (Longenecker et al., 2000), confronting words, signs, models and trying to connect them to the other. Various domains of science may be investigated. This favors the creation of emergent structures and new ways of development thanks to an effective transfer from one domain to another. This process is especially convenient to the 'radical' or 'break' innovation because it questions the traditional reasoning and action modes of company at a given time. Innovation arises from a confrontation between individuals or from organizations, which have strategies and thus different models of action ('variety of points of view' (Morin, 1977)). Companies thus invent new solutions by improving existing answers.

Second, in numerous cases, observation of our panel attests that the emergence of a new technology is bound to the emergence of a *new way of thinking*: this can be induced by networking conditions (Swan et al., this volume). Simplicity and surprising aptness of change of analysis of a problem, or a situation often create success. Weld the manhole onto the tool generally used to remove it: here is the innovation of Firm K. Analyzing the covering of nuts with different layers of compounds as a succession of hydrophilic and hydrophobic coverings before the formulation step: here is the beginning of a new sweet formulation (Firm M). Through the image of the 'frog paradox', Debaig et al. (1999) shows numerous cases where this break in representations drives innovation.

As a consequence, innovation may be considered as a cognitive process, during which individuals show themselves creative and capable of developing several approaches to the same situation. Thus, innovation would be a mechanism of *building new representations of the 'reality' or the object*. A sawing machine became a technological engine and the driver a pilot (Firm J)! The major obstacle consists in going beyond the first reflex: 'to make reality possible'. On the contrary, theory of complexity states that from possible to reality, something else exists, which we can call innovation or progress (Lemoigne, 1994). Innovation consists also in a process of *modification of the ways of reasoning*. For example, certain hypotheses at the origin of innovative knowledge arise from the use of deductive reasoning modes instead of analogical ones.

In other words, creation of new knowledge partly relies on the variation in problem approaches:

- in certain cases, it will be necessary to think globally in order to act locally;
- in other cases, it will be necessary to leave local contexts to act globally (changing representations, reasoning modes and corporate culture).

Finally, innovation is also a knowledge creation process thanks to *action*. In certain cases, only using experience allows new concepts to be discovered: Firm F found a brand new product concept of educational equipment by listening to people criticizing one of its playground prototypes. The process of innovation joins in a dialectic between the certainties of experience and the uncertainties of novelty (Divry, 1998; March, 1991) illustrating the tension between the exploitation of the knowledge already acquired and the exploitation of new ways. In other words, the process of innovation activates a process of learning, which in return feeds innovation. It is similar to a process of creation, even re-creation of knowledge having several modes: socialization, outsourcing, combination, internalization (Nonaka, 1995). Vinck, 1999, states that the construction of a collective service is consubstantial in the construction of objects: innovation appears then through the process of realization of knowledge through action.

As a conclusion, innovation is the outcome of a collective intelligence, based on a mobilization of formal and validated knowledge. It requires an ability to experiment. Innovation thus appears from the interaction between various cognitive territories and a representation shared through action and consists in a stream evolving by construction.

2. Toward a Constructivist Approach to Technological Innovation Piloting

2.1. Assets and Limitations of Project Management Practices

In this section, focusing on the confrontation between innovation management (all the activities promoting innovation within a company) and project management (methods and tools supposed to be adapted to any project), we will refer to the AFITEP recommendations. Founded in 1984, this French industrial association formalizes methodological tools and abilities acquisition processes in the fields of planning, estimation and cost control. Most of its recommendations has been taken into account in the definition of norms by the French Bureau of Certification AFNOR: the X50-400 norm on terminology, the X50-105 and ISO 10006 norms on concepts and the X50-107 norm on the certification method. Historically, the main application fields of AFITEP notions in France were civil and process engineering.

To describe the assets and limitations of project management practices in France when projects are characterized by a high innovativeness level, four factors are considered:

- the objectives and goals;
- the approaches;
- the tools;
- the competences.

Objectives of Project and Innovation Management

Project management is not recent as it was applied within cathedral building during the 12th century. This know-how was the result of a long learning process called 'compagnonnage': a system of training apprentices. The first formalized methods, including PERT, appeared during the 19th century.

Today project management goals can be clearly summed up: to optimize financial results while transforming customer demand into a measurable object (a bridge or a building for example). Through projects characterized by a beginning and an end, companies try to achieve a triple objective: to control costs and set the agenda while improving performance.

Whereas project management focuses on profit, innovation highlights the concept of *value*. This is the consequence of international competition but also of methodological constraints. As a matter of fact, in the case of projects without innovative aspects, planned or referential budgets are generally defined during the early stages. Moreover, contracts with the customer guarantee any occasional over cost (O'Connel, 1999). On the other hand, as ideas and concepts move (some times radically) throughout the innovation process, notions such as risk instead of guarantee, and global impact measurement instead of expense control represent major top management concerns. Innovation is linked with multidimensional goals and as a result with multidimensional evaluation including variables such as: finance, competitiveness, notoriety, novelty and knowledge acquisition.

Approaches

Project management is highly influenced by the concept of a dominant operational logic focusing on a hierarchical succession of phases (Heerkens, 2000). Among others, approaches consist of: needs analysis, feasibility study, development, construction and closing (Bonner, 2002). Cost estimation is developed within study steps while cost control is carried out within realization phases. The end of a phase is specified by a marker and backwards tasks are avoided as much as possible.

On the other hand, innovation managers gain useful insight by using (Bück, 1999):

- a divergent process (enlarging the solution research area) in alternation with a convergent process (selecting more valuable solutions);
- iterative approaches: succession and repetition of steps including regular calling into question of outcomes;
- phases put in parallel.

Moreover, innovation is consistent with experimental approaches. Testing represents a major know-how improvement process. People involved in innovation face new problems, new scientific subjects. But, studies are not the only way to get the new required expertise. Some important information emerges only from a confrontation with actual production or selling tests.

Tools

Literature attests to the existence of many methodological tools in the field of project management. They allow profitability evaluation, market quantification, investment comparison and financial risk and resource planning. Nevertheless, in numerous innovation success cases, project management tools (information processing) lead to decisions different from those taken by top management. Entrepreneurs, when questioning the outcomes of formal analysis, applied the successful strategy: in France market studies were negative before the launching of the 'Espace Car', the future leader in the multipurpose vehicle market. Therefore, project management tools contain evident limitations within contexts characterized by newness. Among others, two explanations can be proposed:

- the reliability of information is often weak in the case of innovation (acceptability degree for example). Consequently, project management tools considered as data processing methods produce unreliable outcomes;
- project management tools describe static situations while innovation is a dynamic process. The launching of an innovative product modifies the characteristics of the environment of the company: for example the market size may be greater thanks to a new image of the product through innovation. The volume of chocolate eaten with coffee in France doubled as Firm M (see Appendix 1) proposed its innovation. Thus, managers lack references within their analysis: positive or negative correlation between the launching and the variables describing the environment being almost impossible to forecast.

This is particularly important when considering risk evaluation tools. Risk analysis approaches (Monte Carlo method for example), suggest that the best solution is linked with the weakest degree of risk. But, examples given previously show that this risk reduction policy presents strong limitations.

Competences

Project managers have a special interest in some basic abilities such as rigor, communication competences and technical skills (Tidd, 2000). Innovation highlights some complementary abilities. Creativity remains fundamental as innovation emerges from projects (Boylston, 2000). Curiosity is an asset to invest in approaches such as technological forecasting or benchmarking. Moreover all individual skills facilitating the cognitive phenomena are important.

In conclusion, project management methods and tools attest of strong limitations in the case of innovative projects. In the following section we will detail some of the characteristics of innovation piloting.

2.2. *Key Aspects of the Emerging Piloting Approach*

Adopting a Constructivist Approach

As described in the previous section, to enter into a process of innovation requires abandoning dynamics of reproducing facts in favor of the dynamic of creating value (Hamel, 2000). The consequences are a renewal and improvement of the ways of thinking, exchanging and working together. This evolution highlights the following needs:

- to intensify the way in which a given system is studied so that a global view may be arrived at;
- to seek a methodology to understand systems rather than to analyse or explain them.

Constructivist approaches tend to be consistent with these fundamental needs. The epistemologies developed by Simon (1974), Morin (1980) and more recently Lemoigne (1995) deal with individual modes of constructing mental approaches to reality, with the models created to read the world thereby rendering it more intelligible. As a consequence they encounter the problem of complexity and hence that of innovation processes.

Constructivist approaches are not based on the definition of strict objectives or provisional strategies or on the management of the subsequent tasks to achieve these goals. However, they enhance the description of a global development direction: what are the technologies the company tries to control, does it focus on services, what is its position in the sector, what are the priorities in terms of value acquisition? As individual or collective strategies integrate this general orientation, projects will emerge. Through constructivism, increased attention is directed to drawing up original solutions, collective acceptability, and a development process co-constructed by all the individuals involved. In Firm J, for example, one major task consisted in discussions within the company about this question: how to become a supplier of people piloting high technology equipment from a previous situation as a producer of cutting machines to be driven? From this, different projects arose aiming at the evolution of the product, the company image and its logistics organization. Few individual obstacles to changes could be observed as the future (rather than a diagnosis of previous performance) led the evolution. Coordination between different projects was facilitated thanks to the socialization element inherent in the development direction.

Constructivism lends importance to the process of knowledge production more than the discovery of stable knowledge. It admits both the constructed and constructive character of any competence created by man. One example is the case of Nippon Roche Company (Nonaka et al., this volume). Its 'Super Skill Transfer' consists in transferring to every medical representatives of the company some tacit knowledge developed by high-performing representatives. Constructivism is therefore an epistemology of invention production by doing. It seeks to invent, build, design and create projective knowledge and could be considered as a representation of phenomena creating meaning, designing the intelligible with reference to a project (Lemoigne, 1994).

Constructivists state that piloting innovation consists in achieving a project building management based on this idea: the future is not written, it has to be constructed through the interactions between individuals and groups of people. It aims at coherence when thinking global and acting local.

In practice, managers' attention is directed toward:

- the definition and diffusion of the global direction for the development of the company;
- the knowledge emergence or acquisition mode: how teams collect and process information, what are the reference models of people (what are the basic individual mental representations, do people develop several reasoning modes when faced with problems, are there common representations of the future within the team?), how individuals transmit new knowledge to the other members of the group, how to improve collective experience by analysing on going activities;
- the actions changing the borders in management practices: multidisciplinary teams, internal mobility, outsourcing design activities;
- the creation and stimulation of networks: co designing, partnership;

Development of a Value-Oriented Strategy

In France, several disciplines (economy, management, and engineering sciences) investigate the notion of value.

Our purpose is not to establish a complete overview of the notion of value. We shall quote only some fundamental elements. Economists measure value according to two references: the utility and the physical quantity. Two things are then considered: the costs of obtaining the necessary quantities of the product and the price that the consumer is ready to pay to acquire it.

Later, by taking into account the tastes and needs of consumers, the notion of 'value in use' appeared: value is bound to the reported needs/satisfied needs. The value of a product is specific to the company, its environment and the type of user: a cutting machine that works with a jet of water does not have the same value for a sawmill operator as for a micromechanics manufacturer.

Value is also personalized and characteristic of the industrial system in the same way as the technology. It is represented by different variables: financial margins, skills, technical objects . . . But also individual aspirations: the product also translates the motivations and the moral values of the individuals who are associated with its production. If the tasks relative to the production of a 'high tech' object motivate the staff, this object will have more value for them even if it generates less profit than a classical product.

In France, the value notion is generally defined as:

- financial: a profit created to allow the survival and\or the development of the company;
- strategic: an advantage over existing competitors or new competitors (an idea defended by Microsoft);
- intellectual: knowledge or new know-how which represents a possibility of future development: sale of licenses, potential reduction of costs (the value owned by Toyota in the 1990s with the development of the 'Just in Time' concept). We can also refer here to Nippon Roche experience (Nonaka et al., this volume);
- commercial: measured in market share;
- functional: the product supplies a supplementary service which is significant for the users (for example, the first multipurpose vehicle of the French car manufacturer, Renault);
- bound to the degree of newness: for the customers, the product appears innovative;
- bound to notoriety: value is attached to image as with Coca-cola or Nike;
- hedonist: it is the pleasure and motivation of people who work. Indeed, the craftsman/woman looks at his/her creation with pride;

The value of a product is therefore the result of these eight constituents. Thus, one can consider it as a spatial extension of the notion of performance. Value gives performance a pluridimensional character. It is objective, subjective and evolutionary over time.

If there is a real value, there is also a beneficial effect for the customer or the user, for the producer and for the various partners (suppliers, financial). The notion of value is thus relative to the receiver. It is evolutionary and conditioned by the context, which surrounds it.

In practice, innovation or new value creation thus supposes (Sarlemindj, 1988):

- the capacity to integrate different dimensions based on an 'organizing principle' called: the vision. This allows the understanding of the phenomena by synthesis or by simple analysis;
- the opening of possibilities which is translated by a greater number of configurations;
- access to a global vision;
- meaning to be given to individual and collective actions;
- increased openness and flexibility.

In fact, we note a development of value oriented strategies and a switch from a risk reduction objective to an opportunities capturing approach.

A Management Through Three Levers: Strategy, Piloting, and Sparking

All our observations lead to the question 'is there an integrative process for steering innovation?' This integrative process would be based on the following notions:

- a continuous value development dynamic relies on the integration of various physical processes (design, quality) through a steering process (tools, reasoning, organization) that is oriented towards the development of new values;
- changing the way of considering a situation or an object favors the emergence of new concepts (for example the TRIZ method);
- considering disturbances, uncertainty and paradoxical situations as opportunities constitutes a way to spark innovation;
- innovation is linked to the concept of the learning organization (Nembhard, 2000).

Using a systemic model designed by Lemoigne (1995), suggests the following description of the process of piloting innovation:

- the functional pole represents the production of the process: in the case of innovation, the steering process leads to the design of a new industrial activity and its launching. To sum up, we propose the *confirmed strategy* as a result;
- the ontological pole describes the very nature of the process: in the case of innovation, the steering process is made up of reasoning activities and actions. Intellectual operations and human actions make innovation happen. That is the *sparking* pole;
- the genetic pole is represented by basic evolution mechanisms: in the case of innovation, the steering process is controlled through project management. More precisely, this integrates methodological tools, organization, indicators and all aspects of evaluation. This constitutes the *piloting* tasks.

This *strategy/sparking/piloting* model seems helpful for a better understanding of the global constructivist supervision of innovation.

In fact, the systemic approach suggests a two level model (Fig. 5). Each of the three poles can be defined as a *strategy/sparking/piloting* submode.

Considering the *strategy* pole, for example, the following elements may be described:

- *the sparking of strategy*: all tasks stimulating or being a resource to strategy design. This includes: technological forecasting, benchmarking;
- *the piloting of strategy*: all variables structuring the definition of the development strategy (Gavidarayan et al., 2001). This includes: the use of grids (Meyer grids for example) or structured approaches (Porters Value chain for example);

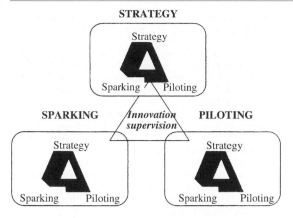

STRATEGY

SPARKING   *Innovation supervision*   **PILOTING**

Figure 5. A nine pole systemic representation of the global supervision of innovation.

• *the strategy of strategy*: the very fundamental decisions concerning the global evolution of the company. Among others: the basic techniques the company invests in, know how to buy.

This model is useful to determine the possible impact of decisions or actions on innovation. It captures a broad spectrum of tasks: technical studies, organizational changes, management practices. It is also consistent with a value improvement objective and/or a new value nature strategy. Thus, managers can analyze the construction and the influence of their decisions and actions on projects. As a result, this model facilitates the constructivist management of innovation. It ensures a certain level of consciousness about the global evolution of value variables throughout the project. It also enables researchers to observe and define the position of any innovation phenomenon thanks to the nine poles.

Taking the example of technological forecasting into account, efficiency depends on:

• *the sparking of strategy*: is the information gathered really at the origin of new project ideas and of successful strategic decisions?
• *the piloting of strategy*: is technological forecasting a way to organize strategic design by structuring information and inducing an agenda?
• *the piloting of sparking*: as a consequence of collecting external information, do people involved in innovation develop their reasoning modes, enrich their mental representations, become more open to change and more creative?

Our team is carrying out on-going research in France to validate this model and to precisely define its strengths and limitations.

Conclusion

The ways in which specialists are considering innovation are numerous and very diversified. And yet is this not the direct consequence of the intrinsic nature of the innovation process?

Of course, one can argue that research in this field is recent, but a consensus seems to be emerging concerning technological innovation. Technical innovation can be considered as:

• complex;
• uncertain;
• a process of technical, cognitive and social transformations;
• formal or not.

Innovation can be considered through various references scales: structures (production or services organizations), teams or individuals. One can also consider all the company's activities or just the innovative projects. Furthermore, the way of acting is a key factor in a process of innovation. We saw that in many cases, new concepts emerge from action, resulting from individual or collective experience.

Finally, an ideal innovation process does not exist. Each technological transformation is characteristic of the project, the finality of the firm, its environment, and the period studied. When one compares the evolution of two projects in an industrial system, one can notice strong differences in most cases. One of the reasons is that the piloting of innovation results from 'engineering', that is to say a capacity to adapt basic knowledge in order to lead a project (for example, the development of an innovative project and/or an innovative industrial system). This leads to the development of new competences and so to new jobs in order to respond to this new way of acting.

The constructivist approach to innovation management shows that the success of an innovative project is based on an organization and piloting methodologies, adapted well through out. Furthermore, one of the key factors of success is located on the team project quality. In this way, a new element emerges: the 'integrating agent', whose role is to help people to develop new ways of acting, the opening up toward a culture of relationships and diversity. This highlights the need for new competences, such as:

• To be able to imagine the possible in the impossible;
• To be able to concretize the 'thing' by passing from 'real' to 'possible' (i.e. the idea), and to return from 'possible' to 'real' (i.e. the action).

Perhaps this is more a question of leading individuals toward the discovery of new possibilities, making them aware that there is always another way of posing a question or a problem and thereby, considering the constraints involved in a different light. This is a real new turn of spirit which may be developed in order to promote a new kind of manager: 'business entrepreneurs', capable of:

- anticipation;
- curiosity;
- facing risk;
- managing uncertainty and the irrational.

References

Afitep (1991). *Le management de projet: principes et pratiques*. Collection gestion. Paris: Afnor édition.

Afnor (1994). *Recueil de normes françaises de Management de projet—Qualité et efficacité des organisations*. Paris: Afnor édition.

Afnor (1988). Recommandations pour obtenir et assurer la qualité en conception. *Norme X50–127*. Paris: Afnor édition.

Beaudoin, P. (1984). *La gestion de projets: aspects stratégiques*. Montréal: Editions agence d'Arc inc.

Benghozi, R. & Charue-Duboc, L. (2001). *Innovation based competition and design system dynamics*. Paris: Harmattan edition.

Bernard, J. & Torre, A. (1994). Les dynamiques d'innovation et de R-D des PMI françaises. *Revue Internationale P.M.E.*, **7** (3–4), 19–40.

Boylston, F. (2000). *Creative solutions for unusual projects*. FW publications.

Boly, V., Morel, L. & Renaud, J. (2000). Innovation in low tech SMB'S: evidence of a necessary constructivist approach. *Technovation*, **20**, 161–169.

Bonner, J. M., Ruckert, R. W. & Walker, O. C. (2002). Upper management control of new product development projects and projects performance. *Journal of Product Innovation Management*, **19**, 233–246.

Bück, J. Y. (1999). *Le management des connaissances—Mettre en oeuvre un projet de knowledge management*. Paris: Editions d'Organisation.

Buckler, S. & Zien, K. (1996). From experience, the spirituality of innovation: Learning from stories. *Journal of Product Innovation Management*, **13**, 391–405.

Debaig, M. & Huete, L. M. (1999). *Le paradoxe de la grenouille*. Paris: Dunod.

Divry, C. (1998). Compétences et formes d'apprentissage: pour une approche dynamique d'innovation. *Revue Française de Gestion*, **21**, 454–464.

Garcia, R. & Calantone, R. (2002). A critical look at technological innovation typology an innovativeness terminology: a literature review. *Journal of Product Innovation Management*, **19**, 110–133.

Gavidarayan, V. & Gupta, A. (2001). Strategic innovation: a conceptual road map. *Business Horizons*, August, 3/12.

Griffin, A. (1997). PDMA research new product development practices, updating trend and benchmarking best practices. *Journal of Product Innovation Management*, **14**, 429–459.

Hamel, G. (2000). *Leading the revolution*. Boston, MA: Harvard Business School Press.

Heerkens, M. (2000). *Project management*. McGraw Hill.

Iansiti, M. & West, J. (1997). Technology and integration: turning great research into great products. *Harvard Business Review*, **75**, 69–79.

Karlsson, C. & Ahlstrom, P. (1997). Perspective changing product development strategy, a managerial challenge. *Journal of Product Innovation Management*, **14**, 473–484.

Lemoigne, J. L. (1995). *Les épistémologies constructivistes*. Paris: PUF Edition.

Lemoigne, J. L. (1977). *La théorie du système général—Théorie de la modélisation*. Paris: P.U.F. Edition.

Lemoigne, J. L. (1994). *Le constructivisme. Tome 1: Des fondements*. Paris: ESF Editeur.

Longnecker, C. & Mitchell, N. (2000). Barriers and gateways to management cooperation and teamwork. *Business Horizons*, September-October, 37–44.

Mack, M. (1997). *Co-évolution, dynamique créatrice*. Paris: Editions Village Mondial.

March, J. G. (1991). Exploration and exploitation in organizational learning. *Organization Science*, **2**, 75–85.

Meyer, M. (1997). Revitalize your product times through continuous platform renewal. *Research Technology Management*, March/April, 17–28.

Michell, J. & Singh, W. (1996). Survival of businesses using collaborative relationships to commercialize complex goods. *Strategic Management Journal*, **16**, 169–195.

Ministère de l'économie/Digitip (2001). *L'innovation en chiffres*. Paris: Ministère de l'Économie, des Finances et de l'Industrie.

Nembhard, D. A. & Uzumeri, M. V. (2000). An individual-based description of learning within an organisation. *IEEE Transactions on Engineering Management*, **47**, 370–379.

Nonaka, I. (1994). A dynamic theory of organizational knowledge creation. *Organisation Science*, **5**, 67–77.

O'Connel, F. (1999). *How to run successful projects*. Pearson edition.

Pascale, R. T. (1999). Surfing the edge of chaos. *Sloan Management Review*, **40**, 83–94.

Porter, M. (1986). *L'avantage concurrentiel*. Paris: Inter-éditions.

Sarlemindj, A. & Kroes, P. (1988). Technological analogies and their logical nature. In: P. T. Durbin (Ed.), *Technology and Contemporary Life*. Dordreicht: Reidel Edition.

Sessi (1999). *L'état des PMI*. Paris: Ministère de l'Économie, des Finances et de l'Industrie.

Simon, S. & March, P. (1974). *Les organisations*. Paris: Dunod.

Song, S. M. & Parry, M. E. (1996). What separates Japanese new product winners from losers. *Journal of Product Innovation Management*, **13**, 422–439.

The Minessota Studies (2000). *Research on the management of innovation*. A. H. Van de Ven & H. L. Angle (Eds). New York: Oxford University Press.

Tidd, J. (2000). *From knowledge management to strategic competence: measuring technological and organisational innovation*. London: Imperial College Press.

Vinck, D. (1999). *Ingénieurs au quotidien*. Paris: P.U.F.

Appendix 1: Panel of Firms

Figm	Activity
FIRM A	Industrial activity is based on an invented process transforming wood waste into insulation. FIRM A manages an experimental production unit and a subsidiary involved in technology trading. Both activities represent twenty employees and an international technology user network working under license.
FIRM B	SME working in the electronics sector. Their new product is a beverage vending machine. FIRM B invests in trading and vending machine stock operating whereas machine production is totally undertaken by external partners.
FIRM C	In the mechanical sector, FIRM C is the largest of our panel : 250 employees, producing turbo-compressors for international automotive customers.
FIRM D	Top management is committed to the launching of the large and innovative product family of the 21st century. The 200 employees of FIRM D manufacture towing trailer trucks.
FIRM E	Here, the business is design, assembly, selling and maintenance of workshop doors. Customers are manufacturing companies as well as supermarkets and hospitals. FIRM E numbers 35 people.
FIRM F	In the wood sector, this firm of 10 persons produces outdoor community playgrounds. The originality of its catalogue is its major competitive advantage.
FIRM G	FIRM G and FIRM E have the same shareholders. Firm G (25 people) produces and installs professional partitions. Top management has targeted an innovative product range.
FIRM H	Firm H (20 employees) is in the furniture sector. Their new product reflects radical innovation in the sports equipment sector.
FIRM I	The 150 employees of FIRM I manufacture agricultural machine components. FIRM I is involved in a continuous innovation strategy (new surface treatment process, new materials, . . .)
FIRM J	Firm J has 80 employees and a national agency network. The central structure is in charge of the design and assembly of road verge cutting machines while agencies sell and maintain these machines.
FIRM K	Firm K (50 people) innovates by integrating the concepts of ergonomy and biomechanics in the design of its products: manholes.
FIRM L	Firm L produces cheese and is involved in the renewal of its product range (45 people).
FIRM M	The confectionary market is changing. The traditional products of Firm L are being replaced by products based on chocolate eaten daily with coffee.

The International Handbook on Innovation
Edited by Larisa V. Shavinina

Promotors and Champions in Innovations: Development of a Research Paradigm

Jürgen Hauschildt

Institute for Research in Innovation Management, University of Kiel, Germany

Abstract: The success of innovations is to a great extent dependent upon the activities and the abilities of individuals who enthusiastically support the new product or process. In the Anglo-Saxon world these persons are called 'champions', in Europe the term 'promotors' (in the Latin version) is in use. This chapter presents 30 years of research in this field on both sides of the Atlantic. To date, we know for sure that a specific division of labor between three types of promotors is the most effective and efficient constellation. First, innovations need a technical expert who acts as 'promotor by expertise'. Second, innovations need top management's sponsorship by a 'power promotor'. Third, innovations need boundary spanning skills of a 'process promotor'. The 'troika'-model itself is influenced by the size and the diversification of the company and by the complexity and newness of the innovation.

Keywords: Innovation; Champions; Promoters; 'Troika' model.

Phases of Development

Phase 1: The Discovery of the Champion

Josef Schumpeter can be credited with being the first to draw attention towards the *entrepreneur* in innovation processes, in a book published in 1912 (Theory of Economic Development, Leipzig). The entrepreneur creates new combinations, on a discontinuous basis, in totally new forms, in an act of creative destruction. He brings forth new products, introduces new production methods, opens up new markets, conquers new sources of supply or reorganizes (Schumpeter, 1931, p. 100). The 'dynamic entrepreneur' was thus characterized and described. Apparently, this was sufficient to incorporate him into economic models. To understand him as a real person or even to analyze him in further detail seemed superfluous.

That changed when Schon (1963) introduced a new term for this creative individual: the *'champion'*. The term which has dominated discussion to date was thereby established for the Anglo-Saxon countries. In contrast, in the German-speaking countries this term was not accepted because of a slightly negative connotation.

About ten years later, this precedent phase came to an end, when almost simultaneously in three different parts of the world, mutually independent studies focused on individuals in innovations in empirical investigations. They confirmed that committed and enthusiastic persons play a decisive role in promoting innovations:

- in *Germany*[1] in 1973, Witte (1973) investigated the first procurement of computers. In his survey he proved that the existence of 'promotors'—as he termed the champions—led to significantly higher levels of innovation and of activity than found for processes in which such individuals were absent ('COLUMBUS' Project);
- in the *USA* in 1974, Chakrabarti (1974) discovered that product champions could be found chiefly in successful cases in the further development of NASA innovations ('NASA Study');
- in *England* in 1974, Rothwell (1974) and his team, conducting research into innovations in chemical processes and scientific instruments, found that the human factor was a key determinant for the success of innovation (SAPPHO Project).

---

[1] For the development of German empirical research on promotors in the last 25 years see Hauschildt & Gemünden (1998).

The breakthrough was thus achieved: the importance of the human factor was established beyond doubt. Many research projects, particularly by Howell & Higgins (1990), have confirmed over and over again that identifying the champions or promotors in innovation processes is not a great problem. They normally stand out clearly because of their original contributions and/or because they make quite deliberate use of their power to push the innovation process. Consequently, it is easy to identify the active individuals in the innovation processes. *Champions are no longer merely literary figures, but empirically observable individuals who can be described using suitable quantification conventions and who are clearly successful.*

Phase 2: Confusion

The three seminal studies were not the only ones, but they were the ones which dealt most clearly with the human side of innovation. Other studies published in the early 1970s by Rogers & Shoemaker (1971), Langrish et al. (1972), Globe et al. (1973), and Havelock (1973) should be mentioned. As a rule, these studies identify more than one outstanding individual simultaneously present in innovation processes. The single champion, however, is the exception. Different terms were thought up to distinguish these committed individuals from another. The initial consequence was a confusing *variety of terms* in the literature (Chakrabarti & Hauschildt, 1989), often colored by the language used in normal practice. The following is a small selection:

> Inventor, initiator, stimulator, legitimizer, decision-maker, executor, catalyst, solution giver, process helper, resource linker, technical innovator, product champion, business innovator, chief executive, technology promotor, power promotor.

This plethora of terms for those who actively promote innovation processes has even increased since then. More terms can also be found in publications being addressed to practitioners:

> Political coordinator, information coordinator, resource coordinator, market coordinator, management champion, decider, planner, user, doer, expert, person affected, process promotor, relationship promotor.

Roberts & Fusfeld (1981) did not just describe this variety of terms, they also had them caricatured in a particularly impressive manner.

But what was the outcome?

On the *negative side*, complaints included confusion, lack of clarity, redundancy of terms and encouragement of different schools of thought. Researchers are not exempt from the ambition to establish 'their' terms and to provide evidence that the distinctions they have

selected are particularly useful either for further research or even for direct application.

On the *positive side*, variety became apparent. The fact that a different number of individuals with different functions would be found in different innovation processes was established as a certainty. As a result, the question of the cause and effect of these differences could be raised.

Phase 3: Order

Alok Chakrabarti & Jürgen Hauschildt stood amazed in front of this bewildering variety. They saw it as their first task to establish order so that further research could follow a systematic concept. This concept had to take into account the two functions of an organization: first, to efficiently regulate the *work* to be done, second, to effectively regulate the *power* relationships between the incumbents. We took the terms for those engaged in innovation processes to express certain *activities* performed by them during the process, or certain *power bases* from which they derived their influence on these innovation processes. This produced the following twofold distinction by contributions and power bases (see Table 1).

Phase 4: Explanation of Success

(1) The first studies on the champions in the Anglo-Saxon countries differed from the German promotor studies in one very significant point: in the publications of Rothwell et al. (1974) and Chakrabarti (1974), the *paired comparison approach* was used. The intention was to identify the characteristics which distinguished successful cases from unsuccessful ones. This approach was undoubtedly focused, drawing on the basis of everyday experience and academic topologies, but was not driven by concise theory and did not test a theory in the strict sense.

This is the most prominent difference from the German research conducted under Witte (1977, 47 ff.). He first of all developed a *theoretical concept*, which explains why the presence of promotors improves the success of the innovation process. Witte worked with the hypothetical construct of *barriers*: resistance to the innovation is due to the barrier of *ignorance* and due to the barrier of *unwillingness*. Promotors commit enthusiastically to the innovation and help to overcome these barriers. Witte's promotor model contains three core theorems:

(1) each type of resistance has to be overcome by a specific type of energy. The barrier of unwillingness is overcome by *hierarchical potential*, the barrier of ignorance is overcome by the use of *specific knowledge in a certain technical field* (correspondency theorem);

(2) these types of energy are provided by different people. The *power promotor* ('Machtpromotor') contributes resources and hierarchical potential and the *technology promotor* ('Fachpromotor')

Table 1. Roles in innovation management.

Activities and roles in innovation management	
Activities	Roles in innovation management
1. Initiation of innovative process	Initiator, catalyst, stimulator
2. Development of a solution	Solution finder, solution giver, idea generator, information source
3. Process management	Process helper, connector, resource linker, idea facilitator, orchestrator
4. Decision making	Decision maker, legitimizer
5. Implementation	Realizer, executor

Source: Hauschildt & Chakrabarti (1988, p. 383)

Power bases and roles in innovation management	
Power bases	Roles in innovation management
1. Knowledge specialty	Technology promotor, technical innovator, technologist, inventor
2. Hierarchical potential	Power promotor, chief executive, executive champion
3. Control of resources	Business innovator, investor, entrepreneur, sponsor
4. Organizational know-how and communication potential	Process promotor, product champion, project champion
5. Network know-how and potential for interaction	Relationship promotor

Source: Chakrabarti & Hauschildt (1989, pp. 165–166); Gemünden & Walter (1995, 973 ff.)

contributes specific technical knowledge to the innovation process (theorem of division of labor);

(3) the innovation process is successful when the power promotor and technology promotor form a *coalition* and are *well coordinated*, i.e. when they really co-operate (theorem of team-interaction).

The promotor model is thus based on the specific use of *power bases*. In addition, however, close cooperation between the promotors is also important. Witte chose the term 'tandem structure' (or 'dyad') for this, in the sense of two horses harnessed to a carriage in tandem.

Using a sample of 233 initial acquisitions (by purchase or lease) of computers, the empirical test showed that not only were much more innovative solutions found, but that the work also proceeded much faster and with greater diligence in those cases where such a tandem structure was present (Witte, 1973).

(2) It is undoubtedly true to say that one significant contribution made by promotors lies in overcoming resistance to an innovation. However, this assumption is also a target of criticism of Witte's concept: the promotors do more than just cope with conflicts. This is particularly true when the opposition, overall, has a loyal attitude, as the findings of Markham et al. (1991) prove. The original promotor model was in essence, a conflict-handling model. However, this view distracts from the informative and creative aspects of innovations. After all, innovations are particularly characterized by the fact that information is newly generated and/or a combination of them. Furthermore,

innovations are processes of problem definition, goal formation, generation and identification of new combinations. When Witte's model was developed, these *cognitive tasks* were given less consideration than the *conflict-handling functions* of the promotors. The cognitive tasks could probably supply a different theoretical base for the interaction among the promotors.

The research by Ancona & Caldwell (1992) went into this *interaction of cognitive and conflict-handling activities by the champions* in more detail. The Ancona/Caldwell study determines four characteristic areas of activity by factor analysis.

- 'Ambassadorial activities': Formation of goals and blocking of opposition, above all conflict-handling activities.
- 'Task coordinator activities': Coordination, negotiation and interface management, also basically conflict handling.
- 'Scouting activities': Obtaining information, building expertise, seeking solutions, clearly cognitive activities.
- 'Guard activities': Prevention of an undesirable leak of ideas and information, activities which are not covered by our concept.

This seems to us to provide sufficient evidence of the cognitive contributions of the promotors. The comprehensive model explaining the human influence on the innovation process must definitely combine cognitive and conflict-handling activities.

Phase 5: Systematic Differentiation of the Division of Labor in Contingency Models

(1) The variety of the incumbents which emerged in the wake of the Witte, Rothwell & Chakrabarti studies raised two questions:

• 'What effect does such variety have?'
• 'What determines it?'

The traditional *contingency view* of organizational theory could thus also be applied to innovation management.

The next question was:

• How do external circumstances affect the number of promotors and, the way in which they approach the division of labor?

(2) Witte's (1973, 1977) findings had already indicated that the division of labor between the technology promotor and the power promotor was clearly a phenomenon of *firm size*. Rothwell (1974) and his research team proved that the *industry* is a determinant of division of labor. The *degree of innovativeness* and the *degree of diffusion* of the innovative products or processes also influence the division of labor. With the increasing diffusion of the innovation in an economy, the importance of the technology promotor declines. Maidique (1980) arrived at similar results.

If we put these findings together, we find *two influences superimposed* which are important for the division of labor in innovation management: *system complexity and problem complexity*. The resulting *overall complexity* has to be compared with the capability of the active individuals. If there is a considerable discrepancy between the complexity of the innovation and personal innovative capability, the two center model of division of labor has to be modified. This assumption was confirmed by Lechler (1997) in his survey of 448 projects.

Maidique (1980) dealt with the type of instance where more extensive division of labor becomes an inevitable result of *high system complexity*. In the simplest case, in a small, entrepreneurial company, a *two-center constellation* of technology promotor and power promotor is found. According to Maidique, a *three-center constellation* is typical in medium-sized companies with a functional structure which are still limited to one product line. A *four-center constellation* is to be found in very large, diversified companies.

(3) The many and varied completed research projects which have been analyzed by Hauschildt & Chakrabarti (1988) contained many references to three-center constellations. To find an explanation, we applied the complexity concept, based our work on Witte's concept and identified a third species of promotor: the *process promotor*. Process promotors are needed when innovations affect a very large number of individuals

personally, in relatively large institutions, and trigger conflicts. Like other promotors, process promotors rely on specific power bases: *on system know-how, organizational and planning power, and on interactive skills*. They, too, overcome characteristic forms of resistance: those of an established organization whose aim is to execute routine procedures as efficiently as possible and which rejects innovations as a disruption of its smooth running. Process promotors do not have the formal authority of the power promotor or the expertise of the technology promotor. They rely on leadership qualities and influencing tactics, and like the other two promotors they are characterized by the fact that they take risk and are prepared to sink or swim with the innovation. The study of Howell & Higgins (1990) demonstrated this side of the process champion in particular. This was confirmed for the German-speaking countries: the successful 'interactive project managers' are particularly characterized by a high level of interactive skills, cooperative leadership, above-average problem-solving capabilities and constructive creativity (Medcof, Hauschildt & Keim, 2000).

Building on Witte's concept of the 'tandem structure' we call the team of three the '*troika' of power, process and technology promotor*. In a study of 133 innovation projects in the mechanical engineering industry, we found that this troika structure achieves better technical results, but above all, better economic results, than any other structure (Hauschildt & Kirchmann 2001). These findings correspond with the investigation of Lechler (1997).

In a case research study of ten major innovations, Folkerts (2001) presents new perspectives of the promotor model. Using a role concept, she interprets the promotors as clusters of functions and not necessarily as individuals. She finds the three clusters, i.e. the types of promotors, that are present in each stage of the innovation process. But their roles can be fulfilled by different persons with changing contributions.

Figure 1 summarizes the conflict handling and cognitive activities in the division of labor between the promotors in the troika structure.

(4) Finally, Gemünden & Walter (1998) indicate a further modification of the troika concept: they point out that more and more innovations require *cooperation with external partners* in the value-chain, i.e. with customers or suppliers. There are barriers in this cooperation, too. Just as promotors are needed to eliminate in-house barriers, they are also called upon to overcome extramural barriers of interaction. The barrier concept is developed similarly to Witte's theoretical approach. In place of the process promotor, who overcomes in-house barriers only, Gemünden & Walter's concept includes the '*relationship promotor*'. In a study of 94 technology transfer projects, they proved that processes are more successful if a person is present who deliberately establishes and maintains relationships with the partners.

In summary, research initially proved that an active, committed champion was the most important factor for success in the management of innovations. However, at the same time a variety of other persons were observed who were also striving to make the innovation successful. These individuals could be distinguished by their contributions and their power bases. The successful impact of these individuals is due to their skills in dealing with conflicts constructively and handling information creatively. Finally, we know that the extent and type of the division of labor among these individuals is determined by the complexity of the innovation problem to be solved and by the complexity of the organization concerned. Accordingly, the troika structure consisting of power promotor, process promotor and technology promotor in particular, is the most successful structure for the management of typical innovation projects within a company. It is possible that the increase in extramural innovation activity will shift the role of the process promotor more towards that of a relationship promotor.

Possible Routes for Future Research

As we ask: "How is the research process likely to move forward?", we are entering the realms of science fiction. One can see four routes in future, three of which are theoretical, addressing the problem of *explanation*. The fourth route is directed at the

application of the promotor or champion concept in practice. Let us first of all go down the theoretical routes:

Route 1: Further Details on the Promotor Model

Any serious academic will have no great difficulty in spontaneously reeling off a list of questions which the current body of research cannot answer, or cannot answer satisfactorily. It follows logically that a whole generation of academics can be occupied with identifying and proving further details of the promotor model. We see the following as the most pressing questions:

- Which *key events* stimulate individuals to act as promotors of innovations?
- *How do promotors come together?* The fuzzy front end of innovation is normally lost in mystic obscurity. This phase, when the promotors come together, is a stage which can obviously only be described in social and psychological categories.
- And once these promotors do actually encounter one another—how is the *personal fit* determined and secured? A good fit is essential for the subsequent innovation project to come to a successful end with all its difficulties, and to get it completed in the face of all resistance. Such teams need considerable group cohesion in order to withstand all the pressures from outside.

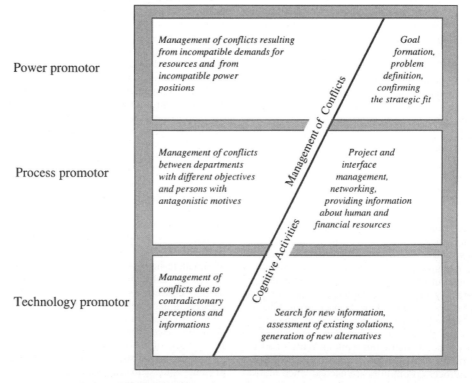

Figure 1. Management of conflicts and of cognitive activities in the troika structure.

- We know very little about the conditions under which promotor structures are *dissolved*. Even if a promising promotor team comes on the scene at the beginning of a process, it is by no means certain that it will see the process through and complete it successfully. On what reefs might the tandem or troika founder?
- The interaction of the promotors and champions is couched in somewhat mysterious terms as '*good cooperation*'. What does that mean specifically? We know very little about whether and how the individual promotors have to take a hand during the innovation process. Do they play changing roles? Do they always appear as a team? What contribution is absolutely essential for which key occurrences?
- Does the promotor model apply *regardless of time and space*? Organizations have changed in the last 20 years: they have become more open, more tolerant of conflict, more process-related, more project-orientated, more targeted, more risk-aware, more informative, and more cooperative. New forms of organization have developed. As a result, the types of resistance have changed. Does the promotor model have to be adapted to these organizational developments?

Let us stop here. The trend is obvious: the deeper one delves into the concepts of leadership and management, the more questions about the details and development of the promotor structures will arise.

Route 2: Extension of the Promotor Model

(1) The promotor model is based on the concept of resistance. Only vague theoretical concepts filed under 'brakers and drivers' or 'devil's advocates' or 'loyal opposition' are currently available to describe the people who embody such resistance. The opposition model by Witte (1973), Chakrabarti & Hauschildt (1988), and Gemünden & Walter (1998) objectivizes the resistance but does not personalize it. It should be asked how *opponent structures* are formed and behave—in accordance with the development of promotor structures. Can an innovative solution with a successful overall outcome be explained through the dialectic of promotors and opponents?

(2) A second corresponding model refers to the firm's partners in the innovation process. If we accept the premise that more and more innovations will be inter-organizational processes, the question of *corresponding promotor structures* arises: The promotor and champion constellation of a firm could be envisaged in that of the cooperating partners, like a mirror image. The hypothesis would be that the success of the inter-organizational depends upon corresponding promotor structures.

Route 3: Focus on Supporting Instruments

Research to date has been based on the tacit assumption that promotors or champions alone determine the success of innovations, without additional supporting instruments or backing. That is, of course, not the case. Actually, success does also depend on other means, which are certainly not neglected by research. However, we know little about the reinforcing or attenuating effects with regard to the human constellation:

- Informal *information and communication* is quite certainly a major factor for success in innovation. And it is also definitely true that promotors quite clearly tend towards an informal interactive and communication style. To this extent, the two effects seem to reinforce one another. But what does informal information and communication mean in the age of electronic media? What type of informality is expected and useful? Is formal information obsolete or even disadvantageous?
- Promotors and champions are active 'temporarily', for a brief period. They have the end of the innovation process in mind. What awaits them then? On the surface, this concerns the question of *incentives*, of rewards, of penalties, of all types of sanctions that firms hold in reserve for successful or unsuccessful managers. The problem is so acute because promotors and champions generally do not only commit themselves because of extrinsic drivers, but intrinsically. They get involved, they commit, they are not called in or appointed. How does a firm react to such self-appointed activists? How does it deal with failing or unsuccessful innovators?
- What role does the change in *organizational or corporate culture* play in the readiness to participate and the success of promotors and champions in innovation processes? Even if we no longer accept the classical dichotomy of 'mechanistic' and 'organic' organization culture, Burns & Stalker (1961) nevertheless show that there is a problem: the more mechanistic an organizational culture is, the more important power promotors are. The more organic it is, the more important process and technology promotors are. The forms of organizational change mentioned above tend more to indicate that organic forms are gaining in importance. Will the role of the power promotor become obsolete?

These three routes may initially be significant more from the theoretical point of view—but we have now learned that there is nothing so practical as a good theory which explains and forecasts reality. To this extent, the contrast between theory and practice—or in our interpretation, between explaining and doing—is much less important than claimed so often.

Route 4: Focus on Application

(1) Innovations are projects, but not all projects are innovations. How far can the practically orientated proposals of *project management* be used in innovation processes? The following striking point emerges from

an analysis of the literature on project management and on innovation management:

- The project management literature emphasizes formal organizational tools for project support, such as matrix management, project controlling, network planning, cost control, information management. In contrast, it devotes much less attention to the human aspects of project management.
- The literature on innovation is quite different. Here, considerable skepticism prevails about formal organizational tools, while at the same time the human perspective on the management of innovation processes is emphasized.

The following question thus arises: Can the domains and overlapping areas of project management and innovation management (promotor concept) be more sharply defined?

An analysis of the available research results prompts the following conclusions:

- Quite clearly, the *degree of innovativeness or the complexity of the innovation problem* is of major importance to the human management of the process. Lechler's (1997) research findings show that, in particular, 'strategic' projects of high complexity with a high degree of innovativeness can be progressed successfully by a troika constellation. It is notable that the formal organizational coordination tools of participation, planning, control, information and communication play a minor role. The promotor concept can be recognized very clearly here: the troika of promotors substitutes for formal coordination;
- The *basic notion* by which the innovation is driven is important: if the *innovation is driven by its end* (demand pull), with new technologies being sought to meet known objectives, this is the domain of the project management concept. However, in *means-driven innovation* (technology push), a known technology is available for which a completely new application is being sought. In this case, the innovators must free themselves of the earlier constraining ties and relationships. They are much more dependent on spontaneous ideas and ad hoc creativity. Collaboration with customers or other external partners in the innovation process play a more important role. Means-driven innovation is the domain of the promotor concept;
- A third influence on the application of certain process management models undoubtedly are the *phases of the innovation process*. Innovation processes generally have a relatively long, relatively fuzzy front end phase in which the problem has to be defined and the objectives have to be generated. According to all the findings available, this phase seems to be a domain of the promotor and champion

concept. After completion of this definition phase it is possible to think about transferring the problem into a 'project', i.e. of institutionalizing it, setting a time frame on it, scheduling it, giving it accountability and responsibility, a formal structure for interaction. Empirical findings advise a certain amount of caution here, since the leap from the 'loose' to the more 'tight' phase is not at all clearly mapped out. Nevertheless, it would be wrong to approach innovative projects from the start with the toolkit of project management, developed for routine situations. It would also be wrong to practice the full openness of self-management in the realization phases, which would allow many sections of the process to be repeated.

(2) A further practical question arises in the light of the frequent observation that champions or promotors occur 'spontaneously' and that their emergence is not amenable to organizational intervention. At first glance, it seems that we have to inquire resignedly whether this is a question which may satisfy our intellectual interest in explanations, but not our practical interest in 'doing'.

It is obvious that the cooperation of promotors or champions cannot be obtained by force. But it can be facilitated. This calls for opportunities or nurturing conditions which improve the chances that these creative spirits will get together. Thus, we are propagating the *idea of 'meeting points'*: a firm should create opportunities for those people who show an enthusiastic interest in a certain technological or market-specific segment to meet, to become acquainted, to evaluate and appreciate one another. We are thus putting in a plea for meeting places, for informal opportunities for communication, for regular, institutionalized, open and non-hierarchical meetings.

Perspectives

Effectivity and effectiveness of innovations are, and will be, dependent upon:

(1) whether or not the promotors will commit themselves to the innovation;
(2) how individually the promotors will react to changing opposition;
(3) how early the promotors will assume their roles and their responsibilities in a conscious division of labor;
(4) whether or not the promotors have some experiences in the management of innovations—even in other roles;
(5) whether or not the promotors are willing and able to make themselves familiar with the domains of the other promotors;
(6) whether or not one of the promotors will accept the additional role of a 'relations promotor' in case of external cooperations;

(7) whether or not the promotors will be supported by the innovation culture of the firm.

References

Ancona, D. G. & Caldwell, D. F. (1992). Bridging the boundary: External activity and performance in organizational teams. *Administrative Science Quarterly*, **37**, 634–665.

Burns, T. & Stalker, G. M. (1961). *The management of innovation*. London: Tavistock Publishing.

Chakrabarti, A. K. (1974). The role of champion in product innovation. *California Management Review*, **XVII**, 58–62.

Chakrabarti, A. K. & Hauschildt, J. (1989). The division of labor in innovation management. *R&D Management*, **19**, 161–171.

Folkerts, L. (2001). *Promotoren in Innovationsprozessen—empirische Untersuchung zur personellen Dynamik*. Wiesbaden: DUV.

Gemünden, H. G. & Walter, A. (1998). Beziehungspromotoren—Schlüsselpersonen für zwischenbetriebliche Innovationsprozesse. In: J. Hauschildt & H. G. Gemünden (Eds), *Promotoren—Champions der Innovation*. Wiesbaden: Gabler.

Globe, S., Levy, G. W. & Schwartz, C. M. (1973). Key factors and events in the innovation process. *Research Management*, **XVI**, 8–15.

Hauschildt, J. & Chakrabarti, A. (1988). Arbeitsteilung im Innovationsmanagement—Forschungsergebnisse, Kriterien und Modelle. *Zeitschrift Führung + Organisation*, **57**, 378–388.

Hauschildt, J. & Gemünden, H. G. (Eds) (1998). *Promotoren—Champions der Innovation*. Wiesbaden: Gabler.

Hauschildt, J. & Kirchmann, E. M. W. (2001). Teamwork for innovation—the 'troika' of promotors. *R&D Management*, **31**, 41–49.

Havelock, R. G. (1973). *The change agent's guide to innovation*. Englewood Cliffs, N.J.: Educational Technology Publishers.

Howell, J. M. & Higgins, C. A. (1990). Champions of technological innovation. *Administrative Science Quarterly* *35*, 317–341.

Langrish, J., Gibbons, M., Evans, W. G. & Jevons, F. R. (1972). *Wealth from knowledge—studies of innovation in industry*. London: Macmillan.

Lechler, T. (1997). *Erfolgsfaktoren des Projektmanagements*. Frankfurt am Main: P. Lang.

Maidique, M. A. (1980). Entrepreneurs, champions, and technological innovation. *Sloan Management Review*, **2**, 59–76.

Markham, S. K., Green, S. G. & Basu, R. (1991). Champions and antagonists: Relationships with R&D project characteristics and management. *Journal of Engineering and Technology Management*, **8**, 217–242.

Medcof, J., Hauschildt, J. & Keim, G. (2000). Realistic criteria for project manager selection and development. *Project Management Journal*, **31**, 23–32.

Roberts, E. B. & Fusfeld, A. R. (1981) Staffing the Innovative Technology-Based Organization. *Sloan Management Review*, **22**, 19–34.

Rogers, E. M. & Shoemaker, F. F. (1971). *Communication of innovations. A cross-cultural approach*. New York: Free Press.

Rothwell, R., Freeman, C., Horsley, A., Jervis, V., Robertson, A. B. & Townsend, J. (1974). SAPPHO updated—project SAPPHO phase II. *Research Policy*, **3**, 258–291.

Schon, D. A. (1963). Champions for radical new inventions. *Harvard Business Review*, **41**, 77–86.

Schumpeter, J. (1931). *Theorie der wirtschaftlichen Entwicklung—Eine Untersuchung über Unternehmergewinn, Kapital, Kredit, Zins und den Konjunkturzyklus* (3rd ed.). Leipzig: Duncker & Humblot (1st ed., 1912).

Witte, E. (1973). *Organisation für Innovationsentscheidungen*. Göttingen: O. Schwartz.

Witte, E. (1977). Power and innovation: A two-center-theory. *International Studies of Management and Organization*, **7**, 47–70.

Part XI

Innovation Leadership

The International Handbook on Innovation
Edited by Larisa V. Shavinina

Innovation and Leadership

Jean Philippe Deschamps

International Management Development Institute (IMD), Switzerland

Abstract: This chapter is about innovation leaders, those critical senior executives which top management sees as the linchpins of its innovation process and the 'evangelists' of an innovation and entrepreneurship culture. It will start with a well-accepted list of generic leadership imperatives as they relate to the specific challenge of innovation. It will continue by listing the common traits of innovation leaders in terms of personal profile and behavioral attributes, also derived from the author's research. The chapter will end with a discussion of the beliefs and management philosophy on innovation leadership adopted by some innovative companies and top managers (CEOs and CTOs).

Keywords: Innovation; Leadership; Innovation management; Innovation champions; Innovation attributes; Leader profile.

Innovation leadership? It is passion; it is learning; it is humility in front of mistakes and errors—understanding that they are necessary elements to learn faster than the others—and it is the target setting . . . yes, stretched targets!

Pekka Ala-Pietilä (2001)
President of Nokia

Introduction

Is innovation just one of the common dimensions of business management? Or is it a special domain that requires a different type of leadership, unique attributes or talents? If the former is true, then most, if not all, business leaders should excel at innovation, provided they pay attention to it, of course. This doesn't seem to fit with practice. Few of the revered leadership icons celebrated by the media for their achievements in shareholder value creation could claim that innovation is their forte. Most would not qualify as innovation leaders. And the opposite also seems to be true: Not all innovation leaders are fully-fledged business leaders. It is, therefore, reasonable to postulate that innovation may require a special leadership profile. If this is true, then we should try to find out how to recognize and develop innovation leaders so we can leverage their talent. This chapter describes some of the main characteristics and attributes of innovation leaders, and also tries to identify what makes them so special.

What do we mean by 'innovation leaders?' We are talking here neither about front-line innovators in R&D or marketing, nor about entrepreneurial middle managers. The general profile of these creative, determined individuals is relatively well known because it has attracted the attention of business scholars. Many stories have described how these mavericks have managed, against all odds, to convince their organizations to allow them to pursue their visions and then pushed them through.

Innovation leaders are senior executives—whatever their functions or positions—who spontaneously instigate, sponsor and steer innovation in their organizations. Even in the face of resistance from their top management colleagues, these executive champions always stand up for innovators and challengers of the status quo. Hauschildt (this volume) calls these champions 'promotors' and distinguishes between three types: 'promotors by expertise', 'power promotors' and 'process promotors', who need to work as a 'troika' to stimulate and support innovation. Whatever their types, true innovation leaders tend to share the same determination, i.e. they are unafraid to risk their credibility with top management in case of failure. Lewis Lehr (1979), the highly charismatic former CEO of 3M, described the behavior of an innovation leader very convincingly when he said: 'We learned to follow the fellow who follows a dream!'

The ideal place for an innovation leader is, obviously, at the head of the company or one of its

815

businesses. The archetype of the innovation leader is the CEO of the company he or she has helped create. Famous names spring to mind: Edwin Land at Polaroid, Robert Noyce at Intel, Steve Job at Apple and, more recently, John Chambers at Cisco or Jeff Bezos at Amazon. But it is restrictive to refer to such charismatic entrepreneurs as the *only* innovation leaders worth considering because it tends to make the management of established 'old-economy' companies feel excluded, and thus authorized to abstain from promoting that special form of innovation leadership. There are, indeed, innovation leaders in most large, mature organizations, although top management does not always properly acknowledge or recognize them. So, we should extend our definition of innovation leaders to include those critical senior executives who—with or without top management blessing—act as the linchpins of the company's innovation process, the 'evangelists' of an innovation and entrepreneurship culture.

Some of the core characteristics of innovative organizations have been amply described in the management literature (Drucker, 1985; Quinn, 1985; Knox, 2002). Jones & Austin (2002, p. 160) highlight five core characteristics of 'innovation leaders' from their research:

- In-depth customer insight;
- Leading-edge technical awareness;
- Inspirational leadership;
- Motivational organizational rewards;
- Sharing knowledge.

But these 'differentiators of enhanced innovation performance', as they call them, relate more to the collective management of innovative companies than to specific individuals. There has been no formal attempt, yet, at brushing a comprehensive portrait of 'innovation leaders', as defined in this chapter.

For the most part, innovation leadership is discussed either by innovation management researchers in the context of the role of top management in innovation (Bessant, this volume; Katz, this volume; Senge, 1999; Tidd et al., 1997; Van De Ven et al., 2000) or by leadership scholars mainly within the topic of 'leadership and organizational change' (Robert, 1991; Schruijer & Vansina, 1999; Shamir, 1999) involving, for example, such concepts as 'transformational leadership' (Bass, 1999) and/or 'effective leadership' (Yukl, 1999). However, not all managers are leaders and not all leaders are innovation leaders. Likewise, not every organizational change leads to innovation as well as not every transformational leadership implies innovation leadership. The phenomenon of innovation leadership thus remains terra incognita from a research viewpoint, being partially known primarily from some (auto) biographies of famous innovation leaders (Grove, 1996; Morita, 1987). This chapter is based on a new direction of research, which tries to analyze specifi-

cally the profiles and attributes of innovation leaders, as defined earlier. This portrait will, by necessity, be more impressionistic than realistic and systematic because it will have to fit a great diversity of characters. It will be built through a succession of brush strokes, each one adding a layer to our description of the special form of leadership that fosters innovation.

We will start with a broad-brush description of the key challenges in innovation, and derive from them a few general leadership imperatives for senior corporate officers. We will also define innovation leaders by what they do specifically to make innovation happen. We will then look at the two main phases of innovation, and identify the various leadership styles and profiles needed to steer them. We will complete our portrait of innovation leaders by listing some of their apparently common personal traits and behavioral attributes. We will end with a discussion of the beliefs and practices of some innovative companies that effectively identify, develop, motivate and retain innovation leaders.

The Innovation Leadership Imperatives

To start, it is useful to recall some of the essential aspects of innovation—the ones that most innovation leaders would probably put on the top of their lists— and reflect on the challenge they raise for business leaders. We will focus on six of these innovation imperatives:

- the urge to do new things;
- an obsession to redefine customer value;
- the courage to take risks;
- an ability to manage risk;
- speed in spotting opportunities and project execution;
- a shift in focus and mindset from business optimization to business creation.

Innovation Requires an Insatiable Urge to Try New Things

It is trite to say that innovation is about challenging the status quo and introducing new and, one hopes, better products, processes, services or management approaches. Fundamentally, innovation requires curiosity, experimentation and openness to change. Innovation leaders are therefore likely to be found among managers who constantly challenge the present state of affairs in their companies, and encourage wild ideas about doing things very differently or even doing new things.

The challenge for innovation leaders is to encourage their subordinates to try new things and experiment, despite the fact that such encouragement is usually not the surest way to move up the management hierarchy. The urge to challenge the status quo, in society as in family life, is normally viewed as the privilege of youth. With maturity, we believe, people normally calm down and become considerably more tolerant of the current state of affairs. This belief applies to companies as well. The desire to change things is an accepted—if not always welcomed—behavioral trait of young managers. To climb the ladder, however, managers are expected to demonstrate maturity and realism. And this often means exhibiting dependability and predictability. Challenging the status quo is encouraged only when the situation is bad enough to call for some fundamental changes. Otherwise, testing the status quo is accepted solely if it does not put the company at risk and if it leads to success. The right to fail is generally limited; not by words, for sure, but in speed of promotion. Despite frequent management denials, many companies still live under the 'If-it-ain't-broken-don't-fix-it' banner. Therefore, under the pressure for order and stability, innovation leaders must have the courage to foster a climate of experimentation and permanent change in their organizations.

It is not surprising, consequently, to find few mavericks and highly charged innovation champions in most top management circles. The career progression rules in mature companies favor managers who deliver results without making waves, not corporate activists. The creators of 'managed or organized chaos', so dear to innovation scholars (Peters, 1987; Quinn, 1985), often meet obstacles on way to the top of their organizations. To stimulate innovation, however, it is incumbent on companies to promote 'challengers', not just 'fixers'.

Innovation Requires an Obsession to Redefine Customer Value

Innovation has to do with adding value, and the way to add value is through leadership, argues Nick Shreiber (2002), CEO of Tetra Pak, an innovative company and world leader in liquid food packaging:

> Adding value is a worthwhile goal in any endeavor, and is particularly true in professional life. One can add value in many ways. The most important, perhaps, is through leadership—a very elusive concept! Just like good judgment, good leadership is hard to define, but you know it when you see it! Leadership can inspire an organization to reach goals it had never dreamed of, and will encourage each employee to reach his or her full potential in pursuit of their objectives. Inspired leadership will encourage new ideas through innovation and entre-preneurship, and will provide the resources to implement them.

In hindsight, highly successful innovators have generally established new standards of value in their industries. For a long time, value creation has come primarily from revolutionary, technology-based products or processes. Michelin redefined the notion of value in tires—as expressed in mileage life—with its radial tire technology, and Sony did something similar with its Trinitron TV screens. So did Pilkington with its revolutionary low-cost flat-glass process. What is new,

perhaps, is the fact that value creation is now felt to come also from introducing radically new business models or management methods. It is no longer necessary to have been a great technical innovator to qualify as an innovation leader. It is by radically changing the economics of the PC industry—some would say forever—not the product itself, that Michael Dell (2001) can arguably be called a great innovation leader:

> People look at Dell and they see the customer-facing aspects of the direct-business model, the one-to-one relationships. What is not really understood is that behind these relationships lies the entire value chain: invention, development, design, manufacturing, logistics, service, delivery, and sales. The value created for our customers is a function of integrating all those things.

W. Chan Kim & Renée Mauborgne (1997) postulate that redefining value starts with questioning current industry assumptions and imagining a new value curve. They advocate asking four probing questions:

- Which of the factors that our industry takes for granted should be eliminated?
- Which factors should be reduced to well below the industry standard?
- Which factors should be raised well above the industry standard?
- Which factors that the industry has never offered should be created?

Consciously or instinctively, innovation leaders challenge industry assumptions in order to unearth opportunities for a quantum jump in customer value. An obsession for customers often fuels this urge to redefine value. Value creators, typically, have an insatiable curiosity for their customers' needs, empathy for their conscious or latent frustrations, and an instinct for what they might need or want in the future. As Akio Morita (1987) eloquently stresses in his story of Sony's legendary *Walkman*™, this type of curiosity is not synonymous with a thirst for traditional market information. No market research, he posits, would have indicated a need for the *Walkman*™. Morita (1987) is referring, rather, to the kind of customer intimacy that comes from a deeply ingrained, quasi-instinctive curiosity. Sony's recent advertising slogan: 'You dreamt it! Sony made it', reflects the company's view of its innovation mission: redefine value constantly by right-guessing the customer's unarticulated desires, and applying its technological expertise to satisfy them.

The challenge for innovation leaders is to encourage this constant reappraisal of value factors, despite the fact that, at times, such an attitude may prove highly destabilizing. Challenging the current ways of delivering value in your industry is, indeed, very difficult when you are an established player and, a fortiori, when you are the market leader. Defying the status quo is much more natural for new entrants looking for ways to challenge incumbents. This is why many innovations have originated with outsiders who forced their way into the market with radically new concepts.

The story of easyJet (Kumar & Rogers, 2000), the latest—and highly successful—European no-frills airline, is a good illustration of this rule. Its founder, Stelios Haji-Ioannou, a typical innovation leader, challenged every single prevailing assumption in the traditional airline industry[1] to come up with a revolutionary business model. This gave him unbeatable low costs and allowed him to redefine the notion of value for Europe's budget-conscious air travelers. Haji-Ioannou carved out a fast-growing share of the European budget airline market. Arguably, it would have been very difficult for any European flag carrier to introduce such radical changes internally. The defenders—companies like British Airways or KLM—had no alternative but to create their own budget airlines by emulating easyJet's business model. But, given the constraints these new airlines inherited from their established owners, imitation proved impossible.

Innovation Requires the Courage to Take Risks

One of the most widely recognized drivers of innovation is management's willingness to take risks. It is, nevertheless, hotly debated because the very concept of risk-taking is subject to all kinds of interpretations. In its classical definition, risk-taking for innovation is related to the concept of entrepreneurship, i.e. being ready to bet one's resources on a new, hence untested, business proposition.

The challenge for innovation leaders is to live up to this principle in the day-to-day reality, and make their managers down the ladder comply with it as well (Perel, 2002). Indeed, if many companies advocate risk-taking as one of their core values, how many fail to change their performance review and reward systems to make them congruent with that belief? We seldom hear stories of managers being penalized for not taking risks, as long as they are meeting their budgets. The right to fail comes up in most innovation speeches, but it is not necessarily applied across the board.

Andy Grove (1996), Intel's legendary former CEO, adds two very interesting dimensions to the risk-taking imperative. First, he claims that innovation leaders must have the courage to focus, which means identifying unambiguously either the things they will *not* do, or the things they will stop doing. The decision to

abandon the fabrication of D-RAM memory chips in order to concentrate on microprocessors was, reportedly, one of the toughest decisions Andy Grove (1996) and his management team ever had to take. Yet, it is this kind of foresight and gamble that enabled Intel to mobilize all its R&D resources for manufacturing microprocessors and, ultimately, to build its success. Second, Grove (1996) believes that innovation leaders must have the courage to 'self-cannibalize', i.e. to make their own business obsolescent before others force obsolescence on them. As we all know, it takes courage to kill one's own products before milking them fully, and replace them with higher-performance, but unproven ones. Arguably, it is this policy, coupled with management's belief in the now-famous Moore's law[2] that enabled Intel to stay on top of its industry for so long. Whereas the willingness to take entrepreneurial risk applies to all managerial echelons, Grove's (1996) observations apply only to the highest level of innovation leaders, say, the CEO, and this is why they are so relevant.

Innovation Also Requires an Ability to Manage Risk

The debate about acceptable levels of risk in an innovation project often pits risk takers (usually the project champions) against risk-containers (typically senior managers). Without risk, argue the former, there will be no innovation. Yes, respond the latter, but without proper risk-management, there will never be any successful or affordable innovation. Innovators often complain that the controlling attitude of their top managers hides a fundamental aversion to risk. And the more conservative proponents of risk-management often suspect risk-takers of being irresponsible. But this debate is fruitless because both arguments are obviously right. Innovation is as much about good risk-containment and management, as it is about risk-taking. This is well known to development engineers, who are increasingly adopting sophisticated approaches to minimizing design risk.

The challenge for innovation leaders is, therefore, to strike the balance between dogged, enterprising risk-taking, and pragmatic, cautious risk-management. The first attitude is necessary for pushing ahead and brushing away objections. In a sense, front-line innovation champions should be so determined and persistent that they can be accused of being both blind and stubborn. Innovation leaders carry the burden of ensuring that all the known risk-factors have been identified at each stage and properly managed. And this needs to be done, one hopes, without discouraging innovators and entrepreneurs.

---

[1] Point-to-point connections using low-cost airports and without prearranged connection possibilities; direct sales (90% through internet), by-passing travel agents; sales staff paid on commission; no fixed prices for tickets (extensive use of 'yield management methods'); no tickets; one type of aircraft; no business class; no meals; etc., but high punctuality level and unbeatable prices.

---

[2] Gordon Moore, one of Intel's founders, predicted in 1965 that the number of transistors inserted on a silicon chip would double every 18 to 24 months. As Intel tried to follow that 'law', it became a kind of self-fulfilling prophecy.

A dilemma arises whenever the CEO or the Business Unit Head is simultaneously both the champion of a particular project *and* the leader, who is supposedly responsible for containing risk. No manager will dare oppose his or her hierarchical head by spotlighting dangerous risk factors on the boss' favorite project. The story of Philips' ill-fated CDi[3] illustrates that danger. It was well known in Philips that its CEO, Jan Timmer, had adopted the CDi as his pet project, as he had successfully championed the CD-Audio years earlier. Many in the company argue today that the CDi concept had inherent flaws, and that its proponents blindly underestimated the competing PC-based technology: CD-ROM. But hardly anyone, it seems, dared openly challenge the notoriously tough Jan Timmer. It took Philips a few years and huge losses to abandon the project.

A similar story can probably be told about Robert Shapiro's energetic pursuit of the market for genetically modified organisms (GMOs) at Monsanto. As CEO, Shapiro was consumed by the vision of Monsanto becoming a life-science powerhouse on the strength of its genetic engineering technology. And he was convinced that realizing his vision meant betting the company's future on GMOs and promoting them aggressively worldwide. But experts are likely to point out that after the controversy over the company's commitment to GMOs erupted in the media, Monsanto's top management failed to grasp the power of the arguments of GMO's detractors. It is hard to be a visionary, risk-taking innovation champion while, at the same time, being a cautious risk-analyzer and container. This, nevertheless, is the challenge of innovation leaders.

Innovation Requires Speed in Spotting Opportunities and Project Execution

Silicon Valley innovators and entrepreneurs have known for a long time that the best idea or the best technology does not necessarily win; the winner is the one that is implemented first (Rogers & Larsen, 1984). Whoever comes first learns fastest. Success with new products comes from launching first, then learning fast to correct mistakes, before others have readied their response, and relaunching a superior product as competitors start coming in. That kind of speed requires three unique skills: (1) an attitude characterized by searching continuously for opportunities; (2) a great deal of management decisiveness at all stages in the process; and (3) speed in execution, typically achieved through a pragmatic reliance on external and internal resources, and, of course, highly effective teams (Cooper, this volume).

---

[3] CDi: Compact Disk interactive, a precursor of CD-ROMs and DVDs, introduced by Philips towards the end of the 1980s and abandoned in the early 1990s due to the growing success of CD-ROMs.

Innovation leaders instinctively create an environment that values the search for opportunities and the generation of ideas to exploit them. They typically encourage people to flag opportunities very early and make their ideas bubble freely upwards for discussion. The challenge lies in the decision process. On what grounds should the project go ahead? What criteria should be met at each stage? When and on what basis should the plug be pulled? This challenge relates directly to our previous discussion on risk-taking versus risk-containment. As the champions of risk-taking entrepreneurs, innovation leaders are bound to allow their staffs both a fair amount of freedom to experiment and the necessary resources. Yet, as advocates of a cautious approach to managing risk, they must ensure that the funding justifications required from their teams are commensurate with the investment amounts requested. Finding an acceptable balance is a challenge, and so is the necessity to decide fast, whatever the decision. In Silicon Valley, innovators usually get the same advice from venture capitalists: if you are going to fail, at least fail fast and fail better.

Innovation Requires a Shift in Focus and Mindset: From Optimizing Business to Creating Business

Most managers see their career progress first through managing a function, and then—if they are of a high enough calibre—through running a business unit and later, a division. Unless they come from R&D, few are asked to manage large projects as their prime job. Their raison d'être and training is managing and optimizing what already exists, not create something new. And if they do happen to innovate, it is as part of their normal business responsibilities. Indeed, business unit heads are generally responsible for new product development in their fields. Nevertheless, even in such cases, innovation is generally pursued to protect and grow the current business, seldom to create new businesses. This is why most companies struggle to exceed the growth rate of their industry. How can Unilever or Nestlé grow in the mature food industry, except by creating entirely new, hence fast-growing product categories? Now that the second-generation mobile phone market is nearing saturation, the same question applies to Nokia and Motorola. Creating new businesses is completely different from tweaking product lines to introduce extensions.

So innovation leaders face a double challenge. The first is to strike the right balance between running the current business and growing new businesses, or as Derek Abell (1993) advocates, between mastering the present and pre-empting the future. The sudden shift in what financial markets demand in the way of share performance—yesterday: growth potential; today: profitability—makes finding the right balance a tough task. Katz (this volume) addresses this dilemma by referring to a series of models to help companies avoid the 'tyranny of success' and learn to 'organize both incremental and disruptive innovative activities'. The second challenge for innovation leaders is sensing market opportunities and choosing promising areas to pursue. Here, innovation leaders must have the ability to shape a vision that will guide them towards new business opportunities.

What Innovation Leaders Really Do

If we follow John Kotter's (1990) research and track what innovation leaders really do, we will note that they excel at six fundamental leadership tasks:

- breeding or attracting and retaining innovators and entrepreneurs;
- formulating a clear innovation vision, and setting innovation priorities;
- charting a roadmap towards their vision, and mobilizing people to implement it;
- accepting the risk of spotting and backing new ideas;
- assembling and nurturing complementary teams of champions; and
- building an innovation process and culture.

The Innovation Leader as a 'Magnet' for Innovators

Marvin Bower (1997), McKinsey's legendary Managing Partner and leadership guru, postulates that '. . . a business should be run by a network of leaders positioned right through the organization' (Bower, 1997, p. 6). This postulate probably applies more to innovation leaders than to any other types of leaders. Indeed, innovation is never the result of a single person's efforts, neither at the project level nor at the sponsoring level. As the well-known saying goes, 'It takes only one "No" coming after nine "Yeses" to kill a project'. Innovation is always in danger if it hangs in the hands of an isolated innovation leader in the top management team, whatever his/her charisma. The first roles of an innovation leader are, therefore, breeding or attracting others to take on leadership roles, propagating innovation values, and supporting concrete projects.

It is relatively easy for innovation leaders to build a team of subordinates who share similar values and behaviors, for two reasons. First, people tend to be attracted by their likes. And second, unless they are authoritarian (which sometimes happens), innovation leaders usually exude a high level of openness and communicate enthusiasm, to say nothing of passion. Working for them is generally exciting.

The situation is more complex at the top management level. Innovation leaders, unless they occupy the top job themselves, may be unable to influence the profile and behavior of their top management colleagues. They can only muster CEO support. If they show growth and results, they can hope to propagate their values through sheer emulation. When they have established a reputed nursery of talent in their organi-

zations, they can also volunteer to transfer some of their best and most motivated staff into other divisions, in the hope of initiating a bottom-up movement of contagion.

The Innovation Leader as a Vision-Builder and Priority Setter

Many people disagree on the very nature of the innovation process. Innovation practitioners, particularly in R&D, often believe in serendipity. They argue that, most of the time, innovation occurs in an unplanned and somewhat erratic fashion, fuelled by randomly generated new ideas and new inventions in search of an application. They claim that the innovation process is often an a posteriori rationalization of a series of trials and errors. It is the result of constructive experimentations by curious and determined individuals who are able to sense a market opportunity behind a serendipitous idea or discovery. The now-famous development of 3M's *Post-it*™ pad provides a near-perfect illustration of this non-linear, 'bubble-up' innovation process. As everyone knows by now, it all started with two determined and inventive engineers trying, on their own initiative, to find applications for a strangely weak glue that had failed all of 3M's classic sticking tests (Nayak & Ketteringham, 1986).

Innovation leaders, in contrast with some of these front-line innovators, do not want to rely only on serendipity. They cannot accept their role in innovation being limited to a kind of benign laissez-faire, i.e. hiring creative people, giving them the freedom to experiment, and hoping that some innovation will emerge! True innovation leaders tend to be more deterministic. They believe that innovation can also result from management ambition and vision. In addition to the traditional 'bottom-up' approach, they strive to create an organized 'top-down' process, starting with a broad vision of environmental changes that are creating new market needs, and selecting priority areas for further exploration. The initial vision, albeit often vague at the outset, determines the boundaries of the search area and triggers a focused pursuit of concrete product or service opportunities. Following such a vision-led process, innovation is no longer random. Would-be front-line innovators in marketing or R&D receive a mandate to explore a number of well-defined areas management has identified and prioritized. Their task is to generate the best ideas, concepts and solutions for implementing the vision.

Tetra Pak's worldwide success in liquid food packaging provides a compelling illustration of such a vision-led process. Ruben Rausing, Tetra Pak's Swedish founder, did not build his company on a serendipitous discovery. Rather, he was animated by a vision of transforming the antiquated milk distribution system that prevailed in 1950s Europe by adapting it to the mass-retailing supermarkets that were emerging.

For the first time, someone tried to optimize a package by creating value throughout the chain, from producers to users, via transporters and retailers. It was Rausing's vision that a package should save more than it costs that led him to search systematically for rational container shapes and efficient filling systems. That search led him, in turn, to invent the first tetrahedron package, now called *Tetra-Classic*™, and a few years later, the ubiquitous *Tetra-Brik*™. And it was this very same vision and ambition to add value to his dairy customers, retailers and consumers that brought him, after a few years, to introduce the first aseptic filling systems and packages. This innovation created an entirely new product category: UHT[4] milk and juice.

Even if they are not visionaries personally, top managers cannot shirk the task of specifying where they expect to see innovation happen in their company. If they want innovation to support their business strategy and priorities, instead of occurring randomly, they need to point towards specific privileged directions. Where do we need to put our innovation priorities and for what purpose? Which market or segment do we want to rejuvenate? On what aspect of our value chain do we want to concentrate our efforts? To what unmet or ill-met customer needs do we want to give priority?

The Innovation Leader as 'Vision Roadmapper' and Implementation Planner

Visions alone do not change things. If they remain vague dreams or ambitions, without any concrete implementation activity, they do not mobilize people either. Innovation leaders get personally involved in charting a roadmap towards their vision; they direct their implementation teams. The involvement of innovation leaders in implementation will, obviously, vary according to the nature of the challenge and the management level at which they operate.

When they occupy the top job, innovation leaders are usually keen to choose, or at least influence the choice of the leaders of their critical corporate innovation projects. They typically see their role as providing strategic direction, empowering, releasing resources, and ensuring that obstacles to implementation are removed. When Jürgen Schremp, DaimlerChrysler's CEO, formulated the vision of his group becoming one of the first automotive manufacturers to offer fuel-cell powered cars, he clearly left it to others in his management team to plan how to achieve that ambitious objective. Nevertheless, as CEO, he surely played an important role in assigning the mission to trusted aides, and in clearing the way for

extending the scope of the company's technology and business alliances, notably with Ballard Power Systems, the Canadian fuel-cell engineering specialist, and with Ford Motor Company, DaimlerChrysler's strategic partners.

When they sit at the supervisory level, such as the Executive Committee, innovation leaders may actually get much more involved, for example, in supervising innovation teams. Many companies—Tetra Pak is one of them—have adopted the practice of having a member of their top management team personally coach each major corporate innovation project. This involvement brings about two major benefits. First, it provides the team with a high-level supporter and protector. A senior coach can shield a team from the natural tendency of any corporate hierarchy to encroach on its autonomy, and thus reducing its level of empowerment. Second, the senior coach is exposed to the day-to-day reality of the project, and hence shares the team's experience and learning.

Nearer the level of operations, the innovation leader may actually direct the overall project, taking responsibility for charting the project path from vision to implementation. Years ago, when Hiroshi Tanaka, Managing Director of Canon's Office Products Development Center, became the official champion of the company's ambitious 'family copier' project, his challenging mission could be summed up in very few words: develop a very small, service-free personal copier to retail under $1,000. His first role as innovation leader was to develop a game plan that would meet his CEO's vision. This meant planning the project management organization, the project itself, the way to fence competitors out, and the launch and rollout. As the anointed innovation leader for this undertaking, his task was not only to chart a path for a breakthrough mission, but also to convince his sceptical team that the job was feasible, and ultimately, to lead them to success (Deschamps & Nayak, 1995).

The Innovation Leader as a 'Spotter and Backer' of Good Ideas

As noted earlier, innovation calls for a delicate balance between risk-taking and risk-containment. Innovation leaders are constantly confronted with this difficult challenge since they are usually the ones who are in contact with, and manage front-line innovators. Consequently, they tend to get involved very early in evaluating new product, process, service or business ideas. They decide, alone or as part of a group, which ideas they should bet on, shelve for a while, or diplomatically turn down.

Obviously, it is critical for the innovation leader to take the right decision. A good flair is needed to ensure that interesting opportunities are not passed over, and that business resources are not squandered on useless pursuits. The difficulty comes from the fact that creative ideas cannot be evaluated with the same

---

[4] UHT: Ultra High Temperature, a conservation process that maintains most of the qualities of fresh (i.e. pasteurized) milk or juice while allowing it to keep, unopened, for months in ambient conditions (i.e. unrefrigerated).

methods and certainty as other types of management decisions (Mannarelli, 2001). They cannot follow the traditional approach through which most companies analyze investment opportunities, e.g. with a detailed justification analysis leading to a single 'Go/No Go' decision point. They need to go through a process of progressive refinement and risk reduction, with funding in phases and in proportion to the removal of major uncertainty factors. This always involves a high degree of management ambiguity that innovation leaders need to accept and explain to their management colleagues.

To preserve the motivation of their creative staff, and hence maintain a positive innovation climate in the future, innovation leaders must ensure that decisions on new ideas are discussed, taken, justified and communicated through a transparent process. This is why having a single person playing the role of 'judge of ideas' is ill advised, particularly if the judge happens to be the 'idea submitter's' direct supervisor.

Some companies have publicly identified (with contact numbers) a number of influential 'idea sponsors' or 'idea advocates' over their corporate intranets. These special types of innovation leader have multiple roles:

• identify the kernels of opportunities behind the raw idea;
• help the originator of the idea to argument it;
• assist in 'packaging' the idea for a presentation to management;
• defend it against hasty negative judgments in management discussions; and
• coach the initial idea validation phase until the next review point.

Increasingly, innovative companies entrust this important screening and backing of ideas to a collective body, which some call an 'Innovation Council'. Whatever its name, this innovation management mechanism bears the important responsibility of selecting the best opportunities for company or business unit funding. There are many advantages in delegating this important task to a management group, instead of a single manager:

• broader and more formal review of the idea's merits and risks;
• more objective assessment through multiple perspectives;
• more credible justification for decisions (positive or negative);
• a pool of resources for coaching initial projects;
• more visibility and transparency in evaluating and selecting ideas.

But the key advantage is the possibility for top management to bring innovation leaders together— traditional and conservative managers should be kept

out of such a body—to give them a chance to influence the flow of new corporate projects.

The Innovation Leader as 'Assembler and Composer' of High-Performance Teams

The team is at the heart of most, if not all innovations. Few advocates of innovation teams are more vocal than the folks at Ideo, America's leading design studio and innovation culture evangelists (Kelley, 2001). A commonly heard saying at Ideo is "Enlightened (team) trial and errors succeeds over the efforts of the lone genius". To reflect the sense of passion that animates design teams as they work on an innovation project, Ideo calls them 'hot groups'.

As the firm's ultimate innovation leader, Dave Kelley—Ideo's founder and CEO—devotes considerable attention to the process through which teams are assembled and formed. Because he believes that teams perform better when they are made of volunteers, he has instituted a very original approach to forming teams. The company is organized around 'hot studios', a Hollywood-like system for quickly building teams around projects and disciplines (Kelley, 2001). Under this approach, studio heads—Ideo's second-level innovation leaders—do not pick the teams they need. They simply describe the project for which they are responsible and the location they will be using; designers then select the projects they want to work on. . . and their leader. Ideo managers are also keenly aware of the fact that it takes a lot of positive reinforcement to turn a group of inspired individuals into a 'hot team', even if it is made of volunteers. This is why they devote so much time and effort thinking about how to make their managerial and physical environment friendlier for teams.

Traditional companies will object that Ideo is a maverick organization. They are right. But while the management style at Ideo may be unorthodox, when it comes to the attention its leaders bring to composing their teams, it is certainly not an isolated case. The experience of large Japanese technology-based companies shows a similar pattern. In firms like Canon or Toshiba, the highest-ranking innovation leader is usually the Chief Technology Officer (CTO), or Chief Engineer, whatever his title. These senior managers are usually not hierarchically responsible for R&D departments, which often report to divisional or plant management. Nevertheless, they consider it one of their key tasks to advise on the leadership and composition of important new product development teams. Japanese innovation leaders actually make up their own 'hot teams' with the same devotion and care as a barman mixing an exotic cocktail. They look for balance in age, seniority, experience, skills, personality and even mindset. This contrasts sharply with many of their Western counterparts who sometimes assemble teams rather rapidly on the basis of staff availability. So, if great products come from great teams, it is undoubtedly a key task of the innovation leader to build such teams very carefully.

The Innovation Leader as Builder of an Innovation Process and Culture

At their inception, most companies were innovators. They were born from the vision and ambition of their owners, for sure, but this vision usually focused on a market opportunity and an innovative idea for exploiting it. At their early stage, innovative companies typically benefit from a strong, almost instinctive entrepreneurial culture. Most probably, very few spend much time building and formalizing what we would today call an innovation process, i.e. a systematic and repeatable approach to generating innovations.

Over time, as they grow in size and complexity and as new management replaces the founding team, many of these early innovators lose some of their natural entrepreneurial spirit. In exchange, though not always, they develop more formal methods, procedures and mechanisms for managing their innovation process, which Bessant (this volume) calls 'routines'. Such processes are certainly helpful and at times indispensable, but they do not completely substitute for culture. World-class innovators such as 3M—the most frequently cited innovation archetype—typically combine culture *and* process. Their innovation culture, which promotes individual freedom, learning and sharing, favours creativity, experimentation, risk-taking and teamwork. Their innovation process adds a market-, technology-operational and economic discipline in the way they finance and manage projects and, ultimately, go to market. Lewis Lehr (1979), former CEO of 3M, expresses well how this combination of culture and process creates a challenge for management:

> Innovation can be a disorderly process, but it needs to be carried out in an orderly way. The truly good manager finds the means to manage a disorderly innovative program in an orderly way without inhibiting disorderly effectiveness.

Innovation leaders tend to recognize these two complementary dimensions and work proactively to improve them, inasmuch as their position in the company allows them to do so, of course. Actually, it is often easy to identify innovation leaders in a top management team: they are they ones who volunteer to set up and animate taskforces to work on innovation improvement tasks. Since it is generally easier to build processes than to change cultures, such taskforces often focus on streamlining or speeding up the innovation process. Unless the effort leads to excessively rigid or bureaucratic rules, this emphasis on process generally conveys a positive message throughout the organization—that management cares about innovation and is determined to enhance it. This, by itself, often helps improve the innovation culture in the company, thus creating a virtuous circle.

Combining Different Styles and Profiles to Lead Innovation

In the past, leadership seemed to come in one flavour. It was an absolute trait of character. You had leadership or you did not. If they had 'the right stuff', leaders could take over almost any company and address any challenge. This simple belief in the universal application of leadership is now under question. Different kinds of leadership, we feel intuitively, may be required for different kinds of objectives. The determined leader needed for business restructuring, cost-cutting and profit improvement may not have what it takes for a growth or globalization strategy, and even less for innovation. The common denominator in leadership is the ability to mobilize, motivate and direct a group of people towards a worthwhile goal. But as the nature of the objective changes, different types of leadership, or at least different leadership styles and attitudes, are required. And this also applies within the very domain of innovation.

Innovation Leadership: The Art of Combining Complementary Styles and Profiles

If we define innovation as the process by which an invention is successfully brought to market, there are clearly two complementary facets or broad phases in this process. At its 'fuzzy front-end', as it is sometimes called, innovation requires a number of 'soft' leadership qualities: strong curiosity, sense of observation, urge for exploration, ability to detect patterns in weak signals, willingness to experiment and learn, openness to new ideas, determination to pursue risky avenues, etc. However, implementing ideas and getting them to market requires a set of very different and 'harder' qualities: rigor in analysis, speed in decision making, clarity in objectives, willingness to dedicate resources, ability to manage risk, skills in problem solving, sense of urgency, faculty to coordinate multiple functions, etc.

This duality of requirements is what makes innovation so difficult to lead from A to Z. At its front-end, innovation requires all the qualities related to creativity. At its back-end, it demands a considerable degree of discipline. Daniel Borel (2001), Chairman of the Board of Logitech, the innovative leader in computer pointing devices, defines the needed capability as 'a mix of emotion and realism'.

What kind of leaders and leadership styles are needed to manage such a complex and multi-faceted process? Can leaders be found with the skills and qualities required to steer both the creative front-end and the disciplined back-end of innovation? Asking these questions to selected leaders of world-class, innovative companies like Logitech, Medtronic, Philips or Nokia serves as an interesting starting point for further discussion.

Logitech's Borel (2001) highlights this challenge by giving his personal definition of the innovation leader:

> Innovation leaders are those unique people who are able to motivate the full set of qualities you need from A to Z to deliver an innovative product. (. . .) They start from the pure innovative aspect and reward creative people, but they do not put too much highlight on them, because you also need the other person who is going to extract every penny out of a design to make it a viable product in the marketplace, so that it is affordable and profitable as well. The real leader is the one who appreciates the passion aspect, the emotion, yet is able to put it in a framework where execution is going to eventually deliver to the customer and end-user a product which is profitable for the company. (. . .) If you look into companies, it is very hard to find someone who is great in execution and at the same time great at the purely creative part of innovation. As a matter of fact the great leader is the one who is able to build a team with people who have a different psyche, and get them to work together and share the same language for the sake of the company.

William George (2001), Chairman of the Board and former CEO of Medtronic, the world's leading medical technology company, uses a sports metaphor to underscore this need for a plurality of innovation talents:

> You might think of it like a (football) team. You need somebody who is going to score. You need someone who can defend. You need somebody who is disciplined, someone who can make the brilliant move!

Ad Huijser (2001), Chief Technology Officer and Management Board Member of Philips, the innovative Dutch electronic giant, concurs with this emphasis on team building:

> On one hand, innovation leaders quite often express the vision behind which the troops align, but at the same time, they are very much team players because they cannot do it themselves. Innovation is absolutely a team effort, and innovation leaders know how to make and build teams, because you need a number of capabilities at the same time to make it happen. And balancing that team is a capability that you quite often see as the strong point of an innovation leader.

Pekka Ala-Pietilä (2001), Nokia's President, extends the team building scope of the innovation leader beyond the boundaries of his company. In his business—mobile telecommunications—different players for hardware, software, services and content need to come together and understand how to contribute in the best possible way to expand the market:

> We have to make sure that there are companies which can come together and win together. We feel that this is not a 'win-win' world! It is a 'win-win-win-win' world, because there are so many partners.

If innovation is a multi-faceted process requiring a diversity of complementary talents and attitudes, then we need innovation leaders who can draw the best from a diverse team. They need not only to be good team integrators, but they also need to have a deep understanding of what it takes to steer the different phases in the process. The leadership profile needed at the creative front-end is, indeed, quite different from the profile needed to support the disciplined back-end. Since very few senior innovation leaders combine these two sets of qualities, one of their key tasks is to develop other innovation leaders who will focus on and excel either at the front-end or at the back-end of the innovation process.

Leading the Creative Front-End of Innovation

The creative-variety of innovation leaders has a relatively well-known profile because it is the one most often described in the innovation literature. But how do the senior business leaders quoted earlier characterize such innovation leaders? William George (2001) sees in Medtronic's Vice-Chairman and Chief Innovation Officer, the man he dealt with in his previous COO job, an archetype of the creative innovation leader:

> He was a Medical Doctor, always open to new ideas. He was always going to try something. He was always willing to put some money aside to fund a new venture that came along, having no idea whether it was going to work or not.

For William George (2001), the main qualities of this purely creative innovation leader are curiosity and tolerance:

> He is really intrigued by the technology; he is very hands-on; he very much knows the products. He can take an idea and has a vision that maybe this is a kernel of an idea. It is like a needle in a haystack, but he is always looking for that needle and says, 'How can you make this work?' and not, 'Oh, that'll never work!'. Sometimes, the business leader may say, 'That'll never work! Look at all the flaws in it!' and he will say, 'No! Look, there is potential in there!' It's like taking a diamond in the rough and polishing it up to make it into something. (. . .) There is also, in that very creative innovation leader, a high tolerance for failure. In fact, he/she realizes, as I do, that most of the great breakthroughs come through failure, through an experiment that does not go as you thought it would. The experiments that go

as you think they would, all they do is confirm previous knowledge. The experiment that doesn't go that way leads you to say, 'Oh! What can I learn from that?' and then you apply that to making it better.

Philips Research, claims Ad Huijser (2001), is trying to encourage its creative innovation leaders to strike the right balance between risk-taking and risk management:

> A creative environment, for me, is an atmosphere in which creativity can flourish, but at the same time is constrained by budgets, manpower, etc. So, if you want to create new things, you have to stop other things. That stimulates people to do both: to push new ideas, but also to be very critical of their own ideas, not at start, but during the course of the action.

Logitech's Daniel Borel (2001) adds an extra dimension to this portrait of the 'front-end' innovation leader: an openness to go outside and 'borrow' technology from whatever source without any trace of 'not-invented-here' syndrome:

> One of the characteristics of the innovation leader is to be open-minded, but I would say open-minded in a way that goes much beyond what we call naturally open-minded, as in accepting any ideas. He/she is able to get out of the box, to look outside the company. (. . .) He/she has this ability to take input from inside, from outside, to take technology here and there, and do the equation that will bring a unique product for the user at the end of the day.

Ad Huijser (2001) concurs with these qualities—curiosity, tolerance for failure and openness to go outside—but adds two nuances. Even at the front-end, innovation leaders need to show a good dose of realism:

> Innovation leaders are creative, but in a balanced way. They are not creative everyday with a new idea, because you cannot lead an organization towards innovation if you change the direction everyday. (. . .) Leadership is also about knowing when to 'pull the plug'. Starting is easier than stopping in a research environment. Stopping requires making choices and taking the enormous risk of stopping something of value. Therefore, stopping projects asks for more leadership than starting projects. Real leaders dare to make choices and say 'No', if they don't believe things will have added value for the company.

Leading the Disciplined Back-End of Innovation

Innovation obviously does not stop with the generation of good market-oriented ideas and product or service concepts. Concepts need to be fleshed out and turned into business propositions and products that can be developed, engineered and produced time- and cost-effectively, then launched into the market. These processes constitute the critical back-end of innovation. Whereas the front-end deals with exploring and inventing, the back-end deals with planning superbly and then—and only then—'running like hell'. Surprisingly, the innovation literature is a lot less loquacious about the leadership traits required for steering these critical back-end activities. Are such leaders only good 'executors'? Is there something specific about leading the implementation of an innovation project? On this point, again, it is interesting to hear what top managers in innovative companies have to say.

The first characteristic of these implementation-oriented innovation leaders, according to William George (2001), is their urge to get new products to market:

> It is the disciplined person that is going to ensure you get the new products to market, because he or she knows that it is only when you get to market that the rubber meets the road, so to speak, and creates the innovation that generates the revenues for the next round of innovation. (. . .) He or she would take a little bit less of a product, accept a less perfect product knowing that, well, we can improve it the next time around! This person is driven to get it to market.

The second characteristic, again according to George (2001), is a strong sense of discipline and speed:

> In the old culture, a timetable was a goal to be shot for, and not a requirement to be met. So, if you missed it, maybe your products got delayed six months, then twelve months, then eighteen months, and everyone accepted it. (. . .) So we had to change that and that meant putting disciplines in place. (. . .) Scientists resented this at first. They kept saying, 'Are we no longer interested in real creativity and breakthroughs?' And we said, 'Oh, Yes. We are very interested, but you have to follow the discipline too (. . .) and if you can't, you are not going to be punished for it; you just can't get on board this product. The product is going to market. You need to be there'! (. . .) The business- and execution-oriented innovation leader also knows what it takes to go through the regulatory process, the quality insurance process, and the production, i.e. gearing up a production line so you are not producing only a hundred, you are producing a hundred thousand.

Logitech's chairman argues that everyone in his company has to be execution-oriented (Borel, 2001):

> Execution is, at the end of the day, what will make a huge difference at the bottom line. In my career, the most productive people I have seen are the ones who, eventually, came from an angle of passion and

emotion, but were able to contain this passion and emotion through experiences that, sometimes, have been extremely expensive. (. . .) Once you have burnt yourself with a huge inventory of the wrong thing, then you learn how to integrate the value of the two sides of the equation.

Passion: The Common Trait of All Innovation Leaders
Leaders generally tend to demonstrate a high level of emotional involvement in the mission they assign themselves. This is particularly true with innovation leaders. Whether they are of the creative type and work at the front-end of innovation, or belong to the disciplined, execution-oriented group, all innovation leaders share one thing in common: a high level of energy and passion. At Medtronic, William George (2001) reckons, all senior managers share that passion:

> People who had no passion for the patients, the doctors, the actual process of the company, did not fare very well in the Medtronic culture. The execution-oriented innovation leaders share the same passion as the creative-types, but it is a different way of looking at the world. In a way, they say, 'What good is your idea if it is in a lab and never helps a patient? I want to drive it and get it to market to help patients, because, you know, these people are out there, dying every day'.

Nokia is so convinced of the need for passion that it actually selects its new entrants, whether they deal with the front-end or the back-end of innovation, on that basis as well. Ala-Pietilä (2001) expresses this conviction unambiguously:

> Whatever you do, if you don't have a passion, then you have lost the biggest source of energy. If you have teams and individuals who don't have the passion—the passion to change the world, the passion to make things better, the passion to always strive for better results and always excel—then you will end up with mediocre results.

Developing and Retaining a Cadre of Innovation Leaders

The general debate about the origin of leadership—whether it is an inborn talent or an acquired skill—has not spared the domain of innovation. Not surprisingly, the answer to the question is the same for leadership *in general* and leadership *for innovation*. There seems to be a consensus that both result from a combination of natural aptitudes that are developed and enhanced through specific leadership development experiences. It is worth considering these two facets in some more detail: the appointment of people with the right skills and attitudes, and the personal development path that turns them into experienced innovation leaders. It is also worth asking, to round out this discussion, what

motivates, and hence helps retain innovation leaders in the company.

Identifying Innovation Leaders
Few companies seem to have developed an explicit, formal process for screening new hires on the basis of their specific potential for innovation leadership. But most classical processes for selecting candidates, be it for R&D or junior business positions, try, somehow, to detect indicators of innate or potential leadership. These provide a first set of clues on innovation leadership, of course. But other more specific pointers are worth taking into account. Not surprisingly, the various business leaders quoted in this chapter share views on what to look for when interviewing candidates for future innovation positions.

Besides checking for compatibility with their culture—a must, since most innovative companies have a very strong culture—they seem to particularly value, and look in their new hiring candidates for at least six main personality traits:

- A high level of passion, or at least energy for what they do;
- A propensity to take risks (and a track record proving that they have taken risk);
- An ability to see the big picture and think 'out of the box';
- An urge to keep learning and to broaden their interests;
- A sense of humility, or at least modesty; and
- A commitment to performance or excellence in whatever they undertake.

The ability to identify potential innovation leaders at entry level, as new hires, is obviously quite valuable. But it becomes much more critical later on, when the new managers make their first important career moves. Actually, the task is not too complicated. Innovation leaders are, indeed, easier to identify in practice than to describe in theory because they usually stand out from the crowd, even in generally innovative companies. It is not necessary to list all their inherent skills and attitudes in order to identify them. It suffices to highlight some of the traits that distinguish them at first glance from other equally competent business leaders. Anyone who has been exposed to innovation leaders will recognize them through their combination of unusual attributes. Not surprisingly, most of these attributes have to do with a high level of emotional intelligence and well-developed 'right-brain' capacity, and that is, perhaps, what makes them so special.

The first and often most distinctive trait of innovation leaders is a strong focus on customers and products. Innovation leaders share the same passion, whether they come from the purely creative side of the business—R&D, for example—or from the business side. But that passion is not disembodied. It is embedded in their products or services. Innovation leaders love their products or services, not for themselves, but for what they do for their customers. And because they instinctively adopt the customer's viewpoint, they are never fully content with their offering. Whereas other business leaders talk daily about strategies, performance, processes and organization, innovation leaders constantly refer to their customers and the products or services they have bought or might want. And when they are part of the top management team, or even sit on the Board of Directors, innovation leaders ensure that discussions on technologies, products and customers get a fair share of all executive meeting agendas.

Those who have known him claim that Akio Morita, Sony's legendary President, was one of those typical 'product-nuts'. History has it that he always came back from trips to his research labs with plenty of miniature electronic 'gizmos' in his pocket to keep there to play with and show his management colleagues. He couldn't help communicating his love for his products, and that love became contagious.

Innovation leaders are also recognizable by their ability to 'fire-up' people at all levels with their enthusiasm. The innovation leaders we tend to notice have enthusiastic followers because they exude a sense of fun, adventure, challenge and self-fulfillment. Their ability to communicate their passion upwards to their bosses—hence to get support for their risky undertakings—is probably just as important, if not more so, than their talent to mobilize the best in their own disciples. This reflects an innate sense of communications, which certain academics would qualify as Aristotelian, since it brings to bear the three classical elements of communications: *logos*, reaching peoples' sense of rationality and logic; *pathos*, touching peoples' emotions; and *ethos*, addressing people's sense of values and beliefs (Eccles et al., 1992). Rosabeth Moss Kanter, of the Harvard Business School, postulates that leaders are characterized by the energy and persistence with which they communicate their aspirations (Blagg & Young, 2001):

> Leaders must pick causes they won't abandon easily, remain committed despite setbacks, and communicate their big ideas over and over again in every encounter. (. . .) Leaders must wake people out of inertia. They must get people excited about something they have never seen before, something that does not yet exist.

This certainly applies to innovation leaders, and it is often what makes them so special in the eyes of their more conservative colleagues, to the point of being sometimes perceived as 'corporate agitators' or 'innovation zealots'.

Innovation leaders often stand out from the crowd through their healthy disrespect for organizational hierarchies, corporate norms and rules. They tend to be non-conformists and are sometimes considered mavericks by their peers. They are so mission-driven or task-oriented that they resent all forms of organizational or bureaucratic encroachment on their freedom and initiatives. This liberty is more easily reached at the top of the pyramid than in the middle, of course, but at any level of management, the urge remains.

Jack Seery (1997), the former head of the Overseas Business Group of Philips Consumer Electronics, is a good example of a high-level 'rule buster'. When asked how he managed to halve TV development lead-times in his overseas business group, he answered:

> The product creation process in our company was getting undue attention, without due results. So, I started at the other end. I said, 'I'm not interested in how long it takes. I'm telling you how long we can stand, and it's 12 months by coincidence, and the next time it will probably be shorter. So that's our target! And the only element you have to worry about is the timing. Nothing else. (. . .) I want it done in 12 months. And I will give you the possibility to see why it has to be 12 months. I'll send you all out in the market place if necessary. No problem. You can have all the exposure you need, but you have to do it'. (. . .) I wanted our development crew, also, to show what they were worth, because I always had the feeling that because of our organization split, the development crew never realized their real potential. They were frustrated. They were always being hammered for what they didn't do. They could tell you how they could do things much better, but the system didn't permit it. So, I got the system out of their way. The rules that I banned were all the rules that were restricting them. 'Write your own rules', I said, 'but you have to guarantee me the integrity of the product at the end of the day, and you're responsible . . . to me.

Innovation leaders differ from traditional managers in their outward-orientation. They stimulate their staffs to go out and broaden their horizon. Ideo's Dave Kelley is adamant that the people who create the most value for the company are not the ones management sees all the time at their desks. They are the ones who go out in the market place to meet customers, users, competitors and suppliers. So, innovation leaders force people out into the field. Jack Seery shares this philosophy. He loves to tell how he pushed his staff to dive into the market and visit consumer electronic shops when they were traveling overseas, not just meet expatriates in the local Philips office. To give his admonition some teeth, he adopted the habit of reimbursing overseas travel expenses of his staff only upon presentation of a field report indicating what they had learned from the market.

Innovation leaders often contrast with their colleagues by daring to defend a long-term view of the business. Companies that conduct surveys to measure their internal innovation climate often come to the same conclusion: one of the greatest perceived obstacles to innovation is management's excessive short-term orientation. Many business managers are reluctant to make a commitment to projects or undertakings with a risky profile or a long-term payback. In the eyes of innovative companies like DuPont, this common reluctance justifies continued heavy spending on central research. If business managers hesitate to invest in the long-term renewal of their products, applications and markets, the corporation must show the way. Of course, the corporation must have earmarked resources for such long-term undertakings.

Innovation leaders tend to resent the 'short-termism' of their business colleagues and will typically lobby hard with top management to be allowed to continue exploring new opportunities and keeping some 'risk-money' at hand. Medtronic's Chief Innovation Officer, alluded to earlier in this chapter, was one of those leaders with a longer-term orientation. His rank as Vice-Chairman, his personal prestige and the support he had from his CEO allowed him to operate in this long-horizon mode (George, 2001):

> When an organization said, 'Well, we don't really believe in this idea!' he said, 'Fine! I'll just set up a little team over here and spend a million dollars a year, and we will finance this little team and we will let them go work on it. If you don't want to do it, fine'. And then, when it worked, all of a sudden the organization would say, 'I want it back'.

Finally, innovation leaders can be identified by their instinctive desire to keep trying harder. They never pause and are never satisfied. Richard Teerlink (1997), the charismatic CEO of motorbike manufacturer Harley-Davidson, likes to refer to what he calls the three enemies of sustained success: arrogance, complacency and greed. Innovation leaders behave very much as if they shared the same belief. The Nokia culture and values, according to its President (Ala-Pietilä, 2001), address these 'innovation killers' almost frontally:

> The key elements or key attitudes derived from our values are 120% target setting, which means that we are not content with 100% target setting, because we already know how to do that. Why bother? So, in everything we do, we try to stretch the target setting to the 120% level, which means that we don't know today how to achieve that goal, but we have confidence in individuals and in teams that they will

come up with the innovative solutions which will then lead to that set target level. That is probably one of the most important and most distinct attitudes: A non-arrogant, non-complacent way of looking at the future.

Tetra Pak's Nick Shreiber (2002) stresses the same point in the leadership values he advocates:

> Delivering value also requires drive—giving energy to others, bringing excitement and enthusiasm—and never allowing complacency to set in. Remember the old English saying, 'Even if you are on the right track, you will get run over if you just sit there'.

Michael Dell (2001) sees his personal role as the CEO of a maverick company very much in this way, too:

> When I see things that are important, I will push on. And also by nurturing things when they are in the skunk works stage and don't necessarily have the organization's support. I am the agitator for progress and change.

Developing Innovation Leaders

The abundance of leadership programs on offer in business schools all over the world is the best evidence that innate leadership talents can be further developed. This development potential applies to *leadership for innovation* as well. In all cases, it seems, developing leadership talents and good reflexes comes from a combination of practical field experiences, coaching and feedback, and periods of reflection and self-examination. For innovation, nothing seems to replace an early exposure to the uncertainties of new product or business development projects. Innovation leadership cannot be taught; it has to be experienced. Venture capitalists know the lesson well: nothing replaces the painful experience of earlier failures and the learning that goes with them. As the Economist noted recently (Micklethwait, 1997), "In Silicon Valley, bankruptcy is treated like a duelling scar in a Prussian officers' mess". William George (2001) also emphasizes the importance of giving people the right experience:

> The best development is achieved by putting people on the right road, saying to them, 'Would you take over a venture? We will put together a little team of ten to twelve people, and let's see how it comes out. Would you take over this project and run it? Start this business from scratch and see if you can create something there?' That is the real test. And then, we would give the young engineers or scientists who joined the company succeeding levels of challenge to see if they could take it on. We would give them a chance, a small budget, a small risk, to see how it comes out. And we would take the creative ones and

give them more and more responsibility, bigger and bigger projects, more and more challenge.

Ad Huijser (2001) concurs with giving scientists and engineers this type of progressive experience-building assignments. But he maintains that research group leaders will not be able to become fully-fledged innovation leaders if they stay in the cozy confines of their research environment. He believes in the learning value of exposing them to the market:

> It is my conviction that, even in research, we have to build on those leaders who have business experience. They have to have been, for a certain while, in one of our business units. If not, I do not believe that they can feel the heat of the market and understand the constraints of the business area. Because in research you can say, 'The sky is the limit'. In business it is not. It is not just understanding the business context, but also being able to interface with colleagues in the business.

As suggested earlier, coaching by senior managers is generally felt to be the necessary complement to this type of experiential leadership development. Through coaching, senior managers can help would-be innovation leaders reflect on their experience. As Harvard's Nitin Nohria discovered in his course (Blagg & Young, 2001), you cannot 'teach leadership':

> When I teach leadership to MBAs, I don't believe that in thirty class sessions I will immediately make them better leaders. What I hope, however, is that I have taught them the capacity for deeper and more thoughtful reflection on their experiences so that they can learn from them and therefore become better leaders.

Coaching is particularly important for R&D staff when they find themselves at the traditional crossroads of having to choose a career orientation. Do they want to go in the direction of management and take on more and more responsibilities and become innovation leaders? Or do they prefer to go in the direction of pure science and technical work, and stay as innovators? Most CTOs recognize that this choice is highly personal, and because they need both types of talents, they tend to avoid influencing their staff one way or another.

Motivating and Retaining Innovation Leaders

The obvious conclusion from the above discussion it that true innovation leaders—managers with technical knowledge and strong leadership capabilities—are rare and precious birds in any company. So one last question remains: How can a company keep its rare innovation leaders loyal and motivated, and avoid their being poached by aggressive competitors? The chal-

lenge is, of course, particularly acute in environments with a high degree of industry competition and staff volatility, like the Silicon Valley, where people shift employers very easily. One possible answer is that companies can keep their innovation leaders with compensation packages and stock options. Money is, undoubtedly, an important motivator and probably still one of the key drivers in Silicon Valley. But even in the Valley, money alone cannot buy talents and maintain loyalty. People want to work on 'cool projects', and innovation leaders tend to be achievement-oriented. They will stay if they are doing something meaningful, if they are allowed to break new ground, and if the company they work for is successful. Even in the highly competitive field of medical technology, as William George (2001) reminds us, financial incentives are not the prime driver:

> I even know engineers who are offended, literally, by financial incentive programs that almost deny them the importance of their work. In many ways they want to feel that they are really doing something important, and the financial follows rather than leads.

Ad Huijser (2001) agrees:

> Mobility—going to work where there is nice work—has increased tremendously. People move particularly to the areas or the places where the excitement is. I believe that, certainly in a technical domain, people want to work at the leading edge. They want to be with the winners. So, as a company, you have to stay at the leading edge. If you fall behind, the first thing you will see is that the good people, your best people, will move out. So, we have bonuses, we have stock option plans, etc. But the best motivators for me are two things: One is the challenge of the work, and second, it is the leadership that people want to work for.

Conclusion: The Key Role of Top Management

Despite all their talents and qualities, innovation leaders will always remain vulnerable, even if they are part of senior management. Their more conservative management colleagues will challenge them for the risks they are taking. Some will not lose any opportunity to remind them, for years, of the failed projects the innovation leaders have launched or supported. Again, William George (2001):

> Innovation leaders need to feel personally secure, and also supported by the organization, from the top. Otherwise, the organization will grind them out. When it comes to budget time, they will get pushed aside. If their projects don't get along, then their budget will be cut. Because, what happens in

organizations (is this): the short-term tends to overtake the real opportunities, the project budgets, and the new products.

Even within a highly innovative culture like 3M, top managers have to be reminded of their personal responsibility in championing innovation leaders. The company cherishes this concluding quote, expressed in 1944 by one of the 3M historical CEOs (Lehr, 1979):

> As our business grows, it becomes increasingly necessary to delegate responsibility and to encourage men and women to exercise their initiative. This requires considerable tolerance. Those men and women, to whom we delegate authority and responsibility, if they are good people, are going to want to do their jobs in their own way. These are characteristics we want, and people should be encouraged as long as their way conforms to our general pattern of operation. Mistakes will be made, but if a person is essentially right, the mistakes he or she makes are not as serious in the long run as the mistakes management will make if it is dictatorial and undertakes to tell those under its authority exactly how they must do their job. Management that is destructively critical when mistakes are made kills initiative, and it's essential that we have many people with initiative if we are to continue to grow.

References

Abell, D. (1993). *Managing with dual strategies*. New York: The Free Press.

Ala-Pietilä, P. (2001). Leadership and innovation. Videotaped interview by J. P. Deschamps, IMD, Lausanne.

Bass, B. M. (1999). Two decades of research and development in transformational leadership. *European Journal of Work and Organizational Psychology*, 8 (1), 9–32.

Bessant, J. (2003). Challenges in innovation management. In: L. V. Shavinina (Ed.), *International Handbook on Innovation*. Oxford: Elsevier Science.

Blagg, D. & Young, S. (2001). What makes a good leader. *Harvard Business School Bulletin*, February, 31–36.

Borel, D. (2001). Leadership and innovation. Videotaped interview by J. P. Deschamps, IMD, Lausanne.

Bower, M. (1997). Developing leaders in business. *The McKinsey Quarterly*, 4, 4–17.

Cooper, R. (2003). Profitable product innovation. In: L. V. Shavinina (Ed.), *International Handbook on Innovation*. Oxford: Elsevier Science.

Dell, M. (2001). Direct from Dell, Q&A with Michael Dell. *Technology Review—Emerging Technologies and Their Impact*, On-line article from MIT Enterprise, July-August, available on http://www.technologyreview.com/articles/qa0701.asp.

Deschamps, J-P. & Nayak, P. (1995). *Product juggernauts—how companies mobilize to generate streams of market winners*. Cambridge, MA: Harvard Business School Press.

Drucker, P. F. (1985). *Innovation and entrepreneurship*. New York: Harper & Row.

Eccles, R., Nohria, N. & Berkley, J. (1992). *Beyond the hype: Rediscovering the essence of management*. Cambridge, MA: Harvard Business School Press.

George, W. (2001). Leadership and innovation. Videotaped interview by J. P. Deschamps, IMD, Lausanne.

Grove, A. S. (1996). *Only the paranoid survive*. New York: Bantam Doubleday.

Hauschildt, J. (2003). Promotors and champions in innovations—Development of a research paradigm. In: L. V. Shavinina (Ed.), *International Handbook on Innovation*. Oxford: Elsevier Science.

Huijser, A. (2001). Leadership and innovation. Videotaped interview by J. P. Deschamps, IMD, Lausanne.

Jones, T. & Austin, S. (2002). *Innovation leadership*. Management Report. London: Datamonitor PLC.

Katz, R. (2003). Managing technological innovation in business organizations. In: L. V. Shavinina (Ed.), *International Handbook on Innovation*. Oxford: Elsevier Science.

Kelley, T. (2001). *The art of innovation*. London: Harper Collins Business.

Kim, W. C. & Mauborgne, R. (1997). Value innovation—The strategic logic of high growth. *Harvard Business Review*, January-February, 103–112.

Knox, S. (2002). The boardroom agenda: Developing the innovative organization. *Corporate Governance*, January, 27–39.

Kotter, J. P. (1990). What leaders really do. *Harvard Business Review*, May-June, 103–111.

Kumar, N. & Rogers, B. (2000). *EasyJet 2000*. Teaching case No. GM 873. Lausanne: IMD.

Lehr, L. W. (1979). Stimulating technological innovation—The role of top management. *Research Management*, November, 23–25.

Mannarelli, T. (2001). Unlocking the secrets of business innovation, mastering people management/managing creativity. *Financial Times*, October 29, 10–11.

Micklethwait, J. (1997). The valley of money's delight. *The Economist*, London, March 29.

Morita, A. (1987). *Made in Japan*. London: William Collins.

Nayak, P. R. & Ketteringham, J. (1986). *Breakthroughs* (pp. 50–73). New York: Rawson Associates.

Perel, M. (2002). Corporate courage: Breaking the barrier to innovation. *Industrial Research Institute*, May-June, 9–17.

Peters, T. (1987). *Thriving on chaos*. New York: Alfred A. Knopf.

Quinn, J. B. (1985). Managing innovation: Controlled chaos. *Harvard Business Review*, May-June, 73–83.

Robert, M. (1991). *The essence of leadership*. New York: Quorum Books.

Rogers, E. M. & Larsen, J. K. (1984). *Silicon Valley fever—growth of high technology culture*. New York: Basic Books.

Schruijer, S. & Vansina, L. (1999). Leadership and organizational change. *European Journal of Work and Organizational Psychology*, 8 (1), 1–8.

Seery, J. (1997). Videotaped interview by J. P. Deschamps, IMD, Lausanne.

Senge, P. M. (1999). The discipline of innovation. *Executive Excellence*, 16 (6), 10–12.

Shamir, B. (1999). Leadership in boundaryless organizations. *Journal of Work and Organizational Psychology*, **8** (1), 49–72.

Shreiber, N. (2002). 2001 MBA graduation speech, published on IMD's Web Letter @imd.ch, January.

Teerlink, R. (1997). Unpublished speech delivered at IMD: Lausanne. October 20.

Tidd, J., Bessant, J. & Pavitt, K. (1997). *Managing innovation*. Chichester, U.K.: John Wiley.

Van de Ven, A. et al. (1999). *Innovation journey*. New York: Oxford University Press.

Yukl, G. (1999). An evaluative essay on current conceptions of effective leadership. *European Journal of Work and Organizational Psychology*, **8** (1), 33–48.

Part XII

Innovation and Marketing

The International Handbook on Innovation
Edited by Larisa V. Shavinina

Innovation and Market Research

Paul Trott

Business School, University of Portsmouth, U.K.

Abstract: Market research results frequently produce negative reactions to discontinuous new products (innovative products) that later become profitable for the innovating company. Famous examples such as the fax machine, the VCR and Dyson's bagless vacuum cleaner are often cited to support this view. Despite this, companies continue to seek the views of consumers on their new product ideas. The debate about the use of market research in the development of new products is long-standing and controversial. Against a backcloth of models of innovation, this chapter examines the extent to which market research is justified and whether companies should sometimes ignore their customers.

Keywords: Innovation; Innovative products; Market research; Marketing; New product development.

Introduction

There is much agreement in the literature that innovation[1] occurs through the interaction of the science base (dominated by universities and industry), technological development (dominated by industry) and the needs of the market (Christensen, 1997; Cooper, 1999; Tidd, Bessant & Pavitt, 2001). Indeed, it is the explanation of the interaction of these activities that forms the basis of models of innovation today. There is, however, much debate and disagreement about precisely what activities influence innovation and, more importantly, the internal processes that affect a company's ability to innovate. Market research results frequently produce negative reactions to discontinuous[2] new products that later become profitable for the innovating company. Famous examples such as the fax machine, the VCR and Dyson's bagless vacuum cleaner are often cited to support this view. Despite this, companies continue to seek the views of consumers on their new product ideas. The debate about the use of market research in the development of new products is long-standing and controversial. Against a backcloth of models of innovation, this chapter examines the extent to which market research is justified and whether companies should sometimes ignore their customers. The first section reviews the various models of innovation that have been used to try to delineate the activities that need to be in place if innovation is to succeed. The section concludes by arguing that it is necessary to view innovation as a series of linked activities, which can be best, described as a management process. The next section looks at the role of market research in innovation. For many years this was seen as controversial and has once again been identified in the literature as troublesome, especially when discontinuous new products are considered (Christensen, 1997; Kumar, Scheer & Kotler, 2000; Lukas & Ferrell, 2000). The final section suggests that organizations may be able to improve their new product development process by emphasizing the internal and external linkages within the management of innovation.

Background

Traditional arguments about innovation have centred around two schools of thought. On the one hand the social deterministic school argued that innovations were the result of a combination of external social

---

[1] Innovation is not a single action but a total process of interrelated sub-processes. It is not just the conception of a new idea, nor the invention of a new device, nor the development of a new market. The process is all these things acting in an integrated fashion' (Myers & Marquis, 1969).

[2] Discontinuous innovations often launch a new generation of technology; whereas continuous product innovations involve improving existing technology. Discontinuous products are generally used to refer to new products that involve dramatic leaps in terms of customer benefits relative to previous product offerings (Chandy & Tellis, 2000; Meyers & Tucker, 1989).

factors and influences, such as demographic changes, economic influences and cultural changes (Drucker, 1985). The argument was that when the conditions were 'right' innovations would occur. Today this view is incorporated and referred to as the *market-based view of innovation*, which argues that market conditions provide the context that facilitates or constrains a firm's innovative activity (Porter, 1985; Slater & Narver, 1994). However, the individualistic school argued that innovations were the result of unique individual talents and such innovators are born (Drucker, 1985). Intertwined with this view is the important role played by serendipity (Boden, 1991; Cannon, 1940; Kantorovich & Ne'eman, 1989; Roberts, 1989; Simonton, 1979). Today this view is incorporated within the *resource-based view of innovation*, which considers that the volatile nature of society and markets is insufficiently stable for a firm to develop long-term technology and innovation strategies. And that a firm's own knowledge and capabilities (resources) better enable firms to cultivate its own markets (Cohen & Levinthal, 1990; Grant, 1997; Hamel & Prahalad, 1994).

Many studies of historical cases of innovation have highlighted the importance of the unexpected discovery (Gallouj, 2002; Sundbo, 2002; Tidd, Bessant & Pavitt, 2001). The role of serendipity or luck is offered as an explanation. This view is also reinforced in the popular media. It is, after all, every persons dream that they will accidentally uncover a major new invention leading to fame and fortune. On closer inspection of these historical cases, such serendipity is rare indeed. After all, in order to recognize the significance of an advance one would need to have some prior knowledge in that area (Cohen & Levinthal, 1990; Simonton, 2002). Hence, most discoveries are the result of people who have had a fascination with a particular area of science or technology and it is following extended efforts on their part that advances are made. Discoveries may not be expected, but in the words of Louis Pasteur, 'chance favors the prepared mind'.

It was U.S. economists after the Second World War who championed the linear model of science and innovation. Since then, largely because of its simplicity, this model has taken a firm grip on people's views on how innovation occurs. Indeed, it dominated science and industrial policy for 40 years. It was only in the 1980s that management schools around the world seriously began to challenge the sequential linear process. The recognition that innovation occurs through the interaction of the science base (dominated by universities and industry), technological development (dominated by industry) and the needs of the market was a significant step forward (see Fig. 1). The explanation of the interaction of these activities forms the basis of models of innovation today.

There is, of course, much debate and disagreement about precisely what activities influence innovation and, more importantly, the internal processes that affect a company's ability to innovate.

Nonetheless there is broad agreement that it is the linkages between these key components that will produce successful innovation. From a European perspective an area that requires particular attention is the linkage between the science base and technological development. The European Union (EU) believes that, compared to the USA, European universities have not established effective links with industry. Whereas in the U.S. universities have been working closely with industry for many years (Miyata, 2003).

Traditionally the innovation process has been viewed as a sequence of separable stages or activities (Gallouj, 2002; Sundbo, 2003). There are two basic variations of this model for product innovation. First, and most crudely, there is the technology driven model (often referred to as 'technology push') where it is assumed that scientists make expected and unexpected discoveries, technologists apply them to develop product ideas and engineers and designers turn them into prototypes for testing. It is left to manufacturing to devise ways of producing the products efficiently. Finally, marketing and sales will promote the product to the potential consumer. In this model the market place was a passive recipient for the fruits of R&D. This so-called 'technology-push' model dominated industrial policy after the Second World War. While

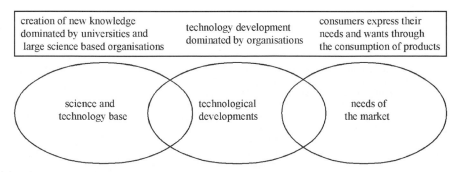

Figure 1. Conceptual framework of innovation.

this model of innovation can be applied to a few cases, most notably in technology-intensive industries such as the pharmaceutical industry, it is not applicable in many other instances. In particular where the innovation process follows a different route.

It was not until the 1970s, that new studies of actual innovations suggested the role of the market place was influential in the innovation process (von Hippel, 1978). This led to the second linear model, the 'market-pull' model of innovation. The customer 'need-driven' model emphasizes the role of marketing, as an initiator of new ideas as a result of close interactions with customers. These, in turn, are conveyed to R&D for design and engineering and then to manufacturing for production.

Whether innovations are stimulated by technology, customer need, manufacturing and a host of other factors, including competition, misses the point. The models above concentrate on what is driving the downstream efforts rather than on how innovations occur (Galbraith, 1982). Hence, the linear model is only able to offer an explanation of where the initial stimulus for innovation was born. That is, where the trigger for the idea or need was initiated. The simultaneous coupling model suggests that it is the result of the simultaneous coupling of the knowledge within all three functions (i.e. marketing, production and R&D) that will foster innovations (Rothwell & Zigweld, 1985; Sundbo, 2003). Furthermore, the point of commencement for innovation is not known in advance.

The interactive model develops this idea further (see Fig. 2) and links together the technology push and market pull models of innovation (Rothwell & Zigweld, 1985; Sundbo, 2003). It emphasizes that innovations occur as the result of the interaction of the market place, the science base and the organization's capabilities. Like the coupling model there is no explicit starting point. The use of information flows is used to explain how innovations transpire and that they can arise from a wide variety of points.

While still oversimplified, it is a more comprehensive representation of the innovation process. It can be regarded as a logically sequential, though not necessarily continuous process, that can be divided into a series of functionally distinct but interacting and interdependent stages (Rothwell & Zigweld, 1985). The overall innovation process can be thought of as a complex set of communication paths whereby knowledge is transferred. These paths include internal and external linkages. The innovation process outlined in Fig. 2 represents the organization's capabilities and its linkages with both the market place and the science base. It is argued that organizations that are to manage this process effectively will be successful at innovation.

At the centre of the model are the organizational functions of R&D, engineering and design, manufacturing and marketing and sales. While at first this may appear as a linear model, the flow of communication is not necessarily in a linear fashion. There is provision for feedback. Also, linkages with the science base and the market place occur between all functions not just with R&D or marketing. For example, as often happens, it may be the manufacturing function which initiates a design improvement that leads to the introduction of either a different material or the eventual development by R&D of a new material. Finally, the generation of ideas is shown to be dependent upon inputs from three basic components (as outlined in Fig. 1): organization capabilities; the needs of the market place; the science and technology base (Bessant, 2003; Katz, 2003).

The preceding discussions have revealed that innovation is not a singular event, but a series of activities that are linked in some way to the others. This may be described as a process and involves (Kelly & Kranzberg, 1978):

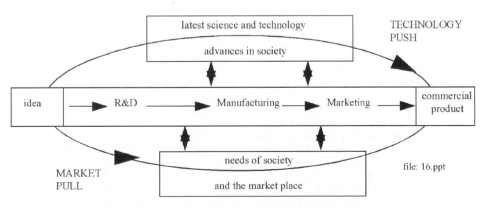

Figure 2. Interactive model of innovation.

(1) a response to either a need or an opportunity that is context dependent;

(2) a creative effort and if successful results in the introduction of novelty;

(3) the need for further changes.

There have been a plethora of models to help us understand the innovation process (see: Gallouj, 2002; Rothwell, 1992; Trott, 1998). Usually in trying to capture this complex process the simplification has led to misunderstandings. The simple linear model of innovation can be applied only to a few innovations and is more applicable to certain industries than others. As previously stated, the pharmaceutical industry characterizes much of the technology push model or resource-based view. Other industries, however, like fast moving consumer goods (FMCG) industries are better represented by the market pull model or market-based view. For most industries and organizations innovations are the result of a mixture of the two. Managers working within these organizations have the difficult task of trying to manage this complex process.

Market Research

Christensen (1997) investigated why well run companies that were admired by many, failed to stay on top of their industry. His research showed that in the cases of well managed firms such as Digital, IBM, Apple and Xerox, 'good management' was the most powerful reason why they failed to remain market leaders (sic). It was precisely because these firms listened to their customers and provided more and better products of the sort they wanted that they lost their position of leadership. He argues that there are times when it is right not to listen to customers. Recent research by Ovans (1998) supports this claim. He suggests that purchase-intention surveys are not effective predicators of sales of new products. The research revealed that people aren't generally reliable predictors of their own long-term purchasing behavior. The type of question used and whether or not the question is placed in context greatly affects the reliability of such market research. James Dyson has good reason to be suspicious of the role of market research in new product development. Not only did he struggle for many years to get anyone in the U.K. to believe it was worth manufacturing his bagless vacuum cleaner, he faced the same scepticism when he launched in the U.S. (Thrift, 1997).

Many industry analysts and business consultants are now arguing that the devotion to focus groups and market research has gone too far (Chandy & Tellis, 2000; Christensen, 1997; Francis, 1994; Martin, 1995). Indeed, the traditional new product development (NPD) process of market research, segmentation, competitive analysis and forecasting, prior to passing the resultant information to the research and development (R&D) department, leads to commonality and

bland new products. This is largely because the process constrains, rather than, facilitates innovative thinking and creativity. Furthermore, and more alarming, is that these techniques are well known and used by virtually all companies operating in consumer markets. In many of these markets the effect is an over-emphasis on minor product modifications and on competition that tends to focus on price (Veryzer, 2003). Indeed, critics of the market-orientated approach to new product development argue that the traditional marketing activities of branding, advertising, positioning, market research and consumer research act as an expensive obstacle course to product development rather than facilitating the development of new product ideas (Cooper, 2003).

For many large multi-product companies it seems the use of market research is based upon accepted practice in addition to being an insurance policy. Many large companies are not short of new product ideas, the problem lies in deciding in which ones to invest substantial sums of money (Trott, Cordey-Hayes & Seaton, 1995), and then justifying this decision to senior managers. Against this background one can see why market research is so frequently used without hesitation, as decisions can be justified and defended. Small companies in general, and small single product companies in particular, are in a different situation. Very often new product ideas are scarce; hence, such companies frequently support ideas based upon their intuition and personal knowledge of the product. This is clearly the situation with James Dyson's bagless vacuum cleaner (Dyson, 1998).

Morone's (1993) study of successful U.S. product innovations suggests that success was achieved through a combination of discontinuous product innovations[3] and incremental improvements. Indeed, Lynn, Morone & Paulson (1997) argue that: in competitive, technology-intensive industries success is achieved with discontinuous product innovations through the creation of entirely new products and businesses, whereas product line extensions and incremental improvements are necessary for maintaining leadership. This, however, is only after leadership has been established through a discontinuous product innovation. This may appear to be at variance with accepted thinking that Japan secured success in the 1980s through copying and improving U.S. and European technology. This argument is difficult to sustain on close examination of the evidence (Lynn et al., 1997; Morone, 1993; Nonaka & Takeuchi, 1995). The most successful Japanese firms

---

[3] Discontinuous innovations often launch a new generation of technology; whereas continuous product innovations involve improving existing technology. Discontinuous products are generally used to refer to new products that involve dramatic leaps in terms of customer benefits relative to previous product offerings (Chandy & Tellis, 2000; Meyers & Tucker, 1989).

have also been leaders in research and development (see: Deshpande, Farley & Webster, 1993; Lyons, 1976; Nonaka, 1991; Nonaka & Kenney, 1991; Nonaka & Takeuchi, 1995). Furthermore, as Cohen & Levinthal have continually argued (1990, 1994) access to technology is dependent on one's understanding of that technology.

Adopting a resource-based view or technology push approach to product innovations can allow a company to target and control premium market segments, establish its technology as the industry standard, build a favorable market reputation, determine the industry's future evolution, and achieve high profits. It can become the centerpiece in a company's strategy for market leadership. It is, however, costly and risky; this is the dilemma facing firms. Such an approach requires a company to develop and commercialize an emerging technology in pursuit of growth and profits. To be successful, a company needs to ensure its technology is at the heart of its competitive strategy. Merck, Microsoft and Dyson have created competitive advantage by offering unique products, lower costs or both by making technology the focal point in their strategies. These companies have understood the role of technology in differentiating their products in the marketplace. They have used their respective technologies to offer a distinct bundle of products, services and price ranges that have appealed to different market segments. Such products revolutionize product categories or define new categories, such as Hewlett-Packard's Laser-jet printers and Apple's (then IBM) personal computer. These products shift market structures, require consumer learning and induce behavior changes, hence, the difficulties for consumers when they are asked to pass judgment.

It seems the dilemma faced by companies when using market research findings is twofold:

- At the policy level: to what extent should companies pursue a strategy of providing more room for technology development of products and less for market research that will *surely* increase the likelihood of failure, but will also increase the chance of a major innovative product.
- At the operational level: to what extent should Product and Brand Managers make decisions based upon market research findings.

Market Research and Discontinuous New Products

In one of the most comprehensive reviews of the literature on product development Brown & Eisenhardt (1995) develop a model of factors affecting the success of product development. This model highlights the distinction between process performance and product effectiveness and the importance of agents, including team members, project leaders, senior management, customers, and suppliers, whose behavior affects these outcomes. The issue of whether customers can hinder

the product development process is not, however, discussed.

It is argued by many from within the market research industry that only extensive market research can help to avoid large scale losses such as those experienced by RCA with its Videodisc, Procter and Gamble with its Pringles and General Motors with its rotary engine (see: Barrett, 1996; Kotler, 1999; Urban & Hauser, 1993). Sceptics may point to the issue of vested interests in the industry, and that it is merely promoting itself. It is, however, widely accepted that most new products fail in the market because consumer needs and wants are not satisfied. Study results show that 80% of newly introduced products fail to establish market presence after two years (Barrett, 1996). Indeed, cases involving international high profile companies are frequently cited to warn of the dangers of failing to utilize market research (e.g. Unilever's Persil Power and R J Reynold's Smokeless cigarette).

Given the inherent risk and complexity, managers have asked for many years whether this could be reduced by market research. Not surprisingly, the marketing literature takes a market driven view, which has extensive market research as its key driver (Booz, Allen & Hamilton, 1982). The benefits of this approach to the new product development process have been widely articulated and are commonly understood (Cooper, 1990; Kotler, 1998). Partly because of its simplicity this view now dominates management thinking beyond the marketing department. Advocates of market research argue that such activities ensure that companies are consumer-orientated. In practice, this means that new products are more successful if they are designed to satisfy a perceived need rather than if they are designed simply to take advantage of a new technology (Ortt & Schoormans, 1993). The approach taken by many companies with regard to market research is that if sufficient research is undertaken the chances of failure are reduced (Barrett, 1996). Indeed, the danger that many companies wish to avoid is the development of products without any consideration of the market. Moreover, once a product has been carried through the early stages of development it is sometimes painful to raise questions about it once money has been spent. The problem then spirals out of control, taking the company with it.

The issue of market research in the development of new products is controversial. The debate will continue for the foreseeable future about whether product innovations are caused by resource-based or market-based factors. The issue is most evident with discontinuous product innovations, where no market exists. First, if potential customers are unable adequately to understand the product, then market research can only provide negative answers (Brown, 1991; Chandy & Tellis, 2000). Second, consumers frequently have difficulty articulating their needs. Hamel and Prahald (1994) argue that customers lack foresight; they refer to Akio Morita, Sony's influential leader:

> Our plan is to lead the public with new products rather than ask them what kind of products they want. The public does not know what is possible, but we do.

This leads many scientists and technologists to view marketing departments with skepticism. As they have seen their exciting new technology frequently rejected due to market research findings produced by their marketing department. Market research specialists would argue that such problems could be overcome with the use of 'benefits research'. The problem here is that the benefits may not be clearly understood, or even perceived as a benefit by respondents. King (1985) sums up the research dilemma neatly:

> Consumer research can tell you what people did and thought at one point in time: it can't tell you directly what they might do in a new set of circumstances.

This is particularly the case if the circumstances relate to an entirely new product that is unknown to the respondent. New information is always interpreted in light of one's prior knowledge and experience. Roger's (1995) studies on the diffusion of innovations as a social process argue that it requires time for societies to learn and experiment with new products. This raises the problem of how to deal with consumers with limited prior knowledge and how to conduct market research on a totally new product or a major product innovation. In their research analyzing successful cases of discontinuous product innovations, Lynn et al. (1997) argue that firms adopt a process of probing and learning. Valuable experience is gained with every step taken and modifications are made to the product and the approach to the market based on that learning. This is not trial and error but careful experimental design and exploration of the market often using the heritage of the organization (Nonaka, 2002). This type of new product development is very different from traditional techniques and methods described in marketing texts (see Kardes, 1998; Kotler, 1999).

Knowing what the customer thinks is still very important, especially when it comes to product modifications or additional attributes. There is, however, a distinction between additional features and the core product benefit or technology (Levitt, 1980). Also, emphasis needs to be placed on the buyer rather than the consumer, for the buyer may be the end user but equally may not, as is the case with industrial markets. Furthermore, for a product to be successful it has to be accepted by a variety of actors such as fellow channel members. In industrial markets the level of information symmetry about the core technology is usually very high indeed (hence the limited use of market research),

but in consumer markets this is not always the case. For example, industrial markets are characterized by:

- Relatively few (information rich) buyers;
- Products are often customized and can involve protracted negotiations regarding specifications;
- And, most importantly, the buyers are usually expert in the technology of the new product (i.e. high information symmetry about the core technology).

A Framework for the Management of Innovation

Industrial innovation[4] and new product development[5] today has evolved considerably from their early beginnings. Indeed, as we have seen, innovation is extremely complex and involves the effective management of a variety of different activities within the organization. It is precisely how the process is managed that needs to be examined. A framework is presented in Fig. 3 that helps to illustrate innovation as a management process. This is simply an aid in describing the main factors that need to be considered if innovation is to be successfully managed by organizations. It helps to show that while the interactions of the functions[6] inside the organization are important, so too are the interactions of the functions with the external environment. Scientists and engineers within the firm will be continually interacting with fellow scientists in universities and other firms about scientific and technological developments (Major & Cordey-Hayes, 2000; Rothwell, 1992). Similarly the marketing function will need to interact with suppliers, distributors, customers and competitors to ensure the day-to-day activities of understanding customer needs and getting products to customers is achieved. Business planners and senior management will likewise communicate with a wide variety of firms and other institutions external to the firm, such as government departments, suppliers, customers, etc. All these information flows contribute to the wealth of knowledge held by the organization (Major & Cordey-Hayes, 2003; Woolgar, Vaux, Gomes, Ezingeard & Grieve, 1998). Recognizing this, capturing and utilizing it to develop successful new products is the difficult management process of innovation.

Within any organization there are likely to be many different functions. Depending on the nature of the business, some functions will be more influential than others. For example in fast moving consumer product

---

[4] Industrial innovation refers to innovation occurring within competitive industrial markets as opposed to research conducted within academia and government funded agencies.
[5] New product development is referred to as the introduction of a new or improved product (Booz, Allen & Hamilton, 1981).
[6] Function here refers to a business function such as marketing, manufacturing or Research and Development (R&D).

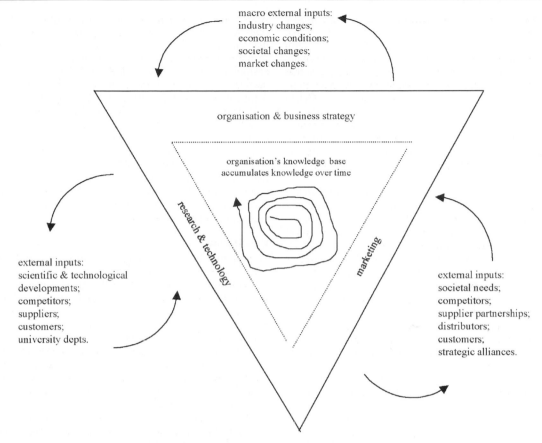

Figure 3. Innovation management framework.

firms Brand Managers are frequently very influential, whereas in technology intensive firms the R&D function is often dominant. The framework shown in Fig. 3 identifies three main functions: marketing, R&D and manufacturing, and business planning. Historical studies of innovations have identified these functions as the most influential in the innovation process (Gallouj, 2002; Sundbo, 2003; Tidd, Bessant & Pavitt, 2001). Whether one lists three or seven functions, misses the point, which is that it is the interaction of these internal functions and the flow of knowledge between them that needs to be facilitated (Nonaka & Takeuchi, 1995; Nonaka, Toyama & Konnon, 2000; Trott, Cordey-Hayes & Seaton, 1995). Similarly, as shown on the framework, effective communication with the external environment also requires encouragement and support (Oliver & Blakeborough, 1998).

The Need to Share and Exchange Knowledge

The framework in Fig. 3 emphasizes the importance placed of interaction (both formal and informal) within the innovation process. Indeed, innovation has been described as an information creation process that arises

out of social interaction (Trott et al., 1995). In effect, the firm provides a structure within which the creative process is located (Nonaka & Kenny, 1991; Nonaka, Sasaki & Ahmed, 2003).

These interactions provide the opportunity for thoughts, potential ideas and views to be shared and exchanged. However, we are often unable to explain what we normally do; we can be competent without being able to offer a theoretical account of our actions (Polyani, 1966). This is referred to as 'tacit knowledge' (Sternberg, Forsythe, Hedlund, Horvath, Wagner, Williams, Snook & Grigorenko, 2000). A great deal of technical skill is 'know-how' and much industrial innovation occurs through on the spot experiments, a kind of action oriented research with ad hoc modifications during step-by-step processes, through which existing repertoires are extended. Such knowledge can only be learned through practice and experience. This view has found significant support from a study of Japanese firms (Nonaka, 1991); where the creation of new knowledge within an organization depends on tapping the tacit and often highly subjective insights, intuitions and hunches of individual employees and making those insights available for testing and use by the organization as a whole. Hence, this implies that certain knowledge and skills, embodied in the term 'know-how', are not easily understood, moreover are less able to be communicated. This would suggest that to gain access to such knowledge one may have to be practising in this or related areas of knowledge. Cohen and Levinthal (1990) refer to this condition as 'lockout' suggesting that failure to invest in research and technology will limit an organization's ability to capture technological opportunities. 'Once off the technological escalator it's difficult to get back on' Cohen & Levinthal (1990, p. 128).

In addition to informal interactions, the importance of formal interactions is also highlighted. There is a substantial amount of research stressing the need for a 'shared language' within organizations to facilitate internal communication (Allen, 1977; Rothwell, 1991; Tushman, 1978; Woolgar et al., 1998). The arguments are presented along the following lines: if all actors in the organization share the same specialized language, they will be effective in their communication. Hence, there needs to be an overlap of knowledge in order for communication to occur. Such arguments have led to developments in cross-function interfaces, for example between R&D, design, manufacturing and marketing (Nonaka & Takeuchi, 1995; Souder, 1988). Concurrent Engineering is an extension of this; in this particular case a small team consisting of a member from each of the various functional departments manage the design, development, manufacture and marketing of a product.

Such thinking is captured in the framework outlined in Fig. 3. It stresses the importance of interaction and communication within and between functions and with the external environment. This networking structure allows lateral communication, helping managers and their staff unleash creativity. This framework emphasizes the importance of informal and formal networking, across all functions.

This introduces a tension between the need for diversity, on the one hand, in order to generate novel linkages and associations, and the need for commonalty on the other, to facilitate effective internal communication. Clearly there will be an organizational trade-off between diversity and commonalty of knowledge across individuals (Beveridge, 1957; Martindale, 1995; McCrae, 1987; Shadish, 1989).

Organizational Heritage
Finally the center of the framework is represented as organizational heritage, sometimes referred to as the organizational knowledge base. This does not mean the culture of the organization. It represents a combination of the organization's knowledge base (established and built up over the years of operating) and the organization's architecture that is unique to a firm (Freeman, 1991; Kay, 1993; Teece & Pisano, 1994; Tushman &

Anderson, 1986). This organizational heritage represents for many firms a powerful competitive advantage that enables them to compete with other firms. For 'American Express' it is its customer service and customer relations, developed and built up over decades, that provides the company with a powerful competitive advantage. 'Bayer, BASF and Siemens' organizational heritage is dominated by their continual investment over almost a hundred years in science and technology and the high profile given to science and technology within their businesses. For 'Unilever', its organizational heritage can be said to lie in its brand management skills and know-how developed over many years. These heritages cannot be ignored or dismissed as irrelevant when trying to understand how companies manage their innovative effort.

Conclusions
Very often product innovation is viewed from purely a marketing perspective with little, if any, consideration of the R&D function and the difficulties of managing science and technology. Likewise many manufacturing and technology approaches to product innovation have previously not taken sufficient notice of the needs of the customer. Finally, the organizational heritage of the firm will influence its future decisions regarding the markets in which it will operate. The point here is that firms do not have a completely free choice. What it does in the future will depend to some extent on what it has done in the past.

Market research can provide a valuable contribution to the development of innovative products. The difficulties lie in the selection and implementation of research methods. It may be that market research has become a victim of its own success, that is, business and product managers now expect it to provide solutions to all difficult product management decisions. Practitioners need to view market research as a collection of techniques that can help to inform the decision process. The conceptual framework outlined in this chapter should help Product and Brand Managers to consider when and under what circumstances market research is most effective. The right sort of market research can be invaluable. The problem is that within consumer markets there are technology intensive and technology vacant industries. In many of the technology intensive industries such as telecommunications, computer hardware and software, these firms are able to utilize their industrial market heritage to balance the need for technology and listen to the needs of consumers. In technology vacant consumer markets, such as food and personal care, the danger is that the technology agenda is completely dominated by market research findings. Minor product modifications may keep a product and brand competitive in the short-term, but if long-term growth is sought then more free-thinking and creativity needs to be afforded to the R&D department.

References

Allen, T. J. (1977). *Managing the flow of technology.* Cambridge MA: MIT Press.

Barrett, P. (1996). The good and bad die young. *Marketing,* July 11, 16.

Bessant, J. (2003). Challenges in innovation management. In: L. V. Shavinina (Ed.), *International Handbook on Innovation.* Oxford: Elsevier Science.

Beveridge, W. I. B. (1957). *The art of scientific investigation* (3rd ed.). New York: Vintage.

Boden, M. A. (1991). *The creative mind: Myths & mechanisms.* New York: BasicBooks.

Booz, Allen & Hamilton (1982). *New product management for the 1980s.* New York: Booz, Allen & Hamilton.

Brown, S. L. & Eisenhardt, K. M. (1995). Product development: Past research, present findings and future directions. *Academy of Management Review,* **20** (2), 343–378.

Cannon, W. B. (1940). The role of chance in discovery. *Scientific Monthly,* **50**, 204–209.

Chandy, R. K. & Tellis, G. J. (2000). The incumbent's curse? Incumbency, size and radical product innovation. *Journal of Marketing,* **64**, July, 1–17.

Christensen, C. M. (1997). *The innovator's dilemma: When new technologies cause great firms to fail.* Cambridge, MA: HBS Press.

Cohen, W. M. & Levinthal, D. A. (1990). A new perspective on learning and innovation. *Administrative Science Quarterly,* **35** (1), 128–152.

Cohen, W. M. & Levinthal, D. A. (1994). Fortune favours the prepared firm. *Management Science,* **40** (3), 227–251.

Cooper, R. (2003). Profitable product innovation. In: L. V. Shavinina (Ed.), *International Handbook on Innovation.* Oxford: Elsevier Science.

Cooper, R. G. (1990). New products: What distinguishes the winners. *Research and Technology Management,* Nov-Dec, 27–31.

Cooper, R. G. (1999). The invisible success factors in product innovation. *Journal of Product Innovation Management,* **16** (2), April 1999, 115–133.

Cooper, R. G. (1999). New product leadership: building in the success factors. *New Product Development & Innovation Management,* **1** (2), 125–140.

Cooper, R. G. & Kleinschmidt, E. J. (1993). Major new products: What distingusihes the winners in the chemical industry? *Journal of Product Innovation Management,* **10** (1), 90–111.

Deshpande, R., Farley, J. U. & Webster, F. E. (1993). Corporate culture, customer orientation, and innovativeness in Japanese firms: a quadrad analysis. *Journal of Marketing,* **57** (January), 23–27.

Dhebar, A. (1996). Speeding high-tech producer, meet the balking consumer. *Sloan Management Review,* **37** (2), 37–50.

Drucker, P. (1985). The discipline of innovation. *Harvard Business Review,* May-June (3), 67–72.

Dyson, J. (1998). *Against the odds.* London: Orion Books.

Francis, J. (1994). Rethinking NPD; giving full rein to the innovator. *Marketing,* May 26, 6.

Freeman, C. (1991). Networks of innovators: A synthesis of research issues. *Research Policy,* **20** (5), 499–514.

Galbraith, J. R. (1982). Designing the innovative organisation. *Organisational Dynamics,* Winter, 3–24.

Gallouj, F. (2002). Interactional innovation: a neo-Schumpeterian model. In: J. Sundbo & L. Fuglsang (Eds), *Innovation as Strategic Reflexivity.* London: Routledge.

Grant, R. M. (1997). Contemporary strategic analysis: Concepts, techniques, applications. Oxford, U.K.: Blackwell.

Hamel, G. & Prahalad, C. K. (1994). Competing for the future. *Harvard Business Review,* **72** (4), 122–128.

Kantorovich, A. & Ne'eman, Y. (1989). Serendipity as a source of evolutionary progress in science. *Studies in History and Philosophy of Science,* **20**, 505–529.

Kardes, F. R. (1999). *Consumer behaviour: Managerial decision making.* New York: Addison-Wesley.

Katz, R. (2003). Managing technological innovation in business organizations. In: L. V. Shavinina (Ed.), *International Handbook on Innovation.* Oxford: Elsevier Science.

Kelly, P. & Kranzberg, M. (1978). *Technological innovation: A critical review of current knowledge.* San Francisco: San Francisco Press.

Kotler, P. (1998). *Marketing management.* London: Prentice Hall.

Kotler, P. (1999). *Kotler on marketing: How to create, win and dominate markets.* New York: The Free Press.

Kumar, N., Scheer, L. & Kotler, P. (2000). From market driven to market driving. *European Management Journal,* **18** (2), 129–141.

Levitt, T. (1980). Marketing success through the differentiation of anything. *Harvard Business Review* (58), Jan-Feb, 56–66.

Lukas, B. A. & Ferrell, O. C. (2000). The effect of market orientation on product innovation. *Journal of Academy of Marketing Science,* **28** (2), 239–247.

Lynn, G. S., Morone, J. G. & Paulson, A. S. (1997). Marketing and discontinuous innovation: The probe and learn process. In: M. L. Tushman & P. Anderson (Eds), *Managing Strategic Innovation and Change, A Collection of Readings* (pp. 353–375). New York: Oxford University Press.

Lyons, N. (1976). *The Sony vision.* New York: Crown Publishers.

Major, E. & Cordey-Hayes, M. (2003). Encouraging innovation in small firms through externally generated knowledge. In: L. V. Shavinina (Ed.), *International Handbook on Innovation.* Oxford: Elsevier Science.

Martin, J. (1995). Ignore your customer. *Fortune,* **8**, May 1, 121–125.

Martindale, C. (1995). Creativity and connectionism. In: S. M. Smith, T. B. Ward & R. A. Finke (Eds), *The Creative Cognition Approach* (pp. 249–268). Cambridge, MA: MIT Press.

McCrae, R. R. (1987). Creativity, divergent thinking, and openness to experience. *Journal of Personality and Social Psychology,* **52**, 1258–1265.

Meyers, P. W. & Tucker, F. G. (1989). Defining roles for logistics during routine and radical technological innovation. *Journal of the Academy of Marketing Science,* **17** (1), 73–82.

Miyata, Y. (2002). An analysis of research and innovative activities of U.S. universities. In: L. V. Shavinina (Ed.), *International Handbook on Innovation.* Oxford: Elsevier Science.

Morone, J. (1993). *Winning in high-tech markets.* Cambridge, MA.: Harvard Business School Press.

Nonaka, I. & Kenney, M. (1991). Towards a new theory of innovation management: A case study comparing Canon, Inc. and Apple Computer, Inc. *Journal of Engineering and Technology Management*, **8**, 67–83.

Nonaka, I. (1991). The knowledge creating company. *Harvard Business Review*, Nov-Dec (6).

Nonaka, I., Sasaki, K. & Ahmed, M. (2003). Continuous innovation: The power of tacit knowledge. In: L. V. Shavinina (Ed.), *International Handbook on Innovation*. Oxford: Elsevier Science.

Nonaka, I. & Takeuchi, H. (1995). *The knowledge-creating company*. New York: Oxford University Press.

Oliver, N. & Blakeborough, M. (1998). Innovation networks: The view from inside. In: J. Greive Smith & J. Michie (Eds), *Innovation, Co-operation and Growth*. Oxford: Oxford University Press.

Ortt, R. J. & Schoormans, P. L. (1993). Consumer research in the development process of a major innovation. *Journal of the Market Research Society*, **35** (4), 375–389.

Ovans (1998). The customer doesn't always know best. *Market Research*, **7** (3), May–June, 12–14.

Polanyi, M. (1966). *The tacit dimension*. London: Rotledge and Kegan Paul.

Roberts, R. M. (1989). *Serendipity: Accidental discoveries in science*. New York: John Wiley.

Rogers, E. (1995). *The diffusion of innovation*. New York: Free Press.

Rothwell, R. (1992). Successful industrial innovation: critical factors for the 1990's. *R&D Management*, **22** (3), 64–84.

Rothwell, R. & Zigweld, W. (1985). *Reindustrialisation and technology*. London: Longman.

Rothwell, R. & Zegvelt, W. (1982). *Innovation and the small and medium sized firm*. London: Frances Pinter.

Shadish, W. R., Jr. (1989). The perception and evaluation of quality in science. In: B. Gholson, W. R. Shadish, Jr., R. A. Neimeyer & A. C. Houts (Eds), *The Psychology of Science: Contributions to Metascience* (pp. 383–426). Cambridge: Cambridge University Press.

Simonton, D. K. (2003). Exceptional creativity and chance: Creative thought as a stochastic combinatorial process. In: L. V. Shavinina & M. Ferrari (Ed.), *Beyond Knowledge* (in Press). Mahwah, NJ: Erlbaum Publishers.

Simonton, D. K. (1979). Multiple discovery and invention: Zeitgeist, genius, or chance? *Journal of Personality and Social Psychology*, **37**, 1603–1616.

Souder, W. E. (1988). Managing relations between R&D and marketing in new product development projects. *Journal of Product Innovation*, **5** (1), 6–19.

Sternberg, R. J., Forsythe, G. B., Hedlund, J., Horvath, J. A., Wagner, R. K., Williams, W. M., Snook, S. A. & Grigorenko, E. L. (2000). *Practical intelligence in everyday life*. New York: Cambridge University Press.

Sundbo, J. (2002). Innovation as strategic process. In: J. Sundbo & L. Fuglsang (Eds), *Innovation as Strategic Reflexivity*. London: Routledge.

Teece, D. & Pisano, G. (1994). The dynamic capabilities of firms: an introduction. *Industrial and corporate change*, **3** (3), 537–555.

Thrift, J. (1997). Too much good advice. *Marketing*, **3**, April 3, 18.

Tidd, J., Bessant, J. & Pavitt, K. (2001). *Managing innovation*. Chichester, U.K.: John Wiley.

Trott, P. (1998). *Innovation management and new product development*. London: FT Management.

Trott, P., Cordey-Hayes, M. & Seaton, R. A. F. (1995). Inward Technology Transfer as an interactive process: A case study of ICI. *Technovation*, **15** (1), 25–43.

Tushman, M. L. (1978). Task characteristics and technical communication in research and development. *Academy of Management Review Journal*, **20** (2), 75–86.

Tushman, M. & Anderson, P. (1986). Technological discontinuities and organizational environments. *Administrative Science Quarterly*, **31**, 439–465.

Urban, G. L. & Hauser, J. R. (1993). *Design and marketing of new products* (2nd ed.). Englewood Cliffs, NJ: Prentice Hall.

von Hippel, E. (1986). Lead users: A source of novel product concepts. *Management Science*, **32** (7), July, 791–805.

Veryzer, R. W. (2003). Marketing and the development of innovative products. In: L. V. Shavinina (Ed.), *International Handbook on Innovation*. Oxford: Elsevier Science.

von Hippel, E. (1978). Cooperation between rivals: Information know-how trading. *Research Policy*, **16**, 291–302.

von Hippel, E. & Thomke, S. (1999). Creating breakthroughs at 3M. *Harvard Business Review*, **77** (5), Sept-Oct, 47–57.

The International Handbook on Innovation
Edited by Larisa V. Shavinina

Marketing and the Development of Innovative New Products

Robert W. Veryzer

Rensselaer Polytechnic Institute, Lally School of Management and Technology, USA

Abstract: This chapter examines innovation from the perspective of marketing concerns and challenges. The role of marketing in the development of highly innovative products is discussed as are a number of the relevant key questions and concerns inherent in this type of product development. As part of this discussion, a framework consisting of three dimensions useful in conceptualizing innovation is presented. This is extended to include a number of influencers of product adoption that are important to consider during the design and development process (and also later on with respect to product launch) for innovative products.

Keywords: Innovation; Discontinuous innovation; Marketing; Consumer behavior; Product adoption.

Introduction

Innovation, or the introducing of new and more effective ways of doing things, is critical to the success of new products. As the competition inherent in a global marketplace increases, product (and service) innovation is the driving force for maintaining company viability and better satisfying customer needs. Really new or discontinuous new products play an important role in building competitive advantage and can contribute significantly to a firm's growth and profitability (Ali, 1994; Cooper, 2002; Kleinschmidt & Cooper, 1991; Robertson, 1967). However, while innovation plays a crucial role in the success (or continued success) of most companies, it does not come easily. As Wheelwright & Clark (1992, pp. 28–29) point out: 'Perhaps no activity in business is more heralded for its promise and approached with more justified optimism than new-product and new-process development. The anticipated benefits almost defy description ... Unfortunately, in most firms the promise is seldom fully realized. Even in many very successful companies, new product development is tinged with significant disappointment and disillusionment, often falling short of both its full potential in general and its specific opportunities on individual projects'. This is true for new product development in general and is especially true for discontinuous new product development.

New product development in the context of high innovation usually involves greater uncertainty than is present for the development of more incremental innovation products. Developing highly innovative product offerings involves considerable risk along with requiring both insight and foresight. The implications of technological advances are often obscured by the high levels of technical and market uncertainty that surround them. Often, it is difficult to know what direction to take for an emerging technology on the path to commercialization as an innovative new product. Appreciating the potential of any disruptive technology for development into a useful product form is a large part of the innovation challenge. Even though innovation at the high end of the spectrum usually entails more 'degrees of freedom' in one sense, it also requires either a great deal of luck or an ability to effectively envision what the market will respond to and embrace in order to produce a successful outcome. The key to this is the ability to link advanced technologies to market opportunities of the future so that the project can be guided through the uncertainties inherent in the development of these types of products. The task of bridging the uncertainty between technological capability and market need is critical for the effectiveness of the development effort if it is to yield a useful and commercially viable new product.

The potential rewards and risks from developing successful new products are high, and many factors can impel organizations to consider new product development activities. Although some companies may survive by trying one product after another in the market until success is achieved, this can be both costly and risky (Urban & Hauser, 1993). Studies of high-tech firms have found that a critical factor for success is an orientation toward marketing (Cooper, 1990, 2002; Gupta, Raj & Wileman, 1985; Kotler, 1999; Souder, 1988; Trott, 2002; Urban & Hauser, 1993).

Although one can look at 'innovation' in many ways, in the context of marketing there are specific aspects that need to be taken into consideration. The role of marketing usually begins with the inception of a product development project. In cases when the project is more than simply technology push, marketing's role may precede the inception phase in the form of explorations of possible customer needs and wants, and usually continues through introducing a new product into the marketplace as well as supporting it with various marketing programs. The heart of marketing concerns and challenges lies in the development of the product itself, for it is here that information (e.g. customer needs, product specifications, market trends, product tests, price points) is most needed if the ultimate product produced is to be one that both benefits consumers (and society at large) and is commercially viable. The key requirement is the alignment of the marketing and market research approaches with the product development task so that they may be used effectively to help develop the product and understand market potentials—in many cases for a market that does not yet exist! Proper alignment is crucial, since without it marketing inputs may seriously undermine innovation efforts.

Marketing's Role in Innovation

Innovation, and especially radical or 'high' innovation, involves fundamental questions that must be answered at some point during the new product development process. Whether these questions are addressed formally by marketing personnel or touched on by people like R&D scientists or engineers, it is inherent to innovation that inputs are needed in order to effectively broach a new commercial frontier (Veryzer, 1998a). Either these questions are considered and examined prior to or as an innovative new product is being formulated, or they will be answered when the product is judged (perhaps harshly) by the marketplace if the product even reaches the introduction stage. Critical questions center around understanding issues such as: what are the potential applications of a technology as a product?, and which application(s) should be pursued first?; what benefits can the proposed product offer to potential customers?; what is a gross estimate of potential market size?; Will the market be large enough to justify moving forward with the project? (Leifer,

McDermott, O'Connor, Peters, Rice & Veryzer, 2000). In addition, in the development of an innovative new product there is always a need to be sensitive to two further issues: (1) the possibility of a product lacking distinctiveness, that is, its being less than unique or the ease with which it may be quickly imitated by competitors; and (2) the possibility of a 'moving target' in terms of matching the product that is actually developed to a set of customers for which it is most beneficial and appropriate (Wheelwright & Clark, 1992). As Wheelwright & Clark (1992, p. 29) point out:

> ... too often the basic product concept misses a shifting technology or market, resulting in a mismatch. This can be caused by locking into a technology before it is sufficiently stable, targeting a market that changes unexpectedly, or making assumptions that just do not hold. In each of the cases the project gets into trouble because of inadequate consistency of focus throughout its duration and an eventual misalignment with reality. Once the target starts to shift, the problem compounds itself: the project lengthens, and longer projects invariably drift as the target continues to shift.

Despite the need to answer these sorts of questions, there has been some debate among marketing scholars concerning the nature of the relationship between marketing and innovation (e.g. Lukas & Ferrell, 2000; Trott, 2002). One view holds that the effect of a market orientation on product innovation yields 'superior innovation and greater new product success' (Lukas & Ferrell, 2000, p. 239; see also: Cooper, 1990, 2002; Gupta, Raj & Wileman, 1985; Kotler, 1999; Montoya-Weiss & Calantone, 1994; Nonaka, Sasaki & Ahmed, 2002; Ortt & Schoormans, 1993; Souder, 1988; Trott, 2002; Urban & Hauser, 1993). *Market orientation* has been defined as the process of generating and disseminating market intelligence (i.e. a 'deep understanding' of the intended customers and product environment) for the purpose of creating superior buyer value (Kohli & Jaworski, 1990; Narver & Slater, 1990). It encompasses both a focus on customers and competitors along with the synthesis and dissemination of market intelligence to various functional areas within a company that may benefit from such information (Lukas & Ferrell, 2000; Narver & Slater, 1990). There are a number of studies that support the view that a market orientation yields superior innovation as well as greater new product success (e.g. Cooper, 1990, 2002; Deshpande', Farley & Webster, 1993; Kohli & Jaworski, 1990; Kotler, 1999; Montoya-Weiss & Calantone, 1994; Trott, 2002). Some researchers go even further in their view of the degree to which customer information should be incorporated into the development of innovative new products. For example, von Hippel

(1988) suggests that focusing on 'lead users'—those customers that identify and craft a solution to a problem being experienced in advance of the rest of the market—is the key to innovation. He suggests that "the locus of almost the entire innovation process is centered on the user and offers a model of the user as the primary actor" (von Hippel, 1988, p. 25). Von Hippel (1988) reports that lead users accounted for 100% of the semiconductor and PC board innovations.

At the same time, a number of researchers have argued that a strong market orientation may result in reduced innovation (e.g. Chandy & Tellis, 2000; Christensen, 1997; Cooper, 2002; Hamel & Prahalad, 1994; Kumar, Scheer & Kotler, 2000; Leonard-Barton & Doyle, 1996; Lukas & Ferrell, 2000). The primary concern involves listening too closely to customers who are focused on current as opposed to future needs. Christensen (1997) has argued that the very decision-making and resource-allocation processes that are key to the success of established companies are the same processes that lead firms to reject the disruptive technologies that they should be embracing—and chief among these is 'listening carefully to customers' (Christensen, 1997, p. 98). He points out the highly negative impact in the disk drive, steel mini-mill, and numerous other industries of a firm's focus on current customers to the exclusion of considering longer-term technological innovations that offer no apparent immediate benefit to the loyal customer base (Christensen, 1997). The result is often new products that are marginally innovative. Concern has also been expressed about the impact of focusing too closely on the competition. A number of researchers have suggested that being overly focused on competitors can lead to a greater introduction of imitations or 'me-too' products (Chandy & Tellis, 2000; Christensen, 1997; Cooper, 2002; Hayes & Abernathy, 1980; Kumar, Scheer & Kotler, 2000; Lukas & Ferrell, 2000; Trott, 2002; Zahra, Nash & Brickford, 1995).

Underlying each of the concerns about market-orientation is the notion that 'an obsessive focus on customers or competitors encourages research and development (R&D) to develop more line extensions and me-too products at the expense of new-to-the-world products' (Lukas & Ferrell, 2000, p. 241). However, regardless of the difference in views concerning the amount of emphasis that should be placed on customer and marketplace input, there is little disagreement about the need to link technology development to potential markets—and this is a critical component for successful innovations (Jolly, 1997; Leifer et al., 2000). Certainly such things as customers being bounded by the familiar, being unaware of emerging new technologies, or having difficulty comprehending or appreciating radical new products and the implications of these innovations for their businesses or their lives can severely undermine marketing research efforts (Christensen, 1997; Jolly, 1997; Ver-

yzer 1998b). Nevertheless, despite problems in procuring and applying appropriate market-based inputs for the development of highly innovative products, inherent in innovation is a fundamental need to address customer needs or to uncover latent needs (Leonard-Barton, 1995; Veryzer, 1998b). Although the innovation context presents significant challenges for incorporating customer or market information that is helpful to the radical new product development process, it does not preclude valid and meaningful customer understanding and input (Veryzer, 1998b).

As Jolly (1997, p. 42) notes, 'deep research, without any context or problem to solve, is inevitably hostage to serendipity alone'. Thus, while he sees customers as neither the sole source of ideas nor the best arbitrators for how ideas should be pursued, they provide a necessary context. A number of researchers (Hamel & Prahalad, 1994; Jolly, 1997, Leifer et al., 2000; Leonard-Barton, 1995) suggest that the imagination underlying all successful technology-based innovations—the 'techno-market insight'—comes from how a problem is approached technically and an ability to identify compelling benefits of that technology and characterize these in terms of a market that may not presently exist. Marketing's role is to both aid in identifying or deriving these benefits as well as in keeping the innovation process grounded in producing a new product that truly provides benefits to the intended users.

Innovation: Context and Dimensions

How we understand innovation is dependent on the perspective from which it is viewed and applied. From the vantage point of business and marketing the primary concerns and challenges center around identifying viable new product directions and then executing the development of the projects so as to produce an offering that provides both a significant benefit for customers and is commercially feasible. This orientation shapes both how innovation is viewed as well as influencing the implications that one draws concerning the best course with respect to any particular product development project.

In the context of marketing, 'innovation' refers to the creation of a new product, service, or process that may be either offered to a market, customer, or group of customers. Innovations may be thought of as falling on a continuum from evolutionary or 'continuous' to revolutionary or 'discontinuous'. The vast majority of new products that are launched—some 25,000 a year (Fellman, 1998)—are incremental innovations or what are referred to in the literature as continuous or dynamically continuous products (Engel, Blackwell & Miniard, 1986). These represent limited changes or improvements to existing products or product forms. The phrase 'discontinuous innovation' is generally used to refer to radically new products that involve dramatic leaps in terms of customer benefits relative to previous product offerings (e.g. Chandy & Tellis, 2000; Garcia & Calantone, 2002; Meyers & Tucker, 1989). Frequently these types of products involve the development or application of significant new core technologies (Ali, 1994; Lee & Na, 1994; Tushman & Nadler, 1986).

S-curves provide a theoretical background for understanding the evolution of these types of innovations as driven by companies seeking to maximize their position in a particular market (Chandy & Tellis, 2000; Narayanan, 2001, p. 76; Utterback & Abernathy, 1975). Technologies evolve along what appears to be a series of these S-shaped curves which occur because initially the new product—based on a new technology—offers limited benefits to customers, but offers an increased number of benefits as the technology underlying the product matures and the product offering is refined (see Fig. 1). Eventually, a product class reaches a mature phase where it levels off in terms of the rate of increase in new benefits offered (Chandy & Tellis, 2000; Narayanan, 2001, p. 76; Utterback & Abernathy, 1975). This remains the case, unless and until, a new technology is applied to create a new product form that offers significantly enhanced and/or new benefits to consumers. In instances when consumers are able to appreciate the value of the new product and the new capabilities that it offers, the new form usually supplants the mature product in the marketplace. This displacement process may occur relatively quickly as in the case of calculators displacing slide rules, slowly

as in the case of steamships displacing sailing vessels, achieve co-existence as in the case of microwaves and conventional ovens, or it may not occur at all as in the case of BetaMax.

In addition to the issue concerning the degree of 'newness' of a technology or its application, there are also questions concerning issues such as the range and scope of innovation. For example, Lee & Na (1994) distinguish between 'incrementally improving innovativeness' and 'radical innovativeness' and explicitly exclude commercial performance as a basis for classifying innovation. However, researchers such as Meyers & Tucker (1989) hold that discontinuous innovation, in addition to being based on new technology and aimed at a market that is unfamiliar with the product class, encompasses both the development and the introduction of the product into the market.

Two critical dimensions may be used to delineate the various levels or degrees of innovation with respect to the application of technology in the form of a product offered to a market (see Fig. 2); product innovation may be viewed as lying along a 'Technological Capability' dimension and a 'Product Capability' dimension (Veryzer, 1998a). The technological capability dimension refers to the degree to which the product involves expanding capabilities—the way product functions are performed—beyond existing boundaries. Discontinuous products involve advanced capabilities that do not exist in current products and cannot be achieved through the extension of existing

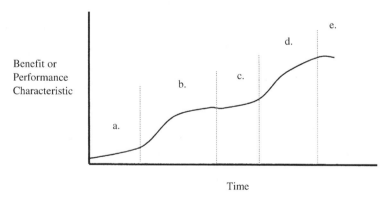

Figure 1. General form of S-curves representing technological evolution.

Stages in the evolution of a performance characteristic:

(a) Emergence phase, product offers limited benefits;
(b) Rapid Improvement phase, product offers increased benefits and/or performance at an accelerating rate;
(c) Plateau phase with limited continued improvement, also convergence in product standards and form;
(d) Successive Generational Improvement phase, product refinement and additional improvement—widespread market acceptance;
(e) Maturity and Possible Decline phase, new applications or improvements may be sought that result in extending the curve or lead to development of a new product. The new technology may begin to challenge an existing technology (already in the Maturity phase) during the Rapid Improvement phase. Efforts to make the older technology competitive with the new technology may result in improvements sufficient to maintain market viability, otherwise products based on the older technology will be replaced by those stemming from the new technology.

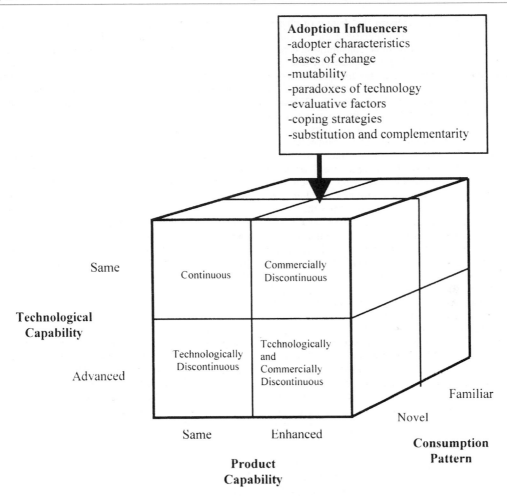

Figure 2. Dimensions of product innovation and adoption influencers.
Adapted from Veryzer (1998a).

technologies. The Product Capability dimension refers to the benefit(s) of the product as perceived and experienced by the customer or user.

In this view of innovation there are essentially four types or levels of innovation (excluding moderately innovative products). The first type encompasses products that utilize existing technology and provide the same benefits as existing products. Such products are continuous in terms of the technology employed and the way they are experienced by customers. Although they may be new, they are not very innovative. In addition to continuous new products, new products may be discontinuous with respect to technology, the benefits perceived by the customer, or both. Products that are perceived by customers as being really new regardless of whether or not they utilize new technology are commercially discontinuous. For example, the SONY Walkman offered new benefits (i.e.

functionality) utilizing available technology. In cases when the delivery of new benefits involves the application of a significant new technology, the product is technologically discontinuous in addition to being commercially discontinuous. PCs and pagers were examples of this type of innovation when they were first introduced. In some cases, products may be perceived as being essentially the same as existing products even though they utilize highly advanced technology. For example, the switch from vacuum tube televisions to televisions utilizing solid-state technology had little impact on consumers in terms of product benefits or use. Thus, even though solid-state circuitry represented a dramatic change in technology (Technologically Discontinuous), the product as perceived by consumers had changed little (Veryzer, 1998a).

To this conceptualization of innovation, a third relevant dimension involving change may be added.

'Consumption Pattern' refers to the degree of change required in the thinking and behavior of the consumer in using the product. Although the consumption pattern dimension is not often explicitly discussed by researchers, it seems to be implied in many discussions of new products (cf. Ali, 1994; Hughes & Chafin, 1996; Mahajan & Wind, 1986; Millison, Raj & Wilemon, 1992; Robertson, 1967; Rogers, 1995). A product can be familiar or novel in the way it requires users to interact with it. The nature of the change involved with respect to this aspect of a new product can play a significant role in product evaluation and adoption (Veryzer, 1998b). Products involving consumption pattern changes can require customers to alter their thinking and habits and this may affect their willingness to embrace a new product.

In considering highly innovative products it is important to take the customer's view and experience of the product into account. A technology-driven view of innovation that does not consider the customer's perspective might result in a product that is at odds with the market's perception of it (Veryzer, 1998b). Even though technology may be the means for enabling an innovation, new products are more than simply (technical) innovations—they must benefit and be used by people who can appreciate them and the impact that the innovative new products will have on their lives. The different dimensions of innovation are also useful in terms of understanding marketing's role in innovation. The role of marketing during the product development process is likely to be especially important for the success of products that fall into the commercially discontinuous portion of the grid. Certainly, marketing's role in ensuring acceptance when a product is introduced into a market could be significant for virtually any product (regardless of which part of the grid into which it fits), particularly if the benefits offered are not readily apparent.

Marketing, Innovation, and Diffusion

Marketing can play a key role both in the development of an innovative new product and in the successful diffusion of an innovative new product. In both cases it is usually the insights with respect to understanding potential customers that marketing supplies (Cooper, 1990, 2002; Kohli & Jaworski, 1990; Kotler, 1999; Trott, 2002). Such insights do not necessarily come directly from customers. In fact, quite often, deep insights relevant for the development of truly innovative products require that much of the information that may be easily ascertained from customers be ignored (Leifer et al., 2000; Veryzer, 1998a, 1998b). In this 'high' or radical innovation product development domain, one has to delve well beyond the superficial and readily apparent in order to gain an understanding of the deeper influences that drive customers' thinking and choices. This requirement to 'delve deeper' can be applicable to new product development in general, but

is particularly necessary in cases involving radical innovations where there is little, if any, precedent to follow and high uncertainty concerning market reaction.

One of the most dramatic illustrations of this need to delve deeply is the outcome of the market research conducted by the Coca-Cola Company for the development of what became known as 'New Coke'. As Pepsi Cola, with its sweeter taste and 'Pepsi Challenge' taste tests, continued to successfully chip away at Coke's market share in the 1980s, the Coca-Cola Company began the largest new product research project in the company's long history (Honomichl, 1985; Kotler & Armstrong, 2001, p. 130). In an effort that spanned over two years, the company spent in excess of \$4 million exploring new, sweeter product formulations and conducting extensive testing with customers to develop a taste that customers would prefer over both Pepsi Cola and the Coke formulation it was to replace (Honomichl, 1985; Kotler & Armstrong, 2001, p. 130). Some 200,000 taste tests were conducted, 30,000 of these on the final new Coke formula (Kotler & Armstrong, 2001, p. 130). The blind taste tests showed that 60% of consumers chose the new Coke product over the old, and 52% chose the new Coke product over Pepsi (Kotler & Armstrong, 2001, p. 130). However, in researching customers' stated preferences for the new Coke taste, important questions concerning other drivers behind Coke drinkers' consumption behavior were overlooked. Among these were factors such as brand identity, brand familiarity, 'intangibles' relating to the product's history and peoples' emotional ties to and memories of it, symbolic meaning, and so on (Honomichl, 1985; Kotler & Armstrong, 2001, p. 130). When the new Coke product was introduced into the market as a replacement for the original formula Coke (what is now known as 'Coke Classic'), customers quickly rejected the new product, clamoring for reinstatement of original or 'real' Coke. Within three months the Coca-Cola Company was forced to bring back the old formula, which quickly eclipsed the highly researched, 'sure-winner' that the company had introduced with such confidence. Thus, although marketing research with customers did reflect taste perceptions, that is customers were asked questions and provided accurate answers, the research conducted did not delve deeply enough to sufficiently uncover other important influences (e.g. brand loyalty, emotional attachment) that act to shape customer purchase and consumption behavior. Ultimately, these influences proved to be the determinants of the product's fate. This need to push toward a deeper and more complete understanding of the influences driving customer thinking and choices is especially true for uncovering the insights that are relevant for various product development phases such as market research and concept testing as well as the market introduction phase.

Discussions of the diffusion process for innovations have drawn on ideas presented by sociologist Everett M. Rogers (1995), who has suggested that diffusion curves are essentially normal curves made up of various adopter categories (innovators, early adopters, early majority, late majority, and laggards). The marketing discipline has relied heavily on this approach to categorizing adopters (e.g. Kotler & Armstrong, 2001, p. 200). Although it is tempting to construct ever more complicated typologies of adopters for highly innovative products, Weil & Rosen (1997) relate a simple three-category break-down oriented toward technological products that is basic but effective. 'Eager Adopters' view technology as fun and challenging and are drawn to it. These sorts expect to have problems with technology, so when a problem arises they are not daunted by the challenge. 'Hesitant "Prove Its" ' do not think technology is fun and prefer to wait until a new technology is proven before trying it. Even though they may take some convincing, if you can show these types how something will make life easier, they will consider it. Diffusion among this group is affected by ease of operation since they do not enjoy solving the problems that may come with the new product. Finally, 'Resisters' want little, if anything, to do with innovative new technology products, no matter what anyone says or does to convince them—technology is absolutely no fun for these people. Awareness of these types of adopter '(pre)dispositions' can provide useful insights that help guide the development effort (e.g. the design of user-product interfaces) as well as marketing efforts (e.g. product positioning and advertising) during the product introduction phase.

In assessing the market or commercial potential of an innovative new product there are a number of important questions to consider: Does or will diffusion for a particular new product occur in a predictable manner?; Is there such a person as a consumer-innovator?, and can he or she be identified?; and What role will advertising and personal influence play in terms of their effectiveness for new product diffusion? (Robertson, 1967, p. 16; Wilson, Kennedy & Trammell, 1996, p. 62; Urban & Hauser, 1993, p. 222). Beyond these questions, are other less direct questions that relate to the diffusion of innovations. Questions such as, 'how are consumers likely to use this new product?', 'will it be as the product developers envisioned or are there latent needs or social trends that may affect how customers apply or use the new technology offered in a new product?' These questions are central to a new innovation's viability and acceptance in the marketplace. Consider, for example, the well-known case of Sony's BetaMax. Sony launched its BetaMax video cassette recorder for the mass consumer market in 1975. At the time, Sony generally positioned its products by deciding what the best uses would be and then selling those uses to customers (Cooper, 2000). In the words of Akio Morita, legendary cofounder of the firm that became Sony Corporation, ''We don't believe in market research for a new product unknown to the public. . . so we never do any. We are the experts'' (Lyons, 1976, p. 110). Morita, who had been responsible for many successful product launches for the firm, regarded the primary function of the BetaMax product as freeing people from a preset television programming schedule—thus allowing people to 'time shift' in order to watch their favorite programs when it suited them rather than at a time that they could not control (Cooper, 2000, p. 4). Unfortunately, Sony missed, or at least underestimated consumer desire to watch full-length feature films at home and therefore did not adjust its product offering to accommodate this desire (e.g. tape length, licensing agreements) (Cooper, 2000). JVC's product offering was better able to meet consumer needs and thus it eventually won the 'VCR war' despite some of the superior performance characteristics of the BetaMax product (Clark & Fujimoto, 1991, p. 4; Cooper, 2000).

Innovation is essentially about change, and diffusion is essentially consumer willingness for change. Fundamental to the change brought by product innovation is how the new offering will interact with the actual needs and desires of consumers. Marketing works across these two elements of the innovation equation (i.e. Innovation Change and Consumer Desire for Change) in order to help produce an optimal result that maximizes everyone's utility (at least that is the goal for good marketing). 'Change' is not, however, a simple uni-dimensional concept. For example, Golembiewski, Billingsley and Yeager (1976) discussed three varieties of change—Alpha change, Beta change, and Gamma change. Alpha change, or what might be called 'perceptual change' involves shifts in peoples' understanding or interpretation of a product or brand (Golembiewski et al., 1976; Wilson et al., 1996, p. 53). This is similar to repositioning a product or brand in an existing framework such as a perceptual map (a representation of the positions of products on a set of primary customer needs; see Urban & Hauser, 1993, p. 205 for a discussion of perceptual maps) and might be accomplished through an advertising campaign (Cooper, 2000), or this type of change could be accomplished via changes in elements of a product's design (Veryzer, 1997).

Beta change involves a shift in consumer values. Here the frame of reference or context circumscribing a consumer's consideration of products (or some set or class of products) has changed. For example, when a person receives a significant increase in income the types of products that are considered for purchase may change dramatically (e.g. shift to expensive, high quality products). Thus, the consumer's 'ideal points' have changed rather than this simply involving a change in a product's position with respect to already established ideal points.

Gamma change involves adding a new dimension for consideration. This has the effect of both defining (or redefining) a new product and is likely to lead to a shift in the ideal points that a market has for products. For example, the Internet, with its ability to make a virtually unlimited supply of information readily available, has changed how people access information and to some extent how they use it. This type of change is what tends to occur in the case of highly innovative products. Here, consumers are faced with more than just a shift in perceptions or values (although these may accompany such changes), rather the very nature of the product and its use are changed and new (Cooper, 2000; Golembiewski et al., 1976; Wilson et al., 1996, p. 53).

Customers' abilities to comprehend the change inherent in innovations is affected by their existing knowledge. The notion of 'mutability' is useful for understanding consumers' abilities for transforming existing knowledge structures (that they perceive as being relevant to an innovative new product they have encountered) so as to accommodate the discrepant information presented by an innovation (Moreau, Lehmann & Markman, 2001). Mutability refers to the conceptual transformability of different features or attributes in a category schema (Love, 1996; Love & Sloman, 1995; Sloman, Love & Ahn, 1998). Mutability depends on both the variability of a feature across category members as well as the number of other features in the (product) category that depend on that feature (Love & Sloman, 1995). As Moreau et al. (2001) point out, in the context of innovative products, the more immutable the feature change, the greater the perceived discontinuity of a product. This has implications for how people process and perceive innovative new products. For example, since people are likely to draw on or transfer knowledge from the product domain they perceive as being closest to the new innovation encountered, this alignment—which may or may not be appropriate—will exert a significant influence on customer assessments of the new product. Further, as Moreau et al. (2001) point out, for discontinuous innovations, expertise in a primary base domain (i.e. an existing product category that is most similar to the innovation in terms of benefits provided) will be negatively related to both comprehension and perception of net-benefits of the innovative new product(s). In their study, these researchers found this to be the case. For the most part, with respect to discontinuous innovations, experts' entrenched knowledge was related to lower comprehension, fewer perceived net benefits, and lower preferences compared to that of novices (Moreau et al., 2001).

Consumers' reactions to innovative new products and their willingness to embrace them are also, of course, driven by the benefit they expect to derive from the products. For discontinuous innovations, however, there may be subtle yet critical currents underlying

peoples' reactions. Such products, which often involve new technologies, frequently require changes in thinking and usage behavior (Veryzer, 1998b). Moreover, by their nature, innovative new products tend to entail a certain paradox in that the trade-off in what they offer (in terms of speed, convenience, etc.) and what they cost (learning time, frustration, complexity) is more apparent to users given the increase across the various product discontinuity dimensions (benefit, technological, consumption pattern) discussed earlier. Such products involve a certain 'paradox of technology'— the very technology that can make us more efficient, may also foster feelings of incompetence and frustration (LaPorte & Metlay, 1975; Mick & Fournier, 1998; Rosen & Weil, 1997; Winner, 1994). Moreover, products like appliances purchased in order to save time, often end up wasting time (Goodman, 1988; Mick & Fournier, 1998). In their codification of the various paradoxes discussed across the technology literature, Mick & Fournier (1998) present eight central paradoxes of technological products. In brief, these are: Control/Chaos (technology can facilitate order, and lead to upheaval or disorder); Freedom/Enslavement (facilitate independence, and lead to dependence); New/Obsolete (provide new benefits, and are rapidly outmoded); Competence/Incompetence (promote feelings of intelligence or efficacy, and lead to feelings of ignorance or ineptitude); Efficiency/Inefficiency (reduce time needed, and increase time required); Fulfills/Creates Needs (help fulfill needs or desires, and reveal unrealized needs or desires); Assimilation/ Isolation (facilitate human togetherness, and lead to human separation); and Engaging/Disengaging (facilitate involvement, and lead to disconnection) (Mick & Fournier, 1998, p. 126). These paradoxes play an important role in shaping consumers' perceptions of innovations as well as determining their willingness to adopt new products.

In addition to the various trade-offs or paradoxes that affect consumers' willingness to embrace innovative products, an important aspect of a new product offering—that should be considered in the design stage as well as the later product launch stage—is that consumers develop their own ways of coping with innovations and these can impact diffusion as well. Potential customers may ignore a new technology altogether, delay obtaining the new product, attempt to try an innovative new product without the risk of outright purchase, embrace the product and master it, and so on (Carver, Scheier & Weintraub, 1989; Mick & Fournier, 1998). Furthermore, in evaluating discontinuous new products, there are certain factors that are likely to come into play more than they do for less innovative products. Lack of familiarity, 'irrationality', user-product interaction problems, uncertainty and risk, and accordance or compatibility issues may play a decisive role in customers' evaluations of products in either the development and testing stages or once the product is introduced into a market (Veryzer, 1998b, p. 144). For example, during the course of one radical innovation development project, managers were struck by how 'irrational' customers were in that they often focused on things that the product development team thought to be unimportant. Test customers ignored aspects of a prototype product on which the team had expended a great deal of effort and money. Even though this type of 'irrationality' may frustrate product development teams, in the domain of highly innovative products, assumptions must be checked and measured against the metric of those who will be the final arbitrators of success (Veryzer, 1998b).

Beyond consumer concerns that are relevant to the development and marketing of innovative products are more 'macro' influences that can affect adoption and thus need to be considered. The substitution of one technology for another is an obvious concern (recall the earlier discussion of S-curves). Along with this, the issue of product complementarity, or when there is a positive interrelationship between products (e.g. a computer printer and a computer), can also be important with respect to product adoption. Thus, in addition to displacing products, new technological innovations often modify or complement existing products that may still be diffusing throughout a given market. This has significant implications for market planning decisions for both products since their diffusion processes are interlinked (Norton & Bass, 1987, 1992). In such cases (e.g. new computer products, electric vehicles) consideration of: (1) whether there is a positive interdependence between a new product and existing products; (2) whether the old technology will be fully replaced by a newer product; and (3) how the size of the old technology's installed base will affect the speed of diffusion of the new product or product generation, represent critical marketing concerns (Dekimpe, Parker & Sarvary, 2000; Norton & Bass, 1987, 1992).

Diffusion may also be examined from an even more macro-perspective, and in some instances it can be particularly important to do so. For example, researchers like Dekimpe, Parker & Sarvary (2000) have investigated the global diffusion of technological innovations. In their work, they focus on: (1) issues concerning the two-stage (implementation stage and confirmation stage) nature of the global diffusion process as defined by Rogers (1995); (2) the irregularity of a diffusion pattern due to network externalities and/or central decision makers; and (3) the role of the installed base of older generation technologies that an innovation replaces (Dekimpe et al., 2000). As they point out, "for most innovations, the adoption process of each country starts with the implementation stage, which is followed by the confirmation stage" (Dekimpe et al., 2000, p. 49). However, they suggest that for technological innovations within-country diffusion might be instantaneous—due to network externalities

(e.g. established standards) or central decision makers, and as such, the confirmation stage for certain countries may have a zero duration.

'Innovation' inherently involves change and thereby at least some degree of disruption and uncertainty. The forces underlying customers' decisions to embrace an innovative product run deep and are complex in that they occur in both individual as well as societal contexts simultaneously. Adopter dispositions (e.g. 'techno adopter' categories) as well as the various paradoxes, evaluation factors, coping strategies, and substitution/complementarity concerns indicate some of the innovation and diffusion challenges that the marketing function must address in order to ensure innovative products that people will choose to purchase and use. It is in understanding these deeper influences that shape customers' reactions to innovative new products that marketing makes its greatest contribution to the potential that 'innovations' represent. This it does by acting as a sort of mediator (or interpreter) between new ideas/technical possibilities and peoples' needs and wants.

Conclusions

Market vision, or the ability to bridge between technological capability and market need or opportunity, is a crucial ability. It is particularly important for high innovation products because such projects usually involve a significant degree of uncertainty about exactly how an emerging technology may be formulated into a usable product and what the final product application will be. The conceptual framework presented here can help to clarify the innovation and adoption context with respect to the marketing challenge(s). Since such products are often difficult to convey to customers—especially early on in the development process—many traditional market research techniques (e.g. concept tests, focus groups, surveys) may not be useful or valid (Leifer et al., 2000; Leonard-Barton, 1995; Veryzer, 1998b). In this domain, particular care must be taken to make certain that the research that is conducted provides valid input that does not lead the development effort astray. In the case of highly innovative products this means relying more heavily on techniques such as prototype reaction studies. Prototypes are essential instruments for market learning during the course of radical innovation, even beginning with the first formative prototype pushed out by the R&D lab (Leifer et al., 2000; Veryzer, 1998a). They provide a critical means for product developers to assess the product direction and test critical assumptions (Veryzer, 1998a). Such efforts play a key role in the 'probe and learn' process that serves to accelerate market learning (Leifer et al., 2000; Lynn, Morone & Paulson, 1996). Apart from these sorts of methods for gaining information to help direct the product development effort, customers can be studied (using various techniques such as laddering which involves successive questioning to uncover the values or motivations linked to a product, and observational study which essentially involves watching or video taping customers using products in settings of interest) in order to identify and understand latent needs that may suggest product ideas or serve to revise the design direction that a product under development is taking. Customer research also provides a means for determining product specifications since for many discontinuous products established benchmarks do not exist (Leifer et al., 2000). In addition, customer and market research is needed in order to determine optimal levels of various aspects of the product offering such as the price or pricing structure, product positioning, advertising approach, promotion(s), and so on.

Given the challenges inherent in innovation, gaining an understanding of the factors that are likely to influence customer evaluations of a new product and how customers are likely to react to the product is necessary for ensuring a successful market outcome. Marketing provides a necessary and useful function in helping to shape an innovative idea into a product offering that meets the needs and desires of the people who are intended to use it.

References

Abernathy, W. J. & Utterback, J. M. (1988). Patterns of industrial innovation. In: M. Tushman & W. Moore (Eds), *Readings in the Management of Innovation* (2nd ed., pp. 25–36). New York: Harper Business.

Ali, A. (1994). Pioneering vs. incremental innovation: Review and research propositions. *Journal of Product Innovation Management*, **11**, 46–61.

Bennett, R. C. & Cooper, R. G. (1979). Beyond the marketing concept. *Business Horizons*, **24**, 51–61.

Carver, C. S., Scheier, M. F. & Weintraub, J. K. (1989). Assessing coping strategies: A theoretically based approach. *Journal of Personality and Social Psychology*, **56** (2), 267–283.

Chandy, R. K. & Tellis, G. J. (2000). The incumbent's curse? Incumbency, size, and radical product innovation. *Journal of Marketing*, **64** (July), 1–17.

Christensen, C. M. (1997). *The innovator's dilemma*. Boston, MA: Harvard Business School Press.

Clark, K. B. & Fujimoto, T. (1991). *Product development performance*. Boston, MA: Harvard Business School Press.

Cooper, R. (2003). Profitable product innovation. In: L. V. Shavinina (Ed.), *International Handbook on Innovation*. Elsevier Science.

Cooper, R. G. (2000). Strategic marketing planning for radically new products. *Journal of Marketing*, **64** (January), 1–16.

Cooper, R. G. (1990). New products: What Distinguishes The Winners. *Research and Technology Management*, Nov.-Dec., 27–31.

Dekimpe, M. G., Parker, P. P. & Sarvary, M. (2000). Global diffusion of technological innovations: A coupled-hazard approach. *Journal of Marketing Research*, **37** (February), 47–59.

Deshpande', R., Farley, J. U. & Webster, F. E. (1993). Corporate culture, customer orientation, and innovativeness in Japanese firms: A quadrad analysis. *Journal of Marketing*, **57** (January), 23–27.

Engle, J. F., Blackwell, R. D. & Miniard, P. W. (1986). *Consumer behavior*. Hinsdale, IL: Dryden Press.

Fellman, M. W. (1998). Forecast: New products storm subsides. *Marketing News*, **32** (March 30), 1.

Garcia, R. & Calantone, R. (2002). A critical look at technological innovation typology and innovativeness terminology: A literature review. *Journal of Product Innovation Management*, **19**, 110–132.

Golembiewski, R. T., Billingsley, K. & Yeager, S. (1976). Measuring change and persistence in human affairs: Types of change generated by OD designs. *Journal of Applied Behavioral Science*, **12** (2), 133–157.

Goodman, E. (1988). Time bandits in the machine age. *Chicago Tribune*, Tempo section (January 10), 2.

Gupta, A. K., Raj, S. P. & Wilemon, D. (1986). A model for studying R&D—Marketing interface in the product innovation process. *Journal of Marketing*, **50**, 7–17.

Hamel, G. & Prahalad, C. K. (1994). *Competing for the future*. Boston, MA: Harvard Business School Press.

Hayes, R. H. & Abernathy, W. J. (1980). Managing our way to economic decline. *Harvard Business Review*, **58** (July/August), 67–77.

Honomichl, J. (1985). Missing ingredients in 'new' Coke's research. *Advertising Age*, (July), 1.

Hughes, G. D. & Chafin, D. C. (1996). Turning new product development into a continuous learning process. *Journal of Product Innovation Management*, **13**, 89–104.

Jolly, V. K., (1997). *Commercializing new technologies*. Boston, MA: Harvard Business School Press.

Kleinschmidt, E. J. & Cooper, R. G. (1991). The impact of product innovativeness on performance. *Journal of Product Innovation Management*, **8**, 240–251.

Kohli, A. K. & Jaworski, B. J. (1990). Market orientation: The construct, research propositions, and managerial implications. *Journal of Marketing*, **54** (April), 1–8.

Kotler, P. (1999). *Kotler on marketing: How to create, win and dominate markets*. New York: The Free Press.

Kotler, P. & Armstrong, G. (2001). *Principles of marketing*. Upper Saddle River, NJ: Prentice Hall.

Kumar, N., Scheer L. & Kotler, P. (2000). From market driven to market driving. *European Management Journal*, **18** (2), 129–141.

LaPorte, T. R. & Metlay, D. (1975). Technology observed: Attitudes of a wary public. *Science*, **188** (April), 121–127.

Lee, M. & Na, D. (1994). Determinants of technical success in product development when innovative radicalness is considered. *Journal of Product Innovation Management*, **11**, 62–68.

Leifer, R., McDermott, M., O'Connor, G., Peters, L., Rice, M. & Veryzer, R. (2000). *Radical innovation: How mature companies can outsmart upstarts*. Boston, MA: Harvard Business School Press.

Leonard-Barton, D. (1995). *Wellsprings of knowledge*. Boston, MA: Harvard Business School Press.

Leonard-Barton, D. & Doyle, J. L. (1996). Commercializing technology: Imaginative understanding of user needs. In: R. S. Rosenbloom & W. J. Spencer (Eds), *Engines of Innovation* (pp. 177–207). Boston, MA: Harvard Business School Press.

Love, B. C. (1996). Mutability, conceptual transformation, and context. In: *Proceedings of the Eighteenth Annual Conference of Cognitive Science Society* (pp. 459–463). Mahwah, NJ: Lawrence Erlbaum Associates.

Love, B. C. & Sloman, S. A. (1995). Mutability and the determinants of conceptual transformability. In: *Proceedings of the Seventeenth Annual Conference of the Cognitive Science Society* (pp. 654–659). Mahwah, NJ: Lawrence Erlbaum Associates.

Lukas, B. A. & Ferrell, O. C. (2000). The effect of market orientation on product innovation. *Journal of the Academy of Marketing Science*, **28** (2), 239–247.

Lynn, G. S., Morone, J. G. & Paulson, A. S. (1996). Marketing and discontinuous innovation: The probe and learn process. *California Management Review*, **38**, 8–37.

Lyons, N. (1976). *The Sony vision*. New York: Crown Publishers.

Mahajan, V. & Wind, J. (1986). New product models: Practice, shortcomings, and desired improvements. *Journal of Product Innovation Management*, **9**, 128–139.

Meyers, P. W. & Tucker, F. G. (1989). Defining roles for logistics during routine and radical technological innovation. *Journal of the Academy of Marketing Science*, **17** (1), 73–82.

Mick, D. G. & Fournier, S. (1998). Paradoxes of technology: Consumer cognizance, emotions, and coping strategies. *Journal of Consumer Research*, **25** (September), 123–143.

Millison, M. R., Raj, S. P. & Wilemon, D. (1992). A survey of major approaches for accelerating new product development. *Journal of Product Innovation Management*, **9**, 53–69.

Montoya-Weiss, M. & Calantone, R. (1994). Determinants of new product performance: A review and meta-analysis. *Journal of Product Innovation Management*, **11** (5), 397–417.

Moreau, C. P., Lehmann, D. R. & Markman, A. B. (2001). Entrenched knowledge structures and consumer response to new products. *Journal of Marketing Research*, **38**, 14–29.

Narayanan, V. K., (2001). *Managing technology and innovation for competitive advantage*. Upper Saddle River, NJ: Prentice Hall.

Narver, J. C. & Slater, S. F. (1990). The effect of a market orientation on business profitability. *Journal of Marketing*, **62** (October), 20–35.

Nonaka, I., Sasaki, K. & Ahmed, M. (2003). Continuous innovation: The power of tacit knowledge. In: L. V. Shavinina (Ed.), *International Handbook on Innovation*. Elsevier Science.

Norton, J. A. & Bass, F. M. (1992). Evolution of technological generations: The law of capture. *Sloan Management Review*, **22** (Winter), 66–77.

Norton, J. A. & Bass, F. M. (1987). A diffusion theory model of adoption and substitution for successive generations of high-technology products. *Management Science*, **33** (September), 1069–1086.

Ortt, R. J. & Schoormans, J. P. L. (1993). Consumer research in the development process of a major innovation. *Journal of the Market Research Society*, **35** (4), 375–389.

Robertson, T. S. (1967). The process of innovation and the diffusion of innovation. *Journal of Marketing*, **31** (January), 14–19.

Rogers, E. M. (1995). *Diffusion of innovations* (4th ed.). New York: Free Press.

Sloman, S. A., Love, B. C. & Ahn, W. (1998). Feature centrality and conceptual coherence. *Cognitive Science*, **22** (2), 189–228.

Souder, W. E. (1988). Managing relations between R&D and marketing in new product development projects. *Journal of Product Innovation*, **5** (1), 6–19.

Trott, P. (2003). Innovation and market research. In: L. V. Shavinina (Ed.), *International Handbook on Innovation*. Elsevier Science.

Tushman, M. L. & Nadler, D. (1986). Organizing for innovation. *California Management Review*, **28** (3), 74–92.

Urban, G. L. & Hauser, J. R. (1993). *Design and marketing of new products* (2nd ed.). Englewood Cliffs, NJ: Prentice Hall.

Utterback, J. M. & Abernathy, W. J. (1975). A dynamic model of process and product innovation. *Omega*, **3** (6), 639–656.

Veryzer, R. W. (1998a). Discontinuous innovation and the new product development process. *Journal of Product Innovation Management*, **15**, 304–321.

Veryzer, R. W. (1998b). Key factors affecting customer evaluation of discontinuous new products. *Journal of Product Innovation Management*, **15**, 136–150.

Veryzer, R. W. (1997). Measuring consumer perceptions in the product development process. *Design Management Journal*, **8** (2), 66–71.

von Hippel, E. (1988). *The sources of innovation*. New York: Oxford University Press.

Weil, M. M. & Rosen, L. D. (1997). *Technostress*. New York: John Wiley.

Wilson, C. C., Kennedy, M. E. & Trammell, C. J. (1996). *Superior product development: Managing the process for innovative products*. Cambridge, MA: Blackwell Publishers.

Winner, L. (1994). Three paradoxes of the information age. In: G. Bender & T. Druckery (Eds), *Culture on the Brink: Ideologies of Technology* (pp. 191–197). Seattle: Bay.

Wheelwright, S. C. & Clark, K. B. (1992). *Revolutionizing product development*. New York: The Free Press.

Zahra, S. A., Nash, S. & Brickford, D. J. (1995). Transforming technological pioneering into competitive advantage. *Academy of Management Executive*, **9** (February), 17–31.

Part XIII

Innovation Around the World:
Examples of Country Efforts, Policies,
Practices and Issues

The International Handbook on Innovation
Edited by Larisa V. Shavinina

Innovation Process in Hungary

Annamária Inzelt

IKU Innovation Research Centre, BUESPA, Hungary

Man, I have Spoken
Struggle and Have Face
Imre Madách (Tragedy of Humankind)

Abstract: The pioneering Hungarian pilot innovation survey reveals that the level of innovation in the Hungarian economy is low. This can largely be attributed to the financial stringency imposed by the transition process. Hungarian firms are aware of the importance of developing new products and addressing new markets—though, in practice, they do this largely within the limited Hungarian context. The firms pay little attention to the kind of long-term, strategic thinking about technological change which, alone, can provide the basis for sustained innovatory dynamism.

Keywords: Innovation; Measurement of innovation; Transition economy.

Introduction

'To be or not to be innovative' is a fundamental question for the entire world. National economic stability and future prosperity can be secured only if a country has an innovative manufacturing base. The importance of innovation in economic development and competition theory is increasingly acknowledged (Dosi et al., 1988 as the central work), and technological change and innovation have become central topics in economic analysis and policy discussion on economic performance in the industrialized countries (Lundvall, 1992; Nelson, 1993; Porter, 1990; Tidd et al., 1997). In recent years the relationship between competitive performance and technological strategy has been widely explored, to see whether it can account for success and failure, at both micro and macro levels (Carlsson & Jacobson, 1993; Foray & Freeman, 1993; Patel & Pavitt, 1994: Scherer, 1989; Smith, 1991; Soete & Arundel, 1993).

Transforming a less innovative economy into a more innovative one is a great challenge for policy-makers. This general transformation task is much more complicated for the former socialist countries. For any redeployment crucial to understand how systems of innovation work. Theoretical and empirical knowledge are critically important, and tremendous efforts need to be made to develop that knowledge. Empirical evidence on the transition of Hungarian innovation process is also needed.

This chapter concentrates on the introduction of a metric tool of innovations in a post-socialist transition economy, Hungary. Measuring innovation means to measure something, which is extremely complex, a moving target, where many impacts of innovations are hardly measurable.

Conceptual Framework

The common goal of all activities is to measure innovations and provide innovation indicators that can meet the theoretical and policy-related requirements. Policies and innovation fostering measures may be broadly based, touching many aspects of the innovation process, or they may be precisely targeted in order to tackle a particular problem that needs attention.

There are several different approaches and methodologies that seek to measure innovation activities. Looking at international innovation literature as a whole, we find a variety of views on the process of innovation and, the factors that create innovation. Some parts of the literature focus on technological development, technical research and R&D functions (Chesnais, 1995; Dosi et al., 1988; Freeman & Soete, 1997; Grupp, 1998; Rosenberg, 1994; Scherer, 1989). Information on technical innovation activities maybe based on case-studies, semi-structured interviews, literature-based innovation indicators and innovation surveys. One crucial element in these methods is that of innovation survey. This instrument has advantages and disadvantages to measure the performance of national innovation capacity—as previous literature has discussed (Archibugi & Pianta, 1996; Arundel & Garffels, 1997; Arundel et al., 1998; Sirilli & Evangelista, 1998; Smith, 1999).

These surveys may employ the '*subject approach*' to collect data on the innovative behavior and activities of the economic unit as a whole, whether or not a unit is a legal entity.

There are several needs for indicators, e.g. technological significance of innovation, quality of innovation, technological nature of the innovations, which cannot be measured with a survey at the economic unit (firm, corporate, factory) level, where each of them can be active in a wide range of innovation projects. Another type of survey may focus on a targeted technology, such as biotechnology, laser-technology and so on. These surveys employ the '*object approach*' to collect data about specific innovations *from* firm/business unit (OECD, 1997, p. 103) that help to identify possible future innovation patterns.

The present chapter highlights one way of collecting information regarding innovation, namely the *innovation survey* by subject approach (Archibugi & Pianta, 1996; Arundel et al., 1998; Evangelista et al., 1998; Smith, 1992).

The Oslo Manual, developed by OECD member countries,[1] is a guidebook for data collection on technological innovations. It describes definitions, criteria and classifications which are relevant for studies of industrial innovation. Description of measuring aspects, alternative solutions make the *Manual* a starting point of any technological product and process innovation survey design. It contains suggestions and recommendations for national and international innovation surveys. The Manual is the required framework to measure innovations in an internationally comparable way. Common core indicators and common definitions help to avoid unnecessary differences in measurement of different groups of countries.

Internationally comparable indicators need to measure and evaluate national innovation capabilities, and performances. Parallel with the development of the

Oslo Manual, a harmonized innovation survey was developed. (Harmonization has included a questionnaire with set of core questions, survey population, sampling methods, time frame and so on.) The name of this common survey: Community Innovation Survey (CIS), drawn up by the OECD and EU Statistical Office (Eurostat) as a basis for collecting statistically comparable data across countries. The CIS questionnaire has to run in all European Union member countries (survey cost has been partially covered by the European Union) and some other OECD member countries. This accepted international standard furnishes a very sound background, even if many methodological problems and interpretations of innovation surveys have not yet been resolved.

The so-called Community Innovation Survey (CIS) questionnaire is revised for each survey period. The different versions of the questionnaire are named: CIS-1 (surveying period: 1990–1992), CIS-2 (surveying period 1994–1996), CIS-3 (surveying period (1998–2000).

Not only the reference period was different among CIS-1, CIS-2 and CIS-3 (see also Wilhelm in this volume) but the questionnaires and coverage of sectors and size of surveyed units were modified taking into consideration the findings the previous surveys. The CIS-1 focused only on the manufacturing sector, CIS-2 also included the service sector. More attention was devoted to include random samples of small firms, not only to cover the innovative but also non-innovative firms. Each modification has involved a considerable amount of intellectual effort.[2]

The accumulated knowledge by surveys proved that the Oslo Manual guided innovation survey contributed to an analysis of the dynamics of technological change in the business sphere and, enable policy-makers to investigate more effective ways to maximize the socio-economic and industrial development potential and, to support competitiveness and productivity in market economies.

Considerations on Adaptation in a Transition Economy

There are many newcomers in the surveying innovation arena. Among them, the specificity of transition economies has been originated from the legacy of socialism.

The conundrum of socialist economies lies between relatively well-developed science and inefficient, ad-hoc business enterprises.

The socialist system was very weak in diffusing knowledge and commercialization. The firms in the

---

[1] Oslo Manual belongs to the OECD Frascati Family of Manuals. The development of Oslo Manual started in the 1980s and the first version of Oslo Manual was published in 1992 (OECD, 1992). Following the experiences the (internationally harmonized) national surveys Manual was thoroughly updated and the revised version was published in 1997 (OECD/Eurostat, 1997). The revision of the Manual is still ongoing incorporating the new knowledge and expertise that have been accumulated about innovation and innovation process through innovation surveys and other investigations of innovation activities.

[2] Limited part of these efforts are available as scientific publications. Most of them were prepared as discussion papers for Eurostat and OECD working party meetings, European Union project reports and studies for national governments.

market economy sense did not exist in socialist economies. In the context of command economies literature use 'enterprise' which differs from profit-oriented micro-economic organizations of market economy. Situated in a socialist system, this type of late-comer periphery industry was cut-off from the major international sources of technology. They operated in isolation from the world centers of science and innovation and consequently were technologically disadvantaged. The surrounding technological and industrial infrastructure was underdeveloped. The economic actors had little interests in commercializing R&D results (Dyker & Radosevic, 1999; Hanson & Pavitt, 1987; Inzelt, 1988, 1991, 1994a; Jasinski, 2002; Radosevic, 2002). Also, the information system did not contain relevant indicators on innovation. If any science and technology measures existed, they were very different from market economies. The main aims of these measures were to support macro-economic decisions of command economies and, prove the prestige of the socialist world, in bi-polar world system. These interests had a strong influence on the collection, production and use of statistical data, indicators, evaluation, and so on.

In the structural transformation phase of CEECs economic and development, there is a burning need to explore the condition of innovation, and R&D, besides establishing economic stability.

During the process of developing a modern information system for innovation transition economies two different but parallel tasks of modernization require solving: (a) to adopt the traditional metrics of innovation and of related areas through; and (b) to be involved in the development of modern innovation indicators. The involvement in the development of modern indicators is important for all countries of the world.

This chapter deals with the first task, the constructive adaptation of innovation survey for the manufacturing sector in a transition economy. The adoption procedure of innovation indicators is a great challenge for all transition economies.[3] Survey attempts are crucial to build internationally comparable information for deci-sion-makers about product and process technology innovation. Only such exercises can prove the relevancy of innovation surveys (existing standardization of concepts and statistical methodologies) outside the advanced OECD/EU countries and can highlight the need for further development.

Hungary has carried out several important steps for transforming her science and technology (S&T) information system during the past 12 years. One of the very important attempts has been employing the Oslo Manual guided innovation survey. This chapter summarizes the findings of the first Hungarian pilot innovation survey. The Hungarian pilot innovation survey, based on a standard questionnaire, largely operates with the concepts and questions proposed in the Oslo Manual and follows the internationally harmonized questionnaire. This was the pioneering survey in a transition economy. Following several knowledge dissemination seminars (OECD, 1993, Paris; OECD-OMFB, 1996, Budapest) several other attempts have been carried out.

From the mid-1990s CEE transition economies started to attempt surveying innovations in the manufacturing sector (parallel to, revision work to R&D surveys). These exercises on innovation data collection in the manufacturing sector, was based on conception, definitions, classifications of the Oslo Manual. The OECD (Organization for Economic Cooperation and Development) and Eurostat are working in tandem and they involve CEECs in the harmonization of innovation indicators, using the Community Innovation Survey (CIS) questionnaire guided by Oslo Manual standards. The national innovation surveys guided by the Oslo Manual are penetrating into CEECs. After 10 years of transition, there are some fruits of this learning process, however, the level of implementation of OECD/Eurostat standards and the adoption of CIS-1 or CIS-2 questionnaires in national environments of CEECs are different.[4] EUROSTAT invovles CEECs into CIS-4 as full-fledged partners.

Following sections of this chapter highlight some of the findings of the first Hungarian pilot innovation survey, carried out in 1994 and based on 1990–1993

---

[3] In this process the international community has been playing an important role which may shorten the adoption-development phase. Foreign governments and international organizations have been supporting such knowledge-transfer through different channels. The first milestone of knowledge transfer was OECD Vienna/Bratislava Conference, in 1991. It was followed by a series of international training seminars (OECD) for Central and Eastern European experts (in 1993 Paris; 1996 Budapest). The OECD has involved transition economies into several activities (workshops, conferences, on-the-job training). The OECD supported the translation and dissemination of the OECD Frascati Family Manuals as the theoretical framework containing the definition of this process. Many countries started the adoption process by means of a translation and dissemination of the OECD Frascati Family Manuals. The Frascati Manual has been

published in several CEECs. And the availability of these manuals in national languages closed the gap between existing and employed knowledge. In the late 1990s Eurostat involved transition economies (called also newly associated countries) in indicator development process. However, penetration by newly acquired knowledge is not a very rapid process.. For example, the Frascati Manual was published in Hungarian in 1996 and the first citations outside Central Statistical Office and academic circles were observed in 1999.

[4] Data and analytical reports on CEE innovation survey attempts see in: Auriol & Radosevic, 1998; Bazhal, 2002; Csobanova, 2002; EC-Eurostat, 2000; Inzelt, 1991, 1993, 1994a, 2002; Radosevic, 2002; Sandu, 2002).

data,[5] the first section of this chapter, 'The Main Objectives of Innovation', summarizes the main objectives of innovation. The second section discusses the 'inputs into innovation'. Then the next section goes into the details of the 'outputs of the innovation process'. The fourth section investigates the factors that hamper the whole process. The fifth section makes some concluding remarks. Appendix 1 shows the modifications that were carried out to the OECD/EU harmonized questionnaire to suit Hungarian conditions. Appendix 2 details the main features of the pilot survey, as based on the harmonized questionnaire.[6]

On the scope of the pilot survey, it should be mentioned that most of the 110 respondents were the successors of former state-owned enterprises or co-operatives. There are very few newly emerged small businesses and companies in the sample.

Before presenting the main findings of this pioneering pilot survey it has to be emphasized that the surveying period belonged to the first phase of transition. In transformation of the former socialist economies into market ones, two different phases may define, which are typical in the CEE region. All CEE countries were exposed to the phenomena of Phase I for some years and suffered its consequences. *Phase I is* characterized in the following way: declining economy, decreasing R&D expenditure and R&D staff, the inherited structures that still exist or are under reorganization, strong brain-drain from the country. Many old partnerships were ceasing. At the same time there are many far-reaching changes in the legal system, governmental structure and policies. Institutional building process is going ahead. However, system of institutions is still gap-toothed, enforcement of new laws are still weak. In other words the countries have to overcome on 'transition crisis' for stabilization, development and transformation.

By the late 1990s some of Central and Eastern European countries, including Hungary seem to have reached Phase II and show a consolidated picture. Phase II is characterized: the economy is growing, the R&D expenditures grow again, new structures began to work, company R&D develops again, the brain-drain balance is nearly neutral. The overall macro economic situation provides for R&D and innovation activities.

The Main Objectives of Innovation

The level of innovation in Hungarian industry is relatively low. There is a desperate need for investment

Table 1. Ranking of objectives of innovation (Reference period 1990–1993).

Objective	Number of firms answering 'very significant' and 'crucial'
Improve product quality	71
Extend product range within main field	66
Create new domestic market	62
Extend product range outside main field	56
Improve production flexibility	43
Create new markets within EU	41
Reduce material consumption	41
Energy consumption	37
Environmental damage	35
Other objectives	30

in product design, in packaging, and in the application of technology. In order to improve market positions, you need competitive products.

In this connection, firms were asked to rate the significance of each of 19 possible objectives underlying their innovation activities during the investigated period. The pattern of ranking of these main aims illustrates well the changes in economic environment (see Table 1 above).

It is particularly striking that the highest ranked aim was to improve product quality. The next most crucial objectives show how firms need, at the same time, new markets and changes in their product ranges. Innovating with the objective of opening up new markets abroad was judged to be less crucial than for the domestic market. Multiplying the product range is important, not only in the main field of production activity, but also outside them. Redeployment may lead the firms out of their main fields. The market sectors targeted for innovations are, however, generally very close to the original markets. This allows us to assume that innovations have been more incremental than fundamental. What does come through clearly is that innovative firms are market-oriented, which is a positive phenomenon in a former socialist country. Reducing costs in terms of material and energy consumption is also important, but far less important than the aims mentioned in Table 1. Other objectives, such as reduce production and design consumption, improve working conditions, replace outgoing products were not important for firms during the investigated period.

Inputs into Innovation

The costs of innovation and the breakdown of costs (R&D, acquisition of patents and licenses, and product design) depend largely on the strategy of the firms involved. A firm might be innovative even if it does not invest in resources, for example, in R&D. But it is not

---

[5] The project was basically financed by OMFB and organized by IKU. The first English version was written during my stay as a Leverhulme visiting fellow at SPRU.

[6] Until mid-2002, this survey was followed by a feasibility survey in the service sector and by Central Statistical Office pilot survey in the manufacturing sector. Both samples are very different from this one. So they do not allow statistical comparison over time.

Table 2. Ratio of R&D expenditures to sales.

	0%	<1%	1–4%	4–8%	>8%
Number of companies	38	29	21	12	8

possible to avoid spending on introduction, adaptation, market analyzes, and so on. Among the inputs of innovations, the main proportion of expenditure made by the firms questioned went on R&D (see Table 2). Table 2 shows the proportion across the distribution of companies.

It is very rare for companies to spend more than 8% of their sales revenue on R&D (the nation-wide aggregate R&D to GDP ratio was 1.59% in 1990 and just 0.99% in 1993. This declining trend had continued by 1999. The lowest fraction was 0.7%). The absolute scale of their expenditures is modest because these are not large firms. Three quarters of the respondents employed between 100 and 1,000, the remainder of respondents were above 1,000, but less than 5,000. Their average net sale was HUF 650 million (in USD = 2.2 million). The surveying period was 1990–1993. These years were the initial years of the first phase of transition: the redeployment of so-called socialist enterprises into firms meant shrinking sizes, changing ownership if the firms have been privatized by multinational companies.

R&D expenditure have usually financed product innovation rather than process development. Only electricity, water, chemical and printing industries spend relatively more on process innovation. The first three cases hardly require any explanation. The case of the printing industry is interesting because it reflects the reconstruction of the industry through the application of new techniques of the last few decades.

The small number of companies expenditure on R&D means that the enterprises have not been compelled by competition to introduce new products and to apply new technologies for the sake of their survival and profitability. Of course, low spending on R&D does not necessarily mean a low rate of innovativeness. The firms were asked to estimate the break-down of their innovation costs. To calculate the sector average of the proportion of different expenditures on innovation, we have to know total innovation costs. Two thirds of the respondents were willing to give total current expenditure on innovation activity. Their resultant data are shown in Table 3, with aggregate figures for manufacturing added for comparative purposes.

After R&D, patents and licenses are the biggest form of expenditure. They are followed by trial production, training and tooling up, which have grown in importance over the transition period. Product design accounts for 8% of innovation expenditure in the manufacturing industry, with the remainder being spent on market analysis. From the point of view of transformation enterprises to firms and, redeployment of sectors, some interesting differences in the way different sectors spend their innovation budgets are observable. For example, rates of expenditure on R&D are relatively low in fabricated metal products, vehicles and transport-equipment industry. The former is a fast-shrinking industry in Hungary, and where it undertakes innovations those are based on licenses. The transport industry is in a mixed position—production of traditional Hungarian products (buses, vans etc.) has been declining, with the firms in the pre-privatization phase in the investigated period. These were reluctant to invest in innovation, and concentrate their efforts rather on seeking new markets and investors. The newly re-emergent automobile industry is a fast-growing one. Its

Table 3. Breakdown of expenditures on innovation, by sectors (%).

Sectors	R&D	Patents and licenses	Product design	Training tooling up trial products	Market analysis
Manufacturing	42.8	27.0	8.0	16.9	4.3
Food, Beverages, Tobacco	74.5	1.0	10.3	9.0	5.2
Textile, Apparel, Furniture	56.8	—	20.9	10.8	11.4
Wood, Paper, Print	84.6	—	—	13.1	2.3
Chemicals	59.8	4.3	8.3	18.0	9.6
Drugs	57.6	8.3	4.3	25.5	4.4
Rubber, Plastic	39.7	8.7	9.1	38.5	3.9
Fabricated metals	9.2	86.1	1.1	2.4	1.2
Machinery	49.4	1.1	15.7	25.0	8.3
Equipment, Instruments	42.2	9.5	32.5	18.8	6.0
Vehicles	33.4	5.2	19.8	23.9	5.9
Furniture	—	30.0	40.0	30.0	—
Construction	—	100.0	—	—	—
Others	92.4	4.1	0.9	2.0	0.6
Taken together	48.2	23.7	6.8	14.6	3.8

innovations (new product design, new instruments etc.) are based on foreign laboratories at assembler companies, and suppliers usually get technical support for innovation from their assemblers. This observation is confirmed if we compare these results of the innovation survey with information from other sources relating to the present situation of the former Hungarian industry R&D laboratories that used to belong to different ministries (Goldman & Ergas, 1997). These applied research, experimental development and design institutes have virtually lost their former customers.

The high proportion of trial production, training and tooling-up in some sectors represent a good example of changes in innovation strategies and adaptation to a new environment. Market analysis shares are relatively high in those sectors that used to have high export shares toward CMEA (Council for Mutual Economic Aid) countries.[7]

An important aim of innovations was to extend the product range outside the main product fields. So, it is worth investigating which sectors are chosen by firms when they do this. Respondents seeking to diversify their product range have tended to invest in upstream or downstream sectors (For example food processing company into packaging). That time restructuring strategies have generally followed the old way of thinking, diversification and in-house production of components and instruments, instead of buying outsourcing. These strategies are usually based on short-term profit-seeking activities, and they have less strategic redeployment elements.

Postal surveys can pick out these tendencies, but cannot give enough information for drawing definite conclusions.

---

[7] CMEA (also used abbreviation was COMECON) used to be an economic organization. It was established in 1949 to encourage trade and friendly relation among 10 command economies, including the Soviet Union. The organization was dissolved in 1991.

Cooperation Arrangements in R&D Activities

Cooperation in R&D is growing in importance all over the world. Both formal and informal networks are playing an important role in innovation (Freeman, 1991; Meyer-Krahmer & Schmoch, 1998). For knowledge-based economies the key of the whole process is the partnership between universities, the business sector, and among firms. The relationship between innovation and networking is discussed in details by Swan & Robertson in this volume. Innovation survey can focus on few dimensions of this relationship.

The questionnaire tried to investigate cooperation by type of actor and by their geographical origin. From 110 respondents just 55 firms engaged in cooperation agreements in R&D activities with other parties in 1993. If we concentrate on the parties with which such arrangements were formed, we see that research institutes of the Academy of Sciences, universities and sector research institutes are at the top of the list. Customers and suppliers are also frequent collaboration partners, but they come far behind that of knowledge-based institutes, as is usual in market economies among medium-sized tech companies.

Table 4 shows the number of innovators using particular cooperation partners at different regional levels.

Table 4 shows clearly that where universities and R&D institutes and industry R&D laboratories were the cooperation partners, they were most likely to be located in Hungary. In the case of customers and suppliers there was no such tendency.

One of the outstanding characteristics of domestic industry-academy linkages, is that regional cooperation is weaker than nationwide. To some extent this may reflect a pattern of seeking excellence anywhere it is to be had. The distance of potential partners are much less important than their complementary capabilities, the quality of their knowledge (similar phenomenon was observed by Faulkner & Senker, 1994).

Table 4. Entering into co-operation in R&D activities.

Base	Location of Partners			
	Domestic		Foreign	
	Regional	National	European Community	Elsewhere
R&D institutes	7	24	3	0
Universities/higher education	8	23	2	1
Suppliers	7	6	10	2
Clients/customers	3	7	7	7
Industry R&D laboratories	3	16	0	0
Consultants	2	6	3	2
Parent/subsidiary company	4	1	1	0
Joint ventures	0	3	2	1
Government laboratories	1	5	0	0
Competitors	1	0	0	0

But taken into account that innovation activities in the sample were usually incremental rather than fundamental, the tendency to weak regional cooperation may have been mainly inherited from the former socialist system. During the years when the industrial structure was centralized, sectors were commonly concentrated in single firms, and even where many units existed in the enterprise, their in-house R&D laboratories were situated at the headquarters. Until the 1970s, universities, Academy of Science, and industry R&D institutes were concentrated in very few regions, mainly in Budapest. The headquarters of enterprises usually established industry-university linkages close to their location. From the beginning of the transition period, until the late 1990s the overwhelming tendency had been towards a decline in university (Academy)-industry relations. At best, companies have been able to preserve inherited links, mostly in Budapest, and have not yet started to seek new partners on a more dispersed basis. Regionalization of the Hungarian innovation system started in the late 1990s however, it still remains one of the main tasks for the near future.

The internationalization of the economy is strongly dependent on the possibilities and results of co-operation in the field of innovation. According to the responses, companies have been able to establish co-operation (in development, testing, and quality control; that is, mainly in R&D services) with foreign companies, but not with laboratories, which reflects international practice.

Sources of Information for Innovation

Among the inputs, the sources of information are very important. As Boly et al. emphasized in this volume, innovation activity is associated with uncertainty. Quality and volume of available information are reducing the risks. Innovation requires managers' ability to select the sources of relevant information.

The questionnaire of innovation survey asked the importance of different kinds of information sources such as market-oriented, R&D related and general information. If we look only at the pattern of responses by sources, that itself indicates a problem. Internal sources (within the enterprise or group of enterprises) were of key importance for 49 out of 89 responding firms (Table 5).

The external sources form three logical groupings: market/commercial sources, educational/research establishment, and generally available information.

In the case of external sources of information, two subgroups are distinguished: national and international. It is reasonable to make this distinction because Hungarian firms are much less internationalized than West European ones. Lack of world-wide communication, sporadic foreign information were important hampering factors of innovations, and lack of competitiveness in the socialist system. As mentioned

Table 5. Ranking of sources of information for innovation by types.

Factors answer	Number of firms which 'very significant' and 'crucial'	
In-house	**49 [89]**	
External	**Domestic**	**Foreign**
	Information sources	
Market/commercial	**107 [107]**	**104 [107]**
Clients/customers	55	48
Competitors in your line of business	30	28
Suppliers of materials and components	19	28
Suppliers of equipment	7	25
Consulting firms	1	2
Educational/research establishments	**76 [76]**	**57 [76]**
Universities	14	1
Government laboratories (Academy)	6	—
Science and innovation parks	1	3
Generally available information	**84 [84]**	**75 [84]**
Fairs/exhibitions	41	32
Professionals conferences, meetings	25	17
Professional journals	24	32
Patent disclosures	14	10
Professional associations, chambers	15	2

Note: Figures in square brackets = number of responses.

Table 6. *Share of sales accounted for by products with incremental or significant changes (numbers of respondents).*

Share of Sales	Domestic sales		Export sales	
%	Significant	Incremental	Significant	Incremental
Below 1	0	0	1	0
1.0–5.0	15	10	9	8
5.1–25.0	26	25	13	18
25.1–50.0	10	16	6	6
50.1–75.0	4	6	5	8
75.1–100	3	4	7	8
Total	58	61	41	48

earlier the enterprises operated in isolation from the world centers of science and innovation and were behind technologically. The surrounding technological and industrial infrastructure was underdeveloped. The internationalization of the economy was an urgent need. Openness of thinking could help to improve competitiveness. The question is whether political 'open up' and governmental measures for internationalization can be found in employing foreign sources of information?

The distinction by national and international sources is more important for periphery countries than advanced ones because the former firms are much less internationalized than West European ones.

Firms usually rated the importance of domestic information higher than foreign, which is a sign that minds are opening too slowly. An important reason for this slowness is the problem of knowledge of foreign languages. Table 5 summarizes the ranking of the various types of information required in the development and introduction of new products and processes.[8]

It is a positive phenomenon that the companies evaluated market information (clients are first, and competitors third) so highly. As regards to domestic sources of information in the rank two of the generally available information are on the second and fourth places. Fairs, exhibitions are second and professional meetings fourth. These events make it possible to meet people and to start building networks (see Swan et al., this volume).

Between competitors, interaction is considered the best way to share information among each other. Almost one third of respondents evaluated competitors in their line of business significantly or crucially important source for their innovation activities.

At the other end of the scale, with very low rankings, are consulting firms, science and innovation parks,

government laboratories and universities. Consulting firms and parks were their own initial phases, building their own capabilities. The government laboratories-university-enterprise linkages hardly existed before the transition period and companies during their own transformation process have not become more hungry for the inventions, novelties. If we take into account the knowledge is becoming the core capability in that competition, the information on new knowledge that is needed for long-term strategy, large changes, and technological breakthroughs, are regarded, at best, as moderately significant sources. The role of professional journals seems to be the most important source of new knowledge.

Comparing the importance of cross-country information sources to domestic ones clearly, observable firms usually rated the importance of domestic information higher than foreign. A small number of companies evaluated foreign market sources 'very significant' and 'crucial' than domestic ones. The difference is wider in the case of generally available information. An important reason for this slowness is the problem of knowledge of foreign languages. Foreign education and research establishments was a negligible information source for the investigated companies.

There are two important features of employing foreign sources: role of professional journals and technology transfer. Technology transfer (suppliers of materials and components and suppliers of equipment) is becoming an important source of information.

Outputs of the Innovation Process

It is not an easy task to measure innovation outputs. The questionnaire targets the share of sales accounted for by innovative products in a number of ways, namely:

- sales of products by different stages of the product life cycle;
- breakdown of domestic/export sales by three types of products (unchanged, incrementally and significantly changed) (see Table 6);
- breakdown of total sales by newness of innovative products (new globally; new in Hungary; and new or

---

[8] The main findings of different innovation surveys show many similarities in relation to sources of information and objectives of innovation. Particularly striking are the similarities between the results of the Irish Innovation Survey and the Hungarian one. (For the Irish results see FitzGerald & Breathnach, 1994) Systematic comparison among smaller European economies seems fruitful.

Table 7. Share of sales accounted for by new products (number of respondents).

Proportion of total sales %	New to		
	Globally	Hungary	Firm
Below 1	2	0	0
1.0–5.0	4	3	1
5.1–25.0	7	10	9
25.1–50.0	6	14	5
50.1–75.0	1	5	5
75.1–100	3	14	28

substantially improved for enterprise only) (Table 7).

It was assumed by the OECD/EU questionnaire that the share of sales accounted for products in the introductory phase would indicate the output of innovation. The product life cycle is not a clearly understood category for many respondents. On the basis of experience with the pilot survey, we shall not use this indicator widely as an output measurement of innovation, because the information based on this indicator is rather uncertain. The quality of responses on the other two indicators is more reliable.

Out of the 110 firms, 82 developed or introduced technologically altered products, and 62 technologically changed processes, during 1990–1993 and 88 intend to develop or introduce technologically changed products or processes in the years 1994–1996.

The data confirms that the main objective of innovation activity is to develop and introduce new products on the domestic market. Breakdown by domestic and export sales of innovative and non-innovative products are very important indicators for less advanced countries. The number of companies which introduced significantly or incrementally changed products is higher in the case of domestic sales than in that of export.

More than 75% of the changed products came from the food, beverage, tobacco and electronic equipment, medical, precision and optical industries. Although a small sample contains many uncertainties, it is worth mentioning these industries because they have a common characteristic—the firms, or their upstream firms (such as assembler and trading companies), have been privatized, and their new owners have started to redeploy production assets to introduce new products on the domestic market. The food industry reacted to new challenges on the domestic market emanating from import liberalization by innovating, to limit the displacement of their products by foreign competition. But they were not able to introduce their new products on (new) export markets. Maybe the reasons are export quotas (European Union trade barriers), licensing restrictions in the case of foreign-owned companies, or weak competitiveness of products. To answer this very crucial question will require more detailed investigation.

Changed products account for a relatively high proportion of exports in the metal and machinery industries. These markets are shrinking, and it is no longer possible to sell many of the old metal and machinery products.

Incrementally and significantly changed products can also be investigated on the basis of their newness. In the survey, the newness of innovative products was categorized into three groups—'globally', 'in Hungary', and 'enterprise/group only'. The ratio of sales of innovations is differentiated between three groups of innovations (Table 7).

Half of the respondents sold altered products that were new only in Hungary or in the firm. If we combine this answer with responses on the product life cycle—in those cases where information seems reliable—most of the products new to firm or only to Hungary were in their maturing or declining phases. These innovations may suffice for short-term survival—they can help the companies avoid bankruptcy, quickly replace lost markets and so on. But they cannot improve international competitiveness, which require breaking out of the old mould.

Factors Hampering Innovation

A review of the innovation survey gives some indications as to why innovation activities and their results are so very limited. A range of 21 possible key impediments can be subsumed under two major factors, with a range of miscellaneous issues completing the picture. Table 8 shows what firms thought were the central factors constraining innovative activity.

Apart from the many changes in the banking system and financial situation in Hungary, the greatest hindrance for firms was the lack of appropriate sources of finance. There was no proper financial system for innovation and development activities. In a matured market economy the financial factors also hamper innovation. However inappropriateness of financial sources is very different between capitalized and undercapitalized countries. In undercapitalized transition economies, the financial market is semi-developed, meaning that many sophisticated financial

Table 8. Ranking of factors hampering innovative activity.

Economic factors	Number of firms with answers of 'very significant' and 'crucial'
Lack of financial sources	80
Innovation costs too high	48
Pay-off period too long	40
Perception of excessive risk	20
Firm factors	
Innovation potential too small	36
Lack of information on markets	18
Innovation costs hard to control	18
Organisational structure of firm	18
Lack or weakness of innovation management	15
Poor availability of external technical services	15
Lack of skilled personnel	14
Lack of information on technologies	8
Resistance to change in firm	6
Lack of opportunities for co-operation	6
Other reasons	
Legislation, norms, regulations, standards, taxation	28
Lack of technological opportunities	21
Uncertainty in timing of innovation	15
Lack of customer responsiveness	14
Innovation too easy to copy	8
No need, due to earlier innovations	2

techniques and measures are missing. During the investigated period very few financial measures have existed in Hungary, which promoted realization of innovation. In the first decade of the transition period, interest rates were high, little funding was available, venture capital had been non-existent. Tax benefits and other state measures directly or indirectly were introduced in the late 1990s after the surveying period. Before that time, public funds contributing to the financing of innovation in Hungary were limited.

All other factors were picked out by a substantially smaller number of firms. The next most important issue—judged to be significant by roughly every second respondent—were excessively high innovation costs and excessively long pay-off period.

But policy-makers must understand that even if a company has enough financial resources, it may remain non-innovative. *Supplying funds for research does not automatically increase the intensity of innovation activities.* In many countries, lack of skilled personnel is becoming an important hampering factor of innovation activity. (For example information and communication technology specialists in Germany, computer illiterate blue-collar workers in the car accessory industry.)

A second block of constraints relates to the potential of the firm to engage in innovation. It is worth mentioning that interviewed firms were much more critical of these firm-level factors than respondents to the postal survey. Almost every person interviewed

face-to-face emphasized: lack or weakness of innovation and inappropriate organizational structure within the firm. Only one in four respondents to the postal survey judged these factors to be a serious impediment. This was the biggest discrepancy between interview and postal survey responses. It may be that self-evaluation in written form was much less critical (or more careful) than in oral form. This highlights one of the limits of the postal survey as a tool of investigation.

A third block of factors hampering innovation can be summarized as general external issues. It is a positive feature that earlier innovations (2) or lack of customer responses to new products and processes (14)—i.e. demand-side problems—do not greatly constrain the innovation process.

Some Concluding Remarks

By the late 1990s some of Central and Eastern European countries, including Hungary seem to have reached Phase II and shows a consolidated picture. According to some experiences in Central Europe, there is a shift of about 2 to 4 years between the recovery of the economy and the revitalization of the R&D. Overall the Hungarian macro economic situation provides an opportunity for R&D and innovation activities.

These results show that the effort put into the Hungarian innovation survey was not in vain. In the new environment of the transition economy it is rational to collect innovation data according to the Oslo Manual. Internationally comparable indicators are extremely important in the age of globalization. They allow inter-country comparisons.

At least the first phase of the transition period from command economy towards a market economy is a very special environment for measuring such typical market economic phenomenon as innovation. (The period of such transition offers different environment for any type of surveys.) The first period did not offer many grounds for optimism regarding modernizing and improving the competitiveness of the Hungarian economy. But analysis of the factors that hamper innovation may help to prepare a better innovation policy for the future.

Summing up the main conclusions of the survey carried in the first phase of transition:

• the main objectives of innovation are the improvement of product quality, the opening-up of new markets, and changes in products;
• the sources of information rated highly by the greatest number of innovators are those which are closest to the firm: clients and customers, internal sources and even competitors. Educational and research establishments are judged to be significant as sources of information by only a small number of firms. Traditional external sources—fairs and exhibitions, professional conferences, meetings, journals—are judged to be more significant;
• financial factors, including lack of appropriate sources of finance, are evaluated as the most significant impediments to innovation. Another major complex of factors hampering innovation relates to company-specific issues such as inadequate innovation potential, lack of information on markets, problems of controlling innovation costs, and deficiencies in the existing organizational structure of enterprise;
• the pilot survey experiences, reveal that respondents were able correctly to identify the declared market economic definitions.

This pilot survey was one of the first attempts to take a wide-ranging look at the actual levels of product and process innovation in Hungary. This survey has provided a much more comprehensive picture of what has been happening than periodic surveys of R&D performance. Policy-makers need to have reliable information on R&D inputs and outputs, as well as on factors hindering diffusion.

In conditions of transition, initial gambits may lead in many different directions. Government policy has to find ways to create an environment friendly toward innovation and entrepreneurs, and to build a system of guidelines to help locate the best solution for the whole economy. Business firms must create demand for R&D results, and must be capable of utilizing them. They need to possess an intellectual base that can promote inventions and innovations introduced in other countries. International experiences suggests that the market cannot by itself solve this problem. It must be supported by a general economic policy, as well as by science, technology and innovation policies. The present research can help to lay the foundation for those policies.

The new indicators—time-series and international comparison—can make policy-makers and businessmen better informed, to provide them up-to-date knowledge. The innovation indicators may back up policy-making process and help to change the way of thinking on innovation policy matters gradually in coming years. Transition economies have to go beyond the traditional innovation indicators developed by advanced economies to learn the reasons of low level of collaboration, lower share of new products in export sales, factors constraining innovation which relate to firm potential.

The innovation survey has highlighted some of the key policy questions which have to be answered. The time series on Hungarian innovation activities are still missing. Development of reliable, timing indicators is also a part of the learning process in transformation of policy-making. The pioneering pilot survey was followed by several other pilot surveys in Hungary that brought several problems into even sharper relief, such as market economy is not a guarantee for improvement in innovativeness, financial conditions are crucial for innovations but they are not enough to encourage innovations. Government has to facilitate in many ways companies to upgrade innovative capabilities.

Acknowledgments

I am grateful for helpful comments and suggestions to Sándor Bottka, József Imre, István Lévai at OMFB, Budapest, to Péter Vince at IE-HAS, Budapest, to Katalin György and Ildikó Poden at IKU, Budapest, to David Dyker, Chris Freeman, Toshiro Hiroto, Pari Patel, Slavo Radosevic, Margaret Sharp, Jacky Senker and Nick von Tunzelmann, at SPRU, Brighton, to Roberto Simeonetti at the Open University, to Gernot Hutschenreiter at WIFO Vienna, and to co-workers at OECD DSTI, Paris and several participants at the Six Countries Programme Conference (Vienna, Dec. 1994) and the 3rd Euro-Japanese Conference, (Sunderland, Dec. 1994). All responsibility for any remaining errors is mine.

References

Archibugi, D. & Pianta, M. (1996). Innovation Surveys and Patents as Technology Indicators: The State of the Art. In: *Innovation, Patents and Technological Strategies*. Paris: OECD.

Arundel, A., Smith, K., Patel, P. & Sirilli, G. (1998). The future of innovation measurement in Europe. *IDEA paper*, **3**. STEP Group.

Arundel, A. & Garffels, R. (Eds) (1997). Innovation measurement and policies. *European Commission, EIMS publication,* **50**. Luxembourg.

Auriol, L. & Radosevic, S. (1998). Measuring S&T activities in the former socialist economies of Central and Eastern Europe: conceptual and methodological issues in linking past with present. *Scientometrics,* **42** (3), July-August, 273–297.

Bazhal, Y. M. (2002). Contemporary issues of innovation activities in Ukraine economy. In: A. Inzelt & L. Auriol (Eds), *Innovation in Promising Economies* (pp. 65–82). Budapest: Aula Publisher.

Carlsson, B. & Jacobson, S. (1993). Technological systems and economic performance: the diffusion of factory automation in "Sweden techological system and future development potential" project, financed by the national board for Industrial and Technical Development and the Swedish Council for Planning and Coordination of Research, November 1992.

Chesnais, F. (1995). Some relationships between foreign direct investment, technology, trade and competitiveness. In: J. Hagedorn (Ed.), *Technical Change and the World Economy. Convergence and Divergence in Technology Strategies.* Aldershot: Edward Elgar.

Csobanova R. (2002). Market for innovation in Bulgaria. In: A. Inzelt and L. Auriol (Eds), *Innovation in Promising Economies* (pp. 43–64). Budapest: Aula Publisher.

Dosi, G., Freeman, C., Nelson, R., Silverberg, G. & Soete, L. (1988). *Technical change and economic theory.* London: Pinter Publishers.

Dyker. D. & Radosevic, S. (Eds) (1999). *Innovation and structural change in post- Socialist Countries: A quantitative approach.* The Netherlands: Kluwer Academic Publishers.

EC-Eurostat (2000). R&D and innovation statistics in candidate countries and Russian Federation, Data 1996–1997, European Communities, 2000. Luxembourg.

Evangelista, R., Sandven, T., Sirilli, G. & Smith, K. (1998). Measuring innovation in European industry. *International Journal of the Economics of Business,* **5** (3), 311–333.

Faulkner, W. & Senker, J. (1994). Making sense of diversity: public-private sector research linkage in three technologies. *Research Policy,* **23**, 673–695.

Freeman, C. (1991). Networks of Innovators: a synthesis of research issues. *Research Policy,* **20**, 499–514.

Freeman, C. & Soete, L. (1997). *The economics of industrial innovation* (3rd ed.). Cambridge, MA: MIT Press.

FitzGerald, A. & Breathnach, M. (1994). *Technological innovation in Irish manufacturing industry.* Dublin: Forfás.

Foray, D. & Freeman, C. (Eds) (1993). *Technology and the wealth of nations.* London: Pinter Publishers.

Goldman, M. & Ergas, H. (1997). Technology institutions and policies. World Bank Technical Papers, No. 383. Washington, D.C.: The World Bank.

Grupp, H. (1998). *Foundations of the economics of innovation—Theory, measurement and practice.* Cheltenham: Edward Elgar.

Hanson, P. & Pavitt, K. (1987). *The comparative economics of research and development and innovation in East and West: A survey.* New York: Harwood Publishers.

Inzelt, A. (1988). *Rendellenességek az ipar szervezetében* (Abnormalities in the structure of industry). Budapest: Közgazdasági és Jogi Könyvkiadó.

Inzelt, A. (1991). Certains problemes de la recherche-developement et de l'innovation. *Revue d'Etudes Comparatives Est-Ouest,* **22** (1), 21–36.

Inzelt, A. (1993). The objective, the method and the process of the 1993 Innovation Survey in Hungary. Paris: OECD Working Paper.

Inzelt, A. (1994a). Restructuring of the Hungarian manufacturing industry. *Technology in Society,* **16** (1), 35–63.

Inzelt, A. (1994b). Preliminary lessons of Hungarian Pilot Innovation Survey. Paris: OECD Working Paper.

Inzelt, A. (2002). Attempts to survey innovation in the Hungarian service sector. *Science and Public Policy,* November. Guilford, U.K.

Jasinski, H. A. (Eds) (2002). *Innovation in transition: The case of Poland.* Warsaw.

Lundvall, B. A. (1992). *National systems of innovation.* London: Pinter Publishers.

Meyer-Krahmer, F. & Schmoch, U. (1998). Science-based technologies: university-industry interactions in four fields. *Research Policy,* **27**, 835–851.

Nelson, R. (1993). *National innovation systems, a comparative analysis.* Oxford: Oxford University Press.

OECD (1992). *Oslo manual, proposed guidelines for collecting and interpreting technological innovation data.* Paris.

OECD and Eurostat (1997). *Oslo manual, proposed guidelines for collecting and interpreting technological innovation data.* Paris.

Patel, P. & Pavitt, K. (1994). National innovation systems: why they are important and how they might be measured and compared. *Economics of Innovation and New Technology,* **3**, 77–95.

Porter, M. (1990). *The competitive advantage of nations.* New York: The Free Press.

Radosevic, S. (2002). Understanding patterns of innovative activities in countries of CEE: A comparison with EU innovation survey. In: A. Inzelt & L. Auriol (Eds), *Innovation in Promising Economies* (pp. 1–27). Budapest: Aula Publisher.

Radosevic, S. & Auriol, L. (1998). Patterns of restructuring in research, development and innovation activities in Central and Eastern European countries: analysis based on S&T indicators. Brighton, SPRU, 1998, 50 pp (SPRU Electronic Working Papers Series; No. 16).

Rosenberg, N. (1994). *Exploring the black box: Technology, economics and history.* Cambridge: Cambridge University Press.

Sandu, S. (2002). Innovativeness of Romanian manufacturing firms. In: A. Inzelt & L. Auriol (Eds), *Innovation in Promising Economies* (pp. 29–42). Budapest: Aula Publisher.

Scherer, F. M. (1989). *Innovation and growth: Schumpeterian perspectives* (pp. 32–58). Cambridge, MA: MIT Press.

Sirilli, G. & Evangelista, R. (1998). *International and high-technology competition.* Cambridge: Harvard University Press.

Smith, K. (1991). Innovation policy in an evolutionary context. In: J. S. Metcalfe & P. P. Saviotti (Eds), *Evolutionary Theories and Technological Change* (pp. 256–275). London: Harwood.

Smith, K. (1992). Technological innovation indicators: experience and prospects. *Science and Public Policy,* **19** (16), 24–34.

Smith, K. (1998). Summary and results from a recent indicators research project: 'IDEA: Indicators and Data for European Analysis'. Room document for OECD Seminar New science and Technology Indicators for the Knowledge-Based Economy. Development Issues. Canberra, Australia.

Soete, L. & Arundel, A. (Eds) (1993). *An integrated approach to European innovation and technology diffusion policy.* Maastricht: European Commission.

Tidd, J., Bessant, J. & Pavitt, K. (1997). *Managing innovation: Integrating technological, market and organisational change.* Chichester, U.K.: John Wiley.

Appendix 1

Comparing the Structure of the Hungarian Questionnaire to CIS-1

OECD/EU HARMONISED (CIS-1)	HUNGARIAN MODIFICATION (Additional/Deleted Questions)
I. General Information	+ ownership structure
II. Sources of Information for Innovation	More subgroups about: • educational/research establishments • professional associations, chambers
III. Objectives of Innovation	Three additional objectives. • increasing or maintaining market share • creating new markets + within former CMEA countries • improving production flexibility
IV. Acquisition/Transfer of Technology	—
V. R&D Activity	—
VI. Factors Hampering Innovation	Two added factors • lack or weakness of innovation management • organizational structure of enterprise
VII. Costs of Innovation	estimated breakdown of total current innovation expenditures by branches
VIII. Impact of Innovation Activities	One more group: innovative products were new to only the enterprise/group

Appendix 2

A Profile of the Hungarian Pilot Survey

The process of developing a Hungarian innovation survey questionnaire was a time-consuming process.[9] The first step was the translation and dissemination of the document *OECD Proposed Guidelines for Collecting and Interpreting Technological Innovation Data* (Oslo Manual, 1992), the theoretical framework containing the definition of the process.[10]

Then we developed a Hungarian questionnaire on the basis of an internationally developed, harmonized one for postal innovation survey work in the OECD (EC) area.

It was decided to run a pre-test of the innovation survey in 1993. We visited some firms and asked members of top management to fill in a draft questionnaire. The pre-testing period was followed by a pilot survey.

The pre-test was done through interview. Every researcher had to fill out the pilot questionnaire and prepare a written report on his or her experiences with the interviews. Thirteen firms filled in the questionnaire and seven others gave valuable, detailed comments on questionnaire design etc.

After the preliminary questionnaires filled in by the firms had been examined and collated, the pilot postal survey questionnaire was developed. That apart, it was necessary to find a suitable register for choosing firms.

---

[9] Appendix 2 is based on Inzelt (1993, 1994).
[10] Translation and publication was supported by the Hungarian Science Policy Committee.

Pilot innovation postal survey questionnaires were sent out in January 1994. Enormous methodological experience was built up through the completed questionnaires, the 24-hour hotline, and follow-up phone calls.[11]

I. The Lifelines of the Hungarian Pilot Survey

Methodology

Kind of survey:	Pilot survey
Survey unit:	Enterprise (mainly innovative firms involved in R&D activities)
Classification:	ISIC Rev. 3 (with some variations below the two-digit level)
Obligatory/voluntary survey:	Voluntary
Size of survey (number of responses):	110
Cut-off-point:	Employees above 100 and/or net sales above HUF 300 million and/or total sum of balance sheet above HUF 150 million
Questionnaire:	Modified OECD/EC harmonized
Combination with other survey:	No
Population and coverage:	All R&D performing enterprises from the 4000-strong sample of the Hungarian Central Statistical Office (478 in number) were selected. These 478 enterprises operate in various industries Their ownership structure is also diverse (private, state-owned, domestic and foreign joint ventures)
Reference period:	1990 to 1993
Survey method and implementation:	postal survey/phone calls for those missing the deadline
Response rate:	23%

Timing

Started mailing the questionnaire:	01.02.1994
Finished collecting/ processing data:	May 1994
Results available:	November 1994

II. Sampling Method and Response Rate

At the time of sampling, there was not an up-to-date listing of Hungarian companies available. Several ideas were put forward as to where to choose the population for the pilot survey from.

(1) The original idea was to choose them from the list of companies that filed a research and development project with OMFB between 1990 and 1993, i.e. over the period since the new project evaluation system was created following the systemic change (Inzelt, 1993). Unfortunately this list was not available at the time of starting the pre-test.

(2) In the autumn of 1993 the list of those firms that had filled in compulsory R&D statistical survey forms became available. It was not our specific aim to test the composition of the list, but while pre-testing the questionnaire it became clear that many of the companies on the list had disappeared or redeployed fundamentally (e.g. they had gone into bankruptcy, split up, privatized and regrouped, etc.) Only 50% of the list seemed correct at the end of 1993. (A very common problem with registers under conditions of transition is that they quickly go out of date.) It would clearly have been unreasonable to use such a list for a postal survey. We had to find something else.

(3) At the end of the pre-test period the list of respondents in the new Hungarian Board of Statistics business survey (3,600 responses from the sample of 4,000) was ready.

This new business survey contained some questions about R&D activities. On this basis we were able to pick out from the list all firms involved in any type of R&D.[12] Their number was 478. This was our target group.[13] Just 110 firms sent back questionnaires amenable to statistical analysis. (Another 30 firms gave valuable information in letters or by phone.)[14] The response rate of 23% is not very high. But, considering that this was a non-mandatory survey in a transition economy in which trade is not flourishing, and where the key question for many business units is just how to survive from day to day.

---

[11] Reminder letters were largely a waste of money. If firms did not answer the first letter we tried to push them by phone calls in order to achieve a higher response rate. This is a more time-consuming and costly process, but it was the only workable method.

[12] Sales from R&D activities, non-intangible assets, gross fixed capital from R&D, direct cost of own production R&D, indirect cost of own production R&D, cost of bought-in R&D activities.

[13] It would have been useful to investigate not only these, but also a similar number of firms from among those that did not report any R&D activities. Unfortunately, because of financial limitations, we were not able to send out more than 500 questionnaires.

[14] R&D firms as a rule did not give information, but wrote subsequently to ask for the results of the survey.

The International Handbook on Innovation
Edited by Larisa V. Shavinina

Innovation under Constraints: The Case of Singapore

Hung-Kei Tang and Khim-Teck Yeo

Nanyang Technological University, Republic of Singapore

Abstract: Singapore is a small island state in Southeast Asia. Historically it was a British colony and trading was its primary economic activity. In the last four decades Singapore underwent rapid industrialization, progressing from being a center for labor-intensive operations to a pioneer in the application of information technology. However, indigenous attempts to innovate so far have had mixed results. This chapter attempts to explore, through four case studies, how constraints of different types could give rise to innovation as well as cause innovation efforts to fail.

Keywords: Innovation; Constraints; Barriers; Path-dependency; Competence; Types of innovation; Singapore.

Introduction

Innovation has been extolled as the ultimate means to achieve competitiveness for companies and nations alike. Companies are warned that either they innovate or evaporate. It is understandable, therefore, with constant admonishments from the eager business gurus and concerned government officials, many executives would feel left out and be afraid to be seen as unenthusiastic if they do not actively initiate, support and fund what appear to be attractive programs and projects for innovation. Hence innovation has become an increasingly popular buzzword.

Innovation can be defined as the process of applying novel but not necessarily brand new ideas for purposeful gains such as productivity, profit or the elimination of problems. It can also be defined as the product of such a process. While innovation can be simply defined, the process of innovation especially in an organizational context tends to be complex (Tang, 1998). And it is even more difficult to quantify and measure innovation (Buderi, 1999). Many companies say they expect innovations from their employees but ironically not every organization is ready or able to exploit the potential of their employees to innovate (Amabile, 1998). Unlike quality control or supply-chain management, innovation is not a management function that can be standardized or operationalized. Nevertheless, a cornucopia of concepts and methods such as those found in this handbook can help to give

some structures and directions to what inherently is a probabilistic and often chaotic process at its core. When an innovation such as the 3M Post-It finally materializes, it is not uncommon that people would say, with the benefit of hindsight, that the idea is simple. The often ignored fact is that, behind every innovation, there is a long history of preparation and struggles. Innovation is seldom easy or instantaneous.

Like entrepreneurship, innovation is often interpreted as pursuing an idea or opportunity, despite or because of, constraints faced. Often opportunities are described optimistically as limitless and project goals are set ambitiously. However, in this chapter it is argued that an organization's optimism and ambition should be tempered by the recognition that there are constraints to its ability to innovate. These constraints, also called barriers, could be internal such as staff and the work environment (Amabile et al., 1996; Roberts & Fusfeld, 1981), or external such as societal and economic factors (Nelson, 1993; Patel & Pavitt, 1994; Shane, 1993). In this handbook, Hadjimanolis gives a comprehensive review and analysis on the barriers against innovation, whereas Major & Cordey-Hayes describe the special barriers faced by small firms. It should be emphasized up front that constraints must not be interpreted completely as final or negative. Many constraints in fact can be interpreted positively because they give impetus to innovations that result in the removal or mitigation of the constraints themselves. For example, the vast distance between communities in Canada and Finland make them pioneers in satellite and wireless telecommunications respectively. The lack of space is the constraint that propels Japan into miniaturizing many consumer products. The microscooter was created by Gino Tsai who described himself having short legs but wanted to move around fast in his vast bicycle factory in Taiwan. However, it would be wrong to assume that there are viable or fast solutions to all constraints. The location of a country cannot be moved. A person could be very creative but without working knowledge in a field he could not be innovative in that field. Hence, it is important first to recognize constraints that can impede or jeopardize efforts in innovation, then remove them, make them irrelevant or turn them into opportunities if possible. Not recognizing the constraints and going against them unknowingly account for many companies' failures in their efforts to produce innovations. Internal and external constraints can also limit the types of innovation a company is capable of producing. Hence knowing how the different types of innovation demand different organizational capabilities is another step toward producing innovations successfully.

There are different ways of classifying innovations and they are used for different purposes (Chesbrough & Teece, 1996; Henderson & Clark, 1990; Hobday, 1998; Tushman & O'Reilly, 1997; see also Sternberg et al., this handbook). For the purpose of this chapter it is useful to classify innovations according to two criteria. The first classifies innovations according to whether it is incremental or radical, the second whether it is autonomous or systemic. Incremental innovations refer to the refinement of existing products, processes and services or adapting them to different applications. Radical innovations are those that offer brand new application, user experience and can potentially set a new standard or even give rise to a new industry. Innovators aiming for radical innovation face great uncertainty in many areas e.g. unforeseen technical difficulties, costing, performance reliability, market acceptance, sustaining initial success. Incremental innovations have far less barriers to overcome. They usually depend on proven technologies or the market and customer requirements are quite well defined.

The autonomous-systemic classification of innovation is determined by the degree of need for complementary assets to develop the technology or the market. Take the case of the introduction of the audio compact disk (CD) in the early 1980s (Nayak & Ketteringham, 1993). The audio CD player would be useless to the consumers if there were no CDs to buy. The CD-player manufacturers and the record companies that provided the contents and recorded the CDs were thus providing complementary assets. Furthermore, to overcome initial resistance to this revolutionary product they recruited leading artists to record on the new medium. In terms of technologies, Philips was the pioneer in using laser in the playback of contents recorded on plastic disks. However, they were stronger in analog electronics than digital electronics. Hence Sony, which was stronger in digital electronics and solid-state laser, was engaged as a joint developer to give the technology the final and decisive push. This example illustrates the systemic nature of many technological innovations that are the results of collaborators who contribute complementary capabilities. Two other examples are the GSM cellular phone and the personal computer. The former was an innovation developed by many European partners. The latter was developed by IBM with major components from Microsoft, Intel and others. In comparison, an autonomous innovation can be realized by a firm largely with its internal resources. The firm may still need to buy services and parts from other companies but these are widely available and are not core to the innovation.

The Case of Singapore

Singapore is one of the newly industrialized economies and is also notable for its small size, a dot on the map in Southeast Asia. It became a British colony in 1819 and an independent nation in 1965. Situated in the middle of Southeast Asia, it has an area of little more than 640 square kilometers and a small population. The British acquired Singapore as a colony primarily for its potential as a port and a trading post. To date Singapore remains an important trading and transportation hub. However, manufacturing, banking and finance now contribute most to the nation's economy. Singapore is in the ranks of the developed nations in terms GNP per capita and considered very competitive economically according to the rankings of global competitiveness published by organizations such as the World Economic Forum. PSA (Port of Singapore Authority) Corp is the largest container port operator and Singapore Airline is one of the most profitable airlines in the world. For many years Singapore produced more than 40% of the world's hard disk drives and still maintains its position as an important hard disk drive manufacturing hub.

The strategic location of Singapore straddling the Indian and Pacific Oceans and its deep-water harbor pre-destined it to be a great port for the British Empire in the nineteenth and first half of the twentieth century. Its central location in Southeast Asia made it a logical choice as a hub for trans-shipment of goods and intra-regional trade. Hence, Singapore for a long time (except during the Second World War when it was occupied by Japan) was primarily a society of traders and clerks who performed fairly well defined and slowly changing tasks that trading and shipping entailed. There was little incentive for innovation or to be a pioneer. Moreover, the great distances that separate Singapore from the great centers of business and technological development in the developed world was an impediment for innovation activities in Singapore.

The societal upheaval around the time of independence in 1965 and the subsequent withdrawal of the British military bases produced a sharp increase in unemployment. Concurrently the economic boom in Western economies marked the beginning of the trend of outsourcing labor-intensive manufacturing operations for consumer goods such as electronics and apparels. Faced with the situation, the government adopted the policy of actively engaging the foreign multinational corporations (MNCs) to set up manufacturing facilities in Singapore. Unlike many developing countries that became newly independent at the time, Singapore maintained a free trade policy and generously provided tax incentives for foreign firms that pioneered manufacturing in Singapore, which continues to date.

The early economic policy of Singapore changed the bleak employment situation to almost full employment within just a few short years. However, the majority of the workforce was still lowly educated. The scope for work widened to include manufacturing but the nature of labor-intensive tasks in the factories, which were almost totally transplanted from other countries, did not provide much scope for innovation. The situation gradually changed. Learning from the success of Singapore, other countries in the region also implemented similar MNC-friendly economic policies. That meant the MNCs had more choices for the relocation of their factories. The spread and rise of manufacturing brought economic growth to the region as a whole and thus demand for consumer goods in the regional economies as well. The new development brought both competition and opportunity.

New policies to upgrade the economy were implemented at the end of the 1970s. There was the unique high-wage policy designed to force up wages and drive out low value-added operations. Tertiary education in engineering was quickly expanded to supply the manpower to satisfy anticipated demand for technical staff. There were also the policies to give incentives for companies that computerize, automate or widen the scope of work to include more value-adding functions such as product design. As a result of steady economic development in Southeast Asia, consumer demands for goods increased. At the same time, China was also reforming its economy and opted for a more open-door policy. Hence the regional market for goods became more attractive to the MNCs from the West and Japan manufacturing in the region. Opportunities thus arose for adapting their products for local consumption. In order to encourage foreign MNCs to set up research and development activities in Singapore, the government established research institutes in IT, microelectronics and life sciences. In 1991 the National Science and Technology Board (now Agency for

Science, Technology and Research) was set up and charged with establishing Singapore not only as a manufacturing hub but also as a center of research and development.

The Cases

PSA Corporation

For many years Singapore has been the second busiest container port in the world after Hong Kong. However, PSA has been the single biggest container operator in the world. It handled 17.04 million TEUs in 2000 (twenty-foot equivalent units of containers) with 6,200 staff. In 1972 the then Port of Singapore Authority became one of the early adopters of container technology, in anticipation of the rising trend of containerization in sea cargo transportation. In the same year PSA's electronic data processing department joined the operations departments to run round the clock, seven days a week. These two events heralded the beginning of the intimate joint development of operational and informational technological capabilities at PSA. Contrary to the commonly perceived image of a government bureaucracy being inefficient, PSA has been a productivity champion and consistently innovative.

Its subsequent corporatization in 1997 was not so much an attempt to boost efficiency, but as an effort to facilitate business expansion, particularly overseas. PSA had to be innovative for two very important reasons—to stay competitive and to maintain a lean workforce. PSA has to stay competitive, despite Singapore's strategic location, because there are up and coming ports nearby that have equal claim to be strategically located. Port Klang, Port Tanjung Pelepas in neighboring Malaysia are eager competitors. PSA has to stay lean with its workforce because for decades Singapore had very tight labor supply, which was a major constraint. In fact while the number of TEUs handled increased from 4.36 to 17 million from 1989 to 1998, the staff strength actually decreased from 7,500 to 6,200. The increase of labor productivity from 583 to 2,746 TEUs per staff was due to technological innovations pioneered by PSA. This upward trend could be traced to a series of research and development projects undertaken by PSA.

In 1987 PSA started collaborating in the application development of expert system (a branch of computer-based Artificial Intelligence) with the recently established Information Technology Institute. The objective was to capture the expertise of a group of operation planners who had a wealth of tacit knowledge among them, which helped them to map out the optimal sequence and positioning of the containers to be loaded or unloaded in as short a time as possible. For a few months the software developers from the Institute and PSA, and the planners were co-located in order to lower the cultural barrier between the two groups. The software development project was a success. The result was the first version of CITOS (Computer Integrated Terminal Operations System). Subsequent development by PSA's IT department improved and expanded the functions of CITOS to include berth allocation, stowage planning, yard planning and resource allocation (Tung & Turban, 1996). CITOS was later augmented by electronics and communication systems that included entry/exit gate control on the movement of container trucks, container identification by computer vision, automatic container locating system and communication system with drivers. These were also developed internally with assistance from external experts. In parallel to operations and planning, PSA's IT development teams developed the computer network systems called Portnet and Boxnet that link PSA's computer with that of shipping agents and other business associates to automate various business transactions and information exchange.

LTA (Land Transport Authority, Singapore)

Singapore is one of the most densely populated countries in the world. The lack of land space is a major constraint to Singapore. Controlling the growth of traffic and regulating the traffic patterns in order to reduce road congestion have been high-priority objectives for Singapore's road transport planners. One example of an innovative policy is the Certificate of Entitlement (COE) scheme that requires new car buyers to bid for a limited supply of COEs that entitle them to buy. Another example is the road-pricing scheme, which is a combination of both policy and technological innovation. From 1975 a rudimentary road-pricing scheme had been in place that required motorists entering into the central business district during peak hours to buy a pass at roadside kiosks or designated petrol stations, and to display it prominently on the windscreen. While this scheme was effective in restricting traffic it caused great inconveniences and relied on visual inspection from a distance by traffic police who stood on the roadside to spot offenders driving by at normal speed. So in 1989 the LTA began to explore the feasibility of an electronic road pricing (ERP) system to be implemented not only for the central business district but also on busy sections of expressways. In order not to slow down traffic flow, the introduction of separation barriers between lanes was ruled out.

In 1993, after four years of study the LTA put up a conceptual system design and preliminary specifications and invited proposals from potential suppliers. LTA also engaged experts on electronics and communication from the Nanyang Technological University to be its consultants in the evaluation of the technical merits of the proposals.

Three proposals were received from the vendors but none met the required specifications. The invitation to propose was repeated in 1995. Again none met the specifications fully, attesting to the gap between state of the art and the stringent requirements. However, this time the consortium led by Mitsubishi-Philips, whose proposed system most closely met the specifications, was commissioned to build a demonstration system. The results were encouraging enough that the supplier was asked to made improvements and be ready for qualification test under realistic traffic conditions in May 1996. The test revealed the prototype road pricing system's susceptibility to electromagnetic interference generated particularly by motorcycles. Further modifications were made and the susceptibility problem was solved. Electronic road pricing is now widely implemented in Singapore. The final ERP system consists of: (a) an in-vehicle, cash-card enabled, electronic transponder that is installed on every road vehicle; and (b) an automatic vehicle identification and debiting sub-system and an enforcement sub-system installed above and beside the road. The system makes clever use of electronics, communications and image processing technology. During designated operational hours, toll is deducted electronically from the cash-card in the in-vehicle unit when a vehicle passes under an ERP gantry. Although Singapore was not the first one to come up with the idea of road pricing, it is the first one that has implemented the most advanced form of electronic road pricing to date in a large scale (Do, 2000).

Creative Technology

Creative Technology is a leading Singapore electronics company that has carved itself a niche in the digital audio player and PC sound card market with a turnover of US$1.2 billion in fiscal year 2001. Its brand *SoundBlaster* is well known amongst consumers worldwide. However, its takeoff in 1989 was only possible after it expanded its marketing and sales function from Singapore to the USA in the heart of the Silicon Valley. The move was designed to overcome the constraints of having only a small domestic market and the lack of reach to the major consumer markets outside of Singapore.

For many years Creative Technology was a struggling small company like many others trying to develop into a viable enterprise. However, unlike others, it had the ambition of creating its own product with its own brand name from the very beginning. Besides making add-on cards for PCs, it also designed and made its own PC with unique features such as good quality sound. So in terms of the scope of operations, it essentially covered the full range, from product development to manufacturing and marketing, just like the bigger and vertically integrated firms. However after investing very heavily on a new PC, which failed to win customers despite having good sound and Chinese character handling capability, the company was in a quandary. The founder then took the audacious step of expanding to the USA and set up a one-man operation initially.

Soon after the move, the founder discovered the nascent market for PC sound card. Opportunely, the company had the people with the design and manufacturing experience of a similar but simpler product earlier. So technically, the company was up to the challenge. However, it was a major hurdle to secure the supply of a hard-to-get sound synthesizing IC chip in large quantity. It was only available from the Japanese company Yamaha, which was quite reluctant to sell to the then unknown Creative Technology. This hurdle was cleared after Singapore's Trade Development Board Office in Japan was able to convince the manufacturer that Creative Technology was a bona fide Singaporean company. Another major hurdle was to convince established computer-game software companies to develop games that supported its sound card. After pursuing them tenaciously Creative Technology finally got their support and so customers could buy computer game software that ran on Creative's sound card. In 1989 Creative's *SoundBlaster* sound card became a runaway success in the USA. Subsequently, it nurtured its research and development capability in the USA through a series of acquisitions and thus laid the technological foundation for sustained product innovation (Tang, 1996). Subsequently it also developed its own sound synthesizing IC and thus cut off its reliance on others for this key component.

Goldtron

Goldtron is a medium-size indigenous electronics company in Singapore with a turnover of about US$100 million in fiscal year 2001. It was traditionally a distributor of electronics components and a contract manufacturer. In the early 1990s the company dabbled in developing its own products. It established the Proteq division to develop home automation electronics. In mid-1990s the market for personal wireless communication began its high growth phase. Sales of pagers increased rapidly because the newer models had more compact design and increased functionality. GSM cellular phones sales began to take off and became the standard in Europe and most of Asia. GSM is the so-called second generation cellular phone system introduced in the early 1990s. GSM phones use digital rather than the analog technology found in the first generation of cellular phones. Seeing the opportunity, Goldtron embarked on an ambitious diversification plan. It aimed to compete internationally with pagers and cellular phones to be designed and manufactured by itself. A new subsidiary, Goldtron Telecommunications, was established and a group of engineers with experience in pager design and radio frequency technology were hired. Consequently the pager was developed successfully. But compared to the cellular phone, the pager was simple and mature technologically.

The building-block technologies in a GSM cellular phone are the GSM protocol, radio frequency, digital signal processing (DSP), audio frequency, microcontroller, battery and power supply. Of these, DSP and the GSM protocol are considered the core technologies. When Goldtron undertook the task of developing its own GSM phones, these core technologies were new to it. Great technical difficulties were encountered and at the end the product cost was too high for Goldtron to compete effectively with the leading brands. Goldtron tried to compensate by introducing what could be considered an innovative product: a cellular phone with a built-in pager. Its phone function would be normally switched off, leaving only the pager turned on to receive incoming message. This way the battery power would last longer. However, battery technology improved rapidly, making the benefit irrelevant almost as soon as the product was launched into the market. Consequently Goldtron's cellular phones found few buyers. So after spending tens of million of dollars the company stopped its product development efforts and abandoned the telecommunication business in 1998. Goldtron became a much weaker company after its failed attempt. After Goldtron's exit, the cellular phone market continued to grow rapidly. In order to capture more market share, some semiconductor manufacturers began to sell DSP chips customized for GSM application thus lowering the technological barrier significantly for newer and smaller players. However, it was too late for Goldtron.

Discussions

Of the four cases covered, PSA and LTA illustrate innovations that resulted from systematic and purposeful exploitation of state-of-the-art technologies successes. The challenges were clear and well defined, and the goals are unambiguous and credible. On the other hand, the remaining two cases are examples of innovation projects undertaken with uncertainty in terms of the market or internal technological capability. The PSA and LTA cases illustrate two fundamental *resource constraints* faced in Singapore, shortage of land and manpower. These constraints are inherent to Singapore's size, which are hard to overcome. However these constraints are exactly the same reasons that drove PSA and LTA to innovate. The nature of the innovations, CITOS for PSA, ERP for LTA, is systemic and infrastructural. The technological innovation of key components contributed to system innovation and breakthroughs. In both cases IT, networking, artificial intelligence and electronics were the key technologies that must be integrated to work together as a system, which leads to dramatically improved performance, productivity and reliability. PSA, because of its internal technological capability plays the role of an innovative system integrator and benefited in the process as it accumulates knowledge in developing and operating such complex systems in the container port environment. PSA's technological capability also puts it in an advantageous position in expanding its container port operating business to other ports in Asia and Europe. In contrast, LTA lacks internal techno-logical development capability. Despite this *competence constraint*, it succeeded through out-sourcing the technology development and system integration. LTA is only a policy and not a techno-logical innovator despite its success with implementing the ERP.

Managing stakeholders' interest is both a constraint and opportunity for PSA. The stakeholders for PSA are the shipping agents, shipping lines and their customers. When there are many stakeholders involved, there are bound to be potential conflict of interest. Conflicts between stakeholders constitute a constraint to innova-tors of complex systems in that not all stakeholders necessarily perceive an innovation as beneficial to their interest and hence some stakeholders resisted the innovation. Hence in order for an innovation to be accepted, this *stakeholder constraint* needs to be dealt with skillfully. In this regard, the stronger is the innovator's position the better is the chance of the innovation being adopted. Internally PSA's innovations in automation were readily accepted by the workers at home because PSA's business was growing, whereas the supply of port workers was shrinking. Hence job security was not an issue. Externally PSA's reputation as being one the most technologically advanced as well as one of the biggest and most profitable, gives it a lot of credence when it exports its innovations overseas.

In the LTA case, the stakeholders are, on the one hand, the government agencies that want to minimize traffic congestion and maximize land usage for the transport system, and on the other hand the motorists. Experience in many countries shows that road pricing is not easy to implement, not because of the lack of technology but because of the resistance from the motorists. Singapore is unique because it is a small city-state with a strong government and a citizenry that is used to tough policies for the sake of efficiency. Rolling out road pricing hence is not much a political problem. This is not the case in other countries where social and political desirability weigh heavier than technological or economic feasibility (Button & Ver-hoef, 1998; Evans & Oswald, 2000). Hence to date, Singapore is still the only country with the most advanced electronic road pricing system in the world. Because of the *social-political constraint* this innova-tion adopted by Singapore has not yet diffused to other countries. In this sense LTA's innovation has less impact than PSA's.

The above two cases deal with infrastructural innovations in Singapore's public sector and both achieved considerable success. In the private sector, it is more difficult to achieve success due to the lack of experience in innovation both in terms of scale and depth. The Creative Technology and Goldtron cases are examples of small and medium sized Singaporean companies' attempts in product innovations. The former was a case of success that took the company to another plateau and the latter a debilitating failure that set back the company. The products in question, a sound card for the PC and a cellular phone/pager, are not as big and complex as the PSA and LTA's systems. They were nevertheless very challenging given the limited resources that Creative Technology and Gold-tron could command. In fact, the chance of success for product innovation is arguably lower than infra-structural innovations like those of PSA and LTA because there are a diverse range of factors that determine the eventual success or failure of a new product (see Cooper, this handbook).

Spotting the market opportunity for the sound cards occurred only after Creative Technology expanded into the USA. With this strategic move Creative overcame the *market constraint* it was facing in Singapore, a small domestic market which tended to follow trend-setting big markets. Creative also had to overcome the *credibility constraint*. In order to achieve impact upon launch market, i.e. having sufficient supplies of hardware and software, Creative must enlist the support of the sound synthesizer chip manufacturer and the computer game software developers. One of the ironies of innovation is that before an innovation becomes a proven success, stakeholders are sometimes reluctant to accept or support it, even though the innovation will benefit them if it succeeds. This is the dilemma Creative faced before the launch of the *Soundblaster*. Because Creative was a small and unknown company the founder had to put in extra efforts to win the support of the bigger stakeholders: the chip manufacturer and the software developers. In comparison, Goldtron was not a small start-up com-pany when it decided to develop the cellular phone on its own. Nor was it totally unfamiliar with developing new products. However, it did face the competence constraint. The gap between its technological compe-tence and the task requirements was not bridged. This failure underscores the need for innovators to recognize the core competence required for the intended innova-tion and be prepared to cultivate it over the long haul.

Creative Technology is so far the only Singaporean company that has grown successfully from a small start-up to a billion-dollar sales technology-based company in less than two decades. It is not because Singaporean companies have not been trying. The foreign MNCs driven industrialization in Singapore provided many opportunities for local start-up com-panies that provide contract manufacturing services, machined parts, jigs and tools. Many of these start-ups have grown into small and medium sized companies. Some have ventured into designing and manufacturing products under their own brands. However, when these companies, such as Goldtron, attempted to diversify by taking on product development or expand overseas, many failed to replicate the initial success they had enjoyed. Creative Technology's success has been the exception rather than the rule.

The need for Singapore to start up and grow indigenous companies that can compete internationally has been well recognized. But despite the government's efforts to encourage start-ups, help indigenous companies to upgrade and promote spin-off companies from the public research institutes in the last decade, there has been no success comparable to Creative Technology. This lack of success is attributed to the shortage of entrepreneurship, innovations and creative people in Singapore. And the shortage is attributed to the education system and ironically Singapore's own success in creating an economy that had been providing practically full employment in the past. Moreover, there is evidence that the work environment in many local companies and government organizations are not particularly conducive to innovation (Tang, 1999; Thomson, 1980).

One key dimension of the work environment is the type and level of tasks that people are required to perform on their jobs. The types of task people perform determine the scope and the opportunities for innovation. It is through performing challenging tasks that new knowledge and skills are acquired and developed. Over time, this virtuous process builds up an organization's innovative competence (Leonard-Barton, 1995; Nonaka, 1998). However, the gap between the firm's present level of competence and what is required in the task may be too wide to bridge relative to the time, people and other resources available. In such a case the firm would be wise not to undertake the new and unfamiliar task by itself. Instead it should consider various forms of acquisitions of collaboration with firms that have complementary competence (Roberts & Berry, 1985).

A new task can challenge an organization in three ways. First, it can be challenging by requiring the acquisition of a new field of knowledge or a new level of mastery of an existing field of knowledge. Second, it can be challenging because of the need to integrate the work or satisfy the needs of multiple parties, in other words the challenge of dealing with complexity. Third, it can be challenging because the problem to be solved and the solutions to be found are ambiguous and uncertain (Schrader et al., 1993; Yeo, 1995), in other words the challenge of dealing with fuzziness. Thus it can be said that a challenging task has three dimensions: knowledge, complexity and fuzziness.

In the introduction, innovations are described as incremental or radical, autonomous or systemic. The challenges of radical innovation in terms of defining the scope, framing the problem, finding the solutions are full of ambiguity and uncertainty. Systemic innovation calls for the performance of tasks that involve complexity in terms of attending to multiple stakeholders' interest and the integration of different parts

that constitute the innovation. Furthermore, every innovation is characterized by the fields of knowledge it embodies. Without a minimum level of competence in the requisite fields of knowledge, an organization will be greatly impaired in producing the intended innovation. Hence, the ability of an organization to undertake a certain type of innovation is constrained by its ability to perform tasks that are associated with the type of innovation. In order to succeed, an organization therefore needs to align its competence in task performance with the type of innovation intended. This observation is supported by the four cases presented.

PSA succeeded because it has intimate knowledge of container port operations and the internal technological competence to absorb new knowledge. It has also a long history of dealing with tasks of complexity: integrating diverse systems into its daily operations and accommodating the interests of many stakeholders. LTA succeeded in implementing an innovation that was first of its kind, despite its lack of in-depth knowledge of technical systems. It achieved this by outsourcing the tasks of system integration and reduction of technical uncertainty through experimentation. Creative Technology already had sufficient technical knowledge to develop the sound card which was only an incremental and architectural innovation. Nevertheless the systemic nature of the innovation that required the support of an IC chip supplier and software developers was a big challenge to the small company. Goldtron failed primarily because it did not foresee or acquire the new knowledge needed.

Recognizing innovation is a necessity for the next stage of economic development, the government has recently intensified its strategic thrust to promote innovation and entrepreneurship in Singapore (SEDB, 2002). In 1996 the Singapore Economic Development initiated the Innovation Programme with the objective of developing a wide base of indigenous creative capabilities in Singapore. This is to be achieved through strategic initiatives that promote investments in innovation projects and promising entrepreneurial startups. The government itself invests substantial amount of financial commitment to schemes such as the Innovation Development Scheme, Patent Application Fund and Startup Enterprise Development Scheme. However, investing government funds into worthwhile innovation projects and startup enterprises is the only an extension to the government's past practices of channeling resources, particularly money, to what it sees fit. However, trying to nudge the culture at work and study to favor innovation is a fundamentally new endeavor. There are four notable initiatives driving for cultural change:

- the first is for high-level recognition of innovators through the giving out of Singapore Innovation Awards starting 2001;

- the second is for bold attempt to nudge the educational system toward emphasizing creative and critical thinking skills. The opening address (Goh, 1997) delivered by the Prime Minister at the Seventh International Conference on Thinking was a symbolic gesture to signify the intention of the government to overhaul its education system so that students will more likely think creatively and critically in the future;
- the third is to supplement and add diversity to Singapore's limited human capital. Scholarships are given to foreign students especially from the neighboring region to enroll in Singapore's schools and universities. The Technopreneur Pass Scheme is aimed at foreign entrepreneurs who want to start high-tech ventures in Singapore;
- the fourth is a joint government private-sector review of government rules and regulations with the aim of removing obstacles to entrepreneurship in Singapore.

The above initiatives should help to ameliorate the resource and competence constraints on innovation in Singapore. To a certain extent the market constraint is also addressed through the import of foreign talents whose familiarity of the countries of origin should broaden Singapore's market reach. However, these initiatives are still new and the eventual outcome will depend on their implementation.

Conclusion

The case of Singapore and four Singaporean companies have shown that the challenge of innovation includes not only identifying opportunities and committing to innovation but also identifying and mitigating the internal and external constraints that impede the process of innovation. Singapore as a young nation has achieved outstanding infrastructural innovations that are prime examples of system integration. Economically, Singapore succeeded in its rapid industrialization within the last few decades, despite its early competence constraint, by relying heavily on the foreign multinational companies to bring in the management, technology and market expertise. In contrast, many local companies are still constrained by the lack of the same expertise. It continues to be a major challenge for local companies to become innovators and compete globally. The government has recognized the need for a better environment for innovation and entrepreneurship in Singapore and has embarked on several strategic thrusts aimed at making it better.

References

Amabile, T. M. (1998). How to kill creativity. *Harvard Business Review*, **76** (5), 76–87.

Amabile, T. M., Conti, R., Coon, H., Lazenby, J. & Herron, M. (1996). Assessing the work environment for creativity. *Academy of Management Journal*, **39**, 1154–1184.

Button, K. J. & Verhoef, E. T. (Ed.) (1998). *Road pricing, traffic congestion and the environment: Issues of efficiency and social feasibility*. Cheltenham, U.K.: Edward Elgar.

Chesbrough, H. W. & Teece, D. J. (1996). When is virtual virtuous? Organising for innovation. *Harvard Business Review*, **74** (1), 65–73.

Do, M. A. (2000). An electronic road pricing system designed for the busy multi-lane road environment. *ITS Journal*, **5**, 327–341.

Evans, S. & Oswald, A. (2000). The price of freedom: A non-technical explanation of the case for road pricing. *Transport Review*, **Winter**, 28–29.

Goh, C. T. (1997). Shaping our future: Thinking schools, learning nation. http://www1.moe.edu.sg/speeches/1997/020697.htm.

Henderson, R. M. & Clark, K. B. (1990). Architectural innovation: The reconfiguration of existing product technologies and the failure of established firms. *Administrative Science Quarterly*, **35**, 9–30.

Hobday, M. (1998). Product complexity, innovation and industrial organisation. *Research Policy*, **26** (6), 689–710.

Leonard-Barton, D. (1995). *Wellsprings of knowledge: Building and sustaining the sources of innovation*. Boston, MA: Harvard Business School Press.

Nelson, R. R. (Ed.) (1993). *National innovation systems: A comparative analysis*. New York: Oxford University Press.

Nonaka, I. (1998). The concept of 'Ba': Building a foundation of knowledge creation. *California Management Review*, **40** (3), 40–54.

Patel, P. & Pavitt, K. (1994). National innovation systems: Why they are important, and how they might be measured and compared. *Economics of Innovation and New Technology*, **3**, 77–95.

Roberts, E. B. & Berry, C. A. (1985). Entering new businesses: Selecting strategies for success. *Sloan Management Review*, **26** (3), 3–17.

Roberts, E. B. & Fusfeld, A. R. (1981). Staffing the innovative technology-based organization. *Sloan Management Review*, **22** (3), 19–34.

Schrader, S., Riggs, W. M. & Smith, R. P. (1993). Choice over uncertainty and ambiguity in technical problem solving. *Journal of Engineering and Technology Management*, **10**, 73–99.

SEDB (2002). *Annual report 2002*. Singapore Economic Development Board, http://www.sedb.com

Shane, S. (1993). Cultural influences on national rates of innovation. *Journal of Business Venturing*, **8**, 59–73.

Tang, H. K. (1996). The ascent of creative technology: A case study of technology entrepreneurship. In: B. S. Neo (Ed.), *Exploiting Information Technology for Business Competitiveness* (pp. 295–307). Singapore: Addison-Wesley.

Tang, H. K. (1998). An integrative model of innovation in organizations. *Technovation*, **18** (5), 297–310.

Tang, H. K. (1999). An inventory of organizational innovativeness. *Technovation*, **19** (1), 41–51.

Thomson, D. (1980). Adaptors and innovators: A replication study on managers in Singapore and Malaysia. *Psychological Reports*, **47**, 383–387.

Tung, L. L. & Turban, E. (1996). Port of Singapore Authority: Using AI technology for strategic positioning. In: B. S. Neo

(Ed.), *Exploiting Information Technology for Business Competitiveness* (pp. 141–158). Singapore: Addison-Wesley.

Yeo, K. T. (1995). Strategy for risk management through problem framing in technology acquisition. *International Journal of Project Management*, **13**, 219–224.

Continuous Innovation in Japan: The Power of Tacit Knowledge*

Ikujiro Nonaka, Keigo Sasaki and Mohi Ahmed

Graduate School of International Corporate Strategy, Hitotsubashi University, Japan

Abstract: Management practices of innovative Japanese companies, the 'knowledge creating companies', has been widely researched and extensively documented. We have begun to look 'behind the scenes', and examine more closely the question of 'what is the basic pattern of innovation in the knowledge-creating companies?' In this chapter, we will introduce the basic pattern of innovation at Nippon Roche. We argue that learning only or just breaking rules are not enough for continuous innovation. Rather individuals as well as organizations need to possess tacit knowledge.

Keywords: Innovation; Continuous innovation; Tacit knowledge; Knowledge creation.

Introduction

While most companies around the world are still trying to manage the explicit dimension of knowledge using various tools and techniques, Nippon Roche has succeeded in creating knowledge through capturing high-quality tacit knowledge (HQTK), synthesizing tacit and explicit knowledge, and incorporating synthesized knowledge into organizational activities. Nippon Roche is a part of Roche Group, a multinational healthcare company based in Switzerland. The group focuses on discovering, manufacturing, and marketing products and services aimed at addressing the prevention, diagnosis, and treatment of diseases. Nippon Roche, established in 1924 and now employing more than 1,600 workers in Japan, went through a difficult time in the late 1990s, primarily due to low market growth, fierce competition, and institutional changes in the healthcare industry.

To deal with the situation, the company developed a concept in early 1998 called 'consulting promotion'. This replaced the 'push sales' concept previously dominating the healthcare industry. Through this new concept, Nippon Roche made efforts to realize the needs of its customers and offer timely solutions. To learn about customer needs and create better solutions, Nippon Roche initiated the 'Super Skill Transfer' (SST) project in 1998, which focused on people, the creators of knowledge. In naming the project, Nippon Roche considered 'supersonic transport' an appropriate metaphor, as the company needed to increase sales productivity at supersonic speed.

The medical representatives (MRs)—the key sales people at Nippon Roche—often contribute to product innovation, as they work in the frontline and improvise with key customers (medical doctors) in the sales process. To succeed in such a competitive marketplace, the MRs need to be able to provide the latest information about healthcare products to their key customers. To deal with the diverse needs of its customers, the development of different levels of MRs was deemed crucial for the company.

Before initiating the SST project, Nippon Roche made efforts to enhance its sales productivity several times by executing conventional training programs, but to no avail. In the past, the company had primarily used different forms of media for communicating best practice, together with role playing at individual branches, but realized that existing tools and techniques were inadequate to facilitate the sharing of the tacit knowledge embedded in individuals.

Nippon Roche initiated and executed the SST project to encourage sales process innovation. In categorizing the sales processes of MRs, the company found that high-performing MRs employed unique

---

* This chapter is an extended version of an article originally published in Ark Group's *Knowledge Management* magazine (Vol. 4, Issue 10, July/August 2001, www.kmmagazine. com).

sales processes, and developed their skills through their experiences over time. These skills and experiences were embedded within themselves. Recognizing the importance of this tacit knowledge, Nippon Roche initiated and executed the project in an attempt to transfer such skills to those MRs who were performing less effectively. The passionate efforts of the company have not only contributed to sales process innovation, but also to continuous innovation in the company as a whole. In this article, we will introduce the practices developed at Nippon Roche. We will also briefly discuss how the Nippon Roche case relates to and explains the theory of knowledge creation.

The SST Project

Nippon Roche executed the SST project in an attempt to capture and transfer the HQTK of high-performing MRs, the key sales force of the company, to other MRs within the company. Hiroaki Shigeta, the president of Nippon Roche, championed the project, and his leadership role was critical to its success. At the beginning of the SST project, the company made efforts to analyse the skill gaps between the higher-performing MRs and the average MRs (see Fig. 1). In this process, it categorized the required knowledge and skills, along with the selling processes of the MRs, as follows:

• Product and medical knowledge.

• Targeting the right customers.
• Process of accessing the potential customers.
• Detailing skills.

From the in-depth analyzes, it became clear that the high-performing MRs generally relied on experience (leaning by doing), while the average MRs relied on existing information (learning by manual). The critical differences between the high-performing MRs and the average MRs were in terms of 'access' to potential customers. The high-performing MRs generally employed the most effective timing to access potential customers, and the access skills of the high-performing MRs were considered as HQTK. Furthermore, the high-performing MRs proved to be very good at contextual practice and improvisation with their customers, and they continuously developed such skills through their personal experiences.

In its journey towards finding an effective mechanism to capture and transfer these skills to other MRs, Nippon Roche initiated the SST project, in an attempt to capture and transfer the HQTK embedded in the high-performing MRs and make it explicit (as far as possible), and incorporate this knowledge into broader corporate activities. As President Shigeta, the knowledge leader of the company, says: "The tacit knowledge that a high-performing MR has should be regarded as a defining factor. Based on his experiences, he knows everything about timing and has the knack of

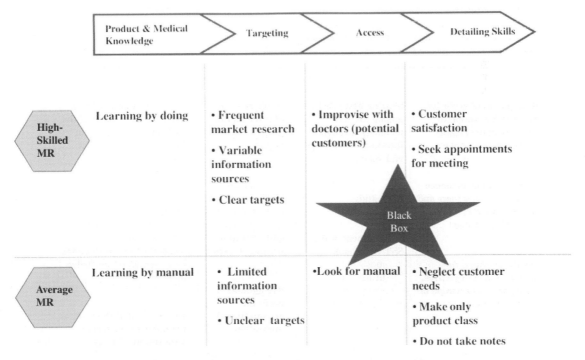

Figure 1. Skill gaps between high-skilled MR and average MR.

(Adapted from Corporate Document of Japan Roche).

getting to know individual doctors. He approaches doctors at the right time, by choosing the moment when his competitors are not around. Considering these models of behavior and action, I do not believe our established methods of training were having enough impact. It was hard to provide MRs with this kind of tacit knowledge through traditional forms of communication, such as training videos. As such, I decided to start up the SST project. I thought we would never be able to improve productivity unless we drastically improved the level of sales skills".

Instead of requesting help from any of the external consulting firms providing services in the area of knowledge management, Shigeta focused on the theory of knowledge creation, and decided to implement hands-on efforts in the creation of knowledge (see Fig. 2). The collaborative efforts of the participants of the SST project and their distributed leadership made it possible for them to develop an original methodology for creating knowledge that ultimately contributed to continuous innovation within the company. The marketing division of Nippon Roche was assigned to manage the SST project, and an executive of the company, Nakajima, became the co-ordinator of the project. The following section introduces what they did, and how.

The First Phase of the Project

After analyzing the skill gaps, 24 high-performing MRs from different units of Nippon Roche were carefully selected and gathered at the company's headquarters. All 24 MRs (the SST members) worked directly under the leadership of the president, Shigata. They discussed their skills and shared their experiences with other MRs over the course eight weeks. At the very beginning of the program, this process revolved around asking and responding to fundamental questions such as: what is our mission? For what do we exist? What is the ideal role of an MR? The participants were required to devote themselves to thinking about

Knowledge Creating Activities
(Nippon Roche)

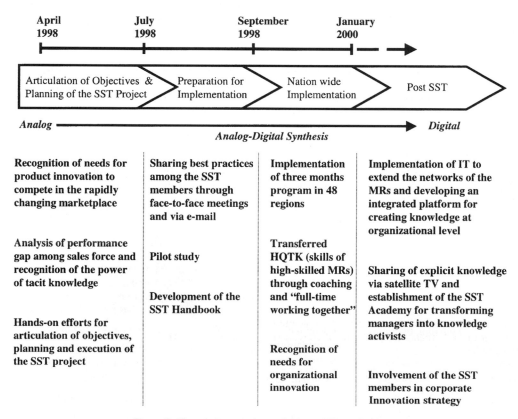

Figure 2. Knowledge creating activities at Nippon Roche.

(Adapted from Corporate Document of Japan Roche).

884

their own beliefs, articulating their knowledge (particularly tacit knowledge), and sharing with others. They were authorized to review highly classified corporate documents, so that they understood the background of the project and could contribute to achieving its goals.

At the end of the first phase, the project members scrapped the existing sales manual and created a new one by synthesizing knowledge (tacit and explicit) in their own words, by using metaphors and stories. From their experiences in the past, they realized that the use of metaphors and stories was critical for sharing knowledge, particularly tacit knowledge, as such approaches helped people to get inside their own minds and capture context.

Once they had created the new handbook, they experimented with it in the Tokyo area, as a part of a pilot program aimed at improving the handbook. After the experimentation, they refined the handbook further. The meaning of SST, the description of an ideal MR, the methodology to become one, together with evaluation criteria, were all clearly described in the handbook. The MRs considered the handbook to be a guideline for doing the right thing, rather than a manual for doing things right.

The Second Phase of the Project

Teams of three were seconded to different branches of the company for a period of three months. In their study, the SST project participants had found that a team of three was the optimal size to allow critical decisions to be made quickly. During the three-month nation-wide implementation period, the major activities of the SST members included:

- gathering preliminary information;
- prioritizing critical issues;
- interviewing managers;
- interviewing frontline employees;
- planning activities with average MRs;
- receiving confirmation from branch managers;
- visiting potential customers with the average MRs;
- attending meetings in which the results of the initiatives were reported;
- follow-ups after four months and six months.

The SST members also attended monthly meetings at the headquarters of the company and discussed the most desirable characteristics of an MR, as well as ways to ensure the continuous improvement of the program. It was recognized that the whole-hearted commitment of participants is critical for the creation of new knowledge.

The SST members (the selected MRs) went to different branches of the company without having any kind of legitimate power. They were not given any administrative staff to take as support, extra facilities to use, or any spare money to spend. All they had with them was their 'expert power'—the skills and experiences embedded within themselves. They helped other

MRs to capture HQTK, making it explicit as much as possible. The high-performing MRs transferred their tacit knowledge (skills and experiences) to other MRs by utilizing a variety of different mechanisms (for example, on-the-job-training, workshops, exchange experiences meetings, storytelling, etc.). The major purpose of their visits to the individual branches was not to temporarily improve performance, but rather to transfer the HQTK they had developed through their own experience.

Tacit knowledge needs care, love, trust, and commitment for it to be transferred. The project members shared tacit knowledge through socializing with other MRs at the branches, not by forcing them to learn, but by working closely with them (for example, by visiting potential customers together). The creed of the SST was 'you can't move people unless you do first, convince them, let them try, and then praise them', a well-known quote from Admiral Isoroku Yamamoto. The project members accompanied the MRs for one full week. They decided to work together full-time, realizing that socialization on a full-time basis allowed the observation, sharing, and experiencing of realities together, and such efforts were considered critical for the capturing and transferring of tacit knowledge. Also, they knew that it would take time to understand the particular context in which each MR worked; it takes time to build trust and it takes time and requires passionate effort to transfer tacit knowledge. Through working closely with the high-performing MRs, the average MRs were able to capture the knowledge (particularly tacit knowledge) needed to enhance their performance.

Evaluation of the Project

The SST project ended in January 2000, and the sales performance of the branches soared after its completion. Nippon Roche concluded that the participants of the project had become 'more valuable assets to the company'. The project motivated MRs and facilitated the capturing of the HQTK embedded in individuals, as well as the conversion of tacit knowledge into explicit, and the incorporation of synthesized knowledge into key organizational activities within the company. After coaching by the more experienced MRs, the less experienced MRs started looking within themselves with a different perspective. They started contributing to the entire organization through the creation of new knowledge.

Another remarkable outcome of the SST project was the further development of the SST members themselves. Since they were assigned to help the average MRs to capture their skills and experiences, they were also able to look at themselves from the viewpoint of others. The coaching activities helped them to redefine the knowledge (tacit) embedded within themselves. 'We brought our own experiences to the discussion and created a sort of standard for our oral presentations. We collaborated and kept updating our presentations, and these continued to become more effective. I think we could improve our level of skills further, to the extent of fundamentally altering the entire company ... I knew that the top management of the company would like to achieve this' (SST member).

Although the SST project was designed and executed to help average MRs to capture the HQTK of high-performing MRs for sales process innovation, the power of tacit knowledge captured the attention of those at Nippon Roche. The SST project converted individual knowledge into organizational knowledge, and the experience suggested the need for enhancing the management capabilities of managers in the separate branches. The project also clearly indicated the need for improving the links between corporate headquarters and the branches. Through initiating and executing the SST project, people involved in the project realized the importance of changing the organizational and management infrastructure of the company.

Following the successful completion of the SST project, Nippon Roche recently established the SST Academy. The goal of the academy is to develop the skills of sales managers, exploiting the know-how that has emerged from SST experiences in the past. Its target is to transform managers into leaders—knowledge activists—to provide distributed leadership in knowledge-creating processes. The SST members who contributed to organizational innovation at Nippon Roche are now playing the role of key knowledge activists within the company, and their next assignment is to contribute to corporate strategy innovation, based on knowledge creation. Nippon Roche recently announced its 'best value provider' vision, which places an emphasis on product innovation with customers.

To provide best value to its customers, Nippon Roche is now attempting to implement information technology (IT) to extend the networks of the MRs. Along with tacit knowledge, the company is now also focusing on sharing explicit knowledge via satellite TV, with the ultimate goal of developing an integrated platform for creating knowledge throughout the branches across the country. Nippon Roche has also recently formed a separate department, the Oncology Area Management Group (OAM), the mandate of which is not only to provide product knowledge, but also knowledge about the latest trends in the relevant fields. The company has also established a team, the eNR, which focuses on innovation relating to the customer relationship management (CRM) systems of the organization.

The SST Academy, OAM group, and the eNR team are three major initiatives executed by Nippon Roche after the remarkable success of the SST project. In our view, recognizing the power of tacit knowledge and making efforts to create knowledge through capturing

HQTK, synthesizing tacit and explicit knowledge, and incorporating the synthesized knowledge into corporate activities have contributed not only to sales process innovation, but also to overall organizational innovation at Nippon Roche.

The Theory of Knowledge Creation and the Nippon Roche Case

To create knowledge for continuous innovation, organizations need to adopt a holistic approach. The critical components of the theory of knowledge creation include:

- the SECI model;
- the concept of *Ba*;
- knowledge assets;
- leadership issues.

We know that tacit knowledge is subjective and experience-based. Tacit knowledge is highly personal and is deeply rooted in action and in an individual's commitment to a specific context (Nonaka, 1991). It is hard to express in words, sentences, and numbers. In the words of Michael Polyani, "we can know more than we can tell" (Polyani, 1966). However, explicit knowledge is objective. It can be expressed in words, sentences, and numbers. We understand that continuous innovation is the product of new knowledge that is generated from synthesizing tacit and explicit knowledge, and this synthesis depends on SECI, *Ba*, knowledge assets, and leadership (see Fig. 3).

The SECI Model

The SECI model describes knowledge creation as a spiral process of interactions between explicit and tacit knowledge (Nonaka & Takeuchi, 1995). In this model, S stands for socialization; E for externalization; C for combination; and I for internalization.

Socialization may start in different forms. It can occur within or outside an organizational boundary. In this process, through interactions between individuals, tacit knowledge can be created and shared. In the externalization process, tacit knowledge is made explicit through dialogue and reflections among individuals. When tacit knowledge is made explicit, knowledge is crystallized. But the successful conversion of tacit knowledge into explicit knowledge depends upon the use of metaphor, analogy, and model. Combination is the process of converting explicit knowledge into more complex and systematic sets of explicit knowledge, through interactions. And in the internalization process, explicit knowledge is converted into tacit knowledge. In this process, explicit knowledge is shared throughout the groups and organization, and converted back into tacit knowledge within individuals. This tacit knowledge becomes the valued asset of the organization (see Fig. 4).

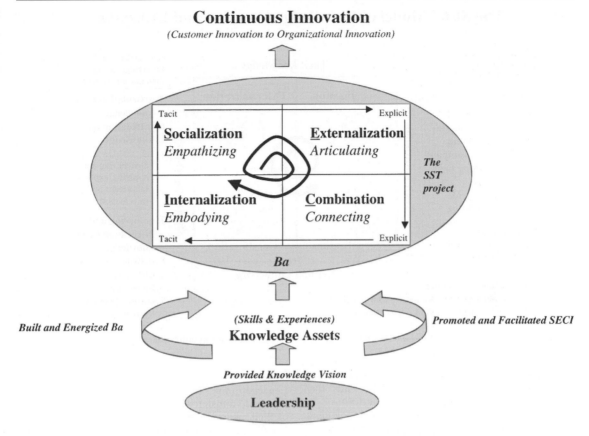

Figure 3. SECI, *Ba*, knowledge assets, and leadership in the Nippon Roche Case.

Knowledge creation is a continuous process of dynamic interactions between tacit and explicit knowledge. Articulating (converting tacit knowledge into explicit) in the externalization stage of the SECI process and embodying (converting explicit knowledge into tacit) in the internalization of the SECI process are the critical steps in the spiral of knowledge. "Knowledge creation is a craft, not a science" (von Krogh, Ichijo & Nonaka, 2000). In the Nippon Roche case, after recognizing the power of tacit knowledge, the company started making hands-on efforts towards the creation of knowledge. In the knowledge-creating process, knowledge conversion was carried into practice, and the converted knowledge made personal skills and experiences (tacit knowledge) more rich and contributed to creating new knowledge at an organizational level.

The Concept of *Ba*

Knowledge creation needs a context. The Japanese word *Ba*, which roughly means a place, provides a shared context for knowledge conversion (Nonaka & Konno, 1998). *Ba* is not necessarily just a physical

space; it can be equally a mental, a physical, or a virtual space. We understand *Ba* as a shared context in motion, a dynamic place where knowledge emerges. The most transcendental characteristics of *Ba* include synchronicity, resonance, kinetics, empathy, and sharing body knowledge. These characteristics of *Ba* are critical for sharing HQTK, embedded in individuals, and creating knowledge collaboratively.

Ba exists at many levels, and these levels may be connected to form a greater *Ba* that provides energy and quality in knowledge-creating processes. In a good *Ba*, participants get involved with whole-hearted commitment. When they get involved with such passion, they can see realities from a different perspective, deeply rooted in their own beliefs. Such commitment and involvement of participants of *Ba* are critical for sharing HQTK and creating knowledge.

The SST project itself was considered to be a *Ba* in which the medical needs of customers were redefined at the beginning. Then, monthly section meetings were also considered to be *Ba*, in which the SST members shared knowledge. *Ba* at different levels helped to create the contexts that were shared among individuals

The SECI Model of Knowledge Creation and Utilization

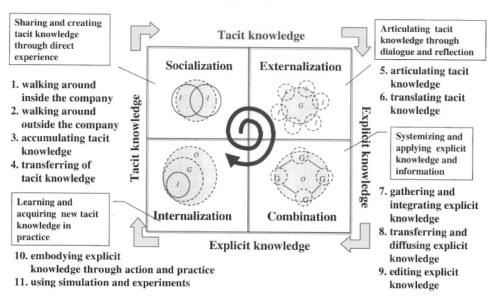

Figure 4. The SECI model of knowledge creation and utilization.

and teams at Nippon Roche. Tacit knowledge was first shared among individuals (the MRs) and then converted into explicit knowledge in the form of the SST handbook. The project helped expand individual, as well as organizational knowledge, through enriching the knowledge of individuals and teams. As a part of the SST project, various *Ba* were connected and expanded. The outcomes of the efforts went beyond knowledge creation at project level and contributed not only to sales process innovation, but also to organizational innovation.

Knowledge Assets

Knowledge assets are both inputs and outputs of an organization's knowledge-creating activities (Nonaka, Toyama & Konno, 2000). Different kinds of knowledge assets are continuously generated and utilized in knowledge-creating processes. Knowledge assets that are generated from *Ba* include: love; conviction; energy; business concepts; product concepts; design concepts; documents; manuals; specifications; intellectual property; skills; experiences; and organizational culture.

Recognizing the value of tacit knowledge as an asset, and synthesizing such knowledge (particularly skills and experiences) are critical for continuous innovation. The Nippon Roche case clearly shows the importance of recognizing such knowledge assets (particularly the tacit knowledge embedded in the high-performing MRs) and of making an effort to capture and synthesize the HQTK embedded in people.

Leadership

Leadership roles that contribute to recognizing, capturing and transferring HQTK, synthesize tacit and explicit knowledge, and incorporate synthesized knowledge into organizational activities, are now one of the major management issues in the knowledge-creating company. In knowledge-creating processes, a single charismatic leader is not enough. A team of knowledge activists—in other words, distributed leadership of knowledge activists—and their collaborative efforts are critical for creating knowledge. The team of knowledge activists includes knowledge leaders and knowledge producers.

The knowledge leaders provide vision for knowledge creation (Nonaka, Toyama & Konno, 2000). They promote the capture and synthesis of knowledge assets (tacit as well as explicit); promote and facilitate the knowledge spiral (SECI); and build and energize *Ba*. In our view, the collaborative efforts of knowledge leaders and knowledge producers are critical in knowledge creating processes. In the Nippon Roche case, Shigeta played the role of knowledge leader. He provided knowledge vision, recognized knowledge assets, built and energized *Ba*, and promoted and facilitated SECI, the knowledge spiral.

The knowledge producers—Nakajima and the SST members—also played critical roles at Nippon Roche. In the SST project, Nakajima's leadership role as co-ordinator of the project was very important. The SST members played the role of coach for other MRs. They transferred their skills and experiences to other MRs through coaching and by working closely with them. They helped the average MRs to capture HQTK, and synthesize tacit and explicit knowledge for sales process innovation. In our view, the distributed leadership and collaborative efforts of Shigeta, Nakajima, the SST members, the average MRs, and the members of the support team and the follow-up team made it possible for the company to create knowledge that contributed not only to sales process innovation but also to continuous innovation at Nippon Roche.

Conclusions

Knowledge is now widely acknowledged to be a source of competitive advantage, and a great deal of attention surrounds knowledge management research and practice. We believe that creating knowledge through the synthesis of tacit and explicit knowledge is more important than managing knowledge (explicit knowledge). Nippon Roche has made it clear that organizations can benefit from recognizing the power of tacit knowledge, creating knowledge through capturing the HQTK embedded in people, making it explicit, and incorporating synthesized knowledge into key organizational activities.

We know that individuals create knowledge through collaborating with others in groups/teams in an organizational context. In today's competitive and complex business environment, helping individuals to achieve their full potential and contribute new knowledge is a critical management issue. Although coaching, mentoring, and storytelling are receiving increasing attention from KM researchers and practitioners, alone these techniques are not enough to ensure the creation of knowledge.

Through introducing the SST project, Nippon Roche facilitated the knowledge spiral (the SECI model), implemented the concept of *Ba* (the SST project), and exploited knowledge assets (skills and experiences of the high-performing MRs) to create new knowledge. The outcome of the SST project went beyond sales process innovation at the project level. It contributed to continuous innovation at an organizational level. In our view, recognizing the power of tacit knowledge and making efforts to create knowledge through capturing HQTK, synthesizing tacit and explicit knowledge, and incorporating the synthesized knowledge into corporate activities, distributed leadership and collaborative efforts among the people involved have contributed to continuous innovation at Nippon Roche.

The key lessons for researchers and practitioners in the field of knowledge management are quite simple: just managing knowledge (explicit knowledge) by using existing tools and techniques is not enough for continuous innovation. Organizations seeking continuous innovation must recognize the power of tacit knowledge, make efforts to capture HQTK, synthesize tacit and explicit knowledge and incorporate synthesized knowledge into key organizational activities. They also must nurture and facilitate knowledge activists to play distributed leadership roles as synthesizers of knowledge. In short, to enhance competitive advantage in today's rapidly changing business environment, organizations must recognize the power of tacit knowledge and make efforts towards creating knowledge, rather than just managing it.

References

Nonaka, I. (1991). The knowledge-creating company. *Harvard Business Review*, (Nov.–Dec.), 96–104.

Nonaka, I. & Takeuchi H. (1995). *The knowledge-creating company*. Oxford: Oxford University Press.

Nonaka, I. & Konno, N. (1998). The concept of Ba. *California Management Review*, **40** (3), 40–54.

Nonaka, I., Toyama, R. & Konno, N. (2000). SECI, Ba, and leadership: A unified model of dynamic knowledge creation. *Long Range Planning*, **33**, 5–34.

Nippon (Japan) Roche Website: www.nipponroche.co.jp

Polyani, M. (1966). *The tacit dimension*. Routledge & Kegan Paul.

Von Krogh, G, Ichijo, K. & Nonaka, I. (2000). *Enabling knowledge creation: How to unlock the mystery of tacit knowledge and release the power of innovation*. Oxford: Oxford University Press.

The International Handbook on Innovation
Edited by Larisa V. Shavinina

Innovation in Korea

Sunyang Chung

*Institute for Technological Innovation (ITI), School of Business Administration,
Sejong University, South Korea*

Abstract: Korea's development has been based on its innovation capabilities. Since the end of the 1960s, it has been making great efforts to increase its innovation potential. In three decades, Korea has been able to formulate and implement a competent national innovation system that is composed of academia, public research sector and industry. Korea has increased national R&D resources in order to improve the national innovation system. The economic crisis in the fall of 1997 did not have a severe impact on the country's innovation potential. Instead, it has become a driving force to increase the innovation potential of the Korean national innovation system.

Keywords: Innovation; National innovation system; Korea; R&D; Innovation policy.

Introduction

Korea has developed remarkably since the beginning of its history of economic development. Many excellent products in international markets, e.g. CDMA, semiconductors, automobiles, and steel, are made in Korea. Success has been based on its capabilities of technological innovation. Korea has accumulated its strong innovation capabilities since the end of the 1960s, and, especially in the 1980s and hereinafter, such innovation capabilities have been transformed into new products and services that are competent in international markets. Korea's efforts to enhance its innovation capabilities are characterized by the very close cooperation among major actors. Here, the government has played an important role. The role of government in innovation promotion has been called its innovation policy, and it has made a great contribution to the development of Korea's national economy.

In many countries, the role of government in enhancing innovation capabilities has been a focus of innovation studies and practices. There are several definitions of innovation. Schumpeter (1934) defines innovation comprehensively as something new in product, process, organization and market. In this chapter, we refer to innovation as technological innovation, so that science, technology and innovation will be used interchangeably. Since the beginning of governmental activities in the areas of science, technology, and innovation the discussion of the legitimation of governmental intervention has been a focus of research (e.g. Ewers, 1990; Fritsch et al., 1993; Gielow

et al., 1985; Meyer-Krahmer & Kuntze, 1992). Many attempts have been made to prove positive or negative results by evaluating specific innovation public measures. In economic reality, there has been a strong resemblance between the innovation policies of different countries (Chung & Lay, 1997; Roobeek, 1990). The hectic competition in the area of technological innovation has led countries to imitate each others' innovation policies. There are numerous instances of countries having modeled their innovation policies or policy contents according to those of other countries, which have been considered successful.

It seems that one of the important reasons for this imitation of innovation policies lies in insufficient, unsystematic research in the area of innovation policies. In particular, there have not been sufficient studies on those of developing countries like Korea. If we recognize the role of innovation in Korea's economic development, there would be a strong demand to learn Korea's innovation policy. In view of the radical change taking place in the technological and economic environment and the strong competition between countries, it is very important to satisfy this learning demand. A rational innovation policy could be defined as a policy that met this demand on institutional learning.

However, it is very important to remember that such learning demand should not be confined to innovation policy *per se*, but, it should extend to institutional frame conditions around innovation policy. If we want to learn successful innovation policy, we should understand why this policy has been successful. Any successful and rational innovation policy should be based on country-specific S&T frame conditions (e.g. Chung & Lay, 1997; Porter, 1990; Vickery & Campbell, 1989). The emphasis on national innovation frame conditions stems from the institutional approach to technological innovation. It is a relatively recent phenomenon to emphasize the importance of the institutional frame conditions surrounding science, technology, and innovation activities (see e.g. Freeman, 1987, 1988, 1989; Lundvall, 1988; Majer, 1977; Nelson, 1988; Nelson & Winter, 1977, 1982). Now, these national institutional conditions for technological innovation are referred to as a *national innovation system*.

Even though there have been many studies on national systems of innovation, they have been carried out at a general level. Detailed analysis on specific actor groups in a national innovation system has not been systematically analyzed. More importantly, they have concentrated on analyzing national innovation systems of developed countries (e.g. Nelson, 1993). Insufficient investigation has been made specifically on national innovation systems of developing and less developed countries. As developing countries have been trying to establish more efficient national innovation systems in order to catch-up to developed countries, it would be also very interesting to look into developing countries' national innovation systems. Also, from successful developing countries we can identify important policy implications that might be useful for not only developing but also developed countries.

One of most interesting countries to learn from would be Korea, which has been one of the fastest growing economies in the world. Korea's economy has been developed, based on its accumulation of innovation capabilities. Many innovation actors could accumulate their innovation capabilities to a large scale, especially based on the active innovation policy of the Korean government. However, it was confronted with an economic crisis in the second half of the 1990s so that it had to be under the IMF jurisdiction in the fall of 1997. During the IMF jurisdiction period, Korea's economy experienced a deep recession. However, Korea could overcome this crisis much faster than expected, and it could regain a strong economic driving force, which is comparable to, and even much stronger than, that before the IMF jurisdiction. In this period of overcoming the crisis, science, technology, and innovation have played an essential role.

The purpose of this chapter is to analyze Korea's efforts to enhance innovation capabilities since the beginning of its industrialization. Special attention is paid to how Korea has overcome the IMF jurisdiction in the late 1990s and what kind of role innovation has played in this overcoming process. As a research framework we will adopt a concept of a national

innovation system. We will analyze the Korean national innovation system according to its major innovation actor groups, i.e. government, academia, public research sector, and industry. As there have been only a few researches on Korean national innovation system (e.g. Chung, 1996; Chung & Lay, 1997; Kim, 1993), this chapter would be very helpful for understanding the dynamism of the Korean national innovation system.

This chapter consists of six major parts. An introduction, 'National Innovation System' deals with the concept of a national innovation system as a theoretical framework of the chapter. Here, we will briefly discuss definition, purpose, and compositions of national innovation system. In 'Role of Government', we will analyze the role of government, which is also a major actor group in the national innovation system, in accumulating innovation capabilities since the beginning of Korea's industrialization. The strong role of the government in enhancing innovation capabilities is an important characteristic of Korean innovation system (Chung & Lay, 1997; Kim, 1993). We look into how Korea's innovation policy has been developed in the process of its industrialization. 'Division of Labor Among Innovation Actors' is concerned with the division of labor among actual innovation actor groups, i.e. academia, public research sector, and industry. Statistics on national R&D resources are used in this analysis. Special attention will be placed on what kind of role of these innovation actors played in overcoming the IMF jurisdiction in the late 1990s. 'IMF Jurisdiction and Korea's Innovation Potential', based on previous analysis, deals with the impact of the economic crisis during the IMF jurisdiction on the Korea's national innovation system. In 'Conclusions', we will summarize the characteristics of the Korean national innovation system and draw some meaningful implications for other countries.

National Innovation System

At present the concept of a national innovation system (NIS) is a frequent topic of discussion in innovation policy research (see e.g. Chung, 1996; Chung & Lay, 1997; Freeman, 1987; Lundvall, 1992; Nelson, 1988, 1993; Patel & Pavitt, 1994). The scholars in this area emphasize interactive learning between knowledge producers and users for the generation of innovations and the role of nation state for it. They also argue that an institutional framework plays an important role for interactive learning which leads to innovation. As a relevant institutional framework for innovation they concentrate on the analysis on the national level. If we agree that S&T and innovation policy targets at enhancing national competitiveness, this concept will be very relevant for this purpose and also for an analysis on competitiveness at national level.

There are several definitions of national innovation system. As Lundvall (1992) illustrated, there can be not only a broad definition, which encompasses all inter-related institutional and non-institutional factors which are concerned with generating, diffusing, and exploiting innovations, but also a narrow one, which includes organizations and institutions in searching and exploring, e.g. R&D departments, technical institutes and universities. In this chapter we follow a narrow definition for the effective analysis of national innovation system. We define a national innovation system as a complex of institutions, i.e. actors, in a nation, which are directly related with the generation, diffusion, and appropriation of technological innovation.

Under this definition we can identify four groups of actors in a national innovation system, i.e. business firms, public research institutes, universities, and government. The first three categories are actual research producers who carry out R&D activities while government can play the role of coordinator between the research producers in terms of its policy instruments, visions and perspectives for the future. The relative importance of these four groups in a national innovation system differs according to the history and country-specific frame conditions of the national innovation system. In general, the government plays a more important role in emerging or developing national innovation systems than in existing ones. Innovation policy is crucial in formulating a new national innovation system and improving its innovation performance.

In the concept of a national innovation system, the inter-relationship or interaction between innovation actors is very important. Most countries prepare for important policy measures to promote these interactions. Such efforts could be measured by the R&D resources of a national innovation system. Looking into the relative distribution of the R&D resources among innovation actor groups, we can identify their relative importance in a national innovation system.

From the terminology *national*, we can easily assume that national innovation systems will differ between countries. Some authors have already performed international comparisons of various national innovation systems with explicitly using terminology of national innovation system (e.g. Edquist & Lundvall, 1993; Nelson, 1993) and without using this terminology (Martin & Irvine, 1989). Historically developed, national innovation systems vary and should vary greatly from one country to another (e.g. Chung, 1996; Chung & Lay, 1997; Ergas, 1987). Therefore, creative learning from other national innovation systems is needed to refine and improve any national innovation system. This implies that, in order to gain a reasonable understanding of a national innovation system, a kind of historical analysis should be applied. In general, the target of learning has been confined in the national innovation systems of advanced countries. However, we should make an effort to learn from the successful innovation systems of developing countries. Under the rapidly changing economic and technological environment, only a national innovation system that has a strong demand on learning could be competent and successful.

Although broadly divisible into four groups of innovation actors, a national innovation system is not easy to grasp because they consist of numerous and diverse institutions and organizations. However, the concept of national innovation system is very helpful to analyze them, because it is based on the institutional theory, which emphasizes the possibility of institutional learning. We argue that the analysis of a national system of innovation should at least examine the role of the government and the division of labor among research producers, i.e. academia, public research sector, and industry. Based on the careful analysis on these actors and institutions, we can identify some important characteristics of a national innovation system.

Role of Government

In Korea, the government has been playing a very important role in almost every aspect of society. As a centralized country, the Korean government has increased its efforts to enhance innovation capabilities. Korea has specific innovation policies that could be interesting to foreign countries. Several methods might be adopted to classify and describe the historical role of the central government in the area of science and technology. It could be described in terms of changes in the administration of a central government or in terms of decades. According to our studies, there has been a tendency of changing policy directions in many countries with the change in decades (Bruder & Dose, 1986; Chung & Lay, 1997). The Korean government also has had a tendency to adopt some major new policy programs with the turn of decade. There have been some studies and reports on the historical development of Korean innovation policies, e.g. MOST (1990), OECD (1996), and Chung & Lay (1997). They discuss the role of government in Korea's accumulating innovation capabilities in terms of decades. Table 1 shows the role played by the Korean government in science, technology, and innovation in terms of decades.

In the 1960s

It was not until the beginning of the 1960s that national efforts for the promotion of science, technology and innovation were initiated in line with the *First Five Year Plan for Economic Development*, which was introduced in 1962. Since then, the Korean government has intervened very strongly in the areas of science, technology, and innovation. The government has applied a strong technology-push approach in the construction and improvement of the national system of innovation. Korean economic policy in the 1960s was characterized by import substitution and export

Table 1. Role of Korean government in national innovation system.

Periods	Characteristics of Korean innovation policies
1960s	• beginning of scientific education • beginning of S&T infrastructure construction
1970s	• construction of government-sponsored research institutes • technical, scientific and further education • beginning of industrial R&D
1980s	• promotion of key technologies through *National R&D Program* • activation of industrial R&D • mass production of highly-qualified R&D personnel • expansion of S&T-related ministries
1990s	• expansion of R&D resources and their efficient utilization • promotion of academic innovation potentials • introduction of regional innovation policy • introduction of Research Council system
2000s	• enactment of *Basic Law of Science and Technology* • selection and concentration on major technologies (5T) • coordination of innovation policies • basic research and welfare technologies

orientation. At this time, automobile production (1960), ship-building (1967), mechanical engineering (1967), and electronics industry (1967) were the central concern of governmental promotion (see Byun, 1989; Song, 1990). In order to activate this economic policy effectively, an institutional framework in the area of science and technology began to be established, for example:

- the foundation of the Korea Institute of Science and technology (KIST) in 1966, which carries out R&D activities, especially in the technology areas mentioned above;
- the passing of the Science and Technology Promotion Act in 1967;
- the establishment of the Ministry of Science and Technology (MOST) in 1967, which has the task of formulating and implementing science and technology policy.

At the same time Korean universities attempted to produce as many engineers as possible, because there was a great shortage of qualified engineers, who were indispensable for the development of the Korean economy. Korea's national innovation system concentrated on the digestion and imitation of imported technologies from advanced countries. There were no concrete S&T policy measures and programs in Korean system. Most technological and innovation needs were covered by KIST, which was the only research institute in Korea.

In the 1970s

In the 1970s the Korean government placed the main emphasis of its industrial policy on the establishment and expansion of heavy, chemical, and export-oriented industries (Byun, 1989; Song, 1990). These industries were technology-oriented and needed a certain level of domestic technological and innovation capabilities. With a view to meeting the needs of these industries, the Korean government founded several corresponding government-sponsored research institutes, e.g. Korea Institute of Machinery and Materials (KIMM), Korea Research Institute of Chemical Technology (KRICT), and Electronics and Telecommunication Research Institute (ETRI). These government-sponsored research institutes in addition to KIST were the grounding stones of the Korean national innovation system.

During this period, the major emphasis of the innovation policy shifted from the simple imitation of imported technologies to their complex adoption and the domestic development of simple, less complex technologies. Creative imitation started in this period (Kim, 1993). The Korean government implemented a series of strong policies for producing researchers and engineers as many as possible, who were needed by these strategic industries. Some policy measures were initiated to further train these engineers. Some big Korean industrial enterprises, especially those in the industries mentioned above, began to carry out their own R&D activities.

In the 1980s

The 1980s were characterized by a very strong increase of industrial R&D activities within the Korean national innovation system. Using several policy instruments, the Korean government motivated industrial enterprises to establish their own R&D institutes, so that the number of private research institutes rose dramatically from 53 in 1981 to 966 in 1990 (KITA, each year). In line with this strong increase of industrial R&D capabilities, the government tried to shift the Korean industrial structure away from traditional branches towards high technology areas.

In 1982 the Korean government initiated the first big project in the areas of science, technology, and innovation, the *National R&D Program*. This program aimed at developing not only high technologies but also large technologies (MOST, 1987). In this program the industrial key technologies that industrial companies could not deal with alone were developed through joint projects, especially between industrial companies and government-sponsored research institutes. As a result of strong R&D efforts in the public and private sectors in this period, Korea could attain a certain level of innovation capabilities to compete with advanced countries in some advanced technology areas like semiconductor (STEPI, 1991).

Since the end of the 1980s, several ministries, the Ministry of Commerce, Industry and Energy (MOCIE), the Ministry of Environment (MOE), and the Ministry of Information and Telecommunications (MOTI), became concerned with science, technology and innovation. In 1987, the Ministry of Commerce, Industry and Energy (MOCIE) initiated the *Program for Nurturing Industrial Technology Base* for the first time among S&T-related ministries except the Ministry of Science and Technology (MOST). Following the MOCIE, other ministries started to initiate their own programs. It made a great contribution to enhancing innovation capabilities in many industrial sectors. However, the problem of resource duplication had started during this period, as these ministries competed very strongly with each other to collect more innovation resources.

In the 1990s

Despite the greatly increased importance of industry, in the 1990s, the Korean government intervened in the areas of science, technology and innovation more actively than before. Based on some successes in the last decade, the Korean government recognized the importance of S&T and innovation in economic development. The government tried to step up national R&D expenditures, with the result that in 1991 the share of national R&D expenditures in Gross Domestic Product (GDP) exceeded 2% for the first time in the Korea's history (MOST, each year).

Based on the strong increase in industrial R&D capabilities, in the 1990s, industrial companies took over major areas of R&D activities, which were previously performed by government-sponsored institutes. As a result, during the 1990s there were frequent reorganization, merger and disorganization of Korean public research institutes. Criticism on the role of public research institutes rose in this period (Chung, 2002; Chung et al., 2001; Kim et al., 2002; Song et al., 2001). Therefore, in March 1999, the Korean government introduced a new public research system, the

Research Council system, by benchmarking the Germany's *Gesellschaft* system.

In this period, the Korean government very strongly promoted R&D and innovation capabilities of Korean universities, which had been the weakest point of the Korean innovation system until this period (Chung, 1996; Chung & Lay, 1997; OECD, 1996). In order to strengthen academic R&D capabilities, the government has initiated the Excellent Research Center (ERC) program for the most advanced research centers in universities in 1990. This program consists of Science Research Centers (SRCs) in the area of basic science and Engineering Research Centers (ERCs) in the area of engineering and applied research. When a center in a university is accepted as an excellent research center, it can be supported by a very large amount of money for ten years. As there was a hierarchy in the level of research capabilities in Korean universities, a few best universities, especially in Seoul, dominated the excellent research centers. Therefore, the Korean government initiated the Regional Research Center (RRC) Program in 1995 in order to strengthen the R&D and innovation capabilities of universities in other regions outside of Seoul. As of 2001 there are 25 SRCs, 34 ERCs, and 45 RRCs (MOST, 2001). These centers have played an important role in enhancing R&D capabilities of Korean universities.

In the middle of the 1990s, a new policy direction arose in Korean innovation policy. The Korean government initiated a regional innovation policy. Korea developed in the middle of capital city, Seoul, and its outskirts. The development of politics, economy, society and culture was centered in these areas, so that the regional level of industry, science and technology was still very low. The central government was always a dominant player in innovation policy and there had not been a regional S&T policy in Korea. Even in 1999, the R&D budget of the total regional governments represents only 6.8% of the national S&T budget (MOST, 1999). Korean research organizations are located in and around Seoul and in Dae-Duck Science Park, which is about 200 Km south of Seoul. Nowadays, however, regional governments have recognized the importance of S&T for the economic development of their regions, especially since the inauguration of the *Local Government System* in March 1995. As of 2000, eight among 16 regional governments established an independent organization for promoting technological and innovation capabilities in their regional administrations (Chung, 2002; MOST, 1999). We can say that Korea is in the early stages of its regional innovation policy.

In the 2000s

Turning to the 21st century, Korea initiated a very ambitious plan to enhance technological and innovation capabilities more systematically. The Korean government enacted a comprehensive law, the *Basic*

Law of Science and Technology, in January 2001, which aimed at more systematic promotion of science and technology. According to this law, the *Basic Plan of Science and Technology* should be formulated and implemented every five years (MOST, 2001). This plan comprised of detailed S&T plans of all S&T-related ministries. Based on the law, the first *Basic Plan for Science and Technology* was formulated in December 2001. This plan had a comprehensive goal and implementation strategies for enhancing technological capabilities for next five years, e.g. from 2002 until 2006. According to this plan, Korea aims at reaching the top 10 countries in the areas of science, technology, and innovation (MOST, 2002). In particular, Korea selected six technology areas, i.e. information technologies (IT), biotechnologies (BT), nanotechnologies (NT), space technologies (ST), environmental technologies (ET), and cultural technologies (CT), which will be essential to the knowledge-based 21st century.

As technological innovation was promoted by many ministries and large amounts of money was invested in science, technology and innovation, it was necessary to coordinate innovation policies among major S&T-related ministries. Therefore, the *Presidential Committee on Science and Technology* was established in 1999 to better coordinate innovation policies among ministries. This committee is originated from the *Committee of S&T-Related Ministers*, whose chairman was Minister of Science and Technology. Because there had been strong competition in innovation policies, especially sector-specific ones, between ministries, it was impossible to attain effective policy coordination under this old committee. However, much better coordination was anticipated, as the chairman of this new committee was the President of Korea.

In addition, regional innovation policy gained an important priority in the Korean innovation policy, especially those of the Ministry of Science and Technology (MOST) and the Ministry of Commerce, Industry and Energy (MOCIE). Special attention was placed on enhancing region-specific technological capabilities that could be transformed into a regional comparative advantage. With regard to the policy goals, future-oriented goals, e.g. enhancing the quality of life in terms of science, technology and innovation, were seriously pursued for the first time in the Korean innovation policy. That implies that the Korean government fully recognized the importance of science, technology and innovation in the development of Korean economy and society.

Division of Labor Among Innovation Actors

A national innovation system can be described in terms of their national R&D resources, which reflect not only the history of the development of the system itself, but also the division of labor between research producers, i.e. innovation actors. In this section, we will discuss national R&D resources in Korea since the beginning

of the 1980s. Special focus is placed on the development of Korea's national R&D resources since the middle of the 1990s. This section describes the dynamics of the Korean national innovation system. There are two kinds of statistics on national R&D resources, i.e. national R&D expenditures and number of researchers. The former can be classified into sources and uses of R&D expenditures. In this chapter we will analyze these statistics in depth in order to grasp the dynamism of the Korean national innovation system.

(1) R&D Expenditures

The short history of the Korean national innovation system is confirmed by examining its national R&D resources. Until the beginning of the 1980s, Korea invested very little in R&D activities. Korean national R&D expenditures in 1980 only amounted to 0.58% of GDP. However, national R&D expenditures rose dramatically over the 1980s, so that in 1990 about 1.91% of GDP was spent on national R&D activities. This is equivalent to an annual increase rate of 31.2% over this period. In the 1990s, the importance of technological innovation had been widely diffused in Korean society (Chung, 2001b). The total R&D expenditures in 1995 were 94 billion won, which amounted to 2.50% of Korea's GDP. Korea continues to be one of the countries that make a strong investment in R&D activities. There are not many countries, even advanced ones, that are comparable to Korea.

In the 1990s, until the IMF jurisdiction in the fall of 1997, the total R&D expenditures had increased dramatically. The years 1995, 1996, and 1997 showed an annual increase rate of over 10%, especially 1995 that showed a 19.6% increase rate. However, 1998, showed a 7% decrease rate, just after the IMF jurisdiction. Such a decrease in national R&D investment was for the first time in the history of Korean S&T and innovation development. However, the decrease rate was not so severe, as it was only 7%. Only 1998 showed a decrease rate, and the year 1999 turned to an annual increase rate of 5.2%. In particular, the year 2000 showed 16.2% of increase rate, which was very comparable to the year 1996 before the IMF jurisdiction. This implies that Korea has not been influenced very much by the IMF jurisdiction. As a result, Korea's total national R&D expenditures in 2000 were 138 billion won. The share of total national R&D expenditures in GDP in 2000 was 2.78%. This implies the Korea was one of the most significant R&D investing countries in the world, even after the deep recession.

When we look at the sources of national R&D expenditures, the private sector has played a far more significant role in R&D financing than the public sector, especially since the beginning of the 1990s. However, until the end of the 1970s, Korean national R&D expenditure had been financed predominantly by

Table 2. Sources of Korea's national R&D expenditures.

(Unit: hundred million Won)

	1980	1990	1995	1996	1997	1998	1999	2000
Total R&D expenditures	2,117	32,105	94,406	108,780	121,858	113,366	119,218	138,485
(Increase rate)	(21.7%)	(18.7%)	(19.6%)	(15.2%)	(12.0%)	(−7.0%)	(5.2%)	(16.2%)
• Public source	1,054	5,108	17,809	23,977	28,507	30,518	32,031	34,518
(Increase rate)	(15.5%)	(10.6%)	(41.3%)	(34.7%)	(18.9%)	(7.1%)	(5.0%)	(7.8%)
• Private source	1,024	26,989	76,597	84,667	93,233	82,764	87,117	103,967
(Increase rate)	(29.3%)	(20.4%)	(15.4%)	(10.5%)	(10.1%)	(−11.2%)	(5.3%)	(19.2%)
• Foreign source	39	8	13	136	118	84	70	–
Public : Private	49.8 : 50.2	15.9 : 84.1	18.9 : 81.1	22.0 : 78.0	23.4 : 76.6	26.9 : 73.1	26.9 : 73.1	24.9 : 75.1
Share of GDP	0.58%	1.91%	2.50%	2.60%	2.69%	2.55%	2.47%	2.78%

Source: MOST (Each Year), *Report on the Survey of Research and Development in Science and Technology,* Seoul.

the public sector. In 1980, about 50% of national R&D expenditures were still publicly financed. However, the role of the public sector diminished in the 1980s as a result of the strong increase in financing by the private sector. Thus, the ratio of public to private financing was about 15.9% vs. 84.1% in 1990. This implies that the private sector recognized the importance of technological innovation more strongly than the public sector did. In fact, industrial companies established their own R&D institutes and increased their R&D expenditures to a large degree.

The sharing between the public and private sector in R&D financing changed from 19% vs. 81% in 1995 through 23% vs. 77% in 1997 to 25% vs. 75% in 2000. This indicates that in the 1990s the public sector increased its financing more than the private sector did. Private companies could not increase its R&D expenditures because the recession resulted from the IMF jurisdiction. They were influenced more strongly by the recession than the public sector. When we look at the trend of R&D expenditures according to sectors, the public sector has always increased its financing in R&D activities. In particular, it increased annual R&D investment in the middle of the 1990s: 41.3% in 1995 and 34.7% in 1996. Even in 1998 it increased R&D investment at a rate of 7.1%. This implies that the public sector in Korea recognized the importance of science, technology, and innovation in overcoming the crisis in this period. However, it is safe to say that the public sector was also influenced by the IMF jurisdiction. The annual increase rate declined sharply during the IMF jurisdiction period and it did not recover after the jurisdiction, i.e. in 1999 and 2000.

The private sector was strongly influenced by the IMF jurisdiction. Before the IMF jurisdiction the Korean private sector had shown over 10% of annual increase rates of R&D investment. However, in 1998, the private sector showed a decrease rate of 11.2%. This suggests that the Korean private companies

decreased a very significant portion of their R&D investment because of the recession in the IMF jurisdiction period. We could also see the declining trend of R&D investment in the private sector since 1995. The rate declined from 15.4% in 1995 to 10.1% in 1997. This indicates that the Korean economy was in recession even before the IMF jurisdiction. Also, Korean industrial companies did not take technological innovation seriously enough to overcome the recession in this period.

However, in 2000, the private sector showed a much higher level of increase rate. Showing a 19.2% annual increase rate, the private sector made greater R&D investment in 2000 than in the middle of the 1990s. In means that Korean companies, having experienced the economic crisis in the late 1990s, recognized the importance of R&D activities to overcome the recession and to enhance their competitive advantage in globalized international markets. Korean companies implemented an aggressive R&D and innovation strategies after the IMF jurisdiction period.

Looking at the national R&D expenditures according to the performing sectors in the national innovation system, we can see the changes in the division of labor in the Korean national innovation system since the beginning of the 1980s (see Table 3). Until the end of the 1970s almost all Korean R&D activities were carried out by public research institutes. At this time the industrial R&D infrastructure was not in existence in Korea.

Even in 1980, the public research institutes were still absorbing 49.4% of national R&D expenditures, while the share of industry was only 38.4%. However, as in the financing of national R&D resources, the role of the private sector in national R&D activities increased continuously over the 1980s. As a result, Korean industry was utilizing 74.0% of national R&D expenditures in 1990. This confirms the strong increase of R&D capabilities of Korean industry. For example,

Table 3. National R&D expenditures by sector of performance.

(Unit: hundred million Won)

	Total R&D exp.	Public research sector			Academia			Industry		
		Amount	Increase rate	Share	Amount	Increase rate	Share	Amount	Increase rate	Share
1980	2,117	1,045	6.4%	49.4%	259	56.6%	12.2%	813	37.2%	38.4%
1990	32,105	5,917	23.9%	18.4%	2,443	6.6%	7.6%	23,745	18.8%	74.0%
1995	94,406	17,667	15.3%	18.7%	7,709	26.6%	8.2%	69,030	20.2%	73.1%
1996	108,780	18,956	7.3%	17.4%	10,188	32.2%	9.4%	79,636	15.4%	73.2%
1997	121,858	20,689	9.1%	17.0%	12,716	24.8%	10.4%	88,453	11.1%	72.6%
1998	113,366	20,994	1.5%	18.5%	12,651	−0.5%	11.2%	79,721	−9.9%	70.3%
1999	119,218	19,792	−5.7%	16.6%	14,314	13.1%	12.0%	85,112	6.8%	71.4%
2000	138,485	20,320	2.7%	14.7%	15,619	9.1%	11.3%	102,547	20.5%	74.0%

Source: MOST (Each Year), *Report on the Survey of Research and Development in Science and Technology*, Seoul.

the number of Korean industrial private research institutes increased from 53 institutes in 1981 to approximately 1,200 institutes in 1991 (KITA, each year). As a result, the industry took over the leading position in the Korean national system of innovation from the public research institutes. The Korean universities, however, always played a very minor role in the national innovation system in the 1980s. Only 12.2% of national R&D expenditures were utilized by the universities in 1980 and their role diminished further in the 1980s. In 1990, the share of universities in the total R&D expenditures was only 7.6%. As a result, the share of national R&D expenditures in 1990 was 18.4% (public research institutes) vs. 7.6% (universities) vs. 74.0% (industrial companies).

During the 1990s the share of innovation actor groups in national R&D expenditures changed. The share of R&D expenditures among public institutes, universities, and companies changed from 18.7% vs. 8.2% vs. 73.1% in 1995 to 14.7% vs. 11.3% vs. 74.0% in 2000. Private companies were the greatest R&D performing actors and their role had not changed during this period. However, private companies decreased their R&D expenditures in 1998 (−9.9%) and, by having 6.8%, the increase rate in 1999, just after the IMF jurisdiction, was much smaller compared to that before the IMF jurisdiction. Korean private companies were strongly affected by the recession in the IMF jurisdiction period. However, Korean private companies showed a very high level of annual increase rate in R&D expenditures, i.e. 20.5% in 2000, which was bigger than that of 1995. This suggests that Korean companies reached again the level of R&D intensity in the middle of the 1990s and showed even stronger R&D activities. As a result, the portion of private companies in national R&D expenditures in 2000 was 74.0%. It was greater than that of 1995 (73.1%).

In the 1990s, the role of public research institutes declined. The public research institutes' portion in

national R&D expenditures declined from 18.4% in 1990 through 18.7% in 1995 to 14.7% in 2000. The role of public research sector declined especially in the second half of the 1990s. In fact there was a strong structural reform in the Korean public research sector in the beginning of 1998, which resulted in the reduction of research potential of public research institutes (Chung, 2001a). It is confirmed by the decrease rate of this sector's R&D expenditures in 1999 (−5.7%) and also by the very small increase rate in 1998 (1.5%). In 2000 the public research sector showed only 2.7% of annual increase rate, which was far smaller than that of 1995, i.e. 15.3%. All these statistics on the public research sector showed that its role has declined remarkably during the 1990s and such a trend is still continuing. Considering the successful role of the Korean public research institutes in the history of the Korean economic development, it is a challenging issue for Korea how to enhance the role of public research institutes in the Korean national innovation system from quantitative and qualitative perspectives.

By contrast, Korean universities increased their R&D activities during the 1990s. As mentioned above, Korean academia had been the weakest area in the Korean national innovation system, especially before the beginning of the 1990s (Chung & Lay, 1997; OECD, 1996). However, Korean universities increased their R&D potential on a very large scale especially during the second half of the 1990s (see Table 3). The portion of Korean universities' total national R&D expenditures increased from 8.2% in 1995 to 11.3% in 2000. In particular, more than 25% of the annual increase rates were shown before the IMF jurisdiction. As a result, Korean academia could play a relevant role in the Korean national innovation system since the middle of the 1990s. The impact of the recession on academic research was not so severe, as only a 0.5% decrease rate was shown in 1998, shortly after the year of IMF jurisdiction. However, Korean academia showed a 13.1% of annual increase rate in 1999 and 9.1% in 2000. Nevertheless, Korean universities were influenced by the recession to a degree, because the annual increase rates after the recession were much smaller than those before the IMF jurisdiction period.

(2) R&D Manpower

We can also see the transformation of technological innovation activities in Korea in terms of R&D manpower. Table 4 shows the trend in the number of researchers of the Korean national innovation system. In 1980, the Korean system of innovation comprised only 18,434 researchers. During the 1980s the number of researchers rose rapidly to 70,503 researchers in 1990. This corresponds to an annual increase rate of 14.4%. In particular, the number of highly-qualified researchers increased during this period. In 1980, academia was employing 47.0% of the total researchers, so that it was the biggest employer of Korean researchers. The second biggest employer was the industry by having 27.9% of total researchers. It is worth noting that public research institutes were still employing about 24.9% of researchers in 1980.

During the 1980s, as shown in the financing of national R&D resources, the role of the private sector in national R&D activities increased continuously. As a result, in 1990, Korean industry became the biggest employer for researchers by having 54.9% of total researchers. It was due to the strongly increased establishment of private research institutes. In fact, the Korean government implemented several policy measures to induce private companies to employ researchers. For example, a researcher who had a master degree could be exempt from military service if he had been employed in a private research institute for

five years. However, Korean universities were also very important employers for researchers during the 1980s, especially for researchers with a Ph.D. degree. But their role had been diminished in the 1980s, and in 1990 universities employed 30.3% of total researchers in Korea. Still public research institutes had a significant proportion of Korea's researchers by having 14.8% of the total researchers in 1990.

During the 1990s the number of Korean researchers more than doubled. In particular, there was an increase of more than 30,000 researchers in the second half of the 1990s. However, 1998 showed a 6.3% decrease compared to the previous year. This implies that the IMF jurisdiction had a severe impact on the number of researchers in the Korean national innovation system. However, 2000 showed a very strong increase in researchers. Compared to the previous year, there was an 18.9% increase in the number of researchers in the Korean national innovation system as a whole. During this period, there were some changes in the share of researchers among major innovation actor groups in Korea. The share of researchers among public research institutes, universities, and private companies changed from 14.8% vs. 30.3% vs. 54.9% in 1990 through 11.7% vs. 34.8% vs. 53.5% in 1995 to 8.7% vs. 32.3% vs. 59.0% in 2000. This shows that private companies increased the number of researchers to a large scale, while public research institutes decreased dramatically.

Private industry has been the biggest employers of researchers in the Korean national innovation system. Especially in the first half of the 1990s, Korean companies increased the number of researchers. The number increased from 38,737 researchers in 1990 to 68,625 researcher in 1995. However, during the second half of the 1990s, Korean companies were not so

Table 4. Number of researchers by year.

(Unit: person)

	1980	1990	1995	1996	1997	1998	1999	2000
Total number of researchers	18,434	70,503	128,315	132,023	138,438	129,767	134,568	159,973
(Increase rate)	(17.3%)	(6.5%)	(9.3%)	(2.9%)	(4.5%)	(-6.3%)	(3.7%)	(18.9%)
(Share)	(100%)	(100%)	(100%)	(100%)	(100%)	(100%)	(100%)	(100%)
Public institutes	4,598	10,434	15,007	15,503	15,185	12,587	13,982	13,913
(Increase rate)	(8.0%)	(2.3%)	(-3.0%)	(3.3%)	(-2.2%)	(-17.1%)	(11.1%)	(-0.5%)
(Share)	(24.9%)	(14.8%)	(11.7%)	(11.7%)	(11.0%)	(9.7%)	(10.4%)	(8.7%)
Universities	8,659	21,332	44,683	45,327	48,588	51,162	50,155	51,727
(Increase rate)	(22.8%)	(2.3%)	(4.6%)	(1.4%)	(7.2%)	(5.3%)	(-2.0%)	(3.1%)
(Share)	(47.0%)	(30.3%)	(34.8%)	(34.3%)	(35.1%)	(39.4%)	(37.3%)	(32.3%)
(Share of Ph.D)	n.a.	(76.9%)	(77.1%)	(76.0%)	(75.4%)	(78.2%)	(76.8%)	(76.2%)
Companies	5,141	38,737	68,625	71,193	74,665	66,018	70,431	94,333
(Increase rate)	(16.7%)	(10.2%)	(15.8%)	(3.7%)	(4.9%)	(-11.6%)	(6.7%)	(33.9%)
(Share)	(27.9%)	(54.9%)	(53.5%)	(54.0%)	(53.9%)	(50.9%)	(52.3%)	(59.0%)

Source: MOST (Each Year), *Report on the Survey of Research and Development in Science and Technology*, Seoul.

aggressive in employing researchers because of the recession in this period. In particular, in 1998, the year just after the IMF jurisdiction, there was a remarkable decrease in researchers by having an 11.6% decrease rate. However, around the end of the 1990s, Korean companies were very aggressively recruiting researchers in trying to overcome the recession. In particular, the year 2000 showed a 33.9% annual increase rate of researchers. As a result, in 2000, 59% of the total researchers were employed by the industry. This implies that Korean companies recognized the importance of science, technology and innovation in overcoming the recession and enhancing their competitiveness. In line with the increase in R&D investment, Korean companies increased well-qualified researchers to a remarkable scale.

In contrast, Korean public research institutes decreased the number of researchers in the 1990s. There were several years that the number of researchers in this sector decreased, e.g. 1995, 1997, and 1998, compared to the previous year. In the same period, other sectors, i.e. academia and industry, increased researchers except the IMF jurisdiction period. Between 1995 and 2000, it showed a reduction of about 10,000 researchers in the public research group. In particular, 1998 shows a decrease of 2,598 researchers, which was about 17% of the total researchers of this sector. We can ascribe two reasons for this. First, in this period, the Korean government demanded a strong and radical structural transformation in the public research sector. During this transformation period, many researchers had to leave public research institutes (Chung, 2001a). Second, there was a strong gale of start-ups in Korean society, so that many well-qualified scientists and engineers left institutes and started their own venture companies. In 1999, however, there was an increase (11.1%) of researchers in this sector, but 2000 showed a weak decrease. This indicates that the Korean public research sector was under a strong transformation process during the IMF jurisdiction.

Finally, Korean universities were always important employers for researchers in the 1990s. Korean academia increased researchers steadily and, in 2000, 32.3% of the total researchers were employed by this sector. In particular, academia was always the biggest employer for researchers with Ph.D. degrees far outnumbering other actor groups. Over 75% of these well-qualified researchers were employed by academia. Considering that academia expended just slightly over 10% share of the total national R&D expenditures at least in the second half of the 1990s, Korean universities had always possessed too many researchers, especially those with Ph.D. degrees. Therefore, it would be desirable that academia reduces the number of well-qualified researchers, and additionally more R&D investment should be made to academia. In summary, the Korean government should prepare for relevant measures for the virtuous flow of researchers

from academia to industry in order to strengthen the national innovation system.

IMF Jurisdiction and Korea's Innovation Potential

It is interesting to investigate more systematically whether, and how much, the economic crisis during the IMF jurisdiction influenced the Korean national innovation system. Just after the IMF jurisdiction, many experts expected the Korean economic crisis to last a long time, because the main reason being there was a structural problem in the Korean economic system. However, Korea overcame the economic crisis in a much shorter period than most expected and regained its economic vitality by the turn of the new century. We could ascribe the main reason for this to the continuous increase in innovation potential of the Korean national innovation system.

Figure 1 summarizes the trend of innovation capabilities of the Korean national innovation system since the middle of the 1990s. It depicts both national R&D expenditures and the number of researchers according to major innovation actor groups. It also shows the similar trend between R&D expenditures and number of researchers before and after the IMF jurisdiction. According to Fig. 1, the unexpected economic crisis towards the end of the 1990s did not have a severe impact on the Korean national innovation system. The recession under the IMF jurisdiction did influence innovation activities of the Korean national innovation system. In 1998, just after the IMF jurisdiction, the total national R&D investment recorded a decrease of 7.0% compared to the previous year for the first time in the history of the Korean national innovation system. There was also a 6.3% of decrease in the number of researchers in the same year.

Looking into major actor groups, the industry was most severely influenced by the IMF jurisdiction. Compared to 1997, it decreased R&D investment by approximately 9.9% and research investment by about 11.6% in 1998 also for the first time in Korea's industrial development. The Korean public research sector was also affected by the deep recession, but not as severe as in industry. However, it had to lay off many researchers in this period so that there was a 17.1% of decrease in the number of researchers in 1998. However, academia was not affected by the recession, even though there was a slight decrease in R&D expenditures and manpower. Overall, the deep recession under the IMF jurisdiction made an influence on innovation activities of the Korean industry and public research sector.

However, the impact of the deep recession on the Korean national innovation system lasted just one year, i.e. 1998 or at the longest, two years until 1999. By the turn of the new century, Korea regained its dynamism and innovation capabilities. In 1999, just less than 2 years after the IMF jurisdiction, the Korean national innovation system regained its innovation vitality to the

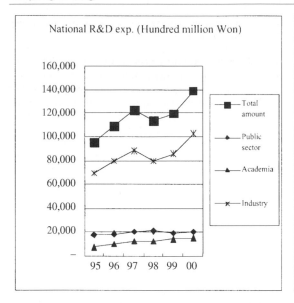 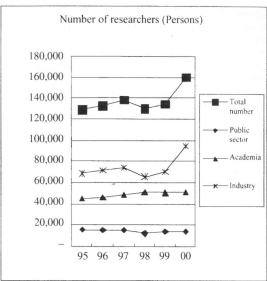

Figure 1. Impact of IMF jurisdiction on the Korean national innovation system.

level of 1997 before the IMF jurisdiction. In particular, 2000 showed much stronger innovation capabilities compared to those before the IMF jurisdiction. This implies that the unexpectedly quick overcoming of the recession was mainly due to the strong increase in innovation capabilities in the Korean national innovation system.

Many studies argue that in the recession period organizations including the *nation* have a tendency to decrease their R&D and innovation capabilities that are not directly related to short-term performance. However, this was not the case for Korea. Korea did not decrease its investment in R&D and innovation activities and even increased it to a remarkable degree. For example, the Korean government increased its annual R&D budget approximately 15% per year after the IMF jurisdiction, even though it had to retrench its annual budget significantly. In fact, just after the IMF jurisdiction, there was a consensus in Korean society that the deep recession in that period came from the low level of innovativeness of the Korean economy, i.e. the low efficiency of the Korean national innovation system and the boosting technological innovation is the only way to overcome the economic crisis. As a result, Korean society made a great effort to improve the efficiency and effectiveness of the Korean innovation system by increasing R&D investment and refining institutional frame conditions of the innovation system.

It is interesting that the Korean industry, which had been most strongly influenced by the recession among innovation actor groups, increased its innovation capabilities to a remarkable degree after the IMF

jurisdiction. In 2000, Korean industry reported a 20.5% of increase in R&D expenditures and 33.9% increase in the number of researchers compared to the previous year. This indicates the high level of dynamism and innovativeness of Korean industrial companies. Having experienced the difficulties in international markets in the middle of the 1990s, Korean companies recognized the importance of technological innovation in enhancing their competitiveness and made a greater effort to increase their innovation potential. Such efforts to increase innovativeness have resulted in the strong competitiveness of Korean industrial companies in international markets nowadays.

Conclusions

Korea's development has been based on its innovation capabilities. In this chapter, we investigated how Korea has accumulated its innovation potential since the beginning of its industrialization. Special focus was placed on the second half of the 1990s because the Korean economy was under the deep recession in this period. We analyzed whether innovation capabilities were diminished in this period and how important a role innovation played in Korea's overcoming the recession. Some of results can be summarized as follows.

First, in about three decades, Korea could formulate and implement a competent national innovation system comparable to that of advanced countries. Due to the short history of its economic development, Korea's national innovation system had many structural weaknesses. But Korea has overcome these weaknesses and now can implement a relatively competent and dynamic national innovation system. Figure 2 shows the historical development of Korea's national innovation system according to major actor groups, i.e. academia, public research sector and industry. In the 1960s, Korea did not have a national innovation system at all, and only one public research institute, the Korea Institute of Science and Technology (KIST), established in 1966. In the 1970s, only the public research sector was operating in the national innovation system as there were 13 major public research institutes in major technological and industrial areas. There were no industrial innovation activities and academia had no R&D capabilities.

However, Korea could formulate a national innovation system in the 1980s because industrial companies started to increase its R&D and innovation capabilities. In this period, the Korean government initiated a *National R&D Program* in order to promote interactive learning between industrial companies and public research institutes. However, it was not until the 1990s that Korea could have a satisfactory national innovation system. As Korean academia secured its R&D capabilities in the 1990s, Korea's national innovation system had three major actor groups: industry, public research sector and academia. It took about three decades for Korea to have a relatively competent national innovation system. However, it is unbelievably quick because an institutional setting, especially at national level as a whole, takes an enormous amount of time and resources. This prompt establishment of a national innovation system has played a key role in enhancing Korea's economic competitiveness.

By the turn of the new millennium, Korea made a great effort to enhance the efficiency and effectiveness of its national innovation system. The Korean government has refined legal and institutional frame conditions to further accelerate R&D and innovation

activities of major innovation actors and to activate interaction between innovation actor groups. Korea has also increased its R&D and innovation investment to a remarkable degree. As a result, Korea is now one of the most R&D-investing and innovative countries in the world. This means that Korea's national innovation system has been enlarged and its efficiency has increased greatly, compared to the 1990s. Figure 2 depicts this trend in the 2000s.

Second, in the development of the Korean national innovation system, the government has played an essential role. Since the beginning of its industrialization the Korean government has implemented a series of policy measures to enhance innovation capabilities of Korean innovation actors and to activate an interaction between them in the national innovation system. This strong involvement by the Korean government has led to the remarkably quick establishment of the Korean national innovation system. Such strong involvement of the government has been possible because Korea has been historically a centralized country. Since the mid-1990s, Korean regional governments have participated in promoting innovation activities in their regions (Chung, 2002; MOST, 1999). As efficient regional innovation systems constitute a competent national innovation system (Chung, 2002), this strong involvement by Korean regional governments in addition to the central government will make a great contribution in enhancing the performance of the Korean national innovation system.

Third, public research institutes have played a very important role in the development of the Korean national innovation system. As discussed earlier, Korea has established its innovation system based on the public research sector. The public research sector has been a very important policy tool for Korea as a developing country that tries to establish its own

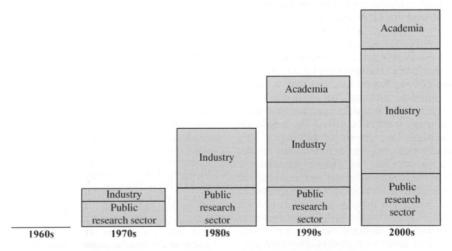

Figure 2. Development of Korean national innovation system.

national innovation system. However, the Korean public research sector has had to re-orient its mission, and role, since the beginning of the 1990s because industry has increased its innovation capabilities since the 1980s and academia has accumulated its R&D capabilities since the beginning of the 1990s. Considering that the public research sector has been playing a pivotal role in the national innovation system, its strategic re-orientation based on rapidly changing technological and economic environment is indispensable for the competent national innovation system for Korea. Based on the strong innovation capabilities of public research institutes, Korea could formulate and implement an excellent national innovation system, which consists of academia, public research sector, and industry (see Fig. 2). Only a few advanced countries, e.g. Germany, the Unites States, and France, have such a balanced national innovation system.

Finally, the unexpected economic crisis at the end of the 1990s did not have a severe impact on the Korean national innovation system. The recession under the IMF jurisdiction in the fall of 1997 did influence innovation potential of the Korean national innovation system. However, the deep recession affected the Korean national innovation system for just one year, i.e. 1998 or at the longest, two years until 1999. By the turn of the new century, Korea regained its dynamism and innovation capabilities. In 1999 just one and half years after the IMF jurisdiction, the Korean national innovation system regained its innovation capabilities to the level of 1997 before the IMF jurisdiction. In particular, 2000 showed much stronger innovation capabilities compared to those before the IMF jurisdiction. This implies that the unexpected quick overcoming of the recession was due to the strong increase in innovation capabilities in the Korean national innovation system. Korean industry, which was most strongly influenced by the recession among innovation actor groups, increased its innovation capabilities to a remarkable degree after the IMF jurisdiction.

As a whole, Korea is a very innovative and dynamic country. It has established and implemented a relatively competent and dynamic national innovation system in a very short period. Korea has invested a lot of resources in order to enhance the efficiency of its national innovation system and increase the innovation capabilities of major innovation actors. The innovation system as a whole and its major actors are very innovative and produce a lot of innovative results, products and so on. We can ascribe the dynamic development of the Korean economy in this globalized economy to the efficient national innovation system. In the meantime, Korea has made a great effort to learn from the successful national innovation systems of advanced countries, and finally it can have a relatively competent national innovation system. However, considering the hectic competition between countries to attain and implement a better national innovation system in the world, Korea should continue to learn from other advanced national innovation systems.

References

Bruder, W. & Dose, N. (1986). Forschungs- und technologiepolitik in der bundesrepublik Deutschland. In: W. Bruder (Ed.), *Forschungs- und Technologiepolitik in der Bundesrepublik Deutschland* (pp. 11–75). Opladen: Westdeutscher Verlag.

Byun, H. Y. (1989). Industry. In: H. Y. Byun (Ed.), *The Korean Economy* (pp. 263–290). Seoul: Yoopoong Publishing Co. (Korean).

Chung, S. (1996). *Technologiepolitik für neue produktionstechnologien in Korea und Deutschland.* Heidelberg: Physica-Verlag.

Chung, S. (2002). Building a national innovation system through regional innovation systems. *Technovation,* **22**, 485–491.

Chung, S. et al. (2001a). *Improvement of the role of government-sponsored research institutes and the operation of research councils.* Seoul: Presidential Council on Science and Technology (Korean).

Chung, S. (2001b). Knowledge Diffusion in Korean society. In: P. Banerjee & F–J. Richter (Eds), *Intangibles in Competition and Cooperation: Euro-Asian Perspectives* (pp. 239–258). Hampshire, New York: Palgrave.

Edquist, C. & Lundvall, B-A. (1993). Comparing the Danish and Swedish systems of innovation. In: R. R. Nelson (Ed.), *National Innovation Systems: A Comparative Analysis* (pp. 265–298). Oxford: Oxford University Press.

Ergas, H. (1987). The importance of technology policy. In: P. Dasgupta & P. Stoneman (Eds), *Economic Policy and Technological Performance* (pp. 51–96). Cambridge: Cambridge University Press.

Ewers, H-J. (1990). Marktversagen und politikversagen als legitimation staatlicher forschungs- und technologiepolitik. In: H. Krupp (Ed.), *Technikpolitik Angesichts der Umweltkatastrophe* (pp. 147–160). Heidelberg: Physica-Verlag.

Freeman, C. (1987). *Technology policy and economic performance: Lessons from Japan.* London, New York: Pinter Publishers.

Freeman, C. (1988). Japan: A new national system of innovation? In: G. Dosi et al. (Eds), *Technical Change and Economic Theory* (pp. 330–348). London, New York: Pinter Publishers.

Fritsch, M., Wein, T. & Ewers, H-J. (1993). *Marktversagen und wirtschaftspolitik: Mikroökonomische grudlagen staatlichen handels.* München: Verlag Franz Vahlen.

Gielow, G., Krist, H. & Meyer-Krahmer, F. (1985). *Industrielle forschungs- und technologieförderung—diskussion theoretischer ansätze und ihrer empirischen evidenz.* Karlsruhe: FhG-ISI.

Johnson, B. (1992). Institutional learning. In: B-A. Lundvall (Ed.), *National Systems of Innovation: Towards a Theory of Innovation and Interactive Learning* (pp. 23–44). London, New York: Pinter Publishers.

Kim, L. (1993). National system of industrial innovation: Dynamics of capability building in Korea. In: R. R. Nelson (Ed.), *National System of Innovation: A Comparative Analysis* (pp. 357–383). Oxford: Oxford University Press.

Kim, S. et al. (2001). *New roles of national R&D programs and a shift of funding pattern.* Seoul: Ministry of Science and Technology (Korean).

Korea Institute of Science and Technology (KIST) (1994). *The 25 Years' history of KIST*. Seoul: KIST (Korean).

Korea Industrial Technology Association (KITA) (Each Year). *A white paper of industrial technology*. Seoul: KITA (Korean).

Lundvall, B-A. (1988). Innovation as an interactive process: user-producer relations. In: G. Dosi et al. (Eds), *Technical change and economic theory* (pp. 349–369). London, New York: Pinter Publishers.

Lundvall, B-A. (Ed.) (1992). *National systems of innovation: Towards a theory of innovation and interactive learning*. London, New York: Pinter Publishers.

Majer, H. (1973). *Die 'technologische lücke' zwischen der bundesrepublik Deutschland und den vereinigten staaten von Amerika: Eine empirische analyse*. Tübingen: J. C. B. Mohr.

Martin, B. R. & Irvine, J. (1989). *Research foresight: Priority-setting in science*. London, New York: Pinter Publishers.

Meyer-Krahmer, F. & Kuntze, U. (1992). Bestandsaufnahme der forschungs- und technologiepolitik. In: K. Grimmer, J. Häusler, S. Kuhlmann & G. Simonis (Eds), *Politische Techniksteuerung* (pp. 95–118). Opladen: Leske und Budrich.

Ministry of Science and Technology (MOST) (Each Year). *Report on the survey of research and development in science and technology*. Seoul (Korean).

Ministry of Science and Technology (MOST) (1990). *Introduction to Science and technology Republic of Korea*. Seoul.

Ministry of Science and Technology (MOST) (1991). *Annual report*. Seoul (Korean).

Ministry of Science and Technology (MOST) (1987). *Five years report of the 'National R&D Program'*. Seoul (Korean).

Ministry of Science and Technology (MOST) (1999). *Regional S&T annual report*. Seoul: STEPI/MOST (Korean).

Nelson, R. R. & Winter, S. G. (1977). In search of useful theory of innovation. *Research Policy*, **6**, 312–329.

Nelson, R. R. & Winter, S. G. (1982). *An evolutionary theory of economic change*. Cambridge, MA: Harvard University Press.

Nelson, R. R. (1988). Institutional supporting technical change in the United States. In: G. Dosi et al. (Eds), *Technical Change and Economic Theory* (pp. 312–329). London, New York: Pinter Publishers.

Nelson, R. R. (Ed.) (1993). *National system of innovation: A comparative analysis*. Oxford: Oxford University Press.

OECD (1996). *Reviews of national science and technology policy: Republic of Korea*. Paris.

Patel, P. & Pavitt, K. (1994). The nature and economic importance of national innovations systems. *STI Review*, 9–32.

Porter, M. E. (1990). *The competitive advantage of nations*. New York: The Free Press.

Roobeek, A. J. M. (1990). *Beyond the technology race: An analysis of technology policy in seven industrial countries*. Amsterdam: Elsevier Science Publishers.

Schumpeter, J. A. (1934). *The theory of economic development*, Cambridge, MA: Harvard University Press.

Science and Technology Policy Institute (STEPI) (1991). *Evaluation of national R&D program (1982–1989)*. Seoul (Korean).

Song, H. et al. (2001). *Survey on government-sponsored research institutes and improvement of their operations*. Seoul: Office for Government Policy Coordination (Korean).

Vickery, G. & Campbell, D. (1989). Advanced Manufacturing Technology and the Organization of Work. *STI Review*, (December), 105–146.

The International Handbook on Innovation
Edited by Larisa V. Shavinina

Regional Innovations and the Economic Competitiveness in India

Kavita Mehra

National Institute of Science, Technology and Development Studies (CSIR), India

Abstract: In today's world of global economy with no territorial boundaries, the regional resources and capabilities are also playing a crucial role. The tacit knowledge based innovations are territorial specific because of its embodiments in individuals, in its social and cultural context. This fact may be responsible for not making the globe a homogenous entity and thus, let the developing nations retain their identity. The chapter highlights the cases of regional innovations from diverse locations in India, true to the region and bound to local culture and resources. Innovations make the enterprises economically competitive and sustainable and thus, in specific cases, preserve local art and craft as national heritage.

Keywords: Innovation; Tacit knowledge; Regional innovation; Learning by doing; Economic competitiveness; Entrepreneurial innovation; R&D based innovations.

Introduction

In the present era of global economy, national boundaries are diminishing leading to movement of resources not only across regions of a nation but also across countries. However, the aspect of regional capabilities has too gained importance in the spatial framework of economic competition. The awareness of the social embeddedness of economic interaction (Grabher, 1993) has given a further impetus to the recognition of the regions as a main territorial framework for learning and knowledge based economic growth. The core argument is that tacit knowledge—with its crucial role in innovation—is highly territorial specific because of its embodiments in individuals, its social and cultural context and therefore it requires proximity (Geenhuizen & Nijkamp, 2000; Nonaka et al., this volume; Pan & Scarbrough, 1999). Badaracco (1991) refers to the tacit knowledge as embedded knowledge, embedded in the psyche of the individual or the culture of organization (or region) and relationships amongst individuals. Hildreth et al. (2000) describes the tacit knowledge as soft knowledge comprising experience and work knowledge and that resides only within individuals. Nonaka (1991) considers tacit knowledge as highly personal and that it is deeply rooted in action and in individual's commitment to a specific context

(see Nonaka et al., this volume). The exploitation of that tacit knowledge along with its interaction with the new knowledge (flowing into through various channels) in a specific region leads to economic competence of that region. The realization of this very fact perhaps may be responsible for not making the globe a homogenous entity in the coming period. This way it may help developing nations retain their identity for which they are striving very hard.

The economic gaps among regions do reflect differences in the regions ability to compete, which increasingly depends upon the innovative capacity of production units and regional systems as a whole (Fagerberg & Verspagen, 1996; Fagerberg et al., 1997; Neven & Gouyette, 1994; Quah, 1996). The process of technological accumulation takes place at local or regional level, even in the era of globalization, and the technological spillovers tend to be highly concentrated at geographical level (Evangelista et al., 2001, see Gibson et al., this volume). This is why regions have become fundamental units of analysis in the cost/benefit evaluation of the economic integration and in the studies, which look at the process of economic convergence (or divergence) in Europe (Evangelista, 1996; Evangelista et al., 2001).

Concept of Innovation and its Relevance in Regional Economic Development

This section elaborates upon the concept of innovation and related issues. Innovations are new ideas in the business of producing, distributing and consuming products or services (Beije, 2000). The most general definition of innovation is 'the process to undertake a change in one or more of many aspects of production, distribution, and consumption of economic goods' (Beije, 2000, p. 22). Schumpeter (1961) has made a classification of innovation that is more practical, consisting of: (1) new products; (2) new processes; (3) ways to penetrate new markets; (4) new supply sources or distribution methods; and (5) industry. It contains all basic categories of ways in which the entrepreneur can earn money by undertaking new activities. In addition to that, there are innovations in other areas beyond these five categories, such as use of new management practices and organization structure, developing and retaining skilled personnel, organization culture, securing financial resources, managing interface with government and other external agencies (Mehta & Joshi, 2002). Wilhelm (this volume) notes that innovations are the results of innovation processes and innovation activities; innovations exist as products, processes and organizational structures.

The importance of innovation at micro level lies in the increase of profits or market share by individual firms. The importance of innovation at macro level goes further. Technical change is initiated by innovations, but its effect ultimately depends on the extent to which one innovation may stimulate a whole chain of subsequent innovations. An important role in this respect is played by the relationship between the creator and users of new technology (Beije, 2000). Innovator can be a person or an organization. The change of any type (e.g. technological) he brings forward is basically a new knowledge.

Economists usually put the emphasis on innovation, not on invention. Innovations have a larger sphere to show the newness through so many activities of the production unit. It may be the combining of the existing activities in new ways in a more beneficial way (Beije, 2000). Therefore, innovations largely reflect the role of innovator, which is more important in terms of economic benefits. Schumpeter (1961) first emphasized the role of innovator, considering it to be crucial for the economy in the process of technological change. Inzlet (this volume) through pilot innovation surveys in Hungary, found that the role of innovations in the Hungarian economy is low. She concluded that the reason for this is the type of role played by innovators. Innovators rely more on market needs than on R&D inputs through links with academia or other firms.

Drucker (1985) discusses innovation in relation to entrepreneurs. Innovation is a specific tool of entrepre-

neurs, the means by which they exploit change as an opportunity for different business or different service. Entrepreneurs need to search purposefully for sources of innovation, the change and their symptoms that indicate opportunities for successful innovation. They need to know and to apply the principles of successful innovation (Drucker, 1985).

Innovation is the act that endows resources with a new capacity to create wealth (Drucker, 1985). Innovation creates resources and also changes the value and satisfaction obtained from the resources by the consumer. Drucker (1985) discussed this in the case of Japan wherein the country became the big economic power by adopting creative imitation, importing the low cost technology and adopting it instead of undergoing increased R&D and new product development. He considered this phenomenon as social innovation and concluded that the Japanese success was primarily based on social innovation.

With poor infrastructure in R&D and limited scope for higher education, entrepreneurs in developing countries—even though they may be imitators—play an important role. They provide a distinct thrust to economic growth. Regional disparities create social and political problems that must be addressed, especially in countries with federal constitution where 'regions' and the gaps among them correspond fairly closely to states or provinces. National economies are aggregations of regional economies, which vary widely in the degree of integration among them. In some countries some regions are more closely integrated with the world economy than with other regions in the same national economy (Higgins & Savoie, 1988). One can substantiate this fact by taking the case of India as nation and Bangalore & Hyderabad as the regions within India, which are very well linked to the world economy through information technology. Similarly, Ahmedabad is specialized in gems, jewelry and in textiles also, the region is well linked to the world economy. There are numerous other examples of regional developments in India. Regions are integral part of the structure of national economy and thus ultimately of the global economy.

National economic systems cannot be understood nor effective policies formulated and plans can be made, without understanding the regional structure. If one considers regions as such regional economic systems, one may assert that regions can only survive if their participating systems (such as political system and legal system) are also able to survive. All participating systems are actors in the whole system (the society) and are influenced by its environment (Brandt et al., 2002).

In parallel, the worldwide economic system is made up of many sub-systems that are connected with each other by customer-supplier chains (Brandts et al., 2002; Fig. 1). The system's ability to survive depends on the ability of sub-systems to survive. The nation wide

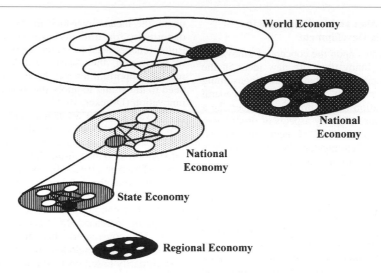

Figure 1. Role of the regional economy in the global economy

Source: Brandt et al., 2002.

economic systems have started to cooperate with each other through international networks. As a matter of principle, these different (nationwide) systems and their sub-entities are to be understood in their relations as of equal weight or equal importance for their survival. In this context, the overall aim of economic politics should be to assist regions and regional enterprises to survive.

According to Higgins & Savoie (1988), accelerating growth of national economy as a whole requires an attack on the problems of retarded regions. Some kinds of resource management—natural and human—are best studied and executed at the regional level because the package of resources involved are best defined in terms of space. Examples are river valleys, mining and forest areas, drainage systems, transport systems etc. As a specific example, Ganga river basin (in India) management is best understood by the regions in the north of India along the river. Each region (as space) has its enrichment or scarcity of resources and that region over the period develop specialization to utilize or manage the uniqueness.

There could be various approaches to regional development. Broadly, it could be regional resources based development, more relevant for present period of global economy, when one can capture the resources from anywhere and make use of it for the regional development. Example of the later situation is the information technology based regional developments (e.g. Bangalore & Hyderabad in India) when the technological achievements at global level have been put together with the local infrastructure development so that those cities have become linked to the global economy. The regional comparative advantage approach has been popular among policy makers

(Higgins & Savoie, 1988). It represents, potentiality at least, a no-cost regional development policy. Regions will grow and specialize according to what they can do best, and government will not waste funds attempting to reproduce the industrial structure of highly developed regions in all slow growth areas. It promises self-sustaining economic activity (Higgins & Savoie, 1988). The same concept has been used in the present chapter in discussing various models of innovations in India.

In today's world, the need remains that to fully realize the regional potentialities and to make them economically competitive, even if such regions are not directly linked with the global economy, they have to be linked to the national economy in a more impact-full way. For that, regional innovations—specific to local resources and skills embodied in the culture—play a crucial role. The following account discusses a few of the many models of regional enterprise innovations experienced in India.

Models of Regional Innovations in India

India is a vast country comprising varied regions with a variety of resources. Particularly art, craft and culture as well as the natural resources, are specific to each region. Regional innovations are thus, an outcome of culture, resource-based activities and these have evolved due to limitation in supply of resources, sustenance in the market, diffusion of culture and art across regions and learning by doing. There are cases of both types: entrepreneurial efforts, as well as organized attempts due to the flow of R&D from public funded research institutions to upgrade regional capabilities. It is the blending of traditional capabilities with modern technologies to make the regional occupations more remunerative ones by providing efficient production, innovative products or processes, new designs and new markets. The cases discussed for entrepreneurial type of innovations reflect the way in which enterprises have been benefited by the changes (i.e. innovations) introduced by entrepreneurs (see Conceicao & Heitor, this volume). The chapter discusses three types of regional innovations in India as cases building towards regional economic competitiveness.

(1) Innovations in small scale enterprises set up in the Gujarat region, with and without any formal R&D efforts.
(2) Innovations in the informal sector—the agriculture, in and around Delhi (Uttar Pradesh)—non-planned and non-R&D based activity.
(3) Innovations based on the inputs from public funded R&D laboratories (Council of Scientific and Industrial Research—CSIR). Modernization and up-gradation of traditional art and craft.

Innovations in Small Scale Enterprises—Gujarat Region

Gujarat is a major centre for innovations happening at the grass root level with strong cultural influence. The enterprising behavior is largely attributed to typical culture comprising of traditions, values beliefs and attitudes of the region. The growth of the Indian textile industry without any tariff protection, which is competitive with British textile industry, is proof of the vigor of Indian entrepreneurs from Gujarat. Gujarati culture has put a prestige value on business. These factors, together with Western ideas and Western contacts, created a new ferment in Gujarat's economic development in the 19th century, which still continues today. They provide a distinct thrust to economic growth: for example, the modern textile industry in Gujarat—and especially in Ahmedabad—was developed as an imitation of a British textile industry. It has contributed a great deal to the Indian economic growth for the past 150 years. The entrepreneurial class consolidated itself socially, as well as economically, and relatively small, close knit agencies, often supplemented by caste ties, made further expansion possible. As managing agents, they applied the profits of one mill to the requirements of new ones and solicited public and private deposits to cover the major capital requirements.

The creative genius of Indian entrepreneurs does not lie in inventing new products as much as in inventing new forms of organization and management (such as managing agency system) and in creative imitation and modification of foreign products. It is the same principle in Japan, the imitation of relevant outside R&D—as mentioned earlier—has been considered as social innovations by Drucker (1985). Recognizing the problem, Bert & Hoselitz (1963) remarked that in an underdeveloped economy, in addition to the Schumpetarian innovators, even imitator-entrepreneurs have a distinct role to play. They provide, he feels, a fillip to the process of economic growth, sometimes having as strong or perhaps even a stronger impact on economic growth as real or alleged innovators. Even the multiplicity of entrepreneurs—lacking in creative genius of inventors, but possessing all other characteristics of successful entrepreneurs—is desirable for underdeveloped economies.

In a broader definition of innovation, relatively novel ideas, which are novel in local context, can also create competitive advantage and large-scale economic and social change in the region. In Gujarat, there are innovations of such kind, which fit in Schumpetarian (1961) framework of five categories of innovations, namely product innovation, process innovation, market innovation, supply source innovation, and innovation regarding industry organization. Entrepreneurial ventures in Gujarat rarely have formal R&D set-ups, but a collaborative relationship is established with research and academic institutes or laboratories to make available testing facilities and information available. Some entrepreneurs have used semi-finished and unsuccessful research of others. Gujarat, however, needs to develop innovative ways of managing the R&D efforts without having formal departments.

Entrepreneurs have tested the feasibility of new ideas before investing in them through their informal networks. It was observed further that since childhood their family members, who are in this particular business, train them and skills are passed on from generation to generation. Mehta & Joshi (2002) describe the culture and region specific characteristics of entrepreneurship and innovativeness.

Cases of Enterprise Innovations in Small and Medium Enterprises in Gujrat

Gems and jewelry. Diamond processing is a traditional craft and one of the oldest industries in India. The main features of this industry are low investment, low level technology used and high employment. Gujarat accounts for more than 80% of diamond processing business. The processing is done through indigenously manufactured and manually operated machines. R&D efforts during the last six years by two Indian Companies, Sahjanand Laser Machines and Gitanjali Laser House, have resulted in the development of laser processing machinery for diamond processing. Use of these machines will reduce manual error considerably and will also decrease the chance of weight loss. Therefore, this technology has been attracting more and more people as it is eventually reflected in better prices. Indian R&D in this field has been able to give the diamond an advantage on the export front by giving an edge in quality, product mix and price. In the last ten years alone, the diamond industry exports have gone up from US$830 million to US$4400 million.

The diamond and gems entrepreneurs undertook process innovations to make use of locally available skilled labour or different quality of locally available raw material, to reduce costs or improve quality. They developed laser kerfing and sawing machines to make it labour-intensive in place of capital-intensive machines imported from abroad. This is a cost-effective, low capital process of diamond polishing in a local situation where skilled labour is easily available and cheap. Israel may use fully automatic, innovative diamond machinery, while India uses non-perfect diamonds like macles, distorted crystals, 'near gems' or 'near industrials' processing which are not easily amenable to automation. Gujarat entrepreneurs have developed machinery to suit these applications and also to take advantage of cheap skilled labour.

Gujarat entrepreneurs also succeeded in structural innovations. Structural innovations are how work is organized, which can enhance competitive advantage of an enterprise. For example, entrepreneurs in Surat Diamond Industry have a lean organization structure with labor on contract allowing them greater flexibility in terms of workload and skills. Also, owner-manager role of the entrepreneurs allowed innovation in cost control, quality assurance and market development.

Rajkot Oil Engine Industry. The first manufacturing unit of the diesel engine, was created and developed by Umakant Pandit in 1951. Today, Rajkot has found a place among the leading industrial centers of the country. Over the last 40 years, it has become one of the main centers for production of diesel engines and its spare contribution is substantial to the total production of oil engines in India as a whole. Rajkot has made progress in the export of diesel engines and spares worth more than Rs. 500 million to the Middle East and several African countries.

Forty years ago, there were approximately 25 units in the field of engineering, There was an acute shortage of diesel-engine spare parts. Those engineering units which were engaged mostly in repairing and over-hauling of the different types of diesel engines besides casting and machining of iron and non-iron metal. Many entrepreneurs were also importing certain parts of diesel engines, manufacturing some of the parts here and assembling the slow-speed vertical diesel engine. The machines at that time were imported from Lister, U.K. In time, gradually, the production of almost all the different parts of the oil-engines commenced in Rajkot itself. This was the beginning of production of diesel-engines as a whole in Rajkot. Umakant Pandit was the pioneer in establishing the foundation stone of this industry in Rajkot. Then a spree of entrepreneurs came into the picture, like P. M. Diesel Ltd., which is now one of the leading diesel engine manufacturers in the world. Today there are approximately 4,000 small-scale industrial units engaged in the production of diesel engines, spares and engineering goods.

Rajkot contributes more than 70% to the production of slow-speed diesel engines (Lister type). The annual production of 150 units engaged in the manufacturing of Lister type of diesel engines is more than 240,000 of slow-speed diesel engines. The Lister type engines are mainly used in agriculture. Of late, the production of Peter-type engines has also been increasing, with the annual turnover of about US$25 million. The diesel engine industry provides employment to more than forty thousand workers.

Entrepreneurship in Rajkot today is a combination of two styles. While the astute and tenacious spirit of the first generation entrepreneurs is unmistakable, there is a new wave of professionalism amongst the second generation of entrepreneurs who are opening new windows of opportunity. The people here are by nature entrepreneurial, hard working and have keen engineering skills. And added with this also comes the commercial sense, which creates a critical combination of success. All these skills have enabled Rajkot to produce and sell at competitive prices all over the country as well as to overseas market.

A major form of supply source innovation is vertical integration where an enterprise manufactures its own raw material. This is prompted by shortage of supplies. In case of Rajkot Diesel Engine Manufacturers, a unique system for supply sources is developed. In Rajkot, when the industry started, everything was sourced from outside by way of imports or purchased in other parts of India. Today Rajkot has more than 600 ancillary units for the supply of raw material, components and assemblies. This is a classic case of ancilliarization and vendor development.

The product innovation identified in Gujarat mainly relates to product modifications of existing products and introduction of new products locally. Rajkot Diesel Engine manufacturers are other examples where they adopted Peter type high speed diesel engine designs from Lister type designs to suit local changing environment, which helped them to capture 70% share of diesel engine markets in India. Besides this, entrepreneurs in the diesel engine industry at Rajkot are in touch with the latest information, they maintain quality and they also know the market pulse too. The market needs new and innovative products every day. Modernization of products and plant is the only alternative to that. And the entrepreneurs take all the steps to modernize their plants and machinery.

The structural innovation made by Gujarati entrepreneurs in engineering, gems and jewelry, ceramics, brass parts and cooperative entrepreneurship is the formation of clusters and networking. In both industrialized and developing countries there is increasing evidence that clustering and networking can help small and medium scale enterprises (SMEs) to boost their competitiveness (see Swan et al., this volume). Groups of firms working together are better able than isolated firms, to rapidly develop product niches, access export markets and offer new employment opportunities. The cluster development concept is important in the establishment of cooperation linkages between firms, their raw material and equipment suppliers, sub-contractors, customers and service providers. The cooperation of specialized firms enables collective efficiency and enhances the ability to innovate new processes and products.

Innovation in the Agriculture Sector (Delhi and Nearby Areas, Uttar Pradesh): Emergence of Floriculture in Open Field Conditions

This case is to illustrate entrepreneurial skills of Indian farmers and their innovativeness in capturing the market demand for good quality cut flowers. It depicts the diffusion of new technology in a farming community, so as to lead to a technological change, is dependent on culture-based informal channels of communication and the tacit-knowledge driven entrepreneurial spirit of a few. Adoption of floriculture amongst traditional crops like wheat and sugarcane growing farmers was an imitation (innovation) of 'green house technology' for their open fields which was based on the exploitation of tacit-knowledge of farming, learning by doing, awareness amongst the farmers and other similar reasons. All these developments are proving farmers as innovators and responsible for bringing farm level technical change.

Entrepreneurial Skills of Farmers

Agriculture is the backbone of developing economies. In India, agriculture is the largest economic activity and in the top position in providing work and jobs to people. The 'New Strategy' for agricultural development in India was initiated in 1966, which in essence called for the implementation of High Yielding Varieties Programme (HYVP). These technological breakthroughs brought spectacular changes in agricultural production. The large increase in production of food grains recorded after 1966–1967 is described as 'Green Revolution' (Sadhu & Mahajan, 1985).

The entrepreneurial spirit of farmers can be regarded as a key to this revolution; perhaps they knew the potential of their lands and need for the adoption of new technology. The response of farmers in the use of HYVs and other inputs had been enthusiastic. They were confident of their tacit-knowledge of farming and were willing to capture the opportunity for change.

The technological change brought a shift in the minds of illiterate but 'scientifically informed' farmers, making them realize the importance of high quality seeds and other farming inputs for increasing the productivity of same land. In the Indian history of agriculture, the role of 'green revolution' is a landmark as it has made India more than self-sufficient in wheat grain production in spite of continuous population growth. However, the higher productivity of land gained by adopting HYV seeds of wheat has now produced a saturation in the benefit level to farmers. The 'alert farmers' of India have thus remained on the lookout for other opportunities for higher economic returns. Popularity of the floriculture sector amongst traditional crop (wheat and sugarcane) growing farmers appears to be one such case during the early 1990s.

During those years, the focus of the Indian government was to promote green house cultivation of flowers for export purpose so as to capture the international demand of cut flowers and add to India's foreign exchange earnings. At the same time, policies towards liberalization of economy were also introduced in the country. As a result, a number of multinational companies were selling superior quality planting material for flowers and also the greenhouse technologies to start the green house enterprises by Indian entrepreneurs. Volumes of greenhouse based produce, meant for export, started appearing in domestic markets too and Delhi market showed high demand. The moment when some traditional farmers of a village Khatoli—nearly 100 km away from Delhi—learnt of the availability of imported bulbs of superior quality flowers and the demand of superior quality cut flowers in the market, they decided to venture into it. It was not for 'green house' based production. It was purely to replace their traditional crops (wheat and sugarcane) in the open fields with floral cultivation. Indian farmers went ahead in the business of flower cultivation by themselves without any guidance and experience of growing exotic bulbs in open field conditions. They started with small plots of cultivation for experimentation. Once they succeeded in growing the crop and selling it, they expanded the area of open field cultivation. For marketing, initially farmers faced exploitation by the middlemen as cut flowers are perishable commodity and they do not have any controlled temperature arrangements to store flowers. So, the middlemen were taking away the produce at throw away prices for sale to retailers or bulk buyers and were earning substantial profits. Ultimately, farmers succeeded in locating and establishing the market for direct sale in Delhi. It was a newly created market by the flower growers from Delhi *and* the nearby region, Uttar Pradesh. Achievement in this front had also followed an innovative route (Mehra, 2002). Gradual success *in establishing the market for the sale of cut flowers*, encouraged the adoption of floriculture by more and more farmers. By now, most of them have devoted a major portion of their land to floriculture activity, and some of them have even gone ahead by taking more land on lease for flower cultivation. Over this period, they exhibited 'learning by doing' innovations through several examples such as regulating the flowering time of the crop, manipulation of more number of crops of tuberose per year or in storage of bulbs etc. We can thus say that farmers have acted as innovative entrepreneurs.

Technical Change in Socio-Cultural Context

It was not a government plan to promote open field cultivation of superior (from imported bulbs) cut flowers for the domestic market. Only the alertness of the scientifically informed farmers led to the change. They came to know about the availability of planting material through each other. The fear of competition is not felt amongst them for marketing of their agricultural produce. Marketing of cut flowers is a difficult task as it is a perishable commodity and they do not have any air-conditioned storage facility. Even then the farmers of the same village were sharing their knowledge. This is the reflection of the culture and the practice of villages of being together and sharing. Most of the growers cum sellers of Khatoli village have gained information/knowledge for floriculture only from their colleagues. As a result, today the Khatoli village is the largest producer of tuberose for Delhi and nearby markets. Khatoli is a village in Muzafar Nagar district of the state Uttar Pradesh in the North of India. Like industrial clusters, Khotoli has emerged as the tuberose cluster during the 1990s. In the context of agriculture, the term cluster has perhaps been used for the first time describing cultivation of flowers in Khatoli (Mehra, 2002) due to the following reasons. First, in Khatoli, almost all farmers are engaged in floriculture activity, mainly specializing in tuberose. Second, farmers have become such experts in growing tuberose that now a substantial supply of flower from Khatoli is available during the winter season. Before, vitually no tuberose from Khatoli or nearby areas was seen in the winter months in the Delhi market. Finally, until the beginning of 1990s, Calcutta was a big supplier of tuberose to Delhi market. At present, the enormous production of tuberose in Khatoli has replaced the supply from Calcutta to Delhi and, in addition, the flowers are also sent to other regions for sale. This all happened due to the rapid diffusion of the popularity of the flower cultivation amongst the farmers. Reason for the popularity was the more profitability in adoption of floriculture over growing conventional crops—wheat and sugarcane. An indicative comparative economic analysis of the investments/returns in floriculture versus wheat and sugarcane crops clearly explains the reasons of adoption of floriculture by farmers (Mehra, 1999). It is the exploitation of tacit-knowledge of farming, learning by doing, awareness amongst the farmers, informal channels of communication, the faith of the community, and in each other are some reasons to prove farmers are innovators and responsible for bringing technical change to farming.

Thus, an Indian farmer, though not formally educated, is alert and sensitive to opportunities leading to higher returns. This is another type of innovation model observed in the informal sector. After green revolution, the boom in the domestic floriculture market offered the opportunities to innovative farmers to discover entrepreneurial skills, building upon their tacit-knowledge of the framing practice. They proved themselves to be successful entrepreneurs, and in some cases innovators, too. This type of enterprise is possible due to the socio-cultural system of the village. A closed-knit system of informal communication amongst farmers makes it possible to launch such a technical change at the community level. Such changes definitely have a bearing on the socio-economic status for the betterment of the region.

Modernization and Upgrading of Traditional Art and Craft Skills: Interventions of Public R&D Institutions

In a country like India, one finds great diversity in art and culture. Each region depicts a culture with specific art and craft. Performers of that art are called artisans or craftsmen. The main occupation of the rural population in India is either agriculture or art/craft work. Artisans are practitioners of tacit-knowledge— uncodified knowledge, which is passed on from generation to generation. Historically, artisans can be considered as the institutions of excellence of human skills as they have been utilizing the local resources in the best possible way for making mostly utility items showing the skill and the sense of aesthetics of artisans. The creative art of the artisans is the source of meeting their meager needs. Artisans had been playing a significant role in the rural economy. However, in recent years, one notices a declining trend among the regions. There could be various reasons for this, but the prime factor appears to be without innovation the traditional modes of operations are still being used. This has resulted in low returns, which dissuades the artisans to follow the traditional practices. Perhaps the solution to this problem lies in the incorporation of modern Science and Technology into the traditional practices. The process should happen in such a way that a blending of the new technology with the traditional one takes place. It would serve: (a) to preserve the national heritage (the human skill, the tacit-knowledge of artisans); and (b) to enhance the earnings of artisans for better living by bringing efficiency in the processes, cost effectiveness, diversity in the product lines and better opportunities for commercialization. This way it would lead to regional economic development.

There is a number of public funded R&D laboratories such as Council of Scientific and Industrial Research (CSIR) and many others, which have made conscious efforts to look into the problems of regional nature, technological or other *types*, to cater to the needs of production units of various industries.

Cases from West Bengal—Bankura

Bankura is a district of West Bengal and, apart from agriculture, its vast population is engaged in the tradition of producing handicraft products like fishing hooks, brass and bell metal products, bell mala, pottery (terra cotta), dokra, conch shells, baluchari saree, stonewares, slate carving and bamboo products. The preponderance of artisan oriented small-scale industries constitutes about 60% of the registered Small Scale Industries.

National Institute of Science Technology and Development Studies (NISTADS) set up a S&T field station at Bankura during 1982. A number of projects are being handled by this center to upgrade the technology of artisans and to promote the commercialization of artisan products. The field station of NISTADS is trying to provide help to artisans not only in terms of newer and better technology (with the help of different CSIR laboratories like Central Mechanical Engineering Research Institute in Durgapur, National Metallurgical Laboratory in Jamshedpur, and Cental Glass and Ceramic Research Institute in Calcutta) but also in terms of procurement of materials and assistance in marketing of their products.

Baluchari Saree weaving—towards modernization. Baluchari is traditional form of handloom silk saree woven in Bishnupur in Bankura district. Saree is an Indian woman dress, which is a designed cloth of 6.0 m × 1.25 m. A baluchari is characterized by artistic designs on the borders of the saree. These designs consist of motifs of different sizes depicting the sculptures made on the historical temples or monuments in India. For making one saree, 15–18 weeks time is required. The traditional process is thus very labor intensive and expensive with the limitation of modifying or rectifying the design during the process or sampling of different color combinations. The practice of weaving of baluchari type originated at 'Baluchar' in Murshidabad of West Bengal on the patronage of the nababs of Murshidadabad (Das & Mukhopadyay, 1995). They tried to portray imageries of muslim rulers, the courts and 'harems' on the spread of these sarees. During the 18th century, designs were biased by Persian miniatures. During the 20th century, designer Sri Das introduced paintings of famous Ajanta-Ellora caves as designs on baluchari sarees in Bankura. These days, paintings and sculptures of various famous temples and historical places are used as designs for weaving baluchari saree. The border of a saree even speaks the story of an epic. These days Jacquard looms are used in which intricate designs are coded and punched on long chains of Jacquard cards. As these punched cards control the movement of the warp on the loom, designs come up on the saree with greater ease.

The process of Baluchari Saree weaving. The process of baluchari making and the stage-wise labor costs involved are as following:

- *Designing stage*—Cost involved varies from Rs.3000 to 6,000 (US$600 to US$120).

- *Preparation of card-deck for Jacquard loom*: Cost involved (labor + material)—Rs.12,000 to 15,000 (US$240—300). For an exclusive design, the cost may go beyond Rs.20,000 (US$400).
- *Preparation of silk threads*—coloring, plying, sizing and polishing. The process costs around Rs.100 (US$2).
- *Weaving:* Done on hand looms normally by using two shuttles. Cost of weaving one saree—Rs.400–500 (US$8–10).

In all the four stages described above, maximum labor is involved in designing and punching of cards. The process of Baluchri saree making has the following limitations:

- Labor intensive process; limited designs; defects in the designing cannot be detected in the initial stages of sarce weaving; testing of color combinations without weaving can not be done *and* limited colors are offered to the customer. Product variation (other than saree such as scarf, neck tie or napkin etc.) is not there because the designer is tuned for designing a saree. He never thought of demand for any other product than traditional product (saree).

Since the same card deck is used a number of times (approximately 100 times), the designing cost is divided by the number of pieces produced. On average, a baluchari saree costs Rs.2,500. Every year, six to eight designs are released in the market.

Steps towards modernization of the Baluchari Saree weaving. Participation of NISTADS project team at Bankura with the artisans brought all limitations into the limelight. The survey conducted, revealed that due to the above limitations, the art was diminishing. Designers were going towards simpler motifs. The younger generation were also found not too willing to follow the family tradition due to low returns. Thus, the need for modernizing the process was realized at two levels: designing and card punching (Mehra & Mukhopadhyay, 2000):

- *At designing level:* With the advent of computerization and extensive use of CAD by professionals in designing, NISTADS made an attempt to get a software, MADHU, developed for baluchari designing by Entrepreneurship Development Centre at Indian Institute of Technology (IIT) in Kharagpur. It was to reduce the designing time and make the process more flexible in terms of the size of motifs, their arrangements on the saree and also to use designs for making diverse products. With the help of the MADHU package, it may be possible to design a full saree within two to three days. Once the software at the designing level, has been developed, a program for the sensitization of users is going on in the field. The feedback of users has gone into the improvement of the software package. At present, the new technology is showing to be acceptable by the users.

- *At card punching level:* If Auto-card punching is introduced; the cost would substantially come down as efficiency of card punching would increase by 20 times or more. Council of Scientific & Industrial Research and Department of Science & Technology, Ministry of Science and Technology are also looking into this aspect.

Brass metal craft—Dokra casting. Dokra artisans of Bankura, where 36 families in all and now the largest cluster in West Bengal, live in a small village called Bikna just outside Bankura town. In the Dokra metal casting process, craftsmen use bees wax and hot metal that permits them to give a metal shape to the creative images visualized by them as an artist. Dokra work is carried out by both, men and women. Skills by children are acquired through imitation and instruction. CSIR has found it important to improve the skills and processes for efficient and cost effective production. NISTADS field station has helped these craftsmen in the following way (Roy & Mohapatra, 2002).

- Skill up-gradation for making of alloys of different non-ferrous compositions.
- Development and application of high collapsible molding material for making the master mould.
- Introduction of direct pouring techniques that facilitates production of a bigger casting in a single firing.
- Introduction of improved finishing tools.

All these inputs are the outcome of the participation of R&D laboratories of CSIR in identifying the problems and providing solutions for the economic betterment of the craft work in the region.

Dokra craftsmen were facing a specific problem in firing of the artifacts in open furnaces that they use. There was evaporation of metal and resulted in a 20% increase in the input costs. During 2001, NISTADS in consultation with senior artists from the village, designed and commissioned five pucca furnaces in three sizes for community use. This step has been effective in creating social, economic and creative impact. New furnaces have brought a noticeable social change—better interaction amongst individuals in the community (Kochhar, 2001).

EU-India Project on *Cross Cultural Innovation Networks*

India is a developing country with limited financial resources but rich in natural resources along with diversity of cultures across various regions in the country. The European Union is a group of countries with diverse cultures and resources. India's vastness as a country is comparable to EU. To look into these issues, a special EU-India project was initiated during October 1999, sponsored by EU. The project is 'Cross

Cultural Innovation Network' and is being managed by Project Director, Prof. Karamjit Gill (University of Brighton, U.K.). The project partners include scholars from U.K. (Brighton and Wales), Germany (Aachen), Italy (Bologna), Denmark (Lyndby), NISTADS (CSIR, India), Delhi University (Delhi), Gujarat (Ahmedabad, India), and Punjab (Ludhiana, India). The project focuses on the value-added applications of university and R&D institutions research into the area of social and economical change, regional models of innovations, role of cultural aspects in innovation process, entrepreneurship skills development, and their transferability between cultures and regions. Cultural aspects in this context emphasize the social, economic and communication environments in which technology or regional art is designed and applied to solve problems. The observations made by the project partners across countries (Gill, in press) reflect remarkable variations in innovations and development across regions due to influence of regional specific cultures and knowledge based practices within India, as well as in Europe. The project is developing a 'virtual network' consisting of distributed knowledge base residing within the partner institutions. EU-India network website has been launched to bring the synergy of the knowledge of different partners and to act as a tool for sustainability and future EU-India collaborations. Through multimedia, traditional knowledge, which is in practice at the grass root level across regions in India, has been captured for the diffusion purpose across cultures. The project has resulted in bringing out a special issue of the International journal *Artificial Intelligence (AI) & Society* on entrepreneurial innovations, mainly from India for which I was a guest editor. The special issue, dwells upon some aspects like exploring the ways how to be innovative to compete in the international market—may it be educational innovations to create qualitative and competitive manpower; or industrial enterprises/clusters to bring out new products with customer orientation; or cooperatives with new organizational innovations creating new products as well as opportunities for employment at the grass root level. The percolation and diffusion of new coded knowledge and its symbiosis with tacit knowledge in the small and medium enterprises of the Indian traditional sector has been a unique factor and has been mainly responsible for the innovation and technical change in India (Mehra, 2002).

Conclusion

India is an ancient civilization with a rich heritage. It has a vast storehouse of knowledge in various fields, particularly in tacit form, but general awareness of which is inadequate. In the present scenario of globalization and liberalization, international competition is compelling the enterprises and organizations to be dynamically innovative which is only possible through a knowledge society—a learning society committed to innovation (Conceicao & Heitor, this volume; Kharbanda & Mehra, 2002). The slogan must be—innovate at all levels of human activity, be it economical, social or cultural.

The focus of this chapter was on three different models of innovations—At SME level, where the role of entrepreneur remained important. The case described regional (Gujarat) specific culture of entre-preneurship encompassing all Schumpeter's (1961) types of innovations. Innovation is associated with 'creativity' but also with initiative and risk taking (see Conceicao & Heitor, this volume) and that qualifies to be as 'entrepreneurial innovation', well observed in the first two types discussed (SME and the unorganized sector—the floriculture). The imitation of the British cotton industry (which can been referred to as social innovation) has also been observed in Gujarat and led to the establishment of a successful textile industry in that region, where the role of entrepreneur-innovator has been very significant. The ability to learn and put knowledge to work by entrepreneurs has been well observed in the process of the development of Gujarat region (see Gibson & Conceicao, this volume).

The second case highlighted, how the process of innovation in an unorganized sector took place and led to a technological change. In this case, innovators-farmers 'imitated' new technology (that can be considered as a social innovation) for greenhouse production. It reflects their entrepreneurial skills, alertness and the application of tacit-knowledge. Farmers for growing flowers mainly relied on their farming experience. Their tacit-knowledge regarding farming was the determining factor for their adoption of floriculture. They succeeded by combining that knowledge with new knowledge (information about new technological inputs such as high quality planting material etc.) the way Nonaka et al. (this volume) have described in the case of Nippon Roche. Gibson and Conceicao (this volume) have discussed the combination of tacit-knowledge with the new one for creation of wealth in the form of combining 'skills' and 'ideas'. The diffusion of change is bound to the social structure and culture, informal channels of communication, faith of the community in each other. This depicts the flavor of culture of rural farming communities that is true to India.

The third case discussed the role of formal R&D of public R&D laboratories in up-grading the skills of artisans. It was based on the targeted role of R&D laboratories of those regions to look for the resources of the regions and the problems faced to be economically competitive. Solutions to the identified problems have been provided from time to time and training and modernization have been carried out to make the occupations more profitable for the better living of crafts men and also to preserve the regional art and craft as national heritage. There are countless examples to show the way *in which* small enterprises producing

ethnic goods have undergone change—in terms of product diversification or variations in the product design and color to capture the larger markets or in the usage of raw material due to shortage of resources. All these developments happen through various channels of information (public agencies or non government organizations (NGOs) so as to make the enterprises more competitive and sustainable.

One can conclude from the above cases that in the age of global economy, the described regions also function as collectors of and repositories of knowledge and ideas and provide the underlying environment or infrastructure, which facilitate flow of learning and knowledge. Van Geenhuizen & Nijkamp (2000) discuss learning capabilities of regions as their ability to create, attract, absorb and act upon new knowledge. The cases described in the chapter reflect these issues, each region has different ways to innovate depending on its knowledge base, circumstances and available resources. Innovations and economic competitiveness in the discussed cases had no direct link with the global economy. It is tacit-knowledge, along with some new knowledge, and availability of lower order competitive advantage (e.g. use of technologies, economies of scales easy to duplicate or imitate, as described by Porter (1990) help in building up the economic competitiveness and thus, the sustainability of the region. These peculiar regional and cultural based, enterprise innovations distinguish one part of India from the other. Such patterns allow the world as a whole to maintain national identities in the era of global economy.

Acknowledgments

I would like to thanks Professor Dhawal Mehta, the Director of the GLS Institute of Business Management in Ahmedabad for sharing information on Gujarat cases discussed in the chapter. I also wish to thank the European Union for its financial support, Professor Karamjit Gill, Chief Coordinator of the EU-India Project 'EU-India Cross Cultural Innovation Network' at the University of Brighton for initiating the project and creating the network of European and Indian partners in the project, making it possible to look into various models of innovations and the entrepreneurship. I extend my thanks to all the partners from India and Europe for collectively looking into the regional culture based aspects of enterprise innovation and sharing observations and experiences. It has provided an insight into important aspects of building competitiveness of regional economy. Thanks are also due to Professor Rajesh Kochhar, Director NISTADS for providing all support to undertake the work.

References

Badaracco, J. (1991). *The knowledge link: How firms compete through strategic alliances*. Boston, MA: Harvard Business School Press.

Beije, P. (2000). *Technological change in the modern economy*. Cheltenham, U.K.: Edward Elgar.

Bert, F. & Hoselitz, A. (1963). A review of entrepreneurs of Lebanon: The role of business leaders in a developing economy. *Business History Review*, **37** (3), 300.

Brandt, D., Rose, C. & Olbertz, E. (2002). The regional networking of industry and craft enterprises and the role of university. *Nistads News*, **4** (1), 25–39. (NISTADS News Letter, Pusa Gate, K. S. Krishnan Marg, New Delhi).

Das, P. P. & Mukhopadhyay, A. K. (1995). *Balucharee of Mallabhum*. India: NISTADS—CSIR.

Drucker, P. F. (1985). *Innovation and entrepreneurship—Practice and principles*. London: Heinmann.

Evangelista, R. (1996). Embodied and disembodied innovative activities: Evidence from Italian innovation survey. In: *Innovation, Patent and Technological Strategies*. Paris: OECD.

Evangelista, R., Iammarino, S., Mastrostefano, V. & Silvani, A. (2001). Measuring the regional dimension of innovation. Lessons from Italian Innovation Survey, *Technovation*, **21**, 731–745.

Fagerberg, J. & Verspagen, B. (1996). Heading for divergence? Regional growth in Europe reconsidered. *Journal of Common Market Studies*, **34**, 431–448.

Fagerberg, J., Verspagen, B. & Caniels, M. (1997). Technology gaps, growth and unemployment across European regions. *Regional Studies*, **31**.

Geenhuizen, M. & Nijkamp, P. (2000). The learning capabilities of regions—Conceptual policies and patterns. In: F. Boekema, K. Morgan, S. Bakkers & R. Rutten (Eds), *Knowledge, Innovation and Economic Growth* (pp. 38–56). U.K.: Edward Elgar.

Grabher, G. (Ed.) (1993). *The embedded firm. On the socioeconomics of industrial networks*. London: Routledge.

Higgins, B. & Savoie, D. (1988). The economics and policies of regional development. In: B. Higgins & D. Savoie (Eds), *Regional Economic Development* (pp. 1–20). Boston, MA: Hyman.

Hildreth, P., Kimble, C. & Wright, P. (2000). Communities of practice in the distributed international environment. *MCB Journal of Knowledge Management*, **4** (1), 13.

Kharbada, V. P. & Mehra, K. (2002). Editorial. *AI & Society*, **16** (1), 1–3.

Kochhar, R. (2001). Brass workers of Bankura: Current activity report. *Nistads News*, **3** (1), 1–2. (NISTADS News Letter, Pusa Gate, K. S. Krishnan Marg, New Delhi).

Mehra, K. (1999). *Farm level technical change: A case of floriculture in India*. Paper presented in an International Seminar on 'Global knowledge partnership: Creating value for the 21st century held from 30th August—2nd September, 1999 at Austin (Texas), USA. In press as 'Entrepreneur Innovativeness & Farm Level Technical Change'. In: P. Coneicao, D. Gibson, M. Heitor & C. Stolp. (Eds), '*Systems and Policies for the Globalized Learning Economy*'—Series on '*Technology Policy & Innovation*' Volume 3, USA Quorum Books (in press).

Mehra, K. (2002). Entrepreneurial spirit of the Indian farmers. *AI & Society*, **16** (1), 112–118.

Mehra, K. & Mukhopadhyay, A. K. (2000). Transition of artisan enterprises to new industrial culture—Towards moderanisation of traditional Saree weaving in India. Preprints of *7th Automated systems based on human skills—Joint Design of Technology & Organization* (pp. 337–340). Aachen: University of Technology.

Mehta, D. & Joshi, B. (2002). Entrepreneurial innovations in Gujarat. *AI & Society*, **16** (1), 73–88.

Neven, D. J., Gouyette, C. (1994). *Regional convergence in the european community*. CERP discussion paper.

Nonaka, I. (1991). The knowledge-creating company. *Harvard Business school Review*, **6** (8), 96–104.

Pan, S. L. & Scarbrough, H. (1999). Knowledge management in practice: An exploratory case study of Buckman labs. *Technology analysis and Strategic Management*, **11** (3), 359–374.

Porter, M. (1990). *The comparative advantage of nations*. London: The Macmillan Press.

Roy, S. & Mohapatra, P. K. J. (2002). Regional specialization for technological innovation in R&D laboratories: A strategic perspective. *AI & Society*, **16** (1), 100–111.

Sadhu, A. N. & Mahajan, R. K. (1985). *Technological change and agriculture development in India*. Bombay: Himalaya Publishing House.

Schumpeter, J. A. (1961). *The theory of economic development*. Boston: MA: Harvard Business Press.

Quah, D. T. (1996). *Regional convergence clusters across Europe*. CEPLSE Discussion paper.

The International Handbook on Innovation
Edited by Larisa V. Shavinina

Innovation Process in Switzerland

Beate E. Wilhelm

z-link, Zurich, Switzerland

Abstract: Many politicians and scientists in Europe declare that today's innovation process is inefficient, which is reflected in a lack of transformation of scientific knowledge into commercialized products and processes. They hold responsible a kind of bottleneck for the missing exchange between science and economy. Fostering the utilization of scientific knowledge via organized technology transfer from universities to industry is the most popular aim of technology policy in Switzerland. But does this really solve the problem—and is the real problem based on the lack of that kind of transfer?

Keywords: Innovation system; Innovation process; Knowledge and technology transfer; Switzerland.

Introduction

Politicians in Europe are still complaining about the innovation gap, caused by the insufficient valorization of scientific knowledge into innovative products: they call it the 'European Paradox' (CEU, 1993, 1995). Its implicit message is as follows: scientific knowledge is the basis for innovative products, which in turn are the basis for economic development at regional, national, and international levels. While scientific knowledge reaches a high competitive level, mainly in natural and technical sciences, the output of new, global high competitive products is assessed to be poor (EC, 1995; Federal Council, 1997, 1998). How does that happen? They hold responsible a kind of 'bottleneck' for the missing exchange, and transfer from science to economy (CEU, 1993, 1995; Ellwein & Bruder 1982; Ewers et al., 1980; Federal Council 1997, 1998; basics: Grimmer et al., 1992; Hauff & Scharpf 1975). This missing exchange results in much less innovation as the output of innovation process. But does fostering this kind of transfer really solve the problem? The findings presented in this chapter discuss the asymmetries and misunderstandings with respect to today's 'knowledge and technology transfer' as an instrument for pushing the innovation process in the case of Switzerland.

For a better understanding of the so-called 'European Paradox', it is helpful to discuss the following issues. First, definitions of innovation and innovation processes in firms are considered. Second, the innovation process on a macro-level regarding all systems involved (like the system of higher education and public research, i.e. 'science system', the industries and political system) is described and explained in light of recent findings in innovation research. Third, the related question about how government supports innovation is discussed. Finally, conclusions are drawn and recommendations are given about further needs in the development of national innovation systems.

Conceptual Issues

The attempt to define innovation has to be regarded as a permanently developing and adapting process, which is not finished yet, and perhaps never will be. There is a wide range of definitions of innovation—from very complex to very minimalist definitions about innovations, always containing something new (Kuhn, 1967). But up to now, there is no generally accepted single definition of innovation.

Innovations exist as *products* as well as *processes* and as *organizational structures*. In terms of *products* and *organizational structures* innovations are regarded as a result. In distinction to this meaning *innovation process* has a twofold meaning: First, innovation process as a procedure intended to result in innovative products[1] in firms and organizations. Therefore often different kinds of innovation activities, interaction and feedback-loops are needed. In this sense innovation

---

[1] The term 'product' is used to cover both: goods and services.

process or innovation activities are a main precondition for innovative products. Besides that micro-level-perspective innovation process is also considered on a macro-level where innovation activities and interaction between and among scientific,[2] industrial, and political systems is analyzed and explained. Second, innovation process in firms or among firms and other organizations means, to do a process in an innovative way, for example a new production process such as 'just-in-time'. In this sense, innovation process can be regarded as an implemented innovation. There exist close connections and interdependencies between both aspects: the making of innovations has impacts on the results of innovations, as well as intended innovation results shape the way of making innovations.

During the last 15 years some progress was made in understanding innovation process in Europe, particularly by the experience gained from the first round of national innovation surveys conducted in 1991 to 1993. The Organization of Economic Co-Operation and Development (OECD), the European Commission (EC), and Eurostat offer a definition of technological innovation as follows: 'technological product and process innovations comprise implemented technologically new products and processes and significant technological improvements in products and processes' (OECD, EC & Eurostat, 1996, p. 31). Innovations always have to be new or significantly improved to be counted as innovation. But up to now, a clear-cut definition is not possible because of the complexity of innovation and because of the variations in the way it occurs. A minimalist basic definition of innovation process (on a micro-level) is that 'innovation is a complex, diversified activity with many interacting components' (OECD, EC & Eurostat, 1996, p. 24).

Throughout economic literature, technological aspects of innovations are focused. This is due to findings of macroeconomic studies and technology portfolios showing direct connections between export of high-technology products and gained export position of nations (in detail see Section 3; Grupp & Legler, 1987; Soete, 1987). However, the growing influence of non-technical innovations and of the service sector is leading to more awareness and knowledge about its impact and contribution on and activities in innovations. OECD, EC and Eurostat are well aware, that organizational, social, and non-technical innovations are innovations too. Therefore, they take steps to improve the measurement basis to grasp these kinds of innovations in future national innovation surveys too (OECD, EC & Eurostat, 1996).

Innovation as Result

Process and product innovations include a very broad spectrum of phenomena, which appear at the level of enterprises, sectors of industry, or even at regional level. Besides that, innovation can appear in all fields of human endeavor. In this sense the 'disc player' is a product innovation, whereas the implementation of 'new public management' in local government is a process innovation, which might be also considered as a structural innovation.

Innovations are differentiated by their *degree of novelty* from 'radical' to 'incremental' innovations (OECD, EC & Eurostat, 1996). The common understanding of radical innovations is to shape large changes whereas incremental innovations have to be seen as step-by-step improvements and by doing so proceeding the change continuously. History of technology shows, that radical innovations like the steam engine, automobile, and personal computers, did not emerge from one day to another, but were results of a long-term research and development process, often lasting decades and always bearing high risks and open ends.[3] But once the breakthrough succeeded, the impact of radical innovations has the potential to be very extensive, namely in their spatial reach and in their permeation of society and economy. Closely connected with their degree of newness, innovations are also discriminated by their *spatial reach* as well as by their *organizational-structural reach*. In the sense of *spatial reach*, innovations can be new just for a single person or an enterprise; it can be new for a region, for a whole nation or for an international alliance, or even for the whole world (see Fig. 1).

However, findings in national and international innovation surveys show, that incremental innovations are the most common, most widespread and the usual way to innovate (Arvanitis et al., 2001, 1998; Bosworth, Stoneman & Sinha, 1996; Eurostat, 2001). Product innovations new to the world, average about 4% of sales revenues in the manufacturing industry in Switzerland, whereas, innovations significantly improved and new to the firm average approximately 34% (Arvanitits et al., 2001). Products, which are not innovative because they were not significantly improved, come to about 63% of sales revenues. For the service industry, this share is about 89%, compared to 11% for innovative products. In the EU the share of innovative products new to the international markets is about 7% in average (Eurostat, 2001).

Organizational-structural reach means innovations, which are new for sectors of industry, specific technology communities, organizational systems, and institutional settings. Innovations can also contribute to

---

[2] Scientific system as it is used in Europe includes all universitarian disciplines as well as the universities themselves and public funded R&D institutions.

---

[3] For computers see for example Rojas & Hashagen (2000).

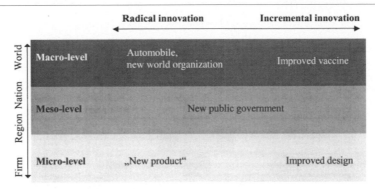

Figure 1. Spatial reach and degree of novelty of innovations.

improving the landscape of institutional and organizational structures. For example, networks between different kind of protagonists and institutions are an innovation for regions and their development, changing the organizational and may be also the institutional structure. This kind of innovation, namely networks, is situated between the market system and the hierarchical system (i.e. government). The profile, structure, and organization of networks differ between each unit analyzed. Networks are able to support innovation processes and innovative activities in enterprises, in government, and in non-profit organizations by developing regional competitiveness and more innovative structures (see Swan et al., this volume). However, networks are also able to hinder the emergence of innovative structures, namely if organization or codes of conduct of networks are deadlocked and established (for example Grabher, 1993). In this light, the link between innovation as results and innovation process is obvious: innovative structures are regarded as a favorable precondition in fostering innovations. This is described in detail in the National Innovation System of OECD (1997).

Both kinds of differentiation, namely by spatial and organizational-structural reach, bear asymmetries between reach and localization of their own 'making'. That makes it hard to grasp, identify, and allocate these innovations and their respectively impacts.

In categories of radical innovation to incremental innovation, products and processes are only counted, if they were implemented or commercialized (OECD, EC & Eurostat, 1996, p. 31): 'technological product and process (TPP) innovation has been implemented if it has been introduced on the market (product innovation) or used within a production process (process innovation). TPP innovations involve a series of scientific, technological, organizational, financial and commercial activities. The TPP innovating firm is one that has implemented technologically new or significantly technologically improved products or processes during the period under review'. The notion of 'implementation' in the definition given above is a very important

condition. It also says that innovations exist before and after their implementation respectively their valorization. In other words, innovation activities do not automatically lead to innovation results. Internal factors like, for example, managerial decisions, financial cuts, organizational restrictions, insufficient technological information, lack of specialists, as well as external factors like changed market situations may emerge as obstacles and ultimately hinder the completion and implementation of innovations. Regarding the difference between failed and successful innovation, there is little knowledge about the ratio to what extend firms have invested in innovations, and what is the gained success out of these efforts. Up to now data about the impact of commercialized innovations is only available by their sales revenues.

Innovation Process on a Macro-Level: About the 'Making of Innovations'

Previously, *innovation process* (on a macro-level) was described as a cascade of scientific knowledge: starting from a (technical) invention made by scientists at universities, being improved or/and transformed by applied scientists and marketing specialists then being implemented and produced by enterprises (see Fig. 2) (Mensch, 1975). The inventory power was attributed to single persons, mostly professors (type 1). Up to now the Swiss education, research, and technology policy program for the years 2000 to 2003 is based on this theoretical background (Federal Council, 1998). In this process there is a clear distinction between the liabilities of universities and universities of applied science (UAS). Universities are in charge of excellence in basic research and in teaching, and regarded as the main input factor for generating new knowledge (Federal Council, 1998). UAS are assigned to a sandwich position as a 'mediator and valorizator' between basic non-oriented science and the commercial side of firms. They have to develop applied problem solutions especially for small and medium sized enterprises (SME) in their very regions—about 99% of all enterprises fall in this category, not only in

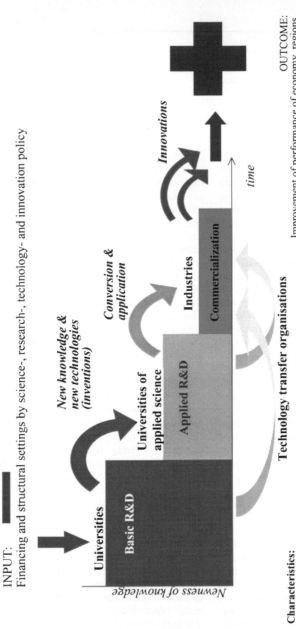

INPUT:
Financing and structural settings by science-, research-, technology- and innovation policy

New knowledge & new technologies (inventions)

Conversion & application

Innovations

Universities

Basic R&D

Universities of applied science

Applied R&D

Industries

Commercialization

Newness of knowledge

time

Technology transfer organisations

Characteristics:
• Strongly oriented towards scientific disciplines
• Knowledge production process alike a cascade
• Hierarchie of knowledge

Process results:
• Self-preservation of science systems and disciplines by sharing of tasks
• Excellence in specific tasks
• Distinguishing single persons, scientific theories and doctrines

OUTCOME:
Improvement of performance of economy, regions, and national economy
➔ **Excellence in innovation, by:**
- rapid implementation of knowledge in innovative products and processes
- structure of economy is designed to foster high-technology and innovations

Figure 2. Linear model of knowledge production and innovation.

Switzerland but also in the European Union.[4] Therefore universities of applied science are supposed to use their wide-spread, exhaustive structure to reach each SME in all regions of Switzerland. Finally this applied knowledge shall be utilized by enterprises buying these applied problem solutions, produce new innovative products, and diffuse by selling it. To fulfil their sandwich position, UAS have to be engaged in networking with universities to ensure the flow of basic knowledge to themselves.

To unify both types of universities in their efforts to foster innovation processes, they are regarded as 'equivalent but different'—different with respect to their main tasks (Federal Council, 1998). Politicians suppose the time distance between gaining results of research from universities and competitive new products on the markets (i.e. 'time-to-market') being much too long (Federal Council, 1998). Therefore, they intend to foster technology and knowledge transfer from universities to industries but also from universities to universities of applied science. That process is supported by technology transfer institutions not only established in universities, but also in UAS.

The model, as shown in Fig. 2, is based on the assumption, that innovations rely on scientific knowledge and know-how, which again is regarded as result of R&D. In the end innovation process is equated with knowledge production process, and simultaneously also equated with scientific knowledge production process.

Nowadays, innovation process (on a macro-level) is understood as a process of complex interactions between different kind of protagonists and institutions (type 2). The commission of the European Union describes the innovation process as follows (CEU, 1995, p. 8): "It is no linear process with strictly separated steps, which are connected by an automatically step by step procedure, but it is more a system of interchanges, a system of to- and from-movements among single functions and protagonists, whose knowledge and experience increase and complements one another" (see Fig. 3). This definition and perception of innovation process will be used further in this chapter. For a better understanding of innovation processes, it is important to know for each innovation system, based on regions, industrial branches and so on, which protagonists are involved at which time, what are their specific contributions and why it is so.

Although, no specific pre-conditions for starting innovation process are listed in the Oslo Manual or Frascati Manual, the OECD therefore claims other sources, namely the book about national innovation systems (OECD, 1997). One definition (besides others)

given there about a national innovation system is: ". . . the elements and relationships which interact in the production, diffusion, and use of new, and economically useful knowledge . . . are either located within or rooted inside the borders of a nation state" (Lundvall, 1992 in OECD 1997, p. 10). All definitions listed there, point out the role of interactions or networks between different institutions—like university institutes, research laboratories, enterprises, trade associations, business promotion institutions and so on. Differences among these definitions appear in focusing specific aspects of interactions or results out of these interactions. It is striking that virtually no differentiation of innovation systems exists, for example describing structured procedures in different constellations. Besides, each definition equates innovations with technologies, being explicit or implicit.

Also, recently, studies about new production forms of knowledge have appeared. Mostly looking from the point of science, authors located in fields of social science have analyzed and stated changing structures of interactions between science on the one hand and technology and economy on the other hand. This structural change is expressed twofold: science is opening up to needs and demands from economy, technology and society, whereas innovations in technology, economy, and society are more and more based on scientific knowledge (Caracostas & Muldur, 1998; OECD, 1998; Stokes, 1997; Weingart, 1995). Some authors even describe a shift towards more interactive and more dependent and tighter relations between science, industry, and politics (Leydesdorff & Etzkowitz, 1998). A new form of knowledge production emerges as a result of these structural changes: knowledge is no longer seen as isolated part embedded in the linear sequence of basic research, applied research and diffusion, but emerges in the context of application and application demands (see Fig. 3) (Gibbons et al., 1994; Nowotny, 1996; Nowotny, Scott & Gibbons, 2001). This new application oriented knowledge production is called 'mode 2' in contrary to the traditional, linear academic knowledge production, called 'mode 1'. As science intends to contribute in problem-solving in future, science activities have to be much more transdisciplinary to even be able to do so (Gibbons et al., 1994). The part of unscientific knowledge production is 'somehow' implicitly woven in the network approach of 'mode 2'. Further, the authors assume, that the old model of innovation process is disintegrating for the benefit of the new model (Gibbons et al., 1994; Nowotny, 1996; Nowotny, Scott & Gibbons, 2001). But this shift in knowledge production has first to be empirically tested. Do both types already exist, and if yes, since when and to what extent, are there patterns of precondition for their respectively existence, and so on.

The expansion of knowledge production as 'mode 2' also brings up an important aspect of task sharing

---

[4] For Switzerland see: http://www.statistik.admin.ch/stat_ch/ber06/bz01/actuel/dact01.htm 'BFS: Betriebszählung 2001'; for the European Union see: http://europa.eu.int/comm/eurostat 'SME statistics'.

INPUT:
Financing and structural settings only partly by science-, research-,
technology- and innovation policy

Universities

**Universities of
applied science**

Politics & NGOs

Economy/enterprises

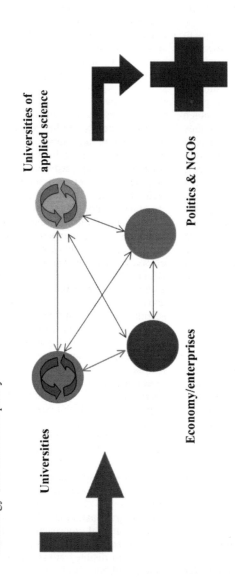

OUTCOME:

- Improvement of performance of economy, regions,
  and national economy
  ➔ **Excellence in innovation**
  - Improvement of ability in problem solving
    together with localized protagonists

➔ **Excellence in ability for problem solving, by:**
  - common development of problem solutions
  - innovation orientiented instead of purely
    tehnology-oriented economy structure

Characteristics:
• Oriented towards transdisciplinarity
• Knowledge production by networking
• Equality of knowledge
• Implicit and explicite knowledge is being considered

Process results:
• Flexible system in problem solving by networking
• Oriented towards problem solving
• Distinguishing of problem-solving methods and their respectively
  teams and know-how bearer

Figure 3. Network model of knowledge production and innovation.

between universities and universities of applied science in Switzerland. As soon as universities are also doing application-oriented research as universities of applied science, this factor is not discriminating both types of universities sufficiently. For universities, it is supposable, to put the focus more on application oriented basic research whereas, universities of applied science put their focus on applied research and its implementation. But this kind of job sharing implies a continuum and a co-ordinated interface between both types of universities, which up to now is unlikely to happen. Last, but not least, this is due to the fact that universities of applied science do not offer the same 'disciplines' and fields of activity as universities. However, the difference in the respective fields of activity is a chance for differentiation. For universities, the emerging approach of transdisciplinarity can be a starting point for problem-solving and oriented basic research.

In Europe scientific knowledge and know-how production is supposed to be based on a strict neutral, objective and perception-oriented R&D-process at universities. The outcome of this process shapes out in different kinds of artefacts and is reflected in knowledge carriers. Carriers of scientific knowledge are graduates and employees, who bear explicit (or codified) knowledge as well as implicit (or tacit) knowledge. Therefore, knowledge and technology transfer is always a people business, where people are involved and affected directly or indirectly by different forms of structures and frameworks, such as incentive systems (e.g. 'publish or perish' fixed with a minimum of publications dedicated for selected scientific communities in appropriate journals; newness; originality). This process of scientific knowledge production is examined in fields of science and technology studies,[5] especially in the field of social construction of scientific perception. Although, this process is not examined exhaustively, the emergence of scientific knowledge so far can be considered as a product of purely neutral, objective and perception-oriented scientific processes (Felt, Nowotny & Taschwer, 1995). As scientists are driven by doctrines, theories, trends, and personal attitudes there is always a risk in lack of 'neutrality' and 'objectivity'. This has to be born in mind whenever looking at scientific knowledge, knowledge and technology transfer from science to industry and others.

About Diffusion of Innovation: Knowledge and Technology Transfer

The most common known type of knowledge transfer from universities and R&D institutions to industry is the transfer of graduates to industries. Of course, graduates do not exclusively switch over to industry: In Switzerland after one year of graduation, approximately 43% of university graduates (without UAS) are working in industry, whereas 41% are working in public services, which includes universities[6] (Diem, 2000). All types of knowledge transfer regarded as typical in the linear cascade model of knowledge transfer (type 1) are shown in Fig. 4. The common understanding of knowledge transfer focuses on R&D studies, consulting and information exchange, which is regarded as one-way transfer flow from science to industry. Instruments for these forms of knowledge transfer are R&D cooperations between science and industry, and innovation cooperations, which additionally include other forms of input and transfer necessary for innovation processes.

About Measuring Innovations

Ministers for science and technology policies of the OECD call for improved indicators to measure better innovation performance and other related output of a knowledge-based economy. In order to do so, the Oslo Manual was developed and first published in 1992 (second edition in 1996 and further updates). The Oslo Manual is mainly intended to develop the measurement of innovations and innovative activities across countries for a better understanding of innovation processes. By doing so it is necessary to understand what innovation and innovation processes are about before being able to measure it. Therefore a great deal of (scientific) knowledge and expertise resulting from the first and second Community Innovation Survey (CIS 1: 1993 to 1994 and CIS 2: 1996 to 1997) of the EU about innovation process was taken into consideration.

Innovation surveys and technology portfolios are the main instruments to measure the performance of the specific national innovation and technology capacity. Up to now innovation surveys are mainly measuring input, output, and throughput factors for innovations in single enterprises. Innovations are only counted and regarded as useful if they have entered the market and/or the enterprises and—in general—have been valorized (see sub-section 'Innovation as Results'). Enterprises, that still have not finished their innovative activities or only have failed innovations, were not counted as 'innovative' in innovation surveys. It is obvious, that this exclusion covers only parts of innovations and especially innovation activities that happen in total. Innovation activities as preparative activities for technological innovations are listed in the Oslo Manual as follows (OECD, EC & Eurostat, 1996):

---

[5] For an overview of science and technology studies see for example Felt, Nowotny & Taschwer (1995); Edge (1994); Jasanoff et al. (1994).

[6] Another 2% start up their own business, another 2% are working in NGOs, 10% remain unemployed and about 3% have a job prospective. But to know more about the destination of graduates, either a long-term survey or a survey across different years of lag towards graduation would be necessary.

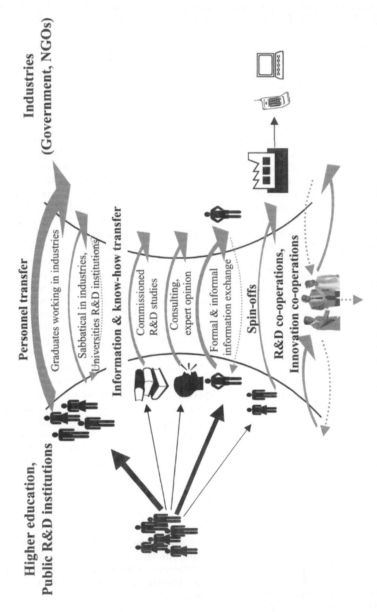

Figure 4. Types of transfer in linear knowledge production and innovation.

Notes:
(1) Thickness of arrows sketches each transfer flow roughly in its extend (for Europe).
(2) Dotted lines indicate transfer flows additionally considered in network model of knowledge and technology transfer.

- Acquisition and generation of relevant knowledge new to the firm, including: R&D; construction and testing of prototype; software development; acquisition of embodied technology; acquisition of disembodied technology and know-how.
- Other preparations for production, including: tooling-up and industrial engineering; industrial design; other capital acquisition (e.g. acquisition of machinery, tools); production start-up.
- Marketing for new or improved products.

Some of these activities are borderline cases, for example design, training, marketing, software, which were only counted, if they were carried out only for innovation purposes. Furthermore, even the new edition of the Oslo Manual only recommends measures regarding 'technological innovation' like products and processes, for the measurement of 'organizational innovation' is classified as very difficult, highly firm-specific, hard to summarize and therefore hard to compare. Being aware that measurement of innovative activities mainly focuses on technological innovations it is obvious that the service industry is not addressed ideally. Up to now, there are still no exhaustive nation-wide surveys in the wide field of social, organizational and structural innovations in middle Europe, neither considering industries nor services, public services or non-governmental organizations.[7] Up to now, innovations are also not measured on an exhaustive macro-economic level on their respectively impact. For the first time service industries have been considered in the EU innovation survey of 1996–1997, in the Swiss innovation survey 1996 (Arvanitis et al., 1998; and further in 1999; Arvanitis et al., 2001), in the Netherlands (Brouwer & Kleinknecht, 1996), and in Germany (Ebling & Janz, 1999). the EU innovation survey of 1996–1997 has considered only a few service industries, namely wholesale trade, traffic, telecommunication, credit and insurance industry, data processing and database, and architecture and engineering companies. Facing these limited service industries considered, the Swiss innovation survey from 1999 covers nearly all private service industries, including service industries for enterprises and R&D for example (Arvanitis et al., 2001).

Data in the innovation surveys shows, that enterprises in the service sector analyzed do have less expenses for innovations (2.8% of total turnover) compared to enterprises in industrial sector (3.5% of total turnover) (EC, 2001). Of course there is a wide spectrum of different expenses between the EU-countries and between both sectors. However, this finding must not lead to the general conclusion of service industries being less innovative. It has to be interpreted carefully: first, the service sector analyzed

is not complete. Second, the importance of single innovative activities in service and industrial sector cannot be equated or directly measured with the amount of expenses, because some innovation activities are not just more expensive than others but not necessarily more important. This fact holds also true for comparisons between the different kind of innovative activities between the service and industrial sector.

Based on new knowledge about innovations, additional needs arise for analysing innovations by their impact, by the role of diffusion, by the sources of information for innovation and their obstacles, by input factors, as well as the role of public policy in innovation and the gained output (OECD, EC & Eurostat, 1996). And there also exist different innovation strategies by industrial branches and industrial clusters, which raises a need for analysing these differencies (Arvanitis et al., 2001; Wilhelm, 2000). All these discriminating factors help in a better understanding of innovations and their emergence.

Focusing on output measurement is also not sufficient for a better understanding of how innovations develop, who is involved and so on. OECD and Eurostat are also aware of the necessity of feeding the debate about innovation policy and gathering much more information and analysis of many aspects of innovation for better supporting innovation activities: 'ideally, a comprehensive information system should be constructed that covers all types of factors within the innovation policy terrain' (OECD, EC & Eurostat, 1996, p. 25). Therefore, in future, it is intended to do both: collecting data based on indicators, which will be able to cover some more parts of innovation activities as well as collecting more qualitative information by case studies. Hopefully, procedural aspects of innovation activities are one of the main aims of future investigations in case studies. Case studies therefore are useful if they are conducted, co-ordinated and adjusted, for example by conducting cross-industries studies or cross-countries studies.

Although a great deal about innovation processes still remains unknown or just with a poor understanding, the today knowledge about it is much more sophisticated than 10 to 20 years ago. For a better understanding of the 'making of innovations', processes of technical and especially non-technical innovations have to be analyzed and differentiated. With respect to the implementation of this knowledge in R&D-, technology-, innovation-, and education policy other important aspects arise: how does knowledge about innovation processes diffuse into governance, namely to the main policy decision-makers, and what happens in the 'black box' of political decision-making especially in innovation policy. Standard definitions about innovation and R&D as given in the Oslo Manual, Frascati Manual or the Green Paper are basic orientation guidelines for

---

[7] For a review of the state of the art in service innovation research see Küpper (2001).

government in single member states. But it does not mean, that the design of the respective policies, especially of innovation policy, is automatically based on this knowledge! This is important, because measures and programs as 'output' of this black box are often based on implicit assumptions about innovations far from current (scientific) knowledge (Hofmann, 1993). The understanding of innovation can be differentiated as follows: (a) innovation processes as they objectively happen;[8] (b) innovation processes as they are partly described and reflected in innovation surveys, case studies and so on and as they are interpreted, described, and explained by researchers, scientists, politicians, and others; and (c) assumptions about innovation processes reflected in activities and measures to foster innovation processes by politics.

In the following sections, innovations are looked at closer to find some answers about the role of innovations, what is known about innovations, and by whom.

Why Innovation? The Role of Innovations for National Economies

For the general purpose in a better understanding of innovation it may help, to embed the complex innovation phenomena as a whole in the context of research and innovation policy and national economic development. Therefore, it is necessary to analyze and evaluate political statements, programs and measures regarding to foster innovation.

Innovation is widely regarded as the most important factor for developing economies, to raise the competitiveness of enterprises up to national economies, and in the long-run to raise the welfare for the inhabitants of national economies. This 'chain reaction' is roughly sketched in Fig. 5.

First, a general distinction has to be made: the 'direction of effects' in Fig. 5 shows the innovation process: (1) and its diffusion; and (2) as it is understood and partly supported by research and innovation policy programs measures in Switzerland. The main reason for putting R&D at the beginning of this 'chain reaction' as shown in Fig. 5 lies in statistical relevant findings of positive connections between technical standards and gained export position (Grupp & Legler, 1987; Soete, 1987). Technical standards again were found being mainly based on R&D in technological fields. Studies have identified different branches characteristic in high amounts of R&D and ranked on top positions in export (BMBF, 2002 (annual volumes); Grupp & Legler, 1987; Soete, 1987). As the average value of R&D in all branches was found to be 3.5%, all

branches under this level were called 'low-tech industries', branches with an amount between 3.5% and 8.5% were called 'higher-tech industries', and such with more than 8.5% were called 'high- or top-tech industries'. That is the reason why technological based R&D was supposed to be the main source of good performing industries and therefore it was assumed to be good for the whole nation. As a consequence of this finding key industries with high levels of R&D have been identified and classified to be worth being supported by technology and innovation politics.

The finding of the connection between 'technical standard and gained export position' described above is correct to some extend but not sufficiently exact. R&D can lead to technological innovative products and processes in an enterprise, which gain high commissions and therefore, have positive effects on its employees; but it does not always happen or automatically—which makes an important difference! This short-gripping understanding of innovation expresses a lack of knowledge of what happens in the 'black-box', i.e. in 'the making of innovation'. As we know today R&D is not the only source for innovations (Acs & Audretsch, 1993; Eurostat, 2001; Felder et al., 1994; Grupp, 1997): approximately 57% of all expenses for innovation activities are spent on their own R&D in large sized industrial enterprises, whereas the share is only about 25% of all expenses for innovation in small and medium sized enterprises (SMEs) (Eurostat, 2001). Especially buying and implementing new machines and new equipment is taking about 50% of all innovation costs. For SMEs new machines, design, experimental testing, and construction development are more important factors for being innovative than R&D (Biegelbauer, 1997; Bosworth, Stoneman & Sinha, 1996; Eurostat, 2001; Licht & Stahl, 1997). As innovative products mainly result out of step-by-step improvements of existing products or processes or as results of combinations of already existing products they are often products of design, marketing, customer-relationships or a combination of these factors, especially in small and medium sized enterprises. Therefore, factors like design, experimental development, marketing, and other innovation activities have also to be included in a definition of innovation activities, as it is done in the Oslo Manual (OECD, EC & Eurostat, 1996).

Up to now, no theory is able to prove the neo-classical assumption between well-being enterprises and well-being nations exactly (Ehrenberg, 1995). One relevant indicator for breaking off this simple linear logic is the phenomena of job-less growth. Thurow claims a threat of industrial growth by simultaneously stagnating job opportunities in the USA (Thurow, 1996). Economic growth between 1991 and 1995 in Europe did not lead to a decline of the unemployment rate either, contrarily, the unemployment rate rose from 8.2% to 10.9% during this time (Eurostat, 1997b).

---

[8] Of course, 'the real innovation process' has always to be regarded as a social construct which can never absolutely be objective. Here, it is meant as the infinite search for 'objectivity' or 'truth' of at least some researchers.

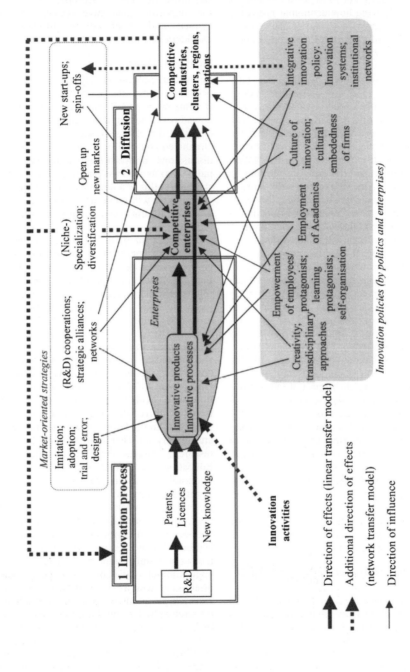

Figure 5. Influences on innovations and effects of innovations.

However, job-less growth is again no phenomena that can be observed neither in all industrialized countries nor in all countries in Europe (Christoffel, 1995; OECD, 1996a; Weeber, 1995, Wolter & Curti, 1996). It is realized that there does not exist an automatic or mechanistic way of showing how economic growth is effecting employment (or vice versa) (CEU, 1993b; Ehrenberg, 1995; Oppenländer, 1994; Walterskirchen, 1994; Wolter & Curti, 1996).

The findings of 'knowledge as the main source of innovation' and at least national welfare has only recently emerged, and led to innovation policy regarded as amalgam of science, technology, and industry policy (OECD, EC & Eurostat, 1996). The logic of good performing industries that benefit the whole nation is common place, but enlightens just parts of its underlying connections: strategies of innovation processes in enterprises are changing by becoming internationally oriented and are therefore dis-embedding from their former innovation network (Sauer & Lang, 1999). This may lead to a phase out of personnel and/or institutional responsibilities of single enterprises and at last to an erosion of innovation networks.

A main aspect of impact of innovations is, that innovations bear two sides of a medal or in other words innovations are a paradoxical phenomenon: not each progress for enterprises means progress for its employees or its region. For example: on the one hand a technologically new production process in an enterprise produces an enormous raise of sales; on the other hand it costs a lot of people their jobs. But, innovations also can improve both: the performance of the enterprise as well as the job-situation of employees. In all cases it is necessary to put the impact of innovations into context and therefore, take several aspects of impact into account. However, this depends on the point of view one takes, and it also depends on the demarcation regarding space, time, and which type of costs and effects of innovation are taken into account. This means the 'location' of impact, and—closely connected—the duration of impact, are regarded the main discriminating factors for impact of innovations in general.

The following section sheds some light on the involved systems and protagonists and their role in innovation process in Switzerland.

The Swiss Landscape of Tertiary Education, R&D and Innovation

Compared to other states in Europe, measures and instruments for supporting innovation and innovation activities are less differentiated and much less multifaceted in Switzerland. Innovation policy in Switzerland is even formally in-existent up to now and also the awareness about the necessity of innovation policy is hardly developed (Freiburghaus et al., 1991). Instead, education, research, and technology policy offers measures and instruments also supporting inno-

vation activities. This has to be considered, when using the term of 'Swiss innovation policy' later. Swiss innovation policy is still focusing on R&D as the main input factor for innovation. Whereas in most middle and northern states of Europe, the network approach for supporting innovative activities has been implemented and supported in different ways for a considerable time.

Innovation Policy

Innovation process is regarded to take place mainly between science and economy being directly involved (see arrow B and B′ in Fig. 6). Civil society claims nowadays to having more influence and more rights to a say in directions of scientific activities like R&D. Organized by political activities like 'round table discussions' between science and society (partly with industries) as well as 'platforms for society' have been started. As this is a new measure it is not sure, whether government will continue to support it, and whether it will have an impact on innovation activities in science in future.

The role of politics is trying to influence the direction of innovations by setting up the required general conditions, also by financing parts of innovation activities, virtually all in science (see arrow A in Fig. 6). Swiss research and technology policy makes it its business in mainly strengthening R&D in science (arrow A) which is expected to help industries and economy most (arrow B). In Swiss politics mainly two departments are responsible for R&D and technology and innovation policy, namely the Department of Inner Affairs and the Department of Economic Affairs. In the Department of Inner Affairs the Federal Office for Education and Science and the council of both technical universities and its four annex-institutions are connected in the Group of Science and Research. They are mainly in charge of universities and scientific research, whereas the Federal Office for Vocational Education and Technology as part of the Department of Economic Affairs is mainly responsible for the universities of applied science, for innovation, and for technology transfer. Its main instruments therefore are the Commission of Technology and Innovation and the Section for Universities of Applied Science. The Swiss National Science Foundation (SNF) is another instrument for mainly supporting basic research. Therefore approximately 80% of its budget is spent on non-oriented as well as oriented basic research projects by research programs, whereas the remainder is spent on research associates and different activities in the fields of basic science, e.g. support of science conferences. Although organized as a foundation the SNF is fulfilling its tasks by order of the federal state, represented by the federal office for education and science.

The Swiss Council of Science and Technology (SWTR) is the main organ of science, research, and technology policy for the Swiss Federal Council and for federal government, namely for the Department of Inner Affairs and the Department of Economic Affairs. However, members of SWTR are determined by the Federal Council and not by the universities or universities of applied science itself. Although, the SWTR is supposed to be an independent organ, neither influenced by interests resulting from faculty nor by industrial affiliations, each member of the SWTR council is simultaneously either a professor at university, or a decision-maker, or a person with wide influence in one or more industrial enterprises. As national universities and universities of applied science are mainly financed and determined by the cantons, the SWTR could be of great significance for co-ordination of scientific concerns among science institutions and the federal government and its respective departments. However, up to now issues dealt by SWTR are dominated by basic aspects on scientific systems, but neither by aspects of applied science, nor by elaborating concepts or exhaustive consulting activities for politicians.

Industries are also influenced and determined by general set-up of policies, like policy programs and instruments, taxes, laws about venture capital and so on (C in Fig. 6). Industries again are able to influence innovation policy by their specific pressure groups and

by militia system which here means direct influences of leaders in economy who are at the same time leaders of parties, leaders in the national or cantonal government system or members of scientific councils (C').

Education, research, and technology policy ('innovation policy') in Switzerland is based on a diffusion-oriented approach (Federal Council, 1997), whereas innovation policy in the EU is program-oriented manifested in five years termed research and technology frame programs. In Switzerland all governmental measures aim to directly or indirectly influence the emergence and diffusion of new and technology-relevant knowledge (BFK, 1992; Federal Council, 1997, 1998). Therefore mainly universities, federal technical universities[9] and research institutions conducting basic research[10] are supported. Additionally, federal government also conducts research in its own governmental institutions (or resorts) but it also delegates liability of research support to different

---

[9] As described in the section on 'R&D Structure and Expenses' in more detail.

[10] This includes four research institutions, so called annex-organizations—which are closely connected to the two federal technical universities, as well as 19 institutions named in the law about research, article 16, which are assigned to fill gaps in the Swiss 'basic research landscape'.

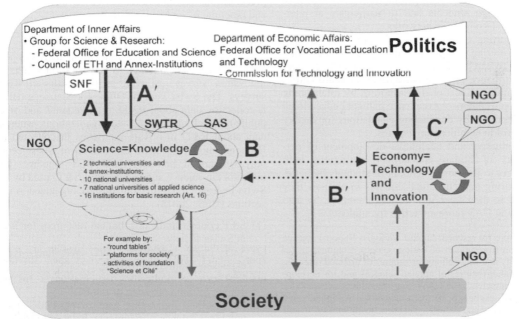

Figure 6. Involved subsystems in innovation processes in Switzerland.

institutions.[11] Between federal and cantonal government there is a kind of job-sharing, as universities and universities of applied science are in charge of the cantons and mainly financed by them. Federal government takes the role of strategic thinking and developing concepts about science, R&D, and technology development, whereas the cantons are in charge of its operative implementation. As Switzerland is strictly based on principles of federalism and subsidiarity, the cantons have huge autonomy in the respectively implementation.

The approved education, research, and technology policy program for 2000 until 2003 intends to spend about 11.8 billion CHF[12] to reach the goal of fostering research and technology. From this amount about 1.92 billion CHF or 16% are dedicated for applied research, for supporting research and technology in SMEs, and for exchange between science and society. With the education, research, and technology policy program mentioned above five main aims are intended to succeed (Federal Council, 1998).

(1) Development of Swiss higher education networks between universities, technical universities, and universities of applied science (UAS) to mainly foster the interactions among them for at least a better performance of technology transfer.
(2) Integration of these networks into international co-operation for also being competitive and attractive for foreign students and to participate at international knowledge networks.
(3) Supporting excellence in education and research by fostering competition among higher education and research institutions, by establishing more performance-oriented basic financial contributions for higher educational institutions, by establishing national centers of excellence in higher educational institutions, by conducting evaluations for quality control in technical universities.
(4) Quantitative and qualitative development of networks of higher educational institutions with respect to the raising amount of students by better preparing and educating students in abilities they need in their later job-life and therefore, by also making UAS more attractive for students.

---

[11] Institutions for research support are Swiss National Science Foundation (for support of basic research), Commission for Technology and Innovation (for support of oriented research, mainly for universities of applied sciences and for development of competences in UAS), four scientific academies (for support of exchange between scientific disciplines and between science and society), and other institutions to support a few selected fields of research.

[12] Without expenses for education on the secondary level, for education programs of the EU (secondary level), for contributions to international organizations, and without contributions for the European Space Agency (federal council, 1998). All these expenses added makes a total amount of 14.12 billion CHF for the period 2000 to 2003.

(5) Valorization of knowledge of these institutions, mainly by fostering the efficiency of technology transfer of UAS, by fostering the generation of spin-offs, and by supporting the Swiss Network on Innovation which again supports to raise the efficiency of technology transfer at Swiss higher educational institutions.

The Federal Council states a large gap between gained research results from universities and having this knowledge transformed into competitive products. For this reason it aims to foster the exploration of the potential which it recognizes being inherent in these research results (Federal Council, 1998). Therefore the Federal Council intends to improve the return on investment (ROI) of publicly funded research activities in universities and to improve the transfer processes in UAS (see Fig. 2).

In the following section, general R&D input data is considered, before financial structure of R&D financing and realization is regarded for Switzerland. After this a closer look on innovation activities and knowledge transfer is taken based on results of R&D—and innovation surveys as well as on studies about exchange and networks between science and industry. As this contribution focuses on exchange, knowledge, and technology transfer between science and enterprises, it emphasizes more on input factors as preconditions for innovative activities, than on output and outcome of innovative activities.

R&D Structure and Expenses

As one intends to localize innovation activities and R&D as part of it, one has to fall back on results of innovation surveys as well as on results of R&D surveys. The Swiss Federal Office for Statistics (BFS) is collecting data about R&D in universities and other public research institutions and organizations, whereas data about R&D in private industries in collected by Economiesuisse, the main association of Swiss economy (organized privately) and BFS together. The indicators in science and technology (S&T) used by the Swiss Federal Office for Statistics are organized in five sections (BFS, 2001a):

(1) S&T context: indicator: human resources for S&T in Swiss society;
(2) S&T input: indicators: R&D personnel—R&D expenses—sources for R&D;
(3) S&T process: indicator: participation at 3rd and 4th (. . .) frame program of the EU;
(4) S&T output: indicators: patents registered—technological balance of payments; balance of trade of high-tech industries;
(5) S&T impact: indicators: none.

One of the most popular indicators is input-oriented, like R&D financing and R&D personnel. With respect to a better understanding of innovation processes, indicators as listed above, are relevant but insufficient. Therefore, results of innovation surveys as well as qualitative studies have to be added, in order to complete the knowledge about innovation processes, which is indispensable for policy making.

R&D Expenditure in General

Switzerland spends approximately 2.7% of its gross domestic expenditure in R&D during 1996 which is much higher than the average value of OECD with 2.2% and of EU with 1.8% in 1997. Sweden spends much more on R&D, namely 3.9% in 1997 and the USA about 2.8% in 1998 of its gross domestic expenditure (OECD, 1999: MSTI database). At the same time Switzerland has 55 researchers per 10,000 labour force, which is equal to the average value of OECD and a higher than average value of EU with 50. The highest amount of R&D labour force are in countries like Japan (92), Sweden (86), Finland (83), Norway and Iceland (both 76), and the USA (74). Although, these data used here, do neither help in comparison nor tell anything about the efficiency and the outcome of any of these expenses. Therefore, they should have to be set in its specific context, and compared much more in detail with respect to its impact and efficiency, which should be done between specific (technological) industries or other issue clusters. Although Switzerland has one of the highest rates of patent application, and one of the highest rates of industrial R&D, the share of high-technology industrial goods in export is comparatively small (Schmoch, Grupp & Laube, 1996). In general, Switzerland has gained good basics and export positions in medium-technology industries but has not reached a high position in specific high-technology industries, which are regarded as key-industries (Hotz-Hart & Küchler, 1996).

Where does all the money come from that is spent on R&D in Switzerland? Private industry bears about 67% of all R&D expenses in Switzerland, whereas federal government contributes about 19%, and cantonal government about 8% of all expenses for R&D, 3% are contributed by abroad, and 3% are contributed from other sources (data for 1996; BFS, 2001a). Hence, the share of R&D expenses is virtually equal to the share of R&D personnel: about 69% of all R&D personnel are working in the private industry, and 29% in higher education institutions. The rate of industrial financed R&D in Switzerland is above-average, compared to other OECD countries where the total average value is approximately 62%. In Japan the share of industrial financed R&D is about 68% and in the USA it is about 59% (data for 1997; OECD, 2000).

Financing R&D at Universities (Without Universities of Applied Science)

In Switzerland three kinds of universities are existing, which are financed, regulated, and controlled different:

two federal technical universities (FTH), one in Zurich and one in Lausanne, which are mainly financed by the federal government. Both federal technical universities additionally have funds from projects and acquisitions, which again are mostly public funds. Ten cantonal universities and seven universities of applied science (UAS) have mixed financial sources, most contributed by cantons where the respective universities are located, by federal funds as well as by projects. UAS have been transformed of former technical, economical, and governmental higher educational schools. But up to now, UAS still suffer from these old structures based rather on a more school-oriented institution than oriented towards an applied-science and education institution: Most of the lecturers in UAS are obligated to participate in a high share of teaching and extended vocational training. Institutional structures, rules, and incentives for applied R&D still are missing at UAS. Compared to universities, UAS have a flat hierarchy, being reflected in a lack of institutional structure and assistants, which also prevent activities besides teaching. Also comparing the law about universities (here without technical universities) and UAS differences between liabilities of both are shrinking: both types have the task to offer services to enterprises or government, both have to be actively doing transformation of knowledge. Although the last mentioned aspect is not formulated for universities, the transformation of knowledge has to be carried out in the field of research[13] (also it is not clear how fields of research are defined).

All universities are mainly financed by public funds (see Fig. 7): about 80% of the whole university expenses are covered by the regular university budget (in 2000), which is public money from the federal government and the cantons. Of which about 2% are fees from students (BFS, 2001b). The balance of 20% of university expenses is derived from acquisition activities in a competitive way for public and private funded money.[14] This 20% are composed of 6% from Swiss National Science Foundation (SNF) and 14% from other public sources, private industries, and non-profit organizations. Universities are financed by commissioned work from industry and public services

---

[13] Federal law about supporting universities and about co-operation between higher education institutions; effective date 1.4.2000.

[14] There is only one exception within the share of financial sources between universities: The highest share of acquisition funds is featured by the University of St. Gall: 43% of its total budget in 1999 consists of 34% from private enterprises and 9% from other private and non-private sources. All other universities reach a maximum of nearly half of it (BFS, 2001a, 2001b). Although, one has to be aware that the highest amount of funds from private enterprises is acquired by the University of Geneva with about 75.9 millions of CHF compared to 38.5 millions of CHF at the University of St. Gall.

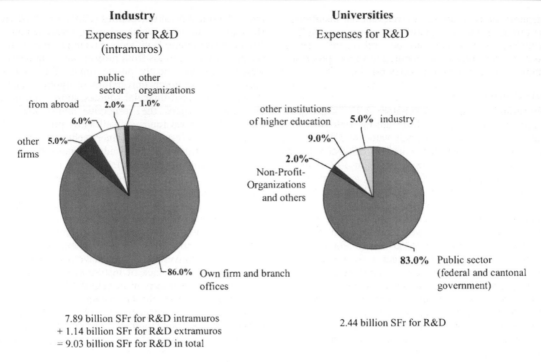

Industry
Expenses for R&D
(intramuros)

Universities
Expenses for R&D

7.89 billion SFr for R&D intramuros
+ 1.14 billion SFr for R&D extramuros
= 9.03 billion SFr for R&D in total

2.44 billion SFr for R&D

Figure 7. Expenses for R&D by source of financing in Switzerland, 2000.

Source: Data: BFS 2002; BFS 2001b; BFS/Economiesuisse 2001.

for 14% whereas this share is about 8% for UAS[15] (average value for 2000; BFS, 2001a). So at the moment the situation is more the other way round than it is supposed to be: universities are also active in applied research while UAS are still unable to conduct the main part of applied research.

In the year 2000 about 94% of expenses of FTHs are financed by the federal government, compared to about 24% of expenses financed by the federal government and 58% financed by cantonal government (18% by others) at cantonal universities (BFS, 2001a). These differences in financing regulatories between cantonal universities and UAS on the one hand, and FTHs and their annex-institutions on the other hand appear to be unbalanced. Cantonal universities have to negotiate their main budgets with cantonal governments, whereas FTHs are financed automatically by federal funds. This also implies, that FTHs are financed by taxes from all inhabitants, whereas, cantonal universities and UAS have to be financed mainly by inhabitants of the respective cantons plus some countervailing duties from other cantons. Some politicians and executives of cantonal universities and of UAS therefore, strive to maintain fairness through adaption

of the existing financing regulatories among all universities.

In general, these differences in financing are an expression of esteem for natural and engineering sciences being the most important science basic for Swiss industries. Therefore a few highly promising technology fields are promoted under the condition to reach excellence and to keep or reach world leadership in these fields.

Measuring R&D Activities in Universities

As universities are non-profit organizations their income flow should be equal to the output flow— although universities are able to accumulate reserves. This general standard can be applied on the financial flows as well as on the activities financed by the specific financial sources. In other words, money from acquisition from enterprises is supposed to be spent for applied R&D or services as instructed. Existing statistics are about time expense in universities, in fields of teaching and further education, R&D, and other activities. The average values of work time by fields of activities are listed as average, as they do not differ significantly between cantonal universities and FTHs, but mainly among each single university: the

---

[15] Financial data between universities and UAS is hard to compare, as it is structured and aggregated in different ways.

average values of work time is about 35% for teaching and further education, 48% for R&D, and about 17% for 'other activities' (data for 2000; BFS, 2001c). Although big efforts were made to gain these data, some crucial weaknesses still remain. First, the method of gaining this data is kind of 'rough sketching': as a lack of detailed working hours registration in most universities, the data about working hours are based on assumptions and guesses conducted by some personnel working in university administration. Second, the shares of work time spent by fields of activities are insufficient: They do not allow to draw connections between the sources of funds and the respective and accurate activities therefore conducted. For example, definitions of R&D and 'other activities' do not distinguish between basic or free R&D and oriented and applied R&D conducted for enterprises or public services. Also the definition of 'other activities' contains 'services' (like expertise) conducted for enterprises and other customers. For this reason it is virtually unknown, how much oriented and applied R&D is already conducted in universities, and how much basic or free research is left at universities and its faculties or disciplines.

In other words, applied research and services are already existing in universities but not sufficiently reflected in research, technology, and innovation policy with respect to differentiation between universities and UAS. Data does not exist about innovative activities in universities which may be in cooperation with external partners. Universities are still not in charge for proving their efficiency towards public money spent. No longer can this be regarded as a 'pièce de resistance' of old linear-transfer-model-thinking, wherein universities and their activities in basic research and teaching are 'per se' supposed to be the most useful precondition for innovations following.

Financing R&D in Industries

Also not surprisingly, private industry finances its R&D mainly itself: In the year 2000 Swiss industry spent 7.71 billions Swiss Francs for R&D conducted in their own enterprises (intramuros) (BFS & Economiesuisse, 2001). The sources of R&D financing in industry are composed of 91% from industry itself, 6% from foreign countries, mainly through participation in EU programs, 2% from the public sector, and about 1% from other organizations in Switzerland. Additionally, Swiss industry has spent another 1.7 billions Swiss Francs for mission oriented R&D, for supporting R&D, and for gaining know-how for further R&D (extramuros). These R&D mandates are spent for external institutions located in Switzerland and abroad (extramuros). Approximately 7%, or 125 million Swiss Francs, of these industrial expenses extramuros are spent on universities, which makes about 3% of all expenses of universities—without UAS. About 23% are spent on other enterprises for R&D, and about 65%

are spent for R&D on different institutions[16] abroad. Regarding all R&D expenses of Swiss industry of about 9.5 billions Swiss Francs (intra- and extramuros) the amount spent on universities in Switzerland is shrinking to 1%.

Research activities in industry are focused on two industrial branches, namely machine industry, pharmaceutical and chemical industry. They include 70% of all industrial expenses on R&D (intramuros) and about the same share of industrial R&D personnel (BFS & Economiesuisse, 2001). Compared to 1996 the amount of R&D personnel in 2000 has risen by 11 percentage points, which is virtually the same as the amount of R&D expenditure, risen by 12 percentage points.

As seen, there exist very little financial flows for R&D between private industry and universities: Private industry in Switzerland spends about 125 million SFr for mission-oriented R&D conducted in universities (in 2000). These flows have to be regarded as one indicator for interactions and knowledge transfer between both sub-systems. Financial flows based on R&D—as described above—make clear, that there is a huge industrial R&D system and a smaller scientific based R&D system. Both systems are separated with very little cross-overs: there are only a few R&D commissions from industry to universities or vice versa. Altogether these data are important hints that industry prefers to conduct R&D on its own and for its specific needs and purposes. It would be interesting to know, whether, and if yes, how the rate of commissioned R&D from industry to universities—mainly UAS—will change, as soon as UAS have developed their R&D potential.

Resulting in what is required: (1) Defining R&D and innovative activities at universities and at UAS is essential. This could be carried out in a sense of an Oslo Manual for universities and UAS; (2) to create a basis of measuring innovative activities and R&D at universities, better data about researcher's activities should be collected and assigned to their funding better than before. Therefore, a system of working-hours registration should be implemented; and (3) contacts and links between universities, their institutes and enterprises should also be measured and mapped more systematically for example by acquired funds, by kinds of contacts, and by regions.

To find out, whether industry or science between them has a need in interactions, knowledge and technology transfer and who really uses them, these forms of contact are discussed the following section.

Innovations: Structure and Expenses

In addition to the structure of financing R&D, innovation surveys are another source for interactions and knowledge transfer between science and economy. It is

---

[16] These institutions may contain universities as well as enterprises and others.

not intended to describe the whole structure of innovation landscape in Switzerland here. Instead only a few remarks about innovations in Switzerland shall be made. Therefore, the innovation survey from Switzerland with data from 1997 to 1999 (Arvanitis et al., 2001) is compared with the innovation survey from EU with data from 1996 to 1997 (Eurostat, 2001). In all states compared, large differences can be seen between manufacturing and service industries being innovative (Table 1). But as the measurement of innovative activities focuses on technological innovations, the question still remains, whether or to what extent are service industries less innovative than manufacturing industries or whether service industries have to be measured differently.

Manufacturing industry in the EU spends about 3.7% of its total revenue for product and process innovations, in the service industry it is about 2.8%. Again, the range between single member states is huge: it differs in the manufacturing industry between 1.8% in Portugal to 7% in Sweden, in service industry it differs between 1.1% in Portugal and 4.7% in Denmark. Companies in manufacturing industries in Switzerland have the second highest expenses as a share of total revenue, namely about 6% (data for 1994–1996), after Sweden with 7%, followed by Denmark with about 5% (data for 1996–1997) (OECD, 2001). In the services sector two different data exist for companies in Switzerland: one far below average, namely 1.7% of innovation expenses as a share of total sales (OECD, 2001) and one significantly above average, namely 3.5% (Arvanitis et al., 2001). The United Kingdom with 4% and Sweden with 3.8% have the highest shares of expenses for innovation activities (Arvanitis et al., 2001).

How many innovators are there in Switzerland internationally compared? All the following data contain average values, bearing the potential of large differences between different industries and between single enterprises.

In general, Switzerland ranks among top three innovative states in Europe. The share of innovative enterprises rises with the size of enterprises: For example in the manufacturing industry in Europe about 43% of small enterprises, 58% of medium sized

enterprises and about 79% of all large enterprises are innovators (Eurostat, 2001). This pattern can also be identified for service industry but on a lower level, namely 36%, 48% up to 73%. For Swiss manufacturing and service industries there is also a dependency between the size of enterprise and being an innovator, although it is a little less pronounced than the average of EU, especially with respect to the manufacturing industry. The ranges are from 77% for small enterprises up to about 82% for big enterprises in manufacturing industry, and from 58% up to 87% for large enterprises in the service industry. Very small enterprises from 5 to 19 employees are in average less innovative in both kinds of industry.

Compared to the less pronounced dependency between size of enterprise and being innovative there is a very pronounced dependency between size of enterprise and carrying out R&D: large enterprises are much more often conducting their own R&D than small and medium sized enterprises (Arvanitis et al., 2001; Eurostat, 2001). This is valid for Switzerland, for the whole EU and also for manufacturing and service industry in the EU, although the differences among service industries are not that distinctive. In Swiss manufacturing industry about 49% of all firms are doing their own R&D whereas about 18% of all service industries conduct R&D (Arvanitis et al., 2001). These data differ between firm sizes as well as between single industries and between countries.

Obstacles claimed to oppose (more) innovation activities by firms are: costs and risks for innovations, which are considered as being too high, problems in financing innovation activities, mostly based on a lack of own capital, lack of qualified personnel, and governmental regulations (Arvanitis et al., 2001). Universities and other R&D institutions are affected here only in the fields of 'qualified personnel', which is an important obstacle for firms. In other words, firms hardly address obstacles in a lack of knowledge and technology exchange with firms.[17] This is an average

---

[17] However, this kind of obstacle was not listed in the questionnaire of the Swiss innovation survey directly, but there was open space for further factors and acknowledgments.

Table 1. Share of innovative enterprises.

Share of innovators (in per cent)	Manufacturing industries				Service industries			
Size of enterprises	total	small	medium	big	total	small	medium	big
Switzerland (1999)[1]	71	*	*	*	54	*	*	*
EU (1996–1997)[2]	51	43	58	79	40	36	48	73
—Ireland	73	68	78	85	58	60	49	87
—Denmark	71	64	76	91	30	24	45	71
—Germany	69	63	70	85	46	41	60	83

Source: [1] Arvanitis et al., 2001; [2] Eurostat, 2001; * no data available because of different arrangement in groups.

opinion of innovative as well as non-innovative firms. That means first, obstacles can differ in their relevance for single industries and firms as well as for SMEs in general, as they suffer more in financing and in a lack of technology support than others. Second, it also means, that non-innovators are probably not able to fully incorporate these obstacles, because they do not know the relevance of these factors for innovations. This aspect is indeed reflected in obstacles like 'high costs and high risk', which is considered less important of non-innovators than of innovators. This leads to the assumption that the more experience firms have on innovations, the more their perception and knowledge about obstacles of innovation rise.

Innovation surveys conducted prove, that innovations cannot be equated with R&D (Arvanitis et al., 2001; Eurostat, 2001; Felder et al., 1994). As in the Swiss manufacturing industry about 71% of all enterprises are innovators, the share of enterprises doing R&D is about 49% (Arvanitis et al., 2001). In the service sector the share of innovators is 55%, whereas the share of enterprises doing R&D is 18% on average. That means many innovations in both industrial sectors are based on non-R&D activities. In the EU about 53% of all enterprises in service industry are innovative without doing R&D, whereas this share is about 30% in manufacturing industries.

With respect to the average values of innovators in Swiss industries as listed in Table 1, the highest share of innovators in manufacturing industry is in paper and wood industry, in electronic and electrotechnical industry, in textile industry, in chemical industry, in engineering, and in food industry (namely between 74% and 98%). In service industry electronic data processing and R&D companies, banks and insurance companies, hotel and restaurant industry, and service industry for companies are innovative above average, namely between 57% and 66%. These shares are early results of the decision each enterprise has to make when answering the questionnaire of innovation surveys, whether it has implemented innovations in a certain period under review: yes or no. That answer does not tell anything about the amount or intensity of innovative activities nor does it tell anything about the level of implemented innovations (like for example regarding the newness of innovation). Therefore, the intensity of innovation activities should be taken into account. While Eurostat considers the amount of expenses spent for innovation activities (Eurostat, 2001), the Swiss innovation survey constructs a complex indicator for innovation performance of each single industry, considering innovation intensity and non-innovative enterprises by each industry (Arvanitis et al., 2001). Therefore, about 23 single indicators for manufacturing industry and 19 for service industry are taken into account. Measured by this complex indicator in the manufacturing industry, vehicle construction industry, electronic industry, textile industry, engineer-

ing, and chemical industry have the highest innovation performance. In the service industry, the ranking as mentioned previously in this section is still valid except for the hotel and restaurant industry which does not any longer, perform above average.

Innovation Output

The ratio of sales of innovations has to be differentiated between innovations, which are new for the world and innovations, which are new or substantially improved for the enterprise. In contrast to this definition used in the innovation survey in Switzerland, Eurostat combines the category 'new or substantially improved for the market' and adds another category 'new or substantially improved for the enterprise' (Eurostat, 2001, 111 ff.). So the ratio of sales of innovations new to the world is 3.6% in Switzerland, and the one of innovations new or substantially improved for the enterprise is 37%. The total average in ratio of sales of innovations 'new or substantially improved for the market' is about 7%, and the one of innovations 'new or substantially improved for the enterprise' is about 26% in the EU (Arvanitis et al., 2001).

In general, a positive correlation is stated between being innovative and growth of sales. It is found to be more significant for product innovations than for process innovations (Arvanitis et al., 2001). Furthermore, firms being more innovative are performing better than firms being less innovative. Innovation activities are also found to correlate positive with highly qualified personnel and negative with low or unqualified personnel. For input factors no correlation could be found, whether positive or negative. This also means, that R&D in general does not play an all decisive role for innovations. But one of the most crucial findings is, that innovation activities are strongly affected by business cycles (Arvanitis et al., 2001). That means innovation activities depend more on positive business activities—or at least expected positive business activities—instead of generating them.

Contacts and Knowledge-Flows between Science and Economy

As already mentioned in the section 'R&D Structures and Expenses' industries as well as science each have its own research system. Universities are financed up to an average of 14% maximum from economy. Most of these contract activities conducted by universities are supposed to be applied R&D. Besides R&D-projects different forms of contacts and knowledge flows between both systems exist, which not necessarily are paid for or paid fully for and therefore cannot be measured by expenses sufficiently. In the EU innovation survey two main forms of information and knowledge sources are distinguished: (1) cooperation for innovation, composed of R&D-project cooperations, both sides being actively involved, and other innovation project cooperations; and (2) Using different sources for obtaining relevant knowledge for being innovative, for example, the own enterprise, clients and competitors, universities, and exhibitions (Eurostat, 2001). In contrast to point 1, Swiss innovation survey in 1999 is just measuring cooperation in R&D-projects and –activities but no other innovation project cooperations (Arvanitis et al., 2001).

Considering all enterprises in Switzerland carrying out R&D, which is on average 49% in manufacturing and 18% in service industry, together about 31% have R&D cooperation partners. The share of having R&D cooperations is approximately 27% in manufacturing and 37% in service industry. These shares of cooperation partners just refer to companies with their own R&D activities, which is less than one third of all enterprises. The respective partners for these R&D cooperations are listed in Fig. 8.

Most partners for R&D cooperation are suppliers of material and customers, followed by university-level institutions. The most pronounced R&D cooperations between firms and scientific institutions can be stated in electrotechnical industry, where 86% of all enterprises have cooperations. The highest rate of R&D cooperations in the service industry with scientific institutions have services for enterprises with about 52%, whereas in the service sector banks and insurance industry have the lowest rate. Although there is no differentiation made between different forms of R&D cooperation with the respective partners, some forms of cooperation exclude some kinds of partners. Here it is obvious to consider universities and other university-level institutions as partners for typical forms of R&D

contracts, in order to conduct R&D projects in common by sharing resources, as well as for informal information and for technology exchange, or more exactly: for exchange of technological know-how. R&D cooperation like joint-ventures as well as minority stakes with universities are not imaginable. These forms of cooperation are confirmed by the reasons for choosing these cooperations: for firms, the main reasons for R&D cooperations with universities are 'having access to specialised technology', 'gain complementary', and 'shorten development processes' (Arvanitis et al., 2001).

The low rate of industry-partners in other R&D institutions in Switzerland is also due to the fact, that besides universities only a few research institutions exist. Even they are mainly engaged in basic research, and only very few institutions are specialised in applied R&D. That means firms in Switzerland almost solely have a choice between university-level institutions and other enterprises, like consultants, for R&D Partnership. Second, firms have little choice between institutions specialised in applied, non-basic-oriented R&D, which are not university institutions. In other words, the gap between R&D knowledge of companies and R&D knowledge of science and higher education institutions is not bridged sufficiently in Switzerland because of the absence of institutional thickness.[18] In

---

[18] For the term 'institutional thickness' see Amin & Thrift (1993). In short, it means the existance of diversified institutions, being preliminary and downstream oriented along the chain of value added as well as supporting institutions by providing knowledge, information, consulting and assistance.

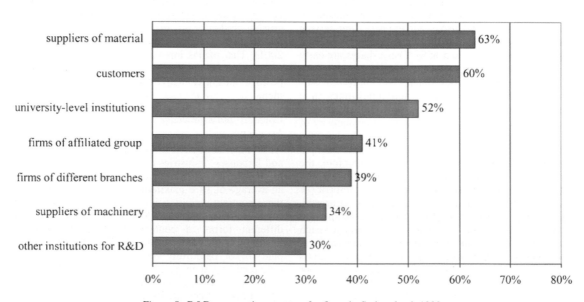

Figure 8. R&D cooperation partners for firms in Switzerland, 1999.

Source: According to Arvanitis et al. (2001).

general, networks of innovation between science and economy are not supported intensively by Swiss politics. Therefore, the connections between economy and science in Switzerland are regarded as improvable (Schmoch, Grupp & Laube, 1996).

Besides the forms of R&D cooperation listed above, there are some more forms existing, regarding R&D as well as innovation activities. For example technology exchange does not include other possible forms of technology exchange between universities and enterprises for example, loan or donation of machinery from industry and conducting studies by students in return. These forms of cooperation are already existing, being explicitly formulated or not, as studies for Switzerland, Germany, and Austria have proven (Balthasar et al., 2000; Wilhelm, 2000). Other forms of cooperation between firms and universities are already existing, like financing or sponsoring chairs or institutes at universities, but are still a single case phenomena in Switzerland.[19]

By default of data about cooperation partners in innovation activities for Swiss industries, some results from CIS 2 survey in Europe are presented (Eurostat, 2001). This is done to enlighten the meaning and importance of cooperation partners in innovation activities compared to cooperation partners for R&D (see Fig. 8). In Europe about 51% of all enterprises in manufacturing industries and 40% of all enterprises in services are innovative, about 27% of all innovative enterprises in manufacturing industry and 24% in service industry have cooperation in innovation processes (Eurostat, 2001). These shares differ somewhat between the EU member states. The most important groups of cooperation partners for innovative activities are located in the own affiliated group, followed by partners at clients and suppliers. Universities and other R&D institutions are also accepted as innovation partners on fourth and fifth rank. In all industries the five main aims are intended to reach by innovative activities, namely to improve product quality, to extend the variety of products and to open up new markets, to reduce labour costs, and to make production more flexible.

Information Sources

The innovation process is a multifaceted process, which is influenced by different kinds of information sources. The questionnaire of innovation surveys in Switzerland as well as in EU asks for the importance of different information and know-how sources for the respectively innovation activities in each company. However, this does neither correspond exactly with the use of these sources nor with the contentment about

this source nor the contentment resulting from the use of these sources. With respect to the real use of different information or knowledge sources their respective importance can be over- or underestimated. It is important to consider the respective question regarding information sources in both innovation surveys is different and therefore hard to compare directly.[20]

Information sources first have to be distinguished between such as the own firm (internal) and such being external (including firms of affiliated group). The Swiss innovation survey of 1999 does not consider internal information sources in its questionnaire, which is hard to understand, as internal information source is known to be the most important information source. Considering external information sources, three components have to be differentiated: knowledge and information from other firms (like customers, suppliers, competitors, joint ventures), science-based knowledge and information from universities and other R&D institutions, patent specifications etc, and information open for the public (e.g. exhibitions, conferences, literature, computer-supported information systems). In all cases implicit (or: embodied) and explicit (or: disembodied) knowledge is asked for and used by firms.

In Switzerland, companies along the value-added chain are regarded as most important information sources for innovation activities (see Fig. 9). Also general information sources open for the public, for example exhibitions, are regarded as an important source, although this explicit named source is more relevant to the manufacturing industry rather than to the service industry. Universities are considered to be less important than issue conferences and literature, but more important than other R&D institutions. This may also be due to the fact of few R&D institutions existing in Switzerland.

For every other firm in the EU the intern knowledge source in the own company is very important (Eurostat, 2001; Fig. 9). The most important external source is the one of customers. Generally said, companies along the value-added chain and competitors are playing a crucial role as information source for innovations. Universities and other R&D institutions, patent specifi-

---

[19] For example the chair of Entrepreneurship at the University of Lausanne.

[20] In the CIS 2 survey, information sources are judged on a scale of 1 to 4, where 4 is considered as 'very important source', and only rank 4 is taken into account. Information sources are judged on a scale of 1 to 5 in the Swiss innovation survey, where 4 is considered as 'important source' and 5 is considered as 'very important'. Here both ranks (4 and 5) are taken into account. Furthermore, Swiss innovation survey regards companies with five employees or more, whereas CIS 2 considers only companies with 20 employees or more. It is assumed, that these very small sized companies use less information sources and therefore results for Switzerland could be underestimated.

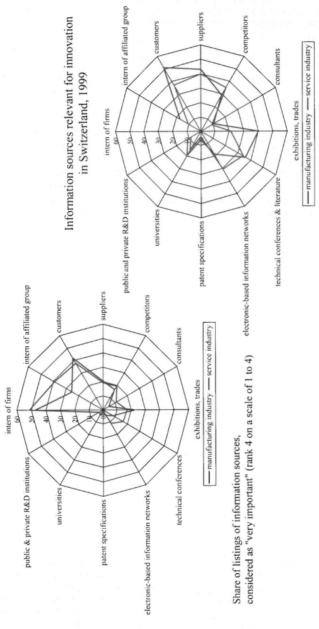

Figure 9. Information sources relevant for innovation activities in Switzerland, 1999, and in EU 1996–1997.

Source: Data based on Eurostat, 2001; Arvanitis et al. (2001).

cations, and even consultants are regarded as far less important. It is obvious, that these institutions are not able to provide much information and know-how necessary for direct and short-term implementation for innovation activities. Results from Swiss innovation survey prove, that universities are accepted as important R&D cooperation partner although they are not a very important source for information in innovation activities (Arvanitis et al., 2001). In general, universities, R&D institutions, and patent specifications are regarded as important but not the most important source for R&D. They offer complementary scientific and technological know-how and access to know-how to shorten development processes in companies. But they are not regarded as important knowledge source for innovation activities. But there are some large differences in the ranking and importance between different external information sources, among Swiss innovation survey in 1996 and 1999. These differences may be caused by cyclical variations. In any case they make it hard to compare and to come to conclusions about the real importance of these sources.

As there are only a few indicators in innovation surveys measuring innovation activities mainly at the interface between science and industry other relevant studies like case studies have to be drawn on. Empirical surveys concentrating on different forms of cooperation and information exchange between universities (and other R&D institutions) and firms or networks of different protagonists in R&D and innovation activities are rare. There are a few results from these surveys presented here in short, always focused on interaction between universities, other R&D institutions and firms or networks.[21]

Case Studies About Knowledge Transfer and Interactions Between Universities, R&D Institutions and Firms

Audretsch & Stephan (1996) focused on the importance of spatial closeness between universities, scientists and biotechnology firms in the USA for cooperation. They state a positive influence of spatial closeness on the emergence of cooperations, but this influence is far from being overwhelming. Also, the importance of closeness depends on the kind of collaboration, which is for example much more important for very intensive collaboration being the co-founder of a new start-up than for project cooperation.

Fritsch & Schwirten (1998) analyzed forms and extend of collaboration between universities, universities of applied science (UAS), other R&D institutions, like Fraunhofer-Institutes, Max-Planck-institutes, institutes of so called 'blue list', and firms in

three regions in Germany.[22] Therefore, in total 246 chairs or institutes were examined only in disciplines supposed to being able to foster innovation processes in industries. Asked about whether each chair (at universities and UAS) or institute (other R&D institutes) ever had some collaboration with firms in the period between 1993 to 1995 about 78% of all institutions agreed. However, this does not tell anything about quality and intensity of these collaborations. Among all institutes asked, R&D institutions outside universities work together with firms most frequently (91%), followed by chairs of UAS (83%) and by chairs of universities (74%). However, about 34% of all firms in these respective regions asked did work together with universities, UAS, and other R&D institutions in the period under review. For research institutions the most important sources to initiate collaboration with firms are 'personnel contacts of employees' (39%), 'approach of firms' (29%), 'contacts on exhibitions and congresses' (14%), whereas 'place of contacts by transfer institutions' are only 4%. The use of electronic databases is marginally less than 1%. So the common organized technology transfer institutions and their acting as a broker shows only little potential impact in this field. The role of R&D institutions is to support firms in the early stages of innovation activities, namely in the development of new ideas and concepts. Here, in all phases of innovation activities other R&D institutions play a crucial role in supporting firms. Spatial closeness is beneficial for collaboration but not crucial.

Czarnitzki, Rammer & Spielkamp (2000) have a similar survey concept like Fritsch & Schwirten (1998): They asked approximately 850 research institutions at universities, technical oriented universities (TOU), universities of applied science (UAS), and 4 R&D institutions outside of universities ('others') in natural and engineering science in Germany about their interactions with industry. These interactions are commonly understood as knowledge and technology transfer. The authors concentrated on intensity of interactions, the forms of transfer used, obstacles and problems of knowledge and technology transfer. Figure 10 shows the main finding of this survey, namely a

---

[21] For a sample of studies about structure and intensity of mainly technology-oriented transfer in Germany see also Schmoch, Licht & Reinhard (2000).

[22] The regions analyzed were Hannover, Baden, and Saxony. To even be able to asses the extend of cooperation of 'other R&D institutions', it is important to know about their financial structure: universities and UAS in Germany are principally financed similar as in Switzerland: about 14% of their total expenses of 35.7 billion DM (about 18 billion Euro) are financed by acquired third-party-funds. However looking at other R&D institutions, only institutes of the Fraunhofer-Association are designed to conduct applied R&D and therefore have to finance themselves mostly by acquired third-party-funds, which up to now have a share of about 60% of their whole expenses (this share does not consider four Fraunhofer-institutes operative in defense-oriented R&D; BMBF 2000).

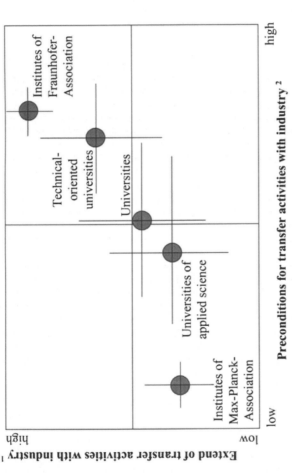

Figure 10. Typology of research institutions in Germany in terms of interactions with firms.

Source: According to Czarnitzki, Rammer & Spielkamp (2000).

typology of all research institutions analyzed with respect to their interactions with industry. There, Fraunhofer-institutes and TOU show the highest rates in interaction with firms as well as the most favorable preconditions for these interactions. Although in Germany (as well as in Switzerland), UAS are supposed to play a crucial part in knowledge transfer, but as can be seen here, they have worse preconditions and less interactions with firms than universities. Obviously UAS are still unable to fulfil their task in applied R&D and intensive interactions with firms, although they have been in operation for more than 30 years. The main obstacles for this weak link between UAS and firms are indicated by UAS in too much teaching, which plays a very crucial role therefore, and in a lack of qualified personnel. Additionally the payment structure of UAS personnel is discriminating compared to the one of universities, which may also cause the weak position of UAS stated[23] (Schmalholz in: Schmoch, Licht & Reinhard, 2000). Institutes busy in applied and industry-oriented R&D and therefore significantly financed by acquired third-party funds have much more interactions with industries than other institutes. The connection between R&D orientation, financial structure of R&D institutes and their extend of interaction activities with industry is also stated by further studies[24] (Balthasar et al., 2000; Wilhelm, 2000).

Hellmer, Friese, Kollros & Krumbein (1999) examined innovation activities of SMEs in two regions in north Germany and their use of, and relation to, cooperations for gaining innovations in three industries.[25] They found that on the one hand SMEs rely on cooperation in innovation activities, because of their small resources. But on the other hand the risk of failure of resource intensive cooperation is much too high for SMEs. Therefore, SMEs have less coopera-

tions than large sized enterprises. Most of their product innovations are incremental, based on development of existing products, partly as a result of collaboration with clients. The few existing cooperations are oriented towards technological excellence and competence of the respective partners but they are not primarily oriented towards partners located in their specific region. This is also valid for cooperations with universities and other R&D institutions. Regional cooperations among firms in their innovation processes are stated to be much less pronounced than theoretical frameworks about regional innovation networks and cooperation suggests. For the few cases of spin-offs from R&D institutions examined here, the authors also found out, that they are located close to their former employer, if further cooperation or exchange is wanted. Again, the excellence in technological or other forms of know-how is a decisive factor for cooperation, as well as trustful personnel contacts to well-known institutes and colleagues.

Peters & Becker (1998) observed the effects of academic research on innovation activities of firms in the German automobile supply industry. They found that the contribution of academic research to innovation activities in firms is less relevant than industrial sources, but for R&D cooperations, universities are the most likely partners. However, they conclude that R&D cooperations, between universities and firms seem to have a positive impact on the improvement of existing products and processes rather than the development of innovations.

In total, these studies about interaction and collaboration between universities, R&D institutions and firms revolve around structures and intensity of interactions. Yet, they do not allow general conclusions about the extent, the efficiency and impact of these collaborations and interactions. Aspects of impact are mainly considered as economic impact in appropriate studies.

Economic Impact of Public Funded Basic R&D for Innovation Processes in Industry and for Economic Growth

The discussion about knowledge-based society shows an increasing need for knowledge and for know-how about how to handle this knowledge. Therefore, it seems obvious, that there is an increasing need in academics and their scientific knowledge—although the knowledge of the knowledge based society cannot be automatically equated with scientific and academic knowledge. However, the return on investment in higher education and scientific R&D is fostered by politics by fostering knowledge transfer. Knowledge transfer is politically and economically intended to result in improving and fostering innovation activities in firms may also be in other institutions of government NGOs. But all gained effects and impacts resulting out of knowledge transfer are yet to be examined. Different

---

[23] Schmalholz concludes, that UAS are still a young kind of R&D infrastructure, only in operation since 1969 (in the western part of Germany), which is too short in time to be able to have established an equivalent but different kind of university, and being accepted in industry as much as universities. But it is not comprehensible, why 30 years should not be enough to gain this position.

[24] For Switzerland up to now there are no extensive and basic studies about knowledge and technology transfer available. Interaction between engineers in firms and universities and different kinds of R&D and transfer institutions were examined for two branches, namely mechanical engineering industry and plastics processing industry for Switzerland, Austria and Baden-Wurttemberg (in Germany) (see Balthasar et al., 2000; Wilhelm, 2000).

[25] The two regions examined were the region of southern Lower Saxony and the region of ‚Oldenburg county'. Therein co-operation structure of SMEs in mechanical engineering industry, electrical industry, and precision engineering and optical industry were surveyed. 37 expert interviews conducted with member of the executive board of these SMEs are empirical basis of this survey, which therefore has to be judged as more qualitative survey.

approaches of impact studies revolve around these issues in economic literature: (1) studies about economic impact of public funded basic research (e.g. Cohen, Nelson & Walsh, 2000; Jaffe, 1989); (2) studies about economic impact of public funded infrastructure (e.g. Conrad & Seitz, 1994; Mintz & Preston, 1993); (3) studies about economic and regional impact of public funded infrastructure such as universities (most of them are case studies: e.g. Fischer & Wilhelm, 2001; Pfähler, 1997, 1999); (4) studies about economic and regional impact of knowledge-spillovers and technology transfer (e.g. Acs, Audretsch & Feldman, 1999; Beise & Spielkamp, 1996; Feldman, 1999; Frey & Brugger, 1984; Griliches, 1992; Saxenian, 1994); and (5) studies about private and social benefit of higher education (e.g. Bassanini & Scarpetta, 2001; Borland et al., 2001; Temple, 2000; Wolter & Weber, 1998).

Salter & Martin (1999) present a review on economic benefits of publicly funded basic research as mentioned in point (1) above. In the literature three main methodological approaches can be stated: first, econometric studies, second, economic surveys, and third, case studies. Econometric studies of course, always bear methodological limitations, based on specific assumptions about impact correlation and based on demarcations involved protagonists and correlations are regarded. In these studies assumptions about R&D impacts on industry have mostly been supposed to be very substantial. Therefore, the results of these studies show huge ranges in their social rate of return, which means the benefits of public funded basic R&D, which accrue to the society: The ranges are between 10% to 160% with a concentration around 50% (Salter & Martin, 1999). However, in some spatial economics there is only weak evidence of positive externalities for university research (Acs, Audretsch & Feldmann, 1992; Anselin, Varga & Acs, 1997; Jaffe, 1989). Anselin, Varga& Acs (2000) found evidence of sectoral and regional differences of positive externalities from R&D.

Mansfield made a survey about the benefits of recent academic research for firms, published in 1991, and a follow-up study published in 1998. In the first study a sample of R&D managers from 76 U.S. firms were asked for the proportion of a firm's products and processes over a ten-year period, which could not have been developed without academic research. The sample of the second study was 70 U.S. firms. In 1991 he found, that about 11% of new products and 9% of new processes could not have been developed without academic research within this time period (Mansfield, 1991). That share of new products and processes, based on academic research account for 3 respectively 1% of sales. The follow-up study shows an increasing importance or influence of academic research: Now about 15% of all new products and 11% of all new processes could not have been developed without academic R&D within this time period (Mansfield, 1998). These

products and processes account for 5% of total sales for the firms examined. These are very optimistic findings compared to a survey conducted by Beise & Stahl for 2,300 firms in manufacturing industries (Beise & Stahl, 1999). They found about 5% of all new products and processes are based on support of academic R&D. They also stated, that small firms are less likely to use academic support and R&D, which is also a main result in nearly each innovation survey conducted.

Economic surveys and case studies show that economic benefits of public funded R&D can take a variety of forms—they all vary within the scientific field, technology and industrial sector. The emergence of new and perhaps useful knowledge, out of public funded basic R&D is one type of benefit, but not necessarily the most important one. Also, new instrumentation and methodologies, skills incorporated in graduates, like tacit and codified knowledge, access to networks, to experts and information, as well, as gaining the ability to solve complex problems by participation in basic research. Finally, the creation of spin-off companies is considered as a further form of benefit resulting from basic research (Salter & Martin, 1999). All these benefits are also listed and partly incorporated in the types of knowledge transfer as seen in Fig. 4. In general, a great heterogeneity in the relationship between basic research and innovation is stated, so that no simple model of economic benefits from basic research is available (Salter & Martin, 1999).

Regarding the impact and benefits of universities for their specific regions, there is strong evidence found in the literature of local academic technology transfers. But regarding the effects of university technology transfer on local economic development the evidence is still vague (Varga, 1997). The impact study about knowledge transfer between the University of St. Gall, their region and up to international regions and protagonists even intensifies this vagueness (Fischer & Wilhelm, 2001[26]): The region directly and significantly benefits from expenses of students and employees and of university administration for office equipment and so on. But the benefits from knowledge transfer are vague. Measured by commissioned R&D studies, R&D cooperations, consulting and other services for customers, proximately 5% to 10% of this commissioned work is carried out for clients in the region of this university. This also holds true for further education

---

[26] The study was conducted for the accounting year 1999, therefore, data is based on this one year. Compared to other regional impact studies about universities this study is based on origin data in important issues and not only on estimations. Additionally, a study about spin-offs of graduates and former employees of universities in the region on St. Gall was conducted (Thierstein/Wilhelm/Behrendt, 2002). This is the first one revolving around this issue, which was conducted in German speaking countries—as far as the authors found out.

seminars, which are demanded by about 10% from people of the region. Also seminar papers, diploma and doctoral theses rarely have a regional context and are seldom conducted in regional localized firms or institutions (about 3%). However, spin-offs and personnel transfer shed another light on transfer structure. An additional study was conducted in order to measure the total amount of spin-offs, generated since the existence of the university within today's structure, and since the operational existence of its Alumni organization in 1963 (Thierstein, Wilhelm & Behrendt, 2002). Although, the total amount of spin-offs could not be calculated from the survey, the number of spin-offs founded in St. Gallen, where the university is located, was surprisingly high: It was virtually as high as in Zurich the largest city in Switzerland. But translated to inhabitants, the most spin-offs were founded in St. Gallen.[27]

Summary

To sum up, the equation of scientific knowledge production process with innovation process bears an asymmetry in the understanding of innovations: scientific knowledge production is not primarily intended to result in innovations. Innovations are not only based on scientific knowledge but often by innovation activities in firms and by interactions and feedback-loops between the 'unscientific' and the scientific knowledge production systems. As there is a need in bridging the gap between science, industries, and governmental systems in order to meet future demands in problem-solving knowledge generation, the innovation system in Switzerland has to be improved. First and generally accepted is a need for developing innovation policy, which includes better coordination between science, industrial, and government systems and also society demands. Coordination among these systems has to be supported by appropriate incentives for protagonists involved, mainly within the science system. Still there is a lack of robust knowledge about innovation processes, innovation systems, and knowledge transfer including, basic aspects of its structure, impacts and flows. Also the implementation of structures of intermediates, designed to bridge the gap between science and industry (as carried out in the Fraunhofer-institutions in Germany), should be considered in this context. Therefore, basic and action-oriented studies, political concepts, strategies, and knowledge about developing innovation systems are required. Last but not least, the results of these studies then have to be secured so as to have an impact on developing innovation policy.

The existence of two large but separated R&D systems in science and industries connected with very little financial flows, leads to reflections about their respectively structure, knowledge flows and their aims. The enormous R&D system in industries is hardly known by its fields, its know-how and findings, and its structure. But to know this is crucial under the aspect of the generation of new knowledge and know-how in networks of relevant protagonists. However, this requires studies about the whole R&D landscape and structure in Switzerland as well as conceptual thinking and studies about how to connect both systems more intensively and effectively. Therefore, a kind of 'R&D database' should be created and established on a national level being oriented internationally to offer matching processes between demand and offer. That also would ease self-organized interactions among both R&D systems on a national but also on an international level. The need for conceptual studies and implementation is required under the aim of 'problem-solving' as completion—not replacement—for generating epistemological knowledge.

In the context of knowledge production, another question is raised about job-sharing between universities and UAS and their respective tasks. The Swiss concept about 'equally good but different' bears some unsolved aspects. As UAS are also supposed to do applied R&D in order to mainly support SMEs in their respective region, reality shows that UAS in Switzerland are still focused on job-oriented teaching. Swiss politics will have to reconsider the task of UAS and universities or it will have to raise the support for UAS, enabling them to fulfil their tasks in applied R&D, knowledge and technology transfer.

As R&D are only part of innovative activities, the focus has to be broadened to other forms of interaction between universities, firms, government, and other kind of institutions. Innovation surveys, as they are conducted up to now in Switzerland, reflect the linear model of knowledge and technology transfer, by asking firms as receiver and for utilization of scientific knowledge. Therefore this shows a limited insight of all innovation activities and processes taking place. Innovation surveys therefore, have to be broadened to all kinds of relevant protagonists in an innovation system (e.g. universities, other R&D institutions, government, NGOs) and their network of information and knowledge flows.

Nowadays university institutes are more and more dependent on third-party funds. Reductions in public basic budgets for higher education are used as an incentive to rise acquisition activities of universities and their interaction with firms, as new financing source and may be as sponsor of university chairs. But this measure cannot be applied on all disciplines equally, as industrial partners or just markets, are not available for all disciplines in the same dimension. However, that approach causes strong initiative and self-organisation of university institutes and its employees and of course the abilities and skills to do

---

[27] Of course, this spread of spin-offs cannot be considered as representative: from 10,430 graduates addressed, about 1,396 answered; therefore 29% indicated to have founded a firm.

so. However, universities need markets for their knowledge or may even have to develop them first. New fields of disciplines and markets may arise in the intersection of different disciplines. There new markets could open up for oriented basic as well as for applied R&D and even for implementation. Therefore, alternative validation rules and career paths in universities and in UAS have to be developed. Up to now validation rules and the career paths in universities hinder interactions between science and industries as they are marked by teaching and 'publish or perish'. This rule even holds true for some UAS, which is contradictory to the tasks and aims of these universities. In both types of universities the development of new forms or at least broadening the spectrum of career paths is essential.

As innovation processes are known not to proceed in a linear way, the task of transfer institutions has to be reconsidered. The old concept of commercialization of scientific knowledge has to be completed and partly replaced. Given that universities want, or have to, raise their third-party funds, transfer institutions need a profile more in initiate and moderate interactions between university, industry, government, and other institutions. At the same time, transfer institutions should keep their supportive role in juridical and business consulting and coaching, accompanying project partners along their activities.

References

Acs, Z. & Audretsch, D. B. (1993). Analysing innovation output indicators: The U.S. experience. In: A. Kleinknecht & D. Bain (Eds), *New concepts in innovation output measurement* (pp. 10–41). Houndsmill: The Macmillan Press.

Acs, Z., Audretsch, D. & Feldman, M. (1992). Real effects of academic research: A comment. *American Economic Review*, **82**, 363–367.

Amin, A. & Thrift, N. (1993). Globalisation, institutional thickness and local prospects. *Revue d' Economie Regionale et Urbaine*, (3), 405–427.

Anselin, L., Varga, A. & Acs, Z. (2000). Geographical spillovers and university research: A spatial econometric approach. *Growth and Change*, **31** (4), 501–515.

Anselin, L., Varga, A. & Acs, Z. (1997). Local geographic spillovers between university research and high technology innovations. *Journal of Urban Economics*, **42**, 422–448.

Arvanitis, S., Bezzola, M., Donzé, L., Hollenstein, H. & Marmet, D. (2001). Innovationsaktivitäten in der Schweizer Wirtschaft. Eine Analyse der Ergebnisse der Innovationserhebung 1999. Strukturberichterstattung Nr. 5. Bern: Staatssekretariat für Wirtschaft (seco).

Arvanitits, S., Donzé, L., Hollenstein, H. & Lenz, S. (1998). Innovationstätigkeit in der Schweizer Wirtschaft. Teil 1: Industrie; Teil 2: Bauwirtschaft und Dienstleistungssektor. Eine Analyse der Innovationserhebung 1996. Strukturberichterstattung. Bern: Bundesamt für Wirtschaft (BWA).

Audretsch, D. B. & Stephan, P. E. (1996). Company-scientist locational links: The case of biotechnology. *American Economic Review*, **86**, 641–652.

Balthasar, A., Bättig, C., Thierstein, A. & Wilhelm, B. (2000). Developers: Key actors of the innovation process. Types of developers and their contacts to institutions involved in research and development, continuing education and training, and the transfer of technology. *Technovation*, **20**, 523–538.

Bassanini, A. & Scarpetta, S. (2001). Does human capital matter for growth in OECD countries? Evidence from pooled mean-group estimates. Economics Department Working Paper. No. 282. ECO/WKP (2001) 8. Paris: OECD.

Beise, M. & Spielkamp, A. (1996). Technologietransfer von Hochschulen. Ein Insider-Outsider-Effekt. Discussion Paper. No. 96–10. Mannheim: ZEW.

Beise, M. & Stahl, H. (1999). Public research and industrial innovations in Germany. *Research Policy*, **28**, 397–422.

Biegelbauer, P. (1997). To be innovative or not to be. Innovative behaviour of Austrian companies. *Wirtschaftspolitische Blätter*, (5), 517–526.

Borland, J., Dawkins, P., Johnson, D. & Williams, R. (2001). Rates of return to investment in higher education. *Australian Social Monitor*, **4** (2), 33–40.

Bosworth, D., Stoneman, P. & Sinha, U. (1996). Technology transfer, information flows and collaboration: An analysis of the CIS 1. European Innovation Monitoring System. Publ. No. 36.

Brouwer, E. & Kleinknecht, A. (1996). Determinants of innovation: A microeconometric analysis of three alternative innovation output measures. In: A. Kleinknecht (Ed.), *Determinants of Innovation and Diffusion* (pp. 89–128). London: Macmillan.

Bundesamt für Konjunkturfragen (BFK) (1992). *Technologiepolitik des Bundes*. Bern. Bundesamt für Konjunkturfragen.

Bundesamt für Statistik (BFS) (2001a). *Indikatoren 'Wissenschaft und Technologie'* (http://www.statistik.admin.ch/stat_ch/ber15/dber15.htm).

Bundesamt für Statistik (BFS) (2001b). *Finanzen der universitären Hochschulen 2000*. Reihe 15: Bildung und Wissenschaft. Neuchâtel.

Bundesamt für Statistik (BFS) (2001c). *Personelle und finanzielle Ressourcen der Hochschulen 2000*. Reihe 15: Bildung und Wissenschaft. Neuchâtel.

Bundesamt für Statistik (BFS); Economiesuisse (2001). Indikatoren 'Wissenschaft und Technologie'. Erhebung über die Forschung und Entwicklung (F+E) in der schweizerischen Privatwirtschaft. Pressekonferenz, 20. Dezember 2001. Reihe 15: Bildung und Wissenschaft. Neuchâtel.

Bundesministerium für Bildung und Forschung (BMBF; Deutschland) (2002). Zur technologischen Leistungsfähigkeit Deutschlands 2001 (anually conducted). Bonn.

Bundesministerium für Bildung und Forschung (BMBF; Deutschland) (2000). Bundesbericht Forschung 2000. Bonn.

Caracostas, P. & Muldur, U. (1998). Die Gesellschaft. Letzte Grenze. Eine europäische Vision der Forschungs- und Innovationspolitik im 21. Jahrhundert. Luxemburg.

Christoffel, J. (1995). Unproduktive Schweizer Wirtschaft? *Die Volkswirtschaft*, (8), 36–41.

Cohen, W., Nelson, R. & Walsh, J. (2000). *Links and impacts: Survey results on the influence of public research on industrial R&D*. Carnegie: mimeo.

Commission of the European Union (CEU) (1995). *The green paper on innovation*. Brussels.

Commission of the European Union (CEU) (1993). Growth, competitiveness and employment—challenges of the present and ways towards 21st century. Delors White Paper. Brussels.

Conrad, K. & Seitz, H. (1994). The economic benefits of public infrastructure. *Applied Economics*, **26**, 303–311.

Czarnitzki, D., Rammer, C. & Spielkamp, A. (2000). Interaktionen zwischen Wissenschaft und Wirtschaft in Deutschland. Ergebnisse einer Umfrage bei Hochschulen und öffentlichen Einrichtungen. Dokumentation. Nr. 00–14. Mannheim: ZEW.

Diem, M. (2000). *Von der universitären Hochschule ins Berufsleben. Absolventenbefragung*. Neuchâtel: BFS.

Ebling, G. & Janz, N. (1999). Export an innovation activities in the german service sector: Empirical evidence at the firm level. Discussion Paper No. 99–53. Mannheim: ZEW.

Edge, D. (1994). reinventing the wheel (pp. 2–23). In: S. Jasanoff et al. (Eds), *Handbook of Science and Technology Studies*. Thousand Oaks/London/New Delhi: Sage.

Ehrenberg, H. (1995). Die Standortdebatte jenseits der gesamtwirtschaftlichen Fakten. *Wirtschaftsdienst*, **7**, 366–370.

Ellwein, T. & Bruder, W. (1982). *Innovationsorientierte Regionalpolitik*. Opladen: Westdeutscher Verlag.

European Commission (EC); Eurostat (2001). *Innovationsstatistik in Europa*. Daten 1996–1997. Luxembourg.

Eurostat (1997). *Jahrbuch 1997*. Luxemburg.

Ewers, H-J. et al. (1980). *Innovationsorientierte Regionalpolitik. Schriftenreihe des Bundesministeriums für Raumordnung, Bauwesen und Städtebau*. Bonn.

Federal Council (1998). *Botschaft über die Förderung von Bildung, Forschung und Technologie in den Jahren 2000–2003*. Bern.

Federal Council (1997). *Bericht des Bundesrates über die Umsetzung der Technologiepolitik des Bundes*. Bern.

Felder, J., Harhoff, D., Licht, G., Nerlinger, E. A. & Stahl, H. (1994). *Innovationsverhalten der deutschen Wirtschaft. Ergebnisse der Innovationserhebung 1993*. Mannheim: ZEW.

Feldman, M. P. (1999). The new economics of innovation, spillovers, and agglomeration: A review of empirical studies. *The Economics of Innovation and New Technology*, **8**, 5–25.

Felt, U., Nowotny, H. & Taschwer, K. (1995). *Wissenschaftsforschung*. Frankfurt A.M./New York: Campus.

Fischer, G. & Wilhelm, B. E. (2001). Die Universität St. Gallen als Wirtschafts- und Standortfaktor. Ergebnisse einer regionalen Inzidenzanalyse. Schriftenreihe des IDT-HSG. Beiträge zur Regionalwirtschaft. Nr. 3. Bern; Stuttgart; Wien: Paul Haupt.

Freiburghaus, D., Balthasar, A., Zimmermann, W. & Knöpfel, C. (1991). Technik-Standort Schweiz. Von der Forschungs- zur Technologiepolitik. Bern; Stuttgart; Wien: Paul Haupt.

Frey, R. L. & Brugger, E. A. (1984). Infrastruktur, Spillovers und Regionalpolitik. Methode und praktische Anwendung der Inzidenzanalyse in der Schweiz. NFP 'Regionalprobleme'. Diessenhofen: Rüegger.

Fritsch, M. & Schwirten, C. (1998). öffentliche Forschungseinrichtungen im regionalen Innovationssystem. Ergebnisse einer Untersuchung in drei deutschen Regionen. *Raumforschung und Raumordnung*, **56** (4), 253–263.

Gibbons, M., Limoges, C., Nowotny, H., Schartzman, S., Scott, P. & Trow, M. (1994). *The new production of knowledge. The dynamics of science and research in contemporary societies*. London.

Grabher, G. (1993). The weakness of strong ties. The lock-in of regional development in the Ruhr area. In: G. Grabher (Ed.), *The Embedded Firm. On the Socioeconomics of Industrial Networks* (pp. 255–277). London and New York: Routledge.

Griliches, Z. (1992). The search for R&D spillovers. *Scandinavian Journal of Economics*, **94** (Supplement), S29–S47.

Grimmer, K., Häusler, J., Kuhlmann, S. & Simonis, G. (Eds) (1992). Politische Technikgestaltung. Schriften des Institut Arbeit und Technik. Bd. 5. Opladen: Leske + Budrich.

Grupp, H. (1997). Messung und Erklärung des technischen Wandels. Grundzüge einer empirischen Innovationsökonomik. Berlin et al.: Springer.

Grupp, H. & Legler, H. (1987). Spitzentechnik, Gebrauchstechnik, Innovationspotential und Preise. Trends, Positionen und Spezialisierung der westdeutschen Wirtschaft im internationalen Wettbewerb. Köln: TüV Rheinland.

Hauff, V. & Scharpf, F. W. (1975). Modernisierung der Volkswirtschaft. Technologiepolitik als Strukturpolitik. Frankfurt A.M.: Verlagsanstalt.

Hellmer, F., Friese, C., Kollros, H. & Krumbein, W. (1999). Mythos Netzwerke. Regionale Innovationsprozesse zwischen Kontinuität und Wandel. Berlin: edition sigma.

Hotz-Hart, B. & Küchler, C. (1996). Das Technologieportfolio der Schweizer Industrie im In- und Ausland. *Schweizerische Zeitschrift für Volkswirtschaft und Statistik*, **3**, 317–333.

Jaffe, A. (1989). Real effects of academic research. *American Economic Review*, **79**, 957–970.

Jasanoff, S., Markle, G. E., Petersen, J. C. & Pinch, T. (Eds) (1994). *Handbook of science and technology studies*. Thousand Oaks/London/New Delhi: Sage.

Küpper, C. (2001). *Service innovation—A review of the state of the art*. Münchner Betriebswirtschaftliche Beiträge, No. 2001–06. München.

Kuhn, T. (1967). *Die Struktur wissenschaftlicher revolution*. Frankfurt A.M.

Leydesdorff, L. & Etzkowitz, H. (1998). Triple helix of innovation: Introduction. *Science and Public Policy*, **25** (6), 358–364.

Licht, G. & Stahl, H. (1997). *Ergebnisse der Innovationserhebung 1996*. Dokumentation No. 97–07. Mannheim: ZEW.

Lundvall, B. (Ed.) (1992). *National systems of innovation—Towards a theory of innovation and interactive learning*. London, New York: Pinter.

Mansfield, E. (1998). Academic research and industrial innovation: An update of empirical findings. *Research Policy*, **26**, 773–776.

Mansfield, E. (1991). Academic research and industrial innovation. *Research Policy*, **20**, 1–12.

Mensch, E. (1975). *Das technologische Patt*. Innovationen überwinden die Depression. Frankfurt A.M.

Mintz, J. & Preston, R. (Eds) (1993). *Infrastructure and competitiveness*. Ottawa: John Deutsch Institute for the Study of Economic Policy.

Nowotny, H. (1996). Mechanismen und Bedingungen der Wissensproduktion. Zur gegenwärtigen Umstrukturierung des Wissenschaftssystems. *Neue Zürcher Zeitung*, **6** (7)..

Nowotny, H., Scott, P. & Gibbons, M. (2001). *Re-thinking science. Knowledge and the public in an age of uncertainty*. Cambridge U.K.: Polity Press.

OECD (2001). *Science, technology, and innovation score-board 2001*. Paris.

OECD (2000). *OECD in figures* (Edition 2000). Paris.

OECD (1999). *Science, technology and industry scoreboard 1999*. Benchmarking knowledge-based economies.

OECD (1998). *University research in transition*. Paris.

OECD (1997). *National innovation systems*. Paris.

OECD (1996). *Politiques du marché du travail en Suisse*. Paris.

OECD; EC; Eurostat (1996). *Oslo Manual. The measurement of scientific and technological activities*. Paris.

Oppenländer, K. H. (1994). Wirtschaftswachstum, Beschäftigung und Arbeitslosigkeit. Theoretische und empirische Zusammenhänge. *Wirtschaft und Gesellschaft*, **20** (4), 361–388.

Oppenländer, K. H. (1975). Das Verhalten kleiner und mittlerer Unternehmen im industriellen Innovationsprozess. In: K. H. Oppenländer, (Hg), *Die gesamtwirtschaftliche Funktion kleiner und mittlerer Unternehmen*. München.

Peters, J. & Becker, W. (1998). Technological opportunities, academic research, and innovation activities in the German automobile supply industry. Paper (see: www.wiso.uni-augsburg.de/vwl/institut/paper/175.pdf).

Pfähler, W., Bönte, W., Gabriel, C. & Kettner, A. (1999). Wirtschaftsfaktor Bildung und Wissenschaftssenschaftseinrichtungen. Die regionalwirtschaftliche Bedeutung der Hochschulbildungs- und Wissenschaftseinrichtungen in Bremen. Frankfurt A.M. u.a.: Lang.

Pfähler, W., Clermont, C., Gabriel, C. & Hofmann, U. (1997). *Bildung und Wissenschaft als Wirtschafts- und Standortfaktor*. Baden-Baden: Nomos.

Rojas, R. & Hashagen, U. (Eds) (2000). *The first computers. History and architectures*. Cambridge, MA: MIT Press.

Salter, A. J. & Martin, B. R. (1999). *The economic benefit of publicly funded basic research: A critical review*. SPRU Electronic Working Paper Series. No. 34.

Sauer, D. & Lang, C. (1999). *Paradoxien der Innovation. Perspektiven sozialwissenschaftlicher Forschung*. Frankfurt A.M. New York: Campus.

Saxenian, A. (1994). *Regional advantage: Culture and competition in Silicon Valley and Route 128*. Cambridge/MA: Harvard University Press.

Schmoch, U., Grupp, H. & Laube, T. (1996). Standortvoraussetzungen und technologische Trends (pp. 55–156). In: Bundesamt für Konjunkturfragen (Ed.), *Modernisierung am Technikstandort Schweiz*. Zürich: vdf.

Schmoch, U., Licht, G. & Reinhard, M. (Hg.) (2000). *Wissens- und Technologietransfer in Deutschland*. Stuttgart: IRB.

Soete, L. (1987). The impact of technological innovation on international trade patterns: The evidence reconsidered. *Research Policy*, (16), 101–130.

Stokes, D. E. (1997). *Pasteurs quadrant: Basic science and technological innovation*. Washington.

Temple, J. (2000). Growth effects of education and social capital in the OECD countries. Economics Department Working Paper. No. 263. ECO/WKP (2000) 36. Paris: OECD.

Thierstein, A., Wilhelm, B. E. & Behrendt, H. (2002). *Gründerzeit. Schriftenreihe des IDT-HSG. Beiträge zur Regionalwirtschaft* (No. 2). Bern; Stuttgart; Wien: Paul Haupt.

Thurow, L. (1996). *The future of capitalism*. London: Brealey Publishing.

Varga, A. (1997). *Regional economic effects of university research: A survey*. Working Paper. Regional Research Institute, West Virginia University (see: www.rri.wvu.edu/wpapers/pdffiles/surveyattila.pdf).

Walterskirchen, E. (1994). Wirtschaftswachstum und Arbeitslosigkeit in Westeuropa. *Wirtschaft und Gesellschaft*, **3**, 377–388.

Weeber, J. (1995). Wachstum ohne Beschäftigung? *WSI Mitteilungen*, (9, 598–603.

Weingart, P. (1995). Konzepte einer neuen gesellschaftsorientierten Wissenschaftspolitik. *Wirtschaft & Wissenschaft* (3), 44–52.

Wilhelm, B. (2000). *Systemversagen im Innovationsprozess. Zur Reorganisation des Wissens- und Technologietransfers*. Wiesbaden: DUV.

Wolter, S. C. & Curti, M. (1996). Wachstum ohne Beschäftigung? *Die Volkswirtschaft*, (8), 48–51.

Wolter, S. C. & Weber, B. A. (1998). Der monetäre Nutzen von Bildung. *Wirtschaftspolitik*, **71** (9), 10–15.

The International Handbook on Innovation
Edited by Larisa V. Shavinina
© 2003 Published by Elsevier Science Ltd.

Systems of Innovation and Competence Building Across Diversity: Learning from the Portuguese Path in the European Context

Pedro Conceição and Manuel V. Heitor

*Center for Innovation, Technology and Policy Research, IN +, Instituto Superior Técnico,
Technical University of Lisbon, Portugal*

Abstract: Innovation, a broad social and economic activity within the emerging learning societies, transcends any specific technology, even if revolutionary, and is tied to attitudes and behaviors oriented towards the exploitation of change by adding value. We analyze the ongoing Portuguese path towards an innovative society, characterizing innovation in Portugal within the European context. We conceptualize 'learning' and the process of knowledge accumulation, as a framework to understand the new demands for being innovative. We conclude by suggesting elements for innovation policies for Portugal, arguing for the need to promote *systems of innovation and competence building*.

Keywords: Systems of innovation; Learning society; Competence building; Techno-economic paradigms.

Introduction: Framing Innovation Practices and Theory Building

The understanding of innovation adopted in this chapter encompasses the way in which firms and entrepreneurs create value by exploiting change. Change can be associated with technological advances, but also with modifications of the regulatory framework of industry, shifts in consumers tastes, changes in demographic makeover, or even major alterations of global geopolitics. Further, in the current socio-economic context, innovation increasingly means the ability to cope with uncertainty in diversified environments, which are particularly influenced by social and institutional factors (see, for example, Conceição, Heitor & Lundvall, 2003; Smith, 2002). To choose such an ambitious definition of innovation presents important challenges. First, it calls for an analysis of many economic, social and institutional issues. Our effort cannot attempt to deal with these issues comprehensively. We will rather attempt, throughout the chapter, to discuss important trends that are likely to influence innovative performance in the presence of diversity, looking at the specific environment in which Portuguese firms conduct their business, and conse-

quently, determine the conditions and opportunities for innovation in the European context. The choice of such an ambitious definition of innovation limits equally the extent to which clear-cut solutions and recommendations to enhance the innovative performance of a country or region can be provided. Our hope is that by raising and discussing some selected questions, and concerns, we contribute to a better awareness of possible weaknesses and potential strengths of the Portuguese system of innovation within a diversified European environment.

It should be noted that innovation is a shared goal of countries within the European Union and even beyond, including other European countries, namely those that are candidates to becoming members of the EU in the coming years. We argue that this unified goal requires policies that are designed in an integrated and systemic way, but that are implemented with diversified actions. 'Policy integration' should occur across a 'portfolio dimension', since innovation policies require coordination across several areas: science and education policies; social and health policies; environmental and industrial policies; employment and market regulation policies. However, the implementation of policies designed in an integrated way need, in a multi-country and multi-cultural context, to consider differences across countries, regions and cultures, thus requiring action diversification. In fact, balancing action diversification with policy integration involves significant problems that extend into the very systemic nature of the relationships between country governments and the role and mission of multi-national political institutions, apart from specific regional and local contexts.

Many contributions in recent years have confirmed the perception that the success of developing systems of innovation, either at national or regional levels, depend on the creation, dissemination and accumulation of knowledge, which per se are fundamental factors for the promotion of economic growth (Conceição, Heitor & Lundvall, 2003; Swan et al., this volume). However, the scarcity of empirical data on intangible economic factors makes it extremely difficult to demonstrate the growing importance of knowledge. Economic growth has traditionally been explained as being the result of increases in the labor and capital factors, with technological change contributing as an exogenous—that is, 'outside' of the realm of economic modeling-factor (Solow, 1997, for a recent review). However, the challenge posed by the endogenous growth theories to this traditional approach (see, for example, the review in Aghion & Howitt, 1997) have led to a need to rethink how these three factors influence the process of economic development. This rethinking has been taking place, in part, by bringing together other perspectives on the process of the relationship between technological change and economic growth, such as the evolutionary theory (Nelson & Winter, 1982) and the perspective of techno-economic paradigms (see discussion below).

Our inspiration to frame the process of knowledge accumulation comes from the contribution of Lundvall & Johnson (1994), who introduced the simple, but powerful idea of *learning*. Lundvall & Johnson (1994) suggest that a 'learning economy', rather than a 'knowledge economy', describes better the way in which knowledge contributes to economic development, promoting innovation. The fundamental difference between the two terms is associated with the fact that the former considers a dynamic perspective. According to Lundvall & Johnson (1994), some types of knowledge do indeed become more important, but there is also knowledge that becomes *less* important. There is both knowledge creation *and* knowledge destruction. By forcing us to look at the process, rather than at the mere accumulation of knowledge, Lundvall & Johnson (1994) add a dimension that makes the discussion more complex and more uncertain, but also more interesting and intellectually fertile.

Following the concept of the learning economy, which is further demonstrated in the volumes edited by Archibugi & Lundvall (2001) and Conceição, Heitor &

Lundvall (2002), innovation is the key process that characterizes a knowledge economy understood from a dynamic perspective. Lundvall & Johnson's (1994) learning economy is about new knowledge replacing old knowledge. This dynamics is very close to Schumpeter's concept of 'creative destruction', which is a standard description of the innovation process. Innovation is associated with creativity, with the generation of new ideas, but also with initiative and risk-taking. Innovation entails bringing new ideas to fruition in the marketplace, satisfying demands or creating new needs, in a process that improves overall welfare.

Beyond innovation, we also consider in this chapter the need to look at competence, as the foundation from which innovation emerges, and which allows many innovations to be enjoyed. In other words, competence contributes both to the 'generation' of innovations (on the supply side of the knowledge economy) and to the 'utilization' of innovations (on the consumptions side of the knowledge economy). Competence is also fueled by innovation itself. Competence is associated with skills and capacities, both individual and collective ones. When we consider competence, we focus on 'higher order of skills' Carneiro (2003). These generic skills include higher levels of education, but also capacities that are more generic, such as creativity, risk-taking, and initiative.

This chapter is organized in seven sections. Following this introduction, we consider innovation in Portugal within the European context. To better understand this case study, we introduce, in 'Looking at Innovation Over Time and Across Space: The Techno-Economic Paradigms Approach', the analysis of innovation over time and across space, looking at the techno-economic paradigms approach. This leads us to conceptualise, in 'The Learning Society: A Framework to Understand the New Demands for Being Innovative', 'learning' as a framework to understand the new demands for being innovative. Then, in 'Deepening our Understanding of Learning towards Innovation: Building on the Economics of Knowledge', we briefly discuss the fundamentals of the economics of knowledge in order to attempt to deepen our understanding of learning towards innovation. In 'Fostering Systems of Innovation and Competence Building: The Challenges of Inclusiveness' we build on the conceptual framework of the previous sections to frame innovation policy, namely in the Portuguese context within a diversified European environment. 'Summary and Conclusions' provides a brief summary and conclusion.

Characteristics of Innovation in Portugal

The measurement of innovative performance of an entire country—in a way, comparable across the diverse realities of many countries—is a demanding challenge, which has been addressed in Europe by a joint effort of the OECD and the Eurostat through the development of innovation surveys according to a set of criteria that values cross-country comparability of results (see Conceição & Ávila, 2001). Portugal has been an integral part of this effort. This European effort is designated by Community Innovation Surveys (CIS), and its framework of enquire has been adopted both in official and autonomous research surveys in many countries, from Eastern European countries to Latin America (Inzelt, in this volume, describe the Hungarian experience, for example).

By giving more importance to cross-country comparability, the CIS looses somewhat of its potential ability to probe into the dynamics of innovation within each country, since it only asks broad and generic questions, which can be accepted to have similar meanings in different economies. However, it provides a reliable way to compare national innovative performance across countries. Figure 1 shows the overall innovative performance of countries in Europe measured by the shares of firms that have introduced innovations over a

two-year period (with reference to the period 1995–1997, as quantified through the CIS-2 example). The horizontal axis indicates innovative performance in manufacturing, and the vertical axis—in services. The results show a general close relationship between innovation in services and in manufacturing, since countries are located across a 45-degree diagonal. In general, innovation rates are lower in services than in manufacturing.

Portugal appears towards the bottom of performance, being the least innovative country in manufacturing. However, in services, Portugal innovates more than Belgium, Finland and Norway. Slightly more than a quarter of Portuguese manufacturing firms are innovative, while almost 30% of service firms are innovative. Here again we can see an indication of the duality: unlike other countries, services in Portugal—which have grown as a share of the economy at rates higher than the EU average—are more innovative than manufacturing firms, which are still largely dominated by traditional sectors of the

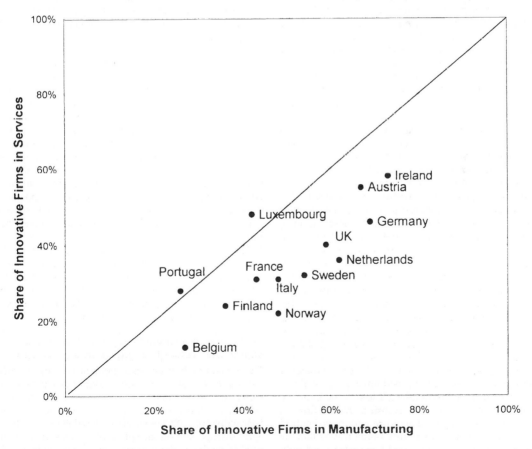

Figure 1. Innovative performance of EU countries for the period 1995–1997.

Source: Conceição & Ávila (2001).

Portuguese economy (such as the textile sector, for example).

Knowledge of the process of innovation in Portugal, and of the way in which it contrasts with the innovation process in Europe can be gathered from other aspects of CIS. Thus, Fig. 2 shows that Portuguese firms rely much more on resources external to the firm (as, for example, information sources for innovation process) than European firms (on average).

Figure 3, however, shows that issues related to high costs and difficulties in funding are much more prominent in Portugal than on average in Europe.

Even though there is a large consensus across Europe that the lack of qualified personnel is the most important factor hampering innovation (Fig. 4), still this factor pales behind high innovation costs and lack of financing as a deterrent of innovation in Portugal.

However, it is important to look at the diversity that exists within Portugal. We concentrate on manufacturing only. Even within manufacturing, though, there are substantial differences across sectors (Fig. 5). The machinery, electrical and optical equipment sector exhibits almost 50% of innovative firms (the rate of innovation in this sector is comparable to the average rate in countries such as Italy and Norway).

Innovation in Portugal seems to be associated with a number of characteristics of the firms in a way that conforms both with theory, and to results in other countries. A descriptive analysis of the results of CIS show that size classes of large firms have a higher share of innovative firms than size classes composed of small firms. A descriptive analysis also shows that firms that are part of a group of companies, show higher rates of innovation. Combining these two variables in a multivariate model, with the dependent variable being dichotomous (1 if the firm has innovated, 0 otherwise) shows, without any other conditioning variables) that large firms and firms that are part of a group do have higher probability to innovate (that is, actually introducing an innovation) than small firms and firms that are not part of a group of firms (first column in).

However, as we saw above, there is large diversity of innovative performance across manufacturing sectors. Still, when industry dummies are added to the model (second column in) none shows up as significant. This can be interpreted by saying that the sector effects are not strong determinants of innovation (when the size of the firm and whether the firm is part of the group are included).

However, when we consider only two groups of firms—those that are high or medium high technology, on the one hand, and those that are low or medium low technology, on the other—the results show that firms in the high/medium-high technology group do indeed exhibit a much higher probability of innovating that the average firm (note that the coefficient associated with the dummy for the low/medium-low technology firms is not significant).

The results indicate the existence of duality, as explained in further detail by Conceição & Heitor (forthcoming). Note how large and statistically significant the coefficient associated with high/medium-high technology is, even after controlling for the size of the firm and the fact that it may belong to a group. Thus, more sophisticated firms in markets with higher demands seem to have a substantially higher probability of innovating than other firms. This is not tied, one should stress, to a mere 'sector effect' (the sector dummies were not significant), it is really a characteristic of a large group of sectors that have in common belonging to the high/medium-high technology category. The duality here is clearly substantiated.

Naturally, other factors, beyond size and belonging to a group, influence innovation and Conceição & Heitor (forthcoming) report also on the effect of the firm level of productivity and the importance of exports. Both of these variables are known to have important effects on innovation and the results tell exactly the same story: when the differentiation is made according to the technological intensity, the duality comes up again, not as strong as before (part of the variation is now picked-up by productivity), but it is still present.

Of course, the models above have merely descriptive value; we do not make any claims in terms of causality, much less explanation. They are understood as showing the correlations among the variables included. It is known, for example, that several of the variables are simultaneously determined (namely innovation and productivity; on this see Conceição & Veloso, 2002). Thus, the point we make is that, even controlling for a number of characteristics that influence innovation, there is a clear duality in terms of probability of innovating when considers technology intensity as a criteria for differentiating firms. This duality appears to be a clear characteristic of the Portuguese society, which in turn must be understood within a diversified European context.

The question which does arise is related with the ability of Portugal to cope with the accelerated rate of technological change, in a way that will allow fostering innovation. This is a complex and evolving question, which requires a better understanding of the process of technological change, as described below.

Looking at Innovation Over Time and Across Space: The Techno-Economic Paradigms Approach

The interaction between the emergence of new technologies and the larger economic and social patterns of behavior can be understood, following Schumpeter (1934), as a process of *creative destruction*. At a first approximation, this statement is obvious: new technologies disrupt, and often replace older ones. At a higher level of analysis, the implications of new technologies are broader. The impact is often felt not only as a replacement of old for new technologies, but brings

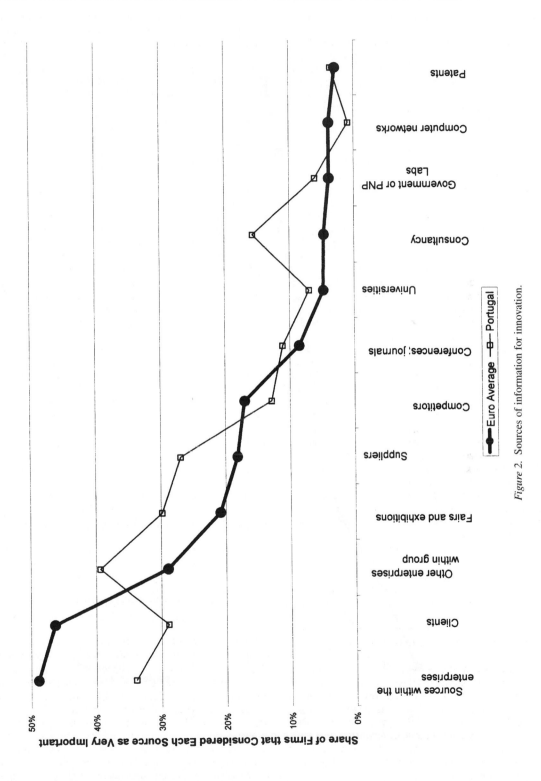

Figure 2. Sources of information for innovation.

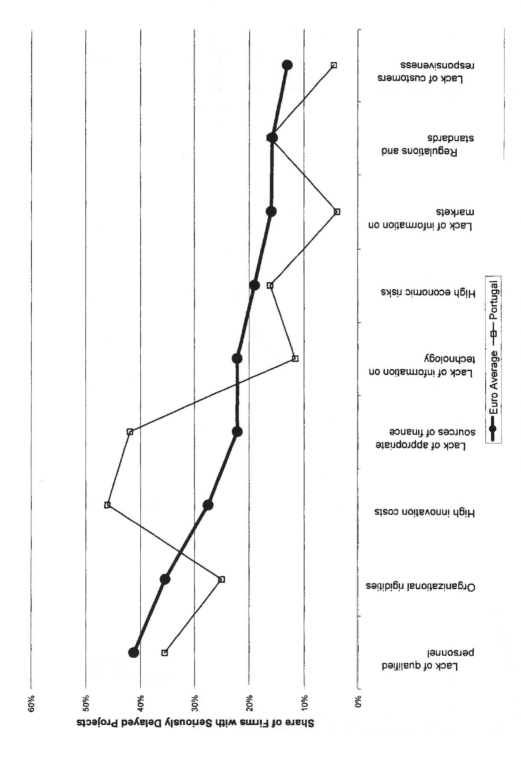

Figure 3. Factors hampering innovation.

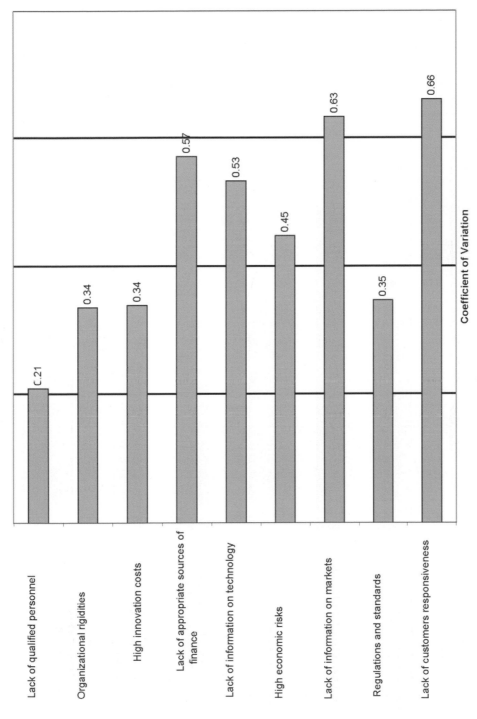

Figure 4. Degree of consensus across Europe on the hampering factors. Adapted from Conceição & Ávila (2001).

Coefficient of Variation

- Lack of qualified personnel — 0.21
- Organizational rigidities — 0.34
- High innovation costs — 0.34
- Lack of appropriate sources of finance — 0.57
- Lack of information on technology — 0.53
- High economic risks — 0.45
- Lack of information on markets — 0.63
- Regulations and standards — 0.35
- Lack of customers responsiveness — 0.66

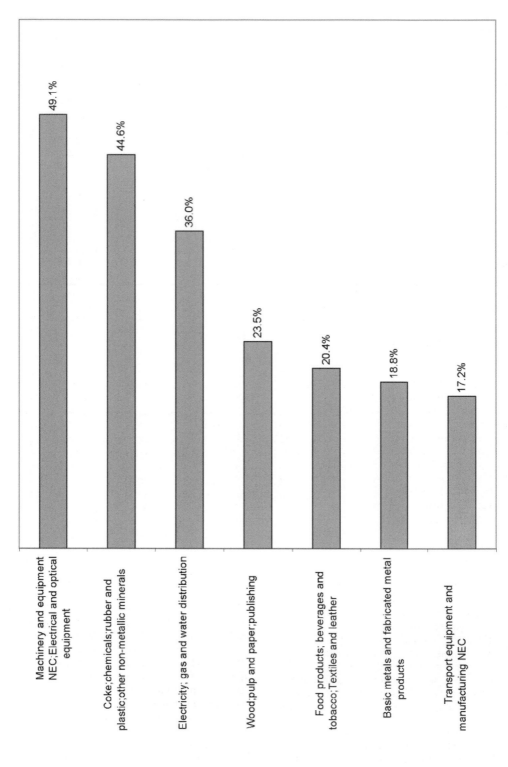

Figure 5. Innovation rates (% of innovative firms) in the Portuguese manufacturing sector. Adapted from Conceição & Ávila (2001).

Table 1. Regression results on the characteristics of innovative firms—first model*.

	Unconditioned (1)	Industries Dummies (2)	Technological Intensity (3)
Intercept	−1.576** (0.2448) [0.0000]	−9.104 [−0.0001] [1.0000]	−1.773** [0.2562] [0.0000]
Firm is part of Group	0.529** (0.1423) [0.0002]	0.318* [2.0834] [00372]	0.474** [0.1435] [0.0009]
Log of Number of Employees	0.213** (0.0613) [0.0005]	0.262** [4.0365] [0.0001]	0.224** [0.0625] [0.0003]
High/ Medium High Technology		Conditioning Industries Dummies None is significant	0.757** [0.138] [0.0000]
Medium Low Technology			0.163 [0.1163] [0.1614]
Concordant	84%	87%	85%
Observations	820	820	820

* Dependent variable: 1 if the firm has introduced any type of innovation, 0 otherwise. Standard errors in brackets, p-values in square brackets. ** significant at 1% or less. Logistic regression. Results with a normally distributed link function (Probit) were not dramatically different. Manufacturing only.

with it opportunities for new firms and difficulties for existing firms, the obsolescence of some occupations and shifts in the structure of employment, changes in the terms of trade between regions and countries. However, it is clear that not all advances in technology are disruptive to the point of creating substantial changes in economic and social conditions. In fact, most technological advances and innovations make their impact felt in a relatively smooth way, when analyzed from a macro perspective.

One way to conceptualize the interaction between technological change and shifts in economic conditions, together with the process of sometimes disruptive innovations, but most often smooth adoption and diffusion of new technologies, is the idea of techno-economic paradigms. The discussion that follows is based on Freeman, Clark & Soete (1982), Freeman & Perez (1986), and Dosi (1988). A techno-economic paradigm embodies a relatively stable cluster of core technologies, around which innovation and economic activity take place. The core technologies have a strong impact in the economy and society, being defined as core given their potential for generalization and penetration across a wide number of products and processes, across all sectors of economic, and often human, activity.

Within a paradigm, the core technologies are virtually unchanged over time, but this does not mean that there is not economic and technological progress.

On the contrary, these core technologies provide a positive heuristic that defines the knowledge and incentives for innovation and economic activity to occur. At the same time, this progress in inherently limited by the conditions set by the interaction of the core technologies with the dominant modes of economic activity, from the organization of firms, to the distribution of employment. Therefore, progress exists within a certain techno-economic paradigm, but occurs within a framework defined by a set of core technologies and modes of organizing economic activity.

Thus, within a paradigm, innovation occurs as the core technologies become more and more pervasive and influence ever-wider realms of production and distribution. When a major technological advance occurs, disrupting the existing core technologies and modes of economic operation, then a new techno-economic paradigm emerges. The displacement of the core technologies of the old paradigm creates a new wave of invention and innovation and is no longer tied to the previous paradigm core technologies. The emergence of a new core technology requires, and creates the opportunity for, an entire new set of small and incremental innovations that permit the widespread usage of the new core technologies. Thus, when a shift in techno-economic paradigm occurs, we have not only a 'substitution effect', but also an expansion of the creative frontier that allows the emergence of new technologies and enables, in the end, a shift to yet another techno-economic paradigm.

Additionally, beyond the technological and purely economic factors, the social and institutional frameworks that fit a certain techno-economic paradigm may not be adequate for a new one. Indeed, the process of emergence of a new techno-economic paradigm results from the interaction of the technological, economic, institutional and social spheres. Having just a new technology coming in may not have any effect if a set of changes in the other dimensions does not accompany the technological novelty. A certain set of institutions and social features may provide enough contexts for innovation within a certain paradigm; in other words, it is not necessarily needed to create institutions and social rules at the same pace that technological innovation progresses. But when there is a shift in techno-economic paradigm, a new institutional framework may be needed.

A number of authors, working together and independently, developed the theory of techno-economic paradigms beginning with Schumpeter, who argued that the expectations of profits would drive the 'entrepreneur' to innovate (see Freeman, Clark & Soete (1982), Freeman & Perez (1986), Dosi (1988)). The entrepreneur's drive towards innovation is motivated by the temporary monopolistic position from which the innovator would benefit. Schumpeter (1934) regarded this position as temporary because the advantages from this privileged position would eventually 'perish in the vortex of the competition which streams after them', since other firms would copy the innovator. Schumpeter (1934) called this process *creative destruction*. Therefore, for Schumpeter, innovation appears at the forefront of economic progress, driving prosperity. In a later version of these same fundamental ideas, Schumpeter refined this earlier simplistic version of an entrepreneur in a perfect market composed by a multitude of competing firms that destroy any persistent market advantage. In his final work Schumpeter (1942) acknowledged that some large corporations could sustain a market advantage by an institutionalization of the effort to innovate through the establishment of large R&D facilities.

The reinterpretation of Schumpeter's fundamental ideas of innovation as a process of disequilibrium in the broader context of techno-economic paradigm is due primarily to Christopher Freeman and his co-authors. Often called a 'neo-Schumpeterian' approach, this perspective is articulated, as mentioned, in Freeman, Clark & Soete (1982), Freeman & Perez (1986), Dosi (1988) and, more recently, in McKnight et al. (2000), to cite a few representative examples. Freeman and his co-authors generalized the concept of Schumpeterian innovation to the national level, making an analogy between innovation at the firm level and a

change in a techno-economic paradigm at the country level (Freeman, 1988; Freeman & Soete, 1997).

This macroeconomic definition of innovation corresponds to what is, at the firm level, a radical innovation. Under this extreme, there are milder types of innovation, such as incremental innovations, that correspond, at the micro level, to improvements in existing products and processes. Freeman (1988) builds a similar hierarchy for his macro analysis of innovation, leading to a conceptual framework that has some similarity to the evolutionary perspective of Nelson & Winter (1982).

It is important to stress two important dimensions of the techno-economic paradigm theory: *time* and *space*. Time is, indeed, crucial, as we saw, since the process of technological change and its economic and social impact is seen as a progress, more stable within a certain techno-economic paradigm, and very different across techno-economic paradigms, which differ over time. Space is equally important, since it is not clear that a certain techno-economic paradigm will not affect all the regions of the world similarly. Certainly there will be different rates of adoption of new core technologies when there is a paradigm shift, or even, within a paradigm, different ways in which specific innovations and modes of economic organization develop in different countries and different regions. Some countries may originate or lead the development of a new techno-economic paradigm, and others may lag behind, or even stay closer to the 'older' rather than the new techno-economic paradigm.

An important idea joining the time and space dimensions of the techno-economic paradigm theory is that of technological trajectories within national innovation systems. The idea of trajectories in national innovation systems (developed, with a comparative analysis across countries, in Nelson, 1991, for example) is because each country follows its own developmental path, within the general framework of the existing techno-economic paradigm, but also—and this is crucially important—influenced by the past history and specific conditions of the local context.

This brings to the discussion the asymmetries in country performance, which according to our interpretation advanced in earlier papers, can be seen as being dependent on what we could call with generality knowledge accumulation through 'learning' processes. Conceptually, the foundations for the relationship between learning and economic growth are well established in the recent literature (Bruton, 1998), and stem from a combination of the pure neo-classical perspective of growth with the Schumpeterian view. Learning is reflected in improved skills in people and in the generation, diffusion, and usage of new ideas. Likewise, organizational learning reflects social processes driven by collective cultures and appropriate management attitudes. The ability to continuously generate skills and ideas (which is to say, to accumulate knowledge through learning) is the ultimate driver of an economy long-run prospects (World Bank, 1997).

The fact that countries have different levels of income is clearly self-evident. Therefore, it is equally obvious that each country has followed its own trajectory, within the context of an existing techno-economic paradigm and the specific innovation system of the nation. We look here at some evidence on the translation of different paths in the economic performance of countries. But we begin with an interpretation of the major techno-economic paradigms, illustrated in Table 2.

The table shows five important techno-economic paradigms. While the paradigms presented result from one interpretation, they serve now to illustrate with some empirical evidence the features of techno-economic paradigms presented before. Let us consider, for example, the first techno-economic paradigm. This corresponds to the emergence of the Industrial Revolution, as mechanization was increasingly incorporated in manufacturing, especially in some industries such as textiles. However, the technologies well diffused and used within this paradigm presented some important limitations for the increase of the scale and output of the productive activity. Most firms remained small and local. Process control was poor and hand operated machines did not allow for output of reliable quality. Naturally, advances in steam engine technologies and

machinery were already taking place, but it took a long time until they were ready for fruition. When these important technologies matured to the level that made their economic utilization possible, they became the core technologies of the second techno-economic paradigm. The new techno-economic paradigm based on steam engine and on machinery ameliorated some of the previous limitations, and created in itself the germ for new types of economic organization, as the table details.

If we cross the techno-economic paradigms with geography, then we start joining together the ideas of technological trajectory and national innovation system. The two first techno-economic paradigms were led by Britain. In this context, the U.S. and Germany, for example, were 'latecomers'. Still, they became leaders in the third techno-economic paradigm, with Japan also leading in the fourth and the U.S. arguably retaining the lead alone in the fifth, although we will be looking at this claim in more detail later.

Still, the manifestations of the current differences in the paths followed by different countries are dramatic. Even taking a set of relatively homogeneous countries, such as the OECD, shows great disparities in income per capita and productivity. Productivity, in a way, is probably the best indicator of the extent to which a nation is taking full advantage of the conditions provided by the existing techno-economic paradigm. A recent study by Ark & McGuckin (1999) tackles international comparisons of productivity and income

Table 2. Tentative sketch of major techno-economic paradigms.

Approximate Period	Description	Key Sectors	Economic Organization
1770s to 1840s	Early Mechanization	Textiles, Canals, Turnpike Roads	Individual entrepreneurs and small firms; local capital and individual wealth
1830s to 1890s	Steam Power and Railway	Steam Engines, Railway, World Shipping	Small firm competition, but emergence of large firms with unprecedented size; limited liability corporations and joint stock ownership
1880s to 1940s	Electrical and Heavy Engineering	Electrical Engineering, Chemical Process Industries, Steel ships, Heavy Armaments	Giant firms, cartels, trusts; mergers and acquisitions; state regulation and enforcement of anti-trust; professional management teams
1930s to 1980s	Fordist Mass Production	Automobiles, Aircraft, Consumer Durables, Synthetic Materials	Oligopolistic competition; emergence of multinational corporations; rise of foreign direct investment; vertical integration; technocratic management styles and approaches
1970s to . . .	Information and Communication	Computers, Software, Telecommunications, Digital Technologies	Networks of large and small firms based increasingly on computer networks; wave of entrepreneurial activity associated with new technologies; strong regional clusters of innovative and entrepreneurial firms

Source: Adapted from Freeman & Soete (1997), Table 3.5.

in a particularly careful way, especially in finding comparable measures across countries. They also link labor productivity with output per capita following a common decomposition procedure. While the relationship between these two variables may seem obvious, in fact there are many subtleties involved. For example, a country that is very productive but where workers engage in productive activities fewer hours than a less productive country can result in an output per capita that is higher in the second country. Table 3 shows the results presented in this work. Column (1) indicates labor productivity and column (8) provides the level of GDP per capita.

Portugal and Turkey have the lowest hourly labor productivity rate of the OECD. Portuguese hourly productivity is about half of the OECD average. Productivity in Greece is 19 points above Portugal's and Spain's productivity is 28 points above the Portuguese hourly labor productivity. Still, when one looks at column (8), Greece's GDP per capita is actually lower than Portugal's by two points and Spain's GDP is only 11 points above Portugal's.

The decomposition of the table shows the variety of effects involved. Column (2) shows the impact of the number of hours worked. The summation of columns (1) and (2) produces the GDP per person employed. We see that Spanish and Japanese workers work longer hours than in most of the other countries. Per worker productivity in Spain, measured as GDP per worker, raises almost to the OECD level. Portuguese workers also work long hours, adding 2 points to the per hour productivity measures. In Italy, France, The Netherlands, Norway and the United Kingdom less hours of work reduce per employee productivity. Standards of living are determined not only by the number of hours worked and the productivity of each hour of work, but also by the 'number of mouths to feed'. The effect of the labor force participation connects per worker productivity and GDP per person. It is the effect of the labor force participation, for example, that brings down the income per capita of the productive and hard working Spanish workers: the combined effect of unemployment and the low level of labor force in the working age population take 26 points to the per worker productivity. The same happens in Greece, where 12 points are taken to the per worker GDP. In Portugal, both the effects of hours worked and labor force participation are small and positive. It is, therefore, clear that the real challenge to increase the level of GDP per capita in Portugal is not so much a reduction of unemployment or, more generally, an increase in labor force participation (as in Spain, for example), but that it is really the increase in the fundamental hourly labor productivity. To understand impact of these differences on innovative performance and, consequently, to derive innovation policies, it is important to look at the new demands for being innovative, to which we turn in the next section.

The Learning Society: A Framework to Understand the New Demands for Being Innovative

Recent models of long-term economic growth have been able to explain the increase in per capita income in developed countries (see Johnson, 2000, for a summary perspective, and Landes, 1988, for a broader treatment) with extremely parsimonious models based exclusively on the growth of knowledge. The factors behind the increase of knowledge are equally simple: the increase in population and the emergence of specialization in the production of knowledge. Kremer (1993) uses a model exclusively based on population growth, where more people means that there are more individuals capable of making a significant discovery and that the larger the population the larger the benefits from those discoveries. In other words, technological improvements make population growth possible which, in turn, creates more possibilities for new discoveries. A slightly more complex model by Hall & Jones (1999) includes also the effect of the specialization of growing proportion of the population in activities associated exclusively with the creation and transmission of knowledge. This entails the need to include *institutions* and *policies*—a combination that the authors call *social infrastructure*—which, according to this model, explain difference across countries in their level of knowledge generation and income per capita.

The gradual transition towards knowledge-based economies has intensified in the last part of the 20th century. According to the OECD (1999) more than 50% of the OECD countries' GDP is associated with knowledge-based industries.[1] Lundvall (2000) asserts that the intensity of the acceleration of knowledge creation and diffusion requires a more dynamic characterization. In Lundvall's opinion, we should speak about the emergence of a *learning society*.

In summary, while much attention has been devoted to specific technologies, namely to digital technologies in recent years, the association between information technologies and augments in productivity remains ambiguous. Still, it is undeniable that the spread of the computer and the Internet is changing in profound ways the way people and firms behave and interact, with important consequences for policy and strategy. A more fundamental change at the start of the new millennium is the increasing importance of knowledge for economic prosperity. This feature of current developed countries corresponds to the continuing of a trend of acceleration of the importance of the creation and diffusion of knowledge throughout the century. Beyond digital technologies, other technological breakthroughs, in many areas from the life sciences to

---

[1] Even if the definition of knowledge-based industries is rather generous, including a large part of services and the high and medium-high technology manufacturing.

Table 3. Decomposition of GDP per hour worked into effects of working hours, labor force participation and GDP per capita, 1997.

	GDP per hour worked as a % of the OECD Average (1)	Effect of working hours (2)	GDP per person employed as a % of the OECD Average (3) = (1) + (2)	Effect of unemployment (4)	Effect of labor force as a % of the working age population (5)	Effect of working age population as a % of the total population (6)	Total effect of labor force participation (7) = (4) + (5) + (6)	GDP per person as a % of the OECD Average (8) = (3) + (7)
Australia	96	0	96	−1	2	0	1	97
Austria	102	−4	98	3	−2	1	2	100
Belgium	128	−5	123	−3	−19	−1	−22	101
Canada	97	2	98	−2	2	2	2	100
Denmark	92	0	92	1	9	1	11	103
Finland	93	0	94	−7	2	0	−5	88
France	123	−9	113	−6	−9	−2	−17	97
Germany	105	−5	100	−3	4	2	4	96
Greece	75	4	71	−2	−11	1	−12	58
Ireland	108	5	113	4	−12	−3	−18	95
Italy	106	−11	96	−5	−1	2	−5	91
Japan	82	10	92	4	6	4	14	106
The Netherlands	121	−26	95	2	−4	2	0	96
New Zealand	69	8	77	1	3	−1	2	79
Norway	126	−17	109	4	12	4	12	122
Portugal	56	2	58	0	1	1	2	60
Spain	84	13	97	−14	−13	2	−26	71
Sweden	93	−3	89	−3	6	4	−1	88
Switzerland	94	0	94	3	12	1	17	111
Turkey	36	2	38	0	−8	−1	−9	29
United Kingdom	100	−9	91	0	3	−2	0	92
United States	120	−1	118	3	9	−2	10	128
EU-14	103	−5	98	−4	−4	0	−8	90

Source: Ark and McGuckin (1999); summations may not add exactly due to rounding errors.

the many fields of engineering, are likely to be seen in the future (see Coates, this volume).

In this context it is important to look both at the *level* of the measures that indicate the extent to which a country is engaged in the knowledge economy and to the *growth* in recent years. Figure 6 provides a first illustration, with the horizontal axis representing the intensity of knowledge-based industries in the mid-1990s and the vertical axis the growth rate of these industries in the previous decade.

Most countries are clustered at the bottom of the figure, with growth rates between 2% and 4% a year. The horizontal distribution of the countries shows Germany, the U.S., Japan and other leading developed countries to the right, with Spain and Greece to the left. In this context, Portugal and Korea stand out. The intensity of the knowledge-based industries in these countries is relatively low, especially for Portugal, which has the lowest level of knowledge-based industries. However, the growth rates for Portugal and Korea are remarkably higher, with the knowledge-based industries in Portugal growing close to 7% a year, and Korean knowledge-based industries at more than 12% a year. The rate of growth of knowledge-based industries in comparable periods was of 3.1% for the European Union and of 3.5% for the entire OECD.

The difference between the growth rates of Portugal and Korea is not as extraordinary as it may seem. In fact, the business sector as whole rose in Korea at 9.1% a year, while in Portugal the growth rate of the entire business sector was 4.6%. Consequently, the difference between knowledge-industries growth rate and the entire business sector growth was of 2.3% for Portugal (or 50% of the business sector growth rate) while in Korea the difference was 3.4% (a higher difference, but only 37% of the entire business growth rate). The case of Portugal and Korea are relevant because they are illustrative of latecomer industrialization and may represent indications of the process through which these latecomer countries become engaged in the new techno economic paradigm.

Turning our attention only to information and communication technologies (ICT), Fig. 7 presents essentially the same framework of the previous figure, but now with the intensity of ICT expenditure in 1997 on the horizontal axis and the growth rate of this intensity from 1992 to 1997. Again, most countries are clustered in the bottom of the figure, with growth rates below 4%. The levels, as indicated by the horizontal distribution of countries, confirm the perception that the U.S. is a leading country. The expenditures on ICT as a percentage of GDP in the U.S. are about 2% above the European average. Individual countries, such as Sweden, outperform the U.S., but most countries lag behind.

But, as with knowledge-based industries, the growth rate in expenditures provides a different picture. In fact, Portugal is the leading OECD country in the growth rate of ICT expenditure from 1992 to 1997, with a growth rate of more than 10%. Most of this growth rate can be accounted for by increases in expenditures in telecommunications (about 9%). Expenditures in IT services and software are particularly low, below 1%. Only Turkey, Greece and Poland have shares of expenditure on IT software and services below the Portuguese value. The growth in this category has been equally dismal, below 2% a year.

Returning to the conceptualization of the knowledge-based or learning economy that we presented earlier, it can be said that, fundamentally, the performance in this knowledge-rich competitive environments in terms of innovative performance depend on the **quality of human resources** (their skills, competencies, education level, learning capability) and on the activities and incentives that are oriented towards the generation and diffusion of knowledge. But beyond human capital, which corresponds to the **aggregation** of an individual capacity for knowledge accumulation, developing a collective capacity for learning—as suggested by Wright (1999) in the context of the U.S.—is as, if not more important, than individual learning. Instead of individual or even aggregated human capital, a further important concept for learning seems to be *social capital*, as analyzed by Conceição et al. (2000), among others.

The importance of social capital, while still controversial, is increasingly being seen as an important determinant of economic performance and, especially, of innovation and creativity. Temple (2000) discusses the impact of education and social capital together as determinants of growth. Temple (2000) argues that there is a growing number of works suggesting that social capital is at least as important as education as a driver of economic growth.

Education is often used as a proxy for human capital. For social capital, the equivalent indicator is the level of 'trust'. Figure 8 shows the results of a survey conducted in the early 1990s on each country's citizens' perception of the internal level of trust. Respondents in each country were asked if their countrymen could be trusted, and the percentage that replied yes is reported in the chart.

The next question is, then, to find out what are the determinants of social capital. Glaeser (2000) suggests that education is strongly associated with social capital, which indicates that an important component of policies aimed at increasing social capital necessarily needs to go hand in hand with policies aimed at increasing the educational level. The reason is not only the fact that there is an association between human and social capital, but also the fact that being in school provides a context for social interaction and learning that has important spillover effects in strengthening social relationships and networks. Alesina & Ferrara (2000) confirm the important role of education as a determinant of social capital, but show also that beyond

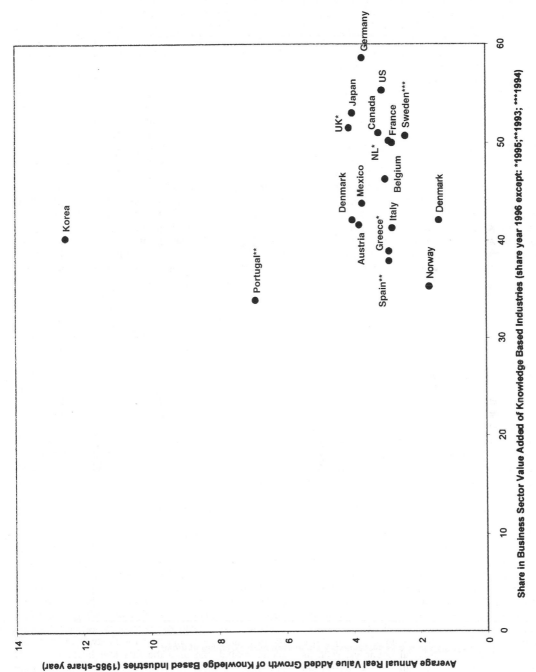

Figure 6. Knowledge based industries intensity and growth.

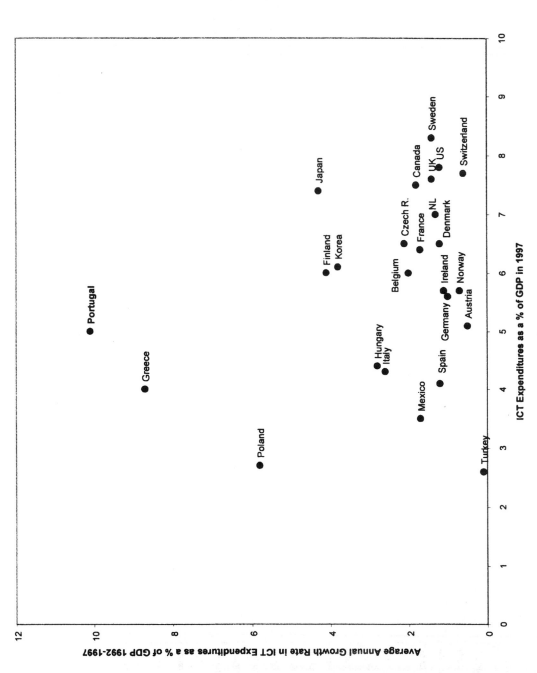

Figure 7. Information and communication technology (ICT) intensity and growth.

Source: OECD (2000).

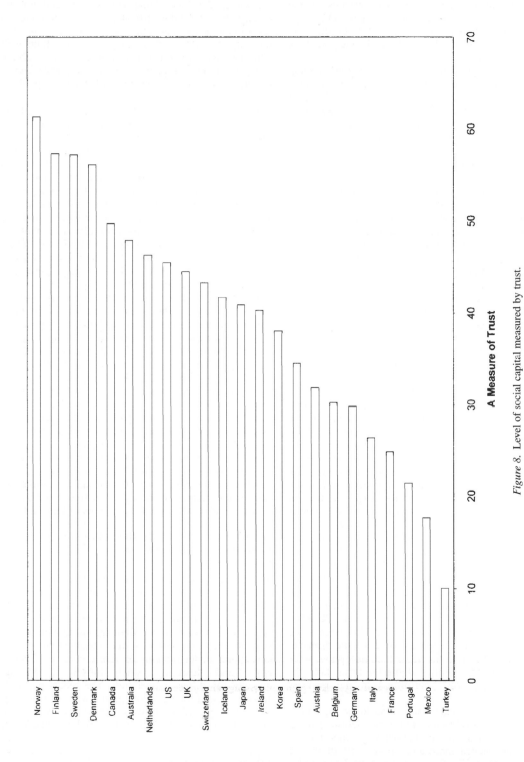

Figure 8. Level of social capital measured by trust.

Source: World values survey. Percentage of people who responded in the affirmative to the question: "Generally speaking would you say that most people can be trusted?"

individual characteristics, the characteristics of the community are equally important. These characteristics include dimensions associated with the way people compare themselves with each other, such as income inequality.

One other important dimension of the learning society includes the activities expressly oriented towards the generation and diffusion of knowledge. It is, as with education, risky to reduce a complex set of activities to a single educator, but the national effort on research and development provides an indication of the commitment, at the country level, to activities explicitly oriented towards the generation of new knowledge. These activities tend to occur in institutions, such as universities and research labs, or within institutional settings, such as the R&D unit within a firm, that provide incentives that foster the specialization on exploration and discovery, as well as exchange of knowledge (Conceição & Heitor, 1999).

Figure 9 shows both the scale and the intensity of national expenditures on R&D for several OECD countries, with the horizontal axis, representing the scale of the expenditure, having a logarithmic scale. The relationship between scale and intensity shows decreasing returns: as the scale of the investment grows, the increase in intensity also grows but at a decreasing (in fact, logarithmic) rate. The results also suggest that there are **three different** 'paths' in which this relationship is expressed.

In the lower left-hand corner of the figure we identify a line that includes the Southern European countries. The thick line in represents a simple fitting of the position of most countries. Nordic countries have a path of their own, with a much higher responsive intensity to increases in scale. For Ireland the scale of R&D expenditure is almost the same as for Portugal, but the intensity for Ireland is comparatively much higher. The large intensity of R&D expenditures in Ireland is largely due to the fact that the R&D that is performed in the business sector, which in 1997 accounted for almost three-quarters of the total R&D expenditure in Ireland. Ireland showed the largest increase in business R&D expenditure of all OECD countries in the 1990s, at an annual growth rate of close to 20%. However, most of this growth is being driven by foreign affiliates doing business in Ireland. The share of foreign affiliates in manufacturing R&D in Ireland in 1995 was close to 70%. This large share indicates a very low capacity of domestic firms to innovate. Ireland is, in this regard, an exception, since for most OECD countries domestic firms take the largest share of R&D performed in the business sector.

R&D efforts are understood as an input; an important outcome of R&D expenditures is scientific papers. Scientific articles are, in themselves, important to diffuse and deepen innovation. Figure 10 shows the same countries as Fig. 9, and the horizontal axis is also

the same: the logarithmic absolute expenditure by country. In Fig. 10 the vertical axis is also presented in logarithmic scale. As when we analyzed scale and intensity of R&D, we fit a straight line, which fits well with the data.[2] Given that both axes are in logarithmic form, scientific production follows a power law, a feature known to be associated with scientific publications.

R&D expenditure is an important indication of the commitment and resources a country devotes to knowledge production and diffusion, but the growing importance of knowledge extends beyond those activities traditionally associated with creativity and learning. Innovative performance, in particular, depends on conditions that foster technology-based entrepreneurship. Mechanisms such as venture capital and high growth start-up stock markets (like the NASDAQ) are ways to mobilize private capital for investment in knowledge economies (Soete, 2000). Gompers & Lerner (1999) show that venture-capital backed start-ups appear to have a disproportionate positive impact on innovation.

However, following Antonelli & Calderini (1999), 'the internal bottom-up learning process based upon the improvement of design and technological processes plays a major role in feeding the continual introduction of technological and organizational innovations'. In this respect, the authors conclude that technological knowledge is embedded in the specific circumstances in which the firm operates, and its generation is the result of a joint process of production, learning and communication, of which R&D activities are only a part (Conceição & Ávila, 2000; Evangelista, Sandven, Sirilli & Smith, 1998). In more general terms, the analysis of the innovative performance of countries in the learning society calls for the need to consider all the processes of learning (both 'formal' and 'informal', in the nomenclature of Conceição and Heitor, 1999) and to better understanding the economics of knowledge.

Deepening our Understanding of Learning towards Innovation: Building on the Economics of Knowledge

The paragraphs above show that, from a systemic perspective of innovation, *learning* is understood, broadly, as *knowledge accumulation*. There are different levels of 'learning entities', from individuals, to organizations, to whole economies. A first important step in our discussion is the clarification of our conceptual understanding of terms such as 'knowledge' and 'learning', often loosely used with dramatically different meanings. The recent paper by Johnson et al. (2002), following the work of Cowan et al. (2000), provides further evidence for the need to clarify these concepts. This conceptual clarification of

---

[2] The R-squared is 0.95.

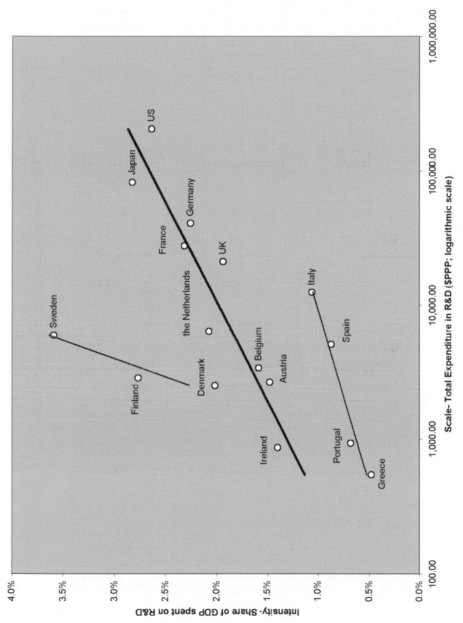

Figure 9. Intensity and scale of R&D expenditure in the OECD (1997).

Source: OECD (2000).

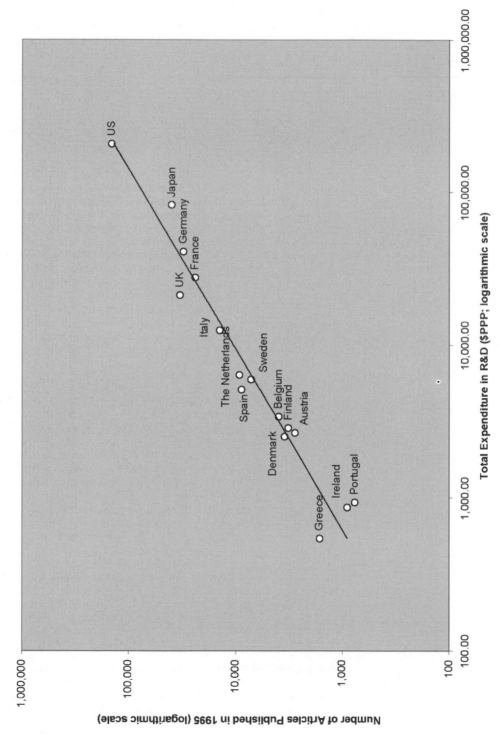

Figure 10. Absolute R&D expenditures and scientific production in the OECD (1997).

Source: OECD (2000).

our understanding of learning as knowledge accumulation is the objective of this section.

We find it useful, as developed in more detail in Conceição & Heitor (1999), to follow Nelson & Romer's (1996) differentiation between ideas and skills, or software and wetware, to use these authors' nomenclature. The conceptual difference between *software* and *wetware* lies in the level of codification. While ideas correspond to knowledge that can be articulated in words, symbols, or other means of expression, skills cannot be formalized, but always remain in tacit form (see Nonaka et al., this volume). Under this taxonomy, knowledge may be divided into two worlds (Johnson et al., 2002): the world of codified ideas (*software*) and the world of non-codified skills (*wetware*).

The difference in the level of codification has implications in terms of the 'economic properties' of the two types of knowledge that we consider. The most important implication is associated with the differences in the rivalry associated with the consumption of each type of knowledge. Since the knowledge underlying *software* is codified, it is easily articulated and reproduced by simple, inexpensive means. Consequently, rivalry in the consumption of software is low. By contrast, the transmission of skills (*wetware*) is complex, expensive, and slow. Skills result from a combination of factors, ranging from their largely innate quality, through individual experience, to formal training. Thus, rivalry is comparatively higher in the consumption of wetware.

The differences in rivalry between *software* and *wetware* have important implications for knowledge production. Dasgupta & David (1994) suggest that there are basically two alternatives for the production of software. The first consists of **intervention by the state** in the production of ideas, by means of direct production, or by subsidizing production, such as funding of university R&D. The second alternative consists of granting property rights for the creation of ideas, that is by defining regulations for **intellectual property** specific instruments that include patents, registered trade marks and copyright (see Conceição & Heitor, 2001; Conceição, Heitor & Oliveira, 1998, for a more comprehensive analysis). Therefore, the production of ideas requires more complex institutional mechanisms than those provided by the market. As for skills, the market provides a large proportion of the incentives needed for their production, at least when these are analyzed in isolation, although with important limitations (see, again, Conceição & Heitor, 2001).

We bring our own understanding to the process of knowledge accumulation when the interaction between software and wetware is explored. The idea of **interaction** between ideas (software) and skills (wetware) is what, in our understanding, defines learning. Analysis of the interaction between ideas and skills leads us to explore the learning processes associated with the generation of each type of knowledge in a more integrated and dynamic way, beyond the mere accumulation of ideas and skills, each in isolation. Our view is yet another perspective on the ongoing debate between the complex and multifaceted interaction between different types of knowledge. Recent manifestations of this debate include Johnson et al. (2002), in which they contest the implicit assumption of Cowan et al. (2000) that codification always represents progress.

Indeed, according to Freeman & Soete (1997), ideas and skills are no more than two sides of the same coin, two essential aspects of the accumulation of knowledge. New ideas spur the development of the skills required to use those new ideas. The bridge from the production of ideas to the usage of ideas is established by producing new skills. Increased use of an idea, which requires its diffusion, will lead to a constellation of other ideas, aimed at improving and extending the initial idea, which will lead to the need for further skills and so on, in a self-reinforcing cycle that leads to the accumulation of knowledge. The accumulation of knowledge results from the production, usage, and diffusion of both software and wetware, in an interactive learning process that leads to knowledge accumulation, as initially proposed by Conceição & Heitor (1999).

Learning Processes and the Accumulation of Knowledge: The Interaction between Software and Wetware

According to Solow (1997), the formalization of the process of economic development in the new growth theories follows the conceptual structure originally proposed by Arrow (1962). It is worth looking briefly at Arrow's analysis, as it contains the kernel of the reasoning behind the idea of economic development as a learning process. Instead of following the orthodox thinking of his time, which attributed to technological change the component of growth that could not be explained by the accumulation of labour and capital factors, Arrow argued that experience in the use of capital led to an increase in the knowledge used in production. In plainer terms, Arrow drew up a relatively simple model in which workers in a company learn by using the means of production, thereby increasing the company's productivity.

In this way learning, that is the accumulation of knowledge, appears as the driving force behind the increases in efficiency, which lead to economic growth. It is interesting to note that Arrow chose an informal way of learning, learning by doing, as the basis for his reasoning. It should also be noted that in this model knowledge is accumulated only in the form of skills. The contribution of the new economic growth theories has been precisely to extend this reasoning to other types of learning, as well as to the accumulation of ideas, starting from when Romer (1986) showed the wider implications of Arrow's arguments.

Table 4. Accumulation of knowledge and learning processes in the new growth theories.

		Learning by			
		Formal processes		Informal processes	
		Education	R&D	Experience (by-doing)	Interaction
Accumulation of	Software (Ideas)		Romer (1990) Grossman & Helpman (1991)		
	Wetware (Skills)	Lucas (1988)		Arrow (1962) Romer (1986)	

Thus, Lucas (1988) also analysed the accumulation of knowledge in the form of skills, but this time putting forward education as a formal learning process. In turn, Romer (1990) and Grossman & Helpman (1991) constructed models in which the accumulation of ideas results from effort put into research, another formal learning process. In this context, Table 4 summarizes how these contributions fit into a framework of possibilities which relates the accumulation of knowledge to the different kinds of learning that can lead to this accumulation. The construction of this table was also inspired by Foray & Lundvall's analysis (1996), in which they placed particular emphasis on the formation of networks of personal and professional contacts, which result from processes of social interaction, the fourth process in Table 4.

This table also illustrates three other points. First, is the analysis that remains to be made in respect of the empty boxes. Second, examination of the dates of the contributions reveals that the emphasis at the beginning of the 1990s was on the study of the accumulation of ideas through R&D, a tendency that has become stronger in recent work (see Romer, 1991, 1993, 1994).

There are at least two reasons for this. On one hand, the study of informal learning processes is more complex and less amenable to empirical testing. We are accordingly left with the study of the accumulation of ideas through R&D, since the role of education has already been extensively researched since the theories of human capital appeared in the 1960s. On the other hand, the really striking aspect of the times in which we live is the increasing codification of knowledge, and the potential of the 'digital economy' and the 'information society' (Romer, 1996; Foray & Lundvall, 1996).

It is important to note that the potential of the 'digital economy' is strongly reflected in the existence of increasing returns, which leads to phenomena such as the apparently unstoppable growth of companies that trade in ideas, such as Microsoft. Indeed, the economic value of an idea is associated with its market potential

(Romer, 1996). As has been seen, it can be extremely expensive to produce ideas, but they are cheap to distribute. The first disk containing the Windows operating system cost Microsoft several million dollars (the entire cost of development), but all the rest cost less than a dollar each. Since there is a vast market and costs, after initial development, are low, the only limit to Microsoft's growth is the size of the market itself. Arthur (1994) points out that the fact of increasing returns, besides being linked to the non-rivalry of ideas, is reinforced by the phenomenon, originally explored by David (1986), known as 'lock-in'. In the case of Microsoft, 'lock-in' took place when the Windows operating system became established as the virtual industry standard. As can be seen, there is much to explore concerning the impact on growth of the accumulation of ideas, but our concern at the moment is to examine the boxes in Table 4 that remain empty, particularly the interaction between ideas and skills.

It is thus time to begin moving into territory that is still being explored, which requires reference to contributions from other groups of economists concentrating on the study of economic growth. Before pursuing this theme, we should note the difficulties that have beset the new economic growth theories. The main criticism is linked to their lack of empirical evidence, despite the intellectual validity of their arguments (Pack, 1994). Mankiw (1995), in a relatively recent assessment, even suggested a return to Solow's traditional formulation. However, according to Soete (1996), empirical difficulties should lead not to a reduction in efforts to pursue the new concepts further, but rather to a recognition that new indicators and quantitative methods must be found that are more appropriate for the knowledge-based economy.

One crucial aspect of the accumulation of knowledge is the interaction between ideas and skills, which gives rise to the learning processes in Table 4. Indeed, according to Soete (1996), ideas and skills are no more than two sides of the same coin, two essential aspects of the accumulation of knowledge. Herbert Simon, quoted by Varian (1995), puts the argument as follows:

> "What information (in the sense of ideas, according to our terminology) consumes is rather obvious: it consumes the attention of its recipients. Hence, a wealth of information (that is, of ideas) creates a poverty of attention, and a need to allocate that attention efficiently among the overabundance of information sources that might consume it".

In other words, many good ideas are useless if the skills needed to use them do not exist. Studies by Pavitt (1987), Nelson (1996), and Rosenberg (1990) follow the same line of thinking. Nelson (1997) describes various circumstances, in which individuals, companies, universities, and other institutions have made use of their skills in order to increase their accumulation of knowledge, acquiring further skills as well as ideas. The main implication of this argument is that the interdependence between ideas and skills casts doubt on the idea that the market supplies the necessary incentives for the production of skills, where these were analyzed in isolation. It seems, therefore, that there is greater scope in the knowledge-based economy for institutional arrangements and public policies that

go beyond the logic of the market (World Bank, 1998).

Although to a great extent skills result from the innate characteristics of an individual or from the history of an institution or a country, they also depend on the learning processes (education, research, experience, social interaction) in which these entities are involved (North, 1990). Without skills, ideas may be irrelevant, and without ideas, there is no need for new and better skills. Analysis of the interaction between ideas and skills understandably brings us to explore learning processes in a more integrated and dynamic way, beyond the mere individual accumulation of ideas and skills set out in Table 4. To illustrate the close and complex interdependence between ideas and skills, Fig. 11 shows the interactions between these two kinds of knowledge.

At this point we should stress that our analysis would be enriched by drawing from the large output of scholarship that originated from the cognitive sciences and from the education sciences on learning. However, this project lays outside the aim of this paper, since we do not intend to contribute to a theory of learning. Our purpose is rather to propose a simplified framework to model the dependency between software and hardware,

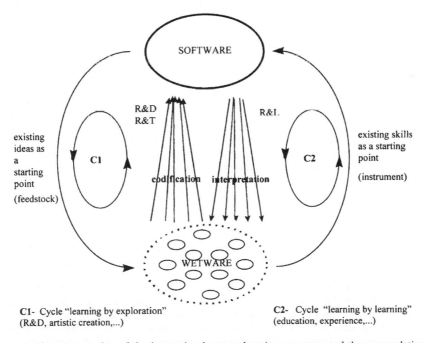

C1- Cycle "learning by exploration"
(R&D, artistic creation,...)

C2- Cycle "learning by learning"
(education, experience,...)

Figure 11. Diagrammatic representation of the interaction between learning processes and the accumulation of knowledge, identifying the various aspects of university research (notably Research & Development, Research & Training, and Research & Learning, as analysed by Conceição & Heitor, 1999).

Notes: R&D—Research and Development; R&T—Research and Training: R&L—Research and Learning.

Source: Figure 4.2 (p. 93) from The Globalizing Learning Economy by D. Archibugi and Bengt-Åke Lundvall (2001). Reprinted by permission of Oxford University Press.

suggesting that it is through this interaction that new knowledge is generated, that is, learning occurs.

From Fig. 11 it can be seen that while skills appear as a cluster of small ovals, reflecting the individual nature of the skills of people and of institutions, ideas appear as a single oval. This represents the indivisibility of ideas (David, 1993), meaning that, once created, an idea remains at least potentially accessible everywhere, and there is no need to rediscover it—hence the common expression 'There's no need to re-invent the wheel'.

Figure 11 shows several learning processes that have been analyzed in various places in the literature. Again we should stress that we have been selective in the way we chose the types of learning processes that are depicted in the figure. Our objective is not to be exhaustive, but rather to emphasize the learning mechanisms that are more directly related to the functions of the university, as will become clear in the next section.

Thus, there are two main cycles:

cycle 1—codification of knowledge (Foray & Lundvall, 1996), the result of progress in information technology, telecommunications and the scientific and technological base; that is, the great number of existing ideas that are the starting point or 'feedstock' for new ideas to be constructed using existing skills;

cycle 2—interpretation of codified knowledge (OECD, 1997), using existing skills as a starting point or instrument to decode the ideas which are being studied or used, leading to improved skills.

Cycle 1 covers learning processes that result in the codification of knowledge, which is the generation of new ideas. Specific examples include R&D and artistic creation. In both cases, ideas are generated as a result of a process of exploration, in science or in search of a form of expression. This type of learning is convergent, meaning that on the basis of different and unique skills, ideas are generated that have the potential for common use.

Cycle 2, however, relates to learning by assimilation of knowledge, which results from activities such as education, experience, and social interaction. Through interpretation of these ideas, different skills emerge. Imagine a mathematics class: all the students are using the same book, they attend the same classes, they do the same exercises. However, the ways in which they assimilate and interpret these are different, meaning that the learning process is divergent. Schon (1987) and others expand on the inner workings of this type of learning, but we keep our discussion at a more superficial level.

The main conclusion of this section, as shown in Fig. 11, is that the accumulation of knowledge, which is the basis for economic growth, is the result of a series of complex processes, in which there is considerable interdependence between the accumulation of ideas and of skills. It is necessary to examine the role of the principal institutions of contemporary society and to attempt to determine how they fit into these processes. Conceição & Heitor (1999) show how this conceptual understanding can be used to analyze broad historical interactions between knowledge and development (such as in the evolution of China and Europe, Landes, 1998) as well as the adoption and diffusion of specific technological innovations (such as standards of videotape recorders). The model also acknowledges the indivisibility of ideas, as proposed by David (1993) (once created, an idea remains at least potentially accessible everywhere, and there is no need to rediscover it).

This conceptual understanding of the learning processes is critical to draw implications in terms of the complex relations associated with the building-up of innovation systems (Christensen, 2002). In this chapter we next develop, the importance of stimulating innovation (generation of ideas) and the parallel importance of developing competencies, within an overall framework looking at inclusiveness.

Fostering Systems of Innovation and Competence Building: The Challenges of Inclusiveness

The analysis presented above considers a context in which the wealth and well being of individuals, organizations and nations is increasingly based on the creation, dissemination, and use of knowledge in a way to foster innovation. This fact is reflected in the trend in developed economies towards an increasing investment in advanced technology, research and development, education, and culture. As a consequence, concepts such as learning ability, creativity and sustainable flexibility gain greater importance as guiding principles for the conduct of individuals, institutions, nations and regions. Against this background, and emphasizing concepts such as the non-rivalry of information and the externalities associated with education and research and development, this section builds on the notion of localized technological change and the need to develop an agenda to promote the inclusive development. This is particularly appropriated to understanding the dynamics of innovation in much of Portuguese industry, which is heavily characterized by the so-called 'traditional sectors'.

Following Antonelli & Calderini (1999), 'the internal bottom-up learning process based upon the improvement of design and technological processes plays a major role in feeding the continual introduction of technological and organizational innovations'. In this respect, the authors conclude that technological knowledge is embedded in the specific circumstances in which the firm operate, and its generation is the result of a joint process of production, learning and communication, of which R&D activities are only a part. In these terms, current evolutionary economics has shown the importance of path dependence of economic processes, in that it is at the core of selection mechanisms between competitive firms and technologies (Metcalfe, 1997). Competition is therefore the result of the rate of change of market share, apart from being dependent on differences in the rates of growth of individual firms. The result is a fully endogenous process, which, in the presence of increasing returns, gives rise to a strong interdependence between specialization and diversification. The direct implication for innovation policies in Portugal is the important, but limited role of demand at the firm level in assessing the amount of incentives for firms to introduce technological innovations. In more general terms, the analysis call for the need to feeding all the processes of learning, implementing technological cooperation among firms and between firms and research institutions, and on the process of on-job-training of the workforce. Technological centers specifically designed to sustain localized processes of technological change might play an important role in this context. However, it is important to clearly emphasize the important role of the science and technology system, S&T, in fostering innovation, as well as the related implications for public policy.

In these terms, although there is an emerging set of literature on technological innovation and industrial economics looking at the distinctive features and institutional characteristics of European regions (e.g. Gambardella & Malerba, 1999), there have been few attempts to build analytical frameworks to improve understanding and to allow the development of well-sustained technology policies for less favored zones and late industrialized European regions, such as those of Portugal. In fact, the neo-classical approaches in industrial economics have emphasized the analysis of the microeconomic behavior of firms and built theories specialized in the American and Anglo-Saxon systems and related market dynamics. However, evolutionary economics have attempted to improve our understanding of learning processes and the role of institutions in economic development, but have not specialized on the specific historical context of European regions, namely those characterized by late industrialization (e.g. Cooke & Morgan, 1998). Building on the evolutionary approaches and system theory, the concept of 'national system of innovation' (e.g. Edquist, 1997; Lundvall, 1992; Nelson, 1993) has led to numerous studies of individual European countries, but there is still a long way to go in order to assess the specificity of transition economies and late industrialized regions and countries, including Portugal.

The various aspects above include heterogeneous approaches to innovation, but consider 'change' at the center of the analysis. This has been considered throughout the entire chapter, but taking into account that firms' competencies are characterized by stability and inertia and, therefore, lock-ins and competence

traps are expected to occur, in that successful firms may be driven by their success in existing technologies to disregard new alternatives. Other important aspect to take into consideration is that the phenomena of increasing returns and path-dependence affect the nature of the innovation processes and the dynamics of industries in Portugal, and Europe.

Among the various aspects raised earlier, it should also be noted that the sectoral specificity in the organization of innovative activities, on one hand, and the specific characteristics of local systems of innovation, on the other hand, are expected to play a significant role in shaping the organization of innovative activity in Portugal. The prevalence of one effect over another dependes on history and competitiveness of firms and their degree of internationalization.

The Importance of Stimulating Innovation

The section above made explicit the way in which we understand learning as knowledge accumulation, which is a result of a complex set of learning processes where there is considerable interdependence between the accumulation of ideas and of skills. We now turn for the analysis of innovation as the concept that best fits with the idea of the knowledge economy understood from a dynamic perspective.

It is by now well understood that the early conceptualizations of innovation as a linear process were clearly insufficient to describe the complexity and contingency of the innovative effort of people, firms and countries (Dosi, 1988; Kline & Rosenberg, 1986; Nelson, 1993). Still, what is surprising is the extent to which the linear perspective still informs much of today's public perceptions about innovation, as well as policy design and implementation. The reliance on simple and direct indicators such as expenditure of R&D by the private sector, and the obsession in some circles associated with improving these types of indicators, reflects the dominance of the linear perspective.

We do not question the importance of these and other indicators, but it should also have become clear by now that they provide an incomplete description of the innovation process and are tied to the linear perspective (see, for the continuation of the linear perspective, Guellec & Pottelsberghe, 2000). Romer (1990, 1993) recognizes the importance of what he calls *appreciative theories* of growth and innovation in helping more formal approaches to better describe the richness of the innovation process, but somehow the link has been hard to accomplish.

The link between the complexity of the innovation process and the special economic characteristics of knowledge, and of conceptualizations of the learning process such as the one advanced in 'Deepening our Understanding of Learning towards Innovation: Building on the Economics of Knowledge', could be a bridge. In fact, Romer (1990) and Nelson & Romer

969

(1996) construct a theory of endogenous growth drawing on the non-rival nature of ideas. Dasgupta & David (1994) advance new ideas about the economics of science building also on the same principles associated with the special characteristics of knowledge. Thus, the conceptual understanding of learning advanced in 'Deepening our Understanding of Learning towards Innovation: Building on the Economics of Knowledge' of this chapter could serve more than just being an interesting modelling tool, allowing the development of new conceptual approaches. It could also become a useful guide for policy, especially in light of the still predominant domination of the linear model. In a series of papers, Conceição & Heitor (2001) have explored the implications of the conceptual model presented in 'Deepening our Understanding of Learning towards Innovation: Building on the Economics of Knowledge' to advance policies associated with innovation (that is, the generation of ideas, or software). We turn, next, to the other side of our conceptual model of learning: the importance of wetware.

The Relevance of Competence Building

Competence is the foundation on which innovation is generated and diffused. Competence is associated with individual skills, but also with collective capacities. It is also on competence that a learning society can be constructed and sustained in order to foster innovation. Some suggest that technological change is (or has become) skill-biased (Autor, Katz & Krueger, 1997). Empirical work supporting the skill-biased technological change conjecture includes studies such as Krueger (1993). Thus, for some, the connection between innovation and competences is primarily understood as being related with this hypothesis.

However, the skill-biased technological change hypothesis is far from being uncontroversial, as the discussion that follows shows. From a conceptual point of view, critics note that the treatment of technological change rarely goes beyond asserting that new technologies, and especially computers, are responsible for a steady increase in the demand for skills (Galbraith, 1998). Technology is conceptualized as in the linear models of innovation. Criticisms based on empirical analysis include DiNardo & Pischke (1997) and the realization that there is a mismatch in the timing of the increase in inequality and the spread in the diffusion of computers, and the fact that the increased adoption of information technology has not noticeably contributed to increased productivity (see Galbraith, 1998, for a comprehensive review). Alternatives to the skill-biased technological change include the perspective advanced by Bresnahan (1999), who proposes an organizational complementarity between information technologies

and telecommunications (ICTs) and highly skilled workers.

But the relationship between competences and innovation is not only seen through the skill-biased technological change perspective. And competence building also entails much more than formal skills. For example, Dore (1976) differentiates 'education' from 'schooling', which refers to 'mere qualification-earning', leading to an 'educational inflation' spiral. Several other authors (e.g. Boudon, 1973; Bourdieu & Passeron, 1970; Bowles & Gintis, 1976; Jencks, 1972) are similarly skeptical about a direct relationship between increases in the level of education and economic performance. The differences between the economists of human capital and these other authors, who come primarily from sociology, remain until today. In fact, some of the critiques have important parallels with economic perspectives, such as Boudieu & Passeron's (1970) theory of the social filter, whereby schools work as filters to preserve and maintain social and educational differences, and the 'inheritance of inequality' perspective of Meade (1964).

However, if one is ready to accept the existence of a labor market where wages reward, at least partially, productivity and skill, Katz & Murphy (1992) provide strong evidence that supply and demand go a long way in explaining the patterns in the evolution of inequality. Most of the recent studies on inequality focus on a single-country longitudinal analysis of the evolution of the dispersion of income. Examples of the same methodology applied to other single country studies include Schmitt (1995) for the U.K., and Edin & Holmlund (1995) for Sweden.

This discussion clearly highlights the link between competence (skills, education), and innovation (technological change) towards inclusive learning. The connection between education, skills and competence, on the one hand, and the learning society, on the other, must consider the manifold interconnections between competence and the learning society and links them with the broader context of the anxieties and concerns, hopes and expectations that we live with today.

An important issue is to know what it takes to be part of the learning society. We may not know exactly what the learning society is, but we do know that there are requirements to be part of it. We need, in particular, to build competence, of which skills are a part. However, for some cases, the need for new skills is not associated with technological change, but with an organizational change, and the new skills provided are not particularly intensive in specialized knowledge. It is important to stress this point because the discussion can easily be drawn into the skill-biased technological change discussion. Naturally, technological change does indeed play a role in increasing the demand for 'a higher order of skills', but there are other elements of change driving this demand. What is hardly questionable is that those that do not possess the skills nor the ability or possibility to acquire them become excluded.

The Need for a Dynamic National Science Base

Pavitt (1998) noted that innovation studies confirm Tocqueville's idea that technological change would require the development of publicly funded basic research and associated training. In this context, analysis has shown that the main practical benefits of academic-based research are not 'easily transmissible information', but it involves the transmission of tacit and non-codifiable knowledge, with tendency for geographically localized benefits (e.g. Katz, 1994). Furthermore, following Hicks (1995), countries and firms benefit academically and economically from basic research performed elsewhere only if they belong to the international professional networks that exchange knowledge. This requires high quality foreign research training and a strong presence in basic research, mainly because academic research is certainly not a 'free good', although it has some attributes of a 'public good'. In this context, Pavitt, among others (e.g. Mowery & Rosenberg, 1998; Narin et al., 1997), conclude that 'public expenditure on academic research is a necessary investment in a modern country's capacity for technical change'.

It is also clear that one must consider the nature and extent of the influence of national patterns of technological change on the national science base. The analysis suggests the co-evolution of scientific performance with national technology and economy (Pavitt, 1996).

Casual observations have, however, shown that patterns of scientific strength and weakness are strongly influenced by the nature of the societal and technological problems to be solved. In any case, current understanding of the complexities of the knowledge bases that underlie future technological knowledge base is very limited.

If any conclusion can be taken with direct application to Portugal, it is that allocation to resources between broad fields of science should remain incremental, and that inadequacies in the rate of technological change should not be claimed to academic research. However, important questions remain to be solved, mainly in terms of the way academic governance influence the performance of basic research activities, and the linkages between basic and applied disciplines. Also, the way the demands for knowledge influence research policies remain to be examined.

It is clear today that one important dimension of the knowledge economy includes the activities expressly oriented towards the generation and diffusion of knowledge. It is, as with education, risky to reduce a complex set of activities to a single educator, but the national effort on research and development provides an indication of the commitment, at the country level, to activities explicitly oriented towards the generation of new knowledge. These activities tend to occur in institutions, such as universities and research labs, or within institutional settings, such as the R&D unit within a firm, that provide incentives that foster the specialization on exploration and discovery, as well as exchange of knowledge. If it is unquestionable today the critical role of the national S&T systems, it is also clear that they do not represent by themselves a true measure of innovation, namely in socio-economic terms. This has led us to broaden our analysis and to attempt to relate current practices for the evaluation of S&T with innovation measurements and other social measures.

A Policy Exercise: Promoting Innovation in Portugal

Recent work within the framework of the OECD International Futures Program suggests two broad policy-related conclusions which apply not only to OECD countries in general, but to a large extent also to the case of Portuguese regions. The first is that if one is to build on the opportunities offered by the considerable progress that has been made in key technological sectors, if one is to reap to the full the economic benefits of rapidly integrating markets and the emerging knowledge society; and if solutions are to be found to tackling the challenges that the management of such a rapidly changing world raises, then what is needed are innovative, creative societies. The second is that in achieving that higher degree of innovativeness and creativity, policy will matter. The way ahead does not necessarily mean less government, not less policy but—certainly in some key areas—different policy.

The reservation 'in some key areas' is important. Just because we are heading into a rapidly changing world in the coming decades does not mean that we have to throw out all policies and make a completely fresh start. Indeed, some policies that have proved their worth in the past may well continue to do so in the future. However, it is clear that in other policy areas at least incremental adjustments are called for, and in yet others some radical new thinking is required. This provides, in fact, a simple but convenient framework for looking at the role of general policies in the future and their implications for innovation: (1) policy continuity; (2) policy reform; and (3) policy breakthroughs.

In this context, we present below four main groups of strategies to be considered for Portugal, which, per se, reinforce the need to develop innovation policies:

(1) *Human Capital for Innovation*: Substantial investments in human capital, and mainly at the basic and secondary levels, will continue to be a main target to promote and nurture innovation if the skill and qualification requirements of future jobs are to be met. This will require imaginative new ways of organizing education and validating people's

knowledge. Regarding the Higher Education System, our work suggests two important ideas. First, we propose that the institutional integrity of the university needs to be preserved. Universities are a special type of learning organization specialized in producing and diffusing knowledge in unique ways. Second, we argue that, important as universities are, they are not enough to guarantee prosperity, and there is a need to promote a diversity of organizational arrangements, even at the higher education level. Indeed, this organizational diversity could be a major contributor to ensure the institutional integrity of the university. In addition, it is concluded that the allocation of resources between broad fields of science should remain incremental, in a way that the aim of policy should be to create a broad and productive science base.

(2) *Institutional Renewal for Innovation*: The evidence from OECD suggests the value of structural and regulatory reforms in supporting the development of innovative and creative societies and economic growth. Among dominant factors, we envisage the role of market liberalization, and market opening, including the privatization of critical infrastructures. The process is to be implemented together a comprehensive program of organizational renewal, namely at the State level, and in a way to promote the establishment of cooperative agreements towards the establishment of social capital. Fiscal incentives for network organizations and a new regulatory framework for employment protection and market regulation should be attempted.

(3) *Networking and Corporate Strategies for Innovation*: a framework for devising and implementing strategies in business environments typical of transitional economies, such as those in the Portugal, is to be considered taking into account clustering effects. The low level of 'thrust' typical of the Portuguese society is a major barrier, that is to be overcome along the enterprise chain value and making use of aggressive 'product development strategies', together with specific factors as: time to market; market and technology; product and process innovation; increasing returns markets; managing environmental complexity; managing organizational change; devising knowledge strategies.

(4) *Alternative Forms of Financing Innovation*: different funding forms to be used in Portugal, including offset and countertrade tools, are conceived in order to promote and develop different approaches to innovation within national companies. Traditional means in financial innovation tends to be 'outdated' on the 'new' economy context. Although national security is not a priority, activities such coast inspection, citizen protection and rescue, and humanitarian programs, are some

examples of the existing need for the country and, at the same time, to consider the use of offsets to foster economic development. Beyond offsets in processes for buying military equipment, countertrade should be considered as well for purchase of civil goods and critical infrastructures, such as the new Lisbon international airport. The research carried out aims to launch guidelines the benefits for the Portuguese economy of the innovative use of tools as offset and countertrade to increment new forms of cooperation between existing firms and new technology based firms creating multipolar, interdisciplinary and market driven networks.

Summary and Conclusions

This chapter addresses complementary aspects of relevance towards improved understanding of innovation in an emerging learning society. It focus on Portugal within a European scene, considering a context increasingly characterized by uncertainty and diversified environments, which are particularly influenced by social and institutional factors. Under this scope, our understanding of innovation encompasses the way in which firms and entrepreneurs create value by exploiting change. This leads us to question the traditional way of viewing the role that contemporary institutions play in the process of economic development and to argue for the need to promote *systems of innovation and competence building* based on learning and knowledge networks.

We describe a conceptual understanding of the relationship between *learning* and *knowledge accumulation*, leading to *innovation*. Our analysis led us to suggest that while the role of institutions needs to be re-examined, the variety of demands and the continuously changing social and economic environment is calling for **diversified systems** able to cope with the need to produce policies that nurture and enhance innovation in the emergent learning society.

In addition to the various arguments used in this chapter derived from emerging concepts associated with the economics of knowledge, a growing body of literature illustrates the importance of demand conditions to allow for technological diffusion in the network society. It is through the diffusion process that technological innovations are translated into wide economic impact, as more and more people and firms consume and use the new products or processes. And if we accept that this increasingly generalized usage of technological innovations fuels, not only increases in well being, but also the conditions to generate further innovations, one cannot escape the importance of demand conditions for economic and technological prosperity in the emerging learning society.

In fact, historians of economic evolution have shown that demand conditions were crucial in the process of early industrialization in the U.S. For example,

Rosenberg (1994) describes the demand conditions that were conducive to the earliest stages of industrialization in the 19th century. In fact, in Rosenberg's (1994) argument, they were crucial to create a new industrial system out of an agricultural society. An important component of the demand conditions was a relatively high level of income per capita and, equally crucial, a relatively egalitarian distribution of the marginal income available beyond the one needed for subsistence. Inspired by this analysis of the interaction between inequality and technology, we believe the concept of *system of innovation and competence building* discussed in this chapter should be further analysed to improve understanding whether, with the current wave of technological innovations, there is also a relationship between levels of inequality and the rates of diffusion of technology. The argument we are advancing here is that social cohesion, beyond the issues associated with ethical judgement and justice, may also be of importance to the learning society.

Innovation should then be understood as a **broad** social and economic activity within the framework of the learning society. It should transcend any specific technology, even a revolutionary one, and should be tied to attitudes and behaviors oriented towards the exploitation of change by adding value. Recent work within the framework of the OECD International Futures Program suggests two broad policy-related conclusions. The first is that if one is to build on the opportunities offered by the considerable progress that has been made in key technological sectors, if one is to reap to the full the economic benefits of rapidly integrating markets and the emerging knowledge society; and if solutions are to be found to tackling the challenges that the management of such a rapidly changing world raises, then what is needed are **innovative**, creative societies. The second is that in achieving that higher degree of innovativeness and creativity, policy will matter. The way ahead does not necessarily mean less government, not less policy but—certainly in some key areas—**different** policy.

References

Aghion, P. & Howitt, P. W. (1997). *Endogenous growth theory*. Cambridge, MA: MIT Press.

Alesina, A. & Ferrara, E. (2000). *The Determinants of Trust*. NBER Working Paper 7621. Cambridge, MA: NBER.

Antonelli, C. & Calderini M. (1999). The dynamics of localized technological change. In: A. Gambardella & F. Malerba (Eds), *The Organization of Economic Innovation in Europe* (pp. 158–176). Cambridge: Cambridge University Press.

Archibugi, D., Lundvall, B.-A. (Eds) (2001). *The globalizing learning economy*. Oxford: Oxford University Press.

Ark, B. van & McGuckin, R. H. (1999). International comparisons of labor productivity and per capita income. *Monthly Labor Review*, (July), 33–41.

Arrow, K. (1962). The economic implications of learning by doing. *Review of Economic Studies*, **28**, 155–173.

Arthur, W. B. (1994). *Increasing returns and path dependency in the economy*. Ann Arbor: University of Michigan Press.

Autor, D., Katz, L. & Krueger, A. (1997). *Computing inequality: Have computers changed the labor market?* NBER Working Paper 5956.

Boudon, R. (1973). *L'Inegalité des chances*, Paris: Libraire Armand Collin.

Bourdieu P. & Passeron J-C. (1970). *La réproduction. Éléments pour une théorie du système d'ensignement*. Paris: Éditions du Minuit.

Bowles, B. & Gintis, H. (1976). *Schooling in capitalist America*. London: Routledge.

Bresnahan, T. (1999). Computerisation and wage dispersion: An analytical reinterpretation. *Economic Journal*, **109** (456), 390–415.

Bruton, H. J. (1998). A reconsideration of import substitution. *The Journal of Economic Literature*, **XXXVI** (June), 903–936.

Carneiro, R. (2003). On knowledge and learning for the new millennium. In: P. Conceição, M. Heitor & B.-A. Lundvall (Eds), *Innovation, Competence Building and Social Cohesion in Europe – Towards a Learning Society*. London: Edward Elgar.

Christensen, J. F. (2002). Corporate strategy and the management of innovation and technology. *Industrial and Corporate Change*, **11** (2), 263–288.

Coates, J. F. (2003). Future innovations in science and technology. In: L. V. Shavinina (Ed.), *International Handbook on Innovation*, Oxford: Elsevier Science Publishers.

Conceição, P. & Ávila, P. (2001). *Community innovation survey in Portugal—1999* (in Portuguese). Lisboa: CELTA Publishers. Also available through http://www.oct.mct.pt

Conceição, P., Gibson, D. V., Heitor, M. V. & Sirilli, G. (2000). Knowledge for inclusive development: The challenge of globally integrated learning and implications for science and technology policy. *Technological Forecasting and Social Change*, **66** (1), 1–29.

Conceição, P. & Heitor, M. V. (1999). On the role of the university in the knowledge economy. *Science and Public Policy*, **26** (1), 37–51.

Conceição, P. & Heitor, M. V. (2001). Universities in the learning economy: Balancing institutional integrity with organizational diversity. In: D. Archibugi & B. Lundvall (Eds), *The Globalizing Learning Economy* (pp. 83–107). Oxford: Oxford University Press.

Conceição, P. & Heitor, M. V. (forthcoming). *Innovation and competence building in the presence of diversity: Learning from the Portuguese path in the European context*. Westport, CT: Praeger Publishers.

Conceição, P., Heitor, M. V. & Lundvall, B.-A. (2003). *Innovation, competence building and social cohesion in Europe – Towards a learning society*. London: Edward Elgar.

Conceição, P., Heitor, M. V. & Oliveira, P. (1998). Expectations for the university in the knowledge based economy. *Technological Forecasting and Social Change*, **58** (3), 203–214.

Conceição, P. & Veloso, F. (2002). Is investing in innovation unproductive? A time to sow and a time to reap. Academy of Management Annual Conference, 11–14 August. Denver, Colorado.

Cooke, P. & Morgan, K. (1998). *The associational economy.* Oxford: Oxford University Press.

Cowan, R., David, P. A. & Foray, D. (2000). The explicit economics ok knowledge codification and tacitness. *Industrial and Corporate Change*, **9**, 211–253.

Dasgupta, P. & David, P. (1994). Toward a new economics of science. *Research Policy*, **23**, 487–521.

David, P. (1986). Clio and the economics of QWERTY. *American Economic Review*, **75** (2), 332–337.

David, P. (1993). Knowledge, property, and the system dynamics of technological change. In: L. H. Summers & S. Shah (Eds), *Proceedings of the World Bank Annual Conference on Development Economics 1992* (Supplement to The World Bank Economic Review).

DiNardo, J. & Pischke, J. (1997). The returns to computer use revisited: Have pencils changed the wage structure too? *Quarterly Journal of Economics*, **112** (1), 291–303.

Dore, R. (1976). *The diploma disease—education, qualification, and development.* Berkeley, CA: University of California Press.

Dosi, G. (1988). Sources, procedures and microeconomic effects of innovation. *Journal of Economic Literature*, **26** (3), 1120–1171.

Edin, P. & Holmlund, B. (1995). The Swedish wage structure: The rise and fall of solidarity wage policy? In: R. Freeman & L. Katz (Eds), *Differences and Changes in Wage Structures.* Chicago: Chicago University Press.

Edquist, C. (1997). *Systems of innovation—technologies, institutions and organizations.* London: Pinter Publishers.

Evangelista, R., Sandven, T., Sirilli, G. & Smith, K. (1998). Measuring innovation in european industry. *International Journal of the Economics of Business*, **5** (3), 311–333.

Foray, D., Lundvall, B.-A. (1996). The Knowledge-based economy: From the economics of knowledge to the learning economy. *Employment and Growth in the Knowledge-based Economy.* Paris: OCDE.

Freeman, C. (1988). Diffusion: The spread of new technology to firms, sectors and nations. In: A. Heetje (Ed.), *Innovation, Technology and Finance.* Oxford: Blackwell.

Freeman, C., Clark, J. & Soete, L. (1982). *Unemployment and technical change.* London: Frances Pinter.

Freeman, C. & Perez, C. (1986). *The diffusion of technical innovations and changes in techno-economic paradigm.* Mimeo.

Freeman, C. & Soete, L. (1997). *The economics of industrial innovation* (3rd ed.). Cambridge, MA: MIT Press.

Galbraith, J. K. (1998). *Created unequal.* New York and London: Free Press.

Gambardella, A. & Malerba, F. (1999). *The organization of economic innovation in Europe.* Cambridge: Cambridge University Press.

Glaeser, E. L. (2000). The formation of social capital. Presented at the *International Symposium on the Contribution of Human and Social Capital to Sustained Economic Growth and Well-being.* Québec City, Canada, 19–21 March.

Gompers, P. A. & Lerner, J. (1999). *The venture capital cycle.* Cambridge, MA: MIT Press.

Grossman, G. M. & Helpman, E. (1991). *Innovation and growth in the global economy*, Cambridge, MA: MIT Press.

Guellec, D. & Pottelsberghe, B. V. (2000). *The impact of public R&D expenditure on business R&D.* STI Working Paper 2000/4. Paris: OECD.

Hall, R. E. & Jones, C. I. (1999). Why do some countries produce so much more output per worker than others? *Quarterly Journal of Economics*, **114** (1), 83–116.

Hicks, D. (1995). Published papers. Tacit competencies and corporate management of the public/private character of knowledge. *Industrial and Corporate Change*, **4**, 401–424.

Inzelt, A. (2003). For a better understanding of the innovation process in Hungary. In: L. V. Shavinina (Ed.), *International Handbook on Innovation.* Oxford: Elsevier Science Publishers.

Jencks, C. (1972). *Inequality.* New York, N.Y.: Basic Books.

Johnson, D. G. (2000). Population, Food, and knowledge. *American Economic Review*, **90** (1), 1–14.

Johnson, B., Lorenz, E. & Lundvall, B.-A. (2002). Why all this about codified and tacit knowledge? *Industrial and Corporate Change*, **11** (2), 245–262.

Katz, J. (1994). Geographical proximity and scientific collaboration. *Scientometrics*, **31** (1), 31–43.

Katz, L. & Murphy, K. (1992). Changes in relative wages, 1963–1987: Supply and demand factors. *Quarterly Journal of Economics*, **107** (1), 35–78.

Kline, S. J. & Rosenberg, N. (1986). An Overview of Innovation. In: R. Landau & N. Rosenberg (Eds), *The Positive Sum Strategy. Harnessing Technology for Economic Growth.* Washington: National Academy Press.

Kremer, M. (1993). Population growth and technological change: One million B.C. to 1990. *Quarterly Journal of Economics*, **108** (3), 681–716.

Krueger, A. (1993). How computers have changed the wage structure? Evidence from micro data. *Quarterly Journal of Economics*, **108** (1), 33–60.

Landes, D. (1998). *The wealth and poverty of nations: Why some are so rich and some so poor.* New York: W. W. Norton & Company.

Lucas, R. E. (1988). On the mechanics of economic development. *Journal of Monetary Economics*, **22**, 3–42.

Lundvall, B.-A. (1992). *National system of innovation—towards a theory of innovation and interactive learning.* London: Pinter Publishers.

Lundvall, B.-A. (2000). Towards the Learning Society: challenges and opportunities for Europe. Presentation in the Research Seminar *Towards a Learning Society: Innovation and Competence Building with Social Cohesion for Europe.* Lisbon, May 28–30. Available at http://in3.dem.ist.utl.pt/learning2000/default.htm.

Lundvall, B-A. & Johnson, B. (1994). The learning economy. *Journal of Industry Studies*, **1** (2), 23–42.

Mankiw, N. G. (1995). The growth of nations. In: W. C. Brainard & G. L. Perry (Eds), *Brookings Papers on Economic Activity* (Vol. 1). Washington, D.C.: Brookings Institution.

McKnight, L., Vaaler, P. & Katz, R. L. (2000). *Creative destruction—Business survival strategies in the global internet economy.* Cambridge, MA: MIT Press.

Meade, J. E. (1964). *Efficiency, equality and the ownership of property.* London: Allen and Unwin.

Metcalfe, J. S. (1997). Science policy and technology policy in a competitive economy. *Intl. J. Social Economics*, **24** (7/8/9), 723–740.

Mowery, D. C. & Rosenberg, N. (1998). *Paths of innovation—technological change in the 20th Century America.* Cambridge: Cambridge University Press.

Narin, F., Hamilton, K. & Olivastro, D. (1997). The increase linkage between us technology and public science. *Research Policy*, **26**, 317–330.

Nelson, R. (1991). *National innovation systems*. Oxford: Oxford University Press.

Nelson, R. (1993). *National innovation systems: A comparative analysis*. Oxford: Oxford University Press.

Nelson, R. R. (1996). What is "commercial" and what is "public" about technology, and what should be?. In: N. Rosenberg, R. Landau & D. C. Mowery (Eds), *Technology and the Wealth of Nations*. Stanford, CA: Stanford University Press.

Nelson, R. R. (1997). How new is new growth theory? *Challenge*, **40** (5), 29–58.

Nelson, R. R. & Romer, P. (1996). Science, economic growth, and public policy. In: B. L. R. Smith & C. E. Barfield (Eds), *Technology, R&D, and the Economy*. Washington, D.C.: Brookings.

Nelson, R. R. & Winter, S. G. (1982). *An evolutionary theory of economic change*. Cambridge, MA: The Belknap Press of Harvard University Press.

Nonaka, I., Sasaki, K. & Ahmed, M. (2003). Continuous innovation: The power of tacit knowledge. In: L. V. Shavinina (Ed.), *International Handbook on Innovation*. Oxford: Elsevier Science Publishers.

North, D. C. (1990). *Institutions, institutional change and economic performance*. Cambridge: Cambridge University Press.

OCDE (1997). *Technology and industrial performance*. Paris: OCDE.

OECD (1999). *Science, technology and industry scoreboard-Benchmarking knowledge-based economies*. Paris: OECD.

OECD (2000). *Information technology outlook*. Paris: OECD.

Pack, H. (1994). Endogenous growth theory: Intellectual appeal and empirical shortcomings. *Journal of Economic Perspectives*, **8** (1), 55–72.

Pavitt, K. (1987). The objectives of technology policy. *Science and Public Policy*, **14**, 182–188.

Pavitt, K. (1996). National policies for technical change: Where are the increasing returns to economic research? *Proceedings of the National Academy of Sciences*, **23** (Nov. 12). Washington, D.C.

Pavitt, K. (1998). The social shaping of the national science base. *Research Policy*, **27** (8), 793–805.

Romer, P. (1986). Increasing returns and long-run growth. *Journal of Political Economy*, **98** (5), 1002–1037.

Romer, P. (1990). Endogenous technological growth. *Journal of Political Economy*, **98** (5), s71–s102.

Romer, P. (1993). Idea gaps and object gaps in economic development. *Journal of Monetary Economics*, **32**, 543–573.

Romer, P. (1994). The origins of endogenous growth. *Journal of Economic Perspectives*, **8** (1), 3–22.

Romer, P. (1996); Why, indeed, in America? Theory, history, and the origins of modern economic growth. *American Economic Review*, **86** (2), 202–206.

Rosenberg, N. (1990). Why do companies do basic research with their own money? *Research Policy*, **19**, 165–174.

Rosenberg, N. (1994). *Exploring the black box: Technology, economics and history*. Cambridge: Cambridge University Press.

Schmitt, J. (1995). The changing structure of male earnings in Britain 1974–1988. In: R. Freeman & L. Katz (Eds), *Differences and Changes in Wage Structures*. Chicago: Chicago University Press.

Schon, D. (1987). *Educating the reflective practitioner*. San Francisco: Jossey-Bass Publishers.

Schumpeter, J. (1934). *The Theory of Economic Development*. Cambridge, MA: Harvard University Press.

Schumpeter, J. (1942). *Capitalism, Socialism and Democracy*. New York: Harper & Row.

Smith, K. (2000). What is the knowledge economy? Knowledge-intensive industries and distributed knowledge bases. UNU-INTECH Discussion Paper 2002-6; also available from http://www.intech.unu.edu/publications/index.htm.

Soete, L. (1996). The challenges of innovation. *IPTS Report*, **7**, 7–13.

Soete, L. (2000). The Challenges and the Potential of the Knowledge Based Economy in a Globalised World, presented at the *International Hearing within the Portuguese Presidency of the European Union*. Mimeo.

Solow, R. M. (1997). *Learning from learning by doing*. Stanford, CA: Stanford University Press.

Swan, J., Scarbrough, H. & Robertson, M. (2003). Linking knowledge, networking and innovation processes: A conceptual model. In: L. V. Shavinina (Ed.), *International Handbook on Innovation*; Oxford: Elsevier Science Publishers.

Temple, J. (2000). Growth Effects of Education and Social Capital in the OECD. presented at the *International Symposium on the Contribution of Human and Social Capital to Sustained Economic Growth and Well-being*, Québec City, Canada, 19–21 March.

Varian, H. R. (1995). The information economy. *Scientific American*, **273** (3), 200–202.

World Bank (1997). *World Development Report 1998: Knowledge for Development*. (Annotated Outline). Mimeo.

World Bank (1998). *World development report 1998: Knowledge for development*. Oxford: Oxford University Press.

Wright, G. (1999). Can a Nation Learn? American Technology as a Network Phenomenon. In: N. Lamoreaux, D. M. G. Raff & P. Temin (Eds), *Learning by Doing in Markets, Firms, and Countries*. Chicago: The University of Chicago Press.

The International Handbook on Innovation
Edited by Larisa V. Shavinina

The Taiwan Innovation System

Chiung-Wen Hsu and Hsing-Hsiung Chen

Industrial Economics and Knowledge Center, Industrial Technology Research Institute, Taiwan, R.O.C.

Abstract: The Taiwan innovation system is continually contributing to technology-based industrial development in Taiwan. This chapter discusses the most representative characteristics of the innovation system, that is: (1) the Technology Development Program of the Ministry of Economic Affairs illustrating the industry innovation policy; (2) the R&D and technology diffusion strategy of the Industrial Technology Research Institutes; (3) the Hsinchu Science-Based Industrial Park's method of technology commercialization; and (4) the recruitment of overseas experts and the cultivation of talent to supply human resources to academia. Finally, we outline the practical achievements of the formation of the Taiwan integrated circuit industry.

Keywords: Innovation system; Government-supported R&D project; Technology transfer; Industrial technology development.

Introduction

Industrial development is the motive force behind economic growth. Industrial development is also a result of international economic competition. Development of technology-intensive industries is often the guiding principle of government for a developing country with limited natural resources and a small-scale domestic market. To assist technology-intensive industries, it is often necessary to subsidize research institutes and provide tax incentives. In addition to encouraging universities to diffuse knowledge for industrial development, the cooperation of industry, government, universities and non-profit research institutes in order to develop technology-intensive industries is critical for a national innovation system in an economy such as Taiwan's, with limited scientific and technological resources. Innovation almost always means the creation of a product, service, or process that is new to an organization. It is the introduction into the marketplace, either by utilization or by commercialization, of a new or improved product, service, or process. In this chapter we define innovation as not being new to the world, that is, it is viewed as the first use of an idea and its exploitation or commercialization within Taiwan.

The Taiwan economy has progressed from the agricultural society of the 1940s to light industries, such as consumer goods, in the 1950s, and then on to heavy industries, such as the petrochemical industry, in the 1960s. At the beginning of the 1970s, the government along with local and foreign scholars recognized that Taiwan, an island nation with scarce natural resources and a limited domestic market, should set up an export-oriented strategy for economic development and should develop high-technology industries so that economic development could be maintained. However, in the 1970s, it was anticipated that no existing industry in Taiwan could lead the way in developing high-technology industries into the future for more than ten years. A consensus was achieved among representatives of industry, government, and academia, that the government should assist in the initial development of high-technology industries. The strategies used to build up an innovation system for high-tech developments were as follows:

(1) in order to enable domestic industry to obtain a foundation for high-tech development, government, enterprises and experts and scholars from universities and non-profit research institutions pooled their talents, trying to build an innovation system for the development of high-tech industry;
(2) a high-tech industry with development potential and wide inter-industry interdependence was selected as the target to be developed. And then, the key technology in the industry to be built up was selected as the subject to be developed. The key technology was the one likeliest to increase

greatly the industrial competitiveness of Taiwan. After an effort of some 20 years, the success of the innovation system for high-tech industrial development in Taiwan has been widely recognized. Among many other achievements, the computer industry here has grown to be the third largest exporter in the world, surpassed only by the USA and Japan. In addition, the production volumes of more than ten products now rank first in the world, e.g. notebook personal computers (PCs), hand-held scanners, and modems, to mention a few examples. Taiwan has also become the world's fourth-largest producer in the semiconductor industry. Taiwan is ranked first in the integrated circuit (IC) design sub-industry. In addition, the liquid crystal display (LCD) industry has grown rapidly in Taiwan. The electronics and information industries have now become the leading industries in Taiwan.

The main purpose of this chapter is to introduce the Taiwan innovation system's main characteristics. This chapter is organized as follows: first, the innovation system in Taiwan will be described; then, the most representative characteristics of the innovation system will be examined, including: (1) the industry innovation policy of government; (2) the technology research and transfer strategy of the research institutes; (3) the technology commercialization method of the Hsinchu Science-Based Industrial Park (HSIP); and (4) the recruitment of overseas scholars and the cultivation of talent in universities; and finally, the IC industry in Taiwan, in terms of the Taiwan innovation system, will be discussed as a case study.

The Innovation System in Taiwan

The Concept of a National System of Innovation

The concept of a national innovation system (NIS) originated from Schumpeter's discussion on the importance of innovation to economic development (OECD, 1997). Among scholars who have studied this topic, some of the more renowned academics are Freeman (1987), Lundvall (1992), and Nelson (1993). They each define a national innovation system from different research points.

Freeman (1987) defines a national innovation system as '. . . the network of institutions in the public and private sectors whose activities and interactions initiate, import, modify and diffuse new technologies' (Freeman, 1987, p. 1). He discusses the interplay between activities that lead to advances in technology and the organizations in the public and private sector that comprise the national system. He applies the concept of national innovation to show how Japan at that time maintained its superiority in some industries over other countries.

Lundvall says, ". . . A system of innovation is constituted by the elements and relationships which interact in the production, diffusion and use of new, and economically useful, knowledge . . . and are either located within or rooted inside the borders of a nation state" (Lundvall, 1992, p. 2). He studies this subject from the perspective of institutions that, taken together, make up the national innovation system, as well as from the microcosmic point of view of the individual.

Nelson calls the national innovation system ". . . a set of institutions whose interactions determine the innovative performance . . . of national firms" (Nelson, 1993, p. 4). Basically, the focus of his research is to compare innovation systems adopted by various countries.

The Organization for Economic Cooperation and Development (OECD) discusses and compares interactions among the various players in the national innovation system, in order to gauge how effectively those interactions are at facilitating the flow of knowledge in the greater technology sector. Four types of interaction between different constituents of the innovation system are compared: (1) interactions among enterprises; (2) interactions among enterprises, universities and public research laboratories; (3) diffusion of knowledge and technology to firms; and (4) movement of personnel (OECD, 1997).

Although there are several definitions of a national innovation system (see Grupp et al., this volume), they can be classified into general and mission-oriented definitions based on formulating goals (see Conceição & Heitor, this volume). The general definition encompasses all interrelated institutional facilitators that create, diffuse, and exploit innovation in the nation. The mission-oriented definition includes organizations and institutions directly related to research and exploring technological innovations, such as enterprises' R&D departments, universities, and public research institutions, enabling policy-makers to develop policies and measures that promote successful innovation (see Mehra, this volume).

In this chapter, the mission-oriented definition is followed. It understands innovation to mean technological innovation and defines a NIS as a complex of innovation facilitators and institutions that are directly related to the generation, diffusion, and commercialization of technological innovation and also the interrelationship between innovation facilitators. The major concern in this concept is how Taiwan can formulate an effective national setting of major innovation facilitators and how to motivate networking among them in order to generate and commercial innovation effectively.

The Framework of the Taiwan Innovation System

In order to build up the foundation of high-tech industry for domestic industry, a joint effort by government, industry, university and research institutes is required to operate the innovation system for stimulating the development of high-tech industry. The primary elements in the Taiwan innovation system are as follows.

(1) *Promoting the Establishment of Applied Research Institutions*

When considering R&D organization, the Taiwan government referred to the approach of foreign countries such as Japan's Agency of Industrial Science and Technology (AIST). This was established in 1952 (renamed the National Institute of Advanced Industrial Science and Technology in 2001) as a governmental organization. Korea's Institute of Science and Technology (KIST) was established in 1966 as the first public research institute, but not as a government agent, in order to attract overseas Korean experts to return to the country. To insure the operational flexibility of research institutes, and to free research institutes from government control, Taiwan decided to establish non-profit organizations to carry out technological innovation tasks. The first applied research institute, the Industrial Technology Research Institute (ITRI), was established under a special act, whose primary purpose was to enhance industrial technologies. A number of organizations were established based on the requirement of the development of domestic industry, like the Institute for Information Industry (III) in 1979, and the Development Center for Biotechnology (DCB) in 1984.

(2) *Government Subsidized R&D Funds*

Applied research institutions rely on government funding to conduct R&D. The 'Science and Technology Development Program' was formulated at the First National Conference on Science and Technology in 1979. (The Conference is a nationwide meeting held every four years by delegates from industry, government, academic and research institutes. It focuses on major issues concerning the development of science and technology.) In the program, it was mandated that the Ministry of Economic Affairs (MOEA) should allocate budgets for the Technology Development Program (TDP) and subsidize the non-profit applied research institutes to carry out the R&D of industrial technologies. The requirement was to achieve the optimal goal of stimulating the development of high-tech industry and upgrade the industry's competitiveness.

(3) *Setting up a Science-Based Park*

Since Taiwan had no experience in developing a high-tech industry, copying strategies from abroad was thought the best course of action. One of the measures proposed by the government was to set up a science-based park, and Silicon Valley, the most successful science-based park in the world, was taken as a model. The Government built up HSIP near ITRI, National Tsing Hua University and National Chiao Tung University, to be the main arena for realizing the industrialization of science and technology (S&T).

(4) *Recruitment of Overseas Academics*

After the founding of R&D institutions, government budgets and the science-based park, there was only one remaining critical element: human resources. Although Taiwan had always placed emphasis on education and had excellent human resources, compared to other advanced countries, Taiwan was still far behind at that time. To make it worse, Taiwan had no practical experience in areas of R&D of S&T, fabrication and operations, and information gathering. To recruit overseas academics and professionals was the most effective way to achieve these goals in a short term.

From the above analysis, we formulate a framework for the innovation system in Taiwan, as shown in Fig. 1. Four primary elements can best represent the characteristics of the Taiwan innovation system.

(1) TDP (of MOEA) will be used as an example to explain the industrial innovation policy of the government.

(2) ITRI will be used to gain insight into the R&D and technology diffusion strategies of research institutions.

(3) HSIP will be used to give examples of the application and commercialization of S&T.

(4) Recruitment of overseas scholars, experts, and cultivation of professional talent will be used to explain how academic institutions supply human resources.

This chapter first examines in broad outline the development mechanism of the Taiwan innovation technology in 'The Taiwan Innovation Development Mechanism'. Then the focus shifts to a microcosmic analysis of the way in which the individual institutions making up the innovation system interact with one another. Interactions between the key players of technological innovation (the industry, universities and research institutes) that lead to technology diffusion, the actions of talented professionals, international knowledge exchanges, and other kinds of interactions are examined in detail from 'The Industrial Innovation Policy' through to 'The Technology Commercialization Method Utilized in HSIP'. Finally, 'Innovation in the Taiwan IC Industry' will use the IC industry in Taiwan as a case study to show that the Taiwan innovation system at various stages of development may require varying degrees of assistance and input from industry, government, academia, and research institutions.

The Taiwan Innovation Development Mechanism

How S&T Policies are Formulated

Science and technology strategies in Taiwan are based on the consensus achieved at National Science and

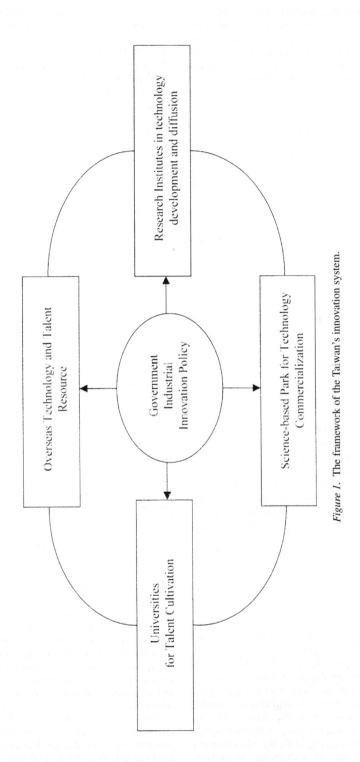

Figure 1. The framework of the Taiwan's innovation system.

Technology Conferences, Science and Technology Advisory Board Meetings, high-tech focus strategy Review Boards, and other major conferences. The following section introduces the formulation of S&T polices in Taiwan from the point of view of major conferences and action plans (NSC, 2000a).

(1) *National Science and Technology Conferences (NSTC)*
The NSTC has been held every four years since 1978. The proceedings of the conference are used to form national policy, as was the case in the sixth NSTC held in 2001, which led to the *National Science and Technology Development Plan (2001–2004)*.

Headed by the Executive Yuan Premier, the sixth NSTC preparatory committee consisted of 47 agency heads, Executive Yuan S&T advisors, and representatives of industry, legislature, universities, and research institutions. Overall responsibility for planning, preparation, and execution was put in the hands of the National Science Council (NSC), Executive Yuan. The preparation for the conference took the form of the following four stages:
(a) formulation of overall planning concepts by the NSC. A Coordination meeting of relevant agencies was held, to set the main directions and focal points of the conference;
(b) the content of the various topics, subtopics, and discussions outlines was decided;
(c) three coordination meetings with the deputy heads of relevant agencies by the NSC, and the first preparatory committee conference by the Executive Yuan were arranged;
(d) at the conclusion stage of the conference, after each of the topics had been discussed, specialists and scholars were instructed to jointly devise specific, feasible programs or suggestions on the basis of preparatory conference conclusions.

(2) *The Science and Technology Advisory Board Meeting of the Executive Yuan (Cabinet)*
Held annually since 1980, the Science and Technology Advisory Board Meeting gives the Science and Technology Advisory Group a chance to make policy recommendations concerning major issues involving the nation's S&T development plans and inter-agency R&D operations.

(3) *The Strategy Review Board (SRB) Meeting on Important Technology, Executive Yuan*
The Executive Yuan has held annual electronics, information, and telecommunications strategy review board meetings since 1992. Since 1997, the Strategy Review Board Meeting on Biotechnology has been held annually. In 2002, the SRB meeting included discussions of technology considered important for national development. It designed a flexible development mechanism for the requirements of industrial development in Taiwan.

The Organizational Framework of S&T Development
The Taiwan S&T development policy was set up on the principle of comprehensive planning and the assigning of tasks to the relevant agencies. The NSC set up by the Executive Yuan, is responsible for the overall implementation of S&T development, while the Ministry of Defense (MOD) looks after defense technology. The Ministry of Economic Affairs (MOEA) takes care of industrial technology and the Ministry of Transportation and Communication (MOTC) is concerned with transportation, telecommunication, and meteorological technology. In addition, various colleges and universities under the administration of the Ministry of Education (MOE) are engaged in basic and applied R&D. A science and technology advisory office or similar office of the Executive Yuan has been founded within each ministry in order to plan and control science and technology development activities (NSC, 2000a).

Measures aimed at fostering the development of technology have already been implemented. For example, the NSC promoted the establishment of the Hsinchu Science-Based Industrial Park so as to create a beneficial R&D environment for the private sector.

Research and development organizations can be classified on the basis of whether they perform basic research, applied research, technology development, or commercial/applications realization. See Fig. 2 for a schematic view of the assignment of tasks between promoting and implementing organizations.

The Industrial Technology Development System
Being the main agency responsible for industrial technology development in Taiwan, the MOEA transfers the results of research to the private sector for product development and commercialization via technical assistance, information diffusion, and manpower training. Industrial technology development is conducted primarily via in-house R&D and secondarily via technology acquisition. To promote the technological upgrading of industry, the MOEA is working to strengthen interaction between industry, government, universities, and research institutions, as Fig. 3 shows.

The role of the research institutions is to interact with the private sector via technology transfer and collaboration with the private sector; they also assist the private sector in technology development by providing technical assistance, technical information, and personnel training. As for academia, the NSC has sponsored specific topics for research projects that enable universities to assist the private sector in technical assistance and personnel training. On the government's part, the development of applied research has been carried out by: (1) programs sponsored by the NSC, e.g. joint research project within industry-

Responding agency / Level of R&D	Promotion	Implementation		
	Government	Universities and Research Institutes	Non-Profit Research Institutes	Industry (Including Science-Based Industrial Park)
Basic Research	• Academia Sinica • Ministry of Education • National Science Council	Universities Labs of Academia Sinica		
Applied Research	• Ministry of Economic Affairs • Ministry of Defense • Ministry of Communications • Council of Agriculture • Atomic Energy Council	• Chunghwa Telecom Labs • Institute of Transportation • Provincial Agriculture Research Institute, etc.	• Industrial Technology Research Institute • Development Center of Biotechnology • Institute for Information Industry • Food Industry Research & Development Institute, etc.	
Technological Development	• Department of Health • Environmental Protection Administration			
Commercialization				• Public Industry • Private Industry

Figure 2. The framework for the allocation of S&T development tasks in Taiwan.

Source: The Annual Report of Science and Technology by NSC.

981

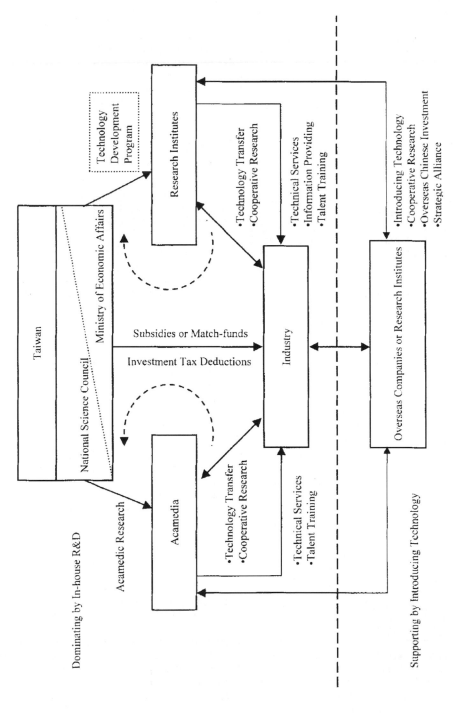

Figure 3. The framework for industrial technology development in Taiwan.

Source: The Annual Report of Science and Technology by NSC.

universities, innovative technology research programs and target-oriented research programs; and (2) programs administered by the MOEA, e.g. importing technology from foreign projects, and TDP. To aid industry, measures for commercialization and applications have been drawn up, such as *the Regulation Governing Assistance in the Development of New Products* by the MOEA.

Universities and research institutes transfer the results of their R&D to the private sector in various forms. In addition, interaction between overseas corporations and research organizations benefit industry through the introduction of new technologies, joint research, overseas investments, and strategic alliances. The MOEA provides subsidies, matching funds, and tax deductions to encourage the private sector to engage in R&D. The Taiwan industrial technology development focuses mainly on the in-house R&D efforts augmented by introduced technologies. The emphasis is on positive interactions among industry, government, universities and research institutes, with the goal of optimizing the facilitation of industrial technology innovation.

The Industrial Innovation Policy

Taiwan industry consists mainly of small and medium enterprises (SMEs) with weak R&D capability, and because of the small scale of the domestic market, the government assumes the role of developing new technologies (see Inzelt, this volume). An annual budget is allocated for the Technology Development Program (TDP) to assist the technology development effort in industrial innovation. The budget grew from NT$10 billion in 1992 to NT$15 billion in 2000, accounting for 20.2% of the total government R&D budget, and 12.1% of private R&D funds. This shows the effect of the government budget on industrial innovation (NSC, 2000b).

TDP plays an important role in the development of key technologies for industrial needs. Its objectives are the establishment of new industries by cultivating professional human resources, developing markets and capital, as well as the upgrading of existing industries. Therefore, TDP has been established for the following reasons: first, and primarily, to support the development of new technologies such as biotechnology and pharmaceuticals; second, to support growing industries such as software engineering and communication and thirdly, to support mature industries such as textiles.

The scope of TDP can be categorized by five areas:

(1) the development of key technologies, parts and components to promote new industries;
(2) the automation and modernization of conventional industries to speed up transformation;
(3) the establishment of inspection facilities;
(4) the promotion of technology for the conservation and efficient use of water and energy;

(5) the promotion of technology for environmental safety and protection to improve the quality of life (MOEA, 2001a).

To enhance the effects of TDP, the results from R&D are passed on to the private sector through various channels. During the process of an R&D project, the private sector receives the R&D results by attending the early stages of the project, collaborating, or receiving subcontracts. At the end of the program, the R&D results are transferred to the private sector through patent licensing and technology transfer. the exchange of information, seminars and conferences, talent diffusion, and various forms of industrial services, are ways of building up interface channels between research institutes and the private sector. These channels also serve as the means of diffusing results from R&D to the private sector. In particular, the diffusion of talent is a direct way of providing technology to the private sector.

The TDP Mechanism

In order to ensure success in its technology development efforts, the strategy for TDP projects is not only to entrust the project to non-profit research institutes, which can result in the industry not knowing much about the particular technology, but also to employ a dual-participation process between the public and private sectors. Once private industry is capable of developing the technology, projects are entrusted to both private and public corporations. It is expected that more than 20% of the TDP budget will be for projects assigned to private industry. Once private industry is in full control of the new technology, an application under *the Regulation Governing Assistance in the Development of New Products* is submitted to the Industrial Development Bureau of the MOEA for the acceleration of commercialization. For crucial technology with significant influence on industrial development, the government becomes a partner for such acceleration of commercialization in order to establish a new industry if private enterprises cannot afford the needed investment.

For the purposes of applying research results to industry, the selection of research topics goes through a number of stages. The Department of Industrial Technology (DOIT) of the MOEA periodically entrusts industrial associations with investigating the needs of the industry. Research topics must be in accordance with industry policy. For each research field, DOIT invites experts from each particular industry, government and academic sectors and research institutes, to discuss long-term planning and policy formulation for industrial development. If necessary, possible plans and policies are proposed to the Strategy Review Board (SRB) of the Executive Yuan. After discussion, development directions and policies for each research field are identified. Then, based on these policies, the research institutes submit their annual R&D proposals. During the review of the R&D proposals by DOIT, the Review Board of Industrial Technology and Planning of MOEA has the power to invite experts to undertake a review of the proposals. The evaluation is based on many considerations, including potential contributions to industry, government policy, industrial needs, and the capability of undertaking and completing projects as well as the estimated budget.

There are many effective ways to promote research results (see Wilhelm, this volume). During project execution, results are passed on to industry whenever possible. After the project is finished, every possible method is used to promote the research results. These include seminars, publications, expositions, technology transfers, patent licensing and the spinning off of new companies if possible. The research institutes provide technical consultations and other services to industries, in order to fully utilize the results of TDP projects. By performing TDP projects and promoting the project results, the capability of industries of developing advanced products is enhanced, so that the effectiveness of *the Regulation Governing Assistance in the Development of New Products* managed by the Industrial Development Bureau is maximized. The management mechanism of the TDP is shown in Fig. 4 (MOEA, 2001a).

Major Measures Undertaken to Upgrade Industrial Technologies

To achieve the goal of upgrading industry technology, DOIT adopted strategies for varying industrial sector needs (Hsu & Chiang, 2001). In addition, adequate measures were implemented in different stages according to the current and future needs of the private sector. In the final stages, the technological development was undertaken by non-profit organizations like ITRI, DCB, and III; the results from the R&D were diffused to the private sector by way of technology transfers, seminars, technical services, and spinning-off companies. Consequently, high-tech has become the main force in the Taiwan industry today.

The measures undertaken to attract cooperation from the private sector in 1994 lured the private sector to engage in other R&D activities to upgrade their technology. The industrial sector joined the TDP R&D projects in the early stages so that the projects could meet the true needs of the private sector. The industrial sector also often attended to research work during development to expedite the commercialization of products and to lessen the risk that the technology developed would have a short life cycle. To accelerate the founding of technological enterprises, resources accumulated from the project, like research capabilities, facilities, human resources, and knowledge, have been made available to the private sector since 1997 in the form of 'open laboratories'.

To meet the requirements from the private sector to upgrade and broaden the technology, DOIT sponsored TDP for the industrial sector in 1997 that entrusted various companies from industry to carry out TDP. This was in order to ride the trends of the coming knowledge-based era in which the creation of knowledge and human resources will be the essential factors. The first priority was to direct abundant R&D and human resources to universities in the private sector. For this reason, DOIT directed TDP to universities in order to assist them in carrying out technological development. The relationship of government industrial policies to various sectors from industry, government, academic and research institutes is depicted in Fig. 5 (MOEA, 2001a). The following describes the five main development strategies for industrial technology development.

The Underpinning of Non-Profit Research Institutes Performing TDP—Interactions between Government, Research Organizations and the Private Sector

In order to encourage the domestic industrial sector to invest in R&D activities the major form of government technical assistance to industry was to contract non-profit research institutes to perform generic technological development and then to transfer their research results to the domestic and private sectors. Some efforts also dealt with innovative and pioneering fields. In the meantime, to encourage research of technologies common to both civil and military use, Chung-Shan Institute of Science and Technology (CSIST), a subordinate to the MOD, actively transferred technologies applicable to civil use to the private sector. Each research institute assisted the private sector in bringing the research results into production. Depending on the needs of the manufacturers, the technology transferring methods included training of professional personnel, technical services, spinning-off new companies, or contracting out jobs for a specific product production for the use of technology testing. Most effective were the spin-off companies that systematically transferred human and technological resources to the private sector. This enabled the rapid emergence of the Taiwan semiconductor industry and the overall accelerated development of Taiwan industry. Take for example the integrated circuit industry that accounts for over half of the production of HSIP, at least half of the manufacturers have come about as the result of TDP (Chang & Hsu, 2001).

Allocating TDP Projects to the Private Sector— Interactions between the Government and the Private Sector

To encourage the private sector to contribute to R&D activities, DOIT has entrusted the private sector to the TDP since 1997 that has attracted the private sector to carry out tasks in TDP. The project has been broadened

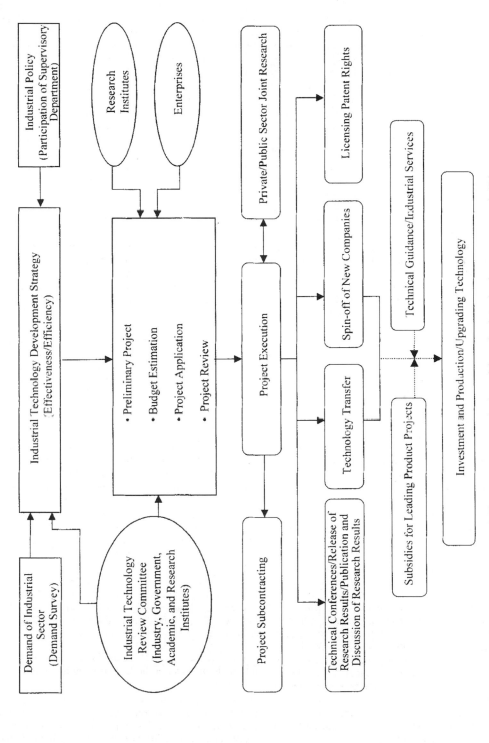

Figure 4. The management mechanism of the technology development program in Taiwan.

Source: MOEA Technology Department Program 2001.

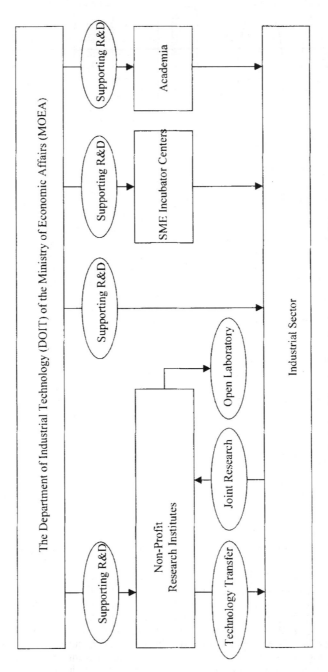

Figure 5. Industrial technology development strategy in Taiwan.

since 1999 to encourage manufacturers cooperating with R&D in industrial technology, subsidies to be limited to no more than 50% of the total expenditure of the program. The following describes the measures taken (MOEA, 2001b).

(1) *Supporting the Private Sector in Developing Industrial Technologies*
    After the project to develop pioneering, critical or aggregate technologies was approved, the private sector was granted subsidies, so that it could invest in R&D and improve its stock of technical personnel. This measure resulted in the enhancement of R&D capability. Furthermore, in order to promote a sound overall R&D capability, a program for integrating technology development was formed. It was formulated in 'Notes for subsidizing of preliminary research for an Integrated Industrial Technology Development Project', in order to stimulate cooperation among manufacturers.

(2) *Encouraging SMEs to Develop New Technologies*
    For the purposes of preliminary research for pioneering innovations or improving technology, or for R&D in some specific innovative techniques, subsidies were granted to those small or medium enterprises that carried out R&D. The grants also encouraged qualified professionals or experts to join innovative research or even set up their own companies. Besides luring the small to medium-sized enterprises to engage in the R&D efforts, the program also stimulated the founding of new high-tech companies.

(3) *Developing Demonstrative Information Applications*
    It was decided to approve and subsidize companies developing pioneer application structures or advanced applied information systems that were innovative domestically, effective, or might stimulate interest in their applications. The goal was to improve the overall capabilities of information applications.

Granting TDP Projects to Universities—Interactions between Government and Academic Institutions

The utilization of academic resources by the government can be traced back to 1992. The MOEA, in its execution of TDP, channelled the results from basic research by the academic organizations to non-profit research institutes, which served as the medium for further applications and test productions. The results from the non-profit organizations were then transferred to manufacturers cooperating in the program.

In order to utilize the valuable R&D capabilities of academia effectively, the program began to plan for direct commissioning of research to universities in 2001. The strong capabilities of the academic teams in R&D is expected to work toward the development of pioneering and innovative industrial technologies. Through the process of innovating high-tech industry, all sectors in the development of industrial technologies it is hoped, may be integrated.

Stimulate Cooperation between the Private Sector and Research Institutes—Interactions between the Research Institutes and the Private Sector

Starting in 1990, DOIT drew up promotional plans, which encouraged research institutes to get together with industrial entities. This joint research strategy has resulted in a better matching of TDP R&D projects with the needs of industry.

Enterprises participating in TDP projects must inject some of their R&D resources, such as manpower, to identify technology research specifications, since this also promotes national R&D investment. Furthermore, by integrating resources from industries, resource utilization efficiency is enhanced. This industrial cooperation strategy has businesses actively participating in R&D projects early-on in the technology development stages, which not only helps steer R&D projects in the right direction, but also achieves gains from the research results. Overall, industry benefits from the new technologies developed, and can discover talented staff early on, to be absorbed into the organization once the person starts his/her professional career.

To sum up, the TDP's cooperative strategy is based on the following objectives: to integrate R&D resources; to accelerate R&D efficacy; to communicate with relevant industries on their needs and requirements; and to accumulate learning experiences. These four critical objectives yield concrete resources and increase application performance for both individual firms and the whole industry.

Establishing Open Laboratories and Incubators—Incubating of High-Tech Companies

Interactions between the research institutes and the private sector will utilize the R&D resources of the research institutes more effectively. The open laboratories resulting from this will breed new technology industries in the fields of communications, opto-electronics, semiconductors, medicine, precision machinery, biotechnology, and computer software. The elements essential for R&D and the starting of new companies will be readily available, such as human resources, technology, information, communications, space, facilities, laboratory equipment, technology management, consultation facilities for starting up businesses, industrial safety, and amenities. With this supportive environment for R&D, entrepreneurs invest and create new industries. This brings about a boost for R&D capability, improves competitiveness, and creates opportunities for manufacturers.

The functions of open laboratories can be divided into two activities: (1) cooperative research; and (2) incubating of new businesses. Cooperative research is cooperation between manufacturers and research organizations on various specific topics. Open laboratories provide necessary resources for research work. The research personnel work closely together to achieve higher levels and standards of quality. Incubating new business programs provides a sound research environment to people or enterprises wanting to create new careers or pioneer new businesses. The program also renders services for the commercialization of products, managerial training, and capital assistance. The goal is to help reduce the risks associated with the founding of new businesses or new careers so that the ventures can succeed.

The research institutes established the open laboratories and the new-business incubator centers for the private sector to effectively utilize R&D facilities and to enhance the industrialization of the technologies developed. Participating members include ITRI, DCB, the China Textile Institute, the Metal Industries Development Center, the Food Industry Research & Testing Institute, and CSIST. There were approximately 50 new-business incubator centers in 2000, spread around the country with the active promotion of the government. In addition, the program was extending to include national laboratories to help cultivate professional personnel for the private sector and to develop pioneering technologies.

R&D and the Technology Diffusion Strategy of the Research Institutions

In order to develop higher-added value and technology-intensive industry, to promote the establishment of emerging high-tech industry, and improve the present industry structure so as to reach sustainable development, government helped the establishment of research institutions in phases and also took the responsibility to improve industry technology based on the requirements of industrial development. For instance, ITRI was established in 1973, with a predominant role in innovation technology. It was to be the stimulus for starting up domestic industry technology; III was established in 1979, to promote the development of information software/hardware industry in Taiwan. In addition, in order to advance biotechnology, the government also set up DCB, with the expectation of improving the Taiwan biotechnology. Since the industrial structure of Taiwan is still dominated by SMEs, seven non-profit research institutions were set up in 1992 and 1993, including printing, plastics, shoemaking, bicycles, precision machinery, stone materials processing, and pharmaceutical technology research center.

Among them, ITRI is the largest non-profit industry technology application research institute in Taiwan, and has engaged in R&D for nearly 30 years. Starting from the establishment of TDP by MOEA, it built up its own R&D resources, then encouraged enterprises to

engage in technological development, further propelling the Taiwan industrial structure to advance toward technology-intensive development. There are around 6,000 employees in ITRI, with more than half holding Masters or Ph.D. degrees. They range over various extensive fields of technological research, and are distributed in seven research laboratories and four research centers. These are the Electronic Research & Service Organization (ERSO), the Computer & Communication Research Laboratories (CCL), the Opto-Electronic & System Laboratories (OES), the Mechanical Industry & Research Laboratories (MIRL), the Union Chemical Laboratories (UCL), the Material Research Laboratories (MRL), the Energy & Resources Laboratories (ERL), and the Center for Measurement Standards (CMS), the Center for Aviation & Space Technology (CAST), the Biomedical Engineering Center (BMEC) and the Center for Industrial Safety and Health Technology (CISH).

ITRI is actively involved in the development of industrial technology, to assist industry to improve technology, to establish new technology-based industries and upgrade existing industries. In order to improve industrial technology, the R&D departments of ITRI complies flexibly with the requirements of industries. For instance, it emphasized R&D planning from the very beginning, selecting technology sectors suitable for development in Taiwan, and focusing on introducing technology from abroad. Following urgent requests from industrial circles, it made great efforts to diffuse its R&D achievements, and to transfer experts to appropriate areas so as to equally emphasize R&D for and implementation of, emerging technology-based industries (Chang et al., 1993).

At present, it strongly encourages industry to engage in R&D activities to actively create economic benefits, and to balance industrial development. For the task of improving industrial technology, ITRI will continue to implement actively R&D projects for enterprises in the future. It will also open laboratories to accelerate the establishment of innovative enterprises. It will also engage in R&D of cutting edge advanced technology for Taiwan's domestic industries, in order to open up new technological possibilities (including cutting edge) for enterprises. It will also sustain an unfailing supply of technology resources to create technology required by enterprises at opportune times. At the same time, it will accelerate internationalization of R&D and combine with the development of the science-based industrial park.

The technological development strategy of ITRI is the R&D strategy stimulated by the goal of establishing core technologies, R&D construction of the appropriate environment for technological innovation, and the improvement of enterprises' R&D capability. With respect to establishing core technologies, it is to integrate technological fields and create the following benefits.

- As far as R&D environment construction is concerned, it is to build up information networks and promote exchange of specialists from different fields.
- Regarding technological innovation, it is to strengthen state of the art technology R&D, promote internationalization and also recreation of innovation culture, to enable ITRI to own an unfailing supply of motive power of technology.
- With respect to improving R&D efficiency of enterprises, it is to use active methods of putting R&D outcomes into use by enterprises such as cooperation among enterprises, technology transfer, R&D strategic alliances, and public laboratories etc., to accelerate the improvement of R&D efficiency by enterprises.

The Recruitment and Training of Talented Professionals

The recruitment and training of talent has been one of the most critical strategies in the Taiwan technology innovation program. Since 1980, the influx of returning overseas scholars has created a mushrooming effect on high-tech industry.

The Recruitment of S&T Talented Professionals

To accelerate the upgrading of domestic industries, the government has adopted a strategy to recruit overseas professionals needed by the country. These people are needed for the improvement of R&D capabilities to accelerate the increasing involvement of the private sector in the development of cutting-edge technology. Measures have been taken to enhance the effort to recruit overseas researchers in science and technology. In 1983, the government to build up technology personnel framed the 'Program for the enhancement, cultivation and recruitment of talented individuals in advanced technology'. This provided the basic human resources for domestic industries in developing high-tech industries. The government also increased its support for the graduate schools of major universities in fields of science, technology, medical science, and agriculture. Overseas scholars were recruited as university faculty members to improve research capabilities. In 1995, another program, 'Program for the enhancement in utilization of talent in advanced technology', was implemented to recruit high-tech talent from Silicon Valley in the U.S. and other overseas areas.

The Cultivation of Talented Technology Researchers

The government pushed for the cultivation and use of talented high-tech researchers to meet the needs of high-tech industry for manpower in advanced technology. After the review of effects from the 'Program for the enhancement in utilization of talent in advanced

technology', a 'Program for the cultivation and utilization of technology talent' was brought about by the Executive Yuan. The program was aimed at building up a sound foundation for the cultivation and utilization of domestic scientific talent in advanced technology. This would greatly improve the national development of science and technology. In 2001, the program was further amended to a high degree to meet the needs of the knowledge-based economy. Seven major talent enhancing strategies were proposed:

(1) The cultivation of high-tech researchers in universities.
(2) The training of industrial technology personnel.
(3) The recruitment of high-tech professionals.
(4) The recruitment of overseas experts in technology.
(5) The recruitment of technical staff from Mainland China.
(6) The attracting of talented graduates.
(7) The promotion, circulation and utilization of talented scientists in the private sector, government, universities, and research organizations.

To meet the demand for high-tech engineering talent in the new technology industries, the 'Project for the cultivation of important high-tech industrial sector talented individuals' was put into effect in July 1998, for a period of three years. The 'Project for the cultivation of advanced talented researchers throughout various technology fields' was implemented in 2000. The project screened talented people with technical backgrounds in R&D or management from domestic high-tech industries and research institutes. These people were recommended by the organizations where they worked and were sent abroad for practical training or short-term workshops. The government subsidized their training abroad for a term of six to nine months of practical training in areas of investment appraisal, technology transfer, and intellectual property rights.

In addition, to meet the development trends in integrating 3C (consumer electronics, computer, and communication) technology and the government's subsidies to R&D efforts in the 3C technologies, the Ministry of Education has urged schools to open classes of 3C integrated technologies since 2000. Due to the rapid development of the electronics, information, and engineering industries, there is a severe shortage of qualified individuals, particularly in the fields of wireless communications, Internet, and software. In addition, such individuals were badly needed for businesses in the electronic sector. As a result, plans to cultivate talented researchers in electronics technology, industrial computers, and information electronics, were put into effect. The projects included on-the-job-training and preparation classes. The goal is to resolve the shortage of talent in the domestic technology industries.

The Technology Commercialization Method Utilized in HSIP

The Taiwan government has adopted a policy of establishing science-based parks to develop high-tech industries in Taiwan. Since the main conditions for developing high-tech industry are professional staff, the most advanced technology and fast acquirement and digestion of information, Hsinchu was chosen as the place to establish the first science-based park in Taiwan. The reason for this was that Tsing Hua University, Chiao Tung University and ITRI are located in Hsinchu, and provide excellent human resources, assisting manufacturers to break through technology bottlenecks. The establishment of HSIP is an important measure taken by the Taiwan government to promote the development of high-tech industries. After 20 years of such development, it is indeed an inspiring symbol for the Taiwan high-tech industry.

Since HSIP was founded in 1980, up to the end of 2000, it comprises 289 manufacturers of high-tech products, employing 102,840 workers, with an invested capital of NT$694.5 billion. There are six major sectors: integrated circuit, computer and peripherals, communications, opto-electronics, precision machinery, and biotechnology. Since the main objective for establishing the science-based park in Taiwan is to promote the development of high technology and industry, and the government uses the necessary resources for high-tech industrial development as the means of carrying out its policies, this section explains the ways of innovation by HSIP based on different measures used by the government to promote the use of the resources to benefit manufacturers or entrepreneurs who engage in the development of high-tech industry (see Boly et al., this volume). Technology sources, human resources, funds requirements, preferential land tax and other levies, and the management structure are five important resources provided by the government for manufacturers in HSIP. There are five major factors to be discussed are:

(1) technical resources;
(2) human resources;
(3) capital resources;
(4) risk sharing of new business; and
(5) the operating environment (Chang & Hsu, 2001).

Technical Resources

Technical sources for manufacturers in HSIP come from R&D projects by manufacturers or subsidies from the Science-based Industrial Park Administration (SIPA) for innovative projects: technology transfer or collaboration with ITRI, subsidies from the TDP sponsored by MOEA; cooperative programs with NSC or universities; import from or cooperation with, foreign sources; and national laboratories like the National Nano Device Laboratories, the Synchronous Radiation Center, and the Precision Instrument Center.

The relations between HSIP manufacturers and nearby ITRI, Tsing Hua University, and Chiao Tung University can be discussed on three technological levels: basic research, applied research, and technology development. The universities are mainly engaged in the basic research, subsidized by research grants from the NSC. ITRI and other Research Institutes mainly receive contracts from the TDP to carry out R&D of applications of advanced and generic technologies. ITRI ensures the realization of the commercialization of technology by the private sector. The industrial sector is mainly involved in the development of specific technologies and the commercialization process. Besides research work done by the private sector itself, results from the basic and applied research work of the government are also adopted by the private sector. For manufacturers in HSIP close to the universities and ITRI, effective cooperation is easy and an efficient network of operations is achieved.

There was a great need for imported technology in the early days of HSIP. ITRI served as the main source of technology. Manufacturers could also contract ITRI to develop technology for new products or to resolve technical problems. The mission for ITRI was to accelerate the development of Taiwan domestic industry by passing on its achievements in R&D to the private sector to generate maximum benefits. For this reason, ITRI utilized all available channels to diffuse technology to the private sector.

ITRI has served HSIP very well. In advanced countries, universities work very closely with the private sector in R&D, and the way HSIP has been set up also enables the practical application of research work by the universities. Back in the 1970s when Taiwan was planning to develop high-tech industries, there were only a few people experienced in R&D, even in academic circles. For this reason, ITRI was founded in 1973 with government subsidies to link basic research by universities and the development of technology by the private sector. By 1980 when HSIP was established, ITRI had become the critical and main source of technology for HSIP manufacturers. With the growth of HSIP, the main sources of technology were from abroad or from technology transfer and cooperation, or from the nearby universities.

Human Resources

Human resources are the motive power of creation. The gathering together of advanced manpower is the greatest success of the HSIP. Many university graduates went into ITRI to sharpen their professional skills before moving into the private sector. HSIP is right next to National Tsing Hua University, Chiao Tung University, and ITRI (Guan Fu Division). Another ITRI division, the Chung Hsin Division, is only ten minutes away. This has greatly assisted the exchange of talented managers and researchers among the four parties.

This occurred in the following ways:

(1) college graduates working directly in ITRI or the factories inside HSIP;
(2) technical training courses sponsored by ITRI;
(3) workshops given by universities targeted for employees inside HSIP or ITRI, with emphasis on theory, advanced technology, or technical management;
(4) employees of ITRI and HSIP manufacturers attending classes given by universities;
(5) seminars and speeches inviting scholars or experts sponsored by HSIP, ITRI, or universities;
(6) review boards to review master theses, doctoral dissertations, and research projects; and
(7) research projects commissioned out to school faculties for consultation or collaboration.

The experience from HSIP showed the private sector that investment in advanced technology personnel often provides maximum returns. There were four measures taken by the government for cultivating talented people in advanced technology:

(1) providing advanced training on the job;
(2) providing training for career change;
(3) recruiting overseas talents to work and train domestic personnel;
(4) sending research personnel for training abroad.

Taiwan began its development in advanced technologies in 1980. With the shortage of domestic R&D capacity, overseas experts and scholars were recruited, especially those from Silicon Valley in the USA. Besides bringing in knowledge of advanced technology, market information, equipment procurement information, these people expanded technology transfer and cooperation using their overseas experience, some even established subsidiaries here. From their attendance at overseas seminars and conferences, information about advanced technology was at hand.

In order to meet the demand for qualified personnel, to improve the quality of manpower, and to shorten new personnel training time, SIPA sponsored training programs with ITRI and nearby universities. Workers in HSIP actively attended the programs in order to keep up with the rapid advances in the technology field and to maintain competitiveness. Furthermore, the recent policy of the Ministry of Education to acknowledge Master's degrees earned from on-the-job programs (studying during spare time for a Master's degree) should provide another incentive for personnel to undertake further qualifications.

Capital Resources

In the early stage of HSIP, the capital needed for development was insufficient. The government provided loans to the private sector or invested directly

through the Chiao Tung Bank. Development Funds from the Executive Yuan were granted by the administration to provide for investment in high-tech industries. In order to implant technology into the private sector, ITRI also set up a venture capital company. Besides assisting ITRI in technology transfer, this company helped manufacturers of promising technology to acquire the needed capital. The company also invested directly in high-risk technologies, which had strong market potential. Measures to encourage the founding of venture capital enterprises to invest in high-technology companies proved to be very effective in starting up new companies in the HSIP.

With the growth of the domestic capital market, manufacturers in HSIP issued common stocks, and distributed stocks to employees. This enabled the manufacturers to directly acquire capital from the public and employees, from entrepreneurs, venture capitalists, banks, and other related enterprises.

The government took several measures to assist manufacturers gain the necessary operating funds. Rapid depreciation measures allowed the investors of large equipment in the park to accelerate depreciation of the equipment and thus delay tax. This gave the manufacturers the advantage of gaining operating capital without paying interest. The measures were very helpful for semiconductor and opto-electronics industries that engage mainly in manufacturing. The government also provided mid-to-long term business loans to makers of important technologies for industrial development. The policy attracted manufacturers to invest in high-risk advanced products and met the manufacturers' need for expanding capital.

Diverted investing was another major source of capital. Because the electronics industry is a major manufacturer, investing in the makers of important components or products was common. With the readiness of capital, market, production, top managers, and managerial systems, shortening the preparation time to start up a business reduced the costs of innovation. Manpower, capital, technology, market, marketing, regulations, managerial systems, all these formed a tight interrelated network between existing or new manufacturers. When the timing was right, these factors generated another manufacturer. After more than twenty years of fostering new enterprises, HSIP has created a sound environment for new businesses.

New Business—Risk Sharing

To cope with the insufficient investment in high technology in 1980s, the government drew up several measures to stimulate investment. Some examples are:

• four- to five-year period of profit-seeking enterprise income tax exemption;
• accelerated depreciation;
• an extended tax-free period;
• capital gains credits; and

• broad expenditure listing.

Further measures on exports included credit and tariff free periods on commodity taxes of exported goods. These overall reductions on tax and tariffs increased manufacturers' net profits.

Two important characteristics of the investment in HSIP are that it is high risk and concentrates on high-technology. This kind of enterprise will contribute most to accelerate the upgrading of domestic industry because it stimulates the development of related technologies. Due to the need for high investment in R&D, it requires the government to share the risk and cost. The tax reduction measures encourage and help the start-up enterprises in high technology in their initial phase of development.

The Operating Environment

HSIP is a community of abundant information. Besides periodicals issued by the SIPA and information on the Internet, employee organizations and associations sponsor various gatherings, such as training courses, seminars, speeches, and cultural activities. Being alumni or alumnae, room-mates, or colleagues, generating a free flow of information brings many people in the park together through their interactions. This also influences the way business operates.

So as to provide a supportive environment for high-tech industry, the government implemented several measures, including a one-shop window service and simplified administrative procedures. Constant efforts were made to improve the basic facilities in order to achieve an appropriate environment. A system to automate customs clearance was put into effect. All these were aimed at giving advantages to manufacturers inside HSIP to increase their competitiveness in the world market.

Innovation in the Taiwan IC Industry

This section describes one important area of the Taiwan high-tech industrial scene—the integrated circuit (IC) industry. The innovation process involves four major sectors: government, research institutes, academia, and the private sector. The Taiwan IC industry in 1973 consisted of assembly plants but had then grown to comprise a complete industrial infrastructure, including design, mask, fabrication, and assembly capabilities by 1990. By 1995, Taiwan had caught up with the advanced nations and reached 0.5-micron ultra-large-sized IC technology. Taiwan has also become the world's fourth largest producing country (Chang et al., 1994).

The selection of the IC industry as a target industry by the government, the introduction of foreign technology from the Radio Corporation of America (RCA) in the USA, the assimilation and improvement of imported technology by ITRI, and the transfer of this technology to local industry all helped in the formation of this industry. ITRI collaborated further with the private sector to develop sub-micro technology, and advanced into the independent making of memory products such as dynamic random access memory/static random access memory (DRAM/SRAM). During the evolution of these developments over more than twenty years, the government subsidized ITRI's R&D funds while ITRI played the role of stimulating technology research and diffusion. Together they have assisted the establishment of new technologies and generated the present prosperity of the Taiwan semiconductor industry.

The Taiwan IC industry, under government supervision, has developed from rough beginnings into the flourishing industry it is today. Throughout this development process, ITRI has been involved in the planning and performing of R&D activities, and the transferring of technology to the IC industry. There are mainly four stages in the development of the Taiwan semiconductor industry: medium scale IC (MSI), large scale IC (LSI), very large scale IC (VLSI), and ultra large scale IC (ULSI) technology. The following describes the measures takes by the interaction of government, ITRI, universities, and the industry in each stage of the process (Chang & Hsu, 1998).

The Initiation Stage—MSI Transfer from Abroad

The Taiwan semiconductor industry began with the founding of the U.S.-owned Kaohsiung Electronics Company assembling transistors in 1966. Not until 1973 did Taiwan-owned Wonban Electronics begin its production of transistors. In the academic world of Taiwan, only National Chiao Tung University had a semiconductor laboratory to cultivate this field.

The Taiwan government were hoping that the local electronics industry would move in the direction of technology-intensive products, and government advisors suggested that it should develop IC design and manufacturing technology in order to stimulate innovation throughout the island's electronics industry, which would also spin off related industries in the process. A task force was funded by the MOEA to discuss how to carry out the development strategy of this high-tech industry. The strategy is described below:

(1) The Technology Advisory Committee (TAC) was established to take responsibility for planning.
(2) The technology transfer was adopted as an initial strategy for the quick development of the industrial base in Taiwan.
(3) ITRI was chosen to undertake the introduction, assimilation, improvement and initial feasibility studies of manufacturing ICs in Taiwan.
(4) Over a period of four years, around NT$410 million (this was a significantly large figure for Taiwan in the economic climate of the 1970s) was invested in purchasing the required manufacturing

technology, product designing and testing, purchasing required instruments and equipment, and building a pilot IC plant.

Since at that time advanced countries had reached the level of large scale IC technology and because there were no domestic IC manufacturers, ITRI decided to attempt to gain a leading role in medium scale IC technology. The resulting success was later transferred to the private sector.

The MSI Stage-Introduction of Foreign Technology

The strategies for introducing foreign technology mainly included decisions regarding technology specifications and technology partners, and evaluation of the progress of technology transfer and the results of its introduction. These items will be described below.

(1) *Decisions Concerning Technology Specifications*
Many types of IC technology were developed in the 1970s. The members of the TAC discussed the technologies available and then decided to obtain low power, high-density technology that would provide sub-micron development potential. The complementary metal-oxide semiconductor (CMOS) was thus selected as the technology to be developed. This decision successfully led Taiwan to the sub-micron technology level in 1992.

In the mid-1970s, the IC process technology of the USA had advanced to 3.0-micron, but only 7.0-micron technology was available for technology transfer. What made Taiwan purchase 7.0-micron technology? Different opinions on this subject are summarized as follows:
(a) on the basis of the principles of technology transfer, it was highly unlikely that advanced countries would be willing to transfer new technologies to other countries; instead, mostly technologies with a lower comparative advantage or obsolescent technologies would be transferred. Therefore, it would either be impossible for Taiwan to purchase the most advanced technology, or Taiwan would have to pay a very high price to purchase the technology;
(b) in advanced countries, 7.0-micron technology was a mature technology with the advantages of higher consistency, complete technical documents, many skilled technicians, and effectiveness in the operation of equipment. This technology was thus very suitable to be transferred to Taiwan and redeveloped by the inexperienced organizations there;
(c) because products manufactured with 7.0-micron technology had already been introduced into the market, feedback was available concerning process technologies, product development or design technology, and mar-

keting channels. This meant that Taiwan would easily be able to learn about all aspects of IC technology, from R&D to commercialization of products;
(d) Process improvements were to be achieved from technical personnel, equipment, and clean room environments. Therefore, although the technology introduced greatly lagged behind that of advanced countries, it was still possible to upgrade it through investment and R&D after introduction. Clauses stressing personnel training were thus essential in the introduction contract so that capabilities for assimilating and improving the technology could be built up.

(2) *Selecting the Technology Partners*
IC technology originated in the USA (see Gutmann, this volume), and American companies not only had experience in technology transfer but also were willing to transfer technology to developing countries. American companies were thus selected as the source of the technology transfer and requests for proposals (RFPs) were sent to more than 20 companies in the USA. The evaluation team visited more than ten IC companies who responded to the RFP, and RCA and another company (referred to here as company X) entered the final round of selection.

The budget required by RCA was twice as high as that for company X, but the technology provided by RCA was far more complete, comprising design, process, manufacturing management and cost accounting, whereas company X proposed to provide only process and design technologies. Moreover, RCA could provide training for 35–40 people for a period of six months to one year; in contrast, company X suggested that training for three to four persons for a period of three months would be enough.

The objective of establishing IC technology in Taiwan could be achieved only through extensive personnel training. The technology content and personnel training provided by company X could not be expected to meet the requirements of the plan. Although the winning-big principle under the budget system of the Taiwan government was to select the bid with the lowest price, since company X could not offer a package with a technology content similar to that offered by RCA, RCA was finally selected as the technology partner.

(3) *Decisions Concerning Technology Specifications*
RCA would provide 7.0-micron CMOS process technology and the product specifications and design and testing technology for a 3.5 inch digital electronic clock that was to serve as the product vehicle. Assistance in building an IC pilot plant, suggestions for equipment specifications, and training of personnel were also included in the

technology transfer. In 1974, the Electronics Research and Service Organization (ERSO) was established by ITRI to implement the technology transfer. A team of young engineers was recruited and trained by ERSO for a period of time. These personnel were then sent to RCA for one year of technical training at various sites in the USA.

(4) *Successful Introduction*

The schedules for building the IC pilot plant, purchasing equipment, and training personnel were coordinated well by ERSO/ITRI, and after the engineers had completed their training and returned to Taiwan, the pilot plant and equipment were ready for pilot runs. In 1977, the first IC was produced by the pilot plant and the functions of this IC conformed to the evaluation standards, this being a criterion for the success of the technology introduction stipulated in the introduction contract signed with RCA. The following describes the four main factors affecting the success of this endeavor:

(a) development direction: The development direction was determined and planned by the technology advisory committee composed of experts and scholars from the government, industry, and academic and research institutes. The committee made suggestions and decided on appraisals for the long-term strategy to develop IC technology;

(b) collaboration in design, manufacturing, perfecting of products and subsequent marketing: ITRI obtained the 7.0-micron technology with the 3.5 inch digital electronic clock needed to verify the technology. The products were then exported to Southeast Asia countries. From the planning of the technology, ITRI was positioned to control all links to the making, design, perfecting of products, and marketing. ITRI also opened channels to gather related information. This helped ITRI in its planning and integrating of technology transfer and the commercialization of the products;

(c) human resources: Recruitment from overseas and cooperation with local universities in cultivating talented professionals;

(d) funds: Government subsidizes to support R&D.

The Transfer of Technology to the Private Sector

After the process technology introduced from RCA was verified through the output from the pilot plant of ITRI, the results were transferred from the research institute (i.e. ITRI) to local industry. As no related operation had been set up at that time, there was no receiver for this technology. Therefore, a team of personnel was transferred from ITRI to spin off a new business, the United Microelectronics Corporation (UMC).

In October 1979, UMC became the first company to apply officially to the Science-Based Industrial Park to establish a plant there. In order to assist in the upgrading of domestic industries and the development of technology-intensive industries, the Science-Based Industrial Park was established in Hsinchu by the Taiwan government in July 1979. Preferential loans, tax reductions, administration services and other incentives were granted to companies established in HSIP. UMC, with capital of NT$500 million, commenced pilot runs in April 1982 and had reached breakeven point by November 1982. UMC's sales exceeded NT$100 million per month in June 1983, and the company was marketing its products in Taiwan, Hong Kong, Korea and the USA.

There are five aspects to consider in understanding how this result was achieved:

(1) the process technologies transferred from ITRI to UMC;
(2) the transferred product technology;
(3) the project planned by ITRI in order to establish UMC, including the transfer and training of personnel;
(4) the coordination provided by ITRI for the competitive operation of UMC in the future; and
(5) the successful future transfer of technologies developed by ITRI to UMC.

The Rapid Expansion Stage-Upgrade to LSI, VLSI

After UMC was spun off, Taiwan had a private enterprise capable of producing IC although compared to America and Japan, the standards were still very low. There was a growing need for higher-level semiconductor technology. A four-year project (1979–1983) was proposed by ITRI to the government. The major goal of this project was to upgrade the technology to large scale IC level, and to build up related development tools, particularly the making of masks and computer aided design (CAD) capability. The developed technology was to be transferred to the private sector, after verification by the pilot plant.

The LSI Stage-Technology Development

From 1979 to 1983, ITRI invested more than NT$670 million to upgrade IC technology. The production capacity was raised from 7.0-micron to 3.0-micron, and the making of a bipolar metal gate was achieved. As regards IC design technology, a computer simulation program was obtained from abroad and a logic simulation program was also independently developed by ITRI. An automatic mask design system was developed to increase the speed of product design. To better the masking technology, a mask duplicating technology was obtained and improved to perform the masking locally. This enabled the domestic IC makers to control the timing of the production process, and eliminated any dependence on foreign services.

The LSI Stage-Technology Transfer

In IC design, the Syntek Semiconductor Co. Ltd. was established by staff from the Digital Circuit Design Department in ERSO/ITRI in 1982. Many other companies were also established in a similar way, such as Wel Trend Semiconductor Inc. and Silicon Integrated Systems Corp. (SIS).

ITRI planted the first seeds of the domestic IC industry by engaging in the design and fabrication of IC. ITRI also made use of every possible method of technology diffusion to transfer new technologies to industry. These methods included: spinning off new companies, offering seminars, accepting design commissions from manufacturers, technical staff training, newsletters, etc.

The VLSI Stage-Technology Development

After these achievements, there was still a great gap of technical levels between the domestic IC industry and industries in the U.S. and Japan. It was unlikely that domestic manufacturers could catch up with foreign advanced countries by their own efforts in research and development. Because of this, ITRI proposed to the government a five-year plan (1983–1988) for technology development. The major goal was to raise the IC technology to very large scale IC design and manufacturing. From 1983 to 1988 a total of NT$245 million was invested in upgrading process technology from the 3.0-micron to the 1.0-micron level. In addition a VLSI laboratory was built. A common design center was promoted to diffuse ASIC (application-specific integrated circuit) design technology. Furthermore, optical masking capability was upgraded to incorporate electron-beam masking technology.

The VLSI Stage-Technology Transfer

(1) *IC Design Technology*

ITRI adopted direct training to expedite the transfer of IC technology, and to shorten the R&D time for enterprises. A joint design small scale center was set up with mature developed computer-aided design programs and circuit design methods. Engineers from various enterprises were trained to be able to utilize the computer workstations to design and develop customized IC products.

After careful evaluation, ITRI recognized the high-risk involved for domestic enterprises in applying special purpose IC products to the market. Due to the lack of IC design talent, domestic enterprises were limited to technology development, and were lukewarm in accepting technology transfers. To generate an environment for the enterprises to acquire necessary talents, ITRI turned to domestic universities to foster IC design engineers. The following describes the measures taken by ITRI to transfer technologies to the academic and industrial sectors (Chang et al., 1993).

(a) Transfers to Academia

Because of the general scarcity of specialists in the IC design field in Taiwan at that time and the relative immaturity of ITRI's IC design technology, there was a lack of interest in such technology in the electronics industry. Therefore, ITRI looked to educational institutions for help. By joint training with academic institutions, it was expected that more specialists in this field would become available to industry. In addition, ITRI continued to perform R&D activities in order to develop more sophisticated IC design technology.

Computer-aided, very large scale integration design techniques had not been introduced to academic institutions in Taiwan until July 1983, when specialists in this field were still extremely scarce. The National Science Council and Ministry of Education, together with ITRI, initiated the Multi Project Chip (MPC) project. The main objective of the project was to improve students' and professors' own practical ability in IC design. The students and professors taking part in the MPC were encouraged to participate in computer-aided design training at ITRI. A basic logic cell library and computer-aided equipment were provided to different universities, so that the universities could design their own ICs. Also, a chip foundry service was provided, so that the universities could manufacture their IC designs.

Chiao Tung University, National Taiwan University, Tsing Hua University, Cheng Kung University, National Taiwan Institute of Technology, Central University, Tatung Institute of Technology, Chung Cheng University and Tamkang University were the nine major universities participating in the MPC project. There were 12 departments involved and approximately 150–200 specialists were trained each year.

(b) Technology Transfer to the Private Sector

To achieve effective transferral of technology to industry, ITRI adopted a dissemination approach to allow firms to design their own ICs according to their needs. A joint design center was established which offered training courses and CAD tools. Thus, IC design technology could be transferred directly to industries.

The joint design center was founded in Hsinchu in March 1985. By promoting ASIC design concepts, publishing IC design handbooks, organizing conferences and training classes, and issuing newsletters about ASIC, the center aimed to consolidate the design techniques of system engineers and familiarize them with CAD applications, so that the idea of specialists designing their own ICs could be realized. The center provided basic design methodology and CAD tools. This proved to be an effective method, because corporations were able to find out exactly what they required. At the same time, this approach reduced the time and effort which the research institute formerly might have spent learning about the client's requirements. The advantages of such a method of transferring technology were that a high degree of security in the design of an IC could be maintained and the time required for both parties to communicate was shortened considerably.

The Taiwan domestic firms' IC design technology was developed in 1976. However, the number of design companies increased rapidly to 30 only after 1987, when the Taiwan Semiconductor Manufacturing Company (TSMC) was founded. TSMC's main business was VLSI foundry and ASIC design services. TSMC allowed IC design and manufacturing to be separated. Consequently, design specialists no longer had to worry about the substantial investment needed for manufacturing. Some design specialists now come from the Joint Design Center at ITRI and some from overseas. The number of professional design companies has increased from 30 in 1987 to 55 in 1990; and marketing revenues have increased from US$32 million to US$236 million.

(2) *VLSI Fabrication Technology*

TSMC is a global enterprise specializing in IC foundries. This was a joint venture by the government, research institutes, domestic and foreign enterprises. Major technical and supporting staffs composed of more than 150 persons, transferred from ITRI. ITRI further assisted by leasing its VLSI plant to TSMC. As a result, TSMC achieved world class VLSI fabrication technology, and gained independent key technology for manufacturing IC products.

(3) *Masking Technology*

After the founding of TSMC, many domestic enterprises invested in the IC design, and many ASIC design companies emerged. As a result, the demand for masking services grew tremendously. The Industrial Technology Investment Corporation (established by ITRI in order to transfer ITRI's technology to local industry) invited relevant entrepreneurs to jointly found the Taiwan Mask Corporation (TMC), and transferred staff, technology, and business from ITRI to the company. TMC was established in 1988 with most masking

techniques transferred from ITRI, and signed a ten-year contract with ITRI for technology transfer and joint development.

The Growth Stage- Joining ULSI

By 1990, Taiwan had the beginnings of an industrial infrastructure for the IC industry including design, masking, fabrication, and assembly facilities. In order to upgrade the technical ability of the Taiwan IC industry, the government contracted ITRI to perform a five-year (1990–1995) project aimed at the development of sub-micron fabrication technology, with DRAM/SRAM products to be used as the primary vehicles for testing the feasibility of the fabrication technology. The success of the sub-micron project makes Taiwan only the fifth country in the world (following the U.S., Japan, Germany, and Korea) able to independently develop 16 megabit (Mb) DRAMs. This also attracted a record (1994–1997) three-year investment of over NT$200 billion in 8-inch memory-product-wafer IC manufacturing plants.

The ULSI Stage-Technology Development

After ITRI had completed the transfer of about 150 technical personnel to the newly established TSMC the following question arose: how would ITRI be able to proceed efficiently with the development of sub-micron technology? ITRI's upper management then decided that since most of the technical personnel had been shifted to private industry, and the purpose of ITRI's technology development was to transfer it to industry, there was no reason companies should not participate in ITRI development projects in order to avoid the manpower shortage problems. At the same time, this would also facilitate and speed-up the technology transfer process.

The funds appropriated for the development of sub-micron technology, and its effect on domestic IC technology, were unprecedented. The following will describe the administration of the project.

(1) *Direction Guidance*

During the execution of the project, a 'Sub-micron Advisory Committee' was formed by MOEA and was composed of representatives from the industry itself, government, academic and research institutions. A meeting was held two to four times a year to provide guidance and consultation for the sub-micron project on targets, contents, progress, technologies, achievements, and ongoing plans.

(2) *Execution of the Project by ITRI*

The sub-micron project undertaken by ERSO/ ITRI. ITRI was able to attain its goal by the expected date by recruiting capable persons as project leaders, as well as completely allocating all responsibility and authority of the ERSO director to the project leader. The following describes the main guidelines for the project.

(a) Setting up Specific and Achievable Goals

After careful analysis, the project leader aimed to raise the technique level from 1.0-micron to 0.7-micron within two years. This specific goal gave the project members a clear target to work for and in the meantime, it enabled the project leader to manage the progress of the project.

(b) Modifying Specifications According to the Needs of the Project

During the sub-micron project's planning stage the wafer size was set at 6 inches. In addition all necessary equipment was devoted to producing 6-inch wafers. Although IBM at the time, possessed 8-inch wafer production technology, the equipment used in Taiwan was modified from the original 6-inch wafer production equipment. Thus, Taiwan decided 6-inch wafers to be the target for development. By the beginning of the second year of the project, with the Japanese IC equipment manufacturers ready to accept 8-inch wafer production equipment purchase orders, the project leader applied to the committee for a change of target from 6-inch to 8-inch wafers. After reviewing the following points, the government decided to upgrade the project to making 8-inch wafers with a reapportionment of NT$1.2 billion more and a six-month extension of the project.

  (i) Rapid progression in technology: IC equipment manufacturers had already begun making 8-inch wafers. It was estimated that by 1995, there would be limited 6-inch wafer equipment available; thus reducing the willingness of the private sector to invest.

 (ii) Lack of competitiveness: For instance, the cost of producing DRAM using 6-inch wafers would be 180% that of 8-inch wafers.

(iii) The project leader promised to speed up the process and limit the delay to six months.

(iv) To increase the budget at this time would be less expensive than to change to 8-inch wafers at the end of the project.

Since the primary concern was to develop fully-competitive sub-micron technology by 1995, technologies developed by ITRI would still require technology transfer and investment for production-site build-ups. Therefore, the industry might be reluctant to accept a transfer of less up-to-date technology; besides, converting a completed sub-micron lab to 8-inch wafer production would require a far greater investment than converting it at the initial stages. With these considerations in mind, the committee agreed to shift the project goal to 8-inch wafers, add an additional NT$1.2 billion to the current budget (making the total budget NT$7 billion in all), and extend the schedule by six months. These changes were later approved by MOEA.

(c) A Supportive Environment for R&D

ITRI gave the project director total authority in the technical area, and provided maximum assistance in administration. The project leader was able to turn all the participants' attention to the project goals, set precise milestones, and give incentives to participants to enhance the team spirit by continuous revision and timely modification of the project where necessary.

(3) *Monitoring by the Private Sector*

The firms' investments and participation in the sub-micron R&D process not only assisted the government in monitoring the implementation of the project, but also enabled the companies to obtain the newest technology developed by ITRI in the shortest possible time. Therefore, the project advanced without undue difficulty and was completed on time.

The ULSI Stage-Technology Transfer

ITRI formed a 'Sub-micron Working Consortium' to transfer the technology and products to member firms so that the domestic IC industry could catch up with the advanced countries. The 'Sub-micron Technology Application Association' was then formed to assist IC design firms in applying advanced fabrication technology and design of prototypes for the 8-inch wafer. This greatly reduced the time needed to develop highly added value novelties.

At each stage of the implementation of the sub-micron project, ITRI worked to transfer all technical data to the firms in the Sub-micron Working Consortium in the shortest amount of time possible. Therefore, even the most advanced 0.5-micron fabrication technology was immediately transferred for use by firms, enabling UMC and TSMC, for example, to make immediate use of ITRI technology transfers to develop 0.6-micron fabrication technology. ITRI also provided design rules for DRAM and SRAM logic to design companies in the Sub-micron Working Consortium for advanced product designs such as 16Mb DRAMs and 4Mb SRAMs.

Meanwhile, MOEA made use of the developed sub-micron fabrication technology and related purchase of equipment and personnel, and targeted the much-needed key component—the DRAM—as the choice for assisting local firms to raise their competitive levels through production in the domestic private sector. Thus MOEA used the sub-micron lab equipment, technology, and human resources in collaboration with private companies to form professional production companies devoted to DRAMs. This was carried out in the form of a practical plan to apply sub-micron fabrication

Table 1. The development strategy of industry.

Development Factors \ Industry Development Stages	Initiation	Burgeoning	Growth
Technology Selection	Technology level is far behind those of leading countries	Technology level closely follows those of most countries	Technology level is synchronous with those leading countries
R&D Activities	The research institute transfers the technology from abroad, then assimilates and progresses by in-house R&D activities	The industry participates in the research institute's R&D projects	The industry collaborates with the research institute in implementing the R&D projects
Technology Transfer	The main method used by the research institute to transfer the R&D results is to spin-off new companies after the R&D project is finished	Technology diffusion is used during the R&D implementation stage. And the new spin-off company is used after the R&D project is finished	The research institute will mainly use technology diffusion to industry during the R&D implementation stage and after the R&D project is finished
Industrial Development Result	A rough and ready industry is established	Upgrades the industrial technology ability and expands marketing areas	Increases the industry's R&D capabilities to improve competitiveness in international markets

technology, commencing with the selection of suitable private corporations to participate in the plan.

The spin-off of a new company from the sub-micron project was done in an open and fair manner, by publicly seeking investment partners. The open bidding attracted 13 local companies, headed by TSMC in September 1994. These 13 companies included suppliers, at various levels of the information and electronics industries, as well as funding institutions. The new company was named Vanguard International Semiconductor Corp. (VISC) and its primary goal was the production of DRAMs. The initial capitalization was NT$18 billion. MOEA, possessing the sub-micron lab, equipment, and technology, took 32% of the company stock right away, making this the fastest return on investment ever among MOEA R&D projects. ITRI completed the transfer by the end of 1994, and transferred a total of 330 people to VISC.

Summary of the Taiwan IC Industry Development Strategy

To summarize, the progress of the Taiwan the semiconductor industry from the 1970s to the end of 1990s can be divided into three stages: the initiation stage, the burgeoning stage, and the growth stage. During the evolution of the Taiwan IC industry, the government made conscious choices about which industries would be targeted for technology transfer from abroad, contracted non-profit research institutes to develop interim technologies until the technology was obtained,

then transferred the technology to industry to facilitate setting up corporations for commercial purposes. Furthermore, the process assisted the firms in establishing R&D capabilities, and increased the industry's international competitiveness. In each of these three stages, e.g. the initiation stage, consisting of obtaining technology and facilitating setting up domestic companies, the burgeoning stage, consisting of the formation of manufacturers' R&D facilities, and the growth stage consisting of further raising the industry's international competitive levels, the strategies employed were rather different from one another. The following is a summary of the differences between the three stages of technology selection, R&D activities, technology transfer, and industry development results. These are also listed in Table 1.

9. Conclusions

In order to promote the development of the Taiwan high-tech industry, based on the development environment in the 1970s and considering all factors needed to develop high-tech industries, various measures were implemented, to advance the germination and development of high technology. After development lasting thirty years, these seeds have grown up to become large trees, and today's high-tech forest in Taiwan. These seeds have brought new vitality to this forest but their function has gradually been superseded.

The Taiwan innovation system is gathering strength from industry, government, academic and research institutes. Technology-based enterprises have become the core of the whole industry. Technology-based enterprises from the industry, government, academic and research institutes have come together to form Taiwan's own industrial innovation system. This system will continue to instill competitiveness into the Taiwan technology-based enterprises, and serve as a guideline for government to consider which policies to follow. With the coming era of the knowledge-based economy, the innovation system shall play an even greater role when Taiwan is forging ahead as a knowledge-based nation.

To develop high added value knowledge-based industry, the emphasis should be on services as well as manufacturing. In addition, the academic and research organizations must become more efficient in transferring their technological innovations to the private sector. As regards the hardware structure, the Taiwan innovation system will continue the collaboration between industry, government, academic and research organizations and a knowledge management system will be introduced to increase operational efficiency. As for the software aspect, the effort will be on the synchronous diffusion of talented personnel among the private sector, academic and research circles while research work progresses.

The Taiwan attempt to develop an innovation system has resulted in a successful technology industry. With the coming of the knowledge economy era, considering the current structure of the Taiwan industry, relying on innovation to maintain economic growth is a feasible strategy. Future innovation systems based on the existing system will be developed, but will be more open, integrated, and advanced. The new measures will include government policies to utilize global resources; incentives for collaboration or alliances among industrial, academic and research sectors; and the pressure from the government for more successful innovative results. The goal now is to advance the Taiwan innovation system to a more comprehensive and efficient stage.

References

Chang, P. L. & Hsu, C. W. (1998). The development strategies for Taiwan's semiconductor industry. *IEEE Trans. Eng. Manag.*, **45** (4), 349–356.

Chang, P. L. & Hsu, C. W. (2001). Made by Taiwan: Booming in the information technology era. In: C. Y. Chang & P. L. Yu (Ed.), *The Industrial Park: Government's Gift to Industrial Development* (pp. 273–297). World Scientific.

Chang, P. L., Shih, C. T. & Hsu, C. W. (1993). Linking technology development to commercial applications. *Int. J. Technology Management*, **8** (6/7/8), 697–712.

Chang, P. L., Shih, C. T. & Hsu, C. W. (1993). Taiwan's approach to technological change: The case of integrated circuit design. *Technology Analysis & Strategic Management*, **5** (2), 173–177.

Chang, P. L., Shih, C. T. & Hsu, C. W. (1994). The formation process of Taiwan's IC industry-method of technology. *Technovation*, **14** (3), 161–171.

Freeman, C. (1987). *Technology and economic performance: Lessons from Japan*. London and New York: Pinter Publishers.

Hsu, C. W. & Chiang, H. C. (2001). The government strategy for the upgrading of industrial technology in Taiwan. *Technovation*, **21** (2), 123–131.

Lundvall, B-Å. (Ed.) (1992). *National systems of innovation: Towards a theory of innovation and interactive learning*. London and New York: Pinter Publishers.

Ministry of Economic Affairs (MOEA), Taiwan, R.O.C. (2001a). *MOEA technology development program 2001*. MOEA Press.

Ministry of Economic Affairs (MOEA), Taiwan, R.O.C. (2001b). *White book of industrial technology 2001* (in Chinese). MOEA Press.

National Science Council (NSC), Executive Yuan, Taiwan, R.O.C. (2000a). *Indicators of science and technology R.O.C. 2000*. NSC Press.

National Science Council (NSC), Executive Yuan, Taiwan, R.O.C. (2000b). *The annual report of science and technology*. In: R.O.C. NSC Press.

Nelson, R. R. (Ed.). (1993) *National innovation systems: A comparative analysis*. New York: Oxford University Press.

Organisation for Economic Co-operation and Development (OECD) (1997). *National innovation systems*. Paris: OECD.

The International Handbook on Innovation
Edited by Larisa V. Shavinina

Innovation in the Upstream Oil and Gas Sector: A Strategic Sector of Canada's Economy

A. Jai Persaud[1], Uma Kumar[2] and Vinod Kumar[2]

[1] *Natural Resources Canada (NRCan)*
[2] *Sprott School of Business, Carleton University, Canada*

Abstract: The upstream oil and gas sector is a strategic part of Canada's economy and an important driving force of innovation. It relies on partnerships to develop new ideas and bring them to market, adopt new knowledge and technologies to improve productivity, gain competitive advantage, and deal with environmental problems. Significant interests exist in understanding innovation in the sector, given Canada's innovation agenda, recent Speeches From the Throne, and climate change concerns. This chapter will address the meaning of innovation, the importance of the sector, key aspects of innovation in the sector and their relationships to Canada's innovation strategy.

Keywords: Oil; Gas; Energy; Technology; Innovation Policy; Sustainable Development; Climate Change.

Introduction

Studying innovation in the sector is crucial to understanding Canada's innovation system. Innovation underlies Canada's quality of life and provides environmental, economic, and social benefits. Innovation has always been considered as a driving force in economic growth and social development but in today's knowledge-based economy, the importance of innovation has increased (IC, 2002). The Government of Canada is interested in making Canada a highly innovative economy as reflected in the launch of *Canada's Innovation Strategy* in February 2002 and the commitments made in the Speech From the Throne (SFT) in 2001 and in 2002.

The oil and gas sector is a major part of Canada's energy and resources industries and a strategic component of Canada's economy and innovation system. With $51 billion in new capital expenditures, the oil sands is expected to be Canada's largest natural resource development opportunity over the next decade. Significant interest exists in innovation from the standpoint of reducing costs, arresting the depletion of oil and gas reserves, successfully adding to reserves, exploiting non-conventional more expensive and less

accessible sources and supplies of oil, and reducing the environmental impacts. The fields of energy and environment are inherently difficult to separate, especially because there seems to be a strong relationship between environmental awareness and energy production and use.

One of the most serious contemporary environmental problems is climate change (global warming). It is considered as one of the most important aspects of sustainable development, which calls for the integration of economic and environmental objectives. The signing, in 1997, of the Kyoto Protocol,[1] to limit greenhouse gas (GHG) emissions thereby reducing the risks of global climate change, has brought to the forefront the importance of the environment and sustainable development. The Protocol is considered

---

[1] In 1997, in Kyoto, Japan, 159 countries negotiated a treaty setting out legally binding reduction targets for six greenhouse gases (carbon dioxide (CO_2), methane (CH_4), nitrous oxide (N_2O), hydrofluorocarbons (HFCs), perfluorocarbons (PFCs), and sulphur hexaflouride (SF_6)) averaging 5% below 1990 levels for industrialized countries. The timetable agreed to is 2008–2012.

one of the most important international agreements of the 21st century (Grubb et al., 1999). The Kyoto Protocol is likely to have significant implications for business, industry, governments, regions, countries and all other aspects of our lives. The Government of Canada just released its *Climate Change Plan for Canada*, which underscores the importance of innovation in reducing GHGs and achieving other economic objectives, noting:

> Through innovation, we will be able to maintain our strong economic growth, create additional export opportunities and reduce greenhouse gas emissions.[2]

For Canada, the oil and gas sector will figure prominently.

This chapter will discuss the role of innovation in the upstream oil and natural gas sector in Canada and against key aspects of Canada's innovation strategy. It will also draw on results of the recent innovation consultation exercises and the National Roundtable on Innovation and Skills in the Natural Resource Sectors and Allied Industries. The chapter is organized as follows: 'Technology and Innovation' discusses the meaning of technology and innovation; 'Canada's Innovation Strategy' highlights key aspects of Canada's innovation strategy; 'The Natural Resources Sector in Canada' discusses the strategic importance of the resource sector and innovation system; 'Overview of Upstream Sector' provides an overview of the upstream sector; 'Key Oil and Gas Innovation and Applications' addresses innovations in the sector and their impacts on several key concepts of Canada's innovation strategy. The final section provides a brief conclusion.

Technology and Innovation

Technology can be considered as an ensemble of theoretical and practical knowledge, know-how, skills and artifacts[3] that are used by the firm to develop, produce and deliver its products and services (Burgleman et al., 1996; Gerwin & Kolodny, 1992). In a business firm, operating activities such as purchasing, manufacturing and distribution; support activities such as R&D; maintenance activities such as personnel; and control activities such as accounting, all have their own technologies (Gerwin & Kolodny, 1992). Technology is responsible for many of the important changes in our society and for determining corporate success, profitability and growth (Chakravarti et al., 1998). Emerging technologies bring not only new opportunities to grasp, but also threaten to replace old ones (Burgleman et al., 1996).

The critical importance of technology is lodged in the dynamic aspect of the concept, i.e. technological change. Technological changes include small improvements in machines and the organization of labor, arising from slow processes of learning by doing, learning by using, and from major or radical changes. Incremental changes (or minor innovations) are innumerable and occur more or less continuously in any industry (Rosenberg, 1994). Major or radical changes are usually termed discontinuous events, which spawn other changes (Steele, 1989). Some major technological advances which will change the boundaries or rules of competition (Dror, 1988; Hart & Milstein, 1999; Schumpeter, 1934). Among process inventions that have greatly increased productivity was the Hall process for producing aluminum, the Bessemer process for producing steel, and automatic looms that replaced earlier hand looms (Samuelson & Nordhaus, 1989). As a result of technological progress, capital, output and real wages per worker grow over time. The importance of technology is widely acknowledged is considered critically important to the success and survival of individual companies and economic well-being and growth and one of the principal drivers of competition in industries (Burgleman et al., 1996; Porter, 1985).

The word innovation can have different meanings in different context and the one chosen will depend on the particular analysis of the measurement. The *Oslo Manual*[4] defines innovation with respect to technological product and process innovation. The technological product and process innovating firm is one that has implemented technologically new or significantly improved products and/or processes during the period under review.[5] The innovation system involves a number of key elements concerned with the generation, dissemination and application of new knowledge: research and development to provide new ideas; education and information services to develop the required personnel; and design engineering and marketing services to incorporate the new ideas into production and distribution systems (SC, 2001).

---

[2] Posted on Government of Canada's Website:
http/www.climatechange.gc.ca/plan_for_Canada/index.html
[3] By 1987, when Random House released its completed updated unabridged dictionary, the word technology had grown to include 'interaction with life society and the environment'.

[4] The *Oslo Manual* considered as a textbook on innovation and national systems of innovation, and a compendium of socio-economic questions on the nature of innovation in market oriented economies.
[5] Holbrook & Hughes (2001) argues that the Oslo Manual uses the first two of five types of innovation presented in the classic work of Joseph Schumpeter. The five types of innovation defined by Joseph Schumpeter (1934) are: introduction of a new product or qualitative change in an existing product; process innovation new to an industry; the opening of a new market; development of new sources of supply for raw materials or other inputs; and changes in industrial organization.

What does innovation include?

The term 'innovation' encompasses much more than R&D or technological change. Innovation makes knowledge useful and turns it into wealth and prosperity. However, innovation does not come out of the blue. It requires investment in a variety of activities, such as bright ideas, learning systems, training, R&D, technology commercialization, corporate culture and entrepreneurial spirit.

Source: CBOC, *Investing in Innovation: 3rd Annual Innovation Report*, November 2001a.

According to IC (2002), through innovation, knowledge is applied to the development of new product and services or to new ways of designing, producing or marketing an existing product or service for private and public markets. The term 'innovation' refers to the world first, new to Canada or simply new to the organization that applies them. Innovation is seen as the process through which new economic and social benefits are extracted from knowledge. Innovation is about the means to achieve a higher standard of living, economic prosperity, international competitiveness, environmental improvement and social well-being that lower production costs, utilize less energy and minimize environmental impacts. It is also the pathway leading to sustainable development in Canada.

Research and development (R&D) is at the heart of the innovation process. R&D is defined as creative work undertaken on a systematic basis in order to increase the stock of knowledge.[6] R&D, however, is not in itself considered as innovation. Companies that perform R&D are far more likely to report innovations. The popular approach to assess the adequacy of R&D is to quantify the amount of money spent on R&D as a percentage of sales. Roberts (2001) found a strong correlation between R&D intensity and sales from new products as well as between R&D intensity and the overall newness of a firm's technology emphasizing R&D as a vital area for company success. Also, most successful economies invest heavily in R&D. Current real incomes in Canada are 30% below those in the United States and 90% of this gap is considered due to productivity differences. The CBOC (2001a) argues that given that investment in technology boosts productivity, it is plausible to argue that the productivity gap can partially be explained by under investment in R&D in Canada's private sector. Firms can benefit by investing in R&D in two ways (Globerman et al.,

1999). First they can reduce their production costs or improve their product quality relative to other firms. Second, they can introduce new products into the market place. Although these are the economic arguments, there are also other factors involved in the decision of firms to innovate, e.g. corporate culture (see Carayannis & Gonzalez, this handbook).

Canada's Innovation Strategy

Government action is of vital importance to the technological progress and economic development of a country.[7] On February 12, 2002, the Government of Canada launched an innovation strategy for Canada by releasing two papers—*Achieving Excellence: Investing in People, Knowledge and Opportunity* and *Knowledge Matters: Skills and Learning for Canadians*. The two papers build on the 2001 Speech from the Throne (SFT), which committed to improve our environment, create an innovative economy and share opportunities with all Canadians from coast to coast. The SFT states that the Government of Canada will:

> promote innovation, growth and development in all parts of the economy, including the agriculture and resource sectors.

It commits to spurring innovation through new federal investments in strategically targeted research in natural resource management, coordinated with partners. The 2002 SFT reaffirms the government commitment to innovation.

From the launch to October 2002, the Government of Canada engaged key stakeholders from a wide range of large and small businesses; academia; governments; industry, business, and labour organizations; voluntary sector organizations and other stakeholders and partners in a series of regional, national, and sectoral meetings expert round tables and best practice events to solicit feedback and commitment to *Canada's Innovation Strategy*. *Achieving Excellence* lays four major areas (*Challenges*), which presents ideas and explores opportunities in to improve Canada's innovation system:

• new knowledge and bringing it to market;
• skills needed;
• business and regulatory environment; and
• elements at the community level.

---

[6] R&D is defined by the Frascati Manual, OECD as creative work undertaken on a systematic basis in order to increase the stock of knowledge, including the knowledge of man, culture and society, and the use of this stock of knowledge to devise new applications.

[7] Industrialized countries have recognized the importance of innovation. Some countries like the U.K. and Australia have launched formal strategies while others such as the U.S. and Sweden without formal strategies have vigorously supported innovation. The U.S., the most innovative country in the world, by almost all measures, has a plethora of methods such as increasing R&D, funding basic research in universities, funding government laboratories and conducting large amounts of defence research, investment in schools, the development of partnerships.

Table 1 also provides targets in these four main areas. Over 10,000 Canadians participated in the national engagement process on innovation and learning.[8] These consultations led up to the National Summit on Innovation on November 18–19, 2002, which called for short term and longer commitments to improve innovation in Canada.

According to *Achieving Excellence*, the private sector needs to strengthen its ability to develop innovations for world markets and adopt leading-edge innovations from around the world. Relatively low levels of investment in R&D, too few strategic technology alliances and limited pools of capital contribute to Canada's relatively poorer innovation performance. The two major goals of the innovation strategy are to vastly increase public and private investments in knowledge to improve Canada's R&D, and to ensure that a growing number of firms benefit from commercial application of knowledge. The Government's targets by 2010 are to: at least double the Government of Canada's current investment in R&D, place Canada among the top five countries in the world in terms of R&D performance, make Canada a world leader in the share of private sector sales attributable to new innovations, and raise venture capital investments per capita to prevailing U.S. levels.

As part of the consultations, sectors within the economy including the oil and gas sector and associations provided input to the Government innovation strategy. The natural resources sector and allied industries provided input in several broad areas during the engagement process and the Round Table on

---

[8] http://www.innovationstrategy.gc.ca/

Innovation, held on October 3–4, 2002, in Vancouver. The resource sector and allied industries which included the upstream oil and gas sector commented on six broad cross-cutting areas: Investment Climate, Access to Resources, Sustainable Development, Market, R&D, and Skills (NRCan, 2002). The oil and gas and resources sector views will be highlighted in the context of the four broad challenges, including, the key roles the Government of Canada plays in promoting and fostering innovation in the oil and gas sector through several ways including partnerships with industry, providing a conducive environment to spur innovation.

The Natural Resources Sector in Canada

Canada's natural resources sector include the energy, minerals and metals and forest sectors, and are closely linked to a broad group of allied industries, including supplier and service industries. Canada's natural resources industries have long been a foundation of its economic strength. Figure 1 shows the economic contribution of the sector.

Canada has a natural advantage in its rich endowment of natural resources and state of the art—knowledge and technology base. Canada has built a strong economy and a high quality of life for all Canadians on its natural resource endowment and innovations, which improve productivity and competitiveness. The sector is part of the 'knowledge-based economy' by consuming and developing leading edge technologies such as sophisticated electronics, robotics, computers lasers and sensors. In 2001, natural resources contributed 13% of the GDP ($129 billion), and almost $150 billion in exports, accounting for $73 billion to Canada's positive trade balance. It employs some 1.5 million Canadians in direct and indirect jobs

---

Table 1. Building a more innovative Canada.

Knowledge Performance Challenge (new knowledge and bringing it to market)	Address key challenges for the university research environment
	Renew the Government of Canada science and technology capacity to respond to emerging public policy, stewardship and economic challenges
	Encourage innovation and the commercialization of knowledge in the private sector
Skills Challenge (skills needed)	Produce new graduates
	Modernize the Canadian immigration system
Innovation Environment Challenge (business and regulatory environment)	Ensure effective decision making for new and existing policies and regulatory priorities
	Ensure that Canada's business taxation regime is internationally competitive
	Brand Canada as a location of choice
Community Based Innovation Challenges (elements at the community level)	Support the development of globally competitive structures
	Strengthen the innovation performance of communities

Source: Extracted from *Achieving Excellence*, 2002.

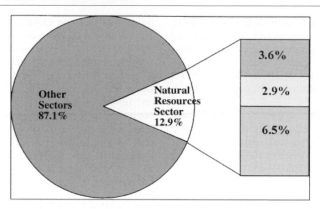

Figure 1. Proportion of Canada's economy based on natural resources in 2001.

Notes: 1. The figure was created with data from Natural Resources Canada, http://www.nrcan.gc.ca/statistics/factsheet.htm. 2. Mining and mineral processing industries (or minerals industries) include mineral extraction and concentrating, smelting and refining, non-metals and metals-based semi-fabricating industries, and metals fabricating industries. Minerals include uranium mining; energy includes coal mining..

(well paid and highly skilled) and is the lifeblood of over 650 communities from coast to coast[9] Canada's high quality of life and a significant portion of its economy is linked to natural resources (NRCan, 2002).

Natural Resource System of Innovation

The sector's system of innovation is unique and more complex than other sectors in the economy. The system is multidimensional with linkages along the supply and value chain (CBOC, 2001c; NRCan, 2002a). The firms often buy technologies rather than developing them with in-house R&D; innovation is usually undertaken through collaboration; the sector focuses more on improving processes than developing value added products.

Canada's natural resources sector operates in a competitive, national and international environment and must compete for investment dollars with other sectors (e.g. information and communication technologies) within Canada and internationally. In Canada and around the world, business confidence is ensured by a stable and competitive business environment that encourages investment in the natural resources sector and promotes sustainable development and wise use of resources. Investment in science is not just for technology development but also for new knowledge acquisition. New knowledge leads to informed decisions on regulations, codes and standards. More importantly, knowledge contributes to increased public and business confidence, which fuels the innovation engine.

> The natural resources industry is . . . an important contributor to the economic wealth of Canada. The industry is very capital intensive, highly competitive and global in nature. The resource industry is also highly innovative. It is a world leader in 'spawning' high-technology services industries.
>
> Conference Board of Canada: *Investing in Innovation in the Resource Sector*, February 2001b, p. 4.

Energy Sector

The energy sector is an important part of the resources sector. Canada has a secure, reliable and diverse source of energy including fossil fuels (oil and gas, and coal), hydro electricity, and nuclear power. Canada is also a leader in energy efficiency technologies and R&D of renewable and alternate energy sources, including hydraulic, solar, wind, biomass and other innovative technologies such as the fuel cell.

The primary source of energy has changed over time. In the 19th century, wood was the primary source of energy. At the turn of the 20th century, coal replaced wood as the primary source for about 50 years. With the proliferation of the automobile and the growing demand for gasoline, petroleum and its associated products have become the primary source of energy. Today, energy is produced as a mix of all energy sources. That mix changed in the past, and is changing now and will most likely change in the future. Sustainable energy development challenges us to examine the present mix of how energy development would change and how new environmental

---

[9] http://www.nrcan.gc.ca/inter/innrndtbl_e.html

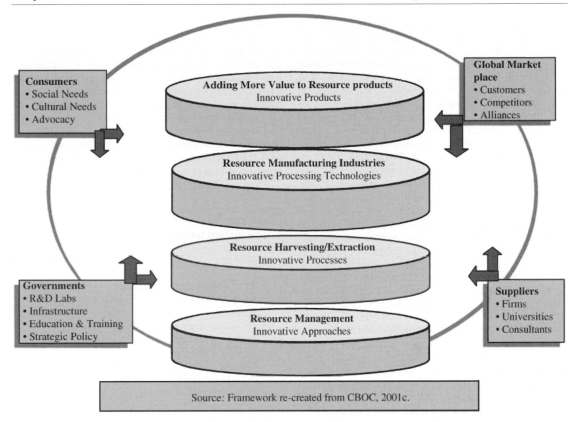

Consumers
• Social Needs
• Cultural Needs
• Advocacy

Global Market place
• Customers
• Competitors
• Alliances

Adding More Value to Resource products
Innovative Products

Resource Manufacturing Industries
Innovative Processing Technologies

Resource Harvesting/Extraction
Innovative Processes

Governments
• R&D Labs
• Infrastructure
• Education & Training
• Strategic Policy

Resource Management
Innovative Approaches

Suppliers
• Firms
• Universities
• Consultants

Source: Framework re-created from CBOC, 2001c.

Figure 2. Natural resource system of innovation.

technologies will impact both production and energy use (NRCan, 2000).

The availability of different sources of energy at reasonable cost has been a key factor in the attainment of a high standard of living in Canada. This advantage led to the development of industries having particularly strong demands for energy. This abundance of cheap energy resources has helped Canada develop a strong industrial base and has given Canada a competitive edge over other countries. It has also helped Canada to deal with the economic disadvantages of small domestic markets, long distances, rugged geography and relatively harsh climate. In 2001, energy provided 6.5% of Canada's GDP ($64.5 billion), direct employment of 222 jobs, $37 new capital investments and $57 billion in exports (NRCan, 2002).

There are three broad underlying energy policy goals in Canada (NRCan, 2000). First, is energy security— ensuring secure and reliable access for present and future generations while increasing flexibility and diversity of the Canadian energy supply system. The second policy objective is to ensure economic growth in terms of investment, regional development, and employment. The third and most recent policy objective is to reduce and manage atmospheric emissions

effluents and wastes resulting from energy development and help Canada to meet its climate change commitments and its environmental, health and safety goals. Energy policy in Canada must reflect a balance of these issues and Canada's responsibility acting within the community of nations to resolve global issues. The issue of climate change will be discussed later as it will be a major variable that firms will have to address in maintaining competitiveness. The IEA notes that the major factors that bring about energy technology innovations include:

energy security requirements, environmental concerns, economic competitiveness, the availability of finance and prevailing social values (IEA, 1997, p. 17).

Overview of Upstream Sector

The petroleum industry includes the oil and gas sectors and is usually discussed in terms of upstream and downstream operations. According to the Petroleum Communications Foundation (2001), the upstream petroleum industry finds, produces and processes oil and natural gas. Exploration and development activities refer to all the upstream activities that are required to find new oil and natural gas resources and develop them to the point where they are ready to be processed. Exploration involves the search for petroleum including seismic surveys and drilling. Production describes the steps of getting petroleum out of the ground. The upstream sector includes more than 1,000 exploration and production companies as well as hundreds of associated businesses such as seismic and drilling contractors, service rig operators, engineering firms and various scientific, technical, service and supply companies (Canadian Petroleum Foundation, 1999).

Upstream expenditures include geological and geophysical activities (e.g. seismic, geological and geophysical), land costs (operating lease, and bonus bids costs), costs of drilling oil and gas wells and oil sands expenditures. Development wells are drilled to produce the oil and natural gas reserves and properly assess the level of reserves.[10] In the case of oil sands in Canada, the reserves are known to exist and no exploration activity is required to find the resources. It should be noted that both Canada and the U.S. are the most extensively drilled areas in the world.

Economic and Strategic Importance of the Oil and Gas Sector

The oil and gas sector is an important contributor to the Canadian economy and energy security. In 2001, oil and gas production accounted for $24.1 billion, 32% of the energy sector GDP of $64.5, and $36 billion trade surplus. Alberta accounted for 76% of Canada's total oil and gas production in 2001 (valued at $49 billion; different from GDP estimates which are value added, reflecting intermediate inputs). The industry provided a secure oil and gas supply for North America. Canada produces oil and gas well in excess of its own requirements and, as a result, is a significant net oil and gas exporter. Canada is the largest oil and natural gas exporter to the U.S. In 2001, Canada exported about 1.4 million barrels of oil per day with 99% going to the U.S. Canada also exported 3.8 Tcf (valued at $25.6 billion) of natural gas to the U.S., meeting 94% of the U.S. import requirements, and 15% of total U.S. demand. In 2001, oil and natural gas accounted for 73% of Canada' energy supply. The upstream petroleum industry has operations in seven provinces and two territories. Most Canadian oil and gas production is concentrated in the western provinces of Alberta, Saskatchewan and British Columbia.

Oil and Gas Reserves

Canada has significant oil and gas resources. However, only a small percentage is considered low-cost as established reserves (economic at today's prices).

Established conventional reserves are in the range of 3.4 billion barrels. Canada has vast high-cost non-conventional resources and oil sands (very heavy oil) and offshore supplies. Oil sands reserve is upgraded into synthetic oil at the Suncor Energy and Syncrude Canada plants, or produced and exported in the form of bitumen. Canada's oil sands reserve is the third largest oil resource basin in the world, after the Persian Gulf and Venezuela. On a world scale, Canada like the U.S., is a high-cost oil producer. Similarly, Canada has about 61 Tcf of established reserves (lower cost supplies) as opposed to higher costs gas resources, which total 630 Tcf.

Oil and Natural Gas Key reserves and production parameters

Crude Oil

- Established conventional reserves—3.4 billion barrels, oil sands 6.8 billion barrels, 1.4 billion frontier
- Oil sands ultimate recoverable potential represent over 300 billion barrels greater than Saudi Arabia's reserves
- Canada current oil production of 800 million barrels annually
- 62% exported to the US, 17% of U.S. Demand
- Canada first as crude oil exporter to the U.S. ahead of Saudi Arabia

Natural Gas

- 61 Tcf established reserves
- undiscovered potential of 301 Tcf in Western Canada
- Third largest producer in the World
- Current production of 6.1 Tcf
- 63% exported representing 15% of U.S. demand

Sources: Natural Resources Canada, U.S. EIA 2002.

Canada produced 2,200 thousand barrels per day of crude oil and 6.1 trillion cubic feet of natural gas in 2001. Canada's domestic oil supply includes various sources: conventional (light and heavy), oil sands, and offshore production. Canada's oil supply is projected to increase significantly due to production from oil sands. Most of Canada's natural gas production comes from conventional sources of the Western Sedimentary Basin (WSCB). Production has increased by 50% over the past 10 years due mainly to higher exports to the U.S. Supplies from the east coast offshore are expected to play an increasing role in the future.

---

[10] The wells drilled in new virgin territories are called wildcat wells. Those in already explored areas are called step out wells.

The Innovation Environment Challenge

Developing Canada's resources, with due consideration for the business environment, requires an approach to decision-making that is based on sound science, Canadian values, public involvement, corporate social responsibility, and new governance models (NRCan, 2000). Finding balanced approaches that achieve economic prosperity and environmental performance will also help to maintain access to global markets. *Achieving Excellence*'s target is for Canada to be recognized internationally as an innovative country in order to attract required talent and capital. The Government's goals are to address public and business confidences, Canada's stewardship regimes, incentives for innovation, and to ensure that Canada is recognized as a leading innovative country.

Drivers of the Application of Advanced Technologies

The substantial declines in real oil and gas prices since the mid-1980s and a dim prospect for the sustained recovery over the long-term, provided significant motivation for the use of advanced technology to reduce costs and to improve profit margins. Firms also turned to rationalizing their operations by downsizing, and focusing more on core their competencies. The requirement to reduce costs had to be achieved in spite of the resource depletion working in the opposite direction. As reserves are found and extracted over time, it is argued that it becomes more expensive to replace them. It is assumed that firms exploit their best resources first.

Unless technology improves to make it easier to find and produce the next generation of resources, the average cost of production will inevitably rise over time. The pressure to lower costs also underscores the need to undertake the development of new technologies. Emerson (1998) notes that investment will take place as long as prices exceed costs. Both Canada and the U.S. are price takers. If oil cannot be produced at less than the prevailing market prices, it could be imported into both countries from other producing nations. Governments have a strong interest in competitiveness because companies will relocate to lower costs producing regions.

Significant motivation in the use of more advanced technologies in the upstream sector stems largely from competitive pressures to reduce costs in the face of lower oil and natural gas prices. Most oil and gas companies today stress the importance of technology in achieving the strategic objectives of their organizations. The recent concern over the environment, in particular, climate change, is likely to provide another impetus to implement and develop new technologies in the energy sector. This is captured by the U.S. National Petroleum Council (NPC, 1995), which notes that the competitive edge of the industry will increasingly depend on the ability to manage and apply technology effectively and rapidly. This will include leadership in technology development and environmentally acceptable hydrocarbon fuels.

Oil and Gas Prices

The behavior of oil and gas prices are key determinants of oil and gas supply (Persaud & Kumar, 2001). A well known fact of the resources sector is that commodity prices are cyclical and volatile. Prior to 1973, crude oil prices, expressed in WTI at Cushing, averaged less than US$10/barrel (bbl). The following 29 years saw significant volatility in prices due to a number of major events. The first oil shock took place in 1973 due to the Arab oil embargo which caused oil prices to triple and brought profound economic disruption in the world for the ensuing years (Martin et al., 1996). The Arab oil embargo was followed by a number of destabilizing events in the Middle East including the Iran Revolution, Iran/Iraq War and the Gulf War, the Asian economic crisis in 1998 and 1999 and the tragic incident of September 11, 2001. The EIA (2002) notes that crude oil prices in 2002 began at roughly $16 per barrel and were between $25 and $30 per barrel by the fall. The EIA is projecting a slight decline in real oil prices ($2000) to 2005 increasing slightly thereafter to $26.50/bbl by 2025. One of the key factors that will impact oil prices is technology.

Unlike oil prices which are determined in the world market, natural gas prices are determined in the North American market. Several factors that have affect gas prices in recent years including increased competition, improved pipeline accessibility, the commoditization of natural gas and the greater use of storage capacity. This point will serve to illustrate the gas price cyclicality. In the United States, for instance, when prices increased in 1992 and 1993, after several years of decline, industry responded by bringing additional supplies to the market thus forcing prices downwards in 1994 and 1995. Average natural gas prices in the U.S. are expected to drop from $4.12 per thousand cubic feet (Mcf) including purchases on the spot market to $2.75 in 2002. Prices are expected to increase to about $3.90/Mcf in 2025 (IEA, 2002), with technology being a key determining factor in accessing and producing resources competitively in an environmentally sound manner.

Key Oil and Gas Innovation and Applications

Historically, until very recently, the technology used in the upstream sector of the petroleum industry has remained relatively unsophisticated. Only over the past 15–20 years, have there been rapid developments and applications of technology in the industry (Bohi, 1997). Significant motivation in the use of advanced technologies in the upstream sector stems largely from competitive pressures to reduce costs in the face of a declining commercial resource base, and relatively low oil and gas prices (the recent run up in prices are not expected to be sustainable according to most forecasters). Advanced technologies, including enhanced oil recovery techniques, are required to recover a large proportion of oil and gas in place. Only about 10–20% of oil in place from a producing reservoir is recovered. Thus, much of the oil remained in abandoned reservoirs. This is not the case for natural gas, which has a higher recovery factor. New technologies are required to further reduce the costs of frontier, Arctic, and oil sands supplies.

With the recent concern for climate change, technology is considered crucial in reducing emissions in the upstream sector. Companies may reduce their emissions per unit of production through increased efficiency of their energy use, reduced flaring and reduced non-fuel emissions of GHGs. The Upstream Oil and Gas Working Group (UOGWG)[11] of the NCCP argued that in the absence of technology to supply alternative energy sources or greatly increase energy efficiency, meeting Canada's Kyoto Target would put severe constraints on households and lifestyles, as well as all Canadian industry, imposing severe costs on the whole economy.

A broad range of technologies is being forecast for upstream oil and gas, such as, reservoir evaluations, reservoir engineering, and drilling techniques. These technologies span the entire value-chain of the industry—finding, development and producing oil and gas. Also, they cover a wide range of deposits including conventional, offshore, Arctic oil and heavy oil,[12] and oil sands (see Table A3, Appendix).

There are four major technologies in the industry, these are: 3-D seismic, horizontal drilling, offshore technology, and oil sands technology. Horizontal drilling and 3-D seismic may also be considered Schumpeterian changes given the number of other developments being spawned. The development of these technologies was due to a culmination of other major developments in computing and drilling. These four technologies could be considered to be the same for the oil and gas industry as the Bessemer process was perceived to be for producing steel and the Hall-process for producing aluminum. Overall, the four major technologies are fairly complementary in many respects and are adaptable to many situations. Provided in Table 2 is a summary of the description, application, and impact of these technologies.

These technologies are expected to play a key role in reducing costs of adding reserves by improving knowledge of the size of the resource base, extending

the life of the reserves, improving success ratios, the targeting of drilling of oil and gas deposits, influencing project development of frontier and oil sands projects (e.g. Adelman & Lynch, 1997; Fagan, 1997; Fisher, 1994; IEA, 1997). In terms of labour productivity, IC (2002) shows that the crude petroleum and natural gas sector in Canada to exceed that of the U.S. Several papers (e.g. Saklou, 2000; Williams, 2000) presented at the 16th World Petroleum Congress (WPC), stressed the importance of technology in the petroleum industry.[13] Emerson (1997) makes a similar point but goes even further by noting that it is time to abandon the concept of fixed stock of oil in favor of a more subtle characterization of oil reserves as a flow—a flow which is dependent on continual adaptation and innovation to keep replacement costs below price and this will lead to continued investment.

Forecasting Upstream Oil and Gas Technology

In the oil and gas supply sector, technology progress for exploration and production activities is represented by trend-based improvement in finding rates, success rates, costs, and the size of the resource base. The view of new technologies and their impact is obtained mainly from experts in industry and government. The U.S. National Energy Modelling System (NEMS),[14] which is considered a highly sophisticated and 'technology rich model', assumes the impact of technologies through the cost function based on expert opinions (U.S. DOE/EIA, 2000).

The UOGWG developed a list of a broad range of technologies to reduce GHG emissions (see Table A3). The UOGWG considered available, emerging and embryonic technologies.[15] These primarily include opportunities to improve the energy efficiency of operations, reduce methane losses, and either conserve or utilize waste stream. These options were chosen through several brainstorming sessions. A key requirement was that the analysis considers only technologies or incremental penetration levels that would reasonably be excluded from a business as usual scenario (i.e. based on current forthcoming requirements).

A Paradigm for Oil and Gas Research: Collaboration and Outsourcing

The U.S. NPC (1995) found that there is a 'new paradigm' for oil and gas R&D and demonstration

---

[11] The Upstream Oil and Gas Working Group (UOGWG) was one of the seven working groups of the Industry Table to carry out analysis to address climate change issues for the Climate Change Secretariat. The UOGWG comprised of several upstream industry organizations such as the CAPP, Canadian Gas Association, Canadian Energy Pipeline Association, Small Explorers and Producers Association of Canada.
[12] Term is used to refer to conventional heavy crude and sometimes bitumen unless explicitly stated.

[13] The first author was a participant at the WPC which was held in Calgary, Alberta Canada, June 11–15, 2000.
[14] NRCan is in the process of implementing a Canadian version of this model.
[15] Each technology is classified as near term, medium term, or future option. Near term technologies are deemed practicable to implement during the period 1999–2008. Emerging technologies or high capital costs, would be more applicable during 2008–2012. Future options are embryonic technologies that are still at the conceptual stage or early stages of development.

Table 2. Four major oil and gas technologies.

Types of Technology	Description and Application	Impact
Three-dimensional seismology	Uses sound waves to provide inference on the structure and properties of sub-surface rock layers	Improves the ability to locate new hydrocarbon deposits (success ratios) thereby reduces the amount of drilling;
(3-d seismic)	Yield higher quality images than two-dimensional 2-D but more expensive and involves processing of a significant amount of data	Allows for monitoring production of reserves; and
		Increases productivity and lowers costs, and energy required and thus lower emissions.
Horizontal wells (directional wells)	Intersect the reservoirs at the side rather than from above as compared to vertical wells	Increases productivity, reserves, and flexibility particularly in offshore and environmentally sensitive areas; and
	Provides more contact with a pay zone (area containing oil and/or gas)	Increasingly important in oil sands recovery, reduces energy consumption and emissions.
	Much more expensive, and riskier than a vertical well and needs highly trained professionals	
Deep-water Exploration	A wide variety of methods exists: including drill ships, directional drilling methods, production platforms, remote-controlled subsea wells, and subsea pipelines	Makes exploration and development possible in water deeper than 1000 feet;
		Increases flexibility in production; and
	High cost and risky	Makes offshore (e.g. U.S. Gulf, U.K. and Canada important sources of production.
Oil sands Technology	Several techniques exist to undertake surface and in-situ extraction, e.g. Steam Assisted Gravity Drainage System (SAGD)—recent innovation which uses two horizontal wells (one for steam and the other for producing)	Increase recovery and lowers costs significantly (e.g. Suncor and Syncrude, and bitumen projects); lowers emissions.

(R&D) are evolving because intense competition that contributes to low oil and gas prices, and the need to re-allocate scarce funds for large investments related to environmental compliance, have forced the industry to re-evaluate every aspect of its business, including their R&D investments. In general, the U.S. Petroleum Council found that this re-evaluation brought better efficiency and cost-effectiveness to private sector (R&D) activities. The IEA (1997) argues that such a new paradigm for oil and gas R&D may also suggest significant opportunities for greater international collaboration among member countries and the oil and natural gas industry. The new paradigm suggests a greater on user-driven collaborative R&D

(including government-industry collaboration) to compensate for smaller in-house industry programs.

The NPC's (1995) survey responses indicate that only a few companies would carry out in-house research, where the outcomes of those programs are vital to their future business interests and/or their long term viability depended on. The results of the 1999 Statistics Canada Survey in the resources sector show that a high percentage of firms rely on external sources of technologies. The CBOC (2000) in its Report, *Collaborating on Innovation* noted that firms with a greater propensity of inter-organizational technological collaboration are more likely to produce innovations than those that collaborate less.

Table 3. Paradigms of resource development.

	Old paradigm	New paradigm
Source of technology	In-house	Leverage and collaborate
Project prioritization	Technology push	User needs
Motive	Own it	Use it

Source: NPC *Research Development and Demonstration Needs of the Oil and Gas Industry, Volume 1: Summary and Discussion* (1995) p. 7.

In Canada, financial pressures in the mid-1990s have led to an emphasis on collaborative R&D to reduce costs and to achieve a critical mass of funding in the energy sector. Experience has shown that the prospects for promoting and accelerating innovation are optimized when the generation of expertise, knowledge and technology and the fostering of a business environment are met concurrently—through targeted partnerships among governments, business and academia (NRCan, 2002). There are many reasons for collaborative research including: cost reduction, sharing risks, gaining expertise, internalizing economic externalities, and co-ordinating strategies. Consortia have grown in the resource sector and are likely to play an increasingly important role in the future (Globerman et al., 1999). One of the key drivers of consortia has been the globalization of world markets (Kumar & Magun, 1995).

Achieving Excellence, Canada's innovation strategy, highlights the increasing involvement of government laboratories in partnerships. NTCan's National Round Table on Innovation, October 2002, underscored the importance of national collaborative efforts in addressing innovation and sustainable development challenges.

Sustainable Development of Canada's Oil Sands

Canada's oil sands are a world class resource and considered unique to Canada. The technology to find safe and environmental sustainable ways to recovering the oil, and to create tens of thousands of jobs is continuously being improved. This is taking place in cooperation with governments, academics and industry, researchers who have helped reduce economic and environmental barriers to development this enormous and important resource. With $51 billion in new capital expenditures, the oil sands resource is expected to be Canada's largest natural resources development in the next decade (IC, 2002; NRCan, 2002a).

Industry has undertaken R&D for innovation, technological improvement, and increased productivity in the exploration development, extraction, upgrading, distribution and sustainable management of resources. During the consultation process on the government innovation strategy, industry argued that commercialization of R&D was critical—Canada cannot afford to 'leave good ideas on the shelf'. Industry argued for improvement in government funding for pilots and prototypes, improvement in the tax system for R&D, find good international technologies that could be imported, developed and adopted in Canada, and seek out international partnerships. Industry argued for

better links with government laboratories, universities and private sector.

A substantial amount of research collaboration takes place in the upstream oil and gas industry and this has become more popular in recent years. Since the mid-1990s, two non-traditional organizations (Canadian Oil Sands Network for Research and Development (CONRAD); and the Petroleum Technology Alliance Canada (PTAC) have emerged that have catalysed innovation in the upstream oil and gas industry, while encouraging collaboration, free information exchange and shared intellectual property ownership use rights. NRCan (CANMET's Western Research Centre) is and active member and preferred research provider to these two organizations. The Government of Canada, through NRCan and other departments and agencies, play several key roles in developing and disseminating the knowledge infrastructure that is critical for innovation, sustainable development decision-making and the generation of new ideas and technologies. NRCan (2002, p. 3) notes that:

> NRCan has a key role to play in the innovation system for the natural resources sector across Canada, as both a catalyst for research and development (R&D) and innovation, and a major science and technology (S&T) performer and funding organization.

Syncrude and Suncor collaborate with other companies, governments, universities and research centres. They are industry leaders and are involved in research areas that cover all parts of the value-chain of extraction, upgrading, reusing material, of energy usage. Syncrude was a key player in the formation of Canadian Oil Sands Network of Research and Development (CONRAD). Companies are involved in collaborative efforts such as the national Centre for Upgrading Technology, Alberta Research Council, the National Research Council, and the Canada Centre for Mineral and Energy Technology (CANMET).

Government research is becoming increasingly partnership focused. Through Program for Energy Research and Development (PERD), NRCan funds R&D projects in collaboration with other federal departments. One such initiative has looked at the impact of ice on offshore floating oil production and storage platforms. The outcome of this research will help companies make better informed decisions about offshore platform operations, thus increasing production, reducing expenses incurred from unnecessary shutdowns and improving the safety of personnel.

The National Centre for Upgrading Technology (NCUT) is a joint venture partnership between NRCan and the Government of Alberta. Its mission is to make oil sands bitumen and heavy oil upgrading in Canada more economically viable and environmentally responsible by conducting R&D leading to technologies that have commercial potential.[16] A technique, known, as Steam Assisted Gravity Drainage (SAGD), as noted in Table 2, took a decade of successful piloting to give operators the confidence to build independent large field demonstrations, could become a technology of choice for in situ recovery of bitumen. Developing the SAGD process was a collaborative efforts in which NRCan was involved at the early stages of development. This technique holds promise for development of Canada's oil sands resources.

NRCan is involved in research in both capturing CO_2, and developing ways to use and/or store it. In Weyburn, Saskatchewan, NRCan is a partner in an International Energy Agency monitoring project where CO_2 is injected into an oil reservoir. Capturing CO_2 emissions and storing them underground is a process with significant potential environmental and economic benefits. The CO_2 helps the oil flow to the surface; the project studies the impact of storing the CO_2 underground.

Oil and gas extraction is one of the few natural resource industry sectors whose spending has been robust over the decade. Several energy companies including Imperial Oil, Syncrude, Alberta Energy rank among the to 50 research spenders in Canada (Research Infosource Inc., 2002). It can be argued that R&D expenditure may not take full account of the level of innovation in the resources sector given the links with other sectors in which it drives innovation through massive capital expenditures and high capital intensity (Persaud & Norris, 2002).

Advanced Technologies and The Resource Base

The literature is rich with discussion on the role of the resource base, the way it changes over time, and its ultimate development (e.g. Adelman & Lynch, 1997; Cleland & Kaufmann, 1997; Dasgupta & Heal, 1979; Fisher, 1994; Hotelling, 1931; Kneese & Sweeny, 1993; NEB, 1999). A key point in the literature is that resource estimates do not take adequate account of technological developments. Further, the usefulness of the estimates of ultimate reserves from an economic standpoint is frequently questioned (Emerson, 1997). Technological advances have allowed industry to find and produce more and more oil in mature regions at lower costs, and resources inaccessible 20 years ago are now routinely being developed and produced. Adelman & Lynch (1997) note that ultimate recoverable oil potential, based on Hubbert's 'bell curve' approach (Hubbert, 1967), in the U.S. was estimated at 170 billion barrels in 1974. Production to date has already exceeded this amount, proven reserves are 20

billion barrels, and discoveries continue to be made. A parallel story exists for natural gas.

To simply extrapolate supply from a given reserve base, therefore, would underestimate supply unless sufficient allowance is made for technological impact. The impact of technology is also bringing into question the theory of depletable resources. Fagan (1997) maintains that technology is mitigating resource declines and reducing costs. Cleland & Kaufmann (1991) note that much of the debate between resource economists, on one hand, and physical scientists, on the other, is with respect to the relative impact of technical change vs. depletion. Resource economists are usually more optimistic than the physical scientists. Adelman & Lynch (1997) argue that the issue of resource shortage has led to misguided policies for several years. Emerson (1997) makes a similar point but goes even further by noting that it is time to abandon *Hotelling's* (1931) assumption of an operator's fixed stock of oil in favor of a more subtle characterization of oil reserves as a flow—a flow which is dependent on continual adaptation and innovation to keep replacement costs below price and this will lead to continued investment.

Forecasting Oil and Gas Supply and Emissions

Forecasting oil and gas supply is a fairly complex exercise involving a large number of variables (NRCan, 1999; Persaud & Kumar, 2001). Projecting oil and gas supply cannot be done by one single approach and, unlike energy demand for which econometric techniques are available, the approach to projecting oil and gas supply is more eclectic. Available information on the geology, technology, engineering and economics of various potential sources of supply is assembled to arrive at oil and gas resource estimates. With continued impact of innovation in the sector, forecasting supply models will have to reflect the underlying change on reserves and costs.

Sustainable Development and Innovations

Canada is well endowed with natural resources and strongly subscribes to the concept of sustainable development.[17] Canada subscribes to the World Commission on the Environment Sustainable Development (WCED, 1987, p. 8), which called for 'development that meets the needs of the present without compromis-

---

[16] The extraction process currently used at Fort McMurray project produces tailings that consist of fine clays and water that are stored in tailings ponds until the fines can settle out and be discarded.

[17] In Rio de Janeiro, Brazil, 10 years ago, at the United Nations Conference on Environment and Development, Canada and more than 175 other nations committed to a comprehensive plan of action for socially, economically, and environmentally sustainable development—a plan known as Agenda 21.

ing the ability of future generations to meet their own needs'. Sustainable development is not considered a fixed state but a process of change in which the exploitation of resources, investments, technological development, and institutional change are made consistent with future as well as present needs (The Brundtland Commission, 1987).

According to NRCan's Sustainable Development Strategy:

Sustainable development is absolutely essential in the resources sector to fulfil its role as a major engine of growth and job creation in this country, helping to provide the highest standard of living and quality of life to Canadians for the 21st century (NRCan, 2001, p. 3).

In the natural resources sector sustainable development is seen as a constantly evolving process of identifying and seizing new opportunities to improve the environmental and economic performance to maximize social and economic benefits.

There are several environmental problems associated with the production and consumption of energy. Exploration, production and transportation activity impact the natural environment in different ways. Tens of thousands of wells and thousands of kilometers of pipeline and roads have a large cumulative effect over a wide area of the landscape. Major industrial operation such as oil sand mines, and gas-processing plants have major effects on relatively small areas. Offshore development poses threats to marine life and birds. Cold water temperatures and frequent storms also make it difficult to clean up oil spills further.

Low water temperatures also slow down the processes of evaporation and bacterial action that eventually break down oil.

The most significant concern with regard to energy and the environment is with greenhouse gases and the impact on global warming. This issue has attracted attention for the past 25 years and intensified in recent years. The combustion of fossil fuels results in emissions of all three of the major greenhouse gases: carbon dioxide (CO_2), nitrous oxide (N_2O) and methane (CH4). CO_2, which has received the most attention, represent the largest share of global greenhouse gas emissions and most of it comes from energy consumption. Scientists warn that continued increase in greenhouse gases concentration could cause irreversible global warming and climate change, entailing severe detrimental economic and ecological effects.

The application of new technologies is considered as an important instrument for reducing greenhouse gas emissions and dealing with other sustainable development issues across all sectors. GHG emissions in the upstream oil and gas industry have increased since 1980 (Environment Canada, 1999). In the case of conventional oil pools, emissions have increased per unit of production due to declining pool pressure and the need for rising fuel consumption because of higher compression requirements. Advances in oil sands technologies have steadily improved the energy efficiency of the operations. However, oil sands mining and upgrading is the most emission intensive, followed by bitumen, conventional heavy, conventional light oil[18] production. In addition, oil sands production has increased continuously because of improved technology.

Emissions from natural gas production account for 35% of CO_2 equivalent of total GHG emissions of the upstream oil and gas industry. The sector consists of several source categories such as wells, gathering systems, and field facilities (e.g. compressor stations, metering stations, etc.). In addition, emissions result from gas flaring. New technologies are helping to significantly reduce the flaring of solution gas (by-product of oil production).

Skills and Innovation

In order to accelerate innovation and boost R&D performance, the public and private sector research facilities will need to be able to recruit and retain world-leading scientists, researchers, engineers and technicians, and skilled trade people. *Achieving Excellence* sets several targets in terms of developing a skilled and talented workforce, including increasing graduate admissions at Canadian universities and recruiting foreign talent.

The oil and gas sector is highly capital intensive, relying on a highly skilled and educated labor force. Training can determine the difference between successes and failures in the field. It is argued that anyone could purchase 3-D data and process the data using dedicated computer workstations, but interpreting the results for oil and gas potential is a combination of art, science, and experience. The view of several companies in the oil and gas industry is that there is a significant effort to train employees in acquiring the new skills because of the high costs associated with equipment and processes. For instance, without skilled employees significant investment can be lost in using horizontal well technology.

The oil and gas sector and other parts of the natural resources sector and allied industries face an increasing number of immediate and longer-term skill challenges. There are several reasons for these shortages including the overall aging of the labor force and stiff competition from other sectors for increasingly skilled workers. The unique characteristics of the resource sectors add significantly to the challenge of recruiting, retaining and developing sufficient pools of skilled

---

[18] Includes conventional light oil, upgraded heavy crude oil, synthetic crude and pentanes plus.

workers. The Innovation Summit, held in Toronto, November 18–19, 2002, stressed the need for lifelong learning and flexibility to Canada's formal education system. Canadian Association of Petroleum Producers (CAPP) notes that companies increasingly demand experienced workers who can be instantly productive. Therefore, more needs to be done to help new graduates overcome the job-experience hurdle, e.g. government and industry need to collaborate on a system of first-job internships. CAPP also notes that sector associations must do more to get companies involved in such programs (NRCan, 2002b).

Innovation at the Community Level

Achieving Excellence calls for cooperation at all levels of governments to unleash the full innovation potential of communities across Canada. One major target is to create at least ten internationally recognized clusters. The Competitive Institute (1998) notes that clusters are related to core capabilities within a certain group of geographically concentrated firms. Clusters are geographically proximate groups of interconnected firms, their customers, suppliers, research and training institutions, and other physical infrastructure. They are intended to stimulate competitiveness and foster strategic alliances that combine to encourage firms to constantly upgrade and innovate.

Given the regional concentration of the oil and gas industry, it offers significant opportunities for innovation at the community level. An innovation cluster to support the work and address major challenges on development of oil and gas resources on the east coast offshore was recommended by stakeholders during the consultation process on the Government innovation strategy. It was argued that cooperative efforts are required by industry, government and academics to address the challenges of this sector and focus expertise in the most promising areas of the east coast.

Conclusion

The upstream oil and gas sector makes significant economic contributions to Canada and has a high-level of productivity. Innovation in the sector have resulted in improved success rates, increased recovery of oil and gas, expansions of the commercial resource base, and reduction of costs, and reduced environmental stress. Technology is now permitting access to frontier and non-conventional sources of supply, particularly oil sands production. The outlook is for continued favorable impact of technology and further development of these sources of supplies, which will make an increasingly larger contribution to Canada's policy and economic objectives. The favorable impact of technology across the value-chain of the industry and the view for continuous improvement is bringing into question the conventional view of a fixed resource base and rising costs. At the organizational level, the application

of technologies is improving flexibility, reducing hierarchical levels and facilitating the implementation of team work and creating the recognition that the development of skill sets is very important.

Canada recognizes the importance of innovation on economic growth, productivity and well-being. Innovation leads to a higher standard of living for Canadians. Innovations in the upstream oil and gas sector can make greater contributions to sustainable development, play a crucial role in reducing greenhouse gas emissions, and un-tapping Canada's vast resources. The Government has played a key role in several ways in the industry for example through partnerships, policies, and support for R&D, as shown through Canada's innovation strategy intends to vigorously pursue making Canada a highly innovative economy. The sector recognizes that the government can play an important role in creating the appropriate environment for innovations through smart regulations, the tax system, fostering R&D, contributing to skills development, and continued partnerships with industry (NRCan, 2002b).

Given *Canada's Innovation Strategy*, it would be worthwhile undertaking further research in these areas. Of great interest, will be studies on R&D collaboration and smart regulations. Very little information is available on the organizational and individual impacts of technology probably because of the capital intensity of this industry. This is also another worthwhile area of further research. Important, and timely studies of the impact on technologies, and the forecasting of technologies to reduce CO_2 emissions will be of increasing interests, given Canada's Commitment to Climate Change.

Acknowledgments

The authors acknowledge the use of statistics, information and projections from NRCan including *Canada's Emissions Outlook: An Update*; The results of NRCan's Round Table on Innovation and Skills (NRCan's Webpage—http://www.nrcan.gc.ca, and drawing from the results of the 1999 Statistics Canada Survey on Innovation. The authors wish to acknowledge the contributions of Frances Anderson and Susan Schaan of the Science, Innovation and Electronic Information Division to the background analysis on the 1999 Survey of Innovation; and Tim Norris, Rob Dunn, and John Hector of NRCan; and Hugh Deng (Agriculture Canada), to the underlying the analysis of the 1999 Survey of Innovation relating to the resources sector. The authors also wish to express their thanks to Dr. Hertsel Labib of the Energy Sector for his insightful comments, Bert Plaus of Statistics Canada for providing R&D estimates and to Jacek Warda of the Conference Board of Canada for providing useful thoughts on innovation. Errors and omissions are the sole responsibility of the authors and views expressed in this chapter are not necessarily those of the respective organizations to which they belong.

References

Adelman, M. A. & Lynch, M. (1997). Fixed view of resource limits creates undue pessimism. *Oil and Gas Journal* (Special Issue), 51–54.

Bohi, D. (1997). *Technology change and productivity in petroleum exploration and development*. Washington, D.C.: Charles River and Associates—Consulting Firm.

Burgleman, R. A., Maidique, M. A. & Wheelwright, S. (1996). *Strategic management of technology and innovation*. Chicago, USA: Irwin.

Canadian Petroleum Foundation (1999). *Our petroleum challenge: Exploring Canada's oil and gas industry*. Calgary, Alberta.

CBOC—Conference Board of Canada (2000). *Collaborating for innovation, 2nd annual innovation report*. Ottawa, Ontario.

CBOC (2001a). *Investing in innovation: 3rd annual innovation report*. Ottawa, Ontario.

CBOC (2001b). *Investing in innovation in the resources sector*. Ottawa, Ontario.

CBOC (2001c). *Members' briefing: Investing I innovation in the resource sector*. Ottawa, Ontario.

Chakravarti, A. K., Vasanta, B., Krishnan, A. S. A. & Dubash, R. K. (1998). Modified Delphi methodology for technology forecasting: Case study of electronics and information technology in India. *Technological Forecasting and Social Change*, **58** (1), 155–165.

Cleland, C. J. & Kaufmann, R. K. (1997). Natural gas in the U.S.: How far can technology stretch the resource base? *The Energy Journal*, **12** (2), 17–46.

Cleland, C. J. & Kaufmann, R. K. (1991). Forecasting ultimate oil recovery and its rate of production: Incorporating economic forces into the models of M. King Hubbert. *The Energy Journal*, **12** (2), 17–46.

Competitive Institute (1998). Nuts and bolts of cluster development, summary of first annual conference proceedings (Workgroup A-1)—(http://www.competitiveness.org/activities/firstconfe/workshop.htm#). Barcelona, Spain.

Dasgupta, P. S. & Heal, G. M. (1979). *Economic theory and exhaustible resources*. Oxford: James Nisbet and Co., Ltd. and Cambridge: Cambridge University Press.

Dror, I. (1988). Forecasting technologies within their socioeconomic framework. *Technological Forecasting and Social Change*, **34** (1), 69–80.

Emerson, S. A. (1997). Why fears of an oil crisis is misinformed. *Harvard International Review* (19), 12–18.

Energy Information Administration (EIA) (2002). *Early Release of the Annual Energy Outlook 2003*. Washington, D.C.

Environment Canada (1999). *Canada's greenhouse gas inventory: 1997 emissions and removals with trends*. Ottawa, Canada.

Fagan, M. N. (1997). Resource depletion and technical change: Effects on U.S. crude oil finding costs from 1977 to 1994. *Energy Journal*, **18** (4), 99–105.

Fisher, W. S. (1994). How technology has confounded U.S. gas resource estimators. *Oil and Gas Journal*, **92** (43), 100–107.

Gerwin, D. & Kolodny, H. (1992). *Management of advanced manufacturing technology*. New York: John Wiley.

Globerman, S., Nakamura, M., Ruckman, K. & Vertinsky, I. (1999). *Technological progress and competitiveness in the Canadian forest products industry*. Ottawa, Ontario: Natural Resources Canada, Canadian Forest Service, Science Branch, and Industry Economics and Programs Branch.

Government of Canada (2002). *Climate change plan for Canada*. Ottawa, Canada.

Grubb, M., Vrolijk C. & Brack, D. (1999). *The Kyoto protocol*. United Kingdom: The Royal Institute of International Affairs, Energy and Environmental Program.

Hart, S. L. & Milstein, M. B.(1999). Global sustainability and creative destruction. *Sloan Management Review*, **41**, 23–34.

Hotelling, H. (1931). The economics of exhaustible resources. *Journal of Political Economy*, **39** (2), 137–175.

Hubbert, M. K. (1967). Degree of advancement of petroleum exploration in the United States. *American Association of Petroleum Geologists Bulletin*, **51**, 2207–2227.

Human Resources Development Canada (2002). Knowledge matters: Skills and learning for Canadians. Ottawa, Ontario.

IEA—International Energy Agency (1997). *Energy technologies for the 21st century*. Paris: OECD Head Office Publications.

Industry Canada (IC) (2002). *Achieving excellence: Investing in people, knowledge and opportunity*. Ottawa, Ontario

Kneese, A. V. & Sweeny, J. (1993). Economic theory of depletable resources: An introduction. In: A. V. Kneese & J Sweeny (Eds), *Handbook of Natural Resource and Energy Economics* (Vol. 11). Elsevier Science Publishers.

Kumar, V. & Magun, S. (1995). *The role of consortia in technology development*. Ottawa, Ontario: Industry Canada, Micro-economic Policy Analysis Branch.

NEB—National Energy Board (1999). Canadian energy: Supply and demand to 2025. Calgary, Alberta.

NPC—U.S. National Petroleum Council (1995). *Research development and demonstration needs of the oil and gas industry*. Washington, D.C.

NRCan—Natural Resources Canada (2000). *Energy in Canada*. Ottawa, Ontario.

NRCan (2002a). Innovation and skills in Canada's natural resources sector. Ottawa, Ontario.

NRCan (2002b). Round table on innovation and skills (NRCan's Webpage—http://www.nrcan.gc).

NRCan (2001). *Sustainable development strategy*. Ottawa, Ontario.

NRCan (1999). *Canada's emissions outlook: An update*. Ottawa, Ontario.

Persaud, A. J. & Kumar, U. (2001). An eclectic approach in energy forecasting: A case of natural resources Canada's (NRCan's) oil and gas outlook. *Energy Policy*, **29** (4) 303–313.

Persaud, A. J. & Norris, T. (2002). R&D and innovation expenditures in Canada's natural resources sector. A paper presented at the Statistics Canada Economic Conference (May 6–7). Ottawa, Ontario.

Porter, M. (1983). The technological dimension of competitive strategy. *Research on Technological Innovation: Management and Policy*, **1**, 1–33, JAI Press.

Research Infosource Inc (2002). *Canada's top 100 corporate R&D spenders*. (www.researchinfosource.com).

Roberts, E. B. (2001). Benchmarking global strategic management of technology, http://www.onlinejournal.net/iri/RTM/allpdfs/44_2_25.html.

Rosenberg, N. (1994). *Exploring the black box: Technology, economics and history*. Cambridge: Cambridge University Press.

Samuelson, P. A. & Nordhaus, W. (1989). *Economics.* McGraw Hill Company of Canada Limited.

Saklou, T. H. (2000). Technology transfer as a competitive advantage: Aligning companies competencies with host countries needs. A paper presented at the 16th World Petroleum Congress. Calgary, Alberta.

Schumpeter, J. (1934). *The theory of economic development.* Cambridge MA: Harvard University Press.

Statistics Canada (SC) (2001). *Industrial research and development* (With 2001 intentions and 2000 Preliminary Etimates). Catalogue no. 88–202-XIB. Ottawa, Ontario.

Steele, L. W. (1989). *Managing technology: The strategic view.* McGraw Hill.

Suncor Energy (April, 1998). Project Millennium: Taking Suncor to the twenty first century. A presentation to representatives from the International Energy Agency, and NRCan employees, Fort McMurray, Alberta.

SFT-Government of Canada (2002). *Speech from the throne, the Canada we want.* Ottawa, Canada.

SFT-Government of Canada (2001). *Speech from the throne (SFT) to open the first sessions of the 37th Parliament of Canada.* Ottawa, Canada.

U.S. DOE/EIA (2000). National energy modeling system, oil and gas supply module documentation. Washington, D.C.

WCED—World Commission on the Environment and Development (1987). *Our common future.* Oxford.

Williams, K. C. (2000). Technology evolution and commercial development at Imperial's Cold Lake production project. A paper presented at the 16th World Petroleum Congress. Calgary, Alberta.

Appendices

Table A1. Canada's oil and gas production and forecast.

	1997	2000	2010	2020
	Thousands of Barrels Per Day (mb/d)			
Conventional Light	1,040	994	965	885
Conventional Heavy	565	575	495	425
Oil Sands—Synthetic	290	360	770	1,030
Oil Sands—Bitumen	235	300	460	735
Frontier	10	135	250	250
Total	2,140	2,364	2,940	3,325
Natural Gas Production (Billions of Cubic Feet—Bcf)	5,505	5,870	7,050	7,850

Source: NRCan, 1999.

Table A2. Technologies and reasons for application.

Technology	Reasons for Technology
Exploration • Geophysics • Geology • Geochemistry • Information technology	The discovery rate is declining because the hydrocarbon traps remaining are complex, and the volume of hydrocarbons they contain will be generally smaller than in known reservoirs. R&D into new and improved techniques is necessary to mitigate this trend.
Reservoir Evaluation • Fluid/rock interactions • Structure assessment	The goal is to preserve access to identified deposits while developing and testing technologies designed to overcome specifically problems associated with reservoir structure and potential.
Reservoir Engineering • Enhanced recovery • Artificial Lift	Much of the oil in a reservoir is left in place when a field is abandoned. Advanced secondary recovery and enhanced oil and gas recovery make valuable contributions to the security of supply.
Drilling and Well Completion	Improvements in drilling technology will greatly reduce the cost of exploiting discoveries. Advances in a number of technology areas will be crucial for developing currently uneconomic reserves.
Offshore Field Development	R&D will reduce capital and operating costs. Significant technology progress could radically change production installations.
Field Operations	R&D is necessary to increase efficiency of operations. Cost increases are expected for activities related to safety or environmental impact mitigation.
Transport Systems	Development of marginal fields and the production of severe fluids will require the development of multi phase flow technology and the application of advanced materials. Work on natural gas conversion to liquid products for transport should be advanced.
Technology for the Arctic Regions	Large supplies will be obtainable from the Arctic regions with the development and deployment of appropriate technology.
Extra Heavy Oils, Natural Bitumens and Shale Oils	There are very large reserves of these sources of hydrocarbons. Appropriate R&D will make them accessible and competitive.

Source: IEA, (International Energy Agency), *Energy technologies for the 21st century*. Paris: OECD Head Office Publications, 1997.

Table A3. Key Upstream Technologies for Reducing GHG Emissions.

Seismic Drilling	Drilling Fluids, Drill-Stem Tests, Drilling Rigs	3D seismic, horizontal wells, new and more efficient motors
Well Servicing and Testing	Venting Activities, Service Rigs, Pumping Units, Wireless Units	implementation of leak detection program, metre regulator, propylene pipes, valves that reduce venting or flaring
Gas Production	Wells, Gathering Systems, Field Facilities	3D seismic, horizontal wells
Light Conventional Oil Production	Wells, Flow lines	3D seismic, horizontal wells
Heavy Oil Conventional Production	Wells, Flow Lines	3D seismic, horizontal wells
Crude Bitumen	Wells, Flow lines, Trucks,	SAGD
Synthetic Production	Upgrading and Processing Equipment	truck and shovel mining, double roll crushers, hydro transport pipelines, low temperature extraction; water recycling, to reduce energy use
Gas Processing	Gas Plants	
Product Transmission and Transportation	Natural Gas Systems	new welding technologies, new leak detection programs, low Nox engines, new fuel efficiency engines, new pipeline pipes
Waste Oil Reclaiming and Disposal	Oil field waste, Transporters, Land Treatment, Road oiling	small scale electrical turbines
Accident and equipment Failures	Pipeline ruptures, Well blowouts, spills, Gas Migration	valves and pipelines

Source: UOGWG, 1998.

The International Handbook on Innovation
Edited by Larisa V. Shavinina

The National German Innovation System: Its Development in Different Governmental and Territorial Structures

Hariolf Grupp[1,2], Icíar Domingue-Lacasa[1] and Monika Friedrich-Nishio[2]

[1] *Fraunhofer Institute for Systems and Innovation Research (ISI), Germany*
[2] *Institute of Economic Policy Research (IWW), University of Karlsruhe, Germany*

Abstract: This chapter provides several long-time series of innovation indicators bridging more than 100 years of German economic history. The ups and downs are put into perspectives and compared to important events in the global innovation system. The pertinence found for the national German innovation system seems to rest in educational, cultural and language traditions rather than in governance and territorial coverage, which changed many times. The chapter represents an analysis of the national level and two long-term studies on electrical engineering and chemistry. Thereby, the path dependence of technological paradigms within such a national innovation system becomes obvious.

Keywords: Innovation; Economic history; Germany; Science system; R&D; Patents; Path-dependence; Evolution.

It is always difficult to record the history of events that have not yet run their course and whose outstanding players are all still living . . . . Events appear different, once they are concluded; different again, while they are still developing. In both instances, the aims of the reporter also differ.

Gustav Struve (1849, p. 290).

Methodological Introduction

The general appreciation of innovation corresponds with a typically European method of thinking which is not found in all cultures. "The positive evaluation of new findings, the esteem for innovation, the idolization of inventors, as well as inventions and patents, are achievements of the modern world dominated by European-American influence, which, from a historical point of view, are relatively young" (Dohrn-van Rossum, 1999, p. 39). However, even in the Christian Occident, the present predominant emphasis on innovation results from the manifold historical changes of the past centuries. Initially, inventions and discoveries were not considered as an act of creation but only

represented the re-discovery of natural phenomena created by God. This change of consciousness—which took place prior to the period investigated by this chapter (1850–2000)—should be dealt with in order to better localize innovation-critical opinions in the present; however, this cannot be done here.

A practicable way to measure innovation could be the elaboration of definitions and measurement methods by starting from a historical point of view, with the objective of recording the enormous change characterising innovation activities. However, this chapter takes the opposite point of departure: starting from today's definitions, an investigation of the comprehensive statistical material including related indicators is carried out, followed by the attempt to trace and complete these back before the foundation of the German Empire. This means, the presently achieved level of theory and methodology serves as a point of departure for the following retrospective.

Consequently, this chapter tries to include a considerable number of quantitative variables, preferentially in the form of *time series*. Therefore, this analysis can be included in the field of *cliometrics*, the 'new' kind of economic history, which is based on quantitative methods including econometrics, aimed at reconstructing and interpreting the past (Bannock et al., 1998, p. 61). This method is regularly criticized since *indicators* cannot be *facts*; however, according to some points of view, narrative historiography cannot distinguish facts from interpretation either. According to these, no fundamental difference exists between the description of facts and the interpretation of these, since every description already represents a certain interpretation, which, moreover, depends on the definitions presently available to the describing person (Lorenz, 1997, p. 32).[1]

Even if there is no basic cognitive difference between the (widespread) narrative approach and the (less common) quantitative approach, the objective of the quantitative approach is the statement of a relationship between variables, based on many cases, and thus generalization (Fogel, 1994, p. 237 onwards). In contrast, the qualitative approach aims to compare case by case, i.e. this contribution has an analytical or holistic character (Lorenz, *loc cit.*, p. 238). It seems that most historians prefer an approach by case (qualitative approach), whereas most social scientists choose the variable approach, verifying hypotheses for a whole series of cases.

To be provided with variables for many cases on the aggregated level, conceptional ideas are needed which regulate inclusion and exclusion. Whereas, in general, brief experiences can be clearly delimited, the application of specific selection criteria is often difficult, especially in the case of long time series, due to the fact that the individual investigation of all relevant indications is absolutely impossible. Moreover, indicators have a *selective* rather than objective quality, so their undeniable status only applies to a specific disciplinary context: according to the present status of empirical economic research, more or less ideal indicators are available for theoretical constructs. The process of 'statistical adequation' means the 'tailor-made adaptation' of measurement concepts, the result of which is not entirely satisfactory in view of the theoretical constructs but which at least corresponds with the descriptive framework used as a basic structure for measurement.[2]

This is aggravated by the fact that the theoretical constructs of innovation research are not clearly defined. Up to the present, rival and unconciliatory innovation theories still exist in several disciplines (Grupp, 1998). In addition, linear models are widespread, presuming a sequential succession of innovation-oriented phases, the point of departure of which is an unpredictable serendipity in basic research or exogenous technical progress which falls like manna from heaven. Orthodox approaches have developed,

which try to subordinate or marginalize alternative approaches in order to find the 'truth' with the help of their own theory. From an empirical perspective such attempts must be considered with scepticism since modern epistemology actually tries to erase any efforts to find the real truth (Hoyningen-Huene, 1999). Therefore, the research on science and technology indicators should ideally begin from an heterogeneous level of theory and definitions in order to find historically solid indicators. Empirical standardization of theoretically heterogeneous constructs must take the individual contexts into consideration, however, it should lead to 'adequate' indicators; herewith, empirical statistical adaptation often remains incomplete, and, yet, only a minimum of discrepancy should be left (only refer to the discussion about real and ideal concepts, Machlup, 1960). In view of an economic-historical approach, definitions must be fixed as an orientation structure for both theoretical and empirical analyses; these definitions should stand iterative modification, they should apply on an inter-disciplinary level and be valid over time.

The standard German textbook for 'businessmen and students' by Roscher (1886), which was reprinted about 20 times by the second half of the 1880s, and which also shaped Schumpeter's theory of economic development (1911), distinguishes six different economic activities, listing *Invention* and *Discovery* first (sic!)—ahead of mining, agriculture, the processing industry, the distribution of goods, and the service sector last (excluding wholesale). Based on this, the result-oriented concept by Schumpeter (1942, p. 136 cont.) defines criteria according to which innovation represents all that yields a profit for the entrepreneur by being the first (the so-called *quasi-rent* or *innovation rent*). Quasi-rents of innovation tend to be neutralized by time due to the effects of concurrent processes. Innovations can be seen as new goods or services, new methods of production or transportation, new markets or new organizations.

In Germany, the term '*novelty*' was used as a definition of innovation for a long period of time. The word 'innovation' was unknown for a long-time: it only reached the German speaking regions after Schumpeter's emigration to the United States, where he published English texts; here, the English word 'innovation' was maintained as the germanized 'innovation' instead of being translated back to the term 'novelty' originally used by Schumpeter. Probably the definition of innovation was adopted around 1960. Consequently, it is evident that the definition of innovation as it is used nowadays cannot be considered as an anchor for the investigation period since 1850. Prior to the 1960s, innovation phenomena were described using other definitions: archives, libraries, research institutions as well as documents from management, personnel departments, or from production centers show terms which are different from those

---

[1] See also the contributions included in Müller & Rüsen (1997).

[2] Cf. Machlup (1960), Grohmann (1988) and Grupp (1998).

used according to present standards (like laboratory, try establishment, experimental factory).

According to today's view, the concept of a specific research process which leads to measurable innovation and which requires personnel and financial expenditure, is based on Bernal's (1939) farsighted and clear analyzing work. Bernal distinguished the role of public research expenditure from that of civil research and—as things stood—from that of the war industry. The first statistics on expenditure for 'industrial research' by British companies are found in the annexes of his works. As reported by Freeman (1992, p. 3), the definitions used by Bernal during his lectures at the London School of Economics were brought to international committees (by Freeman himself as well as by others), which, in the 1960s, worked for another standardization of definitions, which led to a first paper about the measurement of output of research and development (Freeman, 1969).

Consequently, the empirical framework underlying this chapter will be determined using the current definitions and concepts. These may have had other meanings in the past, however, this '*anachronism*' of long-living indicators' definitions must be accepted. Language is used by any historiography including the hermeneutic one; however, as language develops over time, definitions may arise, disappear, or change their meanings. Consequently and independently from the method applied, any historiography is anachronistic to a certain degree (Lorenz, 1997, p. 364). Former innovators were not masters of today's historical knowledge, and neither of information about present innovation processes—knowledge and information, which we have acquired from our observation post.

Because of this, the definitions found in the leading OECD manuals[3] from the 1990s will be used. Cliometrics are also more concerned with anomalies than with constantly ongoing, inconspicuous processes. If a structural breakage is found in time series, this could point to a statistical artifact arisen by the change of used definitions and conventions. Consequently, found structural breakages[4] must always be interpreted and categorized in a qualitative way. The problem of anachronism—if not avoided—can thus at least be moderated. Inter-temporal shifts of emphasis as an explanation of structural breakage are all the more permissible since a functional innovation model serves as an additional basis (cf. Grupp, 1998), working on the assumption that different innovation-oriented processes can be influenced by all types of research and development (R&D).

---

[3] According to these (OECD, 1992, 1993), technological innovations comprise new products and processes and significant technological changes of products and processes.
[4] Maddison (1982, pp. 2, 3) talks about 'system shocks'. Gerschenkron (1943) points to the pervasive institutional powers that may overcome external shocks for decades.

National Innovation System

This chapter is divided into a national (this section) and a sectorial level (see the next section; for the definitions of 'national innovation systems', see Grupp, 1998, p. 244). From a historical point of view, modifications of the territorial situation (population, etc.) may not be ignored. In this way, from an empirical point of view it is essential to consider, for example, the size of the Empire or of each federal territory. Not only is the German Democratic Republic considered here, but also Saarland, the Corridor, East Prussia, and others. Territorial changes which took place can be considered on the basis of today's statistical procedures, so that such data series, a priori, do not have to be absolutely consistent with a territory (also refer to Hoffmann, 1965, 2f). However, it must be pointed out that the omission of *smaller* districts (such as Alsace-Lorraine from 1871 to 1917) in most cases brings in its train less important errors of estimation than the *big variances* in the series of the whole territory of the German Reich (same paper, p. 3).

Public Expenditure for National Science and Technology

Traditionally, the development of science and technology is measured by the number of *scholars*. In this way, for example Gascoigne submitted a historical demography (1992) of the scientific community between 1450 and 1900, by listing the nationality and age of all the scientists. According to this study, Italy was the leading scientific country at the beginning of modern times in the late 15th century, representing about half of all the scientists in the world. This has remained almost unchanged during the entire Middle Ages before that century; then exponential growth with a doubling period of approximately 50 years took place.

Detailed and complete statistics are available about *scientific staff* in Germany since the foundation of the Empire, accessible via today's electronic means. However, generally accessible statistical material about *R&D personnel* in Germany has only been recorded since the 1960s (in the framework of the Federal Research Reports which have been published since 1965).

Another traditional access to the empirical definition of the importance of an innovation system is scientific expenditure (the sum of R&D funds and those for training, teaching, maintenance and diffusion of knowledge). Whereas the evaluation of expenditure for pure educational and R&D institutions is rather simple, this is more complicated in the case of institutions engaged in *both* research *and* teaching. Quota were adopted to cope with the individual fields of specialization as well as with the individual types of universities. However, it is questionable whether these reflect the right proportions between the percentage of research and that of teaching at all historical points in

time. In addition, not only is the historical consideration problematic but also the consideration of the
present time. Nevertheless, it is common statistical
practice in all OECD countries to work with such quota
(Hetmeier, 1990; Irvine et al., 1990).

Pfetsch (1982) undertook adding up scientific expenditure between 1850 and 1975, so that rough estimates
about the degree of R&D financing can be derived from
this; however, these data records only include *public*
expenditure, disregarding the private sector. Consequently, industrial innovation indicators must be
researched separately (see below).

In order to avoid dealing with the difficulty of
different currencies, the development of scientific
expenditure can be best evaluated by its percentage of
the *total expenditure* of public budgets. According to
this, scientific expenditure in the German regions prior
to the foundation of the Empire was approximately 1%
(see Fig. 1). Linked to the foundation of the Empire,
this percentage reached more than 2%, but dropped to
almost 1.5% between the 1880s and the first World
War. The Republic of Weimar attained a doubling of
scientific financing which, however, was lost again due
to the worldwide economic crisis. In West Germany,
the support of science was pushed dramatically to
reach a proportion of 6.5% of all public budgets by the
1970s (university expansion), in order to fall off to
approximately 5% by German reunification. Finally,
due to reunification, the level dropped even further.
These indications are based on the numbers of Empire
or Federal institutions and those of regions and states.

Besides the above-mentioned data records, and also
based on an analysis of the older part of these, Pfetsch
(1974) submitted an extensive analysis of German
scientific policy between 1850 and 1914. For example,
Pfetsch (same paper, p. 60 and p. 171) cannot confirm
the thesis that the state spends more money on science
and technology in times of increasing economic
wealth. In spite of some references to anti-cyclical
research policy, overall an *irregular economic attitude*
is shown by government *policy.*

Surprisingly, the post-World War II expenditures
start at around the same level as at the beginning of the
last century (after World War I) which is at the same
level as the endpoint of records in the 1940s war times,
and increases in a similar way after World War II as
after World War I. This points to quite *stable* and
persistent institutional structures underlying the financial totals.

The financial support of *Research* and *Development*
is typical for post-war Germany. Until 1945, the
financing share for R&D only played a subordinate role
in total scientific expenditure. Although the research
share[5] was between 20% to 30% during the first period
after the foundation of the Empire, it dropped to less

than 20% by the beginning of World War I (cf. Fig. 2).
In addition, it is important to know that a great deal of
scientific expenditure by the Empire was used for
defence tasks shortly after the foundation of the Reich.
During the Weimar Republic and the Third Empire, the
R&D share of the total of scientific expenditure
continued to fluctuate around 20% (industrial research
not included).

An increase in the R&D share of scientific expenditure was the case when research in certain areas was
admitted again in the young Federal Republic, after the
signing of the Treaty of Paris in 1955: at times it
reached 70% and has only declined due to the recent
reunification.

Prior to World War II, the relatively insignificant role
played by R&D within the scientific world is also
shown by the *distribution* of funds to the different
institutions. During the decades preceding and following the foundation of the Empire, the lion's share of
approximately 70% of the total public scientific
expenditure is accounted for by all types of universities. Empire agencies and other institutions were
established over time, so that the universities' share of
scientific expenditure decreased to about 35% at the
beginning of the 20th century. Now, between 20% to
30% goes to these institutions charged with varied
tasks, and almost 10% is accounted for by pure R&D
institutions.

During the entire period before and between the two
World Wars, a small but significant and strongly
varying amount of public expenditure went to missions
other than institutional support. The non-institutional
support is described as *'project-specific scientific
expenditure'* by Pfetsch (1982, p. 113); this could be
misunderstood, since part of it consists of public grants
and support given to a wide range of projects and not
only R&D projects. They range from 'Scientific efforts
for opening up Central Africa', the publication of
archives, international contributions to the surveying of
earth, measures to combat typhoid fever or infantile
mortality, and to financial support for congresses.

Until World War I, more attention is generally given
to scientific support on an industrially relevant level, as
well as to scientific application (Pfetsch, 1974).
However, it would be incorrect to conclude that a major
part of these funds were granted to private companies,
as can be shown for the period following World War II.
Industry was more interested in a proportional increase
in public support of production-relevant branches of
science (consequently, in the creation of external
effects upon science-based industries, same paper,
p. 107) than in the support of their own R&D.

Of course scientific expenditure was borne exclusively by the German states until the foundation of the
Empire; afterwards, the central power moderately
supported the total public science budget by 20%. Only
during the Republic of Weimar did this share grow
considerably (see Fig. 3). After the occupation of

---

[5] More precisely, 'research share' means the 'R&D share' of
the total expenditure for science and technology.

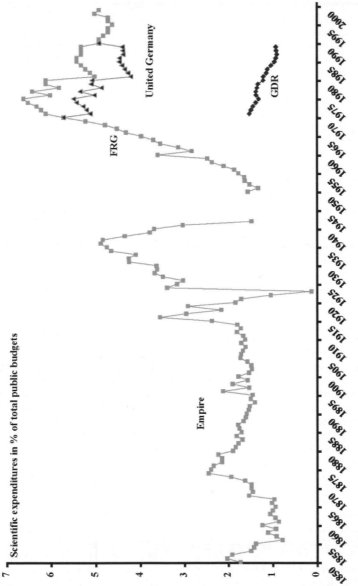

Figure 1. Development of scientific expenditure in proportion to the total expenditure of public budgets.

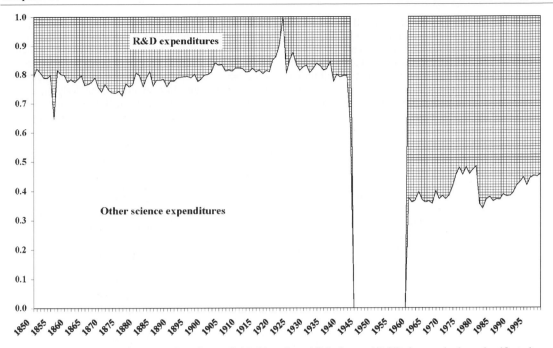

Figure 2. Government expenditure for science, divided into the publicly financed R&D share and other scientific tasks.

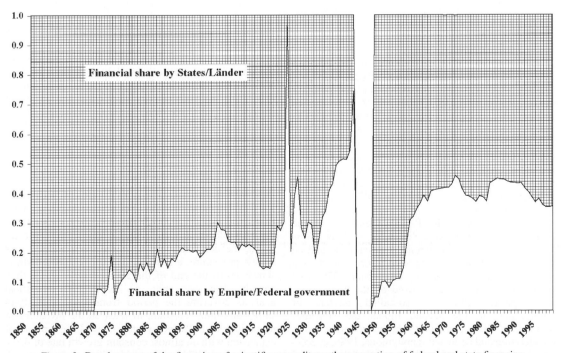

Figure 3. Development of the financing of scientific expenditure: the proportion of federal and state financing.

Germany at the end of World War II the Federal Government could not play its prior role again, all the more so since several research areas were prohibited (research in the fields of armament, nuclear science, chemistry and aviation). However, federal institutions systematically increased their influence on science until the level during the Third Reich was reached again. Since the recent reunification, federal administration has been slowly but surely withdrawing from scientific support.

Consequently, it must be noted that, following critical upheavals such as the foundation of the Empire, the First and the Second World Wars, the Federal states have always begun to take over important tasks of scientific support and are later relieved by the *central power*. The events following the foundation of the Empire, which were called '*getting empired*' by Pfetsch (1974, p. 105), occurred in a hesitant way on complex scientific-political levels. The results were manifold forms of cooperation between private, mixed and individual government institutions and Empire authorities. The same can be observed since 1945 under completely different political circumstances: the Federal government only slowly became a dominant and central supporter and organizer of the scientific system, its scientific expenditure diminishing both on an absolute and on a relative level only after the reunification. Again this points to *persistent* basic structures in the national innovation system. The only historical exception is the reunification of 1990: whereas the Federal government played a dominant role as a central supporter and organizer during the reunification of the two scientific systems, its scientific expenditure diminished afterwards both on an absolute and on a relative level.

Regarding the R&D expenditure of the *German Democratic Republic* (cf. Fig. 4), note that the individual statistics were centrally maintained and are comprehensive. However, the conditions which were applied do not fully comply with those used by OECD countries and often show exaggerated values. Following reunifcation the relevant statistics were revised and adapted to Western standards; however, the conversion problem of the East German bank's Mark persists. Due to the non-convertibility of this currency the reliable purchasing-power parity values of OECD countries cannot be applied.

Figure 4 compares the R&D expenditure of the German Democratic Republic with that of West Germany. In order to be certain, a pessimistic and an optimistic variation can be applied in order to show a range of uncertainty due to conversion. The first possibility of conversion is based on the purchasing-power parity (PPP) of so-called baskets of commodities, in the second model the subsidies included in GDR commodity prices are taken into consideration and deducted (anonymous, 1986, p. 259–268). It is shown in both estimations that the national

R&D expenditure of the German Democratic Republic could not equal the West German level (per head of the population) but the general upward trends somehow resemble each other. This may come as a surprise to those who point to the inefficiencies of the communist part of Germany, but, again, the underlying institutional structures remained basically the same as before the war requiring similar amounts of public support.

Development of Scientific Activities

It is impossible to achieve an insight into the development of non-codified and thus 'tacit' experienced knowledge of the scientific staff. For this reason the historical development of an innovation system is often shown by the personnel statistics, or by statistics showing monetary expenditure. However, only expenditure is measured by this method, instead of the fruit of scientific activities. Efficiency measurements are particularly impossible. Consequently, modern innovation statistics make regular use of yield measures; regarding scientific work, statistics of publications are a typical output indicator. Analyses about the degree of *publication activities* have been maintained for centuries; however, it must be noted that the publication media chosen by scientists may differ from one faculty to another, as well as over time (Wagner-Döbler & Berg, 1996, p. 289). Only during the 19th century did scientific magazines achieve the same degree of significance as books, the dominating publication media until then.

Regarding publication activities in selected areas, only a few but informative historical time series for selected areas are available (cf. below). However, in this section we consider totals first. Due to the known difficulties of aggregation, only limited sources of publication activities are available on this level of analysis. An analysis of the Catalogue of Scientific Papers for the 19th century shows that the output of scientific papers had been growing constantly since 1800; and it accelerated tremendously from 1884 onwards (anonymous, 1925, p. 129 onwards). This analysis is not limited to Germany but refers to worldwide publication output.

From 1900 onwards the availability of data improved worldwide. The growth rates of periodicals were evaluated in 'Ulrichs' Periodical Database' (CD-ROM version) by Mabe & Amin (2001); if one takes the example of academic magazines from this catalogue, the contributions of which are submitted to a review procedure, a remarkable exponential growth is shown up to the end of the 20th century, which then slightly decreases (Fig. 5). A detailed statistical analysis covering the period from 1900 until 1944 shows an almost constant increase in the inventory of magazines of 3.2% per year, followed by a phase of expansion with an almost constant growth rate of 4.8% per year until 1974, ending with a lower rate of growth

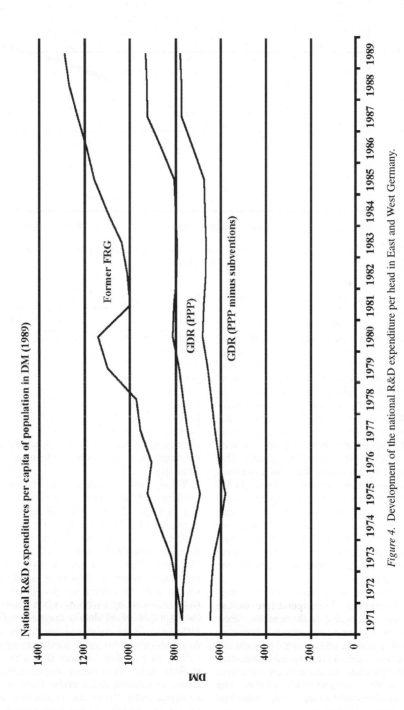

Figure 4. Development of the national R&D expenditure per head in East and West Germany.

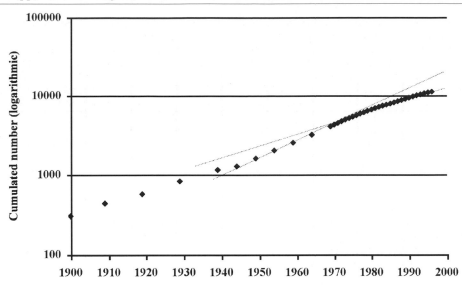

Figure 5. Accumulated number of worldwide inventory of referred scientific magazines (shown as a semi-logarithmic figure) from 1900–2000.

of 3.7% per year, similar to the first half of the century ('century standard').

This development seems to be linked with historical structural breakages on a worldwide level. Following its academization during the 19th century, *science* was characterized by constant expansion until the Second World War. This can be called the *normal development*. After the Second World War, the growth rate of scientific output increased and reached almost 5% for almost three decades (the so-called boom phase), due to the economic miracle, the armament race, reinforced industrial research and development activities, as well as an expansion of scientific activities in general. This phase was followed by a normal phase of expansion similar to pre-war conditions; it can be linked to the ending of university expansion, the consequences of a severe recession due to the oil crisis, as well as with a general decrease of economic growth rates ('limits to growth'). Using completely different time series, Maddison (1982, p. 92) found phases showing a high degree of temporary similarity (1930–1950, 1950–1973, following).

Analyzing the situation in *Germany*, the 'Bibliography of German Periodicals' Literature' represents an alternative regarding historical data. This bibliography covers the German-speaking periodicals' literature from 1896 until 1964, and the international periodicals' literature from 1965 onwards. However, no information about the author's nationality or the institute's location is found in this source; probably most of the older records come from German authors, but quota and changes of these are also unknown.

The Science Citation Index (SCI), which has been available as an online version as early as 1964 (see

below), has a printed version listing the publications from 1945 until 1974. Although no indications are found regarding the authors' nationalities or the institutes' locations, the listing of periodicals is classified by the countries editing and printing them. Since Mabe & Amin (2000) have proved a highly positive correlation between the number of periodicals and that of the journals' articles, the number of German-speaking articles listed in this data record can be counted for the period since 1945. The repeatedly written announcement by the SCI that records would be completed back to 1900 was withdrawn,[6] so there is no hope for the soon publication of a *century's inventory*.

Taking into account that the total volume of publications has grown enormously, it is surprising to see that the share of German periodicals has been constant since 1945, i.e. that it has augmented in line with the worldwide volume. The average proportion of German periodicals, which is 8.9% of the total SCI inventory, shows only minimal fluctuation. Since 1974 an equally constant proportion of German authors is shown in the SCI online version so that it is obvious to link these two data records, all the more so since they overlap each other over a long period (from 1975 to 1984) so that the corresponding factor of extrapolation from periodicals to publications can be determined.

The SCI online version shows the *nationality*, a corresponding field being encoded where a German author or an institute located in Germany are concerned (independently from the medium's language). The strongest growth of the (extrapolated) publication

---

[6] Personal communication Garfield, 14 October 2000.

numbers is stated from the middle of the 1960s, and during the 1970s (Fig. 6). This matches perfectly with the observation of a worldwide expansion of the scientific system, even though the German scientific world showed a *delay* of almost 20 years resulting from the special situation of reconstruction as well as from the Allied Forces' restrictions regarding certain research areas.

At the end of the 1980s the growth rate decreases not only in Germany but on a worldwide level; after 1989 the total number of German articles shows a dramatic decrease. It must be noted that the statistics cover both West and East Germany, and that this decrease in publication activities could principally represent the decay or dissolution of the East German scientific system. The publication level of 1987 was only reached again during the publication year 1993, which was characterized by strong growth in an anti-cyclical course to the worldwide slowdown. A comparison of research activities both on a disciplinary and on a qualitative level is suggested below. Figure 6 also shows a simulation of the course of the three constant growth rates, which point out the exponential growth.

From 1974 to 1990, SCI publications from West and East Germany can be compared electronically.[7] This period is characterized by moderate growth of publication activities ('century standard'). In the 1970s, the share of East German publications was approximately 16% to 17%. However, if one compares East and West Germany, both the proportional shares of population and the proportion of R&D staff is almost 30%, so that scientific publications from the German Democratic Republic are less represented in the U.S.-based database. The proportion of East German publications had constantly diminished to reach 13% by the end of the 1980s; and there is no answer to the question as to whether the *representation* in the database was even worse or if the output efficiency of East German research activities continued declining until the end of this state.

Measured by its publication output, the *profile* of GDR research resembles that of the former Federal Republic. In proportion to worldwide average shares, researchers of both parts of Germany published much more than a pro rata share in the research areas of energy and nuclear technology, chemistry, solid-state physics and microbiology. A weaker level than the worldwide average was shown in information science, engineering, environmental research, the area of public health, as well as in other biomedical subjects. According to our estimation, this *structural similarity* could be the reason for such a strong diminution of

publication activities on an all-German level following reunification. Integration did not concern differently specialized East and West research systems but research systems with the same principal orientation, which led to the deplorable 'reallocation and consolidation' in East Germany. Independently from a political evaluation of the organization of GDR research institutes this structural similarity must be pointed out; obviously 40 years of division were not sufficient for a differentiated development of the basic specialization patterns of research in both parts of German. To a great extent, and in the sense of *path dependency,* research is still based on the (common) preferences which existed prior to the division. This unique historical situation could be understood as an unintended experiment: basic patterns of scientific specialization change only slowly, even in times of great political system change (Hinze & Grupp, 1995, p. 65).[8]

Another method of comparing the two *scientific systems* is to observe the frequency of citations. More frequently cited scientific studies are considered to be more significant (in some way), for example because they include either important methods to which many successive authors return, or especially important results (or errors which are repudiated later). GDR literature shows a lower rate of citations per scientific publication than West German literature; there are two possible reasons for this: on the one hand, the scientific value of GDR publications could have been minor; on the other hand, the periodicals in which these articles appeared were received to a lesser degree on an international level, in particular in the English-speaking realm. This could also be due to the lower circulation figure of the corresponding periodicals (partly of Soviet origin).

In the bibliometric statistics these two effects can be viewed separately, by referring the rate of citations either to the worldwide average or to the average of the individually selected periodicals (Grupp et al., 2001). The following analytical results emerge: GDR research publications show an almost general *lack* of *worldwide communication* of results (particularly in the Anglo-Saxon linguistic area). This explains the low rate of citations. However, a correlation between the frequency of citations and the publication organs (which are typically less read in the Anglo-Saxon area) selected by GDR authors shows that a favourable rate of citations is found in comparison with articles from countries other than the GDR which are also published in these periodicals. Consequently, GDR publications are considered to an above-average degree once the citation quota due to the lack of international spread is mathematically corrected. The *highest regard* (quota of

---

[7] Due to the delay in appearance of scientific publications following submission, no quantitative cutback in literature production by the researchers of German Democratic Republic institutes can be perceived until the end of 1990 (Weingart et al., 1991, p. 4).

---

[8] Recently, surprisingly similar results were also found using other methods.

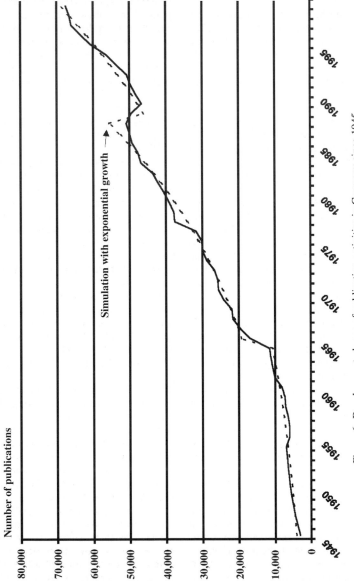

Figure 6. Development phases of publication activities in Germany since 1945.

citations) is found in the areas of neuro science, internal medicine, as well as dietetics and agriculture. Following the correction of the quota of citations, the results of GDR environmental research were the least noticed by the professional world.

A division of science into 27 sub-sections (Hinze & Grupp, *loc. cit.*) leads to a surprising correlation between the degree of internationalization and that of attention received: it was found that those research areas which are *not* internationally spread are significantly more noticed by the professional world than the publications with a higher degree of internationalization. No such correlation can be found for West German research. It is possible that hierarchy and cadre selection mechanisms play a role regarding access to Western periodicals.

Industrial Research and Development in Germany

Since the foundation of the Empire, economic growth of industrial countries, in particular in Europe, has increasingly been based on the innovation energy of the *knowledge-based industry*. 'This is undeniably true for the impulses of growth immediately released by these industries, starting with carbon chemistry and electrical technology' (Wengenroth, 1997). There is hardly a clearer and more distinct way to describe the conducive effects of industrial research on the culture and efficiency of innovation.

It is still difficult to prove the companies' increasing R&D expenditure for such an undeniable success. In particular, no complete data records are available about monetary expenditure or research personnel prior to the end of the Second World War, i.e. the data record established by Pfetsch (1982) regarding public scientific expenditure has no counterpart for industry. Today's statistics about R&D expenditure and personnel of the Federal Republic systematically start from the year 1962; certain presumptions allow the reconstruction of the corresponding indicators starting from 1948/1949 (Fig. 7). According to this, industry has continuously increased its R&D budgets to a higher degree than government, the share of which is presently approximately 40%.

The reconstruction of the corresponding indicators prior to World War II is only possible if one starts from the individual companies or branches. This was tried for in the areas of chemistry and electrical technology (cf. next section). The investigation of the history of R&D in leading companies of two selected industrial branches shows that the execution of case studies in order to represent time series is principally possible and, moreover, makes sense. During the whole investigation period many large and leading companies in Germany maintained archives which can be considered as complete in spite of difficult conditions at times in German history.

Presently the representativity of company-related R&D time series cannot be plausibly proved for the development of a whole branch and that of all branches for the whole national economy. A more detailed

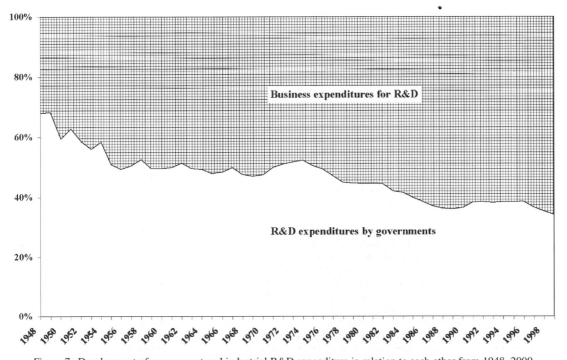

Figure 7. Development of government and industrial R&D expenditure in relation to each other from 1948–2000.

analysis should include other important branches, in particular, the metal industry. However, the results found so far show that the existence of different 'cultures' of R&D is possible in the different branches, which influences not only the forms of organization but also researchers' liberties and practical work.

Development of Invention Activity in Germany

The observation of the development of innovation activity is important in itself in order to establish R&D results, mostly on a technological or application level. Adopted methods are statistics on *patent applications* (a figure representing successful innovation activity seen from the innovators' or applicants' subjective perspective) on the one hand, and, on the other hand, statistics on the number of *granted patents* (as a figure representing successful innovation activity, seen from the objective perspective of patent examiners). Statistics on patents make even more sense if one takes into consideration that only fragments of industrial R&D expenditure are known prior to the Second World War. Instead of inputs, industrial R&D activities can be measured by their patent outputs, and this even more precisely from a technological perspective than by monetary indicators. This also explains our interest in both patent grants and patent applications: if no patent is granted after verification of the novelty, the inventive step and its commercial usefulness, for example due to a lack of novelty, the applying company had nevertheless invested R&D efforts—even if these led to an objectively already known result. Consequently, the 'subjective' perspective of a successful invention is closely linked to the R&D performance, which was in fact realized. Statistics on patent applications as a *proxy variable* for R&D expenditure may ignore whether the object of the invention was a world novelty or not. R&D expenditure also includes the costs of unsuccessful or belated inventions in comparison with competitors (imitations).

The period to be considered is fully included in the statistics of patents. In some German regions, patents were applied for as early as 1820, starting from the South, due to the influence of the Napoleonic legislation. From July 1st, 1877, a *patent act* for the German Empire standardized procedures. Thus, the creation of patent acts in Germany follows the scientific-technological innovation push of the 19th century, at the end of which Germany was one of the leading industrial nations. In about the middle of the century the local, largely secluded markets were dissolved, and the German economy was integrated into the quickly expanding world economy (North, 2000, p. 13; Ziegler, 2000, p. 198).

Since 1879, *patent statistics* have been available using *machine readable* methods. On the one hand, electronic data records since 1970 are more informative than those of former periods, leading to a largely increased importance and use of these patent data records by modern studies in science and technology. On the other hand, if one makes the effort to bring the individually valid patent classifications together with technical know-how, and to chain together different patent data records for the appropriate historical sequences, assembled patent statistics can be established for the whole period. Moreover, regarding the assignment of priority years prior to 1969, these must be established according to the reference system of the individual patent authority.

Considering global patent activities in Germany (Fig. 8), an obvious difference is found between the dynamics of chronological inventions and that of scientific activities (publication statistics). The strongest growth on a low level takes place from 1820 to the foundation of the German Empire; the total growth rate for German regions is shown to be constant with the setback due to the war of 1870/1871. Following the introduction of the countrywide German patent act, the number of applications and grants rose rapidly within a few years, and continued growing at a constant rate, which, although on a considerably higher level, is lower compared with that of the period preceding 1870. This growth, which had lasted for almost one century, was abruptly stopped by the First World War, the annual patent production being halved. From approximately 1920 to 2000, an eventful development pattern nevertheless shows nearly zero growth. During almost a century, the number of annual patent applications is approximately 50,000 to 60,000. Nevertheless, German patent productivity per person reaches one of the highest degrees in comparison with the United States, Japan, and the European Union.

Diverging from this rough rule, growth is observed during the Weimar Republic phase until the beginning of the Third Reich, followed by an immense setback during the Second World War, which is distinctly more serious than that of the First World War, and a return to the secular quota by approximately 1960. Another boom follows until 1975 when a deep recession takes place, which is only overcome in the mid-1990s.

No investigation has yet discovered whether these growth cycles have *only* economic causes. The economic boom after the foundation of the Empire is well-known (Ziegler, 2000, p. 201); the same is true for the serious recession following the oil crisis in 1973 straight after the economic miracle. The question remains as to whether the reduction of innovation activities at the beginning of the Third Reich was only due to economic reasons or to a modified practice of patenting (for example by stronger observance of secrecy due to the early war economy, by expulsion or migration of Jewish scientists). Further, the question is asked as to why the growing R&D budgets granted after the Second World War did not lead to an increase in patent activities. Obviously this decrease of patent efficiency is due to an economic calculation, which is not exclusively driven by R&D inputs.

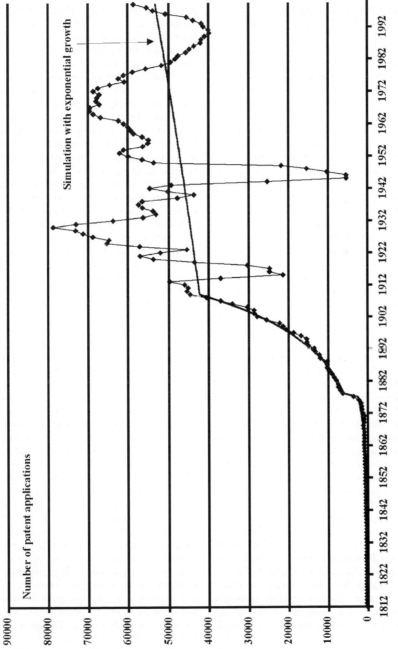

Figure 8. Development of patent applications from 1812 to the present.

Up to now, global indications have been made about patent documents of national and international actors. By *international actor* it is understood that either the inventors' residence or the applying company is located abroad. From 1881 to 1913, the share of foreign patent grants was extremely high showing an average of 35%: until 1933, Germany's reputation as the leading scientific country attracted many young scientifists from abroad. Especially Americans who came to the German Empire in order to benefit from practice-oriented education for their degree, and possibly even to experience some years of active industrial research (Erker, 1990; Smith, 1990). After the First World War and the efforts to achieve self-sufficiency in the 1930s, the share of foreigners was reduced to almost 10% but remained a significant figure in spite of all war speculations. Since the reconstruction of the German patent administration following the Second World War, the share of foreign patent grants has constantly increased reaching more than 60% at this so-called globalization time.

Although most patent applications originate from industry, universities and other public research institutions are increasingly also involved in patent production. This can be linked to their stronger orientation towards applied research, as well as to their acceptance of outside resources from private industry. Prior to the reunification in 1990, universities attained a share of 3% of all West German patent applications, which grew to more than 4% after reunification. Similar contributions are shown by all other public research institutions and companies, above all the Helmholtz Association (big national labs) and the Fraunhofer Institutes. In view of their limited human resources the patent productivity especially of the Fraunhofer Society must be more highly evaluated than that of universities (Schmoch et al., 2000).

Compared with Western conditions, certain deviations in the patent law conditions of the *former GDR* were ruled by the socialist spirit of ownership. Consequently, the national patent applications at the former GDR Authority of Invention and Patent Administration (AfEP) can hardly be compared with those submitted in the West (Hinze & Grupp, 1995, p. 42 onwards). Therefore, another method was chosen for this analysis, which is now based on GDR patent activities in the West-European foreign countries. With the help of this method all the particularities related to patent law specifications are circumvented, enabling comparison with Western countries. GDR inventors were mostly interested in the economic market of the former Federal Republic, so that the foreign applications submitted for this target market can be referred to (independently of whether the application was submitted to the German Patent and Trademark Office, to the European Patent Office, or to the International Patent Authority WIPO designating the Federal Republic of Germany).

The basic framework conditions for GDR activities in view of industrial property rights are fixed in the *patent law* of 6 September 1950 (Albrecht et al., 1991, p. 4). Nevertheless, GDR patent activities according to Western legislation are hard to ascertain during the first years. This is linked to the various forms of recognition of the GDR as an autonomous state by different nations. Some GDR inventors operated from Federal Republic addresses. In spite of these imponderabilities, an increase in patent activities by GDR inventors is seen until approximately 1983/1984. Then the figures are characterized by stagnation, and since 1987 they have been decreasing. The same tendency is shown on a higher level by national patent applications in the German Democratic Republic, so that the *drop* of innovation activities prior to reunification is undoubtedly proved (Hinze & Grupp, same paper, p. 47). The causal and significant explanation of this development is the fact that since 1981 the share of R&D personnel in the economic sector of the GDR had continuously declined, and financial resources for R&D in the economic sector were also reduced.

A comparison of the *specialization* of GDR patent portfolios with those of West Germany is very interesting. According to a division of the whole GDR technology area into 28 fields, particular strength is found in the fields of paper and print, textiles, machine-tool manufacture, handling, optical instruments, and metrology. Distinct weakness is shown in the fields of chemistry, electrical technology and electronics, information technology, as well as traffic and transportation. This specialization profile was constant over time. In particular, the eastern regions' patent profile of the 1990s (including East Berlin) corresponds largely with that of the GDR of the 1980s (Schmoch & Saß, 2000). In addition, there is an amazing correlation with that of West Germany. In spite of completely different economic conditions (Stolper et al., 1964), large fields of technology show a *correspondence* between East and West Germany until reunification (Grupp & Schmoch, 1992, p. 118 onwards). This was also found for the area of basic research (publication statistics) and explained by path dependencies and *persistent* structures in both parts of Germany despite their different political regimes (see above). Since reunification, the new federal regions have expanded their top level technology (semi-conductors, biotechnology, surface technology) starting from a low level.

Consequently, the question is whether a similar influence of scientific development on technology can be found in both parts of Germany. An evaluation of the *scientific dependancy of technology* is usually based on the fact that patent specifications include replies to former inventions in the form of citations to patent documents. If an invention is directly based on *science*, which was published but not patented, the patent examiners annotate references to the corresponding scientific *literature*. It could be shown that the frequency of such scientific hints included in patent specifications is a valid measure of the dependency on science inherent in a field of technology (Grupp & Schmoch, *loc. cit.*).

Regarding the degree to which technology was based on science, the German Democratic Republic remained constantly behind the *world average*. Although the same is true for West Germany, the gap was distinctly smaller. During the 1980s, both West and East Germany significantly increased their orientation to science-based technology—the GDR on a lower level, West Germany showing a reduced growth rate. Comparison between the two countries is interesting in that the East—West distance was reduced in this concern.

When Germany was reunified in 1990, two almost identically *specialized* technology systems came together. It was not possible to integrate the strength of one side and the weakness of the other one; instead, the fields characterized by strength were the same on both sides and the weaker fields were equally neglected.

The limited use of science by GDR technology is clear in international comparison; the same is true for the Federal Republic, however, on a different level. In view of the extended scientific activities, which were most significant in proportion to the size of the GDR, it is surprising to find that technological development did not benefit there from the use of science and therefore, remained on a relatively science-poor level.

Sectorial Innovation Systems: Electrical Technology and Chemistry

So far, the *national* German innovation activities were analyzed in the varying territories. In the framework of this chapter, it is not possible to consider all the university and industrial sectors individually. Therefore, sectorial analysis of the two areas *chemistry* and *electrical technology* will be presented as representative studies. In literature, chemistry and electrical technology are considered perfect examples of the science-based industries, which came into existence in the second half of the 19th century. These industries are characterized by a rapid transfer of research results to production. Due to the intense exchange activities on both sides one can also talk of 'industry-based science' (König, 1995, p. 283), since science probably benefits more from industry than inversely in certain phases (e.g. in the case of the newly emerging academic electrical engineering).

Another motive for the choice of these two specific sectors is the power of innovation in German industry *today*, which, roughly speaking, is considered as contrary: chemistry with its brilliant innovation and export performance compares with international standards, and electrical technology (or its sub-sumption into information technology) is considered as a weak point in the German economy. According to our thesis,

an exemplary analysis of these two sectors within the national German innovation system should reveal essential facts to explain their different characters at the *end* of the 20th century.

Sectorial Expenditure for Science

Statistics on science expenditure until 1945 include a classification by *sectors*. However, only science and engineering are identified in totals, so that these data records cannot be used for the study of electrical technology and chemistry. The establishment of a data base with sufficient capacity from an institutional level (ministries, universities, etc.) also seems impossible. This has remained unchanged since the Second World War. However, personnel figures from universities are available for selected years in both chemistry and electrical technology.

Development of Scientific Activities

At first, the publication history of the two selected areas, electrical technology and chemistry, will be summarized. Another problem remains, the tracking of publication activities in the fields of electrical technology and chemistry prior to 1974, the period for which no online data base is available. To resolve this problem, the corresponding monographs were submitted to manual analyses, which made the period from approximately 1924 onwards accessible. Due to the existence of backup periods outlasting the year 1974, the manually obtained figures can be extrapolated to the database level. First analyses support the earlier hypothesis: during the 1960s and 1970s, the growth of publication activities of chemistry was higher than in later periods, so that a synchronism of scientific publication activities and the economic success of mass chemistry (base) could be confirmed.

An extension of the historical period shows rapid growth of publication activities in the field of electrical technology since the 1950s, reaching its peak from the middle to the end of the 1960s, then again in the mid-1970s. This development is followed by a decrease of publications to reach a relative low level in 1981, then followed by another period of growth. Consequently, the present study of data records is substantially supported and completed by manual tracking up to the Weimar Republic. The manual method probably has an even higher explanatory power, since the continuous completion of inventory data records could lead to artefacts, which cannot be recognized and erased during analysis. Although the SCI strives for transparency with the selection procedures of periodicals entering the database, quantitative conclusions for individual subjects are not always possible (Testa, 1997).

Publication activities in the fields of electrical technology and chemistry show interesting subject-specific deviations from the average trends of

German publications

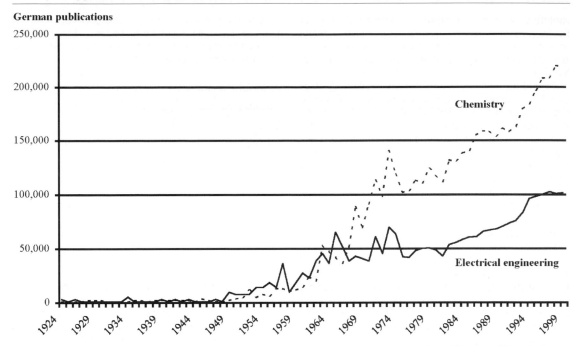

Figure 9. Development of publications in the fields of electrical technology and chemistry, from 1924 to 1999 (early figures extrapolated).

publication activities in all scientific disciplines in Germany. In contrast to the upward development shown by electrical technology prior to, and following, reunification, which takes an anti-cyclical course compared with global publication activities, the growth of publication activities in chemistry is reduced synchronously with the publication activities in all scientific fields in the middle of the 1970s. Presumably the anti-cyclical development of publication activities in the area of chemistry can be linked to the division of basic material chemistry into specialized and fine chemistry which took place in the 1970s, followed by biotechnology. The anti-cyclic boom of publication activities in the area of electrical technology could be linked to the appearance of present information and communication technology, and could be interpreted as its scientific predecessor. First analyses of the development of R&D expenditure in chemistry and electrical technology industries support this thesis.

Industrial R&D Expenditure and R&D Personnel

In the scope of investigations about industrial R&D expenditure, special attention was given to the two industrial branches selected in this analysis. In the fields of chemistry and electrical technology, a small sample of large industrial companies could be personally interviewed regarding the existence of corresponding data records. Sometimes the relevant archives could be accessed. The corresponding indus-

trial federations were involved. Interesting series of figures are available to the federations, however, no figures were found regarding R&D expenditure or R&D personnel. In contrast to this, time series can be found about the status of memberships, about lectures and others.

Comprehensive archive material concerning the number of chemists and physicists is also available from the successor organizations of IG Farben such as BASF. These indications correlate strikingly with the patents held by these companies; the hypothesis according to which a lack of input figures can be substituted by patent statistics is thus supported. Besides this, employment figures of other technical professions can be constructed. A comparison between the total number of chemists employed by BASF, Hoechst and Bayer, and the chemists employed by universities and technological universities (cf. Pfetsch, 1974, p. 158), shows that, since approximately 1880, the contribution of industry can no longer be ignored. Prior to the foundation of the Empire, only a single-digit percentage of all chemists were employed by industry; equality was already reached in 1885, and in 1890 the number of industrial personnel exceeded that of the university staff. Prior to the beginning of the First World War, the three companies employed already three times as many chemists as German universities.

Seen from a quantitative point of view, it is found that even prior to the First World War industrial R&D

efforts parallel to public engagement could *not* be *neglected* either.

The most important German companies in the electrical technology branch were recorded on the basis of the list of patents granted in 1928, which was established by the German Empire Patent Office (in total 1,121 patents, classified by companies). According to this record, the contribution to patents in the area of electrical technology was 20% by the leading company Siemens-Schuckert-Werke, 15% by AEG, 9% by Siemens und Halske, and 5% by the subsidiary of the latter, Telefunken. However, the *large scale* to which electrical technology was embodied in the German economy during this period is shown by the fact that the ten most important patent assignees produced only little more than 60% of all patents; almost 40% is accounted for by other companies. Such a limited degree of concentration means that it would be almost hopeless to register complete R&D input figures by individually interviewing all companies of this branch. Since only 80% of all patents are attained by the total of the 50 most 'patent-active' companies so that the number of 'patent-active' companies, which were also involved in R&D, could have exceeded 100 according to a rough estimate.[9]

The number of scientists employed in research institutes by the above-mentioned large companies can also be ascertained; however, no indications about monetary expenditure for R&D are found in the business reports. Even the annuals of the research institutes were not informative. In the area of electrical technology it seems *almost impossible* to take an inventory of relevant R&D input indicators, covering both the broad industry structure and the corresponding period of time. However, as in the case of chemistry, a study based on a selection of companies could be carried out to find a possible correlation between personnel data and patent data. In any case, statistics on patents will serve as a significant method of investigation.

Development of Sectorial Invention Activities

Due to very detailed patent classifications, the inventory of patents in the area of electrical technology and chemistry can be recorded. Although partly contestable delimitations are necessary for this investigation, these can be documented; moreover, a study based on different delimitations is possible.

It is shown that high-voltage technology (electricity, electrical energy) and low-voltage technology (audio-visual technology and telecommunications or communication engineering) as parts of electrical technology can be divided for the whole time. On an

analytical level, a strong growth of high-voltage engineering is shown since approximately 1900 (the more modern subject in times of electrification), as well as its stagnation after 1950, whereas low-voltage engineering shows moderate growth until 1944, and becomes stronger after 1950 (Fig. 10).

In total, the very different development dynamics are reflected in global statistics by a 4% share of all German patents held by electrical technology until approximately the turn of the century; then the share shows an almost exponential growth to more than 20% during the Second World War (Fig. 11). In spite of an increased growth of low-voltage engineering, the sum of patent shares is reduced from almost 20% (1958) to approximately 13% (1998); this is in accordance with the internationally common image of a lack of specialization in the area of German information technology.

Already at the point of departure of the observation period, *chemistry* had achieved a patent share of 10% which remained quasi unchanged until 1945 (except for short-term cyclical fluctuations in a few years). During the period of reconstruction after World War II, the number of patents in the area of chemistry increased reaching more than 20%, but was hit by a severe crisis following 1967. Since approximately 1970—the proportion had dropped to less than 10%—the share of the total number of patents held showed moderate growth of approximately 12%. These high shares of national patent inventory reveal that the fields of electrical engineering and chemistry were and remain *leading industrial sectors*. They show above-average expansion, and their global economic importance is characterized by continuous growth (Ziegler, 2000, p. 240).

Analogous with electrical technology, chemistry can also be divided into patent-statistical classifications (for example organic chemistry, plastics, pharmaceutical substances, biochemistry, detergents and agrarian chemistry). Innovation activities in these six fields show extremely different degrees of development. Both of the first two fields show a continuous increase in the total number of chemistry patents (in spite of cyclical difference in detail), whereas the remaining category 'other fields', which was very significant from approximately 1880, started to grow from about 1970. This only points to different definitions concerning 'other fields'.

An attempt to subdivide the smaller fields of chemistry (the so-called 'other fields') shows the spectacular development of biochemistry and its high share of the total number of chemistry patent applications prior to the turn of the century. This surprising development is explained by a comparison of patent classes included in the definition *biochemistry* in the years 1900 and 1998: according to our definition, *biochemistry* was dedicated to the food industry in 1900. Between the turn of the century and

---

[9] In 1928, 358 member companies were registered in the central association of electrical technology and electrical industry (Zentralverband Elektrotechnik und Elektroindustrie ZVEI).

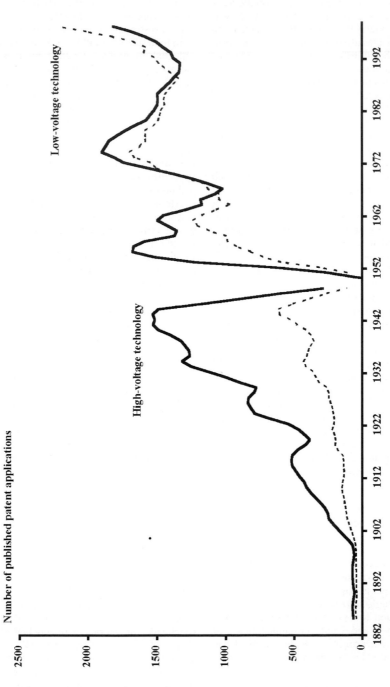

Figure 10. Moving average line of the first publications of national patent applications submitted to the German or European Patent Office (with destination Germany).

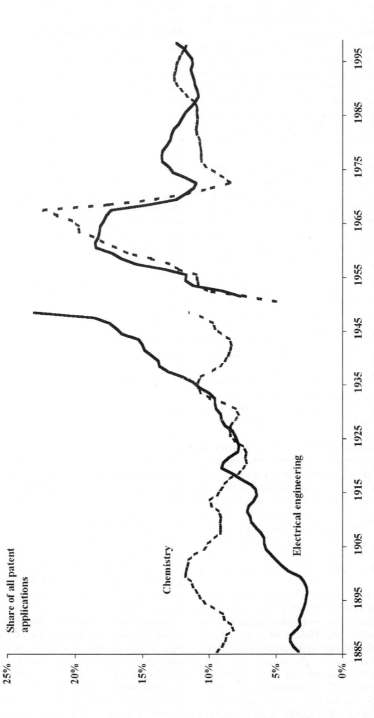

Figure 11. Share of the total patent publications about national patent applications held by electrical technology and chemistry (for Patent Offices and averaging cf. Fig. 10).

the First World War, the food industry based on biochemistry reached its peak: living organisms were used in the most varied ways. At that time, biotechnological procedures competed with chemical-synthetic procedures in the field of production.

According to Marschall (1999, p. 280), prior to the First World War a '*path dependency*' arose in German chemistry, which supported the evolutionary choice for chemical synthesis. This path dependency was based on subsidized investments in high pressure technology, as well as on its use by the chemical industry. Therefore, inventions in biochemistry were repressed. Nevertheless, it must be noted that brewing beer accounted for almost 50% of all inventions in the area of biochemistry around 1900 (only a few percent today), whereas second place was held by the extraction, refining and conservation of fat (today, also only a few percent). In 1998, however, the dominating area includes devices for enzymology or microbiology, second place is held by microorganisms or enzymes.

The difference between electrical technology and chemistry also comes from the significance of foreign inventors. According to the same criteria as above, patents can also be classified in correspondence with their *country of origin*. The results show that more important foreign inventions were realized in the area of electrical technology than in chemistry. The proportion of foreign inventors in the area of electrical technology corresponds approximately with the average of all technical areas; i.e. it regressed from more than 35% (1902) to approximately 20% by 1928 and to 10% by 1942. In the area of chemistry, the situation is different: foreign inventors always played a minor role (1902, 20%), however, this quota was slightly reduced (1942, 13%).

Regarding the historical statistics of the selected sectors, it is concluded that, under conditions of a relatively low level of knowledge exchange with foreign countries, a branch suffers less from periods of crises and autarchy and that, on the other hand, intense knowledge exchange with foreign countries leads to correspondingly sensible change. This will certainly have important effects on the organization of innovation activities which should be analyzed on the level of individual institutions (companies as well as research institutes).

The patent statistics of the *German Democratic Republic* can also be divided both on a technical and on an institutional level. For example, studies exist showing to what extent the Humboldt university of East Berlin applied for patents in the areas of chemistry and electrical technology. According to this, chemistry had become one of the most prominent research areas of the GDR in approximately 1990, whereas, electrical technology lost significance (Albrecht et al., 1991, p. 107). Identical analyses were submitted for central institutes, other centers and companies (*loc cit.*, Hinze & Grupp, 1995, p. 59). The

function of universities in view of technological development was very important not only for West Germany but also in the GDR.

Discussion and Conclusions

The view back into historical times of innovation reveals many interesting perspectives: for instance the present globalization trends in R&D may now be interpreted as a renaissance of the times around 1900: before the autarky and war situations in national socialist Germany the innovation system was internationalized in a similar way as today but possibly not to the same quantitative extent. Yet, at present, the logistic and travel possibilities for exchange of knowledge are much better than hundred years ago.

Most astonishingly, the German innovation system was very *stable*, although it witnessed several political system changes in the past century. The total amount of government spending on science and innovation followed similar quantitative tracks after its formation in the 19th century, the First World War and after the Second World War. The respective central power was not a strong pillar in science and technology. Contrary, the science and technology operation was maintained and was always reconstructed by the German states before the central power found ways to establish itself as dominating. However, considerable differences are observed when regarding the strong role of enterprises on innovation after the Second World War, which was—in pecuniar terms—not as visible before. Only after reunification in 1990, the acting power was the federal government at a time when enterprises were largely dominating the financing of R&D. This was definitely different hundred years ago.

In terms of the basic sectorial structures in science and technology, the strong and the weak sides were almost the same whatever regime and territorial boundaries existed. This *persistence* of the innovation system points to a *resistant innovation culture* in and around Germany, which may not be influenced too much by external shocks or incentives, be it in monetary or institutional form. If technology and innovation policy intends to change the German innovation culture in its basics, one probably needs other government methods than those being used up to today. Even the isolation of the former GDR and its subjection under the communist regime could not change much.

There seems to be a specific German understanding of the opening and prosecution of technology trajectories. The industrial research system in Germany was one of the first in the world to be formed and developed. Other countries followed that pattern more or less closely. Yet the subjects of research seemed to be different between the countries and remained largely constant over long periods. Obviously the technical and scientific elites in Germany succeeded to follow their interests in any political system col-

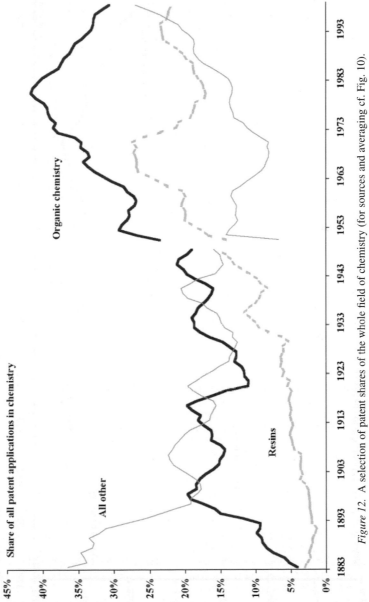

Figure 12. A selection of patent shares of the whole field of chemistry (for sources and averaging cf. Fig. 10).

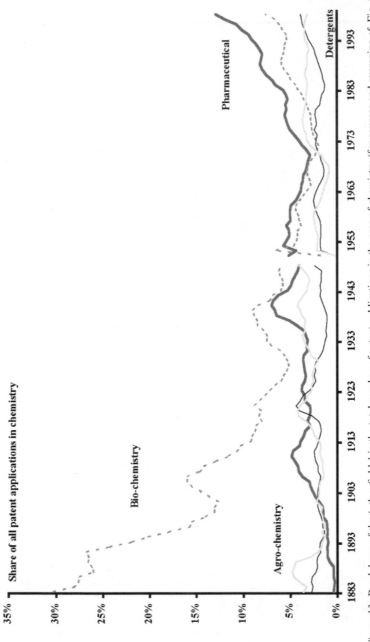

Figure 13. Breakdown of the 'other fields' in the total number of patent publications in the area of chemistry (for sources and averaging cf. Fig. 10).

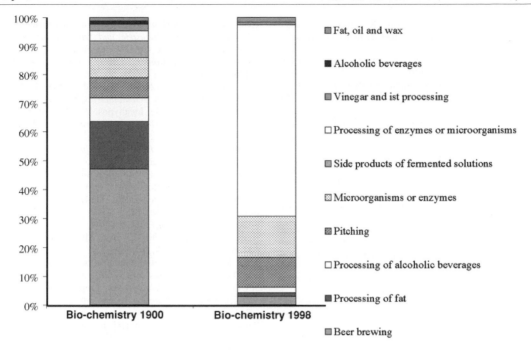

Figure 14. Breakup of patent shares within biochemistry in 1900 and 1998.

lectively. For the research and education policy this means that soft factors like group identity, schools of thought and personal exchange are more reliable and more efficient government instruments than the traditionally monetary incentive systems. This sustainable culture imprint can only be analyzed and detected in historical time series.

Innovation is no 'reality' whatsoever, and it is not considered as existent 'in themselves'; instead, innovation activities are only constituted by the specific method of scientific approach. An historical object of investigation about innovation activities is not 'given', instead, it is 'imposed' on the researcher, i.e. he or she must establish a measurement concept. An operationalization considered as ideal could lead to an exorbitant cost of collection, whereas, a less adequate method could have favorable effects on both the process and the results of measurement. This applies particularly to evolutionary innovation research, where results are not always shown by a formal mathematical model. Here, the keyword of an 'appreciative' theory was created. Based on such an approach, absolutely no immediate, constituent measuring directives can be derived either. Consequently, the task of adequation is the construction of corresponding indicators for insufficient 'tailor-made' theoretical constructs. Although this procedure could be considered wearisome, there is no other option.

The suggested range of indicators on a national or sectorial level gives a detailed impression of both the extent and the contents of innovation activities during more than the past hundred years. The empirical base, which evolutionary researchers interested in innovation and economic history-related questions can rely on, was broadened to a large extent, so that there is no longer the possibility of a serious *empirical gap*. On the one hand, many of the questions raised by evolutionary theory cannot yet be studied on an empirical level. On the other hand, however, the inverse fact is equally correct: empirical findings have become available discovering structures, which could be picked up by the theoretical side for further construction of the theoretical thought.

On the future research agenda there should be many more such studies of innovation systems. The basic findings for Germany should be compared to other countries, which possibly suffered less from territorial and political changes. The data used in this chapter should exist in other countries as well, and maybe brought to the surface. Also, we need more sectoral studies in order to work out typologies of innovation development over long periods. Altogether results achieved so far should encourage more cliometric research all over the world.

References

Albrecht, E., Dohnert, O., Schneider, M. & Bourcevet, H. (1991). *DDR-Forschung im internationalen Vergleich unter*

Zugrundelegung der Patentstatistik. Report for the Science Council, Sektion Wissenschaftstheorie und Wissenschaftsorganisation. Berlin: Humboldt University.

Anonymous, E. W. H. (1925). Output of scientific papers. *Nature*, **116** (2908), 129–130.

Anonymous (1985). Das Kaufkraftverhältnis Zwischen D-Mark und Mark in der DDR. *DIW-Wochenbericht 21/86*, **53**, 259–268.

Bannock, G., Baxter, R. E. & Davis, E. (1998). *Dictionary of Economics* (6th ed.). London: Penguin.

Bernal, J. D. (1939). *The social function of science*. London: George Routledge & Sons.

Boch, R. (Hrsg.) (1999). *Patentschutz und Innovation in Geschichte und Gegenwart*. Frankfurt A.M.: Peter Lang.

Dohrn-van Rossum, G. (1999). Erfinder und Erfinderschutz im Spätmittelalter und in der frühen Neuzeit. In: Boch (pp. 39–50).

Erker, P. (1990). Die Verwissenschaftlichung der Industrie. Zur Geschichte der Industrieforschung in den Europäischen und Amerikanischen Elektrokonzernen 1890–1930. *Zeitschrift für Unternehmensgeschichte*, **35**, 73–94.

Fogel, R. W. (1964). *Railroads and American economic growth*. Baltimore: The Johns Hopkins Press.

Freeman, C. (1969). Measurement of output of research and experimental development. Statistical reports and studies. Paris: UNESCO.

Freeman, C. (1992). *The economics of hope*. London: Pinter Publishers.

Freimann, K.-D. & Ott, A. E. (Eds) (1988). *Theorie und Empirie in der Wirtschaftsforschung*. Tübingen: J. C. B. Mohr.

Gascoigne, R. (1992). The historical demography of the scientific community, 1450–1900. *Social Studies of Science*, **22**, 545–573.

Gerschenkron, A. (1943). *Bread and democracy in Germany*. Ithaca: Cornell University Press.

Grohmann, H. (1988). Die statistische Adäquation als Postulat einer sachgerechten Abstimmung zwischen Theorie und Empirie. In: Freimann & Ott (pp. 25–42).

Grupp, H. (1998). *Foundations of the economics of innovation—theory, measurement and practice*. Cheltenham: Edward Elgar.

Grupp, H. & Schmoch, U. (1992). *Wissenschaftsbindung der Technik*. Heidelberg: Physica-Verlag.

Grupp, H., Schmoch, U. & Hinze, S. (2001). International alignment and scientific regard as macroindicators for international comparisons of publications. *Scientometrics*, **51** (2), 359–380.

Hetmeier, H.-W. (1990). Öffentliche Ausgaben für Forschung und experimentelle Entwicklung 1987. *Wirtschaft und Statistik*, **2**, 123–129.

Hinze, S. & Grupp, H. (1995). Ein Rückblick auf Wissenschaft und Technik in der ehemaligen DDR: Ostdeutschlands Forschungs- und Entwicklungspotential. In: Holland & Kuhlmann (pp. 41–86).

Hoffmann, W. G. (1965). *Das Wachstum der deutschen Wirtschaft seit der Mitte des 19. Jahrhunderts*. Berlin: Springer.

Holland, D. & Kuhlmann, S. (1995). *Systemwandel und industrielle Innovation*. Heidelberg: Physica.

Hoyningen-Huene, P. (1999). Kommt die Physik der Wahrheit immer näher? *Physikalische Blätter*, **55** (3), 56–58.

Irvine, J., Martin, B. R. & Isard, Ph. A. (1990). *Investing in the future*. Aldershot: Edward Elgar.

König, W. (1995). *Technikwissenschaften–Die Entstehung der Elektrotechnik aus Industrie und Wissenschaft zwischen 1880 und 1914*. Chur: G + B Verlag Fakultas.

Lorenz, Ch. (1997). *Konstruktion der Vergangenheit*. Köln: Böhlau.

Mabe, M. & Amin, M. (2001). Growth dynamics of scholarly and scientific journals, *Scientometrics*, **51** (1), 147–162.

Machlup, F. (1960/1961). Idealtypus, Wirklichkeit und Konstruktion. *Ordo—Jahrbuch für die Ordnung von Wirtschaft und Gesellschaft*, **12**, 21–57.

Maddison, A. (1982). *Phases of Capitalist Development*. Oxford: Oxford University Press.

Marschall, L. (1999). Industrielle Biotechnologie im 20. Jahrhundert—Technologische Alternative oder Nischentechnologie? *Technikgeschichte*, **66** (4), 277–293.

Müller, K. E. & Rüsen, J. (Hrsg.) (1997). *Historische Sinnbildung*. Reinbek: Rowohlt.

North, M. (Hrsg.) (2000). *Deutsche Wirtschaftsgeschichte*. München: C. H. Beck.

North, M. (2000). Einleitung, in derselbe, 11–14.

OECD (Ed.) (1993). *Frascati Manual 1992* (5th rev.). Paris.

OECD (Ed.) (1992). *Oslo Manual*. Proposed Guidelines for Collecting and Interpreting Technological Innovation Data, Paris.

Pfetsch, F. R. (1974). *Zur Entwicklung der Wissenschaftspolitik in Deutschland 1750–1914*. Berlin: Duncker & Humblodt.

Pfetsch, F. R. (1982). *Datenhandbuch zur Wissenschaftsentwicklung*. Köln, Zentrum für historische Sozialforschung.

Roscher, W. (1886). Grundlagen der Nationalökonomie (18th ed.). Stuttgart: J. G. Cotta'sche Buchhandlung.

Schmoch, U., Licht, G. & Reinhard, M. (2000). *Wissens- und Technologietransfer in Deutschland*. Stuttgart: Fraunhofer IRB Verlag.

Schmoch, U. & Saß, U. (2000). Erfassung der technologischen Leistungsfähigkeit der östlichen Bundesländer mit Hilfe von Patentindikatoren, Report for the Federal Ministry BMBF. Karlsruhe: FhG-ISI.

Schumpeter, J. A. (1964). *Theorie der wirtschaftlichen Entwicklung*. München and Leipzig, 1911, quoted after the 6th ed. Berlin.

Schumpeter, J. A. (1975). *Capitalism, socialism and demogracy*. New York, 1942, quoted after the German translation, 4th ed. München.

Smith, J. K., Jr. (1990). The scientific tradition in American industrial research. *Technology & Culture*, **31**, 121–131.

Stolper, G., Häuser, K., Borchardt, K. (1964). *Deutsche Wirtschaft seit 1870*. Tübingen: J. C. B. Mohr.

Struve, G. (1980). *Geschichte der drei Volkserhebungen in Baden 1848/1849*. Bern, von Jenni, 1849; quoted after the revised reprint. Freiburg: Rombach.

Testa, J. (1997). The ISI database: The journal selection process, Essay No. 1. Philadelphia: Institute of Scientific Information.

Wagner-Döbler, R. & Berg, J. (1996). Nineteenth-century mathematics in the mirror of its literature: A quantitative approach. *Historia Mathematica*, **23**, 288–318.

Weingart, P., Strate, J. & Winterhager, M. (1991). Bibliometrisches profil der DDR, report for the Science Council. Universitätsschwerpunkt Wissenschaftsforschung, University of Bielefeld.

Wengenroth, U. (1997). Der Beitrag des Staates zu Forschung und Entwicklung in Deutschland seit der Reichsgründung, Typescript.

Ziegler, D. (2000). Das Zeitalter der Industrialisierung. In: North (pp. 102–281).

The International Handbook on Innovation
Edited by Larisa V. Shavinina
© 2003 Published by Elsevier Science Ltd.

Frankenstein Futures? German and British Biotechnology Compared

Rebecca Harding

The Work Foundation, U.K.

Abstract: This chapter compares and contrasts the market based biotechnology policies of the United Kingdom with the regionally engineered biotechnology policies of Germany in the light of the national and regional systems of innovation literature. It shows that innovation systems generally and regional innovation systems in particular are still useful concepts in explaining the clustering of biotechnology. It compares regional clusters in the two countries and argues that German policy has been particularly successful in stimulating rapid catch-up with U.K. and global levels of biotechnology research.

Keywords: National systems of innovation; Biotechnology policy; University-industry links; Clusters; Entrepreneurship.

Introduction

Of all the areas of innovation policy within a government's remit, the area of biotechnology may seem the least obvious as a focus for an Anglo-German comparison to a lay audience. To begin with, the whole area of biotechnology is fraught with ethical difficulties, not least because biotechnology is often seen as synonymous with 'Frankenstein' pictures of chickens without feathers, or mice with ears grotesquely sprouting from their backs. Second, although at an individual level people can clearly see the link between genetic research and their own health prospects, other areas of biotechnology, for example agricultural biotechnology, have less obvious benefits either to society or to the economy. And finally, for all the millions that have been poured into biotechnology across the world, and the fact that biotechnology accounts for 41% of all world wide patents (ZEW, 20002), there is still no long term 'cure' for cancer, no end to world poverty and hunger, and indeed, very few profitable biotechnology companies providing employment in their regional or local communities.

This chapter is an attempt to answer the question as to why governments across the world commit so many resources to biotechnology by comparing German and British policy development over the ten year period since 1992. Much of this policy has been predicated upon a belief that European countries are 'behind' the U.S. in biotechnology and that audacious measures need to be taken to ensure that we 'catch up' with the performance of the United States economy. Policy makers across Europe have admired the strong university–industry links in the U.S. around bio-medical research, have seen the impressive biotechnology 'clusters' of entrepreneurial start-ups and research spin-outs, venture capital, specialist support services and large pharmaceutical companies around leading U.S. universities, and have tried to mimic these structures in their countries. At an EU level the whole area of biotechnology and research-led innovation and entrepreneurship was central to the Lisbon Summit's 2000 agenda.

To a large extent, much of this policy was based on evidence from the United States throughout the 1990s that research-led entrepreneurship generally would lead automatically to increases in productivity and, hence, economic growth. Europe has a productivity gap with the U.S. and studies of innovation would suggest that, by addressing the 'innovation gap' this would be closed (Fagerberg, 1987; Freeman, 1995; Lundvall et al., 1992) and, inevitably, higher employment and wealth generation would follow (Bygrave, Camp, Hay & Reynolds, 1998).

For most people, however, the interest in biotechnology may still seem perplexing. 'Why', puzzled a senior U.K. trade unionist during an interview, 'does the government want to pour money into science? How on earth can a few professors with test tubes and microscopes create jobs?' Given the furore surrounding genetically modified foods, alongside the highly specialized and skilled types of employment that new biotechnology firms create, this question is quite justified and it is the aim of this paper to address that issue by means of an Anglo-German comparison.

The chapter is constructed as follows. It first looks at the features of a specifically Anglo-German comparison and argues that, as the countries with the most highly developed biotechnology structures in Europe, as well as quite distinctive policy structures, they are interesting cases in their own rights. The chapter goes on to examine some of the literature that has informed policy towards biotechnology. By examining the literature on national, regional and sectoral systems of innovation in particular, the specific mechanisms that policy makers in each country have employed can be evaluated. Of course, in the end the real way in which the success of any policy to stimulate or redirect the focus of a national or regional innovation system rests on the effectiveness with which the sustainability of any changes can be measured. The next section takes the discussion of innovation systems on to examine the unique characteristics of biotechnology allowing a further layer to the understanding of policy instruments. It is argued that the real reason why biotechnology has formed such a central part of government science policy in both countries is because of the intrinsic regional and 'clustered' nature of the industry that allows it to fit neatly alongside existing programmes for regional regeneration and economic growth.

The chapter then goes on to look at policy structures in the two countries in terms of basic expenditure on research and development and policies to stimulate private sector involvement, commericialization and clustering. It constructs a map of policies in the two countries and highlights the strong regional dimension to policy in both countries. The section concludes by pulling together the literature search and the overview of policy to highlight the key 'critical success framework' in each country.

The next section uses the critical success framework to examine documentary and attitudinal evidence in the two countries. It looks first at the evidence in terms of clustering effects (large companies, business angel activity, specialized consultancies and venture capital firms) and goes on to examine attitudes towards university-industry links and academic venturing in the area of biotechnology.

The chapter concludes by attempting to answer the question, 'why bother with biotech?' It argues that biotechnology is a central part of any government's science strategy. European policy makers do need to ensure that we do not fall behind the U.S. either in research or in commercialization for the simple reason that a national monopoly over one particular aspect of scientific endeavor would distort future competitiveness as and when commercial biotechnology products become part of everyday life in the way that computers have. Further, there is evidence from the Anglo-German comparison that jobs are created through a strong biotechnology policy and that this leads to a strong sense of regional renewal around science-based industry. However, the critical stage in a biotechnology company's development comes when it is moving from a largely public sector supported, research and prototype base to actual commercialization and sustainable growth. The issue of how to manage this process effectively both at the level of the individual firm and at the level of policy is not clearly understood and there is scope for more research in this area.

Why an Anglo-German comparison?

Britain and Germany represent interesting cases within this wider European effort to catch up with United States. They have, respectively, the largest and the second largest biotechnology sectors in Europe. Both have strong regional biotechnology clusters around world class universities and both countries invest the largest amounts in Europe of private and public sector money in biotechnology. Their biotechnology policies at a government level are strongly supportive of establishing and developing robust biotechnology research and, accordingly have the most policy and infrastructural attributes that are either supportive or strongly supportive of biotechnology (for example, risk capital, public sector support for R&D etc.) (Senker et al., 2001).

What makes the comparison especially interesting, however, is the way in which the two countries have moved towards this position during the last decade. Britain's biotechnology sector is strongly established and has emerged out of high quality research and strong commercial involvement in that research in and around the Cambridge area in particular. Awareness of the significance of biotechnology as a scientific area was first raised in the 1970s and 1980s by a government-commissioned report and subsequent policy sought to increase general awareness, particularly amongst private sector companies about its potential. Government policy has worked over the last five years to extend the cluster-based activity beyond Cambridge to other regions of the U.K. Thus, for example, policies like University Challenge and the Higher Education Innovation Fund have attempted to stimulate the flow of seed corn finance to university research-based spin-outs and encourage academic entrepreneurship generally and, as a corollary, biotechnology in particular.

In contrast, Germany's biotechnology sector was substantially less developed than U.K. or U.S. biotechnology at the beginning of the 1990s. Although the capacity to audit trends in biotechnology did exist within the Fraunhofer Society's technology foresight program, Delphi, research activity was limited to three

Gene Research Centres at Cologne, Heidelberg and Munich. Real awareness of the importance of bio-technology as an area of research and commercial activity did not really become widespread until the latter half of the 1980s (Wörner et al., 2000). Historically-based mistrust of biotechnology because of its association with genetic manipulation and legal restrictions on R&D in this area contributed significantly to its relatively backward stage of development in comparison to the U.K. in the early 1990s (Harding, 1999, 2001). With the amendment of the Genetic Engineering Act in 1993, the legal barriers to biotechnology research were removed. The subsequent government-led BioRegio program was set up in 1995 and has arguably been a central driver behind the development of a systematic and positive approach to biotechnology research and commercialization. It was supported by a strong commitment at a policy level to increasing funds for basic research in biotechnology.

The two countries provide examples of two different approaches to the development of biotechnology then. Both countries have marked similarities according to Senker et al. (2001). For example:

- both have favorable knowledge and skills regimes and strong science bases with an active emphasis on technology transfer between the research base and commercial application;
- both have strong multinationals operating in the pharmaceuticals and chemicals areas;
- in both countries the public is generally positive about biopharmaceutical research and development although more wary of the application of bio-technology techniques such as genetic modification in the agro-food areas.

Yet in the U.K. the structures are largely market led. Policy works to *facilitate* market operations where any gaps are apparent, for example in funding or in research and development. In Germany, biotechnology as a legitimate area of scientific investigation and of commercial enterprise has been developed to a significant extent by policy effort to *create* markets where none existed. This has been done through support for research network development and for biotechnology start-up finance.

There is one final way in which the different policy perspectives is interesting as an area of examination. Within the economic literature comparing and contrasting German and British economic performance there is a tendency to characterize the two economies at different ends of a management spectrum. The British economy is seen as exemplifying the 'Anglo–Saxon' model of market-based flexibility and radical innovation-led growth while the much more rigid, structured and ponderous Rhineland-Capitalist model, of which the German economy is viewed as paradigmatic, is reported to be far less successful in creating structures that support and sustain radical innovation led development (see, for example, Soskice, 1996). Entrepreneurship generally is a test case of this polemic approach to analysis since, by definition, it would be more dominant and more widespread as an expression of labor market activity. Bio-entrepreneurship as a particular case is interesting as it combines both the requirement for radical innovation and regulatory flexibility, especially in labor markets, in order for it to be able to thrive.

Against this background, then, Germany would seem on the face of it to be at a disadvantage. Its labor market inflexibility is well documented (Funk, 2000, 2001) and, arguably manifests itself in the rigidly high levels of unemployment that the economy is currently experiencing. Further, some economists have noted that the German economy has a tendency to produce incremental innovations rather than the radical paradigm shifting innovations that characterize the U.S. or the U.K. economy (Casper et al., 2000; Hall & Soskice 2001; Soskice, 1996). Germany would apparently have a comparative disadvantage in biotechnology innovation, then because of these two structural weaknesses.

The evidence does not support this as a hypothesis, however. In 1995 German policy makers set themselves the target of overtaking the U.K. as the leading country in Europe for biotechnology start-ups. By 2001 it had achieved this with 332 Core Entrepreneurial Life Science Companies (ELISCO's) (BioM, 2002) compared to the U.K.'s 250 (www.dti.gov.uk). Also, although the U.K.'s bioscience sector, including all biotechnology companies, remains larger in terms of longer-established firms and, hence, total turnover, the absolute size of the German industry in terms of the number of firms is marginally bigger at 465 compared to the U.K.'s 450. German biotechnology now accounts for the third highest number of patents per employee in the world after the U.S. and Japan. And finally, in terms of levels of total entrepreneurial activity (TEA) and, indeed, attitudes towards entrepreneurship, Germany and the U.K. are remarkably similar (Harding, 2002; Reynolds et al., 2002; Sternberg & Bergmann, 2002).

Much of the development in Germany is still embryonic—it has a much larger number of start-up companies established for less than three years, for example, where in the U.K. the biotechnology sector is more established. However, the sheer pace with which Germany has caught up in terms of biotechnology research and entrepreneurship warrants further investigation in its own right as it arguably represents an extension of the base of the 'business system' to incorporate this type of activity and, hence, to allow its industrial structures to adapt (Casper, 2000).

Why Support Biotechnology?

The Financing Gap

Within the context of the whole European biotechnology sector and its comparison with the more advanced state of its counterpart in the U.S., the U.K. and Germany are big players with the largest and second largest number of firms respectively. However, the European market is still substantially under-developed, heavily reliant on independent start-up firms and with a relative under-representation of university spin-outs. The risk capital market is still weak with less than 3% of total venture capital investment across the whole of Europe going to biotechnology ventures and less than 1% going towards biotechnology in the U.K.

The financing gap for biotechnology is illustrated in Table 1. In simple terms, potential investors see biotechnology as risky. Biotechnology is research intensive and this is expensive. Thus the amounts of money required even to prove that a concept is viable (let alone commercializable) are very high. It can take anything between ten and twenty years to get to the stage clinical trials have been completed and the development of a commercially viable product at the end of this is by no means assured. Indeed, it has been estimated that for every 100 biotechnology research ideas, only one is likely to have any commercial potential at all (Harding & Lissenburgh, 2000). This means that the risks faced by potential investors are high relative to the rate of return within the 'normal' lifespan of a venture capital investment (where exit is usually planned within three to five years). This manifests itself in a clear under-representation in biotechnology financing relative to total equity financing.

Table 1 clearly shows that, as a proportion of total investment, or even of total high technology investment, biotechnology is a poor relation. Amounts invested across the whole of Europe have remained relatively static since 1996, while total investment has gone up by nearly five times, high-tech investment has nearly doubled and seed and start up investment for all sectors has almost trebled. The EVCA reports that in Germany total investment in biotechnology rose between 2000 and 2001 and in the U.K. they fell over the same period. Across the whole of Europe the trend

for biotechnology investments was downwards, although the level of investment still remains slightly higher than in 1999.

Table 2 looks at the relative structures of venture capital in Germany and Britain and compares it to the EU average.

The table shows quite clearly the different foci of the German and the British venture capital industry. In Britain, investments are heavily focused at the large scale end of the market with management buy-outs (MBOs) and management buy ins (MBIs) comprising the majority of venture capital funding irrespective of sector. In Germany the seed and early stages of start up development account for higher levels of investment than in the U.K. and are substantially higher than the European average. The industrial sectors which attract substantial investment focuses are biotechnology and ICT while in the U.K. the sectors tend to be much more general and not science based.

The reason for this difference in financing focus is arguably because of the different policy emphases in the two countries (for further discussion of this, see Harding, 2001). In Germany, enterprise and innovation policy generally has concentrated on stimulating industrial systems and networks by encouraging public-private sector partnerships, technology transfer and private finance capital at a regional level. The BioRegio program is an example of this but others include the InnoRegio program for the eastern States to develop innovation structures there. In order to ensure that private sector investment has gone into biotechnology and ICT, the government has implemented a system parallel investment through the *Kreditanstalt fuer Wiederaufbau* (KfW) and the *Deutsche Ausgleichsbank* (DtA). Under this system any risk capital investment fund prioritizing early stage technology investments can construct a 'fund of funds' (broadly, a venture capital fund consisting of more than one independently managed fund) and thereby draw down on a further, proportionate, fund of money from the KfW as leverage to 25% extra investment in a specific project. Alongside this, the KfW will also provide up to a 50% guarantee on any investment made by the main 'fund of funds', thus the risk to the investor is on just 25% of the total amount invested. Clearly this acts as a major stimulus to private sector investment in riskier

Table 1. Financing gap for biotechnology.

	1996	1997	1998	1999	2000	2001
Biotech	2.70	2.60	2.40	2.60	2.90	2.65
Hi-tech	19.60	23.90	27.80	25.60	31.40	28.00
Seed and start-up	6.50	7.40	11.40	12.90	19.00	15.00
Total VC funding (m)	6,788	9,655	14,460	25,116	34,986	24,331

Source: EVCA various.

Table 2. *Relative structures for venture capital in Germany and Britain.*

Country	Investment focus of lower level investors	% at each stage (2001)	% total in technology (2000)
Germany	Biotechnology and Information and Communications Technology	Early stage 26 Expansion 22 MBO/I 50	46.8 (but 99% of all early stage is in technology businesses)
U.K.	General investments and cyclical services, increasing general technology focus	Seed/early stage 12 Expansion 32 MBO/I 56	19.3
EU average	Consumer products and services, manufacturing and industrial production	Seed/early 15 Replacement 4 Expansion 35 MBO/I 40	31.4

Source: updated from Harding, 2000.

sectors, and the higher numbers of investors in biotechnology start-ups.

Indeed, it is this desire to mitigate the risk faced by potential biotechnology investors relative to the returns they might get that underpins much of government policy towards the financing of biotechnology entrepreneurship. There is clearly a gap between biotechnology investments and total venture capital investments as Tables 1 and 2 show. What German policy has attempted to do through the KfW and the DtA is to leverage private sector money into riskier areas of investment by guaranteeing any losses that may be made on projects that do not deliver.

That the financing gap is more acute in the U.K. is also apparent from the data presented here. The reasons for this are complex, rooted in the nature of the private equity market in the U.K. (which tends to favor larger investments) and extend beyond the scope of this chapter (for further detail, see Harding, 2000, 2002a). Suffice it to say that successive research has shown U.K. investors to be averse to technology investments both in terms of the amounts of money they put in (Murray & Lott, 1995; Murray et al., 2002) and in terms of their attitudes towards technology investments (Harding, 2000).

The Knowledge Gap

Governments support biotechnology for reasons other than the finance gap, however. Biotechnology concepts are complex and heavily reliant on multi-disciplinary research teams. The research is itself expensive and need to build 'critical mass' in terms of research quantity before any of the concepts are likely to become commercializable. Once an idea with commercial potential is developed, there is still a very long time to full commercialization and many points along the route to market that might end in failure. For example, the public outcry around GM foods has presented many very large biotechnology companies

such as Monsanto with problems in terms of share price and, hence, access to further research funding. Huntingdon Life Sciences is an example of another company that, even when it appeared to be relatively successful, found difficulties in gaining further funding once public opinion swung against it.

This points to the importance of the basic research function within the 'typical' biotechnology firm. Until a product is developed, and this can take some time, the company is organized around a concept or a set of concepts that have commercial potential but that require further research in order to progress them further. This requires a strong research base and highly qualified personnel. It also means that a biotechnology company is high reliant on the research institutions (in particular universities) operating in related fields of research in close proximity.

It is this 'networked' nature of the biotechnology firm that makes it unique and that makes it an attractive vehicle for delivering government policy. The true value of the firm lies in the quality of the research team that 'own' the initial concept and their national and, critically, international research networks. Often these are groups of scientists from related disciplines who have worked together on research and often they do not have any developed knowledge of management, or even of presenting their ideas to a lay audience in an approachable manner. They will always require access to sophisticated research base (including laboratory facilities) and often require access to specialist legal, administrative and financial support as well (OECD, 2001; Sohet & Prevezer, 1996).

Biotechnology as a Policy Focus

There are clearly some factors associated with biotechnology that warrant public sector support. In particular, the above section has highlighted a financing gap and a knowledge gap (the 'personal networked' nature of activity) that mean that it cannot survive alone. This in itself, however, may not justify the large policy emphasis that is being put on biotechnology research in all industrialized economies. In order to examine this, it is necessary to see biotechnology in the context of competitiveness in wider, science-based industries.

Biotechnology is an area of research and commercial activity is derived from the public sector research base in life sciences on the one hand and the research activities of the chemical and pharmaceutical sectors on the other. Increasingly the pharmaceutical industry is the core sector underpinning R&D in biotechnology. These pharmaceutical companies spend very high proportions of their total turnover on R&D as this is where the competitive advantage in that sector originates (Sharp, 1996).

The research intensity of the sector has increased with combinatorial chemistry (inter-disciplinary chemistry) and molecular biology now working together to develop new diagnostic and drug delivery mechanisms as well as long term gene-based solutions to diseases or, in the case of agri-environmental biotechnology, crop development. These tools and techniques are developed in-house in large company research laboratories but the risks associated with the research are high since the 'critical mass' in research effort discussed above applies equally to large companies as it does to small. As a result, brand new areas of biotechnology research, which are not likely to lead to new products or processes in the immediate future, but which may have commercial potential in the future are often strategically 'outsourced' to smaller research-based companies with strong links to research institutions in the public sector science base (Gambaradella et al., 2001; Sharp, 1996; Walsh 1997). In order that the pharmaceutical industry within a country remains competitive, it is essential that structures to support this type of activity exist.

Gambaradella et al. (2001) point to the declining competitiveness of the European pharmaceutical sector and argue that one of the core reasons for this is the underdevelopment of its networks and support infrastructures in the science-based relative to the United States. Germany and the United Kingdom are leaders in the European market, but there is still a competitive gap between them and the U.S. which is potentially extremely damaging to the future competitiveness of the whole sector. They point to the global mobility of researchers in this area since all belong to international networks of scientists and experts. Increasingly, they argue, that local 'innovation clusters' are globally competitive with one another as locations for science-based businesses and it is here that the scale of the U.S. science system as well as its links with the commercial base far exceeds that of Europe (Gambaradella et al., 2001; see also Cooke, 2001).

Biotechnology is core to the competitive success in pharmaceuticals because of this reliance on research outsourcing and excellence in the science base. It is no one clearly defined area (Webber, 1995) and instead appeals to three major policy areas.

University-Industry Links

The first area is science policy generally and university-industry links in particular. Biotechnology requires a strong basic science research base in the life sciences if concepts or ideas with any commercial potential at all are to be developed. This means that firms tend to cluster in 'knowledge sources' (Cooke, 2002). And, since the research and commercial effort in biotechnology is so internationalized, these 'knowledge sources' compete *and* collaborate with one another nationally and globally.

This 'symbiotic tension' where research and industry compete for and collaborate in research projects but are ultimately mutually interdependent in the transfer of technology from the science base through to industrial application is well documented for the German system and regarded as a source of competitive advantage for German innovators (Harding, 2000, 2001). Gambaradella et al measure the extent and scope of networks between the public and the private sector science base for pharmaceuticals in the U.S. and Europe and argue that the concepts of competition and collaboration in technology transfer are important in understanding competitive advantage in the pharmaceutical industry (see also Cooke, 2001; Kaufmann & Todtling, 2001; Love & Roper, 2001; Senker et al., 2001). Further, the transfer of knowledge in biotechnology is tacit in nature and relies substantially on the relationships between scientists in research institutions and private sector laboratories making the spatial concentration in technology transfer at a regional level especially significant in driving the propensity to innovate (Todtling & Kaufmann, 2001; Zeller, 2001).

What policy makers in Europe generally and in Germany and Britain in particular should be aiming to stimulate in the biotechnology sector, therefore, is the development of strong university-industry research networks in the interests of enhancing their attractiveness as locations for global R&D, either by large companies or as the home of international research-based firms. This is the source of national comparative advantage in technology transfer. 'Symbiotic tension', from studies of biotechnology clusters and university-industry links, is key to understanding the viability, sustainability and competitiveness of the biotechnology sector. It is this processes that enhances the development of vibrant university-industry links nationally and, hence, that facilitates the location of international R&D in one country as opposed to another. It is no longer possible to regard the national and the global innovation system or network as independent of one another (Archibugi et al., 1999) since research is internationally mobile and will locate around specialized centres of excellence. Policy has to ensure that research specialization is enhanced by strong university-industry networks and linkages if it is to create attractive locations for global biotechnology R&D.

Regional Policy

There is a substantial body of literature to suggest that, in a world where R&D is mobile internationally, competitive innovation advantage is generated at the regional rather than at the national level (Cantwell & Iammarino, 2000; Cooke et al., 2000; Edquist, 1997; Harding, 1999; Sachsenian, 1997). This is because technological specialization that is so critical to the symbiotic tension relationship within technology transfer is best developed at a regional level. Regional universities have scientific specialization within specific areas and resources to support that focus, any spin-out companies from university research are likely to be within the areas of scientific excellence developed within the university and large companies are more likely to locate and, hence, to transfer knowledge where such excellence exists. Learning and adaptation to changing market and technological conditions is more likely to be effective and sustainable at a regional level since tacit knowledge transfers more easily between actors in close spatial proximity with clear links to the cumulative skills and attributes of the regional labor market (Cooke et al., 2000; Dodgson, 2001; Michie & Oughton, 2001; Porter, 2000). As expertise starts to build, specialist financiers, accountants and lawyers are established to support the science base and any start-up businesses are provided with appropriate and readily accessible advice and consultancy. The evolution of this type of regional 'industrial system' is argued to go some way to explaining the development of Silicon Valley and Route 128 in the U.S. (Sachsenian, 1997).

The attractiveness of the 'cluster' approach (Porter, 1998) to policy makers is clear, especially for biotechnology. Since biotechnology research and commercial activity is interdependent with scientific and commercial networks, since tacit knowledge transfer is behind the symbiotic tension at the heart of competitive success in this industry and since firms cluster close to knowledge sources, it makes sense to operationalize biotechnology policy at a regional level. Universities, as identified above, are 'magnets' of biotechnology activity, but a true innovation system at a regional level is created for biotechnology through the combination of research hospitals and 'chains of transactions between scientists, entrepreneurs and various intermediaries including inventors and lawyers' (Cooke, 2002). Only by systematizing this set of interactions will regionally generated knowledge add value through the cumulative learning process to create the specialization that is so important to international

competitive advantage in research led sectors such as biotechnology.

Evidence suggests that such regional 'centers of excellence' or 'clusters' and their intra-regional links (both within a country and globally) are necessary preconditions for creating attractive locations for global biotechnology R&D. Interestingly, the national, regional and sectoral systems of innovation are peculiarly interdependent for biotechnology because of its knowledge intensive and research-led nature (Freeman, 2002; Gambaradella et al., 2001; Malerba, 2002; Owen-Smith et al., 2002; Senker et al., 2001). For policy makers this is a complex message—that regions are important as the point of delivery but that the sources of learning and added value actually rest in the networks that individual researchers have nationally and internationally. In other words, national science policy and regional cluster policy should be mutually reinforcing and formulated to 'promote network building among firms and other actors of a regional innovation system and to interlink these intra-regional networks with national and international knowledge sources (Koschatsky & Sternberg, 2001).

Finance Policy

Technology-based firms are both more suited to venture capital investment *and* more likely to seek venture capital investment. They require significant amounts of capital but, because their business is based on an innovation rather than a proven business concept, investments in them are inherently more risky. In theory, at least, this ought to be the domain of risk-takers and, hence, also the domain of venture capitalists. Yet the figures presented in Tables 1 and 2 above suggest that both Germany and the U.K. are behind the U.S. in terms of venture capital investments, particularly in biotechnology and this is a clear challenge for policy.

Linking venture capital with bio-innovation through policy is, at best, complicated. Yet, as one German venture capitalist argued, "Venture capital investments are for technology-based companies. I am a financier, but I have had to learn about (bio)technology—quite simply, this is where the money is". The reasons for this are as follows:

- *Returns to technology investments are high.* The Bank of England estimates average returns on technology investments to be around 23% (Bank of England, 2000). But one technology investor claimed return rates of 45% in the U.K. and rates in the U.S. are certainly higher at 33.7% (www.nvca.com). This return rate is evidence of the high growth and wealth creation potential of technology-based firms as much as evidence of their suitability for venture capital funding. Yet venture capitalists themselves will not be able to take advantage of these potential returns unless they can be encouraged into riskier, technology-based investments.

- *The growth potential that these companies have is embedded in the value that they add to their initial concept.* All technology-based companies start with a commercially unproven innovative idea at the seed stage—this is the risk. The growth process is the cumulative 'proof' of the idea or concept's commercial viability. The value at the end is the return. But, especially in science-based industries like biotechnology, this growth process requires substantial development funding. This funding can be necessary over a long period of time—as long as ten years. This is significantly longer than most venture capitalists will invest without a clearly defined exit route, thus there is a clear role for government support at the seed stage and even at the start-up stage to leverage in informal and formal venture capital.

- *The acquisition of substantial capital investments allows the technology-based firms to attract key scientists and innovators into their business.* The value of the company is embodied in the personnel that are employed within the organization. As one Dutch biotechnology fund manager said, "We don't invest in profit, we invest in value. Biotechnology companies never make a profit but their ideas can be worth millions. One company came to us suggesting that the size of the workforce should be reduced in order to show a working profit. This would have been a disaster as we were investing in the high potential value of their scientists. That's what we can sell on".

- *It is important therefore that such companies can access easily the high net worth individuals that add value to an innovative concept.* This is primarily a function of the supply of such people from universities, colleges and industry. In turn, this is a function of the capacity of the education and training

system, the higher education system and the industrial system to create, develop and, critically, keep, these individuals. The role for policy here is in creating an infrastructure that creates such high value 'human capital' in which venture capital can invest.

- *Finally, in order that the rate of return is fully realized and venture capitalists continue to invest in technology projects, there has to be a good supply of investment opportunities for venture capitalists.* This deal flow stems from universities and colleges through academic entrepreneurs and from indigenous and overseas hi-tech companies with research capacity. Governments can do much to stimulate a culture of science and technology-based entrepreneurship through funding for basic science, significant funding for university-business partnerships, science parks, incubators and programs to stimulate high technology investments. Yet there is evidence that there is a weakness in the commercialization of science from the research base across Europe (DG-Enterprise, 2000), but in the U.K. in particular (Bank of England, 2002).

The biotechnology financing life-cycle is presented in Fig. 1.

At the seed stage of concept development, relatively small amounts of money, since much of the activity is research-based rather than commercially-driven. At this stage, the role of public sector funding of the science base is clear in order to provide adequate resources for centres of research excellence to develop.

The second, early stage, of development is where the commercial potential of a concept has been proved and a patent registered. Here there is far greater potential for private sector investors to participate. In the U.K., for example, business angels are particularly important

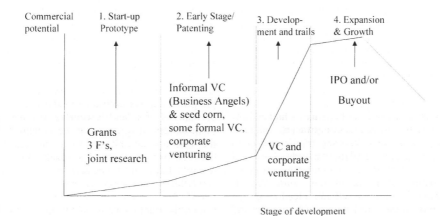

Figure 1. A biotechnology finance life cycle.

in funding biotechnology companies at this stage, while in Germany, corporate venturing (where firms fund research in return for an equity stake in the emerging business) is more common. Government guarantees for private sector investments at this stage can be used to encourage investors to take risks with biotechnology based businesses.

Corporate venturing and private equity are most important at the third, development and trials, stage of development. Here the potential for the value of the company to grow is substantial—not least because the networks and research expertise of the personnel within the company become valuable assets in their own right.

The final stage to initial public offering (IPO) or trade sale (sale to a large pharmaceutical business, for example) is the point at which the venture capitalist or private investor will exit and realize their investment.

Each stage is fraught with risks for the private investor. Biotechnology projects can fail at each stage either because insufficient research has been conducted or because the concept is not viable. Even after clinical trials (a process which can take up to ten years), there is still a risk that the route to market will be blocked by adverse or inconclusive results. And, as the example of the agri-environmental biotechnology firm Monsanto showed, if a public outcry threatens the acceptability of the product, then investments are not secure even once the company is floated on a stock exchange.

This perceived risk goes some way to explaining the reluctance of the venture capital industry to make investments in biotechnology (Murray, 1999). However, biotechnology is unique, not just in its *modus operandi* but also in the source of value in the business. This value, as stated above, rests in the people in the business. Their research and their networks are assets to potential investors, especially large pharmaceutical companies seeking to spread the risk of financing non-core but strategically important R&D. Policy has to have two strands to its approach to biotechnology financing therefore—first, to mitigate risks of private sector investors, either by providing support at the seed and early stages of development or by providing parallel investments for riskier projects, and second, to increase awareness of the inherent value in the intangible assets of the bio-business.

Implications for Analysis

The difficulty of assessing the benefits of biotechnology in its own right and of assessing the effectiveness of policy to alter national systems of innovation to stimulate developments in a particular technology or a particular area of scientific entrepreneurship rests in the 'intangible' nature of many of the 'third stream' activities that are so dominant within biotechnology. In the broadest terms these 'third stream' activities are the relationships between the science base and the industrial base. As such, they are extraordinarily difficult to

measure within the conventional toolbox of national systems of innovation theory.

The relationship between the science and engineering base (knowledge generation) and the institutions involved in the commercialization of science (knowledge transfer) can be co-ordinated through policy to produce an innovation system which is capable of adapting to new competitive and technological pressures. However, measuring the effectiveness of policies to develop the commercialization of science remains a challenge for academics and policy makers alike.

The role of nation state generally and of governments in particular is, as it always has been, to ensure that the institutions of industrial society and economy adapt to the pressures of any new 'global' or technological paradigm. It is not the change in itself that is new or, even, interesting. Thus the interesting question is not whether nation states should adapt but how.

It is the means by which institutions adapt that is key to understanding the nature of innovation generally and technology and knowledge transfer in particular (Harding, 2001). The evidence of history suggests that it is not change in itself that is interesting in the context of the path-dependent nature of innovation (Duysters & Hagedoorn, 1995). Change in itself is endemic. What is interesting, however, is the extent to which policy can influence the way and the rate at which a national system of innovation can itself change in response to the exogenous imperatives of global competitiveness. In the U.S., this has taken the form of the interdependent development of the institutions of innovation to create scientific structures that are, "less and less a matter of the independent unfolding of knowledge and more a response to technological progress in the development of a practical means to produce goods and services" (Mowery & Rosenberg, 1999). The key, then, is to establish the source of this adaptiveness within a particular system of innovation.

Within the national systems of innovation literature there are three broad categories of writers. The first largely take as their starting point the importance of relationships between institutions in the science base and indigenous industry, which, they argue can be measured and evaluated using economic/econometric methodologies. Researchers use, for example patenting statistics (Patel & Pavitt, 1994) or citation indices (Hicks et al., 1996) or industry-based matched sampling to establish both the existence and effectiveness of national systems (Mason & Wagner, 1998). The methodological approach of these authors is to narrow the 'institutional linkages' to quantifiable relationships between two variables. The results tend to be biased towards large firms because of the techniques used. Nevertheless this approach can:

- identify distinct national systems which, arguably, are the reason why 'growth rates differ' (Fagerberg, 1987);

- establish the tendency of firms in a particular country to utilize their own, domestic, science base as a source of blue sky research.

A broader methodological approach looks at national systems of innovation through the perspective of corporate governance (Casper et al., 1999; Freeman, 1995; Lundvall et al., 1992; Prais, 1981; Soskice, 1996; Tylecote & Conessa, 1999; Vittols, 1997). This approach identifies key institutional features of national systems of innovation, for example, incentive structures, education and training, industrial relations, finance or government policies. Authors seek to establish the influences that particular aspects of the system have on the capacity of an economy to innovate often using qualitative and historical techniques. Using this approach, it is possible to isolate the key features of a 'national system' which make countries different. Thus, for example, the U.S. has particular structural strengths which mean it has advantages in radical innovation, while Germany and Japan have institutional strengths which favor incremental innovations (Soskice, 1994; Tylecote & Conesa, 1999).

A third school of writers can also be identified who take a socio-historical perspective on the reaction of national structures generally to major exogenous changes (see for example Albert, 1993; Giddens, 1998; Sorge, 1999; Zysman, 1996). Some of these authors take the change itself as the determining variable and attempt to define the features of the 'global' system, while others take as their perspective the need to understand the institutional adaptiveness of national systems of innovation. What is common amongst these authors, however, is the critique of the neo-classical economic approach to globalization—that national economic and business structures will inevitably be subsumed in a 'global' market order. Instead, these authors point to the intrinsic adaptiveness of systems. Their views on the eventual sustainability of systems are different—they range from theories of systemic convergence under the supremacy of the Anglo-Saxon model (Streek, 1997) to hybridization whereby national systems adopt and adapt eclectically aspects of models from elsewhere to suit their own structures (Lane, 1999).

The contributions made by all of these authors cannot, indeed should not, be under-played. Whatever approach is taken, unique national systems of innovation can be defined. All view NSIs from a macro institutional level and derive specific national strengths and weaknesses accordingly. This has an obvious appeal for policy makers seeking to strengthen the science base of an economy and, accordingly, 'maximize national returns for public and private investments in R&D in terms of efficiency and competitiveness' (Patel & Pavitt, 1999, p. 20). Intriguing, however, is the quite different standpoints on the sustainability of national systems that the different

methodological approaches derive. The more quantitative perspective leads to the conclusion that national systems are 'under strain' (Patel & Pavitt, 1999, p. 23). This conclusion derives from the fact that large company R&D is itself internationalizing and there is thus a discernible weakening in the links between corporate R&D and the national science base. Corporate governance and socio-historical writers, however, are often (although by no means uniformly) more optimistic about the sustainability of national or even regional systems. They regard change in itself as inevitable and see the interesting question as being how national systems are adapting (Duysters & Hagedoorn, 1996; Lane, 1999; Whitely, 1999).

That the jury is still out on the sustainability of national systems is largely a question of methodology. Arguably, the burden of proof rests with the institutional writers who stress the importance of historical processes in understanding institutional (and hence national) adaptiveness. But conversely, econometric techniques are criticized by institutional and socio-historical writers on the grounds that the narrowing down to measurable variables mitigates against a full understanding of how national systems work and, hence, adapt.

While accepting that quantitative techniques can underpin qualitative research by providing trend indicators, in order to understand the process of adaptation, it is necessary also to take as broad an institutional perspective as possible. All of the approaches summarized above tend to view national systems and institutional linkages at a macro-level. But this mitigates against a full understanding of the social mechanisms and practices by which these linkages are effected generally and by which technology transfers in particular. In order fully to comprehend this it is necessary also to examine national systems from two perspectives:

- first, a 'bottom up' approach: in other words, to examine the source of innovation within research institutes, within universities and within industry. The aim here is to establish the means by which technology transfers and, hence, to find the mechanisms through which the innovation system adapts;
- second, using a policy-interface approach: in other words, to establish the interface between policy mechanisms and the institutions of the innovation system. The aim here is to establish the role of policy in altering the rate at which the system adapts.

The research for this chapter has taken the 'bottom up' perspective to establish the innovation dynamic that creates both the resilience and the adaptiveness of technology transfer systems. It supplements this by arguing that it is the policy-interface between national policies to stimulate 'leading edge' technologies and the institutions within the system of innovation that are key to ensuring that this adaptiveness is a permanent and dynamic feature of any technology transfer system.

German and U.K. Policy Compared

The previous section identified three areas around which biotechnology policy is built: university–industry links (particularly support for basic research and for technology transfer), regional development ('clusters' or 'centres of excellence') and finance. None of these areas are mutually exclusive, however and, given the multi-disciplinary and networked nature of biotechnology itself, it would not be appropriate to construct a single policy towards this sector. Further, since the whole area biotechnology is interwoven with the Public Understanding of Science as well as with strong ethical considerations, any policies tend to cross-departmental and legislative boundaries.

To this end, this section compares policy formulation and policy implementation and delivery in terms of three areas:

The legislative, regulatory and policy system: this, broadly, includes departmental responsibilities to construct legislation, regulate biotechnology R&D and commercialization and increase awareness and understanding of biotechnology (and ethical considerations around biotechnology) in the wider public. It also includes an analysis of the policy initiatives towards biotechnology in both countries.

The support for university-industry links: including the funding of the science base and technology transfer, as well as the institutional structure of the science system that delivers biotechnology research and commercial activity.

'Cluster' or 'Centres of Excellence' at a regional level: both German and U.K. policy makers have been strongly influenced by the 'regional systems of innovation' and 'cluster' literature surveyed above that suggests a link between innovation at a regional level and strong national biotechnology performance.

Finance policy in both countries is woven into policies towards support for technology transfer and regional cluster development. Thus this area of policy is not examined in its own right but is instead integrated into the wider discussion and analysis (for further reference, see Harding, 2000, 2002).

The Legislative, Regulatory and Policy Framework

Nowhere is the complexity of biotechnology more obvious than in the legislative and regulatory framework that underpins policy formulation and delivery. This complexity is apparent in both countries and is illustrated in Table 3.

Table 3 breaks the framework for biotechnology policy into five areas, legislative responsibility, guidance and advice (including on ethical matters), monitoring (including regulation, intellectual property and technology/impact assessment), access to funding (for R&D and commercialization) and research and expert services.

Table 3. The biotechnology legislative and regulatory framework in Germany and Britain.

	Legislation	Guidance and advice	Monitoring	Access to funding	Research and expert services
Germany	BMU BML BMG	BMU BML BMBF BMG RKI	RKI ZKBS	BMBF BMWi BML BMG Länder DFG	Dedicated govt institutes RKI DFG funded institutes Universities MPG HGF FhG WGL
United Kingdom	DTI Culture HSE DETR MAFF DoH Home Office European Standards	DTI Culture HSE DETR MAFF DoH Home Office Research Councils Trade Associations	HSE MAFF DoH	DTI DoH Home Office Research Councils Wellcome Trust	Culture DETR DoH Home Office Research Councils Universities

Germany and the U.K. have one central department broadly responsible for the research and training agendas—the Bundesministerium fuer Bildung und Forschung (BMBF) and the Department of Trade and Industry (DTI) respectively. In both countries the remit of these departments is extensive and covers the guidance and advice, access to funding and general policy formulation (see Table 4). In addition to this, the

Table 4. The biotechnology policy framework in Germany and Britain.

	Germany	U.K.
Basic Research	*Deutsche Forschungsgemeinschaft*: research council funding prioritizing biotechnology research (14% increase in funding) *Max-Planck Gesellschaft*: national networks of basic research institutes. Research specialisms at regional level *Blaue Liste Institutes*: regional research institutes with Länder funding *Gene Centers* *National initiatives* include specific funding for Nanotechnology, proteomics, bioinformatics, German Human Genome project and sustainable bioproduction *BioFuture*: competition to provide young scientists with resource-base to develop high powered research and commercial careers in applied biotech research	*Biotechnology and Biological Sciences Research Council, Medical Research Council* and the *Natural and Environmental Research Council* provide funding for basic science in universities
Applied Research	*BioFuture*: also provides support for commercialization and incentivization to stay in Germany *Fraunhofer Institutes* work with companies on applying biotechnology research	*Biotechnology Exploitation Platform Challenge* *DTI funding for partnership research*
Technology Transfer	*INSTI*: national network of patenting search organizations linked with technology transfer structures like AN-Institutes and Fraunhofer *Kompetenzzentren*: tech transfer centers within the BioRegio structures to facilitate university industry links Various other programs including *Innovation-Market, Innovationspartner & Deutsche Wirtschaft*	*Biotechnology Exploitation Platform Challenge*: to encourage universities and businesses to work together *University Challenge:* not specific to biotech but a seed fund for university technology spin outs
Commercialization	*BioChance*: competitive access to development finance for established start-up biotech firms conducting high-risk R&D	*Bioscience Unit* (DTI) champions commercial exploitation including IPR agreements, regulation and tech transfer *Biotechnology Mentoring and Incubator Challenge* Fund to create high quality sustainable biotech companies *Trade Partners U.K.*: to encourage exports in biotechnology
Cluster Development	*BioRegio*: Competition between Länder to develop clusters around biotechnology generally *BioProfile*: Competition-based extension finance for BioRegio Regions to develop focus/specialization in dedicated area of research. Designed to give new drive to BioRegio initiative	*Public-private sector partnerships* to stimulate biotechnology R&D and commercialization in the regions following report by Minister for Science in 1999 *University Challenge* to stimulate science and commercial networks through universities

Table 4. *Continued.*

	Germany	U.K.
Regulatory Framework	Regulation falls under three categories: *national environmental policy* (BMU), *agricultural biotechnology* (BML) (including animal testing and research) and *health* (BMG) that governs Genetic Engineering. The *Robert Koch Institute and the Central Advisory Committee for Biological Safety* monitor health and safety issues and develops guidelines	Responsibility spread across a number of different government departments and guidance notes are prepared accordingly. Advisory Committees provide health and safety and ethical advice. Department of Health has responsibility for medicine licensing
Public Understanding	*Science Live*: 'science touring truck' equipped with research facilities to allow scientists to run experiments where resources might otherwise not exist + trained personnel to raise profile and understanding of biotechnology. Web-based reference center for bioethics *Safety research and Monitoring*	BIO-WISE: explaining the commercial potential of biotechnology to businesses *Department of culture* provides information courses *Public understanding of science* (e.g. through Science Museum)

DTI is also responsible for biotechnology legislation since the biotechnology directorate and the Office of Science and Technology sit within that department.

Germany's legislative and regulatory framework is far more embedded within a wider departmental structure than in the U.K. There are 17 different departments or organizations involved with biotechnology at a national level (compared to 11 in the U.K.). The Environment Ministry (Bundesministerium fuer Umwelt—BMU), the Federal Ministry of Food, Agriculture and Forestry (BML) and the Federal Ministry of Health (BMG) control legislation, for example, while the BMBF is responsible for policies that enable researchers and small businesses to get guidance and advice on ethical, regulatory and research matters. The Federal Economics Ministry (BMWi) also provides funding for biotechnology. The State governments also provided research and commercialization funding, as does the KfW/DtA while research and consultancy is provided by a plethora of establishments within the science base of the German economy including the Max Planck Institutes (MPG), the Fraunhofer Institutes (FhG) and public and private sector research laboratories, universities and research establishments (some of which are dedicated to biotechnology—for example the Gene Centres in Cologne, Heidelberg and Munich). Monitoring and ethical guidance is a clear responsibility of the Robert Koch Institute and the Central Advisory Committee for Biological Safety (ZKBS) and forms part of Germany's wider policy to ensure that Technology Assessment is fully integrated into any R&D activity (Harding & Harding, 2001a).

Departmental responsibility for biotechnology within the U.K. rests with the DTI, the Department of Culture, the Ministry for Agriculture, Fisheries and Food, the Department of Health (DoH), the Depart-

ment for the Environment, Transport and the Regions (DETR—now the Deputy Prime Minister's office) and the Home Office. It is only the DTI that has an explicit role towards biotechnology in the form of legislation, funding or regulation—for the other departments biotechnology is integrated into wider policy frameworks. The Health and Safety Executive (HSE) and the European Standards Office play a strong role in monitoring and regulation while the General Medical Council (GMC) has a bioethics committee which assesses the ethical implications of any biotechnology research. R&D is funded by the DTI and the DoH. By far the largest budget is with the DTI since it controls the research council budgets as well. Interestingly, charity funding for research (for example the Wellcome Trust) forms an important, if small, part of the total funding for biotechnology R&D.

There are three marked differences between the German and the U.K. legislative and regulatory structures. First, the German framework is to a large extent embedded within its wider science and technology system. This means that the responsibilities for pure (basic or blue sky) research as opposed to applied or commercial research and development are clearly delineated at the point of delivery (Harding, 2001). Thus, for example, a Max Planck Institute would not be involved in the front line of commercialization research since this extends beyond its remit although the Fraunhofer Institutes may well be. There is a strong 'blue sky' element within the dedicated 'Blaue Liste' research institutes like the Hermann von Helmholtz Gemeinschaften (HGF) and its post-1995 successor in the eastern States, the Wissenschaftgemeinschaft Gottfried Wilhelm Leibnitz (WPL). All this institutes are predominantly supported by the federal government, but a token 10% of funding comes from the regional

States to ensure that long term public interest projects are fully investigated.

This regional role in the biotechnology framework is a second key difference between the U.K. and Germany. The role for regional level governance and funding for research is not clearly defined in the U.K. while in Germany a critical part of the remit of State (Laender) governments is to formulate regionally based science policy based on regionally defined interests and needs.

Finally, the third key difference is in the financing of biotechnology. Again, in Germany, this is embedded within the existing institutional framework of the social market economy which means a clear delineation between national, regional and local responsibilities in the funding of both of public interest research and of commercial activity. Pure research is largely funded by the Deutsche Forschungsgemeinschaft which had in 1999 a total research budget of DM 2,278 billion, 36% of which was dedicated to life science research. Alongside this, the KfW and the Deutsche Ausgleichsbank (now a merged organization) have responsibility for facilitating loans, mezzanine and equity type finance and work with regional and local banks and venture capital firms to operationalize this. The U.K. funding system, as it is depicted in Table 3 is largely for basic research in biotechnology and comes predominantly from the research councils, charitable trusts (around 7%) or from government departments. Where research is applied and likely to lead to commercial exploitation of the science base, finance is through policy initiatives like University Challenge and has a strong private sector dimension to it.

The spread of biotechnology across so many departments and non-governmental organizations (like for example the GMC) is illustrative of a wide dissemination of biotechnology awareness at a policy level. It is interesting to note, however, that, although both governments do pay some attention to the importance of raising public awareness and understanding of biotechnology, the extent to which this is explicit within the above framework is limited.

The policy framework to support biotechnology policy in the two countries is described in Table 4.

There are three things that are immediately obvious from Table 4. The first is the sheer size of the German effort in the area of biotechnology policy in relation to the U.K. Policy has sought to raise the profile of the technology amongst scientists and businesses alike and, simultaneously has put in place a framework for informed public debate about the issues in biotechnology research. The second feature of the German system is its heavy reliance on competitions as a way of providing funding to research (basic, applied and technology transfer) as well as to network building at a local, regional and national level. Applicants for funding through these routes have to demonstrate a clear and established track record and evidence that

they are already following the strategies they propose in their bid for funding. In other words, the structures and systems for delivery have to be in place and some progress has already to have been made if a bid is to be successful. The final feature of the system is its embeddedness within the overall framework of the German science system. That is, the clear delineation between basic scientific and applied research, the integration of technology transfer and commercialization and, as a logical extension, the capacity to build clusters relatively easily on the back of existing institutional structures that support competition and collaboration in R&D.

The U.K. system similarly reflects the intrinsic nature of its science system. It relies heavily on a competitive process for funding of any kind and, similarly, has sought to engage private sector money on a matched basis at all stages of research and commercialization beyond pure, or 'basic' research. This is especially the case for any product development work as well as for cluster development. Collaboration in the system comes from specific policies to support partnership (for example the biotechnology exploitation platform) and from broader policies to support university-industry partnership. These latter policies, such as University Challenge, The Higher Education Innovation Fund and the Science Enterprise Challenge are not purely for biotechnology however. Public understanding is facilitated through BIO-WISE (although this is technically to explain the commercial potential of biotechnology and not to widen public understanding).

Support for University-Industry links

Biotechnology relies heavily on the efficiency and effectiveness of the science base to develop products with any commercial potential at all. And, similarly, the process of biotechnology development which transfers pure science know-how into industrial application (the technology transfer process) is dependent upon collaborative and communication channels with business. So, if government is to be successful in promoting the industry, it has both to ensure the adequate funding of the science base and, critically, develop support structures to facilitate the effective transfer of knowledge from basic scientific research into product development.

The first thing to examine, then, is the overall level of science funding in Germany and the U.K. in order to understand the scale of differences between the two countries. Funding for the science base as a percentage of GDP is given in Table 5. For comparative purposes, the U.S. is included in the next two tables.

Germany spends more as a percentage of GDP than the U.K., although does not spend as much as the U.S. However, Germany has a much larger GDP than the U.K. and this translates into a higher level of overall expenditure on Science, Engineering and Technology

Table 5. Overall levels of science spending in Germany and the U.K. as a percentage of GDP (U.S. as comparitor).

	1993	1994	1995	1996	1997	1998	1999	2000
Germany	2.42	2.32	2.31	2.3	2.31	2.32	2.44	2.46
U.K.	2.15	2.11	2.02	1.95	1.87	1.83	1.87	—
U.S.	2.62	2.52	2.61	2.66	2.7	2.77	2.64	—

Source: Main OECD Science and Technology Indicators, 2001.

(SET). For example, the U.K. SET budget expanded by 7.5% to £6,734 million between 1999 and 2000, but Germany still spends more than twice the amount in real terms on its science base than the U.K. and expanded its funding by 14% over the same period. This is shown in Table 6 which shows government budget allocations for R&D (GBAORD) in current dollar prices for comparative purposes.

Another point is worthy of note here. German reliance on the private and governmental sectors for funding of R&D in comparison to the U.K. This is illustrated in Table 7 which shows the sources and modes of funding in the two countries.

Germany's funding for R&D is largely from government or business. The U.K. in contrast has a lower level of private expenditure on R&D and lower levels of public expenditure on R&D. Funding from abroad as

well as funding from other U.K. organizations, often charities, is a sizeable proportion of total funding. This is particularly important for biotechnology since much of the 'other national sources' category is accounted for by large national medical charities such as the Wellcome Trust.

Actual expenditure on biotechnology is hard to derive on a comparative basis (see also Senker, 2001). The reason for this is that, as can be seen from Tables 3 and 4 above, the reach of biotechnology research and application extends far beyond one government department and is interwoven with the structure of the science system itself. However, the German government claimed to spend something in the region of £750m on biotechnology in 2001 across all government departments. In the U.K. the three research councils with the most explicit remit for funding

Table 6. Total government budget allocations to R&D (million current PPP $).

	1994	1995	1996	1997	1998	1999	2000
Germany	14,952.4	15,696.9	15,879.4	15,595.7	15,625.0	15,991.5	16,224.6
U.K.	8,058.4	8,628.1	8,942.7	9,055.7	8,603.7	8,879.6	—
U.S.	68,331.0	68,791.0	69,049.0	71,653.0	73,569.0	76,886.0	75,415.0

Source: OECD Main Science and Technology Indicators, 2001.

Table 7. Overview of different sources and modes of funding for research and development in selected countries, 1999.

	Aus	Can	Fin	Fra	D	J	Sw	U.K.	U.S.
R&D Performer									
Business Enterprise	45.1	59.8	71.1	63.1	70.0	70.7	75.1	67.8	75.7
Government	23.4	12.0	11.1	17.9	13.7	9.9	3.4	10.7	7.2
Higher Education	29.4	26.9	17.8	17.6	16.3	14.8	21.4	20.0	14.1
Private Non-Profit	2.1	1.2	0.7	1.5	0.7	4.6	0.1	1.4	2.9
Source of Funding									
Business Enterprise	39.7	44.7	66.9	53.5	65.1	72.2	67.8	49.4	66.8
Government	47.8	31.2	29.2	37.3	32.3	19.5	24.5	27.9	29.2
Abroad	2.5	16.7	3.0	7.4	2.3	0.4	3.5	17.6	—
Other National Sources	4.7	7.4	0.9	1.8	0.3	7.9	4.2	5.1	4.0

Source: OECD + National documentation.

biotechnology are the Biotechnology and Biological Sciences Research Council, the Natural and Environmental Research Council and the Medical Research Council. Their combined budget is £567.1m, although this is a general budget allocation and is not for biotechnology specifically.

Since it is so difficult to establish exactly how much is being put into biotechnology in the two countries, and since much of the effectiveness of biotechnology as a vehicle for commercial application and, hence, innovation-led growth, rests in the relationship between universities and industry, it makes sense to dwell on this area a little longer. This is broadly 'technology transfer', although the relationships between universities and industry at a local or regional level is core to specialist cluster development too.

Table 8 examines policy priorities in Germany and the U.K. in this area.

There are a number of points that can be drawn out from this table.

(1) In Germany, funding for teaching and research in Higher Education establishments comes from regional and national level governments. Thus teaching, for example, is broadly funded by regional governments beyond a token 'core' funding from the national government. However, research is funded by both the regional and the national governments (through the DFG and Blaue Liste Institutes in the case of national interest research). This means that regional governments can set research funding priorities to reflect regional economic priorities and that cluster development policies can build on this to develop sectoral specialisms and networks;

(2) The U.K. government has prioritized funding for the science base generally and for biotechnology in particular and there are more resources available for research in this area. The commercialization strategies are reliant on the engagement of private sector businesses through 'matched funding';

(3) Both countries have mechanisms for anticipating technological changes and formulating policies and strategies accordingly. In Germany the mechanism for evaluating biotechnology developments through the Delphi program is based in the Fraunhofer Society. The U.K.'s Foresight Program is run through the Office of Science and Technology and is based on committees of scientists and business people who evaluate the commercial potential of technological change as they occur;

(4) At the point of implementation, the German policy mechanism reflects the institutional structure and responsibilities of the science system generally. The U.K. in contrast has a much more decentralized delivery system alongside a centralized funding system and is reliant on private sector partner involvement.

The highly contested research market in the U.K. makes collaboration, even in an area where it is essential to collaborate, more difficult. In contrast, the strongly collaborative nature of German science system means that collaborative science is easier—this may go some way to explaining the speed with which Germany has caught up in terms of patents.

University–Industry Links and Academic Entrepreneurship

Both governments have put a large effort in to raising the profile of academic entrepreneurship as a driver for technology transfer and commercialization. Policies are similar in both countries and include strategies to stimulate incubators, science parks, venture capital

Table 8. University–industry policy priorities in Germany and the U.K.

	Policy Priority (2001/2)	Formulation Mechanism	Implementation Mechanism
Germany	Enhancing efficiency of science system; ICT, biotechnology; health research, sustainable development, physics chemistry and materials sciences, nanotechnology, energy, transport and mobility, space, marine technology	Federal Government, BLK[1] and Science Council *Delphi program to advise on future scientific trends (through Fraunhofer but also in conjunction with MITI)*	Federal and regional funding initiatives and programs; foundations and institutional structure VC Innoregio and Bioregio programs
U.K.	Increased infrastructure funding, research in key technologies, boost to science budget to build on university research; commercialization of public sector research	DTI, POST, OST, Chief Scientist, Foresight Program and Foresight fund	Government departments, research councils, universities, research and technology organizations in private sector, Faraday Partnerships Programs and initiatives; VC through University challenge and HEIF, R&D tax credits

Source: Harding & Harding, 2001.

and, critically, to streamline intellectual property agreements so that both universities, researchers and businesses can profit from research. Evaluating the effectiveness of these types of policy in any rigorous sense is extremely difficult since there area multitude of different ways in which the relationships between academics, academic entrepreneurs and business are built.

Within the context of this research it was neither necessary nor appropriate to attempt such an evaluation. However the Global Entrepreneurship Monitor study interviews entrepreneurial experts across a number of countries of which Germany and the U.K. are two. This study uses an identical methodology to speak to these experts, and asks them questions around 'entrepreneurial framework conditions', including R&D transfer, education and training, culture, policy, government programs and finance. For the purposes of this paper, the German and the U.K. expert surveys have been used to draw out any general messages on university-industry links generally and biotechnology in particular. The results are shown in Table 9.

What is clear from this table is that problems in the relationship between universities and business exist in both countries. Specifically:

* there is too much regulation. In Germany the experts focused specifically on patenting requirements and technology assessment regulations in biotechnology while in the U.K. the regulation was seen in the broader context of labor market regulations and taxation;
* the support structures that help R&D to commercialize are seen as too expensive in both countries. This includes patent searches and access to professional business support (for example accountancy firms and legal practices);
* the R&D transfer system doesn't always work as effectively as it might—there is still mistrust between industry and the science base in both countries;
* the patenting and IPR systems in both countries are viewed as unwieldy or ineffective.

There are a number of stark differences between the two however.

* *Government Programs*: In Germany interviewees were very positive while in the U.K. University Challenge was seen to have excluded non-university bio-innovators. It was also pointed out that U.K. policies are not focused explicitly on biotechnology and that this might restrict the potential for biotechnology exploitation;
* *Finance*: Seed and early stage funding for biotechnology in Germany was seen as good. Respondents in the U.K. argued that there is still a shortfall in equity-based funding for biotechnology;

* *Universities*: these were seen in the U.K. as still have real problems in dealing with spin outs as well as with small and medium sized businesses in their local communities. In Germany the attitudes were generally more positive—that universities were developing along the right lines but that there is still under-utilized potential;
* *Physical infrastructure*: This was seen by experts as 'awful' in the U.K. but excellent in Germany.

The Regions

The U.K.'s market-based policy contrasts with Germany's 'engineered' cluster development policy through BioRegio. BioRegio rests on an analysis by the German government in the early 1990s that concluded first, that biotechnology was likely to be central to future economic growth (prompted by the Delphi program) and second that mechanisms had to be established to facilitate a quick and effective catch up. The best way of doing this, was seen as being through the regions.

Regions with established biotechnology sectors (through the Gene Centers) along with other regions with strong biological or biomedical research universities competed for funding in a competition launched in 1995. The BioRegio program assessed proposals against four criteria:

* that the networks would create a motor for biotechnology 'catch up';
* that the proposal would stimulate biotechnology start-ups;
* that the proposal would grow existing biotechnology R&D;
* that venture capital provision would be an integral part of the cluster design.

The overall aim of the program in 1997 was to make Germany 'number one' in Europe by the year 2000 (for further reference on BioRegio, see Dohse, 2000). Seventeen projects were approved, although three were selected as 'models': Munich, Rhineland and Rhine-Neckar. These model regions received more public money and priority access to future competitions. None of the regions received more than a maximum of 50% of public sector funding, however being a model region provided greater leverage to private resources.

Dohse (2000) argues that cluster development in Germany was strongly influenced at a policy level by the literature on regional innovation systems as drivers for national technological specialism and competitiveness. Similarly, U.K. policy has been influenced by the literature and by policy and practice in other countries—especially the U.S. and Germany (DTI, 1999). The theory behind cluster development in the two countries is very similar, therefore, and, as shown in Table 8, this translates into a very similar set of critical success factors against which the policies can be assessed.

Table 9. Attitudes towards university–industry links in Germany and the U.K.

Entrepreneurial Framework Condition	Germany	U.K.
Finance	• Banks seen as not having skills to evaluate research-based business proposals • Neuer Markt was important in getting culture of technology-based businesses going • Access to venture capital for university projects good but this is less the case in the eastern States and there is a perception that the money is 'public' money and therefore not commercial • Development finance for university projects is good	• Persistent risk aversion on behalf of U.K. investors towards university start-ups • Equity gap for university projects because financiers do not have the skills to evaluate, especially in biotech • Finance is hard to come by unless it matches with a priority area
Government Policy	• Too many regulations from government. This is particularly severe for biotechnology businesses	• Regulation, especially in areas of employment law, make growth very hard, especially for science start ups
Government Programs	• Programs effective and logical • Incubators work well to transfer technology • Finance measures are used well • Programs have increased awareness of science venturing • Regional policies excellent—especially BioRegio	• Programs tend to favor entrepreneurs within universities and not those from outside the university sector • Incubators work well
Education and Training	• Lack of business education in schools, especially for the life sciences • There is a strong supply of well qualified people • Germans prefer not to work across scientific disciplines which is an issue for biotechnology	• 'Anti-science' culture in schools • Lack of business education throughout the system • Major skills gap in critical scientific areas
R&D Transfer	• Patent protection is not always effective and is over-complex especially for biotechnology • It is not easy to find the best support for patenting searches as there are so many of them • Underutilized potential in research base • Entrepreneurship in universities is increasing but more is necessary	• Universities have real problems with SMEs • There is more entrepreneurship at universities but there is still too little • University scientists have no concept of what it means to set up a business
Commercial Professional Infrastructure	• There is a tight network of support agencies • Commercial support is expensive	• Variable quality across the company • Duplication is an issue • Commercial support expensive
Physical Infrastructure	• Excellent	• Major source of competitive disadvantage—'world class scientists need a world class infrastructure. We can't offer them this'
Market Openness	• When big pharmaceutical companies are involved they are strongly supportive of start-up biotech companies	• Flexibility in labour market • Strong support from large pharma companies
Culture	• Scientific entrepreneurship is a popular career choice increasingly because of the intellectual freedom it gives researchers • Negative attitude to failure • Working hours culture is changing in Germany and this will be positive • Public understanding of science could be improved • Heavy reliance on government programs	• Persistent 'anti-science' culture in the general population made worse by media coverage • University-industry links still generally weak and not based on mutual understanding

Source: Bergmann & Sternberg, 2002; Harding, 2002.

Table 9, then, looks at biotechnology clusters in four regions—two in Germany and two in the U.K. Cambridge and Munich are compared as models of 'best practice' in the two countries. Alongside this Jena and Manchester are compared as examples of regions with a strong historical research base but weaker economic and infrastructural support at the outset. The material in these tables is based on publicly available material and further research would be necessary to assess or evaluate the actual performance of these biotechnology clusters.

In all regions, the biotechnology cluster strategy appears to have created jobs, attracted private capital, stimulated HEI spin-outs and created research specialisms.

Critical Success Factor—Assessment of Regional Cluster Initiatives

Regional level delivery especially has been an especially important policy instrument in both countries. The evidence of dynamic development is clear from Table 10 in both countries.

Combining the 'critical success factors' outlined in Table 8 alongside a number of conditions for successful regeneration provides some initial analysis of the success of the strategies in the two countries against five criteria:

Actual research and patents: this gives and indication of the strength of science base and its potential for production of the critical mass of research necessary for developing commercial products in the future. All four of the regions have strong research universities and specialisms with active patenting activity in core biotechnology areas. Cambridge, Manchester and

Table 10. Critical success factors in regional policy.

Germany	U.K.
Strong research base	Strong science base
Entrepreneurial Culture	Entrepreneurial Culture
Role Models	Increasing corporate base
Integrating management culture with scientific research	Capacity to attract key staff
Incubation	Premises and infrastructure
Supported up start-phase	Business support services and related international large companies
Guidance to market	Skilled Work Force
Access to finance and investment	Effective Networks
Corporate involvement	Supportive policy

Source: DTI 1999 and BioRegio.com.

Munich have attracted R&D capacity from large multinational firms, while Jena has developed its own commercial R&D strength though Jenoptik.

Numbers of large companies: This gives an idea of the private sector networks and investment that has been leveraged through an initial public sector investment. The market is most developed in Cambridge, although Munich also has a strong track record in recent years for attracting private investment. Manchester and Jena have also been successful in attracting some large company investment, especially in related technological areas.

Private finance raised and numbers of VC firms: Venture capital is seen by policy makers as a means of stimulating start-ups and science based entrepreneurship and, although it is by itself, not enough to guarantee this, evidence from the U.S. suggests that it is a necessary if not sufficient condition. All regions have been successful in attracting large amounts of venture capital funds. The key difference between Germany and the U.K. however, is that these funds have been leveraged by strong policy efforts through the KfW and DtA while in the U.K. the government has played a minimal role.

Numbers of start-ups and SMEs: this gives an idea of the 'lead generation' of growth businesses in the cluster. All regions have been successful in creating spin-outs and start-ups. Cambridge is the most established region and has the largest number of publicly listed biotech businesses. The other regions are still in the 'catch-up' phase and have more embryonic life science businesses (ELISCOS). Evidence on the sustainability of these tiny businesses is sparse.

Jobs created: All regions record job creation through life science and biotechnology based businesses. Cambridge, where the cluster is arguably most developed, attributes 10,000 jobs to the biotechnology sector and Munich has a similar number. Jena is slightly different to Manchester in that it already had a large and established life-science based business before BioRegio and hence claims that 6,000 jobs have been created as a direct consequence of growth in biotechnology.

Is there a Frankenstein Future?

Is there a Frankenstein future? The short answer to this is no—whether we like it or not, Europe (and in this case, Germany and U.K.) have to have a biotechnology industry because this underpins the competitiveness of the pharmaceutical industry as well as the long term viability of the life-science sector. The analysis of the above sections is arguably only scratching at the surface of what is actually happening on the ground, but it nevertheless points to the potential of the sector to provide jobs, create innovation and, hence, to stimulate economic growth.

Table 11. Regional clusters in Germany and the U.K.

	Size of the Biotech Community	Other Regional Facts and Figures			Market Area	
		Biotech companies	Employment	Research	Biotech Companies	Specialist service providers
Munich BioRegio Munich	• BioM AG: number of biotech companies grown from 36 in 1996 to 107 in 2000 • 2001: 130 biotech and pharma companies (of which 110 are SMEs) • 10 international pharma companies including Glaxo-SmithKline, AGFA, AUDI, LINOS, Rodenstock, OSRAM • Munich BioTech Region: 120 Pharma and biotech companies; Martinsried—growth from 10 in 1996 to 50 now	• 50 VC financed biotech companies • 5 Neuer Markt listings • BioTech Region Munich: 85 start-ups • 500% growth in direct employment	• BioM AG 2500 employed in biotech SMEs in Munich region • Bayern Photonics: Global turnover of DM 2.5bn + 5,500 employees • BioTech Region Munich: 1800 jobs created	• 82,000 students • Ludwig-Maximilians University • Technical university • 2 Teaching hospitals • 2 applied science universities • 13 non-university research centres • 3 biotech oriented Max Planck Institutes Society for Health and the Environment	• Microoptometry • Materials • Optical communications • Photonics	• Kapitalgesellschaft for seed financing of biotech start ups • Hub of Munich biotech network (includes VC fund) • 10 dedicated VC firms • 1 dedicated consulting co. • 4 knowledge transfer consultant • Boston consulting group • Fraunhofer Management • KPMG • McKinsey • 3 kompetenznetze • Munich Business Angel network • Investors include: 3i, Apax Partners, Atlas ventures

Table 11. Continued.

	Size of the Biotech Community	Other Regional Facts and Figures			Market Area	
		Biotech companies	Employment	Research	Biotech Companies	Specialist service providers
Cambridge	• 175 biotech companies • 250 specialist service providers • 30 research institutes • 20 multinationals (pharma, agbio and food) • 4 leading hospitals	• 1995: 5 quoted companies (£400m market cap) • 2000: 20 quoted companies (£7bn market cap) • 20% Europe's publicly traded cos • 7 of top 15 LSE quoted biotech cos • 25% of Europe's top 50 publicly quoted cos • £1bn in VC funds • 900,000 sq ft utilised by lab-based biobusiness • 29 publicly quoted cos (17 UK, 8 US, 2 Canadian, 2 Euro)	• 10,000 employed directly related to biotech • 20,000 in life sciences • 20,000 in network membership	• 11 Nobel Prize winners • 3500 students • 350 research groups • 6 of top US biotech cos with operations in region • Large company research—AstraZeneca, GlaxoSmithKline, Dohme	• 30% develop biopharma products • 28% pharma services • 15% diagnostics and reagent supplies • 11% with agbio development • 12% biotech instrumentation and equipment	• 40% offer technical services • 9% offer financial services • 5% offer legal services • 15% offer dedicated consulting services • 31% offer other related services (e.g. biotech centre of excellence)

Table 11. Continued.

	Size of the Biotech Community	Other Regional Facts and Figures			Market Area	
		Biotech companies	Employment	Research	Biotech Companies	Specialist service providers
Jena	• Large firms: Jenoptik; Carl Zeiss, ABS, AGFA, H&W optical instruments, OSRAM semiconductors • 56 members of BioRegio Jena • 50 members of Bildverarbeitung Thüringen (training oriented) • 34 BioInstruments start-ups • Opthalmoinnovation Thüringen • 60 members of OptoNet Jena	• 31 new biotech companies since 1995 from BioRegio	• Bildverarbeitung Thüringen: worldwide turnover of companies – DM 80m + 850 jobs • BioInstruments: 350 jobs; 170 patent registrations; DM 98m in Jena biotech companies • DM 270m investment in university campus + DM 30million from BMBF • Turnover of DM 1bn worldwide in OptoNet + 6000 direct jobs created	• Jenaer Friedrich-Schiller Universitaet • Erfurter Universitaet • 2 FE Colleges of applied science (focus on medical technology, neurology, fibroptics, optometry and data analysis) • 11 non-university research centres including Fraunhofer, Steinbeiss & 2 Max Planck Institutes • 1 government laboratory	• BioInstruments (platform technologies) • Optometry and opthalmics • Cellular & molecular biology • Drug targeting • Materials	• 4 venture capital firms • 4 banks • 1 consulting firm • 4 kompetenznetze—networking structures to provide mentoring and support as well as international links.

Table 11. Continued.

Size of the Biotech Community	Other Regional Facts and Figures			Market Area	
	Biotech companies	Employment	Research	Biotech Companies	Specialist service providers
Manchester & NorthWest Public funding package for BioNow: £24.5m (DTI, NWDA & ERDF) Manchester Incubator: £15.4m total project funding from ERDF, University of Manchester, Wellcome Trust and Hulme Regeneration Ltd	• 120 biotech & biomed companies in region • 60 dedicated biotech in region • 15 listed companies in NW • 9 funded companies in Manchester incubator • 5 companies in Manchester Science Park • 5 multinationals (AstraZeneca, Aventis, Bristol Myers Squibb, Eli Lilly, Novartis Powderjet)	• 8 biotech companies in total	• £25m VC funds • 75,000 sq ft incubator building (fully occupied May 2001)	• 9 'biotech related' depts in NW universities (at 5 or 5* 2001) • 8000 S&T graduates from Uni of Manchester • AstraZeneca's largest world R&D centre • NHS networks • DTI networks (MerseyBio, BioNow)	• Vaccines, immunotherapy and gene therapy • Molecular diagnostics • Sensor technology • Speciality Chemicals • Instrumentation and spectrometry • Pharma companies • Wound healing and tissue engineering	• Specialism in biomanufacture

Source: Rebecca Harding fieldwork, 2002.

We can learn a lot from the analysis here. First, the market based strategy in the U.K. and the more 'engineered' strategy in Germany cannot be compared directly in terms of their effectiveness or suitability outside their national context. Germany has a networked science system that is characterized by the 'symbiotic tension' under which firms and research institutions compete for and collaborate in research projects. The BioRegio contest and the spread of other related initiatives through the institutional system of German R&D and technology transfer has produced a rapid catch up in biotechnology. This in itself has been impressive to watch—especially for those who judged the German system incapable of rapid change!

In contrast the U.K.'s more market based system would not be effective in Germany but has merits within the context of the U.K. economy. Universities are used to competition in research and this ensures that the quality of research conducted remains high. There are issues around the extent to which the system can be adapted to further technology transfer and maybe some of the reduction in the competitiveness of the U.K. biotechnology sector relative to Germany in the last couple of years stems from the difficulties that U.K. scientists and businesses have in collaboration—there is 'tension' but no 'symbiosis' between the users and the producers of science.

The issue of sustainability is key, especially for Germany where criticisms of its strongly public sector approach center around the small size of many of the biotechnology start-ups. Where the U.K.'s structures are more established, for example in Cambridge, the sustainability of the sector can be taken much more for granted.

However we can learn three key points from the speed with which Germany has caught up. Specifically these are:

- regions are important as vehicles for appropriate policy formulation and delivery;
- substantial funding is critical;
- funding is key—needs a lot of money because biotech is expensive and networked.

In conclusion, then, there is no Frankenstein Future as such, but there is a role for strong and careful regulation and monitoring of the technology. Along the way, in order to ensure that the public keeps abreast of the pace of change in this sector, it is also critical that the public understanding of the science itself is increased.

Finally, there is scope for understanding much more about the way biotechnology works from the standpoint of a more detailed Anglo-German comparison and further research should concentrate on addressing the following issues:

- first, policy has been a 'leap of faith' and measuring effectiveness has been hard. We need new measurements that incorporate the role of the tacit knowledge transfer and network development intrinsic to biotechnology research. In short, we need to be able to measure 'symbiotic tension' and its effect on the development of biotechnology;
- second, Germany has a higher number of 'platform technologies'—i.e. equipment and supplies or drug delivery systems that have clear commercial potential as opposed to U.K. which is still more research oriented. This may be because of differences in the applied research funding structure and in particular the use of equity-based finance in the early stages of biotechnology start ups. The area of biotechnology finance warrants further investigation since it may well be that the form this takes fundamentally alters the trajectory along which biotechnology research develops;
- finally, the management of small biotechnology firms is an interesting area for further comparative research. This has been conducted for Germany in some detail (ISI, 2001) but there is scope for expanding this on to a much more extensive level in order to examine the impact of networks on the trajectories along which biotech develops.

References

Archibugi, D., Howells, J. & Michie, J. (1999). Innovation systems in a global economy. *Technology Analysis and Strategic Management*. Basingstoke: Carfax.

Bundesministerium fuer Bildung und Forschung (2002). *Studierende und selbstaendigkeit ergebnisse der exist—studierendenbefragung*. Bonn: BMBF.

Cantwell, J. & Iammarino, S. (2000). Multinational corporations and the Location of Technological Innovation in the U.K. Regions. *Regional Studies*, **34** (4), 317–332.

Casper, S., Lehrer, M. & Soskice, D. (1999). Can high technology industries prosper in Germany? Institutional frameworks and the evolution of the German software and biotechnology industries. In: *Industry and Innovation*, **6** (1), 5–24.

Casper, S. (2000). Institutional adaptiveness, technology policy and the diffusion of new business models: The case of German biotechnology. *Organisation Studies*, **21** (5), 887–914.

Cooke, P., Gomez-Uranga, M. & Etzebarria, G. (1997). Regional innovation systems—Institutional and organisational dimensions. *Research Policy*, **26** (4), Issue 5, 475–492.

Cooke, P. (2001). Biotechnology clusters in the U.K.: Lessons from localisation in the commercialisation of science. *Small Business Economics*, (1–2), 43–59.

Cooke, P. (2002). Biotechnology clusters as regional, sectoral innovation systems. *International Regional Science Review*, **25** (1), 8–37.

De la Mothe, J. & Paquet, J. (Eds) (1998). *Local and regional systems of innovation*. London: Pinter.

Department for Trade and Industry (1999). *Biotechnology clusters*. London: DTI.

Diez, M. A. (2001). The evaluation of regional innovation and cluster policies—towards a participatory approach. *European Planning Studies*, **9** (7), 907–923.

Dohse, D. (2000). Technology Policy and the Regions—the Case of the BioRegio Contest. *Research Policy*, **29**, 1111–1133.

Duysters, G. & Hagedoorn, J. (1996). Internationalisation of corporate technology through strategic partnering: An emprical investigation. *Research Policy*, **25** (1), 1–2.

European Commission (1998). *Risk capital—A key to job creation in the EU*. Brussels: European Commission.

European Union/SME Concerted Action (2000). Proceedings of the SME Concerted Action Workshop, Copenhagen, Denmark, 17th–18th January 2000.

Eurostat (2001). *Statistics in focus—Science and technology: 'How much do governments budget for R&D activities?'* European Commission, Brussels.

Fagerberg, J. (1987). A Technology-Gap Approach to Why Growth Rates Differ. In: *Research Policy*, **16**.

Freeman, C. (1995). The 'national system of innovation' in historical perspective. *Cambridge Journal of Economics*, **19**, February, 4–24.

Freeman, C. (2002). Continental, national and sub-national innovation systems—complementarity and economic growth. *Research Policy*, **31** (2), 191–211.

Furman, J. L., Porter, M. E. & Stern, S. (2002). The determinants of national innovative capacity. *Research Policy*, **31** (6), 899–933.

Gambardella, A., Orsenigo, L., Pammolli, F. (2000). *Global competitiveness in pharmaceuticals—A European perspective*. Report prepared for the Enterprise D-G of the European Commission. Brussels: DG Enterprise.

Geuna, A. & Martin, B. (2001). University research evaluation and funding: An international comparison. SPRU electronic working paper 71.

Global Entrepreneurship Monitor (GEM) annual reports, 2000, 2001 published by London Business School/Babson College.

Giddens, A. (1998). *The third way*. Cambridge: Polity Press.

Harding, R. (1999). *Venture capital and regional development*. London: IPPR.

Harding, R. (2000). Resilience in german technology policy: Innovation through institutional symbiotic tension. *Industry and Innovation*, **7** (2), 223–244.

Harding, R. (2001). Competition and collaboration in German R&D *Industry and Corporate Change*, **10** (2), June, 389–417.

Hicks, D., Izard, P. & Martin, B. (1996). A morphology of Japanese and European corporate networks. *Research Policy*, **25**, 359–378.

Kulicke, M., Menrad, K. & Wörner, S. (2002). Innovationsmanagement in kleinen und mittleren unternehmen. Karlsruhe: Fraunhofer ISI.

Koschatsky, K. (2000). A river is a river—cross-border networking between Baden and Alsace. *European Planning Studies*, **8** (4), 429–449.

Koschatsky, K. & Sternberg, R. (2000). R&D cooperation in innovation systems—some lessons from the European Regional Innovation Survey. *European Planning Studies*, **8** (4) 487–501.

Leydesdorff, L. & Heimeriks, G. (2001). The self-organisation of the European Information Society: The case of biotechnology. *Journal of the American Society for Information Science and Technology*, **52** (14), 1262–1274.

Love, J. H. & Roper, S. (2001). Location and network effects on innovation success: Evidence for U.K., German and Irish manufacturing plants. *Research Policy*, **30** (4), 643–661.

Lundvall, B. Å. (Ed.) (1992). *National systems of innovation—Towards a theory of innovation and interactive learning*. London: Pinter.

Kaufmann, A. & Todtling, F. (2001). Science-Industry interaction in the process of innovation: The importance of boundary-crossing between systems. *Research Policy*, **30** (5), 791–804.

Malerba, F. (2002). Sectoral systems of innovation and production. *Research Policy*, **31** (2), 247–264.

Martin, B. & Salter, A. (1996). *The relationship between publicly funded basic research and economic performance: A SPRU review*. A report for HM Treasury, SPRU University of Sussex.

Molas Gerrat, J. (2002). Measuring third stream activities. Final report to the Russell Group of Universities. Brighton: SPRU, University of Sussex.

Mowery, D. & Rosenberg, N. (1999). *Paths of innovation: Technological change in 20th century America*. Cambridge: Cambridge University Press.

Mowery, D. C. & Oxley, J. E. (1995). Inward technology transfer and competitiveness—The role of national innovation systems. *Cambridge Journal of Economics*, **19** (1), 67–93.

OECD (2000a). Main science and technology indicators, Number 2, 2000. Paris: OECD.

OECD (2000b). Basic science and technology statistics, 2000. Paris: OECD.

OECD (2000c). *Innovative clusters: Drivers of national innovation systems*. Paris: OECD.

OECD (2001). Main science and technology indicators, Volume 1, 2001. Paris: OECD.

Owen-Smith, J., Richardson, M., Pammolli, F. & Powell, W. (2002). A comparison of U.S. and European University-Industry relations in the life sciences. *Management Science*, **48** (1), 24–43.

Patel, P. & Pavitt, K. (1999). National systems of innovation under strain: The internationalisation of corporate R&D. In: R. Barrel, G. Mason & M. Mahony (Eds), *Productivity, Innovation and Economic Performance*. CUP.

Porter, M. (1998). Clusters and the new economy of competition. *Harvard Business Review*, Nov.-Dec., 70–90.

Porter, M. (2002). Regions and policy. Unpublished paper to the DTI Regional Policy Seminars, April 2002.

Prais, S. (1981). *The evolution of giant firms in Britain, Germany and America*. Cambridge: Cambridge University Press.

Sachsenian, A. (1997). *Regional advantage: Culture and competition in Silicon Valley and Route 128*. Cambridge, MA: Harvard University Press.

Salter, A., D'Este, P., Martin, B., Geuna, A., Scott, A., Pavitt, K., Patel, P. & Nightingale, P. (2000). *Talent, not technology: publicly funded research and innovation in the U.K.* Report to the CVCP and HEFCE. Brighton: SPRU, University of Sussex.

Senker, J. & Zwangenberg, P. (2001). European biotechnology systems. October 2001, TSER project no. SOE1-CT98–1117. WP 5. Brighton: SPRU, University of Sussex.

Smith, H. L. (1998). Barriers to technology transfer: Local impediments in Oxfordshire. *Environment and Planning, C-Government and Policy.* London: Pion.

Sohet, S. & Prevezer, M. (1996). Institutional linkages, technology transfer and the role of Intermediaries. *R&D Management*, **26** (3) 283–298, July. Oxford: Basil Blackwell.

Sorge, A. (1999). Mitbestimmung, arbeitsorganisation und technikanwendung. In: W. Streek & N. Kluge (Hg.) *Mitbestimmung in Deutschland: Tradition und Effizienz* (Ch. 1). Frankfurt: Campus Verlag.

Soskice, D. (1996). German technology policy, innovation and national institutional frameworks. WZB Discussion Paper, FS 1–96–319. Berlin: Wissenschaftszentrum.

Strombach, S. (2002). Change in the innovation process: New knowledge production and competitive cities—The case of Stuttgart. *European Planning Studies.* Basingstoke: Carfax.

Todtling, F. & Kaufmann, A. (2001). The role of the region for innovation activities of SMEs. *European Urban and Regional Studies*, **8** (3), 203–215.

Vickers, I. & North, D. (1999). Regional technology initiatives: Some insights from the English Regions. *European Planning Studies*, **8** (3), 301–318.

Webber, D. J. (1995). The emerging federalism of U.S. biotechnology policy. *Politics and the Life Sciences*, **14** (1), 65–72. Guildford, U.K.: Beech Tree Publishing.

Wörner, S., Reiss, T., Menrad, M. & Menrad, K. (2001). European biotechnology innovation systems—The case of Germany. WP2 SOE1-CT98–1117. Karlsruhe: Fraunhofer ISI.

Zeller, C. (2001). Clustering biotechnology: A recipe for success? Spatial pattern of growth of biotechnology in Munich, Hamburg, and the Rhineland. *Small Business Economics*, **17** (1–2), 123–141.

Part XIV

Innovations of the Future

The International Handbook on Innovation
Edited by Larisa V. Shavinina

Future Innovations in Science and Technology

Joseph F. Coates

Washington, D.C., USA

Abstract: This chapter describes future innovations in genetics, brain science, information technology, nanotechnology, materials science, space technology, energy, and transportation. This 25 year look into the future in Part I considers scientific developments and in Part II their practical technological applications. The work depends upon two assumptions: first, that we have the capability to see the future to an extent that is useful; second, we can take action to promote the desirable and to discourage and even prevent the undesirable. We of course have the moral obligation to use these abilities to anticipate and to influence.

Keywords: Innovation; Future; Genetics; Brain science; Information technology; Medicine; Ecology; Environment.

Introduction

Over the next quarter century there will be an indeterminately large number of advances in science and technology, which will affect our personal, family, group, work, organizational and governmental lives and behavior. Most of these effects will be good, but others will adversely affect us. The purpose of this chapter is to lay out that broad range of developments, using as the guiding term 'innovation'.

That, however, creates a problem: innovation has two quite distinct meanings. The first meaning is that of invention, creation, or discovery, to bring forth something truly new and useful. The other meaning is adoption of what is new to you, whether 'you' are an individual or an organization. For example, somebody invented the word processor (innovation, meaning no. 1) and that having been invented in different forms, it was then adopted (innovation, meaning no. 2) by literally millions of people and organizations. This chapter concentrates on the creative sense of innovation, and only lightly touches on applications. Sometimes the movement of uses or applications from institution to institution or person to person is referred to as 'technology transfer'. The reason for discussing both of these meanings of innovation is that each is essential to our future prosperity and well-being. A great invention that doesn't propagate might as well not exist, while the capability to propagate something not worth propagating is just an empty game. See for background, the U.S. Dept. of Energy (1984).

The two meanings of innovation are growing more intimate. Corporations, nonprofit groups, and government are getting more integrated in their perspective on the two meanings. For example, common in corporate research these days is the question of what the market is. In other words, if something does come forth from research and development (R&D), how extensively is it likely to be adopted?

The government is increasingly engaged in programs to support basic research that will later develop into practical applications. Clear examples of that are the programs of The National Institutes of Health and the broad sweep of basic research sponsored by The National Science Foundation. On the other hand the government is increasingly attempting to anticipate how the basic research it supports will it be used, that is, what its applications will be. A recent example is by Roco (2002). The government is also lifting the antimonopoly constraints on organizations, allowing them to come together and engage in 'pre-competitive' joint research—that is, research that is of such broad generic value that any organization could use it without necessarily revealing what the organization's detailed proprietary developments or plans are.

New inventions bring new capabilities. The capabilities can be in any dimension: color, size, shape, material, stability, durability, physical or biological properties, or scope of effectiveness. Capabilities determine the future applications or uses and hence determine how likely the new development is to be important.

Time Horizon

The study of the future is now a well-established procedure in business, government, and other large organizations. Though short-term, one to three year forecasts, are usually dependent on mathematical models and masses of data, they deal with business concerns about commodities, markets, and production. Other forecasts, such as this present one, are more strategic in purpose, by helping the reader or user to understand the forces at play which may substantially alter his or her enterprise over a longer period, usually ten to thirty years. These strategic forecasts are based on three assumptions. First is that we can see the future to an extent that is useful. Second is that we have the capability to intervene, to promote desirable and reduce the likelihood of undesirable outcomes. Third, we have the moral obligation to use the capabilities to anticipate and to influence. While forecasting does describe future outcomes, its primary purpose is to draw the user into an awareness and examination of his or her assumptions about the future. After all, it is the case that most organizational failure results from faulty assumptions about the future.

In a study of the future more detailed than this chapter can be, one would define the system under consideration, and then for each component of the system, identify trends and forces at play. With this in hand, one would define a variety of alternative futures from which one would then draw implications for managing the future. This chapter can only sample from a wide range of such studies conducted by the author (Coates, 1997b) and others.

In anticipations of big change, it is almost useless to look out only five years or so. Such a short term finds many technologies frozen because of long planning cycles and long R&D requirements. The exceptions include minor innovations that have to do with eye appeal or the combinations of features that are frequent in electronics and in food. To go out 25 years provides an opportunity to look at today's seminal and emerging developments and anticipate how they may mature over the next generation. Empirical research shows that seminal developments take about fifteen to forty years to enter the market or come into use, and to reach a point where they are of any significance. See the U.S. Dept. of Energy (1984). We strike a middle ground here by looking out about a quarter of a century.

Seven fields are discussed below as the most important ones for innovation in the sense of new inventions, discoveries, and devices. The emphasis is on physical and biological sciences. Relatively little attention is given to the social sciences, although they will be increasingly important as they move from being advisory to being more definitive. Each of the seven areas covers basic developments and some of their implications. Part II of this chapter pulls together practical applications in some exemplary areas to illustrate the complexity of innovations affecting complex systems.

Part I: Innovations in Science and Technology

This part discusses seven areas of scientific and technological developments likely to have broad practical applications. They are genetics, brain science, information technology, materials science, energy, spac and ecology and the environment. While information technology is the most widespread and dynamic in bringing about successive waves of innovation and change, two biological sciences are discussed first, genetics and the brain, because each are at the seminal stage. Each will in the next quarter century blossom into hundreds, if not thousands, of practical applications, delivering unheralded and until recently, unanticipated capabilities to humankind. As a rough analogy, they are each at the stage of information technology in 1950.

Genetics

In the last 50 years, research has established the following:

- all heritable characteristics of living things are carried by a class of chemicals called deoxyribonucleic acid (DNA);
- this is a long chain made up of basically four components, which one can consider as A, B, C, and D;
- those components comprise a code;
- the code forms units that are called genes, which represent the heritable characteristics of the organism;
- that code has been deciphered. It leads, in the egg, to the production of proteins. Those special proteins are catalysts, or more properly, enzymes, which working with the material in the immediate environment proceed to restructure those materials into the organism that the DNA is programmed to produce;
- we have learned to synthesize DNA;
- we have developed means for taking DNA apart, for putting it back together, and for combining synthetic DNA with natural DNA;
- we have learned that we can take DNA from any organism and put it into any other organism, and if circumstances permit the resulting organism will manifest the newly transferred characteristics.

In brief, we have developed a technology of DNA. A good basic reference is Atherly (1999). For a brief view of the future of genetics, see Coates (1997a).

The Human Genome Project, which is the most important biological project now under way, had its origins in medical concerns and has consequently been primarily focused on diseases and disorders. See Shapiro (1991). We should soon be in a better position to identify and relate the structure of DNA to specific diseases and susceptibilities. It is virtually daily news that a connection has been made between some disease and its genetic base. All of the thousands of diseases and disorders fall somewhere on a spectrum from absolute certainty of occurrence to probable occurrence to unlikely occurrence. Hawley (2001) has written a useful introduction to the human genome, as has Tudge (2000).

What are the consequences of this new knowledge? First and most obviously, will be easier and earlier diagnosis. If a disease is known to be genetic and is in your family, it is now fairly straightforward to determine whether or not you carry the gene, for those diseases whose gene locus on the DNA has been identified. Following diagnosis, but not close behind, will be attempts at prevention; that is, to intervene in some way or another to prevent the genetically programmed disorder from manifesting itself. Following that will be therapy.

Therapy will come in two primary forms: gene therapy and pharmacology. First, strategies involving replacing, neutralizing, or eliminating the defective gene, generically called gene therapy, will undoubtedly dominate the future of the treatment of disorders of genetic origin. Today, there have been no outstanding, unequivocal, complete successes, but this is the earliest stage of a true biomedical revolution. One has to be able to see the longer-term future, not just focus on the partial successes and failures of short-term basic research and experimentation.

Gene therapy might work by several different mechanisms. The easiest one to understand would use an organism, such as the influenza virus, which attacks a specific tissue—e.g. the lungs. If the genetic defect were one that affected the lungs, such as cystic fibrosis, one would remove the disease-causing portion of the influenza virus and attach to the remaining, now benign, virus the gene that was absent or defective in the lungs. Then, one would literally attempt to infect the person with that benign new virus, and thereby deliver to the somatic cells—the body cells—the genes necessary to effect correction in the specific biological target, the lungs. Many variations on this, as well as other strategies, are under extensive investigation as explained by Howard Hughes Medical Institute (1999).

The second strategy is less obvious, but in the short run—the next 10 or 15 years—may have bigger consequences. There are many relatively unexplored opportunities to intervene pharmacologically to prevent, arrest, or reduce the potential intensity of a disorder. Historically, most diseases have been treated at their beginning or their end points. For example, we give vaccines to prevent diseases. We dose with antibiotics when pneumonia occurs, Pharmacology

will expand over the next one or two decades, as it searches for and finds remedies at the intermediate biochemical stages in genetically based conditions. Again, see Howard Hughes Medical Insitute (1999).

The most interesting long-term consequence of genetics research is human enhancement. The genetic knowledge that will permit us to identify diseases and disorders will also allow us to identify the means and mechanisms for enhancing human capabilities. Unfortunately, this is an area that has tended to receive thoughtless, automatic, knee jerk, negative responses, as if the capability to enhance people's function must lead to that genetic horror movie, 'The Boys from Brazil', or to the rise of fascistic armies of clones prepared over a generation to sweep the world with their great strength and power.

Almost all of the ethical and conjectural discussion fails to address the single most important and obvious factor in the development of new genetic capabilities— what ordinary people will do when confronted with the opportunity to use specific genetics technologies.

Let me illustrate the kind of problem that might apply to. When Banting and Best discovered the role of insulin in preventing diabetes, that led to the survival of a large number of people who otherwise would have died before they procreated. Now those people survive and they reproduce, and continue to add their defective genes in larger and larger numbers to the human genome, i.e. our collective gene pool. That having happened over the last 70 years, makes the defective insulin gene or genes attractive candidates for genomic correction. We will be able to prevent the defective gene from passing from one generation to the next by replacing it in diabetic's germinal cells with a sound gene.

Inevitably, there will be some group of people who will reject the emerging capabilities to virtually eliminate disease and disorders and to enhance the human condition. It will be for a variety of reasons— fear, resistance to novelty, ideological indisposition and religious beliefs. In the long pull, there will be new genetically-based differentiation among people, but not necessarily in the sense of Aldous Huxley's Alphas, Betas, and Gammas (which were roughly the super intelligent, the intelligent and the dullards). It could come along a variety of different dimensions. We already have people who are genetically differentiated into what, if we were thinking of dogs, chickens and cattle, we would call varieties. We have tall, thin, dark-skinned people from East Africa. We have short-legged, squat, stout people in the Arctic. We have light-skinned, mesomorphic people in Northern Europe. We have shorter, dark-skinned people in the Mediterranean region. We have people with characteristic Asian features throughout the largest continent. We have a variety of distinctly different people among the Amerindians. Those genetic differences came about in response to natural forces shaping the preferential survival of the most environmentally appropriate people. The reality must be faced squarely, that we are the first species to be able to directly intervene in shaping its own evolution. That Lamarckian capability will be uniquely ours, and it will effectively be irrepressible. Our choice is not yes or no. Our choice is whether we intelligently or stupidly manage that capability.

The stages of application of medical genetics will first be diagnosis. Next will be correction or prevention of a disorder. Following that will be human improvement to help some people reach a common norm. Then there will be enhancement to raise people to competencies and capabilities far above the norm.

Let me suggest a plausible, hypothetical calendar of events on the immediate genetic horizon.

Human Genetics Development 2000 to 2040
(Primarily the United States)

2000 to 2015	Expanding knowledge of multiple gene interactions and the associated disorders.
2000	Human genome fully sequenced; genomics gives way to proteomics— protein mapping.
2000 to 2020	Upper middle class people primarily interested in genetic diagnosis and therapy.
2005 to 2025	Declining costs of genetic interventions.
2005 to 2035	Pharmacological and genetic therapy compete vigorously to treat genetically based disorders.
2005 →	Expanding market.
2006	Exponential growth in the practice of genetic enhancement.
2007	Two percent of children in the United States are genetically serviced perinatally, 3% more before age twelve.
2007 to 2025	Upper middle class people are first into genetic enhancement, that is, elimination of nonmedical or nuisance conditions such as overweight, short stature, etc.
2009 to 2016	Public policymakers set genetic enhancement goals in five countries and provide incentives for meeting them. They are North Korea, China, Finland, Israel, and Singapore.
2010	Genetic testing and interventions to influence outcomes become routine for 175 diseases and disorders
2011, 2013, 2022	Law defines parameters of genetic intervention.
2012	First legal genetic intervention to alter

2018

the germ line instigated by the pressure to correct the epidemic of Type 1 diabetics resulting from Banting and Best's discovery in 1922 that the antidiabetic hormone (insulin) was produced in the islets of the pancreas.

2023
2025

The Nobel Prize goes to the largest number of individuals (12) ever, all engaged in a multinational, highly integrated neurogenetics research program. The team established that between 67% and 84% of mental characteristics could be attributed to genes, and the rest largely to environmental circumstances ranging from intrauterine environments to home and school life. This triggers widespread public demand for genetic interventions to deal with mental characteristics. The resulting demand for interventions causes a quintupling of the federal research budget in mental genetics.

Mandatory premarital genetic counseling required in 14 states.

Growth in genetic servicing doubles every three years in the United States. By 2022, 65% of children are genetically serviced, 95% of them for diseases or disorders and 5% for enhancement. Twelve percent of interventions alter the germ line.

2028

Enhanced people are starting to form social groups and affinity groups on the electronic networks. MENSA has a rapidly growing subgroup called MENSA-E for those with IQs above 160.

The International Olympics Committee is discussing a special Olympics for the physically enhanced. Tryouts in three countries show a consistent 2% to 7% improvement by enhanced contenders.

2030

Decades of experience with hundreds of thousands of people define risks and previously unanticipated side effects of genetic interventions, leading to codified intervention strategies and higher safety and confidence in outcomes.

2030

Forty-three percent of American adults and 84% of children have at least a partial genetic profile.

2035

The anti-genetics movement (Americans for God's Way) has 16 million registered members and

	enjoys support by 27% of adults.
2035	Surveys show that 83% of members of Congress have had a genetic intervention in the immediate family.
2036	The American Medical Association Committee on Medical Genetics reports that since 1998 there has been a 35% decrease in genetically based disorders that manifest themselves in the first three years of life. Epidemiological studies have shown that genetically based disorders and conditions in adults 18 to 45 have declined 22%. The studies attribute 26% of this to premarital genetics counseling and the rest to interventions through pharmacological or gene therapeutic routes.
2038	Under the Olsen Act, passed by the U.S. Congress, citizens refusing to have a genetic inventory of their minor children will have their taxes raised by 30%, and will be double-billed for the treatment of their children's genetically-based disorders.

Recent treatments of the future of human genetics are by Stock (2002) and Dawkins (2002). Frankel (2000) deals with issues of inheritable genetic modifications.

Animal Genetics

The genetics of animals other than human beings is extremely important for commercial, scientific, bio-medical, environmental and other reasons. We are able to do things with animals which are unacceptable with people, from a research, experimental and applications point of view, but which will have pay-off in under-standing human biology, in the prevention of human diseases and disorders, and in human enhancement.

With rapidly reproducing animals like mice, it is now routine to identify a gene that has a function in people as well as in the mouse and 'knock out' that gene, that is, eliminate it from the organism. One can then observe what the effects are, and in turn observe what the consequences are of various interventions or treatments. It is also practical to go in the other direction, that is, add a gene to the mouse so that Its effects and consequences can be studied. Obviously these techniques have profound implications for the identification, treatment, and understanding the sources of diseases and disorders, by revealing details of the complex interaction among genes. Few or no genes do a single thing. Many of them interact at several stages in the development of the individual organism to regulate the scope and penetration of different func-tions or structures. See Howard Hughes Medical Institute (1999).

Modern genetics is widespread in animal husbandry. If one has a highly productive or valuable bovine, horse, pig, or other cultivated animal, it is extremely desirable to conserve those characteristics and repro-duce them in progeny with high fidelity. This goal amounts in many cases to preferring cloning. The cloning of agricultural animals has high economic value, leading to continuous herd improvement. That will provide important carry over of techniques to people.

Cloning research is going apace in animal hus-bandry. Cloning techniques form an elaborate process both biologically and technically. The individual steps and stages in that process will have direct transfer value to understanding every step and stage in human reproduction, in particular the use of stem cells for the reproduction of tissue and whole organs, and ulti-mately to the safe and successful cloning of people

The preservation of endangered species can be enormously assisted by contemporary genetics. As a species becomes scarce, one problem is the tendency in zoos to inbreed them. That brings out genetic defects and thereby weakens the whole surviving strain. Modern genetics will now allow more effective cross-breeding and reduces the genetic weakening of survivors.

Not far down the road will be the resurrection of extinct species. Mammoths, dodos, passenger pigeons, and scores of other extinct animals will walk the earth and fly the air again, in the next decades. Museums are vast repositories of tissue to provide the DNA for those resurrections. No Jurassic Park, but revival of recently-made-extinct species from which we have any tissue containing DNA, e.g. dodos, passenger pigeons, and mammoths. See Coates (1992a).

Genetic research on animals is by no means limited to mammals. The whole phylogenetic scale is under investigation in order to understand not only the characteristics of individual groups of animals, but also the parallel functions that genes perform as they appear in different genera. Gene maps of exemplar species are being prepared in the same way the human genome was recently mapped. As we better understand the genome of insects we will be in a better position to deal with them if they are pests or to enhance or promote them if they are, as are our bees and butterflies, economically or socially desirable. A fine description of how genes shape animal design is given by Lawrence (1992).

Another genetically important use of mammals will be as factories for the production of valuable bio-medical or health-promoting materials. By transferring the appropriate genes into a cow or a goat or another relatively large animal, one can produce complex materials, often too difficult in terms of present knowledge of manufacturing, too expensive to manu-facture, or too difficult to produce even in the laboratory.

Plant Genetics

In the same way that all of the animal kingdom is being explored and mapped genetically, similar mappings are going on in the plant kingdom. Objectives are to improve food and other commercial crops, build into plants better resistance to pests, and allow plants to grow in hostile or inadequate environments such as arid zones and saline water. The crossbreeding of plants genetically, that is, the introduction of genetic material from one plant into another, will expand the human diet enormously, improve foods nutritionally, and make new, better-tasting, and more-attractive foods with longer shelf-life. For example, in the Mexican diet rice and beans are critical. Each provides essential proteins, but neither supplies the full complement. Together they do. Genetic manipulation will make it practical for transgenic rice and beans each to be a fully balanced protein source. Mauseth (1998) in his textbook, *Botany,* provides an excellent extended discussion of plant genetics.

At the gustatory level we can anticipate the restoration of taste and improved flavor to many plant crops and fruits. More exciting is the mixing of genes from different plants to create new transgenic varieties of foods that are unknown today. We may have potatoes with the taste of bananas, or strawberries that have the taste of a fruit salad.

In forestry similar genetic developments are leading to pest-resistant trees and to trees that have better commercial characteristics. Genetics will offer better remedies and better preventatives for tree crops threatened by diseases and pests.

Microorganisms

Moving down to microorganisms, there is the great potential for them becoming little factories producing all kinds of commercially valuable materials. The genes introduced into a microorganism representing any characteristic or product of a donor organism will show up, that is, be expressed, unless something specific in the environment precludes it. Consequently microorganisms will continue to expand in their use for the production of exotic and expensive chemicals. Eventually they may produce commodity chemicals with unprecedented degrees of purity, because the microorganism can only produce what its genes instruct it to produce. If temperatures are too hot or too cold, rather than produce flawed products or undesirable byproducts as occurs in a conventional factory, the microorganism will just stop producing. Microorganisms will also be used to produce enzymes, that is, chemical catalysts that can be harvested and used in chemical processing or to treat diseases and disorders.

There is recent awareness of an extensive number of extremophiles among microorganisms and larger organisms, which thrive in very cold environments, very hot environments, in saline environments, or under high pressure. Those characteristics transferred into commercially useful species will enhance their capabilities as small factories. We are well aware of what microorganisms can do in industrial production: they are the basis of beer and wine productions and the backbone of the cheese industry. The industrialization of microorganisms is not new, but the expansion into new domains via genetic manipulation could be revolutionary.

Brain Science and Technology

Scientific developments over the past decade portend effective, reliable, safe, technologies for altering the brain and all its mental and physical functions. Prozac has enjoyed wide enthusiasm. It is the closest drug ever to Aldous Huxley's Soma—a make-you-feel-good, make-you-perform-better pill. Prozac is just the opening wedge for families of new drugs achieving the same or related objectives.

Genetics, particularly developments in molecular biology and the understanding of the human genome, will for the first time give our species direct control over its own mental evolution. Genetic research is establishing beyond any reasonable doubt that mental abilities, mental disorders, and cognitive short-falls and deficiencies are genetically-based if not fully genetically-determined. Genes set potential boundaries on performance, and the person's total environment at all ages adjusts how fully the potential is met, or diminished. A good introduction to the brain is by Restak (1995).

Psychologists have known for many years that there are numerous independent mental functions. Gardner (2000) is a current leading exponent of multiple intelligences. Genetics will tease out the loci or pattern of loci of those abilities in ever-more-refined degrees of detail. Tools for looking at the brain include imaging and tools for seeing the brain's structure and biochemistry, genetic and molecular probes, electrodes, and electromagnetic field detectors. We will be able to see the brain at work in real time and in three dimensions. The consequences of discovered linkages, whether to mental abilities or to disorders, will stimulate research on mechanisms of intervention to prevent, correct, cure, or enhance them.

We are already witnessing physical interventions in some of the outer reaches of the brain. Eyeglasses have been around for so long that we rarely think of them as brain technology, but we are moving to the point where it is becoming plausible that people will have truly artificial eyes; that is, at least light sensors that will directly affect the brain. We already have such technology for hearing—the artificial cochlea—that is steadily improving.

Neurosurgery is better able than ever to identify sites of mental pathology and to intervene positively. We have known of the brain's pleasure center for decades. It is manipulable in rats. It has also been stimulated in people, although that ability has not yet been converted into any practical use. The earliest developments in brain technology will be directed at the relief of diseases and disorders such as depression, schizophrenia, phobias, compulsions, addictions, and destructive stress responses. But the long-term avenues of development point to human enhancement by acoustic, photonic, electronic, biochemical, genetic, and other means of intervention, as well as through improved modes of training in the technologies of meditation and thought. Vandervert in this volume illustrates the rapid pace of brain research and how it is upsetting very long held beliefs. His work is breaking new ground in the relationship of memory to the cerebellum and how that relationship he is exploring stimulates and implements innovative thinking and action.

Hypnosis, as its psychophysiological basis becomes better understood, could go well beyond stage acts and become a powerful route to influencing our behavior. We know little or nothing about the biochemical and neurological bases of hypnosis, which has within its established boundaries demonstrated repeated capability to enhance human performance.

Research Promises in the Coming Decades

- A complete structural, functional and biochemical map of the brain—a 'geographic information system'—of the brain. Overlaid 3-D coordinated, computerized databases will correlate structure, connectivity, neurochemistry, metabolic, and electromagnetic activity to thinking, feeling and all brain activities.
- A genetic map for all brain disorders, along with a guide to the type and extent of environmental and psychosocial influences, which interact with that genetic template.
- The ability to scan and visualize any portion of the brain from outside the head in real time, simulate it and model brain activity and function on a computer, to test the effects of biochemical or behavioral, physical, or surgical therapies.
- The ability to grow neurons in culture, biochemically alter their development potential or function, and genetically engineer them.
- Much more discriminating, objective, accurate diagnosis of mental and neurological conditions based on biochemical and imaging signatures for each mental health state.
- Prevention and successful treatment of many more disorders and conditions—aggression, learning problems, depression, degenerative disease, dementia, and other conditions associated with the physical and social environment, such as aging, stress, eating disorders, and pain.
- Treatment through the brain of conditions not considered primarily neurological, such as infertility, diabetes, asthma, obesity, and hypertension.

- Treatments to enhance cognition, learning and other brain activities.
- Possibly, control of cognitive ability, mood and memory.
- Also possible: direct control of equipment and computers through actuation by brain signals.
- Understanding the genetic and biological bases of cell death and the consequences for brain functions, which could alter the results of injury and aging.

An excellent review of brain science is by Carter (1998).

Information Technology

Information technology has the highest probability of having the most dramatic effects over the next 25 years. While one can see continuous changes over the past 50 years, the outcomes in the next century will be even more dramatic. Information technology will affect personal life, business life, social life, government activities, international commerce and foreign relations. (For convenience, we include under information technology: telecommunications, computers, virtual reality and technologies promoting smartness, that is, intelligence in devices and systems, including robots.)

Among the principal effects of the widespread use of these technologies are the following. First, every business is now primarily an information machine, irrespective of what a company thinks its business is. It may sell shoes, food, or clothing, or manufacture gadgets, but the reality is that information technology has come to dominate every aspect of the business enterprise, to the extent that businesses are more alike than different by being information machines. Accommodating to that will be important, as businesses reset their priorities to keep that information machine humming and exploit its capabilities. They will recognize the tremendous power of the information machine by reaching out to customers, suppliers, public interest groups, government, competitors, and other organizations. Information technology will also affect the traditional internal elements of the firm.

Second, information technology first allows and now demands a total systems orientation in all institutional enterprises. All enterprises for the indefinite past have recognized that they are a part of a larger system, but the practicalities of life in business or government allowed only limited attention to most of the elements in the system. The failure to take the total system into account has left many organizations at a tremendous disadvantage when they find that the world has changed while they were not watching. Information technology now makes it practical to have a truly holistic approach to all information and knowledge relevant to the enterprise.

Third, information technology fundamentally contracts time, moving things at a faster pace. By contracting time, it also in many regards eliminates distance. Aside from any trivial inconvenience in the fraction of a second it takes to transmit an electronic message or image one can now communicate visually, verbally or with data, from any place to any place at any time.

Fourth, the consequences of all of these capabilities lie in two distinct domains, with a fuzzy interface. Almost all of information technology is introduced because it promises to deliver greater effectiveness and efficiency. It is the second level of effects, the transformational ones, that cause fundamental changes in the organization, its functions and operation. One can see a transformation in institutions in the recent enthusiasm for knowledge-management, which is intended to cast over the whole organization a new network of communication that allows the organization to know what is known inside the organization and have timely access to it. That point and that need was nicely put by Lew Platt, a former CEO of Hewlett-Packard: "If HP knew what HP knows, we would be three times more profitable". Similarly electronic commerce is radically changing companies as they try to come to grips with the use of information networks to substitute and replace or augment traditional marketing and sales.

Fifth, the formation of networks is an absolutely central characteristic of information technology. Networks may be inside of organizations, to facilitate communication and the flow of work. They may be outside of organizations, in public interest group or in affinity groups to share concerns about a disease, environmental problems, films, car and travel or anything else. Those networks widen people's awareness and give them more satisfying and rich communications, with regard to their specific interests and concerns.

Sixth, information technology can often substitute for materials and for human resources and create new enterprises. Information technology creates new businesses, not necessarily big ones. Every innovation creates new needs. Those new needs create niche markets and opportunities. The globalization of every commercial enterprise is now well underway. Global networking makes it first attractive and later necessary to be sure that one has a full grasp of what is going on around the world. Going along with globalization and pushing it hard is the need for standardization. Standardization has the general effect of making parts more compatible with each other, systems more interchangeable, and linkages cheaper and more effective. Illustrations throughout this chapter highlight these and other capabilities coming about from information technology.

Seventh, technologies can coalesce to create striking new capabilities. To illustrate, telecommunications and computation have now come together to create a capability called virtual reality. One can create images, scenes, and interactions that do not exist in any place

except in cyberspace. The computational capabilities bring reality to the images of devices and scenes, while telecommunications carry them wherever they need to go.

Virtual reality and associated experiences will drastically alter education by making it tailored to what one knows, what one need to learn, and one's particular preferred learning strategies. Virtual reality will optimize and accelerate remote learning. Virtual reality will also allow all kinds of simulation, so much so that no devices, whether as simple as a new wine bottle opener or as complex as a new housing development or cruise ship, will even begin to be built until it is completely planned, designed, built, tested, evaluated, and modified in cyberspace.

Even today, with relatively primitive virtual reality, one can walk through the offices of a not yet built building, test it for comfort and size, and even begin to place furniture in it and decorate it to one's taste. Virtual reality will have great effects on the modeling of social, economic, political, physical, and infrastructural systems. The model can then be tried out to test modifications, intrusions, inventions, innovations, and habitability.

Jacky Swann and her associates have in this volume been exploring and codifying the connections between knowledge, networking, and innovation. This triadic relationship is crucial to new developments and their applications. The model they are developing linking and comparing invention, diffusion, and implementation promises to shed beneficial light on behavior of all kinds of institutions.

Materials

The science of materials is one of the most dynamic, complex, and important areas of emerging developments. After all, our whole world is made up of materials. All products of industry, all the things that we have as food, clothing, and shelter, are materials. The physical instruments that control devices and regulate and operate the artifacts of our world are made of materials.

Information technology is ubiquitous in its effects on materials. It allows us now through computer programs to picture individual molecules and the molecular structure of materials, including surfaces. We are beginning to understand more deeply, at atomic and molecular levels, interactions among materials and the design changes that influence those interactions. For example, catalysts, which are extremely important in so many sectors, are chemicals that remain unchanged themselves but bring about changes in other chemicals that pass by them or over them or interact with them. Now it is routine to study the structure of a catalytic surface on a computer screen and design and redesign that structure to interact with molecules to be catalysized. This is an enormous advantage in subsequent laboratory work because it provides clear guidance to research goals.

The ultimate uses of the materials are having a dramatic effect on quality control and the production of more uniform, reliable, and longer-life artifacts. Computers are leading to smartness in artifacts and devices. By smartness, we mean that any device will be able to sense and respond to three things. First, is it operating well internally? Second, is it performing its tasks effectively and efficiently? Third, if the answer to either of the first two questions is no, the device will initiate repair or call for help.

Smartness is often accomplished by the introduction of microsensors, microcomputers, and actuators, but increasingly materials themselves are becoming smart. Examples are materials that have a built in memory, and under certain conditions of temperature, pressure or stress will revert to a previous shape or condition. This property opens up a potential for many kinds of applications, including but by no means limited to alarms. Just imagine the consequences when a surface finish or paint can announce that it has holes in it, or when the beams in a building or bridge can announce that they are weakening. Smartness will ultimately become universal. For a good overview of materials, see Forester (1988) and the U.S. Congress Office of Technology Assessment (1988), as well as Ball (1997). For an authoritative account of some leading edge developments, see Thompson (2001). Psaras (1987) has edited an authoritative volume for the National Academy of Sciences on materials research.

Biomimetics

Plants and animals do marvelous things, not all of which are fully understood. Increasing understanding of the chemistry, biology, dynamics and physiology of plants and animals is opening up opportunities to mimic their characteristics with man-made materials. The incredible strength of some kinds of seashells, the dynamic characteristics of animal flight, and insect communication by chemical receptors suggest new materials and new designs. The responses of plants to degrees of sunlight by their own movement create possibilities for imitation, duplication, and even enhancement in terms of material science. Understanding the special things living things do and how they do it is leading to great new things done with materials. See Benyus (1997).

Composites

Historically, most of the materials in the world, while sometimes complex—such as concrete or alloys—have been relatively simple scientifically. Now, through better understanding of fundamental characteristics some material can be given unprecedented strength and resistance to environmental factors, such as heat. For example, the new materials generally referred to as composites often have very high strength and a great deal of environmental resistance so that they become competitive with more traditional materials for special applications. They promise to be able to allow structures of equivalent strength but much lower total weight than those from traditional materials, steel and concrete.

Composites are not limited to use in new super effective golf club and tennis racquets but are likely to become preferred structural material for automobiles, aircraft, other vehicles, and for containers. They are likely to become more durable parts in the structure of buildings too, and could lead to structures that can be assembled, modified or dismantled much more conveniently and cheaply then is practical today.

Traditional Materials

By no means are all of the developments in material science directed at the new. Familiar materials like glass, concrete, and wood are all subjects of a substantial infusion of scientific research to improve their characteristics. One of the wonders of glass technology is optical fibers, so pure that light waves can travel for hundreds of miles through fibers before they have to be amplified. The technology of wood has developed so much in terms of its manipulation that the current theme through the industry is 'Anything can be made from any part of the tree'. All of the wood is useable in various forms to make almost any kind of useful product. In addition, genetics—as already discussed—will improve the quality and durability of wood, enhance the productivity of commercial forests, and lead to better resistance to pests.

Concrete is one of the most widely researched materials in the world as everywhere people search for concrete made out of local material. The U.S. Chemical Industry (1996) has offered its technology vision for 2020.

Surface Science

As devices get smaller the important factors influencing their behavior are less their bulk characteristics and more their surface characteristics. There is a surge of R&D to understand surfaces of materials. Much of that is that directed at microelectromechanical systems (MEMS) and nanoscale devices, but it is not limited to that. As we understand surfaces better, it will also influence the structure of objects of normal scale, add new characteristics to them, improve their finishes, and give them greater durability.

A long term trend in the use of all kinds of materials is the shrinkage in size of devices. The microchip is probably the most widely recognized example of this, in which literally hundreds of thousands, even a million and soon a billion or more transistors are built into a chip no bigger than the surface of a small fingernail. But, that is not the only area of development in shrinking devices.

MEMS (microelectromechanical) involve devices nominally at the size of a cross section of the hair, that can be used to sense, to respond, and to activate and or to act themselves. For example, MEMS are now the smart device that operates the safety bags in an automobile.

MEMS technology is only part way down the scale where a great deal more exciting research is going on. That is at the so-called nanoscale, which has devices and things built at the level of a billionth of a meter (approximately a billionth of a yard), or on the scale of molecules. Nanoscale devices do not now exist in any practical form, but this is a dynamic area of research and it holds the promise of being able in an ordered systematic way to plan, design, structure, and build up to useful macro-scale devices. Nature already does through genes and their interaction with the micro-environment. The goal of nanoscale research is to do with a wider range of materials what their genes are already doing with organic materials—to build-up of atoms and molecules to human scale devices. There is of course a substantial potential for the interaction of genetics technology with nanoscale technology, but the significance of that remains in the future because the research is at such an early stage. Malanowski (2001) has done a comprehensive brief review of nano-technology.

In terms of more traditional chemistry, manipulation at the nanoscale also goes on. Current speculation sees nanoscale devices that could be swallowed or injected into the body and would have the capability to treat sick cells, or attack foreign bodies such as bacteria and neutralize them. The anticipated medical applications are developed in great detail by Freitas (2002) as part of an ongoing trilogy. Taking advantage of scare tactics, Crichton (2002) has written a novel on how nanoparticles attack us.

Energy

Greenhouse warming and the significance of its effects will primarily determine the future sources and use of energy. If one accepts the current scientific consensus that greenhouse warming is real the primary open question is the extent and severity of its consequences. The forecasts anticipate continued global warming from the accumulated carbon dioxide and some other industrial gases in the atmosphere, the earth will continue to retain more of the sun's heat than in the past with less heat and energy reradiated into space. That is the basis of the term greenhouse warming. See Watson (2001).

The critical consequences of greenhouse warming are anticipated to be, first, a smearing out of the seasons. The boundaries between winter and spring, spring and summer, will become more fuzzy. Second, overlaying the smearing out will be much more spiky weather—deeper snow falls, heavier rains, and longer droughts. Warming ecological and agricultural zones

will move toward each pole. That has big implications because agricultural productivity in the last seven or eight decades has been determined by a balance between plant type, soil type, water availability, and other additives in the forms of fertilizers and pesticides. Consequently, worldwide food production will have to be geographically rebalanced in terms of those four variables.

An important factor accelerating that accommodation will be genetics technology as discussed above. A further consequence will be ocean rise, initially the order of a couple of inches or a few centimeters. The net effect of that will be to make hurricanes, tornadoes and other ocean storms more severe by carrying the damaging water farther inland. Lying a few decades further in the future, but perhaps already accelerating at a greater than anticipated pace, will be the melting of the Antarctic ice cap, which will cause ocean rise on the order of meters or yards. This will wipe out a large number of island republics and wreak havoc on low level mainlands such as coastal Bangladesh and southern Florida. O'Neill (2001) discusses climate change and population. Also see U.S, Global Change Research Program (2000).

The implication for energy is the need for massive energy conservation through the use of less carbon-based fuel, but also more energy-efficient and effective devices such as automobiles, improved insulation against both heat and cold, and energy conservation in manufacturing. Those measures will not be enough in the terms of the most scientifically sound forecasts. There must also be a push for non-carbon energy alternatives.

The most attractive general alternative will be nuclear energy, that is, the wider use of fission technology. Fusion technology, while holding great theoretical promise, seems to be continually 50 years in the future. Fission technology may achieve a broader acceptance in Europe and the United States than it has recently had as a new generation of adult citizens comes along who do not make the strong association between nuclear energy for power and nuclear energy for war.

A second, extremely attractive new energy choice is photovotaics, the direct conversion of sunlight into electricity. That is a well-developed technology, based upon silicon, that can be used wherever the sun shines. The patterns of insolation are well worked out. One can anticipate therefore the total amount of solar electrical energy that would be available in most areas. Because of nighttime and bad weather, photovoltaics would have to be backed up by some forms of energy storage. That could be done in a dozen different ways, most obviously with batteries, but also by pumped water storage, pumping water uphill to operate a hydro-electric facility downhill.

At the moment photovoltaics in many parts of the world is not quite competitive with central power station electrically. As technological improvements continue it will have a great appeal everywhere for the operation of remote machinery, in rural villages for the operation of water pumps and television sets and electric lights that are essential to small communities. Recent scientific developments in making organically based photovotaics materials suggest great promise for further reducing price of photovotaic energy.

Other widely discussed energy sources such as biomass have intrinsic limitations. In order to grow enough biomass (beyond harvesting waste left over from crop production) would itself require energy, chemicals, and other treatments. Biomass is not likely to meet more than 3%–5% of total energy production. Geothermal energy has enormous potential prospects wherever there is geothermal activity. It is well developed in Iceland and Italy. The difficulty with it is not technological. High investments for long lifetime facilities do not make economic sense against a volatile price for fossil fuels which could force a geothermal facility into bankruptcy.

Ocean energy coming from tides, or more likely from the thermal gradient where the warm Gulf Stream passes through colder water, can be used to make fantastically large amounts of electrical energy, at least based on scientific principles and detailed analyses. While studies show the great appeal in using ocean energy, one again runs up against unwillingness to give it a try. No government and no corporate consortium have been willing to even pay for a field demonstration.

An attractive and growing possibility for broad scale energy generation is wind. The technology has continually improved, particularly over the last seven or eight years, and will undoubtedly find widespread use either in wind farm generation or in a more distributed way similar to the historic generation of energy in the Netherlands. There are limitations on wind energy. It calls for backup storage. There also is the matter of noise—the thump, thump, thump associated with windmill blades.

Hydrogen is a widely discussed fuel, but in reality hydrogen is not a fuel but a mean of transmitting energy created in some other place. The parallel to hydrogen is electricity. While electricity powers many things it is not a prime energy resource but comes from energy produced in hydroelectric plants, coal- based power plants, and so on. Similarly hydrogen will not be the basic fuel, but will be produced in a variety of different ways.

Hydrogen, while extremely attractive because it produces no undesirable wastes, creates fundamentally new technical demands. It often attacks and embrittles metals; this calls for new high-tech pipe, storage and use. It also has very low density and therefore requires larger and stronger facilities to contain it, for an energy amount equal to a liquid fuel. Hydrogen of course is

particularly attractive in the long pull for powering automobiles, using fuel cells to produce electricity to drive motors on each wheel. See Hoffman (2001).

Fuel cells ideally would use hydrogen as fuel, but they get by using methane or methyl alcohol, while efficiencies are lower and byproducts are not simply water. Fuel cells in automobiles will be getting a basic test over the next decades. One general advantage of fuel cells is that they can be used at almost any scale from something as small as a hand-held flashlight or the activation of a digital camera, to a size approaching backup for a central power station.

On the other hand, assuming the unlikely—that greenhouse warming is not the key driver of the future of energy—then the central driver will be the price structure of petroleum. The limiting factor on the use of fossil fuels will then become environmental effects, particularly associated with sulfur-containing components and the production of nitrogen oxides. Here there will be strong pressures again for more efficient and effective use of fossil fuels. As it stands now the supply of fossil fuels are surprisingly generous around the world, and are growing. The known petroleum reserves are matched by known natural gas reserves of equivalent size. Natural gas of course has a great deal of appeal in that it is more simple and straightforward in fuel processing and has fewer troublesome components to it. It also is a much more effective base for petrochemical production. The rich and complex studies of petroleum and energy in the future are highly conflicted. Three solid studies, for orientation, are by Chen (2001), Deffeyes (2001), and Nakicenovic (1998).

Beyond these two portable fuels lies coal, which is available in the United States for 500 years at the present rate of fossil fuel consumption. Coal has many environmental difficulties, not only associated with its harvesting but also with undesirable impurities and byproducts.

Lying further in the background are methane hydrates. These are weak chemical complexes formed under proper conditions of pressure and cool environment. The complexes consist of a central core molecule of methane and a cage of water molecules surrounding it. The methane can be easily released by injecting heat into a deposit and methane will come forward much like natural gas. Not all deposits are pure; many have other materials that have to be sorted out. The critical factor is that the estimated reserves of methane hydrates are several times the total combined oil and natural gas reserves. If we can continue to use fossil fuels without environmental injury, we have as many as seven or eight decades in which to develop sources of energy not dependent upon fossil fuels. The move from oil to gas to gas hydrates implies a steady increase in the cost of fossil fuel energy. That in turn will be a powerful incentive for conservation and for developing alternatives.

Space

The scientific and technological developments likely with regard to space fall into two categories. One looks to earth itself. The other looks out to space beyond the earth to explore and understand the physical dynamics of the universe, to establish whether there is intelligent life elsewhere in the universe, and to begin plans for active human exploration and perhaps even the habitable development of our or other planetary systems.

Telecommunications depend upon scores of earth-orbiting satellites. They will become more numerous, more sophisticated and more specialized. A low cost, dense, worldwide net for communications of voice, data and any other kind of information will soon girdle the earth. Closely linked to communication satellites is the development of global positioning technology, which will allow the identification of a spot on the ground with increasingly greater accuracy, first down to 100 meters, then down to 10 meters, then to a half-meter and ultimately even closer. As militarily classified technology is civilianized, unprecedented geographical detail will become available to aid the movements of ships, trucks and automobiles. It will have effects on exploration; on outdoor recreation and tourism, making it almost impossible for anyone to become truly lost, that is, without human contact, anywhere on the face of the earth.

An established technology that is flourishing and becoming more commercially as well as governmentally important is remote sensing. The most obvious form of that is photography that can create terrestrial images from space. Less obvious are other forms of scanning in other bands of the spectrum to pick up specialized information, such as temperature. Today it is possible in most of the inhabited world to find a photograph of your dwelling. The images will become more available as souvenirs of what remote sensing can do.

Remote sensing, linked to ground based information will benefit commerce, research, and government with integrated information on virtually any spot on earth. Consider what geographic information systems can do in linking the number of people living in an apartment house: their social economic data, their mobility and movement in and out of the building, the environmental pollutants coming out of the building, the amounts of the goods going in. Those linkages will be a powerful factor in future planning and the management of the environment. Use of remote sensing will also grow in archeological exploration, seeing through ground cover to identify otherwise obscure traces of previous human occupation. Remote sensing and GPS will trace the formation and movement of pollutants and contraband material and other things, which the global community finds unacceptable to transport. Rees (1996) gives an

excellent overview of remote sensing. Current developments are covered on a Web site (U.S. Geological Survey).

War and violent conflict must not be ignored as a factor that will involve space, but that is outside the scope of this chapter. As the technologies of war are refined, whether they involve remote sensing, the use of lasers for communications or as destructive weapons, or the use of explosives for various kinds of space combat, they are all likely ultimately to transfer specific positive capabilities into the civilian sector.

Cosmology has gotten a tremendous shot in the arm in the last decades as new means of exploring space across the full range of the electromagnetic spectrum are coupled to telescopes outside the earth's atmosphere giving greater depth of field and clarity to images distant in space and time. Space-based telescopes are continually revealing wonders about the universe. There has never been a period when cosmological theory has progressed at a more comprehensive and rapid pace.

We know now that there are at least 70 planets outside our solar system. All of them are large like Jupiter and Saturn, but that in part is because of the limitations of observational technology. The fact that those planets exist and the ability to sort them into different kinds of motions and relationships to their parent star is strong evidence that some stars have solar-like planetary systems. The next stage in galactic astronomy will cost enormous amounts of money, but will allow us to search for Earth-, Venus-, and Mars-sized planets. The fact that other solar systems exist is one of the triumphs of knowledge, wiping out any sense of uniqueness to our solar system and questioning any belief about our uniqueness as intelligent beings in the universe.

Other more tangible advantages of both active and passive space exploration will be better understanding of the dynamics of the sun and the comparable dynamics of stars like the sun. The sun goes through cycles that influence weather, communication, and other factors of both short and long-term consequences to earth.

The scanning of space for signals or signs of extraterrestrial life will continue to a large extent as an informal activity. One of the most interesting aspects of that is SETI (Search for Extra-Terrestrial Intelligence) which links together thousands of small business- and home-based computers during their slack hours to process information. The fact that one can have tens of thousands of people voluntarily linked in that way suggests the potential for future cooperation when more serious or urgent issues arise.

The active exploration of space by sending people, robots or other automated devices into space will continue. The early American space program put men in space largely as a propaganda measure to outshine the Russians and to show that we could play the game better than they could. The technological and economic logic of space exploration is to send out robotic equipment first. It is relatively cheap and can be sent in greater numbers, also it is more resistant to adverse conditions and less likely to fail. We can anticipate robotic exploration of our solar system will expand in the next decades. We may send people in interesting numbers to the moon, Mars, and elsewhere, and begin intragalactic exploration by robots. It is difficult to see that happening in less than a quarter of a century, but by no means is it impossible to see that schedule either accelerated or deferred.

On the outer edge of the next quarter century we may give serious consideration to terraforming, that is, a long-term program to alter the atmosphere of Venus, which is rich in carbon dioxide, to make it habitable for people or plant life. That is the kind of plan that would operate on a hundred to thousands of years schedule. There should be no rush to do it.

Bases on the backside of the moon are attractive for many uses. They could be the launching platform for exploration. The space station that is now being assembled is a first step to more elaborate stations. The space station is the first stage of manned and unmanned long-term activities in space. It is a learning exercise. We can anticipate that in the next quarter century newer and larger space stations will be built in the region of the earth or on the moon.

Ecology and the Environment

Ecology is likely to continue its emergence as the central science of national and global environmentalism. The strength of ecology will depend heavily on massive daily data collection through both direct ground-based operations and remote sensing. Mathematical modeling and other quantitative tools are increasingly part of the ecologist's toolkit. The concept of ecology is no longer limited to the so-called natural environment. Close attention is being given to agro-ecology, that is, making the concept central to farms and managed forests.

Genetic manipulation is entering into ecology in interesting ways. As the ability to create transgenic species increases and managed environments in the ocean or on the land become more important, the survivability of species of fish, mammals, and other animals and plants will depend on direct genetic intervention.

Within the environmental movement, particularly as reflected in government, there has been a steady shift from a focus on individual species to biomes—large complex biological units such as forests. Rather than care for a particular tree or animals species, the unit of care will be ecologically broader, sounder, and more complex.

Sustainability is a now worldwide concept shaping the conduct of public and private policy toward the environment. Ecology has to become the core science underlining any practical sustainability. Remote sensing will provide close to real-time monitoring of the quality of the environment. Crops in Russia, China and elsewhere will be monitored, allowing for better world planning against food shortages or surpluses. On the other hand remote sensing will be a better way to identify effluent from factories producing excessive amounts of CO_2. Pollution patterns in oceanic or fresh water when coupled with more traditional ground-based information will enhance the control and management of the environment.

Ecology has such broad sweep that it has spun off sub-areas and related concepts, not all of which are scientifically or technically solid. Among the more solid are urban ecology, ecological risk management, restorative ecology, and conservation ecology. Of less clear scientific merit, but strongly ideologically based, are deep ecology, radical ecology and spiritual ecology. The capstone of sub-areas is human ecology. The general concept of ecology is the total system as the unit of discourse. Human ecology emphasizes the total human enterprise as the unit of discourse.

It is a central point of all ecology that one can never do just one thing. Side effects are universal and are often more important than the primary effect of any action or intervention. Ecology will become more important as a public policy focus as greenhouse warming causes the unstoppable migration of agriculture and forest zones throughout the world. Ayres (2002) has edited a comprehensive review of industrial ecology.

Environmentalism and public interest groups have not come to grips with greenhouse warming, the concept of sustainability, or with a total systems approach, but over the next decade they will, because they must.

Historically environmental issues in the United States have been framed around three core concerns: carcinogenesis, mutagenesis and aesthetic effects. There may be future issues of equivalent importance, among them effects of environmental pollutants on brain function, on the human reproductive apparatus, and on human and animal immune systems.

Part II: Special Application Areas

This section draws together ideas developed in the scientific areas mentioned above to indicate how they may introduce innovations into our lives and our society. Obviously some technologies such as information technology will be universal in their effects. Others such as genetics will be more directly connected to the biology and ecology of plants, animals and agriculture. New materials technology will be very pervasive. The point of this section is to sample some of the most important applications of the scientific and technical innovations already discussed. The extent to which these applications are fulfilled depends less on scientific developments than on other factors determining the speed and extent to which adaptations occur in business and personal life. This section deals with what the author judges to be high probability outcomes.

The seven exemplary topics are transportation, civil engineering and construction, health and medicine, manufacturing, residences, retail marketing, and education.

Forecasts on all of the topics in this Part II are made by Coates (1997b).

Transportation

Transportation involves both people and goods. While they both use the same modes of travel they do not use them the same way or to the same extent. With regard to the movement of goods one of the most important trends is the integration of different modes of transportation to facilitate the hands-off transportation of things from the time that they leave the factory assembly line until the time they reach the hands of the consumer. Information technology affecting logistics is making this continually more practical. One can anticipate for example, that an item manufactured in Silicon Valley, California might be delivered to Bangladesh or to Poland, without ever being touched by the human hand until it reaches its destination. More standardization of packaging sizes will facilitate this. The rapid development of RFID (Radio Frequency Identification) tags will allow the item in transit to literally speak for itself as to what it is and where it is headed.

Coupled to logistics will be global information systems. Global positioning will allow goods in transit to be identified down to a rail car and even the physical location within the car. This will tighten scheduling.

In the movement of people, the role of the automobile throughout the world will continue to increase into the next decades. One can take as a baseline consideration that as economies grow and people prosper they have three generally increasing social goals. One is more meat in their diet. Second is more automotive transportation and third is more recreational travel. Automotive transportation may be as simple as a motorcycle or a motor scooter all the way through light trucks and automobiles. The reality is that the appeal of personal transportation is independent of location and culture.

Greenhouse warming and the cost of fossil fuels will effect the development of automotive transportation. There will be a broader fanning out of vehicle sizes simultaneously with the greater increase in efficiency of all engines and, as discussed above, shifts to new and improved fuels. Overlaying the appeal of the car itself is the strong trend toward more automation. The ordinary item that today is the most information-technology dense is the automobile. The typical automobile may contain as many as fifty to two hundred electronic devices and the associated communication links.

The movement toward automated driving of vehicles will continue. In its initial stages communications will be from off the road sources to the vehicle or to the driver, and interrogation by the vehicle or the driver for information from off the road sources. The direct control of the highway over the movement of the vehicle is three or four decades into the future, but more information technology introduced into the vehicle will improve traffic efficiency and effectiveness, as well as the density of travel. Short distance radar, for example, will allow cars to communicate with each other about speed and distance. More importantly, automation's shorter response time will reduce the likelihood of multiple car accidents so common in a chain of vehicles in dense traffic. Cars will be able to receive information from other vehicles and sense the conditions of other vehicles and in many regards respond more efficiently, effectively, and faster than the human driver.

Autonomous automobiles will find their first mark in wide-open spaces on relatively lightly trafficked roads. As experience evolves this will move into dense suburbs, then into even more dense cities. Overlaying the technology will be geo-positioning and global information systems, allowing the more effective transmission of information to the vehicle and information from the vehicle. In the latter case information will be about breakdowns, accidents, help needed, disasters, criminal activity and so on.

The movement towards smaller vehicles will not eliminate the need or the use of larger vehicles, but it is likely to accelerate an already existing trend in the United States—about 25% of new cars are not purchased by individuals but are purchased by fleets. Those fleets operate by criteria different from that of individual owners. Fleet owners want high resale value, low maintenance and repair costs and high efficiencies. That pattern will gradually effect all car production and accelerate the trend to longer and longer car life.

Also affecting the vehicle will be design for demolition or dismantlement. Environmental pressure will not push for merely a higher percentage of reuse of high value-added components, since the car is already 90% or more recycled. Rather, a different trend affecting vehicles will be the three Rs—recycling, reclamation, and remanufacturing. Recycling is now familiar. Reclamation is the recovery of vehicle parts to be reintroduced into manufacturing. More important in the next decade or so will be remanufacturing, in which the vehicle is returned to the manufacture for refurbishment—into basically new condition at only a fraction of the cost of a new car. That will not come about spontaneously from the operation of the market, but will call for governmental incentives or mandating.

Rail travel by people will undergo some enhancements. It will be increasingly important in developing countries. In the advanced nations it will also continue to play an important role. In North America, where distances are greater than within most other countries, competition from airlines and automobiles will be significant. Current data suggests that Americans prefer to drive up to about 300–325 miles rather than take either a plane or railroad. Economic, social, and environmental regulatory factors may shift that balance. Light rail systems are becoming popular for moving people from the far suburbs into central business districts. The pattern of cities is changing evolving from a present pattern of expanding rings of activities, occupations, and residencies around the central business district, to a much more complex network pattern called a polycentric city. The metropolitan area will have a number of business centers with only one of them retaining the government and the arts functions.

Air travel will continue to grow as global prosperity increases. More people will be engaged in both business and tourism travel, while some of the business travel for economic reasons may be effectively taken over by information technology, i.e. video conferencing. Tourism will continue to boom and sustain the airline industry, after a slowdown associated with recession and the multinational 'war' on terrorism. Information media are delivering information to more and more people about places in the world. The natural desire to be fulfilled is to visit those places that one has learned about from the media.

How big aircraft will get is an open question, particularly in view of two factors. First is the potential risk of large number of lives from accidents, although it is clear that we are ready to accept one to two hundred deaths from aircraft accidents without the industry going into paralysis. More important will be the question of loading and unloading time. Already the arrival and departure times from airports consume an increasing portion of the total travel time from start to destination. That may prove to be a severe limiting factor on very large aircraft. However, for long distance trips such as transpacific travel, very larger aircraft may offer a great reduction in price and faster travel.

The technologies of balloons, both soft and rigid, are improving. Whether they again become modes of human transportation is open to question. The availability of helium rather than combustible hydrogen relieves one big anxiety. In any case, balloons will play an increasing part in construction, movements of heavy goods and loads, the harvesting of trees and anything requiring access to remote places.

Seagoing transport technologies will continue to evolve. Ocean ships are a big time activity in terms of logistics. As global populations and economics become more tightly knit those goods that do not require immediate delivery or are bulky or heavy will not be shipped by air but by sea. More manufacturing processing may occur at sea. For example, the movement of logs across the Pacific may involve some trimming and cutting or final finishing along the way.

The central technologies influencing transportation are information technologies for control, regulation and automation, energy technology to cope with greenhouse warming, the high cost of long distance travel and the new materials which may offer cheaper, lighter, better, faster, and more durable vehicles and vessels.

Civil Engineering and Construction

Civil engineering gives us society's infrastructure, sewers, highways, waterways, rail, etc.. Changes in construction are driven technologically primarily by information technology and new materials. On the other hand, social concerns will have dramatic effects. The events in New York City on September 11th have raised worldwide questions about structures which become symbols to attack or through accidents or natural disasters, death traps for thousands of people. Another driver of change is the continuing and probably unstoppable trend toward worldwide metropolitization. The USA today is about 81% metropolitan. In the next quarter century there will be 15 to 20 megacities with 20 million or more people, many of them in what are now considered developing countries. The scale alone will raise issues of heath and safety in terms of water and sewage, and economic issues in terms of mobility and transportation in prosaic matters such as getting to and from work.

Smartness

The ability to link microprocessors, that is, computer capabilities, sensors, and actuators into devices to move or change or start something, is becoming universal. Microprocessors are in all kinds of physical equipment like cars, trucks, trains, and, building equipment. The opportunity to operate more devices remotely, safely and reliably allows doing things that would be unacceptable if people were put in that same environment. For example, automated mining can now replace numbers of people in a mine and go further and deeper, and even remove more of the valuable ore, where one would not dare do it with people present.

Smartness is a tool in the administrative and organizational aspects of building and construction, which means that it is moving to unprecedented degrees of coordination of labor and supplies at the work site. Today all large construction firms, dealing with hundreds of million-dollar projects are rapidly moving towards becoming totally smart in all their internal and external operations. All of this means that the smart devices and the smart organization will allow us to do things more rapidly and more effectively and make more durable and higher quality products and structures than we have been able to do in the past.

Smartness will allow us to build infrastructures that are self-diagnostic and self-announcing. If this bridge is overstressed, if this concrete weakens, if there is too much paint flaking off of these surfaces, or if tremors having an adverse effect on joints and linkages occur, this will be announced. We will have new tools to maintain structures more reliably, effectively, and cheaply.

Heavy rain often floods sewers. They overflow, simply because the excess run-off can not be accommodated. Millions of tons of beautifully fresh clean water may be dropped on the landscape, thoroughly polluted by overflowing sewers. In part this results from the fact that sewers are dumb. Information technology now makes it possible to develop smart sewers, with the ability to respond to surges and to shut down if necessary to reduce the likelihood of mass scale pollution.

Infrastructure may of course take on new burdens in the future with regard to scarce resources. Infrastructure may have to automatically account for rationing of water, rationing of electricity, rationing of natural gas or other fuels. The ability to do this automatically lends a degree of reliability and fairness to the process and is a net benefit to both those who are inconvenienced by and those who administer the rationing. Infrastructure, housing, and the city are discussed by Coates (1992b, 1992c, 1994, 1995, 1997b, 1999, 2001).

Macroengineering

More grandiose than the mere improvement of present-day infrastructure is the opening up of the concept of macroengineering. Macroengineering has to do with the kinds of projects which are so large that they will run over the boundaries of individual countries either in their structures or in their effects, and will often outstrip the budget of any individual country to accomplish them. The key feature of macroengineering is that it does not improve something by 10%, 15%, or 20%. Instead it fundamentally changes the situation.

The Suez and Panama Canals and their effects are historically well-known and establish beyond question the global effects of macroengineering. There are 70 or so plausible macroengineering projects that have been shown in varying degrees of detail to be practical. Talked about in recent years (but nothing has been done about it) is the movement of fresh water in the form of giant icebergs from Antarctica to the west coast of Latin America, to Baja California, to Saudi Arabia, or to the Australian Outback.

More macroengineering projects will be considered in the next 25 years, and hopefully several will be undertaken. A possible one mentioned earlier in this chapter is harvesting of energy by tapping into the thermal gradient between the Gulf Stream and the surrounding colder water. Another interesting water project would take surplus water from Southern France and pipe it under the Mediterranean Ocean to supply agricultural water to North Africa. Other North Africa projects would include laying down black top, perhaps 300 yards wide, to begin to change the microclimate

along the coast and induce rainfall; then take advantage of that rainfall by plantings to change soil composition.

Mega Cities

Dealing with the problems of mega cities is high on everyone's list of world public health problems—for example, the contamination of water supply by sewage. The high cost of providing modern fresh water and sewage facilities may be inevitable, but the scope and size of these impending problems raise an alternative possibility. What is now routine in large parts of the rural world may be taken advantage of in cities: collecting human waste as a valuable resource for urban gardening. This could convert a serious health problem to a benefit in large cities and at the same time relieve the visual tedium of the vegetation-free urban environment.

The infrastructure also calls for other changes. If you consider a city of 20 million people the places to work should be located closer to the places where people live. Will we have to get away from the concept of a central business district and surrounding rings of other economic functions which have marked the classical cities of North America, Europe, and Asia, in order to deal with the mass populations of mega-cities and the problems of ground transportation?

Smartness will offer many benefits, and materials science will offer additional ones. One of the long-term trends in civil engineering and construction is the recyclablity of materials. For example in road building in the United States it is customary to replace highways with a newer wider highway, or replace the road surface periodically with a new surface. In doing that one removes the old surface and throws it onto the scrap heap. The black top and perhaps even the concrete disposed of this way are resources that have economic value. In the future, there will be more recycling of highways, buildings, and structures. That in turn will call for new road designs for easier dismantlement and recycling.

Construction's new objective is in building information technology into structures. In the United States some housing developers are striving to build that capability into their new homes. The advantage as with most other built-in construction is that it is substantially cheaper than retrofit. Soon all construction will be wired or cabled for information technology or it will be channeled with hollow tubes that can receive fiber optics for cable over the lifetime of the building.

Consider the kind of housing needed by the poorest 20% of the world population. For four people that construction would involve an average total cash allocation of $300 or $400 per person, along with large amounts of sweat equity, that is, the human labor the owners put in. This situation is about as remote a building model as one can get from models familiar in North America and Europe. But suppose one could allocate a quarter of that $300 or $400 to a high tech package to create an enormous jump in the quality of lives of people living there. Would it be solar cells? Would it be a bicycle electricity generator? a TV set? or some kind of piping for sewage? These types of questions have been barely asked, much less answered, but the technological contributions from the advanced nations to the lowest level of housing in the rest of the world offers the potential for great enhancement for quality of life, health and safety for literally billions of people. A thoughtful collection of Manuel Castells' writings on the city has been compiled by Susser (2002).

Health and Medicine

Aging

The process of aging will undergo even more dramatic changes than occurred in the last century. Every species has an average set point for death. In people, it is about 85–90 years. Of course there are people who will live much longer or die much sooner, but that is the average set point for our species. In the short run, a quarter century, the most likely benefits of health and medicine is to 'square off the death curve', meaning that people will die prematurely less frequently from accidents, injuries, disease and the other conventional things. Therefore more and more people will live to the biological set point. We already see strong evidence of this in the rapid growth of the aging population and the forecast of a boom in centenarians.

The crucial question is, will that increased survival to old age be accompanied by vigor and good health? The answer is uncertain, but three things will help reshape the death curve. First, is giving up smoking. Second, is getting more exercise. Third, is having a well-balanced diet to avoid overweight. Medicine may help in promoting each of the objectives but the decision ultimately lies with people. We have made far less progress than we could simply because people tend to choose short-term gratification of smoking, eating, and being a couch potato over the vague, uncertain and distant promise of longer life. Hayflick (1994) outlines implicitly the limitations on moderating aging and life expectancy over the next quarter century.

Over the long-term medicine will be less intrusive in diagnosis and therapy. Early interventions of all sorts will enhance the quality of life of aging people. A big concern is mental health and freedom from dementia. While mental deterioration still threatens the aging, research now underway is likely to develop technologies to arrest mental decline and offer ways to prevent or correct the most important mental disorders.

Life extension by moving the biological set point is not likely to be of much significance to most people living today, because it will most likely involve genetic intervention. The earlier that intervention is made, the greater will be the likelihood of set-point extension. There are several specific avenues for life extension. First is better understanding of cell death or apoptosis. The second depends on the fact that there is a tail on the DNA called a telomere that shortens with each cell division. The shortening is associated with cell death. Third, is the introduction of genes that lead to life extension, most likely by generating endogenous antioxidants.

Substantial progress has made in research along these lines with lower forms of life, and eventually we may reach the point where these findings will be established and proven in primates including ourselves. The long-term implications of a genetic intervention and the limited understanding of side effects make it most unlikely that human experiments in life extension by gene intervention will even begin on a significant scale in the next quarter century. However, along the way large numbers of people will undertake informal experimentation of their own to shift the set point, using growing knowledge of the role of antioxidants in biological systems and improved techniques for their delivery.

Human Enhancement

Human enhancement lies in our future. Normal people are on a distribution of talent and skills, whether of vision, neuromuscular control, fine motor control, stamina, strength, or mental abilities. Most biological and genetic research on people has been directed at diseases and disorders. The same knowledge, however which reveals the sources and leads to the treatments of diseases and disorders usually will carry with it the capabilities for human enhancement. It is unlikely that much of that will occur on a broad scale in the next quarter century by direct genetic interventions. But as knowledge accrues and work with other species proceeds the genetic approach will become widely accepted for enhancement as we will have seen in the treatment of diseases and disorders.

The technologies of life extension and human enhancement both raise substantial issues of public policy. Are they in the public interest? Will they benefit the overall society? How will the choices be made? What will happen to people who suffer adverse effects? What will be the effects on society of life extension and human enhancement in whatever forms they take. Those issues can only be clarified as scientific research and clinical practice unfolds. See Stock (2002).

New Techniques

The quite different line of development in health and medicine is toward more nonintrusive techniques. The use of information technology and the associated rapid development of the ability to detect all kinds of phenomenon electromagnetically, chemically, and genetically from samples of bodily waste, saliva, fingernails, skin, hair and so on, will allow each person to have a more complete inventory of his or her biological capabilities and potentials. A genetic map will show people the likelihood for them of various diseases and disorders and the possibilities of various interventions or compensatory activities. All of this will go a long way in enhancing our over all health and becoming an important factor in squaring off the death curve.

Accompanying new information technology and techniques will be life long records of one's genetics, diseases, disorders, treatments, and their effects. The problems of balancing the benefits of record keeping with the need for privacy are not insuperable. No significant attention has been given to this issue in a positive way because the unfolding of those records has not yet begun on a broad scale, but it soon will. It is quite conceivable that the individual person will by law have exclusive rights to his or her total health records, perhaps on a portable chip with a duplicate in a safe deposit box. None of these records will be in the hands of physician or hospitals unless one voluntarily chooses to have them deposited there. Life-long record keeping will soon begin, started by one's parents and carried on through one's life. For a brief look at the treatment of disease in the present century, see Coates (2000a).

Manufacturing

The making of things in factories is being transformed by information technology as more processes are automated and markets globalized. This in turn is leading to more of the soft form of customization that is so familiar in ordering an automobile, or a hamburger at a fast food place. You have a limited number of choices from which you select, to create the product that you want. Technology however has become more sophisticated in the manipulation of materials and we may move in manufacturing to a deeper, harder form of customization in which the manufacturing processes themselves will change to meet the individual needs of a customer.

Most of the manufacturing tools, whether now automated or not, are sophisticated versions of the tools that were available at the end of eighteenth and the beginning of the nineteenth centuries. New scientific developments are opening up new capabilities for manipulating materials. For example, the automated sculpture of material in a factory rather than molded casts may come about.

Also affecting manufacturing will be the pressures on energy use coming from greenhouse warming. This will lead to higher efficiency in manufacturing and to a shift to materials that will have a net energy saving in the manufacture and use of products. Manufacturing will also have to expand to take other environmental situations into account, including for example manufacturing for refurbishment and manufacturing for dismantlement.

Information technology will provide new levels of feedback from customers. That will be unstoppable as customers are able to develop websites to attack or to applaud various manufacturers and their products. That pressure will eventually have bad effects on companies laggard in responding to the consumer.

Smartness in manufacturing will lead to further declines in the manufacturing labor force. Ultimately some factories will be unmanned except for maintenance and repair crews. Manufacturing in the 21st century is discussed by Coates (2000b).

Residences

Whether one lives in a private house or in an apartment, emerging technology will make that residence safer and more secure. It will be able to announce its own condition and call for help or repairs. The residence will be secure against interlopers and will be effectively smart in all regards. Smart kitchens will allow the householder to make a 15 second transit through the kitchen, announce who will be coming to dinner, scan the menu with the smart appliances and pick out a desirable menu from what is in the pantry and freezer. In 20 to 30 minutes a full four-course meal for four will be ready followed by 7 minutes of clean up.

The most important effects on residences will be invisible in that they will be behind the walls and in infrastructure, giving greater security against environmental mishaps, better containment of heat, a better buffer against cold, and more control of utilities flow into the home by tying utilities to the various forms of price scheduling. It may be preferable to do laundry or showering at certain times of the day or to have a house partially or totally heated or cooled at different times of day.

Throughout the home there will be extensive large or small flat screens, much like television, ranging in size from 3×3 inches to 6×3 feet. The central feature of the home will be the 'electronic home or study center' where mom and dad do some of their work if not all of it, where junior and his sister carry out their school assignments, and where the whole family seeks entertainment, recreation and a large part of socialization through information technology. 'Center' is only a convenient term, since the information tools will be everywhere throughout the residence.

Work at home of course will call for at least one additional room in every house or apartment to accommodate that work. That in turn will have ripple effects on how people dress, where they eat and how they socialize.

New materials' effects on housing will be more diversity on the one hand and more durability on the other. The ability to change the decorations on the wall or windows electronically lies in the future. Smart appliances have already been discussed.

Schooling at home for many will merely be an augmentation of what is going on in traditional schools, but for others it will be a preferred mode. Entertainment at home will not just be with electronic games, although they will become very popular and more sophisticated, but often real time interaction with people through wall-sized screens will become common. Being able to attend a wedding, funeral, or anniversary remotely and slap hands together on each wall and see each other and share the joys or sorrows of the occasion, will be common.

Furniture will be smart. You will announce who you are and it will adjust itself to fit you. Central lighting, rather than the more expensive and wasteful defuse lighting we now have, will use light pipes to carry bright sunshine throughout the home or use a 2,000 watt lamp at night or on cloudy days.

Retail Marketing

Retail marketing is already experiencing big changes in competition among local shops, chain stores, catalogs, and the Internet. E-commerce affects what we choose to buy, when we choose to buy it, who we choose to buy it from, under what conditions, and with what modes of payment. Travel, convenience, safety, reliability, economic and financial security and ease of returns will all shape the balance among those four retailers. Many people still love to shop in the old fashioned way, to feel the goods, to march through the store and to see thousands of items. Others will see the shops only as a place to go to test out or try something and see if they like it, and then order it either through a catalog or over the Internet. Others will limit themselves mostly to remote purchases because of the high value of their own time and the increased reliability of service. See Westland (1999) for a global perspective on electronic commerce.

Right now catalogs have an edge in that they can show things in full color in many choices. But the electronic systems will gradually catch up. Then one or both systems will have full data about your choices, your preferences, your body dimensions, and your previous purchases. They will not merely respond when you call on them, but will alert you to what might especially satisfy your needs, whether those needs are for home furnishing, books, clothing, drugs, medicine, or any of the hundreds of thousands of other things in commerce for daily life hobbies, entertainment, and recreation.

As mentioned earlier under transportation, logistics systems working with radio frequency identification (RFID) will more quickly and reliably bring things to you. Electronic marketing systems will show more things to you and give you wider choices. For example, looking at an electronic catalog you might see fifty different outfits that you like, and you will be able to see yourself in outfits No. 6, 17, 24, and 39, because when a picture of you and your electronic data are deposited with the vendor you will be able to see yourself from every side in any outfit and make

appropriate changes. Try the hat from 11 with the dress from 24 along with the shoes from 27. Similar things will apply to home furnishings. You will have sent a diagram, a picture, a photograph, of where you live and the furniture vendor will work with you to remodel, or to just to select a particular new piece—an armchair, table, or lamp or a full redecoration.

Clothing

Clothing is difficult to forecast in the long term because it is so intimately linked to vagaries in the state of the economy and fashion, Nevertheless clothing will become more weather sensitive, sensitive to body temperature, humidity, specific conditions of diseases and disorders. Comfort in any environment and adjustment to your preferences will be common. A second big effect in clothing will be smartness. Clothing will be able to electronically respond to you and record or deliver messages as appropriate about your health and alert you to treatments or therapy you need to take. Clothing will also be able to interact with electronics embedded in your arm, your leg or chest to satisfactorily balance your internal condition with your environment. Chameleon-like clothing will be variable in color and changeable in texture as you wear it. Material science will allow these changes to be built in at acceptable cost to further aesthetic choices.

Clothing will also become a part of our overall health regimen, shoes being designed and made to fit electronically, garments tailored to fit the body to enhance or compensate for whatever one's wishes or dysfunctions may be. For example, if you have a tendency toward poor posture, clothing may sense that and alert you to stand straighter, to draw back your shoulders. In extreme cases it will even force back shoulders to treat poor posture or scoliosis.

Education

Education is undergoing competitive pricing with standard classroom education, from kindergarten through graduate school, at odds with education offered remotely, off-site and ad lib. As it stands now, remote or distance learning is to a large extent being driven by business and corporate needs for effective broad scale training of adults. It is also increasing in universities and extension programs. Electronic and remote graduate programs are steadily moving down into undergraduate education because they offer unique advantages in cost and quality. For example, in a sophomore composition course in English the professor can now review papers electronically and give a degree of criticism that was literally impossible face to face or with hand notations alone.

College soon will become the place to go, with the operative word being 'go', for only two reasons—to take those courses that absolutely require some hands-on work like choreography or engineering, or to find a spouse. The rest of the educational programs will be a trade-off between those who choose to go and those who choose to get an equivalently effective education remotely at greatly lower cost. A star system of the best professors in each field from the universities across the country and later across the world will develop. Questions of credentials and student evaluation present obstacles, but by no means are they insuperable. The technologies of groupware will allow students to engage remotely in collective activities, to do homework, to discuss assignments and to exchange information and opinions.

Beginning at the other end, the dissatisfaction with the terrible state of K-12 is leading many families to augment and even to replace classroom learning with home-based learning. Many children are now adept with computers by ages three to six. For them computer-associated technology will not be appliances, but prosthesis, intimate parts of their being, in the same way that for massive numbers of people eyeglasses are no longer appliances but prosthesis.

An emerging child generation will drive expectations higher and higher in K-12 and into college for ever more sophisticated learning tools. If they are not there, they will be available at home. The most important development at home will be the electronic home-work study center.

Education of course will be increasingly tailored to the individual. Automated learning aids will allow one to achieve things currently unattainable, by evaluating first what the student exactly knows, second what needs to be poured into his or her head, then what the student's preferred learning strategies—acoustic, visual, or tactile—are. Optimization of learning will occur and the ability to learn more and better will be striking.

Related to that, we will see for the first time complete learning. No longer will 85% to 95% in a course be enough, because it means that you have not learned everything. Teaching technologies will make a complete grasp possible. After all the technology can be infinity patient and generous in stimulating rewards.

Further effects of information technology through education will not merely improve thinking but introduce new modes of thinking, as information technology becomes dynamic, with screens automatically changing and moving, as it becomes interactive, and as it becomes multidimensional. These new aspects of learning are so unfamiliar that as they become familiar they will alter the very ways that we think.

In Summary

What lies ahead is continuing enhancement of the quality of human life on a worldwide scale in unprecedented richness. The damage so common in the early phases of the industrial-scientific era will find their analogs in the future. But the expanding capabilities of science are better able to look for earlier indicators of most problems and to propose and effect resolutions. However, science and technology are not autonomous forces, but they and their benefits are most likely to flourish in democratic societies.

References

Note: Many of the items cited below do not make forecasts, but are excellent introductions to the forces and factors shaping the subjects to which they refer.

American Chemical Industry (1996). *Technology vision 2020—U.S. Chemical Industry*. Washington, D.C..

Atherly, A. G., Girton, J. R. & McDonald, J. F. (1999). *The science of genetics*. Fort Worth: Saunders College Publishing, Harcourt Brace College Publishers.

Ayres, R. U. & Ayres, L. W. (2002). *A handbook of industrial ecology*. Cheltenham, U.K.: Edward Elgar.

Ball, P. (1997). *Made to measure: New materials for the 21st century*. Princeton, NJ: Princeton University Press.

Benyus, J. M. (1997). *Biomimicry: Innovation inspired by nature*. New York: William Morrow and Company, Inc.

Carter, R. (1998). *Mapping the mind*. Berkeley, CA: University of California Press.

Chen, N. Y. (2001). Energy in the 21st century. *Chemical Innovation*, 15–20.

Coates, J. F. (2001). Technology and the changing city. *Journal of Urban Technology*, **8** (2), 95–106.

Coates, J. F. (2000a). Treatment of disease in the 21st century. *Vital Speeches of the Day*, **LXVI** (9), 274–278.

Coates, J. F. (2000b). Manufacturing in the 21st century. *International Journal of Manufacturing Technology and Management*, **1** (1), 42–59.

Coates, J. F. (1999). Changing urban development: Housing and community in the 21st century. *Vital Speeches of the Day*, **65** (22), 690–695.

Coates, J. F. (1995). The probable future and its impacts on infrastructure. National Research Council, Board on infrastructure and the constructed environment. *The challenges of providing future infrastructure in an environment of limited resources, new technologies, and changing social paradigms*. Proceedings of a Colloquium, March 24, 1995, Washington, D.C.

Coates, J. F. (1994). Emerging technologies in first and third-world cities. *Journal of Urban Technology*, **1** (1), 31–46.

Coates, J. F. (1992a). Public policy actors and futures: Considerations in wildlife management. In: W. R. Mangun (Ed.), *American Fish and Wildlife Policy: The Human Dimension*. Carbondale, IL: Southern Illinois University Press.

Coates, J. F. (1992b). Third world housing, the great technological opportunity. *Technological Forecasting and Social Change*, **42**, 91–95.

Coates, J. F. (1992c). Preparing for the urban future. *Technological Forecasting and Social Change*, **42**, 309–316.

Coates, J. F., Mahaffe, J. B. & Hines, A. (1997a). The promise of genetics. *The Futurist*, Bethesda, Md.: World Future Society, 19ff.

Coates, J. F., Mahaffe, J. B. & Hines, A. (1997b). *2025: Scenarios of U.S. and global society reshaped by science and technology*. Greensboro, NC: Oakhill Press.

Crichton, M. (2001). *Prey*. New York: HarperCollins.

Dawkins, R. (2002). Son of Moore's Law. In: J. Brockman (Ed.), *The Next Fifty Years* (pp. 145–158). New York: Vintage Books.

Deffeyes, K. S. (2001). *Hubbert's peak: The impending world oil shortage*. New Jersey: Princeton University Press.

Forester, T. (1988). *The materials revolution: Superconductors, new materials, and the Japanese challenge*. Cambridge: MA: The MIT Press.

Frankel, M. S. & Chapman, A. R. (2000). *Human inheritable genetic modifications: Assessing scientific, ethical, religious, and policy issues*. Washington, D.C.: American Association for the Advancement of Science.

Freitas, Jr., R. A. (1999). *Nanomedicine*, vol. 1, *Basic capabilities*. Landes Bioscience.

Gardner, H. (2000). *Intelligence reframed: Multiple intelligences for the 21st century*. New York: Basic Books.

Hayflick, L. (1994). *How and why we age*. New York: Ballantine Books.

Hawley, R. S. & Mori, C. A. (1999). *The human genome: A user's guide*. San Diego: Academic Press.

Howard Hughes Medical Institute (1999). *Exploring the biomedical revolution*. Baltimore, Md.: The Johns Hopkins University Press.

Hoffman, P. (2001). *Tomorrow's energy: Hydrogen, fuel cells, and the prospects for a cleaner planet*. Cambridge, MA: The MIT Press.

Lawrence, P. A. (1992). *The making of a fly: The genetics of animal design*. Oxford, U.K.: Blackwell Scientific Publications.

Malanowski, N. (2001). *Study for an innovations and technological analysis (ITA) on nanotechnology*. Dusseldorf, Germany: VDI-Technology Center, Future Technologies Division.

Mauseth, J. D. (1998). *Botany: An introduction to plant biology*. Sudbury, MA: Jones and Bartlett Publishers.

Nakicenovic, N., Grübler, A. & McDonald, A. (Eds) (1998). *Global energy perspectives*. Report from the International Institute for Applied Systems Analysis. U.K.: Cambridge University Press.

O'Neill, B. C., MacKellar, F. L. & Lutz, W. (2001). *Population and climate change*. Cambridge, U.K.: Cambridge University Press.

Psaras, P. A. & Langford, H. D. (Eds) (1987). *Advancing materials research*. Washington, D.C.: National Academy Press.

Rees, W. G. & Campbell, J. (1996). *Physical principles of remote sensing*. New York: Guilford Press.

Restak, R. M. (1995). *Brainscapes: An introduction to what neuroscience has learned about the structure, function, and abilities of the brain*. New York: Hyperion.

Roco, M. C. & Bainbridge, W. S. (Eds) (2002). *Converging technologies for improving human performance: Nanotechnology, biotechnology, information technology and cognitive science*. Arlington, VA: NSF/DOC-sponsored report.

Shapio, R. (1991). *The human blueprint: The race to unlock the secrets of our genetic script*. New York: St. Martin's Press.

Stock, G. (2002). *Redesigning humans: Our inevitable genetic future*. Boston: Houghton Mifflin Company.

Susser, I. (Ed.) (2002). *The Castells reader on cities and social theory*. Malden, MA: Blackwell Publishers, Inc.

Thompson, J. M. T. (2001). *Visions of the future: Chemistry and life science*. New York: Cambridge University Press.

Tudge, C. (2000). *The impact of the gene: From Mendel's peas to designer babies*. New York: Hill and Wang.

U.S. Geological Survey, Geographic Information Systems. http://www.usgs.gov/research/gis/title.html.

U.S. Global Change Research Program, National Assessment Synthesis Team (2000). *Climate change impacts on the United States: The potential consequences of climate variability and change—overview*. Cambridge, U.K.: Cambridge University Press.

U.S. Congress, Office of Technology Assessment (1988). *Advanced materials by design*. OTA-E-351. Washington, D.C.: U.S. Government Printing Office.

U.S. Dept. of Energy (1984). *A synthesis of technology transfer methodologies*. Proceedings of a Technology Transfer Workshop, May 30, 31, and June 1, 1984, Washington, D.C.

Watson, R. T. (Ed.) (2001). *Climate change 2001: Synthesis report*. Contribution of Working Groups I, II, and III to the Third Assessment Report of the Intergovernmental Panel on Climate Change. Cambridge, U.K.: Cambridge University Press.

Westland, J. C. & Clark, T. H. K. (1999). *Global electronic commerce: Theory and case studies*. Cambridge, MA: The MIT Press.

The Future of Innovation Research

Tudor Rickards

Manchester Business School, U.K.

Abstract: Innovation research will increasingly be oriented towards the challenges presented by environmental complexity and turbulence. More progress may be expected in understanding the relationships between creativity and innovation. Innovation will inform, and be informed by, related fields of theory-informed practice such as knowledge management. The speed of product turnover (the velocity issue) will strengthen the case for practitioner-researcher coalitions. The relationships between the roles of the innovation players, team factors, and outputs will receive more attention. The theoretical stance advocated by critical theorists may also make a contribution in challenging more conventional organizational views of innovation.

Keywords: Creativity; Innovation; Complexity; Diffusion of innovation; Multi-level models.

Introduction

Someone once ironically said that any attempt to predict is foolish, particularly if the predictions are about the future. There is a more serious point to be made if we consider that to predict the future of a research field is to attempt to identify continuities and shifts in highly abstract mental constructions or paradigms. Nevertheless, there is a case for offering some thoughts, particularly if they are buttressed by ideas of the many diligent and distinguished researchers and practitioners. For this, I have relied primarily on an updating of a survey which I carried out recently (Rickards, 1999).

This treatment gives us some slight impression of the contours of the vast and sprawling territory of innovation. In my own survey, I reached the conclusion that since the time of Schumpeter (1912), innovation has attracted an eclectic set of researchers, policy makers and practitioners. Yet they seemed to fit into two reasonably coherent categories, namely those operating from an economics perspective, and those taking an organizational perspective. I will be concentrating more on the organizational perspective (which again is the more dominant in this handbook). We should not ignore, however, the debt that is owed to work from the so-called Austrian school of innovation theorists, later incorporated into the influential historical analysis of the capitalist system by Friedrich von Hayek (see Peters, 1992, for an introduction to Hayek's contributions to contemporary innovation practice).

Innovation retains its fascination as a process through which change may be purposively influenced. The notion that individuals and social groups may influence their world is a legacy of the enlightenment in Western thought. As a concept it grew along with the belief that rational (scientific) methods permit knowledge gains and human dominion over the forces of the natural world. This provides a platform of understanding of innovation at a philosophical level. At a different level of enquiry, we may be interested in the manner in which innovation occurs in the school-room, the artist's studio, or the entrepreneur's fledgling business empire. We need hardly be surprised that innovation, a word forced to stand for activities across such a rich set of categories, resists a unified yet coherent treatment. We are nevertheless forced to ask whether unifying principles and themes are to be detected—either within this handbook or beyond its pages? A negative answer implies that innovation researchers have begun the 21st century embarrassed by intellectual riches in the field, and impoverished with regard to sound integrative principles of their subject matter.

Linarity and Multi-Dimensionality

There remains a gulf between the instant answers offered in popular managerial texts on innovation, and the caution exercised by scholars. The informed view over the last decade has been one of realistic acknowledgement of the distance to go before the quest for integrating theory can be considered a success (Drazin & Schoonhoven, 1996; Rogers, 1995; Van de Ven, 1986; Van de Ven & Poole, 1995). In one of the most influential textbooks, Tidd et al. (1997) made it a central theme that innovation has suffered from perspectives that have been inadequately multi-dimensional. A special edition of the Academy of Management Journal, a few years ago, called for more efforts at developing innovation models that span levels (e.g. individual, organizational, and market/environmental levels). The editors noted ". . . we were disappointed to discover that no dominant theoretical perspective had emerged to integrate the multiple streams of innovation research" (Drazin & Schoonhoven, 1996, p. 1065).

The trend in innovation research, and one which seems likely to continue, is a shift in attention away from relatively simple, linear, and universalistic models to more complex multi-level multi-dimensional ones. It is a trend well exemplified in this handbook: the more focused research contributions are deepened and extended to acknowledge complex environments, and innovation trajectories.

Historically, innovation was for several decades regarded as primarily a technological phenomenon. Later studies were to replace the centrality of technology with the centrality of market forces. That is to say, technology push was replaced by market pull. This 'either-or' swing misses the point that innovation incorporates both technology and market influences. As is also well-known, contingency theories gain in completeness at the expense of simplicity and clarity. This may also occur the expense of failing to provide any convincing normative principles for practitioners ('It all depends' is not the most comforting advice that might be offered to an executive in search of innovation guidelines).

This is not to reject the utility of limited range linear models grounded in empirical and theoretical evidence. Cooper (1988, 1992) has been among the most influential of the normative analysts of success factors for innovation (See also Cooper & Kleinschmidt, 1987). His celebrated stage gate model illustrates the power for innovation practitioners of a simplifying linear treatment as a means of exemplifying and surmounting institutional barriers to effective innovation (Drazin & Schoonhoven, 1996). It offers a means of controlling the risks of innovation and has found widespread acceptance, for example in complex technologically oriented projects as found in pharmaceuticals and engineering industry sectors. Nevertheless, an important issue returns to haunt the practitioner. Success factors are presented for innovation which has been notoriously regarded as having instability of success factors both at the level of comparing specific innovations, and across studies (Downs & Mohr, 1976). In some ways Cooper's methodology typifies the limitations of a linear approach: it offers general factors with scope for specific project managers to change that which needs to be changed (*mutatis mutandis*) in order to adopt the general template for specific conditions. The linearity, as Mintzberg (1994) warns us, also conceals various practical and conceptual difficulties. Consider, for example the issue of novelty. Cooper, as has other researchers, resorted to modeling creativity as residing primarily at the 'fuzzy front end' of the linear process (Khurana & Rosenthal, 1988, 1997).

An evolutionary path for research models can be traced. Early innovation models as well as later versions such as Cooper's, placed the generation of novelty at the start of a simple set of stages. As Rothwell (1992) points out, research into innovation refined the basic linear model, until today we seek to incorporate more uncertainties, (Gleick, 1987; Stacey, 1992, 1996), and indeed seek to address multiple levels (individual, organizational, and market context).

The implications of this for innovation research are, I believe, profound. I suggest that at present, and in the near future, simple linear models of innovation will increasingly be found theoretically inadequate and empirically inconsistent across studies in different contexts. At present we can only regret the lack of solid longitudinal studies, and multi-level contextually rich studies. In the future, this will be increasingly acknowledged, and researchers will be encouraged to address this difficult challenge. The globalizing context will encourage efforts to understand innovation in more complex and contextual ways, in what has been described as the new competitive landscape (Bettis & Hitt, 1995). New institutional forms contribute to the increasingly information-rich innovation context (Child & McGrath, 2001).

Researching Creativity's Role in Innovation

I have observed a growing tendency for the concept of creativity to be assimilated into the innovation process, often at the 'fuzzy front end' of the process. I predict that this will become a focal point for researchers, and that the research will help clarify the relationships between the two constructs.

Conventional economic models of innovation have little to offer regarding the origins of unexpected ideas that challenge rational expectations. We are better seeking insights into creativity from the social scientists, (de Bono, 1971; Gardner, 1983/1993; Koestler, 1964; Perkins, 1983; Sternberg, 1988; Sternberg & Lubart, 1991; Walberg, 1988) and the information theorists (e.g. Boden, 1994; Nonaka & Teguechi, 1995; Price & Shaw, 1988; Simon, 1996; Stacy, 1996).

An early influential model was that of Campbell (1960), a pioneer of an evolutionary theory of creativity. The core idea is that of blind variation and selective retention leading to persistence of unique new innovations. Thus, once again, we have a perspective which seeks to explain the origination of novelty within innovation processes. Campbell offers a way of dealing with the unpredictable aspects of innovation. These have been addressed in other ways: emergence (Mintzberg, 1994, writing from the perspective of strategic planning); unintended consequences (Argyris, 1990); and 'wicked' problem-solving (Rickards, 1990; Rittel & Webber, 1974).

In an earlier commentary, I noted that the two terms (and therefore the concepts) are often conflated (Rickards, 1991). This observation is strongly supported by contributors to this handbook, in which the term individual innovation has been used as an integrative and organising construct. A similar decision was made in an earlier text by West & Farr (1990). In such a treatment, individual innovation has been applied rather interchangeably with the term creativity. Conflation may also diminish the attention paid to the transcendental and psychological aspects now emerging in the domain of creativity research (Andreasen & Canter, 1974; Eysenck, 1994a, 1994b; Jamison, 1993; Sass, 2000–2001; Sass & Schuldberg, 2000–2001). I select the contributions by Sternberg and Kaufman to illustrate the point, and to suggest that this is a shift from the vocabulary of their own earlier work.

Sternberg and coworkers have made many distinguished contributions to our understanding of the 'creative' part of the innovation process (e.g. Sternberg, 1988; Sternberg & Davidson, 1995; Sternberg & Lubert, 1990). Kaufmann has offered a European perspective, (for example, through his authoritative chapter in Gronhaug & Kaufmann, 1988). It is perhaps significant that each of these authorities writing for this handbook seem prepared to depart from the vocabulary the have found more congenial in other publications (namely the vocabulary of creativity) and espouse the vocabulary of individual innovation. We may conclude that the authors do not feel there to be a significant loss in such a shift. In the past I have welcomed the support provided to the creativity field by these distinguished figures. They have been inspiring companions as the journeys of creative exploration have suffered pretty stormy seas of criticism whipped up by others of a more mainstream psychological orientation. Now it appears that they have 'jumped ship', (or changed horse in mid-stream, to offer another aquatic metaphor).

We may perhaps be seeing a possible future trend in innovation research in which creativity is assimilated. This would have some advantages as for many researchers and practitioners in economic fields, creativity seems to be a less credible and operationalizable construct (Rickards, 1999a; Stein, 1987). In contrast, innovation perhaps receives an over-positive bias or halo effect (Abrahamson, 1990).

Yet we should ask whether there a valuable distinction to be drawn between a creative poem and an innovative poem? I believe the answer to be a clear yes. Is there a distinction to be drawn between a creative

novel and an innovative novel? Again, I believe the answer to be in the affirmative. But what about the distinction between a creative design and an innovative design? Or a creative product and an innovative product? My suspicion is that we have some way yet to go to deal with this question.

The issue may begin to resolve itself, if we return to the idea of creativity as mediated by environment (or 'press', as the creativity researchers refer to it). For the initial stages of design and innovation, the novel and 'blockbusting' idea has a greater chance of succeeding. As the innovation proceeds, there are inevitably more constraints, and the creativity required is more of a marginal or incremental kind. In other words, there is need for creativity as long as there are uncertainties (which persist throughout the innovation process). But the creativity needed at the finish line is creativity is not so much blockbusting, as developmental. The issue will overlap with considerations of composition and competences required in innovation teams.

Emerging Fields Impacting on Innovation Research

This handbook signals the exciting new fields that are impacting on innovation research. Writing from the home base of a Graduate School of Business, I am particularly interested in the emerging themes that have enriched management theory over the last few decades, and which are the candidates for influencing innovation research into the 21st century. In general, the boundaries of proximal fields are particularly rich in innovation opportunities (Gryskiewicz, 1999). The trend towards emergence of fresh innovation interfaces will continue, I suspect.

One such interface is that between the producer and consumer of meaning (e.g. Goldsmith & Foxall, 1994; using the innovation style model of Kirton, 1976, 1994) The work has implications for models assuming diffusion process of innovations (Rogers, 1995).

Knowledge management is another area of research which is clearly muscling in on the innovation domain (see, for example, Bierly & Chakrabarti, 1996, and Spender & Grant, 1996 for a strategic treatment). It is hardly surprising, and indeed the intractable complexities (all too apparent in this work) in the field of innovation may encourage more researchers to look to knowledge management to provide a new perspective. The links are clear. Innovation requires the generation of new insights, and the actualization of innovative outputs ('products'). Knowledge management is also concerned deeply about the origin, retention and management of ideas (Helfat, 1997). My suspicion is that some of the old innovation wine will be reappearing in new vintage knowledge management bottles. It is unclear whether the conceptual underpinnings of knowledge management will be any more secure than those underpinning the great edifice of innovation studies.

Other hot interface issues referred to in this work, and also to be found elsewhere in contemporary management theory generally, are network theory (e.g. Nohria & Eccles, 1992); constructivism (Weick, 1995); cultural variations in economic form (Whitley, 1999) and stakeholder theory (Donaldson & Preston, 1995; Jones, 1995). We may see how such fields may contribute to research development of understanding of innovation in a range of areas. One such would be National Systems of Innovation (Dodgson & Bessant, 1996; Lundvall, 1992; Pavitt, 1999). Another would be a revisiting of the extensive literature on innovation diffusion (Rogers, 1995), and the management of complex innovation projects, e.g. through alliances (Gulati, 1999).

The 'Increasing Turbulence' Theme

It has become a commonplace to point to technological forces contributing to ever-increasing environmental turbulence in which rapid technological change and knowledge turnover are critical features (Bettis & Hitt, 1995; Hamel & Prahalad, 1994). This theme seems to be one that will be enriched though studies of high velocity environments, (Eisenhardt, 1989; Gersick, 1989, 1991; Lawless & Anderson, 1996). The themes seem to be contributing to interest in more complex innovation models (Drazin & Schoonhoven, 1996). It accords with the trends in innovation research indicated in this handbook. The shift is towards interest in complex turbulent and even chaotic environments.

There may well be a more considered view in which turbulence may be of a more cyclic nature, as it plays out in competitive survival rates of innovative firms. Whatever, the details, turbulence remains an emerging and important theme.

The nature of complexity has consequences for the old question of diffusion of innovation, which may become revitalized by a generation of neo-Darwinists taking ideas of social genetics from the likes of Dawkins (1976, 1989), Price & Shaw (1998), and Dennett (1995). In this we may detect another consequence of the electronic innovation revolution. Emerging theoretical ideas seem to be assisted through the speed and ease of electronic publishing. The researchers into memetics typify countless and fragmented 'research factories'. Debate is vibrant, and largely uncensored (one from many recent examples would be Edmonds, 2001). The turbulence may be further linked with increasing interest in managing turbulence e.g. Gryskiewicz, 1999). The knowledge industries will bring their own challenges of ethical management.

This matters for example, in competition for resources across possible research fields, and policy decisions for allocating those resources. These competing fields are seeking to accommodate growing interest in ethical dimensions of corporate life, sustainability, leadership, and new institutional forms. Even such innocent terms as modernization are opening up the debate into work/leisure issues as never before. Much of relevance to these and other contemporary vital topics may be found, often hidden away, within the current and potential boundaries of innovation research. Innovation researchers will find many more points of contact with these fields in the future.

Power to the People: Roles of Players in Innovation Research

The role of the corporate entrepreneur as innovator dates to Schumpeter (1912). Pinchott's intrapreneur (1986) has attracted some attention, and may yet have more life in it (e.g. Barringer & Bluedorn, 1999; Birkinshaw, 1997). Other innovation roles have been attributed to product champions (Schon, 1965; Markham, 1999); and a confusing number of inter-related terms (Roberts & Fusfeld, 1981).

Interestingly, leadership *per se* has not been given a great deal of attention in innovation studies. I expect that to change. (Incidentally, leadership has attracted far more attention within the creativity and problem-solving literature, where a special process-orientated facilitator is frequently described and studied).

Within management studies, leadership remains a topic as confusing and multi-facetted as innovation itself. Currently, attempts to integrate the two fields have been limited. It was good to see one contribution here from Deschamps in this work). However, it is the slightest of signals in such a comprehensive collection of work. I suspect that there will be more interest in the next decades. I base my belief on several converging themes. First, the demand for more authoritative understanding of leadership processes from executives and MBA students is being reflected in more courses on offer (as a web search quickly reveals). Our own recent research in this area suggests that links between leadership, team factors and innovative outcomes exist, and have yet to be examined in a satisfactory fashion (see Rickards, Chen & Moger, 2001). The emerging theme may throw light on the role of transformational leadership in innovation, and organizational productivity.

Towards a More Critical Theory of Innovation

Technology advances have in the past been given primacy in the variables relevant to innovation. The handbook shows that such a view is now far too simplistic. Technology matters, yet is not regarded as one among many variables of interest. Taken with other consequences of uncertainties and risks associated with innovation, we should also remind ourselves of unintended consequences of technology, and the increasing social suspicion of a technically-driven change agenda. Some writers have suggested (e.g. Abrahamson, 1991) that innovation is far too widely investigated from a perspective that innovation is of itself a good thing (the economic driver argument). My electronic experience left me wondering what had been gained, and what lost when we fall prey to new technological innovations. To what degree are we benefitting individually and collectively from it? The technological eutopia/dystopia debate will rumble on, as innovation researchers investigate the latest turns in the story. Hard core technophiles should consider the cases in the amusing book by Tenner (1996) on 'why things bite back'.

The study of innovation, despite its various influences, has tended to locate itself towards the more conventional organizational perspective (premised on economic theory of the firm and a unity of purpose or mission (Alvesson, 1993; Willmot & Knight, 1982; Linstead, Small & Jeffcutt, 1996). Critical theory and post-modern analyses represent two emerging approaches that may have new perspectives to offer.

Critical theorists may be seen as scholars seeking to challenge the very core beliefs of prevailing knowledge systems. For example, they are more inclined to regard organizations as battle-grounds. Innovation missions are studied as expressions of rhetoric. Such approaches are being introduced into conferences as interest turns towards the creative industries (conservatively estimated as 10% of GNP in a recent U.K. government survey). The approach seems particularly appropriate for taking a fresh look at the conflicts between creative individuals seeking to gain acceptance for their ideas, and the organizational executives seeking to commercialize the ideas ('the suits').

Another possible trend comes from workers taking a post-modern approach in some business courses (Boje et al., 1996). The post-modernist argues that innovation studies reflects a dominant belief that has successful silenced the alternative voices. A post-modern approach is believed to be particularly relevant for examining the nature of the virtual, which is of increasing interest (virtual innovation teams, intellectual property, invisible networks of interest). In short, critical and post-modern thinkers may be well placed to offer alternative critiques of the innovation phenomenon in its 21st century guises.

References

Abrahamson, E. (1991). Managerial fads and fashions: The diffusion and rejection of innovations. *Academy of Management Review*, **16** (3), 596–610.

Alvesson, M. (1993). *Cultural perspectives on organizations*. Cambridge: Cambridge University Press.

Amabile, T. M. (1996). *Creativity in context*. Boulder, CO: Westview.

Argyris, C. (1990). *Overcoming organizational defenses: Facilitating organisational learning*. Boston, MA: Allyn & Bacon.

Barringer, B. R. & Bluedorn, A. C. (1999). The relationship between corporate entrepreneurship and strategic management. *Strategic Management Journal*, **20** (5), 421–444.

Bass, B. M. & Avolio, B. J. (1990). The implications of transactional and transformational leadership for individual, team, and organizational development. *Research in Organizational Change and Development*, **4**, 231–272.

Bass, B. M. & Avolio, B. J. (Eds) (1994). *Improving organizational effectiveness through transformational leadership*. Thousand Oaks, CA: Sage.

Bazerman, M. H. (1994). *Judgment in managerial decision making* (3rd ed.). New York: John Wiley.

Bettis, R. H. & Hitt, M. A. (1995). The new competitive landscape. *Strategic Management Journal*, **16** (special issue), 7–19.

Bierly, P. & Chakrabarti, A. (1996). Generic knowledge strategies in the U.S. pharmaceutical industries. *Strategic Management Journal*, **17** (special issue), Winter, 123–135.

Birkinshaw, J. (1997). Entrepreneurship in multinational corporations: The characteristics of subsidiary initiatives. *Strategic Management Journal*, **18** (3), 207–229.

Boden, M. A. (Ed.) (1994). *Dimensions of creativity*. Cambridge, MA: The MIT Press.

Boje, D. M., Gephart, R.P. Jr. & Thatchenkery, T. J. (Eds) (1996). *Postmodern management and organization theory*. Thousand Oaks, CA: Sage.

Burns, T. & Stalker, G. M. (1961). *The management of innovation*. London: Tavistock.

Campbell, D. T. (1960). Blind variation and selective retention in creative thought as in other knowledge processes. *Psychological Review*, **67**, 380–400.

Child, J. & McGrath, R. G. (2001). Organizations unfettered: Organizational form in an information-intensive economy. *Academy of Management Journal*, **44**, 1135–1148.

Cooper, R. G. (1988). The new product process: A decision guide for management. *Journal of Marketing Management*, **3** (3), 238–255.

Cooper, R. G. (1992). The newprod system: The industry experience. *Journal of Product Innovation Management*, **9** (2), 113–127.

Cooper, R. G. & Kleinschmidt, E. J. (1987). New products: What separates winners from losers? *Journal of Product Innovation Management*, **4** (3), 169–184.

Crick, F. (1995). *The astonishing hypothesis: The scientific search for the soul*. London: Touchstone, Simon & Schuster.

Dawkins, R. (1989). *The selfish gene* (2nd ed.). Oxford: Oxford University Press.

de Bono, E. (1992). *Serious creativity: Using the power of lateral thinking to create new ideas*. London: Harper Collins.

Dennett, D. (1995). *Darwin's dangerous idea*. London: Penguin.

Dodgson, M. & Bessant, J. (1996). *Effective innovation policy: A new approach*. London: Thompson Business Press.

Donaldson, T. & Preston, L. E. (1995). The stakeholder theory of the corporation: Concepts, evidence and implications. *Academy of Management Review*, **20**, 660–697.

Downs, G. W. R., Jr. & Mohr, L. B. (1976). Conceptual issues in the study of innovation. *Administrative Science Quarterly*, **21**, 700–714.

Drazin, R. & Schoonhoven, C. B. (1996). Community, population, and organization effects on innovation: A multilevel perspective. *Academy of Management Review*, **39** (5), 1065–1083.

Eisenhardt, K. M. (1989). Making fast strategic decisions. *Academy of Management Journal*, **32**, 543–576.

Eysenck, H. J. (1994a). Creativity and personality: Word association, origence, and psychoticism. *Creativity Research Journal*, **7** (2), 209–216.

Eysenck, H. J. (1994b). The measurement of creativity. In: M. A. Boden (Ed.), *Dimensions of Creativity* (pp. 199–243). Cambridge, MA: The MIT Press.

Foxall, G. R. (1994). Consumer initiators: *Both* innovators *and* adaptors. In: M. Kirton (Ed.), *Adaptors and Innovators: Style of Creativity and problem-Solving* (rev ed., pp. 114–136).

Gardner, H. (1983/1993). *Frames of mind*. New York: Basic books.

Gersick, C. J. G. (1989). Marking time: Predictable transitions in work groups. *Academy of Management Journal*, **32**, 274–309.

Gersick, C. J. (1991). Punctuated equilibria as a model for organisational change. *Academy of Management Review*, **16**, 10–23.

Geschka, H. (1983). Creativity techniques in product planning and development: A view from West Germany. *R&D Management*, **13**, 169–183.

Gleick, J. (1987). *Chaos: Making a new science*. New York: Viking.

Grönhaug, K. & Kaufmann, G. (1988). *Innovation: A cross-disciplinary perspective*. Oslo: Norwegian University Press.

Gryskiewicz, S. S. (1999). *Positive turbulence: Developing climates for creativity, innovation and renewal*. San Francisco, CA: Jossey-Bass Publishers.

Gulati, R. (1999). Network location and learning: The influence of network resources and firm capabilities on alliance formation. *Journal of Strategic Management*, **20** (5), 397–420.

Hamel, G. & Prahalad, C. K. (1996). *Competing for the future*. Boston, MA: Harvard Business School Press.

Helfat, C. E. (1997). Know-how and asset complementarity and dynamic capability accumation: The case of R&D. *Strategic Management Journal*, **18** (5), 339–360.

Jones, T. M. (1995). Instrumental stakeholder theory: A synthesis of ethics and economics. *Academy of Management Review*, **20**, 404–437.

Kaufman, G. (1988). Problem-solving and creativity. In: K. Grönhaug & G. Kaufmann (Eds), *Innovation: A Cross-Disciplinary Perspective* (pp. 87–137). Oslo: Norwegian University Press.

King, D. (2002). Brains in Bahrain. *The British Chess Magazine*, **122**, 622–637.

Kirton, M. J. (1976). Adaptors and innovators: A description and a measure. *Journal of Applied Psychology*, **61**, 622–629.

Kirton, M. J. (1994). *Adaptors & innovators: Styles of creativity and innovation* (2nd ed.). London: Routledge.

Koestler, A. (1964). *The act of creation*. London: Hutchinson.

Khurana, A. & Rosenthal, S. R. (1988). Towards holistic 'front ends' in new product development. *Journal of Product Innovation Management*, **15**, 57–74.

Khurana, A. & Rosenthal, S. R. (1997). Integrating the fuzzy front end of new product development. *Sloan Management Review*, Winter, 103–119.

Land, G. A. (1973). *Grow or die: The unifying principle of transformation*. New York: John Wiley.

Land, G. A. & Jarman, G. (1992). *Breakpoint and beyond: Mastering the future today*. New York: Harper Business.

Lawless, M. W. & Anderson, P. C. (1996). Generational technological change: Effects of innovation and local

rivalry on performance. *Academy of Management Journal*, **39** (5), 1185–1127.

Leana, C. R. & Van Buren III, H. J. (1999). Organizational social capital and employment practices. *Academy of Management Review*, **24**, 538–555.

Leonard, D. (1995). *Wellsprings of knowledge: Building and sustaining the sources of innovation*. Cambridge, MA: Harvard Business School Press.

Linstead, S., Small, R. G. & Jeffcutt, P. (Eds) (1996). *Understanding management*. London: Sage.

Lundvall, B. (1992). *National systems of innovation*. London: Francis Pinter.

Markham, S. K. (1999). A longitudinal examination of how champions influence others to support their projects. In: K. Brockhoff, A. K. Chakrabarti & J. Hauschildt (Eds), *The Dynamics of Innovation: Strategic and Managerial Implications* (pp. 187–212). Berlin: Springer.

Mintzberg, H. (1994). *The rise and fall of strategic planning*. Englewood Cliffs, N.J.: Prentice Hall International Editions.

Nohria, N. & Eccles, R. G. (1992). *Networks and organizations*. Boston: Harvard Business Press.

Nonaka, I. & Takeuchi, H. (1995). *The knowledge creating company: How Japanese companies create the dynamics of innovation*. Oxford: Oxford University Press.

Parnes, S. J. (Ed.) (1992). *Sourcebook for creative problem-solving*. Buffalo, New York: Creative Education Foundation Press.

Parnes, S. F., Noller, R. B. & Biondi, A. M. (1977). *A guide to creative action* (revised workbook). New York: Charles Scribner & Sons.

Perkins, D. N. (1981). *The mind's best work*. Cambridge, MA: Harvard University Press.

Peters, T. (1992). *Liberation management: Necessary disorganization for the nanosecond nineties*. New York: Alfred Knopf; London: MacMillan.

Pinchott, G. (1986). *Intrepreneuring: Why you don't have to leave the organization to become an entrepreneur*. New York: Harper Collins.

Price, I. & Shaw, R. (1998). *Shifting the patterns*. Chalford, U.K.: Management Books, 2000.

Rickards, T. (1990). *Creativity and problem-solving at work*. Farnborough, U.K.: Gower.

Rickards, T. (1991). Creativity and innovation: Woods, trees, and pathways. *R & D Management*, **21**, 97–108.

Rickards, T. (1996). The management of innovation: Recasting the role of creativity. *European Journal of Work and Organizational Psychology*, **5** (1), 1–15.

Rickards, T. (1999a). Brainstorming revisited: A question of context. *International Journal of Management Reviews*, **1** (1), 91–110.

Rickards, T. (1999). *Creativity and the management of change*. Oxford: Blackwells.

Rickards, T., Chen M-W. & Moger, S. T. (2001). Development of a self-report instrument for exploring team factor, leadership and performance relationships. *British Journal of Management*, **12**, (3), 243 –250.

Rittel, H. W. J. & Webber, M. M. (1994). Dilemmas in a general theory of planning. *DMG-DRS Journal*, **8**, 31–39.

Roberts, E. B. & Fusfeld, A. R. (1981). Critical function: Needed roles in the innovation process. In: R. Katz (Ed.), *Career Issues in Human Resource Management*. Englewood Cliffs: Prentice-Hall.

Rogers, E. M. (1995). *Diffusion of innovation* (4th ed.). New York: Free Press.

Rothwell, R. (1992). Successful industrial innovation: Critical success factors for the 1990s. *R&D Management*, **22** (3), 221—239.

Sass, L. A. & Schuldberg, D. (2000–2001). Introduction to the special issue: Creativity and the schizophrenic spectrum. *Creativity Research Journal*, **13** (1), 1–4.

Schon, D. A. (1965). Champions for radical new inventions. *Harvard Business Review*, March-April, 77–86.

Schon, D. A. (1983). *The reflective practitioner*. New York: Basic Books.

Schumpeter, J. A. (1912). *The theory of economic development*. Cambridge, MA: Harvard University Press (English Trans, 1934).

Simon, H. A. (1986). What we know about the creative process. In: R. L. Kuhn (Ed.), *Creative and Innovative Management* (pp. 3–22). Cambridge, MA.: Ballinger.

Spender, J.-C. & Grant, R. M. (1996). Knowledge and the firm: Overview. *Strategic Management Journal*, special issue, Winter 1996, 5–9.

Stacy, R. D. (1992). *Managing chaos: Dynamic business strategies in an unpredictable world*. London: Kogan Page.

Stacy, R. D. (1995). The science of complexity: An alternative perspective for strategic change processes. *Strategic Management Journal*, **16**, 477–495.

Stein, M. I. (1987). Creativity research at the crossroads: A 1985 perspective. In: S. C. Isaksen (Ed.), *Frontiers of Creativity Research: Beyond the Basics* (pp. 417–427). Buffalo, New York: Bearly.

Sternberg, R. J. (1988). *The nature of creativity*. Cambridge: Cambridge University Press.

Sternberg, R. J. & Davidson, J. E. (Eds) (1995). *The nature of insight*. Cambridge, MA: MIT Press, Bradford Books.

Sutton, R. I. & Hargadon, A. (1996). Brainstorming groups in context: Effectiveness in a product design firm. *Administrative Science Quarterly*, **41**, 685–718.

Tenner, E. (1996). *Why things bite back: New technology and the revenge effect*. London: 4th Estate.

Tidd, J., Bessant, J. & Pavitt, K. (1997). *Managing innovation: Integrating technological, market and organizational change*. Chichester, U.K.: John Wiley.

Tidd, J., Bessant, J. & Pavitt, K. (2001). *Managing innovation: Integrating technological, market, and organizational change* (2nd ed.). Chichester, U.K.: John Wiley.

Utterback, J. M. (1994). *Mastering the dynamics of innovation*. Cambridge, MA: Harvard Business School Press.

VanGundy, A. B. (1988). *Techniques of structured problem-solving* (2nd ed.). New York: Van Nostrand, Reinhold.

Van de Ven, A. (1986). Central problems in the management of innovation. *Management Science*, **32**, 590–607.

Van De Ven, A. H. & Poole, M. S. (1995). Explaining development and change in organizations. *Academy of Management Review*, **20** (3), 510–540.

Van Gundy, A. B. (1992). *Idea power: Techniques & resources to unleash the creativity of your organization*. New York: Amacom.

Walberg, H. J. (1988). Creativity and talent as learning. In: R. J. Sternberg (Ed.), *The Nature of Creativity: Contemporary Psychological Perspectives* (pp. 340–361). New York: Cambridge.

Weick, K. E. (1995). *Sense making in organizations*. Thousand Oaks, CA: Sage.

West, M. A. & Farr, J. L. (Eds) (1990). *Innovation and creativity at work: Psychological and organizational strategies*. Chichester, U.K.: John Wiley.

Whitley, R. (1999). *Divergent capitalisms: The social structuring and change of business systems*. Oxford: Oxford University Press.

Part XV

Conclusion

The International Handbook on Innovation
Edited by Larisa V. Shavinina

Research on Innovation at the Beginning of the 21st Century: What Do We Know About It?

Larry R. Vandervert

American Nonlinear Systems, USA

The Nature and Development of Innovation: An Arrow of Innovation?

Abstract: Innovation is about producing new and useful ideas and products. In this handbook the nature and development of innovation center on the mental mechanisms that propel it. To provide a framework for conclusions about the broad spectrum of ideas and applications of parts II through VII, an evolutionary 'arrow' of innovation is developed.

Keywords: Evolution and innovation; Innovation; Innovation management; National innovation systems; Social context of innovation.

The Meaning of Innovation

Innovation is about producing new and useful ideas and products. This may seem like a simple, ho-hum description of innovation, but scientists and business scholars interpret this statement in different ways. What exactly is involved in producing something 'new?' What does 'useful' mean?

Sternberg, Pretz & Kaufman (Part II) propose that, 'Innovation is the channeling of creativity so as to produce a creative idea and/or product that people can and wish to use'. For these authors, this notion of innovation inherently includes a social dynamics (forces) ingredient—innovation invariably 'propels' a field in some way. On the other hand, however, for Weisberg (Part II), while innovation is synonymous with a creative product, this product does not necessarily carry value beyond the individual who creates it. Further, Weisberg does not feel that innovation goes beyond the processes of ordinary thinking—for him innovation is 'ordinary thinking writ large'. In other words, creative and innovative people use ordinary thinking in extraordinary ways. This is analogous to saying that Olympic champions, for example, use the same muscles ordinary people use; they simply excel in their use. In my own work on the neurophysiology of innovation (Vandervert, Part II), I agree with Weisberg that innovation involves no special processes beyond elaborated ordinary thinking.

However, Root-Bernstein (Part II) proposes the idea that creativity and technology consist of effective *problem raising*. In other words, creativity and innovation are based first on the adequate formulation of problems. It can easily be shown that the process of evolutionary natural selection supports Root-Bernstein's approach. That is, if an organism does not have the mental machinery to recognize and formulate a problem as a problem, it is not likely to survive. Each niche is all about the organism's special capacities to first recognize what might be problems, and, then, second to solve those problems. But this leads us to a rather deep theoretical question: *How and why would organisms generate hypothetical problems?*

Standing at the end of a long evolutionary trend in survival strategies, human beings don't just respond to problems, they generate countless hypothetical problems. Each person, ordinary or extraordinary, has an immense capacity for generating hypothetical problems—many that they may never be able to solve. People spontaneously wonder about all sorts of things. Why would mental functions have evolved that create such 'surplus' problems, the solutions to some of which could lead to an ever-expanding knowledge of their surroundings?

The answer to this question seems to be based on the fact that, from the organism's point of view, survival is most often based on what will happen in the next few moments. And, these next few moments will probably not consist of a simple, straightforward extrapolation of past problem situations, or even the immediate past moments. As Fox (1988) pointed out, 'the dynamics of organism movements and their interactions with the dynamics of other organisms and with the environment are usually nonlinear phenomena' (pp. 160–161). This nonlinear moment-to-moment problem situation makes predicting the next moves of a prey or predator very difficult, if not impossible. In order to deal with the problem of predicting the outcome of real-world survival-dependent situations, organisms would have to be selected to generate predictive rapid *simulations*[1] of such problem situations. These simulations are experienced as various series of constantly updated hypothetical moves a prey or predator might make. This is the basis, I think, of the generation of 'surplus problems', and, at the human level, it seems it is literally the basis of wide-eyed 'wonderment'. Simonton (in press) has developed arguments that parallel this idea in what he calls *constrained stochastic processes* of creativity. Such quasi-random processes would be the perceptual-cognitive counterparts of the nonlinear survival situation described by Fox.

The fact of the matter, of course, is that all three of the above approaches to innovation are on the right track. There are strong social, mental and behavioral dynamics associated with innovation, and these dynamics exist in everyone. Innovations are characterized by most as new and useful solutions. But, at the same time, innovation must begin with new problems. And, in the large sense, innovation is about some sort of forward movement or anticipation of change. To more clearly understand this fundamental forward-looking dynamic of innovation, we can look briefly at some of the most exotic innovative ideas humans have ever produced.

Innovation and the Arrow of Time

Creativity and innovation are probably the most powerful engines of the human intellect. It is only through them that all new worlds of art, science, and technology are conceived and ultimately realized—and this includes, of course, the rarified mental/technological world in which the theoretical physicist works. In perhaps the most rarified of worlds, Stephen Hawking (1988) described three arrows that give a direction to time, namely, the thermodynamic arrow of time, the psychological arrow of time, and the cosmological arrow of time.[2] For the physicist, it is fundamental that the evolution of virtually everything in the physical universe has some rule of *directionality* associated with it. Hawking points out that that in everyday, real world affairs, and this would include the phenomenon of human innovation, all three of the arrows of time point in the same direction.

Does the developmental course of innovation have an 'arrow', and, if so, does this arrow have an identifiable underlying mechanism? While the contributors of various chapters of the two landmark volumes of this handbook may use differing terminologies, these are the two most basic questions that are addressed. Therefore, I will probe these two questions to develop general conclusions for the volume that will encourage an open interpretation of the great collection of ideas and applications presented, and, at the same time, hopefully create new questions and problems. In keeping with Root-Bernstein's sentiments on the importance of 'problem raising', perhaps the creation of new questions and problems is the best way to further the extraordinary work of the many contributors of this volume.

---

[1] Mathematically, prediction can be accomplished for linear and simple nonlinear situations using formulae which provide immediate closed-end solutions. However, for maximally complex nonlinear situations like those involved in Fox's prey-predator scenario, prediction of families of future potential trajectories can only be accomplished by stepping through iterative *simulations* that run faster than the real-time events are unfolding. Behaviorally, such rapid simulations can be thought of as translating into vicarious trial and error behavior (rapidly running- or looking-back-and-forth in the face of problem situations) which can be observed in a broad variety of 'organisms' from rats in 'T' mazes to NBA basketball players in the heat of a game. Mentally, rapid simulations in vicarious trial and error, going back to even Tolman (1926) have been interpreted as oscillatory representations in imagination.

[2] Briefly, Hawking (1988) described the three arrows of time as follows:

> There are at least three different arrows of time. First, there is the thermodynamic arrow of time, the direction of time in which disorder or entropy increases. Then, there is the psychological arrow of time. This is the direction in which we feel time passes, the direction in which we remember the past but not the future. Finally, there is the cosmological arrow of time. This is the direction of time in which the universe is expanding rather than contracting (p. 145).

The only term in the above quote the reader may be unfamiliar with is *entropy*. Entropy refers to the tendency of things to become disordered or to 'deteriorate' over time if left to themselves. That is, as time passes things will tend to 'fall apart', unless they are maintained. But even the energy used in maintenance reduces the available ordered energy, thus resulting in a net disorder in accordance with the entropy principle (see Hawking's classic and easy-to-understand explanation of how your brain increases entropy by reading this book, pp. 152–153).

To better understand the broader context and fundamental significance of these three time arrows for innovation and how the arrow of all mental processes is conceived to be embedded in matter/energy/information, it is recommended that the reader be familiar with Hawking's entire chapter, 'The Arrow of Time'. As most know, Hawking wrote *A Brief History of Time* for the layperson, and the chapter on the arrows of time is quite understandable for that reader.

Restarting From Older Times

Sometimes in order to formulate a new view and new questions, we need to revisit older knowledge and re-evaluate its promise in terms of more recent thinking (see *Reconstruction/Redirection* type of innovation (Sternberg, Pretz & Kaufman, Part II)). Either explicitly or implicitly, Darwinian evolution is a common backdrop running through most of the chapters of Part II (especially, Bailey & Ford; Georgsdottir, Lubart & Getz; Marinova & Phillimore; Nickles; Vandervert). Nickles (Part II) points out that Darwin's evolutionary natural selection of biological and technological forms is uninformed as to outcome. However, like the directional arrows that Hawking described for the physical universe, both biological and technological evolution do appear to follow an 'arrow' or direction of development. To understand how such an uninformed, evolutionary arrow could work, we must go back to the evolutionary ideas of Ludwig Boltzmann, who without knowing it, provided the initial key. Boltzmann (1905) brought evolutionary processes into the *energy* realm in his classic statement on what natural selection is really all about: '(the) struggle for existence is the struggle for free energy available for work' (p. 23). In other words, selective 'fitness' refers to energy-capturing fitness. When the fox captures the rabbit, the only real, selective payoff is fur-covered *energy*. Later, further elaborating Boltzmann's conception, Alfred Lotka, a brilliant mathematical biologist, gave evolution the following arrow of *direction*: 'Evolution proceeds in such direction as to make the total energy flux (flow) through the system a maximum' (1922, p. 147).

The Arrow of Artistic, Scientific, and Technological Innovation

Lotka subsequently described how the evolutionary arrow of increasing energy flow has made its appearance in the overall history of innovation:

> Man, one of the latest, and in his own judgment the highest product of evolution, has hitherto signally conformed with the principle of increasing energy flux (flow). By ingenious contrivances he has immensely refined and multiplied the operation of his receptor-effector apparatus. The excess of energy /information captured, over the energy /information barely sufficient for mere maintenance, has in his case grown to a wholly unparalleled magnitude. Normally this leaves him with a large balance available for 'play' activities and luxuries. And some of his play activities have turned out to be a most profitable reinvestment. For among them must be classed scientific research (and all forms of artistic endeavor) indulged in primarily out of curiosity, but resulting among other things in that complete recasting of methods of production which is known as the industrial and agricultural revolution. Aside from its direct benefits this has made it possible to

spend relatively large amounts on sanitary improvements, on medical education and research, and, above all, on better living among the masses of the people (1945, p. 188).

Lotka's scenario is a handy way to get an aerial view of how the uniquely human explosion of innovation is embedded within evolution. In the above portrayal of the mechanism and arrow of innovation, Lotka describes how the increasing efficiency of the energy flows provided by innovative contrivances magnifies all human tendencies. It is seen in the many chapters of this volume that there is not a cognitive capacity, field or domain of knowledge that the nervous system does not, as Lotka says, immensely refine and multiply though its contrivances of innovation (compare, Carayannis, Gonzalez & Wetter, Part II). Below, I will link this idea to several of the domains described in Part VII.

Thus, the arrow of innovation adds a needed 'directional content' to the abstract processes of natural selection that many chapters of this volume skillfully describe as underlying innovation.

Innovation in Different Domains

What is it that propels the innovative work of human beings in the various domains of technology, ideas, and art? The answer to this question has at least two appearances. First, how does the mind (or the receptor-effector apparatus in Lotka's terminology) develop innovations in the various domains of human activity, and how does it generate domain-relevant problems? (See Root-Bernstein, Part II.) Second, how might the various domains be related to the types of contributions of innovation that have been identified by, for example, Sternberg, Pretz & Kaufman (Part II)?

The First Question (The Development of Innovations in Various Domains)

It is through innovation, and only innovation, that the human mind connects the structure of its internal world (see the detailed synthesis by Shavinina, Part VII), with that of the external physical world of technological, and scientific and artistic creations. It appears that innovation, arguably the most complex of all mental operations, is orchestrated in every domain though the guiding structure of the mental models that comprise the domains (Chi & Hausmann; Dasgupta; Scripp & Subotnik; Shavinina, Part VII; Shavinina & Ponomarev, Part VI; Shavinina & Seeratan; Vandervert, Part II). The personal and social collections of mental models of a domain 'are' the domain, and an infinite variety of such domain-specific mental models can be constructed from experience.

Many, including myself, have argued that the mental models of each domain *propel* (in the sense proposed by Sternberg, Pretz & Kaufman, Part II) searches for new levels of efficiency (for example, Vandervert,

Part II). How, exactly, mental models guide problem searches for new and useful variations in the models is not yet clearly understood (compare the categories of problem-generation proposed by Root-Bernstein, Part II). However, David Ingvar has offered a very promising explanation based on both experimental and clinical evidence that centers on how the prefrontal cortex of the brain handles action programs or plans for future behavior and cognition:

> It is concluded that the prefrontal cortex is responsible for the temporal organization of behaviour and cognition due to its seemingly specific capacity to handle serial information and to extract causal relations from such information. Possibly the serial action programs which are stored in the prefrontal cortex are also used by the brain as templates for extracting meaningful (serial) information from the enormous, mainly non-serial, random sensory noise to which the brain is constantly exposed (1985, abstract, p. 127).

The serial action programs that Ingvar describes may be viewed as search programs that are used in planning, simulating (imagining) new states of affairs, and gathering information that potentially can lead to the generation new conceptions within the various domains. Thus, serial action programs are the means by which mental models construct future states. Action programs do this, in concert with other brain functions, by guiding the extraction of new patterns of causally linked information. Since this newly extracted serial information may or may not be fully understood by the person, the new patterns would occasionally be experienced as the recognition of new problems in the pertinent domains. The recognition of these new problems could then potentially lead to various types of innovations.

In accordance with Ingvar's findings, then, when mental models are acquired through education and experience, one thing that happens is that new categories of problem searches are established. For example, the mental models inherent in a particular type of curriculum set in motion in the student's mind the extraction of certain categories of serial information from the environment. Consequently, the student begins to generate new (new to the student) kinds of questions (Compare with Scripp & Subotnik, Part VII).

The Second Question (How Different Domains are Related to Types of Innovation)

William Ogburn, a sociologist, proposed that the basic determinants of social change were invention and innovation. Ogburn believed that material innovation generally became part of the fabric of society more rapidly than new ideas. He referred to this idea as *cultural lag*. Is there a cultural lag effect across different domains of innovation, in that, certain domains tend to be related to certain types of innovative contributions (à la Sternberg, Pretz & Kaufman, Part II) and not others? For example, following in the vein of Ogburn's theory, are rapidly diffusing technological innovations more likely to be of Sternberg et al.'s 'smoother' *replication*, *forward incrementation*, and *integration* types of innovative contributions? And, are the more slowly diffusing idea-related innovations more likely to be of the 'idea/value-weighted' *redefinition*, *advanced forward incrementation*, *redirection*, *reconstruction/redirection*, and *reinitiation* types?

To help illustrate this very tentative idea, each of the chapters appearing in, for example, Part VII can be examined with this question in mind. Although each of these chapters involves several types of innovation, here are what appear to be the two dominant types of innovation for each of the nine chapters of Part VII:

Chapter 1. Replication; Redefinition
Chapter 2. Redefinition; Redirection
Chapter 3. Redefinition; Integration
Chapter 4. Redirection; Integration
Chapter 5. Replication; Forward Incrementation
Chapter 6. Redefinition; Forward Incrementation
Chapter 7. Reinitiation; Integration
Chapter 8. Replication; Forward Incrementation
Chapter 9. Replication; Integration.

An interesting issue that this sort of an examination brings to mind is that innovators (including, of course, those who write about and research innovation) might consider casting and tweaking the 'packaging' of new gadgets and ideas with an eye on their probable rates of diffusion. How quickly will this product or idea be become disseminated among target consumers? (Compare in this regard Goldsmith & Foxall, especially, their section on 'Why Measure Inventiveness?' Part V).

The Development of Innovation

About half of the chapters on the development of innovation appearing in this volume address the mental and emotional mechanisms that are involved in the development of innovation (Root-Bernstein & Root-Bernstein, Part VI; Simonton, Part IV). The other half describes how innovation can be encouraged through structured learning environments. Largely due to the applied nature of their subject matter, these latter chapters are, themselves, well-structured learning environments (see Kostoff; Reis & Renzulli; Shavinina & Ponomarev, Part VI). This situation presents the reader with a special opportunity. By bringing together the mental/emotion mechanism chapters on the one hand with the chapters on structured learning environments on the other, the reader will form a powerful literature-based/workshop-based approach to innovation of the type described by Kostoff (Part VI). In this manner, the reader can experience a real world approach to encouraging thinking about the topics of innovation.

Following this idea, Root-Bernstein and Root-Bernstein's (Part VI) 'tool kit' of pre-logical and pre-verbal skills for innovative thinking (namely: (1) observing; (2) imaging; (3) abstracting; (4) pattern recognizing; (5) pattern forming; (6) analogizing; (7) bodily-kinesthetic thinking; (8) empathizing; (9) dimensional thinking; (10) modeling; (11) playing; (12) transforming; and (13) synthesizing) seems to connect itself to David Ingvar's above description of how the prefrontal lobe extracts meaningful patterns from the largely non-serial sensory environment. The thirteen skills of the tool kit may be looked upon as the critical skills for building robust mental models (in any domain). Such mental models will enhance the search for new categories of serial information that will in turn be most likely lead to new questions and problems and thus to innovation.

Innovation and the Social Context Integrated: Where We Are and Where We Are Headed

Abstract: The great diversity of ideas and issues appearing in Parts VIII through XIV are presented in three integrative mappings. From these mappings a bright thread appears that shows a unity among all of the chapters. This bright thread reveals that a common social matrix resides behind all innovation.

The great diversity of ideas and studies of innovation and the social context present a particularly difficult problem for presenting a succinct, conceptually integrated conclusion. Therefore, in this concluding chapter, I provide three simple mappings for the larger parts of the volume. Mappings have the advantage of providing a graspable snapshot of complex issues. The mappings will be discussed in a manner that will both capture the components of the diverse territory of all of the chapters and reveal many of the dynamics that interrelate them.

The three mappings depict overall chapter-to-chapter integrations and help to promote the following inter-disciplinary and international goals.

(1) The continuing investigation of the relationships among concepts and models in various approaches to innovation, and to help in useful transfers of information from one discipline and nation to another.

(2) The continuing encouragement of the development of adequate theory in areas which lack them.

(3) The continuing elimination of the duplication of theoretical efforts in different disciplinary approaches to the study of innovation.

(4) The continuing promotion of a unity of understandings of innovation by improving communication among specialists.

How the Mappings Were Constructed and How To Interpret Them

To construct a mapping of the various chapters in Parts VIII through XIII, concepts and issues had to be identified and then prioritized as to what seemed to be most fundamental. Certainly, these processes of identification and, especially the prioritizing, are difficult and somewhat subjective ordering tasks that I am aware are fraught with casus belli for the community of innovation scholars and practitioners. However, the purpose of the mappings should be looked upon as a first draft of an evolutionary tool that, while perhaps being a source of controversy itself, will help guide the general thinking of both individuals and teams interested in the topics of innovation.

Once the fundamental concepts and issues had been identified, the actual mapping procedure is quite simple and quite intuitive. For each of the three mappings, a central idea or concept was chosen that seemed to typify all of the rest of the concepts and issues for a particular mapping. This central idea was placed in the center of the mapping. Then the *relative* conceptual distance of each sub-concept or issue was indicated in the mapping by the length of the line from the central idea. It is extremely important to recognize that conceptual distance from the central idea does not in any way indicate the relative importance or value of a particular sub-concept. On the contrary, it may turn out in certain circumstances that a 'remotely associated' concept will prove to be the spark that redirects thinking about the central idea of the mapping.

A Mapping of the Basic Approaches To Understanding Innovation in the Social Context

The organizing theme of the chapters in Part VIII of this volume is the impact of social psychological issues on innovation. This issue has been placed in the center of Fig. 1. Each of the 'spokes' represents one of the chapters appearing in Part VIII.

The overall message of these chapters can be captured in the following general principle: The more the social aspects of a person's or group's 'mind' are directly influenced by innovation, the more this mind will, in turn, influence innovation. Figure 1 shows that the chapters on the effects of managers' creativity schemas (Zhou & Woodman), and innovation/identity (King) seem closest to this idea. But, keep in mind the earlier caution about assuming from the mapping that these issues have the greatest potential impact.

The principle that innovation influences innovation may sound a little like a no-brainer, but it reveals at least three things. First, it means that innovation is socially constructed (e.g. Hadjimanolis). Second, it means that innovation is a learnable enterprise (e.g. Carayannis & Gonzalez). Finally, we can conclude that

the many processes involved in the social construction of innovation are subject to manipulation (e.g. Callahan & Muegge; Hadjimanolis; Nag, Corley & Gioia).

A Mapping of Innovations in Social Institutions

The chapters devoted to Innovations in Social Institutions constitute a nice continuation of the social psychological issues discussed above in Part VIII of this volume. The central idea in Part IX is *social synthesis*. This idea has been placed in the large circle at the bottom of Fig. 2. Each of the additional circles represents one of the other chapters in Part IX or X.

It can be seen from the patterns of the arrows in Fig. 2 that the seven chapters of Part IX are aimed at ideas and issues that are tightly knitted together, with many threads interconnecting them. Social synthesis in these chapters is most reliant upon the human *networking* of ideas and activities (Swan, Scarbrough & Robertson). Extremely close to these issues are the chapters on 'promoters' (Hauschildt, Part X) and knowledge translation (Major & Cordey-Hayes, Part IX). It was of great personal interest to me that the fundamental processes of social synthesis described in all of these chapters are quite analogous to the synthesis toward innovation produced by the collaboration of *brain functions* that I have described in Part II

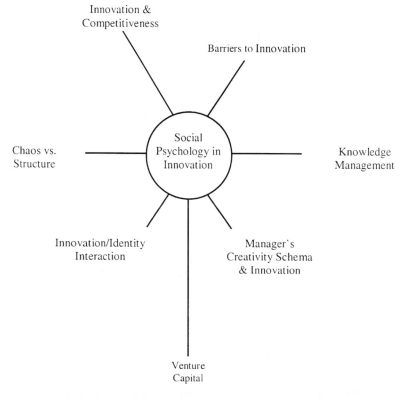

Figure 1. Basic approaches to the understanding of innovation in the social context.

of this Handbook (Vandervert, Part II). For example, Swan, Scarbrough and Robertsons' (Part IX) process model incorporating *invention*, *diffusion*, and *implementation* (see their Fig. 1) has rough counterparts in working memory/cerebellar interaction. And, the neurophysiological model, like theirs, is recursive and nonlinear.

Innovation Management

The main conceptual feature of Part X's four chapters on the management of innovation is the problem of dealing with the contradictory challenges of dualism, that is, 'functioning efficiently today while innovating effectively for tomorrow' (Katz). It seems that the management of innovation must be by nature deeply invested in this dualism, because simultaneous stability and growth are two necessary features of all systems that survive. For example, in the life development of each human being, stages of consolidation are regularly erected while, at the same time, whole new capacities are coming into being.

This 'dualism' that faces innovation management most certainly has the most fundamental of origins. Odum & Odum (1981) laid out the core simultaneous processes that must be taken into account if a system is to survive, and, I think, to *innovate* competitively. These processes comprise the maximum-power principle of *evolution*, and I have restated them very slightly here to help capture their *informational* significance for innovation:

Those organizations that survive in the competition among *alternative choices* are those that develop more energy-information inflow and use it to meet the needs of survival. They do this by:

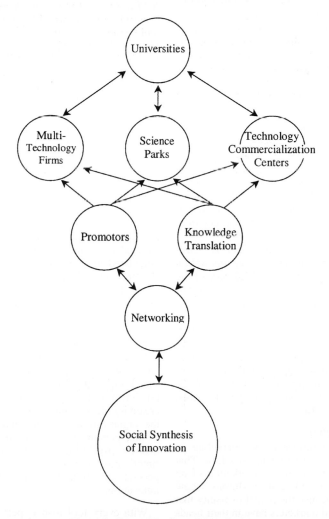

Figure 2. Innovations in social institutions.

(1) developing storages of high-quality information (see Odum, 1988);

(2) feeding forward and feeding back information from the storages to increase inflows;

(3) recycling information as needed;

(4) organizing control mechanisms that keep their systems adapted and stable;

(5) setting up exchanges with other organizations to supply special information needs; and

(6) contributing useful information to surrounding environmental systems that help maintain favorable conditions.

(see Odum, 1988; Odum, 1996; Odum & Odum, 1981)

This, I believe, is the underlying infrastructure for innovation management, and these processes are taken into account by management that squarely confronts the dualism discussed in these chapters. In other words, innovation management requires doing all of these things at once. Seattle-based Amazon.com seems to be surviving well, by doing all of these things right! The maximum-power principle is truly an elaborated constructivist vision of innovation management, where 'The future is not written; it is to be constructed though the interactions between individuals and groups of people' (Boly, Morel & Renaud, Part X).

Innovation and Leadership

Deschamps' Part XI chapter on innovation leadership lists the common traits of leader in terms of personal profile. The reader is urged to connect this leader profile with the story of Taiwan's innovation system that will appear below in the mapping of the 'national vigor of innovation'. It seems that the innovation leader and the national innovation system must be cut from the same cloth.

Although neither Deschamps nor Hsu & Chen (Part XIII) come close to saying it, when innovation leadership is coupled with the *survival mandate* of the maximum-power principle that is stated in the last section, another simple principle emerges: *From an evolutionary standpoint, the vision and energy in innovation is all about kicking some competitor's butt*! Innovation is the selective advantage among alternative choices, and those that don't adequately manage innovation don't survive—period! Jeff Bezos's Amazon.com is a sterling, living demonstration of the features of the maximum-power principle described above in the preceding section.

Innovation and Marketing

The two chapters on innovation and marketing in Part XII somehow remind one of Murphy's Law. That is, if anything can go wrong in the marketing of an innovative product, it will. Of course Murphy's Law comes into play only because the model of reality that planners, researchers and marketers have in their heads doesn't adequately match the details and dynamics that

are out there in the real, physical and social world. Both of the chapters in Part XII offer valuable information that would help bring into alignment the mental models of marketers on the one hand, and the actual dynamics of consumer worlds on the other (see Trott; Veryzer).

Trott offers a model of the interactions of marketing, R&D, and business planning. He proposes that, 'it is the interactions of these internal functions and the flow of knowledge between them that needs to be facilitated'. Why would this be so? It seems that if an organization doesn't do this, it's model of social reality will be off the mark, and some other organization will do precisely what Trott proposes. Then, reflecting back to the evolutionary principle in the above Part XI on Innovation Leadership, it will be your butt that will get kicked.

A Mapping of Innovation Around the World: Examples of Country Efforts, Policies, Practices and Issues

The general question of 'national innovative vigor' is the central theme of the eleven chapters of Part XIII of this Volume. It is the national imperative for the future. The most significant mapping feature in Fig. 3 is how six of the national innovation systems huddle closely around this central theme. Another interesting feature of this mapping is how the 'dynamics analysis' of Germany/United Kingdom (Harding) can be seen to bear on India's regional innovation (Mehra).

The nations tightly huddled around the center of the mapping give a strong sense that, ultimately, innovation has a great deal to do with why societies are formed in the first place. That is, social re-structuring toward higher efficiencies (innovation) bestows greater advantage than any alternative re-structuring. And, as national programs aimed at innovation become more articulated in terms of information flows, as in those clustered around the center of the mapping, the more successful they tend to become in the world marketplace. This general idea has been mentioned many times in this concluding chapter, and is, of course, part and parcel to the messages of all of the chapters of Parts VIII through XIII.

It seems that innovation in the human mind on the one hand and the many information flows necessary to social interaction on the other exist in a dynamic relationship of reciprocal modeling. Thus, ideally, the creativity of the innovative person is embedded in the organization, and the organization is embedded in that person. Sigmund Freud, certainly one of the greatest psychologists who ever lived, noticed this 'inter-embeddedness' in a way that suggested to him a general direction for the future of technological innovation:

> With every tool man is perfecting his own organs, whether motor or sensory, or is removing the limits

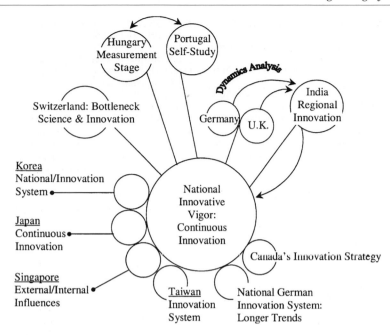

Figure 3. Innovation around the world: examples of country efforts, policies, practices and issues.

to their functioning. Motor power places gigantic forces at his disposal, which, like his muscles, he can employ in any direction; thanks to ships and aircraft neither water nor air can hinder his movements; by means of spectacles he corrects defects in the lens of his own eye; by means of the telescope he sees into the far distance; and by means of the microscope he overcomes the limits of visibility set by the structure of his retina. In the photographic camera he has created an instrument which retains the fleeting visual impressions, just as a gramophone disc retains the equally fleeting auditory ones; both are at bottom materializations of the power he possesses of recollection, his memory. With the help of the telephone he can hear at distances which would be respected as unattainable even in a fairly tale. Writing was in its origin the voice of an absent person (1930/1961, pp. 37–38)

Because distance, time and effort were being reduced essentially to zero, Freud derived from this progression toward 'perfection' that humans had become 'prosthetic Gods'. He expected the trend to continue on into the future.

Future Innovations in Science and Technology

In Part XIV of this volume Joseph Coates provides us with a far-reaching analysis and vision of the future that extends out 25 years.

While the unfolding of the future is no doubt nonlinear in its detail, there are large-scale discernable dynamics that can provide us with more 'linear' arrows

for the future development of innovation (see Rickards concerning the trend toward nonlinear models of innovation). Freud provided us with one important arrow, namely, that future innovations would involve the further refinement and extension of the qualities and capacities of the human senses, mind, and body. Continuing this theme in far more detail, Coates describes how biotechnology and information technology will lead to the greatest predictable changes. The fact that information is far easier to amplify than energy[3] (this fact provides another predictable arrow for innovation) bolsters the soundness of the forecasting judgments Coates makes.

In the second section of his chapter Coates talks about future applications of scientific and technological advances. In his portion on education he mentions that, for children, computers are already not 'equipment', but prostheses. Coates is dead right about this. Although Rickards rightly cautions us on attempting to predict too far down the road, perhaps we should take Coates' idea and run with it a little. Perhaps, along with Freud (and Lotka), we should think of the future of all innovation as a seemingly inescapable tendency toward an increasingly enriched and powerful 'prosthetic' augmentation of ourselves. This can give us one very strong sense of what is likely to come next.

---

[3] For both text discussion and simple calculations of this idea, see Tribus & McIrvine (1971). This article is aimed at a broad audience.

References

Boltzmann, L. (1905). *The second law of thermodynamics*. Dordrecht: Reidel.

Fox, R. (1988). *Energy and the evolution of life*. New York: Freeman.

Freud, S. (1961). *Civilization and its discontents*. New York: W. W. Norton. (Original work published 1930)

Hawking, S. (1988). *A brief history of time*. New York: Bantam Books.

Ingvar, D. (1985). 'Memory of the future': An essay on the temporal organization of conscious awareness. *Human Neurobiology*, **4**, 127–136.

Lotka, A. J. (1922). A contribution to the energetics of evolution. *Proceedings of the National Academy of Science*, **8**, 140–155.

Lotka, A. J. (1945). The law of evolution as a maximal principle. *Human Biology*, **17**, 167–194.

Odum, H. T. (1988). Self-organization, transformity, and information. *Science*, **242**, 1132–1139.

Odum, H. T. (1996). *Environmental accounting: Emergy and environmental decision making*. New York: John Wiley and Sons.

Odum, H. T. & Odum, E. C. (1981). *Energy basis for nature and man*. New York: McGraw-Hill.

Simonton, D. (in press). Scientific creativity as constrained stochastic behavior: The integration of product, process, and person perspectives. *Psychological Bulletin*.

Tolman, E. (1926). A behavioristic theory of ideas. *Psychological Review*, **33**, 352–369.

Tribus, M. & McIrvine, E. (1971, September). Energy and information. *Scientific American*, **225**, 179–188.

Author Index

Page numbers not in brackets refer to the reference lists; those in brackets refer to the citations in the text.

Subject Index

Note: Page numbers in *italic figures* refer to illustrations